Teacher's Edition

PRECALCULUS
A Graphing Approach
Fourth Edition

Teacher's Edition

PRECALCULUS
A Graphing Approach
Fourth Edition

FRANKLIN DEMANA
BERT K. WAITS
The Ohio State University

STANLEY R. CLEMENS
Bluffton College

GREGORY D. FOLEY
Sam Houston State University

ADDISON-WESLEY

Menlo Park, California • Reading, Massachusetts • New York • Don Mills, Ontario
Wokingham, England • Amsterdam • Bonn • Sydney • Singapore • Tokyo
Madrid • San Juan • Paris • Seoul • Milan • Mexico City • Taipei

Reprinted with corrections, April 1997.

ISBN 0-201-87011-8

2 3 4 5 6 7 8 9 10-CRW-99 98 97

The authors dedicate
this book to their wives,
Christine Demana,
Barbara Waits,
Joenita Clemens, and
Jolinda Foley,
without whose patience,
love, and understanding
this book would not
have been possible.

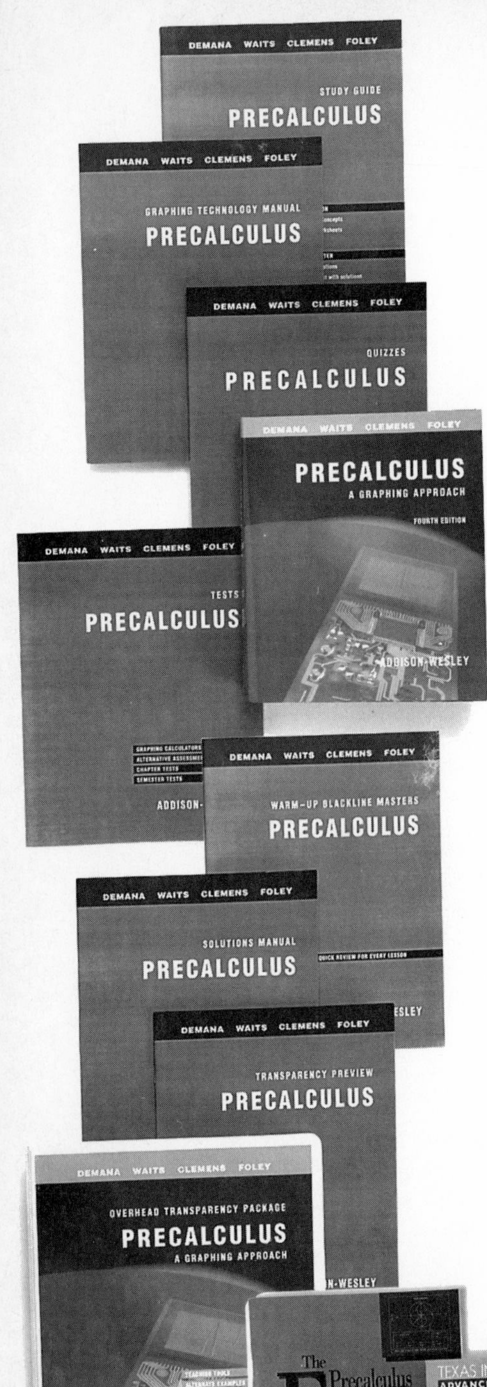

Available for use with PRECALCULUS: A GRAPHING APPROACH

TEACHER'S RESOURCE PACKAGE

- Study Guide
- Graphing Calculator Resource Manual
- Quizzes
- Tests
- Warm Up Masters
- Transparency Preview

ALSO AVAILABLE

- Overhead Transparency Package
- Solutions Manual
- OmniTest
- Mac-81 and PC-81 Emulation Software
- IBM Precalculus Explorer

For more information on these ancillary products, see page x.

To learn about features of this program, see page xiii.

CONTENTS

Precalculus ancillaries provide students with support in preparing for future math courses and exams and provide teachers with flexible teaching options.

TEACHER'S RESOURCE PACKAGE

Study Guide

- Objectives and key terms
- Questions on key concepts
- AP exam-style free response questions
- Sample tests and worked-out solutions

Graphing Calculator Resource Manual

- Texas Instruments, Casio, Sharp, and Hewlett-Packard instructions
- Classroom tips and additional calculator practice

Warm Up Blackline Masters

- Additional Quick Review exercises for use at the beginning of the class

Quizzes

- Two multiple-choice quizzes for each chapter

Tests

- Alternative Assessment
- Two parallel forms of end-of-chapter tests
- Mid-year and end-of-year tests

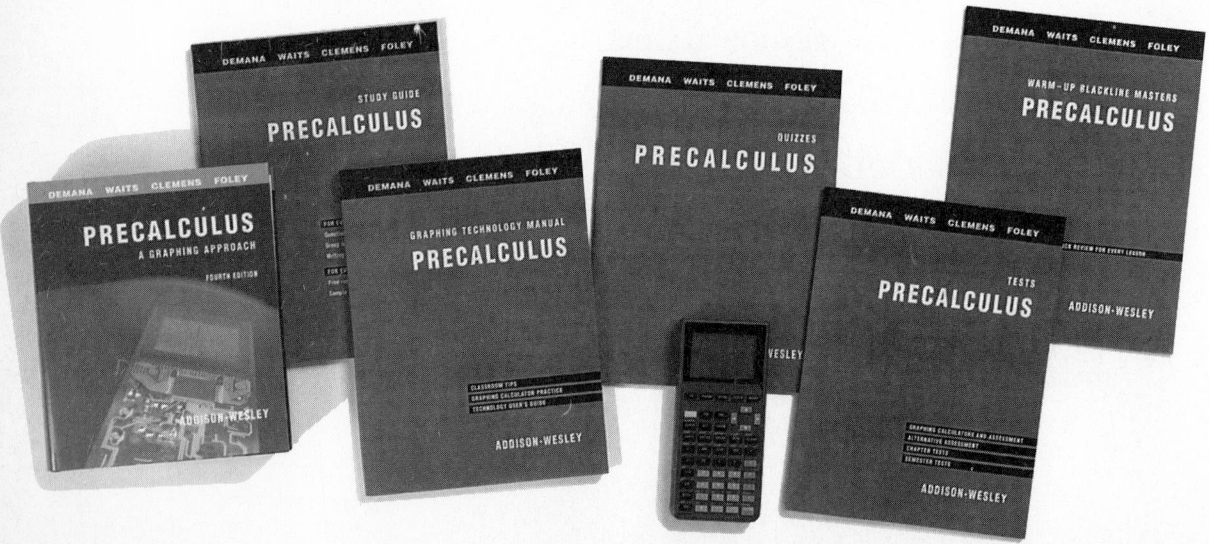

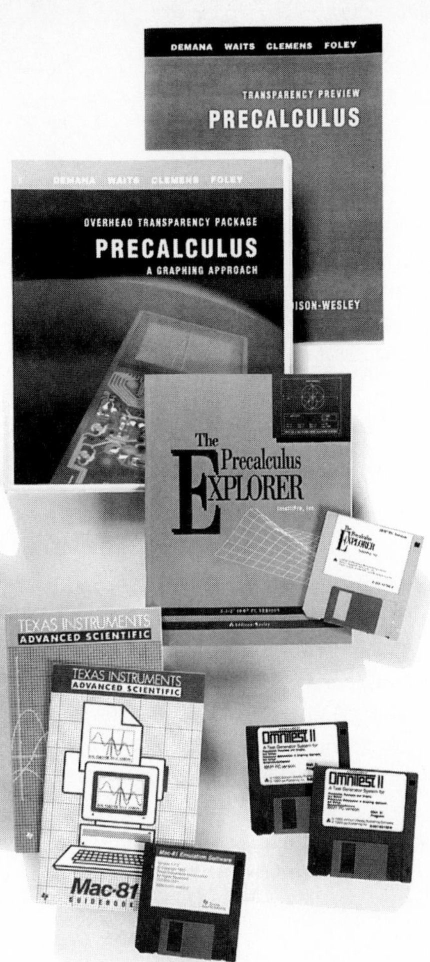

ALSO AVAILABLE

Overhead Transparency Package

- 4-color Examples and Alternate Examples

Solutions Manual

- Step-by-step solutions for all exercises in the Student Edition

OmniTest Software

- Provides options for quizzes, tests, and practice worksheets
- Available for DOS and Macintosh
- Allows customization of items and level of difficulty with add and edit functions

Mac-81 and PC-81 Software

- As products of Texas Instruments, Inc., this software makes all the functions of the TI-81 Graphics Calculator available to Macintosh (Mac-81) or IBM-compatible personal computer (PC-81) users.

Precalculus Explorer Software

- Contains 10 programs that provide the opportunity for the exploration of precalculus concepts. DOS 3.5 format only.

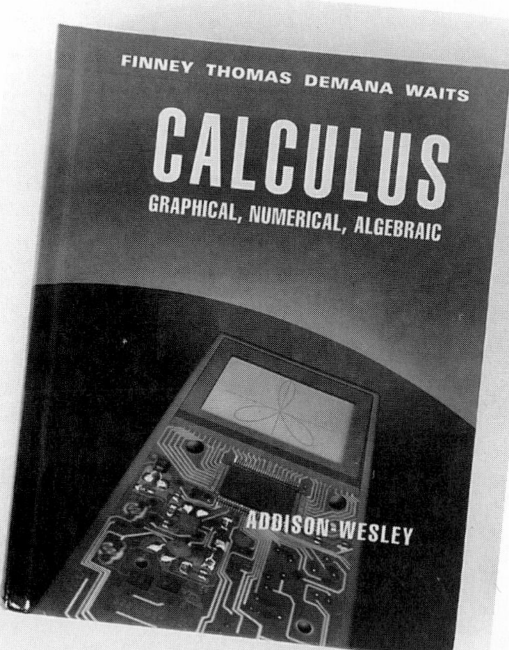

The graphing approach of this book is continued in Calculus: Graphical, Numerical, Algebraic. Contact your local Addison-Wesley sales representative for more information.

ABOUT THE AUTHORS

Franklin Demana

Franklin Demana received his Master's degree in Mathematics and his Ph.D. from Michigan State University. Currently, he is Professor Emeritus of Mathematics at The Ohio State University. As an active supporter of the use of technology to teach and learn mathematics, he is cofounder of the C^2PC (Calculator and Computer Precalculus) technology-enhanced curriculum revision project now operating in over 1000 high schools and colleges. Along with frequent presentations at professional meetings, he has published a variety of articles in the areas of computer- and calculator-enhanced mathematics instruction. Dr. Demana is also cofounder (with Bert Waits) of the annual International Conference on Technology in Collegiate Mathematics (ICTCM) and the T^3 (Teachers Teaching with Technology) professional in-service teacher development program. Dr. Demana coauthored *Essential Algebra: A Calculator Approach; Transition to College Mathematics; Precalculus Mathematics: A Graphing Approach; College Algebra and Trigonometry: A Graphing Approach; College Algebra: A Graphing Approach; Precalculus: Functions and Graphs; Intermediate Algebra: A Graphing Approach;* and *Calculus: Graphical, Numerical, Algebraic.*

Bert K. Waits

Bert Waits is Professor Emeritus of Mathematics at The Ohio State University, where he received his Ph.D. in 1969, and is the Texas Instruments Visiting Professor of Mathematics Education at the University of Texas at Arlington. He is cofounder of the C^2PC project and the T^3 program along with Frank Demana. Dr. Waits has published articles in over 50 nationally recognized professional journals. He frequently gives invited lectures, workshops, and minicourses at national meetings of the Mathematics Association of America (MAA) and the National Council of Teachers of Mathematics (NCTM) on how to use computer technology to enhance the teaching and learning of mathematics. He has given invited presentations at the International Congress on Mathematical Education (ICME 6 and 7) in Budapest (1988) and Quebec (1992), the International Conference on the Teaching of Mathematics by Applications (ICTMA5) in Noordwijkerhout, The Netherlands (1991), and colleges and universities internationally. Dr. Waits coauthored *Precalculus Mathematics: A Graphing Approach; College Algebra and Trigonometry: A Graphing Approach; College Algebra: A Graphing Approach; Precalculus: Functions and Graphs; Intermediate Algebra: A Graphing Approach;* and *Calculus: Graphical, Numerical, Algebraic.*

Stanley R. Clemens

Stanley Clemens received his B.A. from Bluffton College, his Master's degree from Indiana University, and his Ph.D. in Mathematics from the University of North Carolina. He is Professor of Mathematics at Bluffton College. Dr. Clemens has written several journal articles and coauthored *College Algebra and Trigonometry: A Graphing Approach; College Algebra: A Graphing Approach; Precalculus: Functions and Graphs; Intermediate Algebra: A Graphing Approach; Addison-Wesley Pre-Algebra; Access to Algebra and Geometry; Addison-Wesley Mathematics; Precalculus Mathematics: A Graphing Approach; Laboratory Investigations in Geometry; Geometry: An Investigative Approach;* and *Addison-Wesley Geometry.*

Gregory D. Foley

Gregory Foley received his B.A. and M.A. in Mathematics and his Ph.D. in Mathematics Education from the University of Texas at Austin. He is an Associate Professor of Mathematics at Sam Houston State University. Dr. Foley has presented papers, lectures, demonstrations, workshops, minicourses, and summer institutes throughout North America and in Europe, many of these focusing on the effective instructional uses of handheld computers. Dr. Foley coauthored the Instructor's Resource Guides for *Precalculus: Functions and Graphs* and *College Algebra: A Graphing Approach,* and several calculator laboratory manuals for various textbooks.

ABOUT PRECALCULUS

In approaching this revision of *Precalculus: A Graphing Approach,* our principal objectives were to provide students with the best possible understanding of algebra and trigonometry, to show how algebra and trigonometry can be used to model real-life problems, and to address the concerns and incorporate the suggestions of the many students and instructors who have used past editions of the text. We have participated in more than 400 grapher workshops involving some 20,000 teachers since the publication of the third edition. Based on those discussions and recommendations published by AMATYC, MAA, and NCTM, we believe that the changes made in this edition make this course the most effective preparation for future college-level mathematics and science courses.

FEATURES NEW TO THIS EDITION

Balanced Approach

A principal feature of this edition is the balance attained among the algebraic, numerical, graphical, and verbal methods of representing problems: **the rule of four.** We believe that students must learn the value of each of these methods of representation and learn to choose the one most appropriate for solving the particular problem under consideration. Additionally, we believe that students need to master the algebraic techniques necessary to solve problems in this text and to carry that capability forward to their study of calculus.

The Rule of Four

In support of the rule of four, we use a variety of techniques to solve examples. For instance, we obtain solutions algebraically when that is the most appropriate technique to use and obtain solutions graphically or numerically when algebra is difficult or impossible to use. We urge students to solve problems by one method and then support or confirm their solutions by using another method. We encourage students to use the most appropriate technique to solve a particular problem and to decide when use of the grapher may not be the best means to obtain a solution.

Algebraic Skills and Understanding

To ensure that students gain command of the necessary algebra skills, we inserted a new Prerequisite Chapter in this edition. This material gives instructors the option of including algebraic skills review when necessary. Also, new Quick Review exercises precede each section exercise set so that students can review the algebraic skills needed to solve the exercises in that section. An important part of the underlying philosophy of this text is to encourage students to use algebraic methods to solve problems or confirm solutions when the use of such methods is most appropriate. This approach reinforces the idea that to understand a problem fully, students need to understand it algebraically.

Applications

The text includes a rich array of interesting applications that will appeal to students in various fields, including biology, business, chemistry, economics, engineering, finance, physics, the social sciences, and statistics. Many applications are based on actual data from cited sources. As they work through the applications, students are exposed to functions as mechanisms for modeling data and are motivated to learn about how various functions can help model real-life problems. They learn to analyze and model data, represent data graphically, interpret from graphs, and fit curves. Additionally, the tabular representations of data presented in the text highlight the concept that a function is a correspondence between numerical variables, helping students to build the connection between the numbers and the graphs.

Problem-Solving Approach

We introduced the problem-solving process in Section 1.1, where students learn the following variation on Polya's four-step problem-solving process: (1) understand the problem; (2) develop a mathematical model of the problem; (3) solve the mathematical model and support or confirm the solution; and

(4) interpret the solution. Steps 2 through 4 then provide the format for solving applications throughout the remainder of the text.

Graphing Utilities

Students are expected to use a graphing utility (grapher) to visualize and solve problems. The numerous figures in the examples and exercises help students develop graph-viewing skills: *recognize* that a graph is reasonable; *identify* all the important characteristics of a graph; *interpret* those characteristics; and *recognize* "grapher failure." Most graphs either resemble students' actual grapher output or suggest hand-drawn sketches. Appendix A, Grapher Workshop, provides detailed instruction on important grapher features. Technology Notes appear in the chapter margin as needed to help students attain greater grapher proficiency.

Exercise Sets

We revised extensively the exercise sets for this edition and expanded them to include more than 6400 exercises. The different types of exercises include:

- algebraic manipulation
- connecting algebra to geometry
- interpretation of graphs
- graphical representations
- writing to learn
- data analysis
- foundation for calculus
- descriptively titled applications

This variety provides sufficient flexibility to emphasize the skills most needed for each student or class.

Each exercise set in Chapters 1–11 begins with the Quick Review feature, and the traditional exercises that follow are carefully graded from routine to challenging. An additional block of exercises, Extending the Ideas, may be used in a variety of ways, including group work. We also provide Review Exercises at the end of each chapter, except the Prerequisite Chapter. The following sections offer a closer look at some of the new types of material that appear in this edition.

Balanced approach

This application about the growth of Medicare asks students first to solve the problem graphically, then to confirm the results algebraically and finally to interpret the results. Real applications aid the study of functions and help students recognize the importance of a full graphical, numerical, and algebraic understanding of the problem.

For a closer look at this feature see page 154.

Table 2.3 Medicare Data

Year	Medicare Charges (millions of dollars)
1970	2,310
1980	9,011
1985	17,743
1989	26,274
1990	30,447

Source: U.S. Health Care Financing Administration, unpublished data.

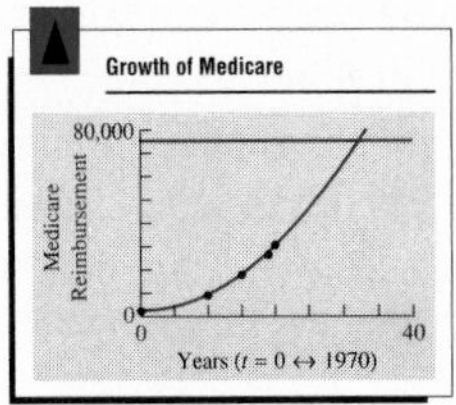

Figure 2.16 Graphs of $y_1 = 71.08x^2 + 47.77x + 2329.3$ and $y_2 = 75{,}000$.

■ EXAMPLE 10 APPLICATION: Modeling the Growth of Medicare

The Medicare charges by physicians between 1970 and 1990 (in millions of dollars) shown in Table 2.3 can be modeled by the quadratic function $f(x) = 71.08x^2 - 47.77x + 2329.3$, where $x = 0$ represents 1970, $x = 1$ represents 1971, and so on. Assuming the suggested trend continues, in what year will physician charges reach \$75 billion (or \$75,000 million)?

Solution

Solve Graphically

Reproduce the graphs in Figure 2.16 on your grapher, and show that $y_1 = 71.08x^2 - 47.77x + 2329.3$ is below $y_2 = 75{,}000$ when $x = 32$, and above 75,000 when $x = 33$.

Confirm Algebraically

Use the quadratic formula to solve $71.08x^2 - 47.77x + 2329.3 = 75{,}000$, or equivalently, $71.08x^2 - 47.77x - 72{,}670.7 = 0$ in standard form.

$$x = \frac{-b \pm \sqrt{b^2 - 4ac}}{2a}$$

$$x = \frac{47.77 \pm \sqrt{(-47.77)^2 - 4(71.08)(-72{,}670.7)}}{2(71.08)}$$

$$x = 32.312 \cdots$$

or $$x = -31.640 \cdots$$

Interpret

If the pattern of charges continues, physicians will charge \$75 billion to Medicare by the 33rd year after 1970, or in the year 2003. ■

Using the grapher as a tool to visualize and solve problems

This graphing activity enables students to use parametric graphing to visualize all the roots of an equation, including its complex roots.

For a closer look at this feature see page 565.

▼ GRAPHING ACTIVITY Visualizing the Roots of Unity

Set your grapher in parametric mode with $0 \leq T \leq 8$, so that Tstep = 1, Xmin = −2.3, Xmax = 2.3, Ymin = −1.5, and Ymax = 1.5.

1. Let $x_1 = \cos(2\pi/8)t$ and $y_1 = \sin(2\pi/8)t$. Use "trace" to visualize the eight eighth roots of unity. We say that $2\pi/8$ *generates* the eighth roots of unity. (Try both dot mode and connected mode.)
2. Replace $2\pi/8$ in Step 1 by other eighth roots of unity. Do any others *generate* the eighth roots of unity?
3. Repeat Steps 1 and 2 for the fifth, sixth, and seventh roots of unity.

What would you conjecture about an nth root of unity that generates all the nth roots of unity in the sense of Step 1? ▲

■ **EXAMPLE 8 APPLICATION: Modeling Women and Work**

Table 3.4 Women and Work

Year	Outside Job (%)
1974	35
1980	46
1985	51
1991	43
1994	45

Source: Roper Starch Worldwide as reported by Mary Cadden and Nick Galifianakis in *USA Today,* May 12, 1995.

Table 3.4 shows the percentage of U.S. women who say they would choose an outside job over being a homemaker. Find both a linear and a cubic regression equation for the data. Which regression equation appears to be the better model of the data?

Solution

Using $x = 0$ for 1970, $x = 1$ for 1971, and so on, we get the linear regression equation

$$y = 0.350x + 38.819$$

that is shown in Figure 3.26a. Figure 3.26b shows the cubic regression[3] equation

$$y = 0.0064x^3 - 0.3637x^2 + 6.2425x + 15.1147.$$

Each regression equation is superimposed on a scatter plot of the data.

The cubic regression equation appears to model the data better than does the linear regression. ■

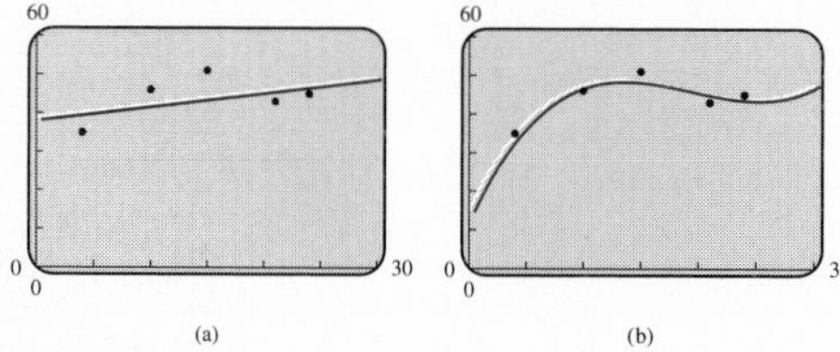

Figure 3.26

Follow-up

In Example 8, have students discuss which regression would be expected to provide a better model for the data over an extended period of time, for example, 1890 to 2050. **(Linear)**

Assignment Guide

Day 1: Ex. 3–45, multiples of 3
Day 2: Ex. 47, 50, 53, 56, 58, 59, 60, 62, 64, 65

Cooperative Learning

Group Activity: Ex. 59, 61

Notes on Exercises

Ex. 47–56 require students to find a polynomial function with particular

■ **EXAMPLE 9 APPLICATION: Modeling Women and Work**

Data tables

Tabular representation of data helps students build the connection between the numbers and the graphs.

Modeling real data problems

Students are given a data set and asked to use regression as a tool to model the data, fit curves to the data, and then use the curves to make predictions.

For a closer look at these features see page 245.

TECHNOLOGY TIP

When you zoom in at a local maximum or minimum point, the graph often appears flat or nearly flat. To prevent this tendency do any one of the following:

1. Change *Y*Fact so that it is larger than *X*Fact.
2. Use a low flat rectangle with *zoom-box.*
3. Manually set *Y*max and *Y*min to be closer together.

The graph of $f(x) = x^4 - 7x^2 + 6x$ (Figure 1.43) suggests that f has two local minimum values and one local maximum value.

Using grapher methods,[6] we find that the local minimum values occur at $x \approx -2.06$ (where $y \approx -24.06$) and $x \approx 1.60$ (where $y \approx -1.77$). The local maximum occurs at $x \approx 0.46$ (where $y \approx 1.32$).

Support Numerically

Trace across the graph. Watch the y-values decrease as you approach a local minimum from the left, then increase as you move away on the right. ■

[6] One grapher method is based on *zoom-in.* Another method (available on some graphers) uses the built-in programs for finding local maximum and local minimum values.

Technology tips

Tips throughout the text offer advice to students on using their grapher to obtain the best, most accurate results.

For a closer look at this feature see page 109.

City B
City A
Temperature
0
24
Time

Figure 1.44

■ **EXAMPLE 3 Interpreting Graphs**

The graph in Figure 1.44 shows the air temperature changes over a 24-hour period in two different cities. All of the following are true.

a. The temperature in city B decreases, then increases, and finally decreases.
b. The largest rate of temperature decrease occurs in city A during the first few hours.
c. The largest rate of temperature increase occurs in city B in the middle of the 24-hour period.
d. In city B a local minimum temperature occurs in the first half and a local maximum temperature occurs in the second half of the 24-hour period.
e. The temperature reaches a local minimum twice during the 24-hour period in city A. ■

To describe the increasing and decreasing behavior of a function, we list the intervals on which it is doing each. To describe the local maximum and minimum values of a function, we list the points at which they occur. Finding these points first can help identify the intervals of increase and decrease.

Reading from and interpreting graphs

Throughout the text, students are encouraged to visualize problems graphically. Examples are presented in realistic settings to encourage students to investigate and understand mathematical concepts.

For a closer look at this feature see page 110.

Remarks provide additional insight and help students avoid common pitfalls

These provide guidance for students, helping them avoid potential algebraic and grapher errors.

For a closer look at this feature see pages 601–602.

sider that conic sections need not have an axis of symmetry parallel to either axis.

REMARK

It is not necessary to simplify y_1 and y_2 as we did in Example 7. The trade-off is fewer pencil and paper errors at the risk of more keystroking errors.

■ **EXAMPLE 7** **Graphing a Second-Degree Equation**

Identify and graph the conic defined by the equation

$$x^2 + 4xy + 4y^2 - 30x - 90y + 450 = 0.$$

Solution

$A = 1$, $B = 4$, $C = 4$, so $B^2 - 4AC = 0$ and the graph is a parabola. To use a grapher we first rewrite the equation as a quadratic in y:

$$4y^2 + (4x - 90)y + (x^2 - 30x + 450) = 0$$

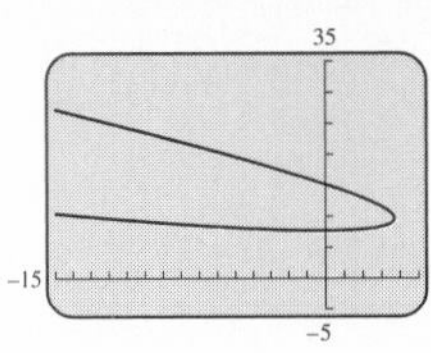

Figure 9.34

Then, we solve for y using the quadratic formula:

$$y = \frac{-(4x - 90) \pm \sqrt{(4x - 90)^2 - 16(x^2 - 30x + 450)}}{8}$$

or

$$y_1 = \frac{45 - 2x + 8\sqrt{225 - 60x}}{4}$$

and

$$y_2 = \frac{45 - 2x - 8\sqrt{225 - 60x}}{4}.$$

The graph of this parabola is shown in Figure 9.34. Notice that its axis is not parallel to either axis (because of the $4xy$ term). ■

Application exercises

Descriptively titled application exercises ask students to model data and use their problem-solving skills in determining how to reach a solution.

Data analysis

Students are asked to use data analysis skills to reach a solution.

For a closer look at these features see page 328.

In Exercises 53–58, solve the problem.

53. *Atmospheric Pressure* Use the model from Example 7 to find the altitude above Atlantic City with an atmospheric pressure of 2 lb/in.^{2}.

54. *Atmospheric Pressure* Use the model from Example 7 to find the atmospheric pressure 40 mi above Colorado Springs (1 mi above sea level).

55. *Penicillin Use* The use of penicillin became so widespread in the 1980s in Hungary that it became practically useless against common sinus and ear infections. Now the use of more effective antibiotics has caused a decline in penicillin resistance. The bar graph shows the use of penicillin in Hungary for selected years.

a. From the bar graph we read the data pairs to be approximately (1, 11), (8, 6), (15, 4.8), (16, 4), and (17, 2.5), using $t = 1$ for 1976, $t = 8$ for 1983, and so on. Complete a scatter plot for these data.

b. **Writing to Learn** Discuss whether the bar graph shown or the scatter plot that you completed best represents the data and why.

Nationwide Consumption of Penicillin

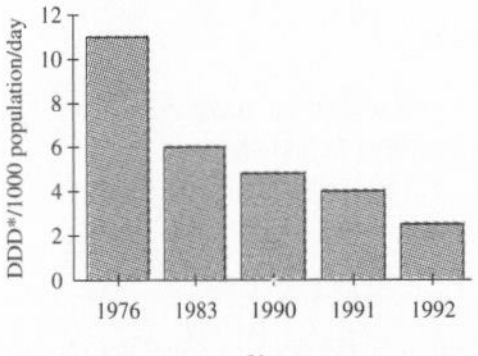

*Defined Daily Dose
Source: Science, vol. 264, April 15, 1994, American Association for the Advancement of Science.

For Exercise 55

56. **Writing to Learn** Find both a logarithmic and a linear regression equation for the data in Exercise 55. Discuss which equation provides the best fit for the data. Which equation would you use to estimate the daily dose for 1993? Why?

57. *Estimating Population Growth* The population of Alaska (in millions) for three known years is indicated in Table 4.12.

a. Find an exponential regression equation for these data. (Use $t = 0$ to represent 1960.)

b. Use the regression to predict the population in the year 2000.

Table 4.12 Population of Alaska

Year	Population (millions)
1960	0.23
1980	0.40
1990	0.55

Source: The Statesman's Yearbook, 129th ed. (London: The Macmillan Press, Ltd., 1992).

c. Estimate when the population of the state was twice the 1960 figure.

58. *Estimating Population Growth* The population of California (in millions) for three known years is indicated in Table 4.13.

Table 4.13 Population of California

Year	Population (millions)
1960	15.72
1980	23.67
1990	29.76

Source: The Statesman's Yearbook, 129th ed. (London: The Macmillan Press, Ltd., 1992).

a. Find an exponential regression equation for these data. (Use $t = 0$ to represent 1960.)

b. Use the regression to predict the population in the year 2000.

c. Predict when the population of the state would be twice the 1960 figure.

EXTENDING THE IDEAS

In Exercises 59–62, solve the problem.

59. *Normal Distribution* The function defined by

$$f(x) = ke^{-cx^2},$$

where c and k are constants, is a bell-shaped curve useful in probability. Find the maximum value of f in the case where $k = 1.30$ and $c = 5.31$.

60. *Normal Distribution* For the function f in Exercise 59 solve the inequality $f(x) \geq 0.5$.

$$P(t) = \frac{800}{1 + 49e^{-0.2t}}.$$

a. What was the initial number of infected students?
b. When will the number of infected students be 200?
c. The school will close when 300 of the 800-student body are infected. When will the school close?

48. *Population of Deer* The population of deer after t years in Cedar State Park is modeled by the function

$$P(t) = \frac{1001}{1 + 90e^{-0.2t}}.$$

a. What was the initial population of deer?
b. When will the number of deer be 600?
c. What is the maximum number of deer possible in the park?

49. *Newton's Law of Cooling* A cup of coffee has cooled from 92° C to 50° C after 12 min in a room at 22° C. How long will the cup take to cool to 30° C?

50. *Newton's Law of Cooling* A cake is removed from an oven at 350° F and cools to 120° F after 20 min in a room at 65° F. How long will the cake take to cool to 90° F?

52. *Newton's Law of Cooling Experiment* A thermometer was removed from a cup of hot chocolate and placed in water whose temperature $T_m = 0°$ C. The data in Table 4.11 were collected over the next 30 sec.

a. Find a scatter plot of the data $T - T_m$.
b. Find an exponential regression equation for the $T - T_m$ data. Superimpose its graph on the scatter plot.
c. Estimate the thermometer reading when it was removed from the hot chocolate.

Table 4.11 Experimental Data

Time t	Temp T	$T - T_m$
2	74.68	74.68
5	61.99	61.99
10	34.89	34.89
15	21.95	21.95
20	15.36	15.36
25	11.89	11.89
30	10.02	10.02

Foundation for calculus

These exercises build intuitive foundation for calculus.

For a closer look at this feature see page 327.

34.

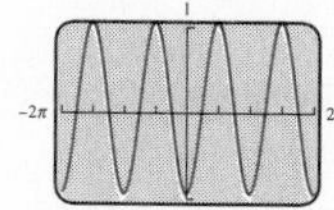

In Exercises 35 and 36, use sum or difference identities to solve the equation exactly.

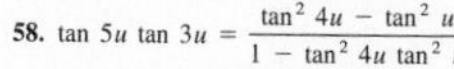

35. $\sin 2x \cos x = \cos 2x \sin x$

36. $\cos 3x \cos x = \sin 3x \sin x$

EXTENDING THE IDEAS

In Exercises 51–59, confirm the identity.

51. $\sin(x - y) + \sin(x + y) = 2 \sin x \cos y$

52. $\cos(x - y) + \cos(x + y) = 2 \cos x \cos y$

53. $\cos 3x = \cos^3 x - 3 \sin^2 x \cos x$

54. $\sin 3u = 3 \cos^2 u \sin u - \sin^3 u$

55. $\cos 3x + \cos x = 2 \cos 2x \cos x$

56. $\sin 4x + \sin 2x = 2 \sin 3x \cos x$

57. $\tan(x + y) \tan(x - y) = \dfrac{\tan^2 x - \tan^2 y}{1 - \tan^2 x \tan^2 y}$

Extending the Ideas

Extending the Ideas are blocks of exercises that can be used in a variety of ways in class, either as groupwork or to challenge better students.

58. $\tan 5u \tan 3u = \dfrac{\tan^2 4u - \tan^2 u}{1 - \tan^2 4u \tan^2 u}$

59. $\cos(x + y + z) = \cos x \cos y \cos z - \sin x \sin y \cos z - \sin x \cos y \sin z - \cos x \sin y \sin z$

In Exercises 60–63, solve the problem.

60. Writing to Learn The figure shows graphs of $y_1 = \cos 5x \cos 4x$ and $y_2 = -\sin 5x \sin 4x$ in one viewing window. Discuss the question, "How many solutions are there to the equation $\cos 5x \cos 4x = -\sin 5x \sin 4x$ in the interval $[-2\pi, 2\pi]$?" Give an algebraic argument that answers the question more convincingly than the graph does. Then support your argument with an *appropriate* graph.

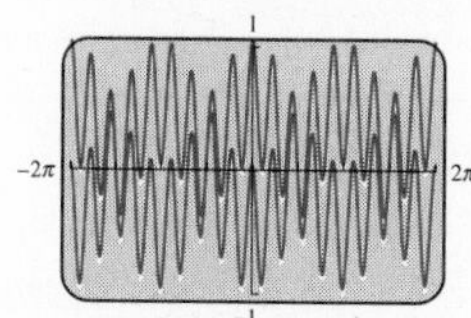

For Exercise 60

61. Writing to Learn Solve $\sin 3x \cos x = \cos 3x \sin x$ both algebraically and graphically. Then discuss the advantages and disadvantages of each method, and describe which method you prefer in this instance and why.

62. *Harmonic Motion* Alternating electric current, an oscillating spring, or any other harmonic oscillator can be modeled by the equation

$$x = a \cos\left(\frac{2\pi}{T}t + \delta\right),$$

where T is the time for one period and δ is the phase constant. Show that this motion can also be modeled by the following sum of cosine and sine, each with zero phase constant:

$$a_1 \cos\left(\frac{2\pi}{T}\right)t + a_2 \sin\left(\frac{2\pi}{T}\right)t,$$

where $a_1 = a \cos \delta$ and $a_2 = -a \sin \delta$.

63. *Magnetic Fields* A magnetic field B can sometimes be modeled as the sum of an incident and a reflective field as

$$B = B_{in} + B_{ref},$$

where
$$B_{in} = \frac{E_0}{c}\cos\left(\omega t - \frac{\omega x}{c}\right), \text{ and}$$
$$B_{ref} = \frac{E_0}{c}\cos\left(\omega t + \frac{\omega x}{c}\right).$$

Show that
$$B = 2\frac{E_0}{c}\cos \omega t \cos \frac{\omega x}{c}.$$

Writing to Learn

Writing to Learn exercises give students practice at communicating about mathematics and at demonstrating their understanding of important ideas.

For a closer look at these features see pages 510–511.

Review Exercises

The Review Exercises contain the full range of exercises covered in the chapter. These exercises give students additional practice with the ideas developed in the chapter.

For a closer look at this feature see page 130.

REVIEW EXERCISES

In Exercises 1 and 2, determine which ordered pair is a solution for the equation.

1. $y = \sqrt{x^2 + 1}$
a. $(0, 1)$
b. $(-1, 0)$

2. $3x - 2y = 4$
a. $(-2, 1)$
b. $(2, 1)$

In Exercises 3–6, complete the table of solutions for the given equation.

3. $y = -3x + 1$

x	-2	-1	?	3	?
y	7	?	-11	?	1

4. $y = 2x - 5$

x	-3	2	?	-4	?
y	-11	?	-7	?	1

5. $f(x) = \dfrac{x}{x - 2}$

x	-1	-2	0	2	?
$f(x)$	$\frac{1}{3}$	?	?	?	-1

6. $g(x) = x^3 - 2x + 1$

x	2	-2	0	1	4
$g(x)$	5	?	?	?	?

In Exercises 7–12, make a table of solutions for the equation for integer values of x from $x = -3$ to $x = 3$. Sketch the graph of the equation.

7. $y = 4x - 1$
8. $y = 2 - 3x$
9. $y = |x - 3|$
10. $y = \sqrt{x - 2}$
11. $y = 3 - x^2$
12. $y = x^2 - 3x$

In Exercises 13–16, match the equation with its graph. Identify the viewing-window dimensions and the values for Xscl and Yscl.

13. $y = 1 + x - x^3$
14. $y = |x - 2| - 1$

15. $f(x) = \begin{cases} \sqrt{1 - x} & \text{if } x < 1 \\ x^2 - 1 & \text{if } x \geq 1 \end{cases}$

16. $f(x) = \begin{cases} -3x + 5 & \text{if } x < 2 \\ x - 3 & \text{if } x \geq 2 \end{cases}$

Problem-solving approach

The following variation of Polya's problem-solving process is used throughout the text: *Develop* a mathematical model; *solve* the mathematical model and support or confirm the solution; and *interpret* the solution. Students are encouraged to use this process.

For a closer look at this feature see page 120.

Teaching Note

The problems introduced in this section are similar to those that will appear throughout the rest of the book. Business application problems (interest, cost, revenue, break-even point) and problems involving mensuration formulas (perimeter, area, volume) are emphasized in this section. Discuss the terms associated with the business application problems and then work one problem of each type. A mensuration problem should be done to see if students remember key geometric concepts.

■ **EXAMPLE 1 APPLICATION: Modeling Profit**

Dependable Auto Rental has fixed annual overhead costs of \$185,000. Each year it generates revenue of \$8540 per car and has operating costs of \$3050 per car. How many cars are needed to earn a profit of \$1,500,000?

Solution

Model

Word Statement: profit = revenue − total cost

$$\begin{aligned} x &= \text{number of cars} \\ 8540x &= \text{revenue} \\ 3050x &= \text{operating cost} \\ 185{,}000 &= \text{overhead cost} \\ 185{,}000 + 3050x &= \text{total cost} \\ \text{Profit} &= 8540x - (185{,}000 + 3050x) \end{aligned}$$

Solve Algebraically

$$\begin{aligned} 1{,}500{,}000 &= 8540x - (185{,}000 + 3050x) \quad \text{Equation to be solved} \\ 1{,}500{,}000 &= 8540x - 185{,}000 - 3050x \\ 1{,}685{,}000 &= 8540x - 3050x \\ 5490x &= 1{,}685{,}000 \\ x &= \frac{1{,}685{,}000}{5490} \\ &= 306.921\cdots \end{aligned}$$

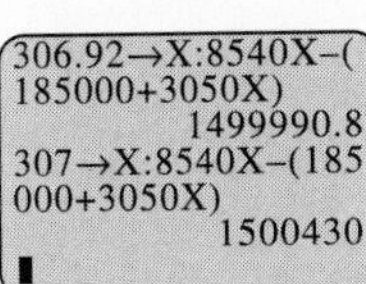

Figure 1.54

Support Numerically

When 306.92 is stored in x and $8540x - (185{,}000 + 3050x)$ is evaluated, we get a value very close to 1.5 million. See the first computation in Figure 1.54.

Interpret

We seek an integer solution. Dependable Auto Rental needs to have 307 cars to make a profit of at least \$1,500,000. See the second computation in Figure 1.54, and check that 306 is too small. ■

In Example 2, drawing a picture of the situation helps to formulate the word statement of the problem.

Quick Review 1.3

In Exercises 1–4, solve for x.

1. $2x - 3 = 3(4x - 1)$ **2.** $5 - 3x = 2(x + 3)$

3. $3(1 - 2x) + 4(2x - 5) = 7$

4. $2(7x + 1) = 5(1 - 3x)$

In Exercises 5–8, solve for y when $x = -2$.

5. $3x - 4y = 22$ **6.** $2x + 5y = 19$

7. $5x^2 + 7y = 48$ **8.** $3x^2 = 14 - 3y$

In Exercises 9 and 10, simplify the fraction.

9. $\dfrac{9 - 5}{-2 - (-8)}$ **10.** $\dfrac{-4 - 6}{-14 - (-2)}$

In Exercises 11–14, simplify the expression.

11. $4 - [x - (5 - 3x)]$ **12.** $-\{3 - [-(x + 1)]\}$

13. $-\dfrac{-(x - 1)}{1 - x}$ **14.** $\dfrac{x - 4}{8 - 2x}$

In Exercises 15–18, solve for y.

15. $2x - 5y = 21$

16. $2x + y = 17 + 2(x - 2y)$

17. $x^2 + y = 3x - 2y$

18. $y(x^2 + 3) = 3y + 2$

Quick Review exercises

All sections except those in the Prerequisite Chapter begin with a set of Quick Review exercises. These exercises help students review and reinforce previously learned algebraic techniques necessary to complete the section exercise set.

In Exercises 9 and 10, solve the problem.

9. *Real Estate Appreciation* Bob Michaels purchased a house 8 years ago for \$42,000. This year it was appraised at \$67,500.

a. A linear equation $V = mt + b$, $0 \le t \le 15$, models the value of this house for 15 years after it was purchased. Find the slope and y-intercept, and then write the equation.

b. Graph the equation and trace to estimate in how many years after purchase will this house be worth \$72,500.

c. Write and solve an equation to determine how many years after purchase this house will be worth \$74,000.

d. Determine how many years after purchase this house will be worth \$80,250.

10. *Investment Planning* Mary Ellen plans to invest \$18,000, putting part of the money (x) into a savings account that pays 5% annually and the rest into an account that pays 8%

In Exercises 23–30, find an equation in point-slope form of the line through the given point with the given slope. Simplify your answer to slope-intercept form.

	Point	*Slope*		*Point*	*Slope*
23.	$(1, 4)$	$m = 2$	**24.**	$(-3, -5)$	$m = \frac{1}{2}$
25.	$(-4, 3)$	$m = -\frac{2}{3}$	**26.**	$(2, 7)$	$m = -\frac{3}{4}$
27.	$(7, 12)$	$m = \frac{3}{5}$	**28.**	$(-3, 4)$	$m = 3$
29.	$(-1, -5)$	$m = \frac{3}{7}$	**30.**	$(5, -4)$	$m = -2$

In Exercises 31–36, find an equation of the line that contains the two given points. Write the equation in the general form $Ax + By + C = 0$.

31. $(-1, 2)$ and $(2, 5)$ **32.** $(4, -1)$ and $(1, 5)$

33. $(-7, -2)$ and $(1, 6)$ **34.** $(-3, -8)$ and $(4, -1)$

35. $(1, -3)$ and $(5, -3)$ **36.** $(-1, -5)$ and $(-4, -2)$

Algebra skills

These exercises require traditional algebraic manipulation.

For a closer look at these features *see pages 90–92.*

40. The line through the points $(4, 2)$ and $(-3, 1)$

41. The line $2x + 5y = 12$

42. The line $7x - 12y = 96$

In Exercises 43 and 44, the lines contain the origin and the point in the upper right corner of the grapher screen.

43. Writing to Learn Which line shown here has the greater slope? Explain.

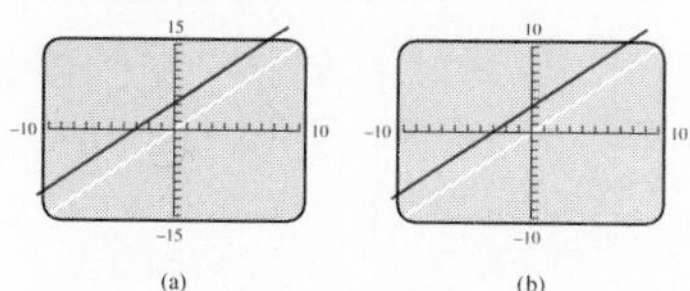

For Exercise 43

44. Writing to Learn Which line shown here has the greater slope? Explain.

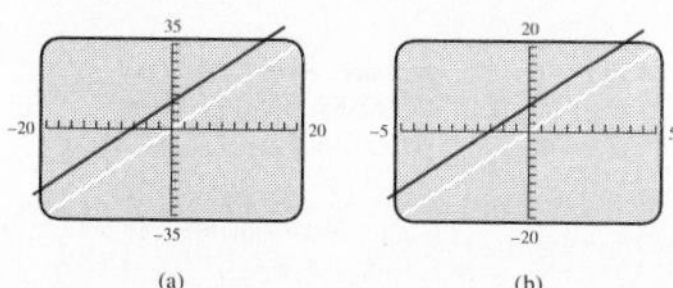

For Exercise 44

In Exercises 45–48, match the equation with one of the graphs in Exercises 43 and 44.

45. $y = x$ **46.** $y = \frac{7}{4}x$

47. $y = 4x$ **48.** $y = \frac{3}{2}x$

In Exercises 49–54, find the value of x and the value of y for which $(x, 14)$ and $(18, y)$ are points on the graph.

49. $y = 0.5x + 12$ **50.** $y = -2x + 18$

51. $x + 2y = 12$ **52.** $3x - 2y = 14$

In Exercises 55 and 56, use the graph provided. Estimate the slope of the line.

55.

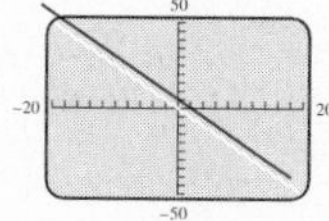

56.

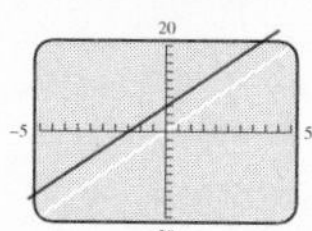

In Exercises 57–60, find the values for *Y*min, *Y*max, and *Y*scl that will make the graph of the equation appear in the viewing window as shown here.

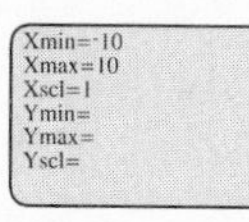

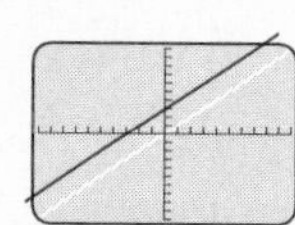

For Exercises 57–60

57. $y = 3x$ **58.** $y = 5x$

59. $y = \frac{2}{3}x$ **60.** $y = \frac{5}{4}x$

In Exercises 61–64, write two equations: one for the line passing through the given point and parallel to the given line, and another for the line passing through the given point and perpendicular to the given line. Support your work graphically.

	Given Point	*Given Line*
61.	$(1, 2)$	$y = 3x - 2$
62.	$(-2, 3)$	$y = -2x + 4$
63.	$(3, 1)$	$2x + 3y = 12$
64.	$(6, 1)$	$3x - 5y = 15$

In Exercises 65–71, solve the given problem.

Graphical representations

These exercises develop students' skills at recognizing graphical representation of algebra.

For a closer look at this feature *see page 92.*

PREREQUISITE CHAPTER: ALGEBRA REVIEW

BIBLIOGRAPHY

For students: *Graphing Calculator Activities for Algebra,* Miller et al. Brooks/Cole Publishing Co., 1992.

For teachers: *Algebra in a Technological World, Addenda Series, Grades 9–12.* National Council of Teachers of Mathematics, 1995.

Innumeracy: Mathematical Illiteracy and Its Consequences, John Allen Paulos. Hill and Wang, 1988. Available through Dale Seymour Publications (XS02001).

Objective

Students will be able to classify real numbers, convert between decimals and fractions, write inequalities, and use the distance formula and the midpoint formula.

Motivate

Ask students how real numbers can be classified.

1 REAL NUMBER SYSTEM

Real Numbers • Real Number Line • Ordering Real Numbers • Intervals of Real Numbers • Absolute Value of a Real Number • Distance between Two Real Numbers • Midpoint Formula

Real Numbers

A **real number** is any number that can be written as a decimal. Real numbers are used in mathematics, science, and daily life to represent quantities or measurements such as the balance in your bank account, your height, age, distance, temperature, miles per gallon, and so forth. Here are some of the symbols we use to represent real numbers:

$$6, \quad -8, \quad 0, \quad 5.25, \quad 2.333\ldots, \quad \tfrac{8}{5},$$
$$\sqrt{3}, \quad \sqrt[3]{16}, \quad e, \quad \text{and} \quad \pi$$

The set of real numbers contains several important subsets:

The **natural numbers:**	$\{1, 2, 3, \ldots\}$
The **whole numbers:**	$\{0, 1, 2, 3, \ldots\}$
The **integers:**	$\{\ldots, -3\ -2, -1, 0, 1, 2, 3, \ldots\}$

The rational numbers are another important subset of the real numbers. A **rational number** is any number that can be written as a ratio a/b of two integers, where $b \neq 0$. The **numerator** of a/b is a, and the **denominator** is b. Here

REMARK

For any rational number a/b,

$$\frac{-a}{b} = \frac{a}{-b} = -\frac{a}{b}.$$

Also, a/b, $a \div b$, and $\frac{a}{b}$ all have the same meaning.

is a convenient way to represent the set of all rational numbers. (The vertical bar that follows a/b is read "such that.")

$$\text{The rational numbers:} \qquad \left\{ \frac{a}{b} \,\middle|\, a,\ b \text{ are integers, and } b \neq 0 \right\}$$

For example, here is the list of all rational numbers with denominator 3:

$$\left\{ \ldots, \frac{-3}{3}, \frac{-2}{3}, \frac{-1}{3}, \frac{0}{3}, \frac{1}{3}, \frac{2}{3}, \frac{3}{3}, \ldots \right\}$$

We can use division to find the decimal form of a rational number. For example,

$$\frac{2}{3} = 0.666\cdots, \qquad \frac{43}{40} = 1.075, \qquad \text{and} \qquad \frac{21}{37} = 0.567567\cdots.$$

The decimal form of a rational number either terminates like $\frac{1}{4} = 0.25$ or is infinite repeating like $\frac{4}{11} = 0.3636\ldots = 0.\overline{36}$. The bar over the 36 indicates the block of digits that repeats.

Sometimes we can use a calculator to determine whether the decimal form of a rational number repeats or terminates, as illustrated in Example 1.

■ EXAMPLE 1 Determining Decimal Forms of Rational Numbers

Determine whether the decimal form of 1/16 repeats or terminates. Do likewise for 55/27.

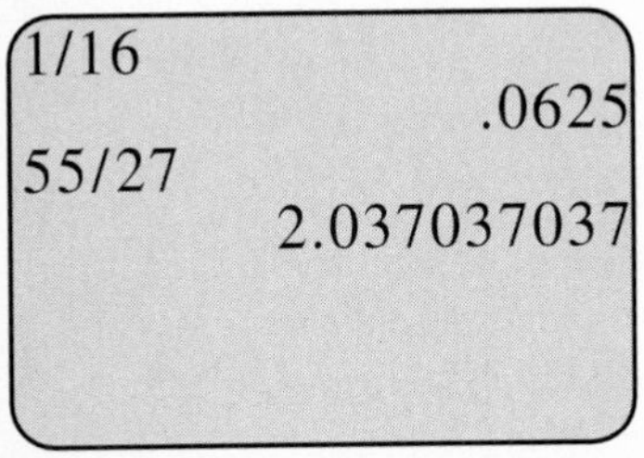

Solution

Figure 1 suggests that the decimal form for 1/16 terminates and that of 55/27 repeats in blocks of 037.

$$\frac{1}{16} = 0.0625 \qquad \text{and} \qquad \frac{55}{27} = 2.\overline{037}$$ ■

```
1/16
                .0625
55/27
         2.037037037
```

Figure 1

Every real number has a decimal form. A real number is **irrational** if it is not rational. Thus, an irrational number is *not* a ratio of two integers and its decimal form *neither* terminates *nor* repeats; its decimal representation is *infinite nonrepeating.*

Two familiar examples of irrational numbers are π and $\sqrt{2}$. By using a calculator we find

$$\pi \approx 3.14159265 \qquad \text{and} \qquad \sqrt{2} \approx 1.41421356.$$

Notice that a calculator gives only a few of the digits of the infinite decimal form. The symbol $\approx$ means "is approximately equal to." To indicate the exact value of π, we write $\pi = 3.14159265 \cdots$.

Real Number Line

The real numbers and the points of a line can be matched *one-to-one* to form a **real number line.** We start with a horizontal line and match the real number zero with a point O, the **origin. Positive numbers** are assigned to the right of the origin, and **negative numbers** to the left, as shown in Figure 2.

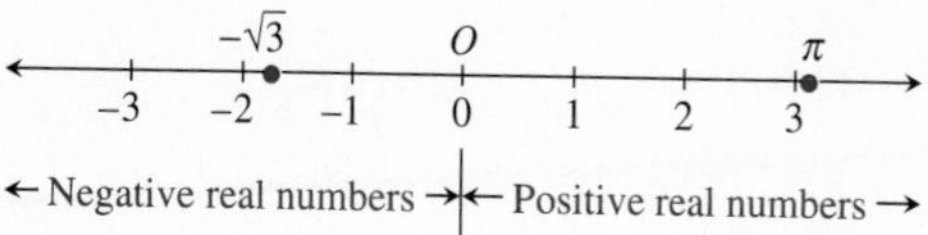

Figure 2 The real number line.

Every real number corresponds to one and only one point on the real number line, and every point on the real number line corresponds to one and only one real number. The number associated with a point is the **coordinate of the point.** The standard convention is to use the real number for both the name of the point and its coordinate. As long as the context is clear, we will follow this convention.

Ordering Real Numbers

The set of real numbers is **ordered**. This means we can compare two numbers using **inequalities** and say that one is "less than" or "greater than" the other.

Order of the Real Numbers

Let a and b be any two real numbers.

Symbol	Definition	Read
$a > b$	$a - b$ is positive	a is greater than b.
$a < b$	$a - b$ is negative	a is less than b.
$a \geq b$	$a - b$ is positive or zero	a is greater than or equal to b.
$a \leq b$	$a - b$ is negative or zero	a is less than or equal to b.

The symbols $>$, $<$, $\geq$, and $\leq$ are **inequality symbols.**

REMARK
If $a < 0$, then a is to the left of 0 on the number line, and $-a$ is to the right of 0. Thus, $-a > 0$.

Geometrically, $a > b$ means that a is to the right of b (equivalently b is to the left of a) on the real number line. For example, since $6 > 3$, 6 is to the right of 3 on the real number line. Note also that $a > 0$ means that $a - 0$, or simply a, is positive and $a < 0$ means that a is negative.

We are able to compare any two real numbers because of the following important property of the real numbers.

Trichotomy Property

Let a and b be any two real numbers. Exactly one of the following is true:

$$a < b, \qquad a = b, \qquad \text{or} \qquad a > b.$$

■ **EXAMPLE 2 Ordering Real Numbers**

Replace ○ with <, =, or > to make a true statement.

a. $4 \bigcirc -1$ **b.** $-3 \bigcirc 5$ **c.** $\frac{-6}{3} \bigcirc -2$

Solution

a. $4 > -1$ because $4 - (-1) = 5$ is positive.
b. $-3 < 5$ because $-3 - 5 = -8$ is negative. Notice also that -3 is to the left of 5 on the number line.

c. $\frac{-6}{3} = -\frac{6}{3} = -2$ ■

Intervals of Real Numbers

Inequalities can be used to describe **intervals** of the real numbers, as illustrated by Example 3.

■ **EXAMPLE 3 Interpreting Inequalities**

Describe and graph the interval of real numbers for the inequality.

a. $x < 3$ **b.** $x \geq -2$ **c.** $-1 < x \leq 4$

Solution

a. The inequality $x < 3$ describes all real numbers less than 3 (Figure 3a).
b. The inequality $x \geq -2$ describes all real numbers greater than or equal to -2 (Figure 3b).

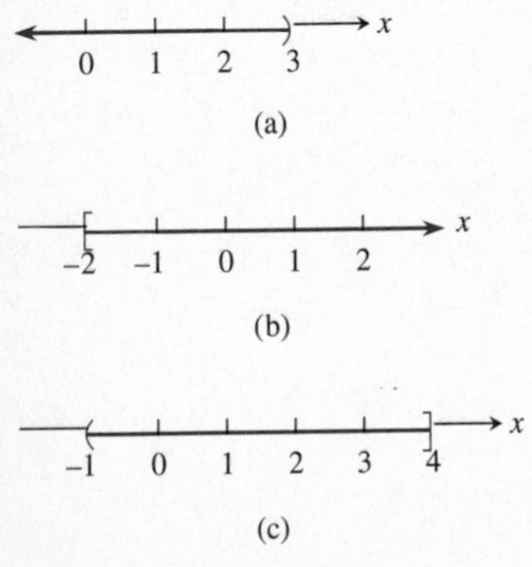

Figure 3 Notice in graphs of inequalities, that the parentheses correspond to < and > and the brackets to ≤ and ≥.

Teaching Note

A mnemonic device that may help students remember the notation for open and closed intervals is the shape of the square bracket, which includes a "ledge" that prevents the endpoint from "falling out" of the interval.

c. The *double inequality* $-1 < x \leq 4$ represents all real numbers between -1 and 4, excluding -1 and including 4 (Figure 3c). ■

EXAMPLE 4 Writing Inequalities

Describe the interval of real numbers using an inequality.

a. The real numbers less than or equal to 3.6
b. The real numbers between -4 and -0.5
c. The real numbers greater than or equal to zero

Solution

a. $x \leq 3.6$
b. $-4 < x < -0.5$
c. These are the **nonnegative** real numbers. The corresponding inequality is $x \geq 0$. ■

As shown in Example 3, inequalities define *intervals* on the real number line. We often find it convenient to use the notation [2, 5] to describe the *bounded interval* determined by $2 \leq x \leq 5$. This interval is **closed** because it contains its *endpoints* 2 and 5. Here are four types of **bounded intervals.**

Bounded Intervals of Real Numbers

Let a and b be real numbers with $a < b$.

Interval Notation	Interval Type	Inequality Notation	Graph
$[a, b]$	Closed	$a \leq x \leq b$	[a ... b]
(a, b)	Open	$a < x < b$	(a ... b)
$[a, b)$	Half-open	$a \leq x < b$	[a ... b)
$(a, b]$	Half-open	$a < x \leq b$	(a ... b]

The numbers a and b are the **endpoints** of each interval.

The interval of real numbers determined by the inequality $x < 2$ can be described by the *unbounded interval* $(-\infty, 2)$. This interval is **open** because it does not contain its endpoint 2.

We may also use the interval notation $(-\infty, \infty)$ to represent the entire set of real numbers. This interval is both *open and closed.* The symbols $-\infty$ (*negative infinity*) and ∞ (*positive infinity*) allow us to use interval notation for unbounded intervals and do *not* represent real numbers. Here are four types of **unbounded intervals.**

Unbounded Intervals of Real Numbers

Let a and b be real numbers.

Interval Notation	Interval Type	Inequality Notation	Graph
$[a, \infty)$	Closed	$x \geq a$	
(a, ∞)	Open	$x > a$	
$(-\infty, b]$	Closed	$x \leq b$	
$(-\infty, b)$	Open	$x < b$	

Each of these intervals has exactly one endpoint, namely a or b.

■ EXAMPLE 5 Converting between Intervals and Inequalities

Convert inequality notation to interval notation or vice versa. State the interval type.

a. $[-6, 3)$ **b.** $(-\infty, 2)$ **c.** $1 \leq x \leq 5$

Solution

a. The interval $[-6, 3)$ corresponds to $-6 \leq x < 3$, and is half-open.
b. The interval $(-\infty, 2)$ corresponds to $x < 2$, and is open.
c. The inequality $1 \leq x \leq 5$ corresponds to the closed interval $[1, 5]$. ■

Absolute Value of a Real Number

The *absolute value* of a real number suggests its **magnitude** (size). For example, the absolute value of 3 is 3, and the absolute value of -5 is 5.

Alert

Many students think that $|-a|$ is equal to a. Show them that $|-(-5)| \neq -5$.

Definition: Absolute Value

The **absolute value** of a real number a is

$$|a| = \begin{cases} a & \text{if } a > 0 \\ -a & \text{if } a < 0 \\ 0 & \text{if } a = 0 \end{cases}$$

■ **EXAMPLE 6** **Using the Definition of Absolute Value**

Evaluate.

a. $|-4|$
b. $|\pi - 6|$

Solution

a. Because $-4 < 0$, $|-4| = -(-4) = 4$.
b. Because $\pi \approx 3.14$, $\pi - 6$ is negative, so $\pi - 6 < 0$. Thus, $|\pi - 6| = -(\pi - 6) = 6 - \pi$. ■

It follows from the above definition that the absolute value of any nonzero real number a is positive. And, since $|a| = 0$ when $a = 0$, we conclude that $|a| \geq 0$. This and other important properties of absolute value are summarized here.

Properties of Absolute Value

Let a and b be real numbers.

1. $|a| \geq 0$ **2.** $|-a| = |a|$

3. $|ab| = |a||b|$ **4.** $\left|\dfrac{a}{b}\right| = \dfrac{|a|}{|b|}, \quad b \neq 0$

Distance between Two Real Numbers

The *distance* between -1 and 4 on the number line is 5 (see Figure 4). This distance may be found by subtracting the smaller number from the larger: $4 - (-1) = 5$. If we use absolute value, the order of the subtraction does not matter: $|4 - (-1)| = |-1 - 4| = 5$.

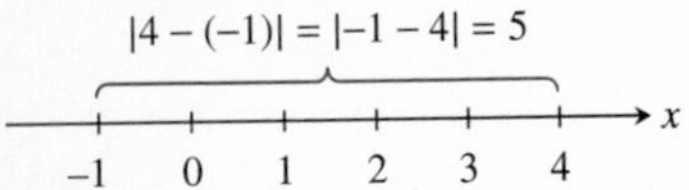

Figure 4

> **Distance Formula (Number Line)**
>
> Let a and b be real numbers. The **distance** between a and b is
>
> $$|a - b|.$$
>
> Note that $|a - b| = |b - a|$.

Teaching Note

It is sometimes helpful for students to see the distance between two points by plotting the points on a number line. For example, in Example 6b students can plot a point for 6 on the number line and then plot a point representing π, somewhere between 3 and 4. If a and b are the two points to be plotted on the number line and $a < b$, then the distance can always be written without absolute value notation as $b - a$. Since 6 is to the right of π on the number line, $6 - \pi$ is the correct representation.

Follow-up

Ask students whether $\dfrac{\pi}{\sqrt{3}}$ is a rational number. (**No**)

Assignment Guide

Day 1: Ex. 2, 5, 12, 16, 23, 27, 37, 41, 47, 54, 57, 64, 68, 71, 81, 84, 85, 88

Notes on Exercises

Ex. 15–20 and 90 require the student to order real numbers.
Ex. 23, 24, 27, 28, 33–38, 41, 42, and 45–54 provide practice with interval notation.
Ex. 25 and 85–88 relate to classifying real numbers.
Ex. 55–70 and 77–82 provide practice with absolute values.

Cooperative Learning

Group Activity: Ex. 83

■ EXAMPLE 7 Using Absolute Value to Find Distance

a. The distance between -2 and -9 is

$$|-2 - (-9)| = |-2 + 9| = 7.$$

b. The distance between $\sqrt{5}$ and 14 is

$$|\sqrt{5} - 14| = 14 - \sqrt{5} \approx 11.76.$$

c. "The distance between x and -3 is less than 9" is stated

$$|x - (-3)| < 9 \quad \text{or} \quad |x + 3| < 9.$$ ■

Midpoint Formula

When the endpoints of a segment in the number line are known, we take the average of their coordinates to find the midpoint of the segment, as suggested by the following formula.

> **Midpoint Formula (Number Line)**
>
> The **midpoint** of the line segment with endpoints a and b is
>
> $$\frac{a + b}{2}.$$

Ongoing Assessment

Self-Assessment: Ex. 7, 17, 33, 49, 55, 63, 73, 79

Embedded Assessment: 84, 88

■ **EXAMPLE 8 Finding the Midpoint of a Line Segment**

The midpoint of the line segment with endpoints −9 and 3 in a number line is

$$\frac{(-9) + 3}{2} = \frac{-6}{2} = -3.$$ ■

EXERCISES FOR SECTION 1

In Exercises 1–8, find the decimal form for the fraction. State whether it repeats or terminates.

1. −37/8 **2.** 25/16

3. 15/99 **4.** −5/11

5. −13/6 **6.** 49/12

7. 5/37 **8.** −5/27

In Exercises 9–14, plot the two numbers on the number line. Then find the distance between them.

9. −3/2, 5 **10.** 7/3, −5/3

11. $\sqrt{7}, \sqrt{2}$ **12.** −4.5, −0.5

13. −5/3, −9/5 **14.** 3/4, 5/7

In Exercises 15–20, replace ○ with <, =, or > to make a true statement.

15. −11/6 ○ −7/6 **16.** 12/3 ○ 20/5

17. −4.1 ○ −4.2 **18.** 3.2 ○ 3.3

19. −(−2) ○ 2 **20.** 1/6 ○ −1

In Exercises 21–26, graph the interval.

21. $x \le 2$ **22.** $-2 \le x < 5$

23. $(-\infty, 7)$ **24.** $[-3, 3]$

25. x is negative.

26. x is greater than or equal to 2 and less than or equal to 6.

In Exercises 27–32, write the inequality.

27. $[-1, 1)$ **28.** $(-\infty, 4]$

29.
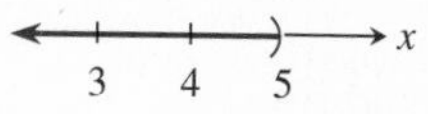

30.
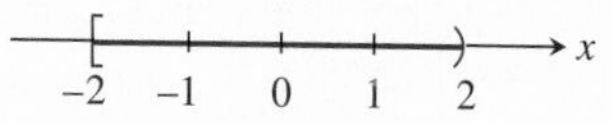

31. x is between −1 and 2.

32. x is greater than or equal to 5.

In Exercises 33–38, write interval notation.

33. $x > -3$ **34.** $-7 < x < -2$

35.
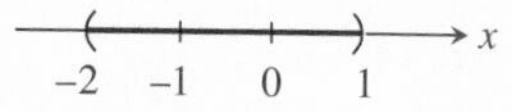

36.
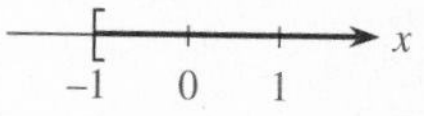

37. x is greater than −3 and less than or equal to 4.

38. x is positive.

In Exercises 39–44, describe the interval in words.

39. $4 < x \le 9$ **40.** $x \ge -1$

41. $[-3, \infty)$ **42.** $(-5, 7)$

43.
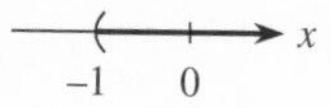

44.
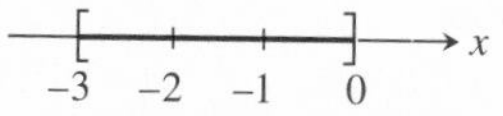

In Exercises 45–50, find the endpoints and state whether the interval is bounded or unbounded.

45. $(-3, 4]$ **46.** $(-3, -1)$

47. $[0, 5]$ **48.** $(2, \infty)$

49. $(-\infty, 5)$ **50.** $[-6, \infty)$

In Exercises 51–54, use both inequality and interval notation to describe the set of numbers.

51. Bill is at least 29 years old.

52. No item at Sarah's Variety Store costs more than \$2.00.

53. The price of a gallon of gasoline varies from \$1.099 to \$1.399.

54. Salary raises at the State University of California at Chico will average between 2% and 6.5%.

In Exercises 55–60, evaluate the expression.

55. $3 + |-3|$

56. $2 - |-2|$

57. $|(-2)3|$

58. $3|-4|$

59. $\dfrac{-2}{|-2|}$

60. $|1.5| + |-1.5|$

In Exercises 61–64, rewrite the expression without using absolute value symbols.

61. $|\pi - 4|$

62. $|5 - \pi|$

63. $|\sqrt{2} - \sqrt{7}|$

64. $|\sqrt{5} - 5/2|$

In Exercises 65–70, replace ○ with <, =, or > to make a true statement.

65. $|-4| \bigcirc -|4|$

66. $-3 \bigcirc -|-3|$

67. $|-2| \bigcirc |2|$

68. $-|-5| \bigcirc |-5|$

69. $-|-6| \bigcirc 0$

70. $|-1| \bigcirc 0$

In Exercises 71–76, find the distance between a and b. Also find the coordinate of the midpoint of the line segment with endpoints a and b.

71. $a = 10.6,\ b = -9.3$

72. $a = -21.5,\ b = 32.6$

73. $a = -17,\ b = -5$

74. $a = 89,\ b = 34$

75. $a = \sqrt{17},\ b = \sqrt{5}$

76. $a = -\sqrt{21},\ b = \sqrt{7}$

In Exercises 77–82, write the statement using absolute value notation.

77. The distance between x and 4 is 3.

78. The distance between y and -2 is greater than or equal to 4.

79. The distance between z and -3 is less than or equal to 1.

80. The distance between x and c is less than d units.

81. y is more than d units from c.

82. z is at most d units from c.

In Exercises 83–90, solve the problem.

83. If a is a negative number, show that $|a^2| = |a|^2 = a^2$.

84. Find the block of decimals that repeat in the decimal form for 2/17.

85. List the whole numbers whose magnitudes are less than 7.

86. List the natural numbers whose magnitudes are less than 7.

87. List the integers whose magnitudes are less than 7.

88. Describe the rational numbers whose magnitudes are less than 7.

89. Describe the real numbers whose magnitudes are less than 7.

90. Place the numbers in order from least to greatest:

$$0,\ -\sqrt{7},\ \sqrt{2},\ 7/5,\ 3/5,\ 2/3,\ -8/3$$

2 PROPERTIES OF REAL NUMBERS

Algebraic Expressions • Basic Properties of Algebra • Equations • Fractions • Exponents • Scientific Notation

Objective

Students will be able to apply the basic properties of algebra and to work with equations, fractions, exponents, and scientific notation.

Motivate

Have students discuss ways to display very large or very small numbers without using a lot of zeros.

Algebraic Expressions

Algebra involves the use of letters and other symbols to represent real numbers. A **variable** is a letter or symbol (for example, x, y, t, θ) that represents an unspecified real number. A **constant** is a letter or symbol (for example, $-2, 0, \sqrt{3}, \pi$) that represents a specific real number.

An **algebraic expression** is a combination of variables and constants involving addition, subtraction, multiplication, division, powers, and roots.

To **evaluate an algebraic expression,** we substitute real number values for each of the variables in the expression, as illustrated in Example 1.

EXAMPLE 1 Evaluating Algebraic Expressions

Evaluate $3x^2 + 4x - 5$ for the value of x.

a. $x = 3$
b. $x = -2.6$

```
-2.6→X
                -2.6
3X²+4X−5
                4.88
```

Figure 5 The calculation shows that $3(-2.6)^2 + 4(-2.6) - 5 = 4.88$.

Solution

a. Using Paper and Pencil

$$3x^2 + 4x - 5 = 3(3)^2 + 4(3) - 5 \quad \text{Substitute } x = 3.$$
$$= 27 + 12 - 5 = 34 \quad \text{Simplify.}$$

Thus, for $x = 3$ the value of the expression is 34.

b. Using a Calculator Figure 5 shows the value -2.6 stored in the variable x, and $3x^2 + 4x - 5$ yielding 4.88 for $x = -2.6$. ■

Basic Properties of Algebra

We begin by investigating some of the properties of the arithmetic operations addition, subtraction, multiplication, and division, represented by the symbols $+$, $-$, $\times$ (or $\cdot$), and $\div$ (or $/$), respectively. Addition and multiplication are the primary operations. Subtraction and division are defined in terms of addition and multiplication.

Subtraction: $a - b = a + (-b)$

Division: $\frac{a}{b} = a\left(\frac{1}{b}\right), \quad b \neq 0$

In the above definitions, $-b$ is the **additive inverse** or **opposite** of b, and $1/b$ is the **multiplicative inverse** or **reciprocal** of b. Do not confuse the additive inverse with negative numbers. The additive inverse of 5 is the negative number -5. However, the additive inverse of -3 is the positive number 3.

The following properties hold for real numbers, variables, and algebraic expressions.

TECHNOLOGY NOTE

On many calculators, there are two "−" keys, one for subtraction and one for negative numbers or opposites. Be sure you know how to use both keys correctly. Misuse can lead to incorrect results.

Properties of Algebra

Let u, v, and w be real numbers, variables, or algebraic expressions.

1. Commutative property

Addition: $u + v = v + u$

Multiplication: $uv = vu$

2. Associative property

Addition: $(u + v) + w = u + (v + w)$

Multiplication: $(uv)w = u(vw)$

3. Identity property

Addition: $u + 0 = u$

Multiplication: $u \cdot 1 = u$

4. Inverse property

Addition: $u + (-u) = 0$

Multiplication: $u \cdot \frac{1}{u} = 1,\ u \neq 0$

5. Distributive property

Multiplication over addition

$u(v + w) = uv + uw$
$(u + v)w = uw + vw$

Multiplication over subtraction

$u(v - w) = uv - uw$
$(u - v)w = uw - vw$

The left-hand sides of the equations for the distributive property show the **factored forms** of the algebraic expressions, and the right-hand sides show the **expanded forms.**

EXAMPLE 2 Using the Distributive Property

a. Write the expanded form of $(a + 2)x$.
b. Write the factored form of $3y - by$.

Solution

a. $(a + 2)x = ax + 2x$
b. $3y - by = (3 - b)y$ ■

You will see that the additive inverse is very useful when solving equations and inequalities. Here are some properties of the additive inverse together with examples that help illustrate their meanings.

Teaching Note

Remind students that $-u$ can be either positive or negative, depending on the sign of u.

Properties of the Additive Inverse

Let u and v be real numbers, variables, or algebraic expressions.

Property	Example
1. $-(-u) = u$	$-(-3) = 3$
2. $(-u)v = u(-v) = -(uv)$	$(-4)3 = 4(-3)$ $= -(4 \cdot 3) = -12$
3. $(-u)(-v) = uv$	$(-6)(-7) = 6 \cdot 7 = 42$
4. $(-1)u = -u$	$(-1)5 = -5$
5. $-(u + v) = (-u) + (-v)$	$-(7 + 9) = (-7) + (-9)$ $= -16$

Equations

An **equation** is a statement of equality between two expressions. Here are some properties of equality.

Properties of Equality

Let u, v, w, and z be real numbers, variables, or algebraic expressions.

Reflexive	$u = u$
Symmetric	If $u = v$, then $v = u$.
Transitive	If $u = v$ and $v = w$, then $u = w$.
Addition	If $u = v$ and $w = z$, then $u + w = v + z$.
Multiplication	If $u = v$ and $w = z$, then $uw = vz$.

In Section 7 we will use the addition and multiplication properties of equality to solve equations. The addition property allows us to add (or subtract) an expression to (from) both sides of an equation. The multiplication property allows us to multiply (or divide) both sides of an equation by a nonzero expression.

Fractions

The following forms are equivalent for fractions:

$$a/b, \quad \frac{a}{b}, \quad a \div b$$

We must have the denominator $b \neq 0$ because *division by zero is not defined.*

Two fractions are **equal,** $a/b = c/d$, if and only if $ad = bc$. We can write fractions *in simpler form* using $ac/bc = a/b$. Here is how we operate with fractions.

Operations with Fractions

Let u, v, w, and z be real numbers, variables, or algebraic expressions. All of the denominators are assumed to be different from zero.

Operation	Example
1. $\dfrac{u}{v} + \dfrac{w}{v} = \dfrac{u + w}{v}$	$\dfrac{2}{3} + \dfrac{5}{3} = \dfrac{2 + 5}{3} = \dfrac{7}{3}$
2. $\dfrac{u}{v} + \dfrac{w}{z} = \dfrac{uz + vw}{vz}$	$\dfrac{2}{3} + \dfrac{4}{5} = \dfrac{2 \cdot 5 + 3 \cdot 4}{3 \cdot 5} = \dfrac{22}{15}$
3. $\dfrac{u}{v} \cdot \dfrac{w}{z} = \dfrac{uw}{vz}$	$\dfrac{2}{3} \cdot \dfrac{4}{5} = \dfrac{2 \cdot 4}{3 \cdot 5} = \dfrac{8}{15}$
4. $\dfrac{u}{v} \div \dfrac{w}{z} = \dfrac{u}{v} \cdot \dfrac{z}{w} = \dfrac{uz}{vw}$	$\dfrac{2}{3} \div \dfrac{4}{5} = \dfrac{2}{3} \cdot \dfrac{5}{4} = \dfrac{10}{12}$

5. For subtraction, replace "+" by "−" in Properties 1 and 2.

The second addition formula gives one way to add or subtract fractions with unlike denominators. We can also use the *least common denominator* (*LCD*), which may require factoring integers into their *prime factors.*

If a, b, and c are integers such that $a = bc$, then b and c are **factors** or **divisors** of a. A **prime number** is an integer greater than 1 whose only natural number factors are itself and 1. The prime numbers less than 20 are

$$2,\quad 3,\quad 5,\quad 7,\quad 11,\quad 13,\quad 17,\quad \text{and}\quad 19.$$

An integer greater than 1 that is not prime is **composite.** Composite numbers can be factored into a product of two or more primes. For example, $4 = 2 \cdot 2$ and $6 = 2 \cdot 3$. Notice that 1 is neither prime nor composite.

The **Fundamental Theorem of Arithmetic** states that every integer greater than 1 can be factored into a product of prime numbers in exactly one way except for the order in which the factors appear. For example, $36 = 2 \cdot 2 \cdot 3 \cdot 3$.

If the numerator and denominator of a fraction have no common prime factors, the fraction is in **reduced form.**

■ EXAMPLE 3 Using the LCD

Rewrite the expression as a single fraction in reduced form.

$$\frac{5}{12} + \frac{3}{10} - \frac{2}{15}$$

Solution

First factor each denominator into its prime factors:

$$12 = 2 \cdot 2 \cdot 3, \qquad 10 = 2 \cdot 5, \qquad 15 = 3 \cdot 5$$

The LCD is $2 \cdot 2 \cdot 3 \cdot 5 = 60$. Each prime must occur the largest number of times it appears in any one denominator.

Next, carry out the operations:

$$\frac{5}{12} + \frac{3}{10} - \frac{2}{15} = \frac{5 \cdot 5}{12 \cdot 5} + \frac{3 \cdot 6}{10 \cdot 6} - \frac{2 \cdot 4}{15 \cdot 4} \qquad \frac{a}{b} = \frac{ac}{bc}.$$

$$= \frac{25 + 18 - 8}{60} \qquad \text{Add numerators.}$$

$$= \frac{35}{60} = \frac{7 \cdot 5}{12 \cdot 5} \qquad \text{Simplify.}$$

$$= \frac{7}{12} \qquad \frac{ac}{bc} = \frac{a}{b}.$$

Notice that 7/12 is in reduced form because 7 and 12 have no common prime factors. ■

Exponents

Exponential notation is used to shorten products of factors that repeat.

Exponential Notation

Let a be a real number, variable, or algebraic expression, and n a positive integer. Then

$$a^n = \underbrace{a \cdot a \cdot \cdots \cdot a}_{n \text{ factors}},$$

where n is the **exponent,** a is the **base,** and a^n is the ***n*th power of *a*,** read as "a to the nth power."

Using exponential notation we can write,

$$(-3)(-3)(-3)(-3) = (-3)^4 \qquad \text{and} \qquad (2x+1)(2x+1) = (2x+1)^2.$$

The two exponential expressions in Example 4 have the same value but have different bases. Be sure you understand the difference.

EXAMPLE 4 Identifying the Base

a. In $(-3)^5$, the base is -3.
b. In -3^5, the base is 3. ■

Here are the basic properties of exponents together with examples that help illustrate their meanings.

Teaching Note

Students should get plenty of practice in applying the properties of exponents.

Alert

Some students may attempt to simplify an expression such as $x^4 + x^3$ by adding or multiplying the exponents. Remind them that some expressions cannot be simplified.

Properties of Exponents

Let u and v be real numbers, variables, or algebraic expressions, and m and n be integers. All bases are assumed to be nonzero.

Property	Example
1. $u^m u^n = u^{m+n}$	$5^3 5^4 = 5^{3+4} = 5^7$
2. $\dfrac{u^m}{u^n} = u^{m-n}$	$\dfrac{x^9}{x^4} = x^{9-4} = x^5$
3. $u^0 = 1$	$8^0 = 1$
4. $u^{-n} = \dfrac{1}{u^n}$	$y^{-3} = \dfrac{1}{y^3}$
5. $(uv)^m = u^m v^m$	$(2z)^5 = 2^5 z^5 = 32z^5$
6. $(u^m)^n = u^{mn}$	$(x^2)^3 = x^{2 \cdot 3} = x^6$
7. $\left(\dfrac{u}{v}\right)^m = \dfrac{u^m}{v^m}$	$\left(\dfrac{a}{b}\right)^7 = \dfrac{a^7}{b^7}$

To simplify an expression involving powers means to rewrite it so that each factor appears only once, all exponents are positive, and exponents and constants are combined as much as possible.

EXAMPLE 5 Simplifying Expressions with Positive Exponents

a. $(2ab^3)(5a^2b^5) = 10(aa^2)(b^3b^5) = 10a^3b^8$

b. $\dfrac{(u^2v)^3}{v^2} = \dfrac{(u^2)^3v^3}{v^2} = \dfrac{u^6v^3}{v^2} = u^6v$

c. $\left(\dfrac{x^3}{y^2}\right)^4 = \dfrac{(x^3)^4}{(y^2)^4} = \dfrac{x^{12}}{y^8}$ ■

Compare the way negative exponents are handled in parts a and b of the next example.

EXAMPLE 6 Simplifying Expressions with Negative Exponents

a. $\dfrac{x^2}{x^{-3}} = x^{2-(-3)} = x^5$

b. $\dfrac{u^2v^{-2}}{u^{-1}v^3} = \dfrac{u^2u^1}{v^2v^3} = \dfrac{u^3}{v^5}$

c. $\left(\dfrac{x^2}{2}\right)^{-3} = \dfrac{(x^2)^{-3}}{2^{-3}} = \dfrac{x^{-6}}{2^{-3}} = \dfrac{2^3}{x^6} = \dfrac{8}{x^6}$ ■

REMARK

Be sure you understand how exponent Property 4 permits us to move factors from the numerator to the denominator and vice versa:

$$\frac{u^n}{v^m} = \frac{v^{-m}}{u^{-n}}$$

Teaching Note

Note that a number like 0.63×10^4 is *not* in scientific notation, because in scientific notation the first digit must be nonzero.

Alert

Some students may try to determine negative exponents in scientific notation by counting the number of zeros that occur after the decimal point. Show them that $0.06 \neq 6 \times 10^{-1}$.

Follow-up

Ask whether -91 can be classified as prime or composite. **(No)**

Assignment Guide

Day 1: Ex. 3–60, multiples of 3, 65, 66, 68, 71, 74, 79

Scientific Notation

Any positive number can be written in **scientific notation,**

$$c \times 10^m, \quad \text{where} \quad 1 \le c < 10 \quad \text{and} \quad m \text{ is an integer.}$$

This notation provides us a way to work with very large and very small numbers. For example, the distance between the earth and the sun is about 93,000,000 miles. In scientific notation,

$$93{,}000{,}000 = 9.3 \times 10^7.$$

The *positive exponent* 7 indicates that moving the decimal point in 9.3 to the right 7 places produces the number on the left side of the equation.

The weight of an oxygen molecule is about 0.000 000 000 000 000 000 000 053 gram. In scientific notation,

$$0.000\,000\,000\,000\,000\,000\,000\,053 = 5.3 \times 10^{-23}.$$

The *negative exponent* -23 indicates that moving the decimal point in 5.3 to the left 23 places produces the number on the left side of the equation.

Notes on Exercises

Ex. 1–22 provide review of basic algebra skills and properties.
Ex. 27–40 are routine arithmetic problems using fractions.
Ex. 55–64 provide practice with the properties of exponents.
Ex. 65–80 provide practice with scientific notation.

Ongoing Assessment

Self-Assessment: Ex. 17, 37, 63
Embedded Assessment: Ex. 48, 60

EXAMPLE 7 Converting Scientific Notation

a. $2.375 \times 10^8 = 237{,}500{,}000$
b. $0.000000349 = 3.49 \times 10^{-7}$ ■

EXAMPLE 8 Using Scientific Notation

Simplify

$$\frac{(370{,}000)(4{,}500{,}000{,}000)}{18{,}000}.$$

Solution

Using Paper and Pencil

$$\frac{(370{,}000)(4{,}500{,}000{,}000)}{18{,}000} = \frac{(3.7 \times 10^5)(4.5 \times 10^9)}{1.8 \times 10^4}$$

$$= \frac{(3.7)(4.5)}{1.8} \times 10^{5+9-4}$$

$$= 9.25 \times 10^{10}$$

$$= 92{,}500{,}000{,}000$$

```
(370000)(4500000
000)/(18000)
           9.25E10
(3.7E5)(4.5E9)/(
1.8E4)
           9.25E10
▮
```

Figure 6 Be sure you understand how your calculator displays scientific notation.

Using a Calculator

Figure 6 shows two ways to perform the computation. In the first, the numbers are entered in decimal form. In the second, the numbers are entered in scientific notation. This calculator uses "9.25E10" to stand for 9.25×10^{10}. ■

EXERCISES FOR SECTION 2

In Exercises 1–8, evaluate the expression for the given values of x.

1. $3x - 4,\ x = -2, 0$

2. $5 - 2x,\ x = 0, 3$

3. $2x^2 - 3x - 1,\ x = 2, 3.2$

4. $x^3 - 2x + 1,\ x = -2,\ 1.5$

5. $x - 4|x| + 2,\ x = -2, 2$

6. $2|x| + x - 1,\ x = -3, 3$

7. $\dfrac{2x^2 - x}{x - 3},\ x = -2, 0$

8. $\dfrac{3x^2 + x}{x - 4},\ x = -1, 0$

In Exercises 9–18, name the algebraic property or properties illustrated by the equation.

9. $(3x)y = 3(xy)$

10. $a^2b = ba^2$

11. $a^2b + (-a^2b) = 0$

12. $(x + 3)^2 + 0 = (x + 3)^2$

13. $a(x + y) = ax + ay$

14. $(x + 2)\dfrac{1}{x + 2} = 1$

15. $1 \cdot (x + y) = x + y$

16. $2(x - y) = 2x - 2y$

17. $2x + (y - z) = 2x + (y + (-z)) = (2x + y) + (-z) = (2x + y) - z$

18. $\dfrac{1}{a}(ab) = \left(\dfrac{1}{a}a\right)b = 1 \cdot b = b$

In Exercises 19 and 20, use the distributive property to write the expanded form of the expression.

19. $a(x^2 + b)$ **20.** $(y - z^2)c$

In Exercises 21 and 22, use the distributive property to write the factored form of the expression.

21. $ax^2 + dx^2$ **22.** $a^3z + a^3w$

In Exercises 23–26, factor the integer into a product of prime numbers.

23. 36 **24.** 100

25. 56 **26.** 88

In Exercises 27–40, rewrite as a single fraction.

27. $\frac{5}{9} + \frac{10}{9}$ **28.** $\frac{17}{32} - \frac{9}{32}$

29. $\frac{20}{21} \cdot \frac{9}{22}$ **30.** $\frac{33}{25} \cdot \frac{20}{77}$

31. $\frac{2}{3} \div \frac{4}{5}$ **32.** $\frac{9}{4} \div \frac{15}{10}$

33. $\frac{5}{12} + \frac{7}{18}$ **34.** $\frac{3}{20} + \frac{7}{50}$

35. $\frac{3}{4} - \frac{5}{49}$ **36.** $\frac{2}{9} - \frac{3}{25}$

37. $\frac{1}{15} + \frac{1}{21}$ **38.** $\frac{1}{10} + \frac{1}{6}$

39. $\frac{1}{14} + \frac{4}{15} - \frac{5}{21}$ **40.** $\frac{1}{6} + \frac{6}{35} - \frac{4}{15}$

In Exercises 41–44, find the additive inverse of the number.

41. $\frac{7}{2}$ **42.** -7

43. $6 - \pi$ **44.** $\sqrt{7} - 4$

In Exercises 45–54, use a calculator to evaluate the expression. Round the value to two decimal places.

45. $3(-5.6) - (-4.1) + 7$ **46.** $\frac{2(-5.5) - 6}{7.4 - 3.8}$

47. $4(-3.1)^3 - (-4.2)^5$

48. $5[3(-1.1)^2 - 4(-0.5)^3]$

49. $\frac{12}{5} + \frac{13}{8}$ **50.** $7\left(\frac{2}{3} - \frac{4}{5}\right)$

51. $(0.3)^{-4} + (0.1)^{-3}$ **52.** $5^{-2} + 2^{-4}$

53. $(0.5)^8$ **54.** $(2.5)^{-6}$

In Exercises 55–64, simplify the expression. Assume that denominators are not zero.

55. $\frac{x^4y^3}{x^2y^5}$ **56.** $\frac{(uv^2)^3}{v^2u^3}$

57. $\frac{(3x^2)^2y^4}{3y^2}$ **58.** $\left(\frac{4}{x^2}\right)^2$

59. $\frac{(2x^2y)^{-1}}{xy^2}$ **60.** $(3x^2y^3)^{-2}$

61. $\frac{x^{-3}y^3}{x^{-5}y^2}$ **62.** $\left(\frac{2}{xy}\right)^{-3}$

63. $\frac{(x^{-3}y^2)^{-4}}{(y^6x^{-4})^{-2}}$ **64.** $\left(\frac{4a^3b}{a^2b^3}\right)\left(\frac{3b^2}{2a^2b^4}\right)$

In Exercises 65–70, write the number in scientific notation.

65. 9,210,000,000 **66.** 0.000 000 000 2

67. The mean distance from Jupiter to the sun is about 483,900,000 miles.

68. The mean distance from Pluto to the sun is about 3,680,000,000 miles.

69. The diameter of a red blood corpuscle is about 0.000007 meter.

70. The electric charge, in coulombs, of an electron is about

$$-0.000\,000\,000\,000\,000\,000\,16.$$

In Exercises 71–76, write the number in decimal form.

71. 3.33×10^{-8} **72.** 6.73×10^{11}

73. The distance that light travels in 1 year (one *light-year*) is about 5.87×10^{12} mi.

74. Our solar system is about 5×10^9 years old.

75. The mass of an electron is about 9.1066×10^{-28} g (gram).

76. The mass of a neutron is about 1.6747×10^{-24} g.

In Exercises 77–80, use scientific notation to simplify.

77. (186,000)(31,000,000) **78.** $\frac{0.0000008}{0.000005}$

79. $\frac{(1.35 \times 10^{-7})(2.41 \times 10^8)}{1.25 \times 10^9}$

80. $\frac{(3.7 \times 10^8)(4.3 \times 10^6)}{2.5 \times 10^7}$

3
RADICALS AND RATIONAL EXPONENTS

Radicals • Simplifying Radical Expressions • Rationalizing the Denominator • Rational Exponents

Objective

Students will be able to apply the properties of radicals and rational exponents and simplify radical expressions by removing factors from radicands or rationalizing the denominator.

Motivate

Ask . . .

If $x^{1/2} \cdot x^{1/2} = x^{1/2+1/2}$, what do you suppose is the meaning of $x^{1/2}$? ($\sqrt{x}$)

Radicals

If $b^2 = a$, then b is a **square root** of a. Both 2 and -2 are square roots of 4 because $2^2 = (-2)^2 = 4$. Similarly, b is a **cube root** of a if $b^3 = a$. Notice that 2 is a cube root of 8 because $2^3 = 8$.

Every real number has exactly one real nth root whenever n is odd. For example, 2 is the only cube root of 8. When n is even, positive real numbers have two real nth roots and negative real numbers have no real nth roots. For example, the fourth roots of 16 are ± 2, and -16 has no real fourth roots. The *principal* fourth root of 16 is 2.

Teaching Note

Stress that the radical symbol denotes the *principal* nth root of the radicand.

Definition: *n*th Root of a Number

Let n be an integer greater than 1 and a and b real numbers.

1. If $b^n = a$, then b is an ***n*th root** of a.
2. If a has an nth root, the **principal *n*th root** of a is the nth root having the same sign as a.

The principal nth root of a is denoted by the **radical expression** $\sqrt[n]{a}$. The positive integer n is the **index** of the radical and a is the **radicand.**

When $n = 2$, special notation is used for the roots. We omit the index and write $\sqrt{a}$ instead of $\sqrt[2]{a}$. If a is a positive real number and n a positive even integer, its two nth roots are denoted by $\sqrt[n]{a}$ and $-\sqrt[n]{a}$.

GRAPHER NOTE

Principal *n*th roots

Calculators often have a key for the principal nth root. Use this feature of your calculator to check the computations in Example 1.

■ **EXAMPLE 1 Finding Principal *n*th Roots**

a. $\sqrt{36} = 6$ because $6^2 = 36$.

b. $\sqrt[3]{\dfrac{27}{8}} = \dfrac{3}{2}$ because $\left(\dfrac{3}{2}\right)^3 = \dfrac{27}{8}$.

c. $\sqrt[3]{-\dfrac{27}{8}} = -\dfrac{3}{2}$ because $\left(-\dfrac{3}{2}\right)^3 = -\dfrac{27}{8}$.

d. $\sqrt[4]{-625}$ is not a real number because 4 is even and -625 is negative. Note that the fourth power of any real number is never negative. ■

Here are some properties of radicals together with examples that help illustrate their meaning.

Teaching Notes

Note that the phrase "assume that all roots are real numbers" means that these properties cannot be applied to expressions involving even roots of negative numbers.

Alert

Some students may try to simplify $\sqrt{x + y}$ by taking the square roots of x and y separately. Show them that $\sqrt{4 + 9} \neq 2 + 3$.

NOTE

Without the restriction that precedes the list, Property 5 would need to be given special attention. For example,

$$\sqrt{(-3)^4} \neq (\sqrt{-3})^4$$

because $\sqrt{-3}$ on the right has no real number value.

Properties of Radicals

Let u and v be real numbers, variables, or algebraic expressions, and m and n be positive integers greater than 1. We assume that all of the roots are real numbers and all of the denominators are not zero.

Property	Example
1. $\sqrt[n]{uv} = \sqrt[n]{u} \cdot \sqrt[n]{v}$	$\sqrt{75} = \sqrt{25 \cdot 3} = \sqrt{25} \cdot \sqrt{3} = 5\sqrt{3}$
2. $\sqrt[n]{\dfrac{u}{v}} = \dfrac{\sqrt[n]{u}}{\sqrt[n]{v}}$	$\dfrac{\sqrt[4]{96}}{\sqrt[4]{6}} = \sqrt[4]{\dfrac{96}{6}} = \sqrt[4]{16} = 2$
3. $\sqrt[m]{\sqrt[n]{u}} = \sqrt[m \cdot n]{u}$	$\sqrt{\sqrt[3]{7}} = \sqrt[2\times 3]{7} = \sqrt[6]{7}$
4. $(\sqrt[n]{u})^n = u$	$(\sqrt[4]{5})^4 = 5$
5. $\sqrt[n]{u^m} = (\sqrt[n]{u})^m$	$\sqrt[3]{27^2} = (\sqrt[3]{27})^2 = 3^2 = 9$
6. $\sqrt[n]{u^n} = \begin{cases} \lvert u \rvert, & n \text{ even} \\ u, & n \text{ odd} \end{cases}$	$\sqrt{(-6)^2} = \lvert -6 \rvert = 6$ $\sqrt[3]{(-6)^3} = -6$

Simplifying Radical Expressions

Many simplifying techniques for roots of real numbers have been rendered obsolete because of calculators. For example, when determining the decimal equivalent of $1/\sqrt{2}$, it was once very common first to change the fraction so that the radical was in the numerator:

$$\frac{1}{\sqrt{2}} = \frac{1}{\sqrt{2}} \cdot \frac{\sqrt{2}}{\sqrt{2}} = \frac{\sqrt{2}}{2}$$

Using pencil and paper, it was then easier to divide a decimal for $\sqrt{2}$ by 2 than to divide the decimal into 1. Now either form is quickly computed with a calculator. However, these techniques are still valid for radicals involving algebraic expressions and for numerical computations when you need exact answers. Examples 2 and 3 illustrate the technique of *removing factors from radicands.*

Alert

Many students have experience in simplifying *square* roots by removing factors from radicands. They may attempt to simplify nth roots without paying attention to the value of n.

EXAMPLE 2 Removing Factors from Radicands

a. $\sqrt[4]{80} = \sqrt[4]{16 \cdot 5}$ Find greatest fourth-power factor.

$= \sqrt[4]{2^4 \cdot 5}$

$= \sqrt[4]{2^4} \cdot \sqrt[4]{5}$ Property 1

$= 2\sqrt[4]{5}$ Property 6

b. $\sqrt{18x^5} = \sqrt{9x^4 \cdot 2x}$ Find greatest square factor.

$= \sqrt{(3x^2)^2 \cdot 2x}$

$= 3x^2\sqrt{2x}$ Properties 1 and 6

c. $\sqrt[4]{x^4y^4} = \sqrt[4]{(xy)^4}$ Find greatest fourth-power factor.

$= |xy|$ Property 6

d. $\sqrt[3]{-24y^6} = \sqrt[3]{(-2y^2)^3 \cdot 3}$ Find greatest cube factor.

$= -2y^2\sqrt[3]{3}$ Properties 1 and 6 ■

Rationalizing the Denominator

The simplifying process of rewriting fractions containing radicals so that the denominator is free of radicals is **rationalizing the denominator.** When the denominator has the form $\sqrt[n]{u^k}$, multiplying numerator and denominator by $\sqrt[n]{u^{n-k}}$ and using Property 6 will eliminate the radical from the denominator because

$$\sqrt[n]{u^k} \cdot \sqrt[n]{u^{n-k}} = \sqrt[n]{u^{k+n-k}} = \sqrt[n]{u^n}.$$

Example 3 illustrates this process.

EXAMPLE 3 Rationalizing the Denominator

a. $\sqrt{\dfrac{2}{3}} = \dfrac{\sqrt{2}}{\sqrt{3}} \cdot \dfrac{\sqrt{3}}{\sqrt{3}} = \dfrac{\sqrt{6}}{3}$

b. $\dfrac{1}{\sqrt[4]{x}} = \dfrac{1}{\sqrt[4]{x}} \cdot \dfrac{\sqrt[4]{x^3}}{\sqrt[4]{x^3}} = \dfrac{\sqrt[4]{x^3}}{\sqrt[4]{x^4}} = \dfrac{\sqrt[4]{x^3}}{|x|}$

c. $\sqrt[5]{\dfrac{x^2}{y^3}} = \dfrac{\sqrt[5]{x^2}}{\sqrt[5]{y^3}} \cdot \dfrac{\sqrt[5]{y^2}}{\sqrt[5]{y^2}} = \dfrac{\sqrt[5]{x^2y^2}}{\sqrt[5]{y^5}} = \dfrac{\sqrt[5]{x^2y^2}}{y}$ ■

Rational Exponents

Exponents can also be rational numbers. It is desirable that the properties of exponents given in Section 2 also hold for rational exponents. It would be helpful, for example, if $x^{1/2} \cdot x^{1/2} = \sqrt{x} \cdot \sqrt{x} = x$, or $x^{1/2} \cdot x^{1/2} = x^{1/2+1/2} = x^1$. This suggests the following definition.

> **Definition:** Rational Exponents
>
> Let u be a real number, variable, or algebraic expression, and n an integer greater than 1. Then
>
> $$u^{1/n} = \sqrt[n]{u}.$$
>
> If m is a positive integer, m/n is in reduced form, and all roots are real numbers, then
>
> $$u^{m/n} = (u^{1/n})^m = (\sqrt[n]{u})^m \quad \text{and} \quad u^{m/n} = (u^m)^{1/n} = \sqrt[n]{u^m}.$$

The numerator of a rational exponent is the *power* to which the base is raised, and the denominator is the *root* to be taken. The fraction m/n needs to be in reduced form because $u^{2/3} = (\sqrt[3]{u})^2$ is defined for all real numbers (since every real number has a cube root), but $u^{4/6} = (\sqrt[6]{u})^4$ is defined only for $u \geq 0$ (since only nonnegative real numbers have sixth roots).

REMARK
If you also want the radical form in Example 4d to be simplified, then continue as follows:

$$\frac{1}{\sqrt{z^3}} = \frac{1}{\sqrt{z^3}} \cdot \frac{\sqrt{z}}{\sqrt{z}} = \frac{\sqrt{z}}{z^2}$$

■ EXAMPLE 4 Converting Radical and Exponential Notation

a. $\sqrt{(x+y)^3} = (x+y)^{3/2}$ **b.** $3x\sqrt[5]{x^2} = 3x \cdot x^{2/5} = 3x^{7/5}$

c. $x^{2/3}y^{1/3} = (x^2y)^{1/3} = \sqrt[3]{x^2y}$ **d.** $z^{-3/2} = \dfrac{1}{z^{3/2}} = \dfrac{1}{\sqrt{z^3}}$ ■

Recall that each factor appears only once and all exponents are positive in a simplified expression involving powers.

Follow-up
Ask . . .
If uv is positive, is it always true that $\sqrt{uv} = \sqrt{u}\sqrt{v}$?
(No, not if u and v are negative.)

■ EXAMPLE 5 Simplifying Exponential Expressions

a. $(x^2y^9)^{1/3}(xy^2) = (x^{2/3}y^3)(xy^2) = x^{(2/3)+1}y^5 = x^{5/3}y^5$

b. $\left(\dfrac{3x^{2/3}}{y^{1/2}}\right)\left(\dfrac{2x^{-1/2}}{y^{2/5}}\right) = \dfrac{6x^{2/3-1/2}}{y^{1/2+2/5}} = \dfrac{6x^{1/6}}{y^{9/10}}$ ■

The last example of this section suggests how to simplify a sum or difference of radicals.

Assignment Guide

Day 1: Ex. 3–81, multiples of 3, 85–87

Notes on Exercises

Ex. 1–24 require students to evaluate *n*th roots.
Ex. 29–44 provide practice with basic simplification and rationalizing denominators.
Ex. 59–66 require students to apply the properties of exponents.
Ex. 67–76 are more difficult simplification problems.

Ongoing Assessment

Self-Assessment: Ex. 11, 23, 41, 65
Embedded Assessment: Ex. 87

EXAMPLE 6 Combining Radicals

a. $2\sqrt{80} - \sqrt{125} = 2\sqrt{16 \cdot 5} - \sqrt{25 \cdot 5}$ Find greatest square factors.

$= 8\sqrt{5} - 5\sqrt{5}$ Remove factors from radicands.

$= 3\sqrt{5}$ Distributive property

b. $\sqrt{4x^2y} - \sqrt{y^3} = \sqrt{(2x)^2y} - \sqrt{y^2y}$ Find greatest square factors.

$= 2|x|\sqrt{y} - |y|\sqrt{y}$ Remove factors from radicands.

$= (2|x| - |y|)\sqrt{y}$ Distributive property ■

Here is a summary of the procedures we use to simplify expressions involving radicals.

1. Remove factors from the radicand (see Example 2).
2. Eliminate radicals from denominator and denominators from radicands (see Example 3).
3. Combine sums and differences, if possible (see Example 6).

EXERCISES FOR SECTION 3

In Exercises 1–6, find the indicated real roots.

1. Square roots of 81
2. Fourth roots of 81
3. Cube roots of 64
4. Fifth roots of 243
5. Square roots of 16/9
6. Cube roots of $-27/8$

In Exercises 7–12, evaluate the expression without using a calculator.

7. $\sqrt{144}$
8. $\sqrt{-16}$
9. $\sqrt[3]{-216}$
10. $\sqrt[3]{216}$
11. $\sqrt[3]{\dfrac{125}{8}}$
12. $\sqrt{\dfrac{64}{25}}$

In Exercises 13–24, use a calculator to evaluate the expression.

13. $\sqrt[4]{256}$
14. $\sqrt[5]{3125}$
15. $\sqrt[3]{15.625}$
16. $\sqrt{12.25}$
17. $\sqrt[3]{-1520.875}$
18. $\sqrt{506.25}$
19. $81^{3/2}$
20. $16^{5/4}$
21. $32^{-2/5}$
22. $27^{-4/3}$
23. $\left(-\dfrac{1}{8}\right)^{-1/3}$
24. $\left(-\dfrac{125}{64}\right)^{-1/3}$

In Exercises 25–28, use information from the grapher screen below to evaluate the expression.

25. $\sqrt{1.69}$
26. $\sqrt{19.4481}$
27. $\sqrt[4]{19.4481}$
28. $\sqrt[3]{3.375}$

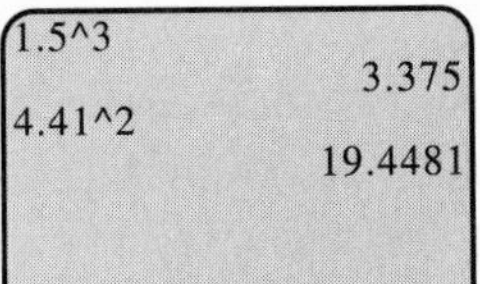

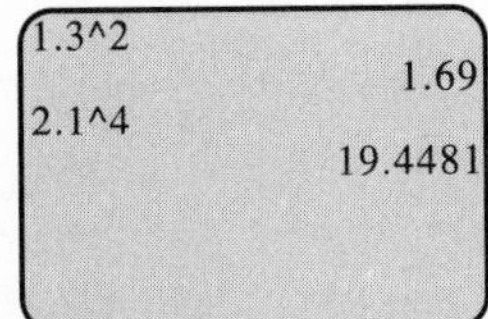

For Exercises 25–28

In Exercises 29–38, simplify by removing factors from the radicand.

29. $\sqrt{288}$
30. $\sqrt[3]{500}$
31. $\sqrt[3]{-250}$
32. $\sqrt[4]{192}$
33. $\sqrt{2x^3y^4}$
34. $\sqrt[3]{-27x^3y^6}$
35. $\sqrt[4]{3x^8y^6}$
36. $\sqrt[3]{8x^6y^4}$
37. $\sqrt[5]{96x^{10}}$
38. $\sqrt{108x^4y^9}$

In Exercises 39–44, rationalize the denominator.

39. $\dfrac{4}{\sqrt[3]{2}}$ **40.** $\dfrac{1}{\sqrt{5}}$

41. $\dfrac{1}{\sqrt[5]{x^2}}$ **42.** $\dfrac{2}{\sqrt[4]{y}}$

43. $\sqrt[3]{\dfrac{x^2}{y}}$ **44.** $\sqrt[5]{\dfrac{a^3}{b^2}}$

In Exercises 45–48, convert to exponential form.

45. $\sqrt[3]{(a + 2b)^2}$ **46.** $\sqrt[5]{x^2y^3}$

47. $2x\sqrt[3]{x^2y}$ **48.** $xy\sqrt[4]{xy^3}$

In Exercises 49–52, convert to radical form.

49. $a^{3/4}b^{1/4}$ **50.** $x^{2/3}y^{1/3}$

51. $x^{-5/3}$ **52.** $(xy)^{-3/4}$

In Exercises 53–58, write using a single radical.

53. $\sqrt{\sqrt{2x}}$ **54.** $\sqrt{\sqrt[3]{3x^2}}$

55. $\sqrt[4]{\sqrt{xy}}$ **56.** $\sqrt[3]{\sqrt{ab}}$

57. $\dfrac{\sqrt[5]{a^2}}{\sqrt[3]{a}}$ **58.** $\sqrt{a}\sqrt[3]{a^2}$

In Exercises 59–66, simplify the exponential expression.

59. $\dfrac{a^{3/5}a^{1/3}}{a^{3/2}}$ **60.** $(x^2y^4)^{1/2}$

61. $(a^{5/3}b^{3/4})(3a^{1/3}b^{5/4})$ **62.** $\left(\dfrac{x^{1/2}}{y^{2/3}}\right)^6$

63. $\left(\dfrac{-8x^6}{y^{-3}}\right)^{2/3}$ **64.** $\dfrac{(p^2q^4)^{1/2}}{(27q^3p^6)^{1/3}}$

65. $\dfrac{(x^9y^6)^{-1/3}}{(x^6y^2)^{-1/2}}$ **66.** $\left(\dfrac{2x^{1/2}}{y^{2/3}}\right)\left(\dfrac{3x^{-2/3}}{y^{1/2}}\right)$

In Exercises 67–76, simplify the radical expression.

67. $\sqrt{9x^{-6}y^4}$ **68.** $\sqrt{16y^8z^{-2}}$

69. $\sqrt[4]{\dfrac{3x^8y^2}{8x^2}}$ **70.** $\sqrt[5]{\dfrac{4x^6y}{9x^3}}$

71. $\sqrt[3]{\dfrac{4x^2}{y^2}} \cdot \sqrt[3]{\dfrac{2x^2}{y}}$ **72.** $\sqrt[5]{9ab^6} \cdot \sqrt[5]{27a^2b^{-1}}$

73. $3\sqrt{48} - 2\sqrt{108}$ **74.** $2\sqrt{175} - 4\sqrt{28}$

75. $\sqrt{x^3} - \sqrt{4xy^2}$ **76.** $\sqrt{18x^2y} + \sqrt{2y^3}$

In Exercises 77–84, replace ○ with <, =, or > to make a true statement.

77. $\sqrt{2 + 6}$ ○ $\sqrt{2} + \sqrt{6}$ **78.** $\sqrt{4} + \sqrt{9}$ ○ $\sqrt{4 + 9}$

79. $(3^{-2})^{-1/2}$ ○ 3 **80.** $(2^{-3})^{1/3}$ ○ 2

81. $\sqrt[4]{(-2)^4}$ ○ -2 **82.** $\sqrt[3]{(-2)^3}$ ○ -2

83. $2^{2/3}$ ○ $3^{3/4}$ **84.** $4^{-2/3}$ ○ $3^{-3/4}$

In Exercises 85–87, solve the problem.

85. The time t (in seconds) that it takes for a pendulum to complete one period is approximately $t = 1.1\sqrt{L}$, where L is the length (in feet) of the pendulum. How long is the period of a pendulum of length 10 ft?

86. The time t (in seconds) that it takes for a rock to fall a distance d (in meters) is approximately $t = 0.45\sqrt{d}$. How long does it take for the rock to fall a distance of 200 m?

87. Explain why $\sqrt[n]{a}$ and a real nth root of a need not have the same value.

Objective

Students will be able to identify, add, subtract, and multiply polynomials.

Motivate

Discuss whether the distributive property can be used to simplify an expression such as $(2x + 3)(x^2 - 4)$. **(Yes)**

4 POLYNOMIALS AND SPECIAL PRODUCTS

Polynomials • Adding and Subtracting Polynomials • Multiplying Polynomials • Special Products

Polynomials

Polynomials are the most familiar type of algebraic expressions. Expressions like the following are *polynomials in x:*

$$3x - 5, \qquad 3x^2 + 4x - 6, \qquad 2x^4 - 5x^3 + x^2 + 7x - 12$$

Expressions like these are *polynomials in x and y:*

$$2x + 3y - 1, \qquad x^2 - y^2, \qquad 2x^2 - 12xy + 6y^2$$

The following expressions are *not* polynomials:

$$\frac{2x + 1}{x - 3}, \qquad x^2 - x + x^{1/2}, \qquad \sqrt{2x^2 - x}$$

Teaching Note

Note that the sentence "it has no degree" does *not* mean that the degree of the zero polynomial is zero. A polynomial of degree zero is a *nonzero* constant.

The **terms** of a polynomial in x have the form ax^k, where the real number a is the **coefficient** and the nonnegative integer k is the **degree of the term.** The terms of $3x^2 + 4x - 6$ are $3x^2$, $4x$, and -6. The corresponding coefficients are 3, 4, and -6, respectively. Polynomials with one, two, and three terms are **monomials, binomials,** and **trinomials,** respectively. The constant 0 is given a special name, the **zero polynomial.** It has no degree.

Definition: Polynomial in x

A nonzero **polynomial in *x*** is any expression that can be written in the form

$$a_n x^n + a_{n-1}x^{n-1} + \cdots + a_1 x + a_0,$$

where n is a nonnegative integer, $a_n \neq 0$, $a_{n-1}, \ldots, a_1, a_0$ are real numbers. The **degree of the polynomial** is n, the **leading coefficient** is a_n, and the **constant term** is a_0.

A polynomial written with powers of x in *descending order* is in **standard form.**

■ EXAMPLE 1 Finding the Standard Form and Degree of a Polynomial

Write the polynomial in standard form and state its degree.

a. $3x^2 - x + x^5 - 1$ **b.** $3 - 2x$ **c.** -4

Solution

a. $x^5 + 3x^2 - x - 1$; the degree is 5.
b. $-2x + 3$; the degree is 1.
c. -4; the degree is 0. ■

A polynomial in several variables may have two variables in the same term. The degree of such a term is the sum of the exponents of the variables in that term, and the degree of the polynomial is the largest degree of its terms. For example, the polynomial

$$2x^4y^2 - y^6 + x^2 - xy + 5$$

has two terms of degree 6, two terms of degree 2, and one term of degree 0. The degree of the polynomial is 6.

Adding and Subtracting Polynomials

To add or subtract polynomials we add or subtract *like terms* using the distributive property. Terms of polynomials that have the same variables each raised to the same powers are **like terms.**

EXAMPLE 2 Adding and Subtracting Polynomials

a. $(2x^3 - 3x^2 + 4x - 1) + (x^3 + 2x^2 - 5x + 3)$

$= (2x^3 + x^3) + (-3x^2 + 2x^2)$

$+ (4x + (-5x)) + (-1 + 3)$ Group like terms.

$= 3x^3 - x^2 - x + 2$ Combine like terms.

b. $(4x^2 + 3x - 4) - (2x^3 + x^2 - x + 2)$

$= (0 - 2x^3) + (4x^2 - x^2)$

$+ (3x - (-x)) + (-4 - 2)$ Group like terms.

$= -2x^3 + 3x^2 + 4x - 6$ Combine like terms. ■

Multiplying Polynomials

To **expand the product of two polynomials** we use the distributive property. Here is what the procedure looks like when we multiply the binomials $3x + 2$ and $4x - 5$.

$(3x + 2)(4x - 5)$

$= 3x(4x - 5) + 2(4x - 5)$ Distributive property

$= (3x)(4x) - (3x)(5) + (2)(4x) - (2)(5)$ Distributive property

$= 12x^2 - 15x + 8x - 10$

Product of *F*irst terms	Product of *O*uter terms	Product of *I*nner terms	Product of *L*ast terms

In the above **FOIL method** for products of binomials, the outer (O) and inner (I) terms are like terms and can be added to give

$$(3x + 2)(4x - 5) = 12x^2 - 7x - 10.$$

Multiplying two polynomials requires multiplying each term of one polynomial by every term of the other polynomial. A convenient way to compute a product is to arrange the polynomials in standard form one on top of another so that their terms align vertically, as illustrated in Example 3.

Notes on Examples

In Example 3, point out that since $0x^3 = 0$, it is not shown in the answer.

EXAMPLE 3 Multiplying Polynomials in Vertical Form

Write the product $(x^2 - 4x + 3)(x^2 + 4x + 5)$ in standard form.

Solution

$$\begin{array}{rrrrrcl}
x^2 & - 4x & + 3 & & & & \\
x^2 & + 4x & + 5 & & & & \\
\hline
x^4 - 4x^3 & + 3x^2 & & & & = & x^2\,(x^2 - 4x + 3) \\
4x^3 & - 16x^2 & + 12x & & & = & 4x\,(x^2 - 4x + 3) \\
 & 5x^2 & - 20x & + 15 & & = & 5\,(x^2 - 4x + 3) \\
\hline
x^4 + 0x^3 & - 8x^2 & - 8x & + 15 & & & \text{Add.}
\end{array}$$

Thus,

$$(x^2 - 4x + 3)(x^2 + 4x + 5) = x^4 - 8x^2 - 8x + 15.$$

Special Products

Certain products provide patterns that will be useful when we factor polynomials in the next section. Here is a list of some special products for binomials.

Teaching Note

You may wish to have students derive some of the special products using the distributive property.

Special Binomial Products

Let u and v be real numbers, variables, or algebraic expressions.

1. **Product of a sum and a difference:** $(u + v)(u - v) = u^2 - v^2$
2. **Square of a sum:** $(u + v)^2 = u^2 + 2uv + v^2$
3. **Square of a difference:** $(u - v)^2 = u^2 - 2uv + v^2$
4. **Cube of a sum:** $(u + v)^3 = u^3 + 3u^2v + 3uv^2 + v^3$
5. **Cube of a difference:** $(u - v)^3 = u^3 - 3u^2v + 3uv^2 - v^3$

Identifying a product as one of the special types simplifies the multiplication process. The next two examples illustrate.

Follow-up

Ask whether a sum or product of two polynomials is always a polynomial. **(Yes)** Discuss why or why not.

Assignment Guide

Day 1: Ex. 3–72, multiples of 3

EXAMPLE 4 Using Special Products

Expand the products.

a. $(3x + 8)(3x - 8) = (3x)^2 - 8^2$ Product of sum and difference

$= 9x^2 - 64$ Simplify.

b. $(5y - 4)^2 = (5y)^2 - 2(5y)(4) + 4^2$ Square of difference
$= 25y^2 - 40y + 16$ Simplify.

c. $(2x - 3y)^3 = (2x)^3 - 3(2x)^2(3y)$
$+ 3(2x)(3y)^2 - (3y)^3$ Cube of difference
$= 8x^3 - 36x^2y + 54xy^2 - 27y^3$ Simplify. ■

Notes on Exercises

Ex. 1–16 are simple exercises that familiarize students with the language of polynomials.
Ex. 17–64 require students to simplify expressions.
Ex. 67–69 are applications of polynomials.

Ongoing Assessment

Self-Assessment: Ex. 25, 37, 59
Embedded Assessment: Ex. 72

Sometimes the special product formulas can be applied in special ways.

■ EXAMPLE 5 Expanding the Square of a Trinomial

Expand the product $(x + y - 1)^2$.

Solution

$(x + y - 1)^2 = [(x + y) - 1]^2$ Consider as a binomial.
$= (x + y)^2 - 2(x + y) + 1$ Square of difference
$= x^2 + 2xy + y^2 - 2x - 2y + 1$ Square of sum ■

EXERCISES FOR SECTION 4

In Exercises 1–6, find the degree, leading coefficient, and constant term of the polynomial.

1. $3x^2 - x + 1$
2. $-2x^4 - x^3 + x^2 + 3$
3. $-x^3 + x^2 - 2x$
4. $4x - 2$
5. $x^5 - 2x^4 + x - 1$
6. -9

In Exercises 7–10, write the polynomial in standard form and state its degree.

7. $2x - 1 + 3x^2$
8. $x^2 - 2x - 2x^3 + 1$
9. $1 - x^7$
10. $x^2 - x^4 + x - 3$

In Exercises 11–16, state whether the expression is a polynomial.

11. $x^3 - 2x^2 + x^{-1}$
12. $\dfrac{2x - 4}{x}$
13. $2xy + y^2 - x^2$
14. $\sqrt{x^2 - y^2}$
15. $(x^2 + x + 1)^2$
16. $1 - 3x + x^4$

In Exercises 17–30, simplify the expression. Write your result in standard form.

17. $(3x^2 + 4) - (5x - 2)$
18. $(x^2 - 3x + 7) + (3x^2 + 5x - 3)$
19. $(-3x^2 - 5) - (x^2 + 7x + 12)$
20. $(4x^3 - x^2 + 3x) - (x^3 + 12x - 3)$
21. $-(y^2 + 2y - 3) + (5y^2 + 3y + 4)$
22. $3y - [7 - (3y + 5)]$
23. $3z + (z^2 + 4z - 3) - (z^2 + 7)$
24. $-(z^4 + z^2 - 3z + 7) - (2z^4 + 2z^2 + 4z - 8)$
25. $2x(x^2 - x + 3)$
26. $y^2(2y^2 + 3y - 4)$
27. $-3u(4u - 1)$
28. $-4v(2 - 3v^3)$
29. $(2 - x - 3x^2)(5x)$
30. $(1 - x^2 + x^4)(2x)$

In Exercises 31–64, expand the product.

31. $(x + 3)(x + 5)$
32. $(x - 2)(x + 5)$
33. $(2x + 3)(4x + 1)$
34. $(3x - 5)(x + 2)$
35. $(x + 6)(x - 6)$
36. $(2x - 3)(2x + 3)$
37. $(3x - y)(3x + y)$
38. $(x + 2y)(x - 2y)$
39. $(x + 2)^2$
40. $(5x - 1)^2$
41. $(3 - 5x)^2$
42. $(3x + 4y)^2$
43. $(u - v + 2)^2$
44. $(x - y - 3)^2$
45. $(a - 2 + b)(a - 2 - b)$
46. $[a + (b + 1)][a - (b + 1)]$

47. $(x - 1)^3$

48. $(x + 2)^3$

49. $(2u - v)^3$

50. $(u + 3v)^3$

51. $(x^2 - 2)(x^2 + 2)$

52. $(2x^3 - 3y)(2x^3 + 3y)$

53. $(5x^3 - 1)^2$

54. $(2 - x^3)^2$

55. $(x^2 - 2x + 3)(x + 4)$

56. $(x^2 + 3x - 2)(x - 3)$

57. $(x^2 + x - 3)(x^2 + x + 1)$

58. $(2x^2 - 3x + 1)(x^2 - x + 2)$

59. $(x - \sqrt{2})(x + \sqrt{2})$

60. $(x^{1/2} - y^{1/2})(x^{1/2} + y^{1/2})$

61. $(\sqrt{u} + \sqrt{v})(\sqrt{u} - \sqrt{v})$

62. $(x^2 - \sqrt{3})(x^2 + \sqrt{3})$

63. $(x - 2)(x^2 + 2x + 4)$

64. $(x + 1)(x^2 - x + 1)$

In Exercises 65–66, evaluate the polynomial for the given values of the variables.

65. $x^2 + y^2 - xy, \quad x = 2, y = -2$

66. $2x^3y - x^2y^2 + y, \quad x = -1, y = 3$

In Exercises 67–72, solve the problem.

67. The polynomial $\pi r^2 h$ gives the volume of a cylinder with radius r and height h. Find the volume of the cylinder if the radius is 2 in. and the height is 5 in.

68. The polynomial $\pi(R^2 - r^2)$ gives the area of one side of a washer with outer radius R and inner radius r. Find the area of the ring if the outer radius is 25 centimeters (cm) and the inner radius is 10 cm.

69. The polynomial $16t^2$ approximates the distance (in feet) fallen by an object in free fall where t is time in seconds (sec). Approximate the distance a skydiver falls during the first 5 sec of free fall.

70. Write a polynomial in x of degree 3 with leading coefficient 2 and constant term -3.

71. Write a polynomial in x of degree 2 with leading coefficient -2 and constant term 5.

72. How does the degree of the product of two polynomials relate to the degree of the two polynomials?

5 FACTORING POLYNOMIALS

Introduction • Common Factors • Factoring Using Special Forms • Factoring Trinomials • Factoring by Grouping

Objective

Students will be able to factor simple polynomials by various methods, including grouping and using special forms.

Motivate

Ask how $x^2 + 6x + 9$ can be written as a product of two binomials. ($(x + 3)(x + 3)$)

Introduction

In Section 4 we multiplied polynomials. Now we reverse the process and **factor a polynomial** into the product of two or more **polynomial factors.** We will later use polynomial factoring to simplify algebraic fractions (Section 6) and to solve polynomial equations (Chapter 3).

Unless specified otherwise, we factor polynomials into factors of lesser degree and with integer coefficients. A polynomial that cannot be factored using integer coefficients is a **prime polynomial.**

A polynomial is **completely factored** if it is written as a product of its prime factors. For example,

$$2x^2 + 7x - 4 = (2x - 1)(x + 4)$$

and

$$x^3 + x^2 + x + 1 = (x + 1)(x^2 + 1)$$

are completely factored. However,

$$x^3 - 9x = x(x^2 - 9)$$

is *not* completely factored because $x^2 - 9$ is *not* prime. Notice that $x^2 - 9 = (x - 3)(x + 3)$ and that

$$x^3 - 9x = x(x - 3)(x + 3)$$

is completely factored.

Teaching Note

Point out that if a polynomial in x has no constant term, then the lowest power of x can be removed as a common factor.

Common Factors

The first step in factoring a polynomial is to remove common factors from its terms using the distributive property.

EXAMPLE 1 Removing Common Factors

a. $2x^3 + 2x^2 - 6x = 2x(x^2 + x - 3)$ $2x$ is the common factor.
b. $u^3v + uv^3 = uv(u^2 + v^2)$ uv is the common factor. ■

Factoring Using Special Forms

In Section 4 we made a list of special polynomial products. Recognizing the expanded forms of such products will help us factor them. Here are some special expanded polynomial forms to look for.

Teaching Note

Point out that $u^2 + v^2$ cannot be factored.

Alert

Some students will not look for common factors before applying a special polynomial form. For example, they may factor $4x^2 - 36$ as $(2x + 6)(2x - 6)$, which is not completely factored.

Special Polynomial Forms

Let u and v be real numbers, variables, or algebraic expressions.

1. **Difference of two squares:** $u^2 - v^2 = (u + v)(u - v)$
2. **Perfect square (sum) trinomial:** $u^2 + 2uv + v^2 = (u + v)^2$
3. **Perfect square (difference) trinomial:** $u^2 - 2uv + v^2 = (u - v)^2$
4. **Sum of two cubes:** $u^3 + v^3 = (u + v)(u^2 - uv + v^2)$
5. **Difference of two cubes:** $u^3 - v^3 = (u - v)(u^2 + uv + v^2)$

The special polynomial form that is easiest to identify is the difference of two squares. The two binomial factors have opposite signs:

$$u^2 - v^2 = (u + v)(u - v)$$

Two squares — Square roots — Difference — Opposite signs

■ EXAMPLE 2 Factoring the Difference of Two Squares

a. $25x^2 - 36 = (5x)^2 - 6^2$ — Difference of two squares

$= (5x + 6)(5x - 6)$ — Factors are prime.

b. $4x^2 - (y + 3)^2 = (2x)^2 - (y + 3)^2$ — Difference of two squares

$= [2x + (y + 3)][2x - (y + 3)]$ — Factors are prime.

$= (2x + y + 3)(2x - y - 3)$ ■

A perfect square trinomial is the square of a binomial and has one of the two forms shown here. The first and last terms are squares of u and v, and the middle term is twice the product of u and v. Notice that the operation signs before the middle term and in the binomial factors are the same.

Perfect square (sum): $u^2 + 2uv + v^2 = (u + v)^2$ — Same signs

Perfect square (difference): $u^2 - 2uv + v^2 = (u - v)^2$ — Same signs

■ EXAMPLE 3 Factoring Perfect Square Trinomials

a. $9x^2 + 6x + 1 = (3x)^2 + 2(3x)(1) + 1^2$ — $u = 3x,\ v = 1$

$= (3x + 1)^2$

b. $4x^2 - 12xy + 9y^2 = (2x)^2 - 2(2x)(3y) + (3y)^2$ — $u = 2x,\ v = 3y$

$= (2x - 3y)^2$ ■

In the sum and difference of two cubes, notice the patterns of the signs.

$u^3 + v^3 = (u + v)(u^2 - uv + v^2)$ — Same signs; Opposite signs

$u^3 - v^3 = (u - v)(u^2 + uv + v^2)$ — Same signs; Opposite signs

> **Teaching Note**
>
> Remind students that if they forget a formula during a test, they can check their answers by multiplying. This is also a good way to prevent careless errors.

■ EXAMPLE 4 Factoring the Sum and Difference of Two Cubes

a. $x^3 - 64 = x^3 - 4^3$ — Difference of two cubes

$= (x - 4)(x^2 + 4x + 16)$ — Factors are prime.

b. $8x^3 + 27 = (2x)^3 + 3^3$ — Sum of two cubes

$= (2x + 3)(4x^2 - 6x + 9)$ — Factors are prime. ■

Factoring Trinomials

Factoring the trinomial $ax^2 + bx + c$ into a product of binomials with integer coefficients requires factoring the integers a and c.

Factors of a

$$ax^2 + bx + c = (\square x + \square)(\square x + \square)$$

Factors of c

Because the number of integer factors of a and c is finite, we can list all possible binomial factors. Then we begin checking each possibility until we find a pair that works. Example 5 illustrates.

Notes on Examples

In Example 5, only factor pairs containing opposite signs are considered because the 14 in the original polynomial was subtracted.

■ EXAMPLE 5 Factoring a Trinomial—Leading Coefficient = 1

Factor $x^2 + 5x - 14$.

Solution

The only factor pair of the leading coefficient is 1 and 1. The factor pairs of 14 are 1 and 14, and 2 and 7. Here are the four possible factorizations of the trinomial:

$$(x + 1)(x - 14) \qquad (x - 1)(x + 14)$$
$$(x + 2)(x - 7) \qquad (x - 2)(x + 7)$$

If you check the middle term from each factorization you will find that

$$x^2 + 5x - 14 = (x + 7)(x - 2).$$ ■

With practice you will find that it usually is not necessary to list all possible binomial factors. Often you can test the possibilities mentally.

■ EXAMPLE 6 Factoring Trinomials—Leading Coefficient ≠ 1

Factor $35x^2 - x - 12$.

Solution

The factor pairs of the leading coefficient are 1 and 35, and 5 and 7. The factor pairs of 12 are 1 and 12, 2 and 6, and 3 and 4. The possible factorizations must be of the form

$$(x - *)(35x+?), \qquad (x + *)(35x-?),$$
$$(5x - *)(7x+?), \qquad (5x + *)(7x-?),$$

where $*$ and ? are one of the factor pairs of 12. Because the two binomial factors have opposite signs, there are six possibilities for each of the four forms—a total of 24 possibilities in all. If you try them, mentally and systematically, you should find that

$$35x^2 - x - 12 = (5x - 3)(7x + 4).$$ ■

Sometimes the techniques of Examples 5 and 6 can be applied to polynomials in two variables, as illustrated in Example 7.

Notes on Examples

Exercise 87 illustrates another way to solve Example 8b by grouping.

Follow-up

Discuss whether any trinomials can be factored by grouping.

Assignment Guide

Day 1: Ex. 3–84, multiples of 3, 87–89

Notes on Exercises

Ex. 1–28 are straightforward factoring problems that require factoring out a common factor or using a special polynomial form.
Ex. 29–46 require factoring without using a special polynomial form.
Ex. 47–87 are more difficult factoring problems. Most require grouping or using more than one step to completely factor.
Ex. 88–89 are simple applications.

Ongoing Assessment

Self-Assessment: Ex. 37, 61, 81
Embedded Assessment: Ex. 72, 84

■ **EXAMPLE 7 Factoring Trinomials in *x* and *y***

Factor $3x^2 - 7xy + 2y^2$.

Solution

The only way to get $-7xy$ as the middle term is with

$$3x^2 - 7xy + 2y^2 = (3x - ?y)(x - ?y).$$

The signs in the binomials must be negative because the coefficient of y^2 is positive *and* the middle term is negative. Checking the two possibilities, $(3x - y)(x - 2y)$ and $(3x - 2y)(x - y)$, shows that

$$3x^2 - 7xy + 2y^2 = (3x - y)(x - 2y).$$ ■

Factoring by Grouping

Notice that $(a + b)(c + d) = ac + ad + bc + bd$. If a polynomial with four terms is the product of two binomials, we can group terms to factor. There are only three ways to group the terms and two of them work. So, if two of the possibilities fail, then it is not factorable.

■ **EXAMPLE 8 Factoring by Grouping**

a. $3x^3 + x^2 - 6x - 2$

$= (3x^3 + x^2) - (6x + 2)$ Group terms.

$= x^2(3x + 1) - 2(3x + 1)$ Factor each group.

$= (3x + 1)(x^2 - 2)$ Distributive property

b. $2ac - 2ad + bc - bd$

$= (2ac - 2ad) + (bc - bd)$ Group terms.

$= 2a(c - d) + b(c - d)$ Factor each group.

$= (c - d)(2a + b)$ Distributive property

■

Here is a checklist for factoring polynomials.

Factoring Polynomials

1. Look for common factors.
2. Look for special polynomial forms.
3. Use factor pairs.
4. If there are four terms, try grouping.

EXERCISES FOR SECTION 5

In Exercises 1–8, factor out the common factor.

1. $5x - 15$
2. $3y - 24$
3. $2x^3 + 2x$
4. $5x^3 - 20x$
5. $yz^3 - 3yz^2 + 2yz$
6. $2xy^3 + xy^2 - 3xy$
7. $2x(x + 3) - 5(x + 3)$
8. $(y + 1)^2 + 3(y + 1)$

In Exercises 9–14, factor the difference of two squares.

9. $x^2 - 81$
10. $z^2 - 49$
11. $9y^2 - 16$
12. $64 - 25y^2$
13. $16 - (x + 2)^2$
14. $(2x - 1)^2 - 16$

In Exercises 15–20, factor the perfect square trinomial.

15. $x^2 - 6x + 9$
16. $y^2 + 8y + 16$
17. $36y^2 + 12y + 1$
18. $4z^2 - 4z + 1$
19. $9z^2 - 24z + 16$
20. $25x^2 - 20x + 4$

In Exercises 21–28, factor the sum or difference of two cubes.

21. $x^3 - 27$
22. $y^3 - 8$
23. $z^3 + 64$
24. $x^3 + 125$
25. $27y^3 - 8$
26. $64z^3 + 27$
27. $1 - x^3$
28. $27 - y^3$

In Exercises 29–46, factor the trinomial.

29. $x^2 + 9x + 14$
30. $x^2 + 15x + 56$
31. $y^2 - 11y + 30$
32. $y^2 - 13y + 36$
33. $z^2 + z - 20$
34. $z^2 - 5x - 24$
35. $6t^2 + 5t + 1$
36. $12t^2 - 7t + 1$
37. $14u^2 - 33u - 5$
38. $18u^2 - 43u - 5$
39. $10v^2 + 23v + 12$
40. $15v^2 - 26v + 8$
41. $12x^2 + 11x - 15$
42. $8x^2 - 14x - 15$
43. $2x^2 - 3xy + y^2$
44. $3x^2 + 4xy + y^2$
45. $6x^2 + 11xy - 10y^2$
46. $15x^2 + 29xy - 14y^2$

In Exercises 47–54, factor by grouping.

47. $x^3 + x^2 - 3x - 3$
48. $x^3 - 4x^2 + 5x - 20$
49. $2x^3 - 3x^2 + 2x - 3$
50. $3x^3 + x^2 - 15x - 5$
51. $x^6 - 3x^4 + x^2 - 3$
52. $x^6 + 2x^4 + x^2 + 2$
53. $2ac + 6ad - bc - 3bd$
54. $3uw + 12uz - 2vw - 8vz$

In Exercises 55–86, factor completely.

55. $x^3 + x$
56. $x^3 + x^2$
57. $x^3 - 4x$
58. $6x^2 - 54$
59. $9x^2 + 12x + 4$
60. $25x^2 - 40x + 16$
61. $4y^3 - 20y^2 + 25y$
62. $18y^3 + 48y^2 + 32y$
63. $4x^2 + 8x - 60$
64. $2x^3 - 16x^2 + 14x$
65. $y - y^3$
66. $16y - y^3$
67. $3x^4 + 24x$
68. $4x^3 + 108$
69. $8 - 22y - 6y^2$
70. $5y + 3y^2 - 2y^3$
71. $z - 8z^4$
72. $16z - 2z^4$
73. $2(5x + 1)^2 - 18$
74. $5(2x - 3)^2 - 20$
75. $x^3y - xy^3$
76. $x^2y^3 - x^4y$
77. $15x^3 - 22x^2 + 8x$
78. $12x^2 + 22x - 20$
79. $3x^2 + 13xy - 10y^2$
80. $2x^2 + 3xy - 14y^2$
81. $2ac - 2bd + 4ad - bc$
82. $6ac - 2bd + 4bc - 3ad$
83. $x^3 + 2x^2 - x - 2$
84. $x^3 - 3x^2 - 4x + 12$

85. $x^4 + x^3 - 9x^2 - 9x$

86. $x^4 - 4x^3 - x^2 + 4x$

In Exercises 87–89, solve the problem.

87. Show that the grouping

$$(2ac + bc) - (2ad + bd)$$

leads to the same factorization in Example 8b.

88. The polynomial $x(50 - x)$ gives the area of Tom's garden, where x is the length of one side in feet. Find the area of the garden if $x = 20$ ft.

89. The polynomial $x(16 - 2x)(20 - 2x)$ gives the volume of Sue's water tank, where x is the height of the tank in feet. Find the volume of the tank if its height is 4 ft.

6 FRACTIONAL EXPRESSIONS

Introduction • Domain • Reducing Rational Expressions • Operations with Rational Expressions • Compound Rational Expressions

Objective

Students will be able to add, subtract, multiply, divide, and simplify fractional expressions, and will be able to determine the domain of an expression.

Motivate

Ask . . .

Are $x^3 - 2x + 14$ and $\dfrac{(x + 2)}{(x - 3)}$ meaningful expressions for all real numbers x? **(No, $\dfrac{(x + 2)}{(x - 3)}$ is not meaningful if $x = 3$.)**

Introduction

A quotient of two algebraic expressions, besides being another algebraic expression, is called a **fractional expression,** or simply a fraction when understood in context. If the quotient can be written as the ratio of two polynomials, the fractional expression is a **rational expression.** Here are examples of each.

$$\frac{x^2 - 5x + 2}{\sqrt{x^2 + 1}} \qquad \frac{2x^3 - x^2 + 1}{5x^2 - x - 3}$$

The one on the left is a fractional expression but not a rational expression. The other is both a fractional expression and a rational expression.

Teaching Note

If your students are familiar with the language of functions, compare the domain of an expression to the domain of a function.

Domain

Unlike polynomials, which are defined for all real numbers, some algebraic expressions are not defined for some real numbers. The set of numbers for which an expression is defined is the **domain of the expression.**

EXAMPLE 1 Finding Domains of Algebraic Expressions

Find the domain of the expression.

a. $3x^2 - x + 5$ **b.** $\sqrt{x - 1}$ **c.** $\dfrac{x}{x - 2}$

Solution

a. The domain of $3x^2 - x + 5$, like that of any polynomial, is the set of all real numbers.

b. Because only nonnegative numbers have square roots, $x - 1 \geq 0$, or $x \geq 1$. In interval notation the domain is $[1, \infty)$.

c. Because division by zero is undefined, $x - 2 \neq 0$, or $x \neq 2$. The domain is the set of all real numbers except 2. ■

Reducing Rational Expressions

We can write rational expressions in simpler form using $uz/(vz) = u/v$ just as we did with fractions (rational numbers) in Section 2. This requires that we first factor the numerator and denominator into common factors. When all common factors have been removed, the expression is in reduced form.

■ EXAMPLE 2 Reducing Rational Expressions

Write $\dfrac{x^2 - 3x}{x^2 - 9}$ in reduced form.

Solution

$$\frac{x^2 - 3x}{x^2 - 9} = \frac{x(x - 3)}{(x + 3)(x - 3)} \qquad \text{Factor completely.}$$

$$= \frac{x}{x + 3}, \quad x \neq 3 \qquad \text{Remove common factors.}$$

We include $x \neq 3$ as part of the reduced form because 3 is not in the domain of the original expression and thus should not be in the domain of the final expression. ■

Two algebraic expressions are **equivalent** if they have the same domain and have the same value for all numbers in their domain. The reduced form of a rational expression must have the same domain as the original rational expression.

Operations with Rational Expressions

We multiply and divide rational expressions using the fraction operations listed in Section 2:

$$\frac{u}{v} \cdot \frac{w}{z} = \frac{uw}{vz}, \quad v \neq 0,\ z \neq 0 \qquad \text{and} \qquad \frac{u}{v} \div \frac{w}{z} = \frac{uz}{vw}, \quad v \neq 0,\ z \neq 0,\ w \neq 0$$

The division step shown above is often referred to as *invert and multiply.*

EXAMPLE 3 **Multiplying and Dividing Rational Expressions**

a. $\dfrac{2x^2 + 11x - 21}{x^3 + 2x^2 + 4x} \cdot \dfrac{x^3 - 8}{x^2 + 5x - 14}$

$= \dfrac{(2x - 3)(x + 7)}{x(x^2 + 2x + 4)} \cdot \dfrac{(x - 2)(x^2 + 2x + 4)}{(x - 2)(x + 7)}$ Factor completely.

$= \dfrac{2x - 3}{x}, \quad x \neq 2, \quad x \neq -7$ Remove common factors.

b. $\dfrac{x^3 + 1}{x^2 - x - 2} \div \dfrac{x^2 - x + 1}{x^2 - 4x + 4}$

$= \dfrac{(x^3 + 1)(x^2 - 4x + 4)}{(x^2 - x - 2)(x^2 - x + 1)}$ Invert and multiply.

$= \dfrac{(x + 1)(x^2 - x + 1)(x - 2)^{2\,1}}{(x + 1)(x - 2)(x^2 - x + 1)}$ Factor completely.

$= x - 2, \quad x \neq -1, \quad x \neq 2$ Remove common factors. ■

To add or subtract rational expressions, we either use the definition

$$\frac{u}{v} + \frac{w}{z} = \frac{uz + vw}{vz}, \quad v \neq 0, \quad z \neq 0,$$

or we find the LCD and add or subtract the corresponding numerators.

EXAMPLE 4 **Adding Rational Expressions**

$\dfrac{x}{3x - 2} + \dfrac{3}{x - 5} = \dfrac{x(x - 5) + 3(3x - 2)}{(3x - 2)(x - 5)}$ Definition of addition

$= \dfrac{x^2 - 5x + 9x - 6}{(3x - 2)(x - 5)}$ Distributive property

$= \dfrac{x^2 + 4x - 6}{(3x - 2)(x - 5)}$ Combine like terms. ■

NOTE

The numerator, $x^2 + 4x - 6$, of the final expression in Example 4 is a prime polynomial. Thus there are no common factors.

If the denominators of fractions have common factors, then it is often more efficient to find the LCD before adding or subtracting the fractions. The **LCD (least common denominator)** is the product of all the prime factors in the denominators, where each factor is raised to the greatest power found in any one denominator for that factor.

■ EXAMPLE 5 Using the LCD

Write the following expression as a fraction in reduced form.

$$\frac{2}{x^2 - 2x} + \frac{1}{x} - \frac{3}{x^2 - 4}$$

Solution

The factored denominators are $x(x - 2)$, x, and $(x + 2)(x - 2)$, respectively. The LCD is $x(x - 2)(x + 2)$.

$$\frac{2}{x^2 - 2x} + \frac{1}{x} - \frac{3}{x^2 - 4}$$

$$= \frac{2}{x(x - 2)} + \frac{1}{x} - \frac{3}{(x - 2)(x + 2)}$$

$$= \frac{2(x + 2)}{x(x - 2)(x + 2)} + \frac{(x - 2)(x + 2)}{x(x - 2)(x + 2)} - \frac{3x}{x(x - 2)(x + 2)} \qquad u/v = (uz)/(vz)$$

$$= \frac{2(x + 2) + (x - 2)(x + 2) - 3x}{x(x - 2)(x + 2)} \qquad \text{Combine numerators.}$$

$$= \frac{2x + 4 + x^2 - 4 - 3x}{x(x - 2)(x + 2)} \qquad \text{Expand terms.}$$

$$= \frac{x^2 - x}{x(x - 2)(x + 2)} \qquad \text{Simplify.}$$

$$= \frac{x(x - 1)}{x(x - 2)(x + 2)} \qquad \text{Factor the numerator.}$$

$$= \frac{x - 1}{(x - 2)(x + 2)}, \quad x \neq 0 \qquad \text{Remove common factors.}$$

■

Compound Rational Expressions

Sometimes a complicated algebraic expression needs to be changed to a more familiar form before we can work with it. A **compound fraction** (sometimes called a **complex fraction**), in which the numerators and denominators may themselves contain fractions, is such an example. One way to simplify a compound fraction is to write both the numerator and denominator as single fractions and then invert and multiply. If the fraction takes the form of a rational expression, then we write the expression in reduced or simplest form.

Notes on Examples

Examples 6 and 7 illustrate two methods of simplifying complex fractions. Students often wonder how to determine which method to use for a particular problem, so it is worth pointing out that most complex fractions can be simplified either way.

■ **EXAMPLE 6 Simplifying a Compound Fraction**

$$\frac{3 - \dfrac{7}{x+2}}{1 - \dfrac{1}{x-3}} = \frac{\dfrac{3(x+2) - 7}{x+2}}{\dfrac{(x-3) - 1}{x-3}} \qquad \text{Form fractions.}$$

$$= \frac{\dfrac{3x-1}{x+2}}{\dfrac{x-4}{x-3}} \qquad \text{Simplify.}$$

$$= \frac{(3x-1)(x-3)}{(x+2)(x-4)}, \quad x \neq 3 \qquad \text{Invert and multiply.}$$

■

A second way to simplify a compound fraction is to multiply the numerator and denominator by the LCD of all fractions in the numerator and denominator, as Example 7 illustrates.

Alert

When using the LCD to simplify a compound fraction, some students will attempt to find the LCD for the numerator and denominator expressions separately. Remind students that the numerator and denominator must be multiplied by the same factor in order to obtain an equivalent expression.

■ **EXAMPLE 7 Simplifying Another Compound Fraction**

Use the LCD to simplify the compound fraction.

$$\frac{\dfrac{1}{a^2} - \dfrac{1}{b^2}}{\dfrac{1}{a} - \dfrac{1}{b}}$$

Solution

The LCD of the four fractions in the numerator and denominator is a^2b^2.

$$\frac{\dfrac{1}{a^2} - \dfrac{1}{b^2}}{\dfrac{1}{a} - \dfrac{1}{b}} = \frac{\left(\dfrac{1}{a^2} - \dfrac{1}{b^2}\right)a^2b^2}{\left(\dfrac{1}{a} - \dfrac{1}{b}\right)a^2b^2} \qquad \text{Multiply numerator and denominator by LCD.}$$

$$= \frac{b^2 - a^2}{ab^2 - a^2b} \qquad \text{Simplify.}$$

$$= \frac{(b+a)(b-a)}{ab(b-a)} \qquad \text{Factor completely.}$$

$$= \frac{b+a}{ab}, \quad a \neq b \qquad \text{Remove common factors.}$$

■

Follow-up

Discuss whether a polynomial can be written as a rational expression. $\left(\textbf{Yes, } \frac{P(x)}{1}\right)$

Assignment Guide

Day 1: Ex. 3–78, multiples of 3

Notes on Exercises

Ex. 1–10 and 19–24 emphasize the concept of the domain of an expression.
Ex. 25–62 are simplification problems that provide practice with the rules for working with fractional expressions.
Ex. 63–70 involve compound fractions.
Ex. 71–78 require students to apply the properties of exponents or radicals.

Ongoing Assessment

Self-Assessment: Ex. 7, 35, 43, 65
Embedded Assessment: Ex. 24, 66

The techniques of this section can be extended to algebraic expressions involving radicals and fractional exponents, which often occur in calculus.

EXAMPLE 8 Simplifying an Algebraic Expression

Simplify

$$\frac{1}{\sqrt{1-x^2}} + x^2(1-x^2)^{-3/2}.$$

Solution

$$\begin{aligned}
\frac{1}{\sqrt{1-x^2}} + x^2(1-x^2)^{-3/2} \\
&= \frac{1}{(1-x^2)^{1/2}} + \frac{x^2}{(1-x^2)^{3/2}} && \text{Positive exponents} \\
&= \frac{1-x^2}{(1-x^2)^{3/2}} + \frac{x^2}{(1-x^2)^{3/2}} && \text{Form common denominators.} \\
&= \frac{(1-x^2)+x^2}{(1-x^2)^{3/2}} && \text{Combine numerators.} \\
&= \frac{1}{(1-x^2)^{3/2}} && \text{Simplify.}
\end{aligned}$$

■

EXERCISES FOR SECTION 6

In Exercises 1–10, find the domain of the expression.

1. $5x^2 - 3x - 7$
2. $2x - 5$
3. $\sqrt{x-4}$
4. $\dfrac{2}{\sqrt{x+3}}$
5. $\dfrac{2x+1}{x^2+3x}$
6. $\dfrac{x^2-2}{x^2-4}$
7. $\dfrac{x}{x-1}, \quad x \neq 2$
8. $\dfrac{3x-1}{x-2}, \quad x \neq 0$
9. $x^2 + x^{-1}$
10. $x(x+1)^{-2}$

In Exercises 11–18, find the missing numerator or denominator so that the two rational expressions are equal.

11. $\dfrac{2}{3x} = \dfrac{?}{12x^3}$
12. $\dfrac{5}{2y} = \dfrac{15y}{?}$
13. $\dfrac{x-4}{x} = \dfrac{x^2-4x}{?}$
14. $\dfrac{x}{x+2} = \dfrac{?}{x^2-4}$
15. $\dfrac{x+3}{x-2} = \dfrac{?}{x^2+2x-8}$
16. $\dfrac{x-4}{x+5} = \dfrac{x^2-x-12}{?}$
17. $\dfrac{x^2-3x}{?} = \dfrac{x-3}{x^2+2x}$
18. $\dfrac{?}{x^2-9} = \dfrac{x^2+x-6}{x-3}$

In Exercises 19–24, consider the original fraction and its reduced form from the specified example. Explain why the given restriction is needed on the reduced form.

19. Example 3a, $x \neq 2$, $x \neq -7$
20. Example 3b, $x \neq -1$, $x \neq 2$
21. Example 4, none
22. Example 5, $x \neq 0$
23. Example 6, $x \neq 3$
24. Example 7, $a \neq b$

In Exercises 25–38, write the expression in reduced form.

25. $\dfrac{18x^3}{15x}$

26. $\dfrac{75y^2}{9y^4}$

27. $\dfrac{x^3}{x^2 - 2x}$

28. $\dfrac{2y^2 + 6y}{4y + 12}$

29. $\dfrac{z^2 - 25}{z^2 - 5z}$

30. $\dfrac{z^2 - 3z}{9 - z^2}$

31. $\dfrac{x^2 + 6x + 9}{x^2 - x - 12}$

32. $\dfrac{x^2 + 2x - 35}{x^2 - 10x + 25}$

33. $\dfrac{y^2 - y - 30}{y^2 - 3y - 18}$

34. $\dfrac{y^3 + 4y^2 - 21y}{y^2 - 49}$

35. $\dfrac{8z^3 - 1}{2z^2 + 5z - 3}$

36. $\dfrac{2z^3 + 6z^2 + 18z}{z^3 - 27}$

37. $\dfrac{x^3 + 2x^2 - 3x - 6}{x^3 + 2x^2}$

38. $\dfrac{y^2 + 3y}{y^3 + 3y^2 - 5y - 15}$

In Exercises 39–62, simplify.

39. $\dfrac{3}{x - 1} \cdot \dfrac{x^2 - 1}{9}$

40. $\dfrac{x + 3}{7} \cdot \dfrac{14}{2x + 6}$

41. $\dfrac{x + 3}{x - 1} \cdot \dfrac{1 - x}{x^2 - 9}$

42. $\dfrac{18x^2 - 3x}{3xy} \cdot \dfrac{12y^2}{6x - 1}$

43. $\dfrac{x^3 - 1}{2x^2} \cdot \dfrac{4x}{x^2 + x + 1}$

44. $\dfrac{y^3 + 2y^2 + 4y}{y^3 + 2y^2} \cdot \dfrac{y^2 - 4}{y^3 - 8}$

45. $\dfrac{2y^2 + 9y - 5}{y^2 - 25} \cdot \dfrac{y - 5}{2y^2 - y}$

46. $\dfrac{y^2 + 8y + 16}{3y^2 - y - 2} \cdot \dfrac{3y^2 + 2y}{y + 4}$

47. $\dfrac{1}{2x} \div \dfrac{1}{4}$

48. $\dfrac{4x}{y} \div \dfrac{8y}{x}$

49. $\dfrac{x^2 - 3x}{14y} \div \dfrac{2xy}{3y^2}$

50. $\dfrac{7x - 7y}{4y} \div \dfrac{14x - 14y}{3y}$

51. $\dfrac{\dfrac{2x^2y}{(x - 3)^2}}{\dfrac{8xy}{x - 3}}$

52. $\dfrac{\dfrac{x^2 - y^2}{2xy}}{\dfrac{y^2 - x^2}{4x^2y}}$

53. $\dfrac{2x + 1}{x + 5} - \dfrac{3}{x + 5}$

54. $\dfrac{3}{x - 2} + \dfrac{x + 1}{x - 2}$

55. $\dfrac{x}{2x + 1} - \dfrac{2}{x - 3}$

56. $\dfrac{x}{x - 1} + \dfrac{x + 1}{3x - 4}$

57. $\dfrac{2}{x^2 - 1} - \dfrac{5}{x^2 - 3x - 4}$

58. $\dfrac{7}{y^2 - 3y - 10} - \dfrac{4}{y^2 - 4}$

59. $\dfrac{2x - 1}{x^2 - x - 2} + \dfrac{x - 3}{x^2 - 3x + 2}$

60. $\dfrac{x + 1}{x^2 - 5x + 6} - \dfrac{3x + 11}{x^2 - x - 6}$

61. $\dfrac{3}{x^2 + 3x} - \dfrac{1}{x} - \dfrac{6}{x^2 - 9}$

62. $\dfrac{5}{x^2 + x - 6} - \dfrac{2}{x - 2} + \dfrac{4}{x^2 - 4}$

In Exercises 63–70, simplify the compound fraction.

63. $\dfrac{\dfrac{x}{y^2} - \dfrac{y}{x^2}}{\dfrac{1}{y^2} - \dfrac{1}{x^2}}$

64. $\dfrac{\dfrac{1}{x} + \dfrac{1}{y}}{\dfrac{1}{x^2} - \dfrac{1}{y^2}}$

65. $\dfrac{2x + \dfrac{13x - 3}{x - 4}}{2x + \dfrac{x + 3}{x - 4}}$

66. $\dfrac{2 - \dfrac{13}{x + 5}}{2 + \dfrac{3}{x - 3}}$

67. $\dfrac{\dfrac{1}{(x + h)^2} - \dfrac{1}{x^2}}{h}$

68. $\dfrac{\dfrac{x + h}{x + h + 2} - \dfrac{x}{x + 2}}{h}$

69. $\dfrac{\dfrac{b}{a} - \dfrac{a}{b}}{\dfrac{1}{a} - \dfrac{1}{b}}$

70. $\dfrac{\dfrac{1}{a} + \dfrac{1}{b}}{\dfrac{b}{a} - \dfrac{a}{b}}$

In Exercises 71–78, simplify.

71. $\left(\dfrac{1}{x} + \dfrac{1}{y}\right)(x + y)^{-1}$

72. $\dfrac{(x + y)^{-1}}{(x - y)^{-1}}$

73. $x^{-1} + y^{-1}$

74. $(x^{-1} + y^{-1})^{-1}$

75. $\sqrt{1 - x} - \dfrac{x}{2\sqrt{1 - x}}$

76. $\sqrt{1 - x^2} - \dfrac{x^2}{\sqrt{1 - x^2}}$

77. $(3 - x)^{-1/2} + \dfrac{x(3 - x)^{-3/2}}{2}$

78. $(x^2 + 2)^{-1/2} - x^2(x^2 + 2)^{-3/2}$

7
LINEAR EQUATIONS AND INEQUALITIES

Objective

Students will be able to solve linear equations and inequalities in one variable.

Motivate

Ask . . .

If $3x + 2 = 14$, then what is the value of x? **(4)**

Solutions of Equations • Linear Equations in One Variable • Linear Inequalities in One Variable

Solutions of Equations

Recall that an equation is a statement of equality between two expressions. For example,

$$2x - 3 = 7, \qquad 2x^2 + 5x - 3 = 0, \qquad \text{and} \qquad \frac{2}{x + 1} = 7$$

are equations in x. To **solve an equation in x** means to find all values of x for which the equation is true. A **solution of an equation in x** is a value of x for which the equation is true.

■ **EXAMPLE 1** **Confirming a Solution**

Show that $x = -2$ is a solution to the equation $x^3 - x + 6 = 0$.

Solution

$$\begin{aligned} x^3 - x + 6 &= (-2)^3 - (-2) + 6 && \text{Substitute } -2 \text{ for } x. \\ &= -8 + 2 + 6 && \text{Simplify.} \\ &= 0 && \text{The equation checks true.} \end{aligned}$$

■

Linear Equations in One Variable

The most basic equation in algebra is a *linear equation.*

Definition: Linear Equation in x

A **linear equation in x** is one that can be written in the form

$$ax + b = 0,$$

where a and b are real numbers with $a \neq 0$.

The equation $2z - 4 = 0$ is linear in z. It turns out that a linear equation in one variable has exactly one solution. We solve the equation by transforming it into an *equivalent equation* whose solution is obvious. Two or more equations are **equivalent** if they have the same solutions. For example, the equations $2z - 4 = 0$, $2z = 4$, and $z = 2$ are all equivalent. Here are operations that produce equivalent equations.

Teaching Note

Compare an equation to a scale. If the scale balances, it is possible to add or subtract equal weights on each side, and the scale will still balance.

Alert

Point out that multiplying or dividing both sides of an equation by x does *not* always result in an equivalent equation.

Operations for Equivalent Equations

An equivalent equation is obtained if one or more of the following operations are performed.

Operation	Given Equation	Equivalent Equation
1. Combine like terms, reduce fractions, and remove grouping symbols.	$2x + x = \frac{3}{9}$	$3x = \frac{1}{3}$
2. Perform the same operation on both sides.		
a. Add (-3).	$x + 3 = 7$	$x = 4$
b. Subtract $(2x)$.	$5x = 2x + 4$	$3x = 4$
c. Multiply by a nonzero constant (1/3).	$3x = 12$	$x = 4$
d. Divide by a nonzero constant (3).	$3x = 12$	$x = 4$

The next two examples illustrate how to use equivalent equations to solve linear equations.

■ EXAMPLE 2 Solving a Linear Equation

Solve $2(2x - 3) + 3(x + 1) = 5x + 2$.

Solution

$2(2x - 3) + 3(x + 1) = 5x + 2$	The given equation
$4x - 6 + 3x + 3 = 5x + 2$	Remove grouping symbols.
$7x - 3 = 5x + 2$	Combine like terms.
$7x = 5x + 5$	Add 3.
$2x = 5$	Subtract $5x$.
$x = 2.5$	Divide by 2.

To confirm that 2.5 is indeed the solution, you can substitute it for x in the

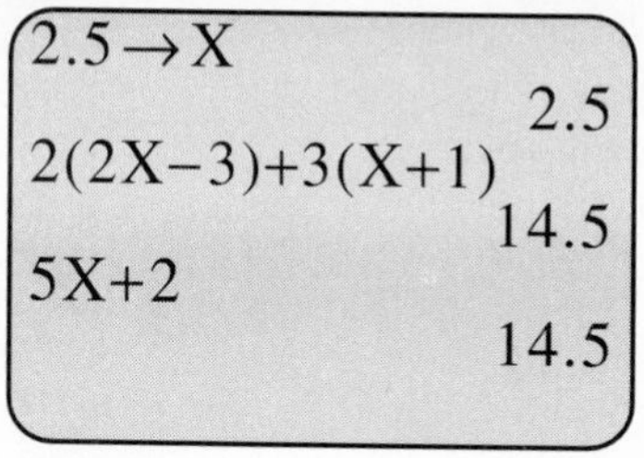

```
2.5→X
                    2.5
2(2X−3)+3(X+1)
                   14.5
5X+2
                   14.5
```

Figure 7

original equation. Figure 7 shows that each side of the original equation is equal to 14.5 if $x = 2.5$. ■

If an equation involves fractions, find the LCD of the fractions and multiply both sides by the LCD. This is sometimes referred to as *clearing the equation of fractions.* Example 3 illustrates.

■ EXAMPLE 3 Solving Another Linear Equation

Solve

$$\frac{5y - 2}{8} = 2 + \frac{y}{4}.$$

Solution

The denominators are 8, 1, and 4. The LCD of the fractions is 8.

$$\frac{5y - 2}{8} = 2 + \frac{y}{4} \qquad \text{The given equation}$$

$$8 \cdot \frac{5y - 2}{8} = 8 \cdot 2 + 8 \cdot \frac{y}{4} \qquad \text{Multiply by the LCD 8.}$$

$$5y - 2 = 16 + 2y \qquad \text{Simplify.}$$

$$5y = 18 + 2y \qquad \text{Add 2.}$$

$$3y = 18 \qquad \text{Subtract } 2y.$$

$$y = 6 \qquad \text{Divide by 3.}$$

We leave it to you to check the solution. ■

Linear Inequalities in One Variable

We used inequalities to describe order on the number line. For example, if x is to the left of 2 on the number line, or x is any real number less than 2, we write $x < 2$. The most basic inequality in algebra is a *linear inequality.*

Definition: Linear Inequality in x

A **linear inequality in x** is one that can be written in the form

$$ax + b < 0, \quad ax + b \le 0, \quad ax + b > 0, \quad \text{or} \quad ax + b \ge 0,$$

where a and b are real numbers with $a \neq 0$.

Teaching Note

Give several numerical examples to demonstrate why the inequality sign is reversed when both sides are multiplied or divided by negative numbers. For example, $2 < 3$, but $-2 > -3$.

REMARK

Multiplying or dividing an inequality by a positive number preserves the direction of the inequality. Multiplying or dividing by a negative number reverses the direction.

To **solve an inequality in x** means to find all values of x for which the inequality is true. A **solution of an inequality in x** is such a value for x. Here is a list of the properties we use to solve inequalities.

Properties of Inequalities

Let u, v, and w be real numbers, variables, or algebraic expressions, and c a real number.

Transitive:	If $u < v$ and $v < w$, then $u < w$.
Addition:	If $u < v$, then $u + w < v + w$.
Multiplication:	If $u < v$ and $c > 0$, then $uc < vc$.
	If $u < v$ and $c < 0$, then $uc > vc$.

The above properties are true if $<$ is replaced by $\leq$. There are similar properties for $>$ and $\geq$.

It turns out that the solutions of a linear inequality in one variable form an interval. We solve the inequality by transforming it into an *equivalent inequality* whose solutions are obvious. Two or more inequalities are **equivalent** if they have the same solutions. Example 4 illustrates.

■ EXAMPLE 4 Solving a Linear Inequality

$$3(x - 1) + 2 \leq 5x + 6 \quad \text{The given inequality}$$

$$3x - 3 + 2 \leq 5x + 6 \quad \text{Distributive property}$$

$$3x - 1 \leq 5x + 6 \quad \text{Simplify.}$$

$$3x \leq 5x + 7 \quad \text{Add 1.}$$

$$-2x \leq 7 \quad \text{Subtract } 5x.$$

$$\left(-\frac{1}{2}\right) \cdot -2x \geq \left(-\frac{1}{2}\right) \cdot 7 \quad \text{Multiply by } -1/2. \text{ Note the inequality reverses.}$$

$$x \geq -3.5$$

The solution of the inequality is the set of all real numbers greater than or equal to -3.5. In interval notation, the solution is $[-3.5, \infty)$. ■

The solution to a linear inequality is an interval of real numbers, as illustrated in Example 4. Thus, we can also show the solution with a **number line graph.** When we want you to graph the solution of an inequality, we will simply say, "graph the inequality."

EXAMPLE 5 Solving Another Linear Inequality

Solve and graph the inequality.

$$\frac{x}{3} + \frac{1}{2} > \frac{x}{4} + \frac{1}{3}$$

Solution

The LCD of the denominators is 12.

$$\frac{x}{3} + \frac{1}{2} > \frac{x}{4} + \frac{1}{3} \quad \text{The given inequality}$$

$$12 \cdot \left(\frac{x}{3} + \frac{1}{2}\right) > 12 \cdot \left(\frac{x}{4} + \frac{1}{3}\right) \quad \text{Multiply by 12.}$$

$$4x + 6 > 3x + 4 \quad \text{Simplify.}$$

$$x + 6 > 4 \quad \text{Subtract } 3x.$$

$$x > -2 \quad \text{Subtract 6.}$$

The solution is the interval $(-2, \infty)$. The graph is shown in Figure 8. ■

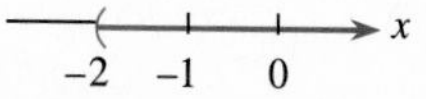

Figure 8

Notes on Examples

In Example 6, point out that the double inequality $-7 < x \leq 5$ means that x must satisfy *both* inequalities, $-7 < x$ and $x \leq 5$.

Follow-up

Ask why one should avoid multiplying both sides of an equation or inequality by 0.

Assignment Guide

Day 1: Ex. 3–60, multiples of 3, 61, 62

Notes on Exercises

Ex. 1–6 and 31–34 emphasize the definition of a solution of an equation or inequality.
Ex. 13–30 and 35–42 require students to solve equations and inequalities.
Ex. 43–54 involve double inequalities.
Ex. 61–62 require students to solve a formula for one of the variables.

Ongoing Assessment

Self-Assessment: Ex. 25, 53, 59
Embedded Assessment: Ex. 30, 62

Sometimes two inequalities are combined in a **double inequality,** whose solution is a double inequality with x isolated as the middle term. Example 6 illustrates.

EXAMPLE 6 Solving a Double Inequality

$$-3 < \frac{2x + 5}{3} \leq 5 \quad \text{The given double inequality}$$

$$-9 < 2x + 5 \leq 15 \quad \text{Multiply by 3.}$$

$$-14 < 2x \leq 10 \quad \text{Subtract 5.}$$

$$-7 < x \leq 5 \quad \text{Divide by 2.}$$

The solution interval is $(-7, 5]$. The graph is shown in Figure 9. ■

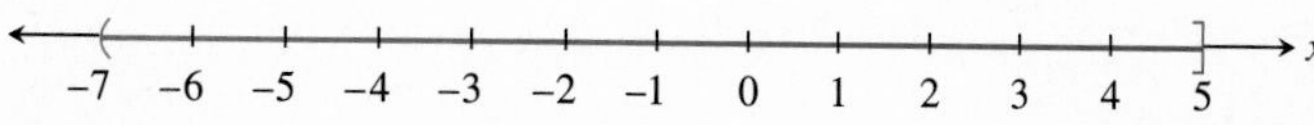

Figure 9

EXERCISES FOR SECTION 7

In Exercises 1–6, find which values of x are solutions of the equation.

1. $7x + 5 = 4x - 7$
a. $x = -4$ **b.** $x = 0$ **c.** $x = 2$

2. $2x^2 + 5x = 3$
a. $x = -3$ **b.** $x = -\frac{1}{2}$ **c.** $x = \frac{1}{2}$

3. $\dfrac{x}{2} + \dfrac{1}{6} = \dfrac{x}{3}$
a. $x = -1$ **b.** $x = 0$ **c.** $x = 1$

4. $x^3 - 2x^2 - 2x + 4 = 0$
a. $x = -2$ **b.** $x = 0$ **c.** $x = 2$

5. $\sqrt{1 - x^2} + 2 = 3$
a. $x = -2$ **b.** $x = 0$ **c.** $x = 2$

6. $(x - 2)^{1/3} = 2$
a. $x = -6$ **b.** $x = 8$ **c.** $x = 10$

In Exercises 7–12, determine whether the equation is linear in x.

7. $5 - 3x = 0$
8. $5 = 10/2$
9. $x + 3 = x - 5$
10. $x - 3 = x^2$
11. $2\sqrt{x} + 5 = 10$
12. $x^3 = x + 1$

In Exercises 13–30, solve the equation.

13. $3x = 24$
14. $4x = -16$
15. $3x - 4 = 8$
16. $2x - 9 = 3$
17. $2x - 3 = 4x - 5$
18. $4 - 2x = 3x - 6$
19. $4 - 3y = 2(y + 4)$
20. $4(y - 2) = 5y$
21. $\frac{1}{2}x = \frac{7}{8}$
22. $\frac{2}{3}x = \frac{4}{5}$
23. $\frac{1}{2}x + \frac{1}{3} = 1$
24. $\frac{1}{3}x + \frac{1}{4} = 1$
25. $2(3 - 4z) - 5(2z + 3) = z - 17$
26. $3(5z - 3) - 4(2z + 1) = 5z - 2$
27. $\dfrac{2x - 3}{4} + 5 = 3x$
28. $2x - 4 = \dfrac{4x - 5}{3}$
29. $\dfrac{x + 5}{8} - \dfrac{x - 2}{2} = \dfrac{1}{3}$
30. $\dfrac{x - 1}{3} + \dfrac{x + 5}{4} = \dfrac{1}{2}$

In Exercises 31–34, find which values of x are solutions of the inequality.

31. $2x - 3 < 7$
a. $x = 0$ **b.** $x = 5$ **c.** $x = 6$

32. $3x - 4 \geq 5$
a. $x = 0$ **b.** $x = 3$ **c.** $x = 4$

33. $-1 < 4x - 1 \leq 11$
a. $x = 0$ **b.** $x = 2$ **c.** $x = 3$

34. $-3 \leq 1 - 2x \leq 3$
a. $x = -1$ **b.** $x = 0$ **c.** $x = 2$

In Exercises 35–42, solve the inequality, and draw a number line graph of the solution.

35. $x - 4 < 2$
36. $x + 3 > 5$
37. $2x - 1 \leq 4x + 3$
38. $3x - 1 \geq 6x + 8$
39. $2 \leq x + 6 < 9$
40. $-1 \leq 3x - 2 < 7$
41. $2(5 - 3x) + 3(2x - 1) \leq 2x + 1$
42. $4(1 - x) + 5(1 + x) > 3x - 1$

In Exercises 43–54, solve the inequality.

43. $\dfrac{5x + 7}{4} \leq -3$
44. $\dfrac{3x - 2}{5} > -1$
45. $4 \geq \dfrac{2y - 5}{3} \geq -2$
46. $1 > \dfrac{3y - 1}{4} > -1$
47. $0 \leq 2z + 5 < 8$
48. $-6 < 5x - 1 < 0$
49. $\dfrac{x - 5}{4} + \dfrac{3 - 2x}{3} < -2$
50. $\dfrac{3 - x}{2} + \dfrac{5x - 2}{3} < -1$
51. $\dfrac{2y - 3}{2} + \dfrac{3y - 1}{5} < y - 1$
52. $\dfrac{3 - 4y}{6} - \dfrac{2y - 3}{8} \geq 2 - y$
53. $\dfrac{1}{2}(x - 4) - 2x \leq 5(3 - x)$
54. $\frac{1}{2}(x + 3) + 2(x - 4) < \frac{1}{3}(x - 3)$

In Exercises 55 and 56, explain how the second equation was obtained from the first.

55. $x - 3 = 2x + 3, \quad 2x - 6 = 4x + 6$

56. $2x - 1 = 2x - 4, \quad x - \frac{1}{2} = x - 2$

In Exercises 57–60, determine whether the two equations are equivalent.

57. $3x = 6x + 9, \quad x = 2x + 9$

58. $6x + 2 = 4x + 10, \quad 3x + 1 = 2x + 5$

59. $3x + 2 = 5x - 7, \quad -2x + 2 = -7$

60. $2x + 5 = x - 7, \quad 2x = x - 7$

In Exercises 61 and 62, solve the problem.

61. The formula for the perimeter of a rectangle is

$$P = 2(L + W).$$

Solve this equation for W. (*Hint:* Treat the other variables as constants.)

62. The formula for the area of a trapezoid is

$$A = \frac{1}{2}h(b_1 + b_2).$$

Solve this equation for b_1. (*Hint:* Treat the other variables as constants.)

8 CARTESIAN COORDINATE SYSTEM

Objective

Students will be able to graph points and find distances and midpoints in a coordinate plane, and to write standard-form equations of circles.

Motivate

Ask . . .

If the post office is 3 miles east and 4 miles north of the bank, what is the straight-line distance from the post office to the bank? **(5 miles)**

Cartesian Coordinate System • Distance Formula • Midpoint Formula • Equations of Circles

Cartesian Coordinate System

The points in a plane can be associated with ordered pairs of real numbers, just as the points on a line can be associated with individual real numbers. Such an association creates the **Cartesian coordinate system** (also called the **rectangular coordinate system**) in the plane.

To construct a rectangular coordinate system, or a **coordinate plane,** draw a pair of perpendicular real number lines, one horizontal and the other vertical, with the lines intersecting at their respective origins (Figure 10). The horizontal line is usually the ***x*-axis** and the vertical line is usually the ***y*-axis.** The positive direction on the *x*-axis is to the right, and the positive direction on the *y*-axis is up. Their point of intersection, *O,* is the **origin of the coordinate plane.**

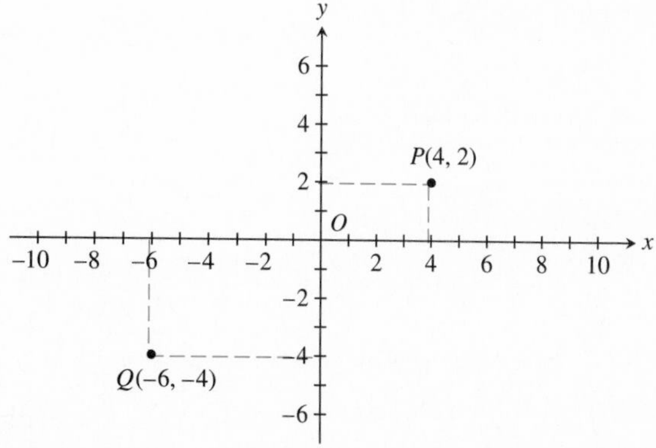

Figure 10 The Cartesian coordinate plane.

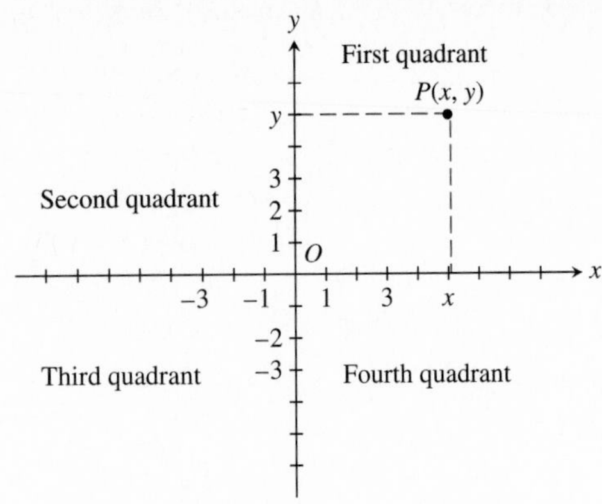

Figure 11 The four quadrants. Points on the x-axis or y-axis are *not* in any quadrant.

The coordinate axes divide the plane into four **quadrants,** as shown in Figure 11. Each point P of the plane is associated with an **ordered pair** (x, y) of real numbers, the **coordinates of the point.** The **x-coordinate** represents the intersection between the x-axis and the perpendicular from P, and the **y-coordinate** represents the intersection between the y-axis and the perpendicular from P. Figure 10 shows the points P and Q with coordinates $(4, 2)$ and $(-6, -4)$, respectively. As with real numbers and a number line, we use the ordered pair (a, b) for both the name of the point and its coordinates.

Table 1 U.S. Exports to Mexico

Year	U.S. Exports (billions of dollars)
1987	14.6
1988	20.6
1989	24.9
1990	28.4
1991	33.3
1992	40.6
1993	41.6

Source: United World Ltd., Inc., as reported in *USA Today,* September 30, 1994.

■ EXAMPLE 1 Plotting Data on U.S. Exports to Mexico

The value in billions of dollars of U.S. exports to Mexico between 1987 and 1993 is given in Table 1. Plot the (year, value) ordered pairs in a rectangular coordinate system.

Solution

The points are plotted in Figure 12. ■

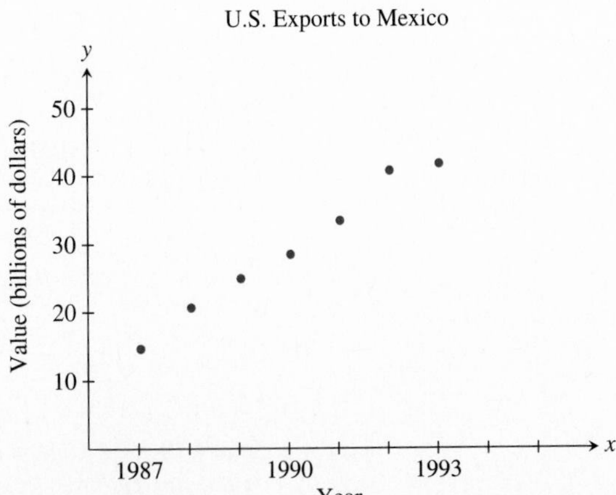

Figure 12

Distance Formula

To find the *distance* between two points that lie on the same horizontal or vertical line in the coordinate plane, we use the distance formula for points on a number line. For example, the distance between the points x_1 and x_2 on the x-axis is $|x_1 - x_2| = |x_2 - x_1|$.

Consider two points $P(x_1, y_1)$ and $Q(x_2, y_2)$ that do not lie on the same horizontal or vertical line, and the right triangle formed by P, Q, and $R(x_2, y_1)$. (See Figure 13.)

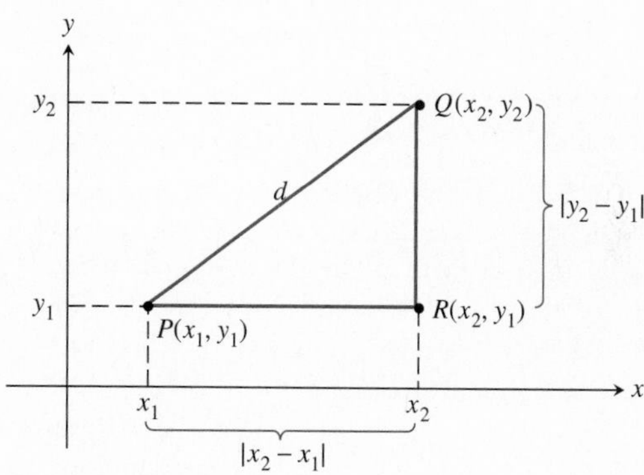

Figure 13

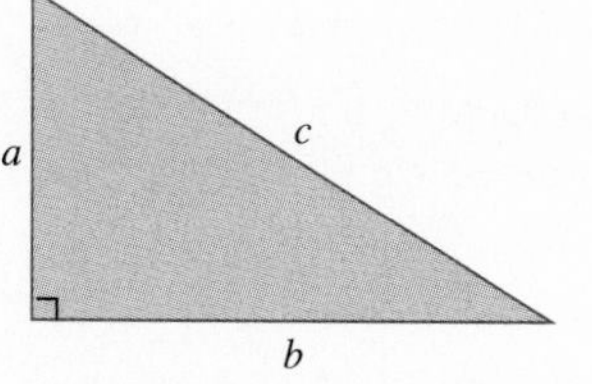

Figure 14 Pythagorean theorem: $c^2 = a^2 + b^2$.

The distance from P to R is $|x_1 - x_2|$, and the distance from R to Q is $|y_1 - y_2|$. By the **Pythagorean theorem** (see Figure 14), the distance d between P and Q is

$$d = \sqrt{|x_1 - x_2|^2 + |y_1 - y_2|^2}.$$

Because $|x_1 - x_2|^2 = (x_1 - x_2)^2$ and $|y_1 - y_2|^2 = (y_1 - y_2)^2$, we obtain the following formula.

Distance Formula (Coordinate Plane)

The **distance** d between points $P(x_1, y_1)$ and $Q(x_2, y_2)$ in the coordinate plane is

$$d = \sqrt{(x_1 - x_2)^2 + (y_1 - y_2)^2}.$$

■ **EXAMPLE 2 Finding the Distance between Two Points**

Find the distance between the points $P(1, 5)$ and $Q(6, 2)$.

Notes on Examples

For Example 2, it may be helpful for students to visualize the distance between the points $P(1, 5)$ and $Q(6, 2)$ by using a pegboard and two golf tees. You may also use your grapher as an "electronic geoboard" and the DRAW menu option to illustrate this concept.

Solution

$$\begin{aligned} d &= \sqrt{(1-6)^2 + (5-2)^2} && \text{The distance formula} \\ &= \sqrt{(-5)^2 + 3^2} \\ &= \sqrt{25 + 9} \\ &= \sqrt{34} \approx 5.83 && \text{Using a calculator} \end{aligned}$$

The distance formula can be used to provide information about geometric figures.

Notes on Examples

In Example 3, the square roots can be simplified. Point out that although this is possible, it is not necessarily desirable because it does not help to solve the problem.

■ **EXAMPLE 3 APPLICATION: Determining Right Triangles**

The converse of the Pythagorean theorem says that if the sum of squares of two sides of a triangle equals the square of the third side, then the triangle is a right triangle. Use this theorem and the distance formula to show that the points $(-3, 4)$, $(1, 0)$, and $(5, 4)$ determine a right triangle.

Solution

We need to show that the lengths of the sides of the triangle satisfy the Pythagorean relationship $a^2 + b^2 = c^2$. The three points are plotted in Figure 15. Applying the distance formula we find that

$$\begin{aligned} a &= \sqrt{(-3-1)^2 + (4-0)^2} = \sqrt{32}, \\ b &= \sqrt{(1-5)^2 + (0-4)^2} = \sqrt{32}, \quad \text{and} \\ c &= \sqrt{(-3-5)^2 + (4-4)^2} = \sqrt{64}. \end{aligned}$$

The triangle is a right triangle because

$$c^2 = 64 = 32 + 32 = a^2 + b^2.$$

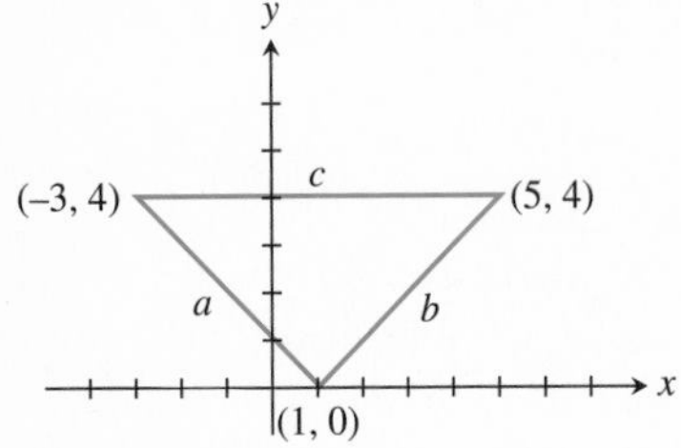

Figure 15

Teaching Notes

The midpoint can be obtained by taking the average of the x-values and the average of the y-values.

Midpoint Formula

When the endpoints of a line segment in the coordinate plane are known, we take the average of their coordinates to find the midpoint of the segment, as suggested by the following formula.

REMARK

If we project the line segment with endpoints (a, b) and (c, d) onto the x-axis and the y-axis, then the midpoints of the segments on the two axes give the coordinates of the midpoint of the line segment in the plane. Draw a figure to support this fact.

Midpoint Formula (Coordinate Plane)

The **midpoint** of the line segment with endpoints (a, b) and (c, d) is

$$\left(\frac{a+c}{2}, \frac{b+d}{2}\right).$$

Teaching Notes

Remind students that the graph of an equation is obtained by drawing all of the points that satisfy the equation.

Follow-up

Ask students to give the center and radius of the circle given by $(x-5)^2 + (y+2)^2 = 9$. **(Center (5, −2), radius 3)**

Assignment Guide

Day 1: Ex. 1–13 odd, 17, 21, 25, 27, 33, 36, 37, 40, 44, 45, 47, 49

Cooperative Learning

Group Activity: Ex. 45–46

EXAMPLE 4 Finding the Midpoint of a Line Segment

Find the midpoint of the line segment with endpoints $(-5, 2)$ and $(3, 7)$.

Solution

The midpoint is

$$(x, y) = \left(\frac{-5+3}{2}, \frac{2+7}{2}\right) = (-1, 4.5).$$

The midpoint formula can be used to confirm properties of geometric figures.

EXAMPLE 5 APPLICATION: Confirming a Parellelogram

It is a fact from geometry that the diagonals of a parallelogram bisect each other. Confirm this with the midpoint formula.

Solution

We can position the parallelogram in the rectangular coordinate plane, as shown in Figure 16. Applying the midpoint formula to segments OB and AC, we find that

$$\text{the midpoint of } OB = \left(\frac{0+a+c}{2}, \frac{0+b}{2}\right) = \left(\frac{a+c}{2}, \frac{b}{2}\right), \quad \text{and}$$

$$\text{the midpoint of } AC = \left(\frac{a+c}{2}, \frac{b+0}{2}\right) = \left(\frac{a+c}{2}, \frac{b}{2}\right).$$

The midpoints of OB and AC are the same, so the diagonals of the parallelogram $OABC$ meet at their midpoints and thus bisect each other.

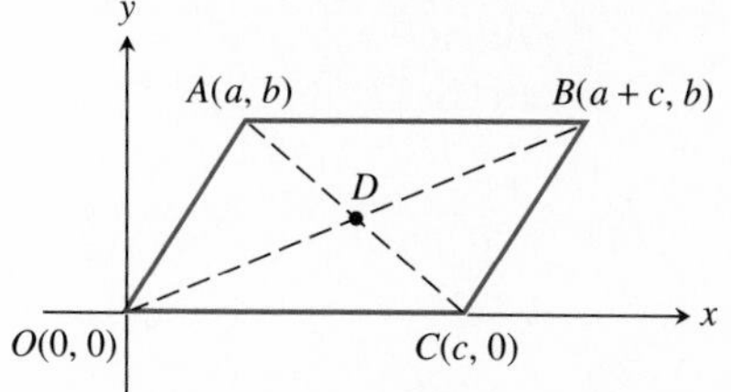

Figure 16 In a parallelogram, opposite sides are parallel.

Equations of Circles

A **circle** is the set of points in a plane at a fixed distance (**radius**) from a fixed point (**center**). Figure 17 shows the circle with center (h, k) and radius r. If (x, y) is any point on the circle, the distance formula gives

$$\sqrt{(x-h)^2 + (y-k)^2} = r.$$

Squaring both sides, we obtain the following equation for a circle.

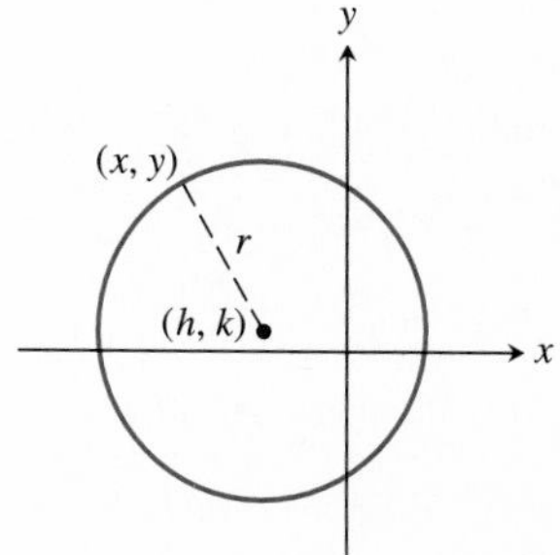

Figure 17

Definition: Equation of a Circle

The **standard form equation of a circle** with center (h, k) and radius r is

$$(x-h)^2 + (y-k)^2 = r^2.$$

Notes on Exercises

Ex. 25–30 require students to find the distances between points to show that figures in the coordinate plane have certain geometric properties. The distances that are found in Ex. 25, 27, 29, and 30 can then be used to solve Ex. 47–50.

Ex. 44–46 are "show" and "prove" problems. A coordinate approach to proof is used here rather than a synthetic approach, as in many geometry books.

Ongoing Assessment

Self-Assessment: Ex. 15, 29, 39

Embedded Assessment: Ex. 44

EXAMPLE 6 Finding Equations of Circles

Find the standard form equation of the circle.

a. center $(-4, 1)$, radius $= 8$.

b. center $(0, 0)$, radius $= 5$.

Solution

a.

$$(x - h)^2 + (y - k)^2 = r^2 \quad \text{Standard form equation}$$
$$(x - (-4))^2 + (y - 1)^2 = 8^2 \quad \text{Substitute } h = -4, k = 1, \text{ and } r = 8.$$
$$(x + 4)^2 + (y - 1)^2 = 64$$

b.

$$(x - h)^2 + (y - k)^2 = r^2 \quad \text{Standard form equation}$$
$$(x - 0)^2 + (y - 0)^2 = 5^2 \quad \text{Substitute } h = 0, k = 0, \text{ and } r = 5.$$
$$x^2 + y^2 = 25$$

EXERCISES FOR SECTION 8

In Exercises 1 and 2, plot the points.

1. $A(3, 5)$, $B(-2, 4)$, $C(3, 0)$, and $D(0, -3)$
2. $A(-3, -5)$, $B(2, -4)$, $C(0, 5)$, and $D(-4, 0)$

In Exercises 3 and 4, find the coordinates of the points.

3.

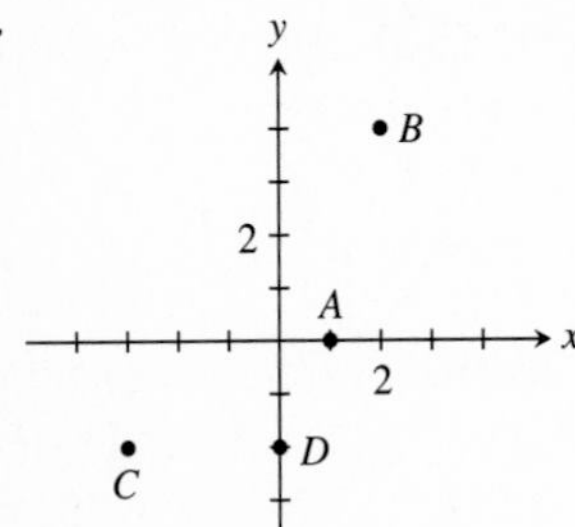

4.

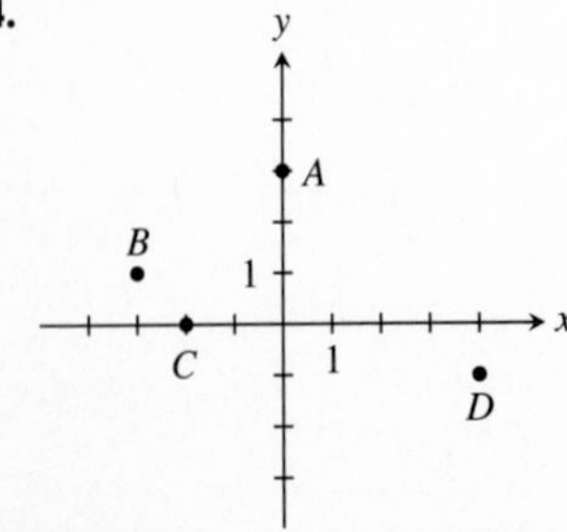

In Exercises 5–8, sketch and identify the figure determined by the points.

5. $(-5, 3)$, $(0, -1)$, $(4, 4)$
6. $(-2, -2)$, $(-2, 2)$, $(2, 2)$, $(2, -2)$
7. $(-3, -1)$, $(-1, 3)$, $(7, 3)$, $(5, -1)$
8. $(-2, 1)$, $(-2, 6)$, $(4, 6)$, $(4, 1)$

In Exercises 9 and 10, name the quadrants containing the points.

9. **a.** $(2, 4)$ **b.** $(0, 3)$ **c.** $(-2, 3)$ **d.** $(-1, -4)$
10. **a.** $(1/2, 3/2)$ **b.** $(-2, 0)$ **c.** $(-1, -2)$ **d.** $(-3/2, -7/3)$

In Exercises 11–16, find the distance between the points.

11. $(-3, -1)$, $(5, -1)$
12. $(-3, -2)$, $(6, -2)$
13. $(0, 0)$, $(3, 4)$
14. $(-3, -4)$, $(0, 0)$
15. $(-1, 2)$, $(2, 3)$
16. $(-4, -3)$, $(1, -1)$

In Exercises 17–24, find the midpoint of the line segment with the given endpoints.

17. $(-1, 3)$, $(5, 9)$
18. $(2, -3)$, $(-1, -1)$
19. $(3, \sqrt{2})$, $(6, 2)$
20. $(-\sqrt{3}, \sqrt{3})$, $(3, -6)$
21. $(-7/3, 3/4)$, $(5/3, 9/4)$
22. $(-2/3, -3/5)$, $(9/4, 7/2)$
23. (a, b), $(2, -3)$
24. (a, b), $(-1, -4)$

In Exercises 25–30, show that the figure determined by the points is the indicated type.

25. Right triangle: $(-2, 1), (3, 11), (7, 9)$

26. Equilateral triangle: $(0, 1), (4, 1), (2, 1 - 2\sqrt{3})$

27. Isoceles triangle: $(1, 3), (4, 7), (8, 4)$

28. Rectangle: $(0, -5/2), (-3, -1), (1, 7), (4, 11/2)$

29. Square: $(-7, -1), (-2, 4), (3, -1), (-2, -6)$

30. Parallelogram: $(-2, -3), (0, 1), (6, 7), (4, 3)$

In Exercises 31–36, find the standard form equation for the circle.

31. Center $(1, 2)$, radius 5

32. Center $(-3, 2)$, radius 1

33. Center $(0, 0)$, radius 2

34. Center $(0, 0)$, radius $\sqrt{3}$

35. Center $(-1, -4)$, radius 3

36. Center $(5, -3)$, radius 4

In Exercises 37–42, find the center and radius of the circle.

37. $(x - 3)^2 + (y - 1)^2 = 36$

38. $(x + 4)^2 + (y - 2)^2 = 121$

39. $(x + 5)^2 + (y + 4)^2 = 9$

40. $(x - 2)^2 + (y + 6)^2 = 25$

41. $x^2 + y^2 = 1$

42. $x^2 + y^2 = 5$

In Exercises 43–46, solve the problem.

43. The total in billions of dollars of U.S. imports from Mexico is given in Table 2. Plot the points in a rectangular coordinate system.

Table 2 U.S. Imports from Mexico

Year	U.S. Imports (billions of dollars)
1987	20.2
1988	23.2
1989	27.2
1990	30.2
1991	31.2
1992	35.2
1993	39.2

Source: United World Ltd., Inc., as reported in *USA Today,* September 30, 1994.

44. Show that the triangle determined by $(3, 0), (-1, 2)$, and $(5, 4)$ is isoceles but not equilateral.

45. Show that the midpoint of the hypotenuse of the right triangle with vertices $(0, 0)$, $(5, 0)$, and $(0, 7)$ is equidistant from the three vertices.

46. Show that the midpoint of the hypotenuse of any right triangle is equidistant from the three vertices.

In Exercises 47 and 48, find the perimeter of the figure from the specified exercise.

47. Exercise 27

48. Exercise 30

In Exercises 49 and 50, find the area of the figure from the specified exercise.

49. Exercise 25

50. Exercise 29

KEY TERMS

The number following each key term indicates the page of its introduction.

AL-KHWARIZMI

Mohammad ibn Musa al-Khwarizmi (ca. A.D. 788–850) was born in what is present-day Iran. He became a professor at the Arab University in Baghdad during the first golden age of Islamic science. His greatest contribution to mathematics was his algebra book, *Kitab al-jabr w'al muqabalah* (*The Book of Integration and Equation*). The book established algebra as an independent discipline for the first time. It showed how to solve equations, manipulate terms, and use the quadratic formula. Al-Khwarizmi also taught the use of Hindu-Arabic numerals, rather than the clumsier Roman numerals then used in Europe.

Al-Khwarizmi's impact on mathematics is reflected in our language to this day. The Arabic word *al-jabr* eventually became the modern word *algebra.* Also, the name Al-Khwarizmi later took on the meaning of "a mathematical procedure," becoming the modern word *algorithm,* by way of Latin.

chapter 1

Objective

Students will be able to apply a problem-solving process involving algebraic, graphical, and/or numerical models.

Motivate

Ask students to name some approaches that they have found helpful in solving word problems.

FUNCTIONS AND GRAPHS

BIBLIOGRAPHY

For students: *How to Solve It: A New Aspect of Mathematical Method,* Second Edition, George Pólya. Princeton Science Library, 1991. Available through Dale Seymour Publications.

For teachers: *The Language of Functions and Graphs,* Shell Centre for Mathematical Education. Available through Dale Seymour Publications.

1.1 PROBLEMS AND THEIR REPRESENTATIONS

Graph of an Equation • Problem Situations • Using Numerical, Algebraic, and Graphical Representations • Using a Problem-Solving Process • Applying the Problem-Solving Process

Graph of an Equation

There are many situations in social service, business, and science enterprises that require analysis of quantitative information. Newspapers are filled with quantitative discussions of inflation, investment alternatives, unemployment, production quantities, balance of payments, and so forth. Often an *equation* is used to describe a relationship between quantities such as these.

We usually give letter names to these quantities, typically x, y, z, and t, but other letters can be used as well. An ordered pair of numbers, written (a, b), is a solution of an equation (in variables x and y) if the substitution of $x = a$ and $y = b$ *satisfies* the equation. For example, (2, 12) is a solution of the equation $y = 3x^2$ because

$$\begin{aligned} y &= 3x^2 && \text{Original equation} \\ 12 &= 3(2)^2 && \text{Substitute } x = 2 \text{ and } y = 12. \\ 12 &= 12. \end{aligned}$$

The **graph** of an equation in x and y consists of all pairs (x, y) that are solutions of the equation. The *point-plotting method* is a traditional way to sketch the graph of an equation using pencil and paper.

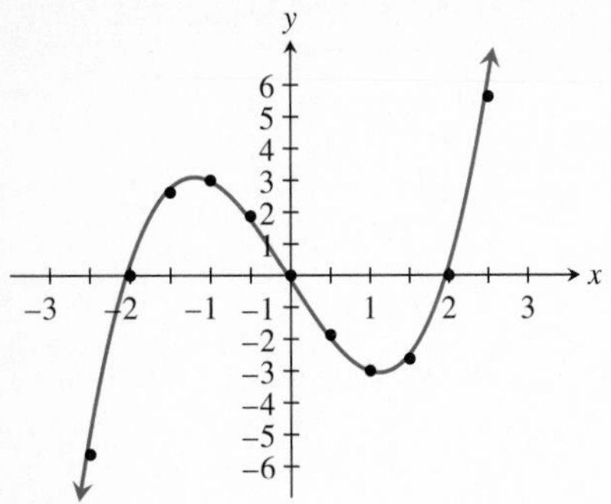

Figure 1.1 A sketch of $y = x^3 - 4x$. First plot (x, y) solutions of the equation. Then join the plotted points with a smooth curve.

Notes on Examples

Students tend to think that setting the scale on a grapher affects the way a graph will be represented on the viewing screen. For Example 2c, have the students set the *X*scl and *Y*scl values to 1. Then the *y*-axis will appear to be a "fat" line. Point out that the *X*scl and *Y*scl values tell how far apart the tick marks will be. Use the example of a 40-foot-long fence. If the fence posts are placed 5 feet apart in a straight line along the 40 feet of fence, then only 9 posts will be needed. On the grapher, the "fence posts" should be far enough apart to clearly distinguish the tick marks.

```
WINDOW FORMAT
Xmin=⁻10
Xmax=10
Xscl=1
Ymin=⁻10
Ymax=10
Yscl=1
```

Figure 1.2 The window dimensions for the *standard window*. The notation "[−10, 10] by [−10, 10]" is used to represent window dimensions like these.

The Point-Plotting Method of Graphing (Paper and Pencil)

To sketch the graph of an equation by point plotting:

1. Make a table of several solutions.
2. Plot these solutions as points in the coordinate plane.
3. Connect these points with a smooth curve.

■ EXAMPLE 1 Sketching a Graph with Pencil and Paper

Graph $y = x^3 - 4x$ using pencil and paper.

Solution

Make a table of solutions.

x	−2.5	−2	−1.5	−1	−0.5	0	0.5	1	1.5	2	2.5
y	−5.625	0	2.625	3	1.875	0	−1.875	−3	−2.625	0	5.625

Plot the (x, y) pairs in the coordinate plane and connect them with a smooth curve (see Figure 1.1). Arrowheads are drawn at the ends of the drawn curve to suggest how the graph continues beyond the portion of the coordinate plane shown. ■

The point-plotting method of graphing is also the basis for the way a graphing utility, often referred to as a *grapher,* does its curve sketching. We simply enter the equation into the grapher. Then we let the graphing utility do the pencil-and-paper work. A grapher is not able to draw a smooth curve. Instead it approximates the curve by drawing short line segments. Also, a grapher does not draw arrowheads at the edge of the **viewing window** to indicate that the graph continues off the screen.

The Point-Plotting Method of Graphing (grapher)

To draw a graph of an equation using a grapher:

1. Rewrite the equation in the form $y =$ (an expression in x).
2. Enter this equation into the grapher.
3. Select an appropriate viewing window (see Figure 1.2).
4. Press the "graph" key.

TECHNOLOGY TIPS

1. *Nothing appears in the viewing window!* Trace to see the coordinates of some points on the graph. Then reset your viewing window dimensions to include those coordinates.
2. *The graph is not what I expected!* Did you enter the function correctly? Check that you have used parentheses correctly—also "minus" versus "negative."

With practice and experience you will learn which *window dimensions* are appropriate for which equations. Example 2 illustrates that some choices are better than others.

EXAMPLE 2 Drawing a Graph with a Grapher

Graph $y = x^3 - 7x^2 + 28$ in each of the viewing windows whose dimensions are given. Then choose the best view of the graph.

a. Dimensions: $[-10, 10]$ by $[-10, 10]$
b. Dimensions: $[-5, 5]$ by $[-50, 10]$
c. Dimensions: $[-5, 10]$ by $[-50, 50]$

Solution

Figure 1.3 shows the equation $y = x^3 - 7x^2 + 28$ entered on the "$Y =$" "edit screen."

Figure 1.4 shows the three desired views of the graph. The window in part (c) appears to be the best view. ■

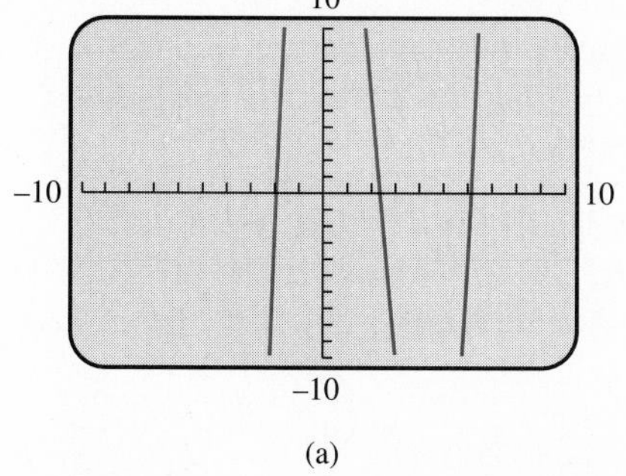

Figure 1.3

With experience you will learn to make good choices for your viewing windows. Choosing a viewing window is a process that often involves some *trial and error* and some *troubleshooting.*

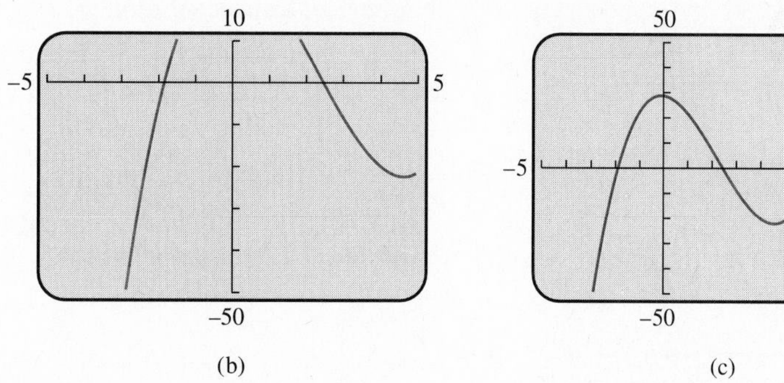

Figure 1.4 Three different views of the same equation $y = x^3 - 7x^2 + 28$.

Alert

Many students think that all graphs of functions exist in the standard $[-10, 10]$ by $[-10, 10]$ viewing window. Give several examples of functions such as $y = x^2 + 10$ that do not exist in this window.

Problem Situations

Workers in this information age of ours need to analyze problems daily. The context of the problem and all of its quantitative constraints are referred to as a *problem situation.* There may be several problems that can be posed within a single problem situation. Example 3 illustrates.

Teaching Notes

Exploration of problem situations using a grapher leads many students to pursue analytical approaches using pencil-and-paper methods.

■ EXAMPLE 3 Manufacturing Automobile Parts

Grant Manufacturing finds that the cost of producing engine mount bolts is \$5,000 in fixed costs plus \$2,500 per ton produced. State two problems that can be posed within this problem situation.

Solution

(There are many correct responses.)

a. How much does it cost to produce 4 tons of bolts?
b. How many tons of bolts can be produced for \$12,500? ■

X	Y_1	
0	5	
1	7.5	
2	10	
3	12.5	
4	15	
5	17.5	
6	20	

$Y_1 = 2.5X+5$

Using Numerical, Algebraic, and Graphical Representations

As you analyze a problem, be a creative and flexible thinker. As you attempt to solve a problem, you will often find it helpful to consider several different representations of the problem. Example 4 illustrates what we mean by numerical, algebraic, and graphical representations.

Bolts (x) (tons)	Cost (y) (\$1000)
0	5.0
1	7.5
2	10.0
3	12.5
4	15.0
5	17.5
6	20.0

Figure 1.5

■ EXAMPLE 4 Manufacturing Automobile Parts (Continued)

Find numerical, algebraic, and graphical representations for the Grant Manufacturing problem situation described in Example 3.

Solution

Numerical Representation

The table in Figure 1.5, generated by hand or by a calculator, shows seven (x, y) pairs of values for x tons of bolts produced at a cost of y thousands of dollars. This table is a *numerical representation* of the problem situation.

Algebraic Representation

Study the pattern in the table in Figure 1.5. To manufacture x tons, the cost in thousands of dollars would be $2.5x + 5$. The equation

$$y = 2.5x + 5$$

is an *algebraic representation* of the problem situation.

Graphical Representation

Figure 1.6 shows a graph of $y = 2.5x + 5$. It is a *graphical representation* of the problem situation. ■

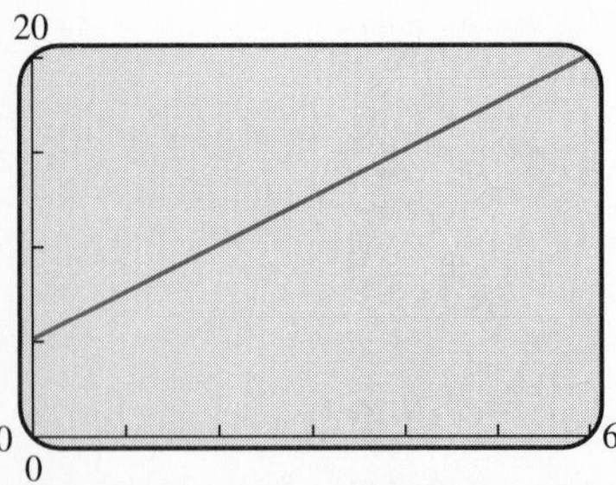

Figure 1.6 A graph of $y = 2.5x + 5$.

Numerical, algebraic, and graphical representations are all important in solving a problem. Each contributes a different approach to the problem's situation. Often there is an interplay between one representation and another that adds an important perspective to the problem.

PÓLYA'S FOUR-STEP PROCESS

1. Understand the problem.
2. Devise a plan.
3. Carry out the plan.
4. Look back.

Using a Problem-Solving Process

George Pólya is sometimes called the father of modern problem solving because of his significant writing and analysis of the mathematical problem-solving process.

Pólya was born in Hungary in 1887 and completed his Ph.D. at the University of Budapest. In 1940 he came to Brown University (Providence, R.I.); he joined the faculty at Stanford University (Palo Alto, Calif.) in 1942. He was scientifically active for 75 years and contributed to many areas of mathematics in addition to problem solving. He died in 1985 at the age of 97. His four-step process continues to be valid in this era using computers and graphing utilities to solve problems.

Throughout this text we use the following variation on Pólya's four steps.

Teaching Note

Many students are either *impulsive* or *reflective* problem solvers. The impulsive problem solver is a brainstormer and a risk taker. The reflective problem solver seems more intent on giving the right answer and may withdraw from class discussions. In using technology in the classroom, it is essential that the teacher observe the problem-solving style of each student and encourage all students to participate in class discussions.

A Problem-Solving Process

Step 1—Understand the problem

- Read the problem as stated, several times if necessary.
- Be sure you understand the meaning of each term used.
- Restate the problem in your own words. Discuss the problem with others.
- Identify clearly the information you need to solve the problem, and decide whether it can be found from the given data.

Step 2—Develop a mathematical model of the problem

- Draw one or more pictures of the problem situation.
- Introduce a variable to represent the quantity requested.
- Translate the word statement of the problem to a mathematical statement, such as an equation or an inequality.

Step 3—Solve the mathematical model and support/confirm the solution

- **Solve algebraically** using traditional algebraic methods and **Support graphically or Support numerically** using a graphing utility.
- **Solve graphically or numerically** using a graphing utility and **confirm algebraically** using traditional algebraic methods.
- **Solve graphically or numerically** because there is no other way possible.

Step 4—Interpret the solution in the problem setting

- Translate your mathematical result into the problem setting and decide whether the result makes sense.

Alert

Some students will complete all of the mathematics of solving a problem but forget to interpret their solution. Stress the importance of Step 4 of the problem-solving process.

Follow-up

Ask . . .

In Example 5,what is the purpose of displaying the graph on the graphing calculator? **(To verify the answer.)**

Assignment Guide

Day 1: Ex. 1, 4, 5, 8, 9, 16, 17, 18, 20, 22, 25, 27, 28
Day 2: Ex. 31, 32, 35, 36, 38–41

Cooperative Learning

Group Activity: Ex. 35–42

Notes on Exercises

The Quick Review exercises in this section and throughout the rest of the book can be assigned if students need to review important concepts.
Ex. 5–22 provide opportunities for students to relate tabular, graphical, and/or algebraic representations.
Ex. 33–34 allow students to explore the effects of changing viewing windows. This gives them an opportunity to explore key concepts that relate to scale.
Ex. 35c and 36c are the first problems in a series of Writing to Learn exercises. Some teachers have students keep journals, and these exercises can be used for journal writing.

Ongoing Assessment

Self-Assessment: Ex. 7, 13, 21, 37
Embedded Assessment: Ex. 32, 36

Applying the Problem-Solving Process

Step 1 of this problem-solving process generally involves oral activity. Consequently, it will not be included in our written solutions. The written part of the process—Steps 2 through 4—will be followed throughout this text.

Example 5 uses the "Solve algebraically and support graphically" strategy.

■ EXAMPLE 5 APPLICATION: Reliability Testing

Ford engineers pay students \$0.08 per mile plus \$25 per day to road test their new vehicles.

a. How much would it cost Ford for Sally to drive 440 miles in one day?
b. How many miles did John test-drive in one day for Ford if he earns \$93?

Solution

Model

Each mile costs \$0.08. Therefore, the cost for x miles is $0.08x$. The total cost y for driving x miles is modeled by $y = 0.08x + 25$.

Solve Algebraically

a. Find y when $x = 440$.

$$y = 0.08x + 25 \quad \text{Original equation}$$
$$y = 0.08(440) + 25 \quad \text{Replace } x \text{ with 440.}$$
$$y = 60.20$$

b. Find x when $y = 93$.

$$y = 0.08x + 25$$
$$93 = 0.08x + 25 \quad \text{Replace } y \text{ with 93.}$$
$$93 - 25 = 0.08x$$
$$\frac{68}{0.08} = \frac{0.08x}{0.08}$$
$$x = \frac{68}{0.08}$$
$$x = 850$$

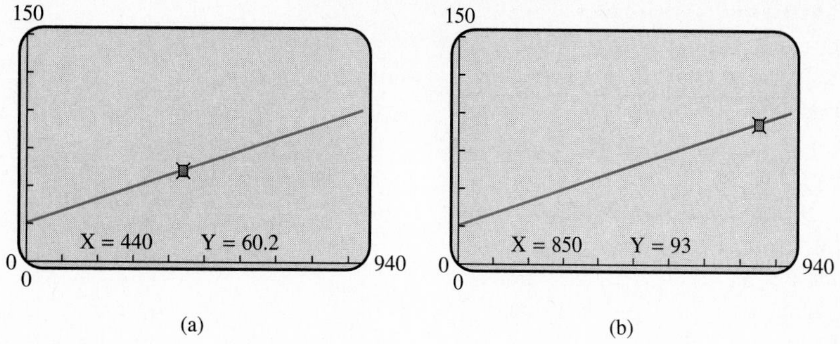

(a)

(b)

Figure 1.7

> **TECHNOLOGY TIP**
>
> Consider the viewing window in Figure 1.7. The problem suggests (as does trial and error) that x should be between 0 and 1000. We chose $X\text{min} = 0$ and $X\text{max} = 940$ so that the trace x-coordinates would be whole numbers. (If that's not true on your grapher, consult the Grapher Workshop for the viewing window that you might use.)

Support Graphically

Figure 1.7a shows that the point (440, 60.2) is on the graph of $y = 0.08x + 25$ to support part (a). Similarly, Figure 1.7b shows that the point (850, 93) is on the graph of $y = 0.08x + 25$ to support part (b).

Interpret

If Sally test-drives 440 mi in one day, she earns \$60.20. If John earns \$93 in one day of test driving, then we know he drove 850 mi. ■

Quick Review 1.1

In Exercises 1–4, evaluate the expression.

1. $(-3.2)^3$

2. $-(-2)^5$

3. $\sqrt{121}$

4. $\sqrt{49^2}$

In Exercises 5–8, substitute the value for x and evaluate.

5. $y = 4x - 19$ for $x = 22$

6. $y = x^2 + 3$ for $x = 12$

7. $y = x^3 - 15x$ for $x = 8$

8. $y = \sqrt{x + 2}$ for $x = 79$

In Exercises 9–12, perform the indicated operations and simplify.

9. $(x^2 + 3)(3x^2 - 2)$

10. $(5st + 2s) - s(t - 2)$

11. $\dfrac{r}{t - 1} + \dfrac{3r}{t}$

12. $\dfrac{2}{x - 2} - \dfrac{3}{x + 1}$

In Exercises 13 and 14, solve the equation.

13. $7(x - 3) + 2 = 8 - 3x$

14. $\dfrac{4}{x - 3} - \dfrac{7}{x + 3} = 0$

SECTION EXERCISES 1.1

In Exercises 1–4, determine which ordered pair is a solution for the given equation.

Equation	*Ordered Pairs*	
1. $y = 2x - x^2$	**a.** (1, 2)	**b.** (2, 0)
2. $y = 3\sqrt{x + 3}$	**a.** (4, 12)	**b.** (6, 9)
3. $2y + 3x = 8$	**a.** (1, 2)	**b.** (2, 1)
4. $2y = x^2 + 1$	**a.** (3, 5)	**b.** (2, 3)

In Exercises 5 and 6, complete the table of solutions for the equation.

5. $y = x + 2$

x	−4	−1	?	4	?
y	−2	?	3	?	11

6. $y = -2x + 3$

x	1	?	3	?	5
y	?	4	?	16	?

In Exercises 7 and 8, choose the equation that is an algebraic representation of the given data.

7.

x	−2	−1	0	1	2
y	−1	1	3	5	7

a. $y = 3x + 1$
b. $y = 2x + 3$
c. $y = 3 - 2x$

8.

x	−2	−1	1	2	3
y	−8	−3	1	0	−3

a. $y = x^2 - 1$
b. $y = -x^2 + 3$
c. $y = 2x - x^2$

In Exercises 9–18, make a table of solutions for the given equation for values of x from $x = -3$ to $x = 3$ in increments of 1. Sketch the graph of the equation.

9. $y = 2x + 5$

10. $y = 3x - 8$

11. $y = x^2 - 7$

12. $y = 2x^2 - 5x - 2$

13. $y = 8x - x^2 + 2$

14. $y = -x^3 + 3x^2 - 4$

15. $y = |x - 2|$

16. $y = |x + 2| - 3$

17. $y = |x^2 - 4|$

18. $y = \sqrt{x + 2}$

In Exercises 19–22, match the equation with its graph. Identify the viewing window dimensions and the values for *X*scl and *Y*scl.

19. $y = \sqrt{x + 3}$

20. $y = \frac{1}{3}x^3 - 2x - 4$

21. $y = 2x^2 - 3x + \frac{1}{2}$

22. $y = |x + 3| - 2$

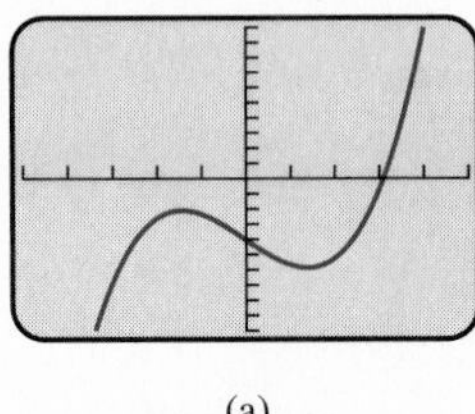
(a)

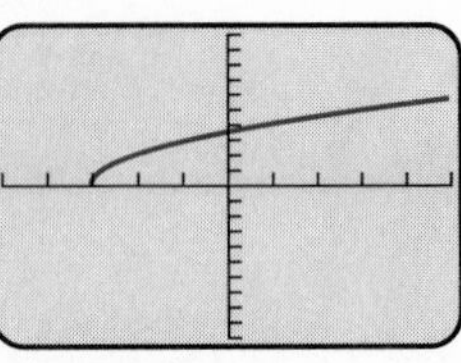
(b)

(c)

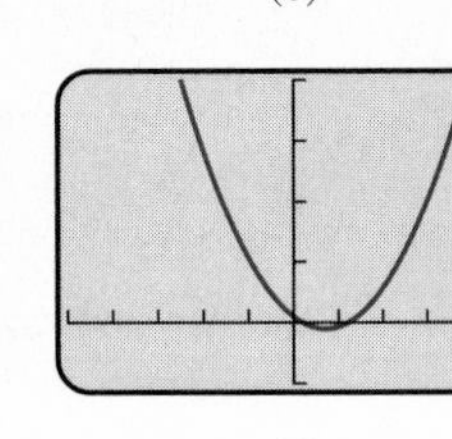
(d)

For Exercises 19–22

In Exercises 23–26, prepare the equation for graphing on a graphing utility by solving for y. Then use a grapher to draw its graph.

23. $2x + 3y = 7$

24. $3x - 4y = 18$

25. $2y + 6x = 4x^2 + 1$

26. $3y + 6x = x^3 - 12$

In Exercises 27–30, graph the equation on a grapher. Then pick one of the following as the best description of the trace cursor moving from left to right on the graph:

a. The y-values decrease, then increase.
b. The y-values decrease, then increase, and then decrease again.
c. The y-values increase.
d. The y-values increase, then decrease, and finally increase again.

27. $y = \frac{1}{3}\sqrt{x + 6}$

28. $y = \frac{2}{7}(x^3 - 13x + 12)$

29. $y = \dfrac{135x}{8(x^2 + 1)}$

30. $y = \frac{1}{5}(x^2 + 2x - 24)$

In Exercises 31 and 32, find an algebraic representation for surface area.

31. Connecting Algebra and Geometry Write an equation for the surface area A of the outside of the box. Assume that the box has a lid.

32. Connecting Algebra and Geometry Write an equation for the surface area A of the box, inside and outside. Assume that the box has no lid.

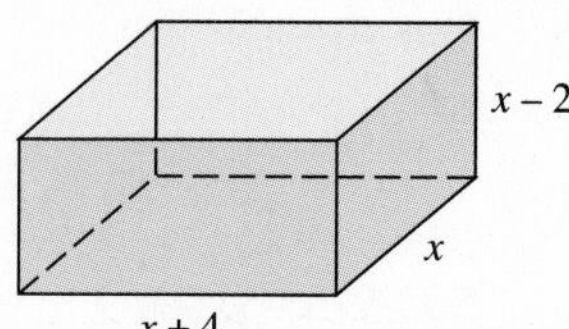

For Exercises 31–32

In Exercises 33 and 34, graph the equation in each of the given windows. Which of the three windows shows the best view of the graph? Explain.

33. $y = 155x - 8$

Xmin=⁻10
Xmax=10
Xscl=1
Ymin=⁻10
Ymax=10
Yscl=1

(a)

Xmin=⁻5
Xmax=5
Xscl=1
Ymin=⁻10,000
Ymax=10,000
Yscl=1000

(b)

Xmin=⁻10
Xmax=10
Xscl=1
Ymin=⁻500
Ymax=500
Yscl=100

(c)

For Exercise 33

34. $y = 0.1x^2 - 12$

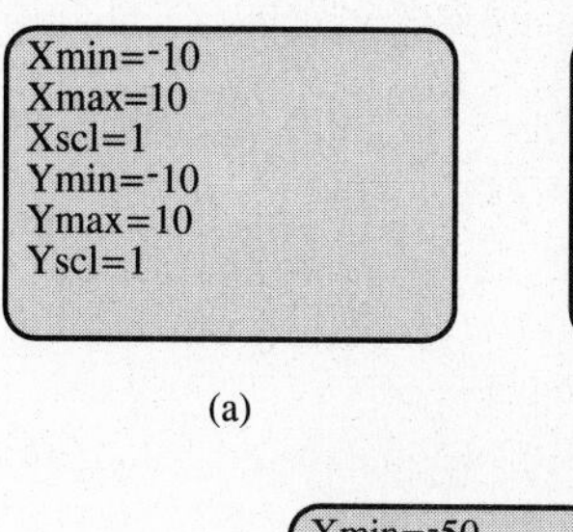

(a)

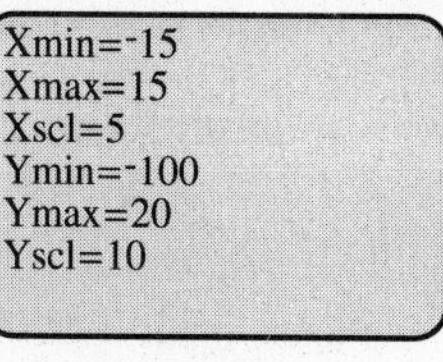

(b)

Xmin=⁻50
Xmax=50
Xscl=5
Ymin=⁻20
Ymax=200
Yscl=50

(c)

For Exercise 34

In Exercises 35–39, solve the problem.

35. *Real Estate Depreciation* Village Green, Inc., purchased an apartment building for \$150,000. It depreciates \$5,000 a year for 30 years.

Model: The value y after x years is

$$y = 150\,000 - 5\,000x, \quad 0 \le x \le 30.$$

a. Graph the modeling equation. Record your viewing-window dimensions.
b. When is the value of the building \$80,000?
c. Writing to Learn Explain why you chose the strategy you used in part (b).

36. *Industrial Depreciation* Sampy, Inc., a tool-and-die manufacturer, purchased a large punch press for \$35,000. It depreciates \$7000 a year for 5 years.

Model: The value y after x years is

$$y = 35\,000 - 7\,000x, \quad 0 \le x \le 5.$$

a. Graph the modeling equation. Record your viewing-window dimensions.
b. When is the value of the press \$14,000?
c. Writing to Learn Explain why you chose the strategy you used in part (b).

37. *Club Fund Raiser* The Ottawa Hills Youth gymnastics club pays \$15.75 to rent a video cassette player to show a movie for a fund-raising project. The club charges $0.75 per ticket for the movie.

a. Write an equation that models the profit y (in dollars), for selling x tickets.
b. How many tickets should be sold for a profit of $8.50?
c. What is the profit for selling 66 students tickets?

38. *Automobile Leasing* The Smithville Motor Company leases a new Cavalier for a one-time payment of $200 plus $180 per month for 48 months.

a. Write an equation that models the cost y (in dollars) after x months of the lease.
b. Graph this modeling equation. Record your viewing-window dimensions.
c. Find the cost of the lease after 42 months.
d. How long will it be until $2540 has been spent?

39. *Archery* An arrow is fired straight up with an initial speed of 200 ft/sec. Its height h after t seconds of flight is modeled by the equation $h = -16t^2 + 200t$.

a. Graph the equation so that it models the flight of the arrow. From your graph and thinking about the problem situation, what window dimensions [Xmin, Xmax] would seem to be most appropriate for the graph?
b. Does the arrow reach a 675-ft height? (*Hint:* Find the graphs of both $y_1 = -16x^2 + 200x$ and $y_2 = 675$ in the same viewing window.)

EXTENDING THE IDEAS

In Exercises 40–42, solve the problem.

40. *Throwing a Baseball* If a baseball is thrown straight up with an initial speed of 88 ft/sec from 5 ft off the ground, its height above the ground t seconds later is modeled by the equation $h = -16t^2 + 88t + 5$.

a. *Solve Numerically (Make a Table)* When is the ball at 96 ft on the way down?
b. *Solve Numerically (Make a Table)* When does the ball reach 121 ft? (*Hint:* Increment the x-values in the table by 0.05.)

41. *Median Family Income* Table 1.1 shows the median family income in the United States (to the nearest hundred dollars).

a. Plot these data points on an (x, y) coordinate system. (Let $x = 1$ represent 1970, $x = 2$ represent 1975, etc.)
b. The graph of the equation $y = 1960 + 6520x$ is the line that comes closest to fitting these data points. We say that the equation *models the data.* Sketch a graph of this equation on the coordinate system used in part (a).

Table 1.1 Median Family Income

Year	Income (dollars)
1970	9,800
1975	13,700
1980	21,000
1985	27,700
1990	35,400

Source: U.S. Bureau of the Census, *Current Population Reports,* series P-60, No. 174.

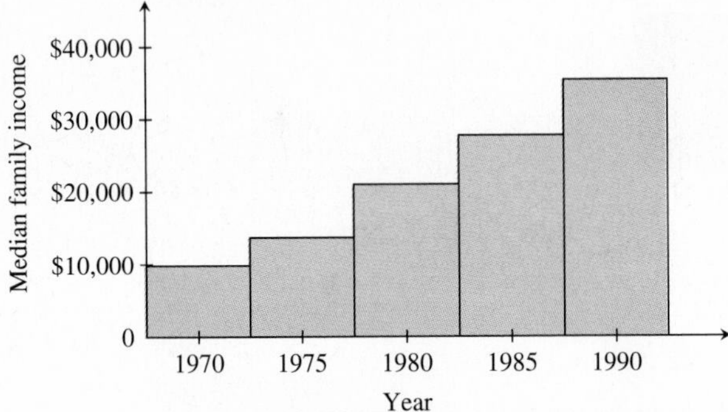

For Exercise 41

c. For 1985 what is the difference between the value given by the (equation) model and the actual value?
d. What does this model predict for the median family income in the year 2000?

42. *Consumer Price Index* The consumer price index (CPI), measures the average price of all items. Compared to a *base price* of 100 set in 1983, it has the values given in Table 1.2.

Table 1.2 Consumer Price Index (CPI)

Year	Income (dollars)
1984	103.9
1985	107.6
1986	109.6
1987	113.6
1988	118.6
1989	124.0
1990	130.7
1991	136.2

Source: U.S. Bureau of Labor Statistics

a. Plot these data points on an (x, y) coordinate system. (Let $x = 1$ represent 1981, $x = 2$ represent 1982, etc.)

b. The equation $y = 4.64x + 83.22$ models these data. Use this model to predict the CPI for 1996.

1.2 FUNCTIONS

Introducing Functions • Function Notation • Finding the Domain of a Function • Finding the Range of a Function • Graph of a Function • Applications

Objective

Students will be able to represent functions numerically, algebraically, and graphically, and determine the domain and range for functions.

Motivate

Ask . . .

If $x = 3$, how many values can $x^2 - 3x + 4$ have? **(One)**

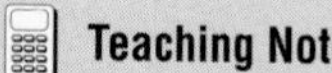

Teaching Note

A familiarity with the concept of functions is very important preparation for calculus. The graphing calculator enables students to understand functions in a way that never existed before.

Introducing Functions

Quantitative information often depicts a relationship between quantities that can be described by a rule, sometimes called a *function.* A function can be thought of as a "machine" that accepts *input* and produces *output.* Here are some examples.

1. If d dollars (input) is invested at 7% interest compounded annually, after 5 years the investment results in a total amount A (output).
2. Producing x copies (input) of a compact disc by The Camarata Singers results in total cost (output) of C dollars.

These two examples fit the following definition of a function.

> **Definition:** Function
>
> A **function** is a rule that assigns each element x (input) to exactly one element y (output). The set X of all inputs is the **domain** of the function, and the set Y of all outputs is the **range** of the function.

Functions occur in business, in science and engineering, in medicine, and in the social sciences. They are important in nearly all aspects of life.

Functions are also found on a graphing utility that contains many built-in functions. Figure 1.8 shows that the "x^{-1}" key on a grapher represents a function. This function takes input value 17 and sends it to output value 0.0588235294. Grapher keys labeled "x^2," "abs," "sin," "cos," "tan," "log," and "ln" each indicate a function. It is also possible to define functions on the "$Y=$" edit screen. (Consult the Grapher Workshop for details.)[1]

A function can be represented in a variety of ways. Example 1 shows that a function can be represented numerically, algebraically, and graphically.

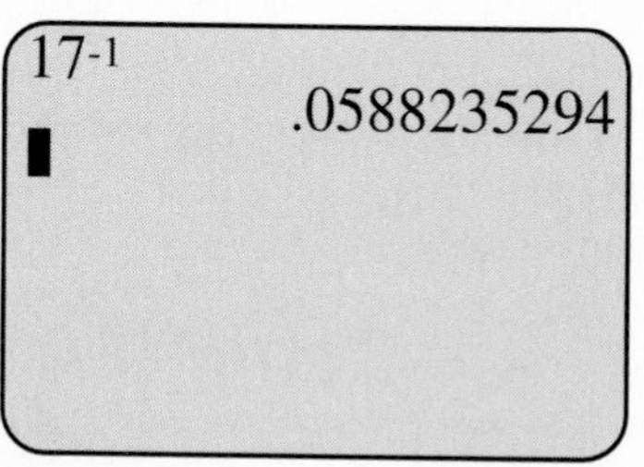

Figure 1.8 Notice that 17 (input) is sent to 0.0588235294 (the output after the x^{-1} key is pressed).

[1] The Grapher Workshop portion of this textbook contains information about grapher terms.

Teaching Note

Many students have difficulty with the concepts of domain and range. Provide opportunities to discuss and write down the domains and ranges of many functions. The examples and exercises in this section lend themselves to using cooperative groups in the classroom. Working in groups allows students to help each other and to take advantage of a variety of exploratory activities.

EXAMPLE 1 Comparing Three Ways to Represent a Function

The relationship between the cost y of Grant Manufacturing producing x tons of engine-mount bolts is a function. Show numerical, algebraic, and graphical representations.

Solution

In Example 4 of Section 1.1, we found numerical, algebraic, and graphical representations for the Grant Manufacturing problem situation. Figure 1.9 reviews these representations. In the table, domain values are in the x-column and range values are in the y-column. In the equation, domain values replace x and range values replace y. In the graph, domain values are on the x-axis and range values are on the y-axis. ■

A function represented as a table

Bolts x (tons)	Cost y ($1000)
0	5.0
1	7.5
2	10.0
3	12.5
4	15.0
5	17.5
6	20.0

A function represented as an equation

$y = 2.5x + 5$

A function represented as a graph

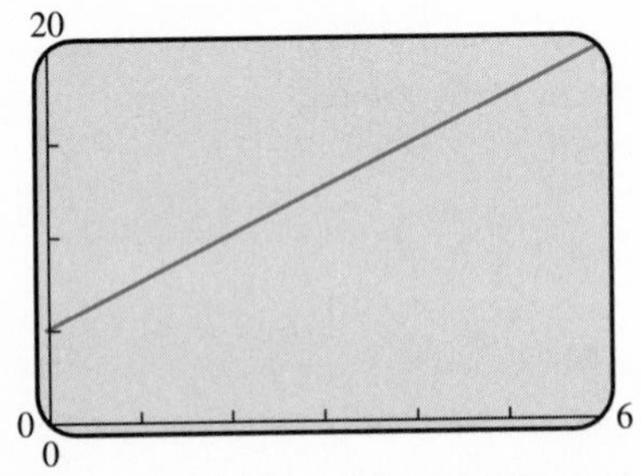

Figure 1.9

Teaching Note

Many students will find it helpful to remember that the value of y *depends* on the value of x, so x is the independent variable and y is the dependent variable.

For each input number (domain element) in a function relationship, there is a *unique* output number (range element). So, a range element (represented by y) has a unique value depending upon what domain value (represented by x) has been chosen. We say that x is the **independent variable** and y is the **dependent variable.**

It is possible for several domain values to be associated with the same range value. For example, the function $y = x^2$ assigns both $x = -2$ and $x = 2$ to the same value, $y = 4$.

A rule that assigns one element x (input) to more than one element y (output), as depicted in Figure 1.10b, is *not* a function.

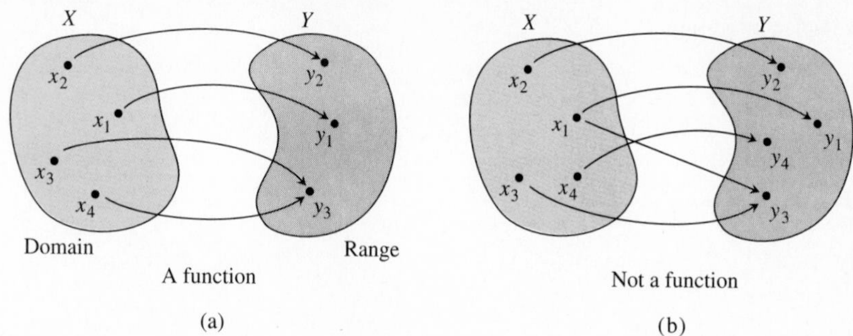

Figure 1.10 The diagram in (a) depicts a function from X to Y. The diagram in (b) depicts a relationship that is not a function from X to Y.

> **Teaching Note**
>
> The graphing calculator allows us to explore many real-world phenomena that can be represented geometrically using this technology. This is important because students should be exposed to many different algebraic representations of functions.

Function Notation

In an equation that defines a function, such as $y = 3x^2 - 5x - 4$, we frequently use the **function notation $f(x)$** (which we read as "f of x") in place of y. Thus, we have two equations that describe the function,

$$y = 3x^2 - 5x - 4 \quad \text{and} \quad f(x) = 3x^2 - 5x - 4,$$

for which we sometimes say $y = f(x)$. This notation is particularly useful because it

- lets us refer to the function as f and lets us say "f is a function of x";
- tells us that the independent variable is x and that the dependent variable is $f(x)$ or y;
- allows us to write an expression like $f(2)$ to communicate "the function value assigned to 2."

■ EXAMPLE 2 Using Function Notation

Suppose $f(x) = 4x^2 + x - 2$. Then

$$\begin{aligned} f(-3) &= 4(-3)^2 + (-3) - 2 && \text{The } x\text{-value is } -3. \\ &= 36 - 3 - 2 \\ &= 31 && \text{The } y\text{-value is } 31. \end{aligned}$$

Thus when we write $f(-3) = 31$ we know that 31 is the range value assigned to the domain value -3. ■

The function notation $y = f(x)$ allows us to do some symbolic manipulation by replacing x with variables or other expressions.

Notes on Examples

If b is a variable, the equation in Example 3b is actually equivalent to the equation $f(x) = x^2 + 1$, and either one could be used to define the function.

EXAMPLE 3 Evaluating a Function

Evaluate the function $f(x) = x^2 + 1$ for the given value of x.

a. $f(5)$ **b.** $f(b)$ **c.** $f(x - 3)$

Solution

a. $f(5) = 5^2 + 1 = 26$ Replace x with 5.

b. $f(b) = b^2 + 1$ Replace x with b.

c. $f(x - 3) = (x - 3)^2 + 1$ Replace x with $x - 3$.

$$= x^2 - 6x + 9 + 1$$
$$= x^2 - 6x + 10$$

Examples 2 and 3 illustrate that there are situations when the $f(x)$ notation is convenient. However, when studying functions on a grapher it is usually necessary to change the form

$$f(x) = 3x^2 - 5x - 4$$

to the form

$$y = 3x^2 - 5x - 4.$$

GRAPHER NOTE

Function Notation

A grapher typically does not use function notation. So the function $f(x) = x^2 + 1$ is entered as $y_1 = x^2 + 1$. On some graphers you can evaluate f at $x = 3$ by entering $y_1(3)$ on the home screen. On the other hand, on other graphers $y_1(3)$ means $y_1 * 3$. Consult the Grapher Workshop.

EXAMPLE 4 Using a Grapher to Evaluate a Function

Evaluate $f(x) = 3x^2 - 5x - 4$ for $x = -2$, $x = -1$, $x = 2$, and $x = 4$.

Solution

Enter the function $y_1 = 3x^2 - 5x - 4$ on the "$Y =$" edit screen. Figure 1.11a shows one way to generate function values one at a time, and Figure 1.11b shows a table that includes y_1 values for $x = -2, -1, 2$, and 4.

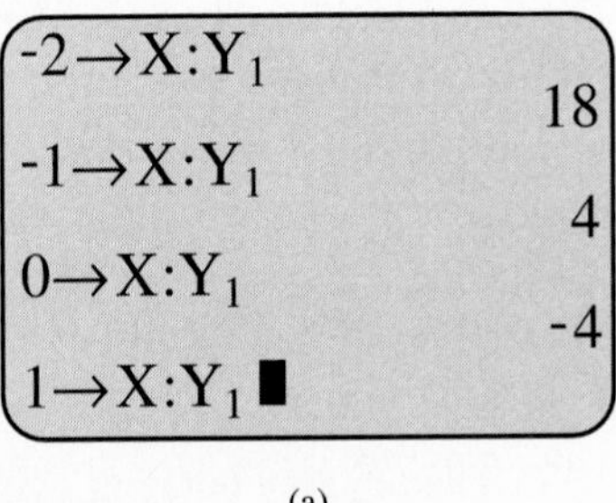

(a)

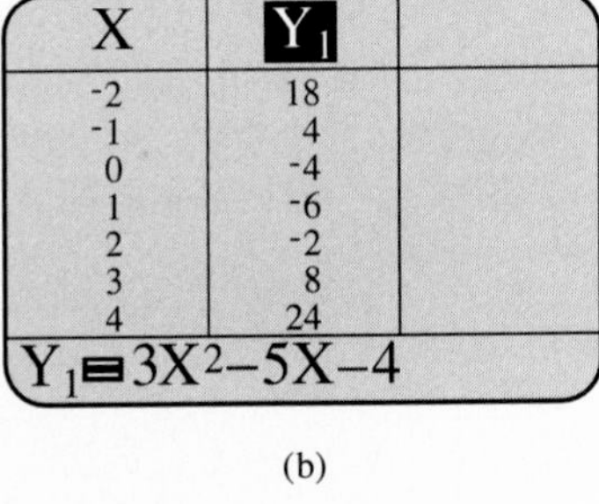

(b)

Figure 1.11

Teaching Note

It may be necessary to review the notation for the union of two sets in the context of intervals.

Finding the Domain of a Function

In the Prerequisite Chapter, we defined the domain of an algebraic expression to be the set of real numbers for which the expression is a real number. When a function is defined by an equation, the domain of the function, unless otherwise restricted, is understood to be all real numbers for which the expression is a real number.

EXAMPLE 5 Finding the Domain of a Function

Find the domain of the function.

a. $y = \sqrt{x + 3}$

b. $y = \dfrac{\sqrt{x}}{x - 5}$

NOTE

The symbol "∪" is read "union." It means that the elements of the two sets are combined to form one set.

Solution

Solve Algebraically

a. The expression under a radical must be nonnegative. Therefore,

$$x + 3 \geq 0 \qquad \text{or} \qquad x \geq -3.$$

The domain of $y = \sqrt{x + 3}$ is the interval $[-3, \infty)$.

b. The expression under a radical must be nonnegative. So for the numerator, $x \geq 0$. A denominator cannot be zero. Thus, for the denominator, $x \neq 5$. The domain is $[0, 5) \cup (5, \infty)$.

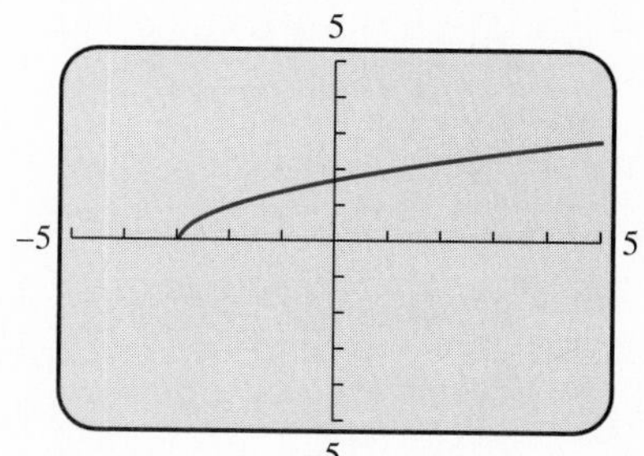

Figure 1.12 The graph of a function can suggest its domain. For example, the graph of $y = \sqrt{x + 3}$ suggests a domain of $x \geq -3$.

Support Graphically

a. The graph in Figure 1.12 shows points on the graph only for $x \geq -3$.

b. View the graph of

$$y = \frac{\sqrt{x}}{x - 5}$$

in the standard viewing window $[-10, 10]$ by $[-10, 10]$. Notice that values of x in the intervals $[0, 5)$ and $(5, \infty)$ appear to be in the domain. Most likely, you cannot tell from the graph that 5 is not in the domain. ■

Finding the Range of a Function

The range of a function can be difficult to find. We need to identify the values that result from evaluating the function for *all* domain values. Often there is no algebraic method for doing this. However, a grapher can be helpful. We can generate y-values in a table or a graph.

EXAMPLE 6 Finding the Range of a Function

Graph the function and make a conjecture about its range.

a. $f(x) = \sqrt{x + 3}$ **b.** $g(x) = |x - 1| - |x + 2|$

Solution

a. Figure 1.13a shows a view of the graph of $f(x) = \sqrt{x + 3}$. Notice that greater values of x yield greater values of $x + 3$, which yield greater square roots. It is reasonable to conjecture that the range of the function is $[0, \infty)$.

b. Figure 1.13b shows a view of the graph of $g(x) = |x - 1| - |x + 2|$. The view suggests that the graph stays flat for values of x outside the viewing window. If this is true, then it is reasonable to conjecture that the range of the function is $[-3, 3]$. ■

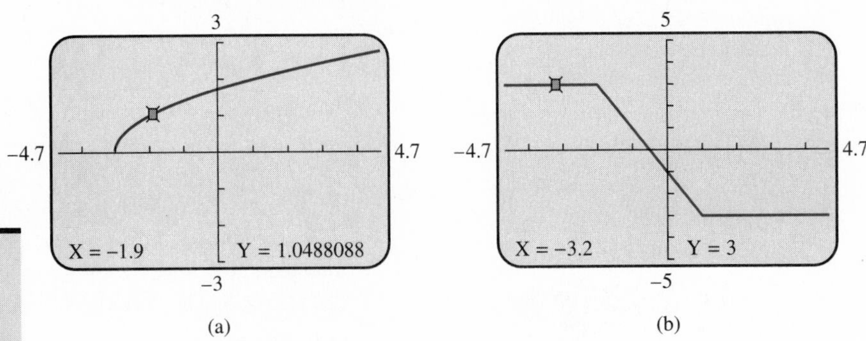

Figure 1.13 (a) A graph of $f(x) = \sqrt{x + 3}$, and (b) a graph of $g(x) = |x - 1| - |x + 2|$.

Graphing Activity Extensions

Sketch the graph of a function that has domain $[-5, -1) \cup (2, 4]$ and range $[-4, -1] \cup [1, \infty)$.

Graph of a Function

The **graph of a function** f consists of all ordered pairs $(x, f(x))$ such that x is in the domain of f. The fact that there is only one value $f(x)$ for each x-value in the domain can be interpreted geometrically by the **vertical line test:** An equation is a function if and only if each vertical line in the coordinate plane intersects the graph of the equation in at most one point.

The following activity focuses on the concepts of domain and range of a function and their relationship to the graph of the function.

Follow-up

Ask . . .

In the first step of Example 7, why is t restricted by $t \geq 0$? **(Arrow was fired at $t = 0$.)** Do you think the equation is valid for *all* nonnegative numbers t? **(No, not after the arrow hits the ground.)**

Assignment Guide

Day 1: Ex. 1, 2, 3–36, multiples of 3
Day 2: Ex. 39–45 odd, 46, 49, 50, 51–57 odd

▼ GRAPHING ACTIVITY Interpreting Graphs of Functions

Figure 1.14 shows the graph of a function $y = f(x)$. Sketch it on a piece of paper.

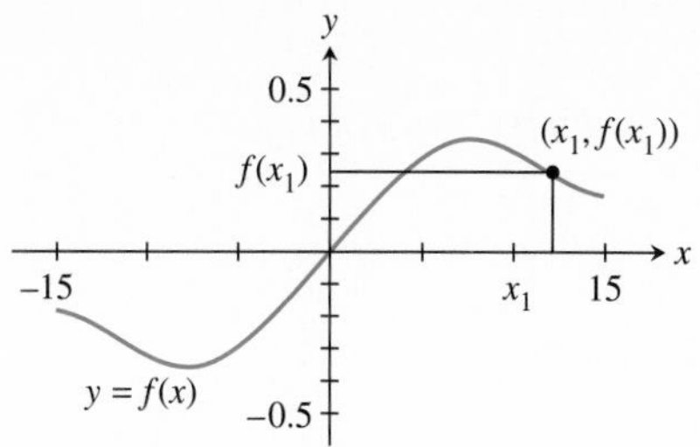

Figure 1.14

Cooperative Learning

Group Activity: Ex. 58, 59

Notes on Exercises

Ex. 11–18 can be used to familiarize students with the language of functions.
Ex. 39–42 illustrate connections between algebra and geometry.
Ex. 50 uses a polynomial function to model statistical data.

Ongoing Assessment

Self-Assessment: Ex. 7, 11, 19, 29, 47
Embedded Assessment: Ex. 46, 50

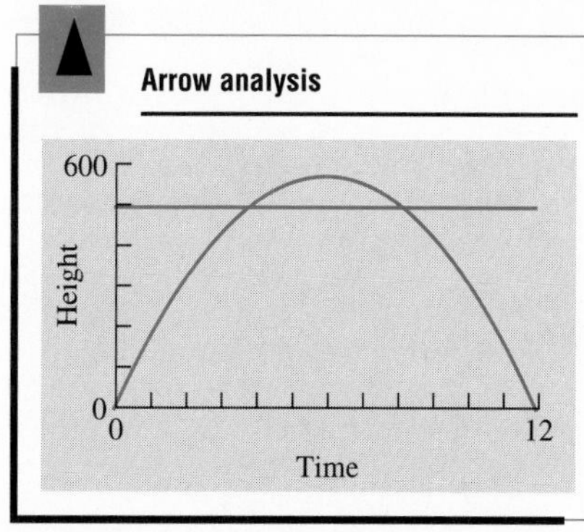

Figure 1.15 The graph of $y_1 = -16x^2 + 192x$ shows the height of the arrow plotted against time. It is not the path of the arrow which was fired straight up.

1. On your sketch, mark and label the domain value -5 on the x-axis. Mark and label the point $(-5, f(-5))$ on the graph. Mark and label the range value $f(-5)$ on the y-axis.
2. Using a different color, shade the domain suggested by the graph. On which axis are you shading?
3. Shade the range suggested by the graph. It is a subset of which axis?
4. For each domain and range given here, sketch an xy-plane and a graph of a function. (You need not include an equation for the function.)

   a. $D: [-2, 2], R: [-1, 5]$
   b. $D: [0, \infty), R: (-\infty, 0]$
   c. $D: (-\infty, \infty), R: [-3, 4]$ ▲

Applications

Sometimes a problem described in words can be reformulated as a mathematical problem. This mathematical form of the problem often involves a function.

■ EXAMPLE 7 APPLICATION: Finding the Height of an Arrow

An arrow is fired straight up from ground level with an initial velocity of 192 ft/sec. At t seconds after firing, its height in feet is $h(t) = -16t^2 + 192t$.

a. How many feet above ground is the arrow 1 sec after firing?
b. Does the arrow reach a height of 500 ft?

Solution

Model

The equation

$$h(t) = -16t^2 + 192t, \quad t \geq 0,$$

models this problem situation by describing the height $h(t)$ of the arrow t seconds after firing.

a. Solve Algebraically

We find $h(1)$:

$$\begin{aligned} h(1) &= -16(1)^2 + 192(1) \\ &= -16 + 192 \\ &= 176 \end{aligned}$$

b. Solve Graphically

Solving part (b) algebraically requires us to decide if the equation $-16t^2 + 192t = 500$ has a solution. A graphical solution allows us to answer the question with visual information. The graphs of $y_1 = -16t^2 + 192t$ and $y_2 = 500$ (see Figure 1.15) suggest that the arrow is at height 500 ft twice.

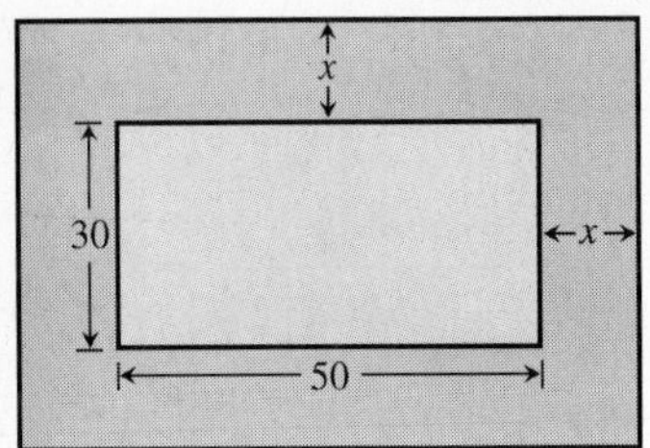

Figure 1.16

Interpret

In 1 sec, the arrow reaches a height of 176 ft. A little later it reaches 500 ft, continues rising to a maximum height, and then falls to 500 ft once more before returning to the earth. ■

■ EXAMPLE 8 APPLICATION: Designing a Lifeline Helicopter Pad

An architect is designing a 30-ft by 50-ft helicopter pad on the roof of a regional trauma center. Regulations require that the "safety apron" around the pad have an area equal to the area of the landing pad. How wide should the apron be if it has a uniform width? (See Figure 1.16.)

Solution

Model

Let $x =$ the width of safety apron. Then

$$\text{Total area} = \text{area of safety apron} + \text{area of pad},$$

$$(50 + 2x)(30 + 2x) = 30(50) + 30(50),$$

$$(50 + 2x)(30 + 2x) = 3000.$$

Solve Graphically

We graph $y_1 = (50 + 2x)(30 + 2x)$ and $y_2 = 3000$. Using the grapher we find that the two graphs intersect at $x \approx 7.8$. (See Figure 1.17.)

Interpret

The apron should be about 7.8 ft wide, which is approximately 7 ft 10 in. To be safe, the architect might design it to be 7 ft 11 in. ■

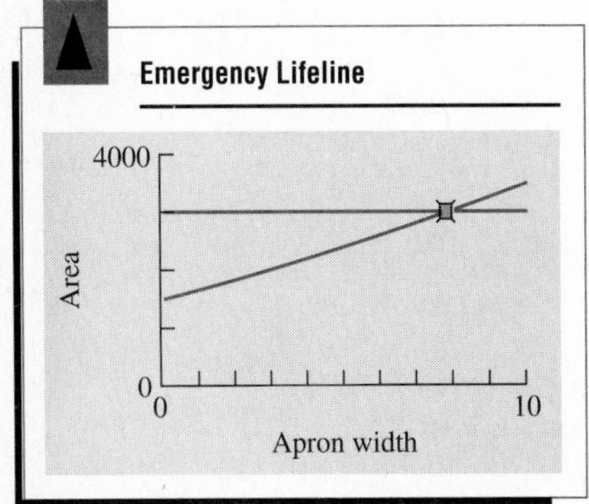

Intersection $x = 7.8388218$; $y = 3000$

Figure 1.17

TECHNOLOGY TIP

Most graphers can find the coordinates of points of intersection of two graphs, as done here for Figure 1.17. Consult your grapher owner's manual.

Quick Review 1.2

In Exercises 1 and 2, evaluate the expression for the given values of x.

1. $x^2 + x$ for $x = -3$, $x = \frac{5}{2}$

2. $x^3 - 3x$ for $x = -2$, $x = -\frac{3}{4}$

In Exercises 3–6, is $\sqrt{x - 1}$ defined for the given value of x?

3. $x = 1$ **4.** $x = -2$

5. $x = 0$ **6.** $x = 3$

In Exercises 7–10, does $|x - 1| = 1 - x$ for the given value of x?

7. $x = 3$ **8.** $x = -1$

9. $x = 0.5$ **10.** $x = -1.3$

In Exercises 11–14, solve for x.

11. $2x + 3 \geq 0$ **12.** $3 - 4x > 0$

13. $3x - 5 < x$ **14.** $1 - 2x \geq 3x$

In Exercises 15–18, for what values of x is the expression undefined?

15. $\sqrt{x}$

16. $2\sqrt{x+3}$

17. $\dfrac{x+1}{x}$

18. $\dfrac{\sqrt{x-1}}{x+3}$

SECTION EXERCISES 1.2

In Exercises 1 and 2, find numerical, algebraic, and graphical representations of the functions described.

1. A "super ball" dropped from a height of x feet bounces to a height y that is 80% of x.

2. The Boston Red Sox great, Ted Williams, had an OBA (on-base percentage) of 48.3%, the highest lifetime OBA in the history of major league baseball. At the time in Ted's career when he had x at bats, the number of times y that he reached base safely was 48.3% of x.

In Exercises 3 and 4, replace the shaded boxes and simplify if possible.

3. $f(x) = x^3 + 2x - 3$

a. $f(2) = (\square)^3 + 2(\square) - 3$
b. $f(-3) = (\square)^3 + 2(\square) - 3$
c. $f(1) = (\square)^3 + 2(\square) - 3$
d. $f(a) = (\square)^3 + 2(\square) - 3$

4. $f(x) = \dfrac{x+1}{x-3}$

a. $f(2) = \dfrac{(\square + 1)}{(\square - 3)}$ **b.** $f(-3) = \dfrac{(\square + 1)}{(\square - 3)}$

c. $f(4) = \dfrac{(\square + 1)}{(\square - 3)}$ **d.** $f(a+1) = \dfrac{(\square + 1)}{(\square - 3)}$

In Exercises 5–10, evaluate the function.

5. $f(2)$ and $f(a)$ for $f(x) = 3x^2 - 5x + 2$

6. $g(-3)$ and $g(b)$ for $g(x) = -2x^3 + 5x^2 - 3x + 2$.

7. $h(5)$ and $h(a)$ for $h(x) = \dfrac{\sqrt{x}}{x-3}$

8. $f(4)$ and $f(c)$ for $f(x) = \dfrac{|x-8|}{x^2}$

9. $h(-2)$ and $h(x+5)$ for $h(x) = \dfrac{1}{x} + \sqrt[3]{x+5}$

10. $g(8)$ and $g(x-3)$ for $g(x) = \dfrac{\sqrt[3]{x}}{x^2+1}$

In Exercises 11 and 12, interpret the graph.

11. Use the graph of $y = f(x)$.

a. $f(8) = ?$ **b.** $f(-1) = ?$
c. The domain of f is ? **d.** The range of f is ?
e. If $f(a) = 3$, then $a = ?$

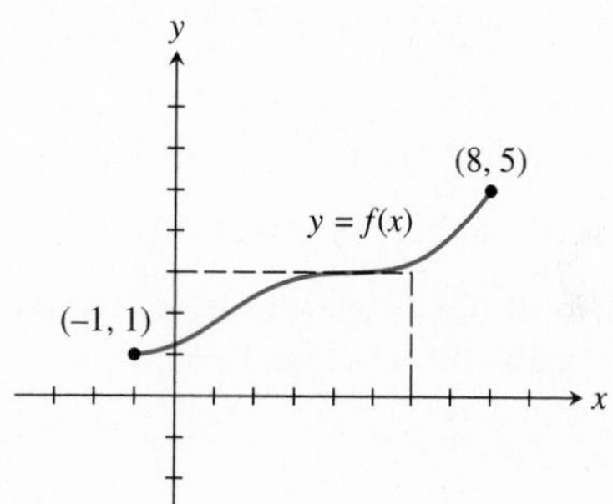

For Exercise 11

12. Use the graph of $y = g(x)$.

a. $g(3) = ?$ **b.** $g(-2) = ?$
c. The domain of g is ? **d.** The range of g is ?
e. If $g(a) = 3$, explain why you cannot conclude that $a = -2$.

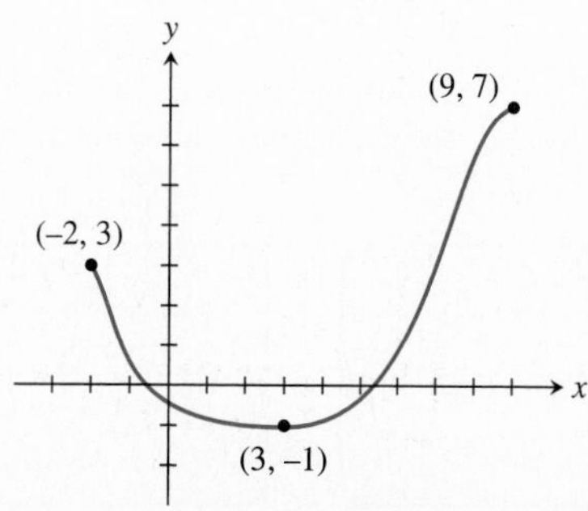

For Exercise 12

In Exercises 13–18, identify the independent variable and the dependent variable of the function.

13. $x = \sqrt{t^2 + 5}$

14. $x^2 + 3x - 2 = y$

15. $t = \dfrac{x^2}{x + 3}$

16. $2r - 5 + r^2 = s$

17. $x = |y + 3|^2$

18. $y = |x + 1|(x^3 - 5)$

In Exercises 19–26, find the domain of the function algebraically, and support your answer graphically.

19. $f(x) = \sqrt{x^2 + 4}$

20. $g(x) = \sqrt{x^2 + 1}$

21. $h(x) = \dfrac{5}{x - 3}$

22. $f(x) = \dfrac{3x - 1}{(x + 3)(x - 1)}$

23. $f(x) = \dfrac{1}{x} + \dfrac{5}{x - 3}$

24. $g(x) = \dfrac{x}{x^2 - 5x}$

25. $h(x) = \dfrac{\sqrt{4 - x^2}}{x - 3}$

26. $h(x) = \dfrac{\sqrt{4 - x}}{(x + 1)(x^2 + 1)}$

In Exercises 27–30, use a grapher to find the value of the function for $x = -3$, $x = -1$, $x = 1$, and $x = 3$.

27. $y = x^3 - 4x^2 + 5x - 3$

28. $y = x^4 - 5x^3 + 7x^2 + x - 12$

29. $y = x^2\sqrt{x + 4}$

30. $y = \dfrac{\sqrt{x^2 + 1}}{x - 5}$

In Exercises 31–34, find a viewing window that shows two intersections with the x-axis and all the points of the graph in between.

31. $f(x) = x^2 - 17x - 22$

32. $g(x) = x^2 + 29x - 119$

33. $h(x) = -x^2 - 18x + 71$

34. $f(x) = -x^2 - 37x + 93$

In Exercises 35–38, examine the graph and use the vertical line test to decide whether the graph represents a function of x.

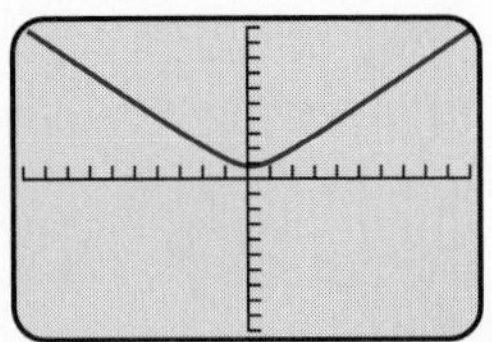

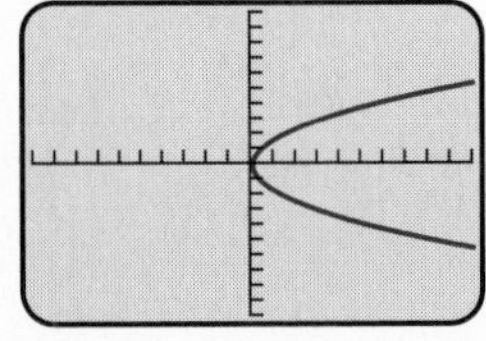

For Exercises 35–36

35. $y = \sqrt{x^2 + 1}$

36. $y^2 = 3x$

37. $2x^2 + 3y^2 = 25$

38. $xy = x - y$

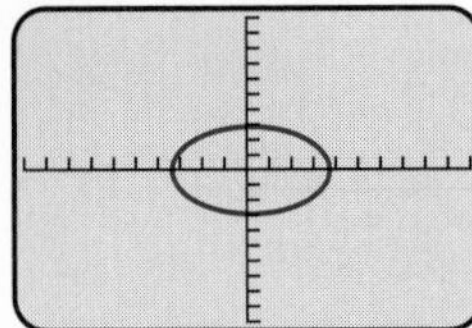

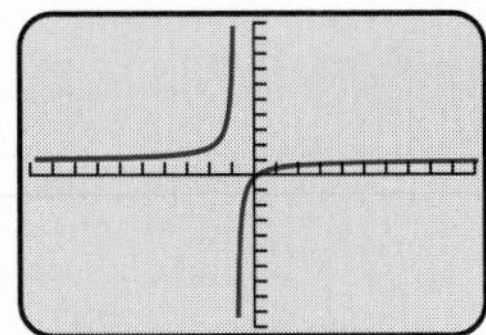

For Exercises 37–38

In Exercises 39 and 40, use the figure of a box that is twice as wide as it is high and 4 times as long as it is wide.

39. *Volume of a Container* If x represents the height of the box, find a function $V(x)$ that represents the volume of the box.

40. *Surface Area of a Container* If x represents the height of the box, find a function $A(x)$ that represents the surface area of the outside of the box assuming that it has a top.

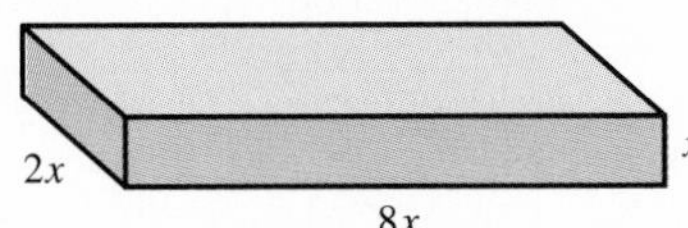

For Exercises 39–40

In Exercises 41 and 42, solve the problems.

41. Connecting Algebra and Geometry Use the function $V = (4/3)\pi r^3$.

a. Find the volume of a sphere with radius 75 cm.
b. Find the radius of a sphere with volume 2000 cm^3.

42. Connecting Algebra and Geometry Use the function $A = 4\pi r^2$.

a. Find the surface area of a sphere with radius 30 cm.
b. Find the radius of a sphere with surface area 150 cm^2.

In Exercises 43–50, solve the given problem.

43. *Area of a Border* A 4-in. by 6.5-in. picture is to be mounted on posterboard so there is a border of area A and uniform width x.

a. Find the area of the border region as $A(x)$, a function of x.

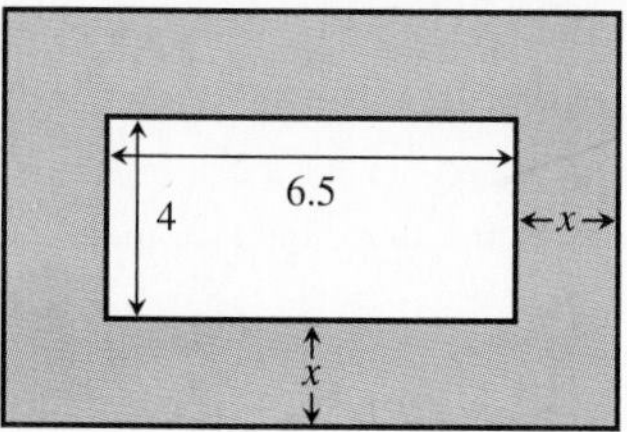

For Exercise 43

b. Graph $y = A(x)$ for those values of x that make sense in this problem. Use it to estimate the width (in inches) of the border if its area is 20 in.^{2}.

44. *Period of a Pendulum* The time (in seconds) for one period of a simple pendulum that is x meters long is $T(x)$, where

$$T(x) = 2\pi\sqrt{\frac{x}{9.807}}.$$

a. What is the period of a simple pendulum that is 2 m long?
b. What is the period of a simple pendulum that is 8 m long?
c. How long is the pendulum if its period is approximately 10 sec?

45. *Throwing a Baseball* Leroy "Satchel" Paige, selected in 1971 as a member of the Baseball Hall of Fame, could throw a baseball straight up with an initial velocity of 88 ft/sec. The height of the ball t seconds later (assuming he released the ball from his hand at a height of 7 ft) would be $h(t) = -16t^2 + 88t + 7$.

a. What would be the height of the ball after 5 sec?
b. Find the graph of $h(t)$ for the values of x that make sense in this problem.
c. Decide whether the ball reaches its peak in less than 3 sec or more than 3 sec.

46. *Shooting Fireworks* The launching pad for Hillville's Fourth of July fireworks is 10 ft above the ground. The final "big boom" is hurled straight up with an initial velocity of 150 ft/sec. The height y in feet of the object t seconds later is $y = -16t^2 + 150t + 10$.

a. What is the height in feet of the firework 3 sec after the explosion that propelled it upwards?
b. Does it reach a height above 350 ft?
c. The manufacturer wants the "big boom" to occur at maximum altitude. What should be the timing for its fuse?

47. *Building a School Track* Salem High School plans to build a new 400-m track around the football field, as shown in the figure.

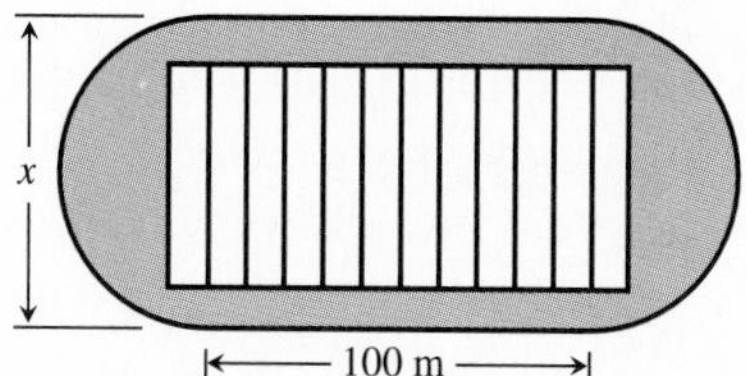

For Exercise 47

a. The track is to have 100 m on a straight side and semicircular turns. Find a function $P(x)$ that describes the length of the track when the perpendicular distance from one side of the infield to the other is x meters. (Recall that the formula for the circumference of a circle is $C = \pi d$.)
b. How far is it across the infield if the track's length is 400 m? That is, what is x if $P(x) = 400$?
c. A football field is 53 yards (yd) wide. What is the perpendicular distance from the edge of the track to the sideline of the football field? (*Hint:* 1 m = 1.0936 yd).
d. Find a function $A(x)$ that describes the total area (in square meters) of the infield.

48. *Industrial Costs* Dayton Power and Light, Inc., has a power plant on the Miami River where the river is 800 ft wide. To lay a new cable from the plant to a location in the city 2 mi downstream on the opposite side costs \$180 per foot across the river and \$100 per foot along the land.

a. Suppose that the cable goes from the plant to a point Q on the opposite side that is x feet from the point P directly opposite the plant. Find a function $C(x)$ that gives the cost of laying the cable in terms of the distance x.
b. Generate a table of values to find if the least expensive

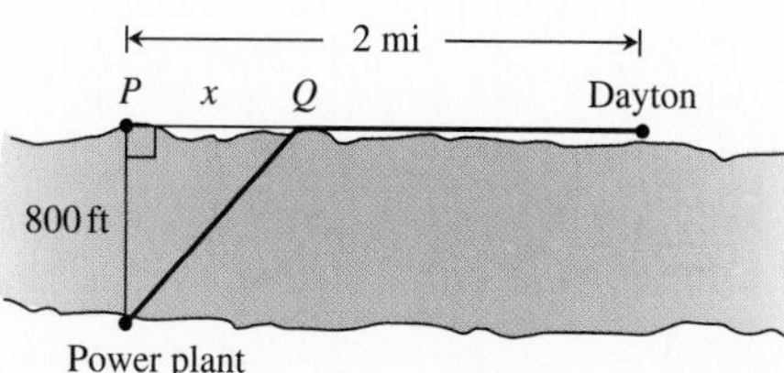

For Exercise 48

location for point Q is less than 2000 ft or greater than 2000 ft from point P.

c. Support your answer in part b graphically.

49. *The Cone Problem* Begin with a circular piece of paper with a 4-in. radius as shown in (a). Cut out a sector with an arc length of x. Join the two edges of the remaining portion to form a cone with radius r and height h, as shown in (b).

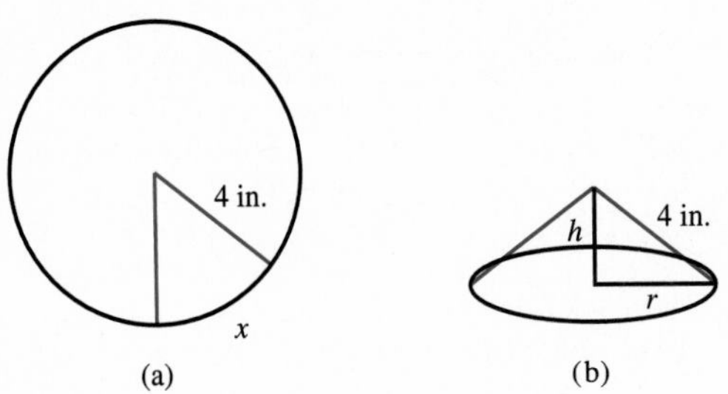

For Exercise 49

a. Explain why the circumference of the base of the cone is $8\pi - x$.

b. Express the radius r as a function of x. (*Hint:* Use the result of part a.)

c. Express the height h as a function of x. (*Hint:* Use the result from part b and the Pythagorean Theorem.)

d. Find a function $V(x)$ that gives the volume of the cone as a function of x. (Recall that the volume of a cone is $V = \frac{1}{3}\pi r^2 h$.)

50. Table 1.3 gives the numbers (in millions) of 16-bit micro-controls shipped worldwide from 1984 to 1990.

Table 1.3 Numbers of Controls Shipped

Year	Number Shipped (millions)
1984	0.1
1985	0.2
1986	0.4
1987	3.4
1988	5.5
1989	9.4
1990	17.4

Source: Dataquest, Inc., San Jose, Calif., unpublished data as reported in U.S. Bureau of the Census, *Statistical Abstract of the United States: 1992* (112th ed.). Washington, D.C., 1992.

a. The function

$$f(x) = 0.0833x^3 - 0.331x^2 + 0.757x - 0.543$$

models these data, where $x = 1$ corresponds to 1984. How closely does the value $f(5)$ match the data for 1988?

b. Use this model to predict the number of controllers shipped in the year 2000.

EXTENDING THE IDEAS

In Exercises 51 and 52, complete the explanation.

51. Writing to Learn Example 6a showed a graph rising to the right, and we conjectured that the range of the function was $[0, \infty)$. Explain how a graph can always rise to the right when its function does not have range $[0, \infty)$.

52. Writing to Learn Explain how a graph could have a view like the one in Example 6b but have a range quite different from $[-3, 3]$.

In Exercises 53–55, use $f(x) = x^2 - 5x + 3$, whose graph is shown here.

53. Use both a numerical method and an algebraic method to show that $f(2 + h) \neq f(2) + f(h)$, where h represents any real number.

54. Simplify $f(2 + h) - f(2 - h)$.

55. a. Simplify the expression

$$\frac{f(2 + h) - f(2 - h)}{2h}.$$

b. Writing to Learn Explain why the expression in part a gives the slope of the line L in the figure.

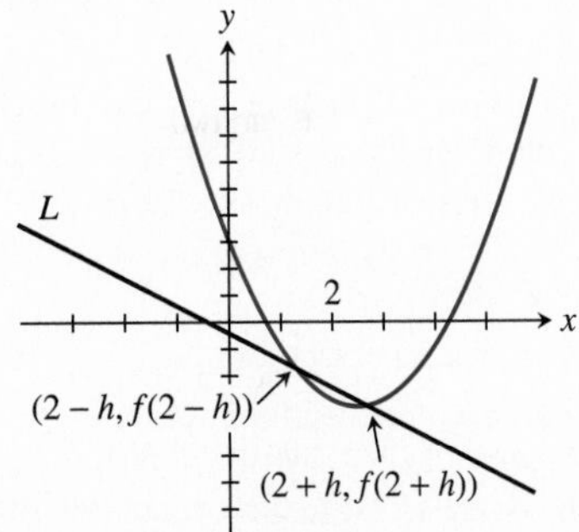

For Exercises 53–55

In Exercises 56 and 57, simplify the expression

$$\frac{f(x+h)-f(x)}{h}$$

for the given function.

56. $f(x) = 3x + 2$

57. $f(x) = x^2 - 3$

58. Group Learning Activity *We suggest that you work in groups of three or four.*
We encounter relationships between sets of numbers routinely in our daily lives. Discuss whether each of the relationships in parts a–e is a function. Then over the next several weeks do part f. Be prepared to defend your conclusions. Remember, to be a function, *one* number in the set you call the domain can associate with *only one* number in the set you call the range.

a. A height-weight chart
b. A sales tax chart
c. The gallon and price gauges on a gasoline pump
d. Taxable income and income tax in the IRS 1040 tax-rate table
e. The won-lost records in major league baseball standings
f. Collect examples of number relationships between two sets of numbers. Classify your examples as functions and nonfunctions, and save them to share with the class. Watch for examples that may be impossible to classify.

59. Group Learning Activity *We suggest that you work in groups of three or four.*
In *function mode* on a grapher we enter equations in the form $y = \cdots$. Consider the relationship $y = \pm\sqrt{x+3}$, which does not define a function. You can enter this relationship on a graphing utility as

$$y_1 = \sqrt{x+3} \quad \text{and} \quad y_2 = -\sqrt{x+3}$$

and graph simultaneously. Do this. Discuss as a group why the relationship fails to be a function of x. In particular, how does graphing simultaneously reinforce this fact?

60. This basketball box score suggests relationships between the number of field goals (FG), the number of free throws (FT) and the total number of points (TP) a player scores. Are any of these relationships functions?

	FG	FT	TP
Renner	3	1	7
Fauver	6	2	14
Harrison	2	6	10
Yount	6	3	15
Lind	6	1	13
Best	5	5	15
Kirkton	1	0	2
Conners	0	3	3

1.3 LINEAR FUNCTIONS AND THEIR GRAPHS

Introducing Linear Functions • Slope of a Line • Interpreting Slope on a Grapher • Point-Slope Form • Slope-Intercept Form • Graphing Linear Equations • Parallel and Perpendicular Lines • Applying Linear Functions

Objective

Students will be able to use the concepts of slope and y-intercept to graph and write linear equations in two variables.

Motivate

Have students graph the functions $y = 2x + 3$, $y = 2x + 1$, and $y = 2x - 2$ on a grapher. Discuss the graphs.

Introducing Linear Functions

A function is **linear** if it can be written in the form

$$f(x) = ax + b \qquad \text{or} \qquad y = ax + b,\ a \neq 0,$$

where a and b are constants. The graph of a linear function is a line. In this section we see the strong connections between these graphs and the algebraic form shown above.

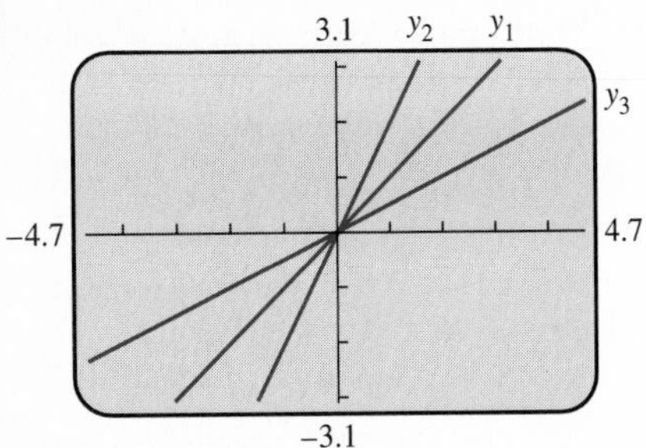

Figure 1.18 $y_1 = x$, $y_2 = 2x$, $y_3 = 0.5x$.

GRAPHER NOTE

Decimal Window

We sometimes call the window used in Figure 1.18 the *decimal window* because the x-coordinates shown on the screen are given in tenths (0.1). If these window dimensions do not result in "nice" x-coordinates on your grapher, consult the Grapher Workshop to find what will work on your machine.

EXPLORATION Introducing Linear Functions

To see one connection between $y = ax + b$ and its graph, first let $b = 0$ and view linear functions of the form $y = ax$.

1. Graph

$$y_1 = x, \qquad y_2 = 2x, \qquad \text{and} \qquad y_3 = 0.5x$$

   in the decimal viewing window. (See Figure 1.18 and the Grapher Note in the margin.)
2. What do the graphs of y_1, y_2, and y_3 have in common? How are they different?
3. List the functions in order of *steepness* from least to greatest.
4. **Describe in your own words:** How do the graphs of

$$f(x) = 3x, \qquad g(x) = 1.5x, \qquad \text{and} \qquad h(x) = 5x$$

   compare in steepness? Support your statement graphically.

Ask and answer a similar set of questions for the functions

$$y_1 = x + 3, \qquad y_2 = x - 4, \qquad y_3 = x + 1.$$

You may have discovered the following while completing the Exploration. The value of a in $y = ax + b$ tells us about the steepness of the line. The value of b tells us where the graph crosses the y-axis. (See Figure 1.19.)

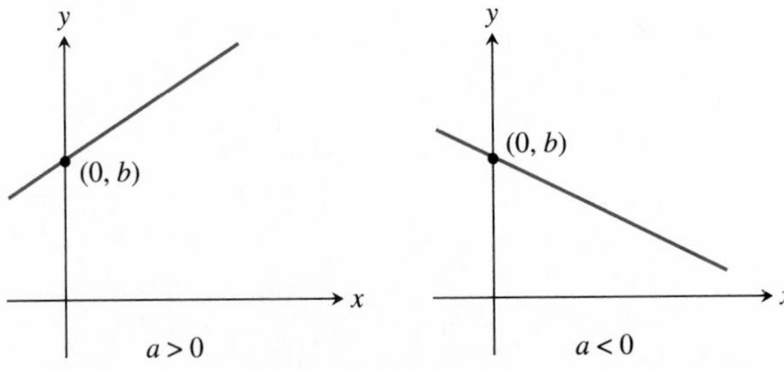

Figure 1.19 When $a > 0$, the line rises from left to right. When $a < 0$, the line falls from left to right. In both cases the line crosses the y-axis at $(0, b)$.

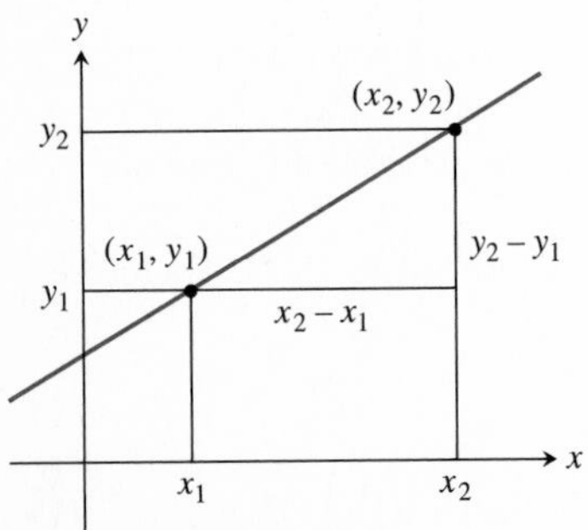

Figure 1.20 The slope of a line can be found from the coordinates of any two points of the line.

Slope of a Line

A key characteristic of a nonvertical line is its *slope.* The slope of a line is the ratio of the amount of vertical change to the amount of horizontal change between points (x_1, y_1) and (x_2, y_2), where the vertical change is $y_2 - y_1$ and the horizontal change is $x_2 - x_1$. See Figure 1.20.

Teaching Note

The Exploration is essential to the lesson. Reviewing the familiar concepts of slope and y-intercept using a grapher gives students an opportunity to make generalizations about the behavior of linear functions.

Exploration Extensions

How does the apparent steepness of a graph change when the viewing window is changed? Try doubling the values for *X*min, *X*max, *Y*min, and *Y*max together. Then see what happens when the values are adjusted separately.

Alert

Some students will attempt to calculate slopes using x in the numerator and y in the denominator. Remind them that slope $= \dfrac{\text{rise}}{\text{run}}$.

Definition: Slope of a Line

The **slope** of the nonvertical line through the points (x_1, y_1) and (x_2, y_2) is

$$m = \frac{y_2 - y_1}{x_2 - x_1}.$$

If the line is vertical, then $x_1 = x_2$ and the slope is undefined.

■ **EXAMPLE 1 Finding the Slope of a Line**

Find the slope of the line through the two points. Sketch a graph of the line.

a. $(-1, 2)$ and $(4, -2)$
b. $(1, 1)$ and $(3, 4)$

Solution

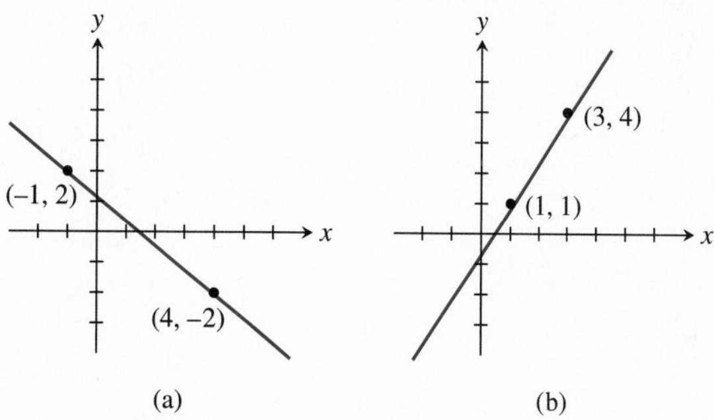

Figure 1.21

a. The two points are $(x_1, y_1) = (-1, 2)$ and $(x_2, y_2) = (4, -2)$. Therefore,

$$m = \frac{y_2 - y_1}{x_2 - x_1} = \frac{-2 - 2}{4 - (-1)} = -\frac{4}{5}.$$

b. The two points are $(x_1, y_1) = (1, 1)$ and $(x_2, y_2) = (3, 4)$. Therefore,

$$m = \frac{y_2 - y_1}{x_2 - x_1} = \frac{4 - 1}{3 - 1} = \frac{3}{2}.$$

The graphs of these two lines are shown in Figure 1.21. ■

REMARK

Notice that the slope does not depend on the order of the points. We could use $(x_1, y_1) = (4, -2)$ and $(x_2, y_2) = (-1, 2)$. Check it out.

Figure 1.22 shows a vertical line through the points (3, 2) and (3, 7). If we try to calculate its slope using the expression $(y_2 - y_1)/(x_2 - x_1)$, we get zero in the denominator. This explains why the slope of a vertical line is undefined.

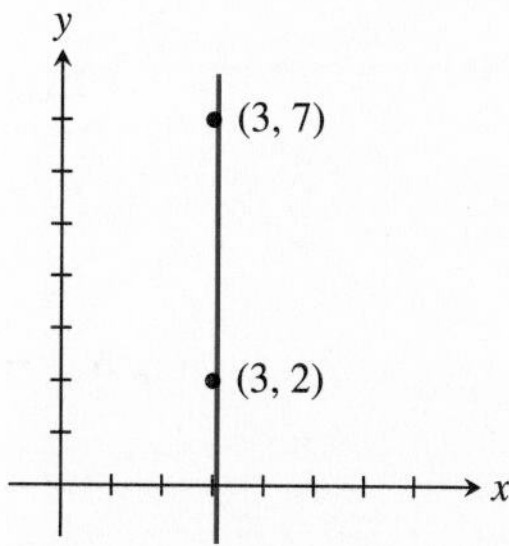

Figure 1.22 Because $m = (7 - 2)/(3 - 3)$ is not defined, the line is vertical and its slope is not defined.

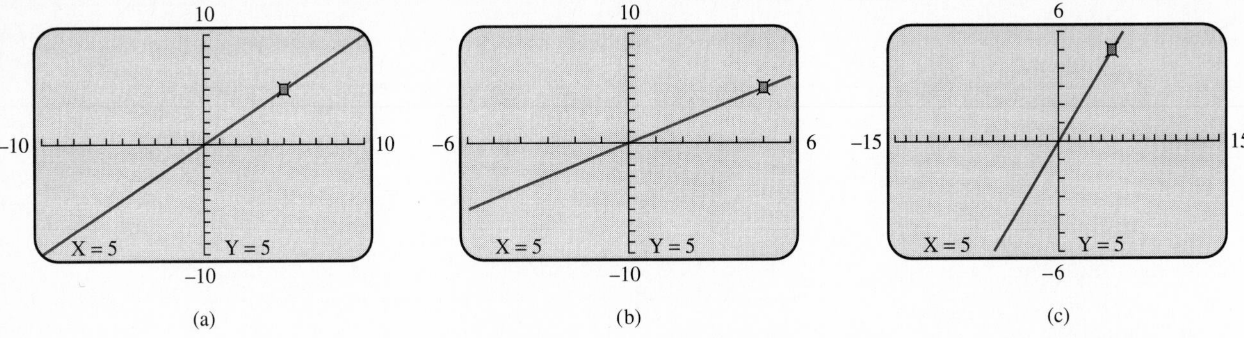

Figure 1.23 Three views of the graph of $y = x$. The slope is 1 in all three views.

Teaching Note

Point out that a graph makes it easy to determine whether a slope is positive or negative.

Interpreting Slope on a Grapher

Figure 1.23 shows that the "steepness" of the graph of $f(x) = x$ appears to differ depending on the viewing window being used.

When you change the y dimension of the viewing window relative to the x dimension, the slope of the graph of f may *appear* to change. But it doesn't. In each window the line passes through both points (0, 0) and (5, 5), so the slope is always

$$m = \frac{5 - 0}{5 - 0} = 1.$$

On the other hand, lines that appear to have the same slope in two viewing windows may in fact have different slopes, as illustrated in Example 2.

REMARK

Example 2 illustrates that to compare slope "visually," it is wise to graph both lines in the same viewing window.

EXAMPLE 2 Observing the Effect of Window Dimension on Slope

Each line in Figure 1.24 passes through the origin and the point in the upper right corner of the screen. Which line has the greater slope?

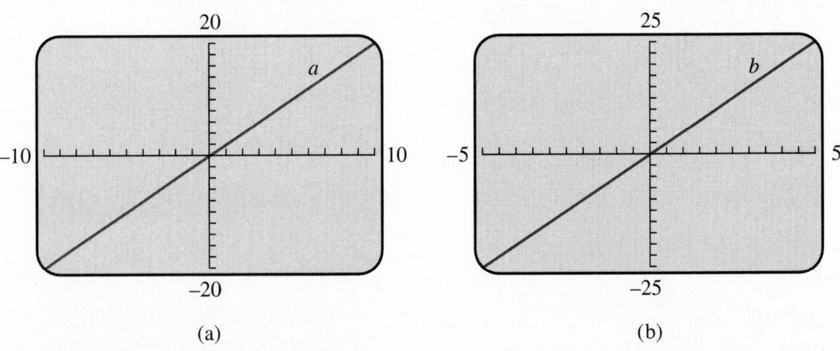

Figure 1.24 Two lines of different slope.

Solution

Line a goes through points (0, 0) and (10, 20). It has the slope

$$m = \frac{20 - 0}{10 - 0}$$

$$= \frac{20}{10} = 2 \quad \text{Steeper than slope 1}$$

Line b goes through points (0, 0) and (5, 25). It has the slope

$$m = \frac{25 - 0}{5 - 0}$$

$$= \frac{25}{5} = 5 \quad \text{Much steeper than slope 1}$$

Line b has the greater slope. ■

Point-Slope Form

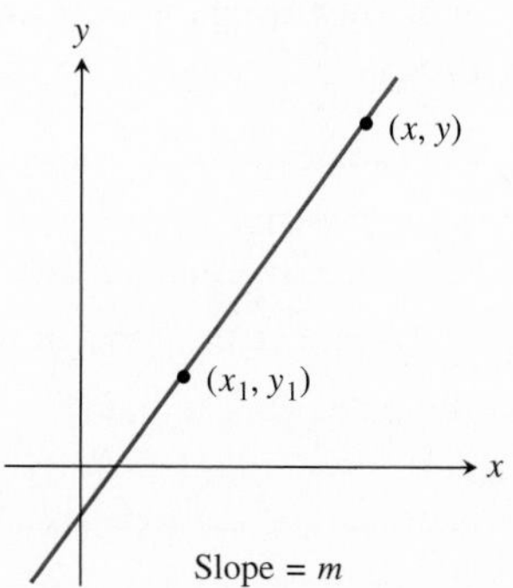

Figure 1.25

If we know the coordinates of one point on a line and the slope of the line, then we can find an equation for that line. For example, the line in Figure 1.25 passes through the point (x_1, y_1) and has slope m. If (x, y) is any other point on this line, the definition of slope yields the equation

$$m = \frac{y - y_1}{x - x_1} \quad \text{or} \quad y - y_1 = m(x - x_1).$$

An equation written this way is in the *point-slope form.*

Teaching Note

You may want to point out that the point-slope form can also be used if two points on the line are known. Ask how the slope of a line through two points can be found.

Definition: Point-Slope Form of an Equation of a Line

The **point-slope form** of an equation of a line that passes through the point (x_1, y_1) and has slope m is

$$y - y_1 = m(x - x_1).$$

■ EXAMPLE 3 Using the Point-Slope Form

Use the point-slope form to find an equation of the line that passes through the point $(-3, -4)$ and has slope 2.

Solution

Substitute $x_1 = -3$, $y_1 = -4$, and $m = 2$ into the point-slope form, and simplify the resulting equation.

$$\begin{aligned} y - y_1 &= m(x - x_1) && \text{Point-slope form} \\ y - (-4) &= 2[x - (-3)] && (x_1, y_1) = (-3, -4) \\ y + 4 &= 2x - 2(-3) \\ y + 4 &= 2x + 6 \\ y &= 2x + 2 && \text{A simplified form} \end{aligned}$$

■

Slope-Intercept Form

The **y-intercept** of a nonvertical line is the point where the line intersects the y-axis. If we know the y-intercept and the slope of the line, we can apply the point-slope form to find an equation of the line.

Figure 1.26 shows a line with slope m and y-intercept $(0, b)$. A point-slope form equation for this line is $y - b = m(x - 0)$. By rewriting this equation we obtain the form known as the *slope-intercept form.*

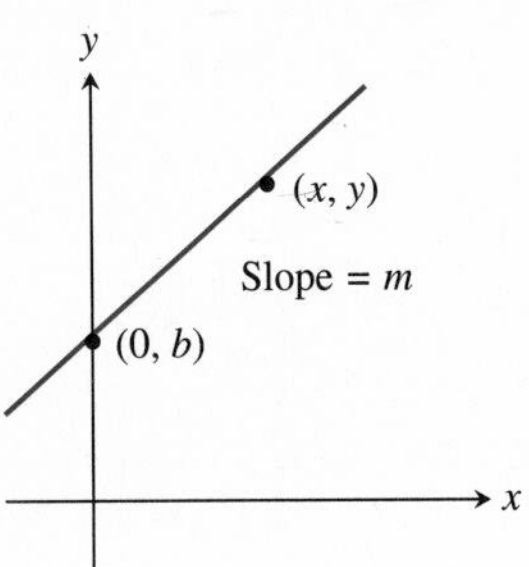

Figure 1.26 A line with slope m and y-intercept $(0, b)$.

> **Definition:** Slope-Intercept Form of an Equation of a Line
>
> The **slope-intercept form** of an equation of a line with slope m and y-intercept $(0, b)$ is
>
> $$y = mx + b.$$

■ EXAMPLE 4 Finding an Equation of a Line

Use the slope-intercept form to find an equation of the line with slope 3 that passes through the point $(-1, 6)$.

Solution

It is given that the line's slope is $m = 3$. To find b we substitute $x = -1$ and $y = 6$ in $y = 3x + b$:

$$\begin{aligned} y &= 3x + b \\ 6 &= 3(-1) + b && (x, y) = (-1, 6) \\ 6 &= -3 + b \\ b &= 9 \end{aligned}$$

The equation is $y = 3x + 9$.

As an alternate solution method you could use the point-slope form and solve for y. ■

We cannot speak of *the* equation of a line because each line has many different equations. Every line has an equation that can be written in the form $Ax + By + C = 0$ where A and B are not both zero. This form is the **general form** for an equation of a line.

If $B \neq 0$, the general form can be changed to the slope-intercept form, as follows:

$$Ax + By + C = 0$$
$$By = -Ax - C$$
$$y = \underbrace{-\frac{A}{B}}_{\text{slope}}x \underbrace{- \frac{C}{B}}_{\text{intercept}}$$

Teaching Note

Point out that in the point-slope form, (x_1, y_1) can represent any point on the line. This form is useful for determining the equation if the slope and one or more points are known.

Forms of Equations of Lines

General form:	$Ax + By + C = 0$, A and B not both zero.
Slope-intercept form:	$y = mx + b$
Point-slope form:	$y - y_1 = m(x - x_1)$
Vertical line:	$x = a$
Horizontal line:	$y = b$

Graphing Linear Equations

A **linear equation in x and y** is one that can be written in the form

$$ax + by = c,$$

where a and b are not both zero. Rewriting this equation in the form $ax + by - c = 0$ we see that it is the general form of an equation of a line. If $b = 0$, the line is vertical, and if $a = 0$, the line is horizontal.

To graph a linear equation on a grapher we first solve the equation for y to obtain a form for the equation that begins $y = \cdots$. Then we enter this equation on a grapher to obtain the graph of the line.

The slope-intercept form fits the form $y = \cdots$. Also, it provides some built-in checks by which we can mentally confirm that the line in the viewing window is correct.

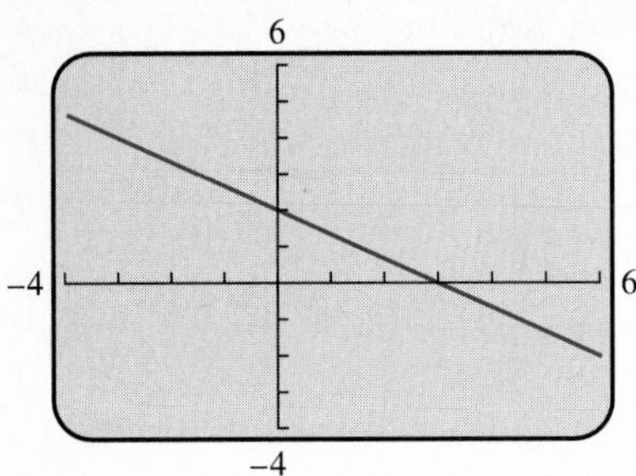

Figure 1.27

■ EXAMPLE 5 Graphing Linear Equations

Draw the graph of $2x + 3y = 6$.

Solution

Solve for y.

$$2x + 3y = 6 \quad \text{Original equation}$$
$$3y = -2x + 6$$
$$y = -\frac{2}{3}x + 2$$

Figure 1.27 gives the graph of $y = -\frac{2}{3}x + 2$, or equivalently, the graph of the linear equation $2x + 3y = 6$. ■

GRAPHER NOTE

Square Window

A square window on a grapher is one in which angles appear to be true. For example, the line $y = x$ will appear to make a 45° angle with the positive x-axis. Furthermore, a distance of 1 on the x- and y-axes will appear to be the same. Consult the Grapher Workshop for more detail. When studying perpendicular lines on a grapher you will find that it is good to use a square window.

Parallel and Perpendicular Lines

There is a useful connection between slopes of lines and whether the lines are parallel or perpendicular, as Example 6 shows.

■ EXAMPLE 6 Comparing Parallel and Perpendicular Lines

a. Graph $y_1 = 0.3x - 1.5$, $y_2 = 0.3x + 1$, and $y_3 = 0.3x + 2.5$ in a square window. Make an observation about these graphs.

b. Graph $y_4 = 2x - 1.5$ and $y_5 = -0.5x + 2.5$ in a square window. Make an observation about these graphs.

Solution

a. The lines in Fig. 1.28a appear to be parallel.

b. The lines in Fig. 1.28b appear to be perpendicular. ■

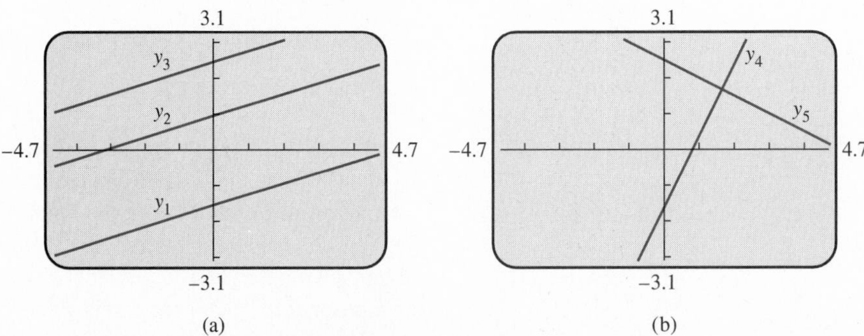

Figure 1.28 (a) Graphs of $y_1 = 0.3x - 1.5$, $y_2 = 0.3x + 1$, and $y_3 = 0.3x + 2.5$. (b) Graphs of $y_4 = 2x - 1.5$ and $y_5 = -0.5x + 2.5$.

Notice that the line $y = x$ in Figure 1.23 appears to make three different angles with the positive x-axis. Try graphing the two lines of Example 6b in other viewing windows, and watch the *angle between the lines* appear to change. Only in a square window will the angle appear to be 90°. Fortunately, we have an algebraic test to determine whether two lines are parallel or perpendicular.

Parallel and Perpendicular Lines

- Two nonvertical lines are parallel if and only if their slopes are equal.
- Two nonvertical lines are perpendicular if and only if their slopes m_1 and m_2 are opposite reciprocals. That is, if and only if

$$m_2 = -\frac{1}{m_1}.$$

In Example 6a the slopes of the three lines y_1, y_2, and y_3 are 0.3, so the lines are parallel. The slopes of the two lines in Example 6b are 2 and $-\frac{1}{2}$, so the lines are perpendicular.

EXAMPLE 7 Finding an Equation of a Parallel Line

Find an equation of the line through $P(1, -2)$ that is parallel to the line ℓ with equation $3x - 2y = 1$.

Solution

Solve Algebraically

By writing the equation in slope-intercept form,

$$3x - 2y = 1,$$
$$-2y = -3x + 1,$$
$$y = \frac{3}{2}x - \frac{1}{2} \quad \text{Slope-intercept form}$$

we find that the slope of ℓ is 3/2.

The line whose equation is being sought has the same slope $m = 3/2$ and contains point $(x_1, y_1) = (1, -2)$. Using the point-slope form, we get

$$y + 2 = \frac{3}{2}(x - 1)$$

$$y = \frac{3}{2}x - \frac{7}{2}$$

■

■ EXAMPLE 8 Finding an Equation of a Perpendicular Line

Find an equation for the line through $P(2, -3)$ that is perpendicular to line ℓ whose equation is $4x + y = 3$.

Solution

Solve Algebraically

By writing the equation in slope-intercept form,

$$4x + y = 3,$$
$$y = -4x + 3,$$

we find that the slope of ℓ is -4.

The line whose equation is being sought has slope 1/4 and passes through point $(2, -3)$. Thus,

$$y - (-3) = \frac{1}{4}(x - 2)$$
$$y + 3 = \frac{1}{4}x - \frac{2}{4}$$
$$y = \frac{1}{4}x - \frac{1}{2} - \frac{6}{2}$$
$$y = \frac{1}{4}x - \frac{7}{2}$$

■

Applying Linear Functions

Linear functions and their graphs occur often in applications. Algebraic solutions to these application problems often require finding an equation of a line and solving a linear equation in one variable. Grapher techniques complement algebraic ones.

■ EXAMPLE 9 APPLICATION: Evaluating the Depreciation of Real Estate

Camelot Apartments purchased a \$50,000 building and depreciates it \$2000 per year over a 25-year period. In how many years will its value be \$24,500?

Solution

Model

This problem has a linear function for a model. Let y be the value of the building in dollars after x years, where $0 \leq x \leq 25$. Notice that $y = 50{,}000$ when $x = 0$, so the line has y-intercept $(0, 50\ 000)$. When $x = 1$, $y = 50{,}000 - 2000$

= 48,000, and the line passes through the point (1, 48 000). The slope of the line is

$$m = \frac{48{,}000 - 50{,}000}{1 - 0} = -2000.$$

The function that models the value of the building is

$$y = -2000x + 50{,}000.$$

To find how many years it will take for the value of the building to be \$24,500, solve the equation

$$24{,}500 = -2000x + 50{,}000.$$

Solve Algebraically

$$2000x + 24{,}500 = 50{,}000 \qquad \text{Add } 2000x.$$

$$2000x = 50{,}000 - 24{,}500 \qquad \text{Subtract } 24{,}500.$$

$$x = \frac{50{,}000 - 24{,}500}{2000} = 12.75 \qquad \text{Divide by } 2000.$$

Support Graphically

The trace coordinates in Figure 1.29a show that (12.75, 24 500) is a solution of $y = 50{,}000 - 2000x$.

Support Numerically

The table of (x, y) values in Figure 1.29b also shows that (12.75, 24 500) is a solution.

Interpret

The value of the building will be \$24,500 in 12.75 years. ■

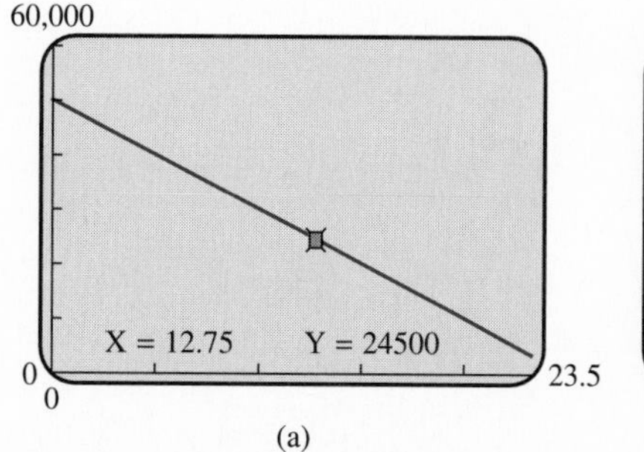

(a)

X	Y_1
12	26,000
12.25	25,500
12.5	25,000
12.75	24,500
13	24,000
13.25	23,500
13.5	23,000

$Y_1 = 50000 - 2000X$

(b)

Figure 1.29 The trace coordinates in (a) and the fourth line of the table in (b) show that (12.75, 24 500) is a solution of $24{,}500 = 50{,}000 - 2000x$.

Follow-up

Ask students how they can tell whether an equation represents a linear function.

Assignment Guide

Day 1: Ex. 1, 3, 6, 9, 10, 12–39, multiples of 3, 40, 42, 44
Day 2: Ex. 43–51 odd, 55–73 odd

Cooperative Learning

Group Activity: Ex. 9, 10, 66, 71

Notes on Exercises

Ex. 3–8 require a grapher for exploration.
Ex. 11–42 include many traditional problems from the analytic geometry of lines.
Ex. 57–60 may lead students to interesting discoveries using a graphing calculator.

Ongoing Assessment

Self-Assessment: Ex. 17, 35, 53, 65
Embedded Assessment: Ex. 10, 44

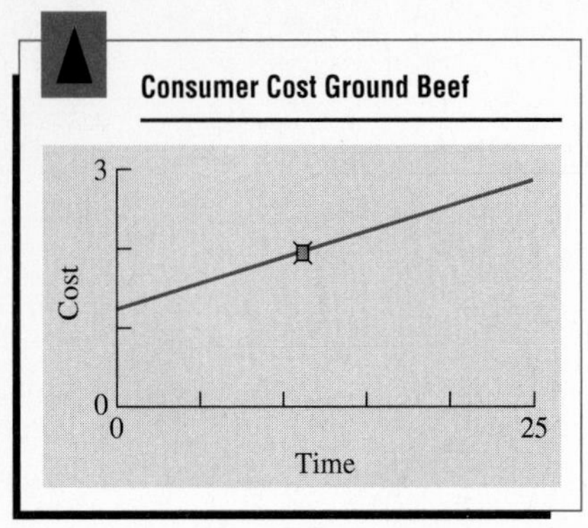

$x = 11.170213; y = 1.9560638$

Figure 1.30 A graph of $y = 0.065x + 1.23$.

EXAMPLE 10 APPLICATION: Cost of Ground Beef

The average consumer cost y of ground beef in dollars per pound between 1985 and 1991 can be modeled by the linear function $y = 0.065x + 1.23$, where $x = 0$ corresponds to 1985.[2] During what year would the cost of ground beef average \$1.95 per pound?

Solution

Model

Because the equation $y = 0.065x + 1.23$ models the cost of ground beef x years after 1985, we need to solve the equation

$$0.065x + 1.23 = 1.95.$$

Solve Algebraically

$$0.065x = 1.95 - 1.23 \quad \text{Subtract 1.23.}$$

$$x = \frac{1.95 - 1.23}{0.065} \quad \text{Divide by 0.065.}$$

$$x = 11.076\cdots$$

Support Graphically

Figure 1.30 shows a graph of $y = 0.065x + 1.23$. It appears that y is approximately 1.95 when x is a little more than 11.

Interpret

Assuming that the model continues to be valid, ground beef will reach an average cost of \$1.95 per pound during 1996. ■

Quick Review 1.3

In Exercises 1–4, solve for x.

1. $2x - 3 = 3(4x - 1)$

2. $5 - 3x = 2(x + 3)$

3. $3(1 - 2x) + 4(2x - 5) = 7$

4. $2(7x + 1) = 5(1 - 3x)$

In Exercises 5–8, solve for y when $x = -2$.

5. $3x - 4y = 22$

6. $2x + 5y = 19$

7. $5x^2 + 7y = 48$

8. $3x^2 = 14 - 3y$

In Exercises 9 and 10, simplify the fraction.

9. $\dfrac{9 - 5}{-2 - (-8)}$

10. $\dfrac{-4 - 6}{-14 - (-2)}$

In Exercises 11–14, simplify the expression.

11. $4 - [x - (5 - 3x)]$

12. $-\{3 - [-(x + 1)]\}$

13. $-\dfrac{-(x - 1)}{1 - x}$

14. $\dfrac{x - 4}{8 - 2x}$

In Exercises 15–18, solve for y.

15. $2x - 5y = 21$

16. $2x + y = 17 + 2(x - 2y)$

17. $x^2 + y = 3x - 2y$

18. $y(x^2 + 3) = 3y + 2$

[2] *Source:* U.S. Bureau of Labor Statistics, *CPI Detailed Report.*

SECTION EXERCISES 1.3

In Exercises 1 and 2, estimate the slope of the line.

1.

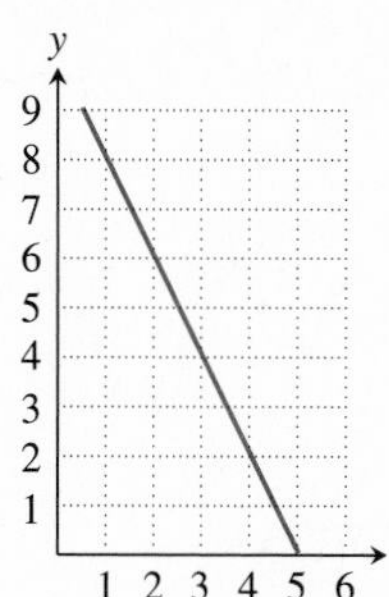

2.

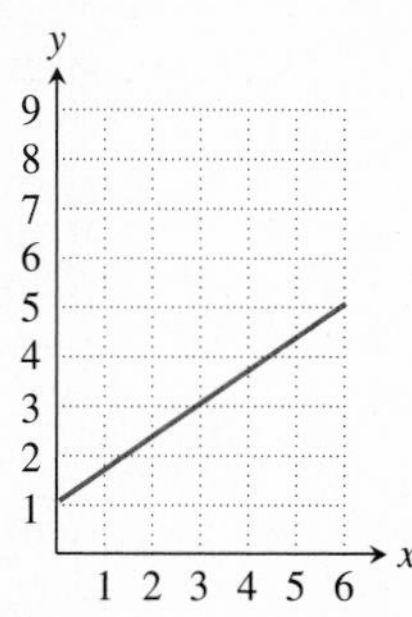

In Exercises 3–8, graph the linear equation on a grapher. Choose a viewing window that shows the line intersecting both the x- and y-axes.

3. $y = 4x + 12$

4. $y = 7x + 41$

5. $8x + y = 49$

6. $2x + y = 35$

7. $123x + 7y = 429$

8. $2100x + 12y = 3540$

In Exercises 9 and 10, solve the problem.

9. *Real Estate Appreciation* Bob Michaels purchased a house 8 years ago for \$42,000. This year it was appraised at \$67,500.

a. A linear equation $V = mt + b$, $0 \le t \le 15$, models the value of this house for 15 years after it was purchased. Find the slope and y-intercept, and then write the equation.
b. Graph the equation and trace to estimate in how many years after purchase will this house be worth \$72,500.
c. Write and solve an equation to determine how many years after purchase this house will be worth \$74,000.
d. Determine how many years after purchase this house will be worth \$80,250.

10. *Investment Planning* Mary Ellen plans to invest \$18,000, putting part of the money (x) into a savings account that pays 5% annually and the rest into an account that pays 8% annually.

a. What are the possible values of x in this situation?
b. If Mary Ellen invests x at 5%, write an equation that describes the total interest I received from both accounts at the end of one year.
c. Graph and trace to estimate how much Mary Ellen invested at 5% if she earned \$1,020 in total interest at the end of the first year.
d. Generate a table to find how much Mary Ellen should invest at 5% to earn \$1,185 in total interest in one year.

In Exercises 11–18, find the slope of the line through the pair of points.

11. $(-3, 5)$ and $(4, 9)$

12. $(-2, 1)$ and $(5, -3)$

13. $(-2, -5)$ and $(-1, 3)$

14. $(12, 23)$ and $(1, -2)$

15. $(-3, 4)$ and $(9, 4)$

16. $(2, -8)$ and $(2, 6)$

17. $(8, 5)$ and $(-3, 7)$

18. $(5, -3)$ and $(-4, 12)$

In Exercises 19–22, find the value of x or y so that the line through the given points has the given slope.

	Points	*Slope*
19.	$(x, 3)$ and $(5, 9)$	$m = 2$
20.	$(-2, 3)$ and $(4, y)$	$m = -3$
21.	$(-3, -5)$ and $(4, y)$	$m = 3$
22.	$(-8, -2)$ and $(x, 2)$	$m = \frac{1}{2}$

In Exercises 23–30, find an equation in point-slope form of the line through the given point with the given slope. Simplify your answer to slope-intercept form.

	Point	*Slope*		*Point*	*Slope*
23.	$(1, 4)$	$m = 2$	**24.**	$(-3, -5)$	$m = \frac{1}{2}$
25.	$(-4, 3)$	$m = -\frac{2}{3}$	**26.**	$(2, 7)$	$m = -\frac{3}{4}$
27.	$(7, 12)$	$m = \frac{3}{5}$	**28.**	$(-3, 4)$	$m = 3$
29.	$(-1, -5)$	$m = \frac{3}{7}$	**30.**	$(5, -4)$	$m = -2$

In Exercises 31–36, find an equation of the line that contains the two given points. Write the equation in the general form $Ax + By + C = 0$.

31. $(-1, 2)$ and $(2, 5)$

32. $(4, -1)$ and $(1, 5)$

33. $(-7, -2)$ and $(1, 6)$

34. $(-3, -8)$ and $(4, -1)$

35. $(1, -3)$ and $(5, -3)$

36. $(-1, -5)$ and $(-4, -2)$

In Exercises 37–42, write an equation for the line in slope-intercept form.

37. The line through $(0, 5)$ with slope $m = -3$

38. The line through $(1, 2)$ with slope $m = \frac{1}{2}$

39. The line through the points $(-4, 5)$ and $(4, 3)$

40. The line through the points (4, 2) and (−3, 1)

41. The line $2x + 5y = 12$

42. The line $7x - 12y = 96$

In Exercises 43 and 44, the lines contain the origin and the point in the upper right corner of the grapher screen.

43. Writing to Learn Which line shown here has the greater slope? Explain.

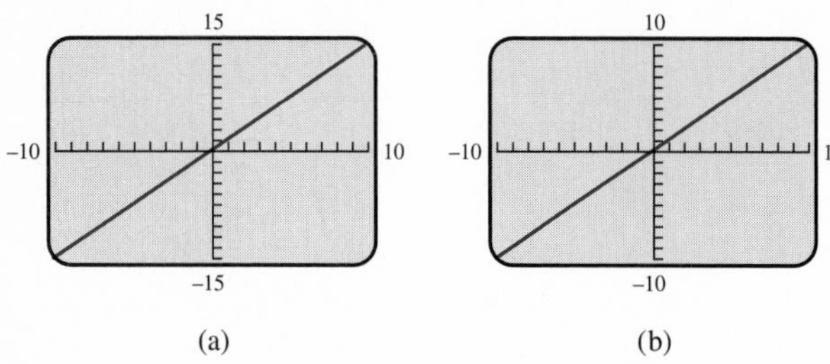

For Exercise 43

44. Writing to Learn Which line shown here has the greater slope? Explain.

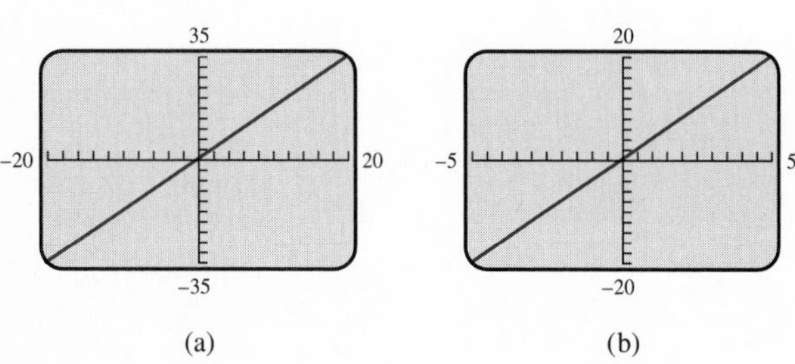

For Exercise 44

In Exercises 45–48, match the equation with one of the graphs in Exercises 43 and 44.

45. $y = x$ **46.** $y = \frac{7}{4}x$

47. $y = 4x$ **48.** $y = \frac{3}{2}x$

In Exercises 49–54, find the value of x and the value of y for which $(x, 14)$ and $(18, y)$ are points on the graph.

49. $y = 0.5x + 12$ **50.** $y = -2x + 18$

51. $x + 2y = 12$ **52.** $3x - 2y = 14$

53. $3x + 4y = 26$ **54.** $2x - 5y = 60$

In Exercises 55 and 56, use the graph provided. Estimate the slope of the line.

55.

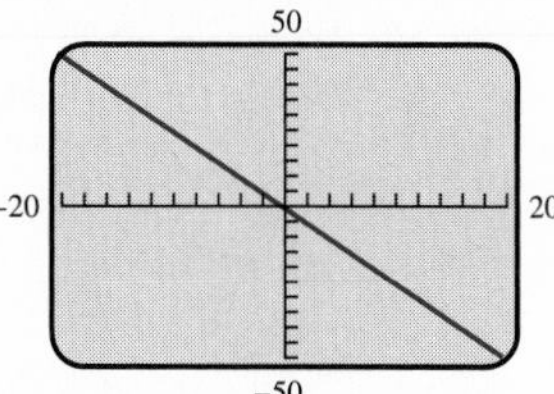

56.

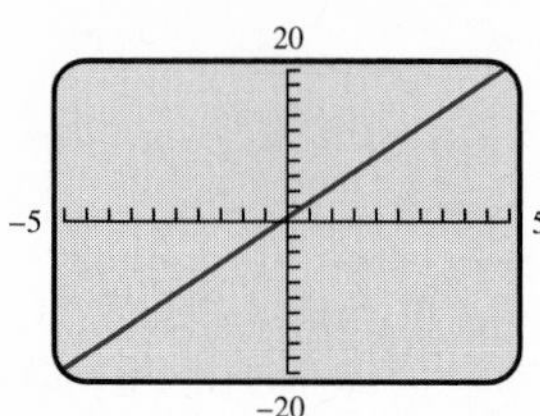

In Exercises 57–60, find the values for Ymin, Ymax and Yscl that will make the graph of the equation appear in the viewing window as shown here.

Xmin=⁻10
Xmax=10
Xscl=1
Ymin=
Ymax=
Yscl=

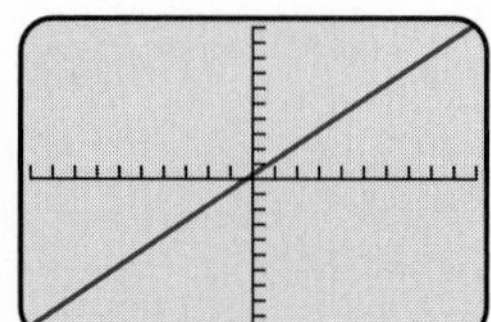

For Exercises 57–60

57. $y = 3x$ **58.** $y = 5x$

59. $y = \frac{2}{3}x$ **60.** $y = \frac{5}{4}x$

In Exercises 61–64, write two equations: one for the line passing through the given point and parallel to the given line, and another for the line passing through the given point and perpendicular to the given line. Support your work graphically.

	Given Point	*Given Line*
61.	(1, 2)	$y = 3x - 2$
62.	(−2, 3)	$y = -2x + 4$
63.	(3, 1)	$2x + 3y = 12$
64.	(6, 1)	$3x - 5y = 15$

In Exercises 65–71, solve the given problem.

65. *Navigation* A commercial jet aircraft climbs at takeoff with slope $m = \frac{3}{8}$. How far (in horizontal distance) will the aircraft fly to reach an altitude of 12,000 ft above the takeoff point?

66. *Highway Engineering* Interstate 70 west of Denver, Colorado, has a section posted as a 6% grade. This means that for a horizontal change of 100 ft there is a 6-ft vertical change.

For Exercise 66

a. Find the slope of this section of the highway.
b. On a highway with a 6% grade what is the horizontal distance required to climb 250 ft?
c. A sign along the highway says 6% grade for the next 7 mi. Estimate how many feet of vertical change there are along those next 7 mi. (There are 5280 ft in 1 mile.)

67. *Roof Building* Asphalt shingles do not meet code specifications on a roof that has less than a 4-12 pitch. (A 4-12 pitch means there are 4 ft of vertical change in 12 ft of horizontal change.) A certain roof has slope $m = \frac{1}{3}$. Could asphalt shingles be used on that roof? Explain.

68. Suppose that $a > 0$ and $b > 0$. Which of these equations describes the line whose x- and y-intercepts are $(a, 0)$ and $(0, b)$ respectively?

a. $y = -\dfrac{b}{a}x + b$ **b.** $y = -\dfrac{a}{b}x + a$

c. $ay - bx = ab$ **d.** $\dfrac{x}{a} + \dfrac{y}{b} = 1$

69. Show that if $c \neq d$, and if a and b are not both zero, then $ax + by = c$ and $ax + by = d$ are parallel lines.

70. Show that if $c \neq d$, and if a and b are not both zero, then $ax + by = c$ and $bx - ay = d$ are perpendicular lines.

71. *World Population* The world population for the years 1986 to 1991 (in millions) is shown in Table 1.4. The grapher screen shows a plot of these data. A line is drawn through the points (1986, 4936) and (1991, 5422).

Table 1.4 World Population

Year	Population (millions)
1986	4936
1987	5023
1988	5111
1989	5201
1990	5329
1991	5422

Source: Statistical Office of the United Nations, *Monthly Bulletin of Statistics, 1991.*

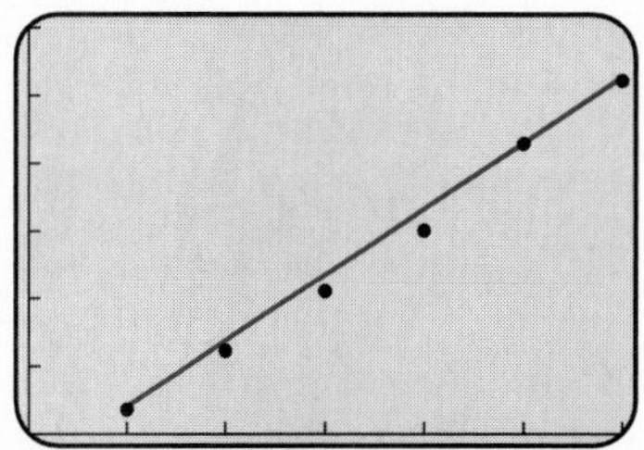

For Exercise 71

a. Find an equation of this line in slope-intercept form.
b. Use this equation to predict what the world population will be in the year 2000.

EXTENDING THE IDEAS

In Exercises 72–78, solve the problem.

72. Connecting Algebra and Geometry Show that if the midpoints of consecutive sides of any quadrilateral are connected, the result is a parallelogram.

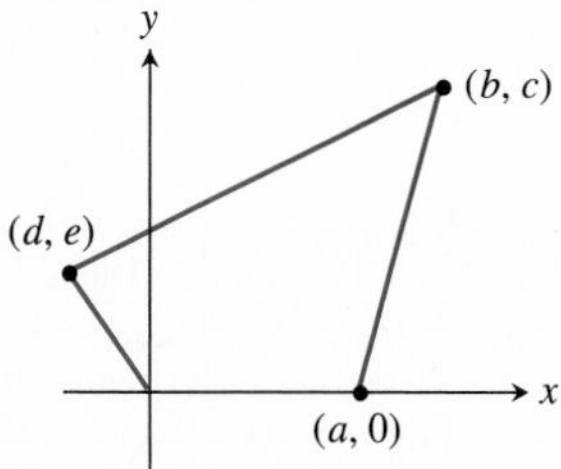

For Exercise 72

73. Connecting Algebra and Geometry Show that the altitudes of a triangle are *coincident* (that is, intersect in a common point). *Hint:* Show that the point $(a, a(c - a)/b)$ is common to the three altitudes ℓ_1, ℓ_2, and ℓ_3.

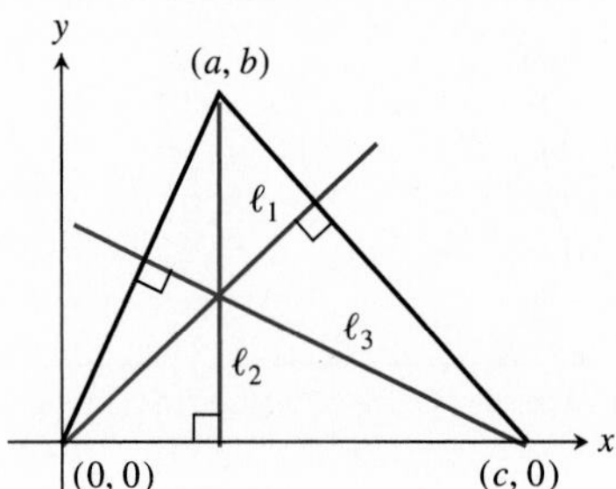

For Exercise 73

74. Connecting Algebra and Geometry Consider the semicircle of radius 5 centered at (0, 0). Find an equation of the line tangent to the semicircle at the point (3, 4). (*Hint:* A line tangent to a circle is perpendicular to the radius at the point of tangency.)

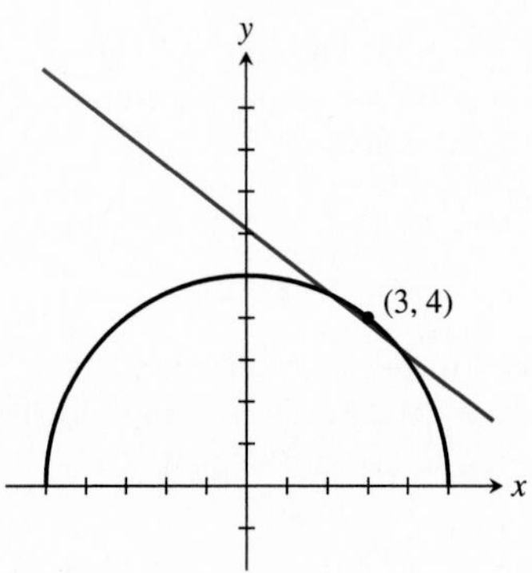

For Exercise 74

75. Writing to Learn The graphs of $y_1 = x^2$ and $y_2 = 4x - 4$ are shown in the figure. Explain how to use "trace" to suggest that the point (2, 4) is the only point common to both graphs. (We say that the line $y = 4x - 4$ is tangent to the graph of $y = x^2$ at (2, 4).)

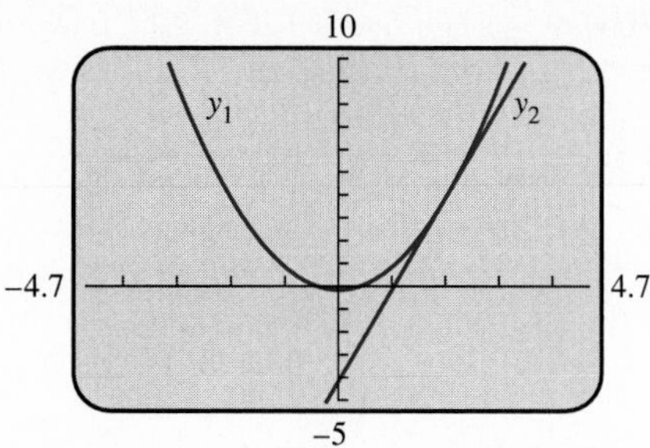

For Exercise 75

76. The line that is tangent to $y = x^2 - 5$ at the point $(1, -4)$ has slope 2. Find an equation of this tangent line.

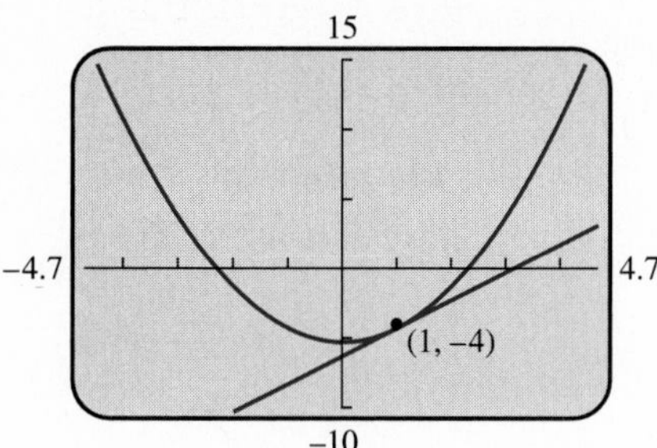

For Exercise 76

77. Writing to Learn Complete the following statement, and explain why you completed it as you did.
"A function is *linear* if a table of values that represents it numerically has the following property: ________________"

78. Connecting Algebra and Geometry Show that in any triangle, the line segment joining the midpoints of two sides is parallel to the third side and half as long.

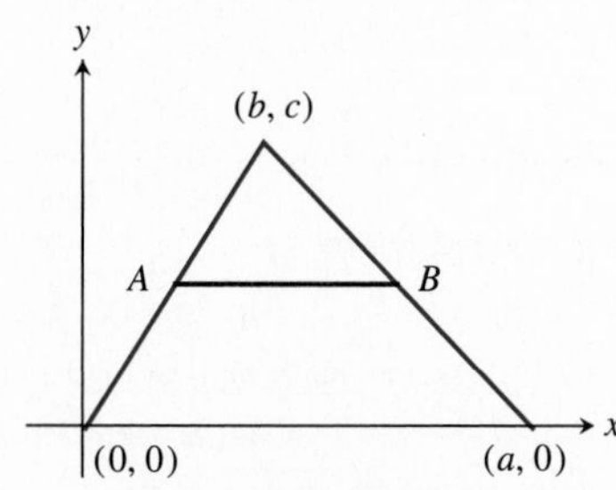

For Exercise 78

1.4 LINES OF BEST FIT FOR DATA

Introducing Data • Making Scatter Plots • Finding Strong and Weak Linear Correlation • Fitting a Line to Data • Modeling Data • Using Linear Regression Lines in Problem Solving

Table 1.5 Renting a Forklift

Hours (x)	Cost ($y = 25x + 18$)
0	18
1	43
2	68
3	93
4	118
5	143
6	168

Introducing Data

Facts collected by statisticians are correctly referred to as **data.**[3] Table 1.5 shows the cost of renting an industrial forklift from Midtown Rent All. It is an example of data generated by a formula.

However, not all data are generated by a formula. **Raw data** are data that are simply collected. Table 1.6, which gives population data, is an example of raw data.

Table 1.6 World Population

Year	Population (millions)
1986	4936
1987	5023
1988	5111
1989	5201
1990	5329
1991	5422

Source: Statistical Office of the United Nations, *Monthly Bulletin of Statistics, 1991.*

Objective

Students will be able to analyze data by making and examining scatter plots, modeling data, and using linear regression.

Motivate

Ask why an equation might be used to approximate the information in Table 1.7.

Teaching Note

Note that scatter plots can be easily produced either on a grapher or by using pencil and paper.

Making Scatter Plots

A state health department worker recorded the ages and heights of 10 children. The data are shown in Table 1.7. Is there any relationship between age and height? It is not evident from a quick glance at Table 1.7. We need to develop some techniques of **data analysis.**

A **scatter plot** is a plotting of the (x, y) data pairs on a rectangular coordinate system. Figure 1.31 shows a scatter plot of the data from Table 1.7. This plot suggests at a quick glance that there is some relationship, perhaps linear, between the two quantities age and height.

Table 1.7 State Health Dept. Age-Height Data

Age (months)	Height (inches)
6	23
9	26
11	17
15	23
25	33
33	36
27	32
46	44
52	43
48	39

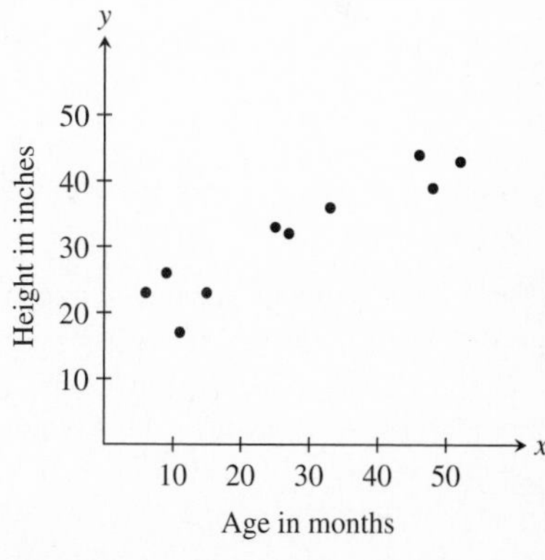

Figure 1.31 A scatter plot of the age-height data in Table 1.7.

[3] The word *data* is the plural of the word *datum,* which means *fact.*

■ EXAMPLE 1 Completing a Scatter Plot

Table 1.8 lists data compiled from 12 randomly chosen students. It pairs their weights with their percentile scores on the PSAT test. Make a scatter plot of these data.[4]

Solution

Table 1.8 Student Weight and Percentile Data

Weight	Percentile Score
78	98
84	34
96	48
97	78
105	28
112	85
123	95
136	34
144	62
148	11
152	78
158	52

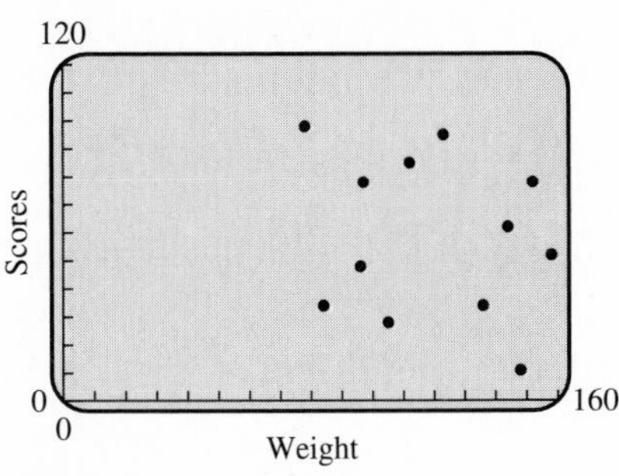

Figure 1.32

■

The scatter plot in Figure 1.32 suggests little or no relationship between student weights and percentile scores on the PSAT test.

Alert

Some students will attempt to determine whether the correlation is strong or weak by calculating a slope. Point out that these terms refer to how well a line can model the data, not to the slope of the line.

Finding Strong and Weak Linear Correlation

Figure 1.33 shows five different scatter plots. When the points in a scatter plot are clustered along a line, we say that there is a **linear correlation** between the quantities represented by the data. When an oval is drawn around the points in a scatter plot, the narrower the oval, the stronger the linear correlation.

When the oval tilts like a line with positive slope (the oval in Figure 1.33a), we say that the data has a **positive linear correlation.** When the oval tilts like a line with negative slope (the oval in Figure 1.33b), we say that the data has a **negative linear correlation.**

[4] Consult the Grapher Workshop to learn how to enter data and make a scatter plot.

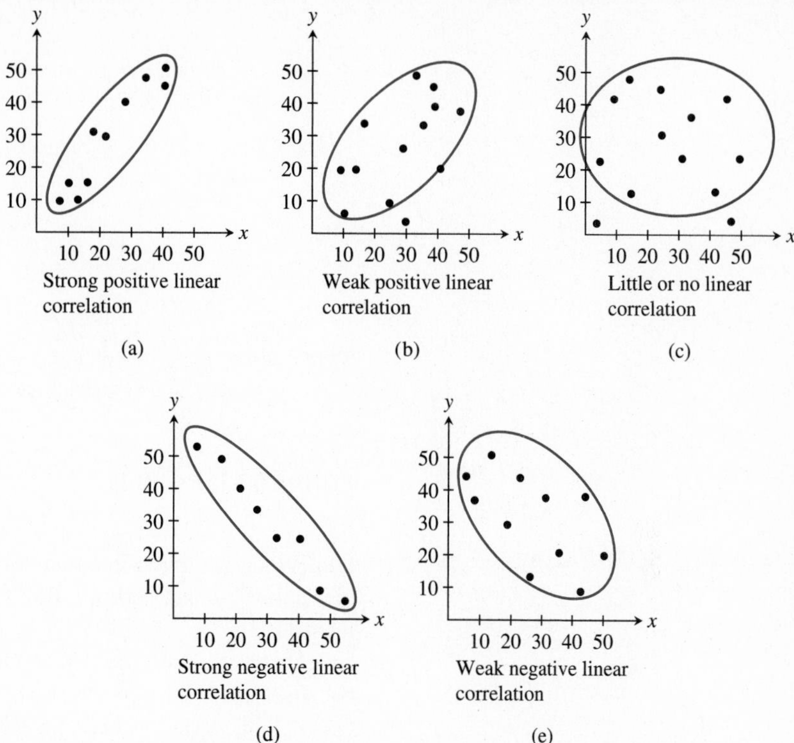

Figure 1.33 Five scatter plots and the types of linear correlation they suggest.

It is also possible for a scatter plot to show a nonlinear pattern, suggesting that there may be some form of nonlinear correlation. This possibility will be thoroughly studied in Chapter 4.

EXAMPLE 2 Determining the Type of Correlation

Make a scatter plot of the (x, y) data below. Determine by visual inspection what type of linear correlation, if any, exists between the x and y quantities.

x	8	12	18	22	31	32	36	40	43	50
y	47	55	36	48	39	26	44	14	32	24

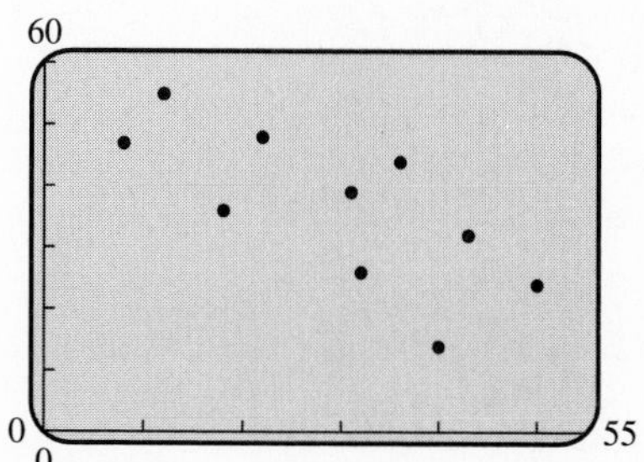

Figure 1.34 Scatter plot of the data in Example 2.

Solution

The scatter plot of these data is shown in Figure 1.34. There appears to be a weak negative linear correlation. ■

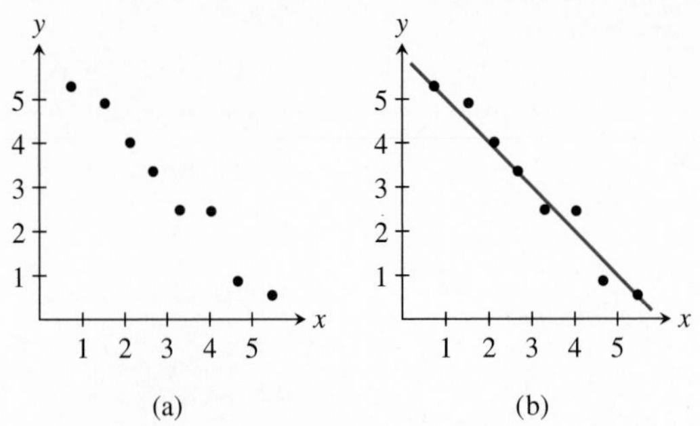

Figure 1.35 (a) A scatter plot of data with a strong negative linear correlation. (b) A *line of fit* superimposed on the data.

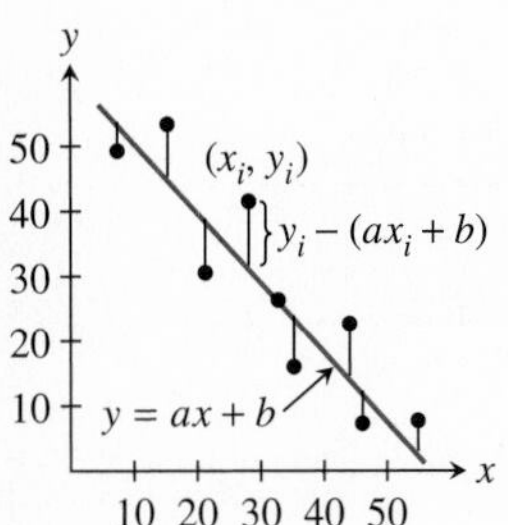

Figure 1.36 A set of data points and a line of best fit with all the residuals indicated by vertical segments.

REMARK

The symbol Σ means "sum." For example

$$\sum_{i=1}^{3} a_i = a_1 + a_2 + a_3.$$

Fitting a Line to Data

Figure 1.35a shows a scatter plot of data with a strong negative linear correlation. When we find a line that comes close to passing through all the scatter-plot points, we are **fitting a line to the data.** Figure 1.35b shows a line fit to the data.

There are many lines that may appear to fit the data. Suppose that the equation $y = ax + b$ is the slope-intercept form of a line that fits a set of data $(x_1, y_1), (x_2, y_2), \ldots, (x_n, y_n)$. (See Figure 1.36.) The difference

$$y_1 - (ax_1 + b)$$

is called the first **residual** and represents the extent to which the first data point (x_1, y_1) misses, or deviates from, the line. As Figure 1.36 suggests, the ith data point (x_i, y_i) has the ith residual $y_i - (ax_i + b)$. (Notice that the subscript identifies which data point.)

The **linear regression line,** sometimes called the **least squares line,** is the one line for which the sum of the squares of the residuals,

$$\sum_{i=1}^{n} [y_i - (ax_i + b)]^2,$$

is the smallest possible. Its equation is called the **linear regression equation.**

Understanding a method for finding the linear regression equation is beyond the scope of this book, but it is built into many graphers. Thus, we are able to find the linear regression equation as shown in Example 3 and to use it as shown in Example 4.

■ EXAMPLE 3 Finding the Linear Regression Equation

Find the linear regression equation for the age-height data shown in Table 1.7 on page 95.

LinReg
y=ax+b
a=.50076
b=17.97933
r=.93652

Figure 1.37 The linear regression screen shows that the linear regression line has slope $a \approx 0.5$ and y-intercept $b \approx 18$.

Solution

Solve Numerically

Using the linear regression feature of a grapher,[5] the *linear regression screen* (Figure 1.37) shows the linear regression equation to be approximately

$$y = 0.50076x + 17.97933, \qquad \text{or} \qquad y = 0.5x + 18.$$

Support Graphically

Figure 1.38 shows a scatter plot of the data with a graph of the linear regression equation $y = 0.5x + 18$ superimposed on it. ■

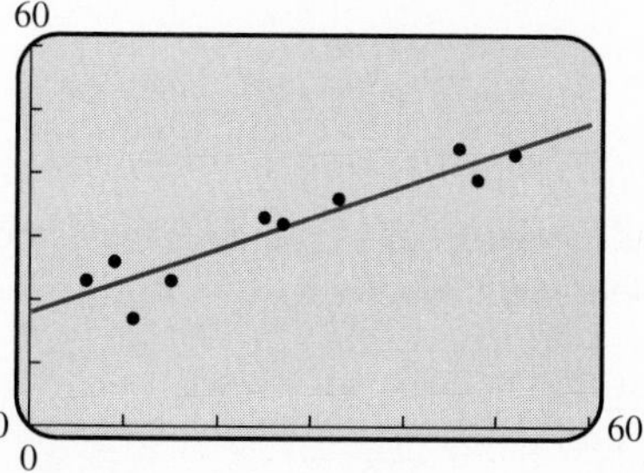

Figure 1.38

Modeling Data

The linear regression screen (Figure 1.37) also shows the number r, called the **correlation coefficient.** This number shows whether the linear correlation is positive or negative, and also indicates its *strength.*

Properties of the Correlation Coefficient r

1. $-1 \le r \le 1$.
2. When $r > 0$, there is a positive linear correlation.
3. When $r < 0$ there is a negative linear correlation.
4. When $|r| \approx 1$, there is a strong linear correlation.
5. When $r \approx 0$, there is weak or no linear correlation.

Alert

Some graphers display a linear regression equation using a and b, where $y = a + bx$ instead of $y = ax + b$. Caution students to be certain that they understand the convention used by the particular graphers they are using.

When there is strong linear correlation between quantities x and y, it may be reasonable to use the linear regression equation to predict unknown values. This line can be used to *estimate* (x, y) pairs that are not included in the data. Example 4 illustrates this modeling method.

■ **EXAMPLE 4 Using a Linear Regression Equation as a Model**

Use the data in Table 1.7 as a basis for estimating a child's height in inches at age 39 months.

Solution

Model

In Example 3 we found the *linear regression equation*

$$y = 0.5x + 18.$$

[5] Consult your grapher owner's manual to learn about these statistical capabilities.

Notes on Examples

In Example 4, students must understand that the linear regression equation provides an *estimate* for *y*. The actual heights of individual children will vary.

Solve Algebraically

Find y when $x = 39$.

$$\begin{aligned} y &= 0.5x + 18 \\ &= 0.5(39) + 18 \\ &= 37.5 \end{aligned}$$

Support Graphically

The linear regression equation is graphed in Figure 1.38. A trace shows that when x is approximately 39, y is approximately 37.5.

Interpret

The estimated height of a 39-month-old child (on the average) is 37.5 in. ■

Using Linear Regression Lines in Problem Solving

We have presented a process for data analysis. We are able to take raw data, make a reasonable decision about whether it shows a relationship between quantities, model the relationship with an equation (using technology), and then use the model to make predictions. Here is a summary.

Using Data Analysis—Linear Regression

1. Complete a scatter plot of the data.
2. Find the linear regression equation $y = ax + b$.
3. Superimpose the graph of the linear regression equation on the scatter plot.
4. Use the linear regression equation to predict y values for x values that are not included in the data.

Example 5 illustrates how linear regression can be used to make a prediction about world population growth.

■ EXAMPLE 5 APPLICATION: Predicting World Population

World Bank officials want to predict world population. Use linear regression with the data in Table 1.6 (page 95) to predict the world population in the year 2010.

REMARK

Population growth data over a short span of time (30 years) can be modeled with a linear function. If the data span a much longer time interval, they would be better modeled by functions studied in Chapter 4.

Solution

Model

Using the linear regression feature of a grapher, we find the linear regression equation to be approximately

$$y = 98.229x - 190{,}157.181.$$

Follow-up

Ask . . .

In Example 5, we assume that the world's population grows linearly with time. Do you think this is a reasonable assumption? **(No, not over a long period of time.)**

Assignment Guide

Day 1: Ex. 1, 4, 8, 9, 10–18, 22
Day 2: Ex. 23, 26, 29, 30, 32, 33, 34, 36

Cooperative Learning

Group Activity: Ex. 37

Notes on Exercises

Ex. 7–8 and 21–22 can be done even if the previous exercises are not assigned.
Ex. 17–22, 29–33, and 35–38 require the student to find a linear regression equation for a set of data.

Ongoing Assessment

Self-Assessment: Ex. 3, 7, 21, 35a, b
Embedded Assessment: Ex. 10, 32

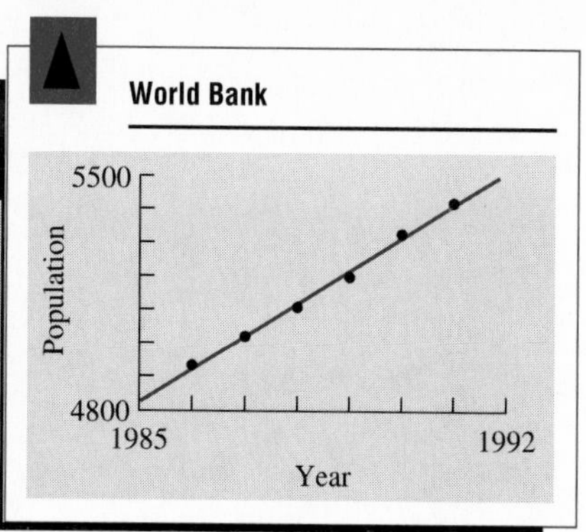

(a)

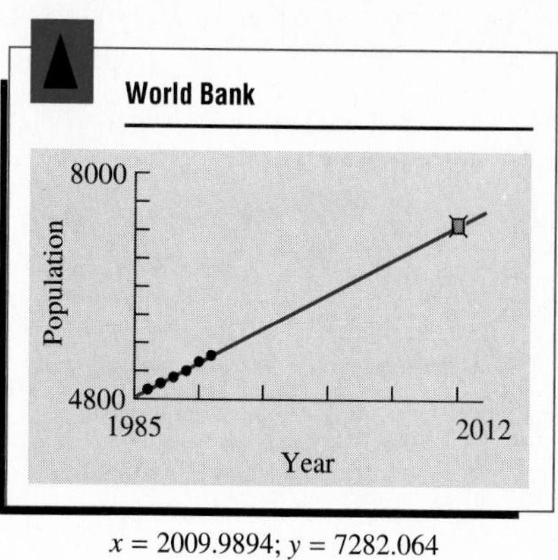

$x = 2009.9894$; $y = 7282.064$

(b)

Figure 1.39

Solve Graphically

Figure 1.39a shows the graph of the linear regression equation superimposed on a scatter plot of the data. Using "trace" (Figure 1.39b) we conclude that y is approximately 7282 when x is 2010.

Confirm Algebraically

Evaluating $y = 98.229x - 190{,}157.181$ for $x = 2010$, we find that

$$\begin{aligned} y &= 98.229x - 190{,}157.181 \\ &= 98.229(2010) - 190{,}157.181 \\ &= 7283.109 \end{aligned}$$

Interpret

The linear regression equation suggests that the world population in the year 2010 will be 7283 million, or approximately 7.3 billion. ■

Quick Review 1.4

In Exercises 1–4, evaluate the expression for the given value of x.

1. $1.4x + 18$ for $x = 12$

2. $-3.2x + 17.1$ for $x = 5$

3. $42{,}000 + 1.5x$ for $x = 7$

4. $18{,}500 - 2.5x$ for $x = 1.1$

In Exercises 5 and 6, write an equation in slope-intercept form whose graph is a line with slope m and y-intercept b.

5. $m = 8,\quad b = 3.6$

6. $m = -1.8,\quad b = -2$

In Exercises 7 and 8, state the slope and the y-intercept of the given line.

7. $y = 2x - (17 + x)$

8. $y = 21 - 2(5x + 3)$

In Exercises 9–12, graph the two points and find an equation of the line containing them.

9. $(-2, 4)$ and $(3, 1)$

10. $(1, 5)$ and $(-2, -3)$

11. $(-5, -1)$ and $(2, 4)$

12. $(-3, 5)$ and $(6, -2)$

SECTION EXERCISES 1.4

In Exercises 1–6, describe the linear correlation as strong positive, weak positive, strong negative, weak negative, or no correlation.

1.
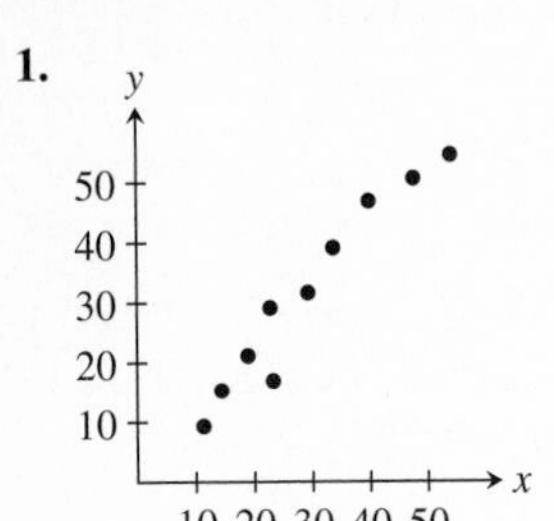

2.
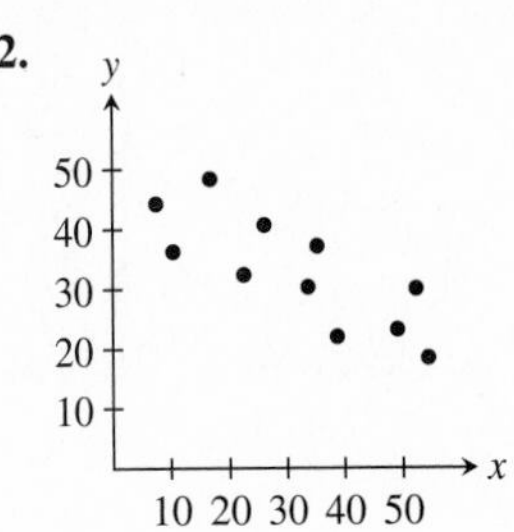

3.
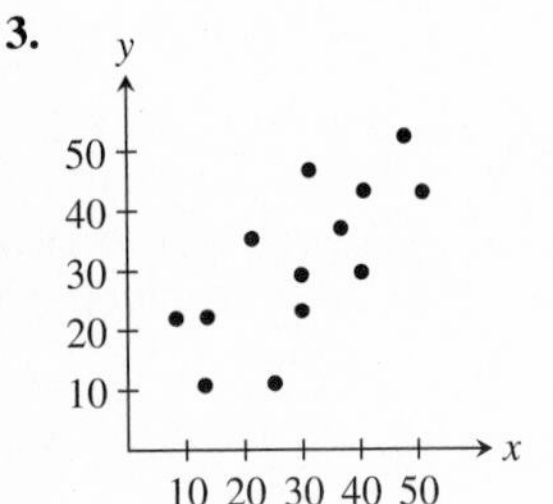

4.
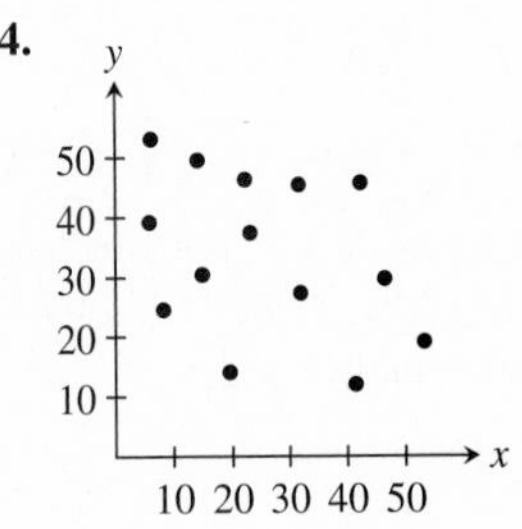

5.
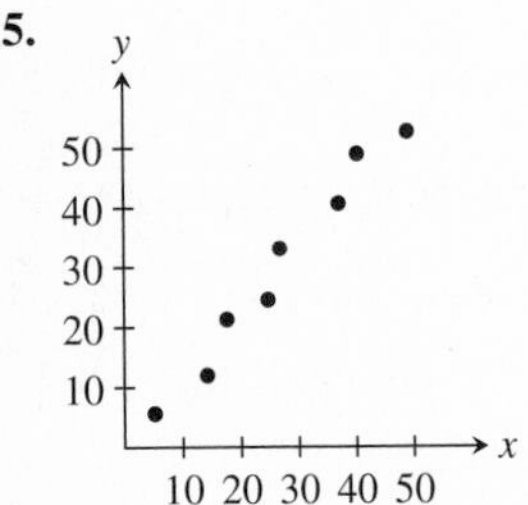

6.
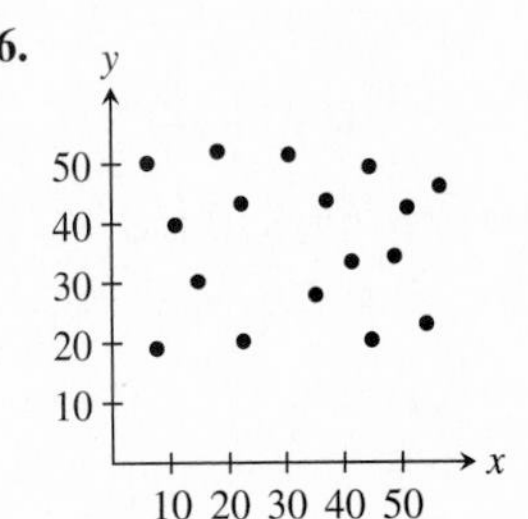

In Exercises 7 and 8, refer to the scatter plots in Exercises 1–6.

7. Select the scatter plot that has a correlation coefficient that is closest to 1.
8. Select the scatter plot that has a correlation coefficient that is closest to −1.

In Exercises 9 and 10, complete a scatter plot and answer the questions.

9. A group of male children were weighed. Their ages and weights are recorded in Table 1.9.

   a. Draw a scatter plot for these data.

Table 1.9 Children's Age and Weight

Age (months)	Weight (pounds)
18	23
20	25
24	24
26	32
27	33
29	29
34	35
39	39
42	44

   b. Does there appear to be any linear correlation? If so, is it positive or negative? Strong or weak?

10. Table 1.10 shows life expectancy data for U.S. citizens as of 1989.

Table 1.10 U.S. Life Expectancy

Age (years)	Life Expectancy (years)
10	66
20	56
30	47
40	37
50	29
60	20
70	14
80	8

Source: U.S. National Center for Health Statistics, *Vital Statistics of the United States.*

   a. Draw a scatter plot for these data.
   b. Does there appear to be any linear correlation? If so, is it positive or negative? Strong or weak?

In Exercises 11–16, decide whether the statement about the scatter plot is true or false.

11. The scatter plot shows (x, y) data pairs.
12. The line shown is a *fitting line* for the data.

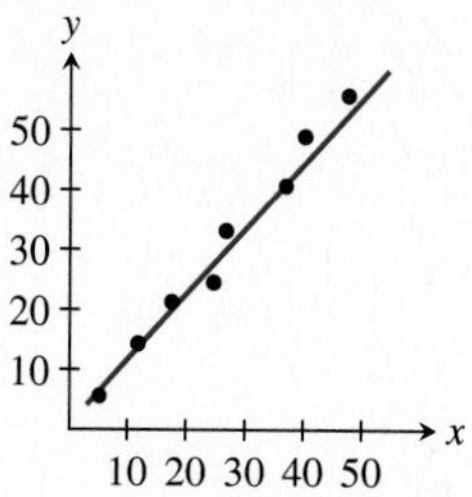

For Exercises 11–16

13. The line shown is a *candidate* for the linear regression line.

14. The equation of the line can be used to predict values for the quantity represented by y.

15. If an equation of the line is $y = 2x + 13$, we can say that if there is a change of 5 units in x then there is a change of 10 units in y.

16. If an equation of the line is $y = 0.9x + 1$, we can say that if there is a change of 10 units in x then there is a change of 9 units in y.

In Exercises 17–20, choose the grapher screen that shows the result of applying the linear regression feature to the data.

17. $\{(1, 4), (2, 8), (3, 9), (4, 12)\}$

18. $\{(1, 3), (2, 5), (3, 8), (4, 9)\}$

LinReg
y=ax+b
a=1.980
b=1.500
r=.864

(a)

LinReg
y=ax+b
a=1.000
b=2.750
r=.958

(b)

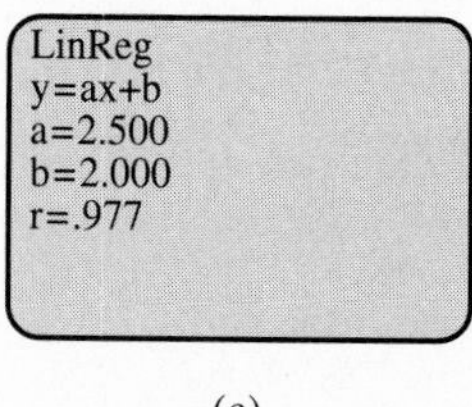

(c)

LinReg
y=ax+b
a=2.100
b=1.000
r=.984

(d)

For Exercises 17–20

19. $\{(1, 3.2), (2, 4.8), (3, 9.6), (4, 8.2)\}$

20. $\{(1, 3.9), (2, 4.8), (3, 5.2), (4, 7.1)\}$

In Exercises 21 and 22, apply steps 1–3 of the linear regression model.

21. Find the linear regression equation using the data from Exercise 19.

22. Find the linear regression equation using the data from Exercise 20.

In Exercises 23–28, the equation is a linear regression equation for some set of (x, y) data. Use the value x_1 to predict a value for y_1.

23. $y = 1.5x + 12$ and $x_1 = 15$

24. $y = -1.8x + 23$ and $x_1 = 6$

25. $y = 9.23x + 5.28$ and $x_1 = 5.5$

26. $y = -12.3x + 91.25$ and $x_1 = 8$

27. $y = -3.21x + 28.9$ and $x_1 = 2$

28. $y = 12.4x + 88$ and $x_1 = 7.5$

In Exercises 29–32, use linear regression to solve the problem.

29. A group of female children were weighed. Their ages and weights are recorded in Table 1.11.

Table 1.11 Children's Ages and Weights

Age (months)	Weight (pounds)
19	22
21	23
24	25
27	28
29	31
31	28
34	32
38	34
43	39

a. Find the linear regression equation.
b. What is the slope of the linear regression line?
c. Draw the scatter plot and the linear regression line.
d. What weight would you expect on the average for a 30-month-old female child?

30. Table 1.12 shows the median base salaries in thousands of dollars of National Football League players.

Table 1.12 National Football League Median Base Salaries

Year	Salary (thousands of dollars)
1985	140
1986	150
1987	175
1988	180
1989	200
1990	236

Source: National Football League, New York, N.Y. Unpublished data as reported by Bureau of the Census, *Statistical Abstract of the United States, 1992* (Washington, D.C., 1992).

a. Find a scatter plot and the linear regression equation for these data.
b. Use the linear regression equation to predict what the median base salary will be in the year 2000.

31. Table 1.13 shows the mean annual compensation of construction workers.

Table 1.13 Construction Worker Average Annual Compensation

Year	Annual Compensation (dollars)
1980	22,033
1985	27,581
1988	30,466
1989	31,465
1990	32,836

Source: U.S. Bureau of Economic Analysis

a. Find the slope of the linear regression line. What does the slope represent?
b. Use linear regression to predict what the construction worker average annual compensation will be in the year 2000.

32. Table 1.14 shows the numbers of phonograph records shipped to stores between 1975 and 1991.

a. Find the linear regression equation.
b. Use the linear regression equation to estimate how many records were shipped in 1988.

Table 1.14 Phonograph Records Shipped

Year	Units Shipped (millions)
1975	531.8
1980	683.7
1985	652.9
1990	865.7
1991	801.0

Source: Recording Industry Association of America, *Inside the Recording Industry: A Statistical Overview* (Washington, D.C., 1991).

EXTENDING THE IDEAS

In Exercises 33–38, solve the problem.

33. Table 1.15 shows the average annual compensation for workers in finance, insurance, and real estate.

a. The viewing window shows the scatter plots and the linear regression line for the data in Table 1.15 and the data in Exercise 31. Which line is for which data?
b. When were the annual compensations of construction workers and financial workers approximately equal?

Table 1.15 Finance, Insurance, and Real Estate Professionals

Year	Annual Compensation (dollars)
1980	18,968
1985	28,014
1988	34,443
1989	35,327
1990	36,679

Source: U.S. Bureau of Economic Analysis

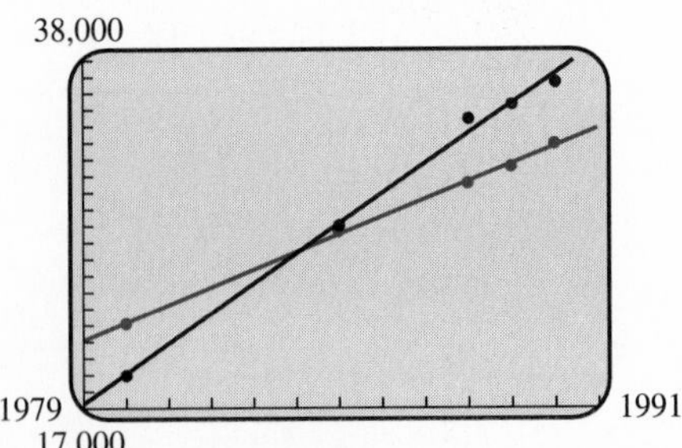

For Exercise 33

34. **Reading to Learn** The linear regression line (least squares regression) is one *line fit* that is often used. Another one is called the median-median line. Use your grapher's owner's manual or do library research to learn how the median-median line is defined.

35. Use the data from Table 1.7.

 a. Find an equation of the median-median line (see the preceding exercise) using a grapher or paper and pencil.
 b. Find the linear regression equation.
 c. Superimpose both lines on a scatter plot. Comment on what you see.
 d. Change the (25, 33) data pair to (25, 62) and repeat parts a and b. (That is, pretend there is a 25 month old who is 5 ft 2 in!) Which line seems to be the better fit?

36. Use this data set:

$$\{(2,8),(3,6),(5,9),(6,8),(8,11),(10,13),(12,14),(15,4)\}$$

 a. Find a scatter plot.
 b. Find and graph the linear regression equation.
 c. Find and graph the median-median regression equation. See Exercise 34.
 d. Which of these two lines would you accept as a *line of best fit?*

37. **Group Learning Activity** This activity investigates the "amount of bounce" to a rubber ball.
We suggest that students work in groups of two or three.

 a. Collecting Data Drop a rubber ball from five to eight different heights. Measure both the height from which it has been dropped and the height of the bounce in each case. Record the data.
 b. Draw a scatter plot.
 c. Superimpose the linear regression line.
 d. Use the linear regression equation to predict the height of the bounce should the ball be dropped from a 300-ft tower.

38. Use the data in Table 1.16. The median price of existing single-family homes has increased consistently during the past two decades. However, the data in the table show that there have been differences in various parts of the country.

 a. Find the linear regression equation for home cost in the Northeast.
 b. Interpret what the slope of the linear regression line means.
 c. Find the linear regression equation for home cost in the Midwest.
 d. Which median price is increasing more rapidly?

Table 1.16 Median Price of Single-Family Homes

Year	Northeast (dollars)	Midwest (dollars)
1970	25,200	20,100
1975	39,300	30,100
1980	60,800	51,900
1985	88,900	58,900
1990	141,200	74,000

Source: National Association of Realtors, *Home Sales Yearbook* (Washington, D.C., 1990).

USING A CBL SYSTEM

Cooperative Learning Activity

Use a Calculator Based Laboratory System (**CBL**) to collect a set of real data: distance compared with time, or temperature compared with time, light intensity compared with time, etc.

1. Draw a scatter plot of the data.
2. Do the data appear to be linearly correlated? If so, describe the correlation as strong positive, weak positive, etc.
3. If not, does any portion of the data suggest a linear relationship?
4. Find a linear regression equation for this subset.

Exercises involving **CBL** will appear throughout the text.

1.5 USING GRAPHS TO STUDY CHARACTERISTICS OF FUNCTIONS

Viewing and Interpreting Graphs • Increasing and Decreasing Functions • Local Maximum and Minimum Values • Piecewise-Defined Functions • Even and Odd Functions • Graphs and Hidden Behavior

Objective

Students will be able to use graphs to determine minimum and maximum values of functions, recognize even and odd functions, and analyze the characteristics of piecewise-defined functions.

Motivate

Ask students why it is important to be able to tell whether a graph on a grapher is reasonable.

Viewing and Interpreting Graphs

Graphing with pencil and paper requires that you develop graph *drawing* skills. Graphing with a grapher requires that you develop graph *viewing* skills.

Graph Viewing Skills

1. Recognize that the graph is reasonable.
2. See all the important characteristics of the graph.
3. Interpret those characteristics.
4. Recognize grapher failure.

Being able to recognize that a graph is reasonable comes with experience. You need to learn the basic functions, their graphs, and how changes in their equations affect the graphs. You have already studied linear functions. Other basic functions are shown in Appendix A.4 and will be studied in later chapters.

Grapher failure refers to those occurrences when the graph produced by a grapher is less than precise, or even incorrect—usually due to the limitations of the screen resolution of the grapher. For example, we know that the graph of $y = 10x$ is a line even though in the standard window we see a graph consisting of a set of vertical segments.

As we proceed, we will study the basic functions and see more interesting examples of grapher failure. First, however, we want to study several general characteristics of functions and learn how to interpret them. This skill of interpreting a graph to understand its *behavior* applies to any graph, whether it is one you draw, one you see in a viewing window, or one you see in print.

Teaching Note

Sometimes it is impossible to show all of the details of a graph in a single window. For example, in Example 10 the graph in Figure 1.53b reveals minute details of the graph, but it hides the overall shape of the graph.

▼ GRAPHING ACTIVITY Interpreting Graphs of Functions

A jogger runs from one end of the Main Street jogging path to the other end and returns. The graph in Figure 1.40 reflects this activity.

Which two of these explanations are valid?

1. The jogger runs for a while, rests for a greater length of time, resumes running, eventually turning around and running all the way home.

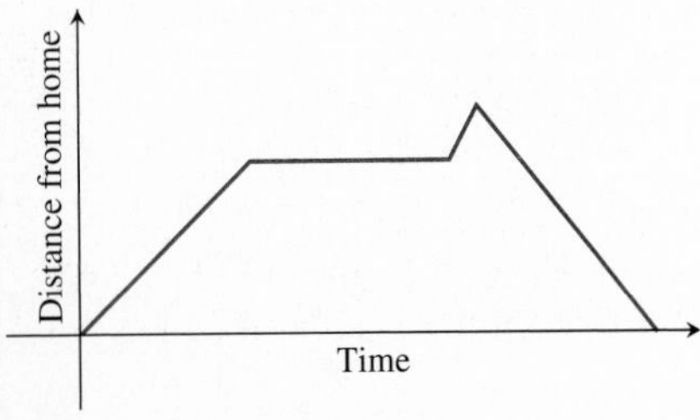

Figure 1.40

Graphing Activity Extensions

Draw a graph that can be explained by the explanation that you did not choose (that is, the explanation that is not a valid explanation of the graph in Fig. 1.40).

2. The jogger speeds up gradually, then runs at a constant rate followed by a short burst of speed before returning to the start.
3. The jogger runs at a constant speed, stops for a long rest, runs to the end and then returns, running faster than at the start. ▲

Increasing and Decreasing Functions

A common characteristic of a graph is its apparent rise or fall from the left to the right over an interval on the x-axis. If the graph of a function rises from the left to the right, the function is *increasing.* If the function falls from the left to the right, the function is *decreasing.* And if the graph is horizontal, the function is *constant.*

Teaching Note

Students must understand that for f to be increasing on an interval, the inequality $f(x_1) < f(x_2)$ must be satisfied for *any* $x_1 < x_2$ in the interval. This means that any line containing two points on the graph within the interval must have positive slope.

Definition: Increasing, Decreasing, and Constant Functions

A function f is **increasing** on an interval if $f(x_1) < f(x_2)$ for any x_1 and x_2 in the interval, where $x_1 < x_2$.
A function f is **decreasing** on an interval if $f(x_1) > f(x_2)$ for any x_1 and x_2 in the interval, where $x_1 < x_2$.
A function f is **constant** on an interval if $f(x_1) = f(x_2)$ for any x_1 and x_2 in the interval, where $x_1 < x_2$.

In Example 1 we investigate the function $f(x) = |x - 1| + |x + 2|$. Recall the following definition of absolute value:

$$|a| = \begin{cases} a, & \text{if } a \geq 0 \\ -a, & \text{if } a < 0. \end{cases}$$

We can rewrite the expression for f using this property.

If $x \leq -2$, then both $x - 1$ and $x + 2$ are either negative or zero. Thus,

$$\begin{aligned} f(x) &= -(x - 1) - (x + 2) \\ &= -x + 1 - x - 2 \\ &= -2x - 1. \end{aligned}$$

If $-2 < x < 1$, then $x - 1$ is negative and $x + 2$ is positive. So

$$\begin{aligned} f(x) &= -(x - 1) + (x + 2) \\ &= -x + 1 + x + 2 \\ &= 3. \end{aligned}$$

Finally, if $x \geq 1$, then both $x - 1$ and $x + 2$ are positive or zero. Thus,

$$\begin{aligned} f(x) &= (x - 1) + (x + 2) \\ &= x - 1 + x + 2 \\ &= 2x + 1. \end{aligned}$$

In Example 1, we give a second expression for f based on this development. In general, **functions f and g are equal** if they have the same domain and if $f(x) = g(x)$ for each x in the common domain.

Notes on Examples

Explain how the steps of the solution are related to the definition of a decreasing function.

EXAMPLE 1 Identifying Decreasing, Constant, and Increasing Behavior

Identify intervals on which the function

$$f(x) = |x - 1| + |x + 2| = \begin{cases} -2x - 1 & \text{if } x \le -2 \\ 3 & \text{if } -2 < x < 1 \\ 2x + 1 & \text{if } 1 \le x \end{cases}$$

is increasing, decreasing, and constant.

Solution

Solve Algebraically

Figure 1.41 suggests that f is decreasing on $(-\infty, -2]$. Suppose that $x_1 < x_2$ in the interval $(-\infty, -2]$. On this interval the function rule is $f(x) = -2x - 1$.

$$x_1 < x_2$$

$$-2x_1 > -2x_2 \quad \text{Multiplication property}$$

$$-2x_1 - 1 > -2x_2 - 1 \quad \text{Addition property}$$

$$f(x_1) > f(x_2)$$

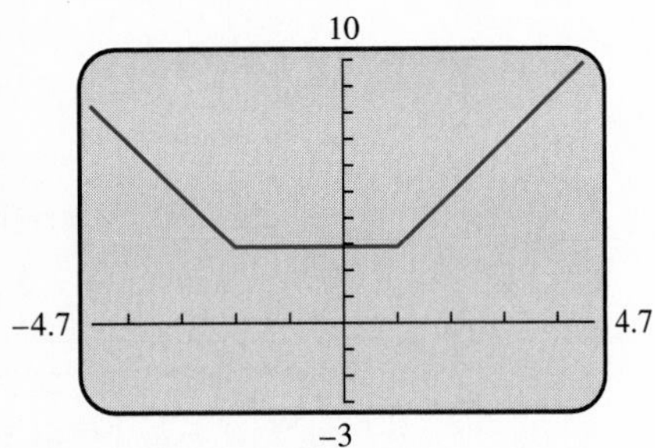

Figure 1.41 A graph of $f(x) = |x - 1| + |x + 2|$.

In a similar way, we can show that if $x_1 < x_2$ in the interval $[-2, 1]$, then $f(x_1) = f(x_2)$, and if $x_1 < x_2$ in the interval $[1, \infty)$, then $f(x_1) < f(x_2)$.

Support Numerically

Use "trace" in Figure 1.41 to observe that the y-values decrease as x traces from the left to -2, remain constant from $x = -2$ to $x = 1$, and increase as x traces to the right of 1.

Interpret

We have shown that f decreases on the interval $(-\infty, -2]$, is constant (that is, equals 3) on the interval $[-2, 1]$, and increases on the interval $[1, \infty)$. ■

Local Maximum and Minimum Values

Other characteristics of many graphs are their peaks and valleys. Figure 1.42 shows a graph of a function $y = f(x)$ with peaks at points P and R and valleys at points Q and S. At P, the y-value is greater than the y-values at nearby points of the graph, so the function has a *local maximum* at P. At Q, the y-value is less than the y-values at nearby points, so the function has a *local minimum* there. Notice that f increases on the intervals $(-\infty, x_1]$, $[x_2, x_3]$, and $[x_4, \infty)$ and decreases on $[x_1, x_2]$ and $[x_3, x_4]$.

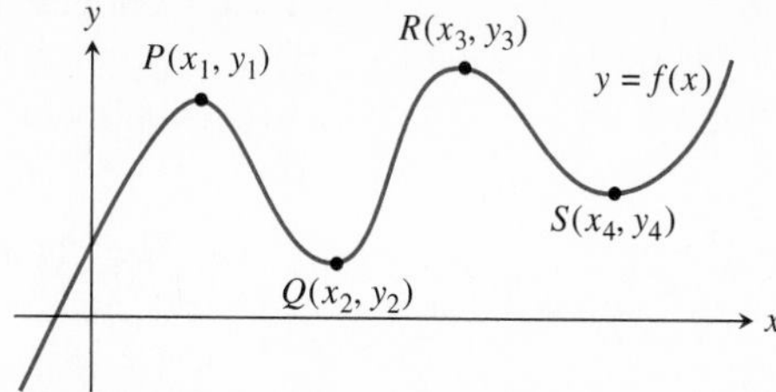

Figure 1.42 The graph suggests that f has local maximum values at P and R and local minimum values at Q and S.

REMARK

We intentionally use the phrase *local maximum* instead of simply *maximum.* There is a point in Figure 1.42 that you would call a *local maximum,* but not a *maximum.* Can you find it?

Definition: Local Maximum and Local Minimum

1. A value $f(a)$ is a **local maximum** of f if there is an open interval (c, d) containing a such that $f(x) \leq f(a)$ for all values of x in (c, d).
2. A value $f(b)$ is a **local minimum** of f if there is an open interval (c, d) containing b such that $f(x) \geq f(b)$ for all values of x in (c, d).

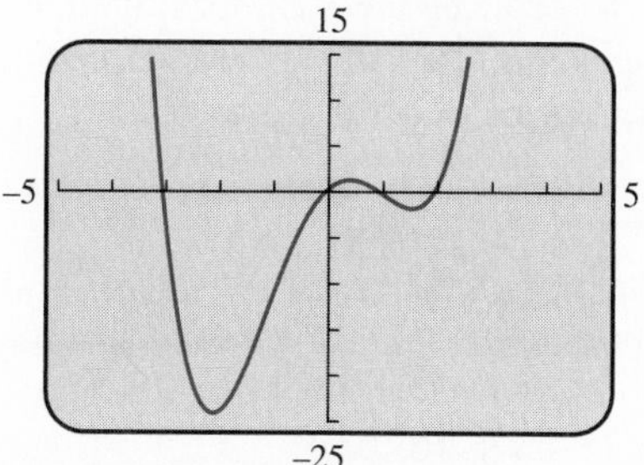

Figure 1.43 A graph of $f(x) = x^4 - 7x^2 + 6x$.

■ EXAMPLE 2 Finding Local Maximum and Minimum Values

Decide whether $f(x) = x^4 - 7x^2 + 6x$ has any local maximum values and local minimum values. If so, find the local maximum values and local minimum values and the value of x where each occurs.

Solution

Solve Graphically

The graph of $f(x) = x^4 - 7x^2 + 6x$ (Figure 1.43) suggests that f has two local minimum values and one local maximum value.

Using grapher methods,[6] we find that the local minimum values occur at $x \approx -2.06$ (where $y \approx -24.06$) and $x \approx 1.60$ (where $y \approx -1.77$). The local maximum occurs at $x \approx 0.46$ (where $y \approx 1.32$).

Support Numerically

Trace across the graph. Watch the y-values decrease as you approach a local minimum from the left, then increase as you move away on the right. ■

TECHNOLOGY TIP

When you zoom in at a local maximum or minimum point, the graph often appears flat or nearly flat. To prevent this tendency do any one of the following:

1. Change *Y*Fact so that it is larger than *X*Fact.
2. Use a low flat rectangle with *zoom-box.*
3. Manually set *Y*max and *Y*min to be closer together.

[6] One grapher method is based on *zoom-in.* Another method (available on some graphers) uses the built-in programs for finding local maximum and local minimum values.

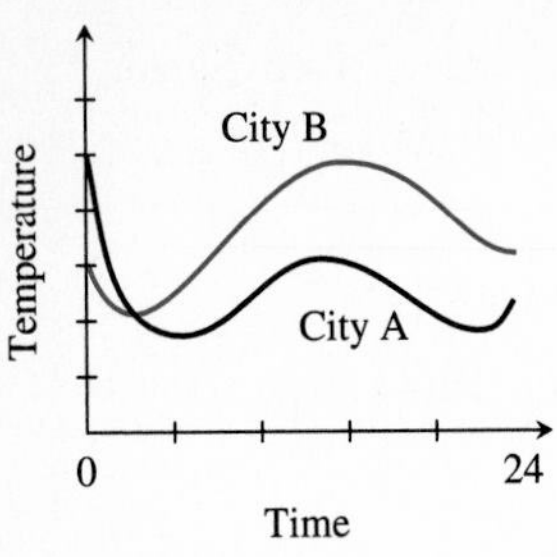

Figure 1.44

■ **EXAMPLE 3 Interpreting Graphs**

The graph in Figure 1.44 shows the air temperature changes over a 24-hour period in two different cities. All of the following are true.

a. The temperature in city B decreases, then increases, and finally decreases.
b. The largest rate of temperature decrease occurs in city A during the first few hours.
c. The largest rate of temperature increase occurs in city B in the middle of the 24-hour period.
d. In city B a local minimum temperature occurs in the first half and a local maximum temperature occurs in the second half of the 24-hour period.
e. The temperature reaches a local minimum twice during the 24-hour period in city A. ■

To describe the increasing and decreasing behavior of a function, we list the intervals on which it is doing each. To describe the local maximum and minimum values of a function, we list the points at which they occur. Finding these points first can help identify the intervals of increase and decrease.

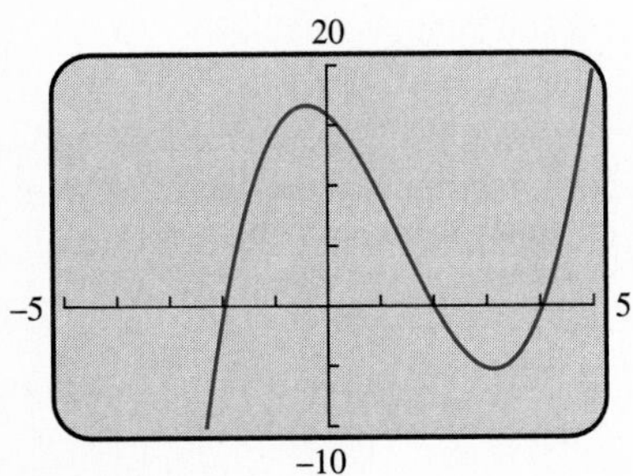

Figure 1.45

■ **EXAMPLE 4 Finding Intervals of Increase and Decrease by Using Local Maximum and Local Minimum**

Describe the behavior of $f(x) = x^3 - 4x^2 - 4x + 16$.

Solution

Solve Graphically

Figure 1.45 shows a graph of f. Using grapher methods, we find that a local maximum occurs at approximately $x = -0.43$ and a local minimum occurs at approximately $x = 3.10$. These values are the approximate endpoints of the intervals on which the function increases and decreases; f increases on the intervals $(-\infty, -0.43]$ and $[3.10, \infty)$ and decreases on the interval $[-0.43, 3.10]$. ■

Alert

When using a grapher, students will sometimes use an automatic zoom-out in order to see all of the features of a graph. Remind students to set the x- and y-scale factors to 0 before beginning the zoom-out procedure. After an appropriate graph has been obtained, students should reset the scale factors.

Examples 2 and 4 illustrate the "solve graphically because there is no other way possible" approach to solving a problem (without using calculus) that is described in the *Problem-Solving Process* in Section 1.1.

■ **EXAMPLE 5 APPLICATION: Finding Efficient Packaging**

The manufacturing industry invests substantial time and money to design efficient packaging for its products. Suppose that a box is made by folding up the

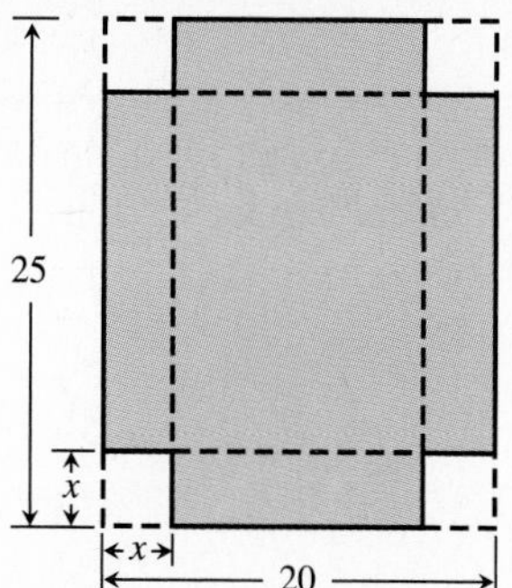

Figure 1.46 Squares cut from a piece of cardboard 20 in. by 25 in.

flaps after squares are cut from a 20-in. by 25-in. piece of cardboard. (See Figure 1.46.) What size of squares will yield a box with greatest volume?

Solution

Model

We want to relate box volume to the cut-out corners and then find the size of the corners that gives the greatest volume.

$$x = \text{edge of cut-out square} = \text{height of box}$$
$$25 - 2x = \text{length of the base}$$
$$20 - 2x = \text{width of the base}$$
$$V(x) = x(20 - 2x)(25 - 2x) \qquad V = \ell wh$$

Solve Graphically

Because x represents the length of one side of a square cut from the cardboard, it is restricted to values from 0 to 10. (Study Figure 1.46 again.) The graph of

$$V(x) = x(20 - 2x)(25 - 2x)$$

shown in Figure 1.47 suggests that the V has a maximum value. Using grapher methods, we show that this maximum occurs at approximately $x = 3.68$.

Interpret

Cutting square corners that are about 3.68 in. on an edge from a piece of cardboard measuring 20 in. by 25 in. will yield a box with maximum volume of about 821 in.^{3}. ■

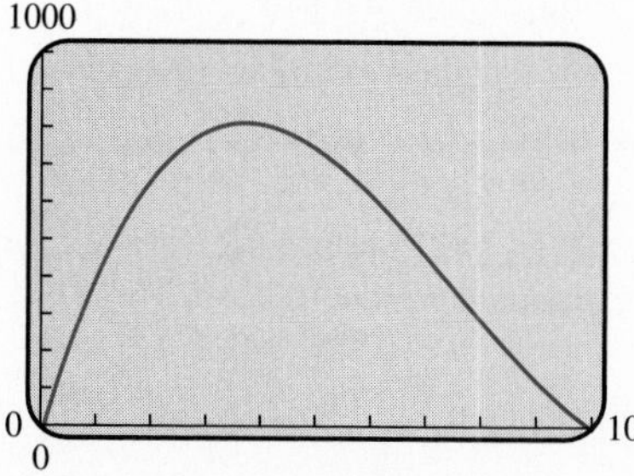

Figure 1.47 The graph of $V(x) = x(20 - 2x)(25 - 2x)$ has a maximum value at $x \approx 3.68$. Confirming this algebraically is not possible without calculus.

Piecewise-Defined Functions

A function can often be defined by a single rule. On the other hand, sometimes in order to model a physical phenomenon or a set of data, it is necessary to define a function in several "pieces" using different rules. Such a function is a **piecewise-defined function.**

■ EXAMPLE 6 Examining a Piecewise-Defined Function

The function

$$f(x) = \begin{cases} 3 - x^2 & \text{if } x < 1 \\ x^3 - 4x & \text{if } x \geq 1 \end{cases}$$

is piecewise-defined. The rule $f(x) = 3 - x^2$ applies only when x is in the interval $(-\infty, 1)$, and the rule $f(x) = x^3 - 4x$ applies only when x is in the interval $[1, \infty)$. The graph of f in Figure 1.48 shows a dramatic break, called a *discontinuity,* at $x = 1$. The open-circle at (1, 2) indicates that the point is *not* part of the graph. The closed-circle at (1, −3) indicates that the point *is* part of the graph. ■

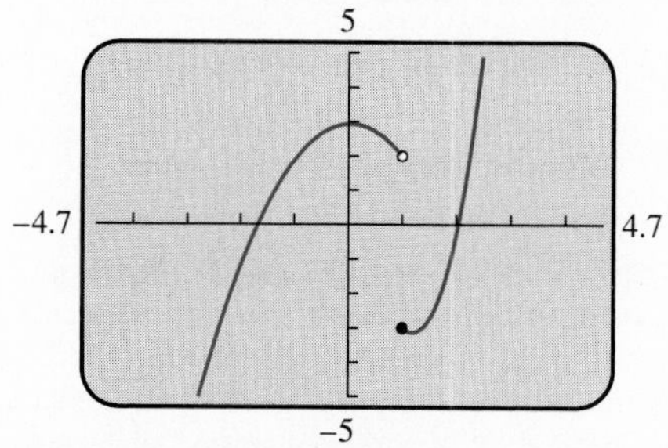

Figure 1.48 A graph of $f(x) = \begin{cases} 3 - x^2 & \text{if } x < 1 \\ x^3 - 4x & \text{if } x \geq 1. \end{cases}$

Table 1.17 GNP in Mining

Year	Dollars (billions)
1980	107.257
1982	132.122
1983	118.351
1984	119.362
1985	114.174
1986	74.289
1987	76.836
1988	80.017
1989	80.254

Source: U.S. Bureau of Economic Analysis

Some data are best modeled by piecewise-defined functions. Consider the data about gross national product (GNP) in mining given in Table 1.17. A scatter plot of the data is shown in Figure 1.49. It appears that something dramatic happened in the industry from 1985 to 1986. In Example 7 we show that these data can be modeled with the piecewise-defined function

$$P(x) = \begin{cases} 1.333x^3 - 12.516x^2 + 30.795x + 107.501 & \text{if } x \leq 5 \\ 2.108x + 62.042 & \text{if } x > 5 \end{cases}$$

where we use $x = 0$ for 1980, $x = 1$ for 1981, and so on.

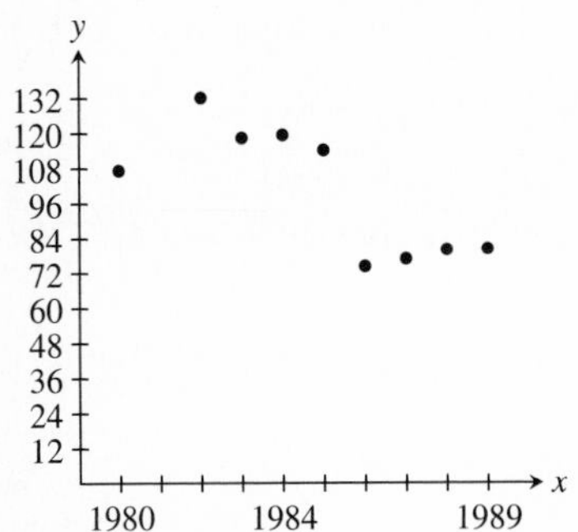

Figure 1.49

■ EXAMPLE 7 APPLICATION: Modeling GNP in Mining

a. Show that the piecewise-defined function $y = P(x)$ models the data in Table 1.17.

b. Use the model to estimate the gross national product in mining for 1983. Compare this value to the actual GNP for 1983.

c. Predict the gross national product in mining for the year 2000.

Solution

Solve Graphically

a. Figure 1.50 shows a graph of $y = P(x)$ superimposed on a scatter plot of the data. It appears that $y = P(x)$ models the data.

Solve Algebraically

b. For 1983, $x = 3$, so we use the model for $x \leq 5$,

$$P(x) = 1.333x^3 - 12.516x^2 + 30.795x + 107.501$$
$$P(3) = 1.333(3)^3 - 12.516(3)^2 + 30.795(3) + 107.501$$
$$= 123.233$$

c. In 2000, $x = 20$, so we use the model for $x \geq 5$,

$$P(x) = 2.108x + 62.042$$
$$P(20) = 2.108(20) + 62.042$$
$$= 104.202$$

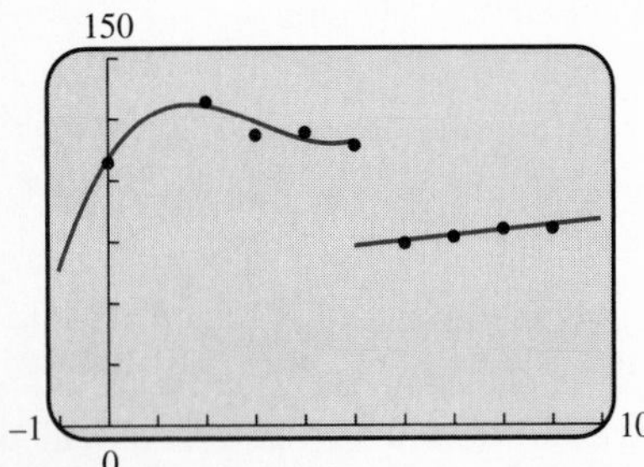

Figure 1.50

Support Graphically

We can use "trace" in Figure 1.50 to show that $P(3) \approx 123$ and $P(20) \approx 104$.

Interpret

According to the model, the gross national product for mining would have been about \$123.233 billion in 1983, while the actual amount was 118.351 billion. The model predicts that the gross national product in mining in the year 2000 should be about \$104 billion. ■

Teaching Note

A piece of wire bent into the shape of an odd function can be used to illustrate the symmetry of the function. If the graph of the function is rotated 180° about the origin, it will coincide with itself.

Alert

Some students may not realize that most functions are neither even nor odd. Give an example of a function such as $y = x^3 - x^2$ that is neither even nor odd.

Follow-up

Ask whether a constant function has a local minimum or local maximum. **(Yes, both)**

Even and Odd Functions

The graph of a function f is **symmetric about the *y*-axis** if $(-x, y)$ is on the graph whenever (x, y) is. This means that $f(-x) = f(x)$ for all x. The y-axis is called the **line of symmetry.**

The graph of a function f is **symmetric about the origin** if $(-x, -y)$ is on the graph whenever (x, y) is. This means that $f(-x) = -f(x)$ for all x. The origin is called the **point of symmetry.**

A function whose graph is symmetric about the y-axis is an *even function;* one whose graph is symmetric about the point (0, 0) is an *odd function.* Here is a more precise algebraic description.

Definition: Even Functions and Odd Functions

- A function f is **even** if $f(-x) = f(x)$ for each value of x in the domain of f. The graph of f is symmetric about the y-axis.
- A function f is **odd** if $f(-x) = -f(x)$ for each value of x in the domain of f. The graph of f is symmetric about the origin.

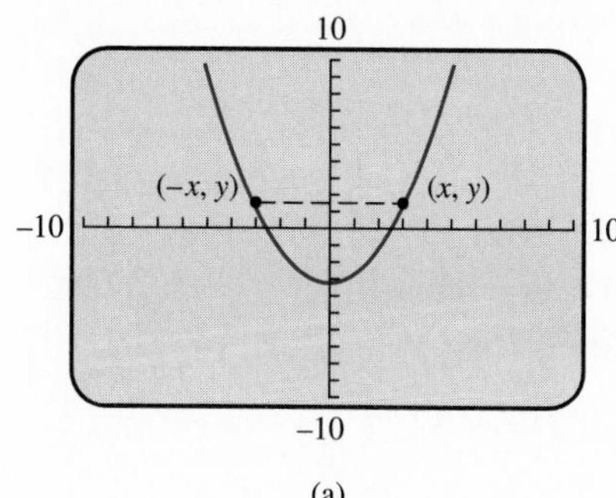

(a)

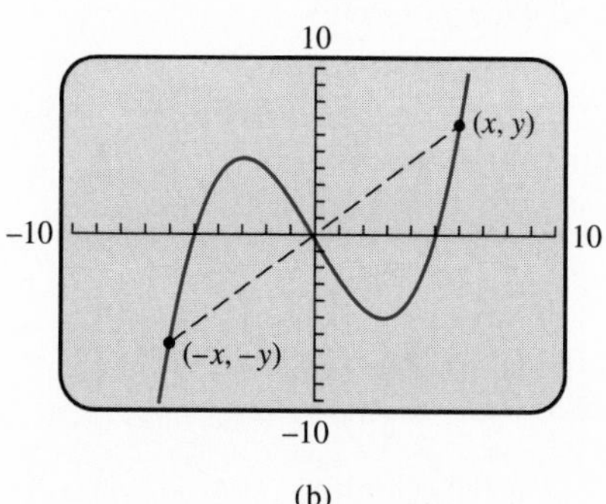

(b)

Figure 1.51 (a) $y_1 = f(x) = 0.5x^2 - 3$. (b) $y_2 = g(x) = 0.1x^3 - 2.5x$.

■ EXAMPLE 8 Determining Even and Odd Functions

Determine whether the function is even or odd.

a. $f(x) = 0.5x^2 - 3$ **b.** $g(x) = 0.1x^3 - 2.5x$

Solution

a.
$$\begin{aligned} f(-x) &= 0.5(-x)^2 - 3 \\ &= 0.5x^2 - 3 \\ &= f(x) \end{aligned}$$

Thus, f is an even function.

b.
$$\begin{aligned} g(-x) &= 0.1(-x)^3 - 2.5(-x) \\ &= -0.1x^3 + 2.5x \\ &= -(0.1x^3 - 2.5x) \\ &= -g(x) \end{aligned}$$

Thus, g is an odd function.

Support Graphically

Figure 1.51 shows that the graph of $y_1 = f(x)$ is symmetric with respect to the y-axis, and the graph of $y_2 = g(x)$ is symmetric with respect to the origin. ■

Example 9 illustrates that algebraic information is sometimes more trustworthy than visual information.

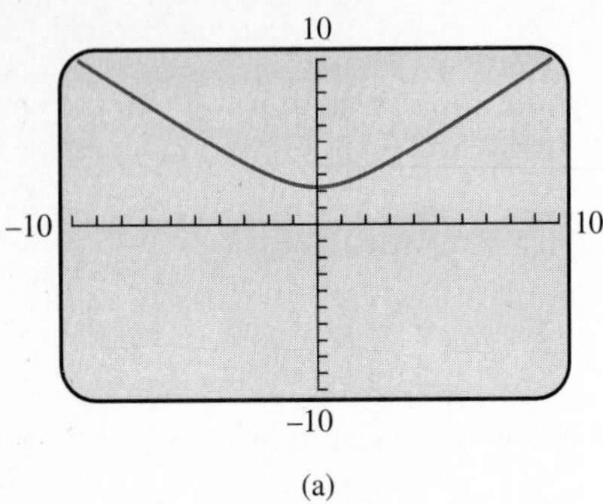

(a)

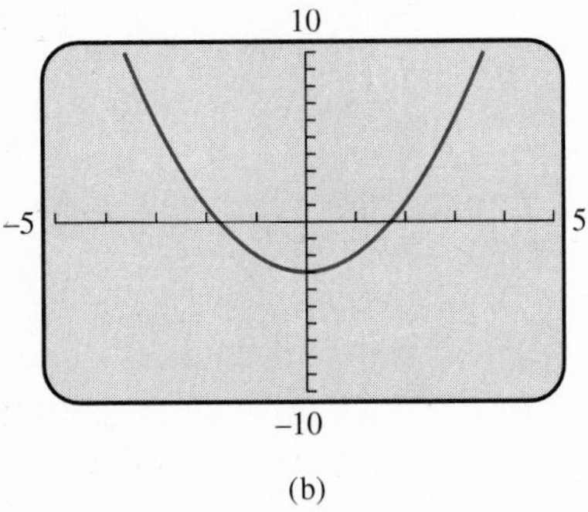

(b)

Figure 1.52 A graph of (a) $f(x) = \sqrt{x^2 + 5}$ and (b) $g(x) = x^2 + 0.01x - 3$.

EXAMPLE 9 Making a Conjecture and Confirming Algebraically

a. The graph of

$$f(x) = \sqrt{x^2 + 5}$$

in Figure 1.52a suggests the conjecture that f is an even function. To confirm we show

$$f(-x) = \sqrt{(-x)^2 + 5} = \sqrt{x^2 + 5} = f(x).$$

b. The graph of $g(x) = x^2 + 0.01x - 3$ in Figure 1.52b suggests the conjecture that g is an even function. To try to confirm, we find

$$\begin{aligned} g(-x) &= (-x)^2 + 0.01(-x) - 3 \\ &= x^2 - 0.01x - 3, \end{aligned}$$

which is not the same as $g(x) = x^2 + 0.01x - 3$. Thus, the conjecture is false, that is, g is not an even function. ■

Try graphing $y_1 = g(x) = x^2 + 0.01x - 3$ and $y_2 = g(-x) = x^2 - 0.01x - 3$ in the same viewing window. The graphs will appear to be the same, thus illustrating the power of algebraic reasoning.

Graphs and Hidden Behavior

If some behavior of a graph is not apparent when the graph is viewed in a window having what we feel are appropriate dimensions, then the behavior is hidden from view. Some **hidden graph behavior** may be viewed by adjusting the window dimensions. A proficient user of a graphing utility is on the alert for hidden behavior, knows to identify portions of a graph that are suspect, and knows ways to modify the viewing window to reveal the hidden behavior when it is possible to do so. Example 10 illustrates.

Assignment Guide

Day 1: Ex. 1, 4, 5, 8, 11, 15, 18, 21, 23, 24, 25, 29, 32
Day 2: Ex. 36–52, multiples of 3, 55–57, 60, 63, 64

Cooperative Learning

Group Activity: Ex. 61–62

Notes on Exercises

Ex. 1–4 introduce the language of increasing and decreasing functions and minimums and maximums.
Ex. 11–22 are designed to give students practice in choosing appropriate ranges on a grapher.
Ex. 41–46 and 56 provide practice with piecewise-defined functions.

EXAMPLE 10 Revealing Hidden Behavior of a Graph

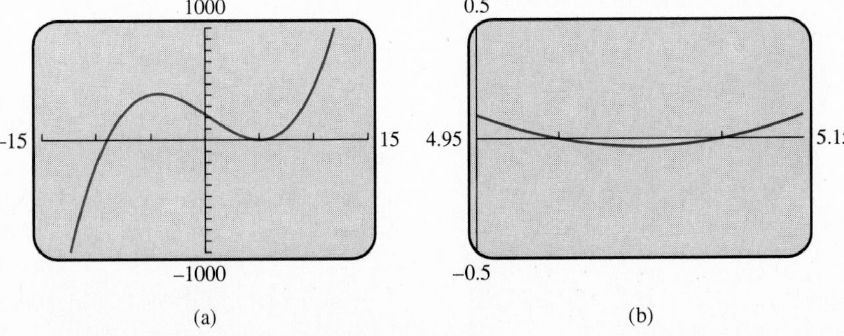

Figure 1.53 Two views of the graph of $f(x) = x^3 - 1.1x^2 - 65.4x + 229.5$.

Ongoing Assessment

Self-Assessment: Ex. 7, 17, 27, 35
Embedded Assessment: Ex. 42, 56

The behavior of the graph (Figure 1.53a) of $f(x) = x^3 - 1.1x^2 - 65.4x + 229.5$ near $x = 5$ is hidden, because it is not clear whether f intersects the x-axis once or twice. Zoom-in reveals that there are two zeros (Figure 1.53b). ■

Quick Review 1.5

In Exercises 1–4, evaluate the expression.

1. $f(x) = 3x^2 - 4x^{-1}$ for $x = -2$.

2. $g(x) = \dfrac{x^2 - 3}{x + 1}$ for $x = 1$.

3. $f(x) = x^2\sqrt{x^2 + 1}$ for $x = -3$.

4. $h(x) = |x - 4|$ for $x = 7$.

In Exercises 5–8, write the inequality in interval form.

5. $x \geq -4$ **6.** $x > \pi$

7. $-3 < x \leq 2$ **8.** $5 \geq x > -2$

In Exercises 9 and 10, solve the double inequality.

9. $-5 \leq 3x - 4 < 8$ **10.** $4 > 1 - 2x \geq 1$

In Exercises 11 and 12, answer the question.

11. Let $f(x) = \begin{cases} x^2 & \text{if } x < 1 \\ 2x + 5 & \text{if } x \geq 1. \end{cases}$
Which expression do you use to find $f(-2)$? $f(1)$?

12. Let $g(x) = \begin{cases} -x^2 + 3 & \text{if } x \leq -2 \\ x^2 + 3 & \text{if } x > -2. \end{cases}$
Which expression do you use to find $g(-2)$? $g(3)$?

SECTION EXERCISES 1.5

In Exercises 1–4, state whether each labeled point identifies a local minimum, a local maximum, or neither. Identify intervals on which the function is decreasing and increasing.

1.

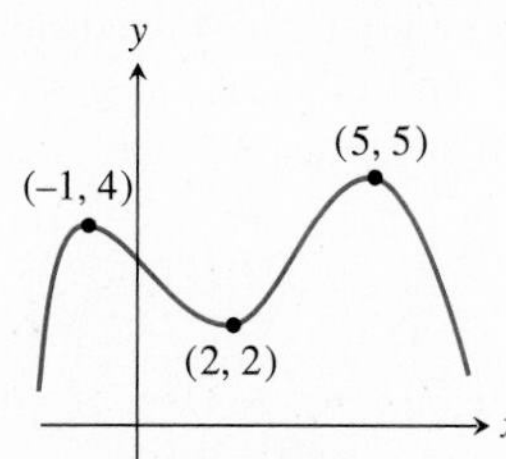

2.

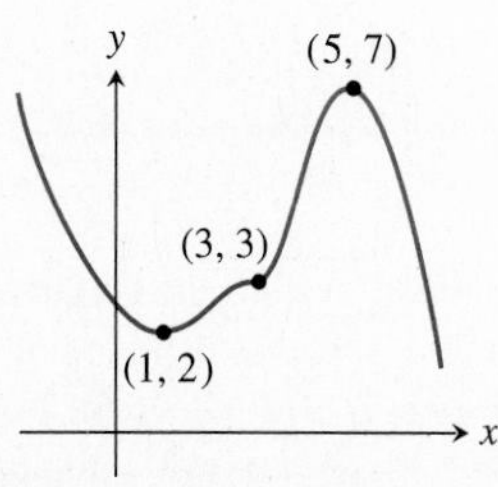

3.

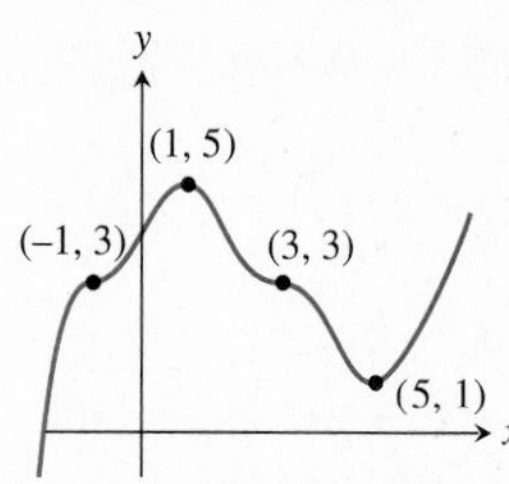

4.

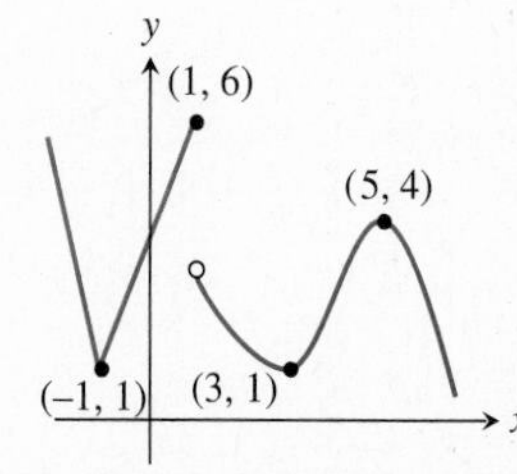

In Exercises 5–10, use a grapher to find all local maximum and minimum values, and the values of x where they occur. Give values rounded to two decimal places.

5. $f(x) = 4 - x + x^2$

6. $g(x) = x^3 - 4x + 1$

7. $h(x) = -x^3 + 2x - 3$

8. $f(x) = (x + 3)(x - 1)^2$

9. $h(x) = x^2\sqrt{x + 4}$

10. $g(x) = x|2x + 5|$

In Exercises 11–14, select the one of the three windows that shows the best view of the graph.

Xmin=$^-$10
Xmax=10
Xscl=1
Ymin=$^-$10
Ymax=10
Yscl=1

(a)

Xmin=$^-$5
Xmax=5
Xscl=1
Ymin=$^-$50
Ymax=50
Yscl=10

(b)

Xmin=$^-$100
Xmax=100
Xscl=10
Ymin=$^-$20000
Ymax=20000
Yscl=10000

(c)

For Exercises 11–14

11. $f(x) = 12x^2 - 5x - 19$

12. $f(x) = x^3 - 10x + 15$

13. $f(x) = 0.05x^3 - 0.5x^2 - 425x - 2625$

14. $f(x) = 0.05x^3 + 0.2x^2 - 0.95x - 2.3$

In Exercises 15–22, find a window that you believe includes the important characteristics of the graph of the function. Begin with the standard window. List the characteristics in the window that you choose.

15. $f(x) = |x^2 - 6x - 12|$

16. $f(x) = x^3 - 12x$

17. $f(x) = (x + 2)^3 - 22x$

18. $k(x) = (x + 15)(x - 20)$

19. $g(x) = \sqrt{x + 12}$

20. $h(x) = \sqrt{|x + 2|} - 3$

21. $f(x) = 0.2\sqrt{x + 8} - 0.12$

22. $h(x) = |0.5x + 1| - 0.5$

In Exercises 23 and 24, which explanation most closely agrees with the information in the graph?

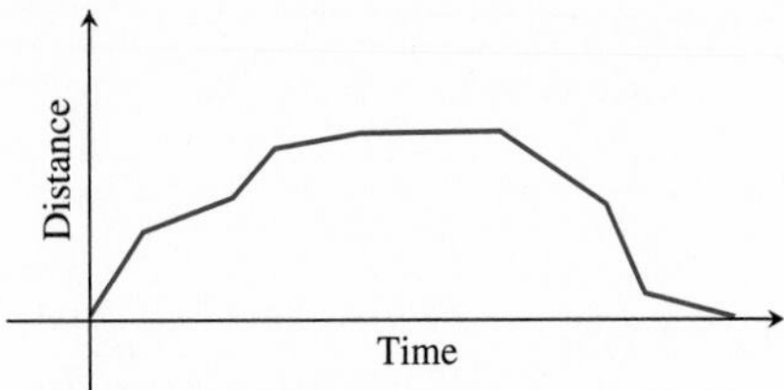

For Exercises 23–24

23. a. The jogger's distance from home is never constant. It first increases rapidly and then decreases rapidly.

b. The jogger's distance from home increases for a period of time, then is constant, and then decreases.

24. a. The jogger begins at a constant rate, slows down a short time later, and then speeds up before slowing down when approaching the far end of the course. After a period of rest, the jogger runs at a steady rate, and then speeds up for a short time before walking the rest of the way home.

b. The jogger's speed increases, first quite rapidly, and then more slowly. Eventually the jogger settles into a constant speed before returning to the starting point. Approaching the starting point, the jogger gradually decreases her speed.

In Exercises 25 and 26, follow the directions.

25. Writing to Learn The graph shows distance from home against time for a jogger. Using information from the graph, write a paragraph describing the jogger's workout.

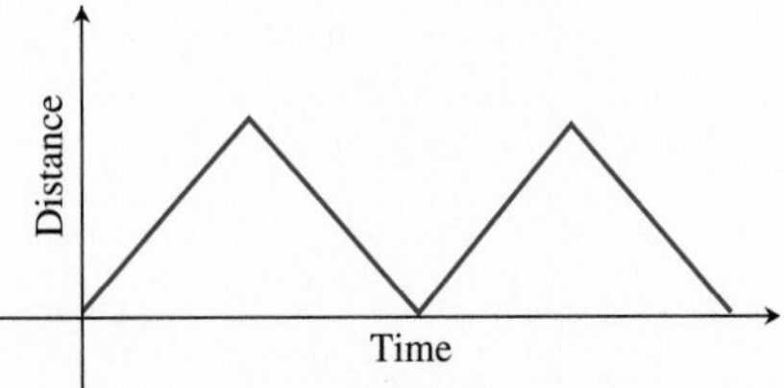

For Exercise 25

26. Sketch a graph that describes the height of a basketball above the floor after it leaves the hand of a free-throw shooter. (The basket is 10 ft above the floor.)

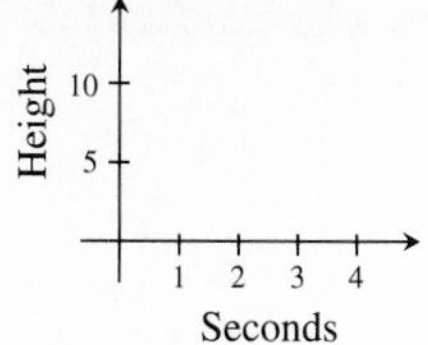

For Exercise 26

In Exercises 27–32, use the decimal window. Identify intervals on which the function is increasing, decreasing, or constant.

27. $f(x) = |x + 2| - 1$

28. $f(x) = |x + 1| + |x - 1| - 3$

29. $g(x) = |x + 2| + |x - 1| - 2$

30. $h(x) = 0.5(x + 2)^2 - 1$

31. $g(x) = 3 - (x - 1)^2$

32. $f(x) = x^3 - x^2 - 2x$

In Exercises 33–40, state whether the function is odd, even, or neither. Support graphically and confirm algebraically.

33. $f(x) = 2x^4$

34. $g(x) = x^3$

35. $f(x) = \sqrt{x^2 + 2}$

36. $g(x) = \dfrac{3}{1 + x^2}$

37. $f(x) = -x^2 + 0.03x + 5$

38. $f(x) = x^3 + 0.04x^2 + 3$

39. $g(x) = 2x^3 - 3x$

40. $h(x) = \dfrac{1}{x}$

In Exercises 41–44, the graph of the function is shown. Match the function with its graph. Use any clues you can find, without using a graphing utility. In each case find $f(-2)$, $f(1)$, and $f(3)$.

41. $f(x) = \begin{cases} |x + 2| & \text{if } x < 2 \\ -2x + 8 & \text{if } x \geq 2 \end{cases}$

42. $f(x) = \begin{cases} x^2 + 1 & \text{if } x < 1 \\ 2x + 3 & \text{if } x \geq 1 \end{cases}$

43. $f(x) = \begin{cases} 2\sqrt{2 - x} & \text{if } x < -2 \\ \frac{1}{2}x + 5 & \text{if } x \geq -2 \end{cases}$

44. $f(x) = \begin{cases} \frac{1}{2}x + 1 & \text{if } x < 2 \\ x^2 - 2 & \text{if } x \geq 2 \end{cases}$

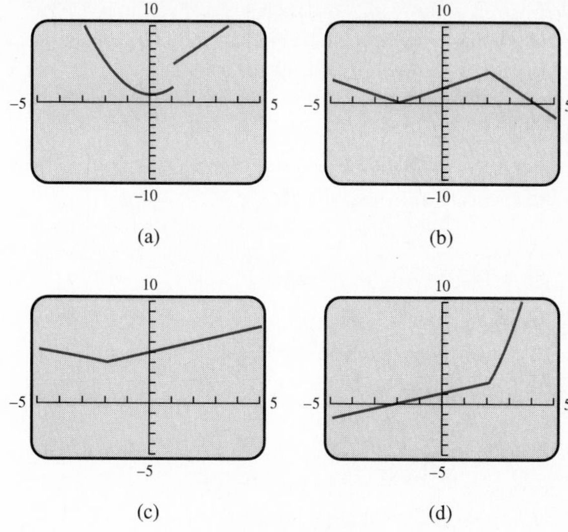

For Exercises 41–44

In Exercises 45 and 46, sketch the graph of the function.

45. $f(x) = \begin{cases} -x + 2 & \text{if } x < -2 \\ x^2 - 2 & \text{if } -2 \leq x \leq 2 \\ -2x + 8 & \text{if } 2 < x \end{cases}$

46. $f(x) = \begin{cases} -x + 1 & \text{if } x < -1 \\ 2 - x^2 & \text{if } -1 \leq x \leq 2 \\ 2x - 2 & \text{if } 2 < x \end{cases}$

In Exercises 47–50, the function has hidden behavior. Find and describe that hidden behavior. State the viewing windows that you used to reveal the hidden behavior.

47. $g(x) = 100x^2 - 203x + 103.02$

48. $h(x) = 1.1 + x - 100x^2$

49. $f(x) = x^3 - 2x^2 + x - 30$

50. $f(x) = x^3 - 13x^2 + 55x$

In Exercises 51–56, solve the problem.

51. *Landscape Design* A decorative water fountain shoots water upward. A drop of water goes straight up with an initial velocity of 60 ft/sec. The height of the drop t seconds later is modeled by $h(t) = -16t^2 + 60t$.

a. What values of t make sense in this problem?

b. Graph h and use "trace" to determine approximately how long until the drop hits the ground.

c. What is the maximum height of the water drop (to the nearest 0.1 ft)?

52. *Throwing a Baseball* A baseball player's hand is 7 ft off the ground as he throws a baseball straight up with an initial velocity of 48 ft/sec. The height of the ball t seconds later is described by $h(t) = -16t^2 + 48t + 7$. Find how high the ball goes.

53. *Constructing Livestock Pens* A farmer uses 800 ft of fencing to construct two square pens.

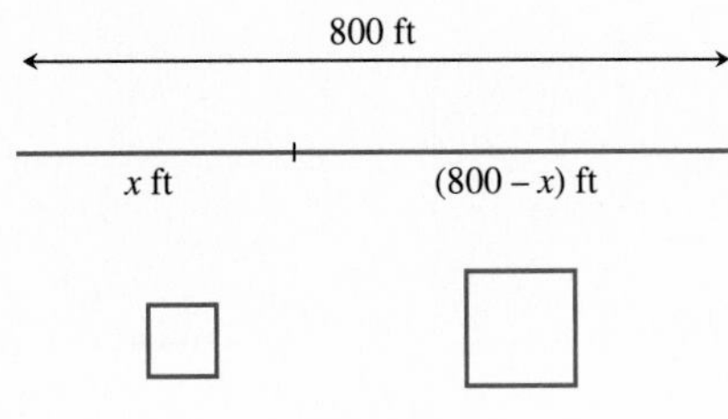

For Exercise 53

a. If the perimeter of one pen is x feet, what is the area of each of the two pens?
b. Express the total area of the two pens as a function $y = A(x)$.
c. What values of x make sense for this problem?
d. Graph $y = A(x)$ for the values of x found in part c.
e. **Writing to Learn** Would you recommend that the farmer use the fencing to build one square pen or two? Explain.

54. *Ranching Management* A rectangular livestock pen with three sides of fencing is to be built against the barn. The fencing is 1050 ft long. Find the dimensions of the maximum area that can be enclosed. What is the maximum area?

For Exercise 54

55. *Cone Problem* In completing the cone problem in Section 1.2 you should have found that

$$V(x) = \frac{\pi}{3}\left(\frac{8\pi - x}{2\pi}\right)^2 \sqrt{16 - \left(\frac{8\pi - x}{2\pi}\right)^2}$$

is the volume of the cone where x is the length of the arc.

a. What values of x make sense for this problem?
b. Graph $V(x)$ for the values of x found in part a.
c. Find the value of x to the nearest 0.01 when the volume is a maximum. What is the maximum volume?

56. *Fuel Prices* The average fuel prices (dollars per million Btu) from 1970 to 1989 are modeled by the function

$$f(x) = \begin{cases} 0.02x^2 + 0.29x + 1.4 & \text{if } 0 \le x \le 13 \\ 0.10x^2 - 3.55x + 38.26 & \text{if } 13 < x \le 19, \end{cases}$$

where $x = 0$ represents the year 1970. (*Source:* U.S. Energy Information Administration.)

a. Use this model to estimate the average cost of fuel in 1975.
b. Estimate the average cost of fuel in 1987.

EXTENDING THE IDEAS

In Exercises 57–64, solve the problem.

57. The figure shows the graphs of $y_1 = 4 - x^2$ and $y_2 = 1$, and a vertical line segment at x.

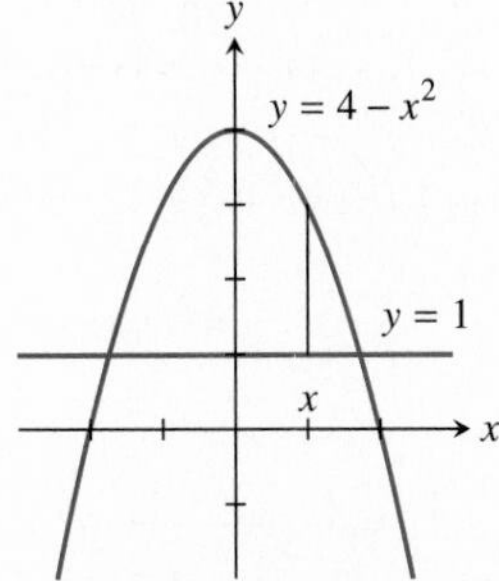

For Exercise 57

a. Find the points of intersection of the graphs of y_1 and y_2.
b. Find the length of the segment as a function of x.

58. The figure shows the graphs of $y_1 = x^2$ and $y_2 = \sqrt{x}$, and a vertical line segment at x.

a. Find the points of intersection of the graphs of y_1 and y_2.
b. Find the length of the segment as a function of x.
c. Find the maximum length of the segment over $[0, 1]$.

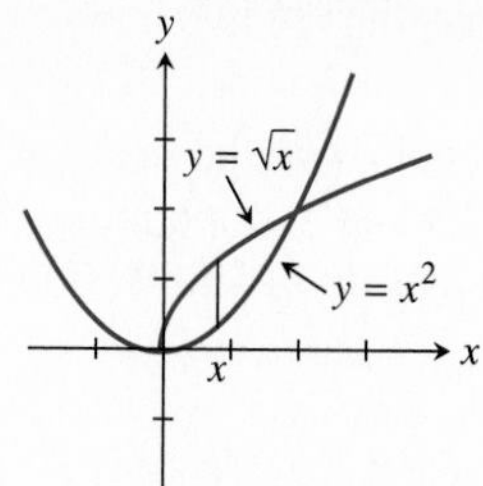

For Exercise 58

In Exercises 59 and 60, note that the domain is restricted so the graph should not go completely across the viewing window.

59. On a graphing utility, graph the function

$$f(x) = x^2 - 2, \quad -3 \leq x \leq 4$$

in the window $[-5, 5]$ by $[-10, 20]$ for the restricted domain.

60. On a graphing utility, graph the function

$$f(x) = |x + 2|, \quad -3 \leq x \leq 2$$

in the window $[-5, 5]$ by $[-3, 5]$ for the restricted domain.

Group Learning Activity In Exercises 61 and 62, *work in groups of three or four* to decide whether or not the given function has hidden behavior.

61. $y = 10x^3 + 7.5x^2 - 54.85x + 37.95$

62. $y = 5x^3 - 30.75x^2 + 63x - 3$

63. The dimensions of rectangles each with a perimeter of 100 in. can have different areas. Suppose x represents the width of such a rectangle.

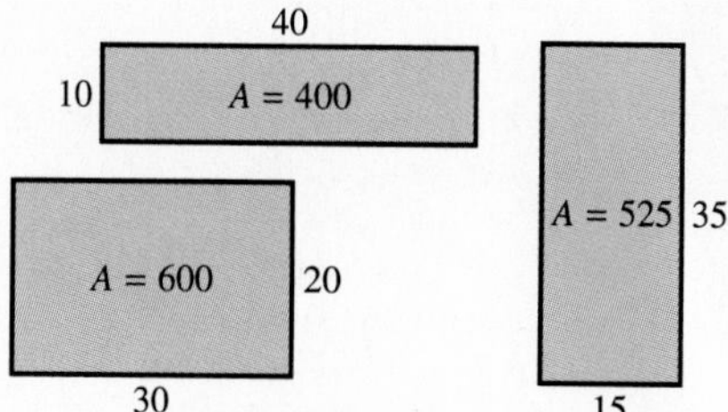

For Exercise 63

a. Find an expression in x for its length.

b. Find an expression $A(x)$ for its area.

c. What are the possible values of x for the rectangles we are considering? Find a graph of $y = A(x)$ for these values of x.

d. Find the local maximum for this function. Interpret your findings in terms of the rectangles we are considering.

64. Writing to Learn Review the definitions of increasing function, decreasing function, local maximum, local minimum, even function, and odd function. Discuss how these ideas apply to the constant function $f(x) = k$.

Objective

Students will be able to use mathematical models in the form of algebraic and graphical representations to solve applied problems.

Motivate

Ask students to name the four steps in Pólya's process for solving problems. **(Understand the problem, devise a plan, carry out the plan, and look back.)**

1.6 APPLYING MATHEMATICS

Using the Problem Solving Process

Using the Problem-Solving Process

All of the problems in this section lead to mathematical models involving linear equations.

As you read these examples, pay particular attention to the process of setting up a *mathematical model* of the problem. The modeling process begins with a word statement that identifies an equation relating the quantities of the problem.

Teaching Note

The problems introduced in this section are similar to those that will appear throughout the rest of the book. Business application problems (interest, cost, revenue, break-even point) and problems involving mensuration formulas (perimeter, area, volume) are emphasized in this section. Discuss the terms associated with the business application problems and then work one problem of each type. A mensuration problem should be done to see if students remember key geometric concepts.

EXAMPLE 1 APPLICATION: Modeling Profit

Dependable Auto Rental has fixed annual overhead costs of \$185,000. Each year it generates revenue of \$8540 per car and has operating costs of \$3050 per car. How many cars are needed to earn a profit of \$1,500,000?

Solution

Model

Word Statement: profit = revenue − total cost

$$\begin{aligned} x &= \text{number of cars} \\ 8540x &= \text{revenue} \\ 3050x &= \text{operating cost} \\ 185{,}000 &= \text{overhead cost} \\ 185{,}000 + 3050x &= \text{total cost} \\ \text{Profit} &= 8540x - (185{,}000 + 3050x) \end{aligned}$$

Solve Algebraically

$$\begin{aligned} 1{,}500{,}000 &= 8540x - (185{,}000 + 3050x) && \text{Equation to be solved} \\ 1{,}500{,}000 &= 8540x - 185{,}000 - 3050x \\ 1{,}685{,}000 &= 8540x - 3050x \\ 5490x &= 1{,}685{,}000 \\ x &= \frac{1{,}685{,}000}{5490} \\ &= 306.921\ldots \end{aligned}$$

```
306.92→X:8540X-(
185000+3050X)
            1499990.8
307→X:8540X-(185
000+3050X)
              1500430
▮
```

Figure 1.54

Support Numerically

When 306.92 is stored in x and $8540x - (185{,}000 + 3050x)$ is evaluated, we get a value very close to 1.5 million. See the first computation in Figure 1.54.

Interpret

We seek an integer solution. Dependable Auto Rental needs to have 307 cars to make a profit of at least \$1,500,000. See the second computation in Figure 1.54, and check that 306 is too small. ■

In Example 2, drawing a picture of the situation helps to formulate the word statement of the problem.

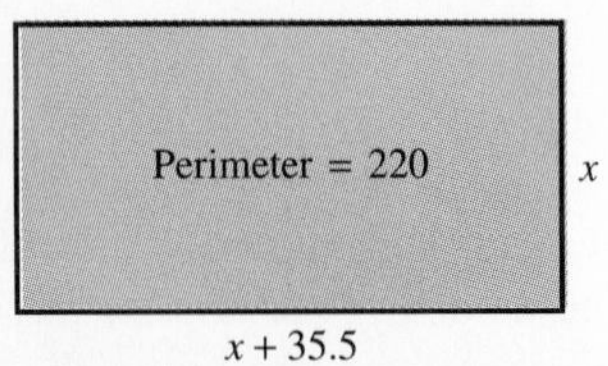

Figure 1.55 Rectangular stage.

EXAMPLE 2 APPLICATION: Finding Stage Dimensions

A concert stage is rectangular. It is 35.5 ft longer than it is wide. (See Figure 1.55.) The stage has a perimeter of 220 ft. What are the dimensions of the stage?

Solution

Model

Word Statement: $2 \times \text{length} + 2 \times \text{width} = \text{perimeter}$

$$\begin{aligned} x &= \text{width of stage} \\ x + 35.5 &= \text{length of stage} \\ 220 &= \text{perimeter} \end{aligned}$$

Solve Algebraically

$$\begin{aligned} 2(x + 35.5) + 2x &= 220 && \text{Equation to be solved} \\ 2x + 71 + 2x &= 220 && \text{Eliminate parentheses.} \\ 4x + 71 &= 220 && \text{Combine } x \text{ terms.} \\ 4x &= 149 && \text{Subtract 71.} \\ x &= 37.25 && \text{Width of stage} \\ x + 35.5 &= 72.75 && \text{Length of stage} \end{aligned}$$

Support Numerically

Check the solution by evaluating $2(x + 35.5) + 2x$ for $x = 37.25$ to obtain 220.

Interpret

The stage is 37.25 ft wide and 72.75 ft long. ■

Example 3 is a problem about an employee's salary and benefits.

■ **EXAMPLE 3 APPLICATION: Finding the Percentage of Benefits**

Sarah received a promotion and pay raise at Universal Advertising Agency. Her new salary is \$31,500 with a total contract, including benefits, of \$34,020. What percentage of her salary are her benefits?

Solution

Multiply by 100 to express a ratio as a percentage.

Model

Word Statement: $\text{benefits percentage} = \dfrac{\text{benefits}}{\text{salary}} \times 100$

$$\begin{aligned} x &= \text{benefits percentage} \\ 31{,}500 &= \text{salary} \\ 34{,}020 &= \text{total contract} \\ 34{,}020 - 31{,}500 &= \text{benefits} \end{aligned}$$

Solve Algebraically

$$x = \frac{34{,}020 - 31{,}500}{31{,}500} \times 100$$

$$x = 8$$

Support Numerically

Check the solution in the context of the problem by evaluating $31{,}500 + 31{,}500x$ for $x = 0.08$ to obtain \$34,020.

Interpret

Sarah's benefits are 8% of her salary. ■

Problems involving at least two of the quantities of distance, rate, and time can often be solved using the equation

$$\text{distance} = \text{rate} \times \text{time}.$$

■ EXAMPLE 4 APPLICATION: Solving a Distance Problem

The Philadelphia Rowing Club practices on the Schuylkill River. One crew travels 5 mi downstream in 1/4 h when the river is flowing 4 mph. How fast would the crew travel in still water?

Solution

Model

Word Statement: rate × time = distance

x = rate of boat in still water (miles per hour)

$x + 4$ = rate of boat downstream (miles per hour)

5 = distance one way (miles)

$\frac{1}{4}$ = time of trip (hours)

Solve Algebraically

$$(x + 4) \cdot \frac{1}{4} = 5 \quad \text{Equation to be solved}$$

$$x + 4 = 20$$

$$x = 16$$

Interpret

The boat travels at 16 mph in still water. ■

Notes on Examples

Examples 5 and 6 illustrate the importance of the fourth step of the problem-solving process. Solving for x alone does not complete the solutions to these problems.

EXAMPLE 5 APPLICATION: Finding an Investment Goal

Bill invests \$20,000. He uses a portion of it for a low-risk investment that earns 6.75% interest and the remainder for a higher-risk investment that earns 8.6% interest. The desired total interest for one year will be \$1,500. How much should Bill invest at each rate?

Solution

Divide by 100 to convert a percent to a decimal.

Model

Word Statement: total interest = interest at 6.75% + interest at 8.6%

$$\begin{aligned} x &= \text{amount invested at } 6.75\% \\ 20{,}000 - x &= \text{amount invested at } 8.6\% \\ 0.0675x &= \text{interest earned at } 6.75\% \\ 0.086(20{,}000 - x) &= \text{interest earned at } 8.6\% \\ 1500 &= \text{total interest in 1 year} \end{aligned}$$

Solve Algebraically

$$\begin{aligned} 0.0675x + 0.086(20{,}000 - x) &= 1500 \quad \text{Equation to be solved} \\ 0.0675x + 1720 - 0.086x &= 1500 \\ (0.0675 - 0.086)x &= 1500 - 1720 \\ x &= \frac{1500 - 1720}{0.0675 - 0.086} \\ x &= 11891.891\ldots \end{aligned}$$

Interpret

Rounding off to whole dollars, and to be sure to reach \$1500, Bill decides to invest \$11,891 at 6.75% and \$8109 at 8.6%.

Support Numerically

Figure 1.56 shows that Bill receives slightly more than \$1500 if he invests \$11,891 at 6.75% and \$8109 at 8.6%. ■

```
.0675*11891+.086
*8109
           1500.0165
```

Figure 1.56

Laboratory workers, custodians, and other industrial workers are involved in diluting solutions. For example, an acid solution is made by taking pure acid and diluting it with water. A 10% solution means that 10% of the solution is pure acid.

EXAMPLE 6 APPLICATION: Finding the Right Mix

Sparks Chemical Supply Co. keeps two acid solutions in stock to fill orders for its customers. One solution is 10% acid and the other is 25% acid. An order is received for 15 liters (15 L) of a 12% acid solution. How much 10% acid solution and how much 25% acid solution should be combined to fill this order?

Teaching Note

Algebra is more important than ever today because its focus has changed. The language for algebra for the twenty-first century is the algebra of representation rather than the algebra of manipulation. Students will be expected to write mathematical models in the language of algebra to describe real problem situations, not contrived ones.

Solution

Model

Word Statement: acid in solution = acid in 10% solution + acid in 25% solution

$$\begin{aligned} x &= \text{number of liters of 10\% acid} \\ 15 - x &= \text{number of liters of 25\% acid} \\ 0.1x &= \text{amount of pure acid in 10\% solution} \\ 0.25(15 - x) &= \text{amount of pure acid in 25\% solution} \\ 0.12(15) &= \text{amount of pure acid in final solution} \end{aligned}$$

Solve Algebraically

$$\begin{aligned} 0.1x + 0.25(15 - x) &= 0.12(15) && \text{Equation to be solved} \\ 0.1x + 3.75 - 0.25x &= 1.8 \\ -0.15x + 3.75 &= 1.8 \\ -0.15x &= -1.95 \\ x &= 13 && \text{Amount of 10\% solution} \\ 15 - x &= 2 && \text{Amount of 25\% solution} \end{aligned}$$

Support Numerically

Thirteen liters of 10% acid is 1.3 L of pure acid and 2 L of 25% acid is 0.5 L of pure acid—a sum of 1.8 L of pure acid in a 15-L solution. On the other hand, 15 L of 12% acid is 1.8 L of pure acid—an identical amount of acid.

Interpret

Thirteen liters of 10% acid and 2 L of 25% acid must be combined for 15 L of 12% acid solution. ■

Follow-up

Ask what assumptions were made in Example 7 about the number of shirts made and the number of shirts sold. **(It was assumed that all shirts made could be sold.)**

Assignment Guide

Day 1: Ex. 3–27, multiples of 3
Day 2: Ex. 29, 30, 32, 35, 37, 40, 43, 44, 46, 49, 51

Cooperative Learning

Group Activity: Ex. 40, 45

Notes on Exercises

Ex. 7–28 are problems that are found in a second-year algebra course. Since the settings are familiar, focus attention on the mathematical models and the relationships between representations.
Ex. 47–50 ask students to pose their own questions. Make sure students read the directions for these exercises.

Ongoing Assessment

Self-Assessment: Ex. 11, 25, 33, 41
Embedded Assessment: Ex. 30, 44

In business and industry, the **breakeven point** occurs when revenue equals cost. An important question is how many items must be produced to reach the breakeven point.

■ EXAMPLE 7 APPLICATION: Finding the Breakeven Point

The Quick Manufacturing Company produces T-shirts. Manufacturing fixed costs are \$200,000 plus \$1.50 for each shirt. The company sells shirts to distributors for \$4 per shirt. Estimate to the nearest 10,000 the number of shirts that must be made and sold for the company to break even.

Solution

Model

Word Statement: revenue = costs

$$\begin{aligned} x &= \text{number of T-shirts made and sold} \\ 1.5x &= \text{cost to produce } x \text{ shirts} \\ 200{,}000 &= \text{fixed costs} \\ 4x &= \text{revenue} \end{aligned}$$

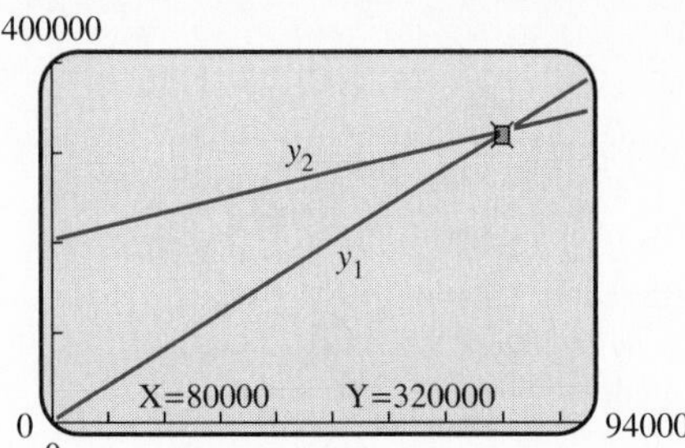

Figure 1.57 Graphs of $y_1 = 4x$ and $y_2 = 1.5x + 200{,}000$.

Solve Graphically

$$4x = 1.5x + 200{,}000 \quad \text{Equation to be solved}$$

Figure 1.57 shows that the graphs of $y_1 = 4x$ and $y_2 = 1.5x + 200{,}000$ intersect near a point with an x-coordinate of 80,000.

Confirm Algebraically

$$4x = 1.5x + 200{,}000 \quad \text{Equation to be solved}$$

$$2.5x = 200{,}000$$

$$x = \frac{200{,}000}{2.5} = 80{,}000$$

Interpret

The Quick Company must sell 80,000 T-shirts for revenue to equal costs—that is, to break even. ■

Quick Review 1.6

In Exercises 1–4, express the number as a percentage.

1. 0.0028 **2.** 0.00087

3. 1.21 **4.** $\dfrac{952}{1157}$

In Exercises 5–10, express all answers to the nearest 0.01.

5. Find 4.8% of 123. **6.** Find 5.3% of 1245.

7. 29 is what percent of 128?

8. 14 is what percent of 3258?

9. 95 is 112% of what number?

10. 1246 is what percent of 23,981?

In Exercises 11–18, the equation is a formula you should recall from previous mathematics courses. Solve the given formula for the given variable.

11. *Area of a Triangle* Solve for b:

$$A = \tfrac{1}{2}bh$$

12. *Area of a Trapezoid* Solve for h:

$$A = \tfrac{1}{2}(b_1 + b_2)h$$

13. *Volume of a Right Circular Cylinder* Solve for h:

$$V = \pi r^2 h$$

14. *Volume of a Right Circular Cone* Solve for h:

$$V = \tfrac{1}{3}\pi r^2 h$$

15. *Volume of a Sphere* Solve for r:

$$V = \tfrac{4}{3}\pi r^3$$

16. *Surface Area of a Right Circular Cylinder* Solve for h:

$$SA = 2\pi rh + 2\pi r^2$$

17. *Simple Interest* Solve for t:

$$I = Prt$$

18. *Compound Interest* Solve for P:

$$A = P(1 + r/n)^{nt}$$

In Exercises 19 and 20, solve for x.

19. $5(7x - 4) + 7(5 - 4x) = 4x + 17$

20. $9(2 - 3x) = 9x + 1 - 2(5x - 2)$

SECTION EXERCISES 1.6

In Exercises 1–12, write a mathematical expression for the given word expression.

1. A number x is decreased by 48.6.

2. Five more than 3 times a number x.

3. Three times 5 more than a number x.

4. A number x is increased by 5, then doubled.

5. Seventeen percent of a number x.

6. Four more than 5% of a number x.

7. *Area of a Rectangle* The area of a rectangle whose length is 12 more than its width x.

8. *Area of a Triangle* The area of a triangle whose altitude is 2 more than its base length x.

9. *Salary Increase* A salary after a 4.5% increase in the current salary of x dollars.

10. *Income Loss* Income after a 3% drop in the current income of x dollars.

11. *Expense Account* An expense account after a 3% decrease in the current amount of x dollars.

12. *Wage Settlement* A salary after a 3.8% increase above the current salary of x dollars together with a bonus of $200.

In Exercises 13–16, choose a variable and write a mathematical expression for the given word expression.

13. *Total Cost* The total cost is $34,500 plus $5.75 for each item produced.

14. *Total Cost* The total cost is $28,000 increased by 9% plus $19.85 for each item produced.

15. *Revenue* The revenue when each item sells for $3.75.

16. *Profit* The profit consists of a franchise fee of $200,000 plus 12% of all sales.

In Exercises 17–28, write an equation for the problem and solve the problem.

17. One positive number is 4 times another positive number. The sum of the two numbers is 620. Find the two numbers.

18. When a number is added to its double and its triple, the sum is 714. Find the three numbers.

19. *Salary Increase* Mark received a 3.5% salary increase. His salary after the raise was $36,432. What was his salary before the raise?

20. *Consumer Price Index* The consumer price index for food in a particular year was 40.4 after a 4.2% increase from the previous year. What was the index the previous year?

21. Connecting Algebra and Geometry A rectangle is 18 ft longer than it is wide and its perimeter is 228 ft. Find the width of the rectangle.

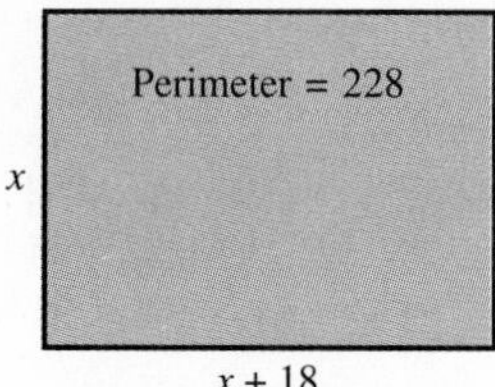

For Exercise 21

22. Connecting Algebra and Geometry One wall of a foundation is x feet long, and the others are as shown in the figure. If the perimeter of the foundation is 166 ft, find the length of each wall.

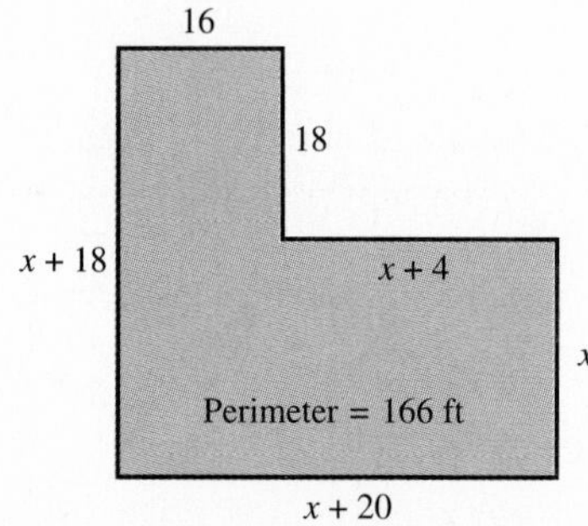

For Exercise 22

23. *Travel Time* A traveler averaged 52 mph on a 182-mi trip. How many hours were spent on the trip?

24. *Travel Time* On their 560-mi trip, the Bruins basketball team spent two more hours on the interstate highway than they did on local highways. They averaged 45 mph on local highways and 55 mph on the interstate highway. How many hours did they spend driving on local highways?

25. *Calculation of Earnings* Smith Bentham had profits in

1994 of \$131,280 and profits in 1995 of \$141,176. What was the percent increase in profits?

26. *Budget Analysis* Centerville School has a total budget of \$3,740,000. Employee salaries amounted to \$2,125,000. What percent of the total budget are salaries?

27. *Flying Time* A United Airlines Boeing 727 makes a trip from Seattle International to Chicago O'Hare Airport when there is a 30 mph tailwind. The 1770-mi trip east takes 3 h and 45 min. What is the average cruising speed of the 727 on a windless day?

28. *Speed Bicycle Racing* Greg Lehman, a competitive racing bicyclist, practices riding to the top of Pike's Peak from a location near Colorado Springs 25 mi from the top of Pike's Peak. The total time for the trip is 2 h. He averages 23 mph slower on the ride up than he does on a level road. What would be his average speed on a level-grade highway?

Exercises 29 and 30 refer to the collection of all rectangles whose lengths are twice their widths.

29. Recall that the perimeter of a rectangle is $P = 2W + 2L$.

a. Write the perimeter P in terms of the width W.
b. Write the perimeter P in terms of the length L.
c. Find the dimensions of the rectangle in this collection whose perimeter is 200.
d. **Writing to Learn** Explain how the graphs below can be used to solve the problem in part c.

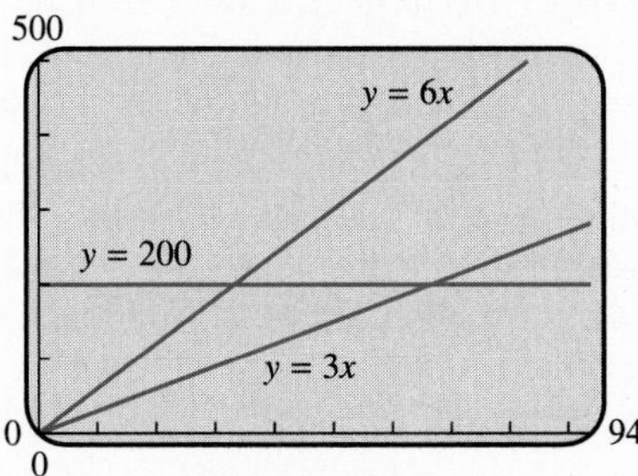

For Exercise 29

30. Recall that the area of a rectangle is $A = LW$.

a. Write the area A in terms of the width W.
b. Write the area A in terms of the length L.
c. Find the dimensions of the rectangle in this collection whose area is 4050 square units.
d. **Writing to Learn** Explain how the graph can be used to solve the problem in part c.

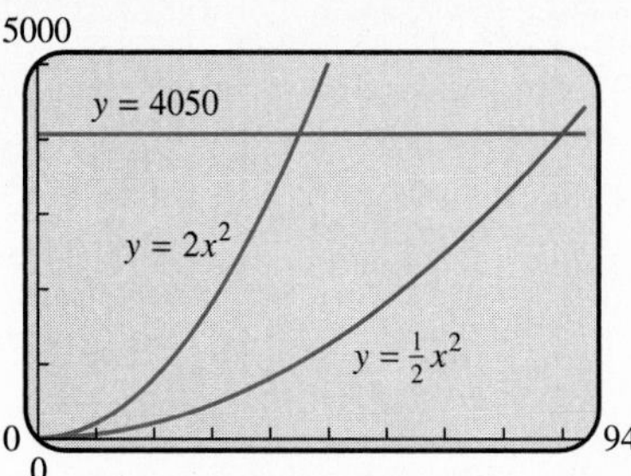

For Exercise 30

In Exercises 31–46, solve the given problem.

31. *Mixing Oil and Gasoline* A Toro two-cycle engine requires that 1 part of oil be mixed with 32 parts of gasoline.

a. Write an equation that models the number of parts of oil, y, to be mixed with x parts of gasoline.
b. What restrictions are there on the variable x?
c. Which of the two graphs is a better graphical representation of the problem, given the restrictions on the variable x?

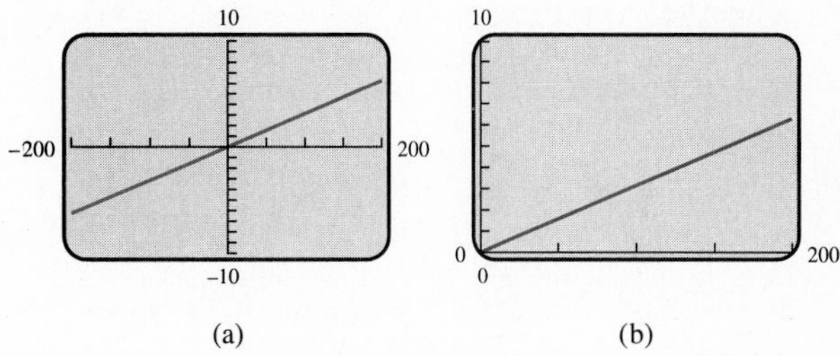

(a) (b)

For Exercise 31

d. How many parts of oil are needed for 125 parts of gasoline?

32. *Fruit Juice Concentrate* Healthful frozen orange juice requires that three parts of water be mixed with one part of frozen concentrate.

a. What do y_1 and y_2 represent in $y_1 = 0.25x$ and $y_2 = 4x$? In each case what does x represent?
b. How much concentrate is needed to make 100 gal (gallons) of orange juice?

33. *Mixture Problem* How much 10% solution and how much 45% solution should be used to make 100 gal of a 25% solution?

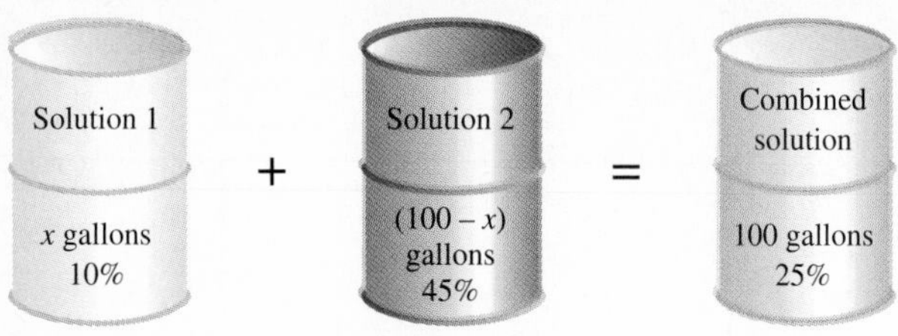

For Exercise 33

a. Write an equation that models this problem.
b. Solve this equation graphically.

34. *Mixture Problem* Show how to change the grapher entries used for Exercise 33 to find the amount of each solution listed here that is required to make 100 gal of mixture.

	Solution 1	*Solution 2*	*Mixture*
a.	10%	25%	15%
b.	22%	48%	30%

35. *Mixture Problem* The chemistry lab at the University of Hardwoods keeps two acid solutions on hand. One is 20% acid and the other is 35% acid. How much 20% acid solution and how much 35% acid solution should be used to prepare 25 L of a 26% acid solution?

36. *Mixing a Cleaning Agent* The Ajax cleaning supply company needs to prepare 100 L of a 14% acid solution. How much 10% acid solution and 25% acid solution must be mixed?

37. *Residential Construction* DDL Construction is building a rectangular house that is 16 ft longer than it is wide. A rain gutter is to be installed in four sections around the 136-ft perimeter of the house. What lengths should be cut for the four sections?

38. *Interior Design* Rene's Decorating Service recommends putting a border around the top of the four walls in a dining room that is 3 ft longer than it is wide. Find the dimensions of the room if its perimeter is 54 ft.

39. *Travel Time* West Coast Logging Co. had one truck leave the tree farm 2 h ahead of a second one. If the first truck travels at an average rate of 55 mph and the second travels at an average rate of 65 mph, how long will it take for the second truck to catch the first truck?

40. *Travel Time* Kim averaged 52 mph on state highways and 62 mph on interstate highways on her 450-mi trip. Her driving time was 7 h 45 min.

a. Write an equation for this problem that relates time and distance, where $t =$ the time Kim traveled on state highways.
b. How long did she travel on interstate highways?
c. How many miles did she travel on state highways?

41. *Investment Decisions* Reggie invests $12,000, part at 7% interest and part at 8.5% interest. How much is invested at each rate if Reggie receives $900 interest in 1 year?

42. *Investment Decisions* Jackie invests $25,000, part at 5.5% interest and the balance at 8.3% interest. How much is invested at each rate if Jackie receives $1571 interest in one year?

43. *Employee Benefits* John's company issues employees a contract that identifies salary and the company's contributions to pension, health insurance premiums, and disability insurance. The company uses the following expressions to calculate these values.

Salary	***x* (dollars)**
Pension	12% of salary
Health Insurance	3% of salary
Disability insurance	0.4% of salary

If John's total contract is worth $48,814.20, what is his salary?

44. *Manufacturing Model* The Buster Green Shoe Co. determines that the annual cost C of making x pairs of one type of shoe is $30 per pair plus $100,000 in fixed overhead costs. Each pair of shoes that is manufactured is sold for $50.

a. Find an equation that models the cost of producing x pairs of shoes.
b. Find an equation that models the revenue produced from selling x pairs of shoes.
c. Use a graphical method to find how many pairs of shoes must be made and sold in order to break even.

45. *Manufacturing Model* Queen, Inc., a tennis racket manufacturer, determines that the annual cost C of making x rackets is $23 per racket plus $125,000 in fixed overhead costs. Each racket is sold for $56 if it is unstrung and $79 if it is strung. It costs the company $8 to string a racket.

a. Find a function $y_1 = u(x)$ that models the cost of producing x unstrung rackets.
b. Find a function $y_2 = s(x)$ that models the cost of producing x strung rackets.
c. Find a function $y_3 = r_u(x)$ that models the revenue generated by selling x unstrung rackets.
d. Find a function $y_4 = r_s(x)$ that models the revenue generated by selling x strung rackets.
e. Graph y_1, y_2, y_3, and y_4 simultaneously in the window [0, 10,000] by [0, 500,000].
f. **Writing to Learn** Write a report to the company, recommending how they should manufacture their rackets,

strung or unstrung. Assume that you can include the viewing window from part e as a graph in the report, and use it to support your recommendation.

46. Table 1.18 gives the total college enrollment in the United States from 1985 to 1991.

Table 1.18 U.S. College Enrollment

Year	Enrollment (thousands)
1985	7799
1986	7613
1987	7932
1988	7973
1989	7987
1990	8742
1991	8304*

*Estimated.
Source: U.S. Bureau of Census

a. Find the linear regression equation that models these data. Let $x = 0$ correspond to 1980. (Find the coefficients to the nearest 0.01.)

b. Use this model to predict when the college enrollment will reach about 9 million students. (9000 in thousands)

EXTENDING THE IDEAS

In Exercises 47–50, you are asked to do some *problem posing.* A problem situation is the situation in which a quantitative problem can be identified. Problems are often posed by asking questions "how much?" or "how many?" or "when?" Often there are several problems that can be posed within the same problem setting. You are asked to pose at least three.

47. Consider the collection of all rectangles with a 42-in. perimeter.

48. Consider the collection of all right triangles.

49. The consumer products division of a large company has general operating costs of \$425,000. The materials for producing one Wally gig cost the manufacturer \$2.43. It can sell each gig to one large distributor for \$6.10 per gig. The distributor in turn sells the Wally gigs for \$9.95.

50. Lisa has \$25,000 to invest. She can invest into an account that carries little risk and pays 5.5% simple interest, or into an account that carries moderate risk and pays 6.8%.

In Exercises 51 and 52, solve the problem.

51. *Height of a Tree* The height of a tree can be found by comparing the length of its shadow to the length of the shadow cast by a stake of a known height. If a 5-ft vertical stake casts a shadow of 8 ft 3 in., and a tree casts a shadow of 78 ft, find the height of the tree.

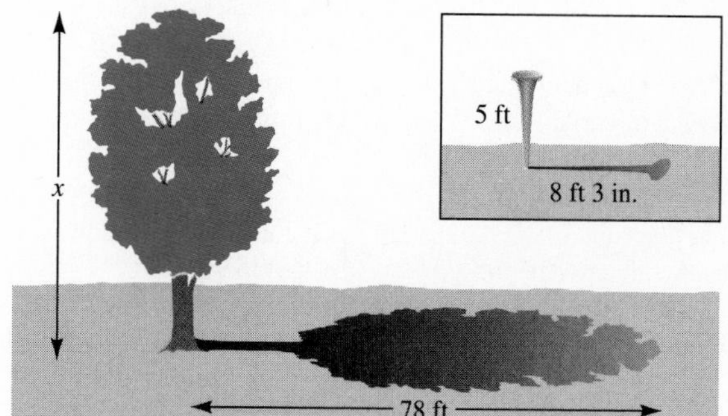

For Exercise 51

52. *Scale Drawing* Jack, who works for Midwest Design Co., usually completes his architectural drawings using a scale of 1 in. per 8 ft.

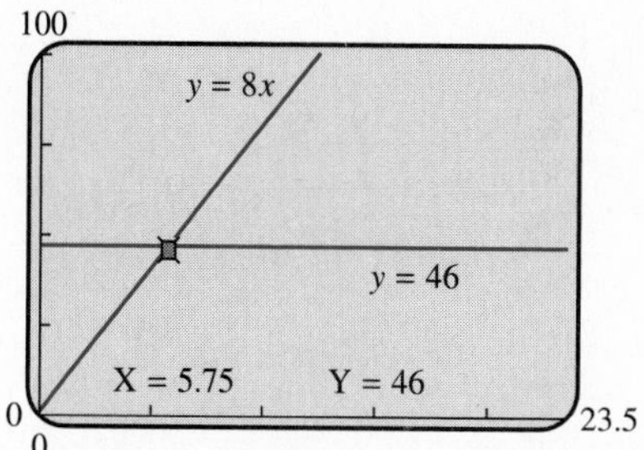

For Exercise 52

a. If the distance *AB* on the drawings is x in. and the corresponding distance *CD* on the building is y in., what is the ratio $\frac{x}{y}$ as a reduced fraction?

b. Writing to Learn Jack claims that both $y = 8x$ and $y = 96x$ can model this scale drawing problem. Explain how to interpret each equation for this to be true.

c. While completing the drawings the architect needed to represent an actual distance of 46 ft. What is the corresponding distance (in inches) on the drawings? Explain how this grapher screen answers the question.

CHAPTER 1 REVIEW

KEY TERMS

The number following each key term indicates the page of its introduction.

REVIEW EXERCISES

In Exercises 1 and 2, determine which ordered pair is a solution for the equation.

1. $y = \sqrt{x^2 + 1}$

a. $(0, 1)$
b. $(-1, 0)$

2. $3x - 2y = 4$

a. $(-2, 1)$
b. $(2, 1)$

In Exercises 3–6, complete the table of solutions for the given equation.

3. $y = -3x + 1$

x	-2	-1	?	3	?
y	7	?	-11	?	1

4. $y = 2x - 5$

x	-3	2	?	-4	?
y	-11	?	-7	?	1

5. $f(x) = \dfrac{x}{x - 2}$

x	-1	-2	0	2	?
$f(x)$	$\frac{1}{3}$	?	?	?	-1

6. $g(x) = x^3 - 2x + 1$

x	2	-2	0	1	4
$g(x)$	5	?	?	?	?

In Exercises 7–12, make a table of solutions for the equation for integer values of x from $x = -3$ to $x = 3$. Sketch the graph of the equation.

7. $y = 4x - 1$

8. $y = 2 - 3x$

9. $y = |x - 3|$

10. $y = \sqrt{x - 2}$

11. $y = 3 - x^2$

12. $y = x^2 - 3x$

In Exercises 13–16, match the equation with its graph. Identify the viewing-window dimensions and the values for Xscl and Yscl.

13. $y = 1 + x - x^3$

14. $y = |x - 2| - 1$

15. $f(x) = \begin{cases} \sqrt{1 - x} & \text{if } x < 1 \\ x^2 - 1 & \text{if } x \geq 1 \end{cases}$

16. $f(x) = \begin{cases} -3x + 5 & \text{if } x < 2 \\ x - 3 & \text{if } x \geq 2 \end{cases}$

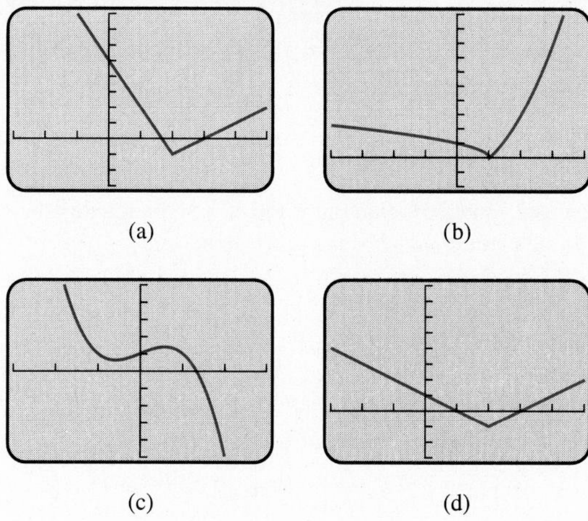

(a) (b)

(c) (d)

For Exercises 13–16

In Exercises 17 and 18, select the one of the three windows below that you feel shows the best view of the graph.

17. $y = 5x^2 - 2x + 6$ **18.** $f(x) = 3x^3 + 5x^2 + 2x - 10$

(a)	(b)	(c)
Xmin=⁻10	Xmin=⁻5	Xmin=⁻5
Xmax=10	Xmax=5	Xmax=5
Xscl=1	Xscl=1	Xscl=1
Ymin=⁻10	Ymin=⁻10	Ymin=⁻20
Ymax=10	Ymax=20	Ymax=10
Yscl=1	Yscl=5	Yscl=5

For Exercises 17 and 18

In Exercises 19 and 20, evaluate the function.

19. $f(-2)$ and $f(a)$ for $f(x) = 2x^2 - x + 5$

20. $f(3)$ and $f(x + 2)$ for $f(x) = \dfrac{\sqrt{x + 1}}{x - 2}$

In Exercises 21 and 22, interpret the graph.

21. Use the graph of $y = f(x)$ to answer the following.

a. $f(-4) = ?$
b. $f(2) = ?$
c. The domain of f is ?
d. The range of f is ?
e. **Writing to Learn** If $f(a) = 2$, explain why we cannot conclude that $a = 2$.

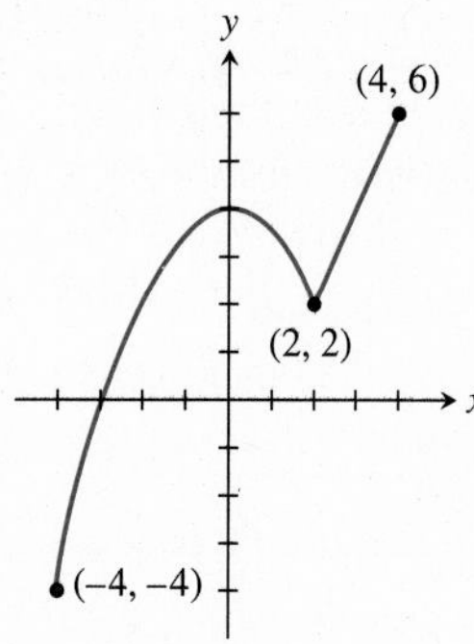

For Exercise 21

22. **Writing to Learn** The graph shows distance from a snack stand plotted against time for a hiker. Using information from the graph, write a paragraph describing the hiker's workout.

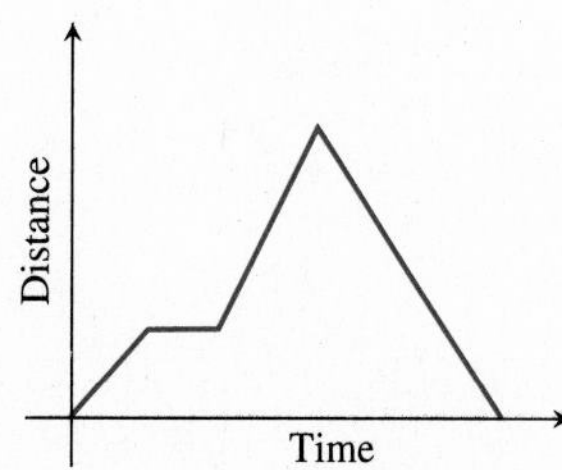

For Exercise 22

In Exercises 23 and 24, identify the independent variable and the dependent variable of the function.

23. $r = \sqrt[3]{s^2 + 1}$ **24.** $y = \dfrac{x^2 - 1}{x + 3}$

In Exercises 25–28, find the domain of the function algebraically. Support your answer graphically.

25. $f(x) = \dfrac{x}{x^2 - 2x}$

26. $g(x) = x^2 + 2x + 1$

27. $h(x) = \sqrt{2 - x}$

28. $k(x) = \sqrt{x^2 + 1}$

In Exercises 29 and 30, use a grapher to find the value of the function for $x = -2$, $x = 1.5$, $x = 2$, and $x = 5$.

29. $f(x) = x^3 - 5x^2 + x - 5$

30. $g(x) = \dfrac{\sqrt{x^2 + 1}}{x^2 + x}$

In Exercises 31 and 32, assume that each graph contains the origin and the point in the upper right-hand corner of the viewing window.

31. Estimate the slope of the line.

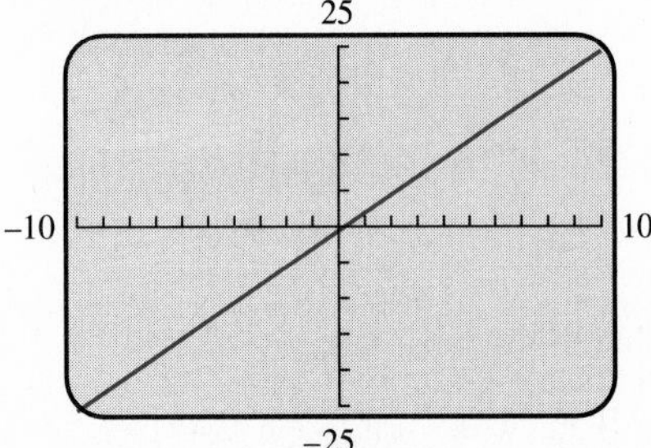

For Exercise 31

32. Writing to Learn Which line has the greater slope? Explain.

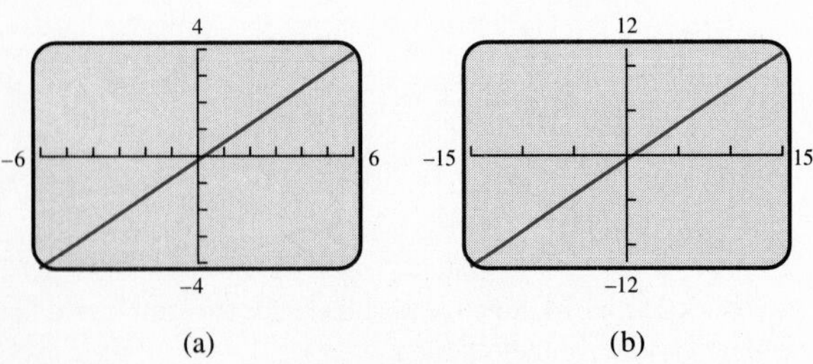

For Exercise 32

In Exercises 33–36, find the slope of the line through the pair of points.

33. (2, −3) and (8, 4)

34. (−1, −2) and (7, 6)

35. (−1, −2) and (4, −5)

36. (−2, 1) and (5, −4)

In Exercises 37 and 38, find the value of x or y such that the line through the given points has the given slope m.

37. $(x, -1)$ and $(4, 7)$; $m = 2$

38. $(-2, y)$ and $(2, -1)$; $m = -1$

In Exercises 39–42, find an equation in point-slope form for the line through the point with the given slope m.

39. $(-3, -5)$; $m = \frac{3}{4}$

40. $(2, -1)$; $m = -\frac{2}{3}$

41. $(-5, 0)$; $m = -2$

42. $(0, 3)$; $m = 2$

In Exercises 43 and 44, find an equation of the line that contains the two points. Write the equation in the general form $Ax + By + C = 0$.

43. (−5, 4) and (2, −5)

44. (−4, −3) and (2, 7)

In Exercises 45–50, find an equation in slope-intercept form for the line.

45. The line through (3, −2) with slope $m = \frac{4}{5}$

46. The line through the points (−1, −4) and (3, 2)

47. The line through (−2, 4) with slope $m = 0$

48. The line $3x - 4y = 7$

49. The line through (2, −3) and parallel to the line $2x + 5y = 3$

50. The line through (2, −3) and perpendicular to the line $2x + 5y = 3$

In Exercises 51 and 52, write two equations: (a) one for the line passing through the given point and parallel to the given line, and (b) another for the line passing through the given point and perpendicular to the given line. Support your work graphically.

51. (1, 3); $y = -2x + 3$

52. (−6, 3); $4x - 3y = 5$

In Exercises 53 and 54, find a viewing window that shows the following.

53. The line $75x - 5y = 500$ intersecting both the x- and y-axes.

54. Two intersections with the x-axis, and all the points on the graph of $f(x) = -x^2 - 2x + 168$ between these two points.

In Exercises 55 and 56, find the values for Ymin, Ymax, and Yscl so that the graph of the equation appears as in the viewing window shown.

55. $y = \frac{7}{4}x$ **56.** $y = 7x$

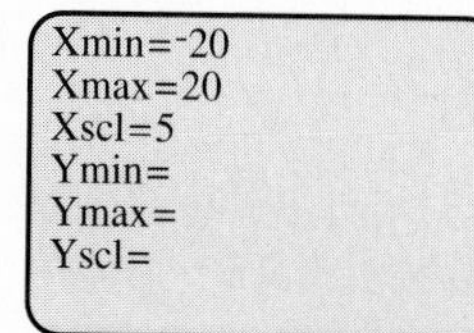

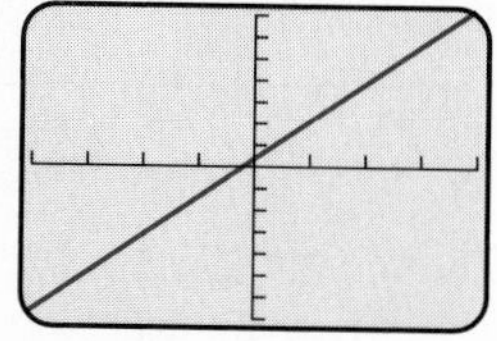

For Exercises 55 and 56

In Exercises 57 and 58, graph the equation. Then pick one of the following as the best description as the "trace" cursor moves from left to right on the graph:

a. The y-values decrease, then increase.
b. The y-values decrease, then increase, and then decrease again.
c. The y-values increase.
d. The y-values increase, then decrease, and then increase again.

57. $y = 2x^2 - 3x + 5$ **58.** $y = -x^3 + 9x + 4$

In Exercises 59 and 60, state whether each specified point identifies a local minimum, a local maximum, or neither. Identify intervals on which the function is decreasing and increasing.

59.

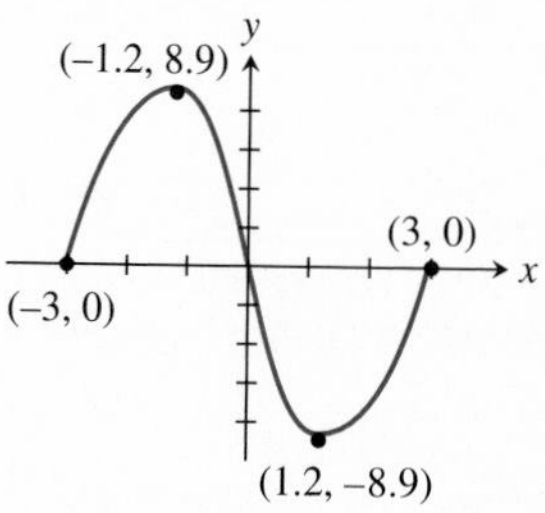

60.

y
(–3, 6)
(4, 2)
x
(2, –1)
(7, –6.4)

In Exercises 61–64, use a grapher to find all local maximum and minimum values, and the values of x where they occur. Give values rounded to two decimal places.

61. $f(x) = x^2|x - 4|$ **62.** $g(x) = |2x - 5| - 6$

63. $f(x) = x^3 - 10x^2 + x + 50$

64. $g(x) = 4 + 2x - x^2$

In Exercises 65–68, find a viewing window that you believe includes the important characteristics of the graph of the function. List the characteristics in the window you choose.

65. $f(x) = |9 + 3x - x^2|$ **66.** $g(x) = x^3 - 4x$

67. $h(x) = 2\sqrt{1 - x} - 5$

68. $k(x) = \begin{cases} -2x - 1 & \text{if } x < -1 \\ x^2 + 1 & \text{if } -1 \le x \le 2 \\ x - 3 & \text{if } x > 2 \end{cases}$

In Exercises 69–72, use the decimal window. Identify intervals on which the function is increasing, decreasing, or constant.

69. $f(x) = 2x^2 - x^3$ **70.** $g(x) = (x + 2)^2 - 3$

71. $f(x) = \frac{1}{2}|x - 2| + \frac{1}{2}|x + 2|$

72. $g(x) = \begin{cases} 2 - x^2 & \text{if } x < 1 \\ x - 1 & \text{if } x \ge 1 \end{cases}$

In Exercises 73 and 74, use the vertical line test on each graph to decide whether it is the graph of a function of x.

73.

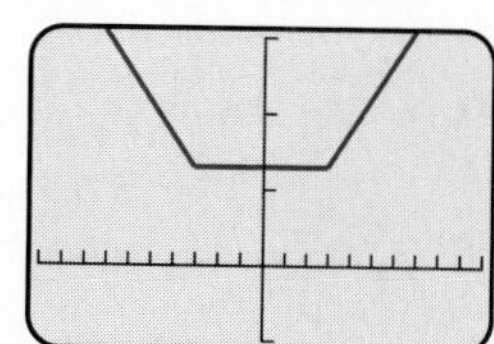

74.

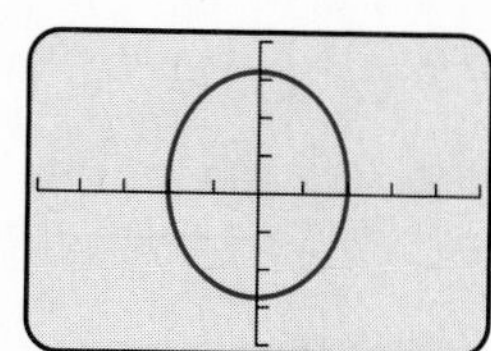

In Exercises 75–78, state whether the function is even, odd, or neither. Support graphically and confirm algebraically.

75. $f(x) = x^3 - 2$ **76.** $g(x) = x^4 - 2x^2$

77. $f(x) = x + \dfrac{1}{x}$ **78.** $g(x) = \dfrac{1}{2 + x^2}$

In Exercises 79–82, describe the linear correlation as strong positive, weak positive, strong negative, weak negative, or no correlation.

79.

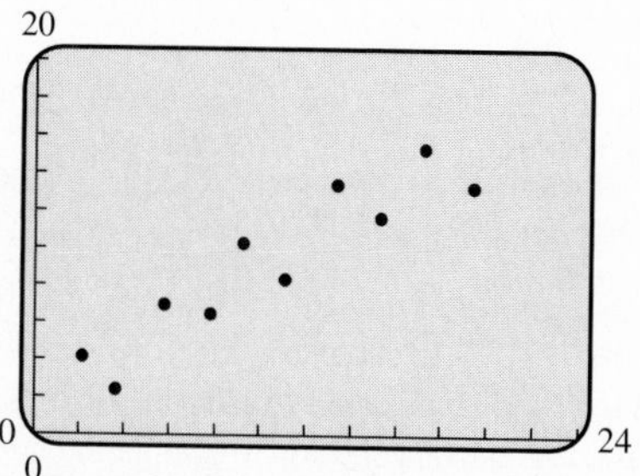

80.

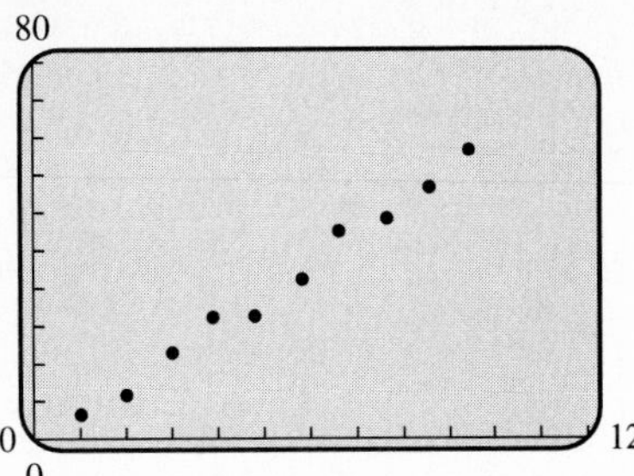

81.

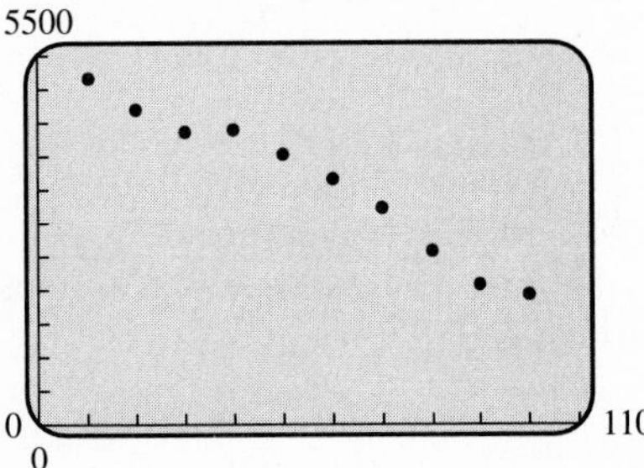

82.

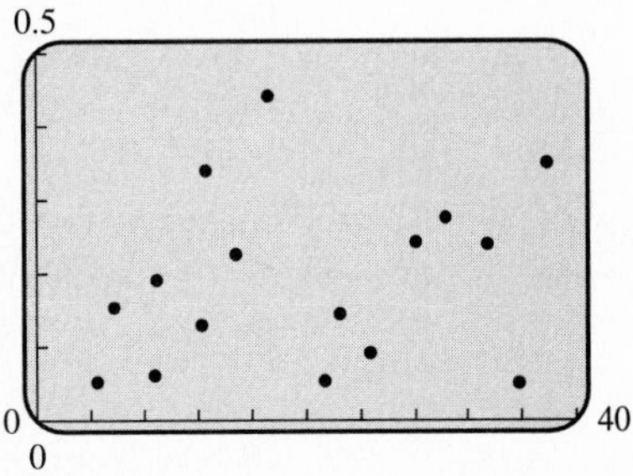

In Exercises 83 and 84, refer to the scatter plots in Exercises 79–82.

83. Select the scatter plot that has a correlation coefficient that is closest to -1.

84. Select the scatter plot that has a correlation coefficient that is closest to 0.

In Exercises 85 and 86:

a. Complete a scatter plot for the data using $x = 0$ to represent the 1970–71 academic year.
b. Find the linear regression equation.
c. Use the regression equation to predict the number of doctoral degrees in the acdemic year 2000–01.

85. *Doctoral Degrees* Table 1.19 shows the number of doctoral degrees earned in the given academic year by Hispanic students.

Table 1.19 Doctorates Earned by Hispanics

Year	Number of Degrees
1976–77	520
1980–81	460
1984–85	680
1988–89	630
1990–91	730
1991–92	810
1992–93	830

Source: U.S. Department of Education, as reported in the *Chronicle of Higher Education,* April 28, 1995.

86. *Doctoral Degrees* Table 1.20 shows the number of doctoral degrees earned in the given academic year by African American students.

Table 1.20 Doctorates Earned by African Americans

Year	Number of Degrees
1976–77	1250
1980–81	1270
1984–85	1150
1988–89	1070
1990–91	1200
1991–92	1220
1992–93	1350

Source: U.S. Department of Education, as reported in the *Chronicle of Higher Education,* April 28, 1995.

In Exercises 87 and 88, complete a scatter plot and solve the problems.

87. *Motor Vehicle Deaths* Table 1.21 shows the number of deaths by motor vehicles per 100,000 population for the given years.

a. Draw a scatter plot and find the linear regression equation.
b. Use the regression equation to predict the number of motor vehicle deaths per 100,000 population in the year 2000.

Table 1.21 Motor Vehicle Deaths in the U.S.

Year	Number of Deaths (100,000)
1980	23.4
1985	19.2
1990	18.8
1992	15.8

Source: USA Today research by Cindy Hall and Web Bryant, as reported on January 16, 1995.

88. *Inflation Rates* Table 1.22 shows the inflation rate as a percentage for the given years.

Table 1.22 U.S. Inflation Rates

Year	Rate (%)
1965	1.9
1970	5.6
1976	4.9
1978	9.0
1982	3.8
1990	6.1
1994	2.7

Source: USA Today research by Cliff Vancura, as reported on January 23, 1995.

a. Draw a scatter plot and find the linear regression equation.
b. Use the regression equation to predict the inflation rate in the year 2000.

In Exercises 89–92, write a mathematical expression for the given word expression.

89. A number is increased by 4, then tripled.

90. *Area of a Rectangle* The area of a rectangle whose width is 5 less than the length x.

91. *Take-Home Pay* The amount of Jim's paycheck after a 35% decrease in his gross salary x for taxes and benefits.

92. *Mixture Problem* The amount of water in x liters of a 20% acid solution.

Exercises 93 and 94 refer to a box (with a lid) that is twice as long as it is wide and whose height is 3 cm less than the width.

93. Connecting Algebra and Geometry

a. Write the volume V of the box in terms of the width x.
b. Graph the equation $y = V(x)$.

94. Connecting Algebra and Geometry

a. Write the surface area S of the box in terms of the width x.
b. Graph the equation $y = S(x)$.

Exercises 95 and 96 refer to the collection of all right circular cylinders whose heights are twice their radii.

95. Connecting Algebra and Geometry

a. Write the volume V of the right circular cylinder in terms of the radius x.
b. Graph the equation $y = V(x)$.

96. Connecting Algebra and Geometry

a. Write the total surface area S of the right circular cylinder in terms of the radius x.
b. Graph the equation $y = S(x)$.

In Exercises 97–100, solve the given problem.

97. Connecting Algebra and Geometry Use the function $A = \pi r^2$ to find the following:

a. The area of a circle with radius 32 cm
b. The radius of a circle with area 2500 cm^2

98. *Truck Depreciation* Jones Hauling purchased an 18-wheel truck for \$100,000. It depreciates \$10,000 per year for 10 years.
Model: The value y after x years is

$$y = 100{,}000 - 10{,}000x,$$

where $0 \leq x \leq 10$.

a. Graph the modeling equation. Record your viewing-window dimensions.
b. When is the value of the truck \$55,000? *Solve graphically and confirm algebraically.*

99. *Highway Engineering* Interstate 70 west of Pittsburgh, Penn., has a section that is posted as a 5% grade. That means that for a horizontal change of 100 ft there is a 5-ft vertical change.

a. Find the slope of this section of highway.
b. On a highway with a 5% grade, what is the horizontal distance required to climb 325 ft?
c. A sign along the highway announces a 5% grade for the next 4 mi. Estimate how many feet of vertical change there is along those 4 mi.

100. Find the value of x and y for which $(x, 2)$ and $(1, y)$ are points on the graph of the equation $5x - 4y = 9$.

In Exercises 101–104, write an equation for the problem and solve the problem.

101. *Salary Increase* Sarah received a 4.2% salary increase. Her salary after the raise was $56,268. What was her salary before the raise? *Solve algebraically and support numerically.*

102. *Mixture Problem* Starkey's Drug Store keeps two acid solutions on hand. One is 24% acid and the other is 45% acid. How much of each solution should be used to fill an order for 30 liters of a 32% acid solution? *Solve algebraically and support graphically.*

103. *Manufacturing Model* The owner of Christine's Crewel Craft Center has monthly fixed expenses of $1500. Craft kits are sold for $22 each, and the cost of the material in each kit is $15. How many kits must be sold to break even in a given month?

104. *Swimming Problem* Sandy can swim 1 mi upstream (against the current) in 20 min. She can swim the same distance downstream in 9 min. Find Sandy's swimming speed and the speed of the current assuming both speeds are constant.

In Exercises 105–112 solve the given problem.

105. *Rectangle Problem* The length of a certain rectangle is 5 in. more than its width. Find the dimensions of the rectangle if its perimeter is 300 in.

106. *Movie Attendance* Ticket prices are $5.50 for adults and $3.00 for children. The revenue collected from one showing was $1500 with 325 people attending. How many tickets of each type were sold?

107. *Coin Problem* The total value of 23 coins consisting of pennies, nickels, and dimes is $1.51. If there are twice as many dimes as pennies, find the number of each coin.

108. *Maximum Area* Ted Saunders has 530 yd (yards) of fence to make a rectangular corral. What is the greatest possible area for the corral?

109. *Designing Pictures* A picture frame's outside dimensions are 16 in. by 20 in. The picture is surrounded by a mat with a uniform border. If the area of the picture is 250 in.^{2}, find the width of the uniform border.

110. *Throwing a Baseball* Pete throws a baseball straight up with an initial velocity of 75 ft/sec from a point 8 ft above level ground.

a. How long is the ball in the air?
b. What is the maximum height of the ball?

111. *Fund Raising* The senior class at Kennedy High School buys a graphing calculator for $65 to use as a raffle prize for a fund-raising project. The class charges $0.75 per raffle ticket.

a. How many tickets must be sold for the class to break even?
b. How many tickets must be sold for the class to earn $500?

112. *Travel Time* Gloria averages 48 mph on a cross-country trip from Boston to Seattle. How long has she been on the road when she has traveled 1200 mi?

GRACE MURRAY HOPPER

Grace Hopper (1906–1992) was a pioneer in the field of technology. With her Ph.D. in mathematics, she spent most of her career with computers and teaching. She started working with computers in 1944 with the Navy. In 1951 she was one of the inventors of COBOL, a programming language that is useful in processing business data. She even discovered the first "bug" in a computer, a real moth that wandered into a Univac computer.

Her career took her in and out of the Navy many times. In 1986 she was the oldest active duty officer. Admiral Hopper was awarded the National Medal of Technology in 1991. She has also received honorary degrees from over 40 colleges and universities. Her life was devoted to an application of mathematics, the development of technology.

chapter 2

SOLVING EQUATIONS AND MORE ON FUNCTIONS

Objective

Students will be able to solve equations by finding x-intercepts or intersections on graphs.

Motivate

Ask . . .

If (2, 3) is a point on the graphs of $f(x)$ and $g(x)$, can you conclude that $f(2) = g(2)$? **(Yes)**

Exploration Extensions

Repeat the Exploration for another function, such as $y = |x| - |2x - 3|$.

BIBLIOGRAPHY

For students: *Aha! Insight,* Martin Gardner. W. H. Freeman, 1978. Available through Dale Seymour Publications (XS02250).

For teachers: *The 7 Ways of Knowing: Understanding Multiple Intelligences,* David Lazear. Skylight Publishing, 1991.

2.1 SOLVING EQUATIONS GRAPHICALLY

Connection between $f(x) = 0$ and $y = f(x)$ • Solving an Equation by Finding x-Intercepts • Solving an Equation by Finding Intersections • Solving Graphically When Solving Algebraically Is Not Possible

Connection between $f(x) = 0$ and $y = f(x)$

There is a fundamental connection between the two equations

$$y = f(x) \quad \text{and} \quad f(x) = 0.$$

We ask that you explore this connection for

$$y = 2x^3 - 3x^2 - 3x + 2 \quad \text{and} \quad 2x^3 - 3x^2 - 3x + 2 = 0.$$

● **EXPLORATION** **Introducing the Concept: Zeros of a Function**

1. Graph $y = 2x^3 - 3x^2 - 3x + 2$ in the decimal window.[1]
2. Trace along the graph. Find the coordinates of the three points where the graph intersects the x-axis.

[1] Recall from Chapter 1 that $[-4.7, 4.7]$ by $[-3.1, 3.1]$ is the decimal window on most graphers. If it is not on your grapher, consult Appendixes A and B.

3. State three solutions of the equation $2x^3 - 3x^2 - 3x + 2 = 0$.
4. Confirm that these values are solutions.

Describe in Your Own Words

What is the relationship between the graph of $y = 2x^3 - 3x^2 - 3x + 2$ and the solutions to the equation $2x^3 - 3x^2 - 3x + 2 = 0$?

For the function $f(x) = 2x^3 - 3x^2 - 3x + 2$ in the Exploration, $f(0.5) = 0$. The function value is zero when $x = 0.5$. We say that $x = 0.5$ is a *zero of the function.* We also say that $x = 0.5$ is a solution to $2x^3 - 3x^2 - 3x + 2 = 0$ and that $(0.5, 0)$ is an *x-intercept* of the graph of f.

If a is a real number and $f(a) = 0$, then a is a **zero of the function** $y = f(x)$, and a **root** (solution) of $f(x) = 0$. The point $(a, 0)$ is an ***x*-intercept** of the graph of f. Sometimes the x-intercept is said to be a rather than $(a, 0)$.

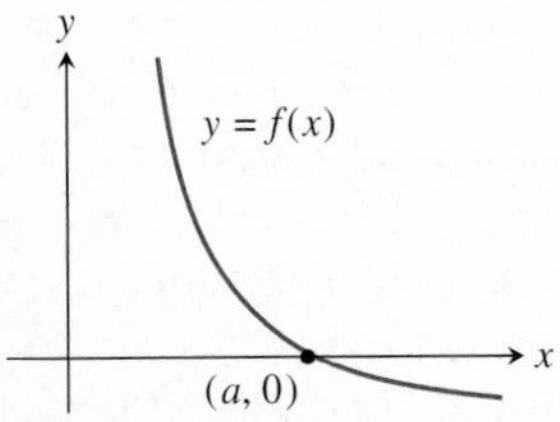

Figure 2.1

Alert

Students may not recognize the difference between a zero and a root. Functions have zeros, while one-variable equations have roots.

Fundamental Connections

Let a be a real number. The following three statements have the same meaning for the equation $f(x) = 0$, the function $y = f(x)$, and the graph of $y = f(x)$. (See Figure 2.1.)

1. The value $x = a$ is a *solution* or a *root* of $f(x) = 0$.
2. The value $x = a$ is a *zero* of the function $y = f(x)$.
3. The point $(a, 0)$ is an *x-intercept* of the graph of f.

Solving an Equation by Finding *x*-Intercepts

An equation in x, like $2x^2 - 6 = x$, has the form

$$g(x) = h(x),$$

where $g(x) = 2x^2 - 6$ and $h(x) = x$. To solve this equation graphically by finding the x-intercepts, rewrite it so that there is a zero on one side, that is,

$$g(x) - h(x) = 0,$$

and find the x-intercepts of the graph of $y = g(x) - h(x)$.

■ EXAMPLE 1 Finding *x*-Intercepts

Solve the equation $2x^2 - 6 = x$.

Solution

Solve Graphically

Rewrite $2x^2 - 6 = x$ as $2x^2 - 6 - x = 0$, and find the x-intercepts of $y = 2x^2 - 6 - x$. Figure 2.2 shows the "trace" cursor located at the two x-intercepts $(-1.5, 0)$ and $(2, 0)$. The two solutions of the equation are $x = -1.5$ and $x = 2$.

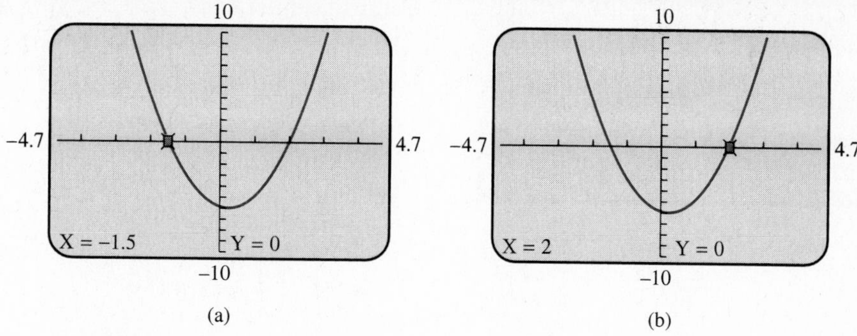

Figure 2.2 $y = 2x^2 - 6 - x.$

Support Numerically

Use a calculator or pencil and paper to check that $2x^2 - 6$ and x have the same value when you substitute $x = -1.5$ or $x = 2$. ■

> **Alert**
>
> Many students will assume that all solutions are contained in their default viewing window. Remind them that it is frequently necessary to zoom-out in order to see the general behavior of the graph, and then zoom-in to find more exact values of x at the intersection points.

More complicated equations, such as those involving absolute values and square roots, are just as easy to solve graphically as the equation in Example 1 even though they can be more difficult, or impossible, to solve algebraically.

■ EXAMPLE 2 Finding x-Intercepts

Solve $|2x + 3| = 2$.

Solution

Solve Graphically

Rewrite $|2x + 3| = 2$ as $|2x + 3| - 2 = 0$, and find the x-intercepts of $y = |2x + 3| - 2$. Figure 2.3 suggests that $x = -2.5$ and $x = -0.5$ are solutions to the equation.

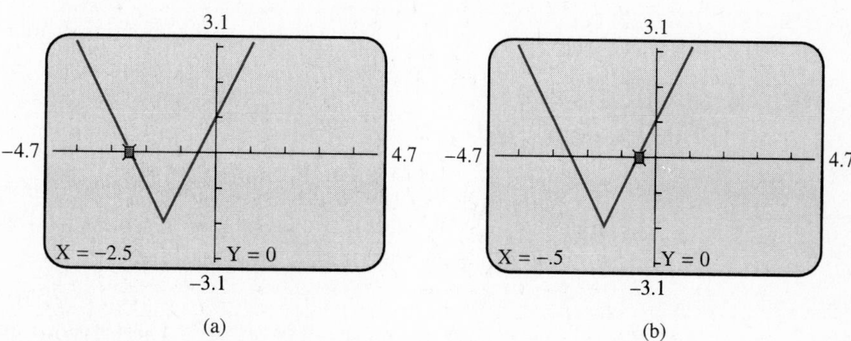

Figure 2.3 $y = |2x + 3| - 2.$

Teaching Note

Point out that when solving an equation by finding intersections, the value of y at an intersection point is irrelevant.

Confirm Numerically

$$|2x + 3| = 2$$
$$|2(-2.5) + 3| = 2 \quad \text{Substitute } x = -2.5.$$
$$2 = 2 \quad -2.5 \text{ checks.}$$
$$|2x + 3| = 2$$
$$|2(-0.5) + 3| = 2 \quad \text{Substitute } x = -0.5.$$
$$2 = 2 \quad -0.5 \text{ checks.}$$

■

Follow-up

Ask how one can determine whether or not a solution obtained by graphing is an exact solution to the equation. **(Substitute the value into the original equation.)**

Solving an Equation by Finding Intersections

To solve the equation

$$g(x) = h(x)$$

graphically by **finding intersections,** graph $y_1 = g(x)$ and $y_2 = h(x)$ in the same viewing window and find values of x that make $y_1 = y_2$. In other words, find the x-coordinates of the points of intersection of the graphs.

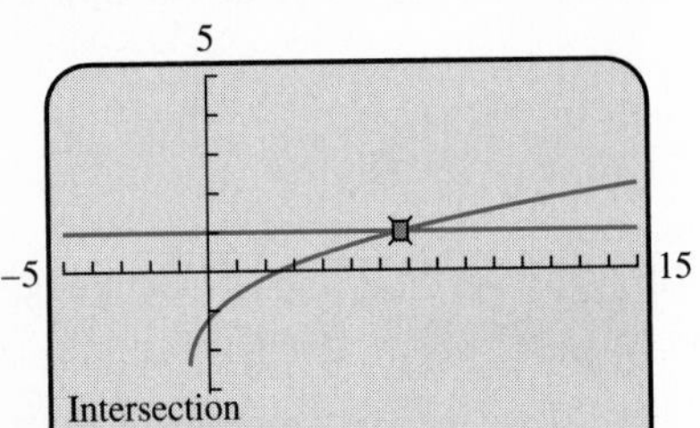

Figure 2.4 $y_1 = \sqrt{3x + 2} - \sqrt{x + 7}$ and $y_2 = 1$.

■ **EXAMPLE 3** **Finding Intersections**

Solve $\sqrt{3x + 2} - \sqrt{x + 7} = 1$.

Solution

Solve Graphically

Graph

$$y_1 = \sqrt{3x + 2} - \sqrt{x + 7} \quad \text{and} \quad y_2 = 1$$

in the same viewing window. There is only one point of intersection of the two graphs, which means that there is only one solution to the equation. (See Figure 2.4.)

A grapher[2] shows that the x-coordinate of the point of intersection is approximately 6.70.

The solution to the equation is approximately $x = 6.70$.

Support Numerically

Figure 2.5 shows that the value of y_1 is very close to 1 when $x = 6.70$. ■

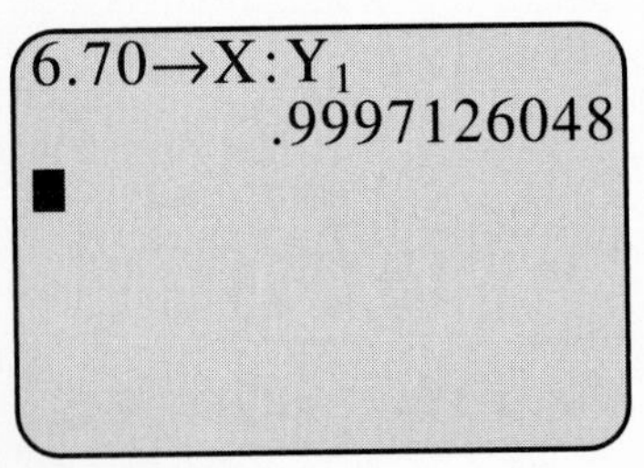

Figure 2.5

For some equations, finding x-intercepts and finding intersections work equally well. For others, one method works better than the other. Experience will suggest which technique to use.

[2]Use any graphical or numerical method you choose. Some graphers have "intersect" keys, as illustrated in Figure 2.4.

Assignment Guide

Day 1: Ex. 3–42, multiples of 3
Day 2: Ex. 43, 44, 47, 50, 51, 54, 57, 60, 61–69 odd

Cooperative Learning

Group Activity: Ex. 58

Notes on Exercises

Ex. 58 relates to Exercise 92 of Section 2.2.
Ex. 59 illustrates the fact that data may be modeled by a function that is not linear.
Ex. 67–70 provide an opportunity to explore the meaning of equal functions.

Ongoing Assessment

Self-Assessment: Ex. 9, 25, 29, 49, 59
Embedded Assessment: Ex. 44, 60

Solving Graphically

To solve $g(x) = h(x)$ graphically,

1. Graph $g(x) - h(x)$ and find all x-intercepts.
2. Graph $g(x)$ and $h(x)$ and find x-coordinates of all points of intersection.

When solutions are integers or certain fraction values, it is possible to select a viewing window so that the graphical solution can be displayed exactly. (See Example 2.) Most often, however, solving graphically or numerically yields solutions that are approximations. In working with approximations, follow this agreement unless directed otherwise.

Agreement about Approximate Solutions

Round approximate solutions to the nearest hundredth (0.01). For applications, round to a value that is reasonable for the context of the problem.

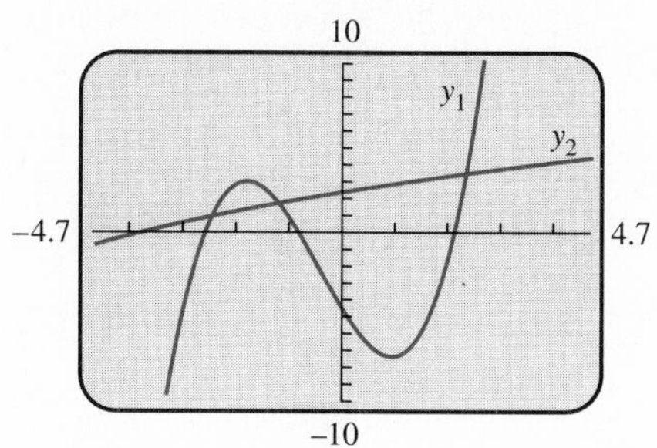

Figure 2.6

Solving Graphically When Solving Algebraically Is Not Possible

The beauty of graphical and numerical methods for solving equations is that they allow us to find sufficiently accurate solutions when a pencil-and-paper solution is difficult, or even impossible!

EXAMPLE 4 Solving Graphically

Solve the equation

$$x^3 + (\sqrt{2})x^2 - 5x - 5 = 5\sqrt{x + 8} - 3\sqrt{x + 16}.$$

Solution

Solve Graphically

Figure 2.6 shows graphs of

$$y_1 = x^3 + (\sqrt{2})x^2 - 5x - 5 \quad \text{and} \quad y_2 = 5\sqrt{x + 8} - 3\sqrt{x + 16}.$$

A grapher shows that $x = -2.50$, $x = -1.24$, and $x = 2.30$ are approximate solutions.

Support Numerically

Figure 2.7 shows that y_1 and y_2 are nearly equal for $x = -1.24$. Showing that $x = -2.50$ and $x = 2.30$ are approximate solutions is left as an exercise.

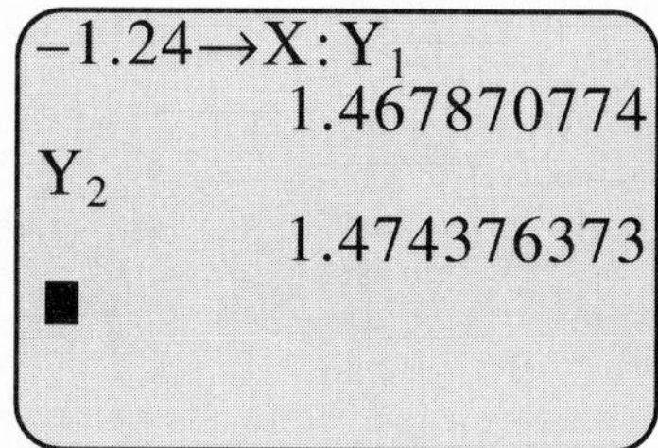

Figure 2.7

Quick Review 2.1

In Exercises 1–6, evaluate the expression.

1. $|2(-3) - 1|$
2. $|-2(-4) + 7|$
3. $\sqrt{(-2)^2}$
4. $\sqrt{(5.1)^2}$
5. $3(-2.5)^2 - 6$
6. $2(-1.5)^3 - 5(-1.5)^2 - 1$

In Exercises 7–10, evaluate the expression for the given values of the variable.

7. $|2x - 3|$, for $x = -2$ and $x = 2$
8. $\sqrt{3x - 2}$, for $x = 0$ and $x = 2$
9. $x + \dfrac{1}{x} - 2$ for $x = -2$ and $x = 2$
10. $(x + 2)^{-1} - x^{-1}$, for $x = -2$ and $x = -1$

In Exercises 11 and 12, write an equivalent expression without using absolute value.

11. $|6 - 3\pi|$
12. $|\sqrt{10} - 2\sqrt{5}|$

SECTION EXERCISES 2.1

In Exercises 1–8, rewrite the equation in the form $f(x) = 0$. Then write the equation $y = f(x)$ whose graph you would use to solve the equation by finding x-intercepts.

1. $|5x - 2| = 3x - 2$
2. $|x - 1| = x^2$
3. $|3x - 1| = \sqrt{x^2 + 1}$
4. $x^3 - 3x = x^2 + 7x$
5. $-4x + 3 = x^2 - 5^3$
6. $7x + x\sqrt{x} = 8$
7. $4x^2 = \sqrt{x^2 + 3}$
8. $|x^3 + 1| = \sqrt{x + 3}$

In Exercises 9–18, solve graphically by finding x-intercepts. Support your results numerically.

9. $|x - 1| - 3 = 0$
10. $|x + 2| - 1 = 0$
11. $|2x - 1| = 3$
12. $|8 - 4x| = 2$
13. $|x + 5| = |x - 3|$
14. $|x + 6| = |2x - 5|$
15. $\sqrt{x + 3} = 1$
16. $\sqrt{x + 3} = \sqrt{2x + 4}$
17. $\sqrt{2x + 3} = \sqrt{x + 8}$
18. $|x - 4| = \sqrt{2x + 3}$

In Exercises 19–26, solve graphically by finding intersections. Support your results numerically.

19. $|t - 8| = 2$
20. $|x + 1| = 4$
21. $|x - 2| = 5$
22. $|2t - 3| = \frac{1}{2}$
23. $|3 - 5x| = |-4|$
24. $6.8 = |2x + 1|$
25. $|0.5x + 3| = x^2 - 4$
26. $\sqrt{x + 7} = -x^2 + 5$

In Exercises 27–38, produce a graph that shows how many solution the equation has. Then find all solutions.

27. $2x - 5 = \sqrt{x + 4}$
28. $|3x - 2| = 2\sqrt{x + 8}$
29. $|2x - 5| = 4 - |x - 3|$
30. $\sqrt{x + 6} = 6 - 2\sqrt{5 - x}$
31. $\sqrt{x + 8} = x^2$
32. $|x| - 2 = x^2 - 5$
33. $2x - 3 = x^3 - 5$
34. $x + 1 = x^3 - 2x - 5$
35. $x + 2 = \dfrac{1}{x} - 2$
36. $2x + 3 = x^{-1} + 4$
37. $x^{-1} + x = 4x + 3$
38. $(x + 1)^{-1} = x^{-1} + x$

In Exercises 39–42, the two candidates for solutions were found algebraically. Does each check as a solution? If not, which one doesn't?

39. $\sqrt{x + 1} = \sqrt{2x} - 1$; $x = 0$, $x = 8$
40. $\sqrt{x} + 1 = \sqrt{3(x - 1)}$; $x = 1$, $x = 4$
41. $\sqrt{x} = x - 2$; $x = 1$, $x = 4$
42. $\sqrt{2x} = x - 4$; $x = 2$, $x = 8$

In Exercises 43 and 44, solve the problems by interpreting the graphs.

43. The graphs in the two viewing windows shown here can be used to solve the equation $3\sqrt{x + 4} = x^2 - 1$ graphically.

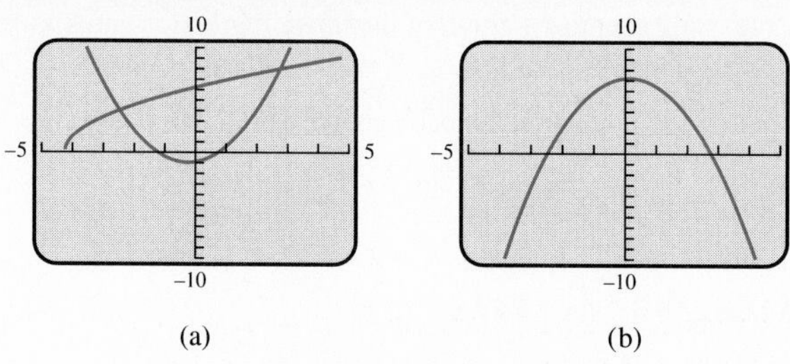

For Exercise 43

a. Viewing window (a) illustrates the intersection method for solving. Identify the two functions that are graphed, and duplicate the graphs on your grapher.

b. Viewing window (b) illustrates the x-intercept method for solving. Identify the function that is graphed, and duplicate the graph on your grapher.

c. How are the intersection points in (a) related to the x-intercepts in (b)?

44. The darker curves in the viewing window shown here are the graphs of $y_1 = 4x + 5$ and $y_2 = x^3 + 2x^2 - x + 3$. The lighter curve is the graph of $y_3 = -x^3 - 2x^2 + 5x + 2$.

a. Write an equation that can be solved by considering the points of intersection of y_1 and y_2.

b. Write an equation that can be solved by considering the x-intercepts of the graph of y_3.

c. The points of intersection of the graphs of y_1 and y_2 appear to be on vertical lines through the x-intercepts of the graph of y_3. Confirm algebraically that this is true.

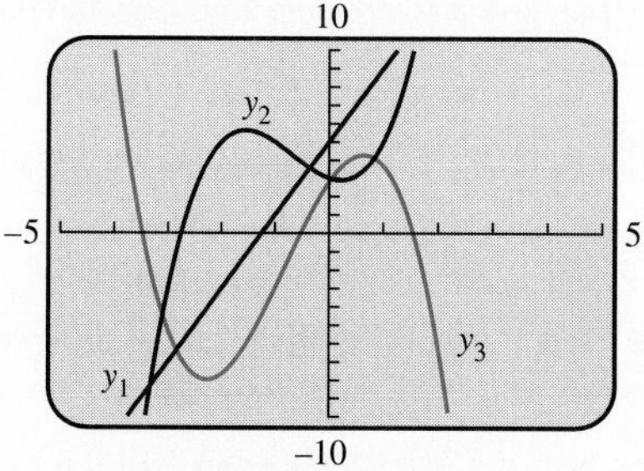

For Exercise 44

In Exercises 45–50, solve graphically for zeros of the function.

45. $f(x) = |2x - 1| - 2$

46. $g(x) = |x^2 - 4| - 3$

47. $h(x) = 2x^2 - x - 1$

48. $f(x) = \sqrt{2x + 3} - 1$

49. $f(x) = 2\sqrt{x + 3} - 0.25(x + 3)^2$

50. $h(x) = x^2 - 1.25x - 3.125$

In Exercises 51–56, solve the equation graphically.

51. $\sqrt{x + 2} = x + 1$

52. $\sqrt{x + 1} = x - 2$

53. $x + 1 - 2\sqrt{x + 4} = 0$

54. $\sqrt{x - 1} = (x - 1)^2$

55. $\sqrt{x} + x = 1$

56. $\sqrt{x} + \sqrt{2x} = 4$

In Exercises 57 and 58, we study free fall. An object in free fall (ignoring air resistance) takes t seconds to fall d feet. An equation that models this situation is $t = \frac{1}{4}\sqrt{d}$.

57. *Free Fall of a Smoke Bomb* At the Oshkosh, Wisc., air show, Jake Trouper drops a smoke bomb to signify the official beginning of the show.

a. How long does it take the smoke bomb to fall 180 ft?

b. If the smoke bomb is in free fall for 12.5 sec after it is dropped, how high was the airplane when the smoke bomb was dropped?

58. *Parachute Jump* Parachute jumpers Beth and Kristin free-fall for 15 sec before linking to complete a formation. They remain in formation for 3 sec and separate for 4 sec before pulling the rip cord.

a. The viewing window shows four graphs related to the situation. Explain each graph.

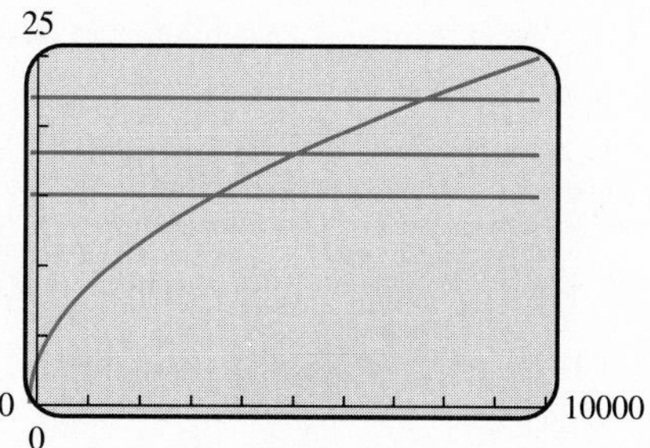

For Exercise 58

b. Writing to Learn Is it safe for Beth and Kristin to jump from 8000 ft and perform the formation routine? Explain.

In Exercises 59 and 60, use the data to solve the problem.

59. *Air Travel* The number of passengers (in millions) that used U.S. planes here and abroad is given in Table 2.1. The equation $y = 2.5x^2 - 45.6x + 651.0$ models the data with $x = 0$ for 1980, $x = 1$ for 1981, etc.

Table 2.1 U.S. Air Travel

Year	Passengers (millions)
1988	441.2
1989	443.6
1990	456.6
1991	445.7
1992	463.0
1993	468.1
1994	509.0

Source: FAA, reported in *USA Today,* February 28, 1995.

a. Graph a scatter plot of the data.

b. Superimpose a graph of $y = 2.5x^2 - 45.6x + 651.0$ on the scatter plot.

c. Use the model to predict when the number of passengers traveling on U.S. planes would be 575 million.

60. *Government Program* The average monthly benefit per person on the food-stamp program is given in Table 2.2.

a. Graph a scatter plot of the data.

b. Find the linear regression equation for the data.

Table 2.2 Food-Stamp Program

Year	Monthly Benefit (dollars)
1975	50
1980	55
1985	60
1990	63
1993	68

Source: Office of Management and Budget, U.S. House Committee on Ways and Means, U.S. Department of Agriculture, reported in *USA Today,* March 7, 1995.

c. Superimpose a graph of the linear regression equation on the scatter plot.

d. Use the linear regression equation to predict when the average monthly benefit per person receiving food stamps would be $75.

EXTENDING THE IDEAS

In Exercises 61 and 62, you may find a grapher helpful.

61. Consider the equation $3\sqrt{2x + 1} - \sqrt{x + 6} = c$.

a. Find a value of c for which this equation has no solution. (There are many such values.)

b. Find a value of c for which this equation has a unique solution. (There are many such values.)

62. Consider the equation $4\sqrt{x + 2} = 2x + c$.

a. Find a value of c for which this equation has two solutions. (There are many such values.)

b. Find a value of c for which this equation has no solution. (There are many such values.)

c. For what values of c do you think there is exactly one solution? Give a convincing argument.

In Exercises 63 and 64, you may find your grapher helpful. Find a value for c for which the equation has **(a)** one solution and **(b)** two solutions. (There are many such values for both (a) and (b).)

63. $\sqrt{x^2 + \sqrt{x + 4}} = c$ **64.** $(\sqrt{x + 3} + x)^2 = c$

In Exercises 65 and 66, find a specific value of c so that the equation $f(x) = c$ has three solutions. (There are many such values.) For your choice of c find the three solutions.

65. $f(x) = x^3 + 2x^2 - x - 2$

66. $f(x) = x^3 + x^2 - 3x - 5$

In Exercises 67 and 68, discuss what it means for two functions to be equal. Then give a convincing argument that the two functions are equal.

67. $f(x) = \begin{cases} x & \text{if } x < 1 \\ 2 - x & \text{if } x \geq 1 \end{cases}$ and $g(x) = -|x - 1| + 1$

68. $f(x) = \sqrt{x^2}$ and $g(x) = |x|$

In Exercises 69 and 70, discuss what is required to show that the two functions are not equal. Then show that the two functions are not equal.

69. $f(x) = \sqrt{x^2 + 2x + 1}$ and $g(x) = x + 1$

70. $f(x) = |-(-x)|$ and $g(x) = -|-x|$

2.2 SOLVING EQUATIONS ALGEBRAICALLY

Objective

Students will be able to recognize quadratic equations and solve them by factoring, completing the square, or using the quadratic formula.

Motivate

Ask . . .
If $(x + 3)(x - 2) = 0$, what are the possible values of x? **($x = -3$ or $x = 2$)**

Standard Form of a Quadratic Equation • Solving by Extracting Square Roots • Solving by Completing the Square • Solving by the Quadratic Formula • Solving Equations by Factoring • The Discriminant • Applications

Standard Form of a Quadratic Equation

A linear equation in x is an equation that can be written in the form $ax + b = 0$. A *quadratic equation* may be defined in a similar way.

Definition: Quadratic Equation in One Variable

A **quadratic equation in x** is an equation that can be written in the form

$$ax^2 + bx + c = 0,$$

where a, b, and c are real numbers and $a \neq 0$. We call $ax^2 + bx + c = 0$ the **standard form of a quadratic equation.**

The Prerequisite Chapter reviews how properties of equality can be used to solve linear equations algebraically. These same properties can be used to solve other types of equations.

Solving by Extracting Square Roots

The simplest quadratic equations are those of the form $x^2 = k$ where k is a positive real number. We solve this type of equation by *extracting square roots.*

Extracting Square Roots

To **extract square roots** observe that the equation $x^2 = k$, where $k > 0$, has two real number solutions:

$$x = \sqrt{k} \quad \text{and} \quad x = -\sqrt{k}.$$

These two solutions can be written as $x = \pm\sqrt{k}$. (See Figure 2.8.)

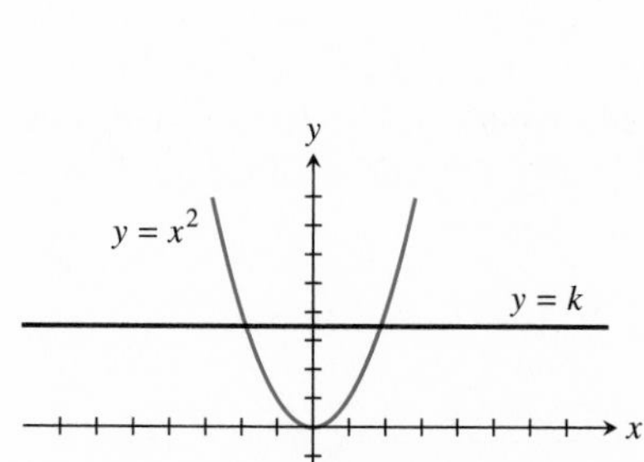

Figure 2.8 To solve the quadratic equation $x^2 = k$, $k > 0$, we find that intersections of $y_1 = x^2$ and $y_2 = k$ occur at $x = -\sqrt{k}$ and $x = \sqrt{k}$.

EXAMPLE 1 Extracting Square Roots

a. If $x^2 = 81$, then $x = 9$ or $x = -9$. We also write $x = \pm 9$.
b. If $10x^2 = 16$, then $x^2 = 1.6$, and $x = \pm\sqrt{1.6}$.

Alert

Many students will forget the ± sign when solving by extracting square roots. Emphasize that these problems usually have *two* solutions and that both are important.

The equation $(2x - 1)^2 = 8$ has the form $u^2 = k$ where $u = 2x - 1$. An equation of this form can also be solved by extracting square roots, but this requires some additional steps.

■ **EXAMPLE 2 Solving by Extracting Square Roots**

Solve $(2x - 1)^2 = 8$ algebraically.

Solution

Solve Algebraically

$$(2x - 1)^2 = 8$$

$$2x - 1 = \pm\sqrt{8} \qquad \text{Extract square roots.}$$

$$2x = 1 \pm \sqrt{8}$$

$$x = \frac{1 \pm \sqrt{8}}{2} \qquad \text{Divide by 2 for exact solutions.}$$

$$x = 1.914 \cdots \quad \text{or} \quad -0.914 \cdots$$

```
(1+√8)/2→X:(2X-1
)²
                 8
(1-√8)/2→X:(2X-1
)²
                 8
```

Figure 2.9

Support Numerically

Figure 2.9 suggests that for

$$x = \frac{1 \pm \sqrt{8}}{2},$$

the expression $(2x - 1)^2$ has the value 8. ■

Solving by Completing the Square

To solve a quadratic equation by extracting square roots we need an equation in the form $u^2 = k$. The expression on the left side contains the variable and has to be a perfect square. If it is not a perfect square, the technique known as *completing the square* might be used.

Completing the Square

To **complete the square** for the expression $x^2 + bx$, add $(b/2)^2$. The completed square is

$$x^2 + bx + \left(\frac{b}{2}\right)^2 = \left(x + \frac{b}{2}\right)^2.$$

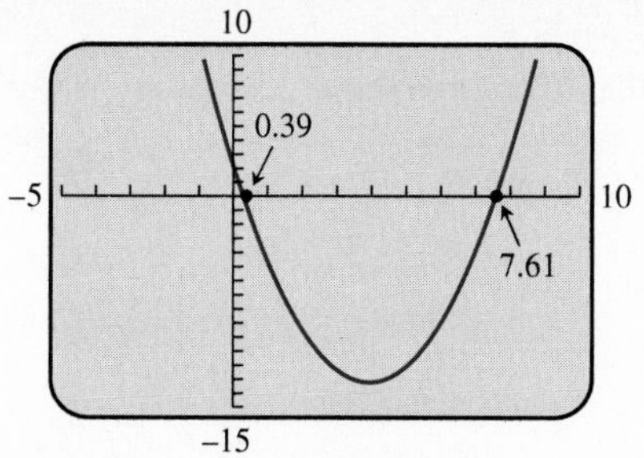

Figure 2.10 The x-intercepts of $y_1 = x^2 - 8x + 3$ are approximately 0.39 and 7.61.

EXAMPLE 3 Solving by Completing the Square

Solve $x^2 - 8x + 3 = 0$ by completing the square.

Solution

Solve Algebraically

$$x^2 - 8x + 3 = 0$$
$$x^2 - 8x = -3$$
$$x^2 - 8x + (-4)^2 = -3 + (-4)^2 \quad \text{Add } (-8/2)^2 \text{ to complete the square.}$$
$$(x - 4)^2 = 13 \quad \text{Factor.}$$
$$x - 4 = \pm\sqrt{13} \quad \text{Extract square roots.}$$
$$x = 4 \pm \sqrt{13} \quad \text{Add 4.}$$
$$x = 7.605\ldots \quad \text{or} \quad 0.394\ldots$$

Support Graphically

Figure 2.10 shows that the x-intercepts of $y = x^2 - 8x + 3$ are approximately 0.39 and 7.61. ■

When the coefficient of x^2 is not 1, we divide both sides of the equation by that number before completing the square. Example 4 illustrates.

EXAMPLE 4 Solving by Completing the Square

Solve $4x^2 - 20x + 17 = 0$ by completing the square.

Solution

Solve Algebraically

$$4x^2 - 20x + 17 = 0$$
$$4x^2 - 20x = -17 \quad \text{Subtract 17.}$$
$$x^2 - 5x = -\frac{17}{4} \quad \text{Divide by 4.}$$
$$x^2 - 5x + \left(-\frac{5}{2}\right)^2 = -\frac{17}{4} + \left(-\frac{5}{2}\right)^2 \quad \text{Add } \left(-\frac{5}{2}\right)^2.$$
$$\left(x - \frac{5}{2}\right)^2 = -\frac{17}{4} + \frac{25}{4} \quad \text{Factor.}$$
$$x - \frac{5}{2} = \pm\sqrt{\frac{8}{4}} \quad \text{Extract square roots.}$$
$$x = \frac{5}{2} \pm \sqrt{2} \quad \text{Add 5/2.}$$
$$x = 3.914\ldots \quad \text{or} \quad 1.085\ldots$$

Support Graphically

Graph $y = 4x^2 - 20x + 17$ and show that the x-intercepts are approximately 1.09 and 3.91. ■

Solving by the Quadratic Formula

When completing the square to solve a general quadratic equation in standard form, we obtain the *quadratic formula.*

$$ax^2 + bx + c = 0$$

$$ax^2 + bx = -c \qquad \text{Subtract } c.$$

$$x^2 + \frac{b}{a}x = -\frac{c}{a} \qquad \text{Divide by } a.$$

$$x^2 + \frac{b}{a}x + \left(\frac{b}{2a}\right)^2 = -\frac{c}{a} + \left(\frac{b}{2a}\right)^2 \qquad \text{Add } \left(\frac{b}{2a}\right)^2.$$

$$\left(x + \frac{b}{2a}\right)^2 = \frac{b^2 - 4ac}{4a^2} \qquad \text{Combine terms (right side).}$$

$$x + \frac{b}{2a} = \pm\sqrt{\frac{b^2 - 4ac}{4a^2}} \qquad \text{Extract square roots.}$$

$$x = -\frac{b}{2a} \pm \sqrt{\frac{b^2 - 4ac}{4a^2}} \qquad \text{Subtract } \frac{b}{2a}.$$

$$x = \frac{-b \pm \sqrt{b^2 - 4ac}}{2a}$$

Alert

Point out that the quadratic *formula* is used to solve quadratic *equations.* Some students confuse these concepts.

Quadratic Formula

The solutions of the quadratic equation $ax^2 + bx + c = 0$, where $a \neq 0$, are given by the **quadratic formula**

$$x = \frac{-b \pm \sqrt{b^2 - 4ac}}{2a}.$$

■ EXAMPLE 5 Using the Quadratic Formula

Solve $3x - x^2 = 1$ using the quadratic formula.

Solution

Solve Algebraically

Write the equation in standard form.

$$-x^2 + 3x - 1 = 0$$

We see that $a = -1$, $b = 3$, and $c = -1$. The quadratic formula yields

$$x = \frac{-b \pm \sqrt{b^2 - 4ac}}{2a}$$

$$x = \frac{-3 \pm \sqrt{3^2 - 4(-1)(-1)}}{2(-1)}$$

$$x = \frac{-3 \pm \sqrt{9 - 4}}{-2}$$

$$x = \frac{3 \pm \sqrt{5}}{2}$$

$$x = 2.618 \cdots \quad \text{or} \quad x = 0.381 \cdots$$

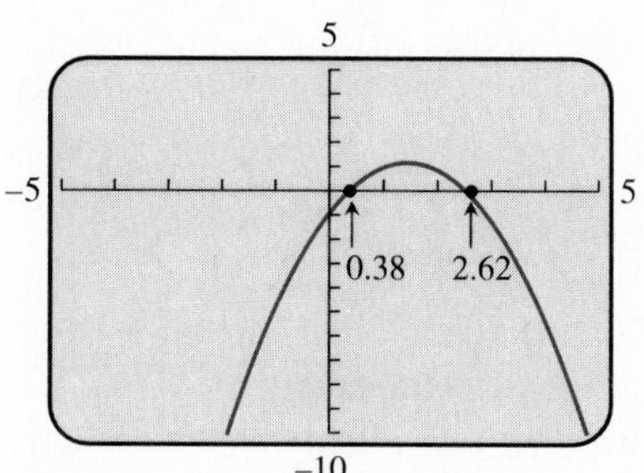

Figure 2.11

Support Graphically

The graph of $y = -x^2 + 3x - 1$ in Figure 2.11 shows that the x-intercepts are approximately 0.38 and 2.62. ■

Solving Equations by Factoring

Another method, often used for solving quadratic equations, but also valid for other equations, is **solving by factoring.** This method uses the **zero factor property,** which says that if the product of two expressions is zero, one of the expressions must be zero.

> **Zero Factor Property**
>
> If $ab = 0$, then $a = 0$ or $b = 0$.

Here is how the zero factor property can be used.

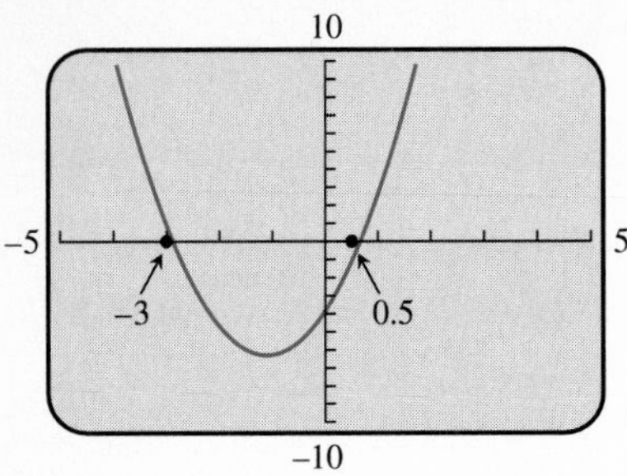

Figure 2.12

EXAMPLE 6 Solving by Factoring

Solve $2x^2 + 5x - 3 = 0$ by factoring.

Solution

Solve Algebraically

$$2x^2 + 5x - 3 = 0$$

$$(2x - 1)(x + 3) = 0 \qquad \text{Factor } 2x^2 + 5x - 3.$$

$$2x - 1 = 0 \quad \text{or} \quad x + 3 = 0 \qquad \text{The zero factor property}$$

$$x = \frac{1}{2} \quad \text{or} \quad x = -3$$

The solutions to the equation are $x = \frac{1}{2} = 0.5$ and $x = -3$.

Support Graphically

The graph of $y = 2x^2 + 5x - 3$ in Figure 2.12 shows that the x-intercepts are $x = -3$ and $x = 0.5$.

The Discriminant

The radicand $b^2 - 4ac$ in the quadratic formula discriminates among the following three cases, which is why it is called the **discriminant.**

When the discriminant is positive, there are two real number solutions. When $b^2 - 4ac$ is negative, there is no real number solution. When it is zero, there is one real number solution, namely

$$\frac{-b \pm \sqrt{b^2 - 4ac}}{2a} = \frac{-b + 0}{2a} = \frac{-b}{2a}.$$

This solution, $x = \dfrac{-b}{2a}$, is called a **double root** of the equation.

EXAMPLE 7 Using the Discriminant

a. The discriminant of $x^2 - 5x + 2$ is

$$b^2 - 4ac = (-5)^2 - 4(1)(2)$$
$$= 25 - 8 = 17.$$

The equation $x^2 - 5x + 2 = 0$ has two real number solutions. (See Figure 2.13a.)

b. The discriminant of $2x^2 - 3x + 5$ is

$$b^2 - 4ac = (-3)^2 - 4(2)(5)$$
$$= 9 - 40 = -31.$$

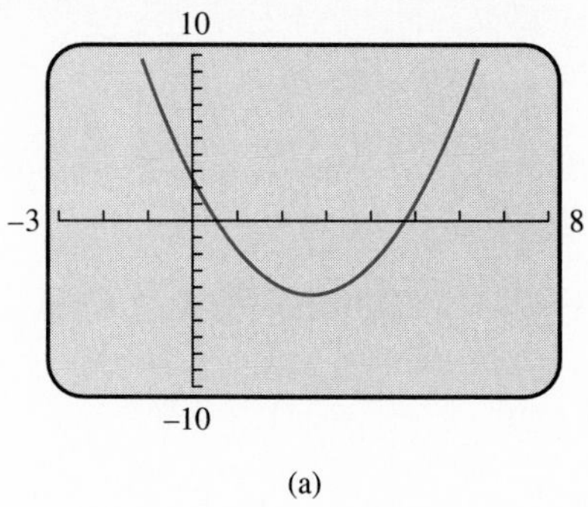

(a)

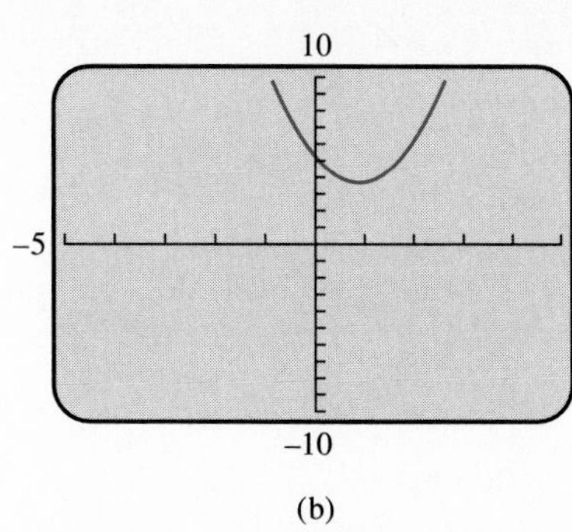

(b)

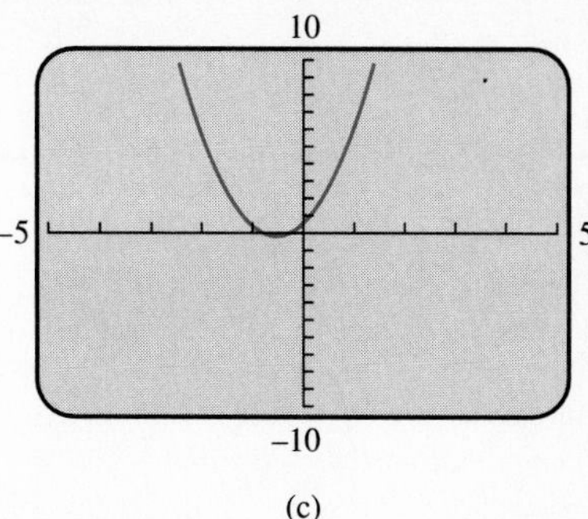

(c)

Figure 2.13 The graphs of (a) $y_1 = x^2 - 5x + 2$, (b) $y_2 = 2x^2 - 3x + 5$, and (c) $y_3 = 2.7x^2 + 3.6x + 1.2$ suggest two, zero, and one real number root, respectively. In (c), the fact that the discriminant is 0 confirms algebraically that there is a double root and not two distinct roots that are too close together to be seen.

The equation $2x^2 - 3x + 5 = 0$ has no real number solutions. (See Figure 2.13b.)

c. The discriminant of $2.7x^2 + 3.6x + 1.2$ is

$$\begin{aligned} b^2 - 4ac &= (3.6)^2 - 4(2.7)(1.2) \\ &= 0. \end{aligned}$$

The equation $2.7x^2 + 3.6x + 1.2 = 0$ has a double root. (See Figure 2.13c.) ■

Applications

The movement of an object that is propelled vertically, but then subject only to the force of gravity, is an example of **projectile motion.**

Teaching Notes

The model for projectile motion is important and is needed in order to solve several of the exercises.

A Model for Projectile Motion

Suppose an object is launched vertically from a point s_0 feet above the ground with an initial velocity of v_0. A model for the vertical position s (in feet) of the object t seconds after it is launched, is

$$s = -16t^2 + v_0 t + s_0.$$

This projectile-motion equation assumes that there is no air resistance. It can be used to model both free fall and vertical launch problems. In a free fall application the initial velocity v_0 is zero.

EXAMPLE 8 APPLICATION: Modeling Projectile Motion in a Throwing Machine

Notes on Examples

Example 9 is important because it illustrates an application of the Pythagorean theorem and because it illustrates a situation where a solution to an equation must be discarded because it does not make sense in the context of the problem.

The Williamstown Little League team has a throwing machine that propels a baseball upward from ground level with an initial velocity of 55 ft/sec.

a. Does the ball reach a height of 52 ft?

b. How many seconds after the ball is propelled is it 42 ft above the ground?

Solution

Model

Use the equation $s = -16t^2 + v_0t + s_0$, where $v_0 = 55$ ft/sec and $s_0 = 0$ ft. The modeling equation becomes

$$s = -16t^2 + 55t.$$

Solve Graphically

a. We need to solve the equation $-16t^2 + 55t = 52$. We graph $y_1 = -16x^2 + 55x$ and $y_2 = 52$ in the same viewing window. Figure 2.14 shows that the graphs of y_1 and y_2 do not intersect. This tells us that the ball does not reach a height of 52 ft.

b. We need to solve the equation $-16t^2 + 55t = 42$. Figure 2.14 shows that the graphs of $y_1 = -16t^2 + 55t$ and $y_3 = 42$ intersect in two points, so there are two solutions. Using the grapher we find that the intersection points occur approximately at $t = 1.15$ and $t = 2.29$.

REMARK

Notice that by viewing one figure, we can answer several different questions. The algebraic work requires an analysis of a separate equation for each question.

Follow-up

Ask whether it is always necessary to discard negative solutions in application problems. **(Usually, but not always.)**

Confirm Algebraically

a. Write the equation $-16t^2 + 55t = 52$ in the standard form

$$-16t^2 + 55t - 52 = 0.$$

Its discriminant

$$b^2 - 4ac = 55^2 - 4(-16)(-52)$$
$$= -303$$

is negative. Thus, there is no solution to the equation.

b. Write the equation $-16t^2 + 55t = 42$ in the standard form

$$16t^2 - 55t + 42 = 0$$

and use the quadratic formula:

$$t = \frac{-(-55) \pm \sqrt{(-55)^2 - 4(16)(42)}}{2(16)}$$

$$t = 2.292\ldots \quad \text{or} \quad t = 1.145\ldots$$

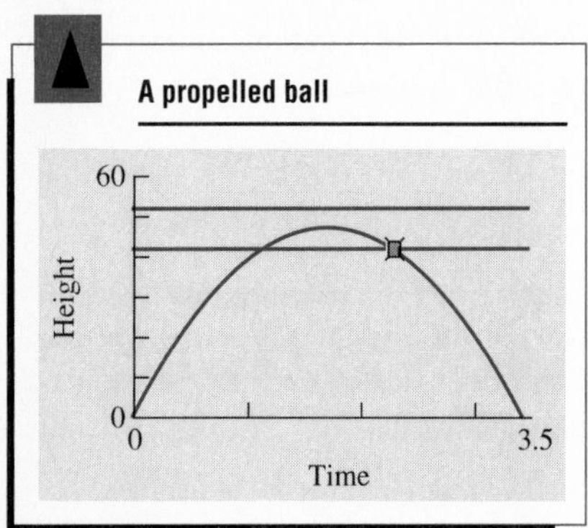

Intersection: $x = 2.2924237$; $y = 42$

Figure 2.14 A graph of $y_1 = -16t^2 + 55t$, $y_2 = 52$, and $y_3 = 42$.

Interpret

The ball does not reach a height of 52 ft. It reaches 42 ft in about 1.15 sec (going up) and again at about 2.29 sec (on the way down).

REMARK

distance = rate × time

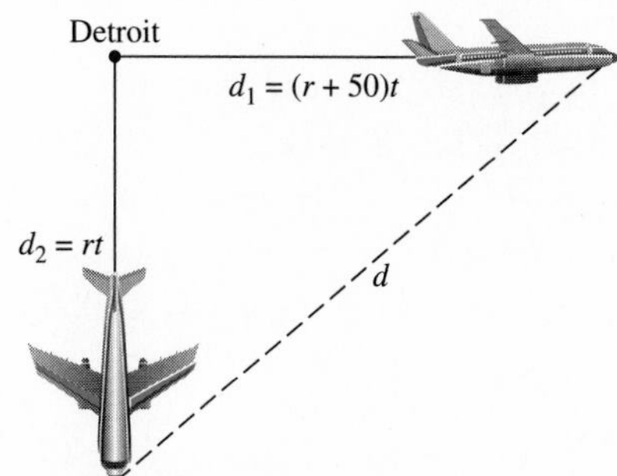

Figure 2.15

Assignment Guide

Day 1: Ex. 3–51, multiples of 3, 16, 32

Day 2: Ex. 53, 56, 59, 62, 65, 68, 71, 73, 76, 77, 79, 83, 86–89, 91

Cooperative Learning

Group Activity: Ex. 84–85

Notes on Exercises

Ex. 15–20 illustrate how a geometric model is related to the concept of completing the square.

Ex. 53–58 provide practice with discriminants.

Ex. 73 and 76 require students to use the Pythagorean theorem.

Ongoing Assessment

Self-Assessment: Ex. 7, 11, 25, 35, 47, 59, 75

Embedded Assessment: Ex. 16, 32

EXAMPLE 9 APPLICATION: Finding Rates of Travel

Two airplanes leave Detroit at the same time. The Boeing 737 flies due east toward New York, and the DC9 flies due south toward Atlanta. The two aircraft travel at constant speeds with the Boeing 737 flying 50 mph faster than the DC9. After flying for 1.5 h, the airplanes are 870 mi apart. How fast are the two airplanes flying?

Solution

Model

The diagram in Figure 2.15 models this problem situation. As time increases, the lengths of all three sides of the triangle increase. Applying the Pythagorean theorem, we get

$$(\text{distance south})^2 + (\text{distance east})^2 = (\text{distance apart})^2$$

$$\begin{aligned} r &= \text{rate of the slower southbound airplane} \\ r + 50 &= \text{rate of the eastbound Boeing 737} \\ 1.5 &= \text{time traveled} \\ 1.5r &= \text{distance of southbound DC9} \\ 1.5(r + 50) &= \text{distance of eastbound Boeing 737} \end{aligned}$$

Solve Algebraically

$$\begin{aligned} (1.5r)^2 + [1.5(r + 50)]^2 &= 870^2 && \text{The equation to be solved} \\ 2.25r^2 + 2.25(r^2 + 100r + 2500) &= 756{,}900 \\ 4.5r^2 + 225r + 5625 &= 756{,}900 \\ 4.5r^2 + 225r - 751{,}275 &= 0 \end{aligned}$$

Using the quadratic formula, we get

$$r = \frac{-b \pm \sqrt{b^2 - 4ac}}{2a}$$

$$r = \frac{-225 \pm \sqrt{(225)^2 - 4(4.5)(-751275)}}{2(4.5)}$$

$$r = 384.359 \cdots$$

$$\text{or } r = -434.359 \cdots \quad \text{Reject } -434.359 \cdots \text{ because it has no meaning in this context.}$$

$$r + 50 = 434.359 \cdots$$

Interpret

The slower DC9 is traveling about 384 mph, and the faster 737 is traveling about 434 mph. ■

Table 2.3 Medicare Data

Year	Medicare Charges (millions of dollars)
1970	2,310
1980	9,011
1985	17,743
1989	26,274
1990	30,447

Source: U.S. Health Care Financing Administration, unpublished data.

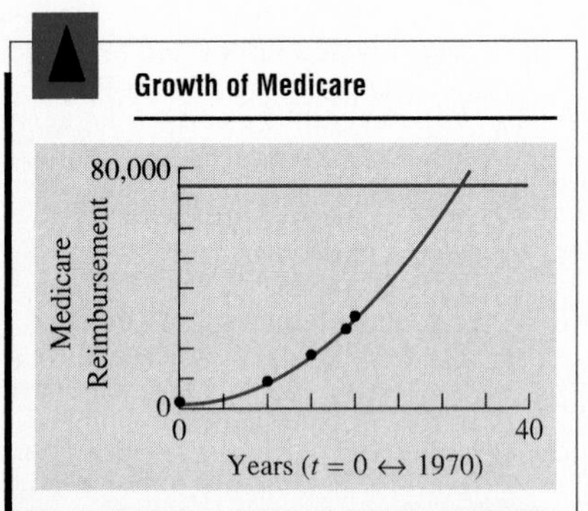

Figure 2.16 Graphs of $y_1 = 71.08x^2 - 47.77x + 2329.3$ and $y_2 = 75{,}000$.

■ EXAMPLE 10 APPLICATION: Modeling the Growth of Medicare

The Medicare charges by physicians between 1970 and 1990 (in millions of dollars) shown in Table 2.3 can be modeled by the quadratic function $f(x) = 71.08x^2 - 47.77x + 2329.3$, where $x = 0$ represents 1970, $x = 1$ represents 1971, and so on. Assuming the suggested trend continues, in what year will physician charges reach \$75 billion (or \$75,000 million)?

Solution

Solve Graphically

Reproduce the graphs in Figure 2.16 on your grapher, and show that $y_1 = 71.08x^2 - 47.77x + 2329.3$ is below $y_2 = 75{,}000$ when $x = 32$, and above 75,000 when $x = 33$.

Confirm Algebraically

Use the quadratic formula to solve $71.08x^2 - 47.77x + 2329.3 = 75{,}000$, or equivalently, $71.08x^2 - 47.77x - 72{,}670.7 = 0$ in standard form.

$$x = \frac{-b \pm \sqrt{b^2 - 4ac}}{2a}$$

$$x = \frac{47.77 \pm \sqrt{(-47.77)^2 - 4(71.08)(-72{,}670.7)}}{2(71.08)}$$

$$x = 32.312 \cdots$$

or $x = -31.640 \cdots$ Reject $-31.640 \cdots$ because it has no meaning in this context.

Interpret

If the pattern of charges continues, physicians will charge \$75 billion to Medicare by the 33rd year after 1970, or in the year 2003. ■

Quick Review 2.2

In Exercises 1–4, recall that $\sqrt{x^2} = |x|$. Use this property to rewrite each of the following expressions.

1. $\sqrt{a^2}$ **2.** $\sqrt{u^2}$

3. $\sqrt{20y^2}$ **4.** $\sqrt{x^2y + 3x^2}$

In Exercises 5–8, simplify the radical.

5. $\sqrt{81}$ **6.** $\sqrt{6^2}$

7. $\sqrt{54}$ **8.** $\sqrt{12}$

In Exercises 9 and 10, confirm each statement.

9. $\sqrt{x^2 + 9} \neq x + 3$

10. $\sqrt{x^2 + 6x + 9} \neq x + 3$

In Exercises 11–14, factor the polynomials completely.

11. $x^2 - 9$ **12.** $x^2 + 4x + 3$

13. $2x^2 + 5x - 3$ **14.** $2x^2 + 7x - 30$

SECTION EXERCISES 2.2

In Exercises 1–10, solve by extracting square roots. Support your work numerically or graphically.

1. $3x^2 = 15$

2. $4x^2 = 25$

3. $2(x - 5)^2 = 17$

4. $3(x + 4)^2 = 8$

5. $4(u + 1)^2 = 18$

6. $[(x + 1)^2 - 8]^2 = 81$

7. $v^2 - 5 = 8 - 2v^2$

8. $(x + 11)^2 = 121$

9. $(x - 2)^2 = (x + 1)^2$

10. $2(x + 3)^2 = (3x - 1)^2$

In Exercises 11–14, write the quadratic equation in standard form. Then solve the equation graphically.

11. $x(x + 3) = 5x + 2$

12. $8 - x^2 = 3x + x(x + 2)$

13. $x^2 - 3x = 12 - 3(x - 2)$

14. $\dfrac{(x - 2)^2}{3} = 3x$

In Exercises 15 and 16, use the diagram to solve the problem.

15. Connecting Algebra and Geometry Explain why the diagram is a model for $(x + \text{b})^2 = x^2 + 2bx + b^2$.

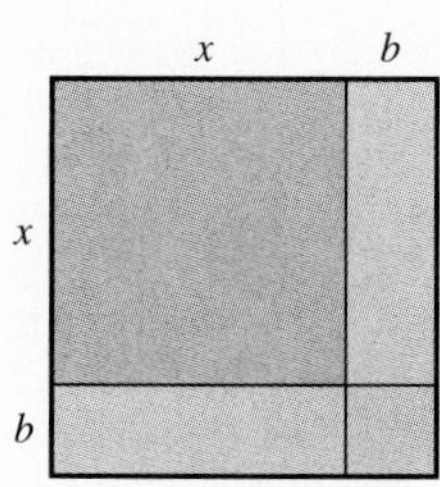

For Exercise 15

16. Connecting Algebra and Geometry Explain why the diagram is a model for $x^2 + bx$. What term must be added to complete the square with side length $x + b/2$?

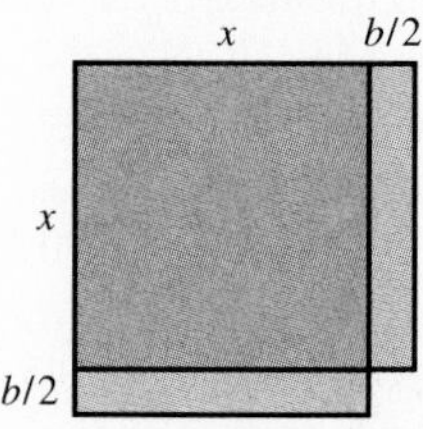

For Exercise 16

In Exercises 17–20, draw a diagram like those in Exercises 15 and 16 to explain completing the square of each expression. Then complete the square of each expression.

17. $x^2 + 8x$

18. $x^2 - 2x$

19. $x^2 - 9x$

20. $x^2 - 15x$

In Exercises 21–28, solve by completing the square.

21. $x^2 + 6x = 7$

22. $x^2 + 4x = 21$

23. $x^2 + 5x - 9 = 0$

24. $x^2 - 7x - \frac{3}{4} = 0$

25. $6x = x^2 - 3$

26. $4 - 6x = x^2$

27. $2x^2 - 5x + 2 = (x + 3)(x - 2) + 3x$

28. $3x^2 - 6x - 7 = x^2 + 3x - x(x + 1) + 3$

In Exercises 29 and 30, select the graph that shows the x-intercept method of solving the equation.

29. $3x^2 - 4x = 3$

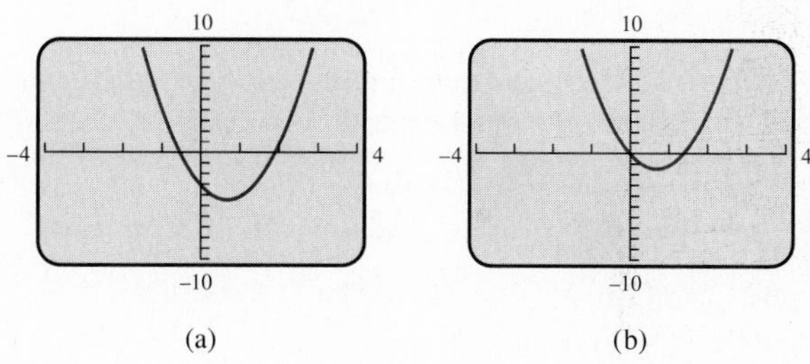

For Exercise 29

30. $2x^2 + 3x - 4 = x^2 + 5$

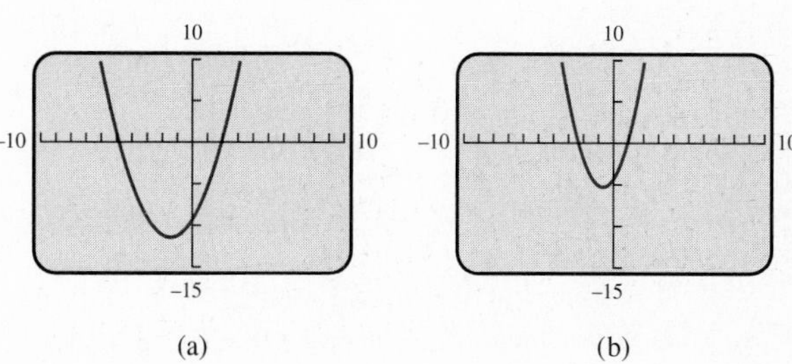

For Exercise 30

31. Writing to Learn Explain why $x^2 = c$ is a quadratic equation.

32. Writing to Learn Explain why $\sqrt{x} = c$ is not a quadratic equation.

In Exercises 33–42, solve using the quadratic formula.

33. $x^2 + x - 1 = 0$

34. $x^2 - 4x + 2 = 0$

35. $x^2 + 8x - 2 = 0$

36. $2x^2 - 3x + 1 = 0$

37. $x^2 - 2x = 7$

38. $3x + 4 = x^2$

39. $5 - x^2 = 8x$

40. $x^2 - 5 = \sqrt{3}x$

41. $x(x + 7) = 14$

42. $x^2 - 3x + 4 = 2x^2 - 7x - 8$

In Exercises 43–52, solve by factoring. Support your work graphically.

43. $x^2 + x - 2 = 0$

44. $x^2 - 5x + 6 = 0$

45. $x^2 - x - 20 = 0$

46. $x^2 - 4x + 3 = 0$

47. $2x^2 + 5x - 3 = 0$

48. $4x^2 - 8x + 3 = 0$

49. $x^2 - 8x = -15$

50. $x^2 + 4x - 3 = 2$

51. $x(2x - 5) = 12$

52. $x(2x - 1) = 10$

In Exercises 53–58, evaluate the discriminant to determine how many real number solutions the equation has. Support your answer graphically.

53. $2x^2 + 5x - 7 = 0$

54. $3x^2 - 2x + 1 = 0$

55. $x^2 = 3 + \sqrt{2}x$

56. $9x^2 + 42x + 50 = 1$

57. $100x^2 - 401x + 403 = 0$

58. $5x^2 + 11x + 9 = 3$

In Exercises 59 to 70, solve by a method of your choice. State the method that you use.

59. $(2x - 3)^2 + 1 = 5$

60. $3x^2 - 14x - 7 = 0$

61. $x^2 + 2x = 3$

62. $4 - (x + 2)^2 = 1$

63. $x^2 - \sqrt{2}x - 9 = 21$

64. $x^2 - 4x = 3$

65. $(x - 3)^2 = 4x^2$

66. $\sqrt{2}x^2 - 5x - 3\sqrt{2} = 0$

67. $x^2 - \sqrt{3}x = 2x^2 - 5$

68. $5t^2 = 7t + 6$

69. $a^2(x - 3)^2 = b$, for x

70. $u^2x^2 - v^2 = 0$, for x

In Exercises 71–81, solve the problem.

71. *Size of a Soccer Field* Several of the World Cup '94 soccer matches were played in Stanford University's stadium in Menlo Park, Calif. The field is 30 yd longer than it is wide, and the area of the field is 8800 yd^2. What are the dimensions of that soccer field?

72. *Size of a Football Field* The playing field at Soldier Field, home of the Chicago Bears, is 14 yd longer (including the end zones) than twice its width. The area of the field is 6360 yd^2. What are the dimensions of the football field?

73. *Height of a Ladder* John's paint crew knows from experience that its 18-ft ladder is particularly stable when the distance from the ground to the top of the ladder is 5 ft more than the distance from the building to the base of the ladder. In this position, how far up the building does the ladder reach?

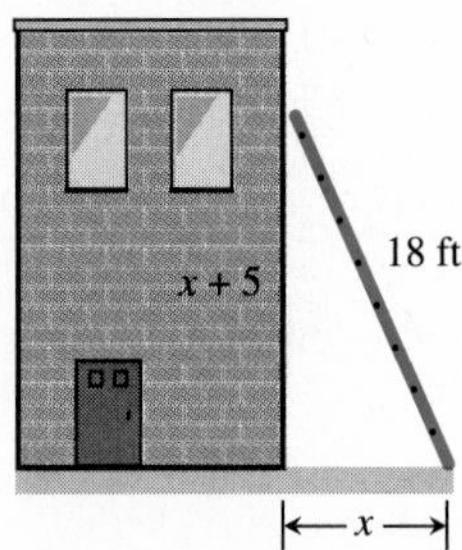

For Exercise 73

74. *Famine Relief Drop* The International Red Cross dropped foodstuffs from an airplane 1500 ft above the earth. How long did it take the food to reach the ground if the parachute did not open? (Hint: $t = \frac{1}{4}d$.)

75. *Sky Diving* Kirstin, Kelly, Michael, and Bill are planning their skydiving jump for the Biloxi Airshow. They need 8 sec from the time they leave the airplane to complete their formation. The judges expect them to hold their formation for 5 sec, and they need 3 sec to separate before pulling their rip cords no later than when they are 1500 ft above the ground. Is an airplane flying 4500 ft above the ground high enough for them to complete a successful jump? (Hint: $t = \frac{1}{4}d$.)

76. Connecting Algebra and Geometry Steve is building a box for strobe lights for his rock band. The end of the box is to be an isosceles right triangle whose hypotenuse is 18 in. How long are the other two sides of the triangle?

a. Solve numerically for the lengths of the other two sides.
b. Confirm the length of the other two sides algebraically.

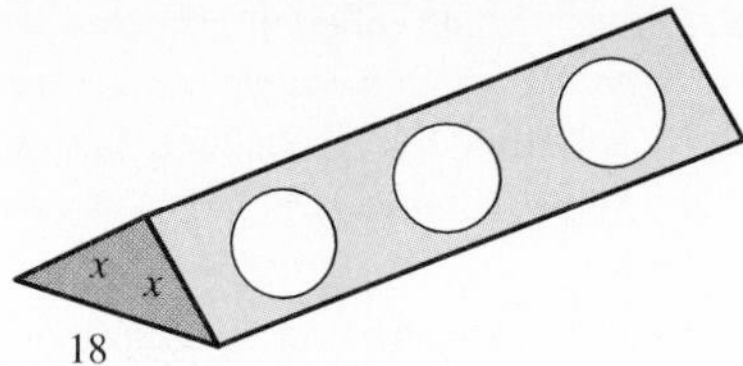

For Exercise 76

77. *Finding the Dimensions of a Norman Window* A Norman window has the shape of a square with a semicircle mounted on it. Find the width of the window if the total area of the square and the semicircle is to be 200 ft^2.

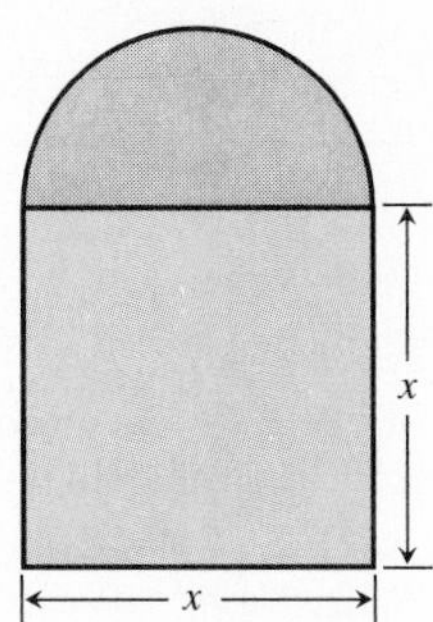

For Exercise 77

78. *Finding Distance* Throughout the Midwest, east-west and north-south county roads are 1 mi apart forming a grid of squares 1 mi on a side. Sarah drove the family Ford pickup from an intersection north at 30 mph at the same time that her father headed west from the same intersection at 10 mph in his John Deere tractor. How long will it be until their CB communication fades? (Their CB radios have a 2-mi range.)

79. *Finding Travel Rates* Two airplanes leave San Francisco at the same time. The Airbus flies due north toward Seattle and the DC10 flies due east toward Chicago. The two airplanes fly at a constant rate and the DC10 flies 30 mph faster than the Airbus due to a tailwind. After 2 h they are 1620 mi apart. How fast are each of the two airplanes flying?

80. *Expenses for Educational Services* The total expenditures for educational services between 1980 and 1989 (in billions of dollars) can be modeled by the equation

$$\text{expenditures} = 0.058x^2 + 1.63x + 16.06,$$

where $x = 0$ represents the year 1980.

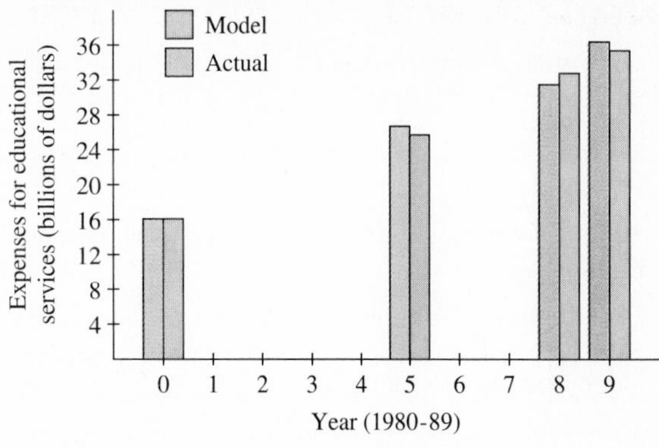

For Exercise 80

a. Draw a graph of this model for the time period 1980–2005.
b. Predict the total expenditures for the year 2005.
c. In what year will the expenditures reach \$278 billion?

81. *Total Revenue* The per unit price p (in dollars) of a popular toy by Tonka when x units (in thousands) are produced is modeled by the function

$$\text{price} = p = 12 - 0.025x.$$

The revenue (in thousands of dollars) is the product of the price per unit and the number of units (in thousands) produced. That is,

$$\text{revenue} = xp = x(12 - 0.025x).$$

a. State the dimensions of a viewing window that shows a graph of the revenue model for producing 0 to 100,000 units.
b. How many units should be produced if the total revenue is to be \$1,000,000.

In Exercises 82 and 83, use the equation $uv^2 - 4u^2v + 3u = 0$, where $u > 0$.

82. Solve the equation for v in terms of u.

83. Solve the equation for u in terms of v.

84. Group Learning Activity *Work in groups of three.* Does the equation $10x^2 - 21x + 5 = -6$ have double roots? One person should use a graphical method, another person an algebraic method, and a third person should check the results. Do the two results agree? Explain.

85. Group Learning Activity *Work in groups of three.* Show how to factor the expression $ax^2 + bx + c$ into a product of two linear factors. Under what circumstances will the two factors be identical?

EXTENDING THE IDEAS

In Exercises 86–88, suppose that $b^2 - 4ac > 0$ for the equation $ax^2 + bx + c = 0$.

86. Show that the sum of the two solutions of this equation is $-(b/a)$.

87. Show that the product of the two solutions of this equation is c/a.

88. The equation $2x^2 + bx + c = 0$ has two solutions, x_1 and x_2. If $x_1 + x_2 = 5$ and $x_1 \cdot x_2 = 3$, find the two solutions. (*Hint:* See Exercises 86 and 87.)

89. Match each equation with the graph that supports graphically the solution of the equation. Defend your match. (You do not need to know the values we used for b.)

a. $x^2 + bx + \dfrac{b^2}{4} = 0$ **b.** $2x^2 - b^2x + \dfrac{b^4}{32} = 0$

c. $3x^2 - bx + \dfrac{b^2}{4} = 0$

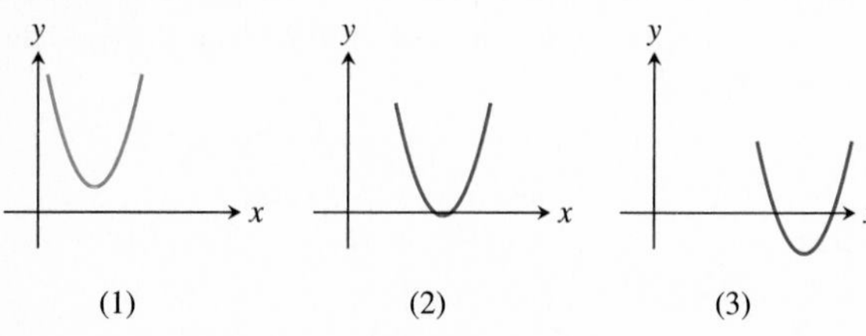

For Exercise 89

90. Writing to Learn Some equations can be solved by *extracting cube roots* in much the same manner that we extract square roots to solve quadratic equations. Explain what you think "solving an equation by extracting cube roots" means. Give an example.

91. Writing to Learn Describe how the graphs of $y_1 = 2x^2 + 3x - 2$ and $y_2 = 0.5x^2 + 5$ shown in the figure relate to solving the equation $2x^2 + 3x - 2 = 0.5x^2 + 5$.

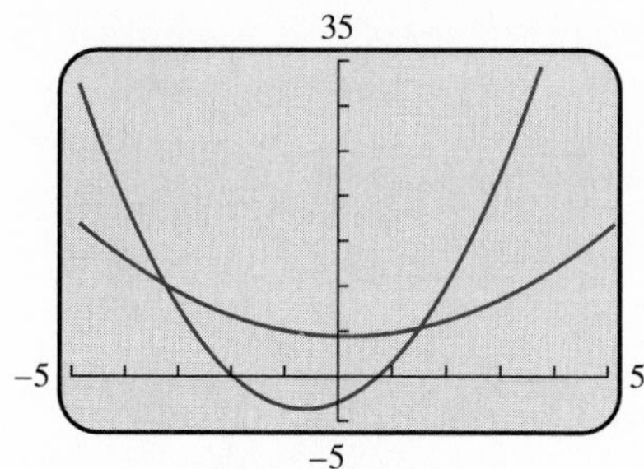

For Exercise 91

92. *Projectile Motion—Free Fall* An object is dropped with no initial velocity from an airplane flying overhead.

a. Explain how the distance the object falls is modeled by the equation $y = 16t^2$.

b. Solve for t in part a, and explain how the resulting equation relates to the parachute-jump problem in Exercise 58 of Section 2.1.

2.3 SOLVING ABSOLUTE VALUE AND RADICAL EQUATIONS AND INEQUALITIES

Objective

Students will be able to solve equations and inequalities involving radical expressions or absolute values.

Motivate

Ask . . .

On a number line, what are the possible values of x if the distance between x and 5 is 2? (**$x = 3$ or $x = 7$**)

Absolute Value Equations • Solving Radical Equations • Extraneous Solutions • Solving Inequalities • Solving Absolute Value Inequalities • Solving Radical Inequalities • Applications

Absolute Value Equations

Solving equations that involve absolute value requires special care. To illustrate we begin with an example that is simple enough to solve mentally.

$$|x| = 5 \quad \text{if and only if} \quad x = 5 \quad \text{or} \quad x = -5$$

There are two solutions—one when the expression "inside" the absolute value is positive and the other when it is negative. In a similar fashion

$$|x - 2| = 5 \quad \text{if and only if} \quad x - 2 = 5 \quad \text{or} \quad x - 2 = -5.$$

Here is a basic principle to use based on the definition of absolute value.

> **Solving Absolute Value Equations**
>
> If u is an expression in x and if a is a positive real number, then
>
> $$|u| = a \quad \text{if and only if} \quad u = a \quad \text{or} \quad u = -a.$$

EXAMPLE 1 **Solving Absolute Value Equations**

a. $|x| = \sqrt{2}$ if and only if $x = \sqrt{2}$ or $x = -\sqrt{2}$.
b. $|x - 2| = 14$ if and only if $x - 2 = 14$ or $x - 2 = -14$. So $x = 16$ and $x = -12$ are solutions to the equation. ■

As you begin solving an absolute value equation, you may be uncertain about the number of solutions to expect. It is often helpful to begin with the visual information shown in graphs.

EXAMPLE 2 **Solving Absolute Value Equations Using Graphs**

Solve $|x^2 + x - 6| = 4$.

Solution

First, we observe in Figure 2.17 that the graphs of $y_1 = |x^2 + x - 6|$ and $y_2 = 4$ appear to intersect four times. Thus, we should expect to find four solutions algebraically.

Solve Algebraically

$$|x^2 + x - 6| = 4 \quad \text{if and only if} \quad x^2 + x - 6 = 4 \qquad \text{Equation 1}$$
$$\text{or} \quad x^2 + x - 6 = -4 \qquad \text{Equation 2}$$

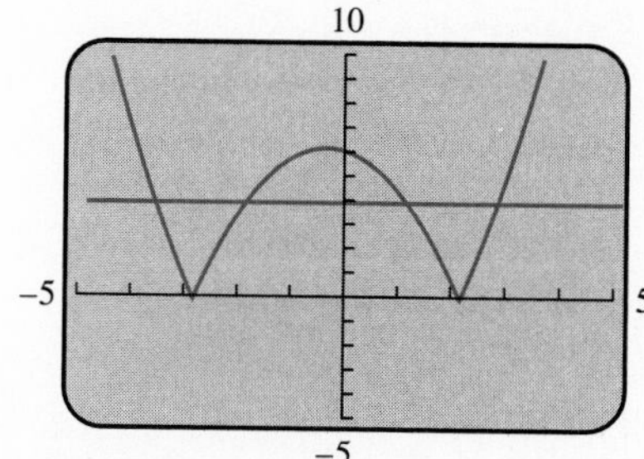

Figure 2.17 Graphs of $y_1 = |x^2 + x - 6|$ and $y_2 = 4$ appear to intersect in four places.

Equation 1:

$$x^2 + x - 6 = 4$$
$$x^2 + x - 10 = 0$$
$$x = \frac{-1 \pm \sqrt{1 + 40}}{2}$$
$$x = \frac{-1 + \sqrt{41}}{2} \quad \text{or} \quad x = \frac{-1 - \sqrt{41}}{2}$$
$$= 2.701 \cdots \quad \text{or} \quad -3.701 \cdots \qquad \text{Two solutions}$$

Equation 2:

$$x^2 + x - 6 = -4$$

$$x^2 + x - 2 = 0$$

$$x = \frac{-1 \pm \sqrt{1 + 8}}{2}$$

$$x = \frac{-1 + 3}{2} \quad \text{or} \quad x = \frac{-1 - 3}{2}$$

$$x = 1 \text{ or } x = -2 \quad \text{Two more solutions}$$

Confirm Numerically

Replace x in $|x^2 + x - 6|$ with each of

$$1, \quad -2, \quad \frac{-1 + \sqrt{41}}{2}, \quad \text{and} \quad \frac{-1 - \sqrt{41}}{2}.$$

Show that the value of $|x^2 + x - 6|$ in each case is 4. ■

Alert

Some students will attempt to solve an equation such as $x + \sqrt{x} = 6$ by squaring both sides without first isolating the radical. Show them that this may not work by illustrating that $(x + \sqrt{x})^2$ contains a radical.

Solving Radical Equations

To solve an equation containing one or more radicals, we sometimes first free the equation of the radical(s) by squaring both sides.

■ EXAMPLE 3 Solving a Radical Equation

Solve $\sqrt{x + 3} = \sqrt{4 - x}$.

Solution

The graphs in Figure 2.18 suggest that there is one solution to the equation.

Solve Algebraically

$$\sqrt{x + 3} = \sqrt{4 - x}$$

$$(\sqrt{x + 3})^2 = (\sqrt{4 - x})^2 \quad \text{Square both sides.}$$

$$x + 3 = 4 - x$$

$$2x = 4 - 3$$

$$x = \frac{1}{2}$$

We conclude that the solution is $x = 1/2$.

Confirm Numerically

Substitute $x = 1/2$ into $\sqrt{x + 3} = \sqrt{4 - x}$, and check that the equation is true. ■

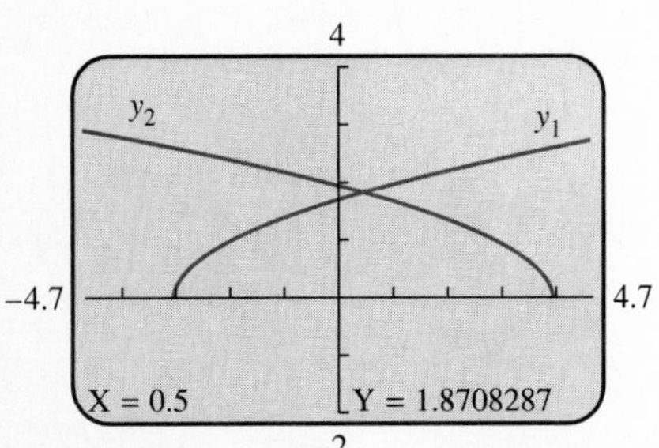

Figure 2.18 Graphs of $y_1 = \sqrt{x + 3}$ and $y_2 = \sqrt{4 - x}$.

Extraneous Solutions

In Examples 2 and 3 each of the values found algebraically, when confirmed, turned out to be a solution of the equation. Sometimes a numerical check will confirm that a value found algebraically is not a solution of the equation. Such a value is called an **extraneous solution.**

Notes on Examples

Point out that the extraneous roots occur because any solution to the equation must satisfy $2x - 4 \geq 0$ (or $x \geq 2$), since $2x - 4$ is equal to an absolute value. Even though the second equation is satisfied, the first one may not be.

■ EXAMPLE 4 Finding an Extraneous Solution

Solve $|x + 3| = 2x - 4$.

Solution

Graph $y_1 = |x + 3|$ and $y_2 = 2x - 4$ to see that there is only one solution. Using the definition of absolute value, we have $|x + 3| = x + 3$ if $x + 3 \geq 0$, and $|x + 3| = -(x + 3)$ if $x + 3 < 0$. Thus, we solve the two equations $x + 3 = 2x - 4$ and $-(x + 3) = 2x - 4$.

Solve Algebraically

$$\begin{aligned} x + 3 &= 2x - 4 \\ -x &= -4 - 3 \\ x &= 7 \end{aligned} \qquad\qquad \begin{aligned} -(x + 3) &= 2x - 4 \\ -x - 3 &= 2x - 4 \\ 3x &= 1 \\ x &= \frac{1}{3} \end{aligned}$$

Confirm Numerically

$$|x + 3| = 2x - 4$$

$$\left|\frac{1}{3} + 3\right| = 2\left(\frac{1}{3}\right) - 4 \qquad \text{Substitute } x = 1/3.$$

$$\frac{10}{3} = -\frac{10}{3} \qquad \text{A false statement}$$

$$|x + 3| = 2x - 4 \qquad \text{Start again.}$$

$$|7 + 3| = 2(7) - 4 \qquad \text{This time substitute } x = 7.$$

$$10 = 10 \qquad \text{A true statement}$$

We say that $x = 1/3$ is an extraneous solution and that $x = 7$ checks. ■

Not only are there extraneous solutions in absolute value equations, but also in equations with radicals.

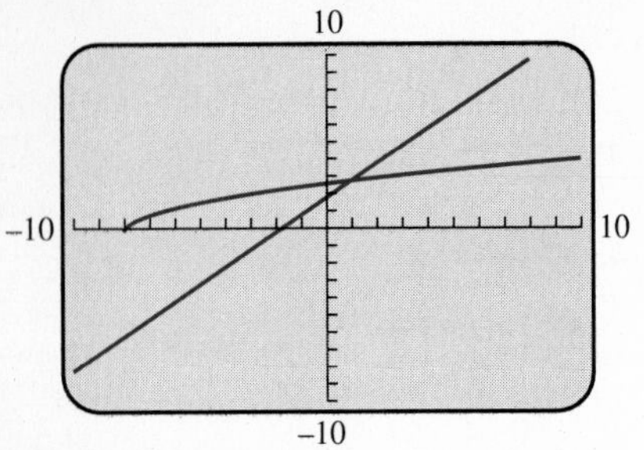

Figure 2.19 Graphs of $y_1 = \sqrt{x + 8}$ and $y_2 = x + 2$. Can you tell which is which?

Notes on Examples

Note that since the equation includes the square root of $x + 8$, any solution to the equation must satisfy $x + 8 \geq 0$ (or $x \geq -8$). Furthermore, since $x + 2$ is equal to a square root, a solution must satisfy $x + 2 > 0$ (or $x > -2$). This may give students some insight into how extraneous solutions can occur in radical equations.

■ EXAMPLE 5 Solving Another Radical Equation

Solve $\sqrt{x + 8} = x + 2$.

Solution

The graphs in Figure 2.19 suggest that there is one solution to the equation.

Solve Algebraically

In order to eliminate the radical we isolate the radical on one side of the equation and then square both sides.

$$\begin{aligned}
\sqrt{x + 8} &= x + 2 \\
x + 8 &= x^2 + 4x + 4 && \text{Square both sides.} \\
x^2 + 3x - 4 &= 0 \\
(x + 4)(x - 1) &= 0 \\
x + 4 = 0 \quad &\text{or} \quad x - 1 = 0 && \text{The zero factor property} \\
x = -4 \quad &\text{or} \quad x = 1
\end{aligned}$$

Confirm Numerically

When x is replaced in the original equation by -4, the equation becomes $\sqrt{4} = -2$, a false statement. Thus, $x = -4$ is an extraneous solution. Replacing x by 1 in the original equation gives a true statement, which confirms that $x = 1$ is a solution. ■

Solving Inequalities

The inequality symbols $<, \leq, >, \geq$ allow us to compare quantities and make statements of comparison. Examples of inequality statements are

$$2x - 3 < 7, \qquad -3x^2 + 1 \leq 4, \qquad 6 + x > -3, \qquad \sqrt{4 - 3x} \geq 0.$$

Recall that a *solution of an inequality* is a value of the variable that makes the inequality true.

Algebraic methods for solving inequalities were illustrated in the Prerequisite Chapter. Inequalities may also be solved graphically.

■ EXAMPLE 6 Solving an Inequality

Solve $x + 1 < 3 - \frac{1}{3}x$.

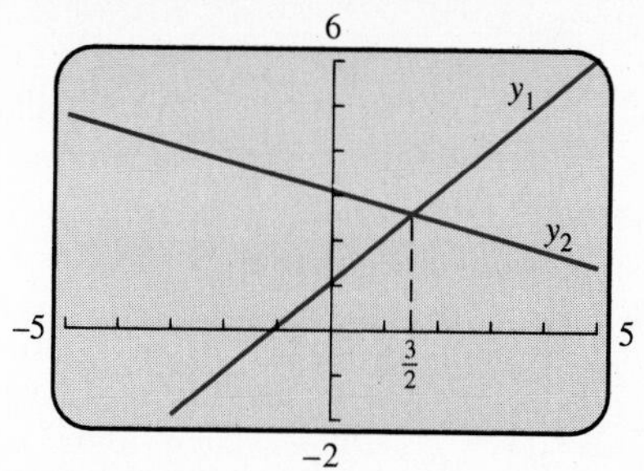

Figure 2.20

Solution

Solve Graphically

Figure 2.20 shows the graphs of $y_1 = x + 1$ and $y_2 = 3 - (1/3)x$. Use "trace" to see that points on the graph of y_1 are below the graph of y_2 when $x < 3/2$.

Confirm Algebraically

$$x + 1 < 3 - \frac{1}{3}x$$

$$x + \frac{1}{3}x < 3 - 1 \qquad \text{Add } \tfrac{1}{3}x - 1.$$

$$\frac{4}{3}x < 2 \qquad \text{Simplify.}$$

$$4x < 6 \qquad \text{Multiply by 3.}$$

$$x < \frac{3}{2} \qquad \text{Divide by 4.}$$

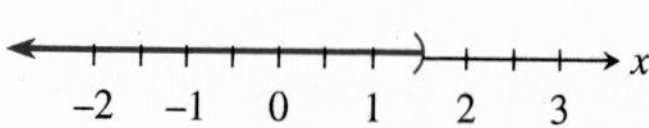

Figure 2.21 The solution to Example 6.

Interpret

The solution is the interval $(-\infty, 3/2)$, pictured in Figure 2.21. ■

Recall from the Prerequisite Chapter that a statement like $-3 \le 4x - 3 < 2$ is a double inequality because it is formed from two inequalities, $-3 \le 4x - 3$ and $4x - 3 < 2$.

■ **EXAMPLE 7 Solving a Double Inequality**

Solve $-3 < \dfrac{2x + 5}{3} \le 5$.

Solution

Solve Algebraically

$$-3 < \frac{2x + 5}{3} \le 5$$

$$-9 < 2x + 5 \le 15 \qquad \text{Multiply by 3.}$$

$$-14 < 2x \le 10 \qquad \text{Add } -5.$$

$$-7 < x \le 5 \qquad \text{Multiply by 1/2.}$$

As an interval the solution is $(-7, 5]$.

Support Graphically

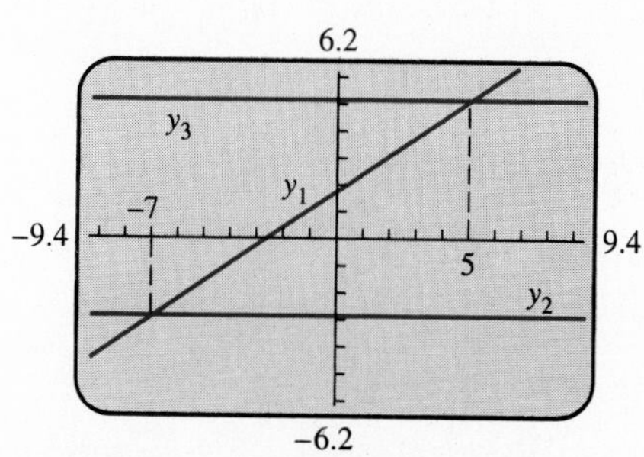

Figure 2.22 The solution is the interval on the x-axis between -7 and 5.

Figure 2.22 shows graphs of $y_1 = (2x + 5)/3$, $y_2 = -3$ and $y_3 = 5$. Observe that the points on the line $y_1 = (2x + 5)/3$ that lie between the lines $y_2 = -3$ and $y_3 = 5$ are those whose x-coordinates are between -7 and 5. ■

Alert

Instead of "$u < -a$ or $u > a$," some students may write "$a < u < -a$." Emphasize the important distinction between *and* and *or* in this context.

Solving Absolute Value Inequalities

Algebraic methods for solving absolute value inequalities parallel the methods used for solving absolute value equations. At the beginning of this section we gave a basic principle to use for solving absolute value equations. Here are two similar principles that we use to solve absolute value inequalities.

Solving Absolute Value Inequalities

Let u be an expression in x and let a be a positive real number.

1. If $|u| < a$, then u is in the interval $(-a, a)$. That is,

$$|u| < a \quad \text{if and only if} \quad -a < u < a.$$

2. If $|u| > a$, then u is in the interval $(-\infty, -a)$ or (a, ∞). That is,

$$|u| > a \quad \text{if and only if} \quad u < -a \quad \text{or} \quad u > a.$$

The inequalities $<$ and $>$ can be replaced with $\le$ and $\ge$ respectively. See Figure 2.23.

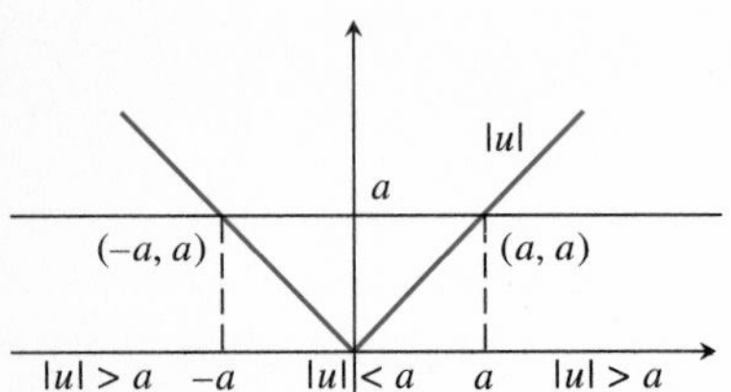

Figure 2.23

■ EXAMPLE 8 Solving an Absolute Value Inequality

Solve $|x - 4| < 8$.

Solution

Solve Algebraically

$$|x - 4| < 8$$

$$-8 < x - 4 < 8 \qquad |u| < 8 \text{ if and only if } -8 < u < 8.$$

$$-4 < x < 12 \qquad \text{Add 4.}$$

As an interval the solution is $(-4, 12)$.

Support Graphically

Figure 2.24 shows that points on $y_1 = |x - 4|$ are below the points on $y_2 = 8$ for values of x between -4 and 12. ■

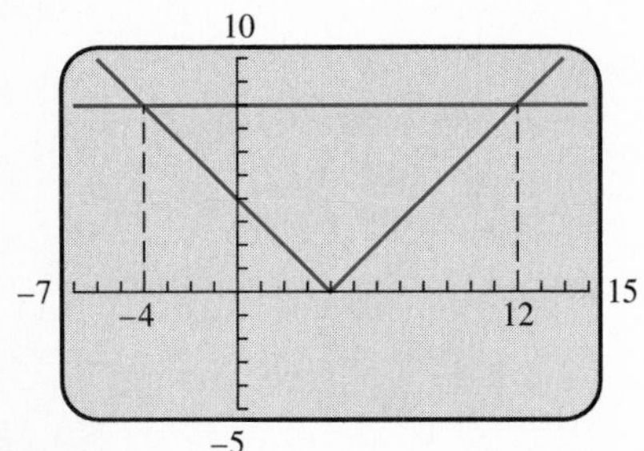

Figure 2.24 $y_1 = |x - 4|$ and $y_2 = 8$.

■ EXAMPLE 9 Solving Another Absolute Value Inequality

Solve $|3x - 2| \ge 1$.

Solution

Solve Algebraically

$$3x - 2 \le -1 \quad \text{or} \quad 3x - 2 \ge 1 \qquad |u| \ge a \text{ if and only if } u \le -a \text{ or } u \ge a.$$

$$3x \le 1 \quad \text{or} \quad 3x \ge 3$$

$$x \le \tfrac{1}{3} \quad \text{or} \quad x \ge 1$$

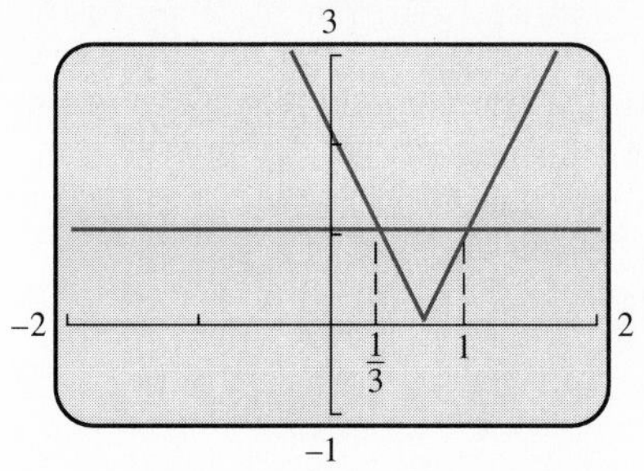

Figure 2.25 $y_1 = |3x - 2|$ and $y_2 = 1$. Note that $y_1 \geq y_2$ to the left of $\frac{1}{3}$ and to the right of 1.

The solution consists of all numbers that are in either one of the intervals $(-\infty, 1/3]$ or $[1, \infty)$, which may be written as $(-\infty, 1/3] \cup [1, \infty)$.

Support Graphically

See Figure 2.25. ■

> **Notes on Examples**
>
> Examples 10 and 11 are important because they emphasize that a value of x must be in the domain of every expression in the inequality in order to be in the solution set.

When faced with an inequality of a type different from those studied so far, any algebraic, graphical, or numerical method can be used. Sometimes it is easier to use a combination of methods as illustrated in Example 10.

■ EXAMPLE 10 Solving Another Inequality

Solve $\dfrac{x + 3}{|x - 2|} > 0$.

Solution

Solve Graphically and Algebraically

Figure 2.26a shows that $x = -3$ is an x-intercept of $y = \dfrac{x + 3}{|x - 2|}$.

When $x > -3$, we can see algebraically that $(x + 3)/|x - 2|$, when defined, is positive. When $x < -3$, we can see algebraically that the values of the expression $(x + 3)/|x - 2|$ are negative.

We note that y is not defined when $x = 2$ (see Figure 2.26b) and conclude that

$$y > 0 \text{ on the intervals } (-3, 2) \text{ and } (2, \infty).$$

The solution consists of all numbers that are in either interval $(-3, 2)$ or interval $(2, \infty)$, that is, $(-3, 2) \cup (2, \infty)$. ■

> **Notes on Examples**
>
> If your students are familiar with set notation, point out that the solution set can be written as $(-3, 2) \cup (2, \infty)$.

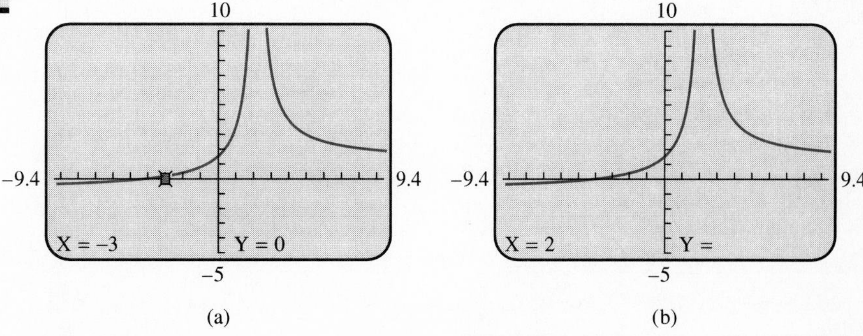

Figure 2.26

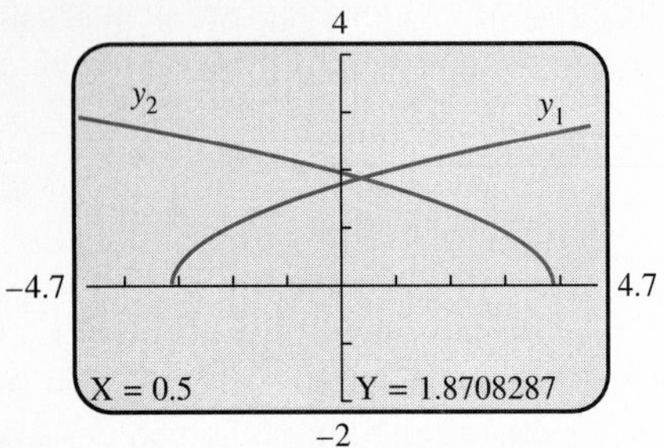

Figure 2.27 Determining where these graphs intersect is an important part of solving the inequality.

Solving Radical Inequalities

To solve an inequality involving radicals it is often helpful to solve the corresponding equation graphically by finding intersections.

■ EXAMPLE 11 Solving an Inequality with a Radical

Solve $\sqrt{x + 3} > \sqrt{4 - x}$.

Solution

Solve Graphically

Figure 2.27 shows graphs of $y_1 = \sqrt{x + 3}$ and $y_2 = \sqrt{4 - x}$. It is evident that $y_1 > y_2$ whenever the graph of y_1 is above the graph of y_2. Show with your grapher that $x = 1/2$ is the x-coordinate of the point of intersection.

We conclude that $y_1 > y_2$ whenever $x > \frac{1}{2}$ and y_2 is defined. Because $\sqrt{4 - x}$ is not defined when $x > 4$, there are no values of x greater than 4 in the solution. Thus, the solution is the interval $(\frac{1}{2}, 4]$. ■

Applications

Linear inequalities often occur in business, consumer, and industrial situations. Whenever quantities are compared, an inequality can often serve as a model.

■ EXAMPLE 12 APPLICATION: Comparing Two Lease Plans

DTQ Auto Parts Company plans to lease a 50-ton punch press from an investment group. Plan 1 charges a flat fee of \$25,000 plus \$2000 per month, and Plan 2 charges a flat fee of \$2000 plus \$3800 per month. When is leasing Plan 1 more economical than leasing Plan 2?

Solution

Model

Word Statement: When is cost of Plan 1 < cost of Plan 2?

t = the time of the lease (in months)

$25{,}000 + 2000t$ = cost of Plan 1 for t months

$2000 + 3800t$ = cost of Plan 2 for t months

Solve Algebraically

$$25{,}000 + 2000t < 2000 + 3800t \quad \text{Inequality to be solved}$$
$$2000t - 3800t < 2000 - 25{,}000$$
$$-1800t < -23{,}000$$
$$t > \frac{230}{18} \approx 12.8$$

Follow-up

Ask . . .

In Example 7, how can you justify the steps used to solve the double inequality algebraically? **(A double inequality is equivalent to a pair of regular inequalities, so the properties of inequalities from Section 0.7 can be applied.)**

Assignment Guide

Day 1: Ex. 3–51, multiples of 3
Day 2: Ex. 57, 59, 63, 66, 71, 72, 74, 77, 81, 87, 88

Cooperative Learning

Group Activity: Ex. 82, 83, 85

Notes on Exercises

Ex. 68 and 73–77 are related to geometry.
Ex. 81 and 84 emphasize the fact that there is often more than one way to solve a problem.
Ex. 85 is related to the definition of the limit of $f(x)$ as x approaches a particular value.

Ongoing Assessment

Self-Assessment: Ex. 11, 23, 43, 69
Embedded Assessment: Ex. 72

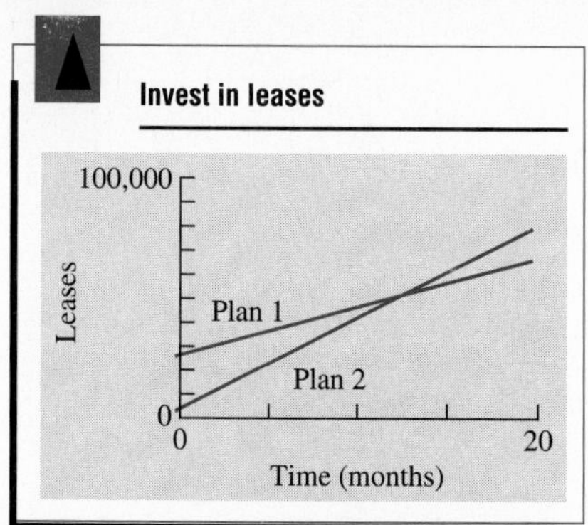

Figure 2.28

Support Graphically

Figure 2.28 shows the graphs for both Plans 1 and 2. Plan 2 is initially less expensive, but catches up and exceeds Plan 1 at a bit before 13 months.

Interpret

Plan 1 will cost less for a leasing period of more than 1 year. ■

Quick Review 2.3

In Exercises 1–4, graph the inequality on a number line.

1. $-2 \le x < 4$ **2.** $-5 < x \le -1$

3. $3 < x < 8$ **4.** $-5 < x \le 4$

In Exercises 5 and 6, describe the number line graphs with interval notation.

5.

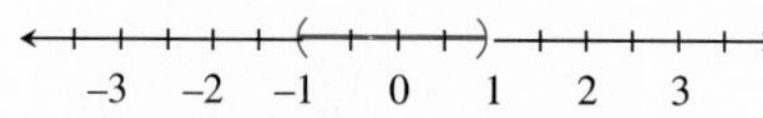

6.

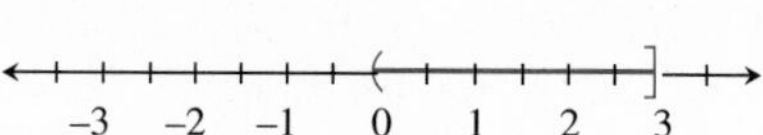

In Exercises 7 and 8, solve for x.

7. $2x + 5 = 7 - 3x$ **8.** $3(x - 4) = 2(5 - 2x)$

In Exercises 9–12, solve the inequality.

9. $3x - 5 < 7$ **10.** $2x + 8 \ge 12$

11. $6 - 5x > 2$ **12.** $8 - 3x \le 4$

In Exercises 13–16, find the domain of the expression.

13. $\dfrac{1}{x - 2}$ **14.** $\dfrac{x + 3}{x^2 - 1}$

15. $\sqrt{x - 3}$ **16.** $\sqrt{x^2 - 4}$

SECTION EXERCISES 2.3

In Exercises 1–14, solve algebraically and support graphically. Identify all extraneous solutions.

1. $|2x| = 8$ **2.** $|3x| = 11$

3. $|2x - 1| = 5$ **4.** $|3x + 4| = 9$

5. $|x^2 + 4x - 1| = 7$ **6.** $|x^2 - 3x - 4| = 2$

7. $|x - 1| = 2x + 1$ **8.** $|x + 2| = 0.5x + 4$

9. $\sqrt{x + 2} = x + 1$ **10.** $\sqrt{x + 1} = x - 2$

11. $x + 1 - 2\sqrt{x + 4} = 0$ **12.** $\sqrt{x + 1} = x - 1$

13. $\sqrt{x} + x = 1$ **14.** $\sqrt{x} + \sqrt{2x} = 4$

In Exercises 15–24, solve the equation by a method of your choice.

15. $x^3 + 4x^2 - 3x - 2 = 0$

16. $24x^3 - 50x^2 + 27x - 4 = 0$

17. $8x^4 + 2x^3 - 11x^2 - 2x + 3 = 0$

18. $16x^4 - 4x^3 - 16x^2 + x + 3 = 0$

19. $5\sqrt{x + 6} - x^2 + 1.5x = 0$

20. $8\sqrt{x + 7} - x^3 - 21 = 0$

21. $3\sqrt{3x + 14} = 15 - 2|x - 2|$

22. $|3x - \pi| = 18 - |2x + 7|$

23. $x^3 - 2\sqrt{12x + 123} = 0$

24. $|x^3 - 5x + 1| - 1 = 0$

In Exercises 25–28, graph f and find the unique x_0 such that $f(x_0) = c$ for the given value of c.

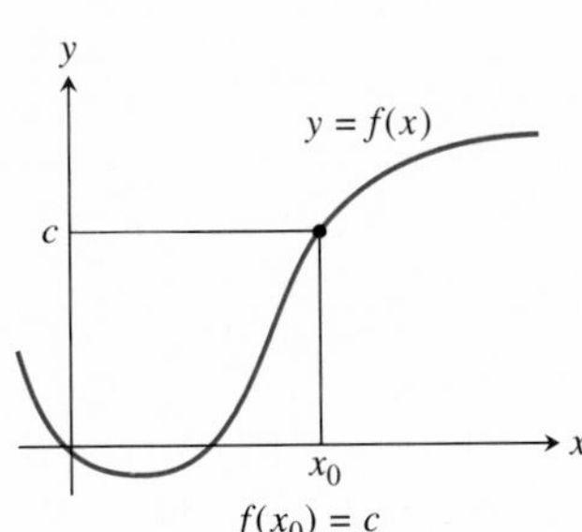

For Exercises 25–28

25. $f(x) = x^3 + 2x^2 - x - 2$ and $c = 2$

26. $f(x) = x^3 + x^2 - 3x - 5$ and $c = 5$

27. $f(x) = 0.5x + 2\sqrt{x + 7}$ and $c = 4$

28. $f(x) = \sqrt{x + 8} + \sqrt{4x + 24}$ and $c = 8$

In Exercises 29–46, solve the inequalities algebraically. Write the solution in interval notation and draw its number line graph.

29. $2x + 3 \leq 7 + 2x$

30. $7x - 9 \geq x + 3$

31. $-2x + 35 < 17x - 4$

32. $4 + 3x > 7 - 0.5x$

33. $-1 < \dfrac{2x + 1}{3} \leq 8$

34. $-4 < \dfrac{3 - x}{4} \leq 7$

35. $5 \geq 2x - 4 > -1$

36. $18 > x + 2 > 5$

37. $|x| < 2$

38. $|x| < 3$

39. $|x + 4| \geq 5$

40. $|2x - 1| > 3.6$

41. $|x - 3| < 2$

42. $|x + 3| \leq 5$

43. $|4 - 3x| - 2 < 4$

44. $\left|\dfrac{x + 2}{3}\right| \geq 3$

45. $\sqrt{x + 4} \geq \sqrt{5 - x}$

46. $\sqrt{2x + 9} < \sqrt{3 - x}$

In Exercises 47–56, solve the inequality by a method of your choice.

47. $\dfrac{3x - 8}{2} > 6$

48. $3|x| - 4 > 0$

49. $x|x - 2| > 0$

50. $\dfrac{x - 3}{|x + 2|} < 0$

51. $\left|\dfrac{1}{x}\right| < 3$

52. $|x| < |x - 3|$

53. $(x + 3)|x - 1| \geq 0$

54. $|x + 3| > |x - 1|$

55. $|x + 3| > |2x - 1|$

56. $|2x + 3| > |x - 1|$

In Exercises 57 and 58, use tracing paper to copy the graph provided. Shade the interval on the x-axis that represents the solution to the inequality.

57.

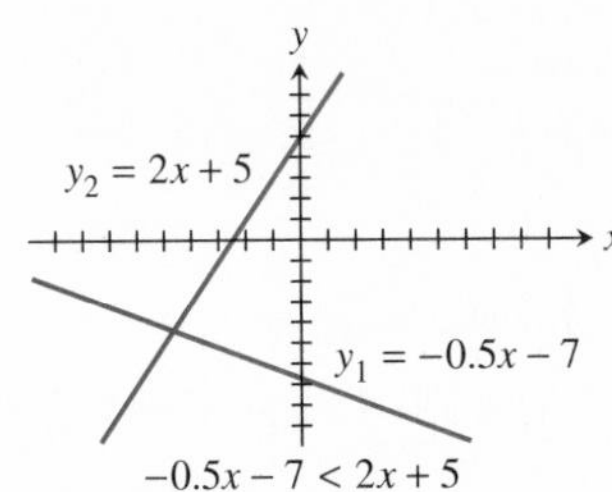

58.

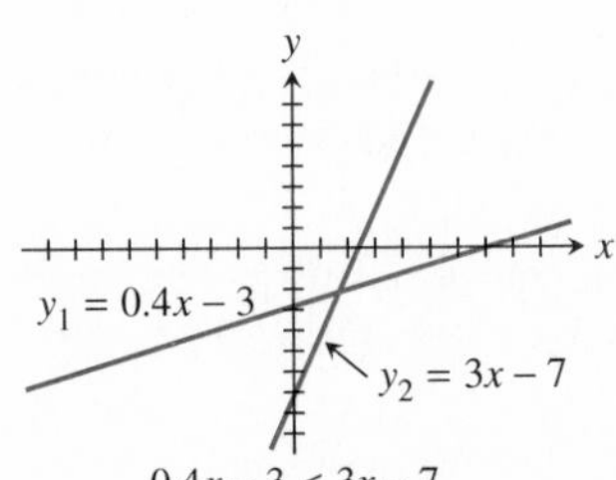

In Exercises 59–62, solve the inequality graphically. Write the solution in interval notation. (*Suggestion:* Use the Integer window.)

59. $1.5x - 18 < 0.5x - 3$

60. $2x + 17 \geq -0.8x - 11$

61. $-2x + 9 < 0.5x - 11$

62. $(3x - 4) - (2.4x + 3) > 0.8x - 11$

In Exercises 63–66, solve the double inequality by a method of your choice.

63. $2x - 10 < -\frac{1}{2}x - 1 < x - 4$

64. $2x < 4 \leq x + 7$

65. $-x < 2x + 3 < -8x + 3$

66. $-x + 4 \leq -3 \leq 3x$

In Exercises 67–72, solve the problem.

67. *Travel Planning* Barb wants to drive to a city 105 mi from her home in no more than 2 h. What is the lowest average speed she must maintain on the drive?

68. Connecting Algebra and Geometry Consider the collection of all rectangles that have lengths 2 in. less than twice their widths. Find the possible widths (in inches) of these rectangles if their perimeters are less than 200 in.

69. *Planning for Profit* The Grovenor Candy Co. finds that the cost of making a certain candy bar is \$0.13 per bar. Fixed costs amount to \$2000 per week. If each bar sells for \$0.35, find the minimum number of candy bars that will earn the company a profit.

70. *Boyle's Law* For a certain gas, $P = 400/V$, where P is pressure and V is volume. If $20 \le V \le 40$, what is the corresponding range for P?

71. *Cash-Flow Planning* A company has current assets (cash, property, inventory, and accounts receivable) of \$200,000 and current liabilities (taxes, loans, and accounts payable) of \$50,000. How much can it borrow if it wants its ratio of assets to liabilities to be no less than 2? Assume the amount borrowed is added to both current assets and current liabilities.

72. *Budget Planning* Sarah has \$65 to spend and wishes to take as many friends as possible to a concert. Parking is \$5.75 and concert tickets are \$9.50 each.

a. Let x represent the number of friends Sarah can take to the concert. Write an inequality that must be satisfied in this situation.

b. Solve the inequality in part a.

c. Interpret your solution.

In Exercises 73–76, assume that an open box is formed by cutting squares from the corners of a rectangular piece of cardboard (see figure) and folding up the flaps.

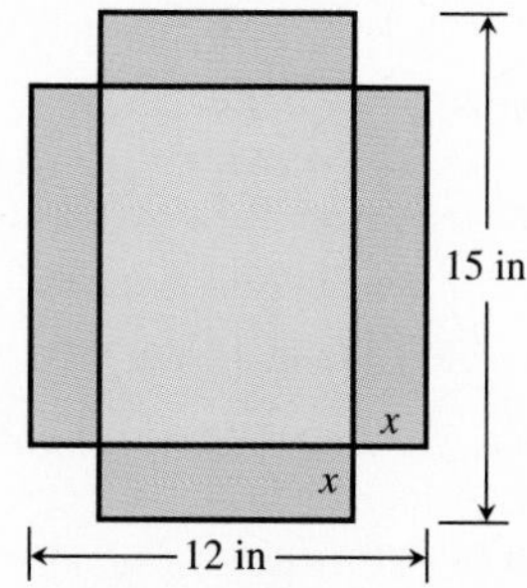

For Exercises 73–76

73. *Package Industry Problem* What sized corner squares should be cut to yield a box with a volume of 125 in.^{3}?

74. *Package Industry Problem* Squares 1.2 in. on a side are cut from the corners of the cardboard. How much smaller should the corners be cut to yield a box volume of 120 in.^{3}?

75. *Varying Package Size* A box is made with a volume of 158 in.^{3}. If the corner squares had been cut 1/2 inch shorter on each side, the box volume would have been greater. What size squares were cut?

76. *Varying Package Size* The volume of the box is 175 in.^{3}. If the squares had been cut 1/10 inch longer, the volume would have increased. What sized squares were cut?

In Exercises 77–83, solve the problem.

77. *A Cone Problem* Begin with a circular piece of paper with a 4 in. radius, as shown in (a). Cut out a sector with an arc length of x. Join the two edges of the remaining portion to form a cone with radius r and height h as shown in (b). What length of arc will produce a cone with a volume of 21 in.^{3}? (See Section 1.2, Exercise 49.)

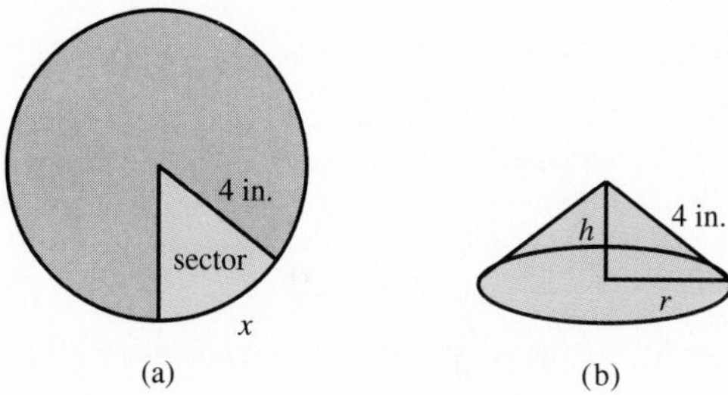

For Exercise 77

78. *Industrial Construction* An electric company has a generating plant alongside the James River where it is 1 mi wide. Richmond is 15 mi downriver on the opposite side. Cable is to be laid from the plant to a point Q on the opposite bank

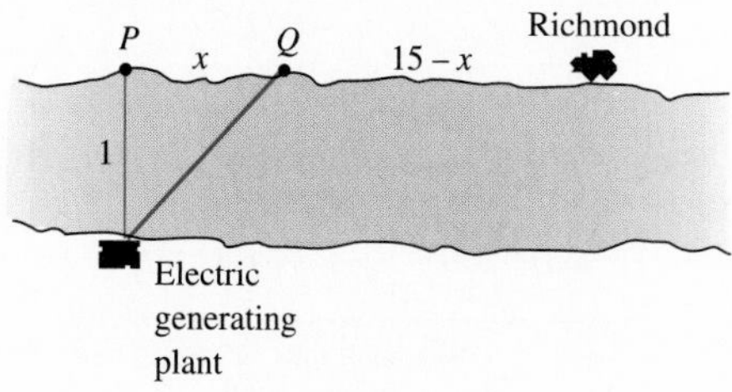

For Exercise 78

x miles from point P, directly opposite the plant. It costs \$15,000 per mile to lay cable underwater and \$7000 per mile to lay cable underground. How far is point Q from P if the total cost of laying the cable is \$135,000?

79. *Modeling R & D Spending* The total number of dollars (in billions) spent in the United States on research and development (R & D) for the 30-year period 1960–89 can be modeled by the equation $y = 0.00377x^3 + 0.168x^2 - 0.926x + 15.907$, where $x = 0$ represents 1960, $x = 1$ represents 1961, and so forth. (*Source:* U.S. National Science Foundation, *National Patterns of R&D Resources,* 1990; and unpublished data.)

a. Use this model to predict what the total expenditures for R & D will be in the year 2000 (to the nearest \$10 billion).

b. Use this model to estimate the year in which expenditures for R & D first reached \$80 billion.

80. *Company Wages* Pederson Electric Co. charges \$25 per service call plus \$18 per hour for home repair work. How long did an electrician work if the charge was less than \$100? Assume the electrician rounds off the time to the nearest quarter hour.

81. Writing to Learn Write a paragraph that explains two ways to solve the inequality $3(x - 1) + 2 \leq 5x + 6$.

82. *Per Capita Income* The U.S. average per capita income (adjusted for inflation) from 1965 to 1990 is given in Table 2.4. These data can be modeled by the linear regression equation

$$\text{income} = 0.23t + 11.3,$$

where $t = 0$ represents 1960, $t = 1$ represents 1961, and so forth.

a. Graph a scatter plot of the data.

Table 2.4 Per Capita Income

Year	Dollars (thousands)
1965	12.7
1970	14.0
1980	14.9
1985	16.6
1989	17.9
1990	19.5

Source: U.S. Bureau of the Census, *Survey of Current Business,* April 1992.

b. Superimpose a graph of $y = 0.23t + 11.3$ on the scatter plot.

c. Use the model to predict when the per capita income exceeds \$25,000.

83. *Weekly Food Cost* The average weekly food costs from 1975 to 1990 for a family with two children between the ages of 6 and 11 is given in Table 2.5.

Table 2.5 Average Weekly Food Cost

Year	Dollars
1975	63.00
1980	88.10
1985	99.80
1990	128.30

Source: U.S. Dept. of Agriculture, Human Nutrition Information Service.

These data can be modeled by the linear regression equation

$$\text{Cost} = 4.15t + 42.9,$$

where $t = 0$ represents 1970, $t = 1$ represents 1971, and so forth.

a. Graph a scatter plot of the data.

b. Superimpose a graph of $y = 4.15t + 42.9$ on the scatter plot.

c. Use the model to predict when the average weekly food cost exceeds \$200 per week.

EXTENDING THE IDEAS

In Exercises 84–88, solve the problem.

84. Writing to Learn Solve the equation $0.5x^4 - 5 = x^2 - 3$ three times—once algebraically, once graphically, and once numerically. Describe the advantages and disadvantages of each method.

85. Looking Ahead to Calculus Let $f(x) = 3x - 5$.

a. Assume that x_0 is in the interval defined by $|x - 3| < \frac{1}{3}$. Give a convincing argument that $|f(x_0) - 4| < 1$.

b. Writing to Learn Explain how part a is modeled by this figure.

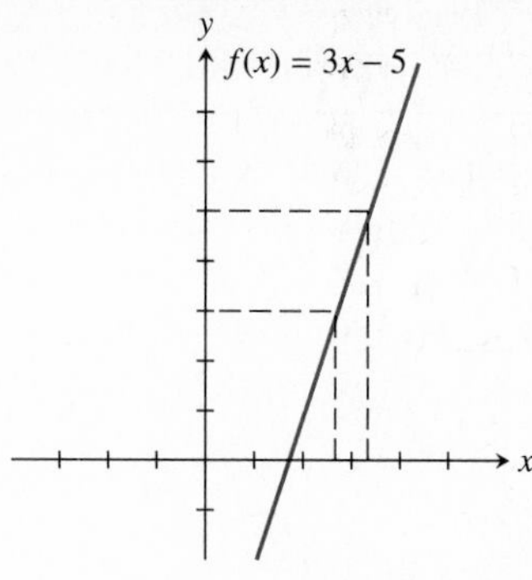

For Exercise 85

c. Show how the algebra used in part a can be modified to show that if $|x - 3| < 0.01$, then $|f(x) - 4| < 0.03$. How would the model in part b change to reflect these inequalities?

86. Here is another type of inequality: $2x + 3 \neq 3x + 4.5$. Discuss how you could solve this inequality graphically.

87. The function $y = f(x)$ is defined piecewise by

$$f(x) = \begin{cases} x^2 & \text{if } x < 2 \\ -2x + 3 & \text{if } x \geq 2. \end{cases}$$

Solve each equation for x.

a. $f(x) = 5$ **b.** $f(x) = 1.5$
c. $f(x) = -0.5$ **d.** $f(x) = -4$

88. *Page Design* A graphic artist designs pages 8-1/2 by 11 in. with a picture centered on the page. Suppose the distance from the outer edge of the page to the picture is x inches on all sides.

a. Find an equation that models the area A of the picture.
b. Estimate the width of the uniform border if the area of the picture is 50 in.^{2}.

89. *Using a Grapher* Read about the "test" menu of your grapher and explain what this grapher screen represents.

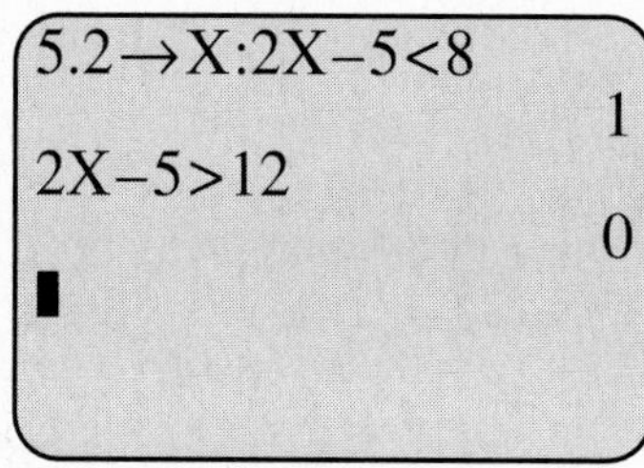

For Exercise 89

Objective

Students will be able to use sum, difference, product, and quotient properties to write the composition of two functions and to perform geometric transformations using composition.

Motivate

Ask . . .

If a person's savings account balance is given by $f(t) = 500 + t$ and their checking account balance is given by $g(t) = 1750$, how can you write a function to give the total balance? ($\boldsymbol{h(t) = 2250 + t}$)

2.4 OPERATIONS ON FUNCTIONS, AND COMPOSITION OF FUNCTIONS

Combining Functions Arithmetically

Just as numbers can be combined (using operations such as addition, subtraction, and so on) to form a new number, so can functions be combined to form a new function. For example, if $f(x) = x^3 - 4x$ and $g(x) = 3x^2 - 2x - 5$, then

$$\begin{aligned} f(x) + g(x) &= (x^3 - 4x) + (3x^2 - 2x - 5) \\ &= x^3 + 3x^2 - 6x - 5. \end{aligned}$$

In a similar way we can subtract, multiply, or divide f and g.

Teaching Note

If your students are familiar with set notation, point out that the domain of the sum, difference, product, or quotient function is the *intersection* of the individual domains (less the zeros of g, in the case of the quotient function).

Alert

Students may expect the domain of the sum, difference, product, or quotient function to be obvious from looking at the new function equation, but this is not always the case. Give the example of $(fg)(x)$, where $f(x) = g(x) = \sqrt{x}$. Then $(fg)(x) = x$, but the domain of fg is the set of nonnegative real numbers only.

Definition: Sum, Difference, Product, and Quotient of Functions

Let f and g be two functions with part of their domains in common. Then for all values of x common to both domains, the arithmetic combinations of f and g are defined as follows:

Sum: $(f + g)(x) = f(x) + g(x)$

Difference: $(f - g)(x) = f(x) - g(x)$

Product: $(fg)(x) = f(x)g(x)$

Quotient: $\left(\dfrac{f}{g}\right)(x) = \dfrac{f(x)}{g(x)}, \quad g(x) \neq 0$

The domain of the new function consists of all numbers that belong to the domains of *both* f and g. For f/g the zeros of g are excluded from the domain.

■ EXAMPLE 1 Defining New Functions

Consider $f(x) = x^2$ and $g(x) = \sqrt{x + 1}$.

a. $\sqrt{x + 1}$ is defined only if $x + 1 \geq 0$ or $x \geq -1$. Therefore,

$$f(x) + g(x) = x^2 + \sqrt{x + 1},$$

and the domain of $f + g$ is the interval $[-1, \infty)$.

b. Similarly,

$$f(x) - g(x) = x^2 - \sqrt{x + 1}$$

$$f(x)g(x) = x^2\sqrt{x + 1}$$

and the domain for both $f - g$ and fg is $[-1, \infty)$.

c. Finally,

$$\frac{f(x)}{g(x)} = \frac{x^2}{\sqrt{x + 1}}$$

and the domain of f/g is $(-1, \infty)$. ■

■ EXAMPLE 2 Evaluating Sum and Product Functions

Let $f(x) = \sqrt{x + 6}$ and $g(x) = \sqrt{x - 2}$. Find

a. $(f + g)(3)$

b. $(fg)(4)$

Solution

a. $(f + g)(x) = \sqrt{x + 6} + \sqrt{x - 2}$, so

$$(f + g)(3) = \sqrt{3 + 6} + \sqrt{3 - 2} = 3 + 1 = 4$$

b. $(fg)(x) = \sqrt{x + 6} \cdot \sqrt{x - 2}$, so

$$\begin{aligned}(fg)(4) &= \sqrt{4 + 6} \cdot \sqrt{4 - 2} \\ &= \sqrt{10} \cdot \sqrt{2} \\ &= \sqrt{20} \\ &= 2\sqrt{5}\end{aligned}$$

■

An Application That Combines Functions

Suppose that a spherical balloon is being inflated. Both the radius r and volume V of the sphere vary with time. Recall that $V = (4/3)\pi r^3$. If $r = f(t)$ then $V(t)$ can be expressed as the *composition*

$$V(t) = \frac{4}{3}\pi[f(t)]^3.$$

■ **EXAMPLE 3 APPLICATION: Studying a Medical Procedure**

In a medical process known as angioplasty, doctors insert a catheter into a heart vein (through a large peripheral vein), and inflate a small spherical-shaped balloon on the tip of the catheter. Suppose the radius r of the balloon increases at the constant rate of 0.5 millimeters per second (mm/sec). (See Figure 2.29.)

a. Find the volume after 5 sec.

b. The balloon is ordinarily inflated to a volume no greater than 400 mm^3. How long would it take to reach 400 mm^3?

Solution

Model

The formula for the volume of a sphere is $V = \frac{4}{3}\pi r^3$.

$$r = 0.5t \qquad r \text{ changes at 0.5 mm/sec.}$$

$$V(r) = \frac{4}{3}\pi r^3 \qquad \text{Volume of a sphere}$$

$$V(t) = \frac{4}{3}\pi(0.5t)^3 \qquad \text{Substitute } r = 0.5t.$$

$$V(t) = \frac{0.5}{3}\pi t^3 \qquad \text{Simplify.}$$

$$= \frac{\pi}{6}t^3$$

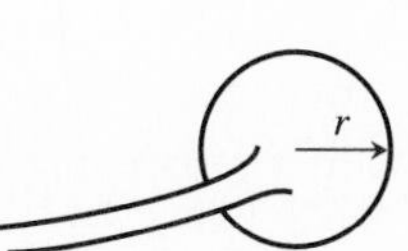

Figure 2.29 A spherical balloon used for angioplasty.

Solve Algebraically

a. Find $V(5)$.

$$V(t) = \frac{\pi}{6}t^3$$

$$V(5) = \frac{\pi}{6}5^3$$

$$= 65.449 \cdots$$

b. $\frac{\pi}{6}t^3 = 400$ Equation to be solved

$$t^3 = \frac{400(6)}{\pi}$$

$t = 9.141 \cdots$ Take cube root.

Interpret

In 5 sec, the balloon's volume is about 65.45 mm^3. It reaches a volume of 400 mm^3 in slightly more than 9.1 sec. ■

Composition of Functions

In Example 3 we began with the function of r, $V = \frac{4}{3}\pi r^3$, and then replaced r with the function of t, $r = 0.5t$, to obtain V as a function of t:

$$V(t) = \frac{4}{3}\pi(0.5t)^3.$$

This is an example of *composition of functions.*

For an additional example let $f(x) = x^2$ and $g(x) = x + \sqrt{x}$. The composition of f and g is the function

$$f(g(x)) = f(x + \sqrt{x}) = (x + \sqrt{x})^2.$$

Definition: The Composition of Two Functions

The **composition of functions** $y_1 = f(x)$ and $y_2 = g(x)$, denoted $f \circ g$ and read "f of g," is the function

$$(f \circ g)(x) = f(g(x)).$$

The domain of $f \circ g$ is the set of all x-values in the domain of g such that $g(x)$ is in the domain of f. (See Figure 2.30.)

The composition $g \circ f$ is defined in a similar way, and usually $f \circ g$ and $g \circ f$ are not the same function.

Alert

Students must understand how the composition notation works. The composition notation $(f \circ g)(x)$ should be compared to the $f(g(x))$ notation so that students can move from one notation to the other with facility. You may wish to read $f \circ g$ as "f following g," to emphasize the order in which the functions are applied. Be aware that not all textbooks use the same convention.

Teaching Note

Point out that the statement "usually $f \circ g$ and $g \circ f$ are not the same function" can also be expressed as "composition of functions is not commutative."

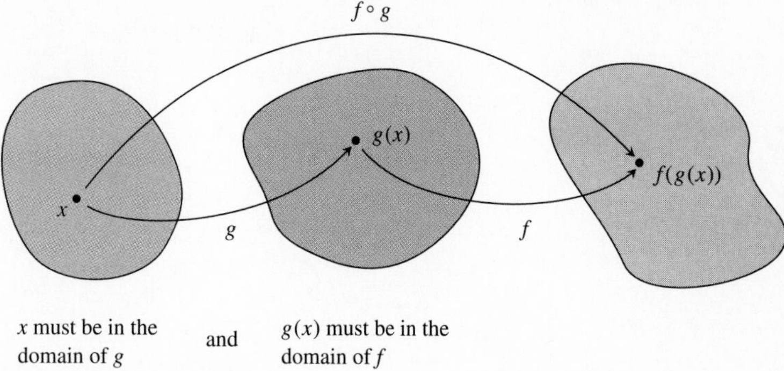

Figure 2.30 In the composition $f \circ g$, the function g is considered first and then f; note this is the reverse order of how we read the symbols.

Notes on Examples

Remind students that two functions are equal if they have the same domain, and for each value of x in the common domain, the two function values are the same. In this example, it is also possible to show that $f \circ g \neq g \circ f$ by demonstrating that they do not have the same domain.

■ EXAMPLE 4 Studying Function Composition

Let $f(x) = x + 1$ and $g(x) = \sqrt{x}$. Find both $f(g(x))$ and $g(f(x))$. Verify that $f \circ g \neq g \circ f$.

Solution

$$f \circ g(x) = f(g(x)) = f(\sqrt{x}) = \sqrt{x} + 1 \quad \text{Replace } g(x) \text{ by } \sqrt{x}.$$

$$g \circ f(x) = g(f(x)) = g(x + 1) = \sqrt{x + 1}$$

We can show that $g(f(x)) \neq f(g(x))$ by showing that they are not equal for at least one value of x. We use $x = 1$.

$$(f \circ g)(1) = f(g(1)) = \sqrt{1} + 1 = 2$$

$$(g \circ f)(1) = g(f(1)) = \sqrt{1 + 1} = \sqrt{2}$$

Because we have shown that $f(g(x)) \neq g(f(x))$ for at least one value ($x = 1$), we can conclude that $f \circ g \neq g \circ f$. ■

In Example 4 two functions were *composed* to form a new function. There are times in calculus when we need to reverse the process. That is, we may begin with a function h and *decompose* it by finding two functions f and g whose composition is h.

Notes on Examples

Example 5 is related to the chain rule for taking derivatives in calculus.

■ EXAMPLE 5 Decomposing Functions: Looking Ahead to Calculus

For the function h, find functions f and g such that $h(x) = f(g(x))$.

a. $h_1(x) = (\sqrt{x + 1})^3$
b. $h_2(x) = (x + 1)^2 - 3(x + 1) + 4$
c. $h_3(x) = \sqrt{x^3 + 1}$

Solution

a. Let $f(x) = x^3$ and $g(x) = \sqrt{x + 1}$. Then

$$f(g(x)) = f(\sqrt{x + 1}) = (\sqrt{x + 1})^3 = h_1(x).$$

b. Let $f(x) = x^2 - 3x + 4$ and $g(x) = x + 1$. Then

$$f(g(x)) = f(x + 1) = (x + 1)^2 - 3(x + 1) + 4 = h_2(x).$$

c. Let $f(x) = \sqrt{x}$ and $g(x) = x^3 + 1$. Then

$$f(g(x)) = f(x^3 + 1) = \sqrt{x^3 + 1} = h_3(x).$$ ■

Function decomposition is not unique. For example, in Example 5c notice that we could let $f(x) = \sqrt{x + 1}$ and let $g(x) = x^3$. Then, $f(g(x)) = f(x^3) = \sqrt{x^3 + 1}$.

Finding the Domain of a Composition

A value $x = a$ is in the domain of a composition $y = f(g(x))$ if two things are true:

1. a is in the domain of g, and
2. $g(a)$ is in the domain of f.

There may be some values of x in the domain of g whose values $g(x)$ are not in the domain of f. For those values, $f(g(x))$ is not defined. In general, to find the domain of $f \circ g$, we first must know the domain of g and how the range of g relates to the domain of f.

Notes on Examples

Example 6b emphasizes that the domain of a composite function cannot always be determined by merely looking at the final function equation.

■ EXAMPLE 6 Finding the Domain of a Composition

Let $f(x) = x^2 - 1$ and $g(x) = \sqrt{x}$. Find the domain of the composite function.

a. $g \circ f$ **b.** $f \circ g$

Solution

a.
$$\begin{aligned}(g \circ f)(x) &= g(f(x)) \\ &= g(x^2 - 1) && x^2 - 1 \text{ must be in the domain of } g. \\ &= \sqrt{x^2 - 1} && x^2 - 1 \text{ must be nonnegative.}\end{aligned}$$

The domain of $y = g(f(x))$ consists of all real numbers such that $x^2 - 1 \geq 0$. The solution of this inequality is

$$x \leq -1 \text{ or } x \geq 1.$$

Therefore, the domain of $g \circ f$ consists of all numbers that belong to the interval $(-\infty, -1]$ or the interval $[1, \infty)$.

Notes on Examples

Examples 7 and 8 anticipate the important calculus topic of related rates.

TECHNOLOGY TIP

Try entering $y_1 = g(x) = \sqrt{x}$ and $y_2 = y_1^2 - 1$ to see if your grapher respects the domain restriction of $f \circ g$ found in Example 6.

Follow-up

Ask . . .

Given two functions f and g, how can you determine the domains of g/f and $g \circ f$?

Assignment Guide

Day 1: Ex. 3–33, multiples of 3, 14
Day 2: Ex. 37–40, 42–45

Cooperative Learning

Group Activity: Ex. 46–49

Notes on Exercises

Ex. 1–10 are routine. Assign enough of them to establish the notation for the sum, difference, product, and quotient of functions.
Ex. 15–24 establish the notation for the composition of functions.
Ex. 46–49 give students a chance to explore the sequence defined recursively by $x_n = f(x_{n-1})$ for a particular function and initial value.

Ongoing Assessment

Self-Assessment: Ex. 5, 7, 17, 23, 41
Embedded Assessment: Ex. 12, 42

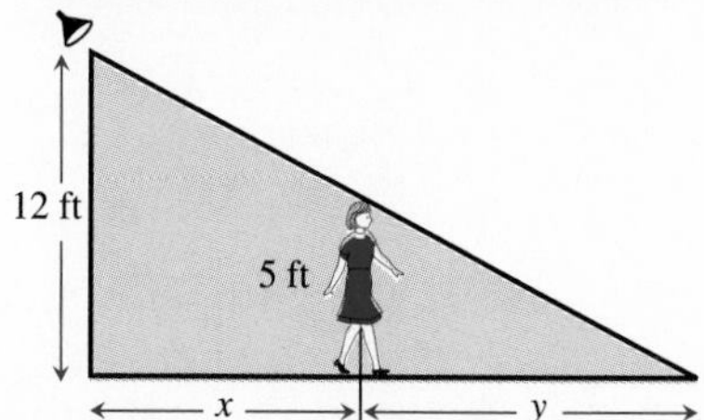

Figure 2.31 Anita casts a shadow y feet long as she walks away from the 12-ft security light.

b. $(f \circ g)(x) = f(g(x))$

$= f(\sqrt{x})$ x must be in the domain of g, so $x \geq 0$.

$= (\sqrt{x})^2 - 1$

$= x - 1$ $(\sqrt{x})^2 = x$ if $x \geq 0$.

Considering only the statement $(f \circ g)(x) = x - 1$, we might be tempted to conclude that the domain of $f \circ g$ is all real numbers. However, we see that x must be nonnegative to be in the domain of g. So the domain of $f \circ g$ is the set $[0, \infty)$. ■

More Applications

■ **EXAMPLE 7 APPLICATION: Finding a Shadow Length**

Anita is 5 ft tall. She is walking at the rate of 4 ft/sec away from a security light. The lamp is 12 ft above level ground. Find the length of her shadow 7 sec after Anita passes directly under the lamp.

Solution

Model

Let t = time (in seconds) since passing under the lamp, y = length (in feet) of Anita's shadow, x = distance (on the ground) of Anita from the lamppost. Then $x = 4t$, and from Figure 2.31 we see that

$$\frac{12}{5} = \frac{x + y}{y} \quad \text{Using similar triangles}$$

$$12y = 5x + 5y \quad \text{Multiply by } 5y.$$

$$7y = 5x \quad \text{Subtract } 5y.$$

$$y = \frac{5}{7}x \quad \text{Divide by 7.}$$

$$y = \frac{5}{7}(4t) \quad \text{Substitute } x = 4t.$$

$$y = \frac{20}{7}t$$

Solve Algebraically

When $t = 7$, $y = \frac{20}{7}(7) = 20$.

Interpret

In 7 sec, Anita's shadow is 20 ft long. ■

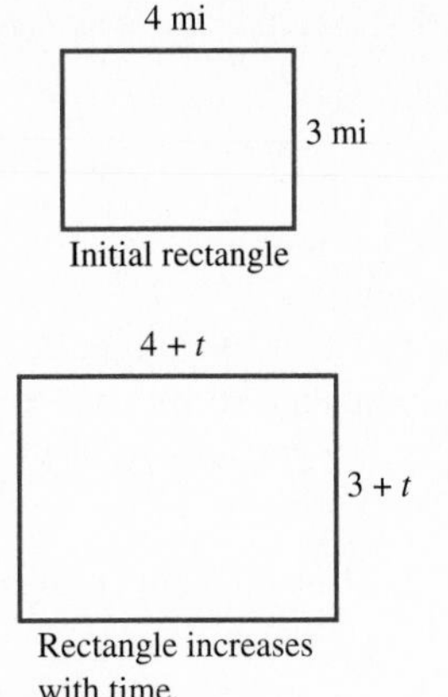

Figure 2.32

Example 8 simulates the zoom-out capability of a camera in space.

■ EXAMPLE 8 APPLICATION: Sizing a Photograph from Space

The dimensions of a rectangular region viewed through a space shuttle camera are 3 mi by 4 mi. As the camera zooms out, the length and width of the rectangle increase at the rate of 1 mi/msec (mile per millisecond). Estimate how long it takes for the area of the rectangle to be at least 120 square miles.

Solution

Model

See Figure 2.32.

$$\begin{aligned} \text{Area of rectangle} &= \text{length} \times \text{width} \\ t &= \text{time (in milliseconds)} \\ 4 + t &= \text{length after } t \text{ milliseconds} \\ 3 + t &= \text{width after } t \text{ milliseconds} \\ \text{Area of rectangle (after } t \text{ milliseconds)} &= (4 + t)(3 + t) \end{aligned}$$

Solve Graphically

$$(4 + t)(3 + t) \geq 120 \quad \text{Inequality to be solved}$$

The graphs of $y_1 = (4 + t)(3 + t)$ and $y_2 = 120$ (see Figure 2.33) show that $y_1 \geq 120$ for $t \approx 7.5$.

Interpret

The area of the rectangle will be at least 120 mi^2 in about 7.5 msec. ■

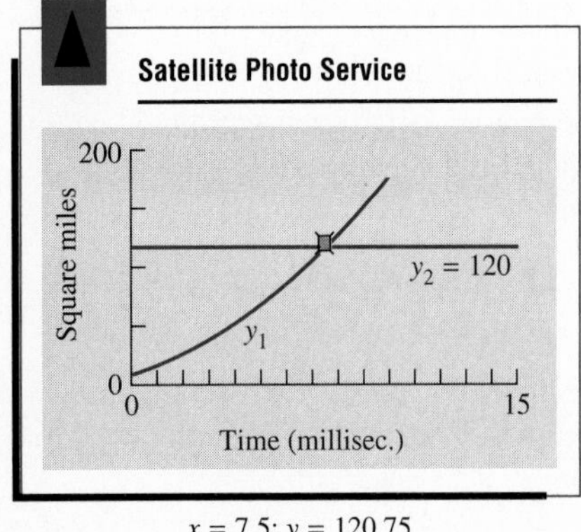

$x = 7.5$; $y = 120.75$

Figure 2.33 Graphs of $y_1 = (4 + x)(3 + x)$ and $y_2 = 120$.

Quick Review 2.4

In Exercises 1–4, perform the operation and simplify the resulting expression.

1. $(3x^2 - 5x^3) - [(x + 3) + x^2]$
2. $(x + 3)^2 - 4$
3. $(2x^2 + 1)(x^2 - 1)$
4. $(x + 1)(x^3 - x)$

In Exercises 5–10, find the domain of the function.

5. $y = \sqrt{x + 1}$
6. $y = \dfrac{1}{x}$
7. $y = x^2 + 3x$
8. $y = \dfrac{1}{(x - 3)(x + 2)}$
9. $y = \dfrac{2x - 3}{\sqrt{x - 1}}$
10. $y = \dfrac{\sqrt{x + 4}}{\sqrt{5 - x}}$

SECTION EXERCISES 2.4

In Exercises 1–6, describe $f + g$, $f - g$, and fg as expressions in x. Also state the domain of each function.

1. $f(x) = 2x - 1;\ g(x) = x^2$

2. $f(x) = (x - 1)^2;\ g(x) = 3 - x$

3. $f(x) = x^2;\ g(x) = 2x$

4. $f(x) = \sqrt{x};\ g(x) = x - 2$

5. $f(x) = x + 3;\ g(x) = \dfrac{2x - 1}{3}$

6. $f(x) = \sqrt{x + 5};\ g(x) = |x + 3|$

In Exercises 7–10, describe f/g and g/f as expressions in x. State the domain of each function.

7. $f(x) = \sqrt{x + 3};\ g(x) = x^2$

8. $f(x) = \sqrt[3]{x + 1};\ g(x) = x^2 + 1$

9. $f(x) = \sqrt{x - 2};\ g(x) = \sqrt{x + 4}$

10. $f(x) = (x + 3)(x - 2);\ g(x) = x^2$

In Exercises 11–14, graphs of f and g are shown in a viewing window. Copy the window on paper, and sketch the graph of h in the window. Then graph h and compare with your sketch.

11. $f(x) = x^2$, $g(x) = 1/x$, and $h = f + g$. (*Hint:* Recall that $(f + g)(x) = f(x) + g(x)$.)

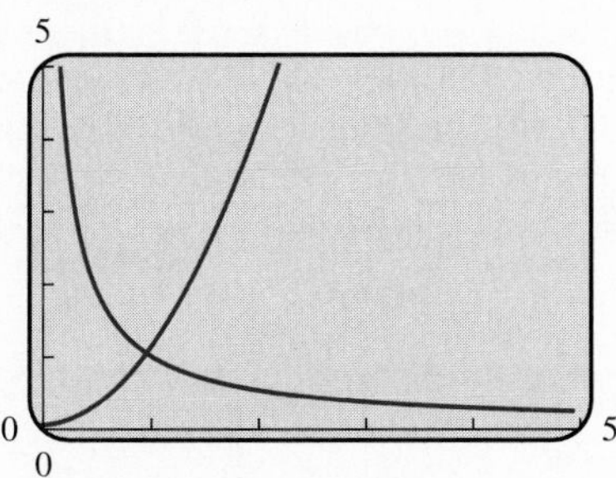

For Exercise 11

12. $f(x) = x^2$, $g(x) = \sqrt{x}$, and $h = f - g$. (*Hint:* Recall that $(f - g)(x) = f(x) - g(x)$.)

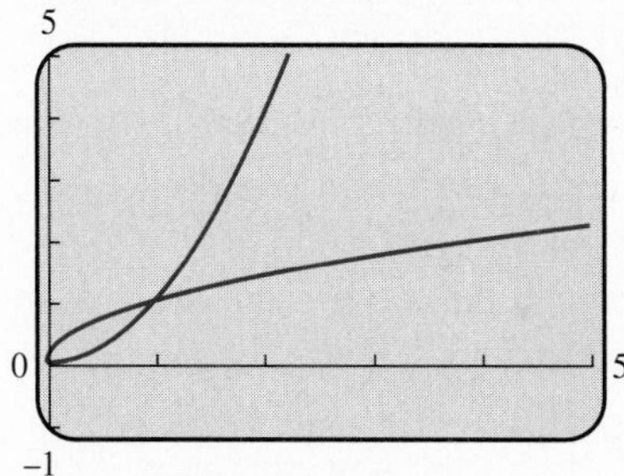

For Exercise 12

13. $f(x) = x^2$, $g(x) = 2x - 3$, and $h = f + g$. Explain why key points in drawing your sketch occur at $x = 0$ and $x = 3/2$.

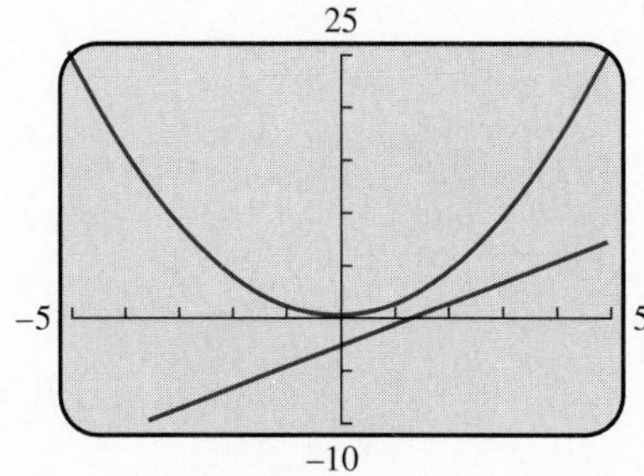

For Exercise 13

14. $f(x) = x^2$, $g(x) = 4 - 3x$, and $h = f - g$. Explain why key points of your sketch occur at $x = -4$ and $x = 1$.

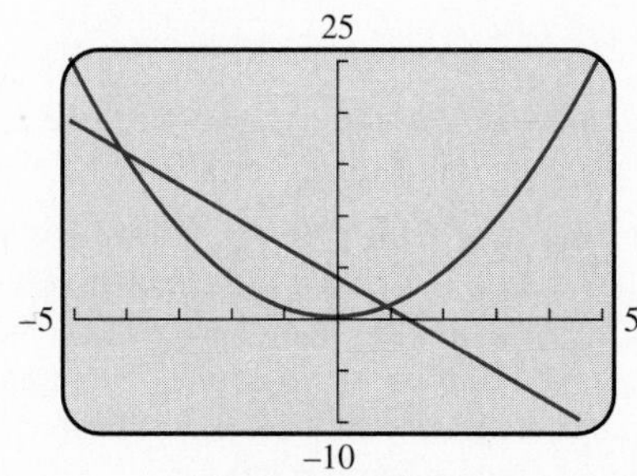

For Exercise 14

In Exercises 15–18, find $(f \circ g)(3)$ and $(g \circ f)(-2)$.

15. $f(x) = 2x - 3;\ g(x) = x + 1$

16. $f(x) = x^2 - 1;\ g(x) = 2x - 3$

17. $f(x) = x^2;\ g(x) = \sqrt{x - 1}$

18. $f(x) = 2x - 3;\ g(x) = x^2 - 2x + 3$

In Exercises 19–24, find $f(g(x))$ and $g(f(x))$. State the domain of each.

19. $f(x) = 3x + 2;\ g(x) = x - 1$

20. $f(x) = x^2 - 1;\ g(x) = \dfrac{1}{x - 1}$

21. $f(x) = 2x - 5;\ g(x) = \dfrac{x + 3}{2}$

22. $f(x) = x^2 - 2;\ g(x) = \sqrt{x + 1}$

23. $f(x) = \dfrac{1}{x - 1};\ g(x) = \sqrt{x}$

24. $f(x) = x^2 - 3;\ g(x) = \sqrt{x + 2}$

In Exercises 25–36, find $f(x)$ and $g(x)$ so that the function can be described as $y = f(g(x))$.

25. $y = \sqrt{x^2 - 5x}$

26. $y = (x^3 + 1)^2$

27. $y = |3x - 2|$

28. $y = \dfrac{1}{x^3 - 5x + 3}$

29. $y = \sqrt[3]{x^2 + 1}$

30. $y = |x^2 + 5|$

31. $y = (x + 3)^2$

32. $y = \left(\dfrac{1}{x + 1}\right)^3$

33. $y = \sqrt{x + 3}$

34. $y = \dfrac{2}{(x - 3)^2}$

35. $y = (x - 3)^4 - 2$

36. $y = 3 - \sqrt{x}$

In Exercises 37–44, solve the given problem.

37. *Angioplasty Treatment* How long does it take the balloon in Example 3 to reach a volume of 900 mm^3?

38. *Weather Balloons* A high-altitude spherical weather balloon expands as it rises due to the drop in atmospheric pressure. Suppose that the radius r increases at the rate of 0.03 in./sec and that $r = 48$ in. at time $t = 0$. Determine an equation that models the volume v of the balloon at time t, and find the volume when $t = 300$ sec.

39. *Lengthening Shadows* For Example 7 express the distance D between the lamp and the tip of Anita's shadow as a function of t. When will that distance D be 100 ft?

40. *Lengthening Shadows* Leon is 6 ft 8 in. tall and walks at the rate of 5 ft/sec away from a streetlight with a lamp 15 ft above ground level. Find an equation that models the length l of Leon's shadow, and find the length of the shadow after 5 sec.

41. *Satellite Photography* A satellite camera takes a rectangular-shaped picture. The smallest region that can be photographed is a 5-km by 7-km rectangle. As the camera zooms out, the length l and width w of the rectangle increase at the rate of 2 km/sec. How long does it take for the area a to be at least 5 times its initial size?

42. *Computer Imaging* New Age Special Effects, Inc., prepares computer software based on specifications prepared by film directors. To simulate an approaching vehicle they begin with a computer image of a 5-cm by 7-cm by 3-cm box. The program increases each dimension at a rate of 2 cm/sec. How long does it take for the volume v of the box to be at least 5 times its initial size?

43. *Wave Effect* A rock is tossed into a pond. The radius of the first circular ripple (wave) increases at the rate of 0.55 ft/sec. Find an equation that models the area a enclosed by the ripple, and find the area when $t = 6$ sec.

44. *Surface Area* A hard candy ball of radius 1.6 cm is placed in a glass of water. As it dissolves, its radius decreases at the rate of 0.0027 cm/sec. Find an equation that models the surface area of the candy ball as a function of t. The surface area of a sphere of radius r is given by $S = 4\pi r^2$. When will the candy be completely dissolved?

45. Writing to Learn Each screen shows part of the graph of $A(t) = (3 + t)(4 + t)$ from Example 8. One graph shows "the important characteristics" of the function $A(t)$. The other shows that portion of the graph "relevant to the problem" discussed in Example 8. Describe the significance of these two phrases in quotation marks.

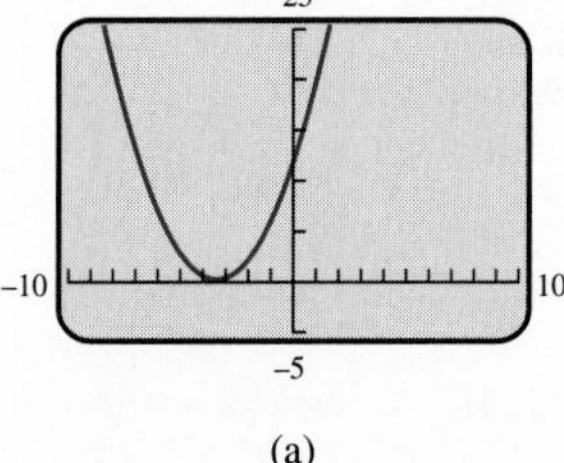

(a)

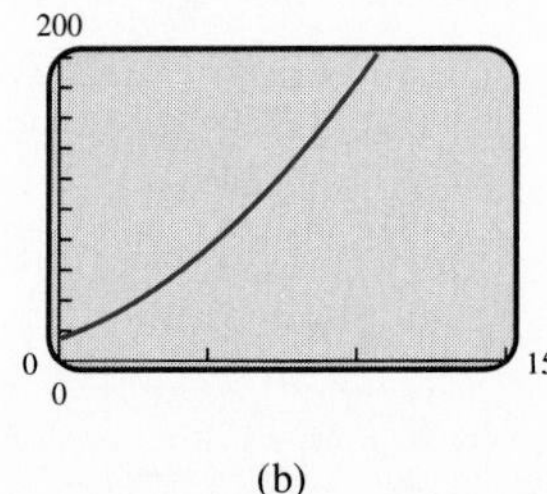

(b)

For Exercise 45

EXTENDING THE IDEAS

In Exercises 46–49, we will use the function defined by $w(x) = 2.8x(1 - x)$ to define a sequence of numbers as follows. Let x_1 be any number such that $0 < x_1 < 1$, and let $x_2 = w(x_1)$, $x_3 = w(x_2)$, $x_4 = w(x_3), \ldots, x_n = w(x_{n-1}), \ldots$.

46. Group Learning Activity *Work in groups of three.* Show that this sequence can be described as the composition of w with w by showing the following.

a. $x_3 = w(w(x_1))$ and $x_4 = w(w(w(x_1)))$.
b. Find x_5 for $x_1 = 0.2$.

47. Group Learning Activity *Work in groups of three.* The figure shows graphs of $y = 2.8x(1 - x)$ and $y = x$.

a. Recall that $x_2 = w(x_1)$. Explain why $x_2 = b = a$.
b. Explain why $(c, d) = (x_2, w(x_2))$.
c. Describe the coordinates of the point (e, f) in terms of this sequence.

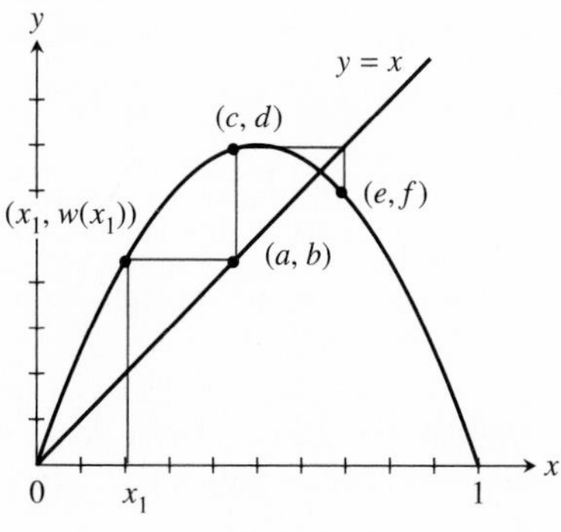

For Exercise 47

48. Group Learning Activity *Work in groups of three.* A grapher with a *sequence mode* and table-generating capabilities can be used to generate the sequence of values $x_1, x_2, \ldots$, where $x_n = w(x_{n-1})$, as follows. In sequence mode, define $u_n = 2.8u_{n-1}(1 - u_{n-1})$. Find the table if $x_1 = 0.2$. In "WINDOW" set "u_nStart = 0.2" and "nStart = 1" and notice the table generated. What numerical value does the sequence $x_n = w(x_{n-1})$ seem to approach? Does this agree with a value that you might infer from the figure for Exercise 47? Explain.

49. Group Learning Activity *Work in groups of three.* The figure shows the graphs of $y = 2.8x(1 - x)$ and $y = x$ and a pattern that is known as a *web pattern.* Discuss how the graph is related to the sequence described in Exercise 48. (This graph is studied in chaos theory.)

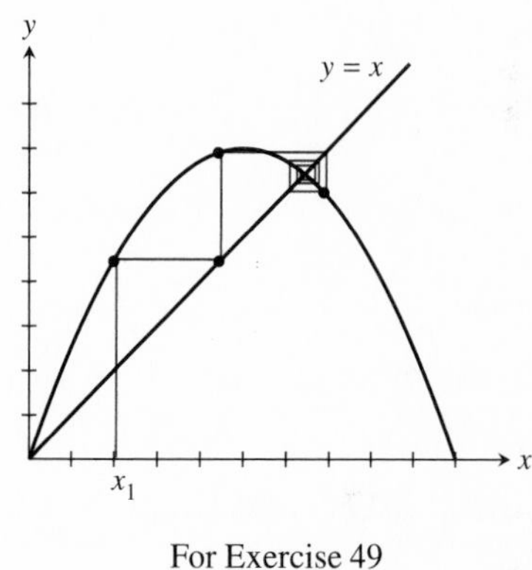

For Exercise 49

2.5 RELATIONS AND PARAMETRIC EQUATIONS

Defining a Relation with an Equation • Introduction to Parametric Equations • Graphing Parametric Equations • Using Parametric Equations for Graphs of Functions

Objective

Students will be able to specify a relation using an equation, inequality, or parametric equations.

Motivate

Ask for an example of an equation in x and y that is easy to graph but does not represent a function.

Defining a Relation with an Equation

Any set of ordered pairs (x, y) of real numbers is a **relation.** The set may be defined to be solutions of an equation (or inequality) in x and y. If a pair (a, b) is a solution, we say that a and b "are related" or that (a, b) is "in the relation." The **graph of a relation** is the set of points in the coordinate plane corresponding to the ordered pairs of the relation.

Alert

Many students will expect the statement "a and b are related" to be equivalent to the statement "b and a are related." You may wish to point out that this is true only if the relation is *symmetric.*

■ **EXAMPLE 1 A Relation Defined by an Equation in x and y**

Determine which of the ordered pairs $(2, -5)$, $(1, 3)$, and $(2, 1)$ are in the relation $x^2y + y^2 = 5$.

Solution

$$x^2y + y^2 = (2^2)(-5) + (-5)^2 \quad \text{Substitute } x = 2 \text{ and } y = -5.$$
$$= -20 + 25 = 5 \quad \text{It checks.}$$
$$x^2y + y^2 = (1)^2(3) + (3)^2 \quad \text{Substitute } x = 1 \text{ and } y = 3.$$
$$= 3 + 9 \neq 5 \quad \text{It doesn't check.}$$
$$x^2y + y^2 = (2)^2(1) + (1)^2 \quad \text{Substitute } x = 2 \text{ and } y = 1.$$
$$= 4 + 1 = 5 \quad \text{It checks.}$$

Thus, $(2, -5)$ and $(2, 1)$ are in the relation, whereas $(1, 3)$ is not. ■

Example 1 shows that the ordered pairs $(2, -5)$ and $(2, 1)$ are both in the relation $x^2y + y^2 = 5$. So the vertical line $x = 2$ must intersect the graph of this relation in at least two points. Thus, we can conclude from the vertical line test that the relation is not a function of x.

Not all relations are defined by equations. For example, the inequality $x^2y + y^2 \geq 5$ defines a relation. Note that all three points $(2, -5)$, $(1, 3)$, and $(2, 1)$ are in this relation.

Alert

Many students always expect the variable t to represent time. Point out that this is frequently the case, but not always.

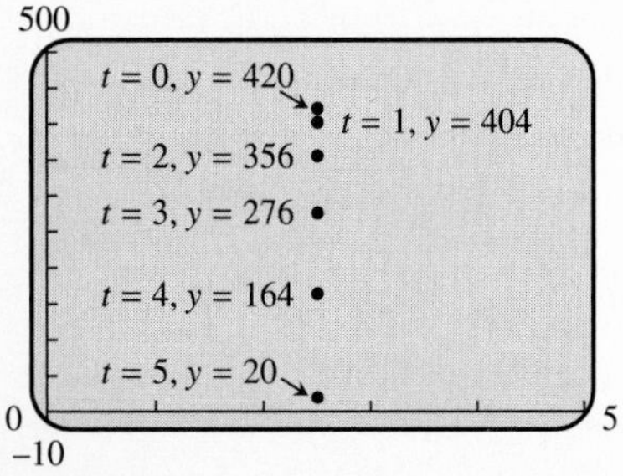

Introduction to Parametric Equations

Imagine that a rock is dropped from a 420-ft tower. The rock's height y feet above the ground t seconds later (ignoring air resistance) is modeled by $y = -16t^2 + 420$. Figure 2.34 shows a coordinate system imposed on the scene so that the line of the rock's fall is on the vertical line $x = 2.5$.

The rock's original position and its position after each of the first 5 sec are the points

$$(2.5, 420), \quad (2.5, 404), \quad (2.5, 356), \quad (2.5, 276), \quad (2.5, 164), \quad (2.5, 20),$$

which are described by the pair of equations

$$x = 2.5, \qquad y = -16t^2 + 420,$$

when $t = 0, 1, 2, 3, 4, 5$. These two equations are an example of *parametric equations.* In general, when two equations in t determine ordered pairs (x, y), the equations

$$x = x(t), \qquad y = y(t),$$

are called **parametric equations,** and the resulting ordered pairs (x, y) form a relation. The variable t is a **parameter.**

Figure 2.34 The position of the rock at 0, 1, 2, 3, 4, and 5 sec.

■ EXAMPLE 2 A Relation Defined with Parametric Equations

Find the ordered pairs in the relation defined by

$$x(t) = t^2 - 2, \qquad y(t) = 3t,$$

for $t = -2$, 1 and 5.

Solution

$$\begin{aligned} x(-2) &= (-2)^2 - 2 = 2 & y(-2) &= 3(-2) = -6 \\ x(1) &= (1)^2 - 2 = -1 & y(1) &= 3(1) = 3 \\ x(5) &= (5)^2 - 2 = 23 & y(5) &= 3(5) = 15 \end{aligned}$$

The ordered pairs are $(2, -6)$, $(-1, 3)$, and $(23, 15)$. ■

For parametric equations that define a relation we sometimes drop the "(t)". Then

$$x(t) = t^2 - 2, \qquad y(t) = 3t,$$

become

$$x = t^2 - 2, \qquad y = 3t.$$

If there is more than one set of parametric equations, we may distinguish between them using subscripts:

$$\begin{aligned} x_1(t) &= t^2 - 2, & y_1(t) &= 3t, \\ x_2(t) &= \sqrt{t + 3}, & y_2(t) &= 5t + 1. \end{aligned}$$

Alert

Many students will confuse range values of t with range values on the function grapher. Point out that while the scale factor does not affect the way a graph is drawn, the *T*step does affect the way the graph is displayed.

Graphing Parametric Equations

The set of all points (x, y) in the coordinate plane for the ordered pairs determined by a pair of parametric equations

$$x = f(t), \qquad y = g(t),$$

is the **graph of the parametric equations.** The *domain of the parameter t* determines exactly which points are in the graph.

The graphs in Example 3 can be produced on a grapher by using *parametric mode.* Notice that different domains of the parameter t result in different relations.

Notes on Examples

Example 3 is important because it shows how the range of values chosen for t affects the graph.

■ EXAMPLE 3 Graphing Parametric Equations

For the given domain of t, graph the parametric equations

$$x(t) = t^2 - 2, \qquad y(t) = 3t.$$

a. $-3 \le t \le 1$ **b.** $-2 \le t \le 3$ **c.** $-3 \le t \le 3$

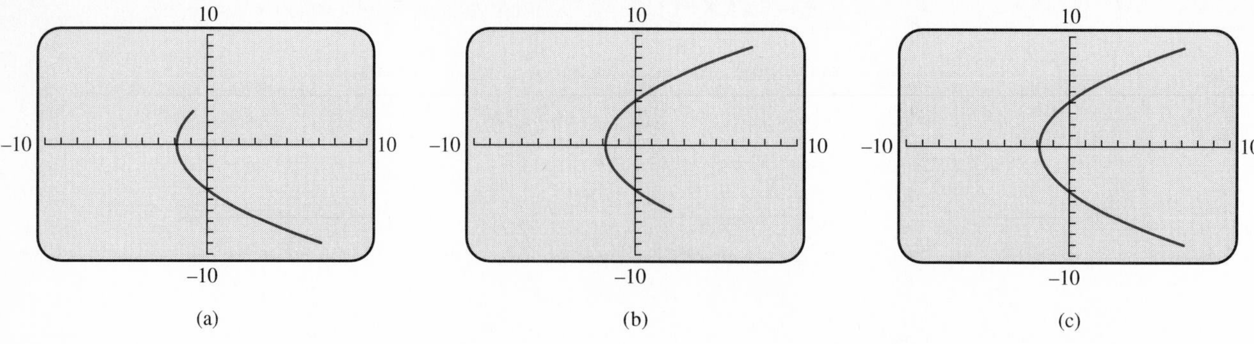

Figure 2.35 Three different relations result from three different parameter intervals: (a) $-3 \le t \le 1$; (b) $-2 \le t \le 3$; and (c) $-3 \le t \le 3$.

GRAPHER NOTE

Note that in *parametric mode* we can control what portion of a curve we graph by the interval we choose for the parameter t. In *function mode* we do not have that flexibility.

Solution

Figure 2.35 shows a graph of the parametric equations for each t domain.[3] The relations are different because the domains of t are different. ■

Using Parametric Equations for Graphs of Functions

The graph of a pair of parametric equations may *not* be the graph of a function. For example, in each of the graphs in Figure 2.35 there are vertical lines that intersect the graph in two points. That implies there are two y-values associated with one x-value—evidence that the graphs don't represent functions of x.

On the other hand, each function $y = f(x)$ can be represented by the parametric equations

$$x = t, \qquad y = f(t).$$

Teaching Note

If students are not familiar with parametric graphing, it might be helpful to show them the graph of the linear function $f(x) = 3x - 2$ and compare it to one defined parametrically as $x = t$ and $y = 3t - 2$, using a trace key to show how t, x, and y are related.

■ EXAMPLE 4 Graphing a Function in Parametric Mode

Graph the function $f(x) = x^2 + 8$ in function mode and in parametric mode.

Solution

Figure 2.36a shows the graph of $y = x^2 + 8$, $-4.7 \le x \le 4.7$, in function mode. Figure 2.36b uses parametric mode for the graph of

$$x = t, \qquad y = t^2 + 8, \qquad -4.7 \le t \le 4.7.$$ ■

Teaching Note

The parametric graphing capabilities of modern technology will greatly enhance a student's ability to visualize the graph of a relation.

[3] In parametric mode the window settings include values Tmin and Tmax that set the domain of the parameter. Consult the Grapher Workshop.

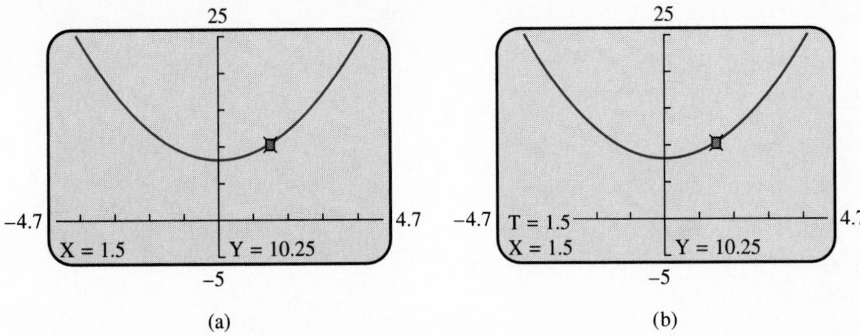

Figure 2.36 In (a) function mode, $X\text{min} = -4.7$ and $X\text{max} = 4.7$. In (b) parametric mode, $T\text{min} = -4.7 = X\text{min}$ and $T\text{max} = 4.7 = X\text{max}$. Both windows show identical portions of the graph of $f(x) = x^2 + 8$.

Example 5 solves a projectile-motion problem. Parametric equations are used in two ways: (a) to find a graph of the modeling equation and (b) to simulate the motion.

■ EXAMPLE 5 APPLICATION: Simulating Projectile Motion

A distress flare is shot straight up from a ship's bridge 75 ft above the water level with an initial velocity of 76 ft/sec. Graph the flare's height against time, and simulate the flare's motion for each length of time.

a. 1 sec **b.** 2 sec **c.** 4 sec **d.** 5 sec

Solution

Model

The equation that models the flare's height above the water t seconds after launch is

$$y = -16t^2 + v_0 t + s_0,$$

where $v_0 = 76$ and $s_0 = 75$.

A graph of the flare's height against time can be found using the parametric equations

$$x_1 = t, \qquad y_1 = -16t^2 + 76t + 75.$$

To simulate the flare's flight straight up and its fall to the water, use the parametric equations

$$x_2 = 5.5, \qquad y_2 = -16t^2 + 76t + 75.$$

(We chose $x_2 = 5.5$ so that the two graphs would not intersect.)

TECHNOLOGY TIP

Parametric mode is powerful because it gives us the ability to stop the motion of the flare (or projectile) at the end of any given interval of time, a capability not possible in function mode. Using $Tstep = 0.1$ allows us to trace along the curve and show positions after each tenth of a second.

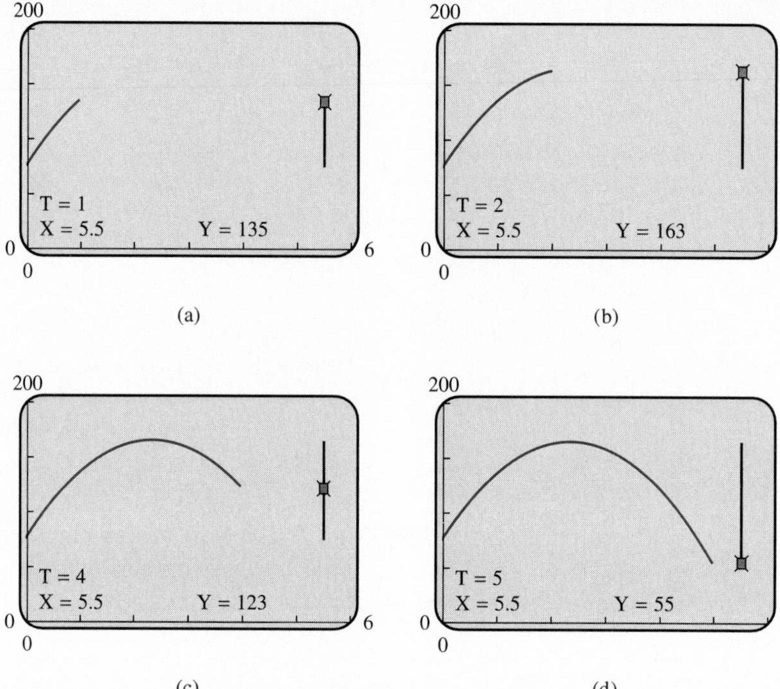

Figure 2.37 Simultaneous graphing of both $x_1 = t$, $y_1 = -16t^2 + 76t + 75$ (height against time) and $x_2 = 5.5$, $y_2 = -16t^2 + 76t + 75$ (the actual path of the flare).

Follow-up

Ask . . .

How can you display the graphs shown in Figure 2.37 on your grapher?

Assignment Guide

Day 1: Ex. 3–33, multiples of 3
Day 2: Ex. 36, 37–41, 44, 45, 48

Cooperative Learning

Group Activity: Ex. 36

Notes on Exercises

Ex. 1–24 give students practice in making the connection between algebraic and numerical representations of relations.
Ex. 35–38 require the use of parametric equations to solve applications.
Ex. 39 and 40 show how to use a grapher to graph a relation that is not a function by solving for *y*.

Ongoing Assessment

Self-Assessment: Ex. 7, 17, 23, 35, 43
Embedded Assessment: Ex. 38

Solve Graphically

Figure 2.37a shows the graphs for $0 \le t \le 1$. Figures 2.37b, c, and d show the graph for t reaching 2, 4, and 5 seconds.

Interpret

The height of the rocket after 1 sec is 135 ft. After 2 sec it is 163 ft. After 4 sec it is 123 ft, and after 5 sec it is 55 ft above the ground. ■

Quick Review 2.5

In Exercises 1–3, evaluate the expression for the given pair of values.

1. $2x^2y + 3xy$ for $x = 3$, $y = -1$

2. $x\sqrt{y + 3} - x^2y^2$ for $x = -3$, $y = 1$

3. $2xy + y^2x - 3$ for $x = 2$, $y = -3$

In Exercises 4–9, solve for y by using the quadratic formula.

4. $y^2 - 4y + 2 = 0$

5. $3y^2 + 5y - 1 = 0$

6. $y^2 + xy - 2 = 0$

7. $y^2 + 2xy - 3 = 0$

8. $y^2 + x^2y - 3 = 0$

9. $y^2 + xy - 1 = 0$

SECTION EXERCISES 2.5

In Exercises 1–4, show that one point is in the relation and the other is not.

1. $2x - y = -2$; (4, 2) and (3, 8)

2. $2x^2 - y = 3$; (2, 5) and (5, 2)

3. $x + y^2 < 33$; (−2, 6) and (−2, 4)

4. $5x^3 - 2y > 1$; (3, 2) and (−3, 2)

In Exercises 5–12, determine whether the ordered pair is in the relation.

5. (3, 5); $x^2 + y^2 = 8$

6. (7, 2); $x - 5y = -3$

7. (1, 2); $x^3 + 3y = -1$

8. (3, 1); $x - 4y^3 = 2$

9. (2, 3); $x^2 - y = 3$

10. (3, 4); $x^2 + y^2 = 25$

11. $(2, \sqrt{2})$; $x^3 - 3y^2 = 2$

12. $(\sqrt{3}, 2)$; $4x - 3y^2 = 5$

In Exercises 13–18, find the (x, y) pair for the value of the parameter.

13. $x = 3t$ and $y = t^2 + 5$ for $t = 2$

14. $x = 5t - 7$ and $y = 17 - 3t$ for $t = -2$

15. $x = 4 - 3t$ and $y = 2t + 5$ for $t = 3$

16. $x = t^2 + 5t$ and $y = 3 - t^2$ for $t = -1$

17. $x = t^3 - 4t$ and $y = \sqrt{t + 1}$ for $t = 3$

18. $x = |t + 3|$ and $y = 1/t$ for $t = -8$

In Exercises 19–24, graph the relation defined by the parametric equations. Use a $[-5, 5]$ by $[-5, 5]$ window and $-2 \leq t \leq 2$.

19. $x = 2t,\ y = 3t - 1$

20. $x = 5t + 1,\ y = 3t - 2$

21. $x = t^2,\ y = t + 1$

22. $x = t,\ y = t^2 - 3$

23. $x = t,\ y = t^3 - 2t + 3$

24. $x = t^3 - 2t + 3,\ y = t$

In Exercises 25–28, refer to the graph of the parametric equations

$$x = 2 - |t|, \quad y = t - 0.5,$$

where $-3 \leq t \leq 3$. Find the values of the parameter t that produces the graph in the indicated quadrant.

25. Quadrant I

26. Quadrant II

27. Quadrant III

28. Quadrant IV

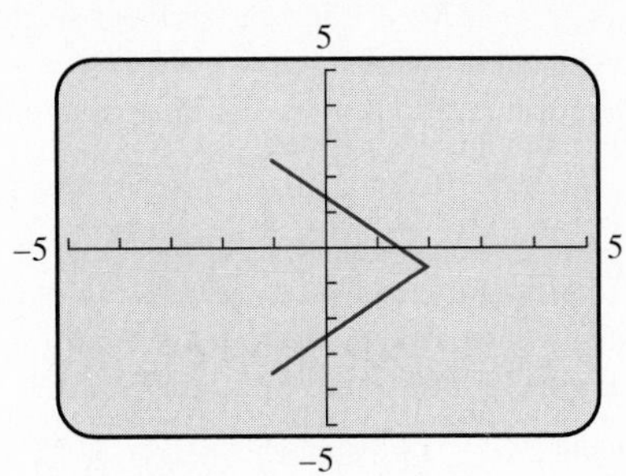

For Exercises 25–28

In Exercises 29–34, use parametric mode and the right choice of Tmin and Tmax to show only the portion of the graph that lies in Quadrant I. Use a standard viewing window.

29. $f(x) = x^2 - 4x + 6$

30. $h(x) = 7 - 2x$

31. $g(x) = 9 - 5x$

32. $f(x) = -x^2 + 2x + 6.84$

33. $h(x) = x^2 - 2x - 3$

34. $g(x) = \sqrt{x + 4}$

In Exercises 35 and 36, solve the projectile-motion problem.

35. *Famine Relief Air Drop* A relief agency drops food containers from an airplane on a war-torn famine area. The drop was made from an altitude of 1000 ft above ground level.

- **a.** Use an equation to model the height of the containers (during free fall) as a function of time t.
- **b.** Use parametric mode to simulate the drop during the first 6 sec.
- **c.** After 4 sec of free fall, parachutes open. How many feet above the ground are the food containers when the parachutes open?

36. *Height of a Pop-up* A baseball is hit straight up from a height of 5 ft with an initial velocity of 80 ft/sec.

- **a.** Write an equation that models the height of the ball as a function of time t.
- **b.** Use parametric mode to simulate the pop-up.
- **c.** Use parametric mode to graph height against time. (*Hint:* Let $x(t) = t$.)
- **d.** How high is the ball after 4 sec?
- **e.** What is the maximum height of the ball? How many seconds does it take to reach its maximum height?

In Exercises 37 and 38, solve the problem.

37. *Simulating a Foot Race* Ben can sprint at the rate of 24 ft/sec. Jerry sprints at 20 ft/sec. Ben gives Jerry a 10-ft head start. The parametric equations can be used to model a race.

$$x_1 = 20t, \qquad y_1 = 3$$
$$x_2 = 24t - 10, \quad y_2 = 5$$

a. Find a viewing window to simulate a 100-yd dash. Graph simultaneously with t starting at $t = 0$ and Tstep $= 0.05$.
b. Who is ahead after 3 sec and by how much?

38. *Capture the Flag* Two opposing players in "Capture the Flag" are 100 ft apart. On a signal, they run to capture a flag that is on the ground midway between them. The faster runner, however, hesitates for 0.1 sec. The following parametric equations model the race to the flag:

$$x_1 = 10(t - 0.1), \quad y_1 = 3$$
$$x_2 = 100 - 9t, \qquad y_2 = 3$$

a. Simulate the game in a [0, 100] by [−1, 10] viewing window with t starting at 0. Graph simultaneously.
b. Who captures the flag and by how many feet?

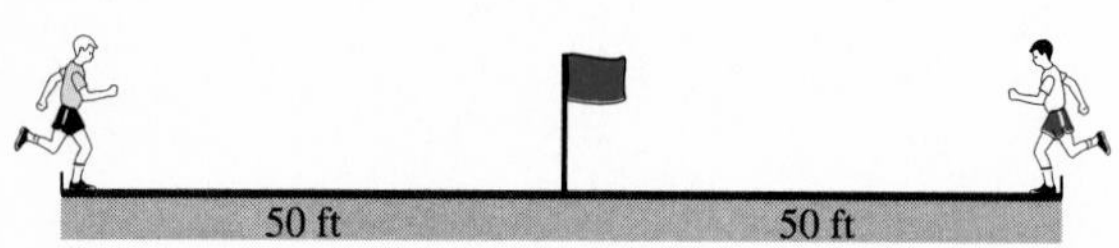

For Exercise 38

EXTENDING THE IDEAS

In Exercises 39 and 40, use parametric equations to graph a relation that is not a function.

39. For the relation $x^2 + y^2 = 25$, solve for $y = \pm\sqrt{25 - x^2}$. Graph both $y_1 = \sqrt{25 - x^2}$ and $y_2 = -\sqrt{25 - x^2}$ in the same viewing window.

40. For the relation $y^2 + x^2y - 5 = 0$, use the quadratic formula to solve for y:

$$y = \frac{-x^2 \pm \sqrt{x^4 + 20}}{2}$$

Graph both

$$y_1 = \frac{-x^2 + \sqrt{x^4 + 20}}{2} \quad \text{and} \quad y_2 = \frac{-x^2 - \sqrt{x^4 + 20}}{2}$$

in the same viewing window.

In Exercises 41–48, graph the relation by using one of the methods of Exercises 39 and 40.

41. $x^2 + y^2 = 36$
42. $y^2 - x^2 = 1$
43. $y^2 - 16 = x^2y^2$
44. $x^2y^2 = 1 - 4y^2$
45. $y^2 + x^2y - 3 = 0$
46. $y^2 + 2xy - 4 = 0$
47. $y^2 - 3xy - x^2 = 0$
48. $y^2 - x^3y - x^2 = 0$

2.6 DESCRIBING GRAPHS USING TRANSFORMATIONS

Objective

Students will be able to algebraically and graphically represent translations, reflections, stretches, and shrinks of functions and parametric relations.

Motivate

Ask . . .

How is the graph of $(x - 2)^2 + (y + 1)^2 = 16$ related to the graph of $x^2 + y^2 = 16$? **(Translated 2 units right and 1 unit down.)**

Introduction • Vertical and Horizontal Translations • Reflections across Axes • Representing Translations and Reflections Parametrically • Vertical and Horizontal Stretch and Shrink • Connecting Geometric and Algebraic Transformations • Combining Transformations

Introduction

A function with domain and range the set of all real numbers is a **transformation.** In this section we relate graphs using transformations. A horizontal translation, vertical translation, or reflection is a **rigid transformation** that leaves the size and shape of a graph unchanged. A horizontal or vertical stretch or shrink is a **nonrigid transformation** that distorts the shape of a graph.

Vertical and Horizontal Translations

A **vertical translation** of the graph of

$$y = f(x)$$

is a shift of the graph up or down in the coordinate plane. A **horizontal translation** is a shift of the graph to the left or right.

> **Teaching Note**
>
> In Explorations 1 and 2, the grapher must be set in a sequential graphing mode rather than a simultaneous mode. This will allow students to see the transformations appear one at a time in the same order they were entered into the calculator.
>
> **Exploration Extensions**
>
> Graph the functions $y_1 = x^2$, $y_2 = (x + 3)^2 - 2$, $y_3 = (x - 1)^2 + 4$, and $y_4 = (x + 1)^2 - 3$. What effect do the added and subtracted constants seem to have?

● EXPLORATION 1 Introducing Translations

1. Graph the functions

$$y_1 = x^2,\quad y_2 = y_1 + 3,\quad y_3 = y_1 + 1,\quad y_4 = y_1 - 2,\quad y_5 = y_1 - 4,$$

in the $[-5, 5]$ by $[-5, 15]$ viewing window. What effect do the $+3$, $+1$, -2, -4 seem to have?

2. Graph the functions

$$y_1 = x^2,\qquad y_2 = (x + 3)^2,\qquad y_3 = (x + 1)^2,$$
$$y_4 = (x - 2)^2,\qquad y_5 = (x - 4)^2.$$

What effect do the $+3$, $+1$, -2, -4 seem to have?

3. Repeat parts 1 and 2 for each of these functions:

$$y_1 = x^3 \qquad y_1 = |x| \qquad y_1 = \sqrt{x}$$

Do your observations agree with those you made after Steps 1 and 2? ●

Translations are caused by adding a number c to $f(x)$ for vertical translations and to x for horizontal translations. The vertical translations are up $(c > 0)$ or down $(c < 0)$, as the value of c might suggest. The horizontal translations, however, are left $(c > 0)$ and right $(c < 0)$, opposite to what c might suggest.

> **Teaching Note**
>
> You may want to explore the connection between transformations and composition of functions. For example, if $g(x) = x + 3$, then $f(x + 3) = f(g(x))$.

Vertical Translations, $c > 0$

- $y = f(x) + c$ — a translation of f up by c units
- $y = f(x) - c$ — a translation of f down by c units

Horizontal Translations, $c > 0$

- $y = f(x + c)$ — a translation of f to the left by c units
- $y = f(x - c)$ — a translation of f to the right by c units

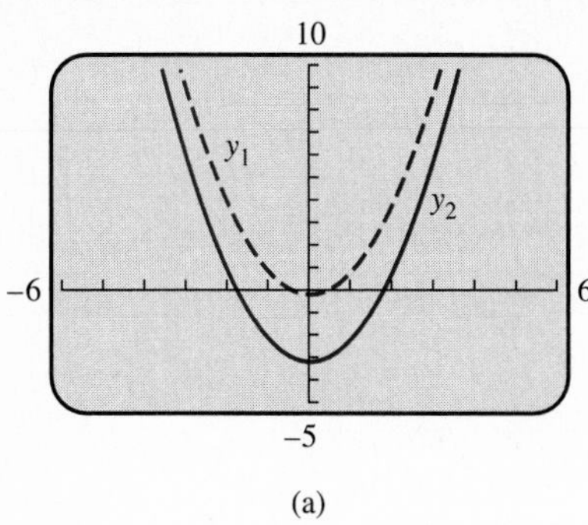

(a)

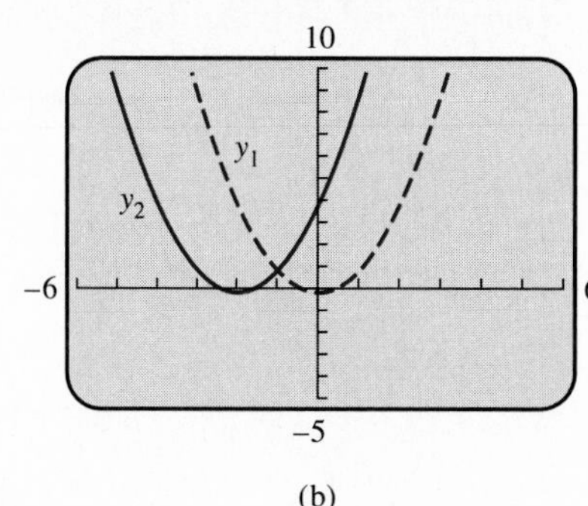

(b)

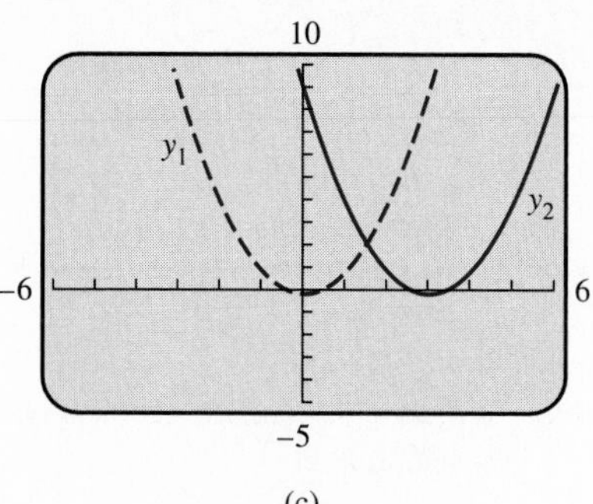

(c)

Figure 2.38 Translations of $y_1 = x^2$.

Teaching Note

Graphing utilities are great tools for showing the transformations applied to parent functions such as $y = x^2$.

■ EXAMPLE 1 Finding Equations for Vertical and Horizontal Translations

Each view in Figure 2.38 shows the graph of $y_1 = x^2$ and a vertical or horizontal translation y_2. Write an equation for y_2 as shown in each graph.

Solution

a. $y_2 = x^2 - 3$, a vertical translation downward by 3 units.
b. $y_2 = (x + 2)^2$, a horizontal translation to the left by 2 units.
c. $y_2 = (x - 3)^2$, a horizontal translation to the right by 3 units. ■

Reflections across Axes

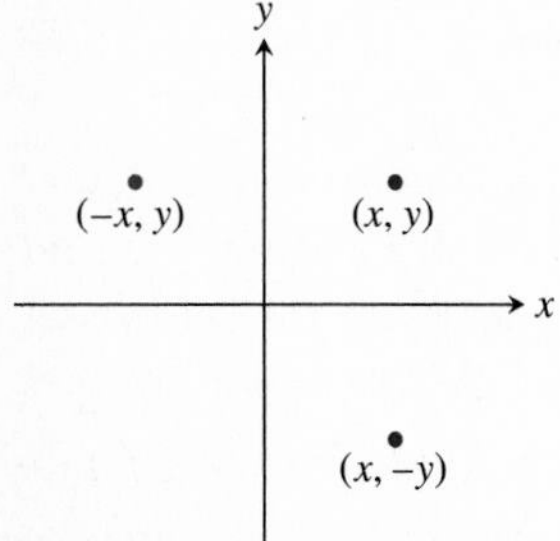

Figure 2.39 The point (x, y) reflected across each line.

Points (x, y) and $(x, -y)$ are **reflections** of each other across the x-axis. Points (x, y) and $(-x, y)$ are **reflections** of each other across the y-axis. See Figure 2.39. Two points (or two graphs) that are symmetric with respect to a line are **reflections** of each other across that line.

● EXPLORATION 2: Introducing Reflections

1. Use a standard viewing window. Graph the functions

$$y_1 = (x - 2)^2 + 1, \qquad y_2 = -y_1.$$

Then graph the functions

$$y_3 = (x - 2)^3 + 4, \qquad y_4 = -y_3.$$

What is the geometrical relationship between the graphs of y_1 and y_2? Between the graphs of y_3 and y_4?

2. Use a standard window. Graph the functions

$$y_1 = x^2 - 5x - 3, \qquad y_2 = y_1(-x) = (-x)^2 - 5(-x) - 3.$$

What is the geometrical relationship between the graphs of y_1 and y_2? ●

REMARK

In part 2 of Exploration 2, let $y_1 = f(x)$. This notation means that $y_1(-x) = f(-x)$.

Exploration 2 suggests that a reflection across the x-axis occurs when y is replaced by $-y$, and a reflection across the y-axis occurs when x is replaced by $-x$.

Exploration Extensions

Consider the function $y_1 = x^3 + 3x^2 + 3x + 2$. Use the conjectures you made in Exploration 2 to write a function y_2 that is a reflection of y_1 across the x-axis, and a function y_3 that is a reflection of y_1 across the y-axis. Use a grapher to check your results.

Reflections of $y = f(x)$

The graphs of

- $y = f(x)$ and $y = -f(x)$ are reflections of each other across the x-axis.
- $y = f(x)$ and $y = f(-x)$ are reflections of each other across the y-axis.

Teaching Note

You can illustrate translations and reflections by taking a piece of stiff wire and bending it into the shape of a particular function. The wire can be used with a grid that has been drawn on the chalkboard.

EXAMPLE 2 Finding Equations for Reflections across the Axes

Find an equation for the reflection of $y_1 = f(x) = x^2 - 5x + 3$ across each axis.

Solution

Solve Algebraically

For a reflection of f across the x-axis, find $y_2 = -f(x)$. Thus,

$$y_2 = -f(x) = -(x^2 - 5x + 3) = -x^2 + 5x - 3$$

For a reflection of f across the y-axis, find $y_3 = f(-x)$. Thus,

$$y_3 = f(-x) = (-x)^2 - 5(-x) + 3 = x^2 + 5x + 3$$ ■

Reflection across a line and symmetry about a line are related ideas. A graph is **symmetric about a line** if it is a reflection of itself across the line. Recall from Section 1.5 that even functions are symmetric about the y-axis.

Representing Translations and Reflections Parametrically

Translations and reflections of graphs of functions can be described using parametric equations.

REMARK

To find the function form of the graph C_2, eliminate the parameter t by solving for t, ($t = x_2 + 2$), and substituting it into the y_2 equation, $y_2 = t^2 = (x_2 + 2)^2$. This algebra explains why, in the function form $y = (x + 2)^2$, the translation is to the left by 2 units.

Translations. Let C_1 and C_2 be the graphs of the following parametric equations.

$$C_1: \quad x_1 = t, \qquad y_1 = t^2$$

$$C_2: \quad x_2 = t - 2, \qquad y_2 = t^2$$

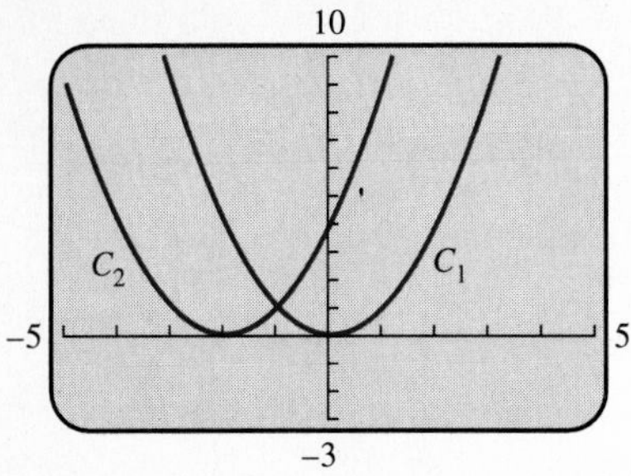

Figure 2.40 A translation of graph C_1 to the left 2 units results in graph C_2.

The point $(t-2, t^2)$ of C_2 can be obtained by moving the corresponding point (t, t^2) of C_1 to the left 2 units. Figure 2.40 shows that C_2 can be obtained from C_1 by a horizontal translation to the left 2 units.

In general, let C_1 and C_2 be the graphs of the following parametric equations.

$$C_1:\quad x_1 = t, \qquad y_1 = f(t)$$
$$C_2:\quad x_2 = t + c, \qquad y_2 = f(t)$$

Then C_2 can be obtained from C_1 by a horizontal translation. If $c > 0$, the translation is to the right, and if $c < 0$, to the left. The magnitude of the translation is $|c|$.

Vertical translations are accomplished in a similar fashion. Let C_1 and C_3 be the graphs of the following parametric equations.

$$C_1:\quad x_1 = t, \qquad y_1 = f(t)$$
$$C_3:\quad x_3 = t, \qquad y_3 = f(t) + c$$

Then C_3 can be obtained from C_1 by a vertical translation. If $c > 0$, the translation is upward, and if $c < 0$, downward. The magnitude of the translation is $|c|$.

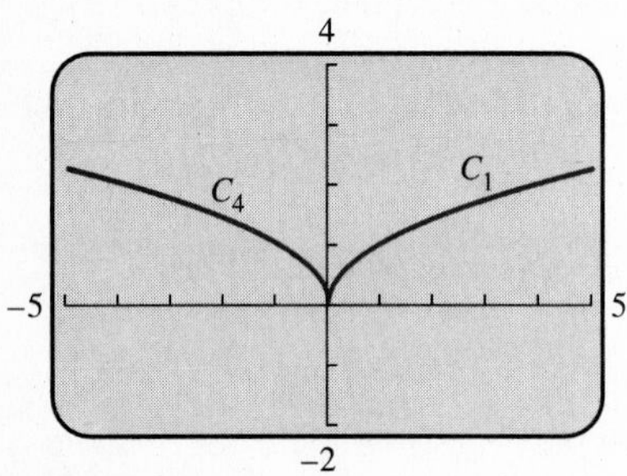

Figure 2.41

Reflections. Let C_1 and C_4 be the graphs of the following parametric equations.

$$C_1:\quad x_1 = t, \qquad y_1 = \sqrt{t}$$
$$C_4:\quad x_4 = -t, \qquad y_4 = \sqrt{t}$$

Figure 2.41 shows that C_4 can be obtained from C_1 by a reflection across the *y*-axis. In Example 3, we show that C_1 is the graph of $f(x) = \sqrt{x}$ and C_4 the graph of $g(x) = \sqrt{-x}$.

■ EXAMPLE 3 Studying Reflections Across the *y*-axis

Eliminate the parameter *t*, and describe each graph in function form.

$$C_1:\quad x_1 = t, \qquad y_1 = \sqrt{t}$$
$$C_4:\quad x_4 = -t, \qquad y_4 = \sqrt{t}$$

Solution

Solve Algebraically

For C_1,

$$y_1 = \sqrt{t} = \sqrt{x_1}.$$

Thus, y_1 is the function $f(x) = \sqrt{x}$.

For C_4,

$$y_4 = \sqrt{t} = \sqrt{-x_4}.$$

Thus, y_4 is the function $g(x) = \sqrt{-x}$.

Support Graphically

Graph f and g in the viewing window of Figure 2.41, and compare with the graphs of the parametric equations that define C_1 and C_4. ■

In general, we let C_1 and C_4 be the graphs of the following parametric equations.

$$C_1:\ x_1 = t, \qquad y_1 = f(t)$$
$$C_4:\ x_4 = -t, \qquad y_4 = f(t)$$

Then C_4 can be obtained from C_1 by a reflection across the y-axis.

Reflections across the x-axis are accomplished in a similar fashion. Let C_1 and C_5 be the graphs of the following parametric equations.

$$C_1:\ x_1 = t, \qquad y_1 = f(t)$$
$$C_5:\ x_5 = t, \qquad y_5 = -f(t)$$

Then C_5 can be obtained from C_1 be a reflection across the x-axis.

Teaching Note

A drawing program on a computer can be used to illustrate the concepts of vertical and horizontal stretch and shrink.

Vertical and Horizontal Stretch and Shrink

We now investigate what happens when we multiply all the y-coordinates (or all the x-coordinates) of the points on a graph by a fixed constant.

Vertical Stretch and Vertical Shrink. Suppose that each point on the graph C_2 of Figure 2.42 is obtained by multiplying the y-coordinate of a point of C_1 directly below it by a constant c that is greater than 1. The graph C_1 appears to "stretch" up to the graph of C_2. We say that C_2 is obtained from C_1 by a **vertical stretch** of factor c, and C_1 is obtained from C_2 by a **vertical shrink** of factor $1/c$. If C_1 is the graph of $y = f(x)$, then C_2 is the graph of $y_2 = cf(x)$.

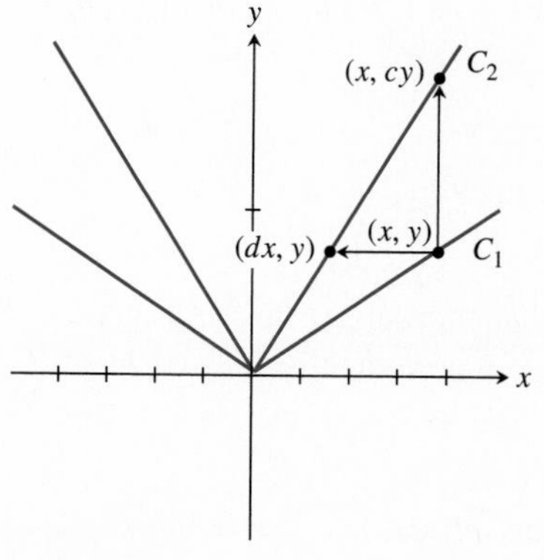

Figure 2.42

Horizontal Shrink and Horizontal Stretch. Suppose that each point on the graph of C_2 in Figure 2.42 is obtained by multiplying the x-coordinate of a point of C_1 directly to the right of it by a constant d such that $0 < d < 1$. The graph of C_1 appears to "shrink" toward the y-axis. We say that C_2 is obtained from C_1 by a **horizontal shrink** of factor d, and C_1 is obtained from C_2 by a **horizontal stretch** of factor $1/d$. If C_1 is the graph of $y = f(x)$, then C_2 is the graph of

$$y_3 = f\left(\frac{1}{d}x\right).$$

Vertical or horizontal stretches or shrinks can also be described using parametric equations, as the next two examples illustrate.

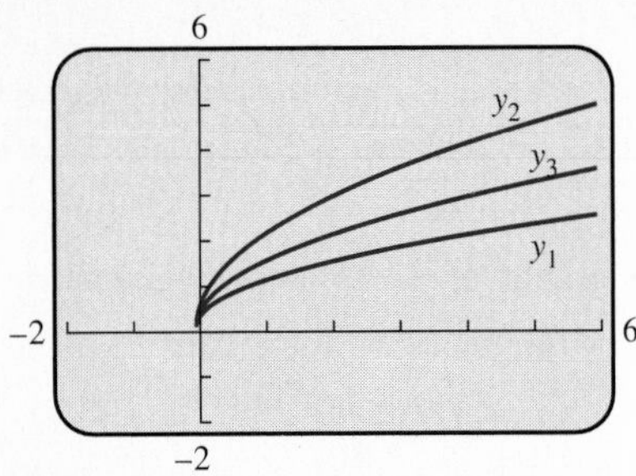

Figure 2.43 The graph of y_2 is a vertical stretch of y_1, and the graph of y_1 is a vertical shrink of y_2. The graph of y_3 is a horizontal shrink of y_1, and the graph of y_1 is a horizontal stretch of y_3.

■ EXAMPLE 4 Studying Vertical Stretch and Shrink

Describe how the graphs C_1 and C_2 can be obtained from each other. Then eliminate the parameter t and describe each graph in function form.

$$C_1: \quad x_1 = t, \qquad y_1 = \sqrt{t}$$
$$C_2: \quad x_2 = t, \qquad y_2 = 2\sqrt{t}$$

Solution

Solve Algebraically

For each value of t, $x_1 = x_2$ and the value of y_2 is twice the value of y_1. Thus, C_2 is obtained from C_1 by a vertical stretch of factor 2. On the other hand, C_1 is obtained from C_2 by a vertical shrink of factor 1/2.

For C_1, $y_1 = \sqrt{t} = \sqrt{x_1}$. Thus, $y_1 = \sqrt{x}$.

For C_2, $y_2 = 2\sqrt{t} = 2\sqrt{x_2}$. Thus, $y_2 = 2\sqrt{x}$.

Support Graphically

Figure 2.43 shows that the graph of $y_2 = 2\sqrt{x}$ is a vertical stretch of the graph of $y_1 = \sqrt{x}$.

Teaching Note

At first, have students operate within a fixed viewing window. Then have them vary the viewing window, keeping the functions fixed. This should lead to an interesting discussion of issues relating to transformations, particularly the effect of the parameter a on the graph of $y = ax^2$.

■ EXAMPLE 5 Studying Horizontal Stretch and Shrink

Describe how the graphs C_1 and C_3 can be obtained from each other. Then eliminate the parameter t and describe each graph in function form.

$$C_1: \quad x_1 = t, \qquad y_1 = \sqrt{t}$$
$$C_3: \quad x_3 = \tfrac{1}{2}t, \qquad y_3 = \sqrt{t}$$

Solution

Solve Algebraically

For each value of t, $y_1 = y_3$ and the value of x_3 is half that of x_1. Thus, C_3 is obtained from C_1 by a horizontal shrink of factor 1/2. On the other hand, C_1 is obtained from C_3 by a horizontal stretch of factor 2.

For C_1, $y_1 = \sqrt{t} = \sqrt{x_1}$. Thus, $y_1 = \sqrt{x}$.

For C_3, $y_3 = \sqrt{t} = \sqrt{2x_3}$. Thus, $y_3 = \sqrt{2x}$.

Support Graphically

Figure 2.43 shows the graph of $y_3 = \sqrt{2x}$ is a horizontal shrink of factor 1/2 of the graph of $y_1 = \sqrt{x}$. The graph of y_1 is a horizontal stretch of the graph of y_3. ■

We have the following summary about vertical and horizontal stretches and shrinks.

TECHNOLOGY TIP

We recommend that you graph all the parametric equations that produced the graphs in Figures 2.40, 2.41, and 2.43 in simultaneous mode. Then "see" how "trace" supports the transformations studied in this section.

Vertical Stretch or Shrink of $y = f(x)$, $c > 0$

- $y = cf(x)$ a stretch of f of factor c, $c > 1$
- $y = cf(x)$ a shrink of f of factor c, $0 < c < 1$

Horizontal Stretch or Shrink of $y = f(x)$, $c > 0$

- $y = f(cx)$ a shrink of f of factor $1/c$, $c > 1$
- $y = f(cx)$ a stretch of f of factor $1/c$, $0 < c < 1$

Connecting Geometric and Algebraic Transformations

We have studied how constants in the algebraic description of a function affect its graph. It is also useful to look at a graph and see it as a possible transformation of a known function.

EXAMPLE 6 Finding Equations for Stretches and Shrinks

Let C_1 be the graph of $y_1 = f(x) = x^3 - 16x$. Find equations for the following transformations of C_1.

a. C_2 is a vertical stretch of factor 3 of C_1 (Figure 2.44a).
b. C_3 is a horizontal shrink of factor 1/2 of C_1 (Figure 2.44b).

Solution

a. If y_2 is an equation for C_2, then

$$\begin{aligned} y_2 &= 3f(x) \\ &= 3(x^3 - 16x) \\ &= 3x^3 - 48x \end{aligned}$$

b. If y_3 is an equation for C_3, then

$$\begin{aligned} y_3 &= f(2x) \\ &= (2x)^3 - 16(2x) \\ &= 8x^3 - 32x \end{aligned}$$

■

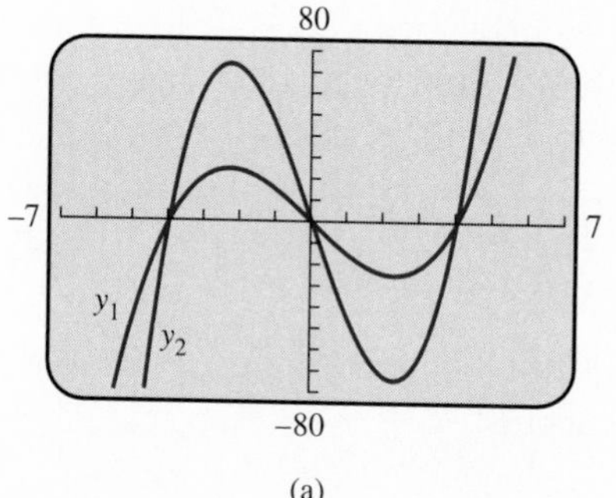

(a)

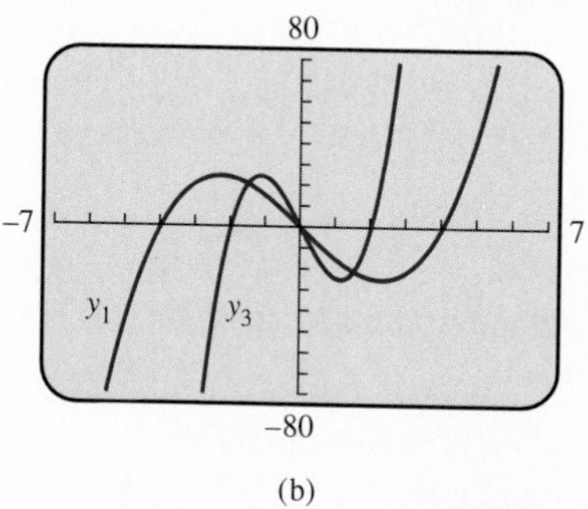

(b)

Figure 2.44 (a) y_2 is a vertical stretch of factor 3 of y_1. (b) y_3 is a horizontal shrink of factor 1/2 of y_1.

Combining Transformations

Transformations may be performed in succession—one after another. If the transformations include stretches, shrinks, or reflections, the order in which the transformations are performed may make a difference. So in those cases, pay particular attention to order.

Alert

Students sometimes make errors in the order in which they apply transformations to a graph. Warn students of the danger of reversing the order of some of these transformations. Point out that, in general, they are not commutative. Explore which transformations are commutative.

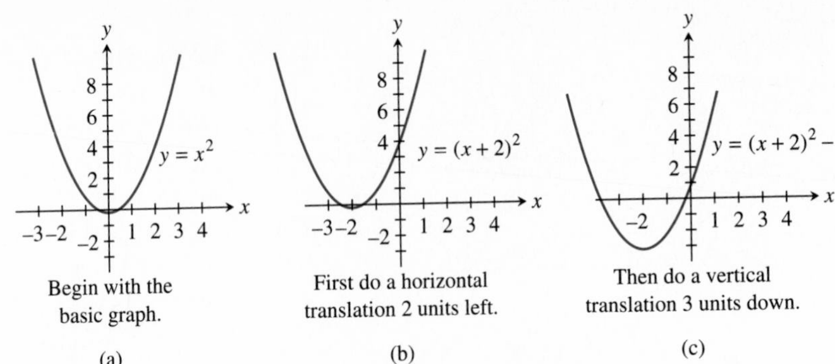

Figure 2.45 Transforming the graph of $y = x^2$ to the graph of $y = (x + 2)^2 - 3$.

■ EXAMPLE 7 Combining Two Translations

Describe how to transform the graph of $y = x^2$ to the graph of $y = (x + 2)^2 - 3$. Sketch graphs to show the result of each transformation.

Solution

The graph of $y = (x + 2)^2 - 3$ can be obtained from $y = x^2$ by the following transformations:

a. a horizontal translation left 2 units to obtain $y = (x + 2)^2$, and

b. a vertical translation down 3 units to obtain $y = (x + 2)^2 - 3$.

These transformations may be performed in either order. (See Figure 2.45.) ■

Follow-up

Discuss how the graphs of even and odd functions are affected by reflections across the x- and y-axes.

Assignment Guide

Day 1: Ex. 3–57, multiples of 3
Day 2: Ex. 59, 60, 63, 64, 67, 68, 73, 76–78, 81–85

Cooperative Learning

Group Activity: Ex. 87

Example 8 uses general function notation instead of a specific function. The rules for transformations can still be applied.

■ EXAMPLE 8 Transforming the Graph of $f(x)$

Describe how to transform the graph of $y = f(x)$ shown in Figure 2.46 to the graph of $y = 2f(x + 1) - 3$. Sketch graphs to show the result of each transformation.

Solution

The graph of $y = 2f(x + 1) - 3$ can be obtained from the graph of $y = f(x)$ by the following transformations:

a. a vertical stretch of factor 2 to get $y = 2f(x)$ (Figure 2.47a);

b. a horizontal translation left 1 unit to get $y = 2f(x + 1)$ (Figure 2.47b);

c. a vertical translation down 3 units to get $y = 2f(x + 1) - 3$ (Figure 2.47c).

The order of the first two transformations can be reversed. ■

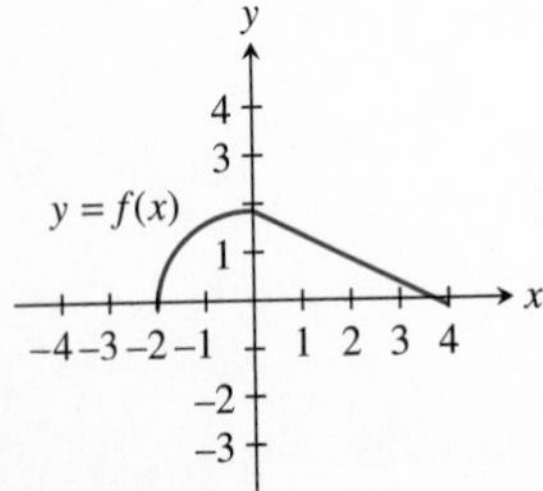

Figure 2.46 Graph of function for Example 8.

Notes on Exercises

Ex. 1–16 give students an opportunity to make and test conjectures about geometric transformations.
Ex. 17–24 illustrate transformations of relations defined parametrically.
Ex. 71–76 involve the use of more than one transformation. Students must consider the question of the commutativity of transformations.

Ongoing Assessment

Self-Assessment: Ex. 13, 19, 37, 47, 55, 75
Embedded Assessment: Ex. 78, 82

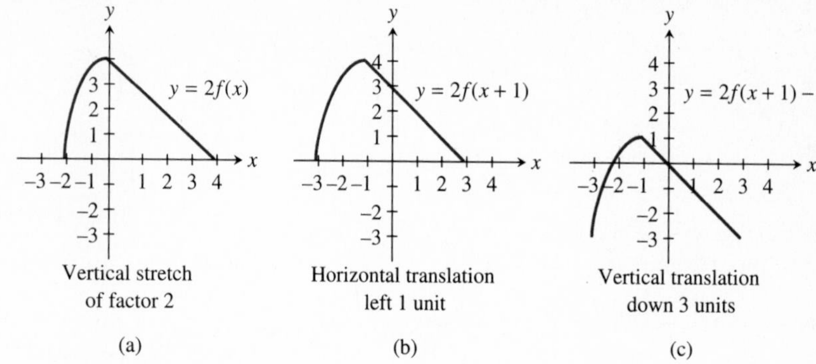

Figure 2.47 Transforming the graph of $y = f(x)$ in Figure 2.46 to get $y = 2f(x + 1) - 3$.

Quick Review 2.6

In Exercises 1–4, write the expression as a binomial squared.

1. $x^2 + 2x + 1$ **2.** $x^2 - 6x + 9$

3. $x^2 + 6x + 9$ **4.** $4x^2 + 4x + 1$

In Exercises 5–8, perform the indicated operations and simplify.

5. $(2x - 1) - (3x^2 + 2x - 3)$

6. $1 - (x^2 - 2x + 5)$

7. $(-2)^3x + (-x)^2 - 3(-x) + 1$

8. $(-2x)^3(-2x)^2$

In Exercises 9–12, solve for t.

9. $x = 2t - 3$ **10.** $y = 3t + 1$

11. $x = \frac{1}{2}t$ **12.** $y = \frac{1}{3}t$

SECTION EXERCISES 2.6

In Exercises 1–10, describe how the graph of $y = x^2$ can be transformed to the graph of the equation.

1. $y = x^2 - 3$ **2.** $y = x^2 + 1$

3. $y = x^2 + 5.2$ **4.** $y = x^2 - 2$

5. $y = (x + 4)^2$ **6.** $y = (x - 3)^2$

7. $y = (1 - x)^2$ **8.** $y = (5 - x)^2$

9. $y = (x - 1)^2 + 3$ **10.** $y = (x + 5)^2 - 3$

In Exercises 11–16, describe how the graph of $f(x) = \sqrt{x}$ can be transformed to the graph of the equation.

11. $y = -\sqrt{x}$ **12.** $y = \sqrt{x - 5}$

13. $y = \sqrt{-x}$ **14.** $y = \sqrt{3 - x}$

15. $y = -\sqrt{x + 4}$ **16.** $y = \sqrt{-(7 - x)}$

In Exercises 17–24, describe how the graph C_2 can be obtained from the graph C_1. Eliminate the parameter t and describe the graphs in function form.

17. C_1: $x_1 = t$, $y_1 = t^3$
C_2: $x_2 = t - 2$, $y_2 = t^3$

18. C_1: $x_1 = t$, $y_1 = \sqrt{t}$
C_2: $x_2 = t + 4$, $y_2 = \sqrt{t}$

19. C_1: $x_1 = t$, $y_1 = t^2 + 2$
C_2: $x_2 = t + 2$, $y_2 = t^2 + 2$

20. C_1: $x_1 = t$, $y_1 = |t| + 3$
C_2: $x_2 = t - 3$, $y_2 = |t| + 3$

21. C_1: $x_1 = t$, $y_1 = t^3$
C_2: $x_2 = t$, $y_2 = 2t^3$

22. C_1: $x_1 = t$, $y_1 = \sqrt{t}$
C_2: $x_2 = t$, $y_2 = 0.5\sqrt{t}$

23. C_1: $x_1 = t$, $y_1 = t^2 + 2$
C_2: $x_2 = 3t$, $y_2 = t^2 + 2$

24. C_1: $x_1 = t$, $y_1 = |t| + 3$
C_2: $x_2 = t$, $y_2 = |(1/2)t| + 3$

In Exercises 25–32, describe how the graph of $y = x^3$ can be transformed to the graph of the equation.

25. $y = 2x^3$
26. $y = \frac{1}{4}x^3$
27. $y = (2x)^3$
28. $y = (0.2x)^3$
29. $y = 0.3x^3$
30. $y = (0.5x)^3$
31. $y = (0.4x)^3$
32. $y = 3x^3$

In Exercises 33–38, starting with f, describe how the graphs of the two functions are related.

33. $f(x) = \sqrt{x + 2}$ and $g(x) = \sqrt{x - 4}$

34. $f(x) = (x - 1)^2$ and $g(x) = -(x + 3)^2$

35. $f(x) = (x - 2)^3$ and $g(x) = -(x + 2)^3$

36. $f(x) = |2x|$ and $g(x) = 4|x|$

37. $f(x) = -2(x + 4)^2 + 3$ and $g(x) = 2(x + 1)^2$

38. $f(x) = x^3$ and $g(x) = (2x)^3$

In Exercises 39–42, do a hand sketch of the graphs of f, g, and h. Support with a grapher.

39. $f(x) = (x + 2)^2$
$g(x) = 3x^2 - 2$
$h(x) = -2(x + 3)^2$

40. $f(x) = x^3 - 2$
$g(x) = (x + 4)^3 - 1$
$h(x) = 2(x - 1)^3$

41. $f(x) = \sqrt[3]{x + 1}$
$g(x) = 2\sqrt[3]{x} - 2$
$h(x) = -\sqrt[3]{x - 3}$

42. $f(x) = -2|x| - 3$
$g(x) = 3|x + 5| + 4$
$h(x) = |3x|$

In Exercises 43–46, the graph is that of a function $y = f(x)$ that can be obtained by transforming the graph of $y = \sqrt{x}$. Write a formula for the function f.

43.

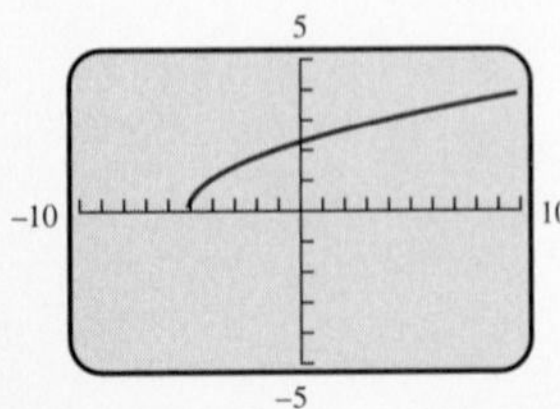

44.

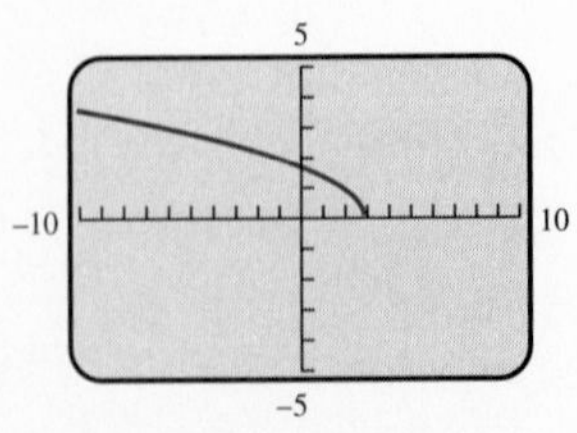

45.

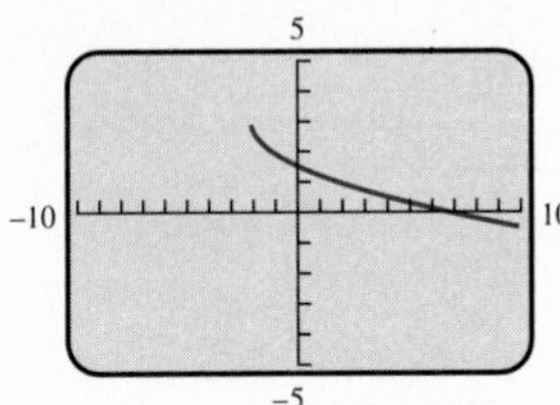

46.

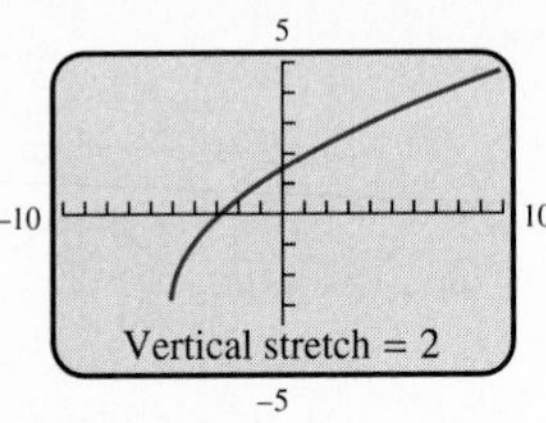

In Exercises 47–50, g is the reflection of f across the y-axis. Find an equation for g.

47. $f(x) = x^3 - 5x^2 - 3x + 2$

48. $f(x) = 2\sqrt{x + 3} - 4$

49. $f(x) = \sqrt[3]{8x}$

50. $f(x) = 3|x + 5|$

In Exercises 51–58, describe a basic graph and a sequence of transformations that can be used to find a graph of the function.

51. $y = 2(x - 3)^2 - 4$
52. $y = -3\sqrt{x + 1}$
53. $y = -(2x)^3$
54. $y = (3x)^2 - 4$
55. $y = 3\sqrt[3]{x + 6} - 7$
56. $y = -2|x + 4| + 1$
57. $y = -2\sqrt{2x}$
58. $y = \sqrt[3]{0.5x}$

In Exercises 59–62, a graph of the function can be obtained from $y = x^2$ by both a vertical stretch (or shrink) and by a horizontal shrink (or stretch). In each case, state the stretch and shrink factors.

59. $y = 9x^2$
60. $y = \frac{1}{4}x^2$
61. $y = 16x^2$
62. $y = 5x^2$

In Exercises 63–66, y_1 is a vertical stretch of y by the factor 2, and y_2 is a vertical shrink of y by the factor 1/3. Find equations for y_1 and y_2.

63. $y = x^3 - 4x$
64. $y = |x + 2|$
65. $y = x^2 + x - 2$
66. $y = 1/(x + 2)$

In Exercises 67–70, y_1 is a horizontal stretch of y by the factor 2, and y_2 is a horizontal shrink of y by the factor 1/3. Find equations for y_1 and y_2.

67. $y = x^2$
68. $y = |x|$
69. $y = |x - 2|$
70. $y = 1/x$

In Exercises 71–76, a graph G is obtained from a graph of y by the transformation indicated. Write an equation whose graph is G.

71. $y = x^2$: a vertical stretch of factor 3, then a shift right 4 units

72. $y = x^2$: a shift right 4 units, then a vertical stretch of factor 3

73. $y = \sqrt[3]{x}$: a vertical stretch of factor 3, then a vertical shift up 4 units

74. $y = \sqrt[3]{x}$: a vertical shift up 4 units, then a vertical stretch of factor 3

75. $y = |x|$: a shift left 2 units, then a vertical stretch of factor 2, and finally a shift down 4 units.

76. $y = |x|$: a shift left 2 units, then a horizontal shrink of factor 1/2, and finally a shift down 4 units.

In Exercises 77–80, a graph of the function f is shown.

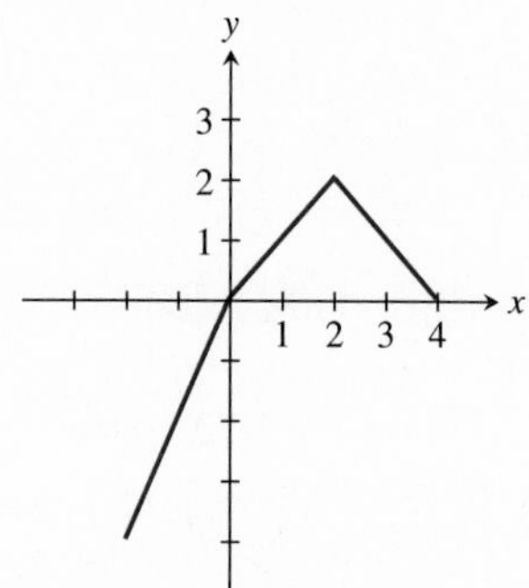

For Exercises 77–80

77. Sketch the graph of $y = 2 + 3f(x + 1)$.

78. Sketch the graph of $y = -f(x + 1) + 1$.

79. Sketch the graph of $y = f(2x)$.

80. Sketch the graph of $y = 2f(x - 1) + 2$.

In Exercises 81–83, solve the problem.

81. *Celsius vs. Fahrenheit* The graph shows the temperature in degrees Celsius in Windsor, Ontario, for one 24-hour period. Describe the transformations that convert this graph to one showing degrees Fahrenheit. (*Hint:* $F(t) = (9/5)C(t) + 32$.)

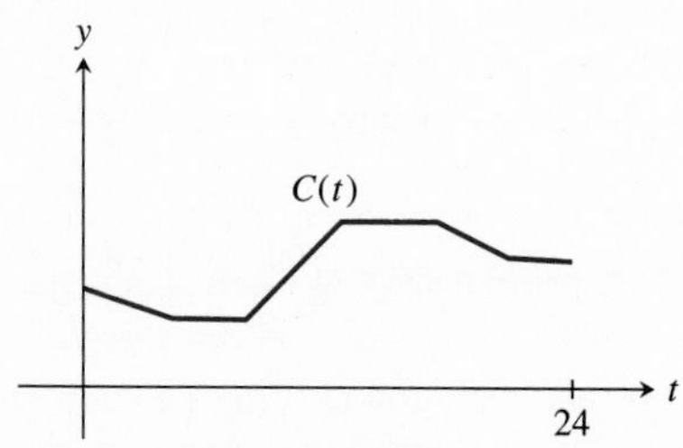

For Exercise 81

82. *Fahrenheit vs. Celsius* The graph shows the temperature in degrees Fahrenheit in Mt. Clemens, Mich., for one 24-hour period. Describe the transformations that convert this graph to one showing degrees Celsius. (*Hint:* Write the equation for converting degrees Fahrenheit to degrees Celsius. See Exercise 81.)

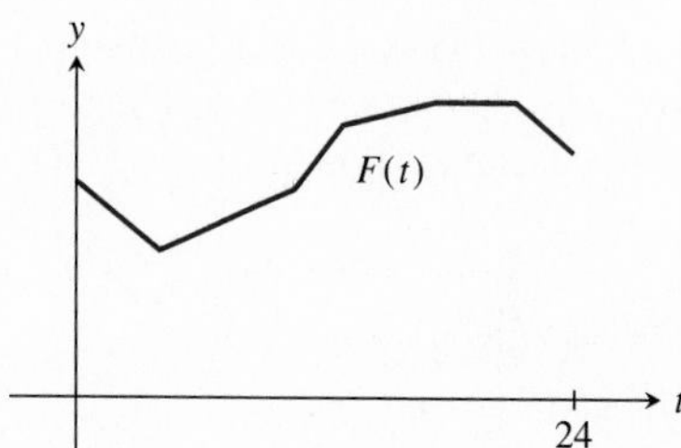

For Exercise 82

83. *Throwing Two Balls* One ball is thrown upward with an initial velocity of 64 ft/sec. A second ball is thrown with the same initial velocity 2 sec later. The equation $y_1 = -16t^2 + 64t + 5$ describes the height h of the first ball assuming that the ball is thrown from a height of 5 ft when $t = 0$. The figure shows the height of each ball plotted against time.

a. Find an equation for the height y_2 of the second ball. (*Hint:* y_2 should be a horizontal shift 2 units to the right of y_1.)

b. Graph y_1 and y_2 to duplicate the graph shown here.

c. Determine the time when the two balls are at the same height.

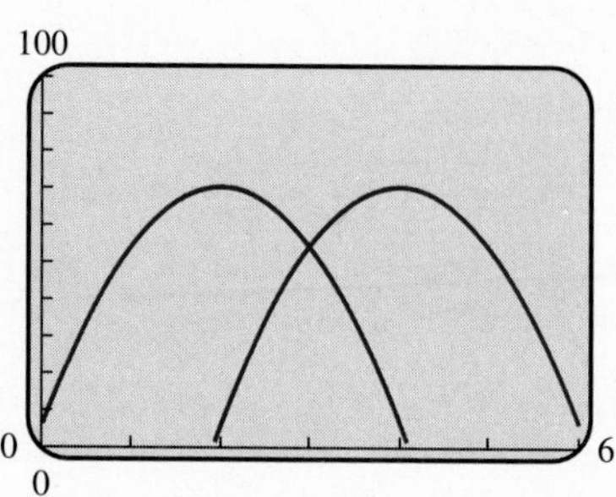

For Exercise 83

EXTENDING THE IDEAS

84. Let G be the graph of the function

$$f(x) = \begin{cases} -x^2 & \text{if } x < 1 \\ x & \text{if } x \geq 1. \end{cases}$$

If G is shifted right 2 units, which graph do you obtain—g or h?

$$g(x) = \begin{cases} -(x-2)^2 & \text{if } x < 1 \\ x - 2 & \text{if } x \geq 1. \end{cases}$$

$$h(x) = \begin{cases} -(x-2)^2 & \text{if } x - 2 < 1 \\ x - 2 & \text{if } x - 2 \geq 1. \end{cases}$$

85. Writing to Learn Does a vertical shift of d units followed by a vertical stretch of factor c give the same result as a vertical stretch of factor c followed by a vertical shift of d units?

86. Writing to Learn Which pairs of transformations studied so far have different outcomes depending on the order in which they are performed?

87. Group Learning Activity *Work in groups of three.* Each of the functions given here is related to $y = \sqrt{x}$ by a transformation or a combination of transformations. Suppose your group is asked to present a "review of transformations" to the rest of your class. Decide the order in which you will present these examples and why.

a. $y = 3\sqrt{2x}$ **b.** $y = 5\sqrt{x}$
c. $y = -3\sqrt{4x}$ **d.** $y = \sqrt{x + 4}$
e. $y = -\sqrt{(-x)}$ **f.** $y = \sqrt{-2x}$
g. $y = \sqrt{x} + 3$ **h.** $y = 7\sqrt{x - 4}$

2.7 INVERSE RELATIONS AND INVERSE FUNCTIONS

Relations and Their Inverses • Graphs of Inverses • A Function and Its Inverse Relation • A Function Whose Inverse Relation Is a Function

Relations and Their Inverses

A relation is a set of ordered pairs. If we reverse the order of each ordered pair we have a new relation called its *inverse.* This idea is embodied in the following definition.

Definition: Inverse Relation

The ordered pair (a, b) is in a relation if and only if (b, a) is in the **inverse relation.**

Table 2.6 Charges versus Time

Time x (hours)	Charges y (\$)
0	5.00
1	7.50
2	10.00
3	12.50
4	15.00
5	17.50

Table 2.7 Time versus Charges

Charges x (\$)	Time y (hours)
5.00	0
7.50	1
10.00	2
12.50	3
15.00	4
17.50	5

Tables 2.6 and 2.7 show relations regarding Wilkens Rent-All charges for a camcorder. Note that they are inverses of each other: Table 2.6 shows the time and the cost. Table 2.7 shows the cost and the time. The ordered pairs in one table are in the reverse order of the ordered pairs in the other table.

Objective

Students will be able to determine the algebraic representation and the geometric representation of a function and its inverse, and to tell if a function is one-to-one.

Motivate

Discuss the relationship between the functions $f(x) = x^3$ and $g(x) = \sqrt[3]{x}$.

When a relation is defined by an equation in two variables, describing its inverse is especially easy.

■ **EXAMPLE 1 Studying the Equations for a Relation and Its Inverse**

The ordered pairs in Table 2.6 of camcorder rental costs are solutions of the equation $y = 2.5x + 5$. Find an equation for Table 2.7.

Solution

If (a, b) satisfies $y = 2.5x + 5$, then (b, a) satisfies $x = 2.5y + 5$. We find the equation for the inverse by simply interchanging x and y. Once this is done, we may solve for y.

$$y = 2.5x + 5 \quad \text{The original relation}$$
$$x = 2.5y + 5 \quad \text{The inverse relation}$$
$$2.5y = x - 5$$
$$y = 0.4x - 2 \quad \text{Solve for } y.$$

■

Graphs of Inverses

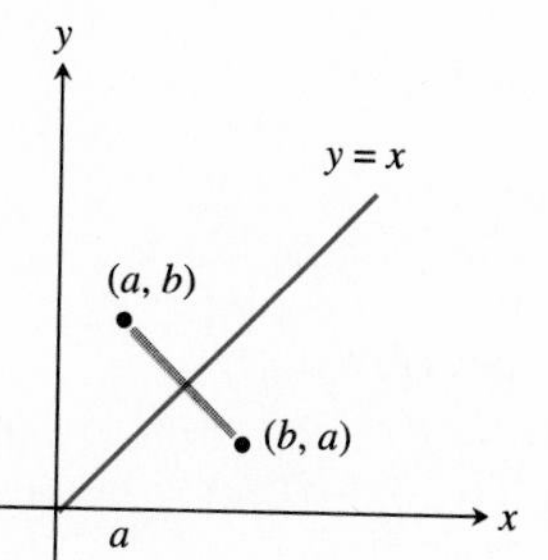

Figure 2.48

The points (a, b) and (b, a) are reflections across the line $y = x$. (See Figure 2.48.) Because (a, b) is in the graph of a relation when (b, a) is in the graph of the inverse relation, the graphs of a relation and its inverse are also reflections of each other across the line $y = x$.

■ **EXAMPLE 2 Observing the Graphs of a Relation and Its Inverse**

Graph $y = x^3 - 4$ and its inverse.

Solution

To find an equation for the inverse, interchange x and y, then solve for y.

$$y = x^3 - 4 \quad \text{The original relation}$$
$$x = y^3 - 4 \quad \text{The inverse relation}$$
$$y^3 = x + 4$$
$$y = \sqrt[3]{x + 4} \quad \text{Solve for } y.$$

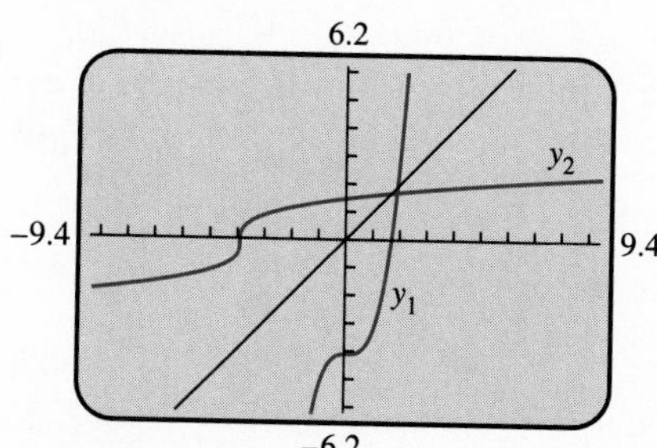

Figure 2.49

Figure 2.49 shows the graph of $y_1 = x^3 - 4$ and $y_2 = \sqrt[3]{x + 4}$. Note that each is the reflection of the other across the line $y = x$. ■

Graphing Activity Extensions

Use the same method to graph the function $f(x) = x^2 + 1, x \geq 0$, and its inverse. Adjust the value of *T*min so that the domain is correct.

▼ **GRAPHING ACTIVITY** **Relations and Their Inverses in Parametric Mode**

Inverse relations are particularly easy to graph in parametric mode. We simply switch the values for x and y. Try it. Graph both (x_1, y_1) and (x_2, y_2) in the same viewing window.[4]

1. $x_1 = t, \quad y_1 = t^2$
$x_2 = t^2, \quad y_2 = t$

2. $x_1 = t, \quad y_1 = t^3 - 4$
$x_2 = t^3 - 4, \quad y_1 = t$

3. $x_1 = 5t, \quad y_1 = 3\sqrt{t + 6}$
$x_2 = y_1, \quad y_2 = x_1$

4. $x_1 = 2t, \quad y_1 = t^3 - 6t$
$x_2 = y_1, \quad y_1 = x_1$

Challenge. To see the symmetry of a relation and its inverse about the line $y = x$, try to include the line $y = x$ in each of these four viewing windows. ▲

Teaching Note

Some students may need to be reminded that in considering inverse relations, it is important to consider what happens to points. Thus, examining (a, b) as an element of a relation and (b, a) as an element of the inverse relation is critical to understanding inverse relations. The concept of symmetry drives much of the thinking about inverses.

A Function and Its Inverse Relation

Because a function is a relation, a function has an inverse. Its inverse relation, however, may not be a function.

■ **EXAMPLE 3** **Observing Graphs of Inverses**

Graph the function $y = x^2 + 3$ and its inverse.

Solution

To find the inverse relation, we interchange x and y, then solve for y.

$$y = x^2 + 3 \qquad \text{The original relation}$$
$$x = y^2 + 3 \qquad \text{The inverse relation}$$
$$y^2 = x - 3$$
$$y = \pm\sqrt{x - 3} \qquad \text{Solve for } y.$$

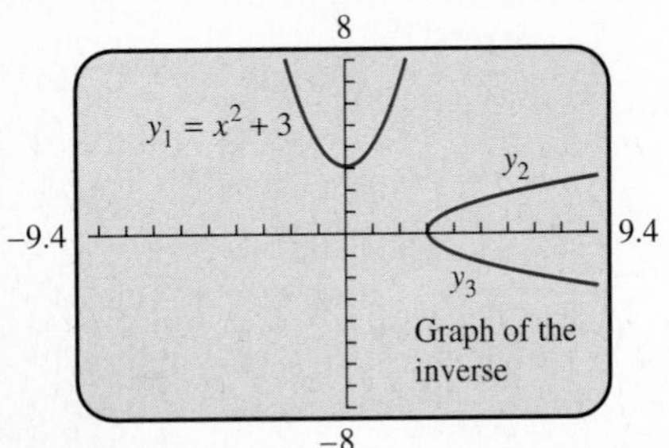

Figure 2.50

Figure 2.50 shows the graphs of $y_1 = x^2 + 3$, $y_2 = \sqrt{x - 3}$, and $y_3 = -\sqrt{x - 3}$. Note that y_2 and y_3 taken together are the inverse of y_1 but are not a function. ■

Alert

Students may think they need to demonstrate all three properties of a function f in order to show it is one-to-one. Point out that the properties are all equivalent.

A Function Whose Inverse Relation Is a Function

Example 3 shows a function whose inverse, y_2 and y_3 taken together, is not a function. To be a function, each element of the domain must match with exactly one element of the range. Thus, for a function to have an inverse that is

[4]Use window settings $T\text{min} = -3$, $T\text{max} = 3$, and $T\text{step} = 0.1$ in a square window.

also a function, each element of the range must also match with exactly one element of the domain. Such functions are **one-to-one.**

Follow-up

Ask . . .

Suppose f and g are inverse functions and $f(3) = 7$. Find $g(3)$ and $g(7)$, if possible. **($g(7) = 3$)**

Assignment Guide

Day 1: Ex. 3–42, multiples of 3, 25, 26

Day 2: Ex. 45, 48, 51, 55–60

Cooperative Learning

Group Activity: Ex. 53–54

Notes on Exercises

Ex. 1–16 can be used to illustrate the relationship between a function and its inverse.

Ex. 17–24 further illustrate inverse functions. It may be worth discussing domain and range issues for these exercises.

Ex. 53 and 54 should be assigned together.

Ongoing Assessment

Self-Assessment: Ex. 11, 23, 37, 47

Embedded Assessment: Ex. 26

Principles of One-to-One Functions

1. A function is one-to-one if and only if its inverse is a function.
2. A function is one-to-one if and only if any horizontal line in the coordinate plane intersects its graph in at most one point.
3. A function f is one-to-one with inverse g if and only if

$$f(g(x)) = x \quad \text{and} \quad g(f(x)) = x$$

for every value of x in the domains of g and f, respectively.

The one-to-one principles suggest two methods, one graphical and one algebraic, for checking that the inverse of a function is also a function.

■ EXAMPLE 4 Verifying Inverse Functions

Show that $f(x) = x^3 + 1$ and $g(x) = \sqrt[3]{x - 1}$ are inverse functions.

Solution

Solve Graphically

The graphs of f, g, and $y = x$ are shown in Figure 2.51. The picture suggests that f and g are reflections of each other across the line $y = x$ and that any horizontal line intersects the graph of f in at most one point. We make the reasonable conjecture that f is one-to-one and its inverse g is a function.

Confirm Algebraically

The domains of both f and g are the set of real numbers. We must show that $f(g(x)) = x$ and $g(f(x)) = x$.

$$f(g(x)) = f(\sqrt[3]{x - 1}) = (\sqrt[3]{x - 1})^3 + 1 = (x - 1) + 1 = x$$

$$g(f(x)) = g(x^3 + 1) = \sqrt[3]{(x^3 + 1) - 1} = \sqrt[3]{x^3} = x$$ ■

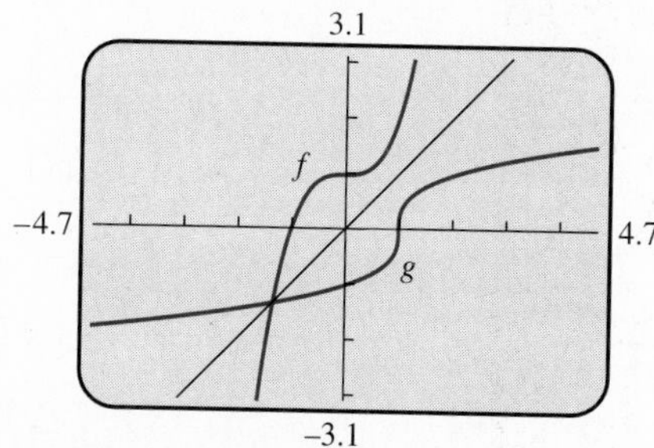

Figure 2.51 The graphs of f and g are reflections of each other across the line $y = x$.

If f is a one-to-one function, its inverse is sometimes written as f^{-1} (read "f inverse"). If (a, b) is in the graph of f, then (b, a) is in the graph of f^{-1}. In other words, a is in the domain of f if and only if a is in the range of f^{-1}, and b is in the range of f if and only if b is in the domain of f^{-1}. The domain and range of f are the range and domain, respectively, of f^{-1}. Here is a summary of this section.

REMARK

f^{-1} refers to the "inverse function" and is not equal to $\frac{1}{f}$. Do not interpret the -1 in the notation f^{-1} as an exponent.

Finding an Inverse Function

Let f be a function.

1. Determine that there is a function f^{-1} by checking that f is one-to-one. State any restrictions on the domain of f.
2. Find f^{-1} by switching x and y in the equation $y = f(x)$ and solving for y to get $y = f^{-1}(x)$. State any restrictions on the domain of f^{-1}.

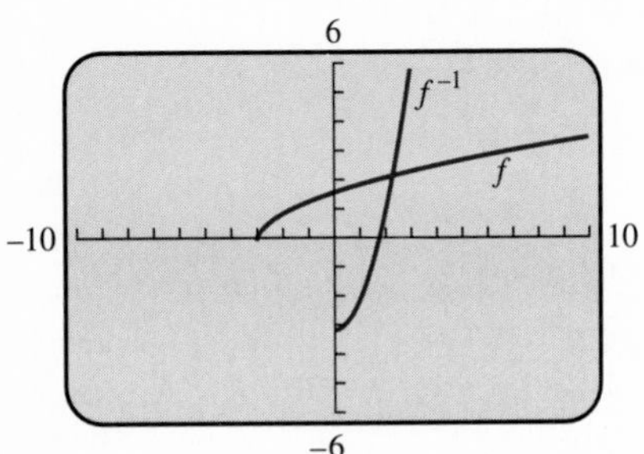

Figure 2.52 A graph of both f and f^{-1}.

EXAMPLE 5 Finding an Inverse Function

Show that $f(x) = \sqrt{x + 3}$ has an inverse function, and find a rule for f^{-1}. State any restrictions on the domains of f and f^{-1}.

Solution

The domain of f is $[-3, \infty)$. To find f^{-1}, we write

$$y = \sqrt{x + 3}, \quad x \geq -3,\ y \geq 0,$$

$$x = \sqrt{y + 3}, \quad y \geq -3,\ x \geq 0, \quad \text{Interchange } x \text{ and } y.$$

$$x^2 = y + 3, \quad y \geq -3,\ x \geq 0, \quad \text{Square both sides.}$$

$$y = x^2 - 3, \quad y \geq -3,\ x \geq 0. \quad \text{Solve for } y.$$

Thus, $f^{-1}(x) = x^2 - 3$, $x \geq 0$.

The domain and range of f are, respectively, $[-3, \infty)$ and $[0, \infty)$. Therefore, the domain and range of f^{-1}, respectively, are $[0, \infty)$ and $[-3, \infty)$. (See Figure 2.52.) ■

Quick Review 2.7

In Exercises 1–10, solve the equation for y.

1. $x = 3y - 6$
2. $x = 0.5y + 1$
3. $x = y^2 + 4$
4. $x = y^2 - 6$
5. $x = \dfrac{y - 2}{y + 3}$
6. $x = \dfrac{3y - 1}{y + 2}$
7. $x = \dfrac{2y + 1}{y - 4}$
8. $x = \dfrac{4y + 3}{3y - 1}$
9. $x = \sqrt{y + 3}$, $y \geq -3$
10. $x = \sqrt{y - 2}$, $y \geq 2$

SECTION EXERCISES 2.7

In Exercises 1–8, graph y and its inverse.

1. $y = 2x + 3$
2. $y = 3x - 4$
3. $y = 1 - 7x$
4. $y = 6 - 2x$
5. $y = 2x^3 - 5$
6. $y = x^3 + 6$
7. $y = \sqrt[3]{x + 1}$
8. $y = \sqrt[3]{x - 3}$

In Exercises 9–12, write a pair of parametric equations that define the inverse relation for the given parametric equations.

9. $x = 2t - 3, \quad y = t^2 - 2$

10. $x = t^3 - 2, \quad y = 2t$

11. $x = 3^t, \quad y = t$

12. $x = t^2, \quad y = t^3$

In Exercises 13–16, graph both the relation (x, y) and its inverse in a square window.

13. $x = 2t, \quad y = t^2$

14. $x = t^2 - 3t + 2, \quad y = 2t - 3$

15. $x = t^2 - 1, \quad y = t^3 + 3$

16. $x = t - t^3, \quad y = 2t$

In Exercises 17–24, show that $f(g(x)) = x$ and $g(f(x)) = x$.

17. $f(x) = 3x - 2; \; g(x) = \frac{1}{3}(x + 2)$

18. $f(x) = \dfrac{x + 3}{4}; \; g(x) = 4x - 3$

19. $f(x) = x^3 + 1; \; g(x) = (x - 1)^{1/3}$

20. $f(x) = \dfrac{1}{x}; \; g(x) = \dfrac{1}{x}$

21. $f(x) = x^2 - 3, \, x \geq 0; \; g(x) = \sqrt{x + 3}$

22. $f(x) = (x + 1)^3; \; g(x) = \sqrt[3]{x} - 1$

23. $f(x) = \dfrac{x + 1}{x}; \; g(x) = \dfrac{1}{x - 1}$

24. $f(x) = \dfrac{x + 3}{x - 2}; \; g(x) = \dfrac{2x + 3}{x - 1}$

In Exercises 25 and 26, interpret the graph.

25. Use a graphical argument to decide whether this function is one-to-one.

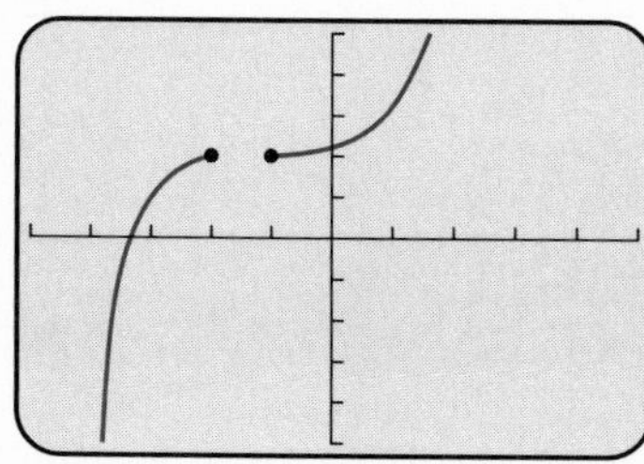

For Exercise 25

26. Use a graphical argument to decide whether this function is one-to-one.

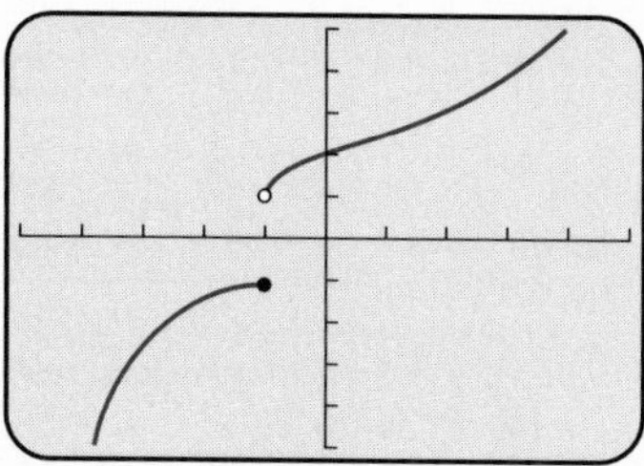

For Exercise 26

In Exercises 27–32, determine whether the function is one-to-one.

27. $f(x) = 3x - 2$

28. $g(x) = \dfrac{3x - 5}{4}$

29. $h(x) = x^2 - 5$

30. $f(x) = x^4 + 3$

31. $f(x) = x^3 - 4$

32. $g(x) = 5 - 0.5x^3$

In Exercises 33–42, determine graphically whether the function is one-to-one.

33. $y = x^2 + 5$

34. $y = x^3 - 4x + 5$

35. $y = x^4 - 5x^2 + 1$

36. $y = 0.001x^3$

37. $f(x) = |2x - 3| - |x + 1|$

38. $h(x) = |x + 4| + |3 - x|$

39. $f(x) = 4 - x^3$

40. $g(x) = \sqrt{x - 4}$

41. $f(x) = \begin{cases} x^3 + 1 & \text{if } x \leq 0 \\ 1 - x & \text{if } 0 < x \end{cases}$

42. $g(x) = \begin{cases} -x^2 - 1 & \text{if } x < 0 \\ 1 - x & \text{if } 0 \leq x \leq 1 \\ x^2 + 2 & \text{if } 1 < x \end{cases}$

In Exercises 43–52, find $y = f^{-1}(x)$.

43. $f(x) = 3x - 6$

44. $f(x) = 2x + 5$

45. $f(x) = \dfrac{2x - 3}{x + 1}$

46. $f(x) = \dfrac{x + 3}{x - 2}$

47. $f(x) = \sqrt{x + 3}$

48. $f(x) = \sqrt{x + 2}$

49. $f(x) = x^3$

50. $f(x) = x^3 + 5$

51. $f(x) = \sqrt[3]{x + 5}$

52. $f(x) = \sqrt[3]{x - 2}$

Table 2.8 B.A. Degrees in Mathematics Awarded in the U.S.

Year	Degrees Awarded
1985	38,445
1986	38,524
1987	38,114
1988	36,775
1989	36,079

Source: Bureau of the Census, *Statistical Abstract of the United States, 1992* (Washington, 1992).

In Exercises 53 and 54, use the data in Table 2.8, which gives the number of B.A. degrees in mathematics awarded in the United States from 1985 to 1989.

53. **a.** Graph a scatter plot of the data in Table 2.8.
b. Find a linear regression equation for the data.
c. Superimpose a graph of the linear regression equation on the scatter plot.

54. **a.** If $y = f(x)$ is the linear regression equation for the data found in Exercise 53, find the inverse function f^{-1}.
b. Use f^{-1} to find the year in which you expect the number of B.A. degrees in mathematics to be approximately 34,000.

EXTENDING THE IDEAS

In Exercises 55 and 56, restrict the domain of the function so that the function on the restricted domain is one-to-one. Find the inverse of this function.

55. $f(x) = (x + 2)^2 - 7$

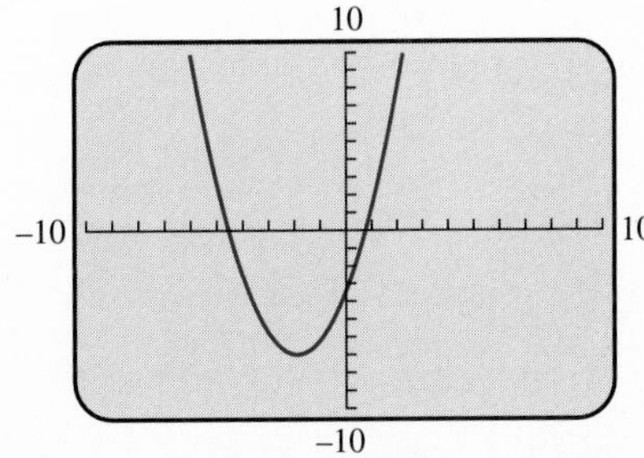

For Exercise 55

56. $f(x) = x^4 - 3$

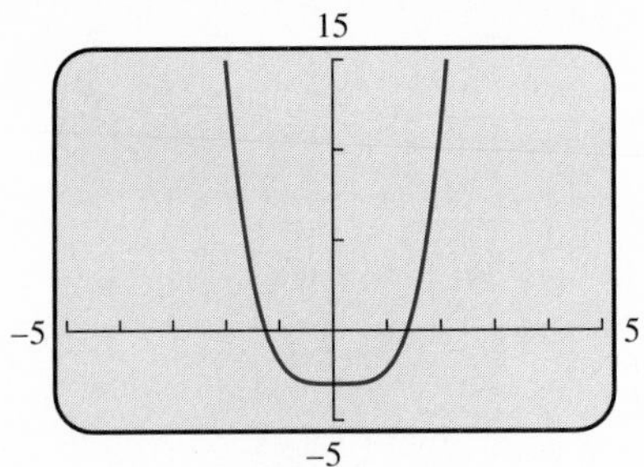

For Exercise 56

In Exercises 57–59, refer to the graphs, each of which shows a function f, its inverse f^{-1}, and the line $y = x$.

57. Suppose a is a number in the domain of f. Point P_2 is on the same horizontal line as point P_1. What are the coordinates of point P_2?

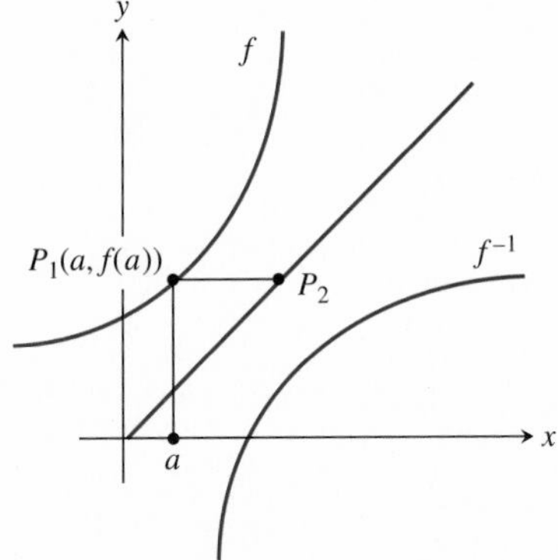

For Exercise 57

58. Point P_3 is on the vertical line through point P_2 and on the

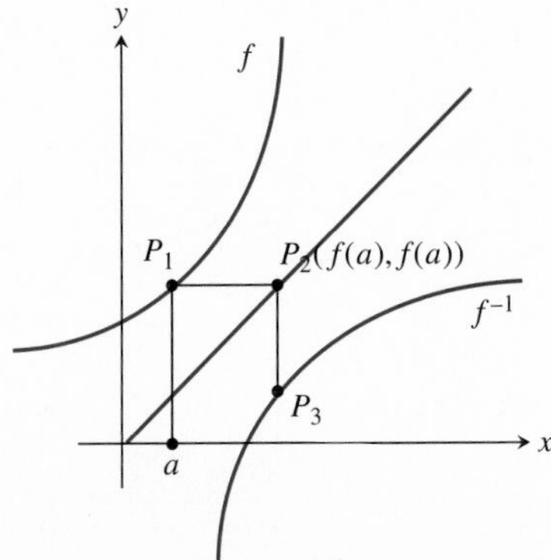

For Exercise 58

graph of f^{-1}. Explain why the coordinates of point P_3 are $(f(a), f^{-1}(f(a)))$.

59. Points (x, y) and (y, x) are reflections of each other about the line $y = x$. How can this fact be used to show that $f^{-1}(f(a)) = a$?

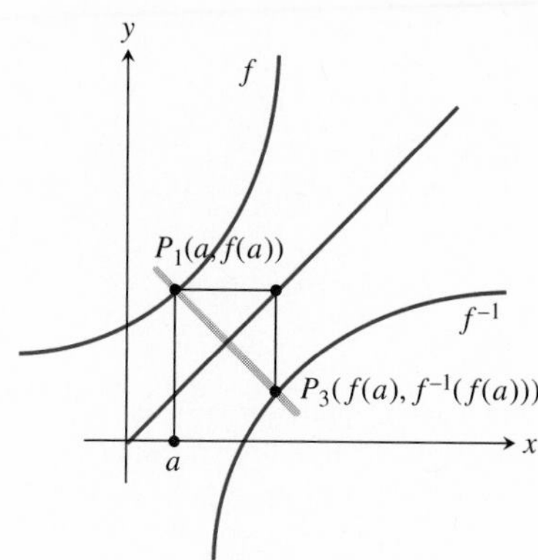

For Exercise 59

60. Writing to Learn Summarize how Exercises 57–59 and their accompanying figures can be used to fill in an explanation that begins "Suppose that f is the graph of a one-to-one function. Then . . . " and ends with the statement " . . . so we see that if f and f^{-1} are functions whose graphs are reflections of each other about the line $y = x$, it must follow that $f(f^{-1}(x)) = x$ for each x in the domain of f."

CHAPTER 2 REVIEW

KEY TERMS

The number following each key term indicates the page of its introduction.

REVIEW EXERCISES

In Exercises 1–4, solve graphically by finding x-intercepts. Confirm algebraically. Identify any extraneous solutions.

1. $6x^2 + 11x = 10$

2. $(x - 2)^2 = 2x^2$

3. $x + 3 = \sqrt{x + 5}$

4. $|3x - 2| = 4$

In Exercises 5–8, solve graphically by finding intersections. Confirm algebraically. Identify any extraneous solutions.

5. $x^2 + x = 2x + 1$

6. $\sqrt{x + 7} = 2$

7. $|x + 3| = 3x + 2$

8. $|x^2 + 2x - 8| = 7$

In Exercises 9–12, solve by completing the square. Support your answer graphically.

9. $x^2 + 1 = 4x$ **10.** $x^2 + 2x = 6$

11. $3x^2 + 4x = \frac{1}{3}$ **12.** $4x^2 + 4x = 3$

In Exercises 13–18, solve by factoring. Support graphically.

13. $(x + 1)^2 = 16$ **14.** $x^2 - 25 = 0$

15. $x^2 + x - 42 = 0$ **16.** $x^2 - 2x - 15 = 0$

17. $2x^2 - x = 6$ **18.** $3x^2 + 11x = 4$

In Exercises 19–22, solve by using the quadratic formula. Support your answer numerically.

19. $2x^2 - 6x + 1 = 0$ **20.** $4x^2 + 8x + 1 = 0$

21. $2x^2 + 3x + 2 = 0$ **22.** $3x^2 + 7x - 20 = 0$

In Exercises 23–38, solve the equation.

23. $4t^2 - 2t + 1 = 0$ **24.** $2u^2 + 5u - 12 = 0$

25. $z(z - 10) = -21$ **26.** $3(s + 2)^2 = 18$

27. $|2x - 3| = 5$ **28.** $|3x + 5| = 7$

29. $|x^2 - 2x - 3| = 5$ **30.** $x^2 - \sqrt{3}x + 5 = 7$

31. $|x + 2| = |2x - 3|$ **32.** $|x - 2| = \sqrt{2x + 5}$

33. $|w - 1| = w^2 - 5$ **34.** $\sqrt{v - 1} = \sqrt{2v - 5}$

35. $x^3 - 4x + 2 = 0$ **36.** $x^4 - 5x^2 + 4 = 0$

37. $\sqrt{x - 3} + \sqrt{x + 2} = 5$

38. $\sqrt{x + 2} = x^2 - 4x + 1$

In Exercises 39 and 40, find a graph that shows how many solutions each equation has. Then find all solutions.

39. $x^2 + x^{-1} = 0.5x + 3$ **40.** $x + x^{-1} = x^2 - 5$

In Exercises 41–44, evaluate the discriminant to determine how many real solutions the equation has. Support your answer graphically.

41. $2x^2 - 3x + 2 = 0$ **42.** $3x^2 + x - 5 = 0$

43. $-x^2 + 4x - 4 = 0$ **44.** $-2x^2 + x + 4 = 0$

In Exercises 45 and 46, solve the problem by interpreting the graphs.

45. The graph shows the x-intercept method of solving the equation

$$|x - 2| = x^2 - 3.$$

Identify the function in the graph.

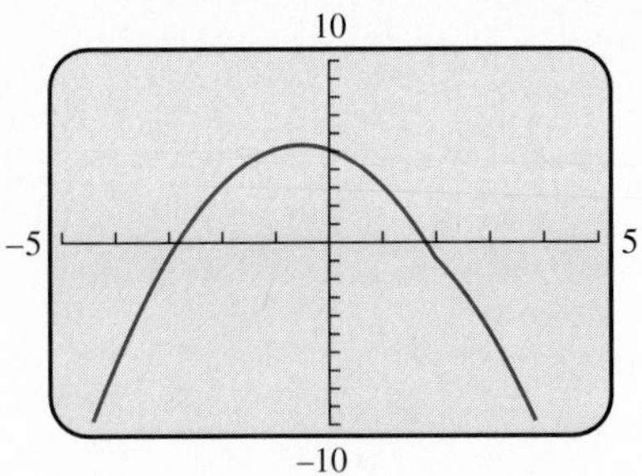

For Exercise 45

46. The solution to an inequality involving $|ax + b|$ is shown on the number line. Identify the inequality.

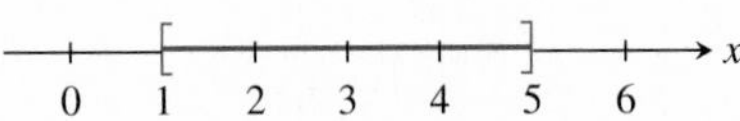

For Exercise 46

In Exercises 47–50, solve the inequality algebraically. Write the solution in interval notation, and draw its number line graph.

47. $3x - 7 < 5x + 1$ **48.** $-3 < \frac{4 - x}{2} \le 8$

49. $|2x + 5| > 3$ **50.** $|3x - 4| \le 2$

In Exercises 51–62, solve the inequality. Write the solution in interval notation.

51. $4x - 3 \ge x + 5$ **52.** $2x - 9 \le 3 - 2x$

53. $7 > 5x - 1 \ge -2$ **54.** $-2 < \frac{2x - 5}{4} < 3$

55. $|5 - 2x| \ge 4$ **56.** $|7x - 1| < 2$

57. $x - 2 \le 2 < x + 3$

58. $2x - 2 \le x + 3 \le 2x + 1$

59. $\sqrt{3x + 5} > \sqrt{3 - x}$ **60.** $|x + 2| < |x - 3|$

61. $\frac{2x - 5}{|x - 4|} \ge 0$ **62.** $(x + 2)|x - 2| \le 0$

In Exercises 63–66, write $f + g$, $f - g$, fg, and f/g as expressions in x. Find the domain of each function.

63. $f(x) = 3x + 5;\ g(x) = x^2$

64. $f(x) = x^2 + x;\ g(x) = x - 1$

65. $f(x) = \sqrt{x - 4};\ g(x) = x^2 - 4$

66. $f(x) = \sqrt{x + 3};\ g(x) = \sqrt{x - 2}$

In Exercises 67 and 68, find $(f \circ g)(2)$ and $(g \circ f)(2)$.

67. $f(x) = x^3 - 1;\ g(x) = x + 3$

68. $f(x) = \sqrt{2x - 1};\ g(x) = 1 - x^2$

In Exercises 69 and 70, find both $(f \circ g)(x)$ and $(g \circ f)(x)$. Find the domain of each.

69. $f(x) = \sqrt{x};\ g(x) = x^2 - 4$

70. $f(x) = x^2 - 3;\ g(x) = \dfrac{1}{x + 2}$

In Exercises 71–74, find $f(x)$ and $g(x)$ so that the given function can be written as $y = f(g(x))$.

71. $y = \sqrt[3]{4x - 3}$

72. $y = |x^2 + x|$

73. $y = \dfrac{2}{(x + 3)^2}$

74. $y = 2\sqrt{x} - 5$

In Exercises 75 and 76, solve the problem by interpreting the graphs.

75. Graphs of f and g are shown. Sketch a graph of $f - g$.

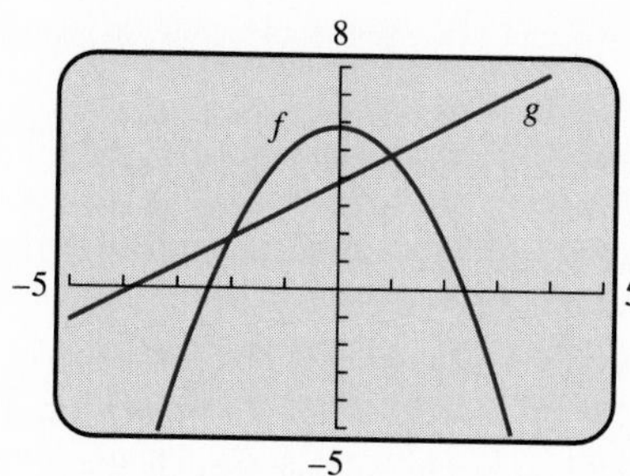

For Exercise 75

76. The graph of the parametric equations

$$x = t - 1, \quad y = 3 - |t|,$$

with $-5 \le t \le 5$ is shown. Find the values of the parameter t for the portion of the graph shown in

a. Quadrant I **b.** Quadrant II
c. Quadrant III **d.** Quadrant IV

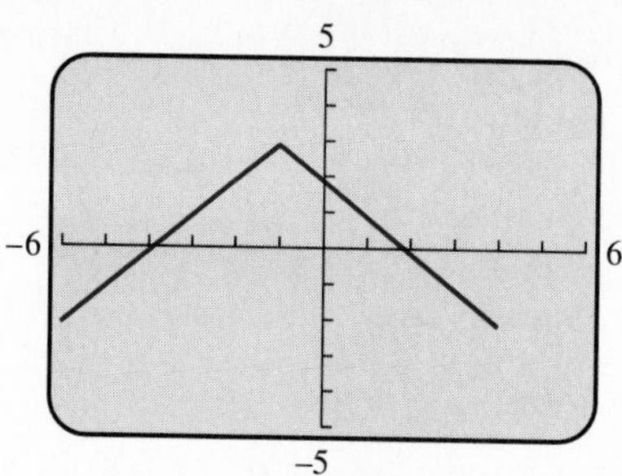

For Exercise 76

In Exercises 77 and 78, determine whether the ordered pair is in the given relation.

77. $x^2 + y^2 < 4$ **a.** $(\sqrt{2}, \sqrt{2})$ **b.** $(1, 1.5)$

78. $x - 2y^2 = 1$ **a.** $(3, -1)$ **b.** $(-1, 1)$

In Exercises 79 and 80, find the (x, y) pair for the given value of the parameter in the relation

$$x = t^2 - 3t, \quad y = 3 - t.$$

79. $t = -1$

80. $t = 2$

In Exercises 81 and 82, find a viewing window that shows a complete graph of the relation defined by the given parametric equations.

81. $x = 2t + 1, \quad y = 1 - 3t, \quad -3 \le t \le 3$

82. $x = t^2 + 1, \quad y = t - 1, \quad -4 \le t \le 4$

In Exercises 83 and 84, find parametric equations that give the graph of the function in the standard viewing window.

83. $f(x) = x^2 - 4x + 1$

84. $g(x) = 5 - 2x$

In Exercises 85–88, describe how the graph of f can be transformed to the graph of g.

85. $f(x) = x^3$ and $g(x) = -2x^3 + 5$

86. $f(x) = x^2$ and $g(x) = (x + 3)^2 - 4$

87. $f(x) = \sqrt{x}$ and $g(x) = \sqrt{4 - x}$

88. $f(x) = |x|$ and $g(x) = 3|x - 2| + 1$

In Exercises 89–92, a graph G is obtained from the graph of y by the transformations indicated. Write an equation whose graph is G.

89. $y = x^2$: a vertical stretch of factor 2, then shift right 3 units.

90. $y = \sqrt[3]{x}$: a horizontal stretch of factor 2, then shift left 3 units.

91. $y = \sqrt{x}$: a reflection across the y-axis, then a horizontal shrink of factor $\frac{1}{3}$, and finally a shift up 3 units.

92. $y = |x|$: a reflection across the x-axis, then a shift left 2 units, and finally a shift down 1 unit.

In Exercises 93 and 94, answer the questions.

93. Describe two ways to transform the graph of $y = x^3$ to the graph of $f(x) = 8x^3$.

94. Describe a basic graph and a sequence of transformations that can be used to find the graph of $y = -3\sqrt[3]{x - 4} - 2$.

In Exercises 95–98, describe how the graph C_2 can be obtained from the graph C_1. Eliminate the parameter t and describe the graphs in function form.

95. C_1: $x_1 = t, \quad y_1 = t^2$
C_2: $x_2 = t + 3, \quad y_2 = t^2$

96. C_1: $x_1 = t, \quad y_1 = |t|$
C_2: $x_2 = t, \quad y_2 = 3|t|$

97. C_1: $x_1 = t, \quad y_1 = t^2 - 1$
C_2: $x_2 = 2t, \quad y_2 = t^2 - 1$

98. C_1: $x_1 = t, \quad y_1 = \sqrt{t}$
C_2: $x_2 = t, \quad y_2 = \sqrt{t} - 4$

In Exercises 99 and 100, solve the problem by interpreting the graphs.

99. Shown is a graph of a function $y = f(x)$ that can be obtained by transforming the graph of $y = \sqrt{x}$. Write a formula for the function f.

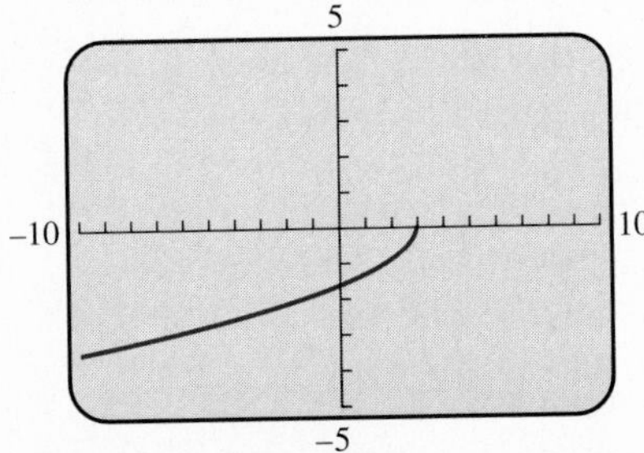

For Exercise 99

100. The graph of a function f is shown. Sketch the graph of $y = -2f(-x) + 1$.

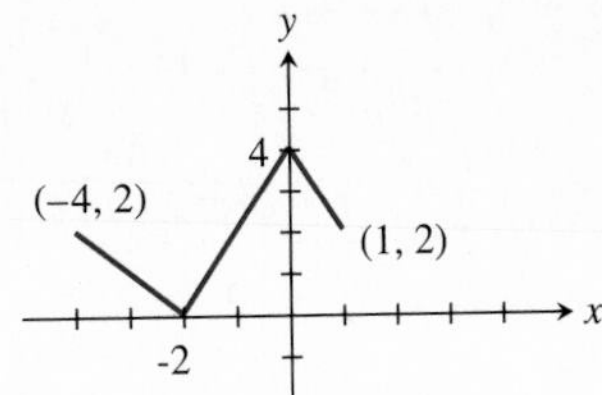

For Exercise 100

In Exercises 101 and 102, write a pair of parametric equations for the inverse relation of the given relation. Graph the relation and inverse relation together in a square viewing window.

101. $x = 3t, \quad y = t^2$ **102.** $x = t, \quad y = 2^t$

In Exercises 103 and 104, show that $f(g(x)) = x$ and $g(f(x)) = x$.

103. $f(x) = 2x + 5; \quad g(x) = \frac{1}{2}(x - 5)$

104. $f(x) = x^2 + 1, x \geq 0; \quad g(x) = \sqrt{x - 1}$

In Exercises 105 and 106, determine whether the function is one-to-one.

105. $f(x) = |x + 6| - |x - 5|$

106. $f(x) = x^3 + x$

In Exercises 107 and 108, find $y = f^{-1}(x)$.

107. $f(x) = 3x - 2$ **108.** $f(x) = \dfrac{x + 1}{x - 3}$

In Exercises 109 and 110, use the given data to answer the questions.

109. *U.S. Deficit* Table 2.9 shows the U.S. parts trade deficit with Japan in billions of dollars for the given years. If $x = 0$ represents 1980, this deficit can be modeled by the equation

$$\text{deficit} = 0.0474x^3 - 1.4217x^2 + 14.324x - 39.019.$$

a. Graph a scatter plot of the data together with the model.
b. Use this model to estimate the deficit in 1996.
c. Use this model to estimate when the deficit will be $30 billion.

110. *PGA Purse* Table 2.10 shows the total annual purses of the Professional Golfer's Association in millions of dollars for the given years. If $x = 0$ represents 1980, this purse can be modeled by the equation

$$\text{purse} = -0.074x^2 + 4.882x + 3.020.$$

a. Graph a scatter plot of the data together with the model.
b. Use this model to estimate the purse in 1997.

Table 2.9 Trade Deficit

Year	Deficit (billions of dollars)
1985	3.1
1986	6.0
1987	7.3
1988	8.8
1989	10.0
1990	9.5
1991	9.1
1992	9.8
1993	11.2
1994	12.8

Source: Commerce Department as reported by Web Bryant in *USA Today,* May 19, 1995.

Table 2.10 PGA Purse

Year	Purse (millions of dollars)
1983	17.6
1985	25.3
1987	32.1
1989	41.3
1991	49.6
1993	53.2
1995	59.5

Source: PGA Tour as reported by Scott Boeck and Julie Stacey in *USA Today,* May 5, 1995.

c. Use this model to estimate when the purse will be \$75 million.

In Exercises 111–118, solve the given problem.

111. *Dimensions of a Window* A window has the shape of a rectangle with a semicircle mounted on a smaller side of it. The rectangle is 5 ft longer than it is wide. Find the width x of the window if the total area A of the window is 150 ft^2.

112. *Finding Travel Rates* Two airplanes leave Wichita at the same time. The Airbus flies due north toward Winnipeg, and the Turbo Prop flies due east toward Raleigh. The two airplanes fly at a constant rate, and the Airbus flies 55 mph less than twice as fast as the Turbo Prop. After 1.5 h they are 765 mi apart. How fast is each of the two airplanes flying?

113. *Planning for Profit* Thrifty Bike finds that the annual cost of making bikes is \$85 per bike plus \$75,000 in fixed overhead costs. If each bike sells for \$190, find the number of bikes, x, that need to be sold to make a profit P of \$40,000.

114. **Connecting Algebra and Geometry** Consider the collection of all rectangles whose length is 6 ft more than the width. Find the possible widths of the rectangle if the perimeter is to be less than 300 ft.

115. *Designing Containers* Equal squares are cut from each corner of a 12-in. by 18-in. piece of cardboard to form an open-top box by folding up the flaps. What length of squares x should be cut to make a box with a volume V of 210 in.^{3}?

116. *Wave Effect* A rock is tossed into a pond. The radius of the first circular ripple (wave) increases at the rate of 0.75 ft/sec. Find when the area will be 250 ft^2.

117. *Height of an Arrow* Stewart shoots an arrow straight up from the top of a building with initial velocity of 245 ft/sec. The arrow leaves from a point 200 ft above level ground.

a. Write an equation that models the height of the arrow as a function of time t.

b. Use parametric equations to simulate the height of the arrow.

c. Use parametric equations to graph height against time.

d. How high is the arrow after 4 sec?

e. What is the maximum height of the arrow? When does it reach its maximum height?

f. How long will it be before the arrow hits the ground?

118. Sarah is 5 ft 6 in. tall and walks at a rate of 6 ft/sec away from a streetlight with a lamp 12 ft above ground level.

a. Find an equation that models the length of Sarah's shadow.

b. Find the length of her shadow after 5 sec.

In Exercises 119 and 120, answer the questions.

119. **Writing to Learn** Explain why applying certain transformations to the graph of $y = f(x)$ leads to the graph of $y = af(bx + c) + d$ for appropriate real numbers a, b, c, and d.

120. **Writing to Learn** Is it possible for an inequality to have exactly one solution? Explain using examples.

CHU SHIH-CHIEH

Chu Shih-chieh (ca. 1280–1303) was the last and greatest mathematician of the golden age of Chinese mathematics during the Sung Dynasty. Very little is known about his personal life. He probably lived in Yen-shan, near modern-day Beijing, but he spent 20 years traveling extensively in China as a renowned mathematician and teacher.

His book *Introduction to Mathematical Studies* was lost for a time, but later became an influential textbook in Korea and Japan. Another book, *Precious Mirror of the Four Elements,* strongly influenced the development of the theory of equations. It discusses four "elements"—heaven, earth, man, and matter—which represent four unknowns in a single equation. Although the book introduces two concepts later named for Western mathematicians (the Horner method and Pascal's triangle), Chu Shih-chieh's contributions to mathematics remain largely anonymous.

chapter 3

POLYNOMIAL FUNCTIONS

BIBLIOGRAPHY

For students: *Conquering Math Anxiety,* Cynthia Arem. Brooks/Cole Publishing Co., 1993.

For teachers: *Explorations in Precalculus Using the TI-82 and the TI-85,* Cochener and Hodge. Brooks/Cole Publishing Co., 1979.

Videos: *Polynomials, Project Mathematics!,* California Institute of Technology. Available through Dale Seymour Publications.

Objective

Students will be able to use transformations to graph quadratic functions and to identify the vertex and line of symmetry of a parabola.

Motivate

Ask students how they think the graph of $y = (x - 2)^2 + 5$ is related to the graph of $y = x^2$. **(Translated 2 units right and 5 units up.)**

3.1 QUADRATIC FUNCTIONS

Introduction to Polynomial Functions • Quadratic Functions and Their Graphs • Using Transformations in Graphing • Applications • Modeling Data with Quadratic Functions

Introduction to Polynomial Functions

Polynomial functions are among the most familiar of all the functions that we shall study. In the Prerequisite chapter we defined the zero polynomial.

Definition: Polynomial Function

A **polynomial function** is one that can be written in the form

$$f(x) = a_n x^n + a_{n-1}x^{n-1} + \cdots + a_1 x + a_0, \quad a_n \neq 0$$

where n is a nonnegative integer and the coefficients $a_0, a_1, \cdots, a_n$ are real numbers. The integer n is the **degree** of the function f, a_n is its **leading coefficient,** and $a_n x^n$ its **leading term.**

Here are some examples.

Function	Degree	Leading Coefficient
$f(x) = 2x^3 - 4x + 3$	3	2
$f(x) = 3x - 4x^5 - 1$	5	−4
$f(x) = (x - 3)^2 + 5x$	2	1

Quadratic Functions and Their Graphs

We have already studied polynomial functions of degree 0 (a constant function) and degree 1 (a linear function). This section focuses on **quadratic functions**—polynomial functions of degree 2. The **standard form equation** of a general quadratic function is

$$f(x) = ax^2 + bx + c, \quad a \neq 0.$$

If $b = 0$, the quadratic function has the form $f(x) = ax^2 + c$. Since

$$f(-x) = a(-x)^2 + c = ax^2 + c = f(x),$$

such quadratic functions are even functions, which means that the y-axis is a line of symmetry of the graph of f.

The graph of a quadratic function is a **parabola,** a line-symmetric curve whose shape is like the graph of $y = x^2$ shown in Figure 3.1. The point of intersection of the parabola and its line of symmetry is the **vertex** of the parabola and is the lowest or highest point of the graph. The graph of a parabola either **opens upward** like $y = x^2$, or **opens downward** like the graph of $y = -x^2$.

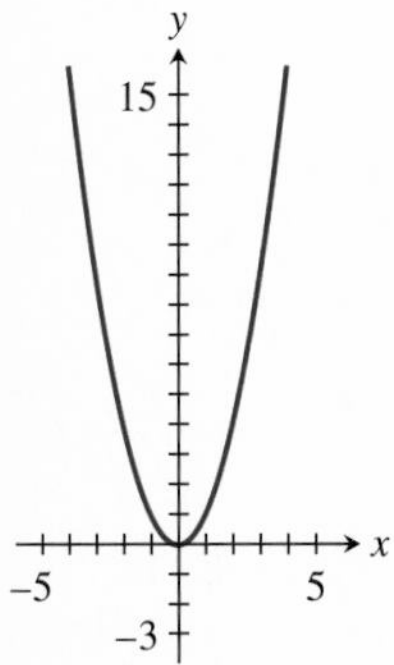

Figure 3.1 The graph of $y = x^2$ is a parabola that is symmetric about the y-axis.

Teaching Note

Note that students have seen some of this material in Section 2.6.

Using Transformations in Graphing

You will learn that the graph of any quadratic function can be described as a transformed graph of $y = x^2$. Example 1 illustrates this property for a function already written in a form convenient for identifying transformations.

■ EXAMPLE 1 Comparing Graphs of Quadratic Functions

Describe how the graph of $f(x) = -x^2 + 2x + 5 = -(x - 1)^2 + 6$ compares with the graph of $y = x^2$. Identify the coordinates of the vertex and the line of symmetry of f.

Solution

The graph of f can be obtained from the graph of $y = x^2$ as follows:

1. Translate $y = x^2$ to the right 1 unit. $\quad y = (x - 1)^2$
2. Reflect the result through the x-axis. $\quad y = -(x - 1)^2$
3. Translate that result up 6 units. $\quad y = -(x - 1)^2 + 6$

In this process, the (0, 0) vertex of the graph of $y = x^2$ moves to the point (1, 6), and the line of symmetry $x = 0$ moves to the line $x = 1$. So the function $f(x) = -(x - 1)^2 + 6$ has the vertex (1, 6) and the line of symmetry $x = 1$.

Support Graphically

Figure 3.2 shows the graph of f is indeed a translation 1 unit right of $y = x^2$ followed by a reflection across the x-axis and then a translation up 6 units. ■

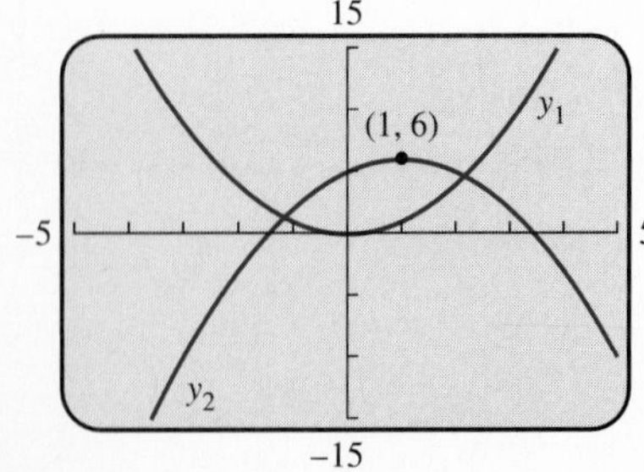

Figure 3.2 $y_1 = x^2$ and $y_2 = -(x - 1)^2 + 6$.

The function f in Example 1 is written in a form convenient for recognizing the transformations that relate the graphs of $y = x^2$ and f. A quadratic function can be written in this convenient form by completing the square (see Section 2.2), as illustrated in Example 2.

■ EXAMPLE 2 Comparing Graphs of Quadratic Functions

Describe how the graph of $f(x) = 6x - 3x^2 - 5$ compares with the graph of $y = x^2$. Find the vertex and the line of symmetry of the graph of f.

Solution

$$\begin{aligned} f(x) &= -3x^2 + 6x - 5 && \text{Standard form} \\ &= -3(x^2 - 2x) - 5 && \text{Factor } -3 \text{ from the first two terms.} \\ &= -3[x^2 - 2x + (\ ) - (\ )] - 5 && \text{Prepare to complete the square.} \\ &= -3[x^2 - 2x + (1) - (1)] - 5 && \text{Complete the square.} \\ &= -3(x^2 - 2x + 1) - (-3)(1) - 5 && \text{Use the distributive property.} \\ &= -3(x^2 - 2x + 1) + (3 - 5) \\ &= -3(x - 1)^2 + (-2) \end{aligned}$$

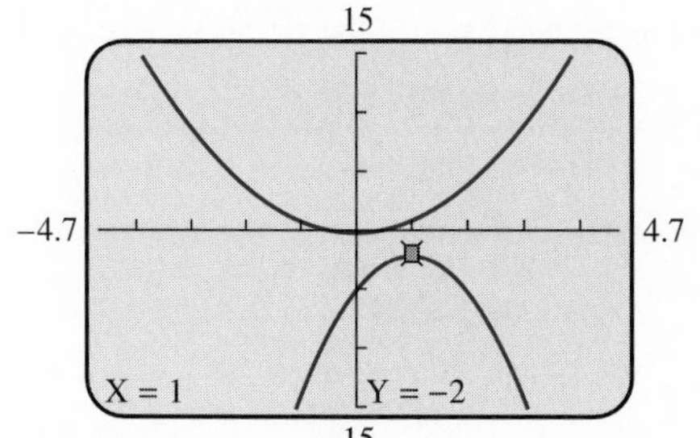

Figure 3.3 $y_1 = x^2$ and $y_2 = 6x - 3x^2 - 5$.

The graph of f can be obtained from the graph of $y = x^2$ by a shift right 1 unit, a vertical stretch of factor 3, a reflection across the x-axis, and finally a shift down 2 units. In this process, the (0, 0) vertex of the graph of $y = x^2$ moves to the point $(1, -2)$, and the line of symmetry $x = 0$ moves to the line $x = 1$. So $f(x) = 6x - 3x^2 - 5$ has the vertex $(1, -2)$ and the line of symmetry $x = 1$, and the parabola opens downward. See Figure 3.3. ■

Example 2 serves as a model for the general quadratic function. By completing the square with $f(x) = ax^2 + bx + c$, we obtain

$$\begin{aligned} f(x) &= ax^2 + bx + c \\ &= a\left[x^2 + \left(\frac{b}{a}\right)x\right] + c && \text{Factor } a \text{ from first two terms.} \\ &= a\left[x^2 + \left(\frac{b}{a}\right)x + (\quad) - (\quad)\right] + c && \text{Prepare to complete the square.} \\ &= a\left[x^2 + \left(\frac{b}{a}\right)x + \left(\frac{b}{2a}\right)^2 - \left(\frac{b}{2a}\right)^2\right] + c && \text{Complete the square.} \\ &= a\left(x + \frac{b}{2a}\right)^2 - a\left(\frac{b^2}{4a^2}\right) + c && \text{Use the distributive property.} \\ &= a\left(x + \frac{b}{2a}\right)^2 + \left(c - \frac{b^2}{4a}\right) \\ &= a\left(x - \frac{-b}{2a}\right)^2 + \left(c - \frac{b^2}{4a}\right) \end{aligned}$$

REMARK

A horizontal translation of 2 means a shift right 2 units, and a horizontal translation of -2 means a shift left 2 units. Similar statements can be made about vertical translations.

This algebra shows that the graph of $f(x) = ax^2 + bx + c$, $a > 0$, can be obtained from the graph of $y = x^2$ by a horizontal translation of $-b/(2a)$ followed by a vertical stretch or shrink of factor a, and then a vertical translation of $c - b^2/(4a)$. If $a < 0$, a reflection across the x-axis is needed before the vertical translation.

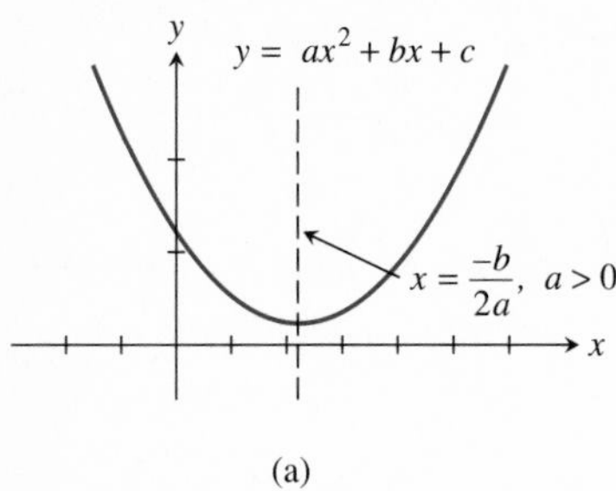

(a)

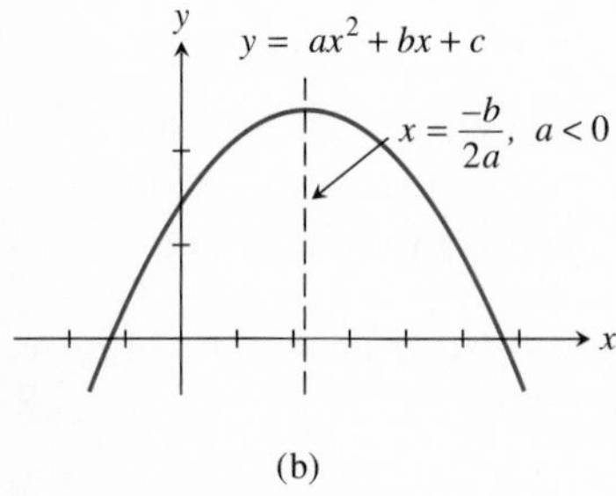

(b)

Figure 3.4 The vertex is at $x = -b/(2a)$, which therefore also describes the line of symmetry. (a) When $a > 0$, the parabola opens upward. (b) When $a < 0$, the parabola opens downward.

Quadratic Function

If $a \neq 0$, $f(x) = ax^2 + bx + c$ can be written in the form

$$f(x) = a(x - h)^2 + k.$$

The vertex of the parabola is (h, k), and its line of symmetry is $x = h$ where $h = \dfrac{-b}{2a}$ and $k = c - \dfrac{b^2}{4a}$.

If $a > 0$, the parabola opens upward, and if $a < 0$ the parabola opens downward. (See Figure 3.4.)

Notice that you need not remember the y-coordinate k of the vertex of a parabola f because $k = f(h)$.

■ EXAMPLE 3 Finding the Line of Symmetry

Find the line of symmetry and the vertex of $f(x) = 7 - 12x - 3x^2$.

Solution

Rewriting f as $f(x) = -3x^2 - 12x + 7$ we see that $a = -3$, $b = -12$, and $c = 7$. Therefore, the line of symmetry is

$$\begin{aligned} x &= \frac{-b}{2a} \\ &= \frac{-(-12)}{2(-3)} = -2 \end{aligned}$$

The vertex is the point $(-2, f(-2)) = (-2, 19)$. ■

Applications

Our model for projectile motion from Section 2.2,

$$s(t) = -16t^2 + v_0 t + s_0$$

is a quadratic function.

Notes on Examples

The formula $s(t) = -16t^2 + v_0 t + s_0$ is important and will be used for several of the exercises.

EXAMPLE 4 APPLICATION: Finding Maximum Height

A flare is launched straight up with an initial velocity of 64 ft/sec from the top of a 200-ft building. What is the maximum height of the flare, and how many seconds after it is launched does it reach its maximum height?

Solution

Model

The function $s(x) = -16x^2 + v_0 x + s_0$, where $v_0 = 64$ and $s_0 = 200$, is a model for the height of the flare at any time x. Because its graph is a parabola that opens downward, we need to find the y-coordinate of the vertex.

Solve Algebraically

In $s(x) = -16x^2 + 64x + 200$ we have $a = -16$ and $b = 64$. The x-coordinate of the vertex is

$$x = \frac{-b}{2a} = \frac{-64}{2(-16)} = 2.$$

The y-coordinate of the vertex is

$$s(2) = -16(2)^2 + 64(2) + 200 = 264.$$

Thus, the vertex of the parabola is (2, 264).

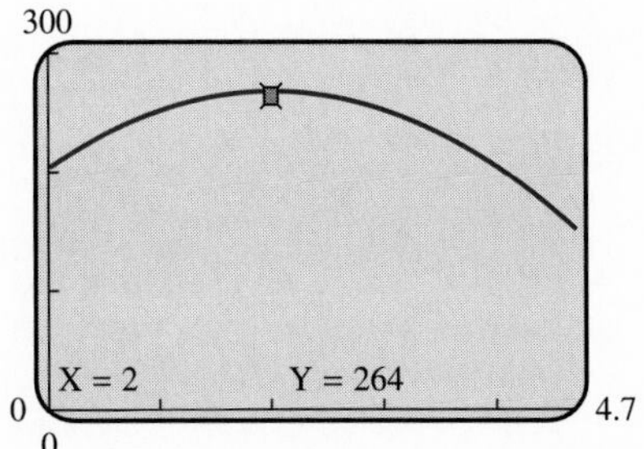

Figure 3.5 The graph of the height of the flare in Example 4 plotted against time. This is not the path of the flare, which was fired straight up. However, the peak of this graph corresponds to the maximum height of the flare.

Support Graphically

A trace on the graph of $s(x) = -16x^2 + 64x + 200$ in Figure 3.5 suggests that the vertex is at (2, 264).[1]

Interpret

The flare reaches a maximum height of 264 ft 2 sec after it is launched. ■

EXAMPLE 5 APPLICATION: Designing a Ventilating System

The ventilating system for the John K. Schertz lecture hall requires ducts with rectangular cross section. These are to be made by folding sheet-metal strips that are 48 in. wide. The volume of air in the ducts can be maximized by designing the ducts to have a maximum cross-sectional area. What are the best dimensions for these ducts?

[1]Experiment with parametric mode on a grapher to simulate the path of the flare. Compare your results with those in Example 5 in Section 2.5.

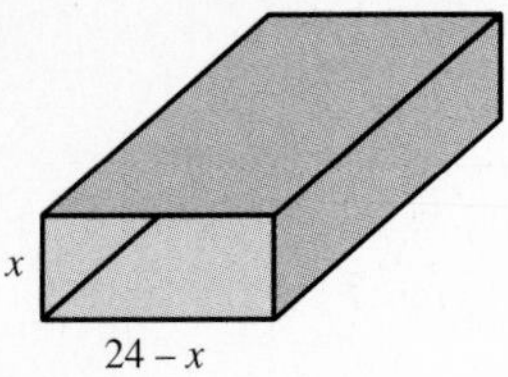

Figure 3.6 A ventilation duct.

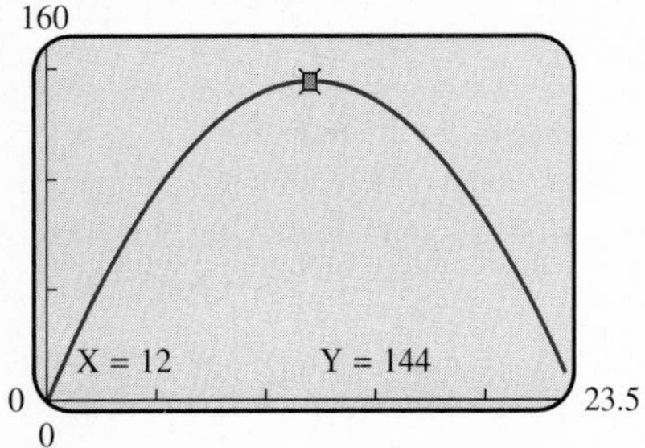

Figure 3.7 $y_1 = A(x) = x(24 - x)$.

Solution

Model

Let x be the width of the rectangular cross section. Then its length is $24 - x$, and the area of the cross section is given by the quadratic function $A(x) = x(24 - x)$. (See Figure 3.6.) The graph of A is a parabola that opens downward, so we need to find the y-coordinate of the vertex.

Solve Graphically

The graph of $y_1 = A(x) = x(24 - x)$ in Figure 3.7 suggests that the vertex is at (12, 144).

Confirm Algebraically

In $A(x) = x(24 - x) = 24x - x^2$ we have $a = -1$ and $b = 24$. The x-coordinate of the vertex is

$$x = \frac{-b}{2a} = \frac{-24}{2(-1)} = 12.$$

The y-coordinate of the vertex is

$$A(12) = 24(12) - (12)^2 = 144.$$

Thus, the vertex of the graph of $A(x)$ is (12, 144).

Interpret

The maximum duct cross section occurs when $x = 12$. That means that the length and width are both 12 in. and the cross section is a square. ■

Modeling Data with Quadratic Functions

A function model for known data can be used to predict future values of a quantity. In Section 1.4 we modeled data with linear functions. Some data are modeled better by a quadratic function than a linear function. Here is a summary of a strategy that you can use.

Using Data Analysis—Quadratic Regression

1. Complete a scatter plot of the data.
2. Find the quadratic regression equation.[2]
3. Superimpose a graph of the quadratic regression equation on the scatter plot.
4. Use the quadratic regression equation to predict y-values for x-values not included in the data.

Follow-up

Ask students whether they think the quadratic regression function found in Example 6 would provide a good estimate of the number of patent applications in 1965. **(No)**

[2]Consult your grapher owner's manual to learn how to find a quadratic regression equation.

In later courses you will learn how to determine how good the fit is through quantitative means. For now, make a visual judgment.

Table 3.1 U.S. Patent Applications

Year	Applications (thousands)
1980	113.0
1981	Not known
1982	118.4
1983	112.4
1984	120.6
1985	127.1
1986	133.0
1987	139.8
1988	151.9
1989	166.3
1990	176.7

Source: U.S. Bureau of the Census, *Statistical Abstract of the United States, 1992* (Washington, D.C., 1992).

Assignment Guide

Day 1: Ex. 3–48, multiples of 3
Day 2: Ex. 51, 54, 55, 58, 59, 61, 62, 63, 64, 65, 68, 70, 75, 76

Cooperative Learning

Group Activity: Ex. 77, 78

Notes on Exercises

Ex. 51–58 require the student to produce a quadratic function with certain qualities.
Ex. 59–70 offer some classical applications of quadratic functions.
Ex. 71–73 are applications of quadratic regression.

Ongoing Assessment

Self-Assessment: Ex. 5, 19, 37, 69
Embedded Assessment: Ex. 58, 62

■ EXAMPLE 6 APPLICATION: Modeling Data on Patent Applications

Table 3.1 shows that the number of patent applications in the United States has been increasing over time in recent years. Find both a linear and a quadratic regression equation for this data. Which appears to be the better model of the data?

Solution

We compute the linear regression equation $y = 6.6x + 100.4$ and the quadratic regression equation $y = 0.8x^2 - 1.65x + 114.1$, with $x = 0$ for 1980, $x = 1$ for 1981, and so forth. Figure 3.8 shows graphs of these linear and quadratic regression equations superimposed on the scatter plot.

The quadratic regression equation appears to model the data better than the linear regression equation. ■

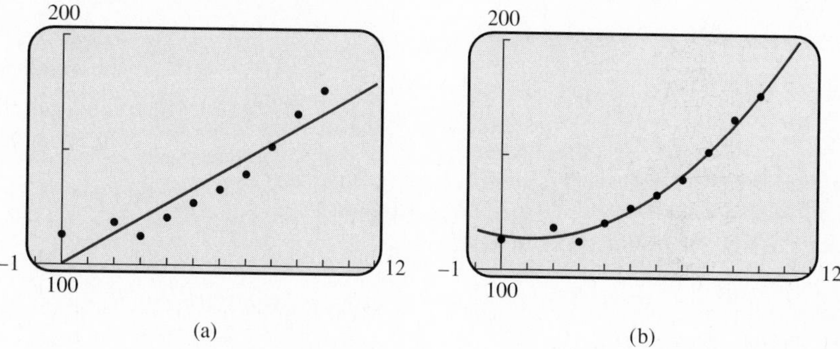

Figure 3.8 (a) Linear regression equation $y = 6.6x + 100.4$. (b) Quadratic regression equation $y = 0.8x^2 - 1.65x + 114.1$.

■ EXAMPLE 7 APPLICATION: Modeling Data on Patent Applications

Use quadratic regression to predict the year in which the number of U.S. applications (see Example 6) would reach 300,000.

Solution

Model

From Example 6, the quadratic regression equation that models the number of U.S. patent applications per year is $y = 0.8x^2 - 1.65x + 114.1$. To predict

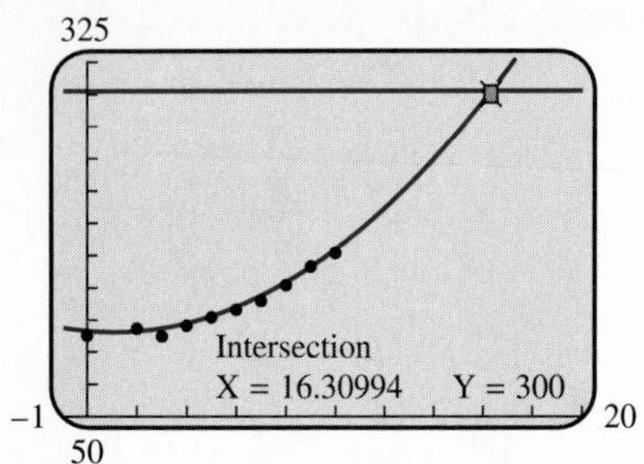

Figure 3.9 A scatter plot of the data, and graphs of $y_1 = 0.8x^2 - 1.65x + 114.1$ and $y_2 = 300$.

when the number of applications would reach 300,000, we need to solve the equation

$$0.8x^2 - 1.65x + 114.1 = 300.$$

Solve Graphically

Figure 3.9 shows that the graphs of $y_1 = 0.8x^2 - 1.65x + 114.1$ and $y_2 = 300$ intersect when x is about 16.3.

Interpret

The number of applications should have reached 300,000 in 1997 (17 years from 1980). ■

Quick Review 3.1

In Exercises 1–4, expand the expression.

1. $(x + 3)^2$ **2.** $(x - 4)^2$

3. $3(x - 6)^2$ **4.** $-3(x + 7)^2$

In Exercises 5–8, factor the trinomial.

5. $x^2 + 10x + 25$ **6.** $x^2 - 4x + 4$

7. $2x^2 - 4x + 2$ **8.** $3x^2 + 12x + 12$

In Exercises 9–12, write the projectile-motion equation that models the situation. Use $h = -16t^2 + v_0 t + s_0$, where h is height in feet, and t is time in seconds.

9. An object is propelled upward from ground level with an initial velocity of 42 ft/sec.

10. A construction explosion hurls a piece of iron upwards from the top of a 125-ft tower with an initial velocity of 82 ft/sec.

11. A stone is thrown downward from the top of a 240-ft tower with an initial velocity of 8 ft/sec.

12. A cork gun shoots a cork from the bottom of a pit 12 ft deep with an initial velocity of 128 ft/sec.

SECTION EXERCISES 3.1

In Exercises 1–6, identify the degree and the leading coefficient of the polynomial function.

1. $f(x) = 4x^2 - 5$

2. $g(x) = 2(x + 1)^2 - 3x + 2$

3. $f(x) = 7 - 2x + (2x + 3)^2$

4. $g(x) = 9 - 4x - x^2$

5. $f(x) = -3(x + 2)^2 - 5$

6. $g(x) = 2(x + 3)^2 - 5(x - 2) + 7$

In Exercises 7–10, match a graph to the function. Include a short explanation for your choice including why dimensions are not needed on the viewing window.

7. $f(x) = 2(x + 1)^2 - 3$ **8.** $f(x) = 3(x + 2)^2 - 7$

9. $f(x) = 4 - 3(x - 1)^2$ **10.** $f(x) = 12 - 2(x - 1)^2$

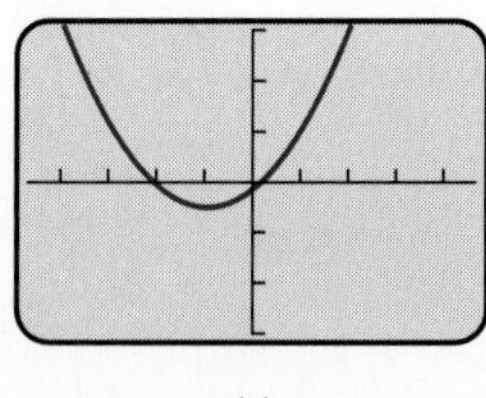

(a)

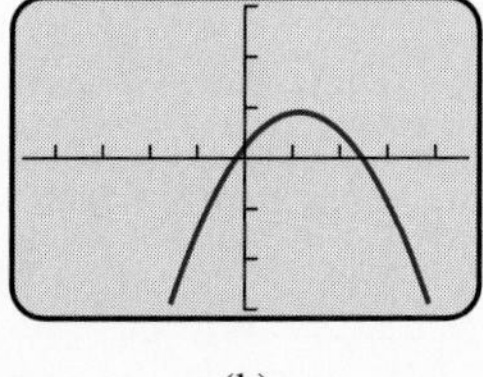

(b)

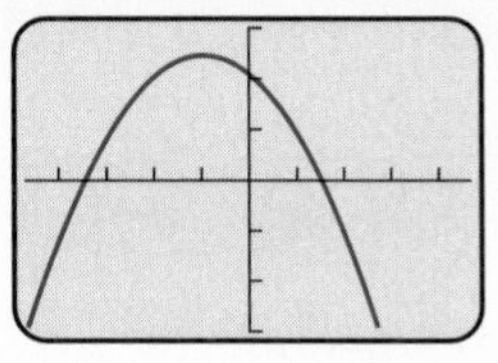
(c)

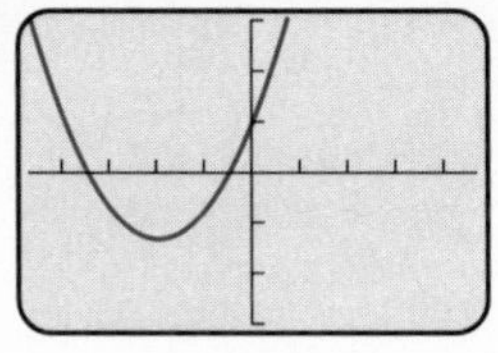
(d)

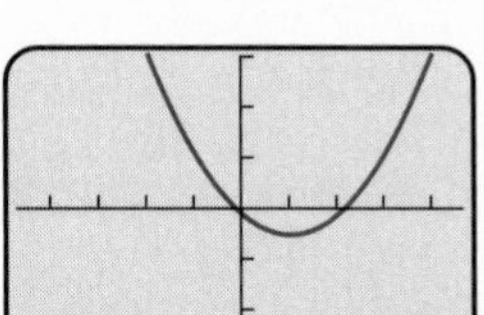
(e)

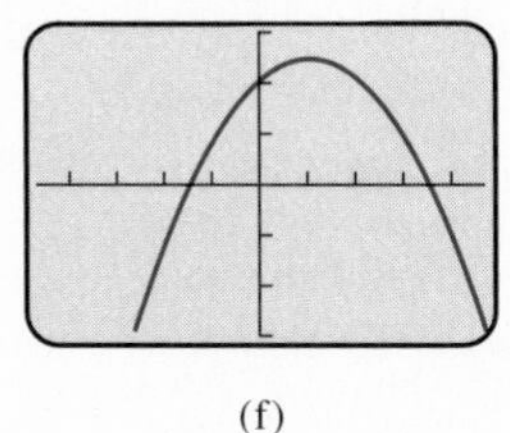
(f)

For Exercises 7–10

In Exercises 11–22, use transformations to explain how the graphs of the given function and of $y = x^2$ compare. Sketch the graph of the function.

11. $f(x) = 4x^2$

12. $k(x) = x^2 - 3$

13. $h(x) = (x - 5)^2$

14. $g(x) = (x + 1)^2$

15. $f(x) = 3(x + 1)^2$

16. $k(x) = 2(x - 4)^2$

17. $g(x) = -(x + 3)^2$

18. $h(x) = -(x + 5)^2$

19. $f(x) = 2(x + 3)^2 + 4$

20. $g(x) = 4(x - 2)^2 - 5$

21. $k(x) = (x + 3)^2 - 2$

22. $f(x) = 2(x - 1)^2 + 7$

In Exercises 23–28, identify the vertex and the line of symmetry of the graph of the function. Support your answer with a grapher.

23. $f(x) = 3(x - 1)^2 + 5$

24. $g(x) = -3(x + 2)^2 - 1$

25. $f(x) = 5(x - 1)^2 - 7$

26. $g(x) = 2(x - \sqrt{3})^2 + 4$

27. $f(x) = \sqrt{2} + (x - 5)^2$

28. $g(x) = \sqrt{3} - (x - \sqrt{2})^2$

In Exercises 29–38, find the vertex and the line of symmetry of the graph of the function.

29. $f(x) = 3x^2 + 5x - 4$

30. $f(x) = -2x^2 + 7x - 3$

31. $f(x) = 8x - x^2 + 3$

32. $f(x) = 6 - 2x + 4x^2$

33. $g(x) = 5x^2 + 4 - 6x$

34. $h(x) = -2x^2 - 7x - 4$

35. $f(x) = x^2 + 5x - 2$

36. $f(x) = x^2 - 4x + 1$

37. $f(x) = 3x^2 - 7x - 12$

38. $f(x) = -2x^2 + 6x + 13$

In Exercises 39–50, use algebraic methods to describe the graph of the function. Include the vertex and the line of symmetry. Support your answer with a grapher.

39. $f(x) = x^2 - 4x + 6$

40. $g(x) = x^2 - 6x + 12$

41. $h(x) = x^2 + 5x + 3$

42. $k(x) = x^2 + 7x - 2$

43. $f(x) = 10 - 16x - x^2$

44. $h(x) = 8 + 2x - x^2$

45. $g(x) = 3x^2 + 12x - 2$

46. $k(x) = 4x^2 - 20x + 5$

47. $f(x) = 2x^2 + 6x + 7$

48. $g(x) = 5x^2 - 25x + 12$

49. $h(x) = 4x^2 - 20x + 23$

50. $k(x) = 3x^2 - 17x + 3$

In Exercises 51–54, find an equation for the parabola shown. One of the given points is the vertex of the parabola.

51.

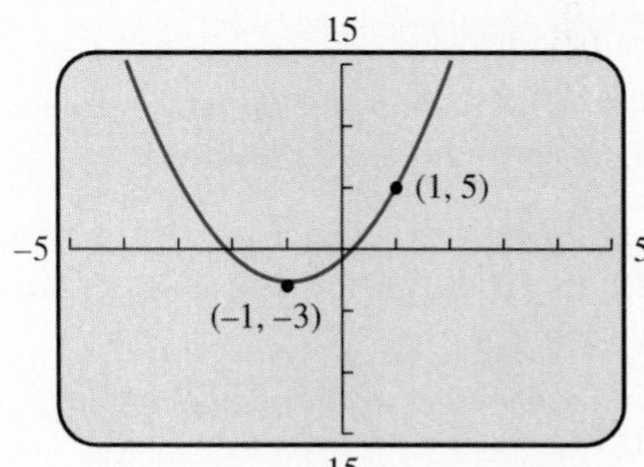

52.

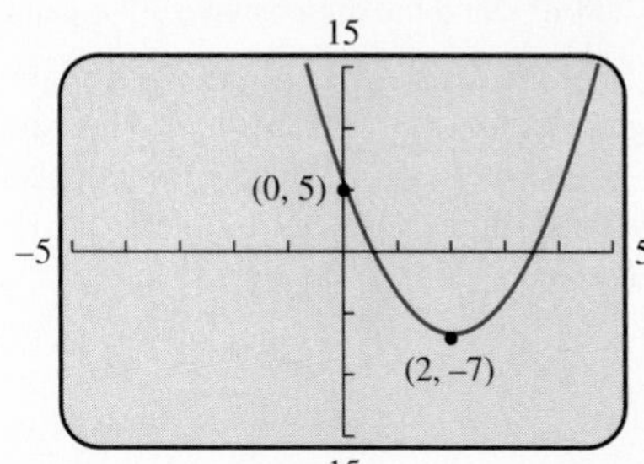

53.

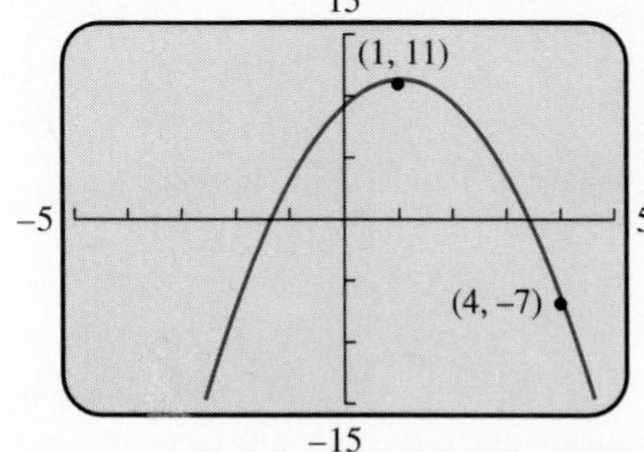

54.

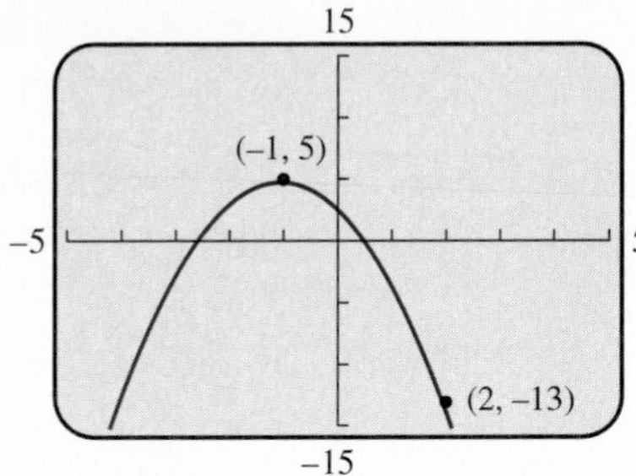

In Exercises 55–58, find an equation for the parabola having the given vertex and passing through the given point.

55. Vertex $(1, 3)$, through the point $(0, 5)$

56. Vertex $(-2, -5)$, through the point $(0, -27)$

57. Vertex $(\sqrt{2}, 5)$, through the point $(1, 0)$

58. Vertex $(3\sqrt{5}, 8)$, through the point $(-\sqrt{5}, 128)$

In Exercises 59–64, solve the problem either algebraically or graphically. Explain your choice of method.

59. *Finding Maximum Area* Among all the rectangles whose perimeters are 100 ft, find the dimensions of the one with maximum area.

60. *Finding the Dimensions of a Painting* A large painting in the style of Rubens is 3 ft longer than it is wide. If the wooden frame is 12 in. wide, the area of the picture and frame is 208 ft^2. Find the dimensions of the painting.

61. *Using Algebra in Landscape Design* Julie Stone designed a rectangular patio that is 25 ft by 40 ft. This patio is surrounded by a terraced strip of uniform width planted with small trees and shrubs. If the area A of this terraced strip is 504 ft^2, find the width x of the strip.

For Exercise 61

62. *Management Planning* The Welcome Home apartment rental company has 1600 units available, of which 800 are currently rented at \$300 per month. A market survey indicates that each \$5 decrease in monthly rent will result in 20 new leases.

a. Determine a function $R(x)$ that models the total rental income realized by Welcome Home, where x is the number of \$5 decreases in monthly rent.

b. Find a graph of $R(x)$ for rent levels between \$175 and \$300 (that is, $0 \le x \le 25$) that clearly shows a maximum for $R(x)$.

c. What rent will yield Welcome Home the maximum monthly income?

63. *Beverage Business* The Sweet Drip Beverage Co. sells cans of soda pop in machines. It finds that sales average 26,000 cans per month when the cans sell for 50¢ each. For each nickel increase in the price, the sales per month drop by 1000 cans.

a. Determine a function $R(x)$ that models the total revenue realized by Sweet Drip, where x is the number of \$0.05 increases in the price of a can.

b. Find a graph of $R(x)$ that clearly shows a maximum for $R(x)$.

c. How much should Sweet Drip charge per can to realize the maximum revenue? What is the maximum revenue?

64. *Sales Manager Planning* Jack was named District Manager of the Month at the Sylvania Wire Co. due to his hiring study. It shows that each of the 30 salespersons he supervises average \$50,000 in sales each month, and that for each additional salesperson he would hire, the average sales would decrease \$1,000 per month. Jack concluded his study by suggesting a number of salespersons that he should hire to maximize sales. What was that number?

In Exercises 65–68, use the formula $h = -16t^2 + v_0 t + s_0$ for the height of an object, where v_0 represents the initial velocity, s_0 represents the initial height, and t represents time.

65. *Baseball Throwing Machine* The Sandusky Little League uses a baseball throwing machine to help train 10-year-old players to catch high pop-ups. It throws the baseball straight up with an initial velocity of 48 ft/sec from a height of 3.5 ft.

a. Find an equation that models the height of the ball t seconds after it is thrown. Graph the equation using both function mode and parametric mode. (See Example 5 in Section 2.5.)

b. What is the maximum height the baseball will reach? How many seconds will it take to reach that height?

66. *Fireworks Planning* At the Bakersville fourth of July celebration, fireworks are shot by remote control into the air from a pit that is 10 ft below the earth's surface.

a. Find an equation that models the height of an aerial bomb t seconds after it is shot upwards with an initial velocity of 80 ft/sec. Graph the equation in both function mode and parametric mode.

b. What is the maximum height above ground level that the aerial bomb will reach? How many seconds will it take to reach that height?

67. *Ballistic Design* Shawna, a jet fighter pilot, ejects from her burning plane. Once safely on the ground she launches a flare to attract the attention of a search plane. The initial velocity of the flare is 1600 ft/sec. What maximum height does the flare achieve, and how long does it take to reach that height?

68. *Landscape Engineering* In her first project after being employed by Land Scapes International, Becky designs a decorative water fountain that will shoot water to a maximum height of 48 ft. What should be the initial velocity of each drop of water to achieve this maximum height? (*Hint:* Use a grapher and a guess-and-check strategy.)

For Exercise 68

In Exercises 69–71, solve the given problem.

69. *Sheet Metal Design* T & P Sheetmetal, Inc., wins a bid for making sheet-metal gutters. T & P begins with a sheet 10 in. wide and then turns up the sides to make a gutter with a rectangular cross section. What must the length and width of the gutter be for the cross section to have maximum area?

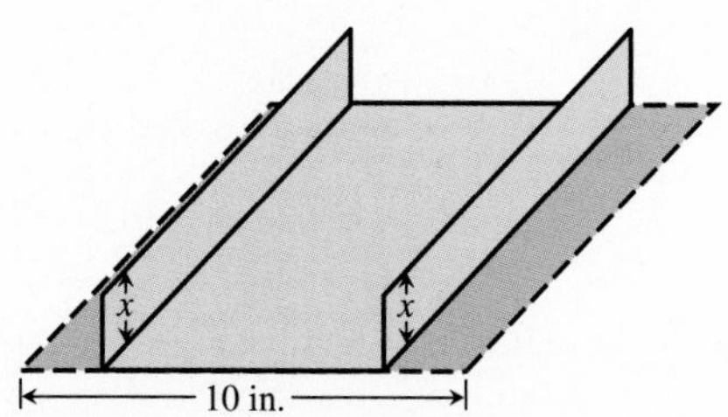

For Exercise 69

70. *Building a Pen* A rectangular puppy-exercise pen is to be made so that one side of the pen is bounded by the wall of a large building. What are the dimensions of the pen with maximum area that can be enclosed if the total length of fencing to be used is 500 ft?

71. *Patent Applications* Using the quadratic regression equation for the data in Example 6, predict the year in which the number of patent applications will reach 250,000.

In Exercises 72 and 73 use the data in Table 3.2 on the amount (in billions of dollars) invested by business in computer equipment.

Table 3.2 Computer Equipment

Year	Amount (billions of dollars)
1989	128
1990	133
1991	139
1992	157
1993	201
1994	249

Source: U.S. Commerce Department as reported by Julie Stacey in *USA Today,* August 9, 1995.

72. *Technology Investment*

a. Find the quadratic regression equation and superimpose its graph on a scatter plot of the data.

b. Find the linear regression equation and superimpose its graph on a scatter plot of the data.

73. Use both the quadratic and the linear regression equation from Exercise 72 to predict the amount that will be invested by business in computer equipment in the year 2000.

EXTENDING THE IDEAS

74. Connecting Algebra and Geometry Show that the line of symmetry for the graph of the quadratic function $f(x) = (x - a)(x - b)$ is $x = (a + b)/2$, where a and b are any real numbers.

75. Connecting Algebra and Geometry Identify the vertex of the graph of $f(x) = (x - a)(x - b)$, where a and b are any real numbers.

76. Connecting Algebra and Geometry Show that if x_1 and x_2 are the zeros of a quadratic function $f(x) = ax^2 + bx + c$ (see Section 2.2), then the line of symmetry of the graph is the line

$$x = \frac{x_1 + x_2}{2}.$$

77. Group Learning Activity *Work in groups of three.* The function f in this exercise is $f(x) = x^2$.

a. Describe the transformation that will produce the graph of $y = f(2x)$.
b. Describe the transformation that will produce the graph of $y = 4f(x)$.
c. Decide whether the transformation in part a produces the same or different result from the transformation in part b when either transformation is applied to the graph of $y = f(x)$.
d. Write a set of activities or exercises that you might use to explain your answer to part c to a group of students.

78. Group Learning Activity *Work in groups of three.* Consider functions of the form $f(x) = x^2 + bx + 1$ where b is a real number.

a. Discuss as a group how the value of b affects the graph of the function.
b. After completing part a, have each member of the group (individually) predict what the graphs of $f(x) = x^2 - 6x + 2$ and $g(x) = x^2 + 4x - 3$ will look like.
c. Compare your predictions with each other. Confirm whether they are correct.

3.2 POLYNOMIAL FUNCTIONS OF HIGHER DEGREE

Objective

Students will be able to graph polynomial functions, predict their end behavior, and find their real zeros using a grapher or an algebraic method.

Motivate

Ask students to predict the characteristics of graphs of degree 4 polynomial functions.

Graphs of Polynomial Functions • Behavior of Polynomial Functions When $|x|$ Is Large • Zeros of Polynomial Functions • Approximating Zeros Using Technology • Applications

Graphs of Polynomial Functions

As we have already seen, a degree 0 polynomial function is a constant function; its graph is a horizontal line. A degree 1 polynomial function is a linear function; its graph is a line that is neither horizontal nor vertical. A degree 2 polynomial function is quadratic; its graph is a parabola.

A **higher-degree polynomial** function is one whose degree is 3 or more. A degree 3 polynomial is commonly called a **cubic,** and a degree 4 polynomial is commonly called a **quartic.**

A polynomial function is **continuous:** its graph is a continuous curve; there are no jumps as we have seen can happen with piecewise-defined functions. A polynomial function often has local maximum and minimum values.

Notes on Examples

Example 1 stresses the importance of using an appropriate viewing window.

EXAMPLE 1 Graphing a Cubic

The cubic function $f(x) = 2x^3 + 8x^2 + 3x - 13$ has a local maximum, a local minimum, and a zero. Graph f, and find a viewing window that shows all three features.

Solution

Figure 3.10a shows a graph of f in the standard viewing window. There appears to be a local minimum that falls outside this viewing window. We can improve the view of this graph by

a. decreasing the horizontal window dimension, and
b. increasing the vertical window dimension. (See Figure 3.10b.)

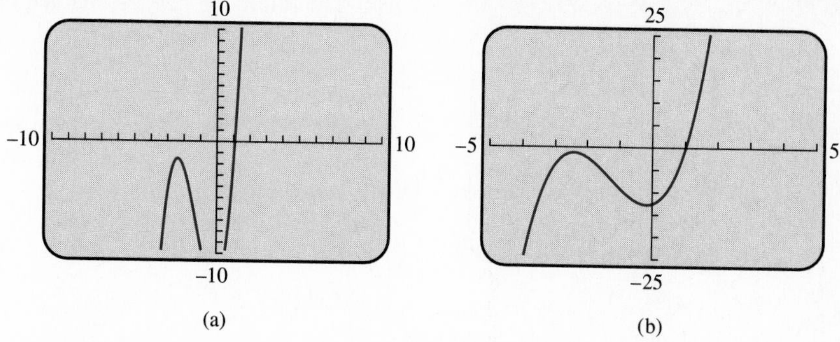

Figure 3.10 Two views of $y_1 = f(x) = 2x^3 + 8x^2 + 3x - 13$.

Step (a) spreads the graph out to the left and right to make better use of the viewing window. Step (b) allows more of the graph to fit into the viewing window from top to bottom. ■

A value $f(a)$ is a **local extremum** if it is a local maximum or a local minimum. The shape of the graph of a polynomial function is related to the number of local extremum values of the polynomial; the position of the graph is related to the number of zeros. And both of these are related to the degree of the polynomial. Figure 3.11 shows several possible shapes for graphs of cubic functions. There can be as many as three zeros and two local extremum values.

Figure 3.12 shows several possible shapes for graphs of quartic functions. There can be as many as four zeros and three local extremum values.

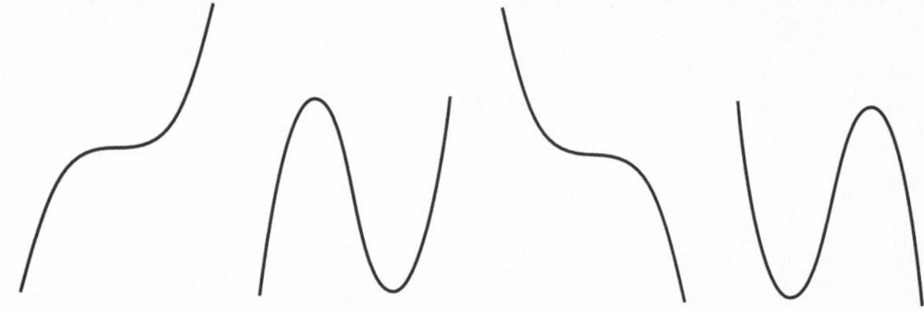

Figure 3.11 Graphs of $f(x) = ax^3 + bx^2 + cx + d$.

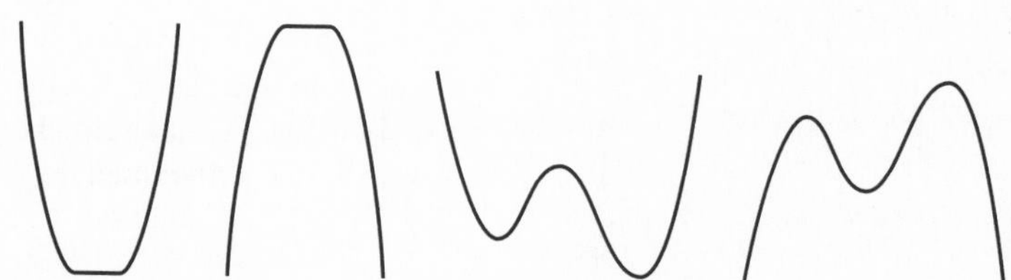

Figure 3.12 Graphs of $f(x) = ax^4 + bx^3 + cx^2 + dx + e$.

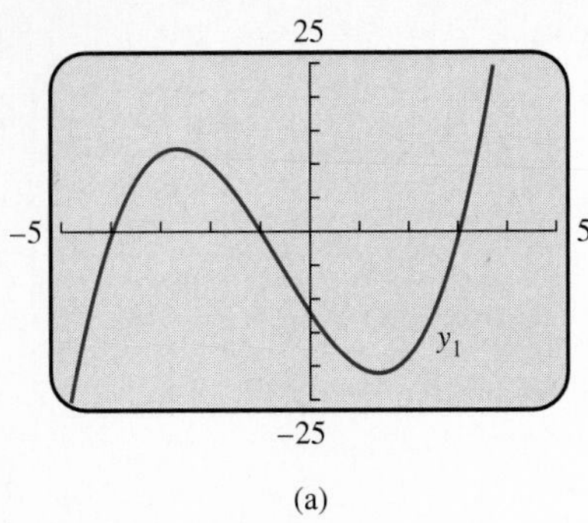

(a)

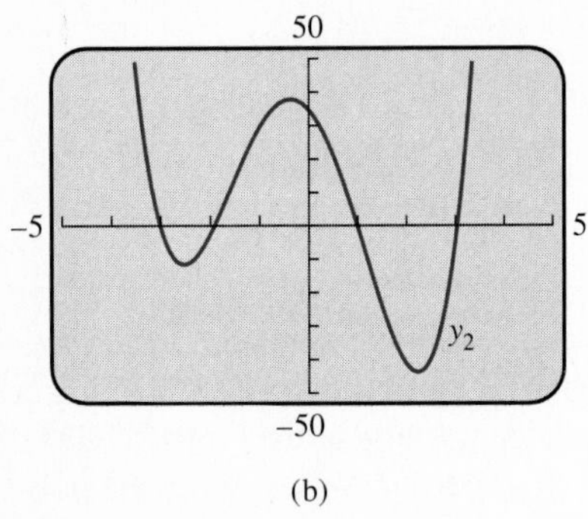

(b)

Figure 3.13
(a) $y_1 = f(x) = x^3 + 2x^2 - 11x - 12$.
(b) $y_2 = g(x) = 2x^4 + 2x^3 - 22x^2 - 18x + 35$.

The following general characteristics of polynomial functions can be confirmed using calculus.

Local Extremum Values and Zeros of Polynomial Functions

A polynomial function

$$f(x) = a_n x^n + a_{n-1}x^{n-1} + \cdots + a_1 x + a_0$$

has at most $n - 1$ local extremum values and n zeros.

■ **EXAMPLE 2 Graphing a Polynomial Function**

Graph the function showing all local extremum values and all zeros.

a. $f(x) = x^3 + 2x^2 - 11x - 12$
b. $g(x) = 2x^4 + 2x^3 - 22x^2 - 18x + 35$

Solution

a. Figure 3.13a shows a graph of f with two local extremum values and three zeros. Since this is the maximum number of local extremum values and zeros for a cubic, we are done.
b. Figure 3.13b shows a graph of g with three local extremum values and four zeros, the maximum possible for a quartic. ■

Behavior of Polynomial Functions When $|x|$ Is Large

An important characteristic of polynomial functions is their behavior when $|x|$ is large. This is the **end behavior** of the function. We begin by introducing notation.

We read

$$f(x) \to \infty$$

as "$f(x)$ increases without bound" and

$$x \to -\infty$$

as "x decreases without bound."

We use these symbols to communicate the nature of graphs on their extreme right and their extreme left (when $x \to \infty$ and $x \to -\infty$, respectively; that is, when $|x|$ is large). For example, the end behavior of the function in Figure 3.13a is described as

$$f(x) \to -\infty \text{ as } x \to -\infty \quad \text{and} \quad f(x) \to \infty \text{ as } x \to \infty.$$

The end behavior of the function in Figure 3.13b is described as

$$f(x) \to \infty \text{ as } x \to -\infty \quad \text{and} \quad f(x) \to \infty \text{ as } x \to \infty.$$

NOTE

Limits at Infinity

In calculus, we use $\lim_{x\to-\infty} f(x) = \infty$ for

$$f(x) \to \infty \text{ as } x \to -\infty$$

and $\lim_{x\to\infty} f(x) = \infty$ for

$$f(x) \to \infty \text{ as } x \to \infty.$$

REMARK

We tell you how to read two of the sentences using "→." We encourage you to use the given phrasing to interpret all sentences that use "→". Also, you will find that sometimes "→" is read as "approaches."

Exploration Extensions

Graph the functions $y = 2x^3 - x^2$, $y = -x^3 + 2$, $y = x^5 - 2x^3$, and $y = -0.5x^7 + 5x$, and compare the end behavior with the functions in Step 1. Adjust the viewing window if necessary.

TECHNOLOGY NOTE

For a cubic, when you change the horizontal window by a factor of k, it usually is a good idea to change the vertical window by a factor of k^3. Similar statements can be made about polynomials of other degrees.

The following Exploration investigates the end behavior of functions of the form $f(x) = ax^n$.

● EXPLORATION End Behavior of Functions $y = ax^n$

Graph each function in the viewing window $[-5, 5]$ by $[-15, 15]$, and describe its end behavior.

1. a. $y = 2x^3$ **b.** $y = -x^3$ **c.** $y = x^5$ **d.** $y = -0.5x^7$
2. a. $y = -3x^4$ **b.** $y = 0.6x^4$ **c.** $y = 2x^6$ **d.** $y = -0.5x^2$
3. a. $y = -0.3x^7$ **b.** $y = -2x^2$ **c.** $y = 3x^4$ **d.** $y = 2x^3$

Describe in Your Own Words

How do the values a and n determine the end behavior of the function $y = ax^n$? ●

Example 3 suggests that there is an interesting link between the end behavior of the polynomials $y = ax^n$ that have one term and the polynomials that have several terms.

■ EXAMPLE 3 Comparing Polynomials of Equal Degree

Compare the graphs of $y = x^3$ and $y = x^3 - 4x^2 - 5x - 3$ in successively larger viewing windows, a process called **zoom-out**. Show that in a sufficiently large window the graphs look nearly identical.

Solution

Figure 3.14 shows three different views of both $y_1 = x^3$ and $y_2 = x^3 - 4x^2 - 5x - 3$. As the viewing dimensions get larger, the graphs become more difficult to distinguish. ■

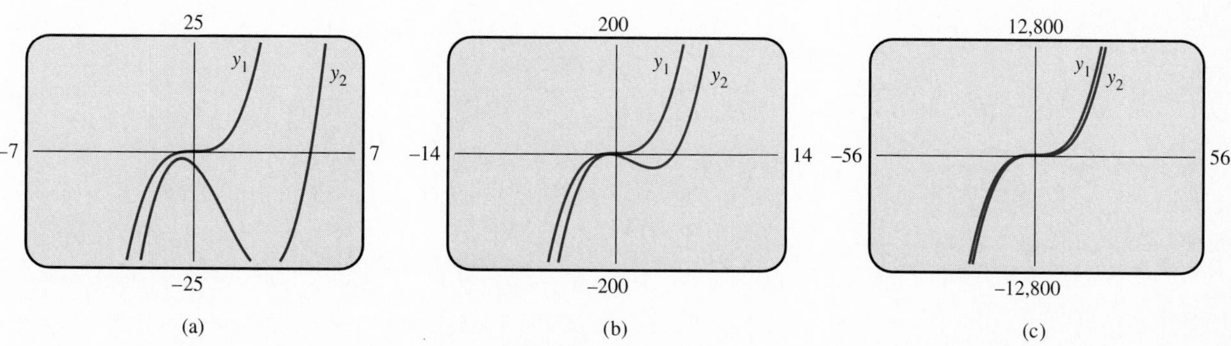

Figure 3.14 As the viewing window gets larger, there is less distinction between the graphs of $y_1 = x^3$ and $y_2 = x^3 - 4x^2 - 5x - 3$.

Teaching Note

The word *limit* and the formal limit notation are purposely avoided in this book. The word *approaches* is used instead. The authors have found that an intuitive approach to the concept of limit is very appropriate for students at the precalculus level, and that formal definitions of limit should be reserved for calculus.

In calculus you will confirm the information in Example 3 by showing that the *limit* as x approaches infinity of the quotient of the two functions is 1.

Example 3 illustrates something that is true for all polynomials: *For sufficiently large viewing dimensions, the graph of a polynomial and the graph of its leading term appear identical.* This fact suggests the following description of polynomial behavior when $|x| \to \infty$.

Polynomial End Behavior

If $f(x)$ is a degree n polynomial function with leading coefficient a_n, then:

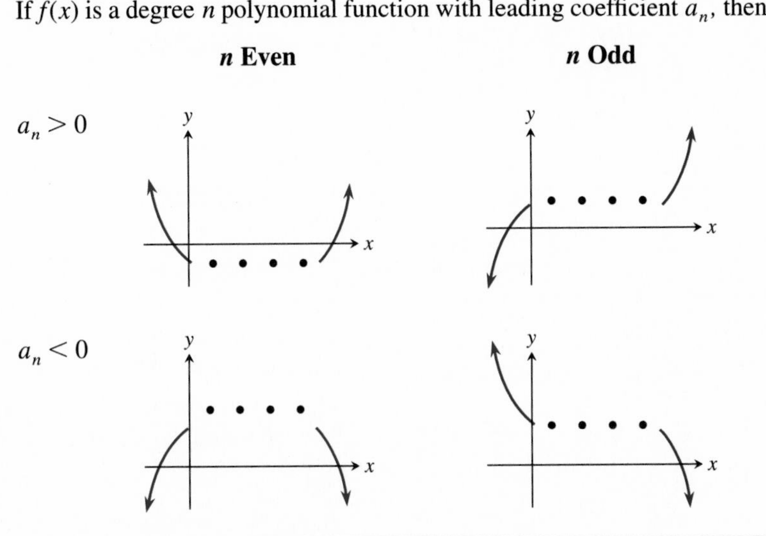

■ EXAMPLE 4 Applying the End Behavior Test

Describe the end behavior of the function.

a. $f(x) = 5x^4 - 3x^2 + x^3 - 8x + 3$
b. $g(x) = 8x + 3x^2 - 3x^5 - 6x^3$

Solution

a. The degree of f is 4 (even), and its leading coefficient is 5 (positive). Therefore,

$$f(x) \to \infty \text{ as } x \to \infty \quad \text{and} \quad f(x) \to \infty \text{ as } x \to -\infty.$$

b. The degree of g is 5 (odd), and its leading coefficient is -3 (negative). Therefore,

$$g(x) \to -\infty \text{ as } x \to \infty \quad \text{and} \quad g(x) \to \infty \text{ as } x \to -\infty.$$

Teaching Note

Discuss the terminology and notation associated with zeros of a function: x-intercept, root, and solution.

Zeros of Polynomial Functions

Recall that a zero of a polynomial function f is a value of x for which $f(x) = 0$. Finding the zeros of polynomial functions is one of the important problems of algebra. It is the key to solving many applied problems. Example 5 reviews a traditional algebraic method of finding zeros by factoring.

EXAMPLE 5 Finding the Zeros of a Polynomial Function

Find the zeros of $f(x) = x^3 - x^2 - 6x$.

Solution

Solve Algebraically

We need to solve the equation $f(x) = 0$.

$$x^3 - x^2 - 6x = 0 \quad \text{Equation to be solved}$$

$$x(x^2 - x - 6) = 0 \quad \text{Factor.}$$

$$x(x - 3)(x + 2) = 0 \quad \text{Factor.}$$

$$x = 0, \quad x - 3 = 0, \quad \text{or} \quad x + 2 = 0 \quad \text{Zero factor property}$$

$$x = 0, \quad x = 3, \quad \text{or} \quad x = -2$$

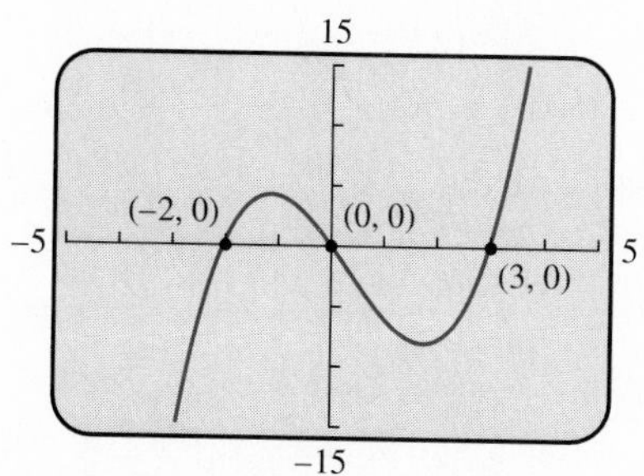

Figure 3.15 A graph of $y = x^3 - x^2 - 6x$ showing the x-intercepts.

Support Graphically

Figure 3.15 shows that the x-intercepts of the graph of $y = f(x)$ are $(0, 0)$, $(3, 0)$, and $(-2, 0)$—coinciding with the zeros found algebraically. ■

In Section 2.1 you learned about the fundamental connections among the equation $f(x) = 0$, the function $y = f(x)$, and the graph of $y = f(x)$. For polynomial functions, the connections extend to the factors of the polynomial.

Fundamental Connections (Polynomial Functions)

Let a be a real number. For a polynomial function of the form $y = f(x)$, the following four statements are equivalent:

1. $x = a$ is a solution or a root of the equation $f(x) = 0$.
2. $x = a$ is a zero of the function $y = f(x)$.
3. $(a, 0)$ is an x-intercept of the graph of $y = f(x)$.
4. $(x - a)$ is a factor of $f(x)$.

We ask that you think in particular about the connection involving Statement 4. It tells us that we can find solutions to $f(x) = 0$ by finding factors of $f(x)$ *and* vice versa.

Alert

Some students fail to make the distinction between exact and approximate answers. They may think that $\sqrt{3}$ (exact) and 1.732 (approximate) are equivalent.

Approximating Zeros Using Technology

The fundamental connections just given state that we can solve the polynomial equation $f(x) = 0$ by finding factors of $f(x)$. Sometimes, however, $f(x)$ is difficult to factor. In such a case, we can approximate the zeros using technology.

Not all polynomials have real number zeros. The following theorem gives us a way to tell when a polynomial has real zeros.

Theorem: Intermediate Value Theorem

If $y = f(x)$ is a polynomial function or any continuous function and $a < b$, then f assumes every value between $f(a)$ and $f(b)$. In other words, if L is any real number such that

$$f(a) < L < f(b) \quad \text{or} \quad f(b) < L < f(a),$$

then there is a number c between a and b such that

$$f(c) = L.$$

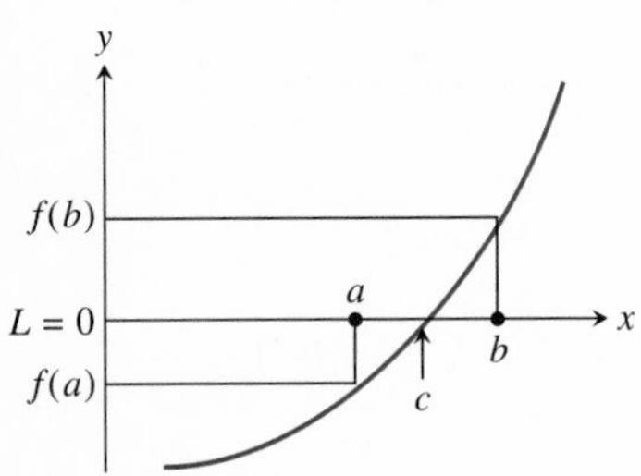

Figure 3.16 If $f(a) < 0 < f(b)$, then there is a zero $x = c$ between a and b.

Suppose that $y = f(x)$ is a polynomial function that takes on positive and negative values. If $f(a) < 0 < f(b)$, then by the Intermediate Value Theorem for polynomials we can conclude that there is a real number c between a and b such that $f(c) = 0$. In other words, there is a zero of f between a and b. (See Figure 3.16.)

■ EXAMPLE 6 Determining Whether Real Zeros Exist

Show that a polynomial of odd degree always has a real zero.

Solution

We can conclude from the end behavior that such a polynomial takes on positive and negative values. Therefore, by the Intermediate Value Theorem a polynomial of odd degree has at least one real zero. ■

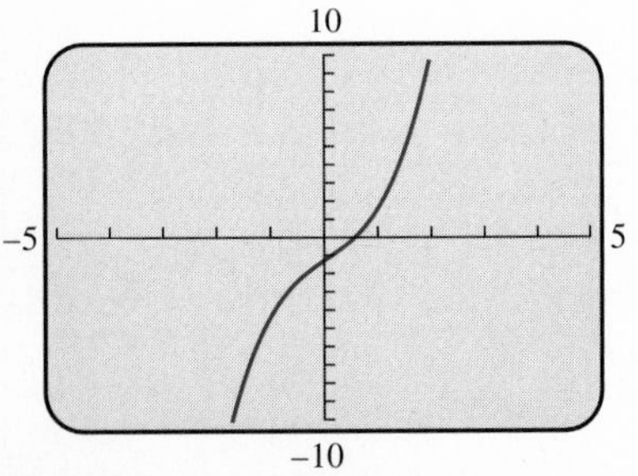

Figure 3.17 A graph of $y = x^3 + 2x - 1$ showing that there is a zero between $x = 0$ and $x = 1$.

Example 6 allows us to conclude, for example, that $f(x) = x^3 + 2x - 1$ has a real zero. Figure 3.17 supports this fact by showing a zero between 0 and 1.

Example 7 illustrates that we must always be alert to the possibility of hidden behavior. (See Section 1.5.)

REMARK

In Section 3.5, we will see that polynomials often have zeros that are not real numbers.

Notes on Examples

In Example 7, students may need to be reminded that if they wish to find solutions with greater accuracy, they can use a very small window on the grapher.

Follow-up

Ask students to discuss the possible number of real zeros for a degree 5 polynomial. **(1, 2, 3, 4, or 5)**

Assignment Guide

Day 1: Ex. 3–51, multiples of 3
Day 2: Ex. 53, 58, 61, 62, 65, 66, 68, 69, 71, 72, 73, 74, 77, 80, 81

Cooperative Learning

Group Activity: Ex. 82

Notes on Exercises

Ex. 29–34 and 71–76 are designed to emphasize the importance of using an appropriate viewing window on the grapher.
Ex. 65–70 offer a variety of applications. Remind students to use the language of models appropriately.
Ex. 79–80 are related to the calculus topic of taking derivatives.

Ongoing Assessment

Self-Assessment: Ex. 19, 37, 47, 63, 67
Embedded Assessment: Ex. 72, 74

EXAMPLE 7 Approximating Zeros with a Grapher

Find all the real zeros of $f(x) = x^4 + 0.1x^3 - 6.5x^2 + 7.9x - 2.4$.

Solution

Solve Graphically

Because f is of degree 4, there are at most four zeros. The graph of f in Figure 3.18a suggests that there is one zero near $x = -3$ and at least one near $x = 1$. Because the graph appears flat near $x = 1$, we should check for other zeros that may be hidden there.

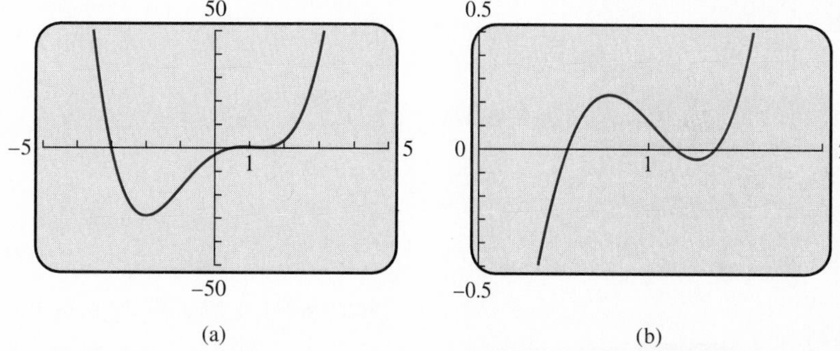

Figure 3.18

The close-up graph of f in Figure 3.18b shows that there are indeed three zeros near $x = 1$. Using the grapher, we find that the four zeros of f, to the nearest hundredth, are $x = -3.10$, $x = 0.50$, $x = 1.13$, and $x = 1.37$. ■

When graphing a function on a grapher, the viewing window may or may not include all of the characteristics of the graph that you need to see. We shall agree that a graph produced on a graphing utility is a *complete graph* if the viewing window clearly shows all the information we need. This might include the y-intercept, any of the x-intercepts, any of the local extremum values, and an indication of the end behavior of the function. By this agreement, neither graph in Figure 3.18 is complete since neither shows all the zeros, the information that we sought.

If we are not sure whether a graph shows all the information we need, we are unable to declare it as complete.

Applications

The power of a grapher makes it possible to solve problems graphically that we presently are unable to solve algebraically.

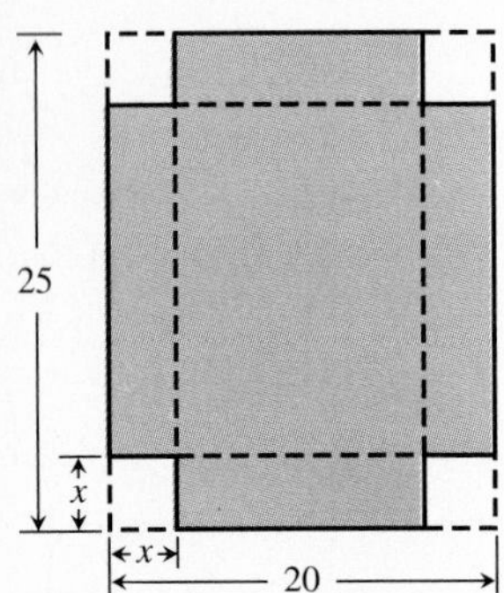

Figure 3.19 Cut square corners from a piece of cardboard, and fold the flaps to make a box.

■ EXAMPLE 8 APPLICATION: Designing Packaging

Dixie Packaging Co. has contracted to make boxes with a volume of about 484 in.^{3} Squares are to be cut from the corners of a 20-in. by 25-in. piece of cardboard, and the flaps folded up to make a box. (See Figure 3.19.) What size squares should be cut from the cardboard?

Solution

Model

Word Statement: Volume (V) = height × width × length

$$\begin{aligned} x &= \text{edge of cut-out square (height of box)} \\ 25 - 2x &= \text{length of the box} \\ 20 - 2x &= \text{width of the box} \\ V &= x(20 - 2x)(25 - 2x) \end{aligned}$$

Solve Graphically

For a volume of 484, we need to solve the equation

$$484 = x(20 - 2x)(25 - 2x).$$

Graph $y_1 = 484$ and $y_2 = x(20 - 2x)(25 - 2x)$. Since the width of the cardboard is 20 in., the context of the problem tells us that x must be between 0 and 10. Consequently we choose $X\min = 0$ and $X\max = 10$ (Figure 3.20), and find that the graphs of y_1 and y_2 intersect at approximately $x = 1.22$ and $x = 6.87$.

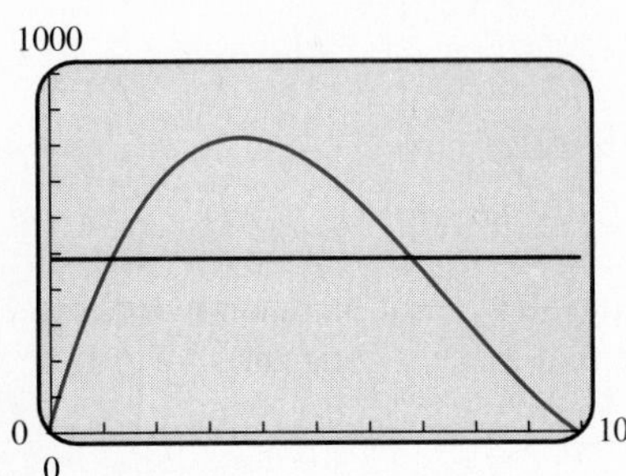

Figure 3.20

Interpret

Squares with length approximately 1.22 in. or 6.87 in. could be cut from the cardboard to produce a box with a volume that is approximately 484 in.^{3} ■

Quick Review 3.2

In Exercises 1–6, factor the polynomial into linear factors.

1. $x^2 - x - 12$
2. $x^2 - 11x + 28$
3. $3x^2 - 11x + 6$
4. $6x^2 - 5x + 1$
5. $3x^3 - 5x^2 + 2x$
6. $6x^3 - 22x^2 + 12x$

In Exercises 7–10, solve the equation mentally.

7. $x|x + 3| = 0$
8. $(x + 1)|x + 1| = 0$
9. $\sqrt{x^2} = 5$
10. $\sqrt{x - 2} = 7$

In Exercises 11–16, complete the exercise without using a grapher. Decide whether the graph of the function is shown among graphs (a) through (d). If it is, select it. All graphs are shown in a standard window.

11. $f(x) = \dfrac{5}{x}$
12. $f(x) = \sqrt{x + 3}$
13. $f(x) = 3\sqrt{x - 4}$
14. $f(x) = |x - 3|$
15. $f(x) = |x^2 + 6|$
16. $f(x) = |x^2 - 6|$

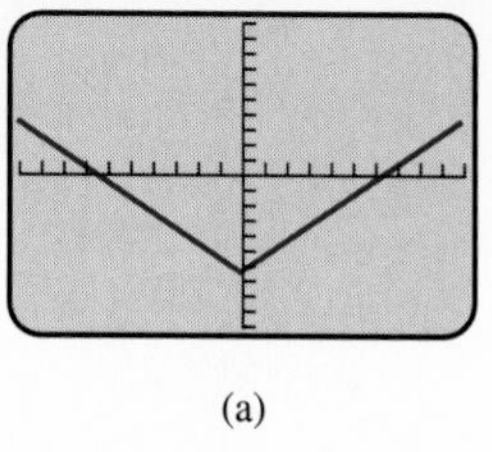

(a)

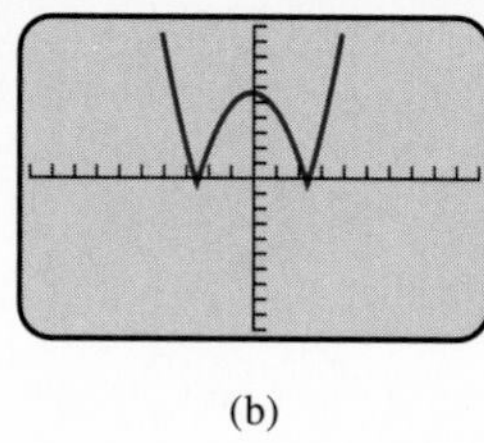

(b)

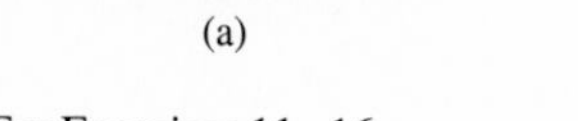

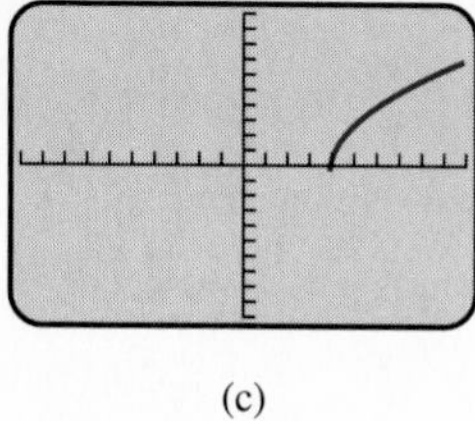

(c)

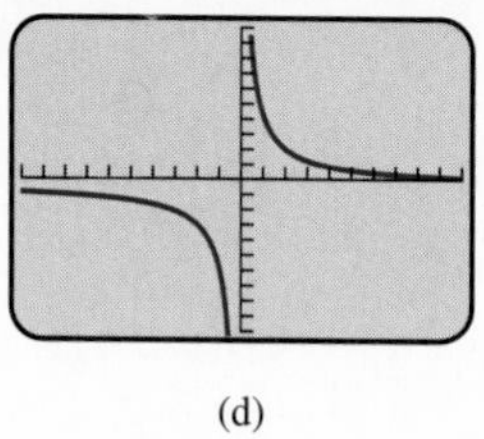

(d)

For Exercises 11–16

SECTION EXERCISES 3.2

In Exercises 1–6, graph the function. Choose a viewing window that shows a local maximum, a local minimum, and all the x-intercepts. Make a sketch of the grapher window, and indicate the viewing-window dimensions.

1. $y = x^3 + 5x^2 - 3x - 2$

2. $y = -2x^3 + 7x^2 + 4x - 6$

3. $y = -x^3 + 4x^2 + 31x - 70$

4. $y = x^3 - 2x^2 - 41x + 42$

5. $y = x^3 + 5x^2 - 58x + 88$

6. $y = x^3 - 8x^2 - 53x + 60$

In Exercises 7–12, graph the function. Choose a viewing window that shows three local extremum values and all the x-intercepts. Make a sketch of the grapher window, and show the viewing-window dimensions.

7. $y = x^4 + 2x^3 - 13x^2 - 14x + 24$

8. $y = 2x^4 + 3x^3 - 40x^2 - 21x + 20$

9. $y = 2x^4 - 5x^3 - 17x^2 + 14x + 41$

10. $y = -3x^4 - 5x^3 + 15x^2 - 5x + 19$

11. $y = -4x^4 - 4x^3 + 7x^2 + 4x - 3$

12. $y = 6x^4 - 23x^3 - 6x^2 + 53x + 18$

In Exercises 13–24, describe the end behavior of the polynomial function.

13. $f(x) = x^3$

14. $f(x) = -x^2$

15. $f(x) = -x^5$

16. $f(x) = x^4$

17. $f(x) = 3x^4 - 5x^2 + 3$

18. $f(x) = -x^3 + 7x^2 - 4x + 3$

19. $f(x) = 7x^2 - x^3 + 3x - 4$

20. $f(x) = x^3 - x^4 + 3x^2 - 2x + 7$

21. $f(x) = x(x^2 - 3x + 2)$

22. $f(x) = (x + 3)(x^3 - 4x + 5)$

23. $f(x) = (x^2 - 1)(7x^2 - x^3 + 2x)$

24. $f(x) = (x - 3)(5 - x)(2x - 1)$

In Exercises 25–28, match the function with its graph. Select the one you think is correct. Then support your answer graphically.

25. $f(x) = 7x^3 - 21x^2 - 91x + 104$

26. $f(x) = -9x^3 + 27x^2 + 54x - 73$

27. $f(x) = x^5 - 8x^4 + 9x^3 + 58x^2 - 164x + 69$

28. $f(x) = -x^5 + 3x^4 + 16x^3 - 2x^2 - 95x - 44$

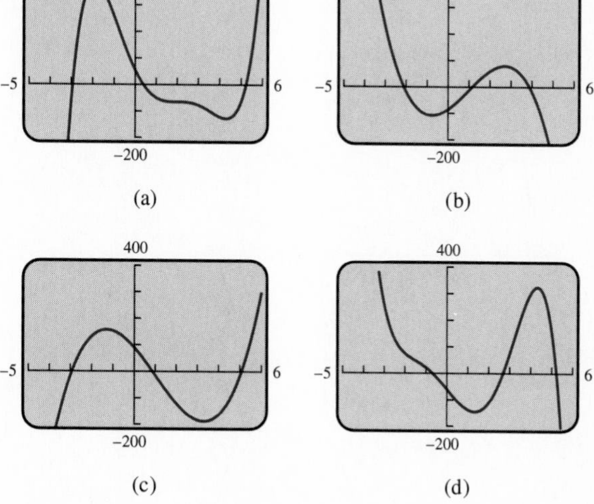

For Exercises 25–28

In Exercises 29–34, graph the function. Which of the three viewing windows in parts a–c shows the best view of the graph? Explain why the view chosen is the best.

a. $-5 \le x \le 5, \quad -10 \le y \le 10$
b. $-5 \le x \le 5, \quad -100 \le y \le 100$
c. $-10 \le x \le 10, \quad -500 \le y \le 500$

29. $f(x) = x^3 - x + 1$

30. $f(x) = x^3 + x^2 - 20x + 13$

31. $f(x) = 4x^4 + 16x^3 + 4x^2 - 24x + 3$

32. $f(x) = x^5 + 2x^4 - 7x^3 - 8x^2 + 12x + 1$

33. $f(x) = x^5 + 6x^4 - 7x^3 - 36x^2 + 36x - 15$

34. $f(x) = 0.5x^4 + 0.5x^3 - 23.5x^2 - 40.5x + 63$

In Exercises 35–40, find all the zeros algebraically.

35. $f(x) = x^2 + 2x - 8$

36. $f(x) = 3x^2 + 4x - 4$

37. $f(x) = 9x^2 + 3x - 2$

38. $f(x) = x^3 - 25x$

39. $f(x) = 3x^3 - x^2 - 2x$

40. $f(x) = 5x^3 - 5x^2 - 10x$

In Exercises 41–44, match the function with its graph. Identify the number of real zeros, and find an approximate value for each.

41. $f(x) = 20x^3 + 8x^2 - 83x + 55$

42. $f(x) = 35x^3 - 134x^2 + 93x - 18$

43. $f(x) = 44x^4 - 65x^3 + x^2 + 17x + 3$

44. $f(x) = 4x^4 - 8x^3 - 19x^2 + 23x - 6$

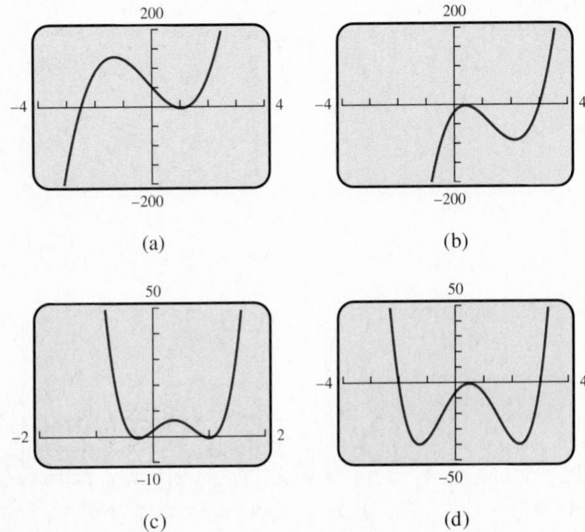

For Exercises 41–44

In Exercises 45–52, find all the real zeros of the function. State whether you solved the problem algebraically or graphically and give reasons for your choice.

45. $f(x) = x^2 - 3x - 10$

46. $f(x) = x^2 + 10x + 24$

47. $f(x) = 1000 - 15x - x^2$

48. $f(x) = x^2 - 14x + 13$

49. $f(x) = x^3 - 36x$

50. $f(x) = x^3 + 2x^2 - 109x - 110$

51. $f(x) = x^3 - 7x^2 - 49x + 55$

52. $f(x) = x^3 - 4x^2 - 44x + 96$

In Exercises 53–58, find a polynomial function with the given zeros.

53. $3, -4, 6$

54. $-2, 3, -5$

55. $\sqrt{3}, -\sqrt{3}, 4$

56. $1, 1 + \sqrt{2}, 1 - \sqrt{2}$

57. $-3, 2 - \sqrt{5}, 2 + \sqrt{5}$

58. $1 - \sqrt{3}, 1 + \sqrt{3}, -5, 3$

In Exercises 59–64, find approximate zeros of the function. Give answers to the nearest hundredth.

59. $f(x) = 2x^3 + 3x^2 - 7x - 6$

60. $f(x) = -x^3 + 3x^2 + 7x - 2$

61. $f(x) = x^3 + 2x^2 - 4x - 7$

62. $f(x) = -x^4 - 3x^3 + 7x^2 + 2x + 8$

63. $f(x) = x^4 + 3x^3 - 9x^2 + 2x + 3$

64. $f(x) = 2x^5 - 11x^4 + 4x^3 + 47x^2 - 42x - 8$

In Exercises 65–70, solve the problem.

65. *Analyzing Profit* Economists for Smith Brothers, Inc., find the company profit P by using the formula $P = R - C$, where R is the total revenue generated by the business and C is the total cost of operating the business.

a. Using data from past years, the economists determined that $R(x) = 0.0125x^2 + 412x$ models total revenue, and $C(x) = 12{,}225 + 0.00135x^3$ models the total cost of doing business, where x is the number of customers patronizing the business. How many customers must Smith Bros. have to be profitable each year?

b. How many customers must there be for Smith Bros. to realize an annual profit of \$60,000?

66. *Stopping Distance* A state Highway Patrol Safety Division collected the "stopping distance" data in Table 3.3 in preparing a demonstration for students. The data are also shown in the scatter plot and can be modeled by the function $D = 0.05x^2 + 0.97x + 0.26$.

Table 3.3 Highway Safety Division

Speed (mph)	Stopping Distance (ft)
10	15.1
20	39.9
30	75.2
40	120.5
50	175.9

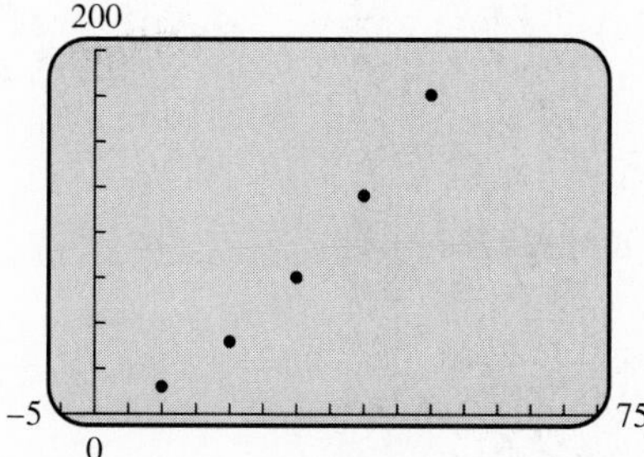

For Exercise 66

a. Find the stopping distance for a vehicle traveling at 25 mph.

b. Find the speed of the car if the stopping distance is 300 ft.

67. *Circulation of Blood* Research conducted at a national health research project shows that the speed at which a blood cell travels in an artery depends on its distance from the center of the artery. The function $v = 1.19 - 1.87r^2$ models the velocity (in centimeters per second) of a cell that is r centimeters from the center of an artery.

a. Find a graph of v that reflects values of v appropriate for this problem. Record the viewing-window dimensions.

b. If a blood cell is traveling at 0.975 cm/sec, estimate the distance the blood cell is from the center of the artery.

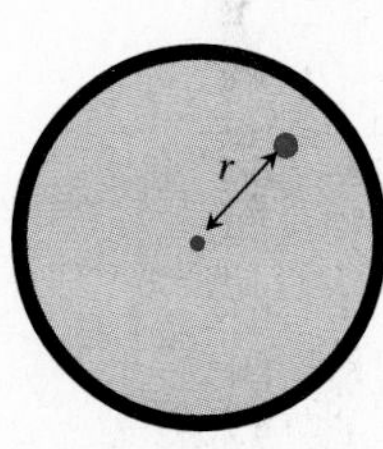

For Exercise 67

68. *Volume of a Box* Dixie Packaging Co. has contracted to manufacture a box with no top that is to be made by removing squares of width x from the corners of a 15-in. by 60-in. piece of cardboard.

a. Show that the volume of the box is modeled by $V(x) = x(60 - 2x)(15 - 2x)$.

b. Determine x so that the volume of the box is at least 450 in.^{3}

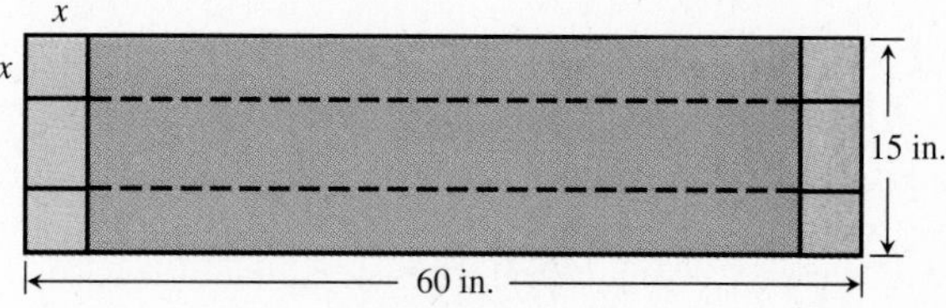

For Exercise 68

69. *Volume of a Box* Squares of width x are removed from a 10-cm by 25-cm piece of cardboard, and the resulting edges are folded up to form a box with no top. Determine all values of x so that the volume of the resulting box is at most 175 cm^3.

70. *Volume of a Box* The function $V = 2666x - 210x^2 + 4x^3$ represents the volume of a box that has been made by removing squares of width x from each corner of a rectangular sheet of material and then folding up the sides. What values are possible for x?

Writing to Learn In Exercises 71 and 72, two views of the function are given. Describe why each view, by itself, may be considered inadequate.

71. $f(x) = x^5 - 10x^4 + 2x^3 + 64x^2 - 3x - 55$

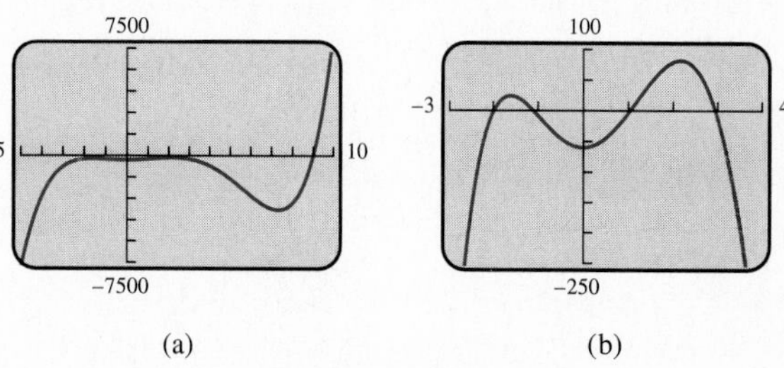

For Exercise 71

72. $f(x) = 10x^4 + 19x^3 - 121x^2 + 143x - 51$

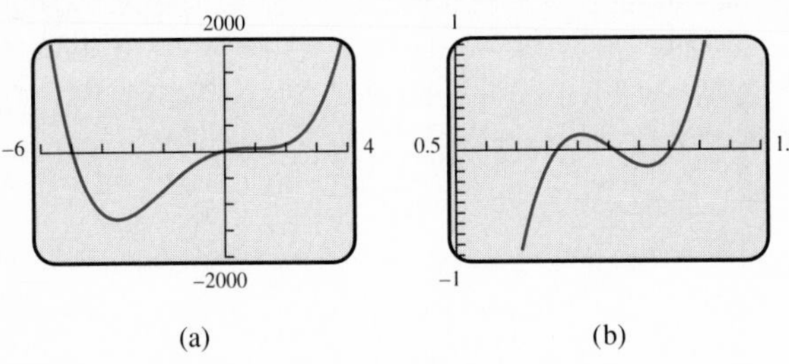

For Exercise 72

In Exercises 73–76, the function has hidden behavior when viewed in the standard window $[-10, 10]$ by $[-10, 10]$. Describe what behavior is hidden, and state the dimensions of a viewing window that reveals the hidden behavior.

73. $f(x) = 10x^3 - 40x^2 + 50x - 20$

74. $f(x) = 0.5(x^3 - 8x^2 + 12.99x - 5.94)$

75. $f(x) = 11x^3 - 10x^2 + 3x + 5$

76. $f(x) = 33x^3 - 100x^2 + 101x - 40$

EXTENDING THE IDEAS

In Exercises 77 and 78, complete a logical argument.

77. Graph the left side of the equation

$$3(x^3 - x) = a(x - b)^3 + c.$$

Then explain why there are no real numbers a, b, and c that make the equation true. (*Hint:* Use your knowledge of $y = x^3$ and transformations.)

78. Graph the left side of the equation

$$x^4 + 3x^3 - 2x - 3 = a(x - b)^4 + c.$$

Then explain why there are no real numbers a, b, and c that make the equation true.

79. Looking Ahead to Calculus The figure shows a graph of both $f(x) = -x^3 + 2x^2 + 9x - 11$ and the line L defined by $y = 5(x - 2) + 7$.

a. Confirm that the point $Q(2, 7)$ is a point of intersection of the two graphs.

b. Zoom in at point Q to develop a visual understanding that $y = 5(x - 2) + 7$ is a *linear approximation* for $y = f(x)$ near $x = 2$.

c. Recall that a line is *tangent* to a circle at a point P if it intersects the circle only at point P. View the two graphs in the window $[-5, 5]$ by $[-25, 25]$, and explain why that definition of tangent line is not valid for the graph of f.

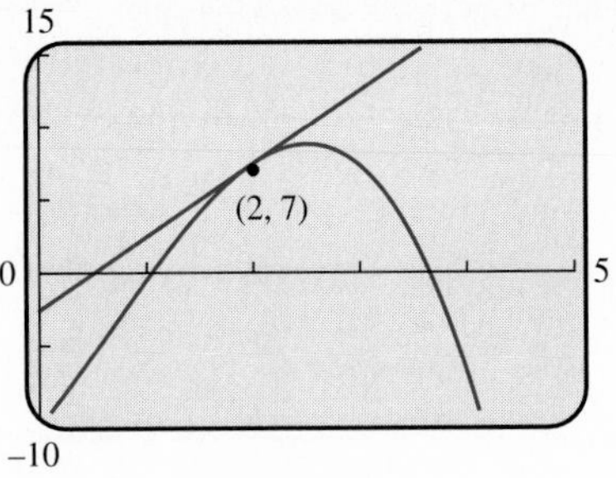

For Exercise 79

80. Looking Ahead to Calculus Consider the function $f(x) = x^n$ where n is an odd integer.

a. Suppose that a is a positive number. Show that the slope of the line through the points $P(a, f(a))$ and $Q(-a, f(-a))$ is a^{n-1}.

b. Let $x_0 = a^{1/(n-1)}$. Find an equation of the line through point $(x_0, f(x_0))$ with the slope a^{n-1}.

c. Consider the special case $n = 3$ and $a = 3$. Show both the graph of f and the line from part b in the window $[-5, 5]$ by $[-30, 30]$.

81. Derive an Algebraic Model of a Problem Show that the distance x in the figure is a solution of the equation $x^4 - 16x^3 + 500x^2 - 8000x + 32000 = 0$ and find the value of D by following these steps.

a. Use the similar triangles in the diagram and the properties of proportions learned in geometry to show that

$$\frac{8}{x} = \frac{y - 8}{y}.$$

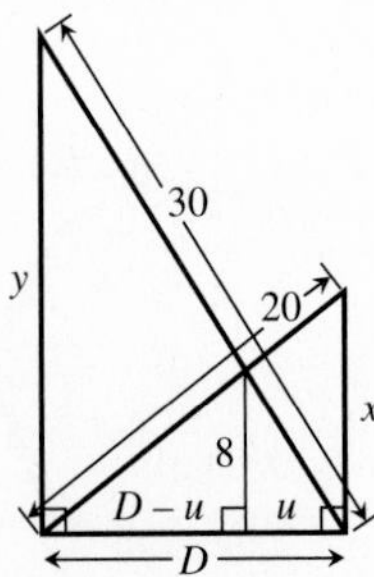

For Exercise 81

b. Show that $y = \dfrac{8x}{x - 8}$.

c. Show that $y^2 - x^2 = 500$. Then substitute for y, and simplify to obtain the desired degree 4 equation in x.

d. Find the distance D.

82. Group Learning Activity *Work in groups of three.* Consider functions of the form $f(x) = x^3 + bx^2 + x + 1$ where b is a nonzero real number.

a. Discuss as a group how the value of b affects the graph of the function.

b. After completing part a, have each member of the group (individually) predict what the graphs of $f(x) = x^3 + 15x^2 + x + 1$ and $g(x) = x^3 - 15x^2 + x + 1$ will look like.

c. Compare your predictions with each other. Confirm whether they are correct.

3.3 POLYNOMIAL DIVISION AND THE FACTOR THEOREM

Factors and Long Division • Synthetic Division • Remainder Theorem • Factor Theorem • Modeling Data with Polynomial Functions

Factors and Long Division

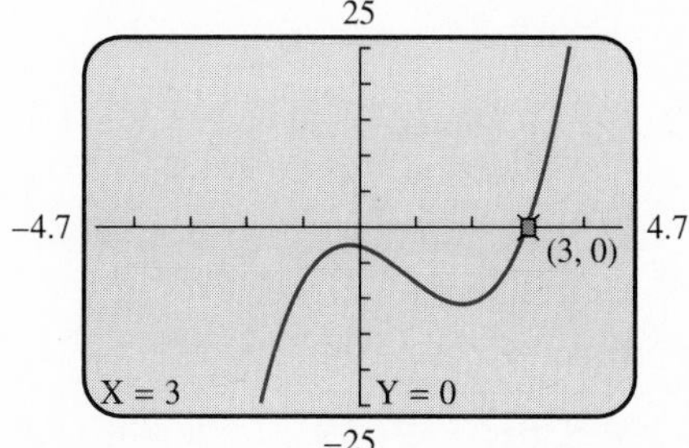

Figure 3.21 $f(x) = 2x^3 - 5x^2 - 2x - 3$.

Objective

Students will be able to divide polynomials using long division or synthetic division, and to apply the Remainder Theorem and the Factor Theorem.

Motivate

Have students discuss the vocabulary of long division: dividend, divisor, quotient, and remainder.

Figure 3.21 shows a graph of $f(x) = 2x^3 - 5x^2 - 2x - 3$. The indicated trace coordinates $x = 3$, $y = 0$ suggest that $x - 3$ is a factor of f. If so, we can write

$$f(x) = (x - 3)q(x),$$

where q is a quadratic factor of f. Long division is one way to find $q(x)$. The long division process for polynomials is similar to the process used for whole numbers.

■ EXAMPLE 1 Using Polynomial Long Division

Divide the polynomial $2x^3 - 5x^2 - 2x - 3$ by $x - 3$.

Solution

First divide $2x^3$ by x to obtain the first term of the quotient, $2x^2$. Then multiply $2x^2$ by the divisor $x - 3$ and subtract the product $2x^3 - 6x^2$. Continue the process until the remainder is zero or its degree is less than that of the divisor.

$$
\begin{array}{rrll}
 & 2x^2 + x + 1 & & \leftarrow \text{Quotient} \\
\text{Divisor} \rightarrow & x - 3 \,\big)\, 2x^3 - 5x^2 - 2x - 3 & & \leftarrow \text{Dividend} \\
 & 2x^3 - 6x^2 & & \leftarrow 2x^2 \text{ is multiplied by } x - 3. \\
 & x^2 - 2x - 3 & & \leftarrow \text{Result of first subtraction} \\
 & x^2 - 3x & & \leftarrow x \text{ multiplied by } x - 3 \\
 & x - 3 & & \\
 & x - 3 & & \\
 & 0 & & \leftarrow \text{Remainder}
\end{array}
$$

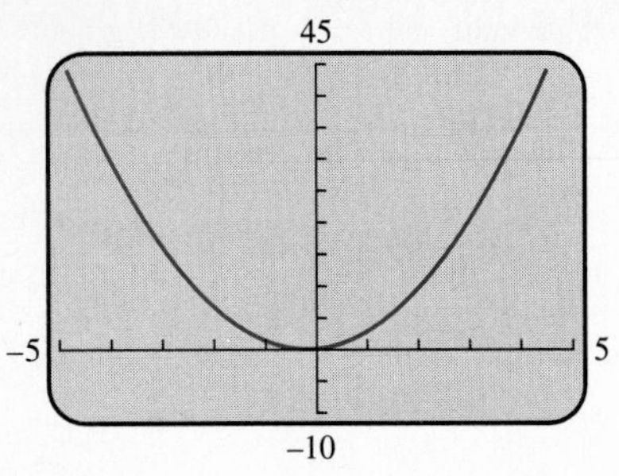

Figure 3.22 $y_1 = \dfrac{2x^3 - 5x^2 - 2x - 3}{x - 3}$ and $y_2 = 2x^2 + x + 1$.

Support Graphically

Figure 3.22 shows a graph of both

$$y_1 = \frac{2x^3 - 5x^2 - 2x - 3}{x - 3} \quad \text{and} \quad y_2 = 2x^2 + x + 1.$$

The graphs appear to be identical.

Confirm Algebraically

Multiplying the factors, we find that

$$(x - 3)(2x^2 + x + 1) = 2x^3 - 5x^2 - 2x - 3.$$ ■

Example 1 shows the long division process by which we found that the function f could be written as $f(x) = (x - 3)(2x^2 + x + 1)$. Example 2 shows that the process sometimes requires writing zero coefficients in the dividend.

■ **EXAMPLE 2** **Using Polynomial Long Division**

Find the quotient and remainder when $2x^4 - x^3 - 2$ is divided by $2x^2 + x + 1$.

Solution

$$\begin{array}{r|l} & x^2 - x \qquad\qquad\qquad\qquad \leftarrow \text{Quotient} \\ 2x^2 + x + 1 & 2x^4 - x^3 + 0x^2 + 0x - 2 \\ & \underline{2x^4 + x^3 + x^2 \qquad\qquad} \\ & \quad -2x^3 - x^2 + 0x - 2 \\ & \quad \underline{-2x^3 - x^2 - x \qquad} \\ & \qquad\qquad\qquad x - 2 \quad \leftarrow \text{Remainder} \end{array}$$

We stop when the remainder is zero or its degree is less than the degree of the divisor. ■

Because the remainder in Example 2 is not 0, the divisor $2x^2 + x + 1$ is not a factor of $2x^4 - x^3 - 2$. However, the following equation is true.

$$\underbrace{2x^4 - x^3 - 2}_{\text{Dividend}} = \underbrace{(2x^2 + x + 1)}_{\text{Divisor}}\,\underbrace{(x^2 - x)}_{\text{Quotient}} + \underbrace{(x - 2)}_{\text{Remainder}}$$

Examples 1 and 2 illustrate the **division algorithm for polynomials.** We can use the algorithm whenever we want to divide one polynomial by another of equal or lesser degree. We get a quotient polynomial and a remainder polynomial that is zero or whose degree is less than that of the divisor.

Teaching Note

It is worth emphasizing the fact that the quotient and remainder are unique. Compare the fact that the degree of the remainder must be less than the degree of the divisor with the corresponding fact for whole number division—the remainder must be less than the divisor.

Division Algorithm for Polynomials

Let $f(x)$ and $h(x)$ be polynomials such that the degree of f is greater than or equal to the degree of h, and $h \neq 0$. Then there are unique polynomials $q(x)$ and $r(x)$, called the **quotient** and **remainder,** such that

$$f(x) = h(x)q(x) + r(x),$$

where either $r = 0$ or the degree of r is less than the degree of h.

Just as with whole numbers, "nice" long divisions have the remainder 0. When this happens,

$$f(x) = h(x)q(x)$$

occurs in a factored form, and we say that "$h(x)$ divides $f(x)$ evenly" and that "$h(x)$ is a factor of $f(x)$." Example 1 shows that $x - 3$ is a factor of $2x^3 - 5x^2 - 2x - 3$.

Synthetic Division

In the long division process shown in Examples 1 and 2, the powers of x did not play much of a role except to indicate the alignment of the terms. The computations involved only the coefficients. So the process could be carried out without writing the powers of x. We illustrate with another division.

With Powers of x

$$\begin{array}{r@{}l}
 & 2x^2 + 3x + 4 \\
x - 3 & \overline{)2x^3 - 3x^2 - 5x - 12} \\
 & \underline{2x^3 - 6x^2} \\
 & \phantom{)2x^3 - {}} 3x^2 - 5x - 12 \\
 & \phantom{)2x^3 - {}} \underline{3x^2 - 9x} \\
 & \phantom{)2x^3 - 3x^2 - {}} 4x - 12 \\
 & \phantom{)2x^3 - 3x^2 - {}} \underline{4x - 12} \\
 & 0
\end{array}$$

Without Powers of x

$$\begin{array}{r|rrrr}
 & 2 & 3 & 4 & \\
\hline
-3 & 2 & -3 & -5 & -12 \\
 & \underline{2} & \underline{-6} & & \\
 & & 3 & -5 & -12 \\
 & & \underline{3} & \underline{-9} & \\
 & & & 4 & -12 \\
 & & & \underline{4} & \underline{-12} \\
 & & & & 0
\end{array}$$

The coefficients in color duplicate some of the coefficients in the quotient or dividend. To streamline our process even more, we eliminate the duplicates and collapse vertically:

$$\begin{array}{r|rrrrl}
 & 2 & 3 & 4 & & \text{Quotient} \\
\hline
-3 & 2 & -3 & -5 & -12 & \text{Dividend} \\
 & & -6 & -9 & -12 & \\
\hline
 & & & & 0 & \text{Remainder}
\end{array}$$

We write the quotient in the bottom row and eliminate the horizontal line across the top:

$$\begin{array}{r|rrrrl} -3 & 2 & -3 & -5 & -12 & \text{Dividend} \\ & & -6 & -9 & -12 & \\ \hline & 2 & 3 & 4 & 0 & \text{Quotient, remainder} \end{array}$$

Alert

Some students will forget to change the sign. Remind them that the number in the divisor position is a zero of the divisor.

Now we change the sign in the divisor position (it thus becomes a zero of the divisor) and change subtraction to addition. (Addition is less likely to result in error than subtraction is.)

$$\begin{array}{lr|rrrrl} \text{Zero of divisor} \rightarrow & 3 & 2 & -3 & -5 & -12 & \text{Dividend} \\ & & & 6 & 9 & 12 & \\ \hline & & 2 & 3 & 4 & 0 & \text{Quotient, remainder} \end{array}$$

This shortcut for long division of polynomials is called **synthetic division,** and is valid only *when the leading coefficient of the divisor is 1 and the divisor is linear.* Example 3 illustrates the process of synthetic division.

■ EXAMPLE 3 Using Synthetic Division

Divide $2x^3 - 3x^2 - 5x - 12$ by $x - 3$ using synthetic division.

Solution

Set Up

The zero of the divisor $x - 3$ is 3. We write 3 in the divisor position followed by the dividend coefficients. Draw a horizontal line as shown.

$$\begin{array}{lr|rrrrl} \text{Zero of divisor} & 3 & 2 & -3 & -5 & -12 & \text{Dividend} \\ & & & & & & \\ \hline \end{array}$$

Calculate

- Write the first coefficient, 2, below the horizontal line.
- Multiply the 3 from the divisor by the most recent number below the line, and enter the product above the line, one column to the right.
- Add that column, and write the sum below the line. Continue multiplying and adding until there is a number below the line in the last column.

$$\begin{array}{r|rrrrl} 3 & 2 & -3 & -5 & -12 & \\ & & 6 & 9 & 12 & \\ \hline & 2 & 3 & 4 & 0 & \text{Quotient, remainder} \end{array}$$

Interpret

The numbers below the line show the coefficients of the quotient. The last number is the remainder. Here the remainder is zero, which indicates that $x - 3$

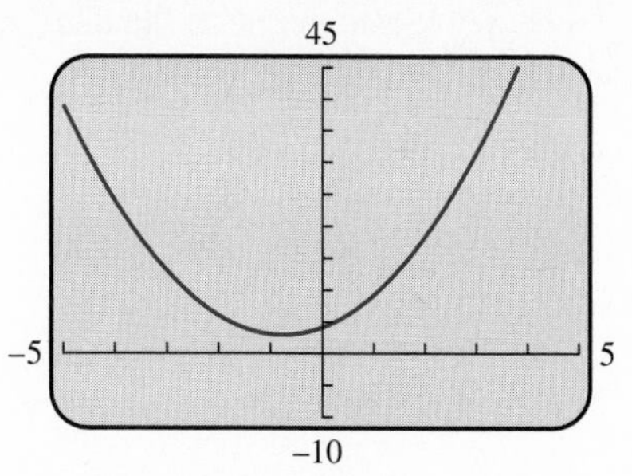

Figure 3.23 The graphs of $y_1 = 2x^2 + 3x + 4$ and $y_2 = \dfrac{2x^3 - 3x^2 - 5x - 12}{x - 3}$.

divides $2x^3 - 3x^2 - 5x - 12$ evenly, and that $x - 3$ is one of its factors. We can write

$$\frac{2x^3 - 3x^2 - 5x - 12}{x - 3} = 2x^2 + 3x + 4, \quad x \neq 3$$

Figure 3.23 provides graphical support for the division. ■

Sometimes we summarize a polynomial division with a fraction equation:

$$f(x) = h(x)q(x) + r(x) \qquad \text{Division algorithm}$$

$$\frac{f(x)}{h(x)} = q(x) + \frac{r(x)}{h(x)} \qquad \text{Divide by } h(x).$$

$$\frac{\text{Dividend}}{\text{Divisor}} = \text{quotient} + \frac{\text{remainder}}{\text{divisor}}$$

Notes on Examples

The dividend in Example 4 does not have an x^2 term. Students may need to be reminded of the importance of using a zero "placeholder" in this situation.

■ **EXAMPLE 4 Using Synthetic Division**

Divide $x^4 - 8x^3 + 11x - 6$ by $x + 3$ using synthetic division. Summarize the results with a fraction equation.

Solution

The zero of $x + 3$ is -3, so we use -3 in the divisor position:

$$\begin{array}{r|rrrrr} -3 & 1 & -8 & 0 & 11 & -6 \\ & & -3 & 33 & -99 & 264 \\ \hline & 1 & -11 & 33 & -88 & 258 \end{array}$$

We conclude that:

$$\underbrace{\overbrace{x^4 - 8x^3 + 11x - 6}^{\text{Dividend}}}_{} \over \underbrace{x + 3}_{\text{Divisor}} = \overbrace{x^3 - 11x^2 + 33x - 88}^{\text{Quotient}} + \frac{\overbrace{258}^{\text{Remainder}}}{\underbrace{x + 3}_{\text{Divisor}}}$$

■

Remainder Theorem

When the divisor is $x - c$, the equation

$$f(x) = h(x)q(x) + r(x)$$

becomes

$$f(x) = (x - c)q(x) + r,$$

where r is a constant because it must have a degree less than that of the divisor, namely zero. Replacing x by c, we see that

$$\begin{aligned} f(c) &= (c - c)q(c) + r \\ &= 0 + r \\ &= r, \end{aligned}$$

or the value of the function at $x = c$ is the remainder r. We call this the **Remainder Theorem.**

Theorem: Remainder Theorem

If a polynomial $f(x)$ is divided by $x - c$, then the remainder is $r = f(c)$.

■ EXAMPLE 5 Applying the Remainder Theorem

REMARK
Synthetic division provides a way to streamline a paper-and-pencil evaluation of a polynomial function. (See Exercise 61.)

Use the Remainder Theorem and synthetic division to find $f(17)$ for $f(x) = 7x^3 - 12x^2 + 6x - 23$.

Solution

$$\begin{array}{r|rrrr} 17 & 7 & -12 & 6 & -23 \\ & & 119 & 1819 & 31{,}025 \\ \hline & 7 & 107 & 1825 & 31{,}002 \end{array}$$

We conclude from the synthetic division that $f(17) = 31{,}002$. ■

Factor Theorem

Suppose that a polynomial function $f(x)$ has $x - c$ as a factor. Then division of $f(x)$ by $x - c$ leaves the remainder 0, so $f(x) = (x - c)q(x)$ and in particular, $f(c) = (c - c)q(c) = 0$.

Conversely, if $f(c) = 0$, then $r = 0$ by the Remainder Theorem, and $f(x) = (x - c)q(x)$. Thus $x - c$ is a factor of $f(x)$. This discussion proves the following theorem.

Theorem: Factor Theorem

A polynomial function $f(x)$ has a factor $x - c$ if and only if $f(c) = 0$.

The Factor Theorem tells us that finding linear factors of a polynomial is equivalent to finding real number zeros. Example 6 illustrates.

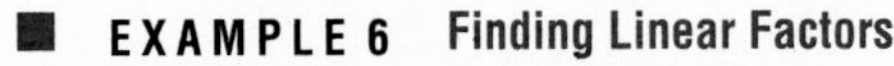

EXAMPLE 6 Finding Linear Factors

Find four linear factors of $f(x) = 2x^4 - 5x^3 - 3x^2 + 13x - 6$.

Solution

We find factors by first finding the zeros of $y = f(x)$. The graphs of $y = f(x)$ in Figure 3.24 suggest that $x = 1.5$ and $x = 2$ are both zeros. The synthetic divisions below both confirm these zeros and show (from the last line) that $2x^2 + 2x - 2$ is also a factor of f.

$$\begin{array}{r|rrrrr}
3/2 & 2 & -5 & -3 & 13 & -6 \\
 & & 3 & -3 & -9 & 6 \\ \hline
2 & 2 & -2 & -6 & 4 & 0 \\
 & & 4 & 4 & -4 & \\ \hline
 & 2 & 2 & -2 & 0 &
\end{array}$$

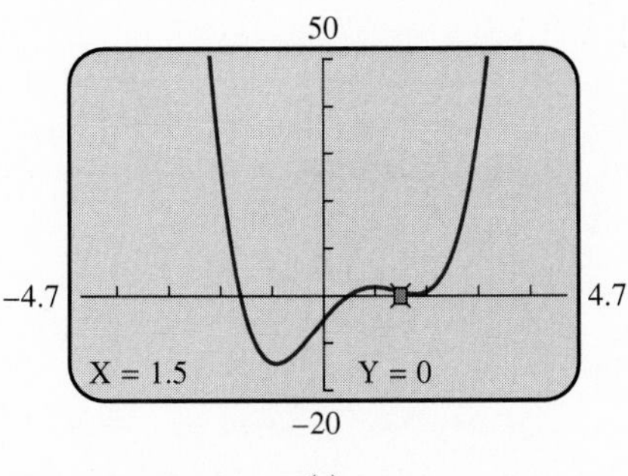

(a)

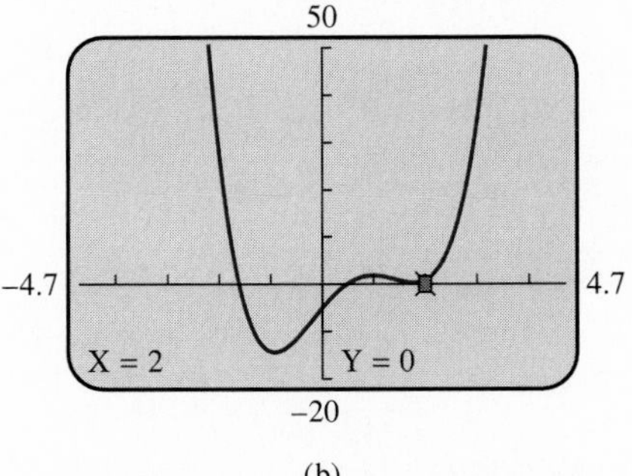

(b)

Figure 3.24 The trace coordinates suggest that $x = 1.5$ and $x = 2$ are zeros of $f(x) = 2x^4 - 5x^3 - 3x^2 + 13x - 6$.

So both $\frac{3}{2}$ and 2 are zeros, which means that $x - \frac{3}{2}$ and $x - 2$ are factors of $f(x)$ and

$$\begin{aligned}
f(x) &= \left(x - \frac{3}{2}\right)(x - 2)(2x^2 + 2x - 2) \\
&= \left(x - \frac{3}{2}\right)(x - 2)(2)(x^2 + x - 1) \\
&= (2x - 3)(x - 2)(x^2 + x - 1).
\end{aligned}$$

After applying the quadratic formula to $x^2 + x - 1 = 0$, we conclude that

$$x = \frac{-1 - \sqrt{5}}{2} \quad \text{and} \quad x = \frac{-1 + \sqrt{5}}{2}$$

are the other zeros of f. Thus, we have four linear factors and can write

$$\begin{aligned}
f(x) &= (2x - 3)(x - 2)\left(x - \frac{-1 - \sqrt{5}}{2}\right)\left(x - \frac{-1 + \sqrt{5}}{2}\right) \\
&= \frac{1}{4}(2x - 3)(x - 2)(2x + 1 + \sqrt{5})(2x + 1 - \sqrt{5}).
\end{aligned}$$

The last line shows the linear factors without fractions. ■

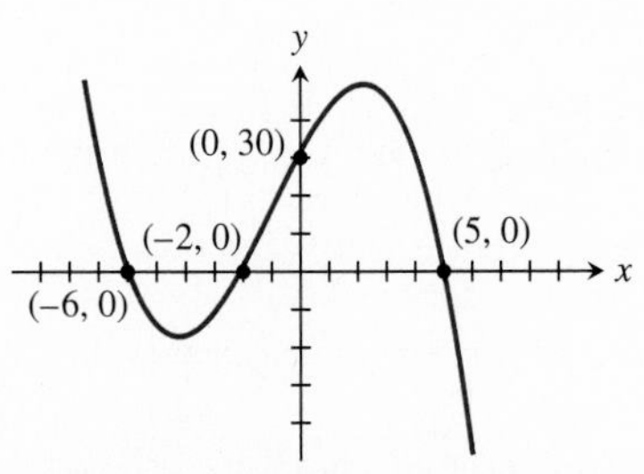

Figure 3.25 A graph of a degree 3 polynomial function f.

Figure 3.25 shows a sketch of a cubic with y-intercept 30 and x-intercepts -6, -2, and 5. In Example 7, we see how to use the Factor Theorem to find a formula for the cubic.

EXAMPLE 7 Finding a Formula for a Function

Find a formula for a cubic $y = f(x)$ whose graph is shown in Figure 3.25.

Solution

The graph shows that $x = -6$, $x = -2$, and $x = 5$ are zeros.

$$-6 \text{ is a zero of } f \text{ if and only if } x + 6 \text{ is a factor of } f$$
$$-2 \text{ is a zero of } f \text{ if and only if } x + 2 \text{ is a factor of } f$$
$$5 \text{ is a zero of } f \text{ if and only if } x - 5 \text{ is a factor of } f$$

Therefore, f has the factorization:

$$f(x) = a(x + 6)(x + 2)(x - 5).$$

To find a, use the fourth point (0, 30):

$$\begin{aligned} 30 &= f(0) \\ &= a(0 + 6)(0 + 2)(0 - 5) \\ &= a(-60) \\ a &= -0.5 \end{aligned}$$

Substituting $a = -0.5$ and multiplying the factors, we find that

$$\begin{aligned} f(x) &= (-0.5)(x + 6)(x + 2)(x - 5) \\ &= -0.5x^3 - 1.5x^2 + 14x + 30 \end{aligned}$$

■

Modeling Data with Polynomial Functions

We modeled data in Section 1.4 with linear functions (polynomials of degree 1), and in Section 3.1 with quadratic functions (polynomials of degree 2). Sometimes data can be modeled better by polynomial functions of higher degree. Here is a summary of a strategy that you can use.

Using Data Analysis—Polynomial Regression

1. Complete a scatter plot and decide what degree of polynomial to use.
2. Find the polynomial regression equation.
3. Superimpose a graph of the polynomial regression equation on the scatter plot.
4. Use the polynomial regression equation to predict y-values for x-values that are not included in the data.

Remember, the choice of which type of polynomial regression to apply depends on what you see in the scatter plot. In addition to lines and parabolas, you have to keep in mind the various cubic and quartic curves. For fun, try finding a cubic regression equation for the data points in Figure 3.25.

Table 3.4 Women and Work

Year	Outside Job (%)
1974	35
1980	46
1985	51
1991	43
1994	45

Source: Roper Starch Worldwide as reported by Mary Cadden and Nick Galifianakis in *USA Today,* May 12, 1995.

Follow-up

In Example 8, have students discuss which regression would be expected to provide a better model for the data over an extended period of time, for example, 1890 to 2050. **(Linear)**

Assignment Guide

Day 1: Ex. 3–45, multiples of 3
Day 2: Ex. 47, 50, 53, 56, 58, 59, 60, 62, 64, 65

Cooperative Learning

Group Activity: Ex. 59, 61

Notes on Exercises

Ex. 47–56 require students to find a polynomial function with particular zeros or function values.
Ex. 57–58 are economics applications using the concepts of supply and demand.
Ex. 61 is an extension of Example 5.
Ex. 66 relates to Example 9.

Ongoing Assessment

Self-Assessment: Ex. 11, 17, 29, 37, 55, 57
Embedded Assessment: Ex. 60

■ EXAMPLE 8 APPLICATION: Modeling Women and Work

Table 3.4 shows the percentage of U.S. women who say they would choose an outside job over being a homemaker. Find both a linear and a cubic regression equation for the data. Which regression equation appears to be the better model of the data?

Solution

Using $x = 0$ for 1970, $x = 1$ for 1971, and so on, we get the linear regression equation

$$y = 0.350x + 38.819$$

that is shown in Figure 3.26a. Figure 3.26b shows the cubic regression[3] equation

$$y = 0.0064x^3 - 0.3637x^2 + 6.2425x + 15.1147.$$

Each regression equation is superimposed on a scatter plot of the data.

The cubic regression equation appears to model the data better than does the linear regression. ■

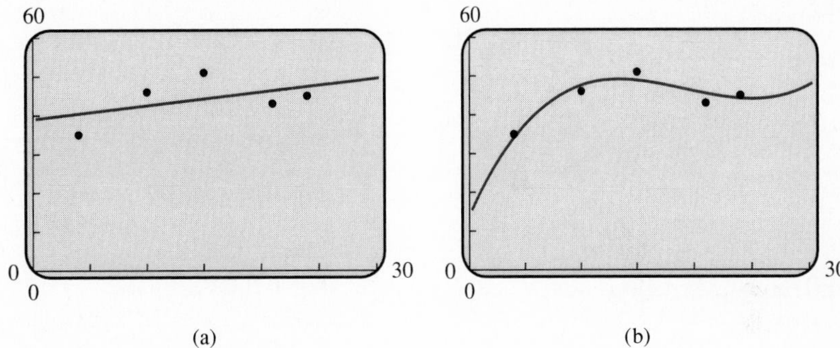

Figure 3.26

■ EXAMPLE 9 APPLICATION: Modeling Women and Work

Use the regression equations from Example 8 to predict the year in which 55% of women would choose an outside job over being a homemaker.

Solution

Model

The equations to be solved are

$$0.350x + 38.819 = 55,$$

$$0.0064x^3 - 0.3637x^2 + 6.2425x + 15.1147 = 55.$$

[3]Consult your grapher owner's manual to learn how to find a cubic regression equation.

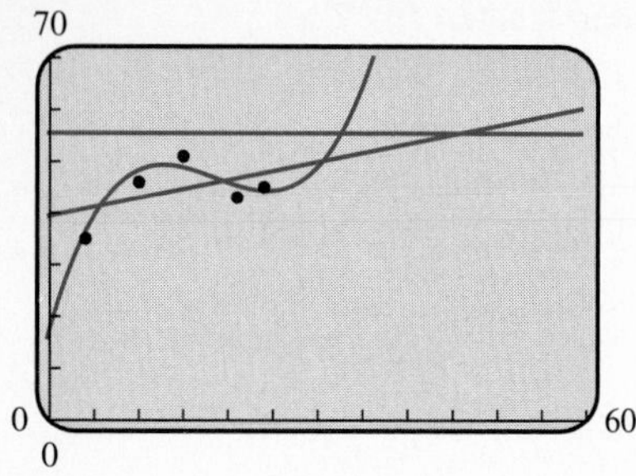

Figure 3.27 A scatter plot of the data and the graphs of $y_1 = 0.350x + 38.819$, $y_2 = 0.0064x^3 - 0.3637x^2 + 6.2425x + 15.1147$, and $y_3 = 55$.

Solve Graphically

The linear equation $y_1 = 0.350x + 38.819$ and the cubic equation $y_2 = 0.0064x^3 - 0.3637x^2 + 6.2425x + 15.1147$ intersect the horizontal line $y_3 = 55$ at about $x = 46.2$ and $x = 33.0$, respectively. (See Figure 3.27.)

Interpret

The linear regression predicts 2016 (= 1970 + 46) to be the year in which 55% of women would choose an outside job. The cubic regression predicts 2003. ■

Quick Review 3.3

In Exercises 1–4, rewrite the expression as a polynomial in standard form.

1. $\dfrac{x^3 - 4x^2 + 7x}{x}$
2. $\dfrac{2x^3 + 5x^2 - 6x}{2x}$
3. $\dfrac{x^4 - 3x^2 + 7x^5}{x^2}$
4. $\dfrac{6x^4 - 2x^3 + 7x^2}{3x^2}$

In Exercises 5–8, write the expression as a polynomial in standard form plus a term in the form a/x.

5. $\dfrac{x^2 + 5x + 3}{x}$
6. $\dfrac{2x^3 + 4x^2 + 7x - 3}{x}$
7. $\dfrac{6x - x^3 + 12x^2 + 4}{2x}$
8. $\dfrac{7x^3 - 3x^2 - 4x}{2x^2}$

SECTION EXERCISES 3.3

In Exercises 1–8, use long division to simplify the quotient. Support your answer graphically.

1. $\dfrac{x^3 - x^2 - 11x + 3}{x + 3}$
2. $\dfrac{x^3 - x^2 - 6x + 8}{x - 2}$
3. $\dfrac{3x^3 + 5x^2 + 3x + 1}{x + 1}$
4. $\dfrac{4x^3 - 4x^2 + 9x + 6}{2x + 1}$
5. $\dfrac{6x^4 + 4x^3 - 15x^2 - 7x + 2}{3x + 2}$
6. $\dfrac{x^4 - 3x^3 + 6x^2 - 3x + 5}{x^2 + 1}$
7. $\dfrac{x^4}{x^2 + 1}$
8. $\dfrac{x^4 + 2x^3 - 5x^2 + x - 11}{x - 1}$

In Exercises 9–14, divide $f(x)$ by $h(x)$. Find the quotient $q(x)$ and remainder $r(x)$, and write the equation $f(x) = q(x)h(x) + r(x)$.

9. $f(x) = x^2 - 2x + 3;\ h(x) = x - 1$
10. $f(x) = x^3 - 1;\ h(x) = x + 1$
11. $f(x) = x^3 + 4x^2 + 7x - 9;\ h(x) = x + 3$
12. $f(x) = x^4 - 7x^3 + 4x^2 + 7x - 9;\ h(x) = x - 2$
13. $f(x) = 4x^3 - 8x^2 + 2x - 1;\ h(x) = 2x + 1$
14. $f(x) = x^4 - 2x^3 + 3x^2 - 4x + 6;\ h(x) = x^2 + 2x - 1$

In Exercises 15–20, divide by synthetic division. Summarize the result by writing a fraction equation (as in Example 4).

15. $\dfrac{x^3 - 5x^2 + 3x - 2}{x + 1}$
16. $\dfrac{2x^4 - 5x^3 + 7x^2 - 3x + 1}{x - 3}$

17. $\dfrac{9x^3 + 7x^2 - 3x}{x - 10}$

18. $\dfrac{3x^4 + x^3 - 4x^2 + 9x - 3}{x + 5}$

19. $\dfrac{5x^4 - 3x + 1}{4 - x}$

20. $\dfrac{x^8 - 1}{x + 2}$

In Exercises 21–24, use the Remainder Theorem to find the remainder when $f(x)$ is divided by $x - c$. Check by using long division or synthetic division.

21. $f(x) = 2x^2 - 3x + 1;\ c = 2$
22. $f(x) = x^4 - 5;\ c = 1$
23. $f(x) = 2x^3 - 3x^2 + 4x - 7;\ c = 2$
24. $f(x) = x^5 - 2x^4 + 3x^2 - 20x + 3;\ c = -1$

In Exercises 25–30, use synthetic division and the Remainder Theorem to find the function value.

25. Find $f(-1)$ for $f(x) = x^3 - 4x^2 - 13x - 2$.
26. Find $f(2)$ for $f(x) = x^4 - 7x^3 + 4x^2 - 3x - 6$.
27. Find $f(-3)$ for $f(x) = 2x^4 + 6x^2 - 3x + 2$.
28. Find $f(-5)$ for $f(x) = x^4 + 2x^3 + 3x + 1$.
29. Find $f(1)$ for $f(x) = 2x^5 - x^4 + x^3 + x^2 - 8x + 3$.
30. Find $f(2)$ for $f(x) = x^4 + 2x^3 + 3x^2 - 4x + 5$.

In Exercises 31–34, use the Factor Theorem to find whether the first polynomial is a factor of the second polynomial.

31. $x - 1;\ x^3 - x^2 + x - 1$
32. $x - 3;\ x^3 - x^2 - x - 15$
33. $x + 2;\ 4x^3 + 9x^2 - 3x - 10$
34. $x + 1;\ 2x^{10} - x^9 + x^8 + x^7 + 2x^6 - 3$

In Exercises 35–40, factor the expression into linear factors.

35. $x^3 + 3x^2 - 10x - 24$
36. $x^3 + 2x^2 - 13x + 10$
37. $2x^4 - 7x^3 - 23x^2 + 43x - 15$
38. $3x^4 - x^3 - 21x^2 - 11x + 6$
39. $3x^5 - 4x^4 - 23x^3 + 14x^2 + 34x - 12$
40. $2x^5 + x^4 - 11x^3 - x^2 + 15x - 6$

In Exercises 41 and 42, use the graph to guess a linear factor for f. Confirm your guess using synthetic division.

41. $f(x) = 5x^3 - 7x^2 - 49x + 51$

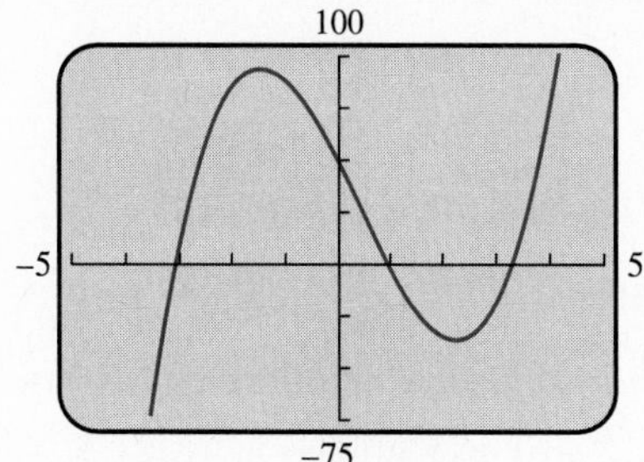

For Exercise 41

42. $f(x) = 5x^3 - 12x^2 - 23x + 42$

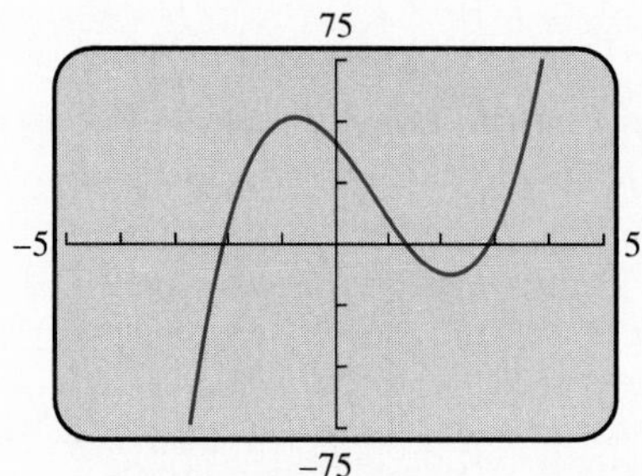

For Exercise 42

In Exercises 43–46, use a graph of the function to help suggest one linear factor. Confirm your guess algebraically.

43. $f(x) = x^3 + 2x^2 - 13x - 6$
44. $f(x) = x^3 - 11x + 6$
45. $f(x) = x^4 + x^3 - 7x^2 - x + 6$
46. $f(x) = 2x^4 + 5x^3 - 2x^2 - 7x + 2$

In Exercises 47–52, find a polynomial with real coefficients, whose leading coefficient is 2, and that satisfies the given conditions.

47. Degree 2, with 3 and -4 as zeros
48. Degree 2, with 2 as the only real zero
49. Degree 3, with -2, 1, and 4 as zeros
50. Degree 3, with -1, 3, and -5 as zeros
51. Degree 3, with 2, $\frac{1}{2}$, and $\frac{3}{2}$ as zeros
52. Degree 4, with -3, -1, 0, and $\frac{5}{2}$ as zeros

In Exercises 53–56, find a polynomial function (whose degree is one less than the number of points) whose graph intersects the given set of points.

53. $(1, 6), (2, 0), (4, 0)$

54. $(2, -30), (-3, 0), (5, 0)$

55. $(0, 180), (3, 0), (5, 0), (-4, 0)$

56. $(-1, 24), (-2, 0), (1, 0), (5, 0)$

Manufacturing Production In Exercises 57 and 58, solve these production schedule problems. Both the supply and demand for a manufactured product is related to its price. As the demand increases, the price also increases. As the supply increases, the price decreases. The manufacturing schedule is *in equilibrium* when the supply is equal to the demand.

57. *Setting Production Schedules* The Sunspot Small Appliance Co. determines that the supply function for their EverCurl hair dryer is $S(p) = 6 + 0.001p^3$ and that its demand function is $D(p) = 80 - 0.02p^2$, where p is the price. Find the price at which the manufacturing level is in equilibrium, and the corresponding demand and supply amounts.

58. *Setting Production Schedules* The Pentkon Camera Co. determines that the supply and demand functions for their 35 mm–70 mm zoom lens are $S(p) = 200 - p + 0.000007p^4$ and $D(p) = 1500 - 0.0004p^3$, where p is the price. Find the price at which the manufacturing level is in equilibrium, and the corresponding demand and supply amounts.

In Exercises 59 and 60, use the data to solve the problem.

59. *Women and Work* Table 3.5 shows the percentage of women who say they would choose being a homemaker over having a job outside the home.

a. Find a cubic regression equation, and graph it together with a scatter plot of the data.

b. Use the regression equation to predict the percentage of women who would choose being a homemaker over an outside job in 2003.

Table 3.5 Women and Work

Year	Homemaker (%)
1974	60
1980	51
1985	45
1991	53
1994	50

Source: Roper Starch Worldwide as reported by Mary Cadden and Nick Galifianakis in *USA Today,* May 12, 1995.

c. Use your estimate from part b together with the cubic regression estimate in Example 9 to estimate the percentage of women in 2003 who would answer "don't know" or "no answer" to this question.

d. **Writing to Learn** Compare the end behavior of the cubic curve in part a with the cubic curve in Example 8. Explain why they are opposites.

60. *Conserving Energy* The aluminum industry reports that its energy use is cut by 95% when it recycles rather than making aluminum from ore. Table 3.6 gives the percentage of aluminum cans that are recycled.

a. Find a quartic regression equation, and graph it together with a scatter plot of the data.

b. Find a quadratic regression equation, and graph it together with a scatter plot of the data.

c. Use the quartic and quadratic regressions from parts a and b to get two estimates of the percentage of aluminum cans that will be recycled in 1997.

d. **Writing to Learn** Give scenarios to justify each of the estimates in part c.

Table 3.6 Recycling Aluminum Cans

Year	Recycled (%)
1974	17
1978	27
1982	56
1986	49
1990	64
1994	65

Source: The Aluminum Association as reported by Anne R. Carey and Suzy Parker in *USA Today,* April 21, 1995.

EXTENDING THE IDEAS

In Exercises 61–66, solve the problem.

61. **Writing to Learn** Synthetic division was used in Example 5 to evaluate $f(17)$ for

$$f(x) = 7x^3 - 12x^2 + 6x - 23.$$

a. Verify algebraically that $f(x) = x[x(7x - 12) + 6] - 23$.

b. Using the result in part a, we see that

$$f(17) = 17[17(7 \cdot 17 - 12) + 6] - 23.$$

Explain how this calculation is related to the synthetic division calculation in Example 5.

c. To evaluate $f(17)$ in part b, how many operations are

necessary? To evaluate $f(17)$ by substituting into $7x^3 - 12x^2 + 6x - 23$, how many operations are necessary?
Which method would likely compute faster if programmed into a calculator?

62. Writing to Learn Explain how to change the form of the divisor so that synthetic division can be used:

$$\frac{4x^3 - 5x^2 + 3x + 1}{2x - 1}$$

63. Find the remainder when $x^{40} - 3$ is divided by $x + 1$.

64. Find the remainder when $x^{63} - 17$ is divided by $x - 1$.

65. Writing to Learn The figure shows a graph of $f(x) = x^4 + 0.1x^3 - 6.5x^2 + 7.9x - 2.4$. Explain how to use a grapher to justify the statement

$$\begin{aligned} f(x) &= x^4 + 0.1x^3 - 6.5x^2 + 7.9x - 2.4 \\ &\approx (x + 3.10)(x - 0.5)(x - 1.13)(x - 1.37) \end{aligned}$$

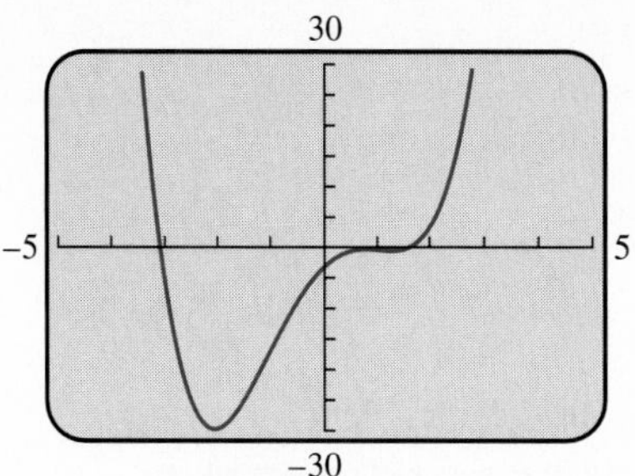

For Exercise 65

66. Writing to Learn Give an argument that defends the use of the linear regression over the cubic regression in Example 9.

Objective

Students will be able to apply Descartes' rule of signs as well as tests for rational zeros and for the upper and lower bounds for real zeros of polynomials.

Motivate

Ask students why it might be useful to know the maximum possible number of positive real zeros a polynomial can have.

3.4 REAL ZEROS OF POLYNOMIAL FUNCTIONS

Descartes' Rule of Signs • Upper and Lower Bounds for Real Zeros • Rational Zeros • Applications

Descartes' Rule of Signs

Recall that a degree n polynomial can have at most n distinct zeros. We have seen examples where this number is less than n. For example, the only real zero of $f(x) = (x - 1)^n$ is $x = 1$.

Be careful when interpreting the graph of a function. Do not draw false conclusions about its zeros. Figure 3.28 shows three views of the graph of

$$f(x) = x^5 - 68x^4 + 1085x^3 + 2450x^2 - 611.$$

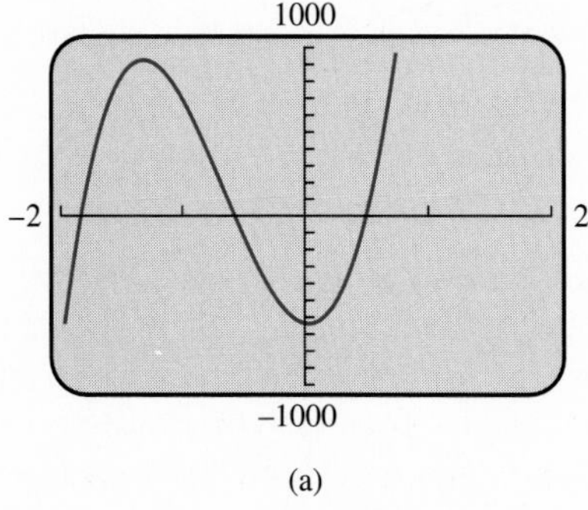

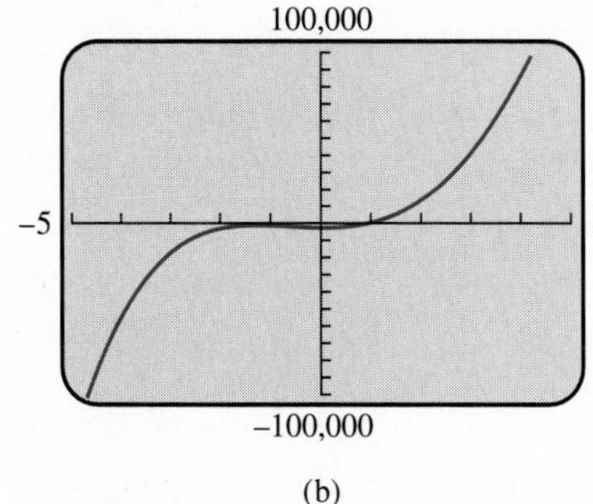

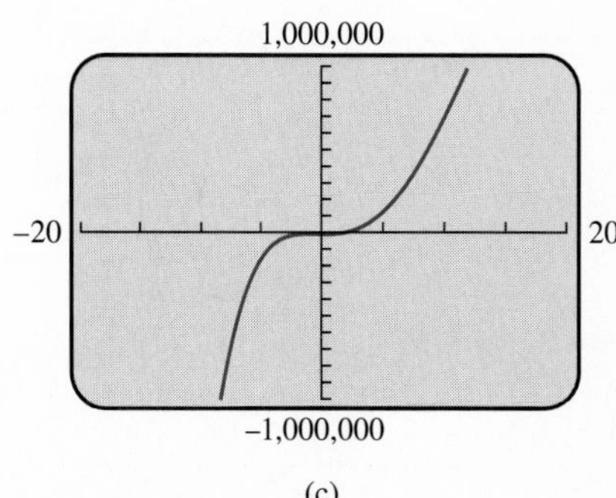

Figure 3.28 Three views of $f(x) = x^5 - 68x^4 + 1085x^3 + 2450x^2 - 611$.

The first view (Figure 3.28a) suggests that there are three real zeros between -2 and 1. The second and third views seem to suggest that the graph has no more x-intercepts no matter how far out we go. Example 2 will show this conclusion to be false.

Descartes' rule of signs and the *upper bound test for real zeros* that follow are two algebraic tools that can be used to gain information about the zeros of a polynomial function.

Alert

Some students may forget to write the terms of the polynomial in descending order before counting the number of variations.

Teaching Note

Using Descartes' rule of signs, note that multiplicity is counted when determining the number of real zeros. Thus $x^2 - 6x + 9$, which has one zero (3) of multiplicity two, is counted as having two positive real zeros. The topic of multiplicity is discussed in Section 3.6.

Descartes' Rule of Signs

Let $f(x) = a_n x^n + a_{n-1}x^{n-1} + \cdots + a_1 x + a_0$ be a polynomial of degree n with real coefficients and $a_n \neq 0$.

1. The number of *positive real zeros* of $f(x)$ is equal to the number of variations in sign of f, or is less than that number by an even integer.
2. The number of *negative real zeros* of $f(x)$ is equal to the number of variations in sign of $f(-x)$, or is less than that number by an even integer.

A variation in sign means that two consecutive coefficients have opposite signs. For example, $x^2 - 3x + 2$ has *two* variations in sign, whereas there is only *one* variation in sign in $x^2 + 3x - 2$.

Notes on Examples

Examples 1 and 2 refer to the function that is graphed in Figure 3.28. These examples emphasize that one should not make assumptions based on the appearance of a function in one or two viewing windows.

■ EXAMPLE 1 Applying the Rule of Signs

Apply the rule of signs to

$$f(x) = x^5 - 68x^4 + 1085x^3 + 2450x^2 - 611.$$

Solution

$$f(x) = x^5 \overset{\downarrow}{-} 68x^4 \overset{\downarrow}{+} 1085x^3 + 2450x^2 \overset{\downarrow}{-} 611 \quad \text{Three variations in sign}$$

$$f(-x) = (-x)^5 - 68(-x)^4 + 1085(-x)^3 + 2450(-x)^2 - 611$$

$$= -x^5 - 68x^4 - 1085x^3 \overset{\downarrow}{+} 2450x^2 \overset{\downarrow}{-} 611 \quad \text{Two variations in sign}$$

By the test of signs there are three or one real zeros to the right of the origin. There are two or no real zeros to the left of the origin. ■

Descartes' rule of signs tells us there are at most two negative real zeros of $f(x) = x^5 - 68x^4 + 1085x^3 + 2450x^2 - 611$. Figure 3.28a shows there are at least two negative zeros; hence, there must be exactly two negative real zeros. Figure 3.28a suggests that there is only one positive zero. However, Example 2 will show that conclusion to be false. The concepts of *upper* and *lower bounds for real zeros* are important in this discussion.

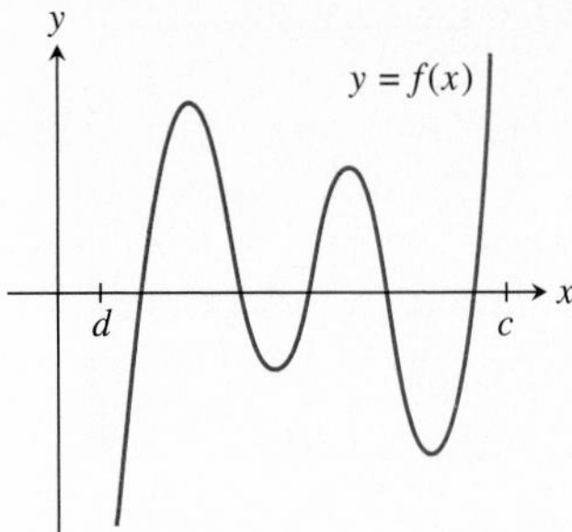

Figure 3.29 c is an upper bound and d is a lower bound for the set of zeros.

Teaching Note

Note that the zeros of f are the same as the zeros of $-f$, so it is sufficient to remember the upper bound for zeros test only for polynomials with positive leading coefficients. For example, the zeros of $f(x) = -2x^3 + x^2$ can be determined by finding the zeros of $g(x) = -f(x) = 2x^3 - x^2$.

Alert

Students may try to apply the upper bound for zeros test to negative values of c. Although it is possible for an upper bound to be negative, it is important to realize that the test does not apply to negative values of c.

Upper and Lower Bounds for Real Zeros

A number c is an **upper bound** for the real zeros of f if $f(x)$ is never zero when x is greater than c. Likewise, a number d is a **lower bound** for the real zeros of f if $f(x)$ is never zero when x is less than d.

Figure 3.29 shows a lower bound d and an upper bound c for a function with five real zeros.

Upper Bound Test for Real Zeros

Let f be a polynomial with a $\begin{pmatrix}\text{positive}\\ \text{negative}\end{pmatrix}$ leading coefficient, and let c be a positive number. Use synthetic division to find the value $f(c)$. If the last line of the synthetic division contains no $\begin{pmatrix}\text{negative}\\ \text{positive}\end{pmatrix}$ numbers, then c is an upper bound for the real zeros of f.

Before explaining why this upper bound test works, we apply it in Example 2 to gain information not evident in Figure 3.28 about the zeros of $f(x) = x^5 - 68x^4 + 1085x^3 + 2450x^2 - 611$.

EXAMPLE 2 Confirming the Number of Real Zeros

Confirm the number of real zeros of

$$f(x) = x^5 - 68x^4 + 1085x^3 + 2450x^2 - 611.$$

Solution

From the discussion following Example 1, we know there are two negative zeros and at least one positive zero.

The coefficients of x^5 and x^4 suggest that we need to try numbers 68 or greater as possible upper bounds using the upper bound test. We test 70 as an upper bound for the real zeros of f.

70	1	−68	1,085	2,450	0	−611
		70	140	85,750	?	?
	1	2	1,225	?	?	?

We see at a glance that all entries in the bottom row will be positive. We conclude that 70 is an upper bound for the positive real zeros.

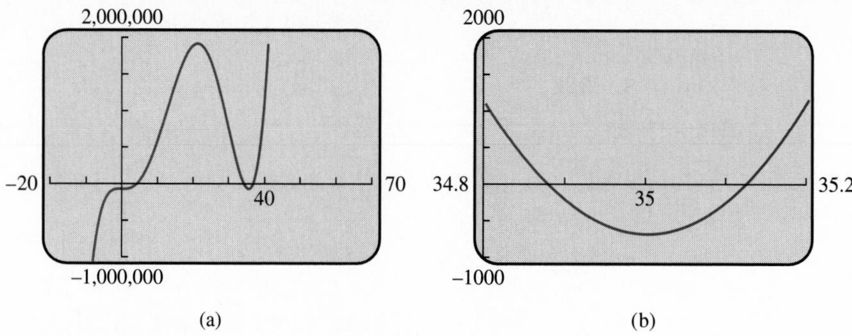

Figure 3.30 $f(x) = x^5 - 68x^4 + 1085x^3 + 2450x^2 - 611$ in two different viewing windows.

REMARK
Notice that Figures 3.28 and 3.30 allow us to conclude that 40 is an upper bound for the zeros of f. However, the upper bound test for real zeros fails to show that 40, or any number less than 68 for that matter, is an upper bound.

The graph of f with $X\max = 70$ (Figure 3.30a) suggests there may be other zeros between 30 and 40. The graph of f in Figure 3.30b shows that indeed two more zeros are hidden near 35. Thus, f has five real zeros, two negative and three positive. ■

The two graphs in Figure 3.30 give a complete picture of the real zeros of $f(x) = x^5 - 68x^4 + 1085x^3 + 2450x^2 - 611$. Without using the information gained from the upper bound test, we may have falsely concluded that all real zeros of f were evident in Figure 3.28.

The steps in the solution of Example 3 explain why the upper bound test for real zeros is valid.

■ EXAMPLE 3 Explaining the Upper Bound Test

Let $f(x) = x^5 - 2x^4 - 24x^3 + 62x^2 + 23x - 60$. Show why the upper bound test for real zeros works for $x = 6$.

Solution

Apply the test with $x = 6$.

$$\begin{array}{r|rrrrrr} 6 & 1 & -2 & -24 & 62 & 23 & -60 \\ & & 6 & 24 & 0 & 372 & 2370 \\ \hline & 1 & 4 & 0 & 62 & 395 & 2310 \end{array}$$

REMARK
Notice that 6 is the smallest number that leads to all nonnegative numbers on the bottom row. Why? Because any smaller number will result in a number less than 24 in the second entry in row 2, which would make the third entry in row 3 negative.

Using synthetic division shows us the quotient and remainder, so we can write

$$\begin{aligned} f(x) &= x^5 - 2x^4 - 24x^3 + 62x^2 + 23x - 60 \\ &= (x - 6)(x^4 + 4x^3 + 62x + 395) + 2310. \end{aligned}$$

a. When x is positive, $x^4 + 4x^3 + 62x + 395$ is positive because substituting a positive number produces a positive number.
b. When x is greater than 6, $x - 6$ is positive.
c. Therefore, when x is greater than 6, $(x - 6)(x^4 + 4x^3 + 62x + 395) + 2310$ is positive because a product of two positive numbers added to a third positive number is positive.

Consequently, $x = 6$ is an upper bound for the real zeros of f. ■

An argument similar to the one given in Example 3 can be used to verify the upper bound test for real zeros for the case of a negative leading coefficient. The test states that the last row in the synthetic division process must be nonpositive. We apply the test in Example 4.

■ EXAMPLE 4 Using the Upper Bound Test with a Negative Leading Coefficient

Let $f(x) = -x^3 + 2x^2 + 21x + 16$. Apply the upper bound for zeros test to show that 6 is an upper bound for the zeros of f.

Solution

Since the leading coefficient of f is negative, we must verify that the last line of the synthetic division with divisor 6 contains no positive numbers.

$$\begin{array}{r|rrrr} 6 & -1 & 2 & 21 & 16 \\ & & -6 & -24 & -18 \\ \hline & -1 & -4 & -3 & -2 \end{array}$$

Because each entry in the last row is a negative number, we have established that 6 is an upper bound for the zeros of f. ■

There is a lower bound test for real zeros of a polynomial function that is somewhat similar to the upper bound test.

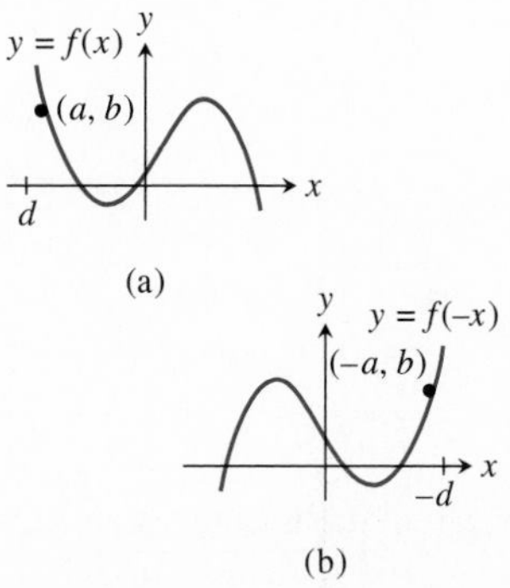

Figure 3.31 The graphs of $f(x)$ and $f(-x)$ are reflections of each other across the y-axis.

> **Lower Bound Test for Real Zeros**
>
> The negative number d is a lower bound for the real zeros of $f(x)$ if and only if the positive number $-d$ is an upper bound for the real zeros of $f(-x)$. (See Figure 3.31.)

EXAMPLE 5 Testing a Lower Bound for Zeros

Show that -6 is a lower bound for $g(x) = x^3 + 2x^2 - 21x + 16$.

Solution

According to the lower bound test for real zeros, we must show that $-(-6) = 6$ is an upper bound for $g(-x)$. (See Figure 3.31 and Exercise 65.)

$$\begin{aligned} g(-x) &= (-x)^3 + 2(-x)^2 - 21(-x) + 16 \\ &= -x^3 + 2x^2 + 21x + 16 \end{aligned}$$

The function $g(-x)$ is identical to the function f in Example 4, that is $g(-x) = f(x)$. So the synthetic division used in Example 4 verifies that 6 is an upper bound for the real zeros of $y = f(x)$ and consequently, -6 is a lower bound for the real zeros of $y = f(-x) = g(x)$. ■

Rational Zeros

Some polynomials have **rational zeros**—zeros that are rational numbers. Some polynomials have **irrational zeros**—zeros that are irrational numbers. For example,

$$f(x) = 4x^2 - 9 = (2x - 3)(2x + 3)$$

has rational zeros 2/3 and $-2/3$, and

$$g(x) = x^2 - 2 = (x - \sqrt{2})(x + \sqrt{2})$$

has irrational zeros $\sqrt{2}$ and $-\sqrt{2}$.

The rational zeros test suggests how to search for rational zeros (or roots), when they exist, for polynomials whose coefficients are integers.

Teaching Note

It is best to see if the terms of $f(x)$ have a common factor that can be factored out before applying the rational zeros test. Factoring out a common factor will reduce the number of extra candidates found.

Rational Zeros Test

Suppose all the coefficients in the polynomial

$$f(x) = a_n x^n + a_{n-1} x^{n-1} + \cdots + a_0, \qquad a_n \neq 0,\ a_0 \neq 0$$

are integers. If $x = p/q$ is a rational zero, where p and q have no common factors, then

- p is a factor of the constant term a_0, and
- q is a factor of the leading coefficient a_n.

EXAMPLE 6 Searching for Rational Zeros

Find all the rational zeros of $f(x) = x^3 - 3x^2 + 1$.

Solution

Because both the leading coefficient and the constant are 1, the only potential rational zeros are the positive and negative factors of 1.

Potential Rational Zeros: ± 1.

$$f(1) = (1)^3 - 3(1)^2 + 1 = -1 \qquad x = 1 \text{ is not a zero of } f.$$

$$f(-1) = (-1)^3 - 3(-1)^2 + 1$$

$$= -1 - 3 + 1 = -3 \qquad x = -1 \text{ is not a zero of } f.$$

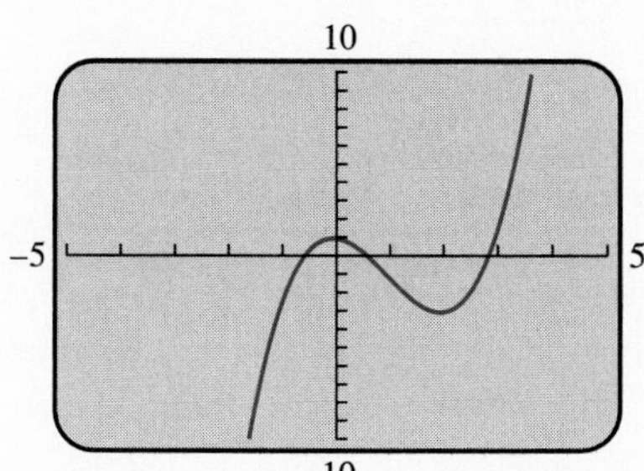

Figure 3.32 $f(x) = x^3 - 3x^2 + 1$.

Therefore, $f(x)$ has no rational zeros. Figure 3.32 shows the graph of f. Each of the three zeros shown must be an irrational number. ■

In Example 6 the rational zeros test suggested only two candidates for rational zeros, and neither of them "checked out." The test often suggests many candidates. When this happens, we can use technology and algebra together to find exact rational zeros when they exist.

EXAMPLE 7 Finding Rational Zeros

Find all the rational zeros of $f(x) = 3x^3 + 4x^2 - 5x - 2$.

Solution

Because the constant is -2, and the leading coefficient is 3, we have:

Potential Rational Zeros:

$$\frac{\text{Factors of } (-2)}{\text{Factors of } 3}: \quad \frac{\pm 1, \pm 2}{\pm 1, \pm 3}: \quad \pm 1, \pm 2, \pm\frac{1}{3}, \pm\frac{2}{3}$$

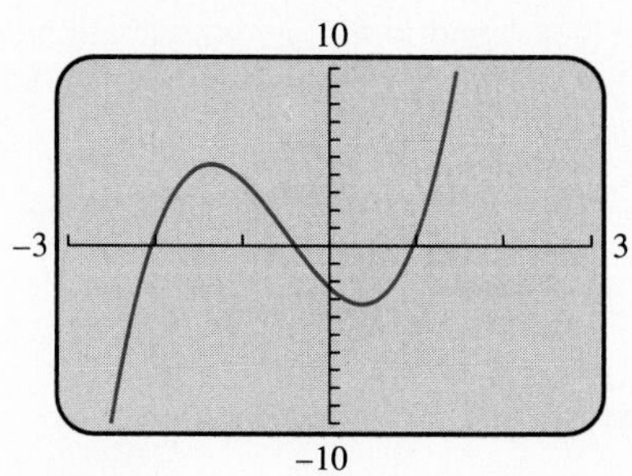

Figure 3.33 $f(x) = 3x^3 + 4x^2 - 5x - 2$.

Figure 3.33 suggests that of the eight potential rational zeros, $x = -2$ and $x = 1$ and perhaps $x = -\frac{1}{3}$ or $-\frac{2}{3}$ are the most likely candidates. Using synthetic division, we find:

$$\begin{array}{r|rrrr} 1 & 3 & 4 & -5 & -2 \\ & & 3 & 7 & 2 \\ \hline & 3 & 7 & 2 & 0 \end{array} \quad \leftarrow f(1) = 0, \text{ so } x - 1 \text{ is a factor.}$$

$$\begin{array}{r|rrr} -2 & 3 & 7 & 2 \\ & & -6 & -2 \\ \hline & 3 & 1 & 0 \end{array} \quad \leftarrow f(-2) = 0, \text{ so } x + 2 \text{ is a factor.}$$

Thus $f(x)$ factors as $f(x) = (x - 1)(x + 2)(3x + 1)$, and we conclude that the zeros are the rational numbers $x = 1$, $x = -2$, and $x = -\frac{1}{3}$. ■

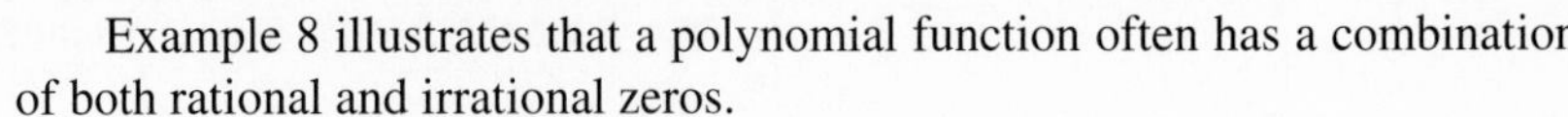

Example 8 illustrates that a polynomial function often has a combination of both rational and irrational zeros.

■ EXAMPLE 8 Finding All Real Zeros

Find all the real zeros of the function $f(x) = 2x^4 - 7x^3 - 8x^2 + 14x + 8$.

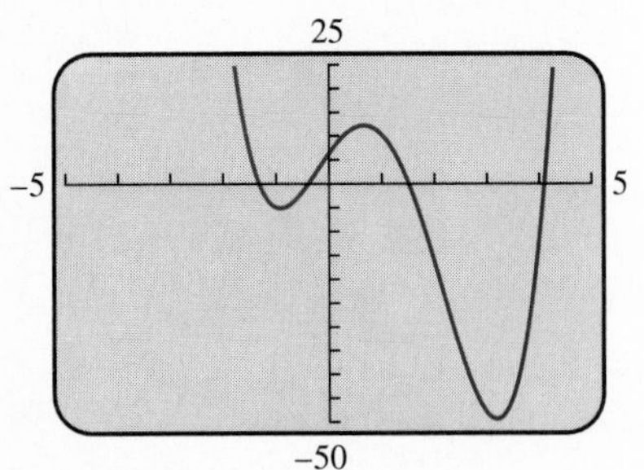

Figure 3.34 $f(x) = 2x^4 - 7x^3 - 8x^2 + 14x + 8$.

Solution

Because the constant is 8 and the leading coefficient is 2 we have:

Potential Rational Zeros:

$$\frac{\text{Factors of } 8}{\text{Factors of } 2}: \quad \frac{\pm 1, \pm 2, \pm 4, \pm 8}{\pm 1, \pm 2}: \quad \pm 1, \pm 2, \pm 4, \pm 8, \pm \frac{1}{2}$$

Figure 3.34 suggests that only $x = -\frac{1}{2}$ and $x = 4$ be considered, and synthetic division confirms that they are indeed both zeros of f. Therefore, $x - 4$ and $x + \frac{1}{2}$ are both factors of f. Another factor (shown using synthetic division) is $2x^2 - 4$. So the factorization of f is

$$\begin{aligned} f(x) &= (x - 4)\left(x + \frac{1}{2}\right)(2x^2 - 4) \\ &= 2(x - 4)\left(x + \frac{1}{2}\right)(x^2 - 2) \\ &= (x - 4)(2x + 1)(x^2 - 2). \end{aligned}$$

The remaining two zeros of f are zeros of $x^2 - 2$, namely $\pm\sqrt{2}$.

Summary

The zeros of f are the rational numbers $x = -\frac{1}{2}$ and $x = 4$, and the irrational numbers $x = \sqrt{2}$ and $x = -\sqrt{2}$. ■

Example 9 illustrates that finding the real zeros of a polynomial function often requires a careful analysis.

■ EXAMPLE 9 Finding All Real Zeros

Find all real zeros of $f(x) = 10x^5 - 3x^2 + x - 6$, and identify them as rational or irrational.

Solution

Potential Rational Zeros:

$$\frac{\text{Factors of } -6}{\text{Factors of } 10}: \quad \pm 1, \pm 2, \pm 3, \pm 6, \pm \frac{1}{2},$$

$$\pm \frac{3}{2}, \pm \frac{1}{5}, \pm \frac{2}{5}, \pm \frac{3}{5}, \pm \frac{6}{5}, \pm \frac{1}{10}, \pm \frac{3}{10}$$

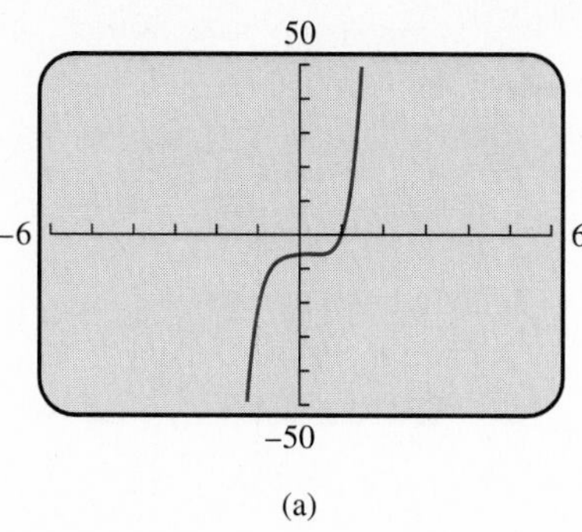

(a)

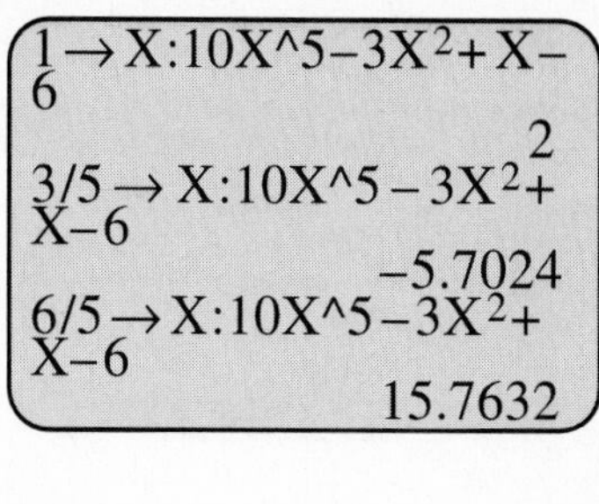

(b)

Figure 3.35 $f(x) = 10x^5 - 3x^2 + x - 6$.

Follow-up

Have students rewrite the lower bound for zeros test using the same kind of language that was used for the upper bound for zeros test.

Assignment Guide

Day 1: Ex. 3–48, multiples of 3
Day 2: Ex. 49, 50, 51, 53, 54, 55, 58, 59, 60, 61, 62, 65

Cooperative Learning

Group Activity: Ex. 66

Notes on Exercises

Ex. 11–30 require students to use tests for upper and lower bounds.
Ex. 49–52 are extensions of the floating-buoy problem of Example 10.
Ex. 59–60 encourage students to think about what kind of regression might be most useful for modeling a particular set of data.

These potential rational zeros all fall in the interval $-6 \leq x \leq 6$. So if the one zero indicated by the graph of f in Figure 3.35a is a rational zero, then it most likely is $x = \frac{3}{5}$, $x = 1$, or $x = \frac{6}{5}$.

Evaluating f for each of these values, however (see Figure 3.35b), shows that $x = 1$, $x = \frac{3}{5}$, and $x = \frac{6}{5}$ are not zeros of f. We conclude that f has *no* rational zeros.

Irrational Zeros: Using grapher methods we find that the one real zero indicated in Figure 3.35a is 0.95, rounded to the nearest hundredth.[4] The upper and lower bound tests confirm that $x = -6$ and $x = 6$ are bounds for the zeros of $f(x)$, so there are no zeros outside the viewing window in Figure 3.35.

Summary

You can show that $x = 1$ is an upper bound and $x = -1$ is a lower bound for the real zeros of f. Thus, the only real zero of f is the irrational number that is approximately 0.95. ■

Applications

■ **EXAMPLE 10 APPLICATION: Finding the Depth of a Floating Buoy**

A spherical floating buoy has a radius of 1 m and a density one-fourth that of seawater. Find the depth to which the buoy sinks in seawater. (See Figure 3.36.)

Solution

Model

We let $x =$ the depth to which the buoy sinks, $d =$ the density of seawater, and define r as shown in Figure 3.36.

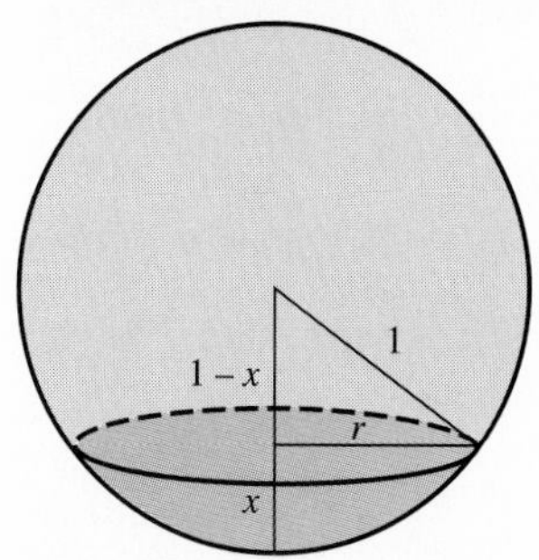

Figure 3.36 A spherical buoy floating in water. Notice that $0 < x < 1$ and that

$$(1 - x)^2 + r^2 = 1,$$
$$r^2 = 1 - (1 - x)^2$$
$$= 2x - x^2.$$

[4] A root-finder grapher routine indicates that the zero of f is approximately $x = 0.950\ 545\ 89$. But we cannot know, without the analysis shown in Example 9, whether the zero is a rational or an irrational number.

Ongoing Assessment

Self-Assessment: Ex. 7, 19, 29, 31, 43, 53
Embedded Assessment: Ex. 60

Archimedes Principle:

Weight of displaced water = weight of buoy

$$\frac{d\pi}{6}x(3r^2 + x^2) = \frac{d\pi}{3}$$ Equation to be solved. See Exercise 49.

Solve Algebraically and Graphically

$$3xr^2 + x^3 = 2$$ Multiply by $6/(d\pi)$.

$$3x(2x - x^2) + x^3 = 2$$ $r^2 = 2x - x^2$. See Figure 3.36.

$$6x^2 - 3x^3 + x^3 = 2$$ Use the distributive property.

$$6x^2 - 2x^3 = 2$$

$$3x^2 - x^3 = 1$$

$$x^3 - 3x^2 + 1 = 0, \quad 0 < x < 1$$

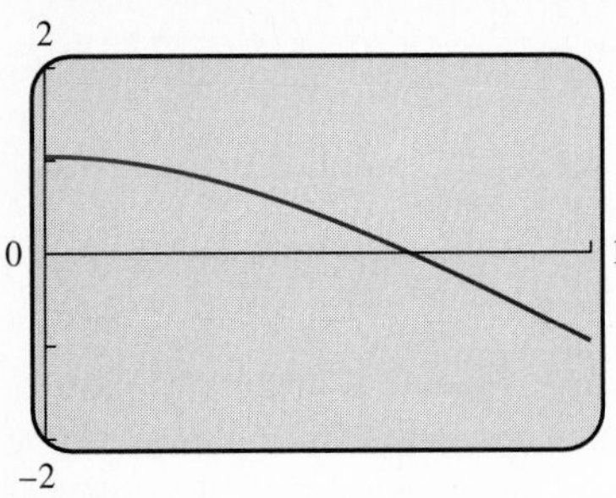

Figure 3.37 $f(x) = x^3 - 3x^2 + 1$.

The graph of $f(x) = x^3 - 3x^2 + 1$ is shown in Figure 3.32 (Example 6), and in Figure 3.37 for $0 < x < 1$. In Example 6 we found that the zeros of f are irrational, so by grapher methods we find the solution $x \approx 0.65$.

Interpret

The buoy sinks approximately 0.65 m into the sea. ■

Quick Review 3.4

In Exercises 1 and 2, use synthetic division to evaluate the function.

1. $f(2)$ for $f(x) = 2x^4 - 7x^3 + 9x^2 - 3x - 7$
2. $g(-1)$ for $g(x) = -3x^4 + 7x^2 - 3x + 2$

In Exercises 3 and 4, write the polynomial $f(-x)$ in standard form for the given polynomial.

3. $f(x) = 3x^3 - 4x^2 + 7x - 2$
4. $f(x) = -x^3 + 3x^2 - 5x + 1$

In Exercises 5 and 6, divide $f(x)$ by $h(x)$ to find a quotient $q(x)$ and remainder $r(x)$.

5. $f(x) = x^4 + 3x^3 - x^2 + 3x - 2;\ h(x) = x^2 + 1$
6. $f(x) = 2x^4 - 2x^3 + 5x^2 - 3x + 3;\ h(x) = 2x^2 + 3$

SECTION EXERCISES 3.4

In Exercises 1–10, use Descartes' rule of signs to find the possible numbers of positive and negative real zeros.

1. $f(x) = x^3 + x^2 - x + 1$
2. $g(t) = t^3 - t^2 - t - 1$
3. $h(s) = 2s^3 - s^2 + s - 1$
4. $f(x) = x^3 + x^2 + x + 1$
5. $f(x) = 2x^3 + x - 3$
6. $g(x) = 5x^4 + x^2 - 3x - 2$
7. $h(x) = x^4 - 5x^3 + x^2 - 3x + 1$
8. $f(x) = -x^4 + 2x^3 + 5x^2 - 7x + 2$
9. $f(t) = -t^5 - t^4 + t^3 + t^2 - 3t + 1$
10. $h(t) = t^4 - t^3 + t^2 - t + 2$

In Exercises 11–16, use synthetic division to show that the given number is an upper bound for the zeros of f.

11. $c = 3;\ f(x) = 2x^3 - 4x^2 + x - 2$
12. $c = 4;\ f(x) = -x^3 + 2x^2 + 3x + 1$

13. $c = 2;\ f(x) = x^4 - x^3 + x^2 + x - 12$

14. $c = 2;\ f(x) = -x^4 - 2x^3 + 3x^2 - x + 1$

15. $c = 5;\ f(x) = -2x^3 + 5x^2 + 5x + 1$

16. $c = 3;\ f(x) = 4x^4 - 6x^3 - 7x^2 + 9x + 2$

In Exercises 17–22, use synthetic division to show that the given number is a lower bound for the zeros of f.

17. $d = -1;\ f(x) = x^2 - 5x - 3$

18. $d = -3;\ f(x) = -2x^2 - 3x + 2$

19. $d = -1;\ f(x) = 3x^3 - 4x^2 + x + 3$

20. $d = -3;\ f(x) = -x^3 - 2x^2 - 2x - 5$

21. $d = -5;\ f(x) = -x^3 + 4x^2 - 7x + 2$

22. $d = -4;\ f(x) = 3x^3 - x^2 - 5x - 3$

In Exercises 23 and 24, explain why the given numbers are an upper bound and a lower bound, respectively, for the zeros of the given function.

23. $x = 2$ and $x = -3;\ f(x) = (x + 3)(x - 2)(x + 1)$

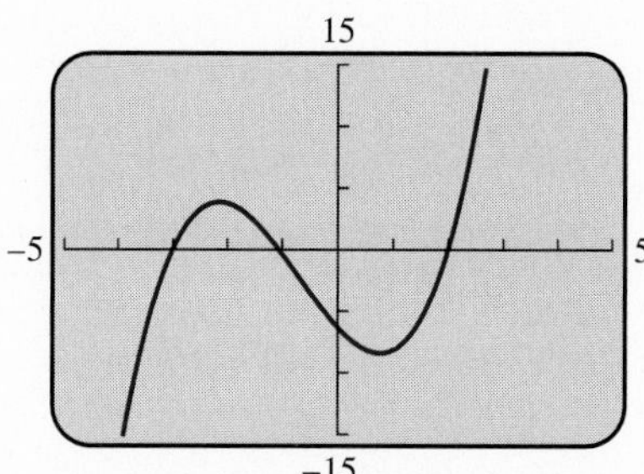

For Exercise 23

24. $x = 2$ and $x = -4;\ f(x) = (x + 4)(x + 1)(x - 2)$

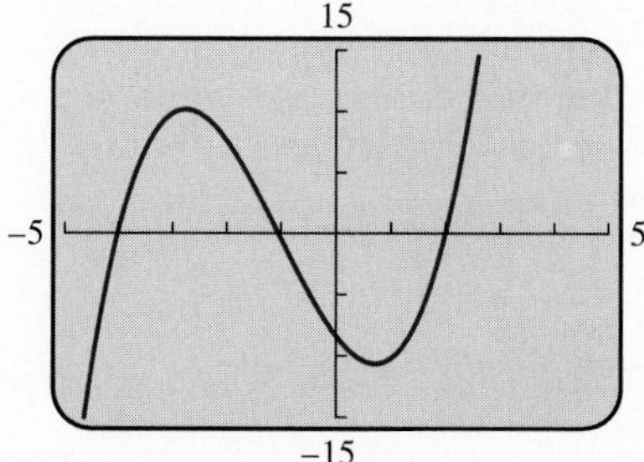

For Exercise 24

In Exercises 25–30, use the upper and lower bound tests to decide whether there is a zero for the given function outside of the viewing window shown.

25. $f(x) = 6x^4 - 11x^3 - 7x^2 + 8x - 34$

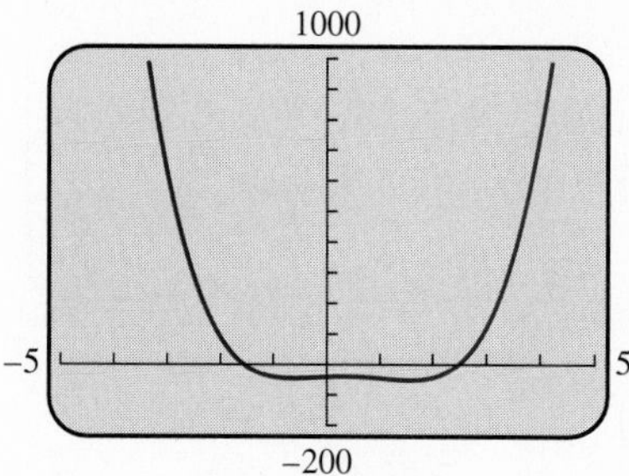

For Exercise 25

26. $f(x) = x^5 - x^4 + 21x^2 + 19x - 3$

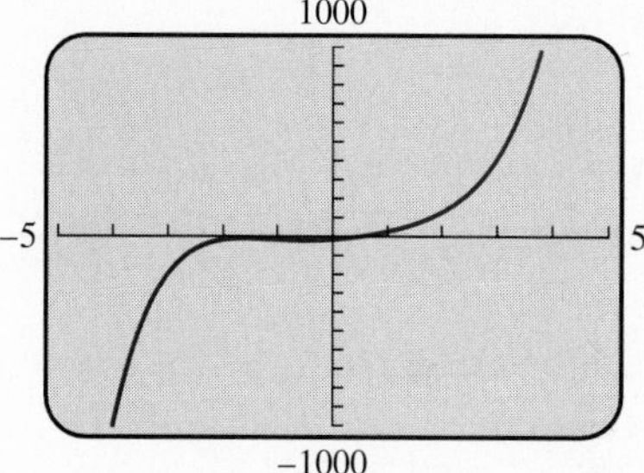

For Exercise 26

27. $f(x) = x^5 + x^4 + 42x^2 + 156x - 32$

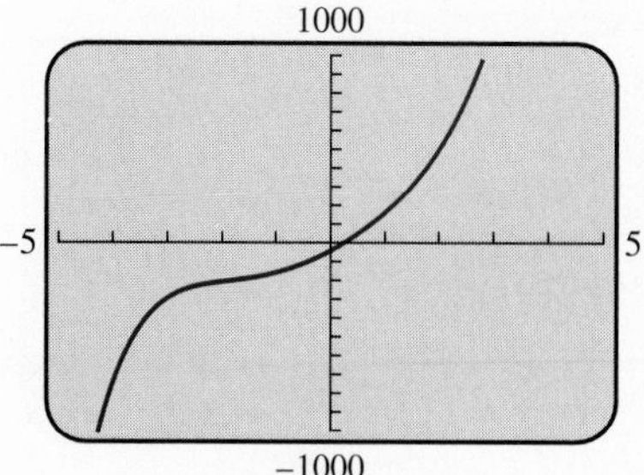

For Exercise 27

28. $f(x) = -x^4 + 75x^2 - 70x - 144$

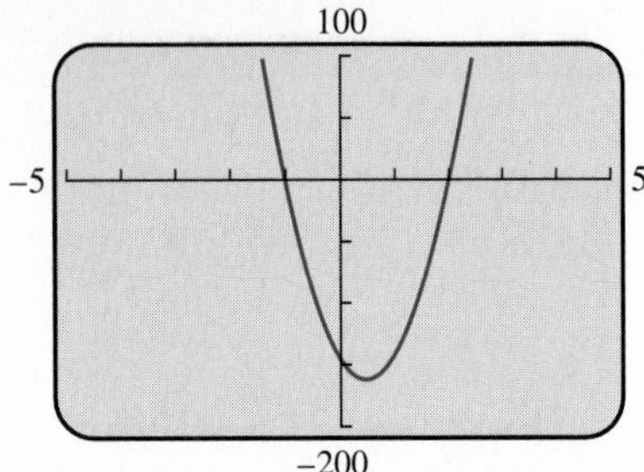

For Exercise 28

29. $f(x) = x^5 - 4x^4 - 129x^3 + 396x^2 - 8x + 3$

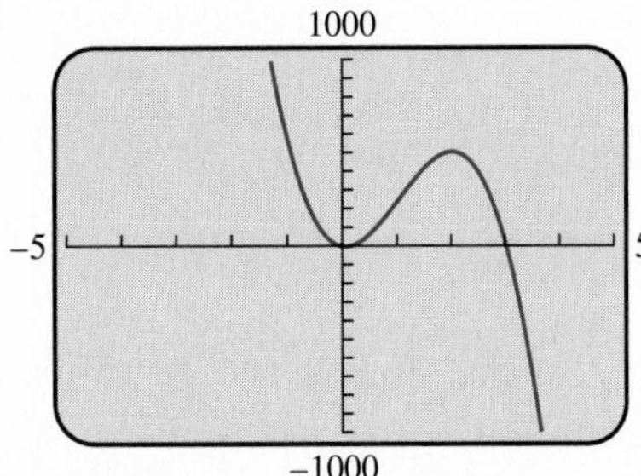

For Exercise 29

30. $f(x) = 2x^5 - 5x^4 - 141x^3 + 216x^2 - 91x + 25$

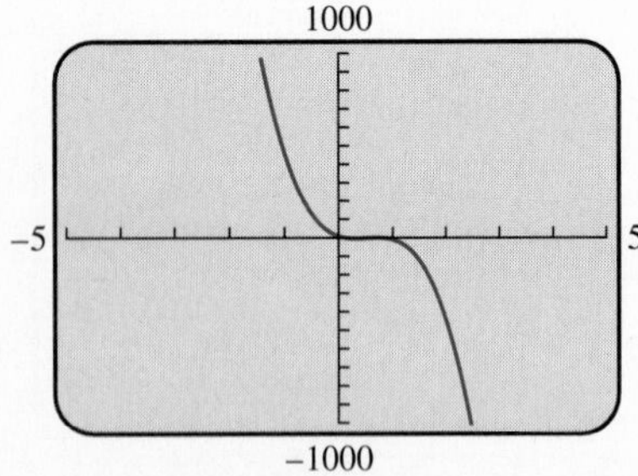

For Exercise 30

In Exercises 31–36, use the rational zeros test to find a list of all potential rational zeros. Then determine which ones, if any, are zeros.

31. $f(x) = x^3 + 4x^2 - 4x - 1$
32. $f(x) = 6x^3 - 5x - 1$
33. $f(x) = 3x^3 - 7x^2 + 6x - 14$
34. $f(x) = 2x^3 - x^2 - 9x + 9$
35. $f(x) = 6x^4 - x^3 - 6x^2 - x - 12$
36. $f(x) = 14x^4 + 11x^3 - 19x^2 + x + 2$

In Exercises 37–48, find all the real zeros of the function. Identify each as rational or irrational.

37. $f(x) = 2x^3 - 3x^2 - 4x + 6$
38. $f(x) = x^3 + 3x^2 - 3x - 9$
39. $f(x) = x^3 + x^2 - 8x - 6$
40. $f(x) = x^3 - 6x^2 + 7x + 4$
41. $f(x) = x^4 - 3x^3 - 6x^2 + 6x + 8$
42. $f(x) = x^4 - x^3 - 7x^2 + 5x + 10$
43. $f(x) = 2x^4 - 7x^3 - 2x^2 - 7x - 4$
44. $f(x) = 3x^4 - 2x^3 + 3x^2 + x - 2$
45. $f(x) = 2x^3 - x^2 - 18x + 9$
46. $f(x) = 3x^3 - x^2 + 27x - 9$
47. $f(x) = x^4 + x^3 - 3x^2 - 4x - 4$
48. $f(x) = x^4 + 3x^2 + 2$

Exercises 49–52 refer to the floating-buoy problem of Example 10.

49. The buoy is a sphere of radius 1 whose density is one-fourth that of seawater.
   a. If d is the density of seawater, show that the weight of the displaced water is $d(\pi/6)x(3r^2 + x^2)$. (*Hint:* the volume of the submerged spherical cap is $(\pi/6)x(3r^2 + x^2)$.)
   b. Show that the volume of the buoy is $4\pi/3$.
   c. Show that the weight of the buoy is $\pi d/3$.
   d. Show that
   $$\frac{d\pi}{6}x(3r^2 + x^2) = \frac{d\pi}{3}.$$
50. Use numerical or graphical methods to show that $x = 0.65$ is an approximate solution to Example 10.
51. Find the depth that the buoy sinks in seawater if its density is one-third that of seawater.

52. Find the depth that the buoy sinks in seawater if its density is one-fifth that of seawater.

In Exercises 53 and 54, solve the problem.

53. *Biological Research* Stephanie, a biologist who does research for the poultry industry, models the population P of wild turkeys, t days after being left to reproduce, with the function

$$P(t) = -0.00001t^3 + 0.002t^2 + 1.5t + 100.$$

a. Graph the function $y = P(t)$ for appropriate values of t.
b. Find what the maximum turkey population is and when it occurs.
c. Assuming that this model continues to be accurate, when will this turkey population become extinct?
d. **Writing to Learn** Create a scenario that could explain the growth exhibited by this turkey population.

54. *Architectural Engineering* Dave, an engineer at the Trumbauer Group, Inc., an architectural firm, completes structural specifications for a 172-ft-long steel beam, anchored at one end to a piling 20 ft above the ground. He knows that when a 200-lb object is placed d feet from the anchored end, the beam bends s feet where

$$s = (3 \times 10^{-7})d^2(550 - d).$$

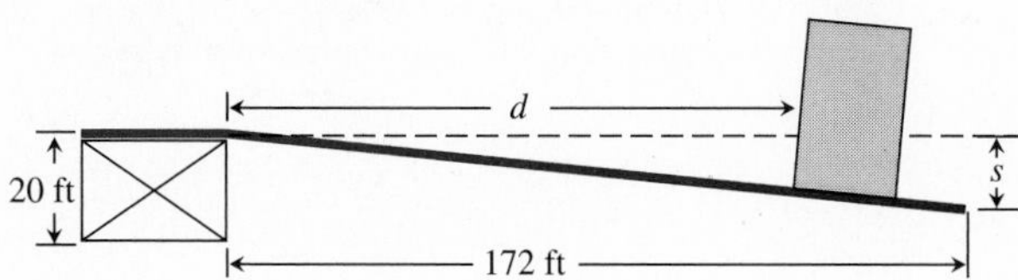

For Exercise 54

a. What is the independent variable in this polynomial function?
b. What are the dimensions of a viewing window that shows a graph for the values that make sense in this problem situation?
c. How far is the 200-lb object from the anchored end if the vertical deflection is 1.25 ft?

In Exercises 55–58, give a convincing argument that each function has no rational zeros.

55. $f(x) = x^3 + 2x^2 + x - 1$

56. $f(x) = x^3 - x + 2$

57. $f(x) = 2x^4 - x^3 + 1$

58. $f(x) = 3x^3 + x^2 - 2x + 1$

In Exercises 59 and 60, use the data on the numbers of personal computers shipped from 1987 to 1991 given in Table 3.7.

Table 3.7 Personal Computers

Year	No. Shipped (thousands)
1987	4178
1988	5493
1989	6147
1990	6638
1991	7052

Source: Dataquest Inc., San Jose, Calif., Consolidated Data Base, November 12, 1991.

59. The quadratic equation

$$y_1 = -140x^2 + 3216x - 11{,}390$$

and the cubic equation

$$y_2 = 48.6x^3 - 1454x^2 + 14{,}876x - 45{,}379$$

both model these data using $x = 7$ for 1987, $x = 8$ for 1988, and so on. The figures show graphs of each modeling equation superimposed on a scatter plot of the data.

a. Reproduce these graphs on your grapher.
b. Match the equation with its graph.
c. Does one graph appear to fit the data better than the other, or do both graphs appear to fit the data equally well? (The formal mathematics for goodness of fit is beyond the scope of this book.)

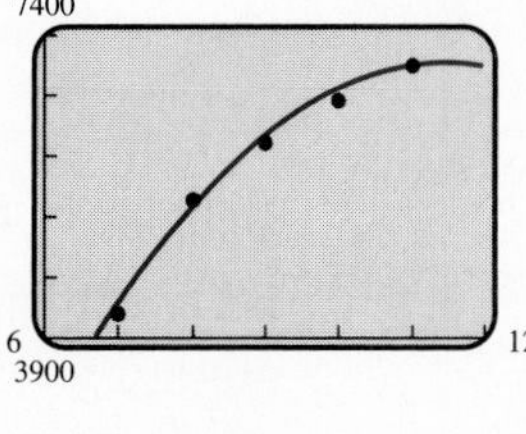

(a)

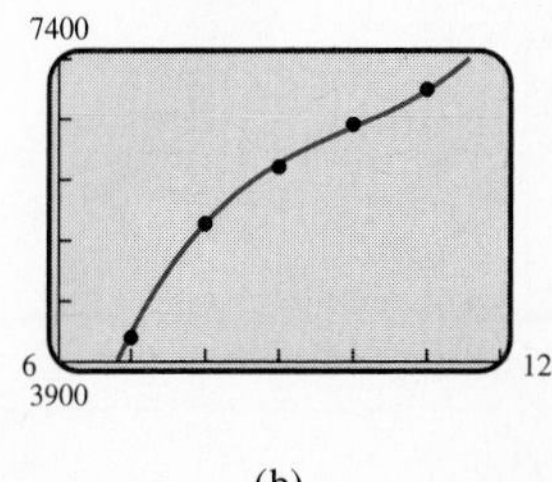

(b)

For Exercise 59

60. Writing to Learn Graphs of the two modeling equations defined in Exercise 59 are shown superimposed on the data in a larger viewing window.

a. What time period is represented by the new viewing windows?

b. In your judgment, which modeling equation most accurately predicts the number of personal computer shipments in the years 1992–96? Explain your decision.

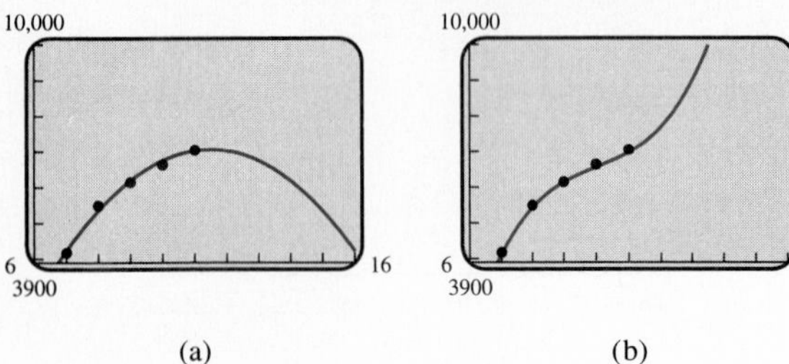

For Exercise 60

EXTENDING THE IDEAS

In Exercises 61–66, solve the problem.

61. Writing to Learn Write a paragraph that describes how the zeros of $f(x) = \frac{1}{3}x^3 + x^2 + 2x - 3$ are related to the zeros of $g(x) = x^3 + 3x^2 + 6x - 9$. In what ways does this example illustrate how the rational zeros test can be applied to find the zeros of a polynomial with *rational* number coefficients?

62. Find the rational zeros of $f(x) = x^3 - \frac{7}{6}x^2 - \frac{20}{3}x + \frac{7}{2}$.

63. Find the rational zeros of $f(x) = x^3 - \frac{5}{2}x^2 - \frac{37}{12}x + \frac{5}{2}$.

64. Writing to Learn Explain how to use the rational zeros test to show that $\sqrt{2}$ is irrational.

65. Writing to Learn Explain how the figure is related to the lower bound for real zeros test.

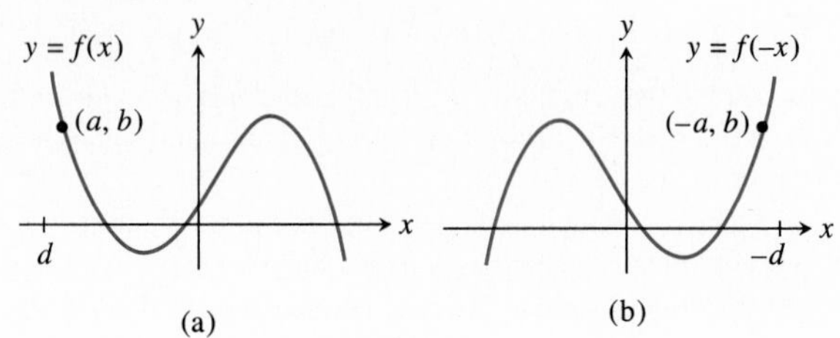

For Exercise 65

66. Group Learning Activity *Work in groups of three.* Graph $f(x) = x^4 + x^3 - 8x^2 - 2x + 7$.

a. Use grapher methods to find approximate real number zeros.

b. Identify a list of four linear factors whose product could be called an *approximate factorization of* $f(x)$.

c. Discuss what graphical and numerical methods you could use to show that the factorization from part b is reasonable.

3.5 COMPLEX NUMBERS

Imaginary Unit *i* • Complex Numbers • Operations with Complex Numbers • Complex Conjugates and Division • Complex Solutions of Quadratic Equations • Complex Plane

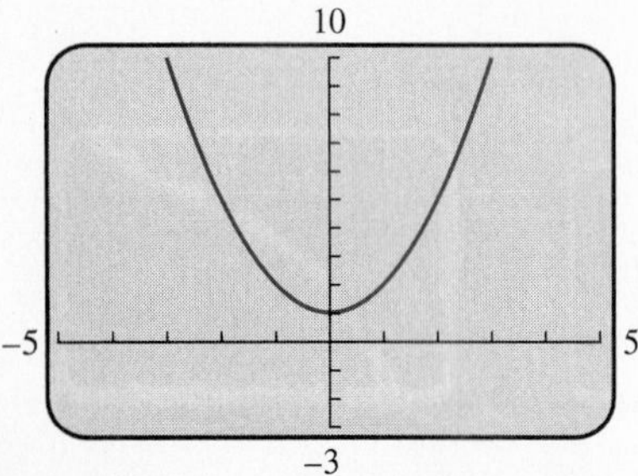

Figure 3.38 There are no real zeros of $f(x) = x^2 + 1$.

Imaginary Unit *i*

Figure 3.38 shows that the function $f(x) = x^2 + 1$ does not have any real zeros, that is, $x^2 + 1 = 0$ has no real number solutions.

To remedy this situation, mathematicians in the 17th century extended the definition of $\sqrt{a}$ to include negative real numbers a. Recall from the Prerequisite Chapter that the meaning of $\sqrt{a}$ was defined only for positive real num-

> **Objective**
>
> Students will be able to add, subtract, multiply, and divide complex numbers, and to find complex zeros of quadratic functions.
>
> **Motivate**
>
> Ask . . .
>
> If i is the square root of -1, what is the square of $-i$? **(-1)**

bers a. The "number" $i = \sqrt{-1}$ is defined by $i^2 + 1 = 0$. Then for any other negative real number a, $\sqrt{a} = \sqrt{|a|} \cdot i$.

Mathematicians were a bit uneasy about creating such a number, so they called i the **imaginary unit.** This number i is the basis for the *complex numbers.*

Complex Numbers

The extended system of numbers, called the complex numbers, consists of all real numbers and sums of real numbers and real number multiples of i. The following are all examples of complex numbers:

$$3i \qquad 5i \qquad \sqrt{5} \qquad -7i \qquad \frac{5}{2}i \qquad \frac{2}{3}$$

$$2 + 3i \qquad 5 - i \qquad \frac{1}{3} + \frac{4}{5}i$$

> **Definition:** Complex Numbers
>
> A **complex number** is any number that can be written in the form
>
> $$a + bi,$$
>
> where a and b are real numbers. The number a is the **real part,** the number b is the **imaginary part,** and $a + bi$ is the **standard form.**

A real number a is the complex number $a + 0i$, so all real numbers are also complex numbers. Two complex numbers are **equal** if their real parts are equal and their imaginary parts are equal. For example,

$$x + yi = 2 + 5i \quad \text{if and only if} \quad x = 2 \text{ and } y = 5.$$

> **Teaching Note**
>
> It is useful to compare operations with complex numbers to operations with binomials, but stress that i is not a variable.

Operations with Complex Numbers

Adding complex numbers involves adding the real parts and the imaginary parts separately. Subtracting complex numbers is also done by parts.

> **Definition:** Adding and Subtracting Complex Numbers
>
> If $a + bi$ and $c + di$ are two complex numbers, then
>
> **Sum:** $(a + bi) + (c + di) = (a + c) + (b + d)i,$
>
> **Difference:** $(a + bi) - (c + di) = (a - c) + (b - d)i.$

The **additive identity** for the complex numbers is $0 + 0i$. The **additive inverse** of $a + bi$ is $-a - bi$ because

$$(a + bi) + (-a - bi) = 0 + 0i.$$

REMARK

Many of the properties of real numbers are also properties for the complex numbers. Here are several:

- Commutative properties of addition and multiplication
- Associative properties of addition and multiplication
- Distributive properties of multiplication over addition and multiplication over subtraction

EXAMPLE 1 Adding and Subtracting Complex Numbers

a. $(7 - 3i) + (4 + 5i) = (7 + 4) + (-3 + 5)i = 11 + 2i$
b. $(2 - i) - (4 + 3i) = (2 - 4) + (-1 - 3)i = -2 - 4i$ ■

Multiplying complex numbers involves treating them as binomials. The equation $i^2 = -1$ helps simplify the resulting expression.

EXAMPLE 2 Multiplying Complex Numbers

Multiply $2 + 3i$ and $5 - i$. Write the answer in standard form.

Solution

$$\begin{aligned}(2 + 3i)(5 - i) &= 2(5) + 2(-i) + (3i)(5) + (3i)(-i)\\ &= 10 - 2i + 15i - 3i^2\\ &= 10 - 2i + 15i + 3 \quad \text{Use } i^2 = -1.\\ &= 13 + 13i\end{aligned}$$

■

Although multiplying complex numbers as binomials is probably the best way to remember the process, we can obtain a general rule by multiplying $a + bi$ and $c + di$.

$$\begin{aligned}(a + bi)(c + di) &= a(c + di) + (bi)(c + di) && \text{Distributive property.}\\ &= ac + adi + bci + bdi^2 && \text{Distributive property.}\\ &= ac + adi + bci - bd && \text{Use } i^2 = -1.\\ &= (ac - bd) + (ad + bc)i\end{aligned}$$

We expand positive integer powers of complex numbers just like we expand binomials.

EXAMPLE 3 Raising Complex Numbers to a Power

If $z = \dfrac{1}{2} + \dfrac{\sqrt{3}}{2}i$, find z^2 and z^3.

Solution

$$
\begin{aligned}
z^2 &= \left(\frac{1}{2} + \frac{\sqrt{3}}{2}i\right)\left(\frac{1}{2} + \frac{\sqrt{3}}{2}i\right) \\
&= \frac{1}{4} + \frac{\sqrt{3}}{4}i + \frac{\sqrt{3}}{4}i + \frac{3}{4}i^2 \\
&= \frac{1}{4} + \frac{\sqrt{3}}{4}i + \frac{\sqrt{3}}{4}i + \frac{3}{4}(-1) \\
&= -\frac{1}{2} + \frac{\sqrt{3}}{2}i
\end{aligned}
$$

$$
\begin{aligned}
z^3 = z^2 \cdot z &= \left(-\frac{1}{2} + \frac{\sqrt{3}}{2}i\right)\left(\frac{1}{2} + \frac{\sqrt{3}}{2}i\right) \\
&= -\frac{1}{4} - \frac{\sqrt{3}}{4}i + \frac{\sqrt{3}}{4}i + \frac{3}{4}i^2 \\
&= -\frac{1}{4} + 0i + \frac{3}{4}(-1) \\
&= -1
\end{aligned}
$$

■

The properties of integer exponents apply to complex numbers. For example, we use properties of exponents together with results from Example 3 to find z^5.

$$
\begin{aligned}
z^5 = z^{(3+2)} &= z^3 \cdot z^2 \\
&= -1 \cdot \left(-\frac{1}{2} + \frac{\sqrt{3}}{2}i\right) \\
&= \frac{1}{2} - \frac{\sqrt{3}}{2}i
\end{aligned}
$$

Teaching Note

Note that the complex conjugate of 3 is 3, and the complex conjugate of $5i$ is $-5i$.

Complex Conjugates and Division

The complex numbers $a + bi$ and $a - bi$ are **complex conjugates.** Notice that the product of complex conjugates $a + bi$ and $a - bi$ is the positive real number $a^2 + b^2$.

$$
\begin{aligned}
(a + bi)(a - bi) &= a^2 - abi + abi - b^2i^2 \\
&= a^2 + b^2
\end{aligned}
$$

A quotient of two complex numbers, written in fraction form, can be simplified to standard form $(a + bi)$ by multiplying the numerator and denominator by the complex conjugate of the denominator. Example 4 illustrates.

Notes on Examples

For students who have limited experience with complex numbers, producing the multiplicative inverse and writing a quotient of complex numbers in standard form are perhaps the only difficult processes.

EXAMPLE 4 Dividing Complex Numbers

Write the complex number in standard form.

a. $\dfrac{2}{3 - i}$ **b.** $\dfrac{5 + i}{2 - 3i}$

Solution

Multiply the numerator and denominator by the complex conjugate of the denominator.

a.
$$\frac{2}{3 - i} = \frac{2}{3 - i} \cdot \left(\frac{3 + i}{3 + i}\right) = \frac{2(3 + i)}{9 - i^2} = \frac{2(3 + i)}{10} = \frac{3}{5} + \frac{1}{5}i$$

b.
$$\frac{5 + i}{2 - 3i} = \frac{(5 + i)(2 + 3i)}{(2 - 3i)(2 + 3i)} = \frac{10 + 15i + 2i - 3}{4 + 9} = \frac{7 + 17i}{13} = \frac{7}{13} + \frac{17}{13}i$$

Complex Solutions of Quadratic Equations

REMARK

The quadratic equations we ask you to solve will have real number coefficients. This means you can use the quadratic formula. It turns out that quadratic equations with complex number coefficients also have two solutions, but study of such equations is beyond the scope of this book.

Consider the quadratic equation $ax^2 + bx + c = 0$, where a, b, and c are real numbers. In Section 2.2 you learned that such an equation has no real number solution whenever the discriminant $b^2 - 4ac$ is a negative real number. That is because the square root of a negative number, for example $\sqrt{-4}$, is not a real number.

On the other hand, $\sqrt{-4}$ *is* defined as a complex number: it is the number $2i$ because

$$\sqrt{-4} = \sqrt{|-4|}\,i = 2i.$$

So in the complex numbers a quadratic equation $ax^2 + bx + c = 0$ with real number coefficients always has two solutions. When the discriminant $b^2 - 4ac$ is positive, the roots are distinct real numbers. When the discriminant is zero, there is a real number double root. When the discriminant is negative, the two roots will be complex numbers with nonzero imaginary parts.

The quadratic formula can always be used to find the solutions of a quadratic equation with real number coefficients.

EXAMPLE 5 Finding Complex Roots

Find the solutions of $x^2 + x + 1 = 0$.

Teaching Note

The complex conjugate is a key idea in thinking about zeros of functions. Students should observe that nonreal roots of quadratic equations with real number coefficients always occur in complex conjugate pairs. In Example 5, lead students to this observation, but show the graph of the function so that students can see that the curve does not touch or cross the x-axis.

Solution

Solve Algebraically

Use the quadratic formula, where $a = b = c = 1$.

$$x = \frac{-1 \pm \sqrt{1 - 4}}{2}$$

$$= \frac{-1 \pm \sqrt{3}i}{2}$$

$$= -\frac{1}{2} \pm \frac{\sqrt{3}}{2}i$$

The solutions are $-\frac{1}{2} + \frac{\sqrt{3}}{2}i$ and $-\frac{1}{2} - \frac{\sqrt{3}}{2}i$.

Confirm Numerically

Substituting

$$x = -\frac{1}{2} + \frac{\sqrt{3}}{2}i \quad \text{and} \quad x = -\frac{1}{2} - \frac{\sqrt{3}}{2}i$$

we find that $x^2 + x + 1 = 0$ for both values of x. ■

Follow-up

Discuss the similarities and differences between the complex plane and the coordinate plane.

Assignment Guide

Day 1: Ex. 3–48, multiples of 3
Day 2: Ex. 50–55, 58–61, 63, 64

Cooperative Learning

Group Activity: Ex. 62

Notes on Exercises

Ex. 1–28 are straightforward computational exercises relating to powers of i and addition, subtraction, and multiplication of complex numbers.
Ex. 61–63 introduce students to a new concept: the modulus of a complex number.

Ongoing Assessment

Self-Assessment: Ex. 7, 11, 31, 47
Embedded Assessment: Ex. 50, 60

Complex Plane

Just as every real number is associated with a point of the real number line, every complex number is associated with a point of the **complex plane.** For example, the complex number $2 + 3i$ is associated with the point 2 units to the right and 3 units up from the origin. (See Figure 3.39.)

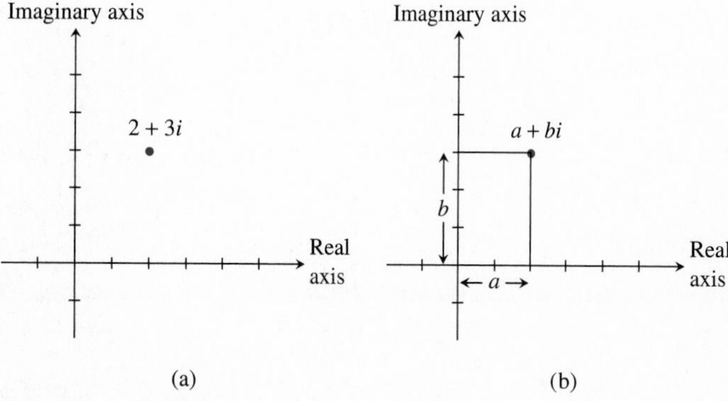

Figure 3.39 The complex plane.

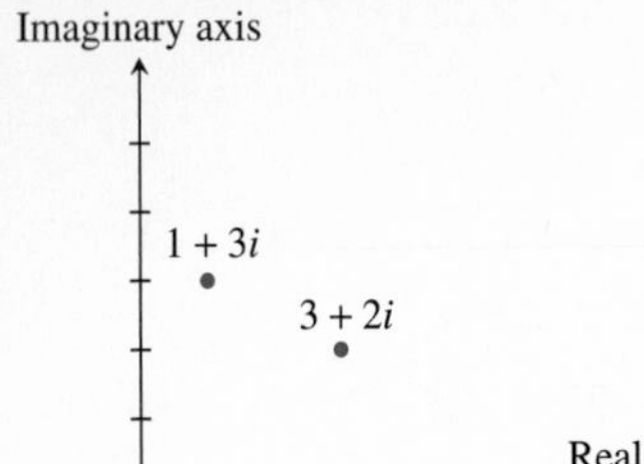

Figure 3.40

EXAMPLE 6 Plotting Complex Numbers

Plot $u = 1 + 3i$, $v = 2 - i$, and $u + v$ in the complex plane.

Solution

$u + v = 3 + 2i$. The numbers are plotted in Figure 3.40. Notice the shape (a parallelogram) suggested by the 3 points and the origin. ■

Quick Review 3.5

In Exercises 1–6, add or subtract, and simplify.

1. $(2x + 3) + (-x + 6)$
2. $(3y - x) + (2x - y)$
3. $(2a + 4d) - (a + 2d)$
4. $(6z - 1) - (z + 3)$
5. $(7x - 3y) - (x - 2y)$
6. $(-x + 3) - (3x + 1)$

In Exercises 7–14, multiply and simplify.

7. $(x - 3)(x + 2)$
8. $(2x - 1)(x + 3)$
9. $(x - \sqrt{2})(x + \sqrt{2})$
10. $(x + 2\sqrt{3})(x - 2\sqrt{3})$
11. $(x - 2y)^2$
12. $(x + 3y)^2$
13. $[x - (1 + \sqrt{2})][x - (1 - \sqrt{2})]$
14. $[x - (2 + \sqrt{3})][x - (2 - \sqrt{3})]$

SECTION EXERCISES 3.5

In Exercises 1–8, write the sum or difference in standard form $a + bi$.

1. $(2 - 3i) + (6 + 5i)$
2. $(2 - 3i) + (3 - 4i)$
3. $(7 - 3i) + (6 - i)$
4. $(2 + i) - (9i - 3)$
5. $(2 - i) + (3 - \sqrt{-3})$
6. $(\sqrt{5} - 3i) + (-2 + \sqrt{-9})$
7. $(i^2 + 3) - (7 + i^3)$
8. $(\sqrt{7} + i^2) - (6 - \sqrt{-81})$

In Exercises 9–16, write the product in standard form.

9. $(2 + 3i)(2 - i)$
10. $(2 - i)(1 + 3i)$
11. $(1 - 4i)(3 - 2i)$
12. $(5i - 3)(2i + 1)$
13. $(7i - 3)(2 + 6i)$
14. $(\sqrt{-4} + i)(6 - 5i)$
15. $(-3 - 4i)(1 + 2i)$
16. $(\sqrt{-2} + 2i)(6 + 5i)$

In Exercises 17–24, write the expression in an equivalent form where the exponent of i is the smallest-possible positive number.

17. i^3
18. i^4
19. i^5
20. i^9
21. i^{11}
22. i^{13}
23. i^{14}
24. i^{16}

In Exercises 25–28, write the expression in the form bi, where b is a real number.

25. $\sqrt{-16}$
26. $\sqrt{-25}$
27. $\sqrt{-3}$
28. $\sqrt{-5}$

In Exercises 29–32, find real numbers x and y that make the equation true.

29. $2 + 3i = x + yi$
30. $3 + yi = x - 7i$

31. $(5 - 2i) - 7 = x - (3 + yi)$

32. $(x + 6i) = (3 - i) + (4 - 2yi)$

In Exercises 33–36, write the expression in standard form.

33. $(3 + 2i)^2$

34. $(1 - i)^3$

35. $\left(\frac{\sqrt{2}}{2} + \frac{\sqrt{2}}{2}i\right)^4$

36. $\left(\frac{\sqrt{3}}{2} + \frac{1}{2}i\right)^3$

In Exercises 37–42, find the product of the complex number and its conjugate.

37. $2 - 3i$

38. $5 - 6i$

39. $-3 + 4i$

40. $-1 - \sqrt{2}i$

41. $i - \sqrt{3}$

42. $1 - (7 - 2i)$

In Exercises 43–50, write the expression in standard form.

43. $\dfrac{1}{2 + i}$

44. $\dfrac{i}{2 - i}$

45. $\dfrac{2 + i}{2 - i}$

46. $\dfrac{2 + i}{3i}$

47. $\dfrac{(2 + i)^2(-i)}{1 + i}$

48. $\dfrac{(2 - i)(1 + 2i)}{5 + 2i}$

49. $\dfrac{(1 - i)(2 - i)}{1 - 2i}$

50. $\dfrac{(1 - \sqrt{2}i)(1 + i)}{(1 + \sqrt{2}i)}$

In Exercises 51–54, plot the complex number in one complex plane.

51. $-3 + 4i$

52. $6i + 1$

53. $2i + (1 - 3i)$

54. $(1 + 2i) - (4 + 6i)$

In Exercises 55–58, find two (complex number) roots for the equation.

55. $x^2 + 2x + 5 = 0$

56. $3x^2 + x + 2 = 0$

57. $4x^2 - 6x + 5 = x + 1$

58. $x^2 + x + 11 = 5x - 8$

EXTENDING THE IDEAS

In Exercises 59–64, solve the problem.

59. Writing to Learn What characteristic does the graph of $f(x) = ax^2 + bx + c$ possess if the equation $ax^2 + bx + c = 0$ has (nonreal) complex number roots?

60. Prove that the difference between a complex number and its conjugate is a complex number whose real part is 0.

61. The *modulus* of a complex number $a + bi$, denoted by $\|a + bi\|$, is defined by:

$$\|a + bi\| = \sqrt{a^2 + b^2}$$

a. Graph $2 + 3i$ in the complex plane.
b. Calculate $\|2 + 3i\|$.
c. Explain how the modulus of $2 + 3i$ is related to the distance between $2 + 3i$ and the origin.

62. Investigate and Conjecture Suppose that $z_1 = a + bi$ and $z_2 = c + di$ are any two complex numbers. Which of the following do you think will always be true? (See Exercise 61 for a definition of $\|z\|$.)

a. $\|z_1 + z_2\| = \|z_1\| + \|z_2\|$
b. $\|z_1 + z_2\| \le \|z_1\| + \|z_2\|$
c. $\|z_1 + z_2\| \ge \|z_1\| + \|z_2\|$

63. Investigate and Conjecture Which of the following do you think will always be true? (See Exercise 61 for a definition of $\|z\|$.)

a. $\|z_1 \cdot z_2\| = \|z_1\| \cdot \|z_2\|$
b. $\|z_1 \cdot z_2\| < \|z_1\| \cdot \|z_2\|$
c. $\|z_1 \cdot z_2\| > \|z_1\| \cdot \|z_2\|$

64. Show that $-(a - bi) = -a + bi$.

3.6 FUNDAMENTAL THEOREM OF ALGEBRA

Objective

Students will be able to factor polynomials with real coefficients.

Motivate

Ask . . .

If a polynomial can be factored as a product of n linear factors, what is the degree of the polynomial? (***n***)

Fundamental Theorem of Algebra and the Linear Factorization Theorem • Complex Conjugate Zeros • Finding Linear Factors • Factoring with Real Number Coefficients • Zeros of Odd-Degree Polynomial Functions • Applications

Fundamental Theorem of Algebra and the Linear Factorization Theorem

We have seen that a degree n polynomial can have from 0 to n real zeros. We have also seen that a polynomial can have nonreal complex number zeros. The

REMARK

The degree 3 polynomial $f(x) = (x - 2)^3$ has the zero $x = 2$ repeated 3 times. A zero is repeated if the corresponding factor is repeated.

Fundamental Theorem of Algebra tells us that a degree n polynomial has n complex zeros (real and nonreal).

Theorem: Fundamental Theorem of Algebra

A polynomial function of degree $n > 0$ has n complex zeros. These zeros may be repeated.

Teaching Note

The Fundamental Theorem of Algebra is very important. It says that another type of number beyond the complex numbers is not needed to solve a polynomial equation. Make an issue about complex and real zeros and the relation of the graph to the x-axis.

The Factor Theorem in Section 3.3 extends to the complex zeros of a polynomial. Thus, if z_1 is a zero of a polynomial, then $x - z_1$ is a factor of the polynomial even if z_1 is nonreal complex. We combine this fact with the Fundamental Theorem to get the following theorem.

Theorem: Linear Factorization Theorem

If $f(x)$ is a polynomial of degree $n > 0$, then $f(x)$ has precisely n linear factors and

$$f(x) = a(x - z_1)(x - z_2) \cdots (x - z_n),$$

where a is the leading coefficient of $f(x)$ and $z_1, z_2, \ldots, z_n$ are complex numbers that are zeros of $f(x)$. The z_i are not necessarily distinct.

The **Linear Factorization Theorem** is an *existence* theorem. It tells us of the existence of linear factors (and zeros), but it does not tell us how to find them. This theorem does suggest, however, how to find a polynomial in standard form whose zeros match a given set of numbers.

When a factor is repeated, we obtain a polynomial with **repeated zeros.** For example, the polynomial

$$\begin{aligned} f(x) &= (x - 2)(x - 2)(x - 2)(x + 1)(x + 1) \\ &= (x - 2)^3(x + 1)^2 \end{aligned}$$

has the factor $x - 2$ occurring 3 times and the factor $x + 1$ occurring twice. We say that 2 is a zero of f of **multiplicity** 3 and -1 is a zero of f of multiplicity 2.

10

-4 4

-10

Figure 3.41 $f(x) = (x - 2)^3(x + 1)^2$.

The graph of f in Figure 3.41 is tangent to the x-axis at $x = -1$ and crosses the x-axis at $x = 2$. In general, a graph of a polynomial is tangent to the x-axis at a zero of even multiplicity and crosses at a zero of odd multiplicity. The more times a factor is repeated, the flatter the graph becomes in the vicinity of the zero.

EXAMPLE 1 Relating Polynomials, Linear Factors, and Zeros

a. The quadratic polynomial function

$$f(x) = (x - 2i)(x + 2i) = x^2 + 4$$

has exactly *two* zeros: $x = 2i$ and $x = -2i$.

b. The cubic polynomial function

$$f(x) = (x - 5)(x - \sqrt{2}i)(x + \sqrt{2}i) = x^3 - 5x^2 + 2x - 10$$

has exactly *three* zeros: $x = 5$, $x = \sqrt{2}i$, and $x = -\sqrt{2}i$.

c. Counting multiplicity, the degree 4 polynomial

$$\begin{aligned} f(x) &= (x - 3)(x - 3)(x - i)(x + i) \\ &= x^4 - 6x^3 + 10x^2 - 6x + 9 \end{aligned}$$

has exactly *four* zeros: $x = 3$, $x = 3$, $x = i$, and $x = -i$. ■

Complex Conjugate Zeros

Recall that the expression $b^2 - 4ac$ is part of the quadratic formula, namely the discriminant. For quadratics $ax^2 + bx + c = 0$ with real coefficients, if the discriminant $b^2 - 4ac$ is negative, the solutions are nonreal complex numbers and the $\pm$ signs in the quadratic formula ensure that these complex zeros occur in conjugate pairs.

Complex Conjugate Zeros

Suppose that $f(x)$ is a polynomial function with real coefficients. If $a + bi$ is a zero of $f(x)$, then its complex conjugate $a - bi$ is also a zero of $f(x)$.

Knowing that nonreal complex zeros occur in pairs is important when trying to find a polynomial with real coefficients for a given set of zeros.

EXAMPLE 2 Finding a Polynomial with Given Zeros

Write the standard form of a degree 5 polynomial function $f(x)$ with real coefficients whose zeros include $x = 1$, $x = 1 + 2i$, $x = 1 - i$.

Solution

Because $1 + 2i$ is a zero, $1 - 2i$ must also be a zero. Therefore, $x - (1 + 2i)$ and $x - (1 - 2i)$ must both be factors of $f(x)$. Likewise, because $1 - i$ is a zero, $1 + i$ must also be a zero. It follows that $x - (1 + i)$ and $x - (1 - i)$

must both be factors of $f(x)$. Therefore,

$$\begin{aligned} f(x) &= (x-1)[x-(1+2i)][x-(1-2i)][x-(1+i)][x-(1-i)] \\ &= (x-1)(x^2-2x+5)(x^2-2x+2) \\ &= (x^3-3x^2+7x-5)(x^2-2x+2) \\ &= x^5-5x^4+15x^3-25x^2+24x-10 \end{aligned}$$

is a polynomial of the type we seek. Any nonzero real number multiple of $f(x)$ will also be such a polynomial. ■

Teaching Note

Students may not be used to seeing a concept expressed as "The following statements . . . are equivalent." Note that this means that any one of the three statements implies the other two, and conversely that if one of them is false, then the other two must be false.

Finding Linear Factors

The following statements about polynomials $f(x)$ are equivalent even if a is a nonreal complex number.

1. $x - a$ is a factor of $f(x)$.
2. $x = a$ is a zero of $f(x)$.
3. $x = a$ is a solution or a root of $f(x) = 0$.

We use these connections together with the quadratic formula, synthetic division, and the rational zeros test to search for linear factors of a polynomial.

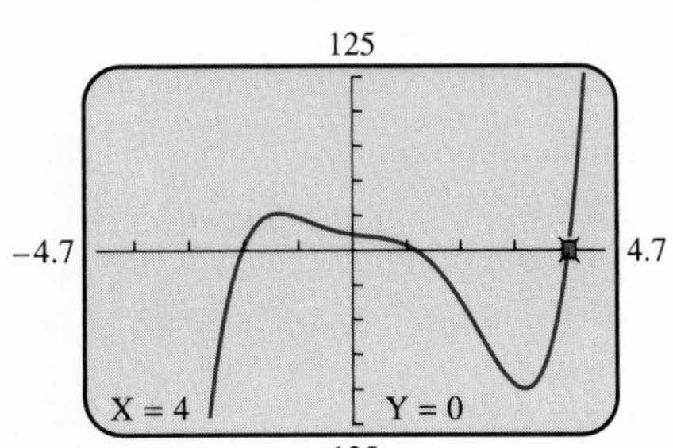

Figure 3.42 $f(x) = x^5 - 3x^4 - 5x^3 + 5x^2 - 6x + 8$.

■ EXAMPLE 3 Finding Linear Factors

Find all zeros of $f(x) = x^5 - 3x^4 - 5x^3 + 5x^2 - 6x + 8$, and write $f(x)$ in its linear factorization.

Solution

Figure 3.42 suggests that the real zeros of f are $x = -2$, $x = 1$, and $x = 4$.

Using synthetic division we can verify these zeros and show that $x^2 + 1$ is also a factor of f. So $\pm i$ are also zeros. Therefore,

$$\begin{aligned} f(x) &= x^5 - 3x^4 - 5x^3 + 5x^2 - 6x + 8 \\ &= (x+2)(x-1)(x-4)(x^2+1) \\ &= (x+2)(x-1)(x-4)(x-i)(x+i). \end{aligned}$$ ■

Synthetic division can be used with complex number divisors in the same way it is used with real number divisors.

■ EXAMPLE 4 Finding Complex Zeros

The complex number $z = 1 - 2i$ is a zero of $f(x) = 4x^4 + 17x^2 + 14x + 65$. Find the remaining zeros of $f(x)$, and write it in its linear factorization.

Solution

We use synthetic division to show that $f(1 - 2i) = 0$:

$$\begin{array}{r|rrrrr} 1-2i & 4 & 0 & 17 & 14 & 65 \\ & & 4-8i & -12-16i & -27-26i & -65 \\ \hline & 4 & 4-8i & 5-16i & -13-26i & 0 \end{array}$$

Thus $1 - 2i$ is a zero of $f(x)$. The complex conjugate $1 + 2i$ must also be a zero. We use synthetic division on the quotient from above to find the remaining quadratic factor:

$$\begin{array}{r|rrrr} 1+2i & 4 & 4-8i & 5-16i & -13-26i \\ & & 4+8i & 8+16i & 13+26i \\ \hline & 4 & 8 & 13 & 0 \end{array}$$

Finally, we use the quadratic formula to find the two zeros of $4x^2 + 8x + 13$:

$$\begin{aligned} x &= \frac{-8 \pm \sqrt{64 - 208}}{8} \\ &= \frac{-8 \pm \sqrt{-144}}{8} \\ &= \frac{-8 \pm 12i}{8} \\ &= -1 \pm \tfrac{3}{2}i. \end{aligned}$$

Thus the four zeros of $f(x)$ are $1 - 2i$, $1 + 2i$, $-1 + \frac{3}{2}i$, and $-1 - \frac{3}{2}i$. Because the leading coefficient of $f(x)$ is 4, we obtain

$$\begin{aligned} f(x) = {} & 4[x - (1 - 2i)][x - (1 + 2i)] \times \\ & \left[x - \left(-1 + \frac{3}{2}i\right)\right]\left[x - \left(-1 - \frac{3}{2}i\right)\right]. \end{aligned}$$

If we want to remove fractions in the factors, we can distribute the 4 to get

$$\begin{aligned} f(x) = {} & [x - (1 - 2i)][x - (1 + 2i)] \times \\ & [2x - (-2 + 3i)][2x - (-2 - 3i)]. \end{aligned}$$

■

Alert

Students may not recognize the difference between the factoring in this section and the factoring they have done in the past. Emphasize that in this section, the factors are not required to have integer coefficients.

Factoring with Real Number Coefficients

Let $f(x)$ be a polynomial with real coefficients. The Linear Factorization Theorem tells us that $f(x)$ can be factored into the form

$$f(x) = a(x - z_1)(x - z_2) \cdots (x - z_n)$$

where the z_i may be (nonreal) complex numbers. Recall, however, that nonreal

Follow-up

Ask students if they can think of another way to justify the statement that an odd-degree polynomial function with real coefficients has at least one real zero. Have them think about end behavior and continuity.

Assignment Guide

Day 1: Ex. 3–42, multiples of 3, 43, 44, 45–48
Day 2: Ex. 50, 51, 54, 56, 57, 58, 61, 62, 63

Cooperative Learning

Group Activity: Ex. 55–57

Notes on Exercises

Ex. 1–12 and 49–50 require that students produce a polynomial with desired characteristics.
Ex. 56–57 require students to use cubic or quadratic regression.
Ex. 58–61 use polynomials with complex coefficients. The nonreal zeros of these polynomials need not occur in complex conjugate pairs.

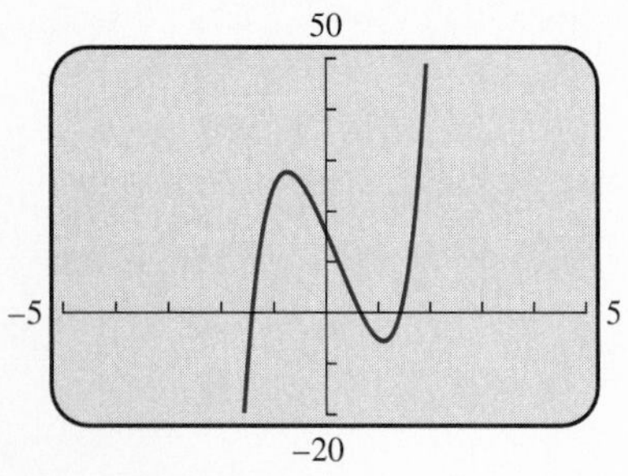

Figure 3.43 $f(x) = 3x^5 - 2x^4 + 6x^3 - 4x^2 - 24x + 16$.

complex number zeros occur in conjugate pairs. The product of $x - (a + bi)$ and $x - (a - bi)$ is

$$\begin{aligned}[x-(a+bi)][x-(a-bi)] &= x^2-(a-bi)x-(a+bi)x+(a+bi)(a-bi)\\ &= x^2-ax+bix-ax-bix+(a^2-b^2i^2)\\ &= x^2-2ax+(a^2+b^2).\end{aligned}$$

The quadratic expression $x^2 - 2ax + (a^2 + b^2)$ is a factor of $f(x)$, and its coefficients are real numbers. A quadratic with no real zeros is **irreducible over the reals.** In other words, if we require that the factors of a polynomial have real coefficients, the factorization can be accomplished with linear factors and irreducible quadratic factors.

> **Factors of a Polynomial with Real Coefficients**
>
> Every polynomial with real coefficients can be written as a product of linear factors and irreducible quadratic factors, each with real coefficients.

■ EXAMPLE 5 Factoring a Polynomial

Write $f(x) = 3x^5 - 2x^4 + 6x^3 - 4x^2 - 24x + 16$ as a product of linear and quadratic factors, each with real coefficients.

Solution

The rational zeros test provides a list of candidates for the rational zeros of f. The graph of f in Figure 3.43 cuts the list down considerably. Using synthetic division on the remaining candidates, we find that $x = \frac{2}{3}$ is a zero. Thus,

$$\begin{aligned} f(x) &= \left(x - \frac{2}{3}\right)(3x^4 + 6x^2 - 24)\\ &= \left(x - \frac{2}{3}\right)(3)(x^4 + 2x^2 - 8)\\ &= (3x-2)(x^2-2)(x^2+4) \quad \text{Factor } x^4+2x^2-8 \text{ as a quadratic in } x^2.\\ &= (3x-2)(x-\sqrt{2})(x+\sqrt{2})(x^2+4).\end{aligned}$$

Because the zeros of $x^2 + 4$ are complex, any further factorization would introduce nonreal complex coefficients. We have taken the factorization of f as far as possible, subject to the condition that *each factor has real coefficients.* ■

Zeros of Odd-Degree Polynomial Functions

We have seen that if a polynomial function has real coefficients, then its nonreal complex zeros occur in conjugate pairs. *Because a polynomial of odd degree has an odd number of zeros, it must have one zero that is real.* This confirms the same information we found in Example 6 of Section 2.2 using the end behavior of polynomials of odd degree.

Polynomial Function of Odd Degree

An odd-degree polynomial function with real coefficients has at least one real zero.

Ongoing Assessment

Self-Assessment: Ex. 11, 19, 25, 35, 49
Embedded Assessment: Ex. 46, 48

The function $f(x) = 3x^5 - 2x^4 + 6x^3 - 4x^2 - 24x + 16$ in Example 5 fits the condition of this theorem, so we know right away that we are on the right track in searching for at least one real zero.

Applications

TECHNOLOGY NOTE

Motion detectors can be connected to a **CBL** (calculator-based laboratory) that was introduced at the end of Section 1.4.

Computers and graphing calculators can be connected to *motion detectors* that record the distance a person is from a sensor as he or she walks back and forth along a straight line.

■ EXAMPLE 6 Using a Motion Detector

Lewis walks back and forth along a straight path. His distance from a sensor is given by the graph in Figure 3.44. Describe his motion.

Solution

The graph gives the distance D that Lewis is from the sensor in meters as a function of time t in seconds. Lewis is moving away from the sensor when the graph is rising and moving toward the sensor when the graph is falling. Initially Lewis is moving away, then towards, and then away again. We can estimate that Lewis is moving away from the motion detector for $0 \le t \le 2.5$ and $5 \le t \le 8$, and moving toward for $2.5 \le t \le 5$. ■

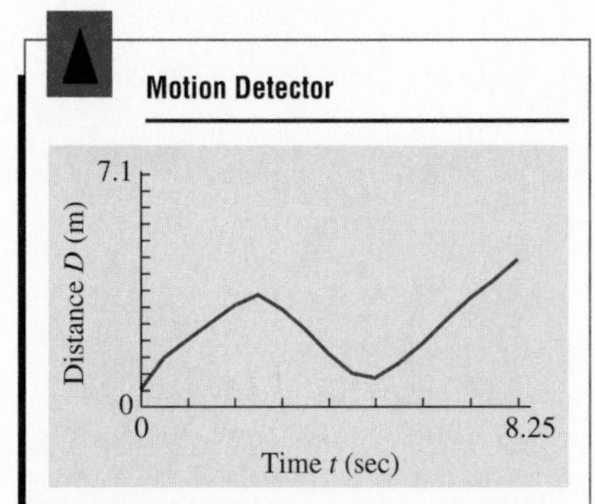

Figure 3.44

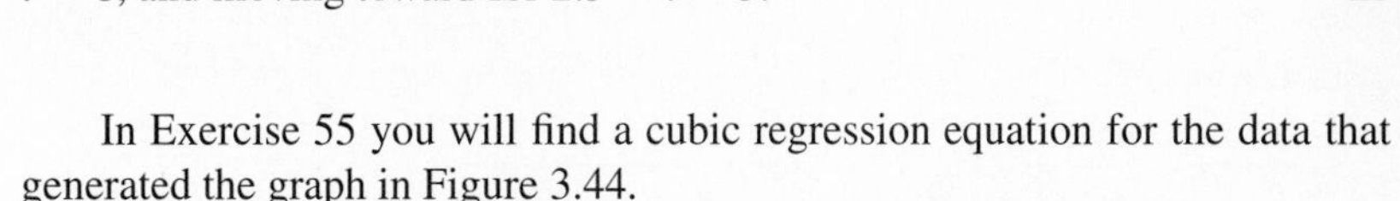

In Exercise 55 you will find a cubic regression equation for the data that generated the graph in Figure 3.44.

Quick Review 3.6

In Exercises 1–4, factor the quadratic expression.

1. $2x^2 - x - 3$

2. $6x^2 - 13x - 5$

3. $10x^2 - 3x - 4$

4. $9x^2 - 6x - 8$

In Exercises 5–8, list all potential rational zeros.

5. $3x^4 - 5x^3 + 3x^2 - 7x + 2$

6. $4x^5 - 7x^2 + x^3 + 13x - 3$

7. $5x^3 + 4x^2 + 3x - 4$

8. $12x^4 + 8x^3 - 7x + 1$

SECTION EXERCISES 3.6

In Exercises 1–8, write a polynomial function with real coefficients in standard form whose zeros are given.

1. i and $-i$

2. $1 - 2i$ and $1 + 2i$

3. 1, $3i$, and $-3i$

4. -4, $1 - i$, and $1 + i$

5. 2, 3, and i

6. -1, 2, and $1 - i$

7. 5 and $3 + 2i$

8. -2 and $1 + 2i$

In Exercises 9–12, write a polynomial function in standard form with the given zeros and their multiplicities.

9. 1 (multiplicity 2), -2 (multiplicity 3)

10. -1 (multiplicity 3), 3 (multiplicity 1)

11. 2 (multiplicity 2), 3 (multiplicity 2)

12. -1 (multiplicity 2), -2 (multiplicity 2)

In Exercises 13–16, for the given zeros and multiplicities, select the matching graph.

13. -1 (multiplicity 2), 3 (multiplicity 1)

14. -1 (multiplicity 1), 3 (multiplicity 2)

15. -3 (multiplicity 2), 1 (multiplicity 1)

16. -3 (multiplicity 1), 1 (multiplicity 2)

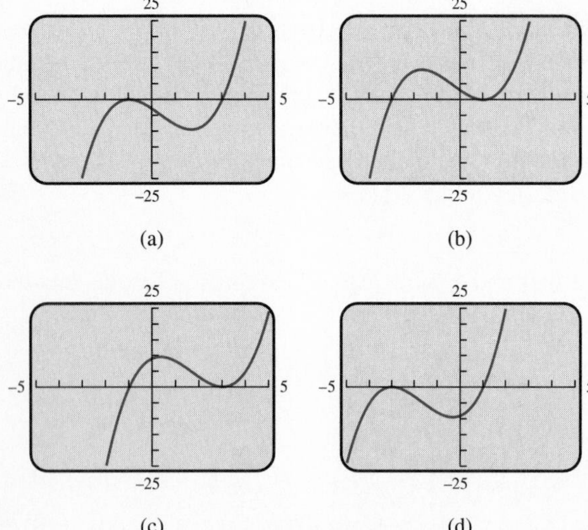

For Exercises 13–16

In Exercises 17–20, for the given zeros and multiplicities, select the matching graph.

17. -3 (multiplicity 2), 2 (multiplicity 3)

18. -3 (multiplicity 3), 2 (multiplicity 2)

19. -1 (multiplicity 4), 3 (multiplicity 3)

20. -1 (multiplicity 3), 3 (multiplicity 4)

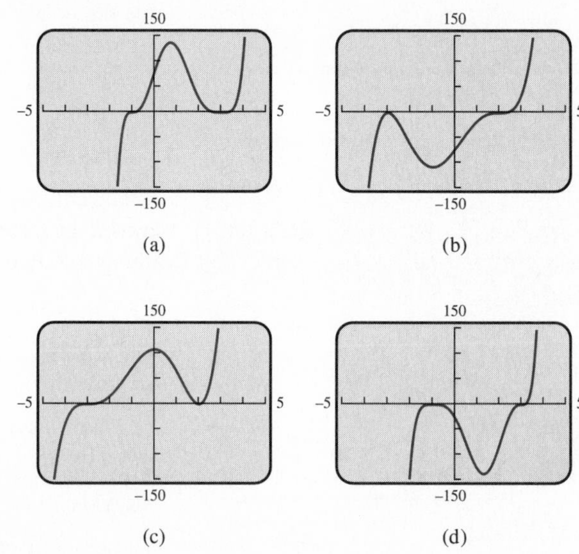

For Exercises 17–20

In Exercises 21–26, find the number of complex zeros that each function has. State how many of these zeros are also real numbers.

21. $f(x) = x^2 - 2x + 7$

22. $f(x) = x^3 - 3x^2 + x + 1$

23. $f(x) = x^3 - x + 3$

24. $f(x) = x^4 - 2x^2 + 3x - 4$

25. $f(x) = x^4 - 5x^3 + x^2 - 3x + 6$

26. $f(x) = x^5 - 2x^2 - 3x + 6$

In Exercises 27–32, find all zeros of $f(x)$, and write a linear factorization of $f(x)$.

27. $f(x) = x^3 + 4x - 5$

28. $f(x) = x^3 - 10x^2 + 44x - 69$

29. $f(x) = x^4 + x^3 + 5x^2 - x - 6$

30. $f(x) = 3x^4 + 8x^3 + 6x^2 + 3x - 2$

31. $f(x) = 6x^4 - 7x^3 - x^2 + 67x - 105$

32. $f(x) = 20x^4 - 148x^3 + 269x^2 - 106x - 195$

In Exercises 33–36, find all zeros of $f(x)$, and write a linear factorization of $f(x)$.

33. $1 + i$ is a zero of $f(x) = x^4 - 2x^3 - x^2 + 6x - 6$.

34. $4i$ is a zero of $f(x) = x^4 + 13x^2 - 48$.

35. $3 - 2i$ is a zero of $f(x) = x^4 - 6x^3 + 11x^2 + 12x - 26$.

36. $1 + 3i$ is a zero of $f(x) = x^4 - 2x^3 + 5x^2 + 10x - 50$.

In Exercises 37–42, write the function as a product of linear and quadratic factors each with real coefficients. Find as many linear factors as possible.

37. $f(x) = x^3 - x^2 - x - 2$

38. $f(x) = x^3 - x^2 + x - 6$

39. $f(x) = 2x^3 - x^2 + 2x - 3$

40. $f(x) = 3x^3 - 2x^2 + x - 2$

41. $f(x) = x^4 + 3x^3 - 3x^2 + 3x - 4$

42. $f(x) = x^4 - 2x^3 + x^2 - 8x - 12$

In Exercises 43 and 44, use *Archimedes' principle,* which states that when a sphere of radius r with density d_S is placed in a liquid of density $d_L = 62.5$ lb/ft^3, it will sink to a depth h where

$$\frac{\pi}{3}(3rh^2 - h^3)d_L = \frac{4}{3}\pi r^3 d_S.$$

Find an approximate value for h if:

43. $r = 5$ ft and $d_S = 20$ lb/ft^3.

44. $r = 5$ ft and $d_S = 45$ lb/ft^3.

In Exercises 45–48, answer yes or no. If yes, include an example. If no, give a reason.

45. Writing to Learn Is it possible to find a polynomial of degree 3 with real number coefficients that has -2 as its only real number zero?

46. Writing to Learn Is it possible to find a polynomial of degree 3 with real number coefficients that has $2i$ as its only nonreal zero?

47. Writing to Learn Is it possible to find a polynomial $f(x)$ of degree 4 with real number coefficients that has zeros -3, $1 + 2i$, and $1 - i$?

48. Writing to Learn Is it possible to find a polynomial $f(x)$ of degree 4 with real number coefficients that has zeros $1 + 3i$, and $1 - i$?

In Exercises 49 and 50, find the unique polynomial with real coefficients that meets these conditions.

49. Degree 4; zeros at $x = 3$, $x = -1$, and $x = 2 - i$; $f(0) = 30$

50. Degree 4; zeros at $x = 1 - 2i$ and $x = 1 + i$; $f(0) = 20$

In Exercises 51–54, match the directed distance-time graph with the description of motion on the x-axis. The distance is from the origin.

51. Dave walks to the right, stops, and then walks to the left.

52. Margaret walks to the left, stops, and then walks to the right.

53. Janet walks to the right, stops, walks to the left, stops, and then walks to the right.

54. Matt walks to the left, stops, walks to the right, stops, and then walks to the left.

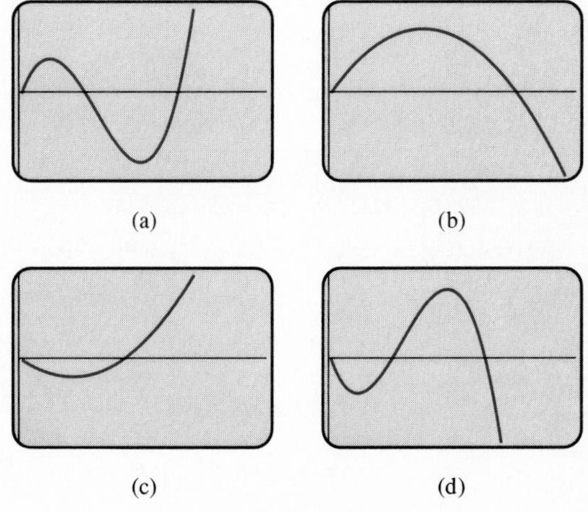

For Exercises 51–54

In Exercises 55–57, use the data in Table 3.8, which were collected with a motion detector.

55. The data in Table 3.8 were used to produce Figure 3.44 of Example 6.

Table 3.8 Motion Detector Data for Exercise 55

t (sec)	D (m)	t (sec)	D (m)
0.0	1.00	4.5	0.99
0.5	1.46	5.0	0.84
1.0	1.99	5.5	1.28
1.5	2.57	6.0	1.87
2.0	3.02	6.5	2.58
2.5	3.34	7.0	3.23
3.0	2.91	7.5	3.78
3.5	2.31	8.0	4.40
4.0	1.57		

a. Find a cubic regression equation, and graph it together with a scatter plot of the data.
b. Use the cubic regression equation to estimate how far Lewis is from the motion detector initially.
c. Use the cubic regression equation to estimate when Lewis changes direction. How far from the motion detector is he when he changes direction?

56. Sally's distance D from a motion detector is given by the data in Table 3.9.

Table 3.9 Motion Detector Data for Exercise 56

t (sec)	D (m)	t (sec)	D (m)
0.0	3.36	4.5	3.59
0.5	2.61	5.0	4.15
1.0	1.86	5.5	3.99
1.5	1.27	6.0	3.37
2.0	0.91	6.5	2.58
2.5	1.14	7.0	1.93
3.0	1.69	7.5	1.25
3.5	2.37	8.0	0.67
4.0	3.01		

a. Find a cubic regression equation, and graph it together with a scatter plot of the data.
b. Describe Sally's motion.
c. Use the cubic regression equation to estimate when Sally changes direction. How far is she from the motion detector when she changes direction?

57. Jacob's distance D from a motion detector is given by the data in Table 3.10.

Table 3.10 Motion Detector Data for Exercise 57

t (sec)	D (m)	t (sec)	D (m)
0.0	4.59	4.5	1.70
0.5	3.92	5.0	2.25
1.0	3.14	5.5	2.84
1.5	2.41	6.0	3.39
2.0	1.73	6.5	4.02
2.5	1.21	7.0	4.54
3.0	0.90	7.5	5.04
3.5	0.99	8.0	5.59
4.0	1.31		

a. Find a quadratic regression equation, and graph it together with a scatter plot of the data.
b. Describe Jacob's motion.
c. Use the quadratic regression equation to estimate when Jacob changes direction. How far is he from the motion detector when he changes direction?

EXTENDING THE IDEAS

In Exercises 58–61, use polynomials whose coefficients are complex numbers.

58. Verify that the complex number i is a zero of the polynomial $f(x) = x^3 - ix^2 + 2ix + 2$.

59. Verify that the complex number $-2i$ is a zero of the polynomial $f(x) = x^3 - (2 - i)x^2 + (2 - 2i)x - 4$.

In Exercises 60 and 61, verify that $g(x)$ is a factor of $f(x)$. Then find $h(x)$ so that $f = gh$.

60. $g(x) = x - i;\ f(x) = x^3 + (3 - i)x^2 - 4ix - 1$

61. $g(x) = x - 1 - i;\ f(x) = x^3 - (1 + i)x^2 + x - 1 - i$

62. **Writing to Learn** Describe a walk along a straight path that will cause a motion detector to produce a graph that appears to be a parabola that opens upward. And again for one that opens downward. Explain.

63. **Writing to Learn** Describe a walk along a straight path that will cause a motion detector to produce a graph that appears to be a straight line with positive slope. And a straight line with negative slope. Explain.

CHAPTER 3 REVIEW

KEY TERMS

The number following each key term indicates the page of its introduction.

additive identity for the complex numbers, 264
additive inverse of a complex number, 264
complex conjugate, 265
complex number, 263
complex plane, 267
continuous function, 224
cubic, 224
degree of a polynomial function, 213
Descartes' rule of signs, 250
division algorithm for polynomials, 238
end behavior, 226
equal complex numbers, 263
Factor Theorem, 242
Fundamental Theorem of Algebra, 270
higher-degree polynomial function, 224
imaginary part of a complex number, 263
imaginary unit, 263
Intermediate Value Theorem, 230
irrational zeros, 254
irreducible over the reals, 274
leading coefficient, 213
leading term, 213
limit, 226
Linear Factorization Theorem, 270
local extremum, 225
lower bound, 251
lower bound test for real zeros, 253
multiplicity, 270
parabola, 214
parabola opening downward, 214
parabola opening upward, 214
polynomial end behavior, 228
polynomial function, 213
polynomial regression, 244
quadratic function, 214
quadratic regression, 218
quartic, 224
rational zeros, 254
rational zeros test, 254
real part of a complex number, 263
Remainder Theorem, 242
repeated zeros, 270
standard form of a complex number, 263
standard form equation of a quadratic function, 214
synthetic division, 240
upper bound, 251
upper bound test for real zeros, 251
vertex of a parabola, 214

REVIEW EXERCISES

In Exercises 1 and 2, use transformations to explain how the graphs of the given function and $y = x^2$ compare. Sketch each graph.

1. $f(x) = 3(x - 2)^2 + 4$
2. $g(x) = -(x + 3)^2 + 1$

In Exercises 3 and 4, identify the vertex and the line of symmetry of the graph. Support your answer with a grapher.

3. $f(x) = -2(x + 3)^2 + 5$
4. $g(x) = 4(x - 5)^2 - 7$

In Exercises 5–8, identify the vertex and the line of symmetry of the graph algebraically. Support your answer graphically.

5. $f(x) = -2x^2 - 16x - 31$
6. $g(x) = 3x^2 - 6x + 2$
7. $h(x) = 3x^2 - 5x - 2$
8. $k(x) = 5x^2 + 3x - 1$

In Exercises 9 and 10, find an equation for the parabola with the given vertex that passes through the given point.

9. Vertex $(-2, -3)$, through the point $(1, 2)$
10. Vertex $(-1, 1)$, through the point $(3, -2)$

In Exercises 11 and 12, find an equation for the parabola shown in the graph. One of the given points is the vertex of the parabola.

11.

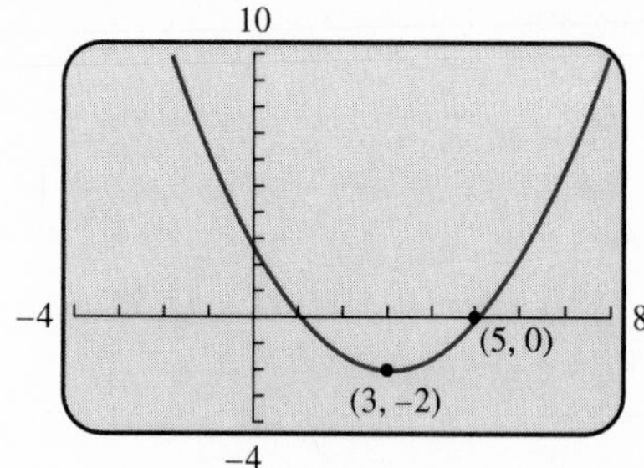

12.

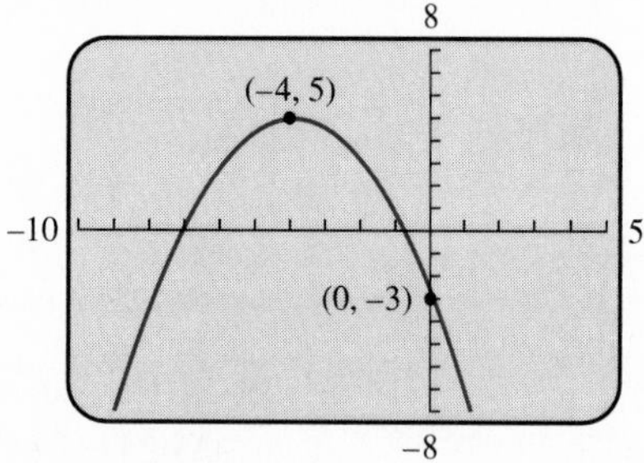

In Exercises 13–18, graph the function. Choose a viewing window that shows all local maximum and minimum values and all x-intercepts. Indicate the viewing-window dimensions.

13. $y = x^2 + 3x - 40$

14. $y = -8x^2 + 16x - 19$

15. $y = x^3 + x^2 + x + 5$

16. $y = x^3 - x^2 - 20x - 2$

17. $y = x^5 + 15x^4 + 87x^3 + 243x^2 + 326x + 156$

18. $y = x^4 - 20x^3 + 137x^2 - 370x + 351$

In Exercises 19–22, describe the end behavior of the function.

19. $f(x) = 12 + 3x - 5x^2$

20. $f(x) = 3x^3 - 2x^2 + 5x - 9$

21. $f(x) = -x^4 - 2x^3 + 19x^2 + 8x - 60$

22. $f(x) = -x^5 + 2x^4 + x^3 - 2x^2 + 72x - 144$

In Exercises 23 and 24, divide by long division. Summarize the result by writing a fraction equation. Support your answer graphically.

23. $\dfrac{2x^4 - 3x^3 + 9x^2 - 14x + 7}{x^2 + 4}$

24. $\dfrac{3x^4 - 5x^3 - 2x^2 + 3x - 6}{3x + 1}$

In Exercises 25 and 26, use synthetic division to divide $f(x)$ by $h(x)$. Find the quotient $q(x)$ and remainder $r(x)$, and write the equation $f(x) = h(x)q(x) + r(x)$.

25. $f(x) = 2x^3 - 7x^2 + 4x - 5;\ h(x) = x - 3$

26. $f(x) = x^4 + 3x^3 + x^2 - 3x + 3;\ h(x) = x + 2$

In Exercises 27 and 28, use the Remainder Theorem to find the remainder when $f(x)$ is divided by $x - c$.

27. $f(x) = 3x^3 - 2x^2 + x - 5;\ c = -2$

28. $f(x) = -x^2 + 4x - 5;\ c = 3$

In Exercises 29 and 30, use synthetic division and the Remainder Theorem.

29. Find $f(-1)$ for $f(x) = 2x^3 - x^2 + 2x - 5$

30. Find $f(4)$ for $f(x) = x^4 - x^3 + 5x^2 - x - 7$

In Exercises 31 and 32, use the Factor Theorem to find whether the first polynomial is a factor of the second polynomial.

31. $x - 2;\ x^3 - 4x^2 + 8x - 8$

32. $x + 3;\ x^3 + 2x^2 - 4x - 2$

In Exercises 33 and 34, use a graph of the function to help suggest one linear factor. Confirm the factor algebraically.

33. $f(x) = 2x^3 + x^2 - 2x - 6$

34. $f(x) = x^4 - 2x^3 - 11x^2 + 12x + 36$

In Exercises 35 and 36, use Descartes' rule of signs to determine the possible number of positive and negative zeros.

35. $f(x) = x^5 - x^4 + 2x^3 + x^2 - x - 1$

36. $g(x) = 2x^4 - 3x^2 + 5x + 1$

In Exercises 37 and 38, use synthetic division to show that the given number is an upper bound for the zeros of f.

37. 5; $f(x) = x^3 - 5x^2 + 3x + 4$

38. 4; $f(x) = -4x^4 + 16x^3 - 8x^2 - 16x + 12$

In Exercises 39 and 40, use synthetic division to show that the given number is a lower bound for the zeros of f.

39. -3; $f(x) = -4x^4 - 4x^3 + 15x^2 + 17x + 2$

40. -3; $f(x) = 2x^3 + 6x^2 + x - 6$

In Exercises 41 and 42, use the rational zeros test to find a list of all potential rational zeros. Then determine which ones, if any, are zeros.

41. $f(x) = 2x^4 - x^3 - 4x^2 - x - 6$

42. $f(x) = 6x^3 - 20x^2 + 11x + 7$

In Exercises 43–52, perform the indicated operation and write your answer in the standard form $a + bi$.

43. $(3 - 2i) + (-2 + 5i)$

44. $(5 - 7i) - (3 - 2i)$

45. $(1 + \sqrt{2}i)(1 - \sqrt{2}i)$

46. $(1 + 2i)(3 - 2i)$

47. $(1 + i)^3$

48. $(1 + 2i)^2(1 - 2i)^2$

49. i^{29}

50. $\sqrt{-16}$

51. $\dfrac{2 + 3i}{1 - 5i}$

52. $\dfrac{3}{2 - i}$

In Exercises 53 and 54, find two complex number solutions of the equation.

53. $x^2 - 6x + 13 = 0$

54. $x^2 - 2x + 4 = 0$

In Exercises 55–58, match a graph to the function.

55. $f(x) = (x - 2)^2$

56. $f(x) = (x - 2)^3$

57. $f(x) = (x - 2)^4$

58. $f(x) = (x - 2)^5$

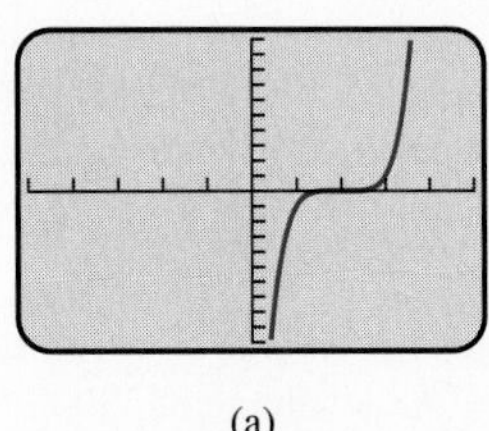

(a)

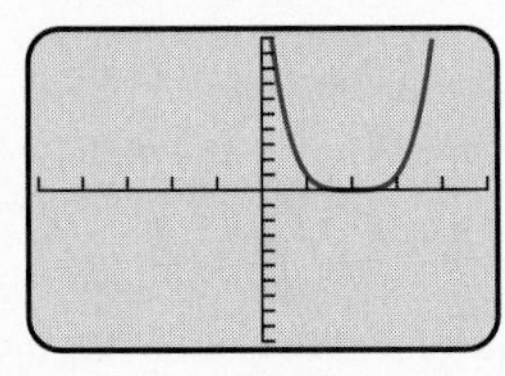

(b)

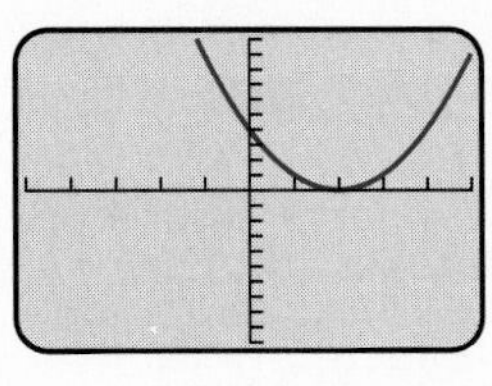

(c)

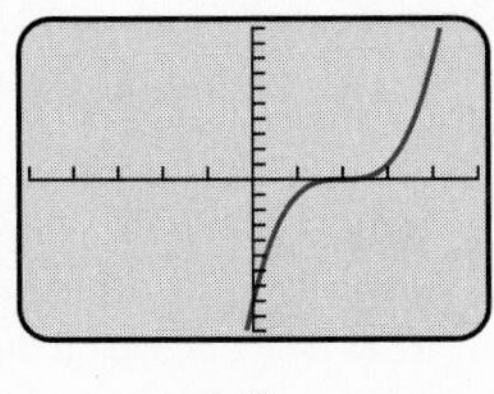

(d)

For Exercises 55–58

In Exercises 59–66, find all real zeros of the function. Identify each as rational or irrational. State the number of nonreal complex zeros.

59. $f(x) = x^4 - 10x^3 + 23x^2$

60. $g(x) = x^2 - 14x + 46$

61. $h(t) = t^3 + 2t - 1$

62. $k(t) = t^4 - 7t^2 + 12$

63. $f(u) = 4u^4 + 5u^3 - 2u^2 + 5u - 6$

64. $g(w) = w^4 - 2w^3 - w^2 - 4w + 12$

65. $h(x) = x^3 - 2x^2 - 8x + 5$

66. $k(x) = x^4 - x^3 - 14x^2 + 24x + 5$

In Exercises 67–72, write the function as a product of linear factors.

67. $f(x) = 2x^3 - 18x$

68. $f(x) = 2x^3 - 2x^2 - x$

69. $f(x) = 2x^3 - 9x^2 + 2x + 30$

70. $f(x) = 5x^3 - 24x^2 + x + 12$

71. $f(x) = 6x^4 + 11x^3 - 16x^2 - 11x + 10$

72. $f(x) = x^4 - 8x^3 + 27x^2 - 50x + 50$, given that $1 + 2i$ is a zero.

In Exercises 73–78, write the function as a product of linear and quadratic factors each with real coefficients. Find as many linear factors as possible.

73. $f(x) = x^3 - x^2 - x - 2$

74. $f(x) = 9x^3 - 3x^2 - 13x - 1$

75. $f(x) = 2x^4 - 9x^3 + 23x^2 - 31x + 15$

76. $f(x) = 3x^4 - 7x^3 - 3x^2 + 17x + 10$

77. $f(x) = 2x^5 - 3x^4 - 15x^3 + 12x^2 + 22x - 12$

78. $f(x) = 3x^5 + 2x^4 - 46x^3 + 6x^2 + 111x - 36$

In Exercises 79–86, find a polynomial function with real coefficients satisfying the given conditions.

79. Degree 3; zeros: $\sqrt{5}$, $-\sqrt{5}$, 3

80. Degree 2; -3 only real zero

81. Degree 4; zeros: 3, -2, $\frac{1}{3}$, $-\frac{1}{2}$

82. Degree 3; zeros: $1 + i$, 2

83. Degree 4; zeros: -2 (multiplicity 2), 4 (multiplicity 2)

84. Degree 3; zeros: $2 - i$, -1, and $f(2) = 6$

85. Degree 3; graph passing through $(-2, 0)$, $(0, 2)$, $(3, 0)$, and $(5, 0)$

86. Degree 4; zeros: $2i$, $3 - i$, and $f(0) = 20$

In Exercises 87 and 88, use the graph to answer the questions.

87. *Motion Detector* Carmen moves in line with a motion detector. The graph gives Carmen's distance from the detector (in meters) as a function of time (in seconds).

a. How far is she from the motion detector initially?
b. Describe her motion.
c. Estimate when she changes direction. How far from the motion detector is she when she changes direction?

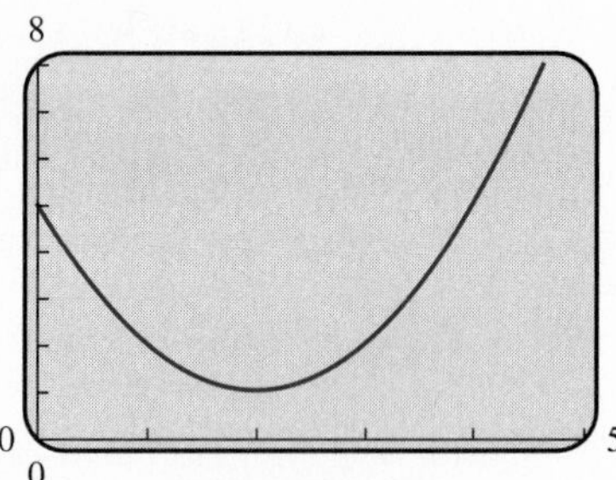

For Exercise 87

88. Writing to Learn Determine whether

$$f(x) = x^5 - 10x^4 - 3x^3 + 28x^2 + 20x - 2$$

has a zero outside the viewing window. Explain.

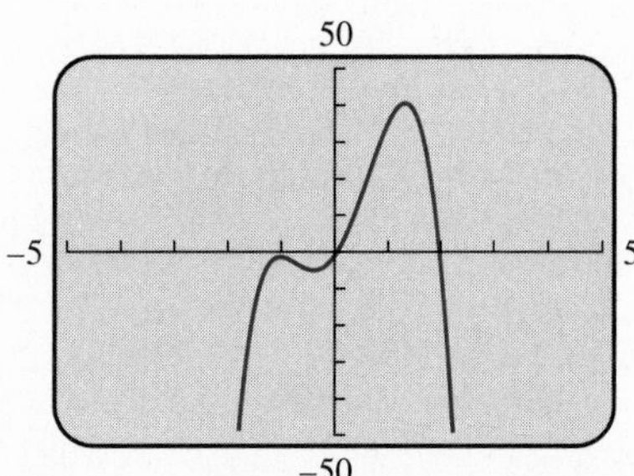

For Exercise 88

In Exercises 89 and 90, plot the complex number on a complex plane.

89. $-3 - 2i$

90. $4 - 3i$

In Exercises 91–98, solve the problem.

91. *Landscape Design* Kuan Min designed a rectangular play area that is 50 yd by 70 yd. The play area is surrounded by a terraced strip of uniform width planted with trees and shrubs. If the area of the terraced strip is 420 yd^2, find the width of the strip.

92. *Setting Production Schedules* Turner Manufacturing determines that the supply and demand functions for their food processor are

$$S(p) = 10 + 0.0001p^3,$$
$$D(p) = 100 - 0.04p^2,$$

where p is the price in dollars. Find the price for which the manufacturing level is in equilibrium, and the corresponding demand and supply amounts.

93. *Launching a Rock* Larry uses a slingshot to launch a rock straight up from a point 6 ft above level ground with an initial velocity of 170 ft/sec.

a. Find an equation that models the height of the rock t seconds after it is launched. Graph the equation using both function mode and parametric mode. (See Example 5 in Section 2.5.)
b. What is the maximum height of the rock? When will it reach that height?
c. When will the rock hit the ground?

94. *Volume of a Box* Edgardo Paper Co. has contracted to manufacture a box with no top that is to be made by removing squares of width x from the corners of a 30-in. by 70-in. piece of cardboard.

a. Find an equation that models the volume of the box.
b. Determine x so that the box has a volume of 5800 in.^{3}.

95. *Temperature Change* The temperature f (in degrees Fahrenheit) in Erie for a 24-hr period starting at 6 A.M. is given by the model

$$f(t) = 0.04t^3 - 1.5t^2 + 13.6t + 48.$$

a. Graph $y = f(t)$ for the 24-hr period.
b. What is the highest temperature, and when does it occur?
c. What is the lowest temperature, and when does it occur?
d. When is the temperature 55°?

96. *Biological Research* Akobundu does biological research for a wildlife protection agency. He has determined that a model for the population P of a certain group of wild pheasants after being left to reproduce for t days is given by

$$P(t) = -0.00005t^3 + 0.003t^2 + 1.2t + 80.$$

a. Graph the function $y = P(t)$.
b. Find the maximum pheasant population and when it occurs.
c. Assuming that this model continues to be accurate, when will this pheasant population become extinct?

97. *Architectural Engineering* Donoma, an engineer at J. P. Cook, Inc., completes structural specifications for a 255-ft-long steel beam anchored between two pilings 50 ft aboveground, as shown in the figure. She knows that when a 250-lb object is placed d feet from the west piling, the beam bends s feet where

$$s = (8.5 \times 10^{-7})d^2(255 - d).$$

a. Graph the function s.
b. What are the dimensions of a viewing window that shows a graph for the values that make sense in this problem situation?
c. What is the greatest amount of vertical deflection s, and where does it occur?
d. **Writing to Learn** Give a possible scenario explaining why the solution to part c does not occur at the halfway point.

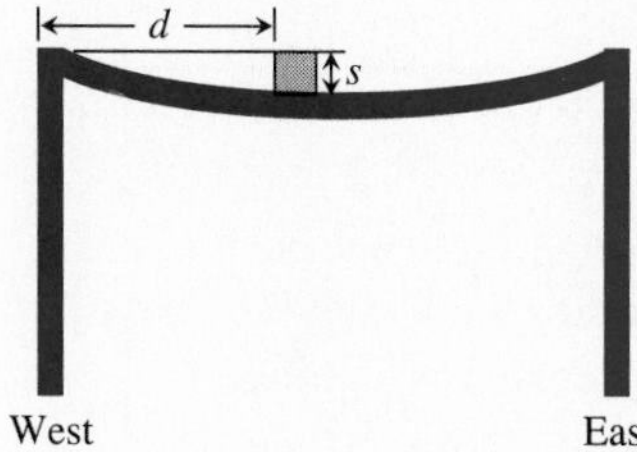

For Exercise 97

98. *Storage Container* A liquid storage container on a truck is in the shape of a cylinder with hemispheres on each end as shown in the figure. The cylinder and hemispheres have the same radius. The total length of the container is 140 ft.

a. Determine the volume V of the container as a function of the radius x.
b. Graph the function $y = V(x)$.
c. What is the radius of the container with the largest possible volume? What is the volume?

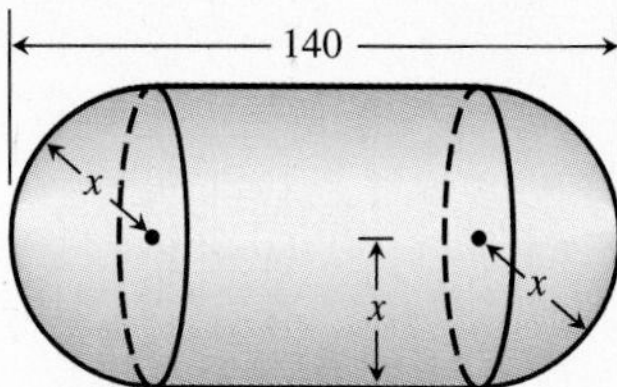

For Exercise 98

In Exercises 99–101, use the data in the tables to solve the problem.

99. Platinum sales are increasing rapidly. Table 3.11 gives the sales to jewelers in ounces for the given years.

Table 3.11 Platinum Sales

Year	Sales (oz)
1990	20,000
1991	20,000
1992	35,000
1993	45,000
1994	55,000

Source: Johnson Matthey as reported by Anne R. Carey and Julie Stacey in *USA Today,* May 25, 1995.

a. Find a linear regression equation, and graph it together with a scatter plot of the data.
b. Find a quadratic regression equation, and graph it together with a scatter plot of the data.
c. Use the linear and quadratic regression equations to estimate the number of ounces of platinum that will be sold to jewelers in 1998.

100. Table 3.12 gives the average amount per grant of emergency financial assistance in Franklin County, Ohio.

Table 3.12 Emergency Help

Year	Avg. Amt. ($)
1986	145.38
1987	155.00
1988	230.43
1989	420.70
1990	494.55
1991	555.00
1992	508.77
1993	460.92
1994	453.77

Source: Franklin County Veterans Service Commission as reported by Mary Stephens in *The Columbus* [Ohio] *Dispatch,* May 31, 1995.

a. Find a cubic regression equation, and graph it together with a scatter plot of the data.
b. Find a quartic regression equation, and graph it together with a scatter plot of the data.
c. **Writing to Learn** Compare the end behavior of the two regression graphs. Create scenarios that could explain the use of one over the other to predict future emergency help.

101. Kibbe's distance from a motion detector is given by Table 3.13.

Table 3.13 Motion Detector Data for Exercise 101

t (sec)	d (m)	t (sec)	d (m)
0.0	2.57	4.5	2.89
0.5	2.62	5.0	3.48
1.0	2.09	5.5	4.12
1.5	1.52	6.0	4.43
2.0	1.00	6.5	4.94
2.5	0.78	7.0	5.23
3.0	1.11	7.5	5.48
3.5	1.63	8.0	5.78
4.0	2.26		

a. Find a cubic regression equation, and graph it together with a scatter plot of the data.
b. Describe Kibbe's motion.
c. Use the regression equation to estimate when Kibbe changes direction. How far from the motion detector is he when he changes direction?

In Exercises 102 and 103, answer the questions.

102. **Writing to Learn** Explain how you can tell from a graph that the multiplicity of a real root is even rather than odd.

103. **Writing to Learn** Suppose that both f and g have a zero of odd multiplicity at $x = c$. Explain how you can tell from a graph which one has a greater multiplicity.

MARIA AGNESI

Maria Gaetana Agnesi (1718–99) was recognized as a genius at an early age. At age 9, her interests included logic, mechanics, chemistry, botany, zoology, mineralogy, and analytic geometry. By age 11, she was fluent in seven languages. Agnesi's first concern was helping the poor and underprivileged people of her native Italy, but her father persuaded her to concentrate on studying mathematics.

Agnesi's masterpiece, *Analytical Institutions for the Use of Italian Youth,* included discoveries in algebra, calculus, and differential equations. Her name is associated with a bell-shaped curve, the *versiera* of Agnesi, which is still used in mathematics and physics. After her father died, she withdrew from her studies to devote herself to charitable work and religious studies.

chapter 4

EXPONENTIAL AND LOGARITHMIC FUNCTIONS

BIBLIOGRAPHY

For students: *Beyond Numeracy,* John Allen Paulos. Alfred A. Knopf, 1991.

For teachers: *e: The Story of a Number,* Eli Maor. Princeton University Press, 1993.

Teaching and Learning Mathematics in the 1990s, 1990 Yearbook, Thomas Cooney and Christian R. Hirsch, eds. NCTM, 1990. Available through Dale Seymour Publications.

4.1 EXPONENTIAL FUNCTIONS; COMPARISONS WITH THE POWER FUNCTION

Objective

Students will be able to identify and understand the differences between exponential and power functions.

Motivate

Ask . . .
If the population of a town increases by 10% every year, what will a graph of the population function look like?

Introduction to Power Functions and Exponential Functions • Graphs of Exponential Functions • Comparing Exponential Functions Having Different Bases • Comparing Power Functions and Exponential Functions • Applications • Modeling Data with Exponential Functions

Introduction to Power Functions and Exponential Functions

It is instructive to compare the functions $y = x^2$ and $y = 2^x$.

$y = x^2$: The base is x, and the exponent is 2.
$y = 2^x$: The base is 2, and the exponent is x.

When the base is a variable and the exponent is a nonzero real number, as in $y = x^2$, the function is a *power function.* When the base is a positive real number and the exponent is a variable, as in $y = 2^x$, the function is *not* a power function; it is an *exponential function.*

Definition: Power Functions and Exponential Functions

Let k and a be nonzero real numbers.

1. A **power function** in x is one that can be written in the form
$$y = k \cdot x^a.$$
2. An **exponential function** in x is one that can be written in the form
$$y = k \cdot a^x$$
where a is positive and $a \neq 1$.

Table 4.1 Function Values

x	$f(x) = 2^x$
−3	0.125
−2	0.25
−1	0.5
0	1
1	2
2	4
3	8
4	16
5	32
6	64

These are examples of power functions:
$$y = x^2 \qquad y = -3x^{-2.1} \qquad y = x^{1/2}$$
These are exponential functions:
$$y = 2^x \qquad y = 1.3^x \qquad y = \left(\frac{1}{3}\right)^x$$

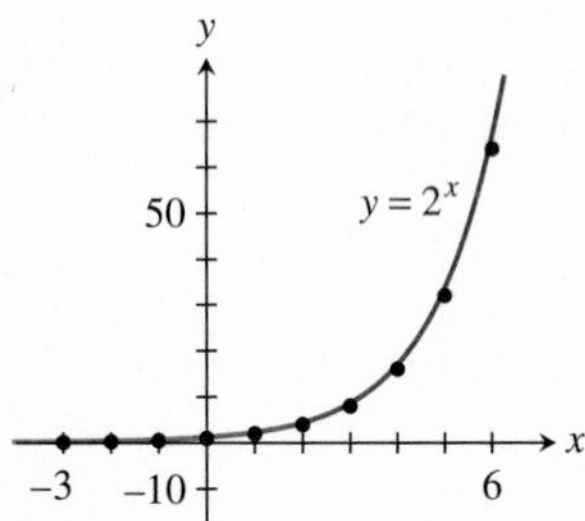

Figure 4.1

Graphs of Exponential Functions

We can evaluate an exponential function like $f(x) = 2^x$ when x is an integer by using properties of exponents. For example,
$$f(-3) = 2^{-3} = \frac{1}{2^3} = \frac{1}{8} = 0.125 \qquad \text{and} \qquad f(4) = 2^4 = 16.$$
In a similar manner, we can find all the function values shown in Table 4.1. To sketch the graph of $y = f(x)$, we plot these points in a coordinate plane and draw a smooth curve through them, as suggested by Figure 4.1.

Because 2^x is defined for all values of x, the domain of f is $(-\infty, \infty)$. Also, $f(x)$ is always positive and the range of f is $(0, \infty)$.

REMARK

You may use a calculator to find approximate values when x is not an integer. For example, confirm that

$2^{\sqrt{3}} \approx 3.322$ and $2^{-0.01} \approx 0.993$.

■ EXAMPLE 1 Comparing Graphs of Exponential Functions

Compare the graphs of $f(x) = 2^x$, $g(x) = 2^{-x}$, and $h(x) = \left(\frac{1}{2}\right)^x$.

Solution

Comparing the Graphs of *f* and *g*

Recall from Section 2.6 that the graphs of $y = f(x)$ and $y = f(-x)$ are reflections of each other across the y-axis for any function f. Thus, the graphs of f and g are reflections of each other across the y-axis (see Figure 4.2) because
$$f(-x) = 2^{-x} = g(x).$$

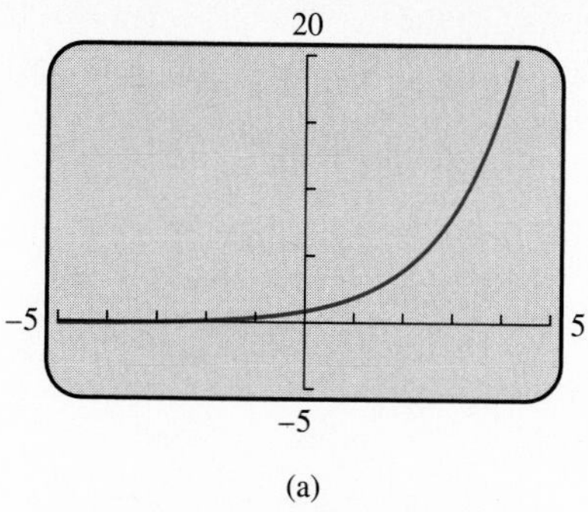

(a)

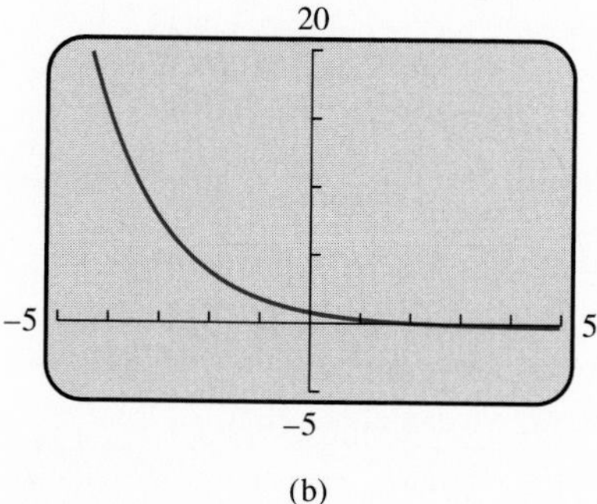

(b)

Figure 4.2 (a) $f(x) = 2^x$. (b) $g(x) = 2^{-x}$ and $h(x) = (\frac{1}{2})^x$.

Exploration Extensions

Graph the functions $y_1 = 2^{-x}$, $y_2 = 3^{-x}$, $y_3 = 4^{-x}$, and $y_4 = 5^{-x}$. Compare to your other graphs. Explain the results.

REMARK

In calculus we will say that

$$\lim_{x\to\infty} 2^x = \infty \quad \text{and}$$

$$\lim_{x\to-\infty} 2^x = 0.$$

Similar statements can be made for 2^{-x}.

Comparing the Graphs of *h* and *g*

$$h(x) = \left(\frac{1}{2}\right)^x = (2^{-1})^x = 2^{-x} = g(x)$$

Since h and g are equal functions, their graphs are identical. ■

Comparing Exponential Functions Having Different Bases

Example 1 compares exponential functions whose bases are reciprocals. The following Exploration asks you to compare exponential functions whose bases are not so nicely related.

● EXPLORATION Graphs of Exponential Functions

1. Graph each function in the viewing window $[-2, 2]$ by $[0, 6]$.
   a. $y_1 = 2^x$ **b.** $y_2 = 3^x$ **c.** $y_3 = 4^x$ **d.** $y_4 = 5^x$
   - What one point lies on all four graphs?
   - How do the end behaviors compare?
2. Graph each function in the viewing window $[-2, 2]$ by $[0, 6]$.
   a. $y_1 = \left(\frac{1}{2}\right)^x$ **b.** $y_2 = \left(\frac{1}{3}\right)^x$ **c.** $y_3 = \left(\frac{1}{4}\right)^x$ **d.** $y_4 = \left(\frac{1}{5}\right)^x$
   - What one point lies on all four graphs?
   - How do the end behaviors compare? ●

Experiences gained from the Exploration support the following characteristics of exponential function graphs.

Graphs of Exponential Functions

Consider an exponential function in the form $f(x) = a^x$.

1. If $a > 1$, then the graph of $f(x) = a^x$
   a. passes through the point (0, 1),
   b. is shaped like the graph of $y = 2^x$, and
   c. has end behavior like $y = 2^x$:
   $$f(x) \to 0 \text{ as } x \to -\infty \quad \text{and} \quad f(x) \to \infty \text{ as } x \to \infty$$
2. If $0 < a < 1$, then the graph of $f(x) = a^x$
   a. passes through the point (0, 1).
   b. is shaped like the graph of $y = \left(\frac{1}{2}\right)^x = 2^{-x}$, and
   c. has end behavior like $y = 2^{-x}$:
   $$f(x) \to 0 \text{ as } x \to \infty \quad \text{and} \quad f(x) \to \infty \text{ as } x \to -\infty$$

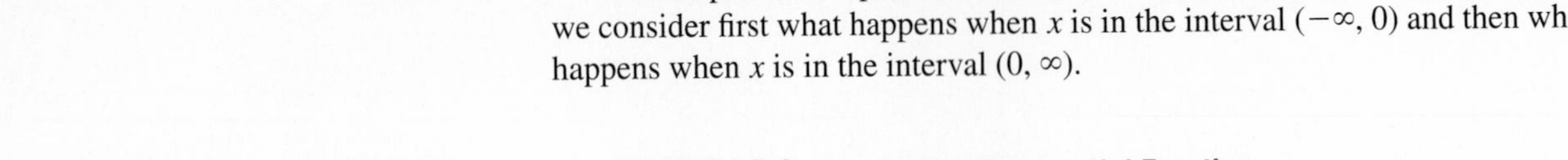

To compare two exponential functions having the form $y = a^x$ when $a > 1$, we consider first what happens when x is in the interval $(-\infty, 0)$ and then what happens when x is in the interval $(0, \infty)$.

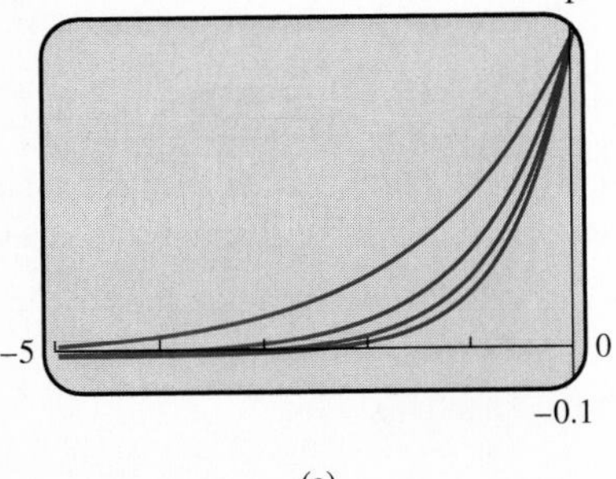

(a)

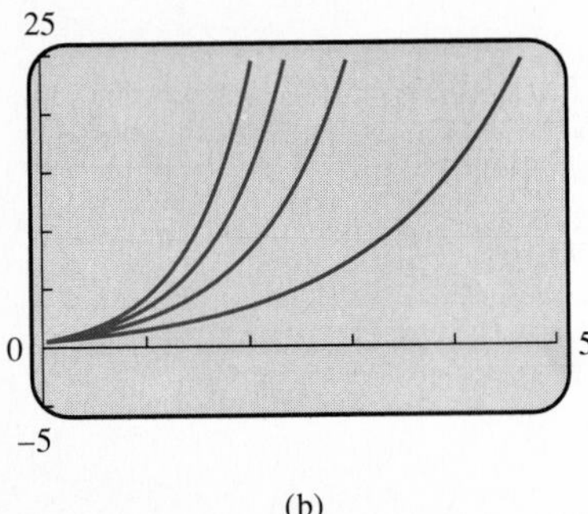

(b)

Figure 4.3 The two graphs may convey a false impression of what happens at $x = 0$. Try graphing $y = 2^x$ for $-1 \le x \le 1$.

Teaching Note

Students may need to be reminded that most of the graphs shown in this section use a large range of y-values in order to show a "global" view of the functions. The functions will have a somewhat different appearance if the x- and y-scales are equal.

■ EXAMPLE 2 Comparing Exponential Functions

The graphs of $y_1 = 2^x$, $y_2 = 3^x$, $y_3 = 4^x$, and $y_4 = 5^x$ are suggested in Figure 4.3a for x-values in the interval $(-\infty, 0)$, and in Figure 4.3b for x-values in the interval $(0, \infty)$. Identify the graph of each.

Solution

a. For x-values in the interval $(-\infty, 0)$, consider what happens at $x = -1$.

$$y_1 = 2^{-1} = \frac{1}{2},\ y_2 = 3^{-1} = \frac{1}{3},$$

$$y_3 = 4^{-1} = \frac{1}{4}, \qquad \text{and} \qquad y_4 = 5^{-1} = \frac{1}{5}$$

so $$2^{-1} > 3^{-1} > 4^{-1} > 5^{-1}.$$

In fact, for any x-value in the interval, we can check that

$$2^x > 3^x > 4^x > 5^x.$$

So the graphs of y_1, y_2, y_3, and y_4 appear from top to bottom, respectively, in Figure 4.3a.

b. For x-values in the interval $(0, \infty)$, consider what happens at $x = 1$.

$$y_1 = 2^1 = 2, \qquad y_2 = 3^1 = 3,$$

$$y_3 = 4^1 = 4, \qquad \text{and} \qquad y_4 = 5^1 = 5,$$

so $$2^1 < 3^1 < 4^1 < 5^1.$$

For any x-value in the interval, we can check that

$$2^x < 3^x < 4^x < 5^x.$$

So the graphs of y_1, y_2, y_3, and y_4 appear from *bottom to top*, respectively, in Figure 4.3b. ■

Comparing Power Functions and Exponential Functions

Figure 4.4 shows the power function $y = x^2$ and the exponential function $y = 2^x$ for positive values of x. It is not obvious at a quick glance which graph is the power function and which graph is the exponential function, so we need to study these two familiar types of functions more carefully.

It usually is difficult to compare a power function $f(x) = x^a$ and an exponential function $g(x) = a^x$ algebraically. It may be easier to compare them using graphical and numerical techniques. We shall illustrate using $y = x^4$ and $y = 2^x$.

Alert

Some students may have trouble entering exponents on a grapher. Many of the exponents are more complex than the students have encountered in previous mathematics courses. Careful attention must be given to the syntax and placement of parentheses.

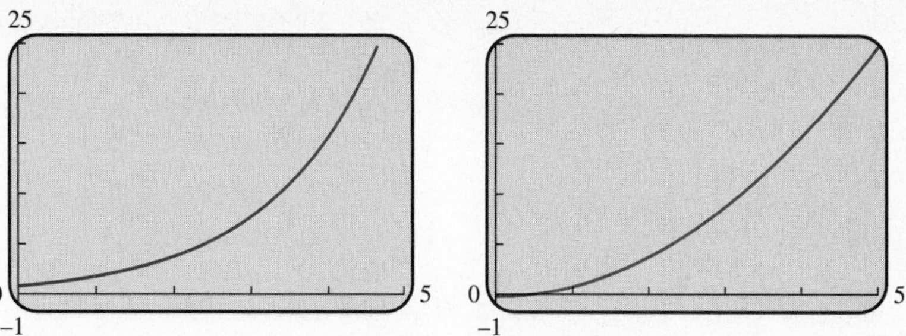

Figure 4.4 $y_1 = x^2$ and $y_2 = 2^x$ using the same viewing-window dimensions.

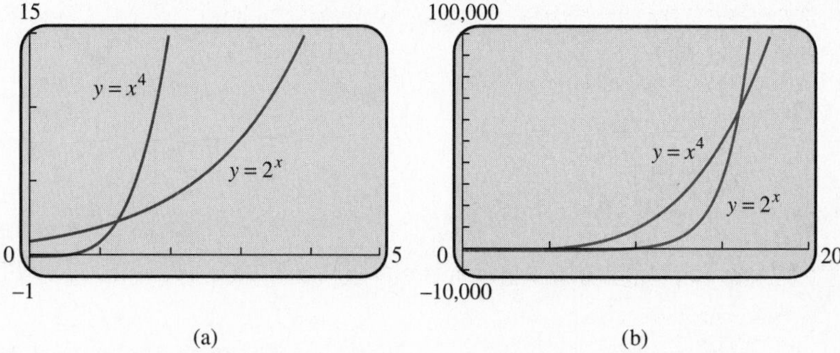

Figure 4.5 The two windows show graphs of $y = x^4$ and $y = 2^x$. Notice that (a) is a magnified view of a small portion of (b).

X	X^4	2^X
1	1	2
2	16	4
3	81	8
4	256	16
5	625	32
6	1296	64
7	2401	128
X=7		

Figure 4.6 Notice that the values of x^4 are greater than the values 2^x starting at $x = 2$.

X	X^4	2^X
12	20736	4096
13	28561	8192
14	38416	16384
15	50625	32768
16	65536	65536
17	83521	131072
18	104976	262144
X=18		

Figure 4.7

Figure 4.5a suggests that the graphs of $y = x^4$ and $y = 2^x$ intersect between $x = 1$ and $x = 2$. In addition, the numerical information in Figure 4.6 suggests that the graph of $y = 2^x$ lies below the graph of $y = x^4$ when $2 < x < 5$.

On the other hand, Figure 4.5b, supported by the numerical data in Figure 4.7, shows that when $x > 16$, the graph of $y = 2^x$ lies above the graph of $y = x^4$. It appears that these graphs intersect at two points in the interval $(0, \infty)$.

The numerical data in Figures 4.6 and 4.7 suggest that 2^x increases much more slowly than does x^4 for small positive values of x, but then speeds up dramatically as x continues to increase; so finally, when x is large enough, 2^x is larger than x^4.

Comparing Exponential Function End Behavior to Power Function End Behavior ($x \to \infty$)

Let a and b be positive integers such that $a > 1$. There is a positive integer n such that $a^x > x^b$ for all values of x in the interval (n, ∞).

EXAMPLE 3 Comparing $y = x^4$ and $y = 2^x$

Find a positive integer n such that $2^x > x^4$ when $x > n$.

Solution

Using grapher methods (in Figure 4.5b), we can show that $y = x^4$ equals $y = 2^x$ when $x = 16$. Thus, $n = 16$. We conclude that if $x > 16$, then $2^x > x^4$. ■

Applications

Exponential functions can be used to model changes in population. Suppose that a population is *increasing* at a constant percentage rate $r > 0$. If r is written in decimal form, we have

$$\begin{aligned} & P && \text{Population now} \\ P + Pr &= P(1 + r) && \text{Population after 1 year} \\ P(1 + r) + P(1 + r)r &= P(1 + r)^2 && \text{Population after 2 years} \\ &\vdots \\ P(1 + r)^{t-1} + P(1 + r)^{t-1}r &= P(1 + r)^t && \text{Population after } t \text{ years} \end{aligned}$$

If the population is *decreasing* at the constant percentage rate $r < 1$, then the population after 1 year is

$$P - Pr = P(1 - r),$$

and the population after t years is $P(1 - r)^t$.

The two functions, $y = P(1 + r)^t$ and $y = P(1 - r)^t$, are exponential functions with bases $1 + r$ and $1 - r$, respectively. Since $1 + r > 1$, the graph of $y = P(1 + r)^t$ is a **growth curve,** shaped like the graph of $y = 2^t$. Since $0 < 1 - r < 1$, the graph of $y = P(1 - r)^t$ is a **decay curve,** shaped like the graph of $y = (\frac{1}{2})^t = 2^{-t}$.

Follow-up

Ask . . .
If $k > 0$, how can you tell whether $y = k \cdot a^x$ represents an increasing or decreasing function?
(The function is increasing if $a > 1$ and decreasing if $0 < a < 1$.)

Assignment Guide

Day 1: Ex. 3–48, multiples of 3
Day 2: Ex. 50, 51, 53–56, 58, 59, 61, 63, 66, 67, 69

Exponential Growth and Decay

The exponential function $y = k \cdot a^x$ $(k > 0)$ is a model for **exponential growth** if $a > 1$, and is a model for **exponential decay** if $0 < a < 1$.

Exponential growth and decay are phenomena that are found frequently in biology, chemistry, business, and the social and physical sciences.

REMARK

We have chosen to use a graphical solution because the algebraic methods needed are not available to us until the next section.

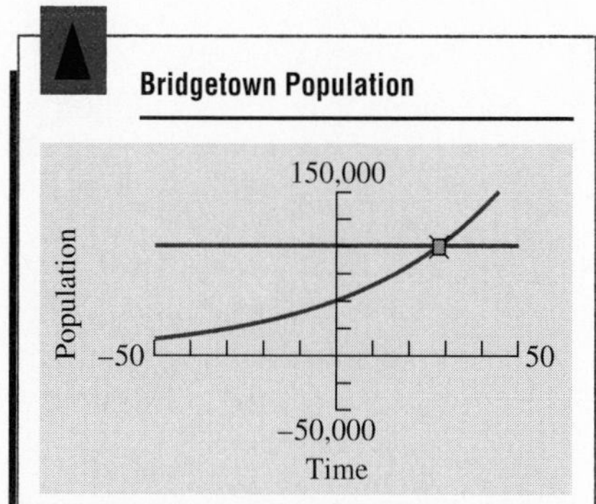

Intersection: $x = 28.071035$; $y = 100{,}000$

Figure 4.8 $y_1 = 50{,}000(1 + 0.025)^t$ and $y_2 = 100{,}000$.

■ EXAMPLE 4 APPLICATION: Modeling Population Growth

The population of Bridgetown is growing at the rate of 2.5% per year. The present population is 50,000. When will the population be 100,000?

Solution

Model

The population now is $P = 50{,}000$ and the growth rate r is 0.025. The function that models the population growth is

$$P(t) = 50{,}000(1 + 0.025)^t.$$

Solve Graphically

To predict when the population will be 100,000, solve the equation

$$100{,}000 = 50{,}000(1 + 0.025)^t.$$

Figure 4.8 shows that the graphs of $y_1 = 50{,}000(1 + 0.025)^t$ and $y_2 = 100{,}000$ intersect when t is about 28.07.

Interpret

The population of Bridgetown will reach 100,000 in a little over 28 years. ■

Modeling Data with Exponential Functions

An exponential function can often model a relationship between population and time. An equation of the model is an **exponential regression equation.** Once a model can be found, that model can be used to predict future population.

Table 4.2 U.S. Population, 1880–1970

Year	Population (millions)
1880	50.2
1890	63.0
1900	76.0
1910	92.0
1920	105.7
1930	122.8
1940	131.7
1950	151.3
1960	179.3
1970	203.3

Source: The Statesman's Yearbook, 129th ed. (London: The Macmillan Press, Ltd., 1992).

Using Data Analysis—Exponential Regression

1. Complete a scatter plot of the data.
2. Find an exponential regression equation $y = a \cdot b^x$.
3. Superimpose a graph of the exponential regression equation on the scatter plot.
4. Use the exponential regression equation to predict y-values for x-values not included in the data.

■ EXAMPLE 5 APPLICATION: Predicting the U.S. Population

Use the population data in Table 4.2 to predict the population for the year 1990. Compare the result with the actual 1990 population of approximately 250 million.

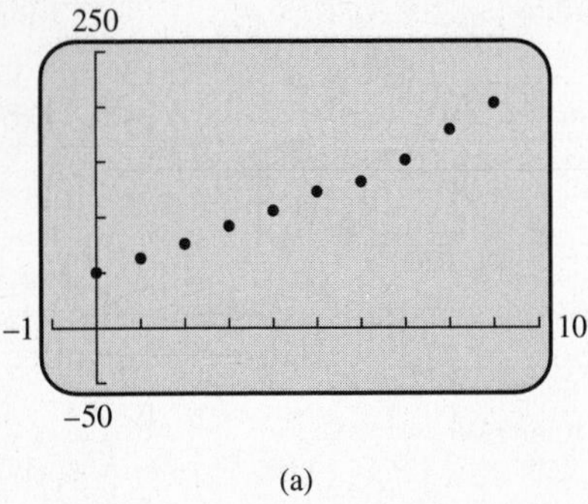

(a)

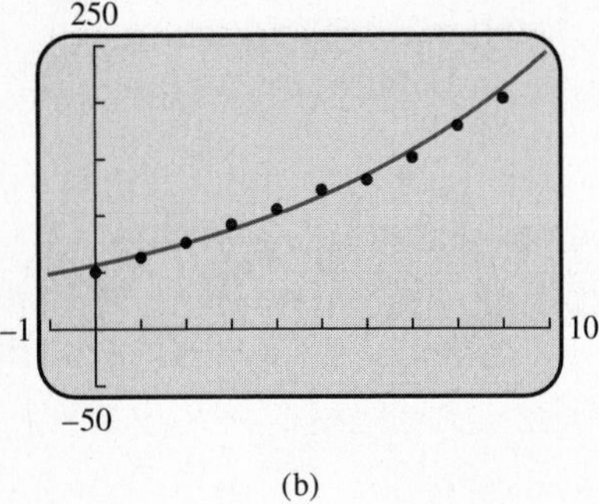

(b)

Figure 4.9

Solution

Figure 4.9a shows a scatter plot for the data, where $x = 0$ represents 1880, $x = 1$ represents 1890, and so on.

We compute the exponential regression equation to be $f(x) = 55.05(1.16^x)$ (see the Grapher Workshop). Its graph is superimposed on the scatter plot in Figure 4.9b.

The year 1990 is represented by $x = 11$, so

$$f(11) \approx 281.71.$$

This exponential model estimates the 1990 population to be 281.71 million, an overestimate of approximately 32 million. ■

In addition to humans, "population growth and decay" can refer to other types of animals, to plants, to bacteria, to radioactive elements, indeed to a population of any sort where the change in the population is directly proportional to the present size of the population. It can even refer to amounts of money that are invested to earn compound interest.

EXAMPLE 6 APPLICATION: Modeling Bacteria Growth

Suppose a culture of 100 bacteria is put into a petri dish and the culture doubles every hour. Predict when the number of bacteria will be 350,000.

Solution

Model

$$200 = 100(2) \quad \text{Total bacteria after 1 h}$$
$$400 = 100(2^2) \quad \text{Total bacteria after 2 h}$$
$$800 = 100(2^3) \quad \text{Total bacteria after 3 h}$$
$$\vdots$$
$$P(t) = 100(2^t) \quad \text{Total bacteria after } t \text{ h}$$

The function $P(t) = 100(2^t)$ models the population at time t.

Solve Graphically

Figure 4.10 shows that the graphs of $y_1 = P(t) = 100(2^t)$ and $y_2 = 350{,}000$ intersect when t is approximately 11.77.

Interpret

The population of the bacteria in the petri dish will be 350,000 in about 11.77 hours, or about 11 h and 46 min. ■

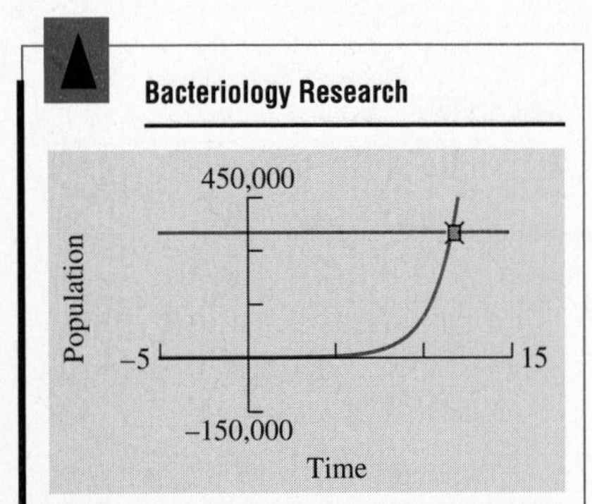

Intersection: $x = 11.773139$; $y = 350{,}000$

Figure 4.10

Exponential functions also model radioactive decay. The **half-life** of a radioactive substance is the amount of time it takes for half of the substance to

Cooperative Learning

Group Activity: Ex. 68

Notes on Exercises

Ex. 15–20 encourage students to think about the appearance of functions without using a grapher.
Ex. 71–74 require students to think about the meaning of different kinds of functions.

Ongoing Assessment

Self-Assessment: Ex. 11, 23, 31, 37, 49, 57
Embedded Assessment: Ex. 18, 56

change from its original radioactive state to a nonradioactive state by emitting energy in the form of radiation.

EXAMPLE 7 APPLICATION: Modeling Radioactive Decay

Suppose the half-life of a certain radioactive substance is 20 days and there are 5 g (grams) present initially. Find the time when there will be 1 g of the substance remaining.

Solution

Model

$$\frac{5}{2} = 5\left(\frac{1}{2}\right) \quad \text{Grams remaining after 20 days}$$

$$\frac{5}{4} = 5\left(\frac{1}{2}\right)^2 \quad \text{Grams remaining after 40 days}$$

$$\vdots$$

$$y = 5\left(\frac{1}{2}\right)^{t/20} \quad \text{Grams remaining after } t \text{ days}$$

The function $y = 5\left(\frac{1}{2}\right)^{t/20}$ models the mass in grams of the radioactive substance at time t.

Solve Graphically

Figure 4.11 shows that the graphs of $y_1 = 5\left(\frac{1}{2}\right)^{t/20}$ and $y_2 = 1$ intersect when t is approximately 46.44.

Interpret

There will be 1 g of the radioactive substance left after approximately 46.44 days, or about 46 days $10\frac{1}{2}$ h. ■

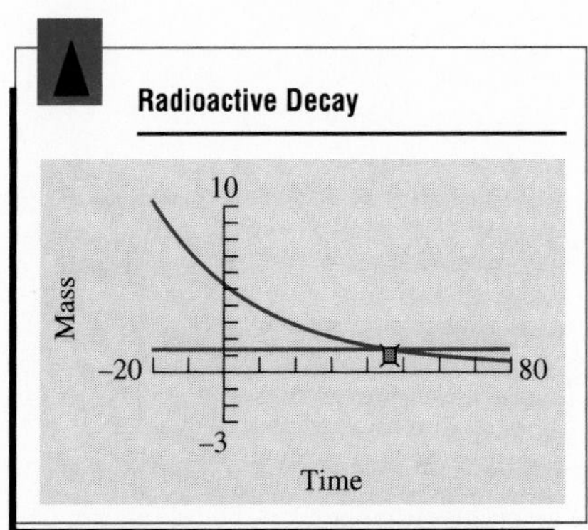

Intersection: $x = 46.438562$; $y = 1$

Figure 4.11

Quick Review 4.1

In Exercises 1–8, evaluate the expression.

1. 7^2 **2.** 2^5

3. 2^{-3} **4.** 4^3

5. 1.2^{-5} **6.** 1.3^{-2}

7. 1.1^7 **8.** 5^5

In Exercises 9–14, rewrite the expression with a single positive exponent.

9. 2^{-5} **10.** 5^{-3}

11. $(2^{-3})^4$ **12.** $(3^4)^{-2}$

13. $(a^{-2})^3$ **14.** $(b^{-3})^{-5}$

SECTION EXERCISES 4.1

In Exercises 1–8, identify the function as an exponential function, a power function, or neither.

1. $y = x^8$ **2.** $y = 3^x$

3. $y = 5^x$ **4.** $y = 4^2$

5. $y = x^{\sqrt{x}}$ **6.** $y = x^{1.3}$

7. $y = 1.7z^{-2.8}$ **8.** $y = 1.9^{-3}$

In Exercises 9–14, graph the function. State its domain, range, and intercepts.

9. $f(x) = 3^x$ **10.** $f(x) = 4^{-x}$

11. $f(x) = 0.3^x$ **12.** $f(x) = 1.6^x$

13. $f(x) = 0.2^{-x}$ **14.** $f(x) = 2.3^{-x}$

In Exercises 15–20, match the given function with its graph. Explain how you can make these choices without using a grapher.

15. $y = 3^x$ **16.** $y = 2^{-x}$

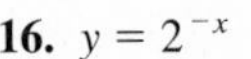

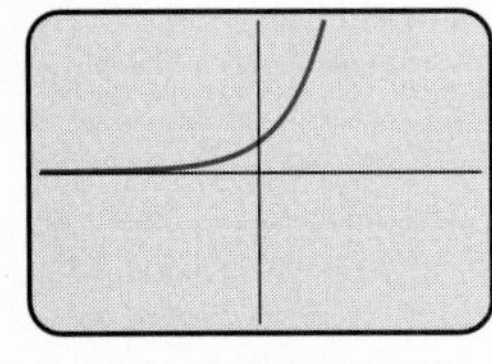

(a)

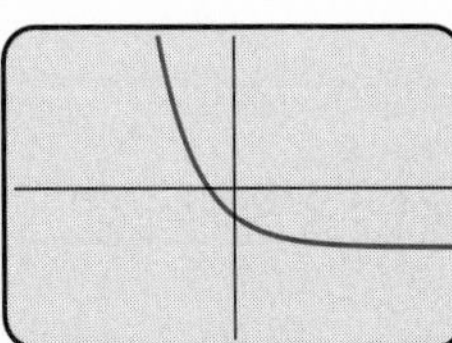

(b)

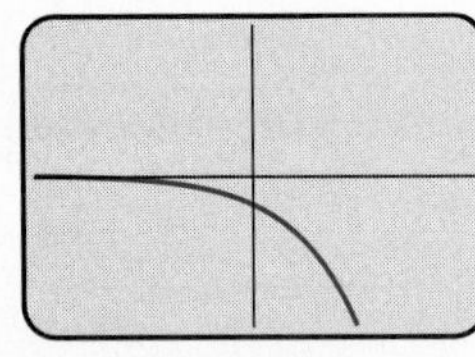

(c)

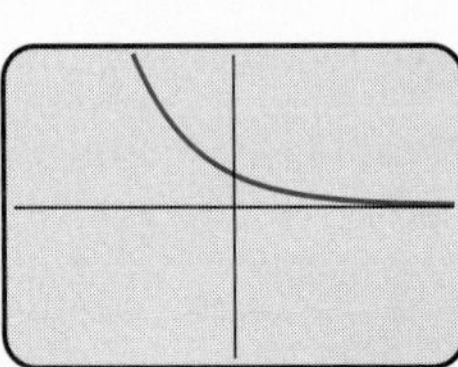

(d)

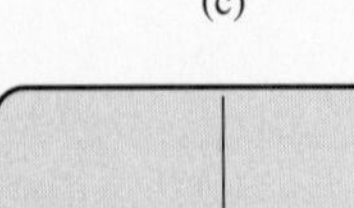

(e)

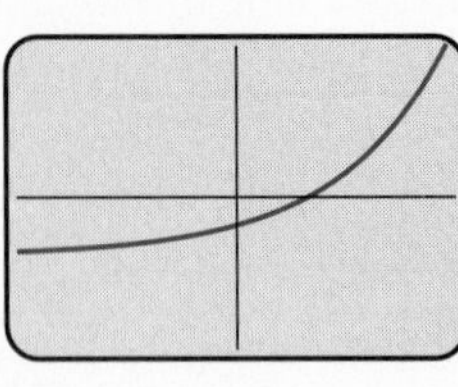

(f)

For Exercises 15–20

17. $y = -2^x$ **18.** $y = -0.5^x$

19. $y = 3^{-x} - 2$ **20.** $y = 1.5^x - 2$

In Exercises 21–24, state whether the graphs of the two functions are identical or are reflections of one another across the y-axis.

21. $f(x) = 3^x, \quad g(x) = 3^{-x}$

22. $f(x) = 1.2^x, \quad g(x) = 1.2^{-x}$

23. $f(x) = 2^x, \quad g(x) = 0.5^{-x}$

24. $f(x) = 4^{-x}, \quad g(x) = 0.25^x$

In Exercises 25–34, state whether the function represents exponential growth or exponential decay. Then describe its end behavior.

25. $y = 7^x$ **26.** $y = 9^x$

27. $y = 6^{-x}$ **28.** $y = 8^{-x}$

29. $y = 3^{-2x}$ **30.** $y = (\frac{1}{e})^x$

31. $y = (\frac{2}{3})^x$ **32.** $y = (\frac{1}{5})^x$

33. $y = 0.5^x$ **34.** $y = 0.75^{-x}$

In Exercises 35–38, rewrite the exponential expression with the specified base.

35. 9^{2x}, base 3 **36.** 16^{3x}, base 2

37. $(\frac{1}{8})^{2x}$, base 2 **38.** $(\frac{1}{27})^x$, base 3

In Exercises 39 and 40, write two different double inequalities that compare the three given functions for $x < 0$ and for $x > 0$. Support your statements with tables and graphs.

39. $y_1 = 4^x, y_2 = 5^x, y_3 = 7^x$

40. $y_1 = 3^{-x}, y_2 = 5^{-x}, y_3 = 8^{-x}$

In Exercises 41–44, solve the inequality. Discuss whether you used an algebraic or graphical method and why.

41. $9^x < 4^x$ **42.** $6^{-x} > 8^{-x}$

43. $(\frac{1}{4})^x > (\frac{1}{3})^x$ **44.** $(\frac{1}{3})^x < (\frac{1}{2})^x$

In Exercises 45 and 46, use rules of exponents to confirm that two of the three given exponential functions are equal. Support your work graphically.

45. **a.** $y_1 = 3^{2x+4}$ **b.** $y_2 = 3^{2x} + 4$ **c.** $y_3 = 9^{x+2}$

46. **a.** $y_1 = 4^{3x-2}$ **b.** $y_2 = 2(2^{3x-2})$ **c.** $y_3 = 2^{3x-1}$

In Exercises 47 and 48, find the smallest positive integer n such that if $x > n$, the exponential function $a^x > x^b$ for the given values of a and b. In each case find a viewing window so that the two graphs look similar to those in the figure.

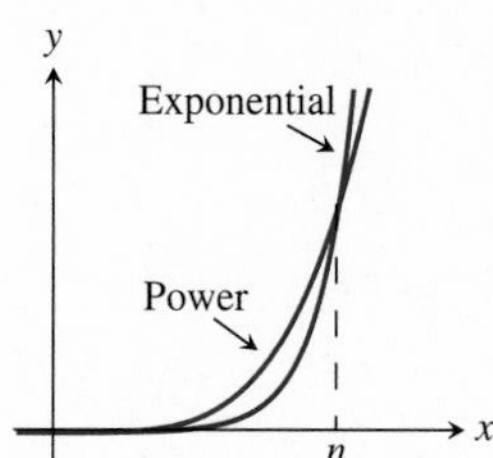

For Exercises 47 and 48

47. $y_1 = 2^x$ and $y_2 = x^4$

48. $y_1 = 3^x$ and $y_2 = x^4$

In Exercises 49–54, use an exponential model to express the population as a function of t.

49. *Population Growth* The population of Knoxville is 475,000 and is increasing at the rate of 3.75% each year. Predict when the population will be 1 million.

50. *Population Growth* The population of Glenbrook is 350,000 and is increasing at the rate of 3.25% each year. Predict when the population will be 1 million.

51. *Population Growth* The population of Silver Run in the year 1890 was 6250. Assume the population increased at a rate of 2.75% per year.

a. Estimate the population in 1915 and 1940.
b. Predict when the population reached 50,000.

52. *Population Growth* The population of Centerville in the year 1910 was 4200. Assume the population increased at a rate of 2.25% per year.

a. Estimate the population in 1930 and 1945.
b. Predict when the population reached 20,000.

53. *Radioactive Decay* The half-life of a certain radioactive substance is 14 days. There are 6.6 g present initially.

a. Express the amount of substance remaining as a function of time t.
b. When will there be less than 1 g remaining?

54. *Radioactive Decay* The half-life of a certain radioactive substance is 65 days. There are 3.5 g present initially.

a. Express the amount of substance remaining as a function of time t.
b. When will there be less than 1 g remaining?

In Exercises 55–58, solve the given problem. Exercises 55–58 focus on the concept of *doubling time.*

55. Writing to Learn Example 4 calculates the length of time for the population of Bridgetown to double—the so-called doubling time. Study Example 4, and explain why this doubling time is the same, no matter what the original population was.

56. Writing to Learn Use the following population growth curve to reason graphically that the time it takes for the population to double (doubling time) is independent of the population size.

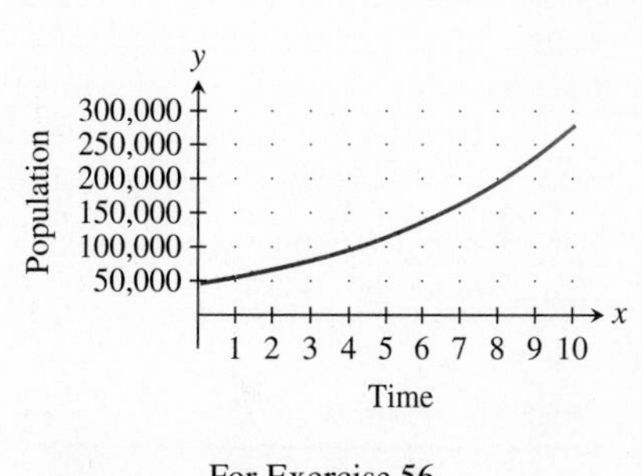

For Exercise 56

57. *Bacteria Growth* The number B of bacteria in a petri dish culture after t hours is given by

$$B = 100e^{0.693t}.$$

a. What was the initial number of bacteria present?
b. How many bacteria are present after 6 hours?
c. When will the number of bacteria be 200? Estimate the doubling time of the bacteria.

58. *Carbon Dating* The amount C in grams of carbon-14 present in a certain substance after t years is given by

$$C = 20e^{-0.0001216t}.$$

a. What was the initial amount of carbon-14 present?
b. How much is left after 10,400 years? When will the amount left be 10 g?
c. Estimate the half-life of carbon-14.

In Exercises 59 and 60, solve the problem.

59. Table 4.3 gives some data about the population of Mexico.

Table 4.3 Population of Mexico

Year	Population (millions)
1950	25.8
1960	34.9
1970	48.2
1980	66.8
1990	81.1

Source: The Statesman's Yearbook, 129th ed. (London: The Macmillan Press, Ltd., 1992).

a. Find an exponential regression equation and superimpose its graph on a scatter plot of the data.

b. Use the exponential regression equation to estimate what the population of Mexico was in 1900. How close is the estimate to the actual population in 1900 of 13,607,272?

60. Table 4.4 gives some data about the population of South Africa.

Table 4.4 Population of South Africa

Year	Population (millions)
1904	5.2
1911	6.0
1921	6.9
1936	9.6
1946	11.4
1951	12.7
1960	16.0
1970	18.3
1980	20.6

Source: The Statesman's Yearbook, 129th ed. (London: The Macmillan Press, Ltd., 1992).

a. Find an exponential regression equation and superimpose its graph on a scatter plot of the data.

b. Use the exponential regression equation to estimate what the population of South Africa was in 1990.

In Exercises 61 and 62, find the interval(s) on which the function is increasing and the interval(s) on which it is decreasing. Also find all local maximum and minimum values, and the domain and range.

61. $f(x) = xe^{-x}$

62. $f(x) = \dfrac{e^{-x}}{x}$

EXTENDING THE IDEAS

In Exercises 63–66, use rules of exponents to solve the equation. Support your answer graphically.

63. $2^x = 4^2$

64. $3^x = 27$

65. $8^{x/2} = 4^{x+1}$

66. $9^x = 3^{x+1}$

Exercises 67 and 68 provide experiences in writing.

67. Writing to Learn Table 4.5 gives function values for $y = f(x)$ and $y = g(x)$. Also, three different graphs are shown.

Table 4.5 Data for Two Functions

x	$f(x)$	$g(x)$
1.0	5.50	7.40
1.5	5.35	6.97
2.0	5.25	6.44
2.5	5.17	5.76
3.0	5.13	4.90
3.5	5.09	3.82
4.0	5.06	2.44
4.5	5.05	0.71

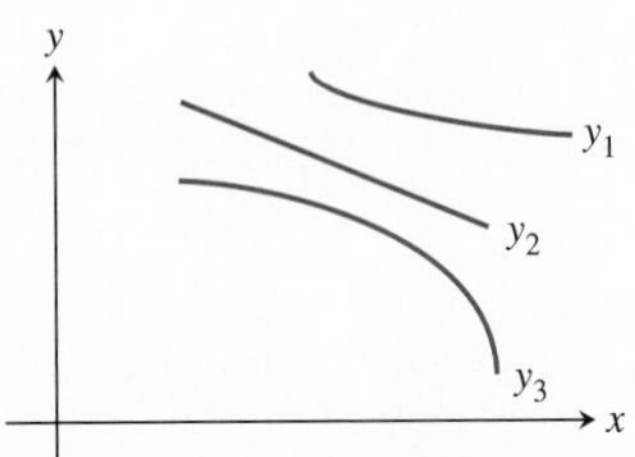

For Exercise 67

a. Which curve of those shown in the graph most closely resembles the graph of $y = f(x)$? Explain your choice.

b. Which curve most closely resembles the graph of $y = g(x)$? Explain your choice.

68. Writing to Learn Let $f(x) = 2^x$. Explain why the graph of $f(ax + b)$ can be obtained by applying one transformation to the graph of $y = c^x$ for an appropriate value of c. What is c?

In Exercises 69 and 70, identify a specific function by finding appropriate values for the constants k and a.

69. The graph of the exponential function $f(x) = ka^x$ passes through the points (1, 6) and (2, 54). Find the values of k and a and specify the function.

70. The power function $g(x) = kx^a$ passes through the points (2, 12) and (3, 27). Find the values of k and a, and specify the function.

Exercises 71–74 refer to the expression $f(a, b, c) = ab^c$. Notice that if $a = 2$, $b = 3$, and $c = x$, the expression is $f(2, 3, x) = 2 \cdot 3^x$, an exponential function.

71. If $b = x$, state conditions on a and c under which the expression $f(a, b, c)$ is a quadratic power function.

72. If $b = x$, state conditions on a and c under which the expression $f(a, b, c)$ is a decreasing linear function.

73. If $c = x$, state conditions on a and b under which the expression $f(a, b, c)$ is an increasing exponential function.

74. If $c = x$, state conditions on a and b under which the expression $f(a, b, c)$ is a decreasing exponential function.

Objective

Students will be able to graph logarithmic functions and convert equations between logarithmic form and exponential form.

Motivate

Graph an exponential function such as $y = 2^x$, and have students discuss the graph of its inverse function.

4.2 LOGARITHMIC FUNCTIONS: INVERSES OF EXPONENTIAL FUNCTIONS

Introduction to Logarithmic Functions • Logarithms with Base 10 • The Number e and Natural Logarithms • Graphs of Logarithmic Functions • Applications • Modeling Data with Logarithmic Functions • Recognizing the Proper Model

Introduction to Logarithmic Functions

The graph of an exponential function $y = a^x$, where $a > 0$ and $a \neq 1$, looks like one of the two graphs in Figure 4.12. We see that these graphs, in either of the two forms, satisfy the horizontal line test. Consequently each exponential function is *one-to-one* and its inverse is a function. The inverse of an exponential function $y = a^x$ with base a is called the **logarithmic function with base *a*,** denoted $y = \log_a x$.

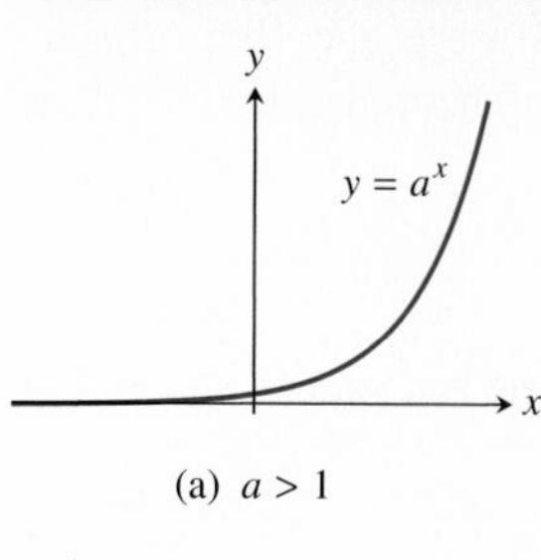

(a) $a > 1$

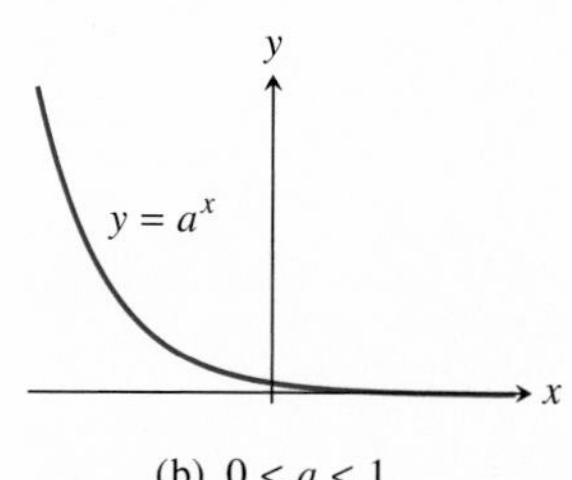

(b) $0 < a < 1$

Figure 4.12

Logarithms with Base 10

The logarithmic function $y = \log_{10} x$ is the inverse of the exponential function $y = 10^x$. As we learned in Section 2.7, another form for the inverse of $y = 10^x$ is $x = 10^y$, so

$$y = \log_{10} x \qquad \text{and} \qquad 10^y = x$$

are equivalent equations for the logarithmic function. In other words, $\log_{10} x$ is nothing more than an exponent!

Definition: Logarithms with Base 10

$$y = \log_{10} x \quad \text{if and only if} \quad 10^y = x.$$

In particular, $10^{\log_{10} x} = x$, and $y = \log_{10} 10^y$.

Applying these relationships, we can convert between the **logarithmic form** and the **exponential form:**

Logarithmic form:

$$\log_{10} 100 = 2 \qquad \log_{10} 1000 = 3 \qquad \log_{10} \frac{1}{1000} = -3$$

Exponential form:

$$10^2 = 100 \qquad 10^3 = 1000 \qquad 10^{-3} = \frac{1}{1000}$$

We shall abbreviate $y = \log_{10} x$ as $y = \log x$, just as your calculator does. Because powers of 10 are positive (that is, $x = 10^y > 0$), $\log x$ is defined only for positive values of x. Logarithms with base 10 are called **common logarithms.** Common logarithms can be evaluated by using the "log" key on a calculator. Here are two logarithmic statements from a calculator and the corresponding exponential statements:

Logarithmic form:	$\log 34.5 = 1.537 \cdots$	$\log 0.43 = -0.366 \cdots$
Exponential form:	$10^{1.537\cdots} = 34.5$	$10^{-0.366\cdots} = 0.43$

Changing from logarithmic form to exponential form and vice versa are important steps in solving equations that involve exponential and logarithmic functions.

■ EXAMPLE 1 Solving Equations with Logarithms and Exponents

Solve.

a. $10^x = 4.6$ **b.** $\log x = -1.3$

Solution

a. $10^x = 4.6$

$x = \log 4.6$ Change to logarithmic form.

$x = 0.662 \cdots$

b. $\log x = -1.3$

$x = 10^{-1.3}$ Change to exponential form.

$x = 0.050 \cdots$

■

X	Y_1	
1000	2.7169	
2000	2.7176	
3000	2.7178	
4000	2.7179	
5000	2.718	
6000	2.7181	
7000	2.7181	
Y_1=(1+1/X)^X		

Figure 4.13 Defining expression for e.

The Number e and Natural Logarithms

Many natural, physical, and economic phenomena are best modeled by an exponential or logarithmic function whose base is the famous number $e = 2.71828\cdots$, defined as the number that the expression $(1 + \frac{1}{x})^x$ approaches as the value of x gets large. (See Figure 4.13.)

The logarithmic function $y = \log_e x$ so frequently models natural phenomena that it is called the **natural logarithm.** It is denoted, without using the subscript e, as $y = \ln x$. The function $y = \ln x$ is defined in a manner analogous to $y = \log x$.

Teaching Note

Using a grapher to explore the concept of the number e is one of the more powerful visualizations in calculus.

Definition: Logarithms with Base e

$$y = \ln x \quad \text{if and only if} \quad e^y = x.$$

In particular, $e^{\ln x} = x$, and $y = \ln e^y$.

Using this meaning of ln x we can find the following values:

$$\ln e^1 = 1 \qquad \ln e^2 = 2 \qquad \ln e^{-2} = -2 \qquad \ln e^{\sqrt{3}} = \sqrt{3}$$

Natural logarithms can be evaluated by using the "ln" key on a calculator. Here are two natural logarithmic statements from a calculator and the corresponding exponential statements.

Logarithmic form:	$\ln 23.5 = 3.157\cdots$	$\ln 0.48 = -0.733\cdots$
Exponential form:	$e^{3.157\cdots} = 23.5$	$e^{-0.733\cdots} = 0.48$

■ EXAMPLE 2 Solving Equations

Solve.

a. $\ln x = 4.62$ **b.** $e^x = 7.68$

Solution

a. $\ln x = 4.62$

$x = e^{4.62}$ Change to exponential form.

$x = 101.494\cdots$

b. $e^x = 7.68$

$x = \ln 7.68$ Change to logarithmic form.

$x = 2.038\cdots$ ■

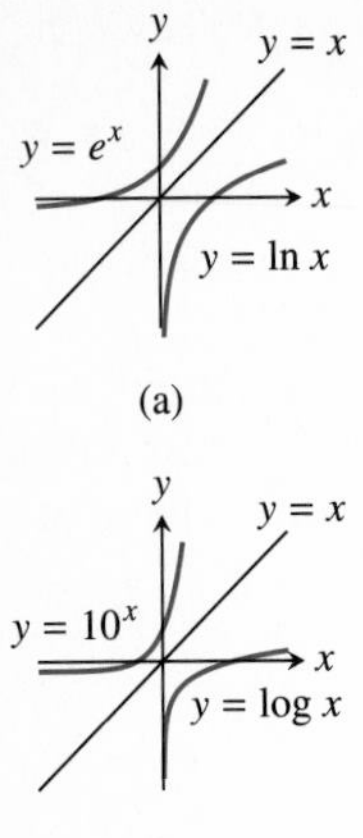

Figure 4.14

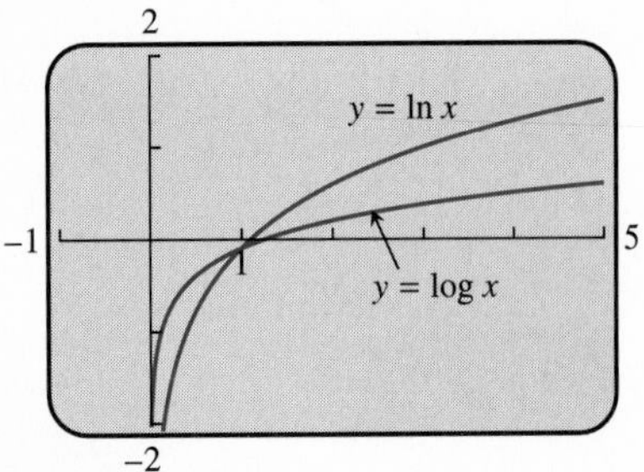

Figure 4.15

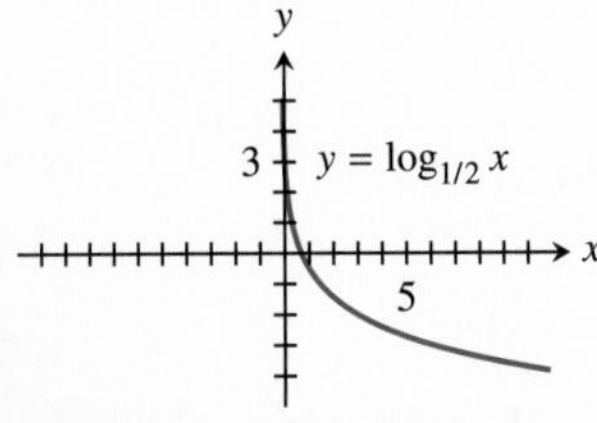

Figure 4.16 Reflect the graph of $y = (\frac{1}{2})^x$ across the line $y = x$ to get $\log_{1/2} x$. You will see how to obtain this graph with your graphing utility in Section 4.3.

Graphs of Logarithmic Functions

The graphs of the logarithmic functions $y = \log x$ and $y = \ln x$ can be found using the fact that they are inverses of $y = 10^x$ and $y = e^x$, respectively. In fact, graphs of logarithmic functions with any base can be found this way. Figure 4.14 shows that the exponential and logarithmic functions of the same bases are reflections of each other across the line $y = x$.

Using the exponential function end behavior description given in Section 4.1 we can conclude that the end behavior of the logarithmic functions $y = \log x$ and $y = \ln x$ is

$$f(x) \to \infty \text{ as } x \to \infty$$

and

$$f(x) \to -\infty \text{ as } x \to 0^+$$

where 0^+ means that x approaches 0 through values greater than 0.

Notice that it is hard to read this information from their graphs in Figure 4.15.

Because the point (0, 1) lies on the graphs of both $y = 10^x$ and $y = e^x$, it follows that the point (1, 0) lies on the graphs of both $y = \log x$ and $y = \ln x$. (See Figure 4.15.)

You will be able to use your graphing utility to graph logarithmic functions with bases other than 10 or e using the change-of-base formula given in Section 4.3. Here are the characteristics of logarithmic function graphs.

Graphs of Logarithmic Functions

Consider a logarithmic function in the form $f(x) = \log_a x$.

1. If $a > 1$, the graph of $f(x) = \log_a x$

a. passes through the point (1, 0),
b. is shaped like the graph of $y = \ln x$, and
c. has end behavior like $y = \ln x$:

$$f(x) \to \infty \text{ as } x \to \infty \quad \text{and}$$
$$f(x) \to -\infty \text{ as } x \to 0^+.$$

2. If $0 < a < 1$, the graph of $f(x) = \log_a x$

a. passes through the point (1, 0),
b. is shaped like the graph of $y = \log_{1/2} x$ (see Figure 4.16), and
c. has end behavior like $y = \log_{1/2} x$:

$$f(x) \to -\infty \text{ as } x \to \infty \quad \text{and}$$
$$f(x) \to \infty \text{ as } x \to 0^+.$$

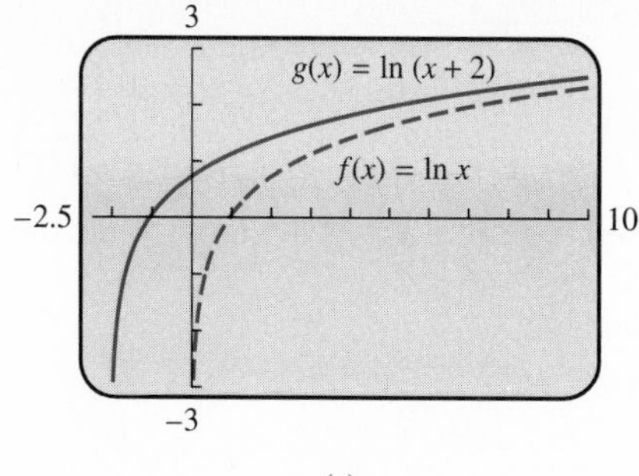

(a)

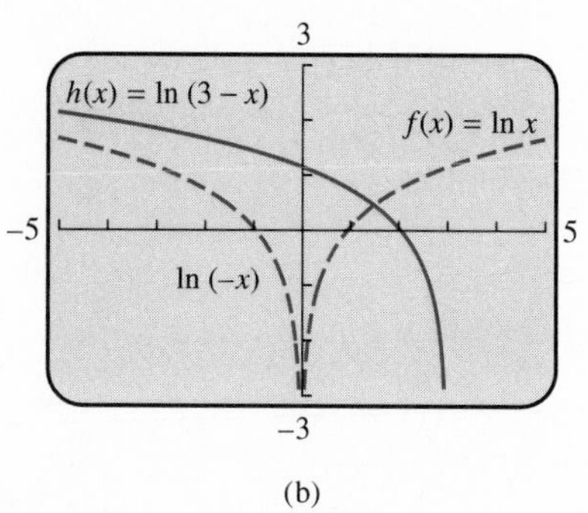

(b)

Figure 4.17 Transforming $f(x) = \ln x$ to get (a) $g(x) = \ln (x + 2)$, and (b) $h(x) = \ln (3 - x)$.

Teaching Note

Note that 1 cannot be the base of a logarithmic function because $1^y = x$ has no solution for most real numbers x.

The transformations studied in Section 2.6 can be applied to any function. Example 3 illustrates their application to logarithmic functions.

EXAMPLE 3 Using Transformations with Logarithmic Functions

Describe transformations that will transform $f(x) = \ln x$ to:

a. $g(x) = \ln (x + 2)$ **b.** $h(x) = \ln (3 - x)$

Solution

Solve Algebraically

a. $g(x) = \ln (x + 2)$
$= f(x + 2)$

So g can be obtained from f by a translation of 2 units to the left.

b. $h(x) = \ln (3 - x)$
$= \ln [-(x - 3)]$
$= f(-(x - 3))$ A reflection followed by a horizontal translation

So h can be obtained from f by applying, in order, a reflection across the y-axis followed by a translation 3 units to the right.

Support Graphically

The graphs of g and h are shown in Figure 4.17. ■

Applications

Logarithmic functions model a variety of scientific and natural phenomena. Example 4 illustrates one such situation.

EXAMPLE 4 APPLICATION: Modeling Drug Absorption

A certain drug is administered intravenously for pain. Beginning with 100 cc (cubic centimeters) of drug, the number of cubic centimeters present in the body after t hours, $f(t)$, is modeled by

$$f(t) = 100 - 43 \ln (1 + t), \qquad 0 \le t \le 6.$$

How long will it be until 25 cc of the drug remain?

Solution

Model

Because $f(t) = 100 - 43 \ln (1 + t)$ models the number of cubic centimeters of drug in the body at time t, we need to solve the equation

$$25 = 100 - 43 \ln (1 + t).$$

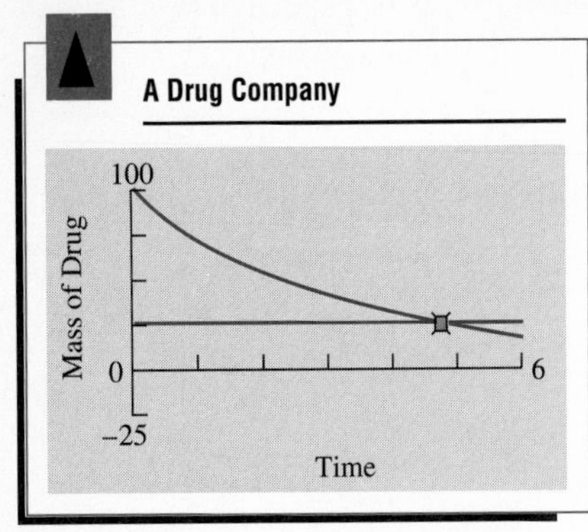

Intersection: $x = 4.7212428$; $y = 25$

Figure 4.18

Solve Algebraically

$$43 \ln(1 + t) + 25 = 100 \qquad \text{Add } 43 \ln(1 + t).$$

$$43 \ln(1 + t) = 100 - 25 \qquad \text{Subtract 25.}$$

$$\ln(1 + t) = \frac{75}{43} \qquad \text{Divide by 43.}$$

$$1 + t = e^{75/43} \qquad \text{Change to exponential form.}$$

$$t = e^{75/43} - 1$$

$$t = 4.721 \cdots$$

Support Graphically

The graphs of $y_1 = 100 - 43 \ln(1 + t)$ and $y_2 = 25$ (see Figure 4.18) intersect at $t = 4.721 \cdots$. ■

Modeling Data with Logarithmic Functions

Sometimes the curve of best fit through a set of data is a logarithmic curve.

Using Data Analysis—Logarithmic Regression

1. Complete a scatter plot of the data.
2. Find a logarithmic regression equation $y = a + b \ln x$ or $y = a + b \log x$.
3. Superimpose a graph of the logarithmic regression equation on the scatter plot.
4. Use the logarithmic regression equation to predict y values for x values not included in the data.

■ EXAMPLE 5 Using Data Analysis

Apply the logarithmic model to the following data. Find a logarithmic regression equation $y = a + b \ln x$ that fits the data. Then predict a value of y for $x = 9$.

x	1	2	3	4	5	6
y	10	17	21	24	26	27

Solution

Solve Graphically

We compute the logarithmic regression equation to be $y = 10.2 + 9.7 \ln x$ (see the Grapher Workshop). Its graph is superimposed on the scatter plot in Figure 4.19.

Using the regression, we find that $y(9) = 10.2 + 9.7 \ln (9) \approx 31.51$. ■

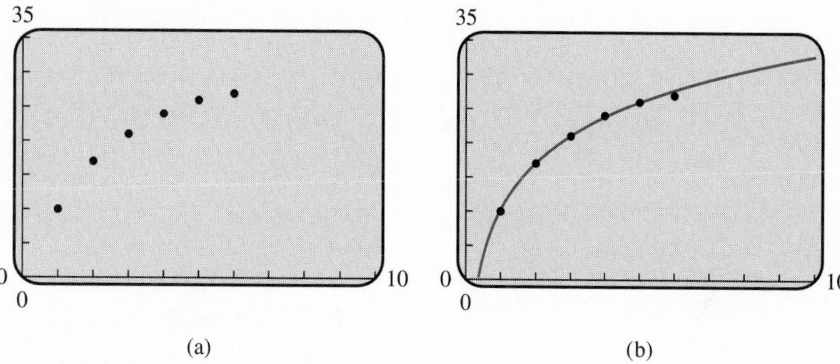

Figure 4.19

Follow-up

Ask students to determine the domain and range of the function $y = \log_a x$. Do the domain and range depend on the value of a?
(No, assuming $a > 0$, $a \neq 1$.)

Assignment Guide

Day 1: Ex. 3–48, multiples of 3
Day 2: Ex. 50–53, 55, 57, 58, 59, 62–64, 66, 68, 69

Cooperative Learning

Group Activity: Ex. 49

Notes on Exercises

Ex. 51–56 have several steps. Students need to read the directions for these problems.
Ex. 57–60 and 63 show how linear regression is related to logarithmic, exponential, and power regression.
Ex. 65–68 require students to determine what kind of regression is best for a particular set of data.

Ongoing Assessment

Self-Assessment: Ex. 13, 19, 37, 47, 61
Embedded Assessment: Ex. 52, 63

Recognizing the Proper Model

When analyzing and studying data, we often choose to find a regression equation of best fit for the data. Here is a summary of four types of regression equations from which we might choose.

Standard Regression Equations

1. **Linear regression:** $y = ax + b$
2. **Natural logarithmic regression:** $y = a + b \ln x$
3. **Exponential regression:** $y = a(b^x)$
4. **Power regression:** $y = a(x^b)$

When we examine a scatter plot of data points (x, y), we should ask which of the four regression equations is the best choice for the data? If the data appear to be linear, a linear regression is the best choice. But when it is visually evident that the data are not linear, the best choice may be a natural logarithmic, exponential, or power regression.

Knowing the shapes of the logarithmic, exponential, and power graphs can suggest the choice. Otherwise it may be helpful to convert the (x, y) data pairs to the pairs ($\ln x$, y), (x, $\ln y$), and ($\ln x$, $\ln y$). If any set of these new data points appear to be linear, then that set suggests which regression equation to use. Figures 4.20–4.22 provide a summary.

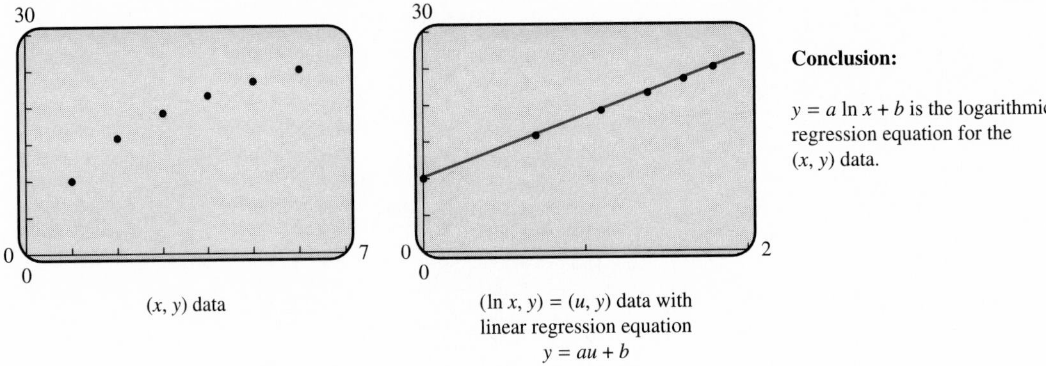

Figure 4.20 Natural logarithmic regression

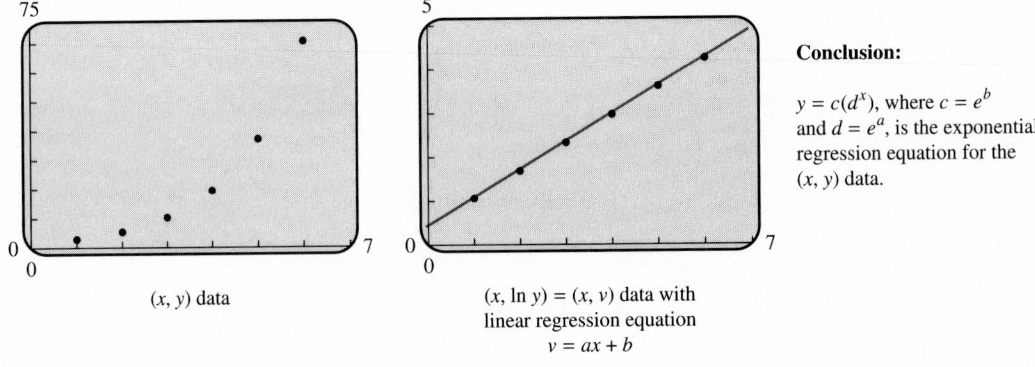

Figure 4.21 Exponential regression

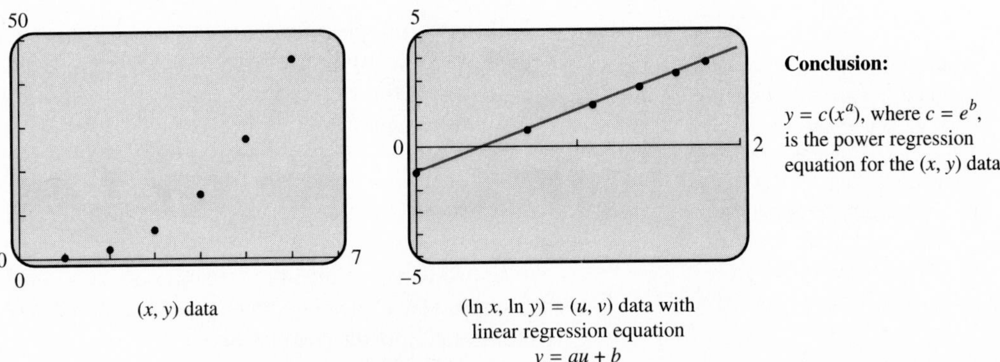

Figure 4.22 Power regression

The three regression equations can be justified algebraically. For exponential regression, for example,

$$
\begin{aligned}
v &= ax + b \\
\ln y &= ax + b && v = \ln y. \\
y &= e^{ax+b} && \text{Change to exponential form.} \\
y &= e^{ax}e^{b} && \text{Use the laws of exponents.} \\
y &= e^{b}(e^{a})^{x} \\
y &= c(d^{x}) && \text{Let } c = e^{b} \text{ and } d = e^{a}.
\end{aligned}
$$

A similar algebraic argument can be given for the other regression equations.

EXAMPLE 6 Selecting a Regression Model

Decide whether these data can be best modeled by logarithmic, exponential, or power regression. Find the appropriate regression equation.

x	1	2	3	4	5	6
y	2	5	10	17	26	38

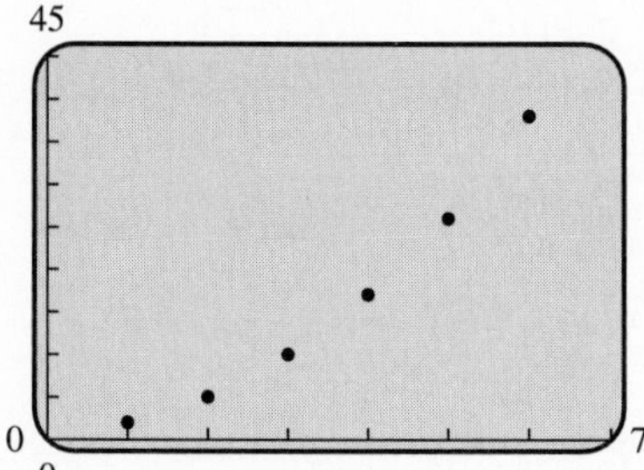

Figure 4.23

Solution

The scatter plot of the data in Figure 4.23 suggests that the data might be modeled by an exponential or power regression equation.

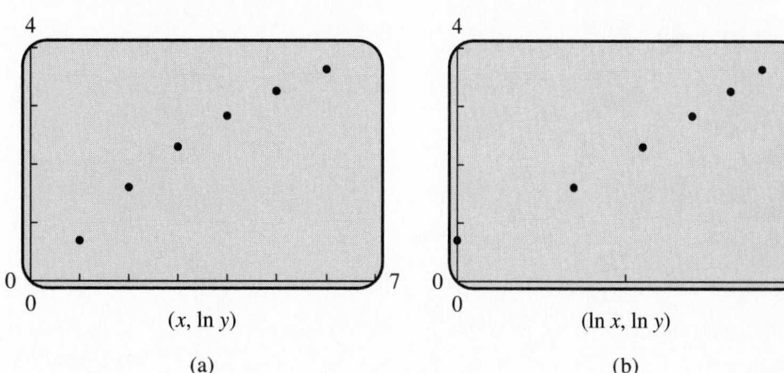

Figure 4.24

Figure 4.24a shows the (x, ln y) scatter plot, and Figure 4.24b shows the (ln x, ln y) scatter plot. Of these two plots, (ln x, ln y) appears to be more linear; so we should find the power regression for the original data.

Figure 4.25 shows the scatter plot of the original (x, y) data with the graph of the power regression equation $y = 1.79x^{1.65}$ superimposed.

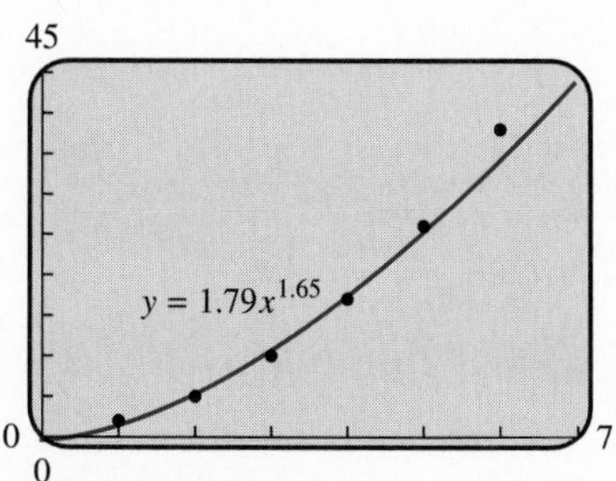

Figure 4.25 Graph of (x, y) and $y = 1.79x^{1.65}$ for Example 6.

Quick Review 4.2

In Exercises 1–6, use the properties and definition of exponents to evaluate the expression.

1. 5^{-2}
2. 10^{-3}
3. $\frac{4^0}{5}$
4. $\frac{1^0}{2}$
5. $\frac{8^{12}}{2^{28}}$
6. $\frac{9^{17}}{27^8}$

In Exercises 7–10, use the horizontal line test (Section 2.7) to determine if the inverse of the function is also a function. If the inverse is a function, sketch its graph.

7. 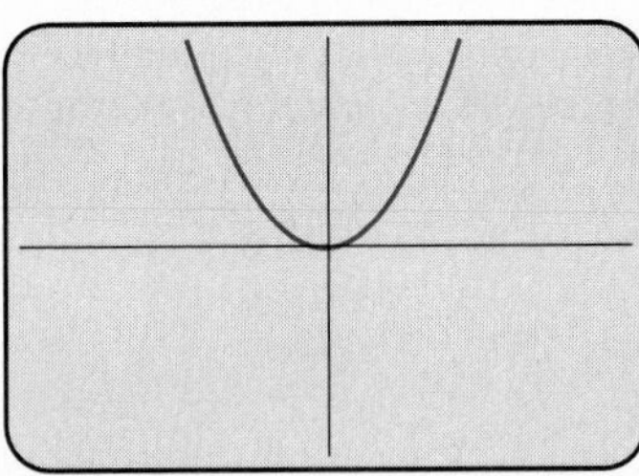

8. 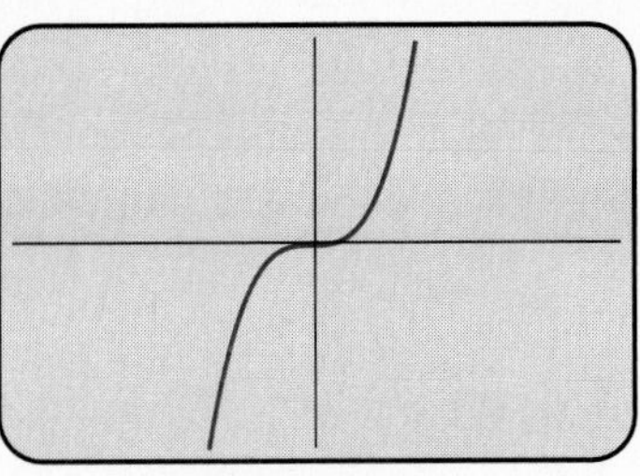

9. 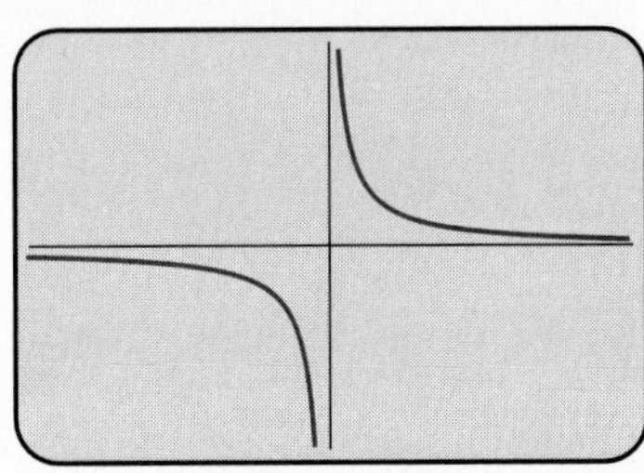

10. 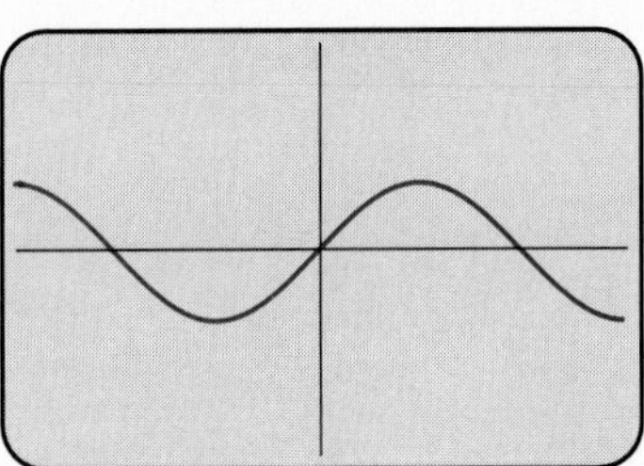

SECTION EXERCISES 4.2

In Exercises 1–8, evaluate the expression.

1. $\log 10^3$
2. $\log 10{,}000$
3. $\log 100{,}000$
4. $\log 10^{-4}$
5. $\ln e^3$
6. $\ln e^{-4}$
7. $\ln \frac{1}{e}$
8. $\ln 1$

In Exercises 9–16, solve the equation by changing to exponential form.

9. $\log x = 2$
10. $\log x = 4$
11. $\log x = -1$
12. $\log x = -3$
13. $\ln x = 3$
14. $\ln x = 5$
15. $\ln x = -2$
16. $\ln x = -1$

In Exercises 17–22, solve by changing the equation to logarithmic form.

17. $10^x = 3$
18. $10^x = 5.1$
19. $e^x = 4.2$
20. $e^x = 7.3$
21. $e^{2x} = 5.3$
22. $10^{3x} = 9.2$

In Exercises 23–26, solve the equation. (Use a calculator if necessary.)

23. $\ln x = 2.8$
24. $\ln x = 3.1$

25. $\log x = 8.23$

26. $\log x = 5.25$

In Exercises 27–30, match the function with its graph.

27. $y = \log(1 - x)$

28. $y = \log(x + 1)$

29. $y = -\ln(x - 3)$

30. $y = -\ln(4 - x)$

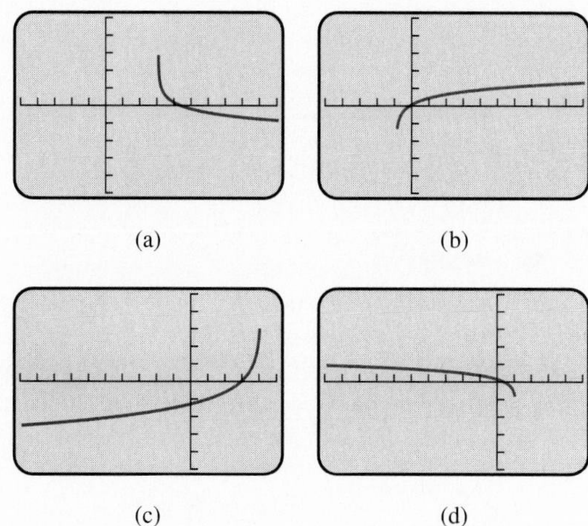

(a) (b) (c) (d)

For Exercises 27–30

In Exercises 31–36, find the intercepts and describe the end behavior of the function.

31. $y = \log(x - 2)$

32. $y = \ln(x + 1)$

33. $y = -\ln(x - 1)$

34. $y = -\log(x + 2)$

35. $y = \ln(3 - x)$

36. $y = \log(2 - x)$

In Exercises 37–40, describe transformations that can be used to transform the graph of one of $y = \ln x$ or $y = \log x$ to a graph of the given function.

37. $f(x) = \ln(x + 3)$

38. $f(x) = \ln(x - 2)$

39. $f(x) = 2\log(x + 1) - 3$

40. $f(x) = 3\log(x - 1) + 2$

In Exercises 41–48, sketch a graph of the given function. Then check your sketch with a grapher.

41. $f(x) = \ln(-x) + 3$

42. $f(x) = \ln(-x) - 2$

43. $f(x) = -2\log(-x)$

44. $f(x) = -3\log(-x)$

45. $f(x) = \ln(2 - x)$

46. $f(x) = \ln(5 - x)$

47. $f(x) = 2\log(3 - x) - 1$

48. $f(x) = -3\log(1 - x) + 1$

In Exercises 49 and 50, solve the given problem.

49. *Drug Absorption* A drug is administered intravenously for pain. The function $f(t) = 80 - 23\ln(1 + t)$, $0 \le t \le 24$, models the amount of the drug in the body after t hours.

a. What was the initial ($t = 0$) number of units of drug administered?
b. How much is present after 2 h?
c. Draw the graph of f.

50. *Light Absorption* The Beer-Lambert law of absorption applied to Lake Erie states that the light intensity I (in lumens), at a depth of x feet, satisfies the equation

$$\log \frac{I}{12} = -0.00235x.$$

Find the intensity of the light at a depth of 30 ft.

In Exercises 51–56, use the given (x, y) data.

a. Find a scatter plot of the data.
b. Use a graphing utility to find the indicated regression equation for this data.
c. Specify a viewing window for a graph of both the scatter plot and the regression equation. Draw the graph.
d. Use the regression equation for a prediction.

51. Estimate the y-value associated with $x = 25$ as predicted by the logarithmic regression equation for the data

x	10	20	30	40
y	1.61	2.99	3.80	4.38

52. Estimate the y-value associated with $x = 13$ as predicted by the logarithmic regression equation for the data.

x	6	12	18	24
y	7.38	9.45	10.67	11.53

53. Estimate the y-value associated with $x = 7$ as predicted by the exponential regression equation for the data.

x	2	4	6	8
y	5.63	12.66	28.48	64.07

54. Estimate the y-value associated with $x = 7$ as predicted by the exponential regression equation for the data.

x	1	2	4	6
y	7.5	3.75	0.94	0.23

55. Estimate the y-value associated with $x = 7.1$ as predicted by the power regression equation for the data.

x	4	6.5	8.5	10
y	2816	31,908	122,019	275,000

56. Estimate the y-value associated with $x = 9.2$ as predicted by the power regression equation for the data.

x	2	3	4.8	7.7
y	7.48	7.14	6.81	6.41

In Exercises 57–60, solve the given problem.

57. Complete an $(\ln x, y)$ scatter plot for the data in Exercise 51. Then find the linear regression equation $y = au + b$ for the $(\ln x, y)$ data. Confirm that $y = a \ln x + b$ is the logarithmic regression equation found in Exercise 51.

58. Complete an $(x, \ln y)$ scatter plot for the data in Exercise 53. Then find the linear regression equation $v = ax + b$ for the $(x, \ln y)$ data. Confirm that $y = e^b \cdot (e^a)^x$ is the exponential regression equation found in Exercise 53.

59. Complete an $(\ln x, \ln y)$ scatter plot for the data in Exercise 55. Then find the linear regression equation $v = au + b$ for the $(\ln x, \ln y)$ data. Confirm that $y = e^b \cdot x^a$ is the power regression equation found in Exercise 55.

60. Complete an $(\ln x, \ln y)$ scatter plot for the data in Exercise 56. Find the linear regression equation for this scatter plot. Use the constants from this equation to find the power equation for the original data.

In Exercises 61 and 62, let $x = 1$ represent 1960, $x = 11$ represent 1970, and $x = 31$ represent 1990 for the given data about oil production.

61. *Algerian Oil Production* Find a natural logarithmic regression equation for the data in Table 4.6. Use it to estimate production for the year 1980.

Table 4.6 Algerian Oil Production

Year	Metric Tons
1960	8.63
1970	47.25
1990	56.67

Source: The Statesman's Yearbook, 129th ed. (London: The Macmillan Press, Ltd., 1992).

62. *Canadian Oil Production* Find a natural logarithmic regression equation for the data in Table 4.7. Use it to estimate production for the year 1985.

Table 4.7 Canadian Oil Production

Year	Metric Tons
1960	27.48
1970	69.95
1990	92.24

Source: The Statesman's Yearbook, 129th ed. (London: The Macmillan Press, Ltd., 1992).

Exercises 63 and 64 refer to the data in Table 4.8 about our sun's planets. The orbit of each planet is an ellipse with the sun at one focus. Let T be the orbit period (that is, the time in days required for one full revolution around the sun), and let x be the length (miles) of the semimajor axis of the planet's orbit.

Table 4.8 Planetary Data

Planet	Period of Revolution, T	Semimajor Axis x
Mercury	88	36,000,000
Venus	225	67,100,000
Earth	365	92,600,000
Mars	687	141,700,000
Jupiter	4,330	483,000,000
Saturn	10,750	886,100,000

63. *Planetary Motion* Confirm that the quantities T and x in the Table 4.8 are related by a power equation of the form $T = ax^m$ for constants a and m by following steps a–c:

a. Complete an $(\ln T, \ln x)$ scatter plot for the data in Table 4.8.

b. Find the linear regression equation for the data in part a.

c. **Writing to Learn** Explain how the slope and the y-intercept of the linear function found in part b give us the values of the constants a and m in $T = ax^m$.

64. Use the equation found in part c of Exercise 63 to estimate the period of Pluto's orbit if the length of its semimajor axis is 3.66 billion mi.

In Exercises 65–68, tables of (x, y) data pairs are given. Determine whether a linear, logarithmic, exponential, or power regression equation is the best choice for the data in each. Explain your method.

65.

x	1	2	3	4
y	3	4.4	5.2	5.8

66.

x	1	2	3	4
y	6	18	54	162

67.

x	1	2	3	4
y	3	6	12	24

68.

x	1	2	3	4
y	5	7	9	11

EXTENDING THE IDEAS

In Exercises 69 and 70, use an exponential function to help you sketch the graph of the logarithmic function.

69. $f(x) = \log_3 x$ **70.** $g(x) = \log_5 x$

In Exercises 71 and 72, answer the questions.

71. Writing to Learn Explain why zero is not in the domain of the logarithmic functions $y = \log_3 x$ and $y = \log_5 x$.

72. Show that for each positive number c such that $0 < c < 1$, $\log c$ is defined.

4.3 PROPERTIES OF LOGARITHMIC FUNCTIONS; EQUATION SOLVING

Graphs of Logarithmic Functions with Base a • Change-of-Base Formula • Properties of Logarithmic Functions • Solving Equations • Application—Earthquake Intensity

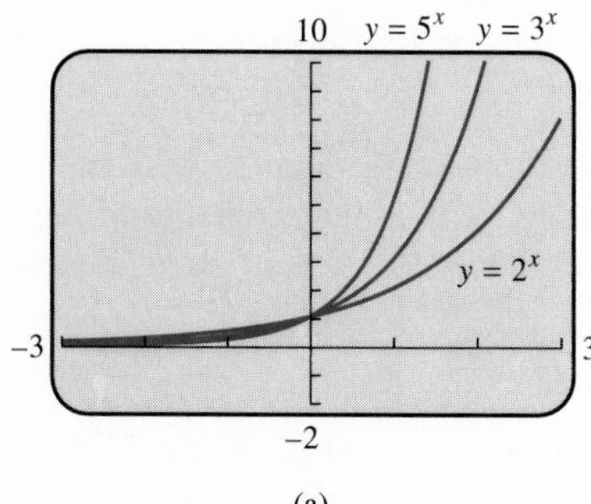

(a)

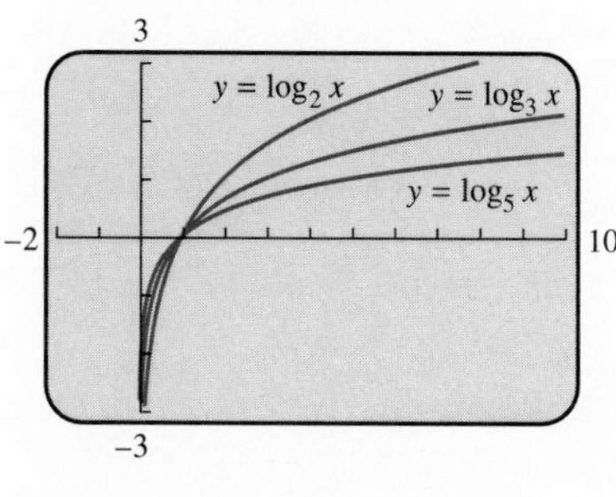

(b)

Figure 4.26

Graphs of Logarithmic Functions with Base *a*

You learned in Section 4.1 that there is an exponential function $f(x) = a^x$ for each positive real number $a \neq 1$. Because the graph of $f(x) = a^x$ satisfies the horizontal line test, the inverse of f is a function. We call it the logarithmic function and represent it as $g(x) = \log_a x$. Certain characteristics of exponential functions are carried over to the corresponding logarithmic functions.

For example, if $a > 1$, the following three properties of $f(x) = a^x$:

a. $f(0) = 1$;
b. the domain of f is $(-\infty, \infty)$, the range of f is $(0, \infty)$; and
c. f is one-to-one and increasing

become these three properties of the logarithmic function $g(x) = \log_a x$:

a. $g(1) = 0$;
b. the domain of g is $(0, \infty)$, the range of g is $(-\infty, \infty)$;
c. g is one-to-one and increasing.

Notice in Figure 4.26 that the rate at which $f(x) = a^x$ increases depends on the value of a. Correspondingly, the rate at which $g(x) = \log_a x$ increases also depends on the value of a.

Objective

Students will be able to apply the properties of logarithms to solve equations algebraically.

Motivate

Have students use a grapher to compare the graphs of $y = \log x$ and $y = \log 10x$. Discuss the graphs.

Change-of-Base Formula

Calculators and graphers have at most two logarithmic keys—"ln" and "log," which correspond to the bases e and 10, respectively. For logarithmic functions with bases a other than e or 10, we can use one of these two equations:

$$y = \log_a x = \frac{\ln x}{\ln a} \qquad y = \log_a x = \frac{\log x}{\log a}$$

We will verify these equations later in this section. (See Example 6 and Exercise 79.)

Teaching Note

As a mnemonic to help students remember the change-of-base formula, tell them to place the "base in the basement."

EXAMPLE 1 Evaluating Logarithmic Functions

Evaluate.

a. $\log_3 16$ **b.** $\log_6 10$.

Solution

We can use either base e or base 10. We somewhat arbitrarily choose base e.

a. $\log_3 16 = \dfrac{\ln 16}{\ln 3} = 2.523 \cdots$ **b.** $\log_6 10 = \dfrac{\ln 10}{\ln 6} = 1.285 \cdots$

Properties of Logarithmic Functions

Your experiences with a wide variety of algebraic expressions may sometimes lead you to draw false conclusions regarding some of the properties of logarithmic functions. The following Exploration may lead you to discover some relationships.

Exploration Extensions

Repeat the Exploration exercise with the following relationships:

9. $\log_3 5 = (\log_3 x)(\log_x 5)$
10. $e^{\ln 6 + \ln x} = 6x$

EXPLORATION Discovering Relationships

Of the eight relationships suggested here, four are true and four are false. Think about what you know from arithmetic, and make a prediction about each. Then test each with some actual numerical values. Finally, compare the graphs of both sides of the equation.

1. $\ln(x + 2) = \ln x + \ln 2$ **2.** $\log_3 7x = 7 \log_3 x$

3. $\log_2 (5x) = \log_2 5 + \log_2 x$ **4.** $\ln \dfrac{x}{5} = \ln x - \ln 5$

5. $\log \dfrac{x}{4} = \dfrac{\log x}{\log 4}$ **6.** $\log_4 x^3 = 3 \log_4 x$

7. $\log_5 x^2 = (\log_5 x)(\log_5 x)$ **8.** $\log |4x| = \log 4 + \log |x|$

Try stating some conclusions about what is not true and some conjectures about what might be true.

You may have been surprised in the Exploration by the "properties" that are false and by those that may be true. In general, logarithms have the following properties, and if you remember that logarithms can be thought of as exponents, then these properties should not seem particularly unreasonable.

Properties of Logarithms

Let a, r, and s be positive real numbers such that $a \neq 1$. Then the following statements are true:

1. **Product rule:** $\log_a rs = \log_a r + \log_a s$
2. **Quotient rule:** $\log_a \frac{r}{s} = \log_a r - \log_a s$
3. **Power rule:** $\log_a r^c = c \log_a r$ for every real number c.

PROPERTIES OF EXPONENTS

1. $a^n \cdot a^m = a^{n+m}$
2. $\frac{a^n}{a^m} = a^{n-m}$
3. $(a^n)^m = a^{nm}$

The properties of exponents lead to the properties of logarithms. For example, the first exponent property listed in the margin is used to verify the logarithmic property called the product rule.

In Exercise 80 you will prove the quotient and power rules for logarithms. Example 1 illustrates techniques that can be used to give the proofs.

■ EXAMPLE 2 Proving the Product Rule for Logarithms

Verify that $\log_a rs = \log_a r + \log_a s$.

Solution

Let $x = \log_a r$ and $y = \log_a s$. The corresponding exponential statements are $a^x = r$ and $a^y = s$. Therefore,

$$\begin{aligned} rs &= a^x \cdot a^y \\ &= a^{x+y} && \text{Use properties of exponents.} \\ \log_a rs &= x + y && \text{Change to logarithmic form.} \\ &= \log_a r + \log_a s && \text{Use the definition of } x \text{ and } y. \end{aligned}$$ ■

When we solve equations algebraically that involve logarithms, we often have to rewrite expressions using properties of logarithms. Sometimes we need to expand as far as possible, and other times we combine as much as possible. The next three examples illustrate how properties of logarithms can be used to change the form of expressions involving logarithms.

■ EXAMPLE 3 Rewriting the Logarithm of a Product

Use properties of logarithms to rewrite $\log 8xy^4$, where $y > 0$, as a sum of logarithms.

Solution

$$\begin{aligned} \log 8xy^4 &= \log 8 + \log x + \log y^4 && \text{Property 1} \\ &= \log 2^3 + \log x + \log y^4 \\ &= 3 \log 2 + \log x + 4 \log y && \text{Property 3} \end{aligned}$$

■

■ EXAMPLE 4 Rewriting the Logarithm of a Quotient

Use properties of logarithms to rewrite $\ln \dfrac{\sqrt{x^2+5}}{x}$ as a sum and/or difference of logarithms.

Solution

$$\begin{aligned} \ln \frac{\sqrt{x^2+5}}{x} &= \ln \frac{(x^2+5)^{1/2}}{x} \\ &= \ln (x^2+5)^{1/2} - \ln x \\ &= \frac{1}{2} \ln (x^2+5) - \ln x \end{aligned}$$

■

■ EXAMPLE 5 Condensing a Logarithmic Expression

Rewrite $\ln|x^2 - 1| - \ln|x+1|$, where $x \neq -1$, as a single logarithm.

Solution

$$\begin{aligned} \ln|x^2-1| - \ln|x+1| &= \ln \frac{|x^2-1|}{|x+1|} \\ &= \ln \left| \frac{x^2-1}{x+1} \right| && \left|\frac{a}{b}\right| = \frac{|a|}{|b|} \\ &= \ln \left| \frac{(x+1)(x-1)}{(x+1)} \right| \\ &= \ln|x-1| \end{aligned}$$

■

Notes on Examples

Example 6 is important because it shows how the change-of-base formula can be derived from the definition of the logarithm.

■ EXAMPLE 6 Developing a Change-of-Base Formula

Suppose that $a > 0$ and $a \neq 1$. Verify that $\log_a x = \dfrac{\ln x}{\ln a}$.

Solution

The exponential statement associated with $y = \log_a x$ is $a^y = x$. We have

$$\begin{aligned} a^y &= x \\ \ln a^y &= \ln x \quad \text{Take ln of both sides.} \\ y \ln a &= \ln x \\ y &= \frac{\ln x}{\ln a} \end{aligned}$$

■

Solving Equations

When a logarithmic or exponential equation involves base 10 or base *e*, as in Section 4.2, it is convenient to solve the equation by changing back and forth between exponential and logarithmic forms, as required. However, when the base is neither 10 nor *e*, it is often convenient to use the two properties listed in the following box.

For any function *f*, if $x = y$, then $f(x) = f(y)$. However, the converse is true only if *f* is one-to-one. Because $f(x) = \log_a x$ is one-to-one, the first property listed here is true. The second property is true because $y = a^x$ and $y = \log_a x$ are inverse functions.

Additional Logarithmic Properties

1. **One-to-one rule:** $x = y$ if and only if $\log_a x = \log_a y$.
2. **Inverse rule:** $a^{\log_a x} = x$ and $\log_a a^x = x$.

■ **EXAMPLE 7** **Solving Exponential Equations**

Solve $2^{x+1} = 3$.

Solution

Solve Algebraically

$$\begin{aligned} 2^{x+1} &= 3 \\ \ln 2^{x+1} &= \ln 3 \quad \text{Take ln of both sides.} \\ (x+1)\ln 2 &= \ln 3 \quad \text{Use the power rule.} \\ x + 1 &= \frac{\ln 3}{\ln 2} \\ x &= \frac{\ln 3}{\ln 2} - 1 \\ x &\approx 0.585 \end{aligned}$$

Support Numerically

Replace x in the equation $2^{x+1} = 3$ with $x = 0.585$. Use your calculator to check that $2^{1.585} \approx 3$.

■ EXAMPLE 8 Solving Logarithmic Equations

Solve $2 \log_3 x - 5 = 2$.

Solution

Solve Algebraically

$$2 \log_3 x - 5 = 2$$

$$2 \log_3 x = 7$$

$$\log_3 x = 3.5$$

$$x = 3^{3.5} \qquad \text{Change to exponential form.}$$

$$x = 46.765 \cdots$$

Support Numerically

Replace x in the equation $2 \log_3 x - 5 = 2$ with $x = 46.765$. Use your calculator to check that $2 \log_3 46.765 - 5 \approx 2$. ■

Application—Earthquake Intensity

The intensity of an earthquake is rated by its measurement R on the **Richter scale.** This scale is actually the logarithmic function

$$R = \log \frac{a}{T} + B,$$

where R is the intensity, a is the amplitude (in micrometers, abbreviated μm) of the vertical ground motion at the receiving station, T is the period of the seismic wave (in seconds), and B is a constant that accounts for the weakening of the seismic wave with increasing distance from the epicenter of the earthquake.

Notes on Examples

Students may be surprised that the answer to Example 9 does not depend on the values of a, T, and B.

■ EXAMPLE 9 APPLICATION: Using the Richter Scale

A particular earthquake had an amplitude a during its early stages, and at later stages the amplitude became $10a$ (10 times more powerful). By how much did the Richter scale intensity R increase as the earthquake progressed?

Solution

Model

The earthquake intensities can be modeled by these equations:

$$R_1 = \log \frac{a}{T} + B$$

$$R_2 = \log \frac{10a}{T} + B$$

Solve Algebraically

We must determine $R_2 - R_1$.

$$\begin{aligned} R_2 - R_1 &= \left(\log \frac{10a}{T} + B\right) - \left(\log \frac{a}{T} + B\right) \\ &= \log \frac{10a}{T} - \log \frac{a}{T} \\ &= \log \frac{10a/T}{a/T} \\ &= \log 10 \\ &= 1 \end{aligned}$$

Interpret

When the amplitude increases by a factor of 10, the intensity increases by 1 on the Richter scale. ■

■ **EXAMPLE 10 APPLICATION: Comparing Earthquake Intensity**

The 1906 San Francisco earthquake measured 8.6 on the Richter scale. The 1989 San Francisco earthquake measured 7.1 on the Richter scale. How many times greater was the amplitude of the ground motion of the 1906 earthquake? (Assume that the Richter scale constants T and B are equal in the two cases.)

Solution

Model

Let a_1 be the amplitude of the ground motion of the 1906 quake and a_2 be the amplitude of the ground motion of the 1989 quake. Then

$$8.6 = \log \frac{a_1}{T} + B$$

and

$$7.1 = \log \frac{a_2}{T} + B$$

We must compare a_1 and a_2.

Follow-up

Ask students how the quotient rule for logarithms can be derived from the product rule and the power rule.

Assignment Guide

Day 1: Ex. 3–60, multiples of 3
Day 2: Ex. 63, 66, 67, 69–71, 74, 75, 78, 79

Cooperative Learning

Group Activity: Ex. 79, 80

Notes on Exercises

Ex. 25–50 provide practice in applying the properties of logarithms.
Ex. 75–78 require students to think about how the domains of functions are affected when the properties of logarithms are applied.

Ongoing Assessment

Self-Assessment: Ex. 7, 11, 19, 35, 49, 57, 65
Embedded Assessment: Ex. 70

Solve Algebraically

$$\left(\log \frac{a_1}{T} + B\right) - \left(\log \frac{a_2}{T} + B\right) = 8.6 - 7.1$$

$$\log \frac{a_1}{T} - \log \frac{a_2}{T} = 1.5$$

$$\log \frac{a_1}{a_2} = 1.5$$

$$\frac{a_1}{a_2} = 10^{1.5} \quad \text{Change to exponential form.}$$

$$\frac{a_1}{a_2} = 31.622 \cdots$$

$$a_1 \approx 31.6a_2$$

Interpret

The ground amplitude of the 1906 earthquake was about 31.6 times greater than that of the 1989 earthquake. ■

Quick Review 4.3

In Exercises 1–6, evaluate the expression without using a calculator.

1. $\log 10^2$ **2.** $\ln e^3$
3. $\ln e^{-2}$ **4.** $\log 10^{-3}$
5. $\log 1000$ **6.** $\log 0.001$

In Exercises 7–14, simplify the expression.

7. $\dfrac{x^5y^{-2}}{x^2y^{-4}}$ **8.** $\dfrac{u^{-3}v^7}{u^{-2}v^2}$

9. $(x^6y^{-2})^{1/2}$ **10.** $(x^{-8}y^{12})^{3/4}$

11. $\left(\dfrac{3}{x^2y}\right)^{-2}$ **12.** $\left(\dfrac{2}{xy^2}\right)^3$

13. $\dfrac{(u^2v^{-4})^{1/2}}{(27u^6v^{-6})^{1/3}}$ **14.** $\dfrac{(x^{-2}y^3)^{-2}}{(x^3y^{-2})^{-3}}$

SECTION EXERCISES 4.3

In Exercises 1–8, evaluate the logarithm.

1. $\log_2 7$ **2.** $\log_5 19$
3. $\log_8 175$ **4.** $\log_{12} 259$
5. $\log_{0.5} 12$ **6.** $\log_{0.2} 29$
7. $\log_{1/3} 14$ **8.** $\log_{1/4} 25$

In Exercises 9–12, rewrite the expression into an equivalent form using only natural logarithms.

9. $\log_3 x$ **10.** $\log_7 x$
11. $\log_2 (a + b)$ **12.** $\log_5 (c - d)$

In Exercises 13–16, rewrite the expression into an equivalent form using only common logarithms.

13. $\log_2 x$ **14.** $\log_4 x$
15. $\log_{1/2} (x + y)$ **16.** $\log_{1/3} (x - y)$

In Exercises 17–20, graph the function. State its domain and range.

17. $f(x) = \log_4 x$

18. $g(x) = \log_5 x$

19. $f(x) = \log_7 (x - 2)$

20. $g(x) = \log_3 (2 - x)$

In Exercises 21–24, match the equation with its graph. Identify window-viewing dimensions and values for Xscl and Yscl.

21. $y = \log_4 (2 - x)$

22. $y = \log_6 (x - 3)$

23. $y = \log_{0.5} (x - 2)$

24. $y = \log_{0.7} (3 - x)$

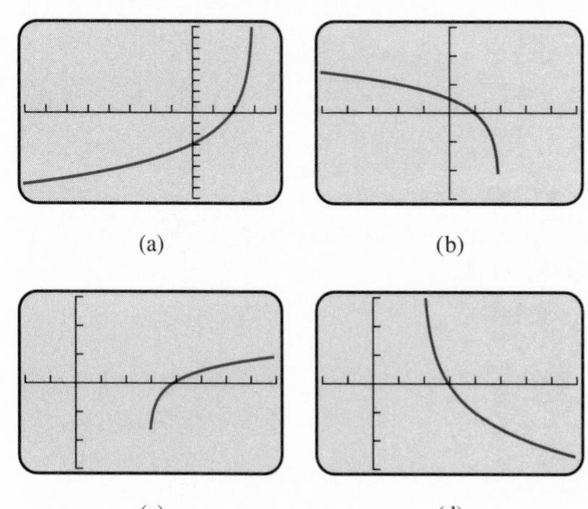

For Exercises 21–24

In Exercises 25–38, use the properties of logarithms to write the expression as a sum, difference, or multiples of logarithms of x, y, or constants.

25. $\ln 8x$

26. $\ln 9y$

27. $\log \dfrac{3}{x}$

28. $\log \dfrac{2}{y}$

29. $\log_2 y^5$

30. $\log_2 x^{-2}$

31. $\log x^3y^2$

32. $\log xy^3$

33. $\ln \dfrac{x^2}{y^3}$

34. $\log 1000x^4$

35. $\log \sqrt[4]{\dfrac{x}{y}}$

36. $\ln \dfrac{\sqrt[3]{x}}{\sqrt[3]{y}}$

37. $\ln 5000x^{360}$

38. $\ln 1000y^{240}$

In Exercises 39–50, use the properties of logarithms to write the expression as a single logarithm.

39. $\log x + \log y$

40. $\log x + \log 5$

41. $\ln y - \ln 3$

42. $\ln x - \ln y$

43. $3 \log_2 (x + 1)$

44. $4 \log_2 (y - 1)$

45. $\frac{1}{3} \log x$

46. $\frac{1}{5} \log z$

47. $2 \ln (x + 3) + 3 \ln (x - 2)$

48. $4 \log (x - 1) + 2 \log (x + 4)$

49. $\frac{1}{2} \log_a (x - 3) - \frac{1}{3} \log_a (x + 3)$

50. $\frac{1}{4} \log_b (x + 2) - \frac{1}{5} \log_b (3 - x)$

In Exercises 51–62, solve the equation algebraically and support the solution numerically.

51. $3^{x-2} = 5$

52. $5^{x+3} = 2$

53. $e^{x-1} = 7$

54. $e^{x+1} = 2$

55. $3 \log x + 1 = 5$

56. $-2 \log x + 3 = 7$

57. $2 - 3 \ln x = 5$

58. $3 - 4 \ln x = 6$

59. $2500 = 1000(1.08)^t$

60. $6000 = 4600(1.05)^t$

61. $10 = 20(\frac{1}{2})^t$

62. $5 = 25(\frac{1}{2})^t$

In Exercises 63 and 64, find the magnitude R of an earthquake on the Richter scale.

$$R = \log \frac{a}{T} + B.$$

63. $a = 250$, $T = 2$, and $B = 4.25$

64. $a = 300$, $T = 4$, and $B = 3.5$

In Exercises 65–68, solve the given problem. (Assume that the Richter scale constants T and B are the same in each case.)

65. *Comparing Earthquakes* How much greater was the amplitude of the ground motion in the 1978 Mexico City earthquake ($R = 7.9$) than in the 1994 Los Angeles earthquake ($R = 6.6$)?

66. *Comparing Earthquakes* How much greater was the amplitude of the ground motion in the 1995 Kobe, Japan, earthquake ($R = 7.2$) than in the 1994 Los Angeles earthquake ($R = 6.6$)?

67. *Light Intensity in Lake Erie* The relationship between intensity I of light (in lumens) at a depth of x feet in Lake Erie is given by

$$\log \frac{I}{12} = -0.00235x.$$

What is the intensity at a depth of 40 ft?

68. *Light Intensity in Lake Superior* The relationship between intensity I of light (in lumens) at a depth of x feet in Lake Superior is given by

$$\log \frac{I}{12} = -0.0125x.$$

What is the intensity at a depth of 10 ft?

In Exercises 69 and 70, answer the question.

69. Writing to Learn Use the change-of-base formula to explain how we know that the graph of $f(x) = \log_3 x$ can be obtained by applying a transformation to the graph of $y = \ln x$.

70. Writing to Learn Explain how we know that the graph of $f(x) = \log_8 x$ can be obtained by applying a transformation to the graph of $y = \log_3 x$.

EXTENDING THE IDEAS

In Exercises 71 and 72, solve the inequality $f(x) > g(x)$.

71. $f(x) = \ln x$ and $g(x) = \sqrt[3]{x}$

72. $f(x) = 7 \ln x$ and $g(x) = 0.5e^x$

In Exercises 73 and 74, solve the inequality $f(x) < g(x)$.

73. $f(x) = 7 \ln x$ and $g(x) = 0.5e^x$

74. $f(x) = e^x$ and $g(x) = 5\sqrt{x}$

In Exercises 75–78, compare the domains of f and g and explain any differences.

75. $f(x) = 2 \ln x + \ln (x - 3)$ and $g(x) = \ln x^2(x - 3)$

76. $f(x) = \ln (x + 5) - \ln (x - 5)$ and $g(x) = \ln \dfrac{x + 5}{x - 5}$

77. $f(x) = \log (x + 3)^2$ and $g(x) = 2 \log (x + 3)$

78. $f(x) = \log (x - 4)^3$ and $g(x) = 3 \log (x - 4)$

In Exercises 79 and 80, solve the problem.

79. Verify the general change-of-base formula

$$\log_b x = \frac{\log_a x}{\log_a b}$$

80. Verify the quotient and power rules for logarithms.

81. Show that $\dfrac{\log x}{\ln x}$ is a constant function.

4.4 MORE EQUATION SOLVING AND APPLICATIONS

Objective

Students will be able to solve exponential and logarithmic equations algebraically, and solve application problems using these equations.

Motivate

Ask students to use a grapher to graph $y = \log x^2$ and $y = 2 \log x$, and to comment on any differences they see. Do the results contradict the power rule for logarithms? **(No)**

Solving Exponential Equations

Some exponential and logarithmic equations can be solved by changing forms between exponential and logarithmic, as needed. Sometimes the one-to-one and inverse properties introduced in Section 4.3 can be used. On occasion, however, equations are complicated enough that a graphical solution may be preferred.

■ EXAMPLE 1 Solving an Exponential Equation

Solve $20\left(\dfrac{1}{2}\right)^{x/3} = 5.$

Solution

Solve Algebraically

$$20\left(\frac{1}{2}\right)^{x/3} = 5$$

$$\left(\frac{1}{2}\right)^{x/3} = \frac{1}{4} \quad \text{Divide by 20.}$$

$$\left(\frac{1}{2}\right)^{x/3} = \left(\frac{1}{2}\right)^{2}$$

$$\frac{x}{3} = 2 \quad \text{Use the inverse property: } a^b = a^c \rightarrow b = c.$$

$$x = 6$$

Support Graphically

Figure 4.27 shows that the point of intersection of the graphs of $y_1 = 20\left(\frac{1}{2}\right)^{x/3}$ and $y_2 = 5$ occurs when $x = 6$. ■

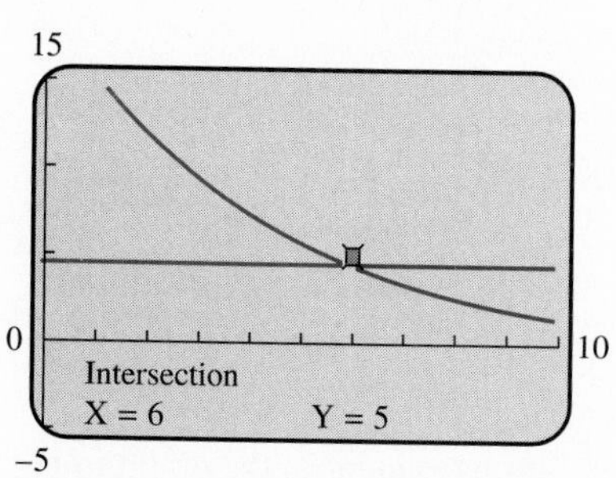

Figure 4.27

■ **EXAMPLE 2 Solving Another Exponential Equation**

Solve $\dfrac{e^x - e^{-x}}{2} = 5.$

Solution

Because this equation includes two exponential functions, it is not as easy to solve algebraically as it is graphically.

Solve Graphically

Figure 4.28 shows that the graphs of $y_1 = \dfrac{e^x - e^{-x}}{2}$ and $y_2 = 5$ intersect when x is approximately 2.31. The solution to the equation is approximately 2.31.

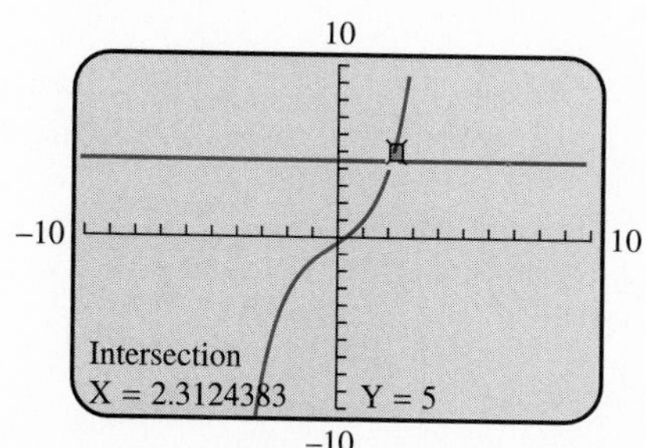

Figure 4.28

Confirm Algebraically

$$\frac{e^x - e^{-x}}{2} = 5$$

$$(e^x)^2 - e^0 = 10e^x \quad \text{Multiply by } 2e^x.$$

$$(e^x)^2 - 10(e^x) - 1 = 0$$

This equation is quadratic in e^x, so the quadratic formula gives

$$e^x = \frac{10 \pm \sqrt{104}}{2} = 5 \pm \sqrt{26}.$$

Because e^x is always positive, we discard the possibility that e^x has the negative value $5 - \sqrt{26}$. Therefore,

$$\begin{aligned} e^x &= 5 + \sqrt{26} \\ \ln e^x &= \ln(5 + \sqrt{26}) && \text{Take ln of both sides.} \\ x &= \ln(5 + \sqrt{26}) && \text{Use the inverse property: } \log_a a^x = x. \\ x &= 2.312 \cdots \end{aligned}$$

■

Teaching Note

Students have probably seen extraneous solutions in previous math courses, but they may not have seen examples where solutions are missed. However, this can also happen with simpler equations. For example, if one attempts to solve $2x = x^2$ by dividing both sides by x, then 0 is deleted from the domain and is missed as a solution.

Solving Logarithmic Equations

When logarithmic equations are solved by algebraic means, it is important to keep track of the domains of each expression. A particular algebraic method may introduce extraneous solutions or perhaps miss some solutions. For example, the equation

$$\log x^2 = 2$$

is solved by the two methods that follow.

Method 1. The first method misses one solution.

$$\begin{aligned} \log x^2 &= 2 \\ 2 \log x &= 2 && \text{Power rule.} \\ \log x &= 1 \\ x &= 10 && \text{Change to exponential form.} \end{aligned}$$

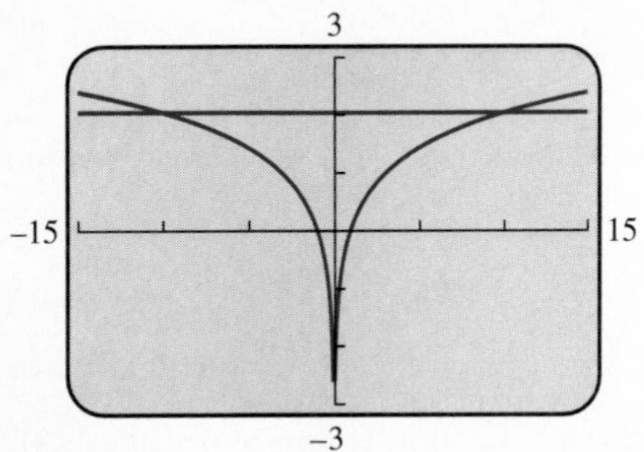

Figure 4.29 Graphs of $y_1 = \log x^2$ and $y_2 = 2$.

Method 2. In the second method both solutions evident in the graph in Figure 4.29 are found.

$$\begin{aligned} \log x^2 &= 2 \\ x^2 &= 10^2 && \text{Change to exponential form.} \\ x^2 &= 100 \\ x &= 10 \text{ or } -10 \end{aligned}$$

Method 1 fails because the domain of $\log x^2$ consists of all nonzero real numbers, whereas the domain of $\log x$ used in the solution consists of only positive real numbers.

Because algebraic manipulation of a logarithmic equation can produce expressions with different "understood domains," a graphical solution is often less prone to error.

■ EXAMPLE 3 Solving Another Logarithmic Equation

Solve

$$\ln(3x - 2) + \ln(x - 1) = 2 \ln x.$$

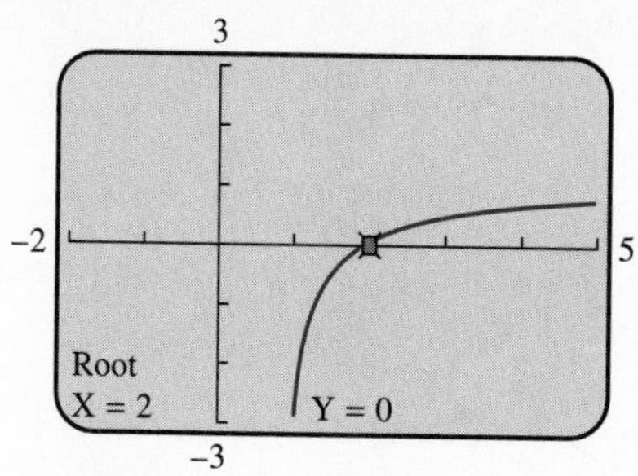

Figure 4.30

Solution

Solve Graphically

To use the x-intercept method of solving, graph

$$y = \ln(3x - 2) + \ln(x - 1) - 2\ln x,$$

as shown in Figure 4.30.

The x-intercept is $x = 2$ which is the solution to the equation.

Confirm Algebraically

$$\begin{aligned} \ln(3x - 2) + \ln(x - 1) &= 2\ln x \\ \ln(3x - 2)(x - 1) &= \ln x^2 \\ (3x - 2)(x - 1) &= x^2 \quad \text{Use the one-to-one rule.} \\ 2x^2 - 5x + 2 &= 0 \\ (2x - 1)(x - 2) &= 0 \end{aligned}$$

$$x = \frac{1}{2} \quad \text{or} \quad x = 2$$

We recognize $x = \frac{1}{2}$ is an extraneous solution of the original equation because $\ln(\frac{1}{2} - 1)$ is not defined.

By substituting $x = 2$, we find that

$$\ln 4 + \ln 1 = \ln(4)(1) = \ln 2^2 = 2\ln 2.$$

Thus, the only solution to the original equation is $x = 2$. ■

Application—Logistic Growth

In Section 4.1 we modeled population growth with the exponential function

$$y_1 = P(t) = ba^t.$$

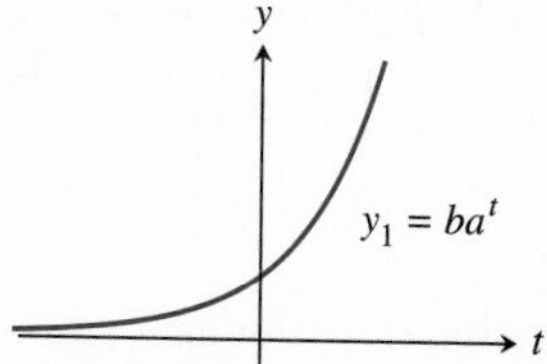

Figure 4.31 An exponential curve.

When $b > 0$ and $a > 1$, the graph is the exponential curve shown in Figure 4.31 with exponential end behavior

$$P(t) \to \infty \text{ as } t \to \infty.$$

This means that eventually there would be an incredible population explosion.

Environmental factors, however, often prevent a population from growing exponentially. For example, we would not expect the number of guppies in a fish bowl to grow exponentially because the number would eventually be too large for the bowl.

A more realistic model, the **logistic growth function,** has the form

$$y_2 = P(t) = \frac{a}{1 + be^{-rt}}$$

where a, b, and r are positive real number constants. The graph of this function is a **logistic curve.**

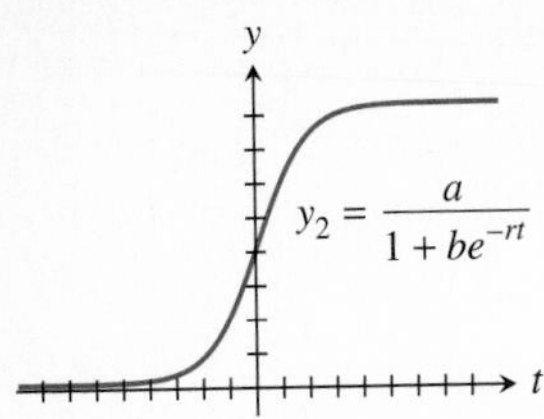

Figure 4.32 A logistic curve.

Logistic curves can have the shape of the graph of

$$P(t) = \frac{10}{1 + e^{-t}}$$

shown in Figure 4.32. In particular, notice that $P(t)$ approaches a constant value as $t \to \infty$. There is some controlling factor that says the population has reached a **saturation value,** and growth will not continue.

EXAMPLE 4 Modeling the Growth of Bears

In one federal Wildlife Reserve the population of bears after t years is modeled by the logistic function

$$P(t) = \frac{300}{1 + 19e^{-0.3t}}.$$

According to this model, when will the bear population be 150?

Solution

Solve Graphically

We must find the value of t for which $P(t) = 150$. To do so we must solve the equation

$$\frac{300}{1 + 19e^{-0.3t}} = 150.$$

Figure 4.33 shows the graphs of $y_1 = \frac{300}{1 + 19e^{-0.3t}}$ and $y_2 = 150$. The graphs intersect when t is approximately 9.81.

Interpret

Predicting from the model, the bear population will be 150 in a little more than 9.8 years. ■

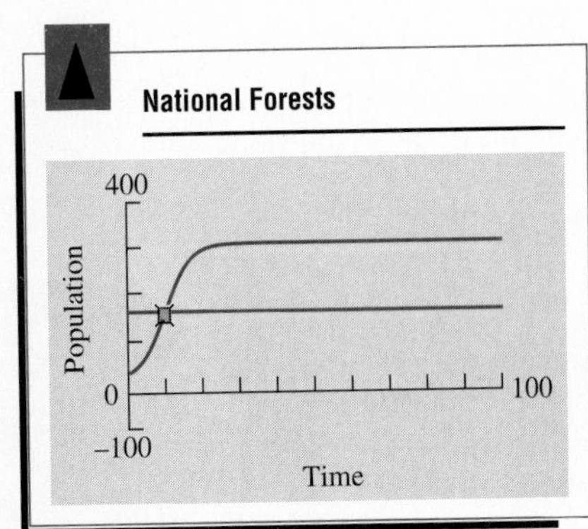

Intersection: $x = 9.8147966$; $y = 150$

Figure 4.33

Application—Newton's Law of Cooling

An object that has been heated will cool to the temperature of the medium in which it is placed such as, for example, the surrounding air or water. The temperature $T(t)$ of the object at time t can be modeled by

$$T(t) = T_m + (T_0 - T_m)e^{-kt}$$

for an appropriate value of k, where

$T(t)$ = the temperature of the object at time t,
T_m = temperature of the surrounding medium,
T_0 = initial temperature of the object.

To keep the situation simple, we assume that the surrounding medium, although taking heat from the object, essentially maintains a constant temperature. Then a Newton cooling curve has the shape of the graph of $y = 1 + 4e^{-t}$ shown in Figure 4.34.

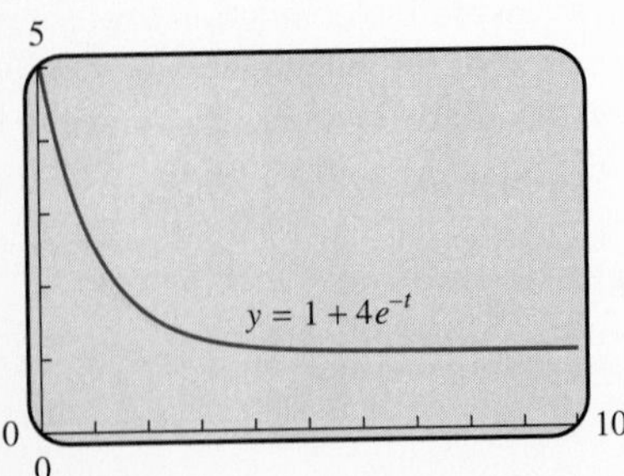

Figure 4.34 A curve modeling Newton's law of cooling.

EXAMPLE 5 Newton's Law of Cooling

A hard-boiled egg at temperature 96° C is placed in 16° C water to cool. Four minutes later the temperature of the egg is 45° C. Use a Newton's cooling law model to determine when the egg will be 20° C.

Solution

Model

We know from the given information that

$$T_0 = 96, \text{ the initial temperature of the egg,}$$
$$T_m = 16, \text{ the temperature of the medium.}$$

Substituting into $T(t) = T_m + (T_0 - T_m)e^{-kt}$, we have

$$T(t) = 16 + (96 - 16)e^{-kt}.$$

To find the value of k we use the fact that $T = 45$ when $t = 4$.

$$45 = 16 + 80e^{-4k}$$
$$45 - 16 = 80e^{-4k}$$
$$\frac{29}{80} = e^{-4k}$$
$$\ln \frac{29}{80} = -4k \qquad \text{Change to logarithmic form.}$$
$$k = \frac{\ln(29/80)}{-4}$$
$$k \approx 0.254$$

A modeling equation, then, for the temperature T of the egg at time t is $T(t) = 16 + 80e^{-0.254t}$. To find t when $T = 20°$ C, we must solve the equation

$$20 = 16 + 80e^{-0.254t}$$

Solve Algebraically

$$20 = 16 + 80e^{-0.254t}$$
$$\frac{4}{80} = e^{-0.254t} \qquad \text{Subtract 16, then divide by 80.}$$
$$\ln \frac{4}{80} = -0.254t \qquad \text{Change to logarithmic form.}$$
$$\ln 0.05 = -0.254t$$
$$t = \frac{\ln 0.05}{-0.254}$$
$$t \approx 11.794$$

Interpret

The temperature of the egg will be 20° C after about 11.8 min (11 min 48 sec). ■

Table 4.9 Experimental Data

Time t	Temp T	$T - T_m$
2	64.8	60.3
5	49.0	44.5
10	31.4	26.9
15	22.0	17.5
20	16.5	12
25	14.2	9.7
30	12.0	7.5

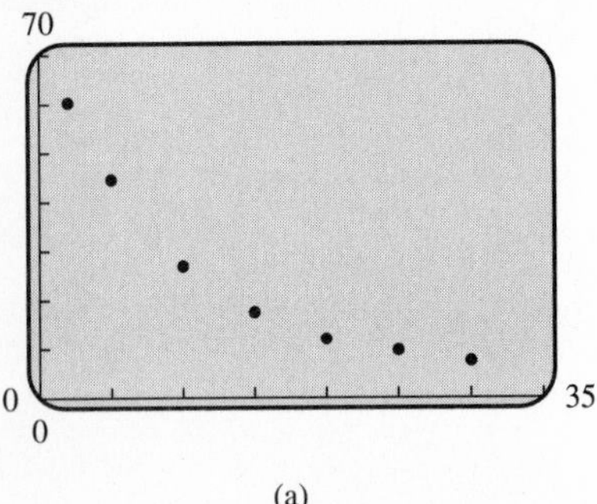

(a)

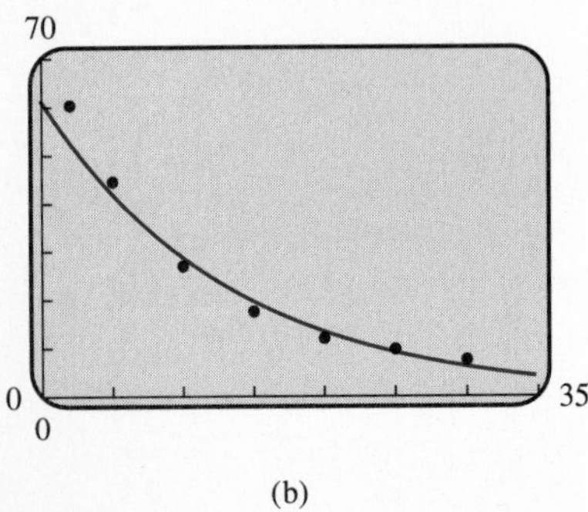

(b)

Figure 4.35

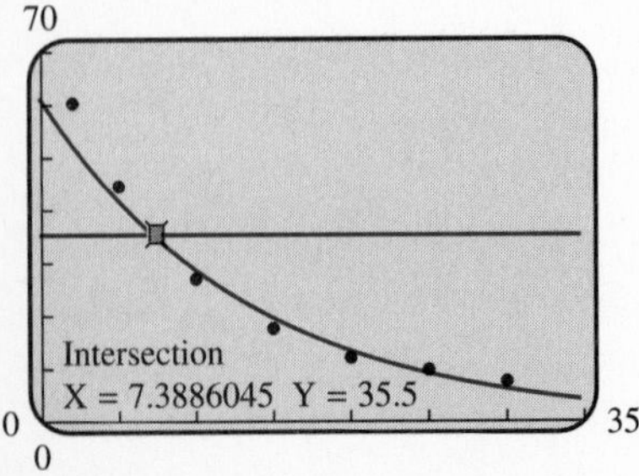

Figure 4.36

By rewriting Newton's law of cooling in the form

$$T(t) - T_m = (T_0 - T_m)e^{-kt}$$

we see that experimental data for the difference $T - T_m$ can be modeled using exponential regression.

■ EXAMPLE 6 APPLICATION: Using Newton's Law of Cooling

A thermometer is removed from a cup of coffee and placed in water that has a temperature of 4.5° C. Temperature readings are taken after 2 sec, 5 sec, and every 5 sec thereafter, as recorded in Table 4.9. Estimate the coffee temperature, and the time when the thermometer reading will be 40° C.

Solution

Model

In this situation the thermometer is the object that is being cooled and the medium is water at 4.5° C. Therefore, $T(t)$ is the temperature of the thermometer at time t and $T_m = 4.5°$ C.

Figure 4.35a shows a scatter plot of the data $T - T_m$. Using a calculator we find the exponential regression equation to be

$$T(t) - 4.5 = 61.66(0.928^t),$$

whose graph is shown in Figure 4.35b.

Solve Graphically

The temperature of the coffee represents the initial temperature, T_0, of the thermometer. So this problem asks us to find $T_0 = T(0)$ and the time t when $T(t) =$ 40° C. To first find T_0, substitute $t = 0$ into the modeling equation.

$$T(0) - 4.5 = 61.66(0.928^0)$$

$$T_0 - 4.5 = 61.66$$

$$T_0 = 61.66 + 4.5 = 66.16$$

To find the value of t for which $T(t) = 40°$ C, we must solve the equation

$$40 - 4.5 = 61.66(0.928^t).$$

Figure 4.36 shows that the graphs of $y_1 = 61.66(0.928^t)$ and $y_2 = 40 - 4.5 = 35.5$ intersect at $t \approx 7.39$.

Support Numerically

Substitute $t = 7.39$ into $61.66(0.928^t)$, and use your calculator to see that its value is approximately 35.5.

Interpret

The temperature of the coffee was approximately 66.2° C. The thermometer reading was 40° C about 7.4 sec after it was placed in the water.

Notes on Examples

Example 6 uses exponential regression to solve a problem related to Newton's law of cooling.

Follow-up

Ask students to explain how extraneous solutions might be introduced by replacing

$\ln x - \ln(x - 2)$ with $\ln\left(\dfrac{x}{x-2}\right)$

in an equation.

Assignment Guide

Day 1: Ex. 3–45, multiples of 3
Day 2: Ex. 48, 49, 51, 54, 55–57, 59, 60, 61, 67, 68

Cooperative Learning

Group Activity: Ex. 61, 62

Notes on Exercises

Ex. 27–38 give students a choice of method and an opportunity to explain their choice. It is possible to solve all of these problems algebraically.
Ex. 39–46 encourage students to think about extraneous solutions.

Ongoing Assessment

Self-Assessment: Ex. 1, 17, 23, 43, 47
Embedded Assessment: Ex. 30, 36, 56

Application—Atmospheric Pressure

Scientists have established that atmospheric pressure at sea level is 14.7 lb/in.^{2}, and the pressure is reduced by half for each 3.6 mi above sea level. For example, the pressure 3.6 mi above sea level is $(\frac{1}{2})(14.7) = 7.35$ lb/in.^{2}. This rule for atmospheric pressure holds for altitudes up to 50 mi.

■ EXAMPLE 7 APPLICATION: Modeling Atmospheric Pressure

Determine the altitude above Atlantic City at which the atmospheric pressure P will be 4 lb/in.^{2}. Atlantic City is at sea level on the Atlantic Ocean coast in New Jersey.

Solution

Model

Study this pattern:

$$14.7\left(\frac{1}{2}\right) = \text{pressure at } 3.6 \text{ mi}$$

$$14.7\left(\frac{1}{2}\right)^2 = \text{pressure at } 2(3.6) = 7.2 \text{ mi}$$

$$\vdots$$

$$14.7\left(\frac{1}{2}\right)^h = \text{pressure at } 3.6h \text{ mi}$$

The equation $P(h) = 14.7(0.5^{h/3.6})$ describes the pressure at h miles of altitude. We must find the value of h that satisfies the equation

$$4 = 14.7(0.5^{h/3.6}).$$

Solve Algebraically

$$\ln 14.7(0.5^{h/3.6}) = \ln 4 \quad \text{Take ln of both sides.}$$

$$\ln 14.7 + \ln 0.5^{h/3.6} = \ln 4 \quad \text{Use the product rule.}$$

$$\frac{h}{3.6} \ln 0.5 = \ln 4 - \ln 14.7$$

$$h = 3.6\left(\frac{\ln 4 - \ln 14.7}{\ln 0.5}\right)$$

$$= 6.759 \ldots$$

Support Numerically

Check that $P(h) = 14.7(\frac{1}{2})^{h/3.6}$ is approximately 4 for $h = 6.76$.

Interpret

The atmospheric pressure will be 4 lb/in.^{2} at an altitude of approximately 6.76 mi above Atlantic City (sea level). ■

Quick Review 4.4

In Exercises 1–4, show that each function in the given pair is the inverse of the other.

1. $f(x) = e^{2x}$ and $g(x) = \ln (x)^{1/2}$
2. $f(x) = 10^{x/2}$ and $g(x) = \log x^2$, $x > 0$
3. $f(x) = \frac{1}{3} \ln x$ and $g(x) = e^{3x}$
4. $f(x) = 3 \log x^2$, $x > 0$ and $g(x) = 10^{x/6}$

In Exercises 5–8, complete the given equality between Fahrenheit and Celsius temperature. Recall that $F = (\frac{9}{5})C + 32$.

5. $25°\text{C} = ?°\text{F}$
6. $5°\text{C} = ?°\text{F}$
7. $75°\text{F} = ?°\text{C}$
8. $10°\text{F} = ?°\text{C}$

In Exercises 9–12, the equation is a quadratic in the first expression. Use factoring or the quadratic formula to solve for x.

9. x^2; $(x^2)^2 - 4(x^2) + 3 = 0$
10. x^2; $(x^2)^2 - 2(x^2) - 4 = 0$
11. $\sqrt{x}$; $(\sqrt{x})^2 - 4(\sqrt{x}) + 1 = 0$
12. $\sqrt{x}$; $(\sqrt{x})^2 - 5(\sqrt{x}) + 6 = 0$

SECTION EXERCISES 4.4

In Exercises 1–8, solve the equation algebraically.

1. $1.06^x = 4.1$
2. $0.98^x = 1.6$
3. $1.09^{-x} = 18.4$
4. $1.12^{-x} = 3.2$
5. $\log x = 4$
6. $\ln x = -1$
7. $\ln (x + 3) = 2$
8. $\log (x - 1) = 3$

In Exercises 9–20, solve the equation algebraically. Support your answer numerically.

9. $50e^{0.035x} = 200$
10. $80e^{0.045x} = 240$
11. $2(10^{-x/3}) = 20$
12. $3(5^{-x/4}) = 15$
13. $20(\frac{1}{2})^{x/4} = 8$
14. $40(\frac{1}{2})^{x/3} = 10$
15. $3 + 2e^{-x} = 6$
16. $7 - 3e^{-x} = 2$
17. $\log_4 (x - 5) = -1$
18. $\log_4 (1 - x) = 1$
19. $3 \ln (x - 3) + 4 = 5$
20. $3 - \log (x + 2) = 5$

In Exercises 21–26, state the domain of the function. Then match the function with its graph. (Each graph is drawn in the decimal window.)

21. $f(x) = \log x(x + 1)$
22. $g(x) = \log x + \log (x + 1)$
23. $f(x) = \ln \dfrac{x}{x + 1}$
24. $g(x) = \ln x - \ln (x + 1)$
25. $f(x) = 2 \ln x$
26. $g(x) = \ln x^2$

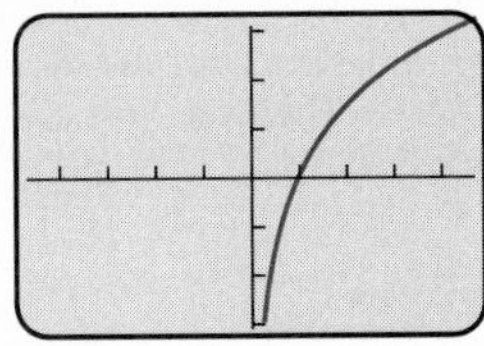

(a)

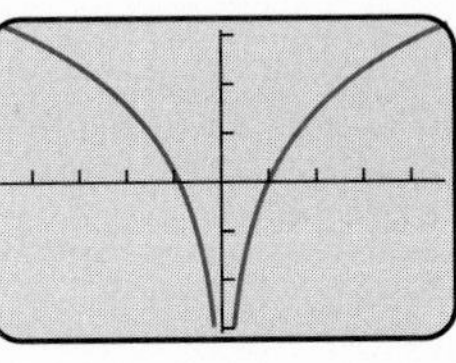

(b)

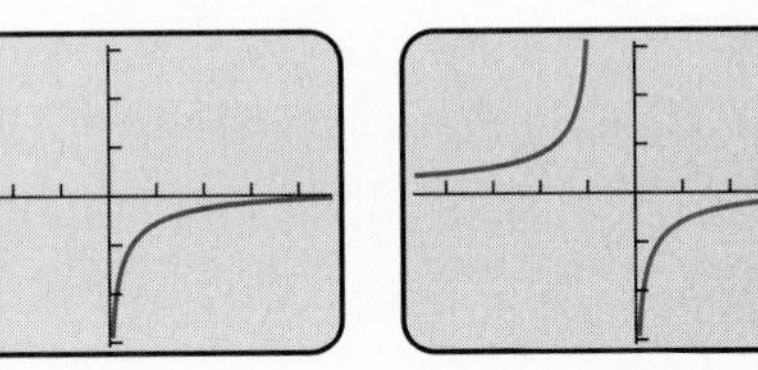

(c) (d)

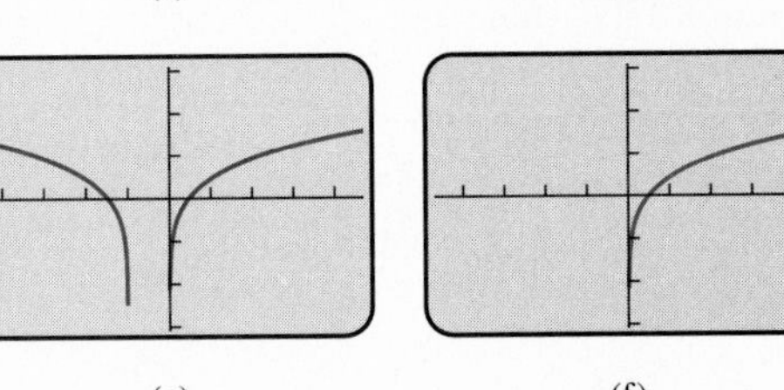

(e) (f)

For Exercises 21–26

In Exercises 27–38, solve the equation algebraically or graphically. Write a sentence explaining your choice of method.

27. $\dfrac{100}{2 + e^x} = 25$
28. $\dfrac{200}{5 + e^{-x}} = 50$
29. $\dfrac{2^x - 2^{-x}}{3} = 4$
30. $\dfrac{2^x + 2^{-x}}{2} = 3$
31. $\dfrac{e^x + e^{-x}}{2} = 4$
32. $2e^{2x} + 5e^x - 3 = 0$

33. $\ln x + \ln 2 = 3$

34. $\log x - \log 5 = 1$

35. $\log (x + 5) - \log x = 1$

36. $\ln x + \ln (x + 2) = 1$

37. $\ln x + \ln (x + 3) = 2$

38. $\ln (x + 1) - \ln x = 1$

In Exercises 39–46, solve the equation and state whether your method introduces an extraneous solution. If so, state which solution is extraneous and explain.

39. $\dfrac{500}{1 + 25e^{0.3x}} = 200$

40. $\dfrac{400}{1 + 95e^{-0.6x}} = 150$

41. $\frac{1}{2} \ln (x + 3) - \ln x = 0$

42. $\log x - \frac{1}{2} \log (x + 4) = 1$

43. $\ln (x - 3) + \ln (x + 4) = 3 \ln 2$

44. $\log (x - 2) + \log (x + 5) = 2 \log 3$

45. $\ln (3x - 1) - \ln (x + 2) = 2 \ln 5$

46. $\log (x + 4) - \log (2x - 3) = \log 2$

In Exercises 47–50, solve the problem.

47. *Spread of Flu* The number of students infected with flu at Springfield High School after t days is modeled by the function

$$P(t) = \frac{800}{1 + 49e^{-0.2t}}.$$

a. What was the initial number of infected students?
b. When will the number of infected students be 200?
c. The school will close when 300 of the 800-student body are infected. When will the school close?

48. *Population of Deer* The population of deer after t years in Cedar State Park is modeled by the function

$$P(t) = \frac{1001}{1 + 90e^{-0.2t}}.$$

a. What was the initial population of deer?
b. When will the number of deer be 600?
c. What is the maximum number of deer possible in the park?

49. *Newton's Law of Cooling* A cup of coffee has cooled from 92° C to 50° C after 12 min in a room at 22° C. How long will the cup take to cool to 30° C?

50. *Newton's Law of Cooling* A cake is removed from an oven at 350° F and cools to 120° F after 20 min in a room at 65° F. How long will the cake take to cool to 90° F?

In Exercises 51 and 52, real data are given. Use the modeling function that you find to predict a temperature value that is not among the data.

51. *Newton's Law of Cooling Experiment* A thermometer is removed from a cup of coffee and placed in water whose temperature (T_m) is 10° C. The data in Table 4.10 were collected over the next 30 sec.

Table 4.10 Experimental Data

Time t	Temp T	$T - T_m$
2	80.47	70.47
5	69.39	59.39
10	49.66	39.66
15	35.26	25.26
20	28.15	18.15
25	23.56	13.56
30	20.62	10.62

a. Find a scatter plot of the data $T - T_m$.
b. Find an exponential regression equation for the $T - T_m$ data. Superimpose its graph on the scatter plot.
c. Estimate the thermometer reading when it was removed from the coffee.

52. *Newton's Law of Cooling Experiment* A thermometer was removed from a cup of hot chocolate and placed in water whose temperature $T_m = 0°$ C. The data in Table 4.11 were collected over the next 30 sec.

a. Find a scatter plot of the data $T - T_m$.
b. Find an exponential regression equation for the $T - T_m$ data. Superimpose its graph on the scatter plot.
c. Estimate the thermometer reading when it was removed from the hot chocolate.

Table 4.11 Experimental Data

Time t	Temp T	$T - T_m$
2	74.68	74.68
5	61.99	61.99
10	34.89	34.89
15	21.95	21.95
20	15.36	15.36
25	11.89	11.89
30	10.02	10.02

In Exercises 53–58, solve the problem.

53. *Atmospheric Pressure* Use the model from Example 7 to find the altitude above Atlantic City with an atmospheric pressure of 2 lb/in.^{2}.

54. *Atmospheric Pressure* Use the model from Example 7 to find the atmospheric pressure 40 mi above Colorado Springs (1 mi above sea level).

55. *Penicillin Use* The use of penicillin became so widespread in the 1980s in Hungary that it became practically useless against common sinus and ear infections. Now the use of more effective antibiotics has caused a decline in penicillin resistance. The bar graph shows the use of penicillin in Hungary for selected years.

a. From the bar graph we read the data pairs to be approximately (1, 11), (8, 6), (15, 4.8), (16, 4), and (17, 2.5), using $t = 1$ for 1976, $t = 8$ for 1983, and so on. Complete a scatter plot for these data.

b. Writing to Learn Discuss whether the bar graph shown or the scatter plot that you completed best represents the data and why.

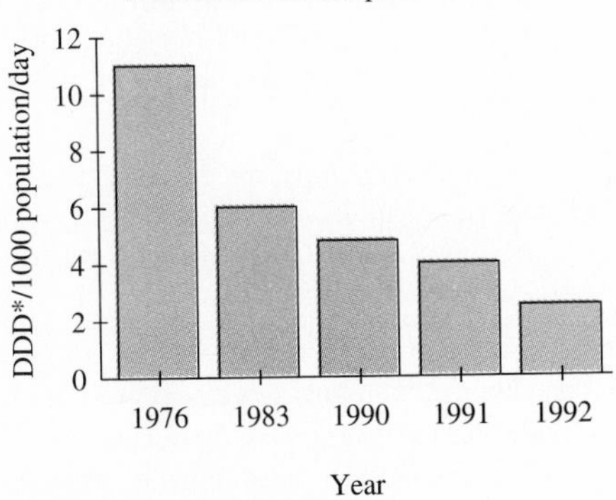

*Defined Daily Dose
Source: Science, vol. 264, April 15, 1994, American Association for the Advancement of Science.

For Exercise 55

56. Writing to Learn Find both a logarithmic and a linear regression equation for the data in Exercise 55. Discuss which equation provides the best fit for the data. Which equation would you use to estimate the daily dose for 1993? Why?

57. *Estimating Population Growth* The population of Alaska (in millions) for three known years is indicated in Table 4.12.

a. Find an exponential regression equation for these data. (Use $t = 0$ to represent 1960.)

b. Use the regression to predict the population in the year 2000.

Table 4.12 Population of Alaska

Year	Population (millions)
1960	0.23
1980	0.40
1990	0.55

Source: The Statesman's Yearbook, 129th ed. (London: The Macmillan Press, Ltd., 1992).

c. Estimate when the population of the state was twice the 1960 figure.

58. *Estimating Population Growth* The population of California (in millions) for three known years is indicated in Table 4.13.

Table 4.13 Population of California

Year	Population (millions)
1960	15.72
1980	23.67
1990	29.76

Source: The Statesman's Yearbook, 129th ed. (London: The Macmillan Press, Ltd., 1992).

a. Find an exponential regression equation for these data. (Use $t = 0$ to represent 1960.)

b. Use the regression to predict the population in the year 2000.

c. Predict when the population of the state would be twice the 1960 figure.

EXTENDING THE IDEAS

In Exercises 59–62, solve the problem.

59. *Normal Distribution* The function defined by

$$f(x) = ke^{-cx^2},$$

where c and k are constants, is a bell-shaped curve useful in probability. Find the maximum value of f in the case where $k = 1.30$ and $c = 5.31$.

60. *Normal Distribution* For the function f in Exercise 59 solve the inequality $f(x) \geq 0.5$.

61. *Potential Energy* The potential energy E (the energy stored for use at a later time) between two ions in a certain molecular structure is modeled by the function

$$E = -\frac{5.6}{r} + 10e^{-r/3}$$

where r is the distance separating the nuclei.

a. Writing to Learn Graph this function in the window $[-10, 10]$ by $[-10, 30]$, and explain which portion of the graph does not represent this potential energy situation.

b. Identify a viewing window that shows that portion of the graph (with $r \leq 10$) which represents this situation, and find the maximum value for E.

62. *Determining Book Value* Suppose that C is the original cost of an automobile and that S is its salvage value at the end of a useful life of n years. At any time t between those two extremes the banking and insurance industries give the automobile a value called its *book value*. The book value B at any time t depends on C, S, n, and an interest rate i, and can be found by using the equation

$$B(t) = C - \left(\frac{C - S}{[(1 + i)^n - 1]/i}\right)\left(\frac{(1 + i)^t - 1}{i}\right).$$

a. If a car costs \$17,000 with a useful life to a company of 6 years and a salvage value of \$1200, find a graph of $B(t)$ if the interest rate is $i = 0.05$.

b. What is the value of the automobile after 4 years and 3 months?

In Exercises 63–68, solve the equation or inequality.

63. $e^x + x = 5$

64. $e^{2x} - 8x + 1 = 0$

65. $e^x < 5 + \ln x$

66. $\ln|x| - e^{2x} \geq 3$

67. $2 \log x - 4 \log 3 > 0$

68. $2 \log (x + 1) - 2 \log 6 < 0$

4.5
INTEREST AND ANNUITIES

Objective

Students will be able to use exponential equations to solve business and finance applications related to simple and compound interest.

Motivate

Ask . . .
How might you determine the interest rate necessary to double your money within 8 years? **(Solve an equation such as $2 = (1 + r)^t$ or $2 = e^{rt}$.)**

Interest Compounded Annually • Interest Compounded k Times per Year • Interest Compounded Continuously • Annual Percentage Yield • Value of an Annuity • Present and Future Value of an Annuity Compared

Interest Compounded Annually

You pay rent for an apartment as a payment for the use of property. Interest is payment for the use of money. When you borrow money, you pay interest, and when you loan money, you receive interest. When you invest in a savings account, you are actually lending money to the bank.

Suppose you invest P dollars at a 5% interest rate with interest calculated at the end of each year. If S_1 represents the total amount after year 1, then

$$S_1 = P + 0.05P = P(1 + 0.05).$$

The amount at the end of year 2, S_2, is

$$\begin{aligned} S_2 &= P(1 + 0.05) + 0.05P(1 + 0.05) \\ &= P(1 + 0.05)(1 + 0.05) \quad \text{Factor out } P(1 + 0.05). \\ &= P(1 + 0.05)^2 \end{aligned}$$

Extending this pattern with an interest rate r, we have

$$\begin{aligned} S_3 &= P(1 + r)^3, \\ &\vdots \\ S_n &= P(1 + r)^n. \end{aligned}$$

Teaching Note

The ideas of business finance are very important in our society. In this section, you may need to teach financial concepts as well as mathematical ideas. This section provides many useful real-world applications.

The total value S of the investment at the end of n years is modeled by the equation

$$S = P(1 + r)^n.$$

Compound interest is interest that becomes part of the investment, as happened in the situation just described. The formula $S = P(1 + r)^n$ is a compound interest formula that is used when interest is **compounded annually,** that is, computed just once a year. Notice that for fixed interest rate r, the formula $S = P(1 + r)^n$ is an exponential function with base $1 + r$ and whose exponent is the variable n. Sometimes it is convenient to use the function $S(t) = P(1 + r)^t$, where t is a real number that represents time.

Notes on Examples

To illustrate the growth that leads to the answer in Example 1, enter 500 on the computation screen of your grapher (key in the number followed by ENTER or EXE). The grapher should return the answer of 500.00. Then key in ANS × 1.07 and press ENTER or EXE ten times, one press for each year of growth. Each time you press the ENTER or EXE key, a new line of text should appear. The screen should eventually show a table of values that culminates in 983.58. This progression of values is a geometric sequence with first term 500.00 and common ratio 1.07.

EXAMPLE 1 Compounding Annually

Suppose Quan Li invests \$500 at 7% interest compounded annually. Find the value of her investment 10 years later.

Solution

Model

Use $P = 500$, $r = 0.07$, and $n = 10$ in $S = P(1 + r)^n$.

Solve Algebraically

$$\begin{aligned} S &= 500(1 + 0.07)^{10} \\ &= 983.575 \cdots \end{aligned}$$

Interpret

The value of Quan Li's investment after 10 years is \$938.58, rounded up. The value is \$938.57, rounded down. ■

Interest Compounded k Times per Year

Suppose P dollars are invested at an annual interest rate r compounded k times a year for t years. Then r/k is the interest rate per compounding period, and kt is the number of compounding periods. The amount S in the account after t years is

$$S = P\left(1 + \frac{r}{k}\right)^{kt}.$$

S is the *value of the investment.*

Teaching Note

Calculating compound and continuous interest using the blank screen of a grapher is one of the main reasons the authors believe a graphing calculator is the best scientific calculator money can buy.

EXAMPLE 2 Compounding Monthly

Suppose Roberto invests \$500 at 9% annual interest **compounded monthly,** that is, compounded 12 times a year. Find the value of his investment 5 years later.

Teaching Note

Developing the concepts of compound interest and continuous interest can be enhanced by using the large-screen display of a grapher, along with its recursive, replay, and editing capabilities. Begin by setting the decimal display to two places to show answers in dollars and cents.

Solution

Model

Use $P = 500$, $r = 0.09$, $k = 12$, and $t = 5$ in the equation

$$S = P\left(1 + \frac{r}{k}\right)^{kt}.$$

Solve Algebraically

$$S = 500\left(1 + \frac{0.09}{12}\right)^{12(5)}$$

$$= 782.840 \cdots$$

Interpret

The value of Roberto's investment after 5 years is \$782.84. ■

The problems in Examples 1 and 2 required that we calculate S. Instead you may want to find the time required for an investment to reach a specific value. In that case, solve for t.

■ EXAMPLE 3 Finding Time for an Investment

Judy has \$500 to invest at 9% annual interest compounded monthly. How long will it take for her investment to grow to \$3000?

Solution

Model

Use $P = 500$, $r = 0.09$, $k = 12$, and $S = 3000$ in the equation

$$S = P\left(1 + \frac{r}{k}\right)^{kt},$$

and solve for t.

Solve Graphically

For

$$3000 = 500\left(1 + \frac{0.09}{12}\right)^{12t},$$

we let

$$y_1 = 500\left(1 + \frac{0.09}{12}\right)^{12t} \quad \text{and} \quad y_2 = 3000,$$

and then find the point of intersection of the graphs of y_1 and y_2. Figure 4.37 shows that it occurs where $t \approx 19.98$.

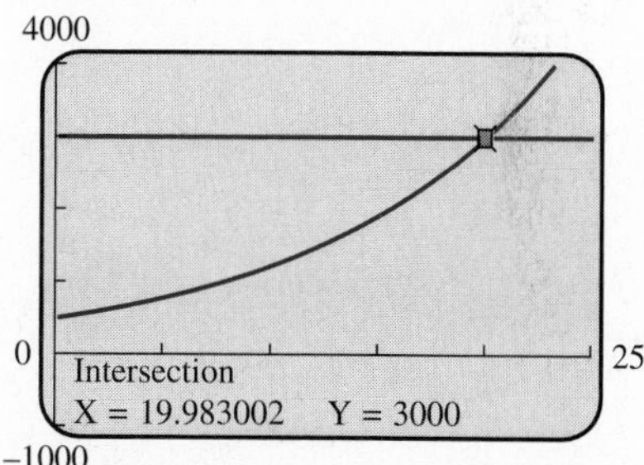

Figure 4.37

Confirm Algebraically

$$3000 = 500\left(1 + \frac{0.09}{12}\right)^{12t}$$

$$6 = 1.0075^{12t} \qquad \text{Divide by 500.}$$

$$\ln 6 = \ln (1.0075^{12t})$$

$$\ln 6 = 12t(\ln 1.0075) \qquad \text{Use the power rule.}$$

$$t = \frac{\ln 6}{12 \ln 1.0075}$$

$$= 19.983 \cdots$$

Interpret

Rounding 19.98 to the next integer, we find that it will take 20 years for the value of the investment to reach (and slightly exceed) $3000. ■

Sometimes you may want to know what interest rate is needed for an investment to reach a certain value in a specified period of time. Example 4 illustrates that a graphical solution to this problem is probably more immediate than an algebraic solution.

■ EXAMPLE 4 Finding an Interest Rate

Steven has $500 to invest. What annual interest rate compounded quarterly is required to double his money in 10 years?

Solution

Model

Use $P = 500$, $k = 4$, $t = 10$, and $S = 1000$ in

$$S = P\left(1 + \frac{r}{k}\right)^{kt}$$

and solve for r.

Solve Graphically

To solve

$$1000 = 500\left(1 + \frac{r}{4}\right)^{4(10)}$$

for r, we let

$$y_1 = 500\left(1 + \frac{r}{4}\right)^{40} \qquad \text{and} \qquad y_2 = 1000$$

and then find the point of intersection of the graphs of y_1 and y_2. Figure 4.38 shows that it occurs at $r \approx 0.07$, or $r = 7\%$.

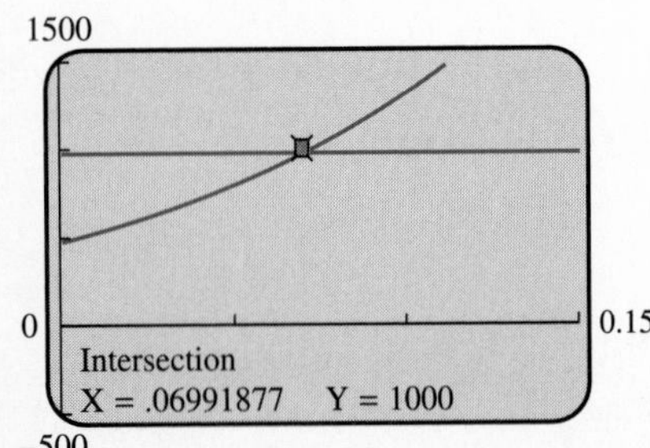

Figure 4.38

Confirm Algebraically

$$
\begin{aligned}
1000 &= 500\left(1 + \frac{r}{4}\right)^{40} \\
2 &= \left(1 + \frac{r}{4}\right)^{40} \\
\ln 2 &= \ln\left(1 + \frac{r}{4}\right)^{40} \\
\ln 2 &= 40 \ln\left(1 + \frac{r}{4}\right) \\
\frac{\ln 2}{40} &= \ln\left(1 + \frac{r}{4}\right) \\
e^{(\ln 2)/40} &= 1 + r/4 \\
\frac{r}{4} &= e^{(\ln 2)/40} - 1 \\
r &= 4e^{(\ln 2)/40} - 4 \\
&= 0.06991 \ldots
\end{aligned}
$$

Interpret

Note that we delay computation until the last step for the best possible accuracy. We find that an investment of \$500 will yield \$1000 in 10 years at an annual interest rate of 7% compounded quarterly. ■

Interest Compounded Continuously

In the following Exploration \$1000 is invested for 1 year at a 10% interest rate. We investigate the value of the investment at the end of 1 year as the number of compounding periods k increases. In other words, we determine the "limiting" value of the expression

$$S = 1000\left(1 + \frac{0.1}{k}\right)^k$$

as k assumes larger and larger integer values.

Exploration Extensions

Repeat the Exploration for an interest rate of 5%. Use an appropriate function for S, and use $y_2 = e^{0.05}$.

● EXPLORATION Compound Interest

Let $S = 1000\left(1 + \dfrac{0.1}{k}\right)^k$.

1. Complete a table of values of S for $k = 10, 20, \ldots, 100$. What do you observe?
2. Figure 4.39 shows the graphs of $y_1 = S$ and $y_2 = 1000e^{0.1}$. Interpret the meanings of these graphs. ●

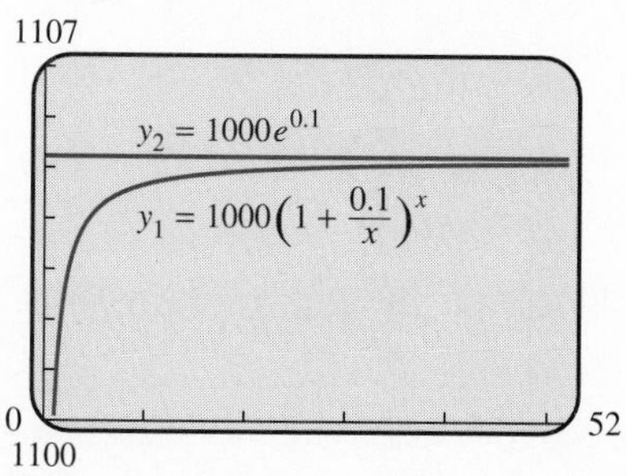

Figure 4.39

Teaching Note

Have students investigate the current rates of interest available for financing a new automobile. Let students determine what auto they want to purchase and how much they would need to finance through a loan. Students should use the formulas from this section to calculate the monthly payment necessary to amortize the loan over a fixed number of months.

Recall that $e = 2.71828 \cdots$. This Exploration and others like it suggest that all of the following are true.

$$\left(1 + \frac{0.1}{k}\right)^k \to e^{0.1} \quad \text{as} \quad k \to \infty$$

$$\left(1 + \frac{1}{k}\right)^k \to e \quad \text{as} \quad k \to \infty$$

$$\left(1 + \frac{r}{k}\right)^k \to e^r \quad \text{as} \quad k \to \infty$$

$$P\left(1 + \frac{r}{k}\right)^{kt} \to Pe^{rt} \quad \text{as} \quad k \to \infty$$

As the number of compounding periods per year increases, the value of $P(1 + r/k)^{kt}$ becomes close to the value of Pe^{rt}. Financial institutions say that interest is **compounded continuously** when they use the formula Pe^{rt} to calculate S, the total value of an account.

Definition: Compound Interest—Value of an Investment

Suppose P dollars are invested at annual interest rate r for t years. The **value of the investment** is

$$S = P\left(1 + \frac{r}{k}\right)^{kt} \quad \text{when interest compounds } k \text{ times per year,}$$

$$S = Pe^{rt} \quad \text{when interest compounds continuously.}$$

■ **EXAMPLE 5 Compounding Continuously**

Suppose Noah invests \$1000 at 8% annual interest compounded continuously. Find the value of his investment at the end of each of the years, 1, 2, . . . , 7.

Solution

Model

Use $y_1 = S = 1000e^{0.08t}$ for continuous compounding.

Solve Algebraically

Figure 4.40 shows the values of $y_1 = 1000e^{0.08x}$ for $x = 1, 2, \ldots, 7$.

Interpret

For example, the value is \$1491.82 at the end of 5 years, and \$1,750.67 at the end of 7 years. ■

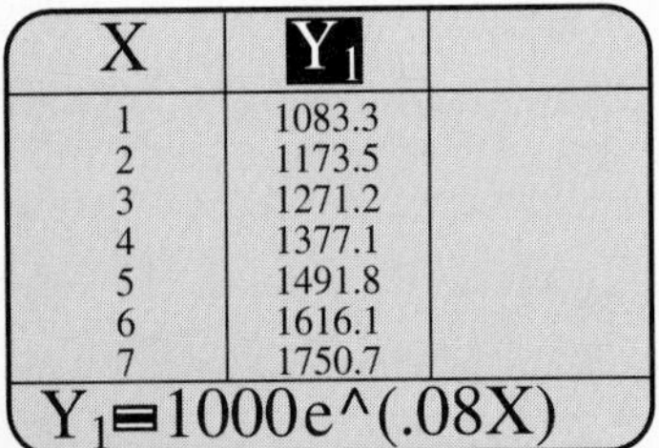

X	Y_1	
1	1083.3	
2	1173.5	
3	1271.2	
4	1377.1	
5	1491.8	
6	1616.1	
7	1750.7	

Y_1=1000e^(.08X)

Figure 4.40

Annual Percentage Yield

With so many different interest rates and methods of compounding it is sometimes difficult for a consumer to compare two different options. For example, would you prefer an investment earning 8.75% annual interest compounded quarterly or one earning 8.7% compounded monthly?

One way to compare two investments is by computing what the financial community calls the **annual percentage yield (APY)**—the rate that would give the same return if interest were computed just once at the end of one year.

■ **EXAMPLE 6 Computing Annual Percentage Yield (APY)**

Ursula invests \$2000 with Household Bank at 5.15% annual interest compounded quarterly. What is the equivalent APY?

Solution

Model

Let $x =$ the equivalent APY. The value of the investment at the end of 1 year using this rate is $S = 2000(1 + x)$. Thus, we have

$$2000(1 + x) = 2000\left(1 + \frac{0.0515}{4}\right)^4.$$

Solve Algebraically

$$(1 + x) = \left(1 + \frac{0.0515}{4}\right)^4 \quad \text{Divide by 2000.}$$

$$x = \left(1 + \frac{0.0515}{4}\right)^4 - 1$$

$$= 0.0525 \cdots$$

Interpret

The APY is 5.25%. In other words \$2000 invested at 5.15% compounded quarterly for 1 year yields the same amount as \$2000 invested at 5.25% compounded once at the end of one year. ■

Example 6 shows that the APY does not depend on the amount of the investment P since both sides of the equation are divided by $P = 2000$. So we can assume that $P = 1$ when comparing investments.

■ EXAMPLE 7 Comparing Annual Percentage Yields (APYs)

Which interest rate is more attractive, 8.75% compounded quarterly or 8.7% compounded monthly?

Solution

Model

Let

$$r_1 = \text{the APY for the } 8.75\% \text{ rate,}$$
$$r_2 = \text{the APY for the } 8.7\% \text{ rate.}$$

Solve Numerically

$$1 + r_1 = \left(1 + \frac{0.0875}{4}\right)^4 \qquad 1 + r_2 = \left(1 + \frac{0.087}{12}\right)^{12}$$
$$r_1 = \left(1 + \frac{0.0875}{4}\right)^4 - 1 \qquad r_2 = \left(1 + \frac{0.087}{12}\right)^{12} - 1$$
$$= 0.09041 \cdots \qquad = 0.09055 \cdots$$

Interpret

The 8.7% rate compounded monthly is more attractive because its APY is 9.055% compared with 9.041% for the 8.75% rate compounded quarterly. ■

Value of an Annuity

So far our investment calculations assume that the investor makes a single, *one-time* deposit. But suppose the investor makes deposits monthly, quarterly, or yearly—the same amount each time. This is an *annuity* situation.

An **annuity** is a sequence of equal periodic payments. Figure 4.41 represents this situation graphically. The annuity is **ordinary** if deposits are made at the same time the interest is posted in the account. We will consider only ordinary annuities in this textbook.

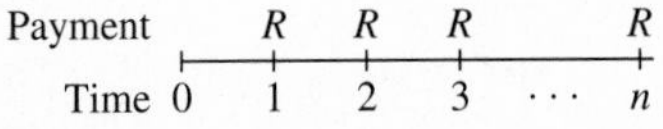

Payment		R	R	R		R
Time	0	1	2	3	$\cdots$	n

Figure 4.41

■ EXAMPLE 8 Computing an Ordinary Annuity

Sarah makes quarterly \$500 payments into a retirement account that pays 8% interest compounded quarterly. How much will be in Sarah's account at the end of the first year?

Solution

If Sarah deposits \$500 on the last day of one quarter and \$500 on the last day of the next quarter, then the total in her account after two quarters will be

\$500 + \$500(1.02). This is the \$500 second-quarter payment plus the principal and interest on the first-quarter deposit. The growth in the value of her account is shown below. Notice the pattern.

End of first quarter:

$$\$500 = \$500$$

End of second quarter:

$$\$500 + \$500(1.02) = \$1010$$

End of third quarter:

$$\$500 + \$500(1.02) + \$500(1.02)^2 = \$1530.20$$

End of fourth quarter:

$$\$500 + \$500(1.02) + \$500(1.02)^2 + \$500(1.02)^3 = \$2060.80$$

The total at the end of the fourth quarter is \$2060.80. ■

Thus the value consists of all the periodic payments together with all the interest. A formula for the value is given here, and it can be proved using mathematical induction (see Section 11.3). Note that the interest rate in the formula is a rate for the payment interval, not an annual rate.

Teaching Note

Show students how an investment in an annuity can accumulate over a 40-year period if they were to begin making monthly deposits of \$50 the first month after graduating from high school. Use several different interest rates.

Notes on Examples

You can challenge your students using Example 9. Have them keep S and R constant and try to solve for i (the quarterly interest rate) using an algebraic method. Then have them graph S as a function of i by letting the variable X replace i in the formula.

Definition: Value of an Annuity

The **value S of an annuity** consisting of n equal payments of R dollars at an interest rate i per compounding period (payment interval) is

$$S = R\,\frac{(1 + i)^n - 1}{i}.$$

■ EXAMPLE 9 Calculating the Value of an Annuity

At the end of each quarter year, Sarah makes a \$500 payment into the Lincoln National Putnam Master Fund. If she earns 7.88% annual interest compounded quarterly, what will be the value of Sarah's annuity in 20 years?

Solution

Model

We use $R = 500$, $i = 0.0788/4$, and $n = 20(4) = 80$ in the equation

$$S = R\,\frac{(1 + i)^n - 1}{i}.$$

Solve Algebraically

$$S = 500 \cdot \frac{(1 + 0.0788/4)^{80} - 1}{0.0788/4}$$

$$S = 95{,}483.389 \cdots$$

Interpret

The value of Sarah's annuity in 20 years will be $95,483.39. ■

Present and Future Value of an Annuity Compared

An annuity is a sequence of equal period payments. The net amount of money put into an annuity is its **present value.** The net amount returned from the annuity is its **future value.** (Future value is what we previously called value.) The periodic and equal payments on a bank loan comprise an annuity. How does the bank determine what the periodic payments should be? It considers what would happen to the present value as an investment with interest compounding over the term of the loan and compares the result to the future value of the loan repayment annuity.

We illustrate this reasoning by assuming that a bank lends you $A = \$50{,}000$ at 6% to purchase a house with the expectation that you will make a loan payment each month (at the monthly interest rate of $0.06/12 = 0.005$).

1. The future value of an investment at 6% compounded monthly for n months is

$$A(1 + i)^n = 50{,}000(1 + 0.005)^n.$$

2. The future value of an annuity of R dollars (the loan payments) is

$$R\,\frac{(1 + i)^n - 1}{i} = R\,\frac{(1 + 0.005)^n - 1}{0.005}.$$

To find R, we would solve the equation

$$50{,}000(1 + 0.005)^n = R\,\frac{(1 + 0.005)^n - 1}{0.005}.$$

In general, the monthly payments of R dollars for a loan of A dollars must satisfy the equation

$$A(1 + i)^n = R\,\frac{(1 + i)^n - 1}{i}.$$

Dividing both sides by $(1 + i)^n$ leads to the following formula for the present value of an annuity.

Definition: Present Value of an Annuity

The **present value A of an annuity** consisting of n equal payments of R dollars earning at an interest rate i per period (payment interval) is

$$A = R\,\frac{1 - (1 + i)^{-n}}{i}.$$

Follow-up

Ask students how the interest rate affects the present and future values of an annuity. **(A higher interest rate gives a lower present value and a higher future value.)**

Assignment Guide

Day 1: Ex. 1, 4, 5, 8, 9, 12, 13, 16–22, 25, 27, 30, 31, 34
Day 2: Ex. 36, 37, 40, 43, 45, 48–50, 52–54

Cooperative Learning

Group Activity: Ex. 51

Notes on Exercises

The exercises in this section should be interesting to students because they deal with real-life financial situations. Students can apply these methods to their own financial planning.
Ex. 45–46 illustrate the results of making accelerated payments on a mortgage.
Ex. 47–48 can be solved by applying exponential regression to the tax data.

Ongoing Assessment

Self-Assessment: Ex. 7, 11, 29, 33, 39
Embedded Assessment: Ex. 50

The annual interest rate charged on consumer loans is the **annual percentage rate (APR)**. The APY for the lender is higher than the APR. See Exercise 50.

■ EXAMPLE 10 Calculating Loan Payments

Carlos purchases a Ford Taurus from Ricart Ford for \$18,500. What are the monthly payments for a 4-year loan with a \$2000 down payment if the annual interest rate (APR) is 2.9%?

Solution

Model

The down payment is \$2000, so the amount borrowed is \$16,500. Since APR = 2.9%, $i = 0.029/12$ and the monthly payment is the solution to

$$16{,}500 = R\,\frac{1 - (1 + 0.029/12)^{-4(12)}}{0.029/12}.$$

Solve Algebraically

$$R\left[1 - \left(1 + \frac{0.029}{12}\right)^{-4(12)}\right] = 16{,}500\left(\frac{0.029}{12}\right) \qquad \text{Multiply.}$$

$$R = \frac{16{,}500(0.029/12)}{1 - (1 + 0.029/12)^{-48}} \qquad \text{Divide.}$$

$$= 364.487\ldots$$

Interpret

Carlos will have to pay \$364.49 per month for 48 months. ■

Quick Review 4.5

In Exercises 1–8, complete the given percentage calculation.

1. Find 3.5% of 200.
2. Find 2.5% of 150.
3. 78 is what percent of 120?
4. 28 is what percent of 80?
5. 48 is 32% of what number?
6. 176.4 is 84% of what number?
7. How much does Jane have at the end of 1 year if she invests \$300 at 5% simple interest?
8. How much does Carlos have at the end of 1 year if he invests \$500 at 4.5% simple interest?

SECTION EXERCISES 4.5

In Exercises 1–4, find the value S of investing P dollars for n years with the interest rate r compounded annually.

1. $P = \$1500, \quad r = 7\%, \quad n = 6$ years

2. $P = \$3200, \quad r = 8\%, \quad n = 4$ years

3. $P = \$12{,}000, \quad r = 7.5\%, \quad n = 7$ years

4. $P = \$15{,}500, \quad r = 9.5\%, \quad n = 12$ years

In Exercises 5–8, find the value S of investing P dollars for n years with the interest rate r compounded k times a year.

5. $P = \$1500, \quad r = 7\%, \quad k = 4, \quad n = 5$ years

6. $P = \$3500, \quad r = 5\%, \quad k = 4, \quad n = 10$ years

7. $P = \$40{,}500, \quad r = 3.8\%, \quad k = 12, \quad n = 20$ years

8. $P = \$25{,}300, \quad r = 4.5\%, \quad k = 12, \quad n = 25$ years

In Exercises 9–16, solve the problem.

9. *Finding Time* If John invests $2300 in a savings account with a 9% interest rate compounded quarterly, how long will it take until John's account has a balance of $4150?

10. *Finding Time* If Joelle invests $8000 into a retirement account with a 9% interest rate compounded monthly, how long will it take until this single payment has grown in her account to $16,000?

11. *Trust Officer* Megan is the trust officer for an estate. If she invests $15,000 into an account that carries an interest rate of 8% compounded monthly, how long will it be until the account has a value of $45,000 for Megan's client?

12. *Chief Financial Officer* Willis is the financial officer of a private university with the responsibility for managing an endowment. If he invests $1.5 million at an interest rate of 8% compounded quarterly, how long will it be until the account exceeds $3.75 million?

13. *Finding the Interest Rate* What interest rate compounded daily (365 days/year) is required for a $22,000 investment to grow to $36,500 in 5 years?

14. *Finding the Interest Rate* What interest rate compounded monthly is required for an $8500 investment to triple in 5 years?

15. *Pension Officer* Jack is an actuary working for a corporate pension fund. He needs to have $14.6 million grow to $22 million in 6 years. What interest rate compounded annually does he need for this investment?

16. *Bank President* The president of a bank has $18 million in his bank's investment portfolio that he wants to grow to $25 million in 8 years. What interest rate compounded annually does he need for this investment?

In Exercises 17 and 18, find the time required for an investment to double or triple. Notice that this time is independent of the amount of the investment.

17. *Doubling Your Money* Determine how much time is required for an investment to double in value if interest is earned at the rate of 5.75% compounded quarterly.

18. *Tripling Your Money* Determine how much time is required for an investment to triple in value if interest is earned at the rate of 6.25% compounded monthly.

In Exercises 19–22, complete the information requested in the table about continuous compounding.

Continuous Compounding

	Initial Investment	APR	Time to Double	Amount in 15 Years
19.	$12,500	9%	?	?
20.	$32,000	8%	?	?
21.	$ 9,500	?	4 years	?
22.	$16,800	?	6 years	?

In Exercises 23–30, calculate the effect of the annual percentage rate and number of compounding periods on the doubling time of an investment.

Doubling Time

	APR	Compounding Periods	Time to Double
23.	4%	Quarterly	?
24.	6%	Quarterly	?
25.	8%	Quarterly	?
26.	10%	Quarterly	?
27.	7%	Annually	?
28.	7%	Quarterly	?
29.	7%	Monthly	?
30.	7%	Continuously	?

In Exercises 31–34, find the annual percentage yield (APY) for the investment.

31. $3000 at 6% compounded quarterly

32. $8000 at 5.75% compounded daily

33. P dollars at 6.3% compounded continuously

34. P dollars at 4.7% compounded monthly

In Exercises 35 and 36, compare the two investments.

35. *Comparing Investments* Which interest rate is more attractive, 5% compounded monthly or 5.1% compounded quarterly?

36. *Comparing Investments* Which interest rate is more attractive, $5\frac{1}{8}\%$ compounded annually or 5% compounded continuously?

In Exercises 37–40, a consumer invests the same amount each month with payments made the same day that interest is posted in the account.

37. *An IRA Account* Amy contributes $50 per month into the Lincoln National Bond Fund that earns 7.26% annual interest. What is the value of Amy's investment after 25 years?

38. *An IRA Account* Matthew contributes $50 per month into the Kaufman Fund that earns 15.5% annual interest. What is the value of Matthew's investment after 20 years?

39. *An Investment Annuity* Betsy contributes to the Fidelity Puritan Fund that earns 12.4% annual interest. What should her monthly payments be if she wants to accumulate $250,000 in 20 years?

40. *An Investment Annuity* Diego contributes to the Commercial National Money Market Account that earns 4.5% annual interest. What should his monthly payments be if he wants to accumulate $120,000 in 30 years?

In Exercises 41–44, you will be finding loan payments.

41. *Car Loan Payment* What is Kim's monthly payment for a 4-year $9000 car loan with an APR of 7.95% from Century Bank?

42. *Car Loan Payment* What is Sarah's monthly payment for a 3-year $4500 car loan with an APR of 10.25% from County Savings Bank?

43. *House Mortgage Payment* Gendo obtains a 30-year $86,000 house loan with an APR of 8.75% from National City Bank. What is her monthly payment?

44. *House Mortgage Payment* Roberta obtains a 25-year $100,000 house loan with an APR of 9.25% from NBD Bank. What is her monthly payment?

In Exercises 45–54, solve the problem.

45. *Mortgage Payment Planning* An $86,000 mortgage for 30 years at 12% APR requires monthly payments of $884.61. Suppose you decided to make monthly payments of $1050.00.

a. When would the mortgage be completely paid?
b. How much do you save with the greater payments compared with the original plan?

46. *Mortgage Payment Planning* Suppose you make payments of $884.61 for the $86,000 mortgage in Exercise 45 for 10 years and then make payments of $1050 until the loan is paid.

a. When will the mortgage be completely paid under these circumstances?
b. How much do you save with the greater payments compared with the original plan?

47. *School Taxes* Table 4.14 gives the real estate taxes and the total taxes (in millions) collected for the Columbus, Ohio Public Schools.

Table 4.14 School Tax Data

Year	Real Estate Taxes	Total Taxes
1988	130.8	169.4
1989	136.8	175.8
1990	139.4	179.5
1991	142.6	182.8
1992	188.2	236.8
1993	193.3	239.0

Source: Franklin County (OH) Auditor's Office as reported by Doug Miller in *The Columbus Dispatch,* July 17, 1994.

a. Graph a scatter plot of the real estate tax data.
b. Compute and graph an exponential regression equation for the real estate tax data. (Let $x = 0$ represent 1988, $x = 1$ represent 1989, and so forth.)
c. Use the regression equation to estimate the real estate taxes the schools will receive in 1998.

48. Use the data in Table 4.14.

a. Graph a scatter plot of the total tax data.
b. Compute and graph an exponential regression equation

for the total tax data. (Let $x = 0$ represent 1988, $x = 1$ represent 1989, and so on.)

c. Use the regression equation to estimate the total taxes the schools will receive in the year 2000.

49. Writing to Learn Explain why computing the APY for an investment does not depend on the actual amount being invested. Give a formula for the APY on a \$1 investment at annual rate r compounded k times a year. How do you extend the result to a \$1000 investment?

50. Writing to Learn Give reasons why banks might not announce their APY on a loan they would make to you at a given APR. What is the bank's APY on a loan that they make at 4.5% APR?

51. Group Learning Activity *Work in groups of three or four.* Consider population growth of humans or other animals, bacterial growth, radioactive decay, and compounded interest. Explain how these problem situations are similar and how they are different. Give examples to support your point of view.

52. *Simple Interest versus Compounding Annually* Steve purchases a \$1000 certificate of deposit and will earn 6% each year. The interest will be mailed to him, so he will not earn interest on his interest.

a. Show that after t years, the total amount of interest he receives from his investment plus the original \$1000 is given by

$$f(t) = 1000(1 + 0.06t).$$

b. Steve invests another \$1000 at 6% compounded annually. Make a table that compares the value of the two investments for $t = 1, 2, \ldots, 10$ years.

EXTENDING THE IDEAS

53. The function

$$f(x) = 100\,\frac{(1 + 0.08/12)^x - 1}{0.08/12}$$

describes the future value of a certain annuity.

a. What is the annual interest rate?
b. How many payments per year are there?
c. What is the amount of each payment?

54. The function

$$f(x) = 200\,\frac{1 - (1 + 0.08/12)^{-x}}{0.08/12}$$

describes the present value of a certain annuity.

a. What is the annual interest rate?
b. How many payments per year are there?
c. What is the amount of each payment?

CHAPTER 4 REVIEW

KEY TERMS

The number following each key term indicates the page of its introduction.

annual percentage rate (APR), 339
annual percentage yield (APY), 335
annuity, 336
common logarithm, 298
compound interest, 330
compounded annually, 330
compounded continuously, 334
compounded k times per year, 330
compounded monthly, 330
exponential decay, 290
exponential form, 298
exponential function, 286
exponential growth, 290
exponential regression, 291, 303
future value of an annuity, 338
half-life, 292
inverse rule of logarithms, 313
linear regression, 303
logarithm, 297
logarithmic form, 298
logarithmic function with base a, 297
logarithmic regression, 302
logistic curve, 321
logistic growth function, 321
natural logarithm, 299
natural logarithmic regression, 303
Newton's law of cooling, 322
one-to-one rule of logarithms, 313
ordinary annuity, 336
power function, 286
power regression, 303
power rule of logarithms, 311
present value of an annuity, 338
product rule of logarithms, 311
quotient rule of logarithms, 311
Richter scale, 314
value of an annuity, 338, 339
value of an investment, 334

REVIEW EXERCISES

In Exercises 1–4, use the definition of logarithmic function to rewrite the equation in exponential form.

1. $\log_3 x = 5$

2. $\log_2 x = y$

3. $\ln \frac{x}{y} = -2$

4. $\log \frac{a}{b} = -3$

In Exercises 5–12, describe how the graph of the function can be obtained from the graph of $y = 2^x$ or $y = e^x$ by using transformations. Support your answer graphically.

5. $f(x) = 4^{-x} + 3$

6. $f(x) = -4^{-x}$

7. $f(x) = -8^{-x} - 3$

8. $f(x) = 8^{-x} + 3$

9. $f(x) = e^{4-x}$

10. $f(x) = e^{1-x}$

11. $f(x) = e^{2x-3}$

12. $f(x) = e^{3x-4}$

In Exercises 13–18, sketch a graph of the function. Support your answer with a grapher.

13. $y = 3^x + 1$

14. $y = 3^x - 1$

15. $y = e^{4-x} + 2$

16. $y = e^{3-x} + 1$

17. $y = 2(5^{x-3}) + 1$

18. $y = 3(4^{x+1}) - 2$

In Exercises 19–24, describe how the graph of the function can be obtained from the graph of $y = \log_2 x$.

19. $f(x) = \log_2 (x + 4)$

20. $f(x) = \log_2 (x - 5)$

21. $g(x) = \log_2 (4 - x)$

22. $g(x) = \log_2 (5 - x)$

23. $h(x) = -\log_2 (x - 1) + 2$

24. $h(x) = -\log_2 (x + 1) + 4$

In Exercises 25–28, find the interval(s) on which the function is increasing, and those on which it is decreasing. Also find all local maximum and minimum values, the domain and range, and describe its end behavior.

25. $f(x) = x \ln x$

26. $f(x) = x^2 \ln x$

27. $f(x) = x^2 \ln|x|$

28. $f(x) = \frac{\ln x}{x}$

In Exercises 29–32, find a positive real number that the logistic function is close to when x is a large positive number.

29. $f(x) = \frac{100}{5 + 3e^{-0.05x}}$

30. $g(x) = \frac{100}{4 + 2e^{-0.01x}}$

31. $f(x) = \frac{50}{5 - 2e^{-0.04x}}$

32. $g(x) = \frac{24}{2 - 5e^{-0.01x}}$

In Exercises 33 and 34, rewrite the exponential expression with the specified base.

33. $\left(\frac{1}{e^2}\right)^{3x}$, base e

34. $\left(\frac{1}{e^{-2}}\right)^{4x}$, base e

In Exercises 35 and 36, write two different double inequalities that compare the three given functions for $x < 0$ and for $x > 0$. Support your statements with tables and graphs.

35. $y_1 = 1.2^x$, $y_2 = 2.1^x$, $y_3 = 3.4^x$

36. $y_1 = 1.2^{-x}$, $y_2 = 2.1^{-x}$, $y_3 = 3.4^{-x}$

In Exercises 37 and 38, use rules of exponents to confirm that two of the three given exponential functions are equal. Support your work graphically.

37. a. $y_1 = 2^{3x-1}$ **b.** $y_2 = 8^{x-1/2}$ **c.** $y_3 = 2^{3x-3/2}$

38. a. $y_1 = 2^{-(x-4)}$ **b.** $y_2 = 0.5^{x-4}$ **c.** $y_3 = 0.25^{2x-8}$

In Exercises 39 and 40, find the smallest positive integer n such that if $x > n$, the given exponential function is greater than the given power function. Also find a viewing window so that the two graphs look similar to those in the figure.

39. $y_1 = e^x$ and $y_2 = 2x^e$

40. $y_1 = 4^x$ and $y_2 = x^6$

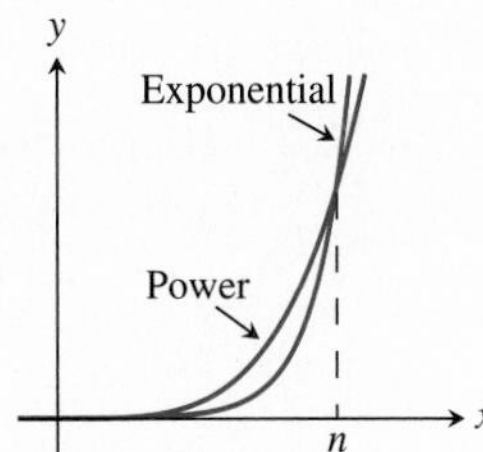

For Exercises 39–40

In Exercises 41–54, solve for x.

41. $10^x = 4$

42. $e^x = 0.25$

43. $1.05^x = 3$

44. $1.045^x = 2$

45. $\ln x = 5.4$

46. $\log x = -7$

47. $3^{x-3} = 5$

48. $2^{x+1} = 0.5$

49. $3 \log_2 x + 1 = 7$

50. $2 \log_3 x - 3 = 4$

51. $\frac{3^x - 3^{-x}}{2} = 5$

52. $\frac{50}{4 + e^{2x}} = 11$

53. $\log(x + 2) + \log(x - 1) = 4$

54. $\ln(3x + 4) - \ln(2x + 1) = 5$

In Exercises 55–57, find the intervals on which the function is increasing and decreasing. Also find all local maximum and minimum values, and the domain and range.

55. $f(x) = x3^{-x}$

56. $g(x) = e^{4-x^2}$

57. $f(x) = \dfrac{1}{\sqrt{2\pi}} e^{-x^2/2}$ (Sometimes called the *normal distribution curve* or a *bell curve.*)

58. Writing to Learn Let $f(x) = e^x$. Explain why the graph of $f(ax + b)$ can be obtained by applying one transformation to the graph of $y = c^x$ for appropriate c. What is c?

In Exercises 59–64, determine the value of k so that the graph of f passes through the given point.

59. $f(x) = 20e^{-kx}$, $(3, 50)$

60. $f(x) = 20e^{-kx}$, $(1, 30)$

61. $f(x) = 30e^{-kx}$, $(1, 20)$

62. $f(x) = 30e^{-kx}$, $(2, 10)$

63. $f(x) = 50e^{-kx}$, $(4, 70)$

64. $f(x) = 50e^{-kx}$, $(1, 70)$

In Exercises 65 and 66, let $x = 1$ represent 1960, $x = 11$ represent 1970, and $x = 31$ represent 1990 for the data.

65. *Oil Production* Find a natural logarithmic regression equation for the data in Table 4.15 and use it to estimate the production for the year 1982.

Table 4.15 Indonesia's Oil Production

Year	Metric Tons
1960	20.56
1970	42.10
1990	70.10

Source: The Statesman's Yearbook, 129th ed. (London: The Macmillan Press, Ltd., 1992).

66. *Oil Production* Find a natural logarithmic regression equation for the data in Table 4.16, and use it to estimate the production for the year 1975.

Table 4.16 Saudi Arabia's Oil Production

Year	Metric Tons
1960	61.09
1970	176.85
1990	321.93

Source: The Statesman's Yearbook, 129th ed. (London: The Macmillan Press, Ltd., 1992).

In Exercises 67–86, solve the problem.

67. *Estimating Population Growth* Use the data given in Table 4.17 about the population of Georgia.

Table 4.17 Population of Georgia

Year	Population (millions)
1960	3.94
1980	5.46
1990	6.48

Source: The Statesman's Yearbook, 129th ed. (London: The Macmillan Press, Ltd., 1992).

a. Find an exponential regression equation for these data. (Use $t = 0$ to represent 1960.)

b. Use the regression equation to predict the population in the year 2000.

c. Predict when the population of the state will be double the 1960 population.

68. *Estimating Population Growth* Use the data in Table 4.18 about the population of New York.

Table 4.18 Population of New York

Year	Population (millions)
1960	16.78
1980	17.56
1990	17.99

Source: The Statesman's Yearbook, 129th ed. (London: The Macmillan Press, Ltd., 1992).

a. Find an exponential regression equation for these data. (Use $t = 0$ to represent 1960.)
b. Use the regression equation to predict the population in the year 2000.
c. Predict when the population of the state will be double the 1960 population.

69. *Drug Absorption* A drug is administered intravenously for pain. The function $f(t) = 90 - 52 \ln(1 + t)$, where $0 \le t \le 4$, gives the amount of the drug in the body after t hours.

a. What was the initial ($t = 0$) number of units of drug administered?
b. How much is present after 2 h?
c. Draw the graph of f.

70. *Population Decrease* The population of Metroville is 123,000 and is decreasing by 2.4% each year.

a. Write a function that models the population as a function of time t.
b. Predict when the population will be 90,000.

71. *Population Decrease* The population of Preston is 89,000 and is decreasing by 1.8% each year.

a. Write a function that models the population as a function of time t.
b. Predict when the population will be 50,000.

72. *Spread of Flu* The number P of students infected with flu at Northridge High School t days after exposure is modeled by

$$P(t) = \frac{300}{1 + e^{4-t}}.$$

a. What was the initial ($t = 0$) number of students infected with the flu?
b. How many students were infected after 3 days?
c. When will 100 students be infected?
d. What would be the maximum number of students infected?

73. *Rabbit Population* The number of rabbits in Elkgrove doubles every month. There are 20 rabbits present initially.

a. Express the number of rabbits as a function of the time t.
b. How many rabbits were present after 1 year? after 5 years?
c. When will there be 10,000 rabbits?

74. *Guppy Population* The number of guppies in Susan's aquarium doubles every day. There are four guppies initially.

a. Express the number of guppies as a function of time t.
b. How many guppies were present after 4 days? after 1 week?
c. When will there be 2000 guppies?

75. *Radioactive Decay* The half-life of a certain radioactive substance is 1.5 sec. The initial amount of substance is S_0 grams.

a. Express the amount of substance S remaining as a function of time t.
b. How much of the substance is left after 1.5 sec? after 3 sec?
c. Determine S_0 if there was 1 g left after 1 min.

76. *Radioactive Decay* The half-life of a certain radioactive substance is 2.5 sec. The initial amount of substance is S_0 grams.

a. Express the amount of substance S remaining as a function of time t.
b. How much of the substance is left after 2.5 sec? after 7.5 sec?
c. Determine S_0 if there was 1 g left after 1 min.

77. *Finding Time* If Joenita invests \$1500 into a retirement account with an 8% interest rate compounded quarterly, how long will it take this single payment to grow to \$3750?

78. *Finding Time* If Juan invests \$12,500 into a retirement account with a 9% interest rate compounded continuously, how long will it take this single payment to triple in value?

79. *Monthly Payments* The time t in months that it takes to pay off a \$60,000 loan at 9% annual interest with monthly payments of x dollars is given by

$$t = 133.83 \ln\left(\frac{x}{x - 450}\right).$$

Estimate the length (term) of the \$60,000 loan if the monthly payments are \$700.

80. *Monthly Payments* Using the equation in Exercise 79, estimate the length (term) of the \$60,000 loan if the monthly payments are \$500.

81. *Finding APY* Find the annual percentage yield for an investment with an interest rate of 8.25% compounded monthly.

82. *Finding APY* Find the annual percentage yield that can be used to advertise an account that pays interest at 7.20% compounded continuously.

83. *Light Absorption* The Beer-Lambert law of absorption applied to Lake Superior states that the light intensity I (in lumens) at a depth of x feet satisfies the equation

$$\log \frac{I}{12} = -0.0125x.$$

Find the light intensity at a depth of 25 ft.

84. For what values of b is $\log_b x$ a vertical stretch of $y = \ln x$? A vertical shrink of $y = \ln x$?

85. For what values of b is $\log_b x$ a vertical stretch of $y = \log x$? A vertical shrink of $y = \log x$?

86. If $f(x) = ab^x$, $a > 0$, $b > 0$, prove that $g(x) = \ln f(x)$ is a linear function. Find its slope and y-intercept.

In Exercises 87–93, solve the given problem.

87. *Spread of Flu* The number of students infected with flu after t days at Springfield High School is modeled by the function

$$P(t) = \frac{1600}{1 + 99e^{-0.4t}}.$$

a. What was the initial number of infected students?
b. When will 800 students be infected?
c. The school will close when 400 of the 1600 student body are infected. When would the school close?

88. *Population of Deer* The population P of deer after t years in Briggs State Park is modeled by the function

$$P(t) = \frac{1200}{1 + 99e^{-0.4t}}.$$

a. What was the initial population of deer?
b. When will there be 1000 deer?
c. What is the maximum number of deer planned for the park?

89. *Newton's Law of Cooling* A cup of coffee cooled from 96° C to 65° C after 8 min in a room at 20° C. When will it cool to 25° C?

90. *Newton's Law of Cooling* A cake is removed from an oven at 220° F and cools to 150° F after 35 min in a room at 75° F. When will it cool to 95° F?

91. The function

$$f(x) = 100\,\frac{(1 + 0.09/4)^x - 1}{0.09/4}$$

describes the future value of a certain annuity.

a. What is the annual interest rate?
b. How many payments per year are there?
c. What is the amount of each payment?

92. The function

$$g(x) = 200\,\frac{1 - (1 + 0.11/4)^{-x}}{0.11/4}$$

describes the present value of a certain annuity.

a. What is the annual interest rate?
b. How many payments per year are there?
c. What is the amount of each payment?

93. *Simple Interest versus Compounding Continuously* Grace purchases a \$1000 certificate of deposit that will earn 5% each year. The interest will be mailed to her, so she will not earn interest on her interest.

a. Show that after t years, the total amount of interest she receives from her investment plus the original \$1000 is given by

$$f(t) = 1000(1 + 0.05t).$$

b. Grace invests another \$1000 at 5% compounded continuously. Make a table that compares the values of the two investments for $t = 1, 2, \ldots, 10$ years.

THE AZTECS

The advanced Aztec civilization in modern-day Mexico was notable for its art, architecture, commerce, and system of justice. To help keep accurate accounts, the Aztecs developed an efficient system of numerals and arithmetic. They borrowed the concepts of place value and a base-20 number system from the earlier Mayan and Olmec cultures, but they had a set of distinctive Aztec numerals. For example, a small ear of corn was used for the zero symbol.

The Aztecs applied mathematics to the records that registered land ownership. These records showed the boundaries, area, and market value of property, and were used to calculate the taxes that the owners had to pay. The measurements included in the Aztec records were more accurate than the measurements of the same farms drawn up by the Spanish conquerors.

chapter 5

RATIONAL FUNCTIONS

BIBLIOGRAPHY

For students: *Career Choices for the 90s for Students of Mathematics.* Walker Publishing Company, 1990.

For teachers: *Algebra Experiments II: Exploring Nonlinear Functions,* Ronald J. Carlson and Mary Jean Winter. Addison-Wesley (81525), 1993. Available through Dale Seymour Publications.

Mathematica, Stephen Wolfram. Addison-Wesley (51502), 1991.

5.1 RATIONAL FUNCTIONS AND ASYMPTOTES

Objective

Students will be able to describe the graphs of rational functions and identify horizontal and vertical asymptotes of rational functions.

Introducing Rational Functions • Vertical Asymptotes • Horizontal Asymptotes • Applications

Introducing Rational Functions

Rational functions are defined using ratios or quotients of polynomials.

REMARK

Recall that 0 is the zero polynomial. (See the Prerequisite Chapter.)

Definition: Rational Function

A **rational function** is one that can be written in the form

$$f(x) = \frac{p(x)}{h(x)} = \frac{a_n x^n + a_{n-1}x^{n-1} + \cdots + a_0}{b_m x^m + b_{m-1}x^{m-1} + \cdots + b_0},$$

where $p(x)$ and $h(x)$ are polynomials and $h(x)$ is not the zero polynomial.

Motivate

Ask students to predict some of the characteristics of the graph of $y = \dfrac{x+2}{x-2}$ and to check their predictions with a grapher.

These are rational functions:

$$f(x) = \frac{3x}{x^2+1} \qquad g(x) = \frac{x^2+1}{x-3} \qquad k(x) = \frac{4-x}{x^2+2x-3}$$

The domain of f consists of *all real numbers,* whereas the domain of g is *all real numbers excluding* $x = 3$ and the domain of k is *all real numbers excluding both* $x = -3$ *and* $x = 1$. Rational functions are not defined for the zeros of the denominator.

None of the next three functions is rational. Do you see why?

$$f(x) = \frac{\sqrt{x^2 - x}}{x + 1} \qquad g(x) = \frac{x^3 + 2}{e^x} \qquad k(x) = \frac{x^2 - 3x + \ln x}{2x^3 + 5x^2 - 4x + 12}$$

Some expressions may need to be rewritten before they can be recognized as rational functions. Example 1 illustrates.

■ EXAMPLE 1 Identifying Rational Functions

Show that the function f defined by

$$f(x) = 1 + \frac{1}{x - \dfrac{1}{x}}$$

is a rational function, and find its domain.

Solution

We must show that f can be rewritten as a quotient of two polynomials.

$$f(x) = 1 + \frac{1}{x - \dfrac{1}{x}} = 1 + \frac{1}{\dfrac{x^2 - 1}{x}}$$

$$= 1 + \frac{x}{x^2 - 1}$$

$$= \frac{(x^2 - 1) + x}{x^2 - 1}$$

$$= \frac{x^2 + x - 1}{x^2 - 1}, \quad x \neq 0$$

We have shown that f can be written as a quotient of two polynomials. The function f in its original form is undefined when $x = 0$ and also when $x - 1/x = 0$. Consequently, the domain of f is all real numbers x except $x = -1, 0, 1$. ■

REMARK

Notice that the original form for f in Example 1 is undefined when $x = 0, 1,$ or -1. However, the final form is undefined only for $x = 1$ or -1. It was necessary to restrict the domain of the final form by excluding $x = 0$ so that the two forms would be *equivalent*.

Vertical Asymptotes

A rational function and its graph have some distinctive characteristics near the zeros of the denominator. To illustrate we shall consider the function

$$f(x) = \frac{1}{x}.$$

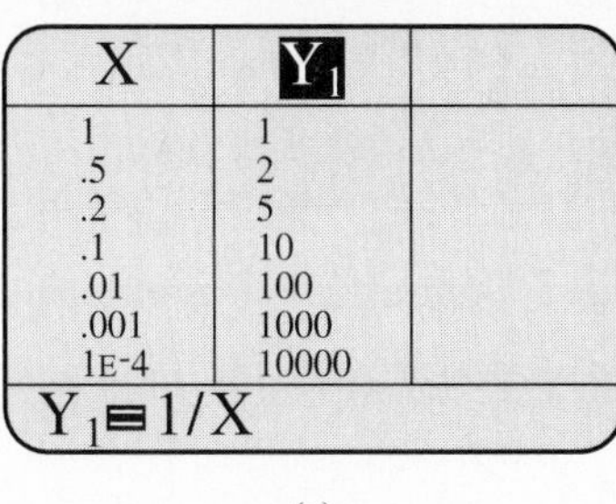

X	Y_1	
1	1	
.5	2	
.2	5	
.1	10	
.01	100	
.001	1000	
1E-4	10000	

Y_1=1/X

(a)

X	Y_1	
-1	-1	
-.5	-2	
-.2	-5	
-.1	-10	
-.01	-100	
-.001	-1000	
-1E-4	-10000	

Y_1=1/X

(b)

Figure 5.1 Notice that 1E^-4 is grapher scientific notation for 0.0001, and -1E^-4 for -0.0001.

REMARK

Notice that the $^+$ and $^-$ superscripts in $x \to 0^+$ and $x \to 0^-$ are read "from the right" and "from the left," respectively.

The table in Figure 5.1a shows that when x has a value greater than but near 0, the expression $1/x$ is a large positive number.[1] Figure 5.1b shows that when x has a value less than but near 0, the expression $1/x$ is negative with a large absolute value. For the behavior to the right of 0, we write

$$f(x) \to \infty \text{ as } x \to 0^+,$$

which is read

"$f(x)$ approaches positive infinity as x approaches 0 from the right."

For the behavior to the left of 0, we write

$$f(x) \to -\infty \text{ as } x \to 0^-,$$

which is read

"$f(x)$ approaches negative infinity as x approaches 0 from the left."

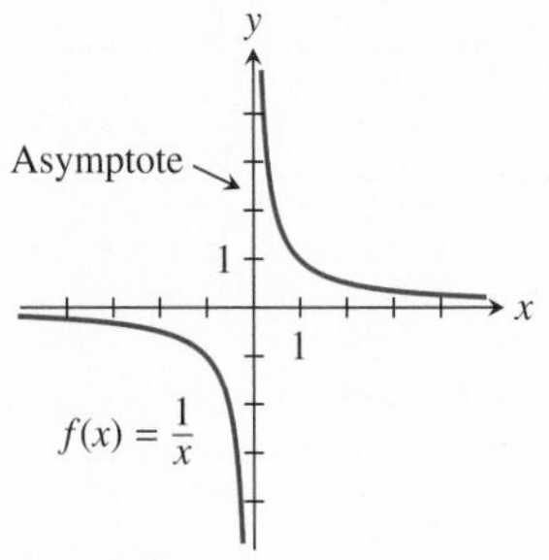

Figure 5.2

Figure 5.2 shows the graph of f near $x = 0$, the zero of the denominator of f. The values of $|f(x)| \to \infty$ as $x \to 0$, and the line $x = 0$ (the y-axis) is a *vertical asymptote*.

Definition: Vertical Asymptotes

The vertical line $x = h$ is a **vertical asymptote** of a function f if, as $x \to h$ from the right or from the left,

$$f(x) \to \infty \quad \text{or} \quad f(x) \to -\infty.$$

[1] Consult your grapher owner's manual to make a table with these x-values.

Teaching Note

If possible, have students compare the graphs of rational functions using both the dot mode and the connected mode. In the connected mode, students may think that the grapher is purposely displaying the vertical asymptotes of the function. The asymptotes that appear on the grapher are actually artifacts of the algorithm used to create the graphical image on the screen. In Example 2, for example, the asymptote shown might be a result of connecting points such as $(-3.05, 21)$ and $(-2.95, -19)$. If a column of pixels corresponds to exactly $x = -3$, the grapher finds that the function is undefined at that value, so the value is skipped and the asymptote does not appear.

EXAMPLE 2 Identifying Vertical Asymptotes

Write statements that describe the behavior of the function $f(x) = 1 - \dfrac{1}{x+3}$ near the zeros of its denominator. Then identify the vertical asymptote(s) of f.

Solution

We write

$$f(x) = 1 - \frac{1}{x+3} = \frac{x+2}{x+3}.$$

We see that the value $x = -3$ is a zero of the denominator of f. Near $x = -3$, the graph of f behaves as shown in Figure 5.3. As x approaches -3 from the right ($x \to -3^+$), it appears that $f(x) \to -\infty$. As x approaches -3 from the left ($x \to -3^-$), it appears that $f(x) \to \infty$. The line $x = -3$ is a vertical asymptote. ■

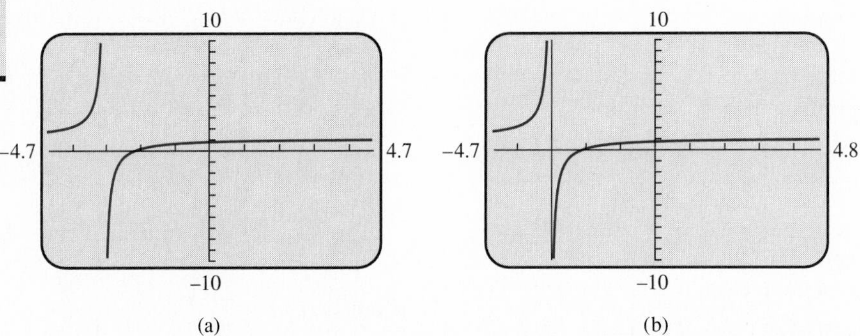

Figure 5.3 The graph of f in slightly different viewing windows. (a) Note that the asymptote $x = -3$ does not show when the grapher window includes a column of pixels whose x-coordinates are exactly $x = -3$. (b) The grapher has to be in connected mode for what appears to be a vertical asymptote to show.

The function $f(x) = 1 + 1/(x - 1/x)$ of Example 1 is not defined at $x = 0$. However, this function does not have $x = 0$ as a vertical asymptote. (See Exercise 64.) For $x = a$ to be a vertical asymptote of a function, the function must be undefined at $x = a$ and the values of $|f(x)|$ must be unbounded as x approaches a from the left or right.

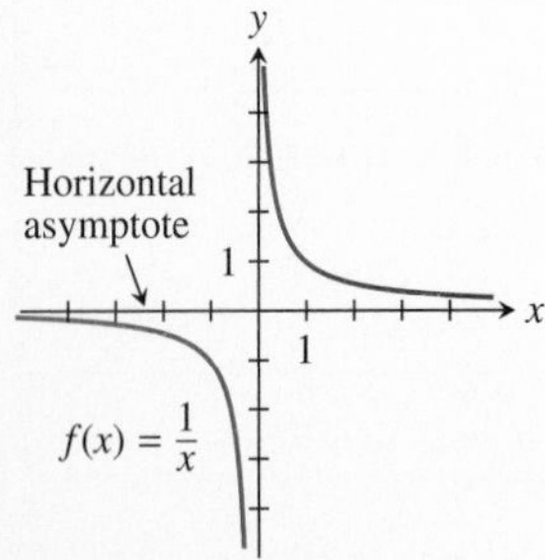

Figure 5.4

Horizontal Asymptotes

Not only does a rational function have distinctive behavior near its vertical asymptotes, it also has distinctive behavior when the value of $|x|$ is large. The graph in Figure 5.4 and the tables in Figure 5.5 indicate that $f(x) = 1/x$ comes close to zero as $|x|$ becomes large.

X	Y_1	
1	1	
5	.2	
10	.1	
50	.02	
100	.01	
1000	.001	
10000	1E-4	

$Y_1 \equiv 1/X$

(a)

X	Y_1	
-1	-1	
-5	-.2	
-10	-.1	
-50	-.02	
-100	-.01	
-1000	-.001	
-10000	-1E-4	

$Y_1 \equiv 1/X$

(b)

Figure 5.5

Because $f(x)$ is never zero, the graph of f never intersects the x-axis. However, the graph gets closer and closer to the x-axis. We write this symbolically as follows:

$$f(x) \to 0 \text{ as } x \to \infty \qquad \text{and} \qquad f(x) \to 0 \text{ as } x \to -\infty,$$

and say that the horizontal line $y = 0$ is a *horizontal asymptote.*

> **Definition:** Horizontal Asymptote
>
> The horizontal line $y = k$ is a **horizontal asymptote** of a function f, if
>
> $$f(x) \to k \text{ as } x \to \infty \qquad \text{or} \qquad f(x) \to k \text{ as } x \to -\infty.$$

We will see that a horizontal asymptote is one possible type of an *end behavior asymptote.*

■ EXAMPLE 3 Finding Horizontal Asymptotes

Identify the horizontal asymptote(s) of the function $f(x) = \dfrac{2x+1}{x}$, and describe the behavior of f when $|x|$ is large.

Solution

Solve Algebraically

Using polynomial division (see Chapter 3), we write

$$f(x) = \frac{2x+1}{x} = 2 + \frac{1}{x}$$

and observe that

$$\frac{1}{x} \to 0 \text{ as } x \to \infty, \quad \text{so} \quad \frac{2x+1}{x} = 2 + \frac{1}{x} \to 2 \text{ as } x \to \infty.$$

Similarly,

$$\frac{1}{x} \to 0 \text{ as } x \to -\infty, \quad \text{so} \quad \frac{2x+1}{x} = 2 + \frac{1}{x} \to 2 \text{ as } x \to \infty.$$

Combining the two observations, we have

$$f(x) \to 2 \text{ as } |x| \to \infty.$$

Support Graphically

Figure 5.6 shows that the graph of $y_1 = f(x)$ appears to approach the graph of $y_2 = 2$ *asymptotically.* ■

> **Teaching Note**
>
> The word *limit* and the formal limit notation are purposely avoided in this book. The word *approaches* is used instead. The authors have found that an intuitive approach to the concept of limit is very appropriate for students at the precalculus level, and that formal definitions of limit should be reserved for calculus.

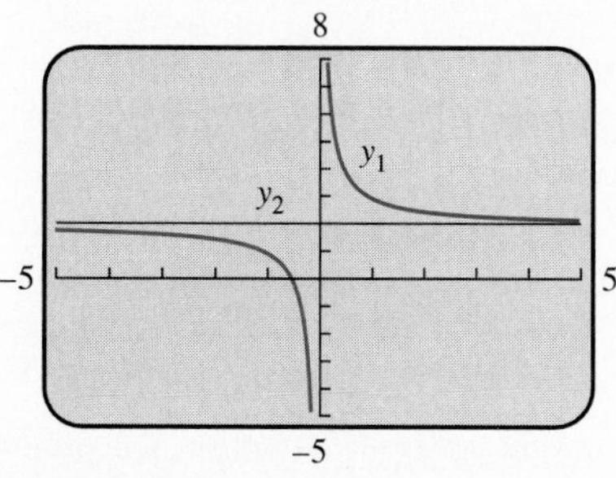

Figure 5.6 Graphs of $y_1 = (2x + 1)/x$ and its horizontal asymptote $y_2 = 2$.

The numerical and graphical behavior of a function can provide clues to horizontal asymptotes.

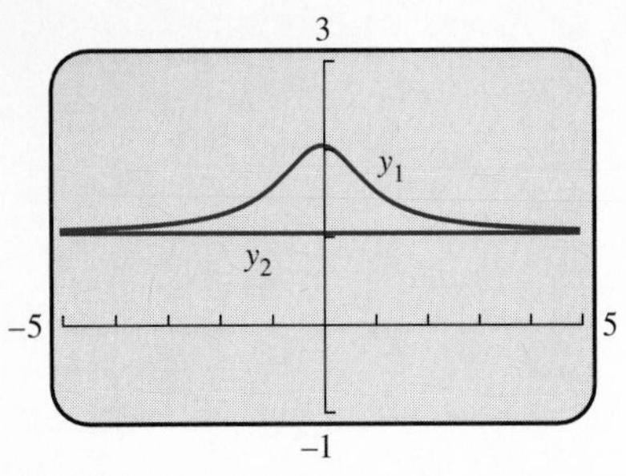

Figure 5.7 Graphs of $y_1 = f(x) = (x^2 + 2)/(x^2 + 1)$ and its horizontal asymptote $y_2 = 1$.

X	Y_1	
10	1.0099	
20	1.0025	
30	1.0011	
40	1.0006	
50	1.0004	
60	1.0003	
70	1.0002	

$Y_1 = (X^2+2)/(X^2+1)$

(a)

X	Y_1	
-10	1.0099	
-20	1.0025	
-30	1.0011	
-40	1.0006	
-50	1.0004	
-60	1.0003	
-70	1.0002	

$Y_1 = (X^2+2)/(X^2+1)$

(b)

Figure 5.8

Teaching Note

The results presented here are further developed and justified in Sections 5.2 and 5.3.

■ EXAMPLE 4 Finding a Horizontal Asymptote

Find a horizontal asymptote of $f(x) = \dfrac{x^2 + 2}{x^2 + 1}$.

Solution

The graph of f in Figure 5.7 and the tables in Figure 5.8 both suggest that $f(x) \to 1$ as $|x| \to \infty$.

Confirm Algebraically

Using polynomial division, we get

$$f(x) = \frac{x^2 + 2}{x^2 + 1} = 1 + \frac{1}{x^2 + 1}.$$

When the value of $|x|$ is large, the denominator $x^2 + 1$ is a large positive number, and $1/(x^2 + 1)$ is a positive number near zero. Therefore,

$$f(x) \to 1 \text{ as } |x| \to \infty,$$

so $y = 1$ is indeed a horizontal asymptote. ■

Examples 3 and 4 illustrate some general facts about horizontal asymptotes that we shall investigate later in the chapter. We summarize the information about asymptotes as follows:

Asymptotes of a Rational Function

Let f be a rational function given by

$$f(x) = \frac{p(x)}{h(x)} = \frac{a_n x^n + a_{n-1}x^{n-1} + \cdots + a_0}{b_m x^m + b_{m-1}x^{m-1} + \cdots + b_0},$$

where $p(x)$ and $h(x)$ have no common factors. Then,

1. The *vertical asymptotes* are at the zeros of $h(x)$.
2. The *end behavior asymptote* is the function that $f(x)$ approaches when the value of $|x|$ is large.
   - **a.** If $n < m$, it is the horizontal asymptote $y = 0$ (x-axis).
   - **b.** If $n = m$, it is the horizontal asymptote $y = a_n/b_m$ (the ratio of leading coefficients).
   - **c.** If $n > m$, it is $y = q(x)$, where $q(x)$ is the quotient when $p(x)$ is divided by $h(x)$; that is, $p(x) = q(x)h(x) + r(x)$.

We can conclude from this discussion that a rational function whose numerator degree is less than or equal to the denominator degree has a *unique* horizontal asymptote. If the degree of the numerator is greater than the degree of the denominator, it has *no* horizontal asymptote. Example 5 shows that a rational function can have more than one vertical asymptote.

EXAMPLE 5 Finding Asymptotes

Find the horizontal and vertical asymptotes for the function $f(x) = \dfrac{x^3}{x^2 - 9}$. Support your answer graphically.

Solution

Solve Algebraically

The exponent of x in the numerator is greater than the exponent of x in the denominator, so there is no horizontal asymptote. Factoring the denominator,

$$x^2 - 9 = (x - 3)(x + 3),$$

shows that the zeros are $x = 3$ or $x = -3$. Consequently, there are two vertical asymptotes for f; these are $x = 3$ and $x = -3$.

Support Graphically

The graph in Figure 5.9 shows the two vertical asymptotes and strongly suggests that there is no horizontal asymptote. ■

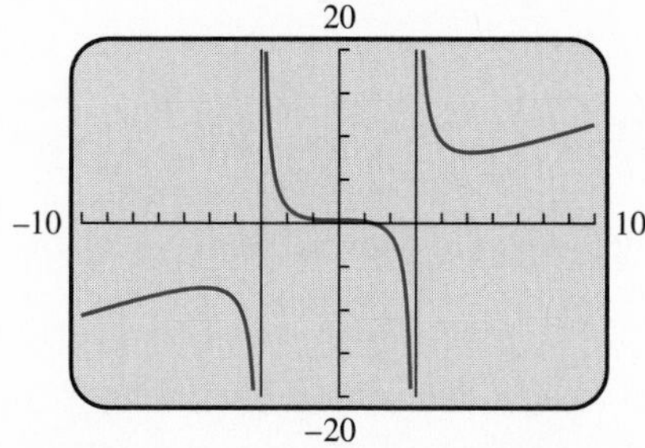

Figure 5.9

Applications

Some problems can be modeled by a rational function.

EXAMPLE 6 APPLICATION: Calculating Acid Mixtures

How much pure acid must be added to 50 oz of a 35% acid solution to produce a mixture that is at least 75% acid? (See Figure 5.10.)

Solution

Model

Word Statement: $\dfrac{\text{ounces of pure acid}}{\text{ounces of mixture}} = \text{concentration of acid}$

$$(0.35)50 = \text{ounces of pure acid in 35\% solution}$$

$$x = \text{ounces of acid added}$$

$$x + (0.35)50 = \text{ounces of pure acid in the resulting mixture}$$

$$x + 50 = \text{ounces of the resulting mixture}$$

$$\frac{x + (0.35)50}{x + 50} = \text{concentration of acid}$$

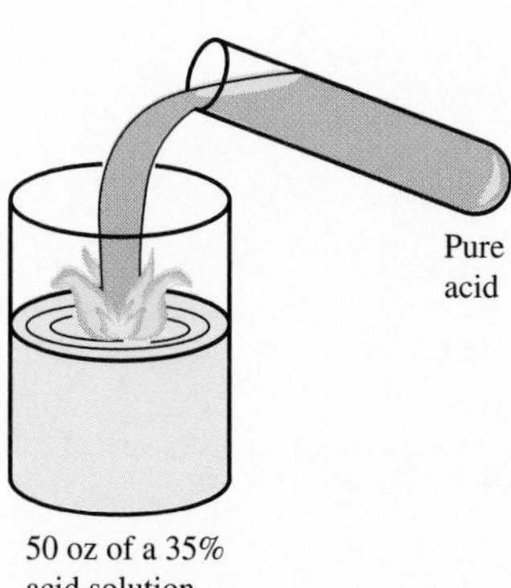

Figure 5.10

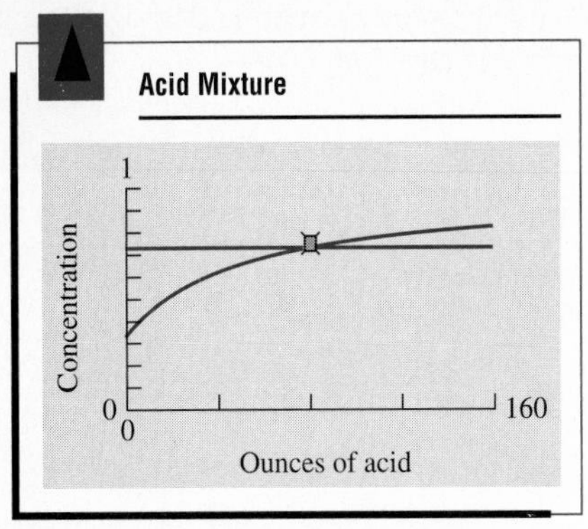

Intersection: $x = 80$; $y = .75$

Figure 5.11

Follow-up

Ask students to discuss whether a polynomial function is a rational function. $\left(\text{Yes, } P(x) = \frac{P(x)}{1}.\right)$

Assignment Guide

Day 1: Ex. 3–45, multiples of 3
Day 2: Ex. 47, 48, 50–53, 56, 57, 59, 62, 63

Cooperative Learning

Group Activity: Ex. 54

Notes on Exercises

Ex. 19–30 give students practice in writing statements using the "approaches" notation.
Ex. 47–51 are straightforward applications of rational functions.
Ex. 55–58 require students to write rational functions with certain properties.

Ongoing Assessment

Self-Assessment: Ex. 7, 23, 31, 37, 49
Embedded Assessment: Ex. 52

Solve Graphically

$$\frac{x + (0.35)50}{x + 50} \geq 0.75 \qquad \text{Inequality to be solved}$$

Figure 5.11 shows graphs of $y_1 = \dfrac{x + (0.35)50}{x + 50}$ and $y_2 = 0.75$. The point of intersection is (80, 0.75). To the right of this point, $y_1 > y_2$.

Interpret

We need to add 80 oz or more of pure acid to the 35% acid solution to make a solution that is at least 75% acid. ■

■ EXAMPLE 7 APPLICATION: Determining a Revenue Goal

Bob, the owner of Landry's Custom Shirts, needs to draw \$2000 per week from the shirt revenue to meet his personal needs. In order to be competitive, he determines that the most he can charge for a T-shirt is \$11.75 each. It costs \$9.25 to produce each shirt. How many T-shirts must be sold each week for Bob to meet his financial goal?

Solution

Model

We add the \$2000 that Bob needs to the weekly production cost. Then we compute a "new" average production cost per T-shirt that includes the \$2000.

Word Statement: Average cost per shirt < 11.75

$$x = \text{number of T-shirts sold per week}$$

$$9.25x + 2000 = \text{total cost per week}$$

$$\frac{9.25x + 2000}{x} = \text{average cost per T-shirt}$$

Solve Algebraically

$$\frac{9.25x + 2000}{x} \leq 11.75 \qquad \text{Inequality to be solved}$$

$$9.25x + 2000 \leq 11.75x \qquad \text{Multiply by } x\ (x > 0).$$

$$2000 \leq 2.5x \qquad \text{Subtract } 9.25x.$$

$$800 \leq x \qquad \text{Divide by } 2.5.$$

Interpret

If Landry's Custom Shirts sells 800 T-shirts in a week, its average cost is

$$\frac{9.25(800) + 2000}{800} = 11.75.$$

By selling 800 or more shirts per week, Bob will reach his revenue goal. ■

Quick Review 5.1

In Exercises 1–4, find the real number zeros of *f*.

1. $f(x) = 3x + 4$

2. $f(x) = 2x - 5$

3. $f(x) = (x - 3)(x + 4)$

4. $f(x) = (x - 1)(x + 5)$

In Exercises 5–10, use factoring to find the real number zeros.

5. $f(x) = 2x^2 + 5x - 3$

6. $f(x) = 3x^2 - 2x - 8$

7. $g(x) = x^2 - 4$

8. $g(x) = x^2 - 1$

9. $h(x) = x^3 - 1$

10. $h(x) = x^2 + 1$

In Exercises 11–14, find the quotient and remainder when $f(x)$ is divided by $g(x)$.

11. $f(x) = 2x + 1, \quad g(x) = x - 3$

12. $f(x) = 4x + 3, \quad g(x) = 2x - 1$

13. $f(x) = 3x - 5, \quad g(x) = x$

14. $f(x) = 5x - 1, \quad g(x) = 2x$

In Exercises 15–18, find the concentration of the solution.

15. 12 oz of pure acid is mixed with 20 oz of water

16. 28 oz of pure acid is mixed with 14 oz of water

17. 10 oz of a 35% acid solution is mixed with 20 oz of an 18% acid solution

18. 15 oz of a 28% acid solution is mixed with 30 oz of a 52% acid solution

SECTION EXERCISES 5.1

In Exercises 1–4, identify the function as a rational function or not a rational function and explain.

1. $f(x) = \dfrac{2x^2 - 3x + 4}{x^3 + x + 1}$

2. $f(x) = \dfrac{3x - 1}{x^2 + \sqrt{x - 1}}$

3. $f(x) = \dfrac{x^2 - 3\sqrt{x} + 1}{2x - 3}$

4. $f(x) = \dfrac{1}{3x + 4}$

In Exercises 5–10, explain why the function is a rational function and state its domain.

5. $f(x) = \dfrac{1}{x} - 4$

6. $f(x) = 2 - \dfrac{1}{x}$

7. $f(x) = 4 + \dfrac{1}{x + 1}$

8. $f(x) = -2 + \dfrac{2}{x + 3}$

9. $f(x) = 1 + \dfrac{2}{x + \dfrac{3}{x}}$

10. $f(x) = 3 - \dfrac{1}{x - \dfrac{2}{x}}$

In Exercises 11–18, complete the given statement about asymptotes as suggested by the graph of *f* shown.

11. As $x \to 3^-$, $f(x) \to ?$

12. As $x \to 3^+$, $f(x) \to ?$

13. As $x \to \infty$, $f(x) \to ?$

14. As $x \to -\infty$, $f(x) \to ?$

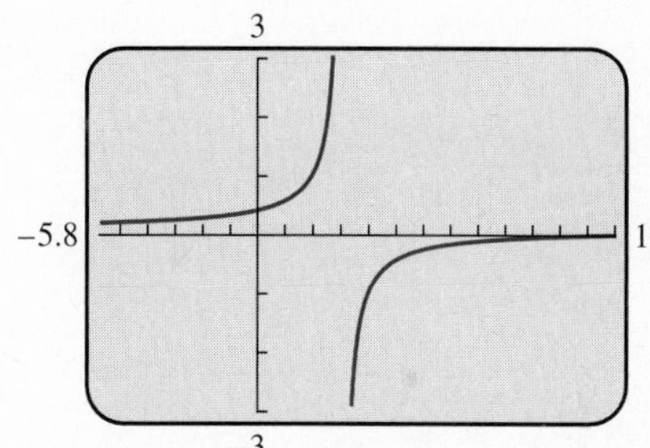

For Exercises 11–14

15. As $x \to -3^+$, $f(x) \to ?$

16. As $x \to -3^-$, $f(x) \to ?$

17. As $x \to -\infty$, $f(x) \to ?$

18. As $x \to \infty$, $f(x) \to ?$

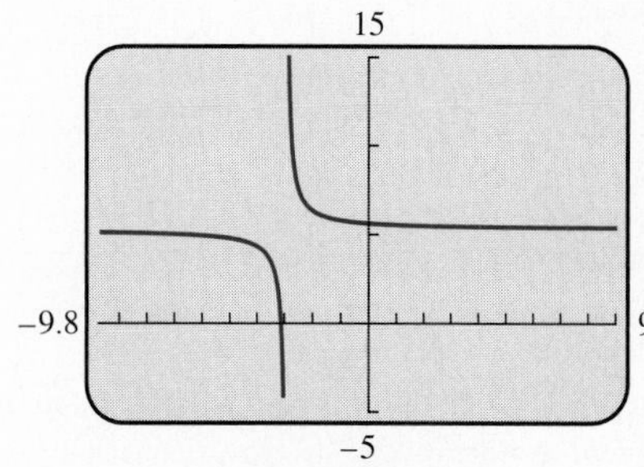

For Exercises 15–18

In Exercises 19–24, identify each vertical asymptote, and write statements that describe the asymptotic behavior of the function near the asymptote.

19. $f(x) = \dfrac{2}{x+1}$

20. $g(x) = -\dfrac{1}{x-2}$

21. $h(x) = 2 + \dfrac{1}{x-1}$

22. $f(x) = x - \dfrac{3}{x+4}$

23. $g(x) = \dfrac{2}{2x^2 - x - 3}$

24. $h(x) = \dfrac{2}{x^2 + 4x + 3}$

In Exercises 25–34, identify the horizontal asymptote, and write statements that describe the asymptotic behavior of f when the value of $|x|$ is large.

25. $f(x) = \dfrac{2}{x} - 3$

26. $f(x) = \dfrac{3}{x} + 4$

27. $f(x) = \dfrac{2x+1}{x+1}$

28. $f(x) = \dfrac{3x-7}{x-2}$

29. $f(x) = \dfrac{x-1}{x^2 - x - 12}$

30. $f(x) = \dfrac{x+1}{x^2 - 3x - 10}$

31. $f(x) = \dfrac{3x^2 - 7x + 2}{5x^2 + 7x - 3}$

32. $f(x) = \dfrac{7x^3 + 5x - 3}{-2x^3 - 3x^2 + 7}$

33. $f(x) = \dfrac{3x^2 + 7x^3 - 4x + 3}{x^3 - 5x^2 + 8x - 3}$

34. $f(x) = \dfrac{x^3 - 5x^4 + 2x - 3}{x^4 + 3x - 2}$

In Exercises 35–40, find the horizontal and vertical asymptotes of the function. Then sketch its graph by hand. Support your sketch graphically.

35. $g(x) = \dfrac{3x-1}{x+2}$

36. $f(x) = \dfrac{8x+6}{2x-4}$

37. $f(x) = \dfrac{2x+4}{x-3}$

38. $h(x) = \dfrac{x-3}{2x+5}$

39. $t(x) = \dfrac{x-1}{x+4}$

40. $k(x) = \dfrac{2x-3}{x+2}$

In Exercises 41–46, match the equation with its graph. Identify the viewing-window dimensions and the values used for Xscl and Yscl.

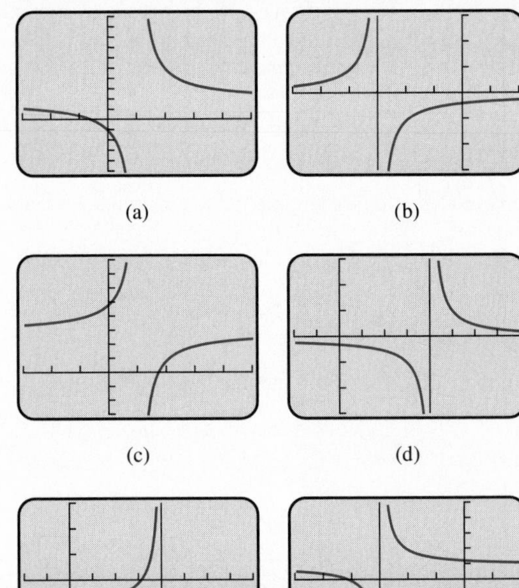

For Exercises 41–46

41. $y = \dfrac{1}{x-4}$

42. $y = -\dfrac{1}{x+3}$

43. $y = 2 + \dfrac{3}{x-1}$

44. $y = 1 + \dfrac{1}{x+3}$

45. $y = -1 + \dfrac{1}{4-x}$

46. $y = 3 - \dfrac{2}{x-1}$

In Exercises 47–54, solve the problem.

47. *Acid Mixture* Suppose that x ounces of pure acid are added to 125 oz of a 60% acid solution. How many ounces of pure acid must be added to obtain a solution of at least 83% acid?

a. Explain why the concentration $C(x)$ of the new mixture is

$$C(x) = \frac{x + 0.6(125)}{x + 125}.$$

b. Suppose the viewing window in the figure is used to find a solution to the problem. What is the equation of the horizontal line?

c. Write and solve an inequality that answers the question of this problem.

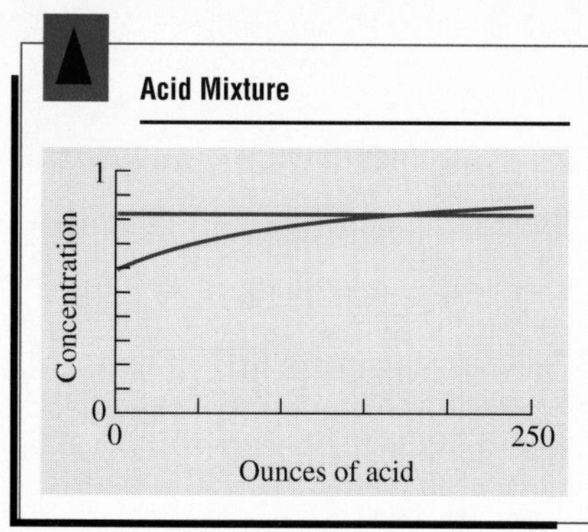

For Exercise 47

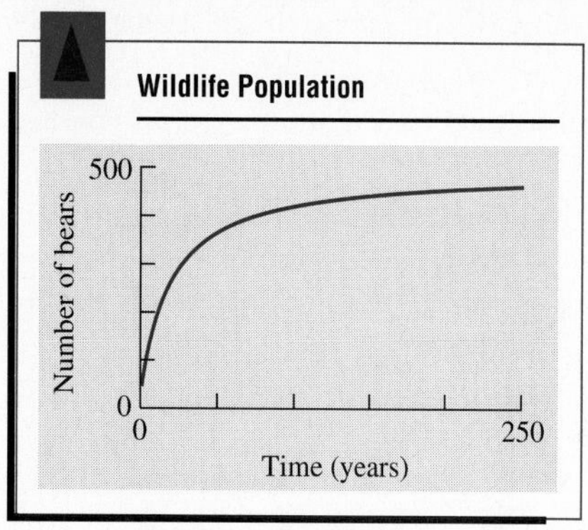

For Exercise 50

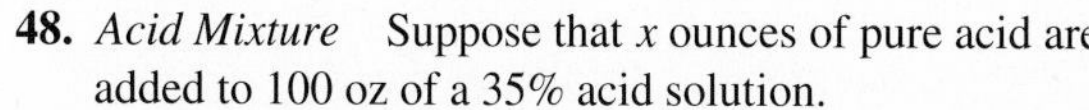

48. *Acid Mixture* Suppose that x ounces of pure acid are added to 100 oz of a 35% acid solution.

a. Express the concentration $C(x)$ of the new mixture as a function of x.
b. Use a graph to determine how much pure acid should be added to the 35% solution to produce a new solution that is less than 75% acid.
c. Solve part b algebraically.

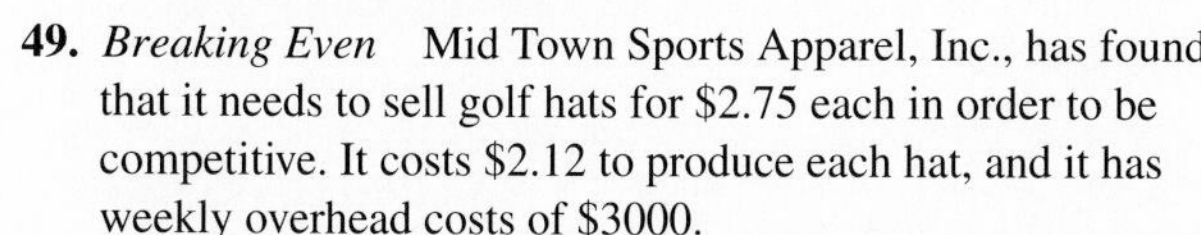

49. *Breaking Even* Mid Town Sports Apparel, Inc., has found that it needs to sell golf hats for \$2.75 each in order to be competitive. It costs \$2.12 to produce each hat, and it has weekly overhead costs of \$3000.

a. Let x be the number of hats produced each week. Express the average cost (including overhead costs) of producing one hat as a function of x.
b. Solve algebraically to find the number of golf hats that must be sold each week to make a profit. Support your answer graphically.
c. How many golf hats must be sold to make a profit of \$1000 in 1 week?

50. *Bear Population* The number of bears at any time t (in years) in a federal game reserve is given by

$$P(t) = \frac{500 + 250t}{10 + 0.5t}.$$

a. Find the population of bears when the value of t is 10, 40, and 100.
b. Does the graph of the bear population have a horizontal asymptote? If so, what is it? If not, why not?
c. According to this model, what is the largest the bear population can become?

51. *Research and Development (R & D)* The graph shows the amounts spent (in billions of dollars) for research and development of prescription drugs. The rational function

$$f(x) = \frac{1}{0.2411 - 0.0186x}$$

is a model for the data (where $x = 1$ represents 1985, $x = 2$ represents 1986, and so forth).

a. Draw a graph of the data and the model f.
b. \$0.2 billion was spent in 1958 for R & D. What does the model f estimate for the amount spent in 1958?

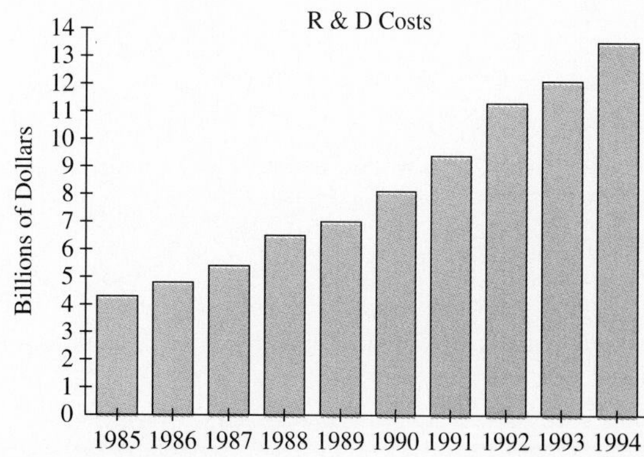

Source: The Pharmaceutical Manufacturers Association, as reported in *Science*, American Association for the Advancement of Science, May 20, 1994.

For Exercises 51–52

52. Use the model given in Exercise 51 for the given data.

a. Find the vertical asymptote of f.

b. Writing to Learn For what years is this model realistic?

53. Writing to Learn Find the horizontal asymptote of the function of Example 7 that represents the average cost of producing one T-shirt. What does it mean for the function to have this asymptote?

54. Group Learning Activity *Work in groups of two.* Compare the functions

$$f(x) = \frac{x^2 - 9}{x - 3} \quad \text{and} \quad g(x) = x + 3.$$

a. Are the domains equal?

b. Does f have a vertical asymptote? Explain.

c. Explain why the graphs appear to be identical.

d. Are the functions identical?

EXTENDING THE IDEAS

In Exercises 55–58, find a rational function of the form

$$f(x) = \frac{ax + b}{cx + d}$$

that satisfies the given conditions.

55. Horizontal asymptote $y = 2$, vertical asymptote $x = 3$.

56. Horizontal asymptote $y = -3$, vertical asymptote $x = -2$.

57. Vertical asymptote $x = 2$ satisfying

$$f(x) \to \infty \text{ as } x \to 2^-$$
$$\text{and } f(x) \to -\infty \text{ as } x \to 2^+.$$

58. Vertical asymptote $x = -4$ satisfying

$$f(x) \to \infty \text{ as } x \to -4^-$$
$$\text{and } f(x) \to -\infty \text{ as } x \to -4^+.$$

In Exercises 59–62, each function has two horizontal asymptotes. Write equations for all asymptotes. Express the function as a piecewise-defined function without absolute value.

59. $h(x) = \dfrac{2x - 3}{|x| + 2}$

60. $h(x) = \dfrac{3x + 5}{|x| + 3}$

61. $f(x) = \dfrac{5 - 3x}{|x| + 4}$

62. $f(x) = \dfrac{3 - 2x}{|x| + 1}$

63. Writing to Learn You learned that each rational function has *at most one* horizontal asymptote. Explain why Exercises 59–62 do not contradict that statement.

64. Consider the function of Example 1.

$$f(x) = 1 + \frac{1}{x - \frac{1}{x}}$$

Show that f does not have a vertical asymptote at $x = 0$ by showing that $f(x) \to 1$ as $x \to 0$. Support your answer graphically.

5.2 GRAPHING RATIONAL FUNCTIONS, $n \leq m$

Confirming Graphs of Rational Functions • Rational Functions Where $n < m$ • Rational Functions Where $n = m$ • Using Transformations to Describe Graphs

Objective

Students will be able to describe the properties of rational functions in which the degree of the numerator is less than or equal to the degree of the denominator, and to identify transformations of the reciprocal function $f(x)$.

Confirming Graphs of Rational Functions

Throughout this section we shall assume that for rational functions

$$f(x) = \frac{p(x)}{h(x)} = \frac{a_n x^n + a_{n-1}x^{n-1} + \cdots}{b_m x^m + b_{m-1}x^{m-1} + \cdots},$$

$p(x)$ and $h(x)$ have no common factors. The situation in which the numerator and denominator have common factors is considered in Exercises 32–36 in Section 5.3.

To confirm the graph of a rational function, it is helpful to follow these five steps.

REMARK

Notice that n is the degree of the *numerator* of the rational function f.

Motivate

Ask students how they might be able to find the zeros of a rational function, assuming that the numerator and denominator have no common factors. **(Find zeros of the numerator.)**

Confirming Graphs of Rational Functions $y = \dfrac{a_n x^n + \cdots}{b_m x^m + \cdots}$

1. **Confirm the end behavior asymptote:**

   If $n < m$, the horizontal asymptote is $y = 0$.
   If $n = m$, the horizontal asymptote is $y = a_n/b_m$.
2. **Confirm vertical asymptotes,** at the zeros of the denominator.
3. **Confirm x-intercepts,** the zeros of the numerator.
4. **Confirm the y-intercept,** the value of $f(0)$, if defined.
5. **Confirm intermediate behavior,** the behavior of f between and beyond the x-intercepts and vertical asymptotes.

Rational Functions Where $n < m$

If the numerator degree of a rational function is less than the denominator degree, then the x-axis is the horizontal asymptote.

To see why this is true, divide numerator and denominator of f by x^m, where m is the degree of the denominator:

$$f(x) = \frac{p(x)}{h(x)} = \frac{a_n x^n + a_{n-1}x^{n-1} + \cdots}{b_m x^m + b_{m-1}x^{m-1} + \cdots}$$

$$= \frac{\dfrac{a_n}{x^{m-n}} + \dfrac{a_{n-1}}{x^{m-n+1}} + \cdots}{b_m + \dfrac{b_{m-1}}{x} + \cdots}$$

Notice that $m - n \ge 1$. When the value of $|x|$ is large, all terms of the numerator and all terms of the denominator except the first are near zero. Consequently, $f(x)$ is near $0/b_m = 0$.

■ **EXAMPLE 1 Confirming the Graph of a Rational Function**

Graph $f(x) = \dfrac{x - 1}{x^2 - x - 6}$, and confirm its behavior.

Solution

The graph of f is shown in Figure 5.12.

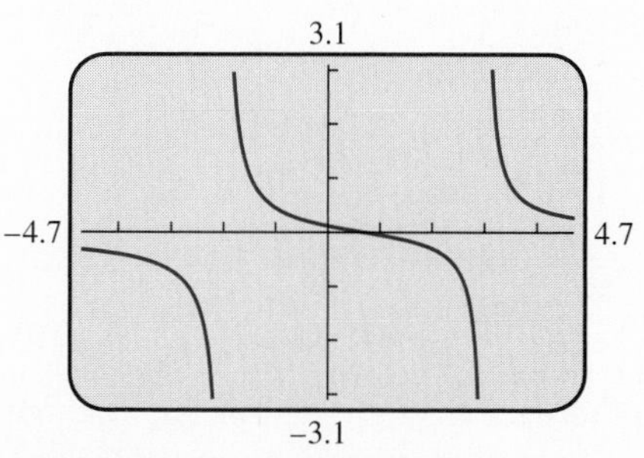

Figure 5.12 $f(x) = (x - 1)/(x^2 - x - 6)$.

End Behavior Asymptote: The degree of the numerator, 1, is less than the degree of the denominator, 2. Thus, there is a horizontal asymptote, $y = 0$.

Vertical Asymptotes: We find the zeros of the denominator:

$$x^2 - x - 6 = (x - 3)(x + 2) = 0 \Rightarrow x = 3 \text{ or } x = -2$$

(The symbol $\Rightarrow$ is shorthand for "implies.") The lines $x = 3$ and $x = -2$ are vertical asymptotes.

x-Intercept: Find the zeros of the numerator.

$$x - 1 = 0 \Rightarrow x = 1$$

The only x-intercept is $x = 1$.

y-Intercept: Because $y = f(0) = \frac{1}{6}$, the y-intercept is $(0, \frac{1}{6})$.

Intermediate Behavior: For the behavior near $x = -2$, the numerical information in Figure 5.13 suggests that

$$f(x) \to -\infty \text{ as } x \to -2^- \quad \text{and} \quad f(x) \to \infty \text{ as } x \to -2^+,$$

and for the behavior near $x = 3$

$$f(x) \to -\infty \text{ as } x \to 3^- \quad \text{and} \quad f(x) \to \infty \text{ as } x \to 3^+. \quad ■$$

X	Y_1	
-2.1	-6.078	
-2.01	-60.08	
-2.001	-600.1	
-2	ERROR	
-1.999	599.92	
-1.99	59.92	
-1.9	5.9184	

Y_1=(X−1)/(X²−X−...

(a)

X	Y_1	
2.9	-3.878	
2.99	-39.88	
2.999	-399.9	
3	ERROR	
3.001	400.12	
3.01	40.12	
3.1	4.1176	

Y_1=(X−1)/(X²−X−...

(b)

Figure 5.13 The behavior of $y_1 = f(x) = (x - 1)/(x^2 - x - 6)$ near (a) $x = -2$ and (b) $x = 3$.

Rational Functions Where $n = m$

If the numerator and denominator of a rational function are of equal degree, then the line $y = a_n/b_n$ is a horizontal asymptote.

To see why this is true, divide the numerator and denominator of f by x^n, where n is the common degree.

$$f(x) = \frac{p(x)}{h(x)} = \frac{a_n x^n + a_{n-1}x^{n-1} + \cdots}{b_n x^n + b_{n-1}x^{n-1} + \cdots}$$

$$= \frac{a_n + \dfrac{a_{n-1}}{x} + \dfrac{a_{n-2}}{x^2} + \cdots}{b_n + \dfrac{b_{n-1}}{x} + \dfrac{b_{n-2}}{x^2} + \cdots}$$

When the value of $|x|$ is large, all terms except the first of both the numerator and denominator are near zero. Consequently, $f(x)$ is near a_n/b_n.

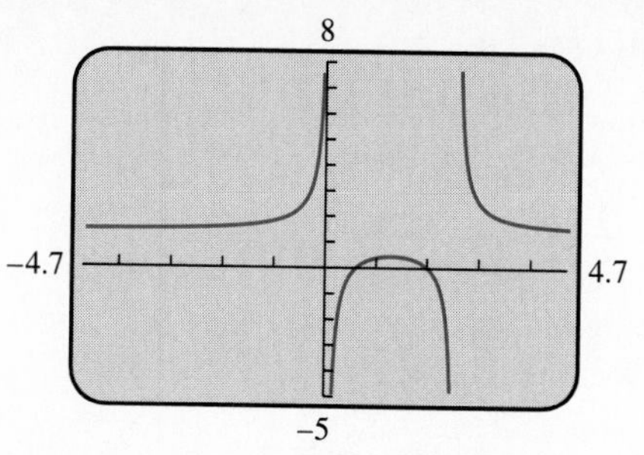

Figure 5.14 $f(x) = \dfrac{6x^2 - 15x + 6}{4x^2 - 10x}$.

■ EXAMPLE 2 Confirming the Graph of Another Rational Function

Graph $f(x) = \dfrac{6x^2 - 15x + 6}{4x^2 - 10x}$, and confirm its behavior.

Solution

The graph is shown in Figure 5.14.

End behavior asymptote: The degree of the numerator equals the degree of denominator, 2, and $a_2 = 6$ and $b_2 = 4$. So there is a horizontal asymptote $y = \frac{6}{4} = \frac{3}{2}$.

Vertical asymptotes: We find the zeros of denominator by factoring:

$$4x^2 - 10x = 2x(2x - 5) = 0 \quad \Rightarrow \quad x = 0 \text{ or } x = \frac{5}{2}$$

The lines $x = 0$ and $x = \frac{5}{2}$ are vertical asymptotes.

***x*-intercepts:** We find the zeros of numerator:

$$\begin{aligned} 6x^2 - 15x + 6 &= 0 \\ 3(2x^2 - 5x + 2) &= 0 \\ 3(2x - 1)(x - 2) &= 0 \\ x = \frac{1}{2} \quad &\text{or} \quad x = 2 \end{aligned}$$

The x-intercepts are $x = \frac{1}{2}$ and $x = 2$.

***y*-intercept:** There are none because $f(0)$ is undefined.

Intermediate behavior: The numerical information in Figure 5.15 suggests, for the behavior near $x = 0$, that

$$f(x) \to \infty \text{ as } x \to 0^- \quad \text{and} \quad f(x) \to -\infty \text{ as } x \to 0^+,$$

and for the behavior near $x = \frac{5}{2}$

$$f(x) \to -\infty \text{ as } x \to \frac{5}{2}^- \quad \text{and} \quad f(x) \to \infty \text{ as } x \to \frac{5}{2}^+.$$ ■

X	Y_1	
-.1	7.2692	
-.01	61.261	
-.001	601.26	
0	ERROR	
.001	-598.7	
.01	-58.74	
.1	-4.75	
Y_1=(6X²−15X+6)/...		

(a)

X	Y_1	
2.4	-4.75	
2.49	-58.74	
2.499	-598.7	
2.5	ERROR	
2.501	601.26	
2.51	61.261	
2.6	7.2692	
Y_1=(6X²−15X+6)/...		

(b)

Figure 5.15 The behavior of $y_1 = f(x) = (6x^2 - 15x + 6)/(4x^2 - 10x)$ near (a) $x = 0$ and (b) $x = \frac{5}{2}$.

Teaching Note

If students have trouble with transformations, have them review Section 2.6.

Using Transformations to Describe Graphs

One of the simplest rational functions is the **reciprocal function,**

$$f(x) = \frac{1}{x},$$

which was graphed in Section 5.1. The graphs of some rational functions are easy to describe as transformations of the graph of the reciprocal function.

EXAMPLE 3 Using Transformations

Describe how the graph of the given function can be obtained by transforming the graph of $f(x) = 1/x$.

a. $g(x) = \dfrac{1}{x - 2}$ **b.** $h(x) = \dfrac{2}{x + 3}$ **c.** $k(x) = \dfrac{1}{2(5 - x)}$

Solution

a. $g(x) = \dfrac{1}{x - 2} = f(x - 2).$

It follows that the graph of g is the graph of f shifted right 2 units.

b.
$$\begin{aligned} h(x) &= \frac{2}{x + 3} \\ &= 2\left(\frac{1}{x + 3}\right) \\ &= 2f(x + 3) \end{aligned}$$

The graph of h is the graph of f shifted left 3 units and then stretched vertically by a factor of 2.

c.
$$\begin{aligned} k(x) &= \frac{1}{2(5 - x)} \\ &= \frac{1}{2}\left(\frac{1}{5 - x}\right) \\ &= -\frac{1}{2}\left(\frac{1}{x - 5}\right) \\ &= -\frac{1}{2}f(x - 5) \end{aligned}$$

The graph of k is the graph of f shifted right 5 units, then shrunk vertically by a factor of $\frac{1}{2}$, and finally reflected across the x-axis. ■

Sometimes the form of a rational function must be changed before we can recognize the required transformations.

Follow-up

Ask students to write a function whose graph is the graph of $f(x) = \dfrac{1}{x}$ shifted right 4 units, reflected across the x-axis, and then translated down 1 unit. $\left(g(x) = \dfrac{3 - x}{x - 4}\right)$

Assignment Guide

Day 1: Ex. 2, 3, 6, 7, 10, 12, 13, 16, 17, 19, 21, 22
Day 2: Ex. 25, 28, 29, 32, 35–40

Cooperative Learning

Group Activity: Ex. 37–40

Notes on Exercises

Ex. 37–40 provide an opportunity for students to make conjectures regarding horizontal and vertical asymptotes, transformations, and the domain and range of functions.

Ongoing Assessment

Self-Assessment: Ex. 1, 5, 11, 15, 23, 31
Embedded Assessment: Ex. 16, 22, 28, 32

■ **EXAMPLE 4** Using Transformations

Describe how the graphs of

$$f(x) = \frac{1}{x} \quad \text{and} \quad g(x) = \frac{3x - 7}{x - 2}$$

are related by transformations. Support your answer graphically.

Solution

Using long division, we get

$$g(x) = \frac{3x - 7}{x - 2}$$
$$= 3 - \frac{1}{x - 2}.$$

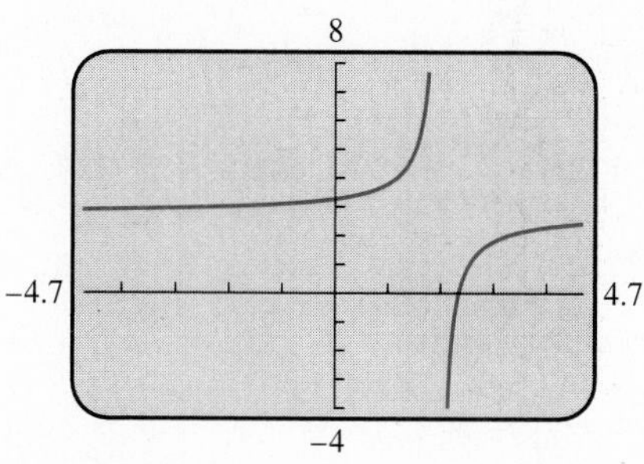

Figure 5.16

The graph of g is the graph of f translated 2 units right, followed by a reflection across the x-axis, and then translated up 3 units up. (Note that the reflection must be executed *before* the vertical translation.) This means that the lines $x = 2$ and $y = 3$ are vertical and horizontal asymptotes, respectively, as supported by Figure 5.16. ■

Quick Review 5.2

In Exercises 1–8, find the x- and y-intercepts.

1. $y = 2x + 3$

2. $y = 7x - 6$

3. $y = 2x^2 - x - 15$

4. $y = 3x^2 + 5x - 28$

5. $y = x^3 - 4x^2 + 4x$

6. $y = x^3 + 2x^2 + x$

7. $y = 2x^2 - 3x - 1$

8. $y = 3x^2 - 2x - 2$

In Exercises 9–14, use transformations to describe how the graph of g can be obtained from the graph of f.

9. $g(x) = x^3 - 4$; $f(x) = x^3$

10. $g(x) = |x + 4|$; $f(x) = |x|$

11. $g(x) = \sqrt{-x}$; $f(x) = \sqrt{x}$

12. $g(x) = -|x|$; $f(x) = |x|$

13. $g(x) = -\sqrt{x + 2}$; $f(x) = \sqrt{x}$

14. $g(x) = 2x^2 + 3$; $f(x) = x^2$

SECTION EXERCISES 5.2

In Exercises 1–4, identify the vertical asymptotes and the x- and y-intercepts.

1. $f(x) = \dfrac{x + 2}{x^2 + 2x - 3}$

2. $f(x) = \dfrac{x - 3}{x^2 - x - 2}$

3. $f(x) = \dfrac{x^2 + x - 2}{x^2 - 9}$

4. $f(x) = \dfrac{x^2 - x - 2}{x^2 - 2x - 8}$

In Exercises 5 and 6, complete the two tables. Write a statement that describes the function behavior suggested by the two tables.

5. $y = \dfrac{x - 5}{x^2 - 2x - 3}$

a.

x	0	−0.9	−0.99	−0.999	−0.9999
y	?	?	?	?	?

b.

x	−2	−1.1	−1.01	−1.001	−1.0001
y	?	?	?	?	?

6. $y = \dfrac{x + 1}{x^2 - x - 6}$

a.

x	2	2.9	2.99	2.999	2.9999
y	?	?	?	?	?

b.

x	4	3.1	3.01	3.001	3.0001
y	?	?	?	?	?

In Exercises 7–10, identify the vertical asymptotes. Describe the behavior of the function on either side of these asymptotes.

7. $f(x) = \dfrac{2x + 5}{x^2 - 5x + 6}$ **8.** $f(x) = \dfrac{7 - x}{x^2 - 5x + 6}$

9. $f(x) = \dfrac{2x^2 - 4x - 5}{x^2 + 2x - 3}$ **10.** $f(x) = \dfrac{4x^2 + x - 1}{x^2 - x - 12}$

In Exercises 11 and 12, complete each table. Write a statement for the table that describes the function behavior when the value of $|x|$ is large.

11. $y = \dfrac{3x^2 + x - 1}{x^2 + 5x - 4}$

a.

x	1	10	100	1000	10,000
y	?	?	?	?	?

b.

x	−1	−10	−100	−1000	−10,000
y	?	?	?	?	?

12. $y = \dfrac{7x - 3x^2 + 2}{2x^2 + 5x - 17}$

a.

x	1	10	100	1000	10,000
y	?	?	?	?	?

b.

x	−1	−10	−100	−1000	−10,000
y	?	?	?	?	?

In Exercises 13–16, identify important function behaviors. Support your work graphically.

13. $f(x) = \dfrac{7 - 3x}{2x^2 + x - 3}$ **14.** $f(x) = \dfrac{3 - x}{2x^2 - 3x - 5}$

15. $f(x) = \dfrac{x^2 + 7x + 10}{x^2 - 4x + 3}$ **16.** $f(x) = \dfrac{3x^2 - 6x - 8}{2x^2 - x - 1}$

In Exercises 17–28, describe how the graph of f can be obtained by transforming the graph of $y = 1/x$. Specify the order of the transformations. Name any vertical or horizontal asymptotes. Support your work graphically.

17. $f(x) = \dfrac{1}{x - 3}$ **18.** $f(x) = \dfrac{2}{x + 2}$

19. $f(x) = -\dfrac{2}{x + 5}$ **20.** $f(x) = \dfrac{1}{x + 3}$

21. $f(x) = -3 + \dfrac{1}{x + 1}$ **22.** $f(x) = -2 + \dfrac{1}{x + 3}$

23. $f(x) = 2 + \dfrac{5}{1 - x}$ **24.** $f(x) = -1 - \dfrac{1}{x + 1}$

25. $f(x) = \dfrac{2x - 1}{x + 3}$ **26.** $f(x) = \dfrac{3x - 2}{x - 1}$

27. $f(x) = \dfrac{5 - 2x}{x + 4}$ **28.** $f(x) = \dfrac{4 - 3x}{x - 5}$

In Exercises 29–36, graph the function. Confirm the asymptotes, intercepts, and intermediate behavior.

29. $f(x) = \dfrac{2}{x - 3}$ **30.** $f(x) = \dfrac{-3}{x + 2}$

31. $g(x) = \dfrac{x - 2}{x^2 - 2x - 3}$ **32.** $g(x) = \dfrac{x + 2}{x^2 + 2x - 3}$

33. $h(x) = \dfrac{2}{x^3 - x}$ **34.** $h(x) = \dfrac{3}{x^3 - 4x}$

35. $k(x) = \dfrac{x - 1}{x^2 + 3}$ **36.** $k(x) = \dfrac{x + 3}{x^2 + 1}$

EXTENDING THE IDEAS

In Exercises 37–40, express your answers in terms of the constants a, b, c, d, h, k, and r. Assume that $r \neq 0$, $c \neq 0$, and $ad \neq bc$.

37. Find the domain and range of

$$f(x) = \frac{r}{x - h} + k.$$

38. Find the domain and range of

$$f(x) = \frac{ax + b}{cx + d}.$$

39. Describe how the graph of $f(x) = \dfrac{r}{x - h} + k$ can be obtained from the graph of $y = 1/x$.

40. Describe how the graph of

$$f(x) = \frac{ax + b}{cx + d}$$

can be obtained from the graph of $y = 1/x$. (*Hint:* Use long division.)

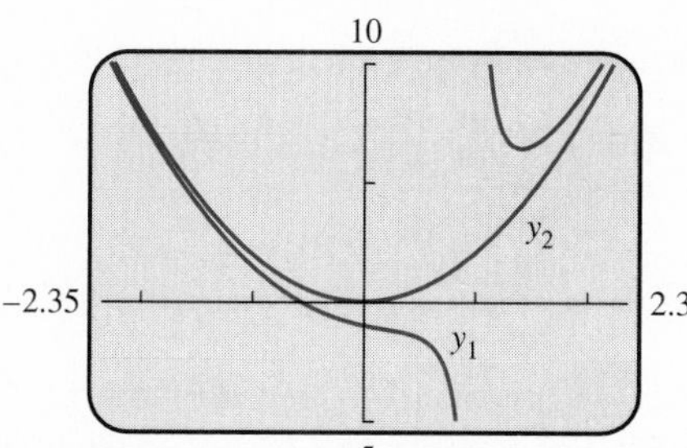

Figure 5.17

X	Y_1	Y_2
-5	49.833	50
-4	31.8	32
-3	17.75	18
-2	7.6667	8
-1	1.5	2
0	-1	0
1	ERROR	2

Y_1=(2X³−2X²+1)/...

(a)

X	Y_1	Y_2
1	ERROR	2
2	9	8
3	18.5	18
4	32.333	32
5	50.25	50
6	72.2	72
7	98.167	98

Y_2=2X²

(b)

Figure 5.18 Some values of $y_1 = (2x^3 - 2x^2 + 1)/(x - 1)$ and $y_2 = 2x^2$ for (a) $x \leq 1$, and (b) $x \geq 1$.

5.3 GRAPHING RATIONAL FUNCTIONS, $n > m$

End Behavior Asymptotes • Rational Functions Where $n > m$ • Using the End Behavior Asymptote to Describe a Graph • Applications

End Behavior Asymptotes

In this section we shall consider rational functions for which the degree of the numerator is greater than the degree of the denominator:

$$f(x) = \frac{p(x)}{h(x)} = \frac{a_n x^n + a_{n-1} x^{n-1} + \cdots}{b_m x^m + b_{m-1} x^{m-1} + \cdots},$$

where $p(x)$ and $h(x)$ have no common factors.

You learned in Section 5.1 that functions in this family do not have a *horizontal* asymptote. We need to investigate what other kinds of end behavior asymptotes are possible for rational functions for which $n > m$. Recall that for $n \leq m$, the horizontal asymptote is the end behavior asymptote.

● **EXPLORATION** **Rational Functions Where $n > m$**

1. Figure 5.17 shows the graphs of

$$y_1 = f(x) = \frac{2x^3 - 2x^2 + 1}{x - 1}$$

and $y_2 = g(x) = 2x^2$ superimposed on one another. Reproduce this figure on a grapher.

2. Zoom out several times. Tell what you see, and make a conjecture about it.
3. Are the tables of values in Figure 5.18 consistent with what you see? Explain. ●

The Exploration should suggest that $g(x) = 2x^2$ closely approximates $f(x) = (2x^3 - 2x^2 + 1)/(x - 1)$ for large values of $|x|$. Saying it another way, as $|x| \to \infty$, the graphs of f and g appear to be identical. As we shall see, $g(x) = 2x^2$ is the *end behavior asymptote* for the function f. The graph of f approaches the graph of g as $|x| \to \infty$.

Objective

Students will be able to identify the end behavior asymptotes and confirm the behavior of rational functions in which the degree of the numerator is greater than the degree of the denominator.

Motivate

Ask students to predict the end behavior of $f(x) = 2x^2 - 3x + 4 - \frac{1}{x}$.

Teaching Note

Note that according to the division algorithm, the degree of $r(x)$ is less than the degree of $h(x)$, so

$$\frac{r(x)}{h(x)} \to 0 \text{ as } |x| \to \infty.$$

Exploration Extensions

Repeat the Exploration for the functions $f(x) = \frac{-x^3 + 5x - 3}{x^2 - 2x}$ and $g(x) = -x - 2$. Produce a table of values similar to Figure 5.18 for these functions.

Teaching Note

Remind students that more than one viewing window may be necessary to reveal all of the important features of a graph. As an example, show them graphs of

$$f(x) = \frac{2x^4 + 7x^3 + 7x^2 + 2x}{x^3 - x + 50},$$

viewed in $[-20, 20]$ by $[-20, 20]$ and $[-2.1, 1.1]$ by $[-0.04, 0.04]$ viewing windows.

Here is a definition of end behavior asymptote that includes the case $n \le m$.

Definition: End Behavior Asymptote

Suppose $q(x)$ and $r(x)$ are the quotient and remainder when the division algorithm is applied to the numerator $p(x)$ and the denominator $h(x)$ of the rational function $f(x)$. Then

$$f(x) = \frac{p(x)}{h(x)} = q(x) + \frac{r(x)}{h(x)},$$

and $q(x)$ is the **end behavior asymptote** of $f(x)$.

Rational Functions Where $n > m$

The Exploration suggested the graphical significance of the end behavior asymptote and why we might want to find it. In Section 3.2 we described the four types of end behavior that a polynomial can have. The **end behavior** of the rational function $f(x)$ is the end behavior of the quotient polynomial $q(x)$.

■ EXAMPLE 1 Finding an End Behavior Asymptote

Find the end behavior asymptote of the rational function

$$f(x) = \frac{x^3 - 3x^2 + 3x + 1}{x - 1}.$$

Solution

Divide $x^3 - 3x^2 + 3x + 1$ by $x - 1$ using long division (or synthetic division).

$$\begin{array}{r|l}
 & x^2 - 2x + 1 \\
\hline
x - 1 & x^3 - 3x^2 + 3x + 1 \\
 & x^3 - x^2 \\
\hline
 & -2x^2 + 3x + 1 \\
 & -2x^2 + 2x \\
\hline
 & x + 1 \\
 & x - 1 \\
\hline
 & 2
\end{array}$$

We see that

$$f(x) = \frac{x^3 - 3x^2 + 3x + 1}{x - 1} = (x^2 - 2x + 1) + \frac{2}{x - 1},$$

so $g(x) = x^2 - 2x + 1$ is the end behavior asymptote of f. ■

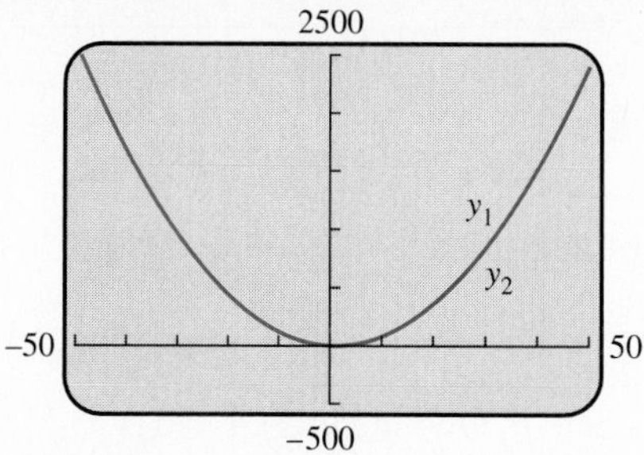

Figure 5.19 The graphs of $y_1 = (x^3 - 3x^2 + 3x + 1)/(x - 1)$ and its end behavior asymptote $y_2 = x^2 - 2x + 1$.

X	Y1	Y2
10	81.222	81
11	100.2	100
12	121.18	121
13	144.17	144
14	169.15	169
15	196.14	196
16	225.13	225

Y2=X2–2X+1

Figure 5.20

■ EXAMPLE 2 Comparing a Function and Its End Behavior Asymptote

Compare

$$f(x) = \frac{x^3 - 3x^2 + 3x + 1}{x - 1}$$

and its end behavior asymptote graphically and numerically.

Solution

From Example 1, we know the end behavior asymptote of $y_1 = f(x)$ is $y_2 = q(x) = x^2 - 2x + 1$. In Figure 5.19 the two graphs appear to be identical. The table in Figure 5.20 shows how very closely the functions approximate each other.

In Figure 5.19 the size of the viewing window hides the behavior of f near its vertical asymptote $x = 1$. If we zoom in on $x = 1$, graphically or numerically, the two functions would appear quite different.

Interpret

$g(x)$ is the end behavior asymptote of $f(x)$ because the values of $f(x)$ get very close to the values of $g(x)$ for large values of $|x|$. ■

Using the End Behavior Asymptote to Describe a Graph

When the degree of the numerator of a rational function is less than or equal to the degree of the denominator, the end behavior asymptote is a horizontal line. However, when the degree of the numerator is greater than the degree of the denominator, the end behavior asymptote can be a curve, as illustrated by the Exploration and Example 2.

The end behavior asymptote is a useful guide to describe or check a graph of a rational function. To illustrate, consider the rational function of Examples 1 and 2.

$$f(x) = \frac{x^3 - 3x^2 + 3x + 1}{x - 1}$$

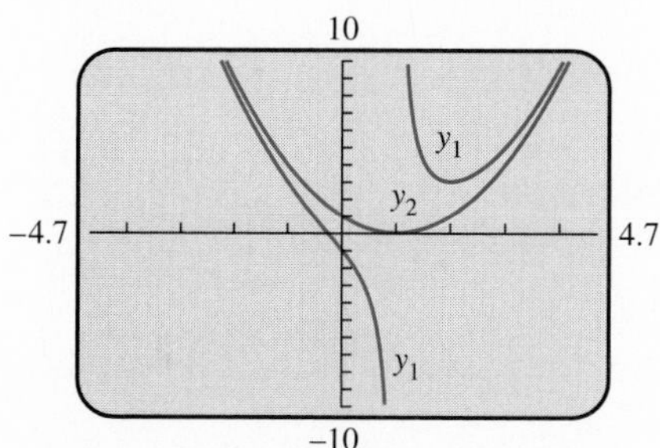

Figure 5.21 The graphs of $y_1 = (x^3 - 3x^2 + 3x + 1)/(x - 1)$ and its end behavior asymptote $y_2 = x^2 - 2x + 1$.

Figure 5.21 shows the graphs of $y_1 = f(x)$ and its end behavior asymptote $y_2 = g(x) = x^2 - 2x + 1$ superimposed on one another.

We can describe the graph of f as follows:

1. Away from the vertical asymptote $x = 1$, the graph of f is very close to the graph of its end behavior asymptote g.
2. Near $x = 1$ the graph of f has the asymptotic behavior

   $$f(x) \to -\infty \text{ as } x \to 1^- \quad \text{and} \quad f(x) \to \infty \text{ as } x \to 1^+,$$

   which can be confirmed algebraically from the equation for f.
3. Combining the information in items 1 and 2, we obtain the graph of f.

To confirm the graph of a rational function it is helpful to follow these five steps.

Teaching Note

Encourage students to develop a true understanding of the behavior of rational functions by attempting to predict the appearance of graphs of rational functions before viewing them on a grapher.

Follow-up

Ask students how one can tell whether or not a rational function has a slant asymptote without doing any division. **(There is a slant asymptote if the degree of the numerator minus the degree of the denominator is one.)**

Confirming Graphs of Rational Functions $y = \dfrac{p(x)}{h(x)} = \dfrac{a_n x^n + \cdots}{b_m x^m + \cdots}$

1. **Confirm the end behavior asymptote:**

   If $n < m$, the horizontal asymptote $y = 0$.
   If $n = m$, the horizontal asymptote $y = a_n/b_m$.
   If $n > m$, the function $y = q(x)$ where $p(x) = q(x)h(x) + r(x)$.

2. **Confirm the vertical asymptotes,** at the zeros of the denominator.
3. **Confirm the x-intercepts,** the zeros of the numerator.
4. **Confirm the y-intercept,** the value of $f(0)$, if defined.
5. **Confirm the intermediate behavior,** the behavior of f between and beyond the x-intercepts and vertical asymptotes. (It sometimes helps to overlay a graph of the end behavior asymptote if $n > m$.)

An end behavior asymptote that is a linear function with a nonzero slope is sometimes called a **slant asymptote.**

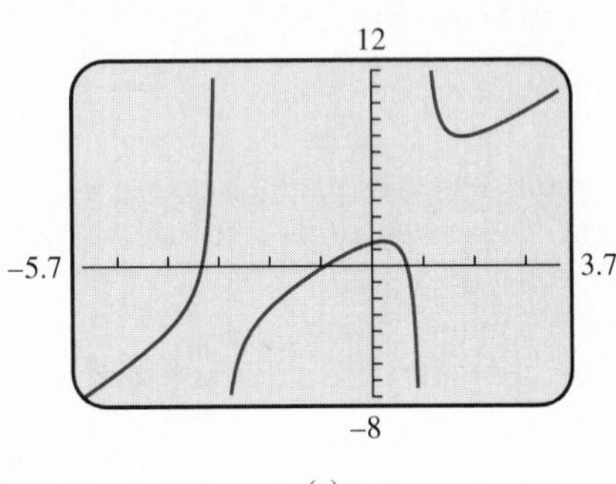

(a)

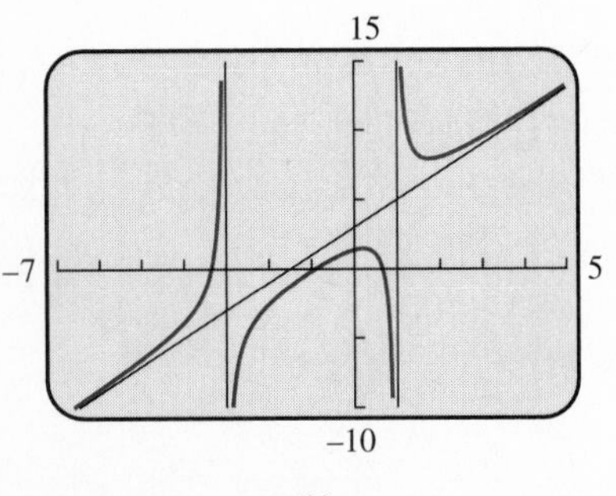

(b)

Figure 5.22 Graphs of $y_1 = (2x^3 + 7x^2 - 4)/(x^2 + 2x - 3)$ showing (a) no asymptotes, and (b) the vertical and slant asymptotes.

EXAMPLE 3 Studying a Slant Asymptote

Graph

$$f(x) = \frac{2x^3 + 7x^2 - 4}{x^2 + 2x - 3}$$

and confirm its behavior.

Solution

The graph is shown in Figure 5.22a.

End Behavior Asymptote: The degree of the numerator, 3, is 1 more than the degree of the denominator, 2. The end behavior asymptote has degree 1, a slant asymptote. Here is the complete long division to find the end behavior asymptote of f.

$$\begin{array}{r} 2x \;+\; 3 \qquad\qquad\quad \\ x^2 + 2x - 3\,\overline{)\,2x^3 + 7x^2 \qquad\quad - 4} \\ \underline{2x^3 + 4x^2 - 6x \quad\;} \\ 3x^2 + 6x - 4 \\ \underline{3x^2 + 6x - 9} \\ 5 \end{array}$$

$y = 2x + 3$ is a slant asymptote of f.

Vertical Asymptotes: We find the zeros of the denominator:

$$x^2 + 2x - 3 = (x + 3)(x - 1) = 0 \quad \Rightarrow \quad x = -3 \text{ or } x = 1$$

The lines $x = -3$ and $x = 1$ are vertical asymptotes.

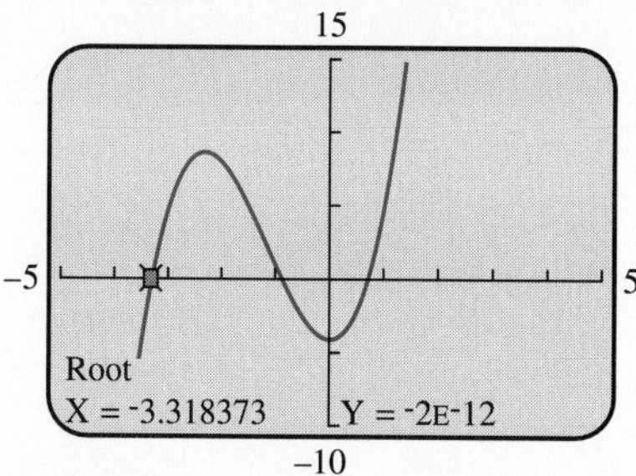

Figure 5.23 A graph of the numerator $y_2 = 2x^3 + 7x^2 - 4$ of the rational function of Example 3.

x-Intercepts: We find the zeros of the numerator using a grapher to be approximately -3.32, -0.87, and 0.69. (See Figure 5.23.)

y-Intercept: Because $f(0) = 4/3$, the y-intercept is $(0, 4/3)$.

Intermediate Behavior: The numerical information in Figure 5.24 suggests, for the behavior near $x = -3$, that

$$f(x) \to \infty \text{ as } x \to -3^- \quad \text{and} \quad f(x) \to -\infty \text{ as } x \to -3^+,$$

and for the behavior near $x = 1$, that

$$f(x) \to -\infty \text{ as } x \to 1^- \quad \text{and} \quad f(x) \to \infty \text{ as } x \to 1^+.$$

Interpret

The above information confirms the information about f shown in Figure 5.22.

■

Assignment Guide

Day 1: Ex. 1, 2, 3, 6, 9, 11, 14, 15, 18, 21, 22, 25, 26

Day 2: Ex. 27, 28, 30, 31, 33, 35, 37, 40

Cooperative Learning

Group Activity: Ex. 32–35, 41–42

Notes on Exercises

Ex. 11–18 illustrate how various viewing windows can show different details about a graph.

Ex. 32–35 illustrate the behavior of rational functions with common factors in the numerator and denominator.

Ongoing Assessment

Self-Assessment: Ex. 7, 23, 29

Embedded Assessment: Ex. 30

X	Y_1	
-3.1	8.9951	
-3.01	121.67	
-3.001	1246.7	
-3	ERROR	
-2.999	-1253	
-2.99	-128.3	
-2.9	-15.62	

Y_1≡(2X³+7X²–4)/…

(a)

X	Y_1	
.9	-8.021	
.99	-120.3	
.999	-1245	
1	ERROR	
1.001	1254.7	
1.01	129.71	
1.1	17.395	

Y_1≡(2X³+7X²–4)/…

(b)

Figure 5.24 The behavior of $y_1 = f(x) = (2x^3 + 7x^2 - 4)/(x^2 + 2x - 3)$ near (a) $x = -3$ and (b) $x = 1$.

Applications

■ EXAMPLE 4 Finding a Minimum Perimeter

Find the dimensions of the rectangle with minimum perimeter if its area is 200 ft^2. Find this least perimeter.

Solution

Model

See Figure 5.25.

Word Statement: Perimeter = 2 × length + 2 × width

$$x = \text{width in feet}$$

$$\frac{200}{x} = \frac{\text{area}}{\text{width}} = \text{length in feet}$$

Function to Be Minimized: $P(x) = 2x + 2\left(\dfrac{200}{x}\right) = 2x + \dfrac{400}{x}.$

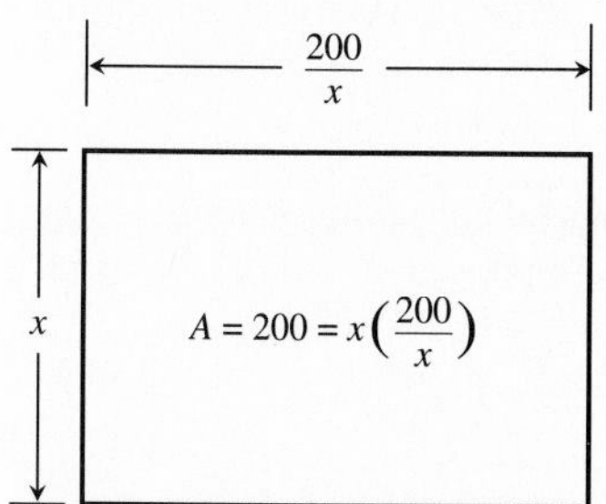

Figure 5.25

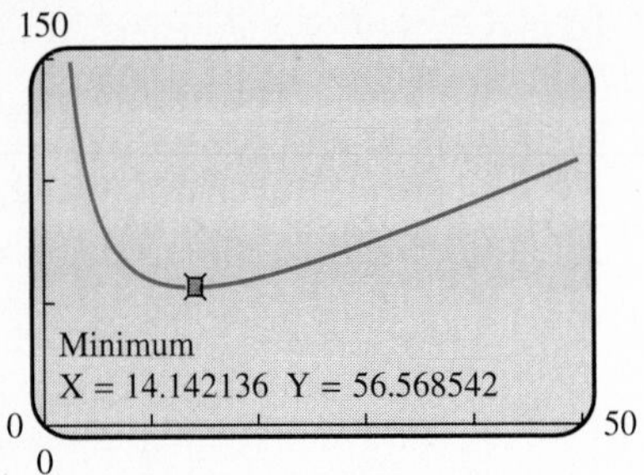

Figure 5.26 We use the feature of the grapher that gives the coordinates of local minimum points.[2]

Solve Graphically

The graph of P in Figure 5.26 shows the minimum point of P to be about (14.14, 56.57).

Interpret

The width is about 14.14 ft and the minimum perimeter is about 56.57 ft. Because $200/14.14 \approx 14.14$, the dimensions of the rectangle with minimum perimeter are 14.14 by 14.14 ft, a square. ■

■ EXAMPLE 5 APPLICATION: Designing a Page

South Dublin Publishing Co. wants to design a page that has a 1-in. left border, a 2-in. top border, and borders on the right and bottom of 0.75 in. They are to surround 36 in.^{2} of print material. Find the dimensions of the page with minimum area. Find the minimum area.

Solution

Model

Let x and y be the dimensions of the print material as suggested in Figure 5.27.

Word Statement: Area of print material = width × length

$$36 = xy, \qquad \text{so } y = \frac{36}{x}$$

$$\begin{aligned} A &= (x + 1.75)(y + 2.75) \\ &= xy + 2.75x + 1.75y + (1.75)(2.75) \\ &= 36 + 2.75x + 1.75\left(\frac{36}{x}\right) + 4.8125 \end{aligned}$$

Function to Be Minimized: $A = 2.75x + 40.8125 + \dfrac{63}{x}$

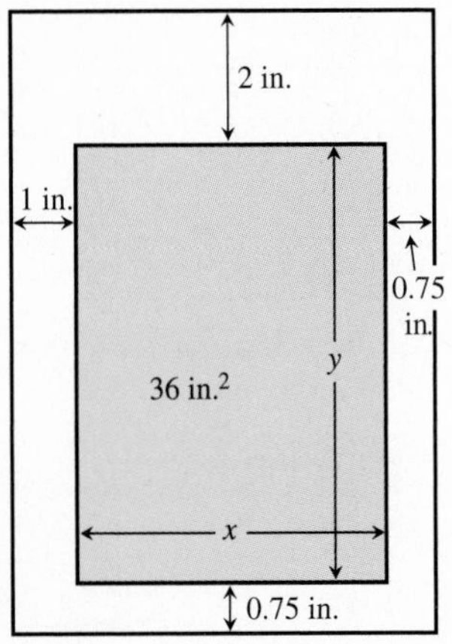

Figure 5.27

Solve Graphically

Notice that A is a rational function in $q(x) + \dfrac{r(x)}{h(x)}$ form. The quotient shows that A has a slant asymptote. The graph of A in Figure 5.28 shows the minimum area is about 67.14 in.^{2}.

Interpret

The page has minimum area when the print material has width of about 4.79 in. and a length of about $36/4.79 \approx 7.52$ in. The minimum area is about 67.14 in.^{2}. The dimensions of the page are approximately 6.54 in. by 10.27 in. ■

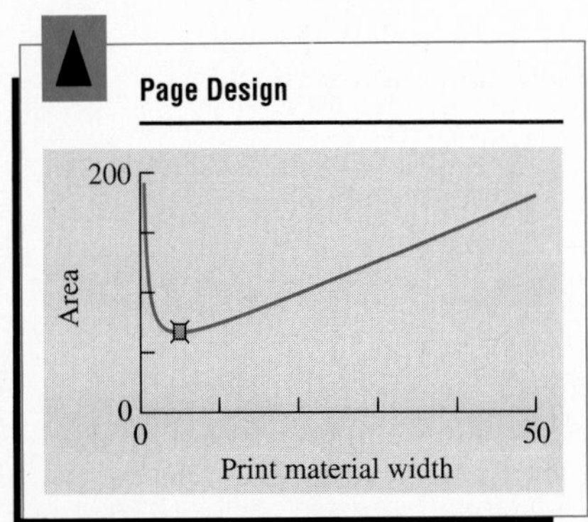

Minimum: $x = 4.7863442$; $y = 67.137393$

Figure 5.28

[2]Consult your owner's manual to see if your grapher has this feature.

Quick Review 5.3

In Exercises 1–4, factor the polynomial.

1. $x^2 + 2x - 15$

2. $2x^2 + 5x - 12$

3. $2x^2 + x - 3$

4. $6x^2 + x - 1$

In Exercises 5–8, find the quotient and remainder using either long division or synthetic division.

5. $\dfrac{x^3 - 5x^2 + x - 4}{x - 2}$

6. $\dfrac{2x^3 + 7x^2 - 3x - 4}{x + 1}$

7. $\dfrac{5x^3 - 7x^2 + 3x - 4}{x^2 + 3}$

8. $\dfrac{3x^4 - 5x^3 + x^2 - 4x + 1}{x^2 - 3}$

SECTION EXERCISES 5.3

In Exercises 1 and 2, complete the tables. Discuss why each table supports the claim that g is the end behavior asymptote of f.

1. $f(x) = \dfrac{2x^3 - 4x^2 + 3}{x - 2}$ and $g(x) = 2x^2$.

a.

x	1	10	100	1000	10,000
$f(x)$	?	?	?	?	?
$g(x)$	?	?	?	?	?

b.

x	−1	−10	−100	−1000	−10,000
$f(x)$	?	?	?	?	?
$g(x)$	?	?	?	?	?

2. $f(x) = \dfrac{3x^3 - 8x^2 - 3x + 2}{x - 3}$ and $g(x) = 3x^2 + x$.

a.

x	1	10	100	1000	10,000
$f(x)$	?	?	?	?	?
$g(x)$	?	?	?	?	?

b.

x	−1	−10	−100	−1000	−10,000
$f(x)$	?	?	?	?	?
$g(x)$	?	?	?	?	?

In Exercises 3–10, find the end behavior asymptote of the function.

3. $f(x) = \dfrac{3x^2 - 2x + 4}{x^2 - 4x + 1}$

4. $f(x) = \dfrac{4x^2 + 2x}{x^2 - 4x + 8}$

5. $f(x) = \dfrac{x^3 - 1}{x^2 + 4}$

6. $f(x) = \dfrac{x^3 - 2}{x^2 - 4}$

7. $f(x) = \dfrac{x^4 + 1}{x + 1}$

8. $\dfrac{2x^5 + x^2 - x + 1}{x^2 - 1}$

9. $f(x) = \dfrac{x^5 - 1}{x + 2}$

10. $f(x) = \dfrac{x^5 + 1}{x - 1}$

In Exercises 11–14, select two viewing windows from those shown. Choose one window that gives the best view of the details around the vertical asymptote of f, and choose the other that shows a graph of f resembling its end behavior asymptote.

11. $y = \dfrac{2x^3 - 3x + 2}{x^3 - 1}$

12. $y = \dfrac{3x^3 + x - 4}{x^3 + 1}$

13. $y = \dfrac{x^2 - 2x + 3}{x - 5}$

14. $y = \dfrac{2x^2 + 2x - 3}{x + 3}$

WINDOW
Xmin=-10
Xmax=10
Xscl=1
Ymin=-10
Ymax=10
Yscl=1

(a)

WINDOW
Xmin=-10
Xmax=10
Xscl=1
Ymin=-100
Ymax=100
Yscl=10

(b)

WINDOW
Xmin=-100
Xmax=100
Xscl=10
Ymin=-10
Ymax=10
Yscl=1

(c)

WINDOW
Xmin=-100
Xmax=100
Xscl=10
Ymin=-100
Ymax=100
Yscl=10

(d)

For Exercises 11–14

In Exercises 15–18, graph the rational function f in two viewing windows, (a) one showing the details around the vertical asymptote and (b) one showing a graph of f that resembles its end behavior asymptote.

15. $y = \dfrac{x^3 - x^2 + 1}{x + 2}$ **16.** $y = \dfrac{x^3 + 1}{x - 1}$

17. $y = \dfrac{x^4 - 2x + 1}{x - 2}$ **18.** $y = \dfrac{x^5 + 1}{x^2 + 1}$

In Exercises 19–26, graph the function and confirm its behavior. Overlay a graph of the end behavior asymptote.

19. $f(x) = \dfrac{2x^2 + x - 2}{x^2 - 1}$ **20.** $g(x) = \dfrac{-3x^2 + x + 12}{x^2 - 4}$

21. $f(x) = \dfrac{x^2 - 2x + 3}{x + 2}$ **22.** $g(x) = \dfrac{x^2 - 3x - 7}{x + 3}$

23. $f(x) = \dfrac{x^3 - 2x^2 + x - 1}{2x - 1}$

24. $g(x) = \dfrac{2x^3 - 2x^2 - x + 5}{x - 2}$

25. $f(x) = \dfrac{2x^4 - x^3 - 16x^2 + 17x - 6}{2x - 5}$

26. $g(x) = \dfrac{2x^5 - 3x^3 + 2x - 4}{x - 1}$

In Exercises 27–31, solve the problem.

27. *Minimizing Perimeter* Consider all rectangles with an area of 182 ft^2. Let x be the length of one side of such a rectangle.

a. Express the perimeter P as a function of x.
b. Find the dimensions of the rectangle that has the least perimeter. What is the least perimeter?

28. *Page Design* Hendrix Publishing Co. wants to design a page that has a 0.75-in. left border, a 1.5-in. top border, and borders on the right and bottom of 1 in. They are to surround 40 in.^{2} of print material. Let x be the width of the print material.

a. Express the area of the page as a function of x.
b. Find the dimensions of the page that has the least area. What is the least area?

29. *Designing a Swimming Pool* Thompson Recreation, Inc., wants to build a rectangular swimming pool with the top of the pool having a surface area 1000 ft^2. The pool is required to have a walk of uniform width 2 ft surrounding it. Let x be length of one side of the swimming pool.

a. Express as a function of x the area of the plot of land needed for the pool and surrounding sidewalk.
b. Find the dimensions of the plot of land that has the least area. What is the least area?

In Exercises 30 and 31, refer to the data in the figure which shows the consumer price indexes and college faculty salaries for the years 1987–1994.

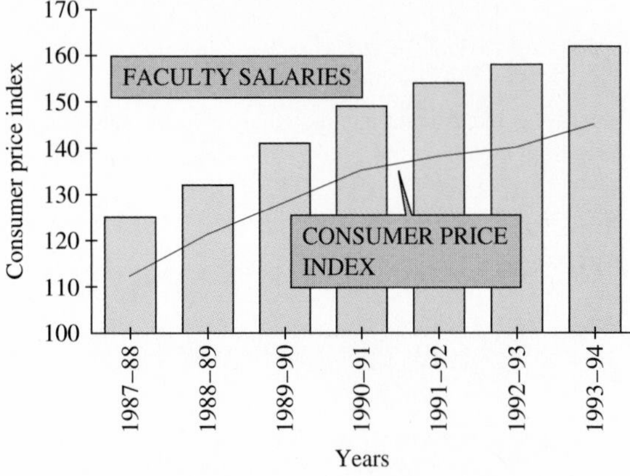

Source: American Association of University Professors; U. S. Department of Labor as reported in *The Chronicle of Higher Education*, May 25, 1994.

For Exercises 30–31

30. *Consumer Price Index (CPI)* The rational function

$$f(x) = \frac{1}{0.008095 - 0.000298x}$$

models the faculty salary data (where $x = 1$ represents the year 1987–88, $x = 2$ represents 1988–89, and so on).

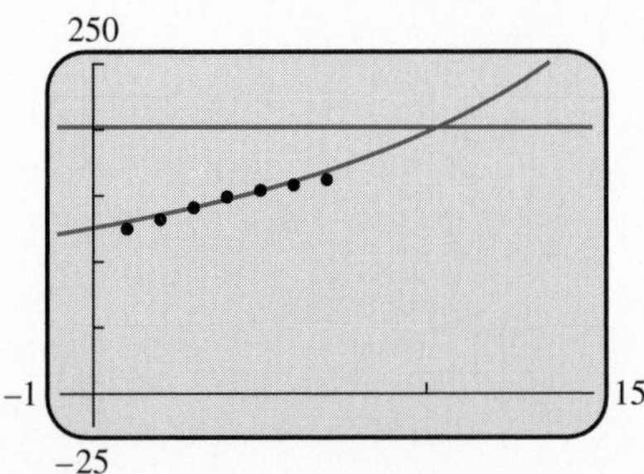

For Exercise 30

a. The figure shows a scatter plot of the data and a graph of the function *f*. It also shows a graph of $y = 200$. Find approximately when the faculty salary will reach 200 on the CPI scale.

b. Explain why this function, while it may be a good model over a short period of time, will not be a good model for faculty salaries over a 50-year period.

31. *Consumer Price Index* The linear regression equation $g(x) = 5.6x + 107.1$ is a model for the CPI data shown in the bar graph.

a. Compare the graph of g with the graph of f from Exercise 30.

b. Discuss reasons why you might expect g to be a better model of the data than f.

EXTENDING THE IDEAS

Group Learning Activity *In Exercises 32 and 33 work in groups of three.* Compare the graphs of the two functions in the decimal window. Explain why the graphs are not identical and why neither one has a vertical asymptote.

32. $y_1 = \dfrac{x^2 + x - 2}{x - 1}$ and $y_2 = x + 2$

33. $y_1 = \dfrac{x^2 - 1}{x + 1}$ and $y_2 = x - 1$

Group Learning Activity *In Exercises 34 and 35 work in groups of three.* Compare the graphs of the two functions in the decimal window. Explain why the graphs are not identical and why each has a vertical asymptote.

34. $y_1 = \dfrac{x^2 - 1}{x^3 - x^2 - x + 1}$ and $y_2 = \dfrac{1}{x - 1}$

35. $y_1 = \dfrac{x^2 + x - 2}{x^3 + 3x^2 - 4}$ and $y_2 = \dfrac{1}{x + 2}$

36. Writing to Learn Suppose $x - a$ is a factor of the numerator and denominator of a rational function f. Under what conditions will f have a vertical asymptote at $x = a$?

In Exercises 37–40, write a rational function with the end behavior $g(x)$ and the given vertical asymptote(s).

37. $g(x) = x^2 + 2x$; vertical asymptote $x = 2$

38. $g(x) = 2x + 1$; vertical asymptote $x = 3$

39. $g(x) = x^2 + 1$; vertical asymptotes $x = -1, x = 1$

40. $g(x) = x^3 - x$; vertical asymptotes $x = -3, x = 2$

Group Learning Activity *Work in groups of three.* In Exercises 41 and 42, find the end behavior asymptote, the vertical asymptote(s), all real zeros, local maximum and minimum values, the intervals on which the function is increasing, and the intervals on which the function is decreasing.

41. $f(x) = \dfrac{2x^4 - 3x^2 + 1}{3x^4 - x^2 + x - 1}$

42. $f(x) = \dfrac{x^4 + 2x^3 - 7x^2 - 26x - 119}{x + 4}$

5.4 RATIONAL EQUATIONS AND INEQUALITIES

Objective

Students will be able to solve equations and inequalities involving rational functions by using both algebraic and graphical techniques.

Motivate

Ask students to suggest ways to solve $\dfrac{3}{x + 2} = \dfrac{2x}{3x + 1}$ algebraically or graphically.

Solving Rational Equations • Extraneous Solutions • Solving Rational Inequalities

Solving Rational Equations

When a rational equation is written in the form

$$\frac{p(x)}{h(x)} = 0,$$

where $p(x)$ and $h(x)$ are polynomial functions with no common factors, the zeros of $p(x)$ are the solutions of the equation.

EXAMPLE 1 Solving Rational Equations

Solve $x + \frac{3}{x} = 4$.

Solution

Solve Algebraically

$$x + \frac{3}{x} = 4$$

$$x^2 + 3 = 4x \qquad \text{Multiply by } x.$$

$$x^2 - 4x + 3 = 0 \qquad \text{Subtract } 4x.$$

$$(x - 1)(x - 3) = 0 \qquad \text{Factor.}$$

$$x - 1 = 0 \quad \text{or} \quad x - 3 = 0 \qquad \text{Zero factor property}$$

$$x = 1 \quad \text{or} \quad x = 3$$

Support Numerically By replacing x with 1 and then with 3 in the original equation, we can see that each value is a solution. ■

Recall that to solve an equation graphically, we find intersections of two graphs or x-intercepts of a graph.

EXAMPLE 2 Solving Rational Equations

Solve $x^2 + \frac{4}{x} = 20$.

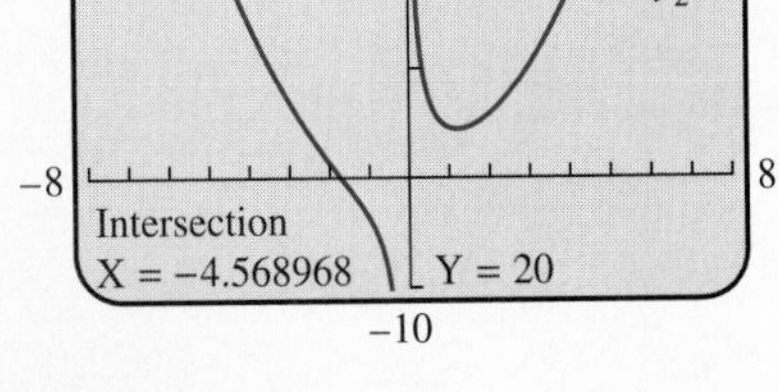

Figure 5.29

Solution

Solve Graphically

Figure 5.29 shows the graphs of $y_1 = x^2 + (4/x)$ and $y_2 = 20$. Using the grapher, we find the x-coordinate of one point of intersection to be approximately $x = -4.57$. The x-coordinates of the other two points of intersection are approximately $x = 0.20$ and $x = 4.37$.

Support Numerically

We replace x by -4.57, then by 0.20, and finally by 4.37 in the expression $x^2 + (4/x)$. In each case the value of $x^2 + (4/x)$ is approximately 20. ■

Figure 5.30

EXAMPLE 3 APPLICATION: Using a Rational Equation in Industrial Design

Stewart Cannery will package tomato juice in 2-liter cylindrical cans. Find the radius and height of the cans (see Figure 5.30) if the cans have a surface area of 1000 cm^2.

Solution

Model

$$S = \text{surface area of can in cm}^3,$$
$$r = \text{radius of can in centimeters},$$
$$h = \text{height of can in centimeters}.$$

Using volume (V) and surface area (S) formulas, and the fact that 1 L = 1000 cm^3, we conclude that

$$V = \pi r^2 h = 2000 \qquad \text{and} \qquad S = 2\pi r^2 + 2\pi r h = 1000.$$

So

$$2\pi r^2 + 2\pi r h = 1000$$

$$2\pi r^2 + 2\pi r\left(\frac{2000}{\pi r^2}\right) = 1000 \qquad \text{Substitute } h = 2000/(\pi r^2).$$

$$2\pi r^2 + \frac{4000}{r} = 1000 \qquad \text{Equation to be solved}$$

Solve Graphically

Figure 5.31a shows the graphs of $y_1 = 2\pi r^2 + 4000/r$ and $y_2 = 1000$. One point of intersection occurs when r is approximately 9.65. The second point of intersection occurs when r is approximately 4.62.

Since $h = 2000/(\pi r^2)$, the corresponding values for h are

$$h = \frac{2000}{\pi(4.619\cdots)^2} \approx 29.83 \qquad \text{and} \qquad h = \frac{2000}{\pi(9.654\cdots)^2} \approx 6.83.$$

REMARK

Remember that it is a good idea not to replace r by its approximation until the value of h is found.

Support Numerically

Figure 5.31b shows that S is about 1000 when $r = 4.62$ and $h = 29.83$, or when $r = 9.65$ and $h = 6.83$.

Interpret

If the radius is 4.62 cm and the height is 29.83 cm, or if the radius is 9.65 cm and the height is 6.83 cm, the can will have volume approximately 2 L and a surface area of about 1000 cm^2. Do either of these dimensions seem reasonable for a tomato juice can? ■

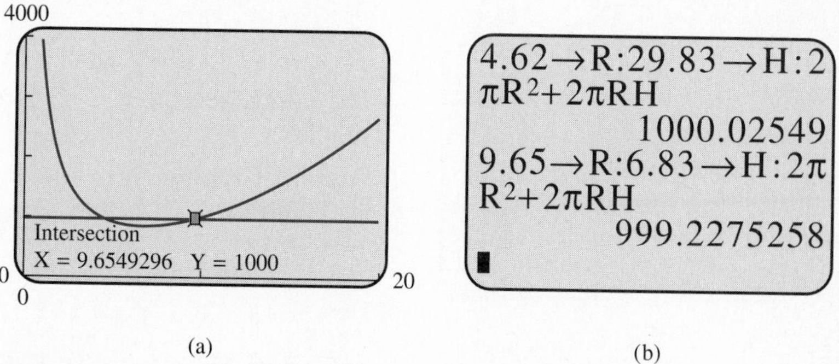

(a) (b)

Figure 5.31

Teaching Note

Encourage students to see if they can identify the step that causes an extraneous solution to be introduced. Emphasize that extraneous solutions reaffirm the importance of checking solutions.

Extraneous Solutions

When both sides of an equation in x are multiplied by an expression containing x, as happened in Example 1, the resulting equation can have an extraneous solution (see Section 2.3). Extraneous solutions can be identified by checking them in the original equation.

EXAMPLE 4 Eliminating Extraneous Solutions

Solve the equation

$$\frac{2x}{x-1} + \frac{1}{x-3} = \frac{2}{x^2 - 4x + 3}.$$

Solution

Solve Algebraically

$x^2 - 4x + 3$ factors into $(x - 1)(x - 3)$. We multiply both sides of the equation by $(x - 1)(x - 3)$:

$$(x-1)(x-3)\left(\frac{2x}{x-1} + \frac{1}{x-3}\right) = (x-1)(x-3)\left(\frac{2}{x^2 - 4x + 3}\right)$$

$$2x(x-3) + (x-1) = 2$$

$$2x^2 - 5x - 3 = 0$$

$$(2x+1)(x-3) = 0$$

$$x = -\frac{1}{2} \quad \text{or} \quad x = 3$$

Teaching Note

The notion of an extraneous solution provides an opportunity for students to investigate the interplay between graphical and algebraic representations of equations.

Confirm Numerically

We replace x by $-\frac{1}{2}$ in the original equation:

$$\frac{2(-\frac{1}{2})}{(-\frac{1}{2}) - 1} + \frac{1}{(-\frac{1}{2}) - 3} \stackrel{?}{=} \frac{2}{(-\frac{1}{2})^2 - 4(-\frac{1}{2}) + 3}$$

$$\frac{2}{3} - \frac{2}{7} = \frac{8}{21} \quad \text{It checks.}$$

The original equation is not defined for $x = 3$, so $x = 3$ is an extraneous solution.

Support Graphically

The graph of

$$f(x) = \frac{2x}{x-1} + \frac{1}{x-3} - \frac{2}{x^2 - 4x + 3}$$

in Figure 5.32 suggests that $x = -\frac{1}{2}$ is an x-intercept and $x = 3$ is not. ■

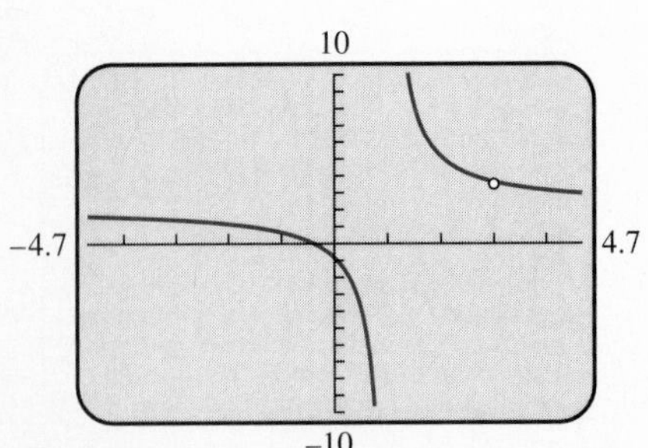

Figure 5.32

Alert

When solving a rational inequality, some students will attempt to multiply both sides by the LCD without regard to the direction of the inequality sign. They need to realize that since the sign of the LCD (and hence the direction of the inequality sign in the resulting inequality) may depend on x, this method will not work.

Solving Rational Inequalities

As you learned in Chapter 2, the solution set of an inequality is an interval or a combination of intervals.

For an inequality in the form

$$\frac{p(x)}{h(x)} > 0 \quad \text{or} \quad \frac{p(x)}{h(x)} < 0$$

we identify intervals that are candidates for the solution set by determining where the quotient $p(x)/h(x)$ changes sign. A sign change may occur in two kinds of places:

a. Where the graph of $f(x) = p(x)/h(x)$ intersects the x-axis [where $p(x) = 0$], or

b. Where vertical asymptotes occur [where $h(x) = 0$].

Solving Rational Inequalities

To solve rational inequalities $f(x) = \dfrac{p(x)}{h(x)} > 0$:

1. Find all real zeros of $p(x)$ and $h(x)$. Arrange them in order:

$$a_1 < a_2 < \cdots < a_n.$$

2. Test one value of x from each interval:

$$(-\infty, a_1), (a_1, a_2), \ldots, (a_{n-1}, a_n), (a_n, \infty).$$

If $f(x) > 0$, then the entire interval belongs to the solution set.

A similar set of steps provides the solution of $f(x) < 0$.

EXAMPLE 5 Solving a Rational Inequality

Solve $f(x) = \dfrac{x - 1}{x + 3} > 0.$

Solution

Solve Algebraically

The zeros of the numerator and denominator, arranged in order, are -3 and 1. The possible solution intervals are

$$(-\infty, -3), \quad (-3, 1), \quad \text{and} \quad (1, \infty).$$

Picking -4, 0, and 2 from the intervals, we find that $f(-4) > 0$, $f(0) < 0$, and $f(2) > 0$. Therefore, the solution set is $(-\infty, -3) \cup (1, \infty)$.

REMARK

The symbol "∪" is read "union" and indicates that two sets have been joined together to form a new set.

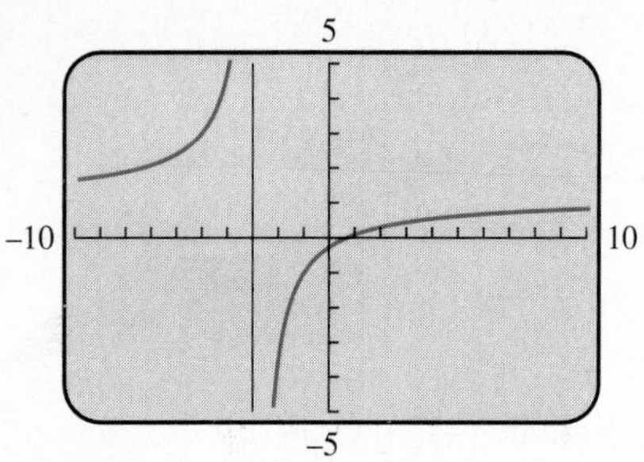

Figure 5.33

Support Graphically

The graph of f in Figure 5.33 suggests that

$$\frac{x-1}{x+3} > 0$$

on the intervals $(-\infty, -3)$ and $(1, \infty)$. ■

Notes on Examples

In Example 6, note that it is important to rewrite the inequality in the form $\frac{p(x)}{h(x)} < 0$ before proceeding.

■ EXAMPLE 6 Solving Another Rational Inequality

Solve

$$\frac{2x^2+6x-8}{2x^2+5x-3} < 1.$$

Solution

Solve Algebraically

$$\frac{2x^2+6x-8}{2x^2+5x-3} < 1$$

$$\frac{2x^2+6x-8}{2x^2+5x-3} - 1 < 0 \quad \text{Subtract 1.}$$

$$\frac{2x^2+6x-8-(2x^2+5x-3)}{2x^2+5x-3} < 0 \quad \text{Write as } \frac{p(x)}{h(x)}.$$

$$\frac{x-5}{(2x-1)(x+3)} < 0 \quad \text{Simplify numerator; factor denominator.}$$

The zeros of the numerator and denominator, arranged in order, are $-3, \frac{1}{2}$, and 5. The possible solution intervals are

$$(-\infty, -3),\ (-3, \tfrac{1}{2}),\ (\tfrac{1}{2}, 5), \text{ and } (5, \infty).$$

Picking -4, 0, 1, and 6 from the intervals, we find that

$$\frac{x-5}{(2x-1)(x+3)} < 0$$

for $x = -4$ and $x = 1$. The solution set is $(-\infty, -3) \cup (\frac{1}{2}, 5)$.

Support Graphically

The graph of

$$f(x) = \frac{2x^2+6x-8}{2x^2+5x-3} - 1$$

in Figure 5.34 appears to be below the x-axis when x is in $(-\infty, -3)$ or in $(\frac{1}{2}, 5)$. ■

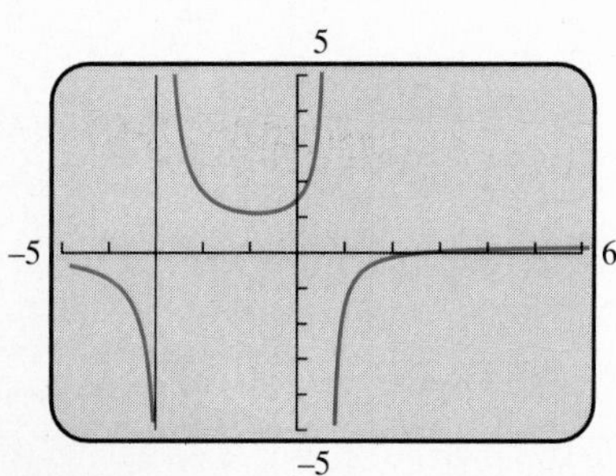

Figure 5.34

EXAMPLE 7 APPLICATION: Using a Rational Inequality in Industrial Design

Stewart Cannery will package tomato juice in a 2-liter (2000 cm^3) cylindrical can. Find the radius and height if the can is to have a surface area that is less than 1000 cm^2. (See Example 3 and Figure 5.30.)

Solution

Model

From Example 3, we know the surface area S is given by

$$S(r) = 2\pi r^2 + \frac{4000}{r}.$$

The inequality to be solved is

$$2\pi r^2 + \frac{4000}{r} < 1000.$$

Solve Graphically

Figure 5.35 shows a graph of $y_1 = S(r) = 2\pi r^2 + 4000/r$ and $y_2 = 1000$. Using grapher methods we find that the two curves intersect at approximately $r = 4.62$ and $r = 9.65$. So the surface area is less than 1000 cm^2 if

$$4.62 < r < 9.65.$$

The volume of a cylindrical can is $V = \pi r^2 h$ and $V = 2000$. Using substitution we see that $h = 2000/(\pi r^2)$. To find the values for h we build a double inequality for $2000/(\pi r^2)$.

$$4.62 < r < 9.65$$

$$4.62^2 < r^2 < 9.65^2 \qquad 0 < a < b \Rightarrow a^2 < b^2 \text{ (see Exercise 48).}$$

$$\pi 4.62^2 < \pi r^2 < \pi 9.65^2 \qquad \text{Multiply by } \pi.$$

$$\frac{1}{\pi(4.62)^2} > \frac{1}{\pi r^2} > \frac{1}{\pi 9.65^2} \qquad 0 < a < b \Rightarrow \frac{1}{a} > \frac{1}{b} \text{ (see Exercise 49).}$$

$$\frac{2000}{\pi(4.62)^2} > \frac{2000}{\pi r^2} > \frac{2000}{\pi 9.65^2} \qquad \text{Multiply by 2000.}$$

$$\frac{2000}{\pi(4.62)^2} > h > \frac{2000}{\pi 9.65^2} \qquad h = 2000/(\pi r^2)$$

$$29.83 > h > 6.84 \qquad \text{Compute.}$$

Interpret

The surface area of the can will be less than 1000 cm^2 if the values of r are between 4.62 cm and 9.95 cm. The corresponding values of h are between 6.84 cm and 29.83 cm.

Follow-up

Ask students to discuss why one must find the zeros of *both* the numerator and denominator to solve a rational inequality of the form $\frac{p(x)}{h(x)} > 0$.

Assignment Guide

Day 1: Ex. 3–33, multiples of 3
Day 2: Ex. 36–38, 41–45

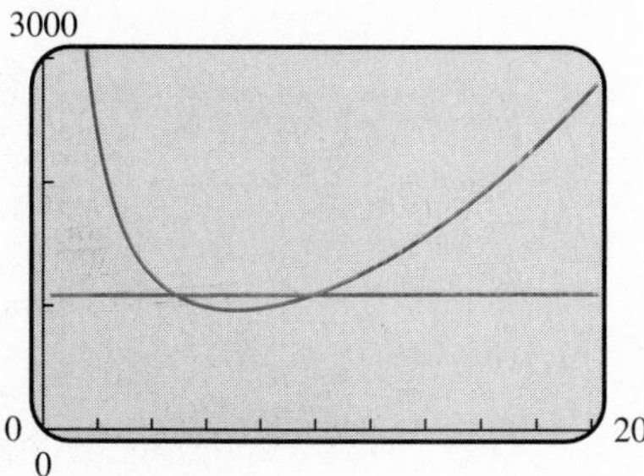

Figure 5.35

Cooperative Learning

Group Activity: Ex. 39

Notes on Exercises

Ex. 35–41 include many opportunities for developing students' problem-solving skills.

Ongoing Assessment

Self-Assessment: Ex. 7, 13, 23, 35
Embedded Assessment: Ex. 38

Quick Review 5.4

In Exercises 1–8, find the missing numerator or denominator.

1. $\frac{3}{4x} = \frac{?}{6x^3}$
2. $\frac{5y}{2z} = \frac{10xy}{?}$
3. $\frac{6x^2}{9xz} = \frac{2x}{?}$
4. $\frac{4xy^2z^3}{6x^2y} = \frac{?}{3x}$
5. $\frac{5}{3x} = \frac{?}{3x^2 - 3x}$
6. $\frac{2x}{x-3} = \frac{?}{x^2 + x - 12}$
7. $\frac{x-1}{x+1} = \frac{x^2-1}{?}$
8. $\frac{x^2+x+1}{x+4} = \frac{x^3-1}{?}$

In Exercises 9–12, use the quadratic formula to find the zeros of the function.

9. $y = 2x^2 - 3x - 1$
10. $y = 2x^2 - 5x - 1$
11. $y = 3x^2 + 2x - 2$
12. $y = x^2 - 3x - 9$

SECTION EXERCISES 5.4

In Exercises 1–8, solve the equation algebraically. Support numerically.

1. $\frac{x-1}{x+2} = 3$
2. $\frac{x+3}{x-4} = 5$
3. $x + \frac{10}{x} = 7$
4. $x + \frac{12}{x} = 7$
5. $x + 2 = \frac{15}{x}$
6. $x + 5 = \frac{14}{x}$
7. $\frac{1}{x} - \frac{2}{x-3} = 4$
8. $\frac{3}{x-1} + \frac{2}{x} = 8$

In Exercises 9–14, solve the equation algebraically. Check for extraneous solutions. Support graphically.

9. $\frac{3x}{x+5} + \frac{1}{x-2} = \frac{7}{x^2+3x-10}$
10. $\frac{4x}{x+4} + \frac{3}{x-1} = \frac{15}{x^2+3x-4}$
11. $\frac{x-3}{x} - \frac{3}{x+1} + \frac{3}{x^2+x} = 0$
12. $\frac{x+2}{x} - \frac{4}{x-1} + \frac{2}{x^2-x} = 0$
13. $\frac{3}{x+2} + \frac{6}{x^2+2x} = \frac{3-x}{x}$
14. $\frac{x+3}{x} - \frac{2}{x+3} = \frac{6}{x^2+3x}$

In Exercises 15–26, solve the inequality algebraically. Support graphically.

15. $\frac{1}{x-3} + 4 > 0$
16. $\frac{x-3}{x+5} - 6 \le 3$
17. $\frac{x+3}{2x-7} < 5$
18. $\frac{x-1}{x+4} > 3$
19. $\frac{x-1}{x^2-4} < 0$
20. $\frac{x+2}{x^2-9} < 0$
21. $\frac{x^2-1}{x^2+1} \le 0$
22. $\frac{x^2-4}{x^2+4} > 0$
23. $\frac{x^2+x-12}{x^2-4x+4} > 0$
24. $\frac{x^2+3x-10}{x^2-6x+9} < 0$
25. $\frac{x^2-x+1}{x+2} < 3$
26. $\frac{x^2+3x-6}{x+3} > 2$

In Exercises 27–34, solve the equation or the inequality.

27. $\frac{2}{x-1} + x = 5$
28. $\frac{x^2-6x+5}{x^2-2} = 3$
29. $\frac{x^3-x}{x^2+1} \ge 0$
30. $\frac{x^3-4x}{x^2+2} \le 0$
31. $x^2 + \frac{5}{x} = 8$
32. $x^2 - \frac{3}{x} = 7$
33. $x^2 - \frac{2}{x} > 0$
34. $x^2 + \frac{4}{x} \ge 0$

In Exercises 35–43, solve the problem.

35. *Resistors* The total electrical resistance R of two resistors connected in parallel with resistances R_1 and R_2 is given by

$$\frac{1}{R} = \frac{1}{R_1} + \frac{1}{R_2}.$$

One resistor has a resistance of 2.3 ohms. Let x be the resistance of the second resistor.

a. Express the total resistance R as a function of x.

b. Find the resistance of the second resistor if the total resistance of the pair is 1.7 ohms.

36. *Designing Rectangles* Consider all rectangles with an area of 200 m^2. Let x be the length of one side of such a rectangle.

a. Express the perimeter P as a function of x.
b. Find the length and width of a rectangle whose perimeter is 70 m.
c. Find the length and width of a rectangle whose perimeter is less than 70 m.

37. *Industrial Design* Drake Cannery will pack peaches in 0.5-L cylindrical cans. Let x be the radius of the can in cm.

a. Express the surface area S of the can as a function of x.
b. Find the radius and height of the can if the surface area is 900 cm^2.

38. *Industrial Design* The graph of the function $y_1 = S(x)$ for the surface area of a can that you found in Exercise 37 is shown. Duplicate this graph on your grapher.

a. Find the possible dimensions for the can if the surface area is to be less than 900 cm^2.
b. Find the least possible surface area for the can.

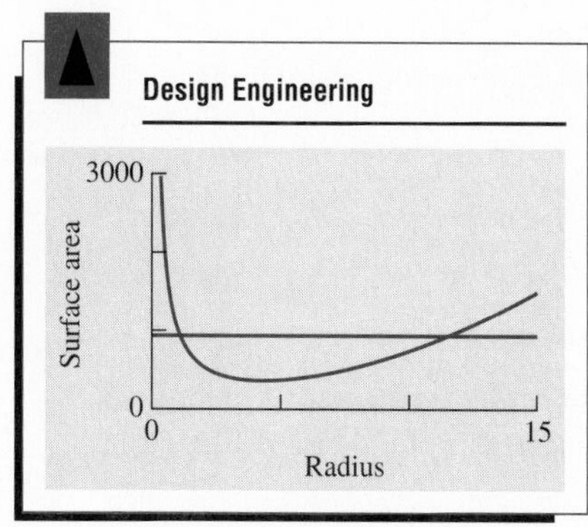

For Exercise 38

39. *Swimming Pool Drainage* Drains A and B are used to empty a swimming pool. Drain A alone can empty the pool in 4.75 h. Let t be the time it takes for drain B alone to empty the pool.

a. Express as a function of t the part D of the drainage that can be done in 1 h with both drains open at the same time.
b. Find graphically the time it takes for drain B alone to empty the pool if both drains, when open at the same time, can empty the pool in 2.6 h. Confirm algebraically.

40. *Time-Rate Problem* Josh rode his bike 17 mi from his home to Columbus, and then traveled 53 mi by car from Columbus to Dayton. Assume the average rate of the car was 43 mph faster than the average rate of the bike.

a. Express the total time t required to complete the 70-mi trip (bike and car) as a function of the rate x of the bike.
b. Find graphically the rate of the bike if the total time of the trip was 1 h 40 min. Confirm algebraically.

41. *Construction Costs* A single-story house with a rectangular base is to contain 900 ft^2 of living area. Local building codes require that both the length L and the width W of the base of the house be greater than 20 ft. To minimize the cost of the foundation, the builder wants to minimize the perimeter of the foundation.

a. Express the perimeter P as a function of L, the length of the base.
b. Find the value of L that minimizes the perimeter. What is the minimum perimeter?

For Exercise 41

EXTENDING THE IDEAS

42. Rewrite the equation of Example 4 in the form $p(x)/h(x) = 0$. Investigate the graph of $f(x) = p(x)/h(x)$, and explain algebraically why the equation has an extraneous solution.

43. Rewrite the equation of Example 2 in the form $p(x)/h(x) = 0$, and let $f(x) = p(x)/h(x)$.

a. Find an upper bound for the number of real zeros of f.
b. Find the zeros of f.

In Exercises 44–47, solve for the specified variable.

44. Solve for x: $y = 1 + \dfrac{1}{1 + x}$

45. Solve for x: $y = 1 - \dfrac{1}{1 - x}$

46. Solve for x: $y = 1 + \dfrac{1}{1 + \dfrac{1}{x}}$

47. Solve for x: $y = 1 + \dfrac{1}{1 + \dfrac{1}{1 - x}}$

In Exercises 48 and 49, use the properties of inequalities to solve the problem.

48. If $0 < a < b$, show that $a^2 < b^2$.

49. If $0 < a < b$, show that $\dfrac{1}{a} > \dfrac{1}{b}$.

5.5
PARTIAL FRACTIONS

Objective

Students will be able to decompose rational expressions into partial fractions.

Motivate

Have students rewrite

$$\frac{3}{x - 4} + \frac{2}{x + 3}$$

as a single rational expression.

$$\left(\frac{5x + 1}{x^2 - x - 12}\right)$$

Explain that this section will give them techniques for "going the other way"—writing a rational expression as a sum of simpler expressions.

Introducing Partial Fractions • Denominators with Linear Factors • Denominators with Irreducible Quadratic Factors

Introducing Partial Fractions

The study of growth rates in calculus requires the ability to decompose a rational function into a sum of simpler rational functions. Each fraction in the sum is called a **partial fraction,** and the sum is called a **partial fraction decomposition** of the original fraction.

For example, the partial fraction decomposition of

$$f(x) = (3x - 4)/(x^2 - 2x)$$

is

$$f(x) = \frac{3x - 4}{x^2 - 2x} = \frac{2}{x} + \frac{1}{x - 2}.$$

Here $2/x$ and $1/(x - 2)$ are the partial fractions of f.

When the degree of the numerator of the rational function

$$f(x) = \frac{p(x)}{h(x)}$$

is equal to or greater than the degree of the denominator, we can find the quotient $q(x)$ and remainder $r(x)$ by long division and write the function in the form

$$f(x) = q(x) + \frac{r(x)}{h(x)},$$

where the degree of $r(x)$ is less than the degree of the denominator $h(x)$. Then we decompose $r(x)/h(x)$. Therefore, in what follows we will consider only rational functions in which the degree of numerator is less than the degree of the denominator.

Denominators with Linear Factors

In Section 3.6, we observed that any polynomial with real coefficients can be factored into linear factors and irreducible quadratic factors, which are factors with no real zeros. We first consider those cases in which the denominator $h(x)$ can be factored completely into linear factors, some of which may repeat.

Teaching Note

Partial fractions are extremely useful for the calculus topic of integration. Point out that certain operations will be easier to accomplish with a sum of simple expressions than with one complicated expression.

Partial Fraction Decomposition of $\dfrac{r(x)}{h(x)}$

For each factor of $h(x)$ in the form $(mx + n)^k$, the partial fraction decomposition must contain the sum of the k terms:

$$\frac{A_1}{mx + n} + \frac{A_2}{(mx + n)^2} + \cdots + \frac{A_k}{(mx + n)^k}$$

Notice that if $k = 1$, there is only one term in the sum.

Examples 1 and 2 demonstrate how the A_i constants can be found.

EXAMPLE 1 Decomposing a Fraction with Nonrepeated Linear Factors

Find the partial fraction decomposition of

$$\frac{5x - 1}{x^2 - 2x - 15}.$$

Solution

Solve Algebraically

The denominator factors into $(x + 3)(x - 5)$. We write

$$\frac{5x - 1}{x^2 - 2x - 15} = \frac{A_1}{x + 3} + \frac{A_2}{x - 5} \qquad \textbf{(1)}$$

and then "clear fractions" by multiplying both sides of equation (1) by $x^2 - 2x - 15$ to obtain

$$5x - 1 = A_1(x - 5) + A_2(x + 3). \qquad \textbf{(2)}$$

Substituting $x = 5$ into equation (2), we obtain

$$\begin{aligned} 5x - 1 &= A_1(x - 5) + A_2(x + 3), \\ 5(5) - 1 &= A_1(5 - 5) + A_2(5 + 3), \\ 24 &= 8A_2, \\ A_2 &= 3. \end{aligned}$$

Alert

When using the method of Examples 1–3, students, by using arbitrary values of x, may get more complicated equations than necessary. To get the simplest results, they should first use the zeros of the original denominator as values of x. Alternatively, they can apply the method of Example 4.

In a similar fashion, we let $x = -3$ and find $A_1 = 2$. Substituting into equation (1) we see that

$$\frac{5x - 1}{x^2 - 2x - 15} = \frac{2}{x + 3} + \frac{3}{x - 5}.$$

(To check, you can add the right side using common denominators and compare to the left side.)

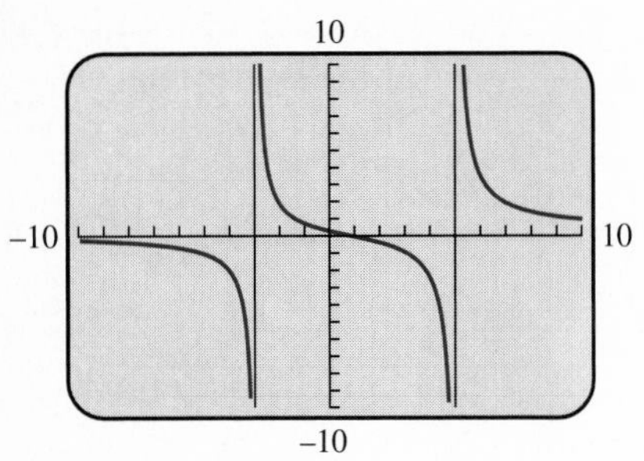

Figure 5.36

Support Graphically

Figure 5.36 shows graphs of both

$$y_1 = \frac{5x - 1}{x^2 - 2x - 15} \quad \text{and} \quad y_2 = \frac{2}{x + 3} + \frac{3}{x - 5}.$$

Use "trace" to verify that the graphs lie one atop the other. ■

Notes on Examples

Students may not understand why they cannot simply use two terms of the form $\frac{A_2}{x - 2}$ and $\frac{A_3}{x - 2}$. They should realize that this would be equivalent to one term of the form $\frac{B}{x - 2}$, where $B = A_2 + A_3$.

■ **EXAMPLE 2 Decomposing a Fraction with a Repeated Linear Factor**

Find the partial fraction decomposition of

$$\frac{-x^2 + 2x + 4}{x^3 - 4x^2 + 4x}.$$

Solution

Solve Algebraically

The denominator factors into $x(x - 2)^2$. Because the factor $x - 2$ is squared, it contributes two terms to the decomposition:

$$\frac{-x^2 + 2x + 4}{x^3 - 4x^2 + 4x} = \frac{A_1}{x} + \frac{A_2}{x - 2} + \frac{A_3}{(x - 2)^2}. \qquad \textbf{(3)}$$

To simplify we multiply both sides of equation (3) by $x^3 - 4x^2 + 4x$ to obtain

$$-x^2 + 2x + 4 = A_1(x - 2)^2 + A_2 x(x - 2) + A_3 x.$$

In this equation we let $x = 2$ to find $A_3 = 2$, and let $x = 0$ to find $A_1 = 1$. Finally, we substitute $A_1 = 1$, $A_3 = 2$, and $x = 1$ to find

$$-1 + 2 + 4 = (-1)^2 + A_2(-1) + 2,$$

$$A_2 = -2.$$

We have $A_1 = 1$, $A_2 = -2$, and $A_3 = 2$. Substitution into equation (3) gives

$$\frac{-x^2 + 2x + 4}{x^3 - 4x^2 + 4x} = \frac{1}{x} + \frac{-2}{x - 2} + \frac{2}{(x - 2)^2}.$$

Support Graphically

Compare the graphs of

$$y_1 = \frac{-x^2 + 2x + 4}{x^3 - 4x^2 + 4x} \quad \text{and} \quad y_2 = \frac{1}{x} + \frac{-2}{x - 2} + \frac{2}{(x - 2)^2}$$

in the same viewing window. ■

Denominators with Irreducible Quadratic Factors

We observed in Section 3.6 that quadratic factors $ax^2 + bx + c$ with real zeros could be factored using real coefficients. Thus, irreducible quadratic factors *cannot* be factored into linear factors with real number constants.

Partial Fraction Decomposition of $\frac{r(x)}{h(x)}$

1. Provide for each linear factor of $h(x)$ as already shown.
2. For each irreducible factor of $h(x)$ in the form $(ax^2 + bx + c)^k$, the partial fraction decomposition must contain the sum of the k terms:

$$\frac{B_1x + C_1}{ax^2 + bx + c} + \frac{B_2x + C_2}{(ax^2 + bx + c)^2} + \cdots + \frac{B_kx + C_k}{(ax^2 + bx + c)^k}$$

■ **EXAMPLE 3 Decomposing a Fraction with a Quadratic Factor**

Find the partial fraction decomposition of

$$\frac{x^2 + 4x + 1}{x^3 - x^2 + x - 1}.$$

Solution

We factor the denominator *by grouping,* as follows:

$$x^3 - x^2 + x - 1 = x^2(x - 1) + (x - 1) = (x - 1)(x^2 + 1)$$

where $x^2 + 1$ has no real zeros. Each factor occurs once, so each one leads to one term in the decomposition:

$$\frac{x^2 + 4x + 1}{x^3 - x^2 + x - 1} = \frac{A}{x - 1} + \frac{Bx + C}{x^2 + 1}. \qquad \textbf{(4)}$$

We multiply both sides of equation (4) by $x^3 - x^2 + x - 1$ to get

$$x^2 + 4x + 1 = A(x^2 + 1) + (Bx + C)(x - 1).$$

Notes on Examples

Example 4 uses the method of comparing coefficients to determine the constants. Note that two polynomials of the same degree in x are equal if and only if the corresponding coefficients are equal.

We let $x = 1$ to find $A = 3$, and let $A = 3$ and $x = 0$ to find

$$1 = 3(0 + 1) - C$$
$$C = 2.$$

Finally, we let $A = 3$, $C = 2$, and $x = -1$ to find

$$\begin{aligned} 1 - 4 + 1 &= 3(1 + 1) + [B(-1) + 2](-1 - 1) \\ -2 &= 6 + (-2)(-B + 2) \\ -2 &= 6 + 2B - 4 \\ -4 &= 2B \\ B &= -2 \end{aligned}$$

Thus,
$$\frac{x^2 + 4x + 1}{x^3 - x^2 + x - 1} = \frac{3}{x - 1} + \frac{-2x + 2}{x^2 + 1}.$$

■

Example 4 illustrates another technique that can be used to determine the constants in a partial fraction decomposition.

■ EXAMPLE 4 Decomposing a Fraction with a Repeated Quadratic Factor

Find the partial fraction decomposition of

$$\frac{2x^3 - x^2 + 5x}{(x^2 + 1)^2}.$$

Solution

The denominator contains only quadratic factors with no real zeros. We write

$$\frac{2x^3 - x^2 + 5x}{(x^2 + 1)^2} = \frac{B_1x + C_1}{x^2 + 1} + \frac{B_2x + C_2}{(x^2 + 1)^2} \tag{5}$$

and then clear fractions by multiplying both sides of equation (5) by $(x^2 + 1)^2$.

$$\begin{aligned} 2x^3 - x^2 + 5x &= (B_1x + C_1)(x^2 + 1) + B_2x + C_2 \\ &= B_1x^3 + C_1x^2 + (B_1 + B_2)x + C_1 + C_2 \end{aligned}$$

Comparing coefficients, we see that

$$\begin{aligned} B_1 &= 2, \\ C_1 &= -1, \\ B_1 + B_2 &= 5; \quad \text{so} \quad B_2 = 3; \\ C_1 + C_2 &= 0; \quad \text{so} \quad C_2 = 1. \end{aligned}$$

Thus,
$$\frac{2x^3 - x^2 + 5x}{(x^2 + 1)^2} = \frac{2x - 1}{x^2 + 1} + \frac{3x + 1}{(x^2 + 1)^2}.$$

■

Follow-up

Ask students to write the form of the partial fraction decomposition (without solving for the constants) for a more complicated expression, such as $\dfrac{x^2 - 3x}{(x^2 + 2x + 5)(x - 4)^4(x + 3)}$.

$$\left(\frac{Ax + B}{x^2 + 2x + 5} + \frac{C}{x - 4} + \frac{D}{(x - 4)^2} + \frac{E}{(x - 4)^3} + \frac{F}{(x - 4)^4} + \frac{G}{(x + 3)}\right)$$

Assignment Guide

Day 1: Ex. 1, 4, 5, 8–10, 12–14
Day 2: Ex. 16–22, 24–26

Cooperative Learning

Group Activity: Ex. 25–26

Notes on Exercises

Ex. 1–8 remind students that there are at least two ways to check their partial fraction decompositions.
Ex. 21–26 require division before the partial fractions can be found.

Ongoing Assessment

Self-Assessment: Ex. 3, 11, 15, 23
Embedded Assessment: Ex. 18, 24

REMARK
Try this technique in Example 3 to see which method you prefer.

Quick Review 5.5

In Exercises 1–4, find the quotient and remainder when $f(x)$ is divided by $g(x)$. Then express $f(x)/g(x)$ as the sum of a polynomial plus a rational function in which the degree of the numerator is less than that of the denominator.

1. $f(x) = x^3 - 3x^2 + 3x - 2;\ g(x) = x^2 + 1$

2. $f(x) = 2x^3 + 5x^2 - 2x - 2;\ g(x) = x^2 - 1$

3. $f(x) = x^5 - 7x^3 + 2x^2 + 12x + 1;\ g(x) = x^3 - 4x$

4. $f(x) = 2x^4 + 3x^3 + x^2 + 3x + 1;\ g(x) = x^3 + x$

In Exercises 5–8, use grouping to completely factor the polynomial.

5. $x^3 - 2x^2 + x - 2$

6. $x^3 + 3x^2 + 4x + 12$

7. $2x^3 + x^2 + 2x + 1$

8. $3x^3 - x^2 + 12x - 4$

In Exercises 9–14, determine the coefficients of $f(x)$ so that the given conditions are satisfied.

9. $f(x) = ax + b,\quad f(0) = 1, f(1) = 2$

10. $f(x) = ax + b,\quad f(0) = -3, f(-1) = 2$

11. $f(x) = ax^2 + bx + c,\quad f(0) = 2, f(1) = f(-1) = 5$

12. $f(x) = ax^2 + bx + c,\quad f(0) = 5, f(2) = f(-2) = -3$

13. $f(x) = ax^2 + bx + c,\quad f(2) = 1, f(5) = f(-1) = 6$

14. $f(x) = ax^2 + bx + c,\quad f(-3) = 4, f(1) = f(-7) = -2$

SECTION EXERCISES 5.5

In Exercises 1–4, find the partial fraction decomposition. Confirm your answer algebraically by combining the partial fractions.

1. $\dfrac{2}{(x-5)(x-3)}$

2. $\dfrac{4}{(x+3)(x+7)}$

3. $\dfrac{4}{x^2-1}$

4. $\dfrac{6}{x^2-9}$

In Exercises 5–8, find the partial fraction decomposition. Support your answer graphically.

5. $\dfrac{1}{x^2+2x}$

6. $\dfrac{-6}{x^2-3x}$

7. $\dfrac{-x+10}{x^2+x-12}$

8. $\dfrac{7x-7}{x^2-3x-10}$

In Exercises 9–20, find the partial fraction decomposition.

9. $\dfrac{x+17}{2x^2+5x-3}$

10. $\dfrac{4x-11}{2x^2-x-3}$

11. $\dfrac{2x^2+5}{(x^2+1)^2}$

12. $\dfrac{3x^2+4}{(x^2+1)^2}$

13. $\dfrac{x^2-x+2}{x^3-2x^2+x}$

14. $\dfrac{-6x+25}{x^3-6x^2+9x}$

15. $\dfrac{3x^2-4x+3}{x^3-3x^2}$

16. $\dfrac{5x^2+7x-4}{x^3+4x^2}$

17. $\dfrac{2x^3+4x-1}{(x^2+2)^2}$

18. $\dfrac{3x^3+6x-1}{(x^2+2)^2}$

19. $\dfrac{x^2+3x+2}{x^3-1}$

20. $\dfrac{2x^2-4x+3}{x^3+1}$

In Exercises 21–24, write the rational function in the form $q(x) + r(x)/h(x)$, where the degree of $r(x)$ is less than the degree of $h(x)$. Then find the partial fraction decomposition of $r(x)/h(x)$.

21. $\dfrac{2x^2+x+3}{x^2-1}$

22. $\dfrac{3x^2+2x}{x^2-4}$

23. $\dfrac{x^3-2}{x^2+x}$

24. $\dfrac{x^3+2}{x^2-x}$

EXTENDING THE IDEAS

In Exercises 25 and 26, solve the problem.

25. Let

$$f(x) = \frac{2x^3 - x^2 - 9x + 14}{x^2 - 4}.$$

a. Write $f(x)$ in the form $q(x) + r(x)/h(x)$, where the degree of $r(x)$ is less than the degree of $h(x)$.
b. Find the partial fraction decomposition of $r(x)/h(x)$.

c. Writing to Learn Explain how the graphs of the end behavior asymptote of f and the partial fractions of $r(x)/h(x)$ could be used to describe the graph of f.

26. Let

$$f(x) = \frac{2x^5 + 2x^2 + x + 3}{x^3 + x}.$$

a. Write $f(x)$ in the form $q(x) + r(x)/h(x)$ where the degree of $r(x)$ is less than the degree of $h(x)$.

b. Find the partial fraction decomposition of $r(x)/h(x)$.

c. Writing to Learn Explain how the graphs of the end behavior asymptote of f and the partial fractions of $r(x)/h(x)$ could be used to describe the graph of f.

CHAPTER 5 REVIEW

KEY TERMS

The number following each key term indicates the page of its introduction.

end behavior, 366
end behavior asymptote, 352, 366
extraneous solution, 376
horizontal asymptote, 351
partial fraction, 382
partial fraction decomposition, 382
rational function, 347
reciprocal function, 362
slant asymptote, 368
vertical asymptote, 349

REVIEW EXERCISES

In Exercises 1–6, state the domain of the function and identify the horizontal and vertical asymptotes (if they exist).

1. $f(x) = \dfrac{x^2 + 2x - 3}{x + 1}$

2. $f(x) = \dfrac{x^3 - 3x^2 + x + 2}{x - 3}$

3. $f(x) = \dfrac{x^2 + x + 1}{x^2 - 1}$

4. $f(x) = \dfrac{2x^2 + 7}{x^2 + x - 6}$

5. $f(x) = \dfrac{4x^3 + 1}{x^2 + 4}$

6. $f(x) = \dfrac{2x^3 - 1}{x^2 + 9}$

In Exercises 7–14, graph the function and confirm its behavior.

7. $f(x) = \dfrac{-x^3 + x + 1}{x^3 - 1}$

8. $g(x) = \dfrac{x^3 - x + 1}{x^3 + 1}$

9. $f(x) = \dfrac{x^2 - 4x + 5}{x + 3}$

10. $g(x) = \dfrac{x^2 - 3x - 7}{x + 3}$

11. $f(x) = \dfrac{x^3 + 1}{x^2 + 1}$

12. $g(x) = \dfrac{2x^3 - 3x + 1}{x^2 + 4}$

13. $f(x) = \dfrac{x^3 - 4x^2 + 3x + 2}{x - 3}$

14. $f(x) = \dfrac{-x^4 + 4x^2 - 4}{x^2 - 1}$

In Exercises 15–18, write a rational function with the given vertical asymptote and end behavior asymptote $g(x)$.

15. Vertical asymptote $x = 2$; $g(x) = x^2 + 2x$

16. Vertical asymptote $x = 3$; $g(x) = 2x + 1$

17. Vertical asymptotes $x = -1, x = 1$; $g(x) = x^2 + 1$

18. Vertical asymptotes $x = -3, x = 2$; $g(x) = x^3 - x$

In Exercises 19 and 20, describe how the graph of f can be obtained by transforming the graph of $y = 1/x$. Specify the order of the transformations. Name any vertical or horizontal asymptotes. Support your work graphically.

19. $f(x) = \dfrac{-x + 7}{x - 5}$

20. $f(x) = \dfrac{3x + 5}{x + 2}$

In Exercises 21–24, solve the equation or inequality algebraically. Support your answer graphically.

21. $2x + \dfrac{12}{x} = 11$

22. $\dfrac{x}{x + 2} + \dfrac{5}{x - 3} = \dfrac{25}{x^2 - x - 6}$

23. $\dfrac{x + 3}{x^2 - 4} \geq 0$

24. $\dfrac{x^2 - 7}{x^2 - x - 6} < 1$

In Exercises 25–30, find the partial fraction decomposition.

25. $\dfrac{3x - 2}{x^2 - 3x - 4}$

26. $\dfrac{x - 16}{x^2 + x - 2}$

27. $\dfrac{3x + 5}{x^3 + 4x^2 + 5x + 2}$

28. $\dfrac{3(3 + 2x + x^2)}{x^3 + 3x^2 - 4}$

29. $\dfrac{5x^2 - x - 2}{x^3 + x^2 + x + 1}$

30. $\dfrac{-x^2 - 5x + 2}{x^3 + 2x^2 + 4x + 8}$

In Exercises 31–45, solve the problem.

31. *Breaking Even* Midtown Sporting Goods has determined that it needs to sell its soccer shinguards for $5.25 a pair in order to be competitive. It costs $4.32 to produce each pair of shinguards, and the weekly overhead cost is $4000.

   a. Express the average cost that includes the overhead of producing one shinguard as a function of the number x of shinguards produced each week.
   b. Solve algebraically to find the number of shinguards that must be sold each week to make $8000 in profit. Support your work graphically.

32. *Deer Population* The number of deer P at any time t (in years) in a federal game reserve is given by

$$P(t) = \frac{800 + 640t}{20 + 0.8t}.$$

   a. Find the number of deer when t is 15, 70, and 100.
   b. Find the horizontal asymptote of the graph of $y = P(t)$.
   c. According to the model, what is the largest possible deer population?

33. *Minimizing Perimeter* Consider all rectangles with an area of 375 ft^2. Let x be the length of one side of one such rectangle.

   a. Express the perimeter P as a function of x.
   b. Find the dimensions of a rectangle that has the least perimeter. What is the least perimeter?

34. *Page Design* Wellsley Publishing Co. wants to design a page that has a 0.5-in. left border, 1.25-in. top and bottom borders, and a 1-in. right border surrounding 40 in.^{2} of print material. Let x be the length of one side of the print material.

   a. Express the area of the page as a function of x.
   b. Find the dimensions of the page that has the least area. What is the least area?

35. *Designing a Swimming Pool* Sunsport Recreation, Inc., wants to build a rectangular swimming pool with a pool surface of 1200 ft^2. They are required to have a walk of uniform width 2.5 ft surrounding the pool. Let x be the length of one side of the swimming pool.

   a. Express the area covered by the pool and sidewalk as a function of x.
   b. Find the dimensions that cover the least area. What is the least area?

36. *Designing Rectangles* Consider all rectangles with an area of 150 m^2. Let x be the length of one side of one such rectangle.

   a. Express the perimeter P as a function of x.
   b. Find the dimensions of a rectangle that has perimeter 60 m. Solve algebraically, and support your answer graphically.
   c. Find the possible dimensions for a rectangle if the perimeter is to be less than 70 m.

37. **Writing to Learn** Find the horizontal asymptote of the function that represents the concentration of acid in Example 6, Section 5.1. What does this asymptote represent?

38. **Group Learning Activity** *Work in groups of two.* Compare the functions $f(x) = \dfrac{x^2 - 4}{x + 2}$ and $g(x) = x - 2$.

   a. Are the domains equal?
   b. Does f have a vertical asymptote? Explain.
   c. Explain why the graphs appear to be identical.
   d. Are the functions identical?

39. *Resistors* The total electrical resistance R of two resistors connected in parallel with resistances R_1 and R_2 is given by

$$\frac{1}{R} = \frac{1}{R_1} + \frac{1}{R_2}.$$

The total resistance is 1.2 ohms. Let $x = R_1$.

   a. Express the second resistance R_2 as a function of x.
   b. Find R_2 if R_1 is 3 ohms.

40. *Acid Mixture* Suppose that x ounces of distilled water are added to 50 oz of pure acid.

   a. Express the concentration $C(x)$ of the new mixture as a function of x.
   b. Use a graph to determine how much distilled water should be added to the pure acid to produce a new solution that is less than 60% acid.
   c. Solve part b algebraically.

41. *Industrial Design* Johnson Cannery will pack peaches in 1-L cylindrical cans. Let x be the radius of the base of the can in centimeters.

   a. Express the surface area S of the can as a function of x.
   b. Find the radius and height of the can if the surface area is 900 cm^2.
   c. What dimensions are possible for the can if the surface area is to be less than 900 cm^2?

42. *Industrial Design* Gilman Construction is hired to build a rectangular tank with a square base and no top. The tank is to hold 1000 ft^3 of water. Let x be a length of the base.

a. Express the outside surface area S of the tank as a function of x.

b. Find the length, width, and height of the tank if the outside surface area is 600 ft^2.

c. What dimensions are possible for the tank if the outside surface area is to be less than 600 ft^2?

43. *Industrial Design* Bryan Construction is hired to build a rectangular tank with a square base and no top. The tank is to hold 600 ft^3 of water. Let x be the side length of the base.

a. Express the outside surface area S of the tank as a function of x.

b. Find the length, width, and height of the tank if the outside surface area is 500 ft^2.

c. What dimensions are possible for the tank if the outside surface area is to be less than 500 ft^2?

d. What is the least possible outside surface area for the tank?

44. *Swimming Pool Drainage* Drains A and B are used to empty a swimming pool. It takes 3 h to drain a swimming pool with both drains open. Let t be the time it takes for drain A alone to empty the pool.

a. Express the part D of the drainage that is done in 1 h by drain B as a function of t.

b. Use a graphical method to find the time it takes for drain B alone to empty the pool if it takes drain A alone 5 h to empty the pool. Confirm your answer algebraically.

45. *Time-Rate Problem* Sarah rode her bike 10 mi from her home in Springfield, Illinois, and then completed a 35-mi trip by car from Springfield to Decatur. Assume the average rate of the car was 40 mph faster than the average rate of the bike.

a. Express the total time t required to complete the 45-mi trip (bike and car) as a function of the rate x of the car.

b. Use a graphical method to find the rate of the car if the total time of the trip was 1 h. Confirm your answer algebraically.

SOPHIE GERMAIN

Sophie Germain (1776–1831) was a middle-class woman from a liberal, educated family in France at the time of the French Revolution. She fought fiercely against the prejudices of the time to become an accomplished mathematician. Starting at age 13, Germain studied secretly at night by candlelight. Although not allowed to attend the university, she obtained lecture notes for many courses and submitted comments to instructors under a male pseudonym.

Even though Germain was denied the stimulation of learning alongside other scientists, she did exchange letters with several prominent mathematicians, including Carl Friedrich Gauss and Joseph-Louis Lagrange. Germain is best known for her work in the areas of number theory and the theory of elasticity. She competed or collaborated with many of the eminent mathematicians and physicists of her day, and she was proud to work at the frontier of nineteenth-century science.

chapter 6

6.1 Angles and Their Measures

6.2 Trigonometric Functions of Acute Angles

6.3 Extending Trigonometric Functions

6.4 Graphs of Sine and Cosine Functions

6.5 Graphs of Other Trigonometric Functions

6.6 Advanced Trigonometric Graphs

6.7 Inverse Trigonometric Functions

6.8 Applications of Trigonometric Functions

Objective

Students will be able to draw angles in standard position using a rotational approach, convert between radians and degrees, and find arc lengths.

Motivate

Ask students how many degrees the second hand of a clock travels in 45 seconds. **(270°)**

TRIGONOMETRY

BIBLIOGRAPHY

For students: *16–19 Mathematics: Modelling with Circular Motion,* The School Mathematics Project. Cambridge University Press, 1990. Available through Dale Seymour Publications.

For teachers: *A History of Mathematics,* Carl B. Boyer (Second Edition revised by Uta C. Merzbach). John Wiley & Sons Inc., 1991.

Videos: *M! Project Mathematics! Sines and Cosines, Part 1,* Tom Apostol. NCTM, 1992.

6.1 ANGLES AND THEIR MEASURES

Introduction • Angles • Degree Measure • Radian Measure • Applications of Radian Measure • Arc Length Formula • Applications of the Arc Length Formula

Introduction

So far in this book, you have studied several types of functions: polynomial, exponential, logarithmic, and rational. Now you will learn about another type—the trigonometric functions. Trigonometric functions are used to model periodic phenomena, from alternating electrical current to the zig-zags of the Dow Jones industrial average.

The Greek mathematician Hipparchus founded trigonometry to explore his interests in astronomy and geography. He lived in Rhodes and Alexandria, dying about 125 B.C. The trigonometry of Hipparchus is what we now call spherical trigonometry. The word *trigonometry* is from the Greek and means "triangle measure." In 1220 Leonardo of Pisa first presented the methods of trigonometry contained in this textbook—plane trigonometry. Throughout the Renaissance, the study of trigonometry was pursued to support work in surveying, astronomy, navigation, and calendar reckoning. With the rise of the function concept in the 18th century, the ideas of trigonometry came to be viewed both in terms of functions and as an outgrowth of geometry.

In this text we develop both perspectives.

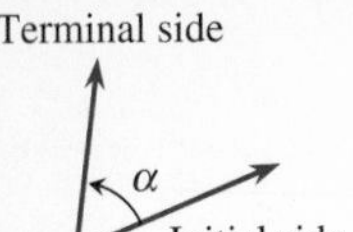

Figure 6.1

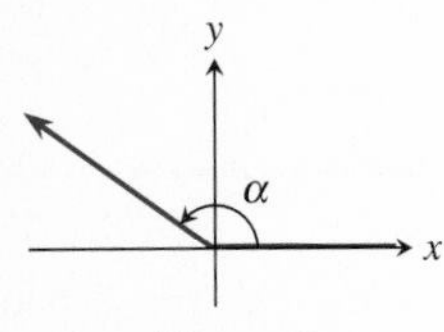

A positive angle (counterclockwise)

(a)

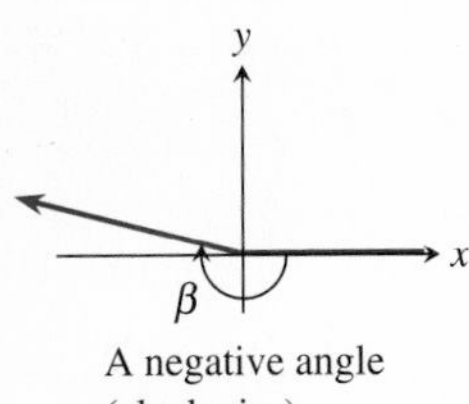

A negative angle (clockwise)

(b)

Figure 6.2

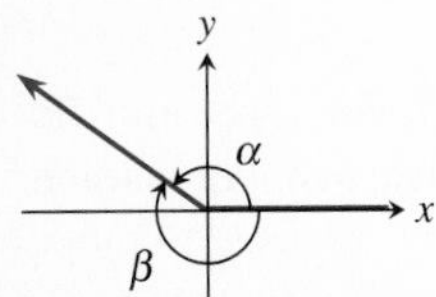

Positive and negative coterminal angle

(a)

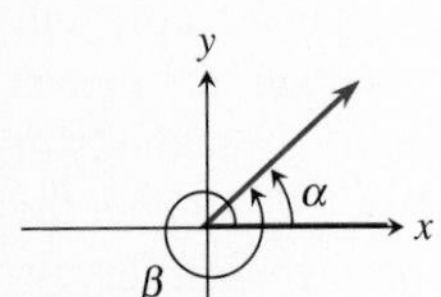

Two positive coterminal angles

(b)

Figure 6.4

Angles

In geometry an **angle** is defined as the union of two rays with a common endpoint called the **vertex.** Trigonometry takes a more dynamic view by describing an angle in terms of a rotating ray. The beginning position of the ray, called the **initial side,** is rotated about its endpoint, also called the vertex. The final position is called the **terminal side.** The curved **directional arrow** indicates the path the rotating ray follows from the initial side to the terminal side. (See Figure 6.1.)

An angle is in **standard position** when its vertex is at the origin of a rectangular coordinate system and the rest of its initial side is the positive x-axis. Figure 6.2 shows two angles in standard position. **Positive angles** are generated by counterclockwise rotation (Figure 6.2a) and **negative angles** by clockwise rotation (Figure 6.2b).

An angle or angle measure is often named with a single Greek letter, such as α (alpha), β (beta), γ (gamma), or θ (theta). We also use the symbol $\angle$ (angle), so that $\angle A$ is read "angle A."

Degree Measure

A **measure of an angle** is a number that describes the amount of rotation from the initial side to the terminal side of the angle. A commonly used unit of angle measure is the *degree,* represented by the symbol °. One **degree** (1°) is 1/360th of a complete revolution about the vertex of an angle.

Figure 6.3a shows the most frequently used positive angles in standard position. Figure 6.3b shows the quadrants of the coordinate plane in which the terminal sides may be found. The terminal side of an angle θ in standard position is in quadrant I if $0° < \theta < 90°$, in quadrant II if $90° < \theta < 180°$, in quadrant III if $180° < \theta < 270°$, and in quadrant IV if $270° < \theta < 360°$.

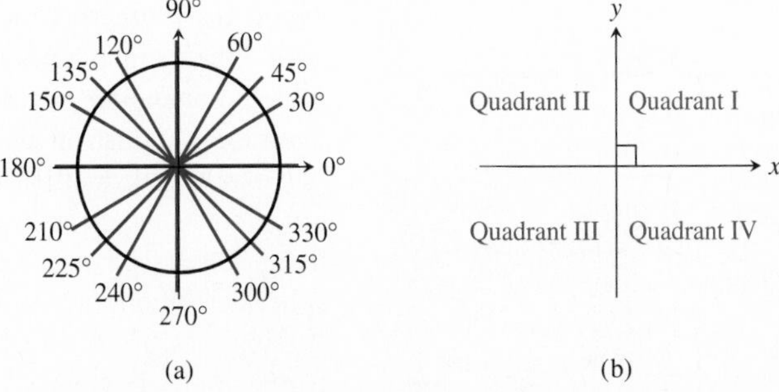

Figure 6.3 This diagram of commonly occurring angles reminds us of the face of a directional compass or of a tool or machine dial that uses angle measurements.

Two angles with the same initial side and terminal side are **coterminal angles.** (See Figure 6.4.) Coterminal angles differ by a multiple of 360°, that is, by $\ldots, -720°, -360°, 0°, 360°, 720°, \ldots$.

EXAMPLE 1 Finding Coterminal Angles

Find and draw a positive angle and a negative angle that are coterminal with the given angle.

a. 30° **b.** −150°

Solution

There are many possible solutions. Here are two for each angle.

a. Add 360°: $30° + 360° = 390°$
Subtract 360°: $30° - 360° = -330°$
Figure 6.5 shows these two angles, which are coterminal with the 30° angle.

b. Add 360°: $-150° + 360° = 210°$
Subtract 360°: $-150° - 360° = -510°$
We leave it to you to draw the two coterminal angles. ■

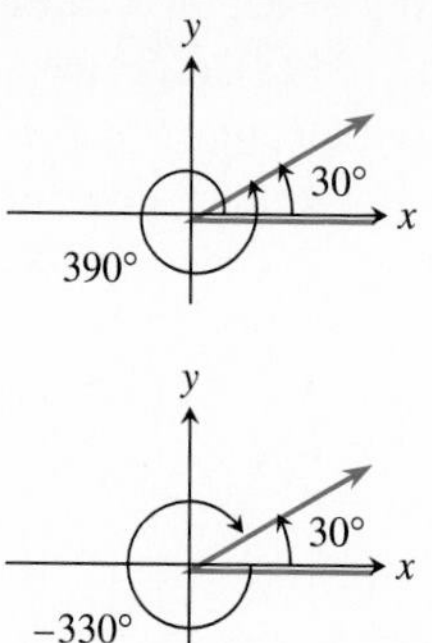

Figure 6.5 To distinguish two coterminal angles, we show two distinct directional arrows from the initial side to the terminal side along with their angle measures.

Two positive angles α and β are **complements** if their sum is 90°, and they are **supplements** if their sum is 180°. (See Figure 6.6.)

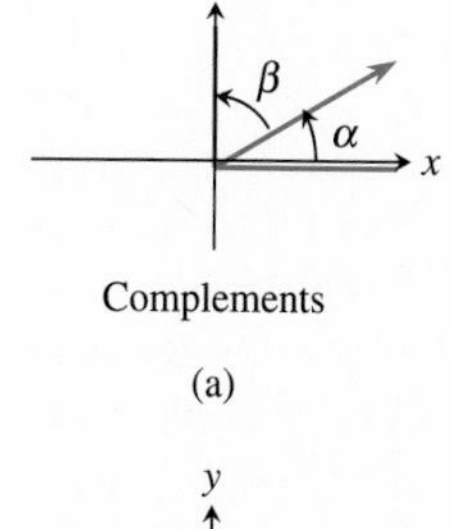

Complements

(a)

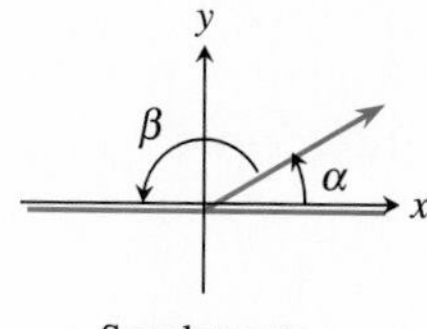

Supplements

(b)

Figure 6.6

EXAMPLE 2 Finding Complements and Supplements

Find the complement and the supplement of the angle, or state that none exists.

a. 48° **b.** 135° **c.** −15°

Solution

a. The complement of 48° is $90° - 48° = 42°$.
The supplement of 48° is $180° - 48° = 132°$.

b. 135° has no complement because it is greater than 90°, and complementary angles must both be positive.
The supplement of 135° is $180° - 135° = 45°$.

c. −15° has neither a complement nor a supplement because an angle must be positive to have a complement or a supplement. ■

Machine dials like those on surveying instruments, express fractional parts of a degree in *minutes* (denoted by ′), and *seconds* (denoted by ″). A **minute** is 1/60th of a degree, and a **second** is 1/60th of a minute. An angle of 83 degrees, 14 minutes, 38 seconds is represented as 83° 14′ 38″. In symbols we have

$$1' = 1 \text{ minute} = \frac{1}{60}(1°) = \left(\frac{1}{60}\right)^{\circ},$$

$$1'' = 1 \text{ second} = \frac{1}{60}(1') = \frac{1}{3600}(1°) = \left(\frac{1}{3600}\right)^{\circ}.$$

Teaching Note

It is common for geometry textbooks to define *angle* as a measure strictly between 0 and 180 degrees. The difference between that definition and the one used here should be pointed out. Focus on the basic idea of rotation and how angle measure can be viewed in terms of rotation.

Teaching Note

Many new terms are introduced in this section. Make sure students understand the vocabulary.

Calculators usually express fractional parts of a degree in decimal form. Example 3 illustrates how to convert degree-minute-second (DMS) angle measure to decimal-degree angle measure.

EXAMPLE 3 Converting from DMS to Decimal Form

Change to decimal form.

a. $83° \, 12' \, 18''$ **b.** $19° \, 24' \, 58''$

Solution

a.

$$83° \; 12' \; 18'' = 83° + \left(\frac{12}{60}\right)^{\circ} + \left(\frac{18}{3600}\right)^{\circ}$$

$$= 83° + 0.2° + 0.005°$$

$$= 83.205°$$

b. Figure 6.7 shows this conversion on a grapher.

$$19° \; 24' \; 58'' = 19.416 \cdots°$$

Figure 6.7 Some graphers have a built-in conversion routine. Consult your owner's manual.

Teaching Note

Students may need help with the radian as a unit of angle measure. You may wish to have students explore the basis for this new measure of angles. In Section 6.3, students will need to think of real numbers as the domain elements of trigonometric functions and to see real numbers as lengths of arcs on a circle.

Radian Measure

The *radian* (abbreviated "rad"), another unit of angle measure, is ordinarily used in scientific applications of trigonometry and for studying the trigonometric functions we mentioned at the start of this section. It is also important because it connects linear measure and angle measure.

To define radian measure, we use a circle, an angle whose vertex is at the circle's center (a **central angle**), and the arc of the circle **intercepted** by the sides of the angle.

Definition: Radian Measure

Suppose θ, a central angle of a circle of radius r, intercepts an arc of length s. Then the **radian measure** of θ is $\frac{s}{r}$.

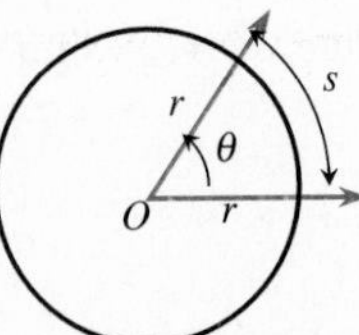

REMARK

Notice that radian measure is free of units because it is a ratio of quantities with the same units,

$$\frac{\text{arc length (linear units)}}{\text{radius (linear units)}},$$

so the units drop out.

Because all circles are similar, the ratio s/r is the same no matter what r is. Thus, the radian measure of central angle θ is independent of the circle used

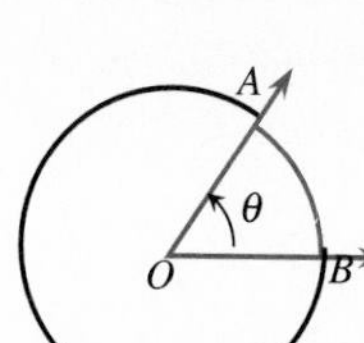

Figure 6.8 A central angle θ and its intercepted arc AB.

to define it. Any circle centered at the vertex of θ could be used. The **unit circle,** with radius $r = 1$, is particularly nice to use because $\theta = s/1 = s$. This associates the angle measure with an arc length that is numerically equal to s.

In the unit circle of Figure 6.8, the central angle θ and its intercepted arc AB both have measure $\theta = s$. Now, imagine θ (and s) increasing in size; that is, imagine the terminal side of θ rotating counterclockwise around O. When it has traveled 360° around to the x-axis, we know that

$$360° = \theta = s = \text{circumference} = 2\pi \text{ (radians)}.$$

From this we can conclude that

$$180° = \pi \text{ rad},$$

which gives us the following conversion equations.

Teaching Note

A piece of string and circular graph paper may be used to demonstrate radian measure. Measure a radius length of string and then show, by wrapping the string around the circle, that it takes a little more than 6 radius lengths to go all the way around the circle.

Degree-Radian Conversion Equations

$$\theta° = \frac{180°}{\pi \text{ rad}}(\theta \text{ radians}) \qquad \theta \text{ radians} = \frac{\pi \text{ rad}}{180°}(\theta°).$$

Because π rad $= 180°$, it follows that

$$\frac{\pi \text{ rad}}{180°} = \frac{180°}{\pi \text{ rad}} = 1.$$

Fractions equal to 1, such as these, are **unit ratios.** From science classes you may be familiar with **dimensional analysis,** a strategy of paying attention to units used to make conversions between measurement systems. Watch how these ratios are used in Example 4.

Teaching Note

Many calculators have built-in functions for converting between degrees and radians. Be aware that students have access to these functions and may not be performing angle-measure conversions as you would expect.

EXAMPLE 4 Making Degree-Radian Conversions

Convert degree measure to radians and radian measure to degrees.

a. 450° **b.** −12° **c.** $5\pi/4$ rad **d.** 7 rad

Solution

a. $450° = 450°\left(\frac{\pi \text{ rad}}{180°}\right) = \frac{5\pi}{2} \text{ rad}$

b. $-12° = -12°\left(\frac{\pi \text{ rad}}{180°}\right) = -\frac{\pi}{15} \text{ rad}$

c. $\frac{5\pi}{4} \text{ rad} = \frac{5\pi}{4}\left(\frac{180°}{\pi \text{ rad}}\right) = 225°$

d. $7 \text{ rad} = 7\left(\frac{180°}{\pi \text{ rad}}\right) \approx 401.07°$ ■

NOTE

Radian measures are often given exactly in terms of π rather than as decimal approximations.

Remembering that 2π rad $= 360°$ or that π rad $= 180°$ will help you remember these common radian sizes. Memorizing the following *benchmarks* for degree-radian equivalence will allow you to focus your attention on deeper matters. (See Figure 6.9.)

NOTE

Here are two other conversions that are often needed:

$$1° = 1° \cdot \frac{\pi \text{ rad}}{180°} = \frac{\pi \text{ rad}}{180}$$

$$\approx 0.0175 \text{ rad}$$

$$1 \text{ rad} = 1 \text{ rad} \cdot \frac{180°}{\pi \text{ rad}} \approx 57.30°$$

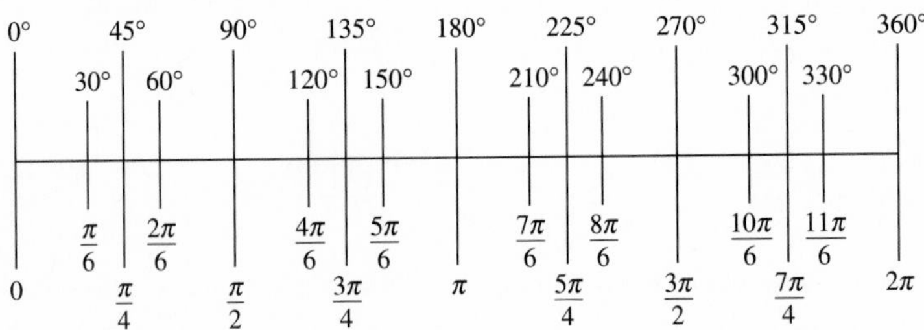

Figure 6.9 Benchmark angles in radians and degrees.

Applications of Radian Measure

Many scientific and mathematical applications are more effectively handled using radian measure. However, historically the world has been measured in degrees rather than radians. Consequently, many industrial and business situations continue to use degree measure.

In navigation, the **course** or **bearing** of an object is sometimes given as the measure of the clockwise angle that the **line of travel** (the terminal side) makes with due north (the initial side). By convention, the angle measure θ is positive even though the angle rotation is clockwise. The line of travel in Figure 6.10 has the bearing 155°.

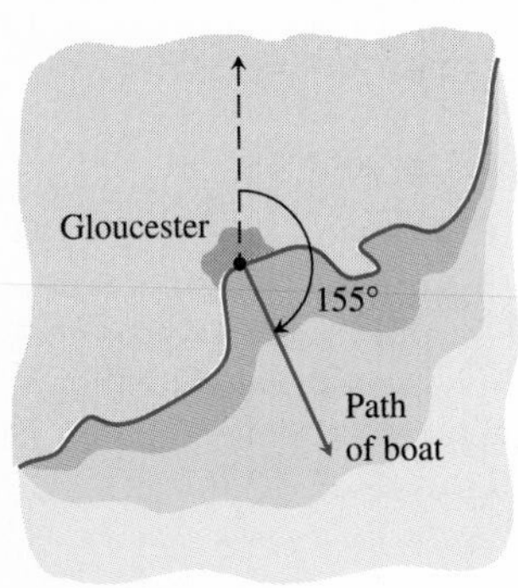

Figure 6.10

■ **EXAMPLE 5 APPLICATION: Using Angles in Navigation**

The captain of a fishing boat steers a bearing of 155° out of Gloucester. Sketch the situation.

Solution

Draw a ray from Gloucester in the due-north direction. Rotate the ray 155° clockwise. (See Figure 6.10.) The boat travels in the direction of this second ray. We say that the bearing of the boat is 155°. ■

Teaching Note

Frequently, as in Example 6, the word *radian* is omitted when giving the measure of an angle in radians. For example, an angle of 2 radians will be reported as an angle whose measure is 2.

Arc Length Formula

The radian measure definition, $\theta = s/r$, is particularly useful for finding the length s of an intercepted arc.

Follow-up

Ask students whether a point on the earth's equator has a higher angular speed than a point near the north pole. **(No)**

Definition: The Arc Length Formula

If θ is a central angle, measured in radians, in a circle of radius r, the directed arc length s of the intercepted arc is

$$s = r\theta.$$

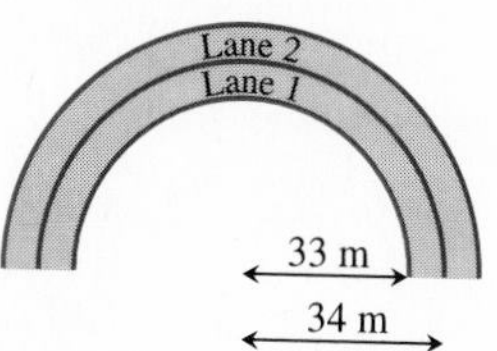

Figure 6.11 A figure that models Example 6.

Applications of the Arc Length Formula

■ EXAMPLE 6 APPLICATION: Designing a Running Track

The running lanes at the Emery Sears track at Bluffton College are 1 m wide. The inside radius of lane 1 is 33 m and the inside radius of lane 2 is 34 m. How much longer is lane 2 than lane 1 around one turn? (See Figure 6.11.)

Solution

Each lane is a semicircle with central angle $\theta = \pi$ and length $s = r\theta = r\pi$. The difference in their lengths, therefore, is $34\pi - 33\pi = \pi \approx 3.14$ m.

Lane 2 is about 3.14 m longer than lane 1. ■

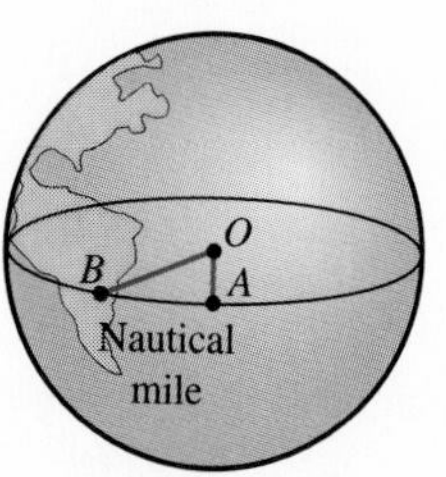

Figure 6.12 Although the earth is not a perfect sphere, its diameter is, on the average, 7912.18 statute miles. A nautical mile is 1′ of earth's circumference at the equator.

A **nautical mile** (naut mi) is the length of 1′ of arc along the earth's equator. Figure 6.12 shows, though not to scale, a central angle AOB of the earth that measures 1′ (1/60 of a degree). It intercepts an arc 1 naut mi long. The arc length formula allows us to convert between nautical miles and **statute miles** (stat mi), the familiar "land mile" of 5280 ft.

■ EXAMPLE 7 APPLICATION: Converting to Nautical Miles

Megan McCarty, a pilot for Western airlines, frequently pilots flights from Boston to San Francisco, a distance of 2698 stat mi. Captain McCarty's calculations of flight time are based on nautical miles. How many nautical miles is it from Boston to San Francisco?

Solution

Because $\left(\frac{1}{60}\right)^\circ = \frac{\pi}{180}\left(\frac{1}{60}\right) = \frac{\pi}{10{,}800}$ rad and the earth's radius ≈ 3956 stat mi, we have by the arc length formula, $s = r\theta$,

$$1 \text{ naut mi} = 3956 \frac{\pi}{10{,}800} \text{ stat mi} = 1.150\cdots \text{ stat mi}.$$

Assignment Guide

Day 1: Ex. 3–60, multiples of 3
Day 2: Ex. 61, 62, 65, 67, 69, 72, 76, 77, 81, 85, 88, 89, 90

Cooperative Learning

Group Activity: Ex. 78–79

Notes on Exercises

Ex. 61 requires students to use arc length to approximate linear distance. Ex. 62–64, 78–79, and 91 are application problems involving angular speed.

Thus, 1 stat mi $= \dfrac{10{,}800}{3956\pi}$ naut mi $= 0.868 \cdots$ naut mi, and, the distance from Boston to San Francisco, 2698 stat mi $= \dfrac{2698 \cdot 10{,}800}{3956\pi}$ naut mi ≈ 2345 naut mi.

Interpret

The distance from Boston to San Francisco is about 2345 naut mi. ■

Arc length equals road length

Figure 6.13

The arc length formula allows us to convert between a vehicle's **linear speed** (highway speed) v and the **angular speed** ω (omega) $= \theta/t$ of one of the vehicle's wheels:

$$v = \frac{s}{t} = \frac{r\theta}{t} = r \cdot \frac{\theta}{t} = r\omega$$

See Figure 6.13. Often in such a situation, v is in miles per hour, r is in inches, and ω is in radians per second, so attention must be paid to the conversion equation and units in use.

Ongoing Assessment

Self-Assessment: Ex. 13, 17, 35, 41, 49, 63

Embedded Assessment: Ex. 88

NOTE

Using dimensional analysis,

1 ft/sec

$$= 1 \frac{\text{ft}}{\text{sec}} \cdot \frac{1 \text{ mi}}{5280 \text{ ft}} \cdot \frac{60 \text{ sec}}{1 \text{ min}} \cdot \frac{60 \text{ min}}{1 \text{ hr}}$$

≈ 0.68 mph

■ EXAMPLE 8 APPLICATION: Using Angular Speed

Albert Juarez's truck has wheels 36 in. in diameter. If the wheels are rotating at 630 rpm (revolutions per minute) find the truck's speed in miles per hour.

Solution

First we convert the angular speed ω of a wheel from revolutions per minute to radians per second:

$$\omega = 630 \text{ rpm} = \frac{630 \text{ rev}}{\text{min}} \cdot \frac{2\pi \text{ rad}}{1 \text{ rev}} \cdot \frac{1 \text{ min}}{60 \text{ sec}} = 21\pi \text{ rad/sec}$$

The radius of a wheel is 18 in. or 1.5 ft, so the linear speed v of the truck is

$$v = r\omega = 1.5(21\pi) \text{ ft/sec} \approx 67.47 \text{ mph}.$$

Interpret

Juarez's truck is traveling at about 67 mph. ■

Quick Review 6.1

In Exercises 1 and 2, find the circumference of the circle with given radius r. State the correct unit.

1. $r = 2.5$ in. **2.** $r = 4.6$ m

In Exercises 3 and 4, find the radius of the circle with the given circumference C. State the correct unit.

3. $C = 12$ m **4.** $C = 8$ ft

In Exercises 5 and 6, evaluate the expression for the given values of the variables. State the correct unit.

5. $s = r\theta$

a. $r = 9.9$ ft, $\theta = 4.8$ rad

b. $r = 4.1$ km, $\theta = 9.7$ rad

6. $v = r\omega$

a. $r = 8.7$ m, $\omega = 3.0$ rad/sec

b. $r = 6.2$ ft, $\omega = 1.3$ rad/sec

In Exercises 7–10, solve the equation. State the correct unit.

7. $2.6 \text{ cm} = 5.9r$

8. $1.6 = \dfrac{93.9 \text{ ft}}{r}$

9. $88.1 = \dfrac{s}{40.4 \text{ km}}$

10. $\dfrac{s}{22.4 \text{ ft}} = \dfrac{1}{2}$

In Exercises 11–14, convert from miles per hour to feet per second or from feet per second to miles per hour.

11. 60 mph

12. 45 mph

13. 8.8 ft/sec

14. 132 ft/sec

In Exercises 15 and 16, state whether the angle is positive or negative.

15.  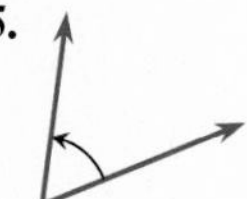

16. 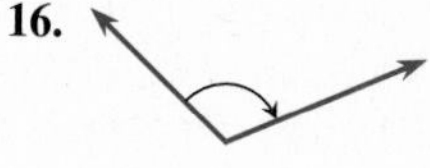

SECTION EXERCISES 6.1

In Exercises 1–8, find a positive angle and a negative angle that are coterminal with the given angle.

1. 40°

2. −30°

3. −157°

4. 457°

5. $-\dfrac{11\pi}{6}$

6. $-\dfrac{3\pi}{4}$

7. $\dfrac{7\pi}{6}$

8. $\dfrac{3\pi}{2}$

In Exercises 9–14, find and draw a positive angle and a negative angle that are coterminal with the angle.

9. 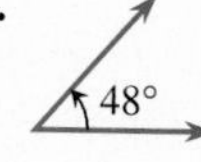

10. 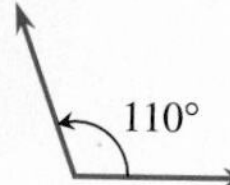

11. −15°

12. 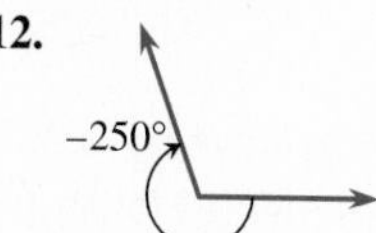

13. $\frac{13\pi}{6}$

14. 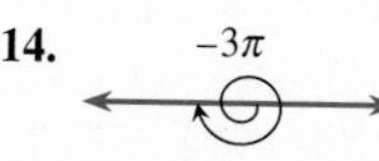

In Exercises 15–24, find the complement and the supplement, or state that none exists.

15. 35°

16. 23°

17. 68°

18. 12°

19. −88°

20. 198°

21. $\dfrac{\pi}{6}$

22. $\dfrac{\pi}{3}$

23. $\dfrac{5\pi}{4}$

24. $-\dfrac{\pi}{6}$

In Exercises 25–28, change to decimal form.

25. 23° 12′

26. 35° 24′

27. 118° 44′ 15″

28. 48° 30′ 36″

In Exercises 29–32, change to degrees, minutes, and seconds (DMS).

29. 21.2°

30. 49.7°

31. 118.32°

32. 99.37°

In Exercises 33–38, use the arc length formula to find the missing information.

	s	r	θ
33.	?	2 in.	25 rad
34.	?	1 cm	70 rad
35.	1.5 ft	?	$\pi/4$ rad
36.	2.5 cm	?	$\pi/3$ rad
37.	3 m	1 m	?
38.	4 in.	7 in.	?

In Exercises 39 and 40, a central angle intercepts arcs s_1 and s_2 on two concentric circles with radii r_1 and r_2 respectively. Find the missing information.

	θ	r_1	s_1	r_2	s_2
39.	?	11 cm	9 cm	44 cm	?
40.	?	8 km	36 km	?	72 km

In Exercises 41–48, express in radians.

41. $60°$ **42.** $90°$

43. $270°$ **44.** $-150°$

45. $71.72°$ **46.** $11.83°$

47. $-61.4°$ **48.** $-75.5°$

In Exercises 49–56, express in degrees.

49. $\pi/6$ **50.** $\pi/4$

51. $3\pi/2$ **52.** $-4\pi/3$

53. -6 **54.** 5

55. 1.9 **56.** -5.4

Exercises 57–60 refer to the 16 compass bearings shown. North corresponds to an angle of 0°, and other angles are measured clockwise from north.

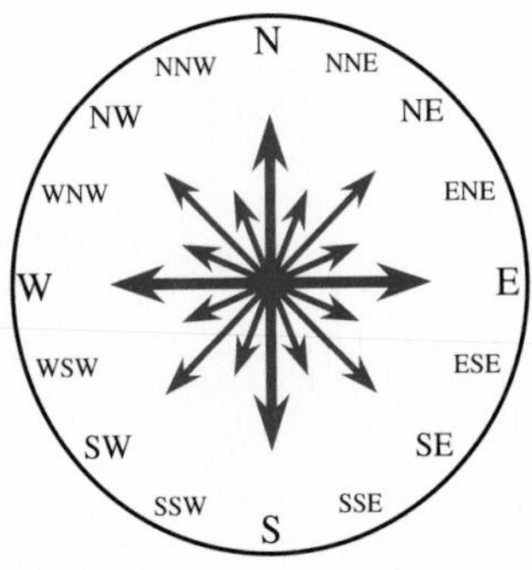

For Exercises 57–60

57. *Compass Reading* Find the angle in degrees that describes the compass bearing.

a. NE (northeast)
b. NNE (north-northeast)
c. WSW (west-southwest)

58. *Compass Reading* Find the angle in degrees that describes the compass bearing.

a. SSW (south-southwest)
b. WNW (west-northwest)
c. NNW (north-northwest)

59. *Compass Reading* Which compass direction is closest to a bearing of 121°?

60. *Compass Reading* Which compass direction is closest to a bearing of 219°?

In Exercises 61–68, solve the problem.

61. *Navigation* Two Coast Guard patrol boats leave Cape May at the same time. One travels with a bearing of 42° 30′ and the other with a bearing of 52° 12′. If they travel at the same speed, approximately how far apart will they be when they are 25 stat mi from Cape May?

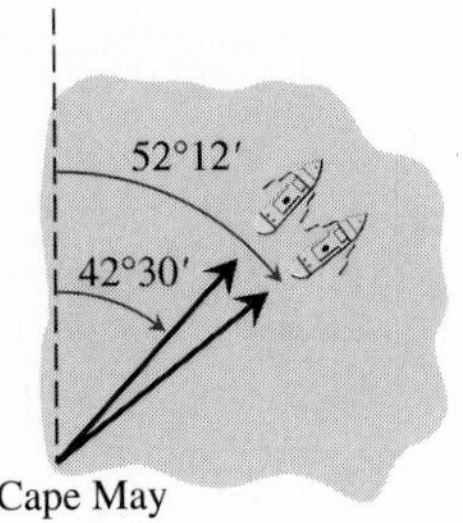

For Exercise 61

62. *Automobile Design* The wheel (including the tire) of a sports car under development by one of the Big Three auto companies has a 11-in. radius. What would the car's speed be in miles per hour if its wheels are turning at 800 rpm?

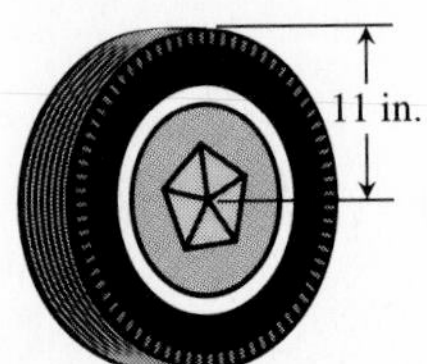

For Exercise 62

63. *Bicycle Racing* Cathy Nguyen races on a bicycle with 13-in.-radius wheels. When she is traveling at a speed of 44 ft/sec, how many revolutions per minute are her wheels making?

64. *Tool Design* A radial arm saw has a circular cutting blade with a diameter of 10 in. It spins at 2000 rpm. If there are 12

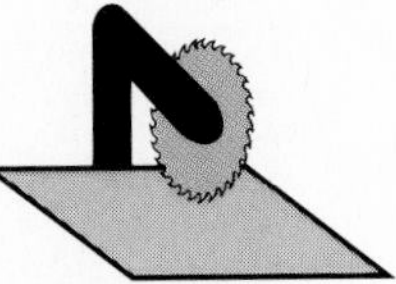

For Exercise 64

cutting teeth per inch on the cutting blade, how many teeth cross the cutting surface each second?

65. *Navigation* Sketch a ship on the given course.

a. 35° **b.** 128° **c.** 310°

66. *Navigation* The captain of the tourist boat *Julia* out of Oak Harbor follows a 38° course for 2 mi and then changes to a 47° course for the next 4 mi. Draw a sketch of this trip.

67. *Navigation* Points A and B are 257 naut mi apart. How far apart are A and B in statute miles?

68. *Navigation* Points C and D are 895 stat mi apart. How far apart are C and D in nautical miles?

In Exercises 69–74, the point is on the terminal side of an angle θ in standard position where $0° \le \theta < 360°$. Find θ in both degrees and radians.

69. $(-1, 0)$ **70.** $(0, 5)$

71. $(3, 3)$ **72.** $(-2, 2)$

73. $(5, -5)$ **74.** $(10, 0)$

In Exercises 75–79, solve the problem.

75. *Designing a Sports Complex* Example 6 describes how lanes 1 and 2 compare in length around one turn of a track. Find the difference in the lengths of these lanes around one turn.

a. Lanes 5 and 6 **b.** Lanes 1 and 6

76. *Mechanical Engineering* A simple pulley used to lift heavy objects is positioned 10 ft above ground level. For the given radius r and rotation θ, determine the height to which the object is lifted.

a. $r = 4$ in., $\theta = 720°$ **b.** $r = 2$ ft, $\theta = 180°$

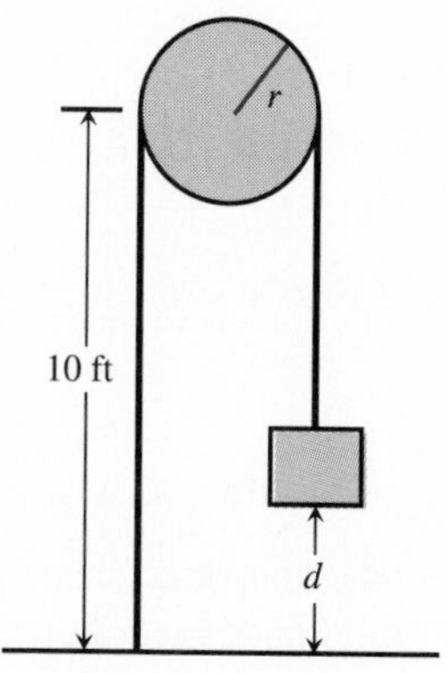

For Exercise 76

77. *Foucault Pendulum* In 1851 the French physicist Jean Foucault used a pendulum to demonstrate the earth's rotation. There are now over 30 Foucault pendulum displays in the United States. The Foucault pendulum at the Smithsonian Institution in Washington, D.C., consists of a large brass ball suspended by a thin 52-ft cable. If the ball swings through an angle of 1°, how far does it travel?

78. *Air Conditioning Belt* The belt on an automobile air conditioner connects metal wheels with radii $r = 4$ cm and $R = 7$ cm. The angular speed of the larger wheel is 120 rpm.

a. What is the angular speed of the larger wheel in radians per second?

b. What is the linear speed of the belt in centimeters per second?

c. What is the angular speed of the smaller wheel in radians per second?

79. *Ship's Propeller* The propellers of the *Amazon Paradise* have a radius of 1.2 m. At full throttle the propellers turn at 135 rpm.

a. What is the angular speed of a propeller blade in radians per second?

b. What is the linear speed at the tip of the propeller blade in meters per second?

c. What is the linear speed (in meters per second) of a point on a blade halfway between the center of the propeller and the tip of the blade?

EXTENDING THE IDEAS

Table 6.1 shows the latitude-longitude locations of U.S. cities. Latitude is measured from the equator. Longitude is measured

Table 6.1 Latitudes and Longitudes of Several U.S. Cities

City	Latitude	Longitude
Atlanta	33° 45′	84° 23′
Chicago	41° 49′	87° 37′
Detroit	42° 22′	83° 10′
Los Angeles	35° 12′	118° 02′
Miami	25° 45′	80° 11′
Minneapolis	44° 58′	93° 15′
New Orleans	30° 00′	90° 05′
New York	40° 40′	73° 58′
San Diego	32° 43′	117° 10′
San Francisco	37° 45′	122° 26′
Seattle	47° 36′	122° 20′

from the Greenwich meridian that passes north-south through London, England.

In Exercises 80–83, find the difference in longitude between the cities.

80. Atlanta and San Francisco

81. New York and San Diego

82. Minneapolis and Chicago

83. Miami and Seattle

In Exercises 84–87, assume that the two cities have the same longitude (that is, assume that one is directly north of the other), and find the distance between them in nautical miles. Use 7912 miles for the diameter of the earth.

84. San Diego and Los Angeles

85. Seattle and San Francisco

86. New Orleans and Minneapolis

87. Detroit and Atlanta

In Exercises 88–91, refer to the figure for help in solving the problem.

88. *Area of a Sector* A *sector of a circle* (shaded in the figure) is the region bounded by a central angle of the circle and its intercepted arc. Use the fact that the areas of sectors are proportional to their central angles to prove that

$$A = \frac{1}{2}r^2\theta,$$

where r is the radius and θ is in radians.

89. *Area of a Sector* Use the formula $A = \frac{1}{2}r^2\theta$ to determine the area of the sector with the given radius r and central angle θ.

a. $r = 5.9$ ft, $\theta = \pi/5$ **b.** $r = 1.6$ km, $\theta = 3.7$

90. *Navigation* Control tower A is 60 mi east of control tower B. At a certain time an airplane is on bearings of 340° from tower A and 37° from tower B. Use a drawing to model the exact location of the airplane.

91. *Bicycle Racing* Ben Scheltz's bike wheels are 28 in. in diameter, and for high gear the pedal sprocket is 9 in. in diameter and the wheel sprocket is 3 in. in diameter. Find the angular speed in radians per second of the wheel and of both sprockets when Ben reaches his peak racing speed of 66 ft/sec in high gear.

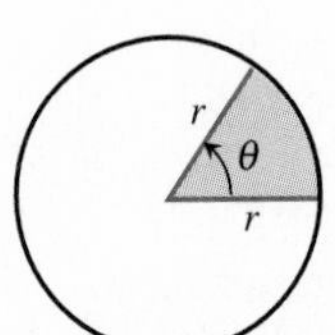

For Exercises 88–89

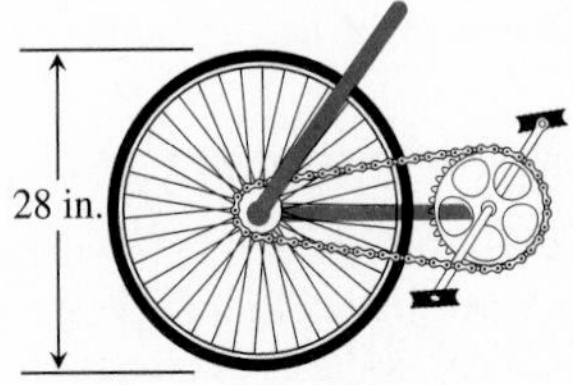

For Exercise 91

6.2 TRIGONOMETRIC FUNCTIONS OF ACUTE ANGLES

Right Triangle Trigonometry • Trigonometric Identities • Evaluating Trigonometric Functions with a Calculator • Solving Right Triangles • Applications Involving Right Triangles

Objective

Students will be able to define the six trigonometric functions using the lengths of the sides of a right triangle.

Motivate

Review the notions of similarity and congruence for triangles. Discuss sufficient conditions for similarity and congruence of right triangles.

Right Triangle Trigonometry

Geometric figures are **similar** if they have the same shape even though they may have different sizes. Having the same shape means that the angles of one are congruent to the angles of the other and their sides are proportional. Similarity is the basis for many applications, including scale drawings, maps, and **right triangle trigonometry.**

It is a theorem of geometry that two triangles are similar if the angles of one are congruent to the angles of the other. For two right triangles we need only know that an acute angle of one is congruent to an angle of the other for the triangles to be similar. Thus one acute angle determines an entire family of similar right triangles.

Similarly, a single acute angle θ of a right triangle determines six distinct ratios of side lengths. Each ratio, therefore, is a function of θ as θ takes on values from 0° to 90°, or from 0 to $\pi/2$ rad.

The six ratios of side lengths in a right triangle are the six *trigonometric functions* of the acute angle θ. (See Figure 6.14.) The names of these trigonometric functions (with abbreviations) are

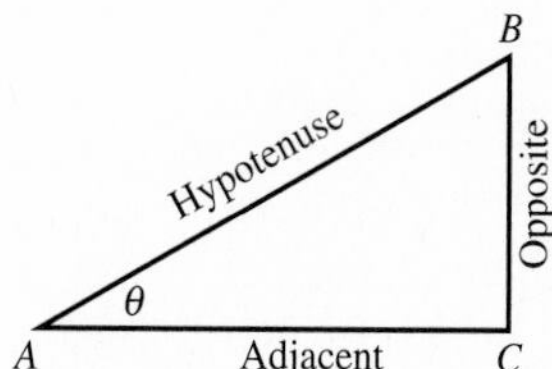

Figure 6.14

sine (sin)	*cosecant* (csc)
cosine (cos)	*secant* (sec)
tangent (tan)	*cotangent* (cot)

The following definitions use the abbreviations *opp, adj,* and *hyp* to refer to the side opposite θ, the side adjacent to θ, and the hypotenuse, respectively, as shown in Figure 6.14. For these functions we streamline the notation a bit and write, for example, $\sin\theta$ instead of $\sin(\theta)$.

Definition: Trigonometric Functions

Let θ be an acute angle in the right $\triangle ABC$.

$$\sin\theta = \frac{\text{opp}}{\text{hyp}} \qquad \cos\theta = \frac{\text{adj}}{\text{hyp}} \qquad \tan\theta = \frac{\text{opp}}{\text{adj}}$$

$$\csc\theta = \frac{\text{hyp}}{\text{opp}} \qquad \sec\theta = \frac{\text{hyp}}{\text{adj}} \qquad \cot\theta = \frac{\text{adj}}{\text{opp}}$$

Exploration Extensions

Label angle ACB as angle γ. Compute the six trigonometric function values for γ. Compare your results with the results for α.

EXPLORATION Trigonometric Functions

1. Refer to Figure 6.15. Draw an acute angle on a sheet of paper using a straightedge. Label the angle α and its vertex A. Measure distances of 4 cm and 7 cm from A along the side adjacent to A, labeling the points obtained B and B', respectively. Using a protractor, carefully draw rays at B and B' that are perpendicular to ray AB and intersect the other side of α at C and C', respectively.
2. Measure the sides opposite α in $\triangle ABC$ and $\triangle AB'C'$ each to the nearest tenth of a centimeter.
3. Compute all six trigonometric function values of α for each triangle. Compare the corresponding values for each triangle.

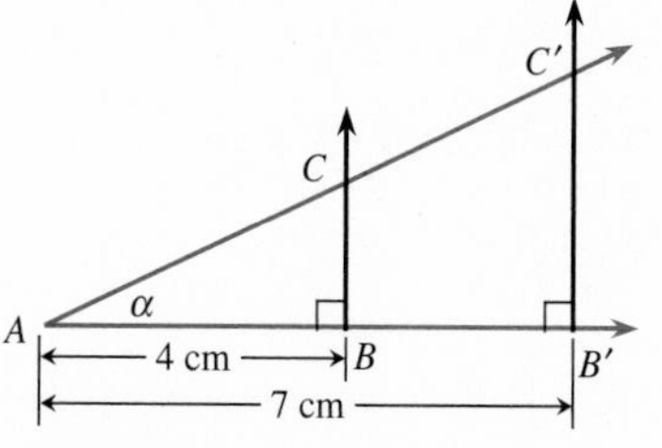

Figure 6.15

Describe in Your Own Words

Based on these activities what would you conjecture for trigonometric functions of acute angles in right triangles?

The results of the Exploration should suggest that indeed the acute angle yields the same six trigonometric function values regardless of the right triangle used.

In Example 1 we find the trigonometric values of one of the benchmark acute angles shown in Figure 6.9 of Section 6.1.

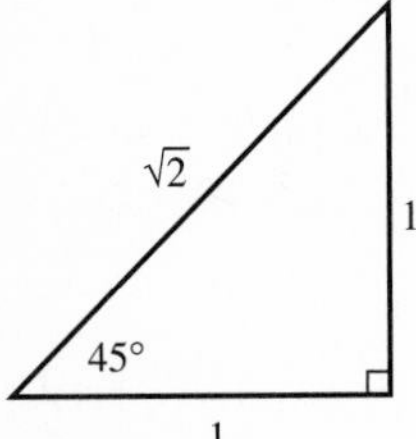

Figure 6.16

■ EXAMPLE 1 Evaluating Trigonometric Functions of 45°

Find the values of the trigonometric functions for a 45° angle.

Solution

A 45° angle occurs in a **45-45 right triangle,** one that is isosceles. (See Figure 6.16.) When the length of a leg is 1, the length of the hypotenuse is $\sqrt{1 + 1} = \sqrt{2}$.

> **NOTE**
>
> The "simplified" radical form for the sine and cosine of 45° is
>
> $$\sin 45° = \cos 45° = \frac{\sqrt{2}}{2}.$$

$$\sin 45° = \frac{1}{\sqrt{2}} = 0.707\ldots \qquad \csc 45° = \frac{\sqrt{2}}{1} = 1.414\ldots$$

$$\cos 45° = \frac{1}{\sqrt{2}} = 0.707\ldots \qquad \sec 45° = \frac{\sqrt{2}}{1} = 1.414\ldots$$

$$\tan 45° = \frac{1}{1} = 1 \qquad \cot 45° = \frac{1}{1} = 1$$

■

Whenever two sides of a right triangle are known, the third side can be found using the Pythagorean theorem, and all six trigonometric ratios can be found for either of the acute angles. This procedure is illustrated in Example 2 with two more benchmark acute angles.

■ EXAMPLE 2 Evaluating Trigonometric Functions of 30° and 60°

Evaluate the sine, cosine, and tangent for the angle.

a. 30° **b.** 60°

Solution

In Figure 6.17, $\triangle ABC$ is equilateral and hence equiangular. Thus, each of its angles has measure 60°. Next, we construct BD perpendicular to AC. It follows that BD also bisects $\angle ABC$ and AC and, therefore, $\angle CBD = 30°$. If $BC = 2$, then $CD = 1$ and $BD = \sqrt{4 - 1} = \sqrt{3}$ by the Pythagorean theorem.

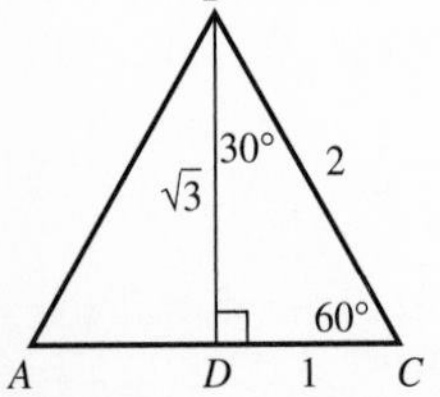

Figure 6.17 $\triangle ABC$ is an equilateral triangle.

> **Alert**
>
> Students will need to pay attention to whether their calculators are in degree or radian mode. A lack of attention to this issue will create computational and graphical results that students will not be able to explain.

a. For $\theta = 30°$, opp $= 1$, adj $= \sqrt{3}$, and hyp $= 2$.

$$\sin 30° = \frac{\text{opp}}{\text{hyp}} = \frac{1}{2} = 0.5 \qquad \cos 30° = \frac{\text{adj}}{\text{hyp}} = \frac{\sqrt{3}}{2} = 0.866\cdots$$

$$\tan 30° = \frac{\text{opp}}{\text{adj}} = \frac{1}{\sqrt{3}} = 0.577\cdots$$

b. For $\theta = 60°$, opp $= \sqrt{3}$, adj $= 1$, and hyp $= 2$.

$$\sin 60° = \frac{\text{opp}}{\text{hyp}} = \frac{\sqrt{3}}{2} = 0.866\cdots \qquad \cos 60° = \frac{\text{adj}}{\text{hyp}} = \frac{1}{2} = 0.5$$

$$\tan 60° = \frac{\text{opp}}{\text{adj}} = \frac{\sqrt{3}}{1} = 1.732\cdots$$

■

We summarize the trigonometric functions for the special angles 30°, 45°, and 60°. You should know these values by memory or be able to recall them with a diagram whenever needed.

Sine, Cosine, and Tangent of 30°, 45°, 60°

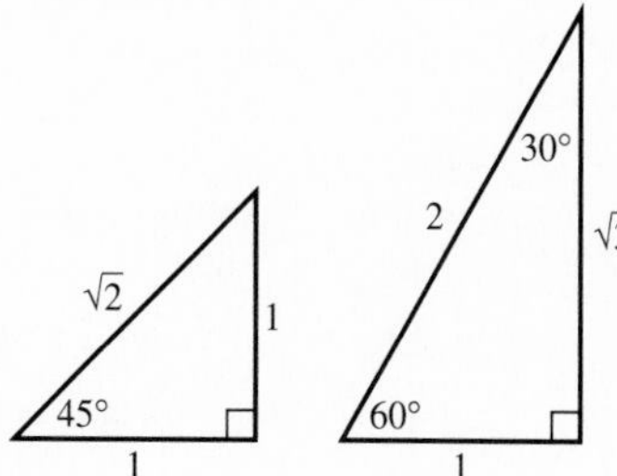

Angle θ	Sin θ	Cos θ	Tan θ
$30° \left(\frac{\pi}{6} \text{ rad}\right)$	$\frac{1}{2} = 0.5$	$\frac{\sqrt{3}}{2} = 0.866\cdots$	$\frac{\sqrt{3}}{3} = 0.577\cdots$
$45° \left(\frac{\pi}{4} \text{ rad}\right)$	$\frac{\sqrt{2}}{2} = 0.707\cdots$	$\frac{\sqrt{2}}{2} = 0.707\cdots$	$\frac{1}{1} = 1$
$60° \left(\frac{\pi}{3} \text{ rad}\right)$	$\frac{\sqrt{3}}{2} = 0.866\cdots$	$\frac{1}{2} = 0.5$	$\frac{\sqrt{3}}{1} = 1.732\cdots$

Example 3 illustrates that whenever one trigonometric ratio is known, all the others can be found.

■ **EXAMPLE 3 Finding All Trigonometric Ratios**

Let θ be an acute angle such that $\sin\theta = \frac{5}{6}$. Evaluate the other trigonometric functions of θ.

Solution

Using the information $\sin\theta = \frac{5}{6}$, label a right triangle as shown in Figure 6.18. From the Pythagorean theorem it follows that

$$b^2 + 5^2 = 6^2,$$
$$b^2 = 6^2 - 5^2 = 11,$$
$$b = \sqrt{11}.$$

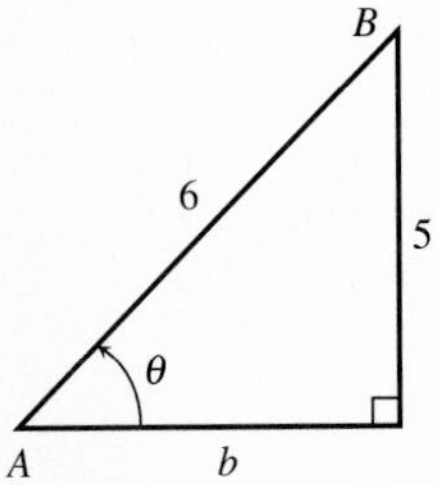

Figure 6.18

$$\sin\theta = \frac{5}{6} = 0.833\cdots \qquad \csc\theta = \frac{6}{5} = 1.2$$

$$\cos\theta = \frac{\sqrt{11}}{6} = 0.552\cdots \qquad \sec\theta = \frac{6}{\sqrt{11}} = 1.809\cdots$$

$$\tan\theta = \frac{5}{\sqrt{11}} = 1.507\cdots \qquad \cot\theta = \frac{\sqrt{11}}{5} = 0.663\cdots$$

■

Trigonometric Identities

An **identity** is an equation that is true for all values of the variable for which each side of the equation is defined. The following trigonometric identities are among the most frequently used relationships.

Fundamental Trigonometric Identities

Reciprocal Identities

$$\sin\theta = \frac{1}{\csc\theta} \qquad \cos\theta = \frac{1}{\sec\theta} \qquad \tan\theta = \frac{1}{\cot\theta}$$

$$\csc\theta = \frac{1}{\sin\theta} \qquad \sec\theta = \frac{1}{\cos\theta} \qquad \cot\theta = \frac{1}{\tan\theta}$$

Quotient Identities

$$\tan\theta = \frac{\sin\theta}{\cos\theta} \qquad \cot\theta = \frac{\cos\theta}{\sin\theta}$$

Pythagorean Identities

$$\sin^2\theta + \cos^2\theta = 1 \qquad \tan^2\theta + 1 = \sec^2\theta \qquad 1 + \cot^2\theta = \csc^2\theta$$

NOTE

We use $\sin^2\theta$ to mean $(\sin\theta)^2$. We follow a similar agreement for the other five trigonometric functions.

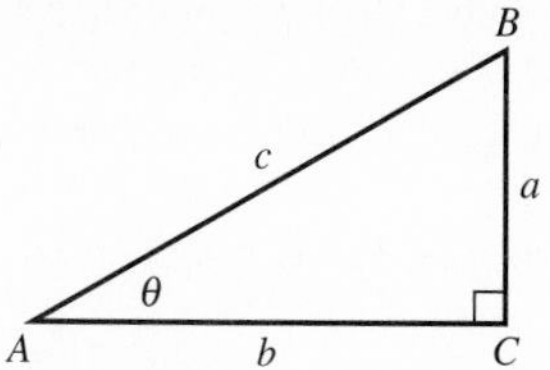

Figure 6.19

■ EXAMPLE 4 Confirming an Identity

Show that if θ is an acute angle, $\sin^2\theta + \cos^2\theta = 1$.

Solution

Sketch and label a right triangle as shown in Figure 6.19.

$$\begin{aligned}\sin^2\theta + \cos^2\theta &= \left(\frac{a}{c}\right)^2 + \left(\frac{b}{c}\right)^2\\ &= \frac{a^2+b^2}{c^2} = \frac{c^2}{c^2} \quad \text{Pythagorean theorem}\\ &= 1\end{aligned}$$

■

Degree mode

sin 25
-.1323517501

Radian mode

Figure 6.20

Evaluating Trigonometric Functions with a Calculator

Historically, trigonometric function values were printed in table form. Now the values are more easily found with calculators. Most calculators have keys for sine, cosine, and tangent. We must also be careful that we have the calculator set in the desired mode—degrees or radians. Figure 6.20 shows that the sine of 25° does not equal the sine of 25 rad.

■ EXAMPLE 5 Finding Trigonometric Functions with a Calculator

Use a calculator to obtain a table of values for sine, cosine, and tangent that includes the values of the special angles 30°, 45°, and 60°.

Solution

Figure 6.21 shows the values of sin, cos, and tan beginning with 0° and incrementing by 15°. Next to the figure we see the sine, cosine, and tangent of 30°, 45°, and 60° found earlier.

Angle θ	Sin θ	Cos θ	Tan θ
30°	$\frac{1}{2} = 0.5$	$\frac{\sqrt{3}}{2} = 0.866\cdots$	$\frac{\sqrt{3}}{3} = 0.577\cdots$
45°	$\frac{\sqrt{2}}{2} = 0.707\cdots$	$\frac{\sqrt{2}}{2} = 0.707\cdots$	$\frac{1}{1} = 1$
60°	$\frac{\sqrt{3}}{2} = 0.866\cdots$	$\frac{1}{2} = 0.5$	$\frac{\sqrt{3}}{1} = 1.732\cdots$

■

X	sin X	cos X	tan X
0	0	1	0
15	.25882	.96593	.26795
30	.5	.86603	.57735
45	.70711	.70711	1
60	.86603	.5	1.7321
75	.96593	.25882	3.7321
90	1	0	ERROR

X=0

Figure 6.21 Trigonometric function values in *degree mode.*

Notice that this special angles table is merely a subset of the grapher table in Figure 6.21. This table also shows what the trigonometric function values of 0° and 90° will be.

Teaching Note

Some attention should be paid to showing students efficient keying sequences for solving right triangles.

Solving Right Triangles

The ancient Greeks established that, given a right triangle, if two sides, or one side and one acute angle, are known, then the other sides and angles can be determined. Finding one or more unknown *parts* in this fashion is known as **solving a right triangle.**

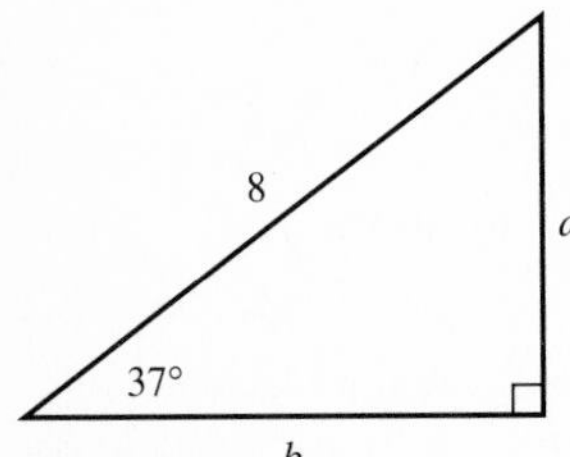

Figure 6.22

EXAMPLE 6 Solving a Right Triangle

A right triangle with a hypotenuse of 8 includes a 37° angle. (See Figure 6.22.) Find the lengths of the other sides.

Solution

We use the definitions of sine and cosine:

$$\sin 37° = \frac{a}{8} \qquad \cos 37° = \frac{b}{8}$$

$$a = 8 \sin 37° \qquad b = 8 \cos 37°$$

$$\approx 4.81 \qquad \approx 6.39$$

Support Numerically

By the Pythagorean theorem, $a^2 + b^2 = 8^2$. Using $a = 4.81$ and $b = 6.39$, we get

$$\sqrt{a^2 + b^2} = \sqrt{4.81^2 + 6.39^2}$$

$$= 7.998 \cdots \approx 8,$$

which gives us no reason to doubt the values we found for a and b. ■

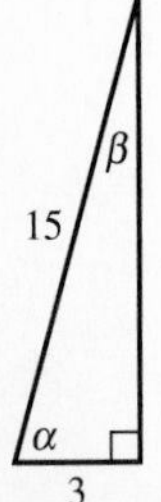

Figure 6.23

EXAMPLE 7 Solving a Right Triangle

The hypotenuse and one side of a right triangle are 15 and 3, respectively. (See Figure 6.23.) Find the two acute angles α and β.

Solution

Solve Numerically

Because the adjacent side of α and the hypotenuse are known parts, we know the value

$$\cos \alpha = \frac{3}{15} = 0.2.$$

Figure 6.23 suggests that the value of α is between 45° and 90°. We use the table-building feature of a grapher to *zoom in numerically* on the value of α. See Figure 6.24.

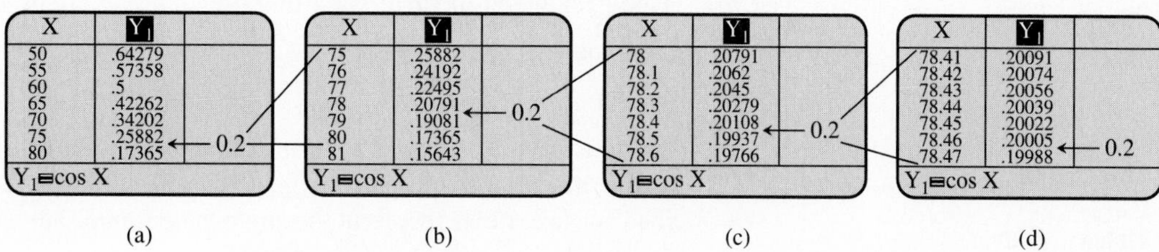

Figure 6.24 Zooming in numerically to find α when $\cos \alpha = 0.2$.

We find that $\alpha \approx 78.46°$.

The other acute angle β is the complement of α, so

$$\beta = 90° - \alpha \approx 90° - 78.46° = 11.54°.$$

■

In Section 6.7 we introduce a quicker way to solve the equation $\cos \alpha = 0.2$. If you already know how to solve for α on your calculator by finding $\cos^{-1} 0.2$, feel free to do so.

Applications Involving Right Triangles

Often a length that cannot be measured directly can be found indirectly by solving a right triangle.

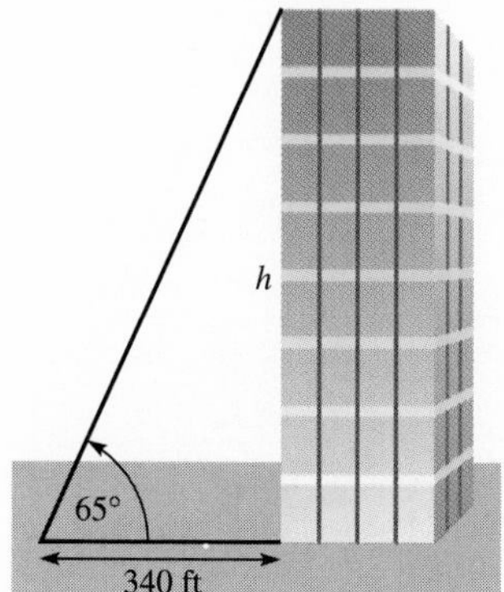

Figure 6.25

■ EXAMPLE 8 APPLICATION: Finding Height

From a point 340 ft away from the base of the Peachtree Center Plaza in Atlanta, Georgia, the angle of elevation of the top is 65°. (See Figure 6.25.) Find the height h of the building.

Solution

$$\tan 65° = \frac{h}{340} \qquad \tan = \frac{\text{opp}}{\text{adj}}$$

$$h = 340 \tan 65°$$

$$= 729.132 \cdots$$

The height of the building is approximately 729 ft. ■

■ EXAMPLE 9 APPLICATION: Finding Angle of Elevation

In one 8-mi stretch of Interstate 70 west of Denver, the highway rises from approximately 6200 ft above sea level to 8700 ft above sea level. Find the average angle of inclination for this stretch of the highway.

Solution

Model

Assume that the road rises at a constant angle of inclination so that the situation can be modeled as a right triangle ABC with the hypotenuse representing the road surface. Let α represent the angle to be found. See Figure 6.26.

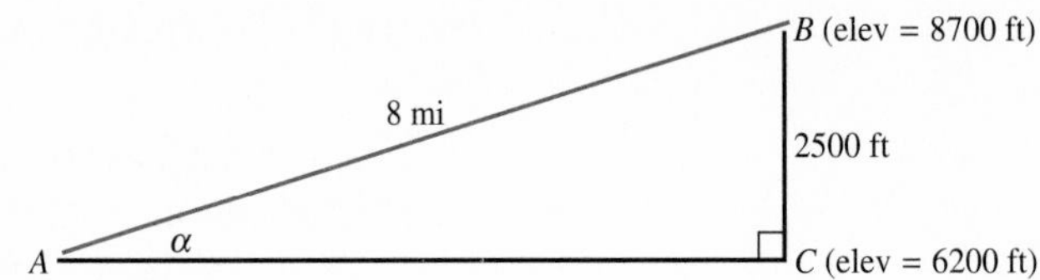

Figure 6.26

Solve Numerically

$$\sin \alpha = \frac{(8700 - 6200)\text{ ft}}{8\text{ mi}} \qquad \sin \alpha = \frac{\text{opp}}{\text{hyp}}$$

$$\sin \alpha = \frac{2500\text{ ft}}{8\text{ mi}}\left(\frac{1\text{ mi}}{5280\text{ ft}}\right) \qquad \text{Freeing of units}$$

$$= \frac{2500}{8 \cdot 5280}$$

$$= 0.05918 \cdots$$

Figure 6.26 suggests that α lies between 0° and 45°. Zooming in numerically (see Figure 6.27) we find that $\alpha \approx 3.39°$.

Interpret

The angle of inclination averages about 3.39°. ■

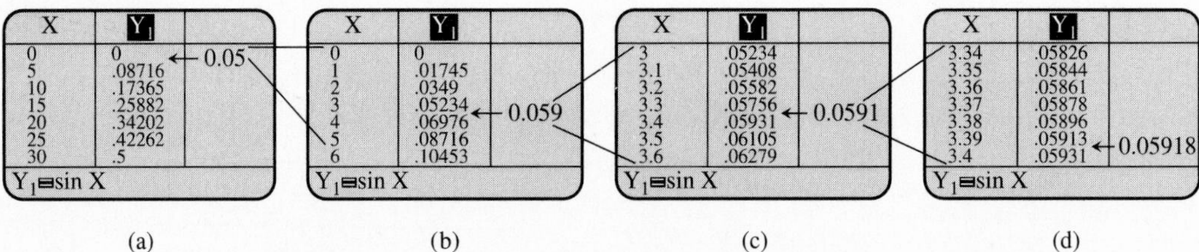

(a)

X	Y_1
0	0
5	.08716
10	.17365
15	.25882
20	.34202
25	.42262
30	.5

Y_1=sin X

(b)

X	Y_1
0	0
1	.01745
2	.0349
3	.05234
4	.06976
5	.08716
6	.10453

Y_1=sin X

(c)

X	Y_1
3	.05234
3.1	.05408
3.2	.05582
3.3	.05756
3.4	.05931
3.5	.06105
3.6	.06279

Y_1=sin X

(d)

X	Y_1
3.34	.05826
3.35	.05844
3.36	.05861
3.37	.05878
3.38	.05896
3.39	.05913
3.4	.05931

Y_1=sin X

Figure 6.27 Zooming in numerically to find α when $\sin \alpha = 0.05918 \cdots$.

Follow-up

Ask students to explain why the identities $\csc \theta = \frac{1}{\sin \theta}$, $\tan \theta = \frac{\sin \theta}{\cos \theta}$, and $\sin^2 \theta + \cos^2 \theta = 1$ must be true for any acute angle θ.

Assignment Guide

Day 1: Ex. 3–69, multiples of 3
Day 2: Ex. 73, 75, 78, 81, 82, 86, 89, 90, 95, 98, 101

Cooperative Learning

Group Activity: Ex. 101, 102

Notes on Exercises

Ex. 25–44 give students practice in using the trigonometric functions on a calculator.
Ex. 51–58 anticipate the concept of inverse trigonometric functions.
Ex. 81–86 are application problems involving right triangles. Students may find Ex. 82 and 84 especially challenging since no illustration is provided.

Ongoing Assessment

Self-Assessment: Ex. 11, 23, 55, 79, 85
Embedded Assessment: Ex. 48, 54, 82

Quick Review 6.2

In Exercises 1–4, use the Pythagorean theorem to solve for x.

1.

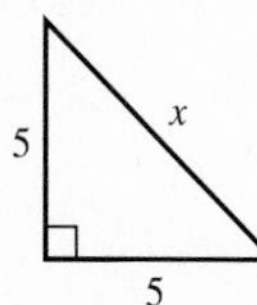

2.

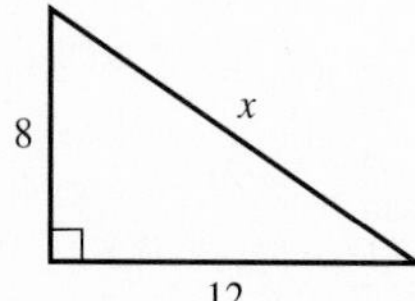

3.

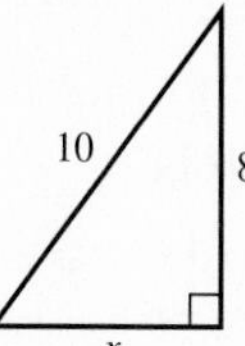

4.

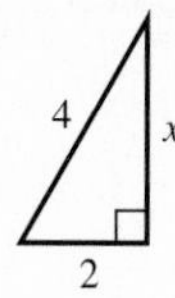

In Exercises 5–8, convert units.

5. 8.3 mi to feet

6. 8.4 ft to inches

7. 940 ft to miles

8. 760 in. to feet

In Exercises 9–12, solve the equation. State the correct unit.

9. $0.388 = \dfrac{a}{20.4\text{ km}}$

10. $1.72 = \dfrac{23.9\text{ ft}}{b}$

11. $\dfrac{2.4\text{ in.}}{31.6\text{ in.}} = \dfrac{\alpha}{13.3}$

12. $\dfrac{5.9}{\beta} = \dfrac{8.66\text{ cm}}{6.15\text{ cm}}$

In Exercises 13 and 14, simplify the radical. (See Section 3 of the Prerequisite Chapter.)

13. $\dfrac{3}{\sqrt{7}}$

14. $\dfrac{1}{\sqrt{13}}$

In Exercises 15 and 16, give the length of the side opposite θ and the side adjacent to θ.

15.

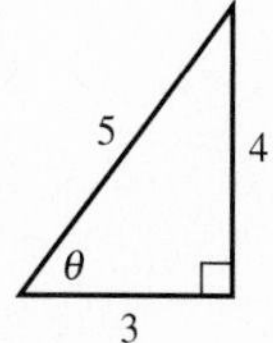

16.

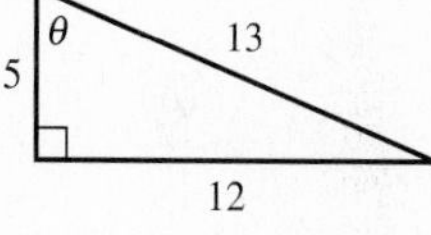

SECTION EXERCISES 6.2

In Exercises 1–4, evaluate all six trigonometric functions of the angle θ.

1.

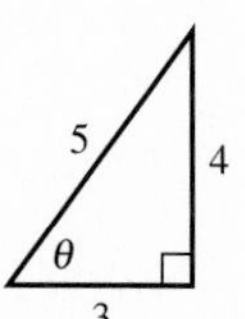

2.

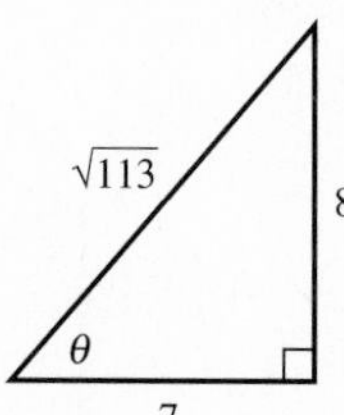

3.

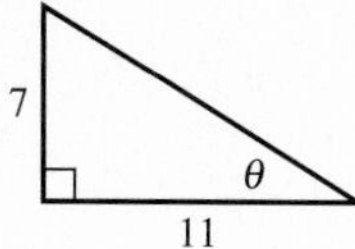

4.

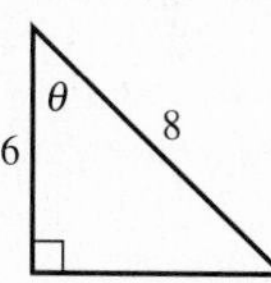

In Exercises 5–8, find all six trigonometric functions of the angle θ. (*Hint:* Use the Pythagorean theorem.)

5.

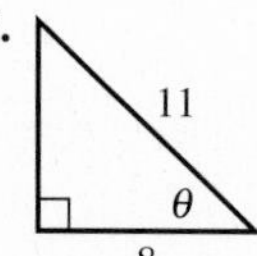

6.

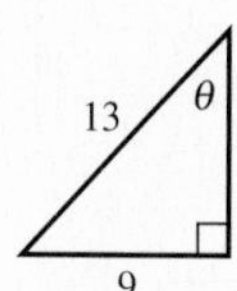

7.

8.

In Exercises 9–18, assume that θ is an acute angle in a right triangle satisfying the given conditions. Evaluate the remaining trigonometric functions.

9. $\sin\theta = \dfrac{3}{7}$

10. $\sin\theta = \dfrac{2}{3}$

11. $\cos\theta = \dfrac{5}{11}$

12. $\cos\theta = \dfrac{5}{8}$

13. $\tan\theta = \dfrac{5}{9}$ **14.** $\tan\theta = \dfrac{12}{13}$

15. $\cot\theta = \dfrac{11}{3}$ **16.** $\csc\theta = \dfrac{12}{5}$

17. $\csc\theta = \dfrac{23}{9}$ **18.** $\sec\theta = \dfrac{17}{5}$

In Exercises 19–24, evaluate without using a calculator.

19. $\sin(\pi/3)$ **20.** $\tan(\pi/4)$

21. $\cot(\pi/6)$ **22.** $\sec(\pi/3)$

23. $\cos(\pi/4)$ **24.** $\csc(\pi/3)$

In Exercises 25–44, evaluate using a calculator. Be sure the calculator is in the correct mode. Give answers accurate to two decimal places.

25. $\sin 74°$ **26.** $\cos 48°$

27. $\tan 8°$ **28.** $\sin 82°$

29. $\cos 19° 23'$ **30.** $\tan 23° 42'$

31. $\sin 0.43$ **32.** $\cos 0.92$

33. $\tan(\pi/12)$ **34.** $\sin(\pi/15)$

35. $\sec 49°$ **36.** $\csc 19°$

37. $\cot 63°$ **38.** $\cot 24°$

39. $\sec 1.24$ **40.** $\cot 0.89$

41. $\csc 0.39$ **42.** $\sec 0.44$

43. $\cot(\pi/8)$ **44.** $\csc(\pi/10)$

In Exercises 45–50, show that the equation is an identity. (Use the letters from this figure.)

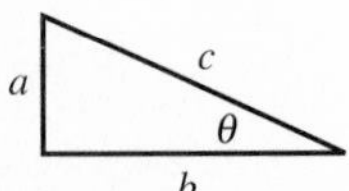

For Exercises 45–50

45. $\sin\theta = \dfrac{1}{\csc\theta}$ **46.** $\tan\theta = \dfrac{1}{\cot\theta}$

47. $\sec\theta = \dfrac{1}{\cos\theta}$ **48.** $\tan\theta = \dfrac{\sin\theta}{\cos\theta}$

49. $1 + \tan^2\theta = \sec^2\theta$ **50.** $1 + \cot^2\theta = \csc^2\theta$

In Exercises 51–58, find an acute angle θ in both degrees and radians without using a calculator.

51. $\sin\theta = \dfrac{1}{2}$ **52.** $\sin\theta = \dfrac{\sqrt{3}}{2}$

53. $\cot\theta = \dfrac{1}{\sqrt{3}}$ **54.** $\cos\theta = \dfrac{\sqrt{2}}{2}$

55. $\sec\theta = 2$ **56.** $\cot\theta = 1$

57. $\tan\theta = \dfrac{\sqrt{3}}{3}$ **58.** $\cos\theta = \dfrac{\sqrt{3}}{2}$

In Exercises 59–66, use the table-building feature of a grapher to find an acute angle θ in degrees rounded to the nearest hundredth.

59. $\sin\theta = 0.39$ **60.** $\sin\theta = 0.97$

61. $\cos\theta = 0.41$ **62.** $\cos\theta = 0.89$

63. $\tan\theta = 3.2$ **64.** $\tan\theta = 1.45$

65. $\sin\theta = 0.85$ **66.** $\cos\theta = 0.72$

In Exercises 67–70, solve for *x, y,* or *z.*

67.

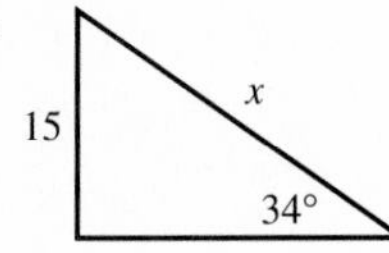

68.

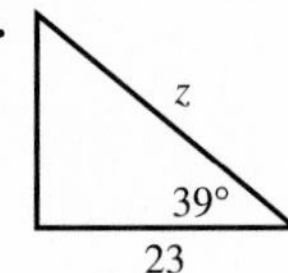

69.

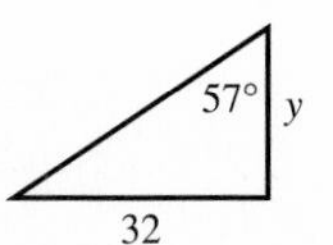

70.

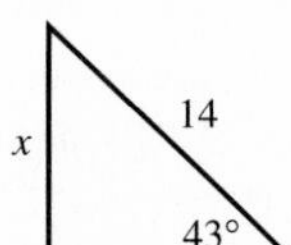

In Exercises 71–80, solve the right triangle $\triangle ABC$ for all of its unknown parts.

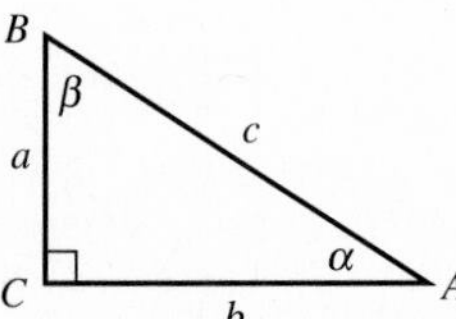

For Exercises 71–80

71. $\alpha = 20°;\ a = 12.3$ **72.** $a = 3;\ b = 4$

73. $\alpha = 41°;\ c = 10$ **74.** $\beta = 55°;\ a = 15.58$

75. $b = 5;\ c = 7$ **76.** $a = 20.2;\ c = 50.75$

77. $a = 2;\ b = 9.25$ **78.** $a = 5;\ \beta = 59°$

79. $c = 12.89;\ \beta = 12.5°$ **80.** $\alpha = 10.2°;\ c = 14.5$

In Exercises 81–86, solve the problem.

81. *Height* A guy wire from the top of the transmission tower at WJBC forms a 75° angle with the ground at a 55 ft distance from the base of the tower. How tall is the tower?

For Exercise 81

82. *Height and Angle Measure* A guy wire 30 m long runs from an antenna to a point on level ground 14 m from the base of the antenna. Determine the angle the guy wire makes with the horizontal, the angle the guy wire makes with the antenna, and the height of the point where the guy wire is attached to the antenna.

83. *Area* For locations between 20° and 60° north latitudes a solar collector panel should be mounted so that its angle with the horizontal is 20 greater than the local latitude. Consequently, the panel mounted on the roof of Solar Energy, Inc., in Atlanta (latitude 34°) forms a 54° angle with the horizontal. The bottom edge of the 12-ft-long panel is resting on the roof, and the high edge is 5 ft above the roof. What is the total area of this collector?

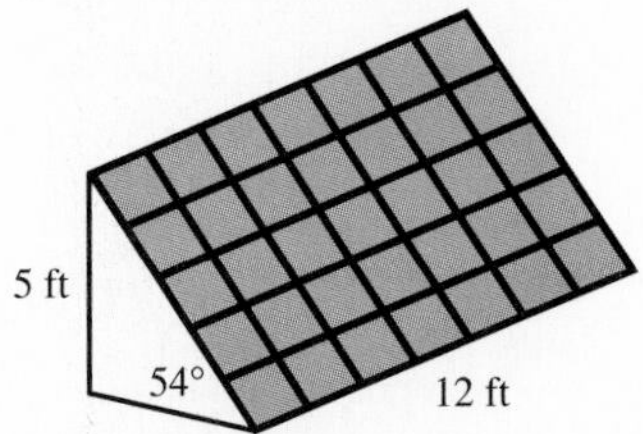

For Exercise 83

84. *Height* The Chrysler Building in New York City was the tallest building in the world at the time it was built. It casts a shadow approximately 130 ft long on the street when the sun's rays form a 82.9° angle with the earth. How tall is the building?

85. *Distance* DaShanda Taylor's team of surveyors had to find the distance AC across the lake at Montgomery County Park. Field assistants positioned themselves at points A and C, and DaShanda set up an angle-measuring instrument at point B, 100 ft from C in a perpendicular direction. DaShanda measured $\angle ABC$ as 75° 12′ 42″. What is the distance AC?

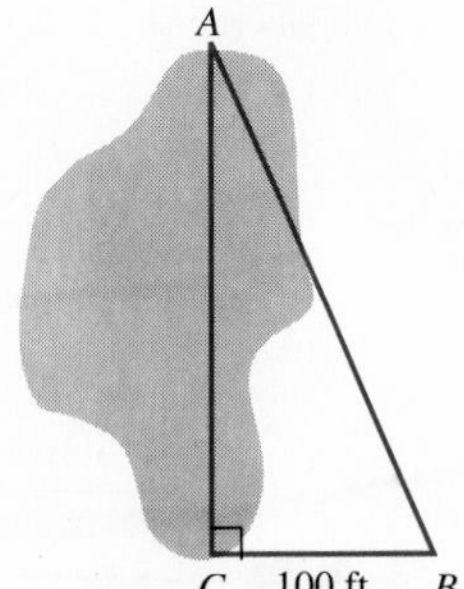

For Exercise 85

86. *Height* Kirsten places her surveyor's telescope on the top of a tripod 5 ft above the ground. She measures an 8° elevation above the horizontal to the top of a tree that is 120 ft away. How tall is the tree?

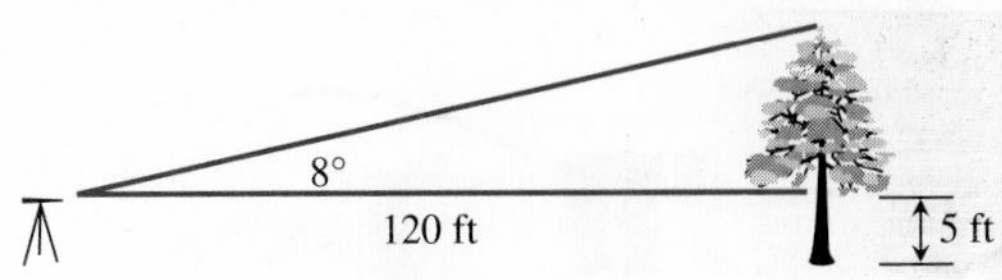

For Exercise 86

EXTENDING THE IDEAS

In Exercises 87–92, solve the problem.

87. **Writing to Learn** This grapher display seems to indicate that sin 1 has values of both 0.841 · · · and 0.017 · · · . How can this be? Explain.

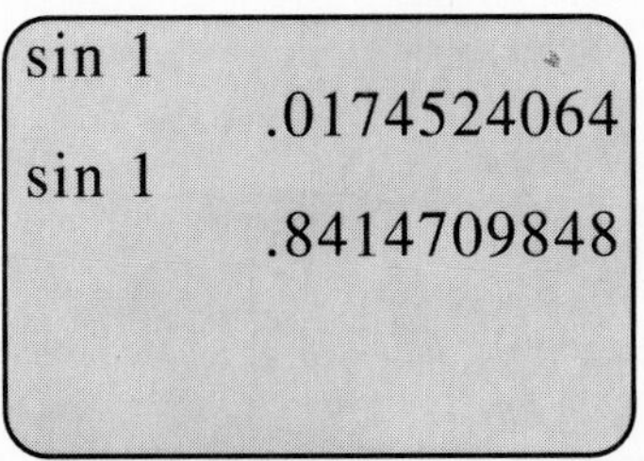

For Exercise 87

88. This grapher display shows both tan 1° + tan 1° and tan 2°. Is $\tan \alpha + \tan \beta = \tan(\alpha + \beta)$ an identity? Explain.

```
tan 1+tan 1
          .0349101299
tan 2
          .0349207695
```

For Exercise 88

89. Find $\sin \theta$ if $\cos \theta = \dfrac{\sqrt{2}}{2}$.

90. Find $\cos \theta$ if $\sin \theta = \dfrac{\sqrt{3}}{2}$.

91. Find $\sin \theta$ if $\tan \theta = 1$.

92. Find $\sin \theta$ if $\cos \theta = 0.25$.

In Exercises 93–100, decide whether the statement is true or false, and justify your choice.

93. $(\cos 56°)(\csc 56°) = 1$

94. $(\cos 22°)(\sec 22°) = 1$

95. $(\cos 34°)(\sin 56°) = 1$

96. $(\cos 19°)(\csc 71°) = 1$

97. $\cos^2 43° + \sin^2 47° = 1$

98. $\cot^2 39° + \tan^2 39° = 1$

99. $\csc^2 43° - \cot^2 43° = 1$

100. $1 + \tan^2 39° = \cot^2 39°$

In Exercises 101 and 102, solve the problem.

101. *Mirrors* In the figure, a light ray shining from point A to point P on the mirror will bounce to point B in such a way that the *angle of incidence* α will equal the *angle of reflection* β. This is the *law of reflection* derived from physical experiments. Both angles are measured from the *normal line,* which is perpendicular to the mirror at the point of reflection P. If A is 2 m farther from the mirror than is B, and if $\alpha = 30°$ and $AP = 5$ m, what is the length PB?

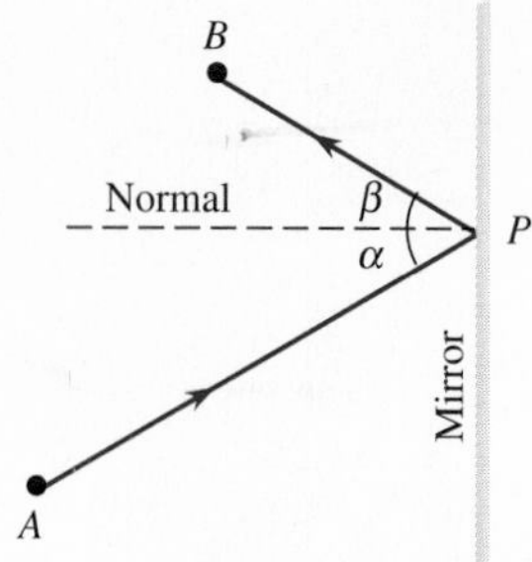

For Exercise 101

102. *Pool* On the pool table shown in the figure, where along the portion CD of the railing should you direct ball A so that it will bounce off CD and strike ball B? Assume that A obeys the law of reflection relative to rail CD.

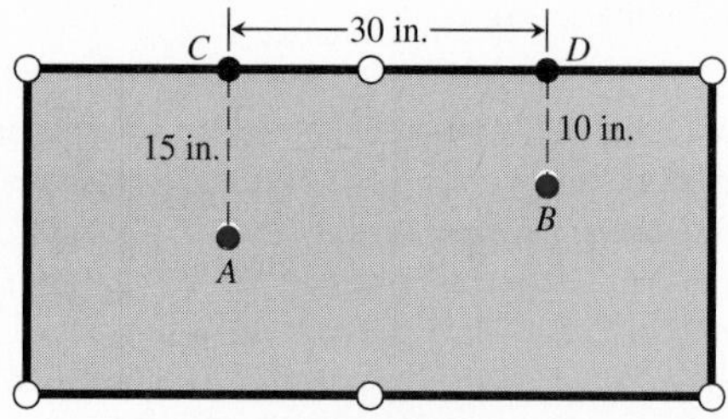

For Exercise 102

6.3 EXTENDING TRIGONOMETRIC FUNCTIONS

Objective

Students will be able to solve problems involving the trigonometric functions of real numbers and the properties of the sine and cosine as periodic functions.

Trigonometric Functions of Any Angle • Trigonometric Functions of Real Numbers • Properties of the Sine and Cosine Functions

Trigonometric Functions of Any Angle

In Section 6.2 you learned to find the trigonometric function values for an acute angle θ using a member of the family of right triangles determined by θ. We now extend the definitions of the six trigonometric functions to any angle regardless of size.

Motivate

Have students use a calculator to evaluate sin 23° and sin (−23°) and see what they observe. Repeat for the cosine function.
(sin 23° = −sin (−23°) ≈ 0.39
cos 23° = cos (−23°) ≈ 0.92)

Definition: Trigonometric Functions

Let θ be any angle in standard position, $P(x, y)$ any point on the terminal side of the angle (except the origin), and $r = \sqrt{x^2 + y^2}$. (See Figure 6.28.)

$$\sin\theta = \frac{y}{r} \qquad \cos\theta = \frac{x}{r} \qquad \tan\theta = \frac{y}{x},\ x \neq 0$$

$$\csc\theta = \frac{r}{y},\ y \neq 0 \qquad \sec\theta = \frac{r}{x},\ x \neq 0 \qquad \cot\theta = \frac{x}{y},\ y \neq 0$$

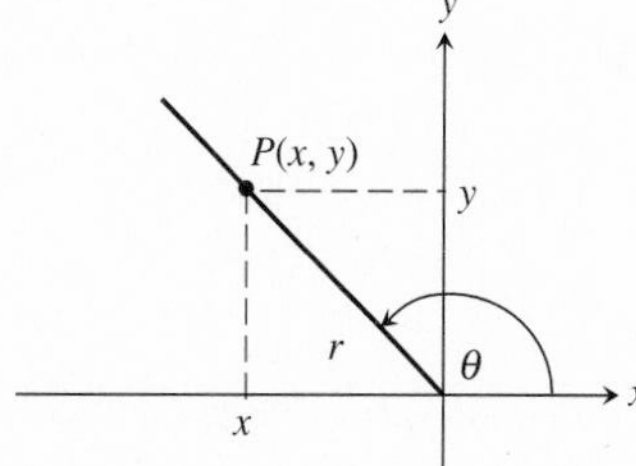

Figure 6.28

NOTE

Note that $r = \sqrt{x^2 + y^2}$ is the radial distance from the origin to the point P and is always positive. So $\sin\theta$ and $\cos\theta$ are defined for all values of θ. The other four trigonometric functions may be undefined if P lies on an axis. For example, if $\theta = \pi = 180°$, $y = 0$ and the cosecant and cotangent functions are undefined.

Observe that these definitions agree with those given in Section 6.2 when $0 < \theta < \pi/2$ (or $0° < \theta < 90°$), that is, for acute angles. For such angles, y = opp, x = adj, and r = hyp. Just as the definitions of the trigonometric functions for any acute angle were independent of the choice of right triangle used, these new definitions are independent of the point chosen on the terminal side of θ (as long as it is not the origin).

■ EXAMPLE 1 Evaluating Trigonometric Functions

Suppose angle θ is in standard position with point $P(-2, 3)$ on its terminal side. Evaluate the trigonometric functions for θ.

Solution

For point $P(-2, 3)$, $x = -2$, $y = 3$, and $r = \sqrt{(-2)^2 + 3^2} = \sqrt{13}$. (See Figure 6.29.) Therefore, we have the following trigonometric values.

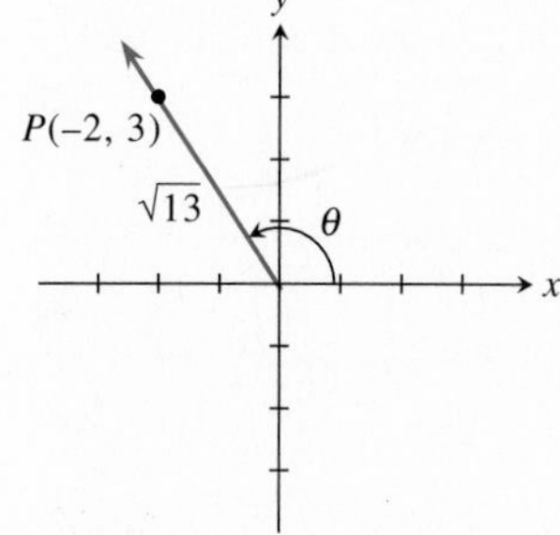

Figure 6.29

$$\sin\theta = \frac{3}{\sqrt{13}} = 0.832\cdots \qquad \cos\theta = \frac{-2}{\sqrt{13}} = -0.554\cdots \qquad \tan\theta = \frac{3}{-2} = -1.5$$

$$\csc\theta = \frac{\sqrt{13}}{3} = 1.201\cdots \qquad \sec\theta = \frac{\sqrt{13}}{-2} = -1.802\cdots \qquad \cot\theta = \frac{-2}{3} = -0.666\cdots$$

■

Example 1 illustrates that when the terminal side of θ is in quadrant II, some trigonometric values are negative. The same is true when the terminal side of θ is in quadrants III or IV.

Because r is always positive, the *sign* of a trigonometric value can be determined from the signs of x and y, as summarized in Figure 6.30.

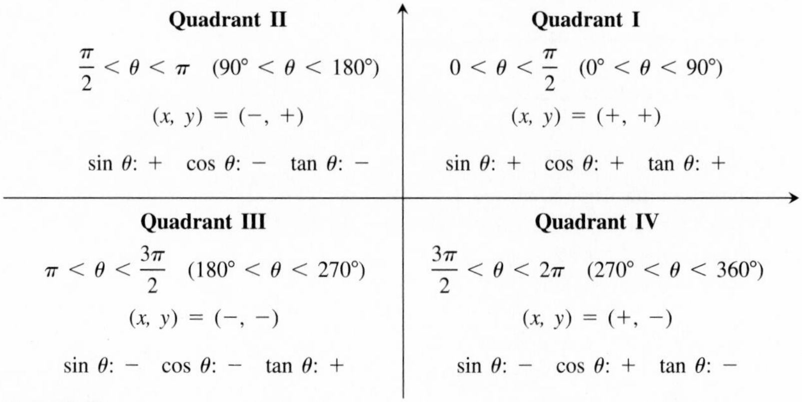

Quadrant II	**Quadrant I**
$\frac{\pi}{2} < \theta < \pi \quad (90° < \theta < 180°)$	$0 < \theta < \frac{\pi}{2} \quad (0° < \theta < 90°)$
$(x, y) = (-, +)$	$(x, y) = (+, +)$
$\sin\theta$: + $\cos\theta$: − $\tan\theta$: −	$\sin\theta$: + $\cos\theta$: + $\tan\theta$: +
Quadrant III	**Quadrant IV**
$\pi < \theta < \frac{3\pi}{2} \quad (180° < \theta < 270°)$	$\frac{3\pi}{2} < \theta < 2\pi \quad (270° < \theta < 360°)$
$(x, y) = (-, -)$	$(x, y) = (+, -)$
$\sin\theta$: − $\cos\theta$: − $\tan\theta$: +	$\sin\theta$: − $\cos\theta$: + $\tan\theta$: −

Figure 6.30

■ EXAMPLE 2 Finding Signs of Trigonometric Functions

State whether the function value is positive or negative. Explain.

a. $\sin 210°$ **b.** $\tan \frac{3\pi}{4}$ **c.** $\cos \frac{-\pi}{3}$

Solution

a. $\sin 210° = \frac{y}{r}$ is negative. Because $180° < 210° < 270°$, the angle $210°$ is in quadrant III, where y-coordinates are negative.

b. $\tan \frac{3\pi}{4} = \frac{y}{x}$ is negative. The angle $\frac{3\pi}{4}$ is in quadrant II, where x-coordinates are negative and y-coordinates are positive.

c. $\cos \frac{-\pi}{3} = \frac{x}{r}$ is positive. The angle $\frac{-\pi}{3}$ is in quadrant IV, where x-coordinates are positive. ■

An angle in standard position is a **quadrantal angle** whenever its terminal side lies on an axis.

■ EXAMPLE 3 Finding Trigonometric Values of Quadrantal Angles

Evaluate the sine, cosine, and tangent for the four quadrantal angles 0, $\pi/2$, π, and $3\pi/2$. (See Figure 6.31.)

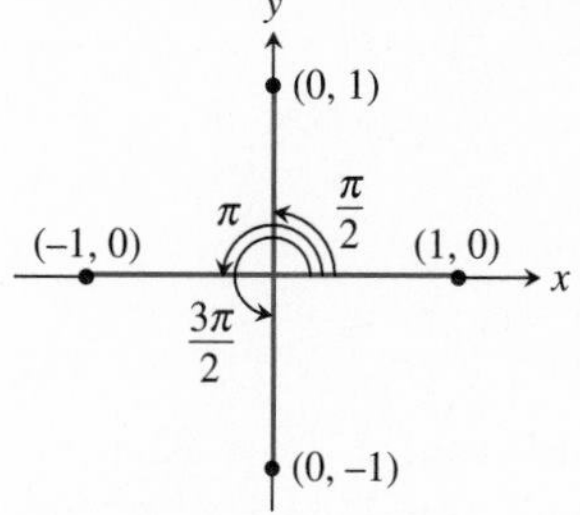

Figure 6.31

Solution

Choose the point on the terminal side of the angle with $r = 1$.

$$\sin 0 = \frac{y}{r} = \frac{0}{1} = 0 \qquad \cos 0 = \frac{x}{r} = \frac{1}{1} = 1 \qquad \tan 0 = \frac{y}{x} = \frac{0}{1} = 0$$

$$\sin \frac{\pi}{2} = \frac{1}{1} = 1 \qquad \cos \frac{\pi}{2} = \frac{0}{1} = 0 \qquad \tan \frac{\pi}{2} = \frac{1}{0} \text{ (undefined)}$$

$$\sin \pi = \frac{0}{1} = 0 \qquad \cos \pi = \frac{-1}{1} = -1 \qquad \tan \pi = \frac{0}{-1} = 0$$

$$\sin \frac{3\pi}{2} = \frac{-1}{1} = -1 \qquad \cos \frac{3\pi}{2} = \frac{0}{1} = 0 \qquad \tan \frac{3\pi}{2} = \frac{-1}{0} \text{ (undefined)}$$

■

The **reference angle** of an angle θ in standard position is the acute angle θ' (read "theta prime") formed by the terminal side of θ and the horizontal axis. (See Figure 6.32.)

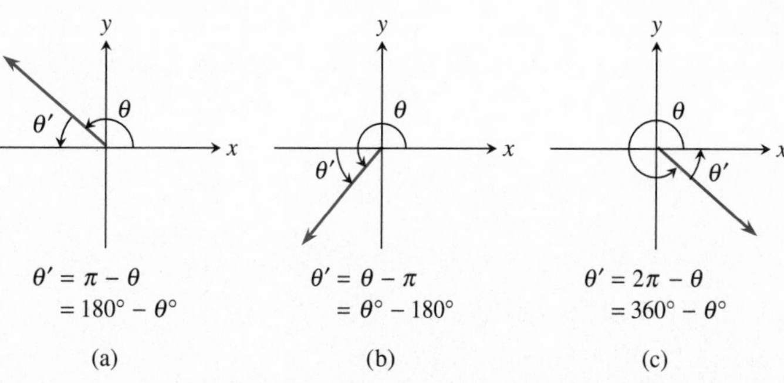

Figure 6.32 Reference angles θ' for three angles θ.

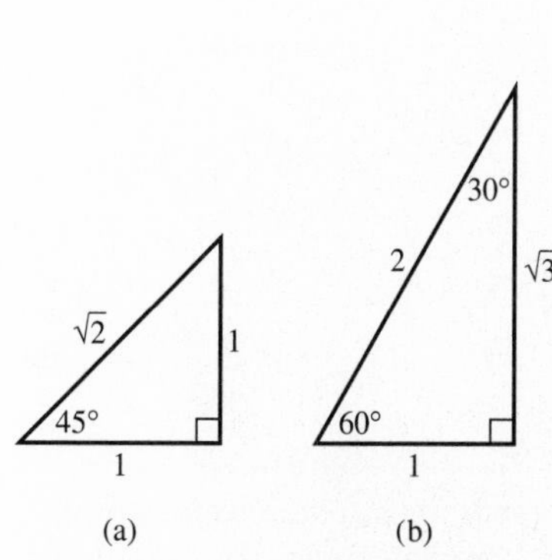

Figure 6.33 Watch how the side measures of these special triangles are used in Example 4.

Example 4 illustrates the use of reference angles in evaluating the trigonometric functions of the special angles without using a calculator. We also use the linear measures of the special 45-45 and 30-60 right triangles (see Figure 6.33) to find coordinates of points needed to evaluate the trigonometric functions.

EXAMPLE 4 Using Reference Angles in Evaluations

Find the sine, cosine, and tangent using the reference angle.

a. $150°$ **b.** $\dfrac{4\pi}{3}$ **c.** $315°$

Solution

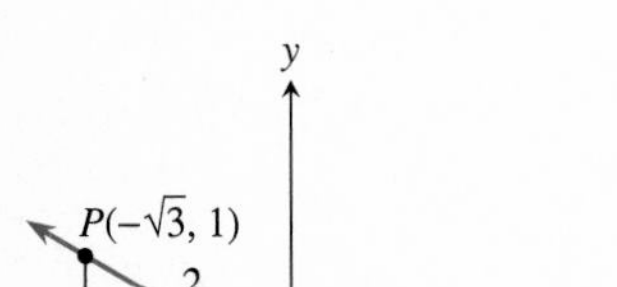

Figure 6.34

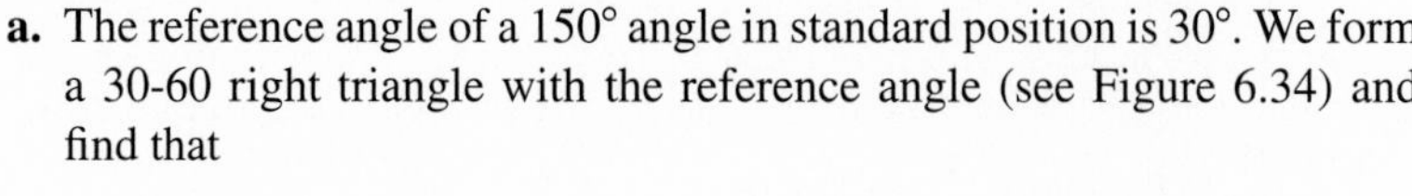

a. The reference angle of a $150°$ angle in standard position is $30°$. We form a 30-60 right triangle with the reference angle (see Figure 6.34) and find that

$$\sin 150° = \frac{y}{r} = \frac{1}{2}, \qquad \cos 150° = \frac{x}{r} = \frac{-\sqrt{3}}{2},$$

$$\tan 150° = \frac{y}{x} = \frac{1}{-\sqrt{3}} = -\frac{\sqrt{3}}{3}.$$

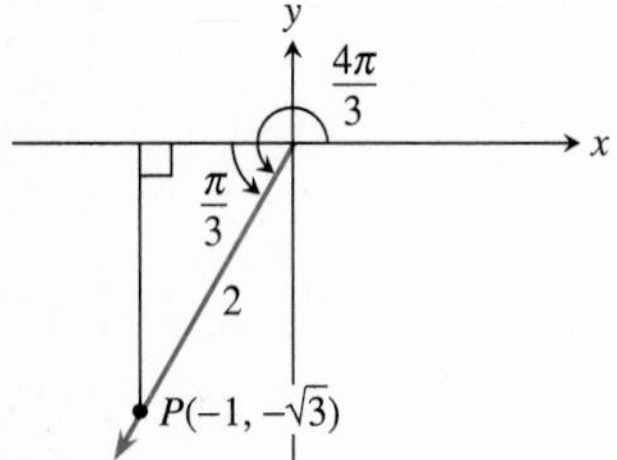

Figure 6.35

b. The reference angle of $4\pi/3$ in standard position is $\pi/3$. We form a 30-60 right triangle ($\pi/3$ radians $= 60°$) with the reference angle (see Figure 6.35) and find that

$$\sin\frac{4\pi}{3} = \frac{y}{r} = \frac{-\sqrt{3}}{2}, \qquad \cos\frac{4\pi}{3} = \frac{x}{r} = \frac{-1}{2},$$

$$\tan\frac{4\pi}{3} = \frac{y}{x} = \frac{-\sqrt{3}}{-1} = \sqrt{3}.$$

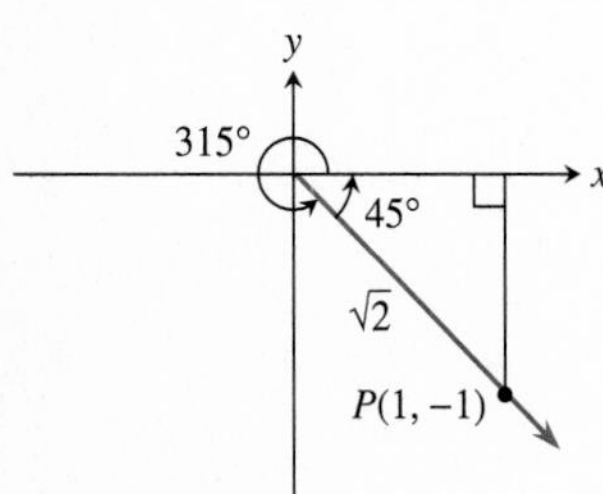

Figure 6.36

c. The reference angle of a $315°$ angle in standard position is $45°$. We form a 45-45 right triangle with the reference angle (see Figure 6.36) and find that

$$\sin 315° = \frac{-1}{\sqrt{2}} = -\frac{\sqrt{2}}{2}, \qquad \cos 315° = \frac{1}{\sqrt{2}} = \frac{\sqrt{2}}{2},$$

$$\tan 315° = \frac{-1}{1} = -1.$$

■

We summarize the exact values of the trigonometric functions for the special angles in $0° \le \theta \le 360°$ including the quadrantal angles. You should be able to recall them for angles in degrees or radians with a diagram whenever needed. Table 6.2 shows the corresponding decimal values obtained using a grapher.

Table 6.2 Special Angles

Angle θ	$\sin\theta$	$\cos\theta$	$\tan\theta$
0	0	1	0
30	.5	.86603	.57735
45	.70711	.70711	1
60	.86603	.5	1.7321
90	1	0	ERROR
120	.86603	−.5	−1.732
135	.70711	−.7071	−1
150	.5	−.866	−.5774
180	0	−1	0
210	−.5	−.866	.57735
225	−.7071	−.7071	1
240	−.866	−.5	1.7321
270	−1	0	ERROR
300	−.866	.5	−1.732
315	−.7071	.70711	−1
330	−.5	.86603	−.5774
360	0	1	0

The sine, cosine, and tangent of the special angles in decimal form.

Sine, Cosine, and Tangent of Special Angles

Angle θ	**$\sin\theta$**	**$\cos\theta$**	**$\tan\theta$**
$0°$ (0)	0	1	0
$30°$ $(\pi/6)$	$1/2$	$\sqrt{3}/2$	$\sqrt{3}/3$
$45°$ $(\pi/4)$	$\sqrt{2}/2$	$\sqrt{2}/2$	1
$60°$ $(\pi/3)$	$\sqrt{3}/2$	$1/2$	$\sqrt{3}$
$90°$ $(\pi/2)$	1	0	Undefined
$120°$ $(2\pi/3)$	$\sqrt{3}/2$	$-1/2$	$-\sqrt{3}$
$135°$ $(3\pi/4)$	$\sqrt{2}/2$	$-\sqrt{2}/2$	-1
$150°$ $(5\pi/6)$	$1/2$	$-\sqrt{3}/2$	$-\sqrt{3}/3$
$180°$ (π)	0	-1	0
$210°$ $(7\pi/6)$	$-1/2$	$-\sqrt{3}/2$	$\sqrt{3}/3$
$225°$ $(5\pi/4)$	$-\sqrt{2}/2$	$-\sqrt{2}/2$	1
$240°$ $(4\pi/3)$	$-\sqrt{3}/2$	$-1/2$	$\sqrt{3}$
$270°$ $(3\pi/2)$	-1	0	Undefined
$300°$ $(5\pi/3)$	$-\sqrt{3}/2$	$1/2$	$-\sqrt{3}$
$315°$ $(7\pi/4)$	$-\sqrt{2}/2$	$\sqrt{2}/2$	-1
$330°$ $(11\pi/6)$	$-1/2$	$\sqrt{3}/2$	$-\sqrt{3}/3$
$360°$ (2π)	0	1	0

Example 5 illustrates that whenever we know one trigonometric ratio and the quadrant in which the terminal side lies then the other ratios can be found.

EXAMPLE 5 Evaluating Trigonometric Functions

Suppose $\sin\theta = \frac{3}{7}$ and $\tan\theta < 0$. Evaluate $\cos\theta$ and $\cot\theta$.

Solution

Because $\sin\theta$ is positive we know that θ is in either quadrant I or II. Knowing that $\tan\theta$ is negative tells us that θ is in quadrant II.

Choose point P in quadrant II, 7 units from the origin with $y = 3$. (See Figure 6.37.) Then x in Figure 6.37 can be found using the Pythagorean theorem:

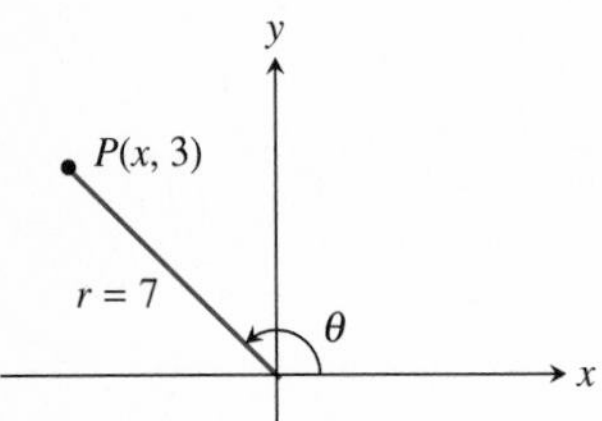

Figure 6.37

$$x = -\sqrt{7^2 - 3^2} = -\sqrt{40}$$

It follows that

$$\cos\theta = \frac{-\sqrt{40}}{7} = -0.903\cdots \quad \text{and} \quad \tan\theta = \frac{3}{-\sqrt{40}} = -0.474\cdots.$$

Teaching Note

The material in this section helps students to see angle measure and arc length as being equivalent. Wrapping the number line around the unit circle opens up a new way of considering trigonometric functions. Ultimately, students will need to think in terms of real numbers as lengths of arcs on the unit circle and hence the domain elements of the trigonometric functions. Most students can easily see the relationship between the unit circle and angle measure. You may wish to show a rope or a string (with the units of the axis marked on it) being wrapped around a unit circle.

Alert

Many students do not think of π as a number. The experience of working with circular functions should help students to correct this misconception. This approach paves the way to graphing these functions.

Trigonometric Functions of Real Numbers (Circular Functions)

We have defined the trigonometric functions on angles. Now we want to define them on real numbers. To do so, we draw a number line vertically with its origin at the point (1, 0) and the positive coordinates above the x-axis. (See Figure 6.38.)

Imagine wrapping the number line around the unit circle $x^2 + y^2 = 1$ with positive numbers corresponding to counterclockwise wrapping (Figure 6.38a),

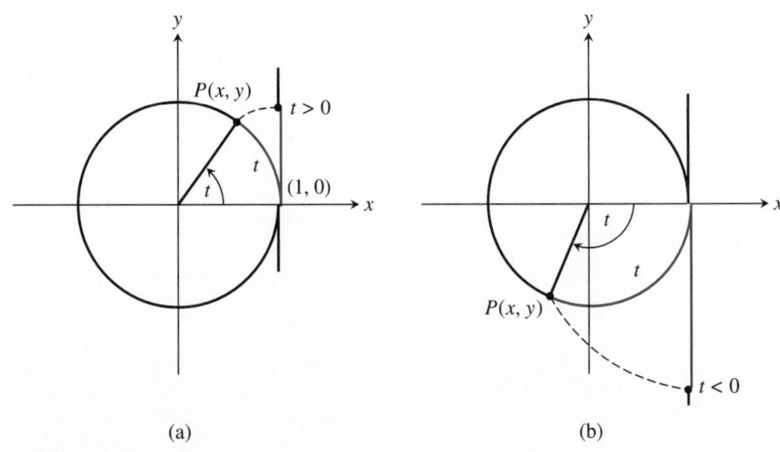

Figure 6.38

REMARK

If t is positive, then t gives the length of the counterclockwise-produced intercepted arc. If t is negative, then t gives the opposite of the clockwise-produced intercepted arc. So t is the directed arc length of the intercepted arc of θ.

and negative numbers to clockwise wrapping (Figure 6.38b). Each real number t corresponds to a central angle θ of the unit circle. Because the circle has radius 1, the number t is also the radian measure of θ and the **directed arc-length** distance around the unit circle from the point (1, 0) ($s = r\theta = 1 \cdot \theta = \theta$). Now we simply define the trigonometric functions of t to be their values on the corresponding central angle $\theta = t$ in radians.

With these definitions, the domain of both $y_1 = \cos t$ and $y_2 = \sin t$ is the set of all real numbers. Because $\cos t$ and $\sin t$ are x- and y-coordinates of points on the unit circle, we have

$$-1 \leq \cos t \leq 1 \quad \text{and} \quad -1 \leq \sin t \leq 1.$$

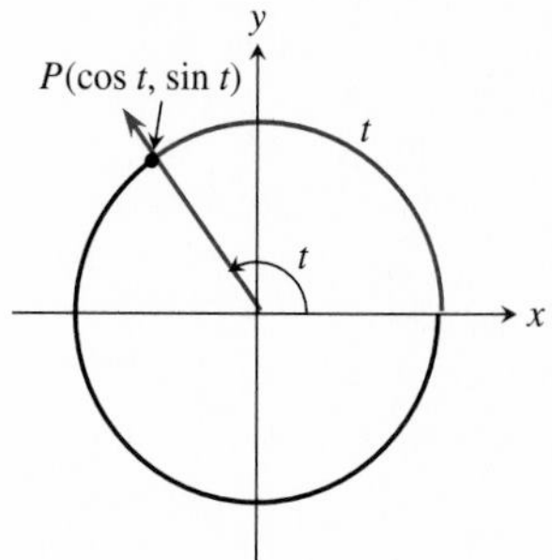

Figure 6.39

Thus, the range of both functions is $[-1, 1]$. We will investigate the domains and ranges of the other trigonometric functions later in the chapter.

If $P(x, y)$ is the point on the unit circle corresponding to t (see Figure 6.39), then

$$\sin t = \frac{y}{r} = \frac{y}{1} = y \quad \text{and} \quad \cos t = \frac{x}{r} = \frac{x}{1} = x.$$

This form for the points on the unit circle allows us to graph it parametrically, as suggested by the following activity.

▼ GRAPHING ACTIVITY Unit Circle and Trigonometric Values in Parametric Mode

Let $x_1 = \cos t$ and $y_1 = \sin t$. Set your grapher in degree mode where $T\text{min} = 0$, $T\text{max} = 360$, and $T\text{step} = 15$.

1. Find a square viewing window that displays a graph of the unit circle.
2. How are the values of x and y found using "trace" connected to Table 6.2?
3. Now set your grapher in radian mode where $T\text{min} = 0$, $T\text{max} = 2\pi$, and $T\text{step}$ is the equivalent of 15° in radians. Does the viewing window found in Step 1 still give a graph of the unit circle?
4. How are the values of x and y found using "trace" (in radian mode) connected to Table 6.2?
5. Set $T\text{max} = 4\pi$. Use "trace," and explain the effect of this change. ▲

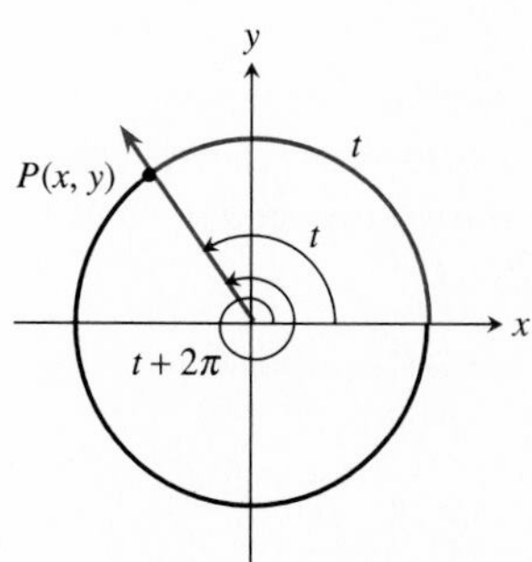

Figure 6.40

Properties of the Sine and Cosine Functions

You may have discovered in this Graphing Activity that the arc length for one complete revolution around a unit circle is 2π. The arc length on a unit circle for an angle greater than 2π is also greater than 2π, and the associated arc overlaps itself. Figure 6.40 shows that angles t and $t + 2\pi$ have the same terminal side, so $(\cos t, \sin t)$ and $(\cos (t + 2\pi), \sin (t + 2\pi))$ are the same point. Thus

$$\cos t = \cos (t + 2\pi) \quad \text{and} \quad \sin t = \sin (t + 2\pi).$$

Functions that behave in such a regularly repetitive fashion are called *periodic functions.*

Definition: Periodic Function

A function $y = f(t)$ is **periodic** if there is a positive number c such that

$$f(t + c) = f(t) \text{ for all values of } t \text{ in the domain of } f.$$

The smallest such number c is the **period** of the function.

X	X+2π	sin X	sin(X+2π)
0	6.2832	0	0
.1	6.3832	.09983	.09983
.2	6.4832	.19867	.19867
.3	6.5832	.29552	.29552
.4	6.6832	.38942	.38942
.5	6.7832	.47943	.47943
.6	6.8832	.56464	.56464

X=0

Figure 6.41

Because angles t and $t + 2\pi$ have the same terminal side,

$$(\cos t, \sin t) = (\cos (t + 2\pi), \sin (t + 2\pi)),$$

and the cosine and sine functions are periodic, as supported by the table in Figure 6.41. In fact, the period of each function is 2π. (See Exercises 71 and 72.)

The behavior of sine and cosine on the interval $[0, 2\pi]$ is repeated on $[2\pi, 4\pi]$, $[4\pi, 6\pi]$, . . . , as well as $[-2\pi, 0]$, $[-4\pi, -2\pi]$, ... This leads to the identities

$$\sin (t \pm 2\pi n) = \sin t \qquad \text{and} \qquad \cos (t \pm 2\pi n) = \cos t,$$

for $n = 1, 2, 3, \ldots$. Example 6 shows how these identities can be used.

EXAMPLE 6 Using Period to Find Trigonometric Values

a. $\sin \dfrac{7\pi}{3} = \sin \left(\dfrac{7\pi}{3} - 2\pi\right) = \sin \dfrac{\pi}{3} = \dfrac{\sqrt{3}}{2} \approx 0.87$

b. $\cos \left(-\dfrac{13\pi}{4}\right) = \cos \left(-\dfrac{13\pi}{4} + 4\pi\right) = \cos \dfrac{3\pi}{4} = -\dfrac{\sqrt{2}}{2} \approx -0.71$

■

Because the domains of the sine and cosine functions are the set of real numbers, you can use your grapher to draw their graphs because they are built-in functions. Confirming these graphs and more detailed investigation of the properties of these two functions will occur in Section 6.4.

EXAMPLE 7 Finding the Domain and Range

Find the domain and range of the function.

a. $f(t) = 3 \cos t$ **b.** $g(t) = 2 \sin 3t$

NOTE

We write $3 \cos t$ to mean $3 \cdot \cos t$ and we write $\sin 3t$ to mean $\sin (3t)$.

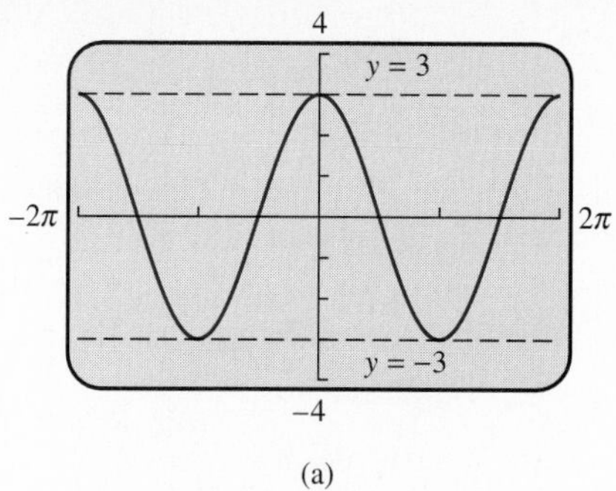

(a)

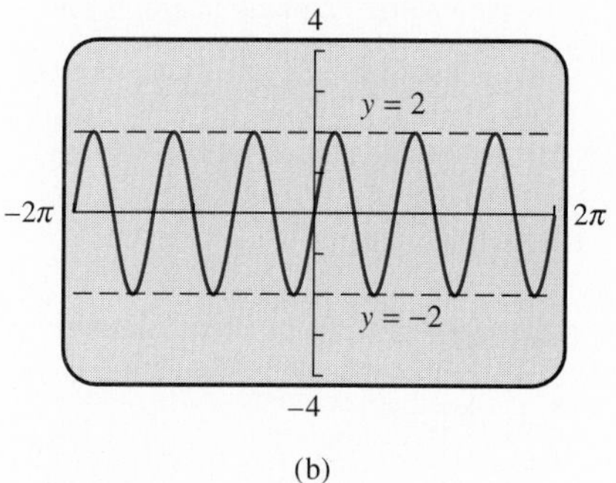

(b)

Figure 6.42 (a) $y_1 = f(t) = 3 \cos t$, and (b) $y_2 = g(t) = 2 \sin 3t$. The window used for these graphs is sometimes called the *trig window.*

Solution

a. Because $\cos t$ is defined for all values of t, so is $3 \cos t$. The domain of $3 \cos t$ is $(-\infty, \infty)$.
Because $-1 \le \cos t \le 1$, it follows that $-3 \le 3 \cos t \le 3$. The range of $3 \cos t$ is $[-3, 3]$.

b. Because $\sin t$ is defined for all values of t, so are $\sin 3t$ and $2 \sin 3t$. The domain of $2 \sin 3t$ is $(-\infty, \infty)$.
Because $-1 \le \sin 3t \le 1$, it follows that $-2 \le 2 \sin 3t \le 2$. The range of $2 \sin 3t$ is $[-2, 2]$.

Support Graphically

The graphs in Figure 6.42 give us no reason to doubt the domains and ranges found. ■

The unit circle helps us understand the domain and range of the sine and cosine functions and the fact that they are periodic. Other properties of these functions can be discovered. For example, Figure 6.43 suggests that the cosine function is an *even* function and the sine function is an *odd* function. That is,

$$\cos(-t) = \cos t \quad \text{and} \quad \sin(-t) = -\sin t.$$

Here is a summary of some key properties of the sine and cosine functions that we have discovered so far.

Properties of Sine and Cosine

Property	Sine	Cosine
Period	2π	2π
Domain	$(-\infty, \infty)$	$(-\infty, \infty)$
Range	$[-1, 1]$	$[-1, 1]$
Even or Odd	Odd	Even

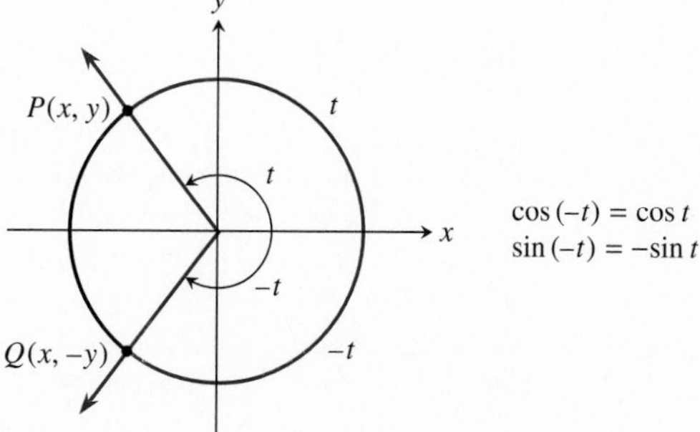

Figure 6.43 P and Q have the same x-coordinate, and have y-coordinates that are opposite in sign.

Quick Review 6.3

In Exercises 1–6, state the quadrant of the terminal side of the angle θ.

1. $\theta = 135°$ **2.** $\theta = 210°$

3. $\theta = \dfrac{-\pi}{6}$ **4.** $\theta = \dfrac{-5\pi}{6}$

5. $\theta = \dfrac{25\pi}{4}$ **6.** $\theta = \dfrac{16\pi}{3}$

In Exercises 7–12, use special triangles to evaluate.

7. $\cos 45°$ **8.** $\sin 60°$

9. $\tan \dfrac{\pi}{6}$ **10.** $\cot \dfrac{\pi}{4}$

11. $\csc \dfrac{\pi}{4}$ **12.** $\sec \dfrac{\pi}{3}$

In Exercises 13–16, use a right triangle to find the values of the other five trigonometric functions of the acute angle θ.

13. $\sin \theta = \dfrac{5}{13}$ **14.** $\cos \theta = \dfrac{15}{17}$

15. $\cot \theta = \dfrac{21}{20}$ **16.** $\sec \theta = \dfrac{29}{21}$

In Exercises 17 and 18, use parametric graphing to graph the function. (See Section 2.5.)

17. $f(x) = x^2 + 3$ **18.** $f(x) = \sqrt{x}$

SECTION EXERCISES 6.3

In Exercises 1–4, evaluate the six trigonometric functions for θ.

1.

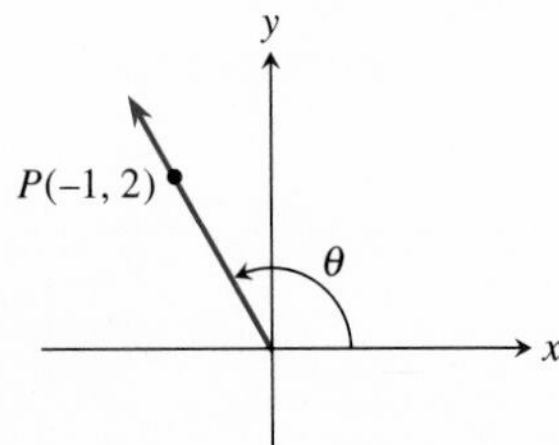

2.

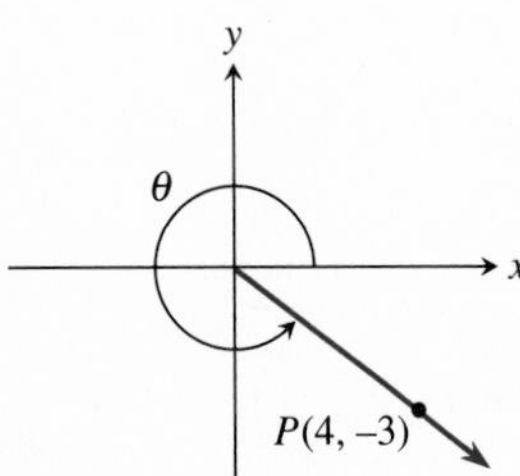

3.

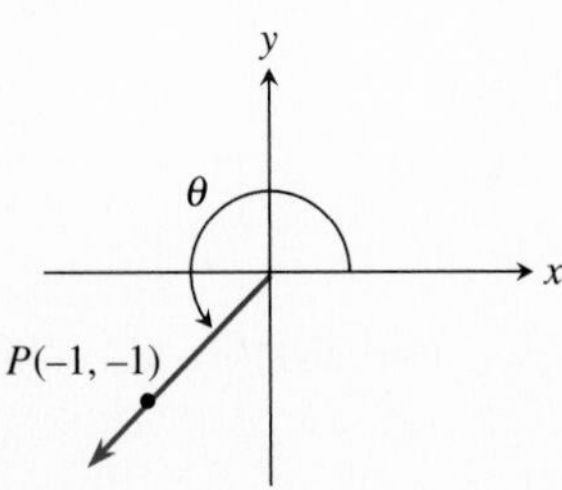

4.

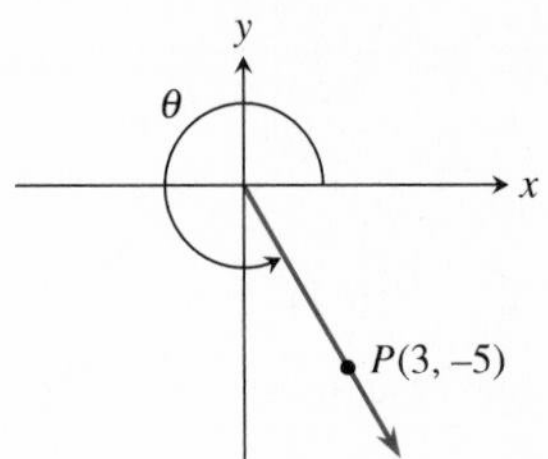

In Exercises 5–12, point P is on the terminal side of angle θ. Evaluate the six trigonometric functions for θ.

5. $P(1, 2)$ **6.** $P(3, 4)$

7. $P(-3, 0)$ **8.** $P(0, 5)$

9. $P(-4, -6)$ **10.** $P(5, -2)$

11. $P(22, -22)$ **12.** $P(-8, -1)$

In Exercises 13–16, state the sign (+ or −) of the function on the given interval of real numbers.

a. $\sin t$ **b.** $\cos t$ **c.** $\tan t$.

13. $\left(0, \dfrac{\pi}{2}\right)$ **14.** $\left(\dfrac{\pi}{2}, \pi\right)$

15. $\left(\pi, \dfrac{3\pi}{2}\right)$ **16.** $\left(\dfrac{3\pi}{2}, 2\pi\right)$

In Exercises 17–24, state whether the value is positive or negative.

17. $\cos 143°$ **18.** $\tan 192°$

19. $\sin \frac{5\pi}{3}$ **20.** $\cos \frac{7\pi}{8}$

21. $\tan \frac{4\pi}{5}$ **22.** $\sin \frac{11\pi}{6}$

23. $\cos \frac{-\pi}{12}$ **24.** $\sin \frac{3\pi}{4}$

In Exercises 25–28, choose the point on the terminal side of θ.

25. $\theta = 45°$

a. $(2, 2)$ **b.** $(1, \sqrt{3})$ **c.** $(\sqrt{3}, 1)$

26. $\theta = \frac{2\pi}{3}$

a. $(-1, 1)$ **b.** $(-1, \sqrt{3})$ **c.** $(-\sqrt{3}, 1)$

27. $\theta = \frac{7\pi}{6}$

a. $(-\sqrt{3}, -1)$ **b.** $(-1, \sqrt{3})$ **c.** $(-\sqrt{3}, 1)$

28. $\theta = -60°$

a. $(-1, -1)$ **b.** $(1, -\sqrt{3})$ **c.** $(-\sqrt{3}, 1)$

In Exercises 29–34, evaluate the function for the given quadrantal angle. (Do not use a calculator.)

a. $\sin \theta$ **b.** $\cos \theta$ **c.** $\tan \theta$

29. $-450°$ **30.** $-270°$

31. 7π **32.** $\frac{11\pi}{2}$

33. $\frac{-7\pi}{2}$ **34.** -4π

In Exercises 35–42, evaluate using the reference angle.

35. $\cos 120°$ **36.** $\tan 300°$

37. $\sec \frac{\pi}{3}$ **38.** $\csc \frac{3\pi}{4}$

39. $\sin \frac{13\pi}{6}$ **40.** $\cos \frac{7\pi}{3}$

41. $\tan \frac{-15\pi}{4}$ **42.** $\cot \frac{13\pi}{4}$

In Exercises 43–48, evaluate without using a calculator.

43. Find $\sin \theta$ and $\tan \theta$ if $\cos \theta = \frac{2}{3}$ and $\cot \theta > 0$.

44. Find $\cos \theta$ and $\cot \theta$ if $\sin \theta = \frac{1}{4}$ and $\tan \theta < 0$.

45. Find $\tan \theta$ and $\sec \theta$ if $\sin \theta = \frac{-2}{5}$ and $\cos \theta > 0$.

46. Find $\sin \theta$ and $\cos \theta$ if $\cot \theta = \frac{3}{7}$ and $\sec \theta < 0$.

47. Find $\sec \theta$ and $\csc \theta$ if $\cot \theta = \frac{-4}{3}$ and $\cos \theta < 0$.

48. Find $\csc \theta$ and $\cot \theta$ if $\tan \theta = \frac{-4}{3}$ and $\sin \theta > 0$.

In Exercises 49–56, evaluate using the periodic property of $\sin t$ and $\cos t$.

49. $\sin \frac{13\pi}{3}$ **50.** $\cos \frac{23\pi}{6}$

51. $\cos \frac{17\pi}{4}$ **52.** $\sin \frac{11\pi}{3}$

53. $\sin \frac{-13\pi}{3}$ **54.** $\cos \frac{-15\pi}{4}$

55. $\tan \frac{25\pi}{6}$ **56.** $\cot \frac{19\pi}{6}$

In Exercises 57–62, find the domain of the function. (Assume that the angle is in radians.)

57. $h(x) = 5 \cos 2x$ **58.** $f(x) = 2 \sin 3x$

59. $f(x) = \frac{\sin x}{x}$ **60.** $g(x) = \frac{1 - \cos x}{1 - x}$

61. $f(x) = \frac{\cos x}{\sqrt{x}}$ **62.** $g(x) = \frac{3}{\sin x}$

In Exercises 63–68, find the range of the function. Use a grapher to support your answer.

63. $g(x) = 3 \sin x$ **64.** $f(x) = 2 \cos 3x$

65. $f(x) = \frac{1}{2} \sin x$ **66.** $h(x) = \frac{1}{3} \cos x$

67. $f(x) = \sqrt{2} \cos x$ **68.** $g(x) = \sqrt{5} \sin 4x$

In Exercises 69 and 70, use parametric graphing to draw a graph of the circle with center $(0, 0)$ and radius r.

69. $r = 3$ **70.** $r = 5$

In Exercises 71–74, solve the problem.

71. Writing to Learn Give a convincing argument that the period of $\sin t$ is 2π. That is, show that there is no smaller positive real number p such that $\sin (t + p) = \sin t$ for all real numbers t.

72. Writing to Learn Give a convincing argument that the period of $\cos t$ is 2π. That is, show that there is no smaller positive real number p such that $\cos (t + p) = \cos t$ for all real numbers t.

73. *Refracted Light* Light is *refracted* (bent) as it passes through glass. In the figure below θ_1 is the angle of incidence and θ_2

is the *angle of refraction.* The *index of refraction* is a constant μ that satisfies the equation

$$\sin \theta_1 = \mu \sin \theta_2.$$

If $\theta_1 = 83°$ and $\theta_2 = 36°$ for a certain piece of flint glass, find the index of refraction.

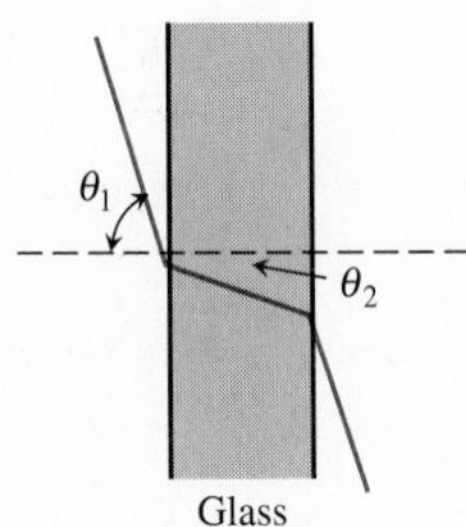

For Exercises 73–74

74. *Refracted Light* A certain piece of crown glass has an index of refraction of 1.52. If a light ray enters the glass at an angle $\theta_1 = 42°$, what is the angle of refraction θ_2?

Exercises 75 and 76 describe two physical situations that can be modeled by the cosine function in which t represents time.

75. *Damped Harmonic Motion* A weight suspended from a spring is set into motion. Its displacement d from equilibrium is modeled by the equation

$$d = 0.4e^{-0.2t} \cos 4t,$$

where d is the displacement in inches and t is the time in seconds. Find the displacement at the given time. Use radian mode.

a. $t = 0$ **b.** $t = 3$

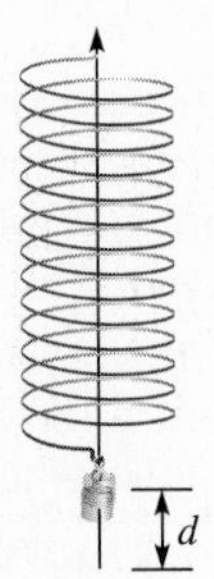

For Exercise 75

76. *Swinging Pendulum* The Columbus Museum of Science and Industry exhibits a Foucault pendulum 32 ft long that swings back and forth on a cable once in approximately 6 sec. The angle (in radians) θ between the cable and an imaginary vertical line is modeled by the equation

$$\theta = 0.25 \cos t.$$

Find the measure of angle θ when $t = 0$ and $t = 2.5$.

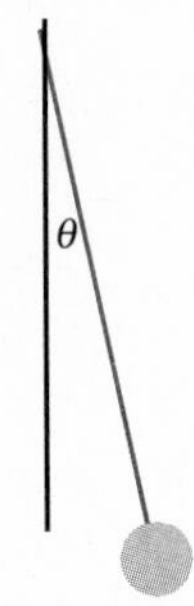

For Exercise 76

In Exercises 77 and 78, solve the problem.

77. *Too Close for Comfort* An F-15 aircraft flying at an altitude of 8000 ft passes directly over a group of vacationers hiking at 7400 ft. If θ is the angle of elevation from the hikers to the F-15, find the distance d from the group to the jet for the given angle.

a. $\theta = 45°$ **b.** $\theta = 90°$ **c.** $\theta = 140°$

78. *Manufacturing Swimwear* Get Wet, Inc., manufactures swimwear, a seasonal product. The monthly sales x (in thousands) for Get Wet swimsuits are modeled by the equation

$$x = 72.4 + 61.7 \sin \frac{\pi t}{6},$$

where $t = 1$ represents January, $t = 2$ February, and so on. Estimate the number of Get Wet swimsuits sold in January, April, June, October, and December. For which two of these months are sales the same? Explain why this might be so.

In Exercises 79 and 80, use the fundamental trigonometric identities to evaluate the expression.

79. $\cos \theta$ if $\sin \theta = 0.36$

80. $\tan \theta$ if $\cos \theta = -0.65$

EXTENDING THE IDEAS

In Exercises 81–84, find the value of a unique real number θ, $0 < \theta < 2\pi$, that satisfies the two conditions.

81. $\sin \theta = \frac{1}{2}$ and $\tan \theta < 0$.

82. $\cos \theta = \frac{\sqrt{3}}{2}$ and $\sin \theta < 0$.

83. $\tan \theta = -1$ and $\sin \theta < 0$.

84. $\sin \theta = \frac{-\sqrt{2}}{2}$ and $\tan \theta > 0$.

Exercises 85–88 refer to the unit circle in this figure. Point P is on the terminal side of an angle t and point Q is on the terminal side of an angle $t + \pi/2$.

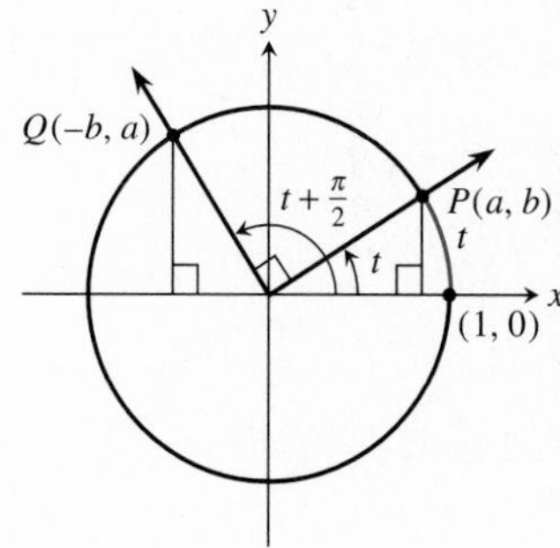

For Exercises 85–88

85. *Using Geometry in Trigonometry* Drop perpendiculars from points P and Q to the x-axis to form two right triangles. Explain how the right triangles are related.

86. *Using Geometry in Trigonometry* If the coordinates of point P are (a, b), explain why the coordinates of point Q are $(-b, a)$.

87. Explain why $\sin\left(t + \dfrac{\pi}{2}\right) = \cos t$.

88. Explain why $\cos\left(t + \dfrac{\pi}{2}\right) = -\sin t$.

In Exercises 89–91, solve the problem.

89. Writing to Learn In the figure for Exercises 85–88, t is an angle with radian measure $0 < t < \pi/2$. Draw a similar figure $\pi/2 < t < \pi$, and use it to explain why

$$\sin\left(t + \frac{\pi}{2}\right) = \cos t.$$

90. Writing to Learn Use the accompanying figure to explain each of the following.

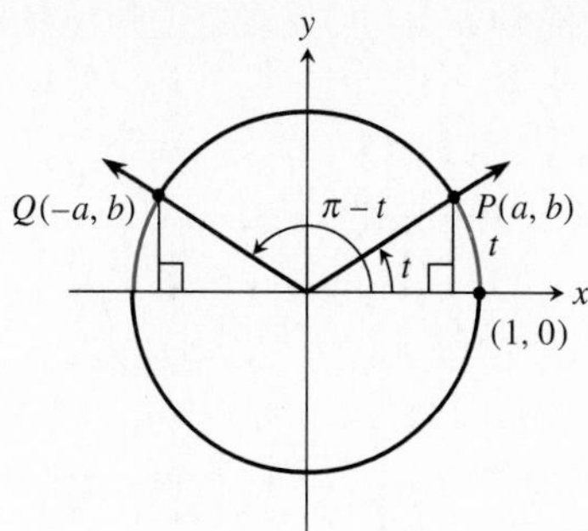

For Exercise 90

a. $\sin(\pi - t) = \sin t$
b. $\cos(\pi - t) = -\cos t$

91. *Approximation and Error Analysis* Use your grapher to complete the table to show that $\sin \theta \approx \theta$ (in radians) when $|\theta|$ is small. Physicists often use the approximation $\sin \theta \approx \theta$ for small values of θ. For what values of θ is the *magnitude of the error* in approximating $\sin \theta$ by θ less than 1% of $\sin \theta$? That is, solve the relation

$$|\sin \theta - \theta| < 0.01|\sin \theta|.$$

(*Hint:* Extend the table to include a column for values of

$$\frac{|\sin \theta - \theta|}{|\sin \theta|}.)$$

θ	$\sin \theta$	$\sin \theta - \theta$
−0.03		
−0.02		
−0.01		
0		
0.01		
0.02		
0.03		

Taylor Polynomials Radian measure allows the trigonometric functions to be approximated by simple polynomial functions. For example, in Exercises 92 and 93, sine and cosine are approximated by *Taylor polynomials,* named after the English mathematician Brook Taylor (1685–1731). Complete each table showing a Taylor polynomial in the third column. Describe the patterns in the table.

92.

θ	$\sin\theta$	$\theta - \dfrac{\theta^3}{6}$	$\sin\theta - \left(\theta - \dfrac{\theta^3}{6}\right)$
−0.3	−0.295 ⋯		
−0.2	−0.198 ⋯		
−0.1	−0.099 ⋯		
0	0		
0.1	0.099 ⋯		
0.2	0.198 ⋯		
0.3	0.295 ⋯		

93.

θ	$\cos\theta$	$1 - \dfrac{\theta^2}{2} + \dfrac{\theta^4}{24}$	$\cos\theta - \left(1 - \dfrac{\theta^2}{2} + \dfrac{\theta^4}{24}\right)$
−0.3			
−0.2			
−0.1			
0			
0.1	0.995 ⋯		
0.2	0.980 ⋯		
0.3	0.955 ⋯		

Objective

Students will be able to generate the graphs of the sine and cosine functions and explore various transformations of these graphs.

Motivate

Have students use a grapher to graph $y = \sin x$ and ask them what observations they can make about the graph.

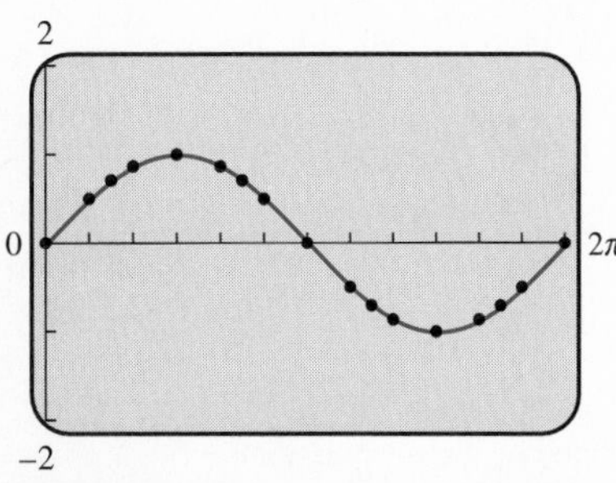

Figure 6.44 A graph of $y = \sin x$ superimposed on a scatter plot of its 17 special-angle values in the interval.

6.4 GRAPHS OF SINE AND COSINE FUNCTIONS

Introduction • Graph of $y = \sin x$ • Graph of $y = \cos x$ • Applying Transformations to Sinusoids • Combining Transformations of Sinusoids • Applications

Introduction

In the first three sections of this chapter you have seen that the trigonometric functions are rooted in the geometry of triangles and circles. It is these connections with geometry that give trigonometric functions their mathematical power and make them widely applicable in many fields.

Using the unit circle in Section 6.3, you learned that trigonometric functions, like polynomial, exponential, logarithmic, and rational functions, can be defined as functions of real numbers. At the end of that section you encountered some properties of the sine and cosine functions. In this section you will see the graphs of these functions, how the graphs relate to the unit circle, and transformations of these graphs.

Graph of $y = \sin x$

Figure 6.44 shows a graph of $y = \sin x$ obtained with a grapher in radian mode superimposed on a scatter plot of the 17 points corresponding to the special angles. This figure correctly suggests that the graph of $y = \sin x$ is a smooth curve through the 17 points.

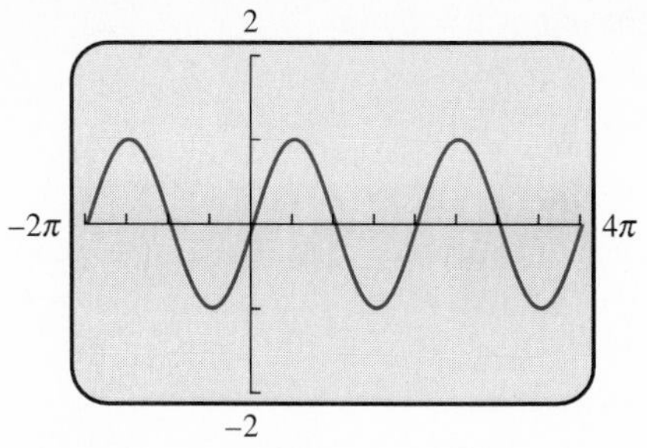

Figure 6.45 A graph of $y_1 = \sin x$ showing three complete periods.

We know that the sine function has the domain $(-\infty, \infty)$ and is periodic with a period of 2π. So to obtain a complete graph of $y = \sin x$ we shift the graph over the interval $[0, 2\pi]$ to the left and to the right 2π units, then 4π units, and so on, obtaining a graph over $(-\infty, \infty)$. This allows us to see that the end behavior, and all other behavior, is apparent in one complete period of the graph. (See Figure 6.45.)

Does it surprise you that knowing the sine function is periodic and knowing what the 17 special function values are together give such a good idea of what the graph looks like? More information is provided in Exercise 66.

The following Graphing Activity shows how the graph of the sine function is related to the unit circle.

Alert

Students should use the simultaneous graphing option in the Graphing Activity.

Graphing Activity Extensions

Discuss the connection between the Graphing Activity and Figure 6.46. A central part of the discussion should be the relationship between the unit circle and the graph of the sine function. Discuss how the arc length on the circle relates conceptually to the length on the x-axis.

▼ **GRAPHING ACTIVITY** **Graphing $y = \sin x$ and Unit Circle in Parametric Mode**

Set your grapher in radian mode and also in simultaneous graphing mode with $T\text{min} = 0$, $T\text{max} = 2\pi$, and $T\text{step} = \pi/24$. Let

$$x_1 = \cos t, \qquad y_1 = \sin t \qquad \text{Unit circle}$$

$$x_2 = t, \qquad y_2 = \sin t \qquad \text{The sine function}$$

1. Explain why the special angles are included in the above T-values.
2. Draw the graphs of the two parametric equations in a square viewing window that contains $[-2, 2\pi]$ by $[-2.7, 2.7]$. Use "trace" to see the connection between the y-coordinates on the unit circle and the y-coordinates on the sine function graph. (See Figure 6.46.) ▲

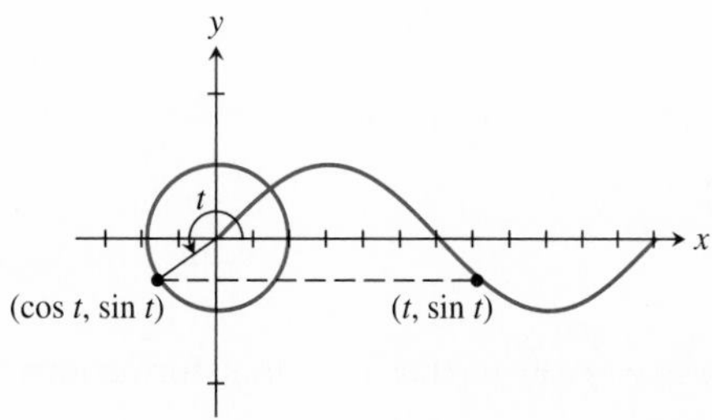

Figure 6.46

Teaching Note

Parametric graphing allows students to explore trigonometry in a way that was not possible before. Seeing the unit circle unwrap on the graphing screen is a powerful visualization that is critical for a deeper understanding of trigonometry.

There are two important ways to investigate the behavior of trigonometric functions. One involves the use of the unit circle, and the other a complete period of the function. Example 1 illustrates both approaches.

■ **EXAMPLE 1** **Describing the Behavior of $y = \sin x$**

Identify x-intercepts, maximum values, and minimum values for $y = \sin x$.

Solution

Because $y = \sin x$ is the second coordinate of the points on the unit circle, $y = 0$ when the angle puts the unit circle points on the x-axis, that is, when $x = 0$, $\pm\pi$, $\pm 2\pi$, and so on. Thus, we have

x-intercepts: $x = \pm n\pi$, for $n = 0, 1, 2, \ldots$.

Looking at the graph of $y = \sin x$ over the interval $[0, 2\pi]$ (see Figure 6.45), we see that there is a local maximum of 1 at $x = \pi/2$. Because the period of the sine function is 2π, all other maximum values occur at values of x obtained by adding (or subtracting) a multiple of 2π to (or from) $\pi/2$. Thus,

Maximum value: equal to 1 and occurs at

$$x = \frac{\pi}{2} \pm 2n\pi, \qquad \text{for } n = 0, 1, 2, \ldots.$$

NOTE

The values of x at which the local maximum values of $\sin x$ occur can also be written in the form

$$\frac{\pi}{2} + 2n\pi,$$

for $n = 0, \pm 1, \pm 2, \ldots$.

From the unit circle, we can see that the smallest value of y is -1. So,

Minimum value: equal to -1 and occurs at

$$x = \frac{3\pi}{2} \pm 2n\pi, \qquad \text{for } n = 0, 1, 2, \ldots.$$

■

Graph of $y = \cos x$

The function $y = \cos x$ also has the domain $(-\infty, \infty)$, the range $[-1, 1]$, and is periodic with period 2π. We can obtain the graph of $y = \cos x$ over $(-\infty, \infty)$ from its graph over $[0, 2\pi]$ just as we did with the sine function.

■ **EXAMPLE 2** **Describing the Behavior of the Cosine Function**

Draw a graph of $f(x) = \cos x$ that shows three periods. Find the x-intercepts, maximum values, and minimum values.

Solution

Because the period of f is 2π, any viewing window with $X\text{max} - X\text{min} = 3(2\pi) = 6\pi$ is acceptable. A graph of $y_1 = \cos x$ is shown in Figure 6.47.

Using the unit circle or one period of the graph of $y = \cos x$, we find the following.

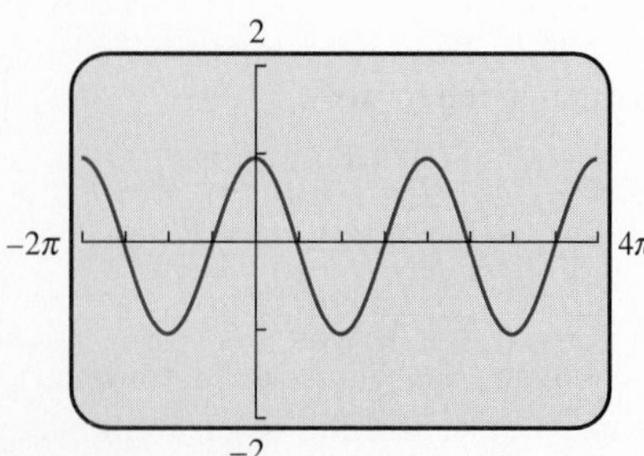

Figure 6.47

x-intercepts: occur at $x = \frac{\pi}{2} \pm n\pi$, for $n = 0, 1, 2, \ldots$.

Maximum value: equal to 1 and occurs at

$$x = \pm 2n\pi, \qquad \text{for } n = 0, 1, 2, \ldots.$$

Minimum value: equal to -1 and occurs at

$$x = \pi \pm 2n\pi, \qquad \text{for } n = 0, 1, 2, \ldots.$$

■

In Section 6.2 several problems required finding an acute angle of a right triangle. In these situations we were faced with solving equations such as $\sin \beta = 21/29$ or $\cos \theta = 0.437$. We used numerical zoom-in to solve these types of equations. We can now solve them graphically as well, because we know the graphs of the sine and cosine functions.

■ EXAMPLE 3 Solving a Trigonometric Equation

Find x, $0 \le x < \pi/2$, such that $\cos x = 3/15$.

Solution

Solve Graphically

The graphs of $y_1 = \cos x$ and $y_2 = 3/15$ intersect at $x = 1.369 \cdots$. (See Figure 6.48.)

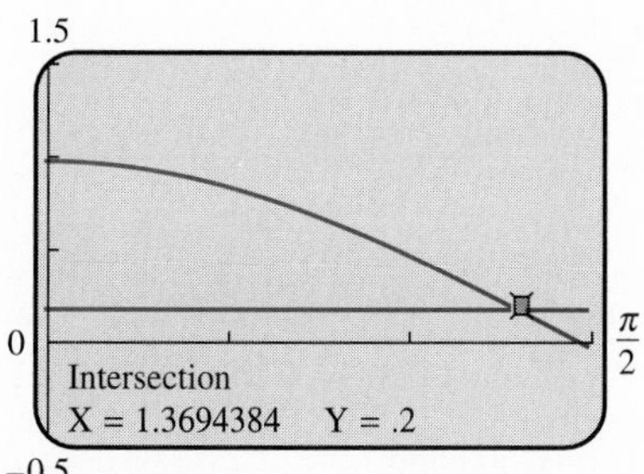

Figure 6.48

Interpret

$\cos 1.37 \approx 3/15$.

Support Numerically

We do this two ways. Our grapher shows that $\cos 1.37 = 0.199 \cdots \approx 3/15$. Also, we convert 1.37 rad to $1.37\left(\dfrac{180°}{\pi}\right) = 78.495 \cdots°$, which compares quite nicely with the solution found by numerical zoom-in in Example 7 of Section 6.2. ■

Teaching Note

Students should be expected to apply previously examined transformations to the graphs of the sine and cosine. Give attention to translation, reflection, stretching, and shrinking. A brief review on how $f(x) = x^2$ is related to $f(x) = ax^2$ and to $f(x) = (x - h)^2$ may be valuable. Relate this discussion to the trigonometric functions.

Applying Transformations to Sinusoids

The transformations studied in Section 2.6 can be applied to any function, including trigonometric functions. The graphs obtained by applying these transformations to the sine and cosine are *sinusoids.*

Definition: Sinusoid

A function is a **sinusoid** if it can be written in the form

$$f(x) = a \sin (bx + c) + d \qquad \text{or} \qquad f(x) = a \cos (bx + c) + d,$$

where a, b, c, and d are constants.

In Example 7a we will show that $\cos x = \sin (x + \pi/2)$. Thus, the graph of the cosine function is the graph of the sine function translated left $\pi/2$ units. Because of this connection, we could rewrite all sinusoids in the form $f(x) = a \sin (bx + c) + d$.

The following box reviews transformation terminology and applies it to the function $y = \sin x$. It also shows how the traditional sinusoid terms *amplitude, period,* and *phase shift* apply. We will introduce these concepts more fully in the next few pages.

Applying Transformations to Sinusoids

Transformation	Applied to $y = f(x)$	Applied to $y = \sin x$	Trigonometric Terminology
Vertical stretch/shrink	$y = af(x)$	$y = a \sin x$	Amplitude
Horizontal stretch/shrink	$y = f(bx)$	$y = \sin bx$	Period
Horizontal translation	$y = f(x - h)$	$y = \sin (x - h)$	Phase shift
Vertical translation	$y = f(x) + k$	$y = \sin x + k$	
Reflection across x-axis	$y = -f(x)$	$y = -\sin x$	
Reflection across y-axis	$y = f(-x)$	$y = \sin (-x)$ $= -\sin x$	

NOTE

We can conclude from the last two lines that reflecting the graph of $y = \sin x$ across the y-axis is the same as reflecting it across the x-axis.

EXAMPLE 4 Using Transformations on Sinusoids

Describe how the graphs of $y_1 = \sin x$ and $y_2 = 3 \sin 2x$ are related.

Solution

The graph of $y = 3 \sin 2x$ can be obtained from the graph of $y = \sin x$ by applying a vertical stretch of factor 3 and a horizontal shrink of factor 1/2, in either order. (See Figure 6.49.) ■

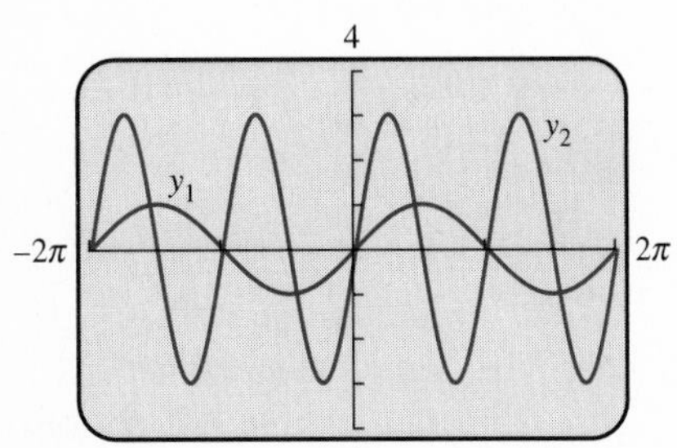

Figure 6.49 $y_1 = \sin x$ and $y_2 = 3 \sin 2x$.

Vertical Stretch or Shrink, Amplitude, and Reflection. When a vertical stretch or shrink is applied to a sinusoid, the characteristic of the graph that changes is called its *amplitude*.

Definition: Amplitude of a Sinusoid

The **amplitude** of

$$f(x) = a \sin (bx + c) + d \quad \text{or} \quad f(x) = a \cos (bx + c) + d,$$

is $|a|$.

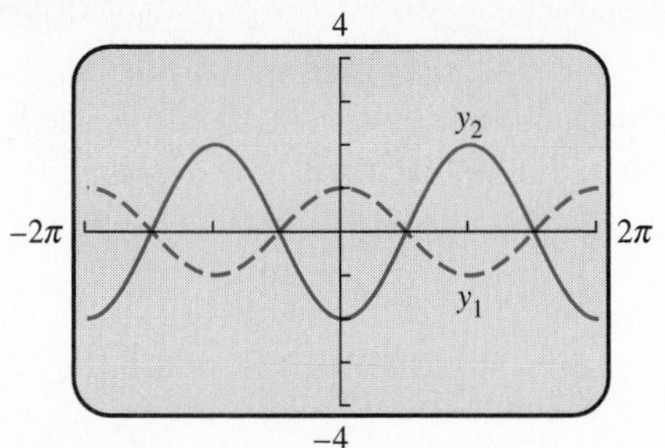

Figure 6.50 Graphs of $y_2 = -2 \cos x$ and $y_1 = \cos x$ over two periods.

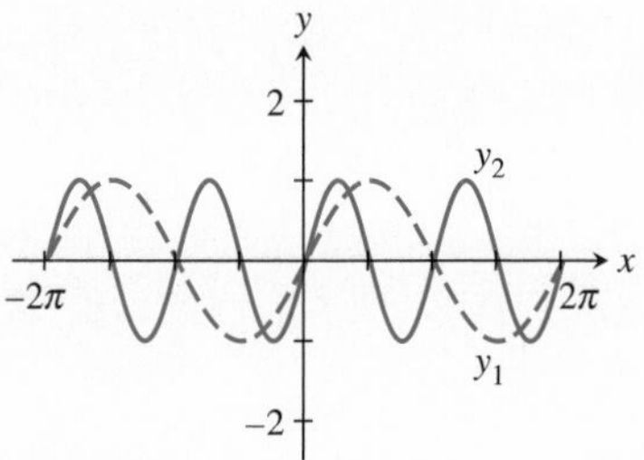

Figure 6.51 $y_1 = \sin x$ and $y_2 = \sin 2x$.

> **Alert**
>
> Students must understand that the period of a periodic function is the *smallest* positive real number h such that $f(x + h) = f(x)$ for all real numbers x in the domain of f.

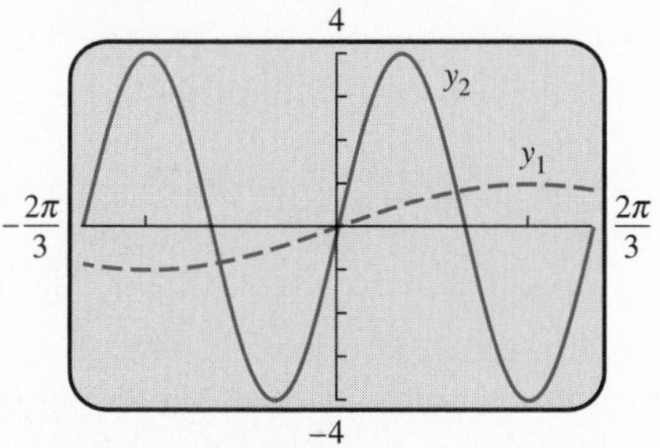

Figure 6.52 The graph of $y_2 = 4 \sin 3x$ has amplitude 4 and period 1/3 that of $y_1 = \sin x$.

EXAMPLE 5 Using Vertical Stretch and Amplitude

Describe how the graphs of $y_2 = -2 \cos x$ and $y_1 = \cos x$ are related. Graph two periods of $y_2 = -2 \cos x$.

Solution

The amplitude of $y_2 = -2 \cos x$ is $|-2| = 2$. The graph of $y_2 = -2 \cos x$ is a vertical stretch of $y_1 = \cos x$ by a factor of 2 and a reflection across the x-axis, done in either order. See Figure 6.50. ■

Horizontal Stretch or Shrink, and Period. The graph of $y_2 = \sin 2x$ is a horizontal shrink of the graph of $y_1 = \sin x$ by a factor of 1/2. This also shrinks the period by a factor of 1/2 from 2π to π. (See Figure 6.51.) More generally, the graph of $\sin x$ is shrunk horizontally by a factor $1/b$ to obtain the graph of $\sin bx$ for $b > 1$. The same is true for $\cos bx$.

Use your grapher to show that the graph of $\sin x$ is stretched horizontally by a factor $2 = 1/(1/2)$ to obtain the graph of $\sin (1/2)x$. This also stretches the period by a factor of 2 from 2π to 4π. In general, the graph of $\sin x$ is stretched horizontally by a factor $1/b$ to obtain the graph of $\sin bx$ for $0 < b < 1$. The same is true for $\cos bx$.

Recall from Section 6.3 that

$$\sin(-\theta) = -\sin\theta \qquad \text{and} \qquad \cos(-\theta) = \cos\theta.$$

So, when $b < 0$, the graph of $\cos bx$ is the same as the graph of $\cos |b|x$, and the graph of $\sin bx$ is the graph of $\sin |b|x$ reflected across the x-axis.

Here is a summary of how the constant b affects the period of the sine or cosine functions.

> **Period of Sine and Cosine Functions**
>
> The period of $y = a \sin bx$ and $y = a \cos bx$ is $\dfrac{2\pi}{|b|}$, $a \neq 0$, $b \neq 0$.

EXAMPLE 6 Using Horizontal Shrink and Period

Describe how the graphs of $y_2 = 4 \sin 3x$ and $y_1 = \sin x$ are related. Graph two periods of $y_2 = 4 \sin 3x$.

Solution

The graph of $y_2 = 4 \sin 3x$ can be obtained from the graph of $y_1 = \sin x$ by a vertical stretch of factor 4 and a horizontal shrink of factor 1/3. The period is $2\pi/3$. To graph two periods we use $X\text{min} = -2\pi/3$ and $X\text{max} = 2\pi/3$. See Figure 6.52. ■

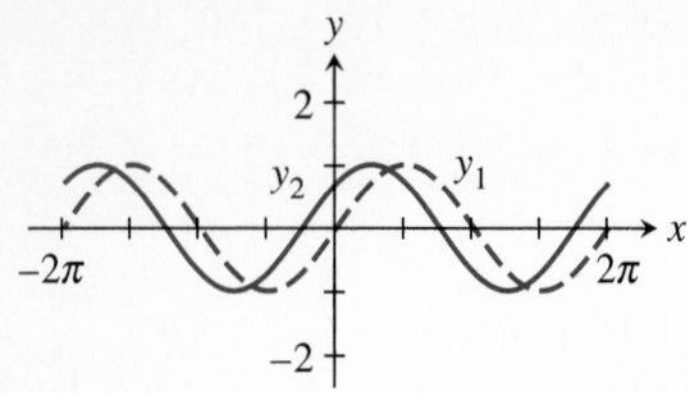

Figure 6.53 $y_2 = \sin(x + \pi/4)$ is a phase shift of $y_1 = \sin x$ of $-\pi/4$.

Teaching Note

It is worth noting that since the sine and cosine functions are periodic, different phase shifts can produce the same graph. For example, the graphs of

$$y = \sin\left(x - \frac{\pi}{2}\right)$$

and

$$y = \sin\left(x + \frac{3\pi}{2}\right)$$

are the same.

Horizontal Translation and Phase Shift. Recall from Section 2.6 that the graph of $y_2 = \sin(x + \pi/4)$ is a translation $\pi/4$ units to the left of $y_1 = \sin x$. (See Figure 6.53.) Using terminology that has its roots in electrical engineering, we say that $\sin(x + \pi/4)$ represents a **phase shift** of $\sin x$ of $-\pi/4$.

■ EXAMPLE 7 Showing Horizontal Translation or Phase Shift

Show that

a. $\cos x$ is a phase shift of $\sin x$ of $-\pi/2$ units; that is,

$$\cos x = \sin\left(x + \frac{\pi}{2}\right).$$

b. $\sin x$ is a phase shift of $\cos x$ of $\pi/2$ units; that is,

$$\sin x = \cos\left(x - \frac{\pi}{2}\right).$$

Solution

See Figure 6.54. ■

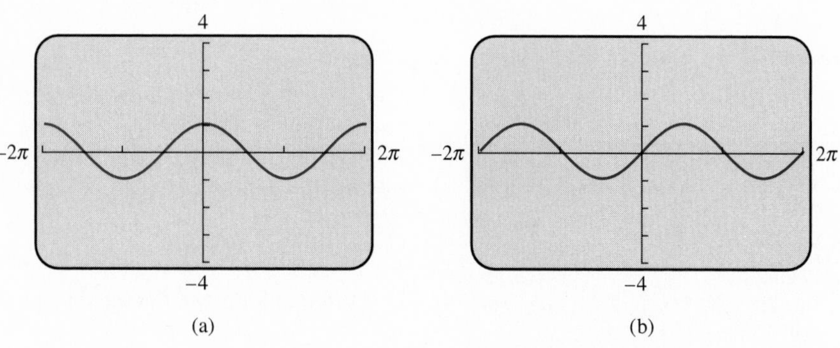

Figure 6.54 (a) Graphs of $\cos x$ and $\sin(x + \pi/2)$ appear to be the same. (b) Graphs of $\sin x$ and $\cos(x - \pi/2)$ appear to be the same.

The observations of Example 7 are among those summarized here.

Horizontal Translation Identities

For all values of x the following are true.

$$\cos\left(x - \frac{\pi}{2}\right) = \sin x \qquad \sin\left(x + \frac{\pi}{2}\right) = \cos x$$

$$\cos(x \pm 2\pi) = \cos x \qquad \sin(x \pm 2\pi) = \sin x$$

The last line of this box can be generalized as follows: A periodic function of period p is a translation of itself p units to the left or right. This is the property of the sine and cosine that we used to describe their graphs over $(-\infty, \infty)$.

Combining Transformations of Sinusoids

First we rewrite the two expressions for sinusoids in the form

$$a \sin [b(x - h)] + k \qquad \text{and} \qquad a \cos [b(x - h)] + k.$$

Here is a summary of the effect the constants a, b, h, and k have on the graphs of $y = \sin x$ or $y = \cos x$.

Characteristics of Sinusoids

The graphs of $y = a \sin [b(x - h)] + k$ and $y = a \cos [b(x - h)] + k$ have the following characteristics:

Terminology	**Transformation**
Amplitude $= \lvert a \rvert$	Vertical stretch or shrink of $\lvert a \rvert$
Period $= \dfrac{2\pi}{\lvert b \rvert}$, $b \neq 0$	Horizontal stretch or shrink of $\dfrac{1}{\lvert b \rvert}$
Phase shift $= h$	Horizontal translation of h
k	Vertical translation of k

There is a reflection across the x-axis for the cosine if $a < 0$, and for the sine if $ab < 0$.

■ **EXAMPLE 8 Describing Sinusoids**

Tell how to obtain the graph of the function from a basic trigonometric graph. Also, find the amplitude and period.

a. $f(x) = 3 \cos \left(\dfrac{x}{2} - \dfrac{\pi}{6} \right) + 1$ **b.** $g(x) = \sin (2 - x) + 4$

Solution

a. $f(x) = 3 \cos \left(\dfrac{x}{2} - \dfrac{\pi}{6} \right) + 1 = 3 \cos \left[\dfrac{1}{2} \left(x - \dfrac{\pi}{3} \right) \right] + 1$. The graph of f is obtained from the graph of $\cos x$ by applying, in order, the following transformations:

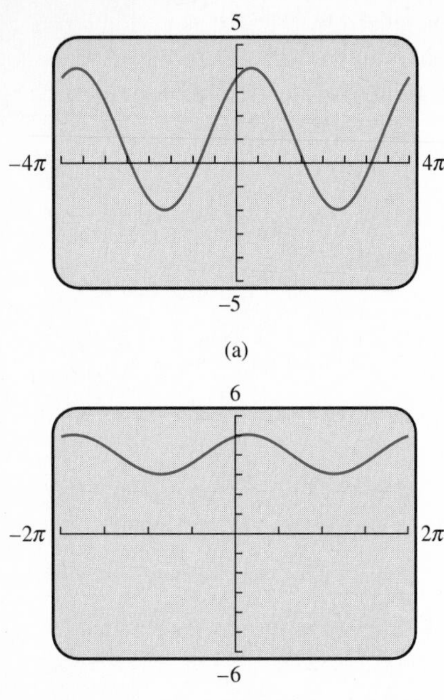

Figure 6.55

(a) $f(x) = 3\cos\left(\frac{x}{2} - \frac{\pi}{6}\right) + 1.$

(b) $g(x) = \sin(2 - x) + 4.$

1. A horizontal stretch of $1/(1/2) = 2$ to get $\cos\frac{1}{2}x$
2. A horizontal translation right $\pi/3$ units to get $\cos\left[\frac{1}{2}\left(x - \frac{\pi}{3}\right)\right]$
3. A vertical stretch of 3 to get $3\cos\left[\frac{1}{2}\left(x - \frac{\pi}{3}\right)\right]$
4. A vertical translation up 1 unit to get $f(x) = 3\cos\left[\frac{1}{2}\left(x - \frac{\pi}{3}\right)\right] + 1$

Comparing with $y = a\cos[b(x - h)] + k$, we see that $a = 3$ and $b = 1/2$. Therefore,

- amplitude $= |a| = |3| = 3$,
- period $\frac{2\pi}{|b|} = 2\pi \div \frac{1}{2} = 4\pi$.

b. $g(x) = \sin(2 - x) + 4 = \sin[-(x - 2)] + 4 = -\sin(x - 2) + 4.$ The graph of g is obtained from the graph of $\sin x$ by applying, in order, the following transformations:

1. A horizontal translation right 2 units to get $\sin(x - 2)$
2. A reflection across the x-axis to get $-\sin(x - 2)$
3. A vertical translation up 4 units to get $g(x) = -\sin(x - 2) + 4$.

Comparing with $y = a\sin[b(x - h)] + k$, we see that $a = -1$ and $b = 1$. Therefore,

- amplitude $= |a| = |-1| = 1$,
- period $2\pi/|b| = 2\pi/1 = 2\pi$.

Support Graphically

Figure 6.55 shows two periods of each graph. ■

Notice that the high and low points of the graph of f in Figure 6.55a are equidistant from the horizontal line $y = k = 1$. For this reason, the horizontal line $y = k$ is sometimes called the **midline of a sinusoid.** Also, the range of the function f is

$$[-2, 4] = [k - a, k + a],$$

which helps you find a viewing window for the graph of a sinusoid.

Applications

A sinusoid f always has a maximum (max f) and a minimum (min f). In Example 9, we use the following form for the amplitude (see Exercise 55)

$$a = \frac{1}{2}(\max f - \min f).$$

EXAMPLE 9 APPLICATION: Calculating Ebb and Flow

On a particular July 4 high tide in Galveston, Texas, occurs at 9:36 A.M. At that time the water at the end of the 61st Street Pier is 2.7 m deep. Low tide occurs at 3:48 P.M., and the water is only 2.1 m deep. Assume the depth of the water is a sinusoidal function of time with a period of 1/2 a lunar day, which is about 12 hours and 24 minutes.

a. At what time on the 4th of July does the first low tide occur?
b. What is the approximate depth of the water at 6:00 A.M. and at 3:00 P.M.?
c. What is the first time on the 4th of July when the water was 2.4 m deep?

Solution

Model

Let $D(t) = a \sin [b(t - h)] + k$ be the depth of the water at any time t in hours after midnight. For example 3:48 P.M. becomes $t = 12 + 3 + 48/60 = 15.8$.

Amplitude:

$$a = \frac{1}{2}(\max D - \min D)$$

$$= \frac{2.7 - 2.1}{2} = 0.3$$

Period: The period is 12 h 24 min. So

$$\frac{2\pi}{b} = 12.4 \quad \text{12 h 24 min = 12.4 h.}$$

$$b = \frac{2\pi}{12.4} = \frac{\pi}{6.2}.$$

Midline:

$$k = \frac{\max D + \min D}{2}$$

$$= 2.4$$

Horizontal Translation h: $D(t)$ reaches a maximum value at 9:36 A.M. (9.6 h after midnight). If we back up 1/4 of a period (3.1 h), we are at a starting point of a sinusoidal wave: $9.6 - 3.1 = 6.5$. The curve is a translation to the right 6.5 units, so $h = 6.5$.

Sinusoid:

$$D(t) = 0.3 \sin \left[\frac{\pi}{6.2}(t - 6.5)\right] + 2.4$$

Follow-up

Ask students what the graphical approach reveals about the sine and cosine functions that was not observed in the representations based on (a) acute angles of right triangles, (b) general angles in standard position, and (c) arcs of the unit circle.

Assignment Guide

Day 1: Ex. 3–51, multiples of 3
Day 2: Ex. 53, 54, 55–57, 59–62, 67–69, 72

Cooperative Learning

Group Activity: Ex. 65, 66, 73

Notes on Exercises

Ex. 41–44 emphasize that any sinusoidal graph has numerous representations.
Ex. 65–66 are related to Taylor polynomials.

Ongoing Assessment

Self-Assessment: Ex. 7, 13, 25, 47, 63
Embedded Assessment: Ex. 30, 56

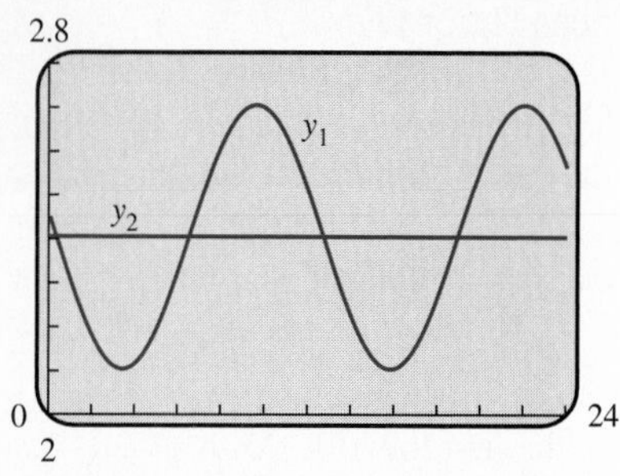

Figure 6.56 The depth of the water graphed as a function of time from midnight to midnight.

Solve Graphically

Figure 6.56 shows the graph of $y_1 = D(t)$ and $y_2 = 2.4$ (the midline) over a 24-h period beginning at midnight on July 4.

a. The first low tide corresponds to the first local minimum on the graph, at $t = 3.4$.

b. The depth at 6:00 A.M. is $D(6) \approx 2.32$, and at 3:00 P.M. is $D(12 + 3) = D(15) \approx 2.12$.

c. The first time the water is 2.4 m deep corresponds to the leftmost intersection of the sinusoid and the midline in Figure 6.56, which is $t = 0.3$.

Interpret

The first low tide is at 3:24 A.M. The water is about 2.32 m deep at 6:00 A.M. and about 2.12 m deep at 3:00 P.M. The water first has a depth of 2.4 m at 12:18 A.M. ■

Quick Review 6.4

In Exercises 1–6, state the sign (positive or negative) of the function in each quadrant.

1. $\sin x$ **2.** $\cos x$

3. $\tan x$ **4.** $\cot x$

5. $\sec x$ **6.** $\csc x$

In Exercises 7–12, give the radian measure of the angle.

7. 135° **8.** 270°

9. −150° **10.** −30°

11. 450° **12.** 510°

In Exercises 13–18, state how the graphs of y_1 and y_2 are related.

13. $y_1 = x^2$ and $y_2 = (x + 2)^2$

14. $y_1 = \sqrt{x}$ and $y_2 = 3\sqrt{x}$

15. $y_1 = e^x$ and $y_2 = e^{-x}$

16. $y_1 = \ln x$ and $y_2 = 0.5 \ln x$

17. $y_1 = x^3$ and $y_2 = x^3 - 2$

18. $y_1 = \dfrac{1}{x}$ and $y_2 = -\dfrac{1}{x}$

SECTION EXERCISES 6.4

In Exercises 1–4, graph one period of the function.

1. $y = 2 \sin x$ **2.** $y = 2.5 \sin x$

3. $y = 3 \cos x$ **4.** $y = -2 \cos x$

In Exercises 5–10, graph three periods of the function.

5. $y = 5 \sin 2x$ **6.** $y = 3 \cos \dfrac{x}{2}$

7. $y = 0.5 \cos 3x$ **8.** $y = 20 \sin 4x$

9. $y = 4 \sin \dfrac{x}{4}$ **10.** $y = 8 \cos 5x$

In Exercises 11–16, specify the period and the amplitude of each function. Then specify the viewing window that is shown.

11. $y = 1.5 \sin 2x$ **12.** $y = 2 \cos 3x$

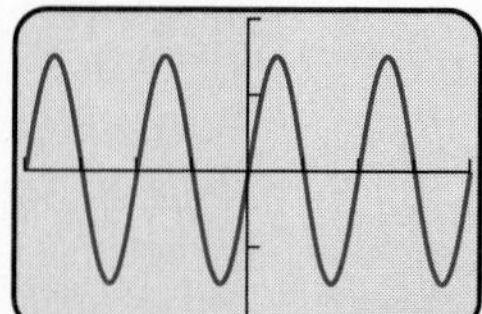

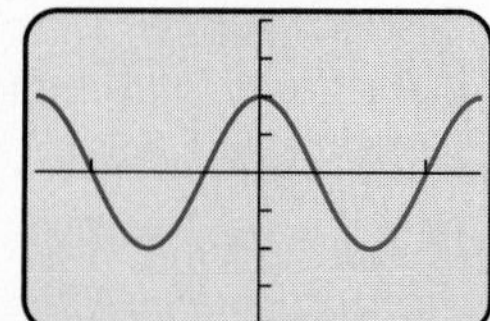

For Exercises 11–12

13. $y = -3 \cos 2x$

14. $y = 5 \sin \frac{x}{2}$

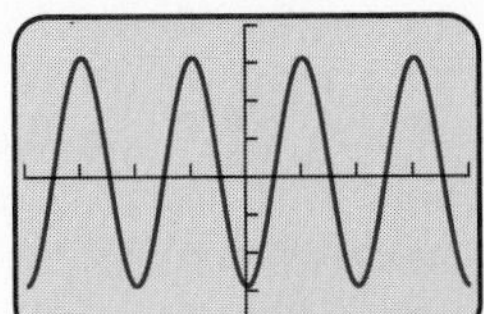

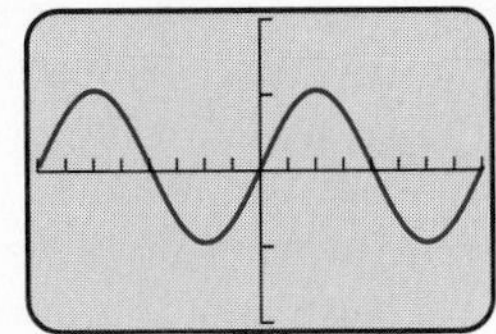

For Exercises 13–14

15. $y = -4 \sin \frac{\pi}{3}x$

16. $y = 3 \cos \pi x$

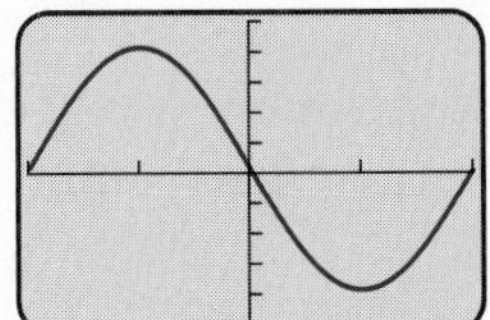

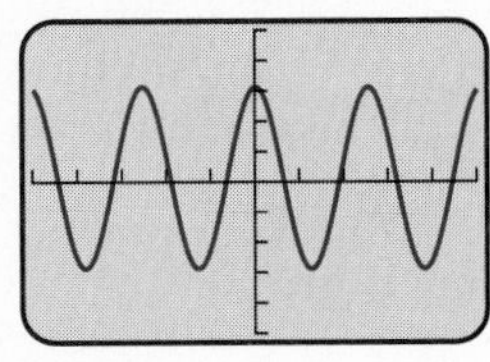

For Exercises 15–16

In Exercises 17–20, identify the maximum and minimum values and the zeros for the function in the interval $[-2\pi, 2\pi]$.

17. $y = 2 \sin x$

18. $3 \cos \frac{x}{2}$

19. $y = \cos 2x$

20. $y = \frac{1}{2} \sin x$

In Exercises 21–26, solve for x in the given interval.

21. $\sin x = 3/4, \quad 0 \le x < \pi/2$

22. $\cos x = 2/3, \quad 0 \le x < \pi/2$

23. $\cos x = 0.4, \quad 3\pi/2 \le x < 2\pi$

24. $\sin x = 0.6, \quad \pi/2 \le x < \pi$

25. $\sin x = -4/5, \quad \pi \le x < 3\pi/2$

26. $\cos x = -1/5, \quad \pi/2 \le x < \pi$

In Exercises 27–32, describe how the graph of the function is related to a basic trigonometric graph. Graph two periods.

27. $y = 0.5 \sin 3x$

28. $y = 1.5 \cos 4x$

29. $y = -\frac{2}{3} \cos \frac{x}{3}$

30. $y = \frac{3}{4} \sin \frac{x}{5}$

31. $y = 3 \cos \frac{2\pi x}{3}$

32. $y = -2 \sin \frac{\pi x}{4}$

In Exercises 33–40, describe how the graphs of f and g are related.

33. $f(x) = \sin x$ and $g(x) = 2 \sin x$

34. $f(x) = \cos 2x$ and $g(x) = \frac{5}{3} \cos 2x$

35. $f(x) = \sin x$ and $g(x) = \sin\left(x + \frac{\pi}{3}\right)$

36. $f(x) = 2 \cos\left(x + \frac{\pi}{3}\right)$ and $g(x) = \cos\left(x + \frac{\pi}{4}\right)$

37. $f(x) = 3 \sin 2x + 1$ and $g(x) = \sin 2x - 1$

38. $f(x) = 0.3 \cos 2x - 4$ and $g(x) = 0.5 \cos 2x + 1$

39. $f(x) = 2 \cos \pi x$ and $g(x) = 2 \cos 2\pi x$

40. $f(x) = 3 \sin \frac{2\pi x}{3}$ and $g(x) = 2 \sin \frac{\pi x}{3}$

In Exercises 41–44, select the pair of functions that have identical graphs.

41. **a.** $y = \cos x$ **b.** $y = \sin\left(x + \frac{\pi}{2}\right)$ **c.** $y = \cos\left(x + \frac{\pi}{2}\right)$

42. **a.** $y = \sin x$ **b.** $y = \cos\left(x - \frac{\pi}{2}\right)$ **c.** $y = \cos x$

43. **a.** $y = \sin\left(x + \frac{\pi}{2}\right)$ **b.** $y = -\cos(x - \pi)$

c. $y = \cos\left(x - \frac{\pi}{2}\right)$

44. **a.** $y = \sin\left(2x + \frac{\pi}{4}\right)$ **b.** $y = \cos\left(2x - \frac{\pi}{2}\right)$

c. $y = \cos\left(2x - \frac{\pi}{4}\right)$

In Exercises 45–52, state the amplitude, period, phase shift, and vertical translation of the sinusoid.

45. $y = -2 \sin\left(x - \frac{\pi}{4}\right) + 1$

46. $y = -3.5 \sin\left(2x - \frac{\pi}{2}\right) - 1$

47. $y = 5 \cos\left(3x - \frac{\pi}{6}\right) + 0.5$

48. $y = 3 \cos(x + 3) - 2$

49. $y = 2 \cos 2\pi x + 1$

50. $y = 4 \cos 3\pi x - 2$

51. $y = \frac{7}{3} \sin\left(\frac{x + 5}{2}\right) - 1$

52. $y = \frac{2}{3} \cos\left(\frac{x - 3}{4}\right) + 1$

In Exercises 53 and 54, find values for *a, b, h* and *k* so that the graph of the function $y = a \sin[b(x - h)] + k$ is the one shown.

53.

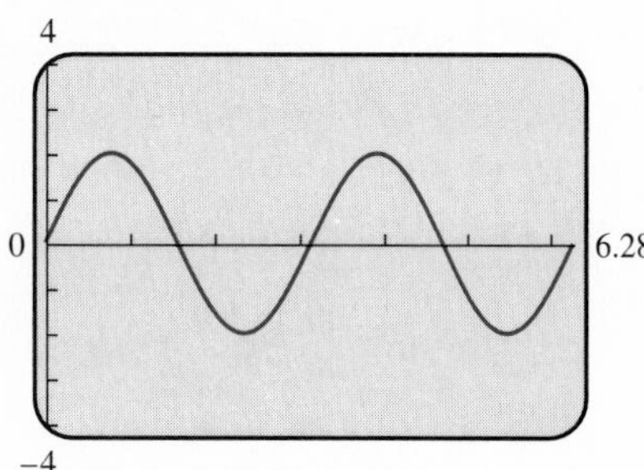

54.

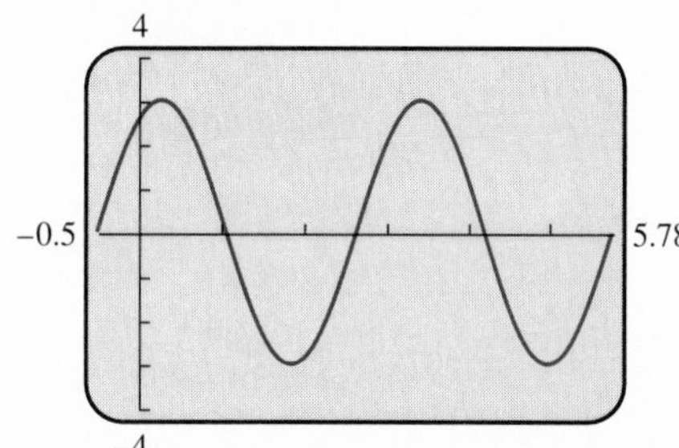

In Exercises 55–68, solve the problem.

55. Let *f* be a sinusoidal function.

a. Show that the maximum value of *f* is $\max f = k + |a|$, and the minimum value is $\min f = k - |a|$.

b. Show that the amplitude of *f* is

$$\frac{1}{2}(\max f - \min f).$$

56. Writing to Learn Explain why shifting the points found for sin *x* in Example 1 to the left $\pi/2$ units gives the corresponding points found for cos *x* in Example 2.

57. *Points of Intersection* Graph $y = \ln x$ and $y = 2 \cos x$.

a. How many points of intersection do there appear to be?

b. Find the coordinates of each point of intersection.

58. *Points of Intersection* Graph $y = 1.3^{-x}$ and $y = 1.3^{-x} \cos x$ for *x* in the interval $[-1, 8]$.

a. How many points of intersection do there appear to be?

b. Find the coordinates of each point of intersection.

59. *Motion of a Buoy* A signal buoy in the Chesapeake Bay bobs up and down with the height *h* of its transmitter (in feet) above sea level modeled by $h = a \sin bt + 5$. During a small squall its height varies from 1 to 9 ft and there are 3.5 sec from one 9-ft height to the next. What are the values of the constants *a* and *b?*

For Exercise 59

60. *Ferris Wheel* A Ferris wheel 50 ft in diameter makes one revolution every 40 sec. If the center of the wheel is 30 ft above the ground, how long after reaching the low point is a rider 50 ft above the ground?

61. *Tsunami Wave* An earthquake occurred at 9:40 A.M. on Nov. 1, 1755, at Lisbon, Portugal, and started a *tsunami* (often called a tidal wave) in the ocean. It produced waves that traveled more than 540 ft/sec (370 mph) and reached a height of 60 ft. If the period of the waves was 30 min or 1800 sec, estimate the length *L* between the crests.

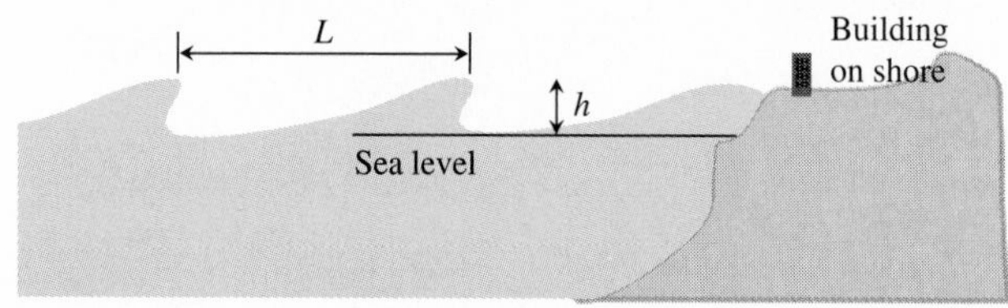

For Exercise 61

62. *Blood Pressure* The function

$$P = 120 + 30 \sin 2\pi t$$

models the blood pressure (in millimeters of mercury) for a person who has a (high) blood pressure of 150/90; *t* represents seconds.

a. What is the period of this function?

b. How many heartbeats are there each minute?

c. Graph this function to model a 10-sec time interval.

63. *Ebb and Flow* On a particular Labor Day, the high tide in southern California occurs at 7:12 A.M. At that time you measure the water at the end of the Santa Monica Pier to be 11 ft deep. At 1:24 P.M. it is low tide, and you measure the water to be only 7 ft deep. Assume the depth of the water is a sinusoidal function of time with a period of 1/2 a lunar day, which is about 12 h 24 min.

a. At what time on that Labor Day does the first low tide occur?

b. What was the approximate depth of the water at 4:00 A.M. and at 9:00 P.M.?

c. What is the first time on that Labor Day that the water is 9 ft deep?

64. *Lunar Month* A lunar month is 29.5 days because it is about 29.5 days from one full moon to the next. The phases of the moon (*full, waning quarter, new, waxing quarter, full, . . .*) are modeled by a periodic function of time. Suppose the area of the lunar surface visible from earth is a sinusoidal function of time with a period of one lunar month. Write an algebraic formula for visible lunar area versus time based on the sine function, and draw its graph. Start ($t = 0$) with a full moon.

65. *Approximating Cosine*

a. Draw a scatter plot (x, cos x) for the 17 special angles x, where $-\pi \leq x \leq \pi$.

b. Find a quartic regression for the data.

c. Compare the approximation to the cosine function given by the quartic regression with the Taylor polynomial approximations given in Exercise 93 of Section 6.3.

66. *Approximating Sine*

a. Draw a scatter plot (x, sin x) for the 17 special angles x, where $-\pi \leq x \leq \pi$.

b. Find a cubic regression for the data.

c. Compare the approximation to the sine function given by the cubic regression with the Taylor polynomial approximations given in Exercise 92 of Section 6.3.

EXTENDING THE IDEAS

67. *Visualizing a Musical Note* A piano tuner strikes a tuning fork for the note middle C and creates a sound wave that can be modeled by

$$y = 1.5 \sin 524\pi t,$$

where t is the time in seconds.

a. What is the period p of this function?

b. What is the frequency $f = 1/p$ of this note?

c. Graph the function.

68. Writing to Learn In a certain video game a cursor bounces back and forth horizontally across the screen at a constant rate. Its distance d from the center of the screen varies with time t and hence can be described as a function of t. Explain why this horizontal distance d from the center of the screen *does not vary* according to an equation $d = a \sin bt$, where t represents seconds. You may find it helpful to include a graph in your explanation.

In Exercises 69–72, the graphs of the sine and cosine functions are waveforms like the figure below. By correctly labeling the coordinates of points A, B, and C, you will get the graph of the function shown here.

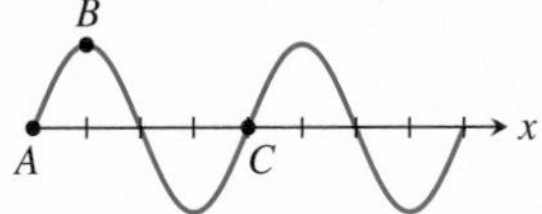

For Exercises 69–72

69. $y = 3 \cos 2x$ and $A = \left(-\frac{\pi}{4}, 0\right)$. Find B and C.

70. $y = 4.5 \sin\left(x - \frac{\pi}{4}\right)$ and $A = \left(\frac{\pi}{4}, 0\right)$. Find B and C.

71. $y = 2 \sin\left(3x - \frac{\pi}{4}\right)$ and $A = \left(\frac{\pi}{12}, 0\right)$. Find B and C.

72. $y = 3 \sin(2x - \pi)$, and A is the first x-intercept on the right of the y-axis. Find A, B, and C.

73. *The Ultimate Sinusoidal Equation*

$$y = a \sin[b(x - h)] + k$$

a. Explain why you can assume b is positive. (*Hint:* Replace b by $-B$ and simplify.)

b. Use one of the horizontal translation identities to prove that the equation

$$y = a \cos[b(x - h)] + k$$

has the same graph as

$$y = a \sin[b(x - H)] + k$$

for a correctly chosen value of H. Explain how to choose H.

c. Give a unit-circle argument for the identity $\sin(\theta + \pi) = -\sin\theta$. Support your unit-circle argument graphically.

d. Use the identity from part c to prove that

$$y = -a \sin[b(x - h)] + k, \quad a > 0,$$

has the same graph as

$$y = a \sin [b(x - H)] + k, \quad a > 0$$

for a correctly chosen value of H. Explain how to choose H.

e. Combine your results from parts a–d to prove that any sinusoid can be represented by the equation

$$y = a \sin [b(x - H)] + k,$$

where a and b are both positive.

Objective

Students will be able to generate the graphs for the tangent, cotangent, secant, and cosecant functions and to explore various transformations of these graphs.

6.5 GRAPHS OF OTHER TRIGONOMETRIC FUNCTIONS

Graph of the Tangent Function • Graph of the Cotangent Function • Graphs of the Secant and Cosecant Functions

Graph of the Tangent Function

We investigate the graph of the tangent function by utilizing its connection with the unit circle (see Figure 6.57).

$$\tan t = \frac{\sin t}{\cos t}.$$

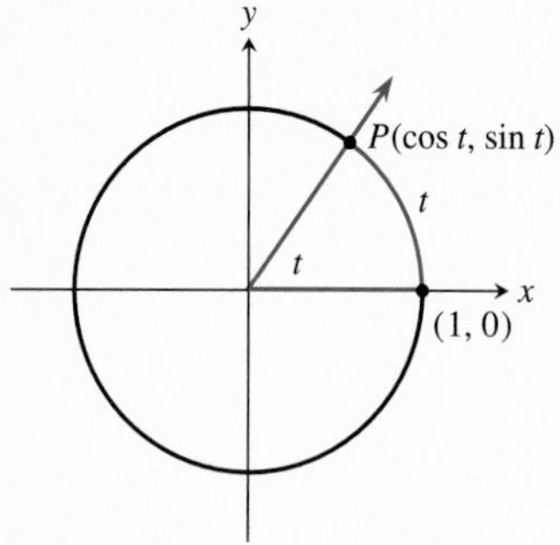

Figure 6.57 Notice t is the measure of the angle and the arc length (see Exercise 34).

Whenever the denominator $\cos t$ is zero, the quotient $\tan t$ is undefined. Consequently, $\tan t$ is undefined whenever t results in a point on the y-axis, that is, when $t = \pi/2 + n\pi$ and n is any integer. Notice also that the zeros of $\tan t$ are the same as the zeros of $\sin t$.

The table below shows the values of $y = \tan t$ for the special angles between $-\pi/2$ and $\pi/2$, plus two angles near these two zeros of the denominator.

t	-1.45	$-\pi/3$	$-\pi/4$	$-\pi/6$	0	$\pi/6$	$\pi/4$	$\pi/3$	1.45
y	-8.24	-1.73	-1	-0.58	0	0.58	1	1.73	8.24

Motivate

Have students use a grapher to graph $y = \tan x$ in both dot mode and connected mode and ask them what observations they can make about the graph.

By plotting these $(t, \tan t)$ pairs and connecting the points with a smooth curve, we obtain the graph of $y = \tan t$ for t in the interval $(-\pi/2, \pi/2)$. (See Figure 6.58.)

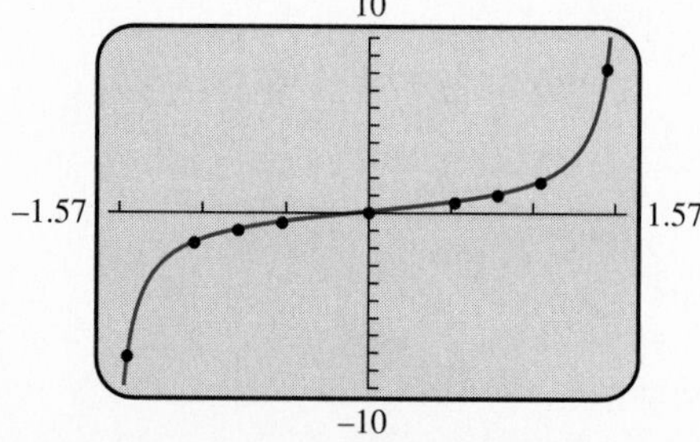

Figure 6.58 This is a graph of the built-in grapher function $y_1 = \tan t$ superimposed on a scatter plot of nine values of $\tan t$.

Teaching Note

The graph of $y = \tan x$ deserves special attention. It is helpful for students to see the graphs of $\sin x$, $\cos x$, and $\tan x$ superimposed in the same viewing window. Discuss the relationships among these three functions.

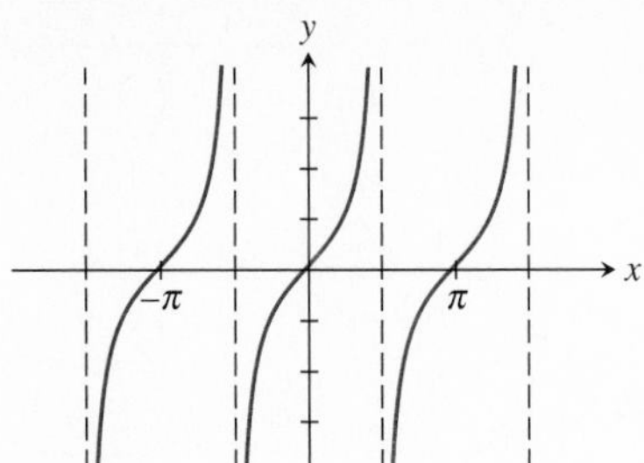

$y = \tan x$
Period: π
Domain: $x \neq \frac{\pi}{2} \pm n\pi$
Range: $(-\infty, \infty)$
Vertical asymptotes: $x = \frac{\pi}{2} \pm n\pi$
Zeros: $\pm n\pi$
$n = 0, 1, 2, 3, \ldots$

Figure 6.59

REMARK

In case you were wondering, once we have the trigonometric functions defined on the real numbers, it does not matter which variable we use (x, t, or some other) for the independent variable.

Because $\sin t$ and $\cos t$ are periodic and have the same period, we expect $\tan t = (\sin t)/(\cos t)$ to be periodic. You may be surprised to find that its period is π. (See Exercise 33.) Figure 6.59 shows three periods of $y = \tan x$ and describes its behavior.

The constants a, b, h, and k influence the graph of $y = a \tan [b(x - h)] + k$ in much the same way that they do for the graph of $y = a \sin [b(x - h)] + k$. The constant a yields a vertical stretch or shrink, b affects the period, h causes a horizontal translation, and k causes a vertical translation. The terms *amplitude* and *phase shift,* however, are not used because they only apply to sinusoids. A grapher is an effective tool for exploring graphs of tangent functions.

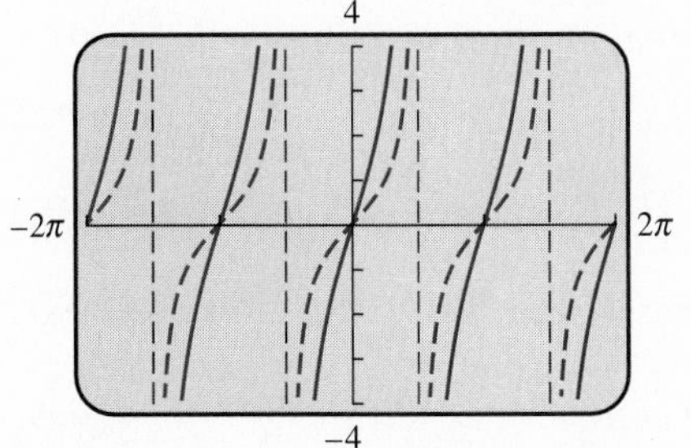

Figure 6.60 Four periods of $y_2 = 3 \tan x$ (solid curves) and $y_1 = \tan x$ (dashed curves).

■ EXAMPLE 1 Graphing Two Tangent Functions

Describe how the graphs of $y_2 = 3 \tan x$ and $y_1 = \tan x$ are related. Graph four periods of each function.

Solution

Because $y_2 = a \tan x$ where $a = 3$, the graph of $y_2 = 3 \tan x$ can be obtained from the graph of $y_1 = \tan x$ by a vertical stretch of factor 3. (See Figure 6.60.) Notice that it is more difficult to "see" the magnitude of the stretch for tangent than it is for sine or cosine. ■

■ EXAMPLE 2 Graphing a Tangent Function

Describe the graph of $y_1 = -\tan 2x$. Graph four periods, and locate the vertical asymptotes.

Solution

Because $y_1 = a \tan bx$ where $a = -1$ and $b = 2$, the period of $y_1 = -\tan 2x$ is $\pi/b = \pi/2$. Any viewing window where

$$X\max - X\min = 4\left(\frac{\pi}{2}\right) = 2\pi$$

will show four periods of $y_1 = -\tan 2x$.

Teaching Note

Remind students that the sine and cosine functions each have a period of 2π, but the tangent and cotangent functions each have a period of π. You may wish to introduce the identities $\cos (x \pm \pi) = -\cos x$ and $\sin (x \pm \pi) = -\sin x$ to explain this fact.

Alert

Remind students to pay close attention to the radian and degree mode settings on their calculators. They should always check to see if the calculator mode is properly set.

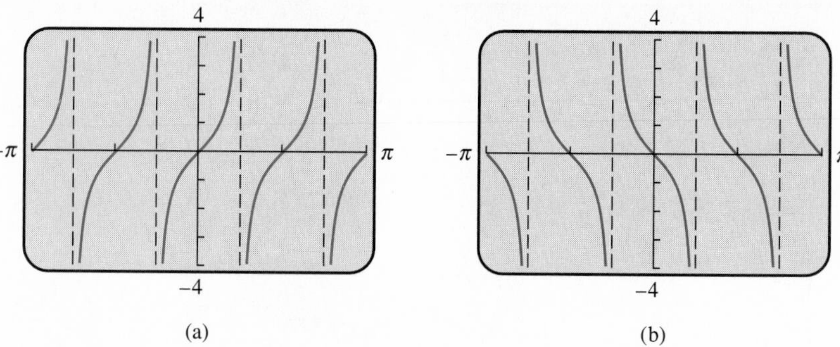

Figure 6.61

The value $a = -1$ causes the graph of $y_2 = \tan 2x$ (Figure 6.61a) to be reflected across the x-axis to obtain the graph of $y_1 = -\tan 2x$ (Figure 6.61b).

The graph of $y_2 = \tan 2x$ is a horizontal shrink of $\tan x$ by a factor of 1/2. Thus, the vertical asymptotes $x = \pi/2 \pm n\pi$ of $\tan x$ are shrunk by 1/2 to $\pi/4 \pm n\pi/2$ to obtain the vertical asymptotes of y_2. And y_1 and y_2 have the same vertical asymptotes. ■

Graph of the Cotangent Function

REMARK

The fundamental trigonometric identities stated in Section 6.2 are also valid for the expanded domains of the trigonometric functions. (See Exercises 35 and 36.)

We learn about the graph of the cotangent function by analyzing the quotient identity

$$\cot x = \frac{\cos x}{\sin x}.$$

Exercise 33 also establishes that the period of $\cot x$ is π. Notice that the zeros of the $\cot x$ are the same as the zeros of $\cos x$.

Whenever the denominator $\sin x$ is zero, that is, whenever x is a multiple of π, the function $\cot x$ is undefined. Also, when $\sin x$ is near zero, $|\cot x|$ is large. There is a vertical asymptote on the graph of $y = \cot x$ whenever $x = \pm n\pi$. Figure 6.62 shows the graph of $y = \cot x$ and describes its behavior.

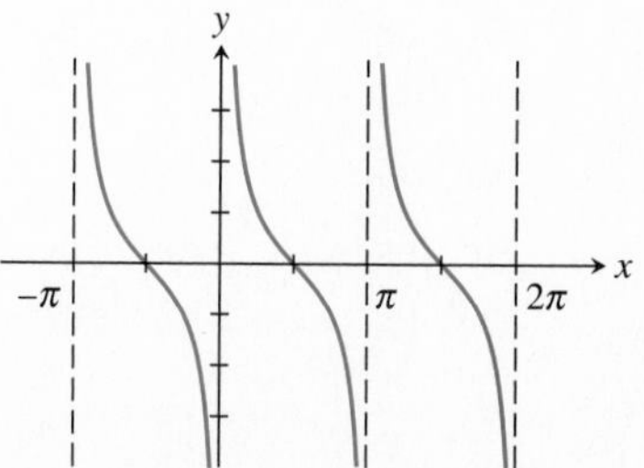

$y = \cot x$
Period: π
Domain: $x \neq \pm n\pi$
Range: $(-\infty, \infty)$
Vertical asymptotes: $x = \pm n\pi$
Zeros: $\frac{\pi}{2} \pm n\pi$
$n = 0, 1, 2, 3, \ldots$

Figure 6.62 Three periods of $y = \cot x$.

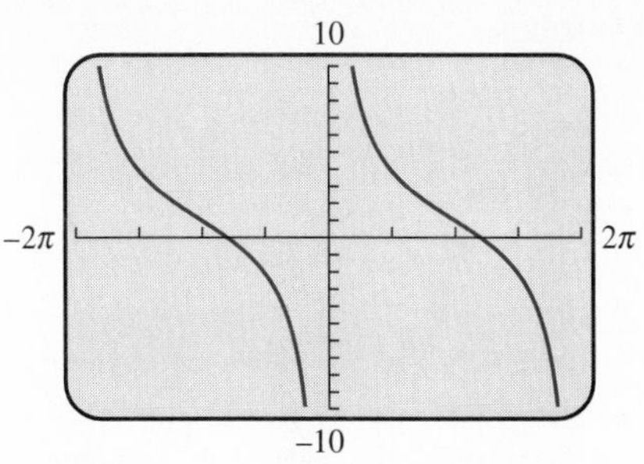

Figure 6.63

Follow-up

Ask what the relationship is between the domain of the tangent and secant functions and the zeros of the cosine function. **(The domain of the tangent and secant functions is the set of all real numbers that are not zeros of the cosine function.)**

Assignment Guide

Day 1: Ex. 3–33, multiples of 3
Day 2: Ex. 34, 35, 37–39, 42, 46, 49, 50, 51, 53, 55

Cooperative Learning

Group Activity: Ex. 56

Notes on Exercises

Ex. 17–32 concern domain, range, period, and asymptotes. Have students think of these concepts in terms of transformations. This will make the sketching procedure much easier.
Ex. 39–44 can also be solved numerically or algebraically. The algebraic approach will be covered in Section 7.3. Help students make the connection between the different solution methods.

Ongoing Assessment

Self-Assessment: Ex. 5, 23, 43, 45
Embedded Assessment: Ex. 34, 38

EXAMPLE 3 Graphing a Cotangent Function

Describe the graph of $f(x) = 3 \cot \frac{x}{2} + 1$. Graph two periods.

Solution

The period of f is 2π, twice the period of $y = \cot x$, as a result of the horizontal stretch caused by the constant $b = 1/2$. The value $a = 3$ causes a vertical stretch with factor 3, and the constant $k = 1$ causes a vertical translation up one unit.

Figure 6.63 shows two periods of the graph of f. ■

Graphs of the Secant and Cosecant Functions

Important characteristics of the graphs of the two remaining trigonometric functions, $y = \sec x$ and $y = \csc x$, can be inferred from the fact that they are reciprocals of the cosine and the sine functions, respectively.

Whenever $\cos x = 1$, its reciprocal $\sec x$ is also 1. Whenever $\cos x$ is close to zero, $|\sec x|$ is large. Whenever $\cos x = 0$, $\sec x$ is undefined.

Similarly, whenever $\sin x = 1$, its reciprocal $\csc x$ is also 1. Whenever, $\sin x$ is close to zero, $|\csc x|$ is large. Whenever $\sin x = 0$, $\csc x$ is undefined.

A handy fact to keep in mind is that the reciprocal of a periodic function is periodic with the same period. (See Exercise 37.) We can use this fact to show that the period of $\sec x$ or $\csc x$ is 2π.

Figure 6.64 shows the graphs of the secant and cosecant functions together with the graphs of $y = \cos x$ and $y = \sin x$ respectively.

Notice in Figure 6.64a that a maximum of $y = \cos x$ corresponds to a local minimum of $y = \sec x$, and a local minimum of $y = \cos x$ corresponds to a local maximum of $y = \sec x$. A similar observation can be made for $y = \sin x$ and $y = \csc x$.

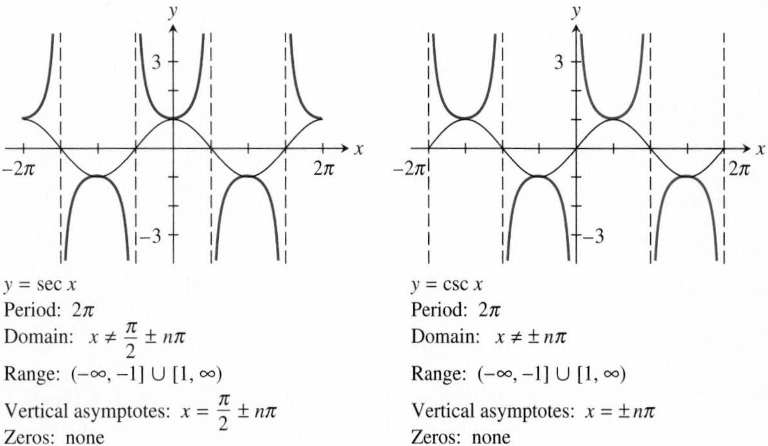

$y = \sec x$
Period: 2π
Domain: $x \neq \frac{\pi}{2} \pm n\pi$
Range: $(-\infty, -1] \cup [1, \infty)$
Vertical asymptotes: $x = \frac{\pi}{2} \pm n\pi$
Zeros: none
$n = 0, 1, 2, 3, \ldots$

(a)

$y = \csc x$
Period: 2π
Domain: $x \neq \pm n\pi$
Range: $(-\infty, -1] \cup [1, \infty)$
Vertical asymptotes: $x = \pm n\pi$
Zeros: none
$n = 0, 1, 2, 3, \ldots$

(b)

Figure 6.64 Four periods each of (a) $y = \sec x$, and (b) $y = \csc x$.

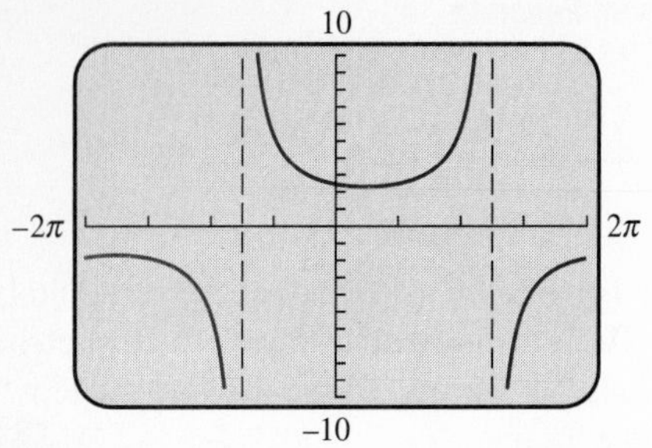

Figure 6.65

■ EXAMPLE 4 Graphing a Secant Function

Describe the graph of $f(x) = 2\sec\left(0.5x - \frac{\pi}{8}\right)$. Graph one period.

Solution

$f(x) = 2\sec\left[0.5\left(x - \frac{\pi}{4}\right)\right]$, so $a = 2$, $b = 0.5$, and $h = \pi/4$.

The period of f is $2\pi/b = 2\pi/0.5 = 4\pi$, twice the period of $y = \sec x$. The value $a = 2$ causes a vertical stretch of factor 2.

Because $h = \pi/4$, the graph of f is a translation of $y = 2\sec(0.5x)$ to the right $\pi/4$ units. Figure 6.65 shows one period of the graph of f. ■

Here is a summary of some key properties of the trigonometric functions.

Properties of Trigonometric Functions

Property	sin x	cos x	tan x	cot x	sec x	csc x
Period	2π	2π	π	π	2π	2π
Domain	$(-\infty, \infty)$	$(-\infty, \infty)$	$x \neq \frac{\pi}{2} \pm n\pi$	$x \neq \pm n\pi$	$x \neq \frac{\pi}{2} \pm n\pi$	$x \neq \pm n\pi$
Range	$[-1, 1]$	$[-1, 1]$	$(-\infty, \infty)$	$(-\infty, \infty)$	$(-\infty, -1] \cup [1, \infty)$	$(-\infty, -1] \cup [1, \infty)$
Asymptotes	None	None	$x = \frac{\pi}{2} \pm n\pi$	$x = \pm n\pi$	$x = \frac{\pi}{2} \pm n\pi$	$x = \pm n\pi$
Zeros	$x = \pm n\pi$	$x = \frac{\pi}{2} \pm n\pi$	$x = \pm n\pi$	$x = \frac{\pi}{2} \pm n\pi$	None	None
Even or Odd	Odd	Even	Odd	Odd	Even	Odd

$(n = 0, 1, 2, \ldots)$

Quick Review 6.5

In Exercises 1–4, state the period of the function.

1. $y = \cos 2x$

2. $y = \sin 3x$

3. $y = \sin \frac{1}{3}x$

4. $y = \cos \frac{1}{2}x$

In Exercises 5–8, find the zeros and vertical asymptotes of the function.

5. $y = \frac{x - 3}{x + 4}$

6. $y = \frac{x + 5}{x - 1}$

7. $y = \frac{x + 1}{(x - 2)(x + 2)}$

8. $y = \frac{x + 2}{x(x - 3)}$

In Exercises 9 and 10, θ is an acute angle satisfying the given condition. Evaluate the remaining trigonometric functions of θ.

9. $\sin\theta = \dfrac{3}{5}$ **10.** $\cos\theta = \dfrac{12}{13}$

In Exercises 11–16, tell whether the function is even, odd, or neither.

11. $y = x^2 + 4$ **12.** $y = \sqrt{x}$

13. $y = x^3 + x$ **14.** $y = \dfrac{1}{x^2 - 4}$

15. $y = x^2 - x$ **16.** $y = \dfrac{1}{x}$

SECTION EXERCISES 6.5

In Exercises 1–4, identify the graph of each function.

1. Graphs of one period of csc x and 2 csc x are shown.

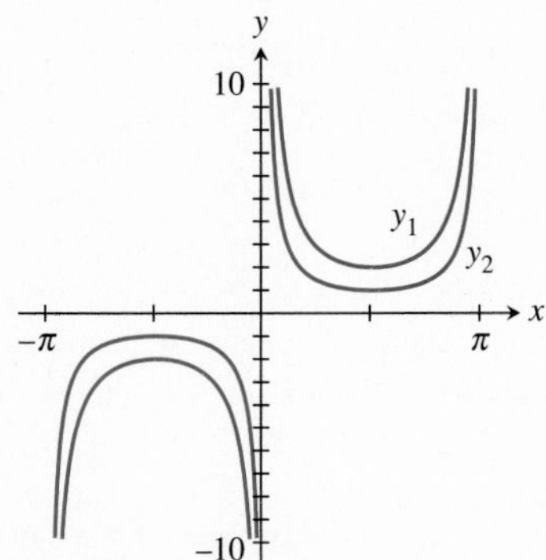

For Exercise 1

2. Graphs of two periods of 0.5 tan x and 5 tan x are shown.

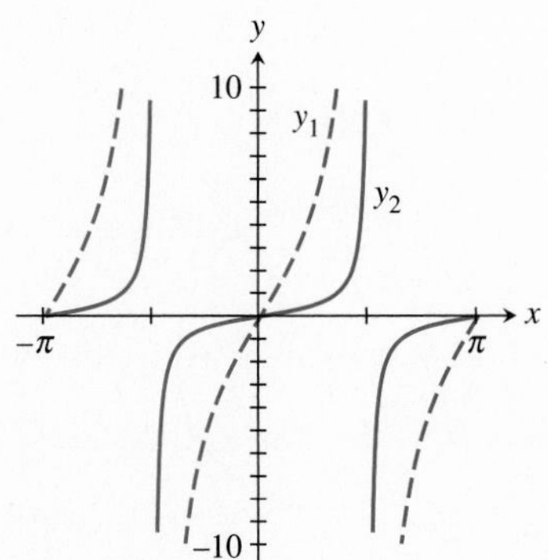

For Exercise 2

3. Graphs of csc x and 3 csc $2x$ are shown.

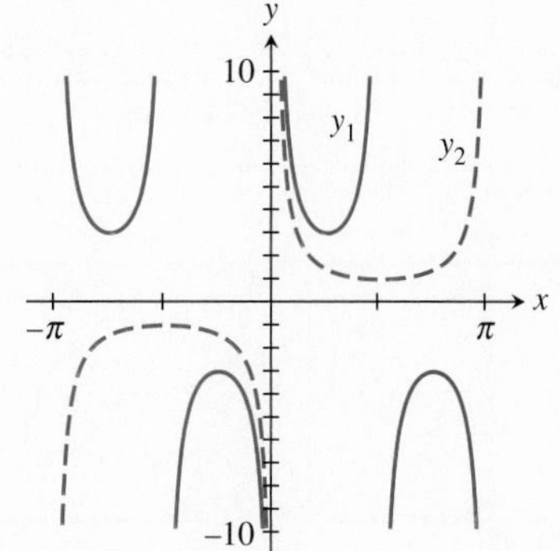

For Exercise 3

4. Graphs of cot x and cot $(x - 0.5) + 3$ are shown.

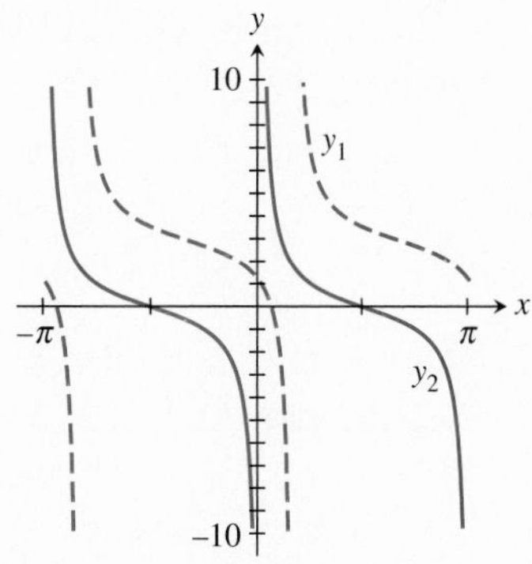

For Exercise 4

In Exercises 5–12, graph two periods of the function.

5. $y = \tan 2x$ **6.** $y = -\cot 3x$

7. $y = \sec 3x$ **8.** $y = \csc 2x$

9. $y = 2 \cot 2x$ **10.** $y = 3 \tan (x/2)$

11. $y = \csc (x/2)$ **12.** $y = 3 \sec 4x$

In Exercises 13–16, match the trigonometric function with its graph. State the Xmin and Xmax values of the viewing window.

13. $y = -2 \tan x$ **14.** $y = \cot x$

15. $y = \sec 2x$ **16.** $y = -\csc x$

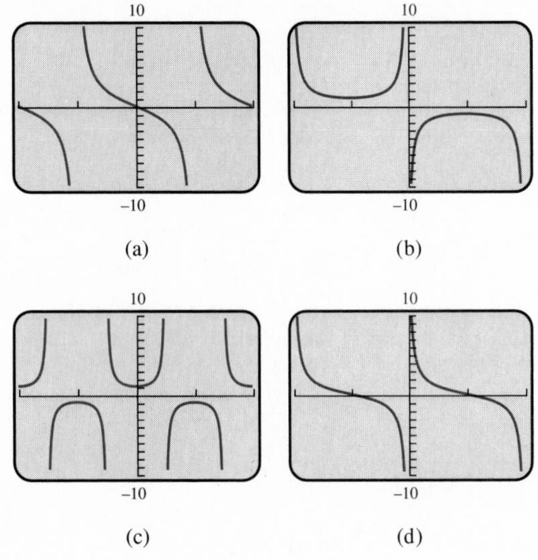

For Exercises 13–16

In Exercises 17–32, explain how the graph of the function is related to a basic trigonometric graph. Determine the period, domain, range, zeros, and asymptotes (if any).

17. $y = 3 \tan x$ **18.** $y = -\tan x$

19. $y = 0.5 \sec x$ **20.** $y = \sec (-x)$

21. $y = 3 \csc x$ **22.** $y = 2 \tan x$

23. $y = -3 \cot \frac{1}{2}x$ **24.** $y = 2 \cot \frac{1}{2}x$

25. $y = 2 \csc x$ **26.** $y = -2 \sec \frac{1}{2}x$

27. $y = 2 \tan 3x$ **28.** $y = \sec -\frac{1}{2}x$

29. $y = -\tan \frac{\pi}{2}x$ **30.** $y = 2 \tan \pi x$

31. $y = 3 \sec 2x$ **32.** $y = 4 \csc \frac{1}{3}x$

In Exercises 33–38, solve the problem.

33. Writing to Learn The figure shows a unit circle and an angle t whose terminal side is in quadrant III.

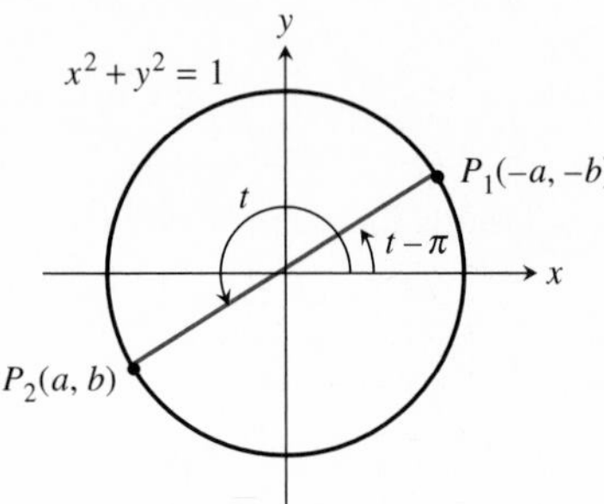

For Exercise 33

a. If the coordinates of point P_2 are (a, b), explain why the coordinates of point P_1 on the circle and the terminal side of angle $t - \pi$ are $(-a, -b)$.

b. Explain why $\tan t = \frac{b}{a}$.

c. Find $\tan (t - \pi)$, and show that $\tan t = \tan (t - \pi)$.

d. Explain why the period of the tangent function is π.

e. Explain why the period of the cotangent function is π.

34. Writing to Learn Explain why it is correct to say $y = \tan x$ is the slope of the terminal side of angle x in standard position. P is on the unit circle.

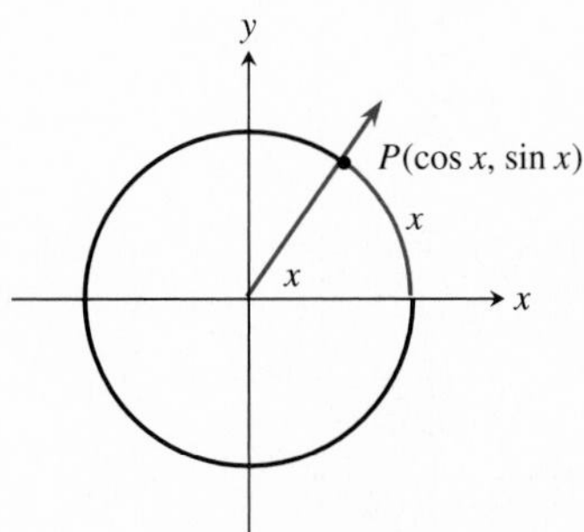

For Exercise 34

35. *Pythagorean Identities* Confirm that the Pythagorean identities stated in Section 6.2 are valid for the trigonometric functions with expanded domains.

36. *Reciprocal and Quotient Identities* Confirm that the reciprocal and quotient identities stated in Section 6.2 are valid for the trigonometric functions with expanded domains.

37. *Periodic Functions* Let f be a periodic function with period p. That is, p is the smallest positive number such that

$$f(x + p) = f(x)$$

for any value of x in the domain of f. Show that the reciprocal $1/f$ is periodic with period p.

38. *Identities* Use the unit circle to give a convincing argument for the identity.

a. $\sin(t + \pi) = -\sin t$

b. $\cos(t + \pi) = -\cos t$

c. Use parts a and b to show that $\tan(t + \pi) = \tan t$. Explain why this is *not* enough to conclude that the period of tangent is π.

In Exercises 39–44, solve for x in the given interval graphically.

39. $\tan x = 1.3, \quad 0 \le x \le \frac{\pi}{2}$

40. $\sec x = 2.4, \quad 0 \le x \le \frac{\pi}{2}$

41. $\cot x = -0.6, \quad \frac{3\pi}{2} \le x \le 2\pi$

42. $\csc x = -1.5, \quad \pi \le x \le \frac{3\pi}{2}$

43. $\csc x = 2, \quad 0 \le x \le 2\pi$

44. $\tan x = 0.3, \quad 0 \le x \le 2\pi$

In Exercises 45 and 46, solve the problem.

45. *Lighthouse Coverage* The Bolivar Lighthouse is located on a small island 350 ft from the shore of the mainland as shown in the figure.

a. Express the distance d as a function of the angle x.

b. If x is 1.55 rad, what is d?

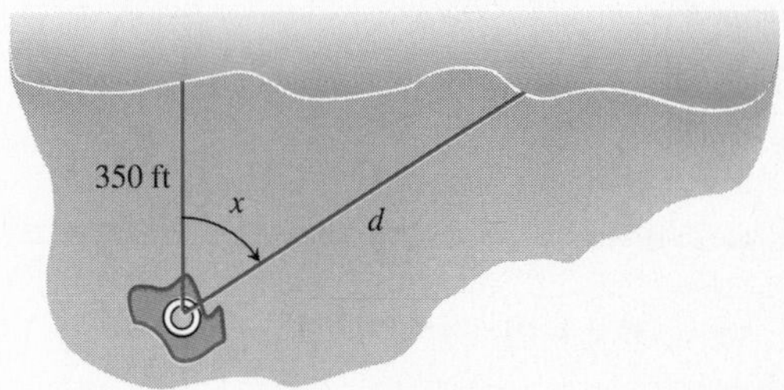

For Exercise 45

46. *Hot-Air Balloon* A hot-air balloon over Albuquerque, New Mexico, is being blown due east from point P and traveling at a constant height of 800 ft. The angle y is formed by the ground and the line of vision from P to the balloon. This angle changes as the balloon travels.

a. Express the horizontal distance x as a function of the angle y.

b. When the angle is $\pi/20$ rad, what is its horizontal distance from P?

c. An angle of $\pi/20$ rad is equivalent to how many degrees?

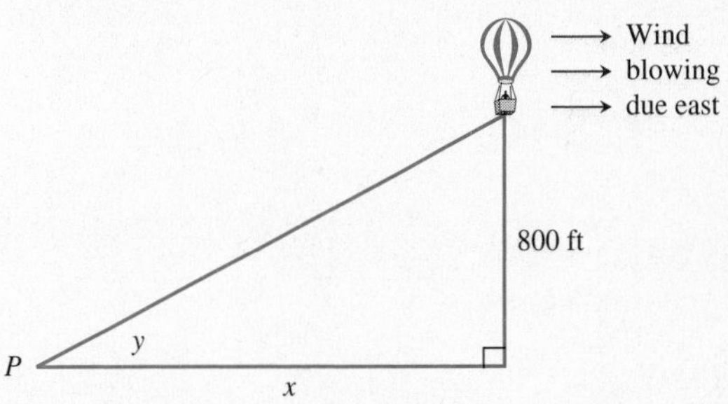

For Exercise 46

EXTENDING THE IDEAS

In Exercises 47–50, find approximate solutions for the equation in the interval $-\pi < x < \pi$.

47. $\tan x = \csc x$

48. $\sec x = \cot x$

49. $\sec x = 5 \cos x$

50. $4 \cos x = \tan x$

In Exercises 51 and 52, graph both f and g in the $[-\pi, \pi]$ by $[-10, 10]$ viewing window. Estimate values in that interval for which $f > g$.

51. $f(x) = 5 \sin x$ and $g(x) = \cot x$

52. $f(x) = -\tan x$ and $g(x) = \csc x$

In Exercises 53–57, solve the problem.

53. Writing to Learn Graph the function $f(x) = -\cot x$ on the interval $(-\pi, \pi)$. Explain why it is correct to say that f is increasing on the interval $(0, \pi)$, but it is not correct to say that f is increasing on the interval $(-\pi, \pi)$.

54. **Writing to Learn** Graph functions $f(x) = -\sec x$ and

$$g(x) = \frac{1}{x - \dfrac{\pi}{2}}$$

simultaneously in the viewing window $[0, \pi]$ by $[-10, 10]$. Discuss whether you think functions f and g are equivalent.

55. *Television Coverage* A television camera is on a platform 30 m from the point on High Street where the Worthington Memorial Day Parade will pass. Express the distance d from the camera to a particular parade float as a function of the angle x, and graph the function over the interval $-\pi/2 < x < \pi/2$.

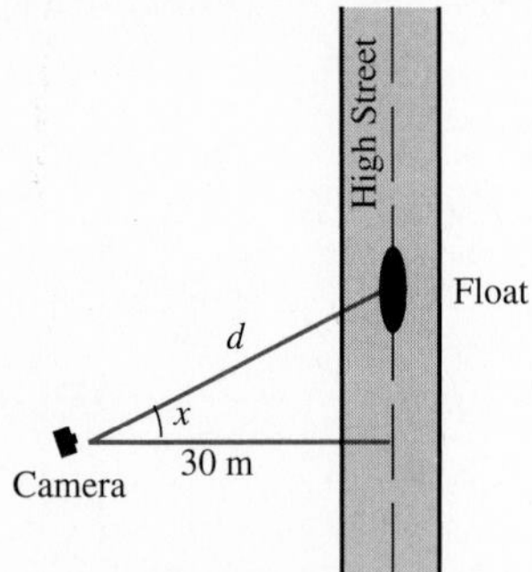

For Exercise 55

56. *Name Game* The word *sine* comes from the Latin *sinus,* which means bend, curve, fold, or hollow. Authorities do not agree on how that came to be a name for the trigonometric function.

  a. *Cosine* means "sine of the complement." Explain why this makes sense.

  b. In the figure, segment CE is perpendicular to OD. That is, CE is tangent to the unit circle at D. Use similar triangles to find the four segment lengths that represent $\tan t$, $\cot t$, $\sec t$, and $\csc t$. (*Hint:* $\sin t = OA$, $\cos t = OB$, and OD has length 1.)

  c. What kind of segments are the ones you found in part a relative to the circle? Use your answer to explain why the names "tangent, cotangent, secant," and "cosecant" make sense.

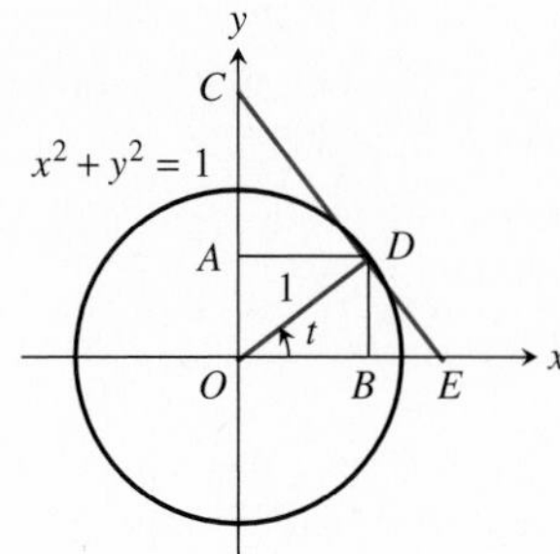

For Exercise 56

57. *Capillary Action* A film of liquid in a thin (capillary) tube has surface tension γ (gamma) given by

$$\gamma = \frac{1}{2}h\rho gr \sec \phi,$$

where h is the height of the liquid in the tube, ρ (rho) is the density of the liquid, $g = 9.8$ m/sec^2 is the acceleration due to gravity, r is the radius of the tube, and ϕ (phi) is the angle of contact between the tube and the liquid's surface. Whole blood has a surface tension of 0.058 N/m (newton per meter) and a density of 1050 kg/m^3. Suppose that blood rises to a height of 1.5 m in a capillary blood vessel of radius 4.7×10^{-6} m. What is the contact angle between the capillary vessel and the blood surface? ($1\text{ N} = 1(\text{kg} \cdot \text{m})/\text{sec}^2$.)

6.6 ADVANCED TRIGONOMETRIC GRAPHS

Objective

Students will be able to graph sums, differences, and other combinations of trigonometric and algebraic functions.

Sums and Differences That Are Sinusoids • Sums and Differences That Are Not Sinusoids • Combining Trigonometric and Algebraic Functions • Applications

Sums and Differences That Are Sinusoids

In Section 6.4 you studied sinusoidal functions, functions that can be written in the form

Motivate

Have students use a grapher to graph $f(x)$, $g(x)$, and the sum $(f + g)(x)$, where

$$f(x) = x^2 \text{ and } g(x) = \frac{x^2}{2} - 3x.$$

Point out that the sum of two quadratic functions is another quadratic function (or a line or a constant) and ask whether students think it is possible for the sum of two sinusoids to be a sinusoid. **(Yes)**

Exploration Extensions

Repeat the exercise for the functions $y = \sin x + \sin 2x$ and

$$y = 2 \sin x - 2 \cos\left(x - \frac{\pi}{2}\right).$$

$$y = a \sin [b(x - h)] + k,$$

and therefore have the shape of a sine curve.

Sinusoids model a variety of physical and social phenomena—such as sound waves, voltage in alternating electric current, the velocity of air flow during the human respiratory cycle, and many others. Sometimes these phenomena interact in an additive fashion. For example, if y_1 models the sound of one tuning fork and y_2 models the sound of a second tuning fork, then $y_1 + y_2$ models the sound when they are struck simultaneously. So we are interested in whether the sums and differences of sinusoids are again sinusoids.

● EXPLORATION Investigating Sinusoids

Graph these functions, one at a time, in the viewing window $[-2\pi, 2\pi]$ by $[-6, 6]$. Which ones appear to be sinusoids?

$y = 3 \sin x + 2 \cos x$ $\qquad$ $y = 2 \sin x - 3 \cos x$

$y = 2 \sin 3x - 4 \cos 2x$ $\qquad$ $y = 3 \sin 5x - 5 \cos 5x$

$y = 4 \sin x - 2 \cos x$ $\qquad$ $y = 3 \cos 2x + 2 \sin 7x$

What relationship between the sine and cosine functions ensures that their sum or difference will again be a sinusoid? Check your guess with a graphing utility. ●

Sums That Are Sinusoid Functions

If $y_1 = a_1 \sin [b(x - h_1)]$ and $y_2 = a_2 \cos [b(x - h_2)]$, then

$$y_3 = a_1 \sin [b(x - h_1)] + a_2 \cos [b(x - h_2)]$$

is a sinusoid with the period $2\pi/b$.

Notice that the period of both y_1 and y_2 is $2\pi/b$. Whenever y_1 and y_2 have the same period, the sum $y_1 + y_2$ is again a sinusoid with that common period.

■ EXAMPLE 1 Finding the Amplitude and Phase Shift

Approximate the amplitude and phase shift (to the nearest hundredth) for the sinusoid $f(x) = 2 \sin x + 5 \cos x$. That is, find a and h such that for all values of x

$$2 \sin x + 5 \cos x \approx a \sin (x - h).$$

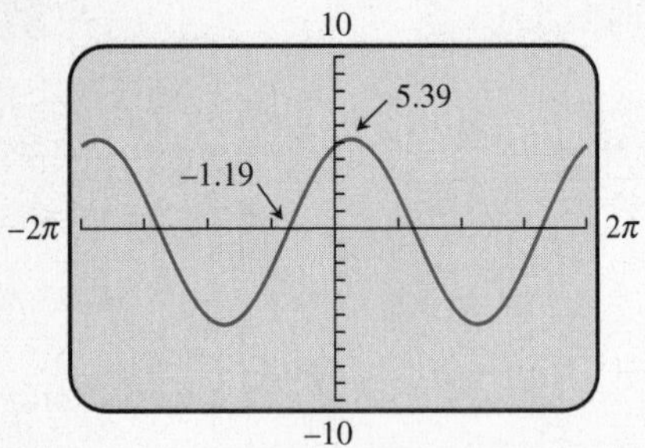

Figure 6.66 $f(x) = 2 \sin x + 5 \cos x$.

Notes on Examples

Example 2 shows how to find the period of a sum of sinusoids when the sum is not a sinusoid.

Follow-up

Ask students whether the sum of two periodic functions is always periodic. **(No, not if the ratio of the periods is irrational.)**

Assignment Guide

Day 1: Ex. 3–63, multiples of 3
Day 2: Ex. 66, 68, 69, 70–73, 78, 81, 83, 84, 87, 95

Cooperative Learning

Group Activity: Ex. 65

Solution

Solve Graphically

The graph in Figure 6.66 suggests that indeed f is a sinusoid. Using the grapher we find the maximum value for f and the x-intercept closest to $x = 0$.

Maximum Value: The maximum value, rounded to the nearest hundredth, is 5.39, so the amplitude of f is about $a = 5.39$.

x intercept: The intercept closest to $x = 0$, rounded to the nearest hundredth, is -1.19, so the phase shift of the sine function is about -1.19.

The period of f is 2π. Putting all this information together, we conclude that

$$f(x) = a \sin (x - h) \approx 5.39 \sin [x - (-1.19)].$$

Support Graphically

Draw the graphs of $f(x) = 2 \sin x + 5 \cos x$ and $y = 5.39 \sin (x + 1.19)$ and see that they appear to be identical. ■

Sums and Differences That Are Not Sinusoids

Example 2 illustrates that if the periods of the individual terms of a sum or difference of sinusoids are unequal, the sum will not be a sinusoid.

Finding the period of a sum of periodic functions can be tricky. Here is an important fact to keep in mind. Let f be a periodic function. Assume $s > 0$ is a candidate for the period of f, that is, suppose $f(x + s) = f(x)$ for all x in the domain of f. Then the period of f divides s exactly—in fact, s could be the period of f. (See Exercise 65.) Watch how we use this fact in Example 2.

■ EXAMPLE 2 Showing a Function Is Periodic but Not a Sinusoid

Show that $f(x) = \sin 2x + \cos 3x$ is periodic but not a sinusoid. Graph one period.

Solution

First we show that 2π is a candidate for the period of f, that is, $f(x + 2\pi) = f(x)$.

$$\begin{aligned} f(x + 2\pi) &= \sin [2(x + 2\pi)] + \cos [3(x + 2\pi)] \\ &= \sin (2x + 4\pi) + \cos (3x + 6\pi) \\ &= \sin 2x + \cos 3x \\ &= f(x) \end{aligned}$$

This means either 2π is the period of f, or the period is an exact divisor of 2π. Figure 6.67 suggests that the period is not smaller than 2π, so it must be 2π. Notice that indeed f is not a sinusoid. ■

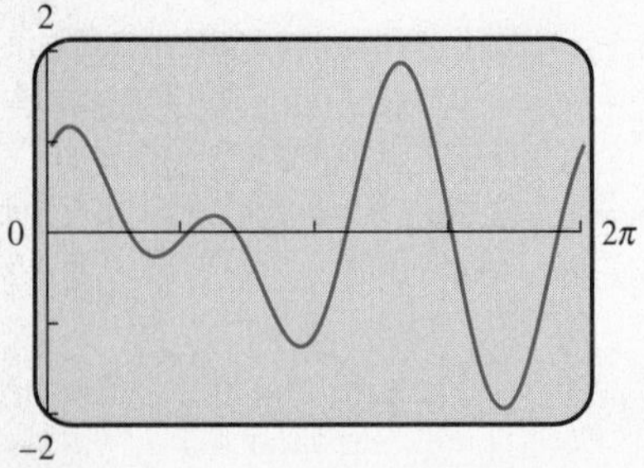

Figure 6.67 One period of $f(x) = \sin 2x + \cos 3x$.

Notes on Exercises

Ex. 7–12 require students to use the graphical solution method of Example 1.

Ex. 33–36 are similar to Example 5. You may wish to specify that y_1 and y_2 should actually be *tangent* to the graph of *y*.

Ex. 65–66 provide practice working with the definition of the period of a function.

NOTE

Recall that $\sin^3 x$ means $(\sin x)^3$.

Combining Trigonometric and Algebraic Functions

A theme of this text has been "families of functions." We have studied polynomial functions, exponential functions, logarithmic functions, rational functions, to name a few, and in this chapter trigonometric functions. Now we consider adding, multiplying, or composing trigonometric functions with functions from these other families.

Periodicity is the property that distinguishes trigonometric functions from all the others. Example 3 shows that when a trigonometric function is composed with a polynomial, the resulting function may continue to be periodic.

■ EXAMPLE 3 Composing $y = \sin x$ and $y = x^3$

Graph $f(x) = \sin^3 x$ and prove that it is a periodic function. Then prove that $|\sin^3 x| \le |\sin x|$.

Solution

The graph of $y_1 = f(x) = \sin^3 x$ in Figure 6.68a suggests that f is periodic with a period no smaller than 2π. To confirm that f is periodic we show that $f(x + 2\pi) = f(x)$.

$$\begin{aligned} f(x + 2\pi) &= \sin^3(x + 2\pi) \\ &= [\sin (x + 2\pi)]^3 \\ &= (\sin x)^3 \\ &= \sin^3 x = f(x) \end{aligned}$$

Figure 6.68b suggests that the graph of $\sin^3 x$ is never farther from the x-axis than is $\sin x$, or that

$$|\sin^3 x| \le |\sin x|.$$

We confirm algebraically:

$\lvert\sin x\rvert \le 1$	Range of $\sin x$ is $[-1, 1]$
$\lvert\sin x\rvert^2 \le \lvert\sin x\rvert$	Multiply by $\lvert\sin x\rvert$.
$\lvert\sin x\rvert^3 \le \lvert\sin x\rvert^2$	Multiply by $\lvert\sin x\rvert$.
$\lvert\sin x\rvert^3 \le \lvert\sin x\rvert$	Transitive property of $\le$ ■

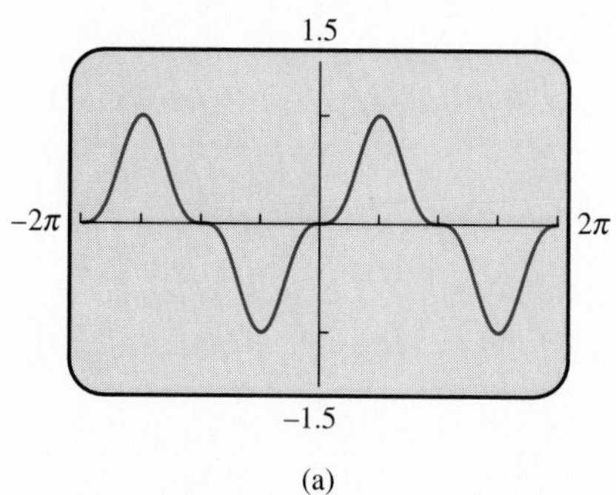

(a)

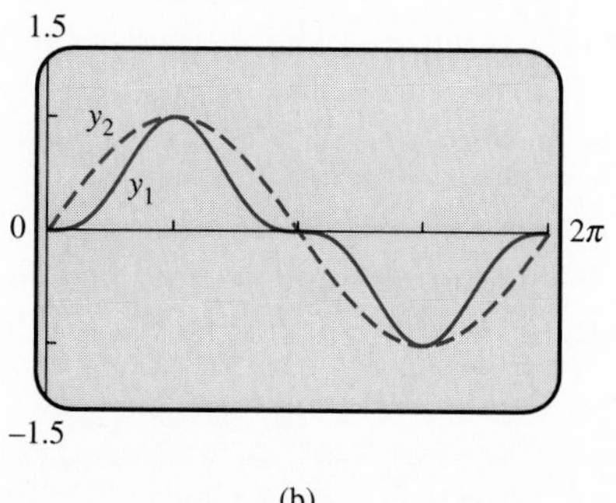

(b)

Figure 6.68 (a) $y_1 = f(x) = \sin^3 x$, and (b) a comparison of $y_1 = \sin^3 x$ and $y_2 = \sin x$.

If $y = f(x)$ is a periodic function, then $y = |f(x)|$ is also a periodic function. Example 4 illustrates this for the function $y = \tan x$.

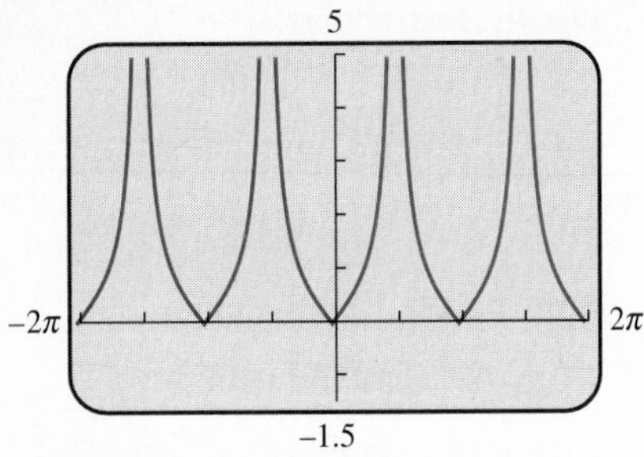

Figure 6.69 $f(x) = |\tan x|$.

Ongoing Assessment

Self-Assessment: Ex. 11, 25, 29, 35, 49, 61, 67
Embedded Assessment: Ex. 70, 72

EXAMPLE 4 Studying a Periodic Function Involving Absolute Value

Find the domain and range of $f(x) = |\tan x|$. Graph four periods of f.

Solution

Whenever $\tan x$ is defined, so is $|\tan x|$. Therefore, the domain of f is the same as the domain of $y = \tan x$, that is, all real numbers except $x = \pi/2 \pm n\pi$.

Because $f(x) = |\tan x| \geq 0$ and the range of $\tan x$ is $(-\infty, \infty)$, the range of f is all real numbers greater than or equal to 0 or $[0, \infty)$.

The period of f, like that of $y = \tan x$, is π. Any viewing window in which $X\text{max} - X\text{min} = 4\pi$ will show four periods of f. See Figure 6.69. ■

When a trigonometric function is combined with an algebraic function by addition or subtraction, the resulting function is, in general, not periodic. However, knowing that a trigonometric term is periodic helps in our analysis of the function.

EXAMPLE 5 Graphing the Sum of $y = \sin x$ and $y = 0.5x$

Graph $f(x) = 0.5x + \sin x$ over the interval $[-2\pi, 2\pi]$. State its domain and range.

Solution

Figure 6.70a shows the graph of f. Because $\sin x$ is defined for any real number x, the domain of f is $(-\infty, \infty)$.

The values of $\sin x$ oscillate between -1 and 1, which means that the values of $f(x) = 0.5x + \sin x$ oscillate between $0.5x - 1$ and $0.5x + 1$. The graph of f falls between the lines $y = 0.5x - 1$ and $y = 0.5x + 1$, as shown in Figure 6.70b, and the range of f is $(-\infty, \infty)$. ■

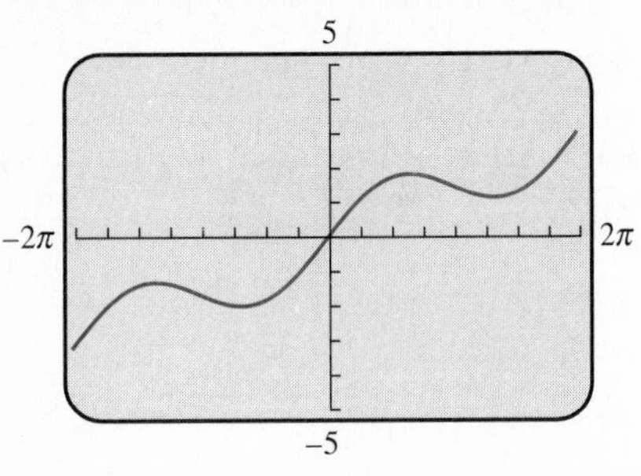

(a)

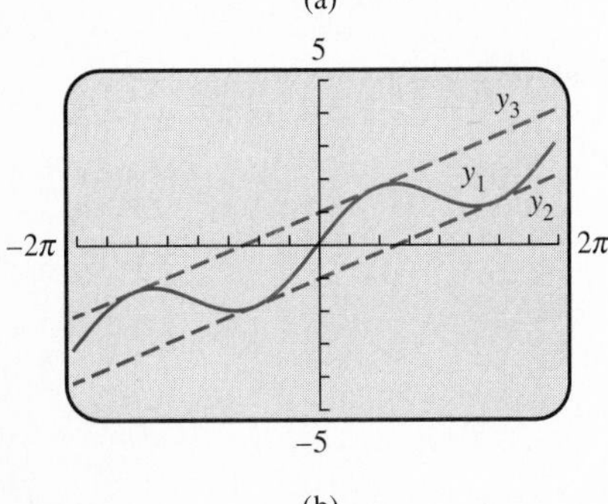

(b)

Figure 6.70 (a) $y_1 = f(x) = 0.5x + \sin x$. (b) $y_1 = f(x) = 0.5x + \sin x$, $y_2 = 0.5x - 1$, and $y_3 = 0.5x + 1$.

Applications

An oscillating spring under the influence of friction and the amount of electrical charge on a capacitor over time are two examples of quantities that can be modeled by an equation that combines an exponential function and a trigonometric function as a product:

$$y = Ae^{-at} \cos bt.$$

The factor Ae^{-at} is called the **damping factor** because the two functions $y = -Ae^{-at}$ and $y = Ae^{-at}$ provide an upper and lower bound for the oscillating quantity.

EXAMPLE 6 APPLICATION: Modeling An Oscillating Spring

Dr. Sanchez's physics class collected data for an air table glider that oscillates between two springs. The class determined that the equation

$$y = 0.22e^{-0.065t} \cos 2.4t$$

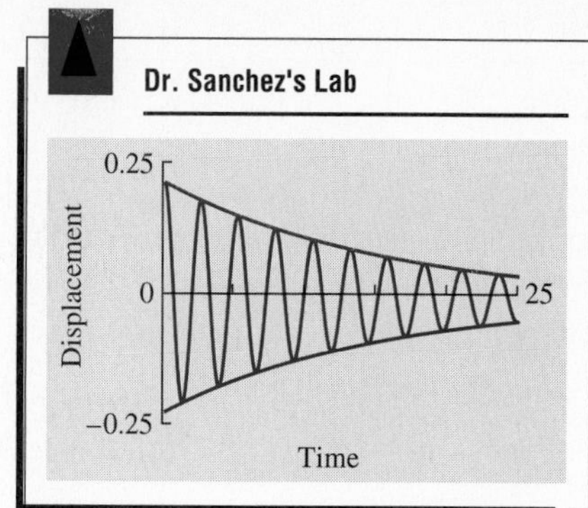

Figure 6.71

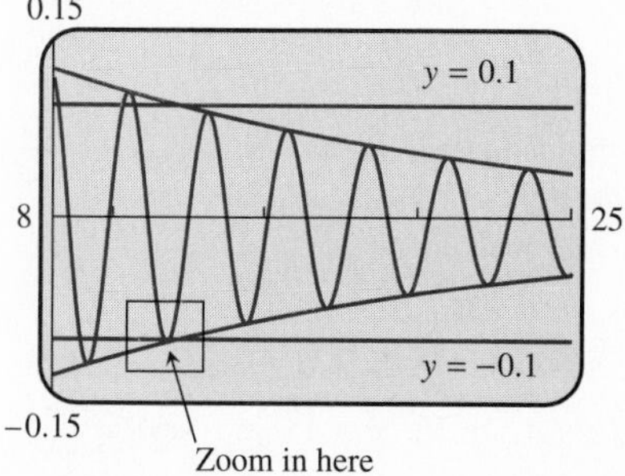

Figure 6.72 The graph of the oscillating air table glider stays between the graphs of its damping factors.

modeled the data where y was the displacement of the spring from its original position.

a. Show algebraically and graphically that the inequality

$$-0.22e^{-0.065t} \le y \le 0.22e^{-0.065t}$$

is valid for $0 \le t \le 25$.

b. Approximately how long does it take for the spring to be damped so that $-0.1 \le y \le 0.1$?

Solution

The equation $y = 0.22e^{-0.065t} \cos 2.4t$ models the action of the air table glider.

a. The cosine function oscillates between -1 and 1. Therefore,

$$-1 \le \cos 2.4t \le 1,$$

$$-0.22e^{-0.065t} \le 0.22e^{-0.065t} \cos 2.4t \le 0.22e^{-0.065t} \quad \text{Multiply by } 0.22e^{-0.065t}.$$

Figure 6.71 shows that the graph of $y = 0.22e^{-0.065t} \cos 2.4t$ oscillates back and forth between the damping equations $y = -0.22e^{-0.065t}$ and $y = 0.22e^{-0.065t}$.

b. We want to find how soon the curve $y = 0.22e^{-0.065t} \cos 2.4t$ falls entirely between the lines $y = -0.1$ and $y = 0.1$. By zooming in on the region indicated in Figure 6.72 and using grapher methods, we find that it takes approximately 11.86 sec until the graph of $y = 0.22e^{-0.065t} \cos 2.4t$ is entirely between $y = -0.1$ and $y = 0.1$.

Interpret

It takes approximately 11.86 sec for the spring to be damped so that $-0.1 \le y \le 0.1$. ■

Quick Review 6.6

In Exercises 1–6, state the domain and range of the function.

1. $f(x) = 3 \sin 2x$

2. $f(x) = -2 \cos 3x$

3. $f(x) = \sqrt{x - 1}$

4. $f(x) = \sqrt{x}$

5. $f(x) = |x| - 2$

6. $f(x) = |x + 2| + 1$

In Exercises 7–10, show that the equation is an identity.

7. $\sin(x + 4\pi) = \sin x$

8. $\cos(x + 4\pi) = \cos x$

9. $\tan(2x + 3\pi) = \tan 2x$

10. $\sin(4x + 6\pi) = \sin 4x$

In Exercises 11 and 12, describe how the graphs of y_1 and y_2 are related.

11. $y_1 = |x^2 - 4|$ and $y_2 = x^2 - 4$

12. $y_1 = |2x - 3|$ and $y_2 = 2x - 3$

In Exercises 13–16, describe the end behavior of the function, that is, the behavior as $|x| \to \infty$.

13. $f(x) = 5e^{-2x}$

14. $g(x) = -3e^{-0.5x}$

15. $f(x) = -0.2(5^{-0.1x})$

16. $g(x) = 5(2^{-0.4x})$

In Exercises 17 and 18, form the compositions $f \circ g$ and $g \circ f$. State the domain of each function.

17. $f(x) = x^2 - 4$ and $g(x) = \sqrt{x}$

18. $f(x) = x^2$ and $g(x) = \cos x$

SECTION EXERCISES 6.6

In Exercises 1–6, determine whether $f(x)$ is a sinusoid.

1. $f(x) = \sin x - 3\cos x$

2. $f(x) = 4\cos x + 2\sin x$

3. $f(x) = 3\sin 2x - 5\cos 3x$

4. $f(x) = 2\cos \pi x + \sin \pi x$

5. $f(x) = \pi \sin 3x - 4\pi \sin 2x$

6. $f(x) = 2\sin x - \tan x$

In Exercises 7–12 find, a, b, and h so that $f(x) \approx a\sin[b(x-h)]$.

7. $f(x) = 2\cos x + \sin x$

8. $f(x) = 3\sin 2x - \cos 2x$

9. $f(x) = 2\sin 2x - 3\cos 2x$

10. $f(x) = \cos 3x + 2\sin 3x$

11. $f(x) = \sin \pi x - 2\cos \pi x$

12. $f(x) = \cos 2\pi x + 3\sin 2\pi x$

In Exercises 13–18, graph one period of the function.

13. $y = \sin 2x - 2\cos x$

14. $y = 2\cos x + \cos 3x$

15. $y = 2\sin 2x + \cos 3x$

16. $y = \cos 3x - 4\sin 2x$

17. $y = \sin 2x + \sin 5x$

18. $y = \cos 3x - 2\cos 4x$

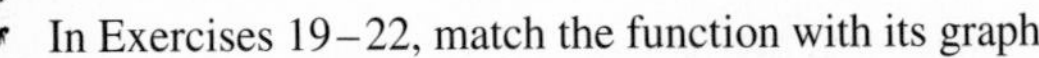

In Exercises 19–22, match the function with its graph.

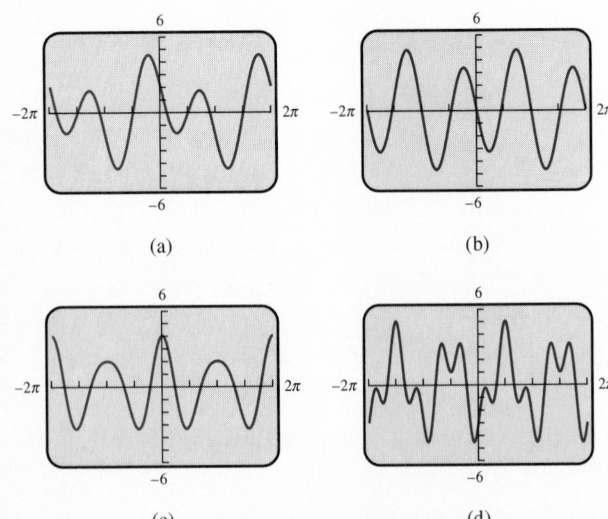

For Exercises 19–22

19. $y = 2\cos x - 3\sin 2x$

20. $y = 2\sin 5x - 3\cos 2x$

21. $y = 3\cos 2x + \cos 3x$

22. $y = \sin x - 4\sin 2x$

In Exercises 23–26, graph two periods of f and compare it to the graph of $y = \cos x$. Show that f is a periodic function. State its period.

23. $f(x) = \cos^2 x$

24. $f(x) = \cos^3 x$

25. $f(x) = \sqrt{\cos^2 x}$

26. $f(x) = |\cos^3 x|$

In Exercises 27–32, state the domain and range of the function and graph four periods.

27. $y = \cos^2 x$

28. $y = |\cos x|$

29. $y = |\cot x|$

30. $y = \cos|x|$

31. $y = -\tan^2 x$

32. $y = -\sin^2 x$

In Exercises 33–36, the function is of the form $y = g(x) + \cos bx$. Find two linear functions y_1 and y_2 related to g such that $y_1 \le y \le y_2$. Graph all three functions y, y_1, and y_2 in the same viewing window.

33. $y = 2x + \cos x$

34. $y = 1 - 0.5x + \cos 2x$

35. $y = 2 - 0.3x + \cos x$

36. $y = 1 + x + \cos 3x$

In Exercises 37–40, graph f over the interval $[-4\pi, 4\pi]$. Describe how the graphs of $y = x$ and $y = -x$ are related to the graph of f.

37. $f(x) = x\sin x$

38. $f(x) = -x|\cos x|$

39. $f(x) = |x\sin x|$

40. $f(x) = x\cos|x|$

In Exercises 41–44, graph both f and plus or minus its damping factor in the same viewing window. Describe the behavior of the function f for $x \ge 0$. What is the end behavior of f?

41. $f(x) = 1.2^{-x}\cos 2x$

42. $f(x) = 2^{-x}\sin 4x$

43. $f(x) = x^{-1}\sin 3x$

44. $f(x) = e^{-x}\cos 3x$

In Exercises 45–50, find the period and graph the function over two periods.

45. $y = \sin 3x + 2\cos 2x$

46. $y = 4\cos 2x - 2\cos(3x - 1)$

47. $y = \dfrac{\sin 2x}{5} - \dfrac{2\sin 4x}{3}$

48. $y = \dfrac{3\cos 3x}{\sqrt{5}} + \dfrac{2\cos(6x - 1)}{\sqrt{3}}$

49. $y = 2\sin(3x + 1) - \cos(5x - 1)$

50. $y = 3\cos(2x - 1) - 4\sin(3x - 2)$

In Exercises 51–56, graph f over the interval $[-4\pi, 4\pi]$. Determine whether the function is periodic. State the period.

51. $f(x) = |\sin \frac{1}{2}x| + 2$

52. $f(x) = 3x + 4\sin 2x$

53. $f(x) = x - \cos x$

54. $f(x) = x + \sin 2x$

55. $f(x) = \frac{1}{2}x + \cos 2x$

56. $f(x) = 3 - x + \sin 3x$

In Exercises 57–64, find the domain and the range of the function.

57. $f(x) = 2x + \cos x$

58. $f(x) = 2 - x + \sin x$

59. $f(x) = |x| + \cos x$

60. $f(x) = -2x + |3\sin x|$

61. $f(x) = \sqrt{\sin x}$

62. $f(x) = \sin |x|$

63. $f(x) = \sqrt{|\sin x|}$

64. $f(x) = \sqrt{\cos x}$

In Exercises 65–70, solve the problem.

65. *Periodic Functions* Let f be a periodic function with period p. Assume $s > 0$ and $f(x + s) = f(x)$ for all values of x in the domain of f. Show that s must be a multiple of p by following these steps.

a. $s \geq p$

b. If $s = pq + r$, where q is an integer and $0 \leq r < p$, then $f(x + r) = f(x)$ for all values of x in the domain of f.

c. In part b, $r = 0$ and $s = pq$.

66. *Periodic Functions* If f is a periodic function with period p, show that $f(x + 2p) = f(x)$ for all x-values in the domain of f. Generalize this result.

67. *Predicting Economic Growth* Stogner Auto Parts, Inc., had total sales of \$75 million in 1995. Historically the industry has an annual growth of 4% subject to an adjustment due to its cyclical nature. The company's economists have found that the equation $y = 75(1.04^x) + 2\sin \pi x$ models the sales (in millions of dollars), where $x = 0$ corresponds to the year 1995.

a. Graph this model to reflect the years from 1995 to 2010.

b. How many years are there in the economic cycle?

c. What does the model predict for sales in 2001?

68. *Oscillating Spring* The oscillations of a spring subject to friction are modeled by the equation $y = 0.43e^{-0.55t}\cos 1.8t$.

a. Graph y and its two damping curves in the same viewing window for $0 \leq t \leq 12$.

b. Approximately how long does it take for the spring to be damped so that $-0.2 \leq y \leq 0.2$?

EXTENDING THE IDEAS

69. Writing to Learn Example 3 shows that the function $y = \sin^3 x$ is periodic. Explain whether you think that $y = \sin x^3$ is periodic and why.

70. Writing to Learn Example 4 shows that $y = |\tan x|$ is periodic. Write a convincing argument that $y = \tan |x|$ is not a periodic function.

In Exercises 71 and 72, select the one correct inequality, **a** or **b.** Give a convincing argument.

71. a. $x - 1 \leq x + \sin x \leq x + 1$ for all x

b. $x - \sin x \leq x + \sin x$ for all x.

72. a. $-x \leq x\sin x \leq x$ for all x

b. $-|x| \leq x\sin x \leq |x|$ for all x

In Exercises 73 and 74, use the graph to answer the question.

73. Writing to Learn Explain why the graph suggests the conjecture that

$$-|x| \leq x\sin x \leq |x|.$$

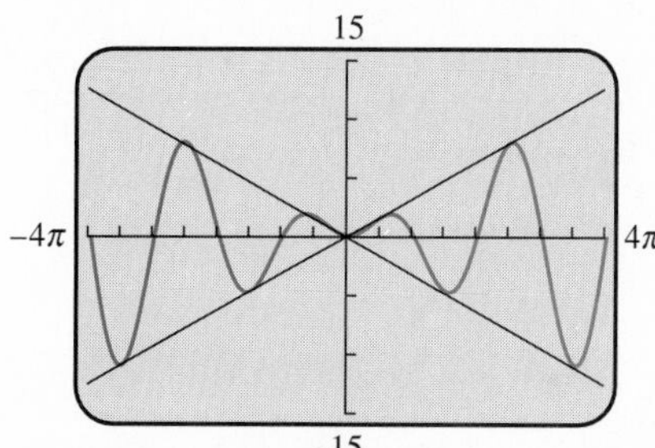

For Exercise 73

74. Writing to Learn Explain why the graph suggests the conjecture that

$$-\left|\frac{3}{x}\right| \leq \frac{3\sin x}{x} \leq \left|\frac{3}{x}\right|.$$

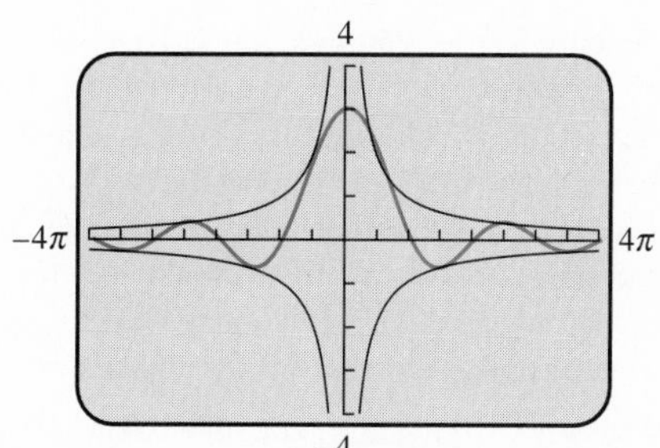

For Exercise 74

In Exercises 75–78, match the function with its graph. In each case state the viewing window.

75. $y = \cos x - \sin 2x - \cos 3x + \sin 4x$

76. $y = \cos x - \sin 2x - \cos 3x + \sin 4x - \cos 5x$

77. $y = \sin x + \cos x - \cos 2x - \sin 3x$

78. $y = \sin x - \cos x - \cos 2x - \cos 3x$

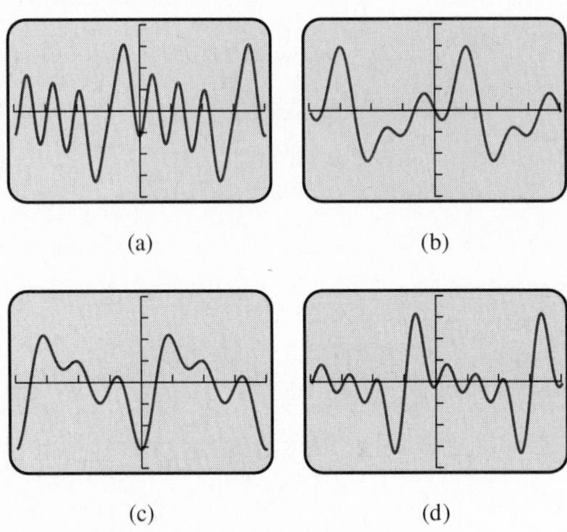

For Exercises 75–78

In Exercises 79–82, draw graphs to support the identity.

79. $\sin x = \cos\left(\frac{\pi}{2} - x\right)$ **80.** $\cos x = \sin\left(\frac{\pi}{2} - x\right)$

81. $\tan x = \cot\left(\frac{\pi}{2} - x\right)$ **82.** $\cot x = \tan\left(\frac{\pi}{2} - x\right)$

In Exercises 83 and 84, solve the problem.

83. *Inaccurate or Misleading Graphs*

a. In function mode, set $X\text{min} = 0$ and $X\text{max} = 2\pi$. Move the cursor along the x-axis. What is the distance between one pixel and the next (to the nearest hundredth)?

b. What is the period of $f(x) = \sin 250x$? Consider that the period is the length of one full cycle of the graph. Approximately how many cycles should there be between two adjacent pixels? Can your grapher produce an accurate graph of this function between 0 and 2π?

84. *Length of Days* The graph shows the number of hours of daylight in Boston as a function of the day of the year, from September 21, 1983, to December 15, 1984. Key points are labeled and other critical information is provided. Write a formula for the sinusoidal function and check it by graphing.

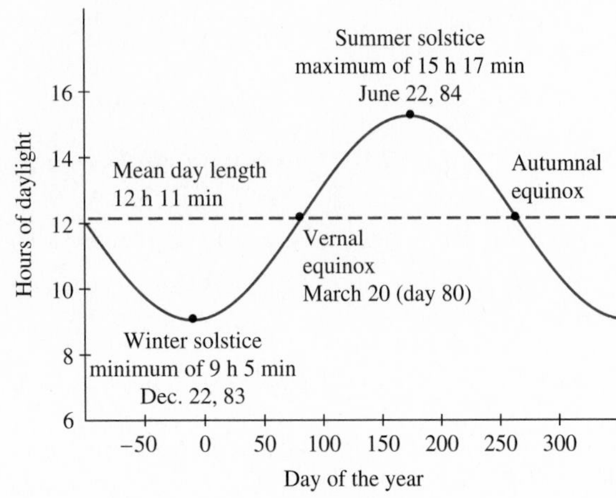

For Exercise 84

(*Source:* Farmer's Almanac.)

In Exercises 85–92, first try to predict what the graph will look like (without too much effort, that is, just for fun). Then graph the function in one or more viewing windows to determine the main features of the graph, and draw a summary sketch. Describe in words the nature of the graph. Where applicable, name the period, amplitude, domain, range, asymptotes, and zeros.

85. $f(x) = \cos e^x$ **86.** $g(x) = e^{\tan x}$

87. $f(x) = \sqrt{x} \sin x$

88. $g(x) = \sin \pi x + \sqrt{4 - x^2}$

89. $f(x) = \dfrac{\sin x}{x}$ **90.** $g(x) = \dfrac{\sin x}{x^2}$

91. $f(x) = x \sin \dfrac{1}{x}$ **92.** $g(x) = x^2 \sin \dfrac{1}{x}$

In Exercises 93–96, confirm the identity.

93. $\ln|\tan x| = \ln|\sin x| - \ln|\cos x|$

94. $\ln|\cot x| = \ln|\cos x| - \ln|\sin x|$

95. $\log|\sec x| = -\log|\cos x|$

96. $\log(\sin^2 x + \cos^2 x) = 0$

6.7 INVERSE TRIGONOMETRIC FUNCTIONS

Inverse Relations • Inverse Sine Function • Inverse Cosine and Tangent Functions • Compositions of Trigonometric and Inverse Trigonometric Functions • Using Inverse Trigonometric Functions in Calculus • Transformations of Inverse Trigonometric Functions • Applications

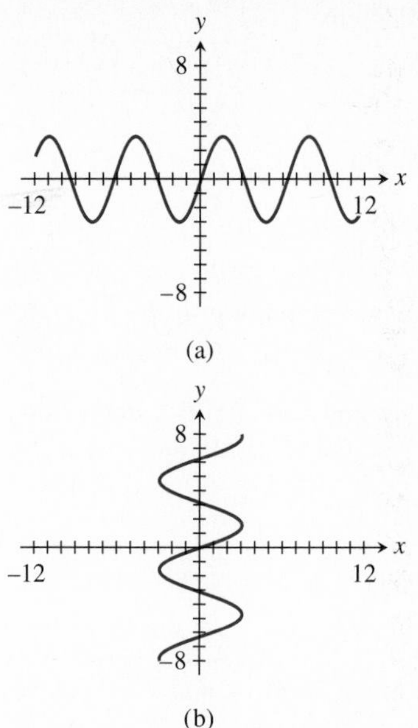

Figure 6.73 (a) The function $y = 3 \sin x$. (b) Its inverse relation, $x = 3 \sin y$.

Inverse Relations

You learned in Section 2.7 that each function has an inverse relation, and that this inverse relation is a function only if the original function is one-to-one.

The graph of $y = 3 \sin x$ in Figure 6.73a indicates that the sine function is not one-to-one because different values of x yield the same value of y. Consequently, its inverse relation is not a function. See Figure 6.73b.

Similar observations can be made about the other five trigonometric functions and their inverse relations. None of the functions is one-to-one, so each of the inverse relations fails to be a function. However, by restricting the domain of a trigonometric function to an interval on which the function *is* one-to-one, it is possible to obtain an inverse function.

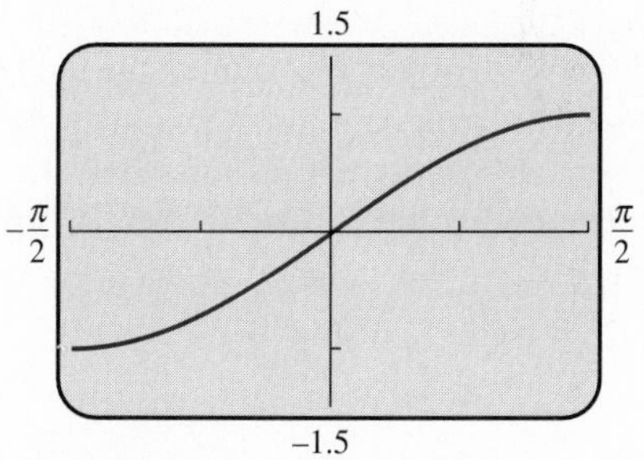

Figure 6.74 $y = \sin x$ restricted to $[-\pi/2, \pi/2]$.

Inverse Sine Function

If you restrict the domain of $y = \sin x$ to the interval $[-\pi/2, \pi/2]$, as shown in Figure 6.74, the function is one-to-one. The **inverse sine function** $y = \sin^{-1} x$ is the inverse of the sine function $y = \sin x$ with the domain restricted to $[-\pi/2, \pi/2]$. We read $\sin^{-1} x$ as "inverse sine of x" or "sine inverse of x." As we learned in Section 2.7, another form for the inverse of $y = \sin x$ is $x = \sin y$, so

$$y = \sin^{-1} x \qquad \text{and} \qquad x = \sin y$$

are equivalent equations with suitable restrictions on the domains. In other words, $\sin^{-1} x$ is nothing more than an angle!

REMARK

In the equation $y = \sin^{-1} x$ in this definition, we can regard y as an angle or as a *directed arc length.* In this context y is the "arc" whose sine is x. In other words, $\sin^{-1} x$ is the arc (in quadrant I or IV) whose sine is x. This is why the function $\sin^{-1}$ is also known as *arcsin.*

Inverse Sine Function

$$y = \sin^{-1} x \qquad \text{if and only if} \qquad x = \sin y$$

for $-1 \le x \le 1$ and $-\pi/2 \le y \le \pi/2$.

In particular, $\sin(\sin^{-1} x) = x$ and $y = \sin^{-1}(\sin y)$.

Applying this relationship, we can convert between inverse sine function and sine function.

Objective

Students will be able to relate the concept of inverse function to trigonometric functions.

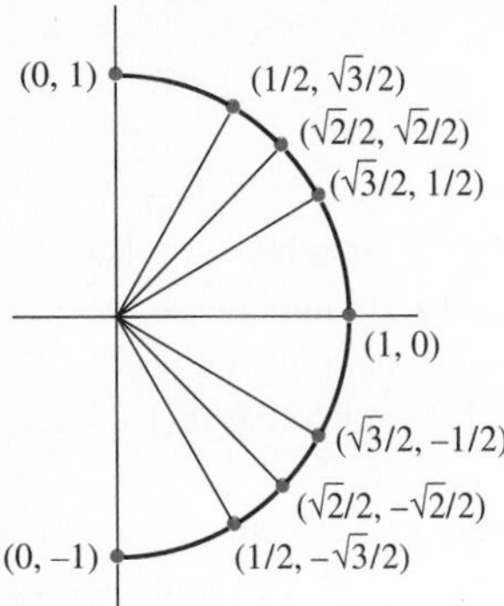

Figure 6.75

EXAMPLE 1 Evaluating Inverse Sine Function

Find the exact value of the expression.

a. $\sin^{-1}\left(\frac{1}{2}\right)$ **b.** $\sin^{-1}\left(-\frac{\sqrt{3}}{2}\right)$

Solution

Figure 6.75 shows the points on the unit circle corresponding to the nine special angles in $[-\pi/2, \pi/2]$: $-\pi/2, -\pi/3, -\pi/4, -\pi/6, 0, \pi/6, \pi/4, \pi/3, \pi/2$.

a. $\sin^{-1}(1/2) = \theta$ if and only if $\sin\theta = 1/2$ and $-\pi/2 \le \theta \le \pi/2$. Since $\sin\theta$ is positive, θ must be in quadrant I. The special angle satisfying this equation is $\theta = \pi/6$. Therefore,

$$\sin^{-1}\left(\frac{1}{2}\right) = \frac{\pi}{6}.$$

b. $\sin^{-1}(-\sqrt{3}/2) = \theta$ if and only if $\sin\theta = -\sqrt{3}/2$ and $-\pi/2 \le \theta \le \pi/2$. Since $\sin\theta$ is negative, θ must be in quadrant IV and $\theta = -\pi/3$ is a solution. Therefore,

$$\sin^{-1}\left(-\frac{\sqrt{3}}{2}\right) = -\frac{\pi}{3}.$$

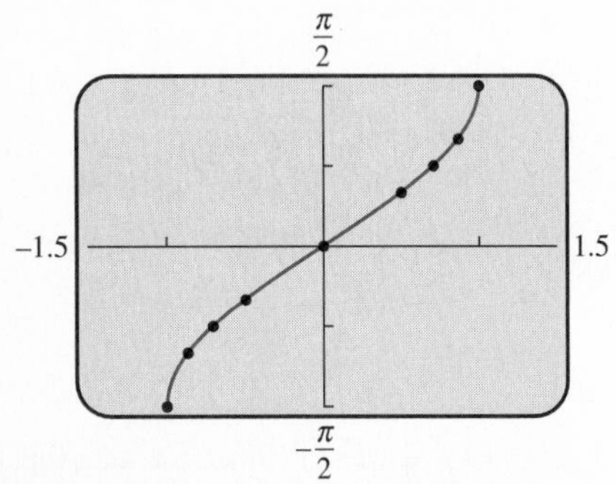

Figure 6.76 $y = \sin^{-1} x$ and a scatter plot of nine of its points corresponding to the special angles in $[-\pi/2, \pi/2]$.

The ordered pairs $(\sin\theta, \theta)$ for the nine special angles in $[-\pi/2, \pi/2]$ give nine points on the graph of $y = \sin^{-1} x$. Figure 6.76 shows a graph of the built-in function $y = \sin^{-1} x$ superimposed on a scatter plot of these nine points.

The keystrokes used to evaluate $\sin^{-1} x$ vary from calculator to calculator. Often you need to press two keys. For example, "2nd," "2ndF," "shift," or "inv" followed by "sin" or "asin." Depending on the model of calculator you are using, enter x and press the keys, or press the keys and enter x.

Motivate

Ask students to state the definition of an inverse function and explain why the sine function (as defined over the real numbers) does not have an inverse function. **(The sine function is not one-to-one.)**

EXAMPLE 2 Evaluating the Inverse Sine Function

Using a calculator in radian mode evaluate these inverse sine values:

a. $\sin^{-1}(-0.81) = -0.944\cdots$ **b.** $\sin^{-1}(0.92) = 1.168\cdots$

Inverse Cosine and Tangent Functions

The **inverse cosine function** $y = \cos^{-1} x$ is the inverse of the cosine function $y = \cos x$. The **inverse tangent function** $y = \tan^{-1} x$ is the inverse of the tangent function $y = \tan x$. The domains of the cosine and tangent functions must be suitably restricted, of course.

Teaching Note

Be sure students understand the distinction between the inverse relation associated with a function and an inverse function. As an example, the inverse relation associated with $y = \sin x$ is not a function because it fails the vertical line test, so domain restrictions *must* be considered and observed. Note that this text does not use an initial capital letter (e.g., Sin^{-1}) to indicate the principal-value function.

Teaching Note

Graphing calculators can graph the function $y = \sin^{-1} x$, but they cannot directly graph the relation $\sin y = x$; hence the name *function grapher.* The parametric graphing techniques introduced in this chapter can be used to graph relations such as $\sin y = x$.

Alert

The symbol used to denote inverse functions, f^{-1}, frequently causes confusion because it is similar to the symbol x^{-1} used for the reciprocal of a number. Stress that $x^{-1} = \frac{1}{x}$, but $f^{-1}(x) \neq \frac{1}{f(x)}$, and $\sin^{-1}(x) \neq \frac{1}{\sin x}$.

Alert

Students often expect the inverse sine, inverse cosine, and inverse tangent functions to have the same range. This is not the case because, for example, the sine and cosine functions are not one-to-one on the same interval.

Inverse Cosine and Tangent Functions

$$y = \cos^{-1} x \quad \text{if and only if} \quad \cos y = x$$

for $-1 \leq x \leq 1$ and $0 \leq y \leq \pi$.

$$y = \tan^{-1} x \quad \text{if and only if} \quad \tan y = x$$

for $-\infty < x < \infty$ and $-\pi/2 < y < \pi/2$.

In Figures 6.77 and 6.78, the graphs of the restricted cosine and tangent functions are shown next to the graphs of their inverse functions.

The graphs in Figures 6.77 and 6.78 remind us of the following relationships we observed in Section 2.7:

$$\text{Domain of } f = \text{range of } f^{-1}$$
$$\text{Range of } f = \text{domain of } f^{-1}$$

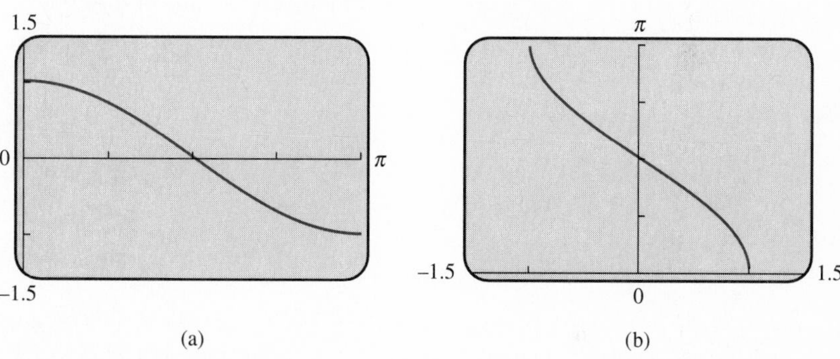

Figure 6.77 (a) The restricted cosine function $y = \cos x$, where $0 \leq x \leq \pi$. (b) Its inverse $y = \cos^{-1} x$.

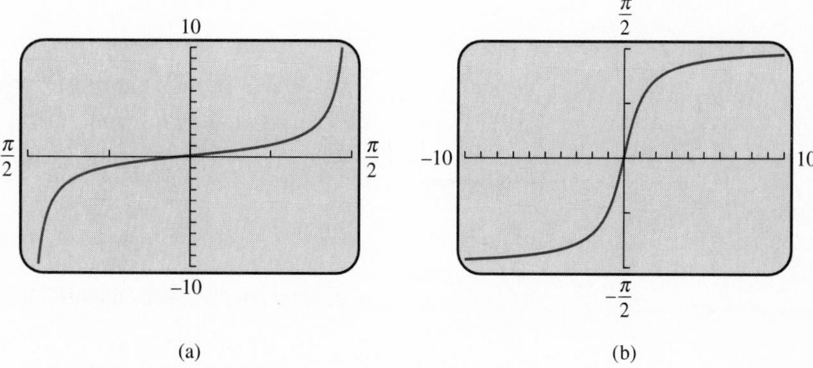

Figure 6.78 (a) The restricted tangent function $y = \tan x$, where $-\pi/2 < x < \pi/2$. (b) Its inverse $y = \tan^{-1} x$.

However, the figures do not give a good picture of the relationship between the graph of a function and its inverse. Recall that the graph of a function and its inverse are reflections of each other across the line $y = x$. Draw graphs of each pair of trigonometric function and inverse trigonometric function, together with the graph of $y = x$, in a square viewing window to see this relationship.

The inverse cosine and inverse tangent functions may also be denoted using the word *arc* in a manner consistent with the use in arcsin.

NOTE

$\cos^{-1} x$ is also read as "the angle whose cosine is x." Similar statements can be made for $\sin^{-1} x$ and $\tan^{-1} x$.

$$\cos^{-1} x = \arccos x = \text{the arc (in quadrant I or II) whose cosine is } x.$$
$$\tan^{-1} x = \arctan x = \text{the arc (in quadrant I or IV) whose tangent is } x.$$

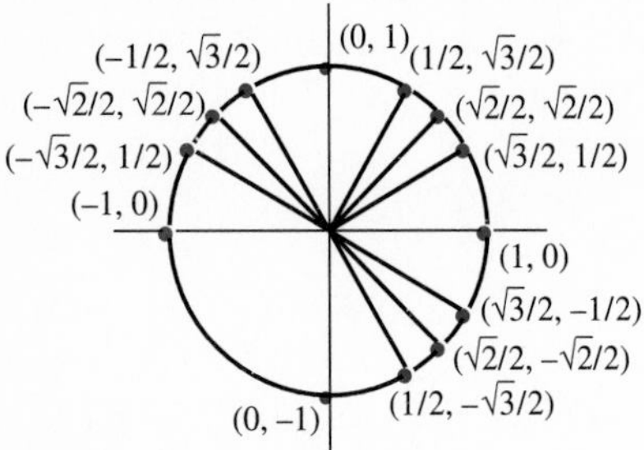

Figure 6.79

Figure 6.79 shows the points on the unit circle corresponding to the special angles in quadrants I, II, and IV. The coordinates of these points help us find exact values for the inverse trigonometric functions, as illustrated in Example 3.

■ **EXAMPLE 3 Evaluating Inverse Trigonometric Functions**

Find the exact value of the expression.

a. $\cos^{-1}\left(-\dfrac{\sqrt{2}}{2}\right)$ **b.** $\tan^{-1} \sqrt{3}$

Solution

a. $\cos^{-1}(-\sqrt{2}/2) = \theta$ if and only if $\cos \theta = -\sqrt{2}/2$ and $0 \le \theta \le \pi$. Since $\cos \theta$ is negative, θ must be in quadrant II, and $\theta = 3\pi/4$ is a solution. Therefore,

$$\cos^{-1}\left(-\frac{\sqrt{2}}{2}\right) = \frac{3\pi}{4}.$$

b. $\tan^{-1} \sqrt{3} = \theta$ if and only if $\tan \theta = \sqrt{3}$ and $-\pi/2 \le \theta \le \pi/2$. Since $\tan \theta$ is positive, θ must be in quadrant I, so $\theta = \pi/3$. Therefore,

$$\tan^{-1} \sqrt{3} = \frac{\pi}{3}.$$

■

Compositions of Trigonometric and Inverse Trigonometric Functions

In Section 2.7 we saw that, if f and f^{-1} are inverse functions, then

$$f(f^{-1}(x)) = x \quad \text{and} \quad f^{-1}(f(x)) = x$$

for each value of x in the domain of f^{-1} and for each value of x in the domain of f, respectively. Consequently, we have the equations

$$\sin^{-1}(\sin x) = x \qquad \cos^{-1}(\cos x) = x \qquad \tan^{-1}(\tan x) = x$$
$$\sin(\sin^{-1} x) = x \qquad \cos(\cos^{-1} x) = x \qquad \tan(\tan^{-1} x) = x$$

with the domains suitably restricted.

We can evaluate $\sin^{-1}(\sin x)$ for all x-values in the domain of sine, but if x falls outside of $[-\pi/2, \pi/2]$, then $\sin^{-1}(\sin x) \neq x$, as Example 4 illustrates. The equation $\sin^{-1}(\sin x) = x$ is true only for x in the restricted domain of sine that allows $\sin^{-1}$ to exist.

Notes on Examples

Example 4 is important because it emphasizes domain and range issues.

NOTE

Notice in Example 4a that

$$\sin^{-1}\left(\sin \frac{7\pi}{6}\right) \neq \frac{7\pi}{6}$$

because $7\pi/6$ is *not* in the restricted domain of sine.

■ **EXAMPLE 4 Evaluating Composite Functions**

a. $$\sin^{-1}\left(\sin \frac{7\pi}{6}\right) = \sin^{-1}\left(-\frac{1}{2}\right) = -\frac{\pi}{6}$$

b. $$\cos^{-1}\left(\cos \frac{-\pi}{4}\right) = \cos^{-1}\left(\frac{\sqrt{2}}{2}\right) = \frac{\pi}{4}$$

c. $$\tan^{-1}\left[\tan\left(\frac{3\pi}{4}\right)\right] = \tan^{-1}(-1) = -\frac{\pi}{4}$$

■

Teaching Note

Check that in using calculators to produce solutions, students are aware of the restrictions for inverses that affect how many solutions they can produce. Proficiency in using the technology is one of the goals of this section. Focus students' attention on what happens to domain and range.

Using Inverse Trigonometric Functions in Calculus

In calculus you will learn to use trigonometric substitutions to accomplish certain mathematical tasks. When using trigonometric substitutions, you will sometimes need to use inverse trigonometric functions as well.

■ **EXAMPLE 5 Using Inverse Trigonometric Functions in Calculus**

Write the expression as an algebraic expression in x.

a. $\cos(\arcsin 2x)$ for $0 \leq x \leq 1/2$
b. $\tan(\arcsin 2x)$ for $0 \leq x \leq 1/2$

Solution

If $u = \arcsin 2x$, then $\sin u = 2x$. This explains the labeling of the triangle in Figure 6.80.

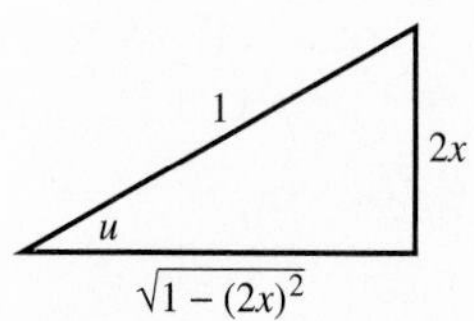

Figure 6.80

a. $\cos(\arcsin 2x) = \cos u = \sqrt{1 - (2x)^2} = \sqrt{1 - 4x^2}$

b. $\tan(\arcsin 2x) = \tan u = \dfrac{2x}{\sqrt{1 - (2x)^2}} = \dfrac{2x}{\sqrt{1 - 4x^2}}$ ■

Transformations of Inverse Trigonometric Functions

The inverse trigonometric functions can be transformed just like the other functions we have studied.

■ EXAMPLE 6 Identifying Transformations

Describe how the graph of

$$y_2 = 3\sin^{-1}\left[\frac{1}{4}(x+8)\right] - 5$$

can be obtained from the graph of $y_1 = \sin^{-1} x$. State the domain and range of y_2.

Solution

The graph of y_2 is obtained from the graph of y_1 by applying, in order, the following transformations:

1. A horizontal stretch of $1/(1/4) = 4$ to get $\sin^{-1}\frac{1}{4}x$
2. A horizontal translation 8 units to the left to get $\sin^{-1}\left[\frac{1}{4}(x+8)\right]$
3. A vertical stretch of factor 3 to get $3\sin^{-1}\left[\frac{1}{4}(x+8)\right]$
4. A vertical translation down 5 units to get $3\sin^{-1}\left[\frac{1}{4}(x+8)\right] - 5$

The first two transformations transform the domain of $y_1 = \sin^{-1} x$ from $[-1, 1]$ to $[-12, -4]$. The third and fourth transformations transform the range of $y_1 = \sin^{-1} x$ from $[-\pi/2, \pi/2]$ to $[-3\pi/2 - 5, 3\pi/2 - 5]$. Therefore, the domain of y_2 is $[-12, -4]$ and the range is $[-3\pi/2 - 5, 3\pi/2 - 5]$.

Draw graphs of y_1 and y_2 to support these conclusions. ■

Follow-up

Ask students to name the values of x for which $x = \cos^{-1}(\cos x)$ and the values for which $x = \cos(\cos^{-1} x)$. **($0 \le x \le \pi$; $-1 \le x \le 1$)**

Assignment Guide

Day 1: Ex. 3–45, multiples of 3
Day 2: Ex. 49, 51, 52, 53, 55, 57, 60, 63

Cooperative Learning

Group Activity: Ex. 48

Notes on Exercises

Ex. 1–12 and 37–46 include many types of problems traditionally associated with the topic of inverse trigonometric functions.

Ongoing Assessment

Self-Assessment: Ex. 7, 17, 29, 37, 43, 47
Embedded Assessment: Ex. 49

Applications

When an application involves an angle as the dependent variable, as in $\theta = f(x)$, then to solve for x, it is natural to use an inverse trigonometric function and find $x = f^{-1}(\theta)$.

■ EXAMPLE 7 APPLICATION: Calculating Viewing Angle

The bottom of the 20-ft replay screen at Dodger Stadium is 45 ft above the playing field. As you move away from the wall, the angle formed by the screen

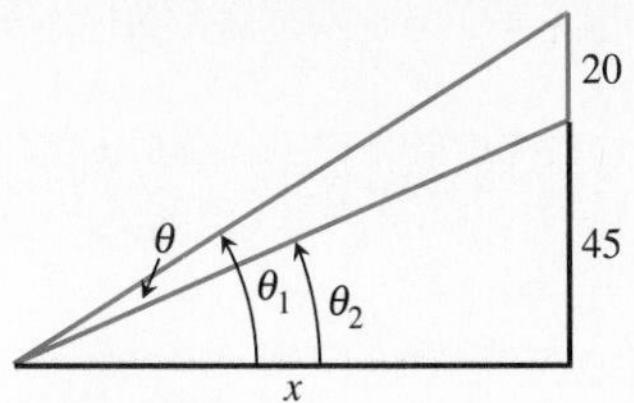

Figure 6.81

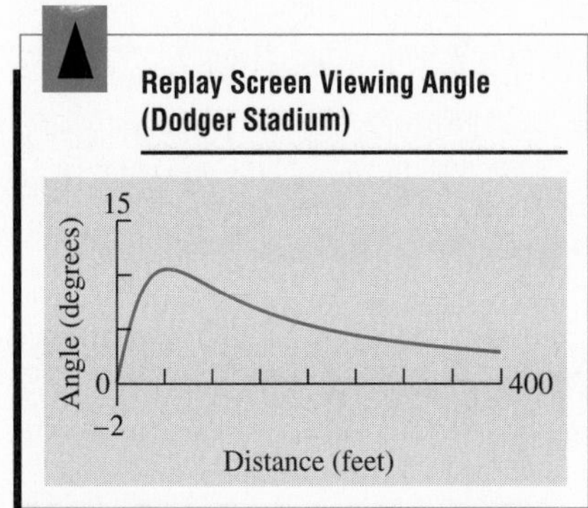

Figure 6.82

at your eye changes. There is a distance from the wall at which the angle is greatest. What is that distance?

Solution

Model

The angle subtended by the screen is represented in Figure 6.81 by θ, and $\theta = \theta_1 - \theta_2$. Since $\tan\theta_1 = 65/x$, it follows that $\theta_1 = \tan^{-1}(65/x)$. Similarly, $\theta_2 = \tan^{-1}(45/x)$. Thus,

$$\theta = \tan^{-1}\frac{65}{x} - \tan^{-1}\frac{45}{x}.$$

Solve Graphically

Figure 6.82 shows a graph of θ that reflects degree mode. The question about distance for maximum viewing angle can be answered by finding the x-coordinate of the maximum point of this graph. Using grapher methods we see that this maximum occurs when $x \approx 54$ ft.

Interpret

The maximum angle subtended by the replay screen occurs about 54 ft from the wall. ■

Quick Review 6.7

In Exercises 1–4, state the sign (positive or negative) of the sine, cosine, and tangent in the quadrant.

1. Quadrant I **2.** Quadrant II

3. Quadrant III **4.** Quadrant IV

In Exercises 5–14, find the exact value.

5. $\sin(\pi/6)$ **6.** $\tan(\pi/4)$

7. $\cos(2\pi/3)$ **8.** $\sin(2\pi/3)$

9. $\sin(-\pi/6)$ **10.** $\cos(-\pi/3)$

11. $\cos(-\pi/4)$ **12.** $\tan(-\pi/4)$

13. $\sin(\pi/2)$ **14.** $\tan\pi$

In Exercises 15 and 16, refer to the unit circle in Figure 6.79.

15. Name the special angles shown for $[-\pi/2, \pi]$.

16. Confirm that the labeled points are on the unit circle.

EXERCISES FOR SECTION 6.7

In Exercises 1–12, find the exact value.

1. $\sin^{-1}\left(\frac{\sqrt{3}}{2}\right)$ **2.** $\sin^{-1}\left(-\frac{1}{2}\right)$

3. $\tan^{-1} 0$ **4.** $\cos^{-1} 1$

5. $\cos^{-1}\left(\frac{1}{2}\right)$ **6.** $\tan^{-1} 1$

7. $\tan^{-1}(-1)$ **8.** $\cos^{-1}\left(-\frac{\sqrt{3}}{2}\right)$

9. $\sin^{-1}\left(-\frac{1}{\sqrt{2}}\right)$

10. $\tan^{-1}(-\sqrt{3})$

11. $\cos^{-1} 0$

12. $\sin^{-1} 1$

In Exercises 13–18, use a calculator to find the approximate value. Express your result in degrees.

13. $\sin^{-1}(0.362)$

14. arcsin 0.67

15. $\tan^{-1}(-12.5)$

16. $\cos^{-1}(-0.23)$

17. arctan 23.8

18. arccos 0.17

In Exercises 19–22, use a calculator to find the approximate value. Express your result in radians.

19. $\tan^{-1}(2.37)$

20. $\tan^{-1}(22.8)$

21. $\sin^{-1}(-0.46)$

22. $\cos^{-1}(-0.853)$

In Exercises 23–30, find the exact value.

23. $\cos[\sin^{-1}(1/2)]$

24. $\sin[\tan^{-1} 1]$

25. $\sin^{-1}[\cos(\pi/4)]$

26. $\cos^{-1}[\cos(7\pi/4)]$

27. $\cos[2\sin^{-1}(1/2)]$

28. $\sin[\tan^{-1}(-1)]$

29. $\arcsin[\cos(\pi/3)]$

30. $\arccos[\tan(\pi/4)]$

In Exercises 31–36, use transformations to describe how the graph of the function is related to a basic inverse trigonometric graph. State the domain and range.

31. $f(x) = \sin^{-1}(2x)$

32. $g(x) = 3\cos^{-1}(2x)$

33. $h(x) = 5\tan^{-1}(x/2)$

34. $f(x) = \sin^{-1}(x + 1)$

35. $g(x) = 3\arccos(x/2)$

36. $h(x) = 0.3\arctan(2x)$

In Exercises 37–40, find an exact solution to the equation.

37. $\sin(\sin^{-1} x) = 1$

38. $\cos^{-1}(\cos x) = 1$

39. $2\sin^{-1} x = 1$

40. $\tan^{-1} x = -1$

In Exercises 41–46, find an algebraic expression equivalent to the given expression. (*Hint:* Form a right triangle as done in Example 5.)

41. $\sin(\tan^{-1} x)$

42. $\cos(\tan^{-1} x)$

43. $\tan(\arcsin x)$

44. $\cot(\arccos x)$

45. $\cos(\arctan 2x)$

46. $\sin(\arccos 3x)$

In Exercises 47–49, solve the problem. (Use degree mode.)

47. *Viewing Angle* You are standing in an art museum viewing a picture. The bottom of the picture is 2 ft above your eye level, and the picture is 12 ft high. Angle θ is formed by the lines of vision to the bottom and to the top of the picture.

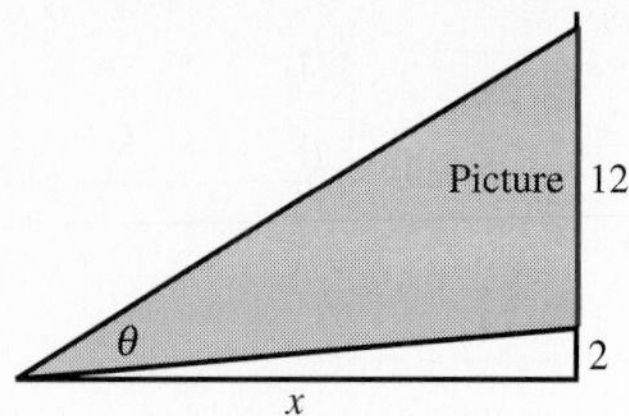

For Exercise 47

a. Show that $\theta = \tan^{-1}\left(\frac{14}{x}\right) - \tan^{-1}\left(\frac{2}{x}\right)$.

b. Graph θ in the [0, 25] by [0, 55] viewing window. Use your grapher to show that the maximum value of θ occurs approximately 5.3 ft from the picture.

c. How far (to the nearest foot) are you standing from the wall if $\theta = 35°$?

48. *Analysis of a Lighthouse* A rotating beacon L stands 3 m across the harbor from the nearest point P along a straight shoreline. As the light rotates, it forms an angle θ as shown in the figure, and illuminates a point Q on the same shoreline as P.

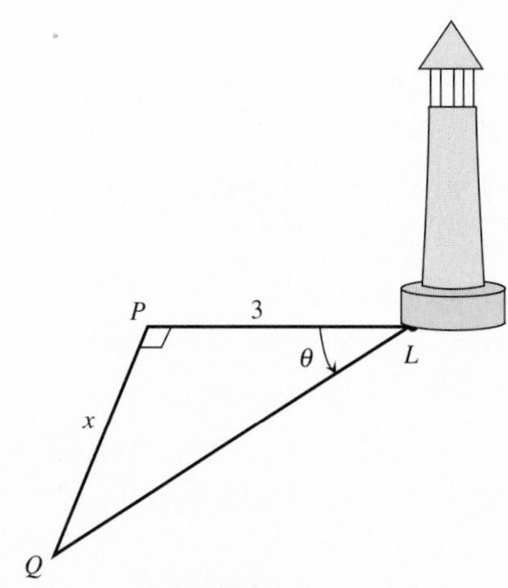

For Exercise 48

a. Show that $\theta = \tan^{-1}\left(\frac{x}{3}\right)$.

b. Graph θ in the viewing window [−20, 20] by [−90, 90]. What do negative values of x represent in the problem? What does a positive angle represent? A negative angle?

c. Find θ when $x = 15$.

49. *Rising Hot-Air Balloon* The hot-air balloon festival held each year in Phoenix, Arizona, is a popular event for photographers. Jo Silver, an award-winning photographer at the event, watches a balloon rising from ground level from a point 500 ft away on level ground.

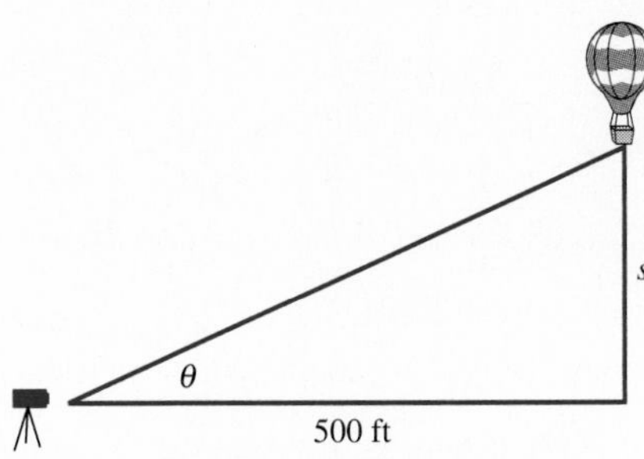

For Exercise 49

a. Write θ as a function of the height s of the balloon.
b. Is the change in θ greater as s changes from 10 ft to 20 ft, or as s changes from 200 ft to 210 ft? Explain.
c. **Writing to Learn** In the graph of this relationship shown here, do you think that the x-axis represents the height s and the y-axis angle θ, or does the x-axis represent angle θ and the y-axis height s? Explain.

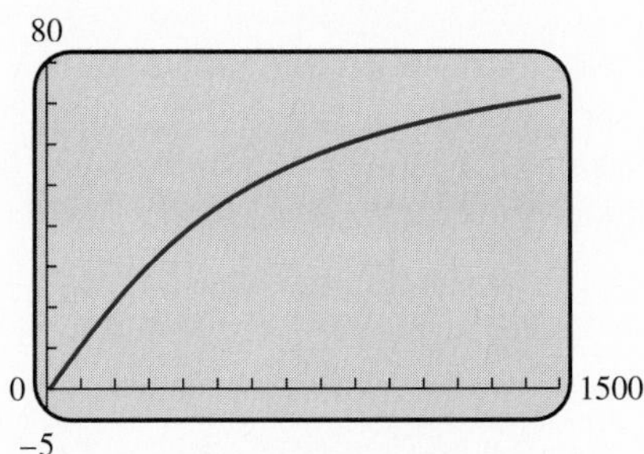

For Exercise 49c

In Exercises 50–52, graph $y = x$ simultaneously with the restricted trigonometric function and its inverse in a square viewing window. Describe the relationship among the graphs.

50. Sine **51.** Cosine

52. Tangent

53. **Writing to Learn** Compare the following exercises. Are the answers identical? Why or why not?

a. Solve $\sin x = 0.5$. **b.** Evaluate $\sin^{-1}(0.5)$.

EXTENDING THE IDEAS

In Exercises 54 and 55, solve the inequality over the interval $[-\pi/2, \pi/2]$.

54. $(\cos x)(\tan^{-1} x) \geq 0$ **55.** $\dfrac{\sin^{-1}(0.5x)}{\sin x} \leq 0.8$

In Exercises 56–63, verify the identity.

56. $\sin^{-1}(-x) = -\sin^{-1} x$

57. $\tan^{-1} x = -\tan^{-1}(-x)$

58. $\sin^{-1} x = \tan^{-1} \dfrac{x}{\sqrt{1 - x^2}}$ for $-1 < x < 1$

59. $\cos^{-1} x = \tan^{-1} \dfrac{\sqrt{1 - x^2}}{x}$ for $0 < x \leq 1$

60. $\arcsin x + \arccos x = \pi/2$ for $-1 \leq x \leq 1$

61. $\arcsin x + \arctan \dfrac{\sqrt{1 - x^2}}{x} = \pi/2$ for $0 < x \leq 1$

62. $\cos(\sin^{-1} x) = \sqrt{1 - x^2}$ for $-1 \leq x \leq 1$

63. $\tan(\cos^{-1} x) = \dfrac{\sqrt{1 - x^2}}{x}$ for $0 < x \leq 1$

Objective

Students will be able to apply the concepts of trigonometry to solve real-world problems.

Motivate

Ask students if they can think of real-world situations that might be approximated by sinusoidal functions.

6.8 APPLICATIONS OF TRIGONOMETRIC FUNCTIONS

Additional Applications Involving Right Triangles • Harmonic Motion

Additional Applications Involving Right Triangles

Since the time of the ancient Greeks, people have solved right triangles as a step in surveying the earth. The process involves superimposing a right triangle on the landscape and then using trigonometry to calculate lengths of sides and

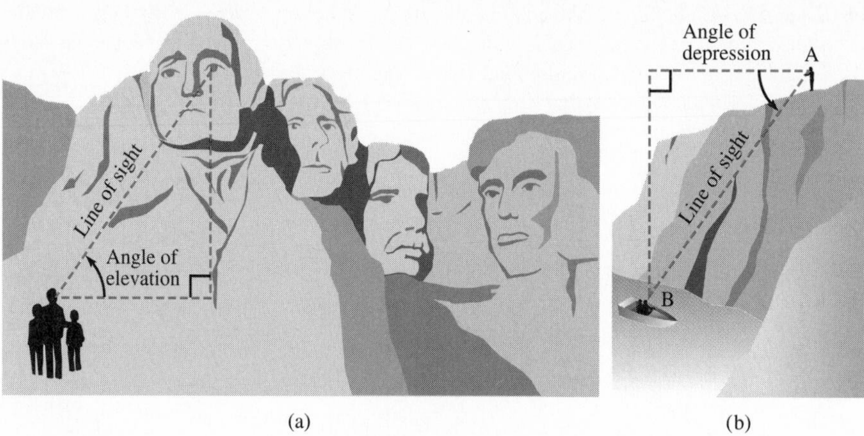

Figure 6.83 (a) Angle of elevation at Mt. Rushmore. (b) Angle of depression at the Grand Canyon.

measures of the acute angles of the triangle. The concepts **angle of elevation** and **angle of depression** are illustrated in Figure 6.83 the way they might be used at Mount Rushmore or at the Grand Canyon.

NOTE

We can conclude that the angles of elevation and depression are equal: Recall from your study of geometry that line of sight is a transversal that intersects a pair of horizontal parallel lines, and these two angles are the resulting equal alternate interior angles.

Teaching Note

The applications in this section illustrate the usefulness of trigonometric functions when viewed as ratios of sides of right triangles.

EXAMPLE 1 APPLICATION: Using Angle of Depression

The angle of depression of a buoy from the top of the Barnegat Bay lighthouse 130 feet above the surface of the water is 6°. Find the distance x from the base of the lighthouse to the buoy.

Solution

Model

Figure 6.84 models the situation. Notice that $\theta = 6°$ using the marginal note above. The height 130 and the distance x are the sides opposite and adjacent to angle θ, respectively.

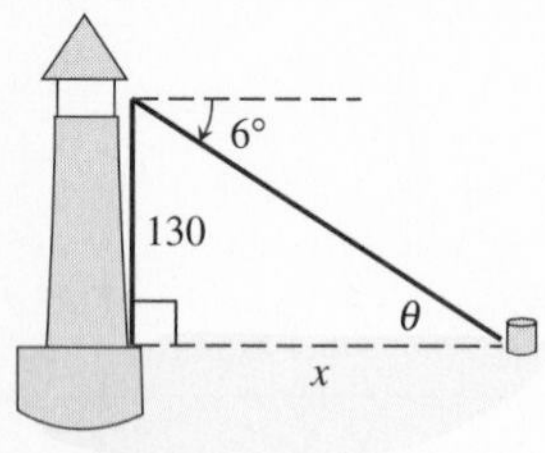

Figure 6.84

Solve Algebraically

Use the tangent function.

$$\tan\theta = \tan 6° = \frac{130}{x}$$

$$x = \frac{130}{\tan 6°} = 1236.867\ldots$$

Interpret

The distance from the base of the lighthouse to the buoy is about 1237 ft. ■

Notes on Examples

Example 2 shows a situation in which it is necessary to draw and use *two* right triangles to solve the problem.

EXAMPLE 2 APPLICATION: Making Indirect Measurements

From the top of the 100-ft-tall Altgelt Hall a man observes a car moving toward the building. If the angle of depression of the car changes from 22° to 46° during the period of observation, how far does the car travel?

Solution

Model

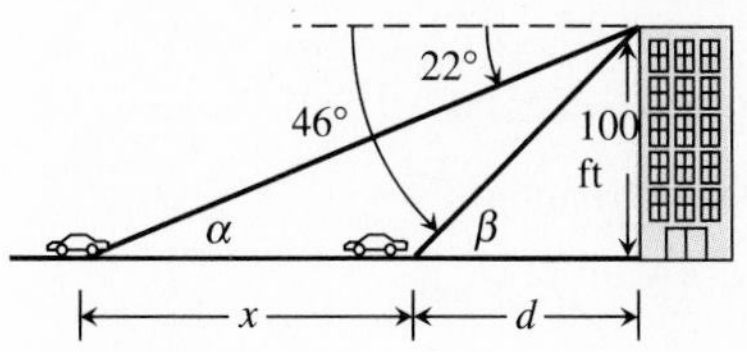

Figure 6.85

Figure 6.85 models the situation. Note the two right triangles with acute angles α and β. We must find length x.

Solve Algebraically

Notice that α and 22° are equal alternate interior angles when the car is first observed, and that β and 46° are equal alternate interior angles when the car is next observed.

From One Right Triangle We Conclude:

$$\tan \beta = \frac{100}{d}$$

$$d = \frac{100}{\tan 46°}$$

From the Other Right Triangle We Conclude:

$$\tan \alpha = \tan 22° = \frac{100}{x + d}$$

$$x + d = \frac{100}{\tan 22°}$$

$$x = \frac{100}{\tan 22°} - d$$

$$x = \frac{100}{\tan 22°} - \frac{100}{\tan 46°}$$

$$x = 150.939 \cdots$$

Interpret

The car travels about 151 ft. ■

EXAMPLE 3 APPLICATION: Using Trigonometry in Navigation

A U.S. Coast Guard patrol boat leaves Port Cleveland and averages 35 knots (naut mph) traveling for 2 h on a course of 53° and then 3 h on a course of 143°. What is the boat's bearing and distance from Port Cleveland?

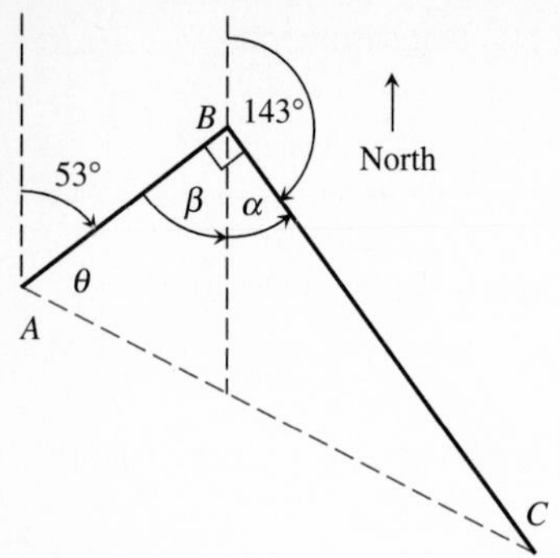

Figure 6.86

Teaching Note

Encourage students to draw diagrams and pictures to help them visualize problem situations.

Solution

Model

Figure 6.86 models the situation.

Line AB in Figure 6.86 is a transversal that intersects a pair of vertical parallel lines. Thus, $\beta = 53°$ because they are alternate interior angles. Angle α, as the supplement of a 143° angle, is 37°. Consequently $\angle ABC = 90°$, and AC is the hypotenuse of right $\triangle ABC$.

Solve Algebraically

Use distance = rate × time to determine distances AB and BC.

$$AB = (35 \text{ knots})(2 \text{ h}) = 70 \text{ naut mi}$$

$$BC = (35 \text{ knots})(3 \text{ h}) = 105 \text{ naut mi}$$

Solve the right triangle for AC and θ.

$$AC = \sqrt{70^2 + 105^2} \quad \text{Pythagorean theorem}$$

$$= 126.194 \ldots$$

$$\theta = \tan^{-1}\left(\frac{105}{70}\right)$$

$$= 56.309 \ldots°$$

Interpret

The Coast Guard boat's bearing from Port Cleveland is about

$$53 + \theta \approx 53 + 56.3 = 109.3°.$$

They are about 126.2 mi out. ■

Harmonic Motion

The sine and cosine functions, because of their periodic nature, are helpful in describing the motion of objects that oscillate, vibrate, or rotate. For example, the linkage in Figure 6.87 converts the rotary motion of a motor to the back-and-forth motion needed for some machines. When the wheel rotates, the piston oscillates back and forth.

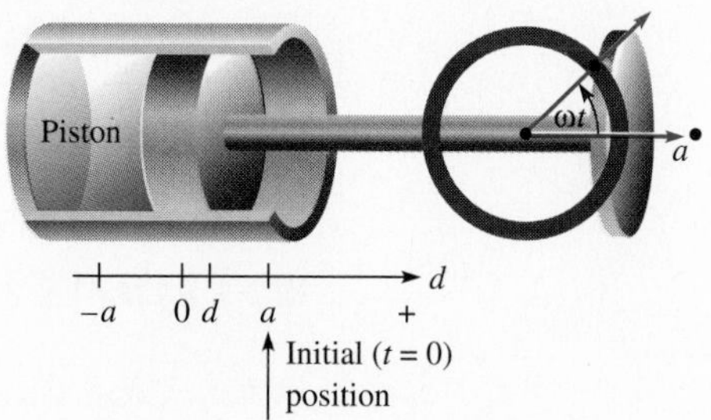

Figure 6.87

If the wheel rotates at a constant rate ω radians per second, the back-and-forth motion of the piston is an example of *simple harmonic motion* and can be modeled by an equation of the form

$$d = a \cos \omega t, \quad a > 0$$

where a is the radius of the wheel and d is the position of the piston on a number line.

NOTE

Notice that harmonic motion is a sinusoidal function with amplitude $|a|$ and period $2\pi/\omega$. The period is the reciprocal of the frequency.

Definition: Simple Harmonic Motion

A point moving on a number line is in **simple harmonic motion** if its directed distance d from the origin at time t is given by either

$$d = a \sin \omega t \qquad \text{or} \qquad d = a \cos \omega t,$$

where a and ω are real numbers and $\omega > 0$. The motion has the **frequency** $\omega/2\pi$, which is the number of oscillations per unit of time.

EXAMPLE 4 Calculating Harmonic Motion

In a mechanical linkage like the one shown in Figure 6.87 a wheel with an 8-cm radius turns with an angular velocity of 8π rad/sec.

a. What is the frequency of the piston?
b. What is the distance from the initial position ($t = 0$) exactly 3.45 sec after starting?

Follow-up

Ask students if they can think of other equations that model the situation in Example 5 on page 472.

Assignment Guide

Day 1: Ex. 1, 2, 5, 6, 7, 10, 12, 13, 14, 16, 17, 19, 20, 21, 25
Day 2: Ex. 29, 31, 33, 34, 36, 37, 38, 40, 42

Solution

Model

The sinusoid $d = 8 \cos 8\pi t$ models the motion of the piston.

Solve Algebraically

a. The frequency of $d = 8 \cos 8\pi t$ is $8\pi/2\pi$, or four complete back-and-forth cycles per second. The graph of $y = 8 \cos 8\pi t$ in Figure 6.88 models the four cycles of the motor or the four complete strokes of the piston over 1 sec. Notice that this sinusoid has a period of 1/4.

b. We must find the distance between the positions at $t = 0$ and $t = 3.45$. The initial position at $t = 0$ is

$$d(0) = 8.$$

The position at $t = 3.45$ is

$$d(3.45) = 8 \cos [8\pi(3.45)] = 2.472 \cdots$$

The distance between the two positions is $8 - 2.472 \cdots = 5.527 \cdots$.

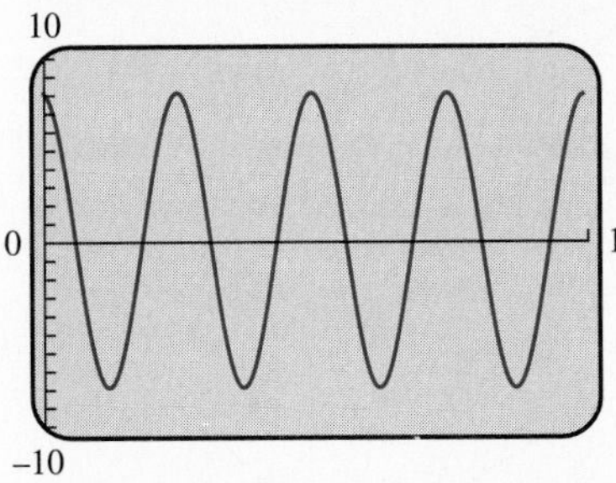

Figure 6.88 The piston completes four complete cycles in 1 sec.

Cooperative Learning

Group Activity: Ex. 43

Interpret

At 3.45 sec, the piston moving at four cycles per second is about 5.53 cm from its initial position. ■

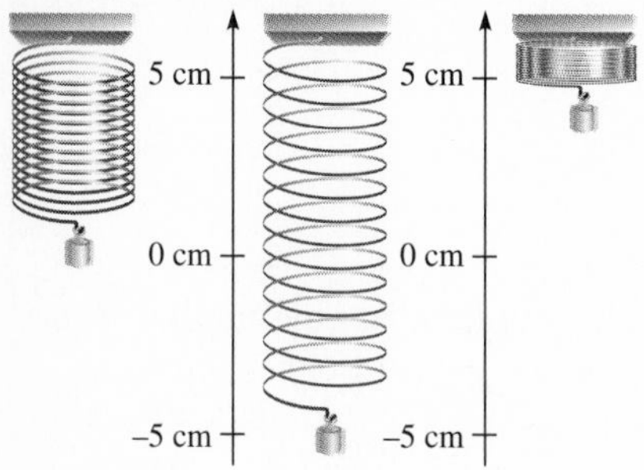

Figure 6.89

■ EXAMPLE 5 Calculating Harmonic Motion

A mass oscillating up and down on the bottom of a spring (assuming perfect elasticity and no friction or air resistance) can be modeled as harmonic motion. If the weight is displaced a maximum of 5 cm, find the modeling equation if it takes 2 sec to complete one cycle. (See Figure 6.89.)

Solution

Model

This harmonic motion can be modeled with the equation $d = a \cos \omega t$ or $d = a \sin \omega t$. Assuming that the spring is at the origin of the coordinate system when $t = 0$, we choose the equation $d = a \sin \omega t$.

Solve Algebraically

Finding Amplitude: Because the maximum displacement is 5 cm, we conclude that the amplitude $a = 5$.

Finding ω: If it takes 2 sec to complete one cycle, it completes 1/2 cycle in 1 sec. Therefore,

$$\frac{\omega}{2\pi} = \frac{1}{2},$$

$$\omega = \pi.$$

Interpret

The modeling equation is $d = 5 \sin \pi t$. ■

Notes on Exercises

Ex. 27–30 deal with simple harmonic motion.

Ex. 33, 35, and 43 require students to find sinusoidal functions that approximate data.

Ex. 37–39 use geometry concepts.

Ongoing Assessment

Self-Assessment: Ex. 3, 9, 15, 23, 41

Embedded Assessment: Ex. 34, 36

Quick Review 6.8

In Exercises 1–4, find the lengths a, b, and c.

1.

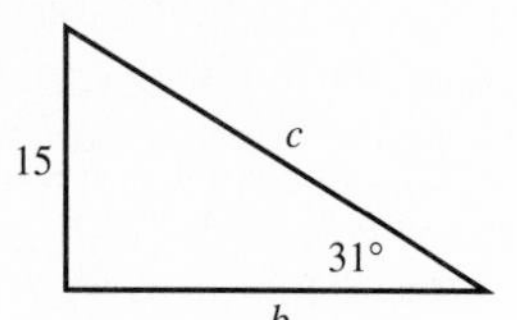

2.

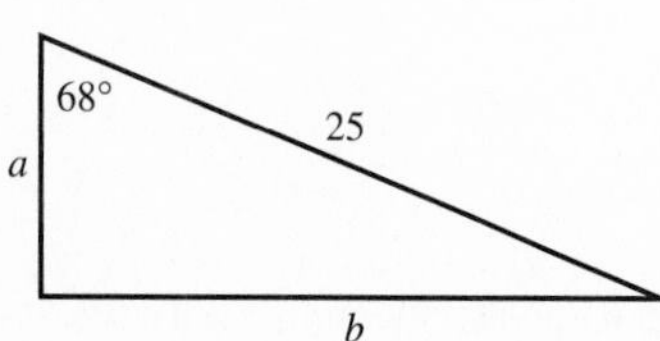

3.

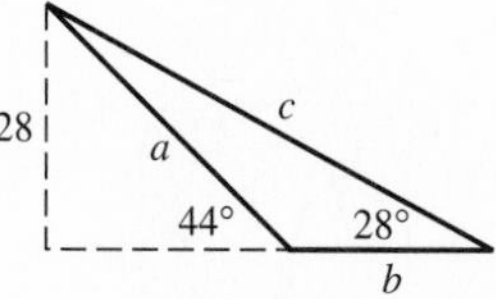

4.

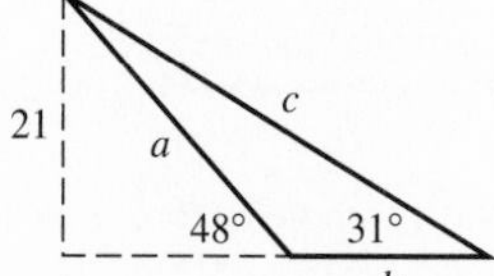

In Exercises 5–8, find the complement and the supplement of the angle, or state that none exists.

5. 32°

6. 73°

7. 95°

8. 120°

In Exercises 9 and 10, state the bearing that describes the direction.

9. NE (northeast)

10. SSW (south-southwest)

In Exercises 11 and 12, state the amplitude and period of the sinusoid.

11. $-3 \sin 2(x - 1)$

12. $4 \cos 4(x + 2)$

SECTION EXERCISES 6.8

In Exercises 1–43, solve the problem. Sketch a figure if one is not provided.

1. *Finding a Cathedral Height* The angle of elevation of the top of the Ulm Cathedral from a point 300 ft away from the base of its steeple on level ground is 60°. Find the height of the cathedral.

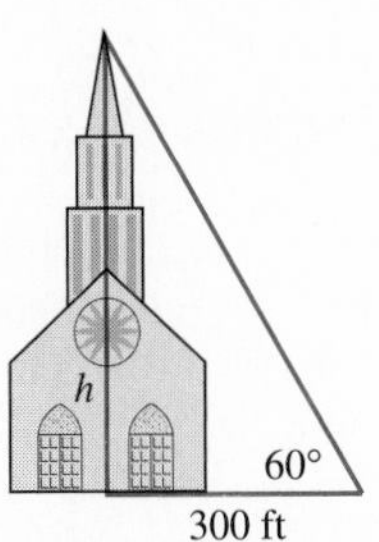

For Exercise 1

2. *Finding a Monument Height* From a point 100 ft from its base the angle of elevation of the top of the Arch of Septimus Severus, in Rome, Italy, is 34° 13′ 12″. How tall is this monument?

3. *Finding a Distance* The angle of depression from the top of the Smoketown Lighthouse 120 ft above the surface of the water to a buoy is 10°. How far is the buoy from the lighthouse?

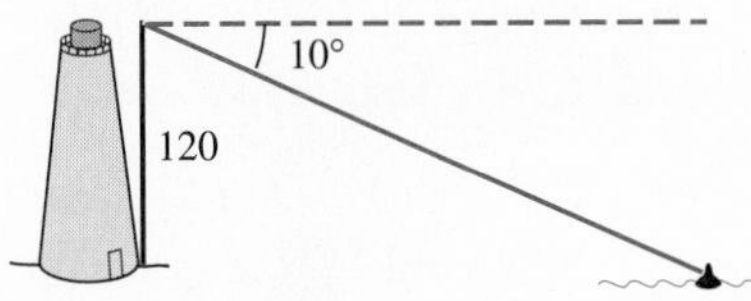

For Exercise 3

4. *Finding a Baseball Stadium Dimension* The top row of the red seats behind home plate at Cincinnati's Riverfront Stadium is 90 ft above the level of the playing field. The angle of depression to the base of the left field wall is 14°. How far is the base of the left field wall from a point on level ground directly below the top row?

5. *Finding a Guy-Wire Length* A guy wire connects the top of an antenna to a point on level ground 5 ft from the base of the antenna. The angle of elevation formed by this wire is 80°. What are the length of the wire and the height of the antenna?

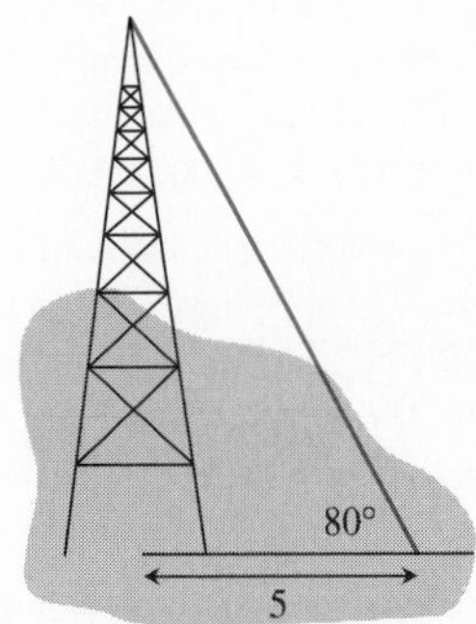

For Exercise 5

6. *Finding a Length* A wire stretches from the top of a vertical pole to a point on level ground 16 ft from the base of the pole. If the wire makes an angle of 62° with the ground, find the height of the pole and the length of the wire.

7. *Height of Eiffel Tower* The angle of elevation of the top of the TV antenna mounted on top of the Eiffel Tower in Paris is measured to be 80° 1′ 12″ at a point 185 ft from the base of the tower. How tall is the tower plus TV antenna?

8. *Finding the Height of Tallest Chimney* The world's tallest smokestack at the International Nickel Co., Sudbury, Ontario, casts a shadow that is approximately 1580 ft long when the sun's angle of elevation (measured from the horizon) is 38°. How tall is the smokestack?

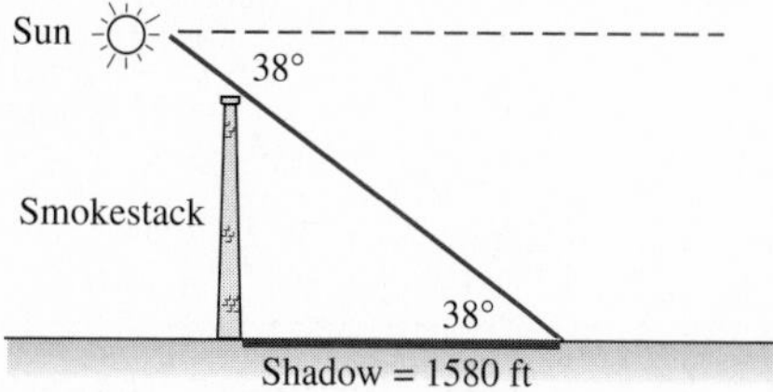

For Exercise 8

9. *Cloud Height* To measure the height of a cloud, you place a bright searchlight directly below the cloud and shine the beam straight up. From a point 100 ft away from the searchlight, you measure the angle of elevation of the cloud to be 83° 12′. How high is the cloud?

10. *Ramping Up* A ramp leading to a freeway overpass is 470 ft long and rises 32 ft. What is the average angle of inclination of the ramp to the nearest tenth of a degree?

11. *Antenna Height* A guy wire attached to the top of the KSAM radio antenna is anchored at a point on the ground 10 m from the antenna's base. If the wire makes an angle of 55° with level ground, how high is the KSAM antenna?

12. *Building Height* To determine the height of the Louisiana-Pacific (LP) Tower, the tallest building in Conroe, Texas, a surveyor stands at a point on the ground, level with the base of the LP building. He measures the point to be 125 ft from the building's base and the angle of elevation to the top of the building to be 29° 48′. Find the height of the building.

13. *Navigation* The *Paz Verde,* a whalewatch boat, is located at point P, and L is the nearest point on the Baja California shore. Point Q is located 4.25 mi down the shoreline from L and $\overline{PL} \perp \overline{LQ}$. Determine the distance that the *Paz Verde* is from the shore if $\angle PQL = 35°$.

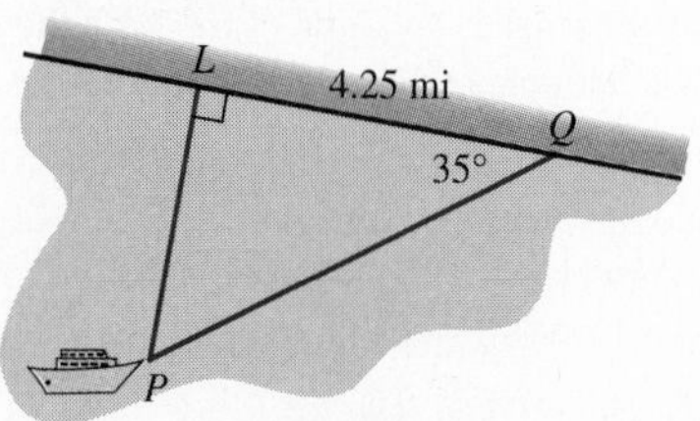

For Exercise 13

14. *Recreational Hiking* While hiking on a level path toward Colorado's front range, Otis Evans determines that the angle of elevation to the top of Long's Peak is 30°. Moving 1000 ft closer to the mountain, Otis determines the angle of elevation to be 35°. How much higher is the top of Long's Peak than Otis's elevation?

15. *Civil Engineering* The angle of elevation from an observer to the bottom edge of the Delaware River drawbridge observation deck located 200 ft from the observer is 30°. The angle of elevation from the observer to the top of the observation deck is 40°. What is the height of the observation deck?

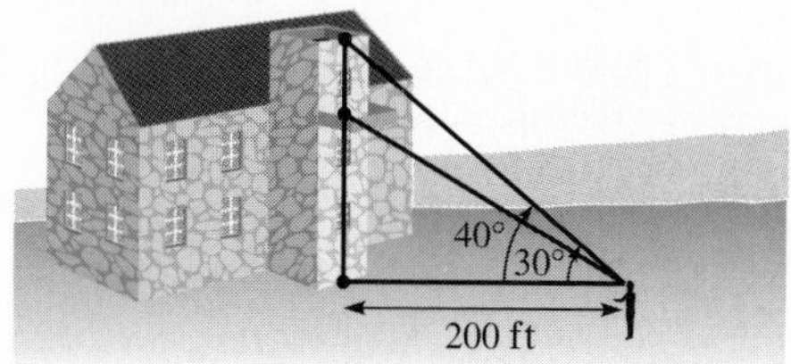

For Exercise 15

16. *Traveling Car* From the top of a 100-ft building a man observes a car moving toward him. If the angle of depression of the car changes from 15° to 33° during the period of observation, how far does the car travel?

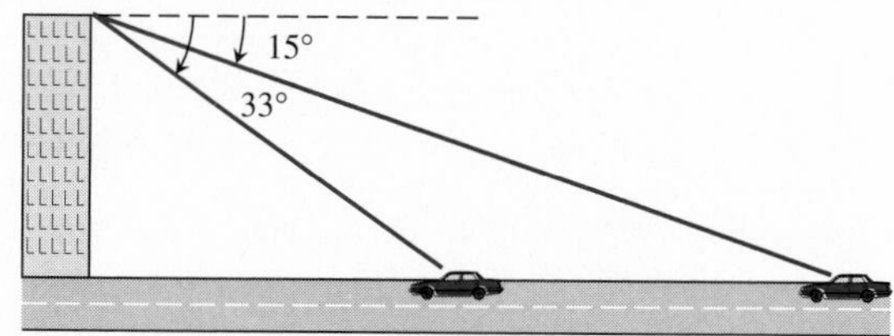

For Exercise 16

17. *Navigation* The Coast Guard cutter *Angelica* travels at

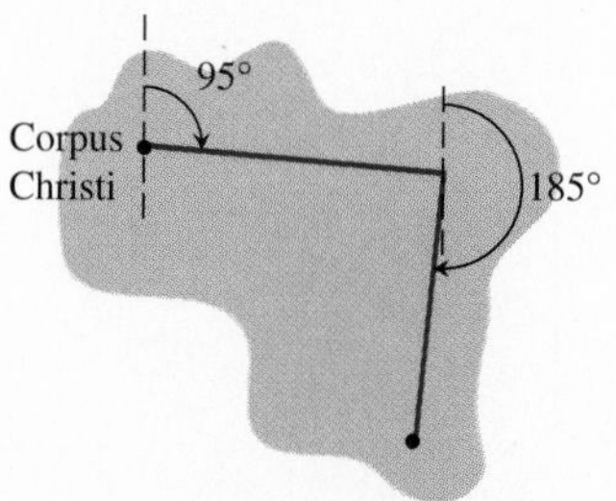

For Exercise 17

30 knots from its home port of Corpus Christi on a course of 95° for 2 h and then changes to a course of 185° for 2 h. Find the distance and the bearing from the Corpus Christi port to the boat.

18. *Navigation* The *Cerrito Lindo* travels at a speed of 40 knots from Fort Lauderdale on a course of 65° for 2 h and then changes to a course of 155° for 4 h. Determine the distance and the bearing from Fort Lauderdale to the boat.

19. *Land Measure* The angle of depression is 19° from a point 7256 ft above sea level on the north rim of the Grand Canyon level to a point 6159 ft above sea level on the south rim. How wide is the canyon at that point?

20. *Ranger Fire Watch* A ranger spots a fire from a 73-ft tower in Yellowstone National Park. She measures the angle of depression to be 1° 20′. How far is the fire from the tower?

21. *Civil Engineering* The bearing of the line of sight to the east end of the Royal Gorge footbridge from a point 325 ft due north of the west end of the footbridge across the Royal Gorge is 117°. What is the length ℓ of the bridge?

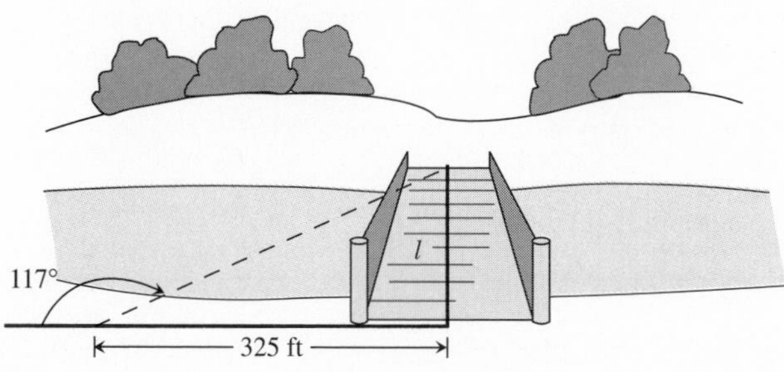

For Exercise 21

22. *Space Flight* The angle of elevation of the space shuttle *Columbia* from Cape Canaveral is 17° when the shuttle is directly over a ship 12 mi downrange. What is the altitude of the shuttle when it is directly over the ship?

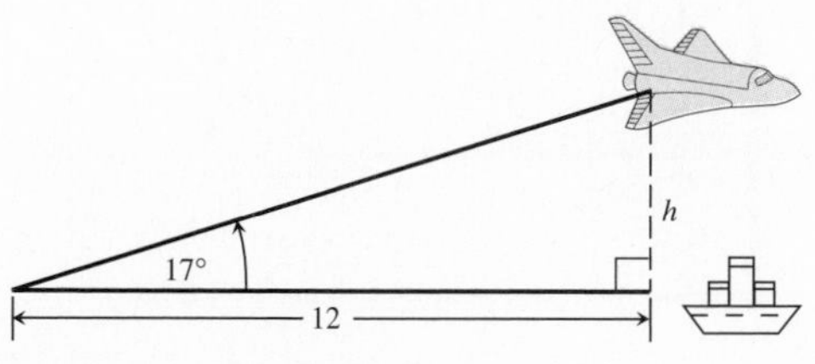

For Exercise 22

23. *Architectural Design* A barn roof is constructed as shown in the figure. What is the height of the vertical center span?

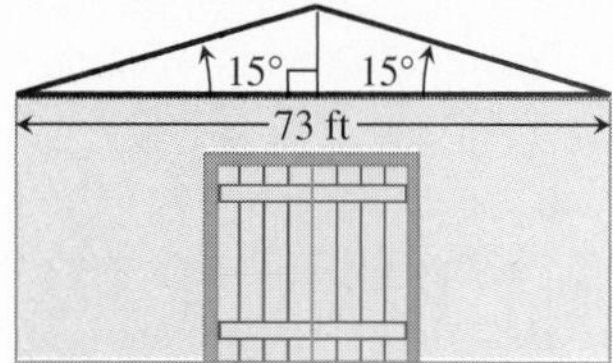

For Exercise 23

24. *Recreational Flying* A hot-air balloon over Park City, Utah, is 760 ft above the ground. The angle of depression from the balloon to an observer is 5.25°. Assuming the ground is relatively flat, how far is the observer from a point on the ground directly under the balloon?

25. *Navigation* A shoreline runs north-south, and a boat is due east of the shoreline. The bearings of the boat from two points on the shore are 110° and 100°. Assume the two points are 550 ft apart. How far is the boat from the shore?

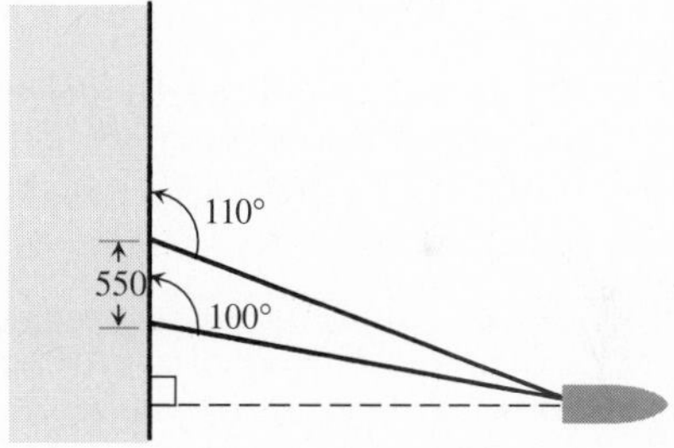

For Exercise 25

26. *Navigation* Milwaukee, Wisconsin, is directly west of Grand Haven, Michigan, on opposite sides of Lake Michigan. On a foggy night, a law enforcement boat leaves from Milwaukee on a course of 105° at the same time that a small smuggling craft steers a course of 195° from Grand Haven. The law enforcement boat averages 23 knots and collides with the smuggling craft. What was the smuggling boat's average speed?

27. *Mechanical Design* *Refer to Figure 6.87.* The wheel in a piston linkage like the one shown in the figure has a radius of 6 in. It turns with an angular velocity of 16π rad/sec. The initial position is the same as that shown in Figure 6.87.

a. What is the frequency of the piston?

b. What equation models the motion of the piston?

c. What is the distance from the initial position 2.85 sec after starting?

28. *Mechanical Design* Suppose the wheel in a piston linkage like the one shown in Figure 6.87 has a radius of 18 cm and turns with an angular velocity of π rad/sec.

a. What is the frequency of the piston?
b. What equation models the motion of the piston?
c. How many cycles does the piston make in 1 min?

29. *Vibrating Spring* A mass on a spring oscillates back and forth and completes one cycle in 0.5 sec. Its maximum displacement is 3 cm. Write an equation that models this motion.

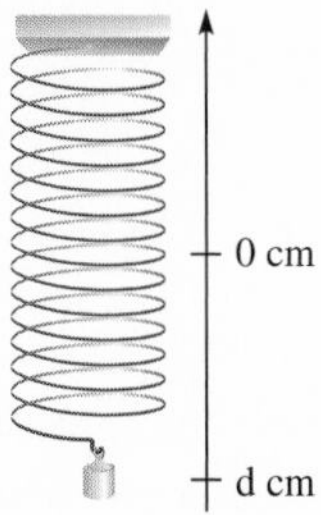

For Exercise 29

30. *Tuning Fork* A point on the tip of a tuning fork vibrates in harmonic motion described by the equation $d = 14 \sin \omega t$. Find ω for a tuning fork that has a frequency of 528 vibrations per second.

31. *Ferris Wheel Motion* The Ferris wheel shown in this figure makes one complete turn every 20 sec. A rider's height, h, above the ground can be modeled by the equation $h = a \sin \omega t + k$, where h and k are given in feet and t is given in seconds.

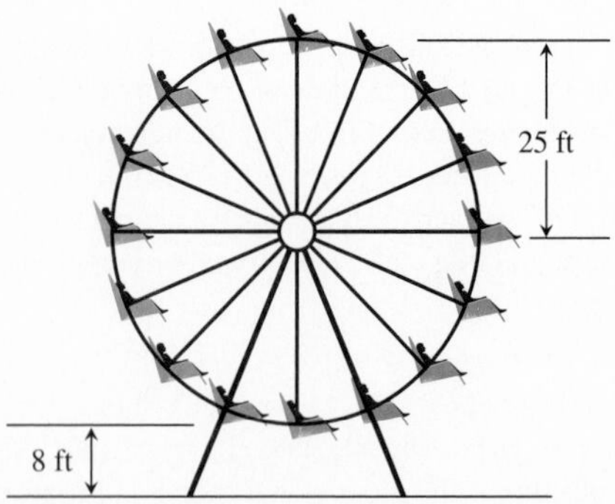

For Exercise 31

a. What is the value of a?
b. What is the value of k?
c. What is the value of ω?

32. *Ferris Wheel Motion* Jacob and Emily ride a Ferris wheel at a carnival in Billings, Montana. The wheel has a 16-m diameter and turns at 3 rpm with its lowest point 1 m above the ground. Assume that Jacob and Emily's height h above the ground is a sinusoidal function of time t (in seconds), where $t = 0$ represents the lowest point of the wheel.

a. Write an equation for h.
b. Draw a graph of h for $0 \le t \le 30$.
c. Use h to estimate Jacob and Emily's height above the ground at $t = 4$ and $t = 10$.

33. Assume that the data shown in Table 6.3 and graphed in the figure repeat each year, and you want to find an equation of the form $y = a \sin [b(t - h)] + k$ that models these data.

Table 6.3 Temperature Data for St. Louis

Time (months)	Average Temperature
1	34°
2	30°
3	39°
4	44°
5	58°
6	67°
7	78°
8	80°
9	72°
10	63°
11	51°
12	40°

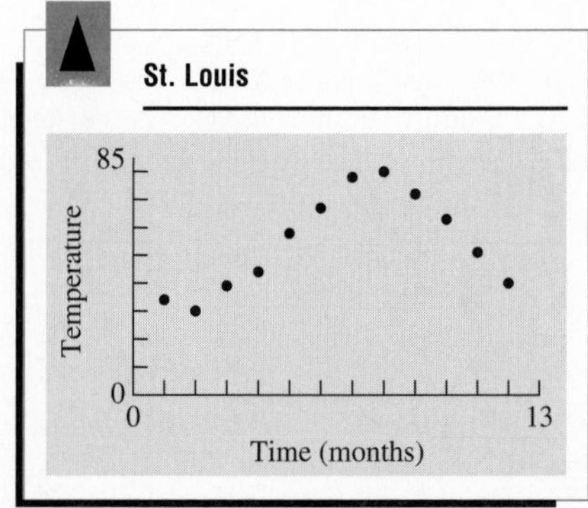

For Exercise 33

a. Given that the period is 12 months, what is the value of b?

b. How is the amplitude a related to the difference $80° - 30°$?

c. Use the information in part b to find k.

d. Find h, and write an equation for y.

e. Superimpose a graph of y on a scatter plot of the data in Table 6.3.

f. Use the modeling equation from part d (with the assumption that $t = 0$ means January 1) to predict dates when the temperature will be 70°.

34. Writing to Learn For the Ferris wheel in Exercise 31, which equation correctly models the height of a rider who begins the ride at the bottom of the wheel when $t = 0$?

a. $h = 25 \sin \frac{\pi t}{10}$

b. $h = 25 \sin \frac{\pi t}{10} + 8$

c. $h = 25 \sin \frac{\pi t}{10} + 33$

d. $h = 25 \sin \left(\frac{\pi t}{10} + \frac{3\pi}{2}\right) + 33$

Explain your thought process, and use of a graphing utility in choosing the correct modeling equation.

35. The data for displacement (pressure) versus time on a tuning fork, shown in Table 6.4, were collected using a **CBL** (see box in Section 1.4 Exercises) and a microphone.

Table 6.4 Tuning Fork Data

Time	Displacement	Time	Displacement
0.00091	−0.080	0.00362	0.217
0.00108	0.200	0.00379	0.480
0.00125	0.480	0.00398	0.681
0.00144	0.693	0.00416	0.810
0.00162	0.816	0.00435	0.827
0.00180	0.844	0.00453	0.749
0.00198	0.771	0.00471	0.581
0.00216	0.603	0.00489	0.346
0.00234	0.368	0.00507	0.077
0.00253	0.099	0.00525	−0.164
0.00271	−0.141	0.00543	−0.320
0.00289	−0.309	0.00562	−0.354
0.00307	−0.348	0.00579	−0.248
0.00325	−0.248	0.00598	−0.035
0.00344	−0.041		

a. Graph a scatter plot of the data in the [0, 0.0062] by [−0.5, 1] viewing window.

b. Select the equation that appears to be the best fit of these data.

i. $y = 0.6 \sin (2464x - 2.84) + 0.25$
ii. $y = 0.6 \sin (1210x - 2) + 0.25$
iii. $y = 0.6 \sin (2440x - 2.1) + 0.15$

c. What is the approximate frequency of the tuning fork?

36. Writing to Learn Human sleep—awake cycles at three different ages are described by the accompanying graphs. The portions of the graphs above the horizontal lines represent times awake, and the portions below represent times asleep.

Newborn

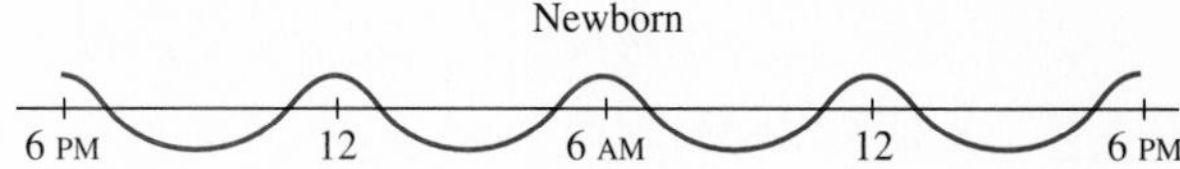

Four years

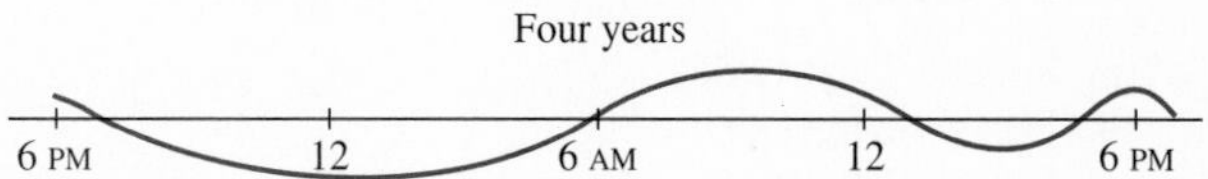

Adult

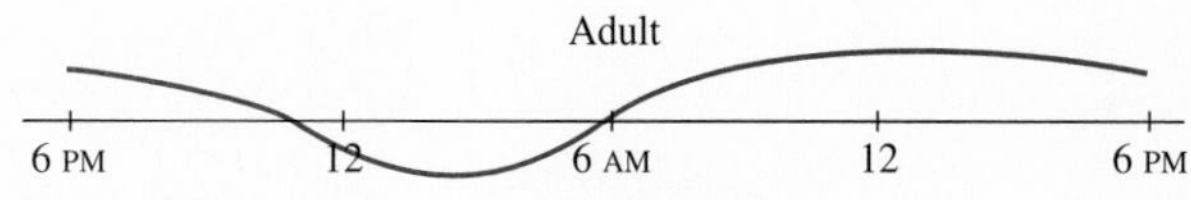

For Exercise 36

a. What is the period of the sleep—awake cycle of a newborn? Of a four year old? Of an adult?

b. Which of these three sleep—awake cycles is the closest to being modeled by a function $y = a \sin bx$?

Using Trigonometry in Geometry In a *regular polygon* all sides have equal length and all angles have equal measure. Consider the regular seven-sided polygon whose sides are 5 cm.

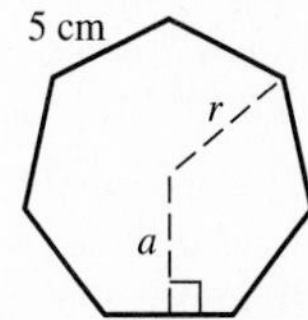

For Exercises 37–38

37. Find the length of the *apothem,* the segment from the center of the seven-sided polygon to the midpoint of a side.

38. Find the radius of the circumscribed circle of the regular seven-sided polygon.

39. A *rhombus* is a quadrilateral with all sides equal in length. Recall that a rhombus is also a parallelogram. Find length AC and length BD in the rhombus shown here.

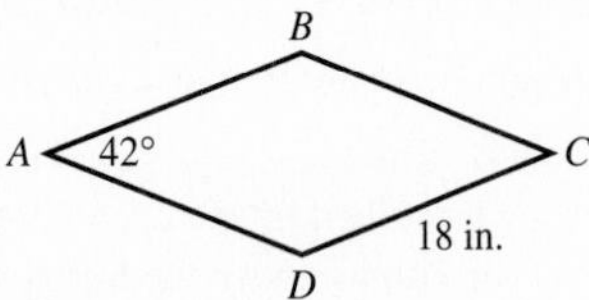

For Exercise 39

EXTENDING THE IDEAS

40. A roof has two sections, one with a 50° elevation and the other with a 20° elevation, as shown in the figure.

a. Find the height BE.
b. Find the height CD.
c. Find the length $AE + ED$, and double it to find the length of the cross section of the roof.

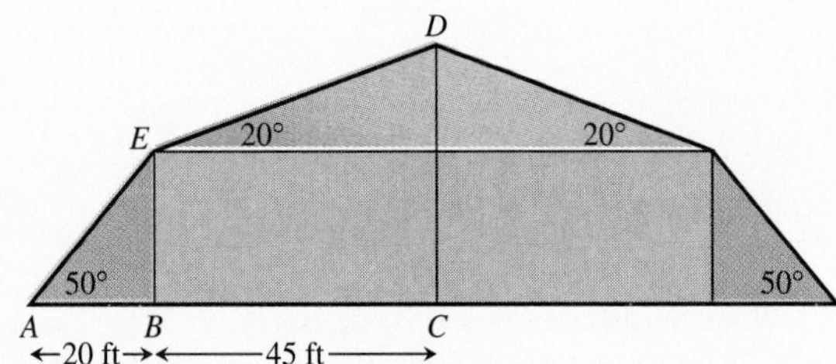

For Exercise 40

41. *Steep Trucking* The *percentage grade* of a road is its slope expressed as a percentage. A tractor-trailer rig passes a sign that reads, "6% grade next 7 miles." What is the average angle of inclination of the road?

42. *Television Coverage* Many satellites travel in *geosynchronous orbits,* which means that the satellite stays over the same point on the earth. A satellite that broadcasts HBO is in geosynchronous orbit 100 mi above the earth. Assume that the earth is a sphere with radius 4000 mi, and find the arc length of coverage area for the HBO satellite on the earth's surface.

43. Group Learning Activity A musical note like that produced with a tuning fork or pitch meter is a pressure wave. Typically, frequency is measured in hertz (1 Hz = 1 cycle per second). Table 6.5 gives frequency (in Hz) of several musical notes. The time-vs.-pressure tuning fork data in Table 6.6 was collected using a **CBL** and a microphone.

Table 6.5 Tuning Fork Data

Note	Frequency (Hz)
C	262
C♯ or D♭	277
D	294
D♯ or E♭	311
E	330
F	349
F♯ or G♭	370
G	392
G♯ or A♭	415
A	440
A♯ or B♭	466
B	494
C (next octave)	524

Table 6.6 Tuning Fork Data

Time (sec)	Pressure	Time (sec)	Pressure
0.0002368	1.29021	0.0049024	−1.06632
0.0005664	1.50851	0.0051520	0.09235
0.0008256	1.51971	0.0054112	1.44694
0.0010752	1.51411	0.0056608	1.51411
0.0013344	1.47493	0.0059200	1.51971
0.0015840	0.45619	0.0061696	1.51411
0.0018432	−0.89280	0.0064288	1.43015
0.0020928	−1.51412	0.0066784	0.19871
0.0023520	−1.15588	0.0069408	−1.06072
0.0026016	−0.04758	0.0071904	−1.51412
0.0028640	1.36858	0.0074496	−0.97116
0.0031136	1.50851	0.0076992	0.23229
0.0033728	1.51971	0.0079584	1.46933
0.0036224	1.51411	0.0082080	1.51411
0.0038816	1.45813	0.0084672	1.51971
0.0041312	0.32185	0.0087168	1.50851
0.0043904	−0.97676	0.0089792	1.36298
0.0046400	−1.51971		

a. Graph a scatter plot of the data.

b. Determine a, b, and h so that the equation $y = a \sin [b(t - h)]$ is a model for the data.

c. Determine the frequency of the sinusoid in part b, and use Table 6.5 to identify the musical note produced by the tuning fork.

d. Identify the musical note produced by the tuning fork used in Exercise 35.

CHAPTER 6 REVIEW

KEY TERMS

The number following each key term indicates the page of its introduction.

REVIEW EXERCISES

In Exercises 1–8, determine the quadrant of the terminal side of the angle in standard position. Express degrees in radians and radians in degrees.

1. $\frac{5\pi}{2}$

2. $\frac{3\pi}{4}$

3. $-135°$

4. $-45°$

5. $78°$

6. $112°$

7. $\frac{\pi}{12}$

8. $\frac{7\pi}{10}$

In Exercises 9 and 10, determine the angle measure in both degrees and radians. Draw the angle in standard position if its terminal side is obtained as described.

9. A three-quarters counterclockwise rotation

10. Two and one-half counterclockwise rotations

In Exercises 11–16, the point is on the terminal side of an angle in standard position. Give the smallest positive angle measure in both degrees and radians.

11. $(\sqrt{3}, 1)$

12. $(-1, 1)$

13. $(-1, \sqrt{3})$

14. $(-3, -3)$

15. $(6, -12)$

16. $(2, 4)$

In Exercises 17–28, evaluate the expression exactly.

17. $\sin 30°$

18. $\cos 330°$

19. $\tan -135°$

20. $\sec -135°$

21. $\sin \frac{5\pi}{6}$

22. $\csc \frac{2\pi}{3}$

23. $\sec -\dfrac{\pi}{3}$

24. $\tan -\dfrac{2\pi}{3}$

25. $\csc 270°$

26. $\sec 180°$

27. $\cot -90°$

28. $\tan 360°$

In Exercises 29–32, evaluate exactly all six trigonometric functions of the angle. Use the reference angle and properties of a 30-60 right triangle and a 45-45 right triangle.

29. $-\dfrac{\pi}{6}$

30. $\dfrac{19\pi}{4}$

31. $-135°$

32. $420°$

In Exercises 33–38, solve the problem.

33. Find all six trigonometric functions of α in $\triangle ABC$.

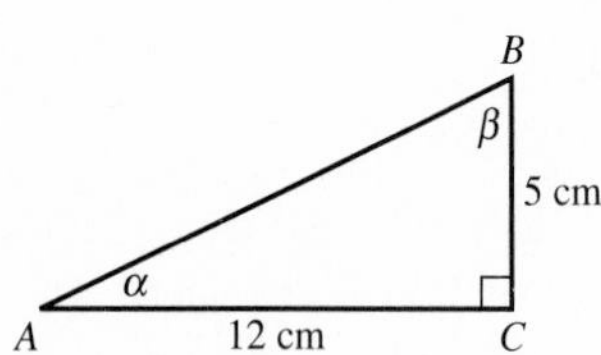

For Exercise 33

34. Use a right triangle to determine the values of all trigonometric functions of θ, where $\cos \theta = 5/7$.

35. Use a right triangle to determine the values of all trigonometric functions of θ, where $\tan \theta = 15/8$.

36. Use a calculator in degree mode to solve $\cos \theta = 3/7$ if $0° \le \theta \le 90°$.

37. Use a calculator in radian mode to solve $\tan x = 1.35$ if $\pi \le x \le 3\pi/2$.

38. Use a calculator to solve $\sin x = 0.218$ if $0 \le x \le 2\pi$.

In Exercises 39–44, solve the right triangle $\triangle ABC$.

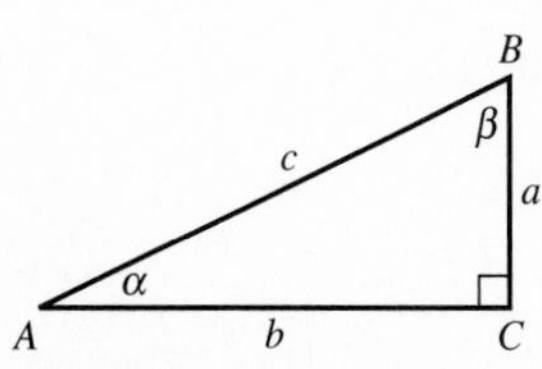

For Exercises 39–44

39. $\alpha = 35°$, $c = 15$

40. $b = 8$, $c = 10$

41. $\beta = 48°$, $a = 7$

42. $\alpha = 28°$, $c = 8$

43. $b = 5$, $c = 7$

44. $a = 2.5$, $b = 7.3$

In Exercises 45–48, x is an angle in standard position with $0 \le x \le 2\pi$. Determine the quadrant of x.

45. $\sin x < 0$ and $\tan x > 0$

46. $\cos x < 0$ and $\csc x > 0$

47. $\tan x < 0$ and $\sin x > 0$

48. $\sec x < 0$ and $\csc x > 0$

In Exercises 49 and 50, use the fundamental trigonometric identities to evaluate the expression. Check your answer with a calculator.

49. $\tan \theta$ if $\cos \theta = 0.82$

50. $\tan \theta$ if $\sin \theta = -0.58$

In Exercises 51–54, point P is on the terminal side of angle θ. Evaluate the six trigonometric functions for θ.

51. $(-3, 6)$

52. $(12, 7)$

53. $(-5, -3)$

54. $(4, 9)$

In Exercises 55–62, use transformations to describe how the graph of the function is related to a basic trigonometric graph. Graph two periods.

55. $y = \sin(x + \pi)$

56. $y = 3 + 2\cos x$

57. $y = -\cos(x + \pi/2) + 4$

58. $y = -2 - 3\sin(x - \pi)$

59. $y = \tan 2x$

60. $y = -2\cot 3x$

61. $y = -2\sec \dfrac{x}{2}$

62. $y = \csc \pi x$

In Exercises 63–68, state the amplitude, period, phase shift, domain, and range for the sinusoid.

63. $f(x) = 2\sin 3x$

64. $g(x) = 3\cos 4x$

65. $f(x) = 1.5\sin(2x - \pi/4)$

66. $g(x) = -2\sin(3x - \pi/3)$

67. $y = 4\cos(2x - 1)$

68. $g(x) = -2\cos(3x + 1)$

In Exercises 69 and 70, graph the function. Then estimate the values of a, b, and h so that $f(x) \approx a\sin[b(x - h)]$.

69. $f(x) = 2\sin x - 4\cos x$

70. $f(x) = 3\cos 2x - 2\sin 2x$

In Exercises 71–74, use a calculator to evaluate the expression. Express your answer in both degrees and radians.

71. $\sin^{-1}(0.766)$

72. $\cos^{-1}(0.479)$

73. $\tan^{-1} 1$

74. $\sin^{-1}\left(\dfrac{\sqrt{3}}{2}\right)$

In Exercises 75–78, describe how the graph of the function is related to a basic inverse trigonometric graph. State the domain and range.

75. $y = \sin^{-1} 3x$ **76.** $y = \tan^{-1} 2x$

77. $y = \sin^{-1}(3x - 1) + 2$ **78.** $y = \cos^{-1}(2x + 1) - 3$

In Exercises 79–82, solve for x.

79. $\sin x = 0.5,\ 0 \le x \le \pi$

80. $\cos x = \sqrt{3}/2,\ 0 \le x \le \pi$

81. $\tan x = -1,\ 0 \le x \le 2\pi$

82. $\sin x = 0.7,\ \pi/2 \le x \le \pi$

In Exercises 83 and 84, describe the end behavior of the function.

83. $\dfrac{\sin x}{x^2}$ **84.** $\dfrac{3}{5}e^{-x/12} \sin(2x - 3)$

In Exercises 85–88, evaluate the expression.

85. $\tan[\tan^{-1} 1]$ **86.** $\cos^{-1}(\cos \pi/3)$

87. $\tan[\sin^{-1} 0.75]$ **88.** $\cos^{-1}(\tan 0.2)$

In Exercises 89–92, determine whether the function is periodic. State the period, domain, and range.

89. $f(x) = |\sec x|$ **90.** $g(x) = \sin|x|$

91. $f(x) = 2x + \tan x$

92. $g(x) = 2 \cos 2x + 3 \sin 5x$

In Exercises 93–104, solve the problem.

93. *Arc Length* Find the length of the arc intercepted by a central angle of $2\pi/3$ rad in a circle with radius 2.

94. *Algebraic Expression* Find an algebraic expression equivalent to $\tan(\cos^{-1} x)$.

95. *Height of Building* The angle of elevation of the top of a building from a point 100 m away from the building on level ground is 78°. Find the height of the building.

96. *Height of Tree* A tree casts a shadow 51 ft long when the angle of elevation of the sun (measured with the horizon) is 25°. How tall is the tree?

97. *Traveling Car* From the top of a 150-ft building Flora observes a car moving toward her. If the angle of depression of the car changes from 18° to 42° during the observation, how far does the car travel?

98. *Finding Distance* A lighthouse L stands 4 mi from the closest point P along a straight shore (see figure). Find the distance from P to a point Q along the shore if $\angle PLQ = 22°$.

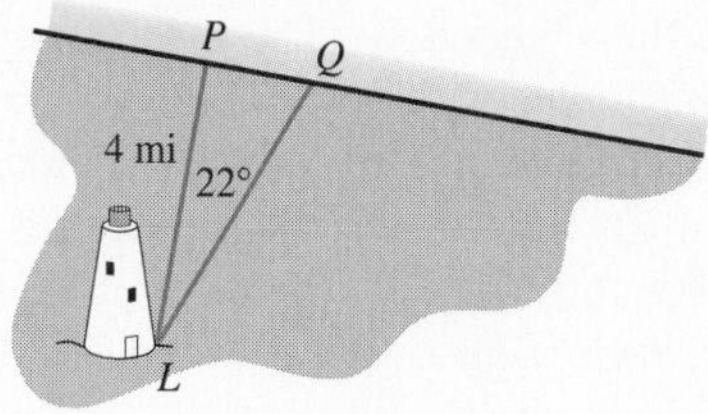

For Exercise 98

99. *Navigation* An airplane is flying due east between two signal towers. One tower is due north of the other. The bearing from the plane to the north tower is 23°, and to the south tower is 128°. Use a drawing to show the exact location of the plane.

100. *Finding Distance* The bearings of two points on the shore from a boat are 115° and 123°. Assume the two points are 855 ft apart. How far is the boat from the nearest point on shore if the shore is straight and runs north-south?

101. *Height of Tree* Dr. Thom Lawson standing on flat ground 62 ft from the base of a Douglas fir measures the angle of elevation to the top of the tree as 72°24′. What is the height of the tree?

102. *Storing Hay* A 75-ft-long conveyor is used at the Lovelady Farm to put hay bales up for winter storage. The conveyor is tilted to an angle of elevation of 22°.

a. To what height can the hay be moved?

b. If the conveyor is repositioned to an angle of 27°, to what height can the hay be moved?

103. *Swinging Pendulum* In the Hardy Boys Adventure *While the Clock Ticked,* the pendulum of the grandfather clock at the Purdy place is 44 in. long and swings through an arc of 6°. Find the length of the arc that the pendulum traces.

104. *Finding Area* A windshield wiper on a Plymouth Acclaim is 20 in. long and has a blade 16 in. long. If the wiper sweeps through an angle of 110°, how large an area does the wiper blade clean?

THE GREEKS

The Greek civilization was the first to utilize trigonometric properties. Trigonometry evolved from the connections between mathematics and astronomy. The Greeks wanted to find the lengths of chords of a circle, and, in actuality, they were building the foundation for trigonometry. The first known table of chords was produced by Hipparchus in about 140 B.C. Although the terms "sine" and "cosine" were not used, these tables represented the basis for the current values of these trigonometric functions. Therefore, Hipparchus may be looked upon as the founder of trigonometry.

Hipparchus' work was followed by other Greek mathematicians, the next most notably being Menelaus. In about 100 B.C., Menelaus produced another table of chords. Another Greek mathematician, Ptolemy, in about 100 A.D., used a form of the well-known trigonometric identity: $\sin^2 a + \cos^2 a = 1$. He later used other trigonometric expressions involving the concepts of sine and cosine. However, these were all in terms of chords.

chapter 7

ANALYTIC TRIGONOMETRY

BIBLIOGRAPHY

For students: *Mathematics Write Now!,* Peggy A. House and Nancy S. Desmond. Janson Publications, 1994.

For teachers: *Encyclopedia of Math Topics and References,* Dale Seymour, ed. Dale Seymour Publications, 1995.

Classic Math: History Topics for the Classroom, Art Johnson. Dale Seymour Publications, 1994.

7.1 APPLICATIONS OF FUNDAMENTAL IDENTITIES

Fundamental Trigonometric Identities • Simplifying Trigonometric Expressions
Factoring Trigonometric Expressions

Objective

Students will be able to use the fundamental identities to simplify and factor trigonometric expressions.

Motivate

Have students graph

$$y = \sin^3 x + \cos^2 x \sin x$$

in a standard trigonometric viewing window. Then discuss why the graph looks as it does and ask students to name the function being viewed.
$(\boldsymbol{y = \sin x})$

Fundamental Trigonometric Identities

Recall that an identity is an equation that is true for all values of the variable for which each side of the equation is defined. For example, $\sin^2 x + \cos^2 x = 1$ is an identity because it is true for all real numbers. On the other hand, an equation like $\sin x = 0.5$, which is true for some real numbers and false for others, is not an identity.

In Chapter 6 you learned the definitions, the graphs, and some fundamental properties of the six trigonometric functions. In this chapter you will use the fundamental identities from Chapter 6 to simplify expressions, solve trigonometric equations, and develop other trigonometric identities.

Teaching Note

The identities and proofs presented in this section are fairly standard. Although the role of identities is different in today's technologically oriented mathematics, you should feel free to explore the fundamental identities using a grapher and to design other approaches to teaching them.

Fundamental Trigonometric Identities

Reciprocal Identities

$$\sin x = \frac{1}{\csc x} \qquad \cos x = \frac{1}{\sec x} \qquad \tan x = \frac{1}{\cot x}$$

$$\csc x = \frac{1}{\sin x} \qquad \sec x = \frac{1}{\cos x} \qquad \cot x = \frac{1}{\tan x}$$

Quotient Identities

$$\tan x = \frac{\sin x}{\cos x} \qquad \cot x = \frac{\cos x}{\sin x}$$

Pythagorean Identities

$$\sin^2 x + \cos^2 x = 1 \qquad \tan^2 x + 1 = \sec^2 x \qquad 1 + \cot^2 x = \csc^2 x$$

Odd-Even Identities

$$\sin(-x) = -\sin x \qquad \cos(-x) = \cos x \qquad \tan(-x) = -\tan x$$

$$\csc(-x) = -\csc x \qquad \sec(-x) = \sec x \qquad \cot(-x) = -\cot x$$

Simplifying Trigonometric Expressions

REMARK

You will use the fundamental identities so often in the work of this chapter that we suggest you commit them to memory.

In calculus it is often necessary to convert a trigonometric expression into an equivalent simpler form. This is what simplifying means. Unfortunately, other than applying the fundamental identities, there are no general methods to apply. We use the techniques that work for us with algebraic expressions. We combine fractions, remove common factors, and use properties and formulas that you learned in algebra.

For example, the equation

$$(a - b)(a + b) = a^2 - b^2,$$

where $a = 1$ and $b = \sin x$, becomes

$$(1 - \sin x)(1 + \sin x) = 1^2 - \sin^2 x.$$

Or the equation

$$(a + b)^2 = a^2 + 2ab + b^2,$$

where $a = \cos x$ and $b = 3$, becomes

$$(\cos x + 3)^2 = \cos^2 x + 6 \cos x + 9.$$

In this fashion trigonometric expressions can be simplified or changed in form just as algebraic expressions can be.

EXAMPLE 1 Simplifying a Trigonometric Expression

Write the expression $\sin x - \sin^3 x$ as a single term.

Solution

Solve Algebraically

$$\begin{aligned} \sin x - \sin^3 x &= (\sin x)(1 - \sin^2 x) && \text{Factor } \sin x. \\ &= (\sin x)(\cos^2 x) && \text{Pythagorean identity} \\ &= \sin x \cos^2 x \end{aligned}$$

Support Graphically

The equation is an identity if the functions y_1 and y_2 determined by the left and right sides of the equation are equal. The graphs of $y_1 = \sin x - \sin^3 x$ and $y_2 = \sin x \cos^2 x$ appear to be identical. (See Figure 7.1.) ■

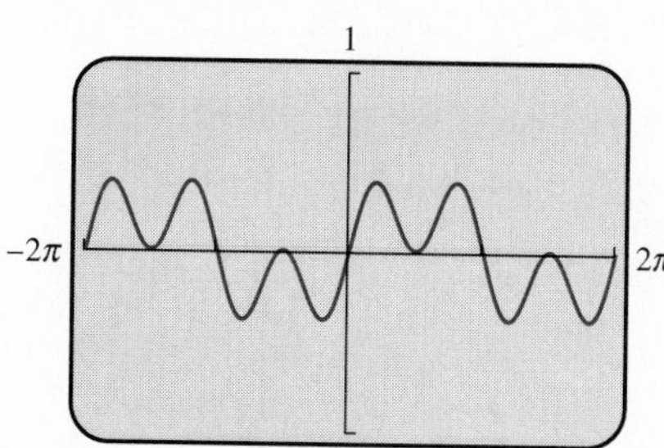

Figure 7.1 Graph of $y_1 = \sin x - \sin^3 x$ and also of $y_2 = \sin x \cos^2 x$.

Teaching Note

Graphers are powerful tools. They will quickly convince students about the validity of identities. Seeing is believing.

Sometimes a sum or difference of fractional expressions can be simplified to a single term.

EXAMPLE 2 Simplifying an Expression Involving Fractions

Simplify this difference:

$$\frac{\cos x}{1 - \sin x} - \frac{\sin x}{\cos x}$$

Solution

$$\frac{\cos x}{1 - \sin x} - \frac{\sin x}{\cos x}$$

$$\begin{aligned} &= \frac{\cos x \cos x}{(1 - \sin x)(\cos x)} - \frac{(\sin x)(1 - \sin x)}{(\cos x)(1 - \sin x)} && \text{Find common denominator.} \\ &= \frac{\cos^2 x - (\sin x)(1 - \sin x)}{(1 - \sin x)(\cos x)} && \text{Subtract numerators.} \\ &= \frac{\cos^2 x - \sin x + \sin^2 x}{(1 - \sin x)(\cos x)} && \text{Distributive property} \\ &= \frac{1 - \sin x}{(1 - \sin x)(\cos x)} && \text{Pythagorean identity} \\ &= \frac{1}{\cos x} = \sec x && \text{Reciprocal identity} \end{aligned}$$

■

One strategy for simplifying an expression is to first change all terms to sines and cosines, as Example 3 illustrates.

TECHNOLOGY NOTE

Your grapher will be a valuable tool in this chapter. Drawing graphs and using trace and tables can help you decide whether two graphs are the same.

EXAMPLE 3 Simplifying by Changing to Sine and Cosine

Simplify

$$\sin x \tan x - \sec x.$$

Solution

$$\begin{aligned} \sin x \tan x - \sec x &= \sin x \cdot \frac{\sin x}{\cos x} - \frac{1}{\cos x} && \text{Quotient and reciprocal identities} \\ &= \frac{\sin^2 x}{\cos x} - \frac{1}{\cos x} \\ &= \frac{\sin^2 x - 1}{\cos x} \\ &= \frac{-\cos^2 x}{\cos x} && \text{Pythagorean identity} \\ &= -\cos x \end{aligned}$$

Support Graphically

Check that the graphs of $y = \sin x \tan x - \sec x$ and $y = -\cos x$ are identical. ■

Factoring Trigonometric Expressions

When solving trigonometric equations algebraically in Section 7.3, it sometimes will be helpful to factor a trigonometric expression. This factoring often requires that the expression be written in terms of just one trigonometric function.

Follow-up

Ask students if they think the factored form of a trigonometric expression is always unique. **(No, $(1 + \sin x)(1 - \sin x)$ and $\cos^2 x$ are both factored.)**

Assignment Guide

Day 1: Ex. 3–36, multiples of 3
Day 2: Ex. 39–60, multiples of 3, 61, 63

Cooperative Learning

Group Activity: Ex. 50, 52

EXAMPLE 4 Factoring a Trigonometric Expression

Write $1 + \cos x - \sin^2 x$ in a factored form.

Solution

No factorization is readily apparent, so we replace $\sin^2 x$ with $1 - \cos^2 x$ to obtain an expression in terms of only $\cos x$.

$$\begin{aligned} 1 + \cos x - \sin^2 x &= 1 + \cos x - (1 - \cos^2 x) && \text{Pythagorean identity} \\ &= 1 + \cos x - 1 + \cos^2 x \\ &= \cos^2 x + \cos x \\ &= (\cos x)(1 + \cos x) && \text{Factor } \cos x. \end{aligned}$$

■

Notes on Exercises

Ex. 31–44 require students to factor trigonometric expressions.
Ex. 53–58 are related to trigonometric substitutions used in calculus.

Ongoing Assessment

Self-Assessment: Ex. 5, 17, 23, 35, 43
Embedded Assessment: Ex. 63

EXAMPLE 5 Factoring a Trigonometric Expression

Write $\sec^2 x + \tan x - 3$ in a factored form.

Solution

Replace $\sec^2 x$ with $\tan^2 x + 1$ so that the expression is given in terms of just one trigonometric function.

$$\sec^2 x + \tan x - 3 = \tan^2 x + 1 + \tan x - 3 \quad \text{Pythagorean identity}$$
$$= \tan^2 x + \tan x - 2 \quad \text{Simplify.}$$
$$= (\tan x - 1)(\tan x + 2) \quad \text{Factor as a trinomial in } \tan x.$$

Quick Review 7.1

In Exercises 1–4, factor the expression into a product of linear factors.

1. $a^2 - 2ab + b^2$
2. $4u^2 + 4u + 1$
3. $2x^2 - 3xy - 2y^2$
4. $2v^2 - 5v - 3$

In Exercises 5–8, simplify the expression.

5. $\dfrac{1}{x} - \dfrac{2}{y}$
6. $\dfrac{a}{x} + \dfrac{b}{y}$
7. $\dfrac{x + y}{\dfrac{1}{x} + \dfrac{1}{y}}$
8. $\dfrac{x}{x - y} - \dfrac{y}{x + y}$

In Exercises 9 and 10, solve the problem.

9. Use degree mode.
   a. Graph $y_1 = \sin x$ in $[-10, 10]$ by $[-2, 2]$. Is the graph correct? Explain.
   b. Now graph y_1 in $[-360, 360]$ by $[-2, 2]$. Why is this a better view?
10. Use radian mode.
   a. Graph $y_1 = \sin x$ in $[-360, 360]$ by $[-2, 2]$. Is the graph correct? Explain.
   b. Now graph y_1 in $[-10, 10]$ by $[-2, 2]$. Why is this a better view?

SECTION EXERCISES 7.1

In Exercises 1–8, use fundamental identities to simplify the expression. Support your result graphically.

1. $\tan x \cos x$
2. $\cot x \tan x$
3. $\sec y \cos y$
4. $\cot u \sin u$
5. $\dfrac{1 + \tan^2 x}{\csc^2 x}$
6. $\dfrac{1 - \cos^2 \theta}{\sin \theta}$
7. $\cos x - \cos^3 x$
8. $\dfrac{\sin^2 u + \tan^2 u + \cos^2 u}{\sec u}$

In Exercises 9–14, simplify the expression to either 1 or −1.

9. $\sin x \csc(-x)$
10. $\sec(-x) \cos(-x)$
11. $\cot(-x) \tan(-x)$
12. $\sin(-x) \csc(-x)$
13. $\sin^2(-x) + \cos^2(-x)$
14. $\sec^2(-x) - \tan^2 x$

In Exercises 15–18, simplify the expression to a single term. Support your result graphically.

15. $\dfrac{\tan x \csc x}{\csc^2 x}$
16. $\dfrac{1 + \tan x}{1 + \cot x}$
17. $(\sec^2 x + \csc^2 x) - (\tan^2 x + \cot^2 x)$
18. $\dfrac{\sec^2 u - \tan^2 u}{\cos^2 v + \sin^2 v}$

In Exercises 19–24, use the fundamental identities to change the expression to one involving only sines and cosines. Then simplify. Support your result graphically.

19. $\tan \alpha + \cot \alpha$
20. $\sin \theta + \tan \theta \cos \theta$

21. $(\sec y + \csc y)^2 \cot y$

22. $(\csc \theta - \sec \theta) \sin \theta \cos \theta$

23. $\dfrac{1}{\csc^2 x} + \dfrac{1}{\sec^2 x}$

24. $\dfrac{\sec x \csc x}{\sec^2 x + \csc^2 x}$

In Exercises 25–30, combine the fractions and simplify. Support your result graphically.

25. $\dfrac{1}{1 - \sin x} + \dfrac{1}{1 + \sin x}$

26. $\dfrac{1}{\sec x - 1} - \dfrac{1}{\sec x + 1}$

27. $\dfrac{1}{\cot^2 x} - \dfrac{1}{\cos^2 x}$

28. $\dfrac{\sec x}{\sin x} - \dfrac{\sin x}{\cos x}$

29. $\dfrac{\sin x}{1 - \cos x} + \dfrac{1 - \cos x}{\sin x}$

30. $\dfrac{1}{\sin x \cos x} - \dfrac{\cos x}{\sin x}$

In Exercises 31–40, write in factored form involving one trigonometric function only. Support your result graphically.

31. $\sin^2 \theta + \sin^2 \theta \tan^2 \theta$

32. $\cos x + \sec^2 x - \tan^2 x$

33. $\cos^2 x + 2 \cos x + 1$

34. $1 - 2 \sin x + \sin^2 x$

35. $1 - 2 \sin x + (1 - \cos^2 x)$

36. $\sin x - \cos^2 x - 1$

37. $\cos x - 2 \sin^2 x + 1$

38. $\sin^2 x + \dfrac{2}{\csc x} + 1$

39. $4 \tan^2 x - \dfrac{4}{\cot x} + \sin x \csc x$

40. $\sec^2 x - \sec x + \tan^2 x$

In Exercises 41–44, simplify to an expression involving one trigonometric function only.

41. $\dfrac{1 - \sin^2 x}{1 + \sin x}$

42. $\dfrac{\tan^2 \alpha - 1}{1 + \tan \alpha}$

43. $\dfrac{\sin^2 x}{1 + \cos x}$

44. $\dfrac{\tan^2 x}{\sec x + 1}$

In Exercises 45–52, use a graphing utility to conjecture whether the equation is an identity. If not, find a counterexample. That is, find a value for x for which the equation is false.

45. $1 - 2 \sin^2 x = 2 \cos^2 x - 1$

46. $\sin x = \sqrt{1 - \cos^2 x}$

47. $\sec x = \sqrt{1 + \tan^2 x}$

48. $\cos^2 x + 1 = 2 \cos^2 x + \sin^2 x$

49. $\sin x + \cos x \cot x = \csc x$

50. $\sin 3x = 3 \sin x$

51. $1 + \cot^2 x = \csc^2 x$

52. $\cos (x - 2) = \cos x - \cos 2$

EXTENDING THE IDEAS

In Exercises 53–58, use the suggested trigonometric substitution to write the algebraic expression as a trigonometric expression, $0 \le \theta \le \pi/2$. Simplify.

53. $\sqrt{1 - x^2}, \quad x = \cos \theta$

54. $\sqrt{x^2 + 1}, \quad x = \tan \theta$

55. $\sqrt{x^2 - 9}, \quad x = 3 \sec \theta$

56. $\sqrt{36 - x^2}, \quad x = 6 \sin \theta$

57. $\sqrt{x^2 + 81}, \quad x = 9 \tan \theta$

58. $\sqrt{x^2 - 100}, \quad x = 10 \sec \theta$

In Exercises 59–63, solve the problem.

59. a. Show that $\cos x = \pm\sqrt{1 - \sin^2 x}$.
b. Use part a to evaluate $\cos x$ if $\sin x = 3/5$ and $\pi/2 \le x \le \pi$.
c. Use part a to evaluate $\cos x$ if $\sin x = -3/5$ and $3\pi/2 \le x \le 2\pi$.

60. a. Show that $\sec x = \pm\sqrt{\tan^2 x + 1}$.
b. Use part a to evaluate $\sec x$ if $\tan x = 5/12$ and $\pi \le x \le 3\pi/2$.
c. Use part a to evaluate $\sec x$ if $\tan x = 5/12$ and $2\pi \le x \le 5\pi/2$.

61. Write the other five trigonometric functions in terms of $\sin x$.

62. Write the other five trigonometric functions in terms of $\cos x$.

63. Writing to Learn Explain how the figure can be used to confirm that

$$\csc \theta = \frac{1}{\sin \theta} \text{ and } \cot \theta = \frac{1}{\tan \theta}.$$

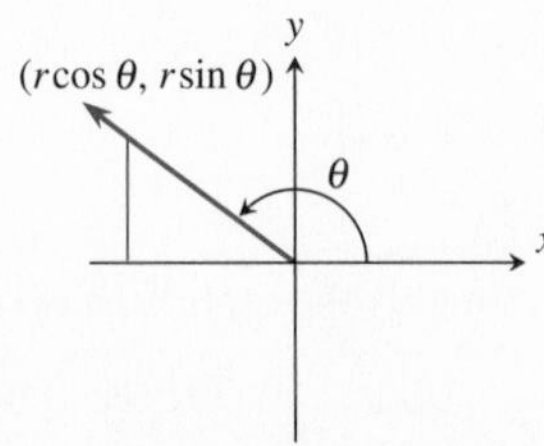

For Exercise 63

7.2 CONFIRMING TRIGONOMETRIC IDENTITIES

Deciding Whether an Equation Is an Identity • Confirming an Identity • An Identity Used in Calculus

Deciding Whether an Equation Is an Identity

When confronted with a trigonometric equation, first decide whether you think the equation is an identity. That is, decide whether the two functions determined by the left and right sides of the equation are equal. For example, consider the equation

$$\sin^2 x + \cos x = \cos^2 x + \cos x + 1. \qquad \textbf{(1)}$$

To decide whether you think this is an identity, graph $y_1 = \sin^2 x + \cos x$ and $y_2 = \cos^2 x + \cos x + 1$ to see if the graphs appear to be identical. Figures 7.2a and 7.2b indicate that they are not identical. Thus, the two functions are not equal and equation (1) is *not* an identity.

Sometimes it is easy to show that the two functions are not equal algebraically. For example, the values of y_1 and y_2 when $x = 0$ are 1 and 3, respectively. Therefore, the two functions are not equal and the equation is not an identity.

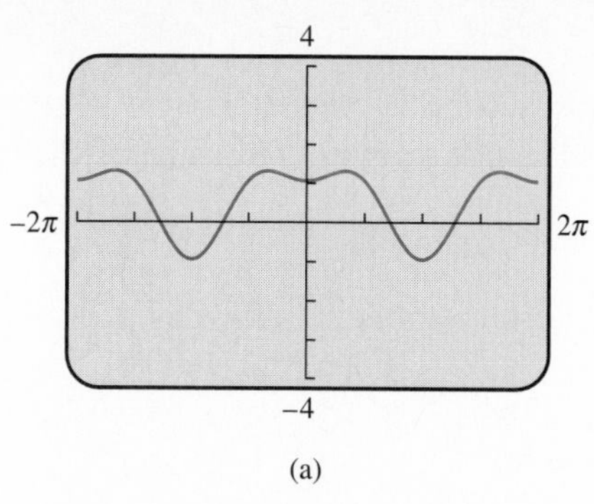

(a)

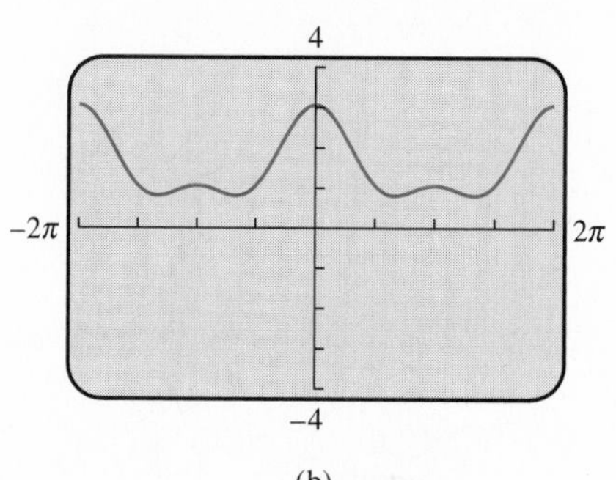

(b)

Figure 7.2 (a) $y_1 = \sin^2 x + \cos x$. (b) $y_2 = \cos^2 x + \cos x + 1$.

REMARK

An equation that is false for at least one value in the domain is sometimes called a *conditional equation.* An equation that is not an identity is conditional.

Objective

Students will be able to decide whether an equation is an identity and to confirm identities analytically.

Motivate

Have students use a grapher to graph the functions $y = \cos 2x$ and $y = \cos^2 x - \sin^2 x$ and make observations about what they see.

EXAMPLE 1 Finding an Identity Graphically

Do the graphs suggest that the equation is an identity?

a. $\cos 2x = 2 \cos x$ **b.** $\sin 2x = 2 \sin x \cos x$

Solution

a. The graphs of $y_1 = \cos 2x$ and $y_2 = 2 \cos x$ (Figure 7.3a) are clearly different. The equation $\cos 2x = 2 \cos x$ is not an identity.

b. The graphs of $y_1 = \sin 2x$ and $y_2 = 2 \sin x \cos x$ (Figure 7.3b) appear to be the same, suggesting that the equation $\sin 2x = 2 \sin x \cos x$ is an identity. ■

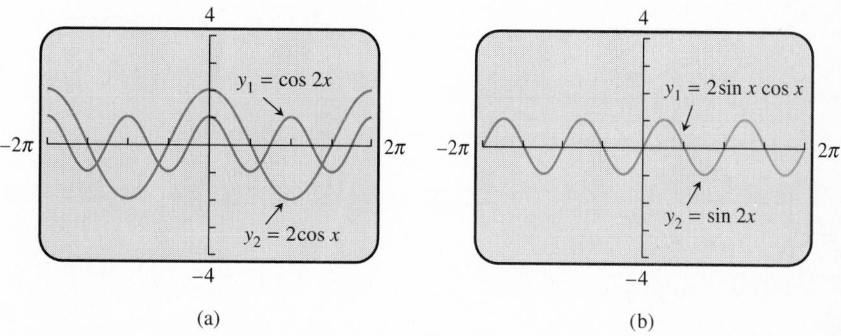

Figure 7.3

Teaching Note

A grapher is an excellent tool for identifying a trigonometric identity. We can say that there is strong visual evidence that the equation in question is an identity, but we cannot say that the visual representation replaces an analytical proof.

Confirming an Identity

To confirm that an equation is an identity, you need to attempt to show that the two expressions have the same values for identical values of the variable. There are several ways to accomplish this:

- Combining terms
- Separating terms
- Multiplying factors
- Factoring
- Using known identities
- Simplifying fractions
- Changing to sines and cosines only

■ EXAMPLE 2 Confirming an Identity

Confirm that

$$\frac{(1 - \cos u)(1 + \cos u)}{\cos^2 u} = \tan^2 u$$

is an identity.

Solution

Figure 7.4 suggests that the equation is indeed an identity. To confirm it, we simplify the more complicated left side to the form of the right side.

$$\begin{aligned} \frac{(1 - \cos u)(1 + \cos u)}{\cos^2 u} &= \frac{1 - \cos^2 u}{\cos^2 u} && \text{Multiply factors.} \\ &= \frac{\sin^2 u}{\cos^2 u} && \text{Pythagorean identity} \\ &= \left(\frac{\sin u}{\cos u}\right)^2 \\ &= \tan^2 u && \text{Quotient identity} \end{aligned}$$

■

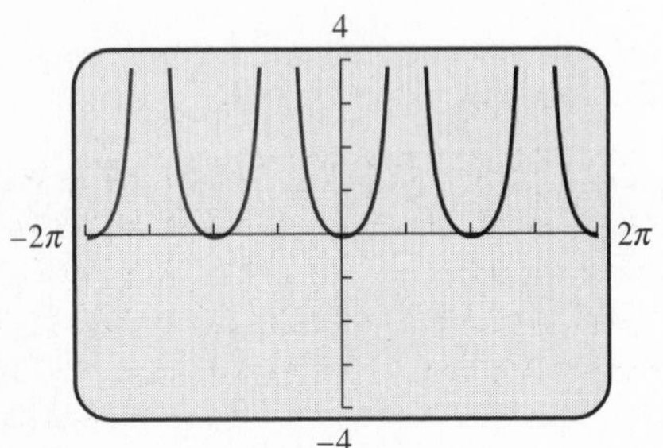

Figure 7.4 $y_1 = \dfrac{(1 - \cos x)(1 + \cos x)}{\cos^2 x}$ and $y_2 = \tan^2 x$.

In Example 3 we change all terms to sines and cosines.

Alert

Some students may not understand that the phrase *confirm an identity* normally means to confirm the identity *analytically,* not graphically.

■ EXAMPLE 3 Confirming an Identity

Confirm that

$$2 \cot x \csc x = \frac{1}{\sec x - 1} + \frac{1}{\sec x + 1}$$

is an identity.

X	Y_1	Y_2
-.3	21.878	21.878
-.2	49.662	49.662
-.1	199.67	199.67
0	ERROR	ERROR
.1	199.67	199.67
.2	49.662	49.662
.3	21.878	21.878
Y_1=2(1/tan X)(1...		

Figure 7.5 A table of values for $y_1 = 2 \cot x \csc x$ and $y_2 = \dfrac{1}{\sec x - 1} + \dfrac{1}{\sec x + 1}$.

Solution

Figure 7.5 suggests that the equation is indeed an identity. To confirm it, we simplify the more complicated right side.

$$\begin{aligned} \frac{1}{\sec x - 1} + \frac{1}{\sec x + 1} &= \frac{(\sec x + 1) + (\sec x - 1)}{(\sec x - 1)(\sec x + 1)} && \text{Find a common denominator.} \\ &= \frac{2 \sec x}{\sec^2 x - 1} && \text{Simplify.} \\ &= \frac{2 \sec x}{\tan^2 x} && \text{Pythagorean identity} \\ &= 2 \sec x \cot^2 x && \text{Reciprocal identity} \\ &= \frac{2}{\cos x} \cdot \frac{\cos^2 x}{\sin^2 x} && \text{Two identities used} \\ &= \frac{2 \cos x}{\sin x} \cdot \frac{1}{\sin x} && \text{Simplify.} \\ &= 2 \cot x \csc x \end{aligned}$$

■

Alert

Some students will test an identity by substituting a number for x. Caution them that it is possible for an equation that is not an identity to work for certain values of x. For example, the equation $\sin x = \cos x$ is true for $x = \dfrac{\pi}{4}$, but it is not an identity.

Sometimes a fraction can be simplified by multiplying the numerator and denominator by the same factor.

■ EXAMPLE 4 Confirming an Identity

Confirm this identity:

$$\frac{\cos t}{1 - \sin t} = \frac{1 + \sin t}{\cos t}$$

Solution

We begin with the left side.

$$\begin{aligned} \frac{\cos t}{1 - \sin t} &= \frac{\cos t}{1 - \sin t} \cdot \frac{1 + \sin t}{1 + \sin t} && \text{Multiply numerator and denominator by } 1 + \sin t. \\ &= \frac{(\cos t)(1 + \sin t)}{1 - \sin^2 t} \\ &= \frac{(\cos t)(1 + \sin t)}{\cos^2 t} && \text{Pythagorean identity} \\ &= \frac{1 + \sin t}{\cos t} && \text{Eliminate common factor.} \end{aligned}$$

■

Sometimes it is easier to simplify each side independently until both sides are identical.

■ EXAMPLE 5 Confirming an Identity

Confirm this identity:

$$\frac{\cot^2 u}{1 + \csc u} = (\cot u)(\sec u - \tan u)$$

Solution

Simplify the Left Side:

$$\frac{\cot^2 u}{1 + \csc u} = \frac{\csc^2 u - 1}{1 + \csc u} \quad \text{Pythagorean identity}$$

$$= \frac{(\csc u - 1)(\csc u + 1)}{1 + \csc u} \quad \text{Factor.}$$

$$= \csc u - 1 \quad \text{Simplify.}$$

Write the Right Side in Sines and Cosines, and Simplify:

$$(\cot u)(\sec u - \tan u) = \left(\frac{\cos u}{\sin u}\right)\left(\frac{1}{\cos u} - \frac{\sin u}{\cos u}\right) \quad \text{Change to sines and cosines.}$$

$$= \frac{1}{\sin u} - 1 \quad \text{Distributive property}$$

$$= \csc u - 1 \quad \text{Reciprocal identity}$$

We have shown that both sides are equivalent to $\csc u - 1$, so we conclude that

$$\frac{\cot^2 u}{1 + \csc u} = (\cot u)(\sec u - \tan u)$$

■

Follow-up

Ask students to confirm the identity in Example 4 by beginning with the right side and multiplying the numerator and denominator by $1 - \sin t$.

Assignment Guide

Day 1: Ex. 3–33, multiples of 3
Day 2: Ex. 36–63, multiples of 3

Cooperative Learning

Group Activity: Ex. 61–62

Notes on Exercises

Ex. 7–44 should be done using algebraic skills; however, feel free to incorporate the use of a grapher to support answers or for further exploration.
Ex. 61 is related to the derivative of the sine function.

An Identity Used in Calculus

Some problems in calculus require that we change a product of powers of sines and cosines to include a factor that is the sum of powers of either sine or cosine. In those instances we develop the identities by beginning from what appears to be the simpler side, contrary to the more conventional approach of changing the more complicated side to the simpler side.

■ EXAMPLE 6 An Example from Calculus

Confirm this identity:

$$\sin^2 x \cos^5 x = (\sin^2 x - 2 \sin^4 x + \sin^6 x)\cos x$$

Ongoing Assessment

Self-Assessment: Ex. 5, 11, 17, 23, 35, 43, 53
Embedded Assessment: Ex. 24, 54

Solution

We begin with the left side and expand using a Pythagorean identity.

$$\begin{aligned}\sin^2 x \cos^5 x &= \sin^2 x \cos^4 x \cos x \\ &= (\sin^2 x)(1 - \sin^2 x)^2 \cos x \\ &= (\sin^2 x)(1 - 2\sin^2 x + \sin^4 x)\cos x \\ &= (\sin^2 x - 2\sin^4 x + \sin^6 x)\cos x\end{aligned}$$

■

Quick Review 7.2

In Exercises 1–6, change to sines and cosines. Write your answer as a single fraction.

1. $\csc x + \sec x$

2. $\tan x + \cot x$

3. $\cos x \csc x + \sin x \sec x$

4. $\sin \theta \cot \theta - \cos \theta \tan \theta$

5. $\dfrac{\sin x}{\csc x} + \dfrac{\cos x}{\sec x}$

6. $\dfrac{\sec \alpha}{\cos \alpha} - \dfrac{\sin \alpha}{\csc \alpha \cos^2 \alpha}$

In Exercises 7–12, determine whether the functions f and g are equal. If so, give a convincing argument. If not, find a value of x for which $f(x) \neq g(x)$.

7. $f(x) = \sqrt{x^2}$, $g(x) = x$

8. $f(x) = \sqrt[3]{x^3}$, $g(x) = x$

9. $f(x) = \sqrt{1 - \cos^2 x}$, $g(x) = \sin x$

10. $f(x) = \sqrt{\sec^2 x - 1}$, $g(x) = \tan x$

11. $f(x) = \ln \dfrac{1}{x}$, $g(x) = -\ln x$

12. $f(x) = \ln x^2$, $g(x) = 2 \ln x$

SECTION EXERCISES 7.2

In Exercises 1–6, use a grapher to conjecture whether the equation is likely to be an identity. If not, give a counterexample, i.e., find a value of x for which the equation is false.

1. $\cos^2 x + \cos x = 1 + \sin x + \cos x$

2. $\cos x \sin x = \sin 2x$

3. $\sin^2 x - \cos^2 x = 2\sin^2 x - 1$

4. $4 \sin x = \sin 4x$

5. $1 + \sin x + \sin^2 x = 1 + \cos x + \cos^2 x$

6. $\sin^2 x - \cos^2 x = \cos 2x$

In Exercises 7–44, confirm the identity.

7. $(\cos x)(\tan x + \sin x \cot x) = \sin x + \cos^2 x$

8. $(\sin x)(\cot x + \cos x \tan x) = \cos x + \sin^2 x$

9. $(1 - \tan x)^2 = \sec^2 x - 2 \tan x$

10. $(\cos x - \sin x)^2 = 1 - 2 \sin x \cos x$

11. $\tan x + \cot x = \sec x \csc x$

12. $\tan x + \sec x = \dfrac{\cos x}{1 - \sin x}$

13. $\dfrac{\cos^2 x - 1}{\cos x} = -\tan x \sin x$

14. $\dfrac{\sec^2 \theta - 1}{\sin \theta} = \dfrac{\sin \theta}{1 - \sin^2 \theta}$

15. $(1 - \sin \beta)(1 + \csc \beta) = 1 - \sin \beta + \csc \beta - \sin \beta \csc \beta$

16. $\dfrac{1}{1 - \cos x} + \dfrac{1}{1 + \cos x} = 2 \csc^2 x$

17. $(\cos t - \sin t)^2 + (\cos t + \sin t)^2 = 2$

18. $\sin^2 \alpha - \cos^2 \alpha = 1 - 2\cos^2 \alpha$

19. $\dfrac{1 + \tan^2 x}{\sin^2 x + \cos^2 x} = \sec^2 x$

20. $\dfrac{1}{\tan \beta} + \tan \beta = \sec \beta \csc \beta$

21. $\dfrac{\cos \beta}{1 + \sin \beta} = \dfrac{1 - \sin \beta}{\cos \beta}$

22. $\dfrac{\sec x + 1}{\tan x} = \dfrac{\tan x}{\sec x - 1}$

23. $\dfrac{\tan^2 x}{\sec x + 1} = \dfrac{1 - \cos x}{\cos x}$

24. $\dfrac{\cot v - 1}{\cot v + 1} = \dfrac{1 - \tan v}{1 + \tan v}$

25. $\cot^2 x - \cos^2 x = \cos^2 x \cot^2 x$

26. $\tan^2 \theta - \sin^2 \theta = \tan^2 \theta \sin^2 \theta$

27. $\cos^4 x - \sin^4 x = \cos^2 x - \sin^2 x$

28. $\tan^4 t + \tan^2 t = \sec^4 t - \sec^2 t$

29. $(x \sin \alpha + y \cos \alpha)^2 + (x \cos \alpha - y \sin \alpha)^2 = x^2 + y^2$

30. $\dfrac{1 - \cos \theta}{\sin \theta} = \dfrac{\sin \theta}{1 + \cos \theta}$

31. $\dfrac{\tan x}{\sec x - 1} = \dfrac{\sec x + 1}{\tan x}$

32. $\dfrac{\sin t}{1 + \cos t} + \dfrac{1 + \cos t}{\sin t} = 2 \csc t$

33. $\dfrac{\sin x - \cos x}{\sin x + \cos x} = \dfrac{2 \sin^2 x - 1}{1 + 2 \sin x \cos x}$

34. $\dfrac{1 + \cos x}{1 - \cos x} = \dfrac{\sec x + 1}{\sec x - 1}$

35. $\dfrac{\sin t}{1 - \cos t} + \dfrac{1 + \cos t}{\sin t} = \dfrac{2(1 + \cos t)}{\sin t}$

36. $\dfrac{\sin A \cos B + \cos A \sin B}{\cos A \cos B - \sin A \sin B} = \dfrac{\tan A + \tan B}{1 - \tan A \tan B}$

37. $\sin^2 x \cos^3 x = (\sin^2 x - \sin^4 x)(\cos x)$

38. $\sin^5 x \cos^2 x = (\cos^2 x - 2 \cos^4 x + \cos^6 x)(\sin x)$

39. $\cos^5 x = (1 - 2 \sin^2 x + \sin^4 x)(\cos x)$

40. $\sin^3 x \cos^3 x = (\sin^3 x - \sin^5 x)(\cos x)$

41. $\dfrac{\tan x}{1 - \cot x} + \dfrac{\cot x}{1 - \tan x} = 1 + \sec x \csc x$

42. $\dfrac{\cos x}{1 + \sin x} + \dfrac{\cos x}{1 - \sin x} = 2 \sec x$

43. $\dfrac{2 \tan x}{1 - \tan^2 x} + \dfrac{1}{2 \cos^2 x - 1} = \dfrac{\cos x + \sin x}{\cos x - \sin x}$

44. $\dfrac{1 - 3 \cos x - 4 \cos^2 x}{\sin^2 x} = \dfrac{1 - 4 \cos x}{1 - \cos x}$

In Exercises 45–50, match the function with one of the following. Then confirm the match.

a. $\sec^2 x \csc^2 x$ **b.** $\sec x + \tan x$ **c.** $2 \sec^2 x$
d. $\tan x \sin x$ **e.** $\sin x \cos x$

45. $\dfrac{1 + \sin x}{\cos x}$

46. $(1 + \sec x)(1 - \cos x)$

47. $\sec^2 x + \csc^2 x$

48. $\dfrac{1}{1 + \sin x} + \dfrac{1}{1 - \sin x}$

49. $\dfrac{1}{\tan x + \cot x}$

50. $\dfrac{1}{\sec x - \tan x}$

In Exercises 51–56, identify a simple function that has the same graph. Then confirm your choice.

51. $\sin x \cot x$

52. $\cos x \tan x$

53. $\dfrac{\sin x}{\csc x} + \dfrac{\cos x}{\sec x}$

54. $\dfrac{\csc x}{\sin x} - \dfrac{\cot x \csc x}{\sec x}$

55. $\dfrac{\sin x}{\tan x}$

56. $(\sec^2 x)(1 - \sin^2 x)$

EXTENDING THE IDEAS

In Exercises 57–60, confirm the identity.

57. $\sqrt{\dfrac{1 - \sin t}{1 + \sin t}} = \dfrac{1 - \sin t}{|\cos t|}$

58. $\sqrt{\dfrac{1 + \cos t}{1 - \cos t}} = \dfrac{1 + \cos t}{|\sin t|}$

59. $\sin^6 x + \cos^6 x = 1 - 3 \sin^2 x \cos^2 x$

60. $\cos^6 x - \sin^6 x = (\cos^2 x - \sin^2 x)(1 - \cos^2 x \sin^2 x)$

In Exercises 61–63, solve the problem.

61. **Writing to Learn** Let $y_1 = \dfrac{\sin (x + 0.001) - \sin x}{0.001}$ and $y_2 = \cos x$.

a. Use graphs and tables to decide whether $y_1 = y_2$.

b. Find a value for h so that the graph of $y_3 = y_1 - y_2$ in $[-2\pi, 2\pi]$ by $[-h, h]$ appears to be a sinusoid. Give a convincing argument that y_3 is a sinusoid.

62. *Hyperbolic Functions* The *hyperbolic trigonometric functions* are defined as follows:

$$\sinh x = \frac{e^x - e^{-x}}{2} \qquad \cosh x = \frac{e^x + e^{-x}}{2} \qquad \tanh x = \frac{\sinh x}{\cosh x}$$

$$\operatorname{csch} x = \frac{1}{\sinh x} \qquad \operatorname{sech} x = \frac{1}{\cosh x} \qquad \coth x = \frac{1}{\tanh x}$$

Confirm the identity.

a. $\cosh^2 x - \sinh^2 x = 1$
b. $1 - \tanh^2 x = \operatorname{sech}^2 x$
c. $\coth^2 x - 1 = \operatorname{csch}^2 x$

63. Writing to Learn Write a paragraph to explain why

$$\cos x = \cos x + \sin(10\pi x)$$

appears to be an identity when the two sides are graphed in a decimal window. Give a convincing argument that it is not an identity.

7.3 SOLVING TRIGONOMETRIC EQUATIONS

Introduction • Equations Involving Special Angles • Equations Involving Multiples of Angles • Finding Approximate Solutions • Applications

Objective

Students will be able to solve trigonometric equations algebraically and graphically and will understand the role that periodicity plays in the equation solving process.

Motivate

Ask students to name a solution of the equation $\cos x = \frac{1}{2}$ and to tell how many solutions the equation has.

($x = \frac{\pi}{3}$; infinitely many solutions.)

Introduction

In Chapter 6 we first solved trigonometric equations numerically, then graphically, and finally algebraically using inverse trigonometric functions. In this section we see how to solve trigonometric equations with the aid of identities.

Figure 7.6 shows that the graphs of $y_1 = \sin x$ and $y_2 = 1/2$ intersect at many points, which means that the equation $\sin x = 1/2$ has many solutions. This is what we expect whenever a trigonometric function is set equal to a constant within its range, because trigonometric functions are periodic. In the case of $\sin x = 1/2$, the two solutions in $[0, 2\pi]$ are $x = \pi/6$ and $x = 5\pi/6$. Because the period of the sine function is 2π, all other solutions are obtained by adding multiples of 2π to these two values.

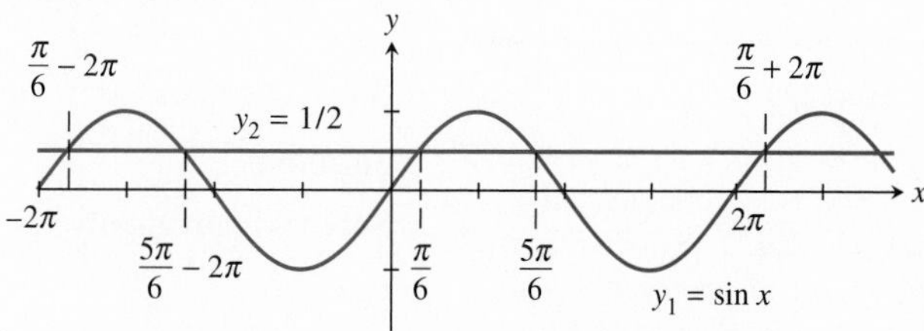

Figure 7.6 $y_1 = \sin x$ and $y_2 = 1/2$.

These infinitely many values can be written in the general form

$$x = \frac{\pi}{6} + 2n\pi \qquad \text{and} \qquad x = \frac{5\pi}{6} + 2n\pi,$$

where n is any integer. The set of all these values is the solution of the trigonometric equation $\sin x = 1/2$.

Notes on Examples

Some students are inclined to think only in terms of a period of 2π. They need to see some examples involving functions with different periods, such as Example 1.

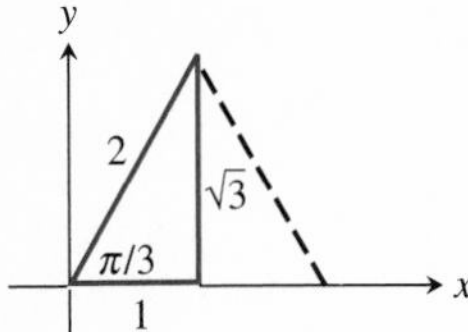

Figure 7.7

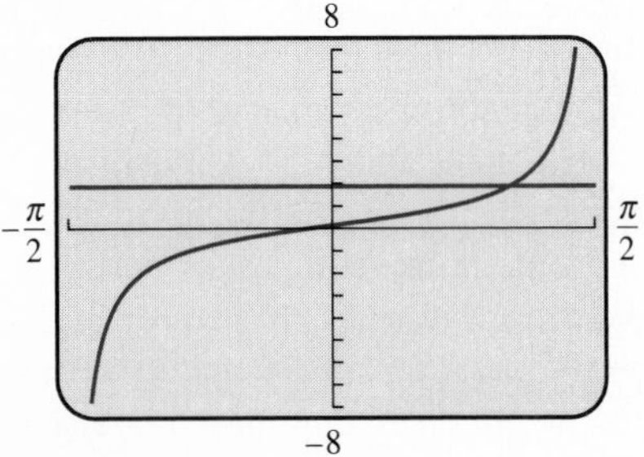

Figure 7.8

Notes on Examples

Example 2 is important because it shows how to find *all* solutions of an equation involving a periodic function, as opposed to finding the solutions within a specified interval. It also shows how various solutions in terms of n can sometimes be combined into a single expression.

Equations Involving Special Angles

Some trigonometric equations can be solved exactly because they involve the special angles.

■ EXAMPLE 1 Finding a Solution

Solve $\tan x = \sqrt{3}$.

Solution

Solve Algebraically

We draw a triangle with $\tan x = \sqrt{3}$ (see Figure 7.7) and see that it is a 30-60 right triangle. Thus, $\tan x = \sqrt{3}$ when $x = \pi/3$. Because $y = \tan x$ has the period π, the general solution of the equation is

$$x = \frac{\pi}{3} + n\pi, \quad n \text{ any integer.}$$

Support Graphically

Figure 7.8 shows that the graphs of $y_1 = \tan x$ and $y_2 = \sqrt{3}$ intersect only once in the interval $(-\pi/2, \pi/2)$. ■

A good strategy to use when solving a trigonometric equation algebraically is to first write the equation in the form

$$f(x) = 0,$$

because sometimes the zero factor property can be used, as shown in Example 2.

■ EXAMPLE 2 Solving by Factoring

Solve $\cot x \cos^2 x = \cot x$.

Solution

Solve Algebraically

$$\cot x \cos^2 x = \cot x$$

$$\cot x \cos^2 x - \cot x = 0 \qquad \text{Subtract } \cot x.$$

$$(\cot x)(\cos^2 x - 1) = 0 \qquad \text{Factor.}$$

$$\cot x = 0 \quad \text{or} \quad \cos^2 x - 1 = 0 \qquad \text{Zero factor property}$$

$$\cos^2 x = 1$$

$$\vdots \qquad \cos x = \pm 1$$

$$x = \frac{\pi}{2} + n\pi \quad \text{or} \quad x = n\pi \qquad n \text{ any integer}$$

The last line above is a satisfactory form of the general solution, but it can be condensed. We start by rewriting it as

$$x = (2n + 1)\frac{\pi}{2} \quad \text{or} \quad x = (2n)\frac{\pi}{2},$$

which reveals the solution to be the odd and even multiples of $\pi/2$. Thus we can write

$$x = n\frac{\pi}{2}, \quad n \text{ any integer.}$$

■

Alert

Students will often fail to list all of the solutions of a periodic equation when only algebraic techniques have been applied. Encourage students to support solutions graphically. The graphical support will also help students know where to look for additional analytical solutions.

Equations Involving Multiples of Angles

An equation like

$$2 \cos 3x = \sqrt{2}$$

is called a **multiple-angle equation** because the angle in the equation, $3x$, is a multiple of x.

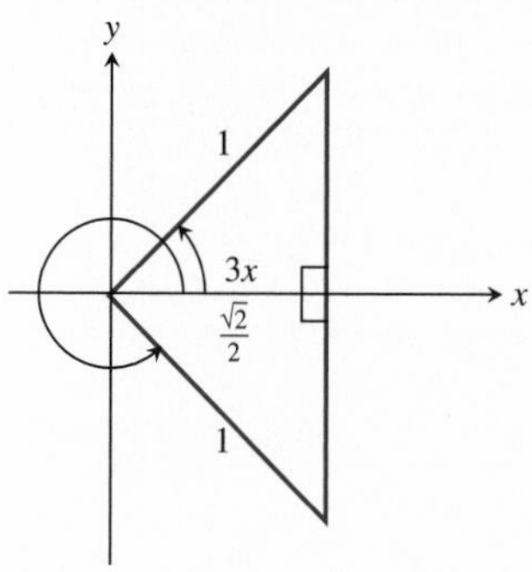

Figure 7.9 The two positions for angle $3x$.

■ EXAMPLE 3 Solving a Multiple-Angle Equation

Find the general solution of $2 \cos 3x = \sqrt{2}$. Then find the solutions in the interval $[0, 2\pi)$.

Solution

Solve Algebraically

$$2 \cos 3x = \sqrt{2}$$

$$\cos 3x = \frac{\sqrt{2}}{2}$$

This equation is satisfied when $3x = \pi/4$ or $3x = 7\pi/4$ (Figure 7.9), or when multiples of 2π are added.

$$3x = \frac{\pi}{4} + 2n\pi \quad \text{or} \quad 3x = \frac{7\pi}{4} + 2n\pi$$

$$x = \frac{\pi}{12} + \frac{2n\pi}{3} \quad \text{or} \quad x = \frac{7\pi}{12} + \frac{2n\pi}{3}, \quad n \text{ an integer}$$

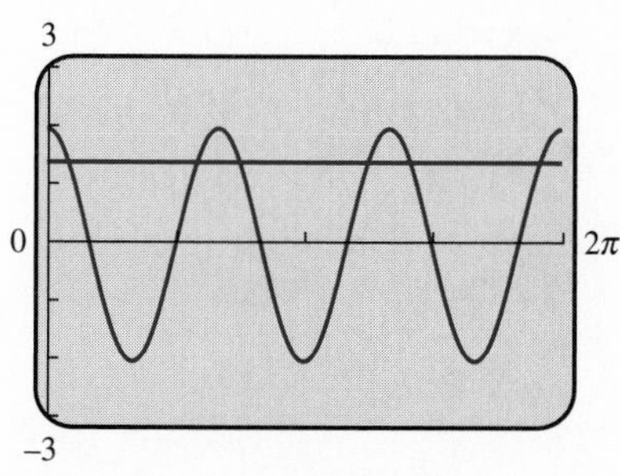

Figure 7.10

The graphs of $y_1 = 2 \cos 3x$ and $y_2 = \sqrt{2}$ in Figure 7.10 suggest that there are six solutions in the interval $[0, 2\pi)$. They can be found by letting $n = 0$, 1, and 2. Notice that if $n \geq 3$, then $2n\pi/3 \geq 2\pi$ and the solutions fall outside the desired interval. The six solutions are

$$\frac{\pi}{12}, \quad \frac{7\pi}{12}, \quad \frac{9\pi}{12}, \quad \frac{15\pi}{12}, \quad \frac{17\pi}{12}, \quad \frac{23\pi}{12}.$$

Teaching Note

Throughout this section, stress the graphical method to support the analytic solutions. "Do algebraically; support graphically."

Support Graphically

Finding intersections of y_1 and y_2 (Figure 7.10) shows that $x = 0.261 \cdots$ (or $\pi/12$), $x = 1.832 \cdots$ $(7\pi/12)$, and so on, are indeed solutions of $2 \cos 3x = \sqrt{2}$. ■

■ **EXAMPLE 4 Solving Another Multiple-Angle Equation**

Solve $4 \sin^2 2x = 1$.

Solution

$$4 \sin^2 2x = 1$$

$$\sin^2 2x = \frac{1}{4}$$

$$\sin 2x = \frac{1}{2} \quad \text{or} \quad \sin 2x = -\frac{1}{2}$$

Knowing that the sine function is positive in quadrants I and II and negative in quadrants III and IV, we use reference angles to draw the special triangles shown in Figure 7.11. They show that there is one value for $2x$ in each of the four quadrants:

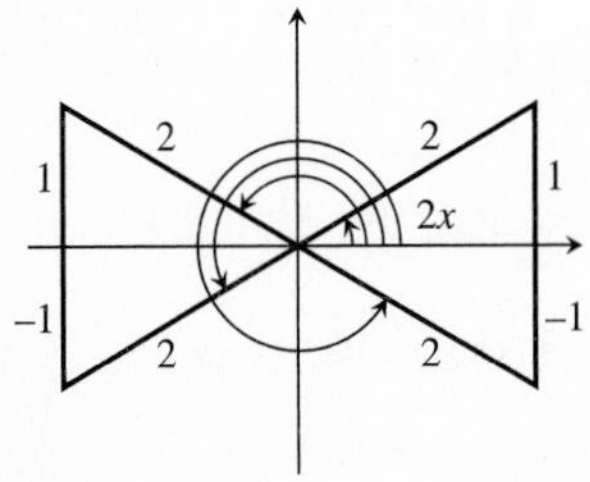

Figure 7.11 The four positions for angle $2x$.

$$2x = \frac{\pi}{6} \qquad 2x = \frac{5\pi}{6}$$

$$2x = \frac{7\pi}{6} \qquad 2x = \frac{11\pi}{6}.$$

Adding multiples of 2π we have

$$2x = \frac{\pi}{6} + 2n\pi, \qquad 2x = \frac{5\pi}{6} + 2n\pi,$$

$$2x = \frac{7\pi}{6} + 2n\pi, \quad \text{or} \quad 2x = \frac{11\pi}{6} + 2n\pi.$$

Dividing by 2 gives us

$$x = \frac{\pi}{12} + n\pi, \qquad x = \frac{5\pi}{12} + n\pi,$$

$$x = \frac{7\pi}{12} + n\pi, \quad \text{or} \quad x = \frac{11\pi}{12} + n\pi.$$ ■

Teaching Note

Students should be cautioned to zoom out far enough to identify the period accurately and then zoom in and trace to identify the solution(s) correctly.

Finding Approximate Solutions

Most trigonometric equations have solutions that are not one of the special angles. These solutions call for the aid of technology.

In Example 5 the trigonometric equation has a quadratic form.

EXAMPLE 5 Using the Quadratic Formula

Solve $\cos^2 x + 3\cos x - 1 = 0$.

Solution

Solve Algebraically

The equation $(\cos x)^2 + 3(\cos x) - 1 = 0$ is a quadratic equation in $\cos x$. The quadratic formula gives

$$\cos x = \frac{-3 + \sqrt{13}}{2} \quad \text{or} \quad \cos x = \frac{-3 - \sqrt{13}}{2}.$$

Because $-1 \le \cos x \le 1$, $\cos x = (-3 - \sqrt{13})/2 = -3.302\cdots$ has no solution. Because $\cos x = (-3 + \sqrt{13})/2 = 0.302\cdots$, we look for solutions in quadrants I and IV. Using $\cos^{-1}$ on a grapher we find

$$x = \cos^{-1}(\cos x) = \cos^{-1}\left(\frac{-3 + \sqrt{13}}{2}\right) = 1.263\cdots.$$

Because the angle $-x$ in quadrant IV has the same cosine value, we have

$$x = 1.263\cdots \quad \text{or} \quad x = -1.263\cdots.$$

Therefore, the general solution is

$$x = 1.263\cdots + 2n\pi \quad \text{or} \quad x = -1.263\cdots + 2n\pi, \quad n \text{ an integer.}$$

Support Graphically

Figure 7.12 shows $y = \cos^2 x + 3\cos x - 1$ with four x-intercepts:

$$x \approx 1.26 - 2\pi \approx -5.02 \quad x \approx -1.26 \quad x \approx 1.26$$
$$x \approx -1.26 + 2\pi \approx 5.02$$

In all of the examples so far it has been possible to change the form of the equation to

$$f(x) = c, \quad c \text{ a constant,}$$

where f is one of the trigonometric functions.

For instance, in Example 5 a numerical solution using grapher keystrokes was possible because the equation we had to solve had the form $\cos x = c$. When that form is not possible, as in Example 6, a graphical solution may be the only available method.

Follow-up

Ask students to condense the answer to Example 4, as was done in Example 2.

$$\left(x = \frac{n\pi}{2} \pm \frac{\pi}{12}\right)$$

Assignment Guide

Day 1: Ex. 3–42, multiples of 3
Day 2: Ex. 43, 46–48, 53–56, 59, 63

Cooperative Learning

Group Activity: Ex. 51–52

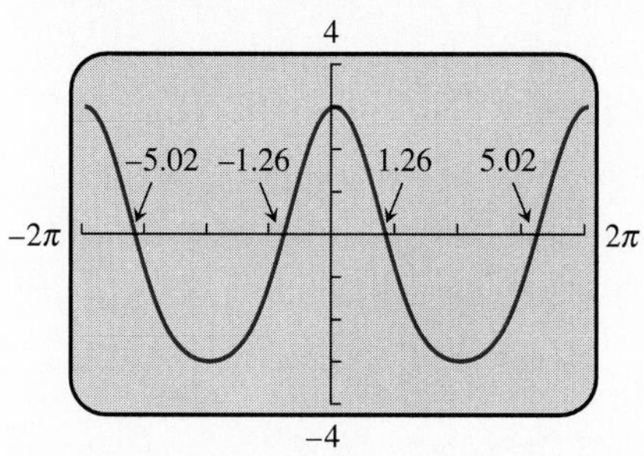

Figure 7.12 Four solutions of $y = \cos^2 x + 3\cos x - 1 = 0$.

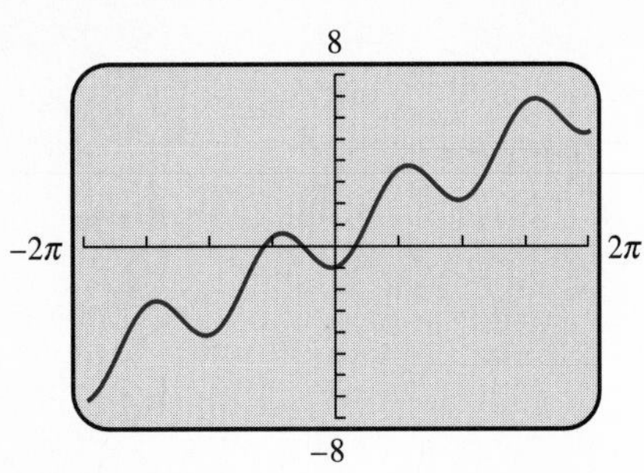

Figure 7.13

■ EXAMPLE 6 Using a Graphical Method

Solve $x = 3\cos^2 x - 2$.

Solution

Solve Graphically

The graph of $y = x - 3\cos^2 x + 2$ in Figure 7.13 suggests that there are three solutions to the equation

$$x - 3\cos^2 x + 2 = 0.$$

Using the grapher we find $x \approx -1.82, -0.93$, or 0.44. ■

Applications

In Examples 7 and 8 we use trigonometric equations to solve real-world problems.

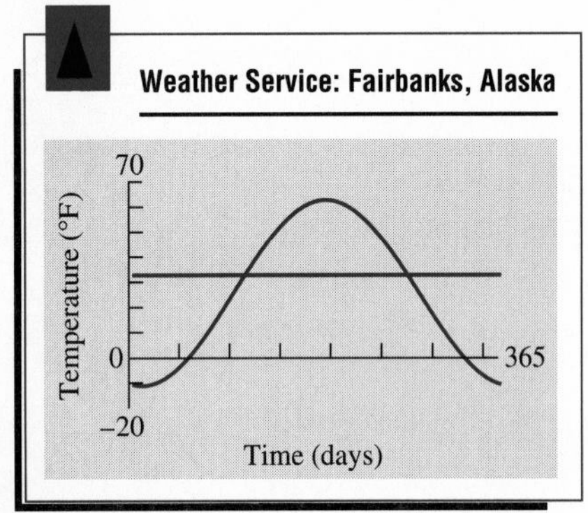

Figure 7.14
$y_1 = 37\sin\left[\frac{2\pi}{365}(x - 101)\right] + 25$
and $y_2 = 32$ for a 1-year period.

■ EXAMPLE 7 APPLICATION: Modeling Mean Temperature

The average daily air temperature (° F) for Fairbanks, Alaska, from 1941 to 1970, can be modeled by the equation

$$T(x) = 37\sin\left[\frac{2\pi}{365}(x - 101)\right] + 25,$$

where x is time in days with $x = 1$ representing January 1. On what days do you expect the average temperature to be 32° F?

Solution

Solve Graphically

The equation to be solved is $T(x) = 32$. Figure 7.14 shows graphs of $y_1 = T(x)$ and $y_2 = 32$. Using the grapher we find that $x \approx 112.1$ or $x = 272.4$.

Interpret

The average temperature in Fairbanks should be about 32° F on about the 112th day (April 22) and the 272nd day (September 29) of the year. ■

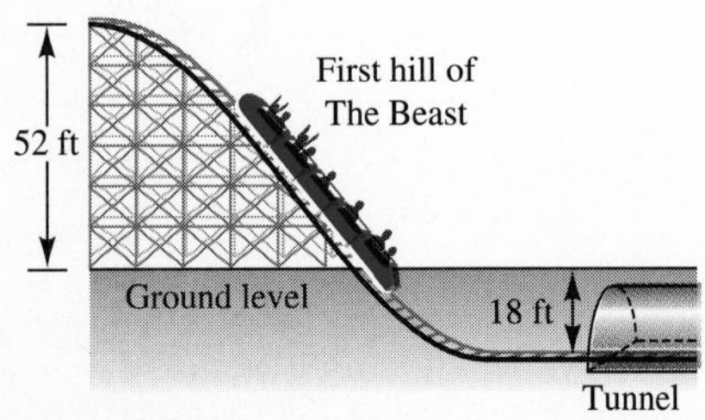

Figure 7.15

■ EXAMPLE 8 APPLICATION: Taming The Beast

The Beast is a featured roller coaster at the King's Island amusement park just north of Cincinnati. On its first and biggest hill, The Beast drops from a height of 52 ft above the ground along a sinusoidal path to a depth 18 ft underground as it enters a frightening tunnel. (See Figure 7.15.) The mathematical model

Notes on Exercises

For Ex. 1–42, graphing should be viewed as a backup strategy when the algebraic approach is too complicated. Students need to develop judgment about when to use each approach.
Ex. 47–54 are application problems related to trigonometric equations.

Ongoing Assessment

Self-Assessment: Ex. 5, 11, 29, 41, 49, 57
Embedded Assessment: Ex. 27

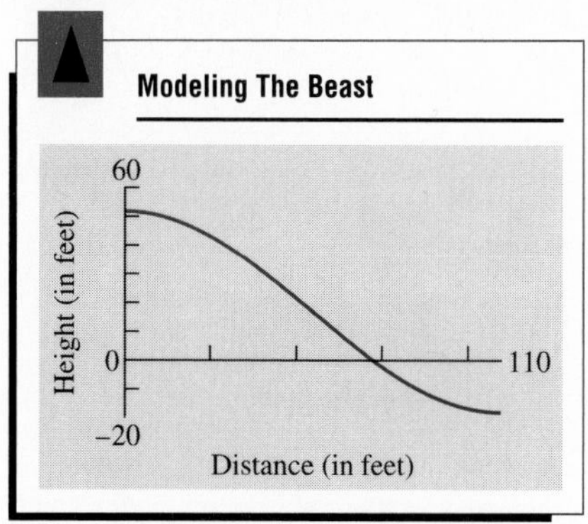

Root: $x = 72.729176$; $y = 0$

Figure 7.16

for this part of track is

$$h(x) = 35 \cos\left(\frac{x}{35}\right) + 17, \quad 0 \le x \le 110,$$

where x is the *horizontal* distance from the top of the hill and $h(x)$ is the vertical position relative to ground level (both in feet). What is the horizontal distance from the top of the hill to the point where the track reaches ground level?

Solution

Solve Algebraically

$$h(x) = 35 \cos\left(\frac{x}{35}\right) + 17 = 0 \quad \text{Equation to be solved}$$

$$35 \cos\left(\frac{x}{35}\right) = -17$$

$$\cos\left(\frac{x}{35}\right) = -\frac{17}{35}$$

$$\frac{x}{35} = \cos^{-1}\left[\cos\left(\frac{x}{35}\right)\right] = \cos^{-1}\left(-\frac{17}{35}\right)$$

$$x = 35 \cos^{-1}\left(-\frac{17}{35}\right) = 72.729\ldots$$

Interpret

The Beast reaches ground level after traveling about 73 ft horizontally from the top of the first hill.

Support Graphically

The graph of h in Figure 7.16 indicates a zero at $x \approx 72.73$. ■

Quick Review 7.3

In Exercises 1–4, evaluate the expression.

1. $\sin^{-1}\left(\frac{12}{13}\right)$ **2.** $\cos^{-1}\left(\frac{3}{5}\right)$

3. $\cos^{-1}\left(-\frac{4}{5}\right)$ **4.** $\sin^{-1}\left(-\frac{5}{13}\right)$

In Exercises 5–14, find all solutions, if any, to the equation in the interval $[0, 2\pi)$.

5. $\sin x = \frac{4}{5}$ **6.** $\cos x = -\frac{12}{13}$

7. $\cos x = -1.5$ **8.** $\sin x = 2.1$

9. $\tan x = -1$ **10.** $\sec x = -2$

11. $\csc x = 5$ **12.** $\cot x = 3$

13. $\csc x = -0.7$ **14.** $\sec x = 0.5$

In Exercises 15–18, list all the angles in the interval $[0, 2\pi)$ that have the given form where n is an integer.

15. $\frac{\pi}{6} + \frac{2n\pi}{3}$ **16.** $\frac{\pi}{2} + \frac{2n\pi}{3}$

17. $\frac{\pi}{8} + n\pi$ **18.** $\frac{\pi}{6} + \frac{n\pi}{2}$

SECTION EXERCISES 7.3

In Exercises 1–8, solve the equation. Support your answer graphically.

1. $\sin x = \dfrac{1}{2}$ **2.** $\cos\theta = \dfrac{\sqrt{3}}{2}$

3. $2\cos x = 1$ **4.** $\sqrt{3}\tan\alpha = 1$

5. $\csc x = 2$ **6.** $\sqrt{2}\sec\beta = 1$

7. $3\cot x = -\sqrt{3}$ **8.** $-2\sin y = 1$

In Exercises 9–18, solve the equation by factoring and/or extracting square roots. Support your work graphically.

9. $2\cos x\sin x - \cos x = 0$

10. $\sqrt{2}\tan x\cos x - \tan x = 0$

11. $\tan x\sin^2 x = \tan x$ **12.** $\sin x\tan^2 x = \sin x$

13. $\tan^2 x = 3$ **14.** $2\sin^2 x = 1$

15. $4\cos^2 x - 4\cos x + 1 = 0$

16. $2\sin^2 x + 3\sin x + 1 = 0$

17. $2\sqrt{2}\sin x\cos x - \sqrt{2}\cos x - 2\sin x = -1$

18. $2\cos x\tan x + \tan x - 2\cos x - 1 = 0$

In Exercises 19–32, find all solutions in the interval $[0, 2\pi)$.

19. $\sin 2x = 1$ **20.** $\sin 3t = 1$

21. $2\cos 3t = 1$ **22.** $\tan 2x = 1$

23. $3\tan 2\theta = 1$ **24.** $3\cos 2t = 2$

25. $\sin^2 3x = 0$ **26.** $\tan^2 2x = 1$

27. $\sin 2x\tan 2x + \sin 2x = 0$

28. $2\cos 2x\sin 2x + \cos 2x = 0$

29. $2\cos^2 x - 3\cos x - 4 = 0$

30. $\sin^2 x - \sqrt{2}\sin x - 1 = 0$

31. $\tan^2 2x - 2\tan 2x - 5 = 0$

32. $3\sin^2 2x + \sin 2x - 1 = 0$

In Exercises 33–42, solve the equation.

33. $\sin^2\theta - 2\sin\theta = 0$ **34.** $2\cos 2t = 0.75$

35. $3\sin t = 2\cos^2 t$ **36.** $\sin 2y = \sin y$

37. $\cos(\sin x) = 1$ **38.** $\tan(\cos x) = 0.75$

39. $2\sin^2 x + 3\sin x = 2$ **40.** $2\cos^2 2x + \cos 2x = 1$

41. $\cos 2x + \cos x = 0$

42. $\tan^2 x\cos x + 5\cos x = 0$

In Exercises 43–46, solve graphically.

43. $x = \cos x - 2$ **44.** $x^2 = \sin x + 1$

45. $x = 1 - 3\sin^2 x$ **46.** $0.5x = \cos^2 x + 0.75$

In Exercises 47 and 48, use the equation

$$T(x) = 37\sin\left[\frac{2\pi}{365}(x - 101)\right] + 25,$$

which models the average air temperature for Fairbanks, Alaska, where x is time in days with $x = 1$ representing January 1.

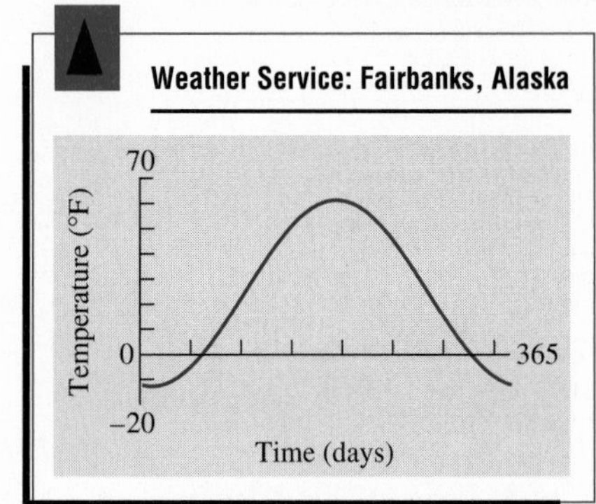

For Exercises 47–48

47. *Average Temperatures* On what days do you expect the average temperature in Fairbanks to be 56° F?

48. *Average Temperatures* On what days do you expect the average temperature in Fairbanks to be less than 22° F?

Exercises 49 and 50 refer to the trajectory as shown in the figure. A cannon is fired with an initial velocity of v_0 (in feet per second) and trajectory angle of θ. The horizontal distance of the

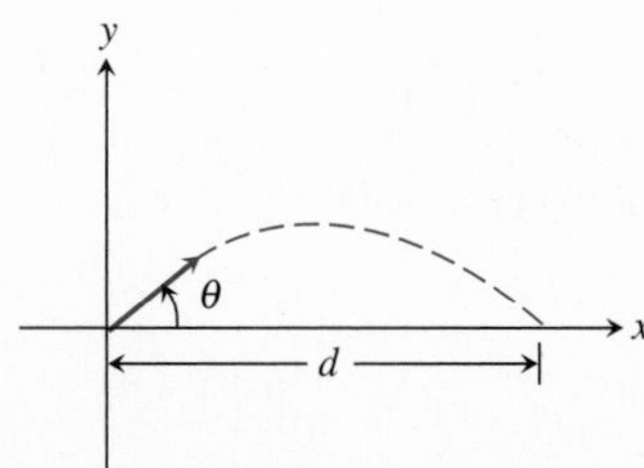

For Exercises 49–50,

cannon ball is modeled by

$$d = \frac{v_0^2}{16} \sin \theta \cos \theta.$$

49. *Cannon Ball Trajectory* Find the angle of trajectory needed to hit a target 2350 ft away if the initial velocity is 500 ft/sec.

50. *Cannon Trajectory* If the initial velocity is 500 ft/sec, what angle of trajectory yields the maximum distance? What is this distance?

In Exercises 51–54, solve the problem.

51. *Analysis of a Roller Coaster* Suppose the first hill of a roller coaster (see Example 8) is 90 ft high and drops to a low point 20 ft below ground level in a horizontal distance of 210 ft.

a. Find a sinusoidal model h for the height as a function of the horizontal distance from the top of the hill. That is, find a, b, and k in the relation

$$h(x) = a \cos bx + k.$$

b. Find the horizontal distance from the top of the hill to the point where the track reaches ground level.

52. *Analysis of a Roller Coaster* Consider the roller coaster described in Exercise 51. After the front car of a train falls a vertical height of 50 ft, what horizontal distance has it traveled?

53. *Thrown Baseball* Major League great Willie Mays was known for his strong throwing arm. He could throw the ball in the air from center field 400 ft to home plate. The horizontal distance d of a thrown ball is modeled by

$$d = \left(\frac{1}{32}\right) v_0^2 \sin 2\theta,$$

where θ is the angle with the horizontal of the throw. If Mays threw with an initial velocity of 118 ft/sec, what was his throwing angle?

54. *Football Punt* A football is kicked from the 35 yard line with an initial velocity of $v_0 = 82$ ft/sec and travels the 195 ft to the opponent's end zone. Use the equation in Exercise 53 to find the angle of the kick.

In Exercises 55–58, solve the inequality.

55. $\sin x < \frac{1}{2}$ for $0 \le x \le 2\pi$

56. $\sin x < \cos x$ for $0 \le x \le 2\pi$

57. $|\sin x| < \frac{1}{\sqrt{2}}$ for $0 \le x \le 2\pi$

58. $\tan x > 0$ for $0 \le x \le 2\pi$

EXTENDING THE IDEAS

In Exercises 59–63, solve the problem.

59. *Zeros* Find the zeros of the function $f(x) = \dfrac{\sin x}{x^2 - 3x}$.

60. *Vertical Asymptotes* Find the vertical asymptotes of $f(x) = \dfrac{x^2 - 2x}{\cos x}$.

61. *Periodic Function* Let f be a periodic function with period p. That is, $f(x + p) = f(x)$ for all values of x in the domain of f. Assume that x_1 is the only zero of f in the interval $[0, p]$. Find the general solution of $f(x) = 0$.

62. *Periodic Function* Consider the function f of Exercise 61. Find the general solution of $f(x) = 0$ if f has exactly two zeros x_1 and x_2 in the interval $[0, p)$.

63. *Transformations* There are two solutions to $\sin x = 0.5$ in the interval $[0, 2\pi)$. Explain how the concepts "period of the function" and "horizontal shrink of the graph" can be used to explain that there are six solutions to the equation $\sin 3x = 0.5$ in the interval $[0, 2\pi)$.

Objective

Students will be able to apply algebraic and trigonometric methods to sum and difference identities.

Motivate

Have students calculate $\sin(60° + 30°)$ and $\sin 60° \cos 30° + \cos 60° \sin 30°$. **(1; 1)**

7.4 SUM AND DIFFERENCE IDENTITIES

Introduction to Sum and Difference Identities • Cosine of a Sum or a Difference • Cofunction Identities • Sine of a Sum or a Difference • Applying Identities

Introduction to Sum and Difference Identities

In this section you will study the *sum* and *difference identities.*

COMMON ERROR

It is tempting to think of sin $(x + a)$ as a product with sin as a variable and apply a distributive property. However, for a trigonometric function f,

$$f(x + a) \neq f(x) + f(a),$$

in general.

EXPLORATION Testing for Identities

Guess whether the equation is an identity. Then check with a grapher.

1. $\sin(x + 2) = \sin x + \sin 2$
2. $\cos(x - 3) = \cos x - \cos 3$
3. $\tan(x + \pi) = \tan x + \tan \pi$
4. $\tan(2 + x) = \tan 2 + \tan x$

You might have guessed that all four of the equations are identities and the Exploration may have changed your mind. In fact the sum and difference formulas for the sine, cosine, or tangent are considerably more complex, as shown here:

Sum and Difference Identities

$$\sin(u + v) = \sin u \cos v + \cos u \sin v \qquad \sin(u - v) = \sin u \cos v - \cos u \sin v$$

$$\cos(u + v) = \cos u \cos v - \sin u \sin v \qquad \cos(u - v) = \cos u \cos v + \sin u \sin v$$

$$\tan(u + v) = \frac{\tan u + \tan v}{1 - \tan u \tan v} \qquad \tan(u - v) = \frac{\tan u - \tan v}{1 + \tan u \tan v}$$

Exploration Extensions

Guess whether the equation $\cos\left(\frac{\pi}{3} + x\right) = \frac{\cos x - \sqrt{3}}{2}$ is an identity, then check with a grapher.

Cosine of a Sum or a Difference

Confirming the sum and difference formulas requires a combination of algebra, geometry, and trigonometry. We begin by confirming the formula for $\cos(u - v)$.

Figure 7.17a shows angles u and v in standard position on a unit circle. Figure 7.17b shows the angle $u - v$ in standard position. Because arcs AB and

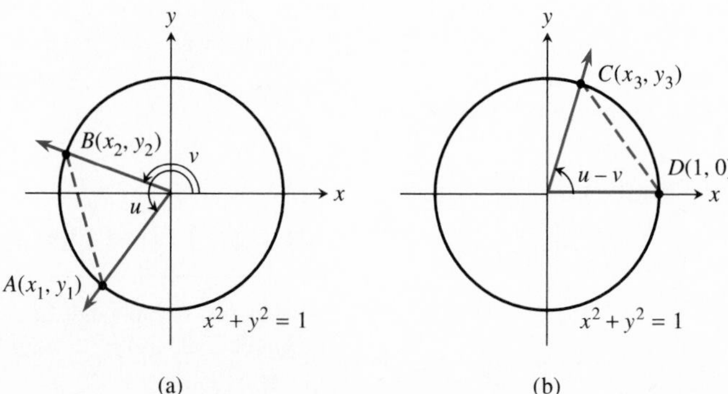

Figure 7.17

CD each have $u - v$ as their central angle, the corresponding line segments AB and CD are equal in length.

Applying the distance formula to $CD = AB$ gives us

$$\sqrt{(x_3 - 1)^2 + (y_3 - 0)^2} = \sqrt{(x_1 - x_2)^2 + (y_1 - y_2)^2}$$
$$[\sqrt{(x_3 - 1)^2 + (y_3 - 0)^2}\,]^2 = [\sqrt{(x_1 - x_2)^2 + (y_1 - y_2)^2}\,]^2$$
$$x_3^2 - 2x_3 + 1 + y_3^2 = x_1^2 - 2x_1x_2 + x_2^2 + y_1^2 - 2y_1y_2 + y_2^2$$
$$1 + (x_3^2 + y_3^2) - 2x_3 = (x_1^2 + y_1^2) + (x_2^2 + y_2^2) - 2x_1x_2 - 2y_1y_2.$$

Teaching Note

Students may need help in understanding the algebraic manipulation used to verify the difference identity for the cosine. Help students make the connection between the first line of the proof and the familiar distance formula
$d = \sqrt{(x_2 - x_1)^2 + (y_2 - y_1)^2}$.

Because A, B, and C lie on the unit circle, their coordinates satisfy the relation $x^2 + y^2 = 1$. So each expression in parentheses can be replaced by 1, which leads to

$$1 + 1 - 2x_3 = 1 + 1 - 2x_1x_2 - 2y_1y_2$$
$$-2x_3 = -2(x_1x_2 + y_1y_2)$$
$$x_3 = x_1x_2 + y_1y_2.$$

Using the coordinates of the points A, B, and C on the unit circle we have

$$x_1 = \cos u, \quad y_1 = \sin u, \quad x_2 = \cos v, \quad y_2 = \sin v, \quad \text{and}$$
$$x_3 = \cos(u - v), \quad y_3 = \sin(u - v).$$

By substitution we conclude that

$$\cos(u - v) = \cos u \cos v + \sin u \sin v.$$

Historically, this identity was used to find cosine values for specific angles, as shown in Example 1.

EXAMPLE 1 Using the Cosine Difference Identity

$$\cos 15^\circ = \cos(45^\circ - 30^\circ)$$
$$= \cos 45^\circ \cos 30^\circ + \sin 45^\circ \sin 30^\circ \quad \text{Difference identity}$$
$$= \frac{\sqrt{2}}{2} \cdot \frac{\sqrt{3}}{2} + \frac{\sqrt{2}}{2} \cdot \frac{1}{2}$$
$$= \frac{\sqrt{6} + \sqrt{2}}{4}$$

■

REMARK

Check the result of Example 1 with your grapher.

The cosine-of-a-difference identity can also be used to easily confirm some other identities we have already encountered. The identity in Example 2 can be viewed as a statement about a transformation of the cosine graph, or a statement about the cosine of the supplement of an angle x.

EXAMPLE 2 Using the Cosine Difference Identity

$$\begin{aligned} \cos(\pi - x) &= \cos \pi \cos x + \sin \pi \sin x && \text{Difference identity} \\ &= -1 \cdot \cos x + 0 \cdot \sin x \\ &= -\cos x \end{aligned}$$

Now we confirm the cosine-of-a-sum identity.

EXAMPLE 3 Confirming the Cosine Sum Identity

$$\begin{aligned} \cos(u + v) &= \cos[u - (-v)] \\ &= \cos u \cos(-v) + \sin u \sin(-v) && \text{Difference identity} \\ &= \cos u \cos v - \sin u \sin v && \text{Odd-even identities} \end{aligned}$$

Cofunction Identities

The **cofunction identities** relate the basic trigonometric functions to their cofunctions (sine to cosine, tangent to cotangent, and secant to cosecant) and relate the functions of an angle to functions of its complement. They also provide a bridge between the cosine of a difference and the sine of a sum.

Cofunction Identities

$$\cos\left(\frac{\pi}{2} - u\right) = \sin u \qquad \sin\left(\frac{\pi}{2} - u\right) = \cos u$$

$$\tan\left(\frac{\pi}{2} - u\right) = \cot u \qquad \cot\left(\frac{\pi}{2} - u\right) = \tan u$$

$$\sec\left(\frac{\pi}{2} - u\right) = \csc u \qquad \csc\left(\frac{\pi}{2} - u\right) = \sec u$$

Here is a confirmation of one of the cofunction identities.

EXAMPLE 4 Confirming a Cofunction Identity

$$\begin{aligned} \cos\left(\frac{\pi}{2} - x\right) &= \cos \frac{\pi}{2} \cos x + \sin \frac{\pi}{2} \sin x && \text{Difference identity} \\ &= 0 \cdot \cos x + 1 \cdot \sin x \\ &= \sin x \end{aligned}$$

Sine of a Sum or a Difference

We confirm the sine of a sum identity using the cosine of a difference.

Notes on Examples

Example 5 is important because it illustrates how one identity can be used to confirm another identity. This method is much simpler than trying to use a geometric approach to confirm the sine-of-a-sum identity.

EXAMPLE 5 Confirming a Cofunction Identity

$$\sin x = \cos\left[\frac{\pi}{2} - x\right]$$

$$\begin{aligned}
\sin(u + v) &= \cos\left[\frac{\pi}{2} - (u + v)\right] && \text{Substitute } u + v \text{ for } x.\\
&= \cos\left[\left(\frac{\pi}{2} - u\right) - v\right]\\
&= \cos\left(\frac{\pi}{2} - u\right)\cos v + \sin\left(\frac{\pi}{2} - u\right)\sin v && \text{Difference identity}\\
&= \sin u \cos v + \cos u \sin v && \text{Cofunction identity}
\end{aligned}$$

■

The remaining sum and difference identities and cofunction identities will be confirmed in Exercises 37–44.

Applying Identities

Example 6 confirms an identity that is needed to prove an important theorem in calculus. You will revisit it in your calculus course.

Notes on Examples

Example 6 is related to the derivative of the sine function.

EXAMPLE 6 Confirming an Identity Used in Calculus

Confirm that

$$\frac{\sin(x + h) - \sin x}{h} = \sin x\left(\frac{\cos h - 1}{h}\right) + \cos x\left(\frac{\sin h}{h}\right).$$

Solution

$$\begin{aligned}
\frac{\sin(x + h) - \sin x}{h} &= \frac{\sin x \cos h + \cos x \sin h - \sin x}{h} && \text{Sum identity}\\
&= \frac{(\sin x)(\cos h - 1) + \cos x \sin h}{h} && \text{Factor.}\\
&= (\sin x)\left(\frac{\cos h - 1}{h}\right) + (\cos x)\left(\frac{\sin h}{h}\right) && \text{Rewrite.}
\end{aligned}$$

■

Follow-up

Ask students if other values of a and h are possible in Example 7. **(Yes, for example, $a = -\sqrt{41}$ and $h = \frac{\pi}{3} + \frac{1}{3}\tan^{-1}\left(-\frac{5}{4}\right)$.)**

Assignment Guide

Day 1: Ex. 3–30, multiples of 3, 31–34
Day 2: Ex. 39–57, multiples of 3, 59, 61, 62

Cooperative Learning

Group Activity: Ex. 61

Notes on Exercises

For Ex. 11–22, after students have rewritten each expression, suggest that they check their work by using a calculator or grapher.
Ex. 49–50 are related to the calculus topic of taking derivatives.

Ongoing Assessment

Self-Assessment: Ex. 7, 17, 29, 37
Embedded Assessment: Ex. 27, 61

We used a graphical method in Example 1 of Section 6.6 to show that $y = 2\sin x + 5\cos x$ is a sinusoid. In particular, we found that

$$2\sin x + 5\cos x \approx 5.39\sin(x + 1.19).$$

We now use an algebraic method to solve a similar problem. Our results will be exact.

■ EXAMPLE 7 Finding the Amplitude and Phase Shift

Find values for a and h so that for all values of x,

$$5\cos 3x + 4\sin 3x = a\sin[3(x - h)].$$

Solution

Solve Algebraically

$$\begin{aligned} 5\cos 3x + 4\sin 3x &= a\sin[3(x - h)] \\ &= a\sin(3x - 3h) \\ &= a\sin 3x\cos 3h - a\cos 3x\sin 3h \quad \text{Difference identity} \\ &= (-a\sin 3h)(\cos 3x) + (a\cos 3h)(\sin 3x) \end{aligned}$$

Equating the coefficients of $\cos 3x$ and $\sin 3x$ we find

$$-a\sin 3h = 5, \qquad a\cos 3h = 4,$$

$$\sin 3h = -\frac{5}{a}, \qquad \cos 3h = \frac{4}{a}.$$

Therefore, $$\tan 3h = \frac{\sin 3h}{\cos 3h} = \frac{-5/a}{4/a} = -\frac{5}{4}.$$

So $$3h = \tan^{-1}\left(-\frac{5}{4}\right) \quad \text{and} \quad h = \frac{1}{3}\tan^{-1}\left(-\frac{5}{4}\right).$$

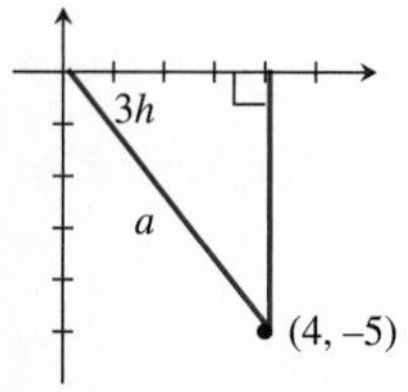

To find a, we use $\sin 3h = -\dfrac{5}{a}$ and $\cos 3h = \dfrac{4}{a}$ as the basis for Figure 7.18. Because a is the hypotenuse of the triangle shown, we find $a = \sqrt{5^2 + 4^2} = \sqrt{41}$. Thus,

$$5\cos 3x + 4\sin 3x = \sqrt{41}\sin\left\{3\left[x - \frac{1}{3}\tan^{-1}\left(-\frac{5}{4}\right)\right]\right\}.$$

Figure 7.18

Support Graphically

We let $y_1 = 5\cos 3x + 4\sin 3x$ and $y_2 = \sqrt{41}\sin\left[3x - \tan^{-1}\left(-\frac{5}{4}\right)\right]$.

Their graphs are identical in Figure 7.19. ■

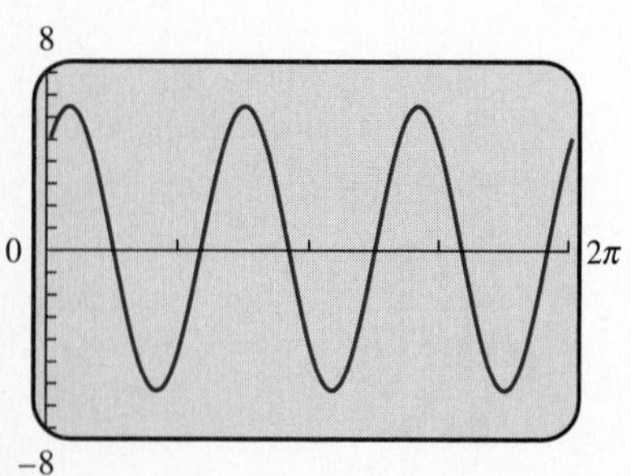

Figure 7.19

Quick Review 7.4

In Exercises 1–8, express the angle as a sum or difference of special angles (multiples of 30°, 45°, $\pi/6$, or $\pi/4$).

1. 15°
2. 75°
3. 165°
4. 255°
5. $\pi/12$
6. $5\pi/12$
7. $7\pi/12$
8. $-\pi/12$

In Exercises 9–12, determine whether the function is a sinusoid. If so, estimate its amplitude, period, and phase shift (based on $y = \sin x$).

9. $f(x) = 3 \sin 2x + 2 \cos 2x$
10. $g(x) = 5 \sin 3x - 4 \cos 3x$
11. $f(x) = \sin 2x + \sin 3x$
12. $g(x) = 2 \sin 4x - \cos 3x$

SECTION EXERCISES 7.4

In Exercises 1–10, use a sum or difference identity to find an exact value.

1. $\sin 15°$
2. $\tan 15°$
3. $\sin 75°$
4. $\cos 75°$
5. $\cos \frac{\pi}{12}$
6. $\sin \frac{7\pi}{12}$
7. $\tan \frac{5\pi}{12}$
8. $\tan \frac{11\pi}{12}$
9. $\cos \frac{7\pi}{12}$
10. $\sin \frac{-\pi}{12}$

In Exercises 11–22, write the expression as the sine, cosine, or tangent of an angle.

11. $\sin 42° \cos 17° - \cos 42° \sin 17°$
12. $\cos 94° \cos 18° + \sin 94° \sin 18°$
13. $\sin \frac{\pi}{5} \cos \frac{\pi}{2} + \sin \frac{\pi}{2} \cos \frac{\pi}{5}$
14. $\sin \frac{\pi}{3} \cos \frac{\pi}{7} - \sin \frac{\pi}{7} \cos \frac{\pi}{3}$
15. $\dfrac{\tan 19° + \tan 47°}{1 - \tan 19° \tan 47°}$
16. $\dfrac{\tan \frac{\pi}{5} - \tan \frac{\pi}{3}}{1 + \tan \frac{\pi}{5} \tan \frac{\pi}{3}}$
17. $\cos \frac{\pi}{7} \cos x + \sin \frac{\pi}{7} \sin x$
18. $\cos x \cos \frac{\pi}{7} - \sin x \sin \frac{\pi}{7}$
19. $\sin 3x \cos x - \cos 3x \sin x$
20. $\cos 7y \cos 3y - \sin 7y \sin 3y$
21. $\dfrac{\tan 2y + \tan 3x}{1 - \tan 2y \tan 3x}$
22. $\dfrac{\tan 3\alpha - \tan 2\beta}{1 + \tan 3\alpha \tan 2\beta}$

In Exercises 23–30, confirm the identity.

23. $\sin\left(x - \frac{\pi}{2}\right) = -\cos x$
24. $\tan\left(x - \frac{\pi}{2}\right) = -\cot x$
25. $\cos\left(x - \frac{\pi}{2}\right) = \sin x$
26. $\cos\left[\left(\frac{\pi}{2} - x\right) - y\right] = \sin(x + y)$
27. $\sin\left(x + \frac{\pi}{6}\right) = \frac{\sqrt{3}}{2} \sin x + \frac{1}{2} \cos x$
28. $\cos\left(x - \frac{\pi}{4}\right) = \frac{\sqrt{2}}{2}(\cos x + \sin x)$
29. $\tan\left(\theta + \frac{\pi}{4}\right) = \dfrac{1 + \tan \theta}{1 - \tan \theta}$
30. $\cos\left(\theta + \frac{\pi}{2}\right) = -\sin \theta$

In Exercises 31–34, match a pair of equations with the graph.

a. $y = \cos(3 - 2x)$
b. $y = \sin x \cos 1 + \cos x \sin 1$
c. $y = \cos(x - 3)$
d. $y = \sin(2x - 5)$
e. $y = \cos x \cos 3 + \sin x \sin 3$
f. $y = \sin(x + 1)$
g. $y = \cos 3 \cos 2x + \sin 3 \sin 2x$
h. $y = \sin 2x \cos 5 - \cos 2x \sin 5$

31.

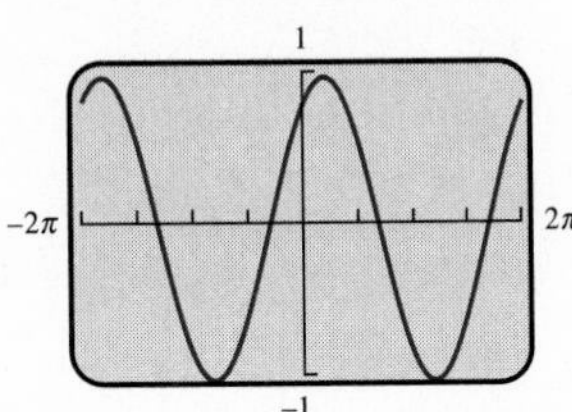

32.

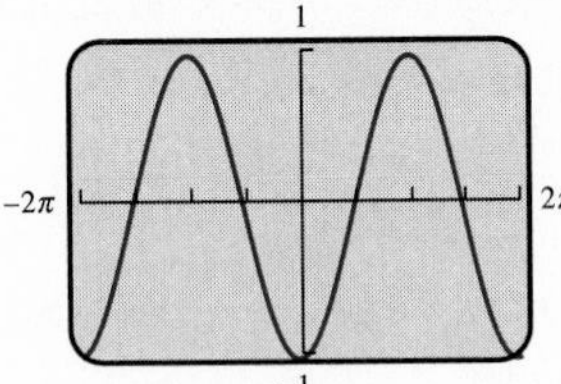

33.

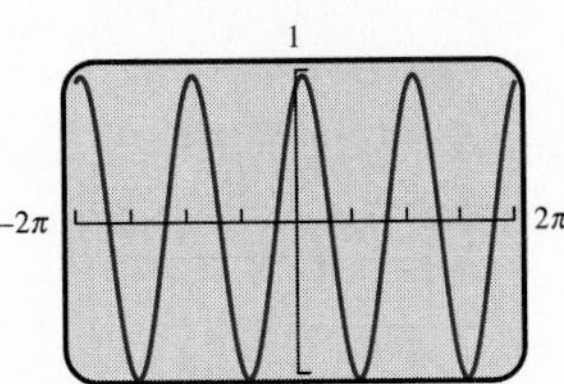

34.

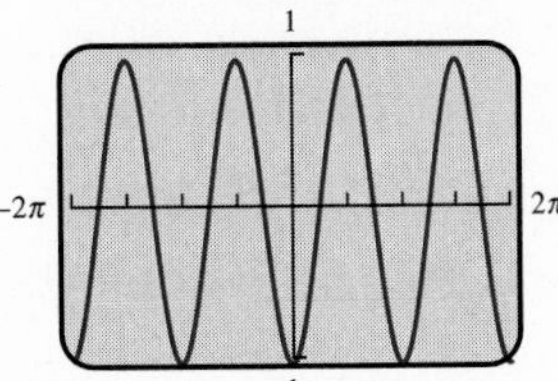

In Exercises 35 and 36, use sum or difference identities to solve the equation exactly.

35. $\sin 2x \cos x = \cos 2x \sin x$

36. $\cos 3x \cos x = \sin 3x \sin x$

In Exercises 37–44, confirm the cofunction or the sum or difference identity.

37. $\sin\left(\frac{\pi}{2} - u\right) = \cos u$

38. $\tan\left(\frac{\pi}{2} - u\right) = \cot u$

39. $\cot\left(\frac{\pi}{2} - u\right) = \tan u$

40. $\sec\left(\frac{\pi}{2} - u\right) = \csc u$

41. $\csc\left(\frac{\pi}{2} - u\right) = \sec u$

42. $\sin(u - v) = \sin u \cos v - \cos u \sin v$

43. $\tan(u + v) = \dfrac{\tan u + \tan v}{1 - \tan u \tan v}$

44. $\tan(u - v) = \dfrac{\tan u - \tan v}{1 + \tan u \tan v}$

In Exercises 45–48, find exact values for the amplitude, period, and phase shift in $a \sin[b(x - h)]$ for the sinusoid.

45. $y = \cos 3x + 2 \sin 3x$

46. $y = 3 \cos 2x - 2 \sin 2x$

47. $y = 3 \sin x + 5 \sin(x + 2)$

48. $y = 3 \sin(2x - 1) + 5 \sin(2x + 3)$

In Exercises 49 and 50, confirm the identity.

49. $\dfrac{\cos(x + h) - \cos x}{h} = \cos x\left(\dfrac{\cos h - 1}{h}\right) - \sin x\left(\dfrac{\sin h}{h}\right)$

50. $\dfrac{\tan(x + h) - \tan x}{h} = \sec^2 x\left(\dfrac{\sin h}{h}\right)\dfrac{1}{\cos h - \sin h \tan x}$

EXTENDING THE IDEAS

In Exercises 51–59, confirm the identity.

51. $\sin(x - y) + \sin(x + y) = 2 \sin x \cos y$

52. $\cos(x - y) + \cos(x + y) = 2 \cos x \cos y$

53. $\cos 3x = \cos^3 x - 3 \sin^2 x \cos x$

54. $\sin 3u = 3 \cos^2 u \sin u - \sin^3 u$

55. $\cos 3x + \cos x = 2 \cos 2x \cos x$

56. $\sin 4x + \sin 2x = 2 \sin 3x \cos x$

57. $\tan(x + y) \tan(x - y) = \dfrac{\tan^2 x - \tan^2 y}{1 - \tan^2 x \tan^2 y}$

58. $\tan 5u \tan 3u = \dfrac{\tan^2 4u - \tan^2 u}{1 - \tan^2 4u \tan^2 u}$

59. $\cos(x + y + z) = \cos x \cos y \cos z - \sin x \sin y \cos z - \sin x \cos y \sin z - \cos x \sin y \sin z$

In Exercises 60–63, solve the problem.

60. Writing to Learn The figure shows graphs of $y_1 = \cos 5x \cos 4x$ and $y_2 = -\sin 5x \sin 4x$ in one viewing window. Discuss the question, "How many solutions are there to the equation $\cos 5x \cos 4x = -\sin 5x \sin 4x$ in the interval $[-2\pi, 2\pi]$?" Give an algebraic argument that answers the question more convincingly than the graph does. Then support your argument with an *appropriate* graph.

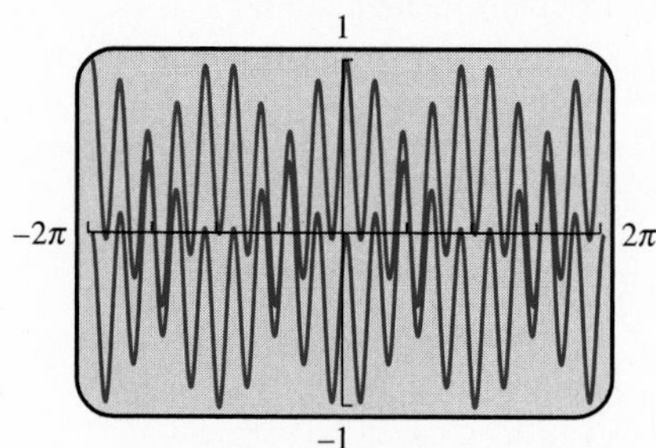

For Exercise 60

61. Writing to Learn Solve $\sin 3x \cos x = \cos 3x \sin x$ both algebraically and graphically. Then discuss the advantages and disadvantages of each method, and describe which method you prefer in this instance and why.

62. *Harmonic Motion* Alternating electric current, an oscillating spring, or any other harmonic oscillator can be modeled by the equation

$$x = a \cos\left(\frac{2\pi}{T}t + \delta\right),$$

where T is the time for one period and δ is the phase constant. Show that this motion can also be modeled by the following sum of cosine and sine, each with zero phase constant:

$$a_1 \cos\left(\frac{2\pi}{T}\right)t + a_2 \sin\left(\frac{2\pi}{T}\right)t,$$

where $a_1 = a \cos \delta$ and $a_2 = -a \sin \delta$.

63. *Magnetic Fields* A magnetic field B can sometimes be modeled as the sum of an incident and a reflective field as

$$B = B_{\text{in}} + B_{\text{ref}},$$

where $B_{\text{in}} = \dfrac{E_0}{c}\cos\left(\omega t - \dfrac{\omega x}{c}\right)$, and

$$B_{\text{ref}} = \frac{E_0}{c}\cos\left(\omega t + \frac{\omega x}{c}\right).$$

Show that $B = 2\dfrac{E_0}{c}\cos \omega t \cos \dfrac{\omega x}{c}$.

7.5 MULTIPLE-ANGLE IDENTITIES

Objective

Students will be able to apply trigonometric and algebraic methods to double-angle identities, power-reducing identities, and half-angle identities.

Motivate

Have students apply the cosine-of-a-sum identity in order to evaluate $\cos(x + x)$. **($\cos^2 x - \sin^2 x$)**

Double-Angle and Power-Reducing Identities • Using Double-Angle Identities to Solve Equations • Half-Angle Identities • Using Half-Angle Identities to Solve Equations • Applying Identities

Double-Angle and Power-Reducing Identities

Double-angle identities include an angle u on one side and double that angle, $2u$, on the other side. The double-angle identities listed here are among the most frequently used identities.

The identity for $\cos 2u$ has three forms because of the Pythagorean identity, $\sin^2 u + \cos^2 u = 1$. We can solve for $\sin^2 u$ and $\cos^2 u$ using the second and third forms for $\cos 2u$, respectively. Then dividing gives an identity for $\tan^2 u$. These three *power-reducing identities* are also listed below.

Double-Angle Identities

$\sin 2u = 2 \sin u \cos u$ $\quad \cos 2u = \cos^2 u - \sin^2 u$ $\quad \tan 2u = \dfrac{2 \tan u}{1 - \tan^2 u}$

$\cos 2u = 2 \cos^2 u - 1$

$\cos 2u = 1 - 2 \sin^2 u$

Power-Reducing Identities

$\sin^2 u = \dfrac{1 - \cos 2u}{2}$ $\quad \cos^2 u = \dfrac{1 + \cos 2u}{2}$ $\quad \tan^2 u = \dfrac{1 - \cos 2u}{1 + \cos 2u}$

EXAMPLE 1 Confirming a Double-Angle Identity

Confirm the double-angle identity $\sin 2u = 2 \sin u \cos u$.

Solution

$$\begin{aligned} \sin 2u &= \sin (u + u) \\ &= \sin u \cos u + \cos u \sin u && \text{Sum identity} \\ &= 2 \sin u \cos u \end{aligned}$$

■

The other double-angle identities can be confirmed in a similar way. See Exercises 1–4.

Using Double-Angle Identities to Solve Equations

Sometimes we rewrite a trigonometric equation in an equivalent form and then use algebraic properties to solve it.

EXAMPLE 2 Using a Double-Angle Identity

Solve $\sin 2x = \sin x$ in the interval $[0, 2\pi)$.

Solution

The graph in Figure 7.20 leads us to expect four solutions in $[0, 2\pi)$.

Solve Algebraically

$$\begin{aligned} \sin 2x &= \sin x \\ \sin 2x - \sin x &= 0 \\ 2 \sin x \cos x - \sin x &= 0 && \text{Double angle identity} \\ (\sin x)(2 \cos x - 1) &= 0 && \text{Factor.} \end{aligned}$$

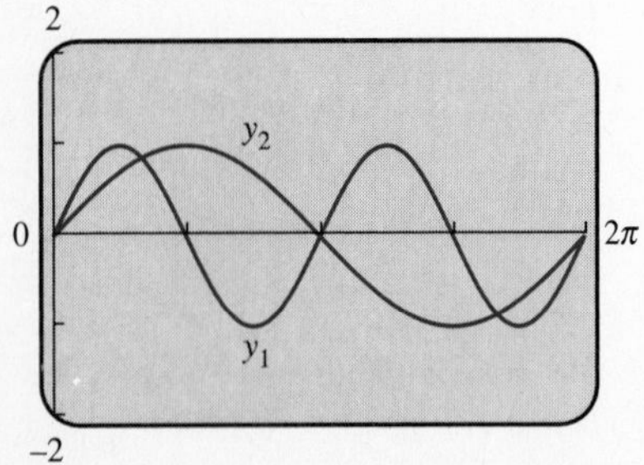

Figure 7.20 $y_1 = \sin 2x$ and $y_2 = \sin x$.

$$\sin x = 0 \quad \text{or} \quad 2\cos x - 1 = 0 \qquad \text{Zero factor property}$$

$$\vdots \qquad \cos x = \frac{1}{2}$$

$$x = 0 \text{ or } \pi \quad \text{or} \quad x = \frac{\pi}{3} \text{ or } \frac{5\pi}{3}$$

So the four solutions of $\sin 2x = \sin x$ in $[0, 2\pi)$ are 0, $\pi/3$, π, and $5\pi/3$. ■

Example 3 shows that several identities may need to be used to solve an equation algebraically. We begin with a graphical solution.

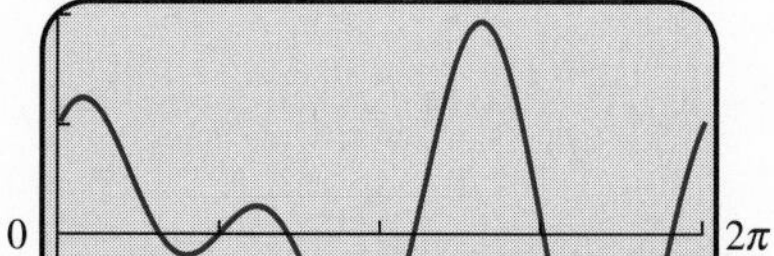

Figure 7.21 $y = \sin 2x + \cos 3x$ has six zeros in $[0, 2\pi)$.

■ EXAMPLE 3 Using Double-Angle Identities

Solve $\sin 2x + \cos 3x = 0$ in the interval $[0, 2\pi)$.

Solution

Solve Graphically

Figure 7.21 suggests that there are six solutions, which we find to be about

$$0.94, \quad 1.57, \quad 2.20, \quad 3.46, \quad 4.71, \quad \text{and} \quad 5.97.$$

Confirm Algebraically

We rewrite the equation into one involving $\sin x$ and $\cos x$ only. To do this we first view $\cos 3x$ as $\cos(2x + x)$ and then use the $\cos(u + v)$ identity and the double-angle identities.

$$\begin{aligned}
\sin 2x + \cos 3x &= 0 \\
\sin 2x + (\cos 2x \cos x - \sin 2x \sin x) &= 0 && \text{Cosine of sum} \\
\sin 2x + (1 - 2\sin^2 x)\cos x - (2\sin x \cos x)(\sin x) &= 0 && \text{Double angle} \\
\sin 2x + \cos x - 2\sin^2 x \cos x - 2\sin^2 x \cos x &= 0 && \text{Multiply.} \\
\sin 2x + \cos x - 4\sin^2 x \cos x &= 0 && \text{Simplify.} \\
2\sin x \cos x + \cos x - 4\sin^2 x \cos x &= 0 && \text{Double angle} \\
(\cos x)(2\sin x + 1 - 4\sin^2 x) &= 0 && \text{Factor.}
\end{aligned}$$

$$\cos x = 0 \quad \text{or} \quad -4\sin^2 x + 2\sin x + 1 = 0 \qquad \text{Zero factor property}$$

$$x = \frac{\pi}{2} \text{ or } \frac{3\pi}{2} \quad \text{or} \quad \sin x = \frac{1 \pm \sqrt{5}}{4} \qquad \text{Quadratic formula}$$

$$x = \sin^{-1}\left(\frac{1 + \sqrt{5}}{4}\right)$$

$$\vdots \qquad \text{or} \quad \sin^{-1}\left(\frac{1 - \sqrt{5}}{4}\right)$$

$$x = \frac{\pi}{2} \text{ or } \frac{3\pi}{2} \quad \text{or} \quad x = 0.942\cdots, 2.199\cdots, 3.455\cdots, \text{ or } 5.969\cdots.$$

REMARK

Notice that it is easy to find the six solutions in Example 3 graphically. The algebraic analysis assures us that we indeed have found all the solutions.

Interpret

Use your grapher to see that these six exact solutions found algebraically agree with the approximations found graphically. ■

Example 4 uses $\cos^4 x$ to illustrate a way to reduce powers of trigonometric functions, something that is sometimes needed in calculus.

■ EXAMPLE 4 Using a Power-Reducing Identity

Rewrite $\cos^4 x$ as a sum including $\cos 2x$ and $\cos 4x$.

Solution

$$\begin{aligned}
\cos^4 x &= (\cos^2 x)^2 \\
&= \left(\frac{1 + \cos 2x}{2}\right)^2 && \text{Power-reducing identity} \\
&= \frac{1 + 2\cos 2x + \cos^2 2x}{4} && \text{Multiply.} \\
&= \frac{1}{4} + \frac{1}{2}\cos 2x + \frac{1}{4}\left(\frac{1 + \cos 4x}{2}\right) && \text{Power-reducing identity} \\
&= \frac{1}{4} + \frac{1}{2}\cos 2x + \frac{1}{8} + \frac{1}{8}\cos 4x && \text{Multiply.} \\
&= \frac{1}{8}(3 + 4\cos 2x + \cos 4x)
\end{aligned}$$

■

Alert

Students are often confused as to which of the signs (+ or −) is correct in the sine and cosine half-angle identities. They must be shown that the signs are chosen according to the quadrant of $\frac{u}{2}$.

Half-Angle Identities

The half-angle identities have an angle u on one side of the equation and an angle $u/2$ on the other side. These identities can be obtained from the power-reducing identities by taking the square root of both sides of the equations. These identities permit us to write $\sin(x/2)$, $\cos(x/2)$, and $\tan(x/2)$ in terms of $\sin x$, $\cos x$, and $\tan x$.

Half-Angle Identities

$$\sin\frac{u}{2} = \pm\sqrt{\frac{1 - \cos u}{2}} \qquad \cos\frac{u}{2} = \pm\sqrt{\frac{1 + \cos u}{2}}$$

$$\tan\frac{u}{2} = \pm\sqrt{\frac{1 - \cos u}{1 + \cos u}} \qquad \tan\frac{u}{2} = \frac{1 - \cos u}{\sin u} \qquad \tan\frac{u}{2} = \frac{\sin u}{1 + \cos u}$$

NOTE

To use the half-angle identities involving radicals you need to determine the quadrant of $u/2$. Then use the correct sign (+ or −) for the trigonometric function in that quadrant.

Historically, the half-angle identities were used to find trigonometric values for specific angles. For example,

$$\tan(\pi/8) = \frac{1 - \cos(\pi/4)}{\sin(\pi/4)} = \frac{1 - \sqrt{2}/2}{\sqrt{2}/2} = \sqrt{2} - 1.$$

They can also be used to confirm other identities.

Using Half-Angle Identities to Solve Equations

When a trigonometric equation includes half angles, you can expect to use half-angle identities when solving the equation.

■ EXAMPLE 5 Using Half-Angle Identities

Solve $\sin^2 x = 2\sin^2 \frac{x}{2}$.

Solution

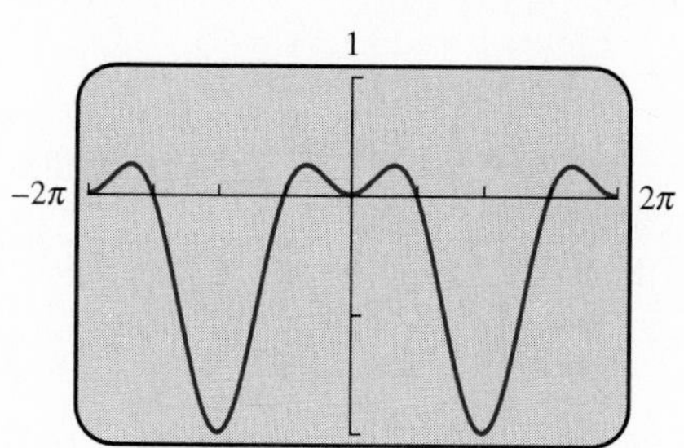

Figure 7.22

The graph of $y = \sin^2 x - 2\sin^2(x/2)$ in Figure 7.22 suggests that this function is periodic with the period 2π, and that the equation $\sin^2 x - 2\sin^2(x/2) = 0$ has three solutions in $[0, 2\pi)$.

Solve Algebraically

$$\sin^2 x = 2\sin^2 \frac{x}{2}$$

$$\sin^2 x = 2\left(\frac{1 - \cos x}{2}\right) \quad \text{Half-angle identity}$$

$$1 - \cos^2 x = 1 - \cos x \quad \text{Pythagorean identity}$$

$$\cos x - \cos^2 x = 0$$

$$(\cos x)(1 - \cos x) = 0 \quad \text{Factor.}$$

$$1 - \cos x = 0 \quad \text{or} \quad \cos x = 0 \quad \text{Zero factor property.}$$

$$x = 0 \quad \text{or} \quad x = \frac{\pi}{2} \quad \text{or} \quad \frac{3\pi}{2} \quad \text{Solutions in } [0, 2\pi)$$

In general,

$$x = 2n\pi, \quad x = \frac{\pi}{2} + 2n\pi, \quad \text{or} \quad \frac{3\pi}{2} + 2n\pi, \quad n \text{ an integer.}$$

■

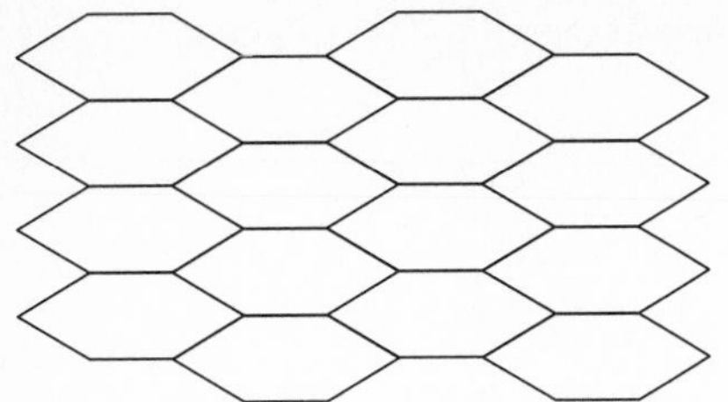

Figure 7.23 Hexagon-shaped wire mesh.

Applying Identities

Mishmash Mesh manufactures a wire mesh used by the construction industry. The individual cells in this mesh are hexagons with a horizontal line of symmetry and whose edges are all the same length, as shown in Figure 7.23. Company design engineers want to find the relationship between the "side" angle α of the hexagon and the area of an individual hexagon shown in Figure 7.24.

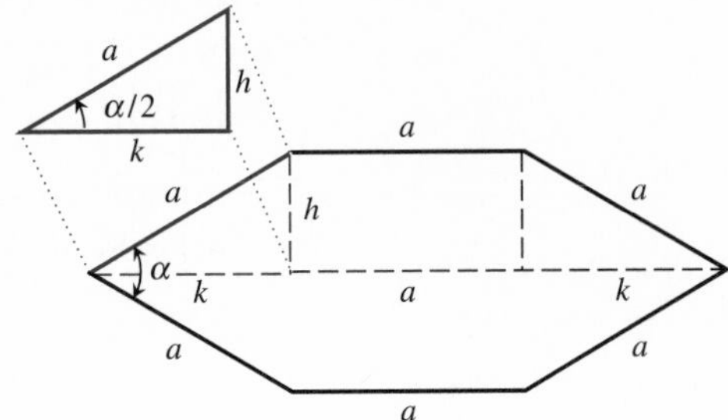

Figure 7.24

■ EXAMPLE 6 APPLICATION: Solving the Mishmash Mesh Problem

Express the area of the hexagon in Figure 7.24 in terms of the angle α and length a of a side.

Solution

The horizontal line of symmetry assures that the horizontal diagonal divides the hexagon into a pair of congruent isosceles trapezoids.

The area of a trapezoid in terms of the bases b_1 and b_2 and altitude h is

$$A = \frac{1}{2}(b_1 + b_2)h.$$

We apply this formula to the hexagon in Figure 7.24, where $b_1 = a$ and $b_2 = a + 2k$:

$$A = 2 \cdot \frac{1}{2}(b_1 + b_2)h$$

$$= (a + a + 2k)h$$

$$= \left(a + a + 2a\cos\frac{\alpha}{2}\right)\left(a\sin\frac{\alpha}{2}\right) \qquad k = a\cos\frac{\alpha}{2},\ h = a\sin\frac{\alpha}{2}.$$

$$= 2a^2\sin\frac{\alpha}{2} + 2a^2\sin\frac{\alpha}{2}\cos\frac{\alpha}{2}$$

$$= a^2\left(2\sin\frac{\alpha}{2} + 2\sin\frac{\alpha}{2}\cos\frac{\alpha}{2}\right) \qquad \text{Factor.}$$

$$= a^2\left(2\sin\frac{\alpha}{2} + \sin\alpha\right) \qquad \text{Double-angle identity}$$

■

Follow-up

Have students discuss why the identity $\tan\left(\frac{u}{2}\right) = \frac{\sin u}{1 + \cos u}$ does not require a $\pm$ sign. **(sin u and tan$\left(\frac{u}{2}\right)$ have the same sign, and 1 + cos u cannot be negative.)**

Assignment Guide

Day 1: Ex. 3–36, multiples of 3
Day 2: Ex. 37, 39, 42, 43, 46, 47, 50–52, 56

Cooperative Learning

Group Activity: Ex. 47–48

Notes on Exercises

Ex. 23–30 require students to solve equations using double-angle or other identities.
Ex. 37–42 relate to the power-reducing identities.

Ongoing Assessment

Self-Assessment: Ex. 7, 11, 19, 25, 35, 41, 45, 49
Embedded Assessment: Ex. 21, 37

■ EXAMPLE 7 APPLICATION: Maximizing Mesh Area

What angle α should be used to maximize the area of the mesh hexagon described in Example 6?

Solution

Solve Graphically

From Example 6, a hexagon with sides of length a and a horizontal line of symmetry has area

$$A = a^2\left(2 \sin \frac{\alpha}{2} + \sin \alpha\right).$$

So the area is maximum whenever the expression

$$2 \sin \frac{\alpha}{2} + \sin \alpha$$

is maximum. Figure 7.25 shows a graph of $y = 2 \sin (x/2) + \sin x$ over the interval $[0, \pi]$. We find that the maximum occurs when x is about 2.09 rad.

Interpret

Notice that 2.09 rad is about 120°. Mishmash Mesh can generate the maximum area for a given amount of wire by using $\alpha = 120°$. ■

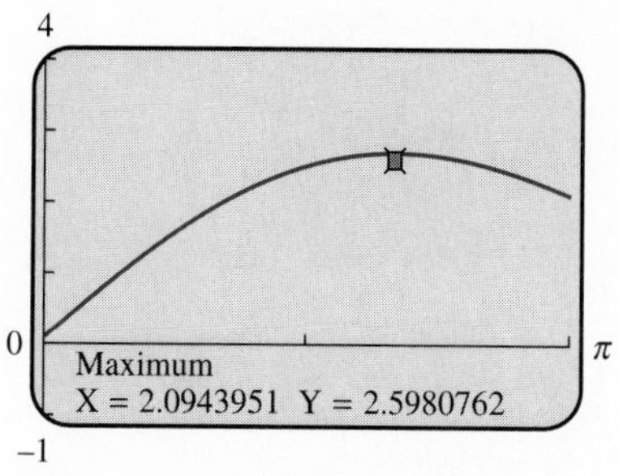

Figure 7.25 $y = 2 \sin \frac{x}{2} + \sin x.$

Quick Review 7.5

In Exercises 1–8, find the general solution of the equation.

1. $\tan x - 1 = 0$
2. $\tan x + 1 = 0$
3. $(\cos x)(1 - \sin x) = 0$
4. $(\sin x)(1 + \cos x) = 0$
5. $\sin x + \cos x = 0$
6. $\sin x - \cos x = 0$
7. $(2 \sin x - 1)(2 \cos x + 1) = 0$
8. $(\sin x + 1)(2 \cos x - \sqrt{2}) = 0$

In Exercises 9 and 10, find the area of the trapezoid. All angles that appear to be right angles are right angles, and all segments that appear to be parallel are parallel.

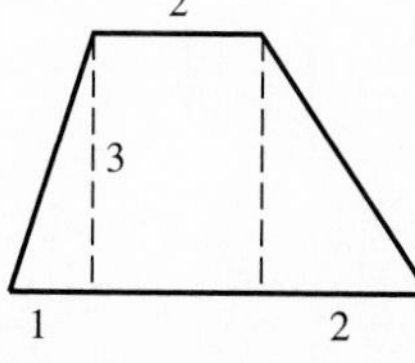

10.

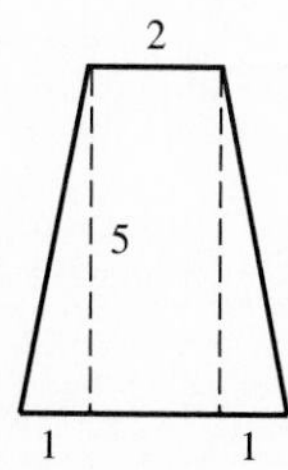

11. Find the height of the isosceles triangle.

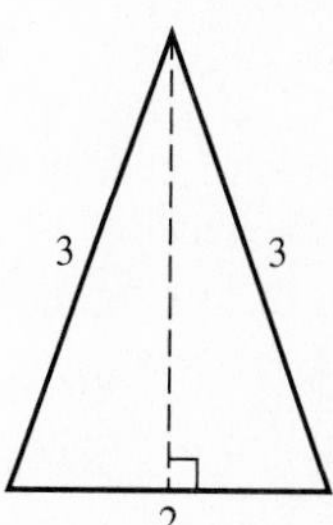

For Exercise 11

SECTION EXERCISES 7.5

In Exercises 1–4, use the appropriate sum or difference identity to confirm the double-angle identity.

1. $\cos 2u = \cos^2 u - \sin^2 u$

2. $\cos 2u = 2\cos^2 u - 1$

3. $\cos 2u = 1 - 2\sin^2 u$

4. $\tan 2u = \dfrac{2\tan u}{1 - \tan^2 u}$

In Exercises 5–10, find all solutions to the equation in the interval $[0, 2\pi)$.

5. $\sin 2x = 2\sin x$

6. $\sin 2x = \cos x$

7. $\cos 2x = \sin x$

8. $\cos 2x = \cos x$

9. $\sin 2x - \tan x = 0$

10. $\cos^2 x + \cos x = \cos 2x$

In Exercises 11–14, write the expression as one involving only $\sin\theta$ and $\cos\theta$.

11. $\sin 2\theta + \cos\theta$

12. $\sin 2\theta + \cos 2\theta$

13. $\sin 2\theta + \cos 3\theta$

14. $\sin 3\theta + \cos 2\theta$

In Exercises 15–22, confirm the identity.

15. $\sin 4x = 2\sin 2x\cos 2x$

16. $\cos 6x = 2\cos^2 3x - 1$

17. $2\csc 2x = \csc^2 x\tan x$

18. $2\cot 2x = \cot x - \tan x$

19. $\sin 3x = (\sin x)(4\cos^2 x - 1)$

20. $\sin 3x = (\sin x)(3 - 4\sin^2 x)$

21. $\cos 4x = 1 - 8\sin^2 x\cos^2 x$

22. $\sin 4x = (4\sin x\cos x)(2\cos^2 x - 1)$

In Exercises 23–30, solve the equation in $[0, 2\pi)$. Find exact values. Support your work graphically.

23. $\cos 2x + \cos x = 0$

24. $\cos 2x + 2\sin x = 0$

25. $\cos x + \cos 3x = 0$

26. $\sin x + \sin 3x = 0$

27. $\sin 2x + \sin 4x = 0$

28. $\cos 2x + \cos 4x = 0$

29. $\sin 2x - \cos 3x = 0$

30. $\sin 3x + \cos 2x = 0$

In Exercises 31–36, use half-angle identities to find an exact value for the expression.

31. $\sin 15°$

32. $\tan 195°$

33. $\cos 75°$

34. $\sin 5\pi/12$

35. $\tan 7\pi/12$

36. $\cos \pi/8$

In Exercises 37 and 38, confirm the power-reducing identities.

37. $\sin^2 u = \dfrac{1 - \cos 2u}{2}$

38. $\cos^2 u = \dfrac{1 + \cos 2u}{2}$

In Exercises 39–42, use the power-reducing identities to confirm the identity.

39. $\sin^4 x = \dfrac{1}{8}(3 - 4\cos 2x + \cos 4x)$

40. $\cos^3 x = \left(\dfrac{1}{2}\cos x\right)(1 + \cos 2x)$

41. $\sin^3 2x = \left(\dfrac{1}{2}\sin 2x\right)(1 - \cos 4x)$

42. $\sin^5 x = \left(\dfrac{1}{8}\sin x\right)(3 - 4\cos 2x + \cos 4x)$

In Exercises 43–46, use the half-angle identities to find all solutions in the interval $[0, 2\pi)$. Then find the general solution.

43. $\cos^2 x = \sin^2\left(\dfrac{x}{2}\right)$

44. $\sin^2 x = \cos^2\left(\dfrac{x}{2}\right)$

45. $\tan\left(\dfrac{x}{2}\right) = \dfrac{1 - \cos x}{1 + \cos x}$

46. $\sin^2\left(\dfrac{x}{2}\right) = 2\cos^2 x - 1$

In Exercises 47–50, solve the problem.

47. Connecting Trigonometry and Geometry In a regular polygon all sides are the same length and all angles are equal in measure.

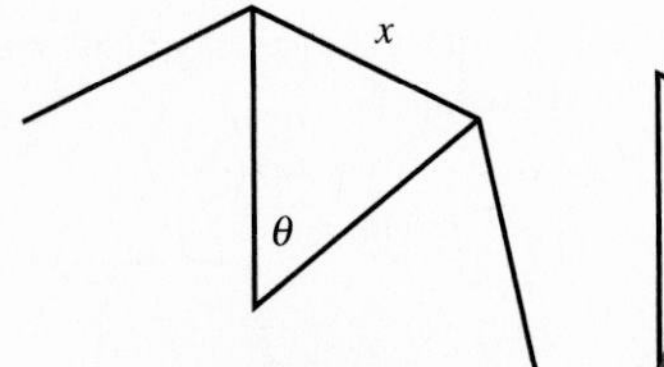

For Exercise 47

a. If the perpendicular distance from the center of the polygon with n sides to the midpoint of a side is R, and if the length of the side of the polygon is x, show that

$$x = 2R \tan \frac{\theta}{2}$$

where $\theta = 2\pi/n$ is the central angle subtended by one side.

b. If the length of one side of a regular 11-sided polygon is approximately 5.87 and R is a whole number, what is the value of R?

48. Connecting Trigonometry and Geometry A rhombus is a quadrilateral with equal sides. The diagonals of a rhombus bisect the angles of the rhombus and are perpendicular bisectors of each other. Let $\angle ABC = \theta$, d_1 = length of AC, and d_2 = length of BD.

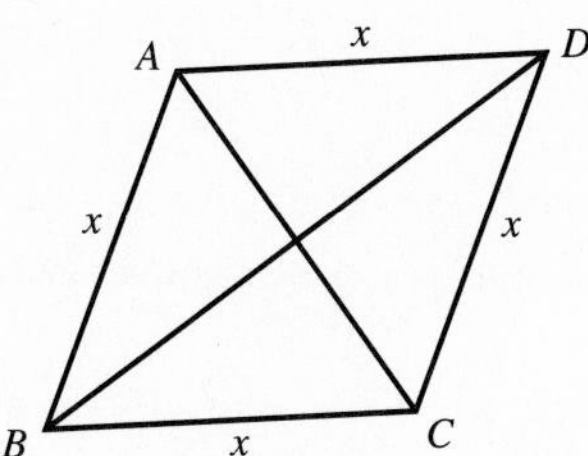

For Exercise 48

a. Show that $\cos \frac{\theta}{2} = \frac{d_2}{2x}$ and $\sin \frac{\theta}{2} = \frac{d_1}{2x}$.

b. Show that $\sin \theta = \frac{d_1 d_2}{2x^2}$.

49. *Maximizing Volume* The ends of a 10-foot-long water trough are isosceles trapezoids as shown in the figure. Find the value of θ that maximizes the volume of the trough and the maximum volume.

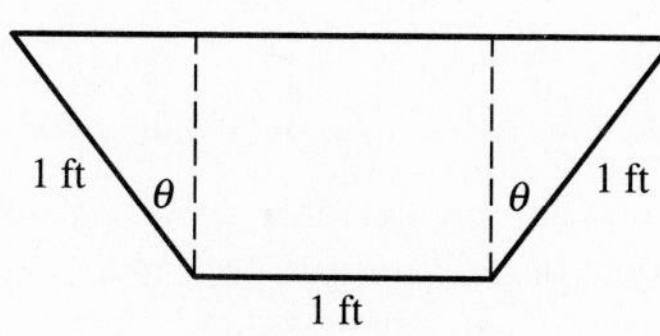

For Exercise 49

50. *Tunnel Problem* A rectangular tunnel is cut through a mountain to make a road. The upper vertices of the rectangle are on the circle $x^2 + y^2 = 400$, as illustrated in the figure.

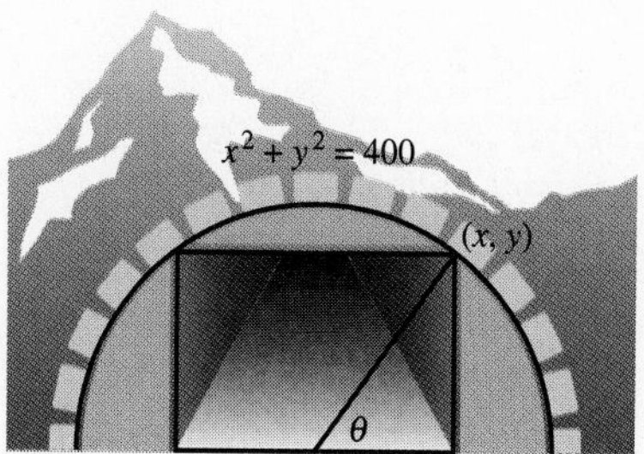

For Exercise 50

a. Show that the cross-sectional area of the end of the tunnel is $400 \sin 2\theta$.

b. Find the dimensions of the rectangular end of the tunnel that maximizes its cross-sectional area.

EXTENDING THE IDEAS

In Exercises 51–55, confirm the double-angle identities.

51. $\csc 2u = \frac{1}{2} \csc u \sec u$

52. $\cot 2u = \frac{\cot^2 u - 1}{2 \cot u}$

53. $\sec 2u = \frac{\csc^2 u}{\csc^2 u - 2}$

54. $\sec 2u = \frac{\sec^2 u}{2 - \sec^2 u}$

55. $\sec 2u = \frac{\sec^2 u \csc^2 u}{\csc^2 u - \sec^2 u}$

56. Writing to Learn Explain why

$$\sqrt{\frac{1 - \cos 2x}{2}} = |\sin x|$$

is an identity but

$$\sqrt{\frac{1 - \cos 2x}{2}} = \sin x$$

is not an identity.

CHAPTER 7 REVIEW

KEY TERMS

The number following each key term indicates the page of its introduction.

cofunction identity, 506
difference identity, 504
double-angle identity, 512
half-angle identity, 514
multiple-angle equation, 497
odd-even identity, 484
power-reducing identity, 512
Pythagorean identity, 484
quotient identity, 484
reciprocal identity, 484
sum identity, 504

REVIEW EXERCISES

In Exercises 1 and 2, write the expression as the sine, cosine, or tangent of an angle.

1. $2 \sin 100° \cos 100°$

2. $\dfrac{2 \tan 40°}{1 - \tan^2 40°}$

In Exercises 3 and 4, simplify the expression to a single term. Support your answer graphically.

3. $(1 - 2 \sin^2 \theta)^2 + 4 \sin^2 \theta \cos^2 \theta$

4. $1 - 4 \sin^2 x \cos^2 x$

In Exercises 5–22, confirm the identity. Support your work graphically.

5. $\cos 3x = 4 \cos^3 x - 3 \cos x$

6. $\cos^2 2x - \cos^2 x = \sin^2 x - \sin^2 2x$

7. $\tan^2 x - \sin^2 x = \sin^2 x \tan^2 x$

8. $2 \sin \theta \cos^3 \theta + 2 \sin^3 \theta \cos \theta = \sin 2\theta$

9. $\csc x - \cos x \cot x = \sin x$

10. $\dfrac{\tan \theta + \sin \theta}{2 \tan \theta} = \cos^2\left(\dfrac{\theta}{2}\right)$

11. $\dfrac{1 + \tan \theta}{1 - \tan \theta} + \dfrac{1 + \cot \theta}{1 - \cot \theta} = 0$

12. $\sin 3\theta = 3 \cos^2 \theta \sin \theta - \sin^3 \theta$

13. $\cos^2\left(\dfrac{t}{2}\right) = \dfrac{1 + \sec t}{2 \sec t}$

14. $\dfrac{\tan^3 \gamma - \cot^3 \gamma}{\tan^2 \gamma + \csc^2 \gamma} = \tan \gamma - \cot \gamma$

15. $\dfrac{\cos \phi}{1 - \tan \phi} + \dfrac{\sin \phi}{1 - \cot \phi} = \cos \phi + \sin \phi$

16. $\dfrac{\cos(-z)}{\sec(-z) + \tan(-z)} = 1 + \sin z$

17. $\sqrt{\dfrac{1 - \cos y}{1 + \cos y}} = \dfrac{1 - \cos y}{|\sin y|}$

18. $\sqrt{\dfrac{1 - \sin \gamma}{1 + \sin \gamma}} = \dfrac{|\cos \gamma|}{1 + \sin \gamma}$

19. $\tan\left(u + \dfrac{3\pi}{4}\right) = \dfrac{\tan u - 1}{1 + \tan u}$

20. $\dfrac{1}{4} \sin 4\gamma = \sin \gamma \cos^3 \gamma - \cos \gamma \sin^3 \gamma$

21. $\tan \dfrac{1}{2}\beta = \csc \beta - \cot \beta$

22. $\arctan t = \dfrac{1}{2} \arctan \dfrac{2t}{1 - t^2}, \quad -1 < t < 1$

In Exercises 23 and 24, use a grapher to conjecture whether the equation is likely to be an identity. Confirm your conjecture.

23. $\sec x - \sin x \tan x = \cos x$

24. $(\sin^2 \alpha - \cos^2 \alpha)(\tan^2 \alpha + 1) = \tan^2 \alpha - 1$

In Exercises 25–28, write the expression in terms of sin x and cos x only.

25. $\sin 3x + \cos 3x$

26. $\sin 2x + \cos 3x$

27. $\cos^2 2x - \sin 2x$

28. $\sin 3x - 3 \sin 2x$

In Exercises 29–38, find the general solution of the equation. Give exact answers.

29. $\sin 2x = 0.5$

30. $\cos x = \frac{\sqrt{3}}{2}$

31. $\tan x = -1$

32. $\sin x = 0.7$

33. $\cos 2x = 0.13$

34. $\cot x = 1.5$

35. $3 \sin x = 0.9$

36. $2 \cos 3x = 0.45$

37. $2 \sin^{-1} x = \sqrt{2}$

38. $\tan^{-1} x = 1$

In Exercises 39–42, solve the equation graphically.

39. $\sin^2 x - 3 \cos x = -0.5$

40. $\cos^3 x - 2 \sin x - 0.7 = 0$

41. $\sin^4 x + x^2 = 2$

42. $\sin 2x = x^3 - 5x^2 + 5x + 1$

In Exercises 43–48, find all solutions in the interval $[0, 2\pi)$. Support graphically.

43. $3 \sin 2x = 1$

44. $2 \cos x = 1$

45. $\sin 3x = \sin x$

46. $\sin^2 x - 2 \sin x - 3 = 0$

47. $\cos 2t = \cos t$

48. $\sin (\cos x) = 1$

In Exercises 49–52, solve the inequality algebraically. Support your answer graphically.

49. $2 \cos 2x > 1$ for $0 \le x < 2\pi$

50. $\sin 2x > 2 \cos x$ for $0 < x \le 2\pi$

51. $2 \cos x < 1$ for $0 \le x < 2\pi$

52. $\tan x < \sin x$ for $-\frac{\pi}{2} < x < \frac{\pi}{2}$

In Exercises 53 and 54, graph the function. Then find an equivalent equation of the form $y = a \sin [b(x - h)]$. Give exact values for a, b, and h. Support your work graphically.

53. $y = 3 \sin 3x + 4 \cos 3x$

54. $y = 5 \cos 2x - 12 \sin 2x$

In Exercises 55–58, solve the problem.

55. *Maximizing Area* A trapezoid is inscribed in the upper half of a unit circle, as shown in the figure.

a. Write the area of the trapezoid as a function of θ.

b. Find the value of θ that maximizes the area of the trapezoid and the maximum area.

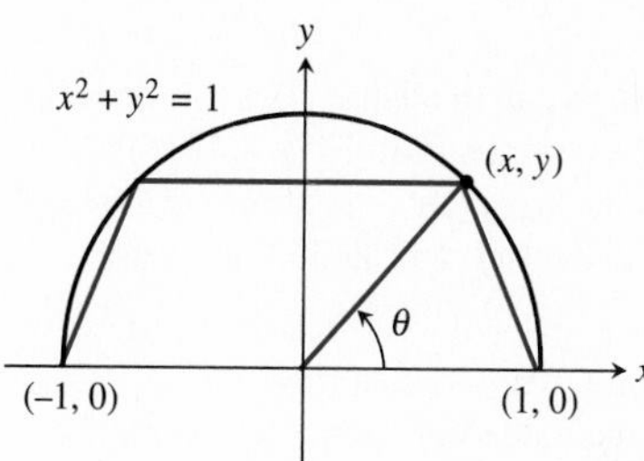

For Exercise 55

56. *Beehive Cells* A single cell in a beehive is a regular hexagonal prism open at the front with a trihedral[1] top cut at the back. It can be shown that the surface area of a cell is given by

$$S(\theta) = 6ab + \frac{3}{2}b^2\left(-\cot \theta + \frac{\sqrt{3}}{\sin \theta}\right),$$

where θ is the trihedral angle, a is the depth of the prism, and $2b$ is the length of the line segment through the center connecting opposite vertices of the hexagonal front. Assume $a = 1.75$ in. and $b = 0.65$ in.

a. Graph the function $y = S(\theta)$.

b. What value of θ gives the minimum surface area? (*Note:* This answer is quite close to the observed angle in nature.)

c. What is the minimum surface area?

57. *Cable Television Coverage* A cable broadcast satellite S orbits a planet at a height h (in miles) above the earth's sur-

[1] *Note:* trihedral refers to an angle formed by three lines each in a different plane.

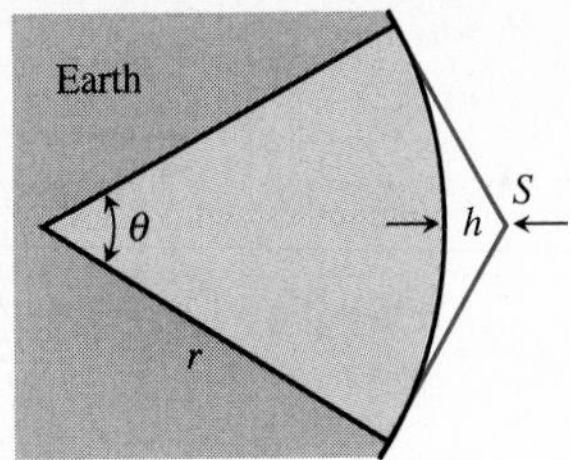

For Exercise 57

face, as shown in the figure. The two lines from S are tangent to the earth's surface. The part of the earth's surface that is in the broadcast area of the satellite is determined by the central angle θ indicated in the figure.

a. Assuming that the earth is spherical with a radius of 4000 mi, write h as a function of θ.

b. Approximate θ for a satellite 200 mi above the surface of the earth.

58. *Finding Extremum Values* The graph of

$$y = \cos x - \frac{1}{2}\cos 2x + \frac{1}{3}\cos 3x$$

is shown in the figure. The x-values that correspond to local maximum and minimum points are solutions of the equation $\sin x - \sin 2x + \sin 3x = 0$. Solve this equation algebraically, and support your solution using the graph of y.

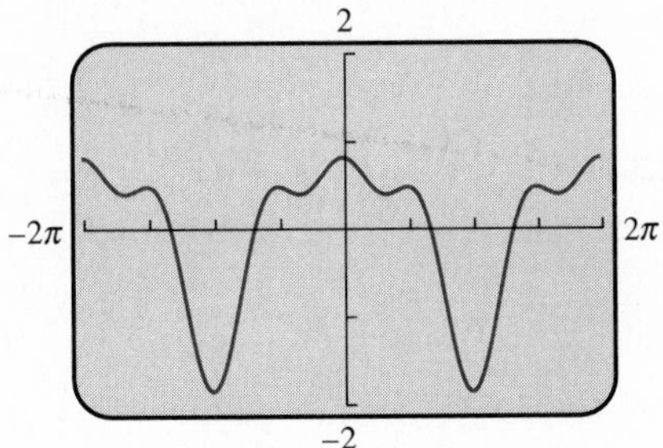

For Exercise 58

THE HINDUS/THE ARABS

The Hindus used the concept of sine of an angle. In about 500 A.D., a Hindu mathematician, Aryabhata, produced tables of half chords that correspond to the sine tables used today. The Hindus used *jya* for what is now referred to as sin. So, sin $2a$ was written *jya* $2a$ in the Hindu society. The next Hindu mathematician to reproduce Aryabhata's work was Brahmagupta in 628 A.D. Later, Bhaskara provided a detailed method for constructing the sine tables for any given angle produced by Aryabhata and Brahmagupta.

The Arabs adopted the Hindu word *jya* for sine, and called it *jiba,* which is literally meaningless. However, jiba later became *jaib* in Arab writings, which means "fold." This is an important characteristic of the sine function. One of the most well-known mathematicians was Abu'l Wafa. He extracted another important trigonometric identity from earlier works: $\sin 2a = 2 \sin a \cos a$ (one of the double-angle identities). Once translated into Latin, the work of the Hindus and the Arabs was the foundation for European mathematics during the Age of Enlightenment.

c h a p t e r 8

ADDITIONAL APPLICATIONS OF TRIGONOMETRY

BIBLIOGRAPHY

For students: *The Sky's the Limit in Math-Related Careers* (Handbook). Women's Educational Equity Publishing Center, 55 Chapel Street, Suite 200, Newton, MA 02160.

For teachers: *Handbook of Research on Math Teaching & Learning,* Douglas A. Grouws, ed. Macmillan, 1992.

Space Mathematics: A Resource for Secondary School Teachers, Bernice Kastner. USGPO, 1986. Available through Dale Seymour Publications.

8.1 LAW OF SINES

Solving Triangles • Law of Sines • Using the Law of Sines • Determining the Number of Triangles and the Ambiguous Case • Applications

Objective

Students will be able to understand the proof of the law of sines and use the computational applications of the law of sines to solve a variety of problems.

Motivate

Ask students to recall the SSS, ASA, and SAS postulates from geometry and have them discuss why there is no SSA postulate.

Solving Triangles

In this section and the next we have two goals. One is to solve a triangle. (Recall that *solving a triangle* means to find one or more of its unknown parts.) The other is to determine how many (if any) triangles exist when three of its six parts are specified. For example, there is *no* right triangle with one side 8 units and a hypotenuse of 5 units. We know this because the measure of the hypotenuse must be greater than the measure of either of the other two sides.

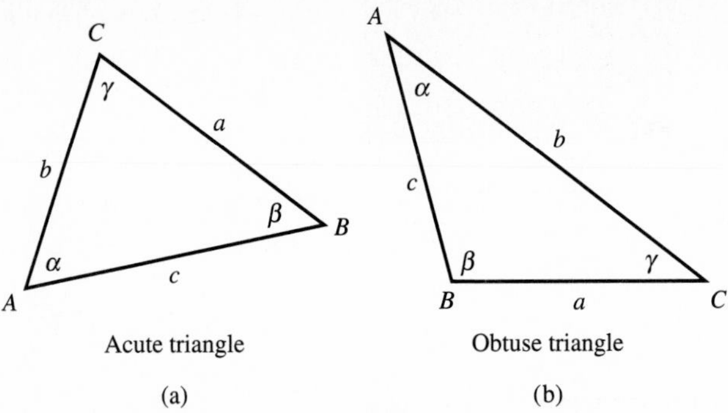

Figure 8.1 In this chapter we will label the vertices of a triangle *A, B, C,* the angles at these vertices α, β, γ, and the sides opposite these angles *a, b, c.*

REMARK

An angle between 90° and 180° is an **obtuse angle.**

We will use the *law of sines* (in this section) and the *law of cosines* (Section 8.2) to solve triangles that are **acute** (all angles less than 90°), and triangles that are **obtuse** (one angle greater than 90°). See Figure 8.1.

Law of Sines

The **law of sines** states that the ratio of the sine of an angle to the side opposite the angle is the same for all three angles of a triangle.

Teaching Note

In discussing the proof of the law of sines, you may wish to label Figure 8.2 with the reference angle *CAD* as α'. This labeling should help students recognize that right triangle $\triangle ACD$ determines $\sin \alpha' = \frac{h}{b}$ and therefore $\sin \alpha = \sin(\pi - \alpha') = \sin(\alpha') = \frac{h}{b}$. Similarly, focus students' attention on right triangle $\triangle BCD$ to determine the ratio for $\sin \beta$.

Law of Sines

In any triangle *ABC* labeled as shown in Figure 8.1,

$$\frac{\sin \alpha}{a} = \frac{\sin \beta}{b} = \frac{\sin \gamma}{c}.$$

Proof. We assume that $\angle \alpha$ and, therefore, $\triangle ABC$ is obtuse. (See Exercise 46 for the other two possibilities.) To confirm the relationship involving α and β, we consider the triangle with *A* at the origin and *B* on the positive *x*-axis. (See Figure 8.2.)

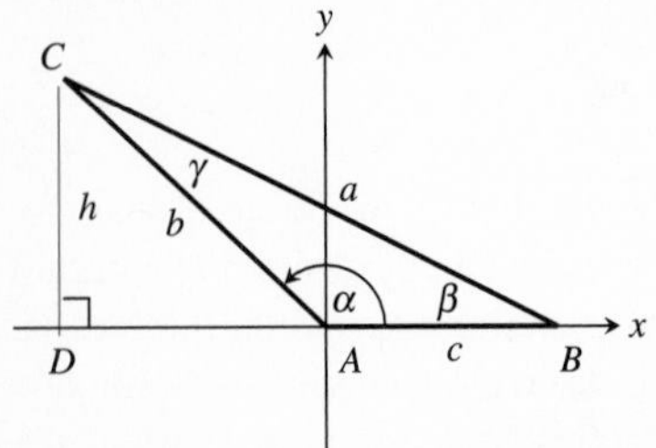

Figure 8.2

$$\sin \alpha = \frac{h}{b} \qquad \sin \beta = \frac{h}{a} \qquad \text{Definitions of sine.}$$

$$h = b \sin \alpha \qquad h = a \sin \beta \qquad \text{Multiply by denominators.}$$

Now we combine these two equations:

$$b \sin \alpha = a \sin \beta$$

$$\frac{\sin \alpha}{a} = \frac{\sin \beta}{b} \qquad \text{Divide by } ab.$$

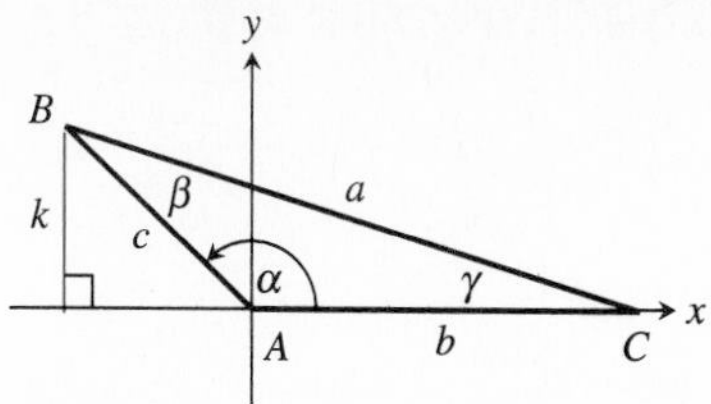

Figure 8.3

To confirm the relationship involving α and γ, we consider the triangle with A at the origin and C on the positive x-axis. (See Figure 8.3.) Equations similar to those above yield

$$c \sin \alpha = a \sin \gamma$$

$$\frac{\sin \alpha}{a} = \frac{\sin \gamma}{c} \quad \text{Divide by } ac.$$

Combining these results gives us the law of sines:

$$\frac{\sin \alpha}{a} = \frac{\sin \beta}{b} = \frac{\sin \gamma}{c}.$$

Notes on Examples

Before discussing Example 1, it is worth discussing ways to form a triangle given any three of its six parts. You may wish to refer to Figure 8.7 on page 528 or use manipulatives or software. Students should recognize that three parts may uniquely determine a triangle, but sometimes zero, two, or infinitely many triangles are possible. This exploration should speed the development of this section.

Using the Law of Sines

The law of sines allows us to solve triangles when one side and two angles (and hence the third angle) are known. Of course, the sum of the two angles must be less than 180°.

■ EXAMPLE 1 Solving a Triangle Given One Side and Two Angles

Solve $\triangle ABC$ given that $a = 8$, $\alpha = 36°$, and $\beta = 48°$.

Solution

We must find b, c, and γ. (See Figure 8.4.)

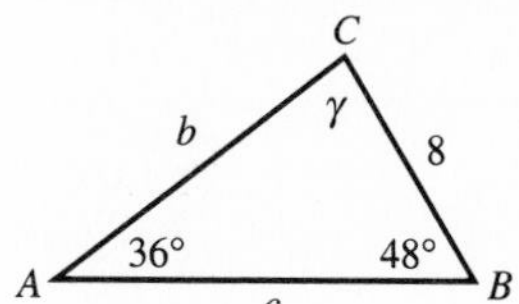

Figure 8.4

Solve Algebraically

$$\gamma = 180° - (36° + 48°) \quad \alpha + \beta + \gamma = 180°.$$
$$= 96°$$

$$\frac{\sin \alpha}{a} = \frac{\sin \beta}{b} \qquad \frac{\sin \alpha}{a} = \frac{\sin \gamma}{c} \quad \text{Law of sines}$$

$$\frac{\sin 36°}{8} = \frac{\sin 48°}{b} \qquad \frac{\sin 36°}{8} = \frac{\sin 96°}{c} \quad \alpha = 36°, \beta = 48°, \gamma = 96°, a = 8.$$

$$b = \frac{8 \sin 48°}{\sin 36°} \qquad c = \frac{8 \sin 96°}{\sin 36°}$$

$$= 10.114 \cdots \qquad = 13.535 \cdots$$

Interpret

In $\triangle ABC$, $b \approx 10.11$, $c \approx 13.54$, and $\gamma = 96°$. ■

The law of sines allows us to solve triangles when two sides and one of the angles *not* formed by the two sides are known.

Figure 8.5

■ EXAMPLE 2 Solving a Triangle Given Two Sides and One Angle

Solve $\triangle ABC$ given that $a = 7$, $b = 6$, and $\alpha = 26.3°$.

Solution

We must find c, β, and γ. (See Figure 8.5.)

Solve Algebraically

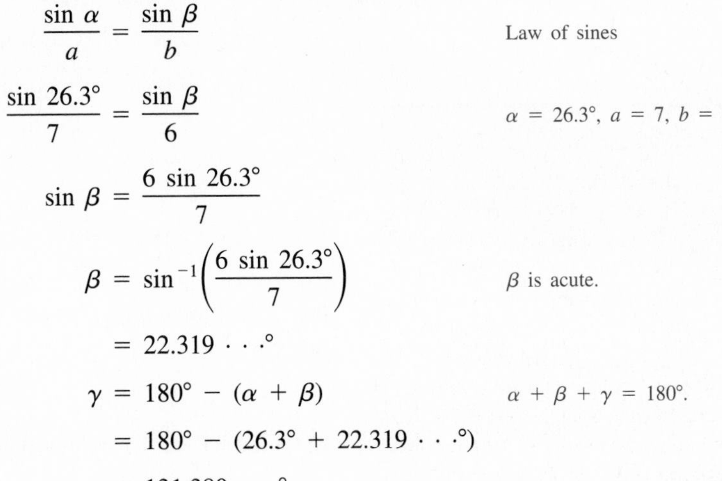

$$\frac{\sin\alpha}{a} = \frac{\sin\beta}{b} \qquad \text{Law of sines}$$

$$\frac{\sin 26.3°}{7} = \frac{\sin\beta}{6} \qquad \alpha = 26.3°,\ a = 7,\ b = 6.$$

$$\sin\beta = \frac{6\sin 26.3°}{7}$$

$$\beta = \sin^{-1}\left(\frac{6\sin 26.3°}{7}\right) \qquad \beta \text{ is acute.}$$

$$= 22.319\ldots°$$

$$\gamma = 180° - (\alpha + \beta) \qquad \alpha + \beta + \gamma = 180°.$$

$$= 180° - (26.3° + 22.319\ldots°)$$

$$= 131.380\ldots°$$

Finally, we find c:

$$\frac{\sin\alpha}{a} = \frac{\sin\gamma}{c} \qquad \text{Law of sines}$$

$$\frac{\sin 26.3°}{7} = \frac{\sin 131.380\ldots°}{c}$$

$$c = \frac{7\sin 131.380\ldots°}{\sin 26.3°}$$

$$= 11.854\ldots$$

REMARK

Recall that the range of the function $\sin^{-1} x$ is the interval $[-90°, 90°]$.

Interpret

In $\triangle ABC$, $c \approx 11.85$, $\beta \approx 22.3°$, and $\gamma \approx 131.4°$. ■

In Example 3, we are also given two sides and one angle. However, this time there are two possible triangles.

■ EXAMPLE 3 Solving a Triangle Given Two Sides and One Angle

Solve $\triangle ABC$ given that $a = 6$, $b = 7$, and $\alpha = 30°$.

Solution

Figure 8.6 shows that there are two possible triangles: in (a) β is acute and in (b) β is obtuse.

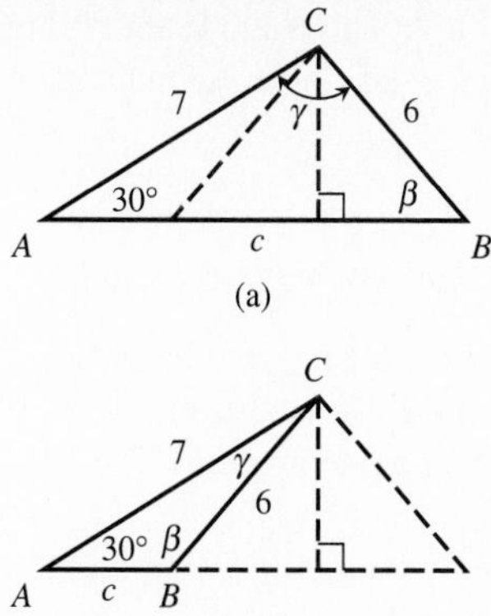

Figure 8.6

Solve Algebraically

Solve $\triangle ABC$ with β Acute:

$$\frac{\sin\alpha}{a} = \frac{\sin\beta}{b} \qquad \text{Law of sines}$$

$$\frac{\sin 30^\circ}{6} = \frac{\sin\beta}{7}$$

$$\sin\beta = \frac{7\sin 30^\circ}{6}$$

$$\beta = \sin^{-1}\left(\frac{7\sin 30^\circ}{6}\right) \qquad 0^\circ < \beta < 90^\circ$$

$$= 35.685\ldots^\circ$$

$$\gamma = 180^\circ - (\alpha + \beta)$$

$$= 114.314\ldots^\circ$$

$$\frac{\sin\alpha}{a} = \frac{\sin\gamma}{c} \qquad \text{Law of sines}$$

$$\frac{\sin 30^\circ}{6} = \frac{\sin 114.314\ldots^\circ}{c}$$

$$c = \frac{6\sin 114.314\ldots^\circ}{\sin 30^\circ}$$

$$= 10.935\ldots$$

Teaching Note

The law of sines and the law of cosines (Section 8.2) provide an opportunity for students with graphers to write programs for solving the various cases of acute and obtuse triangles. A program just to run the ambiguous case of SSA offers an excellent analytical challenge for students.

Solve $\triangle ABC$ with β obtuse: Equations similar to those above yield the following:

$$\sin\beta = \frac{7\sin 30^\circ}{6}$$

$$\beta = 180^\circ - \sin^{-1}\left(\frac{7\sin 30^\circ}{6}\right) \qquad 90^\circ < \beta < 180^\circ.$$

$$= 144.314\ldots^\circ$$

$$\gamma = 180^\circ - (\alpha + \beta)$$

$$= 5.685\ldots^\circ$$

$$c = \frac{a\sin\gamma}{\sin\alpha} = \frac{6\sin 5.685\ldots^\circ}{\sin 30^\circ}$$

$$= 1.188\ldots$$

Interpret

In $\triangle ABC$, $c \approx 10.94$, $\beta \approx 35.69^\circ$, and $\gamma \approx 114.31^\circ$ when $0^\circ < \beta < 90^\circ$. Similarly, $c \approx 1.19$, $\beta \approx 144.31^\circ$, and $\gamma \approx 5.69^\circ$ when $90^\circ < \beta < 180^\circ$. ■

Notice it is not possible in Example 2 to draw a second triangle satisfying the given conditions with $90° < \beta < 180°$ as we do in Example 3. One reason is that a is greater than b.

Teaching Note

Before discussing the ambiguous case, it may be useful to investigate numerically what would happen in Example 3 if side a's length was 2 or 3.5, instead of 6.

Determining the Number of Triangles and the Ambiguous Case

The case of two sides and an angle opposite one of the sides given is **ambiguous** because the number of triangles may be 0, 1, or 2. To analyze these possibilities suppose that the given parts are a, b, and an acute angle α. The situation when α is obtuse is similar. See Exercise 47.

Figure 8.7 shows how these possibilities arise by comparing a, b, and the height $h = b \sin \alpha$ of the triangle. If $a < h$, no triangle is formed. If $a = h$, a unique right triangle is formed. If $h < a < b$, two triangles are possible (see Example 3). And if $a \geq b$, there is a unique triangle (see Example 2).

Notes on Examples

In Example 4b students may need some help in making the connection between the relationship $h = b \sin \alpha$ as the altitude of the triangles in Figure 8.7 and 6 sin 30° as its value for the specific cases in Example 4. Help students to visualize geometrically and analytically how the height can be used to identify the four possible cases.

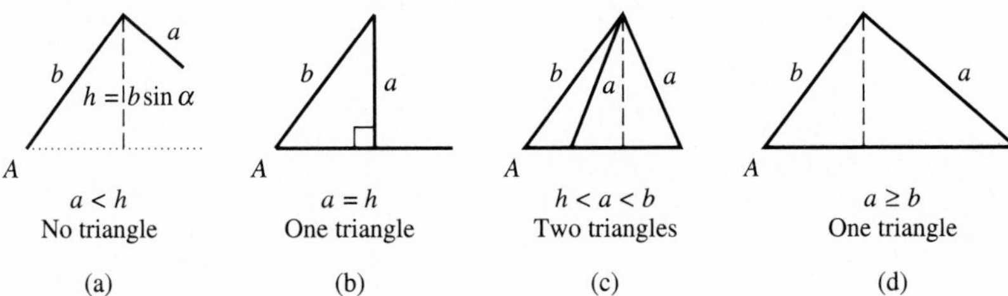

Figure 8.7

■ EXAMPLE 4 Finding the Number of Triangles

Find how many triangles are formed when $\alpha = 30°$, $b = 6$, and the value of a is as given. Then solve the triangle(s). (See Figure 8.8.)

a. $a = 2$ **b.** $a = 3$

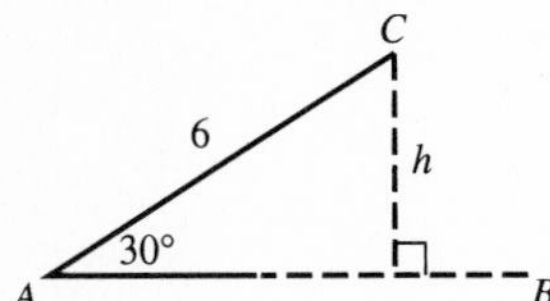

Figure 8.8

Solution

The altitude from C is $h = b \sin \alpha = 6 \sin 30° = 3$.

a. Because $a = 2 < h = 3$, there are no triangles.

b. Because $a = 3 = h$, there is a unique right triangle. Thus, $\beta = 90°$, $\gamma = 60°$, and $c = 3\sqrt{3}$. ■

We can summarize using notation you may recall from geometry. Let A indicate that an angle of a triangle is given, and S that a side is given. For example, ASA means that, as you go around the perimeter of a triangle, you are given an angle, an adjacent side, and the angle that follows. Similarly, the abbreviation AAS means that an angle, the next angle, and the side that follows are given. Be sure you understand why there are only six possibilities: AAA, ASA, SAA, SSA, SAS, and SSS. See Exercise 51.

In Exercise 48 you will show there are infinitely many triangles with AAA given, provided, of course, that the sum of their angles is 180°. Example 1 illustrates that a unique triangle is formed with SAA given, provided the sum of the two given angles is less than 180°. ASA is similar to SAA. Examples 2–4 show that no, one, or two triangles are possible with SAA given.

In Section 8.2 we will see that if there is a triangle with SAS or SSS given, then there is only one such triangle.

Follow-up

Have students show that the law of sines can be applied to a right triangle to obtain the equation

$$\sin \theta = \frac{\text{opposite}}{\text{hypotenuse}}.$$

Assignment Guide

Day 1: Ex. 3–36, multiples of 3
Day 2: Ex. 37, 38, 40, 42, 43, 45–49

Cooperative Learning

Group Activity: Ex. 51–52

Notes on Exercises

Ex. 1–22 and 27–36 allow students to practice solving acute and obtuse triangles before they are asked to solve application problems.
Ex. 23–24 provide a good perspective on applying the ambiguous case of SSA.
Ex. 42–44 may be challenging for students because no illustration is provided. Encourage students to make drawings.

Applications

Many problems involving angles and distances can be solved by superimposing a triangle onto the situation and solving the triangle.

EXAMPLE 5 APPLICATION: Locating a Fire

Forest Ranger LaToya Shoulders at ranger station *A* sights a fire in the direction 32° east of north. Ranger Rick Thorpe at ranger station *B*, 10 mi due east of *A*, sights the same fire 48° west of north. Find the distance from each ranger station to the fire.

Solution

Model

Let *C* represent the location of the fire. (See Figure 8.9.) Because the fire is 32° east of north from *A*, $\alpha = 58°$. Because the fire is 48° west of north from *B*, $\beta = 42°$. We need to find sides *a* and *b* in $\triangle ABC$.

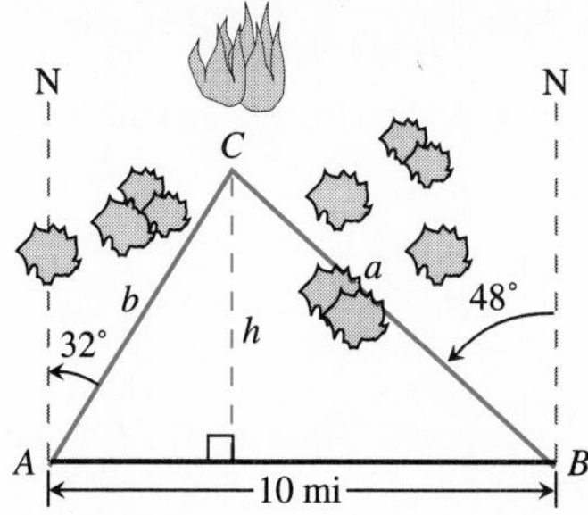

Figure 8.9

Solve Algebraically

$$\gamma = 180° - (\alpha + \beta) = 180° - (58° + 42°) = 80°$$

$$\frac{\sin \alpha}{a} = \frac{\sin \gamma}{c} \qquad \frac{\sin \beta}{b} = \frac{\sin \gamma}{c} \qquad \text{Law of sines}$$

$$\frac{\sin 58°}{a} = \frac{\sin 80°}{10} \qquad \frac{\sin 42°}{b} = \frac{\sin 80°}{10} \qquad \text{Law of sines}$$

$$a = \frac{10 \sin 58°}{\sin 80°} \qquad b = \frac{10 \sin 42°}{\sin 80°}$$

$$= 8.611 \cdots \qquad = 6.794 \cdots$$

Interpret

The fire is about 6.79 mi from ranger station *A* and about 8.61 mi from ranger station *B*. ■

Ongoing Assessment

Self-Assessment: Ex. 5, 11, 17, 19, 29, 39
Embedded Assessment: Ex. 48

EXAMPLE 6 APPLICATION: Finding the Height of a Pole

A road slopes 10° above the horizontal, and a vertical telephone pole stands beside the road. The angle of elevation of the sun is 62°, and the telephone pole casts a 14.5-ft shadow downhill along the road. Find the height of the telephone pole.

Solution

Model

Let AB represent the shadow and a the height of the telephone pole. (See Figure 8.10.) We need to find side a in $\triangle ABC$.

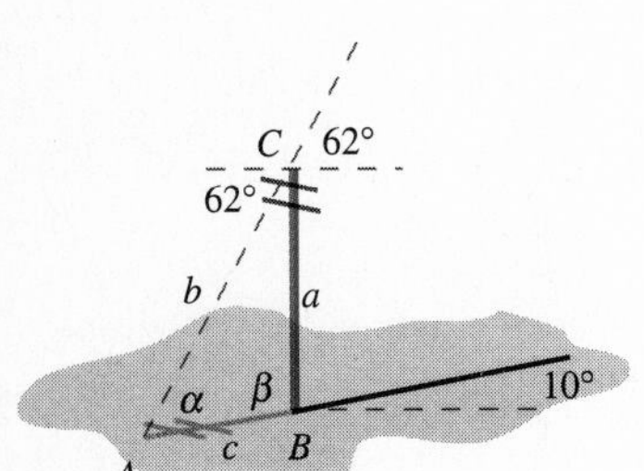

Figure 8.10

Solve Algebraically

Because the roads tilts and AB is not horizontal, we show the angle of elevation with the horizontal at C. From this, we find that $\gamma = 90° - 62° = 28°$. Because the angle of the road with the horizontal is 10°, $\beta = 90° + 10° = 100°$.

$$c = 14.5 \qquad c = \text{length of shadow}$$

$$\alpha = 180° - (\beta + \gamma)$$

$$= 180° - (100° + 28°) = 52°$$

$$\frac{\sin \alpha}{a} = \frac{\sin \gamma}{c} \qquad \text{Law of sines}$$

$$\frac{\sin 52°}{a} = \frac{\sin 28°}{14.5}$$

$$a = \frac{14.5 \sin 52°}{\sin 28°} = 24.338 \cdots$$

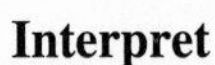

Interpret

The pole is approximately 24.3 ft high. ■

Quick Review 8.1

In Exercises 1–4, solve the equation $\frac{a}{b} = \frac{c}{d}$ for the given variable.

1. a

2. b

3. c

4. d

In Exercises 5 and 6, evaluate the expression.

5. $\frac{7 \sin 48°}{\sin 23°}$

6. $\frac{9 \sin 121°}{\sin 14°}$

In Exercises 7–10, solve for the angle x.

7. $\sin x = 0.3,\quad 0° < x < 90°$

8. $\sin x = 0.3,\quad 90° < x < 180°$

9. $\sin x = -0.7,\quad 180° < x < 270°$

10. $\sin x = -0.7,\quad 270° < x < 360°$

SECTION EXERCISES 8.1

In Exercises 1–4, solve the triangle.

1.

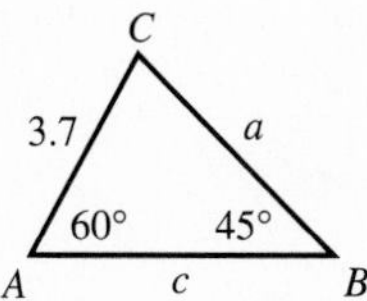

2.

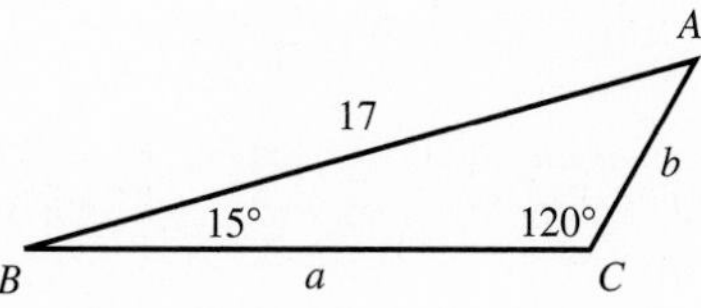

3.

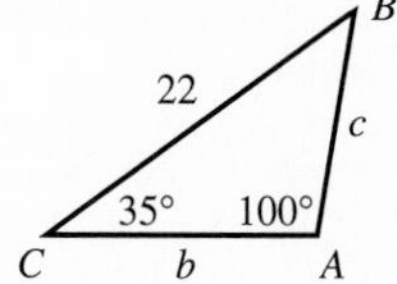

4.

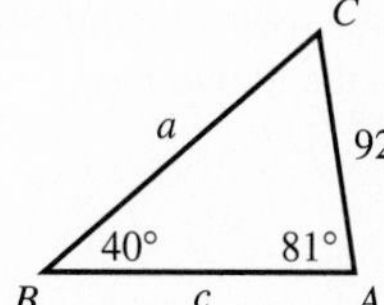

In Exercises 5–8, solve the triangle.

5. $\alpha = 40°$, $\beta = 30°$, $b = 10$

6. $\alpha = 50°$, $\beta = 62°$, $a = 4$

7. $\alpha = 33°$, $\beta = 70°$, $b = 7$

8. $\beta = 16°$, $\gamma = 103°$, $c = 12$

In Exercises 9–12, solve the triangle.

9. $\alpha = 32°$, $a = 17$, $b = 11$

10. $\alpha = 49°$, $a = 32$, $b = 28$

11. $\beta = 70°$, $b = 14$, $c = 9$

12. $\gamma = 103°$, $b = 46$, $c = 61$

In Exercises 13–18, state whether the given measurements determine zero, one, or two triangles.

13. $\alpha = 36°$, $a = 2$, $b = 7$

14. $\beta = 82°$, $b = 17$, $c = 15$

15. $\gamma = 36°$, $a = 17$, $c = 16$

16. $\alpha = 73°$, $a = 24$, $b = 28$

17. $\gamma = 30°$, $a = 18$, $c = 9$

18. $\beta = 88°$, $b = 14$, $c = 62$

In Exercises 19–22, two triangles can be formed using the given measurements. Solve both triangles.

19. $\alpha = 64°$, $a = 16$, $b = 17$

20. $\beta = 38°$, $b = 21$, $c = 25$

21. $\gamma = 68°$, $a = 19$, $c = 18$

22. $\beta = 57°$, $a = 11$, $b = 10$

In Exercises 23 and 24, solve the problem.

23. If $a = 10$ and $\beta = 42°$, determine the values of b that will produce the given number of triangles.

a. Two **b.** One **c.** Zero

24. If $b = 12$ and $\gamma = 53°$, determine the values of c that will produce the given number of triangles.

a. Two **b.** One **c.** Zero

In Exercises 25 and 26, decide whether the triangle can be solved using the law of sines. If so, solve it. If not, explain why not.

25.

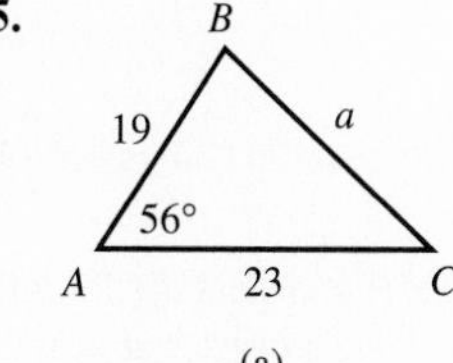

(a)

B
19
a
56°
A
b
C

(b)

26.

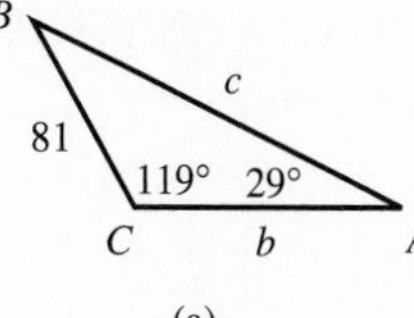

(a)

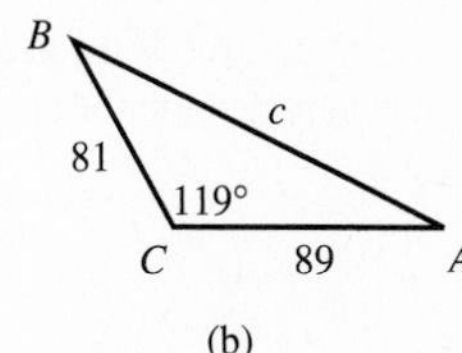

(b)

In Exercises 27–36, respond in one of the following ways:

a. State, "Cannot be solved with law of sines."
b. State, "No triangle is formed."
c. Solve the triangle(s).

27. $\alpha = 61°$, $a = 8$, $b = 21$

28. $\beta = 47°$, $a = 8$, $b = 21$

29. $\alpha = 136°$, $a = 15$, $b = 28$

30. $\gamma = 115°$, $b = 12$, $c = 7$

31. $\beta = 42°$, $c = 18$, $\gamma = 39°$

32. $\alpha = 19°$, $b = 22$, $\beta = 47°$

33. $\gamma = 75°$, $b = 49$, $c = 48$

34. $\alpha = 54°$, $a = 13$, $b = 15$

35. $\beta = 31°$, $a = 8$, $c = 11$

36. $\gamma = 65°$, $a = 19$, $b = 22$

In Exercises 37–52, solve the problem.

37. *Surveying a Canyon* Two markers A and B on the same side of a canyon rim are 56 ft apart. A third marker C, located across the rim, is positioned so that $\angle BAC = 72°$ and $\angle ABC = 53°$.

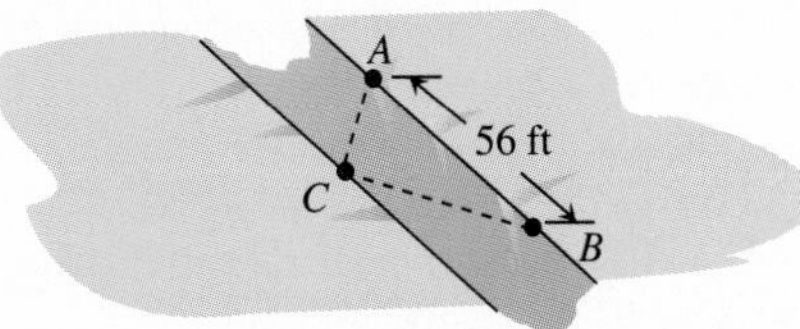

For Exercise 37

a. Find the distance between C and A.
b. Find the distance between the two canyon rims. (Assume they are parallel.)

38. *Weather Forecasting* Two weather forecasters are 25 mi apart located on an east-west road. The forecaster at point A sights a tornado 38° east of north. The forecaster at point B sights the same tornado at 53° west of north. Find the distance from each forecaster to the tornado. Also find the distance between the tornado and the road.

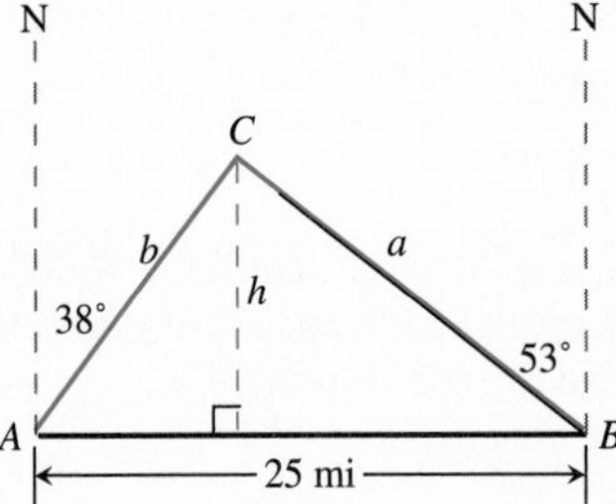

For Exercise 38

39. *Engineering Design* A vertical flagpole stands beside a road that slopes at an angle of 15° with the horizontal. When the angle of elevation of the sun is 62°, the flagpole casts a 16-ft shadow downhill along the road. Find the height of the flagpole.

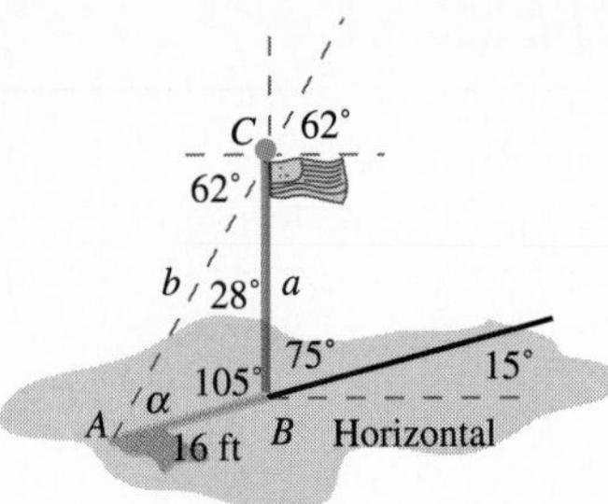

For Exercise 39

40. *Altitude* Observers 2.32 mi apart see a hot-air balloon directly between them but at the angles of elevation shown in the figure. Find the altitude of the balloon.

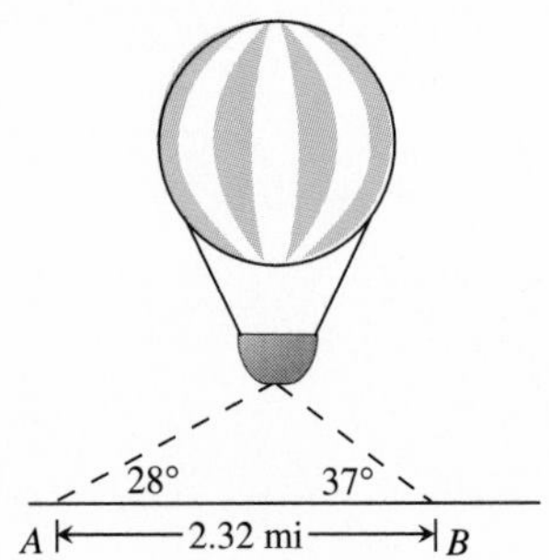

For Exercise 40

41. *Reducing Air Resistance* A 4-ft airfoil attached to the cab of a truck reduces wind resistance. If the angle between the airfoil and the cab top is 18° and angle β is 10°, find the length of a vertical brace positioned as shown in the figure.

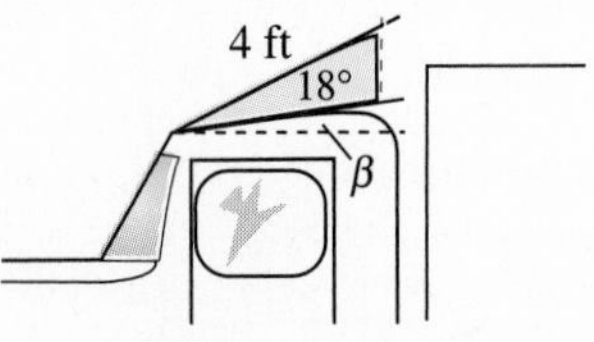

For Exercise 41

42. *Ferris Wheel Design* A Ferris wheel has 16 evenly spaced cars. The distance between adjacent chairs is 15.5 ft. Find the radius of the wheel (to the nearest 0.1 ft.)

43. *Finding Height* Two observers are 600 ft apart on opposite sides of a flagpole. The angles of elevation from the observers to the top of the pole are 19° and 21°. Find the height of the flagpole.

44. *Finding Height* Two observers are 400 ft apart on opposite sides of a tree. The angles of elevation from the observers to the top of the tree are 15° and 20°. Find the height of the tree.

45. *Finding Distance* Two lighthouses A and B are known to be exactly 20 mi apart on a north-south line. A ship's captain at S measures $\angle ASB$ to be 33°. A radio operator at B measures $\angle ABS$ to be 52°. Find the distance from the ship to each lighthouse.

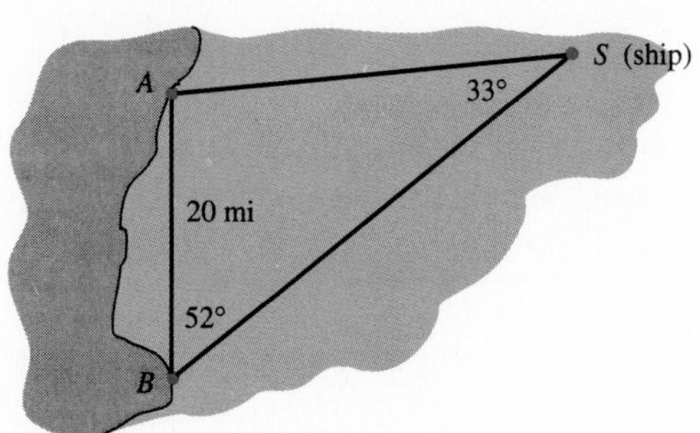

For Exercise 45

46. *Law of Sines* Prove the law of sines if the angle α in Figure 8.2 is

a. a right angle. **b.** an acute angle.

47. *Ambiguous Case, SSA* In $\triangle ABC$, the obtuse angle α and the side b are given. Describe the possible triangles that result when a is also given. (*Hint:* See Figure 8.7.)

48. Writing to Learn

a. Show that there are infinitely many triangles with AAA given if the sum of the three positive angles is 180°.

b. Give three examples of triangles where $\alpha = 30°$, $\beta = 60°$, and $\gamma = 90°$.

c. Give three examples where $\alpha = \beta = \gamma = 60°$.

EXTENDING THE IDEAS

49. Solve this triangle assuming that $\angle\beta$ is obtuse. (*Hint:* Draw a perpendicular from A to the line through B and C.)

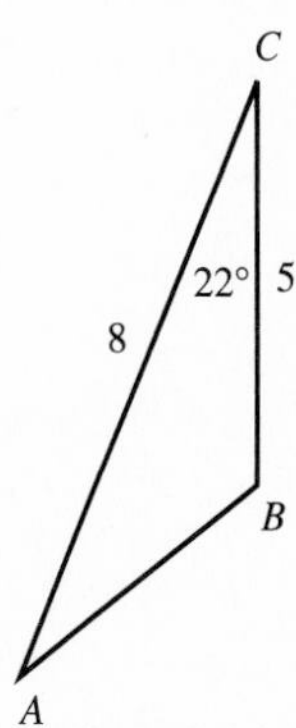

For Exercise 49

50. *Pilot Calculations* Towers A and B are known to be 4.1 mi apart on level ground. A pilot measures the angles of depression to the towers to be 36.5° and 25°, respectively, as shown in the figure. Find distances AC and BC and the height of the airplane.

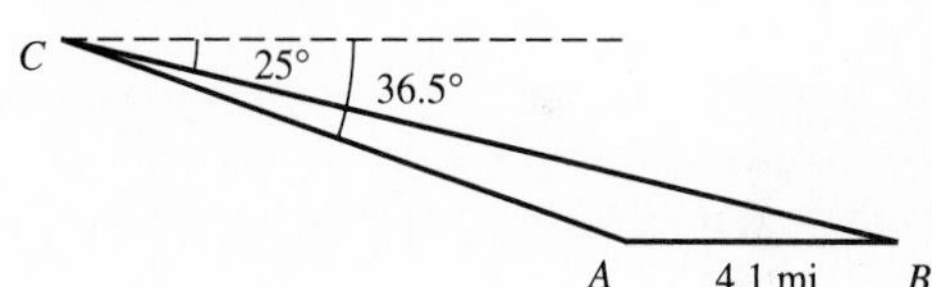

For Exercise 50

51. Writing to Learn Give a convincing argument why the following situations are the same using the notation introduced in this section.

a. AAS and SAA

b. ASS and SSA

Using (a) and (b) we can conclude the following.

c. There are only six different situations to consider when three of the six parts of a triangle are given.

52. *Law of Sines*

a. Prove that the area of $\triangle ABC$ is given by $0.5ac \sin\beta$, $0.5ab \sin\gamma$, or $0.5bc \sin\alpha$. (*Hint:* Use Figures 8.1 and 8.2 for the first two forms.)

b. Use the three area formulas in part a to give another proof of the law of sines.

Objective

Students will be able to apply the law of cosines to solve acute and obtuse triangles and to determine the area of a triangle in terms of the measures of its sides and angles.

Motivate

Have students try to construct a triangle with sides of length 2, 3, and 6. Demonstrate that in any triangle the sum of the lengths of any two sides must exceed the length of the third side.

8.2 LAW OF COSINES

Introduction • Law of Cosines • Using the Law of Cosines • Using the Law of Cosines in Geometry • Area of a Triangle and Heron's Formula • Applications

Introduction

In Section 8.1 we used the law of sines to solve triangles and to determine the number of triangles that exist with ASA, SAA, SSA, or AAA given. In this section we do the same with SAS and SSS using the law of cosines. We also revisit the case SSA.

Law of Cosines

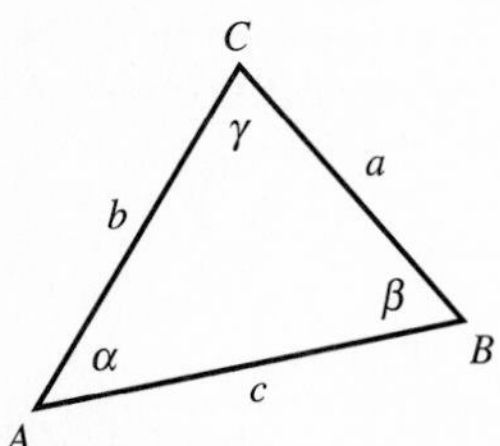

Figure 8.11 An acute $\triangle ABC$. The law of cosines is also true if the triangle is right or obtuse.

Law of Cosines

Let $\triangle ABC$ be labeled in the usual way (see Figure 8.11). Then

$$a^2 = b^2 + c^2 - 2bc\cos\alpha,$$
$$b^2 = a^2 + c^2 - 2ac\cos\beta,$$
$$c^2 = a^2 + b^2 - 2ab\cos\gamma.$$

Proof. To confirm

$$a^2 = b^2 + c^2 - 2bc\cos\alpha,$$

place $\triangle ABC$ in the coordinate plane with α in standard position and B on the positive x-axis. (See Figure 8.12.) Because $\cos\alpha = d/b$ and $\sin\alpha = h/b$, we have $d = b\cos\alpha$ and $h = b\sin\alpha$.

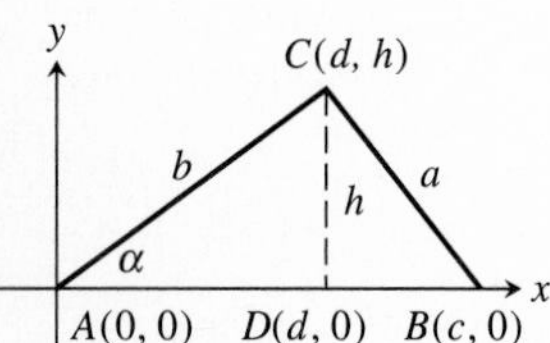

Figure 8.12

$$\begin{aligned} a^2 &= |c - d|^2 + h^2 && \text{Distance formula} \\ &= (c - b\cos\alpha)^2 + (b\sin\alpha)^2 && d = b\cos\alpha,\ h = b\sin\alpha \\ &= c^2 - 2bc\cos\alpha + b^2\cos^2\alpha + b^2\sin^2\alpha \\ &= c^2 - 2bc\cos\alpha + b^2(\cos^2\alpha + \sin^2\alpha) \\ &= b^2 + c^2 - 2bc\cos\alpha && \cos^2\alpha + \sin^2\alpha = 1 \end{aligned}$$

The other two forms of the law of cosines can be found by positioning $\triangle ABC$ so that β and then γ are in standard position.

Using the Law of Cosines

The law of cosines allows us to solve triangles when two sides and the included angle are known (SAS), provided that the angle is less than 180°. There is a unique triangle in this case.

■ **EXAMPLE 1** **Solving a Triangle Given Two Sides and the Included Angle (SAS)**

Solve $\triangle ABC$ given that $a = 11$, $b = 5$, and $\gamma = 20°$.

Solution

We must find c, α, and β. (See Figure 8.13.)

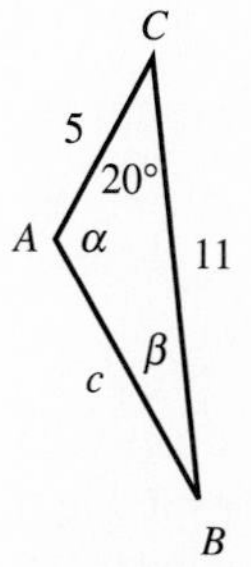

C, 5, 20°, A, α, 11, β, c, B

Figure 8.13

Solve Algebraically

$$c^2 = a^2 + b^2 - 2ab\cos\gamma \qquad \text{Law of cosines}$$

$$= 11^2 + 5^2 - 2(11)(5)\cos 20°$$

$$c = \sqrt{11^2 + 5^2 - 2(11)(5)\cos 20°}$$

$$= 6.529\cdots$$

$$a^2 = b^2 + c^2 - 2bc\cos\alpha \qquad \text{Law of cosines}$$

$$11^2 = 5^2 + (6.529\cdots)^2 - 2(5)(6.529\cdots)\cos\alpha$$

$$\cos\alpha = \frac{5^2 + (6.529\cdots)^2 - 11^2}{2(5)(6.529\cdots)}$$

$$= -0.817\cdots$$

$$\alpha = 144.816\cdots°$$

$$\beta = 180° - (\alpha + \gamma)$$

$$= 15.183\cdots°$$

NOTE

You could also use the law of sines to determine α or β after c is found. However, the advantage of using the law of cosines is that it distinguishes between acute and obtuse angles.

Interpret

In $\triangle ABC$, $c \approx 6.53$, $\alpha \approx 144.82°$, and $\beta \approx 15.18°$. ■

The law of cosines also allows us to solve triangles when three sides are known, provided that the sum of any two of the numbers is greater than the third. In this case there is exactly one such triangle.

■ **EXAMPLE 2 Solving a Triangle Given Three Sides (SSS)**

Solve $\triangle ABC$ if $a = 9$, $b = 7$, and $c = 5$.

Solution

We must find α, β, and γ. (See Figure 8.14.) Using the law of cosines, we find the following:

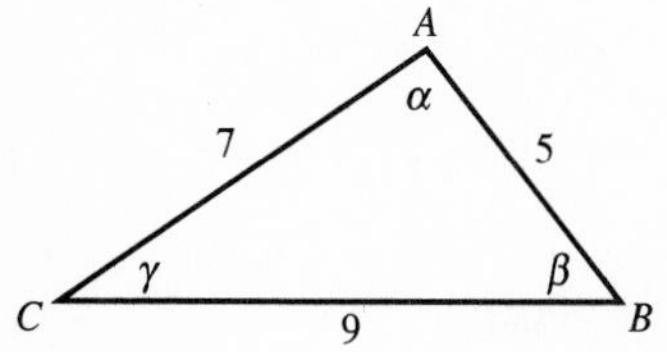

$$a^2 = b^2 + c^2 - 2bc\cos\alpha \qquad b^2 = a^2 + c^2 - 2ac\cos\beta$$

$$9^2 = 7^2 + 5^2 - 2(7)(5)\cos\alpha \qquad 7^2 = 9^2 + 5^2 - 2(9)(5)\cos\beta$$

$$70\cos\alpha = -7 \qquad 90\cos\beta = 57$$

$$\cos\alpha = -0.1 \qquad \cos\beta = 0.633\cdots$$

$$\alpha = 95.739\cdots^\circ \qquad \beta = 50.703\cdots^\circ$$

$$\gamma = 180^\circ - (\alpha + \beta)$$

$$= 33.557\cdots^\circ$$

Figure 8.14

Interpret

In $\triangle ABC$, $\alpha \approx 95.74^\circ$, $\beta \approx 50.70^\circ$, and $\gamma \approx 33.56^\circ$. ■

The law of cosines offers an alternate approach to the ambiguous SSA case studied in Section 8.1. Recall there are zero, one, or two possible triangles with given SSA. Finding the third side using the law of cosines involves solving a quadratic equation. The number of possible triangles equals the number of positive solutions of the quadratic equation. Example 3 uses this approach.

■ **EXAMPLE 3 Solving a Triangle Given Two Sides and One Angle (SSA)**

Solve $\triangle ABC$ given that $a = 6$, $b = 7$, and $\alpha = 30^\circ$.

Solution

We must find c, β, and γ.

Solve Algebraically

$$a^2 = b^2 + c^2 - 2bc\cos\alpha \qquad \text{Law of cosines}$$

$$6^2 = 7^2 + c^2 - 2(7)c\cos 30^\circ$$

$$0 = c^2 - 7\sqrt{3}c + 13 \qquad \cos 30^\circ = \sqrt{3}/2$$

$$c = \frac{7\sqrt{3} \pm \sqrt{(-7\sqrt{3})^2 - 4(1)(13)}}{2} \qquad \text{Quadratic formula}$$

$$= 10.935\cdots \quad \text{or} \quad c = 1.188\cdots$$

Each positive value of c corresponds to a separate triangle, so there are two

> **Alert**
>
> Students need to be cautioned against using intermediate rounding as they progress through a problem. Encourage students to use the full calculator approximations (which may be stored in the calculator's memory) until their last calculation to prevent any loss of accuracy.

triangles. Solving $b^2 = a^2 + c^2 - 2ac\cos\beta$ for $\cos\beta$, we find that

$$\cos\beta = \frac{a^2 + c^2 - b^2}{2ac}.$$

$$\begin{aligned}\cos\beta_1 &= \frac{6^2 + (10.935\cdots)^2 - 7^2}{2(6)(10.935\cdots)} \\ \cos\beta_1 &= 0.812\cdots \\ \beta_1 &= 35.685\cdots^\circ \\ \gamma_1 &= 180^\circ - (\alpha + \beta_1) \\ &= 114.314\cdots^\circ\end{aligned}$$

$$\begin{aligned}\cos\beta_2 &= \frac{6^2 + (1.188\cdots)^2 - 7^2}{2(6)(1.188\cdots)} \\ \cos\beta_2 &= -0.812\cdots \\ \beta_2 &= 144.314\cdots^\circ \\ \gamma_2 &= 180^\circ - (\alpha + \beta_2) \\ &= 5.685\cdots^\circ\end{aligned}$$

Interpret

One triangle has $c \approx 10.94$, $\beta \approx 35.69^\circ$, and $\gamma \approx 114.31^\circ$. The other triangle has $c \approx 1.19$, $\beta \approx 144.31^\circ$, and $\gamma \approx 5.69^\circ$. Compare with Example 3 of Section 8.1, which solves the same triangle using the law of sines. ■

Using the Law of Cosines in Geometry

A regular tetrahedron is a solid that consists of four equilateral triangular faces. We can use the law of cosines to find a dihedral angle of the tetrahedron, that is, an angle formed by two of the solid's faces.

■ EXAMPLE 4 Measuring the Dihedral Angle of a Regular Tetrahedron

Figure 8.15 shows a regular tetrahedron with six edges of length 2. Point B is the midpoint of an edge, and A and C are the vertices of the opposite edge. Find the measure of $\angle ABC$ (which is the measure of the dihedral angle formed by the faces ADE and CDE).

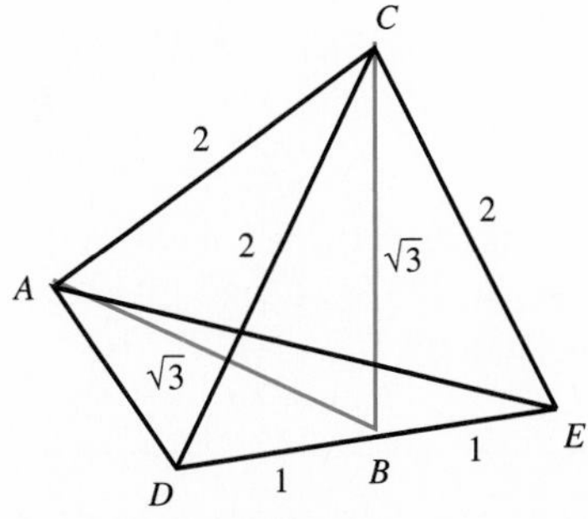

Figure 8.15 The measure of $\angle ABC$ is the same as the measure of any dihedral angle formed by two of the solid's faces.

Solution

The faces of a regular tetrahedron are equilateral triangles. Because B is a midpoint, AB and CB are altitudes of equilateral triangles and thus have length $\sqrt{3}$. Applying the law of cosines to $\triangle ABC$, we have

$$\begin{aligned}b^2 &= a^2 + c^2 - 2ac\cos\beta \\ 2^2 &= (\sqrt{3})^2 + (\sqrt{3})^2 - 2\sqrt{3}\sqrt{3}\cos\beta \\ \cos\beta &= \frac{1}{3} \\ \beta &= \cos^{-1}\left(\frac{1}{3}\right) \approx 70.53^\circ\end{aligned}$$

■

Area of a Triangle and Heron's Formula

Trigonometry allows us to extend the basic and familiar formula

$$\text{area} = \frac{1}{2} \times \text{base} \times \text{altitude}$$

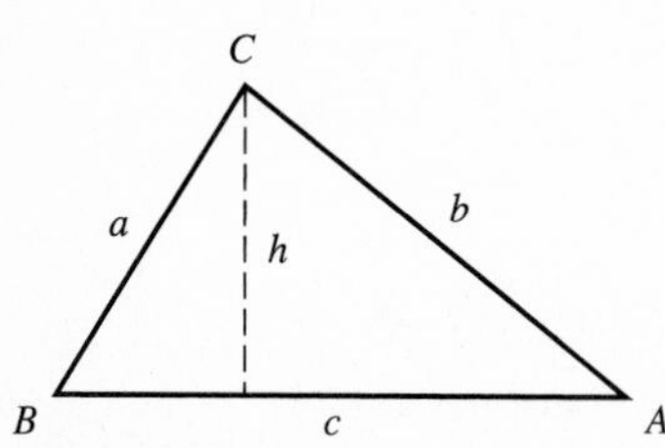

Figure 8.16

for the area of a triangle. In Figure 8.16, c is the base and $h = b \sin \alpha$ is the altitude, so

$$\text{area} = \frac{1}{2}bc \sin \alpha.$$

Using any side as the base, we can find the area in a similar way. Thus we have three trigonometric formulas for the area of a triangle.

Teaching Note

Note that the formula for the area of a triangle holds for obtuse triangles as well as acute and right triangles.

Area of a Triangle

The area of $\triangle ABC$ is given by

$$\text{Area} = \frac{1}{2}bc \sin \alpha = \frac{1}{2}ac \sin \beta = \frac{1}{2}ab \sin \gamma.$$

EXAMPLE 5 Finding the Area of a Triangle

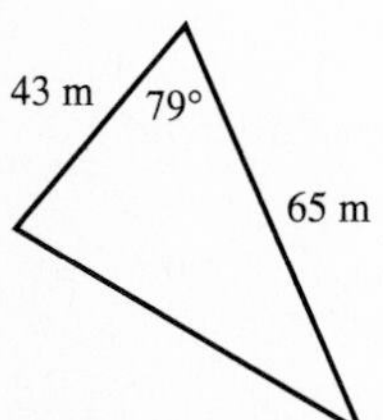

Figure 8.17

The area of a triangular field with sides of 43 m and 65 m and an included angle of 79° (Figure 8.17) is

$$\frac{1}{2}(43)(65)\sin 79° = 1371.823 \cdots,$$

or about 1372 m^2. ■

Heron's formula for the area of a triangle requires only the lengths of the sides. The formula was discovered by Archimedes (287–212 B.C.), Heron's predecessor and the greatest mathematician in antiquity.

Teaching Note

Because Heron's formula can be helpful but not essential, it can be omitted if time is a factor. You may wish to have students write programs for finding areas of triangles.

Heron's Formula

The area of $\triangle ABC$ is given by

$$\text{Area} = \sqrt{s(s - a)(s - b)(s - c)},$$

where $s = \frac{1}{2}(a + b + c)$ is the **semiperimeter** (half the full perimeter).

Proof

$$
\begin{aligned}
\text{Area} &= \frac{1}{2}ab\sin\gamma && \text{Area of } \triangle ABC\\
4(\text{Area}) &= 2ab\sin\gamma\\
16(\text{Area})^2 &= 4a^2b^2\sin^2\gamma && \text{Square both sides.}\\
&= 4a^2b^2(1-\cos^2\gamma) && \text{Pythagorean identity}\\
&= 4a^2b^2 - 4a^2b^2\cos^2\gamma\\
&= 4a^2b^2 - (2ab\cos\gamma)^2\\
&= 4a^2b^2 - (a^2+b^2-c^2)^2 && \text{Law of cosines}\\
&= [2ab-(a^2+b^2-c^2)] && \text{Difference of squares}\\
&\quad\cdot[2ab+(a^2+b^2-c^2)]\\
&= [c^2-(a^2-2ab+b^2)]\\
&\quad\cdot[(a^2+2ab+b^2)-c^2]\\
&= [c^2-(a-b)^2][(a+b)^2-c^2] && \text{Binomial squared}\\
&= [c-(a-b)][c+(a-b)] && \text{Difference of squares}\\
&\quad\cdot[(a+b)-c][(a+b)+c]\\
&= (c-a+b)(c+a-b)(a+b-c)(a+b+c)\\
&= (2s-2a)(2s-2b)(2s-2c)(2s) && 2s=a+b+c\\
16(\text{Area})^2 &= 16s(s-a)(s-b)(s-c)\\
(\text{Area})^2 &= s(s-a)(s-b)(s-c) && \text{Divide by 16.}\\
\text{Area} &= \sqrt{s(s-a)(s-b)(s-c)} && \text{Take square root.} \blacksquare
\end{aligned}
$$

Applications

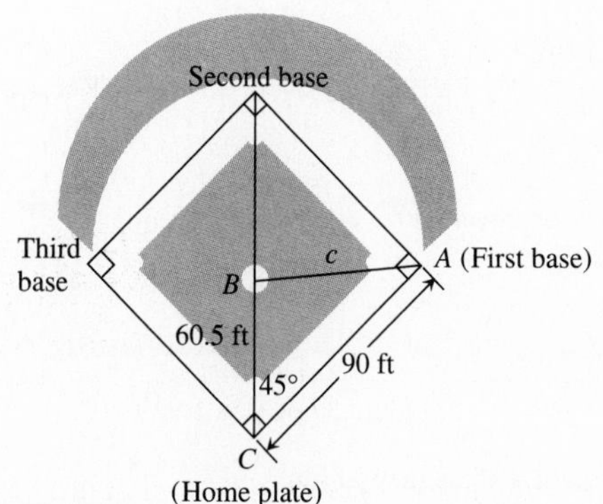

Figure 8.18

■ EXAMPLE 6 APPLICATION: Measuring a Baseball Diamond

The bases on a baseball diamond are 90 ft apart, and the front edge of the pitcher's rubber is 60.5 ft from the back corner of home plate. Find the distance from the center of the front edge of the pitcher's rubber to the far corner of first base.

Solution

Model

Figure 8.18 shows first base as A, the center of the pitcher's rubber as B, and home plate as C. The distance we seek is c in $\triangle ABC$.

Solve Algebraically

In $\triangle ABC$, $a = 60.5$, $b = 90$, and $\gamma = 45°$. Using the law of cosines, we get

$$c^2 = a^2 + b^2 - 2ab\cos\gamma$$
$$= 60.5^2 + 90^2 - 2(60.5)(90)(\cos 45°),$$
$$c = \sqrt{60.5^2 + 90^2 - 2(60.5)(90)(\cos 45°)}$$
$$= 63.717 \cdots .$$

Interpret

The distance from the center of the front edge of the pitcher's rubber to first base is about 63.72 ft, or 63 ft 9 in. ■

Follow-up

Have students discuss whether the law of sines or the law of cosines is easier to use for various kinds of problems (SSS, SSA, SAA, ASA, or SAS).

Assignment Guide

Day 1: Ex. 3–30, multiples of 3
Day 2: Ex. 29, 31, 34, 35, 36, 39–42, 44–47

Cooperative Learning

Group Activity: Ex. 32–33

Notes on Exercises

Ex. 5–16 can be solved by writing a program that applies the law of cosines.
Ex. 21–28 can be assigned even if you choose not to teach Heron's formula. Students can find an angle using the law of cosines and then apply the basic area formula on page 538.

Ongoing Assessment

Self-Assessment: Ex. 1, 13, 19, 25, 37
Embedded Assessment: Ex. 42

■ **EXAMPLE 7** **APPLICATION: Converting Units for Land Appraisal**

Mary McCullough, a certified real estate appraiser, evaluates a triangular lot whose sides are 297 ft by 252 ft by 329 ft, and is required to state the lot's area in acres in her report.

Solution

Model

We use Heron's formula with $a = 297$, $b = 252$, and $c = 329$.

Solve Algebraically

$$s = \frac{a + b + c}{2} = \frac{297 + 252 + 329}{2} = 439$$

By Heron's formula,

$$\text{area} = \sqrt{s(s - a)(s - b)(s - c)}$$
$$= \sqrt{439(439 - 297)(439 - 252)(439 - 329)}$$
$$= 35{,}809.114 \cdots .$$

In acres,

$$\text{area} = (35{,}809.114 \cdots \text{ft}^2)\left(\frac{1 \text{ mi}^2}{5280^2 \text{ ft}^2}\right)\left(\frac{640 \text{ acre}}{1 \text{ mi}^2}\right) \approx 0.822 \text{ acre}.$$

Interpret

The lot has an area of about 35,809 ft^2, or about 0.82 acres. ■

NOTE

There are 640 acres in 1 square mile.

Quick Review 8.2

In Exercises 1–6, find an angle between 0° and 180° that is a solution to the equation.

1. $\cos\alpha = 3/5$ **2.** $\cos\beta = 5/8$

3. $\cos\gamma = -0.23$ **4.** $\cos\alpha = -0.68$

5. $2\cos\beta = 1.38$ **6.** $3\cos\gamma = 1.92$

In Exercises 7 and 8, solve the equation for

a. $\cos\alpha$ **b.** $\alpha, 0 \le \alpha \le 180°$.

7. $9^2 = x^2 + y^2 - 2xy\cos\alpha$ **8.** $y^2 = x^2 + 25 - 10\cos\alpha$

In Exercises 9–12, find a quadratic polynomial with real coefficients that satisfies the given conditions.

9. Has two positive zeros

10. Has one positive and one negative zero

11. Has no real zeros

12. Has exactly one positive zero

SECTION EXERCISES 8.2

In Exercises 1–4, solve the triangle.

1. B, 8, 131°, 13, A, C

2. A, 12, 42°, C, 14, B

3.

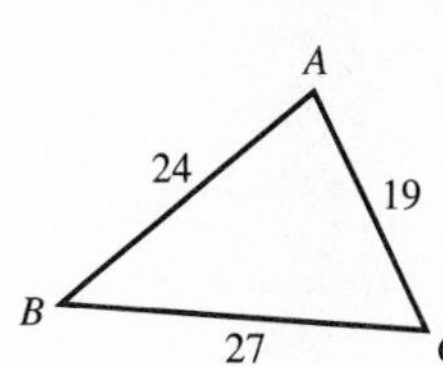

4. C, 35, A, 17, 28, B

In Exercises 5–16, solve the triangle.

5. $\alpha = 55°$, $b = 12$, $c = 7$

6. $\beta = 35°$, $a = 43$, $c = 19$

7. $a = 12$, $b = 21$, $\gamma = 95°$

8. $b = 22$, $c = 31$, $\alpha = 82°$

9. $a = 1$, $b = 5$, $c = 4$

10. $a = 1$, $b = 5$, $c = 8$

11. $a = 3.2$, $b = 7.6$, $c = 6.4$

12. $a = 9.8$, $b = 12$, $c = 23$

13. $\alpha = 42°$, $a = 7$, $b = 10$

14. $\alpha = 57°$, $a = 11$, $b = 10$

15. $\alpha = 63°$, $a = 8.6$, $b = 11.1$

16. $\alpha = 71°$, $a = 9.3$, $b = 8.5$

In Exercises 17–20, find the area of the triangle.

17. $\alpha = 47°$, $b = 32$ ft, $c = 19$ ft

18. $\alpha = 52°$, $b = 14$ m, $c = 21$ m

19. $\beta = 101°$, $a = 10$ cm, $c = 22$ cm

20. $\gamma = 112°$, $a = 1.8$ in., $b = 5.1$ in.

In Exercises 21–28, decide whether a triangle can be formed with the given side lengths. If so, use Heron's formula to find the area of the triangle.

21. $a = 4$, $b = 5$, $c = 8$

22. $a = 5$, $b = 9$, $c = 7$

23. $a = 3$, $b = 5$, $c = 8$

24. $a = 23$, $b = 19$, $c = 12$

25. $a = 19.3$, $b = 22.5$, $c = 31$

26. $a = 8.2$, $b = 12.5$, $c = 28$

27. $a = 33.4$, $b = 28.5$, $c = 22.3$

28. $a = 18.2$, $b = 17.1$, $c = 12.3$

In Exercises 29–42, solve the problem.

29. Find the radian measure of the largest angle in the triangle with sides of 4, 5, and 6.

30. A parallelogram has sides of 18 and 26 ft, and an angle of 39°. Find the shorter diagonal.

31. *Measuring Distance Indirectly* Juan wants to find the distance between two points *A* and *B* on opposite sides of a building. He locates a point *C* that is 110 ft from *A* and 160 ft

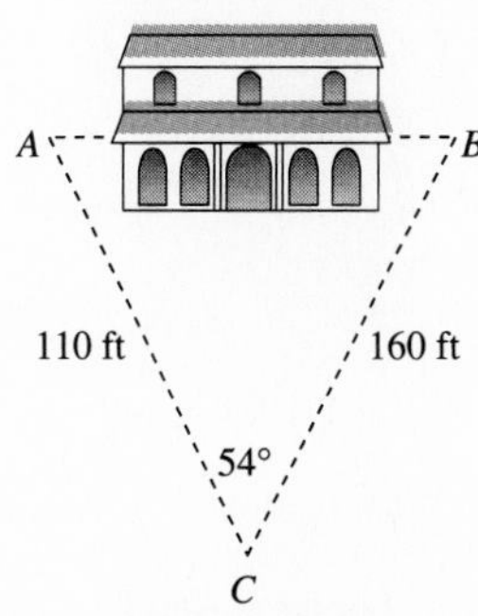

For Exercise 31

from B, as illustrated in the figure. If the angle at C is 54°, find distance AB.

32. *Designing a Baseball Field*

a. Find the distance from the center of the front edge of the pitcher's rubber to the far corner of second base. How does this distance compare with the distance from the pitcher's rubber to first base? (See Example 6.)

b. Find β.

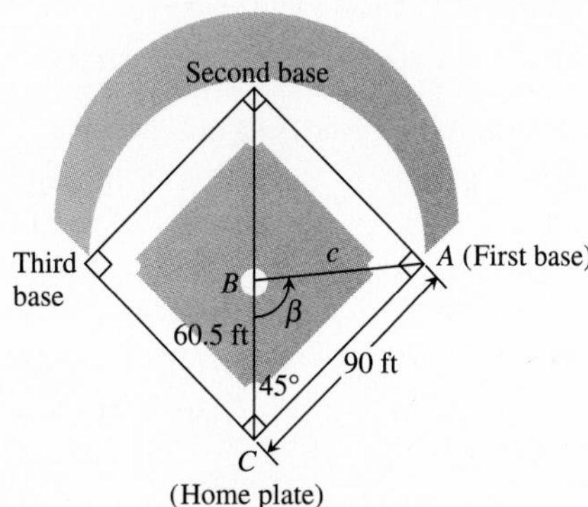

For Exercise 32

33. *Designing a Softball Field* In softball, adjacent bases are 60 ft apart. The distance from the center of the front edge of the pitcher's rubber to the far corner of home plate is 40 ft.

a. Find the distance from the center of the pitcher's rubber to the far corner of first base.

b. Find the distance from the center of the pitcher's rubber to the far corner of second base.

c. Find β.

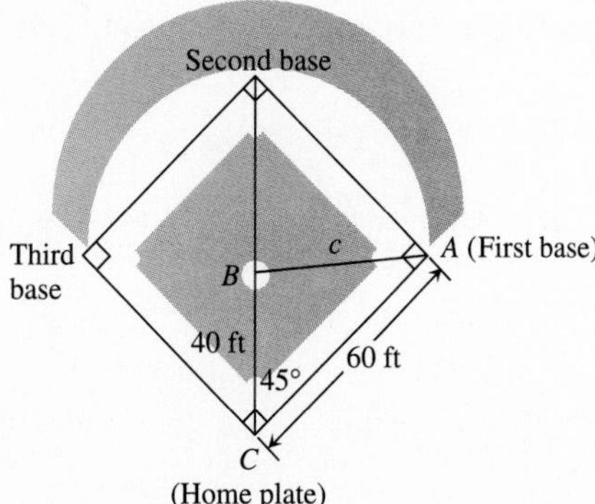

For Exercise 33

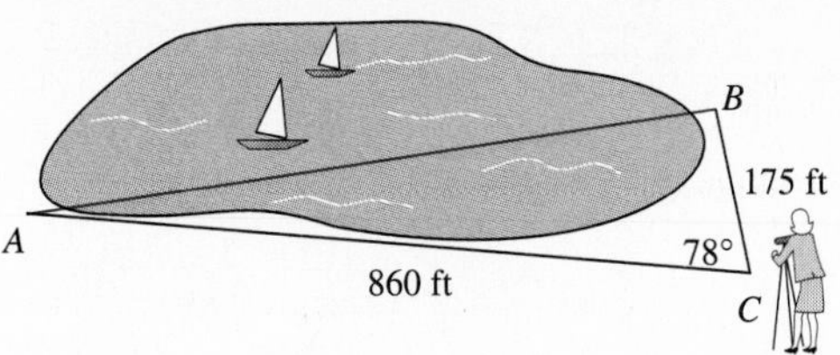

For Exercise 34

34. *Surveyor's Calculations* Tonya must find the distance from A to B on opposite sides of a lake. She locates a point C that is 860 ft from A and 175 ft from B. She measures the angle at C to be 78°. Find distance AB.

35. *Construction Engineering* A manufacturer is designing the roof truss that is modeled in the figure shown below.

a. Find the measure of $\angle CAE$.

b. If $AF = 12$ ft, find the length DF.

c. Find the length EF.

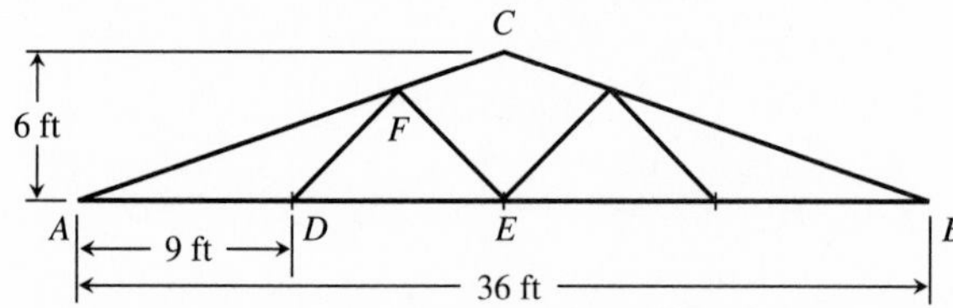

For Exercise 35

36. *Navigation* Two airplanes flying together in formation take off in different directions. One flies due east at 350 mph, and the other flies east-northeast at 380 mph. How far apart are the two airplanes 2 h after they separate, assuming that they fly at the same altitude?

37. *Football Kick* The player waiting to receive a kickoff stands

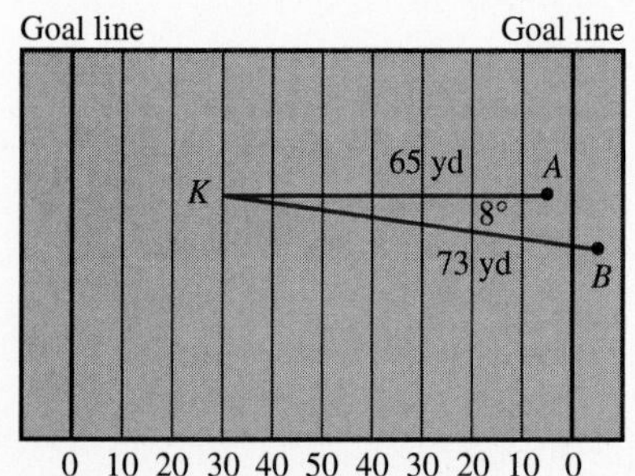

For Exercise 37

at the 5 yard line as the ball is being kicked 65 yd up the field from the opponent's 30 yard line. The kicked ball travels 73 yd at an angle of 8° to the right of the receiver, as shown in the figure. Find the distance the receiver runs to catch the ball.

38. *Architectural Design* Building Inspector Julie Wang checks a building in the shape of a regular octagon, each side 20 ft long. She checks that the contractor has located the corners of the foundation correctly by measuring several of the diagonals. Calculate what the lengths of diagonals *HB, HC,* and *HD* should be.

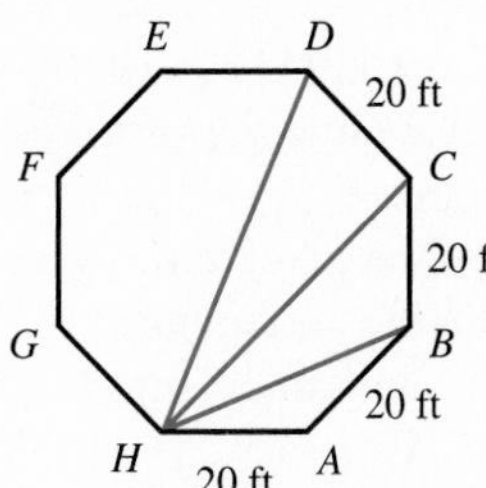

For Exercise 38

39. Connecting Trigonometry and Geometry $\angle CAB$ is inscribed in a rectangular box whose sides are either 1, 2 or 3 ft long. Find the measure of $\angle CAB$.

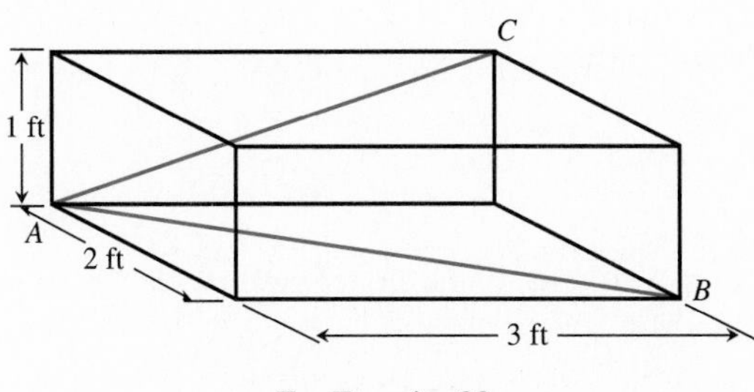

For Exercise 39

40. Connecting Trigonometry and Geometry A cube has edges of length 2 ft. Point *A* is the midpoint of an edge. Find the measure of $\angle ABC$.

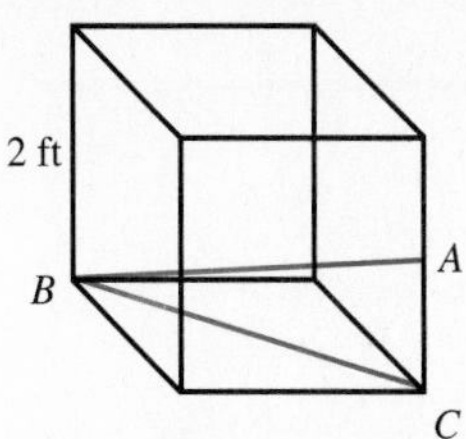

For Exercise 40

41. Writing to Learn State the number of triangles that are possible with AAA, ASA, SAA, SSA, SAS, or SSS given. In each case, state the condition that ensures there is at least one such triangle.

42. Writing to Learn State which of the following cases can be solved using the law of sines, the law of cosines, or both: ASA, SAA, SSA, SAS, or SSS.

EXTENDING THE IDEAS

In Exercises 43 and 44, use the law of sines and/or the law of cosines to solve the problem.

43. *Surveyor's Calculations* Find all the unknown sides and angles of the field shown. (*Hint:* Draw a diagonal.)

44. *Surveyor's Calculations* Find all the unknown sides and angles of the field shown.

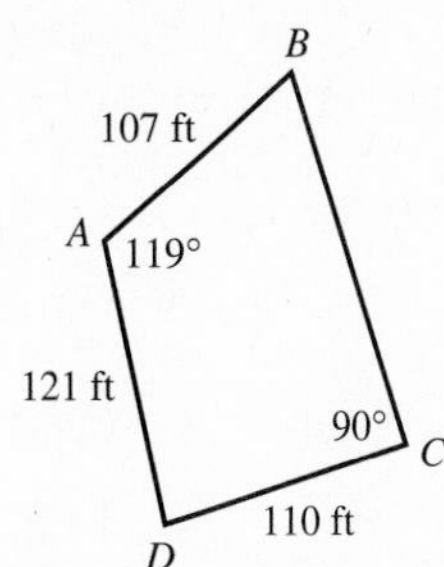

For Exercise 43

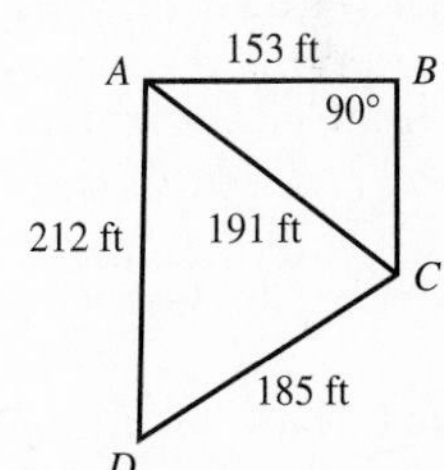

For Exercise 44

In Exercises 45–47, use the following law-of-cosines formula: $c^2 = a^2 + b^2 - 2ab\cos\gamma$.

45. *Extending the Pythagorean Theorem* If $\gamma = 90°$, show that $c^2 = a^2 + b^2$. Draw a figure for this situation.

46. If $\gamma = 0°$, show that $c = |a - b|$. Draw a figure for this situation.

47. If $\gamma = 180°$, show that $c = a + b$. Draw a figure for this situation.

Objective

Students will be able to apply the arithmetic of vectors and use vectors to solve real-world problems.

8.3 VECTORS

Directed Line Segment • Component Form of a Vector • Vector Operations • Unit Vectors • Direction Angle • Applications

Directed Line Segment

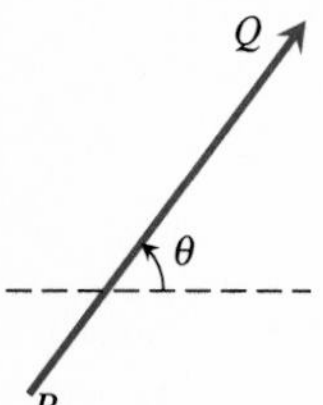

Figure 8.19

Some real-world measurements, like those of temperature, distance, height, area, volume, and money amounts, can be represented by numbers that depict their *magnitude,* or size. Other measurements such as force, velocity, and acceleration, depict both magnitude and *direction.* They need more than a single number as a model. Typically, they are represented by a **directed line segment.**

For example, in Figure 8.19 the directed line segment $\overrightarrow{PQ}$ has **initial point** P and **terminal point** Q. We use the symbol $\|\overrightarrow{PQ}\|$ to represent the **length** or **magnitude** of the directed line segment. The **direction** is given by the angle θ that the directed line segment makes with the horizontal.

Two directed line segments with the same length and direction are **equivalent.**

Motivate

Discuss the difference between the statements "Jose lives 3 miles away from Mary" and "Jose lives 3 miles west of Mary."

■ EXAMPLE 1 Confirming Equivalent Directed Line Segments

Let $R = (-4, 2)$, $S = (-1, 6)$, $O = (0, 0)$, and $P = (3, 4)$. Show that $\overrightarrow{RS}$ and $\overrightarrow{OP}$ are equivalent. (See Figure 8.20.)

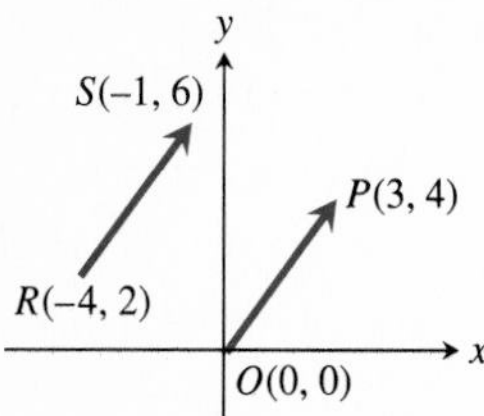

Figure 8.20

Solution

We must show that $\overrightarrow{RS}$ and $\overrightarrow{OP}$ have the same length and direction. Using the distance formula,

$$\|\overrightarrow{RS}\| = \sqrt{[-1-(-4)]^2 + (6-2)^2} = \sqrt{3^2 + 4^2} = 5$$
$$\|\overrightarrow{OP}\| = \sqrt{3^2 + 4^2} = 5$$

Therefore, $\|\overrightarrow{RS}\| = \|\overrightarrow{OP}\|$.

The directions of $\overrightarrow{RS}$ and $\overrightarrow{OP}$ are to the upper right with the slope $\frac{4}{3}$. The angle with the horizontal is

$$\tan^{-1}\left(\frac{4}{3}\right) \approx 53.1°.$$

Interpret

$\overrightarrow{RS}$ and $\overrightarrow{OP}$ have the same length and direction, so

$$\overrightarrow{RS} = \overrightarrow{OP}. \quad ■$$

Teaching Note

It is worth pointing out that although the exact definition of a vector may vary between texts, the general concept is the same—a vector is defined by its direction and magnitude but *not* by its location.

Component Form of a Vector

If P and Q are two points in the coordinate plane, then the set of all directed line segments equivalent to $\overrightarrow{PQ}$ is the **vector** determined by $\overrightarrow{PQ}$, and is denoted by

$$\mathbf{v} = \overrightarrow{PQ}.$$

In particular, there is one directed line segment equivalent to $\overrightarrow{PQ}$ whose initial point is the origin. It is the representative of **v** in **standard position** and, unless otherwise stated, will be the representative of **v** we will use from here on.

This convention allows us to make the extraordinary connection between vectors and the points of the coordinate plane. If the representative of a vector in standard position has terminal point (a, b), then

$$\langle a, b \rangle$$

is the **component form** of the vector, and a and b are called the **components** of the vector. Two vectors $\langle a, b \rangle$ and $\langle c, d \rangle$ are equal if and only if $a = c$ and $b = d$. For simplicity, we will speak of this vector as *vector* $\langle a, b \rangle$. Note that its length is $\sqrt{a^2 + b^2}$ and its direction is given by the angle the line segment from the origin to (a, b) makes with the horizontal. The vector $\langle 0, 0 \rangle$ with length 0 and no direction is the **zero vector.**

Teaching Note

Attention to what constitutes equal vectors will contribute to success later on. Failure to understand the definition of equal vectors results in a surprising number of errors in students' thinking.

Teaching Note

The textbook uses bold print to denote vectors. Suggest to students that they use arrows on homework and tests.

Teaching Note

Note that the magnitude of the vector determined by $\overrightarrow{PQ}$ is simply the distance between P and Q.

Component Form of a Vector

The vector determined by the directed line segment with initial point $P(p_1, p_2)$ and terminal point $Q(q_1, q_2)$ has component form

$$\overrightarrow{PQ} = \langle q_1 - p_1, q_2 - p_2 \rangle = \langle v_1, v_2 \rangle = \mathbf{v}.$$

The magnitude of **v** is

$$\|\mathbf{v}\| = \sqrt{(q_1 - p_1)^2 + (q_2 - p_2)^2} = \sqrt{v_1^2 + v_2^2}.$$

EXAMPLE 2 Finding the Component Form of a Vector

Find the component form, magnitude, and direction of $\mathbf{v} = \overrightarrow{PQ}$ for points $P(-3, 4)$ and $Q(-5, 2)$. (See Figure 8.21.)

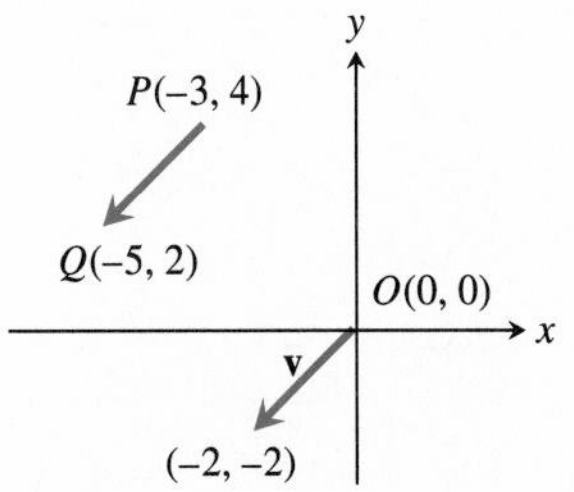

Solution

$$\mathbf{v} = \langle q_1 - p_1, q_2 - p_2 \rangle = \langle -5 - (-3), 2 - 4 \rangle = \langle -2, -2 \rangle$$

and $$\|\mathbf{v}\| = \sqrt{(-2)^2 + (-2)^2} = 2\sqrt{2}$$

The direction is lower left with slope $(-2)/(-2) = 1$.

Interpret

The vector **v** has the length $2\sqrt{2}$ and makes a 225° angle with the positive x-axis. ■

Figure 8.21

Vector Operations

Two key operations involving vectors are *vector addition* and *multiplication of a vector by a* **scalar** (a real number). These operations satisfy many properties similar to those for real numbers. (See Exercise 44.)

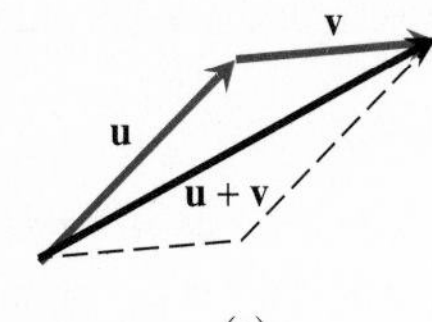

(a)

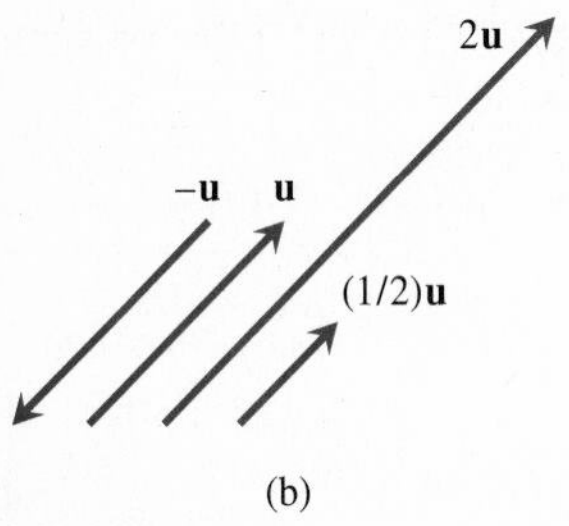

(b)

Figure 8.22

> **Definition:** Vector Addition, and Multiplication by a Scalar
>
> Let $\mathbf{u} = \langle u_1, u_2 \rangle$ and $\mathbf{v} = \langle v_1, v_2 \rangle$ be vectors and k be a real number. Then the **sum of vectors u and v** is the vector
>
> $$\mathbf{u} + \mathbf{v} = \langle u_1, u_2 \rangle + \langle v_1, v_2 \rangle = \langle u_1 + v_1, u_2 + v_2 \rangle,$$
>
> and the **product of the scalar *k* and vector u** is
>
> $$k\mathbf{u} = k\langle u_1, u_2 \rangle = \langle ku_1, ku_2 \rangle.$$

The sum of **u** and **v** can be obtained geometrically by joining the initial point of the second vector **v** with the terminal point of the first vector **u,** as shown in Figure 8.22a. The sum $\mathbf{u} + \mathbf{v}$ is represented by the directed line segment from the initial point of **u** to the terminal point of **v** and is a diagonal of the parallelogram formed by vectors **u** and **v.** So we often say that we add vectors by the **parallelogram law.**

The product $k \cdot \mathbf{u}$ of the scalar k and the vector **u** can be represented geometrically by a stretch or shrink of **u** by the factor k when $k > 0$. (See Figure 8.22b.) If $k < 0$, then there is a stretch or shrink by the factor $|k|$, and the direction of $k \cdot \mathbf{u}$ is opposite that of **u.**

■ EXAMPLE 3 Performing Vector Addition and Scalar Multiplication

Let $\mathbf{u} = \langle -1, 3 \rangle$ and $\mathbf{v} = \langle 4, 7 \rangle$. Find the component form of these vectors:

a. $\mathbf{u} + \mathbf{v}$ **b.** $3\mathbf{u}$ **c.** $2\mathbf{u} + (-1)\mathbf{v}$

Solution

a. Solve Algebraically

$$\begin{aligned} \mathbf{u} + \mathbf{v} &= \langle -1, 3 \rangle + \langle 4, 7 \rangle \\ &= \langle -1 + 4, 3 + 7 \rangle \\ &= \langle 3, 10 \rangle \end{aligned}$$

Support Graphically

Start with the terminal point $(-1, 3)$ of **u.** Move right 4 units and up 7 units (the components of **v**) to arrive at the point $(-1 + 4, 3 + 7) = (3, 10)$, which is the terminal point of $\mathbf{u} + \mathbf{v}$. (See Figure 8.23.)

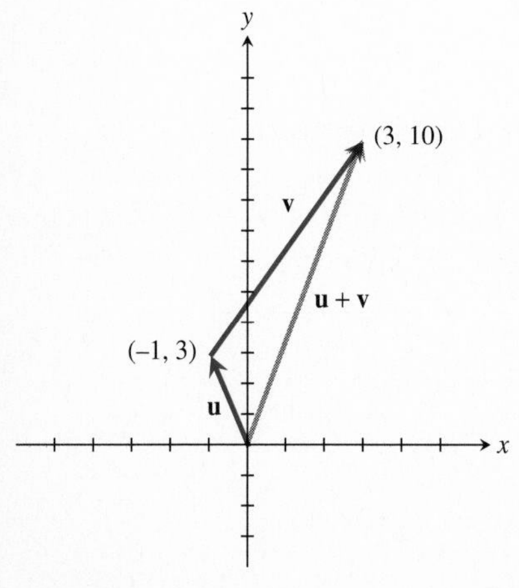

Figure 8.23 Using the parallelogram law to find $\mathbf{u} + \mathbf{v}$ in Example 3a.

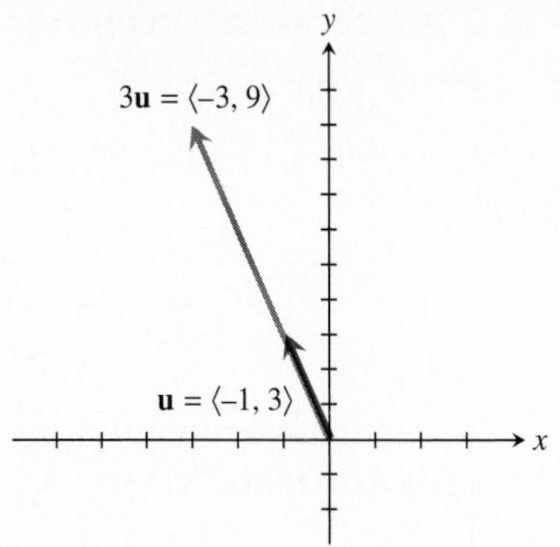

Figure 8.24 Multiplication by the scalar 3 stretches the vector by the factor 3 both horizontally and vertically.

b. Algebraically,

$$3\mathbf{u} = 3\langle -1, 3\rangle = \langle -3, 9\rangle.$$

Figure 8.24 shows that $3\mathbf{u}$ is a stretch of $\mathbf{u}$ by the factor 3.

c.
$$\begin{aligned} 2\mathbf{u} + (-1)\mathbf{v} &= 2\langle -1, 3\rangle + (-1)\langle 4, 7\rangle \\ &= \langle -2, 6\rangle + \langle -4, -7\rangle \\ &= \langle -2 - 4, 6 - 7\rangle = \langle -6, -1\rangle \end{aligned}$$ ■

Unit Vectors

A vector $\mathbf{u}$ is a **unit vector** if its length is 1, that is, if $\|\mathbf{u}\| = 1$. If $\mathbf{v}$ is not the zero vector $\langle 0, 0\rangle$, then the vector

$$\mathbf{u} = \frac{\mathbf{v}}{\|\mathbf{v}\|} = \frac{1}{\|\mathbf{v}\|} \cdot \mathbf{v}$$

is a **unit vector in the direction of v.** Any vector in the direction of **v,** or the opposite direction, is a scalar multiple of this unit vector **u.**

■ EXAMPLE 4 Finding a Unit Vector

Let $\mathbf{v} = \langle -3, 2\rangle$. Find $\frac{\mathbf{v}}{\|\mathbf{v}\|}$ and confirm that its magnitude is 1.

Solution

$$\|\mathbf{v}\| = \|\langle -3, 2\rangle\| = \sqrt{(-3)^2 + 2^2} = \sqrt{13}$$

Therefore,

$$\begin{aligned} \frac{\mathbf{v}}{\|\mathbf{v}\|} &= \frac{1}{\sqrt{13}} \cdot \langle -3, 2\rangle \\ &= \left\langle \frac{-3}{\sqrt{13}}, \frac{2}{\sqrt{13}} \right\rangle. \end{aligned}$$

To confirm that this is a unit vector, we find its magnitude:

$$\begin{aligned} \left\| \left\langle \frac{-3}{\sqrt{13}}, \frac{2}{\sqrt{13}} \right\rangle \right\| &= \sqrt{\left(\frac{-3}{\sqrt{13}}\right)^2 + \left(\frac{2}{\sqrt{13}}\right)^2} \\ &= \sqrt{\frac{9}{13} + \frac{4}{13}} \\ &= \sqrt{\frac{13}{13}} = 1. \end{aligned}$$ ■

Teaching Note

In Figure 8.25 note that a and b are simply the Cartesian coordinates of the terminal point of the vector's representative in standard position.

The two unit vectors $\mathbf{i} = \langle 1, 0\rangle$ and $\mathbf{j} = \langle 0, 1\rangle$ are the **standard unit vectors.** Any vector $\mathbf{v}$ can be rewritten using standard unit vectors, as follows:

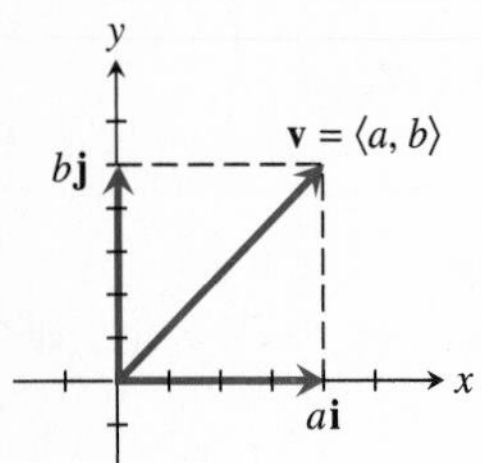

Figure 8.25

$$\begin{aligned}\mathbf{v} &= \langle a, b\rangle \\ &= \langle a, 0\rangle + \langle 0, b\rangle \\ &= a\langle 1, 0\rangle + b\langle 0, 1\rangle \\ &= a\mathbf{i} + b\mathbf{j}\end{aligned}$$

We say that $\mathbf{v} = \langle a, b\rangle$ has been written as the **linear combination** $a\mathbf{i} + b\mathbf{j}$ of the standard unit vectors. The scalars a and b are the **horizontal** and **vertical** components, respectively, of the vector **v.** (See Figure 8.25.)

Direction Angle

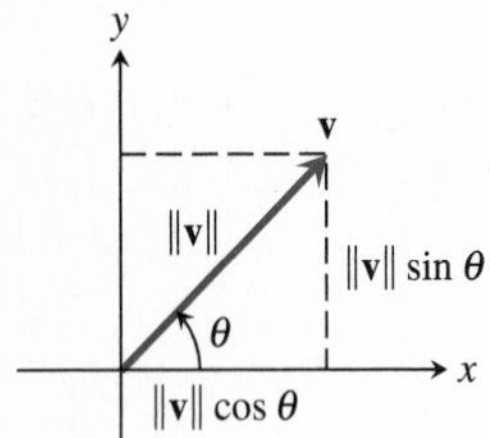

Figure 8.26

If θ is the *direction angle* of **v,** as shown in Figure 8.26, then $\mathbf{v} = \langle \|\mathbf{v}\|\cos\theta, \|\mathbf{v}\|\sin\theta\rangle$, and

$$\mathbf{v} = (\|\mathbf{v}\|\cos\theta)\mathbf{i} + (\|\mathbf{v}\|\sin\theta)\mathbf{j}.$$

The unit vector in the direction of **v,** then, is

$$\mathbf{u} = \frac{\mathbf{v}}{\|\mathbf{v}\|} = (\cos\theta)\mathbf{i} + (\sin\theta)\mathbf{j}.$$

The numbers $\|\mathbf{v}\|\cos\theta$ and $\|\mathbf{v}\|\sin\theta$ in Figure 8.26 are the horizontal component and vertical component of **v,** respectively. The angle θ is the **direction angle** of both **u** and **v.**

■ EXAMPLE 5 Finding the Components of a Vector

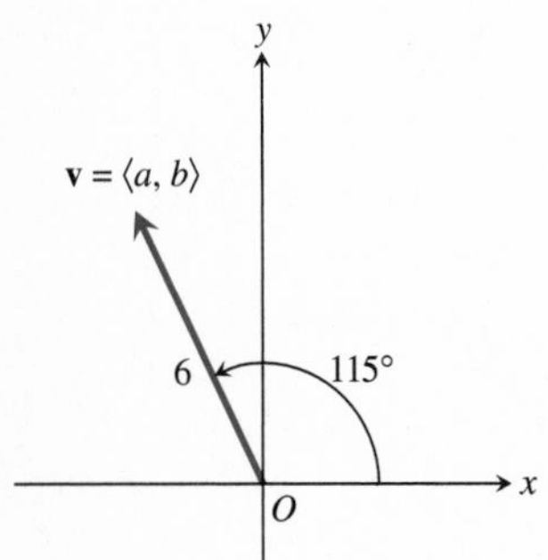

Figure 8.27

Find the components of the vector **v** shown in Figure 8.27.

Solution

$\mathbf{v} = \langle a, b\rangle = \langle 6\cos 115°, 6\sin 115°\rangle$, so the horizontal component is $6\cos 115°$ and the vertical component is $6\sin 115°$. ■

■ EXAMPLE 6 Finding the Direction of a Vector

Let $\mathbf{u} = \langle 3, 2\rangle$ and $\mathbf{v} = \langle -2, 5\rangle$. Find the direction angles of **u** and **v** and the angle between **u** and **v.** (See Figure 8.28.)

Solution

For **u,**

$$\begin{aligned}3 &= \|\mathbf{u}\|\cos\alpha && \text{Horizontal component of } \mathbf{u}. \\ &= \sqrt{3^2 + 2^2}\cos\alpha \\ &= \sqrt{13}\cos\alpha\end{aligned}$$

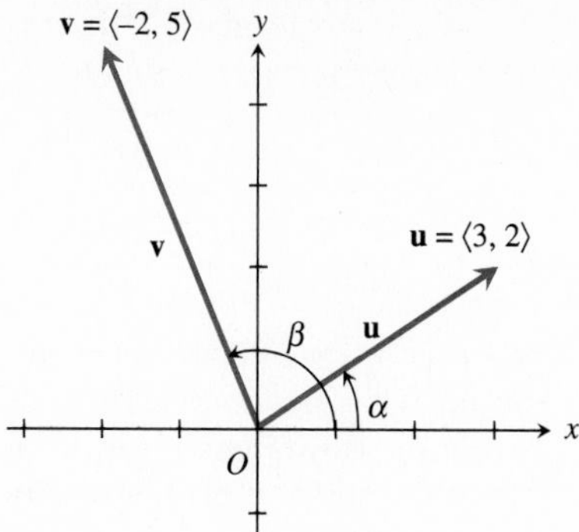

Figure 8.28

$$\cos\alpha = \frac{3}{\sqrt{13}} \qquad \text{Solve for } \cos\alpha.$$

$$\alpha = \cos^{-1}\left(\frac{3}{\sqrt{13}}\right) \qquad \alpha \text{ is acute.}$$

$$\alpha = 33.690\ldots^\circ$$

For **v**,

$$-2 = \|\mathbf{v}\|\cos\beta \qquad \text{Horizontal component of } \mathbf{v}$$

$$= \sqrt{(-2)^2 + 5^2}\cos\beta$$

$$= \sqrt{29}\cos\beta$$

$$\cos\beta = \frac{-2}{\sqrt{29}} \qquad \text{Solve for } \cos\beta.$$

$$\beta = \cos^{-1}\left(\frac{-2}{\sqrt{29}}\right) \qquad 90^\circ < \beta < 180^\circ$$

$$= 111.801\ldots^\circ$$

The angle between the vectors is

$$\beta - \alpha = \cos^{-1}\left(\frac{-2}{\sqrt{29}}\right) - \cos^{-1}\left(\frac{3}{\sqrt{13}}\right)$$

$$= 78.111\ldots^\circ$$

Interpret

The direction angle for **u** is approximately $\alpha = 33.7^\circ$. The direction angle for **v** is approximately $\beta = 111.8^\circ$, and the angle between **u** and **v** is $\beta - \alpha \approx 78.1^\circ$. ■

Applications

The **velocity** of a moving object is a vector because velocity has both magnitude and direction. The magnitude of velocity is **speed.**

■ EXAMPLE 7 APPLICATION: Calculating Velocity

A DC-10 jet aircraft is flying on a bearing of 65° at 500 mph. Find the component form of the velocity of the airplane.

Solution

Model

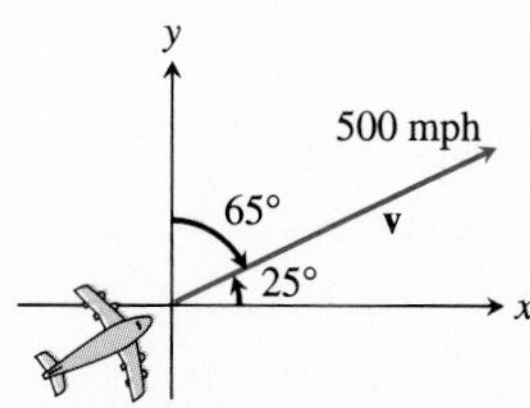

Figure 8.29 The unit vectors **i** and **j** represent 1 mph eastward and 1 mph northward, respectively.

Let **v** be the velocity of the airplane. A bearing of 65° is equivalent to a direction angle of 25°. The plane's speed, 500 mph, is the magnitude of vector **v**; that is, $\|\mathbf{v}\| = 500$. (See Figure 8.29.)

Teaching Note

Encourage students to draw pictures to analyze the geometry of various situations.

Solve Algebraically

$$\begin{aligned}\mathbf{v} &= (500 \cos 25°)\mathbf{i} + (500 \sin 25°)\mathbf{j} \\ &= \langle 500 \cos 25°, 500 \sin 25° \rangle \\ &\approx \langle 453.15, 211.31 \rangle\end{aligned}$$

Interpret

The plane's velocity is approximated by the vector $\langle 453.15, 211.31 \rangle$. The components give the eastward and northward speeds. That is, the plane travels about 453.15 mph eastward and about 211.31 mph northward as it travels at 500 mph on a bearing of 65°. ■

A typical problem for a navigator involves calculating the effect of wind on the direction and speed of the airplane. Example 8 illustrates.

■ EXAMPLE 8 APPLICATION: Calculating Effect of Wind Velocity

Pilot Megan McCarty's flight plan has her leaving San Francisco International Airport and flying a Boeing 727 due east. There is a 65-mph wind with the bearing 60°. Find the compass heading McCarty should follow, and determine what the airplane's ground speed will be (assuming that its speed with no wind is 450 mph).

Solution

Model

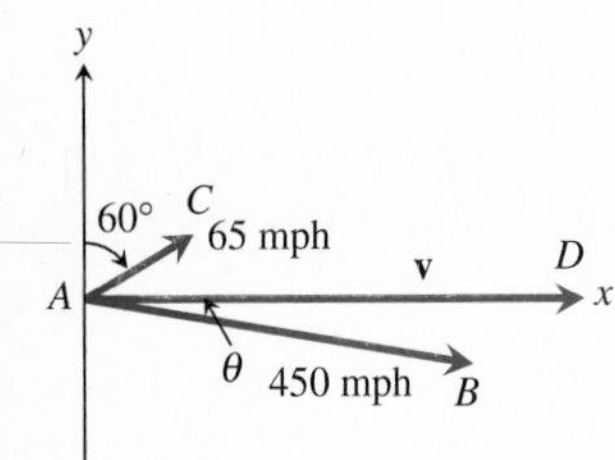

Figure 8.30

See Figure 8.30. Vector $\overrightarrow{AB}$ describes the velocity produced by the airplane alone, and $\overrightarrow{AC}$ represents the velocity of the wind. Vector $\mathbf{v} = \overrightarrow{AD}$ describes the resulting velocity, so

$$\overrightarrow{AD} = \overrightarrow{AC} + \overrightarrow{AB}.$$

We must find the bearing of $\overrightarrow{AB}$ and $\|\mathbf{v}\|$.

Solve Algebraically

$$\begin{aligned}\overrightarrow{AC} &= \langle 65 \cos 30°, 65 \sin 30° \rangle \\ \overrightarrow{AB} &= \langle 450 \cos \theta, 450 \sin \theta \rangle \\ \overrightarrow{AD} &= \overrightarrow{AC} + \overrightarrow{AB} \\ &= \langle 65 \cos 30° + 450 \cos \theta, 65 \sin 30° + 450 \sin \theta \rangle\end{aligned}$$

Because the plane is traveling due east, the second component of $\overrightarrow{AD}$ must be zero.

$$65 \sin 30° + 450 \sin \theta = 0$$

$$\theta = \sin^{-1}\left(\frac{-65 \sin 30°}{450}\right) \qquad -90° < \theta < 90°$$

$$= -4.141 \ldots°$$

Follow-up

Have students discuss why it does not make sense to add a scalar to a vector.

Assignment Guide

Day 1: Ex. 3–30, multiples of 3
Day 2: Ex. 33, 36, 38–40, 42, 43, 45, 47

Cooperative Learning

Group Activity: Ex. 44

Notes on Exercises

Ex. 5–20 deal with the component form of a vector.
Ex. 38–39 are problems that students would typically encounter in a physics course.
Ex. 48 introduces the dot product of two vectors.

Ongoing Assessment

Self-Assessment: Ex. 5, 17, 23, 31, 35
Embedded Assessment: Ex. 36, 38

Thus, the compass heading McCarty should follow is

$$90° + |\theta| \approx 94.14°.$$

The ground speed of the airplane is

$$\begin{aligned}\|\mathbf{v}\| &= \overrightarrow{AD} = \sqrt{(65 \cos 30° + 450 \cos \theta)^2 + 0^2}\\ &= |65 \cos 30° + 450 \cos \theta|\\ &= |65 \cos 30° + 450 \cos(-4.141 \cdots°)|\\ &\approx 505.12\end{aligned}$$

Interpret

McCarty should use a bearing of approximately 94.14°. The airplane will travel due east at approximately 505.12 mph. ■

Quick Review 8.3

In Exercises 1–4, find the values of x and y.

1.

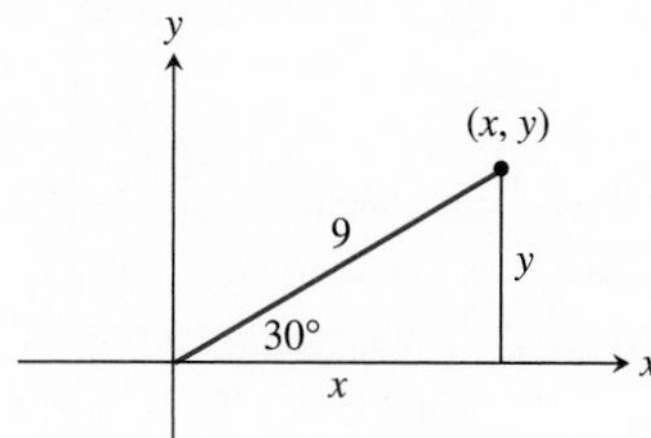

2.

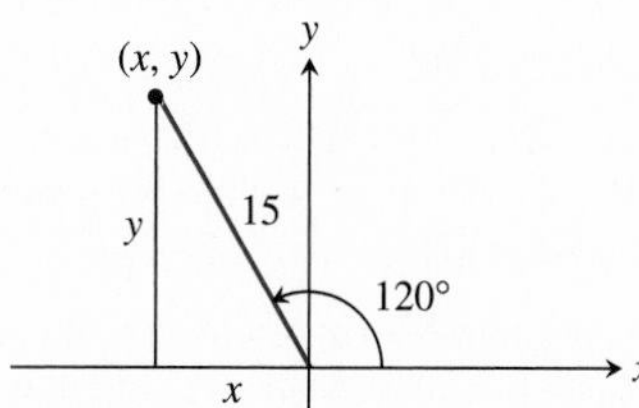

3.

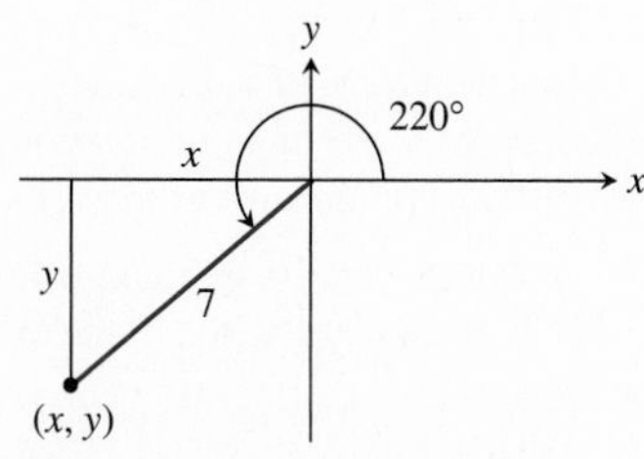

4.

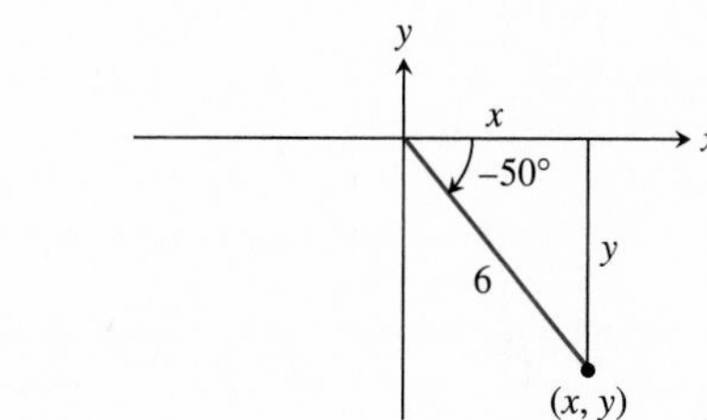

In Exercises 5–8, solve for θ in degrees.

5. $\theta = \cos^{-1}\left(\frac{1}{4}\right)$ **6.** $\theta = \sin^{-1}\left(\frac{5}{8}\right)$

7. $\theta = \sin^{-1}\left(\frac{3}{\sqrt{29}}\right)$ **8.** $\theta = \cos^{-1}\left(\frac{-1}{\sqrt{15}}\right)$

In Exercises 9–12, the point P is on the terminal side of the angle θ. Find the measure of θ if $0° < \theta < 360°$.

9. $P(-3, 4)$ **10.** $P(5, 9)$

11. $P(5, -7)$ **12.** $P(-2, -5)$

13. A naval ship leaves Port Norfolk and averages 42 knots (nautical mph) traveling for 3 h on a course of 40° and then 5 h on a course of 125°. What is the boat's bearing and distance from Port Norfolk after 8 h?

SECTION EXERCISES 8.3

In Exercises 1–4, show that $\overrightarrow{RS}$ and $\overrightarrow{OP}$ are equivalent.

1. $R = (-4, 7)$, $S = (-1, 5)$, $O = (0, 0)$, and $P = (3, -2)$

2. $R = (7, -3)$, $S = (4, -5)$, $O = (0, 0)$, and $P = (-3, -2)$

3. $R = (2, 1)$, $S = (0, -1)$, $O = (1, 4)$, and $P = (-1, 2)$

4. $R = (-2, -1)$, $S = (2, 4)$, $O = (-3, -1)$, and $P = (1, 4)$

In Exercises 5–12, let $P = (-1, 2)$, $Q = (3, 4)$, $R = (-2, 5)$ and $S = (2, -8)$. Write the vector **v** in component form, and find the magnitude of **v**.

5. $\mathbf{v} = \overrightarrow{PQ}$

6. $\mathbf{v} = \overrightarrow{RS}$

7. $\mathbf{v} = \overrightarrow{QR}$

8. $\mathbf{v} = \overrightarrow{PS}$

9. $\mathbf{v} = 2\overrightarrow{QS}$

10. $\mathbf{v} = (\sqrt{2})\overrightarrow{PR}$

11. $\mathbf{v} = 3\overrightarrow{QR} + \overrightarrow{PS}$

12. $\mathbf{v} = \overrightarrow{PS} - 3\overrightarrow{PQ}$

In Exercises 13–20, let $\mathbf{u} = \langle -1, 3\rangle$, $\mathbf{v} = \langle 2, 4\rangle$, and $\mathbf{w} = \langle 2, -5\rangle$. Find the component form of the vector.

13. $\mathbf{u} + \mathbf{v}$

14. $\mathbf{u} + (-1)\mathbf{v}$

15. $\mathbf{u} - \mathbf{w}$

16. $3\mathbf{v}$

17. $2\mathbf{u} + 3\mathbf{w}$

18. $2\mathbf{u} - 4\mathbf{v}$

19. $-2\mathbf{u} - 3\mathbf{v}$

20. $-\mathbf{u} - \mathbf{v}$

In Exercises 21–24, find the unit vector in the direction of the given vector. Write your answer in

a. component form, and
b. as a linear combination of the standard unit vectors **i** and **j**.

21. $\mathbf{u} = \langle 2, 1\rangle$

22. $\mathbf{u} = \langle -3, 2\rangle$

23. $\mathbf{u} = \langle -4, -5\rangle$

24. $\mathbf{u} = \langle 3, -4\rangle$

In Exercises 25–28, write the vector **v** in component form.

25.

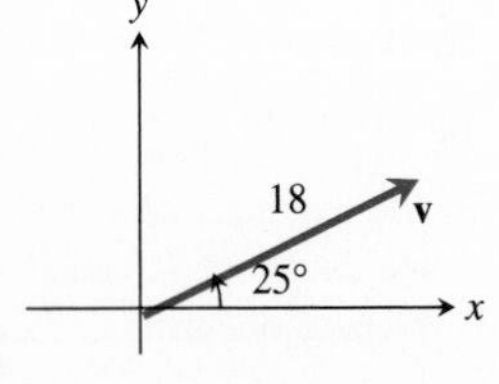

26.

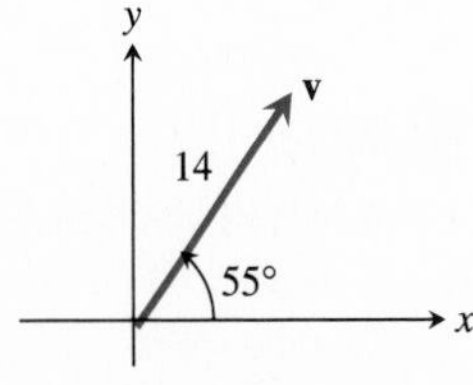

27.

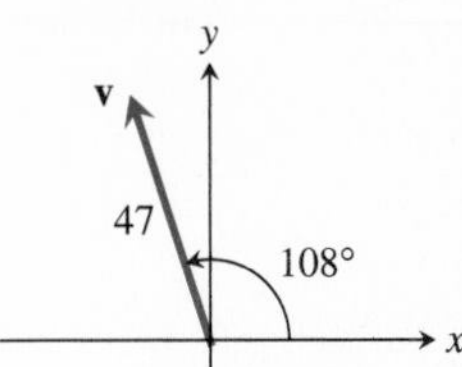

28.

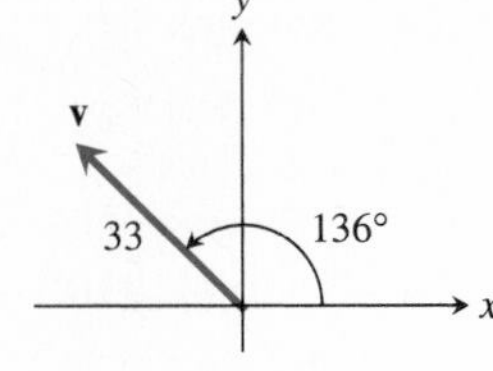

In Exercises 29–32, find the following:

a. The direction angles of **u** and **v**
b. The angle between **u** and **v**

29. $\mathbf{u} = \langle 3, 4\rangle$ and $\mathbf{v} = \langle 1, 0\rangle$

30. $\mathbf{u} = \langle -1, 2\rangle$ and $\mathbf{v} = \langle 3, 2\rangle$

31. $\mathbf{u} = \langle -1, 2\rangle$ and $\mathbf{v} = \langle 3, -4\rangle$

32. $\mathbf{u} = \langle 2, -3\rangle$ and $\mathbf{v} = \langle -3, -5\rangle$

In Exercises 33–40, solve the problem.

33. *Navigation* An airplane is flying on a bearing of 335° at 530 mph. Find the component form of the velocity of the airplane.

34. *Navigation* An airplane is flying on a bearing of 170° at 460 mph. Find the component form of the velocity of the airplane.

35. *Flight Engineering* An airplane is flying on a compass heading of 340° at 325 mph. A wind is blowing with the bearing 320° at 40 mph.

a. Find the component form of the velocity of the airplane.
b. Find the actual ground speed and direction of the plane.

36. *Flight Engineering* An airplane is flying on a compass heading of 170° at 460 mph. A wind is blowing with the bearing 200° at 80 mph.

a. Find the component form of the velocity of the airplane.
b. Find the actual ground speed and direction of the airplane.

37. *Shooting Free Throws* A basketball is shot at a 70° angle with the horizontal direction with an initial speed of 10 m/sec.

a. Find the component form of the velocity.
b. **Writing to Learn** Give an interpretation of the horizontal and vertical components of the velocity.

38. *Moving a Heavy Object* In a warehouse a box is being pushed up a 15° inclined plane with a force of 2.5 lb, as shown in the figure.

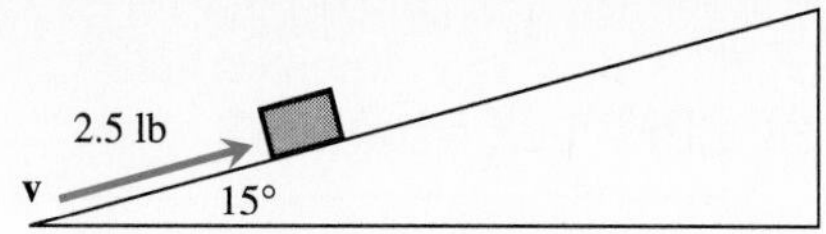

For Exercise 38

a. Find the component form of the force.

b. Writing to Learn Give an interpretation of the horizontal and vertical components of the force.

39. *Moving a Heavy Object* Suppose the box described in Exercise 38 is being towed up the inclined plane, as shown in the figure below. Find the force **w** needed in order for the component of the force parallel to the inclined plane to be 2.5 lb. Give the answer in component form.

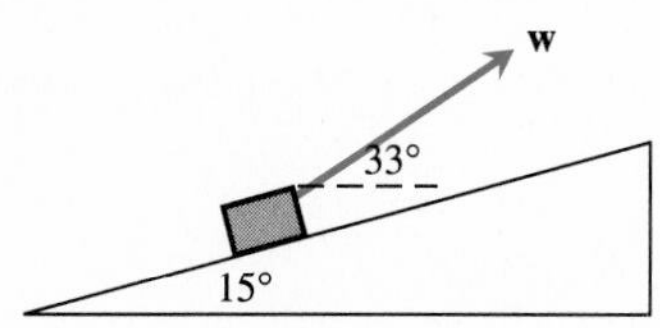

For Exercise 39

40. *Combining Forces* Juana and Diego Gonzales, ages six and four respectively, own a strong and stubborn puppy named Corporal. It is so hard to take Corporal for a walk that they devise a scheme to use two leashes. If Juana and Diego pull with forces of 23 lb and 18 lb at the angles shown in the figure, how hard is Corporal pulling if the puppy holds the children at a standstill?

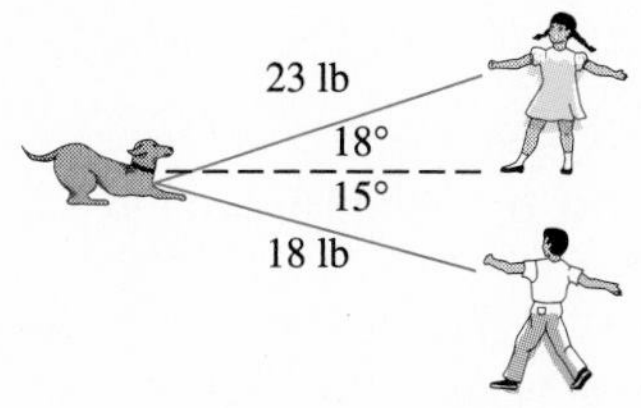

For Exercise 40

In Exercises 41 and 42, find the direction and magnitude of the resultant force.

41. *Combining Forces* A force of 50 lb acts on an object at an angle of 45°. A second force of 75 lb acts on the object at an angle of −30°.

42. *Combining Forces* Three forces with magnitudes 100, 50. and 80 lb, act on an object at angles of 50°, 160°, and −20°, respectively.

In Exercises 43–48, solve the problem.

43. *Unit Vectors* Let **v** be a nonzero vector and

$$\mathbf{u} = \frac{\mathbf{v}}{\|\mathbf{v}\|}.$$

Confirm the following.

a. **u** is a unit vector.

b. **u** and **v** have the same direction.

c. If **w** is in the direction of **u,** then $\mathbf{w} = k\mathbf{u}$. State the value of k.

44. Group Learning Activity *Work in groups of three.* Confirm that the following equations involving vector addition and scalar multiplication are identities.

a. $\mathbf{u} + \mathbf{v} = \mathbf{v} + \mathbf{u}$

b. $(\mathbf{u} + \mathbf{v}) + \mathbf{w} = \mathbf{u} + (\mathbf{v} + \mathbf{w})$

c. $\mathbf{u} + \mathbf{0} = \mathbf{u}$, where $\mathbf{0} = \langle 0, 0\rangle$

d. $\mathbf{u} + (-\mathbf{u}) = \mathbf{0}$ where $-\langle a, b\rangle = \langle -a, -b\rangle$

e. $a(\mathbf{u} + \mathbf{v}) = a\mathbf{u} + a\mathbf{v}$

f. $(a + b)\mathbf{u} = a\mathbf{u} + b\mathbf{u}$

g. $(ab)\mathbf{u} = a(b\mathbf{u})$

h. $a\mathbf{0} = \mathbf{0}, 0\mathbf{u} = \mathbf{0}$

i. $(1)\mathbf{u} = \mathbf{u}, (-1)\mathbf{u} = -\mathbf{u}$

j. $\|a\mathbf{u}\| = |a|\,\|\mathbf{u}\|$

EXTENDING THE IDEAS

45. *Linear Combination* Let $\mathbf{r} = \langle 1, 2\rangle$ and $\mathbf{s} = \langle 2, -1\rangle$. Show that the vector $\mathbf{v} = \langle 5, 7\rangle$ can be expressed as a linear combination of **r** and **s.** Explain what this means geometrically.

46. *Linear Combination* Let $\mathbf{r} = \langle 1, 2\rangle$ and $\mathbf{s} = \langle 2, -1\rangle$. Show that *any* vector $\mathbf{v} = \langle x, y\rangle$ can be expressed as a linear combination of **r** and **s.**

47. *Perpendicular Vectors* Show that the two vectors are perpendicular.

a. $\mathbf{u} = \langle 2, 3\rangle$ and $\mathbf{v} = \langle -3, 2\rangle$

b. $\mathbf{u} = \langle 1, -1\rangle$ and $\mathbf{v} = \langle -1, -1\rangle$

c. $\mathbf{u} = \langle 5, -3\rangle$ and $\mathbf{v} = \langle -3, -5\rangle$

48. *Dot Product* If $\mathbf{u} = \langle u_1, u_2\rangle$ and $\mathbf{v} = \langle v_1, v_2\rangle$, then the **dot product** of **u** and **v** is defined by

$$\mathbf{u} \cdot \mathbf{v} = u_1 v_1 + u_2 v_2.$$

Find the dot product of each pair of vectors in Exercise 47.

8.4 TRIGONOMETRIC FORM OF COMPLEX NUMBERS

Complex Plane • Absolute Value of a Complex Number • Trigonometric Form of a Complex Number • Multiplication and Division of Complex Numbers

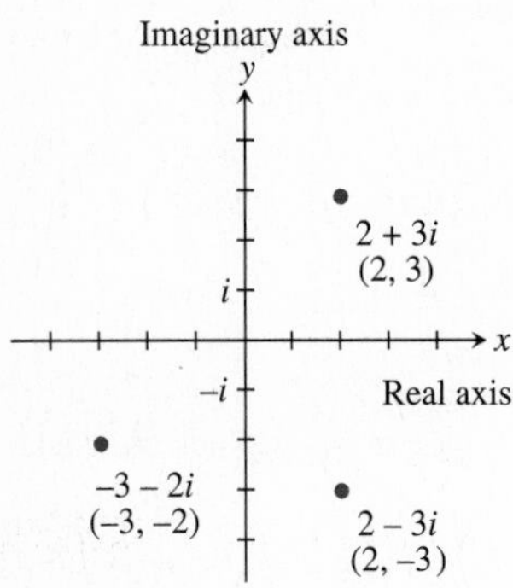

Figure 8.31

Complex Plane

In Section 3.5 we introduced complex numbers. In this section, after a brief review, we will see how complex numbers relate to trigonometry and vectors. Recall that the real numbers are represented geometrically as points on a number line. In a similar way the complex number $a + bi$ is represented geometrically as the point (a, b) in the coordinate plane. The real part of $a + bi$ is a, and the imaginary part is b.

When the coordinate plane is viewed as a model for the complex numbers, it is called the **complex plane** or **Gaussian plane.** Figure 8.31 shows the complex numbers $2 + 3i$, $-3 - 2i$, and $2 - 3i$. The horizontal axis of the complex plane is called the **real axis,** and the vertical axis is called the **imaginary axis.**

Objective

Students will be able to represent complex numbers in trigonometric form and perform operations on them.

Motivate

Have students find all solutions of the equation $z^4 = 1$, where z is a complex number. ($z = \pm 1, z = \pm i$)

Absolute Value of a Complex Number

Just as the absolute value of a real number is its distance from the origin, the *absolute value* of a complex number is also its distance from the origin.

Definition: Absolute Value of a Complex Number

The **absolute value,** or **modulus,** of the complex number $z = a + bi$ is

$$|z| = |a + bi| = \sqrt{a^2 + b^2}.$$

The absolute value of $a + bi$ is the length of the segment from the origin O to $a + bi$.

Notice that the *absolute value of the complex number $a + bi$ is the same as the magnitude of the vector* $\mathbf{v} = \langle a, b \rangle$ studied in Section 8.3.

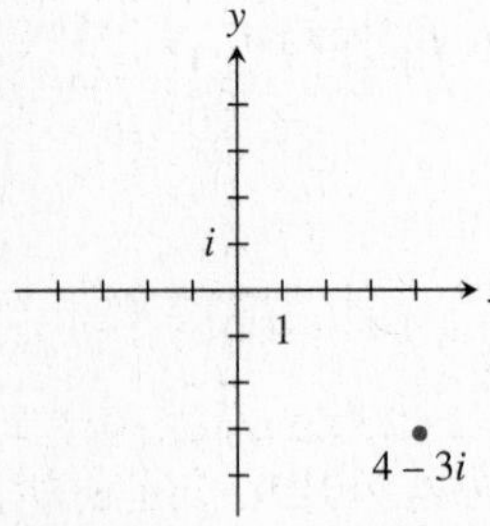

Figure 8.32 Example 1 shows $|4 - 3i|$ is 5, which is also the distance of the point $(4, -3)$ to the origin and the length of the vector $\langle 4, -3 \rangle$.

■ **EXAMPLE 1 Finding Absolute Value**

Plot $4 - 3i$, and find its absolute value.

Solution

See Figure 8.32.

$$|4 - 3i| = \sqrt{4^2 + (-3)^2} = \sqrt{25} = 5$$ ■

Alert

Some students may confuse the standard unit vector **i** with the imaginary unit i. In particular, it is important to recognize that **i** is a horizontal vector, while the imaginary axis is vertical.

Trigonometric Form of a Complex Number

There is a strong link between the component form of a vector and the *trigonometric form* of a complex number. You learned in Section 8.3 that a vector $\mathbf{v} = \langle a, b \rangle$ can be represented in terms of the standard unit vectors as

$$\begin{aligned}\mathbf{v} &= (\|\mathbf{v}\|\cos\theta)\mathbf{i} + (\|\mathbf{v}\|\sin\theta)\mathbf{j}\\ &= \|\mathbf{v}\|(\cos\theta\,\mathbf{i} + \sin\theta\,\mathbf{j}).\end{aligned}$$

We might call this the *trigonometric form* of the vector **v**. It is not surprising that there is a similar form for complex numbers.

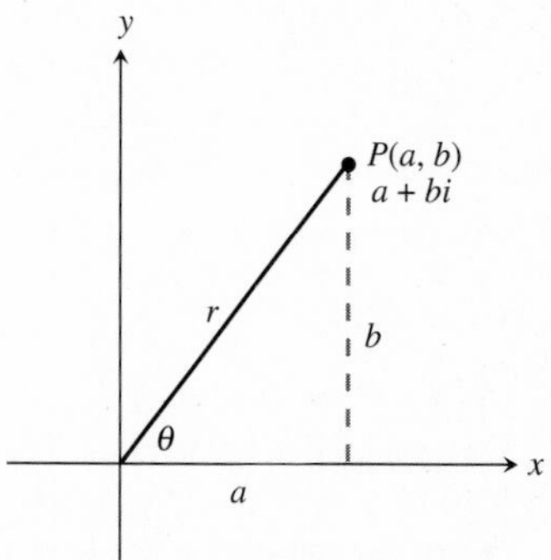

Figure 8.33 The angle θ is the argument of the complex number $a + bi$, and the direction angle of the vector $\langle a, b \rangle$.

Definition: Trigonometric Form of a Complex Number

Let $z = a + bi$ be a complex number where $r = |z| = \sqrt{a^2 + b^2}$ and $\theta =$ the **argument** of z. (See Figure 8.33.) Then $a = r\cos\theta$, $b = r\sin\theta$, and

$$\begin{aligned}z &= (r\cos\theta) + (r\sin\theta)i\\ &= r(\cos\theta + i\sin\theta)\end{aligned}$$

is a **trigonometric form** of z.

An angle θ can always be chosen so that $0 \le \theta < 2\pi$, although any angle coterminal with θ could be used. Consequently, the angle θ is not unique, and hence the trigonometric form of z is not unique.

Teaching Note

It may be useful to review complex numbers as introduced in Section 3.5.

■ EXAMPLE 2 Finding Trigonometric Forms

Find the trigonometric form with $0 \le \theta < 2\pi$ for the complex number.

a. $1 - \sqrt{3}i$ **b.** $-3 - 4i$

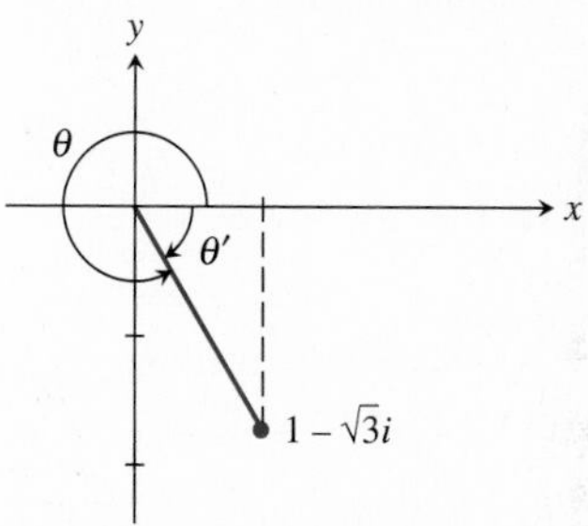

Figure 8.34 Graph for Example 2a.

Solution

a. For $1 - \sqrt{3}i$,

$$r = |1 - \sqrt{3}i| = \sqrt{1^2 + (\sqrt{3})^2} = 2.$$

Because θ' is $-\pi/3$ (Figure 8.34), $\theta = 2\pi + (-\pi/3) = 5\pi/3$.

$$1 - \sqrt{3}i = 2\cos\frac{5\pi}{3} + 2i\sin\frac{5\pi}{3}.$$

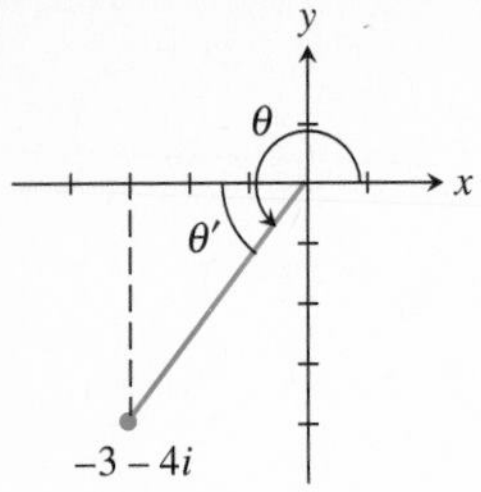

Figure 8.35 Graph for Example 2b.

b. For $-3 - 4i$,

$$|-3 - 4i| = \sqrt{(-3)^2 + (-4)^2} = 5.$$

Because angle θ' is formed with the negative x-axis (Figure 8.35), it satisfies the equation

$$\tan \theta' = \frac{4}{3}, \quad \text{so}$$

$$\theta' = \tan^{-1}\left(\frac{4}{3}\right) = 0.927 \cdots.$$

Because the terminal side of θ is in the third quadrant, we conclude that $\theta = \pi + \theta' = 3.141 \cdots + 0.927 \cdots \approx 4.07$. Therefore,

$$-3 - 4i \approx 5(\cos 4.07 + i \sin 4.07).$$

Teaching Note

Many of the calculations discussed in this section and the next can be simplified using a grapher's built-in functions for converting between polar and rectangular coordinates.

Multiplication and Division of Complex Numbers

The trigonometric form is particularly convenient for remembering the rules for multiplying and dividing complex numbers. The product involves the product of the moduli and the sum of the arguments. (*Moduli* is the plural of *modulus.*) The quotient involves the quotient of the moduli and the difference of the arguments.

Teaching Note

The proofs of the product and quotient formulas are good applications of the sum and difference identities studied in Section 7.4.

Product and Quotient of Complex Numbers

Let $z_1 = r_1(\cos \theta_1 + i \sin \theta_1)$ and $z_2 = r_2(\cos \theta_2 + i \sin \theta_2)$. Then

1. $z_1 \cdot z_2 = r_1 r_2[\cos(\theta_1 + \theta_2) + i \sin(\theta_1 + \theta_2)]$.
2. $\dfrac{z_1}{z_2} = \dfrac{r_1}{r_2}[\cos(\theta_1 - \theta_2) + i \sin(\theta_1 - \theta_2)], \quad (r_2 \neq 0)$.

Proof (of the Product Formula)

$$\begin{aligned} z_1 \cdot z_2 &= r_1(\cos \theta_1 + i \sin \theta_1) \cdot r_2(\cos \theta_2 + i \sin \theta_2) \\ &= r_1 r_2[(\cos \theta_1 \cos \theta_2 - \sin \theta_1 \sin \theta_2) + \\ &\quad i(\sin \theta_1 \cos \theta_2 + \cos \theta_1 \sin \theta_2)] \\ &= r_1 r_2[\cos(\theta_1 + \theta_2) + i \sin(\theta_1 + \theta_2)] \end{aligned}$$

You will be asked to prove the quotient formula in Exercise 47.

Follow-up

Have students discuss similarities and differences between complex numbers and vectors.

Assignment Guide

Day 1: Ex. 3–33, multiples of 3
Day 2: Ex. 35, 38, 39, 42, 43, 46, 47, 50, 51

Cooperative Learning

Group Activity: Ex. 49

Notes on Exercises

Ex. 35–42 provide an opportunity to show how much the product and quotient formulas can simplify computations when applicable. A thorough treatment of these problems will prepare students for the next section.
Ex. 50 provides a preview of graphing polar equations.
Ex. 51–52 give a graphical interpretation of the product of two complex numbers.

Ongoing Assessment

Self-Assessment: Ex. 5, 19, 31, 37, 41
Embedded Assessment: Ex. 46

REMARK

Notice that we can use degrees or radians in the trigonometric form when we apply the product or quotient formulas.

EXAMPLE 3 Multiplying Complex Numbers

Express the product of $z_1 = 25\sqrt{2}\left(\cos\frac{-\pi}{4} + i\sin\frac{-\pi}{4}\right)$ and $z_2 = 14\left(\cos\frac{\pi}{3} + i\sin\frac{\pi}{3}\right)$ in standard form.

Solution

$$\begin{aligned} z_1 \cdot z_2 &= 25\sqrt{2}\left(\cos\frac{-\pi}{4} + i\sin\frac{-\pi}{4}\right) \cdot 14\left(\cos\frac{\pi}{3} + i\sin\frac{\pi}{3}\right) \\ &= 25 \cdot 14\sqrt{2}\left[\cos\left(\frac{-\pi}{4} + \frac{\pi}{3}\right) + i\sin\left(\frac{-\pi}{4} + \frac{\pi}{3}\right)\right] \\ &= 350\sqrt{2}\left(\cos\frac{\pi}{12} + i\sin\frac{\pi}{12}\right) \\ &\approx 478.11 + 128.11i \end{aligned}$$

■

EXAMPLE 4 Dividing Complex Numbers

Divide $z_1 = 2\sqrt{2}(\cos 135° + i\sin 135°)$ by $z_2 = 6(\cos 300° + i\sin 300°)$. Express the answer in standard form.

Solution

$$\begin{aligned} \frac{z_1}{z_2} &= \frac{2\sqrt{2}(\cos 135° + i\sin 135°)}{6(\cos 300° + i\sin 300°)} \\ &= \frac{\sqrt{2}}{3}[\cos(135° - 300°) + i\sin(135° - 300°)] \\ &= \frac{\sqrt{2}}{3}[\cos(-165°) + i\sin(-165°)] \\ &= (-0.455\cdots) - (0.122\cdots)i \end{aligned}$$

■

Quick Review 8.4

In Exercises 1–6, write the complex number in standard form $a + bi$.

1. $(2 + i) + (4 - 2i)$
2. $(5 - 3i) - (7 - i)$
3. $(3 - i)(4 + 2i)$
4. $(2 - 2i)(-1 + 3i)$
5. $\dfrac{3 - 2i}{1 + i}$
6. $\dfrac{2 - 3i}{1 - 2i}$

In Exercises 7–12, find an angle θ where $0 \le \theta < 2\pi$ and which satisfies both equations.

7. $\sin\theta = \dfrac{1}{2}$ and $\cos\theta = \dfrac{-\sqrt{3}}{2}$
8. $\sin\theta = \dfrac{-\sqrt{2}}{2}$ and $\cos\theta = \dfrac{\sqrt{2}}{2}$

9. $\sin\theta = \dfrac{-\sqrt{3}}{2}$ and $\cos\theta = \dfrac{-1}{2}$

10. $\sin\theta = \dfrac{-1}{2}$ and $\cos\theta = \dfrac{\sqrt{3}}{2}$

11. $\sin\theta = \dfrac{-\sqrt{2}}{2}$ and $\cos\theta = \dfrac{-\sqrt{2}}{2}$

12. $\sin\theta = \dfrac{\sqrt{3}}{2}$ and $\cos\theta = \dfrac{-1}{2}$

In Exercises 13–16, a and b are given. Find the following.

a. The magnitude of the vector $\langle a, b\rangle$.
b. The distance of the point (a, b) to the origin.
c. The values of r and θ where $0 \le \theta < 2\pi$, such that $a = r\cos\theta$ and $b = r\sin\theta$.

13. $a = -1$ and $b = 1$ **14.** $a = 2$ and $b = -2$

15. $a = -\sqrt{3}$ and $b = -1$ **16.** $a = 1$ and $b = \sqrt{3}$

SECTION EXERCISES 8.4

In Exercises 1–6, plot the complex number in the complex plane and find its absolute value.

1. $2 + 7i$ **2.** $-4 + i$

3. $6 - 5i$ **4.** $-1 - 7i$

5. $5 - 3i$ **6.** $3 + 2i$

In Exercises 7–14, find the trigonometric form of the complex number where $0 \le \theta < 2\pi$.

7. $3i$ **8.** $-2i$

9. $2 + 2i$ **10.** $\sqrt{3} + i$

11. $-2 + 2i\sqrt{3}$ **12.** $3 - 3i$

13. $\dfrac{1}{2} - \dfrac{\sqrt{3}}{2}i$ **14.** $-\dfrac{\sqrt{3}}{2} - \dfrac{1}{2}i$

In Exercises 15–22, find an approximation of the complex number in trigonometric form with $0 \le \theta < 2\pi$.

15. $3 + 2i$ **16.** $4 - 7i$

17. $-1 + 3i$ **18.** $-9 + i$

19. $13 + 8i$ **20.** $-11 - 17i$

21. $\sqrt{5} - i\sqrt{3}$ **22.** $-3\sqrt{2} - i\sqrt{5}$

In Exercises 23–34, write the complex number in standard form.

23. $3(\cos 30° - i\sin 30°)$ **24.** $2(\cos 45° + i\sin 45°)$

25. $8(\cos 210° + i\sin 210°)$

26. $5[\cos(-60°) + i\sin(-60°)]$

27. $3\left(\cos\dfrac{\pi}{3} + i\sin\dfrac{\pi}{3}\right)$ **28.** $2\left(\cos\dfrac{-\pi}{3} + i\sin\dfrac{-\pi}{3}\right)$

29. $5\left(\cos\dfrac{\pi}{4} + i\sin\dfrac{\pi}{4}\right)$ **30.** $4\left(\cos\dfrac{5\pi}{3} + i\sin\dfrac{5\pi}{3}\right)$

31. $\sqrt{2}\left(\cos\dfrac{7\pi}{6} + i\sin\dfrac{7\pi}{6}\right)$

32. $7\left(\cos\dfrac{11\pi}{6} + i\sin\dfrac{11\pi}{6}\right)$

33. $\sqrt{3}\left(\cos\dfrac{5\pi}{4} + i\sin\dfrac{5\pi}{4}\right)$

34. $\sqrt{7}\left(\cos\dfrac{\pi}{12} + i\sin\dfrac{\pi}{12}\right)$

In Exercises 35–38, find the product of z_1 and z_2. Leave the answer in trigonometric form.

35. $z_1 = 7(\cos 25° + i\sin 25°)$
$z_2 = 2(\cos 130° + i\sin 130°)$

36. $z_1 = \sqrt{2}(\cos 118° + i\sin 118°)$
$z_2 = 0.5[\cos(-19°) + i\sin(-19°)]$

37. $z_1 = 5\left(\cos\dfrac{\pi}{4} + i\sin\dfrac{\pi}{4}\right)$
$z_2 = 3\left(\cos\dfrac{5\pi}{3} + i\sin\dfrac{5\pi}{3}\right)$

38. $z_1 = \sqrt{3}\left(\cos\dfrac{3\pi}{4} + i\sin\dfrac{3\pi}{4}\right)$
$z_2 = \dfrac{1}{3}\left(\cos\dfrac{\pi}{6} + i\sin\dfrac{\pi}{6}\right)$

In Exercises 39–42, find the trigonometric form of the quotient.

39. $\dfrac{2(\cos 30° + i\sin 30°)}{3(\cos 60° + i\sin 60°)}$ **40.** $\dfrac{5(\cos 220° + i\sin 220°)}{2(\cos 115° + i\sin 115°)}$

41. $\dfrac{6(\cos 5\pi + i\sin 5\pi)}{3(\cos 2\pi + i\sin 2\pi)}$ **42.** $\dfrac{\cos(\frac{\pi}{2}) + i\sin(\frac{\pi}{2})}{\cos(\frac{\pi}{4}) + i\sin(\frac{\pi}{4})}$

In Exercises 43–46, find the quotient z_1/z_2 in two ways.

a. Use the trigonometric form for z_1 and z_2.
b. Use the standard form for z_1 and z_2.

43. $z_1 = 3 - 2i$ and $z_2 = 1 + i$

44. $z_1 = 1 - i$ and $z_2 = \sqrt{3} + i$

45. $z_1 = 3 + i$ and $z_2 = 5 - 3i$

46. $z_1 = 2 - 3i$ and $z_2 = 1 - \sqrt{3}i$

EXTENDING THE IDEAS

In Exercises 47–50, solve the problem.

47. *Quotient Formula* Let $z_1 = r_1(\cos \theta_1 + i \sin \theta_1)$ and $z_2 = r_2(\cos \theta_2 + i \sin \theta_2)$, $r_2 \neq 0$. Verify that

$$\frac{z_1}{z_2} = \frac{r_1}{r_2}[\cos(\theta_1 - \theta_2) + i \sin(\theta_1 - \theta_2)].$$

48. Writing to Learn Suppose that $z = r(\cos \theta + i \sin \theta)$. Explain why the *complex conjugate* $\bar{z}$ could be written in the form $r[\cos(-\theta) + i \sin(-\theta)]$.

49. Writing to Learn Let z_1 and z_2 be any two complex numbers.

a. If you are asked to show that $|z_1 z_2| = |z_1| \cdot |z_2|$, will you choose to represent z_1 and z_2 in standard form or trigonometric form? Why?

b. Show that $|z_1 z_2| = |z_1| \cdot |z_2|$.

50. *Identifying Graphs* For the indicated value of r, identify the graph of all complex numbers of the form

$$z = r(\cos \theta + i \sin \theta), \quad 0 \leq \theta \leq 2\pi.$$

a. $r = 1$ **b.** $r = 3$

Exercises 51 and 52 concern the product of two complex numbers z_1 and z_2.

51. Explain why the triangles formed by 0, 1, and z_1 and by 0, z_2 and $z_1 z_2$ shown in the figure are similar triangles.

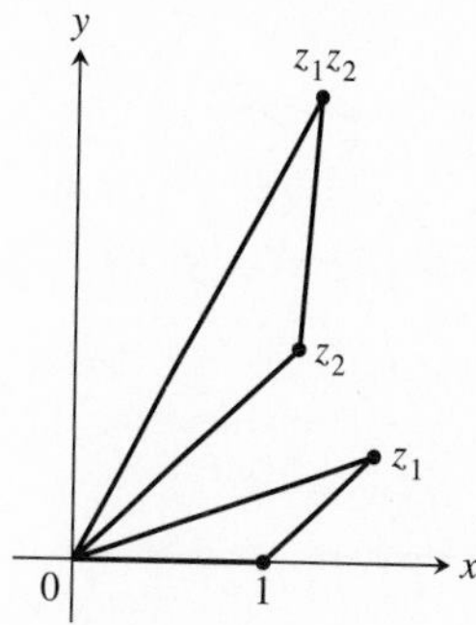

For Exercises 51–52

52. *Compass and Straightedge Construction* Using only a compass and straightedge, construct the location of $z_1 z_2$ given the location of 0, 1, z_1, and z_2.

8.5 DE MOIVRE'S THEOREM AND *n*TH ROOTS

> **Objective**
>
> Students will be able to use De Moivre's theorem and visualize the *n*th roots of a complex number.
>
> **Motivate**
>
> Have students use the product formula to calculate $[3(\cos 30° + i \sin 30°)]^4$ and comment on what they observe. **(81(cos 120° + *i* sin 120°))**

Powers of Complex Numbers • Roots of Complex Numbers • *n*th Roots of Unity

Powers of Complex Numbers

In Section 8.4 you learned that the product of two complex numbers in trigonometric form can be found by multiplying their moduli and adding their arguments. Consequently, the trigonometric form is convenient for raising a complex number to a power.

For example, let $z = r(\cos \theta + i \sin \theta)$. Then

$$\begin{aligned} z^2 &= z \cdot z \\ &= r(\cos \theta + i \sin \theta) \cdot r(\cos \theta + i \sin \theta) \\ &= r^2[\cos(\theta + \theta) + i \sin(\theta + \theta)] \\ &= r^2(\cos 2\theta + i \sin 2\theta). \end{aligned}$$

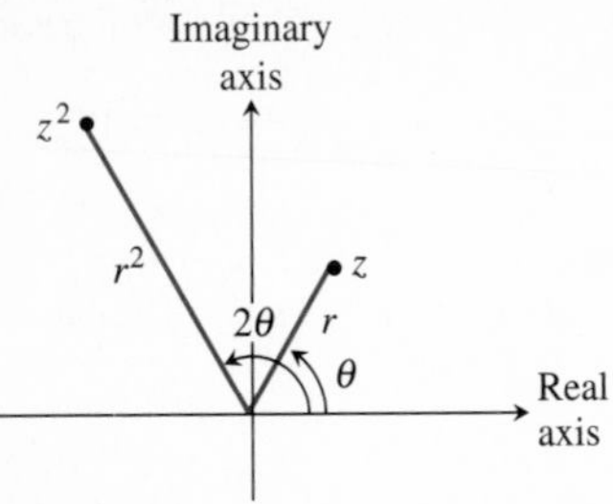

Figure 8.36

> **Teaching Note**
>
> This section provides a nice opportunity to bring geometry and algebra together. Providing geometric motivations to numerical work helps students connect different mathematical ideas.

Figure 8.36 gives a geometric interpretation of squaring a complex number: its argument is doubled, and its modulus is squared. That means its rotational position is doubled and its distance from the origin is changed by a factor r, increased if $r > 1$ or decreased if $r < 1$.

We can find z^3 by multiplying z by z^2:

$$\begin{aligned} z^3 &= z \cdot z^2 \\ &= r(\cos\theta + i\sin\theta) \cdot r^2(\cos 2\theta + i\sin 2\theta) \\ &= r^3[\cos(\theta + 2\theta) + i\sin(\theta + 2\theta)] \\ &= r^3(\cos 3\theta + i\sin 3\theta) \end{aligned}$$

Similarly,

$$\begin{aligned} z^4 &= r^4(\cos 4\theta + i\sin 4\theta) \\ z^5 &= r^5(\cos 5\theta + i\sin 5\theta) \\ &\vdots \end{aligned}$$

This pattern can be generalized to the following theorem, named after the mathematician Abraham De Moivre (1667–1754), who also made major contributions to the field of probability.

> **Theorem:** De Moivre's Theorem
>
> Let $z = r(\cos\theta + i\sin\theta)$ and let n be any positive integer. Then
>
> $$z^n = [r(\cos\theta + i\sin\theta)]^n = r^n(\cos n\theta + i\sin n\theta).$$

■ EXAMPLE 1 Using De Moivre's Theorem

Find $(1 + i\sqrt{3})^3$ using De Moivre's theorem.

Solution

The argument of $z = 1 + i\sqrt{3}$ is $\theta = \pi/3$, and its modulus is $|1 + i\sqrt{3}| = \sqrt{1 + 3} = 2$. (See Figure 8.37.) Therefore,

$$\begin{aligned} z &= 2\left(\cos\frac{\pi}{3} + i\sin\frac{\pi}{3}\right) \\ z^3 &= 2^3\left[\cos\left(3 \cdot \frac{\pi}{3}\right) + i\sin\left(3 \cdot \frac{\pi}{3}\right)\right] \\ &= 8(\cos\pi + i\sin\pi) \\ &= 8(-1 + 0i) \\ &= -8. \end{aligned}$$

■

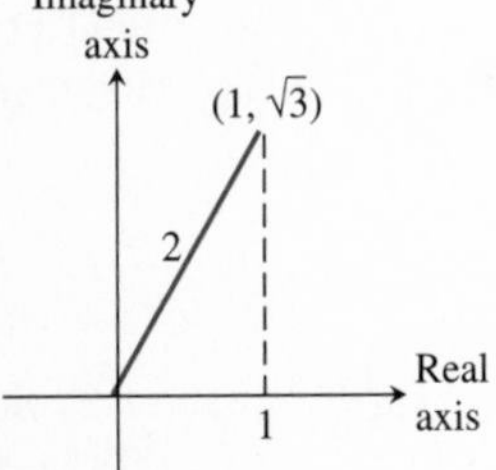

Figure 8.37

Notes on Examples

Problems like Example 2 are frequently found on tests in math contests. They are easy if a student knows De Moivre's theorem.

■ **EXAMPLE 2** Using De Moivre's Theorem

Find $\left(\frac{-\sqrt{2}}{2} + \frac{\sqrt{2}}{2}i\right)^8$ using De Moivre's theorem.

Solution

The argument of $z = (-\sqrt{2}/2) + (\sqrt{2}/2)i$ is $\theta = 3\pi/4$, and its modulus is

$$\left|\frac{-\sqrt{2}}{2} + \frac{\sqrt{2}}{2}i\right| = \sqrt{\frac{1}{2} + \frac{1}{2}} = 1.$$

Therefore,

$$\begin{aligned} z &= \cos\frac{3\pi}{4} + \sin\frac{3\pi}{4}i, \\ z^8 &= \cos\left(8 \cdot \frac{3\pi}{4}\right) + i\sin\left(8 \cdot \frac{3\pi}{4}\right) \\ &= \cos 6\pi + i\sin 6\pi \\ &= 1 + 0i \\ &= 1. \end{aligned}$$

■

REMARK

Finding the eighth power in Example 2 by repeated multiplication would be very tedious. Notice how quick and easy it is using De Moivre's theorem instead.

Roots of Complex Numbers

Notice that the complex number $1 + i\sqrt{3}$ in Example 1 is a solution of $z^3 = -8$ and the complex number $\frac{-\sqrt{2}}{2} + \frac{\sqrt{2}}{2}i$ is a solution of $z^8 = 1$. It is reasonable to call $1 + \sqrt{3}i$ a third root of -8 and $\frac{-\sqrt{2}}{2} + \frac{\sqrt{2}}{2}i$ an eighth root of 1. We will use the following terminology.

Teaching Note

It is worth pointing out that "unity" simply means "one."

nth Root

A complex number $v = a + bi$ is an nth root of z if

$$v^n = z.$$

If $z = 1$, then v is an ***n*th root of unity.**

We can use De Moivre's theorem to develop a general formula for finding the nth roots of a nonzero complex number.

Suppose that $v = s(\cos\alpha + i\sin\alpha)$ is an nth root of $z = r(\cos\theta + i\sin\theta)$. Then

$$\begin{aligned}
v^n &= z \\
(s(\cos\alpha + i\sin\alpha))^n &= r(\cos\theta + i\sin\theta) \\
s^n(\cos n\alpha + i\sin n\alpha) &= r(\cos\theta + i\sin\theta) \\
|s^n(\cos n\alpha + i\sin n\alpha)| &= |r(\cos\theta + i\sin\theta)| && \text{Taking absolute value} \\
s^n &= r && \text{See Exercise 35.} \\
s &= \sqrt[n]{r}
\end{aligned}$$

Substituting $s^n = r$ in the third of the six above equations, we find

$$\cos n\alpha + i\sin n\alpha = \cos\theta + i\sin\theta.$$

Therefore, $n\alpha$ can be any angle coterminal with θ. Consequently, for any integer k, v is an nth root of z if $s = \sqrt[n]{r}$ and

$$\begin{aligned}
n\alpha &= \theta + 2\pi k \\
\alpha &= \frac{\theta + 2\pi k}{n}.
\end{aligned}$$

The expression for **v** takes on n different values for $k = 0, 1, \ldots, n - 1$, and the values start repeating for $k = n, n + 1, \ldots$.

We summarize this result.

nth Roots of a Complex Number

If $z = r(\cos\theta + i\sin\theta)$, then the n distinct complex numbers

$$\sqrt[n]{r}\left(\cos\frac{\theta + 2\pi k}{n} + i\sin\frac{\theta + 2\pi k}{n}\right),$$

where $k = 0, 1, 2, \ldots, n - 1$, are the nth roots of the complex number z.

■ EXAMPLE 3 Finding Fourth Roots

Find the fourth roots of $z = 5\left(\cos\frac{\pi}{3} + i\sin\frac{\pi}{3}\right)$.

Solution

The fourth roots of z are the complex numbers

$$\sqrt[4]{5}\left(\cos\frac{\pi/3 + 2\pi k}{4} + i\sin\frac{\pi/3 + 2\pi k}{4}\right)$$

for $k = 0, 1, 2, 3$. Taking into account that $\dfrac{\pi/3 + 2\pi k}{4} = \dfrac{\pi}{12} + \dfrac{\pi k}{2}$, the list becomes:

$$z_1 = \sqrt[4]{5}\left[\cos\left(\frac{\pi}{12} + \frac{0}{2}\right) + i \sin\left(\frac{\pi}{12} + \frac{0}{2}\right)\right]$$

$$= \sqrt[4]{5}\left(\cos\frac{\pi}{12} + i \sin\frac{\pi}{12}\right)$$

$$z_2 = \sqrt[4]{5}\left[\cos\left(\frac{\pi}{12} + \frac{\pi}{2}\right) + i \sin\left(\frac{\pi}{12} + \frac{\pi}{2}\right)\right]$$

$$= \sqrt[4]{5}\left(\cos\frac{7\pi}{12} + i \sin\frac{7\pi}{12}\right)$$

$$z_3 = \sqrt[4]{5}\left[\cos\left(\frac{\pi}{12} + \frac{2\pi}{2}\right) + i \sin\left(\frac{\pi}{12} + \frac{2\pi}{2}\right)\right]$$

$$= \sqrt[4]{5}\left(\cos\frac{13\pi}{12} + i \sin\frac{13\pi}{12}\right)$$

$$z_4 = \sqrt[4]{5}\left[\cos\left(\frac{\pi}{12} + \frac{3\pi}{2}\right) + i \sin\left(\frac{\pi}{12} + \frac{3\pi}{2}\right)\right]$$

$$= \sqrt[4]{5}\left(\cos\frac{19\pi}{12} + i \sin\frac{19\pi}{12}\right)$$

■

Notes on Examples

Example 4 can also be solved by writing the equation $z^3 + 1 = 0$, factoring, and using the quadratic formula. It is useful for students to see that this method gives the same answer.

■ **EXAMPLE 4 Finding Cube Roots**

Find the cube roots of -1 and plot them.

Solution

Solve Algebraically

The third roots of $z = -1 + 0i = \cos \pi + i \sin \pi$ are the complex numbers

$$\cos\frac{\pi + 2\pi k}{3} + i \sin\frac{\pi + 2\pi k}{3}.$$

for $k = 0, 1, 2$. The three complex numbers are

$$z_1 = \cos\frac{\pi}{3} + i \sin\frac{\pi}{3} = \frac{1}{2} + \frac{\sqrt{3}}{2}i,$$

$$z_2 = \cos\frac{\pi + 2\pi}{3} + i \sin\frac{\pi + 2\pi}{3} = -1 + 0i,$$

$$z_3 = \cos\frac{\pi + 4\pi}{3} + i \sin\frac{\pi + 4\pi}{3} = \frac{1}{2} - \frac{\sqrt{3}}{2}i.$$

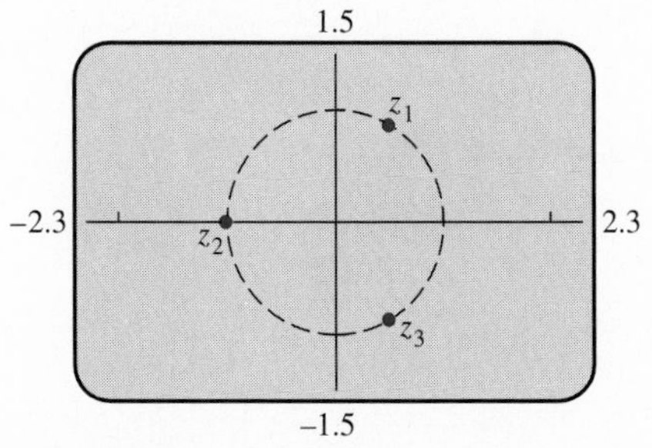

Figure 8.38 z_1, z_2, and z_3 are the three cube roots of -1. The dotted curve is the unit circle.

Figure 8.38 shows the graph of the three cube roots z_1, z_2, and z_3. Notice that they are evenly spaced (with distances of $2\pi/3$ radians) around the unit circle. ■

In Exercise 36, you will show that the nth roots of the complex number $r(\cos\theta + i\sin\theta)$ are evenly spaced around a circle of radius $\sqrt[n]{r}$. The spacing is $2\pi/n$ radians, and one point is

$$\sqrt[n]{r}\left(\cos\frac{\theta}{n} + i\sin\frac{\theta}{n}\right).$$

nth Roots of Unity

To find the nth roots of unity we must find solutions to the equation $z^n = 1$. By writing the number 1 as the complex number $1 + 0i = \cos 0 + i\sin 0$, we can use the formula for finding nth roots. There are n distinct nth roots of $1 + 0i$.

EXAMPLE 5 Finding Roots of Unity

Find the eight eighth roots of unity, and plot them.

Solution

Solve Algebraically

There are eight eighth roots of $1 + 0i = \cos 0 + i\sin 0$. The formula for nth roots when $n = 8$ and $\theta = 0$ is

$$\cos\frac{0 + 2\pi k}{8} + i\sin\frac{0 + 2\pi k}{8}$$

for $k = 0, 1, 2, \ldots, 7$.

$$z_1 = \cos 0 + i\sin 0 = 1 + 0i$$

$$z_2 = \cos\frac{\pi}{4} + i\sin\frac{\pi}{4} = \frac{\sqrt{2}}{2} + \frac{\sqrt{2}}{2}i$$

$$z_3 = \cos\frac{\pi}{2} + i\sin\frac{\pi}{2} = 0 + i$$

$$z_4 = \cos\frac{3\pi}{4} + i\sin\frac{3\pi}{4} = -\frac{\sqrt{2}}{2} + \frac{\sqrt{2}}{2}i$$

$$z_5 = \cos\pi + i\sin\pi = -1 + 0i$$

$$z_6 = \cos\frac{5\pi}{4} + i\sin\frac{5\pi}{4} = -\frac{\sqrt{2}}{2} - \frac{\sqrt{2}}{2}i$$

$$z_7 = \cos\frac{3\pi}{2} + i\sin\frac{3\pi}{2} = 0 - i$$

$$z_8 = \cos\frac{7\pi}{4} + i\sin\frac{7\pi}{4} = \frac{\sqrt{2}}{2} - \frac{\sqrt{2}}{2}i$$

Follow-up

Ask . . .

The complex number $\cos 3\pi + i\sin 3\pi$ is an 8th root of unity. Why is this number not listed in the solution to Example 5 (or is it)? **(It is the same as $\cos\pi + i\sin\pi = 1 + 0i$.)**

Assignment Guide

Day 1: Ex. 3–30, multiples of 3
Day 2: Ex. 33, 34, 36, 38, 39, 42, 45, 48

Cooperative Learning

Group Activity: Ex. 35

Notes on Exercises

Ex. 13–34 require students to use De Moivre's theorem to find nth roots of complex numbers. You might wish to have students think about the radius of the circle on which the roots fall and the angular spacing of the roots.
Ex. 37–38 are related to the Graphing Activity in this section.

Ongoing Assessment

Self-Assessment: Ex. 5, 25, 31
Embedded Assessment: Ex. 36

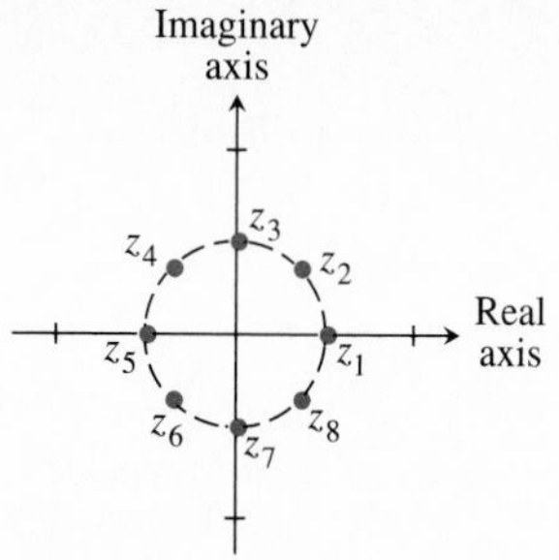

Figure 8.39 The eight eighth roots of unity are equally spaced on a unit circle.

Figure 8.39 shows the eight points. Check that they are spaced $2\pi/8 = \pi/4$ radians apart. ■

▼ GRAPHING ACTIVITY Visualizing the Roots of Unity

Set your grapher in parametric mode with $0 \leq T \leq 8$, so that Tstep $= 1$, Xmin $= -2.3$, Xmax $= 2.3$, Ymin $= -1.5$, and Ymax $= 1.5$.

1. Let $x_1 = \cos(2\pi/8)t$ and $y_1 = \sin(2\pi/8)t$. Use "trace" to visualize the eight eighth roots of unity. We say that $2\pi/8$ *generates* the eighth roots of unity. (Try both dot mode and connected mode.)
2. Replace $2\pi/8$ in Step 1 by the argument of other eighth roots of unity. Do any others *generate* the eighth roots of unity?
3. Repeat Steps 1 and 2 for the fifth, sixth, and seventh roots of unity.

What would you conjecture about an nth root of unity that generates all the nth roots of unity in the sense of Step 1? ▲

Graphing Activity Extensions

See if your conjecture holds for the 15th roots of unity.

Quick Review 8.5

In Exercises 1–6, write the complex number in trigonometric form.

1. $3 - 4i$ **2.** $-2 + 2i$

3. $5 + 3i$ **4.** $8 - 5i$

5. $-2i$ **6.** -10

In Exercises 7–10, write the complex number in standard form.

7. i^{12} **8.** $(-i)^9$

9. $(1 + i)^5$ **10.** $(1 - i)^4$

In Exercises 11–14, find all real solutions.

11. $x^3 - 1 = 0$ **12.** $x^4 - 1 = 0$

13. $x^5 - 1 = 0$ **14.** $x^6 - 1 = 0$

SECTION EXERCISES 8.5

In Exercises 1–12, use De Moivre's theorem to write the complex number in trigonometric form.

1. $\left(\cos\frac{\pi}{4} + i\sin\frac{\pi}{4}\right)^3$ **2.** $\left(\cos\frac{\pi}{3} + \sin\frac{\pi}{3}\right)^4$

3. $2(\cos\pi + i\sin\pi)^6$ **4.** $3\left(\cos\frac{3\pi}{2} + i\sin\frac{3\pi}{2}\right)^5$

5. $2\left(\cos\frac{3\pi}{4} + i\sin\frac{3\pi}{4}\right)^3$ **6.** $5\left(\cos\frac{5\pi}{6} + i\sin\frac{5\pi}{6}\right)^4$

7. $(1 + i)^5$

8. $(3 + 4i)^{20}$

9. $(1 - \sqrt{3}i)^3$

10. $\left(\frac{\sqrt{2}}{2} - i\frac{\sqrt{2}}{2}\right)^4$

11. $-\left(\frac{\sqrt{2}}{2} + i\frac{\sqrt{2}}{2}\right)^4$

12. $\left(\frac{1}{2} + i\frac{\sqrt{3}}{2}\right)^3$

In Exercises 13–18, find the cube roots of the number.

13. $2(\cos 2\pi + i \sin 2\pi)$

14. $2\left(\cos\frac{\pi}{4} + i\sin\frac{\pi}{4}\right)$

15. $3\left(\cos\frac{4\pi}{3} + i\sin\frac{4\pi}{3}\right)$

16. $27\left(\cos\frac{11\pi}{6} + i\sin\frac{11\pi}{6}\right)$

17. $3 - 4i$

18. $-2 + 2i$

In Exercises 19–24, find the fifth roots of the number.

19. $(\cos \pi + i \sin \pi)$

20. $32\left(\cos\frac{\pi}{2} + i\sin\frac{\pi}{2}\right)$

21. $2\left(\cos\frac{\pi}{6} + i\sin\frac{\pi}{6}\right)$

22. $2\left(\cos\frac{\pi}{4} + i\sin\frac{\pi}{4}\right)$

23. $2i$

24. $1 + \sqrt{3}i$

In Exercises 25–30, find the nth roots of the number for the specified value of n.

25. $1 + i, \quad n = 4$

26. $1 - i, \quad n = 6$

27. $2 + 2i, \quad n = 3$

28. $-2 + 2i, \quad n = 4$

29. $-2i, \quad n = 6$

30. $32, \quad n = 5$

In Exercises 31–34, express the roots of unity in the form $a + bi$. Graph each root in the complex plane.

31. Cube roots of unity

32. Fourth roots of unity

33. Sixth roots of unity

34. Square roots of unity

In Exercises 35–38, solve the problem.

35. *Equality of Complex Numbers* Confirm the following where $s(\cos \alpha + i \sin \alpha)$ and $r(\cos \theta + i \sin \theta)$ are trigonometric forms of a complex number.

a. $|s^n(\cos n\alpha + i \sin n\alpha)| = s^n$

b. $|r(\cos \theta + i \sin \theta)| = r$

c. Use the equations in parts a and b to show the following. If

$$s^n(\cos n\alpha + i \sin n\alpha) = r(\cos \theta + i \sin \theta),$$

and $s \neq 0$, $r \neq 0$, then $s = \sqrt[n]{r}$.

36. *nth Roots* Show that the nth roots of the complex number $r(\cos \theta + i \sin \theta)$ are spaced $2\pi/n$ radians apart on a circle with radius $\sqrt[n]{r}$.

37. Connecting Algebra and Geometry Using a graphing utility to draw a regular polygon with seven sides. (*Hint:* See the Graphing Activity in this section.)

38. Connecting Algebra and Geometry Use a graphing utility to draw a regular polygon with five sides. (*Hint:* See the Graphing Activity in this section.)

EXTENDING THE IDEAS

In Exercises 39–44, find all solutions to the equation and graph the solutions.

39. $z^4 - i = 0$

40. $z^4 - (5 - 5i) = 0$

41. $z^6 = 64i$

42. $z^6 = 2i$

43. $z^3 + (1 - i) = 0$

44. $z^3 - (\sqrt{3} - i) = 0$

In Exercises 45–50, solve the problem.

45. Determine z and the three cube roots of z if one cube root of z is $1 + \sqrt{3}i$.

46. Determine z and the four fourth roots of z if one fourth root of z is $-2 - 2i$.

47. Show that $(-1 + i)^{12} = -64$.

48. Show that $(1 + \sqrt{3}i)^8 = 128(-1 + \sqrt{3}i)$.

49. *Parametric Graphing* Write parametric equations that represent $(\sqrt{2} + i)^n$ for $n = t$. Draw and label an *accurate* spiral representing $(\sqrt{2} + i)^n$ for $n = 0, 1, 2, 3, 4$.

50. *Parametric Graphing* Write parametric equations that represent $(-1 + i)^n$ for $n = t$. Draw and label an *accurate* spiral representing $(-1 + i)^n$ for $n = 0, 1, 2, 3, 4$.

CHAPTER 8 REVIEW

KEY TERMS

The number following each key term indicates the page of its introduction.

REVIEW EXERCISES

In Exercises 1–8, solve $\triangle ABC$.

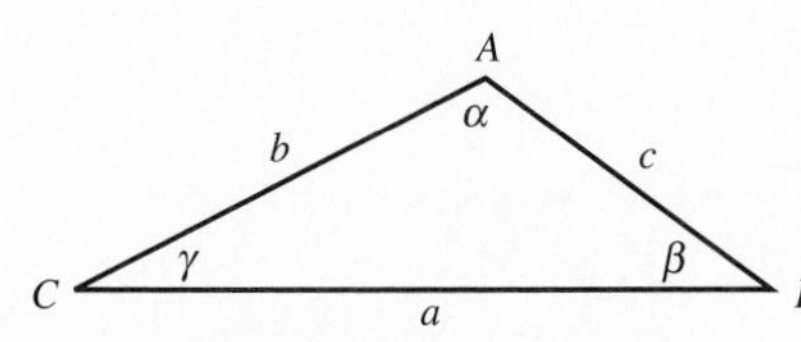

For Exercises 1–11

1. $\alpha = 79°, \quad \beta = 33°, \quad a = 7$

2. $a = 5, \quad b = 8, \quad \beta = 110°$

3. $a = 8, \quad b = 3, \quad \beta = 30°$

4. $a = 14.7, \quad \alpha = 29.3°, \quad \gamma = 33°$

5. $\alpha = 34°, \quad \beta = 74°, \quad c = 5$

6. $c = 41, \quad \alpha = 22.9°, \quad \gamma = 55.1°$

7. $a = 5, \quad b = 7, \quad c = 6$

8. $\alpha = 85°, \quad a = 6, \quad b = 4$

In Exercises 9 and 10, find the area of $\triangle ABC$.

9. $a = 3, \quad b = 5, \quad c = 6$

10. $a = 10, \quad b = 6, \quad \gamma = 50°$

In Exercises 11–16, solve the problem.

11. If $a = 12$ and $\beta = 28°$, determine the values of b that will produce the indicated number of triangles.

a. Two **b.** One **c.** None

12. *Surveying a Canyon* Two markers A and B on the same side of a canyon rim are 80 ft apart, as shown in the figure. A hiker is located across the rim at point C. A surveyor determines that $\angle BAC = 70°$ and $\angle ABC = 65°$.

a. What is the distance between the hiker and point A?
b. What is the distance between the two canyon rims? (Assume they are parallel.)

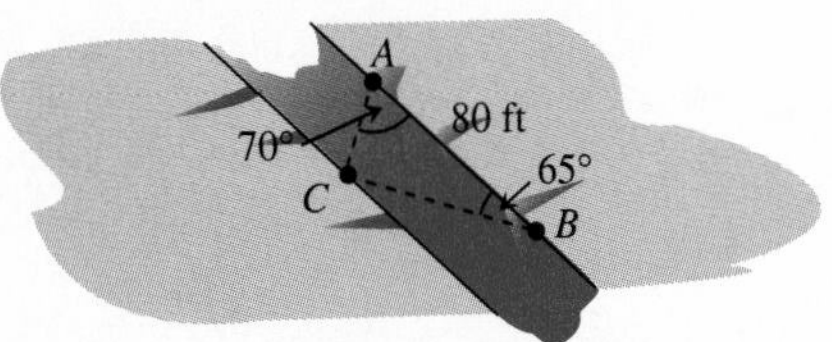

For Exercise 12

13. *Altitude* A hot-air balloon is seen over Tucson, Arizona, simultaneously by two observers at points A and B that are 1.75 mi apart on level ground and in line with the balloon. The angles of elevation are as shown here. How high above ground is the balloon?

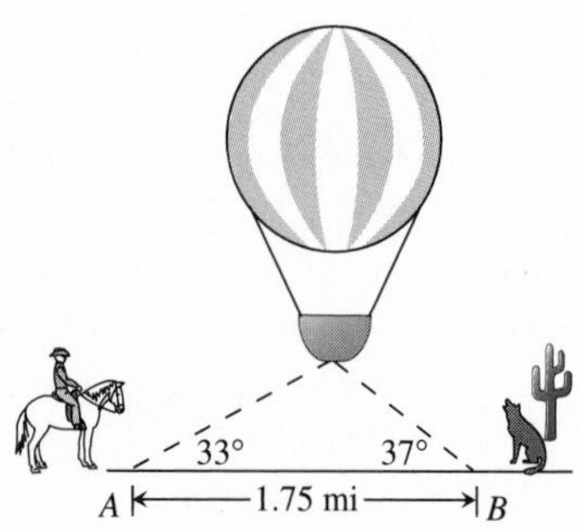

For Exercise 13

14. *Finding Distance* In order to determine the distance between two points A and B on opposite sides of a lake, a surveyor chooses a point C that is 900 ft from A and 225 ft from B, as shown in the figure. If the measure of the angle at C is 70°, find the distance between A and B.

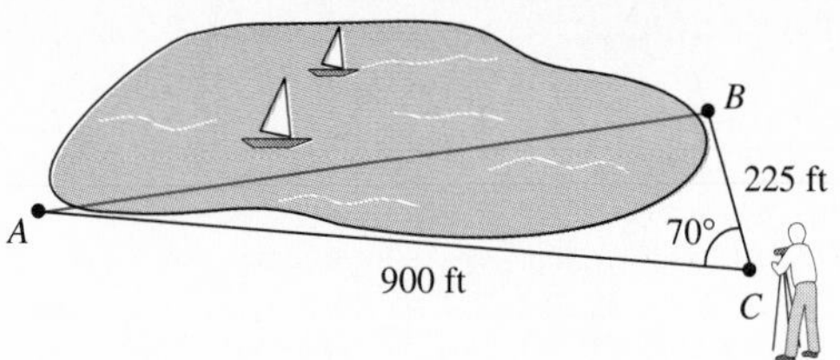

For Exercise 14

15. *Finding Radian Measure* Find the radian measure of the largest angle of the triangle whose sides have lengths 8, 9, and 10.

16. *Finding a Parallelogram* A parallelogram has sides of 15 and 24 ft, and an angle of 40°. Find the diagonals.

In Exercises 17–22, write the complex number in standard $a + bi$ form.

17. $(4 + 3i) - (2 - i)$
18. $(3 + 2i) + (-5 - 7i)$
19. $(1 - i)(3 + 2i)$
20. $(2 - 4i)(3 + i)$
21. $\dfrac{-2 + 4i}{1 - i}$
22. $\dfrac{-1 + 2i}{3 - 7i}$

In Exercises 23–26, graph the complex numbers in the complex plane.

23. $-5 + 2i$
24. $4 - 3i$
25. $2\left(\cos \dfrac{\pi}{3} + i \sin \dfrac{\pi}{3}\right)$
26. $2\left[\cos\left(-\dfrac{\pi}{6}\right) + i \sin\left(-\dfrac{\pi}{6}\right)\right]$

In Exercises 27 and 28, refer to the complex number z_1 shown in the figure.

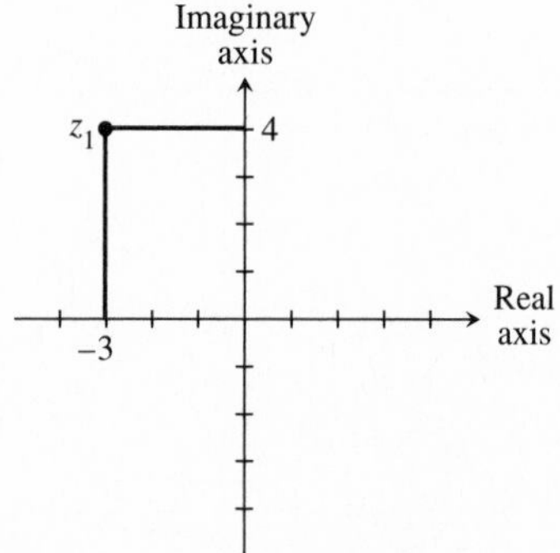

For Exercises 27–28

27. If $z_1 = a + bi$, find a, b, and $|z_1|$.

28. Find the trigonometric form of z_1.

In Exercises 29–32, let $z_1 = -1 + 4i$, $z_2 = 3 + 4i$, and $z_3 = 5 - 2i$. Write the given number in standard form.

29. $z_1 - z_2$

30. $z_2 z_3$

31. $|z_1 z_3|$

32. $z_1(z_2 + z_3)$

In Exercises 33–36, write the complex number in standard form.

33. $6(\cos 30° + i \sin 30°)$

34. $3(\cos 150° + i \sin 150°)$

35. $2.5\left(\cos \frac{4\pi}{3} + i \sin \frac{4\pi}{3}\right)$

36. $4(\cos 2.5 + i \sin 2.5)$

In Exercises 37–40, write the complex number in trigonometric form $r(\cos \theta + i \sin \theta)$ where $0 \le \theta < 2\pi$. Then write three other possible trigonometric forms for the number.

37. $3 - 3i$

38. $-1 + i\sqrt{2}$

39. $3 - 5i$

40. $-2 - 2i$

In Exercises 41 and 42, write the complex numbers $z_1 z_2$ and z_1/z_2 in trigonometric form.

41. $z_1 = 3(\cos 30° + i \sin 30°)$ and $z_2 = 4(\cos 60° + i \sin 60°)$

42. $z_1 = 5(\cos 20° + i \sin 20°)$ and $z_2 = -2(\cos 45° + i \sin 45°)$

In Exercises 43 and 44, solve the problem.

43. Write z_1, z_2, and z_1/z_2 from Exercise 41 in the form $a + bi$. Confirm that the quotient is the same as that obtained using trigonometric form.

44. Write z_1, z_2, and $z_1 z_2$ from Exercise 42 in the form $a + bi$. Confirm that the product is the same as that obtained using trigonometric form.

In Exercises 45–48, use De Moivre's theorem to write the complex number first in trigonometric form and then in $a + bi$ form.

45. $\left[3\left(\cos \frac{\pi}{4} + i \sin \frac{\pi}{4}\right)\right]^5$

46. $\left[2\left(\cos \frac{\pi}{12} + i \sin \frac{\pi}{12}\right)\right]^8$

47. $\left[5\left(\cos \frac{5\pi}{3} + i \sin \frac{5\pi}{3}\right)\right]^3$

48. $\left[7\left(\cos \frac{\pi}{24} + i \sin \frac{\pi}{24}\right)\right]^6$

In Exercises 49–52, find and graph the nth roots of the complex number for the specified value of n.

49. $3 + 3i, \quad n = 4$

50. $8, \quad n = 3$

51. $1, \quad n = 5$

52. $-1, \quad n = 6$

In Exercises 53–56, let $\mathbf{u} = \langle 2, -1 \rangle$, $\mathbf{v} = \langle 4, 2 \rangle$, and $\mathbf{w} = \langle 1, -3 \rangle$ be vectors, and find the indicated expression.

53. $\mathbf{u} - \mathbf{v}$

54. $2\mathbf{u} - 3\mathbf{w}$

55. $\|\mathbf{u} + \mathbf{v}\|$

56. $\|\mathbf{w} - 2\mathbf{u}\|$

In Exercises 57–60, let $A = (2, -1)$, $B = (3, 1)$, $C = (-4, 2)$, and $D = (1, -5)$. Write the expression as a vector whose initial point is at the origin, and find its magnitude.

57. $3\overrightarrow{AB}$

58. $\overrightarrow{AB} + \overrightarrow{CD}$

59. $\overrightarrow{AC} + \overrightarrow{BD}$

60. $\overrightarrow{CD} - \overrightarrow{AB}$

In Exercises 61 and 62, express vector $\overrightarrow{AB}$ as a linear combination of the vectors $i = \langle 1, 0 \rangle$ and $j = \langle 0, 1 \rangle$.

61. $A = (4, 0)$, $B = (2, 1)$

62. $A = (3, 1)$, $B = (5, 1)$

In Exercises 63 and 64, find the following.

a. The direction angles of **u** and **v**.
b. The angle between **u** and **v**.

63. $\mathbf{u} = \langle 4, 3 \rangle, \quad \mathbf{v} = \langle 2, 5 \rangle$

64. $\mathbf{u} = \langle -2, 4 \rangle, \quad \mathbf{v} = \langle 6, 4 \rangle$

In Exercises 65–72, solve the problem.

65. *Flight Engineering* An airplane is flying on a bearing of 80° at 540 mph. A wind is blowing with the bearing 100° at 55 mph.

a. Find the component form of the velocity of the airplane.
b. Find the actual speed and direction of the airplane.

66. *Flight Engineering* An airplane is flying on a bearing of 285° at 480 mph. A wind is blowing with the bearing 265° at 30 mph.

a. Find the component form of the velocity of the airplane.
b. Find the actual speed and direction of the airplane.

67. *Combining Forces* A force of 120 lb acts on an object at an angle of 20°. A second force of 300 lb acts on the object at an angle of −5°. Find the direction and magnitude of the resultant force.

68. Using Trigonometry in Geometry A regular hexagon whose sides are 16 cm is inscribed in a circle. Find the area inside the circle and outside the hexagon.

69. Using Trigonometry in Geometry A circle is inscribed in a regular pentagon whose sides are 12 cm. Find the area inside the pentagon and outside the circle.

70. Using Trigonometry in Geometry A wheel of cheese in the shape of a right circular cylinder is 18 cm in diameter and 5 cm thick. If a wedge of cheese with a central angle of 15° is cut from the wheel, find the volume of the cheese wedge.

71. *Linear Combination* Let $\mathbf{r} = \langle 1, 2 \rangle$ and $\mathbf{s} = \langle 2, -1 \rangle$. Show that any vector $\mathbf{v} = \langle x, y \rangle$ can be expressed as a linear combination of **r** and **s.**

72. *Subtracting Vectors* State a geometric definition of the subtraction of vectors. Then test your definition with the vectors $\mathbf{u} = \langle 2, 4 \rangle$ and $\mathbf{v} = \langle -1, 3 \rangle$ by comparing the result with the component-wise definition.

EVELYN BOYD GRANVILLE

As one of the first African-American women to receive a doctoral degree in mathematics, Evelyn Boyd Granville (b. 1924) paved the way for minorities and women in both education and industry. A graduate of Yale University in 1949, Evelyn used her mathematics background to obtain a position at IBM, working with the space program. Evelyn was closely involved with Vanguard, the first missile program in the United States. She later went on to work at Space Technology Labs as a mathematical analyst who studied rocket trajectories. Evelyn frequently used trigonometry on her job, often concerned with rocket angles and their paths.

In 1967, Evelyn left her industry position to begin teaching. She accepted a position on the faculty of Cal State University at Los Angeles, where she was committed to teaching students mathematics and computer science. Evelyn Boyd Granville retired her professorship in 1984, and is now enjoying a more peaceful lifestyle in Texas. However, she has not given up her love of teaching, as she has devoted some of her time to teaching at one of the local colleges in Texas.

chapter 9

PARAMETRIC EQUATIONS, CONICS, AND POLAR EQUATIONS

BIBLIOGRAPHY

For students: *Practical Conic Sections,* J. W. Downs. Dale Seymour Publications, 1993.

For teachers: *Comic Sections: The Book of Mathematical Jokes, Humour, Wit and Wisdom,* Desmond Machale. Boole Press, 1993.

Multiculturalism in Mathematics, Science, and Technology: Readings and Activities, Miriam Barrios-Chacon et al. Addison-Wesley Publishing, 1993.

Posters: *Conic Sections.* Dale Seymour Publications.

Objective

Students will be able to define parametric equations, graph curves parametrically, and solve application problems using parametric equations.

Motivate

Have students use a grapher to graph the parametric equations $x = t$ and $y = t^2$ for $-5 \leq t \leq 5$. Have them write the equation for this graph in the form $y = f(x)$. ($\boldsymbol{y = x^2}$)

Teaching Note

It is useful to have students plot one or two parametric curves using pencil and paper methods. This will give students an appreciation and an understanding of the work done by a grapher.

9.1 PARAMETRIC EQUATIONS AND MOTION

Graphs of Parametric Equations • Eliminating the Parameter • Motion along a Line • Motion in the Plane • Motion of Wheels

Graphs of Parametric Equations

Parametric equations were introduced in Section 2.5. Example 3 of that section showed that when using parametric equations, you can control what portion of a graph is displayed based on what values of the parameter are allowed.

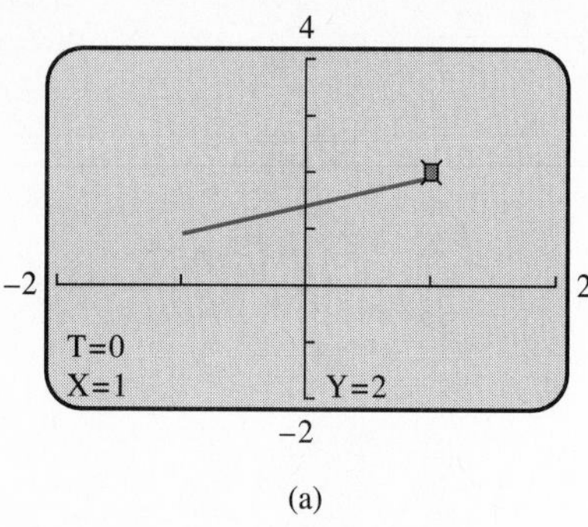

(a)

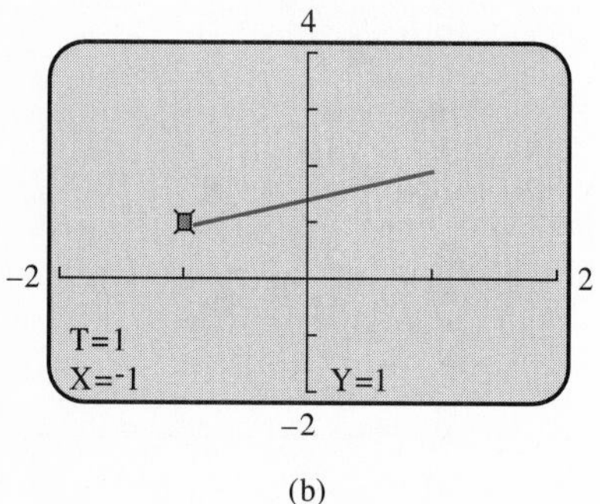

(b)

Figure 9.1 The trace cursor is positioned at the endpoints of the graph of $x = 1 - 2t$, $y = 2 - t$, $0 \leq t \leq 1$.

> **Teaching Note**
>
> Graphing parametric equations was difficult prior to the existence of graphers. To effectively use a grapher, students must attain some measure of control over the range restrictions on the parameter t.

■ EXAMPLE 1 Graphing Parametric Equations

Find the endpoints of and graph the parametric curve

$$x = 1 - 2t, \qquad y = 2 - t, \qquad 0 \leq t \leq 1.$$

Solution

If $t = 0$, then $x = 1$ and $y = 2$ and one endpoint is $(1, 2)$. If $t = 1$, then $x = -1$ and $y = 1$ and the other endpoint is $(-1, 1)$. Figure 9.1 shows a graph of the parametric curve with the trace cursor positioned at the two endpoints. The graph of this curve for $0 \leq t \leq 1$ appears to be a line segment. ■

Eliminating the Parameter

When a curve is defined parametrically using $x(t)$ and $y(t)$, it sometimes is possible to *eliminate the parameter t* and find a familiar x-y relationship. This often helps us identify and be certain that we have a complete understanding of the graph of the parametric curve.

■ EXAMPLE 2 Eliminating the Parameter

Eliminate the parameter and identify the graph of the parametric curve given in Example 1, namely,

$$x = 1 - 2t, \qquad y = 2 - t, \qquad 0 \leq t \leq 1.$$

Solution

First we observe that $-1 \leq x \leq 1$.

$$0 \leq t \leq 1$$

$$0 \geq -2t \geq -2 \qquad \text{Multiply by } -2.$$

$$1 \geq 1 - 2t \geq -1 \qquad \text{Add 1.}$$

Now we solve the first equation for the parameter t:

$$x = 1 - 2t$$

$$2t = 1 - x$$

$$t = \frac{1}{2}(1 - x)$$

Then we substitute this expression for t into the second equation:

$$y = 2 - t$$

$$y = 2 - \frac{1}{2}(1 - x)$$

$$y = 0.5x + 1.5, \quad -1 \leq x \leq 1$$

We see that the parametric relationship between x and y is linear. The line has the slope 0.5 and the y-intercept 1.5. All the information using variables x and y is consistent with Figure 9.1. ■

Alert

Students are sometimes confused by the role of the *T*step on the grapher. They may think it behaves like the *X*scale and *Y*scale. Show them how the *T*step affects the graph produced on the screen. It may be useful to compare graphs of the parametric equations $x = 4\cos t$ and $y = 4\sin t$, $0 < t < 6$, with *T*steps of 1.5 and 0.1.

Teaching Note

Motion problem situations can be effectively represented using the parametric grapher. This method foreshadows the calculus in an important way.

When the parametric equations involve trigonometric functions, the parameter can sometimes be eliminated by using a trigonometric identity.

■ EXAMPLE 3 Eliminating the Parameter

Eliminate the parameter from the parametric equations

$$x = 2\cos t, \qquad y = 2\sin t, \qquad 0 \le t \le 2\pi,$$

and identify the graph.

Solution

The graph in Figure 9.2 appears to be elliptical, and suggests that the x-y relationship is not a function. We will use the familiar trigonometric identity $\cos^2 t + \sin^2 t = 1$ to eliminate the parameter.

$$\begin{aligned} x^2 + y^2 &= 2^2\cos^2 t + 2^2\sin^2 t \\ &= 4(\cos^2 t + \sin^2 t) \\ &= 4(1) \\ &= 4 \end{aligned}$$

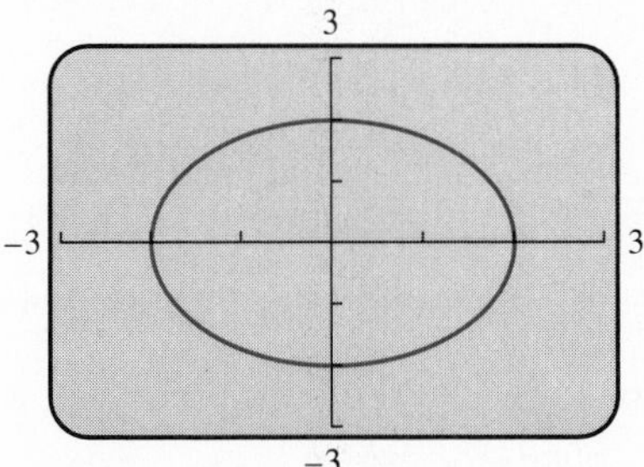

Figure 9.2 A graph of the circle $x = 2\cos t$, $y = 2\sin t$, $0 \le t \le 2\pi$ in a nonsquare viewing window.

We recognize $x^2 + y^2 = 4$ as the equation of the circle with the center (0, 0) and the radius 2. (See Section 8 of the Prerequisite Chapter.) This means our viewing window in Figure 9.2 is not square. Since $x(0) = 2$ and $y(0) = 0$, and $x(2\pi) = 2$ and $y(2\pi) = 0$, the two "endpoints" of this curve are the same point, (2, 0), and the graph is a complete circle. ■

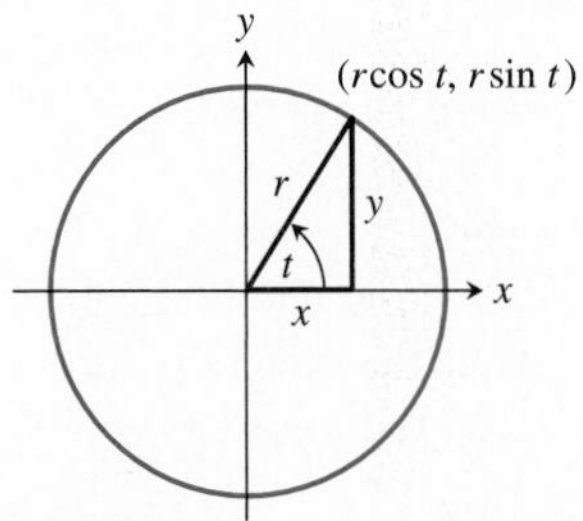

Figure 9.3

We see from Figure 9.3, a figure common to the development of the trigonometric functions, that $x = r\cos t$ and $y = r\sin t$. Consequently,

$$x = r\cos t, \qquad y = r\sin t, \qquad 0 \le t \le 2\pi,$$

are parametric equations for a circle centered at (0, 0) with the radius r. However, parametric equations for a circle are not unique. There are many possible **parametrizations** of a given curve. (See Exercise 30.)

To find parametric equations for a circle with the center point (h, k), apply a horizontal and vertical translation to obtain

$$x = h + r\cos t, \qquad y = k + r\sin t, \qquad 0 \le t \le 2\pi.$$

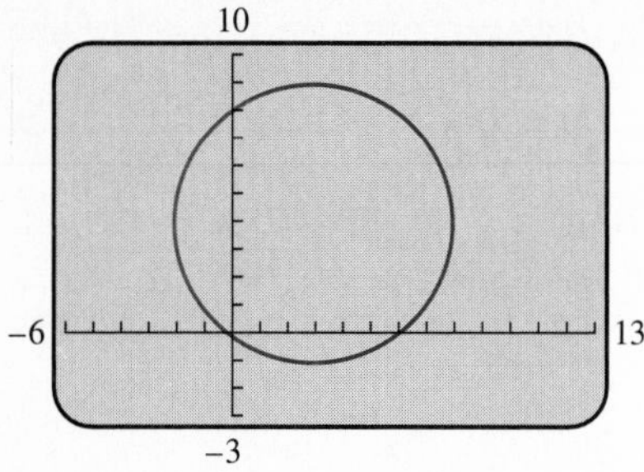

Figure 9.4

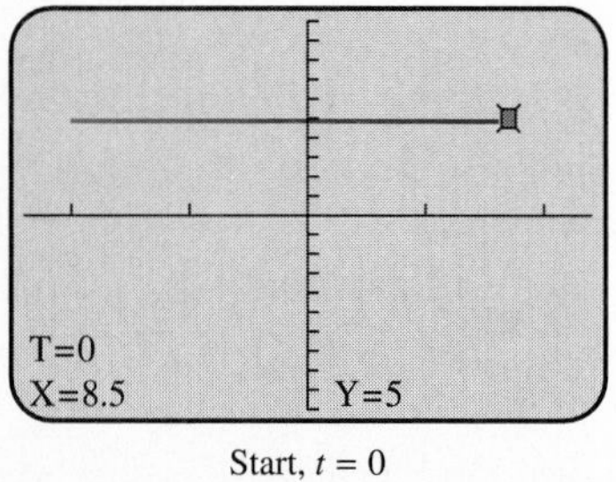

Start, $t = 0$
(a)

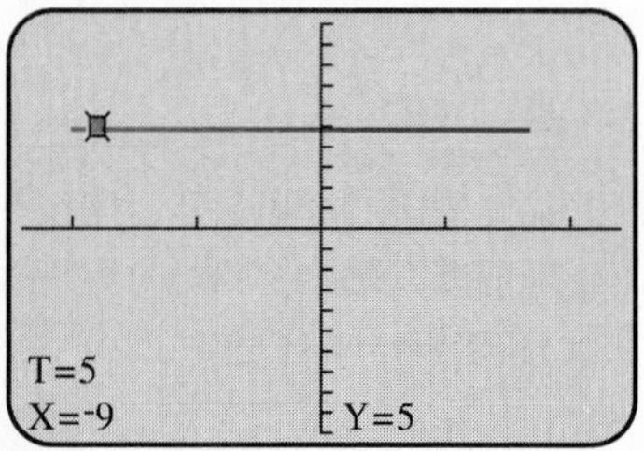

5 sec later, $t = 5$
(b)

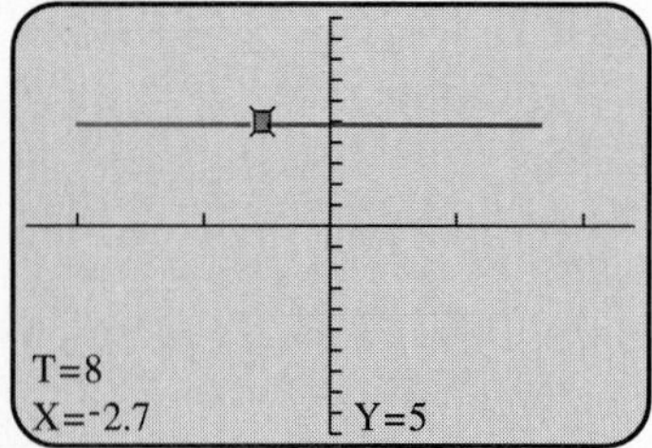

3 sec after that, $t = 8$
(c)

Figure 9.5 Three views of the graph C_1: $x_1 = -0.1(t^3 - 20t^2 + 110t - 85)$, $y_1 = 5, 0 \leq t \leq 12$ in the $[-12, 12]$ by $[-10, 10]$ viewing window.

EXAMPLE 4 Finding Parametric Equations for a Circle

Find parametric equations for the circle whose center is (3, 4) and radius is 5. Use the equations to produce a graph of this circle.

Solution

Parametric equations for the circle with the center (0, 0) and the radius 5 are

$$x = 5 \cos t, \qquad y = 5 \sin t, \qquad 0 \leq t \leq 2\pi.$$

We translate this circle 3 units to the right and 4 units up to obtain

$$x = 3 + 5 \cos t, \qquad y = 4 + 5 \sin t, \qquad 0 \leq t \leq 2\pi.$$

A graph of these parametric equations is shown in Figure 9.4. ■

Motion along a Line

In Example 5 of Section 2.5 we used parametric equations to model the motion of a distress flare shot straight up from a ship's bridge. We can also use parametric equations to simulate motion along a horizontal line. We typically use the variable t for the parameter to represent time.

EXAMPLE 5 Simulating Horizontal Motion

Gary walks along a horizontal line (think of it as a number line) with the coordinate of his position (in meters) given by

$$s = -0.1(t^3 - 20t^2 + 110t - 85),$$

where $0 \leq t \leq 12$. Use parametric equations and a grapher to simulate his motion.

Solution

We arbitrarily choose the horizontal line $y = 5$ to display this motion. The graph C_1 of the parametric equations,

$$C_1: \; x_1 = -0.1(t^3 - 20t^2 + 110t - 85), \quad y_1 = 5, \quad 0 \leq t \leq 12,$$

simulates the motion. His position at any time t is given by the point $(x_1(t), 5)$.

Using "trace" in Figure 9.5 we see that when $t = 0$, Gary is 8.5 m to the right of the y-axis at the point (8.5, 5), and that he initially moves left. Five seconds later he is 9 m to the left of the y-axis at $(-9, 5)$. And after 8 sec he is only 2.7 m to the left of the y-axis. Gary must have changed direction during the walk. The motion of the trace cursor simulates Gary's motion.

A variation in $y(t)$,

$$C_2: \; x_2 = -0.1(t^3 - 20t^2 + 110t - 85), \quad y_2 = -t, \quad 0 \leq t \leq 12,$$

Teaching Note

Focus students' attention on how the variable t (time) plays such an important role in parametric equations. Students will be able to see the direct effects of this variable on motion problems.

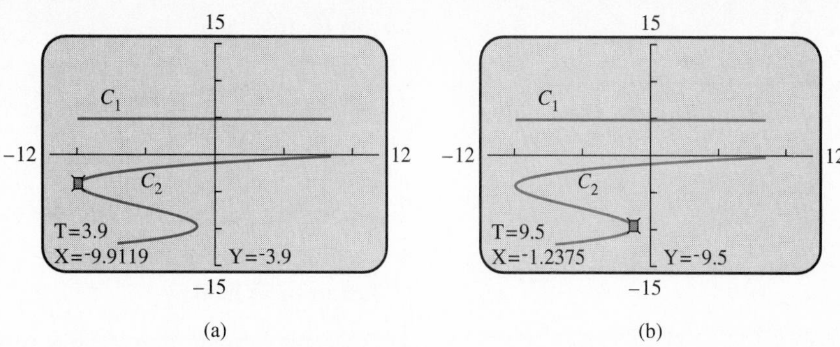

Figure 9.6 Two views of C_1: $x_1 = -0.1(t^3 - 20t^2 + 110t - 85)$, $y_1 = 5$, $0 \le t \le 12$ and C_2: $x_2 = -0.1(t^3 - 20t^2 + 110t - 85)$, $y_2 = -t$, $0 \le t \le 12$.

GRAPHER NOTE

The equation $y_2 = t$ is typically used in this second set of parametric equations. We have chosen $y_2 = -t$ to get two curves in Figure 9.6 that do not overlap. Also notice that the y-coordinates of C_1 are constant ($y_1 = 5$), and that the y-coordinates of C_2 vary with time $t (y_2 = -t)$.

can be used to help visualize where Gary changes direction. The graph C_2 shown in Figure 9.6 suggests that Gary reverses his direction at 3.9 sec and again at 9.5 sec after beginning his walk. ■

Motion in the Plane

In Sections 2.5 and 3.1 we modeled the motion of projectiles that are launched at an angle of 90° (straight up) from the horizontal. Now we investigate the motion of objects, ignoring air friction, that are launched at angles other than 90°.

Suppose that a ball is thrown from a point y_0 feet above ground level with an initial velocity of v_0 ft/sec at an angle θ with the horizontal. (See Figure 9.7.) The velocity in the x-direction is $v_0 \cos \theta$ and in the y-direction $v_0 \sin \theta$. The path of the object is modeled by the parametric equations

$$x = (v_0 \cos \theta)t, \qquad y = -16t^2 + (v_0 \sin \theta)t + y_0.$$

Notice that the x-component is simply distance = velocity × time. The y-component is the familiar vertical projectile-motion equation.

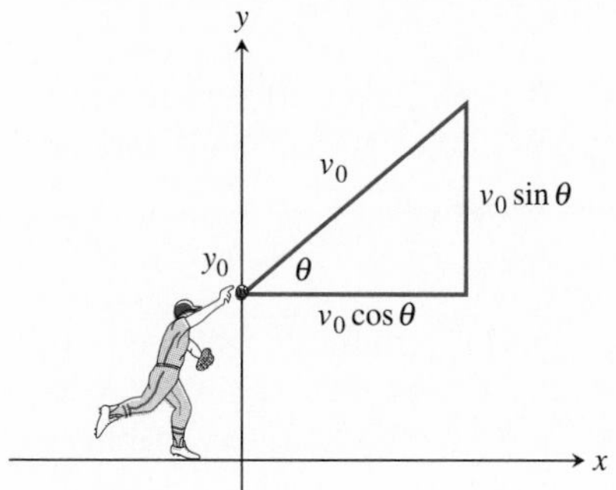

Figure 9.7

■ EXAMPLE 6 APPLICATION: Hitting a Baseball

Kevin hits a baseball at 3 ft above the ground with an initial velocity of 150 ft/sec at an angle of 18° with the horizontal. Will the ball clear a 20-ft wall that is 400 ft away?

Solution

Model

The path of the ball is modeled by the parametric equations

$$x = (150 \cos 18°)t, \quad y = -16t^2 + (150 \sin 18°)t + 3, \quad 0 \le t \le 3.$$

NOTE

Because t stands for time, we choose $t \ge 0$. Also we want to be sure that t is large enough for the ball to get to the fence. After a little trial and error, we found that when $t = 3$ the ball has gone far enough.

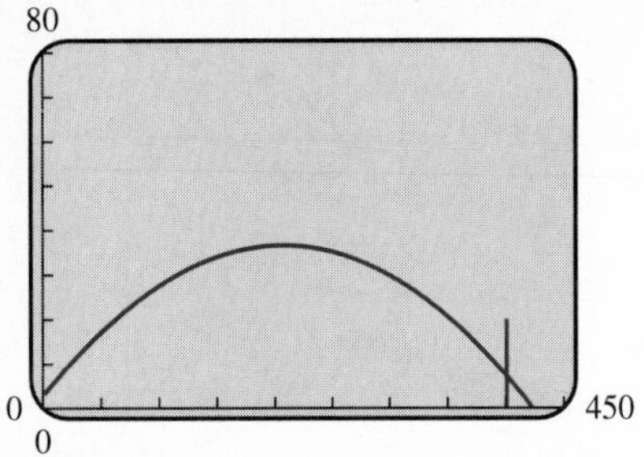

Figure 9.8 Consult your grapher owner's manual to see how to draw the fence.

Solve Graphically

Figure 9.8 shows a graph of both the path of the ball and the 20-ft wall. It is visually evident that the ball's path hits the wall.

Interpret

The ball will not clear the 20-ft wall. You may wish to convince yourself that if the ball leaves the bat at an angle of 21°, the ball will clear the fence. ■

Motion of Wheels

In Example 7 we see how to write parametric equations for position on a moving Ferris wheel using time t as the parameter.

■ EXAMPLE 7 APPLICATION: Riding on a Ferris Wheel

Jane is riding on a Ferris wheel with a radius of 30 ft. The wheel is turning counterclockwise as we view it in Figure 9.9 at the rate of one revolution every 10 sec. Assume the lowest point of the Ferris wheel (6 o'clock) is 10 ft above the ground, and that Jane is at the point marked A (3 o'clock) at time $t = 0$. Use parametric equations to find Jane's position 22 sec into the ride.

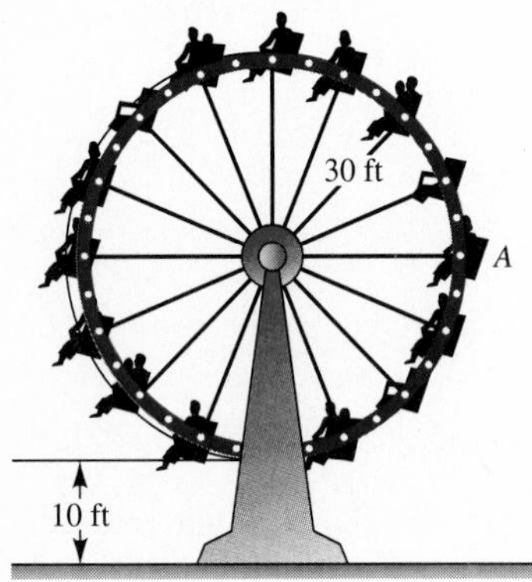

Figure 9.9

Solution

Model

Figure 9.10 shows a circle with center (0, 40) and radius 30 that models the Ferris wheel. The parametric equations for this circle in terms of the parameter θ, the central angle of the circle determined by arc AP, are

$$x = 30 \cos \theta, \qquad y = 40 + 30 \sin \theta, \qquad \theta \geq 0.$$

To take into account the rate at which the wheel is turning we must describe θ as a function of time t in seconds. The wheel is turning at the rate of 2π radians every 10 sec, or $2\pi/10 = \pi/5$ rad/sec. So, $\theta = (\pi/5)t$. Therefore, parametric equations that model Jane's motion are

$$x = 30 \cos \left(\frac{\pi}{5}t\right), \qquad y = 40 + 30 \sin \left(\frac{\pi}{5}t\right), \qquad t \geq 0.$$

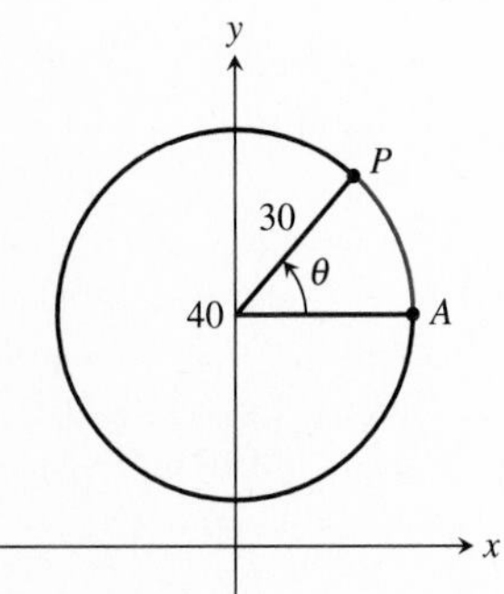

Figure 9.10

Solve Algebraically

To find the desired position we evaluate the parametric equation model for $t =$ 22 sec:

$$x = 30 \cos \left(\frac{\pi}{5} \cdot 22\right) \qquad y = 40 + 30 \sin \left(\frac{\pi}{5} \cdot 22\right)$$

$$x \approx 9.27 \qquad y \approx 68.53$$

Interpret

After riding for 22 sec, Jane is approximately 68.5 ft above the ground and approximately 9.3 ft to the right of the y-axis. Because each revolution takes 10 sec, Jane is into her third time around. ■

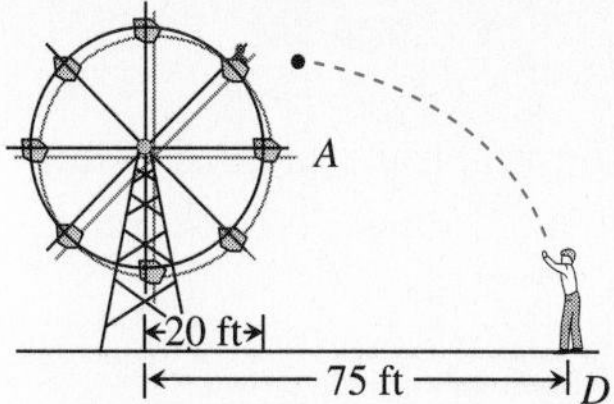

Figure 9.11

Follow-up

Have students explain how the parametric equations in Example 8 were determined.

Assignment Guide

Day 1: Ex. 3–30, multiples of 3
Day 2: Ex. 31, 34–36, 41, 43, 45, 48

Cooperative Learning

Group Activity: Ex. 21–22

Notes on Exercises

Ex. 31–40 and 45–46 include a variety of interesting applications.
Ex. 41–42 relate to cycloids and hypocycloids. A Spirograph can be used to help illustrate these curves.

Ongoing Assessment

Self-Assessment: Ex. 7, 11, 19, 25, 33
Embedded Assessment: Ex. 30

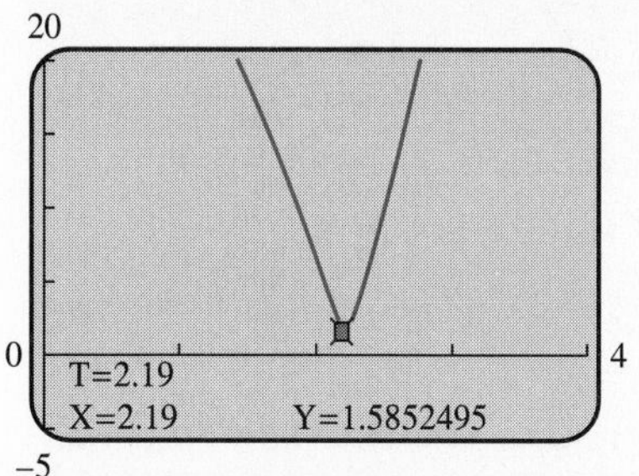

Figure 9.12 Notice that the minimum point of the graph occurs when t is about 2.19, and has coordinates of about $x = 2.19$ and $y = 1.59$.

EXAMPLE 8 APPLICATION: Throwing a Ball at a Ferris Wheel[1]

A 20-ft-radius Ferris wheel turns counterclockwise one revolution every 12 sec. (See Figure 9.11.) Eric stands at point D, 75 ft from the base of the wheel. At the instant Jane is at point A, Eric throws a ball at the Ferris wheel, releasing it from the same height as the bottom of the wheel. If the ball's initial velocity is 60 ft/sec and it is released at an angle of 120° with the horizontal, does Jane have a chance to catch the ball?

Solution

Model

We assign a coordinate system so that the bottom car of the Ferris wheel is at (0, 0) and the center of the wheel is at (0, 20). Then Eric releases the ball at the point (75, 0). Parametric equations for Jane's path, taking into account that the wheel is turning at the rate of $2\pi/12 = \pi/6$ rad/sec, are

$$x_1 = 20\cos\left(\frac{\pi}{6}t\right), \qquad y_1 = 20 + 20\sin\left(\frac{\pi}{6}t\right), \qquad t \ge 0.$$

Equations for the path of the ball are

$$\begin{aligned} x_2 &= (60\cos 120°)t + 75 & y_2 &= -16t^2 + (60\sin 120°)t \\ &= -30t + 75, & &= -16t^2 + 30\sqrt{3}\,t, \qquad t \ge 0. \end{aligned}$$

We graph the two paths simultaneously. Watch closely and you should see that Jane and the ball do not arrive at the point of intersection of the paths at the same moment. Using "trace" you can convince yourself that they come closest to each other between 2.1 and 2.3 sec after the ball is thrown.

The distance d between the ball and Jane at any time t can be found by using the distance formula

$$d(t) = \sqrt{[x_1(t) - x_2(t)]^2 + [y_1(t) - y_2(t)]^2}.$$

Solve Graphically

To find how close the ball gets to Jane we need to find the smallest value for $d(t)$. Figure 9.12 shows the graph of $y = d(t)$ obtained using the parametric equations

$$x_3 = t, \qquad y_3 = d(t).$$

Using "trace" we find that the minimum distance is about 1.59 and occurs when t (or x_3) is about 2.19.

Interpret

The minimum distance between Jane and the ball is approximately 1.6 ft and occurs at approximately $t = 2.2$ sec after the ball is thrown. That is close enough for Jane to have a good chance at catching the ball. An independent check of this result can be obtained by graphing $y = d(t)$ in function mode. (See Exercise 48.) ■

[1]This problem was posed by Neal Koblitz in the March 1988 issue of *The American Mathematical Monthly.*

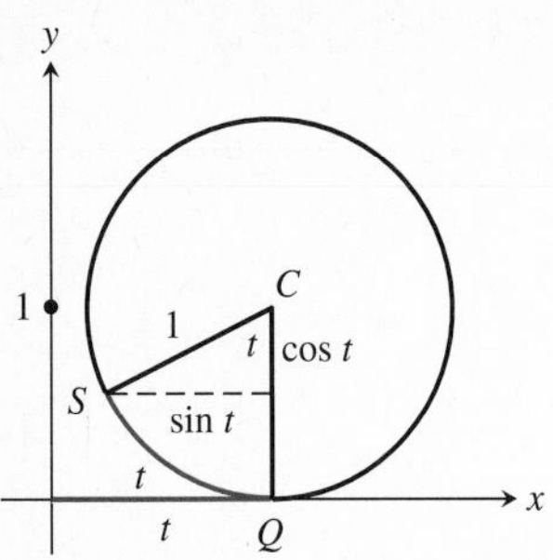

Figure 9.13 Notice that t is the length of the arc SQ, the length of the line segment from the origin to Q, and the radian measure of the central angle of the circle with radius 1.

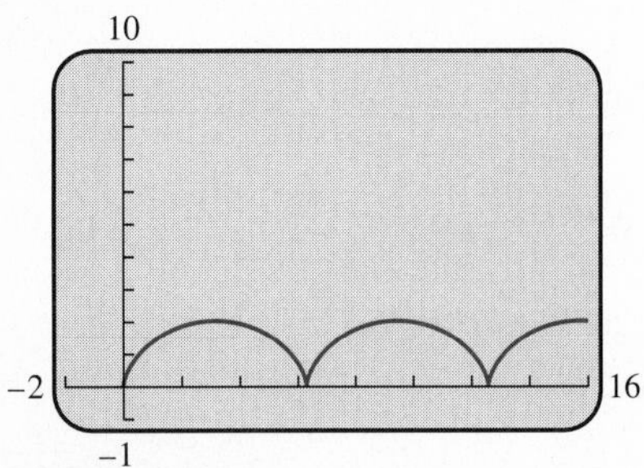

Figure 9.14

EXAMPLE 9 APPLICATION: Doing Industrial Tire Research

The R&D division of Michyear Tire Co. wants to graph the path of a stone stuck in the tread of a tire as the wheel rolls along a flat road. Assume that the radius of the wheel is 1 ft. Find a graph of the stone's path.

Solution

Model

We assign a rectangular coordinate system to agree with the instant the tire picks up the stone. We let (0, 0) be the initial position of the stone and (0, 1) be the center of the wheel, and we assume that the wheel is rolling in the positive x direction. Figure 9.13 shows the wheel after it has rolled a distance t.

$$t = \text{distance wheel rolls after picking up stone } S$$
$$C = (t,\ 1)$$
$$S = (t - \sin t,\ 1 - \cos t)$$

By a careful analysis of Figure 9.13 we see that the path of the stone S can be modeled by the parametric equations

$$x = t - \sin t, \qquad y = 1 - \cos t, \qquad t \geq 0$$

Solve Graphically

Figure 9.14 shows a graph of the parametric equations that model the path of the stone.

Interpret

Whenever t is a multiple of 2π, $\cos t$ has the value 1 and the y-coordinate has the value zero. That is consistent with the fact that the stone in the tire tread touches the road once every revolution. ■

Quick Review 9.1

In Exercises 1 and 2, write an equation in point-slope form for the line through the two points.

1. $(-3, -2), (4, 6)$ **2.** $(-1, 3), (4, -3)$

In Exercises 3 and 4, graph the equation.

3. $y^2 = 8x$ **4.** $y^2 = -5x$

In Exercises 5–8, write an equation for the circle with the given center and radius.

5. (0, 0), 2 **6.** (2, 3), 4

7. (0, 2), 5 **8.** $(-2, 5)$, 3

In Exercises 9 and 10, a wheel with radius r spins at the given rate. Find the angular velocity in radians per second.

9. $r = 13$ in., 600 rpm **10.** $r = 12$ in., 700 rpm

In Exercises 11–14, let θ be an angle in standard position with point P on the terminal side. If P is on the unit circle, find its coordinates.

11. $\theta = 35°$ **12.** $\theta = 160°$

13. $\theta = \dfrac{\pi}{6}$ **14.** $\theta = \dfrac{4\pi}{3}$

In Exercises 15–18, solve for a and b.

15.

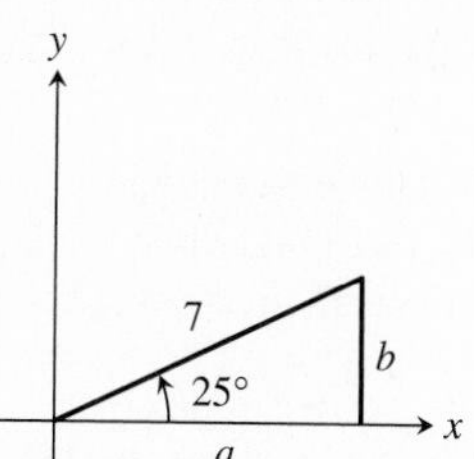

16.

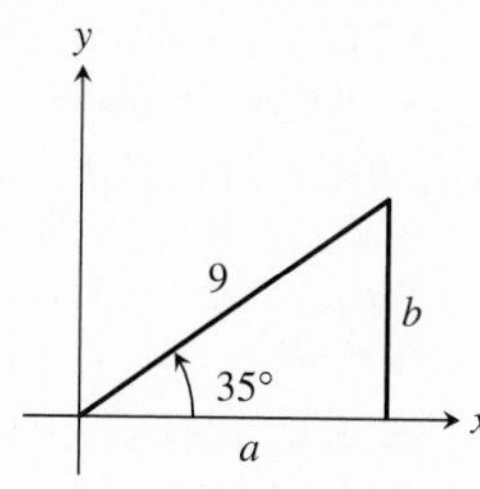

17.

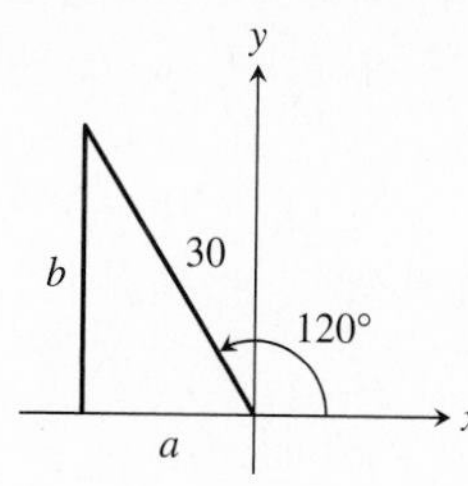

18.

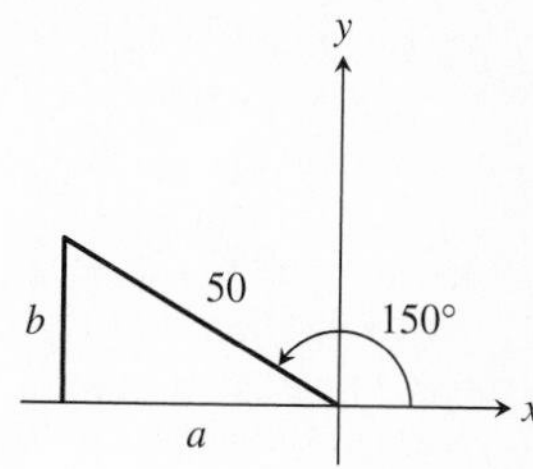

SECTION EXERCISES 9.1

In Exercises 1–4, match the parametric equations with their graph. Identify the viewing window that seems to have been used, and state its dimensions.

1. $x = 2 \cos t, \quad y = 3 \sin 2t$

2. $x = \cos^3 t, \quad y = \sin^3 t$

3. $x = 7 \cos t - \cos 7t, \quad y = 7 \sin t - \sin 7t$

4. $x = 12 \cos t + 3 \cos 6t, \quad y = 12 \sin t - 3 \sin 6t$

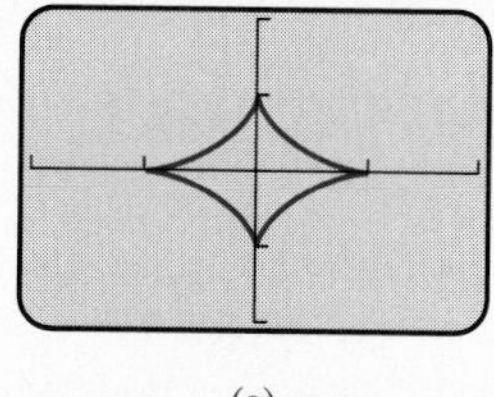

(a)

(b)

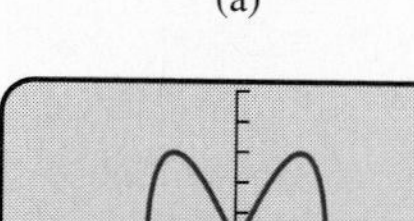

(c)

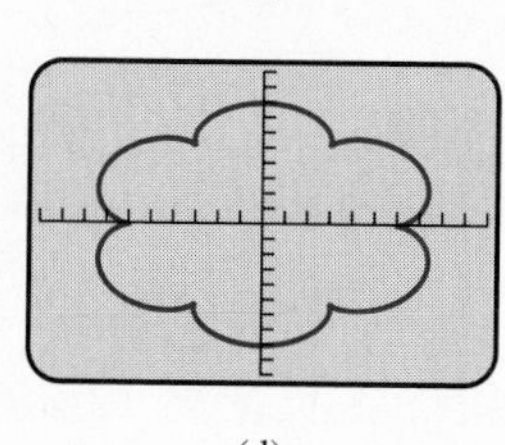

(d)

For Exercises 1–4

In Exercises 5–10, find any endpoints and describe the graph.

5. $x = 1 + t, \quad y = t, \quad 0 \le t \le 5$

6. $x = 1 + t, \quad y = t, \quad -3 \le t \le 2$

7. $x = 2 - 3t, \quad y = 5 + t, \quad -1 \le t \le 3$

8. $x = 2t - 3, \quad y = 9 - 4t, \quad 3 \le t \le 5$

9. $x = t - 3, \quad y = 2/t, \quad -5 \le t \le 5$

10. $x = t + 2, \quad y = 4/t, \quad -8 \le t \le 8$

In Exercises 11–16, eliminate the parameter and name the endpoints.

11. $x = 2 - 3t, \quad y = 5 + t, \quad -2 \le t \le 2$

12. $x = 5 - 3t, \quad y = 2 + t, \quad -1 \le t \le 3$

13. $x = 2 \sin t, \quad y = 2 \cos t, \quad 0 \le t \le 3\pi/2$

14. $x = 3 \cos t, \quad y = 3 \sin t, \quad 0 \le t \le \pi$

15. $x = 8t^3 + 1, \quad y = 2t, \quad -1 \le t \le 1$

16. $x = 0.5t, \quad y = 2t^3 - 3, \quad -2 \le t \le 2$

In Exercises 17–20, find parametric equations for the curve.

17. The line segment with endpoints $(3, 4)$ and $(6, -3)$

18. The line segment with endpoints $(5, 2)$ and $(-2, -4)$

19. The circle with center $(5, 2)$ and radius 3

20. The circle with center $(-2, -4)$ and radius 2

Complete Exercises 21 and 22 working in groups of three.

21. Group Learning Activity Kathy walks back and forth along a number line with her position (in meters) given at any time $t \ge 0$ (in seconds) by $s = f(t) = t^3 - 7t^2 + 10t$. Explain how each parametrization describes her motion.

a. $x = f(t), \quad y = 2$ **b.** $x = f(t), \quad y = t$
c. $x = t, \quad y = f(t)$

22. **Group Learning Activity** Pete moves back and forth along a number line with his position (in meters) given at any time $t \geq 0$ (in seconds) by $s = -t^3 + 8t^2 - 15t$. Explain how each parametrization listed in Exercise 21 describes his motion.

In Exercises 23–28, a particle moves along a horizontal line so that its position at any time t is given by $s(t)$. Use two different pairs of parametric equations and their graphs (as in Example 5) to describe the motion. Write a description of the motion.

23. $s(t) = t^2 - 5t, \quad -1 \leq t \leq 6$
24. $s(t) = t^2 - 6t, \quad -1 \leq t \leq 5$
25. $s(t) = -t^2 + 3t, \quad -2 \leq t \leq 4$
26. $s(t) = -t^2 + 4t, \quad -1 \leq t \leq 5$
27. $s(t) = 0.5(t^3 - 7t^2 + 2t), \quad -1 \leq t \leq 7$
28. $s(t) = t^3 - 5t^2 + 4t, \quad -1 \leq t \leq 5$

In Exercises 29 and 30, solve the problem.

29. The complete graph of the parametric equations $x = 2 \cos t$, $y = 2 \sin t$ is the circle of radius 2 centered at the origin. Find an interval of values for t so that the graph is the given portion of the circle.
    - **a.** The portion in the first quadrant
    - **b.** The portion above the x-axis
    - **c.** The portion to the left of the y-axis

30. **Writing to Learn** Consider the two pairs of parametric equations $x = 3 \cos t$, $y = 3 \sin t$ and $x = 3 \sin t$, $y = 3 \cos t$ for $0 \leq t \leq 2\pi$.
    - **a.** Give a convincing argument that the graphs of the pairs of parametric equations are the same.
    - **b.** Explain how the parametrizations are different.

In Exercises 31–40, solve the problem. Consider both graphical and algebraic solutions.

31. *Hitting a Baseball* Consider Kevin's hit discussed in Example 6.
    - **a.** Approximately how many seconds after the ball is hit does it hit the wall?
    - **b.** How high up the wall does the ball hit?
    - **c.** **Writing to Learn** Explain why Kevin's hit might be caught by an outfielder. Then explain why his hit would likely not be caught by an outfielder, if the ball had been hit at a 20° angle with the horizontal.

32. *Hitting a Baseball* Kirby hits a ball when it is 4 ft above the ground with an initial velocity of 120 ft/sec. The ball leaves the bat at a 30° angle with the horizontal and heads toward a 30-ft fence 350 ft from home plate.
    - **a.** Does the ball clear the fence?
    - **b.** If so, by how much does it clear the fence? If not, could the ball be caught?

33. *Hitting a Baseball* Suppose that the moment Kirby hits the ball in Exercise 32 there is a 5-ft/sec split-second wind gust. Assume the wind acts in the horizontal direction out with the ball.
    - **a.** Does the ball clear the fence?
    - **b.** If so, by how much does it clear the fence? If not, could the ball be caught?

34. *Two-Softball Toss* Chris and Linda warm up in the outfield by tossing softballs to each other. Suppose both tossed a ball at the same time from the same height, as illustrated in the figure. Find the minimum distance between the two balls and when this minimum distance occurs.

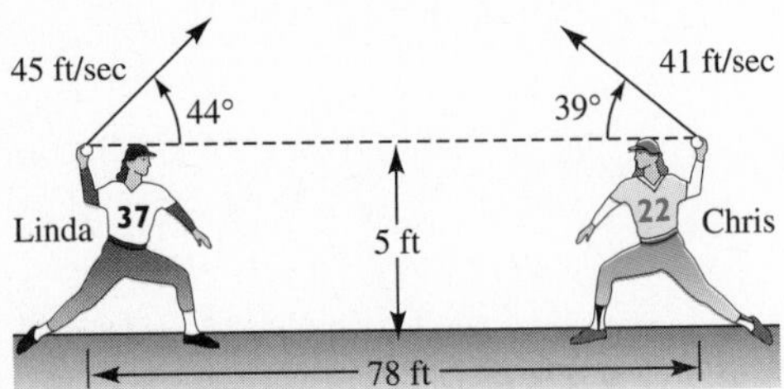

For Exercise 34

35. *Yard Darts* Tony and Sue are launching yard darts 20 ft from the front edge of a circular target of radius 18 in. on the ground. If Tony throws the dart directly at the target, and releases it 3 ft above the ground with an initial velocity of 30 ft/sec at a 70° angle, will the dart hit the target?

36. *Yard Darts* In the game of darts described in Exercise 35 Sue releases the dart 4 ft above the ground with an initial velocity of 25 ft/sec at a 55° angle. Will the dart hit the target?

37. *Hitting a Baseball* Orlando hits a ball when it is 4 ft above ground level with an initial velocity of 160 ft/sec. The ball leaves the bat at a 20° angle with the horizontal and heads toward a 30-ft fence 400 ft from home plate. How strong must a split-second wind gust be (in feet per second) that acts directly with or against the ball in order for the ball to hit within a few inches of the top of the wall? Estimate the answer graphically and solve algebraically.

38. *Hitting Golf Balls* Nancy hits golf balls off the practice tee with an initial velocity of 180 ft/sec with four different clubs. How far down the fairway does the ball hit the ground if it comes off the club making the specified angle with the horizontal?

a. 15° **b.** 20° **c.** 25° **d.** 30°

39. *Analysis of a Ferris Wheel* Ron is on a Ferris wheel of radius 35 ft that turns counterclockwise at the rate of one revolution every 12 sec. The lowest point of the Ferris wheel (6 o'clock) is 15 ft above ground level at the point (0, 15) on a rectangular coordinate system. Find parametric equations for the position of Ron as a function of time t (in seconds) if the Ferris wheel starts ($t = 0$) with Ron at the point (35, 50).

40. *Throwing a Ball at a Ferris Wheel* A 71-ft-radius Ferris wheel turns counterclockwise one revolution every 20 sec. Tony stands at a point 90 ft to the right of the base of the wheel. At the instant Matthew is at point A (3 o'clock), Tony throws a ball toward the Ferris wheel with an initial velocity of 88 ft/sec at an angle with the horizontal of 100°. Find the minimum distance between the ball and Matthew.

Exercises 41 and 42 refer to a graph that could be generated by a point of a circle as the circle rolls along a given curve. The graph can also be described by a pair of parametric equations.

41. *Cycloid* The graph of the parametric equations $x = t - \sin t$, $y = 1 - \cos t$ of Example 9 is a *cycloid.*

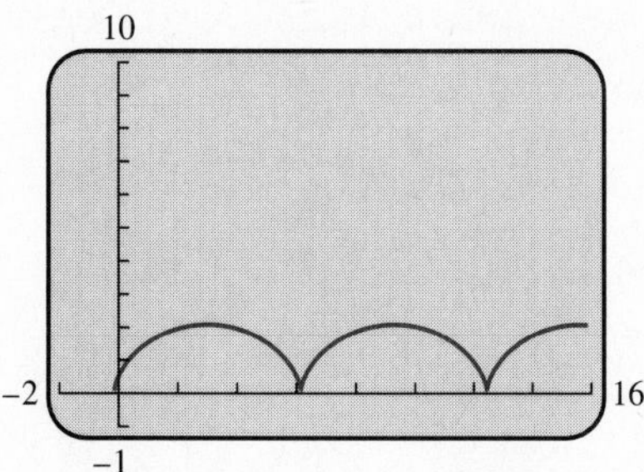

For Exercise 41

a. What is the maximum value of $y = 1 - \cos t$? How is that value related to the graph?

b. What is the distance between neighboring x-intercepts?

42. *Hypocycloid* The graph of the parametric equations $x = 2\cos t + \cos 2t$, $y = 2\sin t - \sin 2t$ is a *hypocycloid.* The graph is the path of a point P on a circle of radius 1 rolling along the inside of a circle of radius 3, as illustrated in the figure.

a. Graph simultaneously this hypocycloid and the circle of radius 3.

b. Suppose the large circle had a radius of 4. Experiment! How do you think the equations in part a should be changed to obtain defining equations? What do you think the hypocycloid would look like in this case? Check your guesses.

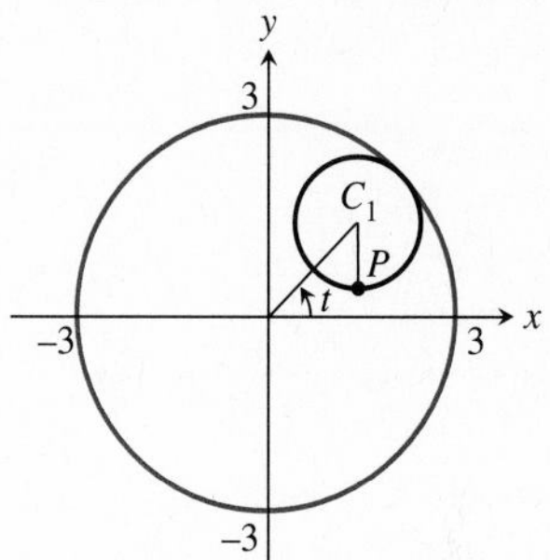

For Exercise 42

EXTENDING THE IDEAS

Exercises 43 and 44 refer to the graph C of the parametric equations

$$x = tc + (1 - t)a, \qquad y = td + (1 - t)b,$$

where $P_1(a, b)$ and $P_2(c, d)$ are two fixed points.

43. *Using Parametric Equations in Geometry* Show that the point $P(x, y)$ on C is equal to

a. $P_1(a, b)$ when $t = 0$. **b.** $P_2(c, d)$ when $t = 1$.

44. *Using Parametric Equations in Geometry* Show that if $t = 0.5$ the corresponding point (x, y) on C is the midpoint of the line segment with endpoints (a, b) and (c, d).

In Exercises 45 and 46, solve the double Ferris wheel problem.

45. *Two Ferris Wheel Problem* Chang is on a Ferris wheel of center (0, 20) and radius 20 ft turning counterclockwise at the rate of one revolution every 12 sec. Kuan is on a Ferris wheel of center (15, 15) and radius 15 turning counterclockwise at the rate of one revolution every 8 sec. Find the minimum distance between Chang and Kuan if both start out ($t = 0$) at 3 o'clock.

46. *Two Ferris Wheel Problem* Chang and Kuan are riding the Ferris wheels described in Exercise 45. Find the minimum distance between Chang and Kuan if Chang starts out ($t = 0$) at 3 o'clock and Kuan at 6 o'clock.

In Exercises 47 and 48, solve the problem.

47. Writing to Learn Consider the three sets of parametric equations.

a. $x = t, \quad y = 3t - 1$ **b.** $x = e^t, \quad y = 3e^t - 1$
c. $x = \sin t, \quad y = 3 \sin t - 1$

Compare and contrast these three parametric equations and their graphs. How are they different? How are they the same?

48. Refer to Example 8. Let $y = d(t)$, and use function mode to show that the minimum value of distance d is about the same as that obtained in Example 8.

Objective

Students will be able to write equations for conic sections in standard form and produce graphs of conic sections by hand and with a grapher.

9.2 CONIC SECTIONS

Introduction to Conic Sections • Parabolas • Applications Using Parabolas • Ellipses • Applications Using Ellipses • Hyperbolas

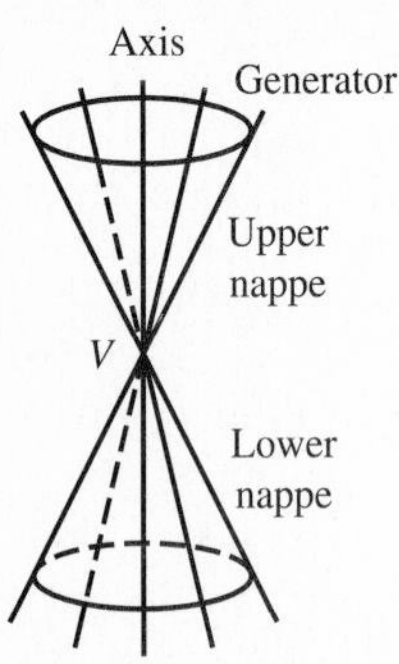

Figure 9.15

Introduction to Conic Sections

Imagine a geometric figure consisting of two nonperpendicular lines intersecting at point V. If we rotate this figure in three-dimensional space using one of the lines as an *axis,* then the other line *generates* a **double-napped right circular cone** with the **vertex** V, as illustrated in Figure 9.15. An ice-cream cone resembles one nappe of a cone.

A *conic section* (or simply a *conic*) is the intersection of a plane with a double-napped right circular cone. The three basic conic sections are the *parabola* (Figure 9.16a), the *ellipse* (Figure 9.16b), and the *hyperbola* (Figure 9.16c). In Exercise 49 you will show that a circle is a special case of an ellipse.

The conic sections can also be defined algebraically in a rectangular coordinate system as the graphs of equations of the form

$$Ax^2 + Bxy + Cy^2 + Dx + Ey + F = 0.$$

Motivate

Discuss the various curves that can be generated as a cross section of a cone.

Teaching Note

Many students have had little experience with solid geometry. It may be useful to construct the nappe of a cone by rolling up a piece of paper. A cone-shaped model can be very effective in illustrating how the various conics can represent a slice of the cone.

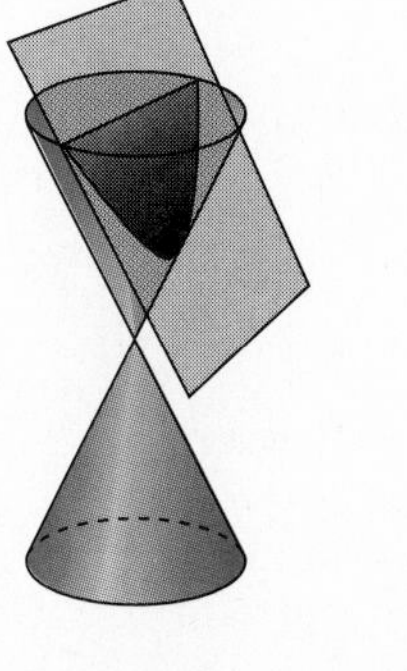

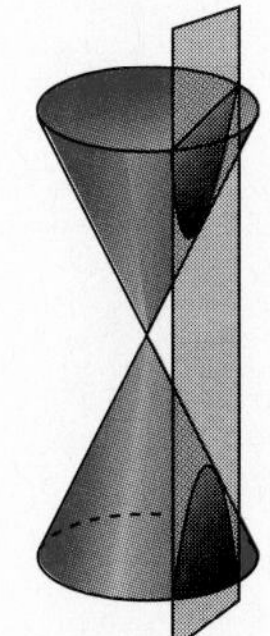

Parabola (a) Ellipse (b) Hyperbola (c)

Figure 9.16

Parabolas

We introduced parabolas in Section 3.1. Since then we have learned more about parabolas, including the three-dimensional description given by Figure 9.16. We summarize what we know with a general algebraic definition, as well as a two-dimensional geometric definition.

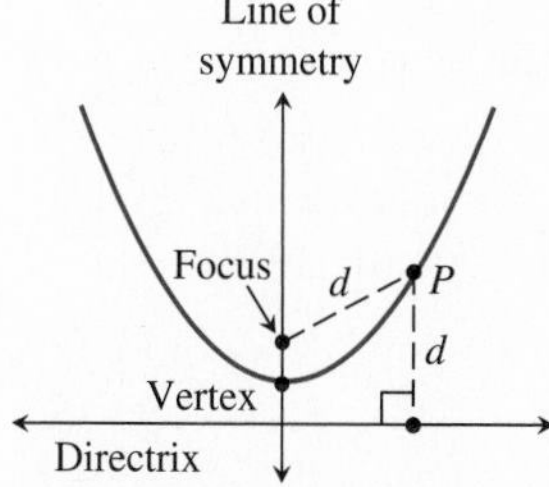

Figure 9.17 Basic structure of a parabola. The distance from P to both the focus and directrix is the same.

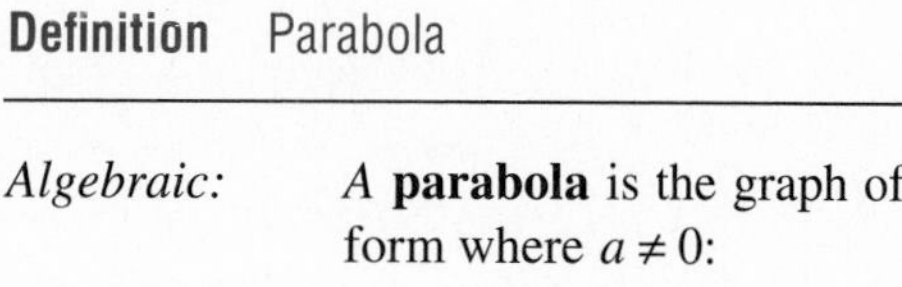

Definition Parabola

Algebraic: A **parabola** is the graph of a quadratic relation of either form where $a \neq 0$:

$$y = ax^2 + bx + c \quad \text{or} \quad x = ay^2 + by + c$$

Geometric: A **parabola** is the set of all points in a plane equidistant from a given point and a given line.

In the geometric description, the given point is the **focus** of the parabola, and the line is its **directrix.** (See Figure 9.17.) It can be shown that the line of symmetry of the parabola is the line perpendicular to the directrix through the focus. The vertex of the parabola is the point of the parabola that is closest to both the focus and directrix.

Example 1 shows a connection between the geometric description of a parabola and the familiar algebraic form for a parabola, $y = ax^2$. See Exercise 50 for $x = ay^2$.

(a)

(b)

Figure 9.18 (a) $p > 0$, and (b) $p < 0$. Notice that the vertex (0, 0) is halfway between the focus and directrix in both cases.

EXAMPLE 1 Connecting Algebra and Geometry

Show that an equation for the parabola with the focus $(0, p)$ and directrix $y = -p$ is

$$y = \frac{1}{4p}x^2.$$

Solution

We must show that a point (x, y) that is equidistant from $(0, p)$ and the line $y = -p$ also satisfies the equation $y = \frac{1}{4p}x^2$. (See Figure 9.18.)

Conversely, we must also show that a point satisfying the equation $y = \frac{1}{4p}x^2$ is equidistant from $(0, p)$ and the line $y = -p$.

Teaching Note

It is assumed that students have substantial knowledge about parabolas and circles from prior courses and from previous sections of this course. The focus of this section is on the use of geometric approaches to conics.

We assume that $p > 0$. The argument is similar for $p < 0$. First, if (x, y) is equidistant from $(0, p)$ and the line $y = -p$, then

$$|y + p| = \text{distance from } (x, y) \text{ to } y = -p,$$
$$\sqrt{x^2 + (y - p)^2} = \text{distance from } (x, y) \text{ to } (0, p).$$

Consequently, we can derive an equation for the parabola as follows:

$$|y + p| = \sqrt{x^2 + (y - p)^2}$$
$$|y + p|^2 = \left(\sqrt{x^2 + (y - p)^2}\right)^2 \quad \text{Square both sides.}$$
$$(y + p)^2 = x^2 + (y - p)^2$$
$$y^2 + 2py + p^2 = x^2 + y^2 - 2py + p^2$$
$$4py = x^2$$
$$y = \frac{1}{4p}x^2$$

By reversing the above steps, we see that a solution (x, y) of $y = \frac{1}{4p}x^2$ is equidistant from $(0, p)$ and the line $y = -p$. ■

The usual algebraic form for the parabola of Example 1 is $y = ax^2$. For this form, $a = \frac{1}{4p}$ or $p = \frac{1}{4a}$.

Characteristics of a Parabola

The **standard forms of a parabola with the vertex (0, 0)** are as follows:

Algebra	**Geometry**
$y = ax^2$	Focus: $\left(0, \frac{1}{4a}\right)$ Directrix: $y = -\frac{1}{4a}$
$x = ay^2$	Focus: $\left(\frac{1}{4a}, 0\right)$ Directrix: $x = -\frac{1}{4a}$

The line of symmetry for $y = ax^2$ is the y-axis. Similarly, the line of symmetry for $x = ay^2$ is the x-axis.

EXAMPLE 2 Finding the Focus and Directrix

Find the focus and directrix for the parabola $y = -\frac{1}{2}x^2$.

Solution

$y = -\frac{1}{2}x^2$ is of the form $y = ax^2$, and $a = -\frac{1}{2}$. Thus,

$$\frac{1}{4a} = \frac{1}{4(-1/2)} = -\frac{1}{2}.$$

The focus is $\left(0, \frac{1}{4a}\right) = \left(0, -\frac{1}{2}\right)$ and the directrix is the line

$$y = -\frac{1}{4a} = \frac{1}{2}.$$

You could support the result of Example 2 by graphing $y = -\frac{1}{2}x^2$ in function mode. However, Example 3 involves the standard form $x = ay^2$. Recall that to graph this relation in function mode you need to enter two expressions:

$$y_1 = \sqrt{\frac{1}{a}x} \quad \text{and} \quad y_2 = -\sqrt{\frac{1}{a}x}$$

EXAMPLE 3 Finding an Equation of a Parabola

Find an equation in standard form for the parabola whose directrix is the line $x = 2$ and focus is the point $(-2, 0)$.

Solution

The directrix, $x = 2$, is a vertical line and the focus $(-2, 0)$ is to the left. Therefore, the parabola is one with a horizontal line of symmetry and opens to the left. Because $x = 2 = -\frac{1}{4a}$, we have

$$-\frac{1}{4a} = 2$$

$$-1 = 8a$$

$$a = -\frac{1}{8}$$

The standard form equation for the parabola is $x = -\frac{1}{8}y^2$.

REMARK

Because of the fraction we often write the equation $x = -\frac{1}{8}y^2$ in the form $-8x = y^2$.

Applications Using Parabolas

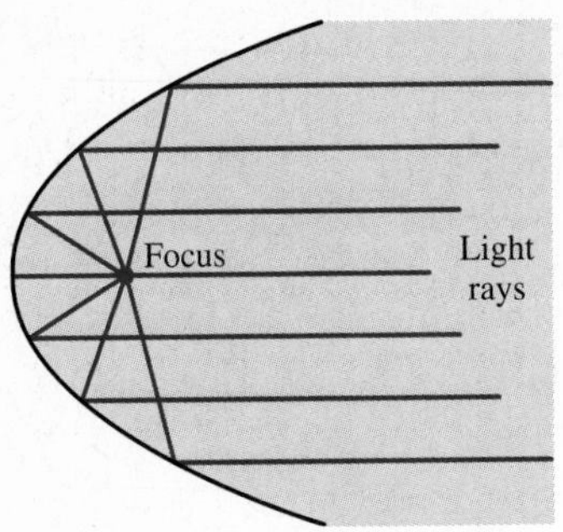

Figure 9.19 A cross section of a paraboloid.

If we rotate a parabola in three-dimensional space using its line of symmetry as an axis, the parabola sweeps out a **parabolic surface** or **paraboloid.** If we place a light source at the **focus** of a reflective paraboloid, the light reflects off the surface in lines parallel to the axis of symmetry, as illustrated in Figure 9.19. This is how car headlights work.

The principle works for signals traveling in the reverse direction—as in microphones (sound signals). When light or sound waves travel parallel to the line of symmetry of a paraboloid, they are reflected through the focus. The effect is to "concentrate" the light or sound at the focus where the electronic receiver is placed. You may have seen parabolic microphones in use at football games.

■ EXAMPLE 4 APPLICATION: Studying a Parabolic Microphone

NFL Films uses a parabolic microphone on the sidelines to capture conversations among football players on the field. If the microphone (paraboloid) is generated by the parabola $x = \frac{1}{15}y^2$, locate the focus (the electronic receiver) of the parabola.

Solution

$x = \frac{1}{15}y^2$ is of the form $x = ay^2$, and $a = \frac{1}{15}$. Thus,

$$\frac{1}{4a} = \frac{1}{4(1/15)} = 3.75.$$

The focus is at

$$\left(\frac{1}{4a}, 0\right) = (3.75, 0).$$

Interpret

If we view the parabolic surface as having the vertex (0, 0) and the line of symmetry the x-axis, the electronic receiver should be placed at the point (3.75, 0). ■

Teaching Note

A loop of string through two pinholes in a piece of cardboard can be used to illustrate the geometric definition of an ellipse.

Ellipses

An ellipse also has both an algebraic and a geometric definition.

Definition Ellipse

Algebraic: An **ellipse** is the graph of a quadratic relation of the form

$$\frac{x^2}{a^2} + \frac{y^2}{b^2} = 1.$$

Geometric: An **ellipse** is the set of all points in a plane the sum of whose distances from two fixed points is constant.

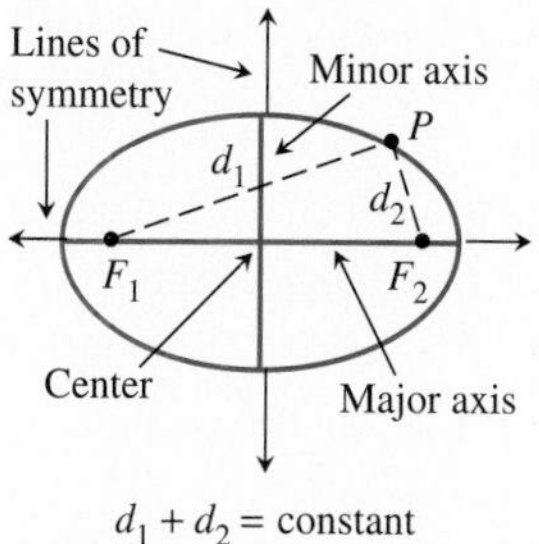

Figure 9.20 Basic structure of an ellipse.

Figure 9.20 shows an ellipse. Each of the fixed points F_1 and F_2 is a **focus** (plural: *foci*) and the distances whose sum is constant are d_1 and d_2. The line segment through F_1 and F_2 with endpoints on the ellipse is the **major axis.** The perpendicular bisector of the major axis with endpoints on the ellipse is the **minor axis.** These axes are lines of symmetry for the ellipse, and their intersection is the **center** of the ellipse.

We can show that the algebraic definition of an ellipse (see Exercise 59),

$$\frac{x^2}{a^2} + \frac{y^2}{b^2} = 1,$$

follows from the geometric definition by using the distance formula to compute $d_1 + d_2$. (See Figure 9.21.) The algebraic and geometric characteristics of an ellipse are summarized here.

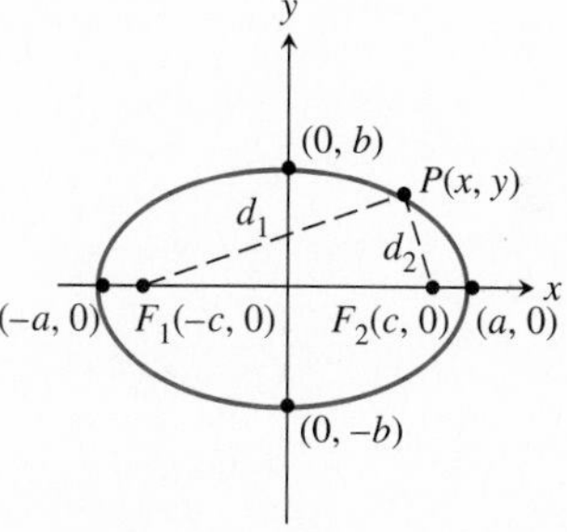

Figure 9.21 The major axis has the length $2a$, the minor axis $2b$, and the foci are at $(c, 0)$ and $(-c, 0)$.

Characteristics of an Ellipse

The **standard forms of an ellipse centered at (0, 0)** are as follows.

Algebra	**Geometry**
$\frac{x^2}{a^2} + \frac{y^2}{b^2} = 1,\ a > b$	Foci: $(-c, 0)$ and $(c, 0)$, where $c^2 = a^2 - b^2$ Major axis from $(-a, 0)$ to $(a, 0)$ Minor axis from $(0, -b)$ to $(0, b)$
$\frac{x^2}{a^2} + \frac{y^2}{b^2} = 1,\ a < b$	Foci: $(0, -c)$ and $(0, c)$, where $c^2 = b^2 - a^2$ Major axis from $(0, -b)$ to $(0, b)$ Minor axis from $(-a, 0)$ to $(a, 0)$

Teaching Note

Help students to see the relationship of the equation for an ellipse to the equation for a circle. Point out that there are only two cases where an ellipse is oriented parallel to the coordinate axes, whereas there are four cases for the parabola.

> **Characteristics of a Hyperbola**
>
> Note that some of the rational functions studied earlier, such as $f(x) = \frac{1}{x}$, were hyperbolas, and some of the radical functions, such as
> $$g(x) = \sqrt{x^2 - 1},$$
> were portions of hyperbolas. Allow students sufficient time to explore the concepts presented in Example 7 on page 590.

To graph an ellipse in function mode, solve for

$$y^2 = f(x)$$

and then enter

$$y_1 = \sqrt{f(x)}, \qquad y_2 = -\sqrt{f(x)}$$

as illustrated in Example 5.

■ EXAMPLE 5 Finding an Equation of an Ellipse

Find an equation of the ellipse with foci $(0, -3)$ and $(0, 3)$ whose minor axis has length 4. Then draw its graph.

Solution

The foci are on the y-axis with $c = 3$. The minor axis is horizontal from $(-2, 0)$ to $(2, 0)$, so $a = 2$. Using $c^2 = b^2 - a^2$, we have

$$\begin{aligned} b^2 &= a^2 + c^2 \\ &= 2^2 + 3^2 \\ &= 13. \end{aligned}$$

Therefore, the standard form equation for the ellipse is

$$\frac{x^2}{4} + \frac{y^2}{13} = 1.$$

To draw its graph we solve for y in terms of x.

$$\begin{aligned} \frac{y^2}{13} &= 1 - \frac{x^2}{4} \\ y^2 &= 13\left(1 - \frac{x^2}{4}\right) \\ y &= \pm\sqrt{13\left(1 - \frac{x^2}{4}\right)} \end{aligned}$$

Figure 9.22 shows two views of the graph of

$$y_1 = \sqrt{13\left(1 - \frac{x^2}{4}\right)} \quad \text{and} \quad y_2 = -\sqrt{13\left(1 - \frac{x^2}{4}\right)}.$$ ■

4
−6.1 6.1
−4
(a)

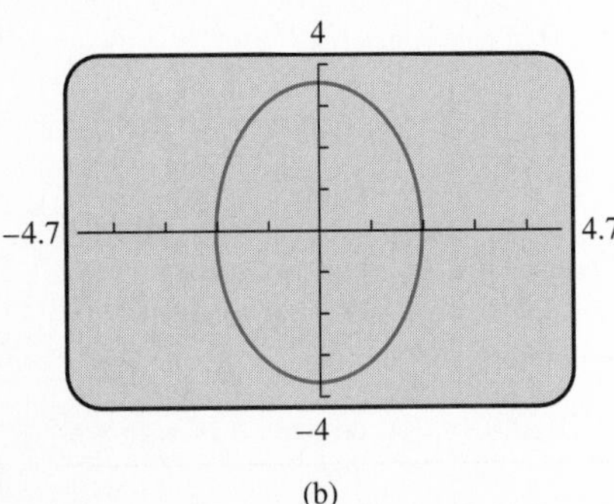

(b)

Figure 9.22 Two views of the ellipse $\frac{x^2}{4} + \frac{y^2}{13} = 1$: (a) An approximately square viewing window. (b) Notice that the gap does not show when the grapher window includes columns of pixels whose x-coordinates are ± 2.

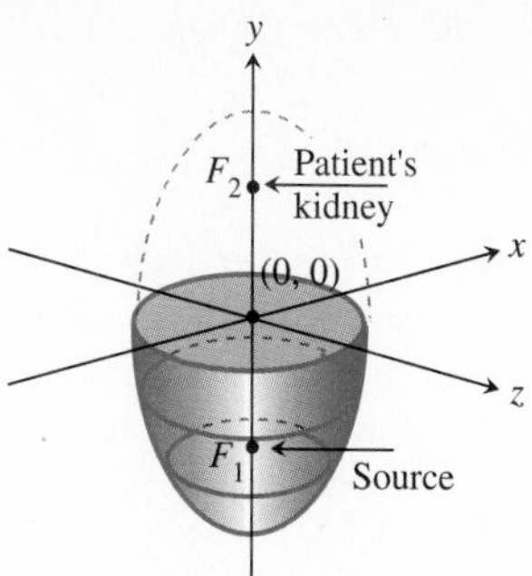

Figure 9.23

Applications Using Ellipses

If we rotate an ellipse in three-dimensional space about its major axis, the ellipse sweeps out an **elliptical surface** or **ellipsoid.** Light or sound emitted at one focus will reflect off the surface and pass through the other focus. Science museums contain "whispering galleries" that work on this principle. The whisper from one person standing at one focus can be heard easily by a second person standing on either focus, even if the speaker is facing away from the hearer.

This also has a medical application in the treatment of kidney stones. An elliptical **lithotripter** emits ultrahigh-frequency (UHF) shock waves from one focus. The shock waves pass through and break up a patient's kidney stone at the other focus. (See Figure 9.23.)

Teaching Note

The asymptotic behavior of hyperbolas is an extension of the ideas about asymptotes that students know from previous experiences with graphing. Allow time for students to explore hyperbolas in detail.

EXAMPLE 6 APPLICATION: Studying a Lithotripter

An ellipse that generates the ellipsoid of a lithotripter has the major axis with endpoints $(-6, 0)$ and $(6, 0)$. One endpoint of the minor axis is $(0, -2.5)$. Find the foci.

Solution

From information about the axes, $a = 6$ and $b = 2.5$. Thus,

$$\begin{aligned} c &= \sqrt{a^2 - b^2} \\ &= \sqrt{6^2 - 2.5^2} \\ &= 5.454 \cdots. \end{aligned}$$

The foci are approximately the points $(-5.45, 0)$ and $(5.45, 0)$. ■

Follow-up

Discuss the differences between ellipses and hyperbolas, including the differences between the equations and the relationship of the foci to the curves.

Assignment Guide

Day 1: Ex. 1, 4, 7, 10, 11, 14, 15, 17, 20, 21, 24, 25, 28, 29, 32
Day 2: Ex. 34–38, 43, 48–52, 55, 58

Cooperative Learning

Group Activity: Ex. 59–60

Notes on Exercises

Ex. 1–42 are traditional problems related to conic sections.
Ex. 51–54 relate to degenerate conics.
Ex. 61 is related to the concept of limits.

Hyperbolas

A hyperbola also has both an algebraic definition and a geometric definition.

Definition Hyperbola

Algebraic: A **hyperbola** is the graph of a quadratic relation of either form:

$$\frac{x^2}{a^2} - \frac{y^2}{b^2} = 1 \quad \text{or} \quad \frac{y^2}{b^2} - \frac{x^2}{a^2} = 1$$

Geometric: A **hyperbola** is the set of all points in a plane the difference of whose distances from two fixed points is constant.

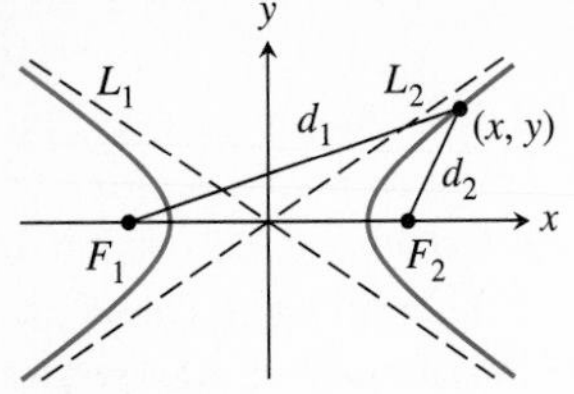

Figure 9.24 Basic structure of a hyperbola.

Figure 9.24 shows a hyperbola. Each of the fixed points F_1 and F_2 is a **focus,** and the distances whose difference is constant are d_1 and d_2. The hyperbola consists of two separate curves called **branches** that have lines L_1 and L_2 as **slant asymptotes.** (See Exercise 61.)

The line through the foci intersects the branches in the **vertices** of the hyperbola. The line segment joining the vertices is the **transverse axis,** and the midpoint of the transverse axis is the **center** of the hyperbola.

As with an ellipse, we can show that the algebraic definition of a hyperbola (see Exercise 60),

$$\frac{x^2}{a^2} - \frac{y^2}{b^2} = 1,$$

follows from the geometric definition by using the distance formula to compute $d_1 - d_2$. (See Figure 9.25.) (A similar figure can be drawn for the other standard form.) The algebraic and geometric characteristics of a hyperbola are summarized here.

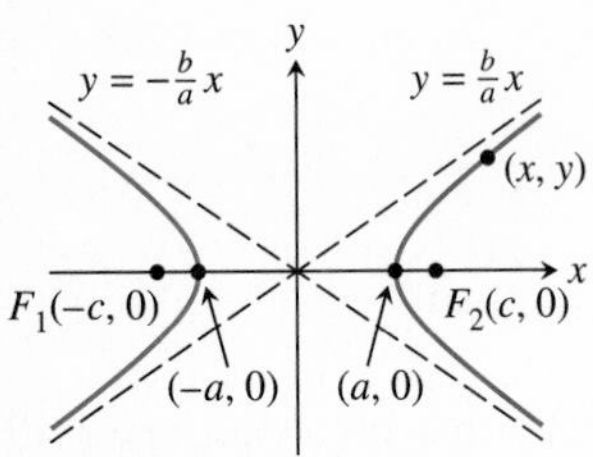

Figure 9.25 The transverse axis has length $2a$, and the foci are at $(c, 0)$ and $(-c, 0)$. The segment on the y-axis between $(0, b)$ and $(0, -b)$ is sometimes called the *conjugate axis.*

REMARK

Notice that a^2 always represents the constant in the denominator below the x^2 term, and b^2 always represents the constant in the denominator below the y^2 term. We also assume that a, b, and c are positive.

Characteristics of a Hyperbola

The **standard forms of a hyperbola centered at (0, 0)** are as follows:

Algebra	**Geometry**
$\frac{x^2}{a^2} - \frac{y^2}{b^2} = 1$	Foci: $(-c, 0)$ and $(c, 0)$, where $c^2 = a^2 + b^2$ Transverse axis from $(-a, 0)$ to $(a, 0)$ Asymptotes: $y = \pm \frac{b}{a}x$
$\frac{y^2}{b^2} - \frac{x^2}{a^2} = 1$	Foci: $(0, -c)$ and $(0, c)$, where $c^2 = a^2 + b^2$ Transverse axis from $(0, -b)$ to $(0, b)$ Asymptotes: $y = \pm \frac{b}{a}x$

Ongoing Assessment

Self-Assessment: Ex. 9, 13, 23, 27, 41, 57
Embedded Assessment: Ex. 50

■ EXAMPLE 7 Finding an Equation of a Hyperbola

Find the standard form equation for the hyperbola with foci $(0, -5)$ and $(0, 5)$, and vertices $(0, -4)$ and $(0, 4)$. Sketch its graph, and support your work with a grapher.

Solution

Because the foci and transverse axis are on the y-axis, we are looking for an equation of the form

$$\frac{y^2}{b^2} - \frac{x^2}{a^2} = 1.$$

One focus $(0, 5) = (0, c)$, so $c = 5$. One vertex $(0, 4) = (0, b)$, so $b = 4$. Therefore, $a^2 = c^2 - b^2 = 5^2 - 4^2 = 25 - 16 = 9$, and $a = 3$. The equation is

$$\frac{y^2}{16} - \frac{x^2}{9} = 1.$$

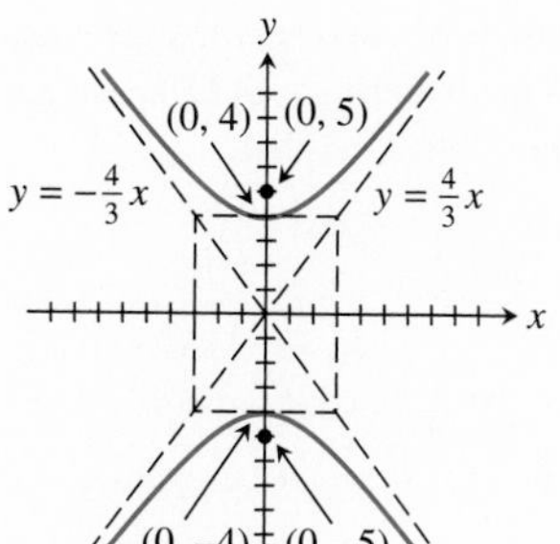

Figure 9.26

The slant asymptotes are the lines $y = \pm\frac{4}{3}x$.

To sketch the hyperbola, we begin by drawing the rectangle with vertices $(\pm a, \pm b) = (\pm 3, \pm 4)$. Then we sketch the slant asymptotes (diagonals of the rectangle). Finally, we sketch the graph of the hyperbola. (See Figure 9.26.)

To use a grapher, we solve for y:

$$\frac{y^2}{16} - \frac{x^2}{9} = 1$$

$$\frac{y^2}{16} = 1 + \frac{x^2}{9}$$

$$y^2 = 16\left(1 + \frac{x^2}{9}\right)$$

$$y = \pm 4\sqrt{1 + \frac{x^2}{9}}$$

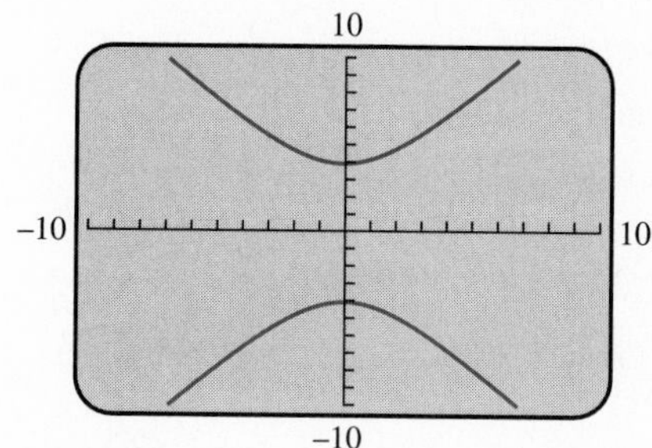

Figure 9.27

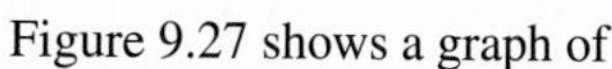

Figure 9.27 shows a graph of

$$y_1 = 4\sqrt{1 + \frac{x^2}{9}} \quad \text{and} \quad y_2 = -4\sqrt{1 + \frac{x^2}{9}}$$

that supports the sketch in Figure 9.26. ■

Quick Review 9.2

In Exercises 1–4, find the distance between the two points.

1. $(-1, 3)$ and $(2, 5)$

2. $(-3, -2)$ and $(2, 4)$

3. $(2, -3)$ and (a, b)

4. $(-3, -4)$ and (a, b)

In Exercises 5–10, solve for y in terms of x.

5. $2x + 3y = 5$

6. $3x - 2y = -1$

7. $2y^2 = 8x$

8. $3y^2 = 15x$

9. $\frac{x^2}{4} + \frac{y^2}{9} = 1$

10. $\frac{y^2}{16} - \frac{x^2}{9} = 1$

SECTION EXERCISES 9.2

In Exercises 1–6, find an equation in standard form for the given parabola.

1. Vertex (0, 0), focus (−3, 0)

2. Vertex (0, 0), focus (0, 2)

3. Vertex (0, 0), directrix $y = 4$

4. Vertex (0, 0), directrix $x = -2$

5. Focus (0, 5), directrix $y = -5$

6. Focus (−4, 0), directrix $x = 4$

In Exercises 7–10, find the focus, directrix, and line of symmetry for the parabola. Sketch a graph of the parabola. Support your work with a grapher.

7. $y = 4x^2$

8. $y = -\frac{1}{6}x^2$

9. $x = -8y^2$

10. $x = 2y^2$

In Exercises 11–14, find the endpoints of the major and minor axes of the ellipse. Sketch a graph of the ellipse. Support your answer with a grapher.

11. $\frac{x^2}{8^2} + \frac{y^2}{6^2} = 1$

12. $\frac{x^2}{9^2} + \frac{y^2}{5^2} = 1$

13. $\frac{y^2}{4^2} + \frac{x^2}{6^2} = 1$

14. $\frac{y^2}{7^2} + \frac{x^2}{5^2} = 1$

In Exercises 15–20, write an equation in standard form for the ellipse with center (0, 0).

15. Endpoints of axes are (±4, 0) and (0, ±5)

16. Endpoints of axes are (±7, 0) and (0, ±4)

17. Major axis length 6 on y-axis, minor axis length 4

18. Major axis length 14 on x-axis, minor axis length 10

19. Foci (±2, 0), major axis length 10

20. Foci (0, ±3), major axis length 10

In Exercises 21–24, write an equation in standard form and find the foci for the ellipse.

21. Major axis endpoints (±5, 0), minor axis length 4

22. Major axis endpoints (0, ±6), minor axis length 8

23. Minor axis endpoints (0, ±4), major axis length 10

24. Minor axis endpoints (±12, 0), major axis length 26

In Exercises 25–28, find the endpoints of the transverse axis and the equations of the asymptotes of the hyperbola. Sketch a graph of the hyperbola. Support your work with a grapher.

25. $\frac{x^2}{7^2} - \frac{y^2}{5^2} = 1$

26. $\frac{y^2}{8^2} - \frac{x^2}{5^2} = 1$

27. $\frac{y^2}{25} - \frac{x^2}{16} = 1$

28. $\frac{x^2}{169} - \frac{y^2}{144} = 1$

In Exercises 29–32, write an equation in standard form for the hyperbola.

29. Foci (±3, 0), transverse axis length 4

30. Foci (0, ±3), transverse axis length 4

31. Foci (0, ±15), transverse axis length 8

32. Foci (±5, 0), transverse axis length 3

In Exercises 33–36, write an equation in standard form for the conic shown.

33.

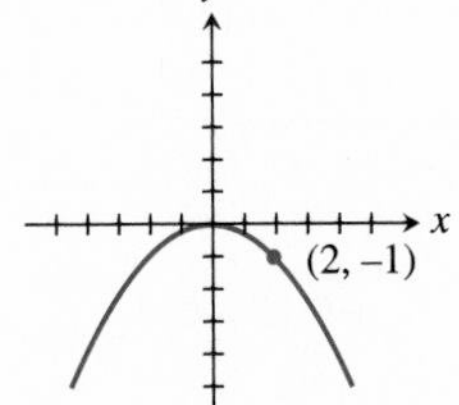

34.

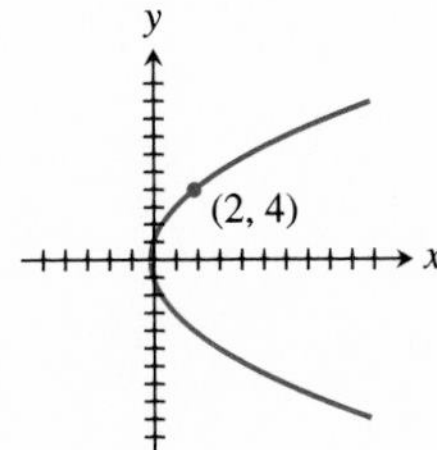

35.

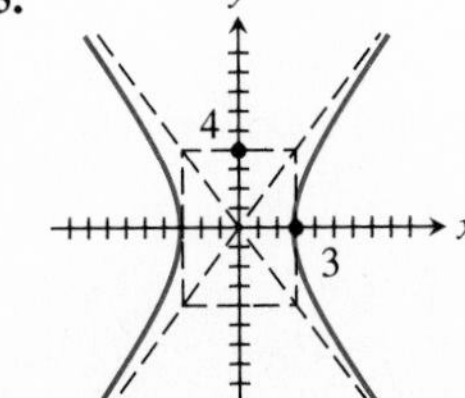

36.

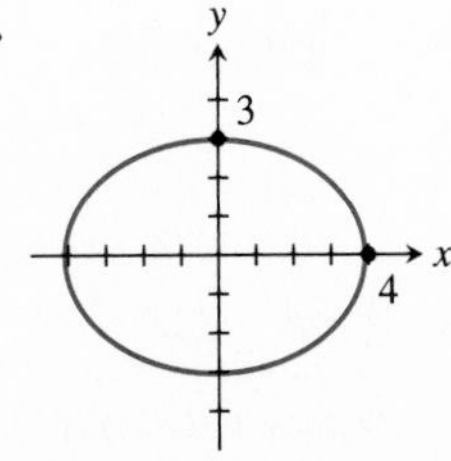

In Exercises 37–42, identify the conic. Then state its characteristics.

37. $\frac{x^2}{16} + \frac{y^2}{5} = 1$

38. $\frac{x^2}{1} - \frac{y^2}{4} = 1$

39. $x^2 = -6y$

40. $\frac{x^2}{7} + \frac{y^2}{9} = 1$

41. $\frac{y^2}{4} - \frac{x^2}{1} = 1$

42. $y^2 = -7x$

In Exercises 43–48, solve for y in terms of x. Then graph the conic.

43. $\frac{x^2}{12^2} + \frac{y^2}{15^2} = 1$

44. $y^2 = 5x$

45. $\frac{x^2}{4} - \frac{y^2}{9} = 1$

46. $\frac{x^2}{18^2} + \frac{y^2}{10^2} = 1$

47. $y^2 = -3x$

48. $\frac{y^2}{16} - \frac{x^2}{9} = 1$

In Exercises 49 and 50, solve the problem.

49. Writing to Learn If $a = b = r$, explain why the ellipse

$$\frac{x^2}{a^2} + \frac{y^2}{b^2} = 1$$

is a circle with center (0, 0) and radius r. (In this case the two foci collapse to the center, so we interpret "sum of whose distance" by "distance" in the geometric definition.)

50. Show that an equation for the parabola with focus $(p, 0)$ and directrix $x = -p$ is

$$x = \left(\frac{1}{4p}\right)y^2.$$

In Exercises 51–54, the graph of the equation is sometimes called a *degenerate conic*. Describe the graph.

51. $\frac{x^2}{4} + \frac{y^2}{9} = 0$

52. $\frac{x^2}{a^2} + \frac{y^2}{b^2} = 0$

53. $\frac{x^2}{4} - \frac{y^2}{9} = 0$

54. $\frac{x^2}{a^2} - \frac{y^2}{b^2} = 0$

In Exercises 55–58, solve the problem.

55. *Parabolic Microphones* Sports Channel uses a parabolic microphone to capture all the sounds from the basketball players and coaches during a regular season game. If one of its microphones has a parabolic surface generated by the parabola $10y = x^2$, locate the focus (the electronic receiver) of the parabola.

56. *Parabolic Headlights* Stein Glass, Inc., makes parabolic headlights for a variety of automobiles. If one of its headlights has a parabolic surface generated by the parabola $x^2 = 12y$, where should its light bulb be placed?

57. *Lithotripter* For an ellipse that generates the ellipsoid of a lithotripter, the major axis has endpoints $(-8, 0)$ and $(8, 0)$. One endpoint of the minor axis is $(0, 3.5)$. Find the coordinates of the foci.

58. *Lithotripter (Refer to Figure 9.23.)* A lithotripter's shape is formed by rotating the portion of an ellipse below its minor axis about its major axis. If the length of the major axis is 26 in., and the length of the minor axis is 10 in., where should the shock-wave source and the patient be placed for maximum effect?

EXTENDING THE IDEAS

In Exercises 59–62, solve the problem.

59. Group Learning Activity Let $c^2 = a^2 - b^2$ with $a > b$ and $d_1 + d_2 = 2a$. *Work in groups of three.* Use the distance formula to show that the standard form equation for the ellipse with foci $(\pm c, 0)$ is

$$\frac{x^2}{a^2} + \frac{y^2}{b^2} = 1.$$

60. Group Learning Activity Let $c^2 = a^2 + b^2$ and $d_1 - d_2 = \pm 2a$. *Work in groups of three.* Show that the standard form equation for the hyperbola with foci $(\pm c, 0)$ is

$$\frac{x^2}{a^2} - \frac{y^2}{b^2} = 1.$$

61. Consider the hyperbola

$$\frac{x^2}{a^2} - \frac{y^2}{b^2} = 1.$$

Show that

$$y^2 \to \frac{b^2x^2}{a^2} \quad \text{as} \quad |x| \to \infty,$$

and conclude that

$$|y| \to \frac{b}{a}|x| \quad \text{as} \quad |x| \to \infty.$$

62. Writing to Learn Show how the standard form equations for the conics are related to

$$Ax^2 + Bxy + Cy^2 + Dx + Ey + F = 0.$$

9.3
CONIC SECTIONS AND TRANSFORMATIONS

Objective

Students will be able to graph transformations of conic sections and recognize conic sections from the general second-degree equation.

Motivate

Have students describe the graph of $(x - 3)^2 + (y + 4)^2 = 25$ according to what they learned in Section 0.8. **(Circle, radius 5, center (3, −4))** How is this graph related to the graph of $x^2 + y^2 = 25$? **(Translated right 3, down 4)**

Teaching Note

Students should be able to relate the transformations of conic sections to the transformations that were studied in Section 2.6.

Translations of Conics • Parabolas • Ellipses • Hyperbolas • Second-Degree Equations • Rotation of Conics • Applications

Translations of Conics

In this section we study conics that are obtained when horizontal or vertical translations are applied to the basic curves studied in Section 9.2. For example, if the ellipse with equation

$$\frac{x^2}{a^2} + \frac{y^2}{b^2} = 1$$

is shifted horizontally h units and vertically k units, the resulting ellipse has the equation

$$\frac{(x - h)^2}{a^2} + \frac{(y - k)^2}{b^2} = 1.$$

We shall explore how the characteristics change when we apply horizontal and vertical translations to parabolas, ellipses, and hyperbolas.

Parabolas

When a parabola with the equation $y = ax^2$ or $x = ay^2$ is translated horizontally by h units and vertically by k units, the vertex of the parabola moves from (0, 0) to (h, k). (See Figure 9.28.)

The connections between this geometry and the associated algebra are highlighted here.

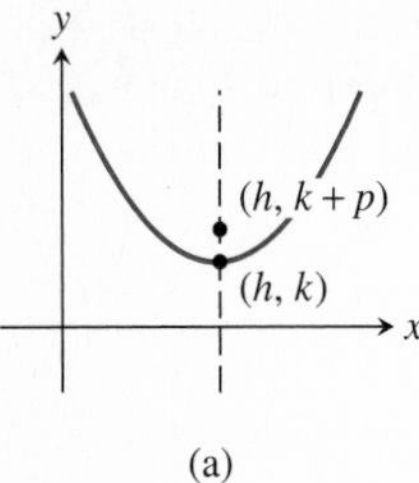

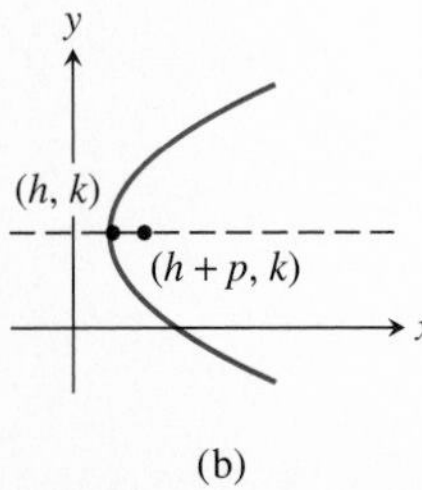

Figure 9.28 Parabolas with the focus on (a) $x = h$ and (b) $y = k$.

Characteristics of a Parabola

The **standard forms of a parabola with the vertex (*h, k*)** are as follows:

Algebra	**Geometry**
$y - k = a(x - h)^2$	Focus: $\left(h,\ k + \frac{1}{4a}\right)$
	Directrix: $y = k - \frac{1}{4a}$
$x - h = a(y - k)^2$	Focus: $\left(h + \frac{1}{4a},\ k\right)$
	Directrix: $x = h - \frac{1}{4a}$

■ EXAMPLE 1 Finding an Equation of a Parabola

Find the standard form equation for the parabola with the vertex (3, 4) and the focus (5, 4).

Solution

Because the vertex and focus are in line horizontally, with the focus to the right of the vertex, we know that the parabola opens to the right and that its equation has the form $x - h = a(y - k)^2$.

The vertex is at $(h, k) = (3, 4)$, which means that $h = 3$ and $k = 4$. The focus is at $(h + 1/(4a), k) = (3 + 1/(4a), 4) = (5, 4)$, which means that

$$3 + \frac{1}{4a} = 5,$$

$$4a = \frac{1}{2},$$

$$a = \frac{1}{8}.$$

The standard form equation for the parabola is $x - 3 = \frac{1}{8}(y - 4)^2$. ■

■ EXAMPLE 2 Using Standard Forms with a Parabola

Show that the graph of $y^2 - 6x + 2y + 13 = 0$ is a parabola, and find its vertex, focus, and directrix.

Solution

Because this equation is quadratic in the variable y, we complete the square with respect to y to obtain a standard form.

$$y^2 - 6x + 2y + 13 = 0$$

$$y^2 + 2y = 6x - 13$$

$$y^2 + 2y + 1 = 6x - 13 + 1 \quad \text{Complete the square.}$$

$$(y + 1)^2 = 6x - 12$$

$$= 6(x - 2)$$

$$x - 2 = \frac{1}{6}(y + 1)^2$$

This equation is in the standard form $x - h = a(y - k)^2$, where $h = 2$, $k = -1$, and $a = 1/6$. It follows that

the vertex is $(2, -1)$,

the focus is $\left(h + \dfrac{1}{4a}, k\right) = \left(2 + \dfrac{1}{4(1/6)}, -1\right) = (3\frac{1}{2}, -1)$, and

the directrix is $x = h - \dfrac{1}{4a} = 2 - 3/2 = 1/2$, or $x = 1/2$. ■

Ellipses

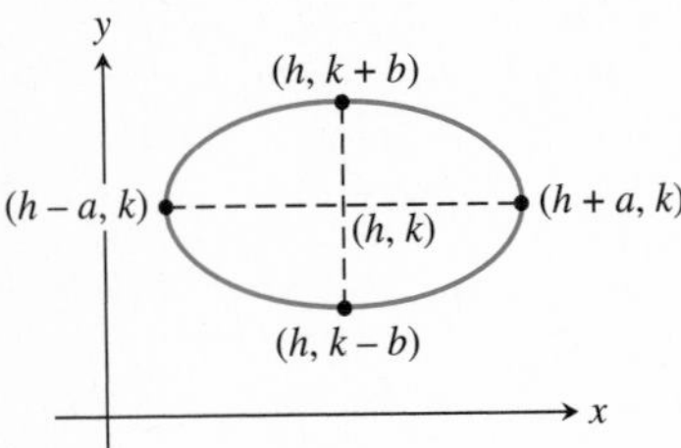

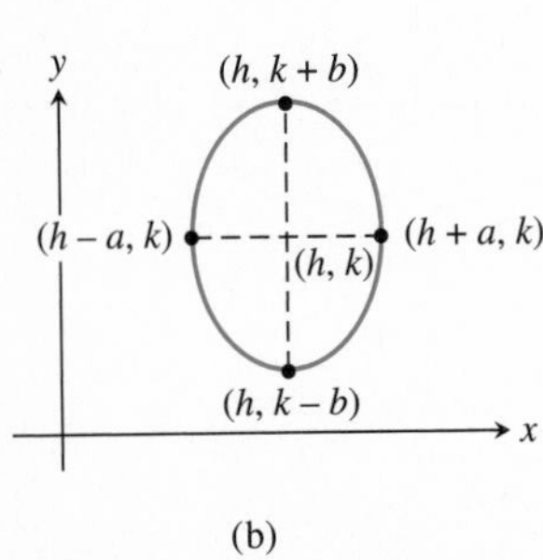

Figure 9.29 Ellipses with foci on (a) $y = k$ and (b) $x = h$.

When an ellipse with center (0, 0) is translated horizontally by h units and vertically by k units, the center of the ellipse moves from (0, 0) to (h, k), as illustrated in Figure 9.29.

The connections between this geometry and the associated algebra are highlighted here.

Characteristics of an Ellipse

The **standard forms of an ellipse centered at (h, k)** are as follows:

Algebra	**Geometry**
$\dfrac{(x - h)^2}{a^2} + \dfrac{(y - k)^2}{b^2} = 1$, $a > b$	Foci: $(h - c, k)$ and $(h + c, k)$, where $c^2 = a^2 - b^2$ Major axis from $(h - a, k)$ to $(h + a, k)$ Minor axis from $(h, k - b)$ to $(h, k + b)$
$\dfrac{(x - h)^2}{a^2} + \dfrac{(y - k)^2}{b^2} = 1$, $a < b$	Foci: $(h, k - c)$ and $(h, k + c)$, where $c^2 = b^2 - a^2$ Major axis from $(h, k - b)$ to $(h, k + b)$ Minor axis from $(h - a, k)$ to $(h + a, k)$

■ EXAMPLE 3 Finding an Equation of an Ellipse

Find the standard form equation of the ellipse whose major axis has the endpoints $(-2, -1)$ and $(8, -1)$, and whose minor axis has length 8.

Solution

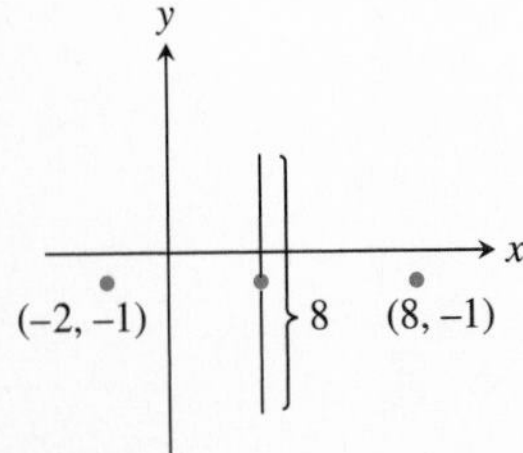

Figure 9.30

Figure 9.30 shows the major axis endpoints and the minor axis. The standard equation of this ellipse has the form

$$\frac{(x-h)^2}{a^2} + \frac{(y-k)^2}{b^2} = 1,$$

where $a > b$. Because $(-2, -1)$ and $(8, -1)$ are endpoints of the major axis, the midpoint

$$\left(\frac{(-2)+8}{2}, \frac{(-1)+(-1)}{2}\right) = (3, -1)$$

is the center of the ellipse. Because a and b are each 1/2 the length of an axis,

$$a = \frac{8-(-2)}{2} = 5 \quad \text{and} \quad b = \frac{8}{2} = 4.$$

The desired equation is

$$\frac{(x-(3))^2}{5^2} + \frac{(y-(-1))^2}{4^2} = 1,$$

$$\frac{(x-3)^2}{25} + \frac{(y+1)^2}{16} = 1.$$

■

To graph an ellipse on a grapher we solve the defining equation for y as follows:

$$\frac{(x-h)^2}{a^2} + \frac{(y-k)^2}{b^2} = 1$$

$$\frac{(y-k)^2}{b^2} = 1 - \frac{(x-h)^2}{a^2}$$

$$(y-k)^2 = \frac{b^2}{a^2}[a^2 - (x-h)^2]$$

$$y - k = \pm\sqrt{\frac{b^2}{a^2}[a^2 - (x-h)^2]}$$

$$y = \pm\frac{b}{a}\sqrt{a^2 - (x-h)^2} + k$$

■ EXAMPLE 4 Graphing an Ellipse

Graph the ellipse

$$\frac{(x-4)^2}{25} + \frac{(y+1)^2}{169} = 1.$$

Confirm the center, the endpoints of the major and minor axes, and the foci.

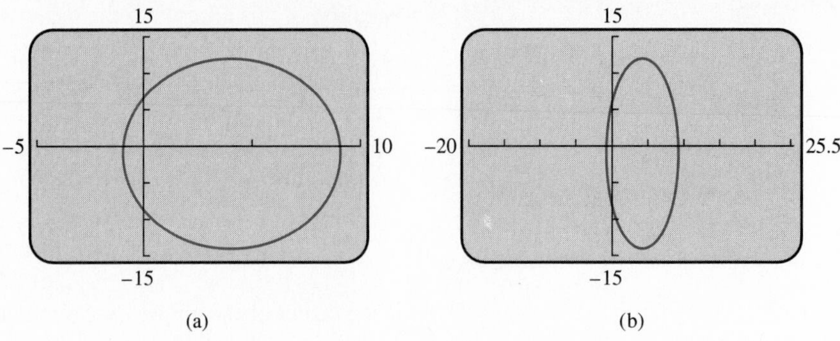

Figure 9.31 Two views of the ellipse of Example 4. Example 5 explains why neither graph has gaps near the x-axis. Also see the Technology Note.

> **TECHNOLOGY NOTE**
>
> The viewing window of Figure 9.31a is large enough to include the complete graph, but gives a distorted view. Notice that the minor axis appears to be larger than the major axis. The viewing window in Figure 9.31b is square, and thus gives a true picture of their relative sizes.

Solution

To graph the ellipse (see Figure 9.31) we use

$$y_1 = \frac{13}{5}\sqrt{25 - (x - 4)^2} - 1$$

and

$$y_2 = -\frac{13}{5}\sqrt{25 - (x - 4)^2} - 1.$$

Center: The standard form of the original equation confirms that the center of the ellipse is $(h, k) = (4, -1)$.

Major Axis: Because $b^2 > a^2$, the major axis is vertical. Because $b^2 = 169$, $b = 13$, which means that the endpoints are 13 units below and above the center at $(4, -14)$ and $(4, 12)$.

Minor Axis: Because $a^2 = 25$, $a = 5$, and the endpoints of the minor axis are 5 units to the left and right of the center, at $(-1, -1)$ and $(9, -1)$.

Foci: $c = \sqrt{169 - 25} = 12$, and the foci are 12 units below and above the center, at $(4, -13)$ and $(4, 11)$. ■

Except for parabolas of one form, the conic sections are not functions. As before, we can describe some nonfunction relations using parametric equations. Example 5 shows parametric equations for an ellipse.

■ EXAMPLE 5 Using Parametric Equations for an Ellipse

The following parametric equations define the same ellipse as given in Example 4. Confirm this algebraically by eliminating the parameter t. Support your work graphically.

$$x = 4 + 5\cos t, \qquad y = -1 + 13\sin t, \qquad 0 \le t \le 2\pi$$

Teaching Note

The graph of an ellipse may be produced on a grapher either by solving the equation for y or by using parametric equations. The use of parametric equations prevents "gaps" in the graph.

Solution

We eliminate the parameter t by using the Pythagorean identity $\cos^2 t + \sin^2 t = 1$:

$$x = 4 + 5\cos t \qquad\qquad y = -1 + 13\sin t$$

$$\cos t = \frac{x-4}{5} \qquad\qquad \sin t = \frac{y+1}{13}$$

Therefore, $$\frac{(x-4)^2}{25} + \frac{(y+1)^2}{169} = \cos^2 t + \sin^2 t = 1.$$

The resulting equation is the standard form for the equation of an ellipse.

Support Graphically

The two graphs in Figure 9.31 were actually produced using these parametric equations. One reason for using parametric equations to graph ellipses is to eliminate the gaps that often show up when graphing in function mode. (Compare with Figure 9.22.) Draw the graphs in Figure 9.31 in function mode to see the gaps near the x-axis. ■

Hyperbolas

When a hyperbola with the center (0, 0) is translated horizontally h units and vertically k units, the center of the hyperbola moves from (0, 0) to (h, k), as illustrated in Figure 9.32.

The connections between this geometry and the associated algebra are highlighted below.

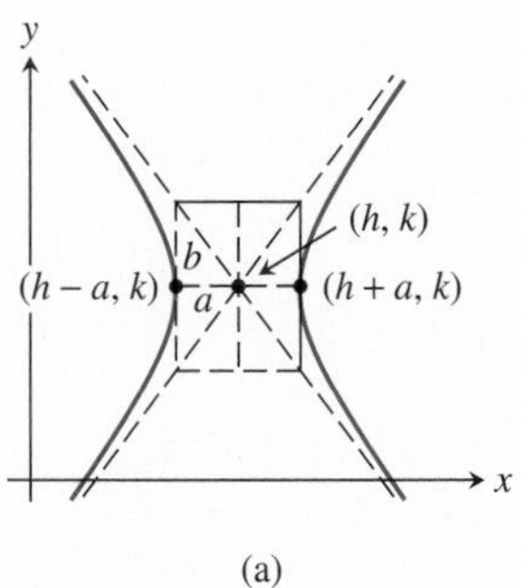

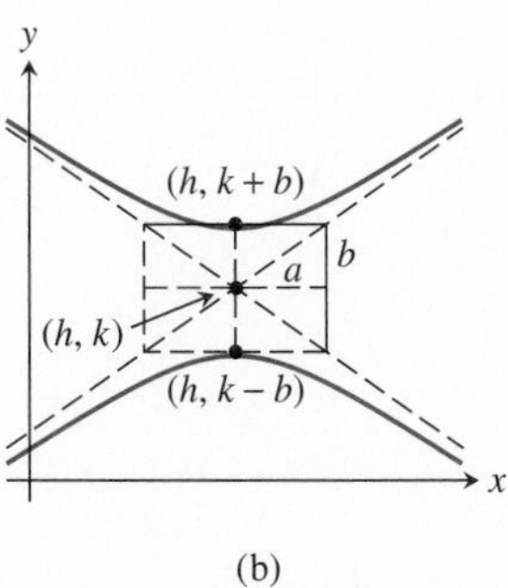

Figure 9.32 Hyperbolas with foci on (a) $y = k$ and (b) $x = h$. These graphs illustrate how to sketch a graph of a hyperbola with paper and pencil.

Characteristics of a Hyperbola

The **standard forms of a hyperbola centered at (h, k)** are as follows:

Algebra	**Geometry**
$\frac{(x-h)^2}{a^2} - \frac{(y-k)^2}{b^2} = 1$	Foci: $(h - c, k)$ and $(h + c, k)$, where $c^2 = a^2 + b^2$ Transverse axis from $(h - a, k)$ to $(h + a, k)$ Asymptotes: $y - k = \pm \frac{b}{a}(x - h)$
$\frac{(y-k)^2}{b^2} - \frac{(x-h)^2}{a^2} = 1$	Foci: $(h, k - c)$ and $(h, k + c)$, where $c^2 = a^2 + b^2$ Transverse axis from $(h, k - b)$ to $(h, k + b)$ Asymptotes: $y - k = \pm \frac{b}{a}(x - h)$

EXAMPLE 6 Using Standard Forms with a Hyperbola

Show that the graph of this equation is a hyperbola. Find the center, the endpoints of the transverse axis, and the asymptotes. Draw its graph.

$$x^2 - 4y^2 + 2x - 24y = 39$$

Solution

To find the standard form, we group the x and y terms and then complete the squares:

$$x^2 - 4y^2 + 2x - 24y = 39$$

$$(x^2 + 2x) - 4(y^2 + 6y) = 39 \qquad \text{Group terms.}$$

$$(x^2 + 2x + 1) - 4(y^2 + 6y + 9) = 39 + 1 - 4(9) \qquad \text{Complete the squares.}$$

$$(x + 1)^2 - 4(y + 3)^2 = 4$$

$$\frac{(x + 1)^2}{4} - \frac{(y + 3)^2}{1} = 1 \qquad \text{Standard form}$$

Thus, the equation is that of a hyperbola (indicated by the "$-$" sign) with a horizontal transverse axis. (Note that the "$-$" sign accompanies the expression in y.) The center is $(-1, -3)$. Because $a^2 = 4$, $a = 2$, so the endpoints of the transverse axis are 2 units to the left and right of the center, at $(-3, -3)$ and $(1, -3)$. Because $b^2 = 1$, $b = 1$, so we can derive the asymptotes as the lines

$$y - k = \pm\frac{b}{a}(x - h),$$

$$y - (-3) = \pm\frac{1}{2}[x - (-1)]$$

$$y + 3 = \pm\frac{1}{2}(x + 1).$$

To graph the hyperbola we solve the standard form equation for y:

$$(y + 3)^2 = \frac{(x + 1)^2}{4} - 1 = \frac{(x + 1)^2 - 4}{4}$$

$$y + 3 = \pm\frac{1}{2}\sqrt{(x + 1)^2 - 4}$$

$$y = \pm\frac{1}{2}\sqrt{(x + 1)^2 - 4} - 3$$

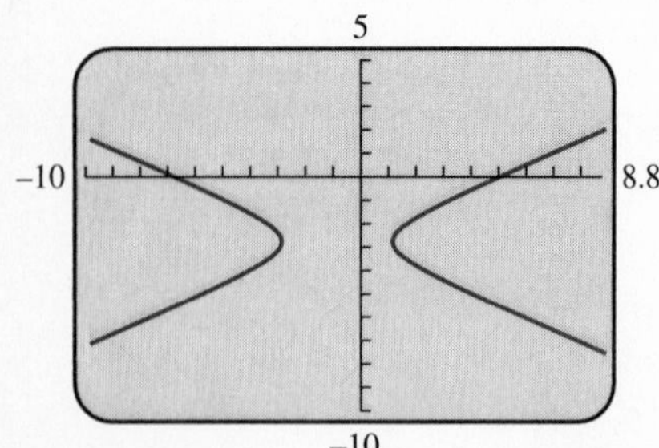

Figure 9.33

Figure 9.33 shows a graph of the hyperbola using

$$y_1 = \frac{1}{2}\sqrt{(x + 1)^2 - 4} - 3 \quad \text{and} \quad y_2 = -\frac{1}{2}\sqrt{(x + 1)^2 - 4} - 3.$$

■

Second-Degree Equations

The equation in Example 6 is equivalent to $x^2 - 4y^2 + 2x - 24y - 39 = 0$, which is given in the form of the **general second-degree equation**

$$Ax^2 + Bxy + Cy^2 + Dx + Ey + F = 0,$$

where $A = 1$, $B = 0$, $C = -4$, $D = 2$, $E = -24$ and $F = -39$. We found it convenient to change its form to the standard form so that we could identify the conic and easily find out information about it. To use a grapher, we solved standard form equations for y, or y_1 and y_2, something that we are also able to do with a second-degree equation in general form.

It can be shown that the graph of a general second-degree equation is a parabola, an ellipse, a hyperbola, or a degenerate conic (a point or a line). It is not a coincidence that the expression $B^2 - 4AC$ discriminates among the conics.

Teaching Note

Students should be able to understand the three cases for the graph of a nondegenerate general second-degree equation where $B = 0$. If $A = 0$ or $C = 0$ (only one second-degree term), the graph is a parabola; if A and C have the same sign, the graph is an ellipse; and if A and C have opposite signs, the graph is a hyperbola. See Exercises 57–58.

Graphs of Second-Degree Equations

A nondegenerate graph of

$$Ax^2 + Bxy + Cy^2 + Dx + Ey + F = 0$$

1. is an ellipse if $B^2 - 4AC < 0$;
2. is a parabola if $B^2 - 4AC = 0$;
3. is a hyperbola if $B^2 - 4AC > 0$.

When $B \neq 0$, the conic is rotated and the axes are not parallel to either of the coordinate axes.

Notes on Examples

Example 7 presents a parabola whose equation has a nonzero rotational term Bxy. It is useful for students to consider that conic sections need not have an axis of symmetry parallel to either axis.

EXAMPLE 7 Graphing a Second-Degree Equation

Identify and graph the conic defined by the equation

$$x^2 + 4xy + 4y^2 - 30x - 90y + 450 = 0.$$

Solution

$A = 1$, $B = 4$, $C = 4$, so $B^2 - 4AC = 0$ and the graph is a parabola. To use a grapher we first rewrite the equation as a quadratic in y:

$$4y^2 + (4x - 90)y + (x^2 - 30x + 450) = 0$$

REMARK

It is not necessary to simplify y_1 and y_2 as we did in Example 7. The trade-off is fewer pencil and paper errors at the risk of more keystroking errors.

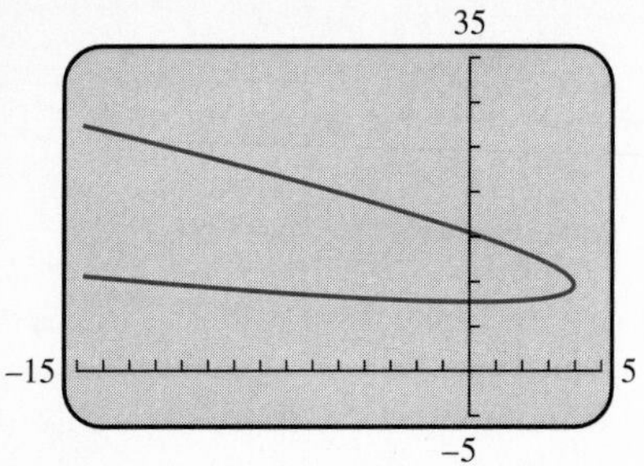

Figure 9.34

Then, we solve for y using the quadratic formula:

$$y = \frac{-(4x - 90) \pm \sqrt{(4x - 90)^2 - 16(x^2 - 30x + 450)}}{8}$$

or

$$y_1 = \frac{45 - 2x + \sqrt{225 - 60x}}{4}$$

and

$$y_2 = \frac{45 - 2x - \sqrt{225 - 60x}}{4}.$$

The graph of this parabola is shown in Figure 9.34. Notice that its axis is not parallel to either axis (because of the $4xy$ term). ■

Rotation of Conics

Suppose two coordinate systems have the same origin and the x'-axis of the $x'y'$ coordinate system makes an angle α with the x-axis of the xy coordinate system. In Exercise 61 you will show that the relationship between the coordinates of the two systems is given by

$$x = x' \cos \alpha - y' \sin \alpha$$

$$y = x' \sin \alpha + y' \cos \alpha$$

A second-degree equation that has an xy term ($B \neq 0$)

$$Ax^2 + Bxy + Cy^2 + Dx + Ey + F = 0$$

can be transformed into one with no such term

$$A'(x')^2 + C'(y')^2 + D'(x') + E'(y') + F' = 0$$

by a rotation of the coordinate axes through an angle α with

$$\cot 2\alpha = \frac{A - C}{B}.$$

■ EXAMPLE 8 Finding an Angle of Rotation

Find an angle of rotation that will eliminate the xy term in

$$x^2 - 3xy + y^2 - 5 = 0.$$

Solution

The angle α of rotation is given by

$$\cot 2\alpha = \frac{A - C}{B}$$

$$\cot 2\alpha = \frac{1 - 1}{(-3)} = 0 \qquad A = 1,\ B = -3,\ C = 1.$$

$$2\alpha = \cot^{-1} 0 = \frac{\pi}{2}$$

$$\alpha = \frac{\pi}{4}$$

You can show that the equations

$$x = x' \cos\frac{\pi}{4} - y' \sin\frac{\pi}{4} = \frac{x' - y'}{\sqrt{2}},$$

$$y = x' \sin\frac{\pi}{4} + y' \cos\frac{\pi}{4} = \frac{x' + y'}{\sqrt{2}}$$

will transform $x^2 - 3xy + y^2 - 5 = 0$ to

$$\frac{(y')^2}{2} - \frac{(x')^2}{10} = 1.$$

Interpret

The graph of $x^2 - 3xy + y^2 - 5 = 0$ is the hyperbola

$$\frac{y^2}{2} - \frac{x^2}{10} = 1$$

rotated through an angle of $\pi/4$ radians with the positive x-axis. ■

Applications

Radio signals travel 980 feet per microsecond (1 μsec = 0.000001 sec). Thus, the time it takes a radio signal to travel from point O to point P is a function of the distance d between these two points. If it takes 5 μsec for the signal to travel from O to P, then

$$OP = 980 \text{ ft}/\mu\text{sec} \times 5\ \mu\text{sec} = 4900 \text{ ft}.$$

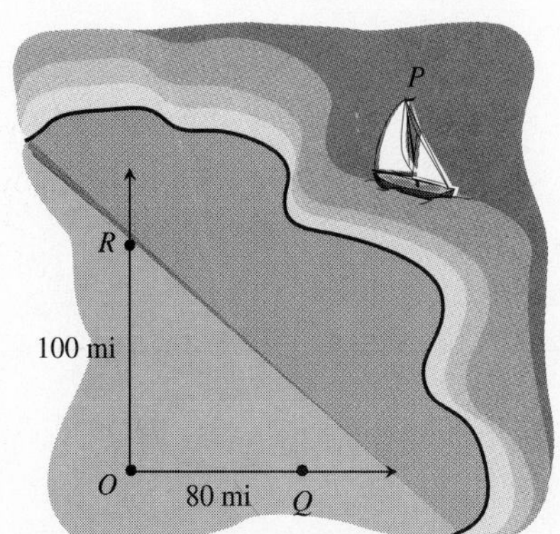

Figure 9.35

If radio signals are transmitted simultaneously from two locations O and R (see Figure 9.35), and if the signal from R reaches a receiver at point P one microsecond before the signal from O, then point P must lie somewhere on the hyperbola with foci O and R defined by the constant difference 980 ft. (Recall that a hyperbola consists of all points on which the difference of whose distances from two fixed points is a constant, which in this case is 980.) The long-range navigation system LORAN uses this fact to locate ships at sea.

■ EXAMPLE 9 APPLICATION: Using the LORAN System

Radio transmitters sending simultaneous signals are located at points O, R, and Q with R due north of O (see Figure 9.35). The LORAN equipment on the sloop *Gloria* receives a radio signal from O, 323.27 μsec after it receives the signal from R, and 258.61 μsec after it receives the signal from Q. Determine the sloop's bearing and distance from point O.

Solution

Model

The *Gloria* is located at a point of intersection of two hyperbolas, one with foci at O and R and on the other with foci at O and Q.

Hyperbola with Foci at O(0, 0) and R(0, 100): The center is (0, 50), so the distance from the center of the hyperbola to the foci is $c = 50$. The length of the transverse axis, $2b$, is the constant difference defining the hyperbola; thus it is the *additional* distance traveled by the signal from O.

$$2b = (323.27\ \mu\text{sec})(980\ \text{ft}/\mu\text{sec}) = 316{,}804.6\ \text{ft} \approx 60\ \text{mi}$$

Thus $b \approx 30$ and $a = \sqrt{c^2 - b^2} \approx \sqrt{50^2 - 30^2} = 40$. The equation is approximately

$$\frac{(y - 50)^2}{30^2} - \frac{x^2}{40^2} = 1.$$

Hyperbola with Foci at O(0, 0) and Q(80, 0): The center is (40, 0), so the distance from the center of the hyperbola to the foci is $c = 40$. As before, the length of the transverse axis, $2a$, is the constant difference defining the hyperbola; thus it is the *additional* distance traveled by the signal from O.

$$2a = (258.61\ \mu\text{sec})(980\ \text{ft}/\mu\text{sec}) = 253{,}437.8\ \text{ft} \approx 48\ \text{mi}$$

Thus $a \approx 24$ and $b = \sqrt{c^2 - a^2} \approx \sqrt{40^2 - 24^2} = 32$. The equation is approximately

$$\frac{(x - 40)^2}{24^2} - \frac{y^2}{32^2} = 1.$$

Solve Graphically

The sloop must be located on the upper branch of the hyperbola with foci O and R, and on the right branch of the hyperbola with foci O and Q. (See Figure 9.36 and Exercises 59 and 60.) We use a grapher to show that the coordinates of this point are approximately (187.09, 193.49). The bearing from point O is approximately

$$90° - \tan^{-1}\left(\frac{193.49}{187.09}\right) = 44.036 \cdots°.$$

The distance from point O is approximately

$$\sqrt{187.09^2 + 193.49^2} = 269.148 \cdots.$$

Follow-up

Ask why it would not be convenient to use the hyperbola with foci at Q and R in Example 9. **(Axes of symmetry are diagonal.)**

Assignment Guide

Day 1: Ex. 3–36, multiples of 3
Day 2: Ex. 39, 41, 44, 47, 50, 54, 55

Cooperative Learning

Group Activity: Ex. 57–60

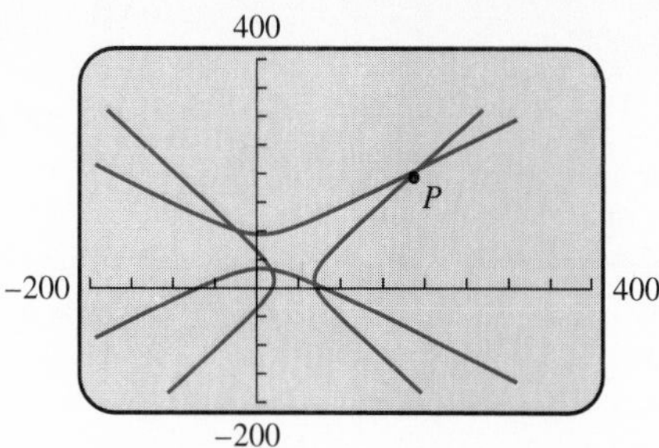

Figure 9.36

Interpret

The *Gloria* is located 187.09 mi east and 193.49 mi north of point O on a bearing of about 44°. Its distance from O is about 269 mi. ■

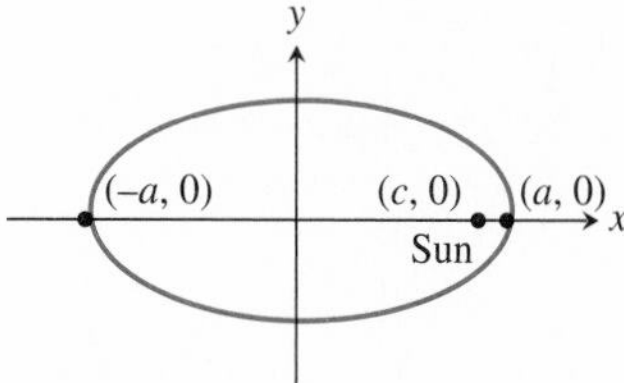

Figure 9.37

Planets and comets travel around the sun in elliptical orbits with the sun as one focus (see Figure 9.37). The earth's orbit is nearly a circle, whereas the orbit of Halley's comet is a very "flat" ellipse. The **eccentricity** e of an ellipse, defined by

$$e = \frac{c}{a} = \frac{\sqrt{a^2 - b^2}}{a} < 1,$$

is a measure of how nearly circular the ellipse is. Table 9.1 gives the eccentricity e for the elliptical orbits of some bodies in our solar system. Also shown for each body is the constant a (in millions of miles). It is called the **semimajor axis** length because it is 1/2 the length of the major axis of the ellipse that models each planet's path.

Table 9.1 Solar System Orbits

Body	Semimajor Axis, a	Eccentricity, e
Earth	93.0	0.02
Jupiter	483.6	0.05
Mars	141.6	0.09
Mercury	36.0	0.21
Venus	67.2	0.01
Halley's comet	1680.0	0.97

■ EXAMPLE 10 APPLICATION: Studying Planetary Orbits

Find the **perihelion** (minimum distance) and **aphelion** (maximum distance) of the earth from the sun.

Solution

Model

Assume that the earth's orbit is modeled as shown in Figure 9.37, with the sun at the focus $(c, 0)$. Then perihelion is $a - c$ and aphelion is $c - (-a) = c + a$.

Solve Algebraically

$$e = \frac{c}{a}$$

$$c = ea$$

$$c \approx (0.02)(93.0) = 1.860 \quad \text{Values from Table 9.1}$$

Notes on Exercises

Ex. 1–46 are traditional exercises on conic sections.
Ex. 59–60 relate the algebraic and geometric definitions of a hyperbola.

Ongoing Assessment

Self-Assessment: Ex. 5, 17, 23, 35, 43, 53

Embedded Assessment: Ex. 54

Therefore,

$$\begin{aligned}\text{Perihelion} &= a - c \approx 93.0 - 1.9 = 91.1,\\ \text{Aphelion} &= c + a \approx 1.9 + 93.0 = 94.9.\end{aligned}$$

Interpret

The minimum distance between the earth and the sun is approximately 91.1 million miles, and the maximum distance is approximately 94.9 million miles. ■

Quick Review 9.3

In Exercises 1–4, use the quadratic formula to solve the equation. Show exact solutions.

1. $2y^2 + 9y - 5 = 0$ **2.** $3y^2 + 13y + 4 = 0$

3. $9y^2 + 6y - 4 = 0$ **4.** $16y^2 + 16y + 1 = 0$

In Exercises 5–8, describe how the graph of the equation can be obtained by applying transformations to the graph of $f(x) = |x|$. Support your answer graphically.

5. $y = |x + 2| + 5$ **6.** $y = |x - 3| + 2$

7. $y = |x - 7| - 3$ **8.** $y = |x + 2| - 5$

In Exercises 9–12, use completing the square to solve the equation. Show exact solutions.

9. $x^2 + 4x - 1 = 0$ **10.** $x^2 - 2x - 2 = 0$

11. $2x^2 - 6x - 3 = 0$ **12.** $2x^2 + 4x - 5 = 0$

SECTION EXERCISES 9.3

In Exercises 1–6, find an equation in standard form for the parabola.

1. Focus $(-2, -4)$ and vertex $(-4, -4)$

2. Focus $(-5, 3)$ and vertex $(-5, 6)$

3. Focus $(3, 4)$ and directrix $y = 1$

4. Focus $(2, -3)$ and directrix $x = 5$

5. Vertex $(4, 3)$ and directrix $x = 6$

6. Vertex $(3, 5)$ and directrix $y = 7$

In Exercises 7–10, find the vertex, line of symmetry, focus, and directrix of the parabola. Sketch the parabola. Support your work graphically.

7. $12(y + 1) = (x - 3)^2$ **8.** $6(y - 3) = (x + 1)^2$

9. $2 - y = 16(x - 3)^2$ **10.** $(x + 4)^2 = -6(y - 1)$

In Exercises 11–14, show that the graph of the equation is a parabola. Find its vertex, focus, and directrix.

11. $x^2 + 2x - y + 3 = 0$

12. $3x^2 - 6x - 6y + 10 = 0$

13. $y^2 - 4y - 8x + 20 = 0$

14. $y^2 - 2y + 4x - 12 = 0$

In Exercises 15–22, find an equation in standard form for the ellipse.

15. The endpoints of one axis are $(-3, 2)$ and $(5, 2)$, and of the other are $(1, -4)$ and $(1, 8)$.

16. The endpoints of one axis are $(-2, -3)$ and $(-2, 7)$, and of the other are $(-4, 2)$ and $(0, 2)$.

17. The foci are $(1, -4)$ and $(5, -4)$; the major axis endpoints are $(0, -4)$ and $(6, -4)$.

18. The foci are $(-2, 1)$ and $(-2, 5)$; the major axis endpoints are $(-2, -1)$ and $(-2, 7)$.

19. The major axis endpoints are $(3, -7)$ and $(3, 3)$; the minor axis length is 6.

20. The major axis endpoints are $(-5, 2)$ and $(3, 2)$; the minor axis length is 6.

21.

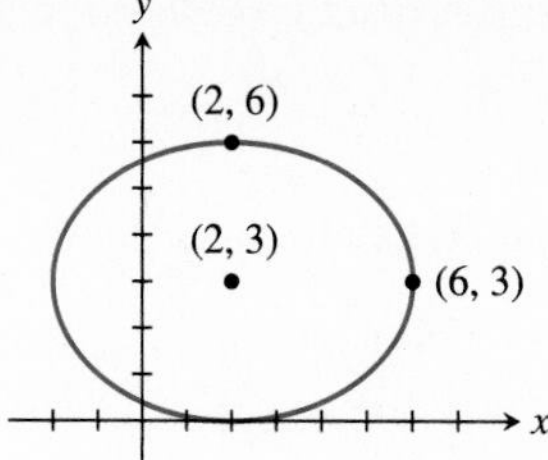

22.

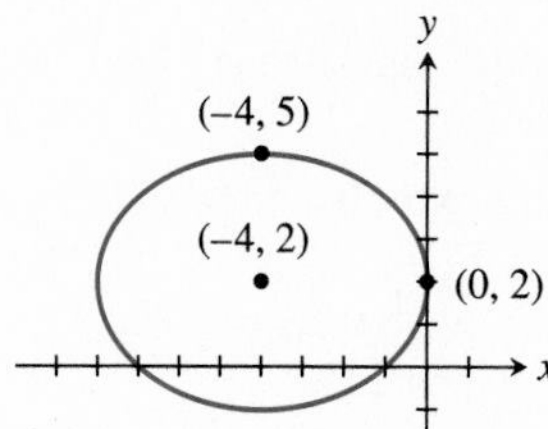

In Exercises 23–26, find the endpoints of the major and minor axes, the center, the foci, and the lines of symmetry of the ellipse. Sketch the ellipse. Support your work graphically.

23. $\frac{(x-1)^2}{2} + \frac{(y+3)^2}{4} = 1$

24. $\frac{(x+3)^2}{16} + \frac{(y-1)^2}{4} = 1$

25. $\frac{(x+2)^2}{5} + 2(y-1)^2 = 1$

26. $\frac{(x-4)^2}{16} + 16(y+4)^2 = 8$

In Exercises 27–30, show that the graph of the equation is an ellipse. Find its major axis endpoints, minor axis endpoints, and foci.

27. $9x^2 + 4y^2 - 18x + 8y - 23 = 0$

28. $3x^2 + 5y^2 - 12x + 30y + 42 = 0$

29. $9x^2 + 16y^2 + 54x - 32y - 47 = 0$

30. $4x^2 + y^2 - 32x + 16y + 124 = 0$

In Exercises 31–34, find an equation in standard form for the hyperbola.

31. The transverse axis endpoints are $(-1, 3)$ and $(5, 3)$, and the slope of one asymptote is $4/3$.

32. The transverse axis endpoints are $(-2, -2)$ and $(-2, 7)$, and the slope of one asymptote is $4/3$.

33. The foci are $(-4, 2)$ and $(2, 2)$, and the transverse axis endpoints are $(-3, 2)$ and $(1, 2)$.

34. The foci are $(-3, -11)$ and $(-3, 0)$, and the transverse axis endpoints are $(-3, -9)$ and $(-3, -2)$.

In Exercises 35–38, find the center, foci, endpoints of the transverse axis, lines of symmetry, and asymptotes of the hyperbola. Sketch the hyperbola. Support your work graphically.

35. $\frac{x^2}{4} - \frac{(y-3)^2}{5} = 1$ **36.** $\frac{(y-3)^2}{9} - \frac{(x+2)^2}{4} = 1$

37. $4(y-1)^2 - 9(x-3)^2 = 36$

38. $4(x-2)^2 - 9(y+4)^2 = 1$

In Exercises 39 and 40, show that the graph of the equation is a hyperbola. Find its transverse axis and its asymptotes.

39. $9x^2 - 4y^2 - 36x + 8y - 4 = 0$

40. $25y^2 - 9x^2 - 50y - 54x - 281 = 0$

In Exercises 41–46, identify the type of conic. Then write the equation in standard form and identify its foci and center.

41. $4y^2 - 9x^2 - 18x - 8y - 41 = 0$

42. $2x^2 + 3y^2 + 12x - 24y + 60 = 0$

43. $9x^2 + 4y^2 - 18x + 16y - 11 = 0$

44. $16x^2 - y^2 - 32x - 6y - 57 = 0$

45. $2x^2 - 4x + y^2 - 6y = 9$

46. $2x^2 - y^2 + 4x + 6 = 0$

In Exercises 47 and 48, solve for y and graph the conic.

47. $3x^2 - 6x + 2y^2 + 8y + 5 = 0$

48. $3y^2 - 5x^2 + 2x - 6y - 9 = 0$

In Exercises 49–52, solve for y and graph the conic. State an angle of rotation that will eliminate the xy term.

49. $2x^2 - xy + 3y^2 - 3x + 4y - 6 = 0$

50. $-x^2 + 3xy + 4y^2 - 5x - 10y - 20 = 0$

51. $2x^2 - 4xy + 8y^2 - 10x + 4y - 13 = 0$

52. $2x^2 - 4xy + 2y^2 - 5x + 6y - 15 = 0$

In Exercises 53 and 54, solve the problem.

53. *Long-Range Navigation* Three LORAN radio transmitters are positioned as shown in the figure, with R due north of O. The cruise ship *Princess Ann* receives simultaneous signals from the three transmitters. The signal from O arrives 323.27 μsec after the signal from R, and 646.53 μsec after

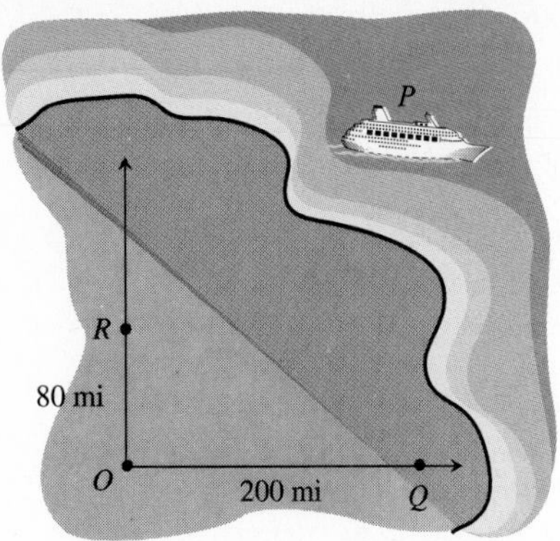

For Exercise 53

the signal from Q. Determine the ship's bearing and distance from point O.

54. *Gun Location* Observers are located at positions A, B, and C with A due north of B. A gun is located somewhere in the first quadrant as illustrated in the figure. A hears the sound of the gun 2 sec before B, and C hears the sound 4 sec before B. Determine the bearing and distance of the gun from point B. (Assume that sound travels at 1100 ft/sec.)

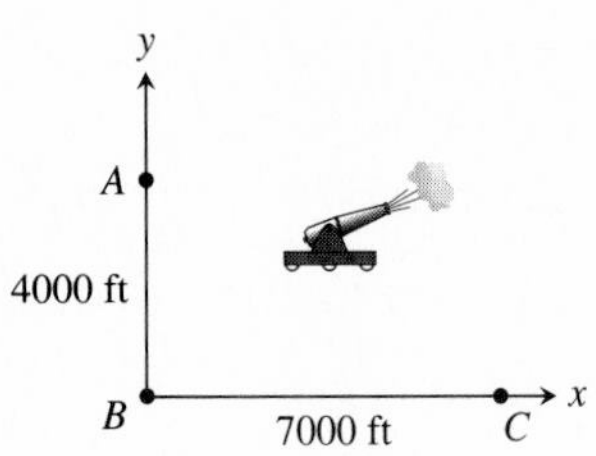

For Exercise 54

In Exercises 55 and 56, use the data in Table 9.1.

55. Find the perihelion and aphelion of Mercury from the sun.

56. Find the perihelion and aphelion of Halley's comet from the sun.

EXTENDING THE IDEAS

In Exercises 57–60, work in groups of three.

57. Group Learning Activity Prove (without using the value of $B^2 - 4AC$) that a nondegenerate graph of the equation

$$Ax^2 + Cy^2 + Dx + Ey + F = 0$$

is a parabola if $A = 0$ and $C \neq 0$.

58. Group Learning Activity Prove (without using the value of $B^2 - 4AC$) that a nondegenerate graph of the equation

$$Ax^2 + Cy^2 + Dx + Ey + F = 0$$

is an ellipse if $AC > 0$.

59. Group Learning Activity Show that the right branch of the hyperbola

$$\frac{x^2}{a^2} - \frac{y^2}{b^2} = 1$$

satisfies the relation $d_1 - d_2 = 2a$, and the left branch satisfies the relation $d_1 - d_2 = -2a$. Recall that d_1 and d_2 are the distances from a point on the hyperbola to the foci F_1 and F_2, respectively.

60. Group Learning Activity Show that the upper branch of the hyperbola

$$\frac{y^2}{b^2} - \frac{x^2}{a^2} = 1$$

satisfies the relation $d_1 - d_2 = 2b$, and the lower branch satisfies the relation $d_1 - d_2 = -2b$.

61. *Rotation of Axes* The xy coordinate system is rotated through an angle α to obtain the $x'y'$ coordinate system as illustrated in the figure.

a. Confirm that

$$x' = r\cos\theta\cos\alpha + r\sin\theta\sin\alpha$$

$$y' = r\sin\theta\cos\alpha - r\cos\theta\sin\alpha$$

(*Hint:* First show that $x' = r\cos(\theta - \alpha)$ and $y' = r\sin(\theta - \alpha)$.)

b. Use the equations in part a to confirm

$$x' = x\cos\alpha + y\sin\alpha$$

$$y' = -x\sin\alpha + y\cos\alpha$$

c. Solve the equations in part b for x and y to confirm that

$$x = x'\cos\alpha - y'\sin\alpha$$

$$y = x'\sin\alpha + y'\cos\alpha$$

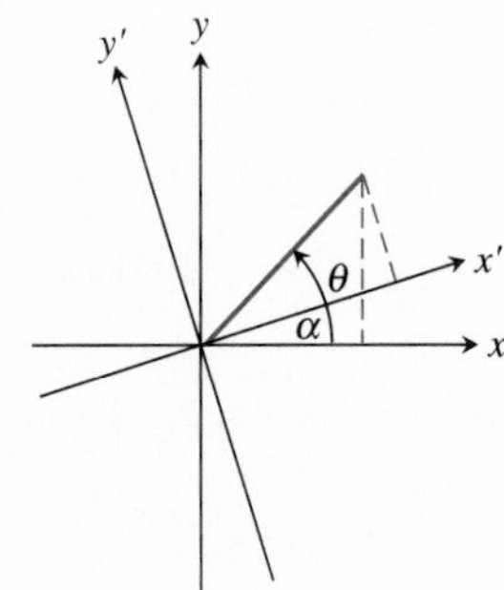

For Exercise 61

9.4 POLAR COORDINATES

Polar Coordinate System • Coordinate Conversion • Graphs of Polar Equations • Equation Conversion • Applications

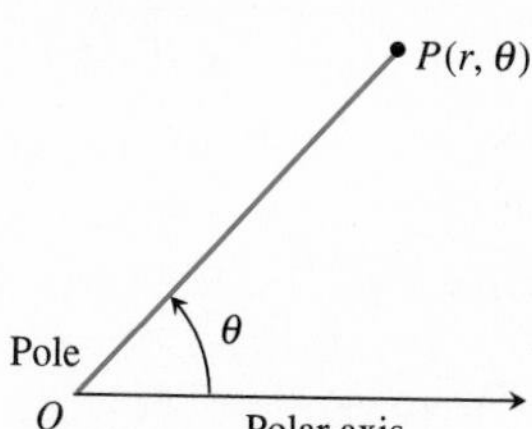

Figure 9.38

Polar Coordinate System

A **polar coordinate system** is a plane with a point O, the **pole,** and a ray from O, the **polar axis,** as shown in Figure 9.38. The polar axis is usually drawn horizontally and to the right like the positive x-axis.

As part of the system, each point P in the plane is assigned **polar coordinates** (r, θ). Here r is the **directed distance** from O to P. And θ is the **directed angle** whose initial side is on the polar axis and whose terminal side is on line OP. As in trigonometry, we measure θ as positive when moving counterclockwise and negative when moving clockwise.

If $r > 0$, then P is on the terminal side of θ. If $r < 0$, then P is on the terminal side of $\theta + \pi$. Normally we use radian measure as the second polar coordinate, but either radians or degrees can be used, as illustrated in Example 1.

REMARK

Notice that when we graph a polar coordinate pair we use the entries in the opposite order: θ first and then r.

EXAMPLE 1 Graphing Points

Graph the point with the given polar coordinates.

a. $P(2, \pi/3)$ **b.** $Q(-1, 3\pi/4)$ **c.** $R(3, -45°)$

Solution

Figure 9.39 shows the three graphs. ■

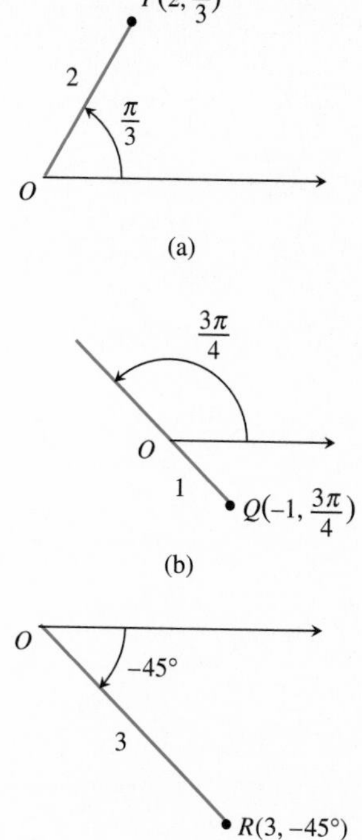

Figure 9.39

Each polar coordinate pair determines a unique point. However, the polar coordinates of a point P are not unique. For example, the coordinates (r, θ), $(r, \theta + 2\pi)$ and $(-r, \theta + \pi)$ all name the same point. In general, the point with polar coordinates (r, θ) also has polar coordinates

$$(r, \theta + 2n\pi) \quad \text{or} \quad (-r, \theta + (2n + 1)\pi),$$

where n is any integer. In particular, the pole has polar coordinates $(0, \theta)$, where θ is any angle.

EXAMPLE 2 Finding Polar Coordinates

One pair of polar coordinates for point P is $(3, \pi/3)$. Find two additional pairs of polar coordinates for P.

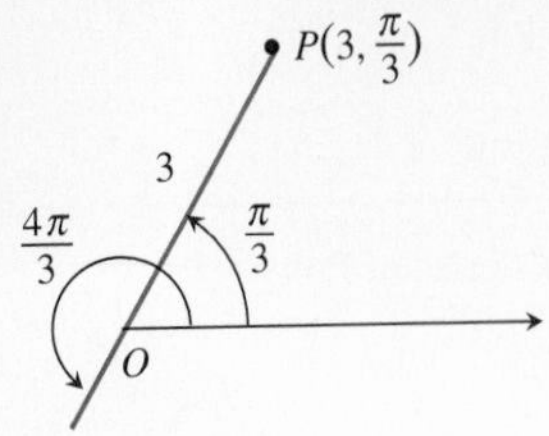

Figure 9.40

Solution

Point P is shown in Figure 9.40. Two additional pairs of polar coordinates for P are

$$\left(3, \frac{\pi}{3} + 2\pi\right) = \left(3, \frac{7\pi}{3}\right) \quad \text{and} \quad \left(-3, \frac{\pi}{3} + \pi\right) = \left(-3, \frac{4\pi}{3}\right).$$

Coordinate Conversion

When we use both polar and Cartesian coordinates, the pole is the origin and the polar axis is the positive x-axis as shown in Figure 9.41. By applying trigonometry we can find equations that relate the polar coordinates (r, θ) and the rectangular coordinates (x, y) of a point P.

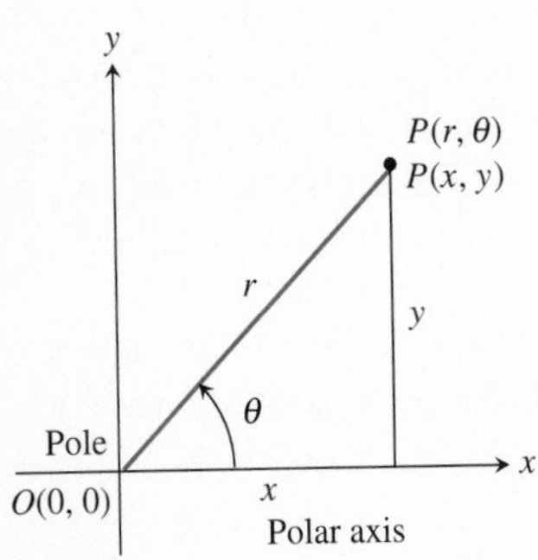

Figure 9.41

Coordinate Conversion Equations

Let (r, θ) be polar coordinates and (x, y) be rectangular coordinates for the point P. Then

$$x = r\cos\theta, \qquad r^2 = x^2 + y^2,$$
$$y = r\sin\theta, \qquad \tan\theta = \frac{y}{x}.$$

These relationships allow us to convert from one coordinate system to the other.

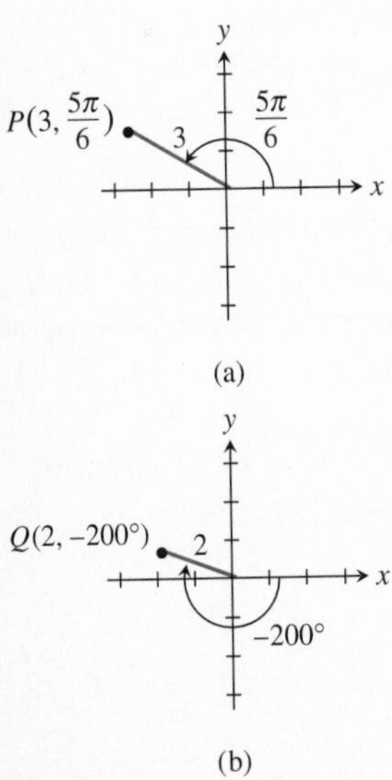

Figure 9.42

EXAMPLE 3 Converting from Polar to Rectangular Coordinates

Find the rectangular coordinates of the points with the given polar coordinates.

a. $P(3, 5\pi/6)$ **b.** $Q(2, -200°)$

Solution

a. For $P(3, 5\pi/6)$, $r = 3$ and $\theta = 5\pi/6$:

$$x = r\cos\theta \qquad y = r\sin\theta$$
$$x = 3\cos\frac{5\pi}{6} \qquad y = 3\sin\frac{5\pi}{6}$$
$$x = -2.598\cdots \qquad y = 1.5$$

The rectangular coordinates for P are approximately $(-2.60, 1.5)$. See Figure 9.42a.

b. For $Q(2, -200°)$, $r = 2$ and $\theta = -200°$:

$$x = r\cos\theta \qquad y = r\sin\theta$$
$$x = 2\cos(-200°) \qquad y = 2\sin(-200°)$$
$$x = -1.879\cdots \qquad y = 0.684\cdots$$

The rectangular coordinates for Q are approximately $(-1.88, 0.68)$. See Figure 9.42b. ■

Objective

Students will be able to convert points and equations from polar to rectangular coordinates and vice versa.

Motivate

Ask students to suggest other methods (besides Cartesian coordinates) of describing the location of a point on a plane.

In converting rectangular coordinates to polar coordinates, we must remember that more than one polar coordinate pair is possible. In Example 4 we are led to two such pairs.

■ EXAMPLE 4 Converting from Rectangular to Polar Coordinates

Find polar coordinates of the point $P(-1, 1)$.

Solution

$$r^2 = x^2 + y^2 \quad \text{and} \quad \tan\theta = \frac{y}{x}$$

$$r^2 = (-1)^2 + (1)^2 \qquad \tan\theta = \frac{-1}{1} = 1$$

$$r = \pm\sqrt{2} \qquad \theta = \tan^{-1}(-1) + n\pi = \frac{-\pi}{4} + n\pi$$

Because P is on the ray opposite the terminal side of $-\pi/4$, the value of r corresponding to $\theta = -\pi/4$ is negative. (See Figure 9.43.) Because P is on the terminal side of $-\pi/4 + \pi = 3\pi/4$, the value of r corresponding to $\theta = 3\pi/4$ is positive. So two polar coordinates of point P are

$$\left(-\sqrt{2}, -\frac{\pi}{4}\right) \quad \text{and} \quad \left(\sqrt{2}, \frac{3\pi}{4}\right).$$ ■

Alert

Because of their extensive use of the Cartesian coordinate system, many students will be surprised that the polar coordinates of a point are not unique. Emphasize the fact that neither r nor θ is uniquely defined.

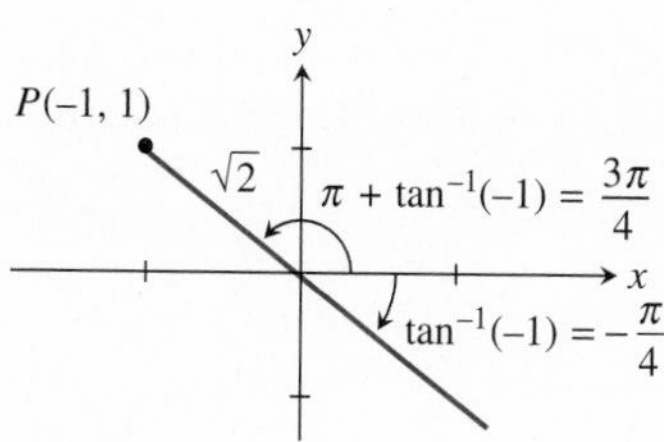

Figure 9.43

Graphs of Polar Equations

An ordered pair (a, α) is a **solution** of an equation in r and θ, a **polar equation,** if the substitution $r = a$ and $\theta = \alpha$ satisfies the equation. The **graph** of a polar equation consists of all points determined by the pairs (a, α) that are solutions of the equation.

Sketching the graph of a polar equation using paper and pencil is similar to sketching the graph of an equation in x and y (see Section 1.1).

1. Make a table of several solutions.
2. Plot the solutions as points in the polar coordinate plane.
3. Connect the points with a smooth curve.

Notes on Examples

Example 4 provides an opportunity to monitor students' use of the inverse keys on their graphers. You may need to assist some students in the correct choice of the quadrant for this example.

TECHNOLOGY NOTE

If your grapher does not have polar mode, Exercise 49 of Section 9.5 shows how to use parametric mode to graph polar equations.

Drawing the graph of a polar equation with a grapher is also similar to drawing the graph of an equation in x and y with a grapher.

1. Rewrite the equation in the form $r =$ (an expression in θ).
2. Enter this equation into the grapher in polar mode.
3. Select an appropriate viewing window.
4. Press the "graph" key.

Example 5 illustrates how to graph the polar equation $r = 4 \cos \theta$.

EXAMPLE 5 Graphing a Polar Equation

Draw the graph of $r = 4 \cos \theta$.

Solution

The table below gives solutions (r, θ) of $r = 4 \cos \theta$ for the special angles in the interval $0 \le \theta \le 2\pi$. Figure 9.44 shows a graph of $r = 4 \cos \theta$ for $0 \le \theta \le 2\pi$ superimposed on a scatter plot of the points corresponding to the table.

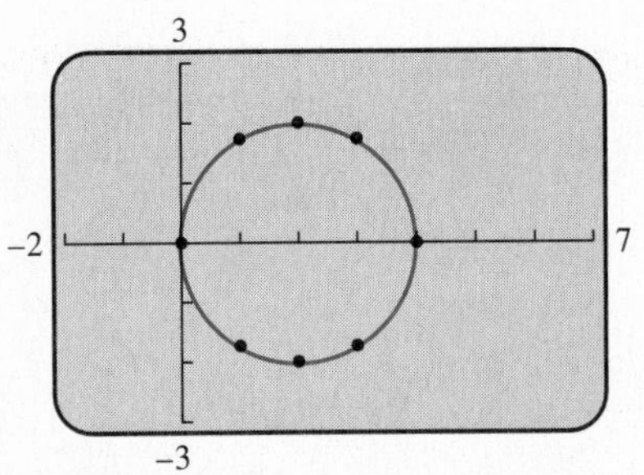

Figure 9.44 The graph of $r = 4 \cos \theta$ appears to be a circle.

θ	0	$\pi/6$	$\pi/4$	$\pi/3$	$\pi/2$	$2\pi/3$	$3\pi/4$	$5\pi/6$	π
r	4	$2\sqrt{3}$	$2\sqrt{2}$	2	0	-2	$-2\sqrt{2}$	$-2\sqrt{3}$	-4

Figure 9.44 correctly suggests that the graph of $r = 4 \cos \theta$ is a smooth curve through these nine points.

Just as with function mode, practice will help you find appropriate viewing windows. Finding an appropriate range for θ in polar mode is similar to finding a range for t in parametric mode. ■

Follow-up

Ask whether it is possible for two polar equations that are not algebraically equivalent to have identical graphs. (Yes)

Assignment Guide

Day 1: Ex. 3–21, multiples of 3, 25–29, 31, 34

Day 2: Ex. 36, 37, 39, 42, 46, 48, 50–52

Equation Conversion

The polar equation $r = 4 \cos \theta$ can be converted to rectangular form by using the coordinate conversion equations, as follows:

$$r = 4 \cos \theta$$

$$r^2 = 4r \cos \theta \quad \text{Multiply by } r.$$

$$x^2 + y^2 = 4x \quad r^2 = x^2 + y^2,\ r \cos \theta = x$$

$$x^2 - 4x + 4 + y^2 = 4$$

$$(x - 2)^2 + y^2 = 4$$

Thus the graph of $r = 4 \cos \theta$ is the circle with the center $(2, 0)$ and radius 2.

You may be surprised by the polar form for a vertical line in Example 6.

EXAMPLE 6 Converting from Polar to Rectangular Coordinates

Graph $r = 4 \sec \theta$, identify the graph, and then confirm algebraically by converting the equation to rectangular form.

Solution

The graph in Figure 9.45 appears to be the vertical line $x = 4$ in rectangular form.

Figure 9.45

Confirm Algebraically

$$r = 4 \sec \theta$$

$$\frac{r}{\sec \theta} = 4$$

$$r \cos \theta = 4$$

$$x = 4 \quad \text{Use } r \cos \theta = x.$$

The graph is indeed the line $x = 4$. ■

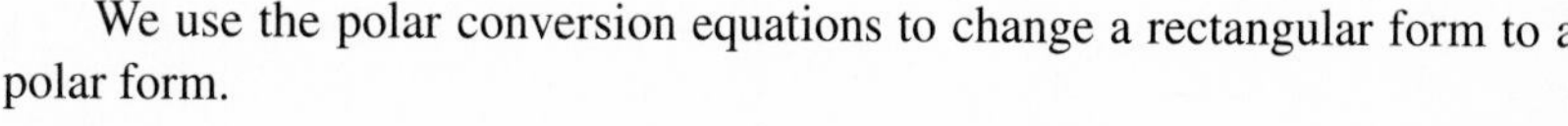

Cooperative Learning

Group Activity: Ex. 49

Notes on Exercises

Ex. 17–24 emphasize the fact that the polar coordinates of a point are not unique.
Ex. 47–48 require students to apply concepts from geometry.

Ongoing Assessment

Self-Assessment: Ex. 7, 11, 17, 33, 41, 45
Embedded Assessment: Ex. 48

We use the polar conversion equations to change a rectangular form to a polar form.

EXAMPLE 7 Converting from Rectangular to Polar Coordinates

Convert the rectangular form $(x - 3)^2 + (y - 2)^2 = 13$ to polar form.

Solution

$$(x - 3)^2 + (y - 2)^2 = 13$$

$$x^2 - 6x + 9 + y^2 - 4y + 4 = 13$$

$$x^2 + y^2 - 6x - 4y = 0$$

$$r^2 - 6r \cos \theta - 4r \sin \theta = 0 \quad r^2 = x^2 + y^2,\ x = r\cos\theta,\ y = r\sin\theta$$

$$r(r - 6 \cos \theta - 4 \sin \theta) = 0$$

$$r = 0 \quad \text{or} \quad r - 6 \cos \theta - 4 \sin \theta = 0$$

Support Graphically

The only solution of $r = 0$ is $(0, \theta)$. It is also a solution of $r - 6 \cos \theta - 4 \sin \theta = 0$, however, so we need only graph the latter equation to see the solution set of the final sentence above. For the grapher, we use the form $r = 6 \cos \theta + 4 \sin \theta$. Its graph is shown in Figure 9.46 and does indeed appear to be the circle $(x - 3)^2 + (y - 2)^2 = 13$. ■

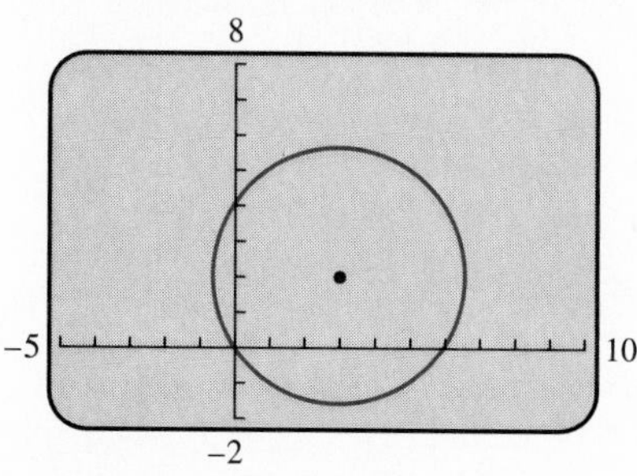

Figure 9.46 Graph of $r = 6 \cos \theta + 4 \sin \theta$.

Applications

A radar tracking system sends out high-frequency radio waves and receives their reflection from an object. The distance and direction of the object from the radar is often given in polar coordinates.

EXAMPLE 8 APPLICATION: Using a Radar Tracking System

Radar detects two airplanes at the same altitude. Their polar coordinates are (8 mi, 110°) and (5 mi, 15°). (See Figure 9.47.) How far apart are the airplanes?

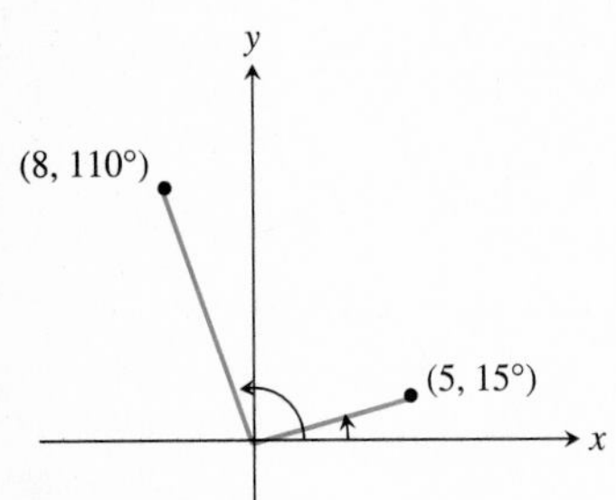

Figure 9.47

Solution

By the law of cosines (Section 8.2),

$$d^2 = 8^2 + 5^2 - 2 \cdot 8 \cdot 5 \cos(110° - 15°)$$

$$d = \sqrt{8^2 + 5^2 - 2 \cdot 8 \cdot 5 \cos(110° - 15°)}$$

$$= 9.796 \cdots .$$

Interpret

The airplanes are about 9.80 mi apart.

Quick Review 9.4

In Exercises 1–4, determine the quadrant containing the terminal side of the angle.

1. $5\pi/6$ **2.** $-3\pi/4$

3. $-300°$ **4.** $210°$

In Exercises 5–8, find a positive and a negative angle coterminal with the angle.

5. $-\pi/4$ **6.** $\pi/3$

7. $160°$ **8.** $-120°$

In Exercises 9 and 10, evaluate the expression. State your answer in degrees and radians.

9. $\tan^{-1} 2$ **10.** $\sin^{-1} 0.5$

In Exercises 11 and 12, solve for θ.

11. $\sin\theta = 0.4, \quad -\pi \le \theta \le \pi$

12. $\tan\theta = 1, \quad 0 \le \theta \le 2\pi$

In Exercises 13 and 14, find the value of all six trigonometric functions of the angle θ.

13.

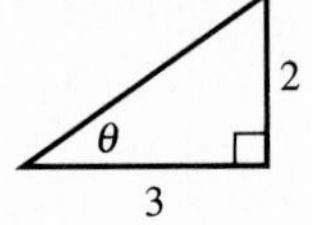

14.

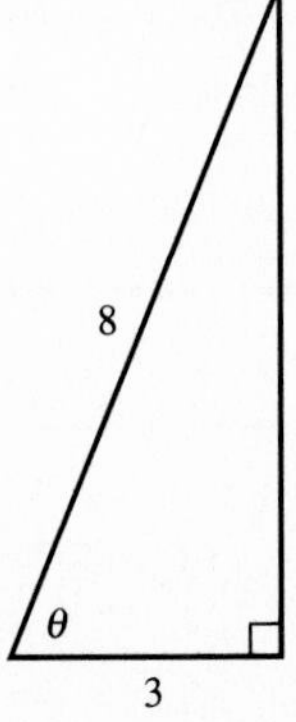

In Exercises 15 and 16, write the equation of the circle in standard form.

15. Center (3, 0) and radius 2

16. Center (0, −4) and radius 3

SECTION EXERCISES 9.4

In Exercises 1–8, graph the point with the given polar coordinates.

1. $(3, 4\pi/3)$ **2.** $(2, 5\pi/6)$

3. $(-1, 2\pi/5)$ **4.** $(-3, 17\pi/10)$

5. $(2, 30°)$ **6.** $(3, 210°)$

7. $(-2, 120°)$ **8.** $(-3, 135°)$

In Exercises 9–16, find the rectangular coordinates of the point.

9. $(1.5, 7\pi/3)$ **10.** $(2.5, 17\pi/4)$

11. $(-3, -29\pi/7)$ **12.** $(-2, -14\pi/5)$

13. $(-2, \pi)$ **14.** $(1, \pi/2)$

15. $(2, 270°)$ **16.** $(-3, 360°)$

In Exercises 17–20, polar coordinates of point P are given. Find all of its polar coordinates.

17. $P = (2, \pi/6)$ **18.** $P = (1, -\pi/4)$

19. $P = (1.5, -20°)$ **20.** $P = (-2.5, 50°)$

In Exercises 21–24, rectangular coordinates of point P are given. Find all polar coordinates of P that satisfy

a. $0 \le \theta \le 2\pi$ **b.** $-\pi \le \theta \le \pi$ **c.** $0 \le \theta \le 4\pi$

21. $P = (1, 1)$ **22.** $P = (1, 3)$

23. $P = (-2, 5)$ **24.** $P = (-1, -2)$

In Exercises 25–28, match the polar equation with its graph.

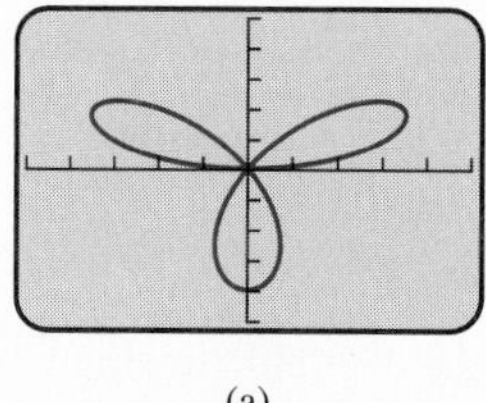
(a)

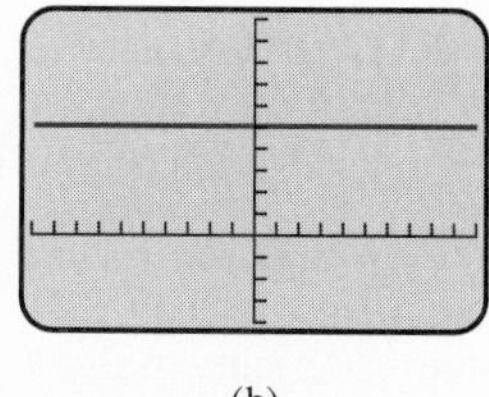
(b)

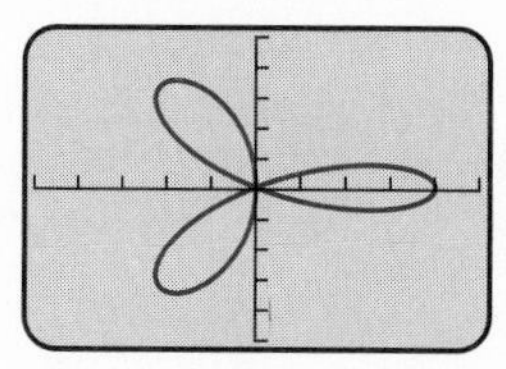
(c)

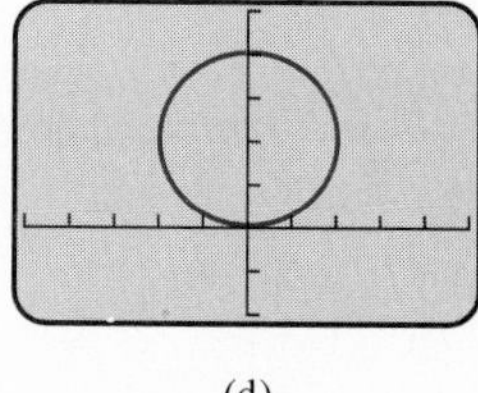
(d)

For Exercises 25–28

25. $r = 5 \csc \theta$ **26.** $r = 4 \sin \theta$

27. $r = 4 \cos 3\theta$ **28.** $r = 4 \sin 3\theta$

In Exercises 29–36, convert the polar equation to rectangular form and identify the graph.

29. $r = 3 \sec \theta$ **30.** $r = -2 \csc \theta$

31. $r = -3 \sin \theta$ **32.** $r = -4 \cos \theta$

33. $r \csc \theta = 1$ **34.** $r \sec \theta = 3$

35. $r = \dfrac{5}{\cos \theta - 2 \sin \theta}$ **36.** $r = \dfrac{2}{3 \cos \theta + \sin \theta}$

In Exercises 37–44, convert the rectangular equation to polar form. Graph the polar equation.

37. $x = 2$ **38.** $x = 5$

39. $2x - 3y = 5$ **40.** $3x + 4y = 2$

41. $(x - 3)^2 + y^2 = 9$ **42.** $x^2 + (y - 1)^2 = 1$

43. $x^2 - y^2 = 4$ **44.** $y^2 - x^2 = 1$

In Exercises 45–48, solve the problem.

45. *Tracking Airplanes* The location, given in polar coordinates, of two planes approaching the Vicksburg airport are (4 mi, 12°) and (2 mi, 72°). Find the distance between the airplanes.

46. *Tracking Ships* The location of two ships from Mays Landing Lighthouse, given in polar coordinates, are (3 mi, 170°) and (5 mi, 150°). Find the distance between the ships.

47. Using Polar Coordinates in Geometry A square with sides of length a and center at the origin has two sides parallel to the x-axis. Find polar coordinates of the vertices.

48. Using Polar Coordinates in Geometry A regular pentagon whose center is at the origin has one vertex on the positive x-axis at a distance a from the center. Find polar coordinates of the vertices.

EXTENDING THE IDEAS

In Exercises 49–52, solve the problem.

49. Let P_1 and P_2 have polar coordinates (r_1, θ_1) and (r_2, θ_2), respectively. Use the law of cosines to show that the distance d between P_1 and P_2 is given by the *polar distance formula*

$$d = \sqrt{r_1^2 + r_2^2 - 2r_1 r_2 \cos(\theta_1 - \theta_2)}.$$

50. Convert $r = 6 \cos \theta + 8 \sin \theta$ to rectangular form. Confirm that the graph is a circle, and find its center and radius.

51. Writing to Learn Explain why the polar equation $\theta = \pi/6$ cannot be drawn on a grapher in polar mode but its rectangular form can be drawn in function mode.

52. Writing to Learn Describe how the graphs of $r = 3$ and $r = -3$ are drawn with a polar grapher for $0 \le \theta \le 2\pi$. Explain any differences.

9.5
GRAPHS OF POLAR EQUATIONS

Introduction • Rose Curves • Limaçon Curves • Maximum r-Values • Determining Symmetry of Polar Graphs

Objective

Students will be able to graph polar equations and determine the maximum r-value and the symmetry of a graph.

Motivate

Ask students to use a grapher to compare the graphs of the polar equations $r = \tan \theta$ and $r = -\tan \theta$ for $0 \le \theta \le 2\pi$. **(The graphs are identical.)**

Exploration Extensions

For the function $r = 5 \sin 3\theta$, determine the sequence in which the petals are generated and the values of θ that generate each petal.

Introduction

In Section 9.4 we used polar mode to graph polar equations like $r = 4 \cos \theta$. If your grapher does not have polar mode, you can graph polar equations using parametric mode, as suggested by Exercise 49.

In this section you will study families of polar curves. These curves are difficult to describe in rectangular form even though they are relatively easy using a polar form.

Rose Curves

In parametric mode, you are able to control what appears in the viewing window by adjusting the interval for the independent variable t. The following Exploration shows that you have the same control in polar mode by adjusting the interval for the independent variable θ.

3.1

−4.7 4.7

−3.1

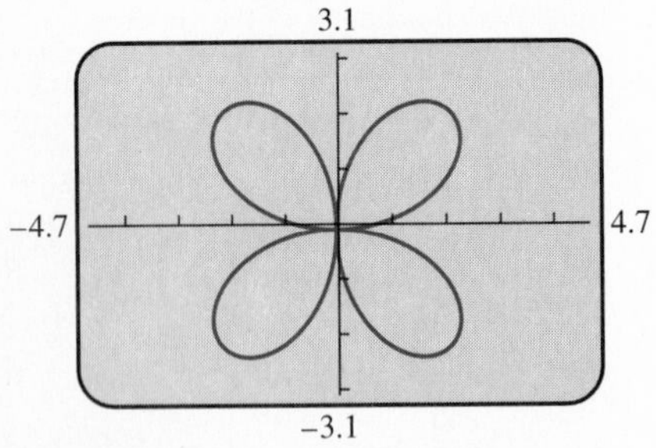

Figure 9.48

EXPLORATION Graphing Polar Equations

With your grapher in polar mode, enter $r = 3 \sin 2\theta$.

1. Set θmin $= 0$, θstep $= 0.1$, and graph r in the decimal viewing window using the given value for θmax.

   a. $\pi/2$ **b.** π **c.** $5\pi/4$ **d.** 2π

2. Describe in writing how parts 1a–d are related to Figure 9.48.
3. State the smallest value for θmax so that $0 \le \theta \le \theta$max gives the complete graph of $r = 5 \sin 3\theta$.

The polar curves $r = a \cos n\theta$ and $r = a \sin n\theta$ are called **rose curves** because, for integer values of n, where $|n| > 1$, their graphs resemble the petals of a flower.

REMARK

In Chapters 6–8 we restricted the range of trigonometric functions by restricting θ. Notice that restricting θ in $r = f(\theta)$ restricts the domain.

■ EXAMPLE 1 Studying an Eight-Petal Rose

Draw the graph of $r = 3 \sin 4\theta$ for θ from 0 to 2π. Identify the sequence in which the petals are generated as θ increases from 0 to 2π. Also determine which petals are generated by positive values and which by negative values of r.

Solution

Watching the viewing window (try it on your grapher), we see the petals drawn in the order shown in Figure 9.49. When θ goes from 0 to 2π, 4θ goes from 0 to 8π and the sine alternates positive and negative in the intervals $(0, \pi)$, $(\pi, 2\pi)$, $(2\pi, 3\pi)$, . . . , $(7\pi, 8\pi)$. Thus petals 1, 3, 5, and 7 draw with r positive and petals 2, 4, 6, and 8 draw with r negative. ■

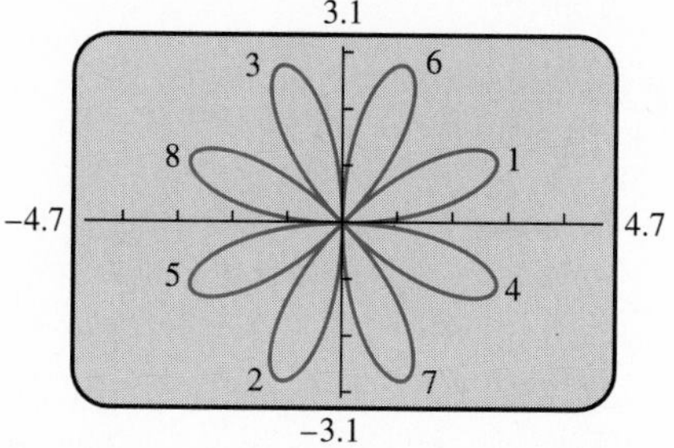

Figure 9.49

■ EXAMPLE 2 Studying a Three-Petal Rose

Draw the graph of $r = 4 \cos 3\theta$ for $0 \le \theta \le 2\pi$.

Solution

The graph of this rose is shown in Figure 9.50. Use "trace" to see that all three petals have been completed when θ reaches the value $\theta = \pi$. Then it is traced a second time as θ increases from π to 2π. ■

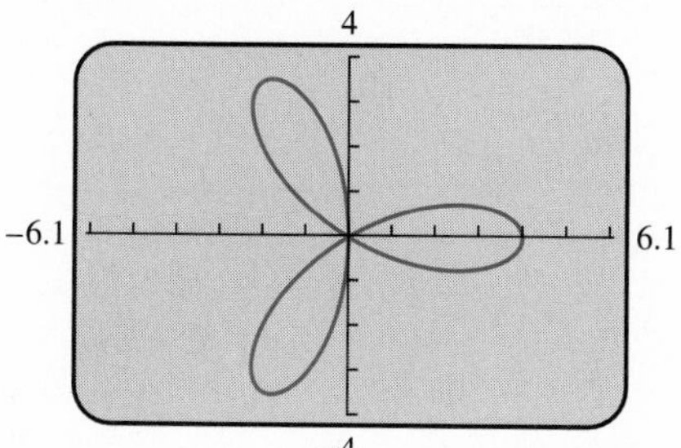

Figure 9.50

The Exploration and Examples 1 and 2 suggest that $2n$ distinct petals are drawn when n is even, and only n petals are drawn, but each is drawn twice, when n is odd.

Limaçon Curves

The **limaçon curves** are graphs of polar equations of the form

$$r = a + b \sin \theta \qquad \text{and} \qquad r = a + b \cos \theta.$$

Limaçon, pronounced "leemasahn," is Old French for "snail." There are three different shapes of limaçons, as illustrated in Figure 9.51.

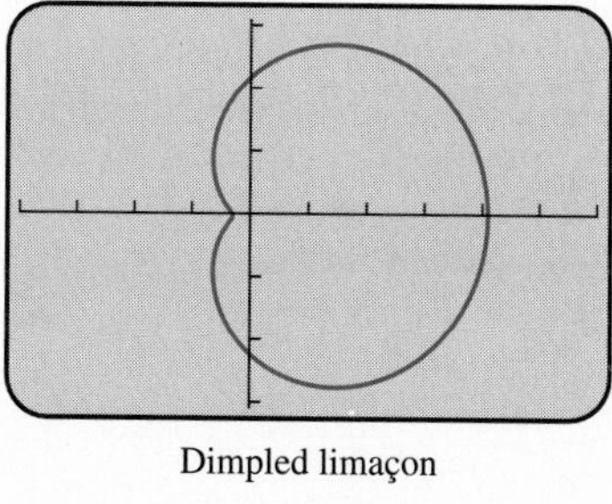

Dimpled limaçon

(a)

Cardiod

(b)

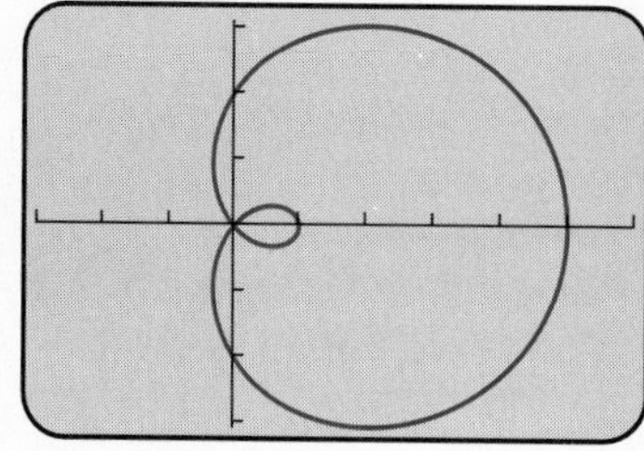

Limaçon with an inner loop

(c)

Figure 9.51

When $|a| = |b|$, the curve is a **cardioid curve,** a subfamily of the limaçons. Example 3 examines two cardioids.

EXAMPLE 3 Studying Cardioid Curves

Graphs of two of these three polar equations

$$r_1 = 2 + 2 \sin \theta, \qquad r_2 = 2 - 2 \sin \theta, \qquad r_3 = 2 + 2 \cos \theta,$$

are shown in Figure 9.52. Use an algebraic method to select the equations whose graphs are shown. Support your work with a grapher.

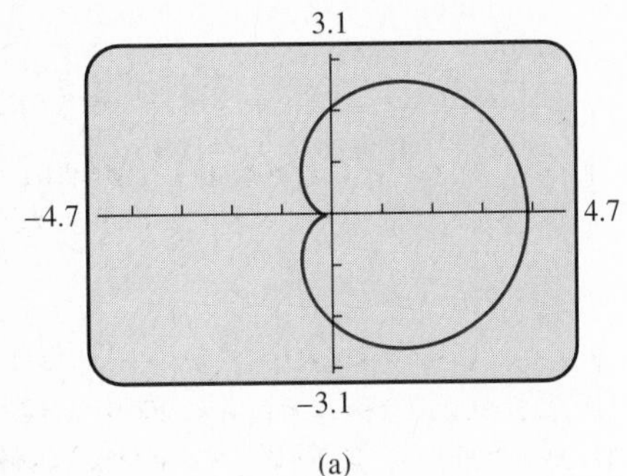

(a)

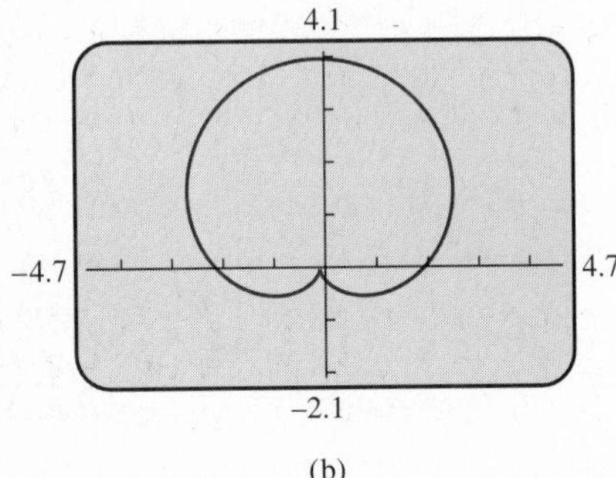

(b)

Figure 9.52

Solution

The graphs have different values at $\theta = 0$ and $\pi/2$. Checking $\theta = 0$, we find

$$r_1(0) = 2 + 2 \sin 0 = 2,$$

$$r_2(0) = 2 - 2 \sin 0 = 2,$$

$$r_3(0) = 2 + 2 \cos 0 = 4.$$

Checking $\theta = \pi/2$, we find

$$r_1\left(\frac{\pi}{2}\right) = 2 + 2 \sin \frac{\pi}{2} = 4,$$

$$r_2\left(\frac{\pi}{2}\right) = 2 - 2 \sin \frac{\pi}{2} = 0,$$

$$r_3\left(\frac{\pi}{2}\right) = 2 + 2 \cos \frac{\pi}{2} = 2.$$

Because $r_2(3\pi/2) = 4$, the graph of r_2 does not appear. Thus, Figure 9.52a is the graph of r_3 only, and Figure 9.52b is the graph of r_1 only. Use your grapher to check these results. ■

Teaching Note

Polar graphs are an invitation for students to explore mathematics. Students will be able to produce elaborate graphs using polar graphing techniques. The graph of the cardioid in Example 3 should be discussed thoroughly. Encourage students to graph different modifications to $r = a + b \cos \theta$ to see the effects of their choices.

If $|a| < |b|$, the graph of the polar equation $r = a + b \sin \theta$ or $r = a + b \cos \theta$ is a limaçon with an inner loop. (If $|a| > |b|$, we get a dimpled limaçon.) We shall use a limaçon with an inner loop to illustrate that two different polar equations may generate the same polar curve, reflecting the fact that one point has many pairs of polar coordinates. Two polar equations that have exactly the same graph are said to be **equivalent.**

EXAMPLE 4 Studying a Limaçon

Show that the graphs of $r_1 = 2 + 3 \cos \theta$ and $r_2 = -2 + 3 \cos \theta$ are the same limaçon with inner loop. As θ increases from 0 to 2π, describe the path around the curve in each case.

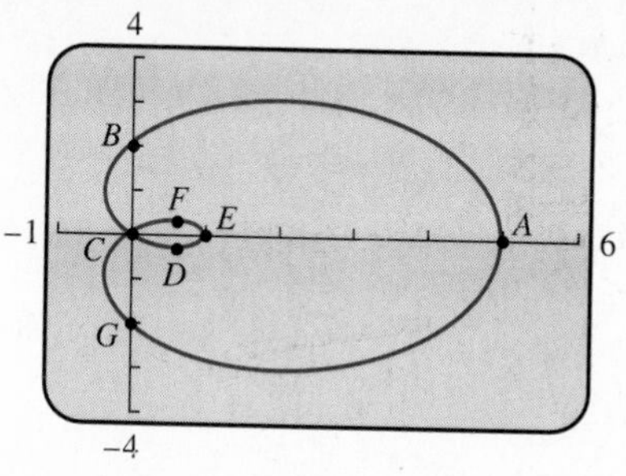

Figure 9.53 A graph of both $r_1 = 2 + 3\cos\theta$ and $r_2 = -2 + 3\cos\theta$.

Solution

The graph is shown in Figure 9.53. Use "trace" to show the following:

r_1: As θ increases from 0 to 2π, the drawing begins at A and moves through points B, C, D, E, F, C, G and back to A, in that order.

r_2: As θ increases from 0 to 2π, the drawing begins at E and moves through F, C, G, A, B, C, D and back to E. ■

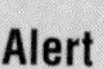 **Alert**

Students are used to specifying different viewing windows to control what they see on the graphing screen. On polar graphs, students should be careful to note their selection of the range values for x, y and θ. Since the environment into which the polar coordinates are patched is the rectangular system, students need to be concerned about controlling both the viewing window and the polar environment. Encourage students to experiment with different input values for both the polar function r and the input variable θ and to observe the effects on the graphs.

Maximum *r*-Values

Recall that r gives the directed distance from the origin to the point having polar coordinates (r, θ). It is easier to find extreme values of r by looking at a graph that displays the directed distances vertically.

For example, Figure 9.54a is the graph of the polar equation $r = 2 + 2\cos\theta$ for $0 \le \theta \le 2\pi$. The directed distances r are given by the corresponding y-values of the graph of $y = 2 + 2\cos x$ in the rectangular coordinate system shown in Figure 9.54b.

Notice that the values of y in Figure 9.54b go from 4 to 0 to 4 as x goes from 0 to 2π. You can use "trace" in Figure 9.54a to see that r goes from 4 to 0 to 4 as θ goes from 0 to 2π.

A **maximum *r*-value** for a polar equation occurs at a point on the curve that is the maximum distance from the pole. To find maximum r-values we must find maximum values of the distance $|r|$ as opposed to the directed distance r.

Example 5 illustrates that there might be several points on a polar curve corresponding to a maximum r-value. Some of them occur at values of θ for which r is negative.

Alert

Some students will find the maximum value of r instead of the maximum value of $|r|$. Emphasize that we are looking for the maximum *distance* from the pole, and r may be either positive or negative at this point.

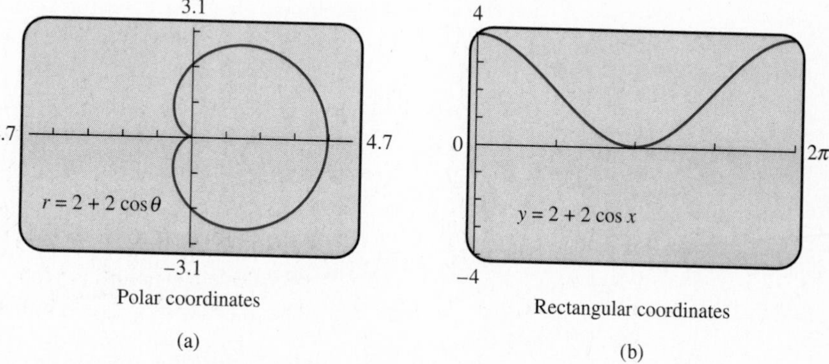

Figure 9.54 When $\theta = x$, the y-values in (b) are the same as the directed distance from the pole to (r, θ) in (a).

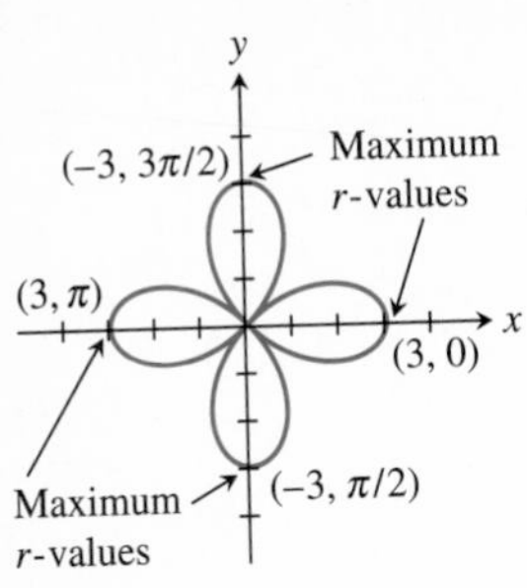

Figure 9.55 $r = 3 \cos 2\theta$.

> **Teaching Note**
>
> Some graphers have no polar mode as such and require the use of the parametric mode to obtain polar graphs. The use of such utilities strongly reinforces the ideas of Section 9.4.

■ EXAMPLE 5 Finding Points for Maximum r-Values

Identify the points on the graph of $r = 3 \cos 2\theta$ for $0 \leq \theta < 2\pi$ that give maximum r-values.

Solution

Using "trace" in Figure 9.55 we can show that there are four points on the graph of $r = 3 \cos 2\theta$ at maximum distance of 3 from the pole:

$$(3, 0) \qquad (-3, \pi/2) \qquad (3, \pi) \qquad (-3, 3\pi/2)$$

Figure 9.56a shows the directed distances r as the y-values of $y_1 = 3 \cos 2x$, and Figure 9.56b shows the distances $|r|$ as the y-values of $y_2 = |3 \cos 2x|$. Notice there are four maximum values of $|r|$ (y_2) in (b). These four maximum values correspond to the four local extremum values of r (y_1) in (a). ■

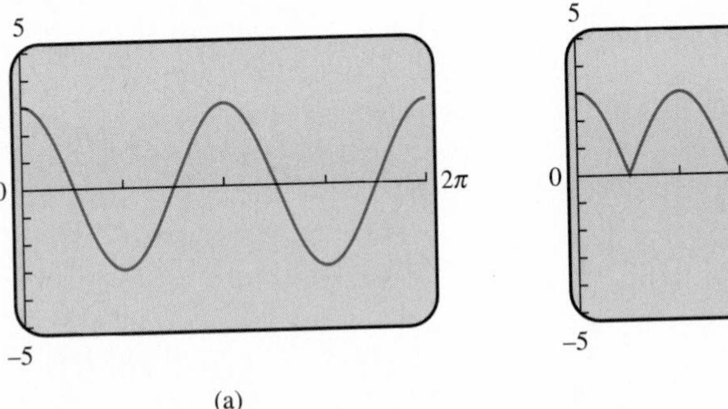

Figure 9.56 (a) $y_1 = 3 \cos 2x$. (b) $y_2 = |3 \cos 2x|$.

> **Teaching Note**
>
> When determining symmetry, sometimes only one of the replacements will appear to produce an equivalent polar equation. For example, the graph of $r = |\theta|$ is symmetric across the polar axis, but at first glance this equation does not appear to be equivalent to $-r = |\pi - \theta|$. Although the pairs (r, θ) that solve each equation are different, the graphs of these two equations are the same.
>
> **Follow-up**
>
> Have students confirm the symmetry of Example 6 by using the replacement $(r, \pi - \theta)$ instead of using $(-r, -\theta)$.

Determining Symmetry of Polar Graphs

You learned algebraic tests for symmetry for equations in rectangular form. Algebraic tests also exist for polar forms.

Figure 9.57 shows a rectangular coordinate system superimposed on a polar coordinate system, with the origin and the pole coinciding and the positive x-axis and the polar axis coinciding.

The three types of symmetry figures to be considered will have:

1. The polar axis (the x-axis) as a line of symmetry (Figure 9.57a)
2. The line perpendicular to the polar axis at the pole (the y-axis) as a line of symmetry (Figure 9.57b)
3. The pole as a point of symmetry (Figure 9.57c)

All three algebraic tests for symmetry in polar forms require replacing the pair (r, θ), which satisfies the polar equation, with another coordinate pair and determining whether it also satisfies the polar equation.

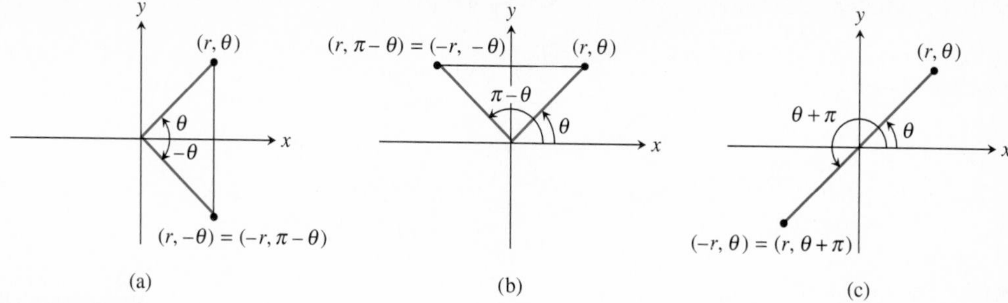

Figure 9.57

Assignment Guide

Day 1: Ex. 1, 4, 5, 8, 9, 10, 12, 14, 15, 18, 19, 22, 23, 26
Day 2: Ex. 27, 31, 34, 35, 38, 41, 44, 47, 48

Cooperative Learning

Group Activity: Ex. 44–49

Notes on Exercises

Ex. 23–26 require students to find the distance from the pole to the furthest point of a petal, not the arc length.
Ex. 44 and 46–49 are a series of Writing to Learn exercises designed to enhance understanding of polar equations. They may be assigned for students to complete individually or in groups.

Ongoing Assessment

Self-Assessment: Ex. 7, 13, 25, 29, 39
Embedded Assessment: Ex. 12, 18

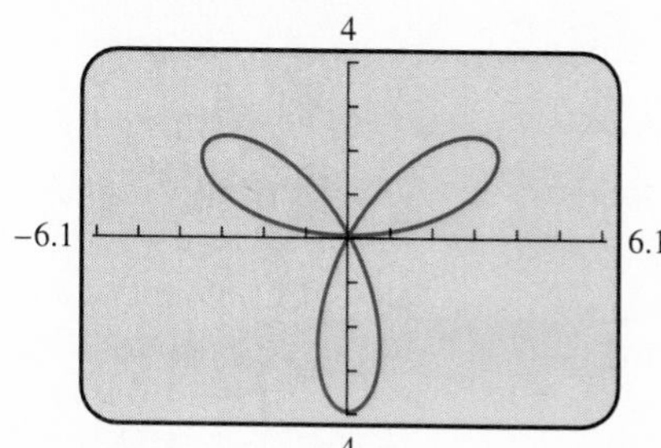

Figure 9.58

Symmetry Tests for Polar Graphs

The graph of a polar equation has the indicated symmetry if either replacement produces an equivalent polar equation.

To Test for Symmetry	Replace	By
1. across the polar axis,	(r, θ)	$(r, -\theta)$ or $(-r, \pi - \theta)$.
2. across the y-axis,	(r, θ)	$(-r, -\theta)$ or $(r, \pi - \theta)$.
3. about the pole,	(r, θ)	$(-r, \theta)$ or $(r, \theta + \pi)$.

We test a rose curve for symmetry in Example 6.

EXAMPLE 6 Testing for Symmetry

The graph of $r = 4 \sin 3\theta$ in Figure 9.58 appears to be symmetric across the y-axis. Confirm this algebraically.

Solution

$$\begin{aligned} r &= 4 \sin 3\theta \\ -r &= 4 \sin 3(-\theta) && \text{Replace } (r, \theta) \text{ by } (-r, -\theta). \\ &= 4 \sin(-3\theta) \\ &= -4 \sin 3\theta \\ r &= 4 \sin 3\theta && \text{(Same as original.)} \end{aligned}$$

Because the equations $-r = 4 \sin 3(-\theta)$ and $r = 4 \sin 3\theta$ are equivalent, there is symmetry about the y-axis. ■

Quick Review 9.5

In Exercises 1–4, complete the square to determine the center and radius of each circle.

1. $x^2 + 4x + y^2 = 0$
2. $x^2 + y^2 - 3y = 0$
3. $x^2 + 6x + y^2 - 4y = 0$
4. $x^2 - 8x + y^2 + 2y = 0$

In Exercises 5–10, simplify the expression.

5. $\sin(\pi - \theta)$
6. $\cos(\pi - \theta)$
7. $\sin(-2\theta)$
8. $\cos(-3\theta)$
9. $\cos 2(\pi + \theta)$
10. $\sin 2(\pi + \theta)$

SECTION EXERCISES 9.5

In Exercises 1–4, predict whether the graph of the equation is a rose curve. Find the graph to support your response.

1. $r = 2 \sin 3\theta$
2. $r = -3 \cos 4\theta$
3. $r = 2 \sin^2 2\theta + \sin 2\theta$
4. $r = 3 \cos 2\theta - \sin 3\theta$

In Exercises 5–8, draw a graph of the rose curve. State the smallest θ-interval $(0 \leq \theta \leq k)$ that will produce a complete graph. Identify the sequence in which the petals are generated as θ increases from 0 to k. State which petals are generated by positive values of r and which by negative values of r.

5. $r = 3 \sin 3\theta$
6. $r = -3 \cos 2\theta$
7. $r = 3 \cos 2\theta$
8. $r = 3 \sin 5\theta$

In Exercises 9 and 10, refer to the curves in the given figure.

9. The graphs of which equations are shown?

   $r_1 = 3 \cos 6\theta \quad r_2 = 3 \sin 8\theta \quad r_3 = 3|\cos 3\theta|$

10. Use trigonometric identities to explain which of these curves is the graph of $r = 6 \cos 2\theta \sin 2\theta$.

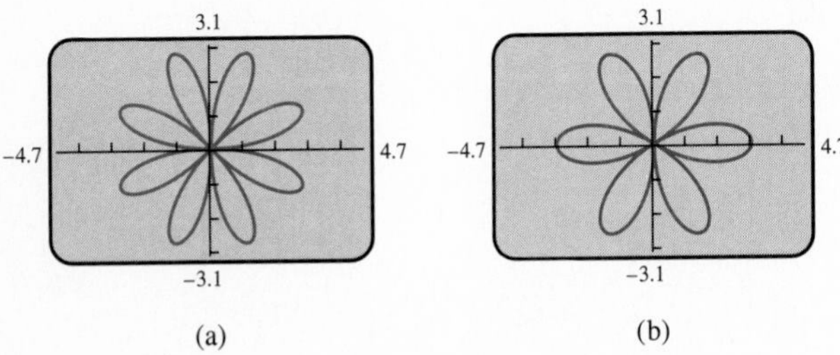

For Exercises 9–10

In Exercises 11–14, match the equation with its graph using algebra.

11. Does the graph of $r = 2 + 2 \sin \theta$ or $r = 2 - 2 \cos \theta$ appear in the figure? Explain.

12. Does the graph of $r = 2 + 3 \cos \theta$ or $r = 2 - 3 \cos \theta$ appear in the figure? Explain.

13. Is the graph in (a) the graph of $r = 2 - 2 \sin \theta$ or $r = 2 + 2 \cos \theta$? Explain.

14. Is the graph in (d) the graph of $r = 2 + 1.5 \cos \theta$ or $r = 2 - 1.5 \sin \theta$? Explain.

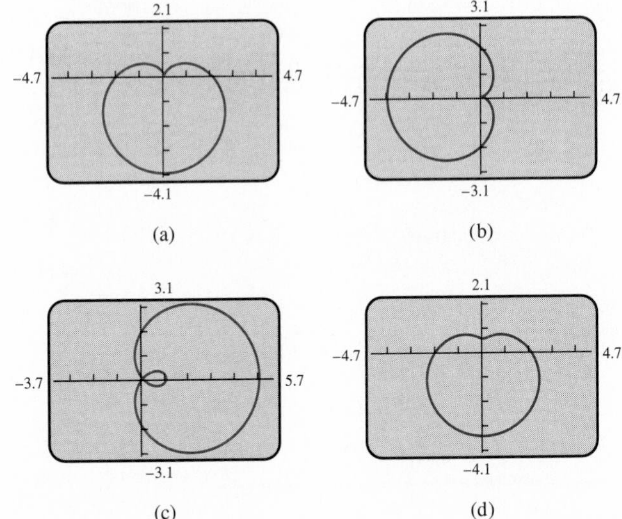

For Exercises 11–14

In Exercises 15–18, select the two equations whose graphs are the same curve. Then, even though the graphs of the equations are identical, describe how the two paths are different as θ increases from 0 to 2π.

15. $r_1 = 1 + 3 \sin \theta, \quad r_2 = -1 + 3 \sin \theta, \quad r_3 = 1 - 3 \sin \theta$
16. $r_1 = 1 + 2 \cos \theta, \quad r_2 = -1 - 2 \cos \theta, \quad r_3 = -1 + 2 \cos \theta$
17. $r_1 = 1 + 2 \cos \theta, \quad r_2 = 1 - 2 \cos \theta, \quad r_3 = -1 - 2 \cos \theta$
18. $r_1 = 2 + 2 \sin \theta, \quad r_2 = -2 + 2 \sin \theta, \quad r_3 = 2 - 2 \sin \theta$

In Exercises 19–22, identify the points where maximum r-values occur on the graph of the polar equation.

19. $r = 3 \cos 3\theta$

20. $r = 4 \sin 2\theta$

21. $r = 2 + 3 \cos \theta$

22. $r = -3 + 2 \sin \theta$

In Exercises 23–26, find the length of each petal of the polar curve.

23. $r = 2 + 4 \sin 2\theta$

24. $r = 3 - 5 \cos 2\theta$

25. $r = 1 - 4 \cos 5\theta$

26. $r = 3 + 4 \sin 5\theta$

In Exercises 27–34, use the polar symmetry tests to determine if the graph is symmetric across the polar axis, across the y-axis, or about the pole.

27. $r = 3 + 3 \sin \theta$

28. $r = 1 + 2 \cos \theta$

29. $r = 4 - 3 \cos \theta$

30. $r = 1 - 3 \sin \theta$

31. $r = 5 \cos 2\theta$

32. $r = 7 \sin 3\theta$

33. $r = \dfrac{3}{1 + \sin \theta}$

34. $r = \dfrac{2}{1 - \cos \theta}$

In Exercises 35–40, draw the graph of the polar curve. Describe and confirm any symmetry.

35. $r = 5 + 4 \sin \theta$

36. $r = 6 - 5 \cos \theta$

37. $r = 2 + 3 \cos \theta$

38. $r = 3 - 4 \sin \theta$

39. $r = 1 - 3 \cos 3\theta$

40. $r = 1 + 3 \sin 3\theta$

EXTENDING THE IDEAS

In Exercises 41–43, graph each polar equation. Describe how they are related to each other.

41. **a.** $r_1 = 3 \sin 3\theta$ **b.** $r_2 = 3 \sin 3\left(\theta + \dfrac{\pi}{12}\right)$
c. $r_3 = 3 \sin 3\left(\theta + \dfrac{\pi}{4}\right)$

42. **a.** $r_1 = 2 \sec \theta$ **b.** $r_2 = 2 \sec\left(\theta - \dfrac{\pi}{4}\right)$
c. $r_3 = 2 \sec\left(\theta - \dfrac{\pi}{3}\right)$

43. **a.** $r_1 = 2 - 2 \cos \theta$ **b.** $r_2 = r_1\left(\theta + \dfrac{\pi}{4}\right)$
c. $r_3 = r_1\left(\theta + \dfrac{\pi}{3}\right)$

In Exercises 44–49, complete a written explanation.

44. **Writing to Learn** Describe how the graphs of $r = f(\theta)$, $r = f(\theta + \alpha)$, and $r = f(\theta - \alpha)$ are related. Explain why you think this generalization is true.

45. *Extended Rose Curves* The graphs of $r_1 = 3 \sin\left(\frac{5}{2}\theta\right)$ and $r_2 = 3 \sin\left(\frac{7}{2}\theta\right)$ may be called rose curves.

a. Determine the smallest θ-interval that will produce a complete graph of r_1; of r_2.
b. How many petals does each graph have?

46. **Writing to Learn** Explain why the number of petals on the rose curve $r = a \sin n\theta$ is $2n$ if n is an even integer.

47. **Writing to Learn** Explain why the graph of $r = a \sec \theta$ is a vertical line for every nonzero real number a.

48. **Writing to Learn** Explain the connection between the amplitude of $y = a \sin x$ and the polar curve $r = a \sin \theta$.

49. **Writing to Learn** Explain why the graph of the parametric equations $x = f(t) \cos t$, $y = f(t) \sin t$ is identical to the graph of the polar equation $r = f(\theta)$ in the polar coordinate plane.

In Exercises 50–53, describe the polar graph. Explain your choice of θ-interval.

50. $r = e^{\theta/10}$

51. $r = \ln \theta$

52. $r = \dfrac{8}{\theta}$

53. $r^2 = \dfrac{100}{\theta}$

9.6 POLAR EQUATIONS OF CONICS

Objective

Students will be able to find the eccentricity of conic sections and explore conic sections using polar equations.

Introduction • Defining a Conic as a Ratio • Polar Equations of Conics • Identifying a Conic • Applications

Introduction

In Sections 9.2 and 9.3 we developed rectangular equations for conics. Polar equations allow us to give a *unified description* of conics. We will see that polar

REMARK

Do not confuse the e used here with the number $e = 2.718 \cdots$. This e is the *eccentricity* ratio first defined in Section 9.3. Its value can vary.

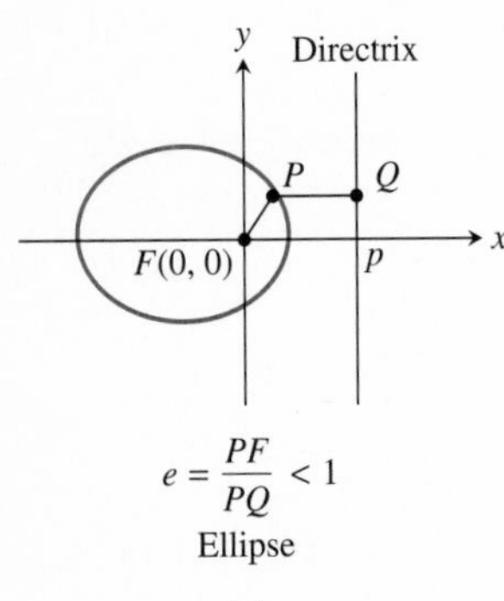

$e = \frac{PF}{PQ} < 1$
Ellipse

(a)

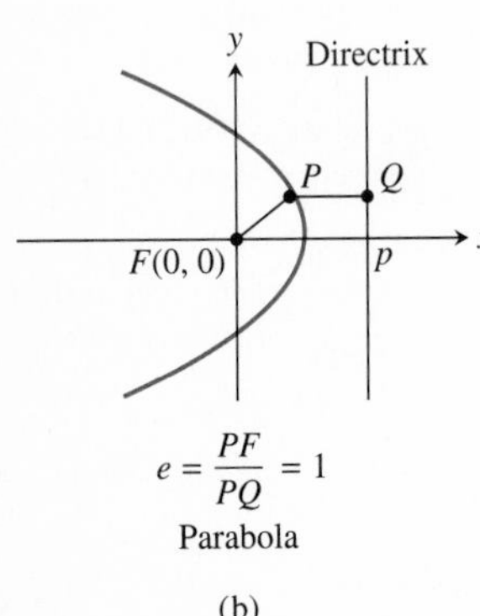

$e = \frac{PF}{PQ} = 1$
Parabola

(b)

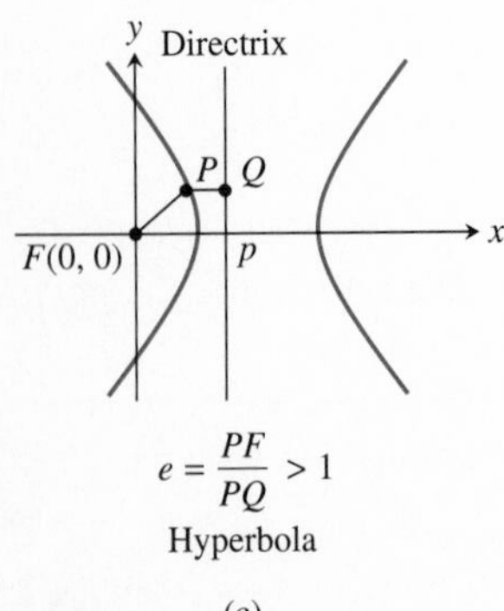

$e = \frac{PF}{PQ} > 1$
Hyperbola

(c)

Figure 9.59

equations for conics have a single geometric definition when one focus is at the pole.

The following Exploration will introduce you to polar equations of conics.

EXPLORATION Polar Equations of Conics

Let $p = -3$. For a particular value of e the graph of

$$r = \frac{ep}{1 - e \cos \theta}$$

is a conic. Graph the equation in the viewing window $[-12, 12]$ by $[-12, 12]$, and identify which type of conic seems to be drawn for each of the values

$$e = 0.7,\ 0.8,\ 1,\ 1.5,\ 2,\ 3.$$

Conjecture

How do you think the value of e is related to the type of conic that is produced?

Defining a Conic as a Ratio

When we defined the conics parabola, ellipse, and hyperbola in a rectangular coordinate setting, we used the concept of directrix only for the parabola. However, there are alternative geometric definitions of ellipse and hyperbola that also use the concept of directrix. The ratio

$$\frac{\text{Distance between a point and a fixed point (focus)}}{\text{Distance between a point and a fixed line (directrix)}}$$

is the basis for the more unified description of the conics. In each part of Figure 9.59, the ratio $e = PF/PQ$ is this defining ratio.

Definition Polar Definition of Conics

A **conic** with focus F and directrix L is the set of all points P such that the ratio of distances from P to F and from P to L is a constant e. (See Figure 9.59.) This ratio,

$$e = \frac{PF}{PQ},$$

is the **eccentricity** of the conic.

If $e < 1$, the conic is an ellipse.
If $e = 1$, the conic is a parabola.
If $e > 1$, the conic is a hyperbola.

Motivate

Have students use a grapher to graph the polar equation $r = \dfrac{5}{1 - \cos\theta}$.

Ask what they observe.

(The graph is a parabola.)

Exploration Extensions

Confirm your guess by trying a few more values of e, such as 0.2, 0.5, 5, and 10. You may even wish to try some negative values.

Teaching Note

This is a difficult section for most students. The connections between the concepts of conic sections and polar coordinates will require some time for students to understand.

In Exercises 39 and 40, you will connect the new definitions of ellipse and hyperbola with their standard form rectangular equations given in Section 9.2. This will enable you, for example, to find equations for the directrix in terms of the familiar parameters a, b, and c.

Polar Equations of Conics

If a polar coordinate system is chosen so that the pole coincides with a focus of a conic and the polar axis is perpendicular to the directrix, then a particularly nice form for an equation of the conic can be found.

EXAMPLE 1 Finding a Polar Equation of a Conic

Show that every point on the conic with eccentricity e, focus $F(0, 0)$, and directrix $x = p$ (where $p > 0$) is a solution of the equation

$$r = \frac{ep}{1 + e\cos\theta}.$$

Solution

Let $P(r, \theta)$ be a point on the conic.

Suppose P and F Are on the Same Side of the Directrix: We see in Figure 9.60a that $PQ = p - r\cos\theta$ and $PF = r$. Therefore,

$$\frac{PF}{PQ} = e,$$

$$\frac{r}{p - r\cos\theta} = e,$$

$$r = e(p - r\cos\theta)$$

$$= ep - er\cos\theta,$$

$$r + er\cos\theta = ep,$$

$$r = \frac{ep}{1 + e\cos\theta}. \qquad (1)$$

Suppose P and F Are on Opposite Sides of the Directrix: We see in Figure 9.60b that $PQ = r\cos\theta - p$ and $PF = r$. We can choose the polar coordinates of P to be $(-r, \theta + \pi)$. In Exercise 41, you will show that $e = PF/PQ$ leads to an equation indicating that P is a solution to the polar equation (1) for r using the coordinates $(-r, \theta + \pi)$. ■

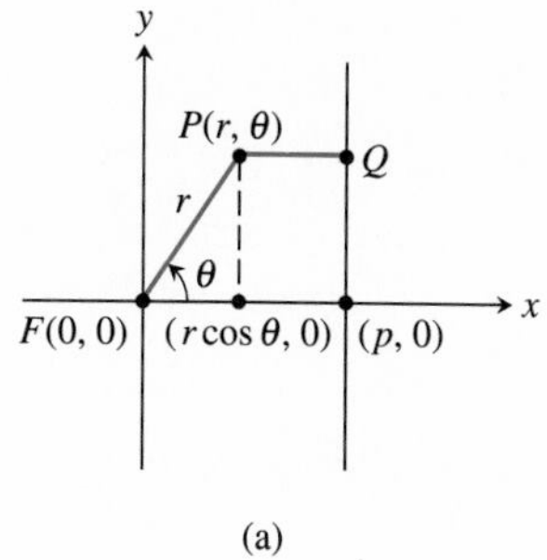

(a)

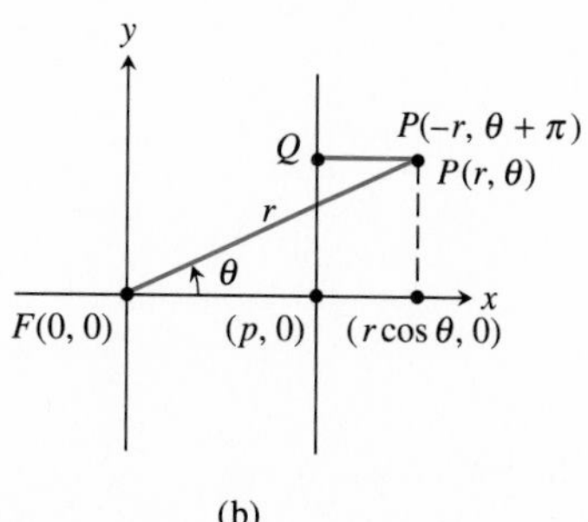

(b)

Figure 9.60

In Example 2, we show that every solution of $r = ep/(1 + e\cos\theta)$ is a point on the conic of Example 1. Again we assume $p > 0$. The arguments for Examples 1 and 2 are similar for $p < 0$. (See Exercises 31 and 33.)

Notes on Examples

If possible, work with students through all of the examples in this section, allowing time for exploration. Consider having some students present a few of the examples in class.

■ EXAMPLE 2 Using the Definition of a Conic

Show that the graph of

$$r = \frac{ep}{1 + e \cos \theta}$$

is a conic with eccentricity e, the focus $(0, 0)$, and the directrix $x = p$.

Solution

Let $P(r, \theta)$ be a point on the graph of the equation.

$$\begin{aligned}
PQ &= |p - r \cos \theta| && \text{From Figure 9.60} \\
&= \left| p - \left(\frac{ep}{1 + e \cos \theta} \right) \cos \theta \right| && r = ep/(1 + e \cos \theta). \\
&= \left| \frac{p + ep \cos \theta - ep \cos \theta}{1 + e \cos \theta} \right| && \text{Combine fractions.} \\
&= \left| \frac{p}{1 + e \cos \theta} \right| \\
&= \left| \frac{1}{e} \cdot \frac{ep}{1 + e \cos \theta} \right| && \text{Multiply by } e/e = 1. \\
&= \left| \frac{r}{e} \right|
\end{aligned}$$

Because $PF = |r|$, we have

$$\frac{PF}{PQ} = \frac{|r|}{|r/e|} = e.$$

Thus, the graph of the original equation is a conic with eccentricity e, the focus $(0, 0)$, and the directrix $x = p$. ■

■ EXAMPLE 3 Finding a Polar Equation for a Conic

Find a polar equation for the conic with the focus $(0, 0)$, eccentricity $e = 0.6$, and the directrix $x = 2$. Identify the type of conic, and graph the equation to support your conclusion.

Solution

Because $e < 1$, the graph is an ellipse. Because the focus is $(0, 0)$ and the directrix is $x = 2$, we see that $p = 2$. Therefore,

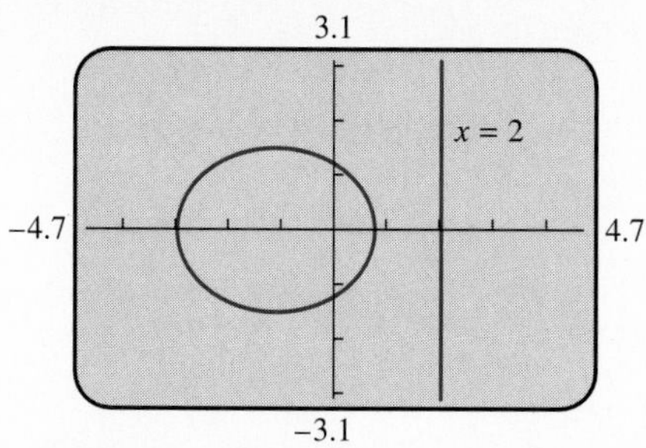

Figure 9.61

Teaching Note

This section provides a good opportunity to observe students' selection of the range variables in the polar setting on the grapher. Students should have a good notion of how the graphs of conics should appear and can now be in a position to refine their ideas about range.

$$r = \frac{ep}{1 + e \cos \theta}$$

$$= \frac{0.6(2)}{1 + 0.6 \cos \theta} \qquad e = 0.6,\ p = 2$$

$$= \frac{1.2}{1 + 0.6 \cos \theta}.$$

Figure 9.61 shows the graphs of the ellipse and its directrix. ■

In Exercises 31–34, you will have a chance to verify the following *standard form polar equations* for conics.

Definition Standard Form Polar Equations for Conics

Suppose that $p > 0$, and that the eccentricity of a conic with a focus at the pole is e.

1. If the directrix of the conic is $x = \pm p$, the **standard form polar equation** for the conic is

$$r = \frac{ep}{1 \pm e \cos \theta}.$$

2. If the directrix of the conic is $y = \pm p$, the **standard form polar equation** for the conic is

$$r = \frac{ep}{1 \pm e \sin \theta}.$$

■ EXAMPLE 4 Finding a Polar Equation for Another Conic

Find a polar equation for the conic with the focus (0, 0), eccentricity $e = 1$, and the directrix $x = -2$. Identify the type of conic, and graph the equation to support your conclusion.

Solution

Because $e = 1$, the graph is a parabola. Because the focus is (0, 0) and the directrix is $x = -2$, we see that $p = 2$. Therefore,

$$r = \frac{ep}{1 - e \cos \theta}$$

$$= \frac{1(2)}{1 - 1 \cos \theta} \qquad e = 1,\ p = 2$$

$$= \frac{2}{1 - \cos \theta}$$

Figure 9.62 shows the graphs of the parabola and its directrix. Notice that when the directrix is to the left of (0, 0), the parabola opens to the right. ■

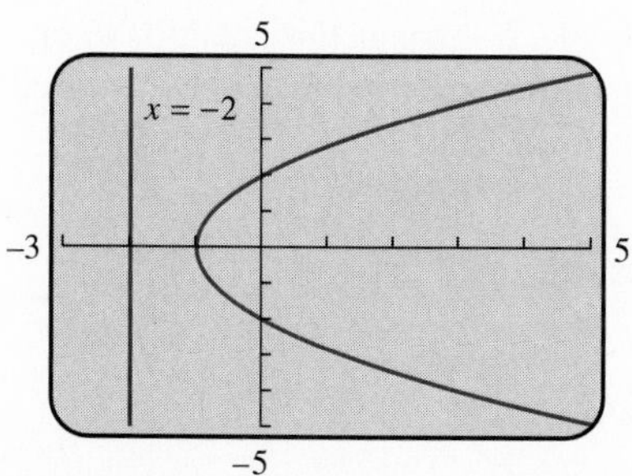

Figure 9.62

EXAMPLE 5 Finding a Polar Equation for Another Conic

Find a polar equation for the conic with the focus (0, 0), eccentricity $e = 2$, and the directrix $y = 4$. Identify the type of conic, and graph the equation to support your conclusion.

Solution

Because $e = 2$, the graph is a hyperbola. Because the focus is (0, 0) and the directrix is $y = 4$, we see that $p = 4$. Therefore,

$$\begin{aligned} r &= \frac{ep}{1 + e \sin \theta} \\ &= \frac{2(4)}{1 + 2 \sin \theta} \qquad e = 2,\ p = 4 \\ &= \frac{8}{1 + 2 \sin \theta} \end{aligned}$$

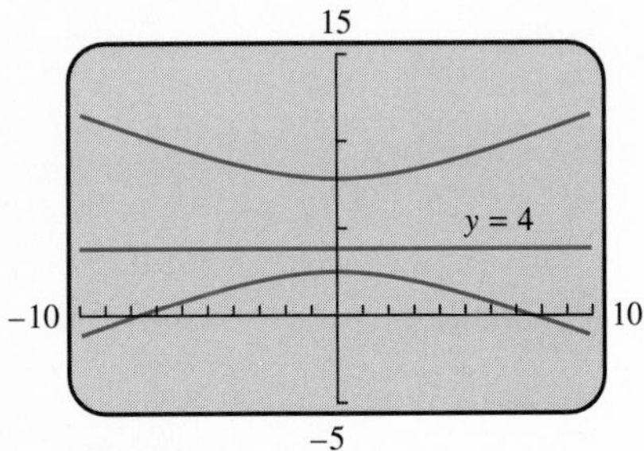

Figure 9.63

Figure 9.63 shows the graphs of the hyperbola and its directrix. Notice that the horizontal directrix results in a hyperbola with a vertical transverse axis. ■

Identifying a Conic

Examples 6 and 7 illustrate how to draw conclusions about a conic directly from its polar equation form.

EXAMPLE 6 Identifying a Conic

Identify the type of conic and a directrix of the equation

$$r = \frac{6}{4 - 3 \sin \theta}.$$

Solution

Figure 9.64 suggests that the conic is an ellipse. To change this equation to one of the standard forms, we divide the numerator and denominator by 4 so that the constant term in the denominator is 1:

$$\begin{aligned} r &= \frac{6}{4 - 3 \sin \theta} \\ &= \frac{1.5}{1 - 0.75 \sin \theta} \end{aligned}$$

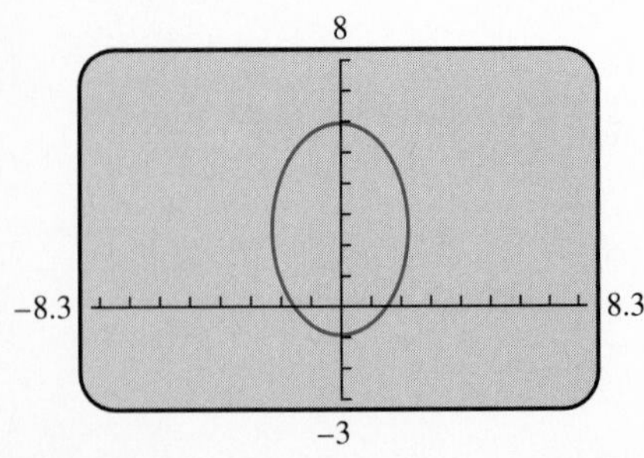

Figure 9.64

The coefficient of $\sin \theta$ in the denominator is the eccentricity $e = 0.75$, which confirms that the graph is an ellipse. From the numerator, we see that $ep = 1.5$, so $p = 1.5/0.75 = 2$. Because the coefficient of $\sin\theta$ is negative, the directrix is $y = -p = -2$. ■

■ **EXAMPLE 7 Identifying a Conic**

Identify the type of conic and a directrix for the equation

$$r = \frac{6}{2 + 3\cos\theta}.$$

Solution

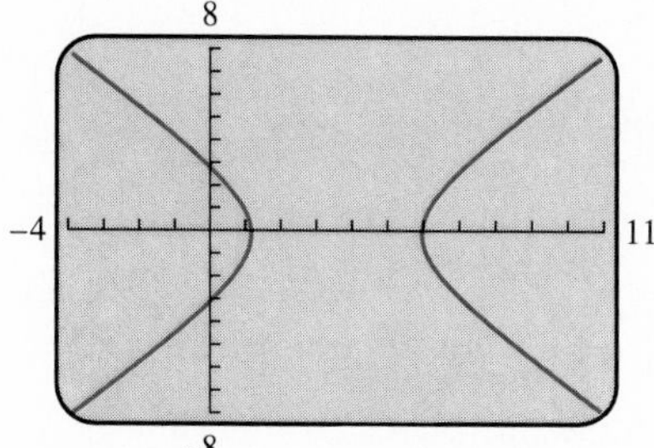

Figure 9.65

Figure 9.65 suggests that the conic is a hyperbola. To change this equation to one of the standard forms, divide numerator and denominator by 2 so that the constant in the denominator is 1:

$$r = \frac{6}{2 + 3\cos\theta}$$

$$= \frac{3}{1 + 1.5\cos\theta}$$

Now the coefficient of $\cos\theta$ in the denominator is the eccentricity $e = 1.5$, which confirms that the graph is a hyperbola. From the numerator we see that $ep = 3$, so $p = 3/1.5 = 2$. Because the coefficient of $\cos\theta$ is positive, the directrix is $x = p = 2$. ■

REMARK

Draw an ellipse or hyperbola with center at the origin, and use the equation $p = a/e - ea$ to locate a directrix. There is a directrix for each focus. Also, if you have drawn the ellipse with a vertical major axis or the hyperbola with a vertical transverse axis, then replace a by b in the equation for p.

Applications

We shall apply polar coordinate equations to an analysis of planetary orbits. The eccentricity of an ellipse or a hyperbola whose center is at the origin and foci are at $(\pm c, 0)$ is c/a. The distance p from a focus to a directrix is given by $p = a/e - ea$. (See Exercises 39 and 40.) Recall that $2a$ is the length of the horizontal major axis of an ellipse or the length of the horizontal transverse axis of a hyperbola.

■ **EXAMPLE 8 Working with Planetary Orbits**

Use polar coordinate equations to find the perihelion (minimum distance) and aphelion (maximum distance) of Jupiter from the sun.

Solution

Model

For Jupiter, $a = 483.6$ and $e = 0.05$ (see Table 9.1 on page 605). Using the

Follow-up

Ask . . .

What do you suppose the eccentricity of a circle is? **(0)**

Assignment Guide

Day 1: Ex. 1–6 all, 9–27, multiples of 3
Day 2: Ex. 29–34, 36, 38, 41

Cooperative Learning

Group Activity: Ex. 39–40

Notes on Exercises

The exercises in this section require that students bring together their knowledge of polar-coordinate graphing and conic sections. This is a challenging set of exercises for most students.
Ex. 37–38 deal with circular orbits.
Ex. 39–40 require students to work with the directrix of an ellipse and hyperbola in rectangular coordinates.

Ongoing Assessment

Self-Assessment: Ex. 11, 17, 23, 25, 35
Embedded Assessment: Ex. 32, 34

equations you derive in Exercises 39 and 40, we conclude that

$$p = \frac{a}{e} - ea$$

$$\approx \frac{483.6}{0.05} - 0.05(483.6) \qquad a = 483.6,\ e = 0.05$$

$$\approx 9647.82.$$

Using these values for e and p we see that a polar equation that models Jupiter's orbit is

$$r = \frac{ep}{1 + e\cos\theta}$$

$$= \frac{0.05(9647.82)}{1 + 0.05\cos\theta} \qquad e = 0.05,\ p = 9647.82$$

$$= \frac{482.391}{1 + 0.05\cos\theta}$$

Solve Algebraically

The distances we seek occur when θ is 0 or π.

$$r(0) = \frac{482.391}{1 + 0.05\cos 0} \approx 459.42$$

$$r(\pi) = \frac{482.391}{1 + 0.05\cos\pi} \approx 507.78$$

Interpret

The minimum distance from Jupiter to the sun is about 459.421 million miles, and the maximum distance is about 507.78 million miles. ■

Quick Review 9.6

In Exercises 1 and 2, find the missing entry so that the two polar coordinates represent the same point.

1. $(3, \theta) = (?, \theta + \pi)$

2. $(-2, \theta) = (?, \theta + \pi)$

In Exercises 3 and 4, solve for θ.

3. $(1.5, \pi/6) = (-1.5, \theta),\ -2\pi \le \theta \le 2\pi$

4. $(-3, 4\pi/3) = (3, \theta),\ -2\pi \le \theta \le 2\pi$

In Exercises 5 and 6, find the foci and the endpoints of the major axis of the ellipse.

5. $\dfrac{x^2}{9} + \dfrac{y^2}{4} = 1$

6. $\dfrac{x^2}{9} + \dfrac{y^2}{25} = 1$

In Exercises 7 and 8, find the foci and the endpoints of the transverse axis of the hyperbola.

7. $\dfrac{x^2}{16} - \dfrac{y^2}{9} = 1$

8. $\dfrac{y^2}{36} - \dfrac{x^2}{4} = 1$

SECTION EXERCISES 9.6

In Exercises 1–6, match the polar equation with its graph. Identify the viewing window.

1. $r = \dfrac{8}{3 - 4\cos\theta}$ **2.** $r = \dfrac{4}{3 + 2\cos\theta}$

3. $r = \dfrac{5}{2 - 2\sin\theta}$ **4.** $r = \dfrac{9}{5 - 3\sin\theta}$

5. $r = \dfrac{15}{2 + 5\sin\theta}$ **6.** $r = \dfrac{15}{4 + 4\cos\theta}$

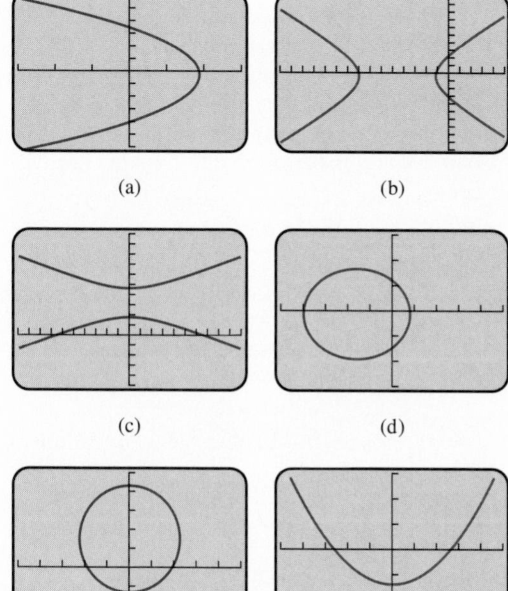

For Exercises 1–6

In Exercises 7–12, find a polar equation for the conic with focus at (0, 0), and the given eccentricity and directrix. Identify the conic and draw its graph.

7. $e = 1, \quad x = -2$ **8.** $e = 5/4, \quad x = 4$

9. $e = 3/5, \quad y = 4$ **10.** $e = 1, \quad y = 2$

11. $e = 7/3, \quad y = -1$ **12.** $e = 2/3, \quad x = -5$

In Exercises 13–20, identify the conic without drawing its graph. State the eccentricity and directrix.

13. $r = \dfrac{2}{1 + \cos\theta}$ **14.** $r = \dfrac{6}{1 + 2\cos\theta}$

15. $r = \dfrac{5}{2 - 2\sin\theta}$ **16.** $r = \dfrac{2}{4 - \cos\theta}$

17. $r = \dfrac{20}{6 + 5\sin\theta}$ **18.** $r = \dfrac{42}{2 - 7\sin\theta}$

19. $r = \dfrac{6}{5 + 2\cos\theta}$ **20.** $r = \dfrac{20}{2 + 5\sin\theta}$

In Exercises 21–24, find a polar equation for an ellipse whose focus is at (0, 0) if the endpoints of its major axis have the given polar coordinates.

21. $(1.5, 0)$ and $(6, \pi)$ **22.** $(1.5, 0)$ and $(1, \pi)$

23. $(1, \pi/2)$ and $(3, 3\pi/2)$

24. $(3, \pi/2)$ and $(0.75, -\pi/2)$

In Exercises 25–28, find a polar equation for a hyperbola with a focus at (0, 0) if the endpoints of its transverse axis have the given polar coordinates.

25. $(3, 0)$ and $(-15, \pi)$ **26.** $(-3, 0)$ and $(1.5, \pi)$

27. $\left(2.4, \dfrac{\pi}{2}\right)$ and $\left(-12, \dfrac{3\pi}{2}\right)$

28. $\left(-6, \dfrac{\pi}{2}\right)$ and $\left(2, \dfrac{3\pi}{2}\right)$

In Exercises 29 and 30, find a polar equation for the conic with focus at (0, 0). The indicated coordinates are rectangular.

29.

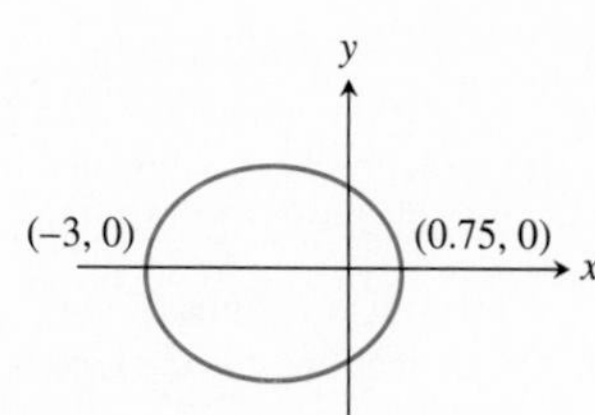

30.

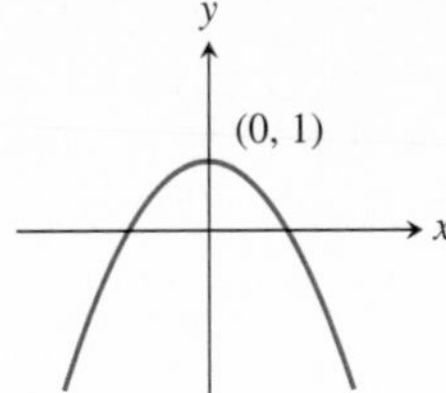

In Exercises 31 and 32, show that every point on the conic with eccentricity e, focus $F(0, 0)$, and the given directrix is a solution to the polar equation.

31. $r = \dfrac{ep}{1 - e \cos \theta}$, $x = -p$ (Assume that $p > 0$.)

32. $r = \dfrac{ep}{1 + e \sin \theta}$, $y = p$ (Assume that $p > 0$.)

In Exercises 33 and 34, use the polar definition of conics. Show that the graph of the polar equation is a conic with eccentricity e, focus $(0, 0)$, and the given directrix.

33. $r = \dfrac{ep}{1 - e \cos \theta}$, $x = -p$ (Assume that $p > 0$.)

34. $r = \dfrac{ep}{1 + e \sin \theta}$, $y = p$ (Assume that $p > 0$.)

In Exercises 35 and 36, use the data in Table 9.1.

35. *Orbit of Comet* Find a polar equation for the orbit of Halley's comet. Find its perihelion and aphelion.

36. *Orbit of Planet* Find a polar equation for the orbit of planet Mercury. Find its perihelion and aphelion.

In Exercises 37 and 38, solve the problem. The velocity of an object traveling in a circular orbit of radius r (distance from center of planet in meters) around a planet is given by

$$v = \sqrt{\frac{3.99 \times 10^{14} k}{r}} \text{ m/sec},$$

where k is a constant related to the masses of the planet and the orbiting object.

37. *Lunar Module* A lunar module is in a circular orbit 250 km above the surface of the moon. Assume that the moon's radius is 1740 km and that $k = 0.012$. Find the following.

a. The velocity of the lunar module

b. The length of time required for the lunar module to circle the moon once

38. *Mars Satellite* A satellite is in a circular orbit 1000 mi above Mars. Assume that the radius of Mars is 2100 mi and that $k = 0.11$. Find the velocity of the satellite.

EXTENDING THE IDEAS

In Exercises 39–41, solve the problem.

39. *Connecting Polar to Rectangular* Consider the ellipse

$$\frac{x^2}{a^2} + \frac{y^2}{b^2} = 1,$$

where half the length of the major axis is a, and the foci are $(\pm c, 0)$ such that $c^2 = a^2 - b^2$. Let L be the vertical line $x = a^2/c$.

a. Show that L is a directrix for the ellipse. (*Hint:* Show that PF/PQ is the constant c/a, where P is a point on the ellipse, and Q is the point on L such that PQ is perpendicular to L.)

b. Show that the eccentricity is $e = c/a$.

c. Show that the distance from F to L is $a/e - ea$.

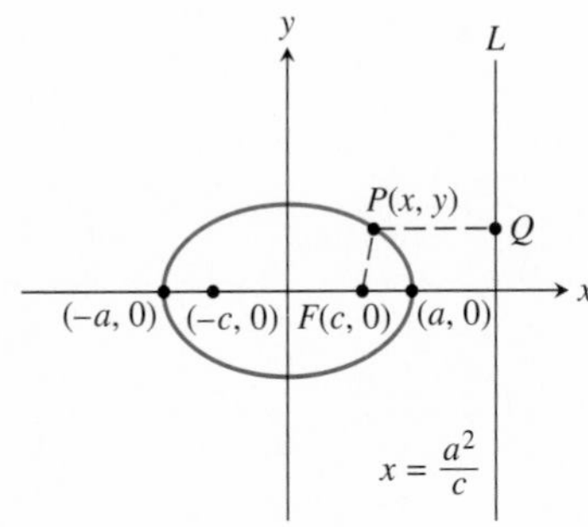

For Exercise 39

40. *Connecting Polar to Rectangular* Consider the hyperbola

$$\frac{x^2}{a^2} - \frac{y^2}{b^2} = 1,$$

where half the length of the transverse axis is a, and the foci are $(\pm c, 0)$ such that $c^2 = a^2 + b^2$. Let L be the vertical line $x = a^2/c$.

a. Show that L is a directrix for the hyperbola. (*Hint:* Show that PF/PQ is the constant c/a, where P is a point on the hyperbola, and Q is the point on L such that PQ is perpendicular to L.)

b. Show that the eccentricity is $e = c/a$.

c. Show that the distance from F to L is $a/e - ea$.

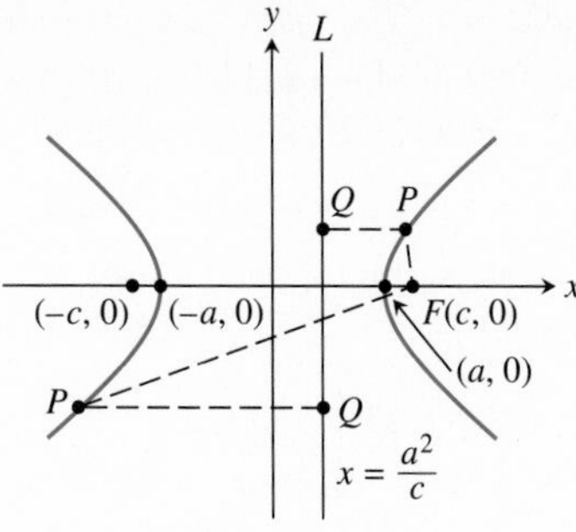

For Exercise 40

41. *Completion of Example 1* Let P and F be defined as in Example 1. If $(-r, \theta + \pi)$ are the polar coordinates of P, show that $PF = r$, $PQ = r\cos\theta - p$, and that P is a solution of

$$r = \frac{ep}{1 + \cos\theta}.$$

CHAPTER 9 REVIEW

KEY TERMS

The number following each key term indicates the page of its introduction.

REVIEW EXERCISES

In Exercises 1–4, describe the graph of the parametric equations. State rectangular coordinates for any endpoints.

1. $x = e^{2t} - 1, \quad y = e^t$
2. $x = 2t^2 + 3, \quad y = t - 1$
3. $x = 3 \cos t, \quad y = 3 \sin t, \quad 0 \le t \le 4\pi$
4. $x = 4 \sin t, \quad y = 4 \cos t, \quad -\pi/2 \le t \le \pi/2$

In Exercises 5–8, graph the point with the given polar coordinates.

5. $(2, -2\pi/3)$
6. $(4, 11\pi/6)$
7. $(-4, -6\pi/5)$
8. $(-5, -17\pi/10)$

In Exercises 9–12, convert the polar coordinates to rectangular coordinates.

9. $(-2.5, 25°)$
10. $(-3.1, 135°)$
11. $(2, -\pi/4)$
12. $(3.6, 3\pi/4)$

In Exercises 13 and 14, polar coordinates of point P are given. Find all of its polar coordinates.

13. $P = (-1, -2\pi/3)$
14. $P = (-2, 5\pi/6)$

In Exercises 15–18, rectangular coordinates of point P are given. Find polar coordinates of P that satisfy these conditions:

a. $0 \le \theta \le 2\pi$ **b.** $-\pi \le \theta \le \pi$ **c.** $0 \le \theta \le 4\pi$

15. $P = (2, -3)$
16. $P = (-10, 0)$
17. $P = (5, 0)$
18. $P = (0, -2)$

In Exercises 19–22, eliminate the parameter t and find any endpoints.

19. $x = 3 - 5t, \quad y = -4 + 3t, \quad -2 \le t \le 3$
20. $x = 4 + t, \quad y = -8 - 5t, \quad -3 \le t \le 5$
21. $x = 2 \ln t, \quad y = t^4, \quad 0 \le t \le 2$
22. $x = t^3, \quad y = \ln t, \quad 0 \le t \le 2$

In Exercises 23 and 24, solve the problem.

23. For $p < 0$, show that an equation in rectangular form for the parabola with focus $(0, p)$ and directrix $y = -p$ is

$$y = \frac{1}{4p}x^2.$$

24. Show that the parabola $x = ay^2$ has the focus $(1/(4a), 0)$ and directrix $x = -1/(4a)$.

In Exercises 25–30, solve for y and graph the conic on a grapher. Use the quadratic formula as needed.

25. $3x^2 - 8xy + 6y^2 - 5x - 5y + 20 = 0$
26. $10x^2 - 8xy + 6y^2 + 8x - 5y - 30 = 0$
27. $3x^2 - 2xy - 5x + 6y - 10 = 0$
28. $5xy - 6y^2 + 10x - 17y + 20 = 0$
29. $-3x^2 + 7xy - 2y^2 - x + 20y - 15 = 0$
30. $-3x^2 + 7xy - 2y^2 - 2x + 3y - 10 = 0$

In Exercises 31–38, decide whether the graph of the given polar equation appears among the four graphs shown.

31. $r = 3 \sin 4\theta$
32. $r = 2 + \sin \theta$
33. $r = 2 + 2 \sin \theta$
34. $r = 3|\sin 3\theta|$
35. $r = 2 - 2 \sin \theta$
36. $r = 1 - 2 \cos \theta$
37. $r = 3 \cos 5\theta$
38. $r = 3 - 2 \tan \theta$

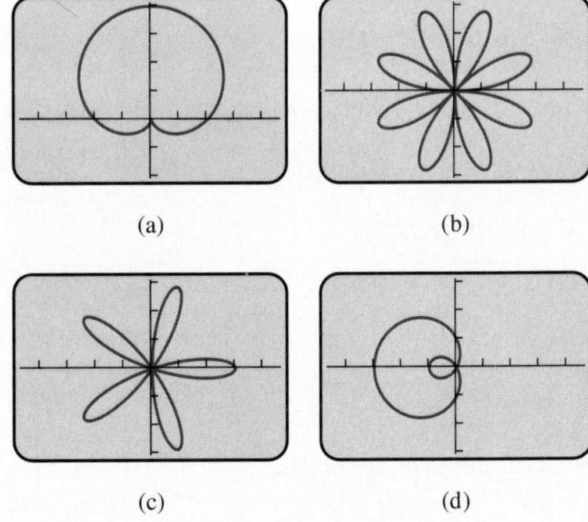

(a) (b) (c) (d)

For Exercises 31–38

In Exercises 39–42, find the center, foci, and endpoints of the axes. Sketch the graph. Support with a grapher.

39. $\dfrac{x^2}{5} + \dfrac{y^2}{8} = 1$
40. $\dfrac{y^2}{16} - \dfrac{x^2}{49} = 1$
41. $\dfrac{(x + 3)^2}{18} - \dfrac{(y - 5)^2}{28} = 1$
42. $\dfrac{(y - 3)^2}{9} - \dfrac{(x - 7)^2}{12} = 1$

In Exercises 43–46, find an equation in standard form for the ellipse.

43. Foci $(2, 0)$ and $(2, 6)$; minor axis endpoints $(0, 3)$ and $(4, 3)$

44. Foci $(-5, -3)$ and $(1, -3)$; minor axis endpoints $(-2, -7)$ and $(-2, 1)$

45. Minor axis endpoints $(-1, 3)$ and $(3, 3)$; major axis length 8

46. Minor axis endpoints $(-2, -4)$ and $(-2, 2)$; major axis length 10

In Exercises 47–52, eliminate t to determine what type of conic is described by the parametric equations.

47. $x = 5 \cos t, \quad y = 2 \sin t, \quad 0 \le t \le 2\pi$

48. $x = 4 \sin t, \quad y = 6 \cos t, \quad 0 \le t \le 4\pi$

49. $x = -2 + \cos t, \quad y = 4 + \sin t, \quad 2\pi \le t \le 4\pi$

50. $x = 5 + 3 \cos t, \quad y = -3 + 3 \sin t, \quad -2\pi \le t \le 0$

51. $x = 3 \sec t, \quad y = 5 \tan t, \quad 0 \le t \le 2\pi$

52. $x = 4 \csc t, \quad y = 3 \cot t, \quad 0 \le t \le 2\pi$

In Exercises 53–64, convert the polar equation to rectangular form and identify its graph.

53. $r = -2$

54. $r = -4$

55. $\theta = \pi/4$

56. $\theta = 2\pi/3$

57. $r^2 = 2 \sin 2\theta$

58. $r^2 = 3 \cos 2\theta$

59. $r^2 \sin 2\theta = 6$

60. $r^2 \cos 2\theta = 9$

61. $r = \dfrac{5}{\cos \theta - 2 \sin \theta}$

62. $r = \dfrac{2}{3 \cos \theta + \sin \theta}$

63. $r^2(4 + 5 \cos^2 \theta) = 36$

64. $r^2(9 + 7 \sin^2 \theta) = 144$

In Exercises 65–70, find the x- and y-intercepts of the conic. Give them in polar coordinates form with $0 \le \theta \le 2\pi$, and in rectangular coordinates.

65. $r = \dfrac{4}{1 + \cos \theta}$

66. $r = \dfrac{5}{1 - \sin \theta}$

67. $r = \dfrac{4}{3 - \cos \theta}$

68. $r = \dfrac{3}{4 + \sin \theta}$

69. $r = \dfrac{35}{2 - 7 \sin \theta}$

70. $r = \dfrac{15}{2 + 5 \cos \theta}$

In Exercises 71–78, convert the rectangular equation to polar form. Graph the polar equation.

71. $y = -4$

72. $y = 3$

73. $xy = 2$

74. $xy = -4$

75. $x^2 + xy + y^2 = 1$

76. $x^2 - xy + y^2 = 2$

77. $(x - 3)^2 + (y + 1)^2 = 10$

78. $(x + 2)^2 + (y - 1)^2 = 5$

In Exercises 79–82, use an algebraic or trigonometric method to find the center and the radius of the circle defined by the equation.

79. $r = 2 \cos \theta + 5 \sin \theta$

80. $r = 3 \sin \theta - \cos \theta$

81. $r = -3 \cos \theta - 2 \sin \theta$

82. $r = \cos \theta - 2 \sin \theta$

In Exercises 83–96, solve the problem.

83. Writing to Learn Consider these three pairs of parametric equations:

a. $x = t, \quad y = 3t - 1$
b. $x = e^t, \quad y = 3e^t - 1$
c. $x = \sin t, \quad y = 3 \sin t - 1$

Explain how they are different. What do they have in common?

84. Writing to Learn Let a and b be real numbers. Explain why the graph of the curve

$$x = a \cos t, \quad y = b \sin t$$

is an ellipse. What is the center? What are the intercepts?

85. *Ferris Wheel Problem* Lucinda is on a Ferris wheel of radius 35 ft that turns at the rate of one revolution every 20 sec. The lowest point of the Ferris wheel (6 o'clock) is 15 ft above ground level at the point $(0, 15)$ of a rectangular coordinate system. Find parametric equations for the position of Lucinda as a function of time t in seconds if Lucinda starts $(t = 0)$ at the point $(35, 50)$.

86. *Ferris Wheel Problem* The lowest point of a Ferris wheel (6 o'clock) of radius 40 ft is 10 ft above the ground, and the center is on the y-axis. Find parametric equations for Henry's position as a function of time t in seconds if his starting position $(t = 0)$ is the point $(0, 10)$ and the wheel turns at the rate of one revolution every 15 sec.

87. *Ferris Wheel Problem* Sarah rides the Ferris wheel described in Exercise 86. Find parametric equations for Sarah's position as a function of time t in seconds if her starting position $(t = 0)$ is the point $(0, 90)$ and the wheel turns at the rate of one revolution every 18 sec.

88. *Parabolic Microphones* Sports Channel uses a parabolic microphone to capture all the sounds from the basketball players and coaches during each regular season game. If one of its microphones has a parabolic surface generated by the parabola $18y = x^2$, locate the focus (the electronic receiver) of the parabola.

89. *Parabolic Headlights* Stein Glass, Inc., makes parabolic headlights for a variety of automobiles. If one of its headlights has a parabolic surface generated by the parabola $y^2 = 15x$ (see figure), where should its lightbulb be placed?

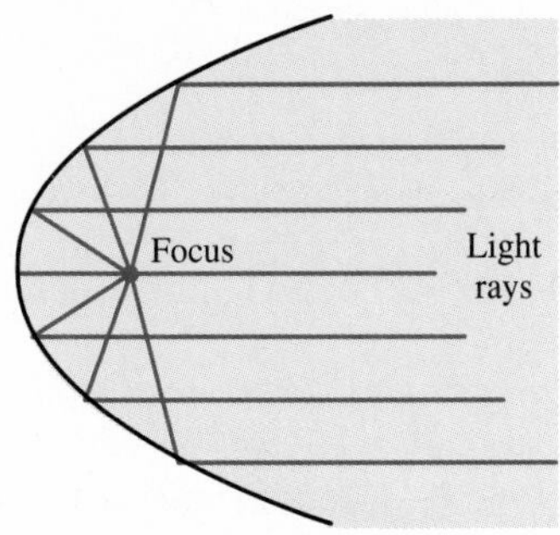

For Exercise 89

90. *Elliptical Pool Table* Elliptical pool tables have been constructed with a single pocket at one focus and a spot marking the other focus. Suppose such a table has a major axis of 6 ft and minor axis of 4 ft.

a. Explain how a "pool shark" who knows conic geometry has a great advantage in a game of pool on this table over a "mark" who knows no conic geometry.

b. How should the ball be hit so that it bounces off the cushion directly into the pocket? Give specific measurements.

91. Suppose $P_1(a, b)$ and $P_2(c, d)$ are two fixed points and consider the parametric equations for the line determined by P_1 and P_2:

$$x = tc + (1 - t)a, \qquad y = td + (1 - t)b.$$

Let $P_t(x, y)$ be the point on the graph of the equations corresponding to t for $0 \leq t \leq 1$. Show that $P_1P_t/P_1P_2 = t$; that is, P_t divides the segment with endpoints P_1 and P_2 into two segments whose lengths have the ratio $t/(1 - t)$.

92. *Epicycloid* The graph of the parametric equations

$$x = 4\cos t - \cos 4t, \quad y = 4\sin t - \sin 4t$$

is an *epicycloid.* The graph is the path of a point P on a circle of radius 1 rolling along the outside of a circle of radius 3, as suggested in the figure.

a. Graph simultaneously this epicycloid and the circle of radius 3.

b. Suppose the large circle has a radius of 4. Experiment! How do you think the equations in part a should be changed to obtain defining equations? What do you think the epicycloid would look like in this case? Check your guesses.

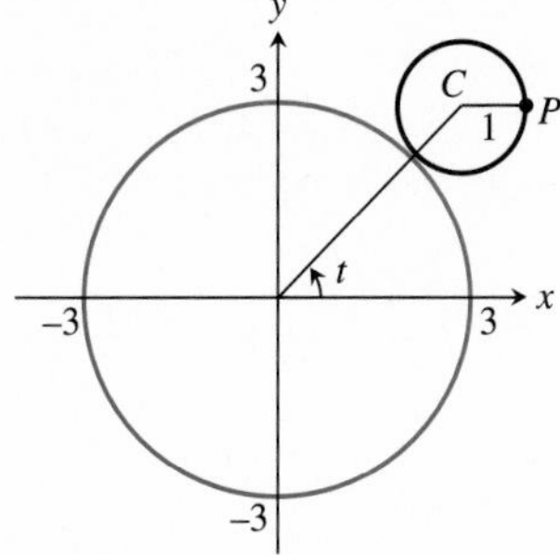

For Exercise 92

93. *Tracking Airplanes* The locations, given in polar coordinates, of two airplanes approaching the Vicksburg airport are (5 mi, 200°) and (6 mi, 170°). Find the distance between the airplanes.

94. *Tracking Ships* The location of two ships from Mays Landing Lighthouse, given in polar coordinates, are (7 mi, 260°) and (9 mi, 280°). Find the distance between the ships.

95. Writing to Learn Explain why the number of petals on the rose curve $r = a\cos n\theta$ is n if n is an odd integer.

96. Writing to Learn Explain why the graph of $r = a\csc\theta$ is a horizontal line for every nonzero real number a.

In Exercises 97–104, draw the graph of the polar curve. Describe and confirm any symmetry.

97. $r = \dfrac{3}{1 - \sin\theta}$ **98.** $r = \dfrac{2}{1 + \cos\theta}$

99. $r = 1 + 2\sin 2\theta$ **100.** $r = 1 - 3\cos 2\theta$

101. $r^2 = -4\cos\theta$ **102.** $r = -4\sin 2\theta$

103. $r^2 = \tan\theta\sec^2\theta$ **104.** $r^2 = \cot\theta\csc^2\theta$

In Exercises 105–112, graph the polar curve. Determine the smallest θ-interval that will produce a complete graph.

105. $r = 4 - 4\cos\theta$ **106.** $r = \dfrac{2}{1 + \cos\theta}$

107. $r = 4 \cos 5\theta$

108. $r = 2 \sin 5\theta$

109. $r = 5 \sin 4\theta$

110. $r = 3 \cos 4\theta$

111. $r = 3 \cos \dfrac{\theta}{2}$

112. $r = -4 \sin \dfrac{\theta}{2}$

In Exercises 113 and 114, solve the problem algebraically and support the solution graphically.

113. *Throwing a Baseball* Sharon releases a baseball 4 ft above the ground with an initial velocity of 66 ft/sec at an angle of 5° with the horizontal. How many seconds after the ball is thrown will it hit the ground? How far from Sharon will the ball be when it hits the ground?

114. *Throwing a Baseball* Diego releases a baseball 3.5 ft above the ground with an initial velocity of 66 ft/sec at an angle of 12° with the horizontal. How many seconds after the ball is thrown will it hit the ground? How far from Diego will the ball be when it hits the ground?

In Exercises 115–125, solve the problem.

115. *Field Goal Kicking* Spencer practices kicking field goals 40 yd from a goal post with a crossbar 10 ft high. If he kicks the ball with an initial velocity of 70 ft/sec at a 45° angle with the horizontal (see figure), will Spencer make the field goal if the kick sails "true?"

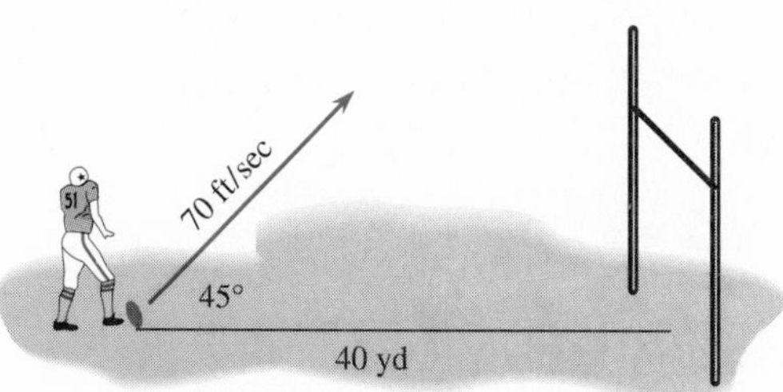

For Exercise 115

116. *Field Goal Kicking* Suppose in Exercise 115 that Spencer kicks the ball at a 30° angle with the horizontal. The kick sails "true." Will it clear the crossbar?

117. *Hang Time* An NFL punter kicks a football downfield with an initial velocity of 85 ft/sec. The ball leaves his foot at the 15 yard line at an angle of 56° with the horizontal. Determine the following:

a. The ball's maximum height above the field.

b. The "hang time" (the total time the football is in the air).

118. *Baseball Hitting* Brian hits a baseball straight toward a 15-ft-high fence that is 400 ft from home plate. The ball is hit when it is 2.5 ft above the ground at an angle of 30° with the horizontal. Find the initial velocity needed for the ball to clear the fence.

119. *Throwing a Ball at a Ferris Wheel* A 60-ft-radius Ferris wheel turns counterclockwise one revolution every 12 sec. Sam stands at a point 80 ft to the left of the bottom (6 o'clock) of the wheel. At the instant Kathy is at point A (3 o'clock), Sam throws a ball with an initial velocity of 100 ft/sec and an angle with the horizontal of 70°. He releases the ball from the same height as the bottom of the Ferris wheel. Find the minimum distance between the ball and Kathy.

120. *Yard Darts* Gretta and Lois are launching yard darts 20 ft from the front edge of a circular target of radius 18 in. If Gretta releases the dart 5 ft above the ground with an initial velocity of 20 ft/sec and at a 50° angle with the horizontal, will the dart hit the target?

121. *Yard Darts* Todd and Mark are launching yard darts 20 ft from the front edge of a circular target of radius 18 in. If Todd releases the dart 6 ft above the ground with an initial velocity of 30 ft/sec and at a 75° angle with the horizontal, will the dart hit the target?

122. *Weather Satellite* The Nimbus weather satellite travels in a north-south circular orbit 500 mi above earth. Find the following. (Assume the earth's radius is 6380 km.)

a. The velocity of the satellite using the formula for velocity v given for Exercises 37 and 38 in Section 9.6 with $k = 1$.

b. The time required for Nimbus to circle the earth once.

123. *Elliptical Orbits* The velocity of a body in an elliptical earth orbit at a distance r (in meters) from the focus (center of earth) is

$$v = \sqrt{3.99 \times 10^{14}\left(\frac{2}{r} - \frac{1}{a}\right)} \text{ m/sec},$$

where a is the length of the semimajor axis of the ellipse. An earth satellite has an *apogee* (maximum distance from earth) of 18,000 km and a *perigee* (minimum distance from earth) of 170 km. Assuming the earth's radius is 6380 km, find the velocity of the satellite at its apogee and perigee.

124. Group Learning Activities *Work in groups of three.* Let $c^2 = b^2 - a^2$ with $a < b$ and $d_1 + d_2 = 2b$. Show that the standard form equation for the ellipse with foci $(0, \pm c)$ is

$$\frac{x^2}{a^2} + \frac{y^2}{b^2} = 1.$$

125. Group Learning Activities *Work in groups of three.* Let $c^2 = a^2 + b^2$ and $d_1 - d_2 = \pm 2b$. Show that the standard form equation for the hyperbola with foci $(0, \pm c)$ is

$$\frac{y^2}{b^2} - \frac{x^2}{a^2} = 1.$$

HYPATIA

Hypatia (A.D. 370–415) was the only known woman professor at the famous university in Alexandria, Egypt. Among her research subjects was the geometry of conic sections. These are the figures formed by the intersection of a plane and a cone. Depending on the angle of the plane, the figure formed is either a circle, an ellipse, a parabola, or a hyperbola.

Hypatia was also interested in science. She wrote descriptions of plans for building an instrument called an astrolabe, which is used to measure the positions of the stars and planets. She also invented several pieces of equipment for working with liquids. Hypatia lived during a time of great social upheaval in Egypt; she was murdered by a mob of fanatics because of her religious convictions. Some historians believe that her death marks the end of ancient mathematics and science.

c h a p t e r 10

SYSTEMS OF EQUATIONS AND INEQUALITIES

BIBLIOGRAPHY

For students: *Calculus: Graphical, Numerical, Algebraic,* Finney, Thomas, Demana, and Waits. Addison-Wesley Publishing Co., 1995.

For teachers: *Math Projects: Organization, Implementation, and Assessment,* Katie DeMeulemeester. Dale Seymour Publications, 1993.

Thinking Connections: Learning to Think & Thinking to Learn, David N. Perkins, Heidi Goodrich, Shari Tishman, and Jill Mirman Owen. Addison-Wesley Publishing Co., 1994.

10.1 SOLVING SYSTEMS OF EQUATIONS

Solutions of Systems of Equations • Solving by Finding Intersections • Solving by Substitution • Solving by Elimination • Number of Solutions • Applications

Objective

Students will be able to solve systems of equations graphically and algebraically.

Solutions of Systems of Equations

Sometimes applications can be modeled by a **system of equations** consisting of two or more equations. For example, suppose that a rectangular garden has perimeter 100 ft and area 300 ft^2. If x and y are lengths of adjacent sides of the garden (see Figure 10.1), then an algebraic model for perimeter and area is the system

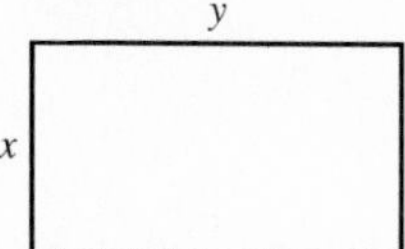

Figure 10.1

$$2x + 2y = 100 \quad \text{Perimeter is 100.}$$

$$xy = 300 \quad \text{Area is 300.}$$

A **solution of a system** of two equations in two variables is an ordered pair of real numbers that satisfies both equations. To *solve* a system means to find *all* solutions of the system. We can solve a system algebraically or graphically.

Motivate

Ask . . .

If a solution of a statement such as "$x > 3$ and $x \le 6$" is a value of x that satisfies *both* inequalities, what do you suppose is meant by a solution of a system such as $2x + 3y = 5$ and $4x - y = 3$? **(An ordered pair (x, y) that satisfies both equations.)**

Solving by Finding Intersections

If we have the system

$$y = f(x)$$
$$y = g(x)$$

we can solve it by solving the equation

$$f(x) = g(x)$$

just as we did in Chapter 2.

EXAMPLE 1 Solving by Finding Intersections

Solve the system.

$$y = \ln x$$
$$y = x^2 - 4x + 2$$

Solution

Solve Graphically

Substituting the expression for y of the first equation into the second equation, we have

$$\ln x = x^2 - 4x + 2.$$

This equation appears difficult to solve algebraically, so instead we graph

$$y_1 = \ln x$$
$$y_2 = x^2 - 4x + 2$$

(Figure 10.2) and see that there are two solutions,

$$(0.711\cdots, -0.340\cdots) \quad \text{and} \quad (3.828\cdots, 1.342\cdots).$$

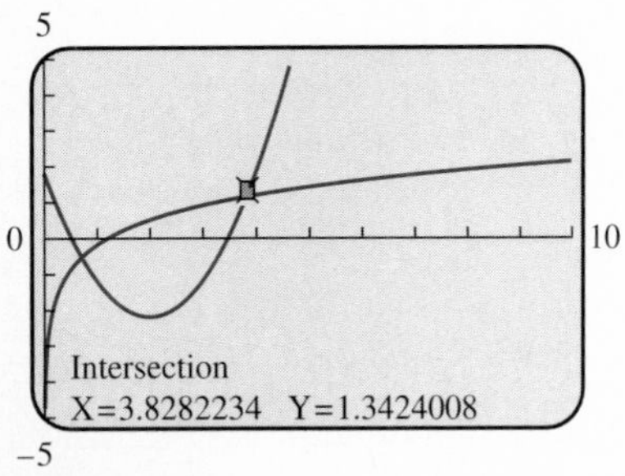

5

0 10

Intersection
X=3.8282234 Y=1.3424008

−5

Figure 10.2

Interpret

Two solutions of the system are approximately $(0.71, -0.34)$ and $(3.83, 1.34)$.

Support Numerically

Substituting $x = 0.71$ yields approximately $y = -0.34$ from both equations. Substituting $x = 3.83$ yields approximately $y = 1.34$ from both equations. ■

Teaching Note

Encourage students to check their solutions by substituting them back into the original equations to make true statements (or "nearly true" statements, in the case of approximate solutions).

We next introduce two algebraic methods that can be used to solve systems of two linear equations in two variables.

> **Teaching Note**
>
> The graph of an equation in two variables is the set of solutions to the equation represented by points in the plane. Every solution to a system of equations in two variables is an *ordered pair.* Geometrically, an ordered pair is a point in the plane. A simultaneous solution to a system of equations is the point at which the graphs of the equations intersect.

Solving by Substitution

Substitution is an algebraic method that we mentioned in Example 1. When we are able to **solve by substitution,** we first solve only one of the equations for one of the variables and then substitute for it in the other equation and solve it.

■ EXAMPLE 2 Solving by Substitution

Solve the system.

$$\begin{aligned} 2x - y &= 10 \\ 3x + 2y &= 1 \end{aligned}$$

Solution

Solve Algebraically

Solving the first equation for y yields $y = 2x - 10$. Then substitute into the second equation.

$$\begin{aligned} 3x + 2y &= 1 && \text{Second equation} \\ 3x + 2(2x - 10) &= 1 && \text{Replace } y \text{ by } 2x - 10. \\ 3x + 4x - 20 &= 1 \\ 7x &= 21 \\ x &= 3 \end{aligned}$$

Substituting $x = 3$ into $y = 2x - 10$ gives $y = -4$.

Interpret

The solution is $x = 3$, $y = -4$, or the ordered pair $(3, -4)$.

Support Graphically

Without even using the grapher it is easy to see that the graphs are two nonparallel lines. Thus there is only one solution.

Confirm Numerically

Substituting $x = 3$ and $y = -4$ into the original equations gives two true statements. ■

> **Teaching Note**
>
> Remind students that the strategy for solving equations using the elimination method involves examining the coefficients. Recall that equivalent equations can be obtained by adding or subtracting the same polynomial to both sides of an equation. Discuss the conditions under which a pair of linear equations may have one, infinitely many, or no solutions (intersecting lines, coincident lines, or parallel lines).

Solving by Elimination

To **solve by elimination** we can avoid solving either equation for one of the variables initially. Instead we rewrite the two equations so that one of the variables has opposite coefficients. Then we add the two equations to eliminate that variable.

Notes on Examples

Example 3 illustrates the elimination method for solving two linear equations. After discussing this example, you may wish to have students solve the system using substitution and graphical methods. Then discuss strategies they can use to choose a method of solution.

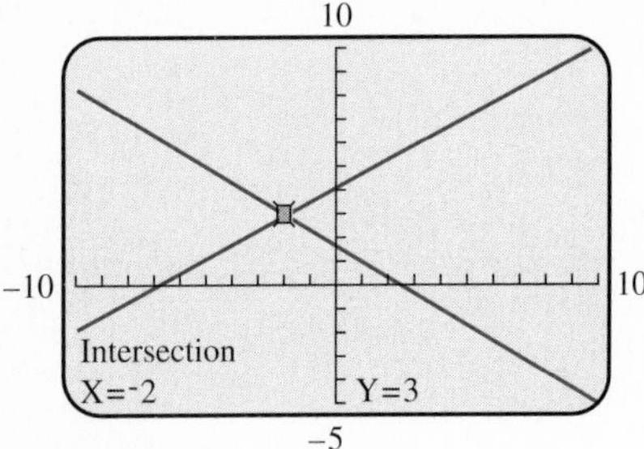

Figure 10.3

■ EXAMPLE 3 Using the Elimination Method

Solve the system.

$$2x + 3y = 5$$
$$-3x + 5y = 21$$

Solution

Solve Algebraically

$6x + 9y = 15$ Multiply first equation by 3.

$-6x + 10y = 42$ Multiply second equation by 2.

$19y = 57$ Add the two equations. The variable x is eliminated.

$y = 3$ Divide by 19.

Substituting $y = 3$ into either equation in the system gives $x = -2$.

Interpret

The solution to the system is $(-2, 3)$.

Support Graphically

Figure 10.3 shows that the graphs of $y_1 = (-2/3)x + 5/3$ and $y_2 = (3/5)x + 21/5$ appear to intersect at $(-2, 3)$. ■

Number of Solutions

The graph of a linear equation in two variables, $ax + by = c$, is a line. Thus there are three possibilities for a system of two linear equations in two variables. There may be no solution (the graphs are parallel lines), there may be one solution (the graphs intersect at one point as in Example 3), or there may be infinitely many solutions (the graphs are the same).

■ EXAMPLE 4 Finding No Solution

Solve the system.

$$x - 3y = -2$$
$$2x - 6y = 4$$

Solution

Solve Algebraically

$-2x + 6y = 4$ Multiply first equation by -2.

$2x - 6y = 4$ Second equation

$0 = 8$ Add.

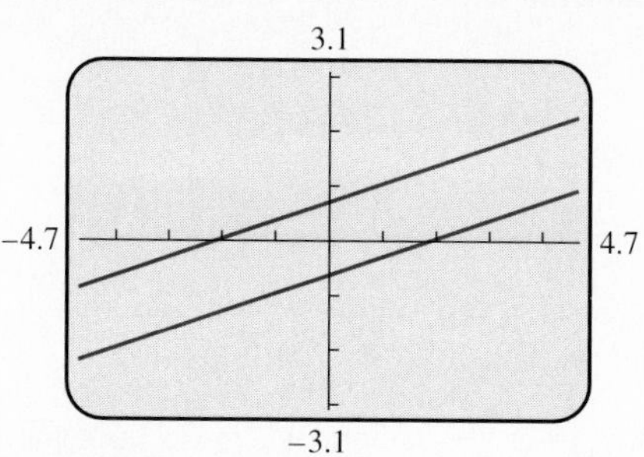

Figure 10.4

Interpret

The last equation is true for *no* values of x and y. The system has no solutions.

Support Graphically

We can rewrite the system as

$$y_1 = \frac{1}{3}x + \frac{2}{3}$$

$$y_2 = \frac{1}{3}x - \frac{2}{3}$$

and see that the two lines are parallel. (See Figure 10.4.) ■

■ **EXAMPLE 5 Finding Infinitely Many Solutions**

Solve the system.

$$4x - 5y = 2$$
$$-12x + 15y = -6$$

Solution

Solve Algebraically

$$12x - 15y = 6 \quad \text{Multiply first equation by 3.}$$
$$-12x + 15y = -6 \quad \text{Second equation}$$
$$0 = 0 \quad \text{Add.}$$

Interpret

The last equation is true for all values of x and y. Thus every ordered pair that satisfies one equation also satisfies the second equation. The system has infinitely many solutions.

Support Graphically

Solving either equation for y yields

$$y = \frac{4}{5}x - \frac{2}{5}.$$

The graphs of the equations are the same line. ■

Systems of equations get particularly interesting when one or both of the equations is not linear. Example 6 involves two functions that you studied in Chapters 1 and 3 and has three solutions.

EXAMPLE 6 Solving a Nonlinear System

Solve the system

$$y = x^3 - 6x$$
$$y = 3x$$

Solution

Solve Graphically

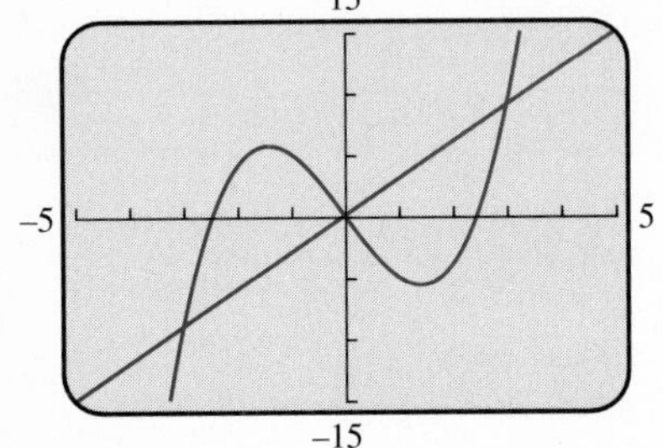

Figure 10.5

Figure 10.5 shows that the graphs of $y_1 = x^3 - 6x$ and $y_2 = 3x$ appear to intersect at $(-3, -9)$, $(0, 0)$, and $(3, 9)$.

Confirm Algebraically

Substituting one value for y into the other equation, we have

$$x^3 - 6x = 3x$$
$$x^3 - 9x = 0$$
$$x(x - 3)(x + 3) = 0 \quad \text{Factor completely.}$$
$$x = 0,\ x = 3,\ \text{or } x = -3 \quad \text{Zero factor property}$$

Interpret

The system of equations has three solutions: $(-3, -9)$, $(0, 0)$, and $(3, 9)$. ■

Applications

According to a report in *USA Today,* the amount of money coming into the Medicare hospital insurance trust fund will someday begin falling behind the money going out. Table 10.1 shows some of the data.

In Exercise 45 you will see that

$$y_R = 6.81x + 79.7$$
$$y_P = 9.24x + 64.86$$

Table 10.1 Medicare Hospital Insurance Trust Fund

Year	Revenues (y_R) (billions of dollars)	Payments (y_P) (billions of dollars)
1990	79.6	66.7
1991	87	71.4
1992	94.3	84.3
1993	97.1	91.4
1994	108.6	102.9

Source: Congressional Budget Office, Office of Management and Budget, and U.S. Census Bureau, as reported by Kevin Rechin in *USA Today* on May 2, 1995.

where $x = 0$ for 1990, are good models for Medicare revenues and payments, respectively. We use these models to find out when the amount coming in will no longer "cover" the amount going out.

Notes on Examples

Examples 7 and 8 show how systems of equations are used in the study of economics. Example 7 relates to payments and revenues, and Example 8 relates to supply and demand.

EXAMPLE 7 APPLICATION: Estimating Medicare Solvency

Estimate when Medicare payments will exceed revenues.

Solution

Model

Solve the system

$$y_R = 6.81x + 79.7$$
$$y_P = 9.24x + 64.86$$

Solve Algebraically

Use substitution.

$$9.24x + 64.86 = 6.81x + 79.7 \qquad \text{Substitute for } y.$$

$$x = \frac{79.7 - 64.86}{9.24 - 6.81} = 6.106 \cdots \qquad \text{Solve for } x.$$

Interpret

Because $x = 0$ stands for 1990, Medicare payments will exceed revenues in about 6.1 years after 1990, that is, sometime during 1996.

Support Graphically

The graphs of y_R and y_P intersect at approximately $x = 6.1$ (Figure 10.6). ■

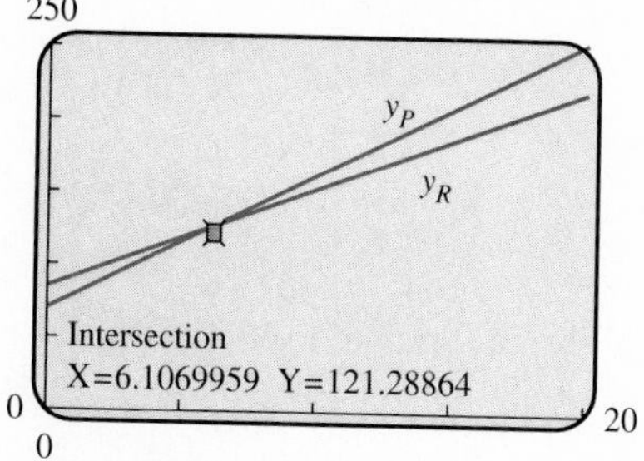

Figure 10.6

Suppliers will usually increase production, x, if they can get higher prices, p, for their products. So, as one variable increases, the other also increases. Normal mathematical practice would be to use p as the independent variable and x as the dependent variable. However, most economists put x on the horizontal axis and p on the vertical axis. In keeping with this practice, we will write $p = f(x)$ for a **supply curve.** On one hand, as the price p increases (vertical axis) so does the willingness for suppliers to increase production x (horizontal axis).

On the other hand, the demand, x, for a product by consumers will decrease as the price, p, goes up. So, as one variable increases, the other decreases. Again economists put x (demand) on the horizontal axis and p (price) on the vertical axis, even though it seems like p should be the dependent variable. In keeping with this practice, we will write $p = g(x)$ for a **demand curve.**

Finally, a point where the supply curve and demand curve intersect is an **equilibrium point.** The corresponding price is the **equilibrium price.**

REMARK

The use of x for the number of units supplied by producers or the number of units demanded by consumers will remind you to put price p on the vertical axis. However, the language may tempt you to put p on the horizontal axis, so be careful.

EXAMPLE 8 APPLICATION: Determining the Equilibrium Price

Nibok Manufacturing has determined that production and price of a new tennis shoe should be geared to the equilibrium point for this system of equations.

$$p = 160 - 5x \quad \text{Demand curve}$$

$$p = 35 + 20x \quad \text{Supply curve}$$

The price, p, is in dollars and the number of shoes, x, is in millions of pairs. Find the equilibrium point.

Solution

Solve Algebraically

We use substitution to solve the system.

$$160 - 5x = 35 + 20x \quad \text{Substitute for } p \text{ in the supply curve.}$$

$$25x = 125$$

$$x = 5$$

$$p = 160 - 5x \quad \text{Demand curve}$$

$$= 160 - 5(5) = 135$$

Interpret

The equilibrium point is (5, 135). The equilibrium price is \$135, the price for which supply and demand will be equal at 5 million pairs of tennis shoes. ■

We return to the garden problem from the first part of the section and solve the system algebraically using substitution.

EXAMPLE 9 APPLICATION: Solving the Garden Problem

A rectangular garden has perimeter 100 ft and area 300 ft^2. Find its dimensions.

Solution

Model

Let x and y be the length and width of the garden. The system of equations that models perimeter and area is

$$2x + 2y = 100 \quad \text{Perimeter}$$

$$xy = 300 \quad \text{Area}$$

Follow-up

Ask students whether it is possible for a system of linear equations to have exactly two solutions. **(No)**

Assignment Guide

Day 1: Ex. 1, 2, 3–36, multiples of 3
Day 2: Ex. 39, 42, 43, 45, 46, 48, 50, 52, 54, 56, 57, 59, 61

Cooperative Learning

Group Activity: Ex. 55

Notes on Exercises

Ex. 43–44 and 60–61 relate to supply and demand.
Ex. 45 and 55–56 require students to find regression models.
Ex. 53–54 require students to use given values of x and y to find the coefficients a and b in an equation.

Ongoing Assessment

Self-Assessment: Ex. 7, 17, 25, 41, 49
Embedded Assessment: Ex. 56

Solve Algebraically

$$y = 50 - x \qquad \text{Solve first equation for } y.$$

$$x(50 - x) = 300 \qquad \text{Substitute for } y \text{ in second equation.}$$

$$x^2 - 50x + 300 = 0$$

$$x = \frac{50 \pm \sqrt{(-50)^2 - 4(300)}}{2} \qquad \text{Quadratic formula}$$

$$x = 6.972\cdots \quad \text{or} \quad x = 43.027\cdots \qquad \text{Use a calculator.}$$

$$y = 43.027\cdots \quad \text{or} \quad y = 6.972\cdots \qquad \text{Use } y = 50 - x.$$

Support Graphically

Use a grapher to see that $y_1 = 50 - x$ and $y_2 = 300/x$ have two points of intersection.

Interpret

The dimensions of the rectangle are approximately 7 ft by 43 ft. ■

Quick Review 10.1

In Exercises 1 and 2, determine whether the ordered pair is a solution of the equation.

1. $5x - 2y = 4$
a. $(2, 3)$ **b.** $(0, 2)$

2. $y = 2x^2 - 3x + 1$
a. $(-2, 3)$ **b.** $(-1, 6)$

In Exercises 3 and 4, solve for y in terms of x.

3. $2x + 3y = 5$ **4.** $xy + x = 4$

In Exercises 5 and 6, use the quadratic formula to solve the equation.

5. $3x^2 - 2x - 2 = 0$ **6.** $2x^2 + 5x - 10 = 0$

In Exercises 7–10, solve the equation algebraically. Support graphically.

7. $2(x + 3) - 4(x - 1) = 7$

8. $2(1 - x) - 2x = 3(x + 2)$

9. $3(2x - 1) - 4(x - 2) = 2(x + 3)$

10. $3(x + 4) + x = 2(2x + 7) - 2$

SECTION EXERCISES 10.1

In Exercises 1 and 2, determine whether the ordered pair is a solution of the system.

1. $5x - 2y = 8$
$2x - 3y = 1$

a. $(0, 4)$ **b.** $(2, 1)$ **c.** $(-2, -9)$

2. $y = x^2 - 6x + 5$
$y = 2x - 7$

a. $(2, -3)$ **b.** $(1, -5)$ **c.** $(6, 5)$

In Exercises 3–12, solve the system by substitution.

3. $\begin{aligned} x + 2y &= 5 \\ y &= -2 \end{aligned}$

4. $\begin{aligned} x &= 3 \\ x - y &= 20 \end{aligned}$

5. $\begin{aligned} 3x + y &= 20 \\ x - 2y &= 10 \end{aligned}$

6. $\begin{aligned} 2x - 3y &= -23 \\ x + y &= 0 \end{aligned}$

7. $\begin{aligned} 2x - 3y &= -7 \\ 4x + 5y &= 8 \end{aligned}$

8. $\begin{aligned} 3x + 2y &= -5 \\ 2x - 5y &= -16 \end{aligned}$

9. $\begin{aligned} x - 3y &= 6 \\ -2x + 6y &= 4 \end{aligned}$

10. $\begin{aligned} 3x - y &= -2 \\ -9x + 3y &= 6 \end{aligned}$

11. $\begin{aligned} y &= x^2 \\ y - 9 &= 0 \end{aligned}$

12. $\begin{aligned} x &= y + 3 \\ x - y^2 &= 3y \end{aligned}$

In Exercises 13–18, solve the system by first eliminating one of the variables.

13. $\begin{aligned} x - y &= 10 \\ x + y &= 6 \end{aligned}$

14. $\begin{aligned} 2x + y &= 10 \\ x - 2y &= -5 \end{aligned}$

15. $\begin{aligned} 3x - 2y &= 8 \\ 5x + 4y &= 28 \end{aligned}$

16. $\begin{aligned} 4x - 5y &= -23 \\ 3x + 4y &= 6 \end{aligned}$

17. $\begin{aligned} 2x - 4y &= 8 \\ -x + 2y &= -4 \end{aligned}$

18. $\begin{aligned} 2x - y &= 3 \\ -4x + 2y &= 5 \end{aligned}$

In Exercises 19–22, use the graph to estimate any solutions to the system.

19. $\begin{aligned} y &= 1 + 2x - x^2 \\ y &= 1 - x \end{aligned}$

20. $\begin{aligned} 6x - 2y &= 7 \\ 2x + y &= 4 \end{aligned}$

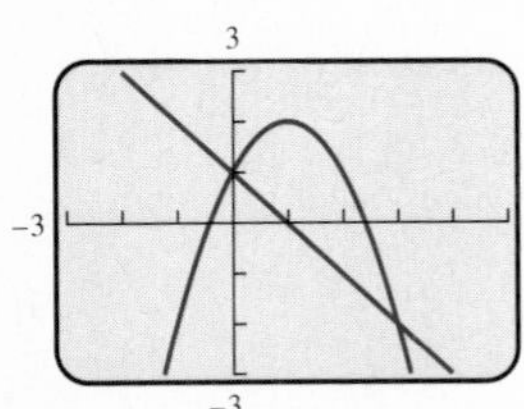

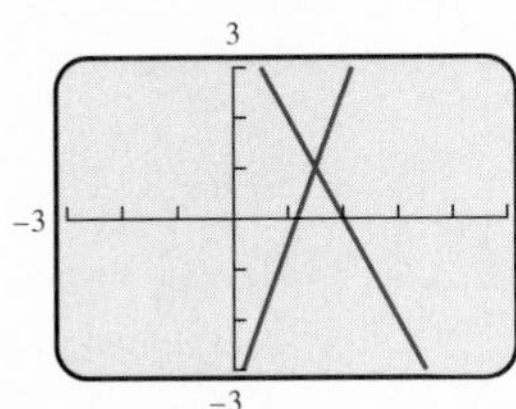

For Exercises 19–20

21. $\begin{aligned} x + 2y &= 0 \\ 0.5x + y &= 2 \end{aligned}$

22. $\begin{aligned} x^2 + y^2 &= 16 \\ y + 4 &= x^2 \end{aligned}$

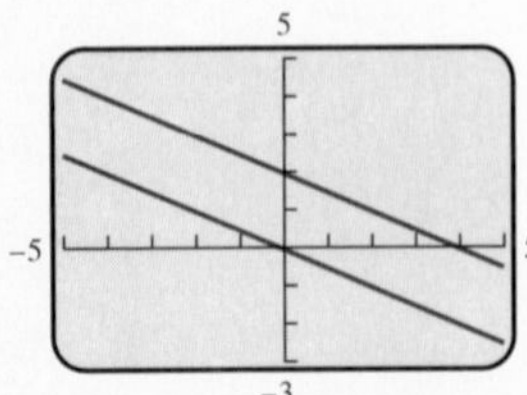

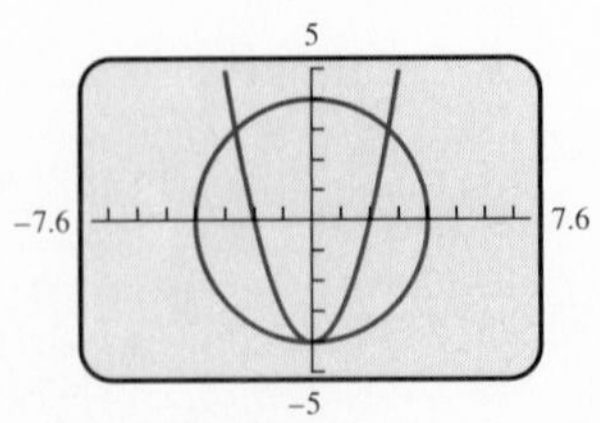

For Exercises 21–22

In Exercises 23–26, use graphs to determine the number of solutions to the system.

23. $\begin{aligned} 3x + 5y &= 7 \\ 4x - 2y &= -3 \end{aligned}$

24. $\begin{aligned} 3x - 9y &= 6 \\ 2x - 6y &= 1 \end{aligned}$

25. $\begin{aligned} 2x - 4y &= 6 \\ 3x - 6y &= 9 \end{aligned}$

26. $\begin{aligned} x - 7y &= 9 \\ 3x + 4y &= 1 \end{aligned}$

In Exercises 27–36, solve the system graphically. Support your answer numerically.

27. $\begin{aligned} y &= \ln x \\ 1 &= 2x + y \end{aligned}$

28. $\begin{aligned} y &= 3 \cos x \\ 1 &= 2x - y \end{aligned}$

29. $\begin{aligned} y &= x^3 - 4x \\ 4 &= x - 2y \end{aligned}$

30. $\begin{aligned} y &= x^2 - 3x - 5 \\ 1 &= 2x - y \end{aligned}$

31. $\begin{aligned} 4x^2 + 9y^2 &= 36 \\ x + 2y &= 2 \end{aligned}$

32. $\begin{aligned} 4x^2 + 9y^2 &= 36 \\ x - 2y &= 2 \end{aligned}$

33. $\begin{aligned} 9x^2 - 4y^2 &= 36 \\ x + 2y &= 4 \end{aligned}$

34. $\begin{aligned} 9x^2 - 4y^2 &= 36 \\ x - 2y &= 4 \end{aligned}$

35. $\begin{aligned} x^2 + 4y^2 &= 4 \\ y &= 2x^2 - 3 \end{aligned}$

36. $\begin{aligned} 9x^2 - 4y^2 &= 36 \\ 4x^2 + 9y^2 &= 36 \end{aligned}$

In Exercises 37–42, solve the system algebraically. Support graphically.

37. $\begin{aligned} y &= 6x^2 \\ 7x + y &= 3 \end{aligned}$

38. $\begin{aligned} y &= 2x^2 + x \\ 2x + y &= 20 \end{aligned}$

39. $\begin{aligned} y &= x^3 - x^2 \\ y &= 2x^2 \end{aligned}$

40. $\begin{aligned} y &= x^3 + x^2 \\ y &= -x^2 \end{aligned}$

41. $\begin{aligned} x^2 + y^2 &= 9 \\ x - 3y &= -1 \end{aligned}$

42. $\begin{aligned} x^2 + y^2 &= 16 \\ 4x + 7y &= 13 \end{aligned}$

In Exercises 43 and 44, find the equilibrium point for the given demand and supply curves.

43. $p = 200 - 15x$ Demand curve

$p = 50 + 25x$ Supply curve

44. $p = 15 - \dfrac{7}{100}x$ Demand curve

$p = 2 + \dfrac{3}{100}x$ Supply curve

In Exercises 45 and 46, use the revenue and payment data for Medicare hospital insurance trust fund given in Table 10.1 of Example 7.

45. **a.** Confirm the linear regression equation $y_R = 6.81x + 79.7$ for the revenue data, and superimpose its graph on a scatter plot of the revenue data.

b. Confirm the linear regression equation $y_P = 9.24x + 64.86$ for the payment data, and superimpose its graph on a scatter plot of the payment data.

c. Use the linear regression equation to estimate Medicare revenues and payments in the year 1998.

46. Assume that the trust fund surplus at the end of 1994 was \$133 billion and that the trend suggested by the two regression lines continues. In what year will the surplus be used up?

In Exercises 47–57, solve the problem.

47. *Garden Problem* Find the dimensions of a rectangle with a perimeter of 200 m and an area of 500 m^2.

48. *Cornfield Dimensions* Find the dimensions of a rectangular cornfield with a perimeter of 220 yd and an area of 3000 yd^2.

49. *Rowing Speed* Hank can row a boat 1 mi upstream (against the current) in 24 min. He can row the same distance downstream in 13 min. If both the rowing speed and current speed are constant, find Hank's rowing speed and the speed of the current.

50. *Airplane Speed* An airplane flying with the wind from Los Angeles to New York City takes 3.75 hr. Flying against the wind, the airplane takes 4.4 hr for the return trip. If the air distance between Los Angeles and New York is 2500 mi and the airplane speed and wind speed are constant, find the airplane speed and the wind speed.

51. *Food Prices* At Philip's convenience store the total cost of one medium and one large soda is \$1.74. The large soda costs \$0.16 more than the medium soda. Find the cost of each soda.

52. *Nut Mixture* A 5-lb nut mixture is worth \$2.80 per pound. The mixture contains peanuts worth \$1.70 per pound and cashews worth \$4.55 per pound. How many pounds of each type of nut are in the mixture?

53. Connecting Algebra and Functions Determine a and b so that the graph of $y = ax + b$ contains the two points $(-1, 4)$ and $(2, 6)$.

54. Connecting Algebra and Functions Determine a and b so that the graph of $ax + by = 8$ contains the two points $(2, -1)$ and $(-4, -6)$.

55. Table 10.2 gives per capita national health expenditures. Use $x = 0$ for 1960.

a. Find the quadratic regression equation and superimpose its graph on a scatter plot of the data.

b. Find the linear regression equation and superimpose its graph on a scatter plot of the data.

Table 10.2 U.S. Health Expenditures

Year	Per Capita (dollars)
1960	143
1965	204
1970	346
1975	591
1980	1,068
1985	1,761
1990	2,686

Source: U.S. Health Care Financing Administration, *Health Care Financing Review,* Fall 1994.

c. Estimate the values given by the two regressions for 1993 and compare them with the actual value of \$3299 per capita.

d. Writing to Learn Give some reasons why one regression appears to be a better model than the other.

56. *Personal Income Data* Table 10.3 gives the disposable personal income per capita for the state of Mississippi. Use $x = 0$ for 1960.

a. Find the linear regression equation and superimpose its graph on a scatter plot of the data.

b. Estimate when the per capita national health expenditures for the data in Exercise 55 will be equal to 20% of the per capita disposable income for the state of Mississippi.

Table 10.3 Disposable Personal Income

Year	Per Capita (dollars)
1980	6,122
1985	8,552
1986	8,985
1987	9,516
1988	10,259
1989	10,880
1990	11,491

Source: U.S. Bureau of Economic Analysis, *Survey of Current Business,* 1993.

57. Writing to Learn Describe all possibilities for the number of solutions to a system of two equations in two variables if the graphs of the two equations are a line and a circle.

EXTENDING THE IDEAS

In Exercises 58 and 59, use the elimination method to solve the given system of equations.

58. $x^2 - 2y = -6$
$x^2 + y = 4$

59. $x^2 + y^2 = 1$
$x^2 - y^2 = 1$

In Exercises 60 and 61, $p(x)$ is the demand curve. The total revenue if x units are sold is $R = px$. Find the number of units sold that gives the maximum revenue.

60. $p = 100 - 4x$

61. $p = 80 - x^2$

10.2
SOLVING SYSTEMS OF LINEAR EQUATIONS USING GAUSSIAN ELIMINATION

Objective

Students will be able to use Gaussian elimination to solve systems of linear equations in n variables.

Motivate

Ask . . .
If $4x - 3y = 7$ is a linear equation in two variables, what would be an example of a linear equation in four variables?

Triangular Form • Gaussian Elimination • Number of Solutions • Applications

Triangular Form

In a **linear system,** the equations are all first-degree polynomial equations. Linear systems in *triangular form* are easily solved using substitution.

EXAMPLE 1 Solving by Substitution

Solve the linear system containing three variables.

$$\begin{aligned} x - 2y + z &= 7 \\ 3y - z &= -5 \\ 2z &= 4 \end{aligned}$$

Solution

The third equation can be used to determine z.

$$\begin{aligned} 2z &= 4 && \text{Third equation} \\ z &= 2 \end{aligned}$$

Substitute the value for z into the second equation to determine y.

$$\begin{aligned} 3y - z &= -5 && \text{Second equation} \\ 3y - 2 &= -5 && \text{Substitute } x = 2. \\ y &= -1 \end{aligned}$$

Finally, substitute the values for y and z into the first equation to determine x.

$$\begin{aligned} x - 2y + z &= 7 && \text{First equation} \\ x - 2(-1) + 2 &= 7 && \text{Substitute } y = -1,\ z = 2. \\ x &= 3 \end{aligned}$$

Interpret

The solution of the system is $x = 3$, $y = -1$, $z = 2$, or $(3, -1, 2)$. ■

The system in Example 1 is in **triangular form** because the third equation has one variable, the second equation has two variables (including the one in the third equation), and the first has three variables.

Gaussian Elimination

In Section 10.1 we solved a system of two linear equations in two variables by eliminating one of the variables. **Gaussian elimination** extends the process to n linear equations in n variables. A **linear equation in $x_1, x_2, \ldots, x_n$** is one that can be written in the form

$$a_1x_1 + a_2x_2 + \cdots + a_nx_n = b,$$

where $a_1, a_2, \ldots, a_n$ and b are real numbers.

The "trick" to Gaussian elimination is to *reduce* the system to triangular form as in Example 1 so that we can then solve for the variables one by one. The key step in reducing the system involves the multiplication and addition properties of equality in the following form:

> The solution of a system of equations is unchanged if a *multiple* of one equation is added to any other equation in the system.

Watch how this idea is used in Example 2.

NOTE

Notice that a linear equation in n variables is also a first-degree polynomial equation.

Notes on Examples

In Example 2, note that when the first equation is multiplied by -2 to be added to the second equation, this is a mental calculation and affects only the second equation. The first equation in the system remains unchanged, keeping the system as simple as possible.

■ EXAMPLE 2 Using Gaussian Elimination

Solve the system.

$$\begin{aligned} x - 2y + z &= -1 \\ 2x + 3y - 2z &= -3 \\ x + 3y - 2z &= -2 \end{aligned}$$

Solution

Multiply the first equation by -2 and add the result to the second equation.

$$\begin{aligned} x - 2y + z &= -1 \\ 7y - 4z &= -1 \\ x + 3y - 2z &= -2 \end{aligned}$$

Multiply the first equation by -1 and add the result to the third equation.

$$\begin{aligned} x - 2y + z &= -1 \\ 7y - 4z &= -1 \\ 5y - 3z &= -1 \end{aligned}$$

Multiply the second equation by $-\frac{5}{7}$ and add the result to the third equation. This system is in triangular form.

$$\begin{aligned} x - 2y + z &= -1 \\ 7y - 4z &= -1 \\ -\frac{1}{7}z &= -\frac{2}{7} \end{aligned}$$

Solve the third equation for z.

$$\begin{aligned} -\frac{1}{7}z &= -\frac{2}{7} \\ z &= 2 \quad \text{Multiply by } -7. \end{aligned}$$

Solve for y by substituting $z = 2$ into the second equation.

$$\begin{aligned} 7y - 4z &= -1 \\ 7y - 4(2) &= -1 \\ y &= 1 \end{aligned}$$

Teaching Note

It is easy to make mistakes when performing Gaussian elimination, so it is particularly important to encourage students to check their answers by substituting them back into the original equations.

Solve for x by substituting $y = 1$ and $z = 2$ into the first equation.

$$\begin{aligned} x - 2y + z &= -1 \\ x - 2(1) + 2 &= -1 \quad \text{Substitute } y = 1,\ z = 2. \\ x &= -1 \end{aligned}$$

The solution to the original system of equations is $x = -1$, $y = 1$, and $z = 2$, or $(-1, 1, 2)$. ■

Number of Solutions

The graph of a linear equation in three variables is a plane in 3-dimensional space. The graphs of three such equations are three planes. The system has a single solution if the three planes intersect in exactly one point, no solution if the three planes have no common intersection point, and infinitely many solutions if the three planes all intersect in a line or are in fact one and the same plane.

A system has no solution if Gaussian elimination produces an equation that is never true.

EXAMPLE 3 Finding No Solution

Solve the system.

$$\begin{aligned} x - 3y + z &= 4 \\ -x + 2y - 5z &= 3 \\ 5x - 13y + 13z &= 8 \end{aligned}$$

Solution

Use Gaussian elimination.

$$\begin{aligned} x - 3y + z &= 4 \\ -y - 4z &= 7 \\ 5x - 13y + 13z &= 8 \end{aligned}$$ Add 1st equation to 2nd equation.

$$\begin{aligned} x - 3y + z &= 4 \\ -y - 4z &= 7 \\ 2y + 8z &= -12 \end{aligned}$$ Multiply 1st equation by -5 and add to 3rd equation.

$$\begin{aligned} x - 3y + z &= 4 \\ -y - 4z &= 7 \\ 0 &= 2 \end{aligned}$$ Multiply 2nd equation by 2 and add to 3rd equation.

Because $0 = 2$ is never true, we conclude that the system has no solution. ■

Notes on Examples

Examples 4 and 5 are important because they show not only how to determine that an equation has infinitely many solutions, but also how to describe these solutions in terms of one or two variables. Although this representation is not unique, the authors consistently solve for the "first" variables in terms of the "last" variables. Encourage students to do the same.

EXAMPLE 4 Finding Infinitely Many Solutions

Solve the system.

$$\begin{aligned} x + y + z &= 3 \\ 2x + y + 4z &= 8 \\ x + 2y - z &= 1 \end{aligned}$$

Solution

Use Gaussian elimination.

$$\begin{aligned} x + y + z &= 3 \\ -y + 2z &= 2 \\ y - 2z &= -2 \end{aligned}$$ Multiply 1st equation by -2 and add to 2nd equation. Multiply 1st equation by -1 and add to 3rd equation.

$$\begin{aligned} x + y + z &= 3 \\ -y + 2z &= 2 \\ 0 &= 0 \end{aligned}$$ Add 2nd equation to 3rd equation.

The last equation is true for all values of x, y, and z. Thus every solution of the original system is also a solution of the following system of two equations in three variables.

$$\begin{aligned} x + y + z &= 3 \\ -y + 2z &= 2 \end{aligned}$$

This system has infinitely many solutions. To describe these solutions, solve the second equation for y in terms of z.

$$y = 2z - 2$$

Then substitute this expression for y into the first equation and solve for x in terms of z.

$$x + y + z = 3$$

$$x + (2z - 2) + z = 3 \qquad \text{Replace } y \text{ by } 2z - 2.$$

$$x = -3z + 5 \qquad \text{Solve for } x.$$

The solutions have the form

$$(-3z + 5,\ 2z - 2,\ z),$$

where z is any real number.

Support Numerically

Substitute several values for z into

$$(-3z + 5,\ 2z - 2,\ z)$$

and show that the resulting ordered triples are all solutions of the original system. ■

Gaussian elimination can be used to solve a system of linear equations, regardless of the number of equations or the number of variables. Example 5 illustrates the process.

■ EXAMPLE 5 Finding Infinitely Many Solutions

Solve the system.

$$\begin{aligned} x + 2y - 3z &= -1 \\ 2x + 3y - 4z + w &= -1 \\ 3x + 5y - 7z + w &= -2 \end{aligned}$$

Solution

Use Gaussian elimination.

$$x + 2y - 3z = -1$$

$$-y + 2z + w = 1 \quad \text{Multiply 1st equation by } -2 \text{ and add to 2nd equation.}$$

$$-y + 2z + w = 1 \quad \text{Multiply 1st equation by } -3 \text{ and add to 3rd equation.}$$

$$x + 2y - 3z = -1$$

$$-y + 2z + w = 1$$

$$0 = 0 \quad \text{Multiply 2nd equation by } -1 \text{ and add to 3rd equation.}$$

The last equation is true for all values of x, y, z, and w. Thus every solution of the original system is also a solution of the following system of two equations in four variables.

$$x + 2y - 3z = -1$$

$$-y + 2z + w = 1$$

This system has infinitely many solutions. To describe these solutions, solve the second equation for y in terms of z and w.

$$y = 2z + w - 1$$

Then substitute this expression for y into the first equation and solve for x in terms of z and w.

$$x + 2y - 3z = -1$$

$$x + 2(2z + w - 1) - 3z = -1 \quad \text{Substitute } y = 2z + w - 1.$$

$$x = -z - 2w + 1$$

The solutions have the form

$$(-z - 2w + 1, 2z + w - 1, z, w),$$

where z and w are any real numbers.

Support Numerically

Substitute several values for z and w in

$$(-z - 2w + 1, 2z + w - 1, z, w)$$

and show that the resulting ordered 4-tuples are solutions of the original system. ■

Applications

■ EXAMPLE 6 APPLICATION: Mixing Acid Solutions

Aileen's Drugstore needs to prepare a 60-L mixture that is 40% acid using three concentrations of acid. The first concentration is 15% acid, the second is 35% acid, and the third is 55% acid. Because of the amounts of acid solution on hand, they need to use twice as much of the 35% solution as the 55% solution. How much of each solution should be used?

Solution

Model

Let

$$x = \text{number of liters of 15\% solution}$$
$$y = \text{number of liters of 35\% solution}$$
$$z = \text{number of liters of 55\% solution}$$

Then

$$x + y + z = 60 \quad \text{New mixture} = 60 \text{ L.}$$
$$0.15x + 0.35y + 0.55z = 24 \quad \text{40\% of 60 is 24.}$$
$$y = 2z \quad \text{35\% amount must be twice 55\% amount.}$$

Solve Algebraically

Use Gaussian elimination.

$$x + y + z = 60$$
$$0.15x + 0.35y + 0.55z = 24$$
$$y - 2z = 0$$

$$x + y + z = 60$$
$$0.20y + 0.40z = 15 \quad \text{Multiply 1st equation by } -0.15 \text{ and add to 2nd equation.}$$
$$y - 2z = 0$$

$$x + y + z = 60$$
$$0.20y + 0.40z = 15$$
$$-4z = -75 \quad \text{Multiply 2nd equation by } -5 \text{ and add to 3rd equation.}$$

$$z = 18.75$$
$$y = 2z = 37.50$$
$$x = 60 - y - z = 3.75$$

```
3.75+37.5+18.75
                60
.15*3.75+.35*37.
5+.55*18.75
                24
2*18.75
              37.5
```

Figure 10.7

Interpret

3.75 L of 15% acid, 37.5 L of 35% acid, and 18.75 L of 55% acid are needed to make 60 L of a 40% acid solution.

Support Numerically

Figure 10.7 shows that $x = 3.75$, $y = 37.5$, $z = 18.75$ is the solution to the problem. ■

We have seen that any two points not in line vertically or horizontally determine exactly one first-degree polynomial. The graph of the first-degree polynomial is the line determined by two points. Similarly, any three noncollinear points with distinct x coordinates determine exactly one second-degree polynomial as illustrated by Example 7. The graph of a second-degree polynomial is a parabola.

Notes on Examples

Example 7 shows how to find a parabola that passes through three points.

Follow-up

Have students explain why a system in triangular form is easy to solve.

Assignment Guide

Day 1: Ex. 3, 6, 9, 12, 14, 15, 18–20
Day 2: Ex. 21–24, 26, 27, 29, 31–34, 37

Cooperative Learning

Group Activity: Ex. 38

Notes on Exercises

Ex. 5–12 each have exactly one solution.
Ex. 13–16 and 20 have infinitely many solutions. The solutions may be described in terms of a single variable.
Ex. 17–18 have infinitely many solutions. The solutions may be described in terms of two variables.
Ex. 19 is an inconsistent system.

■ EXAMPLE 7 APPLICATION: Curve Fitting

Determine a, b, and c so that $(-1, 5)$, $(2, -1)$, and $(3, 13)$ are on the graph of

$$f(x) = ax^2 + bx + c.$$

Solution

Model

We must have $f(-1) = 5$, $f(2) = -1$, and $f(3) = 13$.

$$\begin{aligned} f(-1) &= a - b + c = 5 \\ f(2) &= 4a + 2b + c = -1 \\ f(3) &= 9a + 3b + c = 13 \end{aligned}$$

System to be solved:

$$\begin{aligned} a - b + c &= 5 \\ 4a + 2b + c &= -1 \\ 9a + 3b + c &= 13 \end{aligned}$$

Solve Algebraically

Use Gaussian elimination.

$$a - b + c = 5$$

$$6b - 3c = -21 \quad \text{Multiply 1st equation by } -4 \text{ and add to 2nd equation.}$$

$$12b - 8c = -32 \quad \text{Multiply 1st equation by } -9 \text{ and add to 3rd equation.}$$

$$a - b + c = 5$$

$$6b - 3c = -21$$

$$-2c = 10 \quad \text{Multiply 2nd equation by } -2 \text{ and add to 3rd equation.}$$

Ongoing Assessment

Self-Assessment: Ex. 5, 11, 17, 25
Embedded Assessment: Ex. 29, 32

Solve the third equation for c.

$$c = -5$$

Solve for b by substituting $c = -5$ into the second equation.

$$6b - 3c = -21$$

$$6b - 3(-5) = -21$$

$$b = -6$$

Solve for a by substituting $b = -6$, $c = -5$ into the first equation.

$$a - b + c = 5$$

$$a - (-6) + (-5) = 5$$

$$a = 4$$

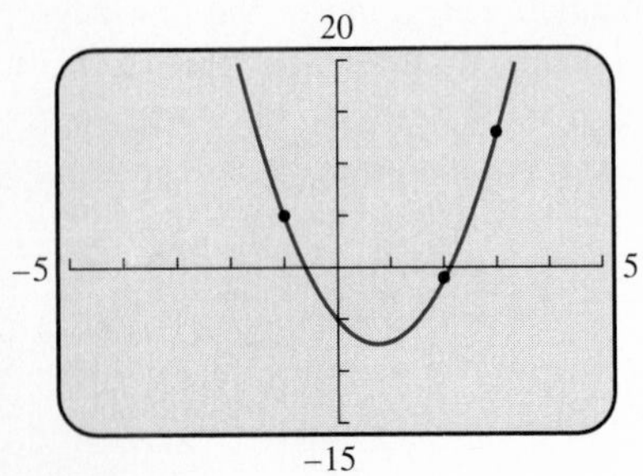

Figure 10.8 A graph of $y_1 = 4x^2 - 6x - 5$ superimposed on a scatter plot of the three points $(-1, 5)$, $(2, -1)$ and $(3, 13)$.

Interpret

The graph of the second-degree polynomial $f(x) = 4x^2 - 6x - 5$ contains the three points $(-1, 5)$, $(2, -1)$, and $(3, 13)$.

Support Graphically

Figure 10.8 suggests that the graph of $y_1 = 4x^2 - 6x - 5$ contains the three points $(-1, 5)$, $(2, -1)$, and $(3, 13)$. ■

By the way, if the graph of f was required to contain only two of the given points in Example 7, there would be infinitely many possibilities for f. See Exercise 29.

Quick Review 10.2

In Exercises 1 and 2, find the amount of pure acid in the solution.

1. 40 L of a 32% acid solution

2. 60 ml of a 14% acid solution

In Exercises 3 and 4, find the amount of water in the solution.

3. 50 L of a 24% acid solution

4. 80 ml of a 70% acid solution

In Exercises 5 and 6, determine which points are on the graph of the function.

5. $f(x) = 2x^2 - 3x + 1$

a. $(-1, 6)$ **b.** $(2, 1)$

6. $f(x) = x^3 - 4x - 1$

a. $(0, -1)$ **b.** $(-2, -17)$

In Exercises 7–10, solve for x or y in terms of the other variables.

7. $x - z = 2$

8. $y - 2w - 3 = 0$

9. $y + z - w = 1$

10. $x - 2z + w = 3$

SECTION EXERCISES 10.2

In Exercises 1–4, determine which ordered triples are solutions to the given system.

1. $\begin{aligned} 2x - y + z &= 4 \\ x - 2y - z &= -4 \\ x - y + 2z &= 5 \end{aligned}$

a. (1, 1, 3) **b.** (1.5, 1.5, 2.5) **c.** (2, 2, 2)

2. $\begin{aligned} x + y + z &= 2 \\ 2x + 3y + z &= 7 \\ x - y + 2z &= 5 \end{aligned}$

a. (−3, 4, 1) **b.** (1, 1, 0) **c.** (1, 2, −1)

3. $\begin{aligned} 2x + 3y + 3z &= 2 \\ 2x + 4y + 4z &= 1 \\ x + 2y + 2z &= 1 \end{aligned}$

a. (2, 1, 1) **b.** (2.5, −1, 0) **c.** (1, 1, −1)

4. $\begin{aligned} x + z &= 1 \\ 2x + 2y &= 6 \\ 2x + y + z &= 4 \end{aligned}$

a. (−1, 4, 2) **b.** (2, 1, −1) **c.** (−2, 5, 3)

In Exercises 5–8, use Gaussian elimination to solve the given system. Support your answer numerically.

5. $\begin{aligned} x - y + z &= 0 \\ 2x - 3z &= -1 \\ -x - y + 2z &= -1 \end{aligned}$

6. $\begin{aligned} 2x - y &= 0 \\ x + 3y - z &= -3 \\ 3y + z &= 8 \end{aligned}$

7. $\begin{aligned} x + y - z &= 4 \\ y + w &= -4 \\ x - y &= 1 \\ x + z + w &= 1 \end{aligned}$

8. $\begin{aligned} \frac{1}{2}x - y + z - w &= 1 \\ -x + y + z + 2w &= -3 \\ x - z &= 2 \\ y + w &= 0 \end{aligned}$

In Exercises 9–20, solve the given system.

9. $\begin{aligned} 2x - y &= 10 \\ x - z &= -1 \\ y + z &= -9 \end{aligned}$

10. $\begin{aligned} 1.25x + z &= 2 \\ y - 5.5z &= -2.75 \\ 3x - 1.5y &= -6 \end{aligned}$

11. $\begin{aligned} x + 2y + 2z + w &= 5 \\ 2x + y + 2z &= 5 \\ 3x + 3y + 3z + 2w &= 12 \\ x + z + w &= 1 \end{aligned}$

12. $\begin{aligned} x - y + w &= -4 \\ -2x + y + z &= 8 \\ 2x - 2y - z &= -10 \\ -2x + z + w &= 5 \end{aligned}$

13. $\begin{aligned} x - y + z &= 6 \\ x + y + 2z &= -2 \end{aligned}$

14. $\begin{aligned} x - 2y + z &= 3 \\ 2x + y - z &= -4 \end{aligned}$

15. $\begin{aligned} 2x + y + z + 4w &= -1 \\ x + 2y + z + w &= 1 \\ x + y + z + 2w &= 0 \end{aligned}$

16. $\begin{aligned} 2x + 3y + 3z + 7w &= 0 \\ x + 2y + 2z + 5w &= 0 \\ x + y + 2z + 3w &= -1 \end{aligned}$

17. $\begin{aligned} 2x + y + z + 2w &= -3.5 \\ x + y + z + w &= -1.5 \end{aligned}$

18. $\begin{aligned} 2x + y + 4w &= 6 \\ x + y + z + w &= 5 \end{aligned}$

19. $\begin{aligned} x + y - z + 2w &= 0 \\ y - z + 2w &= -1 \\ x + y + 3w &= 3 \\ 2x + 2y - z + 5w &= 4 \end{aligned}$

20. $\begin{aligned} x + y + w &= 2 \\ x + 4y + z - 2w &= 3 \\ x + 3y + z - w &= 2 \\ x + y + w &= 2 \end{aligned}$

In Exercises 21–26, solve the problem.

21. *Train Tickets* At the Pittsburgh zoo, children ride a train for 25 cents, adults pay \$1.00, and senior citizens 75 cents. On a given day, 1400 passengers paid a total of \$740 for the rides. There were 250 more children riders than all other riders. Find the number of children, adult, and senior riders.

22. *Manufacturing* Stewart's Metals has three silver alloys on hand. One is 22% silver, another is 30% silver, and the third is 42% silver. How many grams of each alloy is required to produce 80 grams of a new alloy that is 34% silver if the amount of 30% alloy used is twice the amount of 22% alloy used.

23. *Investment* Monica receives an \$80,000 inheritance. She invests part of it in CDs (certificates of deposit) earning 6.7% APY (annual percentage yield), part in bonds earning 9.3% APY, and the remainder in a growth fund earning 15.6% APY. She invests three times as much in the growth fund than in the other two combined. How much does she have in each investment if she receives \$10,843 interest the first year?

24. *Investments* Oscar invests \$20,000 in three investments earning 6% APY, 8% APY, and 10% APY. He invests \$9000 more in the 10% investment than in the 6% investment. How much does he have invested at each rate if he receives \$1780 interest the first year?

25. *Investments* Morgan has \$50,000 to invest and wants to receive \$5000 interest the first year. He puts part in CDs earning 5.75% APY, part in bonds earning 8.7% APY, and the rest in a growth fund earning 14.6% APY. How much should he invest at each rate if he puts the least amount possible in the growth fund?

26. *Mixing Acid Solutions* Simpson's Drugstore needs to prepare a 40-L mixture that is 32% acid from three solutions: a 10% acid solution, a 25% acid solution, and a 50% acid solution. How much of each solution should be used if Simpson's wants to use as little of the 50% solution as possible?

In Exercises 27–30, determine f so that its graph contains the given points.

27. *Curve Fitting* $f(x) = ax^2 + bx + c$
$(-1, 3), (1, -3), (2, 0)$

28. *Curve Fitting* $f(x) = ax^3 + bx^2 + cx + d$
$(-2, -37), (-1, -11), (0, -5), (2, 19)$

29. *Family of Curves* $f(x) = ax^2 + bx + c$
$(-1, -4), (1, -2)$

30. *Family of Curves* $f(x) = ax^3 + bx^2 + cx + d$
$(-1, -6), (0, -1), (1, 2)$

In Exercises 31 and 32, answer the questions.

31. Writing to Learn Can a linear system of three equations with four variables have exactly one solution? Explain.

32. Writing to Learn Describe all possibilities for the number of solutions of a system of two equations in two variables if the graphs are a line and a hyperbola.

EXTENDING THE IDEAS

In Exercises 33–36, use Gaussian elimination to solve the given system of equations

33. $\begin{aligned} x + 2y &= -4 \\ 2x - 3y &= 13 \end{aligned}$

34. $\begin{aligned} 2x - y &= -5 \\ 5x + 4y &= -19 \end{aligned}$

35. $\begin{aligned} x - 3y &= 9 \\ 3x + 2y &= 5 \\ 4x - 5y &= 18 \end{aligned}$

36. $\begin{aligned} 2x - y &= 9 \\ 3x + 4y &= 8 \\ x - 6y &= 10 \end{aligned}$

In Exercises 37–39, answer the questions.

37. Writing to Learn Describe all possible solutions for a system of three linear equations with two variables.

38. Group Learning Activity *We suggest you work in groups of three or four.* Describe all possibilities geometrically for the solution of a system of three linear equations with three variables. Construct physical models if you find that helpful.

39. The solution to the system of equations of Example 4 is the set of all ordered triples $(-3z + 5, 2z - 2, z)$ where z is any real number. Consider these triples as vectors in 3-dimensional space. Describe this collection of vectors.

10.3 SOLVING SYSTEMS OF LINEAR EQUATIONS WITH MATRICES

Matrices • Elementary Row Operations • Row Echelon Form • Reduced Row Echelon Form • Applications

Objective

Students will be able to use matrix methods to solve and interpret systems of linear equations.

Motivate

Ask . . .

Why might it be beneficial to represent a system of linear equations as a rectangular array of numbers?

Matrices

When we solve a system of linear equations using Gaussian elimination all the action is really on the coefficients of the variables. One role of the variables is to keep track of the coefficients. We will use a *matrix* to record the coefficients in an orderly manner without using the variables.

Definition Matrix

Let m and n be positive integers. An $\boldsymbol{m \times n}$ **matrix** (read "m by n matrix") is a rectangular array of m rows and n columns of real numbers.

$$\begin{pmatrix} a_{11} & a_{12} & \cdots & a_{1n} \\ a_{21} & a_{22} & \cdots & a_{2n} \\ \vdots & \vdots & & \vdots \\ a_{m1} & a_{m2} & \cdots & a_{mn} \end{pmatrix}$$

Notice the *double subscript* notation of each element, or entry, a_{ij}. The *row subscript* is the first subscript i, and the *column subscript* is j. The element a_{ij} is said to be in the ith row and jth column. The **order of an** $\boldsymbol{m \times n}$ **matrix** is $m \times n$. If $m = n$, the matrix is a **square matrix.** Two matrices are **equal matrices** if they have the same order and their corresponding elements are equal.

■ **EXAMPLE 1 Determining the Order of a Matrix**

a. The matrix $\begin{pmatrix} 1 & -2 & 3 \\ 2 & 0 & 4 \end{pmatrix}$ has order 2×3.

b. The matrix $\begin{pmatrix} 1 & -1 \\ 0 & 4 \\ 2 & -1 \\ 3 & 2 \end{pmatrix}$ has order 4×2.

c. The matrix $\begin{pmatrix} 1 & 2 & 3 \\ 4 & 5 & 6 \\ 7 & 8 & 9 \end{pmatrix}$ has order 3×3 and is a square matrix. ■

Alert

Emphasize that before the augmented matrix can be created, the equations must be written with the variables in order on the left side and the constant terms on the right side.

Elementary Row Operations

The **coefficient matrix** for the system of equations

$$\begin{aligned} x + 0y + z &= 4 \\ 2x + 2y + 4z &= 10 \\ x + 6y + 8z &= 4 \end{aligned} \quad \text{is} \quad \begin{pmatrix} 1 & 0 & 1 \\ 2 & 2 & 4 \\ 1 & 6 & 8 \end{pmatrix},$$

in which the matrix elements are the coefficients in the equations. The **augmented matrix** for the system of equations

$$\begin{aligned} x + 0y + z &= 4 \\ 2x + 2y + 4z &= 10 \\ x + 6y + 8z &= 4 \end{aligned} \quad \text{is} \quad \begin{pmatrix} 1 & 0 & 1 & 4 \\ 2 & 2 & 4 & 10 \\ 1 & 6 & 8 & 4 \end{pmatrix},$$

in which matrix elements in the last column are the numbers on the right sides of the equations.

When we solve a system of equations using Gaussian elimination, we are essentially manipulating the numbers in the augmented matrix into a certain form from which we can easily pick out the solution of the corresponding system of equations. Watch how this works.

System of Equations **Augmented Matrix**

$$\begin{aligned} x + 0y + z &= 4 \\ 2x + 2y + 4z &= 10 \\ x + 6y + 8z &= 4 \end{aligned} \qquad \begin{pmatrix} 1 & 0 & 1 & 4 \\ 2 & 2 & 4 & 10 \\ 1 & 6 & 8 & 4 \end{pmatrix}$$

Multiply the first equation (row) by -2 and add to the second equation (row).

$$\begin{aligned} x + 0y + z &= 4 \\ 0x + 2y + 2z &= 2 \\ x + 6y + 8z &= 4 \end{aligned} \qquad \begin{pmatrix} 1 & 0 & 1 & 4 \\ 0 & 2 & 2 & 2 \\ 1 & 6 & 8 & 4 \end{pmatrix}$$

Multiply the first equation (row) by -1 and add to the third equation (row).

$$\begin{aligned} x + 0y + z &= 4 \\ 0x + 2y + 2z &= 2 \\ 0x + 6y + 7z &= 0 \end{aligned} \qquad \begin{pmatrix} 1 & 0 & 1 & 4 \\ 0 & 2 & 2 & 2 \\ 0 & 6 & 7 & 0 \end{pmatrix}$$

Multiply the second equation (row) by -3 and add to the third equation (row).

$$\begin{aligned} x + 0y + z &= 4 \\ 0x + 2y + 2z &= 2 \\ 0x + 0y + z &= -6 \end{aligned} \qquad \begin{pmatrix} 1 & 0 & 1 & 4 \\ 0 & 2 & 2 & 2 \\ 0 & 0 & 1 & -6 \end{pmatrix}$$

The third equation now shows that $z = -6$. Substituting this into the second equation, we get $y = 7$. Substituting both values into the first equation, we get $x = 10$. The solution of the system is $(10, 7, -6)$.

Row Echelon Form

Notice that, in the preceding system–matrix pair, we could divide the second equation by 2 and get

System of Equations **Augmented Matrix**

$$\begin{aligned} x + 0y + z &= 4 \\ 0x + y + z &= 1 \\ 0x + 0y + z &= -6 \end{aligned} \qquad \begin{pmatrix} 1 & 0 & 1 & 4 \\ 0 & 1 & 1 & 1 \\ 0 & 0 & 1 & -6 \end{pmatrix}$$

The matrix now is in *row echelon form.*

Definition Row Echelon Form

A matrix is in **row echelon form** if the following conditions are satisfied.

1. Rows consisting entirely of 0's occur at the bottom of the matrix.
2. The first entry in any row with nonzero entries is 1.
3. The column subscript of the leading 1 entries increases as the row subscript increases.

Another way to phrase parts 2 and 3 above is to say that the leading 1's move to the right as we move down the rows.

Our goal is to take a system of equations, write the corresponding augmented matrix, and simplify it, without carrying along the equations, to row echelon form. From there we can pick off the solutions to the system fairly easily.

The steps that we use with equations have corresponding steps for matrices. These are called *elementary row operations.*

Elementary Row Operations

1. Interchange any two rows.
2. Multiply all elements of a row by a nonzero constant.
3. Add a multiple of one row to any other row.

NOTATION

1. R_{ij} indicates interchanging the ith and jth rows of a matrix.
2. kR_i indicates multiplying the ith row of a matrix by the number k.
3. $kR_i + R_j$ indicates adding k times the ith row to the jth row.

Example 2 illustrates how we can manipulate the augmented matrix to solve a system of equations.

Teaching Note

The arrow recording process for describing the operations is provided to help students see clearly what is done at particular steps in the solution process. The process is shown on page 664, and the notation is described above. Encourage students to use this technique.

■ EXAMPLE 2 Simplifying to Row Echelon Form

For this system of equations, simplify the augmented matrix to row echelon form. Then use the row echelon form to solve the system.

$$x - y + 2z = -3$$

$$2x + y - z = 0$$

$$-x + 2y - 3z = 7$$

Solution

We apply elementary row operations to the augmented matrix.

$$\begin{pmatrix} 1 & -1 & 2 & -3 \\ 2 & 1 & -1 & 0 \\ -1 & 2 & -3 & 7 \end{pmatrix} \xrightarrow{(-2)R_1 + R_2} \begin{pmatrix} 1 & -1 & 2 & -3 \\ 0 & 3 & -5 & 6 \\ -1 & 2 & -3 & 7 \end{pmatrix} \xrightarrow{(1)R_1 + R_3}$$

$$\begin{pmatrix} 1 & -1 & 2 & -3 \\ 0 & 3 & -5 & 6 \\ 0 & 1 & -1 & 4 \end{pmatrix} \xrightarrow{R_{23}} \begin{pmatrix} 1 & -1 & 2 & -3 \\ 0 & 1 & -1 & 4 \\ 0 & 3 & -5 & 6 \end{pmatrix} \xrightarrow{(-3)R_2 + R_3}$$

$$\begin{pmatrix} 1 & -1 & 2 & -3 \\ 0 & 1 & -1 & 4 \\ 0 & 0 & -2 & -6 \end{pmatrix} \xrightarrow{(-1/2)R_3} \begin{pmatrix} 1 & -1 & 2 & -3 \\ 0 & 1 & -1 & 4 \\ 0 & 0 & 1 & 3 \end{pmatrix}$$

The last matrix is in row echelon form. Convert each row into equation form and complete the solution:

$$z = 3 \qquad \begin{aligned} y - z &= 4 \\ y - 3 &= 4 \\ y &= 7 \end{aligned} \qquad \begin{aligned} x - y + 2z &= -3 \\ x - 7 + 2(3) &= -3 \\ x &= -3 + 7 - 6 \\ x &= -2 \end{aligned}$$

Interpret

$(-2, 7, 3)$ is the solution of the system. ■

Reduced Row Echelon Form

In Example 2 we obtained the following row echelon form of the augmented matrix to solve the system of equations.

$$\begin{pmatrix} 1 & -1 & 2 & -3 \\ 0 & 1 & -1 & 4 \\ 0 & 0 & 1 & 3 \end{pmatrix}$$

If we apply additional row operations to obtain a matrix in which every column that has a leading 1 has 0's in all other positions, we obtain an augmented matrix that is in **reduced row echelon form.** It is usually easier to read the solution from a reduced row echelon form. Watch what happens when we obtain a reduced row echelon form for the above matrix.

$$\begin{pmatrix} 1 & -1 & 2 & -3 \\ 0 & 1 & -1 & 4 \\ 0 & 0 & 1 & 3 \end{pmatrix} \xrightarrow{(1)R_2 + R_1} \begin{pmatrix} 1 & 0 & 1 & 1 \\ 0 & 1 & -1 & 4 \\ 0 & 0 & 1 & 3 \end{pmatrix} \xrightarrow{(-1)R_3 + R_1}$$

$$\begin{pmatrix} 1 & 0 & 0 & -2 \\ 0 & 1 & -1 & 4 \\ 0 & 0 & 1 & 3 \end{pmatrix} \xrightarrow{(1)R_3 + R_2} \begin{pmatrix} 1 & 0 & 0 & -2 \\ 0 & 1 & 0 & 7 \\ 0 & 0 & 1 & 3 \end{pmatrix}$$

TECHNOLOGY NOTE

Some graphers can perform elementary row operations. Some can even produce a row echelon form or a reduced row echelon form for a matrix. Check your owner's manual to find your grapher's matrix capabilities.

From this reduced row echelon form, we read the solutions to the system of Example 2:

$$x = -2, \qquad y = 7, \qquad \text{and} \qquad z = 3.$$

If you are using pencil and paper to solve a system of equations, then it's usually quicker to use Gaussian elimination. However, if your grapher can produce the row echelon form, or even better, the reduced row echelon, then use it to read the solutions.

Teaching Note

Computer software is available to lessen the computational burden in working with matrices. Many of the newer graphers also have matrix utilities that will produce the desired results. Encourage students to use their calculators to perform matrix operations if possible.

EXAMPLE 3 Obtaining Reduced Row Echelon Form

Find the reduced row echelon form of the augmented matrix and solve the system.

$$\begin{aligned} x + 2y + z + w &= 5 \\ 3x + 2y + 2z - 3w &= 11 \\ x + y + z - w &= 4 \end{aligned}$$

Solution

As you gain experience with elementary row operations you can often perform several steps at one time.

$$\begin{pmatrix} 1 & 2 & 1 & 1 & 5 \\ 3 & 2 & 2 & -3 & 11 \\ 1 & 1 & 1 & -1 & 4 \end{pmatrix} \xrightarrow[(-1)R_1 + R_3]{(-3)R_1 + R_2} \begin{pmatrix} 1 & 2 & 1 & 1 & 5 \\ 0 & -4 & -1 & -6 & -4 \\ 0 & -1 & 0 & -2 & -1 \end{pmatrix} \xrightarrow[(-4)R_3 + R_2]{(2)R_3 + R_1}$$

$$\begin{pmatrix} 1 & 0 & 1 & -3 & 3 \\ 0 & 0 & -1 & 2 & 0 \\ 0 & -1 & 0 & -2 & -1 \end{pmatrix} \xrightarrow[R_{23}]{(-1)R_3} \begin{pmatrix} 1 & 0 & 1 & -3 & 3 \\ 0 & 1 & 0 & 2 & 1 \\ 0 & 0 & -1 & 2 & 0 \end{pmatrix} \xrightarrow{(1)R_3 + R_1}$$

$$\begin{pmatrix} 1 & 0 & 0 & -1 & 3 \\ 0 & 1 & 0 & 2 & 1 \\ 0 & 0 & -1 & 2 & 0 \end{pmatrix} \xrightarrow{(-1)R_3} \begin{pmatrix} 1 & 0 & 0 & -1 & 3 \\ 0 & 1 & 0 & 2 & 1 \\ 0 & 0 & 1 & -2 & 0 \end{pmatrix}$$

From this reduced row echelon form we read the solutions of the original system of equations:

$$x = w + 3, \qquad y = -2w + 1, \qquad \text{and} \qquad z = 2w,$$

where w is any real number. ■

Applications

Mathematics often is used to solve consumer or industrial problems. Mathematics also is applied to solve other mathematical problems.

In Section 5.5 you studied how to decompose a given rational function into a sum of simpler rational functions, an important technique used in calculus. We can use matrices to help find such decompositions, as illustrated in Example 4.

EXAMPLE 4 APPLICATION: Decomposing Partial Fractions

Use matrices to determine the values of A, B, and C so that

$$\frac{x^2 - 2x + 3}{(x-1)(x-2)^2} = \frac{A}{x-1} + \frac{B}{x-2} + \frac{C}{(x-2)^2}.$$

Solution

Clear the equation of fractions by multiplying both sides by the LCD $(x-1)(x-2)^2$.

$$\begin{aligned} x^2 - 2x + 3 &= A(x-2)^2 + B(x-1)(x-2) + C(x-1) \\ &= A(x^2 - 4x + 4) + B(x^2 - 3x + 2) + C(x-1) \\ &= (A+B)x^2 + (-4A - 3B + C)x + (4A + 2B - C) \end{aligned}$$

In order for the quadratic polynomials on the left and right sides of the equation to be equal, we must have

$$\begin{aligned} A + B &= 1 \\ -4A - 3B + C &= -2 \\ 4A + 2B - C &= 3 \end{aligned}$$

Find a reduced row echelon form for the augmented matrix.

$$\begin{pmatrix} 1 & 1 & 0 & 1 \\ -4 & -3 & 1 & -2 \\ 4 & 2 & -1 & 3 \end{pmatrix} \xrightarrow[(-4)R_1 + R_3]{(4)R_1 + R_2} \begin{pmatrix} 1 & 1 & 0 & 1 \\ 0 & 1 & 1 & 2 \\ 0 & -2 & -1 & -1 \end{pmatrix} \xrightarrow[(2)R_2 + R_3]{(-1)R_2 + R_1}$$

$$\begin{pmatrix} 1 & 0 & -1 & -1 \\ 0 & 1 & 1 & 2 \\ 0 & 0 & 1 & 3 \end{pmatrix} \xrightarrow[(-1)R_3 + R_2]{(1)R_3 + R_1} \begin{pmatrix} 1 & 0 & 0 & 2 \\ 0 & 1 & 0 & -1 \\ 0 & 0 & 1 & 3 \end{pmatrix}$$

From the reduced row echelon form, we read

$$A = 2, \quad B = -1, \quad \text{and} \quad C = 3.$$

Follow-up

Ask students to compare the method of using row operations to create a triangular matrix to that of Gaussian elimination.

Assignment Guide

Day 1: Ex. 1–11, 14–20, 23, 26, 27, 30, 31, 34, 35, 38
Day 2: Ex. 39, 42, 43, 44, 46, 49, 52, 53, 56, 58, 60

Cooperative Learning

Group Activity: Ex. 51

Notes on Exercises

Ex. 39–42 and 48 each have exactly one solution.
Ex. 43 and 47 are inconsistent systems.
Ex. 44–46 have infinitely many solutions.
Ex. 49–52 relate to partial fractions.

Ongoing Assessment

Self-Assessment: Ex. 25, 29, 33, 37, 41, 47, 51, 57
Embedded Assessment: Ex. 58

Interpret

The original rational function can be decomposed as

$$\frac{x^2 - 2x + 3}{(x-1)(x-2)^2} = \frac{2}{x-1} - \frac{1}{x-2} + \frac{3}{(x-2)^2}.$$

■

Quick Review 10.3

In Exercises 1–4, perform the indicated operations and combine any like terms.

1. $(-3)(x - 2y + w - 5)$
2. $(-1)(-2x - y + w + 2)$
3. $2(2x + 3y - z + 1) + (-3x + 2y - 3)$
4. $(-2)(x - 3y + z + 2) + (2x + 4y - z - 6)$

In Exercises 5–8, perform the indicated operations and write the answer as a single reduced fraction.

5. $\dfrac{1}{x-1} + \dfrac{2}{x-3}$

6. $\dfrac{5}{x+4} - \dfrac{2}{x+1}$

7. $\dfrac{1}{x} + \dfrac{3}{x+1} + \dfrac{1}{(x+1)^2}$

8. $\dfrac{3}{x^2+1} - \dfrac{x+1}{(x^2+1)^2}$

In Exercises 9 and 10, assume that $f(x) = g(x)$. What can you conclude about A, B, C, and D?

9. $f(x) = Ax^2 + Bx + C + 1$
   $g(x) = 3x^2 - x + 2$

10. $f(x) = (A + 1)x^3 + Bx^2 + Cx + D$
    $g(x) = -x^3 + 2x^2 - x - 5$

SECTION EXERCISES 10.3

In Exercises 1–6, determine the order of the given matrix. Indicate whether the matrix is square.

1. $\begin{pmatrix} 2 & 3 & -1 \\ 1 & 0 & 5 \end{pmatrix}$

2. $\begin{pmatrix} 1 & 3 \\ -1 & 2 \end{pmatrix}$

3. $\begin{pmatrix} 5 & 6 \\ -1 & 2 \\ 0 & 0 \end{pmatrix}$

4. $\begin{pmatrix} -1 & 0 & 6 \end{pmatrix}$

5. $\begin{pmatrix} 2 \\ -1 \\ 0 \end{pmatrix}$

6. (0)

In Exercises 7–10, identify the element specified for the following matrix.

$$\begin{pmatrix} -2 & 0 & 3 & 4 \\ 3 & 1 & 5 & -1 \\ 1 & 4 & -1 & 3 \end{pmatrix}$$

7. a_{13}

8. a_{24}

9. a_{32}

10. a_{33}

In Exercises 11–14, what elementary row operation applied to

$$\begin{pmatrix} -2 & 1 & -1 & 2 \\ 1 & -2 & 3 & 0 \\ 3 & 1 & -1 & 2 \end{pmatrix}$$

will yield the given matrix?

11. $\begin{pmatrix} 1 & -2 & 3 & 0 \\ -2 & 1 & -1 & 2 \\ 3 & 1 & -1 & 2 \end{pmatrix}$

12. $\begin{pmatrix} 0 & -3 & 5 & 2 \\ 1 & -2 & 3 & 0 \\ 3 & 1 & -1 & 2 \end{pmatrix}$

13. $\begin{pmatrix} -2 & 1 & -1 & 2 \\ 1 & -2 & 3 & 0 \\ 0 & 7 & -10 & 2 \end{pmatrix}$

14. $\begin{pmatrix} -2 & 1 & -1 & 2 \\ 1 & -2 & 3 & 0 \\ 0.75 & 0.25 & -0.25 & 0.5 \end{pmatrix}$

In Exercises 15–18, perform the elementary row operation on the given matrix.

$$\begin{pmatrix} 2 & -6 & 4 \\ 1 & 2 & -3 \\ -3 & 1 & -2 \end{pmatrix}$$

15. $(3/2)R_1 + R_3$

16. $(1/2)R_1$

17. $(-2)R_2 + R_1$

18. $(1)R_1 + R_2$

In Exercises 19–22, indicate whether the matrix is

a. in reduced row echelon form.
b. in row echelon form but not reduced.
c. not in row echelon form.

19. $\begin{pmatrix} 1 & 2 & 0 & 2 \\ 0 & 0 & 1 & -1 \\ 0 & 0 & 0 & 0 \end{pmatrix}$

20. $\begin{pmatrix} 1 & -3 & 5 & 0 \\ 0 & 1 & -1 & 2 \\ 0 & 0 & 2 & 1 \end{pmatrix}$

21. $\begin{pmatrix} 1 & 2 & 1 \\ 0 & 1 & 2 \\ 0 & 0 & 0 \end{pmatrix}$

22. $\begin{pmatrix} 1 & 0 & -1 & 4 \\ 0 & 1 & 2 & 3 \\ 0 & 0 & 0 & 0 \end{pmatrix}$

In Exercises 23–26, find a row echelon form for the matrix.

23. $\begin{pmatrix} 1 & 3 & -1 \\ 2 & 1 & 4 \\ -3 & 0 & 1 \end{pmatrix}$

24. $\begin{pmatrix} 1 & 2 & -3 \\ -3 & -6 & 10 \\ -2 & -4 & 7 \end{pmatrix}$

25. $\begin{pmatrix} 1 & 2 & 3 & -4 \\ -2 & 6 & -6 & 2 \\ 3 & 12 & 6 & 12 \end{pmatrix}$

26. $\begin{pmatrix} 3 & 6 & 9 & -6 \\ 2 & 5 & 5 & -3 \end{pmatrix}$

In Exercises 27–30, find a reduced row echelon form for the matrix.

27. $\begin{pmatrix} 1 & 0 & 2 & 1 \\ 3 & 2 & 4 & 7 \\ 2 & 1 & 3 & 4 \end{pmatrix}$

28. $\begin{pmatrix} 1 & -2 & 2 & 1 & 1 \\ 3 & -5 & 6 & 3 & -1 \\ -2 & 4 & -3 & -2 & 5 \\ 3 & -5 & 6 & 4 & -3 \end{pmatrix}$

29. $\begin{pmatrix} 1 & 2 & 3 & 1 \\ -3 & -5 & -7 & -4 \end{pmatrix}$

30. $\begin{pmatrix} 3 & -6 & 3 & -3 \\ 2 & -4 & 2 & -2 \\ -3 & 6 & -3 & 3 \end{pmatrix}$

In Exercises 31–34, write the augmented matrix for the system of equations.

31. $\begin{aligned} 2x - 3y + z &= 1 \\ -x + y - 4z &= -3 \\ 3x - z &= 2 \end{aligned}$

32. $\begin{aligned} 3x - 4y + z - w &= 1 \\ x + z - 2w &= 4 \end{aligned}$

33. $\begin{aligned} 2x - 5y + z - w &= -3 \\ x - 2z + w &= 4 \\ 2y - 3z - w &= 5 \end{aligned}$

34. $\begin{aligned} 3x - 2y &= 5 \\ -x + 5y &= 7 \end{aligned}$

In Exercises 35–38, write the system of equations for the augmented matrix.

35. $\begin{pmatrix} 3 & 2 & -1 \\ -4 & 5 & 2 \end{pmatrix}$

36. $\begin{pmatrix} 1 & 0 & -1 & 2 & -3 \\ 2 & 1 & 0 & -1 & 4 \\ -1 & 1 & 2 & 0 & 0 \end{pmatrix}$

37. $\begin{pmatrix} 2 & 0 & 1 & 3 \\ -1 & 1 & 0 & 2 \\ 0 & 2 & -3 & -1 \end{pmatrix}$

38. $\begin{pmatrix} 2 & 1 & -2 & 4 \\ -3 & 0 & 2 & -1 \end{pmatrix}$

In Exercises 39–48, solve the system of equations.

39. $\begin{aligned} x + y &= -1 \\ 3x + 2y &= 0 \end{aligned}$

40. $\begin{aligned} x - 2y &= 1 \\ 2x - 3y &= 1 \end{aligned}$

41. $\begin{aligned} x + 2y - z &= 3 \\ 3x + 7y - 3z &= 12 \\ -2x - 4y + 3z &= -5 \end{aligned}$

42. $\begin{aligned} x - 2y + z &= -2 \\ 2x - 3y + 2z &= 2 \\ 4x - 8y + 5z &= -5 \end{aligned}$

43. $\begin{aligned} x + y + 3z &= 2 \\ 3x + 4y + 10z &= 5 \\ x + 2y + 4z &= 3 \end{aligned}$

44. $\begin{aligned} x - z &= 2 \\ -2x + y + 3z &= -5 \\ 2x + y - z &= 3 \end{aligned}$

45. $\begin{aligned} x + z &= 2 \\ 2x + y + z &= 5 \end{aligned}$

46. $\begin{aligned} x + 2y - 3z &= 1 \\ -3x - 5y + 8z &= -29 \end{aligned}$

47. $\begin{aligned} x + 2y &= 4 \\ 3x + 4y &= 5 \\ 2x + 3y &= 4 \end{aligned}$

48. $\begin{aligned} x + y &= 3 \\ 2x + 3y &= 8 \\ 2x + 2y &= 6 \end{aligned}$

In Exercises 49–52, use matrices to find the partial fraction decomposition.

49. $\dfrac{x + 22}{(x + 4)(x - 2)} = \dfrac{A}{x + 4} + \dfrac{B}{x - 2}$

50. $\dfrac{x - 3}{x(x + 3)} = \dfrac{A}{x + 3} + \dfrac{B}{x}$

51. $\dfrac{3x^2 + 2x + 2}{(x^2 + 1)^2} = \dfrac{Ax + B}{x^2 + 1} + \dfrac{Cx + D}{(x^2 + 1)^2}$

52. $\dfrac{4x + 4}{x^2(x + 2)} = \dfrac{A}{x} + \dfrac{B}{x^2} + \dfrac{C}{x + 2}$

In Exercises 53 and 54, determine *a, b,* and *c* so that the graph of the function goes through the specified points.

53. *Curve Fitting* $f(x) = \dfrac{a}{bx + c}$
$(-1, -0.5), (0, -2), (1, 1)$

54. *Curve Fitting* $f(x) = ax^2 + bx + c$
$(-1, -12), (1, -2), (2, -6)$

In Exercises 55 and 56, solve the problem.

55. *Loose Change* Matthew has 74 coins consisting of nickels, dimes, and quarters in his coin box. The total value of the coins is $8.85. If the number of nickels and quarters is four more than the number of dimes, find how many of each coin Matthew has in his coin box.

56. *Vacation Money* Heather has saved $177 to take with her on the family vacation. She has 51 bills consisting of $1, $5, and $10 bills. If the number of $5 bills is three times the number of $10 bills, find how many of each bill she has.

In Exercises 57 and 58, answer the questions.

57. Writing to Learn Explain why adding one row to another row in a matrix is an elementary row operation.

58. Writing to Learn Explain why subtracting one row from another row in a matrix is an elementary row operation.

EXTENDING THE IDEAS

In Exercises 59 and 60, answer the questions.

59. Writing to Learn Explain why the three cases in the direction to Exercises 19–22 exhaust all possibilities for a given matrix.

60. Writing to Learn Explain why a row echelon form of a matrix is not unique. That is, show that a matrix can have two unequal row echelon forms. Give an example.

10.4 SOLVING SYSTEMS OF LINEAR EQUATIONS WITH INVERSE MATRICES

Matrix Multiplication • Identity and Inverse Matrices • Systems of Linear Equations

Objective

Students will be able to solve a system of linear equations using matrix multiplication.

Motivate

Have students calculate the sum of matrices $A = \begin{pmatrix} 2 & 4 \\ 5 & -1 \end{pmatrix}$ and $B = \begin{pmatrix} 3 & 0 \\ 1 & 4 \end{pmatrix}$.

$\mathbf{A + B = \begin{pmatrix} 5 & 4 \\ 6 & 3 \end{pmatrix}}$

Ask them to guess what the product *AB* of the matrices might be.

$\mathbf{AB = \begin{pmatrix} 10 & 16 \\ 16 & -4 \end{pmatrix}}$

Matrix Multiplication

It is convenient to write $A = (a_{ij})$ for the matrix A with entry a_{ij} in the ith row and jth column. To form the product AB of two matrices, the number of columns of the matrix A on the left must be equal to the number of rows of the matrix B on the right. Each entry of the product involves multiplying the entries of a row of A by the corresponding entries of a column of B.

Definition Matrix Multiplication

Let $A = (a_{ij})$ be an $m \times r$ matrix and $B = (b_{ij})$ an $r \times n$ matrix. The **product of matrices A and B,** $AB = (c_{ij})$, is an $m \times n$ matrix where

$$c_{ij} = (a_{i1} \quad a_{i2} \quad \cdots \quad a_{ir}) \begin{pmatrix} b_{1j} \\ b_{2j} \\ \vdots \\ b_{rj} \end{pmatrix}$$

$$= a_{i1}b_{1j} + a_{i2}b_{2j} + \cdots + a_{ir}b_{rj}.$$

Teaching Note

Help students remember how to do matrix multiplication by emphasizing that in most contexts with matrices, rows come before columns. For example, an $m \times n$ matrix has m rows and n columns; a_{ij} is in the ith row and the jth column; the rows of the first matrix are multiplied by the columns of the second matrix. Point out that the product AB is not defined unless the number of entries in a row of A is the same as the number of entries in a column of B.

To find c_{ij} we multiply the entries of the ith row of A by the corresponding entries of the jth column of B and then compute their sum.

■ EXAMPLE 1 Finding the Product of Two Matrices

Find the product AB, where

$$A = \begin{pmatrix} 2 & 1 & -3 \\ 0 & 1 & 2 \end{pmatrix} \quad \text{and} \quad B = \begin{pmatrix} 1 & -4 \\ 0 & 2 \\ 1 & 0 \end{pmatrix}.$$

Solution

The number of columns of A is 3 and the number of rows of B is 3, so the product AB is defined. We use the definition of matrix multiplication to find the entries of the 2×2 matrix $AB = (c_{ij})$.

$$c_{11} = (2 \quad 1 \quad -3)\begin{pmatrix} 1 \\ 0 \\ 1 \end{pmatrix} = 2 \cdot 1 + 1 \cdot 0 + (-3) \cdot 1 = -1$$

$$c_{12} = (2 \quad 1 \quad -3)\begin{pmatrix} -4 \\ 2 \\ 0 \end{pmatrix} = 2 \cdot (-4) + 1 \cdot 2 + (-3) \cdot 0 = -6$$

$$c_{21} = (0 \quad 1 \quad 2)\begin{pmatrix} 1 \\ 0 \\ 1 \end{pmatrix} = 0 \cdot 1 + 1 \cdot 0 + 2 \cdot 1 = 2$$

$$c_{22} = (0 \quad 1 \quad 2)\begin{pmatrix} -4 \\ 2 \\ 0 \end{pmatrix} = 0 \cdot (-4) + 1 \cdot 2 + 2 \cdot 0 = 2$$

Thus we have

$$AB = \begin{pmatrix} -1 & -6 \\ 2 & 2 \end{pmatrix}.$$

■

In Example 2 we show that the product BA is not the same as product AB for the matrices of Example 1. This illustrates that, in general, matrix multiplication is *not* commutative.

Notes on Examples

Taken together, Examples 1 and 2 show that matrix multiplication is not commutative.

EXAMPLE 2 Finding the Product of Two Matrices

Find the product BA, where

$$B = \begin{pmatrix} 1 & -4 \\ 0 & 2 \\ 1 & 0 \end{pmatrix} \quad \text{and} \quad A = \begin{pmatrix} 2 & 1 & -3 \\ 0 & 1 & 2 \end{pmatrix}.$$

Solution

The number of columns of B is 2 and the number of rows of A is 2, so the product BA is defined. We use the definition of matrix multiplication to find the entries of the 3×3 matrix BA.

$$BA = \begin{pmatrix} 1 \cdot 2 + (-4) \cdot 0 & 1 \cdot 1 + (-4) \cdot 1 & 1 \cdot (-3) + (-4) \cdot 2 \\ 0 \cdot 2 + 2 \cdot 0 & 0 \cdot 1 + 2 \cdot 1 & 0 \cdot (-3) + 2 \cdot 2 \\ 1 \cdot 2 + 0 \cdot 0 & 1 \cdot 1 + 0 \cdot 1 & 1 \cdot (-3) + 0 \cdot 2 \end{pmatrix}$$

$$= \begin{pmatrix} 2 & -3 & -11 \\ 0 & 2 & 4 \\ 2 & 1 & -3 \end{pmatrix}.$$

■

TECHNOLOGY NOTE

Most graphers can perform matrix multiplication, as well as other matrix operations. Check your owner's manual to find your grapher's matrix capabilities.

Notice that, not only do AB and BA have different entries, they are not even the same size!

Example 3 demonstrates that a system of linear equations can be written as a matrix equation.

Teaching Note

The use of technology in this section is very appropriate. The emphasis is on concepts of matrix operations and the meanings of different terms related to matrix theory. While the concepts of the inverse of a matrix and matrix multiplication are illustrated using paper and pencil methods for small matrices, encourage students to use the grapher to apply these concepts using technology.

EXAMPLE 3 Writing a System as a Matrix Equation

Express the system of equations as a matrix equation.

$$3x - 2y = 0$$
$$-x + y = 5$$

Solution

Let

$$A = \begin{pmatrix} 3 & -2 \\ -1 & 1 \end{pmatrix}, \quad X = \begin{pmatrix} x \\ y \end{pmatrix}, \quad \text{and} \quad C = \begin{pmatrix} 0 \\ 5 \end{pmatrix}.$$

Then

$$AX = \begin{pmatrix} 3 & -2 \\ -1 & 1 \end{pmatrix} \cdot \begin{pmatrix} x \\ y \end{pmatrix} = \begin{pmatrix} 3x - 2y \\ -x + y \end{pmatrix}$$

so that

$$AX = C.$$

■

Identity and Inverse Matrices

Recall that an $n \times n$ matrix, with n rows and n columns, is a square matrix. The square matrix I_n with 1's on the diagonal from the top left to the bottom right (called the *main diagonal*) and 0's everywhere else is the **$n \times n$ identity matrix.** For example,

$$I_2 = \begin{pmatrix} 1 & 0 \\ 0 & 1 \end{pmatrix}, \quad I_3 = \begin{pmatrix} 1 & 0 & 0 \\ 0 & 1 & 0 \\ 0 & 0 & 1 \end{pmatrix}, \quad \text{and} \quad I_4 = \begin{pmatrix} 1 & 0 & 0 & 0 \\ 0 & 1 & 0 & 0 \\ 0 & 0 & 1 & 0 \\ 0 & 0 & 0 & 1 \end{pmatrix}.$$

In Exercise 42 you will show that, if A is any $n \times n$ matrix, then

$$AI_n = I_nA = A.$$

If $3a = 1$ for the real number a, then we know that $a = 1/3$ is the *multiplicative inverse* of 3. Every nonzero real number has a multiplicative inverse. Some square matrices have multiplicative inverses. The $n \times n$ matrix B is the **multiplicative inverse** of the $n \times n$ matrix A if and only if

$$AB = BA = I_n.$$

REMARK
If $AB = BA = I_n$, you can show that both A and B are $n \times n$ matrices. So, only square matrices can have inverses.

If it is clear that the operation is multiplication, then we simply say that B is the *inverse* of A. We write $B = A^{-1}$ (read "A inverse"), and we say that A and B are inverses of each other.

Alert
Some calculators use the x^{-1} key as a reciprocal key in some applications and as an inverse key in others. Students should understand that this key may be used to denote the inverse of a matrix in entering the equation $X = A^{-1}B$. If a matrix does not have an inverse, an error message should occur.

■ EXAMPLE 4 Confirming an Inverse Matrix

Show that

$$A = \begin{pmatrix} 3 & -2 \\ -1 & 1 \end{pmatrix} \quad \text{and} \quad B = \begin{pmatrix} 1 & 2 \\ 1 & 3 \end{pmatrix}$$

are inverse matrices.

Solution

$$AB = \begin{pmatrix} 3 & -2 \\ -1 & 1 \end{pmatrix} \cdot \begin{pmatrix} 1 & 2 \\ 1 & 3 \end{pmatrix} = \begin{pmatrix} (3)(1) + (-2)(1) & (3)(2) + (-2)(3) \\ (-1)(1) + (1)(1) & (-1)(2) + (1)(3) \end{pmatrix}$$

$$= \begin{pmatrix} 1 & 0 \\ 0 & 1 \end{pmatrix}$$

$$BA = \begin{pmatrix} 1 & 2 \\ 1 & 3 \end{pmatrix} \cdot \begin{pmatrix} 3 & -2 \\ -1 & 1 \end{pmatrix} = \begin{pmatrix} (1)(3) + (2)(-1) & (1)(-2) + (2)(1) \\ (1)(3) + (3)(-1) & (1)(-2) + (3)(1) \end{pmatrix}$$

$$= \begin{pmatrix} 1 & 0 \\ 0 & 1 \end{pmatrix}$$

Thus $AB = BA = I_2$. ■

We are interested in inverses of matrices because systems of linear equations that have exactly one solution can be solved by using inverses and matrix multiplication. We illustrate with the system of equations in Example 3.

$$3x - 2y = 0$$
$$-x + y = 5$$

In Example 4, we showed that the coefficient matrix

$$A = \begin{pmatrix} 3 & -2 \\ -1 & 1 \end{pmatrix} \quad \text{has inverse} \quad A^{-1} = B = \begin{pmatrix} 1 & 2 \\ 1 & 3 \end{pmatrix}.$$

Watch what happens when we use this information.

$$3x - 2y = 0$$
$$-x + y = 5$$
$$\begin{pmatrix} 3 & -2 \\ -1 & 1 \end{pmatrix} \cdot \begin{pmatrix} x \\ y \end{pmatrix} = \begin{pmatrix} 0 \\ 5 \end{pmatrix} \qquad \text{Corresponding matrix equation}$$
$$AX = C$$
$$A^{-1}AX = C \qquad \text{Multiply by } A^{-1}.$$
$$I_2X = A^{-1}C \qquad A^{-1}A = I_2.$$
$$X = A^{-1}C$$
$$\begin{pmatrix} x \\ y \end{pmatrix} = \begin{pmatrix} 1 & 2 \\ 1 & 3 \end{pmatrix} \cdot \begin{pmatrix} 0 \\ 5 \end{pmatrix} = \begin{pmatrix} 10 \\ 15 \end{pmatrix}$$

From this matrix equation we read (10, 15) as the solution of the system of equations. (You should verify this.)

The success of this approach depends on our ability to find inverses of matrices. Not all square matrices have inverses (see Exercise 43). A matrix that has an inverse is said to be an **invertible matrix.** There is a simple test that decides if a 2×2 matrix has an inverse.

Teaching Note

Note that A^{-1} is multiplied on the left side of each expression. This is important because matrix multiplication is not commutative.

Inverses of 2×2 Matrices

If $ad - bc \neq 0$, then

$$\begin{pmatrix} a & b \\ c & d \end{pmatrix}^{-1} = \frac{1}{ad - bc} \begin{pmatrix} d & -b \\ -c & a \end{pmatrix}.$$

NOTE

Check that this formula gives the inverse of the matrices in Example 4.

The number $ad - bc$ is the **determinant** of the 2×2 matrix $\begin{pmatrix} a & b \\ c & d \end{pmatrix}$ and is denoted

$$\det A = \begin{vmatrix} a & b \\ c & d \end{vmatrix} = ad - bc.$$

The process we used to solve the system of linear equations $AX = C$, where A is 2×2, generalizes to an $n \times n$ system. Square matrices of higher order also have determinants. However, finding A^{-1} and $\det A$ for $n > 2$ can involve pages of calculations when done by paper and pencil. We expect you to use your grapher or other technology to find inverses and determinants in these cases. Of course, you can also use technology when $n = 2$.

We can now state the condition under which square matrices have inverses.

Theorem Inverses of $n \times n$ Matrices

An $n \times n$ matrix A has an inverse if and only if

$$\det A \neq 0.$$

■ EXAMPLE 5 Determining Inverse Matrices

Determine whether the matrix is invertible. If so, find its inverse matrix.

a. $A = \begin{pmatrix} 3 & 1 \\ 4 & 2 \end{pmatrix}$

b. $B = \begin{pmatrix} 1 & 2 & -1 \\ 2 & -1 & 3 \\ -1 & 0 & 1 \end{pmatrix}$

Solution

a. Because $\det A = ad - bc = 3 \cdot 2 - 1 \cdot 4 = 2 \neq 0$, we conclude that A has an inverse. Using the formula for the inverse of a 2×2 matrix, we have

$$A^{-1} = \frac{1}{2}\begin{pmatrix} 2 & -1 \\ -4 & 3 \end{pmatrix} = \begin{pmatrix} 1 & -0.5 \\ -2 & 1.5 \end{pmatrix}.$$

Support Numerically

Show that $AA^{-1} = A^{-1}A = I_2$.

b. Using our grapher, we find that $\det B = -10 \neq 0$. So B has an inverse.

Again, using our grapher, we find that

$$B^{-1} = \begin{pmatrix} 0.1 & 0.2 & -0.5 \\ 0.5 & 0 & 0.5 \\ 0.1 & 0.2 & 0.5 \end{pmatrix}.$$

Support Numerically

Use your grapher to show that $BB^{-1} = B^{-1}B = I_3$. ■

Systems of Linear Equations

We explain how to use matrix multiplication to solve systems of n linear equations with n variables that have exactly one solution. We illustrate the process with $n = 3$. Consider the following system of three linear equations with variables x_1, x_2, and x_3.

$$\begin{aligned} a_{11}x_1 + a_{12}x_2 + a_{13}x_3 &= b_1 \\ a_{21}x_1 + a_{22}x_2 + a_{23}x_3 &= b_2 \\ a_{31}x_1 + a_{32}x_2 + a_{33}x_3 &= b_3 \end{aligned}$$

We write this system as a *matrix equation.* Recall that two matrices of the same order are equal if their corresponding entries are equal. Thus the following two matrices with three rows and one column are equal.

$$\begin{pmatrix} a_{11}x_1 + a_{12}x_2 + a_{13}x_3 \\ a_{21}x_1 + a_{22}x_2 + a_{23}x_3 \\ a_{31}x_1 + a_{32}x_2 + a_{33}x_3 \end{pmatrix} = \begin{pmatrix} b_1 \\ b_2 \\ b_3 \end{pmatrix}$$

We can write the matrix on the left as a product of the coefficient matrix A and a column matrix X of the variables.

$$\begin{pmatrix} a_{11} & a_{12} & a_{13} \\ a_{21} & a_{22} & a_{23} \\ a_{31} & a_{32} & a_{33} \end{pmatrix} \begin{pmatrix} x_1 \\ x_2 \\ x_3 \end{pmatrix} = \begin{pmatrix} b_1 \\ b_2 \\ b_3 \end{pmatrix}$$

If B represents the matrix on the right side of the equation, we obtain

$$AX = B.$$

If the coefficient matrix A is invertible, then we solve this system of equations in the following way.

$$\begin{aligned} AX &= B && \text{Matrix equation to be solved} \\ A^{-1}AX &= A^{-1}B && \text{Multiply by } A^{-1}. \\ I_3X &= A^{-1}B && A^{-1}A = I_3. \\ X &= A^{-1}B \end{aligned}$$

We can use this last matrix equation to read the values of the variables. This procedure works for general n.

Teaching Note

Note that the system $AX = B$ has exactly one solution if and only if $\det A \neq 0$.

Theorem Square Systems with Exactly One Solution

If A is an invertible square matrix, then the system of equations given by $AX = B$ has the solution

$$X = A^{-1}B.$$

EXAMPLE 6 Using Inverses to Solve Systems

Solve the system.

$$\begin{aligned} 3x - 3y + 6z &= 20 \\ x - 3y + 10z &= 40 \\ -x + 3y - 5z &= 30 \end{aligned}$$

Solution

Solve with a Grapher

Let

$$A = \begin{pmatrix} 3 & -3 & 6 \\ 1 & -3 & 10 \\ -1 & 3 & -5 \end{pmatrix}, \quad X = \begin{pmatrix} x \\ y \\ z \end{pmatrix}, \quad \text{and} \quad B = \begin{pmatrix} 20 \\ 40 \\ 30 \end{pmatrix}.$$

The system of equations can be written as

$$A \cdot X = B.$$

Because $\det A = -30$, A is invertible and

$$A^{-1} = \begin{pmatrix} 0.5 & -0.1 & 0.4 \\ 0.166\cdots & 0.3 & 0.8 \\ 0 & 0.2 & 0.2 \end{pmatrix}.$$

Thus the solution of the system of equations is given by

$$X = A^{-1}B = \begin{pmatrix} 18 \\ 39.333\cdots \\ 14 \end{pmatrix}.$$

Interpret

We read $x = 18$, $y = 39.333\cdots = 39\frac{1}{3}$, and $z = 14$.

Follow-up

Ask . . .

If the system of equations given by $AX = B$ has no solution or infinitely many solutions, what will happen if one tries to solve it using $X = A^{-1}B$? **(A^{-1} does not exist.)**

Assignment Guide

Day 1: Ex. 3–30, multiples of 3
Day 2: Ex. 31, 34, 35, 38–44

Cooperative Learning

Group Activity: Ex. 45–46

Notes on Exercises

Ex. 35–38 require students to find the coefficients of a function that contains given points.
Ex. 41 requires students to find a regression equation.
Ex. 45–46 introduce characteristic polynomials and eigenvalues.

Ongoing Assessment

Self-Assessment: Ex. 11, 19, 29, 33, 37
Embedded Assessment: Ex. 42, 44

```
18→X:(39+1/3)→Y:
14→Z:3X−3Y+6Z
                20
X−3Y+10Z
                40
-X+3Y−5Z
                30
```

Figure 10.9

Support Numerically

Figure 10.9 shows that the values obtained for x, y, and z satisfy the original system of equations. ■

Quick Review 10.4

In Exercises 1–8, solve each equation algebraically. Check your answer.

1. $6x = 1$
2. $13x = 1$
3. $-11x = 1$
4. $-5x = 1$
5. $14x = 14$
6. $7x = 7$
7. $-9x = 9$
8. $-21x = 21$

In Exercises 9 and 10, answer the questions.

9. If $ax = 1$, what can you conclude about the real number a?
10. If $ax = a$, what can you conclude about the real number a?

In Exercises 11–13, let I be the function defined on the real numbers by $I(x) = x$ for all real numbers x.

11. If f is a function with domain the real numbers, show that $f \circ I = I \circ f = f$.
12. Let
$$f(x) = \frac{2x - 1}{x + 2} \quad \text{and} \quad g(x) = -\frac{2x + 1}{x - 2}.$$
Show that $f \circ g = g \circ f = I$.
13. Let
$$f(x) = \frac{1}{x - 3} \quad \text{and} \quad g(x) = 3 + \frac{1}{x}.$$
Show that $f \circ g = g \circ f = I$.

SECTION EXERCISES 10.4

In Exercises 1–6, find (a) AB and (b) BA.

1. $A = \begin{pmatrix} 2 & 3 \\ -1 & 5 \end{pmatrix}$; $B = \begin{pmatrix} 1 & -3 \\ -2 & -4 \end{pmatrix}$
2. $A = \begin{pmatrix} 1 & -4 \\ 2 & 6 \end{pmatrix}$; $B = \begin{pmatrix} 5 & 1 \\ -2 & -3 \end{pmatrix}$
3. $A = \begin{pmatrix} 2 & 0 & 1 \\ 1 & 4 & -3 \end{pmatrix}$; $B = \begin{pmatrix} 1 & 2 \\ -3 & 1 \\ 0 & -2 \end{pmatrix}$
4. $A = \begin{pmatrix} 1 & 0 & -2 & 3 \\ 2 & 1 & 4 & -1 \end{pmatrix}$; $B = \begin{pmatrix} 5 & -1 \\ 0 & 2 \\ -1 & 3 \\ 4 & 2 \end{pmatrix}$
5. $A = \begin{pmatrix} -1 & 0 & 2 \\ 4 & 1 & -1 \\ 2 & 0 & 1 \end{pmatrix}$; $B = \begin{pmatrix} 2 & 1 & 0 \\ -1 & 0 & 2 \\ 4 & -3 & -1 \end{pmatrix}$
6. $A = \begin{pmatrix} -2 & 3 & 0 \\ 1 & -2 & 4 \\ 3 & 2 & 1 \end{pmatrix}$; $B = \begin{pmatrix} 4 & -1 & 2 \\ 0 & 2 & 3 \\ -1 & 3 & -1 \end{pmatrix}$

In Exercises 7–12, find (a) AB and (b) BA, or state that a given product is not possible.

7. $A = (2 \quad -1 \quad 3)$; $B = \begin{pmatrix} -5 \\ 4 \\ 2 \end{pmatrix}$
8. $A = \begin{pmatrix} -2 \\ 3 \\ -4 \end{pmatrix}$; $B = (-1 \quad 2 \quad 4)$
9. $A = \begin{pmatrix} -1 & 2 \\ 3 & 4 \end{pmatrix}$; $B = (-3 \quad 5)$

10. $A = \begin{pmatrix} -1 & 3 \\ 0 & 1 \\ 1 & 0 \\ -3 & -1 \end{pmatrix}$; $B = \begin{pmatrix} 5 & -6 \\ 2 & 3 \end{pmatrix}$

11. $A = \begin{pmatrix} 0 & 0 & 1 \\ 0 & 1 & 0 \\ 1 & 0 & 0 \end{pmatrix}$; $B = \begin{pmatrix} 1 & 2 & 1 \\ 2 & 0 & 1 \\ -1 & 3 & 4 \end{pmatrix}$

12. $A = \begin{pmatrix} 0 & 0 & 1 & 0 \\ 0 & 1 & 0 & 0 \\ 1 & 0 & 0 & 0 \\ 0 & 0 & 0 & 1 \end{pmatrix}$; $B = \begin{pmatrix} -1 & 2 & 3 & -4 \\ 2 & 1 & 0 & -1 \\ -3 & 2 & 1 & 3 \\ 4 & 0 & 2 & -1 \end{pmatrix}$

In Exercises 13–16, solve for a and b.

13. $\begin{pmatrix} a & -3 \\ 4 & 2 \end{pmatrix} = \begin{pmatrix} 5 & -3 \\ 4 & b \end{pmatrix}$

14. $\begin{pmatrix} 1 & -1 & 0 \\ a & -2 & 1 \end{pmatrix} = \begin{pmatrix} 1 & b & 0 \\ 3 & -2 & 1 \end{pmatrix}$

15. $\begin{pmatrix} 2 & a-1 \\ 2 & 3 \\ -1 & 2 \end{pmatrix} = \begin{pmatrix} 2 & -3 \\ b+2 & 3 \\ -1 & 2 \end{pmatrix}$

16. $\begin{pmatrix} a+3 & 2 \\ 0 & 5 \end{pmatrix} = \begin{pmatrix} 4 & 2 \\ 0 & b-1 \end{pmatrix}$

In Exercises 17 and 18, compute AB and BA to verify that the matrices are inverses.

17. $A = \begin{pmatrix} 2 & 1 \\ 3 & 4 \end{pmatrix}$, $B = \begin{pmatrix} 0.8 & -0.2 \\ -0.6 & 0.4 \end{pmatrix}$

18. $A = \begin{pmatrix} -2 & 1 & 3 \\ 1 & 2 & -2 \\ 0 & 1 & -1 \end{pmatrix}$, $B = \begin{pmatrix} 0 & 1 & -2 \\ 0.25 & 0.5 & -0.25 \\ 0.25 & 0.5 & -1.25 \end{pmatrix}$

In Exercises 19–22, find the inverse of the matrix if it has one. If it does, use multiplication to support your result.

19. $\begin{pmatrix} 2 & 3 \\ 2 & 2 \end{pmatrix}$

20. $\begin{pmatrix} 6 & 3 \\ 10 & 5 \end{pmatrix}$

21. $\begin{pmatrix} 1 & 2 & -1 \\ 2 & -1 & 3 \\ 3 & 1 & 2 \end{pmatrix}$

22. $\begin{pmatrix} 2 & 3 & -1 \\ -1 & 0 & 4 \\ 0 & 1 & 1 \end{pmatrix}$

In Exercises 23–26, write each system of equations as a matrix equation $AX = B$, with A as the coefficient matrix.

23. $2x + 5y = -3$
$x - 2y = 1$

24. $x - 2y = 1$
$2x - 5y = 3$

25. $5x - 7y + z = 2$
$2x - 3y - z = 3$
$x + y + z = -3$

26. $2x + 3y - z = 2$
$2x - 3y + 2z = -1$
$-x - y + 3z = -4$

In Exercises 27–30, write each matrix equation as a system of equations.

27. $\begin{pmatrix} 3 & -1 \\ 2 & 4 \end{pmatrix}\begin{pmatrix} x \\ y \end{pmatrix} = \begin{pmatrix} -1 \\ 3 \end{pmatrix}$

28. $\begin{pmatrix} 2 & 4 \\ -1 & -2 \end{pmatrix}\begin{pmatrix} x \\ y \end{pmatrix} = \begin{pmatrix} 5 \\ -2 \end{pmatrix}$

29. $\begin{pmatrix} 1 & 0 & -3 \\ 2 & -1 & 3 \\ -2 & 3 & -4 \end{pmatrix}\begin{pmatrix} x \\ y \\ z \end{pmatrix} = \begin{pmatrix} 3 \\ -1 \\ 2 \end{pmatrix}$

30. $\begin{pmatrix} 1 & -1 & 0 \\ 2 & 1 & -3 \\ -1 & 1 & 2 \end{pmatrix}\begin{pmatrix} x \\ y \\ z \end{pmatrix} = \begin{pmatrix} 3 \\ -1 \\ 4 \end{pmatrix}$

In Exercises 31–34, solve each system by using inverse matrices.

31. $2x - 3y = -13$
$4x + y = -5$

32. $x + 2y = -2$
$3x - 4y = 9$

33. $3x - 2y = 6$
$x + y = 2$

34. $8x - 5y = 16$
$2x + 3y = -13$

In Exercises 35–38, use inverse matrices to determine the coefficients of the function so that its graph contains the given points.

35. *Curve Fitting* $f(x) = ax^2 + bx + c$
(2, 3), (5, 8), (7, 2)

36. *Curve Fitting* $f(x) = ax^2 + bx + c$
(2, 8), (6, 3), (9, 4)

37. *Curve Fitting* $f(x) = ax^3 + bx^2 + cx + d$
(2, 3), (5, 8), (7, 2), (9, 4)

38. *Curve Fitting* $f(x) = ax^4 + bx^3 + cx^2 + dx + e$
(−1, 8), (1, 2), (4, −6), (7, 5), (8, 2)

In Exercises 39 and 40, use inverse matrices to find the equilibrium point for the demand and supply curves.

39. $p = 100 - 5x$ Demand curve
$p = 20 + 10x$ Supply curve

40. $p = 150 - 12x$ Demand curve
$p = 30 + 24x$ Supply curve

In Exercises 41–44, solve the problem.

41. *Personal Income Data* Table 10.4 gives the disposable personal income per capita in constant 1987 dollars for North Dakota and Texas. Use $x = 0$ for 1980.

   a. Find the linear regression equation for the North Dakota data and superimpose its graph on a scatter plot of the data.

   b. Find the linear regression equation for the Texas data and superimpose its graph on a scatter plot of the data.

Table 10.4 Disposable Personal Income

Year	North Dakota per Capita (dollars)	Texas per Capita (dollars)
1988	10,390	12,617
1989	11,158	12,731
1990	12,043	12,868
1991	11,696	12,939
1992	12,432	13,291

Source: U.S. Bureau of Economic Analysis, Survey of Current Business, 1993.

c. Estimate when the per capita disposable income for North Dakota will exceed that of Texas.

42. Let

$$A = \begin{pmatrix} a_{11} & a_{12} & \cdots & a_{1n} \\ a_{21} & a_{22} & \cdots & a_{2n} \\ \vdots & \vdots & & \vdots \\ a_{n1} & a_{n2} & \cdots & a_{nn} \end{pmatrix} \quad \text{and}$$

$$I_n = \begin{pmatrix} 1 & 0 & \cdots & 0 \\ 0 & 1 & \cdots & 0 \\ \vdots & \vdots & & \vdots \\ 0 & 0 & \cdots & 1 \end{pmatrix}$$

a. Show that $AI_n = I_nA = A$ for $n = 2$, 3, and 4.

b. Writing to Learn Give a convincing argument that $AI_n = I_nA = A$ for any integer $n \geq 2$.

43. Show that $A = \begin{pmatrix} 1 & 2 \\ 2 & 4 \end{pmatrix}$ has no inverse by demonstrating that the following matrix equation has no solution. That is, there are no real numbers x, y, z, and w that make the matrix equation true.

$$\begin{pmatrix} 1 & 2 \\ 2 & 4 \end{pmatrix} \begin{pmatrix} x & y \\ z & w \end{pmatrix} = \begin{pmatrix} 1 & 0 \\ 0 & 1 \end{pmatrix}$$

44. Writing to Learn If the product AB is defined for the $n \times n$ matrix A, what can you conclude about the order of the matrix B? Explain.

EXTENDING THE IDEAS

The polynomial

$$C(x) = \begin{vmatrix} a - x & b \\ c & d - x \end{vmatrix} = (a - x)(d - x) - bc$$

is the *characteristic polynomial* of the matrix

$$A = \begin{pmatrix} a & b \\ c & d \end{pmatrix}.$$

The roots of the characteristic polynomial $C(x)$ are the *eigenvalues* of A.

45. Let $A = \begin{pmatrix} 3 & 2 \\ 1 & 5 \end{pmatrix}$.

a. Find the characteristic polynomial $C(x)$ of A.
b. Find the graph of $y = C(x)$.
c. Find the eigenvalues of A.
d. Compare det A with the y-intercept of the graph of $y = C(x)$.
e. Compare the sum of the main diagonal elements of A $(a_{11} + a_{22})$ with the sum of the eigenvalues.

46. Let $A = \begin{pmatrix} 2 & -1 \\ -5 & 2 \end{pmatrix}$.

a. Find the characteristic polynomial $C(x)$ of A.
b. Find the graph of $y = C(x)$.
c. Find the eigenvalues of A.
d. Compare det A with the y-intercept of the graph of $y = C(x)$.
e. Compare the sum of the main diagonal elements of A $(a_{11} + a_{22})$ with the sum of the eigenvalues.

Objective

Students will be able to solve linear programming problems and systems of inequalities using graphical methods.

Motivate

Ask . . .

If the graph of $y = x^2$ is a parabola, how can one describe the graph of $y \leq x^2$? **(It is the parabola and the region below the parabola.)**

10.5 SOLVING SYSTEMS OF INEQUALITIES

Graphs of Inequalities • Systems of Inequalities • Applications

Graphs of Inequalities

An ordered pair (a, b) of real numbers is a **solution of an inequality** in x and y if the substitution $x = a$ and $y = b$ satisfies the inequality. For example, the ordered pair (2, 5) is a solution of $y < 2x + 3$ because $5 < 2(2) + 3 = 7$. However, the ordered pair (2, 8) is not a solution because $8 \not< 2(2) + 3 = 7$.

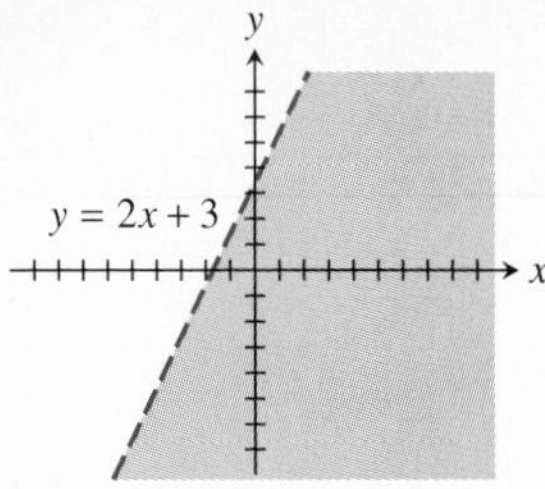

Figure 10.10 A graph of $y = 2x + 3$ (dashed line) and $y < 2x + 3$ (shaded area). The line is dashed to indicate it is not part of the solution of $y < 2x + 3$.

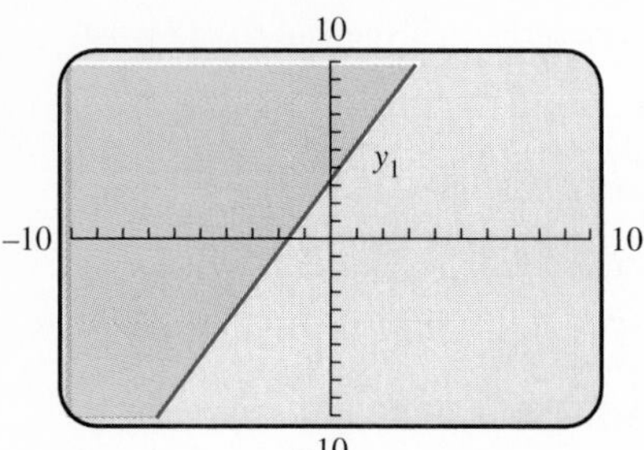

Figure 10.11

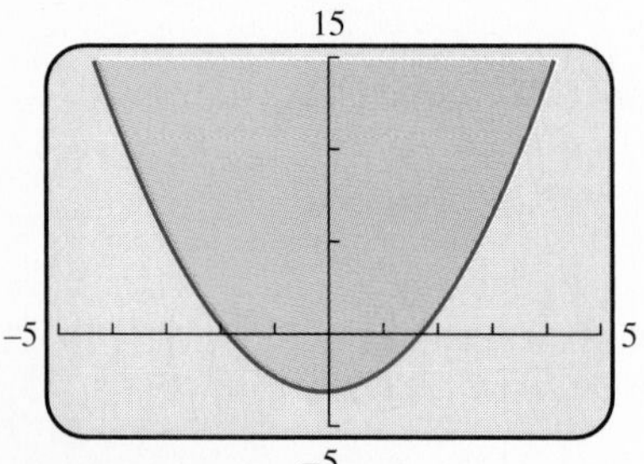

Figure 10.12

Alert

Two common mistakes made in graphing an inequality are using a solid boundary instead of a dashed one (or vice versa) and shading on the wrong side. Emphasize the difference between a dashed line and a solid line and suggest that students test points.

The **graph of an inequality** in x and y consists of all pairs (x, y) that are solutions of the inequality. The graph of an inequality involving two variables typically is a region of the coordinate plane.

The point (2, 7) is on the graph of the line $y = 2x + 3$ but is not a solution of $y < 2x + 3$. A point $(2, y)$ below the line $y = 2x + 3$ is on the graph of $y < 2x + 3$, and those above it are not. The graph of $y < 2x + 3$ is the set of all points below the line $y = 2x + 3$ (Figure 10.10).

■ EXAMPLE 1 Graphing a Linear Inequality

Draw the graph of $y \geq 2x + 3$.

Solution

Because of "$\geq$," the graph of the line $y = 2x + 3$ is part of the graph of the inequality. The points above the line $y = 2x + 3$ are also on the graph of the inequality. Thus the graph of $y \geq 2x + 3$ consists of all points on or above the line $y = 2x + 3$. (See Figure 10.11.) ■

■ EXAMPLE 2 Graphing a Quadratic Inequality

Graph $y \geq x^2 - 3$.

Solution

The graph is shown in Figure 10.12. It includes the points on the parabola $y = x^2 - 3$ and the points above the parabola because, for those points, $y > x^2 - 3$. ■

The graph of $y = x^2 - 3$ is the **boundary** of the region graphed in Figure 10.12.

Systems of Inequalities

The technique for solving a system of inequalities graphically is similar to that for solving a system of equations graphically. First, we graph each individual inequality. Then we determine the points common to the individual graphs.

■ EXAMPLE 3 Graphing a System of Inequalities

Solve the system.

$$y > x^2$$
$$2x + 3y < 4$$

Solution

The graph of $y > x^2$ is shaded in Figure 10.13a. It does not include its boundary $y = x^2$. The graph of $2x + 3y < 4$ is shaded in Figure 10.13b. It

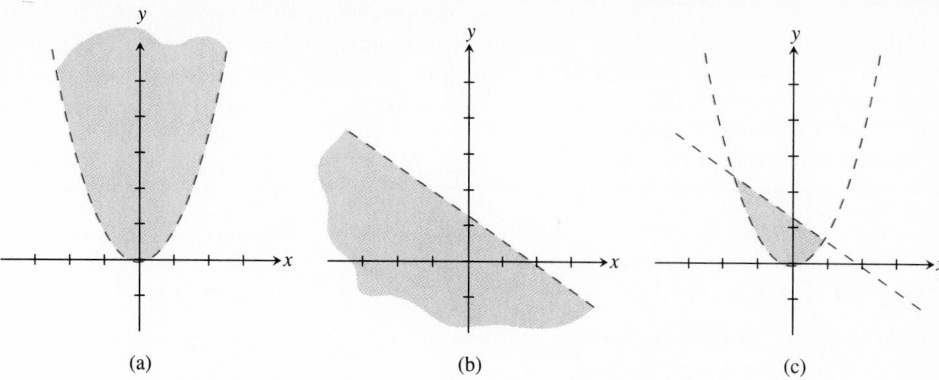

Figure 10.13

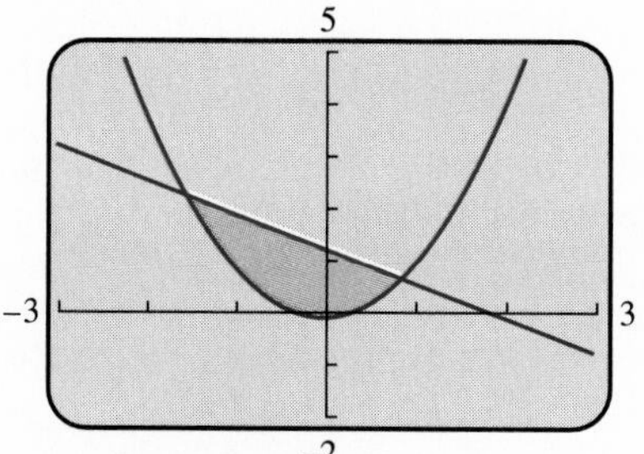

Figure 10.14

does not include its boundary $2x + 3y = 4$. The solution to the system is the intersection of these two graphs, as shaded in Figure 10.13c.

Support Graphically

The graph of the system in Figure 10.14 appears identical to the graph in Figure 10.13c. Note that most graphers cannot distinguish between dashed and solid boundaries. ■

TECHNOLOGY NOTE

Most graphing utilities are capable of shading solutions to inequalities. Check the owner's manual for your grapher.

■ EXAMPLE 4 Graphing a System of Inequalities

Solve the system.

$$\begin{aligned} 2x + y &\le 10 \\ 2x + 3y &\le 14 \\ x &\ge 0 \\ y &\ge 0 \end{aligned}$$

Solution

The solution includes points in the first quadrant that are below the lines $2x + y = 10$ and $2x + 3y = 14$. It also includes all of its boundary points. (See Figure 10.15.) ■

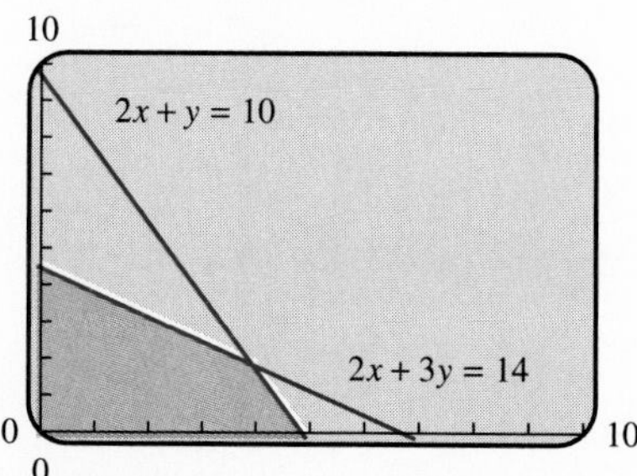

Figure 10.15

Applications

Sometimes a problem situation is modeled by an equation in the form $z = f(x, y)$ whose graph in 3-dimensions lies above a region in the xy-plane such as the one in Figure 10.15 (Example 4). If f is a degree one polynomial in x and y, then the graph of z above the xy-region is a polygonal region that lies at a tilt. A little reflection on the model should convince you that the maximum and minimum values of z will occur above the boundaries of the xy-region, and in particular at its corner points.

NOTE
The first serious use of linear programming was by the United States Air Force. Today these techniques are widely used in business and industry.

Alert

It is particularly easy to make a mistake in shading the feasible region of a linear programming problem. Even when using a grapher, a mistake can be made in solving the inequalities for y. Emphasize the importance of testing a point from the interior of the feasible region in all of the *original* inequalities to be sure the correct region has been shaded.

Follow-up

Ask students to give two additional simple inequalities that could have been used in defining the feasible region of Example 6, but that turned out not to be necessary. ($\mathbf{x \geq 0, y \geq 0}$)

Assignment Guide

Day 1: Ex. 1–7, 9, 12, 14, 16, 17, 20, 21
Day 2: Ex. 23, 26, 27, 30, 31, 32, 34, 35, 36, 37

Cooperative Learning

Group Activity: Ex. 37–40

Notes on Exercises

Ex. 17–26 and 37–40 require students to solve systems of inequalities.
Ex. 31–34 are linear programming problems.

Ongoing Assessment

Self-Assessment: Ex. 11, 19, 25, 29, 33
Embedded Assessment: Ex. 36

The problem then of finding which of the **feasible** $\boldsymbol{xy}$ points gives maximum or minimum values of z is known as a problem in **linear programming,** not to be confused with computer programming. To solve the problem, we simply evaluate z at the corner points of the xy-region.

■ EXAMPLE 5 APPLICATION: Solving a Linear Programming Problem

Find the minimum and maximum values of the expression $z = 5x + 8y$, subject to the following restrictions.

$$2x + y \leq 10$$
$$2x + 3y \leq 14$$
$$x \geq 0$$
$$y \geq 0$$

Solution

Solve Graphically

The feasible xy points are those graphed in Figure 10.15. The following table shows $z = 5x + 8y$ evaluated at the corner points of the xy graph.

(x, y)	$(0, 0)$	$(0, 14/3)$	$(4, 2)$	$(5, 0)$
$z = 5x + 8y$	0	112/3	36	25

Interpret

The maximum value of $z = 5x + 8y$ is 112/3 at $(x, y) = (0, 14/3)$. The minimum value is 0 at $(x, y) = (0, 0)$. ■

■ EXAMPLE 6 APPLICATION: Purchasing Fertilizer

Johnson's Produce is purchasing fertilizer with two nutrients: N (nitrogen) and P (phosphorous). They need at least 180 units of N and 90 units of P. Their supplier has two brands of fertilizer for them to buy. Brand A costs \$10 a bag and has 4 units of N and 1 unit of P. Brand B costs \$5 a bag and has 1 unit of each nutrient. Johnson's can pay at most \$800 for the fertilizer. How many bags of each brand should be purchased to minimize cost?

Solution

Model

Let x = number of bags of Brand A.

Let y = number of bags of Brand B.

Then C = the total cost = $10x + 5y$.

$$4x + y \geq 180 \quad \text{Amount of N is at least 180.}$$
$$x + y \geq 90 \quad \text{Amount of P is at least 90.}$$
$$10x + 5y \leq 800 \quad \text{Total cost is to be at most \$800.}$$

Johnson's Produce

200, 0, 0, 100, $4x + y = 180$, $10x + 5y = 800$, $x + y = 90$

Figure 10.16

Solve Graphically

The region of feasible points is the intersection of the graphs of $4x + y \geq 180$, $x + y \geq 90$, and $10x + 5y \leq 800$ in the first quadrant. (See Figure 10.16.)

The region has three corner points (10, 140), (70, 20), and (30, 60). At the corner points

$$C = 10(10) + 5(140) = 800;$$

$$C = 10(70) + 5(20) = 800;$$

$$C = 10(30) + 5(60) = 600.$$

Interpret

The minimum cost for the fertilizer is \$600 when 30 bags of Brand A and 60 bags of Brand B are purchased. For this purchase, Johnson's Produce gets exactly 180 units of nutrient N and 90 units of nutrient P. ■

Quick Review 10.5

In Exercises 1–4, solve the inequality. Graph the solution on a number line.

1. $5x - 1 < 2x + 3$

2. $2x + 3 \geq 4x - 5$

3. $2(3x - 1) \leq x - 2$

4. $2(x - 6) > 3(x - 4)$

In Exercises 5 and 6, find the x- and y-intercepts of the line. Then sketch a graph of the line.

5. $2x - 3y = 6$

6. $5x + 10y = 30$

In Exercises 7–10, use factoring to solve each equation. Support your results graphically.

7. $x^3 = 4x$

8. $2x^3 = 2x$

9. $x^3 + x^2 - x - 1 = 0$

10. $x^3 + 2x^2 - 4x - 8 = 0$

SECTION EXERCISES 10.5

In Exercises 1–6, match the inequality with its graph. Indicate whether the boundary is included in or excluded from the graph. All graphs are drawn in the decimal window.

1. $x \leq 3$

2. $y > 2$

3. $2x - 5y \geq 2$

4. $y > x^2 - 1$

5. $y \geq 2 - x^2$

6. $x^2 + y^2 < 4$

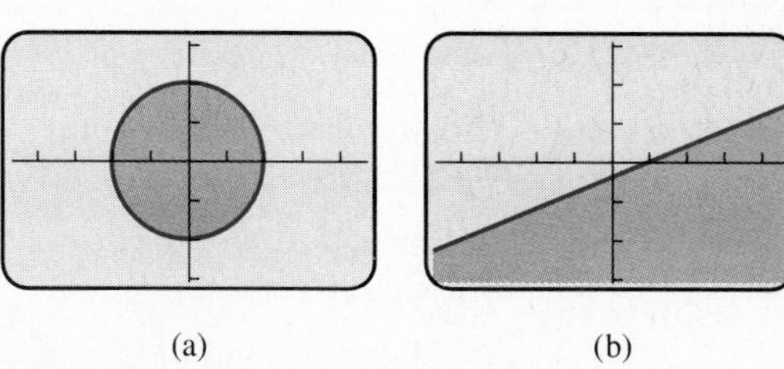

(a) (b)

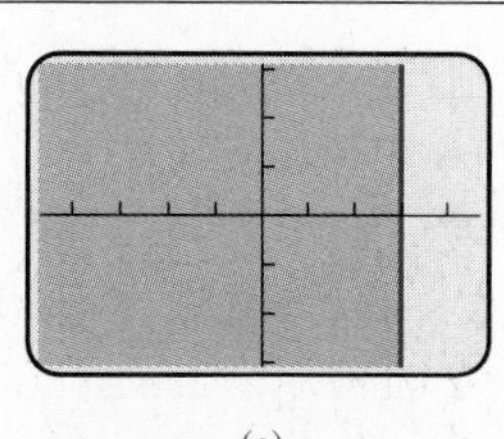

(c)

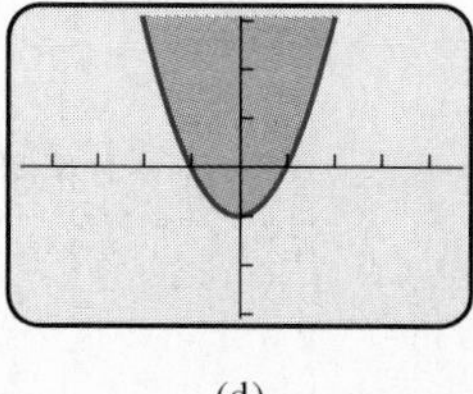

(d)

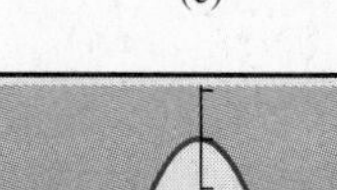

(e)

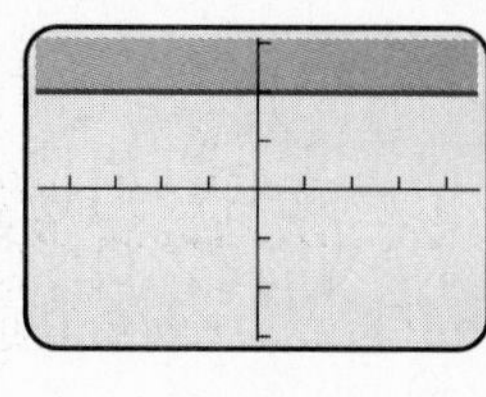

(f)

For Exercises 1–6

In Exercises 7–16, graph the inequality.

7. $x \leq 4$

8. $y \geq -3$

9. $2x + 5y \leq 7$

10. $3x - y > 4$

11. $y < x^2 + 1$

12. $y \geq x^2 - 3$

13. $x^2 + y^2 < 9$

14. $x^2 + y^2 \geq 4$

15. $\dfrac{x^2}{9} + \dfrac{y^2}{4} > 1$

16. $\dfrac{y^2}{4} - \dfrac{x^2}{9} > 1$

In Exercises 17–22, solve the system of inequalities. Give the coordinates of any corner points.

17. $5x - 3y > 1$
$3x + 4y \leq 18$

18. $4x + 3y \leq -6$
$2x - y \leq -8$

19. $y \leq 2x + 3$
$y \geq x^2 - 2$

20. $x - 3y - 6 < 0$
$y > -x^2 - 2x + 2$

21. $y \geq x^2$
$x^2 + y^2 \leq 4$

22. $x^2 + y^2 \leq 9$
$x^2 - y^2 \leq 4$

In Exercises 23–26, solve the system of inequalities. Give the coordinates of any corner points.

23. $2x + y \leq 80$
$x + 2y \leq 80$
$x \geq 0$
$y \geq 0$

24. $3x + 8y \geq 240$
$9x + 4y \geq 360$
$x \geq 0$
$y \geq 0$

25. $5x + 2y \leq 20$
$2x + 3y \leq 18$
$x + y \geq 2$
$x \geq 0$
$y \geq 0$

26. $7x + 3y \leq 210$
$3x + 7y \leq 210$
$x + y \geq 30$

In Exercises 27–30, write a system of inequalities whose solution is the region shaded in the given figure. All boundaries are to be included.

27. decimal window

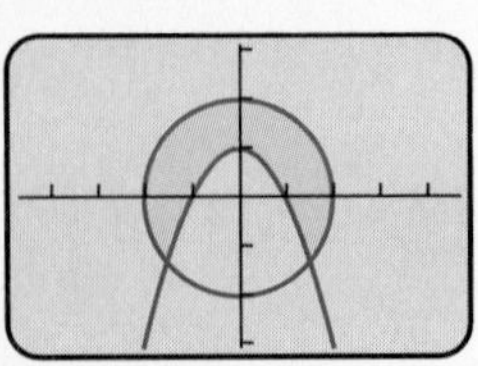

28. decimal window

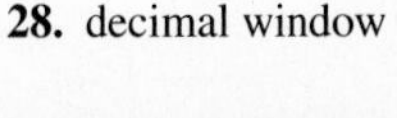

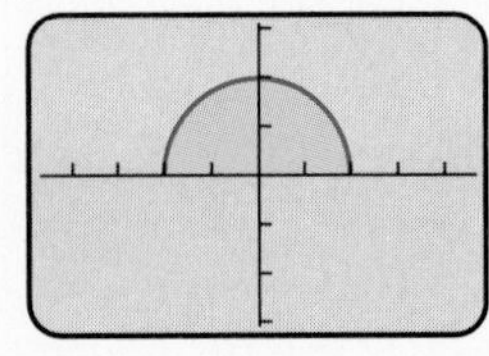

29.

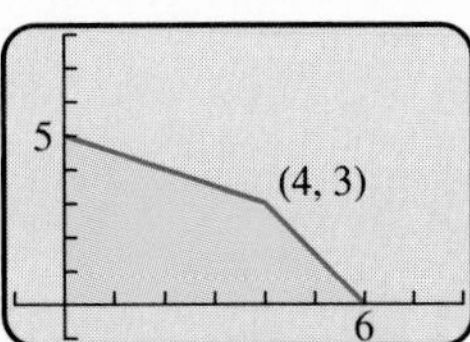

30.

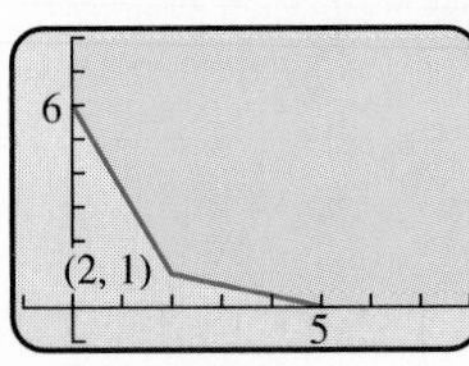

In Exercises 31–34, solve the linear programming problem.

31. Maximize $C = 4x + 3y$, subject to

$$x + y \leq 80$$
$$x - 2y \leq 0$$
$$x \geq 0$$
$$y \geq 0$$

32. Maximize $C = 10x + 11y$, subject to

$$x + y \leq 90$$
$$3x - y \geq 0$$
$$x \geq 0$$
$$y \geq 0$$

33. *Mining Ore* Pearson's Metals mines two ores: R and S. The company extracts minerals A and B from each type of ore. It costs \$50 per ton to extract 80 lb of A and 160 lb of B from ore R. It costs \$60 per ton to extract 140 lb of A and 50 lb of B from ore S. Pearson's must produce at least 4000 lb of A and 3200 lb of B. How much of each ore should be processed to minimize cost? What is the minimum cost?

34. *Planning a Diet* Paul's diet is to contain at least 24 units of carbohydrates and 16 units of protein. Food substance A costs \$1.40 per unit and each unit contains 3 units of carbohydrates and 4 units of protein. Food substance B costs \$0.90 per unit and each unit contains 2 units of carbohydrates and 1 unit of protein. How many units of each food substance should be purchased in order to minimize cost? What is the minimum cost?

In Exercises 35 and 36, answer the questions.

35. Writing to Learn Describe all possible ways that a parabola and an ellipse can intersect. Explain.

36. Writing to Learn Is it possible for an ellipse and a hyperbola to intersect in a single point? Explain.

EXTENDING THE IDEAS

In Exercises 37–40, identify the conics involved and solve the system of inequalities.

37. $$\begin{aligned} x^2 + xy + y^2 + x + y - 6 &\le 0 \\ 2x^2 - 3xy + 2y^2 + x + y - 8 &\ge 0 \end{aligned}$$

38. $$\begin{aligned} 5x^2 - 40xy + 20y^2 - 17x + 25y + 50 &\ge 0 \\ 2x^2 - 3xy + 2y^2 + x + y - 8 &\le 0 \end{aligned}$$

39. $$\begin{aligned} xy - 3 &< 0 \\ 2x^2 - 8xy + 3y^2 + x + y - 10 &< 0 \end{aligned}$$

40. $$\begin{aligned} 3x^2 - 5xy + 6y^2 - 7x + 5y - 9 &> 0 \\ x^2 + y^2 - 2x - 4 &< 0 \end{aligned}$$

CHAPTER 10 REVIEW

KEY TERMS

The number following each key term indicates the page of its introduction.

augmented matrix, 661
boundary of a region, 680
coefficient matrix, 661
demand curve, 645
determinant, 674
elementary row operations, 663
equal matrices, 661
equilibrium point, 645
equilibrium price, 645
feasible points, 682
Gaussian elimination, 651
graph of an inequality, 680
identity matrix, $n \times n$, 672
invertible matrix, 673
linear equation in $x_1, x_2, \ldots, x_n$, 651
linear programming, 682
linear system, 650
matrix, $m \times n$, 661
multiplicative inverse (of a matrix), 672
order of an $m \times n$ matrix, 661
product of matrices A and B, 669
reduced row echelon form, 664
row echelon form, 662, 663
solution of a system, 639
solution of an inequality, 679
solve by elimination, 641
solve by substitution, 641
square matrix, 661
supply curve, 645
system of equations, 639
triangular form of a system, 651

REVIEW EXERCISES

In Exercises 1–4, state whether the system has a solution. If it does, solve the system.

1. $$\begin{aligned} 3x - y &= 1 \\ x + 2y &= 5 \end{aligned}$$

2. $$\begin{aligned} x - 2y &= -1 \\ -2x + y &= 5 \end{aligned}$$

3. $$\begin{aligned} x + 2y &= 1 \\ 4y - 4 &= -2x \end{aligned}$$

4. $$\begin{aligned} x - 2y &= 9 \\ 3y - \frac{3}{2}x &= -9 \end{aligned}$$

In Exercises 5–8, find the reduced row echelon form of the matrix.

5. $$\begin{pmatrix} 1 & 0 & 2 \\ 3 & 1 & 5 \\ 1 & -1 & 3 \end{pmatrix}$$

6. $$\begin{pmatrix} 2 & 1 & 1 & 1 \\ -3 & -1 & -2 & 1 \\ 5 & 2 & 2 & 3 \end{pmatrix}$$

7. $$\begin{pmatrix} 1 & 2 & 3 & 1 \\ 2 & 3 & 3 & -2 \\ 1 & 2 & 4 & 6 \end{pmatrix}$$

8. $$\begin{pmatrix} 1 & -2 & 0 & 4 \\ -2 & 5 & 3 & -6 \\ 2 & -4 & 1 & 9 \end{pmatrix}$$

In Exercises 9–14, use Gaussian elimination to solve the syste...

9. $$\begin{aligned} x + z + w &= 2 \\ x + y + z &= 3 \\ 3x + 2y + 3z + w &= 8 \end{aligned}$$

10. $$\begin{aligned} x + w &= -2 \\ x + y + z + 2w &= -2 \\ -x - 2y - 2z - 3w &= 2 \end{aligned}$$

11. $$\begin{aligned} x + y - 2z &= 2 \\ 3x - y + z &= 4 \\ -2x - 2y + 4z &= 6 \end{aligned}$$

12. $$\begin{aligned} x + y - 2z &= 2 \\ 3x - y + z &= 1 \\ -2x - 2y + 4z &= -4 \end{aligned}$$

13. $$\begin{aligned} -x - 6y + 4z - 5w &= -13 \\ 2x + y + 3z - w &= 4 \\ 2x + 2y + 2z &= 6 \\ -x - 3y + z - 2w &= -7 \end{aligned}$$

14. $$\begin{aligned} -x + 2y + 2z - w &= -4 \\ y + z &= -1 \\ -2x + 2y + 2z - 2w &= -6 \\ -x + 3y + 3z - w &= -5 \end{aligned}$$

In Exercises 15 and 16, find the equilibrium point for the demand and supply curves.

15. $p = 100 - x^2$ Demand curve

$p = 20 + 3x$ Supply curve

16. $p = 80 - \frac{1}{10}x^2$ Demand curve

$p = 5 + 4x$ Supply curve

In Exercises 17 and 18, use the graph to estimate any solutions to the system.

17. $y = x - 1.5$
$y = 0.5x^2 - 3$

18. $y = -0.5x^2 + 3$
$y = 0.5x^2 - 1$

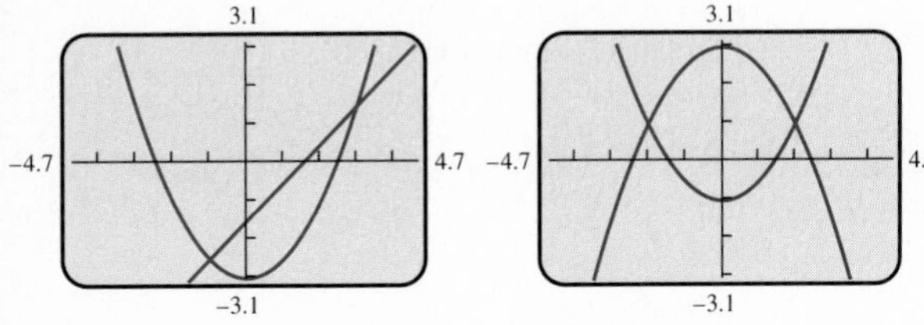

For Exercises 17 and 18

In Exercises 19–24, find the products *AB* and *BA*, or state that a given product is not possible.

19. $A = \begin{pmatrix} -1 & 4 \\ 0 & 6 \end{pmatrix}, \quad B = \begin{pmatrix} 3 & -1 & 5 \\ 0 & -2 & 4 \end{pmatrix}$

20. $A = \begin{pmatrix} -1 & 2 \\ 3 & -1 \\ 4 & 3 \end{pmatrix}, \quad B = \begin{pmatrix} -2 & 3 & 1 \\ 2 & 1 & 0 \\ -1 & 2 & -3 \end{pmatrix}$

21. $A = (-1 \quad 4), \quad B = \begin{pmatrix} 5 & -3 \\ 2 & 1 \end{pmatrix}$

22. $A = \begin{pmatrix} -1 & 1 \\ 0 & 1 \end{pmatrix}, \quad B = \begin{pmatrix} 3 & -4 \\ 1 & 2 \\ 3 & 1 \\ 1 & 1 \end{pmatrix}$

23. $A = \begin{pmatrix} 0 & 1 & 0 \\ 1 & 0 & 0 \\ 0 & 0 & 1 \end{pmatrix}, \quad B = \begin{pmatrix} 2 & -3 & 4 \\ 1 & 2 & -3 \\ -2 & 1 & -1 \end{pmatrix}$

24. $A = \begin{pmatrix} 0 & 1 & 0 & 0 \\ 1 & 0 & 0 & 0 \\ 0 & 0 & 0 & 1 \\ 0 & 0 & 1 & 0 \end{pmatrix}, \quad B = \begin{pmatrix} -2 & 1 & 0 & 1 \\ 3 & 0 & 2 & 1 \\ -1 & 1 & 2 & -1 \\ 3 & -2 & 1 & 0 \end{pmatrix}$

In Exercises 25 and 26, use multiplication to verify that the matrices are inverses.

25. $A = \begin{pmatrix} 1 & -2 & 1 & 1 \\ 1 & -1 & 0 & 3 \\ 1 & -1 & 2 & 2 \\ 2 & -4 & 2 & 3 \end{pmatrix}, \quad B = \begin{pmatrix} 8 & 1.5 & 0.5 & -4.5 \\ 2 & 0.5 & 0.5 & -1.5 \\ -1 & -0.5 & 0.5 & 0.5 \\ -2 & 0 & 0 & 1 \end{pmatrix}$

26. $A = \begin{pmatrix} -1 & 1 & 1 \\ 2 & 1 & 0 \\ -1 & 0 & 2 \end{pmatrix}, \quad B = \begin{pmatrix} -0.4 & 0.4 & 0.2 \\ 0.8 & 0.2 & -0.4 \\ -0.2 & 0.2 & 0.6 \end{pmatrix}$

In Exercises 27 and 28, find the inverse of the matrix if it has one. If it does, use multiplication to support your result.

27. $\begin{pmatrix} 1 & 2 & 0 & -1 \\ 2 & -1 & 1 & 2 \\ 2 & 0 & 1 & 2 \\ -1 & 1 & 1 & 4 \end{pmatrix}$

28. $\begin{pmatrix} -1 & 0 & 1 \\ 2 & -1 & 1 \\ 1 & 1 & 1 \end{pmatrix}$

In Exercises 29–32, solve each system by using inverse matrices.

29. $$\begin{aligned} x + 2y + z &= -1 \\ x - 3y + 2z &= 1 \\ 2x - 3y + z &= 5 \end{aligned}$$

30. $$\begin{aligned} x + 2y - z &= -2 \\ 2x - y + z &= 1 \\ x + y - 2z &= 3 \end{aligned}$$

31. $$\begin{aligned} 2x + y + z - w &= 1 \\ 2x - y + z - w &= -2 \\ -x + y - z + w &= -3 \\ x - 2y + z - w &= 1 \end{aligned}$$

32. $$\begin{aligned} x - 2y + z - w &= 2 \\ 2x + y - z - w &= -1 \\ x - y + 2z - w &= -1 \\ x + 3y - z + w &= 4 \end{aligned}$$

In Exercises 33–40, solve the problem.

33. *Basketball Attendance* At Whetstone High School 452 tickets were sold for the first basketball game There were two ticket prices: $0.75 for students and $2.00 for nonstudents. How many tickets of each type were sold if the total revenue from the sale of tickets was $429?

34. *Investments* Jessica invests \$38,000, part at 7.5% simple interest and the remainder at 6% simple interest. If her annual interest income is \$2600, how much does she have invested at each rate?

35. *Personal Income Data* Table 10.5 gives the disposable personal income per capita in constant 1987 dollars for South Dakota and California. Use $x = 1$ for 1988.

a. Find the linear regression equation for the South Dakota data and superimpose its graph on a scatter plot of the data.

b. Find the linear regression equation for the California data and superimpose its graph on a scatter plot of the data.

c. Estimate when the per capita disposable income for South Dakota will exceed that of California.

Table 10.5 Disposable Personal Income

Year	South Dakota per Capita (dollars)	California per Capita (dollars)
1988	11,191	15,520
1989	11,687	15,342
1990	12,289	15,396
1991	12,351	15,000
1992	12,630	15,015

Source: U.S. Bureau of Economic Analysis, Survey of Current Business, 1993.

36. Writing to Learn Describe all possibilities for the number of solutions of a system of two equations in two variables if the graphs of the two equations are a parabola and a circle.

37. *Truck Deliveries* Brock's Discount TV has three types of television sets on sale: a 13-in. portable, a 27-in. remote, and a 50-in. console. They have three types of vehicles to use for delivery: vans, small trucks, and large trucks. The vans can carry 8 portable, 3 remote, and 2 console TVs; the small trucks, 15 portable, 10 remote, and 6 console TVs; and the large truck, 22 portable, 20 remote, and 5 console TVs. On a given day of the sale they have 115 portable, 85 remote, and 35 console TVs to deliver. How many vehicles of each type are needed to deliver the TVs?

38. *Business Loans* Thompson's Furniture Store borrowed \$650,000 to expand its facilities and extend its product line. Some of the money was borrowed at 4%, some at 6.5%, and the rest at 9%. How much was borrowed at each rate if the annual interest was \$46,250 and the amount borrowed at 9% was twice the amount borrowed at 4%?

39. *Home Remodeling* Sanchez Remodeling has three painters: Sue, Esther, and Murphy. Working together they can paint a large room in 4 hours. Sue and Murphy can paint the same size room in 6 hours. Esther and Murphy can paint the same size room in 7 hours. How long would it take each of them to paint the room alone?

40. *Swimming Pool* Three pipes, A, B, and C, are connected to a swimming pool. When all three pipes are running, the pool can be filled in 3 hr. When only A and B are running, the pool can be filled in 4 hr. When only B and C are running, the pool can be filled in 3.75 hr. How long would it take each pipe running alone to fill the pool?

In Exercises 41–44, solve the system of equations by finding the reduced row echelon form of the augmented matrix.

41.
$$\begin{aligned} x + 2y - 2z + w &= 8 \\ 2x + 3y - 3z + 2w &= 13 \end{aligned}$$

42.
$$\begin{aligned} x + 2y - 2z + w &= 8 \\ 2x + 7y - 7z + 2w &= 25 \\ x + 3y - 3z + w &= 11 \end{aligned}$$

43.
$$\begin{aligned} x + 2y + 4z + 6w &= 6 \\ 3x + 4y + 8z + 11w &= 11 \\ 2x + 4y + 7z + 11w &= 10 \\ 3x + 5y + 10z + 14w &= 15 \end{aligned}$$

44.
$$\begin{aligned} x + 2z - 2w &= 5 \\ 2x + y + 4z - 3w &= 7 \\ 4x + y + 7z - 6w &= 15 \\ 2x + y + 5z - 4w &= 9 \end{aligned}$$

In Exercises 45 and 46, find the coefficients of the function so that its graph goes through the given points.

45. *Curve Fitting* $f(x) = ax^3 + bx^2 + cx + d$

$(2, 8), \quad (4, 5), \quad (6, 3), \quad (9, 4)$

46. *Curve Fitting* $f(x) = ax^4 + bx^3 + cx^2 + dx + e$

$(-2, -4), \quad (1, 2), \quad (3, 6), \quad (4, -2), \quad (7, 8)$

In Exercises 47 and 48, answer the questions.

47. Writing to Learn If the products AB and BA are defined for the $n \times n$ matrix A, what can you conclude about the order of matrix B? Explain.

48. Writing to Learn If A is an $m \times n$ matrix and B is a $p \times q$ matrix, and if AB is defined, what can you conclude about their order? Explain.

In Exercises 49 and 50, solve the system of equations.

49. $x^2 - 4y^2 = 4$
$y = 2x^2 - 3$

50. $9x^2 - 4y^2 = 36$
$9x^2 + 4y^2 = 36$

In Exercises 51 and 52, graph the inequality.

51. $\dfrac{x^2}{25} - \dfrac{y^2}{9} \le 1$

52. $\dfrac{x^2}{4} + \dfrac{y^2}{25} \le 1$

In Exercises 53–58, solve the system of inequalities. Give the coordinates of any corner points.

53. $x - 3y + 6 < 0$
$y > x^2 - 6x + 7$

54. $x + 2y \ge 4$
$y \le 9 - x^2$

55. $x^2 + y^2 \ge 4$
$9x^2 + 4y^2 \le 36$

56. $y \le x^2 + 4$
$x^2 + y^2 \ge 4$

57. $4x + 9y \ge 360$
$9x + 4y \ge 360$
$x + y \le 90$

58. $7x + 10y \le 70$
$2x + y \le 10$
$x + y \ge 3$
$x \ge 0$
$y \ge 0$

In Exercises 59 and 60, solve the linear programming problem.

59. Minimize $C = 7x + 6y$, subject to

$$7x + 5y \ge 100$$
$$2x + 5y \ge 50$$
$$x \ge 0$$
$$y \ge 0$$

60. Minimize $C = 11x + 5y$, subject to

$$5x + 2y \ge 60$$
$$5x + 8y \ge 120$$
$$x \ge 0$$
$$y \ge 0$$

SRINIVASA RAMANUJAN

Born in southern India, Srinivasa Ramanujan (1887–1919) spent much of his life in poverty and obscurity. He never graduated from college, he worked at several menial jobs for low wages, and he spent most of his energies recording his intuitive mathematical discoveries in notebooks.

At the urging of friends, he eventually began to exchange letters with G. H. Hardy of Cambridge University in England. This was the beginning of a 5-year collaboration that produced more than 30 papers on topics such as π, infinite series, prime and composite numbers, integers as the sum of squares, function theory, and combinatorics. More than 70 years after his death, Ramanujan's work is still an important source of new mathematical ideas. In the 1970s it was used by computer programmers to calculate millions of digits in the decimal expansion of π. His notebooks are just now being published, much to the delight of modern-day mathematicians.

chapter 11

DISCRETE ALGEBRA

BIBLIOGRAPHY

For students: *Graphing Calculator Activities with Finite Math,* Miller et al. Brooks/ Cole Publishing Co., 1992.

For teachers: *Chaos: Making a New Science,* James Gleick. Viking, 1987. Available through Dale Seymour Publications.

Discrete Mathematics Through Applications, Nancy Crisler et al. W. H. Freeman, 1994.

Videos: *Discrete Mathematics: Cracking the Code,* COMAP video.

Objective

Students will be able to define sequences explicitly and recursively and solve problems related to arithmetic and geometric sequences.

Motivate

Have students evaluate the expression $(-3)^i$ for $i = 1, 2, 3, 4,$ and 5.
$(-3, 9, -27, 81, -243)$

11.1 SEQUENCES

Finite and Infinite Sequences • Explicitly Defined Sequences • Recursively Defined Sequences • Arithmetic Sequences • Geometric Sequences • Applications

Finite and Infinite Sequences

An **infinite sequence** is a function $f(k) = a_k$ whose domain is the set of natural numbers. In describing a sequence, we usually list the range elements, or *terms* $\{a_k\}$, of a sequence in the order corresponding to their domain elements. For instance, we write the sequence $f(k) = 5k$ as

$$5, 10, 15, 20, \ldots, 5k, \ldots,$$

where $5k$ represents the kth term. Here are two more examples of sequences.

$$-3, 9, -27, 81, -243, \ldots, (-3)^j, \ldots \qquad 1, \frac{1}{2}, \frac{1}{3}, \frac{1}{4}, \frac{1}{5}, \ldots, \frac{1}{k}, \ldots$$

If the domain of a sequence is $\{1, 2, \ldots n\}$ for some natural number n, then there are a finite number of terms in the sequence and the sequence is a **finite sequence.**

Teaching Note

Students often think of a sequence as a set of numbers listed in some order. The secret to understanding sequences and their graphs is to recognize that a sequence is a *function* whose domain is the set of positive integers or $\{1, 2, \cdots, n\}$ for some positive integer n.

Explicitly Defined Sequences

Sometimes a sequence is defined by an explicit rule.

■ **EXAMPLE 1 Evaluating an Explicitly Defined Sequence**

List the first 6 terms and the 100th term of the sequence $\{a_k\}$, where $a_k = k^2 - 7$.

Solution

Evaluating each a_k, we find that

$$a_1 = -6,\ a_2 = -3,\ a_3 = 2,\ a_4 = 9,\ a_5 = 18,\ a_6 = 29, \text{ and } a_{100} = 9993.$$

The first 6 terms are $-6, -3, 2, 9, 18$, and 29 and the 100th term is 9993. ■

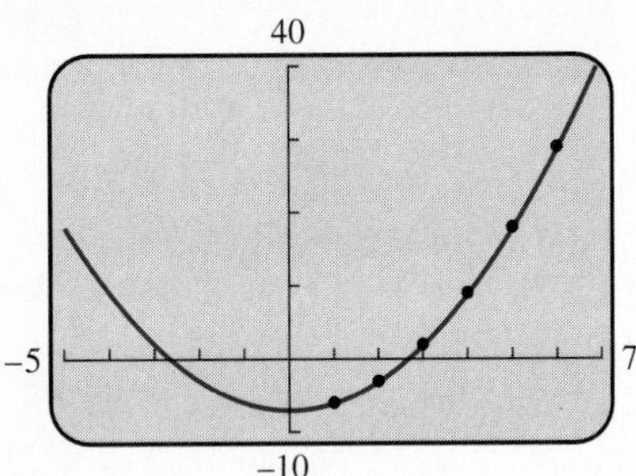

Figure 11.1 A graph of $y_1 = x^2 - 7$ and the first six terms of $a_k = k^2 - 7$.

Because the sequence $a_k = k^2 - 7$ is a function, we can draw its graph as a *discrete* set of points, $(1, -6)$, $(2, -3)$, $(3, 2)$, $(4, 9)$, $(5, 18)$, $(6, 29)$, and so on. We can also draw the graph of $y = x^2 - 7$ as a continuous curve that contains the graph of $a_k = k^2 - 7$. (See Figure 11.1.)

Some graphers have a sequence mode that can be used to draw graphs of sequences. If yours doesn't, Example 2 shows how to graph a sequence in parametric mode.

■ **EXAMPLE 2 Graphing Sequences Parametrically**

Graph the sequence $a_k = k^2 - 7$.

Solution

Choose the dot mode so that the grapher does not connect points. The parametric equations

$$x = t \quad \text{and} \quad y = t^2 - 7,$$

with $T\text{min} = 1$ and $T\text{step} = 1$ gives a graph of the sequence. (See Figure 11.2.) ■

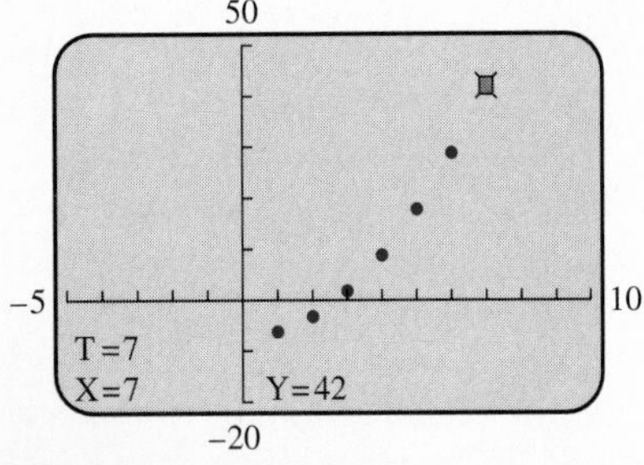

Figure 11.2 Notice that the trace cursor gives the value of the seventh term of the sequence $a_k = k^2 - 7$.

Sometimes we try to find an explicit rule that would define a given sequence.

Teaching Note

Emphasize that a sequence a_i can be defined in either of two ways: (1) explicitly, in terms of the variable i, or (2) recursively, in terms of the previous term a_{i-1}.

■ **EXAMPLE 3 Finding Explicit Rules for Sequences**

Write an explicit rule and find the domain for the sequence.

a. $\{u_k\} = \{5, 10, 15, 20, \ldots, 195, 200\}$

b. $\{v_k\} = \left\{-\frac{1}{3}, \frac{1}{6}, -\frac{1}{9}, \frac{1}{12}, -\frac{1}{15}, \ldots\right\}$

Solution

a. The (k, u_k) pairs, (1, 5), (2, 10), (3, 15), and so on, suggest that $u_k = 5k$. Then, if $u_k = 5k = 200$,

$$k = 40$$

and the domain is $\{1, 2, 3, \ldots, 40\}$.

b. The signs are alternating negative and positive, which are powers of -1, and the denominators are multiples of 3. Thus

$$v_k = \frac{(-1)^k}{3k}, \quad k = 1, 2, 3, \ldots.$$

■

Alert

In giving a recursive definition for a sequence, some students will give only the equation for a_i in terms of the previous term a_{i-1}. Emphasize that the starting point is an essential part of a recursive definition.

Recursively Defined Sequences

If asked to describe the pattern in the sequence $\{a_k\}$ given by

$$81, \quad 9, \quad 3, \quad \sqrt{3}, \quad \sqrt{\sqrt{3}} = \sqrt[4]{3}, \quad \sqrt{\sqrt{\sqrt{3}}} = \sqrt[8]{3}, \quad \ldots,$$

you might say something like, "Start with 81, and then take the square root of a term to find the next term." We can express this mathematically by writing

$$a_1 = 81 \qquad \text{and} \qquad a_n = \sqrt{a_{n-1}}, \qquad \text{for integers } n > 1.$$

This is a *recursive* definition. In a **recursively defined sequence** the first term is given (or the first few terms) along with a procedure for finding the subsequent terms.

■ EXAMPLE 4 Finding Terms in a Recursively Defined Sequence

Find the first four terms and the eighth term of the sequence $\{b_n\}$ defined by $b_1 = 4$ and $b_n = 2b_{n-1} - 3$ for $n > 1$.

Solution

We find the first four terms by using pencil and paper and then show how to use a calculator to find b_8.

Solve Algebraically

$b_1 = 4$ Given

$b_2 = 2b_1 - 3 = 2(4) - 3 = 5$ Replace b_1 by 4.

$b_3 = 2b_2 - 3 = 2(5) - 3 = 7$ Replace b_2 by 5.

$b_4 = 2b_3 - 3 = 2(7) - 3 = 11$ Replace b_3 by 7.

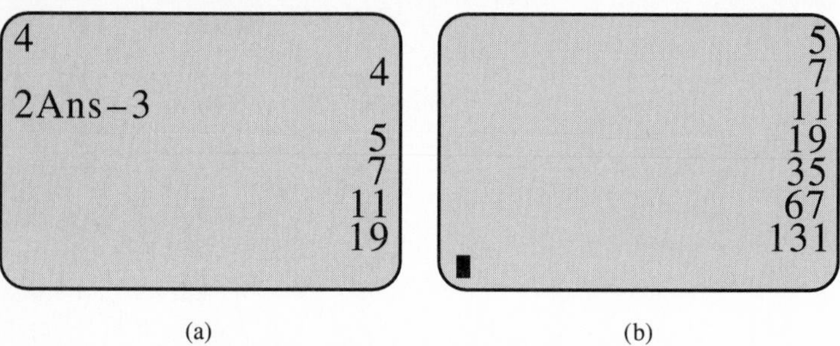

Figure 11.3 In (a) we enter 4 so that our first "Ans" is 4. Then using 2Ans − 3 and pressing "Enter" repeatedly produces the recursive sequence in (b).

TECHNOLOGY NOTE

Graphers have temporary memory locations to store the last result and the expression that produced the result. "Ans" is where a grapher stores the result of the last computation. When "Enter" is pressed after a computation has been completed, the grapher uses the last expression. If the last expression involves Ans, then each time Enter is pressed the grapher uses the last expression with the updated value of Ans.

Notes on Examples

Most calculators can be programmed to calculate automatically the numbers in the Fibonacci sequence.

Solve Numerically

In a calculator the "answer" to a calculation is stored as "Ans." Enter the first term 4, then compute 2Ans − 3 repeatedly to find the subsequent terms. See Figure 11.3. The eighth term is 131. ■

The Ans feature on a grapher greatly eases the calculation load for some, but not all, recursively defined sequences. For example, only the first two terms of the **Fibonacci sequence**—named for the alias of Leonardo of Pisa (ca. 1170–1250), a broadly knowledgeable and well-traveled mathematician who was educated in Africa—are given. Because subsequent terms are based on the two preceding terms and Ans stores only one of them, the Ans feature of a grapher cannot be used to generate this sequence.

■ EXAMPLE 5 Generating the Fibonacci Sequence

Find the first eight terms of the sequence $\{F_n\}$ defined by

$$F_1 = F_2 = 1 \quad \text{and} \quad F_n = F_{n-2} + F_{n-1} \quad \text{for } n \geq 3.$$

Solution

$$F_1 = 1$$

$$F_2 = 1$$

$$F_3 = F_1 + F_2 = 1 + 1 = 2$$

$$F_4 = F_2 + F_3 = 1 + 2 = 3$$

$$F_5 = F_3 + F_4 = 2 + 3 = 5$$

$$F_6 = F_4 + F_5 = 3 + 5 = 8$$

$$F_7 = F_5 + F_6 = 5 + 8 = 13$$

$$F_8 = F_6 + F_7 = 8 + 13 = 21$$

■

The terms in the Fibonacci sequence are the **Fibonacci numbers.** The Fibonacci numbers occur in nature (for example, the number of spirals on a pine cone) and are used in many ways (for example, the dimensions of index cards and the number of lines in stanzas of poetry).

Arithmetic Sequences

A sequence in which consecutive terms differ by a fixed amount is an *arithmetic sequence,* or *arithmetic progression.* Here is a recursive definition of such a sequence.

Definition Arithmetic Sequence

A sequence $\{a_n\}$ is an **arithmetic sequence** (or **arithmetic progression**) if it can be written in the form

$$a_n = a_{n-1} + d, \quad n \geq 2,$$

for some constant d. The number d is the **common difference.**

■ EXAMPLE 6 Recognizing Arithmetic Sequences

Determine whether the sequence could be arithmetic. If so, find the common difference.

a. $-6, \quad -3.5, \quad -1, \quad 1.5, \quad 4, \quad \ldots$
b. $48, \quad 24, \quad 12, \quad 6, \quad 3, \quad \ldots$
c. $\ln 3, \quad \ln 6, \quad \ln 12, \quad \ln 24, \quad \ldots$

Solution

a. We compute the differences between successive terms:

$$a_2 - a_1 = (-3.5) - (-6) = 2.5$$

$$a_3 - a_2 = (-1) - (-3.5) = 2.5$$

$$a_4 - a_3 = (1.5) - (-1) = 2.5$$

$$a_5 - a_4 = (4) - (1.5) = 2.5$$

If this difference pattern continues, the sequence is arithmetic with $d = 2.5$.

b. Here the differences between successive terms vary:

$$a_2 - a_1 = 24 - 48 = -24$$

$$a_3 - a_2 = 12 - 24 = -12$$

The sequence is not arithmetic.

c. $a_2 - a_1 = \ln 6 - \ln 3 = \ln \frac{6}{3} = \ln 2$ $\quad \ln u - \ln v = \ln \frac{u}{v}.$

$$a_3 - a_2 = \ln 12 - \ln 6 = \ln \frac{12}{6} = \ln 2$$

$$a_4 - a_3 = \ln 24 - \ln 12 = \ln \frac{24}{12} = \ln 2$$

If this difference pattern continues, the sequence is arithmetic with $d = \ln 2$. ■

If $\{a_n\}$ is an arithmetic sequence with common difference d, then

$$a_2 = a_1 + d$$

$$a_3 = a_2 + d = a_1 + 2d$$

$$a_4 = a_3 + d = a_1 + 3d$$

We can continue in this way and show that the nth term of an arithmetic sequence has the following form, which we prove in Section 11.3. Also see Exercise 71.

nth Term of an Arithmetic Sequence

The nth term of an arithmetic sequence can be written in the form

$$a_n = a_1 + (n - 1)d,$$

where a_1 is the first term and d is the common difference.

We now have two forms for the nth term of an arithmetic sequence. The definition $a_n = a_{n-1} + d$ gave a_n recursively in terms of the previous term. The new form gives a_n explicitly as a function of n. Example 7 uses this new form for a_n.

Notes on Examples

Examples 7 and 9 demonstrate that by knowing two terms of a sequence and whether the sequence is arithmetic or geometric, the formula that defines the sequence can usually be found. However, it is worth mentioning that for a geometric sequence an ambiguous case is possible where the sign of r is indeterminate.

■ EXAMPLE 7 Determining an Arithmetic Sequence

The third and eighth terms of an arithmetic sequence are 13 and 3, respectively. Find the first term, the common difference, and an explicit rule for the nth term.

Solution

Solve Algebraically

We are given $a_3 = 13$ and $a_8 = 3$, and we know that $a_n = a_1 + (n - 1)d$.

Thus we obtain and solve our system of equations.

$$a_3 = a_1 + 2d = 13$$

$$a_8 = a_1 + 7d = 3$$

$$-5d = 10 \quad \text{Subtract.}$$

$$d = -2 \quad \text{Divide by } -5.$$

$$a_1 + 2(-2) = 13 \quad \text{Substitute } d = -2.$$

$$a_1 = 17 \quad \text{Solve.}$$

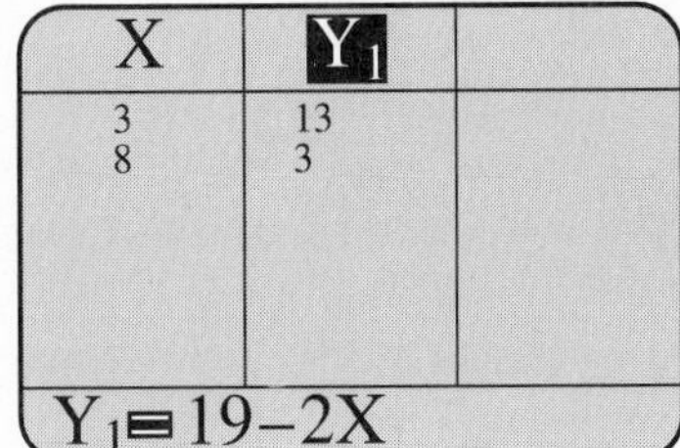

Figure 11.4

Thus $a_n = a_1 + (n-1)d = 17 + (n-1)(-2) = 19 - 2n$.

Interpret

The first term is 17, the common difference is -2, and the nth term is $19 - 2n$.

Support Numerically

Figure 11.4 shows the value of $19 - 2x$ for $x = 3$ and $x = 8$. ■

Geometric Sequences

In an arithmetic sequence, terms are found by adding a constant to the preceding term. A sequence in which terms are found by multiplying the preceding term by a (nonzero) constant is a *geometric sequence,* or *geometric progression.* Here is a recursive definition of such a sequence.

Definition Geometric Sequence

A sequence $\{a_n\}$ is a **geometric sequence** (or **geometric progression**) if it can be written in the form

$$a_n = a_{n-1} \cdot r, \quad n \geq 2,$$

where $r \neq 0$ is the **common ratio.**

■ EXAMPLE 8 Recognizing Geometric Sequences

Determine whether the sequence could be geometric. If so, find the common ratio.

a. $2, \frac{2}{3}, \frac{2}{9}, \frac{2}{15}, \frac{2}{21}, \ldots$

b. $3, 6, 12, 24, 48, \ldots$

c. $10^{-3}, 10^{-1}, 10^{1}, 10^{3}, 10^{5}, \ldots$

Solution

a. We compute the ratio of each term to the preceding term:

$$\frac{a_2}{a_1} = \frac{2/3}{2} = \frac{1}{3}$$

$$\frac{a_3}{a_2} = \frac{2/9}{2/3} = \frac{1}{3}$$

$$\frac{a_4}{a_3} = \frac{2/15}{2/9} = \frac{3}{5}$$

The sequence is not geometric because the ratios vary.

b.

$$\frac{a_2}{a_1} = \frac{6}{3} = 2$$

$$\frac{a_3}{a_2} = \frac{12}{6} = 2$$

$$\frac{a_4}{a_3} = \frac{24}{12} = 2$$

$$\frac{a_5}{a_4} = \frac{48}{24} = 2$$

If this ratio pattern continues, the sequence is geometric with $r = 2$.

c.

$$\frac{a_2}{a_1} = \frac{10^{-1}}{10^{-3}} = 10^{-1-(-3)} = 10^2$$

$$\frac{a_3}{a_2} = \frac{10^1}{10^{-1}} = 10^2$$

$$\frac{a_4}{a_3} = \frac{10^3}{10^1} = 10^2$$

$$\frac{a_5}{a_4} = \frac{10^5}{10^3} = 10^2$$

If this ratio pattern continues, the sequence is geometric with $r = 10^2$. ■

If $\{a_n\}$ is a geometric sequence with common ratio r, then

$$a_2 = a_1 \cdot r$$

$$a_3 = a_2 \cdot r = a_1 \cdot r^2$$

$$a_4 = a_3 \cdot r = a_1 \cdot r^3.$$

We can continue in this way and show that the nth term of a geometric sequence has the following form (see Exercise 72).

nth Term of a Geometric Sequence

The nth term of a geometric sequence can be written in the form

$$a_n = a_1 \cdot r^{n-1},$$

where a_1 is the first term and r is the nonzero common ratio.

As with arithmetic sequences, we now have two forms for the nth term of a geometric sequence—the original form gives a_n recursively, and the new form gives a_n explicitly. In Example 9 we use this new form.

EXAMPLE 9 Determining a Geometric Sequence

The third and eighth terms of a geometric sequence are 20 and -640, respectively. Find the first term, common ratio, and an explicit rule for the nth term.

Solution

Solve Algebraically

We are given $a_3 = 20$ and $a_8 = -640$, and we know that $a_n = a_1 \cdot r^{n-1}$. Thus

$$a_3 = a_1 \cdot r^2 = 20$$

$$a_8 = a_1 \cdot r^7 = -640$$

$$\frac{a_1 r^7}{a_1 r^2} = \frac{-640}{20} \quad \text{Divide.}$$

$$r^5 = -32$$

$$r = -2$$

$$a_1(-2)^2 = 20 \quad \text{Substitute } r = -2.$$

$$a_1 = 5$$

Thus $a_n = a_1 \cdot r^{n-1} = 5(-2)^{n-1}$.

Interpret

The first term is 5, the common ratio is -2, and the nth term is $5(-2)^{n-1}$.

Support Numerically

Show that the value of a_n is 20 for $n = 3$ and -640 for $n = 8$. ■

Follow-up

Ask . . .

A student gave $a_n = 3a_{n-1}$ as a recursive definition for a geometric sequence. What is wrong with this definition? **(a_1 is not defined.)**

Assignment Guide

Day 1: Ex. 3–54, multiples of 3

Day 2: Ex. 55, 59, 63, 64, 66, 69, 71, 74, 84

Cooperative Learning

Group Activity: Ex. 65, 77

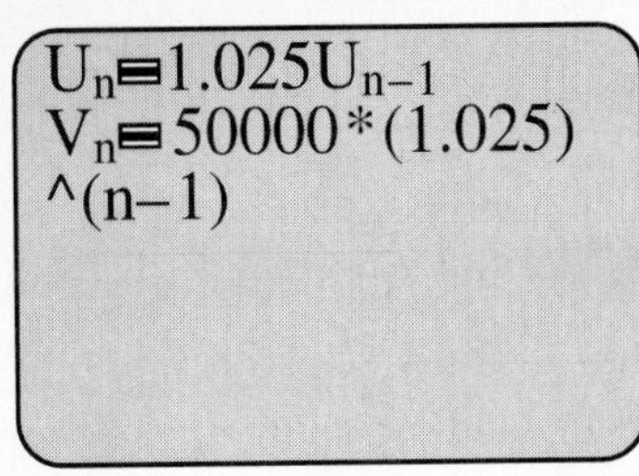

(a)

n	Un	Vn
1	50000	50000
2	51250	51250
3	52531	52531
4	53845	53845
5	55191	55191
6	56570	56570
7	57985	57985

Un=1.025Un−1

(b)

Figure 11.5
(a) Recursive ($u_n = 1.025u_{n-1}$) and explicit ($v_n = 50000(1.025)^{n-1}$) forms for the population of Bridgetown.
(b) Seven resulting values.

Applications

Many of the Chapter 4 applications of exponential functions to problems of growth and decay can be presented in terms of geometric sequences and represented numerically with grapher tables. (See Example 4, Section 4.1.)

■ EXAMPLE 10 APPLICATION: Projecting Population Growth

The population of Bridgetown is growing at the rate of 2.5% per year. The present population is 50,000. Find a sequence that represents Bridgetown's population each year. Represent the nth term of the sequence both explicitly and recursively. Evaluate seven terms of the sequence.

Solution

If $n = 1$ represents the present population, then the population $n - 1$ years later can be represented by the sequence $\{P_n\}$, where P_n is given explicitly by

$$P_n = 50,000(1.025)^{n-1}.$$

The population in any year can also be found by multiplying the previous year's population by 1.025. So this sequence can also be defined recursively by

$$P_1 = 50,000 \quad \text{and} \quad P_n = 1.025P_{n-1}, \quad \text{for } n > 1.$$

See Figure 11.5. ■

On December 29, 1993, the front-page headline of the *Houston Chronicle* boasted, "More folks calling Texas home: State in 1994 may become nation's second most populous." The accompanying article gave U.S. Census Bureau estimates for state populations as of July 1, 1993 (see Table 11.1), and the percentage growth rate based on similar estimates for 1992.

Table 11.1 Population

State	Population (1993)	Growth rate
New York	18,197,000	0.5%
Texas	18,031,000	2.0%

Notes on Exercises

Ex. 81 is related to the concept of limits. Notice how this exercise is related to a horizontal asymptote of a function.

Ongoing Assessment

Self-Assessment: Ex. 11, 19, 23, 25, 37, 41, 43, 61
Embedded Assessment: Ex. 64

■ EXAMPLE 11 APPLICATION: Growing Populations—New York vs. Texas

Assume that the populations of New York and Texas will continue to grow at the annual rates given in Table 11.1.

a. In what year will the population of Texas surpass that of New York?
b. In what year will the population of Texas surpass that of New York by 1 million?

Solution

Model

The population of New York is represented by the geometric sequence

$$u_n = 18,197,000(1.005)^{n-1}$$

n	U_n	V_n
1	18.197	18.031
2	18.288	18.392
3	18.379	18.759
4	18.471	19.135
5	18.564	19.517
6	18.656	19.908
7	18.75	20.306

n=6

Figure 11.6

and the population of Texas by

$$v_n = 18{,}031{,}000(1.02)^{n-1},$$

where $n = 1$ stands for 1993, $n = 2$ for 1994, and so on.

Solve Numerically

Figure 11.6 shows a table of values for u_n and v_n, where the data are in millions. That is, the entry in the first row, second column, stands for 18.197 million, the population of New York in 1993. We can see that

$$v_2 > u_2 \quad \text{and} \quad v_6 > u_6 + 1.$$

Interpret

The headline was right! In 1994, the population of Texas (v_2) was greater than the population of New York (u_2). The population of Texas would first exceed that of New York by 1 million (v_k is 1 more than u_k) in 1998 ($n = 6$). ■

Quick Review 11.1

In Exercises 1–6, evaluate the expression.

1. $(-2)^3$ **2.** $(-2)^4$

3. $(-3)^6$ **4.** $(-4)^4$

5. -3^6 **6.** -4^4

In Exercises 7 and 8, find the function values.

7. $f(x) = 1000(1.045)^x$; **a.** $f(8)$, **b.** $f(20)$

8. $g(x) = 800(1.073)^x$; **a.** $g(9)$, **b.** $g(30)$

In Exercises 9 and 10, find **a.** $f(a + b)$ and **b.** $f(f(x))$.

9. $f(x) = 3x + 2$ **10.** $f(x) = 2x - 3$

In Exercises 11–14, evaluate the expression $2x - 3y - 7$ for the given values of x and y.

11. $x = 0, y = 2$ **12.** $x = 3, y = 0$

13. $x = 3, y = 1$ **14.** $x = 4, y = 5$

In Exercises 15 and 16, find the domain of the function.

15. $f(x) = \sqrt{x - 1}$ **16.** $g(x) = \dfrac{1}{x^2 - 1}$

In Exercises 17 and 18, solve the system of equations algebraically. Support your solution graphically.

17. $\begin{aligned} 2x - 3y &= -8 \\ -3x + 4y &= 11 \end{aligned}$ **18.** $\begin{aligned} 5a + 2b &= 4 \\ a - 3b &= 11 \end{aligned}$

SECTION EXERCISES 11.1

In Exercises 1–4, find the first 6 terms and the 100th term of the sequence whose nth term is given.

1. $u_n = \dfrac{n + 1}{n}$ **2.** $v_n = \dfrac{4}{n + 2}$

3. $c_n = n^3 - n$ **4.** $d_n = n^2 - 5n$

In Exercises 5–12, find the first four terms and the eighth term of the recursive sequence.

5. $a_1 = 8$ and $a_n = a_{n-1} - 4$, for $n \ge 2$

6. $u_1 = -3$ and $u_{k+1} = u_k + 10$, for $k \ge 1$

7. $b_1 = 2$ and $b_{k+1} = 3b_k$, for $k \ge 1$

8. $v_1 = 0.75$ and $v_n = (-2)v_{n-1}$, for $n \ge 2$

9. $a_1 = 2$ and $a_{n+1} = (a_n)^2$, for $n \ge 1$

10. $u_1 = -2$ and $u_m = (u_{m-1})^2$, for $m \ge 2$

11. $c_1 = 2$, $c_2 = -1$, and $c_{k+2} = c_k + c_{k+1}$, for $k \ge 1$

12. $c_1 = -2$, $c_2 = 3$, and $c_k = c_{k-2} + c_{k-1}$, for $k \ge 3$

In Exercises 13–16, determine which table corresponds to the sequence with the given nth term.

13. $a_n = (-2)^n$

14. $b_n = -3n$

15. $a_n = \dfrac{n+2}{n}$

16. $b_n = 5 - \dfrac{2}{n}$

X	Y_1
1	3
2	4
3	4.3333
4	4.5
5	4.6
6	4.6667
7	4.7143

X=7

(a)

X	Y_1
1	-3
2	-6
3	-9
4	-12
5	-15
6	-18
7	-21

X=7

(b)

X	Y_1
1	-2
2	4
3	-8
4	16
5	-32
6	64
7	-128

X=7

(c)

X	Y_1
1	3
2	2
3	1.6667
4	1.5
5	1.4
6	1.3333
7	1.2857

X=7

(d)

For Exercises 13–16

In Exercises 17–20, find the first 2 terms and the 100th term of the sequence.

17. $a_n = (-3)^{n-1}$

18. $b_n = (-2)^{n+1}$

19. $u_n = \dfrac{(-2)^{n-1}}{n+4}$

20. $v_n = \dfrac{(-3)^{n-1}}{n+3}$

In Exercises 21–24, the sequences are arithmetic. Find

a. the common difference,
b. the tenth term,
c. a recursive rule for the nth term, and
d. an explicit rule for the nth term.

21. 6, 10, 14, 18, . . .

22. $-4, 1, 6, 11, \ldots$

23. $-5, -2, 1, 4, \ldots$

24. $-7, 4, 15, 26, \ldots$

In Exercises 25–28, the sequences are geometric. Find

a. the common ratio,
b. the eighth term,
c. a recursive rule for the nth term, and
d. an explicit rule for the nth term.

25. 2, 6, 18, 54, . . .

26. 3, 6, 12, 24, . . .

27. $1, -2, 4, -8, 16, \ldots$

28. $-2, 2, -2, 2, \ldots$

In Exercises 29–32, solve the problem.

29. The fourth and seventh terms of an arithmetic sequence are -8 and 4, respectively. Find the first term and a recursive rule for the nth term.

30. The fifth and ninth terms of an arithmetic sequence are -5 and -17, respectively. Find the first term and a recursive rule for the nth term.

31. The second and eighth terms of a geometric sequence are 3 and 192, respectively. Find the first term, common ratio, and an explicit rule for the nth term.

32. The third and sixth terms of a geometric sequence are -75 and -9375, respectively. Find the first term, common ratio, and an explicit rule for the nth term.

In Exercises 33–38, state whether the sequence could be arithmetic or geometric. For those sequences, find the common difference or common ratio and an explicit rule for the nth term.

33. 5, 10, 20, 40, . . .

34. $-0.25, 1, -4, 16, -64, \ldots$

35. $-16, -9, -2, 5, \ldots$

36. 10.1, 10.201, 10.30301, 10.4060401, . . .

37. $-2, 1, -\dfrac{1}{2}, \dfrac{1}{4}, \ldots$

38. 1, 5, 7, 11, 17, . . .

In Exercises 39–42, write an explicit rule for the nth term of the sequence.

39. $a_1 = 3$ and $a_n = a_{n-1} + 2$, for $n \geq 2$

40. $b_1 = -5$ and $b_n = b_{n-1} + 7$, for $n \geq 2$

41. $u_1 = 2$ and $u_k = 3u_{k-1}$, for $k \geq 2$

42. $v_1 = 3$ and $v_k = (-2)v_{k-1}$, for $k \geq 2$

In Exercises 43–46, write an explicit rule for the sequence and state its domain.

43. $8, 2, \dfrac{1}{2}, \dfrac{1}{8}, \dfrac{1}{32}, \ldots, \dfrac{1}{2048}$

44. $-7, -4, -1, 2, 5, \ldots, 335$

45. $-\dfrac{1}{3}, \dfrac{1}{6}, -\dfrac{1}{9}, \ldots, -\dfrac{1}{999}$

46. $-1, \dfrac{1}{5}, -\dfrac{1}{25}, \dfrac{1}{125}, \ldots, \dfrac{1}{48828125}$

In Exercises 47–50, identify a graph that contains the graph of the sequence with the given nth term. The four viewing window dimensions are the same.

47. $a_n = 2n - 1$ **48.** $b_n = 4 - 3n$

49. $c_n = 4 - n^2$ **50.** $d_n = n^3 - n$

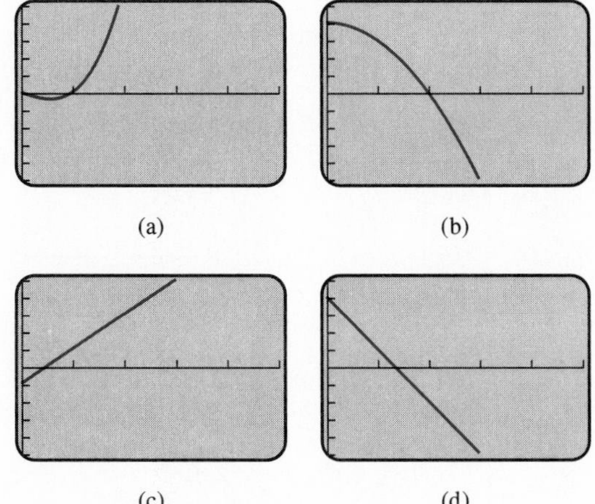

For Exercises 47–50

In Exercises 51–54, graph the sequence.

51. $a_n = 2 - \frac{1}{n}$ **52.** $b_n = \sqrt{n} - 3$

53. $c_n = n^2 - 5$ **54.** $d_n = 3 + 2n$

In Exercises 55–58, use the arithmetic sequence from Exercises 21–24.

a. Draw a scatter plot of the first four terms.
b. Find a linear regression equation for the data.
c. Compare the linear regression equation with the explicit rule for the nth term in the indicated exercise.

55. Exercise 21 **56.** Exercise 22

57. Exercise 23 **58.** Exercise 24

In Exercises 59–66 solve the problem.

59. *Compound Interest* Roberta had \$1250 in a savings account 3 years ago. What will be the value of her account 2 years from now, assuming no deposits or withdrawals are made and the account earns 6.5% interest compounded annually?

60. *Compound Interest* Steve has \$12,876 in a savings account today. He made no deposits or withdrawals during the past 6 years. What was the value of his account 6 years ago if the account earned interest at 5.75% compounded monthly?

61. *Rain Forest Growth* The bungy-bungy tree in the Amazon rain forest grows an average 2.3 cm per week. Write a sequence that represents the weekly height of a bungy-bungy over the course of 1 year if it is 7 meters tall today. Display the first four terms and the last two terms.

62. *Hamburger Inflation* From 1969 through 1994 the average annual inflation rate of a fast-food small hamburger was 5.6%. Although prices vary from location to location, the price of a small hamburger at some McDonald's restaurants was \$0.59 in 1994. What will be the price of a McDonald's hamburger in 2019 if it grows geometrically with the same 5.6% annual inflation rate? What was the price of a McDonald's hamburger in 1969?

63. *Half-Life* (See Section 4.1) Thorium-232 has a half-life of 14 billion years. Make a table showing the half-life decay of a sample of Thorium-232 from 16 grams to 1 gram; list the time (in years, starting with $t = 0$) in the first column and the mass (in grams) in the second column. Which type of sequence is each column of the table?

64. *Population—Conroe vs. Huntsville* The population of Huntsville, Texas, was 25,854 in 1985 and has been increasing by 1.55% each year. The population of Conroe, Texas, was 22,314 in 1985 and has been increasing by 4.35% each year. *Source:* The Chambers of Commerce for the two cities.

a. Make a table, based on the given data, that predicts the population of the two cities for 10 years after 1985.
b. In which year should Conroe's population first exceed Huntsville's?

65. *Rabbit Populations* Assume that rabbits become fertile 1 month after birth and that each male–female pair of fertile rabbits produces one new male–female pair of rabbits each month. Further assume that the rabbit colony begins with one newborn male–female pair of rabbits and no rabbits die for 12 months. Let a_n represent the number of *pairs* of rabbits in the colony after $n - 1$ months.

a. **Writing to Learn** Explain why $a_1 = 1$, $a_2 = 1$, and $a_3 = 2$.
b. Find $a_4, a_5, a_6, \ldots, a_{13}$.
c. **Writing to Learn** Explain why the sequence $\{a_n\}$, $1 \leq n \leq 13$, is a model for the size of the rabbit colony for a 1-year period.

66. *Recursive Sequence* The population of Centerville was 525,000 in 1992 and is growing annually at the rate of

1.75%. Write a recursive sequence $\{P_n\}$ for the population. State the first term P_1 for your sequence.

In Exercises 67–70 find the indicated regression equation for the sequence.

67. $\{2, 0, -2\}$; linear regression equation

68. $\{-4, -1, 2\}$; linear regression equation

69. $\{-2, -3, -2, 1, 6\}$; quadratic regression equation

70. $\{4, 3, 0, -5, -12\}$; quadratic regression equation

EXTENDING THE IDEAS

In Exercises 71–86 solve the problem.

71. Writing to Learn Let $\{a_n\}$ be an arithmetic sequence with common difference d.

a. Give a convincing argument that its nth term can be written in the form

$$a_n = a_1 + (n - 1)d.$$

b. If $f(x) = mx + b$ is the corresponding function determined by the rule of the sequence, show that $m = d$ and that $b = a_1 - d$.

72. Writing to Learn Let $\{a_n\}$ be a geometric sequence with common ratio r. Give a convincing argument that its nth term can be written in the form

$$a_n = a_1 \cdot r^{n-1}.$$

73. *Fibonacci Sequence* Compute the first seven terms of the sequence whose nth term is

$$a_n = \frac{1}{\sqrt{5}}\left(\frac{1+\sqrt{5}}{2}\right)^n - \frac{1}{\sqrt{5}}\left(\frac{1-\sqrt{5}}{2}\right)^n.$$

How do these seven terms compare with the first seven terms of the Fibonacci sequence?

Exercises 74 and 75 concern *constant sequences.* If a is a number, the sequence $a, a, a, \ldots, a, \ldots$ is a *constant sequence.*

74. a. Explain why a constant sequence is both arithmetic and geometric.

b. State why a constant sequence can be considered as both explicitly defined and recursively defined.

75. Determine whether the sequence is a constant sequence.

a. $(-1)^1, (-1)^3, (-1)^5, \ldots$

b. $a_1 = 3$ and $a_k = 2a_{k-1} - 3$, for $k \geq 2$

c. $a_1 = 4$ and $a_k = 2a_{k-1} - 3$, for $k \geq 2$

d. $a_1 = 2$ and $a_k = 2a_{k-1} - 3$, for $k \geq 2$

76. Writing to Learn Based on your understanding of increasing and decreasing functions, write definitions for increasing and decreasing sequences. Give two examples of each type of sequence.

77. *Connecting Geometry and Sequences* In the following sequence of diagrams, regular polygons are inscribed in unit circles with at least one side of each polygon perpendicular to the positive x-axis.

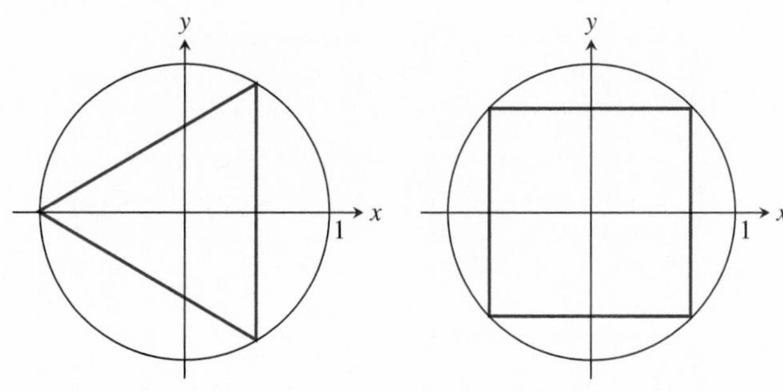

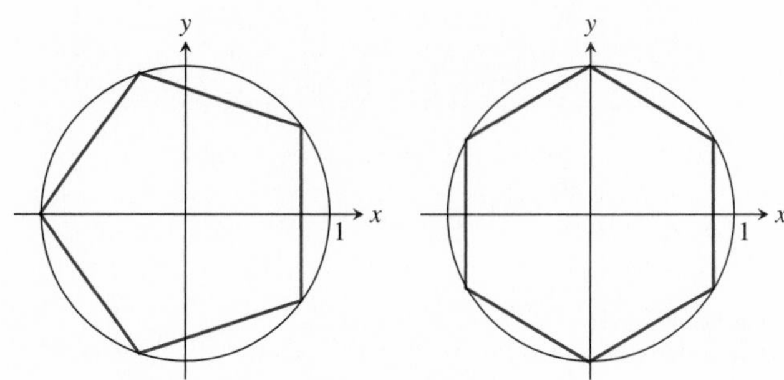

For Exercise 77

a. Prove that the perimeter of each polygon in the sequence is given by $a_n = 2n \sin(\pi/n)$, where n is the number of sides in the polygon.

b. Investigate the value of a_n for $n = 10, 100, 1000$, and 10,000. What conclusion can you draw?

78. If $\{a_n\}$ is a geometric sequence with all positive terms, prove that $\{\log(a_n)\}$ is arithmetic.

79. If $\{b_n\}$ is an arithmetic sequence, prove that $\{10^{b_n}\}$ is geometric.

80. Consider the sequence given by $a_1 = 1$ and $a_n = na_{n-1}$. Compute the first six terms of the sequence and show that in each case $a_n = n(n-1)(n-2)\cdots 3 \cdot 2 \cdot 1$.

81. *Sequence Convergence* A sequence $\{a_n\}$ *converges to a real number K,* denoted $a_n \to K$ as $n \to \infty$, if the line $y = K$ is a horizontal asymptote of the graph of a_n. For each sequence, find K. Then graph a_n and the line $y = K$ in the same viewing window to support the statement that the sequence converges to the number K. State the dimensions of the window.

a. $a_n = \dfrac{2n}{n+1}$ **b.** $a_n = \left(1 + \dfrac{0.05}{n}\right)^n$

c. $a_n = 3 + \dfrac{(-1)^n}{n}$ **d.** $a_n = n \sin\left(\dfrac{\pi}{2n}\right)$

82. *Connecting Recursive Sequences and Equation Solving* Let $f(x) = \left(\frac{1}{2}\right)^x$ and define a sequence $a_n = f(a_{n-1})$, where $a_1 = f(1)$.

a. Compute the first eight terms of this sequence.
b. To what number does this sequence appear to converge? (See Exercise 81.)
c. Graph $y_1 = x$ and $y_2 = f(x)$ in the same viewing rectangle. Solve the equation $f(x) = x$. How does this solution relate to the sequence?

83. *Complex Number Sequences* The terms of a sequence can be complex numbers.

a. Write out the first nine terms of the geometric sequence with the first term 1 and the common ratio the imaginary unit i.
b. Write out the first nine terms of the geometric sequence with the first term $2 + i$ and the common ratio the imaginary unit i.
c. What is the geometric effect of multiplying a complex number by i? Prove your conjecture.

84. *A Sequence of Matrices* Write out the first seven terms of the "geometric sequence" with the first term the matrix $(1 \quad 1)$ and the common ratio the matrix $\begin{pmatrix} 0 & 1 \\ 1 & 1 \end{pmatrix}$. How is this sequence of matrices related to the Fibonacci sequence?

85. *Alternative Notation for Sequences* A notation can be used for sequences that incorporates the domain of the sequence with the rule for the nth term. In this notation a finite sequence is written as an expression $\{a_k\}_{k=1}^{n}$, which equals $\{a_1, a_2, \ldots, a_{n-1}, a_n\}$. An infinite sequence is written as an expression $\{b_k\}_{k=1}^{\infty}$, which equals $\{b_1, b_2, \ldots\}$. State whether the sequence is finite or infinite. Write each sequence in expanded form.

a. $\{3k - 5\}_{k=1}^{7}$ **b.** $\{2^n - 6\}_{n=3}^{8}$

c. $\{5j - 1\}_{j=1}^{\infty}$ **d.** $\{4p^2 + 1\}_{p=1}^{\infty}$

86. *Grapher Activity* Some graphers have a sequence command that allows you to generate a finite sequence as a list given the rule for the nth term, the independent variable, its beginning and ending values, and the step between values. For instance, the first eight terms of the sequence of Example 7 can be displayed by entering the appropriate information.

```
seq(19-2N,N,1,8,
1)
(17 15 13 11 9 ...


Rcl Ans
```

```
seq(19-2N,N,1,8,
1)
(17 15 13 11 9 ...
(17,15,13,11,9,7
,5,3) ▮
```

For Exercise 86

Use the sequence command to write each sequence in expanded form.

a. $\{3k^4 - 5\}_{k=1}^{7}$ **b.** $\{4^n - 6\}_{n=3}^{8}$

Objective

Students will be able to use sigma notation and basic summation formulas to find the sum of a finite series or infinite geometric series.

Motivate

Ask . . .
Can you think of an easy way to add the natural numbers from 1 to 100?
((1 + 100) + (2 + 99) + · · · + (50 + 51) = 50 • 101 = 5050)

11.2 SERIES AND SUMMATION NOTATION

Sequence and Series • Finite Series • Finite Arithmetic Series • Finite Geometric Series • Infinite Series and Sequences of Partial Sums • Summation Notation

Sequence and Series

In everyday language, the words *sequence* and *series* are virtually synonymous. In mathematics, they have distinct but related meanings. A sequence is a list of numbers, whereas a **series** is a sum resulting from adding the terms of a sequence.

Finite Series

REMARK

The third series is the sum of the periods of revolution of the first six planets in the solar system, in years.

The sum of the terms in a finite sequence is a *finite series.* Here are three examples of finite series.

$$1 + 1 + 2 + 3 + 5 + 8 + 13 + 21$$

$$\sqrt{1} + \sqrt{2} + \sqrt{3} + \sqrt{4} + \sqrt{5}$$

$$0.24 + 0.66 + 1 + 1.88 + 11.86 + 29.46$$

Summing a large number of terms caused the ancient Greeks a good deal of trouble. The philosopher Zeno (c. 450 B.C.) contrived the following paradox: A launched arrow will never reach its target because it must first reach the midpoint, then three-quarter point, and so on. Intuition suggests that the sum of infinitely many numbers should be infinite. We will see that such a sum can be finite.

Definition Finite Series

A **finite series** is a sum that can be written in the form

$$a_1 + a_2 + a_3 + \cdots + a_k + \cdots + a_n,$$

where n is a natural number and a_k is the ***k*th term** of the series.

Although it may seem redundant, the numerical value of a finite series is, for convenience, sometimes called the **sum of the series.**

Any time you add a list of numbers, you are summing a finite series. Sometimes the total itself is of interest, as in Example 1.

■ EXAMPLE 1 Investing Money

Orlando invests $100 for his daughter Lawanna the day she is born and again on each of her first five birthdays. If the investments yield 5% compounded annually, how much will Lawanna have on her 21st birthday?

Solution

The first $100 yields $100(1.05^{21})$ dollars, the second $100 yields $100(1.05^{20})$ dollars, and so on. On her 21st birthday, Lawanna will have

$$100(1.05^{21}) + 100(1.05^{20}) + 100(1.05^{19}) + 100(1.05^{18}) + 100(1.05^{17}) + 100(1.05^{16}),$$

or about $1484.77. ■

Finite Arithmetic Series

A series is an **arithmetic series** if its terms form an arithmetic sequence. Let

$$a_1 + a_2 + a_3 + \cdots + a_k + \cdots + a_n$$

be an arithmetic series with a common difference d. Then we know that $a_k = a_1 + (k - 1)d$. If S_n is the sum of this series, then

$$S_n = a_1 + (a_1 + d) + (a_1 + 2d) + \cdots + [a_1 + (k - 1)d] + \cdots + [a_1 + (n - 1)d].$$

If we reverse the order of the terms, then we start with a_n and repeatedly subtract d.

$$S_n = a_n + (a_n - d) + (a_n - 2d) + \cdots + [a_n - (k - 1)d] + \cdots + [a_n - (n - 1)d].$$

Adding both sides of these equations, we get

$$2S_n = (a_1 + a_n) + (a_1 + a_n) + (a_1 + a_n) + \cdots + (a_1 + a_n) + \cdots + (a_1 + a_n).$$

Because the right side of this equation has n terms,

$$2S_n = n(a_1 + a_n) \qquad \text{and thus} \qquad S_n = n\left(\frac{a_1 + a_n}{2}\right).$$

Substituting $a_n = a_1 + (n - 1)d$, we get a second form for S_n,

$$S_n = \frac{n}{2}[2a_1 + (n - 1)d].$$

Sum of a Finite Arithmetic Series

The sum S_n of the arithmetic series $a_1 + a_2 + \cdots + a_n$ with common difference d is

$$S_n = n\left(\frac{a_1 + a_n}{2}\right) \qquad \text{or} \qquad S_n = \frac{n}{2}[2a_1 + (n - 1)d].$$

To find S_n with the formula

$$S_n = n\left(\frac{a_1 + a_n}{2}\right)$$

we need a_1, a_n, and n. To use $S_n = (n/2)[2a_1 + (n - 1)d]$ we need to know a_1, d, and n. Examples 2 and 3 illustrate the use of each form.

EXAMPLE 2 Summing an Arithmetic Series

Find the sum of the first 10 terms of the arithmetic sequence 116, 106, 96, ...

Solution

Because $n = 10$, $a_1 = 116$, and $d = -10$, we use

$$\begin{aligned} S_n &= \frac{n}{2}[2a_1 + (n - 1)d] \\ &= \frac{10}{2}[2(116) + (10 - 1)(-10)] \\ &= 5[232 + 9(-10)] \\ &= 710 \end{aligned}$$

■

EXAMPLE 3 Summing an Arithmetic Series

Sum the arithmetic series $-12 - 7.5 - 3 + 1.5 + \cdots + 60$.

Solution

We know that $a_1 = -12$, $d = 4.5$, and $a_n = 60$. To find n, we use

$$\begin{aligned} a_n &= a_1 + (n - 1)d \\ 60 &= -12 + (n - 1)(4.5) \\ 72 &= 4.5n - 4.5 \\ 76.5 &= 4.5n \\ n &= 17 \end{aligned}$$

We can use either sum formula to find S_n. We use the first one.

$$S_n = n\left(\frac{a_1 + a_n}{2}\right)$$

$$S_{17} = (17)\left(\frac{(-12) + 60}{2}\right) = (17)(24) = 408$$

■

Figure 11.7

The sum T_n of the arithmetic series $1 + 2 + 3 + \cdots + n$ is a **triangular number** because each such sum can be visualized as a triangular array of dots. (See Figure 11.7.)

EXAMPLE 4 Triangular Numbers

Consider the series $1 + 2 + 3 + \cdots + n$.

a. Find an explicit rule for the sum of the series, the triangular number T_n.
b. Use the rule to sum the series $1 + 2 + 3 + \cdots + 100$.

Solution

a. Solve Algebraically

For the arithmetic series $1 + 2 + 3 + \cdots + n$ we know that $a_1 = 1$ and $a_n = n$. Thus the sum of the series is

$$T_n = n\left(\frac{a_1 + a_n}{2}\right) = n\left(\frac{1 + n}{2}\right) = \frac{n(n + 1)}{2}.$$

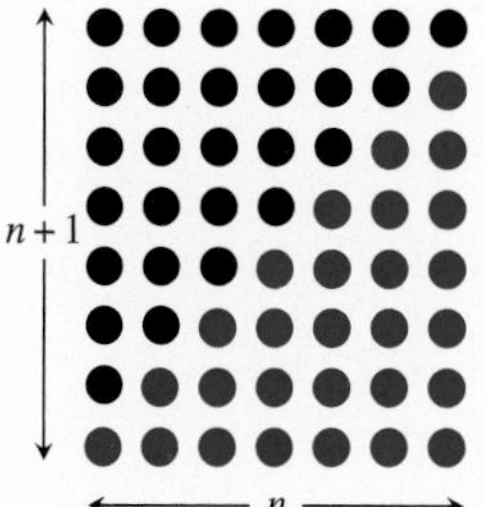

Figure 11.8

Support Geometrically

We can use two copies of the array in Figure 11.7 to form a rectangle, as in Figure 11.8, which suggests that

$$2T_n = n(n + 1) \qquad \text{and thus} \qquad T_n = \frac{n(n + 1)}{2}.$$

b. For $n = 100$, we have

$$T_{100} = 1 + 2 + 3 + \cdots + 100 = \frac{100(100 + 1)}{2} = 5050$$ ■

Finite Geometric Series

A series is a **geometric series** if its terms form a geometric sequence. Let

$$a_1 + a_2 + a_3 + \cdots + a_k + \cdots + a_n$$

be a geometric series with common ratio r. Then we know that $a_k = a_1 r^{k-1}$ for $k > 1$. If S_n is the sum of the series, then

$$S_n = a_1 + a_1 r + a_1 r^2 + \cdots + a_1 r^{k-1} + \cdots + a_1 r^{n-1}$$

$$S_n \cdot r = a_1 r + a_1 r^2 + a_1 r^3 + \cdots + a_1 r^k + \cdots + a_1 r^n$$

Subtracting the second equation from the first, we get

$$S_n - S_n \cdot r = a_1 - a_1 r^n$$

$$S_n(1 - r) = a_1(1 - r^n)$$

$$S_n = \frac{a_1(1 - r^n)}{1 - r}, \qquad r \neq 1$$

Teaching Note

The formula for the sum of a finite geometric series is the basis for the entire section. Make sure students have a firm understanding of the formula and how it is derived.

Sum of a Finite Geometric Series

The sum S_n of the geometric series $a_1 + a_2 + a_3 + \cdots + a_n$ with common ratio $r \neq 1$ is

$$S_n = \frac{a_1(1 - r^n)}{1 - r}.$$

If $r = 1$, the series is a constant series and $S_n = n \cdot a_1$.

■ EXAMPLE 5 Summing a Geometric Series

Sum the geometric series $4 - \frac{4}{3} + \frac{4}{9} - \frac{4}{27} + \cdots + 4\left(-\frac{1}{3}\right)^{10}$.

Solution

We know $a_1 = 4$ and $r = -1/3$. To find n, we use

$$a_n = a_1 r^{n-1}$$

$$4\left(-\frac{1}{3}\right)^{10} = 4\left(-\frac{1}{3}\right)^{n-1}$$

$$10 = n - 1$$

$$n = 11$$

$$S_{11} = \frac{a_1(1 - r^{11})}{1 - r}$$

$$S_{11} = \frac{(4)[1 - (-1/3)^{11}]}{1 - (-1/3)} = 3.0000169 \cdots$$

■

The sum of the series in Example 5 is very close to 3. Example 6 shows that the result in Example 5 was no coincidence.

■ EXAMPLE 6 Finding a Pattern

Let

$$S_n = 4 - \frac{4}{3} + \frac{4}{9} + \frac{4}{27} + \cdots + 4\left(-\frac{1}{3}\right)^{n-1},$$

which is a generalization of the series in Example 5. Graph the sequence $\{S_n\}$ and describe the behavior of S_n as $n \to \infty$.

Solution

Because each S_n is a finite geometric series, we can find an explicit rule for S_n by using $a_1 = 4$ and $r = -1/3$.

$$S_n = \frac{a_1(1 - r^n)}{1 - r}$$

$$= \frac{4[1 - (-1/3)^n]}{1 - (-1/3)}$$

$$= 3[1 - (-1/3)^n]$$

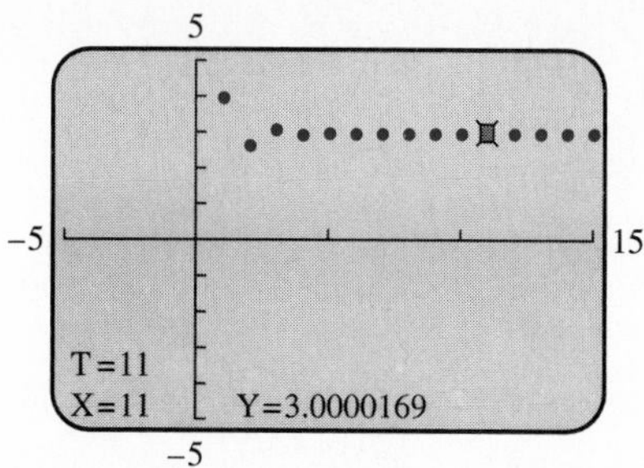

Figure 11.9 Notice that $x = t$ gives a value of n and y the corresponding value of S_n for the series of Example 6. Compare S_{11} with the value found in Example 5.

We can write the sequence $\{S_n\}$ with the parametric equations

$$x = t, \qquad y = S_t = 3[1 - (-1/3)^t],$$

for $T\text{min} = 1$ and $T\text{step} = 1$. The graph of the parametric equations in Figure 11.9 is the graph of the sequence $\{S_n\}$. We can use trace to suggest that $S_n \to 3$ as $n \to \infty$. ■

Alert

Students often fail to grasp the fact that an infinite series is not really a *sum.* Rather, it can be thought of as an infinite sequence, the sequence of partial sums.

Infinite Series and Sequences of Partial Sums

The sequence $\{S_n\}$ in Example 6 is a function whose domain is the natural numbers. Its graph in Figure 11.9 suggests that the horizontal line $y = 3$ is a horizontal asymptote of this function. This example also suggests how to give meaning to the sum of an infinite series

$$a_1 + a_2 + \cdots + a_n + \cdots.$$

First we form the **sequence of partial sums** $\{S_n\}$ defined by

$$S_n = a_1 + a_2 + \cdots + a_n, \qquad \text{for } n = 1, 2, 3, \ldots.$$

If the corresponding function $\{S_n\}$ has the line $y = S$ as a horizontal asymptote, the infinite series

$$a_1 + a_2 + \cdots + a_n + \cdots$$

converges to S, or has **sum** S. If there is no horizontal asymptote (the only other possibility) the series does *not* converge and is said to **diverge.**

If we can find an explicit rule for the sequence of partial sums $\{S_n\}$, then we can graph the sequence and decide if the corresponding infinite series converges. If we cannot find an explicit rule, then deciding whether the series converges may be difficult or impossible. We can always decide whether an infinite geometric series converges, as illustrated by Example 7.

Teaching Note

The grapher allows students to visualize the convergence of an infinite geometric series. The ease of visual understanding was not available before recent technological advances.

■ EXAMPLE 7 Summing Infinite Geometric Series

Determine whether the geometric series converge or diverge.

a. $5 + 10 + 20 + \cdots$ **b.** $1 + \dfrac{1}{2} + \dfrac{1}{4} + \cdots$

Solution

a. We know that $a_1 = 5$ and $r = 2$. The sum of the first n terms of the series is

$$S_n = \frac{a_1(1 - r^n)}{1 - r} = \frac{5(1 - 2^n)}{1 - 2} = 5(2^n - 1)$$

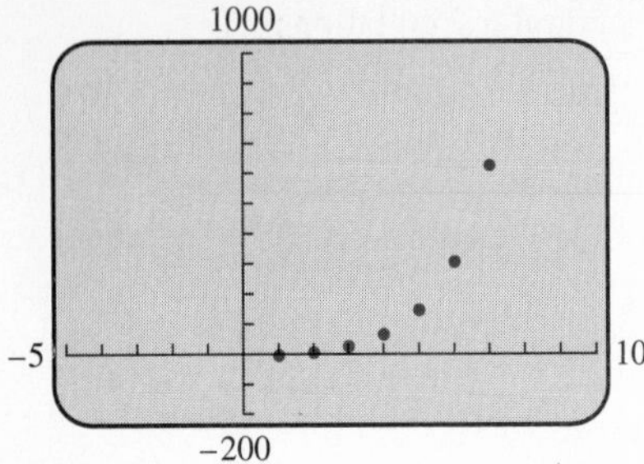

Figure 11.10

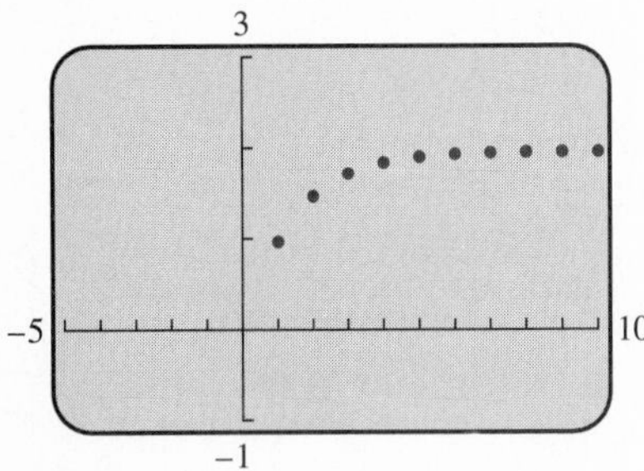

Figure 11.11

The graph of $S_n = 5(2^n - 1)$ in Figure 11.10 can be drawn by using the parametric equations

$$x = t, \qquad y = 5(2^t - 1).$$

The graph does *not* have a horizontal asymptote. In fact, the end behavior is similar to the exponential function $y = 2^x$ presented in Chapter 4. The series diverges.

b. We know that $a_1 = 1$ and $r = 1/2$. The sum of the first n terms of the series is

$$\begin{aligned} S_n &= \frac{a_1(1 - r^n)}{1 - r} \\ &= \frac{1[1 - (1/2)^n]}{1 - (1/2)} \\ &= 2[1 - (1/2)^n] \end{aligned}$$

The graph of $\{S_n\} = \{2[1 - (1/2)^n]\}$ in Figure 11.11 can be drawn by using the parametric equations

$$x = t, \qquad y = 2\left[1 - \left(\frac{1}{2}\right)^t\right].$$

The horizontal line $y = 2$ appears to be a horizontal asymptote. Thus the series converges and has sum 2. ■

We can generalize the results of Example 7. If $\{S_n\}$ is the sequence of partial sums of an infinite geometric series, then

$$S_n = \frac{a_1(1 - r^n)}{1 - r}.$$

If $|r| \geq 1$, $\{S_n\}$ does not have a horizontal asymptote. The series diverges in this case.

If $|r| < 1$, the horizontal asymptote of $\{S_n\}$ is $y = a_1/(1 - r)$. The series converges and has sum $S = a_1/(1 - r)$.

Alert

Some students will expect a series with $r = -1$ to converge because the partial sums do not go to infinity. Explain that the partial sums alternate between two values, so the series does not converge.

Sum of an Infinite Geometric Series

An infinite geometric series converges if and only if $|r| < 1$. In this case the sum is

$$S = \frac{a_1}{1 - r}.$$

■ EXAMPLE 8 Converting a Repeating Decimal to Rational Form

Express 0.234234234 . . . as a fraction.

Solution

Solve Algebraically

$$0.234234234 \cdots = 0.234 + 0.000234 + 0.000000234 + \cdots$$

This is an infinite geometric series, where

$$a_1 = 0.234 \quad \text{and} \quad r = 0.001.$$

Therefore

$$0.234234234 \cdots = \frac{a_1}{1 - r}$$

$$= \frac{0.234}{1 - 0.001}$$

$$= \frac{0.234}{0.999} = \frac{234}{999} = \frac{26}{111}$$

Support Numerically

Check by dividing 26 by 111. ■

Summation Notation

You may be wondering if there is an easier way to write an infinite series. If we have an explicit rule for the nth term, then the series can be expressed in *summation notation* using the capital Greek letter sigma (Σ). This symbol is used for summation because it is the Greek equivalent of the English letter S, for sum.

> **Summation Notation**
>
> The finite series $b_1 + b_2 + \cdots + b_k + \cdots + b_n$ can be written in the form
>
> $$\sum_{k=1}^{n} b_k,$$
>
> which is read "the sum of b_k from $k = 1$ to n."
>
> The infinite series $c_1 + c_2 + \cdots + c_k + \cdots$ can be written in the form
>
> $$\sum_{k=1}^{\infty} c_k,$$
>
> which is read "the sum of c_k from $k = 1$ to infinity."
>
> The variable k is the **index of summation.**

The expression $b_1 + b_2 + \cdots + b_k + \cdots + b_n$ is the **expanded form** of the series $\sum_{k=1}^{n} b_k$. Example 9 illustrates how to convert summation notation to expanded form.

■ EXAMPLE 9 Converting to Expanded Form

Write the series in expanded form and find its sum.

a. $\sum_{k=1}^{3} k^2$ **b.** $\sum_{j=1}^{3} j^2$

Solution

a. $\sum_{k=1}^{3} k^2 = 1^2 + 2^2 + 3^2 = 14$

b. $\sum_{j=1}^{3} j^2 = 1^2 + 2^2 + 3^2 = 14$ ■

In Example 10 watch how the expression written above the summation symbol ($\overset{?}{\sum}$) affects the expanded form of the sum. Also notice that part (b) gives a compact way of expressing a polynomial in x.

■ EXAMPLE 10 Converting to Expanded Form

Write the expanded form and evaluate, if possible.

a. $\sum_{k=1}^{9} (2k - 1)$ **b.** $\sum_{n=3}^{5} nx^n$

c. $\sum_{k=1}^{n} \left(\frac{1}{k+1} - \frac{1}{k}\right)$ **d.** $\sum_{j=1}^{\infty} 3\left(\frac{1}{5}\right)^j$

Solution

a.
$$\begin{aligned}\sum_{k=1}^{9} (2k - 1) &= (2(1) - 1) \\ &\quad + (2(2) - 1) + \cdots + (2(9) - 1) \\ &= 1 + 3 + 5 + 7 + 9 \\ &\quad + 11 + 13 + 15 + 17 && \text{9 terms} \\ &= 9\left(\frac{1 + 17}{2}\right) && \text{Sum of an arithmetic series} \\ &= 81\end{aligned}$$

REMARK

Example 10(a) shows that the sum of the first 9 odd positive integers is 9^2. This is not a coincidence. See Exercise 58.

b. $\sum_{n=3}^{5} nx^n = 3x^3 + 4x^4 + 5x^5$, a fifth-degree polynomial.

c.
$$\sum_{k=1}^{n}\left(\frac{1}{k+1}-\frac{1}{k}\right) = \left(\frac{1}{2}-\frac{1}{1}\right)+\left(\frac{1}{3}-\frac{1}{2}\right) + \left(\frac{1}{4}-\frac{1}{3}\right)+\cdots+\left(\frac{1}{n+1}-\frac{1}{n}\right)$$
$$= \frac{1}{2} - 1 + \frac{1}{3} - \frac{1}{2} + \frac{1}{4} - \frac{1}{3} + \cdots + \frac{1}{n+1} - \frac{1}{n}$$
$$= \frac{1}{n+1} - 1$$

This is a *collapsing* or *telescoping* series.

d. $\sum_{j=1}^{\infty} 3\left(\frac{1}{5}\right)^j = \frac{3}{5} + \frac{3}{25} + \frac{3}{125} + \cdots$ is an infinite geometric series with its first term 3/5 and common ratio 1/5. Its sum is

$$S = \frac{a_1}{1-r} = \frac{3/5}{1-(1/5)} = \frac{3/5}{4/5} = \frac{3}{4} = 0.75.$$ ■

Example 11 illustrates how to convert expanded forms of series to summation notation.

■ EXAMPLE 11 Converting to Summation Notation

Write each series using summation notation assuming that the series in (a) and the denominators in (b) are arithmetic.

a. $-6 - 2 + 2 + 6 + 10 + \cdots + 74$

b. $-\frac{1}{4} + \frac{1}{8} - \frac{1}{12} + \frac{1}{16} - \frac{1}{20} + \cdots$

Solution

a. Because $a_1 = -6$ and $d = 4$, we have

$$a_n = a_1 + (n-1)d$$
$$= -6 + (n-1)(4)$$
$$= 4n - 10$$

Now we can find the number of terms in this series.

$$\begin{aligned} a_n &= 74 \\ 4n - 10 &= 74 \\ 4n &= 84 \\ n &= 21 \end{aligned}$$

In summation notation the series is $\sum_{n=1}^{21} (4n - 10)$.

b. The alternating signs can be generated by the powers of -1. The denominators, being arithmetic, are of the form $4k$. Thus

$$-\frac{1}{4} + \frac{1}{8} - \frac{1}{12} + \frac{1}{16} - \frac{1}{20} + \cdots = \sum_{k=1}^{\infty} (-1)^k \frac{1}{4k}. \quad ■$$

Notes on Examples

Example 11b shows how to use summation notation to express a series in which the terms have alternating signs.

Here are how some of the properties of algebra (see the Prerequisite Chapter, Section 2) are extended to summation notation.

Algebraic Properties of Summation

If $\{a_n\}$ and $\{b_n\}$ are two sequences, n is a positive integer, and c is any real number, then

1. $\sum_{k=1}^{n} ca_k = c \sum_{k=1}^{n} a_k$
2. $\sum_{k=1}^{n} (a_k + b_k) = \sum_{k=1}^{n} a_k + \sum_{k=1}^{n} b_k$
3. $\sum_{k=1}^{n} (a_k - b_k) = \sum_{k=1}^{n} a_k - \sum_{k=1}^{n} b_k$

REMARK

Property 1 is an extension of the distributive property. Properties 2 and 3 involve the commutative, associative, and distributive properties.

These properties allow us to manipulate series written in summation notation, as illustrated in Example 12.

Teaching Note

These properties also apply to infinite series, provided that

$$\sum_{k=1}^{\infty} a_k \text{ and } \sum_{k=1}^{\infty} b_k$$

converge.

■ EXAMPLE 12 APPLICATION: Calculating Total Revenue

From 1990 through 1995, LBH Telecommunications, Inc., approximated its annual revenue by using the formula

$$R_t = 46.39t^2 + 708.71t + 2651.16,$$

where R_t is the annual income (in millions of dollars), $t = 0$ represents 1990, $t = 1$ represents 1991, and so on. Find the total revenue for the 6-year period.

Solution

Solve Algebraically

Use the summation properties and finite series formulas to evaluate the series.

$$\sum_{t=0}^{5} R_t = \sum_{t=0}^{5} (46.39t^2 + 708.71t + 2651.16)$$

$$= \sum_{t=0}^{5} 46.39t^2 + \sum_{t=0}^{5} 708.71t + \sum_{t=0}^{5} 2651.16$$

$$= 46.39 \sum_{t=0}^{5} t^2 + 708.71 \sum_{t=0}^{5} t + 2651.16 \sum_{t=0}^{5} 1$$

Each of the three sums is easy to compute.

$$\sum_{t=0}^{5} t^2 = 0 + 1 + 4 + 9 + 16 + 25 = 55$$

$$\sum_{t=0}^{5} t = 0 + 1 + 2 + 3 + 4 + 5 = 15$$

$$\sum_{t=0}^{5} 1 = 1 + 1 + 1 + 1 + 1 + 1 = 6$$

So the original sum is

$$\sum_{t=0}^{5} R_t = 46.39(55) + 708.71(15) + 2651.16(6)$$

$$= 29{,}089.06.$$

Interpret

The total revenue for LBH Telecommunications from 1990 through 1995 was $29,089,060,000. ■

Follow-up

Ask why a geometric series does not converge when $|r| \geq 1$. **(The terms do not approach zero.)**

Assignment Guide

Day 1: Ex. 1–45, multiples of 3
Day 2: Ex. 47, 48, 50–53, 55, 60, 61

Cooperative Learning

Group Activity: Ex. 57–59

Notes on Exercises

Ex. 33–46, 52–53, and 58–59 focus on summation notation and related summation formulas.
Ex. 49–51 are application problems related to geometric series.

Ongoing Assessment

Self-Assessment: Ex. 13, 17, 23, 31, 37, 43, 49
Embedded Assessment: Ex. 48

Quick Review 11.2

In Exercises 1–4, determine whether the decimal representation of the rational number repeats or terminates.

1. $\frac{3}{4}$

2. $\frac{2}{5}$

3. $\frac{1}{7}$

4. $\frac{2}{3}$

In Exercises 5–8, evaluate the expression for the given values of the variable.

5. $\frac{n(n+1)}{2}$, $n = 5$, $n = 12$

6. $\frac{k(k+3)}{2}$, $k = 6$, $k = 11$

7. $\frac{3(1 - r^6)}{1 - r}$, $r = \frac{1}{2}$, $r = \frac{3}{2}$

8. $\frac{4(1 - r^5)}{1 - r}$, $r = \frac{1}{3}$, $r = \frac{3}{4}$

In Exercises 9–16, determine the end behavior.

9. $\frac{x}{x+1} \to ?$ as $x \to \infty$

10. $\frac{5t}{t+8} \to ?$ as $t \to \infty$

11. $\frac{2x}{3x+1} \to ?$ as $x \to \infty$

12. $\frac{3x}{6x+2} \to ?$ as $x \to \infty$

13. $\frac{2x^3+1}{x^2-5} \to ?$ as $x \to \infty$

14. $\frac{2x+1}{x^2-3x+1} \to ?$ as $x \to \infty$

15. $\frac{x^2+1}{2x^2-x-3} \to ?$ as $x \to \infty$

16. $\frac{5x^2-2}{2x^2+3x-1} \to ?$ as $x \to \infty$

In Exercises 17 and 18, use the distributive property to find the expanded form.

17. $a(b_1 + b_2 + b_3)$

18. $(1 - x)(1 + x + x^2 + x^3 + x^4)$

SECTION EXERCISES 11.2

In Exercises 1–8, find the sum of the arithmetic series.

1. $-7 - 3 + 1 + 5 + 9 + 13$

2. $-8 - 1 + 6 + 13 + 20 + 27$

3. $1 + 2 + 3 + 4 + \cdots + 80$

4. $2 + 4 + 6 + 8 + \cdots + 70$

5. $-3 - 5 - 7 - \cdots - 21$

6. $-9 - 14 - 19 - \cdots - 89$

7. $117 + 110 + 103 + \cdots + 33$

8. $111 + 108 + 105 + \cdots + 27$

In Exercises 9–12, find the sum of the geometric series.

9. $3 + 6 + 12 + \cdots + 12\,288$

10. $5 + 15 + 45 + \cdots + 98\,415$

11. $42 + 7 + \frac{7}{6} + \cdots + 42\left(\frac{1}{6}\right)^8$

12. $42 - 7 + \frac{7}{6} - \cdots + 42\left(-\frac{1}{6}\right)^9$

In Exercises 13–20, find the sum of the first n terms of the sequence. The sequence is either arithmetic or geometric.

13. $2, 5, 8, \ldots; n = 10$

14. $14, 8, 2, \ldots; n = 9$

15. $112, 106, 100, \ldots; n = 15$

16. $96, 88, 80, \ldots; n = 16$

17. $4, -2, 1, -\frac{1}{2}, \ldots; n = 12$

18. $6, -3, \frac{3}{2}, -\frac{3}{4}, \ldots; n = 11$

19. $-1, 11, -121, \ldots; n = 9$

20. $-2, 24, -288, \ldots; n = 8$

In Exercises 21 and 22, graph the sequence of partial sums S_n. What number do the sums appear to be approaching?

21. $6 - \frac{12}{5} + \frac{24}{25} - \frac{48}{125} + \cdots$

22. $5 - \frac{15}{4} + \frac{45}{16} - \frac{135}{64} + \cdots$

In Exercises 23–28, determine whether the infinite geometric series converges, and if so, find its sum. Graph the sequence of partial sums to support your result.

23. $6 + 3 + \frac{3}{2} + \frac{3}{4} + \cdots$

24. $4 + \frac{4}{3} + \frac{4}{9} + \frac{4}{27} + \cdots$

25. $\frac{1}{64} + \frac{1}{32} + \frac{1}{16} + \frac{1}{8} + \cdots$

26. $\frac{1}{48} + \frac{1}{16} + \frac{3}{16} + \frac{9}{16} + \cdots$

27. $\sum_{j=1}^{\infty} 3\left(\frac{1}{4}\right)^j$

28. $\sum_{n=1}^{\infty} 5\left(\frac{2}{3}\right)^n$

In Exercises 29–32, express the rational number as a fraction.

29. $7.14141414\cdots$

30. $5.93939393\cdots$

31. $-17.268268268\cdots$

32. $-12.876876876\cdots$

In Exercises 33 and 34, expand the sums and show that they are equal.

33. $\sum_{j=1}^{n} (2j - j^2)$ and $\sum_{k=1}^{n} (2k - k^2)$

34. $\sum_{k=1}^{n} (3\sqrt{k} - k)$ and $\sum_{i=1}^{n} (3\sqrt{i} - i)$

In Exercises 35–40, write the expanded form and evaluate, if possible.

35. $\sum_{k=1}^{7} (k^2 + 3k + 1)$

36. $\sum_{n=1}^{5} (n^3 + n + 2)$

37. $\sum_{j=1}^{\infty} \left(-\frac{1}{4}\right)^j$

38. $\sum_{k=1}^{\infty} \left(-\frac{2}{5}\right)^k$

39. $\sum_{r=1}^{n+2} r$

40. $\sum_{k=1}^{m+3} k$

In Exercises 41–46, write each series using summation notation, assuming the suggested pattern continues.

41. $-7 - 1 + 5 + 11 + \cdots + 53$

42. $2 + 5 + 8 + 11 + \cdots + 29$

43. $1 + 4 + 9 + \cdots + (n + 1)^2$

44. $1 + 8 + 27 + \cdots + (n + 1)^3$

45. $6 - 12 + 24 - 48 + \cdots$

46. $5 - 15 + 45 - 135 + \cdots$

In Exercises 47–61, solve the problem.

47. *End Behavior* Does

$$f(x) = 2\left(\frac{1 - 1.05^x}{1 - 1.05}\right)$$

have a horizontal asymptote? How does this relate to the convergence or divergence of the series $2 + 2.1 + 2.205 + 2.31525 + \cdots$?

48. *Rainy Days* Table 11.2 lists the average number of days with rain each month for four major U.S. cities. Answer each question by summing the appropriate series.

a. Which of the four cities has the most rainy days? The fewest?

b. Which month is the rainiest for the four cities? The least rainy?

Table 11.2 Number of Rainy Days

Month	Chicago	Los Angeles	New York	Seattle
January	11	6	12	18
February	10	6	10	16
March	12	6	12	16
April	11	4	11	13
May	12	2	11	12
June	11	1	10	9
July	9	0	12	4
August	9	0	10	5
September	9	1	9	8
October	9	2	9	13
November	10	3	9	17
December	11	6	10	19

Source: National Geographic Atlas of the World (rev. 6th ed., 1992).

49. *Follow the Bouncing Ball* When "superballs" sprang upon the scene in the 1960s, kids across the United States were amazed that these hard rubber balls could bounce to 90% of the height from which they dropped. If a superball is dropped from a height of 2 m, how far does it travel by the time it hits the ground for the tenth time? (*Hint:* The ball goes down to the first bounce, then up *and* down thereafter.)

50. *Total Revenue* Assume that the formula given in Example 12 is a valid approximation for annual revenue through the year 2000. What is LBH's total revenue for the 11-year period from 1990 through 2000?

51. *Future Value of an Annuity* Show that the future value of an annuity is the sum of a finite geometric series with the first term $a_1 = R$, the periodic annuity deposit, and the common ratio $r = (1 + i)$, the compounding factor for each period with interest i. (*Hint:* See Example 1.)

EXTENDING THE IDEAS

52. *Finding a Pattern* Write the finite series $-1 + 2 + 7 + 14 + 23 + \cdots + 62$ in summation notation.

53. *Summation Notation* Summation notation allows us to rewrite many formulas compactly. For example, the formula for triangular numbers can be written

$$T_n = \sum_{k=1}^{n} k = \frac{n(n + 1)}{2}.$$

Use summation notation to rewrite each formula.

a. The two forms for the sum of a finite arithmetic series.

b. The form for the sum of an infinite arithmetic series.
c. The sum of a finite geometric series.
d. The sum of a convergent infinite geometric series.

54. *Square Numbers and Triangular Numbers* Prove that the sum of consecutive triangular numbers is a square number; that is, prove

$$T_{n-1} + T_n = n^2$$

for all positive integers $n \geq 2$. Use both a geometric and an algebraic approach.

55. *Fibonacci Sequence and Series* Complete the following table, where F_n is the nth term of the Fibonacci sequence and S_n is the nth partial sum of the Fibonacci series. Make a conjecture based on the numerical evidence in the table.

$$S_n = \sum_{k=1}^{n} F_k.$$

n	F_n	S_n	$F_{n+2} - 1$
1	1		
2	1		
3	2		
4			
5			
6			
7			
8			
9			

56. *Grapher Activity* If your grapher has a sequence mode and sequence and sum commands, try the following. Gather evidence to support the claim that

$$\sum_{k=1}^{n} k^2 = \frac{n(n+1)(2n+1)}{6}$$

by making a table for

$$u_n = \sum_{k=1}^{n} k^2 \quad \text{and} \quad v_n = \frac{n(n+1)(2n+1)}{6},$$

for $n = 1, 2, 3, \ldots, 7$. Verify that you can produce the results shown in the figure.

```
Un=sum seq(K²,K,
1,n,1)
Vn=n(n+1)(2n+1)/
6
```

n	U_n	V_n
1	1	1
2	5	5
3	14	14
4	30	30
5	55	55
6	91	91
7	140	140

Un=sum seq(K²,K...

For Exercise 56

Notice that $\sum_{k=1}^{n} k^2 = 1 + 4 + 9 + \cdots + n^2$, the sum of the first n squares. Use the same method to gather evidence supporting the claim that

$$\sum_{k=1}^{n} k^3 = \left[\frac{n(n+1)}{2}\right]^2.$$

57. *Population Density* The *National Geographic Picture Atlas of Our Fifty States* (1991) groups the states into 10 regions. The two largest groupings are the Heartland (Table 11.3) and the Southeast (Table 11.4). Population and area data for the two regions are given in the tables. The populations are official 1990 U.S. Census figures.

a. What is the total population of each region?
b. What is the total area of each region?
c. What is the population density (in persons per square mile) of each region?
d. **Writing to Learn** For the two regions, compute the population density of each state. What is the average of the seven state population densities for each region? Explain why these answers differ from those found in part (c).

Table 11.3 The Heartland

State	Population	Area (mi^2)
Iowa	2,776,755	56,275
Kansas	2,477,574	82,277
Minnesota	4,375,099	84,402
Missouri	5,117,073	69,697
Nebraska	1,578,385	77,355
North Dakota	638,800	70,703
South Dakota	696,004	77,116

Table 11.4 The Southeast

State	Population	Area (mi^2)
Alabama	4,040,587	51,705
Arkansas	2,350,725	53,187
Florida	12,937,926	58,644
Georgia	6,478,216	58,910
Louisiana	4,219,973	47,751
Mississippi	2,573,216	47,689
S. Carolina	3,486,703	31,113

58. *Sum of Odd Integers* Example 10(a) showed that the sum of the first nine odd positive integers is 9^2. Decide whether the sum of the first n odd positive integers is n^2 for every n, or

$$\sum_{k=1}^{n} (2k - 1) = n^2.$$

If you think it is, try to prove it. If you think it isn't, show why.

59. *Using Summation Properties*

a. Give a convincing argument that $\sum_{k=1}^{n} c = cn$.

b. Given $\sum_{k=1}^{n} k = \dfrac{n(n+1)}{2}$, use the summation properties to prove that

$$\sum_{k=1}^{n} [a_1 + (k-1)d] = \frac{n}{2}[2a_1 + (n-1)d].$$

60. *Harmonic Series* Graph the sequence of partial sums of the *harmonic series:*

$$1 + \frac{1}{2} + \frac{1}{3} + \frac{1}{4} + \cdots + \frac{1}{n} + \cdots.$$

Overlay on it the graph of $f(x) = \ln x$. The resulting picture should support the claim that

$$1 + \frac{1}{2} + \frac{1}{3} + \frac{1}{4} + \cdots + \frac{1}{n} \geq \ln n,$$

for all positive integers n. Make a table of values to further support this claim. Explain why the claim implies that the harmonic series must diverge.

61. Prove that every nonzero infinite arithmetic series diverges.

11.3 MATHEMATICAL INDUCTION

Introduction • Principle of Mathematical Induction • Extended Principle of Mathematical Induction

Objective

Students will be able to use the principle of mathematical induction to prove mathematical generalizations.

Motivate

Ask . . .

Suppose a sequence is defined recursively by $a_1 = 2$ and $a_{n+1} = a_n^2 + 3a_n$. How can you be sure that a_{364} is defined?

Introduction

Mathematical induction is a method of proof used to show that a statement or formula holds for some infinite set of integers, usually the set of positive integers. So mathematical induction is really not induction, but rather *deduction.* In this section, we mainly use mathematical induction to prove formulas for sequences and series. In later sections, we will use it to establish other statements. We begin by examining a problem that cuts across three areas of mathematics: series, geometry, and number theory.

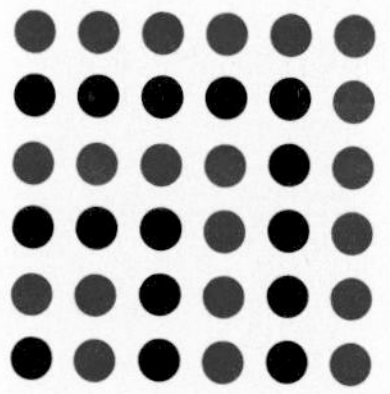

Principle of Mathematical Induction

Figure 11.12 displays the odd integers 1, 3, 5, 7, 9, and 11 as collections of dots of alternating color. Because the total number of dots is 6^2, we have

$$1 + 3 + 5 + 7 + 9 + 11 = 6^2.$$

This statement can be generalized to

$$1 + 3 + 5 + \cdots + (2n - 1) = n^2,$$

for any positive integer n.

1 + 3 + 5 + 7 + 9 + 11

Figure 11.12

With enough time and effort we could draw a diagram like the one in Figure 11.12 for any positive integer n, but we still would not have proved that this relationship holds for all positive integers. To do that we need a mathematical tool such as the following *axiom.*

Axiom of Induction

If S is a set of positive integers that satisfies the conditions

1. S contains the integer 1, and,
2. if S contains the integer k, then S contains the integer $k + 1$,

then S is the entire set of positive integers.

REMARK

You may think that the axiom of induction is obvious. In fact, axioms are accepted as "facts" without proof and generally are intuitively obvious.

The axiom of induction can be illustrated by a domino analogy. (See Figure 11.13.) Imagine lining up an infinite sequence of dominoes so that, if any one is knocked down, the one to its immediate right will also be knocked down. Then, if the first domino is knocked over, they all will fall over.

Teaching Note

The authors have found that it is useful for students to discuss in small groups the axiom of induction. The students should focus on convincing one another that the set S consists of the entire set of positive integers.

Alert

Students need to understand the difference between an implication and its converse when discussing the axiom of induction. Some of them may reason incorrectly as follows: Since 1 is in S and $k + 1 = 1$, $k = 0$. So 0 must be in S.

Figure 11.13

Condition 1 of the axiom of induction corresponds to the first domino being knocked over, and condition 2 corresponds to the positions of the dominoes being such that one falling causes the next one to fall.

Once the axiom of induction is accepted, it can be used to prove that statements are true for all integers n. Pay careful attention to how it is used in Example 1.

■ EXAMPLE 1 Using the Axiom of Induction

Prove that $1 + 3 + 5 + \cdots + (2n - 1) = n^2$ is true for all positive integers n.

Solution

Let S be the set of positive integers for which

$$1 + 3 + 5 + \cdots + (2n - 1) = n^2$$

is true. We show that the two conditions of the axiom of induction are true, so we can conclude that S is the entire set of positive integers.

1. The integer 1 is in S because $1 = 1^2$.
2. We assume that S contains the integer k. We want to deduce, then, that S contains $k + 1$.

$$1 + 3 + 5 + \cdots + (2k - 1) = k^2 \qquad \text{Assumed true}$$

$$\begin{aligned} 1 + 3 + 5 + \cdots + (2k - 1) & \\ + [2(k + 1) - 1] &= k^2 + [2(k + 1) - 1] && \text{Add } [2(k + 1) - 1]. \\ &= k^2 + 2k + 1 && \text{Simplify right side.} \\ &= (k + 1)^2 && \text{Factor right side.} \end{aligned}$$

This shows that the original statement is true if n is replaced by $k + 1$. Thus S contains $k + 1$, and we conclude that the original statement is true for *all n*. ■

We now confirm the formula for the nth term of an arithmetic sequence given in Section 11.1.

■ EXAMPLE 2 Using the Axiom of Induction

Use mathematical induction to prove that the nth term of an arithmetic sequence with its first term a_1 and common difference d is $a_n = a_1 + (n - 1)d$.

Solution

Let S be the set of positive integers for which

$$a_n = a_1 + (n - 1)d$$

is true. We show that the two conditions of the axiom of induction are true so that we can conclude that S is the entire set of positive integers.

1. The integer 1 is in S because $a_1 + (1 - 1)d = a_1$.
2. We assume that S contains the integer k. We want to deduce, then, that S contains $k + 1$.

$$\begin{aligned} a_{k+1} &= a_k + d && \text{Recursive definition} \\ &= a_1 + (k - 1)d + d && \text{Assumed that } a_k = a_1 + (k - 1)d \\ &= a_1 + [(k + 1) - 1]d && \text{Combine terms.} \end{aligned}$$

This shows that the original statement is true if n is replaced by $k + 1$. Thus S contains $k + 1$, and we conclude that the original statement is true for all n. ■

The following principle allows us to bypass the set S used in the axiom of induction.

Principle of Mathematical Induction

Let P_n be a statement about the integer n. Then P_n is true for all positive integers n provided that the following conditions are satisfied:

1. P_1 is true, and,
2. if P_k is true, then P_{k+1} is true.

Examples 3 through 5 illustrate proofs using the principle of mathematical induction.

Alert

Most students find mathematical induction to be a difficult topic. There is a lot of confusion about what is assumed and what needs to be shown. A common mistake is forgetting to show that P_1 is true. Stress that when we assume that P_k is true, we are making an assumption about a particular value of k.

EXAMPLE 3 Using Mathematical Induction

Use mathematical induction to prove that the statement

$$P_n:\quad 1^2 + 3^2 + 5^2 + \cdots + (2n - 1)^2 = \frac{n(2n - 1)(2n + 1)}{3}$$

is true for every positive integer n.

Solution

Is P_1 true?

$$1^2 \stackrel{?}{=} \frac{1[2(1) - 1][2(1) + 1]}{3}$$

$$1 = \frac{1 \cdot 1 \cdot 3}{3} \quad \text{Yes}$$

Assume that P_k is true.

$$1^2 + 3^2 + 5^2 + \cdots + (2k - 1)^2 = \frac{k(2k - 1)(2k + 1)}{3}$$

Add $[2(k + 1) - 1]^2 = (2k + 1)^2$ to both sides.

$$\begin{aligned}
1^2 + 3^2 + 5^2 + \cdots + (2k - 1)^2 + (2k + 1)^2 &= \frac{k(2k - 1)(2k + 1)}{3} + (2k + 1)^2 \\
&= \frac{k(2k - 1)(2k + 1) + 3(2k + 1)^2}{3} \\
&= \frac{(2k + 1)[k(2k - 1) + 3(2k + 1)]}{3} \\
&= \frac{(2k + 1)(2k^2 + 5k + 3)}{3} \\
&= \frac{(2k + 1)(k + 1)(2k + 3)}{3} \\
&= \frac{(k + 1)(2k + 1)(2k + 3)}{3}
\end{aligned}$$

This establishes the statement P_{k+1}. Thus we have shown that, if P_k is true, then P_{k+1} is true. By the principle of mathematical induction, we conclude that P_n is true for all n. ■

■ EXAMPLE 4 Using Mathematical Induction

Use mathematical induction to prove

$$P_n:\quad 1^3 + 2^3 + \cdots + n^3 = \frac{n^2(n+1)^2}{4}.$$

Solution

Is P_1 true?

$$1^3 \stackrel{?}{=} \frac{1^2(1+1)^2}{4}$$

$$1 = \frac{1 \cdot 4}{4} \quad \text{Yes}$$

Assume that P_k is true.

$$1^3 + 2^3 + \cdots + k^3 = \frac{k^2(k+1)^2}{4}$$

$$\begin{aligned} 1^3 + 2^3 + \cdots + k^3 + (k+1)^3 &= \frac{k^2(k+1)^2}{4} + (k+1)^3 \qquad \text{Add } (k+1)^3. \\ &= \frac{k^2(k+1)^2 + 4(k+1)^3}{4} \\ &= \frac{(k+1)^2[k^2 + 4(k+1)]}{4} \\ &= \frac{(k+1)^2(k^2 + 4k + 4)}{4} \\ &= \frac{(k+1)^2(k+2)^2}{4} \end{aligned}$$

This establishes the statement P_{k+1}. Thus we have shown that, if P_k is true, then P_{k+1} is true. By the principle of mathematical induction, we conclude that P_n is true for all n. ■

In Chapter 4 we showed that, if \$200 are invested in a savings account that pays 9% compounded monthly, then the amount in the account n months later is

$$200\left(1 + \frac{0.09}{12}\right)^n.$$

Notice that 0.09/12 is the monthly interest rate.

We then generalized this result in the following way. Suppose A dollars are invested in an account that compounds interest and that the interest rate per compounding period is i. Then the amount in the account after n periods is $A(1 + i)^n$. Now we prove the generalization using mathematical induction.

■ EXAMPLE 5 APPLICATION: Proving the Compound Interest Formula

Use mathematical induction to prove the statement P_n: $A(1 + i)^n$ is the amount in an account after n periods if A dollars are invested at interest rate i per compounding period.

Solution

The amount in the account after one period is

$$A + Ai = A(1 + i).$$

Thus P_1 is true. Assume that P_k is true.

$$\text{Amount after } k \text{ periods} = A(1 + i)^k \qquad \text{Assumed to be true}$$

$$\begin{aligned} \text{Amount after } k + 1 \text{ periods} &= [A(1 + i)^k] + [A(1 + i)^k]i \\ &= A(1 + i)^k(1 + i) \qquad \text{Factor.} \\ &= A(1 + i)^{k+1} \end{aligned}$$

This establishes the statement P_{k+1}. Thus we have shown that, if P_k is true, then P_{k+1} is true. By the principle of mathematical induction, we conclude that P_n is true for all n. ■

Extended Principle of Mathematical Induction

The axiom of induction can be generalized to begin with any fixed integer c.

Follow-up

Ask whether or not mathematical induction can be used to prove the statement P_n:
$\frac{1}{1\cdot 2} + \frac{1}{2\cdot 3} + \cdots + \frac{1}{n(n+1)} = \frac{2n+1}{n+1}$. **(No, P_1 is not true.)**

Assignment Guide

Day 1: Ex. 1, 4, 5, 8, 10, 11, 15, 16, 17, 19, 20, 21, 22
Day 2: Ex. 23, 24, 25, 28, 31, 34, 36, 39, 40

Extended Principle of Mathematical Induction

Let c be an integer and P_n be a statement about the integer n. Then P_n is true for all integers $n \geq c$ provided that the following conditions are satisfied:

1. P_c is true, and,
2. if P_k is true for $k \geq c$, then P_{k+1} is true.

Because a polygon always has at least three sides, Example 6 (from geometry) illustrates the need for the extended principle of mathematical induction.

Cooperative Learning

Group Activity: Ex. 38

Notes on Exercises

Ex. 9–12 require students to write the statements P_1, P_k, and P_{k+1} for a given statement P_n.
Ex. 35–37 require extended mathematical induction.
Ex. 39 is related to Ex. 55 in Section 11.2.

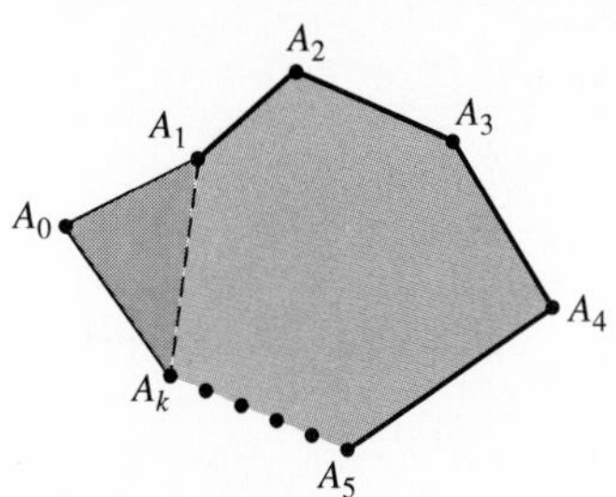

Figure 11.14

Ongoing Assessment

Self-Assessment: Ex. 3, 13, 27, 33, 35
Embedded Assessment: Ex. 8, 16, 20, 22, 24, 28, 36

EXAMPLE 6 Proving the Interior Angle Formula

Use extended mathematical induction to prove P_n:

If $n \geq 3$, the sum of the measures of the interior angles of an n-sided polygon is $180°(n - 2)$.

Solution

P_3 is true because a 3-sided polygon is a triangle with angle sum

$$180° = 180°(3 - 2).$$

Assume that P_k is true. To show that P_{k+1} is true, we take a $(k + 1)$-sided polygon and break it into a k-sided polygon and a triangle in order to find the sum of the measures of its angles.

Let $A_0, A_1, A_2, \ldots, A_k$ $(k \geq c = 3)$ be the vertices of a $(k + 1)$-sided polygon, as in Figure 11.14. Omit vertex A_0 and connect A_1 to A_k to form a k-sided polygon with vertices $A_1, A_2, \ldots, A_k$, whose angle measures we may assume satisfy the formula $180°(k - 2)$. The remaining piece of the original polygon is the triangle whose vertices are A_0, A_1, and A_k and whose angles sum to 180°.

Thus the sum of the measures of the angles of the original $(k + 1)$-sided polygon is

$$180° + 180°(k - 2) = 180°(k - 1) = 180°[(k + 1) - 1].$$

This is the statement P_{k+1}. Thus we have shown that, if P_k is true for $k \geq 3$, then P_{k+1} is true. By the extended principle of mathematical induction, we conclude that P_n is true for all $n \geq 3$. ■

Quick Review 11.3

In Exercises 1–4, expand the products.

1. $n(n + 5)$

2. $(n + 2)(n - 3)$

3. $k(k + 1)(k + 2)$

4. $k(k + 2)(k + 4)$

In Exercises 5–8, factor the polynomial

5. $n^2 + 2n - 3$

6. $k^2 + 5k + 4$

7. $k^3 + 3k^2 + 3k + 1$

8. $n^3 - 3n^2 + 3n - 1$

In Exercises 9–16, evaluate the indicated expressions involving function notation.

9. $f(x) = x + 4;\quad f(1),\ f(t),\ f(t + 1)$

10. $f(x) = x - 3;\quad f(1),\ f(t),\ f(t + 1)$

11. $f(n) = \dfrac{n}{n + 1};\quad f(1),\ f(k),\ f(k + 1)$

12. $f(n) = \dfrac{n}{n - 7};\quad f(a),\ f(k),\ f(k + 1)$

13. $P(n) = \dfrac{2n}{3n + 1};\quad P(1),\ P(k),\ P(k + 1)$

14. $P(n) = \dfrac{3n}{6n + 2};\quad P(1),\ P(k),\ P(k + 1)$

15. $P(n) = 2n^2 - n - 3;\quad P(1),\ P(k),\ P(k + 1)$

16. $P(n) = 3n^2 + 4n - 1;\quad P(1),\ P(k),\ P(k + 1)$

SECTION EXERCISES 11.3

In Exercises 1–4, use mathematical induction to prove that the statement holds for all positive integers.

1. $2 + 4 + 6 + \cdots + 2n = n^2 + n$

2. $8 + 10 + 12 + \cdots + (2n + 6) = n^2 + 7n$

3. $6 + 10 + 14 + \cdots + (4n + 2) = n(2n + 4)$

4. $14 + 18 + 22 + \cdots + (4n + 10) = 2n(n + 6)$

In Exercises 5–8, state an explicit rule for the nth term of the recursively defined sequence. Then use mathematical induction to prove the rule.

5. $a_n = a_{n-1} + 5,\ a_1 = 3$

6. $a_n = a_{n-1} + 2,\ a_1 = 7$

7. $a_n = 3a_{n-1},\ a_1 = 2$

8. $a_n = 5a_{n-1},\ a_1 = 3$

In Exercises 9–12, write the statements P_1, P_k, and P_{k+1}. (Do not write a proof.)

9. P_n: $1 + 2 + \cdots + n = \dfrac{n(n + 1)}{2}$

10. P_n: $1^2 + 2^2 + \cdots + n^2 = \dfrac{n(n + 1)(2n + 1)}{6}$

11. P_n: $\dfrac{1}{1 \cdot 2} + \dfrac{1}{2 \cdot 3} + \cdots + \dfrac{1}{n \cdot (n + 1)} = \dfrac{n}{n + 1}$

12. P_n: $\sum_{k=1}^{n} k^4 = \dfrac{n(n + 1)(2n + 1)(3n^2 + 3n - 1)}{30}$

In Exercises 13–20, use mathematical induction to prove that the statement holds for all positive integers.

13. $1 + 5 + 9 + \cdots + (4n - 3) = n(2n - 1)$

14. $1 + 2 + 2^2 + \cdots + 2^{n-1} = 2^n - 1$

15. $\dfrac{1}{1 \cdot 2} + \dfrac{1}{2 \cdot 3} + \dfrac{1}{3 \cdot 4} + \cdots + \dfrac{1}{n(n + 1)} = \dfrac{n}{n + 1}$

16. $\dfrac{1}{1 \cdot 3} + \dfrac{1}{3 \cdot 5} + \dfrac{1}{5 \cdot 7} + \cdots + \dfrac{1}{(2n - 1)(2n + 1)} = \dfrac{n}{2n + 1}$

17. $2^n \geq 2n$

18. $3^n \geq 3n$

19. 3 is a factor of $n^3 + 2n$.

20. 6 is a factor of $n(n + 1)(n + 2)$.

In Exercises 21 and 22, use mathematical induction to prove that the statement holds for all positive integers.

21. The nth term of a geometric sequence is $a_n = a_1 \cdot r^{n-1}$ where a_1 is the first term and r is the common ratio.

22. The sum of the first n terms of an arithmetic sequence with its first term a_1 and common difference d is

$$S_n = \frac{n}{2}[2a_1 + (n - 1)d].$$

In Exercises 23 and 24, use mathematical induction to confirm that the formula holds for all positive integers.

23. *Triangular Numbers* $\sum_{k=1}^{n} k = \dfrac{n(n + 1)}{2}$

24. *Sum of the First n Squares*

$$\sum_{k=1}^{n} k^2 = \frac{n(n + 1)(2n + 1)}{6}$$

In Exercises 25–30, find the sum.

25. $1 + 2 + 3 + \cdots + 500$

26. $1^2 + 2^2 + \cdots + 250^2$

27. $4 + 5 + 6 + \cdots + n$

28. $1^3 + 2^3 + 3^3 + \cdots + 75^3$

29. $1 + 2 + 4 + 8 + \cdots + 2^{34}$

30. $1 + 8 + 27 + \cdots + 3375$

In Exercises 31–34, find the sum in terms of n.

31. $\sum_{k=1}^{n} (k^2 - 3k + 4)$

32. $\sum_{k=1}^{n} (2k^2 + 5k - 2)$

33. $\sum_{k=1}^{n} (k^3 - 1)$

34. $\sum_{k=1}^{n} (k^3 + 4k - 5)$

In Exercises 35 and 36, use extended mathematical induction to prove the statement.

35. $3n - 4 \geq n$, for all $n \geq 2$.

36. $2^n \geq n^2$, for all $n \geq 4$.

37. Prove that the sum of the measures of the exterior angles of any polygon is 360°.

EXTENDING THE IDEAS

38. *Sum of Cubes* Use previously established summation formulas to confirm that the sum of the first n cubes is given by

$$\sum_{k=1}^{n} k^3 = \left(\sum_{k=1}^{n} k\right)^2 = T_n^2$$

for all positive integers, n. (Do not use mathematical induction.)

In Exercises 39 and 40, use mathematical induction to confirm that the statement holds for all positive integers.

39. *Fibonacci Sequence and Series* $F_{n+2} - 1 = \sum_{k=1}^{n} F_k$, where $\{F_n\}$ is the Fibonacci sequence.

40. If $\{a_n\}$ is the sequence $\sqrt{2}, \sqrt{2 + \sqrt{2}}, \sqrt{2 + \sqrt{2 + \sqrt{2}}}, \ldots$, then $a_n < 2$.

11.4 BINOMIAL THEOREM

Objective

Students will be able to expand a power of a binomial using the binomial theorem or Pascal's triangle. They will also find the coefficient of a given term of a binomial expansion.

Motivate

Have students expand $(2x + y)^4$ using the distributive property.
$(16x^4 + 32x^3y + 24x^2y^2 + 8xy^3 + y^4)$
Explain that this section will give them easier ways to do this kind of calculation.

Introduction • Pascal's Triangle • Binomial Coefficients • Binomial Theorem • Polynomial Expansions

Introduction

A binomial is a polynomial with two terms. In the Prerequisite Chapter you learned how to square and cube binomials. In this section you will learn efficient ways to write the *expanded form* of any positive integer power of a binomial. Our goal is to verify and extend the following *polynomial expansions* for $(a + b)^n$.

$$
\begin{aligned}
(a + b)^1 &= a + b \\
(a + b)^2 &= a^2 + 2ab + b^2 \\
(a + b)^3 &= a^3 + 3a^2b + 3ab^2 + b^3 \\
(a + b)^4 &= a^4 + 4a^3b + 6a^2b^2 + 4ab^3 + b^4 \\
(a + b)^5 &= a^5 + 5a^4b + 10a^3b^2 + 10a^2b^3 + 5ab^4 + b^5
\end{aligned}
$$

Pascal's Triangle

Notice the pattern of the coefficients of the five expanded polynomials. If we insert a 1 at the top, it forms the following triangle. The outer edges consist of 1's. Every other entry is the sum of the two numbers immediately above it.

Row 0						1					
Row 1					1		1				
Row 2				1		2		1			
Row 3			1		3		3		1		
Row 4		1		4		6		4		1	
Row 5	1		5		10		10		5		1
⋮	⋰	⋮		⋮		⋮		⋮		⋮	⋱

REMARK

The 1 at the top of Pascal's triangle makes sense because, if $a + b \neq 0$, $(a + b)^0 = 1$.

Although this triangular number pattern was known to the Chinese mathematician Zhu Shijie (also spelled Chu Shi-kie) as early as 1303, it is called **Pascal's triangle** for the French mathematician Blaise Pascal, who in 1654 used the triangle to find coefficients in polynomial expansions.

Notes on Examples

Note that row 6 of Pascal's triangle is the row that is related to $(a + b)^6$. In some sense, this is the seventh row of the triangle.

EXAMPLE 1 Extending Pascal's Triangle

Find row 6 of Pascal's Triangle.

Solution

The outer numbers are 1's. Each number between them is the sum of the two numbers immediately above it. So row 6 can be found from row 5 as follows:

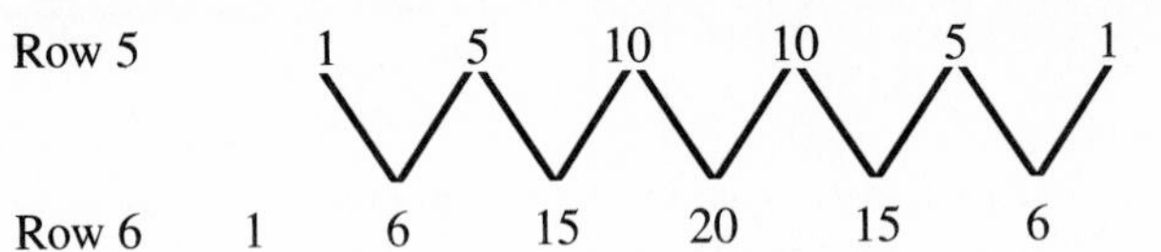

■

EXAMPLE 2 Using a Row of Pascal's Triangle

Expand $(a + b)^6$ and use Pascal's triangle to find the coefficients.

Solution

Use the entries in row 6 as coefficients.

$$(a + b)^6 = a^6 + 6a^5b + 15a^4b^2 + 20a^3b^3 + 15a^2b^4 + 6ab^5 + b^6$$

■

Binomial Coefficients

The numbers in Pascal's triangle are *binomial coefficients*. They can be found recursively row by row with the method in Example 1. Binomial coefficients can also be found by using an explicit formula involving integer products called *factorials*.

Definition Factorial Notation

For any positive integer n, ***n* factorial** is

$$n! = n \cdot (n - 1) \cdot (n - 2) \cdot \cdots \cdot 3 \cdot 2 \cdot 1.$$

Zero factorial is $0! = 1$.

REMARK

The exclamation point was first used for factorials in 1808 by Christian Kramp (1760–1826) to reflect that the factorial of a number can be surprisingly large. For example, $10! = 3{,}628{,}800$ and $70! \geq 10^{100}$. Experiment with your grapher to see the largest factorial that it can compute.

Teaching Note

Note that the binomial coefficient n choose r is denoted on most graphers as ${}_nC_r$.

These definitions yield

$$\begin{aligned} 0! &= 1 \\ 1! &= 1 \\ 2! &= 2 \cdot 1 = 2 \\ 3! &= 3 \cdot 2 \cdot 1 = 6 \\ &\vdots \\ n! &= n \cdot (n-1)! \end{aligned}$$

The following definition provides a direct way to find binomial coefficients.

Definition Binomial Coefficient

If n and r are integers, where $n \geq r \geq 0$, then

$$\binom{n}{r} = \frac{n!}{r!(n-r)!}$$

is a **binomial coefficient.** The symbol ${}_nC_r$ is an alternative for $\binom{n}{r}$.

To become familiar with binomial coefficients you need to compute a few by hand.

■ EXAMPLE 3 Finding Binomial Coefficients

Compute each binomial coefficient by hand. Then use the factorial or binomial coefficient feature of a grapher to support your result.

a. $\binom{6}{4}$ **b.** $\binom{11}{3}$ **c.** $\binom{7}{0}$ **d.** $\binom{5}{5}$

Solution

a. $$\binom{6}{4} = \frac{6!}{4!(6-4)!} = \frac{6!}{4!2!} = \frac{6 \cdot 5 \cdot \cancel{4} \cdot \cancel{3} \cdot \cancel{2} \cdot \cancel{1}}{\cancel{4} \cdot \cancel{3} \cdot \cancel{2} \cdot \cancel{1} \cdot 2 \cdot 1} = \frac{6 \cdot 5}{2 \cdot 1} = 15$$

b. $$\binom{11}{3} = \frac{11!}{3!8!} = \frac{11 \cdot 10 \cdot 9 \cdot \cancel{8!}}{3 \cdot 2 \cdot 1 \cdot \cancel{8!}} = \frac{11 \cdot 10 \cdot 9}{3 \cdot 2 \cdot 1} = 165$$

c. $$\binom{7}{0} = \frac{7!}{0!7!} = \frac{7!}{1 \cdot 7!} = 1$$

d. $$\binom{5}{5} = \frac{5!}{5!0!} = \frac{5!}{5! \cdot 1} = 1$$

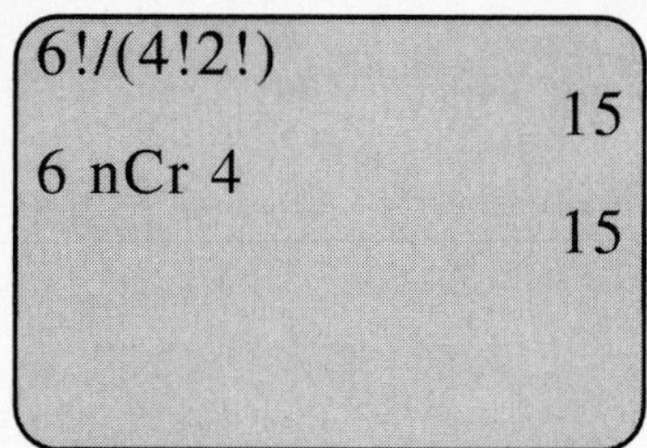

Figure 11.15

Support Numerically

Grapher values of $\binom{6}{4} = {}_6C_4$ are shown in Figure 11.15. Similar support can be used in the other three parts. ■

Example 4 shows that the binomial coefficients $\binom{n}{r}$ form row n,

$$\binom{n}{0}, \binom{n}{1}, \binom{n}{2}, \ldots, \binom{n}{n},$$

of Pascal's triangle. In other words, Example 4 will prove that the following triangle of binomial coefficients is in fact Pascal's triangle.

$$\begin{array}{lccccccccccc}
\text{Row } 0 & & & & & & \binom{0}{0} & & & & & \\
\text{Row } 1 & & & & & \binom{1}{0} & & \binom{1}{1} & & & & \\
\text{Row } 2 & & & & \binom{2}{0} & & \binom{2}{1} & & \binom{2}{2} & & & \\
\text{Row } 3 & & & \binom{3}{0} & & \binom{3}{1} & & \binom{3}{2} & & \binom{3}{3} & & \\
\text{Row } 4 & & \binom{4}{0} & & \binom{4}{1} & & \binom{4}{2} & & \binom{4}{3} & & \binom{4}{4} & \\
\text{Row } 5 & \binom{5}{0} & & \binom{5}{1} & & \binom{5}{2} & & \binom{5}{3} & & \binom{5}{4} & & \binom{5}{5} \\
\vdots & \unicode{x22F0} & \vdots & & \vdots & & \vdots & & \vdots & & \vdots & \ddots
\end{array}$$

Recall that, in Pascal's triangle, (a) the outer numbers are 1's and (b) otherwise a number is the sum of the two numbers immediately above it. Example 4 establishes these two properties for the displayed triangle of binomial coefficients.

■ EXAMPLE 4 Verifying Properties of Binomial Coefficients

Prove the following properties of binomial coefficients:

a. $\binom{n}{0} = \binom{n}{n} = 1, \quad \text{for all integers } n \geq 0.$

b. $\binom{k}{r-1} + \binom{k}{r} = \binom{k+1}{r}$, for all integers $k \geq r \geq 1$.

Solution

a. $\binom{n}{0} = \dfrac{n!}{0! \cdot n!} = \dfrac{n!}{1 \cdot n!} = 1$ and $\binom{n}{n} = \dfrac{n!}{n! \cdot 0!} = \dfrac{n!}{n! \cdot 1} = 1$

b. You might be tempted to use mathematical induction to prove this statement, but there is a fairly straightforward proof using algebraic manipulation. The proof repeatedly uses the idea that $(n + 1)n! = (n + 1)!$ for all integers $n \geq 0$. Also watch how the two fractions on the right side are converted to fractions with common denominators.

$$\begin{aligned}\binom{k}{r-1} + \binom{k}{r} &= \frac{k!}{(r-1)![k-(r-1)]!} + \frac{k!}{r!(k-r)!}\\ &= \frac{rk!}{r(r-1)!(k-r+1)!} + \frac{(k-r+1)k!}{r!(k-r+1)(k-r)!}\\ &= \frac{[r+(k-r+1)]k!}{r!(k-r+1)!}\\ &= \frac{(k+1)k!}{r!(k+1-r)!}\\ &= \frac{(k+1)!}{r![(k+1)-r]!}\\ &= \binom{k+1}{r}\end{aligned}$$

■

Binomial Theorem

Example 4 and mathematical induction can be used to prove the binomial theorem. (See Exercise 42.)

Alert

Students often make sign errors when they apply the binomial theorem to an expression of the form $(a - b)^n$. Notice that this must be interpreted as $(a + (-b))^n$. Example 6 on page 732 illustrates a correct method.

The Binomial Theorem

For any positive integer n,

$$(a+b)^n = \binom{n}{0}a^n + \binom{n}{1}a^{n-1}b + \cdots + \binom{n}{r}a^{n-r}b^r + \cdots + \binom{n}{n}b^n.$$

Polynomial Expansions

We can use the binomial theorem to show a polynomial expansion of any positive integer power of a binomial. Because we have shown the numbers in Pascal's triangle are binomial coefficients, we can either use Pascal's triangle or the formula

$$\binom{n}{r} = \frac{n!}{r!(n-r)!}$$

to obtain the coefficients in these polynomial expansions.

Notes on Examples

Examples 5–7 illustrate that the coefficients of a binomial expansion are not necessarily equal to one of the binomial coefficients. However, if the binomial does not contain fractions, the coefficient of any term is a multiple of a binomial coefficient.

■ EXAMPLE 5 Using the Binomial Theorem

Use the binomial theorem to expand $(x + 2)^6$. Support your result numerically.

Solution

According to the binomial theorem,

$$\begin{aligned}(x+2)^6 &= x^6 + \binom{6}{1}x^5 \cdot 2 + \binom{6}{2}x^4 \cdot 2^2 + \binom{6}{3}x^3 \cdot 2^3 + \binom{6}{4}x^2 \cdot 2^4 + \binom{6}{5}x \cdot 2^5 + 2^6 \\ &= x^6 + 6 \cdot 2x^5 + 15 \cdot 4x^4 + 20 \cdot 8x^3 + 15 \cdot 16x^2 + 6 \cdot 32x + 64 \\ &= x^6 + 12x^5 + 60x^4 + 160x^3 + 240x^2 + 192x + 64\end{aligned}$$

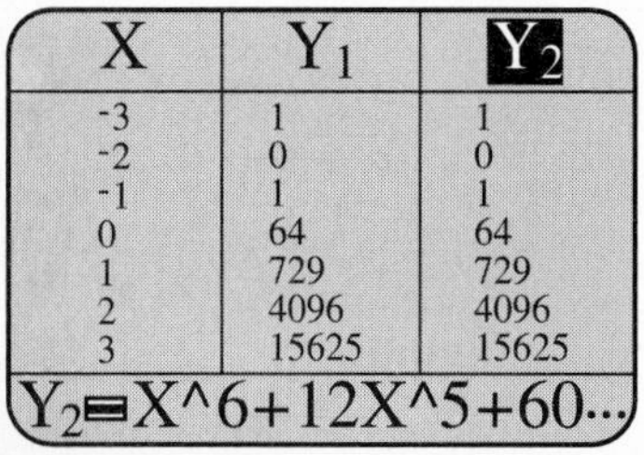

X	Y1	Y2
-3	1	1
-2	0	0
-1	1	1
0	64	64
1	729	729
2	4096	4096
3	15625	15625

Y2=X^6+12X^5+60...

Figure 11.16

Support Numerically

The grapher screen (Figure 11.16) shows identical tables of values for $y_1 = (x + 2)^6$ and $y_2 = x^6 + 12x^5 + 60x^4 + 160x^3 + 240x^2 + 192x + 64$. ■

■ EXAMPLE 6 Finding a Polynomial Expansion

Expand $(2x - 3y^2)^4$.

Solution

The binomial $2x - 3y^2$ can be rewritten as $2x + (-3y^2)$. Let $a = 2x$ and $b = -3y^2$ and substitute into the expression $(a + b)^4$.

$$(a+b)^4 = a^4 + 4a^3b + 6a^2b^2 + 4ab^3 + b^4$$

$$\begin{aligned}(2x - 3y^2)^4 &= (2x)^4 + 4(2x)^3(-3y^2) + 6(2x)^2(-3y^2)^2 + 4(2x)(-3y^2)^3 + (-3y^2)^4 \\ &= 16x^4 + 4 \cdot 8(-3)x^3y^2 + 6 \cdot 4 \cdot 9x^2y^4 + 4 \cdot 2(-27)xy^6 + 81y^8 \\ &= 16x^4 - 96x^3y^2 + 216x^2y^4 - 216xy^6 + 81y^8\end{aligned}$$

■

Follow-up

Ask how many terms the expanded form of $(x + y)^{23}$ has. **(24)**

Assignment Guide

Day 1: Ex. 1, 3, 6, 8, 9, 12, 15, 16, 18, 19, 22
Day 2: Ex. 23, 28, 29, 32, 34–38, 41

Cooperative Learning

Group Activity: Ex. 43

Notes on Exercises

Ex. 9–16 should be done first without a grapher in order to familiarize students with the algebraic manipulations required to calculate binomial coefficients.
Ex. 17, 18, 37, 38, 41, 42, and 44 require students to prove various statements relating to binomial coefficients or the binomial theorem.
Ex. 40 and 43 are related to the fact that for small values of x, $(1 + x)^n \approx 1 + nx$.

Ongoing Assessment

Self-Assessment: Ex. 11, 21, 27, 31
Embedded Assessment: Ex. 18, 34, 36

In summation notation the binomial theorem states that

$$(a + b)^n = \sum_{r=0}^{n} \binom{n}{r} a^{n-r} b^r.$$

This form makes it easy to see that any particular term within a polynomial expansion of $(a + b)^n$ has the form

$$\binom{n}{r} a^{n-r} b^r.$$

This is the key idea in Example 7.

■ EXAMPLE 7 Using the Binomial Theorem

Find the x^3y^{12} term in the polynomial expansion of $(2x + 3y)^{15}$.

Solution

The x^3y^{12} term will have the form

$$\binom{n}{r} a^{n-r} b^r,$$

so let $a = 2x$, $b = 3y$, $n = 15$, and $r = 12$.

$$\begin{aligned}\binom{n}{r} a^{n-r} b^r &= \binom{15}{12} (2x)^{15-12}(3y)^{12} \\ &= 455(2x)^3(3y)^{12} \\ &= 455(2^3)(3^{12})x^3y^{12} \\ &= 1{,}934{,}445{,}240x^3y^{12}\end{aligned}$$

■

Quick Review 11.4

In Exercises 1–4, evaluate the expression.

1. $\dfrac{6 \cdot 5 \cdot 4 \cdot 3 \cdot 2 \cdot 1}{3 \cdot 2 \cdot 1}$

2. $\dfrac{11 \cdot 10 \cdot 9 \cdot 8 \cdot 7 \cdot 6 \cdot 5 \cdot 4 \cdot 3 \cdot 2 \cdot 1}{9 \cdot 8 \cdot 7 \cdot 6 \cdot 5 \cdot 4 \cdot 3 \cdot 2 \cdot 1}$

3. $\dfrac{6 \cdot 5 \cdot 4 \cdot 3 \cdot 2 \cdot 1}{3 \cdot 2 \cdot 1 \cdot 3 \cdot 2 \cdot 1}$

4. $\dfrac{11 \cdot 10 \cdot 9 \cdot 8 \cdot 7 \cdot 6 \cdot 5 \cdot 4 \cdot 3 \cdot 2 \cdot 1}{2 \cdot 1 \cdot 9 \cdot 8 \cdot 7 \cdot 6 \cdot 5 \cdot 4 \cdot 3 \cdot 2 \cdot 1}$

In Exercises 5–14, use the distributive property to expand the binomial.

5. $(x + y)^2$

6. $(a + b)^2$

7. $(5x - y)^2$

8. $(a - 3b)^2$

9. $(3s + 2t)^2$

10. $(3p - 4q)^2$

11. $(u + v)^3$

12. $(b - c)^3$

13. $(2x - 3y)^3$

14. $(4m + 3n)^3$

SECTION EXERCISES 11.4

In Exercises 1 and 2, find the indicated row of Pascal's triangle.

1. Row 8 **2.** Row 10

In Exercises 3–8, expand the binomial and use Pascal's triangle to find the coefficients.

3. $(a + b)^5$ **4.** $(x + y)^6$

5. $(x - y)^7$ **6.** $(a - b)^6$

7. $(3x - y)^4$ **8.** $(a - 2b^2)^4$

In Exercises 9–16, first evaluate the expression without using the factorial or binomial coefficient features of a grapher. Then support your answer by using a grapher.

9. $\binom{7}{3}$ **10.** $\binom{9}{2}$

11. $\binom{11}{9}$ **12.** $\binom{15}{11}$

13. $\binom{12}{6}$ **14.** $\binom{13}{7}$

15. $\binom{166}{166}$ **16.** $\binom{166}{0}$

In Exercises 17 and 18, prove the given property of binomial coefficients.

17. $\binom{n}{1} = \binom{n}{n-1} = n,$ for all integers $n \geq 1$.

18. $\binom{n}{r} = \binom{n}{n-r},$ for all integers $n \geq r \geq 0$.

In Exercises 19–22, use the binomial theorem to find a polynomial expansion for the function. Support your result numerically.

19. $f(x) = (x - 2)^5$ **20.** $g(x) = (x + 3)^6$

21. $h(x) = (2x - 1)^7$ **22.** $f(x) = (3x + 4)^5$

In Exercises 23–28, use the binomial theorem to expand each binomial power.

23. $(2x + y)^4$ **24.** $(2y - 3x)^5$

25. $(\sqrt{x} - \sqrt{y})^6$ **26.** $(\sqrt{x} + \sqrt{3})^4$

27. $(x^{-2} + 3)^5$ **28.** $(a - b^{-3})^7$

In Exercises 29–32, find the coefficient of the given term in the polynomial expansion.

29. $x^{11}y^3$ term, $(x + y)^{14}$ **30.** x^5y^8 term, $(x + y)^{13}$

31. x^4 term, $(x - 2)^{12}$ **32.** x^7 term, $(x - 3)^{11}$

In Exercises 33–44, solve the problem.

33. Determine the largest integer n for which your calculator can compute $n!$.

34. Prove that

$$\binom{n}{2} + \binom{n+1}{2} = n^2,$$

for all integers $n \geq 2$.

35. Find a counterexample to show that each statement is *false*.

a. $(n + m)! = n! + m!$

b. $(nm)! = n!m!$

36. Let the sequence $\{a_n\}$ be given by $a_0 = 1$ and $a_n = n \cdot a_{n-1}$ for $n = 1, 2, 3, \ldots$. Use mathematical induction to prove that $a_n = n!$

EXTENDING THE IDEAS

37. a. Use Pascal's triangle to show that

$$\binom{n}{0} + \binom{n}{1} + \binom{n}{2} + \cdots + \binom{n}{n} = 2^n,$$

for $n = 0, 1, 2, \ldots, 5$.

b. Prove that

$$\binom{n}{0} + \binom{n}{1} + \binom{n}{2} + \cdots + \binom{n}{n} = 2^n$$

is true for all integers $n \geq 0$ by expanding $(1 + 1)^n$.

38. a. Prove that $\binom{n+1}{2}$ is a triangular number (see Section 11.2) for every positive integer n.

b. Make a copy of Pascal's triangle and circle these triangular numbers.

39. Give graphical and numerical justifications for

$$\sum_{k=0}^{\infty} \frac{1}{k!} = e.$$

40. Writing to Learn Use the binomial theorem to write an expansion for each of the following. Think of the expansion as a finite series and make a table for the sequence of partial sums. Use a calculator to evaluate each expression. How well does each term in the sequence of partial sums approximate the value of the expression you calculate?

a. $(1 + 0.03)^6$ **b.** $(1 + 0.08)^5$

41. *Making Mathematical Connections* Explain why $\binom{n}{r}$ is a positive integer for all integers $n \geq r \geq 0$. Then letting $r = 3$, justify that the product of any three consecutive positive integers is divisible by 6.

42. Use mathematical induction to prove the binomial theorem. (*Hint:* To establish the second condition of mathematical induction, write

$$(a + b)^{k+1} = a(a + b)^k + b(a + b)^k$$

and use Example 4.)

43. *Generalizing the Binomial Theorem* Expand $\sqrt{1 + x}$ as an infinite series by thinking of it as $(1 + x)^{1/2}$ and following the pattern of the binomial theorem. Numerically support the validity of your expansion for $|x| < 1$ by trying several such values of x.

44. *Alternative Form of the Binomial Theorem.*

a. Use the summation notation form of the binomial theorem to prove the following alternative form of the theorem:

$$(a + b)^n = \sum_{k=0}^{n} \binom{n}{k} a^k b^{n-k}.$$

(*Hint:* Let $k = n - r$.)

b. Let $0 < x \leq 1$, $a = x$, and $b = 1 - x$. Prove that

$$\sum_{k=0}^{n} \binom{n}{k} x^k (1 - x)^{n-k} = 1.$$

11.5
BASIC COMBINATORICS

Introduction • Fundamental Counting Principle • Permutations • Combinations

Objective

Students will be able to use the fundamental counting principle, permutations, or combinations to count the number of ways that a task can be done.

Motivate

Ask students to find the number of possible ways one can choose two marbles from a bag containing three marbles: (1) if the order is important **(6)**, and (2) if the order is not important. **(3)**

Introduction

We all know how to count. This section investigates more sophisticated kinds of counting. Counting is at the heart of discrete algebra, and we will use it when we discuss probability in Section 11.6.

Many simple and complex problems involve determining the number of elements in a set or the number of ways objects can be arranged and combined. The branch of mathematics that addresses such problems is **combinatorics.**

Fundamental Counting Principle

Some counting problems can be solved by organizing the set to be counted. We begin with two such examples.

■ EXAMPLE 1 Counting by Listing

How many three-letter codes can be formed from the letters A, B, C, and D if no letter can be repeated?

Solution

List the possibilities in an orderly manner:

ABC	ABD	ACB	ACD	ADB	ADC	A first
BAC	BAD	BCA	BCD	BDA	BDC	B first
CAB	CAD	CBA	CBD	CDA	CDB	C first
DAB	DAC	DBA	DBC	DCA	DCB	D first

There are $4 \cdot 6 = 24$ possibilities. ■

If the number of possibilities is small enough, a tree diagram may help to structure the counting task.

■ **EXAMPLE 2 Using a Tree Diagram**

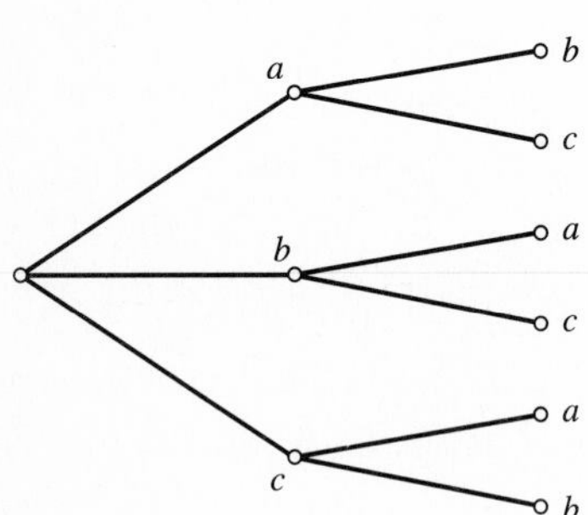

Figure 11.17

In an experiment on animal behavior, a psychologist uses a sequence of two food rewards. Each reward is one of three possible types. How many different sequences are possible if the reward types are not repeated within any sequence?

Solution

Designate the three types of rewards as *a, b,* and *c*. The tree diagram in Figure 11.17 shows all the possibilities. Each pathway along the tree from left to right represents a possible reward sequence. Because there are six pathways, there are six possible sequences. ■

A tree diagram is not practical if the set of possibilities being counted is large. In such cases, a multiplicative method modeled on tree-diagram counting is used.

Teaching Note

The fundamental counting principle refers to breaking a procedure into n successive ordered stages. Encourage students to identify what the stages are and to mentally go through the procedure. For example, in Example 3 the procedure involves choosing two letters and three digits. Draw a strip of five squares to begin the procedure. Then have students choose a letter. Ask them in how many ways this can be done. **(26)** Next, have them choose a different letter and ask them in how many ways this can be done. **(25)** Continue in this way.

Fundamental Counting Principle

If a procedure P has a finite sequence of n stages $S_1, S_2, \ldots, S_n$ and if

$$S_1 \text{ can occur in } r_1 \text{ ways,}$$
$$S_2 \text{ can occur in } r_2 \text{ ways,}$$
$$\vdots$$
$$S_n \text{ can occur in } r_n \text{ ways,}$$

then the number of ways the procedure P can occur is

$$r_1 \cdot r_2 \cdot \cdots \cdot r_n.$$

> **Teaching Note**
>
> Encourage students to ask the following questions prior to actually doing any counting. (1) What is the process that is being completed? Does the order matter (either in terms of completing the problem correctly or simplifying calculations)? (2) What is the first stage? How many ways can it be completed? (3) What is the second stage? How many ways can it be completed? And so on.

The key to solving a counting problem is to identify the stages S_1, $S_2, \ldots, S_n$. Once these stages are identified, next determine how many ways each stage can occur. Then multiply these values to solve the counting problem.

■ EXAMPLE 3 Using the Fundamental Counting Principle

The license plates in South Saskatoba begin with two English letters followed by three digits. How many different license plates are possible? Assume that none of the characters (letters or numbers) repeat within each license plate.

Solution

P: Stamping out a license plate
S_1: Stamping out the first character
S_2: Stamping out the second character

and so on. We have

Stages:	S_1	S_2	S_3	S_4	S_5
Number of ways to complete the stage:	26	25	10	9	8

Number of ways P can occur: $26 \cdot 25 \cdot 10 \cdot 9 \cdot 8 = 468{,}000$

Interpret

There are 468,000 possible license plates in South Saskatoba. ■

■ EXAMPLE 4 Using the Fundamental Counting Principle

There are eight sprinters in the finals of a 100-m dash. How many different outcomes are possible for this race? Assume there are no ties.

Solution

P: Runners list in order of finish
S_1: Sprinter who finishes first
S_2: Sprinter who finishes second

and so on. We have

Stages:	S_1	S_2	S_3	S_4	S_5	S_6	S_7	S_8
Number of ways to complete the stage:	8	7	6	5	4	3	2	1

Number of ways P can occur: $8 \cdot 7 \cdot 6 \cdot 5 \cdot 4 \cdot 3 \cdot 2 \cdot 1 = 8! = 40{,}320$

Interpret

There are 40,320 possible outcomes for this race. ■

Permutations

Order was important in the arrangements considered in Examples 3 and 4. Such ordered arrangements are called **permutations.** In general, the number of permutations of n objects is $n!$, as suggested by Example 4. Many situations dealing with permutations of objects concern only a subset of the objects involved. Example 5 examines one such situation.

EXAMPLE 5 Finding the Number of Permutations

Suppose that the 31 members of your college math club are to select four officers: president, vice-president, secretary, and treasurer. In how many ways can these officers be selected?

Solution

Selection for president:	31 ways
Selection for vice-president:	30 ways
Selection for secretary:	29 ways
Selection for treasurer:	28 ways
Ways to select all four officers:	$31 \cdot 30 \cdot 29 \cdot 28 = 755{,}160$

Example 5 can be generalized to the situation of ordering r objects selected from n objects, where $n \geq r$.

Teaching Note

Note that most graphers are capable of calculating factorials, ${}_nP_r$, and ${}_nC_r$. Encourage students to use these capabilities after they become familiar with the algebraic manipulations involved.

Definition Permutation Formula

If n and r are positive integers with $n \geq r$, then the number of **permutations of n objects taken r at a time** is

$${}_nP_r = \frac{n!}{(n-r)!}.$$

We can see why this formula is valid by applying the fundamental counting principle. Notice that there are n choices for the first object, $n - 1$ for the second, and so on. Because r objects are being ordered, there are r factors. Therefore

$${}_nP_r = n \cdot (n-1) \cdot \cdots \cdot (n-r+1) \qquad r \text{ factors}$$

$$= n \cdot (n-1) \cdot \cdots \cdot (n-r+1)\frac{(n-r) \cdot \cdots \cdot 3 \cdot 2 \cdot 1}{(n-r) \cdot \cdots \cdot 3 \cdot 2 \cdot 1} \qquad \text{Multiply by 1.}$$

$$= \frac{n!}{(n-r)!} \qquad \text{Factorial notation}$$

Notes on Examples

Examples 6, 7, and 9 show some paper-and-pencil methods for calculating permutations and combinations. It is important that students have the algebraic skills to perform these operations since the numbers in some counting problems may exceed the capacity of a calculator.

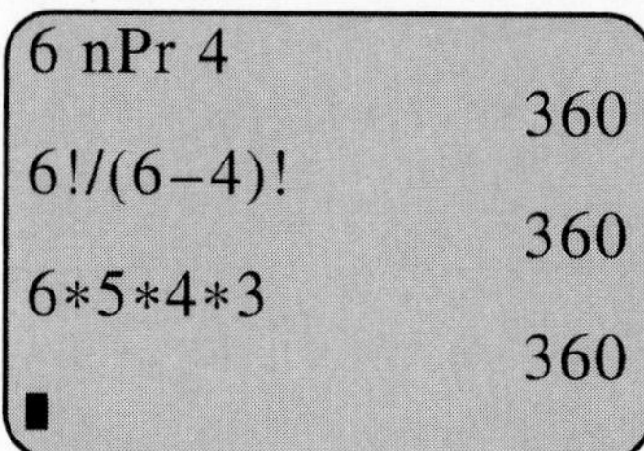

Figure 11.18

EXAMPLE 6 Using the Permutation Formula

Evaluate each expression by hand. Use a grapher to support each result numerically if possible.

a. ${}_6P_4$ **b.** ${}_{11}P_3$ **c.** ${}_nP_3$

Solution

a. $${}_6P_4 = \frac{6!}{(6-4)!} = \frac{6!}{2!} = \frac{6 \cdot 5 \cdot 4 \cdot 3 \cdot 2 \cdot 1}{2 \cdot 1}$$
$$= 6 \cdot 5 \cdot 4 \cdot 3 = 360$$

b. $${}_{11}P_3 = \frac{11!}{8!} = \frac{11 \cdot 10 \cdot 9 \cdot 8!}{8!} = 11 \cdot 10 \cdot 9 = 990$$

c. $${}_nP_3 = \frac{n!}{(n-3)!} = \frac{n \cdot (n-1) \cdot (n-2) \cdot (n-3)!}{(n-3)!}$$
$$= n(n-1)(n-2)$$

Support Numerically

Figure 11.18 shows three ways to evaluate ${}_6P_4$ of part (a). Similar support can be used in (b). Direct numerical support is not possible for (c). ■

EXAMPLE 7 Counting Permutations

How many five-letter code words can be formed from English letters if none are repeated?

Solution

This problem asks us to count the number of permutations of 26 letters taken 5 at a time.

$${}_{26}P_5 = \frac{26!}{(26-5)!} = \frac{26!}{21!} = 26 \cdot 25 \cdot 24 \cdot 23 \cdot 22 = 7{,}893{,}600$$

Interpret

There are 7,893,600 five-letter code words. ■

Combinations

When we want to find the number of subsets of r elements that can be selected from a set of n elements without regard to order, we are counting **combinations.**

Alert

Some students find counting problems very challenging. They often want to apply a formula blindly rather than go through a thinking process. In particular, students frequently forget to pay attention to whether or not the order is important in a particular problem.

■ EXAMPLE 8 Counting Subsets

How many committees of three can be selected from a group of four people?

Solution

Call the four people *A, B, C,* and *D* and consider from Example 1 the listing of the ${}_4P_3 = 24$ orderings possible for selecting three of them.

ABC	ABD	ACB	ACD	ADB	ADC
BAC	BAD	BCA	BCD	BDA	BDC
CAB	CAD	CBA	CBD	CDA	CDB
DAB	DAC	DBA	DBC	DCA	DCB

Notice that a specific committee of three, say, *A, B,* and *C,* appears $3! = 6$ times in the list. Each of these ordered possibilities gives rise to the same committee.

ABC ACB BAC BCA CAB CBA

Therefore the total number of committees is equal to the ${}_4P_3$ orderings possible divided by the 3! different orderings for each committee.

$$\text{Number of committees} = \frac{{}_4P_3}{3!} = \frac{4!/(4-3)!}{3!} = \frac{4!}{(4-3)!3!} = \frac{4!}{1!3!} = 4.$$

■

In general, the number of combinations of r elements selected from any set of n elements is ${}_nP_r$, the number of permutations of the n elements taken r at a time, divided by $r!$, the number of orderings of each set of r elements.

$$\text{Number of combinations} = \frac{{}_nP_r}{r!} = \frac{n!}{(n-r)!r!}$$

Teaching Note

It may be worthwhile to show students the reason why the combination formula is the same as the formula for binomial coefficients. For example, suppose we wish to find the coefficient of xy^2 in $(x + y)^3$ by using the distributive property. The relevant terms in the expansion will be *xyy, yxy,* and *yyx.* There are 3 of these terms, just as there are 3 ways to choose 2 positions out of 3 places, so the coefficient of xy^2 is ${}_3C_2 = 3$.

Definition Combination Formula

If n and r are positive integers with $n \geq r$, then the number of **combinations of *n* objects taken *r* at a time** is

$${}_nC_r = \frac{n!}{r!(n-r)!}.$$

Examples 9 and 10 show two ways this formula can be used.

■ EXAMPLE 9 Finding the Number of Starting Teams

Every NBA basketball team has 12 players. How many different starting 5's are possible?

Solution

We must find ${}_{12}C_5$, the number of different 5-person teams that can be selected from 12 players.

$$ {}_{12}C_5 = \frac{12!}{5!(12-5)!} = \frac{12 \cdot 11 \cdot 10 \cdot 9 \cdot 8}{5 \cdot 4 \cdot 3 \cdot 2 \cdot 1} = 792 $$

Interpret

There are 792 possible starting lineups for each NBA team. ■

Follow-up

Have students explain why ${}_nP_n = n!$.

Assignment Guide

Day 1: Ex. 3–8, 11–14, 19–22
Day 2: Ex. 24, 25, 27, 28, 30, 31, 35, 36, 39

Cooperative Learning

Group Activity: Ex. 41

Notes on Exercises

Ex. 1–12 and 27–29 are based on the fundamental counting principle.
Ex. 21–26 and 30 are based on combinations.

Ongoing Assessment

Self-Assessment: Ex. 1, 9, 15, 17, 23, 29
Embedded Assessment: Ex. 28, 30

Many calculators can compute ${}_{12}C_5$ directly, and the intermediate steps shown are not necessary. However, they are shown to help you handle cases where the numbers may be too large for a calculator to give an exact answer, as in Example 10.

■ EXAMPLE 10 Counting Different Card Hands

In the game of contract bridge, each player is dealt a 13-card hand from a standard deck of 52 cards. How many different card hands are possible?

Solution

The order in which cards are dealt is not important, so this is a combination problem to find ${}_{52}C_{13}$.

$$ {}_{52}C_{13} = \frac{52!}{13!(52-13)!} = 635{,}013{,}559{,}600 $$

Interpret

The number of different bridge hands is 635,013,559,600. ■

The arithmetic simplification in Example 10 is left to you. See if you can find a way to check this exact answer with your grapher. (See Exercise 41.)

Quick Review 11.5

In Exercises 1–14, evaluate the expression.

1. $\dfrac{7 \cdot 6 \cdot 5 \cdot 4 \cdot 3 \cdot 2 \cdot 1}{3 \cdot 2 \cdot 1}$

2. $\dfrac{9 \cdot 8 \cdot 7 \cdot 6 \cdot 5 \cdot 4 \cdot 3 \cdot 2 \cdot 1}{5 \cdot 4 \cdot 3 \cdot 2 \cdot 1}$

3. $\dfrac{11 \cdot 10 \cdot 9 \cdot 8 \cdot 7 \cdot 6 \cdot 5 \cdot 4 \cdot 3 \cdot 2 \cdot 1}{4 \cdot 3 \cdot 2 \cdot 1 \cdot 7 \cdot 6 \cdot 5 \cdot 4 \cdot 3 \cdot 2 \cdot 1}$

4. $\dfrac{10 \cdot 9 \cdot 8 \cdot 7 \cdot 6 \cdot 5 \cdot 4 \cdot 3 \cdot 2 \cdot 1}{2 \cdot 1 \cdot 8 \cdot 7 \cdot 6 \cdot 5 \cdot 4 \cdot 3 \cdot 2 \cdot 1}$

5. $10!$

6. $14!$

7. $\dfrac{7!}{3!}$

8. $\dfrac{6!}{4!}$

9. $\dfrac{5!}{3!2!}$

10. $\dfrac{7}{3!4!}$

11. $\dfrac{52!}{26!26!}$

12. $\dfrac{25!}{13!12!}$

13. $\dfrac{500!}{497!3!}$

14. $\dfrac{500!}{490!10!}$

SECTION EXERCISES 11.5

In Exercises 1–6, list all the possibilities or draw a tree diagram to answer the question.

1. *Three-Letter Code* How many different three-letter code words can be formed, with no repeated letters, from the letters *A, T,* and *X*?

2. *Three-Letter Code* How many different three-letter code words can be formed, with no repeated letters, from the letters *R, S, T,* and *V*?

3. *Possible Routes* There are three roads from town *A* to town *B* and four roads from town *B* to town *C*. How many different routes are there from *A* to *C* by way of *B*?

4. *Possible Routes* Using the information in Exercise 3, how many different routes are there from *A* to *C* and back to *A*, passing through *B* in each direction?

5. *Homecoming King and Queen* There are four candidates for homecoming queen and three candidates for king. How many king–queen pairs are possible?

6. *Airline Tickets* When ordering an airline ticket you can request first class, business class, or coach. You can also choose a window, aisle, or middle seat, except that there are no middle seats in first class. How many different ways can you order a ticket?

In Exercises 7–12, use the fundamental counting principle.

7. *Phone Numbers* How many seven-digit telephone numbers are possible? (A number may not begin with a 0 or a 1. Why?)

8. *Social Security Numbers* How many nine-digit social security numbers are there?

9. *License Plates* How many different license plates begin with two digits, followed by two letters and then three digits if no letters or digits repeat?

10. *License Plates* How many different license plates consist of five symbols, either digits or letters?

11. *Tumbling Dice* Suppose that two dice, one red and one green, are rolled. How many different outcomes are possible for the pair of dice?

12. *Coin Toss* How many different sequences of heads and tails are there if a coin is tossed 10 times?

In Exercises 13–20, calculate the value.

13. ${}_8P_5$
14. ${}_{12}P_7$
15. ${}_{18}P_6$
16. ${}_{11}P_7$
17. ${}_{14}C_5$
18. ${}_8C_6$
19. ${}_{24}C_{15}$
20. ${}_{18}C_8$

In Exercises 21–30, solve the problem.

21. *Forming Committees* A 3-woman committee is to be elected from a 25-member sorority. How many different committees can be elected?

22. *Straight Poker* In the original version of poker known as "straight" poker, a five-card hand is dealt from a standard deck of 52. How many different straight poker hands are possible?

23. *Buying Disks* Juan has money to buy only three of the 48 compact discs available. How many different sets of disks can he purchase?

24. *Coin Toss* A coin is tossed 20 times and the heads and tails sequence is recorded. From among all the possible sequences of heads and tails, how many have exactly seven heads?

25. *Drawing Cards* How many different 13-card hands include the ace and king of spades?

26. The head of the personnel department interviews eight people for three openings. How many different groups of three can be employed?

27. *M & M's Candies* Traditional M & M's "plain" candies come in six colors: brown, green, orange, red, tan, and yellow. If you pick three candies from a bag, how many color possibilities are there?

28. *M & M's Peanut Candies* Traditional M & M's peanut candies have all the colors of plain candies except tan. If you pick three M & M's peanut candies from a bag, how many color possibilities are there?

29. *Yahtzee* In the game of Yahtzee, five dice are tossed simultaneously. How many outcomes can be distinguished if all the dice are different colors?

30. *Indiana Jones and the Final Exam* Professor Indiana Jones gives his class 20 study questions, from which he will select 8 to be answered on the final exam. How many different final exams are possible?

EXTENDING THE IDEAS

In Exercises 31–36, prove the statement algebraically.

31. ${}_nC_n = {}_nC_0 = 1,$ for all integers $n \geq 0$.
32. ${}_nC_1 = {}_nC_{n-1} = n,$ for all integers $n \geq 1$.
33. ${}_nP_n = {}_nP_{n-1},$ for all integers $n \geq 1$.
34. ${}_nP_1 = {}_nC_1,$ for all integers $n \geq 1$.
35. ${}_nC_r = {}_nC_{n-r},$ for all integers $n \geq r \geq 0$.
36. ${}_nC_r + {}_nC_{r+1} = {}_{n+1}C_{r+1}$, where n and r are positive integers and $n \geq r + 1$.

In Exercises 37–41, solve the problem.

37. **Writing to Learn** Suppose that a chain letter (illegal if money is involved) is sent to five people the first week of the year. Each of these five people sends a copy of the letter to five more people during the second week of the year. Assume that everyone who receives a letter participates. Explain how you know with certainty that someone will receive a second copy of this letter later in the year.

38. *Colored Beads* Four beads—red, blue, yellow, and green—are arranged on a string to make a simple necklace as shown in the figure. How many arrangements are possible?

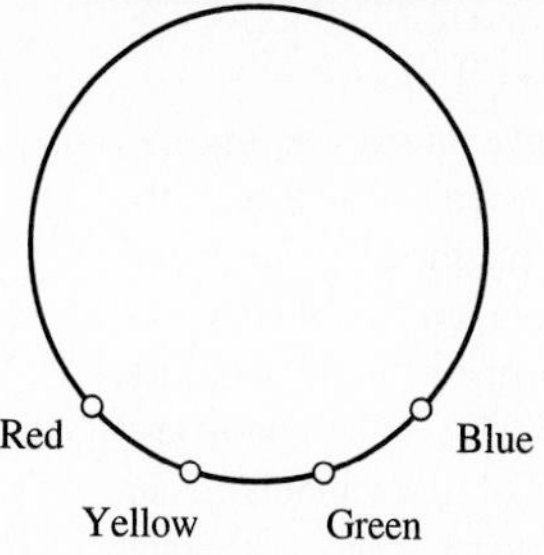

For Exercise 38

39. *Bridge Around the World* Suppose that a contract bridge hand is dealt somewhere in the world every second. What is the fewest number of years required for every possible bridge hand to be dealt?

40. *Basketball Lineups* Following up on Example 9, how many different sets of 10 players can start when two NBA teams play a game?

41. **Writing to Learn** Enter the function $f(x) = {}_{52}C_x$ on your grapher.
    - **a.** Make a table for $x = 0, 1, 2, 3, \ldots, 15$ and record it on paper. Write a paragraph explaining how this table is related to hands that can be dealt from a deck of 52 cards.
    - **b.** Graph the function in the window $[0, 47]$ by $[0, 5 \times 10^{14}]$. Copy the graph onto paper and complete the graph using the idea of symmetry. Write a paragraph explaining how this graph is related to binomial coefficients and how and why it is symmetric.

11.6 PROBABILITY

Objective

Students will be able to identify a sample space and calculate theoretical and experimental probabilities.

Motivate

Ask which is more likely when tossing two dice: a sum of 3 or a sum of 5? **(5)**

Experiment • Sample Spaces • Assigning Probabilities to Events • Mutually Exclusive Events • Complement of an Event • Independent Events • Binomial Probability • Empirical Probability

Experiment

Many everyday situations involve uncertainty. What are the chances of winning the lottery? What are the chances of a newborn baby being a girl? What is the likelihood that a newly assembled calculator is defective? How likely is it that a person holding an airline reservation will be a no-show?

Each of these situations involves an **experiment** because, although the individual results are uncertain, in the long run the results have a regular pat-

REMARK

In everyday conversation, we use *experiment* to describe a process that demonstrates something known or one that discovers the unknown. For example, an experiment might measure the force of gravity, or it could test the effectiveness of a new medicine. In probability, experiment has a broader meaning. It can refer to any process in which the individual results are uncertain.

tern. For example, we don't know what will happen when we toss a fair coin. However, in the long run we expect about half the time to get heads and half the time to get tails.

In everyday conversation we often give ambiguous measures to the likelihood of an event, such as *nearly certain, somewhat likely, not very probable,* or *highly unlikely.* Probability is the branch of mathematics that provides a framework for assigning numerical values to the likelihoods of the various possible results of an experiment.

Sample Spaces

An **outcome** is one of the possible individual results of an experiment. The **sample space** of an experiment is the set of all possible outcomes for the event.

Figure 11.19

■ EXAMPLE 1 Determining Sample Space and Outcomes

Television comedian Red Green tosses two dice (see Figure 11.19), one red and one green. List all the elements in the sample space. How many possible outcomes are there?

Solution

The sample space S can be thought of as a set of ordered pairs, with the first member of each pair representing the number on the red die and the second member the number on the green die. The outcomes in the sample space are

$$
\begin{array}{cccccc}
(1, 1) & (1, 2) & (1, 3) & (1, 4) & (1, 5) & (1, 6) \\
(2, 1) & (2, 2) & (2, 3) & (2, 4) & (2, 5) & (2, 6) \\
(3, 1) & (3, 2) & (3, 3) & (3, 4) & (3, 5) & (3, 6) \\
(4, 1) & (4, 2) & (4, 3) & (4, 4) & (4, 5) & (4, 6) \\
(5, 1) & (5, 2) & (5, 3) & (5, 4) & (5, 5) & (5, 6) \\
(6, 1) & (6, 2) & (6, 3) & (6, 4) & (6, 5) & (6, 6)
\end{array}
$$

There are 36 number pairs and thus 36 possible outcomes for this experiment. This is consistent with the fundamental counting principle because there are 6 outcomes for each die and thus $6 \times 6 = 36$ outcomes for the two dice. ■

Teaching Note

Probability theory got its start as an analysis of games of chance. It has grown to include many industrial and scientific applications.

A subset of a sample space is an **event.** For example, in the experiment of Example 1, we might be interested in the event D, "rolling doubles," or the event N, "rolling (a sum of) nine," or the event L, "rolling less than five." These events can be expressed using set notation:

$$
\begin{aligned}
D &= \{(1, 1), (2, 2), (3, 3), (4, 4), (5, 5), (6, 6)\} \\
N &= \{(3, 6), (4, 5), (5, 4), (6, 3)\} \\
L &= \{(1, 1), (1, 2), (1, 3), (2, 1), (2, 2), (3, 1)\}
\end{aligned}
$$

■ **EXAMPLE 2 Determining Sample Space and an Event**

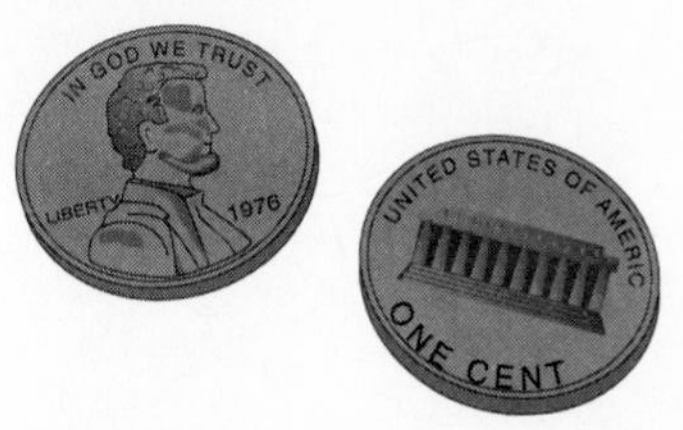

Figure 11.20

Mona Caracruz flips a coin (Figure 11.20) three times in succession, recording the outcomes of heads (H) and tails (T). What is the sample space? What is the event "at least two heads"?

Solution

The sample space is the set

$$S = \{HHH, HHT, HTH, HTT, THH, THT, TTH, TTT\},$$

and the event "at least two heads" is the subset

$$E = \{HHH, HHT, HTH, THH\}.$$

■

Notes on Examples

Example 2 studies the probability experiment of tossing a coin three times in succession. You may want to point out that the experiment of tossing a penny, nickel, and dime at one time has the same sample space. Example 8 on page 750, however, shows that by thinking of the experiment as one coin tossed three times, the theory of independent events can be applied.

Assigning Probabilities to Events

In the experiments considered so far, we assumed that the outcomes in the sample space are **equally likely outcomes.** This condition is described by saying that the dice are "fair" or the coin is "fair." The *mathematical probability* for an experiment with equally likely outcomes is found by dividing the number of outcomes in the event by the number of outcomes in the sample space. We use the notation $n(E)$ for the number of outcomes in event E and $n(S)$ for the number of outcomes in sample space S.

Definition Probability of an Event

If E is an event within a nonempty finite sample space S of equally likely outcomes, then the **probability** of the event E is

$$P(E) = \frac{n(E)}{n(S)}.$$

Because any event is a subset of a sample space, the number of outcomes in an event is always less than or equal to the number of outcomes in the sample space. So the probability of an event is always between 0 and 1.

Properties of the Probability of an Event

Let E be an event within a nonempty finite sample space S.

1. $0 \le P(E) \le 1$
2. If $E = \{\}$, then $P(E) = 0$ and E is an **impossible event.**
3. If $E = S$, then $P(E) = 1$ and E is a **certain event.**

Notes on Examples

The dice were painted two colors to clarify the difference between outcomes of the sample space. Point out that the probability of rolling a sum of 5 would be unchanged if the dice were both the same color.

Alert

Many students, when faced with a new probability experiment, will begin manipulating probability formulas before they understand the experiment. Ask them to list a few elements of the sample space to better understand the problem. Then they can begin calculating the probability of the event.

■ EXAMPLE 3 Calculating the Probability of an Event

What is the probability of event F, "rolling a sum of five," on a toss of the two dice in Example 1?

Solution

As shown in Example 1, the number of outcomes in the sample space is $n(S) = 36$.

Solve Algebraically

$$F = \{(1, 4), (2, 3), (3, 2), (4, 1)\}$$

$$n(F) = 4$$

$$P(F) = \frac{n(F)}{n(S)} = \frac{4}{36} = \frac{1}{9} \approx 0.11$$

Interpret

In the long run, we should expect to "roll a sum of 5" about four of every 36 tries, so the probability of obtaining a 5 on any given roll is 1/9 or about 11%. This does *not* mean we will always get one 5 on nine tosses or four 5's on 36 tosses. ■

Usually, the number of outcomes in events and sample spaces are found by using the counting methods of Section 11.5. Example 4 uses the fundamental counting principle.

■ EXAMPLE 4 Calculating the Probability of an Event

What is the probability that a five-digit telephone number chosen at random has no repeated digits?

Solution

Let E be the event "no repeated digits," and S be the entire sample space. Using the fundamental counting principle and the constraint that a telephone number never begins with a 0 or a 1, we find that

$$n(S) = 8 \cdot 10 \cdot 10 \cdot 10 \cdot 10 = 80{,}000$$

$$n(E) = 8 \cdot 9 \cdot 8 \cdot 7 \cdot 6 = 24{,}192$$

$$P(E) = \frac{n(E)}{n(S)} = \frac{24{,}192}{80{,}000} = 0.3024$$

Interpret

There are 80,000 possible five-digit telephone numbers, and, of these, 24,192 have no repeated digit. So, if a single number is chosen at random, the probability of it having no repeated digits is about 30%. ■

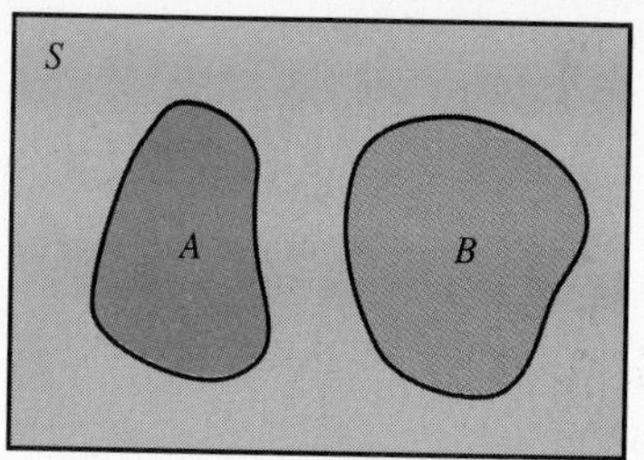

Figure 11.21 A Venn diagram of mutually exclusive events A and B within sample space S.

Mutually Exclusive Events

Events from the same sample space that have no outcomes in common are **disjoint,** or **mutually exclusive, events,** as depicted in Figure 11.21. In the rolling of red and green dice, discussed in Example 1, the events "rolling a sum of 5" and "rolling doubles" are mutually exclusive. By comparison, the events "rolling a sum of 5" and "a 1 on the red die" are not mutually exclusive because the outcome (1, 4) is in both events.

> **Addition Rule for Mutually Exclusive Events**
>
> If events A and B are mutually exclusive, then
>
> $$P(A \text{ or } B) = P(A) + P(B).$$

■ EXAMPLE 5 Using the Addition Rule

Packy randomly chooses a digit between 0 and 9 (inclusive). What is the probability that Packy will choose a digit that is greater than 6 or is a multiple of 6?

Solution

We need to compute $P(G \text{ or } M)$ for the events:

$$G = \text{the digit chosen is greater than 6}$$

$$M = \text{the digit chosen is a multiple of 6}$$

Because $G = \{7, 8, 9\}$ and $M = \{0, 6\}$ have no outcomes in common, the events are disjoint and the addition rule applies.

$$P(G \text{ or } M) = P(G) + P(M) = \frac{3}{10} + \frac{2}{10} = 0.5$$

Interpret

The probability is 0.5 that Packy will choose a digit that is greater than 6 or is a multiple of 6. ■

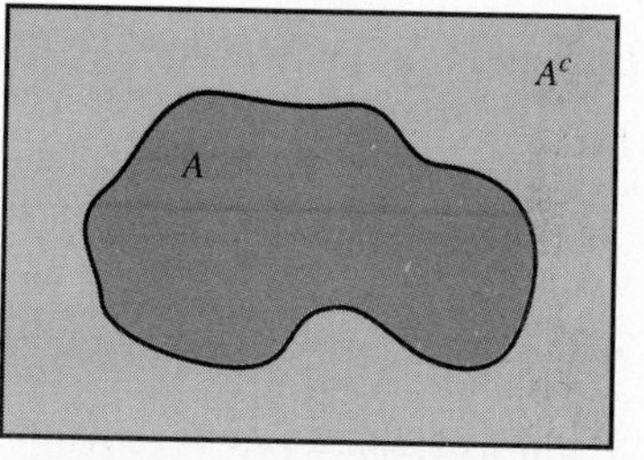

Figure 11.22 A Venn diagram of complementary events within the same sample space.

Complement of an Event

The **complement of the event A,** denoted A^c, is the set of all outcomes in the sample space that are not in event A, as depicted in Figure 11.22. Because events A and A^c taken together constitute the entire sample space, $P(A \text{ or } A^c) = 1$. Moreover, the events A and A^c have no outcomes in common and are thus disjoint. Therefore

$$P(A \text{ or } A^c) = P(A) + P(A^c) = 1.$$

This result can be summarized in the following way.

Complement Rule

If A is an event, then the probability that A does not occur is

$$P(A^c) = 1 - P(A).$$

Example 6 uses the combination formula from Section 11.5 and shows that it is sometimes easier to find the probability of the complement of an event than it is to find the probability of the event itself.

■ EXAMPLE 6 Using the Complement Rule

In the game of contract bridge, each player is dealt a 13-card hand from a standard deck of 52 cards. What is the probability that a bridge hand has at least 1 heart?

Solution

Let H be the event "at least 1 heart," that is, "1 heart, 2 hearts, . . . , or 13 hearts." To find $P(H)$ directly, we would need to sum the probabilities of all 13 possibilities. It is much easier to determine the probability of H^c, the event "no hearts." There are 39 cards that are not hearts.

$${}_{52}C_{13} = \text{the number of 13-card hands}$$

$${}_{39}C_{13} = \text{the number of 13-card hands with no hearts}$$

$$P(H^c) = \frac{{}_{39}C_{13}}{{}_{52}C_{13}} = 0.012 \cdots$$

$$P(H) = 1 - P(H^c) = 0.987 \cdots \quad \text{Complement rule}$$

Interpret

Nearly 99% of bridge hands contain at least 1 heart. ■

Independent Events

Alert

Students tend to confuse the concepts of mutually exclusive events and independent events.

Two events are **independent events** if the occurrence of one has no effect on the probability of the occurrence of the other. For example, when a fair coin is tossed, the event "head on the first toss" has no effect on the likelihood of the event "head on the second toss" because one toss does not influence the other in any way. The two events are independent.

However, when cards are dealt from a deck of 52, the events "heart on the first card" and "heart on the second card" are not independent because the cards remaining in the deck are influenced by the first event.

If we know the individual probabilities of two independent events, we can find the probability of both events happening by using the following formula.

Multiplication Rule for Independent Events

If A and B are independent events, then the probability that both occur is

$$P(A \text{ and } B) = P(A) \cdot P(B).$$

Most graphers can produce random numbers between 0 and 1. Each random number generated is independent of the previous one.

EXAMPLE 7 Using the Multiplication Rule

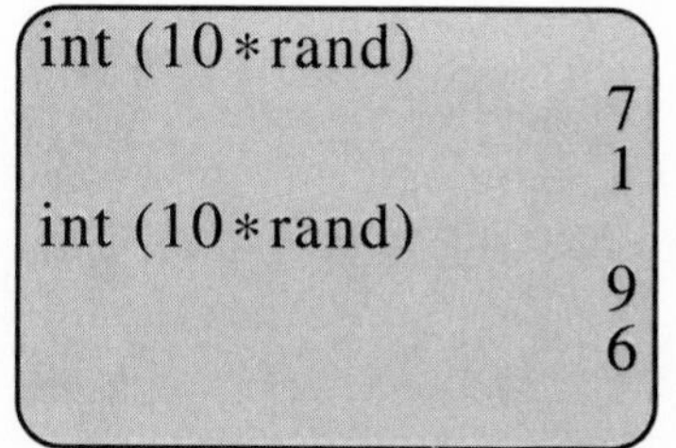

Figure 11.23 "int" is the greatest integer function and "rand" is the random number generator with rand between 0 and 1.

DeShanda develops a grapher formula to generate random digits (see Figure 11.23). What is the probability that the first digit generated is even and the second is a multiple of 3?

Solution

Because the digits are random, the first outcome does not influence the second. The sequence comprises two independent events:

$$E = \text{the digit generated is even}$$

$$T = \text{the digit generated is a multiple of 3}$$

Because $E = \{0, 2, 4, 6, 8\}$ and $T = \{0, 3, 6, 9\}$ are independent, the multiplication rule applies:

$$P(E \text{ and } T) = P(E) \cdot P(T) = \frac{5}{10} \cdot \frac{4}{10} = 0.2$$

Interpret

The probability is 0.2 that the first digit is even and the second is a multiple of 3. ■

Teaching Note

You may need to explain that the greatest integer function is used to find the greatest integer that is less than or equal to a number, which for positive numbers is just the integer part of the number. For example, int (3.72) = 3.

Example 8 illustrates that the multiplication rule can be extended to find the probability of an event that comprises a sequence of more than two independent events.

■ EXAMPLE 8 Using the Multiplication Rule

What is the probability of the outcome HTH when a coin is tossed three times?

Solution

Let:

$$F = \text{head on the first toss}$$
$$S = \text{tail on the second toss}$$
$$T = \text{head on the third toss}$$

The three events are independent, so the multiplication rule applies:

$$P(F \text{ and } S \text{ and } T) = \frac{1}{2} \cdot \frac{1}{2} \cdot \frac{1}{2} = \frac{1}{8}$$

Interpret

The probability of heads, tails, and heads in sequence when a coin is tossed three times is 1/8, or 0.125. ■

Binomial Probability

When a coin is tossed several times, each toss is independent of the others. Suppose that a coin is tossed four times. The binomial theorem provides a way to organize the probabilities of the event of obtaining exactly k heads on four tosses of a fair coin for $k = 0, 1, 2, 3, 4$. Consider the expansion

$$(H + T)^4 = \binom{4}{4}H^4 + \binom{4}{3}H^3T + \binom{4}{2}H^2T^2 + \binom{4}{1}HT^3 + \binom{4}{0}T^4.$$

Each coefficient $\binom{4}{k}$ gives the number of ways exactly k heads can occur when a coin is tossed four times; that is, $\binom{4}{k} = {}_4C_k$.

If we let $H = T = 1/2$, the probability of a head or a tail on one toss of a single coin, then we can rewrite the expansion of $(H + T)^4$ in the following way.

$$\begin{aligned} 1 &= \left(\frac{1}{2} + \frac{1}{2}\right)^4 \\ &= \binom{4}{4}\left(\frac{1}{2}\right)^4 + \binom{4}{3}\left(\frac{1}{2}\right)^4 + \binom{4}{2}\left(\frac{1}{2}\right)^4 + \binom{4}{1}\left(\frac{1}{2}\right)^4 + \binom{4}{0}\left(\frac{1}{2}\right)^4 \\ &= \frac{1}{16} + \frac{4}{16} + \frac{6}{16} + \frac{4}{16} + \frac{1}{16} \end{aligned}$$

Each term ${}_4C_k/16$ represents the probability of exactly k heads occurring on a toss of four coins. This generalizes to tosses of any number of fair coins as illustrated in Example 9. Note that a toss of n fair coins and n tosses of one fair coin are equivalent, in terms of the number of heads or tails that occur.

■ EXAMPLE 9 Using the Binomial Method

A fair coin is tossed seven times, or equivalently, seven fair coins are tossed simultaneously. In either case, what is the probability of obtaining exactly four heads?

Solution

According to the preceding discussion, we obtain the desired probability by substituting $H = T = 1/2$ into ${}_7C_4H^4T^3$. Thus the desired probability is

$${}_7C_4\left(\frac{1}{2}\right)^4\left(\frac{1}{2}\right)^3 = \frac{7!}{4!3!}\left(\frac{1}{2}\right)^7 = 0.273\cdots.$$

■

The *binomial analysis* used in Example 9 applies to any experiment that satisfies the conditions:

1. the experiment consists of a fixed number of observations, n;
2. each observation has exactly two outcomes, H and T;
3. the n observations are independent; and
4. the probability $P(H)$ is the same for each observation.

For example, a coin can be tossed n times with outcomes of "heads" (H) and "tails" (T). Or n light bulbs can be checked as they come off an assembly line with outcomes of "defective" (H) or "nondefective" (T). Or a basketball player can shoot n free throws with outcomes of "made" (H) or "missed" (T). In each case, if $P(H) = p$, where $0 \leq p \leq 1$, then $P(T) = 1 - p$.

Binomial Probability

Suppose that an experiment consists of n independent observations, each with exactly two outcomes H and T. If $P(H) = p$ and E is the event "H occurs exactly k times," then

$$P(E) = {}_nC_k \cdot (P(H))^k \cdot (P(T))^{n-k}$$

$$= \frac{n!}{k!(n-k)!}p^k(1-p)^{n-k}.$$

Alert

Students may have difficulty remembering the difference between empirical probability and theoretical probability. Some textbooks use the term *experimental probability* to describe empirical probability.

Empirical Probability

Throughout this section we have been calculating mathematical probability, sometimes called **theoretical probability** because the computations are done without actually performing an experiment such as tossing coins.

Suppose, however, that a fair coin is physically tossed 100 times and that 48 heads occur. Then the **empirical probability** of heads is the actual **relative frequency,** which is

$$\frac{\text{number of heads}}{\text{number of tosses}} = \frac{48}{100} = 0.48.$$

Follow-up

Ask how the complement rule can be used to find the probability that a bridge hand has at least two hearts. **(1 − *P*(no hearts) − *P*(one heart))**

Assignment Guide

Day 1: Ex. 3, 6, 9, 12, 14, 16, 17, 20, 23, 26, 27, 30, 33, 36
Day 2: Ex. 37, 40, 42, 44, 46, 47, 48, 50, 53, 56, 59, 60, 61, 63

Cooperative Learning

Group Activity: Ex. 64, 65

■ EXAMPLE 10 Calculating Empirical Probability

A 1963 penny from the Denver mint was tossed 55 times, with the results shown in Table 11.5. Heads occurred 26 times. What is the resulting empirical probability of tossing a tail?

Table 11.5 Tossing a 1963D Penny

Trial Number	Result	Relative Frequency of Heads
1	T	$0/1 = 0.00$
2	H	$1/2 = 0.50$
3	T	$1/3 \approx 0.33$
4	T	$1/4 = 0.25$
5	T	$1/5 = 0.20$
6	H	$2/6 \approx 0.33$
7	T	$2/7 \approx 0.29$
8	H	$3/8 = 0.38$
9	H	$4/9 \approx 0.44$
⋮	⋮	⋮
52	H	$25/52 \approx 0.48$
53	H	$26/53 \approx 0.49$
54	T	$26/54 \approx 0.48$
55	T	$26/55 \approx 0.47$

REMARK

Notice that the relative frequency of heads in Table 11.5 changes with each toss of the penny, even though the mathematical probability is a constant of 0.5. In the long run, the relative frequency approaches the mathematical probability value if the coin is fair.

Solution

The number of tails is $55 - 26 = 29$, so the empirical probability of tossing a tail is the relative frequency

$$\frac{\text{number of tails}}{\text{number of tosses}} = \frac{29}{55} \approx 0.53.$$

■

When an actual experiment is conducted, the empirical probability should be consistent with, though not identical to, the mathematical probability. In this sense the mathematical probability is a model of the real-world experiment.

Notes on Exercises

Ex. 11–12, 23–32, 46–47, 53, and 60 relate to independent events.
Ex. 15–18, 48a, and 50 relate to mutually exclusive events.
Ex. 20–22, 48b, and 52 relate to the complement rule.
Ex. 40–44, 51, 54, 56, 58, 62–64, and 66 relate to binomial probabilities.
Ex. 65 introduces the addition rule for nondisjoint events.

Ongoing Assessment

Self-Assessment: Ex. 7, 13, 21, 25, 41, 49

Embedded Assessment: Ex. 50

Suppose that a fair coin is tossed n times and that H_n is the number of times a head occurs. Because the mathematical probability of obtaining a head on a given toss is 1/2, we should expect that the relative frequency of heads would approach a limit of 1/2 as the number of tosses increases without bound. In symbols,

$$\frac{H_n}{n} \to \frac{1}{2} \quad \text{as} \quad n \to \infty.$$

In other words, the observed relative frequency of a given event should tend toward the mathematical probability as its limit if a large number of trials is performed.

Quick Review 11.6

In Exercises 1–12, evaluate the expression. Give exact values.

1. $\frac{1}{3}+\frac{8}{9}-\frac{5}{6}$

2. $\frac{17}{24}+\frac{3}{8}-\frac{5}{2}$

3. $\frac{10}{5!}$

4. $\frac{24}{6!}$

5. $\frac{10 \cdot 9}{4!3!}$

6. $\frac{9 \cdot 8}{4!5!}$

7. $\frac{3!2!}{5!2!}$

8. $\frac{3!4!}{7!4!}$

9. $\frac{{}_4C_2}{16}$

10. $\frac{{}_5C_2}{25}$

11. $\frac{{}_7C_5}{{}_{12}C_5}$

12. $\frac{{}_6C_5}{{}_{10}C_5}$

In Exercises 13–16, evaluate the expression.

13. $(0.99)^{25}$

14. $1-(0.99)^{25}$

15. $\left(\frac{49}{50}\right)^{20}$

16. $1-\left(\frac{49}{50}\right)^{20}$

SECTION EXERCISES 11.6

In Exercises 1–6, list the elements in the sample space of the experiment. How many possible outcomes are there?

1. A single die is rolled.

2. A single fair coin is tossed.

3. A penny, nickel, and dime are tossed at the same time.

4. Ten balls numbered 1 through 10 are in an urn. One ball is selected.

5. Five balls numbered 1 through 5 are in an urn. Two balls are selected from the urn; the second ball is drawn after the first has been replaced.

6. Eight balls numbered 1 through 8 are in an urn. Two balls are selected from the urn without replacement.

In Exercises 7–12, a red die and a green die have been rolled. List the possible outcomes of the event. What is the probability of the event?

7. *Tossing Dice* The sum is 9.

8. *Tossing Dice* The sum is even.

9. *Tossing Dice* The number of the red die is greater than the number on the green die.

10. *Tossing Dice* The sum is less than 10.

11. *Tossing Dice* Both dice are odd.

12. *Tossing Dice* Both dice are even.

In Exercises 13 and 14, solve the problem.

13. *Telephone Numbers* What is the probability that a seven-digit telephone number chosen at random has no repeated digits?

14. *Social Security Numbers* What is the probability that a nine-digit social security number chosen at random has no repeated digits?

In Exercises 15 and 16, what is the probability that one random digit (between 0 and 9 inclusive) satisfies the given conditions?

15. *Random Digit* It is odd or a multiple of 4.

16. *Random Digit* It is a prime number or a multiple of 6.

Exercises 17–22 involve the sample space for traditional M & M's "plain" candies, which is

$$S = \{\text{brown, green, orange, red, tan, yellow}\}.$$

The probability of each event in S is equal to the proportion of all of the M & M's candies produced of that color. The maker of M & M's has released the information about production proportions in the following table.

M & M's Plain Candies

Color	brown	red	yellow	green	orange	tan
Proportion	0.3	0.2	0.2	0.1	0.1	0.1

If you randomly select one M & M's plain candy from a newly opened bag, what is the probability that it has the given color(s)?

17. Brown or tan

18. Red, green, or orange

19. Red

20. Not red

21. Neither orange nor yellow

22. Neither brown nor tan

In Exercises 23–26, what is the probability that two random digits (each between 0 and 9 inclusive) satisfy the given conditions?

23. *Two Random Digits* The first digit is odd and the second is a multiple of 4.

24. *Two Random Digits* The first digit is even and the second is odd.

25. *Two Random Digits* Both digits are less than 7.

26. *Two Random Digits* The second digit is the same as the first.

Exercises 27–32 involve traditional M & M's peanut candies, which have all the colors of plain candies except tan. The production proportions for M & M's peanut candies are given in the following table.

M & M's Peanut Candies

Color	brown	red	yellow	green	orange
Proportion	0.3	0.2	0.2	0.2	0.1

If you randomly select one M & M's peanut candy from each of two newly opened bags, what is the probability that they have the given color(s)?

27. Both are brown.

28. Both are orange.

29. One is red and one is green.

30. The first is brown and the second is yellow.

31. Neither is yellow.

32. The first is not red and the second is not orange.

Exercises 33–36 concern a version of the card game "bid Euchre" that uses a pack of 24 cards, consisting of ace, king, queen, jack, 10, and 9 in each of the four suits spades, hearts, diamonds, and clubs. In bid Euchre a hand contains six cards. What is the probability of the event?

33. *Euchre* A hand is all spades.

34. *Euchre* All six cards are from the same suit.

35. *Euchre* A hand includes all four aces.

36. *Euchre* A hand includes both jacks of the same color (called the right and left bower).

In Exercises 37–44, what is the probability of the given event if 10 dimes, dated 1987 through 1996 are tossed?

37. *Tossing Ten Dimes* Heads on the 1990 dime only.

38. *Tossing Ten Dimes* Heads on the 1987 and 1996 dimes only.

39. *Tossing Ten Dimes* Heads on all 10 dimes.

40. *Tossing Ten Dimes* Heads on all but one dime.

41. *Tossing Ten Dimes* Exactly two heads.

42. *Tossing Ten Dimes* Exactly three heads.

43. *Tossing Ten Dimes* At least one head.

44. *Tossing Ten Dimes* At least two heads.

In Exercises 45–67, solve the problem.

45. *Renting Cars* Floppy Jalopy Rent-a-Car has 25 cars available for rental—20 big bombs and 5 midsize cars. If two cars are selected at random, what is the probability that both are big bombs?

46. *Defective Calculators* Dull Calculators, Inc., knows that a unit coming off an assembly line has a probability of 0.037 of being defective. If four units are selected at random during the course of a workday, what is the probability that none of the units are defective?

47. *You're in the Army Now* Bob, Carol, Ted, Alice, and Cooper join the U.S. Army, and each of them is assigned at random to be either a mechanic or a paratrooper. What is the probability that Ted and Alice will be assigned to the same type of job?

48. *Causes of Death* The government designates a single cause for each death in the United States. The resulting data indicate that 45% of deaths are due to heart and other cardiovascular disease and 22% are due to cancer.

a. What is the probability that the death of a randomly selected person will be due to cardiovascular disease or cancer?
b. What is the probability that the death will be due to some other cause?

49. *Yahtzee* In the game of *Yahtzee,* on the first roll five dice are tossed simultaneously. What is the probability of rolling five of a kind (which is Yahtzee!) on the first roll?

50. **Writing to Learn** Explain why the following statement cannot be true. The probabilities that a computer salesperson will sell zero, one, two, or three computers in any one day are 0.12, 0.45, 0.38, and 0.15, respectively.

51. *Binomial Probability* Decide whether the situation involves binomial probability and explain your decision. Refer to the four conditions given following Example 9.

a. The color of 2 cards that are drawn from a standard deck of 52 cards.
b. Three coins tossed simultaneously.
c. The eye color of five children born of parents who both have brown eyes and both carry the blue eye gene.
d. Seven light bulbs checked for defects at an assembly plant, four from one assembly process and three from another.

52. *HIV Testing* A particular test for HIV, the virus that causes AIDS, is 0.7% likely to produce a false positive result—a result indicating that the human subject has HIV when in fact the person is not carrying the virus. If 60 individuals who are HIV-negative are tested, what is the probability of obtaining at least one false result?

53. *Graduate School Survey* The Earmuff Junction College Alumni Office surveys selected members of the class of 1996. Of the 254 who graduated that year, 172 were women, 124 of whom went on to graduate school. Of the male graduates, 58 went on to graduate school. What is the probability of the given event?

a. The graduate is a woman.
b. The graduate went on to graduate school.
c. The graduate was a woman who went on to graduate school.

54. *Sex of a Child* Assume that the probability that a newborn child is a particular sex is 50%. In a family of four children, what is the probability of the given event?

a. All the children are girls.
b. None of the children are the same sex.
c. All the children are boys or girls.
d. At least two of the children are boys.

55. *Defective Refrigerators* A shipment of 20 refrigerators contains 2 units missing the door handles. An apartment complex has ordered 5 of these 20 units, and because the boxes are identical, the selection will be random. What is the probability of the given event?

a. All five units are good. **b.** Exactly four units are good.
c. At least two units are good.

56. *Indiana Jones and the Final Exam* Professor Indiana Jones gives his class a list of 20 study questions, from which he will select 8 to be answered on the final exam. If a given student knows how to answer 14 of the questions, what is the probability that the student will be able to answer correctly the given number of questions?

a. All 8 questions. **b.** Exactly 5 questions.
c. At least 6 questions.

57. *Graduation Requirement* To complete the kinesiology requirement at Palpitation Tech you must pass two classes chosen from aerobics, aquatics, defense arts, gymnastics, racket sports, recreational activities, rhythmic activities, soccer, and volleyball. If you decide to choose your two classes at random by drawing two class names from a box, what is the probability that you will take racket sports and rhythmic activities?

58. *Expectations* In a mail survey, some questionnaires are printed on green paper and some on yellow. Of the first 15 forms returned, 11 are yellow. If an equal number of green and yellow forms were sent, what is the probability of the event that, of the first 15 returned,

a. 11 are yellow? **b.** 11 or more are yellow?

59. **Writing to Learn** During July in Gunnison, Colorado, the probability of at least 1 hour a day of sunshine is 0.78, the probability of at least 30 minutes of rain is 0.44, and the probability that it will be cloudy all day is 0.22. Write a paragraph explaining whether this statement could be true.

60. *Counting Cars* Suppose that 15% of the cars in the posh River Oaks section of Houston are Lincolns. You decide to stand on the corner of River Oaks and San Felipe and watch the north–south traffic. Assuming that the traffic at this corner is representative of the cars in River Oaks and that the cars pass independently, what is the probability of the given event?

a. The fifth car that passes you is a Lincoln.
b. You have to see four cars go by before seeing the first Lincoln.
c. At least one of the first four cars is a Lincoln.

EXTENDING THE IDEAS

61. *Straight Poker* In the original version of poker known as "straight" poker, a 5-card hand is dealt from a standard deck of 52. What is the probability of the given event?

a. A hand will contain at least one king.
b. A hand will be a "full house" (any three of one kind and a pair of another kind).

62. *Married Students* Suppose that 23% of all college students are married. Answer the following questions for a random sample of eight college students.

a. How many would you expect to be married?
b. Would you regard it as unusual if the sample contained five married students?
c. What is the probability that five or more of the eight students are married?

63. *Blood Type* In a family with 5 children, both parents carry genes for blood types A and B. Because none of the children are identical twins, their blood types are *independent* of each other's. Based on the laws of genetics, the probability of having blood type A is 1/4 for each child. Make a table that shows the probabilities for 0, 1, 2, 3, 4, and 5 of the children having blood type A. What is the sum of the six probability values?

64. *Investigating an Athletic Program* A university widely known for its track and field program claims that 75% of its track athletes get degrees. A journalist investigates what happened to the 32 athletes who began the program over a 6-year period that ended 7 years ago. Of these athletes, 17 have graduated and the remaining 15 are no longer attending any college. If the university's claim is true, the number of athletes who graduate among the 32 examined should have been governed by binomial probability with $p = 0.75$.

a. What is the probability that exactly 17 athletes should have graduated?
b. What is the probability that 17 or fewer athletes should have graduated?
c. If you were the journalist, what would you say in your story on the investigation?

65. *Nondisjoint Events* The addition rule for mutually exclusive (disjoint) events can be modified to accommodate nondisjoint events. Nondisjoint events are events that have common outcomes. If events A and B are in the same sample space, then

$$P(A \text{ or } B) = P(A) + P(B) - P(A \text{ and } B).$$

Use this rule to solve the following problems.

a. If Bonnie chooses a random digit, what is the probability that it is even or is a multiple of 3?
b. If a card is drawn from a standard deck of 52 cards, what is the probability that it is an ace or is a diamond?
c. If you roll a pair of dice, what is the probability of rolling a sum less than 5 or doubles?

66. *Bernoulli's Theorem* The Swiss mathematician Jakob Bernoulli (1654–1705) wrote the first major book on probability, *Ars Conjectandi.* It contained an extension of our binomial probability rule, known to this day as Bernoulli's theorem. The theorem states:

Suppose an experiment consists of n independent observations, each with exactly two outcomes H and T. If $P(H) = p$ and E is the event "H occurs at least k times," then

$$P(E) = \sum_{r=k}^{n} \binom{n}{r} p^r(1-p)^{n-r}.$$

Use the binomial probability rule of this section to prove Bernoulli's theorem.

67. **Writing to Learn** The word *random* is sometimes applied to an irregular or haphazard occurrence of events. But actually random implies regularity in the long run. Write a paragraph that explains this point of view.

Objective

Students will be able to distinguish between categorical and quantitative variables and use various kinds of graphs to display data.

Motivate

Discuss ways to display a set of data graphically.

11.7 DISPLAYING DATA

Statistics • Displaying Categorical Data • Stemplots • Frequency Tables • Histograms • Time Plots

Statistics

Statistics is a major branch of mathematics that draws from discrete algebra. The aim of statistics is to draw meaning from data and communicate it to others.

The objects described by a set of data are **individuals,** which may be people, animals, or things. The characteristic of the individuals being identified or measured is a **variable.** Variables are either *categorical* or *quantitative.* If the variable identifies each individual as belonging to a distinct class, such as male or female, then the variable is a **categorical variable.** If the variable takes on numerical values for the characteristic being measured, then the variable is a **quantitative variable.**

Examples of quantitative variables are annual income (for a company's employees) and shoe size (of the students in your math class). Visual displays of data are often helpful in communicating the pattern of a variable within a group or population of individuals.

Displaying Categorical Data

The National Center of Health Statistics reported that the leading causes of death in 1993 were heart disease, cancer, and stroke. Table 11.6 gives more detailed information.

The "values" for the cause of death are categories, not numbers, so cause of death is a categorical variable. When such variables are presented to an audience in person or on television, or published in a newspaper or magazine, they are often displayed graphically as a **bar chart** (Figure 11.24a), a **pie chart,** or a **circle graph** (Figure 11.24b). The bar chart highlights the rela-

Teaching Note

Note that the *individuals* in this example are individual people who died in 1993.

Table 11.6 Leading Causes of Death in the United States for 1993

Cause of Death	Number (thousands)	Percentage
Heart disease	740	32.6
Cancer	531	23.4
Stroke	150	6.6
Other	848	37.4

Source: U.S. Department of Health and Human Services, as reported in *The World Almanac and Book of Facts* (1995, Mahwah, N.J.: Funk & Wagnalls), p. 959.

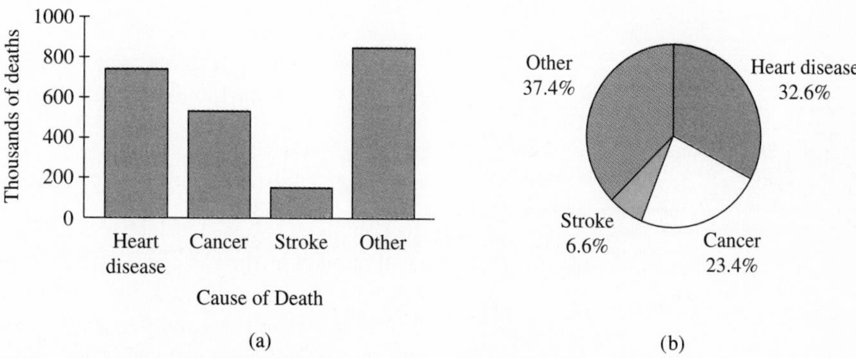

Figure 11.24 Causes of deaths in the United States in 1993 shown in (a) a bar chart and (b) a pie chart, or circle graph.

tive sizes of the categories, and the pie chart emphasizes the part-to-whole relationships.

In bar charts of categorical data the y-axis has a numerical scale, but the x-axis is labeled by category. The rectangular bars are separated by spaces because there is no continuous numerical scale. A pie chart consists of shaded or colored sectors of a circle. The central angles for the sectors are found by multiplying the percentage by 360°. For example, the angle of the sector representing stroke victims is

$$6.6\% \cdot 360° = 0.066 \cdot 360° = 23.76°.$$

In a sense, neither type of chart is crucial for understanding categorical information because a table such as Table 11.6 tells the whole story. Yet, most people like these charts because they provide information at a glance to the trained eye. Using computer spreadsheet programs, you can produce such charts from tabular data with the click of a button.

Stemplots

A quick way to organize and display a small set of quantitative data is with a *stemplot.* In a **stemplot** (short for *stem-and-leaf plot*) each number in the data set is split into a **stem** consisting of its initial digit or digits and a **leaf,** which is the final digit.

■ EXAMPLE 1 Making a Stemplot

Table 11.7 1996 Salaries for Armadillo Biscuit Company (in thousands of dollars)

28.3, 29.7, 28.2, 31.7, 29.8, 31.6,
21.7, 22.3, 23.4, 25.1, 24.5, 26.8,
27.8, 22.9, 23.7, 28.2, 25.6, 29.1,
28.5, 25.7, 29.5, 29.6, 29.7, 28.3,
21.4, 27.4, 23.4, 30.6, 23.8, 31.5,
27.9, 28.6, 25.6, 28.8, 31.7, 22.8,
24.8, 26.7, 30.8, 28.7, 31.6, 21.8

Make a stemplot for the data in Table 11.7. Then answer the following questions about the employees of the Armadillo Biscuit Company.

a. How many employees earn between \$21,000 and \$22,000?
b. How many employees earn between \$23,000 and \$26,000?
c. What is the most common such \$1000 range for employee salaries?

Solution

To complete a stem-and-leaf plot of these data, we use the first two digits of each number as the stem and the last digit as the leaf. Write the stems in order down the first column and, for each number, write the leaf digit in the appropriate stem row. For example, the first salary in Table 11.7, 28.3, has stem 28 and leaf 3. Notice that seven other salaries have the stem 28, so 28 has eight leaves in all.

Stem	Leaf
21	7 4 8
22	3 9 8
23	4 7 4 8
24	5 8
25	1 6 7 6
26	8 7
27	8 4 9
28	3 2 2 5 3 6 8 7
29	7 8 1 5 6 7
30	6 8
31	7 6 5 7 6

a. The first line of the stemplot shows that 3 employees earn between \$21,000 and \$22,000.
b. The third, fourth, and fifth rows show that 10 employees earn between \$23,000 and \$26,000.
c. The row with stem 28 has the most entries, namely, 8, so more employees have salaries between \$28,000 and \$29,000 than any other \$1000 range. ■

NOTE

Usually, leaf digits are written in the order they appear in the data. Sometimes it is useful to list them in increasing magnitude so that the entire table is in order.

The conclusions drawn in Example 1 are not nearly as obvious from the unorganized raw data in Table 11.7.

Table 11.8 Life Expectancy Data for the Nations of Western Africa

Nation	Years
Benin	47
Burkina Faso	46
Cape Verde	61
Côte d'Ivoire	54
Gambia	44
Ghana	54
Guinea	42
Guinea-Bissau	42
Liberia	55
Mali	45
Mauritania	48
Niger	45
Nigeria	49
Senegal	48
Sierra Leone	43
Togo	55

Source: National Geographic Atlas of the World (rev. 6th ed., 1992, Washington, D.C.), plates 128 and 129.

■ EXAMPLE 2 Making a Stemplot

Use the data shown in Table 11.8 to construct a stemplot for the life expectancies for the nations of western Africa. Compare them with the worldwide average of 65 years.

Solution

Because each number has two digits, the stem is the first digit (4, 5, or 6) and the leaf is the second.

4	7 6 4 2 2 5 8 5 9 8 3
5	4 4 5 5
6	1

Interpret

The life expectancies for western African nations are all below the worldwide average of 65 years. It appears that the average is in the 40s. ■

Stemplots show how a quantitative variable is *distributed* over a set of individuals, such as a group of countries. A **back-to-back stemplot** with leaves on either side of a common stem can be used to compare two **distributions.**

Table 11.9 Life Expectancy Data for the Nations of Eastern Africa

Nation	Years
Burundi	52
Comoros	56
Djibouti	48
Ethiopia	47
Kenya	61
Madagascar	55
Malawi	49
Mauritius	69
Mozambique	48
Rwanda	50
Somalia	46
Tanzania	52
Uganda	51
Zambia	53
Zimbabwe	60

Source: National Geographic Atlas of the World (rev. 6th. ed., 1992, Washington, D.C.), plates 128–130.

■ **EXAMPLE 3 Making Back-to-Back Stemplots**

Use the data in Table 11.9 to extend the stemplot of Example 2 to show the life expectancies for eastern and western Africa in a back-to-back stemplot. Compare the distributions.

Solution

We copy the stemplot from Example 2 and place the leaves for eastern Africa to the left of the appropriate stems.

Eastern Africa		Western Africa
8 7 9 8 6	4	7 6 4 2 2 5 8 5 9 8 3
2 6 5 0 2 1 3	5	4 4 5 5
1 9 0	6	1

This side-by-side comparison reveals that life expectancies for eastern Africa, though still generally well below the worldwide average, are longer than for western Africa. ■

Teaching Note

In this context the frequency of an observation is simply the number of times the data value (or any data value in a particular range of values) occurs.

Frequency Tables

In a stemplot, the stems correspond to nonoverlapping intervals of values of equal width for the quantitative variable. The number of leaves for a particular stem is the **frequency** of observations within the stem interval. These counts of observations are often recorded in a **frequency table.** Table 11.10 shows a frequency table for the Armadillo Biscuit Company salary stemplot given in Example 1.

Table 11.10 Frequency Table of Salary Data

Interval	Frequency
21,000–21,999	3
22,000–22,999	3
23,000–23,999	4
24,000–24,999	2
25,000–25,999	4
26,000–26,999	2
27,000–27,999	3
28,000–28,999	8
29,000–29,999	6
30,000–30,999	2
31,000–31,999	5

Histograms

A **histogram** is similar to a *bar chart,* is closely related to a stemplot, and displays the information of a frequency table. A histogram is to quantitative data what a bar chart is to categorical data. Unlike a bar chart, however, both axes have numerical scales and the rectangular bars for adjacent intervals have no gaps.

Figure 11.25 shows a histogram of the information in Table 11.10, where each bar corresponds to an interval in the table (a row) and the height of each bar represents the frequency of observations within the interval.

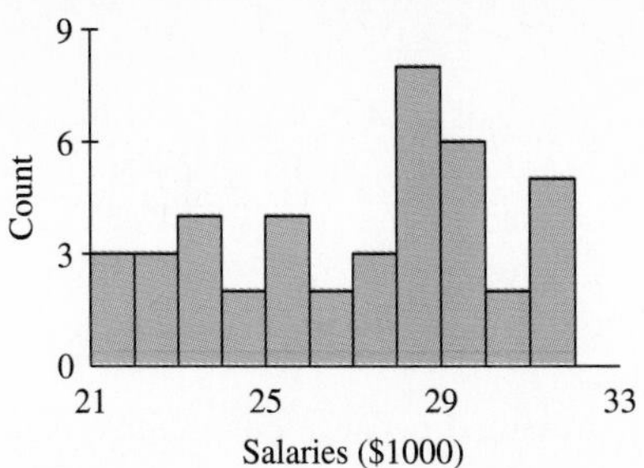

Figure 11.25 Histogram of salaries for the Armadillo Biscuit Company shown in Table 11.10.

Table 11.11 Regular Season Home Run Statistics for Henry Aaron

Year	Home Runs
1954	13
1955	27
1956	26
1957	44
1958	30
1959	39
1960	40
1961	34
1962	45
1963	44
1964	24
1965	32
1966	44
1967	39
1968	29
1969	44
1970	38
1971	47
1972	34
1973	40
1974	20
1975	12
1976	10

Source: The Baseball Encyclopedia (7th ed., 1988, New York: Macmillan) p. 695.

EXAMPLE 4 Graphing a Histogram

Make a histogram of Hank Aaron's annual home run totals given in Table 11.11, using interval widths of five home runs.

Solution

First make a frequency table for the data.

Home Runs	Number of Years
10–14	3
15–19	0
20–24	2
25–29	3
30–34	4
35–39	3
40–44	6
45–49	2

We need to scale the x-axis to be consistent with the row stems (intervals) of the table. Notice that the maximum frequency is 6 years, which occurs for the row width 40–44 home runs, so the y-axis needs to extend to 6. Enter the data from Table 11.11 into the statistics memory of your grapher and plot a histogram in the window [0, 50] by [−2, 6]. See Figure 11.26a. Use trace (Figure 11.26b) and compare the frequency readout on the grapher with the frequency table. ■

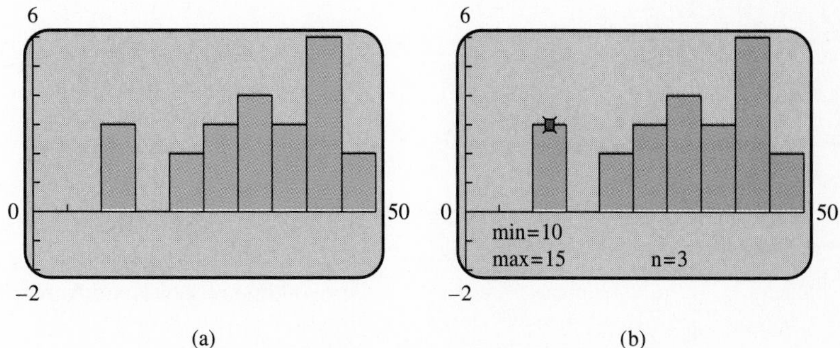

Figure 11.26 (a) Grapher histograms of Hank Aaron's regular season home run statistics, with (b) showing the frequency for the interval [10, 15).

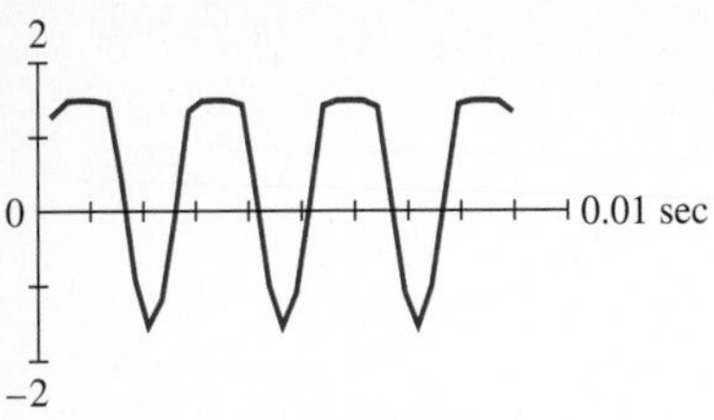

Figure 11.27 Time plot of air pressure caused by a tuning fork.

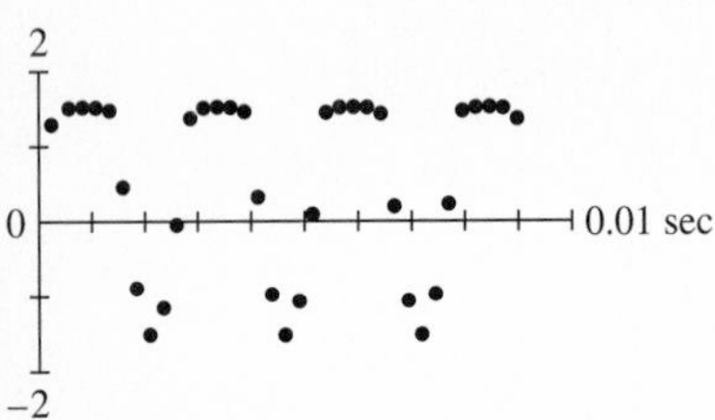

Figure 11.28 The data points of Figure 11.27 without the connecting line segments.

Time Plots

Often the value of a quantitative variable depends on time. For example, in Section 6.8 (page 477), data were collected for the sound wave produced by a tuning fork. We found that the sound was caused by the variable air pressure changing over time. The resulting plot of air pressure versus time is shown in Figure 11.27. This graph is a **time plot** because the variable is plotted against the time at which the variable was measured.

A time plot is a type of **line graph,** so called because consecutive data points are connected by line segments just as graphers do in connected mode. The importance of these connecting line segments is illustrated by Figure 11.28, in which the segments have been removed. The line graph style time plot in Figure 11.27 shows the periodic pattern of the data much better than the scatter plot in Figure 11.28.

Time plots reveal trends in data over time. These plots frequently appear in magazines and newspapers, of which Figure 11.29 is an example.

A time plot is much like the graph of a function with time as the independent variable. In a time plot the horizontal axis represents time and the vertical axis represents the variable that has been measured at various times.

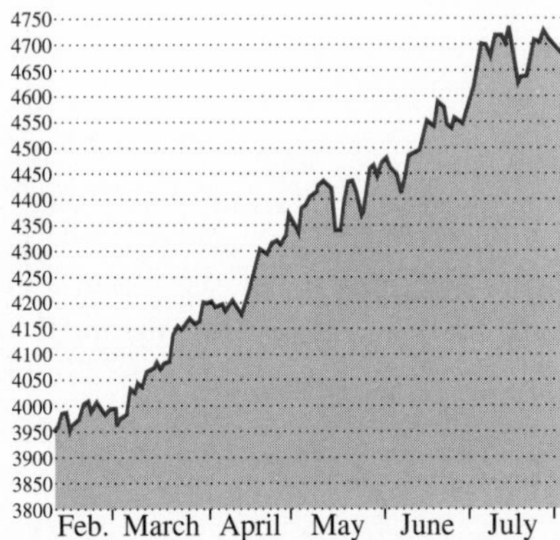

Figure 11.29 Time plot of Dow Jones Industrial Average over a 6-month period in 1995. (*Source: USA Today,* August 4, 1995, p. 3B.)

> **Teaching Note**
>
> When time is one of the varying factors, a time plot is usually the proper way to represent the data.
>
> **Notes on Examples**
>
> The vertical scale in Example 7 on page 764 does not begin at zero. This is frequently the case in line graphs that appear in newspapers and magazines, and if misread they can lead to incorrect conclusions.

> **Follow-up**
>
> Discuss ways to determine whether a bar chart, circle graph, stemplot, histogram, or time plot would be the most effective way to display a set of data.

■ EXAMPLE 5 Drawing a Time Plot

Draw a time plot of Hank Aaron's annual home run totals given in Table 11.11.

Solution

The horizontal axis represents time (in years) from 1954 through 1976. Aaron's highest home run total was 47, so scale the vertical axis from 0 to 50 by 10s.

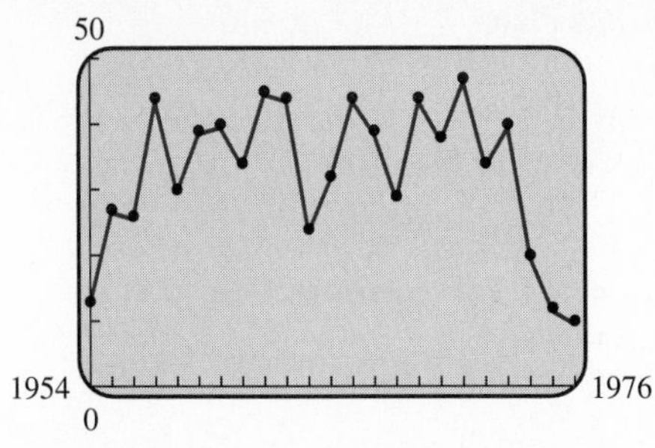

Figure 11.30 Time plot of Hank Aaron's regular season home run totals. Compare this graph with the histogram for the same data shown in Figure 11.26.

Enter the data from Table 11.11 into the statistics memory of your grapher as you did in Example 3. Select the line graph option using the viewing window [1954, 1976] by [0, 50]. See Figure 11.30. ■

Although Figure 11.30 shows some inconsistency in Aaron's home run output, compared to other ballplayers, it also shows that he performed consistently and at an unusually high level over a long career. For example, compare Aaron's performance to that of Roger Maris, given in Table 11.12, who holds the single-season home run record for the Major Leagues.

Table 11.12 Regular Season Home Run Statistics for Roger Maris

Year	Home Runs
1957	14
1958	28
1959	16
1960	39
1961	61
1962	33
1963	23
1964	26
1965	8
1966	13
1967	9
1968	5

Source: The Baseball Encyclopedia (7th ed., 1988, New York: Macmillan) p. 1211.

■ EXAMPLE 6 Overlaying Two Time Plots

Compare Hank Aaron's and Roger Maris's career home run performances by overlaying the time plots for each player.

Solution

Store the data from Tables 11.11 and 11.12 in the statistics memory of your grapher. Make line graph statistical plots for both in the viewing window [1954, 1976] by [0, 65]. See Figure 11.31. The relative consistency and longevity of Aaron's career is apparent. ■

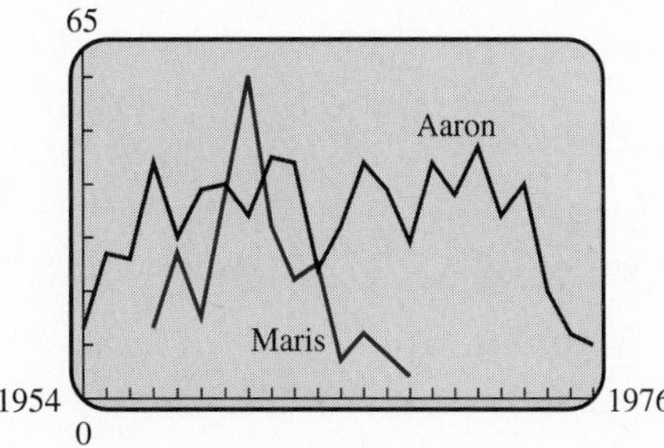

Figure 11.31 Time plots of Hank Aaron's and Roger Maris's regular season home run totals.

Time plots are often used to seek, exhibit, or clarify trends. Example 7 illustrates how a time plot can show a trend even when the time intervals for the data are not uniform.

Assignment Guide

Day 1: Ex. 1, 3, 4, 6, 8, 11, 13
Day 2: Ex. 16, 19, 20, 22, 23, 25

Cooperative Learning

Group Activity: Ex. 28–29

Table 11.13 Life Expectancy in the United States

Year	Male	Female
1920	53.6	54.6
1930	58.1	61.6
1940	60.8	65.2
1950	65.5	71.1
1960	66.6	73.1
1965	66.8	73.7
1970	67.1	74.7
1980	70.0	77.5
1986	71.3	78.3
1990	71.8	78.8
1993	72.1	78.9

Source: National Center of Health Statistics, as reported in *The World Almanac and Book of Facts* (1995, Mahwah, N.J.: Funk & Wagnalls), p. 972.

Notes on Exercises

Most of the exercises in this section are routine and give students practice in the methods of plotting data.

Ongoing Assessment

Self-Assessment: Ex. 5, 7, 17
Embedded Assessment: Ex. 4, 16, 20

EXAMPLE 7 Overlaying Two Time Plots

Use the data in Table 11.13 to overlay time plots of the life expectancy for males and females in the United States from 1920 to 1993. What can you conclude about their life expectancies?

Solution

Enter and plot the data. See Figure 11.32.

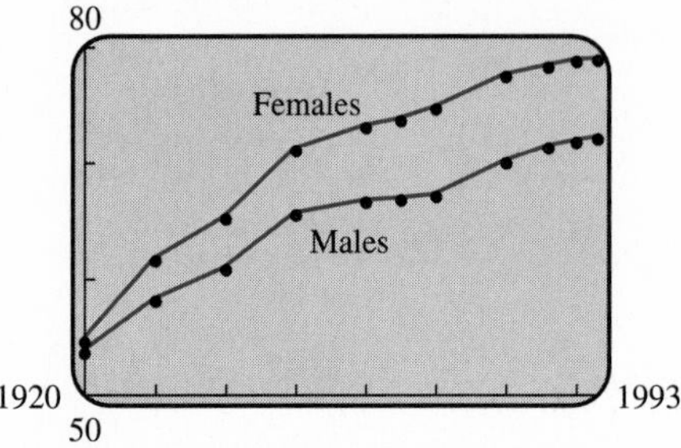

Figure 11.32 Time plots of male and female life expectancies in the United States from 1920 through 1993.

Figure 11.32 shows several trends in the United States from 1920 through 1993:

- The life expectancies for both males and females increased.
- In general, women lived longer than men.
- The gap in life expectancy between men and women widened from 1920 to 1960, then stabilized from 1960 onward. ■

Quick Review 11.7

In Exercises 1–8, solve the problem.

1. 457 is what percent of 2953?
2. 827 is what percent of 3950?
3. 52° is what percent of 360°?
4. 98° is what percent of 360°?
5. 734 is 42.6% of what number?
6. 5106 is 55.5% of what number?
7. Sketch the graph of an increasing function.
8. Sketch the graph of a decreasing function.

SECTION EXERCISES 11.7

In Exercises 1–3, construct the indicated stemplot from the data in Table 11.14. Then write a sentence interpreting the stemplot.

1. Life expectancies of males in the nations of South America.
2. Life expectancies of females in the nations of South America.
3. A back-to-back stemplot for life expectancies of males and females in the nations of South America.

Table 11.14 Life Expectancy by Gender for the Nations of South America

Nation	Male	Female
Argentina	68	75
Bolivia	61	66
Brazil	57	67
Chile	72	78
Columbia	69	75
Ecuador	67	73
Guyana	62	68
Paraguay	72	75
Peru	63	68
Suriname	67	72
Uruguay	71	77
Venezuela	70	76

Source: The World Almanac and Book of Facts (1995, Mahwah, N.J.: Funk & Wagnalls), pp. 743–835.

4. Construct a back-to-back stemplot for the annual home run production of Hank Aaron and Roger Maris using the data in Tables 11.11 and 11.12. Write a sentence interpreting the stemplot.

In Exercises 5 and 6, use the data in Table 11.14 to construct the indicated frequency table, using intervals of 5 years.

5. Life expectancies of males.

6. Life expectancies of females.

In Exercises 7–12, draw a histogram for the given table.

7. The frequency table in Exercise 5.

8. The frequency table in Exercise 6.

9. Table 11.8 of life expectancies for the nations of western Africa, using intervals of 5 years.

10. Table 11.9 of life expectancies for the nations of eastern Africa, using intervals of 5 years.

11. Table 11.15 of Willie Mays's annual home run totals, using intervals of five home runs.

12. Table 11.15 of Mickey Mantle's annual home run totals using intervals of five home runs.

Table 11.15 Regular Season Home Run Statistics for Willie Mays and Mickey Mantle

Year	Mays	Mantle	Year	Mays	Mantle
1951	20	13	1962	38	30
1952	4	23	1963	47	15
1953	41	21	1964	52	35
1954	51	27	1965	37	19
1955	36	37	1966	22	23
1956	35	52	1967	23	22
1957	29	34	1968	13	18
1958	34	42	1969	28	
1959	29	31	1970	18	
1960	40	40	1971	8	
1961	49	54	1972	6	

Source: The Baseball Encyclopedia (7th ed., 1988, New York: Macmillan), pp. 1208–1225.

In Exercises 13–16, make a time plot of the indicated data.

13. Willie Mays's annual home run totals given in Table 11.15.

14. Mickey Mantle's annual home run totals given in Table 11.15.

15. The immunization rates for Houston toddlers shown in the bar graph. Write a sentence interpreting the time plot.

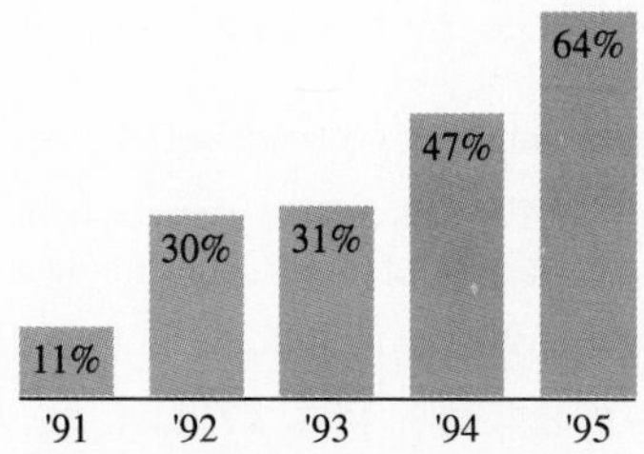

Source: Houston Department of Health and Human Services as published in the *Houston Chronicle*, August 26, 1995, p. 1A.

For Exercise 15

16. The life expectancies for Russian men shown in the bar graph. Write a sentence interpreting the time plot.

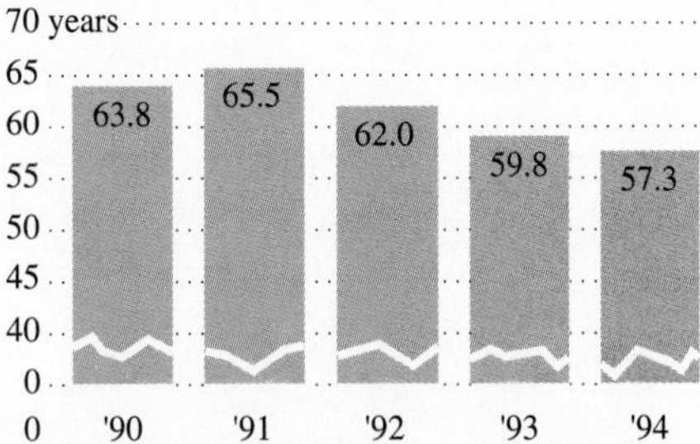

Source: Russian state statistics from NY Times News Service as published in the *Austin American Statesman*, August 6, 1995, p. H5.

For Exercise 16

In Exercises 17 and 18, compare performances by overlaying time plots.

17. The time plots from Exercises 13 and 14 to compare the performances of Mays and Mantle.

18. The time plots from Exercise 13 and Example 4 to compare the performances of Mays and Aaron.

In Exercises 19 and 20, analyze the data as indicated.

19. The salaries of the workers in one department of the Garcia Brothers Company (given in thousands of dollars) are as follows:

33.5, 35.3, 33.8, 29.3, 36.7, 32.8, 31.7, 36.3, 33.5, 28.2, 34.8, 33.5, 35.3, 29.7, 38.5, 32.7, 34.8, 34.2, 31.6, 35.4

a. Complete a stemplot for this data set.
b. Create a frequency table for the data.
c. Draw a histogram for the data. What viewing window did you use?
d. Why does a time plot not work well for these data?

20. The average wind speeds for one year at 44 climatic data centers around the United States are as follows:

9.0, 6.9, 9.1, 9.2, 10.2, 12.5, 12.0, 11.2, 12.9, 10.3,

10.6, 10.9, 8.7, 10.3, 11.0, 7.7, 11.4, 7.9, 9.6, 8.0,

10.7, 9.3, 7.9, 6.2, 8.3, 8.9, 9.3, 11.6, 10.6, 9.0,

8.2, 9.4, 10.6, 9.5, 6.3, 9.1, 7.9, 9.7, 8.8, 6.9,

8.7, 9.0, 8.9, 9.3

a. Complete a stemplot for this data set.
b. Create a frequency table for the data.
c. Draw a histogram for the data. What viewing window did you use?
d. Why does a circle graph not work well for these data?

In Exercises 21 and 22, compare by overlaying time plots for the data in Table 11.16.

Table 11.16 Population (in millions) of California, Florida, Illinois, New York, Pennsylvania, and Texas

Year	CA	FL	IL	NY	PA	TX
1900	1.5	0.5	4.8	7.3	6.3	3.0
1910	2.4	0.8	5.6	9.1	7.7	3.9
1920	3.4	1.0	6.5	10.4	8.7	4.7
1930	5.7	1.5	7.6	12.6	9.6	5.8
1940	6.9	1.9	7.9	13.5	9.9	6.4
1950	10.6	2.7	8.7	14.8	10.5	7.7
1960	15.7	5.0	10.0	16.8	11.3	9.6
1970	20.0	6.8	11.1	18.2	11.8	11.2
1980	23.7	9.7	11.4	17.6	11.9	14.2
1990	29.8	12.9	11.4	18.0	11.9	17.0

Source: U.S. Census Bureau, as reported in *The World Almanac and Books of Facts* (1995, Mahwah, N.J.: Funk & Wagnalls), p. 377.

21. The populations of California, New York, and Texas from 1900 through 1990.

22. The populations of Florida, Illinois, and Pennsylvania from 1900 through 1990.

In Exercises 23 and 24, solve the problem.

23. **Writing to Learn** Consider the Armadillo Biscuit Company salary data in Table 11.7, the associated stemplot in Example 1, frequency table (Table 11.10), and histogram (Figure 11.25). Write a paragraph comparing and contrasting the four types of data displays. Include a discussion of which information is easy to find from each display. For example, which display makes it easiest to find the highest and lowest salaries?

24. Set your calculator to two-place decimal display and generate 50 random numbers between 0 and 1. (*Hint:* See Example 7 and Figure 11.23 in Section 11.6.)

a. Record the 50 numbers generated.
b. Prepare a stemplot.
c. Create a frequency table from the stemplot.
d. Draw a histogram for these data using intervals of 0.1.

EXTENDING THE IDEAS

In Exercises 25 and 26, use the information in Tables 11.17 and 11.18 to make a table in the style of Table 11.6, with an "Other" category and a "Percentage" column. Then construct a bar chart and a circle graph for the data.

25. *Federal Income*

26. *Federal Expenses*

Table 11.17 United States Budget, Net Receipts for Fiscal Year 1993

Source of Revenue	Millions of Dollars
Individual Income Taxes	509,680
Corporation Income Taxes	117,520
Social Insurance Taxes	428,300
Total Receipts	1,153,175

Source: U.S. Department of the Treasury, as reported in *The World Almanac and Book of Facts* (1995, Mahwah, N.J.: Funk & Wagnalls), p. 107.

Table 11.18 United States Budget, Net Outlays for Fiscal Year 1993

Allocation	Millions of Dollars
Defense Department	278,576
Health & Human Services Department	282,774
Interest on the Public Debt	292,502
Total Outlays	1,408,122

Source: U.S. Department of the Treasury, as reported in *The World Almanac and Book of Facts* (1995, Mahwah, N.J.: Funk & Wagnalls), pp. 107–108.

In Exercises 27–29, solve the problem.

27. Use the information in Tables 11.17 and 11.18 to determine the federal deficit for fiscal year 1993.

28. *Time Plot of Periodic Data* Some data are a periodic function of time. If the data vary in an annual cycle, the period is 1 year. Use the information in Table 11.19 and overlay the time plots for the average daily high and the average daily low temperatures of Beijing, China.

Table 11.19 Average Daily High and Low Temperatures in °C for Beijing, China

Month	High	Low
January	2	−9
February	5	−7
March	12	−1
April	20	7
May	27	13
June	31	18
July	32	22
August	31	21
September	27	14
October	21	7
November	10	−1
December	3	−7

Source: National Geographic Atlas of the World (rev. 6th ed., 1992, Washington, D.C.), plate 132.

29. Fit a sinusoid to each time plot in Exercise 28. (See Section 6.4.)

Objective

Students will be able to use measures of center, the five-number summary, a boxplot, and standard deviation to describe quantitative data.

Motivate

Discuss ways to use a number to describe a set of quantitative data.

11.8 DESCRIBING DATA

Descriptive Statistics • Measures of Center • Five-Number Summary • Boxplots • Standard Deviation

Descriptive Statistics

A **statistic** is a number that describes some aspect of a data set. This section examines statistics that measure the *center* or the *spread* of quantitative data sets. A **measure of center** gives a typical, middle, or average value for a data set. A **measure of spread** tells how widely distributed data are.

Teaching Note

Most students should be familiar with the mean, median, and mode from previous mathematics courses.

Measures of Center

The data of the Armadillo Biscuit Company in Table 11.7 reflect employee salaries for 1 year. But suppose that salary trends are studied over several years and that the company president wants a single number that reflects the typical or average salary for a particular year. The *mean* is one such measure of center for a set of data.

NOTE

If the index and range of a summation are understood, we often indicate the summation by Σ alone.

Definition Mean

The **mean** of n numbers $x_1, x_2, x_3, \ldots, x_n$ is

$$\bar{x} = \frac{x_1 + x_2 + \cdots + x_n}{n} = \frac{1}{n}\sum_{i=1}^{n} x_i = \frac{1}{n}\sum x_i.$$

The mean is also called the *arithmetic mean, arithmetic average,* or *average value.*

■ EXAMPLE 1 Computing a Mean

Find the mean salary of Armadillo Biscuit Company employees for 1996.

Solution

Using the data from Table 11.7 gives

$$\bar{x} = \frac{x_1 + x_2 + \cdots + x_n}{n}$$

$$= \frac{28.3 + 29.7 + \cdots + 21.8}{42}$$

$$= \frac{1139.6}{42} \approx 27.1$$

Interpret

The average salary was about \$27,100. ■

When data are collected periodically over time, the mean of the data often represents an average rate. This is the case in Example 2, in which home run production rates are found.

■ EXAMPLE 2 Computing and Comparing Means

Determine the average annual home run production for Hank Aaron and Roger Maris for their career totals of 755 over 23 years and 275 over 12 years, respectively. Who had the higher production rate?

Solution

Using the summation definition of mean, we obtain

$$\text{for Aaron,} \qquad \bar{x} = \frac{1}{n}\sum x_i = \frac{1}{23}(755) \approx 32.83;$$

$$\text{for Maris,} \qquad \bar{x} = \frac{1}{n}\sum x_i = \frac{1}{12}(275) \approx 22.92.$$

Aaron's average production rate was nearly 10 home runs per year higher than Maris's. ■

Most graphers have built-in features to compute the mean of a data set or list of numbers. These features are especially useful for finding the mean of a large data set.

A second key measure of center is the *median.* Whereas the mean is the average of a data set, the median is the middle of the data set.

REMARK

To locate the median, order the data set and count $(n + 1)/2$ numbers from the beginning of the ordered set.

Definition Median

The **median** of n numbers listed in ascending or descending order is

- the middle number of the list if n is odd, and
- the mean of the two middle numbers if n is even.

Another statistic that helps to characterize a data set is the *mode.* The mode is traditionally listed among the measures of *central tendency.* But it does not always exist, and, when it does, it is not always a good measure of center.

Definition Mode

The **mode** of a set of numbers is the number that occurs most frequently.

Notes on Examples

Example 3 gives a situation in which the mode does not exist.

■ EXAMPLE 3 Finding Medians and Modes

Find the median and mode annual home run production for Hank Aaron and for Roger Maris.

Solution

Begin by putting the numbers in ascending order (from smallest to largest).

Aaron: {10, 12, 13, 20, 24, 26, 27, 29, 30, 32, 34, **34,** 38, 39, 39, 40, 40, 44, 44, 44, 44, 45, 47}

Maris: {5, 8, 9, 13, 14, **16, 23,** 26, 28, 33, 39, 61}

Because Aaron's home run data contain 23 observations, the median is located in the $(23 + 1)/2 = 12$th position, the bold 34. Aaron's median production was 34 home runs. His mode was 44 home runs, which he hit in each of four seasons.

To find Maris's median production, we average the two middle numbers (in bold), $(16 + 23)/2 = 19.5$ home runs. Maris had no mode performance because he hit a different number of home runs each year he played in the major leagues. ■

As an alternative to using an ordered list, we can find medians and modes from a sorted stemplot. We could have solved Example 3 by using the following back-to-back stemplot.

Aaron		Maris
	0	5 8 9
0 2 3	1	3 4 6
0 4 6 7 9	2	3 6 8
0 2 4 4 8 9 9	3	3 9
0 0 4 4 4 4 5 7	4	
	5	
	6	1

Depending on the nature of the data, the three measures of center—mean, median, and mode—might all be equal or no two might be equal. As Example 3 shows, a mode may not exist or may be a poor indicator of the center of a data set. For these reasons, the mean and median are the more commonly used measures of center.

The mean is the most traditional measure of center, and it has nice mathematical properties. The median requires no computation, only arrangement. It is a **resistant measure** of center because it changes little due to **outliers,** or data items far from the center.

For example, Maris's 61 home runs in 1961 is an outlier in his performance. Had Maris hit 100 home runs in 1961 instead of 61, his median would not be any different, but his mean would be more than 3 home runs greater.

When there are repeated values in a data set, it can be helpful to complete a frequency table first and then calculate the mean. If the data values x_1, $x_2, \ldots, x_n$ have frequencies $f_1, f_2, \ldots, f_n$, respectively, then the mean of the data set is given by

$$\bar{x} = \frac{x_1 f_1 + x_2 f_2 + \cdots + x_n f_n}{f_1 + f_2 + \cdots + f_n} = \frac{\Sigma\, x_i f_i}{\Sigma\, f_i}.$$

■ **EXAMPLE 4 Finding the Mean from a Frequency Table**

Find the mean of the following data.

Data:	5	6	9	11
Frequency:	2	3	5	2

Solution

$$\bar{x} = \frac{5(2) + 6(3) + 9(5) + 11(2)}{2 + 3 + 5 + 2} = \frac{95}{12} \approx 7.92$$

■

The method of Example 4 is used for **weights** as well as frequencies. If you score 83 and 92 on tests worth 25% each and get a 93 on a final exam worth 50%, your **weighted average** is

$$\bar{x} = \frac{83(25) + 92(25) + 93(50)}{25 + 25 + 50} = \frac{9025}{100} = 90.25.$$

Teaching Note

Note that the weights used can be the values 25, 25, and 50 or the values 0.25, 0.25, and 0.50. Either method will give the same result for the average value.

The weights used to compute meaningful averages are not always percentages; they are whatever quantity makes sense in the situation. Because the life expectancy for a nation is an average over its population, nations' populations are used as weights when averaging across nations.

■ EXAMPLE 5 Finding a Weighted Average

Find the mean life expectancy for the group of western African nations listed in Table 11.20.

Table 11.20 Life Expectancies and Populations for the Nations of Western Africa

Nation	Life Expectancy (years)	Population (millions)
Benin	47	5.0
Burkina Faso	46	9.6
Cape Verde	61	0.4
Côte d'Ivoire	54	13.0
Gambia	44	0.9
Ghana	54	16.0
Guinea	42	7.8
Guinea-Bissau	42	1.0
Liberia	55	2.8
Mali	45	8.5
Mauritania	48	2.1
Niger	45	8.3
Nigeria	49	90.1
Senegal	48	7.9
Sierra Leone	43	4.4
Togo	55	3.8

Source: National Geographic Atlas of the World (rev. 6th ed., 1992, Washington, D.C.), plates 128 and 129.

NOTE

Notice that, because nearly half the people in the 16 nations of western Africa live in Nigeria, Nigeria's life expectancy strongly influences the overall average for western Africa.

Solution

Using the populations as weights yields

$$\bar{x} = \frac{(47)(5.0) + (46)(9.6) + \cdots + (55)(3.8)}{5.0 + 9.6 + \cdots + 3.8}$$

$$= \frac{8879.3}{181.6} \approx 48.89$$

The mean life expectancy among persons living in these 16 nations is about 49 years. ■

Five-Number Summary

REMARK

Don't confuse the range of a data set, which is a number, with the range of a function, which is a set of numbers, often an interval.

Measures of center tell only part of the story of a data set. They do *not* indicate how widely distributed the data are. *Measures of spread* do. They describe the extent of the variability of the data. The simplest and crudest measure of spread is the **range,** which is the difference between the least and greatest values of a data set. That is,

$$\text{range} = \text{maximum} - \text{minimum}.$$

For example, the range of life expectancies for the nations of western Africa given in Table 11.20 is

$$61 - 42 = 19 \text{ years}.$$

The range indicates the full spread of the data, which may include outliers. A single outlier (like Maris's 61 home runs) can greatly increase the range, thus suggesting a much larger spread than generally is the case for the data. To avoid this, we use the *interquartile range,* the spread across the middle half of the data.

To define this measure of spread, we must first define *quartiles.* Just as the median separates the data into halves, **quartiles** separate the data into fourths. The **first quartile** Q_1 is the median of the lower half of the data, the **second quartile** is the median, and the **third quartile** Q_3 is the median of the upper half of the data. The **interquartile range** is

$$IQR = Q_3 - Q_1.$$

Taken together, the maximum, the minimum, and the three quartiles describe both the center and the spread of a data set.

Definition Five-Number Summary

The **five-number summary** of a data set is the collection:

Minimum Q_1 Median Q_3 Maximum

■ EXAMPLE 6 Five-Number Summary and Spread

Determine the five-number summary, the range, and the interquartile range of annual home run production for Hank Aaron and for Roger Maris. Compare the spreads.

Solution

Aaron's home run data in ascending order are

{10, 12, 13, 20, 24, **26,** 27, 29, 30, 32, 34, **34,** 38, 39, 39,
40, 40, **44,** 44, 44, 44, 45, 47}

Because there are 11 numbers below the median, the median of the lower half is in the (11 + 1)/2, or sixth position, the bold 26. So, $Q_1 = 26$. Similarly, counting to the sixth position from the top of the list we obtain $Q_3 = 44$. Together with the minimum of 10 and the maximum of 47, we obtain the five-number summary for Aaron's home run data:

10 26 34 44 47

Thus

$$\text{range} = 47 - 10 = 37 \text{ home runs;}$$

$$IQR = 44 - 26 = 18 \text{ home runs.}$$

REMARK

Many graphers have a built-in feature that will produce the five-number summary, often as part of a comprehensive list of descriptive statistics for a one-variable data set. Check your owner's manual for details about your grapher.

Maris's home run data in ascending order are

{5, 8, 9, 13, 14, 16, 23, 26, 28, 33, 39, 61}

Because there are 6 numbers below the median of 19.5, the median of the lower half is located at the (6 + 1)/2 = 3.5 position, that is between the 3rd and 4th numbers. So to find Q_1, we average the 3rd and 4th smallest numbers.

$$Q_1 = \frac{9 + 13}{2} = 11 \text{ home runs}$$

Similarly, to find Q_3, we average the 3rd and 4th largest numbers.

$$Q_3 = \frac{28 + 33}{2} = 30.5 \text{ home runs}$$

Together with the minimum of 5 and the maximum of 61, we obtain the five-number summary for Maris's home run data:

5 11 19.5 30.5 61

Thus,

$$\text{range} = 61 - 5 = 56 \text{ home runs;}$$

$$IQR = 30.5 - 11 = 19.5 \text{ home runs.}$$

By both measures, Maris's home run production has greater spread, indicating that Aaron's production was more consistent. ■

NOTE

Notice that only two of the five numbers (minimum and maximum) are in Maris's home run data.

Teaching Note

Boxplots provide one of the simplest and visually clearest ways to compare data. Ask students to collect real data and, working in cooperative learning groups, use boxplots to compare the data.

Alert

Some students may interpret the boxplot at the right to mean that there are more data items between 10 and 26 than between 26 and 34. Stress that each quartile has the same number of data items.

Boxplots

A **boxplot** is a graph that depicts the five-number summary of a data set. Boxplot is short for *box-and-whisker plot.* A boxplot is drawn parallel to a number-line scale. The plot consists of a central rectangle (box) that spans the quartiles and line segments (whiskers) that extend to the maximum and minimum values. A segment within the box marks the median. For instance, the boxplot based on the five-number summary for Aaron's home run data is shown in Figure 11.33.

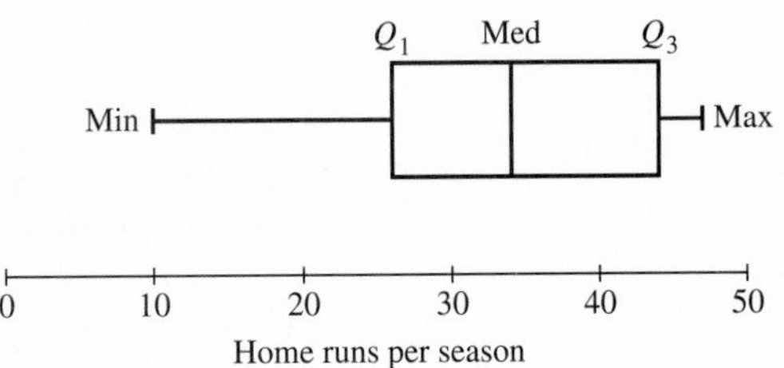

Figure 11.33 Boxplot of Hank Aaron's regular season home run totals.

■ EXAMPLE 7 Drawing a Boxplot

Make a boxplot of the population data for the group of western African nations listed in Table 11.20. Draw the plot by hand and on a grapher.

0 20 40 60 80 100

Population (millions)

Figure 11.34 Boxplot of the populations of western African nations.

Solution

Solve Algebraically

The three quartiles are not in the data set. We indicate their position in the sorted list with a dagger (†), then determine the five-number summary:

{0.4, 0.9, 1.0, 2.1 † 2.8, 3.8, 4.4, 5.0 † 7.8, 7.9, 8.3, 8.5 † 9.6, 13.0, 16.0, 90.1}

Averaging the two numbers next to each quartile split yields the five-number summary.

0.4 2.45 6.4 9.05 90.1

Now make a number-line scale that spans the data, say, from 0 to 100. Place five dots at a common distance above the scale to mark the five numbers in the summary. Then draw the box and whiskers. See Figure 11.34.

Support Graphically

Enter the population data from Table 11.20 into the statistics memory of your grapher. Select the boxplot option from the menu of statistical plots from a window with Xmin = 0 and Xmax = 100. See Figure 11.35. ■

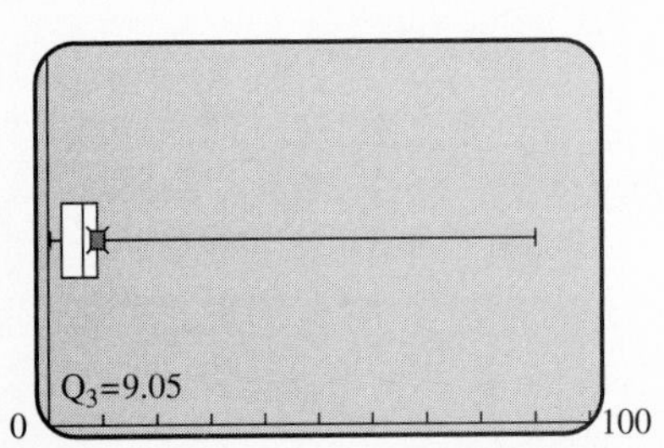

Figure 11.35 Grapher boxplot of western African populations (in millions), with trace activated.

The long whisker extending to the right in Figure 11.34 or Figure 11.35 is caused by an outlier, Nigeria's population of 90.1 million. A boxplot makes it easy to spot outliers. A common criterion is to identify an observation as an outlier if it lies more than $1.5 \times IQR$ below the first quartile or above the third quartile.

In Example 7,

$$IQR = Q_3 - Q_1 = 9.05 - 2.45 = 6.6$$

$$Q_3 + 1.5 \times IQR = 9.05 + 1.5 \times 6.6 = 18.95$$

$$Q_1 - 1.5 \times IQR = 2.45 - 1.5 \times 6.6 = -7.45$$

Because $90.1 > 18.95$, 90.1 is considered an outlier by the $1.5 \times IQR$ rule. In other words, this rule says that a boxplot has an outlier if either whisker is more than 1.5 times the length of the box.

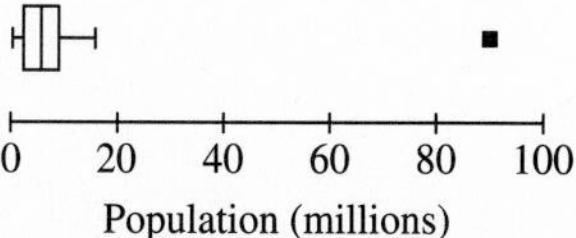

Figure 11.36 Modified boxplot of the populations of western African nations.

Sometimes a boxplot is drawn with outliers shown as points and the whiskers extending to include only nonoutliers. Figure 11.36 is such a modified boxplot.

To compare the center and spread of data sets, their boxplots are drawn side by side along a common scale.

EXAMPLE 8 Drawing Side-by-Side Boxplots

Draw side-by-side modified boxplots of annual home run production of Hank Aaron and Roger Maris.

Solution

See Example 6. Verify that there are no outliers for Aaron. For Maris, we have

$$Q_3 + 1.5 \times IQR = 30.5 + 1.5 \times 19.5 = 59.75.$$

Maris's 61 home runs is an outlier by this criterion, so we modify the right whisker of his boxplot to end at 39, the next highest value, and plot 61 as a separate point (Figure 11.37). ■

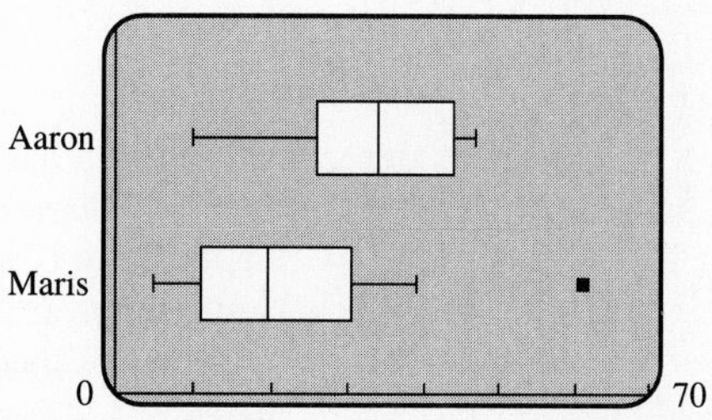

Figure 11.37 Side-by-side modified boxplots of regular season home run totals.

Standard Deviation

The range and interquartile range are measures of spread about the median. We now develop a measure of spread related to the mean. Because the measure is based on an averaging process for the differences $x_i - \bar{x}$, which are deviations from the mean, it is called the *standard deviation.* The standard deviation is denoted by σ, the lower-case Greek letter sigma.

NOTE

This definition is valid when the data set consists of the entire population of interest. If a sample is being used to estimate the standard deviation of a population, you should use

$$s = \sqrt{\frac{1}{n-1} \sum (x_i - \bar{x})^2}$$

as the standard deviation.

Definition Standard Deviation

The **standard deviation** of the numbers $x_1, x_2, \ldots, x_n$ is

$$\sigma = \sqrt{\frac{1}{n} \sum (x_i - \bar{x})^2}.$$

The square of the standard deviation is the **variance,** σ^2.

Alert

Students often have trouble getting an intuitive sense for standard deviation. You may wish to point out that, typically, approximately two-thirds of the data values are within one standard deviation of the mean. The standard deviation is not directly related to the number of values, and a larger standard deviation indicates that the numbers are more spread out.

Follow-up

Ask . . .

In Example 5 on page 771, why is it appropriate to use populations as weights?

Assignment Guide

Day 1: Ex. 3, 6, 8, 9, 12, 15, 17, 21, 24, 25, 28
Day 2: Ex. 31, 32, 34, 35, 38, 39, 40, 44

Cooperative Learning

Group Activity: Ex. 41, 42

To aid in remembering the formula, the standard deviation is sometimes called the *root mean square deviation.* This is because the standard deviation is the square *root* of the *mean* of the *squared deviations.* Calculating the standard deviation by hand is tedious, so it is natural to use a calculator.

■ EXAMPLE 9 Computing a Standard Deviation

Find the standard deviation of employee salaries for the Armadillo Biscuit Company for the year 1996.

Solution

Enter the data from Table 11.7 into the statistics memory of your grapher. Either find the standard deviation specifically or as part of a comprehensive list of descriptive statistics for a one-variable data set. The result should be $\sigma \approx 3.10$, which we interpret as \$3,100. ■

Typically, over half of a data set will lie within one standard deviation of the mean. In Example 1 we found the mean salary for Armadillo Biscuit Company employees to be \$27,100. So a particular salary is within one standard deviation of the average salary if it lies within \$27,100 $\pm$ 3,100, that is, between \$24,000 and \$30,200. Notice that 25 of the 42 salaries are within this interval.

Notes on Exercises

Ex. 1–20 deal with measures of center.
Ex. 21–28 deal with five-number summaries and boxplots.
Ex. 33–38 deal with standard deviation and variance.
Ex. 40 demonstrates that there are no generalizations concerning the order of the mean, median, and mode for a set of data.

Ongoing Assessment

Self-Assessment: Ex. 7, 13, 23, 27, 37
Embedded Assessment: Ex. 32

■ EXAMPLE 10 Computing a Standard Deviation

Find the standard deviation of the population data for the group of western African nations listed in Table 11.20.

Solution

Enter the population data from Table 11.20 into the statistics memory of your grapher and find the standard deviation. It should be $\sigma \approx 20.8$, which we interpret as 20,800,000 people. ■

The outlier (Nigeria's population) accounts for much of the large standard deviation in Example 10. The standard deviation is *not* a resistant measure of spread because it is strongly affected by outliers. The interquartile range, however, is a resistant measure of spread.

Quick Review 11.8

In Exercises 1–6, expand the sum.

1. $\sum_{i=1}^{7} x_i$

2. $\sum_{i=1}^{5} (x_i - \bar{x})$

3. $\frac{1}{7} \sum_{i=1}^{7} x_i$

4. $\frac{1}{5} \sum_{i=1}^{5} (x_i - \bar{x})$

5. $\frac{1}{5} \sum_{i=1}^{5} (x_i - \bar{x})^2$

6. $\sqrt{\frac{1}{5} \sum_{i=1}^{5} (x_i - \bar{x})^2}$

SECTION EXERCISES 11.8

In Exercises 1–4, find the mean of the data set.

1. {12, 23, 15, 48, 36}

2. {4, 8, 11, 6, 21, 7}

3. {32.4, 48.1, 85.3, 67.2, 72.4, 55.3}

4. {27.4, 3.1, 9.7, 32.3, 12.8, 39.4, 73.7}

In Exercises 5 and 6, find the mean population of the six states listed in Table 11.16 (Exercises 11.7) for the indicated year.

5. 1900

6. 1990

In Exercises 7 and 8, find the average of the indicated data.

7. The number of satellites (moons), from the data in Table 11.21.

Table 11.21 Planetary Satellites

Planet	Number of Satellites
Mercury	0
Venus	0
Earth	1
Mars	2
Jupiter	16
Saturn	18
Uranus	15
Neptune	8
Pluto	1

Source: National Geographic Atlas of the World (rev. 6th ed., 1992).

8. The area of the continents, from the data in Table 11.22.

Table 11.22 Size of Continent

Continent	Area (km^2)
Africa	30,269,680
Antarctica	13,209,000
Asia	44,485,900
Australia	7,682,300
Europe	10,530,750
North America	24,235,280
South America	17,820,770

Source: National Geographic Atlas of the World (rev. 6th ed., 1992).

In Exercises 9–12, solve the problem.

9. *Home Run Production* Determine the average annual home run production for Willie Mays and for Mickey Mantle for their career totals of 660 over 22 years and 536 over 18 years, respectively. Who had the greater production rate?

10. *Painting Houses* A painting crew in State College, Pennsylvania, painted 12 houses in 5 days, and a crew in College Station, Texas, painted 15 houses in 7 days. Determine the average number of houses per day each crew painted. Which crew had the greater rate?

11. *Skirt Production* The Hip-Hop House produced 1147 scooter skirts in 4 weeks, and What-Next Fashion produced 1516 scooter skirts in 4 weeks. Which company had the greater production rate?

12. *Per Capita Income* Per capita income (PCI) is an average found by dividing a nation's gross national product (GNP) by its population. India has 882,575,000 people and a GNP of 311 billion dollars, and Mexico has 87,715,000 people and a GNP of 218 billion dollars. Determine the PCI for India and for Mexico. Which nation has the greater income per person?

In Exercises 13–16, use Table 11.2 (Exercises 11.2) to find the median and mode number of rainy days per month for the city.

13. Chicago **14.** Los Angeles

15. New York **16.** Seattle

In Exercises 17 and 18, use the data in Table 11.23 to find the mean life expectancy for the indicated group.

17. South American Women **18.** South American Men

Table 11.23 Life Expectancies and Populations (in millions) for the Nations of South America

Nation	Male	Female	Population
Argentina	68	75	33.9
Bolivia	61	66	7.7
Brazil	57	67	158.7
Chile	72	78	14.0
Columbia	69	75	35.6
Ecuador	67	73	10.7
Guyana	62	68	0.7
Paraguay	72	75	5.2
Peru	63	68	23.7
Suriname	67	72	0.4
Uruguay	71	77	3.2
Venezuela	70	76	20.6

Source: The World Almanac and Book of Facts (1995, Mahwah, N.J.: Funk & Wagnalls), pp. 743–835.

In Exercises 19 and 20, use the data in Table 11.19 (Exercises 11.7).

a. Find the average of the indicated temperatures for Beijing.
b. Find the weighted average using the number of days in the month as the weight. (Assume no leap year.)
c. Compare your results in (a) and (b). Do the weights have an effect on the average? Why or why not? Which average is the better indicator for these temperatures?

19. The monthly high temperatures.

20. The monthly low temperatures.

In Exercises 21–24, determine the five-number summary, the range, and the interquartile range and identify any outliers for the data.

21. The employee salaries for Armadillo Biscuit Company salaries in Table 11.7 (Section 11.7).

22. The annual home run production data for Willie Mays and for Mickey Mantle in Table 11.15 (Exercises 11.7).

23. The following average annual wind speeds at 44 climatic data centers around the United States:
9.0, 6.9, 9.1, 9.2, 10.2, 12.5, 12.0, 11.2, 12.9, 10.3, 10.6, 10.9, 8.7, 10.3, 11.0, 7.7, 11.4, 7.9, 9.6, 8.0, 10.7, 9.3, 7.9, 6.2, 8.3, 8.9, 9.3, 11.6, 10.6, 9.0, 8.2, 9.4, 10.6, 9.5, 6.3, 9.1, 7.9, 9.7, 8.8, 6.9, 8.7, 9.0, 8.9, 9.3

24. The following employees' salaries in one department of the Garcia Brothers Company (in thousands of dollars):
33.5, 35.3, 33.8, 29.3, 36.7, 32.8, 31.7, 37.3, 33.5, 28.2, 34.8, 33.5, 29.7, 38.5, 32.7, 34.8, 34.2, 31.6, 35.4

In Exercises 25–28, make a (a) boxplot and a (b) modified boxplot of the data.

25. The employee salaries for Armadillo Biscuit Company salaries in Table 11.7 (Section 11.7).

26. The annual home run production for Willie Mays in Table 11.15 (Exercises 11.7).

27. The population of the South American nations given in Table 11.23.

28. The population of the western African nations given in Table 11.20.

In Exercises 29 and 30, refer to the wind speed data analyzed in Exercise 23.

29. Some wind turbine generators, to be efficient generators of power, require average wind speeds of at least 10.5 mph. Approximately what fraction of the climatic centers are suited for these wind turbine generators?

30. If technology improves the efficiency of the wind turbines so that they are efficient in winds that average at least 7.5 mph, approximately what fraction of the climatic centers are suited for these improved wind generators?

In Exercises 31 and 32, use side-by-side boxplots of the annual home run production data for Willie Mays and for Mickey Mantle in Table 11.15 (Exercises 11.7) to answer the questions.

31. **a.** Which data set has the greater range?
b. Which data set has the larger interquartile range?

32. **Writing to Learn** Write a paragraph explaining the difference in home run production between Willie Mays and Mickey Mantle.

In Exercises 33–38, find the standard deviation and variance of the data set.

33. $\{23, 45, 29, 34, 39, 41, 19, 22\}$

34. $\{28, 84, 67, 71, 92, 37, 45, 32, 74, 96\}$

35. The employee salaries for Armadillo Biscuit Company salaries in Table 11.7 (Section 11.7).

36. The annual home run production data for Mickey Mantle in Table 11.15 (Section Exercises 11.7).

37. The data in Exercise 23.

38. The data in Exercise 24.

In Exercises 39–44, solve the problem.

39. Is Brazil's population an outlier among the populations of South American nations according to the $1.5 \times IQR$ criterion?

EXTENDING THE IDEAS

40. List a set of data for which the inequality holds.
a. Mode < median < mean
b. Median < mean < mode
c. Mean < mode < median

41. List a set of data for which the inequality holds.
a. Standard deviation < interquartile range
b. Interquartile range < standard deviation
c. Range = interquartile range

42. Is it possible for the standard deviation of a data set to be greater than the range? Explain.

43. Why can we find the mode of categorical data but not the mean or median?

44. Draw a boxplot for which the inequality holds.
a. Median < mean
b. 2 × interquartile range < range
c. Range < 2 × interquartile range

CHAPTER 11 REVIEW

KEY TERMS

The number following each key term indicates the page of its introduction.

arithmetic progression, 693
arithmetic sequence, 693
arithmetic series, 705
axiom of induction, 720
back-to-back stemplot, 759
bar chart, 757
binomial coefficient, 729
binomial probability, 751
binomial theorem, 731
boxplot, 774
categorical variable, 757
certain event, 745
circle graph, 757
combinations, 739
combinations of n objects taken r at a time, 740
combinatorics, 735
common difference, 693
common ratio, 695
complement of an event, 747
converge, 709
disjoint events, 747
distribution, 759
diverge, 709
empirical probability, 752
equally likely outcomes, 745
event, 744
expanded form (of a series), 712
experiment, 743
extended principle of mathematical induction, 724
Fibonacci numbers, 693
Fibonacci sequence, 692
finite sequence, 689
finite series, 704
first quartile, 772
five-number summary, 772
frequency, 760
frequency table, 760
fundamental counting principle, 736
geometric progression, 695
geometric sequence, 695
geometric series, 707
histogram, 760
impossible event, 745
independent events, 748
index of summation, 711
individuals, 757
infinite sequence, 689
interquartile range, 772
kth term (of a series), 704
leaf, 758
line graph, 762
mathematical induction, 719
mean, 768
measure of center, 767
measure of spread, 767
median, 769
mode, 769
mutually exclusive events, 747
n factorial, 728
outcome, 744
outliers, 770
Pascal's triangle, 728
permutations, 738
permutations of n objects taken r at a time, 738
pie chart, 757
principle of mathematical induction, 722
probability (of an event), 745
quantitative variable, 757
quartile, 772
range, 772
recursively defined sequence, 691
relative frequency, 752
resistant measure, 770
sample space, 744
second quartile, 772
sequence, 689
sequence of partial sums, 709
series, 703
standard deviation, 776
statistic, 767
stem, 758
stemplot, 758
sum of a finite arithmetic series, 705
sum of a finite geometric series, 707
sum of an infinite geometric series, 710
summation notation, 711
theoretical probability, 752
third quartile, 772
time plot, 762
triangular number, 706
variable (in statistics), 757
variance, 776
weighted average, 771
weights, 771
zero factorial, 728

REVIEW EXERCISES

In Exercises 1 and 2, find the first 6 terms and the 40th term of the sequence.

1. $a_n = \dfrac{n^2 - 1}{n + 1}$ **2.** $b_k = \dfrac{(-2)^k}{k + 1}$

In Exercises 3–8, find the first 6 terms and the 12th term of the sequence.

3. $a_1 = -1$ and $a_n = a_{n-1} + 3$, for $n \geq 2$

4. $b_1 = 5$ and $b_k = 2b_{k-1}$, for $k \geq 2$

5. Arithmetic sequence, with $a_1 = -5$ and $d = 1.5$

6. Geometric sequence, with $a_1 = 3$ and $r = 1/3$

7. $v_1 = -3$, $v_2 = 1$ and $v_k = v_{k-2} + v_{k-1}$, for $k \geq 3$

8. $w_1 = -3$, $w_2 = 2$ and $w_k = w_{k-2} + w_{k-1}$, for $k \geq 3$

In Exercises 9–16, the sequences are arithmetic or geometric. Find an explicit formula for the nth term. State the common difference or ratio.

9. 12, 9.5, 7, 4.5, . . . **10.** −5, −1, 3, 7, . . .

11. 10, 12, 14.4, 17.28, . . . **12.** $\frac{1}{8}, -\frac{1}{4}, \frac{1}{2}, -1, \ldots$

13. $a_1 = -11$ and $a_n = a_{n-1} + 4.5$ for $n \geq 2$

14. $b_1 = 7$ and $b_n = \frac{1}{4}b_{n-1}$ for $n \geq 2$

15. The fourth and ninth terms of a geometric sequence are −192 and 196 608, respectively.

16. The third and eighth terms of an arithmetic sequence are 14 and −3.5, respectively.

In Exercises 17–20, find the sum of the arithmetic series.

17. $-11 - 8 - 5 - 2 + 1 + 4 + 7 + 10$

18. $13 + 9 + 5 + 1 - 3 - 7 - 11$

19. $2.5 - 0.5 - 3.5 - \cdots - 75.5$

20. $-5 - 3 - 1 + 1 + \cdots + 55$

In Exercises 21–24, find the sum of the geometric series.

21. $4 - 2 + 1 - \frac{1}{2} + \frac{1}{4} - \frac{1}{8}$

22. $-3 - 1 - \frac{1}{3} - \frac{1}{9} - \frac{1}{27}$

23. $2 + 6 + 18 + \cdots + 39\,366$

24. $1 - 2 + 4 - 8 + \cdots - 8192$

In Exercises 25 and 26, find the sum of the first 10 terms of the arithmetic or geometric sequence.

25. 2187, 729, 243, . . . **26.** 94, 91, 88, . . .

In Exercises 27 and 28, use the parametric mode to graph the sequence.

27. $a_n = 1 + \dfrac{(-1)^n}{n}$ **28.** $a_n = 2n^2 - 1$

In Exercises 29–34, determine whether the geometric series converges. If it does, find its sum. Graph the sequence of partial sums to support your answer.

29. $\sum_{j=1}^{\infty} 2\left(\frac{3}{4}\right)^j$ **30.** $\sum_{k=1}^{\infty} 2\left(-\frac{1}{3}\right)^k$

31. $\sum_{j=1}^{\infty} 4\left(-\frac{4}{3}\right)^j$ **32.** $\sum_{k=1}^{\infty} 5\left(\frac{6}{5}\right)^k$

33. $\sum_{k=1}^{\infty} 3(0.5)^k$ **34.** $\sum_{k=1}^{\infty} (1.2)^k$

In Exercises 35–38, write the sum using Σ.

35. $-8 - 3 + 2 + \cdots + 92$

36. $4 - 8 + 16 - 32 + \cdots - 2048$

37. $1^2 + 3^2 + 5^2 + \cdots$

38. $1 + \frac{1}{2} + \frac{1}{2^2} + \frac{1}{2^3} + \cdots$

In Exercises 39–44, use summation formulas to evaluate the expression.

39. $\sum_{k=1}^{n} (3k + 1)$ **40.** $\sum_{k=1}^{n} 3k^2$

41. $\sum_{k=1}^{25} (k^2 - 3k + 4)$ **42.** $\sum_{k=1}^{100} (k^3 - 2k)$

43. $\sum_{k=1}^{175} (3k^2 - 5k + 1)$ **44.** $\sum_{k=1}^{75} (k^3 - k^2 + 2)$

In Exercises 45–50, evaluate the expression.

45. $\binom{12}{5}$

46. $\binom{789}{787}$

47. ${}_{18}C_{12}$

48. ${}_{35}C_{28}$

49. ${}_{12}P_7$

50. ${}_{15}P_8$

In Exercises 51–56, expand the expression.

51. $(2x + y)^5$

52. $(4a - 3b)^7$

53. $(3x^2 + y^3)^5$

54. $\left(1 + \frac{1}{x}\right)^6$

55. $(2a^3 - b^2)^9$

56. $(x^{-2} + y^{-1})^4$

In Exercises 57–60, use mathematical induction to prove that the statement is true for all positive integers n.

57. $1 + 3 + 6 + \cdots + \frac{n(n+1)}{2} = \frac{n(n+1)(n+2)}{6}$

58. $1 \cdot 2 + 2 \cdot 3 + 3 \cdot 4 + \cdots + n(n+1) = \frac{n(n+1)(n+2)}{3}$

59. $2^{n-1} \leq n!$

60. $n^3 + 2n$ is divisible by 3.

In Exercises 61–68, solve the problem.

61. *Code Words* How many five-character code words are there if the first character is always a letter and the other characters are letters and/or digits?

62. *Scheduling Trips* A travel agent is trying to schedule a client's trip from city A to city B. There are three direct flights, three flights from A to a connecting city C, and four flights from this connecting city C to city B. How many trips are possible?

63. *License Plates* How many license plates begin with two letters followed by four digits or begin with three digits followed by three letters? Assume that no letters or digits are repeated.

64. *Forming Committees* A club has 45 members, and its membership committee has three members. How many different membership committees are possible?

65. *Bridge Hands* How many 13-card bridge hands include the ace, king, and queen of spades?

66. *Bridge Hands* How many 13-card bridge hands include all four aces and exactly one king?

67. *Coin Toss* Suppose that a coin is tossed five times. How many different outcomes include at least two heads?

68. *Forming Committees* A certain small business has 35 employees, 21 women and 14 men. How many different employee representative committees are there if the committee must consist of two women and two men?

In Exercises 69–72, list the elements of the sample space.

69. *Spinners* A game spinner numbered 1 through 6 is spun.

70. *Rolling Dice* A red die and a green die are rolled.

71. *Code Words* A two-digit code is selected from the digits $\{1, 3, 6\}$, where no digits are to be repeated.

72. *Production Line* A product is inspected as it comes off the production line and is classified as either defective or nondefective.

In Exercises 73–76, a penny, a nickel, and a dime are tossed.

73. List all possible outcomes.

74. List all outcomes in the event "two heads or two tails."

75. List all outcomes in the complement of the event in Exercise 74.

76. Find the probability of tossing at least one head.

In Exercises 77–82, solve the problem.

77. *Coin Toss* A fair coin is tossed six times. Find the probability of the event "*HHTHTT*."

78. *Coin Toss* A fair coin is tossed four times. Find the probability of obtaining one head and three tails.

79. *Coin Toss* A fair coin is tossed five times. Find the probability of obtaining two heads and three tails.

80. *Assembly Line* In a random check on an assembly line, the probability of finding a defective item is 0.003. Find the probability of a nondefective item occurring 10 times in a row.

81. *Success or Failure* An experiment has only two possible outcomes—success (S) or failure (F)—and repetitions are independent events. If $P(S) = 0.5$, find the probability of obtaining three successes and one failure in four repetitions.

82. *Success or Failure* For the experiment in Exercise 81, explain why the probability of one success and three failures is equal to the probability of three successes and one failure.

In Exercises 83–86, an experiment has only two possible outcomes—success (S) or failure (F)—and repetitions are independent events. The probability of success is 0.4.

83. Find the probability of SF on two repetitions.

84. Find the probability of SFS on three repetitions.

85. Find the probability of at least one success on two repetitions.

86. Explain why the probability of one success and three failures is not equal to the probability of three successes and one failure.

In Exercises 87–90, construct (a) a stemplot, (b) a frequency table, and (c) a histogram for the indicated data.

87. *Faculty Salaries* Use intervals of \$1000. The salaries (in thousands of dollars) of full professors at 4-year U.S. public institutions in 29 selected fields are as follows:

62.3, 53.3, 54.9, 60.7, 53.5, 51.0, 59.5, 57.7, 50.7, 60.0, 51.6, 52.1, 51.4, 50.2, 52.3, 51.1, 53.4, 51.1, 52.6, 53.6, 55.1, 53.1, 53.2, 51.0, 53.2, 51.7, 51.7, 49.9, 48.5

(*Source:* College and University Personnel Association as reported in *The Chronicle of Higher Education Almanac,* August 26, 1992.)

88. *Faculty Salaries* Use intervals of \$5000. The salaries (in thousands of dollars) of full professors at 4-year U.S. private institutions in 29 selected fields are as follows:

59.8, 53.5, 60.4, 71.7, 52.6, 46.7, 58.4, 66.3, 48.8, 71.3, 49.5, 49.7, 50.0, 48.8, 48.6, 49.4, 53.8, 44.9, 47.2, 51.3, 58.6, 54.0, 50.8, 51.8, 47.1, 48.8, 49.2, 42.8, 46.6

(*Source:* College and University Personnel Association as reported in *The Chronicle of Higher Education Almanac,* August 26, 1992.)

89. *Web Sites* Use intervals of 5. The number of schools in the 50 states, the District of Columbia, and the U.S. Virgin Islands having their own World Wide Web sites are as follows. The numbers appear in alphabetical order (name of state).

11, 15, 13, 5, 104, 27, 6, 3, 7, 31, 23, 15, 3, 44, 17, 8, 5, 6, 5, 6, 13, 25, 10, 48, 8, 19, 4, 18, 2, 4, 14, 5, 33, 13, 2, 24, 3, 36, 19, 5, 7, 2, 9, 19, 22, 9, 67, 61, 8, 8, 2, 1

(*Source:* Web 66, as reported in *USA Today* on October 9, 1995.)

90. *Passing Yardage* In 1995, Warren Moon of the Minnesota Vikings became the first pro quarterback to pass for 60,000 total yards. Use intervals of 1000 yards for Moon's regular season passing yards given in Table 11.24.

Table 11.24 Regular Season Passing Yardage Statistics for Warren Moon

Year	Yards	Year	Yards
1978	1112	1987	2806
1979	2382	1988	2327
1980	3127	1989	3631
1981	3959	1990	4689
1982	5000	1991	4690
1983	5648	1992	2521
1984	3338	1993	3485
1985	2709	1994	4264
1986	3489		

Source: The Minnesota Vikings, as reported by Julie Stacey in *USA Today* on September 25, 1995.

In Exercises 91–94, find the five-number summary, the range, the interquartile range, the standard deviation and the variance for the specified data. Identify any outliers.

91. The data in Exercise 87.

92. The data in Exercise 88.

93. The data in Exercise 89.

94. The data in Exercise 90.

In Exercises 95–98, construct (a) a boxplot, and (b) a modified boxplot for the specified data.

95. The data in Exercise 87.

96. The data in Exercise 88.

97. The data in Exercise 89.

98. The data in Exercise 90.

In Exercises 99 and 100, solve the problem.

99. Make a back-to-back stemplot of the data in Exercises 87 and 88. Write a sentence interpreting the stemplot.

100. Make side-by-side boxplots of the data in Exercises 87 and 88.

a. Which set of data has the greater range?

b. Which set of data has the greater interquartile range?

In Exercises 101 and 102, use the data in Table 11.25 and make a timeplot for the indicated data.

Table 11.25 Average Daily High and Low Temperatures in °C for Tokyo

Month	High	Low	Month	High	Low
January	8	−2	July	28	21
February	9	−1	August	30	22
March	12	2	September	26	19
April	17	8	October	21	13
May	22	12	November	16	6
June	24	17	December	11	1

Source: National Geographic Atlas of the World (rev. 6th ed., 1992, Washington, D.C.), plate 132.

101. *Timeplots* The average monthly high temperature for Tokyo.

102. *Timeplots* The average monthly low temperature for Tokyo.

In Exercises 103–108, solve the problem.

103. Use mathematical induction to prove that the sum of the first n terms of a geometric sequence with first term a_1 and common ratio r is

$$\frac{a_1(1 - r^n)}{1 - r}.$$

104. Find the ninth row of Pascal's triangle.

105. Show algebraically that

$${}_nP_k \times {}_{n-k}P_j = {}_nP_{k+j}.$$

106. *Baseball Bats* Suppose that the probability of producing a defective baseball bat is 0.02. Four bats are selected at random. What is the probability that the lot of four bats contains the following?

a. No defective bats. **b.** One defective bat.

107. *Light Bulbs* Suppose that the probability of producing a defective light bulb is 0.0004. Ten light bulbs are selected at random. What is the probability that the lot of 10 contains the following?

a. No defective light bulbs.
b. Two defective light bulbs.

108. *Choosing a Plot* A weather station records the temperature every 15 minutes. If you are to graph the data for a 24-hour period, would you use a histogram, a line graph, or a boxplot graph? Explain your choice.

DAVID BLACKWELL

David Blackwell (b. 1919) is a noted professor of statistics at the University of California, Berkeley. He graduated from high school in Centralia, Illinois, at age 16 and then entered the University of Illinois. Within 6 years he had not only received his baccalaureate degree, but his Ph.D. in mathematics as well, and had become the first African-American to receive a fellowship to the Institute for Advanced Study at Princeton. Professor Blackwell has made contributions to Bayesian statistics, game theory, set theory, information theory, probability, and dynamic programming.

appendix A

GRAPHER WORKSHOP

A.1 INTRODUCTION

We call this appendix a workshop to emphasize that learning the grapher requires active hands-on work. In thousands of workshop hours over the past 10 years we have helped mathematics students and instructors learn how to use the hand-held graphing calculator. Our reward has been the enthusiasm and excitement of the participants as they catch on to the remarkable way that a grapher aids their learning and teaching.

As much as we would like to, we cannot bring that actual workshop directly to you. Still you can learn to use and appreciate this technology with some basic knowledge and your own creativity. A few hours of productive play will allow you to solve mathematical problems in new ways with this powerful tool. Have your grapher out and "on" to work through the examples and activities as you read. Feel free to explore the menus and features of your grapher. We suggest that you refer to this workshop when you encounter mathematics that require unfamiliar aspects of the grapher. The power of the grapher will be what you make it. With play, thought, and practice, many students have found it incredibly exciting.

REMARK

Through the speed and power of the grapher you can investigate ideas quickly and make and test conjectures based on graphical and numerical evidence.

A.2
NUMERICAL COMPUTATION AND EDITING

GRAPHER NOTES

1. In this workshop, boxed items in small caps, such as [TAN], suggest grapher keys. Unboxed words in small caps, such as FUNCTION MODE, suggest on-screen menu items.
2. Most grapher keys have multiple functions. You can access the second function of a key by first pressing the special colored [2ND] or [SHIFT] and its alphabetic function by pressing [ALPHA].

First Steps

Take a minute to study the keyboard of your grapher. The keys are grouped in "zones" according to their function: scientific calculation, graphing, editing, and various menus. Locate [ON]. Not only is it used to turn on your grapher, but it also acts as an [OFF] button as its *second function,* [2ND] [ON].

- Practice turning your calculator on and off.

Next determine how to adjust the screen contrast, something you may need to do as lighting conditions change or battery power weakens. (You may have to check your grapher owner's manual to see how to do this for your grapher.)

- Adjust your screen contrast to make the screen very dark, then very light, and finally to suit your taste.

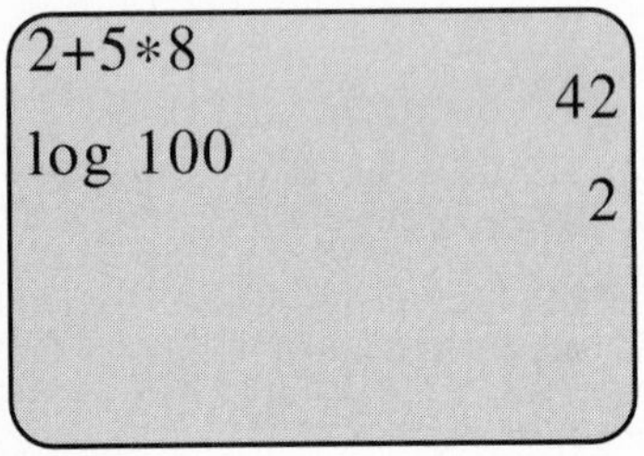

```
2+5*8
                42
log 100
                 2
```

Figure A.1

Performing Calculations

Computation is done on the *Home screen.*

- Try the calculations shown in Figure A.1. Simply key in each expression, followed by [ENTER] (or [EXE]). To find the value of log 100, use [LOG]. Do not enter the individual letters L, O, and G.

Error Messages

Don't be afraid to make a mistake! Just as pencils have erasers, graphers have delete ([DEL]) keys. Let's purposely make a mistake to see what happens.

- Key in 7 [÷] 0, press [ENTER], and observe your grapher's error message.

Error messages vary from grapher to grapher. Take a moment to read your grapher's error message. If a menu appears with a GoTo option, choose this option to "go to" the source of the error.

- Use your cursor keys ([◄], [►], [▲], and [▼]) and [DEL] to change the expression from 7 ÷ 0 to 7 ÷ 2. Then enter this new expression.

Did you obtain the expected answer? If not, check your grapher owner's manual (CYGOM). (See the marginal note.) Take a few minutes to play with the editing features of your grapher. These few minutes could save hours in the long run.

GRAPHER NOTE

Keying sequences and other procedures vary somewhat from grapher to grapher, causing a need for you to "check your grapher owner's manual." We will abbreviate this instruction as CYGOM.

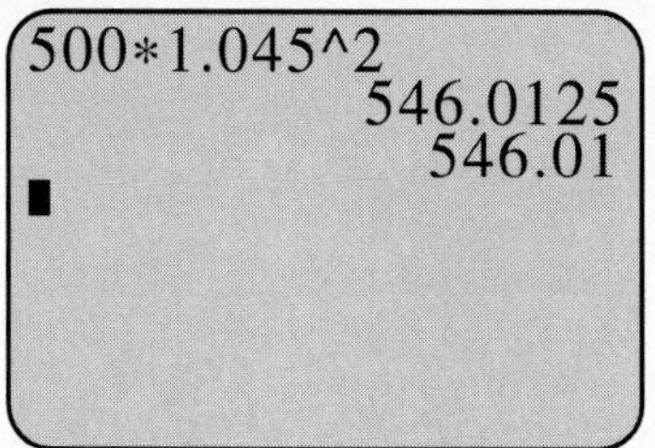

Figure A.2 The compound interest expression computed in *floating point mode,* then in two decimal place display mode.

■ EXAMPLE 1 Replaying and Editing a Computation

If you deposit \$500 in a savings account at a 4.5% interest rate, compounded annually, how much will you have in your account at the end of 2, 4, and 11 years?

Solution

The total value S of the investment at the end of n years is

$$S = P(1 + r)^n,$$

where r is the interest rate. Because $4.5\% = 0.045$, $1 + r = 1.045$.

- Enter the expression 500×1.045^2 on the Home screen of your grapher. If your answer has unwanted decimal places, change your *display mode* to two decimal places (CYGOM if necessary) and reenter the expression, as shown in Figure A.2.
- *Replay* the previous entry. That is, reenter the expression without retyping the entire expression (CYGOM if necessary). Then change the exponent from 2 to 4. Repeat for the exponent 11, as shown in Figure A.3.

500*1.045^2
546.0125
546.01
500*1.045^4
596.26
500*1.045^11
811.43

Figure A.3 The compound interest expression computed for three different time values.

We conclude the values in the account are \$546.01 at the end of 2 years, \$596.26 after 4 years, and \$811.43 after 11 years. ■

Scientific Notation

The U.S. national debt as of 1993 was \$4,351,200,000,000.

- Enter this number of dollars on the Home screen.

Your grapher will return the value in *scientific notation* because the number is so large. (See Figure A.4.) We interpret this result as 4.3512×10^{12} dollars or, because 10^{12} is 1 trillion, as \$4.3512 trillion.

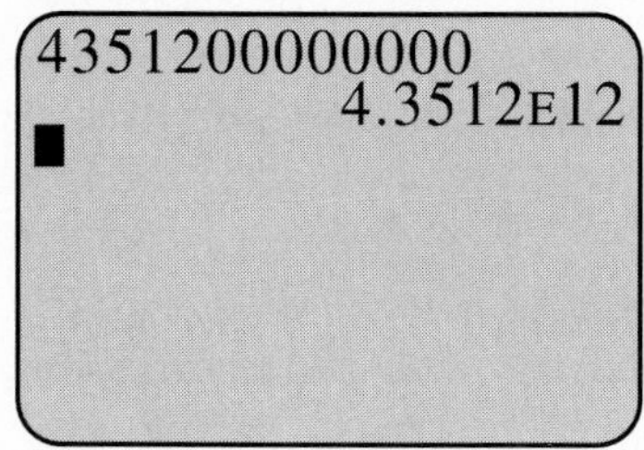

Figure A.4

The ANS Feature

When doing a series of calculations, you can easily use the answer from one calculation in the next calculation.

- Carry out the calculations shown in Figure A.5 by pressing 7 [×] 7 [ENTER], then [×] 9 [ENTER], and finally [√] [ANS] [ENTER]. Note in the second step of the calculation that "Ans" automatically appeared on the grapher screen because the grapher needed a quantity to multiply by 9.

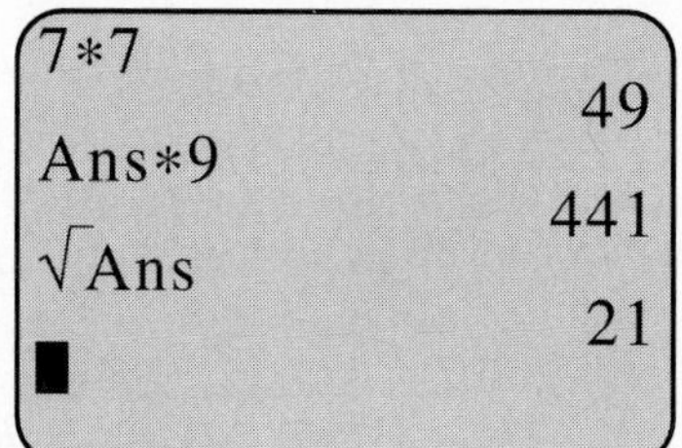

Figure A.5

When repeating a calculation recursively, you can use the ANS feature in an extremely useful way.

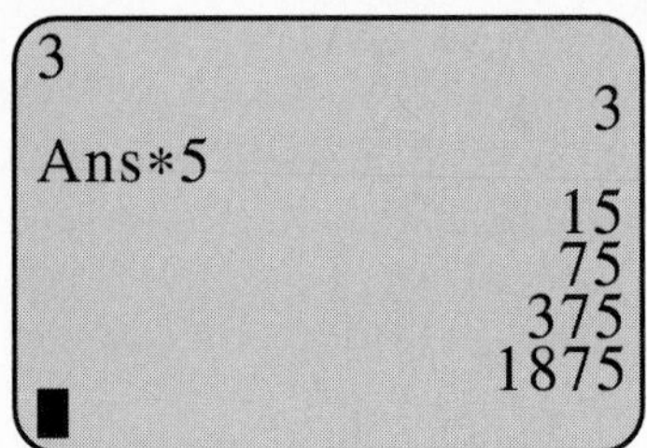

Figure A.6

- Calculate the terms in a geometric sequence that begins with 3 and grows by a constant factor of 5 by carrying out the calculation shown in Figure A.6.

Using Variables

Another extremely helpful computational feature of the grapher is the ability to store and recall numbers as variables. The next activity illustrates using the variable x to evaluate a function at several different values of x. The activity also uses [:] to string together commands. If your grapher does not have [:], use [ENTER] instead.

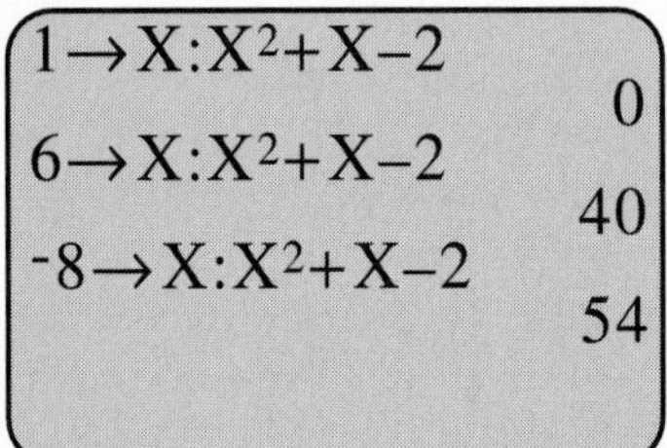

Figure A.7

- Evaluate $f(x) = x^2 + x - 2$ at $x = 1$ by pressing

  1 [STO▶] [X,T,θ] [:] [X,T,θ] [x^2] [+] [X,T,θ] [–] 2 [ENTER].

  Then use the replaying and editing features of your grapher to evaluate $f(6)$ and $f(-8)$, as shown in Figure A.7.

A.3 TABLE BUILDING

A grapher feature even more powerful than the ability to store variables is the ability to store functions. This feature is the basis of graphing and table building. In either case the $Y=$ *edit screen* is used to store the symbolic expressions (rules) for functions. A table can be used to evaluate a function for several different x-values.

- Press [Y=] (or [SYMB]) to go to the $Y=$ screen. Then press [X,T,θ] [x^2] [+] [X,T,θ] [–] 2 [ENTER]. See Figure A.8a.
- Press [TBLSET] 0 [ENTER] 1 [ENTER]. See Figure A.8b. Then press [TABLE]. See Figure A.8c. (This key sequence will vary from grapher to grapher. CYGOM, if necessary.)

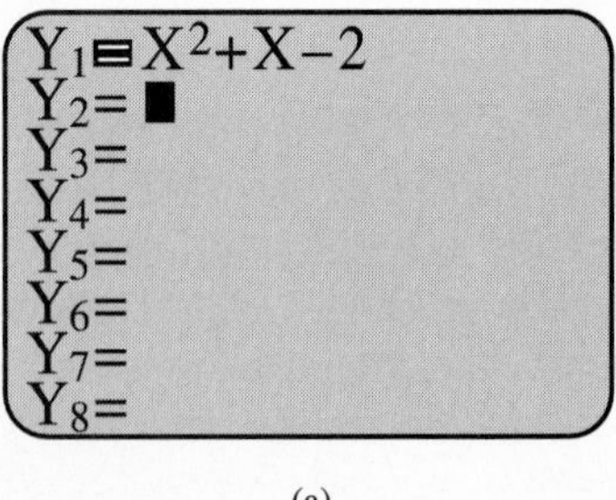

(a)

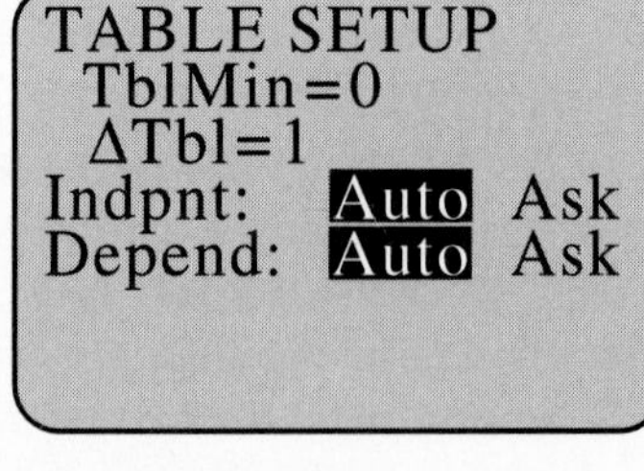

(b)

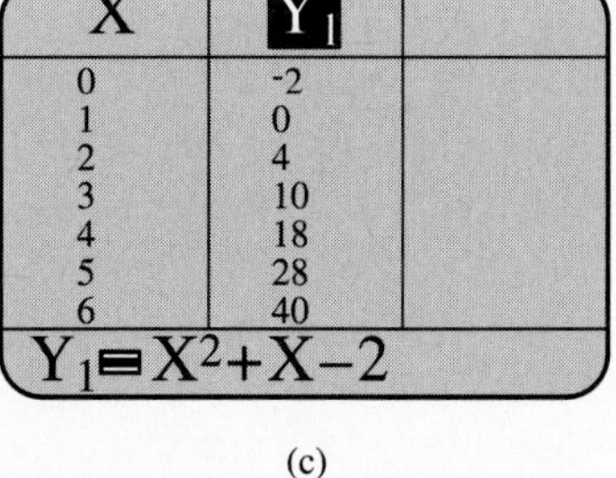

(c)

Figure A.8 The steps in the table-building process on a grapher.

- Use all four cursor keys (▶, ◀, ▲, and ▼) to move around the table and explore. Pay attention to the readout at the bottom of the screen as you move to different "cells" in the table. What happens when you try to move the cursor off the top or bottom of the screen?

A.4 FUNCTION GRAPHING

Graphing and Exploring a Function

Most graphers have several graphing modes. Be sure that your grapher is in FUNCTION mode. The algebraic form for the function needs to be $f(x) = \ldots,$ or $y = \ldots.$

■ EXAMPLE 2 Graphing and Tracing along a Function

Use the FUNCTION mode to graph $f(x) = 2x + 1$. Explore the ordered pairs of the graph with the TRACE feature.

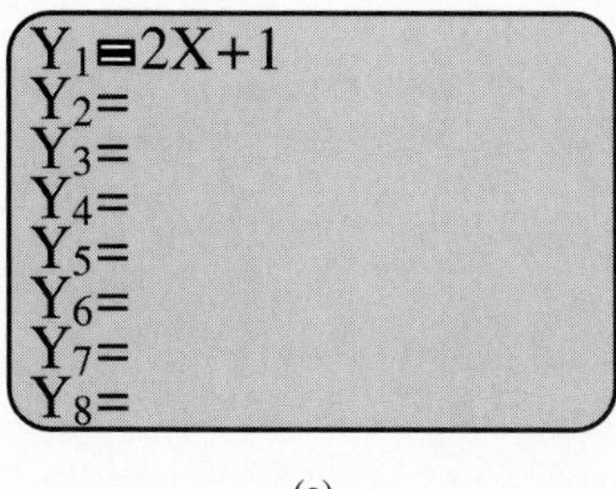

(a)

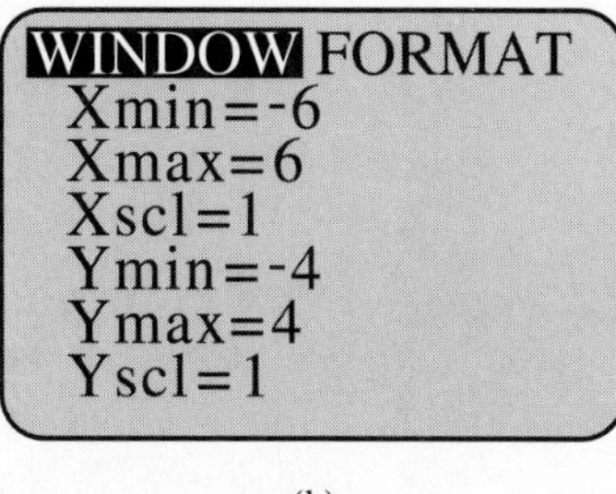

(b)

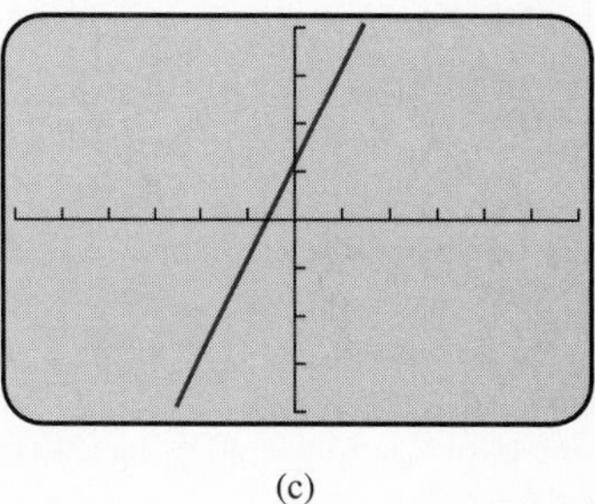

(c)

Figure A.9 The steps in the graphing process on a grapher.

Solution

Figure A.9 illustrates the procedure for graphing f in the window $[-6, 6]$ by $[-4, 4]$. Window is called range on some graphers.

- Enter $y = 2x + 1$ as shown in Figure A.9a and the *window dimensions* as shown in Figure A.9b. Then press GRAPH or PLOT to obtain the graph shown in Figure A.9c.
- Press TRACE to display x- and y-coordinates of points on the graph. Use ▶ or ◀ to move from point to point. See Figure A.10. This *tracing* shows you which points of $f(x) = 2x + 1$ were plotted by the grapher.
- Change the view dimensions to $[-10, 10]$ by $[-10, 10]$, known as the *standard window.* Some graphers have a ZOOM key. If yours does, press

Figure A.10 The graph of $f(x) = 2x + 1$ with the TRACE feature activated.

ZOOM, then choose ZSTANDARD from the menu to set the window $[-10, 10]$ by $[-10, 10]$ automatically. Press TRACE and explore. Are the (x, y) pairs the same as you found using the window $[-6, 6]$ by $[-4, 4]$?

A grapher allows you to obtain several views of the graph of a function. The *X*min and *X*max window dimensions determine which points the grapher plots and hence the coordinate readout when the TRACE feature is activated. ■

Table A.1 The number of columns of pixels on various graphers

Grapher	Columns of Pixels
TI-80	63
Casio, Sharp, TI-82, TI-83	95
TI-81	96
TI-85	127
Hewlett-Packard	131
TI-92	239

Your choice of *X*min and *X*max affects the x-coordinate readout when you trace along a graph. The reason is that the grapher screen is a rectangular array of *pixels,* short for "picture elements." The change in x-value that occurs when tracing is given by

$$\Delta x = \frac{X\text{max} - X\text{min}}{\text{Number of columns of pixels} - 1}.$$

The number of columns of pixels varies from grapher to grapher, as indicated in Table A.1.

GRAPHER NOTE

Some graphers have a ZDECIMAL feature that sets [*X*min, *X*max] to the basic friendly settings and a ZINTEGER feature that can set the x view dimensions to 10 times the basic friendly settings.

Friendly Windows

As we observed in Example 2, the x-values displayed on the screen during tracing can have many decimal places—for example, $x = 0.63829787$, as shown in Figure A.10. Such "unfriendly" x-values can be avoided. You can use the [*X*min, *X*max] settings given in Table A.2, or positive integer multiples of these settings, to guarantee a *friendly* x-coordinate readout when tracing. Windows with friendly x-coordinates are called *friendly windows.*

- Graph the function $f(x) = 2x + 1$ from Example 2 in a friendly window, using the [*X*min, *X*max] settings given for your grapher in Table A.2. Trace along the graph.
- Double the [*X*min, *X*max] and [*Y*min, *Y*max] settings and trace along the new view of the graph. How has the x-coordinate readout changed?
- Enter 10 times the settings given for your grapher in Table A.2 to obtain *integer settings.* Trace to learn the reason for this name.

Table A.2 The [*X*min, *X*max] dimensions for a basic friendly window on various graphers

Grapher	[*X*min, *X*max]
TI-80	$[-3.1, 3.1]$
Casio, Sharp, TI-82, TI-83	$[-4.7, 4.7]$
TI-81	$[-4.8, 4.7]$
TI-85	$[-6.3, 6.3]$
Hewlett-Packard	$[-6.5, 6.5]$
TI-92	$[-11.9, 11.9]$

Square Windows

The graph of $f(x) = 2x + 1$ is a straight line with a slope of 2. You have seen several views of this graph in different windows. The apparent steepness of the graph can be quite different even though the slope is always 2.

- Graph $f(x) = 2x + 1$ in the window $[-9, 9]$ by $[-2, 2]$ and then in the window $[-9, 9]$ by $[-20, 20]$. Compare the apparent steepness of the graph in the two windows.

In general, to obtain a graph that suggests the graph's true shape, you must choose viewing dimensions that are proportional to the dimensions of your grapher screen. Most grapher screens have a width-to-height ratio of roughly

3:2. Windows whose dimensions are proportional to the physical dimensions of the grapher screen are called *square windows.* Square windows yield true shapes: They make perpendicular lines look perpendicular, squares look square, and circles look circular.

■ EXAMPLE 3 Rounding Out a Circle

Use your grapher to plot the circle $x^2 + y^2 = 1$.

Solution

First, you will need to do some algebra:

$$x^2 + y^2 = 1$$
$$y^2 = 1 - x^2$$
$$y = \pm\sqrt{1 - x^2}.$$

So the graph of the circle $x^2 + y^2 = 1$ is the union of the graphs of the functions $y_1 = \sqrt{1 - x^2}$ and $y_2 = -\sqrt{1 - x^2}$.

- Graph $y_1 = \sqrt{(1 - x^2)}$ and $y_2 = -\sqrt{(1 - x^2)}$ in several windows with different x-y dimension ratios. (We used parentheses around $1 - x^2$ because you will need them to enter the functions onto the $Y=$ edit screen.) Continue until you obtain a graph that appears circular. Gaps may appear near the x-axis if the window is not *friendly.*
- Graph $y_1 = \sqrt{1 - x^2}$ and $y_2 = -\sqrt{1 - x^2}$ in a square, friendly window.

GRAPHER NOTE

Some graphers have a ZSQUARE feature that adjusts the window dimensions to make them match the physical proportions of the screen.

Figure A.11 shows three views. Only the view in part (a) looks circular because only in part (a) are the window dimensions proportional to the physical dimensions of the screen. ■

Square, friendly, or standard windows are not always ideal. Example 4 illustrates that window dimensions may need to be exaggerated to reveal the important features of a graph.

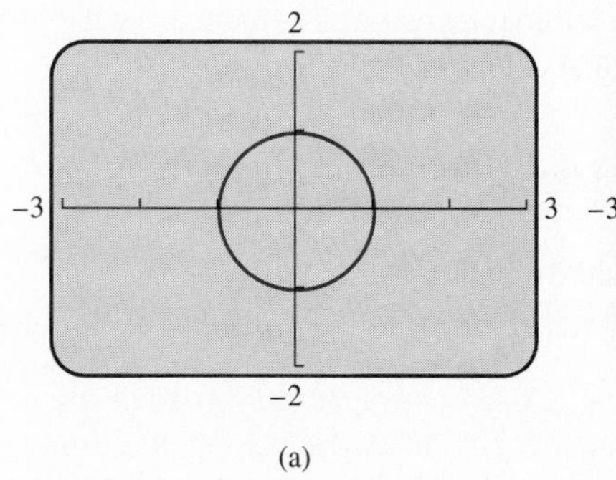

(a)

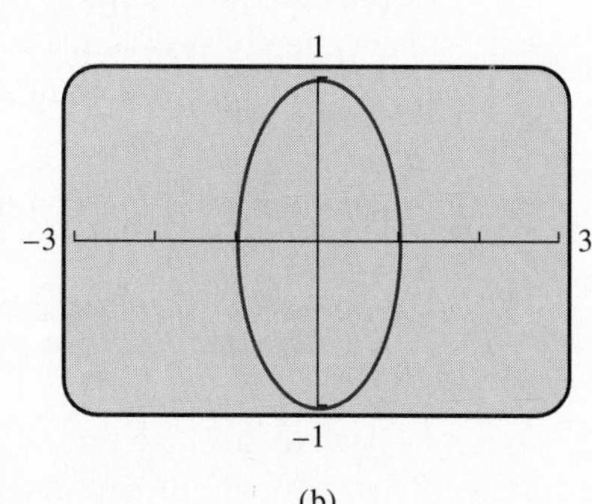

(b)

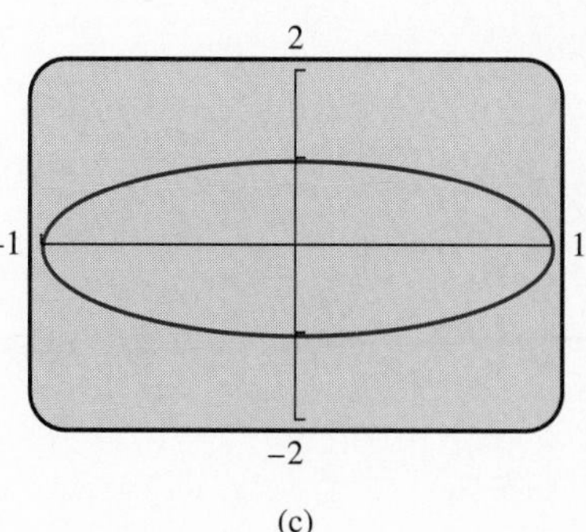

(c)

Figure A.11 The graph of a circle in (a) a square window and (b and c) nonsquare windows.

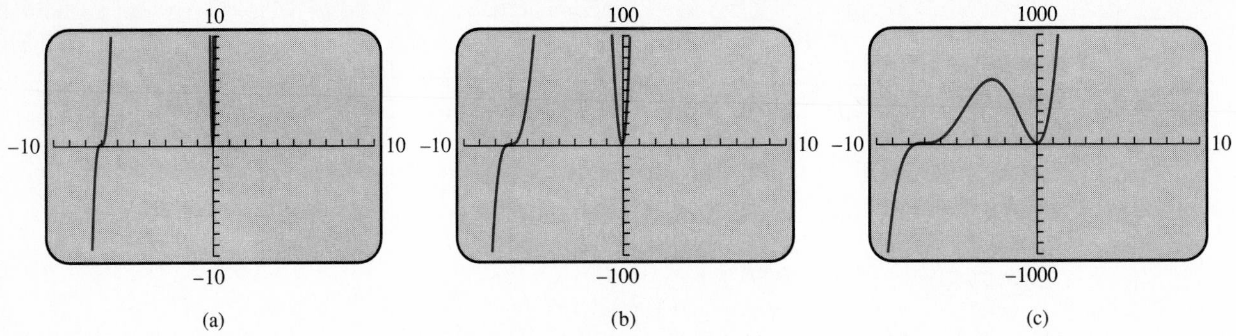

Figure A.12 The graph of the same function in three different viewing windows.

EXAMPLE 4 Finding Key Features of a Graph

Graph $f(x) = x^2(x + 7)^3$, using Xmin $= -10$ and Xmax $= 10$. Try various [Ymin, Ymax] settings.

Solution

Figure A.12 shows three possible views.

- Try the windows shown in Figure A.12 and several others of your own choosing. Continue until you convince yourself that you have observed all the important features of $f(x) = x^2(x + 7)^3$ on the interval $[-10, 10]$.

Note that the view in part (c) shows more features of the graph than either part (a) or part (b) and hence is the best view of the three. Further investigation of the graph in other windows should reveal no other major features on the interval $[-10, 10]$. ■

Example 5 is designed to familiarize you with 12 basic graphs that are used in the textbook and are important as models in many fields of endeavor. You should learn these graphs by heart! That is, you should be able to sketch any of them quickly at any time without a great deal of thought, without doing any hand computation, and without using your grapher.

EXAMPLE 5 Touring a Gallery of Basic Functions

Plot and explore the 12 graphs shown in Figure A.13.

Solution

You may need to do some searching on your grapher keyboard or dig through some menus to find all these functions (CYGOM). You should use function notation—for example, abs (X) rather than abs X—even though your grapher may be forgiving.

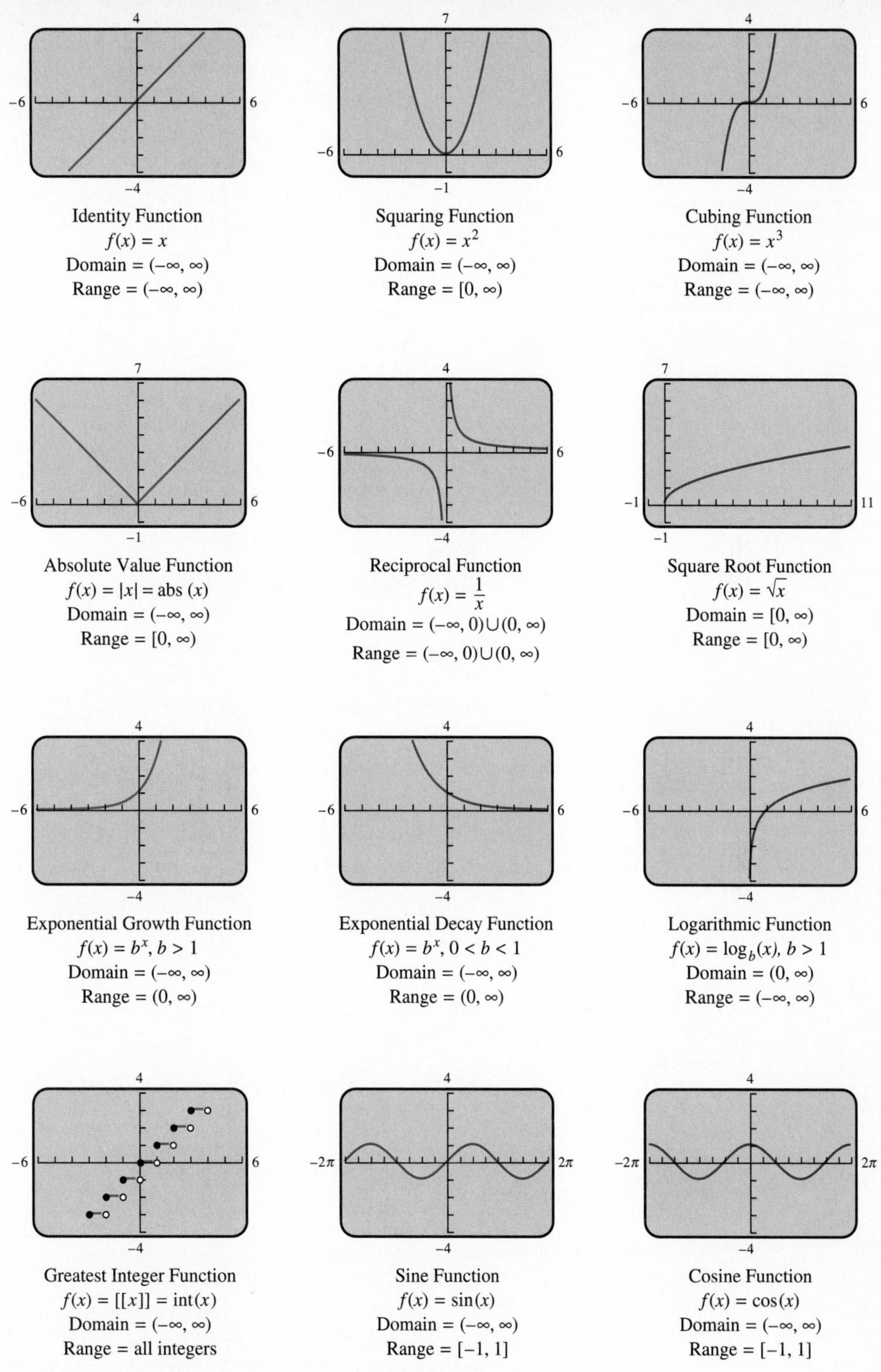

Figure A.13 Gallery of Basic Functions.

- Graph all but the sine and cosine functions with the windows indicated in Figure A.13. Regraph each function for other window dimensions, including standard, friendly, and square. Explore with TRACE.
- Graph the greatest integer function in CONNECTED mode and then in DOT mode. Which produces the better graph? Why?
- Graph the sine and cosine functions in DEGREE mode and then in RADIAN mode. Note that the graphs in Figure A.13 are based on the use of RADIAN mode.

GRAPHER NOTE

Some graphers have ZTRIG that automatically sets desirable window dimensions for viewing trigonometric functions.

You should practice making hand-drawn sketches of the 12 basic functions until you can do them from memory. ■

A.5 GRAPHICAL PROBLEM SOLVING

In this section we explore various grapher methods for solving equations and analyzing the graphical behavior of functions, so you should set your grapher to FUNCTION mode. We begin by showing how to solve equations graphically, using the example

$$|x| = x^2 + x - 2,$$

first by graphing

$$f(x) = |x| \quad \text{and} \quad g(x) = x^2 + x - 2$$

separately and then by investigating the related function

$$h(x) = f(x) - g(x) = |x| - (x^2 + x - 2).$$

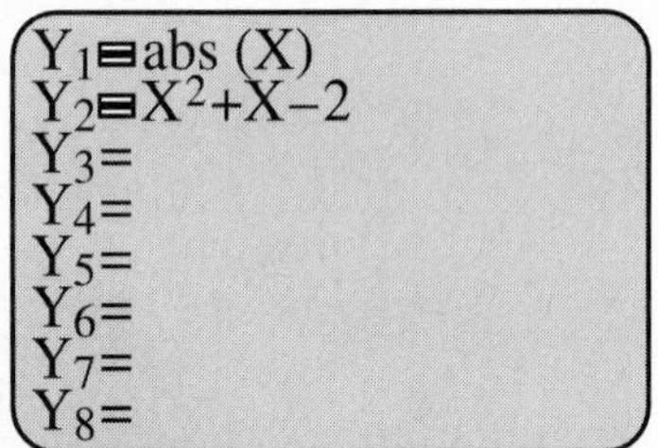

Figure A.14

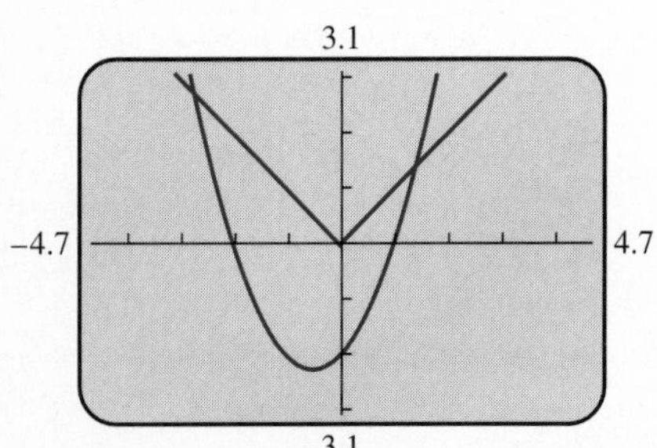

Figure A.15 The dimensions of a friendly window vary from grapher to grapher. Possible x dimensions are given in Table A.2. ZDECIMAL or ZINTEGER yield a square, friendly window on some graphers.

Solving an Equation by Finding Intersections

We can solve an equation by graphing each side as a function and locating the points of intersection.

- Enter each side of the equation $|x| = x^2 + x - 2$ onto the $Y{=}$ edit screen as shown in Figure A.14.
- Graph the equations in a friendly window. See Figure A.15.

The graph of f is V-shaped, the graph of g is an upward opening parabola, and they have two points of intersection. Thus the equation $|x| = x^2 + x - 2$

has two solutions. These solutions are the x-coordinates (one positive and one negative) of the two points of intersection.

- Trace along either graph to approximate the positive solution, that is, to estimate the x-coordinate of the point of intersection in the first quadrant.

To get a better approximation for the positive solution we can *zoom in* by picking smaller and smaller windows that contain the point of intersection in the first quadrant. Three common ways to zoom-in are to

1. change the WINDOW settings manually;
2. use ZOOMBOX, which lets you use the cursor to select the opposite corners of a "box" to define a new window; and
3. use ZOOMIN, which magnifies the graph around the cursor location by a factor that you can set.

- Practice each of the three types of zoom-in (CYGOM if necessary). Trace after each zoom step, as shown in Figure A.16.

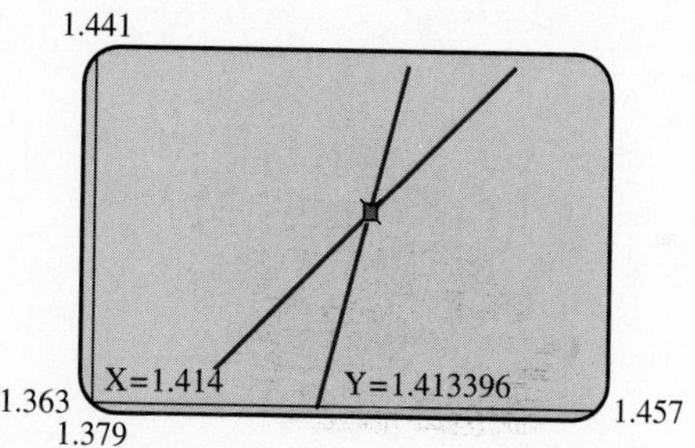

Figure A.16 One possible view of the graphs after some zooming, with the TRACE feature activated.

Most graphers now have an INTERSECTION feature that can be used to automate the process of solving equations graphically without adjusting the viewing window dimensions.

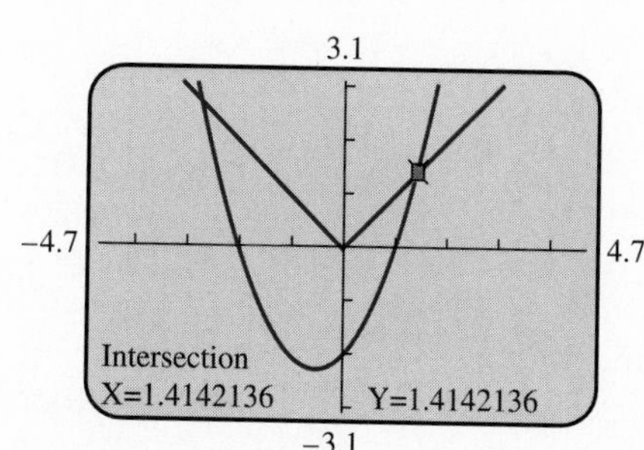

Figure A.17

- Graph the equations of Figure A.14 in a friendly window (Figure A.15). Then use the INTERSECTION feature to locate the point of intersection in the first quadrant, as shown in Figure A.17.

Solving the equation $|x| = x^2 + x - 2$ algebraically reveals that the positive solution is $x = \sqrt{2} = 1.4142\ldots$, which confirms the graphical solution.

Solving by Finding x-Intercepts

To solve an equation of the form $f(x) = g(x)$, we can solve $f(x) - g(x) = 0$. Then the problem becomes one of finding where the functions $y = f - g$ and $y = 0$ intersect, or simply the x-intercepts of $y = f - g$. For example, to solve the equation $|x| = (x^2 + x - 2)$, we can find the x-intercepts of

$$y = h(x) = f(x) - g(x) = |x| - (x^2 + x - 2).$$

- Load the Y= edit screen, as shown in Figure A.18a, selecting only $y_3 = y_1 - y_2 = \text{abs}\,(x) - (x^2 + x - 2)$ to be graphed.
- Graph y_3 in a friendly window, as shown in Figure A.18b.
- The x-intercepts are also the "roots" of the equation. Use the ROOT feature (CYGOM if necessary) to locate the negative x-intercept, as shown in Figure A.18c.

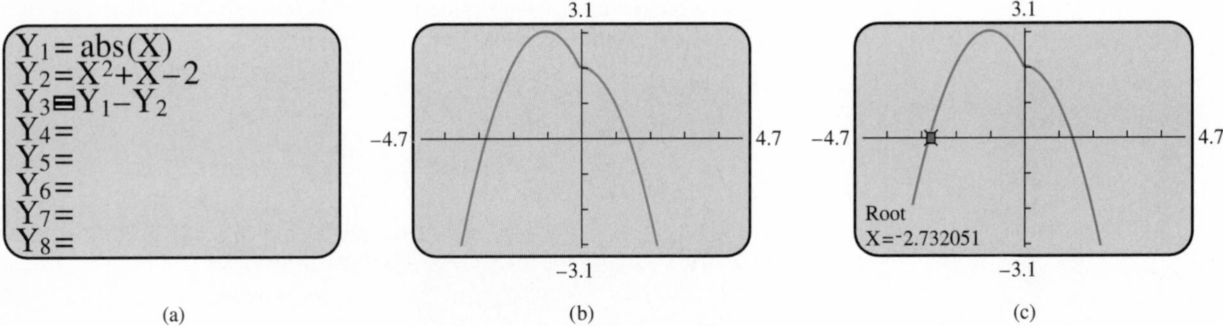

Figure A.18 Finding a solution of an equation by finding an x-intercept of the difference function.

Studying Graph Behavior

As you have seen, TRACE allows you to move from pixel to pixel on a graph with the coordinates of the points displayed to illustrate the numerical behavior of the function. For example, you can see whether the y-coordinate increases, decreases, or remains constant as x increases. ZOOMIN permits a "close-up" examination of the *local behavior* of graphs.

Three other grapher features are useful for investigating graph behavior (CYGOM, if necessary).

1. VALUE evaluates a function for a given domain value, which often avoids the need for a friendly window.

2. MINIMUM finds a local minimum value of a function and the associated domain value.
3. MAXIMUM finds a local maximum value of a function and the associated domain value.

EXAMPLE 6 Investigating Graph Behavior

Graph $f(x) = x^3 + 2x^2 - 5x - 6$ and study its behavior.

Solution

Do the following on your grapher.

- Graph $y = f(x)$ in the standard window. Trace over until $x \approx -3.4$, as shown in Figure A.19a. Then trace from left to right. Observe whether the function values (y-coordinates) increase, decrease, or remain constant as x increases.
- The graph appears to show that $f(2) = 0$, but TRACE fails to give a y for $x = 2$. Use VALUE to find $f(2)$, as shown in Figure A.19b.
- To determine precisely the intervals on which f is increasing or decreasing, locate the domain values at which the local maximums and minimums occur. Use MAXIMUM and MINIMUM to find these values. Figure A.19c shows the result of using the MAXIMUM feature.
- Zoom in around the points that correspond to the local maximum and the local minimum. Describe the graph behavior near each of these points.

What other graph behaviors can you study with the features of your grapher? ■

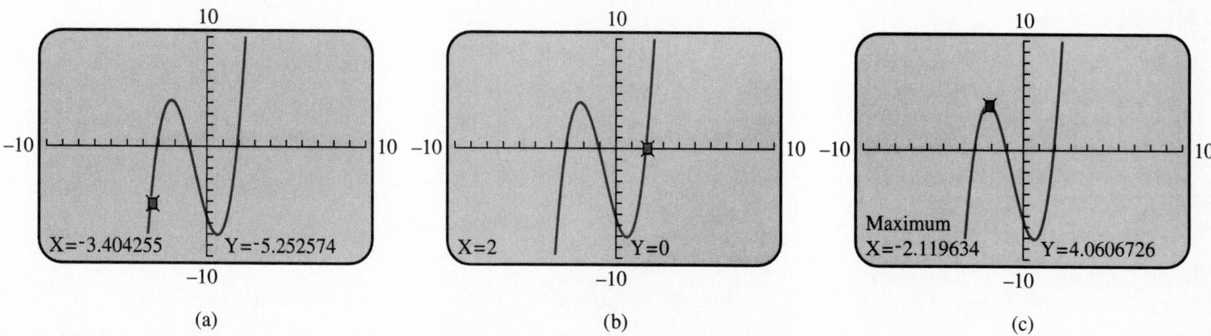

Figure A.19 Exploring a graph with (a) TRACE, (b) VALUE, and (c) MAXIMUM.

A.6
PARAMETRIC AND POLAR GRAPHING

Parametric Graphing

To graph parametric equations, set your grapher to PARAMETRIC mode. In PARAMETRIC mode, pressing X,T,θ will yield the independent variable t. The parametric equations

$$x(t) = \ldots, \qquad y(t) = \ldots$$

are entered in the form $x_1(t) = \ldots, y_1(t) = \ldots$ on the $Y=$ edit screen.

EXAMPLE 7 Graphing Parametric Equations

Graph the parametric equations.

$$x = t^2, \qquad y = t - 1 \qquad \text{for} \quad -2 \leq t \leq 2$$

Solution

Follow these steps:

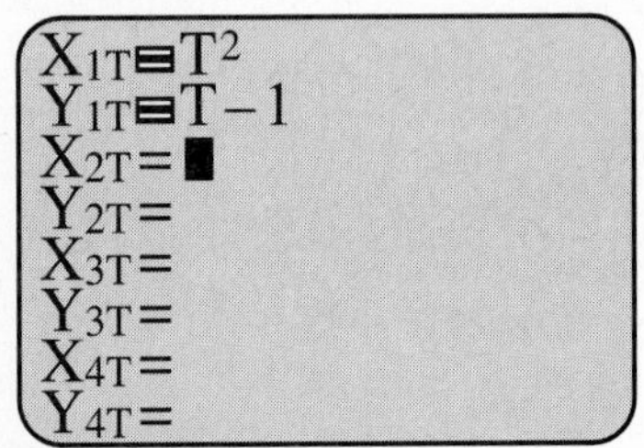

Figure A.20

- Enter the parametric equations on the $Y=$ edit screen, as shown in Figure A.20.
- Set the WINDOW dimensions shown in Figure A.21. You will not be able to see the entire menu given in Figure A.21 at once because it has too many lines.

The *T*step on the parametric WINDOW menu is the change between the successive t-values that the grapher uses to compute and plot (x, y) pairs. In this case, the *T*step of 0.1 will yield 40 steps from the *T*min of -2 to the *T*max of 2. Thus 41 points will be calculated and plotted, with the points corresponding to

$$t = -2.0, -1.9, -1.8, -1.7, \ldots, 1.9, 2.0.$$

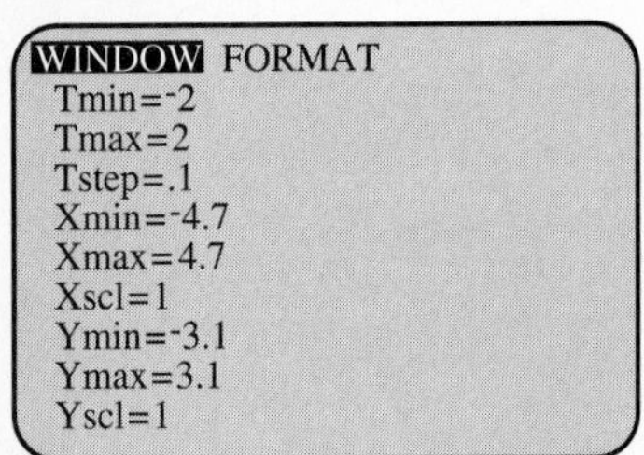

Figure A.21 Facsimile of the WINDOW screen set for plotting the parametric equations.

- Press GRAPH or PLOT to obtain the graph shown in Figure A.22.
- Use TRACE to explore the graph numerically. Note that the values of the parameter t and the coordinates x and y are all shown on the screens in Figure A.23.

Figures A.22 and A.23 show only a piece of the parabola $x = (y + 1)^2$ rather than the complete parabola. The parametric WINDOW menu allows you

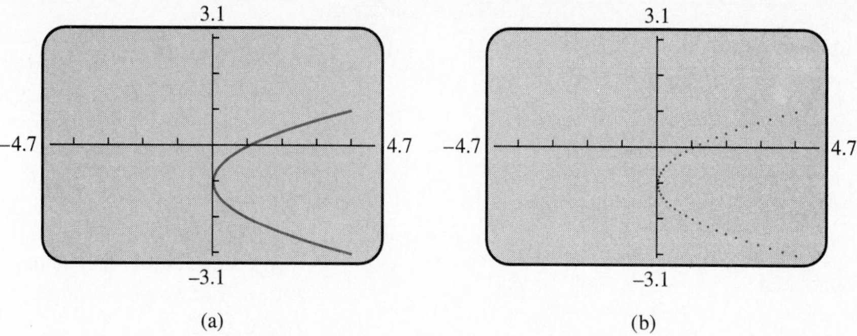

Figure A.22 The graph in (a) CONNECTED mode and (b) DOT mode.

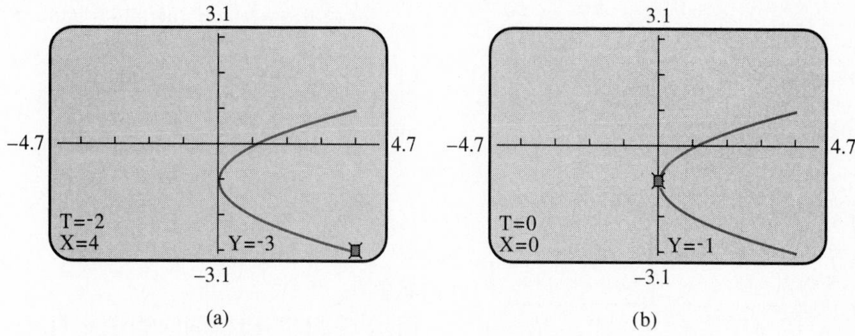

Figure A.23 Two views of the parametric curve with TRACE activated.

to choose a part of the graph by setting Tmin and Tmax. Do you see why? If not, experiment with the Tmin and Tmax settings until you do. ■

Polar Graphing

To graph polar equations set your grapher to POLAR mode. Pressing [X,T,θ] will yield the independent variable θ.

■ EXAMPLE 8 Graphing Polar Equations

Simultaneously graph $r = 9 \sin 5\theta$ and $r = 9$.

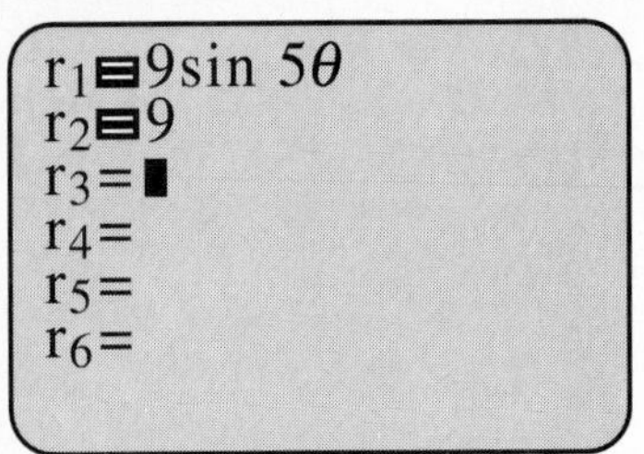

Figure A.24

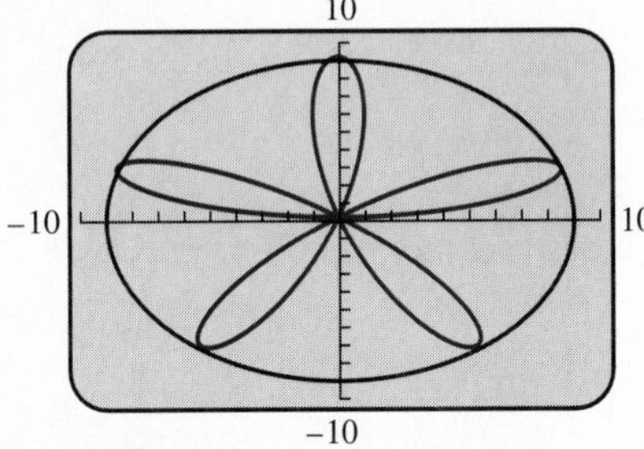

Figure A.25 The circle $r = 9$ and the 5-petaled rose $r = 9 \sin 5\theta$.

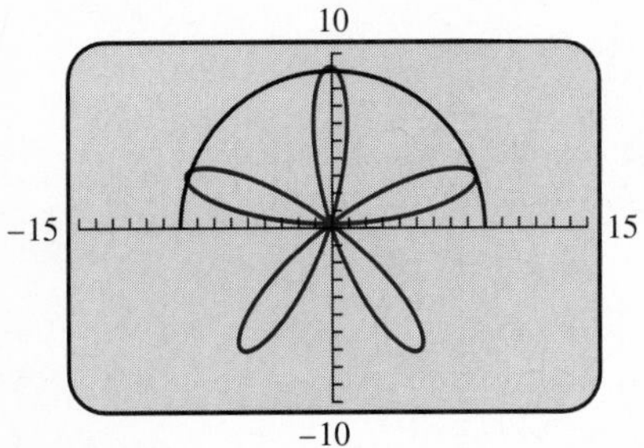

Figure A.26

Solution

Follow these steps:

- Set your grapher to SIMULTANEOUS and RADIAN modes. Then enter the polar equations on the $Y=$ edit screen, as shown in Figure A.24.
- Set the WINDOW dimensions to

$$\theta\text{min} = 0, \qquad \theta\text{max} = 2\pi, \qquad \theta\text{step} = \pi/24,$$

  using the standard dimensions of $[-10, 10]$ by $[-10, 10]$ for x and y. (On some graphers, you can obtain these settings by using ZSTANDARD.)
- Press GRAPH or PLOT to obtain the graphs shown in Figure A.25.
- Trace along the two polar curves.
- Set $\theta\text{max} = \pi$. Then use ZSQUARE to square the viewing window. If no such command is available on your grapher, reset the x-dimensions of the WINDOW menu by hand. Figure A.26 shows the result.

Note that the entire rose curve is plotted for the interval $0 \leq \theta \leq \pi$. Explore the effect of changing the 5 in the equation $r = 9 \sin 5\theta$ to another number. ■

A.7 CURVE FITTING AND STATISTICS

A grapher can help you organize, process, and analyze data, as well as compute and plot models for paired data. The procedures for data analysis and curve fitting vary a great deal from grapher to grapher (CYGOM for details).

■ EXAMPLE 9 Plotting and Fitting Data

Plot the national debt data given in Table A.3, find a model for the data, and then overlay a graph of the model on the scatter plot.

Solution

Follow these steps:

- Enter the data shown in Table A.3 into the statistical memory of your grapher, as shown in Figure A.27a.
- Set an appropriate window for the data, letting x be the year and y be the amount of the debt, as shown in Figure A.27b.
- Make a scatter plot of the data, as shown in Figure A.27c.

Table A.3 U.S. Public Debt, 1950–1990

Year	Debt (Billions of dollars)
1950	256.1
1960	284.1
1970	370.1
1975	533.2
1980	907.7
1985	1,823.1
1990	3,233.3

Source: The World Almanac and Book of Facts (1995, Mahwah, N.J.: Funk & Wagnalls), p. 109.

Most graphers have several regression options. Typically, linear, quadratic, exponential, logarithmic, and power functions are available as regression models. Some graphers offer other options. The following activity illustrates quadratic regression. That is, we find a quadratic function that closely fits the given data.

- Choose the QUADRATIC option from your grapher's regression menu. Use the first column of data for the x-values and the second column as y-values. The grapher should return approximate coefficients for the quadratic regression equation, as shown in Figure A.28a.
- Load the regression equation onto the $Y=$ edit screen, as shown in Figure A.28b.
- Press GRAPH or PLOT to obtain the graph in Figure A.28c, which shows both the scatter plot and the graph of the regression equation.

Repeat the steps to find and graph other regression equations for the data. ■

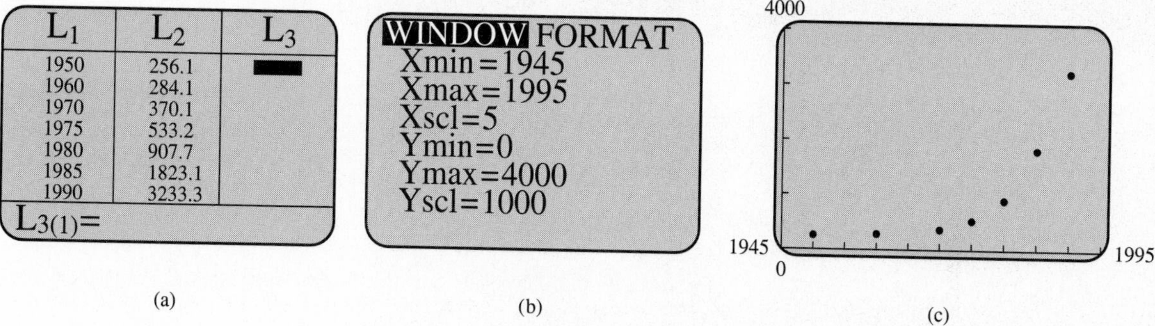

Figure A.27 The steps involved in making a scatter plot on a grapher.

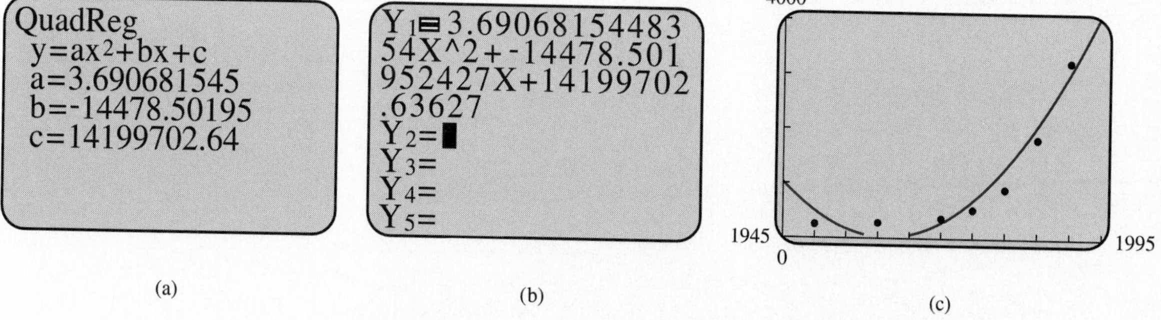

Figure A.28 The steps involved in fitting a quadratic function to a scatter plot.

Most graphers offer a variety of statistical plots, often including histograms, boxplots, and line graphs. In addition, graphers can carry out many types of statistical computations (CYGOM for details).

A.8
MATRIX CALCULATIONS

A grapher can perform many matrix operations, thus avoiding the tedium of hand computation. Matrix procedures vary somewhat from grapher to grapher (CYGOM for details).

- Place matrix A on the Home screen (Figure A.29a) and compute its determinant (Figure A.29b).

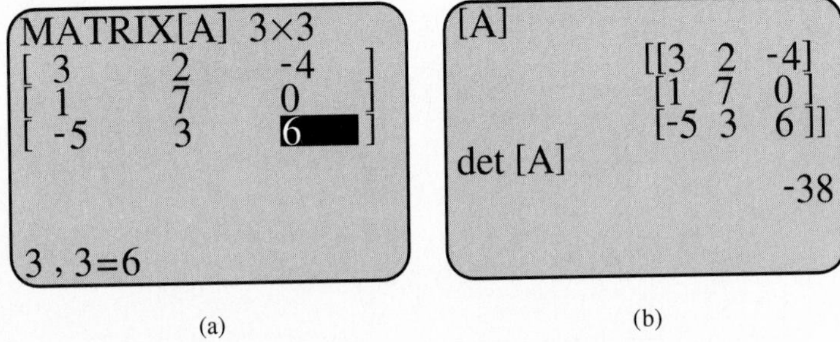

Figure A.29 (a) Matrix entry and (b) matrix display and determinant calculation.

- Compute the inverse of matrix A. You may need to scroll, using ▶ in order to see the entire answer (Figure A.30a).

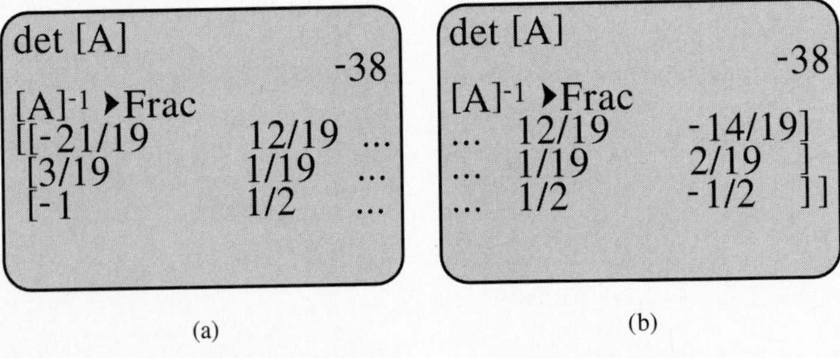

Figure A.30 (a) Computing the inverse of matrix and (b) scrolling.

GRAPHER NOTE

In Figure A.30a, we used the FRAC feature to convert the decimal entries of A^{-1} to fractions.

- Enter the matrix $B = \begin{pmatrix} 12 \\ 9 \\ -13 \end{pmatrix}$ onto the matrix editor of your grapher (Figure A.31a).
- Compute the product $A^{-1} \cdot B$ on the Home screen (Figure A.31b).

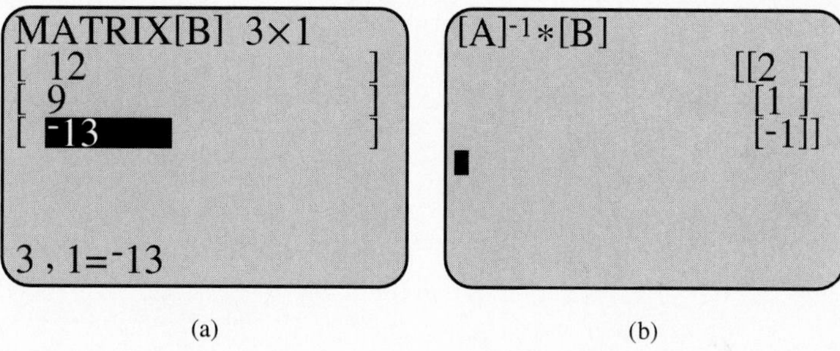

Figure A.31 (a) Entering a column matrix and (b) multiplying matrices.

Graphers have other matrix features. Most importantly, many graphers can perform elementary row operations on a matrix.

A.9 GRAPHER INSIGHTS AND CAVEATS

Limitations of Grapher Computation

Grapher computations are limited in *magnitude* and *relative accuracy.* Numbers less than the lower magnitude limit are rounded to zero. Numbers greater than or equal to the upper magnitude limit yield *overflow errors.*

Regarding relative error, some graphers store only the first 13 significant digits of a number (the rest of the number is rounded off) and they display at most 10 digits. To see this in action,

- Enter (2/3 − 1/3) − 1/3 on your grapher.

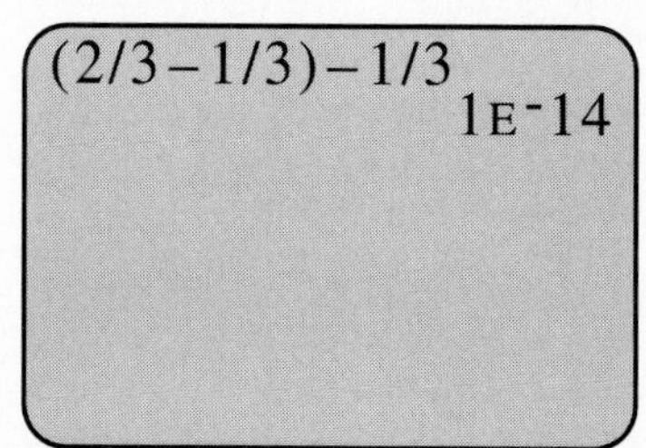

Figure A.32 The surprising result of roundoff error.

We know that (2/3 − 1/3) − 1/3 = 0, but most graphers will not give 0 as the final result, as shown in Figure A.32. What steps in the grapher computation could have produced this result? Explain.

How a Grapher Draws a Function Graph

All graphers produce graphs, or plots, by lighting pixels on a liquid crystal display (LCD). Figure A.33a shows the graph of a line on an LCD, and Figure A.33b shows how it might appear under a magnifying glass. When the grapher is in CONNECTED mode, the plotted pixels are connected by "line segments" of pixels. When in DOT mode, the pixels are left unconnected.

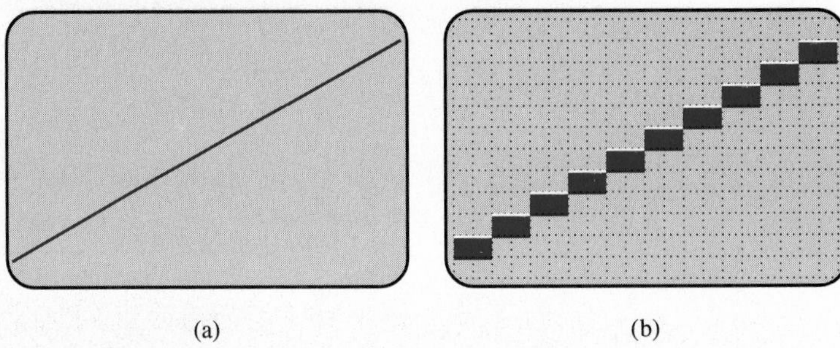

(a) (b)

Figure A.33 (a) The graph of a line on an LCD and (b) how it looks under magnification.

Limitations of Grapher Plotting

Grapher plotting is limited by roundoff error in the calculation of the y-coordinates from the function formula. More significantly, grapher plotting is limited because, for each point plotted, a pixel is rarely centered on the LCD at the exact vertical location corresponding to the y-coordinate of the point.

Despite these limitations, graphers can quickly produce accurate graphs for most functions described in this book if the viewing dimensions are chosen appropriately. It takes mathematical and grapher experience to get good at choosing windows. You will develop this skill over time as you go through the workshop exercises and the rest of the book.

Interpreting Grapher Plots

As Examples 3 and 4 illustrated, a grapher plot may be misleading or incomplete. (See Figures A.11 and A.12.) True visualization occurs in the "mind's eye" when you use the information gained from a grapher together with your mathematical knowledge to obtain a mental image of the mathematical graph. To communicate this mental visualization to another person, however, you must learn to describe and draw graphs accurately. When recording a graph on paper, you should normally add suggestive arrowheads and label axes, key

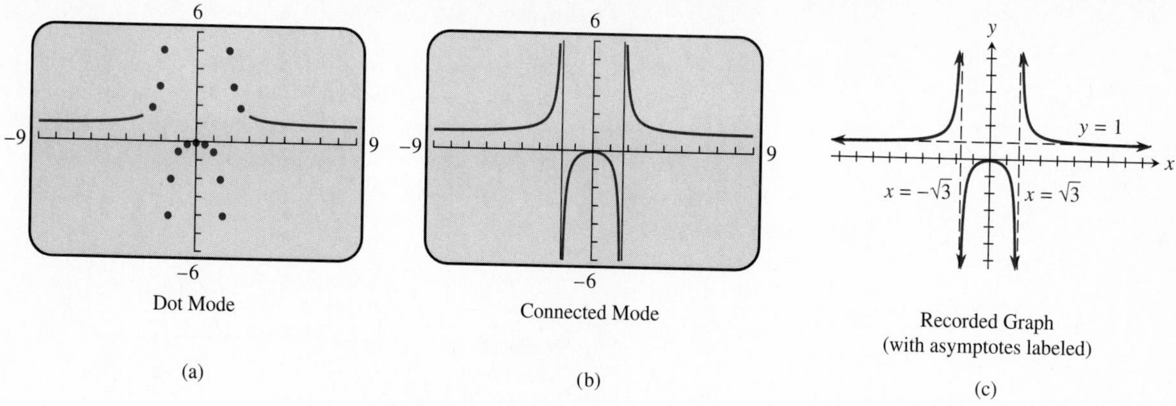

Figure A.34 The graph of $y = x^2/(x^2 - 3)$ as shown (a and b) on a grapher and (c) on paper.

points, and other pertinent information. Figure A.34 shows how two rather misleading views of a complicated graph can be combined into a single recorded graph. This is a skill to strive for as you work with the grapher.

A.10 VIEWING WINDOW SUMMARY

Choosing a Viewing Window

You need experience and mathematical expectations to choose appropriate viewing windows. One approach is to start with the *standard window* of $[-10, 10]$ by $[-10, 10]$ and adjust the y-dimensions.

Some windows may show more features of a graph than others. The view on the right shows the key features of the graph; the other two views do not.

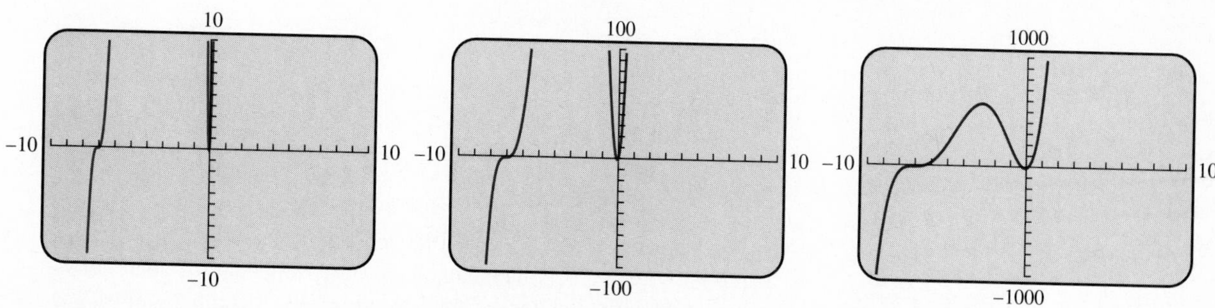

Figure A.35 Three views of $f(x) = x^2(x + 7)^3$.

Friendly Windows

Your choice of Xmin and Xmax affect the x-coordinate readout when you *trace* along a graph. You can use the [Xmin, Xmax] settings given in the table, or positive integer multiples of these settings, to guarantee a *friendly* x-coordinate readout when tracing. Windows with friendly x-coordinates are called *friendly windows.*

Grapher	[Xmin, Xmax]
TI-80	[−3.1, 3.1]
Casio, Sharp, TI-82, TI-83	[−4.7, 4.7]
TI-81	[−4.8, 4.7]
TI-85	[−6.3, 6.3]
Hewlett-Packard	[−6.5, 6.5]
TI-92	[−11.9, 11.9]

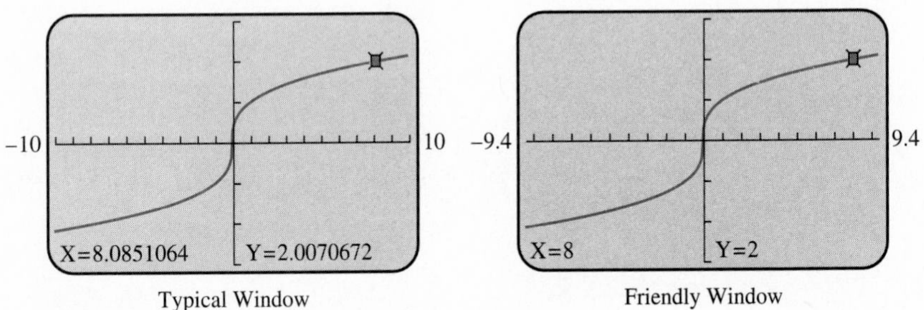

Figure A.36 Graphs of $f(x) = \sqrt[3]{x}$.

Square Windows

A *square window* is a window that shows the true shape of a graph. Such a window makes perpendicular lines look perpendicular, a square look square, and a circle look circular. A square window has the same proportions as your grapher screen. Many grapher screens have a width-to-height ratio of 3:2. Most graphers have a built-in feature for squaring windows.

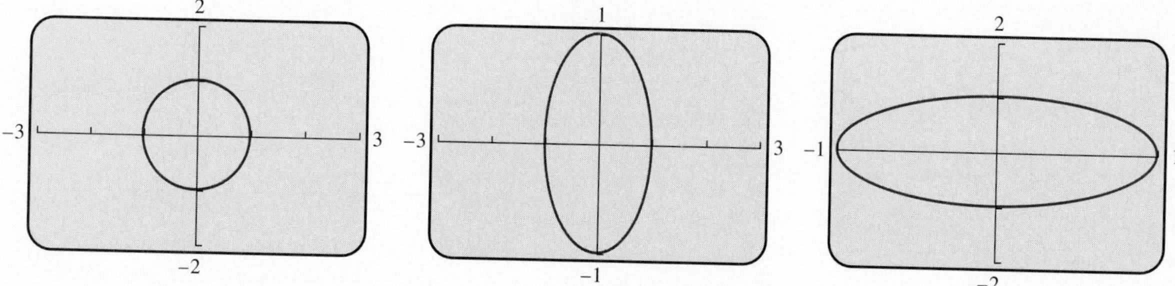

Figure A.37 Three views of the circle $x^2 + y^2 = 1$.

Integer and Decimal Windows

These windows are special types of friendly windows. An *integer window* is a window in which each pixel is centered at a point with integer coordinates and the change in both x and y is 1. A *decimal window* is a window in which each pixel is centered at a point with coordinates of at most one decimal place and the change in both x and y is 0.1. On most graphers, both integer and decimal windows are square and friendly.

EXERCISES FOR APPENDIX A

The exercises are correlated with the sections of the Grapher Workshop. If any give you difficulty, resolve it by consulting either your owner's manual or your instructor.

NUMERICAL COMPUTATION AND EDITING

In Exercises 1–6, use a grapher to evaluate the expression.

1. 600×1.075^2

2. 1.3^5

3. log (1/10)

4. $|-7|$

5. $\sqrt[3]{64}$

6. $\sqrt[4]{81}$

In Exercises 7 and 8, use ANS.

7. Calculate the first seven terms in the geometric sequence that begins with 2 and grows by a constant factor of 6: 2, 12, 72, ...

8. Calculate the first seven terms in the arithmetic sequence that begins with 4 and grows by a constant 9: 4, 13, 22, ...

In Exercises 9 and 10, use replay.

9. Evaluate $f(x) = x^2 + x - 2$ at $x = -3, -2.5, 3$.

10. Evaluate $g(t) = 2t^3 - |5t|$ at $t = -2, 3.4, 7$.

TABLE BUILDING

In Exercises 11–14, make a table with the following inputs.

a. $x = -3, -2, -1, \ldots, 3$
b. $x = 0, 1, 2, \ldots, 6$
c. $x = -10, -5, 0, \ldots, 20$

11. $f(x) = x^2 + 15$ **12.** $f(x) = \ln(x^2 + 1)$

13. $f(x) = |x - 2|$ **14.** $f(x) = \sqrt[3]{x^2 + x - 2}$

FUNCTION GRAPHING

In Exercises 15–20, graph the function in each type of window.

a. The standard window
b. A square window containing the standard window
c. A friendly window using TRACE to support the friendly x-coordinate readout
d. A window that is both square and friendly

15. $y = 3x - 2$ **16.** $y = -\frac{1}{2}x + 3$

17. $y = 1 - x^2$ **18.** $y = 2x^2 - 3x + 1$

19. $y = |x + 2|$ **20.** $y = \sqrt{9 - x^2}$

In Exercises 21 and 22, explain any differences in the two modes.

21. Enter and graph

$$y_1 = 2x + 1;$$
$$y_2 = 2x + 2;$$
$$y_3 = 2x + 3.$$

Switch between SEQUENTIAL and SIMULTANEOUS modes and draw the graphs again.

22. Graph $y = 3x - 1$. Switch between CONNECTED and DOT modes and draw the graph again.

In Exercises 23 and 24, find a window that reveals the key features of the function. How many x-intercepts does the graph have?

23. $y = x^2(x - 12)$ **24.** $y = x(x + 3)^2(x + 14)$

GRAPHICAL PROBLEM SOLVING

In Exercises 25–28, solve the equation by

a. finding intersections.
b. finding x-intercepts.

25. $|x| = \frac{1}{2}x + 1$ **26.** $|x - 3| = -\frac{1}{2}x + 4$

27. $x - 2 = 1 - x^2$ **28.** $x - 3 = x^2 - 5$

In Exercises 29 and 30, **(a)** find a viewing window that reveals the key features of the graph, and **(b)** find the local maximum and minimum values.

29. $y = x^2(x - 12)$ **30.** $y = x(x + 3)^2(x + 14)$

PARAMETRIC AND POLAR GRAPHING

In Exercises 31–34, use PARAMETRIC mode and graph in the viewing window $[-12, 12]$ by $[-12, 12]$.

31. Enter

$$x(t) = t, \qquad y(t) = t.$$

Predict what will happen as a result of the WINDOW settings for t. Then graph and compare with your prediction.

a. Tmin $= -10$, Tmax $= 10$, Tstep $= 0.1$
b. Tmin $= 0$, Tmax $= 10$, Tstep $= 0.1$
c. Tmin $= -10$, Tmax $= 0$, Tstep $= 0.1$
d. Tmin $= -10$, Tmax $= 10$, Tstep $= 1$
e. Tmin $= -5$, Tmax $= 5$, Tstep $= 1$
f. Tmin $= 10$, Tmax $= -10$, Tstep $= -0.1$

32. Enter

$$x_1(t) = t, \qquad y_1(t) = 2t;$$
$$x_2(t) = 2t, \qquad y_2(t) = t,$$

with Tmin $= -10$, Tmax $= 10$, and Tstep $= 0.1$. Predict what the graphs will look like. Then graph and compare with your prediction.

33. Graph

$$x_1(t) = 2t, \qquad y_1(t) = t^2,$$

with Tmin $= -10$, Tmax $= 10$, and Tstep $= 0.1$.

34. Graph

$$x_1(t) = 2t, \qquad y_1(t) = t^2;$$
$$x_2(t) = y_1(t), \qquad y_2(t) = x_1(t),$$

with Tmin $= -10$, Tmax $= 10$, and Tstep $= 0.1$.

In Exercises 35–38, use POLAR and RADIAN modes.

35. Graph $r = 6$ in a square window. Switch between CONNECTED and DOT modes and graph again.

36. Graph $r = 2\cos\theta$ in a square window. Switch between CONNECTED and DOT modes and graph again.

37. Graph $r_1 = 2$ and $r_2 = 5$. Switch between SEQUENTIAL and SIMULTANEOUS modes and graph again.

38. Graph $r_1 = 2$ and $r_2 = 2\cos 3\theta$. Switch between SEQUENTIAL and SIMULTANEOUS modes and graph again.

CURVE FITTING AND STATISTICS

In Exercises 39 and 40, use the data in Table A.4.

Table A.4 Official Census Population (in millions of persons), 1900–1990

Year	Florida	Pennsylvania
1900	0.5	6.3
1910	0.8	7.7
1920	1.0	8.7
1930	1.5	9.6
1940	1.9	9.9
1950	2.7	10.5
1960	5.0	11.3
1970	6.8	11.8
1980	9.7	11.9
1990	12.9	11.9

Source: U.S. Census Bureau as reported in *The World Almanac and Book of Facts* (1995, Mahwah, N.J.: Funk & Wagnalls), p. 377.

39. a. Enter the Florida population data into the statistical memory of your grapher.

b. Set an appropriate window for the data and make a scatter plot.

c. Choose the EXPONENTIAL option from your grapher's regression menu to find the constants in the regression equation.

d. Load the regression equation onto the $Y =$ edit screen and overlay the graph of the regression equation on the scatter plot.

40. a. Enter the Pennsylvania population data into the statistical memory of your grapher.

b. Set an appropriate window for the data and make a scatter plot.

c. Choose the LINEAR option from your grapher's regression menu to find the coefficients for the linear regression equation.

d. Load the regression equation onto the $Y =$ edit screen and overlay the graph of the regression equation on the scatter plot.

MATRIX CALCULATIONS

In Exercises 41 and 42, perform the computations.

41. a. Enter the matrix

$$A = \begin{pmatrix} 1 & -3 \\ 5 & 2 \end{pmatrix}$$

onto the matrix editor of your grapher and compute its determinant.

b. Compute the inverse of matrix A.

c. Enter the matrix

$$B = \begin{pmatrix} -1 \\ 12 \end{pmatrix}$$

onto the matrix editor of your grapher and compute the product $A^{-1} \cdot B$.

42. a. Enter the matrix

$$A = \begin{pmatrix} 4 & 2 & -1 \\ 1 & -1 & 0 \\ 0 & 3 & 5 \end{pmatrix}$$

onto the matrix editor of your grapher and compute its determinant.

b. Compute the inverse of matrix A.

c. Enter the matrix

$$B = \begin{pmatrix} 15 \\ 2 \\ -2 \end{pmatrix}$$

onto the matrix editor of your grapher and compute the product $A^{-1} \cdot B$.

GRAPHER INSIGHTS AND CAVEATS

In Exercises 43 and 44, sketch the graph.

43. Graph

$$y = \frac{(x+3)^2}{x(x+4)}$$

in CONNECTED and DOT modes and then sketch the graph on paper.

44. Graph

$$y = \frac{x(x-3)}{(x-1)(x+4)}$$

in CONNECTED and DOT modes and then sketch the graph on paper.

GRAPHER WORKSHOP EXERCISE ANSWERS

The graphs used in these answers are from the Texas Instruments TI-82 graphing calculator, and are similar to the ones that would appear on the Sharp and most Casio graphing calculators. Use Table A.2 to convert special viewing windows (decimal, friendly, etc.) to appropriate ones for your graphing calculator if it is not one of these.

1. 693.375 **2.** 3.71293

3. −1 **4.** 7

5. 4 **6.** 3

7. 2, 12, 72, 432, 2592, 15552, 93312

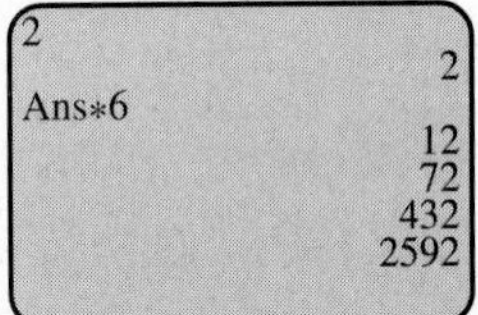

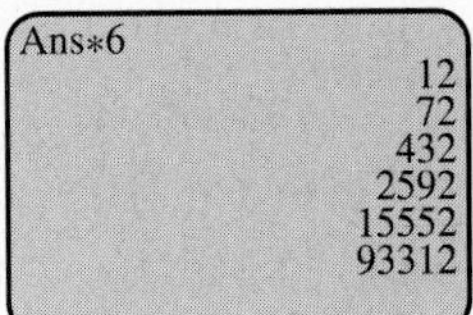

8. 4, 13, 22, 31, 40, 49, 58

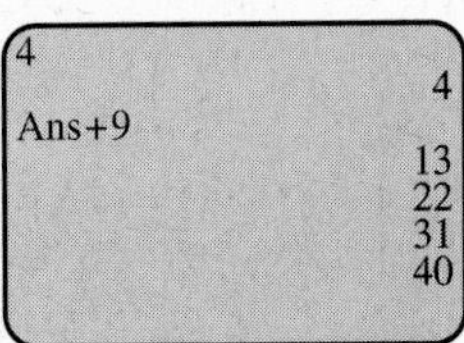

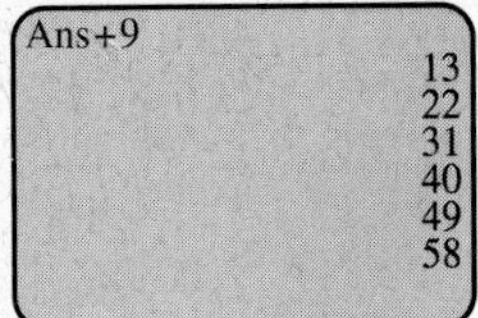

9. 4, 1.75, 10

-3→X:X²+X−2
4
-2.5→X:X²+X−2
1.75
3→X:X²+X−2
10

10. −26, 61.608, 651

-2→T:2T³−abs 5T
-26
3.4→T:2T³−abs 5T
61.608
7→T:2T³−abs 5T
651

11. a.

X	Y1
-3	24
-2	19
-1	16
0	15
1	16
2	19
3	24

Y1≡X²+15

b.

X	Y1
0	15
1	16
2	19
3	24
4	31
5	40
6	51

Y1≡X²+15

c.

X	Y1
-10	115
-5	40
0	15
5	40
10	115
15	240
20	415

Y1≡X²+15

12. a.

X	Y1
-3	2.3026
-2	1.6094
-1	.69315
0	0
1	.69315
2	1.6094
3	2.3026

Y1≡ln (X²+1)

b.

X	Y1
0	0
1	.69315
2	1.6094
3	2.3026
4	2.8332
5	3.2581
6	3.6109

Y1≡ln (X²+1)

c.

X	Y1
-10	4.6151
-5	3.2581
0	0
5	3.2581
10	4.6151
15	5.4205
20	5.994

Y1≡ln (X²+1)

13. a.

X	Y1
-3	5
-2	4
-1	3
0	2
1	1
2	0
3	1

Y1≡abs (X−2)

b.

X	Y1
0	2
1	1
2	0
3	1
4	2
5	3
6	4

Y1≡abs (X−2)

c.

X	Y1
-10	12
-5	7
0	2
5	3
10	8
15	13
20	18

Y1≡abs (X−2)

14. a.

X	Y1
-3	1.5874
-2	0
-1	-1.26
0	-1.26
1	0
2	1.5874
3	2.1544

Y1≡³√(X²+X−2)

b.

X	Y1
0	-1.26
1	0
2	1.5874
3	2.1544
4	2.6207
5	3.0366
6	3.42

Y1≡³√(X²+X−2)

c.

X	Y_1
-10	4.448
-5	2.6207
0	-1.26
5	3.0366
10	4.7622
15	6.1972
20	7.477

$Y_1 = \sqrt[3]{(X^2+X-2)}$

15. a. $[-10, 10]$ by $[-10, 10]$

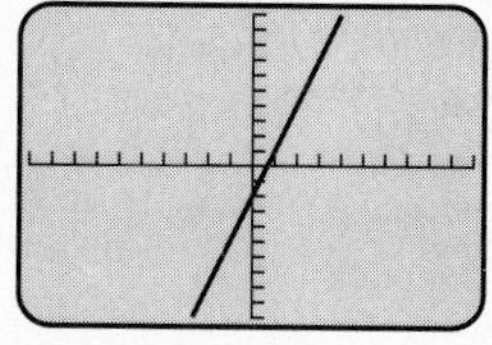

b. $[-15.161\ldots, 15.161\ldots]$ by $[-10, 10]$

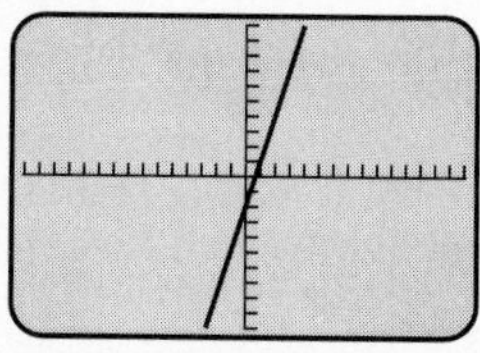

c. & d. $[-4.7, 4.7]$ by $[-3.1, 3.1]$
On the TI-82, Casio, and Sharp graphing calculators, the decimal window is square and friendly. This is not true on all graphers. For example, the decimal window on the TI-85 is friendly but not square.

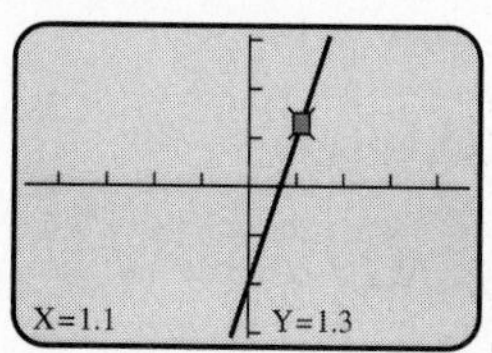

16. a. $[-10, 10]$ by $[-10, 10]$

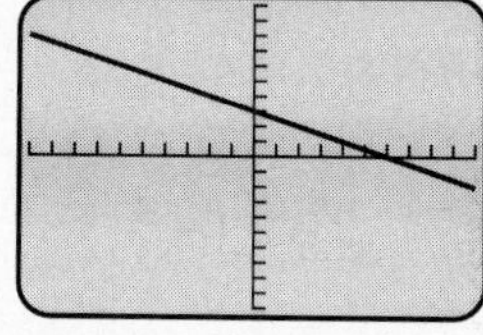

b. $[-15.161\ldots, 15.161\ldots]$ by $[-10, 10]$

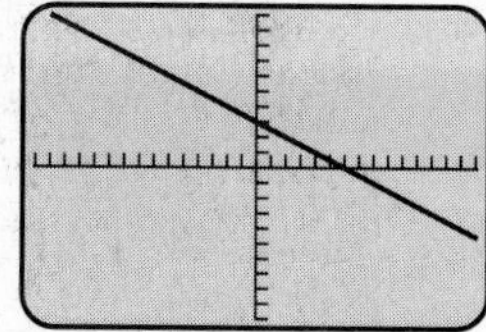

c. $[-9.4, 9.4]$ by $[-5, 10]$

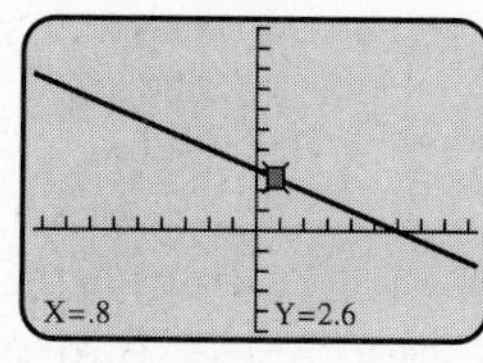

d. $[-9.4, 9.4]$ by $[-6.2, 6.2]$

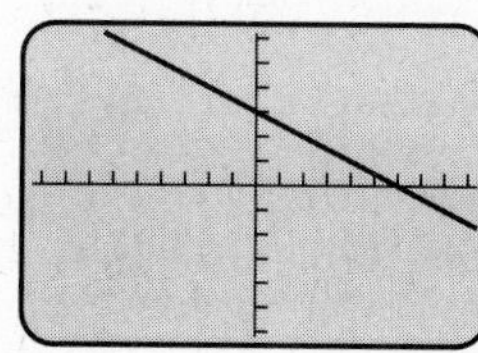

17. a. $[-10, 10]$ by $[-10, 10]$

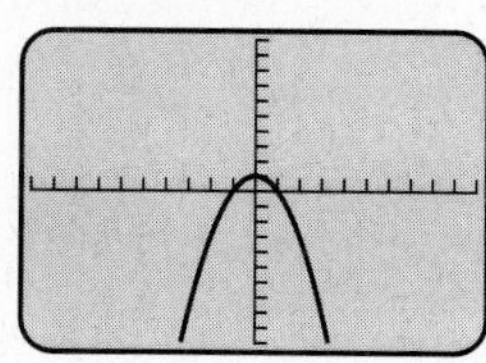

b. $[-15.161\ldots, 15.161\ldots]$ by $[-10, 10]$

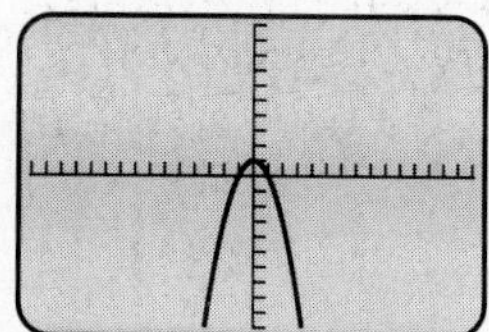

c. & d. $[-4.7, 4.7]$ by $[-3.1, 3.1]$

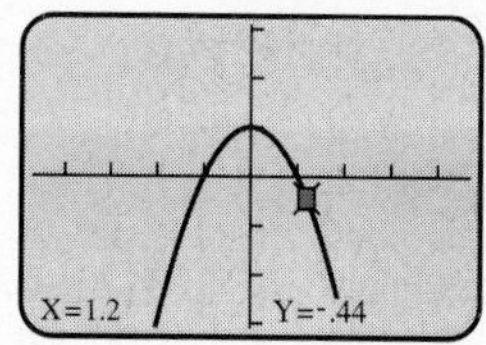

18. a. $[-10, 10]$ by $[-10, 10]$

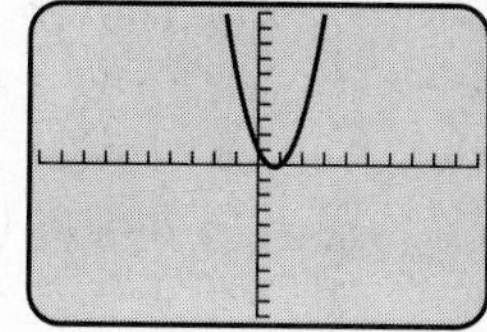

b. $[-15.161\ldots, 15.161\ldots]$ by $[-10, 10]$

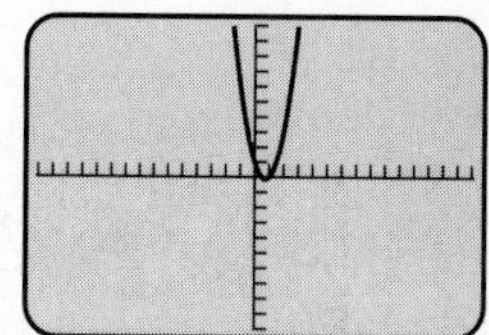

c. & d. $[-4.7, 4.7]$ by $[-3.1, 3.1]$

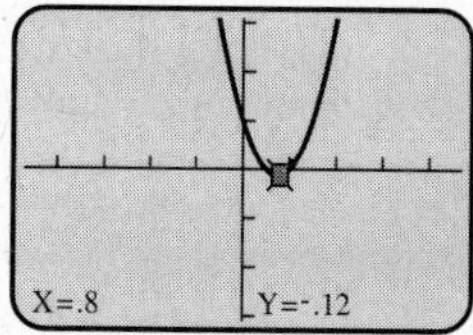

19. a. $[-10, 10]$ by $[-10, 10]$

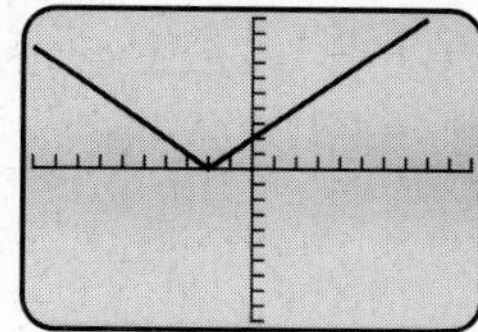

b. $[-15.161\ldots, 15.161\ldots]$ by $[-10, 10]$

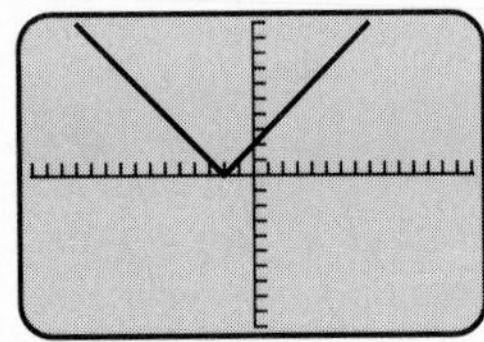

c. & d. $[-4.7, 4.7]$ by $[-3.1, 3.1]$

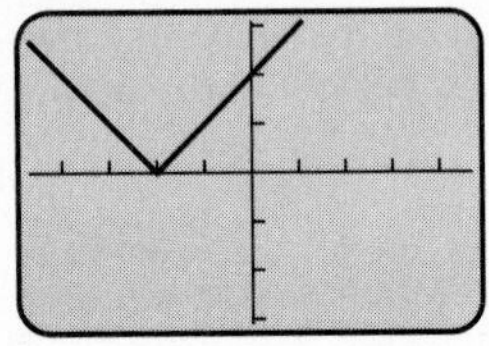

20. a. $[-10, 10]$ by $[-10, 10]$

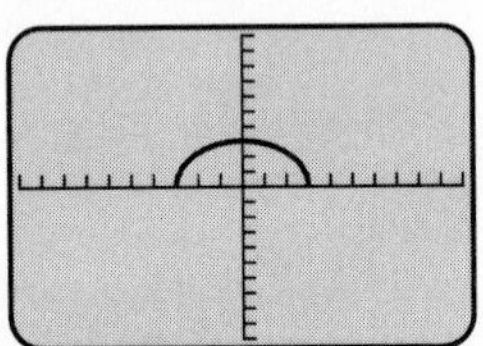

b. $[-15.161\ldots, 15.161\ldots]$ by $[-10, 10]$

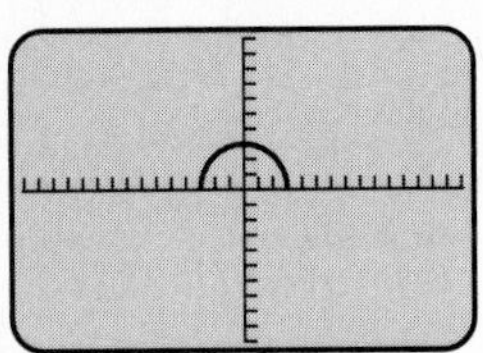

c. & d. $[-4.7, 4.7]$ by $[-3.1, 3.1]$

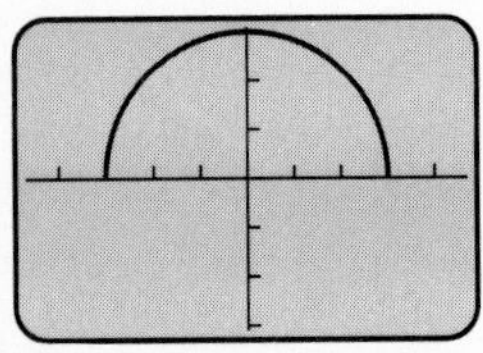

21. In sequential mode, first the graph of y_1 is drawn, then the graph of y_2, and finally the graph of y_3. In simultaneous mode, the three graphs are drawn at the same time.

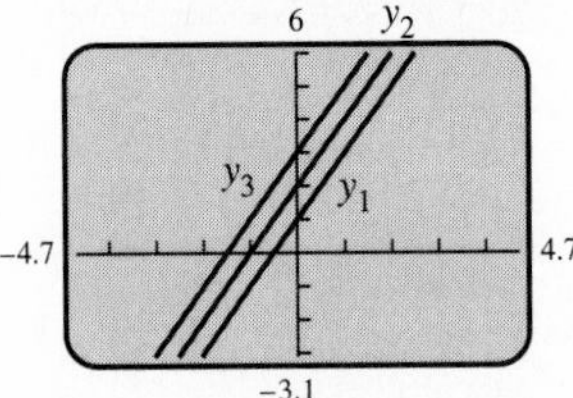

22. In dot mode, one point is plotted in each column of pixels provided the point is in the viewing window. That is, if the corresponding y-value lies between *Y*min and *Y*max. In connected mode, points are plotted and joined together to produce a curve that appears to have no breaks in it.

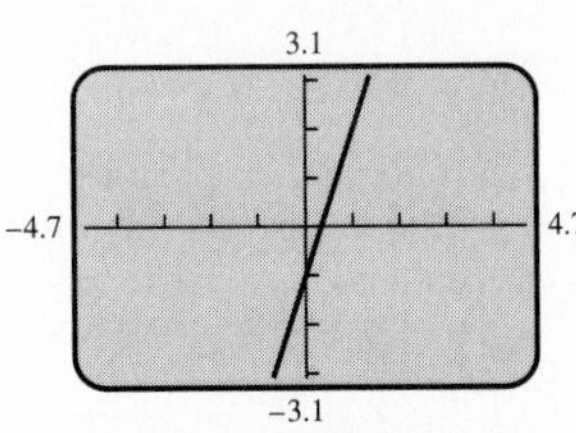

connected

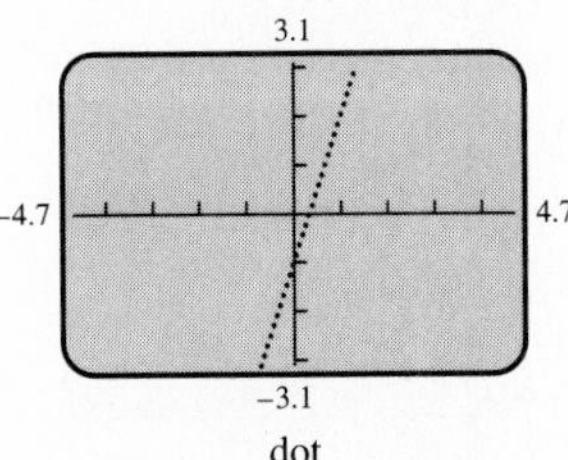

dot

23. One possibility is $[-10, 15]$ by $[-300, 200]$. There are two x-intercepts.

24. One possibility is $[-20, 10]$ by $[-2200, 1000]$. There are three x-intercepts.

25. a. The graphs of $y_1 = |x|$ and $y_2 = (1/2)x + 1$ intersect at $(-0.666\ldots, 0.666\ldots)$ and $(2, 2)$.

b. The x-intercepts of $y_3 = y_1 - y_2$ are $x = -0.666\ldots$ and $x = 2$. The solutions of the equation are $x = -0.666\ldots = -2/3$ and $x = 2$.

26. a. The graphs of $y_1 = |x - 3|$ and $y_2 = -(1/2)x + 4$ intersect at $(-2, 5)$ and $(4.666\ldots, 1.666\ldots)$.

b. The x-intercepts of $y_3 = y_1 - y_2$ are $x = -2$ and $x = 4.666\ldots$. The solutions of the equation are $x = -2$ and $x = 4.666\ldots = 4\frac{2}{3}$.

27. a. The graphs of $y_1 = x - 2$ and $y_2 = 1 - x^2$ intersect at $(-2.302\ldots, -4.302\ldots)$ and $(1.302\ldots, -0.697\ldots)$.

b. The x-intercepts of $y_3 = y_1 - y_2$ are $x = -2.302\ldots$ and $x = 1.302\ldots$. The solutions of the equation are $x \approx -2.30$ and $x \approx 1.30$.

28. a. The graphs of $y_1 = x - 3$ and $y_2 = x^2 - 5$ intersect at $(-1, -4)$ and $(2, -1)$.

b. The x-intercepts of $y_3 = y_1 - y_2$ are $x = -1$ and $x = 2$. The solutions of the equation are $x = -1$ and $x = 2$.

29. a. One possibility is $[-10, 15]$ by $[-300, 200]$.

b. There is a local maximum of 0 at $x = 0$, and a local minimum of -256 at $x = 8$.

30. a. One possibility is $[-20, 10]$ by $[-2200, 1000]$.

b. Accurate to two decimal places there is a local minimum of -2112.40 at $x = -11.05$, a local maximum of -3.00 at $x = 0.00$, and a local minimum of -52.10 at $x = -0.95$. (You may need to zoom-in to find these values.)

31. All graphs are drawn in the $[-12, 12]$ by $[-12, 12]$ viewing window.

a.

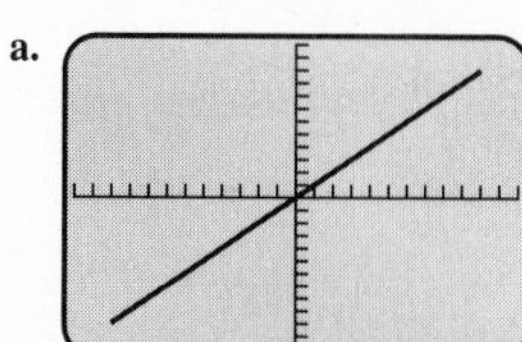

b.

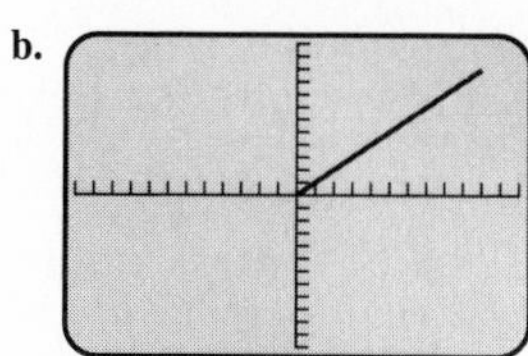

c.

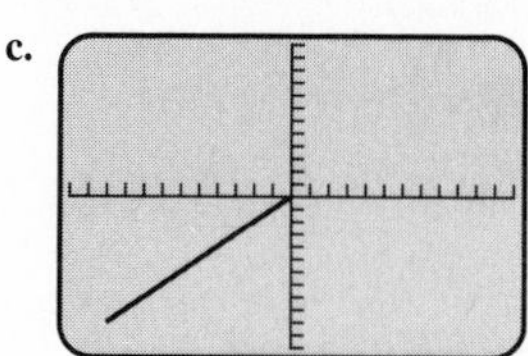

d. This graph may appear to be the same as in (a), but there are fewer points plotted. Also this graph is drawn faster than the one in (a).

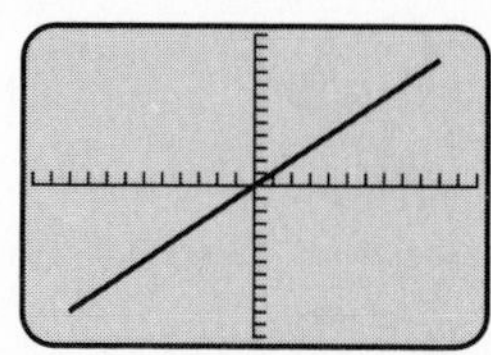

e.

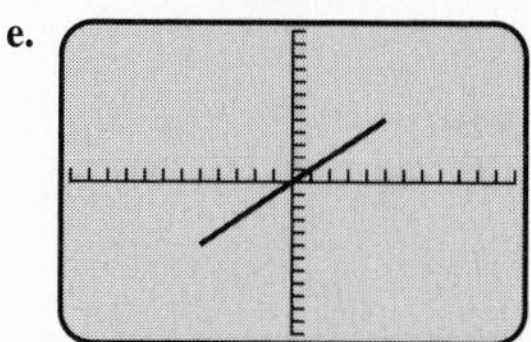

f. Same as (a), except this graph is drawn from upper right to lower left, and the graph in (a) is drawn from lower left to upper right.

32.

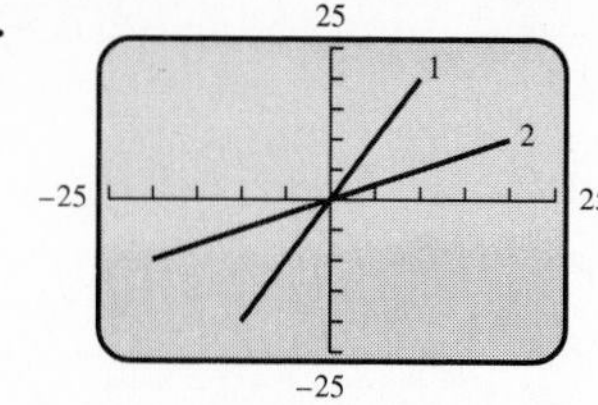

33.

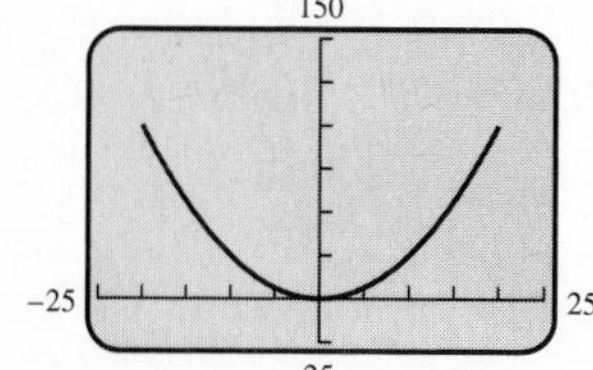

34.

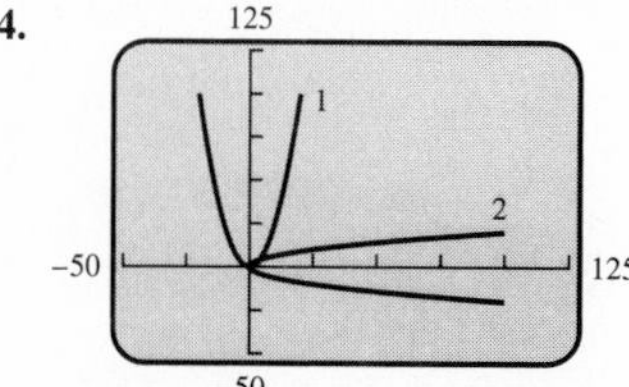

35. Viewing window: $[-9.4, 9.4]$ by $[-6.2, 6.2]$, $\theta\text{min} = 0$, $\theta\text{max} = 2\pi$, $\theta\text{step} = 0.1$.

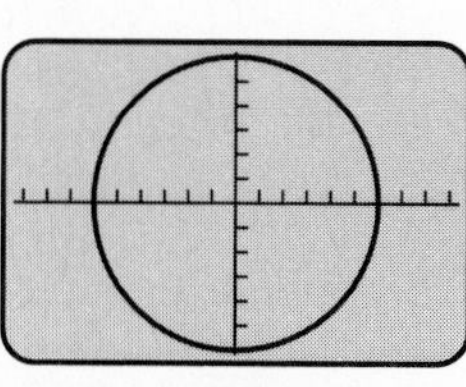

connected

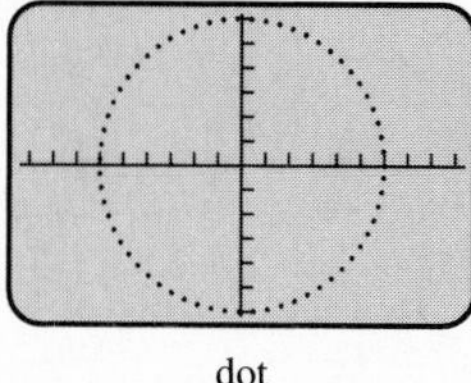

dot

36. Viewing window: $[-4.7, 4.7]$ by $[-3.1, 3.1]$, $\theta\text{min} = 0$, $\theta\text{max} = \pi$, $\theta\text{step} = 0.1$.

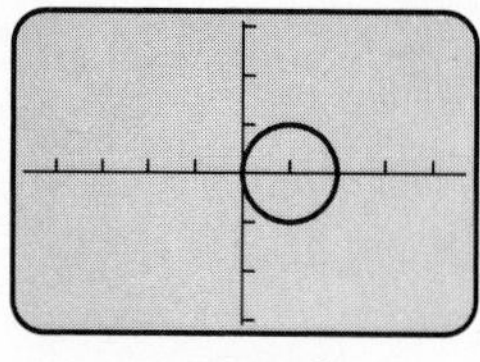

connected

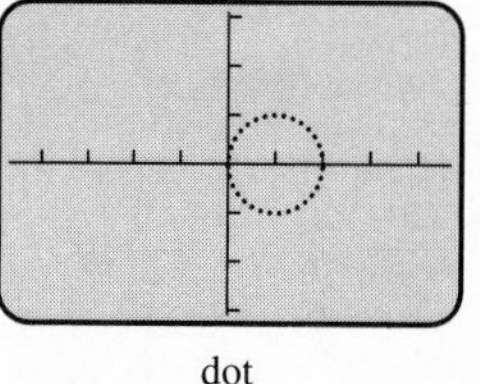

dot

37. Viewing window: $[-9.4, 9.4]$ by $[-6.2, 6.2]$, $\theta\text{min} = 0$, $\theta\text{max} = 2\pi$, $\theta\text{step} = 0.1$.

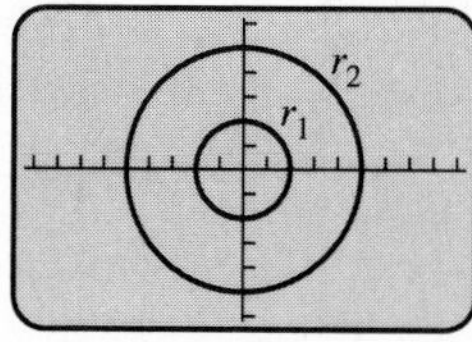

38. Viewing window: $[-4.7, 4.7]$ by $[-3.1, 3.1]$, $\theta\text{min} = 0$, $\theta\text{max} = 2\pi$, $\theta\text{step} = 0.1$.

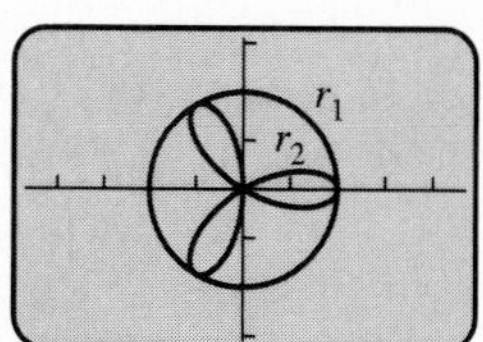

39. b.

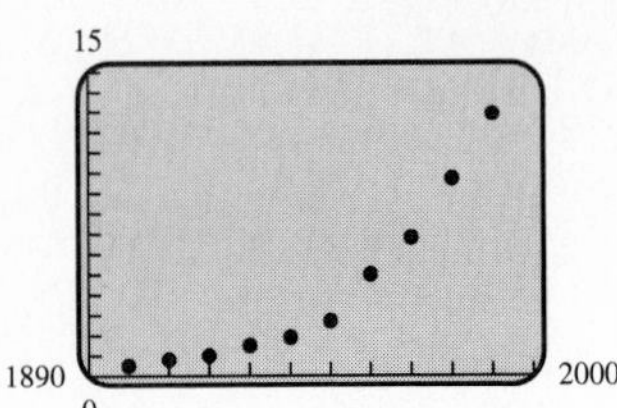

c.

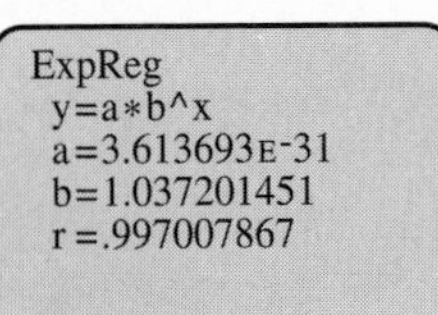

d.

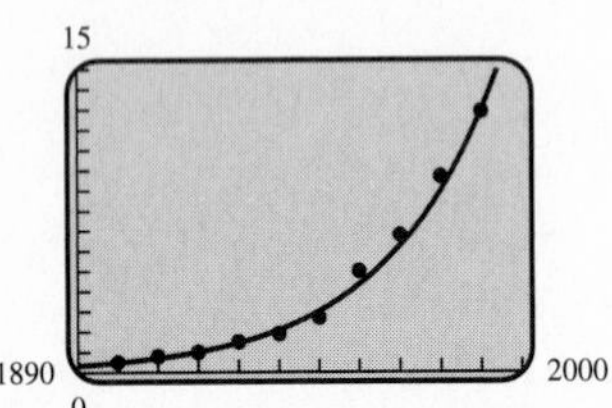

40. b.

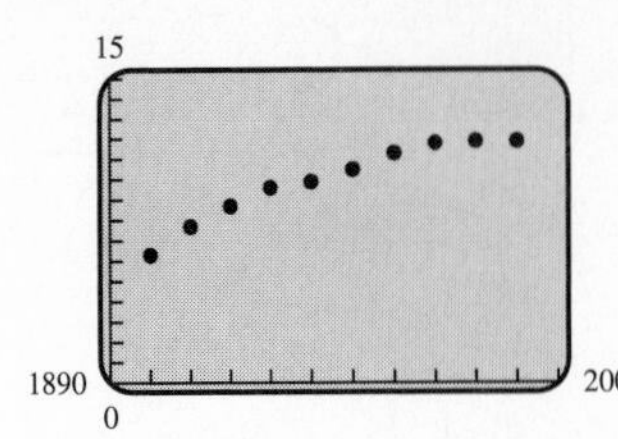

c.

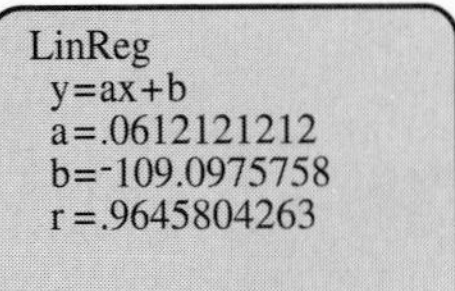

d.

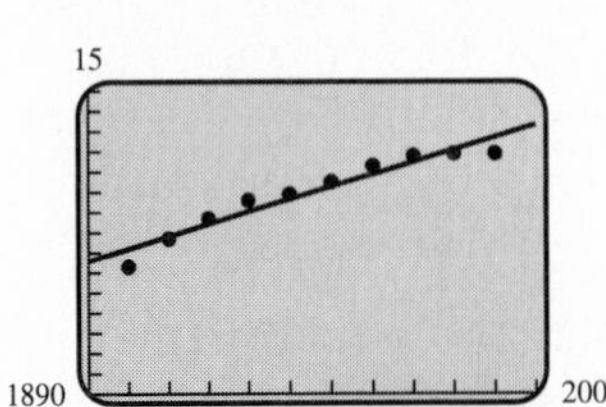

41. a. $\det A = 17$

b. $A^{-1} = \begin{pmatrix} 2/17 & 3/17 \\ -5/17 & 1/17 \end{pmatrix}$

c. $A^{-1}B = \begin{pmatrix} 2 \\ 1 \end{pmatrix}$

42. a. $\det A = -33$

b. $A^{-1} = \begin{pmatrix} 5/33 & 13/33 & 1/33 \\ 5/33 & -20/33 & 1/33 \\ -1/11 & 4/11 & 2/11 \end{pmatrix}$

c. $A^{-1}B = \begin{pmatrix} 3 \\ 1 \\ -1 \end{pmatrix}$

43.

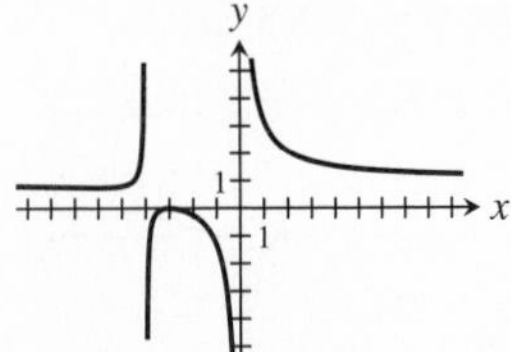

44.

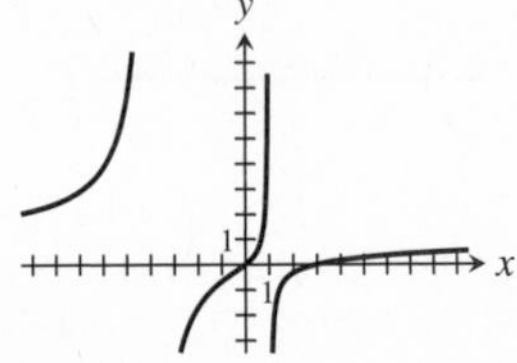

appendix B

KEY FORMULAS

B.1 FORMULAS FROM ALGEBRA

Exponents

If all bases are nonzero:

$u^m u^n = u^{m+n}$

$\dfrac{u^m}{u^n} = u^{m-n}$

$u^0 = 1$

$u^{-n} = \dfrac{1}{u^n}$

$(uv)^m = u^m v^m$

$(u^m)^n = u^{mn}$

$\left(\dfrac{u}{v}\right)^m = \dfrac{u^m}{v^m}$

Radicals and Rational Exponents

If all roots are real numbers:

$\sqrt[n]{uv} = \sqrt[n]{u} \cdot \sqrt[n]{v}$

$\sqrt[n]{\dfrac{u}{v}} = \dfrac{\sqrt[n]{u}}{\sqrt[n]{v}} \quad (v \neq 0)$

$\sqrt[m]{\sqrt[n]{u}} = \sqrt[mn]{u}$

$(\sqrt[n]{u})^n = u$

$\sqrt[n]{u^m} = (\sqrt[n]{u})^m$

$\sqrt[n]{u^n} = \begin{cases} |u|, & n \text{ even} \\ u, & n \text{ odd} \end{cases}$

$u^{1/n} = \sqrt[n]{u}$

$u^{m/n} = (u^{1/n})^m = (\sqrt[n]{u})^m$

$u^{m/n} = (u^m)^{1/n} = \sqrt[n]{u^m}$

Special Products

$(u + v)(u - v) = u^2 - v^2$

$(u + v)^2 = u^2 + 2uv + v^2$

$(u - v)^2 = u^2 - 2uv + v^2$

$(u + v)^3 = u^3 + 3u^2v + 3uv^2 + v^3$

$(u - v)^3 = u^3 - 3u^2v + 3uv^2 - v^3$

Factoring Polynomials

$u^2 - v^2 = (u + v)(u - v)$

$u^2 + 2uv + v^2 = (u + v)^2$

$u^2 - 2uv + v^2 = (u - v)^2$

$u^3 + v^3 = (u + v)(u^2 - uv + v^2)$

$u^3 - v^3 = (u - v)(u^2 + uv + v^2)$

Inequalities

If $u < v$ and $v < w$, then $u < w$.

If $u < v$, then $u + w < v + w$.

If $u < v$ and $c > 0$, then $uc < vc$.

If $u < v$ and $c < 0$, then $uc > vc$.

If $c > 0$, $|u| < c$ is equivalent to $-c < u < c$.

If $c > 0$, $|u| > c$ is equivalent to $u < -c$ or $u > c$.

Quadratic Formula

If $a \neq 0$, the solutions of the equation $ax^2 + bx + c = 0$ are given by

$$x = \frac{-b \pm \sqrt{b^2 - 4ac}}{2a}.$$

Logarithms

If $a > 0$ and $a \neq 1$:

$y = \log_a x$ if and only if $a^y = x$

$\log_a 1 = 0$

$\log_a a = 1$

$\log_a rs = \log_a r + \log_a s \qquad (r, s > 0)$

$\log_a \frac{r}{s} = \log_a r - \log_a s \qquad (r, s > 0)$

$\log_a r^c = c\log_a r \qquad (r > 0)$

$\log_b x = \frac{\log_a x}{\log_a b} \qquad (x > 0,\ b > 0,\ b \neq 1)$

Determinants

$$\begin{vmatrix} a & b \\ c & d \end{vmatrix} = ad - bc$$

Arithmetic Sequences and Series

$a_n = a_1 + (n - 1)d$

$S_n = n\left(\frac{a_1 + a_n}{2}\right)$ or $S_n = \frac{n}{2}[2a_1 + (n - 1)d]$

Geometric Sequences and Series

$a_n = a_1 \cdot r^{n-1}$

$S_n = \frac{a_1(1 - r^n)}{1 - r} \qquad (r \neq 1)$

$S = \frac{a_1}{1 - r} \qquad (|r| < 1)$ infinite geometric series

Factorial

$n! = n \cdot (n - 1) \cdot (n - 2) \cdot \cdots \cdot 3 \cdot 2 \cdot 1$

$n \cdot (n - 1)! = n!, \quad 0! = 1$

Binomial Coefficient

$\binom{n}{r} = \frac{n!}{r!(n - r)!}$ (integers n and r, $n \geq r \geq 0$)

Binomial Theorem

If n is a positive integer:

$$(a+b)^n = \binom{n}{0}a^n + \binom{n}{1}a^{n-1}b + \cdots + \binom{n}{r}a^{n-r}b^r + \cdots + \binom{n}{n}b^n$$

B.2 FORMULAS FROM GEOMETRY

Triangle

$h = a \sin \theta$

$\text{Area} = \frac{1}{2}bh$

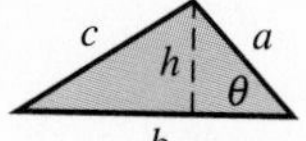

Trapezoid

$\text{Area} = \frac{h}{2}(a + b)$

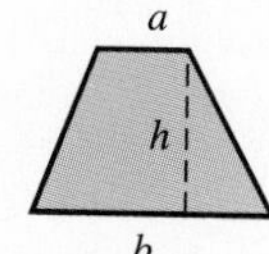

Circle

$\text{Area} = \pi r^2$

$\text{Circumference} = 2\pi r$

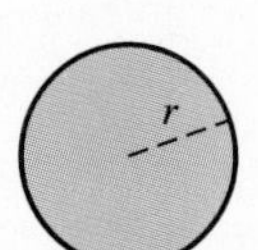

Sector of Circle

$\text{Area} = \frac{\theta r^2}{2}$ (θ in radians)

$s = r\theta$ (θ in radians)

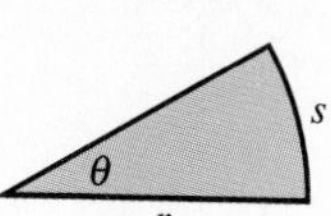

Right Circular Cone

$\text{Volume} = \frac{\pi r^2 h}{3}$

$\text{Lateral surface area} = \pi r\sqrt{r^2 + h^2}$

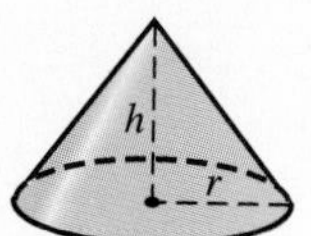

Right Circular Cylinder

$\text{Volume} = \pi r^2 h$

$\text{Lateral surface area} = 2\pi rh$

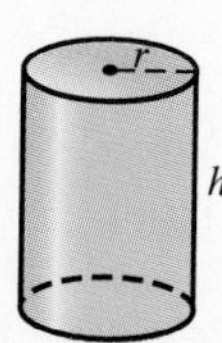

Right Triangle

Pythagorean Theorem:

$c^2 = a^2 + b^2$

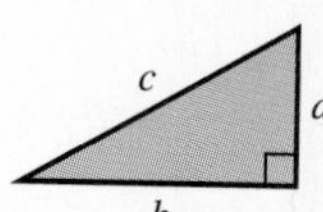

Parallelogram

$\text{Area} = bh$

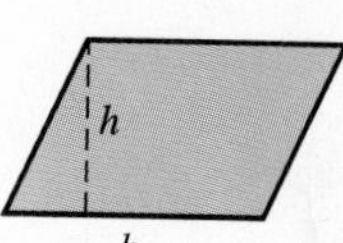

Circular Ring

$\text{Area} = \pi(R^2 - r^2)$

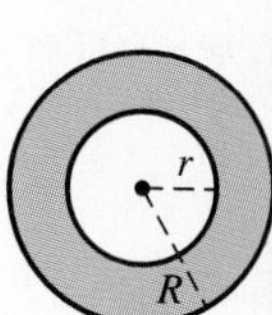

Ellipse

$\text{Area} = \pi ab$

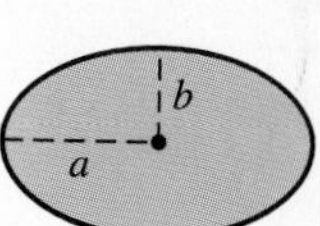

Cone

$\text{Volume} = \frac{Ah}{3}$ (A = Area of base)

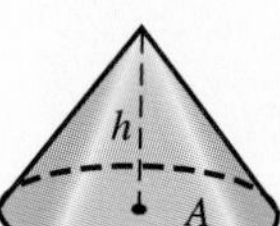

Sphere

$\text{Volume} = \frac{4}{3}\pi r^3$

$\text{Surface area} = 4\pi r^2$

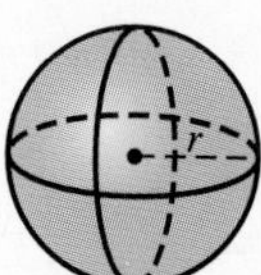

B.3 Formulas from Trigonometry

Angular Measure

π radians = 180°

So, 1 radian = $\dfrac{180}{\pi}$ degrees,

and 1 degree = $\dfrac{\pi}{180}$ radians.

Reciprocal Identities

$\sin x = \dfrac{1}{\csc x}$ $\qquad \csc x = \dfrac{1}{\sin x}$

$\cos x = \dfrac{1}{\sec x}$ $\qquad \sec x = \dfrac{1}{\cos x}$

$\tan x = \dfrac{1}{\cot x}$ $\qquad \cot x = \dfrac{1}{\tan x}$

Quotient Identities

$\tan x = \dfrac{\sin x}{\cos x}$ $\qquad \cot x = \dfrac{\cos x}{\sin x}$

Pythagorean Identities

$\sin^2 x + \cos^2 x = 1$

$\tan^2 x + 1 = \sec^2 x$

$1 + \cot^2 x = \csc^2 x$

Odd–Even Identities

$\sin(-x) = -\sin x \qquad \csc(-x) = -\csc x$

$\cos(-x) = \cos x \qquad \sec(-x) = \sec x$

$\tan(-x) = -\tan x \qquad \cot(-x) = -\cot x$

Sum and Difference Identities

$\sin(u + v) = \sin u \cos v + \cos u \sin v$

$\sin(u - v) = \sin u \cos v - \cos u \sin v$

$\cos(u + v) = \cos u \cos v - \sin u \sin v$

$\cos(u - v) = \cos u \cos v + \sin u \sin v$

$\tan(u + v) = \dfrac{\tan u + \tan v}{1 - \tan u \tan v}$

$\tan(u - v) = \dfrac{\tan u - \tan v}{1 + \tan u \tan v}$

Cofunction Identities

$\cos\left(\dfrac{\pi}{2} - u\right) = \sin u$

$\sin\left(\dfrac{\pi}{2} - u\right) = \cos u$

$\tan\left(\dfrac{\pi}{2} - u\right) = \cot u$

$\cot\left(\dfrac{\pi}{2} - u\right) = \tan u$

$\sec\left(\dfrac{\pi}{2} - u\right) = \csc u$

$\csc\left(\dfrac{\pi}{2} - u\right) = \sec u$

Double-Angle Identities

$\sin 2u = 2 \sin u \cos u$

$$\begin{aligned}\cos 2u &= \cos^2 u - \sin^2 u \\ &= 2\cos^2 u - 1 \\ &= 1 - 2\sin^2 u\end{aligned}$$

$\tan 2u = \dfrac{2 \tan u}{1 - \tan^2 u}$

Power-Reducing Identities

$\sin^2 u = \dfrac{1 - \cos 2u}{2}$

$\cos^2 u = \dfrac{1 + \cos 2u}{2}$

$\tan^2 u = \dfrac{1 - \cos 2u}{1 + \cos 2u}$

Half-Angle Identities

$\sin \dfrac{u}{2} = \pm\sqrt{\dfrac{1 - \cos u}{2}}$

$\cos \dfrac{u}{2} = \pm\sqrt{\dfrac{1 + \cos u}{2}}$

$$\tan \frac{u}{2} = \pm\sqrt{\frac{1 - \cos u}{1 + \cos u}} = \frac{1 - \cos u}{\sin u} = \frac{\sin u}{1 + \cos u}$$

Triangles

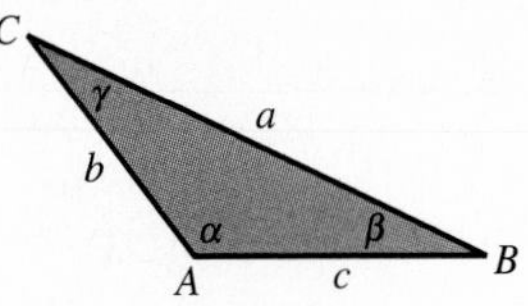

Law of sines:

$$\frac{\sin \alpha}{a} = \frac{\sin \beta}{b} = \frac{\sin \gamma}{c}$$

Law of cosines:

$a^2 = b^2 + c^2 - 2bc \cos \alpha$

$b^2 = a^2 + c^2 - 2ac \cos \beta$

$c^2 = a^2 + b^2 - 2ab \cos \gamma$

Area:

$$\text{Area} = \frac{1}{2} bc \sin \alpha = \frac{1}{2} ac \sin \beta = \frac{1}{2} ab \sin \gamma$$

$\text{Area} = \sqrt{s(s - a)(s - b)(s - c)}$,

where $s = \dfrac{1}{2}(a + b + c)$

Trigonometric Form of a Complex Number

$$\begin{aligned} z = a + bi &= (r \cos \theta) + (r \sin \theta)i \\ &= r(\cos \theta + i \sin \theta) \end{aligned}$$

DeMoivre's Theorem

$$\begin{aligned} z^n &= [r(\cos \theta + i \sin \theta)]^n \\ &= r^n(\cos n\theta + i \sin n\theta) \end{aligned}$$

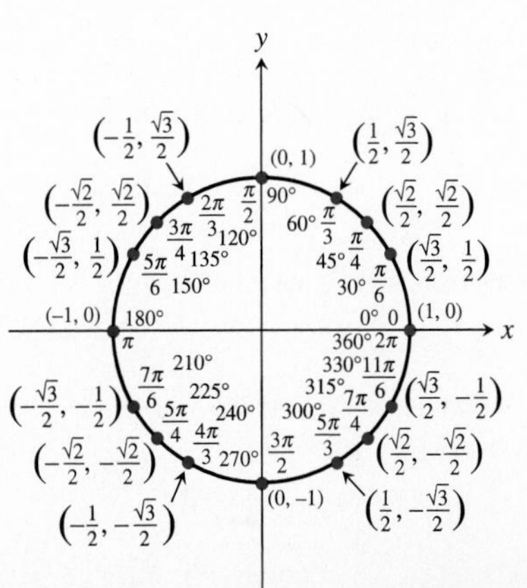

B.4 Formulas from Analytic Geometry

Basic Formulas

Distance d between points $P(x_1, y_1)$ and $Q(x_2, y_2)$:

$$d = \sqrt{(x_1 - x_2)^2 + (y_1 - y_2)^2}$$

Midpoint: $\left(\dfrac{x_1 + x_2}{2}, \dfrac{y_1 + y_2}{2}\right)$

Slope of a line: $m = \dfrac{y_2 - y_1}{x_2 - x_1}$

Condition for parallel lines: $m_1 = m_2$

Condition for perpendicular lines: $m_2 = \dfrac{-1}{m_1}$

Equations of a Line

The point-slope form, slope m and through (x_1, y_1):

$$y - y_1 = m(x - x_1)$$

The slope-intercept form, slope m and y-intercept b: $y = mx + b$

Equation of a Circle

The circle with center (h, k) and radius r: $(x - h)^2 + (y - k)^2 = r^2$

Ellipse

Centered at (0, 0):

$\dfrac{x^2}{a^2} + \dfrac{y^2}{b^2} = 1$ foci: $(-c, 0)$ and $(c, 0)$, where $c^2 = a^2 - b^2$

$a > b$ major axis from $(-a, 0)$ to $(a, 0)$
minor axis from $(0, -b)$ to $(0, b)$

$\dfrac{x^2}{a^2} + \dfrac{y^2}{b^2} = 1$ foci: $(0, -c)$ and $(0, c)$, where $c^2 = b^2 - a^2$

$a < b$ major axis from $(0, -b)$ to $(0, b)$
minor axis from $(-a, 0)$ to $(a, 0)$

Centered at (h, k):

$\dfrac{(x - h)^2}{a^2} + \dfrac{(y - k)^2}{b^2} = 1$ foci: $(h - c, k)$ and $(h + c, k)$, where $c^2 = a^2 - b^2$

$a > b$ major axis from $(h - a, k)$ to $(h + a, k)$
minor axis from $(h, k - b)$ to $(h, k + b)$

$\dfrac{(x - h)^2}{a^2} + \dfrac{(y - k)^2}{b^2} = 1$ foci: $(h, k - c)$ and $(h, k + c)$, where $c^2 = b^2 - a^2$

$a < b$ major axis from $(h, k - b)$ to $(h, k + b)$
minor axis from $(h - a, k)$ to $(h + a, k)$

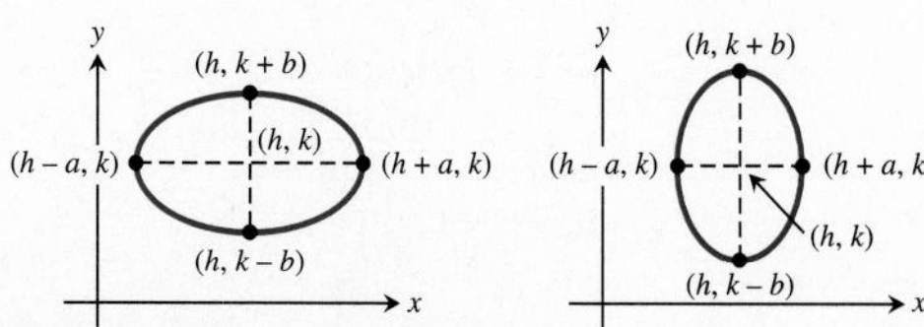

Parabola

Vertex at (0, 0):

$y = ax^2$ focus: $\left(0, \dfrac{1}{4a}\right)$, directrix: $y = -\dfrac{1}{4a}$

$x = ay^2$ focus: $\left(\dfrac{1}{4a}, 0\right)$, directrix: $x = -\dfrac{1}{4a}$

Vertex at (h, k):

$y - k = a(x - h)^2$ focus: $\left(h, k + \dfrac{1}{4a}\right)$, directrix: $y = k - \dfrac{1}{4a}$

$x - h = a(y - k)^2$ focus: $\left(h + \dfrac{1}{4a}, k\right)$, directrix: $x = h - \dfrac{1}{4a}$

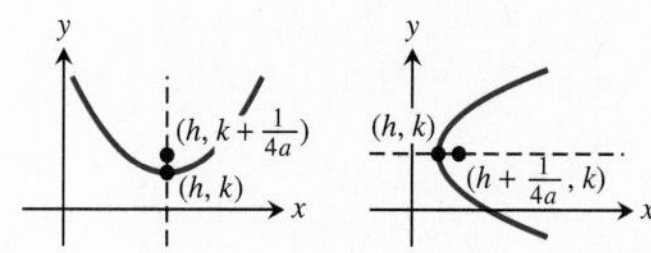

Hyperbola

Centered at (0, 0):

$\dfrac{x^2}{a^2} - \dfrac{y^2}{b^2} = 1$ foci: $(-c, 0)$ and $(c, 0)$, where $c^2 = a^2 + b^2$

transverse axis from $(-a, 0)$ to $(a, 0)$, asymptotes: $y = \pm\dfrac{b}{a}x$

$\dfrac{y^2}{b^2} - \dfrac{x^2}{a^2} = 1$ foci: $(0, -c)$ and $(0, c)$, where $c^2 = a^2 + b^2$

transverse axis from $(0, -b)$ to $(0, b)$, asymptotes: $y = \pm\dfrac{b}{a}x$

Centered at (h, k):

$\dfrac{(x - h)^2}{a^2} - \dfrac{(y - k)^2}{b^2} = 1$ foci: $(h - c, k)$ and $(h + c, k)$, where $c^2 = a^2 + b^2$

transverse axis from $(h - a, k)$ to $(h + a, k)$

asymptotes: $y - k = \pm\dfrac{b}{a}(x - h)$

$\dfrac{(y - k)^2}{b^2} - \dfrac{(x - h)^2}{a^2} = 1$ foci: $(h, k - c)$ and $(h, k + c)$, where $c^2 = a^2 + b^2$

transverse axis from $(h, k - b)$ to $(h, k + b)$

asymptotes: $y - k = \pm\dfrac{b}{a}(x - h)$

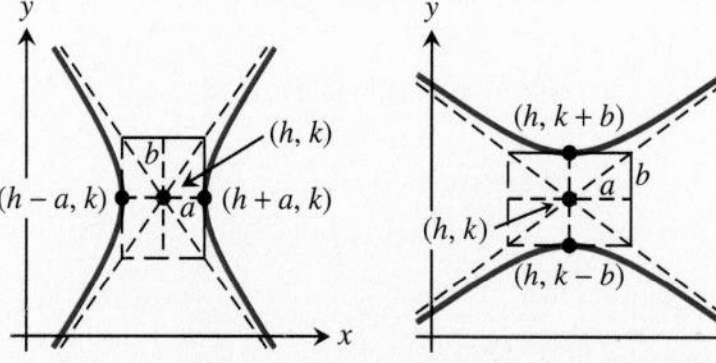

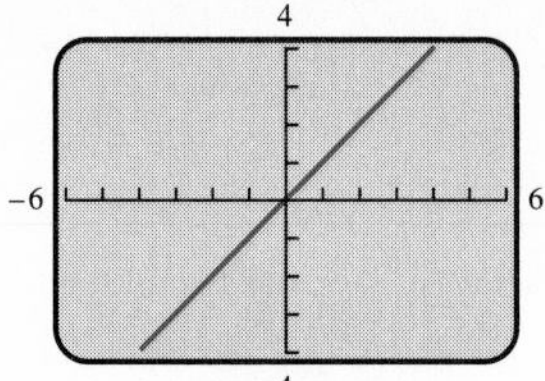

Identity Function
$f(x) = x$
Domain = $(-\infty, \infty)$
Range = $(-\infty, \infty)$

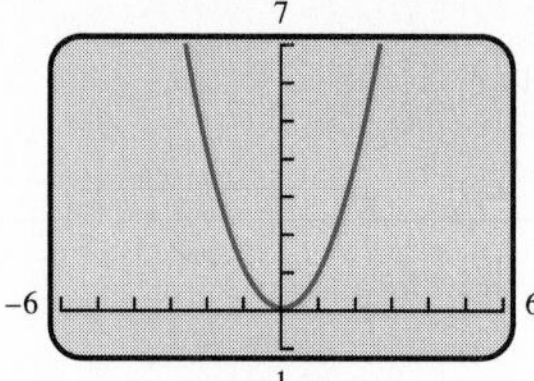

Squaring Function
$f(x) = x^2$
Domain = $(-\infty, \infty)$
Range = $[0, \infty)$

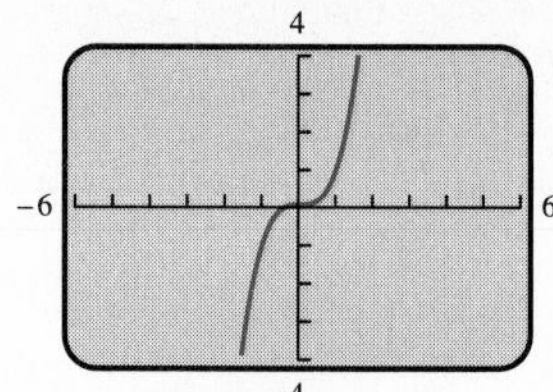

Cubing Function
$f(x) = x^3$
Domain = $(-\infty, \infty)$
Range = $(-\infty, \infty)$

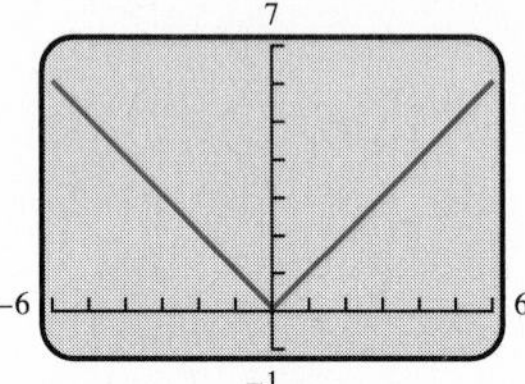

Absolute Value Function
$f(x) = |x| = \text{abs}\ (x)$
Domain = $(-\infty, \infty)$
Range = $[0, \infty)$

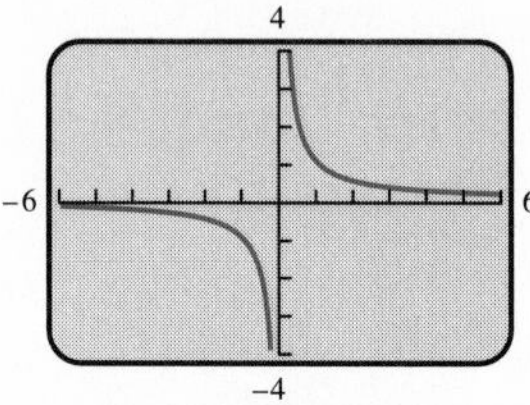

Reciprocal Function
$f(x) = \frac{1}{x}$
Domain = $(-\infty, 0) \cup (0, \infty)$
Range = $(-\infty, 0) \cup (0, \infty)$

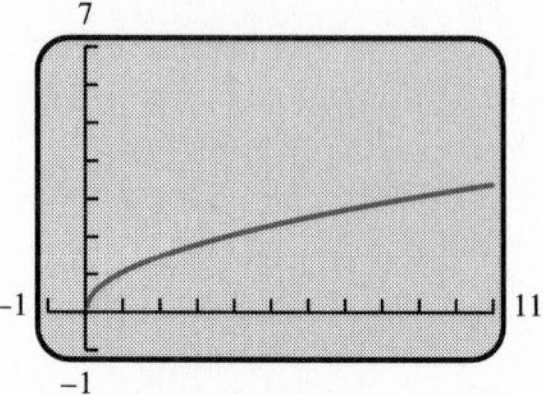

Square Root Function
$f(x) = \sqrt{x}$
Domain = $[0, \infty)$
Range = $[0, \infty)$

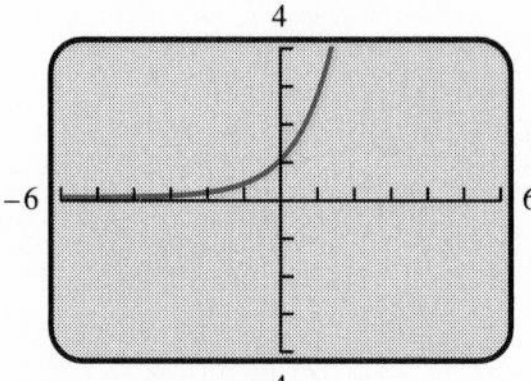

Exponential Growth Function
$f(x) = b^x, b > 1$
Domain = $(-\infty, \infty)$
Range = $(0, \infty)$

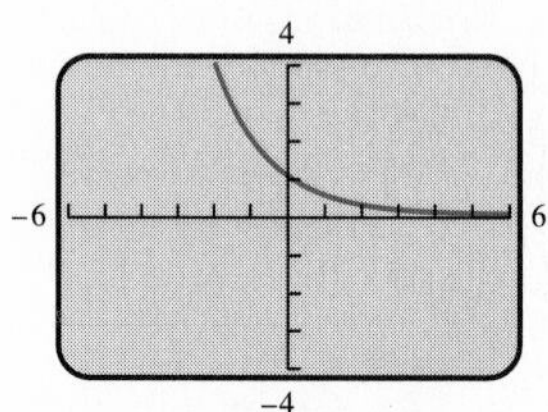

Exponential Decay Function
$f(x) = b^x, 0 < b < 1$
Domain = $(-\infty, \infty)$
Range = $(0, \infty)$

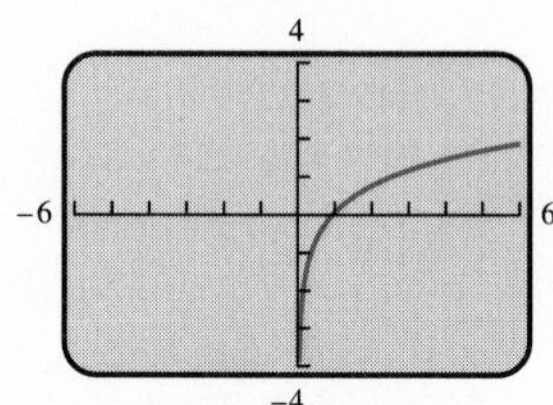

Logarithmic Function
$f(x) = \log_b(x), b > 1$
Domain = $(0, \infty)$
Range = $(-\infty, \infty)$

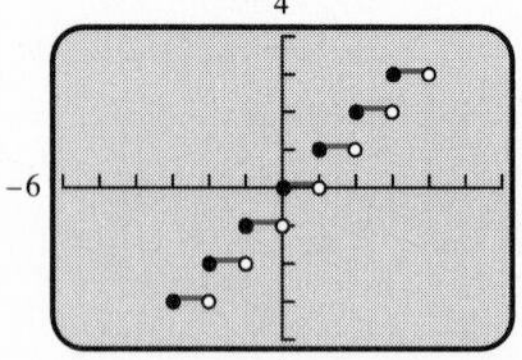

Greatest Integer Function
$f(x) = [[x]] = \text{int}(x)$
Domain = $(-\infty, \infty)$
Range = all integers

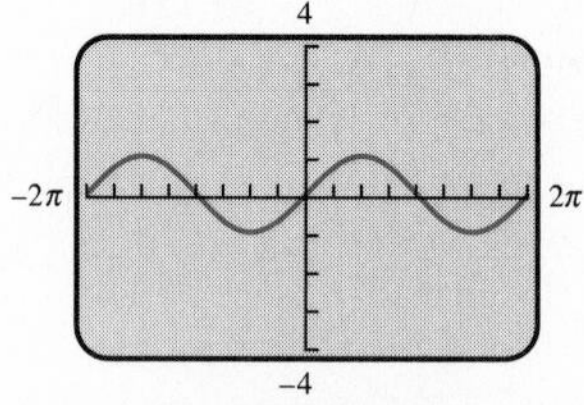

Sine Function
$f(x) = \sin(x)$
Domain = $(-\infty, \infty)$
Range = $[-1, 1]$

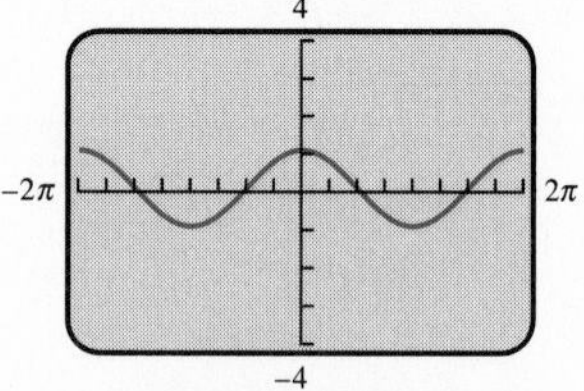

Cosine Function
$f(x) = \cos(x)$
Domain = $(-\infty, \infty)$
Range = $[-1, 1]$

appendix C

C.1
Logic: An Introduction

C.2
Conditionals and Biconditionals

LOGIC[1]

C.1 LOGIC: AN INTRODUCTION

Logic is a tool used in mathematical thinking and problem solving. In logic, a **statement** *is a sentence that is either true or false, but not both.*

The following expressions are not statements because their truth value cannot be determined without more information.

1. She has blue eyes.
2. $x + 7 = 18$
3. $2y + 7 > 1$
4. $2 + 3 = 5$

Expressions (1), (2), and (3) become statements if, for (1), "she" is identified, and for (2) and (3), values are assigned to x and y, respectively. However, an expression involving *he* or *she* or x or y may already be a statement. For example, "If he is over 210 cm tall, then he is over 2 m tall," and "$2(x + y) = 2x + 2y$" are both statements because they are true no matter who *he* is or what the numerical values of x and y are.

From a given statement, it is possible to create a new statement by forming a **negation.** The negation of a statement is a statement with the opposite truth value of the given statement. If a statement is true, its negation is false, and if a statement is false, its negation is true. Consider the statement "It is snowing." The negation of this statement may be stated simply as "It is not snowing."

[1]The authors wish to express their thanks to Rick Billstein, Shlomo Libeskind, and Johnny W. Lott, for permission to copy the material that appears in our Logic Appendix from their text, *A Problem Solving Approach to Mathematics for Elementary School Teachers, fifth edition,* Addison-Wesley Publishing Company, copyright 1993.

EXAMPLE 1

Negate each of the following statements:

a. $2 + 3 = 5$
b. A hexagon has six sides.
c. Today is not Monday.

Solution

a. $2 + 3 \neq 5$
b. A hexagon does not have six sides.
c. Today is Monday.

The statements "The shirt is blue" and "The shirt is green" are not negations of each other. A statement and its negation must have opposite truth values. If the shirt is actually red, then both of the above statements are false and, hence, cannot be negations of each other. However, the statements "The shirt is blue" and "The shirt is not blue" are negations of each other because they have opposite truth values no matter what color the shirt really is.

Some statements involve **quantifiers** and are more complicated to negate. Quantifiers include words such as *all, some, every,* and *there exists.*

The quantifiers *all, every,* and *no* refer to each and every element in a set and are **universal quantifiers.** The quantifiers *some* and *there exists at least one* refer to one or more, or possibly all, of the elements in a set. *Some* and *there exists* are called **existential quantifiers.** Examples with universal and existential quantifiers follow:

1. All roses are red. [universal]
2. Every student is important. [universal]
3. For each counting number x, $x + 0 = x$. [universal]
4. Some roses are red. [existential]
5. There exists at least one even counting number less than 3. [existential]
6. There are women who are taller than 200 cm. [existential]

Venn diagrams can be used to picture statements involving quantifiers. For example, Figure C.1a and b picture statements (1) and (4). The x in Fig-

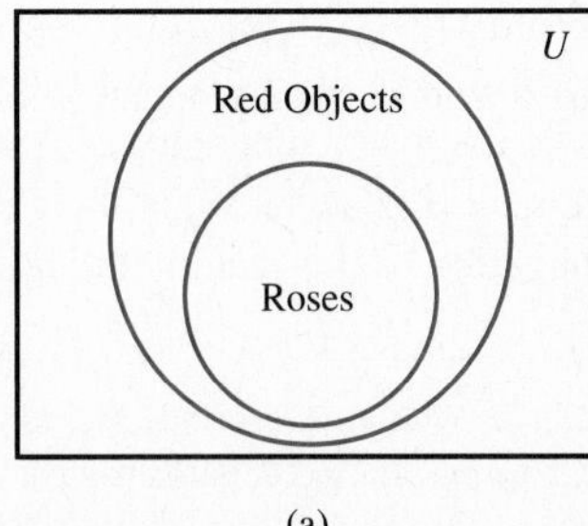

(a)

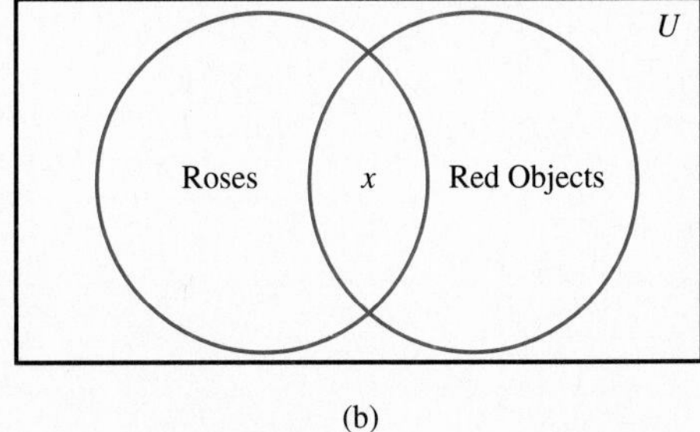

(b)

Figure C.1

ure C.1b can be used to show that there must be at least one element of the set of roses that is red.

Consider the following statement involving the existential quantifier *some*. "Some professors at Paxson University have blue eyes." This means that at least one professor at Paxson University has blue eyes. It does not rule out the possibilities that all the Paxson professors have blue eyes or that some of the Paxson professors do not have blue eyes. Because the negation of a true statement is false, neither "Some professors at Paxson University do not have blue eyes" nor "All professors at Paxson have blue eyes" are negations of the original statement. One possible negation of the original statement is "No professors at Paxson University have blue eyes."

To discover if one statement is a negation of another, we use arguments similar to the preceding one to determine if they have opposite truth values in all possible cases. Some general forms of quantified statements with their negations follow:

Statement	**Negation**
Some *a* are *b*.	No *a* is *b*.
Some *a* are not *b*.	All *a* are *b*.
All *a* are *b*.	Some *a* are not *b*.
No *a* is *b*.	Some *a* are *b*.

■ **EXAMPLE 2**

Negate each of the following statements:

a. All students like hamburgers.
b. Some people like mathematics.
c. There exists a counting number x such that $3x = 6$.
d. For all counting numbers, $3x = 3x$.

Solution

a. Some students do not like hamburgers.
b. No people like mathematics.
c. For all counting numbers x, $3x \neq 6$.
d. There exists a counting number x such that $3x \neq 3x$. ■

Table C.1

Statement p	Negation $\sim p$
T	F
F	T

There is a symbolic system defined to help in the study of logic. If p represents a statement, the negation of the statement p is denoted by $\sim p$. **Truth tables** are often used to show all possible true-false patterns for statements. Table C.1 summarizes the truth tables for p and $\sim p$.

Observe that p and $\sim p$ are analogous to sets P and $\overline{P}$. If x is an element of P, then x is not an element of $\overline{P}$.

Compound Statements

From two given statements, it is possible to create a new, **compound statement** by using a connective such as *and*. For example, "It is snowing" and "The ski run is open" together with *and* give "It is snowing and the ski run is open." Other compound statements can be obtained by using the connective *or.* For example, "It is snowing or the ski run is open."

The symbols $\wedge$ and $\vee$ are used to represent the connectives *and* and *or,* respectively. For example, if p represents "It is snowing," and if q represents "The ski run is open," then "It is snowing and the ski run is open" is denoted by $p \wedge q$. Similarly, "It is snowing or the ski run is open" is denoted by $p \vee q$.

The truth value of any compound statement, such as $p \wedge q$, is defined using the truth table of each of the simple statements. Because each of the statements p and q may be either true or false, there are four distinct possibilities for the truth of $p \wedge q$, as shown in Table C.2. The compound statement $p \wedge q$, is the **conjunction** of p and q and is defined to be true if, and only if, both p and q are true. Otherwise, it is false.

Table C.2

p	q	Conjunction $p \wedge q$
T	T	T
T	F	F
F	T	F
F	F	F

The compound statement $p \vee q$—that is, p *or* q—is a **disjunction.** In everyday language, *or* is not always interpreted in the same way. In logic, we use an *inclusive or.* The statement "I will go to a movie or I will read a book" means that I will either go to a movie, or read a book, or do both. Hence, in logic, p or q, symbolized as $p \vee q$, is defined to be false if both p and q are false and true in all other cases. This is summarized in Table C.3.

Table C.3

p	q	Disjunction $p \vee q$
T	T	T
T	F	T
F	T	T
F	F	F

■ EXAMPLE 3

Given the following statements, classify each of the conjunctions and disjunctions as true or false:

p: $2 + 3 = 5$ $\quad$ r: $5 + 3 = 9$

q: $2 \cdot 3 = 6$ $\quad$ s: $2 \cdot 4 = 9$

a. $p \wedge q$ **b.** $p \wedge r$ **c.** $s \wedge q$ **d.** $r \wedge s$

e. $\sim p \wedge q$ **f.** $\sim(p \wedge q)$ **g.** $p \vee q$ **h.** $p \vee r$

i. $s \vee q$ **j.** $r \vee s$ **k.** $\sim p \vee q$ **l.** $\sim(p \vee q)$

Solution

a. p is true and q is true, so $p \wedge q$ is true.

b. p is true and r is false, so $p \wedge r$ is false.

c. s is false and q is true, so $s \wedge q$ is false.

d. r is false and s is false, so $r \wedge s$ is false.

e. $\sim p$ is false and q is true, so $\sim p \wedge q$ is false.

f. $p \wedge q$ is true [part (a)], so $\sim (p \wedge q)$ is false.

g. p is true and q is true, so $p \vee q$ is true.

h. p is true and r is false, so $p \vee r$ is true.

i. s is false and q is true, so $s \vee q$ is true.

j. r is false and s is false, so $r \vee s$ is false.

k. $\sim p$ is false and q is true, so $\sim p \vee q$ is true.

l. $p \vee q$ is true [part (g)], so $\sim(p \vee q)$ is false. ■

Table C.4

p	q	$p \wedge q$	$q \wedge p$
T	T	T	T
T	F	F	F
F	T	F	F
F	F	F	F

Not only are truth tables used to summarize the truth values of compound statements, they also are used to determine if two statements are logically equivalent. Two statements are **logically equivalent** if, and only if, they have the same truth values. For example, we could show that $p \wedge q$ is logically equivalent to $q \wedge p$ by using truth tables as in Table C.4.

EXAMPLE 4

Use a truth table to determine if $\sim p \vee \sim q$ and $\sim(p \wedge q)$ are logically equivalent.

Solution

Table C.5 shows headings and the four distinct possibilities for p and q. In the column headed $\sim p$, we write the negations of the p column. In the $\sim q$ column, we write the negations of the q column. Next, we use the values in the $\sim p$ and the $\sim q$ columns to construct the $\sim p \vee \sim q$ column. To find the truth values for $\sim(p \wedge q)$, we use the p and q columns to find the truth values for $p \wedge q$ and then negate $p \wedge q$.

Table C.5

p	q	$\sim p$	$\sim q$	$\sim p \vee \sim q$	$p \wedge q$	$\sim(p \wedge q)$
T	T	F	F	F	T	F
T	F	F	T	T	F	T
F	T	T	F	T	F	T
F	F	T	T	T	F	T

APPENDIX EXERCISES C.1

1. Determine which of the following are statements, and then classify each statement as true or false:
   a. $2 + 4 = 8$
   b. Shut the window.
   c. Los Angeles is a state.
   d. He is in town.
   e. What time is it?
   f. $5x = 15$
   g. $3 \cdot 2 = 6$
   h. $2x^2 > x$
   i. This statement is false.
   j. Stay put!
2. Use quantifiers to make each of the following true where x is a natural number:
   a. $x + 8 = 11$
   b. $x + 0 = x$
   c. $x^2 = 4$
   d. $x + 1 = x + 2$
3. Use quantifiers to make each equation in Exercise 2 false.
4. Write the negation for each of the following statements:
   a. The book has 500 pages.
   b. Six is less than eight.
   c. $3 \cdot 5 = 15$
   d. Some people have blond hair.
   e. All dogs have four legs.
   f. Some cats do not have nine lives.
   g. All squares are rectangles.
   h. Not all rectangles are squares.
   i. For all natural numbers x, $x + 3 = 3 + x$.
   j. There exists a natural number x such that $3 \cdot (x + 2) = 12$.
   k. Every counting number is divisible by itself and 1.

l. Not all natural numbers are divisible by 2.
m. For all natural numbers x, $5x + 4x = 9x$.

5. Complete each of the following truth tables:

a.

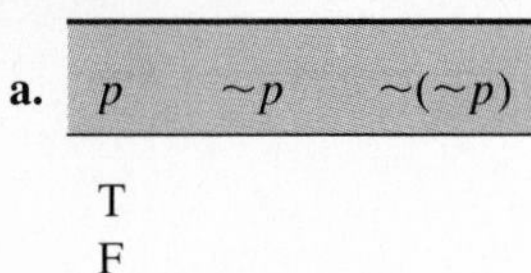

p	$\sim p$	$\sim(\sim p)$
T		
F		

b.

p	$\sim p$	$p \vee \sim p$	$p \wedge \sim p$
T			
F			

c. Based on part a, is p logically equivalent to $\sim(\sim p)$?
d. Based on part b, is $p \vee \sim p$ logically equivalent to $p \wedge \sim p$?

6. If q stands for "This course is easy" and r stands for "Lazy students do not study," write each of the following in symbolic form:

a. This course is easy and lazy students do not study.
b. Lazy students do not study or this course is not easy.
c. It is false that both this course is easy and lazy students do not study.
d. This course is not easy.

7. If p is false and q is true, find the truth values for each of the following:

a. $p \wedge q$ **b.** $p \vee q$
c. $\sim p$ **d.** $\sim q$
e. $\sim(\sim p)$ **f.** $\sim p \vee q$
g. $p \wedge \sim q$ **h.** $\sim(p \vee q)$
i. $\sim(\sim p \wedge q)$ **j.** $\sim q \wedge \sim p$

8. Find the truth value for each statement in Exercise 7 if p is false and q is false.

9. For each of the following, is the pair of statements logically equivalent?

a. $\sim(p \vee q)$ and $\sim p \vee \sim q$
b. $\sim(p \vee q)$ and $\sim p \wedge \sim q$
c. $\sim(p \wedge q)$ and $\sim p \wedge \sim q$
d. $\sim(p \wedge q)$ and $\sim p \vee \sim q$

10. a. Write two logical equivalences discovered in parts 9a–d. These equivalences are called DeMorgan's Laws for "*and*" and "*or*."
b. Write an explanation of the analogy between DeMorgan's Laws for sets and those found in part a.

11. Complete the following truth table:

p	q	$\sim p$	$\sim q$	$\sim p \vee q$
T	T			
T	F			
F	T			
F	F			

12. Restate the following in a logically equivalent form:

a. It is not true that both today is Wednesday and the month is June.
b. It is not true that yesterday I both ate breakfast and watched television.
c. It is not raining or it is not July.

C.2 CONDITIONALS AND BICONDITIONALS

Statements expressed in the form "if p, then q" are called **conditionals,** or **implications,** and are denoted by $p \rightarrow q$. Such statements also can be read "p implies q." The "if" part of a conditional is called the **hypothesis** of the implication and the "then" part is called the **conclusion.**

Many types of statements can be put in "if-then" form; an example follows:

Statement: All first-graders are 6 years old.
If-then form: If a child is a first-grader, then the child is 6 years old.

An implication may also be thought of as a promise. Suppose Betty makes the promise, "If I get a raise, then I will take you to dinner." If Betty keeps her promise, the implication is true; if Betty breaks her promise, the implication is false. Consider the following four possibilities:

	p	q	
(1)	T	T	Betty gets the raise; she takes you to dinner.
(2)	T	F	Betty gets the raise; she does not take you to dinner.
(3)	F	T	Betty does not get the raise; she takes you to dinner.
(4)	F	F	Betty does not get the raise; she does not take you to dinner.

The only case in which Betty breaks her promise is when she gets her raise and fails to take you to dinner, case (2). If she does not get the raise, she can either take you to dinner or not without breaking her promise. The definition of implication is summarized in Table C.6. Observe that the only case for which the implication is false is when p is true and q is false.

Table C.6

p	q	Implication $p \to q$
T	T	T
T	F	F
F	T	T
F	F	T

An implication may be worded in several equivalent ways, as follows:

1. If the sun shines, then the swimming pool is open. (If p, then q.)
2. If the sun shines, the swimming pool is open. (If p, q.)
3. The swimming pool is open if the sun shines. (q if p.)
4. The sun shines implies the swimming pool is open. (p implies q.)
5. The sun is shining only if the pool is open. (p only if q.)
6. The sun's shining is a sufficient condition for the swimming pool to be open. (p is a sufficient condition for q.)
7. The swimming pool's being open is a necessary condition for the sun to be shining. (q is a necessary condition for p.)

Any implication $p \to q$ has three related implication statements, as follows:

Statement:	If p, then q.	$p \to q$
Converse:	If q, then p.	$q \to p$
Inverse:	If not p, then not q.	$\sim p \to \sim q$
Contrapositive:	If not q, then not p.	$\sim q \to \sim p$

■ EXAMPLE 5

Write the converse, the inverse, and the contrapositive for each of the following statements:

a. If $2x = 6$, then $x = 3$.
b. If I am in San Francisco, then I am in California.

Solution

a. *Converse:* If $x = 3$, then $2x = 6$.
Inverse: If $2x \neq 6$, then $x \neq 3$.
Contrapositive: If $x \neq 3$, then $2x \neq 6$.

b. *Converse:* If I am in California, then I am in San Francisco.
Inverse: If I am not in San Francisco, then I am not in California.
Contrapositive: If I am not in California, then I am not in San Francisco. ■

Table C.7 shows that an implication and its converse do not always have the same truth value. However, an implication and its contrapositive do always have the same truth value. Also, the converse and inverse of a conditional statement are logically equivalent.

Table C.7

p	q	$\sim p$	$\sim q$	Implication $p \to q$	Converse $q \to p$	Inverse $\sim p \to \sim q$	Contra-positive $\sim q \to \sim p$
T	T	F	F	T	T	T	T
T	F	F	T	F	T	T	F
F	T	T	F	T	F	F	T
F	F	T	T	T	T	T	T

Connecting a statement and its converse with the connective *and* gives $(p \to q) \wedge (q \to p)$. This compound statement can be written as $p \leftrightarrow q$ and usually is read "p if and only if q." The statement "p if and only if q" is a **biconditional.** A truth table for $p \leftrightarrow q$ is given in Table C.8. Observe that $p \leftrightarrow q$ is true if and only if both statements are true or both are false.

Table C.8

p	q	$p \to q$	$q \to p$	Biconditional $(p \to q) \wedge (q \to p)$ or $p \leftrightarrow q$
T	T	T	T	T
T	F	F	T	F
F	T	T	F	F
F	F	T	T	T

■ EXAMPLE 6

Given the following statements, classify each of the biconditionals as true or false:

p: $2 = 2$ r: $2 = 1$
q: $2 \neq 1$ s: $2 + 3 = 1 + 3$

a. $p \leftrightarrow q$ **b.** $p \leftrightarrow r$ **c.** $s \leftrightarrow q$ **d.** $r \leftrightarrow s$

Solution

a. $p \rightarrow q$ is true and $q \rightarrow p$ is true, so $p \leftrightarrow q$ is true.
b. $p \rightarrow r$ is false and $r \rightarrow p$ is true, so $p \leftrightarrow r$ is false.
c. $s \rightarrow q$ is true and $q \rightarrow s$ is false, so $s \leftrightarrow q$ is false.
d. $r \rightarrow s$ is true and $s \rightarrow r$ is true, so $r \leftrightarrow s$ is true. ■

In the previous section, we discussed analogies between the conjunction $p \wedge q$ and set intersection and between the disjunction $p \vee q$ and set union. Similar analogies exist for implication. Consider the implication "If a flower is a violet, then it is blue." The set of violets is a subset of the set of blue objects. In general, the implication $p \rightarrow q$ is analogous to $P \subseteq Q$. In fact, the definition of $P \subseteq Q$ tells us that $x \in P$ implies $x \in Q$. Thus for every property involving set inclusion, we should have a corresponding property involving implications.

Now consider the following statement:

It is raining or it is not raining.

This statement, which can be modeled as $p \vee (\sim p)$, is always true, as shown in Table C.9. A statement that is always true is called a **tautology.** One way to make a tautology is to take two logically equivalent statements such as $p \rightarrow q$ and $\sim q \rightarrow \sim p$ (from Table C.7) and form them into a biconditional as follows:

Table C.9

p	$\sim p$	$p \vee (\sim p)$
T	F	T
F	T	T

$$(p \rightarrow q) \leftrightarrow (\sim q \rightarrow \sim p)$$

Because $p \rightarrow q$ and $\sim q \rightarrow \sim p$ have the same truth values, then $(p \rightarrow q) \leftrightarrow (\sim q \rightarrow \sim p)$ is a tautology.

Valid Reasoning

In problem solving, the reasoning used is said to be **valid** if the conclusion follows unavoidably from the hypotheses. Consider the following example:

Hypotheses: All roses are red.
This flower is a rose.
Conclusion: Therefore this flower is red.

The statement "All roses are red" can be written as the implication "If a flower is a rose, then it is red" and pictured with the Venn diagram in Figure C.2a.

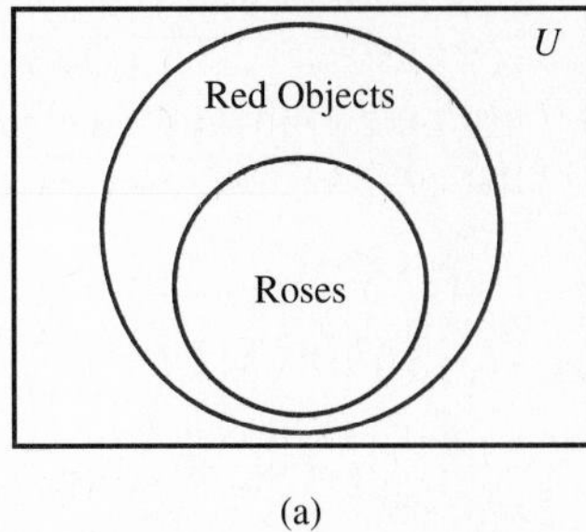

(a)

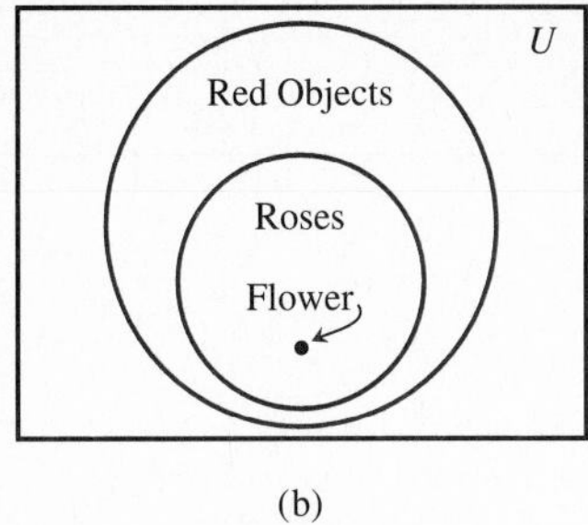

(b)

Figure C.2

The information "This flower is a rose" implies that this flower must belong to the circle containing roses, as pictured in Figure C.2b. This flower also must belong to the circle containing red objects. Thus the reasoning is valid because it is impossible to draw a picture satisfying the hypotheses and contradicting the conclusion.

Consider the following argument:

Hypotheses: All elementary school teachers are mathematically literate.
Some mathematically literate people are not children.

Conclusion: Therefore no elementary school teacher is a child.

Let E be the set of elementary school teachers, M be the set of mathematically literate people, and C be the set of children. Then the statement "All elementary school teachers are mathematically literate" can be pictured as in Figure C.3a. The statement "Some mathematically literate people are not children" can be pictured in several ways. Three of these are illustrated in Figure C.3b–d.

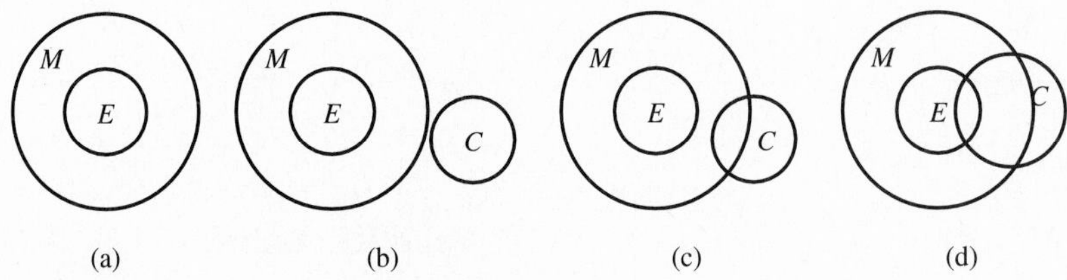

Figure C.3

According to Figure C.3d, it is possible that some elementary school teachers are children, and yet the given statements are satisfied. Therefore the conclusion that "No elementary school teacher is a child" does not follow from the given hypotheses. Hence, the reasoning is not valid.

If a single picture can be drawn to satisfy the hypotheses of an argument and contradict the conclusion, the argument is not valid. However, to show that an argument is valid, *all* possible pictures must be considered to show that there are no contradictions. There must be no way to satisfy the hypotheses and contradict the conclusion if the argument is valid.

EXAMPLE 7

Determine if the following argument is valid:

Hypotheses: In Washington, D.C., all senators wear power ties.
No one in Washington, D.C., over 6 ft tall wears a power tie.
Conclusion: Persons over 6 ft tall are not senators in Washington, D.C.

Solution

If S represents the set of senators and P represents the set of people who wear power ties, the first hypothesis is pictured as shown in Figure C.4a. If T represents the set of people in Washington, D.C., over 6 ft tall, the second hypothesis is pictured in Figure C.4b. Because people over 6 ft tall are outside the circle representing power tie wearers and senators are inside the circle P, the conclusion is valid and no person over 6 ft tall can be a senator in Washington, D.C.

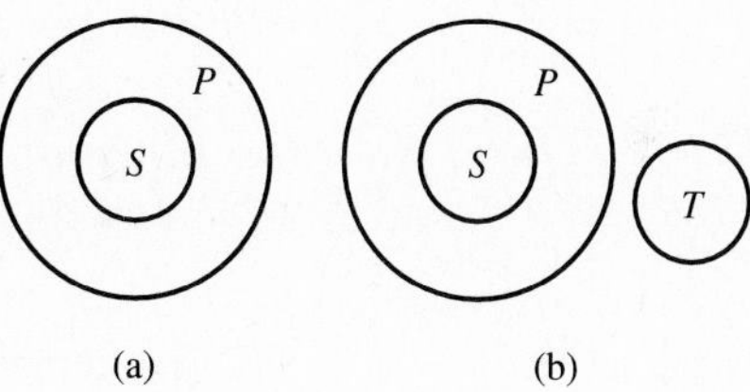

Figure C.4

A different method for determining if an argument is valid uses **direct reasoning** and a form of argument called the Law of Detachment (or **Modus Ponens**). For example, consider the following true statements:

If the sun is shining, then we shall take a trip.
The sun is shining.

Using these two statements, we can conclude that we shall take a trip. In general, the **Law of Detachment** is stated as follows:

If a statement is in the form "if p, then q" is true, and p is true, then q must also be true.

The Law of Detachment is sometimes described schematically as follows, where all statements above the horizontal line are true and the statement below the horizontal line is the conclusion.

$$\begin{array}{l} p \rightarrow q \\ \underline{p \qquad} \\ \quad\; q \end{array}$$

The Law of Detachment follows from the truth table for $p \rightarrow q$ given in Table C.6. The only case in which both p and $p \rightarrow q$ are true is when q is true (line 1 in the table).

■ EXAMPLE 8

Determine if each of the following arguments is valid:

a. Hypotheses: If you eat spinach, then you will be strong.
You eat spinach.
Conclusion: Therefore you will be strong.

b. Hypotheses: If Claude goes skiing, he will break his leg.
If Claude breaks his leg, he cannot enter the dance contest.
Claude goes skiing.
Conclusion: Therefore Claude cannot enter the dance contest.

Solution

a. Using the Law of Detachment, we see that the conclusion is valid.

b. By using the Law of Detachment twice, we see that the conclusion is valid. ■

A different type of reasoning, **indirect reasoning,** uses a form of argument called **Modus Tollens.** For example, consider the following true statements:

If Chicken Little had been hit by a jumping frog, he would have thought the earth was rising.
Chicken Little did not think the earth was rising.

What is the conclusion? The conclusion is that Chicken Little did not get hit by a jumping frog. This leads us to the general form of Modus Tollens:

If we have a conditional accepted as true, and we know the conclusion is false, then the hypothesis must be false.

Modus Tollens is sometimes schematically described as follows:

$$\begin{array}{l} p \rightarrow q \\ \underline{\sim q \qquad} \\ \quad\; \sim p \end{array}$$

The validity of Modus Tollens also follows from the truth table for $p \to q$ given in Table C.6. The only case in which both $p \to q$ is true and q is false is when p is false (line 4 in the table). The validity of Modus Tollens also can be established from the fact that an implication and its contrapositive are equivalent.

EXAMPLE 9

Determine conclusions for each of the following sets of true statements:

a. If an old woman lives in a shoe, then she does not know what to do. Mrs. Pumpkin Eater, an old woman, knows what to do.
b. If Jack is nimble, he will not get burned. Jack was burned.

Solution

a. Mrs. Pumpkin Eater does not live in a shoe.
b. Jack was not nimble. ■

The final reasoning argument to be considered here involves the **Chain Rule.** Consider the following statements:

If my wife works, I will retire early.
If I retire early, I will become lazy.

What is the conclusion? The conclusion is that if my wife works, I will become lazy. In general, the Chain Rule can be stated as follows:

If "if p, then q," and "if q, then r" are true, then "if p, then r" is true.

The Chain Rule is sometimes schematically described as follows:

$$\begin{array}{c} p \to q \\ \underline{q \to r} \\ p \to r \end{array}$$

Notice that the chain rule shows that implication is a transitive relation.

People often make invalid conclusions based on advertising or other information. Consider, for example, the statement "Healthy people eat Super-Bran cereal." Are the following conclusions valid?

If a person eats Super-Bran cereal, then the person is healthy.
If a person is not healthy, the person does not eat Super-Bran cereal.

If the original statement is denoted by $p \to q$, where p is "a person is healthy" and q is "a person eats Super-Bran cereal," then the first conclusion is the converse of $p \to q$—that is, $q \to p$—and the second conclusion is the inverse of $p \to q$—that is, $\sim p \to \sim q$. Table C.7 points out that neither the converse nor the inverse are logically equivalent to the original statement, and consequently the conclusions are not necessarily true.

EXAMPLE 10

Determine conclusions for each of the following sets of true statements.

a. If Alice follows the White Rabbit, she falls into a hole. If she falls into a hole, she goes to a tea party.

b. If Chicken Little is hit by an acorn, we think the sky is falling. If we think the sky is falling, we will go to a fallout shelter. If we go to a fallout shelter, we will stay there a month.

REMARK

Note that in Example 10, the Chain Rule can be extended to contain several implications.

Solution

a. If Alice follows the White Rabbit, she goes to a tea party.

b. If Chicken Little is hit by an acorn, we will stay in a fallout shelter for a month. ■

APPENDIX EXERCISES C.2

1. Write each of the following in symbolic form if p is the statement "It is raining" and q is the statement "The grass is wet."

a. If it is raining, then the grass is wet.
b. If it is not raining, then the grass is wet.
c. If it is raining, then the grass is not wet.
d. The grass is wet if it is raining.
e. The grass is not wet implies that it is not raining.
f. The grass is wet if, and only if, it is raining.

2. For each of the following implications, state the converse, inverse, and contrapositive.

a. If you eat Meaties, then you are good in sports.
b. If you do not like this book, then you do not like mathematics.
c. If you do not use Ultra Brush toothpaste, then you have cavities.
d. If you are good at logic, then your grades are high.

3. Construct a truth table for each of the following:

a. $p \to (p \vee q)$ **b.** $(p \wedge q) \to q$
c. $p \leftrightarrow \sim(\sim p)$ **d.** $\sim(p \to q)$

4. If p is true and q is false, find the truth values for each of the following:

a. $\sim p \to \sim q$ **b.** $\sim(p \to q)$
c. $(p \vee q) \to (p \wedge q)$ **d.** $p \to \sim p$
e. $(p \vee \sim p) \to p$ **f.** $(p \vee q) \leftrightarrow (p \wedge q)$

5. If p is false and q is false, find the truth values for each of the statements in Problem 4.

6. Can an implication and its converse both be false? Explain your answer.

7. Iris makes the true statement, "If it rains, then I am going to the movies." Does it follow logically that if it does not rain, then Iris does not go to the movies?

8. Consider the statement "If every digit of a number is 6, then the number is divisible by 3." Which of the following is logically equivalent to the statement?

a. If every digit of a number is not 6, then the number is not divisible by 3.
b. If a number is not divisible by 3, then some digit of the number is not 6.
c. If a number is divisible by 3, then every digit of the number is 6.

9. Write a statement logically equivalent to the statement "If a number is a multiple of 8, then it is a multiple of 4."

10. Use truth tables to prove that the following are tautologies:

a. $(p \to q) \to [(p \wedge r) \to q]$ Law of Added Hypothesis
b. $[(p \to q) \wedge p] \to q$ Law of Detachment
c. $[(p \to q) \wedge (\sim q)] \to \sim p$ Modus Tollens
d. $[(p \to q) \wedge (q \to r)] \to (p \to r)$ Chain Rule

11. a. If $p \to q$ is true but $q \to p$ is false, what is the analogous relation between sets?
b. Suppose that $p \to q$ and $q \to p$ are true. What is the analogous relation between sets?
c. Suppose that $A \subseteq B$ and $\overline{A} \subseteq \overline{B}$. What are the analogous statements in logic? What can you conclude about A and B?

12. a. Suppose that $p \to q$, $q \to r$, and $r \to s$ are all true, but s is false. What can you conclude about the truth value of p?

b. Suppose that $(p \wedge q) \to r$ is true, r is false, and q is true. What can you conclude about the truth value of p?

c. Suppose that $p \to q$ is true and $q \to p$ is false. Can q be true? Why or why not?

13. Translate the following statements into symbolic form. Give the meanings of the symbols that you use.

a. If Mary's little lamb follows her to school, then its appearance there will break the rules and Mary will be sent home.

b. If it is not the case that Jack is nimble and quick, then Jack will not make it over the candlestick.

c. If the apple had not hit Isaac Newton on the head, then the laws of gravity would not have been discovered.

14. Investigate the validity of each of the following arguments:

a. All women are mortal.
Hypatia was a woman.
Therefore Hypatia was mortal.

b. All squares are quadrilaterals.
All quadrilaterals are polygons.
Therefore all squares are polygons.

c. All teachers are intelligent.
Some teachers are rich.
Therefore some intelligent people are rich.

d. If a student is a freshman, then she takes mathematics.
Jane is a sophomore.
Therefore Jane does not take mathematics.

15. For each of the following, form a conclusion that follows logically from the given statements:

a. All college students are poor.
Helen is a college student.

b. Some freshmen like mathematics.
All people who like mathematics are intelligent.

c. If I study for the final, then I will pass the final.
If I pass the final, then I will pass the course.
If I pass the course, then I will look for a teaching job.

d. Every equilateral triangle is isosceles.
There exist triangles that are isosceles.

16. Write the following in if-then form:

a. Every figure that is a square is a rectangle.

b. All integers are rational numbers.

c. Figures with exactly three sides may be triangles.

d. It rains only if it is cloudy.

ANSWERS TO APPENDIX C EXERCISES

APPENDIX EXERCISES C.1

1. a. False statement **c.** False statement **e.** Not a statement **g.** True statement **i.** Not a statement

2. a. There exists $x = 3$ such that $x + 8 = 11$. **c.** There exists a natural number x such that $x^2 = 4$.

4. a. The book does not have 500 pages. **c.** $3 \cdot 5 \neq 15$
e. There exists a dog which does not have 4 legs.
g. There exists a square which is not a rectangle. **i.** There exists a natural number x such that $x + 3 \neq 3 + x$. **k.** There exists a counting number not divisible by itself and 1.
m. There exists a natural number x such that $5x + 4x \neq 9x$.

5. a.

$\sim p$	$\sim(\sim p)$
F	T
T	F

c. Yes

6. a. $q \wedge r$ **c.** $\sim(q \wedge r)$

7. a. F **c.** T **e.** F **g.** F **i.** F

9. a. No **c.** No

11.

$\sim p$	$\sim q$	$\sim p \vee q$
F	F	T
F	T	F
T	F	T
T	T	T

12. a. Today's not Wednesday or this is not the month of June.
c. It's not true that it's both raining and the month is July.

APPENDIX EXERCISES C.2

1. a. $p \to q$ **c.** $p \to \sim q$ **e.** $\sim q \to \sim p$

2. a. Converse: If you're good in sports, then you eat Meaties.

Inverse: If you don't eat Meaties, then you're not good in sports. Contrapositive: If you're not good in sports, then you don't eat Meaties.

c. Converse: If you have cavities, then you don't use Ultra Brush toothpaste. Inverse: If you use Ultra Brush toothpaste, then you don't have cavities. Contrapositive: If you don't have cavities, then you use Ultra Brush toothpaste.

3. a.

p	q	$p \vee q$	$p \rightarrow (p \vee q)$
T	T	T	T
T	F	T	T
F	T	T	T
F	F	F	T

c.

p	$\sim p$	$\sim(\sim p)$	$p \rightarrow \sim(\sim p)$
T	F	T	T
F	T	F	T

4. a. T **c.** F **e.** T

7. No

8. a. No **c.** No

9. If a number is not a multiple of 4, it is not a multiple of 8.

10. a.

p	q	r	$(p \rightarrow q)$	$(p \wedge r)$	$[(p \wedge r) \rightarrow q]$	$(p \rightarrow q) \rightarrow [(p \wedge r) \rightarrow q]$
T	T	T	T	T	T	T
T	T	F	T	F	T	T
T	F	T	F	T	F	T
T	F	F	F	F	T	T
F	T	T	T	F	T	T
F	T	F	T	F	T	T
F	F	T	T	F	T	T
F	F	F	T	F	T	T

c.

p	q	$\sim p$	$\sim q$	$p \rightarrow q$	$[(p \rightarrow q) \wedge (\sim q)]$	$[(p \rightarrow q) \wedge (\sim q)] \rightarrow (\sim p)$
T	T	F	F	T	F	T
T	F	F	T	F	F	T
F	T	T	F	T	F	T
F	F	T	T	T	T	T

11. a. If A represents p and B represents q, then $A \subset B$.
c. Let A represent p and B represent q. Then $A \subseteq B$ means $p \rightarrow q$, $\overline{A} \subseteq \overline{B}$ means $\sim p \rightarrow \sim q$, so $A = B$ and $p \leftrightarrow q$.

12. a. p is false **c.** Yes

13. b. If p = Jack is nimble and quick, and q = Jack makes it over the candlestick, then $\sim p \rightarrow \sim q$.

14. a. Valid **c.** Valid

15. a. Helen is poor. **c.** If I study for the final, then I will look for a teaching job.

16. a. If a figure is a square, then it is a rectangle.
c. If a figure has three sides, then it may be a triangle.

GLOSSARY

A

ABS Built-in absolute value function on computers and graphing calculators, p. 67, AA-8, AA-10.

absolute value of a complex number See *modulus.*

absolute value of a real number Denoted by $|a|$, represents the number a if $a \geq 0$, or the positive number $-a$ if $a < 0$, p. 7.

acute triangle A triangle in which all angles measure less than 90°, p. 524.

addition property of equality If $u = v$ and $w = z$, then $u + w = v + z$, p. 13.

addition property of inequality If $u < v$, then $u + w < v + w$, p. 46.

additive identity for the complex numbers $0 + 0i$ is the complex number zero, p. 264.

additive inverse of a real number The opposite of b, or $-b$, p. 11.

additive inverse of a complex number The opposite of $a + bi$, or $-a - bi$, p. 264.

algebraic expression A collection of variables and constants that are combined by addition, subtraction, multiplication, division, radicals and rational exponents, p. 11.

ambiguous case A triangle in which two sides and a nonincluded angle are known, p. 528.

amplitude See *sinusoid.*

angle Union of two rays with a common endpoint (the vertex). The beginning ray (the initial side) can be rotated about its endpoint to obtain the final position (the terminal side), p. 392.

angle of depression The acute angle formed by the line of sight (downward) and the horizontal, p. 468.

angle of elevation The acute angle formed by the line of sight (upward) and the horizontal, p. 468.

angular speed Speed of rotation, typically measured in radians or revolutions per unit time, p. 398.

annual percentage rate (APR) The annual interest rate, p. 339.

annual percentage yield (APY) The rate that would give the same return if interest were computed just once a year, p. 335.

annuity A sequence of equal periodic payments, p. 336.

arc length formula The length of an arc in a circle of radius r intercepted by a central angle of θ is $s = r\theta$, p. 397.

argument of a complex number The argument of $a + bi$ is the direction angle of the vector $\langle a, b\rangle$, p. 555.

arithmetic sequence (or arithmetic progression) A sequence $\{a_n\}$ in which $a_n = a_{n-1} + d$ for every integer $n \geq 2$. The number d is the common difference, p. 693.

arithmetic series A series in which the terms form an arithmetic sequence, p. 705.

associative properties $a + (b + c) = (a + b) + c$, $a(bc) = (ab)c$, p. 12.

asymptote A line or curve that a given curve approaches as $|x|$ and/or $|y|$ approaches ∞, p. 349, 351.

augmented matrix A matrix that represents a system of equations, p. 661.

axiom of induction If a set of positive integers contains 1 and the positive integer $k + 1$ whenever it contains the positive integer k, then it is the set of all positive integers, p. 720.

B

back-to-back stemplot A stemplot with leaves on either side used to compare two distributions, p. 759.

bar chart A rectangular graphical display of categorical data, p. 757.

base See *logarithmic function,* n*th power of* a.

bearing Measure of the clockwise angle that the line of travel makes with due north, p. 396.

binomial A polynomial with exactly two terms, p. 26.

binomial coefficients The numbers in Pascal's triangle: $\binom{n}{r} = \frac{n!}{r!(n-r)!}$, p. 729.

binomial probability In an experiment with two possible outcomes, the probability of one outcome occurring k times in n independent trials is $P(E) = \frac{n!}{k!(n-k)!}p^k(1-p)^{n-k}$, where p is the probability of the outcome occurring once, p. 751.

binomial theorem A theorem that gives an expansion formula for $(a+b)^n$, p. 731.

boundary The set of points on the "edge" of a region, p. 680.

bounded interval An interval that has finite length (does not extend to ∞ or $-\infty$), p. 5.

boxplot (or box-and-whisker plot) A graph that displays a five-number summary, p. 774.

branches The two separate curves that make up a hyperbola, p. 590.

break-even point The point at which revenue equals cost, p. 124.

C

cardioid A limaçon whose polar equation is $r = a + b \sin \theta$, or $r = a + b \cos \theta$, where $|a| = |b|$, p. 618.

Cartesian coordinate system An association between the points in a plane and ordered pairs of real numbers, p. 49.

categorical variable A variable (in statistics) that identifies each individual as belonging to a distinct class, p. 757.

center The central point in a circle, ellipse, or hyperbola, pp. 53, 587, 590.

central angle An angle whose vertex is the center of a circle, p. 394.

certain event An event that equals the sample space, p. 745.

circle A set of points in a plane equally distant from a fixed point called the center, p. 53.

circle graph A circular graphical display of categorical data, p. 757.

closed interval An interval that includes its endpoints, p. 5.

coefficient The real number multiplied by the variable(s) in a term, p. 26.

coefficient matrix A matrix whose elements are the coefficients in a system of linear equations, p. 661.

cofunction identity An identity that relates the sine, secant, or tangent to the cosine, cosecant, or cotangent, respectively, p. 506.

combination An arrangement of elements of a set, in which order is not important, p. 739.

combinations of *n* objects taken *r* at a time $_nC_r = \frac{n!}{r!(n-r)!}$, p. 740.

combinatorics A branch of mathematics related to determining the number of elements of a set or the number of ways objects can be arranged or combined, p. 735.

common difference See *arithmetic sequence.*

common logarithm A logarithm with base 10, p. 298.

common ratio See *geometric sequence.*

commutative properties $a + b = b + a$, $ab = ba$, p. 12.

complement of an event The set of outcomes in the sample space that are not in the event, p. 747.

complements or complementary angles Two angles of positive measure whose sum is 90°, p. 393.

completely factored polynomial A polynomial written in factored form with all prime factors, p. 30.

completing the square A method of adding a constant to an expression in order to form a perfect square, p. 146.

complex conjugates Complex numbers $a + bi$ and $a - bi$, p. 265.

complex fraction See *compound fraction.*

complex number An expression $a + bi$, where a (the real part) and b (the imaginary part) are real numbers, p. 263.

complex plane A coordinate plane used to represent the complex numbers. The x-axis of the complex plane is called the real axis and the y-axis is the imaginary axis, p. 267, 554.

component form of a vector If a vector's representative in standard position has a terminal point (a, b), then $\langle a, b \rangle$ is the component form of the vector, and a and b are the horizontal and vertical components of the vector, p. 545.

components of a vector See *component form of a vector.*

composite (number) An integer greater than 1 that is not prime, p. 14.

composition of functions $(f \circ g)(x) = f(g(x))$, p. 174.

compound fraction A fractional expression in which the numerator or denominator may contain fractions, p. 39.

compound interest Interest that becomes part of the investment, p. 330.

compounded annually See *compounded k times per year.*

compounded continuously Interest compounded using the formula $S = Pe^{rt}$, p. 334.

compounded *k* times per year Interest compounded using the formula $S = P\left(1 + \frac{r}{k}\right)^{kt}$ where $k = 1$ is compounded annually, where $k = 4$ is compounded quarterly, where $k = 12$ is compounded monthly, etc., p. 330.

compounded monthly See *compounded k times per year.*

cone See *double-napped right circular cone.*

conic section (or conic) A curve obtained by intersecting a double-napped right circular cone with a plane, p. 582.

constant A letter or symbol that stands for a specific number, p. 10.

constant function (on an interval) $f(x_1) = f(x_2)$ for any x_1 and x_2 (in the interval), p. 107.

constant term See *polynomial in x.*

continuous function A function that is continuous on its entire domain (see *continuous on an interval*), p. 224.

continuous on an interval A function f is continuous on an interval if, for all a and b in the interval, it is possible to trace the graph of the function between a and b without lifting the pencil from the paper, p. 224.

converge An infinite series converges to a number S, or has sum S, if the end behavior of the corresponding sequence of partial sums $\{S_n\}$ is the horizontal line $y = S$ (that is, $S_n \to S$ as $n \to \infty$), p. 709.

coordinate(s) of a point The number associated with a point on a number line, or the ordered pair associated with a point in the Cartesian coordinate plane, p. 3, 50.

coordinate plane See *Cartesian coordinate system.*

correlation coefficient A measure of the strength of the linear relationship between two variables, p. 99.

cosecant The function $y = \csc x$, p. 403.

cosine The function $y = \cos x$, p. 403.

cotangent The function $y = \cot x$, p. 403.

coterminal angles Two angles having the same initial side and the same terminal side, p. 392.

course See *bearing.*

cube root nth root, where $n = 3$ (see *principal* n*th root*), p. 20.

cubic A degree 3 polynomial, p. 224.

D

damping factor The factor Ae^{-at} in an equation such as $y = Ae^{-at} \cos bt$, p. 454.

data Facts collected for statistical purposes (singular form is *datum*), p. 95.

De Moivre's theorem $[r(\cos \theta + i \sin \theta)]^n = r^n(\cos n\theta + i \sin n\theta)$, p. 560.

decreasing function A function whose values $f(x)$ decreases as x increases, p. 107.

degree Unit of measurement (represented by the symbol °) for angles or arcs, equal to 1/360 of a complete revolution, p. 392.

degree of a polynomial (function) The largest exponent on the variable in any of the terms of the polynomial, p. 213.

degree of a term The whole number n in ax^n, p. 26.

demand curve $p = g(x)$, where x represents demand and p represents price, p. 645.

denominator The expression below the line in a fraction, p. 1.

dependent variable Variable representing the range value of a function (usually y), p. 68.

Descartes' rule of signs A rule for determining the possible number of positive and negative zeros of a polynomial function, p. 250.

determinant A number that is associated with a square matrix, p. 674.

difference identity An identity involving a trigonometric function of $u - v$, p. 504.

difference of complex numbers $(a + bi) - (c + di) = (a - c) + (b - d)i$, p. 263.

difference of functions $(f - g)(x) = f(x) - g(x)$, p. 172.

dimensional analysis A strategy that uses unit ratios to make conversions between measurement systems, p. 395.

directed angle See *polar coordinates,* p. 609.

directed arc-length For a central angle of radian measure θ in a circle of radius r, it is the number $r\theta$, p. 421.

directed distance See *polar coordinates.*

directed line segment The notation $\overrightarrow{PQ}$ denoting the directed line segment with initial point P and terminal point Q, p. 544.

direction angle of a vector The direction of one of a vector's representative directed line segments, p. 548.

direction of a directed line segment The angle the line segment makes with the horizontal, p. 544.

directional arrow A curved arrow indicating the path a rotating ray follows from the initial side to the terminal side of an angle, p. 392.

directrix of a parabola, ellipse, or hyperbola A line used to determine the conic, p. 583, 624.

discriminant For the function $f(x) = ax^2 + bx + c$, the expression $b^2 - 4ac$, p. 150.

disjoint events See *mutually exclusive events.*

distance (in a coordinate plane) The distance d between $P(x_1, y_1)$ and $Q(x_2, y_2)$, or $d = \sqrt{(x_1 - x_2)^2 + (y_1 - y_2)^2}$, p. 51.

distance (on a number line) The distance between real numbers a and b, or $|a - b|$, p. 8.

distribution A set of values for a statistical variable, p. 759.

distributive property $a(b + c) = ab + ac$ and related properties, p. 12.

diverge An infinite series diverges if it does not converge, p. 709.

division $\frac{a}{b} = a\left(\frac{1}{b}\right)$, $b \neq 0$, p. 11.

division algorithm for polynomials Given $f(x)$, $h(x) \neq 0$ there are unique polynomials $q(x)$ (quotient) and $r(x)$ (remainder) with $f(x) = q(x)h(x) + r(x)$ with either $r = 0$ or degree of $r <$ degree of h, p. 239.

divisor of a polynomial See *division algorithm for polynomials.*

divisor of an integer See *factor.*

domain of a function The set of all input values for a function, p. 67.

domain of an expression The set of numbers for which an expression is defined, p. 36.

dot product The number found when the corresponding components of two vectors are multiplied and then summed, p. 553.

double-angle identity An identity involving a trigonometric function of $2u$, p. 512.

double inequality A statement that describes a bounded interval, such as $3 \leq x < 5$, p. 47, 163.

double-napped right circular cone A surface generated by rotating a line about an axis that intersects the line (at the vertex of the cone), maintaining a constant angle between the rotating line and the axis, p. 582.

double root Root of multiplicity 2 (see *multiplicity*), p. 150.

E

eccentricity A positive number that specifies the shape of a conic, p. 624.

elementary row operations The following three row operations: Multiply all elements of a row by a nonzero constant; interchange two rows; and add a multiple of one row to another row, p. 663.

elements of a matrix See *matrix element.*

ellipse The set of all points in the plane such that the sum of the distances from a pair of fixed points (the foci), is a constant, p. 587.

elliptical surface or ellipsoid A surface generated by rotating an ellipse about its major axis, p. 589.

empirical probability Probability that has been determined by performing an experiment, p. 752.

end behavior The behavior of a graph of a function as $|x| \to \infty$, p. 226.

end behavior asymptote An asymptote that the function approaches as $|x| \to \infty$, p. 352, 366.

endpoint of an interval A real number that represents one "end" of an interval, p. 5.

equal complex numbers Complex numbers whose real parts are equal and whose imaginary parts are equal, p. 263.

equal fractions $\frac{a}{b} = \frac{c}{d}$ if and only if $ad = bc$ ($b \neq 0$, $c \neq 0$), p. 14.

equal functions f and g are equal if they have the same domain and $f(x) = g(x)$ for each x in the domain, p. 108.

equal matrices Matrices that have the same order and equal corresponding elements, p. 661.

equal vectors Vectors with the same magnitude and direction, p. 544.

equally likely outcomes Outcomes of an experiment that have the same probability of occurring, p. 745.

equation A statement of equality between two expressions, p. 13, 43.

equilibrium point A point where the supply curve and demand curve intersect. The corresponding price is the equilibrium price, p. 645.

equilibrium price See *equilibrium point.*

equivalent (algebraic) expressions Expressions that have the same domain and the same value for each number in the domain, p. 37.

equivalent directed line segments Directed line segments that have the same length and direction, p. 544.

equivalent equations (inequalities) Equations (inequalities) that have the same solutions, p. 44, 46.

even function A function whose graph is symmetric about the y-axis ($f(-x) = f(x)$ for all x in the domain of f), p. 113.

event A subset of a sample space, p. 744.

expanded form The right side of $u(v + w) = uv + uw$, p. 12.

expanded form of a series A series written explicitly as a sum of terms (not in summation notation), p. 712.

experiment Any process in which the individual results are uncertain, p. 743.

exponent See *nth power of* a, p. 15.

exponential decay Decay modeled by $y = k \cdot a^x$ $(0 < a < 1)$, $k > 0$, p. 290.

exponential form An equation written with exponents instead of logarithms, p. 298.

exponential function A function of the form $f(x) = k \cdot a^x$, where $k \neq 0$, $a > 0$, $a \neq 1$, p. 286.

exponential growth Growth modeled by $y = k \cdot a^x$ $(a > 1)$, $k > 0$, p. 290.

exponential regression A procedure for fitting an exponential function to a set of data, p. 291, 303.

extended principle of mathematical induction A principle related to the axiom of mathematical induction, p. 724.

extracting square roots A method for solving equations in the form $x^2 = k$, p. 145.

extraneous solution Any solution of the resulting equation that is not a solution of the original equation, p. 161, 376.

F

factor The numbers or expressions b and c in $a = bc$, p. 14.

Factor Theorem $x - c$ is a factor of a polynomial if and only if c is a zero of the polynomial, p. 242.

factored form The left side of $u(v + w) = uv + uw$, p. 12.

factoring (a polynomial) Writing a polynomial as a product of two or more polynomial factors, p. 30.

feasible points Points which satisfy the constraints in a linear programming problem, p. 682.

Fibonacci numbers The terms of the Fibonacci sequence, p. 693.

Fibonacci sequence The sequence 1, 1, 2, 3, 5, 8, 13, . . . , p. 692.

finite sequence A function whose domain is the first n positive integers for some fixed integer n, p. 689.

finite series Sum of a finite number of terms, p. 704.

first quartile See *quartile.*

fitting a line or curve to data Finding a line or curve that comes close to passing through all the points in a scatter plot, p. 98.

five-number summary The minimum, first quartile, median, third quartile, and maximum of a data set, p. 772.

foci, focus See *ellipse, hyperbola, parabola.*

fractional expression Quotient of two algebraic expressions, p. 36.

frequency Reciprocal of the period, p. 471.

frequency (in statistics) The number of individuals or observations with a certain characteristic, p. 760.

frequency table (in statistics) A table showing frequencies, p. 760.

function A relation that associates each value in the domain with exactly one value in the range, p. 67.

fundamental counting principle A principle used to find the number of ways a sequence of events can occur, p. 736.

Fundamental Theorem of Algebra A polynomial function of degree $n > 0$ has n complex zeros (counting multiplicity), p. 270.

Fundamental Theorem of Arithmetic Every integer greater than 1 has a unique prime factorization, p. 14.

future value of an annuity The net amount of money returned from an annuity, p. 338.

G

Gaussian elimination A method of solving a system of n linear equations in n unknowns, p. 651.

Gaussian plane See *complex plane.*

general form (of a line) $Ax + By + C = 0$, where A and B are not both zero, p. 85.

general second-degree equation $Ax^2 + Bxy + Cy^2 + Dx + Ey + F = 0$, where A, B, and C are not all zero, p. 601.

geometric sequence (or geometric progression) A sequence $\{a_n\}$ in which $a_n = a_{n-1} \cdot r$ for every positive integer $n \geq 2$. The nonzero number r is called the common ratio, p. 695.

geometric series A series whose terms form a geometric sequence, p. 707.

graph of a function *f* The set of all points in the coordinate plane corresponding to the pairs $(x, f(x))$ for x in the domain of f, p. 72.

graph of a polar equation The set of all points in the polar coordinate system corresponding to the ordered pairs (r, θ) that are solutions of the polar equation, p. 611.

graph of a relation The set of all points in the coordinate plane corresponding to the ordered pairs of the relation, p. 181.

graph of an equation in *x* and *y* The set of all points in the coordinate plane corresponding to the pairs (x, y) that are solutions of the equation, p. 57.

graph of an inequality in *x* and *y* The set of all points in the coordinate plane corresponding to the solutions (x, y) of the inequality, p. 680.

graph of parametric equations The set of all points in the coordinate plane corresponding to the ordered pairs determined by the parametric equations, p. 183.

grapher or graphing utility Graphing calculator or a computer with graphing software, p. 58.

H

half-angle identity An identity involving a trigonometric function of $\frac{u}{2}$, p. 514.

half-life The amount of time required for half of a radioactive substance to decay, p. 292.

Heron's formula The area of ΔABC with semiperimeter s is given by $\sqrt{s(s-a)(s-b)(s-c)}$, p. 538.

higher-degree polynomial function A polynomial function whose degree is ≥ 3, p. 224.

histogram A graph with rectangles of equal width that visually represents the information in a frequency table, p. 760.

horizontal asymptote An asymptote that is a horizontal line, p. 351.

horizontal component See *component form of a vector.*

horizontal line $y = b$, p. 85.

horizontal shrink or stretch See *shrink, stretch.*

horizontal translation A shift of a graph to the left or right, p. 189.

hyperbola A set of points in a plane, the absolute value of the difference of whose distances from two fixed points (the foci) is a constant, p. 589.

hypotenuse Side opposite the right angle in a right triangle, p. 403.

I

identity An equation that is always true throughout its domain, p. 406.

identity matrix A square matrix with 1's in the main diagonal and 0's elsewhere, p. 672.

identity properties $a + 0 = a$, $a \cdot 1 = a$, p. 12.

imaginary axis See *complex plane.*

imaginary part of a complex number See *complex number.*

imaginary unit The complex number $i = \sqrt{-1}$, p. 263.

impossible event An event that contains no outcomes, p. 745.

increasing function A function whose values $f(x)$ increase as x increases, p. 107.

independent events The occurrence of one event has no effect on the probability of the occurrence of the other event, p. 748.

independent variable Variable representing the domain value of a function (usually x), p. 68.

index See *radical.*

index of summation See *summation notation.*

individuals The objects described by a set of data, p. 757.

inequality A statement that compares two quantities using an inequality symbol, p. 3.

inequality symbol $<$, $>$, $\leq$, or $\geq$, p. 3.

infinite sequence A function whose domain is the set of all natural numbers, p. 689.

initial point See *directed line segment.*

initial side of an angle See *angle.*

integers The numbers $\ldots, -3, -2, -1, 0, 1, 2, 3, \ldots$, p. 1.

intercept Point where a line crosses the x- or y-axis in a graph, p. 84, 138.

intercepted arc Arc of a circle between the initial side and terminal side of a central angle, p. 394.

intermediate behavior The behavior of a graph of a function between and beyond the x-intercepts and/or vertical asymptotes, p. 359.

Intermediate Value Theorem If f is a polynomial function and $a < b$, then f assumes every value between $f(a)$ and $f(b)$, p. 230.

interquartile range The difference between the third quartile and the first quartile, p. 772.

interval Connected subset of the real number line with at least two points, p. 4.

interval notation Notation used to specify intervals, p. 5.

inverse cosine function The function $y = \cos^{-1} x$, p. 460.

inverse function The inverse relation of a one-to-one function, p. 12.

inverse of a matrix The inverse of matrix *A*, if it exists, is a matrix *B*, such that $AB = BA = I$, where *I* is an identity matrix, p. 672.

inverse properties $a + (-a) = 0$, $a \cdot \frac{1}{a} = 1$ $(a \neq 0)$, p. 12.

inverse relation (of the relation *R*) A relation that consists of all ordered pairs (b, a) for which (a, b) belongs to *R*, p. 200.

inverse rule of logarithms $a^{\log_a x} = x$ and $\log_a a^x = x$, p. 313.

inverse sine function The function $y = \sin^{-1} x$, p. 459.

inverse tangent function The function $y = \tan^{-1} x$, p. 460.

invertible matrix A square matrix with an inverse, p. 673.

irrational numbers Real numbers that are not rational, p. 2.

irrational zeros Zeros of a function that are irrational numbers, p. 254.

irreducible over the reals A polynomial that cannot be factored using real coefficients, p. 274.

K

***k*th term (of a series)** The *k*th expression that is added to form the series, p. 704.

L

law of cosines $a^2 = b^2 + c^2 - 2bc \cos \alpha$, $b^2 = a^2 + c^2 - 2ac \cos \beta$, $c^2 = a^2 + b^2 - 2ab \cos \gamma$, p. 534.

law of sines $\frac{\sin \alpha}{a} = \frac{\sin \beta}{b} = \frac{\sin \gamma}{c}$, p. 524.

leading coefficient See *polynomial in x.*

leading term See *polynomial in x.*

leaf The final digit of a number in a stemplot, p. 758.

least common denominator (LCD) The least common multiple of the denominators of fractions, p. 38.

least squares line See *linear regression line.*

length of a directed line segment The length or magnitude $\|\overrightarrow{PQ}\|$ of the directed line segment $\overrightarrow{PQ}$ is the distance between *P* and *Q*, p. 544.

length of a vector The length of one of a vector's representative directed line segments, p. 545.

like terms Terms in a polynomial that have the same variable part, p. 27.

limaçon A graph of a polar equation $r = a + b \sin \theta$ or $r = a + b \cos \theta$, p. 617.

limit If $f(x) \to a$ as $x \to b$, then *a* is the limit of $f(x)$ as *x* approaches *b* (where *a* and *b* may be real numbers or $\pm\infty$), p. 226.

line graph A graph of data in which consecutive data points are connected by line segments, p. 762.

line of symmetry A line over which a graph is the mirror image of itself, p. 113.

line of travel The path along which an object travels, p. 396.

linear combination of vectors u and v An expression $a\mathbf{u} + b\mathbf{v}$, where *a* and *b* are real numbers, p. 548.

linear correlation A scatter plot with points clustered along a line. Correlation is positive if the slope is positive and negative if the slope is negative, p. 96.

linear equation in *x* A equation that can be written in the form $ax + b = 0$, where *a* and *b* are real numbers and $a \neq 0$, p. 43.

linear equation in $x_1, x_2, \ldots, x_n$ An equation that can be written in the form $a_1x_1 + a_2x_2 + \cdots + a_nx_n = b$, where $a_1, a_2, \ldots, a_n$, and *b* are real numbers, p. 651.

Linear Factorization Theorem A polynomial $f(x)$ of degree $n > 0$ has the factorization $f(x) = a(x - z_1)(x - z_2) \cdots (x - z_n)$ where the z_i are the zeros of *f*, p. 270.

linear function A function that can be written in the form $f(x) = mx + b$, where $m \neq 0$ and *b* are real numbers, p. 79.

linear inequality in two variables *x* and *y* An inequality that can be written in one of the following forms: $y < mx + b$, $y \leq mx + b$, $y > mx + b$, or $y \geq mx + b$ with $m \neq 0$, p. 679.

linear inequality in *x* An inequality that can be written in the form $ax + b < 0$, $ax + b \leq 0$, $ax + b > 0$, or $ax + b \geq 0$, where *a* and *b* are real numbers and $a \neq 0$, p. 45.

linear programming A method of solving certain problems involving maximizing or minimizing a function of two variables with restrictions, p. 682.

linear regression A procedure for finding the straight line that is the best fit for the data, p. 303.

linear regression equation Equation of a linear regression line, p. 98.

linear regression line The line for which the sum of the squares of the residuals is the smallest possible, p. 98.

linear system A system of first-degree polynomial equations, p. 650.

local extremum A local maximum or a local minimum, p. 225.

local maximum A value $f(a)$ is a local maximum of *f* if there is an open interval (c, d) containing *a* such that $f(x) \leq f(a)$ for all values of *x* in (c, d), p. 109.

local minimum A value $f(b)$ is a local minimum of *f* if there is an open interval (c, d) containing *b* such that $f(x) \geq f(b)$ for all values of *x* in (c, d), p. 109.

logarithm An expression of the form $\log_a x$ (see *logarithmic function*), p. 297.

logarithmic form An equation written with logarithms instead of exponents, p. 298.

logarithmic function with base a The inverse of the exponential function $y = a^x$, denoted by $y = \log_a x$, p. 297.

logarithmic regression See *natural logarithmic regression.*

logistic curve The graph of the logistic growth function, p. 321.

logistic growth function A model of population growth:

$$P(t) = \frac{a}{1 + be^{-rt}},$$

p. 321.

lower bound for real zeros A number d is a lower bound for the set of real zeros of f if $f(x) \neq 0$ whenever $x < d$, p. 251.

lower bound test for real zeros A test for finding a lower bound for the real zeros of a polynomial, p. 253.

M

magnitude of a directed line segment See *length of a directed line segment.*

magnitude of a vector See *length of a vector.*

main diagonal The diagonal from the top left to the bottom right of a square matrix, p. 672.

major axis The line segment through the foci of an ellipse with endpoints on the ellipse, p. 587.

mathematical induction A process for proving that a statement is true for all natural numbers n by showing that it is true for $n = 1$ and that, if it is true for $n = k$, then it must be true for $n = k + 1$, p. 719.

matrix, $m \times n$ A rectangular array of m rows and n columns of real numbers, p. 661.

matrix element Any of the real numbers in a matrix, p. 661.

maximum r-value The value of $|r|$ at the point on the graph of a polar equation that has the maximum distance from the pole, p. 619.

mean (of a set of data) The sum of all the data divided by the total number of items, p. 768.

measure of an angle The number of degrees or radians in an angle, p. 392.

measure of center A measure of the typical, middle, or average value for a data set, p. 767.

measure of spread A measure that tells how widely distributed data are, p. 767.

median (of a data set) The middle number (or the mean of the two middle numbers) if the data are listed in order, p. 769.

midline of a sinusoid See *sinusoid.*

midpoint (in a coordinate plane) For the line segment with endpoints (a, b) and (c, d), $\left(\frac{a + c}{2}, \frac{b + d}{2}\right)$, p. 52.

midpoint (on a number line) For the line segment with endpoints a and b, $(a + b)/2$, p. 8.

minor axis of an ellipse The perpendicular bisector of the major axis of an ellipse with endpoints on the ellipse, p. 587.

minute Angle measure equal to $\frac{1}{60}$ of a degree, p. 393.

mode of a data set The category or number that occurs most frequently in the set, p. 769.

modulus The absolute value of the complex number $z = a + bi$ is given by $\sqrt{a^2 + b^2}$; also, the length of the segment from the origin to z in the complex plane, p. 554.

monomial A polynomial with exactly one term, p. 26.

multiple-angle equation An equation involving a trigonometric function of a multiple of the variable, p. 497.

multiplication property of equality If $u = v$ and $w = z$, then $uw = vz$, p. 13.

multiplication property of inequality If $u < v$ and $c > 0$, then $uc < vc$. If $u < v$ and $c < 0$, then $uc > vc$, p. 46.

multiplicative inverse of a matrix See *inverse of a matrix.*

multiplicative inverse of a real number The reciprocal of b, or $\frac{1}{b}$, $b \neq 0$, p. 11.

multiplicity The multiplicity of a zero c of a polynomial $f(x)$ of degree $n > 0$ is the number of times the factor $(x - c)$ occurs in the linear factorization $f(x) = a(x - z_1)(x - z_2) \cdots (x - z_n)$, p. 270.

mutually exclusive events Events from the same sample space that have no outcomes in common, p. 747.

N

natural logarithm A logarithm with base e, p. 299.

natural logarithmic function The inverse of the exponential function $y = e^x$, denoted by $y = \ln x$, p. 299.

natural logarithmic regression A procedure for fitting a curve $y = a + b \ln x$ to a set of data, p. 303.

natural numbers The numbers 1, 2, 3, . . . , p. 1.

nautical mile Length of 1 minute of arc along the earth's equator, p. 397.

negative angle Angle generated by clockwise rotation, p. 392.

negative linear correlation See *linear correlation.*

negative numbers Real numbers shown to the left of the origin on a number line, p. 3.

Newton's law of cooling $T(t) = T_m + (T_0 - T_m)e^{-kt}$, p. 322.

n factorial For any positive integer n, n factorial is $n! = n \cdot (n - 1) \cdot (n - 2) \cdot \cdots \cdot 3 \cdot 2 \cdot 1$; zero factorial is $0! = 1$, p. 729.

nonnegative real number A real number that is ≥ 0, p. 5.

nth power of a The number $a^n = a \cdot a \cdot \cdots \cdot a$ (with n factors of a), where n is the exponent and a is the base, p. 15.

nth root See *principal* n*th root.*

nth root of a complex number z A complex number v such that $v^n = z$, p. 562.

nth root of unity A complex number v such that $v^n = 1$, p. 561.

number line graph of a linear inequality The graph of the solutions of a linear inequality (in x) on a number line, p. 46.

numerator The expression above the line in a fraction, p. 1.

O

obtuse angle An angle whose measure is between 90° and 180°, p. 524.

obtuse triangle A triangle in which one angle is greater than 90°, p. 524.

odd–even identity An identity involving a trigonometric function of $-x$, p. 484.

odd function A function whose graph is symmetric about the origin ($f(-x) = -f(x)$ for all x in the domain of f), p. 113.

one-to-one function A function in which each element of the range corresponds to exactly one element in the domain, p. 203.

one-to-one rule of logarithms $x = y$ if and only if $\log_a x = \log_a y$, p. 313.

open interval An interval that does not include its endpoints, p. 6.

opens upward or downward A parabola $y = ax^2 + bx + c$ opens upward if $a > 0$ and opens downward if $a < 0$, p. 214.

opposite See *additive inverse of a real number* and *additive inverse of a complex number.*

order of an $m \times n$ matrix The order of an $m \times n$ matrix is $m \times n$, p. 661.

ordered pair A pair of real numbers (x, y), p. 50.

ordered set A set is ordered if it is possible to compare any two elements and say that one element is "less than" or "greater than" the other, p. 3.

ordinary annuity An annuity in which deposits are made at the same time interest is posted, p. 336.

origin The number zero on a number line, or the point where the x- and y-axes cross in the Cartesian coordinate system, p. 3, 49.

outcomes The various possible results of an experiment, p. 744.

outliers Data items far from the median, p. 770.

P

parabola The graph of a quadratic function, or the set of points in a plane that are equidistant from a fixed point (the focus) and a fixed line (the directrix), p. 214, 583.

parabolic surface or paraboloid A surface generated by rotating a parabola about its line of symmetry, p. 586.

parallel lines Two lines that are both vertical or have equal slopes, p. 87.

parallelogram law Rule for adding vectors geometrically, p. 546.

parameter The variable used as input in each parametric equation of a relation, p. 182.

parametric equations A method for defining a relation that employs a function for each x and y, p. 182.

parametrization A set of parametric equations for a curve, p. 573.

partial fraction decomposition See *partial fractions.*

partial fractions The process of expanding a fraction into a sum of fractions. The sum is called the partial fraction decomposition of the original fraction, p. 382.

partial sums See *sequence of partial sums.*

Pascal's triangle A number pattern in which row n (beginning with $n = 0$) consists of the coefficients of the expanded form of $(a + b)^n$, p. 728.

period See *periodic function.*

periodic function A function f for which there is a positive number c such that $f(t + c) = f(t)$ for every value t in the domain of f. The smallest such number c is the period of the function, p. 422.

permutation An arrangement of elements of a set, in which order is important, p. 738.

permutations of n objects taken r at a time ${}_nP_r = \dfrac{n!}{(n-r)!}$, p. 738.

perpendicular lines Two lines that are at right angles to each other, p. 87.

phase shift See *sinusoid.*

piecewise-defined function A function whose domain is divided into several parts and a different function rule is applied to each part, p. 111.

pie chart See *circle graph.*

point of symmetry A point about which a graph can be rotated 180° without affecting its appearance, p. 113.

point-slope form (of a line) $y - y_1 = m(x - x_1)$, p. 83.

polar axis See *polar coordinate system.*

polar coordinate system A coordinate system whose ordered pair is based on the directed distance from a central point (the pole) and the angle measured from a ray from the pole (the polar axis), p. 609.

polar coordinates The numbers (r, θ) that determine a point's location in a polar coordinate system. The number r is the directed distance and θ is the directed angle, p. 609.

polar distance formula $d = \sqrt{r_1^2 + r_2^2 - 2r_1r_2\cos(\theta_1 - \theta_2)}$, p. 615.

polar equation An equation in r and θ, p. 611.

pole See *polar coordinate system.*

polynomial factor of $p(x)$ A polynomial that can be multiplied by another polynomial to obtain $p(x)$, p. 30.

polynomial function A function in which $f(x)$ is a polynomial in x, p. 213.

polynomial in x An expression that can be written in the form $a_nx^n + a_{n-1}x^{n-1} + \cdots + a_1x + a_0$, where n is a nonnegative integer, the coefficients are real numbers, and $a_n \neq 0$. The degree of the polynomial is n, the leading coefficient is a_n, the leading term is a_nx^n, and the constant term is a_0. (The number 0 is the zero polynomial), p. 25.

polynomial regression A procedure for fitting a polynomial function to a set of data, p. 244.

positive angle Angle generated by a counterclockwise rotation, p. 392.

positive linear correlation See *linear correlation.*

positive numbers Real numbers shown to the right of the origin on a number line, p. 3.

power function A function of the form $f(x) = k \cdot x^a$, where k and a are nonzero constants, p. 286.

power-reducing identity An identity involving the square of a trigonometric function, p. 512.

power regression A procedure for fitting a curve $y = k \cdot x^a$ to a set of data, p. 303.

power rule of logarithms $\log_a r^c = c \log_a r, \quad r > 0$, p. 311.

present value of an annuity The net amount of your money put into an annuity, p. 338.

prime number An integer greater than 1 whose only natural number factors are itself and 1, p. 14.

prime polynomial A polynomial whose only polynomial factors (with integer coefficients) are itself, its opposite, and ± 1, p. 30.

principal *n*th root If $b^n = a$, then b is an nth root of a. If $b^n = a$ and a and b have the same sign, b is the principal nth root of a (see *radical*), p. 20.

principle of mathematical induction A principle related to the axiom of mathematical induction, p. 722.

probability of an event The number of elements in the event divided by the number of elements in the sample space, p. 745.

product of complex numbers $(a + bi)(c + di) = (ac - bd) + (ad + bc)i$, p. 264.

product of a scalar and a vector The product of scalar k and vector $\mathbf{u} = \langle u_1, u_2 \rangle$ is $k \cdot \mathbf{u} = k\langle u_1, u_2 \rangle = \langle ku_1, ku_2 \rangle$, p. 546.

product of functions $(fg)(x) = f(x)g(x)$, p. 172.

product of matrices *A* and *B* The matrix in which each entry is obtained by multiplying the entries of a row of A by the corresponding entries of a column of B and then adding, p. 669.

product rule of logarithms $\log_a rs = \log_a r + \log_a s,\ r > 0,\ s > 0$, p. 311.

projectile motion The movement of an object that is subject only to the force of gravity, p. 151.

Pythagorean identities $\sin^2 x + \cos^2 x = 1$, $\tan^2 x + 1 = \sec^2 x$, and $1 + \cot^2 x = \csc^2 x$, p. 406, 484.

Pythagorean theorem In a right triangle with sides a and b and hypotenuse c, $c^2 = a^2 + b^2$, p. 51.

Q

quadrant Any one of the four parts into which a plane is divided by the perpendicular coordinate axes, p. 50.

quadrantal angle An angle in standard position whose terminal side lies on an axis, p. 416.

quadratic equation in *x* An equation that can be written in the form

$$ax^2 + bx + c = 0 \qquad (a \neq 0),$$

p. 145.

quadratic formula The formula

$$x = \frac{-b \pm \sqrt{b^2 - 4ac}}{2a}$$

used to solve $ax^2 + bx + c = 0$, p. 148.

quadratic function A function that can be written in the form $f(x) = ax^2 + bx + c$, where a, b, and c are real numbers, and $a \neq 0$, p. 214.

quadratic regression A procedure for fitting a quadratic function to a set of data, p. 218.

quantitative variable A variable (in statistics) that takes on numerical values for a characteristic being measured, p. 757.

quartic A degree 4 polynomial, p. 224.

quartile The first quartile is the median of the lower half of a set of data, the second quartile is the median, and the third quartile is the median of the upper half of the data, p. 772.

quotient identities $\tan x = \sin x/\cos x$ and $\cot x = \cos x/\sin x$, p. 406.

quotient of complex numbers $\dfrac{a + bi}{c + di} = \dfrac{ac + bd}{c^2 + d^2} + \dfrac{bc - ad}{c^2 + d^2}i$, p. 266.

quotient of functions $(f/g)(x) = f(x)/g(x),\ g(x) \neq 0$, p. 172.

quotient rule of logarithms $\log_a(r / s) = \log_a r - \log_a s,\ r > 0,\ s > 0$, p. 311.

R

radian The measure of a central angle whose intercepted arc has a length equal to the circle's radius, p. 394.

radian measure The measure of an angle in radians, or, for a central angle, the ratio of the length of the intercepted arc to the radius of the circle, p. 394.

radical Any expression of the form $\sqrt[n]{a}$ denoting the principal nth root of a $(n > 1)$. The integer n is the index of the radical, and a is the radicand, p. 20.

radicand See *radical.*

radius The distance from a point on a circle to the center of the circle, p. 53.

range of a function The set of all output values corresponding to elements in the domain, p. 67.

range (in statistics) The difference between the greatest and least values in a data set, p. 772.

range screen A special menu of a graphing calculator that allows the user to set the viewing window, p. AA-5.

rational expression An expression that can be written as a ratio of two polynomials, p. 36.

rational function Function of the form $\dfrac{p(x)}{h(x)}$, where $p(x)$ and $h(x)$ are polynomials and $h(x)$ is not the zero polynomial, p. 347.

rational numbers Numbers that can be written as a/b, where a and b are integers, and $b \neq 0$, p. 1.

rational zeros Zeros of a function that are rational numbers, p. 254.

rational zeros test A test for finding the possible rational zeros of a polynomial, p. 254.

rationalizing the denominator Rewriting fractions so that the denominator is free of radicals, p. 22.

raw data Data that are collected (not generated by a formula), p. 95.

real axis See *complex plane.*

real number Any number that can be written as a decimal, p. 1.

real number line A horizontal line that represents the set of real numbers, p. 3.

real part of a complex number See *complex number.*

reciprocal function The function $f(x) = 1/x$, p. 362.

reciprocal identity An identity that equates a trigonometric function with the reciprocal of another trigonometric function, p. 406.

reciprocal of a real number See *multiplicative inverse of a real number.*

rectangular coordinate system See *Cartesian coordinate system.*

recursively defined sequence A sequence defined by giving the first term (or the first few terms) along with a procedure for finding the subsequent terms, p. 691.

reduced form of a fraction The numerator and denominator have no common prime factors, p. 14.

reduced row echelon form A matrix in row echelon form with every column that has a leading 1 having 0's in all other positions, p. 664.

reference angle See *reference triangle.*

reference triangle For an angle θ in standard position, the reference triangle is the triangle formed by the terminal side of angle θ, the x-axis, and a perpendicular dropped from a point on the terminal side to the x-axis. The angle in the reference triangle at the origin is the reference angle, p. 417.

reflection Two points that are symmetric with respect to a line or a point, p. 190.

reflexive property of equality $a = a$, p. 13.

regression model An equation found by regression and can be used to predict unknown values, p. 99.

relation A set of ordered pairs of real numbers, p. 181.

relative frequency Ratio of the number of times an event occurs to the total number of experiments performed, p. 752.

Remainder Theorem If a polynomial $f(x)$ is divided by $x - c$, the remainder is $f(c)$, p. 242.

repeated zeros Zeros of multiplicity ≥ 2 (see *multiplicity*), p. 270.

residual The difference $y_1 - (ax_1 + b)$, where (x_1, y_1) is a point in a scatter plot and $y = ax + b$ is a line that fits the set of data, p. 98.

resistant measure A statistical measure that does not change much in response to outliers, p. 770.

Richter scale A logarithmic scale used in measuring the intensity of an earthquake, p. 314.

right angle A 90° angle, p. 52.

right triangle A triangle with a 90° angle, p. 52.

rigid transformation A transformation that leaves the basic shape of a graph unchanged, p. 188.

root of a number See *principal* n*th root.*

root of an equation A solution, p. 138.

rose curve A graph of a polar equation $r = a \cos n\theta$ or $r = a \sin n\theta$, p. 616.

row echelon form A matrix in which rows consisting of all 0's occur only at the bottom of the matrix, the first nonzero entry in any row with nonzero entries is 1, and the leading 1's move to the right as we move down the rows, p. 662.

row operations See *elementary row operations.*

S

sample space Set of all possible outcomes of an equation, p. 744.

scalar A real number, p. 546.

scalar product See *dot product.*

scatter plot A plot of all the ordered pairs of a two-variable data set on a coordinate plane, p. 95.

scientific notation A positive number written as $c \times 10^m$, where $1 \leq c < 10$ and m is an integer, p. 17.

secant The function $y = \sec x$, p. 403.

second Angle measure equal to $\frac{1}{60}$ of a minute, p. 393.

second quartile See *quartile.*

semimajor axis Line segment with endpoints at the center of and on an ellipse, containing one of the foci, p. 605.

semiperimeter of a triangle One-half of the sum of the lengths of the sides of a triangle, p. 538.

sequence See *finite sequence, infinite sequence.*

sequence of partial sums The sequence $\{S_n\}$, where S_n is the nth partial sum of the series, that is, the sum of the first n terms of the series, p. 709.

series A finite or infinite sum of terms, p. 703.

shrink of factor c A transformation of a graph obtained by multiplying all the x-coordinates (horizontal shrink) or y-coordinates (vertical shrink) of the points by a constant c, $0 < c < 1$, p. 193.

similar figures Figures that have the same shape but not necessarily the same size, p. 402.

simple harmonic motion Motion described by $d = a \sin \omega t$ or $d = a \cos \omega t$, p. 471.

sine The function $y = \sin x$, p. 403.

sinusoid A function that can be written in the form $f(x) = a \sin [b(x - h)] + k$ or $f(x) = a \cos [b(x - h)] + k$. The number a

is the amplitude, the number h is the phase shift, and the line $y = k$ is the midline, p. 431.

slant asymptote An end behavior asymptote that is a line and is not horizontal or vertical, p. 368, 590.

slope Ratio $\frac{\text{change in } y}{\text{change in } x}$, p. 81.

slope-intercept form (of a line) $y = mx + b$, p. 84.

solution of a system in two variables An ordered pair of real numbers that satisfies all of the equations or inequalities in the system, p. 639.

solution of an equation, polar equation, or inequality A value of the variable (or values of the variables) for which the equation, polar equation, or inequality is true, p. 43, 46, 611, 679.

solve an equation or inequality To find all solutions of the equation or inequality, p. 43, 46.

solve a triangle To find one or more unknown sides or angles of a triangle, p. 408.

solve a system To find all solutions of a system, p. 639.

solve by elimination or substitution Methods for solving systems of linear equations, p. 641.

speed The magnitude of the velocity vector, given by distance/time, p. 549.

SQR or SQRT Built-in square root function on computers and graphing calculators, p. 72.

square matrix A matrix whose number of rows equals the number of columns, p. 661.

square root nth root, where $n = 2$ (see *principal* n*th root*), p. 20.

standard deviation A measure of how a data set is spread, p. 776.

standard form:

equation of a circle $(x - h)^2 + (y - k)^2 = r^2$, p. 53.

equation of an ellipse $\frac{(x - h)^2}{a^2} + \frac{(y - k)^2}{b^2} = 1$, p. 587, 596.

equation of a hyperbola $\frac{(x - h)^2}{a^2} - \frac{(y - k)^2}{b^2} = 1$ or $\frac{(y - k)^2}{b^2} - \frac{(x - h)^2}{a^2} = 1$, p. 590, 599.

equation of a parabola $y - k = a(x - h)^2$ or $x - h = a(y - k)^2$ $(a \neq 0)$, p. 584, 594.

of a polynomial $a_n x^n + a_{n-1}x^{n-1} + \cdots + a_1 x + a_0$, p. 26.

equation of a quadratic function $f(x) = ax^2 + bx + c$ $(a \neq 0)$, p. 214.

standard form of a complex number $a + bi$, where a and b are real numbers, p. 263.

standard form polar equation of a conic $r = \frac{ep}{1 \pm e\cos\theta}$ or $r = \frac{ep}{1 \pm e\sin\theta}$, p. 627.

standard position (angle) An angle positioned on a rectangular coordinate system with its vertex at the origin and its initial side on the positive x-axis, p. 392.

standard position (directed line segment) A directed line segment with its initial point at the origin, p. 545.

standard unit vectors $\mathbf{i} = \langle 1, 0\rangle$ and $\mathbf{j} = \langle 0, 1\rangle$, p. 548.

statistic A number that describes some characteristic of a data set, p. 767.

statute mile 5280 feet, p. 397.

stem The initial digit or digits of a number in a stemplot, p. 758.

stemplot (or stem-and-leaf plot) An arrangement of a numerical data set into a specific tabular format, p. 758.

stretch of factor c A transformation of a graph obtained by multiplying all the x-coordinates (horizontal stretch) or y-coordinates (vertical stretch) of the points by a constant $c > 1$, p. 193.

subtraction $a - b = a + (-b)$, p. 11.

sum identity An identity involving a trigonometric function of $u + v$, p. 504.

sum of a finite arithmetic series $S_n = n\left(\frac{a_1 + a_n}{2}\right) = \frac{n}{2}[2a + (n - 1)d]$, p. 705.

sum of a finite geometric series $S_n = \frac{a_1(1 - r^n)}{1 - r}$, p. 707.

sum of a finite series The sum of terms in the series, p. 704.

sum of an infinite geometric series $S_n = \frac{a_1}{1 - r}$, $|r| < 1$, p. 710.

sum of an infinite series See *converge.*

sum of complex numbers $(a + bi) + (c + di) = (a + c) + (b + d)i$, p. 263.

sum of functions $(f + g)(x) = f(x) + g(x)$, p. 172.

sum of two vectors The sum of $\mathbf{u} = \langle u_1, u_2\rangle$ and $\mathbf{v} = \langle v_1, v_2\rangle$ is $\mathbf{u} + \mathbf{v} = \langle u_1 + v_1, u_2 + v_2\rangle$, p. 546.

summation notation The series $\sum_{k=1}^{n} b_k$, where n is a natural number (or ∞), is in summation notation and is read "the sum of b_k from $k = 1$ to n (or infinity)." k is the index of summation and b_k is the kth term of the series, p. 711.

supplements or supplementary angles Two angles of positive measure whose sum is 180°, p. 393.

supply curve $p = f(x)$, where x represents production and p represents price, p. 645.

symmetric about a line A graph that is a reflection of itself across the line, p. 113.

symmetric about the origin A graph in which $(-x, -y)$ is on the graph whenever (x, y) is, p. 113.

symmetric about the pole A graph in which $(-r, \theta)$ or $(r, \pi + \theta)$ is on the graph whenever (r, θ) is, p. 621.

symmetric about the y-axis A graph in which $(-x, y)$ is on the graph whenever (x, y) is; or a graph in which $(-r, -\theta)$ or $(r, \pi - \theta)$ is on the graph whenever (r, θ) is, p. 113, 621.

symmetric across the polar axis A graph in which $(r, -\theta)$ or $(-r, \pi - \theta)$ is on the graph whenever (r, θ) is, p. 621.

symmetric property of equality If $a = b$, then $b = a$, p. 13.

symmetry The exact correspondence of one-half of a graph with the other half of the graph, p. 113.

synthetic division A procedure used to divide a polynomial by a linear factor, $x - a$, p. 240.

system A set of equations or inequalities, p. 639.

T

tangent The function $y = \tan x$, p. 403.

terminal point See *directed line segment.*

terminal side of an angle See *angle.*

term of a polynomial An expression of the form $a_n x^n$ in a polynomial, p. 26.

terms The parts of an expression separated by addition, p. 26.

terms of a sequence The range elements of a sequence, p. 689.

theoretical probability Probability that has been determined without performing an experiment, p. 752.

third quartile See *quartile.*

time plot A graph in which a quantitative variable is plotted against time, p. 762.

transformation A translation, reflection, stretch, shrink, or rotation of a graph, p. 188.

transitive property If $a = b$ and $b = c$, then $a = c$. Similar properties hold for the inequality symbols $<, \leq, >, \geq$, pp. 13, 46.

translation See *horizontal translation, vertical translation.*

transverse axis The line segment whose endpoints are the vertices of a hyperbola, p. 590.

triangular form A special form for a system of linear equations that facilitates finding the solution, p. 651.

triangular number A number that is a sum of the arithmetic series $1 + 2 + 3 + \cdots + n$ for some natural number n, p. 706.

trichotomy property For real numbers a and b, exactly one of the following is true: $a < b$, $a = b$, or $a > b$, p. 4.

trigonometric form of a complex number $r(\cos\theta + i\sin\theta)$, p. 555.

trinomial A polynomial with exactly three terms, p. 26.

U

unbounded interval An interval that extends to $-\infty$ or ∞ (or both), p. 6.

unit circle A circle with radius 1, p. 395.

unit ratio A fraction that equals 1, p. 395.

unit vector Vector of length 1, p. 547.

unit vector in the direction of a vector A unit vector that has the same direction as the given vector, p. 547.

upper bound for real zeros A number c is an upper bound for the set of real zeros of f if $f(x) \neq 0$ whenever $x > c$, p. 251.

upper bound test for real zeros A test for finding an upper bound for the real zeros of a polynomial, p. 251.

V

value of an annuity $S = R\dfrac{(1+i)^n - 1}{i}$, p. 337.

value of an investment $S = P(1 + r/k)^{kt}$ or $S = Pe^{rt}$, p. 334.

variable A letter that represents an unspecified number, p. 10.

variable (in statistics) A characteristic of individuals that is being identified or measured, p. 757.

variance The square of the standard deviation, p. 776.

vector (determined by a directed line segment) The set of all directed line segments equivalent to a particular line segment, p. 545.

velocity A vector that specifies the motion of an object in terms of its speed and direction, p. 549.

vertex of a cone See *double-napped right circular cone.*

vertex of a parabola The point of intersection of a parabola and its line of symmetry, p. 214.

vertex of an angle See *angle.*

vertical asymptote Any asymptote that is a vertical line, p. 349.

vertical component See *component form of a vector.*

vertical line $x = a$, p. 85.

vertical line test A test for determining whether a graph is a function, p. 72.

vertical stretch or shrink See *shrink, stretch.*

vertical translation A shift of a graph up or down, p. 189.

vertices of a hyperbola The points where a hyperbola intersects the line containing its foci, p. 590.

viewing window The rectangular portion of the coordinate plane specified by the dimensions [*X*min, *X*max] by [*Y*min, *Y*max], p. 58.

W

weighted average An average calculated in such a way that some elements of the data set have higher weights (that is, are counted more strongly in determining the average) than others, p. 771.

weights See *weighted average.*

whole numbers The numbers 0, 1, 2, 3, . . . , p. 1.

window dimensions The restrictions on x and y that specify a viewing window. See *viewing window.*

X

***x*-axis** Usually the horizontal coordinate line in a Cartesian coordinate system with positive direction to the right, p. 49.

x-coordinate The directed distance from the y-axis to a point in a plane, or the first number in an ordered pair, p. 50.

x-intercept A point that lies on both the graph and the x-axis, p. 138.

Xmax The x-value of the right side of the viewing window, p. 58.

Xmin The x-value of the left side of the viewing window, p. 58.

Xscl The scale of the tick marks on the x-axis in a viewing window, p. 58.

Y

y-axis Usually the vertical coordinate line in a Cartesian coordinate system with positive direction up, p. 49.

y-coordinate The directed distance from the x-axis to a point in a plane, or the second number in an ordered pair, p. 50.

y-intercept A point that lies on both the graph and the y-axis, p. 84.

Ymax The y-value of the top of the viewing window, p. 58.

Ymin The y-value of the bottom of the viewing window, p. 58.

Yscl The scale of the tick marks on the y-axis in a viewing window, p. 58.

Z

zero factor property If $ab = 0$, then either $a = 0$ or $b = 0$, p. 149.

zero factorial See *factorial.*

zero of a function A value in the domain that makes the function value zero, p. 138.

zero polynomial The constant 0, p. 26.

zero vector The vector $\langle 0, 0 \rangle$, p. 545.

zoom-in A procedure of a graphing utility used to magnify the graph (used, for example, to find solutions to a high degree of accuracy), p. 109, AA-11.

zoom out A procedure of a graphing utility used to view more of the coordinate plane (used, for example, to find the end behavior of a function), p. 227.

ANSWERS

PREREQUISITE CHAPTER

SECTION EXERCISES 1

1. -4.625 (terminating)
2. 1.5625 (terminating)
3. $0.\overline{15}$ (repeating)
4. $-0.\overline{45}$ (repeating)
5. $-2.1\overline{6}$ (repeating)
6. $4.08\overline{3}$ (repeating)
7. $0.\overline{135}$ (repeating)
8. $-0.\overline{185}$ (repeating)
9. $-4\ -3\ -2\ -1\ 0\ 1\ 2\ 3\ 4\ 5\ 6$; distance: 6.5
10. $-5\ -4\ -3\ -2\ -1\ 0\ 1\ 2\ 3\ 4\ 5$; distance: 4
11. $1\ \ 1.5\ \ 2\ \ 2.5\ \ 3$; distance: $\sqrt{7} - \sqrt{2} = 1.231\ldots$
12. $-6\ -5\ -4\ -3\ -2\ -1\ 0\ 1\ 2\ 3\ 4$; distance 4
13. $-2\ \ -1.75\ \ -1.5$; distance: $\frac{2}{15}$ or $0.1\overline{3}$
14. $0.70\ \ 0.75$; distance: $\frac{1}{28}$ or $0.03\overline{571428}$
15. $-\frac{11}{6} < -\frac{7}{6}$
16. $\frac{12}{3} = \frac{20}{5}$
17. $-4.1 > -4.2$
18. $3.2 < 3.3$
19. $-(-2) = 2$
20. $\frac{1}{6} > -1$
21. $-5\ -4\ -3\ -2\ -1\ 0\ 1\ 2\ 3\ 4\ 5$
22. $-3\ -2\ -1\ 0\ 1\ 2\ 3\ 4\ 5\ 6\ 7$
23. $-1\ 0\ 1\ 2\ 3\ 4\ 5\ 6\ 7\ 8\ 9$
24. $-5\ -4\ -3\ -2\ -1\ 0\ 1\ 2\ 3\ 4\ 5$
25. $-5\ -4\ -3\ -2\ -1\ 0\ 1\ 2\ 3\ 4\ 5$
26. $-1\ 0\ 1\ 2\ 3\ 4\ 5\ 6\ 7\ 8\ 9$
27. $-1 \le x < 1$
28. $-\infty < x \le 4$, or $x \le 4$
29. $-\infty < x < 5$, or $x < 5$
30. $-2 \le x < 2$
31. $-1 < x < 2$
32. $x \ge 5$
33. $(-3, \infty)$
34. $(-7, -2)$
35. $(-2, 1)$
36. $[-1, \infty)$
37. $(-3, 4]$
38. $(0, \infty)$
39. The real numbers greater than 4 and less than or equal to 9
40. The real numbers greater than or equal to -1, or the real numbers that are at least -1
41. The real numbers greater than or equal to -3, or the real numbers that are at least -3
42. The real numbers between -5 and 7, or . . . greater than -5 and less than 7
43. The real numbers greater than -1
44. The real numbers between -3 and 0 (inclusive), or . . . greater than or equal to -3 and less than or equal to 0
45. Endpoints -3 and 4; bounded
46. Endpoints -3 and -1; bounded
47. Endpoints 0 and 5; bounded
48. Endpoint 2; unbounded
49. Endpoint 5; unbounded
50. Endpoint -6; unbounded
51. $x \ge 29$; $[29, \infty)$
52. $0 \le x \le 2$; $[0, 2]$
53. $1.099 \le x \le 1.399$; $[1.099, 1.399]$
54. $0.02 \le x \le 0.065$; $[0.02, 0.065]$
55. 6
56. 0
57. 6
58. 12
59. -1
60. 3
61. $4 - \pi$
62. $5 - \pi$
63. $\sqrt{7} - \sqrt{2}$
64. $\frac{5}{2} - \sqrt{5}$, or $2.5 - \sqrt{5}$
65. $|-4| > -|4|$
66. $-3 = -|-3|$
67. $|-2| = |2|$
68. $-|-5| < |-5|$
69. $-|-6| < 0$
70. $|-1| > 0$

71. 19.9; 0.65

72. 54.1; 5.55

73. 12; −11

74. 55; 61.5

75. $\sqrt{17} - \sqrt{5}$, or about 1.887; $(\sqrt{17} + \sqrt{5})/2$, or about 3.18

76. $\sqrt{21} + \sqrt{7}$, or about 7.228; $(\sqrt{7} - \sqrt{21})/2$, or about −0.968

77. $|x - 4| = 3$

78. $|y - (-2)| \geq 4$, or $|y + 2| \geq 4$

79. $|z - (-3)| \leq 1$, or $|z + 3| \leq 1$

80. $|x - c| < d$

81. $|y - c| > d$

82. $|z - c| \leq d$

83. $|a^2| = |aa| = |a| \cdot |a|$ (property 3 of absolute values) $= |a|^2$

84. $0.\overline{1176470588235294}$. These digits can be found by long division.

85. 0, 1, 2, 3, 4, 5, 6

86. 1, 2, 3, 4, 5, 6

87. −6, −5, −4, −3, −2, −1, 0, 1, 2, 3, 4, 5, 6

88. All rational numbers p/q (p, q integers) in the interval $(-7, 7)$

89. The interval $(-7, 7)$

90. $-\frac{8}{3}, -\sqrt{7}, 0, \frac{3}{5}, \frac{2}{3}, \frac{7}{5}, \sqrt{2}$

SECTION EXERCISES 2

1. −10, −4

2. 5, −1

3. 1, 9.88

4. −3, 1.375

5. −8, −4

6. 2, 8

7. −2, 0

8. −0.4 (or $-\frac{2}{5}$), 0

9. Associative property of multiplication

10. Commutative property of multiplication

11. Addition inverse property

12. Addition identity property

13. Distributive property of multiplication over addition

14. Multiplication inverse property

15. Multiplication identity property—or distributive property of multiplication over addition, followed by the multiplication identity property. Note that we also use the multiplicative commutative property to say that $1 \cdot u = u \cdot 1 = u$.

16. Distributive property of multiplication over subtraction

17. Definition of subtraction; associative property of addition; definition of subtraction

18. Associative property of multiplication; multiplicative inverse; multiplicative identity

19. $ax^2 + ab$

20. $yc - z^2c$

21. $(a + d)x^2$

22. $a^3(z + w)$

23. $2 \cdot 2 \cdot 3 \cdot 3$

24. $2 \cdot 2 \cdot 5 \cdot 5$

25. $2 \cdot 2 \cdot 2 \cdot 7$

26. $2 \cdot 2 \cdot 2 \cdot 11$

27. $\frac{5}{3}$

28. $\frac{1}{4}$

29. $\frac{30}{77}$

30. $\frac{12}{35}$

31. $\frac{5}{6}$

32. $\frac{3}{2}$

33. $\frac{29}{36}$

34. $\frac{29}{100}$

35. $\frac{127}{196}$

36. $\frac{23}{225}$

37. $\frac{4}{35}$

38. $\frac{4}{15}$

39. $\frac{1}{10}$

40. $\frac{1}{14}$

41. $-\frac{7}{2}$

42. 7

43. $\pi - 6$

44. $4 - \sqrt{7}$

45. −5.7, or −5.70

46. −4.72

47. 1187.75

48. 20.65

49. 4.03

50. −0.93

51. 1123.46

52. 0.10

53. 0.00

54. 0.00

55. x^2/y^2

56. v^4

57. $3x^4y^2$

58. $16/x^4$

59. $\dfrac{1}{2x^3y^3}$

60. $\dfrac{1}{9x^4y^6}$

61. x^2y

62. $\dfrac{x^3y^3}{8}$

63. x^4y^4

64. $\dfrac{6}{ab^4}$

65. 9.21×10^9

66. 2×10^{-10}

67. 4.839×10^8

68. 3.68×10^9

69. 7×10^{-6}

70. -1.6×10^{-19}

71. 0.000 000 033 3

72. 673,000,000,000

73. 5,870,000,000,000

74. 5,000,000,000

75. 0.000 000 000 000 000 000 000 000 000 910 66 (27 zeros)

76. 0.000 000 000 000 000 000 000 001 674 7 (23 zeros)

77. 5.766×10^{12}

78. $0.16 = 1.6 \times 10^{-1}$

79. 2.6028×10^{-8}

80. 6.364×10^7

SECTION EXERCISES 3

1. 9 or -9

2. 3 or -3

3. 4

4. 3

5. $\frac{4}{3}$ or $-\frac{4}{3}$

6. $-\frac{3}{2}$

7. 12

8. Not a real number

9. -6

10. 6

11. $\frac{5}{2}$

12. $\frac{8}{5}$

13. 4

14. 5

15. 2.5

16. 3.5

17. -11.5

18. 22.5

19. 729

20. 32

21. 0.25

22. $\frac{1}{81}$, or 0.012345679

23. -2

24. -0.8

25. 1.3

26. 4.41

27. 2.1

28. 1.5

29. $12\sqrt{2}$

30. $5\sqrt[3]{4}$

31. $-5\sqrt[3]{2}$

32. $2\sqrt[4]{12}$

33. $xy^2\sqrt{2x}$ (Note: x cannot be negative.)

34. $-3xy^2$

35. $x^2|y|\sqrt[4]{3y^2}$

36. $2x^2y\sqrt[3]{y}$

37. $(2x^2)\sqrt[5]{3}$

38. $6x^2y^4\sqrt{3y}$

39. $2\sqrt[3]{4}$

40. $\dfrac{\sqrt{5}}{5}$

41. $\dfrac{\sqrt[5]{x^3}}{x}$

42. $\dfrac{2\sqrt[4]{y^3}}{y}$

43. $\dfrac{\sqrt[3]{x^2y^2}}{y}$

44. $\dfrac{\sqrt[5]{a^3b^3}}{b}$

45. $(a + 2b)^{2/3}$

46. $x^{2/5}y^{3/5}$

47. $2x^{5/3}y^{1/3}$

48. $x^{5/4}y^{7/4}$

49. $\sqrt[4]{a^3b}$

50. $\sqrt[3]{x^2y}$

51. $\sqrt[3]{x^{-5}} = \dfrac{1}{\sqrt[3]{x^5}}$

52. $\sqrt[4]{x^{-3}y^{-3}} = \dfrac{1}{\sqrt[4]{x^3y^3}}$

53. $\sqrt[4]{2x}$

54. $\sqrt[6]{3x^2}$

55. $\sqrt[8]{xy}$

56. $\sqrt[6]{ab}$

57. $\sqrt[15]{a}$

58. $\sqrt[6]{a^7}$

59. $a^{-17/30}$

60. $|x|\,y^2$

61. $3a^2b^2$ (b must be positive.)

62. $\dfrac{x^3}{y^4}$ (x must be positive.)

63. $4x^4y^2$

64. $\dfrac{q}{3|p|}$

65. $\dfrac{1}{|y|} \cdot \dfrac{|x|}{x}$

66. $\dfrac{6}{x^{1/6}y^{7/6}}$

67. $3y^2/|x^3|$

68. $4y^4/|z|$

69. $\sqrt[4]{\frac{3}{8}}\,|x|^{3/2}\,|y|^{1/2} = \sqrt[4]{\frac{3}{8}}\sqrt{|x^3y|}$

70. $\sqrt[5]{\frac{4}{9}}x^{3/5}y^{1/5} = \sqrt[5]{\frac{4}{9}x^3y}$

71. $\dfrac{2x\sqrt[3]{x}}{y}$

72. $3b\sqrt[5]{a^3} = 3ba^{3/5}$

73. 0

74. $2\sqrt{7}$

75. $(x - 2|y|)\sqrt{x}$

76. $(3|x| + y)\sqrt{2y}$

77. $\sqrt{2+6} < \sqrt{2} + \sqrt{6}$

78. $\sqrt{4} + \sqrt{9} > \sqrt{4+9}$

79. $(3^{-2})^{-1/2} = 3$

80. $(2^{-3})^{1/3} < 2$

81. $\sqrt[4]{(-2)^4} > -2$

82. $\sqrt[3]{(-2)^3} = -2$

83. $2^{2/3} < 3^{3/4}$

84. $4^{-2/3} < 3^{-3/4}$

85. 3.48 sec

86. Approximately 6.36 sec

87. If n is even, then there are two nth roots of a (when $a > 0$): $\sqrt[n]{a}$ and $-\sqrt[n]{a}$.

SECTION EXERCISES 4

1. Degree 2; 3 and 1

2. Degree 4; -2 and 3

3. Degree 3; -1 and 0

4. Degree 1; 4 and -2

5. Degree 5; 1 and -1

6. Degree 0; -9 and -9

7. $3x^2 + 2x - 1$; degree 2

8. $-2x^3 + x^2 - 2x + 1$; degree 3

9. $-x^7 + 1$; degree 7

10. $-x^4 + x^2 + x - 3$; degree 4

11. No (cannot have x^{-1})

12. No (cannot have division)

13. Yes

14. No (cannot have radical)

15. Yes

16. Yes

17. $3x^2 - 5x + 6$

18. $4x^2 + 2x + 4$

19. $-4x^2 - 7x - 17$

20. $3x^3 - x^2 - 9x + 3$

21. $4y^2 + y + 7$

22. $6y - 2$

23. $7z - 10$

24. $-3z^4 - 3z^2 - z + 1$

25. $2x^3 - 2x^2 + 6x$

26. $2y^4 + 3y^3 - 4y^2$

27. $-12u^2 + 3u$

28. $12v^4 - 8v$

29. $-15x^3 - 5x^2 + 10x$

30. $2x^5 - 2x^3 + 2x$

31. $x^2 + 8x + 15$

32. $x^2 + 3x - 10$

33. $8x^2 + 14x + 3$

34. $3x^2 + x - 10$

35. $x^2 - 36$

36. $4x^2 - 9$

37. $9x^2 - y^2$

38. $x^2 - 4y^2$

39. $x^2 + 4x + 4$

40. $25x^2 - 10x + 1$

41. $25x^2 - 30x + 9$

42. $9x^2 + 24xy + 16y^2$

43. $u^2 + v^2 - 2uv + 4u - 4v + 4$

44. $x^2 + y^2 - 2xy - 6x + 6y + 9$

45. $a^2 - 4a + 4 - b^2$

46. $a^2 - b^2 - 2b - 1$

47. $x^3 - 3x^2 + 3x - 1$

48. $x^3 + 6x^2 + 12x + 8$

49. $8u^3 - 12u^2v + 6uv^2 - v^3$

50. $u^3 + 9u^2v + 27uv^2 + 27v^3$

51. $x^4 - 4$

52. $4x^6 - 9y^2$

53. $25x^6 - 10x^3 + 1$

54. $x^6 - 4x^3 + 4$

55. $x^3 + 2x^2 - 5x + 12$

56. $x^3 - 11x + 6$

57. $x^4 + 2x^3 - x^2 - 2x - 3$

58. $2x^4 - 5x^3 + 8x^2 - 7x + 2$

59. $x^2 - 2$

60. $x - y$ (x, y both positive)

61. $u - v$ (u, v both positive)

62. $x^4 - 3$

63. $x^3 - 8$

64. $x^3 + 1$

65. 12

66. -12

67. $20\pi \approx 62.83$ in.^{3}

68. $525\pi \approx 1649.34$ cm^2

69. 400 ft

70. $2x^3 + \_x^2 + \_x - 3$. The blanks can be any number.

71. $-2x^2 + \_x + 5$ (The blank can be any number.)

72. The degree of the product is the sum of the degrees of the two factors.

SECTION EXERCISES 5

1. $5(x - 3)$

2. $3(y - 8)$

3. $2x(x^2 + 1)$

4. $5x(x^2 - 4)$

5. $yz(z^2 - 3z + 2)$

6. $xy(2y^2 + y - 3)$

7. $(x + 3)(2x - 5)$

8. $(y + 1)(y + 4)$

9. $(x + 9)(x - 9)$

10. $(z + 7)(z - 7)$

11. $(3y + 4)(3y - 4)$

12. $(8 + 5y)(8 - 5y)$

13. $(4 + x + 2)(4 - x - 2) = (6 + x)(2 - x)$

14. $(2x - 1 + 4)(2x - 1 - 4) = (2x + 3)(2x - 5)$

15. $(x - 3)^2$

16. $(y + 4)^2$

17. $(6y + 1)^2$

18. $(2z - 1)^2$

19. $(3z - 4)^2$

20. $(5x - 2)^2$

21. $(x - 3)(x^2 + 3x + 9)$

22. $(y - 2)(y^2 + 2y + 4)$

23. $(z + 4)(z^2 - 4z + 16)$

24. $(x + 5)(x^2 - 5x + 25)$

25. $(3y - 2)(9y^2 + 6y + 4)$

26. $(4z + 3)(16z^2 - 12z + 9)$

27. $(1 - x)(x^2 + x + 1)$

28. $(3 - y)(y^2 + 3y + 9)$

29. $(x + 2)(x + 7)$

30. $(x + 7)(x + 8)$

31. $(y - 5)(y - 6)$

32. $(y - 4)(y - 9)$

33. $(z - 4)(z + 5)$

34. $(z - 8)(z + 3)$

35. $(2t + 1)(3t + 1)$

36. $(3t - 1)(4t - 1)$

37. $(2u - 5)(7u + 1)$

38. $(2u - 5)(9u + 1)$

39. $(2v + 3)(5v + 4)$

40. $(3v - 4)(5v - 2)$

41. $(3x + 5)(4x - 3)$

42. $(2x - 5)(4x + 3)$

43. $(x - y)(2x - y)$

44. $(x + y)(3x + y)$

45. $(2x + 5y)(3x - 2y)$

46. $(3x + 7y)(5x - 2y)$

47. $(x + 1)(x^2 - 3)$

48. $(x - 4)(x^2 + 5)$

49. $(2x - 3)(x^2 + 1)$

50. $(3x + 1)(x^2 - 5)$

51. $(x^2 - 3)(x^4 + 1)$

52. $(x^2 + 2)(x^4 + 1)$

53. $(c + 3d)(2a - b)$

54. $(w + 4z)(3u - 2v)$

55. $x(x^2 + 1)$

56. $x^2(x + 1)$

57. $x(x + 2)(x - 2)$

58. $6(x + 3)(x - 3)$

59. $(3x + 2)^2$

60. $(5x - 4)^2$

61. $y(2y - 5)^2$

62. $2y(3y + 4)^2$

63. $4(x + 5)(x - 3)$

64. $2x(x - 1)(x - 7)$

65. $y(1 + y)(1 - y)$

66. $y(4 + y)(4 - y)$

67. $3x(x + 2)(x^2 - 2x + 4)$

68. $4(x + 3)(x^2 - 3x + 9)$

69. $2(y + 4)(1 - 3y)$

70. $y(y + 1)(5 - 2y)$

71. $z(1 - 2z)(4z^2 + 2z + 1)$

72. $2z(2 - z)(z^2 + 2z + 4)$

73. $2(5x + 1 + 3)(5x + 1 - 3) = 2(5x + 4)(5x - 2)$

74. $5(2x - 3 + 2)(2x - 3 - 2) = 5(2x - 1)(2x - 5)$

75. $xy(x + y)(x - y)$

76. $x^2y(y + x)(y - x)$

77. $x(3x - 2)(5x - 4)$

78. $2(2x + 5)(3x - 2)$

79. $(x + 5y)(3x - 2y)$

80. $(x - 2y)(2x + 7y)$

81. $(2a - b)(c + 2d)$

82. $(3a + 2b)(2c - d)$

83. $(x+2)(x+1)(x-1)$

84. $(x-3)(x+2)(x-2)$

85. $x(x+1)(x+3)(x-3)$

86. $x(x-4)(x+1)(x-1)$

87. $(2ac+bc)-(2ad+bd) = c(2a+b)-d(2a+b) = (2a+b)(c-d)$

88. 600 ft^2

89. 384 ft^3

SECTION EXERCISES 6

1. All real numbers

2. All real numbers

3. $x \geq 4$, or $[4, \infty)$

4. $x > -3$, or $(-3, \infty)$

5. $x \neq 0$ and $x \neq -3$

6. $x \neq -2$ and $x \neq 2$

7. $x \neq 1$ and $x \neq 2$

8. $x \neq 0$ and $x \neq 2$

9. $x \neq 0$

10. $x \neq -1$

11. $8x^2$

12. $6y^2$

13. x^2

14. $x^2 - 2x$

15. $(x+3)(x+4) = x^2 + 7x + 12$

16. $(x+3)(x+5) = x^2 + 8x + 15$

17. $x^3 + 2x^2$

18. $(x+3)(x^2+x-6) = x^3 + 4x^2 - 3x - 18$

19. The terms $(x-2)(x+7)$ cancel out during simplification; the restriction indicates that the values 2 and -7 were not valid in the original expression.

20. The terms $(x+1)(x-2)$ in the denominator cancel out during simplification; the restriction indicates that the values -1 and 2 were not valid in the original expression.

21. No factors were removed from the expression; we can see by inspection that 2/3 and 5 are not valid.

22. The variable x cancels out during simplification; the restriction indicates that 0 was not valid in the original expression.

23. The term $(x-3)$ ends up in the numerator of the simplified expression; the restriction reminds us that it began in the denominator, so 3 is not allowed.

24. When $a = b$ in the original, we get division by 0; this is not apparent in the simplified expression because we cancel a factor of $b - a$.

25. $\frac{6x^2}{5}$, $x \neq 0$

26. $\frac{25}{3y^2}$

27. $\frac{x^2}{x-2}$, $x \neq 0$

28. $\frac{y}{2}$, $y \neq -3$

29. $\frac{z+5}{z}$, $z \neq 5$

30. $-\frac{z}{z+3}$, $z \neq 3$

31. $\frac{x+3}{x-4}$, $x \neq -3$

32. $\frac{x+7}{x-5}$

33. $\frac{y+5}{y+3}$, $y \neq 6$

34. $\frac{y(y-3)}{y-7}$, $y \neq -7$

35. $\frac{4z^2+2z+1}{z+3}$, $z \neq \frac{1}{2}$

36. $\frac{2z}{z-3}$

37. $\frac{x^2-3}{x^2}$, $x \neq -2$

38. $\frac{y}{y^2-5}$, $y \neq -3$

39. $\frac{x+1}{3}$, $x \neq 1$

40. 1, $x \neq -3$

41. $-\frac{1}{x-3}$, $x \neq 1$ and $x \neq -3$

42. $12y$, $y \neq 0$, $x \neq 0$ and $x \neq \frac{1}{6}$

43. $\frac{2(x-1)}{x}$

44. $\frac{1}{y}$, $y \neq -2$ and $y \neq 2$

45. $\frac{1}{y}$, $y \neq -5$, $y \neq 5$, and $y \neq \frac{1}{2}$

46. $\frac{y(y+4)}{y-1}$, $y \neq -4$ and $y \neq -\frac{2}{3}$

47. $\frac{2}{x}$

48. $\frac{x^2}{2y^2}$, $x \neq 0$

49. $\frac{3(x-3)}{28}$, $x \neq 0$ and $y \neq 0$

50. $\frac{3}{8}$, $x \neq y$ and $y \neq 0$

51. $\frac{x}{4(x-3)}$, $x \neq 0$ and $y \neq 0$

52. $-2x$, $x \neq 0$, $y \neq 0$, $x \neq -y$ and $x \neq y$

53. $\frac{2x-2}{x+5}$

54. $\frac{x+4}{x-2}$

55. $\frac{x^2-7x-2}{(2x+1)(x-3)} = \frac{x^2-7x-2}{2x^2-5x-3}$

56. $\frac{4x^2-4x-1}{(x-1)(3x-4)} = \frac{4x^2-4x-1}{3x^2-7x+4}$

57. $-\frac{3}{(x-1)(x-4)} = -\frac{3}{x^2-5x+4}$, $x \neq -1$

58. $\dfrac{3}{(y-2)(y-5)} = \dfrac{3}{y^2 - 7y + 10}$, $y \neq -2$

59. $\dfrac{3x+1}{(x+1)(x-1)} = \dfrac{3x+1}{x^2-1}$, $x \neq 2$

60. $-\dfrac{2(x+4)}{(x+2)(x-2)} = -\dfrac{2(x+4)}{x^2-4}$, $x \neq 3$

61. $-\dfrac{1}{x-3} = \dfrac{1}{3-x}$, $x \neq 0$ and $x \neq -3$

62. $-\dfrac{2x+5}{(x+3)(x+2)} = -\dfrac{2x+5}{x^2+5x+6}$, $x \neq 2$

63. $\dfrac{x^2+xy+y^2}{x+y}$, $x \neq 0, y \neq 0$ and $x \neq y$

64. $-\dfrac{xy}{x-y} = \dfrac{xy}{y-x}$, $x \neq 0, y \neq 0$ and $x \neq -y$

65. $\dfrac{x+3}{x-3}$, $x \neq 4$ and $x \neq \frac{1}{2}$

66. $\dfrac{x-3}{x+5}$, $x \neq 3$ and $x \neq \frac{3}{2}$

67. $-\dfrac{2x+h}{x^2(x+h)^2}$, $h \neq 0$

68. $\dfrac{2}{(x+2)(x+h+2)}$, $h \neq 0$

69. $a+b$, $a \neq 0, b \neq 0$ and $a \neq b$

70. $\dfrac{1}{b-a}$, $a \neq 0, b \neq 0$ and $a \neq -b$

71. $\dfrac{1}{xy}$, $x \neq -y$

72. $\dfrac{x-y}{x+y}$, $x \neq y$

73. $\dfrac{x+y}{xy}$

74. $\dfrac{xy}{x+y}$, $x \neq 0$ and $y \neq 0$

75. $\dfrac{2-3x}{2\sqrt{1-x}}$

76. $\dfrac{1-2x^2}{\sqrt{1-x^2}}$

77. $\dfrac{6-x}{2(3-x)^{3/2}}$

78. $\dfrac{2}{(x^2+2)^{3/2}}$

SECTION EXERCISES 7

1. (a)

2. (a) and (c)

3. (a)

4. (c)

5. (b)

6. (c)

7. Yes

8. No

9. No

10. No

11. No

12. No

13. $x = 8$

14. $x = -4$

15. $x = 4$

16. $x = 6$

17. $x = 1$

18. $x = 2$

19. $y = -\frac{4}{5}$, or -0.8

20. $y = -8$

21. $x = \frac{7}{4}$, or 1.75

22. $x = \frac{6}{5} = 1.2$

23. $x = \frac{4}{3}$

24. $x = \frac{9}{4} = 2.25$

25. $z = \frac{8}{19}$

26. $z = \frac{11}{2} = 5.5$

27. $x = \frac{17}{10}$, or 1.7

28. $x = \frac{7}{2} = 3.5$

29. $x = \frac{31}{9}$

30. $x = -\frac{5}{7}$

31. (a)

32. (b) and (c)

33. (b) and (c)

34. (a), (b), and (c)

35. $x < 6$:

−1 0 1 2 3 4 5 6 7 8 9

36. $x > 2$:

−1 0 1 2 3 4 5 6 7 8 9

37. $x \geq -2$:

−6 −5 −4 −3 −2 −1 0 1 2 3 4

38. $x \leq -3$:

−8 −7 −6 −5 −4 −3 −2 −1 0 1 2

39. $-4 \leq x < 3$:

−6 −5 −4 −3 −2 −1 0 1 2 3 4

40. $\frac{1}{3} \leq x < 3$:

−1 0 1 2 3 4

41. $x \geq 3$:

−3 −2 −1 0 1 2 3 4 5 6 7

42. $x < 5$:

−2 −1 0 1 2 3 4 5 6 7 8

43. $x \leq -\frac{19}{5}$

44. $x > -1$

45. $-\frac{1}{2} \leq y \leq \frac{17}{2}$

46. $-1 < y < \frac{5}{3}$

47. $-\frac{5}{2} \leq z < \frac{3}{2}$

48. $-1 < x < \frac{1}{5}$

49. $x > \frac{21}{5}$

50. $x < -\frac{11}{7}$

51. $y < \frac{7}{6}$

52. $y \geq \frac{27}{2}$

53. $x \leq \frac{34}{7}$

54. $x < \frac{33}{13}$

55. Multiply both sides by 2.

56. Divide both sides by 2.

57. No

58. Yes

59. Yes

60. No

61. $W = \frac{1}{2}P - L$

62. $b_1 = \dfrac{2A}{h} - b_2$

SECTION EXERCISES 8

1.

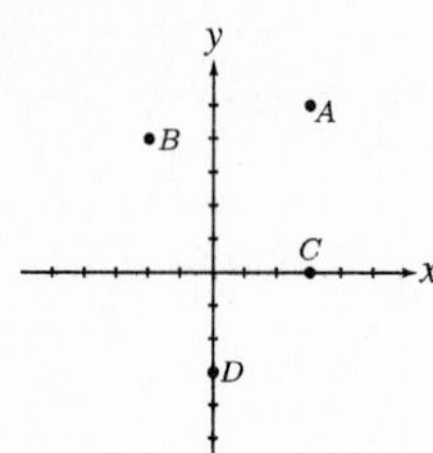

2.

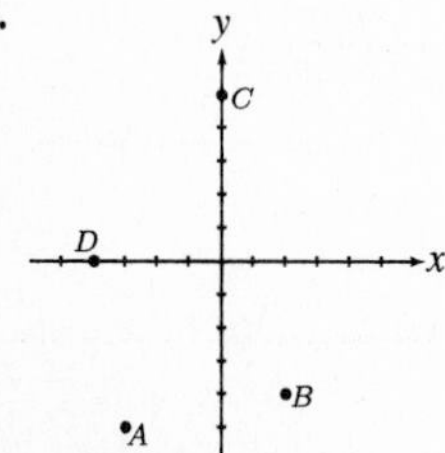

3. $A(1, 0)$; $B(2, 4)$; $C(-3, -2)$; $D(0, -2)$

4. $A(0, 3)$; $B(-3, 1)$; $C(-2, 0)$; $D(4, -1)$

5. An isosceles triangle:

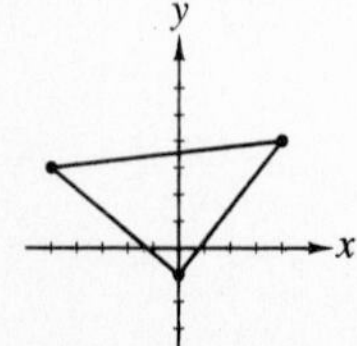

6. A square:

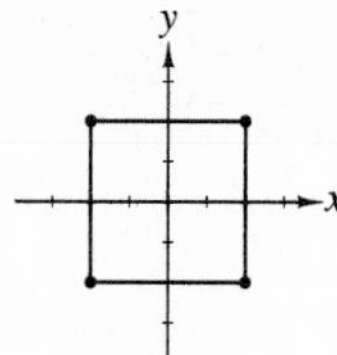

7. A parallelogram:

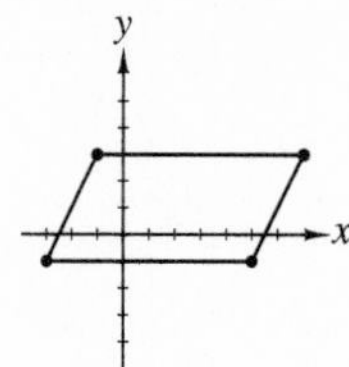

8. A rectangle:

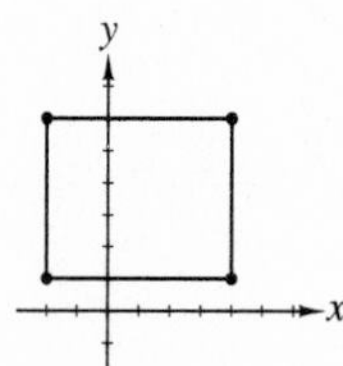

9. **(a)** First quadrant. **(b)** On the y-axis, between quadrants 1 and 2. **(c)** Second quadrant. **(d)** Third quadrant.

10. **(a)** First quadrant. **(b)** On the x-axis, between quadrants 2 and 3. **(c)** Third quadrant. **(d)** Third quadrant.

11. 8

12. 9

13. 5

14. 5

15. $\sqrt{10} = 3.162\cdots$

16. $\sqrt{29} = 5.385\cdots$

17. $(2, 6)$

18. $(\frac{1}{2}, -2)$

19. $\left(\dfrac{9}{2}, \dfrac{2 + \sqrt{2}}{2}\right)$

20. $\left(\dfrac{3 - \sqrt{3}}{2}, \dfrac{\sqrt{3} - 6}{2}\right)$

21. $(-\frac{1}{3}, \frac{3}{2})$

22. $(\frac{19}{24}, \frac{29}{20})$

23. $\left(\dfrac{a + 2}{2}, \dfrac{b - 3}{2}\right)$

24. $\left(\dfrac{a - 1}{2}, \dfrac{b - 4}{2}\right)$

25. The three side lengths are $\sqrt{125} = 5\sqrt{5}$, $\sqrt{20} = 2\sqrt{5}$, and $\sqrt{145}$; the sum of the squares of the two shorter side lengths equals the square of the long side length.

26. All sides are 4 units long.

27. The side lengths are 5, 5, and $\sqrt{50} = 5\sqrt{2}$.

28. The side lengths are $\sqrt{11.25} = \frac{3}{2}\sqrt{5}$, $\sqrt{80} = 4\sqrt{5}$, $\frac{3}{2}\sqrt{5}$, and $4\sqrt{5}$. The diagonals are both $\sqrt{91.25} = \frac{1}{2}\sqrt{365}$ units long.

29. All sides are $5\sqrt{2}$ units long; both diagonals are 10 units long.

30. The side lengths are $2\sqrt{5}$, $6\sqrt{2}$, $2\sqrt{5}$, and $6\sqrt{2}$. The diagonals are different lengths: $2\sqrt{41}$ and $2\sqrt{5}$.

31. $(x - 1)^2 + (y - 2)^2 = 25$

32. $(x + 3)^2 + (y - 2)^2 = 1$

33. $x^2 + y^2 = 4$

34. $x^2 + y^2 = 3$

35. $(x + 1)^2 + (y + 4)^2 = 9$

36. $(x - 5)^2 + (y + 3)^2 = 16$

37. Center $(3, 1)$, radius 6

38. Center $(-4, 2)$, radius 11

39. Center $(-5, -4)$, radius 3

40. Center $(2, -6)$, radius 5

41. Center $(0, 0)$, radius 1

42. Center $(0, 0)$, radius $\sqrt{5}$

43. Shown is the window $X\text{min} = 1986$,

Xmax = 1994, Xscl = 1; Ymin = 18, Ymax = 41, Yscl = 5.

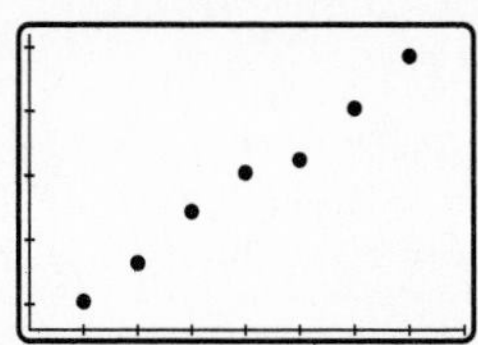

44. The side lengths are $2\sqrt{5}$, $2\sqrt{10}$, and $2\sqrt{5}$.

45. The midpoint of the hypotenuse is (2.5, 3.5); it is $\sqrt{18.5}$ units from all three vertices.

46. If the legs have lengths a and b and the hypotenuse is c units long, then without loss of generality, we can assume that the vertices are (0, 0), (a, 0), and (0, b). Then the midpoint of the hypotenuse is $\left(\frac{a}{2}, \frac{b}{2}\right)$, which is $\sqrt{\frac{a^2}{4} + \frac{b^2}{4}} = \frac{1}{2}c$ from all vertices.

47. $10 + 5\sqrt{2} = 17.071 \cdots$

48. $2(2\sqrt{5} + 6\sqrt{2}) = 2(12.957 \cdots) = 25.914 \cdots$

49. 25

50. 50

CHAPTER 1

QUICK REVIEW 1.1

1. −32.768

2. 32

3. 11

4. 49

5. 69

6. 147

7. 392

8. 9

9. $3x^4 + 7x^2 - 6$

10. $4st + 4s$

11. $\frac{4rt - 3r}{t^2 - t} = \frac{r(4t - 3)}{t(t - 1)}$

12. $-\frac{x - 8}{x^2 - x - 2} = -\frac{x - 8}{(x - 2)(x + 1)}$

13. $x = 2.7$

14. $x = 11$

SECTION EXERCISES 1.1

1. (b)

2. (b)

3. (b)

4. (a)

5.

x	−4	−1	1	4	9
y	−2	1	3	6	11

6.

x	1	$-\frac{1}{2}$	3	$-\frac{13}{2}$	5
y	1	4	−3	16	−7

7. (b)

8. (c)

9.

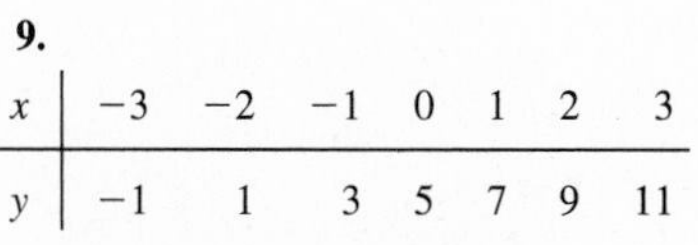

x	−3	−2	−1	0	1	2	3
y	−1	1	3	5	7	9	11

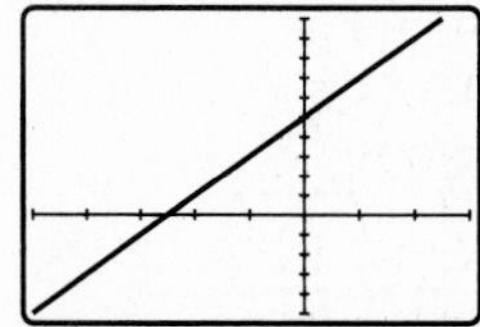

Shown in [−5, 3] × [−5, 10].

10.

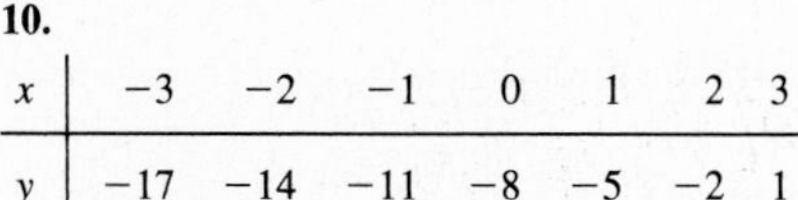

x	−3	−2	−1	0	1	2	3
y	−17	−14	−11	−8	−5	−2	1

Shown in [−3, 5] × [−10, 30].

11.

x	−3	−2	−1	0	1	2	3
y	2	−3	−6	−7	−6	−3	2

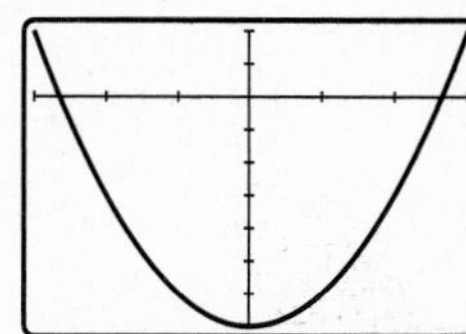

Shown in [−3, 3] × [−7, 2].

12.

x	−3	−2	−1	0	1	2	3
y	31	16	5	−2	−5	−4	1

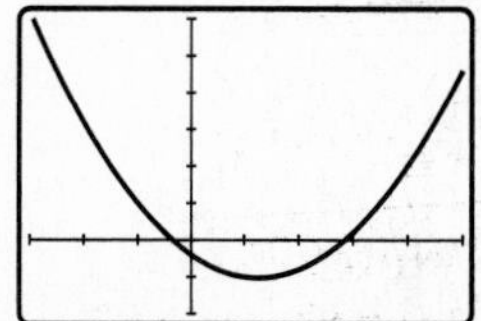

Shown in [−3, 5] × [−10, 30].

13.

x	−3	−2	−1	0	1	2	3
y	−31	−18	−7	2	9	14	17

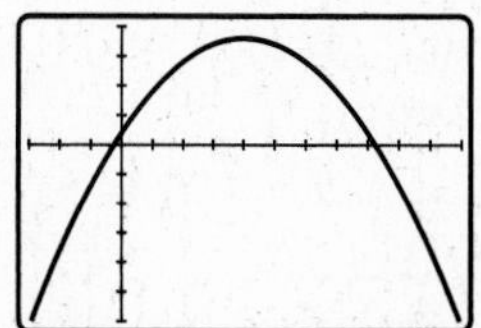

Shown in [−3, 11] × [−30, 20].

14.

x	−3	−2	−1	0	1	2	3
y	50	16	0	−4	−2	0	−4

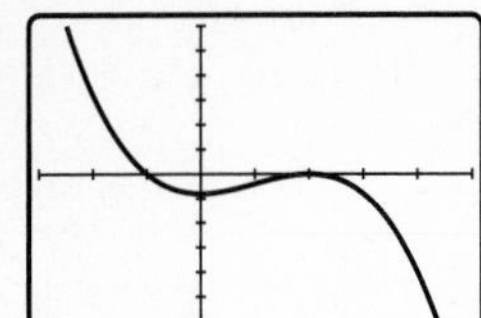

Shown in $[-3, 5] \times [-30, 30]$.

15.

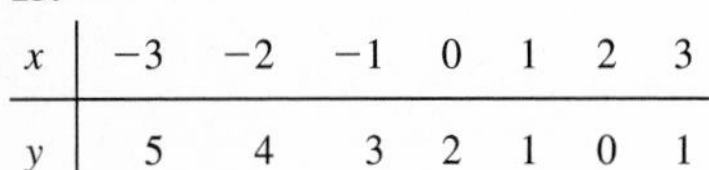

x	-3	-2	-1	0	1	2	3
y	5	4	3	2	1	0	1

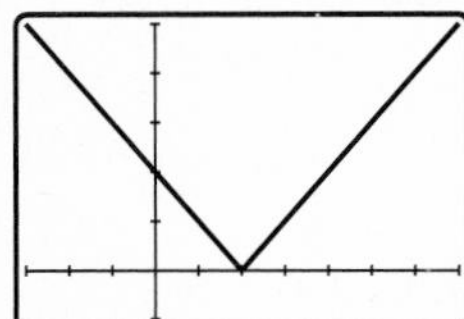

Shown in $[-3, 7] \times [-1, 5]$.

16.

x	-3	-2	-1	0	1	2	3
y	-2	-3	-2	-1	0	1	2

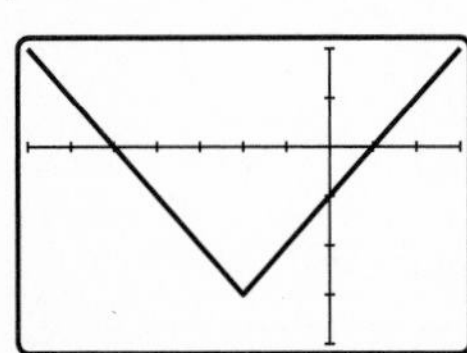

Shown in $[-7, 3] \times [-4, 2]$.

17.

x	-3	-2	-1	0	1	2	3
y	5	0	3	4	3	0	5

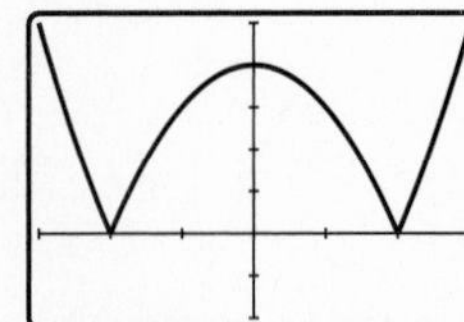

Shown in $[-3, 3] \times [-2, 5]$.

18.

x	-3	-2	-1	0	1	2	3
y	ERR	0	1	1.4142	1.7321	2	2.2361

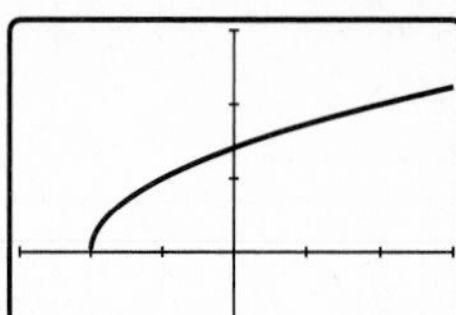

Shown in $[-3, 3] \times [-1, 3]$.

19. The graph in part **(b)**; Xmin $= -5$, Xmax $= 5$, Xscl $= 1$; Ymin $= -5$, Ymax $= 5$, Yscl $= 0.5$

20. The graph in part **(a)**; Xmin $= -5$, Xmax $= 5$, Xscl $= 1$; Ymin $= -10$, Ymax $= 10$, Yscl $= 1$

21. The graph in part **(d)**; Xmin $= -5$, Xmax $= 5$, Xscl $= 1$; Ymin $= -5$, Ymax $= 20$, Yscl $= 5$

22. The graph in part **(c)**; Xmin $= -8$, Xmax $= 3$, Xscl $= 1$; Ymin $= -10$, Ymax $= 10$, Yscl $= 1$

23. $y = (7 - 2x)/3 = 7/3 - 2x/3$. The last part could also be "2/3 * x" or "(2/3)x," but many calculators will not give the correct graph if "2/3x" is used.

24. $y = \dfrac{3x - 18}{4} = \dfrac{3x}{4} - \dfrac{9}{2}$

25. $y = \dfrac{4x^2 - 6x + 1}{2} = 2x^2 - 3x + \dfrac{1}{2}$

26. $y = \dfrac{x^3 - 6x - 12}{3} = \dfrac{x^3}{3} - 2x - 4$

27. (c)

28. (d)

29. (b)

30. (a)

31. $y = 2x(x + 4) + 2x(x - 2) + 2(x + 4)(x - 2) = 6x^2 + 8x - 16$

32. $y = 2x(x + 4) + 4x(x - 2) + 4(x + 4)(x - 2) = 10x^2 + 8x - 32$

33. The third window is best.

34. The third window is best.

35. **(a)** One possibility is Xmin $= 0$, Xmax $= 30$, Xscl $= 10$; Ymin $= 0$, Ymax $= 150000$, Yscl $= 10000$; **(b)** after 14 years.

36. **(a)** One possibility is Xmin $= 0$, Xmax $= 5$, Xscl $= 1$; Ymin $= 0$, Ymax $= 35{,}000$, Yscl $= 5000$. **(b)** After 3 years.

37. **(a)** $y = 0.75x - 15.75$, where $x \geq 0$. **(b)** They cannot earn *exactly* \$8.50 profit; to earn *at least* that much, they should sell 33 tickets (and earn \$9.00). **(c)** \$33.75

38. **(a)** $y = 200 + 180x$, where $0 \leq x \leq 48$. **(b)** One possibility: Xmin $= 0$, Xmax $= 48$, Xscl $= 6$; Ymin $= 0$, Ymax $= 10{,}000$, Yscl $= 1000$. **(c)** \$7760. **(d)** After 13 months.

39. **(a)** One possibility: Xmin $= 0$, Xmax $= 12.5$, Xscl $= 1$; Ymin $= 0$, Ymax $= 625$, Yscl $= 100$. **(b)** No. The maximum height is 625 ft. The graphs of y_1 and y_2 do not intersect.

40. **(a)** Between 4.1 and 4.15 seconds (about 4.12 sec). **(b)** Going up: between 2.15 and 2.2 sec (2.19 sec). Going down: between 3.3 and 3.35 sec (3.31 sec).

41. **(a)** and **(b)** The plot is shown on the window Xmin $= 0.5$, Xmax $= 5.5$, Xscl $= 1$; Ymin $= 7240$, Ymax $= 37960$, Yscl $= 5000$. **(c)** The model gives \$28,040—\$340 different from the actual value. **(d)** \$47,600 (using $x = 7$).

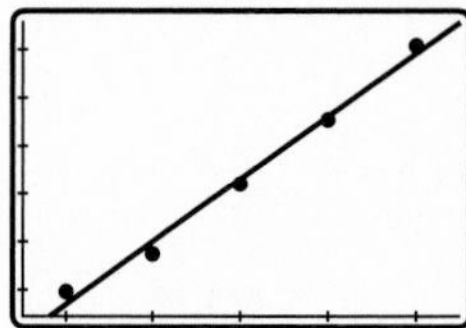

42. **(a)** The plot is shown on the window Xmin $= 3$, Xmax $= 12$, Xscl $= 1$; Ymin $= 100$, Ymax $= 140$, Yscl $= 10$. The plot of the modeling equation is also shown. **(b)** 157.46 (using $x = 16$).

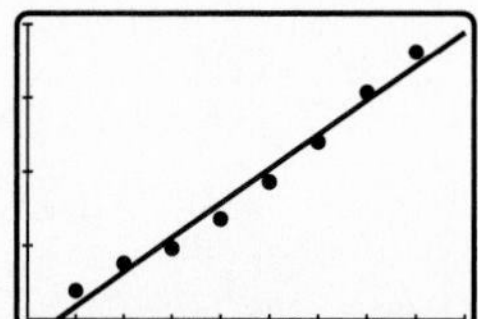

QUICK REVIEW 1.2

1. 6 and $\frac{35}{4}$ (or 8.75)
2. -2 and $\frac{117}{64}$ (or 1.828125)
3. Yes
4. No
5. No
6. Yes
7. No
8. Yes
9. Yes
10. Yes
11. $x \geq -\frac{3}{2}$
12. $x < \frac{3}{4}$
13. $x < \frac{5}{2}$
14. $x \leq \frac{1}{5}$
15. $x < 0$
16. $x < -3$
17. $x = 0$
18. $x < 1$

SECTION EXERCISES 1.2

1. Numerical:

x	1	2	3	4	5
y	0.8	1.6	2.4	3.2	4.0

; algebraic: $y = 0.8x$; graphical: shown is the window $[0, 10] \times [0, 8]$.

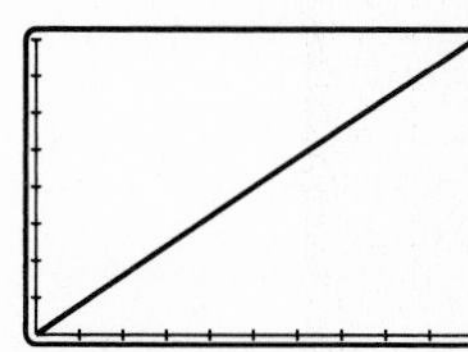

2. Numerical:

x	100	200	500	1000
y	48.3	96.6	241.5	483

; algebraic: $y = 0.483x$; graphical: shown is the window $[0, 500] \times [0, 250]$.

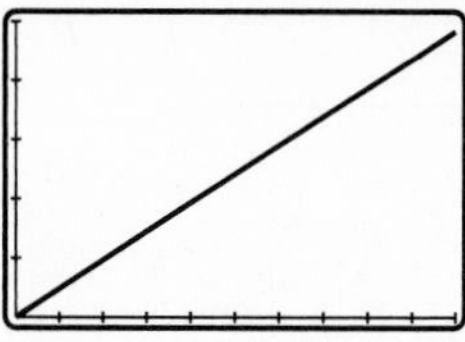

3. **(a)** $f(2) = (2)^3 + 2(2) - 3 = 9$ **(b)** $f(-3) = (-3)^3 + 2(-3) - 3 = -36$ **(c)** $f(1) = (1)^3 + 2(1) - 3 = 0$ **(d)** $f(a) = (a)^3 + 2(a) - 3 = a^3 + 2a - 3$

4. **(a)** $f(2) = \frac{(2+1)}{(2-3)} = \frac{3}{-1} = -3.$ **(b)** $f(-3) = \frac{(-3+1)}{(-3-3)} = \frac{-2}{-6} = \frac{1}{3}.$ **(c)** $f(4) = \frac{(4+1)}{(4-3)} = \frac{5}{1} = 5.$ **(d)** $f(a+1) = \frac{(a+1+1)}{(a+1-3)} = \frac{a+2}{a-2}.$

5. $f(2) = 4$; $f(a) = 3a^2 - 5a + 2$

6. $g(-3) = 110$; $g(b) = -2b^3 + 5b^2 - 3b + 2$

7. $h(5) = \frac{\sqrt{5}}{2} = 1.118\ldots$; $h(a) = \frac{\sqrt{a}}{a-3}$

8. $f(4) = \frac{1}{4}$; $f(c) = \frac{|c-8|}{c^2}$

9. $h(-2) = \sqrt[3]{3} - \frac{1}{2} = 0.942\ldots$; $h(x+5) = \frac{1}{x+5} + \sqrt[3]{x+10}$

10. $g(8) = \frac{2}{65}$; $g(x-3) = \frac{\sqrt[3]{x-3}}{(x-3)^2+1} = \frac{\sqrt[3]{x-3}}{x^2-6x+10}$

11. **(a)** 5. **(b)** 1. **(c)** $[-1, 8]$. **(d)** $[1, 5]$. **(e)** 6 (approximately)

12. **(a)** -1. **(b)** 3. **(c)** $[-2, 9]$. **(d)** $[-1, 7]$ (if we assume that $(3, -1)$ really is the lowest point on the graph). **(e)** The function g takes on the value 3 twice: once at -2, and again somewhere near 7.

13. Independent: t; dependent: x
14. Independent: x; dependent: y
15. Independent: x; dependent: t
16. Independent: r; dependent: s
17. Independent: y; dependent: x
18. Independent: x; dependent: y
19. We need $x^2 + 4 \geq 0$, which is true for all real x. The domain is $(-\infty, \infty)$.
20. We need $x^2 + 1 \geq 0$, which is true for all real x. The domain is $(-\infty, \infty)$.
21. We need $x - 3 \neq 0$. The domain is $(-\infty, 3) \cup (3, \infty)$.
22. We need $x + 3 \neq 0$ and $x - 1 \neq 0$. The domain is $(-\infty, -3) \cup (-3, 1) \cup (1, \infty)$.
23. We need $x \neq 0$ and $x - 3 \neq 0$. The domain is $(-\infty, 0) \cup (0, 3) \cup (3, \infty)$.
24. We need $x^2 - 5x \neq 0$, or $x(x-5) \neq 0$. The domain is $(-\infty, 0) \cup (0, 5) \cup (5, \infty)$.
25. We need $x - 3 \neq 0$ and $4 - x^2 \geq 0$, which means that $x \neq 3$ and $x^2 \leq 4$; the latter implies that $-2 \leq x \leq 2$, so the domain is $[-2, 2]$.
26. We need $x + 1 \neq 0$, $x^2 + 1 \neq 0$, and $4 - x \geq 0$. The first requirement means that $x \neq -1$, the second is true for all x, and the last means that $x \leq 4$. The domain is therefore $(-\infty, -1) \cup (-1, 4]$.

27.

x	-3	-1	1	3
y	-81	-13	-1	3

28.

x	-3	-1	1	3
y	264	0	-8	0

29.

x	-3	-1	1	3
y	9	1.732...	2.236...	23.811...

30.

x	-3	-1	1	3
y	$-0.395\ldots$	$-0.235\ldots$	$-0.353\ldots$	$-1.581\ldots$

31. One must have (approximately) $Xmin < -1.2$, $Xmax > 18.2$, $Ymin < -94.25$, and $Ymax > 0$. A good choice is $[-5, 25] \times [-110, 20]$, with $Xscl = 5$ and $Yscl = 10$.

32. $Xmin < -32.6$, $Xmax > 3.6$, $Ymin < -329.3$, and $Ymax > 0$—for example, $[-40, 10] \times [-400, 100]$, with $Xscl = 10$ and $Yscl = 50$

33. $Xmin < -21.3$, $Xmax > 3.3$, $Ymin < 0$, and $Ymax > 152$—for example, $[-30, 10] \times [-100, 200]$, with $Xscl = 10$ and $Yscl = 50$.

34. $Xmin < -39.4$, $Xmax > 2.4$, $Ymin < 0$, and $Ymax > 435$—for example, $[-50, 10] \times [-100, 500]$, with $Xscl = 10$ and $Yscl = 50$

35. Yes

36. No

37. No

38. Yes

39. $V(x) = x(2x)(8x) = 16x^3$

40. $A(x) = 2(x)(2x) + 2(x)(8x) + 2(8x)(2x) = 52x^2$

41. **(a)** $V = \frac{4}{3}\pi(75 \text{ cm})^3 = 562{,}500\pi \text{ cm}^3 \approx 1{,}767{,}146 \text{ cm}^3$.
(b) $r = \sqrt[3]{\frac{3}{4\pi} \cdot 2000 \text{ cm}^3} = 10\sqrt[3]{\frac{3}{2\pi}} \text{ cm} \approx 7.816 \text{ cm}$.

42. **(a)** $A = 4\pi(30 \text{ cm})^2 = 3600\pi \text{ cm}^2 \approx 11{,}309.7 \text{ cm}^2$.
(b) $r = \sqrt{150 \text{ cm}^2/4\pi} = \frac{5}{2}\sqrt{6/\pi} \text{ cm} \approx 3.455 \text{ cm}$.

43. **(a)** $A(x) = (4 + 2x)(6.5 + 2x) - (4)(6.5) = 4x^2 + 21x \text{ in.}^2$; **(b)** $Xmin$ and $Ymin$ should be 0. There is no *physical* limitation on how large x can be, but it seems reasonable to say that $Xmax$ should be no more than about 3 or 4. In that case, $Ymax$ should be about 150. When $x = 0.823$, $A(x) \approx 20$. So, if the area is 20 in.^{3}, the width is about 0.823 in.

44. **(a)** $T(2) = 2\pi\sqrt{\frac{2}{9.807}} \approx 2.837$ sec. **(b)** $T(8) \approx 5.675$ sec. **(c)** About 24.84 m.

45. **(a)** $h(5) = 47$ ft. **(b)** The x (that is, t) values should range from 0 to about 5.6; $Ymin$ and $Ymax$ should be from 0 to 128 or more. **(c)** The highest point occurs at 2.75 sec, which is less than 3 sec.

46. **(a)** 316 ft. **(b)** Yes. **(c)** The firework reaches it highest point—about 361.6 ft above the ground—after about 4.69 sec.

47. **(a)** $P(x) = 200 + \pi x$ m.
(b) $\frac{200}{\pi} = 63.661\ldots$ m. **(c)** 7.60 m, or 8.31 yards, on each side of the field.
(d) $A(x) = 100x + \pi(x/2)^2 \text{ m}^2$.

48. **(a)** 1 mile is 5280 ft, so 2 miles are 10,560 ft. The segment from the plant to point Q is the hypotenuse of a right triangle. Thus $C(x) = 180\sqrt{800^2 + x^2} + 100(10560 - x)$. **(b)** and **(c)** The minimum—$1,175,733—occurs when x is about 534.5 ft, which is less than 2000 ft. Try the window $[0, 940] \times [1{,}100{,}000, 1{,}200{,}000]$.

49. **(a)** The original circle's circumference was 8π, from which x units have been removed.
(b) $r(x) = (8\pi - x)/2\pi = 4 - x/2\pi$;
(c) $h(x) = \sqrt{16 - (4 - x/2\pi)^2}$
(d) $V(x) = \frac{1}{3}\pi(4 - x/2\pi)^2\sqrt{16 - (4 - x/2\pi)^2}$

50. **(a)** $f(5) = 5.3795$, which rounds to 5.4. $f(5)$ is less than the 1988 value by about 0.1 million. **(b)** $f(17) = 325.9$.

51. The function might never rise above some level—all the function values may be less than 10, for example.

52. The function may appear quite different for x values less than -4.7 or greater than 4.7; it may rise above 3 or fall below -3.

53. Numerical: Trying $h = 1$ yields $f(2 + 1) = -3$, whereas $f(2) + f(1) = -3 + (-1) = -4$; $h = 0$ is even easier: $f(2 + 0) = -3$, whereas $f(2) + f(0) = -3 + 3 = 0$.
Algebraic: $f(2 + h) = (2 + h)^2 - 5(2 + h) + 3 = h^2 - h - 3$, whereas $f(2) + f(h) = -3 + h^2 - 5h + 3 = h^2 - 5h$.

54. $f(2 + h) - f(2 - h) = (h^2 - h - 3) - (h^2 + h - 3) = -2h$

55. **(a)** $\dfrac{f(2 + h) - f(2 - h)}{2h} = \dfrac{-2h}{2h} = -1$;
(b) The expression represents the difference in the y-coordinates divided by the difference in the x-coordinates.

56. $\dfrac{f(x + h) - f(x)}{h} = \dfrac{3(x + h) + 2 - (3x + 2)}{h} = 3$

57. $\dfrac{f(x + h) - f(x)}{h} = \dfrac{(x + h)^2 - 3 - (x^2 - 3)}{h} = \dfrac{x^2 + 2xh + h^2 - x^2}{h} = 2x + h$

58.–59. Answers may vary.

60. No. Fauver, Yount and Lind each made 6 FG, but scored different numbers of points, so TP cannot be a function of FG. Meanwhile, Best and Yount both scored 15 points, but had different numbers of field goals, so FG cannot be a function of TP. None of the other four possibilities are functions.

QUICK REVIEW 1.3

1. $x = 0$

2. $x = -\frac{1}{5}$

3. $x = 12$

4. $x = \frac{3}{29}$

5. $y = -7$

6. $y = \frac{23}{5}$

7. $y = 4$

8. $y = \frac{2}{3}$

9. $\frac{2}{3}$

10. $\frac{5}{6}$

11. $9 - 4x$

12. $-x - 4$

13. $-1\ (x \neq 1)$

14. $-\frac{1}{2}\ (x \neq 4)$

15. $y = (2x - 21)/5 = 2x/5 - 21/5$

16. $y = 17/5$

17. $y = x - x^2/3$

18. $y = 2/x^2$

SECTION EXERCISES 1.3

1. $m = -2$

2. $m = 2/3$

3. The window should include $(-3, 0)$ and $(0, 12)$—for example, $[-5, 2] \times [-5, 15]$.

4. The window should include $(-5.86, 0)$ and $(0, 41)$—for example, $[-10, 5] \times [-10, 50]$.

5. Graph $y = 49 - 8x$; window should include $(6.125, 0)$ and $(0, 49)$—for example, $[-5, 10] \times [-10, 60]$.

6. Graph $y = 35 - 2x$; the window should include $(17.5, 0)$ and $(0, 35)$—for example, $[-5, 20] \times [-10, 40]$.

7. Graph $y = (429 - 123x)/7$; window should include $(3.488, 0)$ and $(0, 61.29)$—for example, $[-1, 5] \times [-10, 75]$.

8. Graph $y = (3540 - 2100x)/12 = 295 - 175x$; the window should include $(1.686, 0)$ and $(0, 295)$—for example, $[-1, 3] \times [-50, 350]$.

9. **(a)** $m = (67{,}500 - 42{,}000)/8 = 3187.5$. The y-intercept is $b = 42000$. The equation is $V = 3{,}187.5x + 42{,}000$. **(b)** The house is worth about \$72,500 after 9.57 years. Window: $[0, 15] \times [40{,}000, 100{,}000]$. **(c)** $3187.5t + 42{,}000 = 74{,}000$; $t = 10.04$; **(d)** $t = 12$ years.

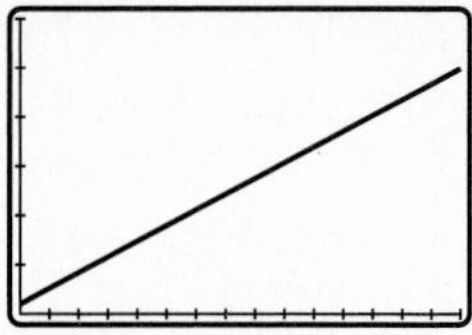

10. **(a)** $0 \le x \le 18{,}000$.
(b) $I = 0.05x + 0.08(18000 - x)$.
(c) $x = \$14{,}000$.
(d) $x = \$8{,}500$.

11. $m = 4/7$

12. $m = -4/7$

13. $m = 8$

14. $m = 25/11$

15. $m = 0$

16. m is undefined.

17. $m = -2/11$

18. $m = -5/3$

19. $x = 2$

20. $y = -15$

21. $y = 16$

22. $x = 0$

23. $y = 2x + 2$

24. $y = \frac{1}{2}x - \frac{7}{2}$

25. $y = -\frac{2}{3}x + \frac{1}{3}$

26. $y = -\frac{3}{4}x + \frac{17}{2}$

27. $y = \frac{3}{5}x + \frac{39}{5}$

28. $y = 3x + 13$

29. $y = \frac{3}{7}x - \frac{32}{7}$

30. $y = -2x + 6$

The coefficients in the answers to Exercises 31–36 can be multiplied by the same nonzero number.

31. $x - y + 3 = 0$

32. $2x + y - 7 = 0$

33. $x - y + 5 = 0$

34. $x - y - 5 = 0$

35. $0x + y + 3 = 0$

36. $x + y + 6 = 0$

37. $y = -3x + 5$

38. $y = \frac{1}{2}x + \frac{3}{2}$

39. $y = -\frac{1}{4}x + 4$

40. $y = \frac{1}{7}x + \frac{10}{7}$

41. $y = -\frac{2}{5}x + \frac{12}{5}$

42. $y = \frac{7}{12}x - 8$

43. The line in part **(a)**: The slope is 1.5, versus 1 in part **(b)**.

44. The line in part **(b)**: The slopes are 7/4 and 4, respectively.

45. The graph in part **(b)** in Exercise 43.

46. The graph in part **(a)** in Exercise 44.

47. The graph in part **(b)** in Exercise 44.

48. The graph in part **(a)** in Exercise 43.

49. $x = 4$, $y = 21$

50. $x = 2$, $y = -18$

51. $x = -16$, $y = -3$

52. $x = 14$, $y = 20$

53. $x = -10$, $y = -7$

54. $x = 65$, $y = -4.8$

55. $m = -5/2$ (assuming that the line passes through the origin and the corners of the window)

56. $m = 4$.

57. $Y\text{min} = -30$, $Y\text{max} = 30$, $Y\text{scl} = 3$

58. $Y\text{min} = -50$, $Y\text{max} = 50$, $Y\text{scl} = 5$

59. $Y\text{min} = -20/3$, $Y\text{max} = 20/3$, $Y\text{scl} = 2/3$

60. $Y\text{min} = -12.5$, $Y\text{max} = 12.5$, $Y\text{scl} = 1.25$

61. Parallel: $y - 2 = 3(x - 1)$, or $y = 3x - 1$; perpendicular: $y - 2 = -\frac{1}{3}(x - 1)$, or $y = -\frac{1}{3}x + \frac{7}{3}$

62. Parallel: $y - 3 = -2(x + 2)$, or $y = -2x - 1$; perpendicular: $y - 3 = \frac{1}{2}(x + 2)$, or $y = \frac{1}{2}x + 4$

63. Parallel: $2x + 3y = 9$, or $y = -\frac{2}{3}x + 3$; perpendicular: $3x - 2y = 7$, or $y = \frac{3}{2}x - \frac{7}{2}$

64. Parallel: $3x - 5y = 13$, or $y = \frac{3}{5}x - \frac{13}{5}$; perpendicular: $5x + 3y = 33$, or $y = -\frac{5}{3}x + 11$

65. $y = \frac{3}{8}x$, where y is altitude and x is horizontal distance. The plane must travel $x = 32{,}000$ ft horizontally—just over 6 miles.

66. **(a)** $m = 0.06$. **(b)** $4166.\overline{6}$ ft, or about 0.79 mile. **(c)** 2217.6 ft.

67. $m = \frac{1}{3}$ corresponds to a 4–12 pitch, so asphalt shingles are acceptable.

68. **(a)** and **(d)**

69. If $b = 0$, both lines are vertical, otherwise, both have slope $m = -a/b$.

70. If either a or b equals 0, then one line is horizontal and the other is vertical. Otherwise, their slopes are $-a/b$ and b/a, respectively.

71. **(a)** $m = (5422 - 4936)/(1991 - 1986) = 97.2$; in point-slope form, this line is $y - 4936 = 97.2(x - 1986)$, which translates to $y = 97.2x - 188103.2$. **(b)** When $x = 2000$, $y = 6296.8$—about 6.3 billion (or 6300 million to Europeans).

72. As in the diagram, we can choose one point to be the origin, and another to be on the x-axis. The midpoints of the sides, starting from the origin and moving counterclockwise in the diagram, are then $A\left(\frac{a}{2}, 0\right)$, $B\left(\frac{a+b}{2}, \frac{c}{2}\right)$, $C\left(\frac{b+d}{2}, \frac{c+e}{2}\right)$, and $D\left(\frac{d}{2}, \frac{e}{2}\right)$. The slopes of the four lines connecting those points are: $m_{AB} = \frac{c}{b}$, $m_{BC} = \frac{e}{d-a}$, $m_{CD} = \frac{c}{b}$, and $m_{DA} = \frac{e}{d-a}$. Thus the opposite sides are parallel.

73. The slopes of the lines containing the sides are (counterclockwise from the origin): $m_1 = 0$, $m_2 = \frac{b}{a-c}$, and $m_3 = \frac{b}{a}$. The altitudes are perpendicular to each of these lines; the altitude to the first side thus has the equation $x = a$. The other two altitudes are $y = \frac{c-a}{b}x$ and $y = -\frac{a}{b}(x - c) = \frac{a}{b}(c - x)$. The values $x = a$ and $y = \frac{a(c-a)}{b}$ satisfy all three equations, so the point $\left(a, \frac{a(c-a)}{b}\right)$ is common to all three altitudes.

74. The line from the origin to point (3, 4) has slope 4/3, so the tangent line is $y - 4 = -\frac{3}{4}(x - 3)$.

75. Tracing anywhere near the point (2, 4), we see *slightly* different y-coordinates when jumping back and forth between the two graphs—except at (2, 4) exactly.

76. $y - (-4) = 2(x - 1)$, or $y = 2x - 6$

77. One possibility: Equally spaced x-values correspond to equally spaced y-values.

78. A has coordinates $\left(\frac{b}{2}, \frac{c}{2}\right)$ and B has coordinates $\left(\frac{a+b}{2}, \frac{c}{2}\right)$, so the line containing A and B is horizontal, and the distance from A to B is $a/2$.

QUICK REVIEW 1.4

1. 34.8

2. 1.1

3. 42010.5

4. 18497.25

5. $y = 8x + 3.6$

6. $y = -1.8x - 2$

7. $m = 1$; $b = -17$

8. $m = -10$; $b = 15$

9. $y - 4 = -\frac{3}{5}(x + 2)$, or $y = -0.6x + 2.8$:

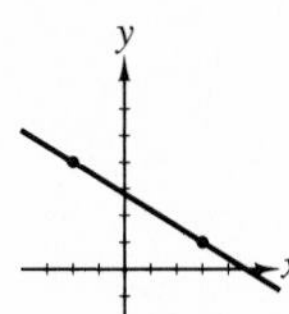

10. $y - 5 = \frac{8}{3}(x - 1)$, or $y = \frac{8}{3}x + \frac{7}{3}$:

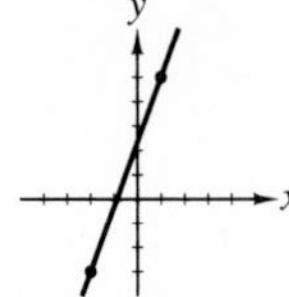

11. $y + 1 = \frac{5}{7}(x + 5)$, or $y = \frac{5}{7}x + \frac{18}{7}$:

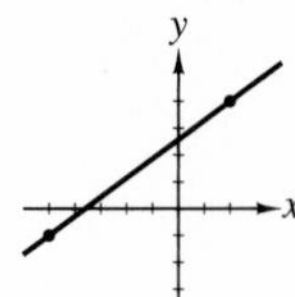

12. $y - 5 = -\frac{7}{9}(x + 3)$, or $y = -\frac{7}{9}x + \frac{8}{3}$:

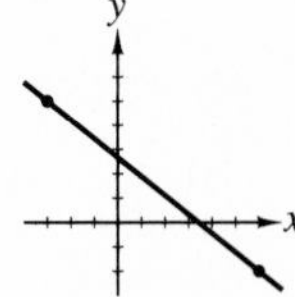

SECTION EXERCISES 1.4

1. Strong positive

2. Strong negative

3. Weak positive

4. Weak negative

5. Strong positive

6. No correlation

7. The scatter plot in Exercise 5

8. The scatter plot in Exercise 2

9. **(a)** Shown in $[15, 45] \times [20, 45]$:

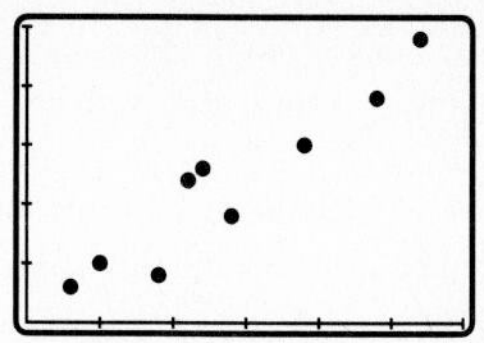

(b) There is a strong positive correlation (in fact, $r = 0.9477$).

10. **(a)** Shown in $[0, 80] \times [0, 70]$:

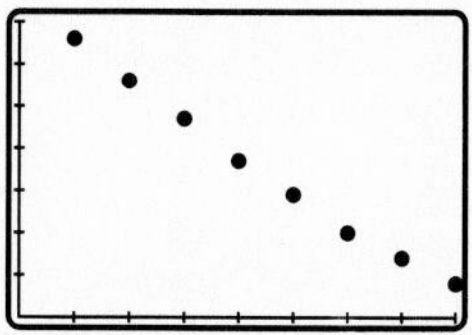

(b) There is a strong negative correlation (in fact, $r = -0.9964$).

11. True

12. True

13. True

14. True

15. True

16. True

17. (c)

18. (d)

19. (a)

20. (b)

21. The output is shown as screen (a), so $y = 1.98x + 1.5$. Note that the calculator is set to display three digits after the decimal.

22. The output is shown as screen (b), so $y = x + 2.75$.

23. $y_1 = 34.5$

24. $y_1 = 12.2$

25. $y_1 = 56.045$

26. $y_1 = -7.15$

27. $y_1 = 22.48$

28. $y_1 = 181$

29. **(a)** $y = 0.68x + 9.0125$; **(b)** $m = 0.68$; **(c)** Shown in $[15, 45] \times [20, 40]$:

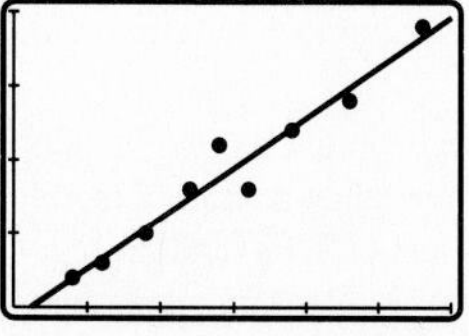

(d) When $x = 30$, $y = 29.4$ lb.

30. **(a)** The plot is shown in $[1984, 1991] \times [130, 245]$. The regression line is approximately $y = 18.1429x - 35878.8$.

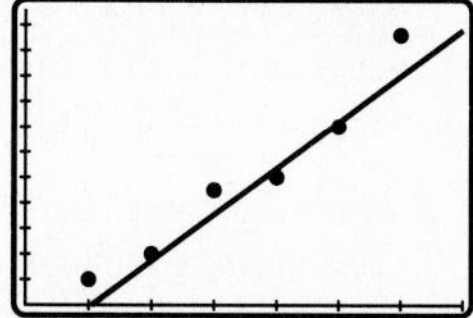

(b) Using this approximation for the regression line, in 2000 the median base salary will be about \$407 thousand.

31. **(a)** $m \approx 1060.42$, which represents the increase in y for an increase by 1 in the x-value. That is, it is the typical annual increase in salary. **(b)** About \$43,298 (the regression equation is approximately $y = 1060.423313x - 2\,077\,548.669$).

32. **(a)** $y = 17.65x - 34310.4$. **(b)** About 777.8 million.

33. **(a)** The line that begins lower and ends higher is for the data in Table 1.15. **(b)** The two lines cross when x is about 1984.

34. Answers may vary.

35. **(a)** The median-median line (computed by a TI-82) is $y = 0.51282x + 18.6453$. **(b)** $y = 0.50076x + 17.97933$. **(c)** The lines shown in $[0, 55] \times [15, 45]$ are about the same—the regression line lies slightly below the median-median line. **(d)** The new median-median line is $y = 0.51282x + 19.1453$—the slope is unchanged, but the intercept has increased by 0.5. The new regression line is $y = 0.47652x + 21.53876$. The median-median line is perhaps slightly better now: The regression line lies above almost all of the data points (except for the extreme one).

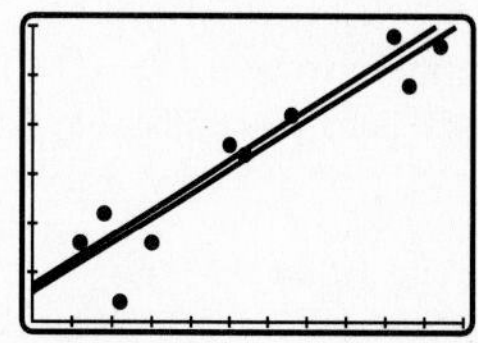

36. **(a)**, **(b)**, and **(c)** Median-median line: $y = 0.5556x + 6.0926$; regression line: $y = 0.1154x + 8.2449$ shown in $[0, 16] \times [0, 15]$. **(d)** The median-median line matches well all of the points, except (15, 4), which deviates from the linear pattern. Meanwhile, the regression line attempts to match up with (15, 4), and as a result does not do a good job overall. Unless we place special importance on matching up with (15, 4), the median-median line seems to be better.

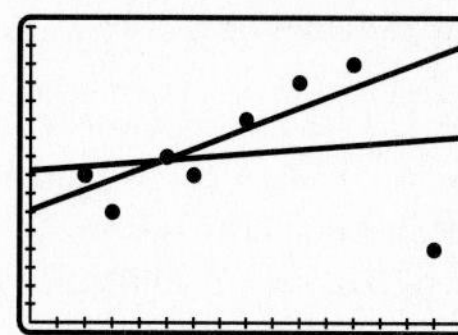

37. Answers may vary.

38. **(a)** $y = 5{,}632x - 11{,}080{,}280$. **(b)** 5632 is the "average" increase in median price each year for homes in the Northeast. **(c)** $y = 2{,}732x - 5{,}362{,}360$. **(d)** The annual increase in the Midwest is only about \$2732 per year; the prices in the Northeast are increasing at over twice that rate (\$2900 *more* per year). Note, however, that the scatter plot of the Northeast data suggests that a line is not the best fit for the data; some other kind of function fits the data better.

QUICK REVIEW 1.5

1. $f(-2) = 14$

2. $g(1) = -1$

3. $f(-3) = 9\sqrt{10} = 28.460\ldots$

4. $h(7) = 3$

5. $[-4, \infty)$

6. (π, ∞)

7. $(-3, 2]$

8. $(-2, 5]$

9. $-\frac{1}{3} \le x < 4$

10. $-\frac{3}{2} < x \le 0$

11. For -2, use the top rule: $f(-2) = (-2)^2$ since $-2 < 1$. For 1, use the bottom rule: $f(1) = 2(1) + 5$ since $1 \ge 1$.

12. Use the top and bottom rules, respectively: $g(-2) = -(-2)^2 + 3$, since $-2 \le -2$; $g(3) = (3)^2 + 3$, since $3 > -2$.

SECTION EXERCISES 1.5

1. Local maximums at $(-1, 4)$ and $(5, 5)$, local minimum at $(2, 2)$. The function increases on $(-\infty, -1]$, decreases on $[-1, 2]$, increases on $[2, 5]$, and decreases on $[5, \infty)$.

2. Local minimum at $(1, 2)$, neither at $(3, 3)$, and local maximum at $(5, 7)$. The function decreases on $(-\infty, 1]$, increases on $[1, 5]$ and decreases on $[5, \infty)$.

3. Neither at $(-1, 3)$ and $(3, 3)$, local maximum at $(1, 5)$, and local minimum at $(5, 1)$. The function increases on $(-\infty, 1]$, decreases on $[1, 5]$, and increases on $[5, \infty)$.

4. Local minima at $(-1, 1)$ and $(3, 1)$, and local maxima at $(1, 6)$ and $(5, 4)$. The function decreases on $(-\infty, -1]$, increases on $[-1, 1]$, decreases on $[1, 3]$, increases on $[3, 5]$, and decreases on $[5, \infty)$.

5. f has a local minimum when $x = 0.5$, where $y = 3.75$. It has no maximum.

6. Local maximum: $y \approx 4.08$ at $x \approx -1.15$; local minimum: $y \approx -2.08$ at $x \approx 1.15$

7. Local minimum: $y \approx -4.09$ at $x \approx -0.82$; local maximum: $y \approx -1.91$ at $x \approx 0.82$

8. Local maximum: $y \approx 9.48$ at $x \approx -1.67$; local minimum: $y = 0$ at $x = 1$

9. Local maximum: $y \approx 9.16$ at $x \approx -3.20$; local minimum: $y = 0$ at $x = 0$

10. Local maximum: $y = 0$ at $x = -2.5$; local minimum: $y = -3.125$ at $x = -1.25$

11. (b)

12. (b)

13. (c)

14. (a)

15. The viewing window should include the minimum values at approximately $(-1.58, 0)$ and $(7.58, 0)$ and the local maximum at $(3, 21)$. Xmin and Xmax should be enough below -1.58 and above 7.58, respectively, to suggest the behavior of the graph to the left and right of the minimum values. An example is $[-6, 12] \times [-5, 45]$.

16. The local maximum is at $(-2, 16)$, and the local minimum is at $(2, -16)$. A possible window is $[-6, 6] \times [-20, 20]$.

17. The local maximum is at approximately $(-4.71, 83.72)$, and the local minimum is at $(0.71, 4.28)$. Another important feature that should be included is at $x = -7.48$, where the graph crosses the x-axis. A possible window is $[-10, 5] \times [-10, 100]$.

18. The minimum is at $(2.5, -306.25)$. The windows should also show $(-15, 0)$ and $(20, 0)$, the two places where the graph crosses the x-axis. A possible window is $[-20, 30] \times [-400, 100]$.

19. Xmin should be less than -12, and Xmax should be more than 0. This graph has no local extreme values, except at $(-12, 0)$. A possible window is $[-15, 2] \times [-1, 5]$.

20. Show the minimum at $(-2, -3)$, and the two x-axis crossings at $(-11, 0)$ and $(7, 0)$. A possible window is $[-15, 10] \times [-4, 2]$.

21. A good window is $[-10, 10] \times [-0.15, 0.8]$, with Yscl $= 0.1$, so that the piece of the graph below the x-axis is visible.

22. The local minimum is at $(-2, -0.5)$, and x-axis crossings are at -3 and -1. A possible window is $[-10, 10] \times [-1, 5]$.

23. (b)

24. (a)

25. One possible story: The jogger travels at an approximately constant speed during her workout. She jogs to the far end of the course, turns around, returns to her starting point, and then goes out again for a second trip. On the second trip, she turns around before reaching the far end of the course.

26. Answers may vary.

27. Decreasing on $(-\infty, -2]$; increasing on $[-2, \infty)$

28. Decreasing on $(-\infty, -1]$; constant on $[-1, 1]$; increasing on $[1, \infty)$

29. Decreasing on $(-\infty, -2]$; constant on $[-2, 1]$; increasing on $[1, \infty)$

30. Decreasing on $(-\infty, -2]$; increasing on $[-2, \infty)$

31. Increasing on $(-\infty, 1]$; decreasing on $[1, \infty)$

32. Increasing on $(-\infty, -0.5]$; decreasing on $[-0.5, 1.2]$, increasing on $[1.2, \infty)$. (Values are approximate.)

33. Even: $f(-x) = 2(-x)^4 = 2x^4 = f(x)$

34. Odd: $g(-x) = (-x)^3 = -x^3 = -g(x)$

35. Even: $f(-x) = \sqrt{(-x)^2 + 2} = \sqrt{x^2 + 2} = f(x)$

36. Even: $g(-x) = \dfrac{3}{1 + (-x)^2} = \dfrac{3}{1 + x^2} = g(x)$

37. Neither: $f(-x) = -(-x)^2 + 0.03(-x) + 5 = -x^2 - 0.03x + 5$, which is neither $f(x)$ nor $-f(x)$

38. Neither: $f(-x) = (-x)^3 + 0.04(-x)^2 + 3 = -x^3 + 0.04x^2 + 3$, which is neither $f(x)$ nor $-f(x)$

39. Odd: $g(-x) = 2(-x)^3 - 3(-x) = -2x^3 + 3x = -g(x)$

40. Odd: $h(-x) = \dfrac{1}{-x} = -\dfrac{1}{x} = -h(x)$

41. The graph in part (b); $f(-2) = 0$, $f(1) = 3$, $f(3) = 2$

42. The graph in part (a). $f(-2) = 5$; $f(1) = 5$; $f(3) = 9$

43. The graph in part (c); $f(-2) = 4$, $f(1) = 5.5, f(3) = 6.5$

44. The graph in part (d). $f(-2) = 0$; $f(1) = 1.5$; $f(3) = 7$

45.

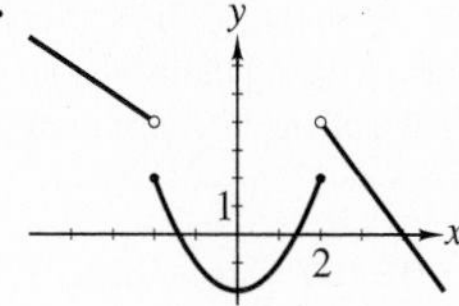

46.

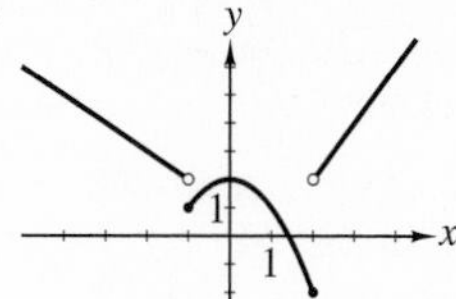

47. This graph crosses the x-axis at 1.01 and 1.02, and has a minimum at $x = 1.015$ and $y = -0.0025$. One possible window is $[1, 1.03] \times [-0.005, 0.005]$.

48. This graph crosses the x-axis at -0.1 and 0.11 and has a maximum at $x = 0.005$ and $y = 1.1025$. A possible window is $[-0.2\ 0.2] \times [-3, 3]$.

49. This graph has a local maximum at approximately $x = 0.333$ ($y \approx -29.85$) and a local minimum at $x = 1$ ($y \approx -30$). One possible window is $[0, 1.5] \times [-30.05, -29.8]$.

50. This graph has a local maximum at approximately $x = 3.667$ ($y = 76.19$) and a local minimum at $x \approx 5$ ($y \approx 75$). A possible window is $[2, 6] \times [72, 78]$.

51. **(a)** Certainly t must be positive, since we are measuring time *after* the drop leaves the fountain. It is also reasonable to require that $t \le 3.75$ (see the next part). **(b)** The height equals 0 (so the drop has returned to the height at which it began, which presumably is ground level) at time $t = 3.75$. **(c)** It is *exactly* 56.25 ft (according to the model); this height may be reported as 56.2 or 56.3 ft.

52. The ball reaches a height of 43 ft after 1.5 seconds.

53. **(a)** The first pen (with side length $x/4$) is $(x/4)^2 = (x^2/16)$ ft^2, and the other is $[(800 - x)/4]^2 = (800 - x)^2/16$ ft^2. **(b)** $A(x) = (x^2/16) + (800 - x)^2/16 = [x^2 + (800 - x)^2]/16$ ft^2. **(c)** $0 < x < 800$. **(d)** The window used for the graph is $Xmin = 0$, $Xmax = 800$, $Xscl\ 50$; $Ymin = 0$, $Ymax = 40000$, $Yscl = 5000$. **(e)** A single pen ($x = 0$ or $x = 800$) gives the greatest total area.

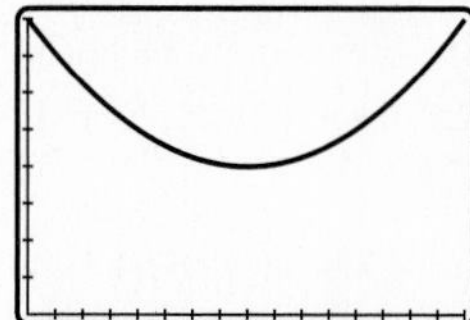

54. The area of the pen is $A(x) = x(1050 - 2x)$, and the values of x that make sense are $0 < x < 525$. The maximum of $A(x)$ occurs when $x = 525/2 = 262.5$ ft, which corresponds to dimensions 262.5×525 and gives an area of 137,812.5 ft^2.

55. **(a)** $0 < x < 8\pi$. **(b)** The window used for the graph is $Xmin = 0$, $Xmax = 8\pi$, $Xscl = \pi$; $Ymin = 0$, $Ymax = 30$, $Yscl = 5$. **(c)** The max occurs when $x = 4.61$ in. The maximum volume is 25.80 in.^{3}

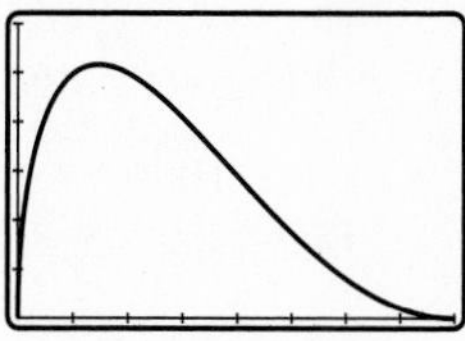

56. **(a)** $f(5) = 0.02(5)^2 + 0.29(5) + 1.4 = 3.35$. **(b)** $f(17) = 0.10(17)^2 - 3.55(17) + 38.26 = 6.81$.

57. **(a)** The graphs intersect at about $(-1.732, 1)$ and $(1.732, 1)$—these x values are $-\sqrt{3}$ and $\sqrt{3}$, respectively. **(b)** The length of the segment is the difference in the y-coordinates. For the value of x shown, the y-coordinate at the top is $4 - x^2$; at the bottom, it is 1; thus the length is $(4 - x^2) - 1 = 3 - x^2$. In general, we could write $|3 - x^2|$ to allow for values of x less than $-\sqrt{3}$ or greater than $\sqrt{3}$.

58. **(a)** The graphs intersect at (0, 0) and (1, 1). **(b)** The segment length is $\sqrt{x} - x^2$ (for x between 0 and 1); in general, we could write $|\sqrt{x} - x^2|$. **(c)** When $x \approx 0.39685$, the maximum length—about 0.47247—is achieved.

59. On TI-8x calculators, entering $Y_1 = (X^2 - 2)/(-3 \le X)(X \le 4)$ or (on a TI-82) $Y_1 = (X^2 - 2)/(-3 \le X$ and $X \le 4)$ produces the result shown. Consult your grapher manual.

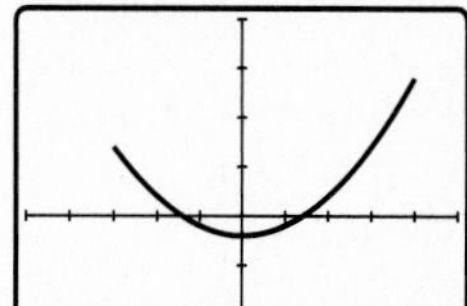

60. On TI-8x calculators, entering $y_1 = \text{abs}(x + 2)/(-3 \le x)(x \le 2)$ or (on a TI-82) $y_1 = \text{abs}(x + 2)/(-3 \le x$ and $x \le 2)$ produces the result shown. Consult your grapher manual.

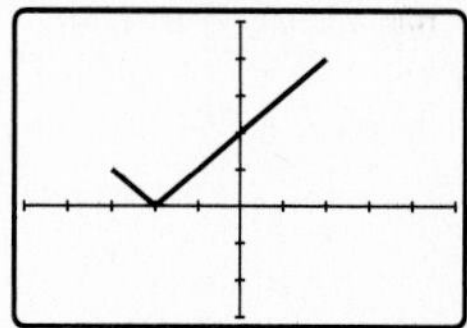

61. It is difficult to see that this graph crosses the x-axis at $x = 1.1$ and $x = 1.15$—just to the left and right of the minimum, which occurs when $x \approx 1.1251$.

62. This function has a local maximum at $x = 2$ and a local minimum at $x = 2.1$ ($y = 40$ and 39.9975, respectively).

63. **(a)** Length: $50 - x$. **(b)** $A(x) = x(50 - x)$. **(c)** $0 \le x \le 50$. **(d)** The maximum occurs when $x = 25$, which corresponds to a square.

64. $f(x) = k$ is neither increasing nor decreasing. According to the definition, k is both a minimum and a maximum value of f; f is even since $f(-x) = k = f(x)$. If $k = 0$, the function $f(x) = 0$ is also odd.

QUICK REVIEW 1.6

1. 0.28%

2. 0.087%

3. 121%

4. Approximately 82.28%

5. 5.90

6. 65.99

7. 22.66%

8. 0.43%

9. 84.82

10. 5.20%

11. $b = 2A/h$

12. $h = 2A/(b_1 + b_2)$

13. $h = V/(\pi r^2)$

14. $h = 3V/(\pi r^2)$

15. $r = \sqrt[3]{3V/(4\pi)}$

16. $h = (SA - 2\pi r^2)/(2\pi r) = SA/(2\pi r) - r$

17. $t = I/(Pr)$

18. $P = A/(1 + r/n)^{nt} = A(1 + r/n)^{-nt}$

19. $x = 2/3$

20. $x = 0.5 = 1/2$

SECTION EXERCISES 1.6

1. $x - 48.6$

2. $3x + 5$

3. $3(x + 5)$

4. $2(x + 5)$

5. $0.17x$

6. $0.05x + 4$

7. $A = x(x + 12)$

8. $A = \frac{1}{2}(x)(x + 2)$

9. $(1 + 0.045)x = 1.045x$

10. $(1 - 0.03)x = 0.97x$

11. $(1 - 0.03)x = 0.97x$

12. $(1 + 0.038)x + 200 = 1.038x + 200$

13. Let C be the total cost and n be the number of items produced; $C = 34{,}500 + 5.75n$.

14. Let C be the total cost and n be the number of items produced; $C = (1.09)28000 + 19.85n$.

15. Let R be the revenue and n be the number of items sold; $R = 3.75n$.

16. Let P be the profit and s be the amount of the sales; $P = 200{,}000 + 0.12s$.

17. $x + 4x = 620$; the numbers are $x = 124$ and $4x = 496$.

18. $x + 2x + 3x = 714$. $x = 119$; the second and third numbers are 238 and 357.

19. $1.035x = 36{,}432$; $x = 35{,}200$

20. $1.042x = 40.4$; $x \approx 38.77$

21. $228 = 2x + 2(x + 18)$; the width is $x = 48$ ft.

22. $166 = x + (x + 4) + 18 + 16 + (x + 18) + (x + 20)$. Then $x = 22.5$ ft, and the other walls (counterclockwise) are 26.5, 18, 16, 40.5, and 42.5 ft.

23. $182 = 52t$; $t = 3.5$ hr

24. $560 = 45t + 55(t + 2)$; $t = 4.5$ hours on local highways

25. $131{,}280(1 + r) = 141{,}176$; $r \approx 0.07538 = 7.538\%$

26. $3{,}740{,}000 \cdot r = 2{,}125{,}000$; $r \approx 0.5682$, or 56.82%

27. $1770 = (s + 30)(3.75)$; the plane's speed is $s = 442$ mph.

28. $25 = (s - 23)(2)$; Greg's speed on level road is $s = 35.5$ mph.

29. **(a)** $P = 2W + 2(2W) = 6W$. **(b)** $P = 2(\frac{1}{2}L) + 2L = 3L$. **(c)** $W = 200/6 = 100/3$, $L = 200/3$. **(d)** The x-coordinate of the intersection of the line $y = 6x$ and $y = 200$ would be the width; the intersection of $y = 3x$ and $y = 200$ would be the length.

30. **(a)** $A = (2W)(W) = 2W^2$. **(b)** $A = L(\frac{1}{2}L) = \frac{1}{2}L^2$. **(c)** $W = 45$; $L = 90$. **(d)** The x-coordinate of the intersection of $y = 2x^2$ and $y = 4050$ would be the width; the intersection of $y = (1/2)x^2$ and $y = 4050$ would be the length.

31. **(a)** $y = x/32$. **(b)** $x \geq 0$. **(c)** the graph in part (b). **(d)** 3.90625, or about 3.9 parts.

32. **(a)** y_1 is the number of gallons of concentrate needed to make x gallons of juice; y_2 is the number of gallons of orange juice made with x gallons of concentrate. **(b)** 25 gallons of concentrate (and 75 gal of water).

33. **(a)** $0.10x + 0.45(100 - x) = 0.25(100)$. **(b)** Graph $y_1 = 0.1X + 0.45(100 - X)$ and $y_2 = 25$; use $x \approx 57.14$ gallons of the 10% solution and about 42.86 gal of the 45% solution.

34. **(a)** Graph $y_1 = 0.1x + 0.25(100 - x)$ and $y_2 = 15$. **(b)** Graph $y_1 = 0.22x + 0.48(100 - x)$ and $y_2 = 30$.

35. $0.20x + 0.35(25 - x) = 0.26(25)$; use $x = 15$ liters of the 20% solution and 10 liters of the 35% solution.

36. $0.10x + 0.25(100 - x) = 0.14(100)$; use $x \approx 73.33$ liters of the 10% solution and about 26.67 liters of the 25% solution.

37. $2x + 2(x + 16) = 136$; two pieces that are $x = 26$ ft long are needed, along with two 42-ft pieces.

38. $2x + 2(x + 3) = 54$. $x = 12$; the room is 12 ft × 15 ft.

39. $55(t + 2) = 65t$; the second truck will catch up with the first $t = 11$ hours after the *second* truck leaves.

40. **(a)** $450 = 52t + 62(7.75 - t)$. **(b)** She spent $t = 3.05$ hours (3 hours, 3 minutes) on state highways and 4 hours, 42 minutes on interstate highways. **(c)** $(3.05)(52) = 158.6$ miles.

41. $900 = 0.07x + 0.085(12{,}000 - x)$; $x = \$8000$ invested at 7%; the other \$4000 was invested at 8.5%.

42. $1571 = 0.055x + 0.083(25{,}000 - x)$; $x = \$18{,}000$ was invested at 5.5%; the other \$7000 was invested at 8.3%.

43. $48{,}814.20 = x + 0.12x + 0.03x + 0.004x$; then $48{,}814.20 = 1.154x$, so $x = \$42{,}300$.

44. **(a)** $C = 100{,}000 + 30x$. **(b)** $R = 50x$. **(c)** Graph $y_1 = 100{,}000 + 30x$ and $y_2 = 50x$; these graphs cross when $x = 5000$ pairs of shoes.

45. **(a)** $u(x) = 125{,}000 + 23x$. **(b)** $s(x) = 125{,}000 + 23x + 8x = 125{,}000 + 31x$. **(c)** $r_u(x) = 56x$. **(d)** $r_s(x) = 79x$. **(e)**

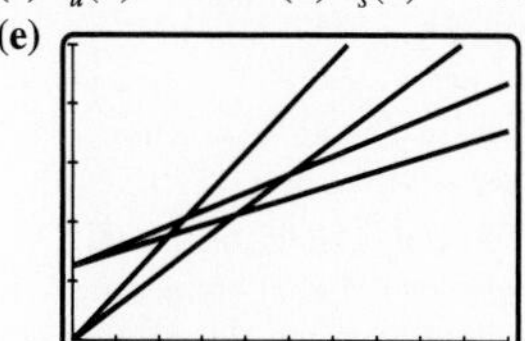

46. **(a)** $y = 136.71x + 6956.29$. **(b)** $y = 9000$ when $x = 14.95$, or about 1995.

47–50. Answers may vary.

51. $\dfrac{5}{8.25} = \dfrac{x}{78}$; $x = 47.\overline{27}$ ft = 47 ft $3.\overline{27}$ in.

52. **(a)** The ratio is 1/96. **(b)** $y = 8x$ gives the "real-world" distance y in feet when the distance on the drawing is x in.; $y = 96x$ gives the "real-world" distance y in *inches* corresponding to a drawing distance of x in. **(c)** The intersection of the two graphs shows the drawing distance: $x = 5.75$ in.

Chapter 1 Review

1. (a)

2. (b)

3.

x	-2	-1	4	3	0
y	7	4	-11	-8	1

4.

x	-3	2	-1	-4	3
y	-11	-1	-7	-13	1

5.

x	-1	-2	0	2	1
y	1/3	1/2	0	ERR	-1

6.

x	2	-2	0	1	4
y	5	-3	1	0	57

7.

x	-3	-2	-1	0	1	2	3
y	-13	-9	-5	-1	3	7	11

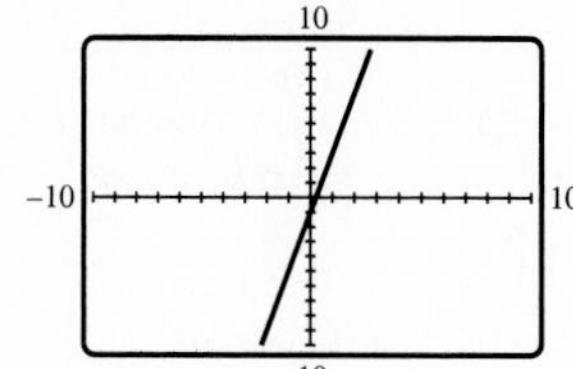

8.

x	-3	-2	-1	0	1	2	3
y	11	8	5	2	-1	-4	-7

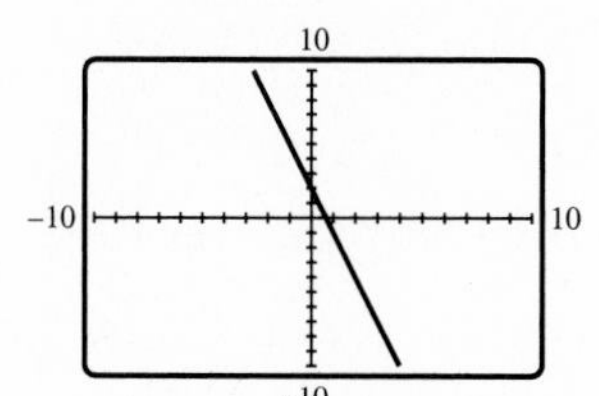

9.

x	-3	-2	-1	0	1	2	3
y	6	5	4	3	2	1	0

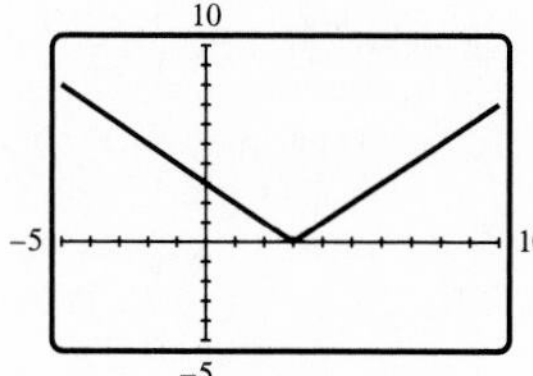

10.

x	-3	-2	-1	0	1	2	3
y	ERR	ERR	ERR	ERR	ERR	0	1

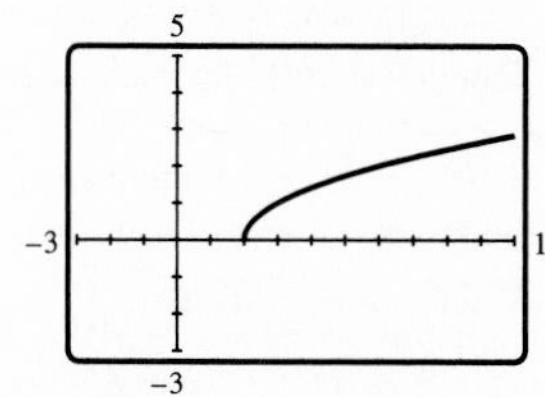

11.

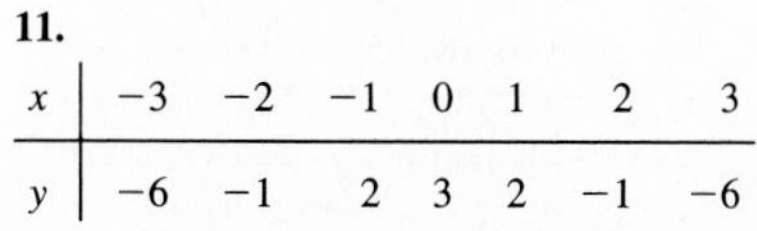

x	-3	-2	-1	0	1	2	3
y	-6	-1	2	3	2	-1	-6

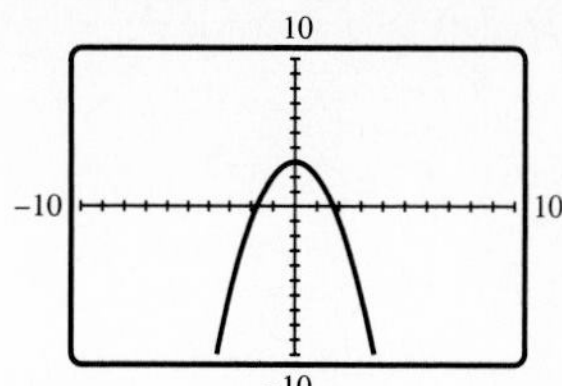

12.

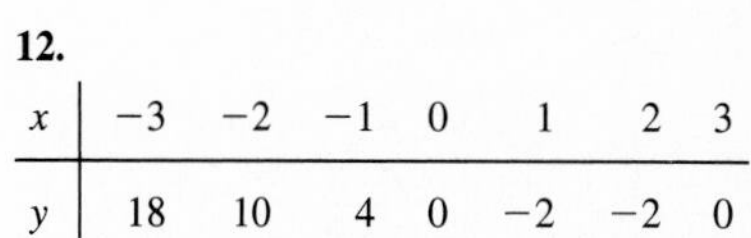

x	-3	-2	-1	0	1	2	3
y	18	10	4	0	-2	-2	0

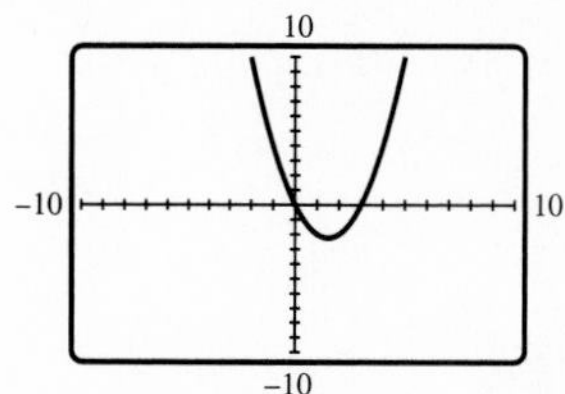

13. The graph in part (c); $X\text{min} = -3$, $X\text{max} = 3$, $X\text{scl} = 1$; $Y\text{min} = -5$, $Y\text{max} = 5$, $Y\text{scl} = 1$

14. The graph in part (d); $X\text{min} = -3$, $X\text{max} = 5$, $X\text{scl} = 1$; $Y\text{min} = -3$, $Y\text{max} = 8$, $Y\text{scl} = 1$

15. The graph in part (b); $X\text{min} = -4$, $X\text{max} = 4$, $X\text{scl} = 1$; $Y\text{min} = -2$, $Y\text{max} = 10$, $Y\text{scl} = 1$

16. The graph in part (a); $X\text{min} = -3$, $X\text{max} = 5$, $X\text{scl} = 1$; $Y\text{min} = -3$, $Y\text{max} = 8$, $Y\text{scl} = 1$

17. (b)

18. (c)

19. $f(-2) = 15$; $f(a) = 2a^2 - a + 5$

20. $f(3) = 2$; $f(x + 2) = \dfrac{\sqrt{x + 3}}{x}$

21. **(a)** $f(-4) = -4$. **(b)** $f(2) = 2$. **(c)** $[-4, 4]$. **(d)** $[-4, 6]$. **(e)** Two input values a give output values of 2, and the other is close to -2.

22. One possibility: The hiker walked away from the snack stand, stopped to rest, and then began walking more rapidly for a while. Finally, the hiker turned around and walked back to the stand.

23. Independent: s; dependent: r

24. Independent: x; dependent: y

25. $x \neq 0$ and $x \neq 2$, or $(-\infty, 0) \cup (0, 2) \cup (2, \infty)$

26. $(-\infty, \infty)$

27. $x \leq 2$, or $(-\infty, 2]$

28. $(-\infty, \infty)$

29.

x	-2	1.5	2	5
y	-35	-11.375	-15	0

30.

x	-2	1.5	2	5
y	$1.118\ldots$	$0.480\ldots$	$0.372\ldots$	$0.169\ldots$

31. $m = 2.5$

32. Both graphs look the same, but the line in the graph in part (a) has slope 2/3, or less than the slope of the line in the graph in part (b), which is $12/15 = 4/5$. The different horizontal and vertical scales make it difficult to judge by looking at the graphs.

33. $m = 7/6$

34. $m = 1$

35. $m = -3/5$

36. $m = -5/7$

37. $x = 0$

38. $y = 3$

39. $y + 5 = \frac{3}{4}(x + 3)$

40. $y + 1 = -\frac{2}{3}(x - 2)$

41. $y = -2(x + 5)$

42. $y - 3 = 2x$

43. $9x + 7y + 17 = 0$

44. $5x - 3y + 11 = 0$

45. $y = \frac{4}{5}x - 4.4$

46. $y = 1.5x - 2.5$

47. $y = 4$

48. $y = \frac{3}{4}x - \frac{7}{4}$

49. $y = -\frac{2}{5}x - \frac{11}{5}$

50. $y = \frac{5}{2}x - 8$

51. **(a)** $y = -2x + 5$. **(b)** $y = \frac{1}{2}x + \frac{5}{2}$.

52. **(a)** $4x - 3y = -33$, or $y = \frac{4}{3}x + 11$.
(b) $3x + 4y = -6$, or $y = -\frac{3}{4}x - \frac{3}{2}$.

53. The intersections occur at $(0, -100)$ and $(6.\overline{6}, 0)$. A possible window is $[-5, 10] \times [-125, 25]$.

54. The intersections occur at $x = -14$ and $x = 12$; between these points it rises to a maximum of $y = 169$. A possible window is $[-20, 20] \times [-50, 200]$.

55. $Y\text{min} = -35$, $Y\text{max} = 35$, $Y\text{scl} = 5$

56. $Y\text{min} = -140$, $Y\text{max} = 140$, $Y\text{scl} = 20$

57. (a)

58. (b)

59. Local minimums at $(-3, 0)$ and $(1.2, -8.9)$, local maximums at $(-1.2, 8.9)$ and $(3, 0)$. Increasing on $[-3, -1.2]$, decreasing on $[-1.2, 1.2]$, increasing on $[1.2\ 3]$.

60. Local maximum at $(-3, 6)$, local minimum at $(2, -1)$ and $(7, -6.4)$, neither at $(4, 2)$. Decreasing on $[-3, 2]$, decreasing on $(2, 7]$.

61. Local minimum: $y = 0$ at $x = 0$ and $y = 0$ at $x = 4$; local maximum: $y \approx 9.48$ at $x \approx 2.67$

62. Local minimum: $y = -6$ at $x = 2.5$

63. Local maximum: $y \approx 50.03$ at $x \approx 0.05$; local minimum: $y \approx -91.51$ at $x \approx 6.62$

64. Local maximum: $y = 5$ at $x = 1$

65. Local minimum: $y = 0$ at $x \approx -1.85$ and $y = 0$ at $x \approx 4.85$; local maximum: $y = 11.25$ at $x = 1.5$. A possible window is $[-10, 10] \times [-5, 15]$.

66. Crosses x-axis at $x = -2$ and $x = 2$; maximum at $x \approx -1.155$ ($y \approx 3.079$) and minimum at $x \approx 1.155$ ($y \approx -3.079$). A possible window is $[-4, 4] \times [-25, 25]$.

67. Crosses x-axis at $x = -5.25$; crosses y-axis at $y = -3$. No maximum or minimum. A possible window is $[-10, 3] \times [-6, 3]$.

68. The window should have $X\text{min} < -1$ and $X\text{max} > 2$ so that the "breaks" in the graph can be seen; also, $Y\text{min} < -1$ and $Y\text{max} > 5$. Local minimum at $x = 0$, $y = 1$; local maxima at $x = -1$, $y = 2$ and $x = 2$, $y = 5$. A possible window is $[-5, 5] \times [-2, 10]$.

69. Decreasing on $(-\infty, 0]$, increasing on $[0, 1.3]$, and decreasing on $[1.3, \infty)$

70. Decreasing on $(-\infty, -2]$ and increasing on $[-2, \infty)$

71. Decreasing on $(-\infty, -2]$, constant on $[-2, 2]$, and increasing on $[2, \infty)$

72. Increasing on $(-\infty, 0]$, decreasing on $[0, 1]$, and increasing on $[1, \infty)$

73. Yes

74. No

75. Neither: $f(-x) = (-x)^3 - 2 = -x^3 - 2$, which is neither $f(x)$ nor $-f(x)$

76. Even: $g(-x) = (-x)^4 - 2(-x)^2 = x^4 - 2x^2 = g(x)$

77. Odd: $f(-x) = (-x) + \dfrac{1}{-x} = -x - \dfrac{1}{x} = -f(x)$

78. Even: $g(-x) = \dfrac{1}{2 + (-x)^2} = \dfrac{1}{2 + x^2} = g(x)$

79. Strong positive

80. Strong positive

81. Strong negative

82. Little or no correlation

83. The scatter plot in Exercise 81

84. The scatter plot in Exercise 82

85. **(a)** Shown is the scatter plot in $[5, 23] \times [400, 850]$.

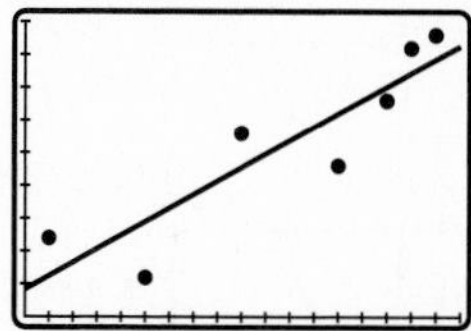

(b) The regression line, included on the scatter plot, is $y = 20.627x + 338.622$. **(c)** When $x = 30$, $y = 957.43$—about 960 doctorates are predicted for Hispanics.

86. **(a)** Shown is the scatter plot (which suggests no particular pattern) in $[5, 23] \times [1050, 1375]$.

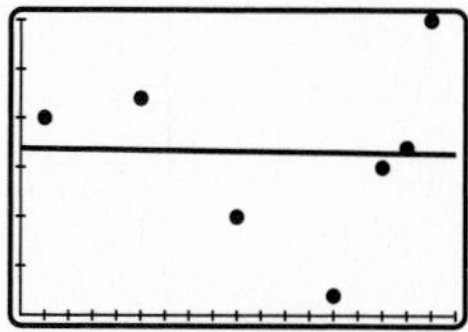

(b) The regression line, included on the scatter plot, is $y = -0.291x + 1220.33$. The correlation is very close to 0: $r = -0.01975$.
(c) When $x = 30$, $y = 1211.6$—about 1210 doctorates earned by African-Americans are predicted. Since the correlation is so small, this prediction is not very reliable.

87. **(a)** The scatter plot and regression line are shown in $[9, 23] \times [14, 24]$. When we use $x = 0$ for 1970, the equation of the line is $y = -0.54755x + 28.4715$. As long as x is measuring years, the slope should be the same, regardless of what $x = 0$ represents.

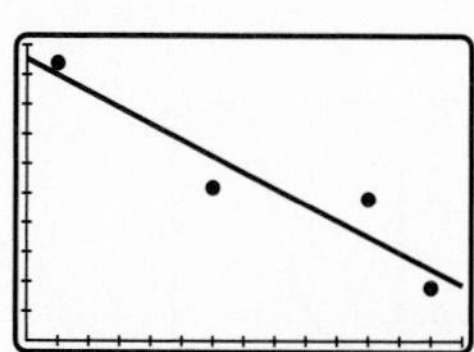

(b) When $x = 30$, $y = 12.05$.

88. **(a)** Shown are the scatter plot and regression line in $[0, 40] \times [0, 10]$. Using $x = 0$ to represent 1960, the equation of the line is $y = 0.01339x + 4.599$. The correlation is small: $r = 0.0583$.

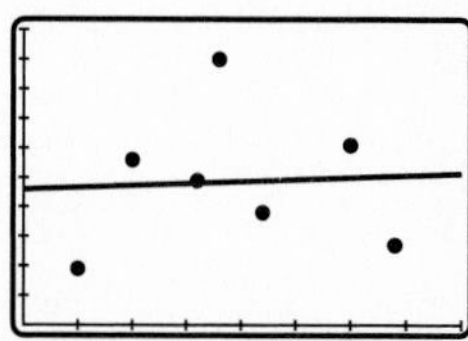

(b) When $x = 40$, $y = 5.13$ percent—note that this prediction is not very reliable.

89. $3(x + 4)$

90. $A = x(x - 5)$

91. $(1 - 0.35)x = 0.65x$

92. $(1 - 0.20)x = 0.80x$

93. **(a)** $V = (2x)(x)(x - 3) = 2x^3 - 6x^2$. **(b)** One possible window is $[3, 8] \times [0, 500]$.

94. **(a)** $S = 2(2x)(x) + 2(2x)(x - 3) + 2(x)(x - 3) = 10x^2 - 18x$. **(b)** One possible window is $[3, 8] \times [0, 500]$.

95. **(a)** $V = \pi x^2(2x) = 2\pi x^3$. **(b)** One possible window is $[0, 5] \times [0, 500]$.

96. **(a)** $S = 2\pi x(2x) + 2\pi x^2 = 6\pi x^2$ **(b)** One possible window is $[0, 5] \times [0, 100]$.

97. **(a)** $A = 1024\pi \approx 3216.99\text{ cm}^2$. **(b)** $r = \sqrt{2500/\pi} = 50/\sqrt{\pi} \approx 28.21$ cm.

98. **(a)** The window used for this graph was $X\text{min} = 0$, $X\text{max} = 10$, $X\text{scl} = 1$; $Y\text{min} = 0$, $Y\text{max} = 100{,}000$, $Y\text{scl} = 10{,}000$. **(b)** After $x = 4.5$ years.

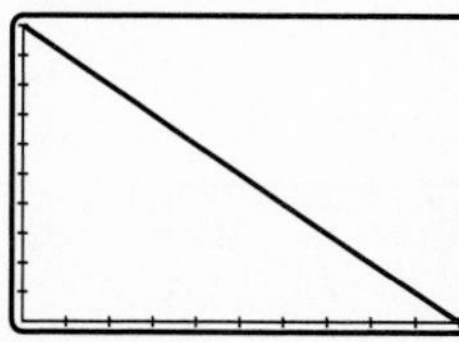

99. **(a)** $m = 5/100 = 1/20 = 0.05$; **(b)** 6500 ft; **(c)** 1056 ft

100. $x = 3.4$, $y = -1$

101. \$54,000

102. $0.24x + 0.45(30 - x) = 0.32(30)$, which gives $x \approx 18.57$ liters of the 24% solution and about 11.43 liters of the 45% solution.

103. The profit from selling x kits is $P(x) = 22x - 1500 - 15x$, which equals 0 when $x \approx 214.29$. The company will begin making money on the 215th kit sold.

104. $1 = (y - x)(20)$ and $1 = (y + x)(9)$, where y is Sandy's swimming speed and x is the speed of the current. Solving these two equations gives $x \approx 0.0306$ and $y \approx 0.0806$ miles per minute, or 1.83 mph and 4.83 mph, respectively.

105. $300 = 2x + 2(x + 5)$. The width is $x = 72.5$ in; the length is 77.5 in.

106. $1500 = 5.50x + 3.00(325 - x)$, so there were $x = 210$ adults and 115 children.

107. $1.51 = 0.01x + 0.05(23 - x - 2x) + 0.10(2x)$. Then there are $x = 6$ pennies, 12 dimes, and 5 nickels.

108. $A = x(\frac{530}{2} - x)$; the maximum value occurs when $x = 132.5$ (a square field), which gives an area of $17{,}556.25\text{ yd}^2$.

109. Picture area: $A = (16 - 2x)(20 - 2x)$. This area equals 250 in^2 when $x \approx 1.03$ in.—and also when $x \approx 16.97$ in., but that isn't reasonable in this situation.

110. **(a)** The ball is at a height of 0 (on the ground) when $t \approx 4.79$. **(b)** The maximum height, at time $t \approx 2.34$, is about 95.89 ft.

111. **(a)** $86.\overline{6}$ tickets will cover their costs; they will begin to make a profit on the 87th ticket. **(b)** 753 tickets will earn them \$499.75, and 754 will earn them \$500.50.

112. 25 hours

CHAPTER 2

QUICK REVIEW 2.1

1. 7
2. 15
3. 2
4. 5.1
5. 12.75
6. -19
7. 7 and 1
8. Undefined and 2
9. -4.5 and 0.5
10. Undefined and 2
11. $3\pi - 6$
12. $2\sqrt{5} - \sqrt{10}$

SECTION EXERCISES 2.1

1. $|5x - 2| - 3x + 2 = 0$; graph $y = |5x - 2| - 3x + 2$ $[y_1 = \text{abs}(5x - 2) - 3x + 2]$.
2. $|x - 1| - x^2 = 0$; graph $y = |x - 1| - x^2$ $[y_1 = \text{abs}(x - 1) - x^2]$.
3. $|3x - 1| - \sqrt{x^2 + 1} = 0$; graph $y = |3x - 1| - \sqrt{x^2 + 1}$ $[y_1 = \text{abs}(3x - 1) - \sqrt{(x^2 + 1)}]$.
4. $x^3 - 3x - x^2 - 7x = 0$, or $x^3 - x^2 - 10x = 0$; graph $y = x^3 - x^2 - 10x$ $(y_1 = x^3 - x^2 - 10x)$.
5. $-4x + 3 - x^2 + 125 = 0$, or $-4x - x^2 + 128 = 0$; graph $y = -4x - x^2 + 128$ $(y_1 = -4x - x^2 + 128)$.
6. $7x + x\sqrt{x} - 8 = 0$; graph $y = 7x + x\sqrt{x} - 8$ $(y_1 = 7x + x\sqrt{x} - 8)$.
7. $4x^2 - \sqrt{x^2 + 3} = 0$; graph $y = 4x^2 - \sqrt{x^2 + 3}$ $[y_1 = 4x^2 - \sqrt{(x^2 + 3)}]$.
8. $|x^3 + 1| - \sqrt{x + 3} = 0$; graph $y = |x^3 + 1| - \sqrt{x + 3}$ $[y_1 = \text{abs}(x^3 + 1) - \sqrt{(x + 3)}]$.
9. $x = -2$ or $x = 4$
10. $x = -3$ or $x = -1$
11. $x = -1$ or $x = 2$
12. $x = 1.5$ or $x = 2.5$
13. $x = -1$
14. $x \approx -0.3333$ or $x = 11$
15. $x = -2$
16. $x = -1$
17. $x = 5$
18. $x \approx 1.54$ or $x \approx 8.46$
19. $t = 6$ or $t = 10$
20. $x = -5$ or $x = 3$
21. $x = -3$ or $x = 7$
22. $t = 1.25$ or $t = 1.75$
23. $x = -0.2$ or $x = 1.4$
24. $x = -3.9$ or $x = 2.9$
25. $x \approx -2.41$ or $x \approx 2.91$
26. $x \approx -1.64$ or $x \approx 1.45$
27. $x \approx 3.91$

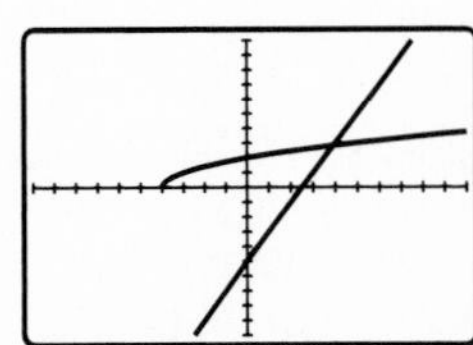

Shown in $[-10, 10] \times [-10, 10]$.

28. $x \approx -1.09$ or $x \approx 2.86$

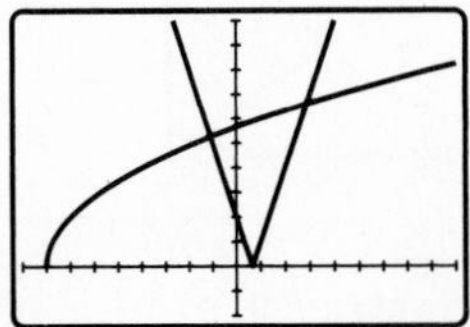

Shown in $[-9, 9] \times [-2, 10]$.

29. $x \approx 1.33$ or $x = 4$

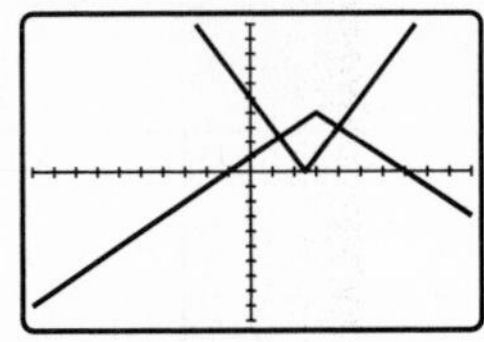

Shown in $[-10, 10] \times [-10, 10]$.

30. $x \approx 2.66$

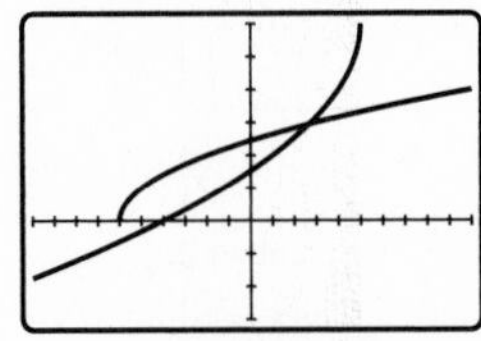

Shown in $[-10, 10] \times [-3, 6]$.

31. $x \approx -1.59$ or $x \approx 1.77$

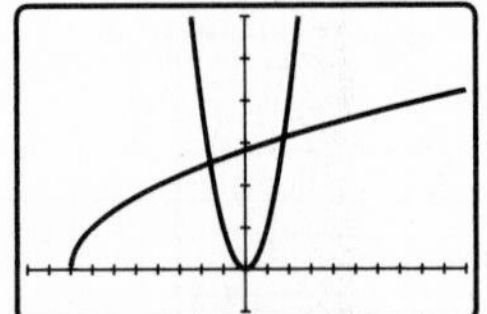

Shown in $[-10, 10] \times [-1, 6]$.

32. $x \approx -2.30$ or $x \approx 2.30$

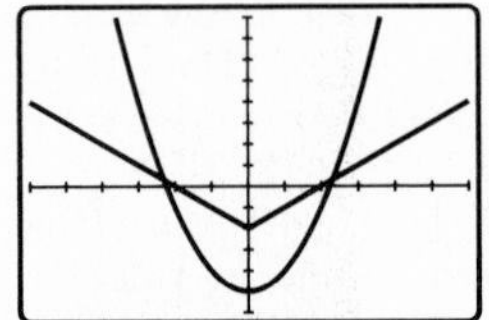

Shown in $[-6, 6] \times [-6, 8]$.

33. $x \approx 1.77$

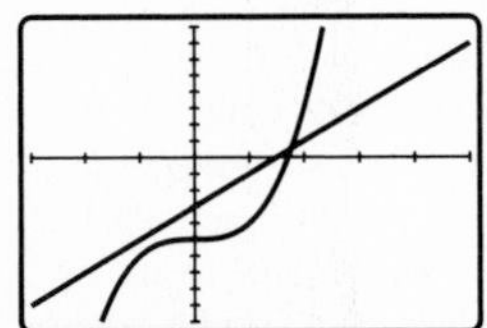

Shown in $[-3, 5] \times [-10, 8]$.

34. $x \approx 2.36$

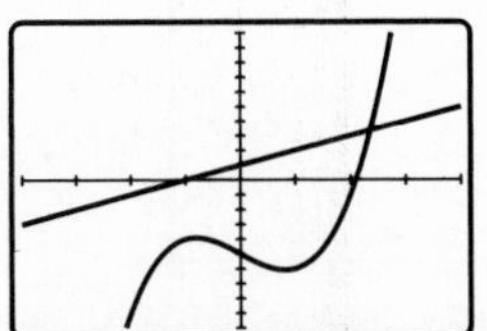

Shown in $[-4, 4] \times [-10, 10]$.

35. $x \approx -4.24$ or $x \approx 0.24$

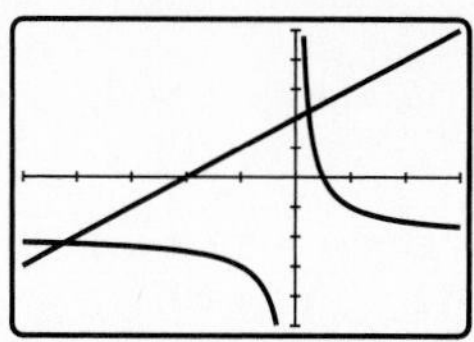

Shown in $[-5, 3] \times [-5, 5]$.

36. $x = -0.5$ or $x = 1$

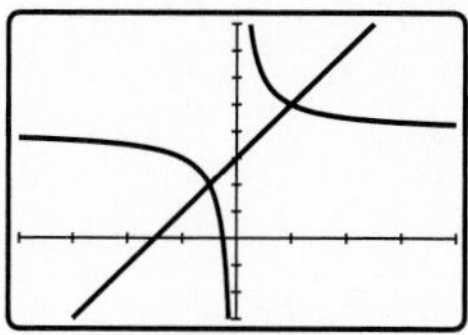

Shown in $[-4, 4] \times [-3, 8]$.

37. $x \approx -1.26$ or $x \approx 0.26$

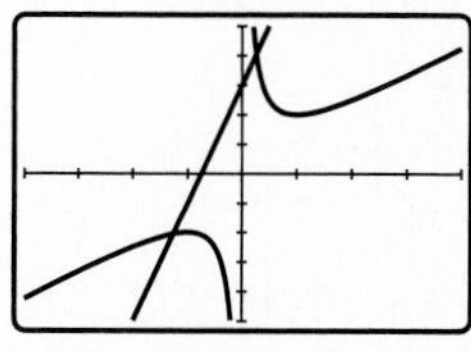

Shown in $[-4, 4] \times [-4, 5]$.

38. $x \approx -1.47$

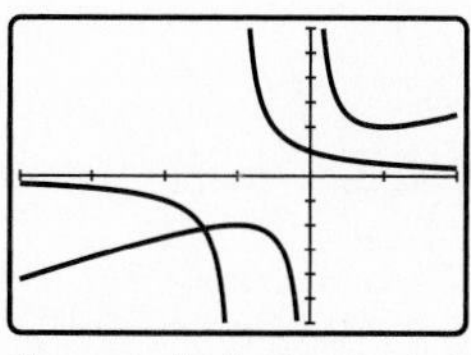

Shown in $[-4, 2] \times [-6, 6]$.

39. $x = 0$ is not a solution.

40. $x = 1$ is not a solution.

41. $x = 1$ is not a solution.

42. $x = 2$ is not a solution.

43. **(a)** The two functions are $y_1 = 3\sqrt{x + 4}$ (the one that begins on the x-axis) and $y_2 = x^2 - 1$. **(b)** This is the graph of $y = 3\sqrt{x + 4} - x^2 + 1$. **(c)** The x-coordinates of the intersections in the first picture are the same as the x-coordinates where the second graph crosses the x-axis.

44. **(a)** $4x + 5 = x^3 + 2x^2 - x + 3$. **(b)** $-x^3 - 2x^2 + 5x + 2 = 0$. **(c)** The two equations are equivalent: We can obtain the second from the first by subtracting y_2 from both sides.

45. $x = -0.5$ or $x = 1.5$

46. $x \approx -2.65$, $x = -1$, $x = 1$, or $x \approx 2.65$

47. $x = -0.5$ or $x = 1$

48. $x = -1$

49. $x = -3$ or $x = 1$

50. $x = -1.25$ or $x = 2.5$

51. $x \approx 0.62$

52. $x \approx 4.30$

53. $x = 5$

54. $x = 1$ or $x = 2$

55. $x \approx 0.38$

56. $x \approx 2.75$

57. **(a)** about 3.35 sec; **(b)** 2500 ft

58. **(a)** The horizontal axis is distance d, and the vertical axis is time t. The curve is the graph of $t = \frac{1}{4}\sqrt{d}$, and the three lines are (bottom to top) $t = 15$, $t = 18$, and $t = 22$. The intersections of these lines with the curve show the distances Beth and Kristin have fallen by the time they get in formation (after 15 sec), when they break formation (after 18 sec), and when they pull their rip cords (after 22 sec). **(b)** If they jump from 8000 ft, they would hit the ground after 22.36 sec (without parachutes). That is too close to the time they plan to pull their rip cords.

59. **(a)** Shown in $[7, 14.5] \times [430, 510]$.

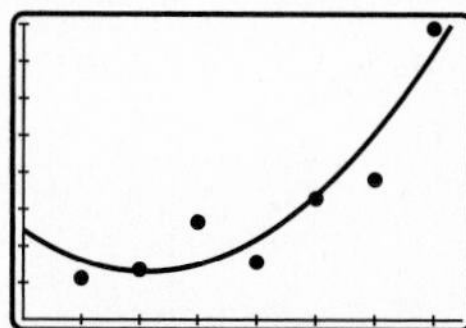

(c) $y = 575$ when $x = 1.86$ or $x = 16.38$. The first is between 1981 and 1982; the second (and the one we are really interested in) is about 1996.

60. **(a)**, **(b)**, and **(c)** The graph is shown in $[0, 25] \times [45, 70]$ along with the linear regression model $y = 0.95x + 45.34$ (using $x = 5$ for 1975, etc.).

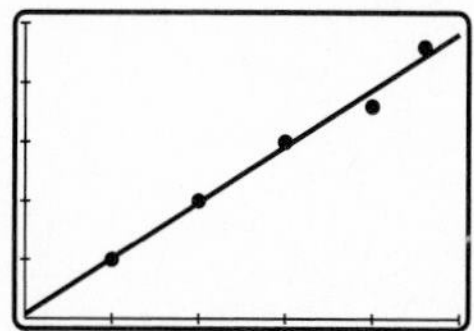

(d) $y = 75$ when $x \approx 31.2$, or in about 2001.

61. **(a)** Any value of $c < -\sqrt{5.5} = -2.345\ldots$ **(b)** Any value of $c \geq -\sqrt{5.5}$.

62. **(a)** Any value of c with $4 \leq c < 6$. **(b)** Any value of $c > 6$. **(c)** $c = 6$ or $c < 4$. **(d)** Arguments may vary.

63. **(a)** $c \approx 1.409$ or $c > 4$. **(b)** $1.409 < c \leq 4$

64. **(a)** $c = 0$ or $c > 9$. **(b)** $0 < c \leq 9$.

65. Any c between approximately -2.113 and 0.631

66. Any c between approximately -6.268 and -1.583

67. Two functions are equal if, given the same input, they return the same output. If $x < 1$, then $|x - 1| = 1 - x$, so $g(x) = -(1 - x) + 1 = x$, the same as $f(x)$. If $x \geq 1$, then $g(x) = -(x - 1) + 1 = 2 - x = f(x)$.

68. If $x < 0$, then $\sqrt{x^2} = -x = |x|$; if $x \geq 0$ then $\sqrt{x^2} = x = |x|$.

69. If, for any input value, the outputs are different, then the functions are different. For any $x < -1$, this is the case. Take $x = -2$: $f(-2) = \sqrt{(-2)^2 + 2(-2) + 1} = 1$ while $g(-2) = -1$.

70. For any $x \neq 0$, $f(x) \neq g(x)$. Take $x = 1$: $f(1) = |-(-1)| = 1$ but $g(1) = -|-1| = -1$.

QUICK REVIEW 2.2

1. $|a|$

2. $|u|$

3. $2\,|y|\,\sqrt{5}$

4. $|x|\,\sqrt{y + 3}$

5. 9

6. 6

7. $3\sqrt{6}$

8. $2\sqrt{3}$

9. Try, for example, $x = 1$: $\sqrt{10} \neq 4$.

10. With $x = -4$: $\sqrt{1} \neq -1$

11. $(x - 3)(x + 3)$

12. $(x + 1)(x + 3)$

13. $(2x - 1)(x + 3)$

14. $(2x - 5)(x + 6)$

SECTION EXERCISES 2.2

1. $x = \pm\sqrt{5}$

2. $x = \pm\frac{5}{2}$

3. $x = 5 \pm \sqrt{8.5}$

4. $x = -4 \pm \sqrt{8/3}$

5. $u = -1 \pm \sqrt{4.5}$

6. $x = -1 \pm \sqrt{17}$

7. $v = \pm\sqrt{13/3}$

8. $x = -11 \pm 11$; $x = -22$ or $x = 0$

9. $x = 0.5$

10. $x = \dfrac{1 - 3\sqrt{2}}{\sqrt{2} + 3} \approx -0.73$ or
$x = \dfrac{-1 - 3\sqrt{2}}{\sqrt{2} - 3} \approx 3.31$

11. $x^2 - 2x - 2 = 0$; $x \approx -0.73$ or $x \approx 2.73$

12. $2x^2 + x - 8 = 0$; $x \approx -2.27$ or $x \approx 1.77$

13. $x^2 - 18 = 0$; $x \approx -4.24$ or $x \approx 4.24$

14. $\frac{1}{3}x^2 - \frac{13}{3}x + \frac{4}{3} = 0$, or $x^2 - 13x + 4 = 0$; $x \approx 0.32$ or $x \approx 12.68$

15. $(x + b)^2$ is the area of the whole square; the three terms on the right side give the areas of the $(x \times x)$ square, the two $(b \times x)$ rectangles, and the $(b \times b)$ square, respectively.

16. The missing piece is a $\frac{b}{2} \times \frac{b}{2}$ square, which has area $b^2/4$.

17. Completed square: $x^2 + 8x + 16 = (x + 4)^2$

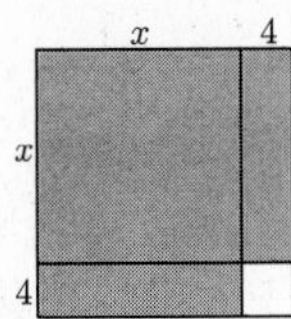

18. Completed square: $x^2 - 2x + 1 = (x - 1)^2$

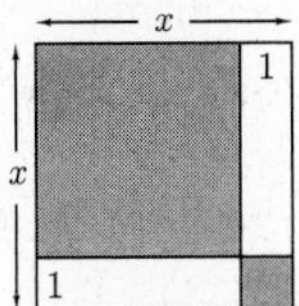

19. Completed square: $x^2 - 9x + 4.5^2 = (x - 4.5)^2$

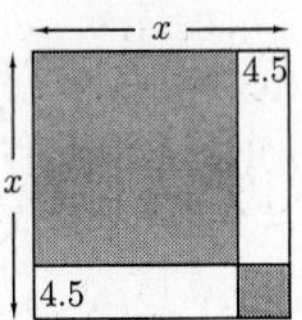

20. Completed square: $x^2 - 15x + 7.5^2 = (x - 7.5)^2$

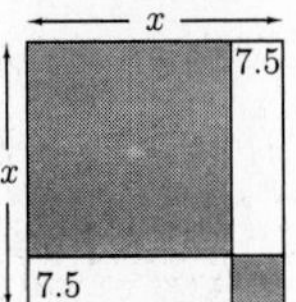

21. $(x + 3)^2 = 16$, so $x = -3 \pm 4$; $x = -7$ or $x = 1$

22. $(x + 2)^2 = 25$, so $x = -2 \pm 5$; $x = -7$ or $x = 3$

23. $(x + 2.5)^2 = 15.25$, so $x = -2.5 \pm \sqrt{15.25}$; $x = -6.405\ldots$ or $x = 1.405\ldots$

24. $(x - 3.5)^2 = 13$, so $x = 3.5 \pm \sqrt{13}$; $x = -0.105\ldots$ or $x = 7.105\ldots$

25. $(x - 3)^2 = 12$, so $x = 3 \pm \sqrt{12}$; $x = -0.464\ldots$ or $x = 6.464\ldots$

26. $(x + 3)^2 = 13$, so $x = -3 \pm \sqrt{13}$; $x = -6.605\ldots$ or $x = 0.605\ldots$

27. $(x - 4.5)^2 = 12.25$, so $x = 4.5 \pm 3.5$; $x = 1$ or $x = 8$

28. $3(x - \frac{4}{3})^2 = \frac{46}{3}$, so $x = \frac{4}{3} \pm \frac{1}{3}\sqrt{46}$; $x = -0.927\ldots$ or $x = 3.594\ldots$

29. (a)

30. (a)

31. Rewritten as $x^2 - c = 0$, this is the standard form with $a = 1$ and $b = 0$.

32. This cannot be rewritten in the standard form.

33. $x = -\frac{1}{2} \pm \frac{1}{2}\sqrt{5}$; $x = -1.618\ldots$ or $x = 0.618\ldots$

34. $x = 2 \pm \sqrt{2}$; $x = 0.585\ldots$ or $x = 3.414\ldots$

35. $x = -4 \pm 3\sqrt{2}$; $x = -8.242\ldots$ or $x = 0.242\ldots$

36. $x = \frac{3}{4} \pm \frac{1}{4}$; $x = 0.5$ or $x = 1$

37. $x = 1 \pm 2\sqrt{2}$; $x = -1.828\ldots$ or $x = 3.828\ldots$

38. $x = \frac{3}{2} \pm \frac{5}{2}$; $x = -1$ or $x = 4$

39. $x = -4 \pm \sqrt{21}$; $x = -8.582\ldots$ or $x = 0.582\ldots$

40. $x = \frac{1}{2}\sqrt{3} \pm \frac{1}{2}\sqrt{23}$; $x = -1.531\ldots$ or $x = 3.263\ldots$

41. $x = -\frac{7}{2} \pm \frac{1}{2}\sqrt{105}$; $x = -8.623\ldots$ or $x = 1.623\ldots$

42. $x = 2 \pm 4$; $x = -2$ or $x = 6$

43. $(x + 2)(x - 1) = 0$; $x = -2$ or $x = 1$

44. $(x - 2)(x - 3) = 0$; $x = 2$ or $x = 3$

45. $(x + 4)(x - 5) = 0$; $x = -4$ or $x = 5$

46. $(x - 1)(x - 3) = 0$; $x = 1$ or $x = 3$

47. $(x + 3)(2x - 1) = 0$; $x = -3$ or $x = 0.5$

48. $(2x - 1)(2x - 3) = 0$; $x = 0.5$ or $x = 1.5$

49. $(x - 3)(x - 5) = 0$; $x = 3$ or $x = 5$

50. $(x + 5)(x - 1) = 0$; $x = -5$ or $x = 1$

51. $(2x + 3)(x - 4) = 0$; $x = -1.5$ or $x = 4$

52. $(x + 2)(2x - 5) = 0$; $x = -2$ or $x = 2.5$

53. $D = 81$; two real solutions

54. $D = -8$; no real solutions

55. $D = 14$; two real solutions

56. $D = 0$; one real solution

57. $D = -399$; no real solutions

58. $D = 1$; two real solutions

59. $x = 0.5$ or $x = 2.5$

60. $x \approx -0.46$ or $x \approx 5.12$ (graphically or quadratic formula)

61. $x = -3$ or $x = 1$

62. $x = -2 \pm \sqrt{3}$; that is, $x = -3.732\ldots$ or $x = -0.267\ldots$ (completing the square)

63. $x \approx -4.82$ or $x \approx 6.23$

64. $x = 2 \pm \sqrt{7}$; that is, $x = -0.645\ldots$ or $x = 4.645\ldots$ (quadratic formula)

65. $x = -3$ or $x = 1$

66. $x \approx -0.71$ or $x = 4.24$ (graphically)

67. $x \approx -3.26$ or $x \approx 1.53$

68. $t = -0.6$ or $t = 2$ (factoring or quadratic formula)

69. $x = 3 \pm \dfrac{\sqrt{b}}{a}$

70. $x = \pm v/u$ (completing the square)

71. 80 yd × 110 yd

72. 53 yd × 120 yd

73. about 14.980 ft

74. $t = 0.25\sqrt{1500} = 9.682\ldots$ sec

75. In 16 seconds, the skydivers will fall 4096 ft, which will not leave them with the 1500 ft they want.

76. $x^2 + x^2 = 18^2$, so $x \approx 12.73$ in.

77. $200 = x^2 + \frac{1}{2}\pi(x/2)^2$, so $x \approx 11.98$ ft

78. The CB will fade out after $(30t)^2 + (10t)^2 = 2^2$, which happens when $t \approx 0.0632$ h, or 3.795 min.

79. $(2r)^2 + (2(r + 30))^2 = 1620^2$, so r, the Airbus's speed, is approximately 557.56 mph, while the DC10 is traveling at about 587.56 mph.

80. **(a)** Shown is the window $[0, 25] \times [0, 100]$.

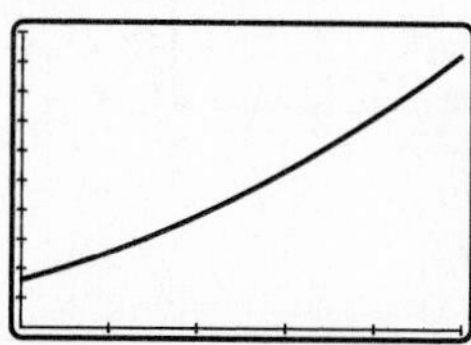

(b) When $x = 25$, the expenditures are predicted to be \$93.06 billion. **(c)** $x = 54.6$, or in about 2035.

81. **(a)** $[0, 100] \times [0, 1000]$ is one possibility. **(b)** either 107,335 units or 372,665 units

82. $v = 2u \pm \sqrt{4u^2 - 3}$

83. $u = \dfrac{v^2 + 3}{4v}$ or $u = 0$

84. There are two distinct real roots.

85. $ax^2 + bx + c =$

$$a\left(x + \frac{b}{2a} - \frac{\sqrt{b^2 - 4ac}}{2a}\right) \cdot \left(x + \frac{b}{2a} + \frac{\sqrt{b^2 - 4ac}}{2a}\right)$$

The two factors are the same if $b^2 - 4ac = 0$.

86. Let $D = b^2 - 4ac$;

$$\frac{-b + \sqrt{D}}{2a} + \frac{-b - \sqrt{D}}{2a} = \frac{-2b + \sqrt{D} - \sqrt{D}}{2a} = -b/a.$$

87. Let $D = b^2 - 4ac$.

$$\frac{-b + \sqrt{D}}{2a} \cdot \frac{-b - \sqrt{D}}{2a} = \frac{(-b)^2 - (\sqrt{D})^2}{4a^2} = \frac{b^2 - (b^2 - 4ac)}{4a^2} = \frac{c}{a}$$

88. $5 = x_1 + x_2 = -b/a = -b/2$, so $b = -10$; $3 = x_1 \cdot x_2 = c/a = c/2$, so $c = 6$. The solutions are $2.5 \pm \frac{1}{2}\sqrt{13}$, or approximately 0.697 and 4.303.

89. **(a)** $D = 0$, so the answer is (2). **(b)** $D = \frac{3}{4}b^4 > 0$, so the answer is (3). **(c)** $D = -2b^2 < 0$, so the answer is (1).

90. Rearrange the cubic polynomial so that it is of the form $(ax + b)^3 = c$; then take cube roots of both sides and solve the resulting linear equation.

91. The x-coordinates of the intersection points correspond to solutions to the equation.

92. **(a)** This is the projectile motion formula with $v_0 = s_0 = 0$, except that we are considering "down" to be a *positive* direction, so we drop the negative. **(b)** $t = \sqrt{\frac{1}{16}y} = \frac{1}{4}\sqrt{y}$, or the same formula used for the earlier problem.

QUICK REVIEW 2.3

1. −5 −4 −3 −2 −1 0 1 2 3 4 5

2. −8 −7 −6 −5 −4 −3 −2 −1 0 1 2

3. −1 0 1 2 3 4 5 6 7 8 9

4. −11 −9 −7 −5 −3 −1 1 3 5 7 9

5. $(-1, 1)$

6. $(0, 3]$

7. $x = 0.4$

8. $x = 22/7$

9. $x < 4$

10. $x \geq 2$

11. $x < 0.8$

12. $x \geq 4/3$

13. $x \neq 2$; $(-\infty, 2) \cup (2, \infty)$

14. $x^2 \neq 1$; $(-\infty, -1) \cup (-1, 1) \cup (1, \infty)$

15. $x \geq 3$; $[3, \infty)$

16. $x^2 \geq 4$; $(-\infty, -2] \cup [2, \infty)$

SECTION EXERCISES 2.3

1. $x = 4$ or $x = -4$

2. $x = 11/3$ or $x = -11/3$

3. $x = -2$ or $x = 3$

4. $x = -13/3$ or $x = 5/3$

5. $x = -2 \pm 2\sqrt{3}$

6. $x = 1.5 \pm \frac{1}{2}\sqrt{17}$ or $x = 1.5 \pm \frac{1}{2}\sqrt{33}$

7. $x = 0$ or $x = -2$ (extraneous)

8. $x = -4$ or $x = 4$

9. $x = -0.5 \pm \frac{1}{2}\sqrt{5}$, so $x = -1.618\ldots$ (extraneous) or $x = 0.618\ldots$

10. $x = 2.5 \pm \frac{1}{2}\sqrt{13}$, so $x = 0.697\ldots$ (extraneous) or $x = 4.302\ldots$

11. $x = -3$ (extraneous) or $x = 5$

12. $x = 0$ (extraneous) or $x = 3$

13. $x = 1.5 \pm \frac{1}{2}\sqrt{5}$, so $x = 0.381\ldots$ or $x = 2.618\ldots$ (extraneous)

14. $x = 48 \pm 32\sqrt{2}$, so $x = 2.745\ldots$ or $x = 93.254\ldots$ (extraneous)

15. $x \approx -4.56$, $x \approx -0.44$, $x = 1$

16. $x = 0.25$, $x = 0.5$, $x = 4/3$

17. $x = -1$, $x = -0.75$, $x = 0.5$, $x = 1$

18. $x = -0.75$, $x = -0.5$, $x = 0.5$, $x = 1$

19. $x \approx -2.42$, $x \approx 4.88$

20. $x \approx -1.20$, $x \approx -0.11$, $x \approx 1.26$

21. $x \approx -4.53$, $x \approx 0.28$, $x \approx 2.53$

22. $x \approx -4.37$, $x \approx 2.83$

23. $x \approx 2.93$

24. $x \approx -2.41$, $x \approx -2.24$, $x = 0$, $x \approx 0.41$, $x = 2$, $x \approx 2.24$

25. $x \approx 1.27$

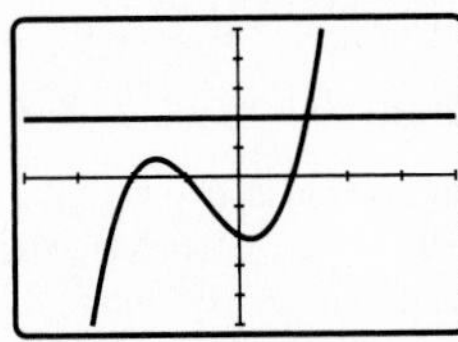

Shown in $[-4, 4] \times [-5, 5]$.

26. $x \approx 2.27$

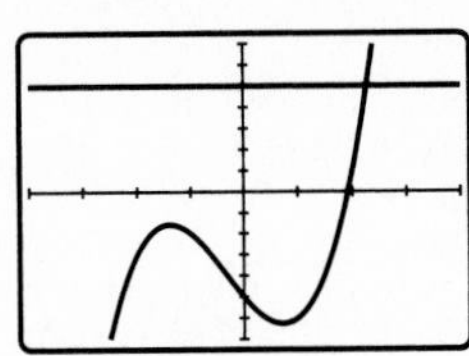

Shown in $[-4, 4] \times [-7, 7]$.

27. $x \approx -1.44$

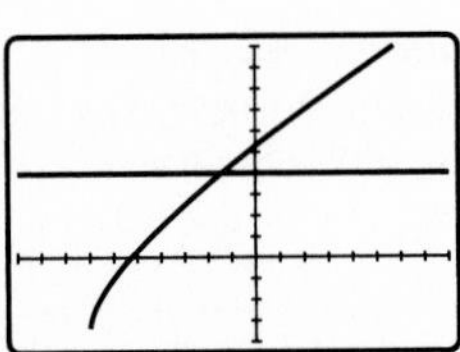

Shown in $[-10, 8] \times [-4, 10]$.

28. $x \approx 0.47$

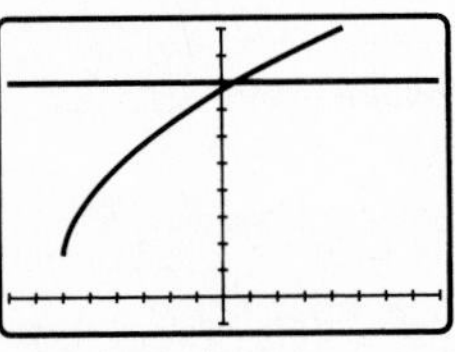

Shown in $[-8, 8] \times [-1, 10]$.

29. All real x, or $(-\infty, \infty)$

30. $[2, \infty)$:

31. $(39/19, \infty)$:

32. $(6/7, \infty)$:

33. $(-2, 23/2]$:

34. $[-25, 19)$:

35. $(3/2, 9/2]$:

36. $(3, 16)$:

37. $(-2, 2)$:

38. $(-3, 3)$:

39. $(-\infty, -9] \cup [1, \infty)$:

40. $(-\infty, -1.3) \cup (2.3, \infty)$:

41. $(1, 5)$:

42. $[-8, 2]$:

43. $(-2/3, 10/3)$:

44. $(-\infty, -11] \cup [7, \infty)$:

45. $[0.5, 5]$:

46. $[-4.5, -2)$:

47. $x > 20/3$, or $(20/3, \infty)$

48. $x < -4/3$ or $x > 4/3$, or $(-\infty, -4/3) \cup (4/3, \infty)$

49. $0 < x < 2$ or $x > 2$, or $(0, 2) \cup (2, \infty)$

50. $x < 3$ and $x \neq -2$, or $(-\infty, -2) \cup (-2, 3)$

51. $x < -\frac{1}{3}$ or $x > \frac{1}{3}$, or $(-\infty, -1/3) \cup (1/3, \infty)$

52. $x < 3/2$, or $(-\infty, 3/2)$

53. $x \geq -3$, or $[-3, \infty)$

54. $x > -1$, or $(-1, \infty)$

55. $-\frac{2}{3} < x < 4$, or $(-2/3, 4)$

56. $x < -4$ or $x > -2/3$, or $(-\infty, -4) \cup (-2/3, \infty)$

57. $(-4.8, \infty)$:

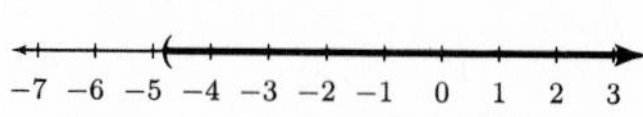

58. $(20/13, \infty)$:

−3 −2 −1 0 1 2 3 4 5 6 7

59. $(-\infty, 15)$

60. $[-10, \infty)$

61. $(8, \infty)$

62. $(-\infty, 20)$

63. $2 < x < 3.6$, or $(2, 3.6)$

64. $-3 \leq x < 2$, or $[-3, 2)$

65. $-1 < x < 0$, or $(-1, 0)$

66. $x \geq 7$, or $[7, \infty)$.

67. 52.5 mph

68. $P = 2[x + (2x - 2)] < 200$ and $2x - 2 > 0$, so 1 in. $< x <$ 34 in.

69. 9091 candy bars per week

70. $10 \leq P \leq 20$

71. They can borrow up to $100,000.

72. **(a)** Assuming that all can ride in one vehicle, $5.75 + 9.50(x + 1) \leq 65$. **(b)** $x \leq 5.236\ldots$ **(c)** Sarah may take no more than 5 friends (6 people altogether) to the concert.

73. Either $x \approx 0.94$ in. or $x \approx 3.78$ in.

74. $x \approx 0.89$ in., or about 0.31 in. smaller.

75. $x \approx 3.11$ in.

76. $x \approx 1.93$ in.

77. Either $x \approx 1.68$ in. or $x \approx 9.10$ in.

78. $x \approx 3.49$ mi

79. **(a)** When $x = 40$, $y = \$488.9$ billion. **(b)** $x \approx 18.48$—between 1978 and 1979.

80. $x < 4.166\ldots$, so the electrician worked no more than 4 hours, 7.5 minutes (which rounds to 4 hours).

81. Answers may vary.

82. Use the window $[0, 40] \times [0, 50]$ to graph **(a)** and **(b)**. **(c)** $t \approx 59.57$, or in about 2020.

83. Use the window $[0, 25] \times [0, 140]$ to graph **(a)** and **(b)**. **(c)** According to the model, when $x \approx 37.85$ the costs equal $200, which occurs between 2007 and 2008.

84. $x \approx \pm 1.80$. Note: To solve algebraically, use the quadratic formula applied to $0.5z^2 - z - 2 = 0$ to determine that $z = x^2 = 1 \pm \sqrt{5}$; then $x = \pm\sqrt{1 + \sqrt{5}}$.

85. **(a)** x_0 falls in the interval $(3 - \frac{1}{3}, 3 + \frac{1}{3}) = (\frac{8}{3}, \frac{10}{3})$, so $3x_0$ is in $(8, 10)$ and $3x_0 - 5$ is in $(3, 5)$. **(b)** The vertical bars mark x values within $\frac{1}{3}$ of 3; the horizontal bars mark y values which are within 1 of 4; the graph shows that x values in the former set lead to y (function) values in the latter set. **(c)** $2.99 < x < 3.01$, so $3(2.99) - 5 < 3x - 5 < 3(3.01) - 5$ and $3.97 < f(x) < 4.03$; the vertical bars in the graph would be closer together, as would the horizontal bars.

86. Graph both functions; any x values for which the two graphs *do not* cross would be solutions to this inequality.

87. **(a)** $x = -\sqrt{5}$. **(b)** $x = \pm\sqrt{1.5}$. **(c)** No real solutions. **(d)** $x = 3.5$

88. **(a)** $A = A(x) = (8.5 - 2x)(11 - 2x)$. **(b)** $x \approx 1.28$ in.

89. The TEST commands return 1 for true statements (for example, $2x - 5 < 8$, when $x = 5.2$) and 0 for false statements (such as $2(5.2) - 5 > 12$).

QUICK REVIEW 2.4

1. $-5x^3 + 2x^2 - x - 3$

2. $x^2 + 6x + 5$

3. $2x^4 - x^2 - 1$

4. $x^4 + x^3 - x^2 - x$

5. $x \geq -1$, or $[-1, \infty)$

6. $x \neq 0$, or $(-\infty, 0) \cup (0, \infty)$

7. All reals, or $(-\infty, \infty)$

8. $x \neq 3$ and $x \neq -2$, or $(-\infty, -2) \cup (-2, 3) \cup (3, \infty)$

9. $x > 1$, or $(1, \infty)$.

10. $x \geq -4$ and $x < 5$, or $[-4, 5)$

SECTION EXERCISES 2.4

1. $(f + g)(x) = 2x - 1 + x^2$, $(f - g)(x) = 2x - 1 - x^2$, and $(fg)(x) = (2x - 1)(x^2) = 2x^3 - x^2$; all three domains are $(-\infty, \infty)$.

2. $(f + g)(x) = (x - 1)^2 + 3 - x = x^2 - 3x + 4$, $(f - g)(x) = (x - 1)^2 - 3 + x = x^2 - x - 2$, and $(fg)(x) = (x - 1)^2(3 - x) = -x^3 + 5x^2 - 7x + 3$; all domains are $(-\infty, \infty)$.

3. $(f + g)(x) = x^2 + 2x$, $(f - g)(x) = x^2 - 2x$, and $(fg)(x) = 2x^3$; all domains: $(-\infty, \infty)$

4. $(f + g)(x) = \sqrt{x} + x - 2$, $(f - g)(x) = \sqrt{x} - x + 2$, and $(fg)(x) = (x - 2)\sqrt{x}$; all domains are $[0, \infty)$.

5. $(f + g)(x) = x + 3 + \dfrac{2x - 1}{3} = \dfrac{5x + 8}{3}$, $(f - g)(x) = x + 3 - \dfrac{2x - 1}{3} = \dfrac{x + 10}{3}$, and $(fg)(x) = (x + 3)\left(\dfrac{2x - 1}{3}\right) = \dfrac{2x^2 + 5x - 3}{3}$; all domains: $(-\infty, \infty)$

6. $(f+g)(x) = \sqrt{x+5} + |x+3|$, $(f-g)(x) = \sqrt{x+5} - |x+3|$, and $(fg)(x) = |x+3|\sqrt{x+5}$; all domains are $[-5, \infty)$.

7. $(f/g)(x) = \dfrac{\sqrt{x+3}}{x^2}$, domain: $[-3, 0) \cup (0, \infty)$; $(g/f)(x) = \dfrac{x^2}{\sqrt{x+3}}$, domain: $(-3, \infty)$

8. $(f/g)(x) = \dfrac{\sqrt[3]{x+1}}{x^2+1}$, domain: $(-\infty, \infty)$; $(g/f)(x) = \dfrac{x^2+1}{\sqrt[3]{x+1}}$, domain: $(-\infty, -1) \cup (-1, \infty)$

9. $(f/g)(x) = \dfrac{\sqrt{x-2}}{\sqrt{x+4}} = \sqrt{\dfrac{x-2}{x+4}}$, domain: $[2, \infty)$; $(g/f)(x) = \dfrac{\sqrt{x+4}}{\sqrt{x-2}} = \sqrt{\dfrac{x+4}{x-2}}$, domain: $(2, \infty)$

10. $(f/g)(x) = \dfrac{(x+3)(x-2)}{x^2} = 1 + \dfrac{1}{x} - \dfrac{6}{x^2}$, domain: $(-\infty, 0) \cup (0, \infty)$
$(g/f)(x) = \dfrac{x^2}{(x+3)(x-2)} = \dfrac{x^2}{x^2+x-6}$, domain: $(-\infty, -3) \cup (-3, 2) \cup (2, \infty)$

11.

12.

13. $x = 0$ and $x = 3/2$ are zeros of f and g; the graph of $f + g$ will cross the graphs of g and f, respectively, at those points.

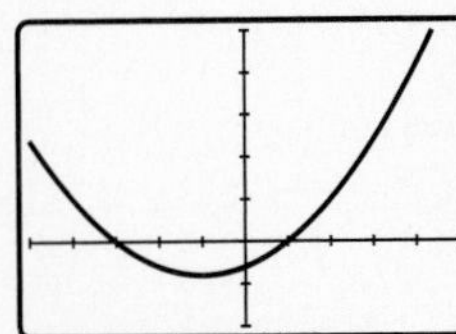

14. At $x = -4$ and $x = 1$, $f(x) = g(x)$, so $(f-g)(x) = 0$

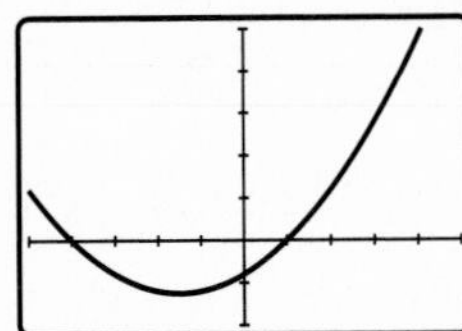

15. $(f \circ g)(3) = 5$; $(g \circ f)(-2) = -6$

16. $(f \circ g)(3) = 8$; $(g \circ f)(-2) = 3$

17. $(f \circ g)(3) = 2$; $(g \circ f)(-2) = \sqrt{3}$

18. $(f \circ g)(3) = 9$; $(g \circ f)(-2) = 66$

19. $f(g(x)) = 3x - 1$, domain: $(-\infty, \infty)$; $g(f(x)) = 3x + 1$, domain: $(-\infty, \infty)$

20. $f(g(x)) = \dfrac{1}{(x-1)^2} - 1$, domain: $(-\infty, 1) \cup (1, \infty)$; $g(f(x)) = \dfrac{1}{(x^2-1)-1} = \dfrac{1}{x^2-2}$, domain: $(-\infty, -\sqrt{2}) \cup (-\sqrt{2}, \sqrt{2}) \cup (\sqrt{2}, \infty)$

21. $f(g(x)) = x - 2$, domain: $(-\infty, \infty)$; $g(f(x)) = x - 1$, domain: $(-\infty, \infty)$

22. $f(g(x)) = (\sqrt{x+1})^2 - 2 = x - 1$, domain: $[-1, \infty)$; $g(f(x)) = \sqrt{(x^2-2)+1} = \sqrt{x^2-1}$, domain: $(-\infty, -1] \cup [1, \infty)$

23. $f(g(x)) = \dfrac{1}{\sqrt{x}-1}$, domain: $[0, 1) \cup (1, \infty)$; $g(f(x)) = \sqrt{\dfrac{1}{x-1}}$, domain: $(1, \infty)$

24. $f(g(x)) = (\sqrt{x+2})^2 - 3 = x - 1$, domain: $[-2, \infty)$; $g(f(x)) = \sqrt{(x^2-3)+2} = \sqrt{x^2-1}$, domain: $(-\infty, -1] \cup [1, \infty)$

25. One possibility: $f(x) = \sqrt{x}$ and $g(x) = x^2 - 5x$

26. One possibility: $f(x) = (x+1)^2$ and $g(x) = x^3$

27. One possibility: $f(x) = |x|$ and $g(x) = 3x - 2$

28. One possibility: $f(x) = 1/x$ and $g(x) = x^3 - 5x + 3$

29. One possibility: $f(x) = \sqrt[3]{x}$ and $g(x) = x^2 + 1$

30. One possibility: $f(x) = |x|$ and $g(x) = x^2 + 5$

31. One possibility: $f(x) = x^2$ and $g(x) = x + 3$

32. One possibility: $f(x) = x^3$ and $g(x) = \dfrac{1}{x+1}$

33. One possibility: $f(x) = \sqrt{x}$ and $g(x) = x + 3$

34. One possibility: $f(x) = 2/x$ and $g(x) = (x-3)^2$

35. One possibility: $f(x) = x^4 - 2$ and $g(x) = x - 3$

36. One possibility: $f(x) = 3 - x$ and $g(x) = \sqrt{x}$

37. $t = \sqrt[3]{5400/\pi} = 11.978\ldots$ sec.

38. $r = 48 + 0.03t$ in., so $v = \frac{4}{3}\pi(48 + 0.03t)^3$; $v = 246{,}924\pi \approx 775{,}734.6$ in.^{3} when $t = 300$.

39. $D(t) = \sqrt{12^2 + \left(\frac{48t}{7}\right)^2}$; $D = 100$ when $t \approx 14.48$ sec

40. Leon's distance from lamp: $d = 5t$; let l be the length of the shadow; then $\dfrac{d+l}{l} = \dfrac{15}{6.6} = 2.25$, so $l = d/1.25 = 4t$; when $t = 5$ sec, $l = 20$ ft.

41. $l = 5 + 2t$, $w = 7 + 2t$, so $a(t) = lw = (5+2t)(7+2t)$; $a = 175$ (5 times its original size) when $t \approx 3.63$ sec

42. $v = (5+2t)(7+2t)(3+2t)$; the volume is at least 525 (5 times the original value of 105) when $t \geq 1.616$ sec (approximately).

43. $a(t) = \pi r^2 = \pi(0 + 0.55t)^2 = 0.3025\pi t^2$ ft^2; $a(6) = 10.89\pi = 34.211\ldots$ ft^2

44. $S = 4\pi(1.6 - 0.0027t)^2$; $r = 0$ (and $S = 0$) when $t \approx 592.59$ sec, or a bit less than 10 min

45. Answers may vary.

46. **(a)** $x_3 = f(x_2)$, by definition, and $x_2 = f(x_1)$, so $x_3 = f(f(x_1))$. Similarly, $x_4 = f(x_3) = f(f(f(x_1)))$. **(b)** $x_2 = 0.448$, $x_3 = 0.69242\ldots$, $x_4 = 0.59631\ldots$, and $x_5 = 0.67402\ldots$

47. **(a)** Since $x_2 = w(x_1)$ and $b = w(x_1)$, $x_2 = b$; also, (a, b) is on the line $y = x$, so $a = b$; therefore $a = b = x_2$. **(b)** From (a), $a = x_2$, and from the graph, $a = c$, so $a = c = x_2$; also from the graph, $d = w(c)$; therefore $(c, d) = (x_2, w(x_2))$. **(c)** The unlabeled point on the line $y = x$ (above (e, f) and to the right of (c, d)) is $(x_3, w(x_2))$; since this point is above (e, f), $e = x_3$, so $f = w(e) = w(x_3)$; therefore, $(e, f) = (x_3, w(x_3))$.

48. The sequence begins 0.2, 0.448, 0.6924 . . . , 0.5963 . . . , 0.6740 . . . , 0.6152 . . . , 0.6628. . . . After a while the initial digits stabilize at 0.6428. . . . The sequence values in fact approach $\frac{9}{14} = 0.6\overline{428571}$.

49. The web is closing in on the intersection point of the parabola and the line; the x coordinate of this point is a solution to the equation $2.8x(1 - x) = x$; this solution is the value found in Exercise 48.

QUICK REVIEW 2.5

1. -27

2. -15

3. 3

4. $y = 2 \pm \sqrt{2}$

5. $y = -\frac{5}{6} \pm \frac{1}{6}\sqrt{37}$

6. $y = -\frac{1}{2}x \pm \frac{1}{2}\sqrt{x^2 + 8}$

7. $y = -x \pm \frac{1}{2}\sqrt{4x^2 + 12} = -x \pm \sqrt{x^2 + 3}$

8. $y = -\frac{1}{2}x^2 \pm \frac{1}{2}\sqrt{x^4 + 12}$

9. $y = -\frac{1}{2}x \pm \frac{1}{2}\sqrt{x^2 + 4}$

SECTION EXERCISES 2.5

1. $2(4) - (2) \neq -2$; $2(3) - 8 = -2$

2. $2(2)^2 - 5 = 3$; $2(5)^2 - 2 \neq 3$

3. $(-2) + (6)^2 \nless 33$; $(-2) + (4)^2 < 33$

4. $5(3)^3 - 2(2) > 1$; $5(-3)^3 - 2(2) \ngtr 1$

5. No: $3^2 + 5^2 \neq 8$

6. Yes: $7 - 5(2) = -3$

7. No: $1^3 + 3(2) \neq -1$

8. No: $3 - 4(1)^3 \neq 2$

9. No: $2^2 - 3 \neq 3$

10. Yes: $3^2 + 4^2 = 25$

11. Yes: $2^3 - 3(\sqrt{2})^2 = 2$

12. No: $4(\sqrt{3}) - 3(2)^2 \neq 5$

13. $x = 6$, $y = 9$

14. $x = -17$, $y = 23$

15. $x = -5$, $y = 11$

16. $x = -4$, $y = 2$

17. $x = 15$, $y = 2$

18. $x = 5$, $y = -1/8$

19.

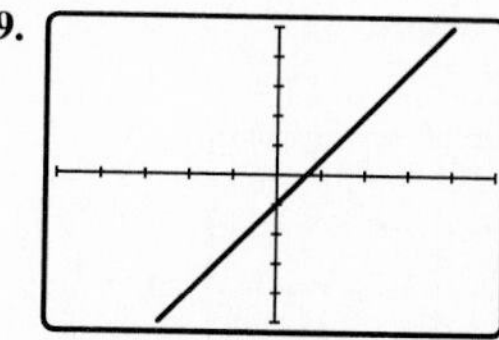

20.

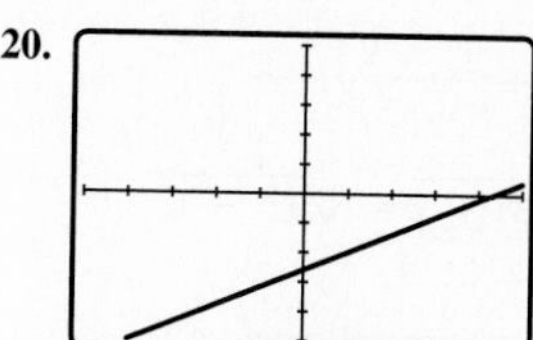

21.

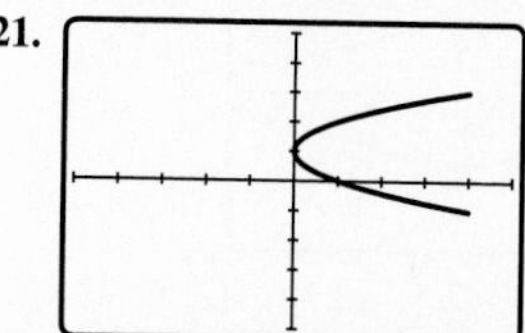

22.

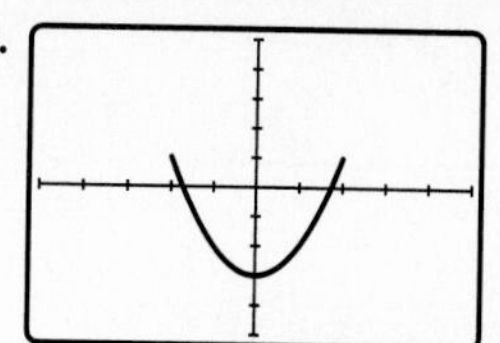

23.

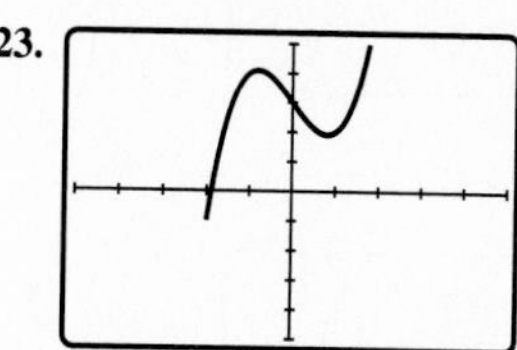

24.

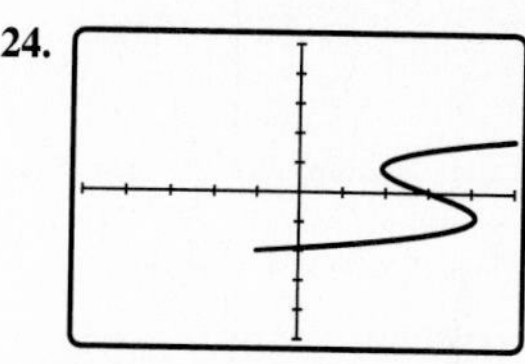

25. $0.5 < t < 2$

26. $2 < t \leq 3$

27. $-3 \leq t < -2$

28. $-2 < t < 0.5$

29. Use $t > 0$, $x = t$, $y = t^2 - 4t + 6$.

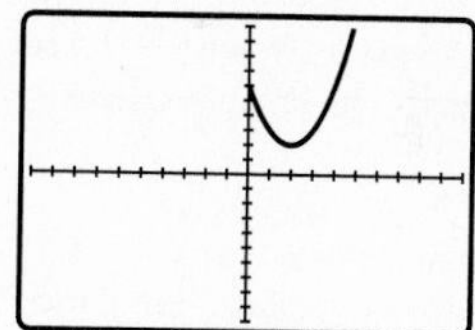

30. Use $0 < t < 3.5$, $x = t$, $y = 7 - 2t$

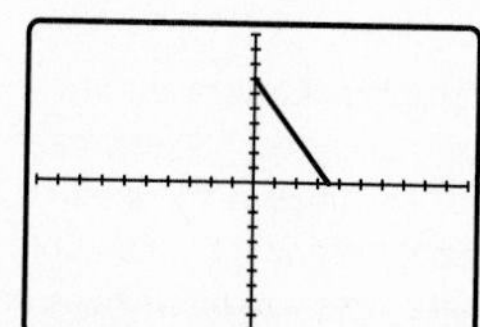

31. Use $0 < t < 1.8$, $x = t$, $y = 9 - 5t$.

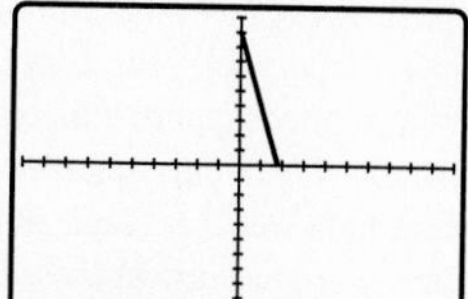

32. Use $0 < t < 3.8$, $x = t$, $y = -t^2 + 2t + 6.84$

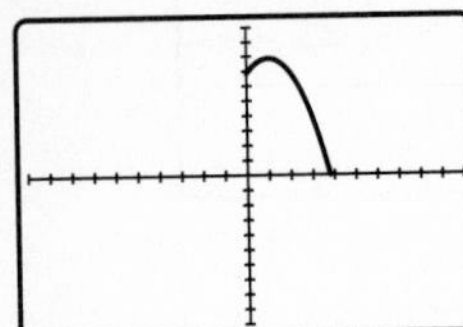

33. Use $t > 3$, $x = t$, $y = t^2 - 2t - 3$.

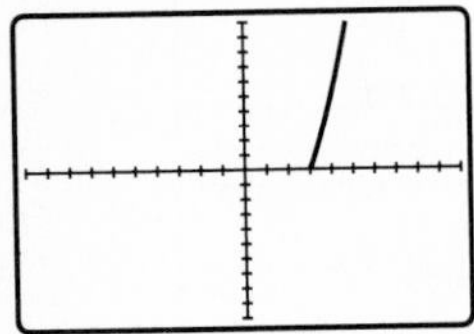

34. Use $t > 0$, $x = t$, $y = \sqrt{t + 4}$

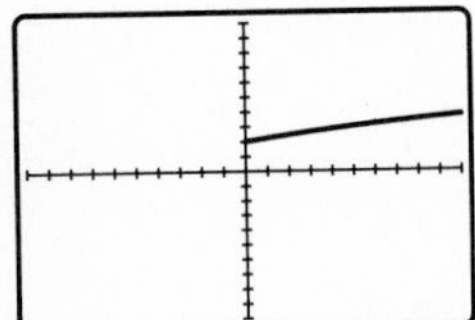

35. **(a)** $y = -16t^2 + 1000$. **(c)** When $t = 4$, $y = 744$ ft.

36. **(a)** $y = -16t^2 + 80t + 5$. **(d)** When $t = 4$, $y = 69$ ft. **(e)** When $t = 2.5$ sec, the ball is at its maximum height of 105 ft.

37. **(a)** One good window is $[-20, 300] \times [-1, 10]$. If your grapher allows, use "Simultaneous" rather than "Sequential" plotting. **(b)** After 3 sec, Ben is ahead by 2 ft.

38. **(a)** If your grapher allows, use "Simultaneous" rather than "Sequential" plotting. **(b)** The faster runner (who is coming from the left in the simulation) arrives at $t = 5.1$ sec. At this instant, the slower runner is 4.1 ft away from the flag.

39. If your graph does not look like a circle, try the "square" option on your ZOOM menu (if your grapher supports this option). Your circle may be missing pieces on the left and right sides. A graph similar to this can be produced with the window $[-9.6, 9.4] \times [-6.4, 6.2]$ (on a TI-81), $[-9.4, 9.4] \times [-6.2, 6.2]$ (TI-82), or $[-12.6, 12.6] \times [-7.4, 7.4]$ (TI-85).

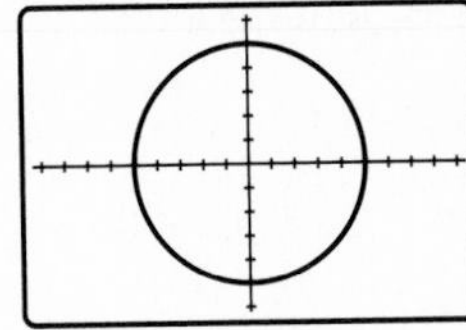

40. Shown is the window $[-8, 8] \times [-5, 3]$. The curve on top is y_1, and the curve on the bottom is y_2.

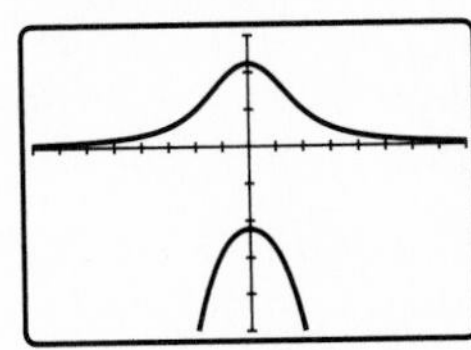

41. $y = \pm\sqrt{36 - x^2}$; this is a circle like that in Exercise 39. Use the same window that you used in Exercise 39.

42. $y = \pm\sqrt{x^2 + 1}$

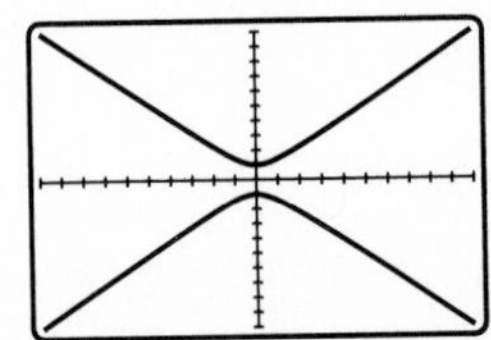

Shown in $[-10, 10] \times [-10, 10]$.

43. $y = \pm \dfrac{4}{\sqrt{1 - x^2}}$

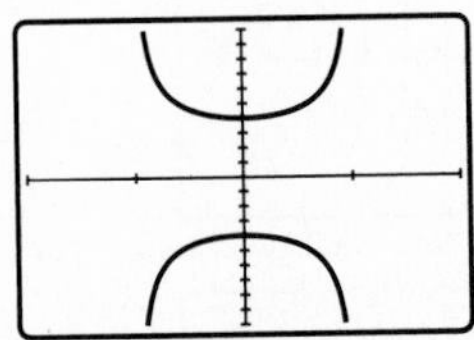

Shown in $[-2, 2] \times [-10, 10]$.

44. $y = \pm \dfrac{1}{\sqrt{x^2 + 4}}$

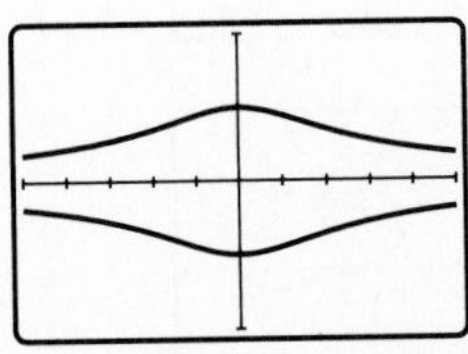

Shown in $[-5, 5] \times [-1, 1]$.

45. $y = -\frac{1}{2}x^2 \pm \frac{1}{2}\sqrt{x^4 + 12}$

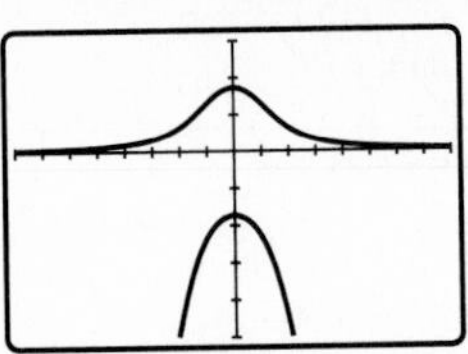

Shown in $[-8, 8] \times [-5, 3]$.

46. $y = -x \pm \sqrt{x^2 + 4}$

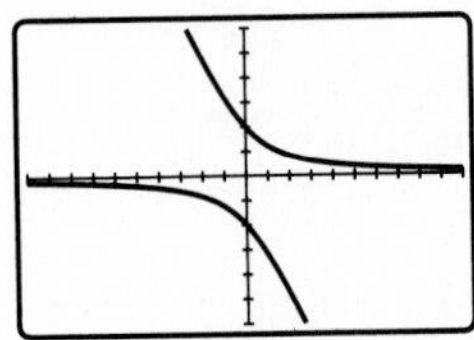

Shown in $[-10, 10] \times [-6, 6]$.

47. $y = \frac{3}{2}x \pm \dfrac{\sqrt{13}}{2}x$

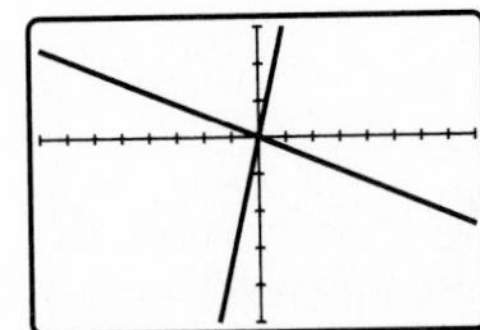

Shown in $[-8, 8] \times [-5, 3]$.

48. $y = \frac{1}{2}x^3 \pm \frac{1}{2}\sqrt{x^6 + 4x^2}$

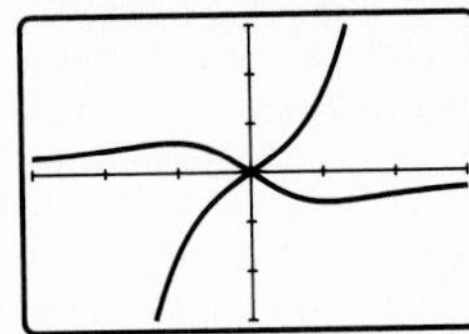

Shown in $[-3, 3] \times [-3, 3]$.

QUICK REVIEW 2.6

1. $(x + 1)^2$
2. $(x - 3)^2$
3. $(x + 3)^2$
4. $(2x + 1)^2$
5. $-3x^2 + 2$
6. $-x^2 + 2x - 4$
7. $x^2 - 5x + 1$
8. $-32x^5$

9. $t = (x + 3)/2$

10. $t = (y - 1)/3$

11. $t = 2x$

12. $t = 3y$

SECTION EXERCISES 2.6

1. Vertical translation downward 3 units

2. Vertical translation upward 1 unit

3. Vertical translation upward 5.2 units

4. Vertical translation downward 2 units

5. Horizontal translation to the left 4 units

6. Horizontal translation to the right 3 units

7. Horizontal translation to the right 1 unit (same as $(x - 1)^2$)

8. Horizontal translation to the right 5 units

9. Horizontal translation to the right 1 unit and vertical translation upward 3 units

10. Horizontal translation to the left 5 units and vertical translation downward 3 units

11. Reflection across x-axis

12. Horizontal translation to the right 5 units

13. Reflection across y-axis

14. Horizontal translation to the left 3 units followed by reflection across y-axis, or reflection across y-axis followed by a horizontal translation right 3 units.

15. Horizontal translation to the left 4 units followed by reflection across x-axis

16. Horizontal translation to the right 7 units

17. $C_1: y = x^3$, $C_2: y = (x + 2)^3$; horizontal translation 2 units left

18. $C_1: y = \sqrt{x}$; $C_2: y = \sqrt{x - 4}$; horizontal translation 4 units right

19. $C_1: y = x^2 + 2$, $C_2: y = (x - 2)^2 + 2$; horizontal translation 2 units right

20. $C_1: y = |x| + 3$; $C_2: y = |x + 3| + 3$; horizontal translation 3 units left

21. $C_1: y = x^3$, $C_2: y = 2x^3$; vertically stretch by a factor of 2

22. $C_1: y = \sqrt{x}$; $C_2: y = 0.5\sqrt{x}$; vertically shrink by a factor of 0.5

23. $C_1: y = x^2 + 2$, $C_2: y = x^2/9 + 2$; horizontally stretch by a factor of 3

24. $C_1: y = |x| + 3$; $C_2: y = |\frac{1}{2}x| + 3$; horizontally stretch by a factor of 2

25. Vertically stretch by 2.

26. Vertically shrink by 1/4.

27. Horizontally shrink by 1/2 or vertically stretch by 8.

28. Horizontally stretch by 5 or vertically shrink by 0.008.

29. Vertically shrink by 0.3.

30. Horizontally stretch by 2 or vertically shrink by $1/8 = 0.125$.

31. Horizontally stretch by 5/2 or vertically shrink by 0.064.

32. Vertically stretch by 3.

33. Starting with f, translate 6 units right to get g.

34. Starting with f, translate 4 units left and reflect across x-axis.

35. Starting with f, translate 4 units left, and reflect through x-axis to get g.

36. Starting with f, vertically stretch by 2.

37. Starting with f, translate 3 units right, reflect through x-axis, and translate 3 units upward to get g.

38. Starting with f, horizontally shrink by 1/2 or vertically stretch by 8.

39.
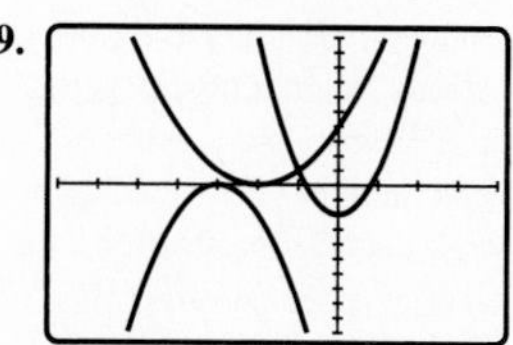

Shown in $[-7, 4] \times [-10, 10]$.

40.
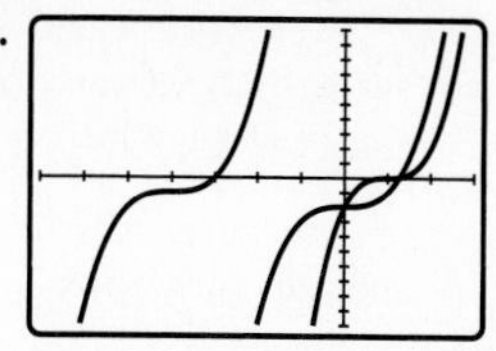

Shown in $[-7, 3] \times [-10, 10]$.

41.
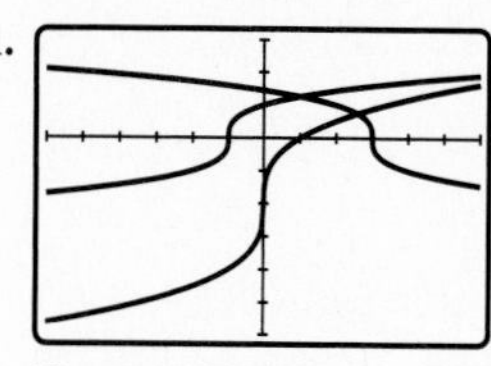

Shown in $[-6, 6] \times [-6, 3]$.

42.
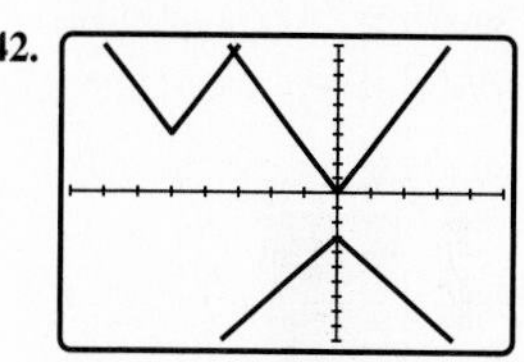

Shown in $[-8, 5] \times [-10, 10]$.

43. $f(x) = \sqrt{x + 5}$

44. $f(x) = \sqrt{3 - x}$

45. $f(x) = 3 - \sqrt{x + 2}$

46. $f(x) = 2\sqrt{x + 5} - 3$

47. $g(x) = f(-x) = -x^3 - 5x^2 + 3x + 2$

48. $g(x) = 2\sqrt{3 - x} - 4$

49. $g(x) = f(-x) = \sqrt[3]{-8x} = -2\sqrt[3]{x}$

50. $g(x) = 3|5 - x|$

51. Starting with $y = x^2$, translate 3 units right, vertically stretch by 2 and translate 4 units downward.

52. Starting with $y = \sqrt{x}$, translate 1 unit left, vertically stretch by 3, and reflect across x-axis.

53. Starting with $y = x^3$, horizontally shrink by 1/2 and reflect across x-axis.

54. Starting with $y = x^2$, horizontally shrink by 1/3 and translate 4 units downward.

55. Starting with $y = \sqrt[3]{x}$, translate 6 units left, vertically stretch by 3 and translate 7 units down.

56. Starting with $y = |x|$, translate 4 units left, vertically stretch by 2, reflect across x-axis, and translate 1 unit upward.

57. Starting with $y = \sqrt{x}$, horizontally shrink by 1/2, vertically stretch by 2 and reflect across x-axis.

58. Starting with $y = \sqrt[3]{x}$, horizontally stretch by 2.

59. Vertically stretch by 9 or horizontally shrink by 1/3.

60. Vertically shrink by 1/4 or horizontally stretch by 2.

61. Vertically stretch by 16 or horizontally shrink by 1/4.

62. Vertically stretch by 5 or horizontally shrink by $1/\sqrt{5}$.

63. $y_1 = 2x^3 - 8x$; $y_2 = \frac{1}{3}x^3 - \frac{4}{3}x$

64. $y_1 = 2|x + 2|$; $y_2 = \frac{1}{3}|x + 2|$

65. $y_1 = 2x^2 + 2x - 4$; $y_2 = \frac{1}{3}x^2 + \frac{1}{3}x - \frac{2}{3}$

66. $y_1 = \dfrac{2}{x + 2}$; $y_2 = \dfrac{1}{3(x + 2)}$

67. $y_1 = (x/2)^2$; $y_2 = (3x)^2$

68. $y_1 = |x/2|$; $y_2 = |3x|$

69. $y_1 = |x/2 - 2|$; $y_2 = |3x - 2|$

70. $y_1 = 2/x$; $y_2 = 1/(3x)$

71. $y = 3(x - 4)^2$

72. $y = 3(x - 4)^2$

73. $y = 3\sqrt[3]{x} + 4$

74. $y = 3(\sqrt[3]{x} + 4) = 3\sqrt[3]{x} + 12$

75. $y = 2|x + 2| - 4$

76. $y = |2x + 2| - 4 = 2|x + 1| - 4$

77.

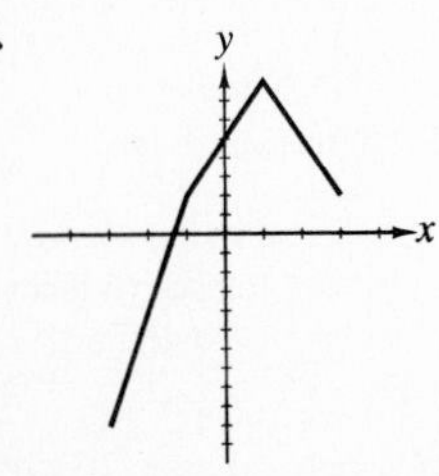

78.

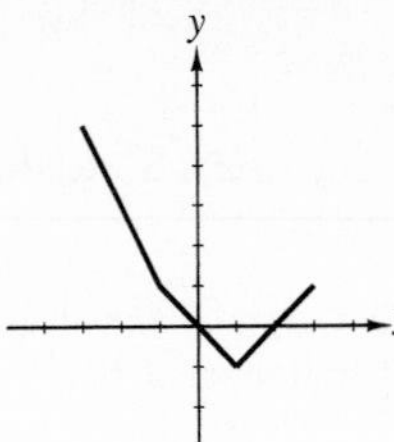

79.

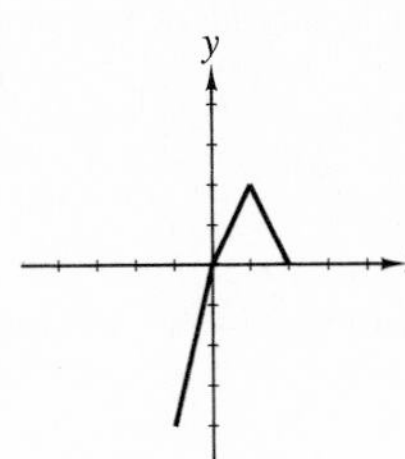

80.

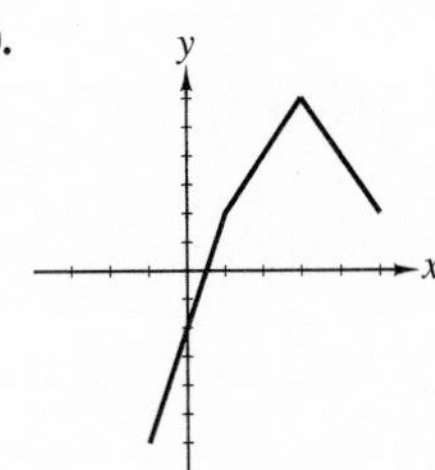

81. Vertically stretch by 9/5 and then translate 32 units upward.

82. $C = \frac{5}{9}(F - 32) = \frac{5}{9}F - \frac{160}{9}$; vertically shrink by 5/9 and then translate downward $160/9 = 17.\overline{7}$ units.

83. **(a)** $y_2 = -16(t - 2)^2 + 64(t - 2) + 5$. **(c)** When $t = 3$ sec, both balls are 53 ft high.

84. $h(x)$. All x's must be replaced with $x - 2$, including those in the inequalities.

85. No. Because $c(f(x) + d) \neq cf(x) + d$ unless $c = 1$ or $d = 0$, which amounts to "no stretch" or "no shift."

86. For a horizontal translation and another horizontal change (stretch, shrink, or y-axis reflection), the outcome depends on the order. Likewise, for a vertical translation and another vertical change (stretch, shrink, or x-axis reflection), the order matters. For other pairs of transformations, the order does not matter.

87. Answers may vary.

QUICK REVIEW 2.7

1. $y = \frac{1}{3}x + 2$

2. $y = 2x - 2$

3. $y = \pm\sqrt{x - 4}$

4. $y = \pm\sqrt{x + 6}$

5. $y = \dfrac{3x + 2}{1 - x}$

6. $y = \dfrac{2x + 1}{3 - x}$

7. $y = \dfrac{4x + 1}{x - 2}$

8. $y = \dfrac{x + 3}{3x - 4}$

9. $y = x^2 - 3, x \geq 0$

10. $y = x^2 + 2, x \geq 0$

SECTION EXERCISES 2.7

1. $y = (x - 3)/2$

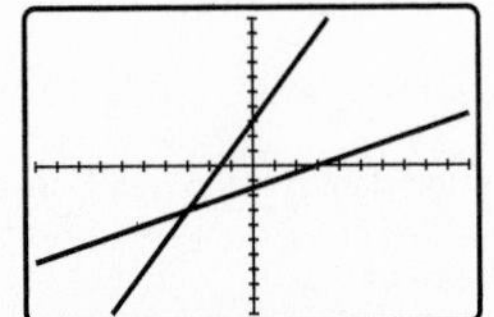

Shown in $[-10, 10] \times [-10, 10]$.

2. $y = (x + 4)/3$

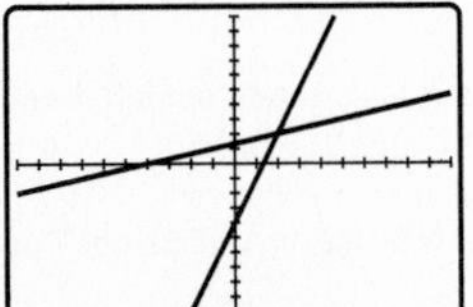

Shown in $[-10, 10] \times [-10, 10]$.

3. $y = (x - 1)/(-7) = (1 - x)/7$

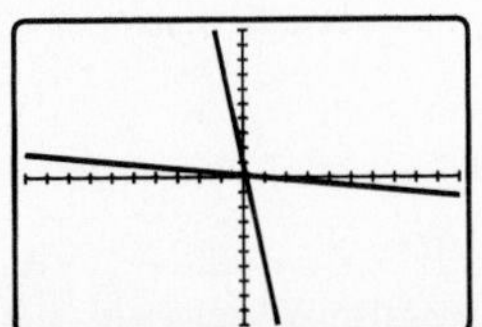

Shown in $[-10, 10] \times [-10, 10]$.

4. $y = 3 - x/2$

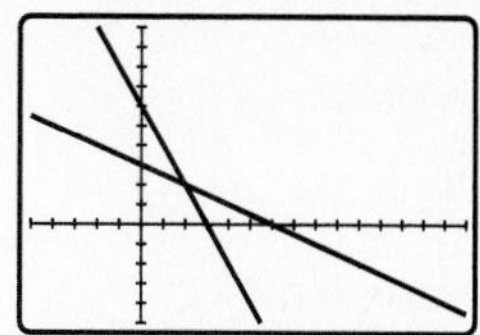

Shown in $[-5, 15] \times [-5, 10]$.

5. $y = \sqrt[3]{(x + 5)/2}$

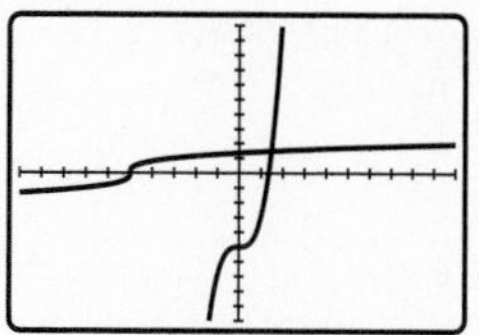

Shown in $[-10, 10] \times [-10, 10]$.

6. $y = \sqrt[3]{x - 6}$

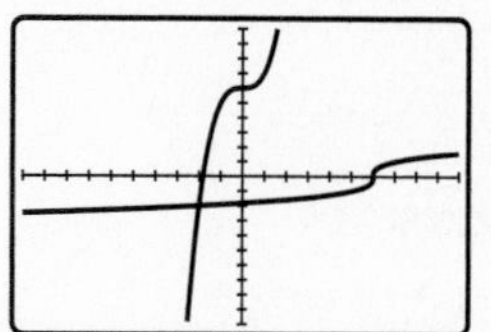

Shown in $[-10, 10] \times [-10, 10]$.

7. $y = x^3 - 1$

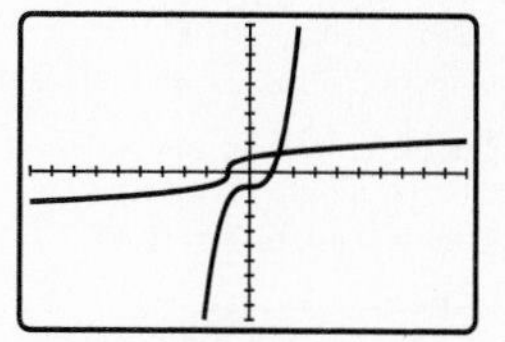

Shown in $[-10, 10] \times [-10, 10]$.

8. $y = x^3 + 3$

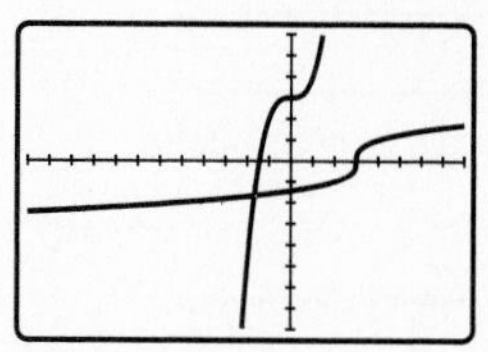

Shown in $[-12, 8] \times [-8, 6]$.

9. $x = t^2 - 2, y = 2t - 3$

10. $x = 2t, y = t^3 - 2$

11. $x = t, y = 3^t$

12. $x = t^3, y = t^2$

13.

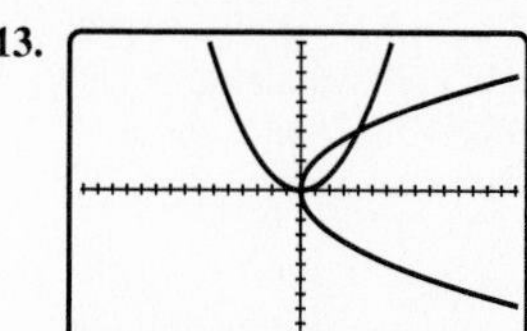

Shown in $[-15.2, 15.2] \times [-10, 10]$.

14.

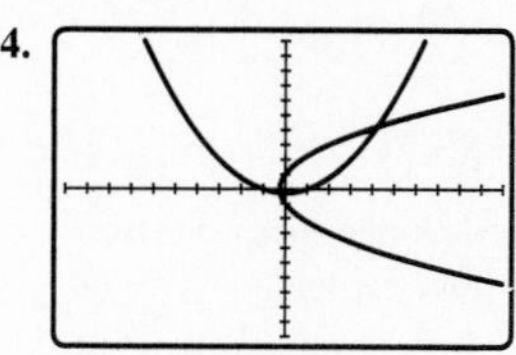

Shown in $[-15.2, 15.2] \times [-10, 10]$.

15.

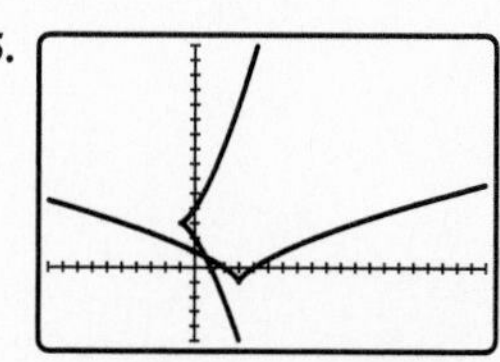

Shown in $[-10.1, 20.2] \times [-5, 15]$.

16.

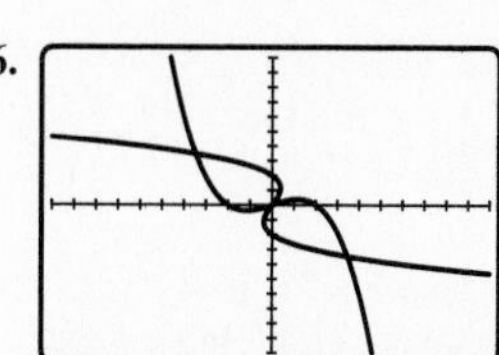

Shown in $[-15.2, 15.2] \times [-10, 10]$.

17. $f(g(x)) = 3\left[\frac{1}{3}(x + 2)\right] - 2 = x + 2 - 2 = x$; $g(f(x)) = \frac{1}{3}[(3x - 2) + 2] = \frac{1}{3}(3x) = x$

18. $f(g(x)) = \frac{1}{4}[(4x - 3) + 3] = x$; $g(f(x)) = 4\left[\frac{1}{4}(x + 3)\right] - 3 = x$

19. $f(g(x)) = [(x - 1)^{1/3}]^3 + 1 = x - 1 + 1 = x$; $g(f(x)) = [(x^3 + 1) - 1]^{1/3} = (x^3)^{1/3} = x$

20. $f(g(x)) = g(f(x)) = \dfrac{1}{1/x} = x$

21. $f(g(x)) = (\sqrt{x + 3})^2 - 3 = x + 3 - 3 = x$; $g(f(x)) = \sqrt{(x^2 - 3) + 3} = \sqrt{x^2} = |x| = x$ since $x \geq 0$

22. $f(g(x)) = [(\sqrt[3]{x} - 1) + 1]^3 = (\sqrt[3]{x})^3 = x$; $g(f(x)) = \sqrt[3]{(x + 1)^3} - 1 = x + 1 - 1 = x$

23. $f(g(x)) = \dfrac{\dfrac{1}{x - 1} + 1}{\dfrac{1}{x - 1}} = (x - 1) \cdot \left(\dfrac{1}{x - 1} + 1\right) = 1 + x - 1 = x$; $g(f(x)) = \dfrac{1}{\dfrac{x + 1}{x} - 1} = \dfrac{x}{x + 1 - x} = \dfrac{x}{1} = x.$

24. $f(g(x)) = \dfrac{\dfrac{2x + 3}{x - 1} + 3}{\dfrac{2x + 3}{x - 1} - 2} = \dfrac{2x + 3 + 3(x - 1)}{2x + 3 - 2(x - 1)} = \dfrac{5x}{5} = x$; $g(f(x)) = \dfrac{2\,\dfrac{x + 3}{x - 2} + 3}{\dfrac{x + 3}{x - 2} - 1} = \dfrac{2(x + 3) + 3(x - 2)}{x + 3 - (x - 2)} = \dfrac{5x}{5} = x$

25. No. A horizontal line at a height of 2 apparently would cross the graph twice.

26. Yes. No horizontal line appears to cross more than once.

27. Yes

28. Yes

29. No

30. No

31. Yes

32. Yes

33. No

34. No

35. No

36. Yes

37. No

38. No

39. Yes

40. Yes

41. No

42. Yes

43. $f^{-1}(x) = \frac{1}{3}x + 2$

44. $f^{-1}(x) = \frac{1}{2}x - \frac{5}{2}$

45. $f^{-1}(x) = -\dfrac{x+3}{x-2}$

46. $f^{-1}(x) = \dfrac{2x+3}{x-1}$

47. $f^{-1}(x) = x^2 - 3, x \geq 0$

48. $f^{-1}(x) = x^2 - 2, \quad x \geq 0$

49. $f^{-1}(x) = \sqrt[3]{x}$

50. $f^{-1}(x) = \sqrt[3]{x-5}$

51. $f^{-1}(x) = x^3 - 5$

52. $f^{-1}(x) = x^3 + 2$

53. The graph is shown in $[4, 9.5] \times [36{,}000, 39{,}000]$, along with the regression line $y = 42124 - 648.1x$ where $x = 5$ represents 1985, $x = 6$ represents 1986, etc.

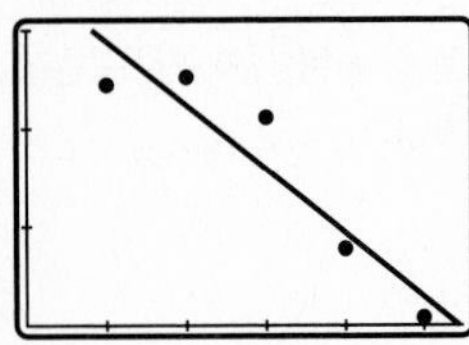

54. **(a)** $f^{-1}(x) = 64.996 - \frac{1}{648.1}x$.
(b) $f^{-1}(34000) = 12.535$, in about 1992 or 1993

55. $x \geq -2$: $f^{-1}(x) = \sqrt{x+7} - 2$; or $x \leq -2$: $f^{-1}(x) = -\sqrt{x+7} - 2$

56. $x \geq 0$: $f^{-1}(x) = \sqrt[4]{x+3}$; or $x \leq 0$: $f^{-1}(x) = -\sqrt[4]{x+3}$

57. $P_2(f(a), f(a))$; all y-coordinates on a horizontal line are the same, and all points on the line $y = x$ have the same x- and y-coordinates.

58. The x-coordinate of P_3 is the same as the x-coordinate of P_2; for any point on the graph of f^{-1}, the y-coordinate is f^{-1} evaluated at the x-coordinate.

59. Since P_1 and P_3 are symmetric about $y = x$, the x-coordinate of P_1 must equal the y-coordinate of P_3.

60. Answers may vary.

Chapter 2 Review

1. $x = -2.5$ or $x = 0.666\ldots$ $\left(\frac{2}{3}\right)$

2. $x \approx -4.828$ or $x \approx 0.828$

3. $x = -1$; an algebraic approach also gives the extraneous solution $x = -4$.

4. $x = -0.666\ldots$ $\left(-\frac{2}{3}\right)$ or $x = 2$

5. $x = -0.618\ldots$ or $x = 1.618\ldots$

6. $x = -3$

7. $x = 0.5$; an algebraic approach also gives the extraneous solution $x = -1.25$.

8. $x = -5$, $x = -2.414\ldots$, $x = 0.414\ldots$, or $x = 3$

9. $(x-2)^2 = 3$, so $x = 2 \pm \sqrt{3}$

10. $(x+1)^2 = 7$, so $x = -1 \pm \sqrt{7}$

11. $\left(x + \frac{2}{3}\right)^2 = \frac{5}{9}$, so $x = -\frac{2}{3} \pm \dfrac{\sqrt{5}}{3}$

12. $(2x+1)^2 = 4$, so $x = -\frac{1}{2} \pm 1$; $x = -1.5$ or $x = 0.5$

13. $(x + 1 + 4)(x + 1 - 4) = 0$, so $x = -5$ or $x = 3$

14. $(x-5)(x+5) = 0$, so $x = \pm 5$

15. $(x-6)(x+7) = 0$, so $x = -7$ or $x = 6$

16. $(x-5)(x+3) = 0$, so $x = -3$ or $x = 5$

17. $(2x+3)(x-2) = 0$, so $x = -\frac{3}{2}$ or $x = 2$

18. $(3x-1)(x+4) = 0$, so $x = -4$ or $x = \frac{1}{3}$

19. $x = 1.5 \pm \frac{1}{2}\sqrt{7}$, so $x = 0.177\ldots$ or $x = 2.822\ldots$

20. $x = -1 \pm \frac{1}{2}\sqrt{3}$; $x = -1.866\ldots$ or $x = -0.133\ldots$

21. No real solutions; $D = -7$

22. $x = -\frac{7}{6} \pm \frac{17}{6}$; $x = -4$ or $x = \frac{5}{3}$

23. No real solutions; $D = -12$

24. $u = -4$ or $u = 1.5$

25. $z = 3$ or $z = 7$

26. $s = -2 \pm \sqrt{6}$; $s = -4.449\ldots$ or $s = 0.449\ldots$

27. $x = -1$ or $x = 4$

28. $x = -4$ or $x = \frac{2}{3}$

29. $x = -2$ or $x = 4$

30. $x \approx -0.79$ or $x \approx 2.52$

31. $x = \frac{1}{3}$ or $x = 5$

32. $x \approx -0.16$ or $x \approx 6.16$

33. $w = -3$ or $w = 2.561\ldots$

34. $v = 4$

35. $x \approx -2.21$, $x \approx 0.54$, or $x \approx 1.68$

36. $x = \pm 1$ or $x = \pm 2$

37. $x = 7$

38. $x \approx -0.09$ or $x \approx 4.35$

39. $x \approx -1.66$, $x \approx 0.33$, or $x \approx 1.84$

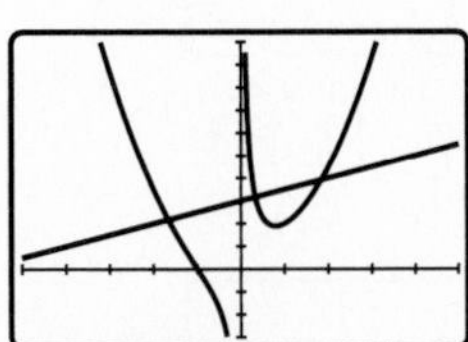

Shown in $[-5, 5] \times [-3, 10]$.

40. $x \approx -1.66$ or $x \approx -0.21$ or $x \approx 2.87$

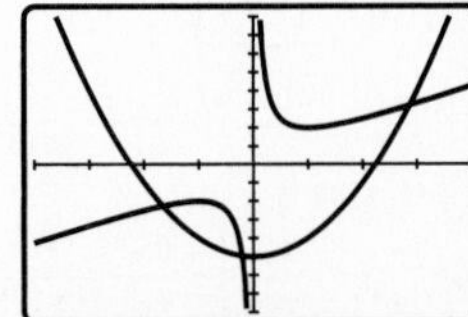

Shown in $[-4, 4] \times [-8, 8]$.

41. $D = -7$; no real solutions

42. $D = 61$; two real solutions

43. $D = 0$; one real solution

44. $D = 33$; two real solutions

45. $y = |x - 2| - x^2 + 3$

46. One possibility is $|x - 3| \le 2$.

47. $x > -4$, or $(-4, \infty)$;

−8 −7 −6 −5 −4 −3 −2 −1 0 1 2

48. $-12 \le x < 10$, or $[-12, 10)$;

−15 −12 −9 −6 −3 0 3 6 9 12 15

49. $x < -4$ or $x > -1$, or $(-\infty, -4) \cup (-1, \infty)$;

−8 −7 −6 −5 −4 −3 −2 −1 0 1 2

50. $\frac{2}{3} \le x \le 2$, or $[\frac{2}{3}, 2]$;

0 1 2

51. $x \ge \frac{8}{3}$, or $[\frac{8}{3}, \infty)$

52. $x \le 3$, or $(-\infty, 3]$

53. $-\frac{1}{5} \le x < \frac{8}{5}$, or $[-\frac{1}{5}, \frac{8}{5})$

54. $-1.5 < x < 8.5$, or $(-1.5, 8.5)$.

55. $x \le 0.5$ or $x \ge 4.5$, or $(-\infty, 0.5] \cup [4.5, \infty)$

56. $-\frac{1}{7} < x < \frac{3}{7}$, or $(-\frac{1}{7}, \frac{3}{7})$.

57. $-1 < x \le 4$, or $(-1, 4]$

58. $2 \le x \le 5$, or $[2, 5]$.

59. $-0.5 < x \le 3$, or $(-0.5, 3]$

60. $x < 0.5$, or $(-\infty, 0.5)$.

61. $x \ge 2.5$ and $x \ne 4$, or $[2.5, 4) \cup (4, \infty)$.

62. $x \le -2$ or $x = 2$, or $(-\infty, -2] \cup \{2\}$.

63. $(f + g)(x) = 3x + 5 + x^2$, $(f - g)(x) = 3x + 5 - x^2$, and $(fg)(x) = (3x + 5)(x^2) = 3x^3 + 5x^2$; all three domains are $(-\infty, \infty)$; $(f/g)(x) = \dfrac{3x + 5}{x^2}$, domain: $(-\infty, 0) \cup (0, \infty)$.

64. $(f + g)(x) = x^2 + x + x - 1 = x^2 + 2x - 1$, $(f - g)(x) = x^2 + x - x + 1 = x^2 + 1$, and $(fg)(x) = (x^2 + x)(x - 1) = x^3 - x$; all three domains are $(-\infty, \infty)$; $(f/g)(x) = \dfrac{x^2 + x}{x - 1}$, domain: $(-\infty, 1) \cup (1, \infty)$.

65. $(f + g)(x) = \sqrt{x - 4} + x^2 - 4$, $(f - g)(x) = \sqrt{x - 4} - x^2 + 4$, $(fg)(x) = (\sqrt{x - 4})(x^2 - 4)$, and $(f/g)(x) = \dfrac{\sqrt{x - 4}}{x^2 - 4}$; all four domains are $[4, \infty)$.

66. $(f + g)(x) = \sqrt{x + 3} + \sqrt{x - 2}$, $(f - g)(x) = \sqrt{x + 3} - \sqrt{x - 2}$, and $(fg)(x) = (\sqrt{x + 3})(\sqrt{x - 2}) = \sqrt{x^2 + x - 6}$; all three domains are $[2, \infty)$; $(f/g)(x) = \dfrac{\sqrt{x + 3}}{\sqrt{x - 2}} = \sqrt{\dfrac{x + 3}{x - 2}}$, domain: $(2, \infty)$.

67. $f(g(2)) = 124$; $g(f(2)) = 10$

68. 2 is not in the domain of $f \circ g$; $g(f(2)) = -2$.

69. $(f \circ g)(x) = \sqrt{x^2 - 4}$; domain: $(-\infty, -2] \cup [2, \infty)$; $(g \circ f)(x) = (\sqrt{x})^2 - 4 = x - 4$; domain: $[0, \infty)$.

70. $(f \circ g)(x) = \left(\dfrac{1}{x + 2}\right)^2 - 3 = \dfrac{1}{(x + 2)^2} - 3$; domain: $(-\infty, -2) \cup (-2, \infty)$; $(g \circ f)(x) = \dfrac{1}{(x^2 - 3) + 2} = \dfrac{1}{x^2 - 1}$; domain: $(-\infty, -1) \cup (-1, 1) \cup (1, \infty)$

71. One possibility is $f(x) = \sqrt[3]{x}$ and $g(x) = 4x - 3$.

72. One possibility is $f(x) = |x|$ and $g(x) = x^2 + x$.

73. One possibility is $f(x) = 2/x$ and $g(x) = (x + 3)^2$.

74. One possibility is $f(x) = 2x - 5$ and $g(x) = \sqrt{x}$.

75. On the same viewing window as the original, the graph of $f - g$ looks like the graph shown here.

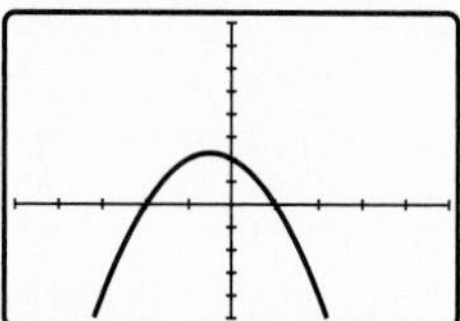

76. **(a)** $1 < t < 3$. **(b)** $-3 < t < 1$. **(c)** $-5 \le t < -3$. **(d)** $3 < t \le 5$.

77. **(a)** No. **(b)** Yes.

78. **(a)** Yes. **(b)** No.

79. $x = 4$, $y = 4$

80. $x = -2$, $y = 1$.

81. $[-10, 10] \times [-10, 12]$ is one possibility.

82. $[0, 20] \times [-10, 10]$ is one possibility.

83. $x = t$, $y = t^2 - 4t + 1$ with $-10 \le t \le 10$

84. $x = t$, $y = 5 - 2t$, with $-10 \le t \le 10$.

85. Vertically stretch by 2, reflect across x-axis, and translate 5 units upward.

86. Translate 3 units left and translate 4 units down.

87. Reflect across y-axis and translate 4 units right or translate 4 units left and reflect across y-axis.

88. Translate 2 units right, vertically stretch by 3, and translate 1 unit upward.

89. $y = 2(x - 3)^2$

90. $y = \sqrt[3]{\frac{1}{2}(x + 3)}$

91. $y = \sqrt{-3x} + 3$

92. $y = -|x + 2| - 1$

93. Vertically stretch by 8 or horizontally shrink by 1/2.

94. Starting from $y = \sqrt[3]{x}$, translate 4 units right, vertically stretch by 3, reflect across the x-axis, and translate 2 units down.

95. $C_1: y = x^2$, $C_2: y = (x - 3)^2$; translate 3 units right.

96. $C_1: y = |x|$; $C_2: y = 3\,|x|$; vertically stretch by 3.

97. $C_1: y = x^2 - 1$, $C_2: y = (x/2)^2 - 1$; horizontally stretch by 2.

98. $C_1: y = \sqrt{x}$; $C_2: y = \sqrt{x} - 4$; translate 4 units downward.

99. $f(x) = -\sqrt{3 - x}$

100.

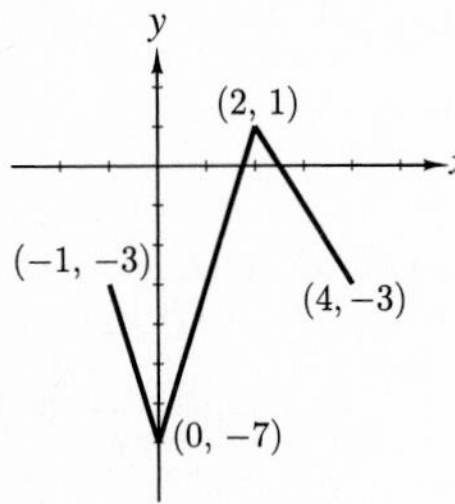

101. $x = t^2$, $y = 3t$

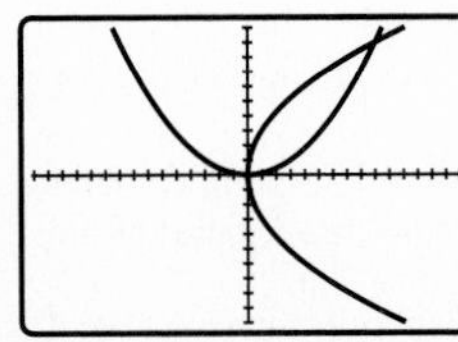

Shown in $[-15.2, 15.2] \times [-10, 10]$.

102. $x = 2^t$, $y = t$

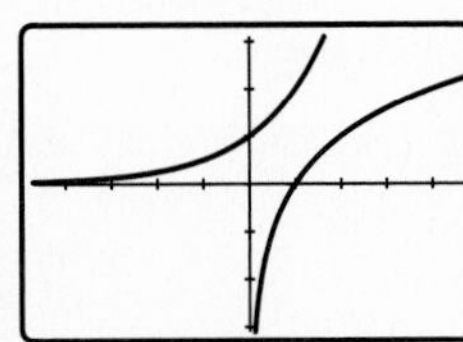

Shown in $[-4.7, 4.7] \times [-3.1, 3.1]$.

103. $f(g(x)) = 2[\frac{1}{2}(x - 5)] + 5 = x - 5 + 5 = x$; $g(f(x)) = \frac{1}{2}[(2x + 5) - 5] = \frac{1}{2}(2x) = x$

104. $f(g(x)) = (\sqrt{x - 1})^2 + 1 = x - 1 + 1 = x$ (for any $x \geq 1$); $g(f(x)) = \sqrt{(x^2 + 1) - 1} = \sqrt{x^2} = |x| = x$ (for any $x \geq 0$)

105. No (though you cannot tell from the standard viewing window)

106. Yes

107. $f^{-1}(x) = (x + 2)/3$

108. $f^{-1}(x) = \dfrac{3x + 1}{x - 1}$

109.

(a) Shown in $[4, 14] \times [2, 13]$.
(b) With $x = 16$, y (the deficit) equals \$18.645 billion. **(c)** $y = 30$ when $x = 17.74$, or about 1997 or 1998.

110.

(a) Shown in $[2, 16] \times [15, 65]$.
(b) When $x = 17$, the purse is about \$64.6 million. **(c)** The first such x is $x \approx 22.24$, in about 2002. If the model continues to hold further into the future, it happens again when $x \approx 43.73$, or in about 2023 or 2024.

111. $A(x) = x(x + 5) + \frac{1}{2}\pi(x/2)^2$; $A = 150$ when $x = 8.737$ ft.

112. The Turbo Prop's speed is $x = 249.8$ mph, and the Airbus's speed is $2x - 55 = 444.6$ mph.

113. $P(x) = 190x - 75000 - 85x$; $P(x)$ equals \$40,000 when $x = 1095.238$, so they must sell 1096 bikes.

114. If x is the width, then $0 < x < 72$.

115. $V(x) = x(12 - 2x)(18 - 2x)$; the volume is 210 in.^{3} when $x = 1.630$ in., or $x = 3.152$ in.

116. $t = \frac{4}{3}\sqrt{250/\pi} \approx 11.894$ sec.

117. **(a)** $h = -16t^2 + 245t + 200$.
(c) Graph $x = t$, $y = -16t + 245t + 200$ in $[0, 16.1] \times [0, 1200]$ for $0 \leq t \leq 16.1$. **(d)** When $t = 4$, $h = 924$ ft. **(e)** When $t \approx 7.66$, the arrow is at its peak: 1137.89 ft. **(f)** The arrow hits the ground after about 16.09 sec.

118. **(a)** Let x be the length of her shadow and $d = 6t$ be her distance from the streetlight; $\dfrac{x}{5.5} = \dfrac{x + d}{12}$, or $\dfrac{x}{5.5} = \dfrac{x + 6t}{12}$, or $x = \dfrac{66t}{13}$.
(b) When $t = 5$, $x = 25.385$ ft.

119. Answers may vary.

120. Yes; take, for example, $|x| \leq 0$.

CHAPTER 3

QUICK REVIEW 3.1

1. $x^2 + 6x + 9$
2. $x^2 - 8x + 16$
3. $3x^2 - 36x + 108$
4. $-3x^2 - 42x - 147$
5. $(x + 5)^2$
6. $(x - 2)^2$
7. $2(x - 1)^2$
8. $3(x + 2)^2$
9. $h = -16t^2 + 42t$
10. $h = -16t^2 + 82t + 125$
11. $h = -16t^2 - 8t + 240$
12. $h = -16t^2 + 128t - 12$

SECTION EXERCISES 3.1

1. Degree 2; leading coefficient 4
2. Degree 2; leading coefficient 2
3. Degree 2; leading coefficient 4
4. Degree 2; leading coefficient −1
5. Degree 2; leading coefficient −3

6. Degree 2; leading coefficient 2

7. The graph in part (a); the vertex is in quadrant III, at $(-1, -3)$, which eliminates all but parts (a) and (d); since $f(0) = -1$, it must be part (a).

8. The graph in part (d); the vertex is in quadrant III, at $(-2, -7)$, which eliminates all but parts (a) and (d); since $f(0) = 5$, it must be part (d).

9. The graph in part (b); the vertex is in quadrant I, at $(1, 4)$, meaning that it must be either part (b) or part (f); since $f(0) = 1$, it cannot be part (f) because if the vertex in part (f) is $(1, 4)$, the intersection with the y-axis would be at about $(0, 3)$.

10. The graph in part (f); the vertex is in quadrant I, at $(1, 12)$, meaning that it must be either part (b) or part (f); since $f(0) = 10$, it cannot be part (b) because, if the vertex in part (b) is $(1, 12)$, the intersection with the y-axis occurs considerably lower than $(0, 10)$.

11. Vertically stretch by 4 (or horizontally shrink by 1/2).

12. Translate downward 3 units.

13. Translate right 5 units.

14. Translate left 1 unit.

15. Translate left 1 unit and vertically stretch by 3 (either order).

16. Translate right 4 units and vertically stretch by 2 (either order).

17. Translate left 3 units and reflect across x-axis (either order).

18. Translate left 5 units and reflect across x-axis (either order).

19. Translate left 3 units and vertically stretch by 2 (either order) and then translate upward 4 units.

20. Translate right 2 units and vertically stretch by 4 (either order) and then translate downward 5 units.

21. Translate left 3 units and translate downward 2 units (either order).

22. Translate right 1 unit and vertically stretch by 2 (either order) and then translate upward 7 units.

23. Vertex: $(1, 5)$; line of symmetry: $x = 1$

24. Vertex: $(-2, -1)$; line of symmetry: $x = -2$

25. Vertex: $(1, -7)$; line of symmetry: $x = 1$

26. Vertex: $(\sqrt{3}, 4)$; line of symmetry: $x = \sqrt{3}$

27. Vertex: $(5, \sqrt{2})$; line of symmetry: $x = 5$

28. Vertex: $(\sqrt{2}, \sqrt{3})$; line of symmetry: $x = \sqrt{2}$

29. Vertex: $\left(-\frac{5}{6}, -\frac{73}{12}\right)$; line of symmetry: $x = -\frac{5}{6}$

30. Vertex: $(1.75, 3.125)$; line of symmetry: $x = 1.75$

31. Vertex: $(4, 19)$; line of symmetry: $x = 4$

32. Vertex: $(0.25, 5.75)$; line of symmetry: $x = 0.25$

33. Vertex: $(0.6, 2.2)$; line of symmetry: $x = 0.6$

34. Vertex: $(-1.75, 2.125)$; line of symmetry: $x = -1.75$

35. Vertex: $(-2.5, -8.25)$; line of symmetry: $x = -2.5$

36. Vertex: $(2, -3)$; line of symmetry: $x = 2$

37. Vertex: $\left(\frac{7}{6}, -\frac{193}{12}\right)$; line of symmetry: $x = \frac{7}{6}$

38. Vertex: $(1.5, 17.5)$; line of symmetry: $x = 1.5$

39. Vertex: $(2, 2)$; line of symmetry: $x = 2$; opens upward; does not intersect x-axis.

40. Vertex: $(3, 3)$; line of symmetry: $x = 3$; opens upward; does not intersect x-axis.

41. Vertex: $(-2.5, -3.25)$; line of symmetry: $x = -2.5$; opens upward; intersects x-axis at about -4.303 and -0.697 $\left(-2.5 \pm \frac{1}{2}\sqrt{13}\right)$.

42. Vertex: $(-3.5, -14.25)$; line of symmetry: $x = -3.5$; opens upward; intersects x-axis at about -7.275 and 0.275 $\left(-3.5 \pm \frac{1}{2}\sqrt{57}\right)$.

43. Vertex: $(-8, 74)$; line of symmetry: $x = -8$; opens downward; intersects x-axis at about -16.602 and 0.602 $(-8 \pm \sqrt{74})$.

44. Vertex: $(1, 9)$; line of symmetry: $x = 1$; opens downward; intersects x-axis at -2 and 4.

45. Vertex: $(-2, -14)$; line of symmetry: $x = -2$; opens upward; intersects x-axis at about -4.160 and 0.160 $\left(-2 \pm \frac{1}{3}\sqrt{42}\right)$; vertically stretched by 3.

46. Vertex: $(2.5, -20)$; line of symmetry: $x = 2.5$; opens upward; intersects x-axis at about 0.264 and 4.736 $(2.5 \pm \sqrt{5})$; vertically stretched by 4.

47. Vertex: $(-1.5, 2.5)$; line of symmetry: $x = -1.5$; opens upward; does not intersect x-axis; vertically stretched by 2.

48. Vertex: $(2.5, -19.25)$; line of symmetry: $x = 2.5$; opens upward; intersects x-axis at about 0.538 and 4.462 $\left(2.5 \pm \frac{1}{10}\sqrt{385}\right)$; vertically stretched by 5.

49. Vertex: $(2.5, -2)$; line of symmetry: $x = 2.5$; opens upward; intersects x-axis at about 1.793 and 3.207 $\left(2.5 \pm \frac{1}{2}\sqrt{2}\right)$; vertically stretched by 4.

50. Vertex: $\left(\frac{17}{6}, -\frac{253}{12}\right)$; line of symmetry: $x = \frac{17}{6}$; opens upward; intersects x-axis at about 0.182 and 5.484 $\left(\frac{17}{6} \pm \frac{1}{6}\sqrt{253}\right)$; vertically stretched by 3.

51. Use the form $y = a(x - h)^2 + k$; you know $h = -1$ and $k = -3$ from the vertex, so substitute $x = 1$, $y = 5$ to find a: $y = 2(x + 1)^2 - 3$.

52. $y = 3(x - 2)^2 - 7$

53. $y = -2(x - 1)^2 + 11$

54. $y = -2(x + 1)^2 + 5$

55. $y = 2(x - 1)^2 + 3$

56. $y = -5.5(x + 2)^2 - 5$

57. $y = -(15 + 10\sqrt{2})(x - \sqrt{2})^2 + 5$, or $y \approx -29.142(x - \sqrt{2})^2 + 5$

58. $y = 1.5(x - 3\sqrt{5})^2 + 8$

59. $A(x) = x(50 - x)$; maximum of 625 ft^2 when $x = 25$. The dimensions are 25 ft × 25 ft.

60. The area of the picture and the frame, if the width of the picture is x ft, is $A(x) = (x + 2)(x + 5)$ ft^2, which equals 208 when $x = 11$, so the painting is 11 ft × 14 ft.

61. If the strip is x feet wide, the area of the strip is $A(x) = (25 + 2x)(40 + 2x) - 1000$ ft^2; $A = 504$ ft^2 when $x = 3.5$ ft.

62. **(a)** $R(x) = (800 + 20x)(300 - 5x)$. **(b)** [0, 25] × [200,000, 250,000] is one possibility (shown). **(c)** The maximum income, $250,000 is achieved when $x = 10$, corresponding to rent of $250 per month.

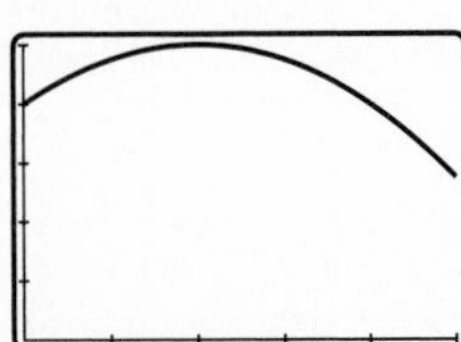

63. **(a)** $R(x) = (26000 - 1000x)(0.50 + 0.05x)$.

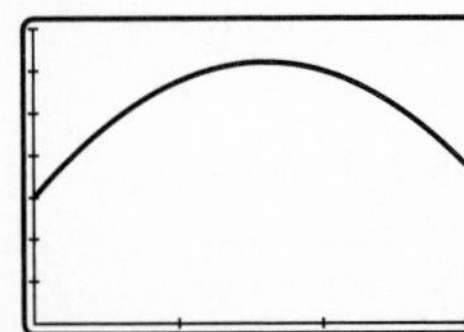

(b) Many choices of Xmax and Ymin are reasonable. Shown is [0, 15] × [10,000, 17,000]. **(c)** The maximum revenue, $16,200, is achieved when $x = 8$; that is, by charging $0.90 per can.

64. Total sales would be $S(x) = (30 + x)(50 - x)$ thousand dollars, when x additional salespeople are hired; the maximum occurs when $x = 10$.

65. **(a)** $h = -16t^2 + 48t + 3.5$; shown is the window [0, 3.5] × [0, 40]; for Parametric mode, graph $x = 1$, $y = -16t^2 + 48t + 3.5$ on the window [0, 2] × [0, 40] and use Tmin = 0 and Tmax = 3.1.

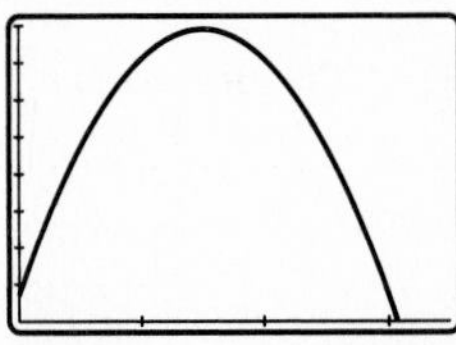

(b) The maximum height is 39.5 ft, 1.5 sec after the ball is thrown.

66. **(a)** $h = -16t^2 + 80t - 10$; shown is the window [0, 5] × [−10, 100]; for parametric mode, graph $x = 1$, $y = -16t^2 + 80t - 10$ on the window [0, 2] × [−10, 100] and use Tmin = 0 and Tmax = 4.9.

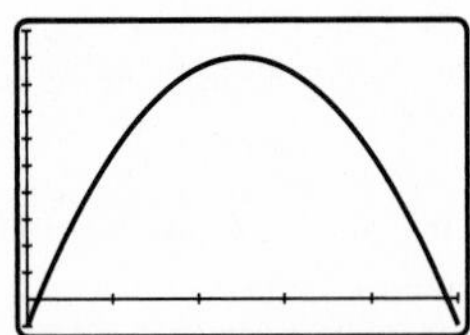

(b) The maximum height is 90 ft, 2.5 sec after the aerial bomb is shot.

67. It reaches a height of 40,000 ft after 50 sec.

68. The exact answer is $32\sqrt{3}$, or about 55.426 ft/sec; it can be found algebraically by noting that the vertex of the parabola $y = ax^2 + bx + c$ with $c = 0$ has y-coordinate $c - \frac{b^2}{4a} = \frac{b^2}{64}$ and setting this equal to 48.

69. Cross-sectional area: $A(x) = x(10 - 2x)$; it is maximized (at 12.5 in.^{2}) when $x = 2.5$ in., so the width is 5 in.

70. Area of pen, when the two sides perpendicular to the wall have length x: $A(x) = x(500 - 2x)$; the largest area (31,250 ft^2) is attained when $x = 125$ and the other dimension is 250 ft.

71. $0.8x^2 - 1.65x + 114.1 = 250$ when $x = 14.1$, or about 1995

72. The formulas shown are using $x = 0$ to represent 1980. If $x = 0$ represents something else, the leading coefficient will be the same in each case, but the others will change. The picture should be the same no matter what $x = 0$ represents. **(a)** $y \approx 6.554x^2 - 127.104x + 743.7$. **(b)** $y \approx 23.629x - 103.895$.

73. The quadratic regression predicts $823 billion, whereas the linear regression predicts $369 billion.

74. One way: Multiply out $f(x)$ to get $x^2 - (a + b)x + ab$. Matching it with the form $Ax^2 + Bx + C$ (using uppercase letters to avoid confusion with a and b in our function), we have $A = 1$ and $B = -(a + b)$, so the axis of symmetry is $x = -B/2A = (a + b)/2$. Another way: $f\left(\frac{a+b}{2} + h\right) = \left(\frac{a+b}{2} + h - a\right)\left(\frac{a+b}{2} + h - b\right) = \left(h - \frac{a-b}{2}\right)\left(h + \frac{a-b}{2}\right) = h^2 - (a - b)^2/4$.

Then compute $f\left(\frac{a+b}{2} - h\right)$ and observe that it is the same; that is, $f\left(\frac{a+b}{2} + h\right) = f\left(\frac{a+b}{2} - h\right)$, so f is symmetric about $x = (a + b)/2$.

75. $x = \frac{(a+b)}{2}$, $y = \frac{-(a-b)^2}{4}$

76. x_1 and x_2 are given by the quadratic formula $\frac{-b \pm \sqrt{b^2 - 4ac}}{2a}$; then $x_1 + x_2 = -b/a$, and the line of symmetry is $x = -\frac{b}{2a}$.

77. **(a)** Horizontally shrink by 1/2. **(b)** Vertically stretch by 4. **(c)** The effect on the graph is the same.

78. **(a)** If $b < 0$, the vertex has a positive x-coordinate; if $b > 0$, the vertex has a negative x-coordinate; in fact, the vertex is $(-b/2, 1 - b^2/4)$; the vertex is above the x-axis if $|b| < 2$, on the axis if $|b| = 2$, and below if $|b| > 2$. **(b)** Use the rules from (a), and note that the graph of $f(x)$ is the graph of $y = x^2 - 6x + 1$

translated upward 1 unit, whereas g is $y = x^2 + 4x + 1$ translated downward 4 units.

QUICK REVIEW 3.2

1. $(x - 4)(x + 3)$
2. $(x - 4)(x - 7)$
3. $(3x - 2)(x - 3)$
4. $(2x - 1)(3x - 1)$
5. $x(3x - 2)(x - 1)$
6. $2x(3x - 2)(x - 3)$
7. $x = 0$ or $x = -3$
8. $x = -1$
9. $x = -5$ or $x = 5$
10. $x = 51$
11. (d)
12. Not shown
13. (c)
14. Not shown
15. Not shown
16. (b)

SECTION EXERCISES 3.2

1. $[-7, 3] \times [-10, 30]$

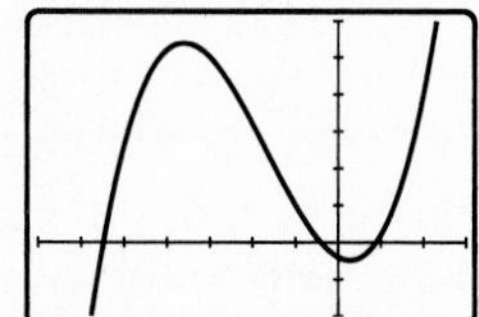

2. $[-4, 6] \times [-10, 20]$

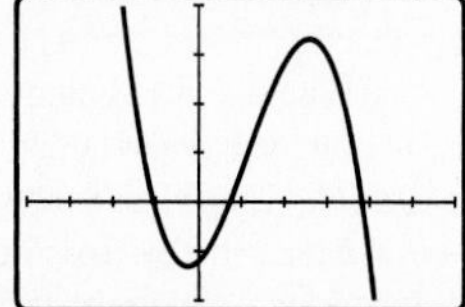

3. $[-8, 10] \times [-120, 100]$

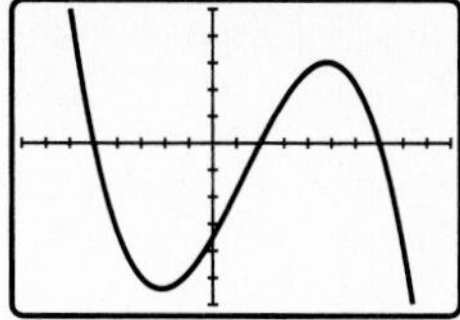

4. $[-10, 10] \times [-100, 130]$

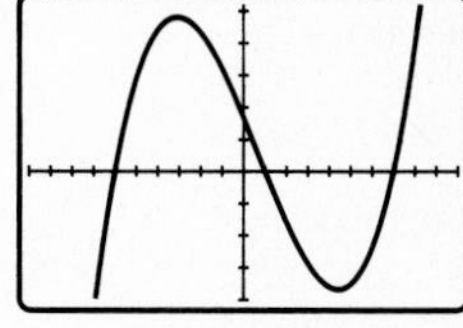

5. $[-12, 8] \times [-50, 425]$

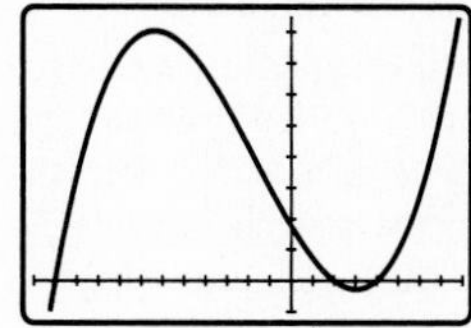

6. $[-8, 15] \times [-400, 150]$

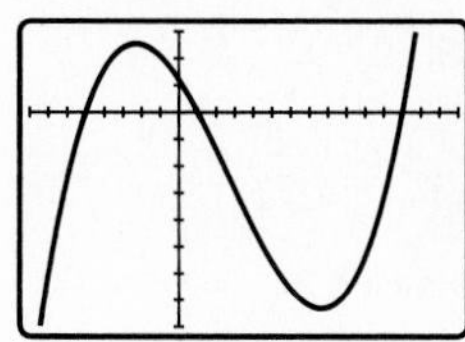

7. $[-5, 5] \times [-30, 30]$

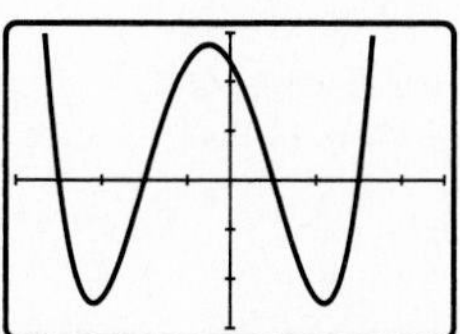

8. $[-6, 5] \times [-250, 50]$

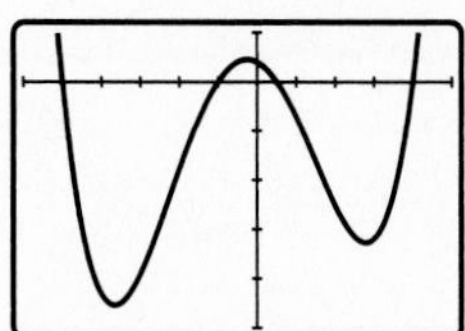

9. $[-3, 5] \times [-50, 50]$

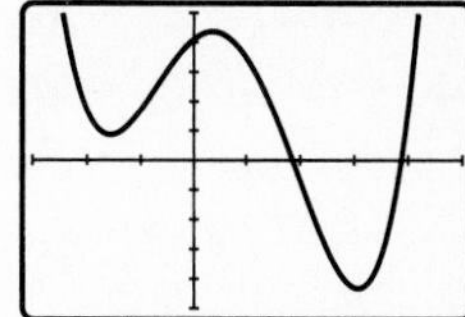

10. $[-4, 3] \times [-20, 90]$

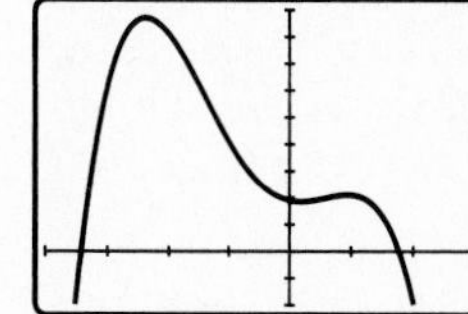

11. $[-2, 2] \times [-4, 2]$

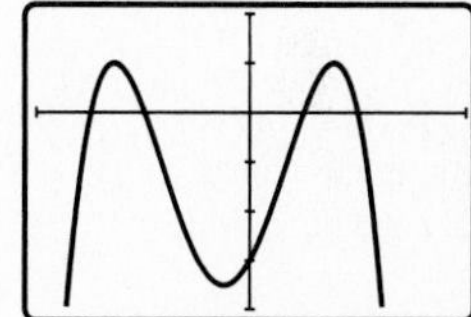

12. $[-3, 5] \times [-20, 50]$

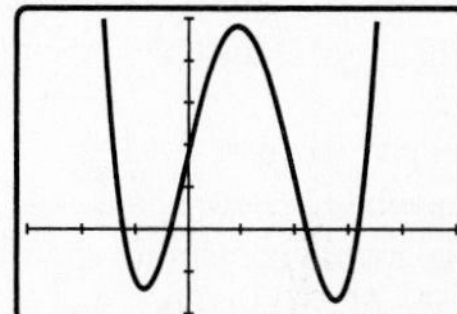

13. $f(x) \to \infty$ as $x \to \infty$ and $f(x) \to -\infty$ as $x \to -\infty$
14. $f(x) \to -\infty$ as $x \to \infty$ and $f(x) \to -\infty$ as $x \to -\infty$
15. $f(x) \to -\infty$ as $x \to \infty$ and $f(x) \to \infty$ as $x \to -\infty$
16. $f(x) \to \infty$ as $x \to \infty$ and $f(x) \to \infty$ as $x \to -\infty$
17. $f(x) \to \infty$ as $x \to \infty$ and $f(x) \to \infty$ as $x \to -\infty$
18. $f(x) \to -\infty$ as $x \to \infty$ and $f(x) \to \infty$ as $x \to -\infty$
19. $f(x) \to -\infty$ as $x \to \infty$ and $f(x) \to \infty$ as $x \to -\infty$

20. $f(x) \to -\infty$ as $x \to \infty$ and $f(x) \to -\infty$ as $x \to -\infty$

21. $f(x) \to \infty$ as $x \to \infty$ and $f(x) \to -\infty$ as $x \to -\infty$

22. $f(x) \to \infty$ as $x \to \infty$ and $f(x) \to \infty$ as $x \to -\infty$

23. $f(x) \to -\infty$ as $x \to \infty$ and $f(x) \to \infty$ as $x \to -\infty$

24. $f(x) \to -\infty$ as $x \to \infty$ and $f(x) \to \infty$ as $x \to -\infty$

25. (c)

26. (b)

27. (a)

28. (d)

29. The graph in part (a); the other windows squeeze the graph too much and obscure the maximum and minimum.

30. The graph in part (c); one of the x-intercepts is less than -5.

31. The graph in part (b); part (a) is not tall enough, and part (c) is too tall.

32. The graph in part (b); part (a) is not quite tall enough to include one maximum and one minimum; part (c) shows them but makes them too short.

33. The graph in part (c); one x-intercept is at about -6.

34. The graph in part (c); one x-intercept is at -6 and another is at 7.

35. -4 and 2

36. -2 and 2/3

37. $-2/3$ and 1/3

38. 0, -5, and 5

39. 0, $-2/3$, and 1

40. 0, -1, and 2

41. The graph in part (a); there are three zeros (though it is difficult to tell on the window shown): -2.5, 1, and 1.1.

42. The graph in part (b); there are three zeros (though it is difficult to tell without zooming in): 0.4, approximately 0.429 (actually 3/7), and 3.

43. The graph in part (c); there are three zeros: approximately -0.273 (actually $-3/11$), -0.25, and 1.

44. The graph in part (d); there are three zeros: -2, 0.5, and 3.

45. -2 and 5; algebraically

46. -6 or -4; algebraically

47. -40 or 25; algebraically

48. 1 and 13; algebraically

49. 0, -6, and 6; algebraically—factor out x first.

50. -11, -1, and 10; graphically. Cubic equations *can* be solved algebraically, but methods for doing so are more complicated than the quadratic formula. A complete graph can be seen on the window $[-15, 15] \times [-800, 800]$.

51. -5, 1, 11; a complete graph can be seen on the window $[-10, 15] \times [-300, 150]$.

52. -6, 2, and 8; graphically, starting from the window $[-10, 15] \times [-500, 500]$

53. $f(x) = (x-3)(x+4)(x-6) = x^3 - 5x^2 - 18x + 72$ is one possibility.

54. $f(x) = (x+2)(x-3)(x+5) = x^3 + 4x^2 - 11x - 30$

55. $f(x) = (x-\sqrt{3})(x+\sqrt{3})(x-4) = (x^2-3)(x-4) = x^3 - 4x^2 - 3x + 12$

56. $f(x) = (x-1)(x-1-\sqrt{2})(x-1+\sqrt{2}) = (x-1)[(x-1)^2 - 2] = x^3 - 3x^2 + x + 1$

57. $f(x) = (x+3)(x-2+\sqrt{5})(x-2-\sqrt{5}) = (x+3)[(x-2)^2 - 5] = x^3 - x^2 - 13x - 3$

58. $f(x) = (x-1+\sqrt{3})(x-1-\sqrt{3}) \cdot (x+5)(x-3) = [(x-1)^2 - 3](x+5)(x-3) = x^4 - 21x^2 + 26x + 30$

59. -2.43, -0.74, and 1.67

60. -1.73, 0.26, and 4.47

61. -2.47, -1.46, and 1.94

62. -4.53 and 2

63. -4.90, -0.45, 1, and 1.35

64. -1.98, -0.16, 1.25, 2.77, and 3.62

65. **(a)** $P(x) = R(x) - C(x)$ is positive if $29.73 < x < 541.74$, so they need at least 30 customers. **(b)** Either 200.49 or 429.73 customers, so 201 or 429 customers.

66. **(a)** When $x = 25$, $D = 55.76$ (or 56.31 if you use the regression equation produced on your calculator). **(b)** $x = 68.3$ mph (or $x = 67.8$ mph if you use the regression equation).

67. **(a)** A good window is $[0, 0.8] \times [0, 1.19]$. **(b)** 0.3391 cm from the center of the artery.

68. **(a)** The height of the box will be x, the width will be $15 - 2x$, and the length $60 - 2x$. **(b)** Any value of x between approximately 0.550 and 6.786 in.

69. $0 < x \leq 0.929$ or $3.644 \leq x < 5$

70. $0 < x < 21.5$. (The side lengths of the rectangle are 43 and 62 units.)

71. The first view shows the end behavior of the function but obscures the fact that there are two local maxima and a local minimum (and four x-axis intersections) between -3 and 4. These are visible in the second view, but missing is the minimum near $x = 7$ and the x-axis intersection near $x = 9$. The second view suggests a degree 4 polynomial rather than degree 5.

72. The end behavior is visible in the first window but not the details of the behavior near $x = 1$. The second view shows those details but at the loss of the end behavior.

73. The exact behavior near $x = 1$ is hard to see. A zoomed-in view around point $(1, 0)$ suggests that the graph just touches the x-axis at 0 without actually crossing it; that is, $(1, 0)$ is a local maximum. One possible window is $[0.9999, 1.0001] \times [-1 \times 10^{-7}, 1 \times 10^{-7}]$.

74. This also has a maximum near $x = 1$, but this time a window such as $[0.6, 1.4] \times [-0.1, 0.1]$ reveals that the graph actually rises above the x-axis and has a maximum at (0.999, 0.025).

75. A maximum and minimum are not visible in the standard window but can be seen on the window $[0.2, 0.4] \times [5.29, 5.3]$.

76. A maximum and minimum are not visible in the standard window but can be seen on the window $[0.95, 1.05] \times [-6.0005, -5.9995]$.

77. The graph of $y = 3(x^3 - x)$, shown on the window $[-2, 2] \times [-5, 5]$, cannot be obtained from the graph of $y = x^3$ by the transformations studied in Chapter 2 (translations, reflections, and stretching/shrinking). Since the right side includes only these transformations, there can be no solution.

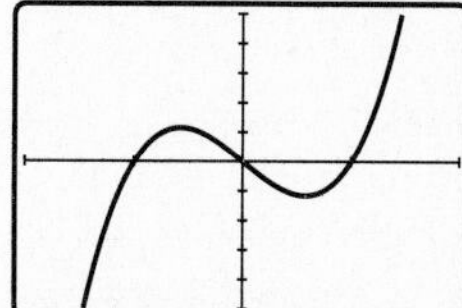

78. The graph of $y = x^4$ has a "flat bottom," whereas the graph of $y = x^4 + 3x^3 - 2x - 3$, shown on $[-4, 2] \times [-8, 5]$, is "bumpy." Therefore this graph cannot be obtained from the graph of $y = x^4$ through the transformations of Chapter 2. Since the right side includes only these transformations, there can be no solution.

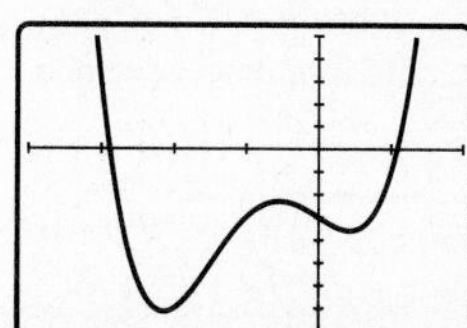

79. **(a)** Substituting $x = 2$, $y = 7$, we find that $7 = 5(2 - 2) + 7$, so Q is on line L and also that $f(2) = -8 + 8 + 18 - 11 = 7$, so Q is on the graph of $f(x)$. **(b)** Shown is a zoomed-in view near $Q(2, 7)$; the viewing window is $[1.8, 2.2] \times [6, 8]$. Calculator output will not show the detail seen here. **(c)** The line L also crosses the graph of $f(x)$ at $(-2, -13)$.

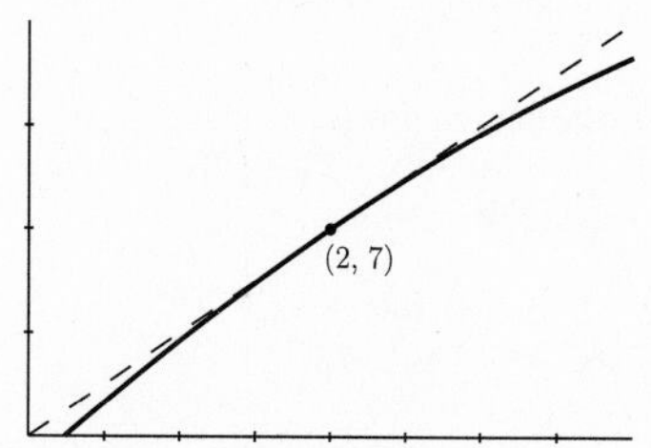

80. **(a)** Note that $f(a) = a^n$ and $f(-a) = -a^n$; $m = \dfrac{y_2 - y_1}{x_2 - x_1} = \dfrac{-a^n - a^n}{-a - a} = \dfrac{-2a^n}{-2a} = a^{n-1}$. **(b)** First observe that $f(x_0) = (a^{1/(n-1)})^n = a^{n/(n-1)}$; using point-slope form gives $y - a^{n/(n-1)} = a^{n-1}(x - a^{1/(n-1)})$. **(c)** With $n = 3$ and $a = 3$, this equation becomes $y - 3^{3/2} = 3^2(x - 3^{1/2})$, or $y = 9(x - \sqrt{3}) + 3\sqrt{3} = 9x - 6\sqrt{3}$.

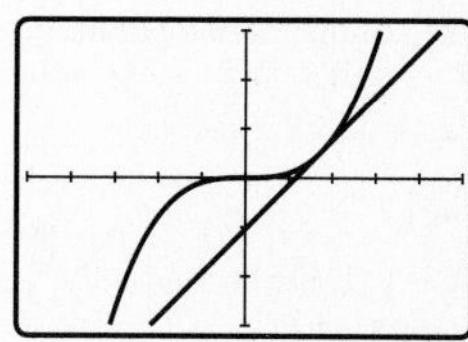

81. **(a)** Consider the right triangle with leg lengths x and D; it is similar (in the geometric sense) to the triangle with leg lengths 8 and $D - u$; therefore, $\dfrac{8}{x} = \dfrac{D - u}{D}$. Now sketch a horizontal segment from the left side to the point where the two diagonal segments cross (note that this segment is $D - u$ units long); the right triangle with leg lengths y and D is similar to the triangle formed by the left side and this new segment, with leg lengths $y - 8$ and $D - u$, so that $\dfrac{y - 8}{y} = \dfrac{D - u}{D}$; therefore $\dfrac{8}{x} = \dfrac{D - u}{D} = \dfrac{y - 8}{y}$. **(b)** Starting with the equation from (a), cross-multiply to get $8y = xy - 8x$, then subtract xy from both sides, factor out y on the left, and divide by $(8 - x)$. **(c)** Applying the Pythagorean theorem to the two large right triangles, we have $x^2 + D^2 = 20^2$ and $y^2 + D^2 = 30^2$, which combine to give $D^2 = 400 - x^2 = 900 - y^2$, or $y^2 - x^2 = 500$; then making the substitution from (b), we get $\left(\dfrac{8x}{x - 8}\right)^2 - x^2 = 500$, so that $\dfrac{64x^2}{(x - 8)^2} - x^2 = 500$, or $64x^2 - x^2(x - 8)^2 = 500(x - 8)^2$; expanding this expression gives us the desired equation. **(d)** The two solutions are about $x = 5.9446$ and $x = 11.7118$; based on the diagram, x must be between 8 and 20 for this problem, so $x \approx 11.7118$, so $D = \sqrt{20^2 - x^2} \approx 16.2121$

82. Methods of calculus can be used to show that the graph has a maximum in Quadrant I and a minimum in Quadrant I or IV if $b < -\sqrt{3}$, a maximum and minimum in Quadrant II if $b > \sqrt{3}$, and has no maximum/minimum if $|b| \leq \sqrt{3}$. Experimentation may not give quite this much detail, but similar conclusions can be reached by trying different values of b.

QUICK REVIEW 3.3

1. $x^2 - 4x + 7$

2. $x^2 + \frac{5}{2}x - 3$

3. $7x^3 + x^2 - 3$

4. $2x^2 - \frac{2}{3}x + \frac{7}{3}$

5. $x + 5 + 3/x$

6. $2x^2 + 4x + 7 - 3/x$

7. $-\frac{1}{2}x^2 + 6x + 3 + 2/x$

8. $\frac{7}{2}x - \frac{3}{2} - 2/x$

SECTION EXERCISES 3.3

1. Quotient $x^2 - 4x + 1$; remainder, 0

2. Quotient: $x^2 + x - 4$; remainder: 0

3. Quotient $3x^2 + 2x + 1$; remainder, 0

4. Quotient: $2x^2 - 3x + 6$; remainder: 0

5. Quotient $2x^3 - 5x + 1$; remainder, 0

6. Quotient: $x^2 - 3x + 5$; remainder: 0

7. Quotient: $x^2 - 1$; remainder: 1

8. Quotient: $x^3 + 3x^2 - 2x - 1$; remainder: -12

9. $f(x) = (x - 1)(x - 1) + (2)$

10. $f(x) = (x^2 - x + 1)(x + 1) + (-2)$

11. $f(x) = (x^2 + x + 4)(x + 3) + (-21)$

12. $f(x) = (x^3 - 5x^2 - 6x - 5)(x - 2) + (-19)$

13. $f(x) = (2x^2 - 5x + 7/2)(2x + 1) + (-9/2)$

14. $f(x) = (x^2 - 4x + 12)(x^2 + 2x - 1) + (-32x + 18)$

15. $\dfrac{x^3 - 5x^2 + 3x - 2}{x + 1} = x^2 - 6x + 9 + \dfrac{-11}{x + 1}$

16. $\dfrac{2x^4 - 5x^3 + 7x^2 - 3x + 1}{x - 3} = 2x^3 + x^2 + 10x + 27 + \dfrac{82}{x - 3}$

17. $\dfrac{9x^3 + 7x^2 - 3x}{x - 10} = 9x^2 + 97x + 967 + \dfrac{9670}{x - 10}$

18. $\dfrac{3x^4 + x^3 - 4x^2 + 9x - 3}{x + 5} = 3x^3 - 14x^2 + 66x - 321 + \dfrac{1602}{x + 5}$

19. $\dfrac{5x^4 - 3x + 1}{4 - x} = -5x^3 - 20x^2 - 80x - 317 + \dfrac{1269}{4 - x}$

20. $\dfrac{x^8 - 1}{x + 2} = x^7 - 2x^6 + 4x^5 - 8x^4 + 16x^3 - 32x^2 + 64x - 128 + \dfrac{255}{x + 2}$

21. The remainder is $f(2) = 3$.

22. The remainder is $f(1) = -4$.

23. The remainder is $f(2) = 5$.

24. The remainder is $f(-1) = 23$.

25. The bottom row is 1 -5 -8 6, so $f(-1) = 6$.

26. The bottom row is 1 -5 -6 -15 -36, so $f(2) = -36$.

27. The bottom row is 2 -6 24 -75 227, so $f(-3) = 227$.

28. The bottom row is 1 -3 15 -72 361, so $f(-5) = 361$.

29. The bottom row is 2 1 2 3 -5 -2, so $f(1) = -2$.

30. The bottom row is 1 4 11 18 41, so $f(2) = 41$.

31. Yes: 1 is a zero of the second polynomial.

32. Yes: 3 is a zero of the second polynomial.

33. Yes: -2 is a zero of the second polynomial.

34. No: when $x = -1$, the second polynomial evaluates to 2.

35. $(x + 4)(x + 2)(x - 3)$

36. $(x + 5)(x - 1)(x - 2)$

37. $(x + 3)(2x - 1)(x - 1)(x - 5)$

38. $(x + 2)(x + 1)(3x - 1)(x - 3)$

39. $(x + 2)(x + \sqrt{2})(3x - 1) \cdot (x - \sqrt{2})(x - 3)$

40. $(x + 2)(x + \sqrt{3})(2x - 1)(x - 1) \cdot (x - \sqrt{3})$

41. Either $(x + 3)$ or $(x - 1)$ can be guessed easily from the graph; $(x - 3.4)$ or $(5x - 17)$ is the other (though it's not so easy to guess).

42. Either $(x + 2)$ or $(x - 3)$ can be guessed easily; the third factor is $(x - 1.4)$ or $(5x - 7)$.

43. The "guessable" factor is $(x - 3)$. The other two crossings occur at $-2.5 \pm \frac{1}{2}\sqrt{17}$ (not very easy to guess).

44. The "guessable" factor is $(x - 3)$; the other two crossings occur at $-1.5 \pm \frac{1}{2}\sqrt{17}$.

45. All four factors—$(x + 3)$, $(x + 1)$, $(x - 1)$, and $(x - 2)$—can be guessed from the graph.

46. $(x - 1)$ can be guessed from the standard window; $(x + 2)$ can be guessed after zooming in; the other two crossings occur at $-0.75 \pm \frac{1}{4}\sqrt{17}$.

47. $2(x - 3)(x + 4) = 2x^2 + 2x - 24$

48. $2(x - 2)^2 = 2x^2 - 8x + 8$

49. $2(x + 2)(x - 1)(x - 4) = 2x^3 - 6x^2 - 12x + 16$

50. $2(x + 1)(x - 3)(x + 5) = 2x^3 + 6x^2 - 26x - 30$

51. $2(x - 2)\left(x - \frac{1}{2}\right)\left(x - \frac{3}{2}\right) = \frac{1}{2}(x - 2)(2x - 1)(2x - 3) = 2x^3 - 8x^2 + \frac{19}{2}x - 3$

52. $2(x + 3)(x + 1)(x)\left(x - \frac{5}{2}\right) = x(x + 3)(x + 1)(2x - 5) = 2x^4 + 3x^3 - 14x^2 - 15x$

53. $2(x - 2)(x - 4) = 2x^2 - 12x + 16$

54. $2(x + 3)(x - 5) = 2x^2 - 4x - 30$

55. $3(x - 3)(x - 5)(x + 4) = 3x^3 - 12x^2 - 51x + 180$

56. $2(x + 2)(x - 1)(x - 5) = 2x^3 - 8x^2 - 14x + 20$

57. The supply and demand graphs are shown on the window $[0, 50] \times [0, 100]$. They intersect when $p = \$36.27$, at which point both supply and demand equal 53.7.

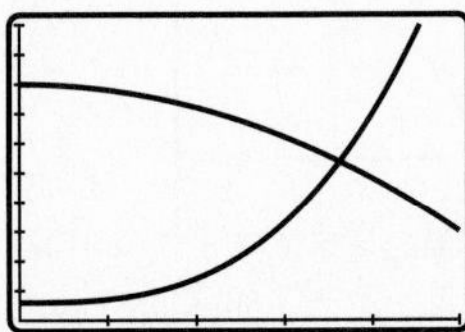

58. The supply and demand graphs, shown on the window $[0, 150] \times [0, 1500]$, intersect when $p = \$106.99$; there $S(p) = D(p) = 1010.15$.

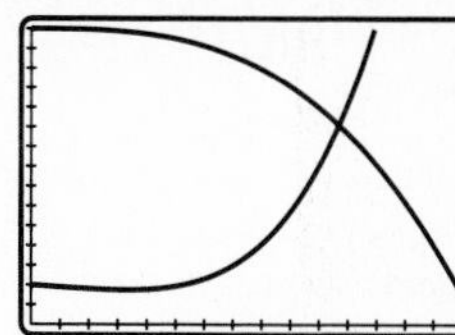

59. **(a)** $y = -0.0054x^3 + 0.3064x^2 - 5.3285x + 77.1473$ (where $x = 0$ represents 1970). Shown in $[0, 25] \times [40, 65]$. **(b)** When $x = 33$, $y \approx 41$ —about 41%. **(c)** If we predict 41% would say "Homemaker" and 55% would say "Outside Job," then we predict no answer from the remaining 4%. **(d)** If the two curves are to be consistent, then when one increases, the other should decrease.

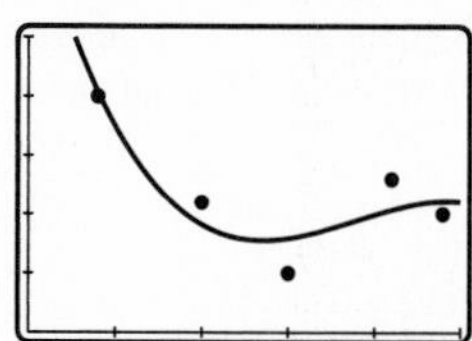

60. **(a)** $y_1 = 0.001546x^4 - 0.08529x^3 + 1.4831x^2 - 5.9970x + 21.5$, where $x = 0$ represents 1970. Shown in $[0, 25] \times [10, 70]$. **(b)** $y_2 = -0.1127x^2 + 5.6134x - 4.9$. Shown in $[0, 25] \times [10, 70]$. **(c)** When $x = 27$, $y_1 = 84\%$ and $y_2 = 65\%$. **(d)** The first curve suggests that recycling increases; the second suggests that it levels off.

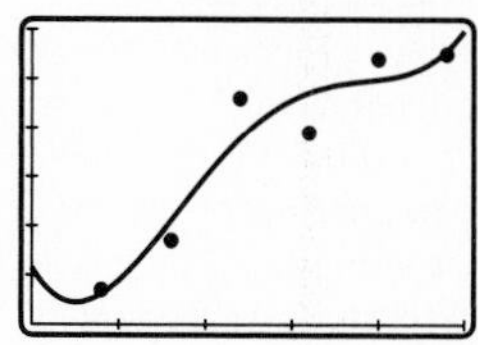

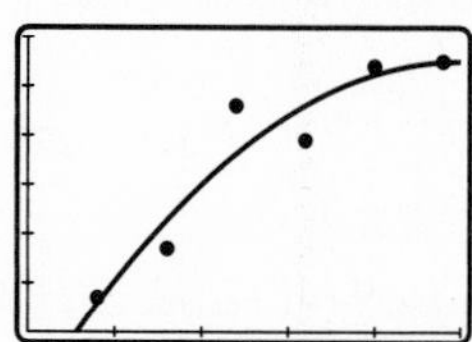

61. **(a)** Multiply out this expression to confirm the equality. **(b)** The innermost operation, $7 \cdot 17 - 12$, is the first addition done in the synthetic division table $(-12 + 119)$. It is followed in both cases by multiplying by 17, then adding 6, then multiplying by 17, and then subtracting 23. **(c)** $f(17) = 31002$; with the expression in (b), three multiplications and three additions/subtractions are required; using the original expression, there are again three additions/subtractions, but six multiplications (for example, $7x^3 = 7 \cdot x \cdot x \cdot x$ is three); presumably, the first would execute faster on a computer since fewer multiplications are required.

62. Synthetic division requires a leading coefficient of 1 in the divisor; if we divide the numerator and denominator by 2 first, we obtain $\dfrac{2x^3 - \frac{5}{2}x^2 + \frac{3}{2}x + \frac{1}{2}}{x - \frac{1}{2}}$, which can be performed synthetically.

63. The remainder theorem gives the remainder as $(-1)^{40} - 3 = -2$.

64. The remainder is $1^{63} - 17 = -16$.

65. The right side of this approximate equality is a degree 4 polynomial, which has zeros -3.10, 0.5, 1.13, and 1.37. If the zeros of degree 4 polynomial $f(x)$ with leading coefficient 1 are r_1, r_2, r_3, and r_4, then $f(x) = (x - r_1)(x - r_2)(x - r_3)(x - r_4)$. Zooming in on the graph of $f(x)$ reveals that $r_1 \approx -3.10$, $r_2 \approx 0.5$ (actually $r_2 = 0.5$), $r_3 \approx 1.13$, and $r_4 \approx 1.37$. Of course, we can also confirm this approximate equation by graphing the two sides and observing that they are nearly identical.

66. One reason is that the cubic model suggests (perhaps unreasonably) rapid growth of the percentage of women choosing outside work: By 2011, it predicts (slightly over) 100%.

More data points might also be useful. With only 5 data points, we can get an exact fit with a quartic polynomial and an *almost* exact fit with a cubic (as can be seen). Would a cubic model fit so much better than the linear model if we had data from *all* the years 1974 to 1994? Perhaps not.

QUICK REVIEW 3.4

1. The bottom row is 2 −3 3 3 −1, so $f(2) = -1$.

2. The bottom row is −3 3 4 −7 9, so $g(-1) = 9$.

3. $f(-x) = -3x^3 - 4x^2 - 7x - 2$

4. $f(-x) = x^3 + 3x^2 + 5x + 1$

5. $q(x) = x^2 + 3x - 2$, $r(x) = 0$

6. $q(x) = x^2 - x + 1$, $r(x) = 0$

SECTION EXERCISES 3.4

1. There are two sign changes in $f(x)$ and one sign change in $f(-x) = -x^3 + x^2 + x + 1$; there are no or two positive zeros and one negative zero.

2. There are one sign change in $g(t)$ and two sign changes in $g(-t) = -t^3 - t^2 + t - 1$; there are one positive zero and no or two negative zeros.

3. One or three positive zeros and no negative zeros

4. No positive zeros and one or three negative zeros

5. One positive zero and no negative zeros

6. One positive zero and one negative zero

7. No, two, or four positive zeros and no negative zeros

8. One or three positive zeros and one negative zero

9. One or three positive zeros and no or two negative zeros

10. No, two, or four positive zeros and no negative zeros

11. The bottom row is 2 2 7 19 (all positive).

12. The bottom row is −1 −2 −5 −19 (all negative).

13. The bottom row is 1 1 3 7 2 (all positive).

14. The bottom row is −1 −4 −5 −11 −21 (all negative).

15. The bottom row is −2 −5 −20 −99 (all negative).

16. The bottom row is 4 6 11 42 128 (all positive).

17. The bottom row is 1 6 3 (all positive).

18. The bottom row is −2 −3 −7 (all negative).

19. The bottom row is −3 −7 −8 −5 (all negative).

20. The bottom row is 1 1 5 10 (all positive).

21. The bottom row is 1 9 52 262 (all positive).

22. The bottom row is −3 −13 −47 −191 (all negative).

23. All three zeros are between −3 and 2 (of course, the endpoints of this interval are two of the three zeros). Therefore, $f(x)$ is never zero when $x > 2$ or when $x < -3$.

24. $f(x)$ is never zero when $x > 2$ or when $x < -4$.

25. Bottom rows: 6 19 88 448 2206 and 6 41 198 982 4876. No zeros are outside window.

26. Bottom rows: 1 4 20 121 624 3117 and −1 −6 −30 −129 −664 −3323. No zeros are outside window.

27. Bottom rows: 1 6 30 192 1116 5548 and −1 −4 −20 −58 −446 −2262. No zeros are outside window.

28. Bottom rows: −1 −5 50 180 756 and −1 −5 50 320 1456. Zeros at −9 and 8 are *not* shown because −5 and 5 are not bounds for zeros of $f(x)$.

29. Bottom rows: 1 1 −124 −224 −1128 −5637 and −1 −9 84 816 4088 20,443. Zeros at −11.002 and 12.003 are *not* shown because −5 and 5 are not bounds for zeros of $f(x)$.

30. Bottom rows: 2 5 −116 −364 −1911 −9530 and −2 −15 66 546 2821 14,130. Zeros at −8.036 and 9.038 are *not* shown because −5 and 5 are not bounds for zeros of $f(x)$.

31. Possible rational zeros: ±1; 1 is a zero.

32. Possible rational zeros: $\dfrac{\pm 1}{\pm 1, \pm 2, \pm 3, \pm 6}$, or $\pm 1, \pm\frac{1}{2}, \pm\frac{1}{3}, \pm\frac{1}{6}$; 1 is a zero.

33. Possible rational zeros: $\dfrac{\pm 1, \pm 2, \pm 7, \pm 14}{\pm 1, \pm 3}$, or $\pm 1, \pm 2, \pm 7, \pm 14, \pm\frac{1}{3}, \pm\frac{2}{3}, \pm\frac{7}{3}, \pm\frac{14}{3}$; $\frac{7}{3}$ is a zero.

34. Possible rational zeros: $\dfrac{\pm 1, \pm 3, \pm 9}{\pm 1, \pm 2}$, or $\pm 1, \pm 3, \pm 9, \pm\frac{1}{2}, \pm\frac{3}{2}, \pm\frac{9}{2}$; $\frac{3}{2}$ is a zero.

35. Possible rational zeros: $\dfrac{\pm 1, \pm 2, \pm 3, \pm 4, \pm 6, \pm 12}{\pm 1, \pm 2, \pm 3, \pm 6}$, or $\pm 1, \pm 2, \pm 3, \pm 4, \pm 6, \pm 12, \pm\frac{1}{2}, \pm\frac{3}{2}, \pm\frac{1}{3}, \pm\frac{2}{3}, \pm\frac{4}{3}, \pm\frac{1}{6}$; $-\frac{4}{3}$ and $\frac{3}{2}$ are zeros.

36. Possible rational zeros: $\dfrac{\pm 1, \pm 2}{\pm 1, \pm 2, \pm 7, \pm 14}$, or $\pm 1, \pm 2, \pm\frac{1}{2}, \pm\frac{1}{7}, \pm\frac{2}{7}, \pm\frac{1}{14}$; $-\frac{2}{7}$ and $\frac{1}{2}$ are zeros.

37. Rational: 1.5; irrational: $\pm\sqrt{2}$

38. Rational: −3; irrational: $\pm\sqrt{3}$

39. Rational: −3; irrational: $1 \pm \sqrt{3}$

40. Rational: 4; irrational: $1 \pm \sqrt{2}$

41. Rational: −1 and 4; irrational: $\pm\sqrt{2}$

42. Rational: −1 and 2; irrational: $\pm\sqrt{5}$

43. Rational: −0.5 and 4; irrational: none

44. Rational: 2/3; irrational: −0.6823

45. Rational: ±3 and 0.5; irrational: none

46. Rational: 1/3; irrational: none

47. Rational: ±2; irrational: none

48. No real zeros

49. **(a)** Weight of water = Density · Volume $= d \cdot \frac{\pi}{6}x(3r^2 + x^2)$. **(b)** $V = \frac{4}{3}\pi r^3 = \frac{4}{3}\pi \cdot 1^3 = \frac{4\pi}{3}$. **(c)** Weight of buoy = Density · Volume $= \frac{d}{4} \cdot \frac{4\pi}{3} = \frac{\pi d}{3}$. **(d)** Archimedes' principle says that the weight of the displaced water equals the weight of the buoy.

50. There are three solutions, but only 0.6527 makes sense in this situation.

51. Following the development in Example 10 and assuming that the buoy has density p times that of water, we find that W_1 = weight of buoy $= \frac{4}{3}pd\pi$ while W_2 = weight of displaced water is unchanged. Therefore we equate W_1 and W_2 and find eventually that $x^3 - 3x^2 + 4p = 0$. When $p = \frac{1}{3}$, the solution to this equation (between 0 and 1) is approximately 0.77. It sinks 0.77 m into the water.

52. If $p = \frac{1}{5}$, the solution to $x^3 - 3x^2 + 4p = 0$ (between 0 and 1) is 0.57. It sinks 0.57 m into the water.

53. **(a)** Shown is one possible view, on the window [0, 600] × [0, 500].

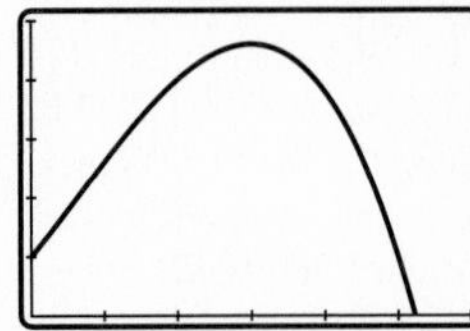

(b) The maximum population, after 300 days, is 460 turkeys. **(c)** $P = 0$ when $t = 523.217$, or about 523 days after release. **(d)** One possibility: After the population increases to a certain point, they begin to compete for food and eventually die of starvation.

54. **(a)** d is the independent variable. **(b)** A good choice is [0, 172] × [0, 5]. **(c)** $s = 1.25$ when $d \approx 95.777$ ft.

55. The only possible rational zeros are ±1, and neither is a zero.

56. The only possible rational zeros are ±1 or ±2; they are not zeros.

57. The only possible rational zeros are ± 1 or $\pm\frac{1}{2}$, and none are zeros.

58. The only possible rational zeros are ± 1 or $\pm\frac{1}{3}$; they are not zeros.

59. **(b)** Graph (a) represents the quadratic equation (y_1) and graph (b) represents the cubic equation (y_2). **(c)** The cubic function is perhaps a slightly better match.

60. **(a)** 1986–1996. **(b)** Computer sales may not have grown *exactly* as predicted by the cubic function, but they certainly have not fallen off as the quadratic model suggests.

61. $g(x) = 3f(x)$, so the zeros of f and the zeros of g are identical. If the coefficients of a polynomial are rational, we may multiply that polynomial by the least common multiple (LCM) of the denominators of the coefficients to obtain a polynomial, with integer coefficients, that has the same zeros as the original.

62. The zeros of $f(x)$ are the same as the zeros of $6f(x) = 6x^3 - 7x^2 - 40x + 21$. Possible rational zeros: $\frac{\pm 1, \pm 3, \pm 7, \pm 21}{\pm 1, \pm 2, \pm 3, \pm 6}$, or $\pm 1, \pm 3, \pm 7, \pm 21, \pm\frac{1}{2}, \pm\frac{3}{2}, \pm\frac{7}{2}, \pm\frac{21}{2}, \pm\frac{1}{3}, \pm\frac{7}{3}, \pm\frac{1}{6}, \pm\frac{7}{6}$. The actual zeros are $-\frac{7}{3}, \frac{1}{2}$, and 3.

63. The zeros of $f(x)$ are the same as the zeros of $12f(x) = 12x^3 - 30x^2 - 37x + 30$. Possible rational zeros: $\frac{\pm 1, \pm 2, \pm 3, \pm 5, \pm 6, \pm 10, \pm 15, \pm 30}{\pm 1, \pm 2, \pm 3, \pm 4, \pm 6, \pm 12}$, or $\pm 1, \pm 2, \pm 3, \pm 5, \pm 6, \pm 10, \pm 15, \pm 30, \pm\frac{1}{2}, \pm\frac{3}{2}, \pm\frac{5}{2}, \pm\frac{15}{2}, \pm\frac{1}{3}, \pm\frac{2}{3}, \pm\frac{5}{3}, \pm\frac{10}{3}, \pm\frac{1}{4}, \pm\frac{3}{4}, \pm\frac{5}{4}, \pm\frac{15}{4}, \pm\frac{1}{6}, \pm\frac{5}{6}, \pm\frac{1}{12}, \pm\frac{5}{12}$. There are no actual rational zeros.

64. $\sqrt{2}$ is a zero of $f(x) = x^2 - 2$; the only possible rational zeros of that polynomial are ± 1 or ± 2.

65. The graph of $f(-x)$ is the reflection across the y-axis of the graph of $f(x)$. An upper bound on the zeros of $f(-x)$ would, after re-reflecting across the y-axis, be a lower bound for the zeros of $f(x)$.

66. **(a)** Approximate zeros: $-3.126, -1.075, 0.910, 2.291$. **(b)** $f(x) \approx f(x) = (x + 3.126)(x + 1.075)(x - 0.910)(x - 2.291)$. **(c)** Graphically, graph the original function and the approximate factorization on a variety of windows and observe their similarity; numerically, compute $f(c)$ and $\hat{f}(c)$ for several values of c.

QUICK REVIEW 3.5

1. $x + 9$

2. $x + 2y$

3. $a + 2d$

4. $5z - 4$

5. $6x - y$

6. $-4x + 2$

7. $x^2 - x - 6$

8. $2x^2 + 5x - 3$

9. $x^2 - 2$

10. $x^2 - 12$

11. $x^2 - 4xy + 4y^2$

12. $x^2 + 6xy + 9y^2$

13. $x^2 - 2x - 1$

14. $x^2 - 4x + 1$

SECTION EXERCISES 3.5

1. $8 + 2i$

2. $5 - 7i$

3. $13 - 4i$

4. $5 - 8i$

5. $5 - (1 + \sqrt{3})i$

6. $\sqrt{5} - 2$

7. $-5 + i$

8. $\sqrt{7} - 7 + 9i$

9. $7 + 4i$

10. $5 + 5i$

11. $-5 - 14i$

12. $-13 - i$

13. $-48 - 4i$

14. $15 + 18i$

15. $5 - 10i$

16. $-(10 + 5\sqrt{2}) + (12 + 6\sqrt{2})i$

17. $-i$

18. 1

19. i

20. i

21. $-i$

22. i

23. -1

24. 1

25. $4i$

26. $5i$

27. $\sqrt{3}i$

28. $\sqrt{5}i$

29. $x = 2, y = 3$

30. $x = 3, y = -7$

31. $x = 1, y = 2$

32. $x = 7, y = -7/2$

33. $5 + 12i$

34. $-2 - 2i$

35. -1

36. i

37. 13

38. 61

39. 25

40. 3

41. 4

42. 40

43. $\frac{2}{5} - \frac{1}{5}i = 0.4 - 0.2i$

44. $-\frac{1}{5} + \frac{2}{5}i = -0.2 + 0.4i$

45. $\frac{3}{5} + \frac{4}{5}i = 0.6 + 0.8i$

46. $\frac{1}{3} - \frac{2}{3}i$

47. $\frac{1}{2} - \frac{7}{2}i = 0.5 - 3.5i$

48. $\frac{26}{29} + \frac{7}{29}i$

49. $\frac{7}{5} - \frac{1}{5}i = 1.4 - 0.2i$

50. $\dfrac{2\sqrt{2} - 1}{3} - \dfrac{2\sqrt{2} + 1}{3}i$

51–54.

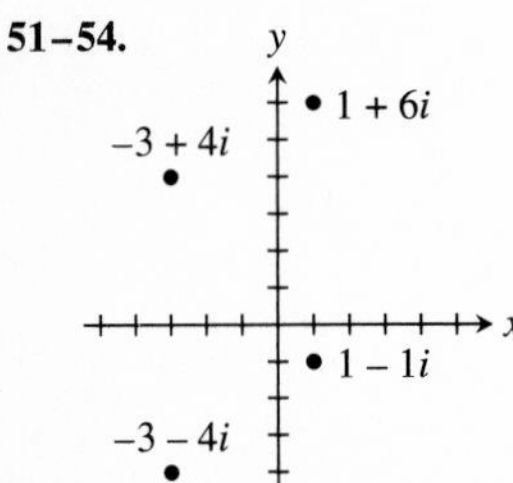

55. $x = -1 \pm 2i$

56. $-\frac{1}{6} \pm \frac{\sqrt{23}}{6}i$

57. $\frac{7}{8} \pm \frac{\sqrt{15}}{8}i$

58. $2 \pm \sqrt{15}i$

59. It either lies entirely above or entirely below the x-axis; it does not touch or cross it.

60. Let a and b be any two real numbers; then $(a + bi) - (a - bi) = (a - a) + (b + b)i = 0 + 2bi = 2bi$.

61. **(a)**

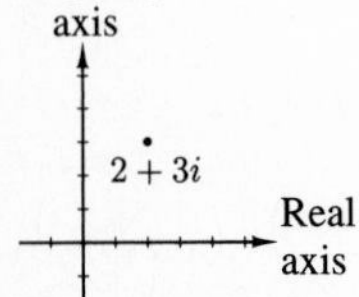

(b) $\|2 + 3i\| = \sqrt{2^2 + 3^2} = \sqrt{13}$.
(c) The formula for the modulus gives the distance from (0, 0) to (2, 3). In general, the modulus of $a + bi$ is the distance from the origin to point (a, b).

62. Part (b) is always true.

63. Part (a) is always true; the modulus of a complex number shares this property with the absolute value of a real number (in fact, if $z = a + 0i$ is actually real, $\|z\| = |a|$).

64. $a - bi + (-a + bi) = (a - a) + (-b + b)i = 0 + 0i = 0$, which is the required property of $-(a - bi)$.

QUICK REVIEW 3.6

1. $(2x - 3)(x + 1)$

2. $(3x + 1)(2x - 5)$

3. $(5x - 4)(2x + 1)$

4. $(3x + 2)(3x - 4)$

5. $\dfrac{\pm 1, \pm 2}{\pm 1, \pm 3}$, or $\pm 1, \pm 2, \pm\frac{1}{3}, \pm\frac{2}{3}$

6. $\dfrac{\pm 1, \pm 3}{\pm 1, \pm 2, \pm 4}$, or $\pm 1, \pm 3, \pm\frac{1}{2}, \pm\frac{3}{2}, \pm\frac{1}{4}, \pm\frac{3}{4}$

7. $\dfrac{\pm 1, \pm 2, \pm 4}{\pm 1, \pm 5}$, or $\pm 1, \pm 2, \pm 4, \pm\frac{1}{5}, \pm\frac{2}{5}, \pm\frac{4}{5}$

8. $\dfrac{\pm 1}{\pm 1, \pm 2, \pm 3, \pm 4, \pm 6, \pm 12}$, or $\pm 1, \pm\frac{1}{2}, \pm\frac{1}{3}, \pm\frac{1}{4}, \pm\frac{1}{6}, \pm\frac{1}{12}$

SECTION EXERCISES 3.6

1. $x^2 + 1$

2. $x^2 - 2x + 5$

3. $x^3 - x^2 + 9x - 9$

4. $x^3 + 2x^2 - 6x + 8$

5. $x^4 - 5x^3 + 7x^2 - 5x + 6$

6. $x^4 - 3x^3 + 2x^2 + 2x - 4$

7. $x^3 - 11x^2 + 43x - 65$

8. $x^3 + x + 10$

9. $x^5 + 4x^4 + x^3 - 10x^2 - 4x + 8$

10. $x^4 - 6x^2 - 8x - 3$

11. $x^4 - 10x^3 + 37x^2 - 60x + 36$

12. $x^4 + 6x^3 + 13x^2 + 12x + 4$

13. (a)

14. (c)

15. (d)

16. (b)

17. (b)

18. (c)

19. (d)

20. (a)

21. Two complex zeros; none real

22. Three complex zeros; all three real

23. Three complex zeros; one real

24. Four complex zeros; two real

25. Four complex zeros; two real

26. Five complex zeros; one real

27. Zeros: $x = 1$, $x = -0.5 \pm \frac{\sqrt{19}}{2}i$; $f(x) = (x - 1)[x - (-0.5 - \frac{\sqrt{19}}{2}i)][x - (-0.5 + \frac{\sqrt{19}}{2}i)]$

28. Zeros: $x = 3$, $x = 3.5 \pm \frac{\sqrt{43}}{2}i$; $f(x) = (x - 3)[x - (3.5 - \frac{\sqrt{43}}{2}i)][x - (3.5 + \frac{\sqrt{43}}{2}i)]$

29. Zeros: $x = \pm 1$, $x = -0.5 \pm \frac{\sqrt{23}}{2}i$; $f(x) = (x - 1)(x + 1)[x - (-0.5 - \frac{\sqrt{23}}{2}i)][x - (-0.5 + \frac{\sqrt{23}}{2}i)]$

30. Zeros: $x = -2$, $x = 1/3$, $x = -0.5 \pm \frac{\sqrt{3}}{2}i$; $f(x) = (x + 2)(3x - 1)[x - (-0.5 - \frac{\sqrt{3}}{2}i)][x - (-0.5 + \frac{\sqrt{3}}{2}i)]$

31. Zeros: $x = -7/3$, $x = 1.5$, $x = 1 \pm 2i$; $f(x) = (3x + 7)(2x - 3)[x - (1 - 2i)][x - (1 + 2i)]$.

32. Zeros: $x = -0.6$, $x = 5$, $x = 1.5 \pm i$; $f(x) = (5x + 3)(x - 5)[2x - (3 - 2i)][2x - (3 + 2i)]$

33. Zeros: $x = \pm\sqrt{3}$, $x = 1 \pm i$; $f(x) = (x - \sqrt{3})(x + \sqrt{3})[x - (1 - i)] \cdot [x - (1 + i)]$

34. Zeros: $x = \pm\sqrt{3}$, $x = \pm 4i$; $f(x) = (x - \sqrt{3})(x + \sqrt{3})(x - 4i)(x + 4i)$

35. Zeros: $x = \pm\sqrt{2}$, $x = 3 \pm 2i$; $f(x) = (x - \sqrt{2})(x + \sqrt{2})[x - (3 - 2i][x - (3 + 2i)]$

36. Zeros: $x = \pm\sqrt{5}$, $x = 1 \pm 3i$; $f(x) = (x - \sqrt{5})(x + \sqrt{5})[x - (1 - 3i][x - (1 + 3i)]$

37. $f(x) = (x - 2)(x^2 + x + 1)$

38. $f(x) = (x - 2)(x^2 + x + 3)$

39. $f(x) = (x - 1)(2x^2 + x + 3)$

40. $f(x) = (x - 1)(3x^2 + x + 2)$

41. $f(x) = (x - 1)(x + 4)(x^2 + 1)$

42. $f(x) = (x - 3)(x + 1)(x^2 + 4)$

43. $h \approx 3.776$ ft

44. $h \approx 6.513$ ft

45. Yes: $(x + 2)(x^2 + 1) = x^3 + 2x^2 + x + 2$ is one such polynomial.

46. No: $-2i$ (the conjugate of $2i$) must also be a zero.

47. No: $1 - 2i$ and $1 + i$ must also be zeros, giving five zeros for a degree 4 polynomial.

48. Yes: $(x - 1 - 3i)(x - 1 + 3i)(x - 1 - i)(x - 1 + i) = x^4 - 4x^3 + 16x^2 - 24x + 20$ is one such polynomial.

49. $-2(x - 3)(x + 1)(x - 2 + i)(x - 2 - i) = -2x^4 + 12x^3 - 20x^2 - 4x + 30$

50. $2(x - 1 - 2i)(x - 1 + 2i)(x - 1 - i)(x - 1 + i) = 2x^4 - 8x^3 + 22x^2 - 28x + 20$

51. (b)

52. (c)

53. (a)

54. (d)

55. **(a)** The model is $d \approx 0.0669t^3 - 0.7420t^2 + 2.1759t + 0.8250$; shown is the window $[0, 8.25] \times [0, 5]$.

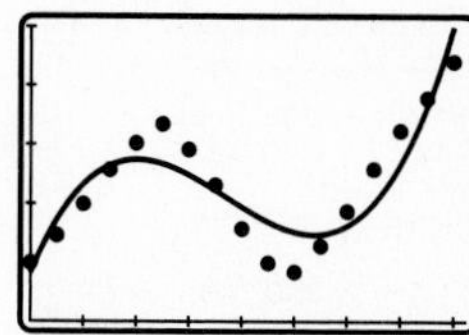

(b) When $t = 0$, the model gives $d = 0.825$. **(c)** The model "changes direction" at approximately $t = 2.02$ (when $d \approx 2.74$ m) and $t \approx 5.38$ (when $d \approx 1.47$ m).

56. **(a)** The model is $d \approx -0.0820t^3 + 0.9162t^2 - 2.5126t + 3.3779$. Shown is the window $[0, 8.25] \times [0, 5]$.

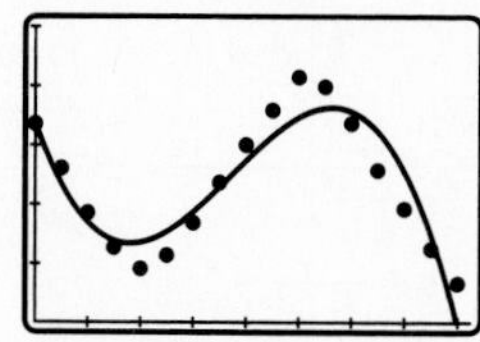

(b) Sally walks toward the detector, turns and walks away (or walks backwards), and then walks toward the detector again.
(c) The model "changes direction" at $t \approx 1.81$ (when $d \approx 1.35$ m) and $t \approx 5.64$ (when $d \approx 3.65$ m).

57. **(a)** The model: $d \approx 0.2434t^2 - 1.7159t + 4.4241$; shown is the window $[0, 8.25] \times [0, 6]$.

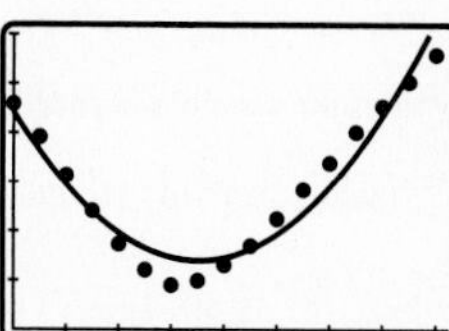

(b) Jacob walks toward the detector, then turns and walks away (or walks backward). **(c)** The model "changes direction" at approximately $t = 3.52$ (when $d \approx 1.40$ m).

58. $f(i) = i^3 - i(i)^2 + 2i(i) + 2 = -i + i - 2 + 2 = 0$

59. $f(-2i) = (-2i)^3 - (2 - i)(-2i)^2 + (2 - 2i)(-2i) - 4 = 8i + (2 - i)(4) - (2 - 2i)(2i) - 4 = 8i + 8 - 4i - 4i - 4 - 4 = 0$

60. $f(i) = i^3 + (3 - i)i^2 - 4ii - 1 = -i - 3 + i + 4 - 1 = 0$—or use synthetic division, which leaves the bottom row as 1 3 $-i$ 0; therefore $f(x) = (x - i)(x^2 + 3x - i)$, and $h(x) = x^2 + 3x - i$.

61. Synthetic division leaves the bottom row: 1 0 1 0, so $f(x) = (x - 1 - i)(x^2 + 1)$, and $h(x) = x^2 + 1$.

62. Up: walking toward the detector and then away (the distance decreases and then increases); down: walking away from the detector and then toward it (the distance increases and then decreases); also, the person should slow down gradually, not just do an abrupt "about face."

63. Positive slope: Walk at a constant rate away from the detector (the distance increases). Negative slope: Walk at a constant rate toward the detector (the distance decreases).

Chapter 3 Review

1. Translate right 2 units and vertically stretch by 3 (either order) and then translate up 4 units.

2. Translate left 3 units and reflect across x-axis (either order) and then translate upward 1 unit.

3. Vertex: $(-3, 5)$; line of symmetry: $x = -3$

4. Vertex: $(5, -7)$; line of symmetry: $x = 5$

5. $f(x) = -2(x + 4)^2 + 1$; vertex: $(-4, 1)$; line of symmetry: $x = -4$

6. $g(x) = 3(x - 1)^2 - 1$; vertex: $(1, -1)$; line of symmetry: $x = 1$

7. $h(x) = 3(x - 5/6)^2 - \frac{49}{12}$; vertex: $(\frac{5}{6}, -\frac{49}{12})$; line of symmetry: $x = \frac{5}{6}$

8. $k(x) = 5(x + 3/10)^2 - 29/20$; vertex: $(-0.3, -1.45)$; line of symmetry: $x = -0.3$

9. $y = \frac{5}{9}(x + 2)^2 - 3$

10. $y = -\frac{3}{16}(x + 1)^2 + 1$

11. $y = \frac{1}{2}(x - 3)^2 - 2$

12. $y = -\frac{1}{2}(x + 4)^2 + 5$

13. $[-10, 7] \times [-50, 10]$

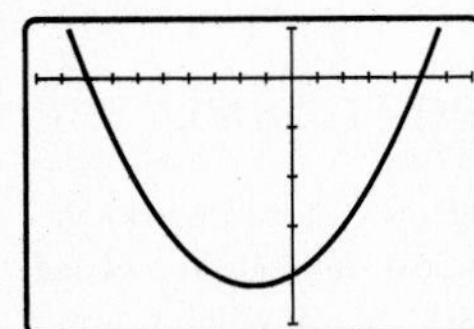

14. $[-3, 5] \times [-50, 10]$

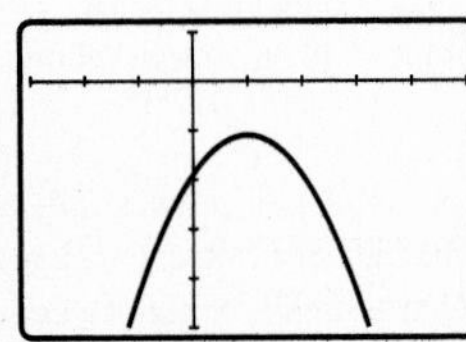

15. $[-5, 3] \times [-30, 30]$

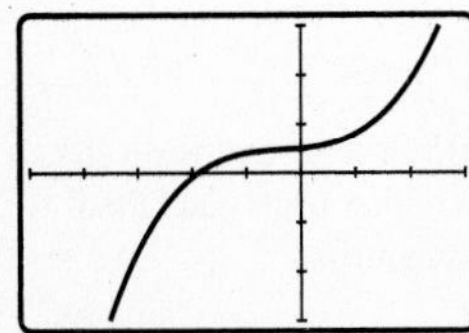

16. $[-6, 7] \times [-50, 30]$

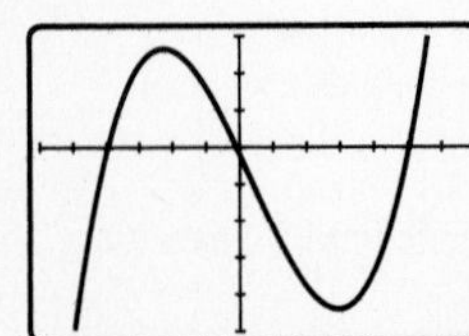

17. $[-5, 0.5] \times [-15, 5]$

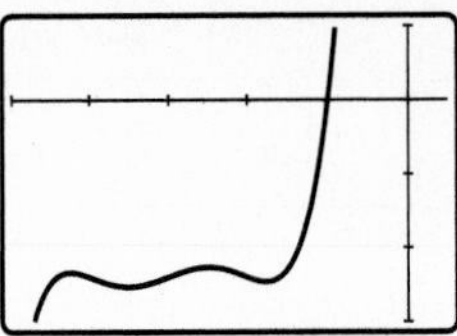

18. $[0, 10] \times [-10, 100]$

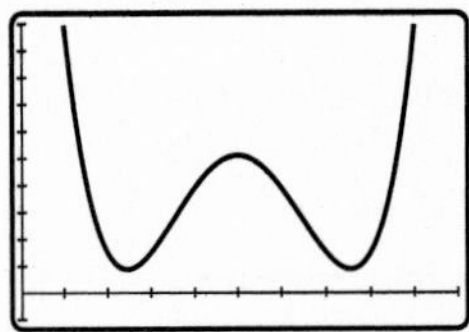

19. $f(x) \to -\infty$ as $x \to \infty$ and $f(x) \to -\infty$ as $x \to -\infty$

20. $f(x) \to \infty$ as $x \to \infty$ and $f(x) \to -\infty$ as $x \to -\infty$

21. $f(x) \to -\infty$ as $x \to \infty$ and $f(x) \to -\infty$ as $x \to -\infty$

22. $f(x) \to -\infty$ as $x \to \infty$ and $f(x) \to \infty$ as $x \to -\infty$

23. $\dfrac{2x^4 - 3x^3 + 9x^2 - 14x + 7}{x^2 + 4} = 2x^2 - 3x + 1 + \dfrac{-2x + 3}{x^2 + 4}$

24. $\dfrac{3x^4 - 5x^3 - 2x^2 + 3x - 6}{3x + 1} = x^3 - 2x^2 + 1 + \dfrac{-7}{3x + 1}$

25. $f(x) = (2x^2 - x + 1)(x - 3) + (-2)$

26. $f(x) = (x^3 + x^2 - x - 1)(x + 2) + (5)$

27. Remainder: $f(-2) = -39$

28. Remainder: $f(3) = -2$

29. Bottom row: 2 −3 5 −10; $f(-1) = -10$

30. Bottom row: 1 3 17 67 261; $f(4) = 261$

31. Yes: 2 is a zero of the second polynomial.

32. No: $x = -3$ yields 1 from the second polynomial.

33. The only real zero is 1.5, so $f(x) = (2x - 3)(x^2 + 2x + 2)$.

34. The zeros (both of multiplicity 2) are -2 and 3, so $f(x) = (x + 2)^2(x - 3)^2$.

35. There are three sign changes in $f(x)$; so there are one or three positive zeros; two sign changes in $f(-x)$, with no or two negative zeros.

36. Two sign changes in $g(x)$ and no or two positive zeros; two in $g(-x)$ and no or two negative zeros.

37. Bottom row: 1 0 3 19 (no negatives)

38. Bottom row: −4 0 −8 −48 −180 (no positives)

39. Bottom row: −4 −8 −9 −44 −130 (no positives)

40. Bottom row: −2 0 −1 −9 (no positives)

41. Possible rational zeros: $\dfrac{\pm 1, \pm 2, \pm 3, \pm 6}{\pm 1, \pm 2}$, or $\pm 1, \pm 2, \pm 3, \pm 6, \pm \frac{1}{2}, \pm \frac{3}{2}$; $-\frac{3}{2}$ and 2 are zeros.

42. Possible rational zeros: $\dfrac{\pm 1, \pm 7}{\pm 1, \pm 2, \pm 3, \pm 6}$, or $\pm 1, \pm 7, \pm \frac{1}{2}, \pm \frac{7}{2}, \pm \frac{1}{3}, \pm \frac{7}{3}, \pm \frac{1}{6}, \pm \frac{7}{6}$; $\frac{7}{3}$ is a zero.

43. $1 + 3i$

44. $2 - 5i$

45. $3 + 0i = 3$

46. $7 + 4i$

47. $-2 + 2i$

48. $25 + 0i = 25$

49. $0 + 1i = i$

50. $0 + 4i = 4i$

51. $-\frac{1}{2} + \frac{1}{2}i = -0.5 + 0.5i$

52. $\frac{6}{5} + \frac{3}{5}i = 1.2 + 0.6i$

53. $3 \pm 2i$

54. $1 \pm \sqrt{3}i$

55. (c)

56. (d)

57. (b)

58. (a)

59. Rational: 0 (multiplicity 2); irrational: $5 \pm \sqrt{2}$; no nonreal zeros

60. Rational: none; irrational: $7 \pm \sqrt{3}$; no nonreal zeros

61. Rational: none; irrational: approximately 0.45; two nonreal zeros

62. Rational: ± 2; irrational: $\pm \sqrt{3}$; no nonreal zeros

63. Rational: $\frac{3}{4}$ and -2; irrational: none; two nonreal zeros

64. Rational: 2 (multiplicity 2); irrational: none; two nonreal zeros

65. Rational: none; irrational: approximately -2.34, 0.57, 3.77; no nonreal zeros

66. Rational: none; irrational: approximately -3.97, -0.19; two nonreal zeros

67. $f(x) = 2x(x - 3)(x + 3)$

68. $f(x) = \frac{1}{2}x[2x - (1 - \sqrt{3})] \cdot [2x - (1 + \sqrt{3})]$

69. $f(x) = (2x + 3)[x - (3 - i)] \cdot [x - (3 + i)]$

70. $f(x) = (5x - 4)[x - (2 + \sqrt{7})] \cdot [x - (2 - \sqrt{7})]$

71. $f(x) = (3x - 2)(2x + 5)(x - 1)(x + 1)$

72. $f(x) = [x - (1 + 2i)][x - (1 - 2i)] \cdot [x - (3 + i)][x - (3 - i)]$

73. $f(x) = (x - 2)(x^2 + x + 1)$

74. $f(x) = (x + 1)[3x - (2 - \sqrt{5})] \cdot [3x - (2 + \sqrt{5})]$

75. $f(x) = (2x - 3)(x - 1)(x^2 - 2x + 5)$

76. $f(x) = (3x + 2)(x + 1)(x^2 - 4x + 5)$

77. $f(x) = (2x - 1)(x + 2)(x - 3) \cdot (x + \sqrt{2})(x - \sqrt{2})$

78. $f(x) = (3x - 1)(x - 3)(x + 4) \cdot (x + \sqrt{3})(x - \sqrt{3})$

79. $(x - \sqrt{5})(x + \sqrt{5})(x - 3) = x^3 - 3x^2 - 5x + 15$

80. $(x + 3)^2 = x^2 + 6x + 9$

81. $(x - 3)(x + 2)(3x - 1)(2x + 1) = 6x^4 - 5x^3 - 38x^2 - 5x + 6$

82. $(x - 2)(x - 1 - i)(x - 1 + i) = x^3 - 4x^2 + 6x - 4$

83. $(x + 2)^2(x - 4)^2 = x^4 - 4x^3 - 12x^2 + 32x + 64$

84. $2(x + 1)(x - 2 - i)(x - 2 + i) = 2x^3 - 6x^2 + 2x + 10$

85. $\frac{1}{15}(x + 2)(x - 3)(x - 5) = \frac{1}{15}(x^3 - 6x^2 - x + 30)$

86. $\frac{1}{2}(x - 2i)(x + 2i)(x - 3 - i)(x - 3 + i) = \frac{1}{2}x^4 - 3x^3 + 7x^2 - 12x + 20$

87. **(a)** When $t = 0$, d appears to be 5 m. **(b)** Carmen moves toward the detector, then turns around and moves away. **(c)** She changes direction at about time $t = 2$, when she appears to be about 1 meter away.

88. Bottom rows from synthetic division: 1 -5 -28 -112 -540 -2702 (so we cannot conclude that 5 is an upper bound) and -1 -15 -72 -332 -1680 -8402 (so -5 *is* a lower bound); there is another zero at $x = 10.0002$.

89.

Imaginary axis

Real axis

$-3 - 2i$

90.

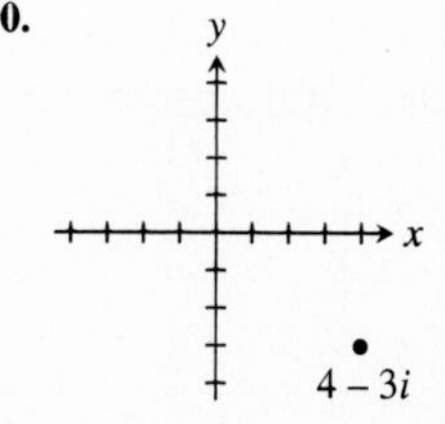

91. $(50 + 2x)(70 + 2x) - 3500 = 420$, so $x \approx 1.7017$ yd

92. When $p \approx \$44.97$, $S(p) = D(p) \approx 19.10$.

93. **(a)** $h = -16t^2 + 170t + 6$; shown is the window $[0, 11] \times [0, 500]$.

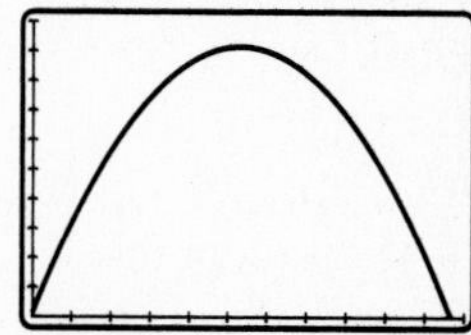

(b) When $t = 5.3125$, $h = 457.5625$. **(c)** The rock hits the ground after about 10.66 sec.

94. **(a)** $V = x(30 - 2x)(70 - 2x)$ in.^{3} **(b)** Either $x \approx 4.57$ or $x \approx 8.63$ in.

95. **(a)** Shown is the window $[0, 24] \times [20, 100]$.

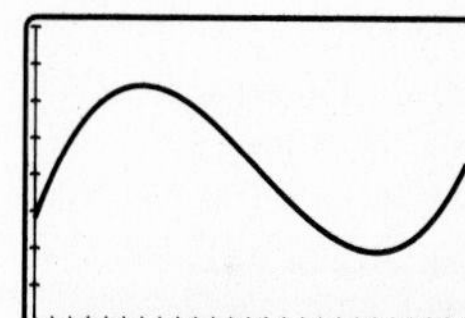

(b) When $t \approx 5.95$ (about 11:57 A.M.), the temperature is about 84.24°. **(c)** When $t \approx 19.05$ (about 1:03 A.M.), the temperature is about 39.26°. **(d)** Three times: $t \approx 0.55$, 13.83, and 23.12, or about 6:33 A.M., 7:50 P.M., and 5:07 A.M.

96. **(a)** Shown is the window $[0, 250] \times [0, 200]$.

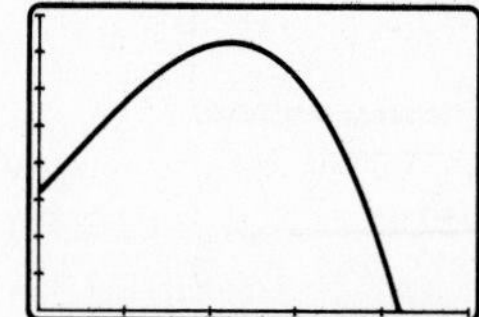

(b) When $t \approx 111.65$ (after about 112 days), the population is about $181.79 \approx 182$. **(c)** $P(t) = 0$ when $t = 210.3$ (after about 210 days).

97. **(a)** and **(b)** Shown is the window $[0, 255] \times [0, 2.5]$.

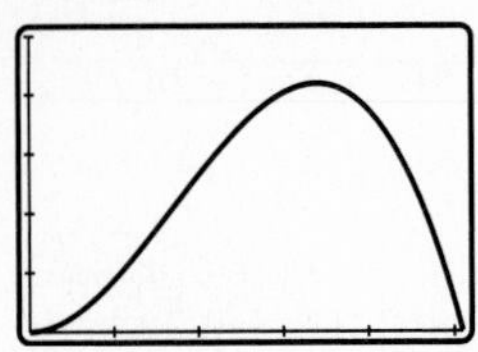

(c) When $d = 170$ ft, $s \approx 2.088$ ft.
(d) Answers may vary.

98. **(a)** $V = \frac{4}{3}\pi x^3 + \pi x^2(140 - 2x)$.
(b) Shown is the window $[0, 70] \times [0, 1{,}500{,}000]$.

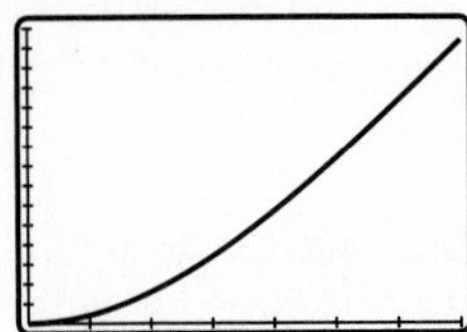

(c) The largest volume occurs when $x = 70$ (so it actually is a sphere) and is $\frac{4}{3}\pi 70^3 \approx 1{,}436{,}755$ ft^3.

99. **(a)** $y = 9500x + 16000$ (where $x = 0$ represents 1990); shown is the window $[-0.5, 4.5] \times [15{,}000, 60{,}000]$.

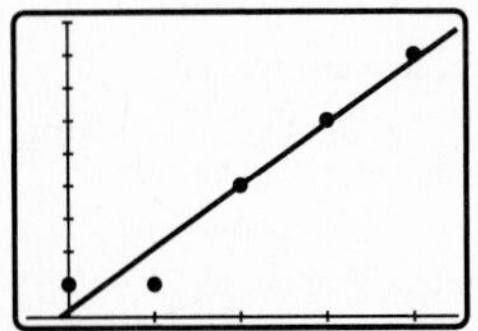

(b) $y = 1071.43x^2 + 5214.29x + 18142.86$; same window.

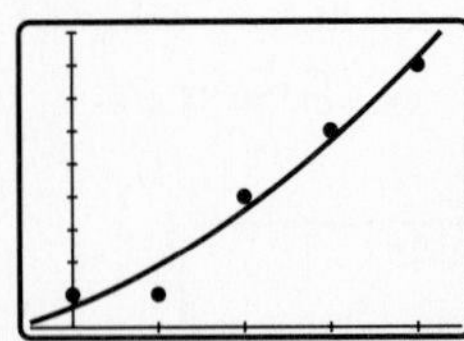

(c) Linear: 92,000 oz when $x = 8$; quadratic: 128,429 oz.

100. **(a)** $y = -2.23x^3 + 54.71x^2 - 351.09x + 733.22$, where $x = 0$ represents 1980; shown is the window $[5, 15] \times [100, 600]$.

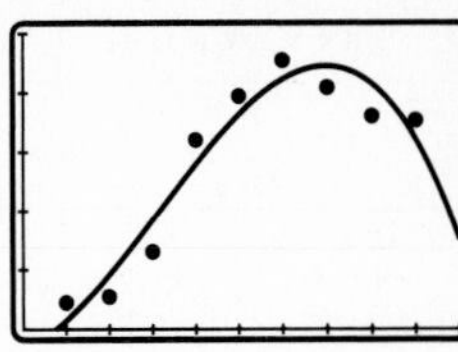

(b) $y = 1.459x^4 - 60.574x^3 + 905.888x^2 - 5706.09x + 12967.629$; same window.

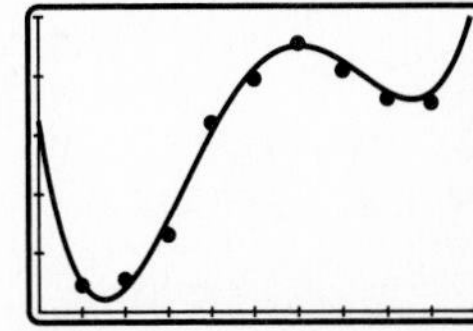

(c) The cubic model shows decreasing support in the future, and the quartic predicts increasing support.

101. **(a)** $y = -0.0418x^3 + 0.6378x^2 - 2.1005x + 3.1102$; shown is the window $[0, 8] \times [0, 6.5]$.

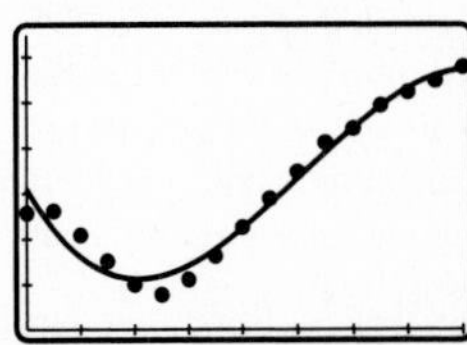

(b) Kibbe walks toward the detector, turns around, and walks away. **(c)** The model shows a minimum distance of 1.12 m when $t = 2.07$ sec.

102. If the multiplicity is odd, the function changes sign (positive to negative or vice versa) at the zero; the graph may "flatten out" at the zero, but it then continues on in the same direction (increasing or decreasing). If the zero has even multiplicity, the graph "turns around" at the zero.

103. The one that is "flatter" around $x = c$ will usually have higher multiplicity.

CHAPTER 4

QUICK REVIEW 4.1

1. 49

2. 32

3. $\frac{1}{8}$

4. 64

5. 0.401 . . .

6. 0.591 . . .

7. 1.948 . . .

8. 3125

9. $\frac{1}{2^5}$

10. $\frac{1}{5^3}$

11. $\frac{1}{2^{12}}$

12. $\frac{1}{3^8}$

13. $\frac{1}{a^6}$

14. b^{15}

SECTION EXERCISES 4.1

1. Power function

2. Exponential function

3. Exponential function

4. Neither

5. Neither

6. Power function

7. Power function

8. Neither

9. Domain: $(-\infty, \infty)$; range: $(0, \infty)$; intercept: $(0, 1)$. Window: $[-3, 3] \times [-2, 5]$.

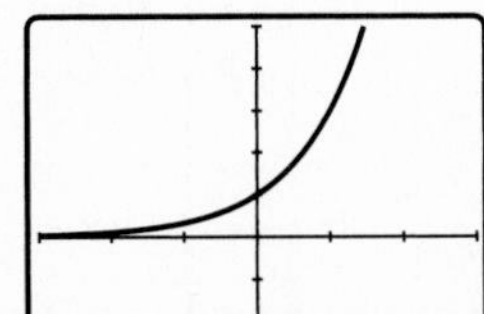

10. Domain: $(-\infty, \infty)$; range: $(0, \infty)$; intercept: (0, 1). Window: $[-3, 3] \times [-2, 5]$.

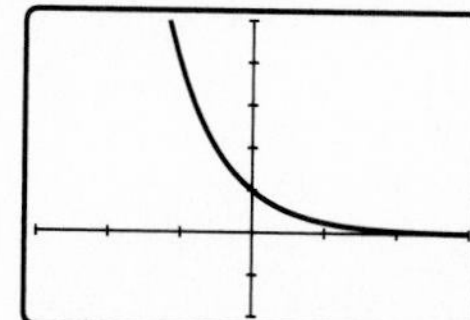

11. Domain: $(-\infty, \infty)$; range: $(0, \infty)$; intercept: (0, 1). Window: $[-3, 3] \times [-2, 5]$.

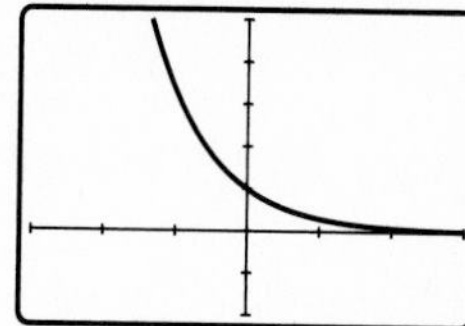

12. Domain: $(-\infty, \infty)$; range: $(0, \infty)$; intercept: (0, 1). Window: $[-3, 3] \times [-2, 5]$.

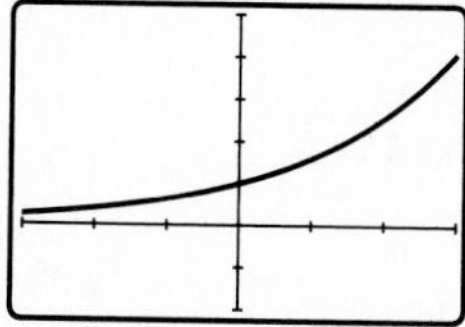

13. Domain: $(-\infty, \infty)$; range: $(0, \infty)$; intercept: (0, 1). Window: $[-3, 3] \times [-2, 5]$.

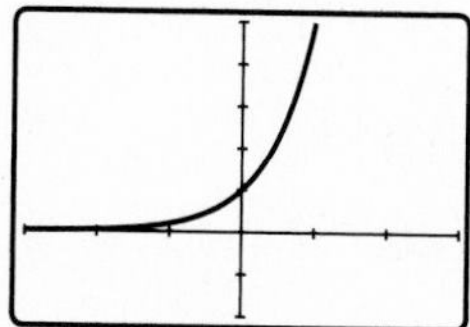

14. Domain: $(-\infty, \infty)$; range: $(0, \infty)$; intercept: (0, 1). Window: $[-3, 3] \times [-2, 5]$.

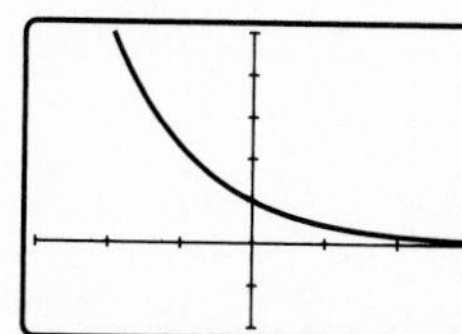

15. (a)

16. (d)

17. (c)

18. (e)

19. (b)

20. (f)

21. Reflections

22. Reflections

23. Identical

24. Identical

25. Exponential growth; $f(x) \to 0$ as $x \to -\infty$ and $f(x) \to \infty$ as $x \to \infty$

26. Exponential growth; $f(x) \to 0$ as $x \to -\infty$ and $f(x) \to \infty$ as $x \to \infty$

27. Exponential decay; $f(x) \to \infty$ as $x \to -\infty$ and $f(x) \to 0$ as $x \to \infty$

28. Exponential decay; $f(x) \to \infty$ as $x \to -\infty$ and $f(x) \to 0$ as $x \to \infty$

29. Exponential decay; $f(x) \to \infty$ as $x \to -\infty$ and $f(x) \to 0$ as $x \to \infty$

30. Exponential decay; $f(x) \to \infty$ as $x \to -\infty$ and $f(x) \to 0$ as $x \to \infty$

31. Exponential decay; $f(x) \to \infty$ as $x \to -\infty$ and $f(x) \to 0$ as $x \to \infty$

32. Exponential decay; $f(x) \to \infty$ as $x \to -\infty$ and $f(x) \to 0$ as $x \to \infty$

33. Exponential decay; $f(x) \to \infty$ as $x \to -\infty$ and $f(x) \to 0$ as $x \to \infty$

34. Exponential growth; $f(x) \to 0$ as $x \to -\infty$ and $f(x) \to \infty$ as $x \to \infty$

35. 3^{4x}

36. 2^{12x}

37. 2^{-6x}

38. 3^{-3x}

39. $x < 0$: $4^x > 5^x > 7^x$; $x > 0$: $4^x < 5^x < 7^x$

40. $x < 0$: $3^{-x} < 5^{-x} < 8^{-x}$; $x > 0$: $3^{-x} > 5^{-x} > 8^{-x}$

41. $x < 0$

42. $x > 0$

43. $x < 0$

44. $x > 0$

45. $y_1 = y_3$, since $3^{2x+4} = 3^{2(x+2)} = (3^2)^{x+2} = 9^{x+2}$

46. $y_2 = y_3$, since $2 \cdot 2^{3x-2} = 2^1 2^{3x-2} = 2^{1+3x-2} = 2^{3x-1}$

47. $n = 16$; one possible window: $[0, 18] \times [0, 80{,}000]$

48. $n = 8$ (in fact, $y_1 = y_2$ when $x \approx 7.175$); one possible window: $[0, 9] \times [0, 5000]$.

49. In about 20.222 years, or about 20 years and 2.7 months

50. In about 32.824 years, or about 32 years, 9.9 months

51. **(a)** In 1915, 12,315; in 1940, 24,265. **(b)** 76.65 years after 1890, in 1966 or 1967.

52. **(a)** In 1930, 6554; in 1945, 9151. **(b)** 70.14 years after 1910, in about 1980.

53. **(a)** $y = 6.6(\frac{1}{2})^{t/14}$, where t is time in days. **(b)** After 38.11 days.

54. **(a)** $y = 3.5(\frac{1}{2})^{t/65}$, where t is time in days. **(b)** After 117.48 days.

55. Starting with initial population P_0, we seek t so that $P_0(1.025)^t = 2P_0$, or $(1.025)^t = 2$, which does not depend on P_0.

56. Choosing any point on the graph—for example, (0, 50,000)—we can double the y (population) value by moving 4 units to the right on the t axis.

57. **(a)** When $t = 0$, $B = 100$. **(b)** When $t = 6$, $B \approx 6394$. **(c)** When $t = 1$, $B \approx 200$—the population doubles every hour.

58. **(a)** When $t = 0$, $C = 20$ grams. **(b)** When $t = 10{,}400$, $C \approx 5.647$. After about 5700.22 years, 10 grams remain. **(c)** The half-life is about 5700 years.

59. **(a)** $y = 26.2366(1.3418)^x$, where $x = 0$ represents 1950, $x = 1$ represents 1960, and so on. Window: $[-1, 4] \times [20, 90]$.

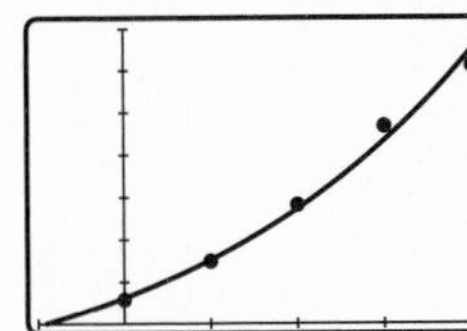

(b) When $x = -5$, $y = 6.0$ (million)—less than half of the actual population.

60. **(a)** $y = 4.8308(1.0190)^x$, where $x = 0$ represents 1900, $x = 1$ represents 1901, and so on. Window: $[-4, 90] \times [0, 25]$.

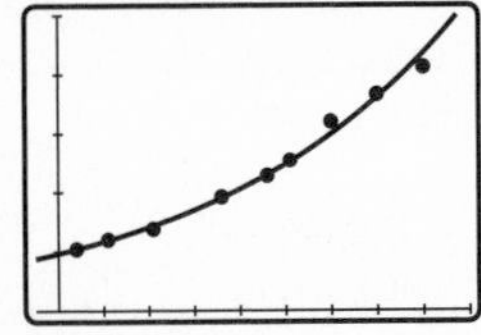

(b) When $x = 90$, $y \approx 26.3$ million.

61. Increasing on $(-\infty, 1]$, decreasing on $[1, \infty)$; maximum: $y = 1/e$ when $x = 1$; domain: $(-\infty, \infty)$; range: $\left(-\infty, \frac{1}{e}\right]$

62. Increasing on $(-\infty, -1]$, decreasing in $[-1, 0)$ and $(0, \infty)$; local maximum: $y = -e$ when $x = -1$; domain: $(-\infty, 0) \cup (0, \infty)$; range: $(-\infty, -e] \cup (0, \infty)$

63. $x = 4$

64. $x = 3$

65. $x = -4$

66. $x = 1$

67. **(a)** y_1—$f(x)$ decreases less rapidly as x decreases. **(b)** y_3; as x increases, $g(x)$ decreases ever more rapidly.

68. Vertically stretch by 2^b the graph of $(2^a)^x$, $c = 2^a$.

69. $a = 9$, $k = 2/3$; $f(x) = \frac{2}{3} \cdot 9^x$.

70. $a = 2$, $k = 3$; $f(x) = 3x^2$

71. $a \neq 0$, $c = 2$.

72. $a < 0$, $c = 1$.

73. $a > 0$, $b > 1$ or $a < 0$, $0 < b < 1$

74. $a > 0$, $0 < b < 1$, or $a < 0$, $b > 1$.

QUICK REVIEW 4.2

1. $\frac{1}{25} = 0.04$

2. $\frac{1}{1000} = 0.001$

3. $\frac{1}{5} = 0.2$

4. $\frac{1}{2} = 0.5$

5. $2^8 = 256$

6. $3^{10} = 59{,}049$

7. No

8. Yes

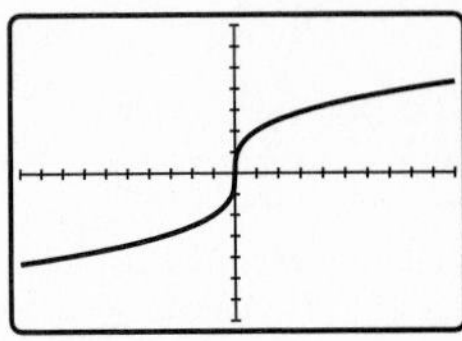

9. Yes

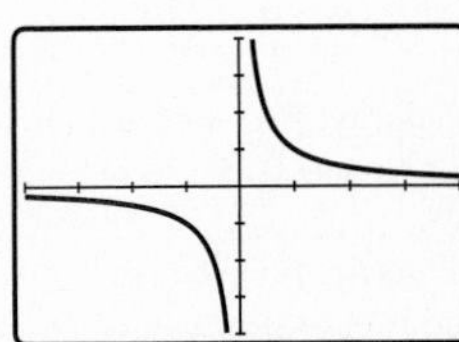

10. No

SECTION EXERCISES 4.2

1. 3

2. 4

3. 5

4. -4

5. 3

6. -4

7. -1

8. 0

9. $x = 100$

10. $x = 10000$

11. $x = \frac{1}{10} = 0.1$

12. $x = \frac{1}{1000} = 0.001$

13. $x\ e^3 = 20.0855\ldots$

14. $x = e^5 = 148.4131\ldots$

15. $x = e^{-2} = 0.1353\ldots$

16. $x = e^{-1} = 0.3678\ldots$

17. $x = \log 3 = 0.4771\ldots$

18. $x = \log 5.1 = 0.7075\ldots$

19. $x = \ln 4.2 = 1.4350\ldots$

20. $x = \ln 7.3 = 1.9878\ldots$

21. $x = \frac{1}{2} \ln 5.3 = 0.8338\ldots$

22. $x = \frac{1}{3} \log 9.2 = 0.3212\ldots$

23. $x = e^{2.8} = 16.4446\ldots$

24. $x = e^{3.1} = 22.1979\ldots$

25. $x = 10^{8.23} = 169{,}824{,}365.2\ldots$

26. $x = 10^{5.25} = 177{,}827.941\ldots$

27. (d)

28. (b)

29. (a)

30. (c)

31. Intercept: (3, 0); $f(x) \to \infty$ as $x \to \infty$ and $f(x) \to -\infty$ as $x \to 2$ from the positive side

32. Intercept: (0, 0); $f(x) \to \infty$ as $x \to \infty$ and $f(x) \to -\infty$ as $x \to -1$ from the positive side

33. Intercept: (2, 0); $f(x) \to -\infty$ as $x \to \infty$ and $f(x) \to \infty$ as $x \to 1$ from the positive side

34. Intercepts: $(-1, 0)$ and $(0, -0.3010)$; $f(x) \to -\infty$ as $x \to \infty$ and $f(x) \to \infty$ as $x \to -2$ from the positive side

35. Intercepts: (2, 0) and (0, 1.0986); $f(x) \to \infty$ as $x \to -\infty$ and $f(x) \to -\infty$ as $x \to 3$ from the negative side

36. Intercepts: (1, 0) and (0, 0.3010); $f(x) \to \infty$ as $x \to -\infty$ and $f(x) \to -\infty$ as $x \to 2$ from the negative side

37. Starting from ln *x*, translate 3 units left.

38. Starting from ln *x*, translate 2 units right.

39. Starting from log *x*, translate 1 unit left, vertically stretch by 2, and translate downward 3 units.

40. Starting from log *x*, translate 1 unit right, vertically stretch by 3, and translate upward 2 units.

41. Window: $[-2, 1] \times [-3, 4]$

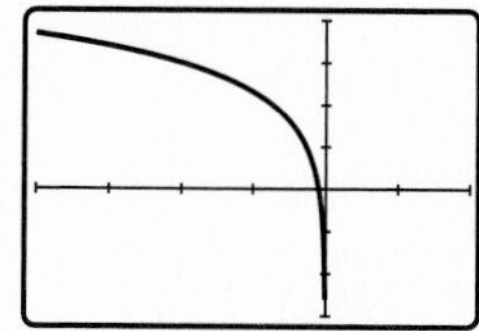

42. Window: $[-4, 1] \times [-5, 1]$

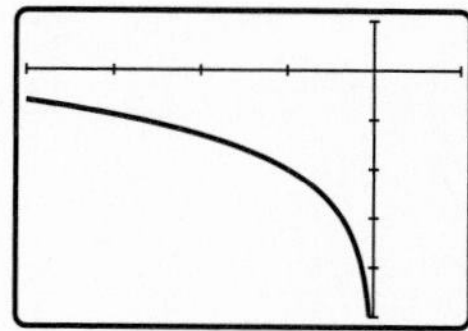

43. Window: $[-2, 1] \times [-2, 7]$

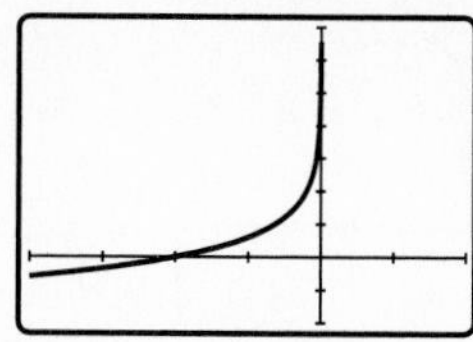

44. Window: $[-8, 1] \times [-4, 3]$

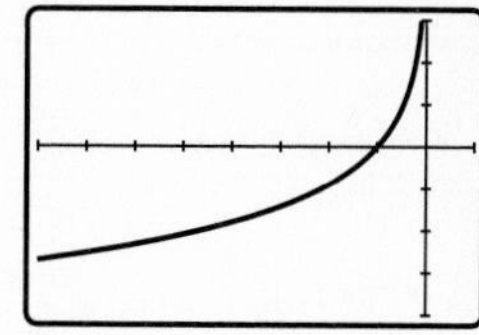

45. Window: $[-4.7, 4.7] \times [-3.1, 3.1]$

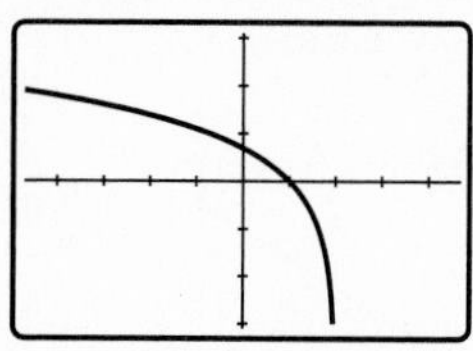

46. Window: $[-8, 8] \times [-3, 3]$

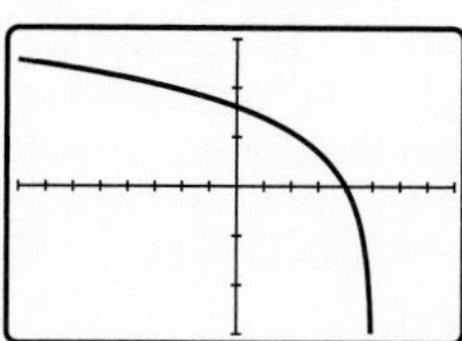

47. Window: $[-4.7, 4.7] \times [-4.1, 2.1]$

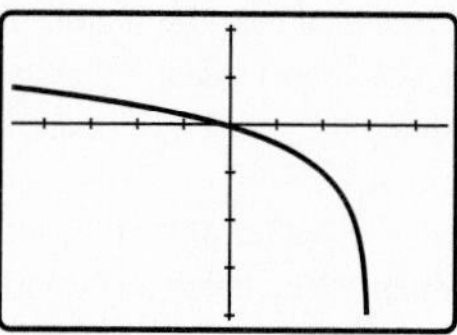

48. Window: $[-6, 2] \times [-2, 3]$

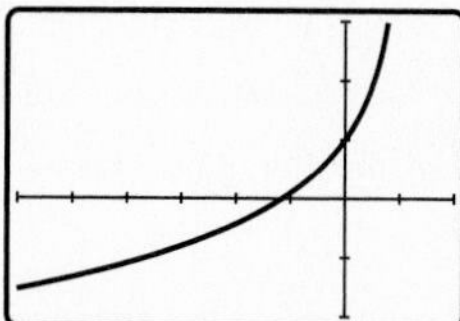

49. **(a)** $f(0) = 80$. **(b)** Approximately 54.7 units; note that $f(2) \approx 54.7319$. **(c)** Window: $[0, 24] \times [0, 80]$.

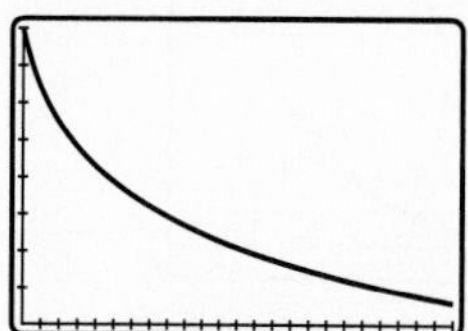

50. $I = 12 \cdot 10^{-0.0705} \approx 10.2019$ lumens.

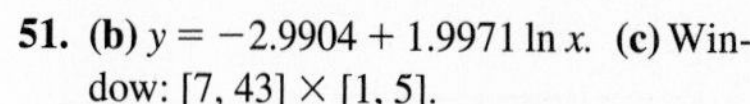

51. **(b)** $y = -2.9904 + 1.9971 \ln x$. **(c)** Window: $[7, 43] \times [1, 5]$.

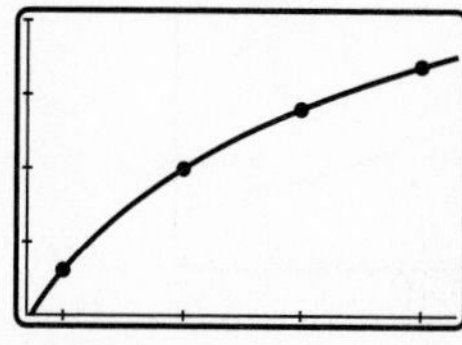

(d) When $x = 25$, $y \approx 3.44$.

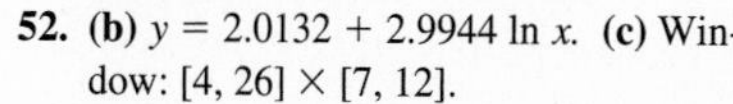

52. **(b)** $y = 2.0132 + 2.9944 \ln x$. **(c)** Window: $[4, 26] \times [7, 12]$.

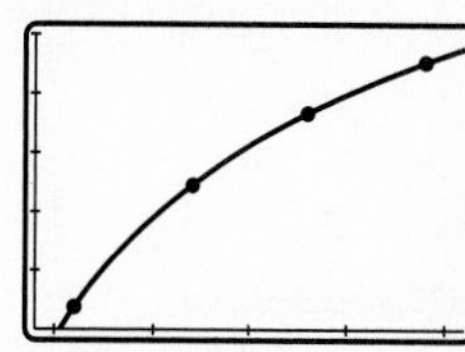

(d) When $x = 13$, $y \approx 9.69$.

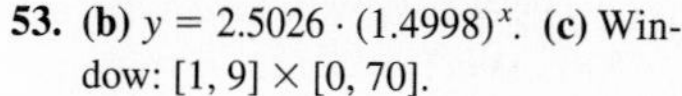

53. **(b)** $y = 2.5026 \cdot (1.4998)^x$. **(c)** Window: $[1, 9] \times [0, 70]$.

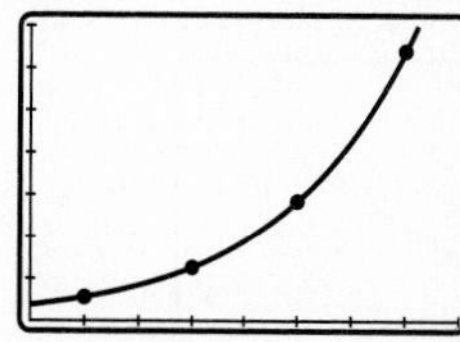

(d) When $x = 7$, $y \approx 42.72$.

54. **(b)** $y = 15.1043 \cdot (0.4983)^x$. **(c)** Window: $[0, 7] \times [0, 9]$.

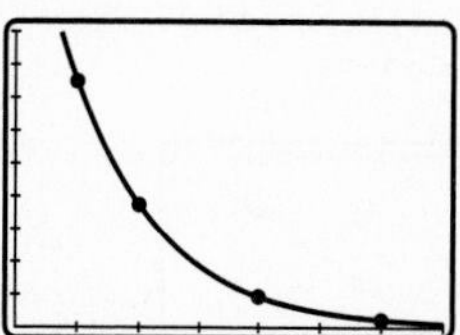

(d) When $x = 7$, $y \approx 0.1152$.

55. **(b)** $y = 2.75x^5$. **(c)** Window: $[3, 11] \times [-24{,}000, 303{,}000]$.

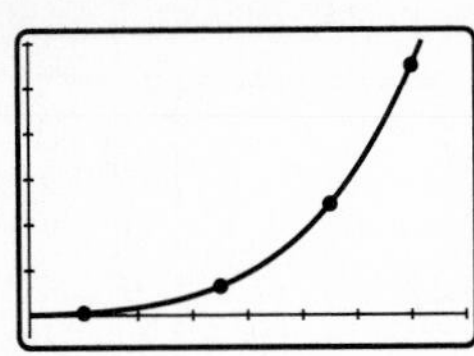

(d) When $x = 7.1$, $y \approx 49616$.

56. **(b)** $y = 8.0955x^{-0.1131}$. **(c)** Window: $[1, 9] \times [6, 8]$.

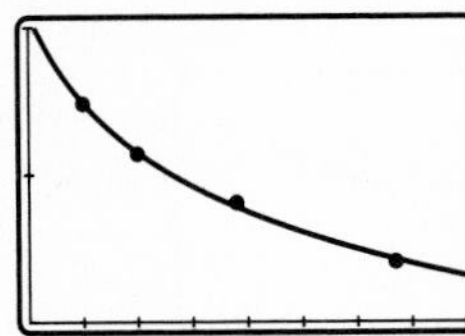

(d) When $x = 9.2$, $y \approx 6.30$.

57. $y = 1.9971u - 2.9904$, where $u = \ln x$. Window: $[2, 4] \times [1, 5]$.

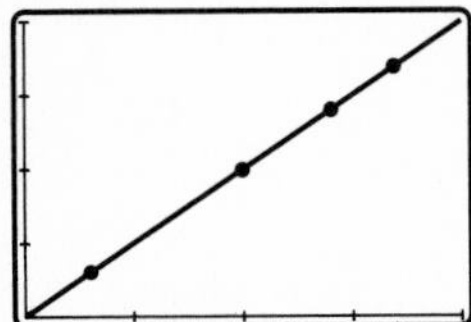

58. $v = 0.4053x + 0.9173$, where $v = \ln y$; then $y = e^{0.9173} \cdot (e^{0.4053})^x = 2.5026 \cdot (1.4998)^x$, as in Exercise 53. Window: $[1.5, 8.5] \times [1.5, 4.5]$.

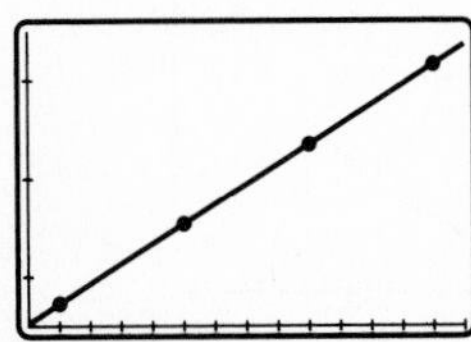

59. $v = 5u + 1.0116$, where $v = \ln y$ and $u = \ln x$; then $y = x^5 \cdot e^{1.0116} = 2.75x^5$, as in Exercise 55. Window: $[1.2, 2.4] \times [7, 13]$.

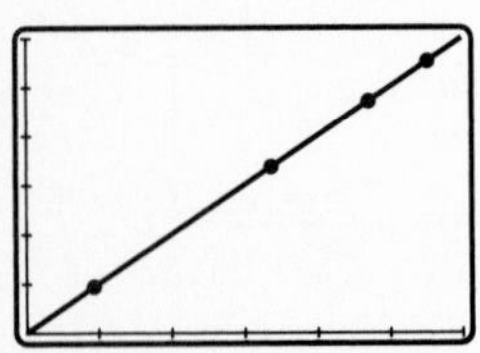

60. $v = -0.1131u + 2.0913$, where $v = \ln y$ and $u = \ln x$; then $y = x^{-0.1131} \cdot e^{2.0913} = 8.0955x^{-0.1131}$, as in Exercise 56. Window: $[0.5, 2.25] \times [1.8, 2.1]$.

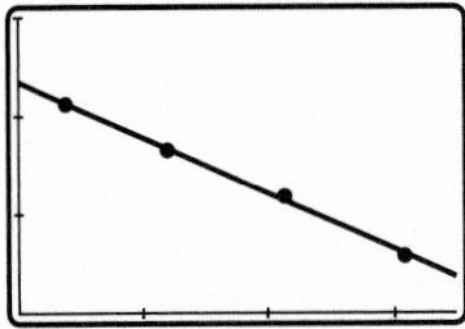

61. $y = 9.6 + 14.36 \ln x$; when $x = 21$, $y \approx 53.32$ metric tons.

62. $y = 26.95 + 18.6573 \ln x$; when $x = 26$, $y \approx 87.74$ metric tons.

63. **(a)** Scatter plot and graph (for part b regression). Window: $[4, 9.5] \times [17, 21]$.

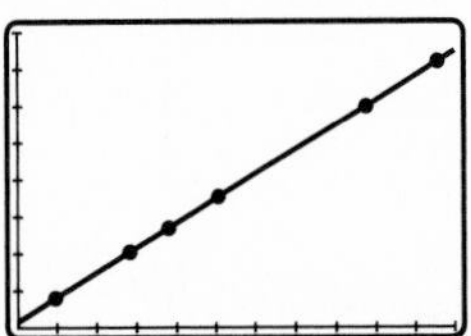

(b) $y = 0.6669u + 14.41$, where $y = \ln x$ and $u = \ln T$. **(c)** Solve for $u = \ln T$, then change to exponential form: $u = 1.5y - 21.61$, so $T = (e^{\ln x})^{1.5} e^{-21.61} \approx (4.12 \times 10^{-10})x^{1.5}$.

64. When $x = 3{,}660{,}000{,}000$ miles, $T \approx 91{,}226$ days.

65. Logarithmic seems best; the scatter plot of $(\ln x, y)$ looks most linear. (The data can be modeled by $y = 3 + 2 \ln x$.)

66. Exponential; the scatter plot of $(x, \ln y)$ is *exactly* linear. (The data can be modeled by $y = 2 \cdot 3^x$.)

67. Exponential; the scatter plot of $(x, \ln y)$ is *exactly* linear. (The data can be modeled by $y = \frac{3}{2} \cdot 2^x$.)

68. Linear; the scatter plot of (x, y) is *exactly* linear.

69. Graph the inverse of $y = 3^x$. Window: $[-3, 5] \times [-3, 3]$.

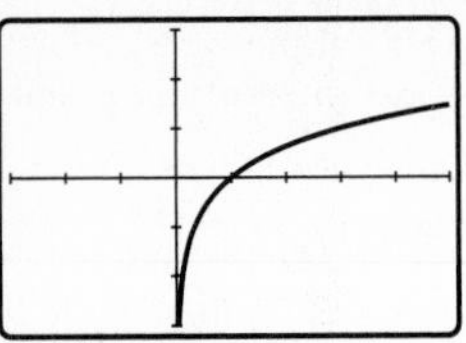

70. Graph the inverse of $y = 5^x$. Window: $[-3, 5] \times [-3, 3]$.

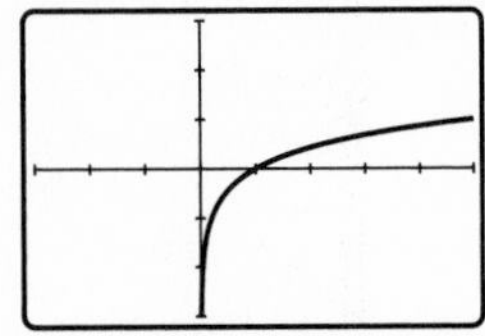

71. 0 is not in the domain of the logarithm functions because 0 is not in the range of exponential functions; that is, a^x never equals 0.

72. For any c between 0 and 1, there is a power of 10 that yields c; that is, there is an x such that $10^x = c$; then $\log c = x$.

QUICK REVIEW 4.3

1. 2

2. 3

3. -2

4. -3

5. 3

6. -3

7. x^3y^2

8. $\frac{v^5}{u}$

9. $\frac{|x|^3}{|y|}$

10. $\frac{|y|^9}{x^6}$

11. $\frac{1}{9}x^4y^2$

12. $\dfrac{8}{x^3y^6}$

13. $\dfrac{1}{3|u|}$

14. $\dfrac{x^{13}}{y^{12}}$

SECTION EXERCISES 4.3

1. 2.8073 . . .

2. 1.8294 . . .

3. 2.4837 . . .

4. 2.2362 . . .

5. −3.5849 . . .

6. −2.0922 . . .

7. −2.4021 . . .

8. −2.3219 . . .

9. $\dfrac{\ln x}{\ln 3}$

10. $\dfrac{\ln x}{\ln 7}$

11. $\dfrac{\ln (a + b)}{\ln 2}$

12. $\dfrac{\ln (c - d)}{\ln 5}$

13. $\dfrac{\log x}{\log 2}$

14. $\dfrac{\log x}{\log 4}$

15. $\dfrac{\log (x + y)}{\log \frac{1}{2}} = -\dfrac{\log (x + y)}{\log 2}$

16. $\dfrac{\log (x - y)}{\log \frac{1}{3}} = -\dfrac{\log (x - y)}{\log 3}$

17. Domain: $(0, \infty)$; range: $(-\infty, \infty)$. Window: $[-1, 10] \times [-2, 2]$.

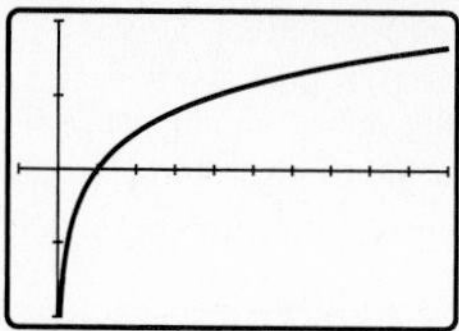

18. Domain: $(0, \infty)$; range: $(-\infty, \infty)$. Window: $[-1, 10] \times [-2, 2]$.

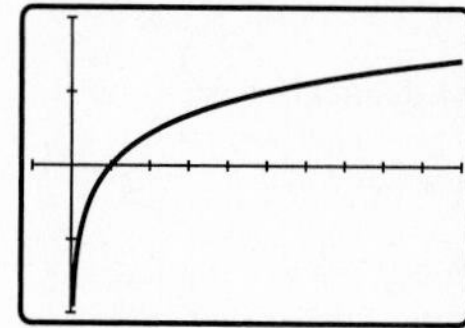

19. Domain: $(2, \infty)$; range: $(-\infty, \infty)$. Window: $[0, 10] \times [-2, 2]$.

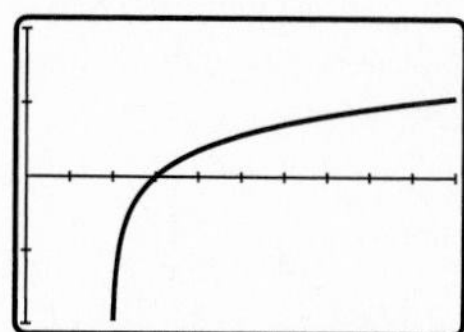

20. Domain: $(-\infty, 2)$; range: $(-\infty, \infty)$. Window: $[-10, 3] \times [-2, 2]$.

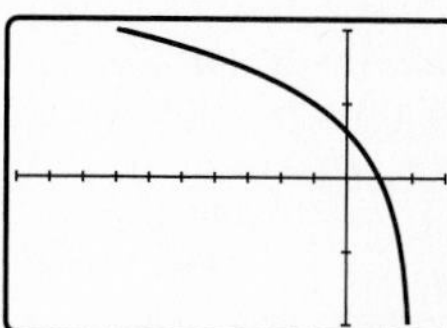

21. The graph in part (b); $X\text{min} = -5$, $X\text{max} = 5$, $X\text{scl} = 1$; $Y\text{min} = -3$, $Y\text{max} = 3$, $Y\text{scl} = 1$

22. The graph in part (c); $X\text{min} = -2$, $X\text{max} = 8$, $X\text{scl} = 1$; $Y\text{min} = -3$, $Y\text{max} = 3$, $Y\text{scl} = 1$

23. The graph in part (d); $X\text{min} = -2$, $X\text{max} = 8$, $X\text{scl} = 1$; $Y\text{min} = -3$, $Y\text{max} = 3$, $Y\text{scl} = 1$

24. The graph in part (a); $X\text{min} = -8$, $X\text{max} = 4$, $X\text{scl} = 1$; $Y\text{min} = -8$, $Y\text{max} = 8$, $Y\text{scl} = 1$

25. $\ln 8 + \ln x = 3 \ln 2 + \ln x$

26. $\ln 9 + \ln y = 2 \ln 3 + \ln y$

27. $\log 3 - \log x$

28. $\log 2 - \log y$

29. $5 \log_2 y$

30. $-2 \log_2 |x|$

31. $3 \log x + 2 \log |y|$

32. $\log |x| + 3 \log |y|$ if $xy > 0$. Undefined if $xy \leq 0$.

33. $2 \ln|x| - 3 \ln y$

34. $3 + 4 \log |x|$

35. $\frac{1}{4}(\log x - \log y)$ if x and y are positive; $\frac{1}{4}(\log |x| - \log |y|)$ if x and y are negative.

36. $\frac{1}{3}(\ln x - \ln y)$

37. $\ln 5000 + 360 \ln |x|$

38. $\ln 1000 + 240 \ln |y| = 3 \ln 10 + 240 \ln |y|$

39. $\log xy$

40. $\log 5x$

41. $\ln (y/3)$

42. $\ln (x/y)$

43. $\log_2 (x + 1)^3$

44. $\log_2 (y - 1)^4$

45. $\log x^{1/3} = \log \sqrt[3]{x}$

46. $\log z^{1/5} = \log \sqrt[5]{z}$

47. $\ln [(x + 3)^2(x - 2)^3]$

48. $\log [(x - 1)^4(x + 4)^2]$

49. $\log_a \dfrac{(x - 3)^{1/2}}{(x + 3)^{1/3}} = \log_a \dfrac{\sqrt{x - 3}}{\sqrt[3]{x + 3}}$

50. $\log_b \dfrac{(x + 2)^{1/4}}{(3 - x)^{1/5}} = \log_b \dfrac{\sqrt[4]{x + 2}}{\sqrt[5]{3 - x}}$

51. $x = 2 + \dfrac{\ln 5}{\ln 3} = 2 + \log_3 5 \approx 3.4650$

52. $x = -3 + \dfrac{\ln 2}{\ln 5} = -3 + \log_5 2 \approx -2.5693$

53. $x = 1 + \ln 7 \approx 2.9459$

54. $x = -1 + \ln 2 \approx -0.3069$

55. $x = 10^{4/3} \approx 21.5443$

56. $x = 10^{-2} = 0.01$

57. $x = e^{-1} = 1/e \approx 0.3679$

58. $x = e^{-3/4} \approx 0.4724$

59. $t = \dfrac{\ln 2.5}{\ln 1.08} = \log_{1.08} 2.5 \approx 11.9059$

60. $t = \dfrac{\ln(30/23)}{\ln 1.05} = \log_{1.05}(30/23) \approx 5.4458$

61. $t = \dfrac{\ln 0.5}{\ln 0.5} = 1$

62. $t = \dfrac{\ln 0.2}{\ln 0.5} = \log_{0.5} 0.2 \approx 2.3219$

63. $R \approx 6.3469$

64. $R \approx 5.3751$

65. If T and B are the same for the two quakes, $a_1/a_2 = 10^{1.3}$, so $a_1 \approx 20a_2$; the Mexico City amplitude was about 20 times greater.

66. If T and B were the same, Kobe's amplitude a_1 was about $10^{0.6}a_2 \approx 4a_2$.

67. $I = 12 \cdot 10^{-0.094} \approx 9.6645$ lumens

68. $I = 12 \cdot 10^{-0.125} \approx 8.9987 \approx 9$ lumens

69. $f(x) = \dfrac{\ln x}{\ln 3}$; vertically shrink by $1/\ln 3 \approx 0.9102$.

70. $f(x) = \dfrac{\ln x}{\ln 8} = \dfrac{\ln 3}{\ln 8} \cdot \dfrac{\ln x}{\ln 3} = \dfrac{\ln 3}{\ln 8} \cdot \log_3 x$; vertically shrink by $\ln 3/\ln 8 \approx 0.5283$.

71. Approximately $6.406 < x < 93.354$

72. Approximately $1.299 < x < 2.589$

73. Approximately $0 < x < 1.299$ or $2.589 < x$

74. Approximately $0.044 < x < 1.941$

75. Both domains are $x > 3$, that is, $(3, \infty)$; using properties of logarithms, we can show that the two functions are equal.

76. The domain of f is $(5, \infty)$, and the domain of g is $(-\infty, -5) \cup (5, \infty)$; neither $\ln(x + 5)$ nor $\ln(x - 5)$—the two parts of $f(x)$—is defined when $x < -5$, since both $x + 5$ and $x - 5$ are negative there; however, the quotient $\dfrac{x+5}{x-5}$ is positive, so $g(x)$ is defined when $x < -5$.

77. The domain of f is $x \neq -3$—that is, $(-\infty, -3) \cup (-3, \infty)$; the domain of g is $x > -3$, or $(-3, \infty)$; for $x < -3$, $x + 3$ is not in the domain of log but $(x + 3)^2$ is.

78. Both domains are $x > 4$; cubing does not change the sign.

79. For any $a, b > 0$, $\log_b x = \dfrac{\ln x}{\ln b} = \dfrac{\ln x/\ln a}{\ln b/\ln a} = \dfrac{\log_a x}{\log_a b}$

80. Let $x = \log_a r$ and $y = \log_a s$. Then $r = a^x$ and $s = a^y$

Quotient Rule: $\dfrac{r}{s} = \dfrac{a^x}{a^y} = a^{x-y}$ so

$\log_a \dfrac{r}{s} = x - y = \log_a r - \log_a s$

Power Rule: $r^c = a^{xc}$ so $\log_a r^c = xc = c \log_a r$

81. $\dfrac{\log x}{\ln x} = \dfrac{\ln x/\ln 10}{\ln x} = \dfrac{1}{\ln 10}$.

QUICK REVIEW 4.4

1. $f(g(x)) = e^{2\ln(x^{1/2})} = e^{\ln x} = x$ and $g(f(x)) = \ln(e^{2x})^{1/2} = \ln(e^x) = x$. Graphical support also is useful.

2. $f(g(x)) = 10^{(\log x^2)/2} = 10^{\log x} = x$ and $g(f(x)) = \log(10^{x/2})^2 = \log(10^x) = x$. Graphical support also is useful.

3. $f(g(x)) = \frac{1}{3}\ln(e^{3x}) = \frac{1}{3}(3x) = x$; $g(f(x)) = e^{3(1/3\ \ln x)} = e^{\ln x} = x$

4. $f(g(x)) = 3\log(10^{x/6})^2 = 6\log(10^{x/6}) = x$; $g(f(x)) = 10^{(3\log x^2)/6} = 10^{\log x} = x$

5. 77°F

6. 41°F

7. $23.\overline{8}$°C

8. $-12.\overline{2}$°C

9. $x^2 = 1$ or $x^2 = 3$, so $x = \pm 1$ or $x = \pm\sqrt{3}$

10. $x^2 = 1 \pm \sqrt{5}$, so $x = \pm\sqrt{1 + \sqrt{5}}$ or $x = \pm\sqrt{1 - \sqrt{5}}$, both of which are nonreal.

11. $\sqrt{x} = 2 \pm \sqrt{3}$, so $x = (2 + \sqrt{3})^2 = 7 + 4\sqrt{3}$ or $x = (2 - \sqrt{3})^2 = 7 - 4\sqrt{3}$

12. $\sqrt{x} = 2$ or $\sqrt{x} = 3$, so $x = 4$ or $x = 9$

SECTION EXERCISES 4.4

1. $x = \log_{1.06} 4.1 \approx 24.2151$

2. $x \approx -23.2644$

3. $x = -\log_{1.09} 18.4 \approx -33.7947$

4. $x \approx -10.2635$

5. $x = 10^4 = 10{,}000$

6. $x = e^{-1} \approx 0.3679$

7. $x = e^2 - 3 \approx 4.3891$

8. $x = 10^3 + 1 = 1001$

9. $x = \frac{1}{0.035}\ln 4 \approx 39.6084$

10. $x = \frac{1}{0.045}\ln 3 \approx 24.4136$

11. $x = -3$

12. $x = -4$

13. $x \approx 5.2877$

14. $x = 6$

15. $x = -\ln\frac{3}{2} \approx -0.4055$

16. $x = -\ln\frac{5}{3} \approx -0.5108$

17. $x = 5 + 4^{-1} = 5.25$

18. $x = -3$

19. $x = 3 + e^{1/3} \approx 4.3956$

20. $x = -2 + 10^{-2} = -1.99$

21. Domain: $(-\infty, -1) \cup (0, \infty)$; the graph in part (e)

22. Domain: $(0, \infty)$; the graph in part (f)

23. Domain: $(-\infty, -1) \cup (0, \infty)$; the graph in part (d)

24. Domain: $(0, \infty)$; the graph in part (c)

25. Domain: $(0, \infty)$; the graph in part (a)

26. Domain: $(-\infty, 0) \cup (0, \infty)$; the graph in part (b)

27. $x = \ln 2$

28. No solution

29. $x = \log_2 (6 + \sqrt{37}) \approx 3.5949$

30. $x = \log_2 (3 \pm 2\sqrt{2}) \approx \pm 2.5431$

31. $x = \ln (4 \pm \sqrt{15}) \approx \pm 2.0634$

32. $x = \ln \frac{1}{2} \approx -0.6931$

33. $x = \frac{1}{2}e^3 \approx 10.0428$

34. $x = 50$

35. $x = \frac{5}{9}$

36. $x = -1 + \sqrt{1 + e} \approx 0.9283$

37. $x = -\frac{3}{2} + \sqrt{2.25 + e^2} \approx 1.6047$

38. $x = \dfrac{1}{e - 1} \approx 0.5820$

39. $x = \frac{1}{0.3} \ln 0.06 \approx -9.3780$

40. $x = \frac{1}{-0.6} \ln \frac{1}{57} \approx 6.7384$

41. $x = 0.5 + \sqrt{3.25} \approx 2.3028$; the algebraic solution also gives $x = 0.5 - \sqrt{3.25}$ (extraneous).

42. $x = 50 + \sqrt{2900} \approx 103.8516$; algebraic solution also gives $x = 50 - \sqrt{2900}$ (extraneous).

43. $x = 4$; the algebraic solution also gives $x = -5$ (extraneous).

44. $x = -1.5 + \sqrt{21.25} \approx 3.1098$; algebraic solution also gives $x = -1.5 - \sqrt{21.25}$ (extraneous).

45. No solution; the algebraic solution gives $x = -51/22$ (extraneous).

46. $x = 10/3$

47. **(a)** $P(0) = 16$. **(b)** When $t \approx 13.97$, or about 14 days. **(c)** When $t \approx 16.90$, or about 17 days.

48. **(a)** $P(0) = 11$. **(b)** When $t \approx 24.51$, after 24 or 25 days. **(c)** As $t \to \infty$, $P(t) \to 1001$.

49. $k = \dfrac{\ln 0.4}{-12} \approx 0.0764$; $t \approx 28.41$ min.

50. $k \approx 0.0823$; $t \approx 29.59$ min

51. **(a)** and **(b)** Modeling function: $T - 10 = 79.4661(0.9322)^t$. Window: $[0, 35] \times [0, 80]$.

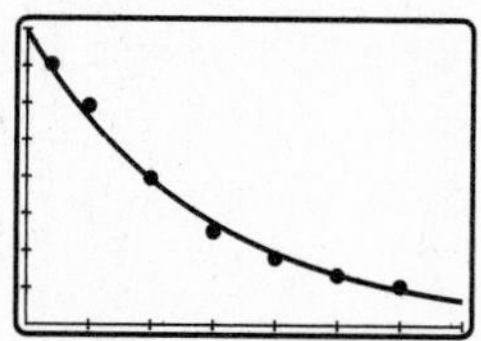

(c) When $t = 0$, $T - 10 = 79.466$, so T was about 89.5° C.

52. **(a)** and **(b)** Modeling function: $T = 79.9607(0.9273)^t$. Window: $[0, 35] \times [0, 80]$.

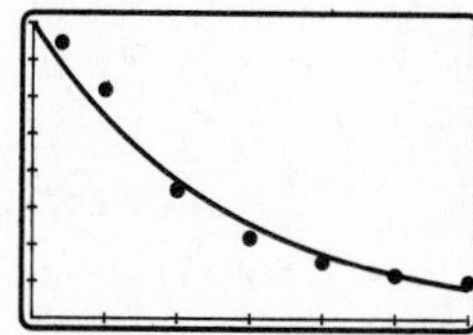

(c) When $t = 0$, T was about 80° C.

53. About 10.36 mi.

54. About 0.0055 lb/in.^{2}

55. **(b)** The scatter plot is better because it accurately represents the times between the measurements. The equal spacing on the bar graph suggests that the measurements were taken at equally spaced intervals.

56. Linear: $y = -0.4596x + 10.8996$; logarithmic: $y = 11.1297 - 2.6314 \ln x$. The logarithmic model matches the first two points very well, but perhaps is not so good for 1993. Window: $[0, 18] \times [2, 12]$.

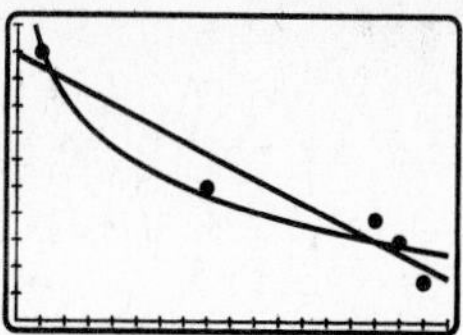

57. **(a)** $P = 0.2286(1.0293)^t$. **(b)** When $t = 40$, P is about 0.73 million. **(c)** P is about 0.46 when $t = 24.22$, or about 1984.

58. **(a)** $P = 15.6655(1.0214)^t$. **(b)** When $t = 40$, P is about 36.52 million. **(c)** P is about 31.44 when $t = 32.92$, in about 1993.

59. The maximum occurs when $x = 0$, where $f(0) = k = 1.30$.

60. $-0.4242 \le x \le 0.4242$

61. **(a)** r cannot be negative since it is a distance. **(b)** $[0, 10] \times [-5, 3]$ is a good choice; the maximum energy, approximately 2.3807, occurs when $r \approx 1.729$.

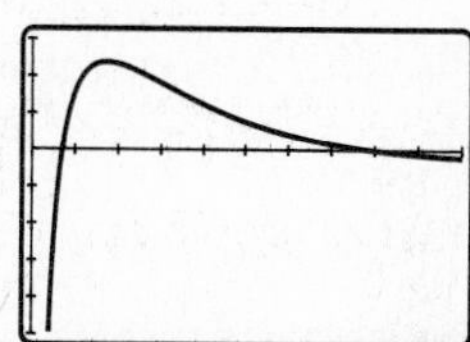

62. **(a)** Window: $[0, 6] \times [0, 17{,}000]$.

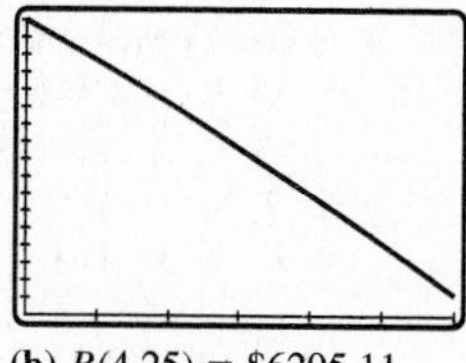

(b) $B(4.25) = \$6295.11$

63. $x \approx 1.3066$

64. $x \approx 0.4073$ or $x \approx 0.9333$

65. $0 < x < 1.7115\ldots$

66. $x \le -20.0855$ (approximate)

67. $x > 9$

68. $-1 < x < 5$

QUICK REVIEW 4.5

1. 7
2. 3.75
3. 65%
4. 35%
5. 150
6. 210
7. $315
8. $522.50

SECTION EXERCISES 4.5

1. $S = 2251.0955$, or $2251.10
2. $S = 4353.5647$, or $4353.56
3. $S = 19{,}908.58968$, or $19,908.59
4. $S = 46{,}057.57814$, or $46,057.58
5. $S = 2122.1673$, or $2122.17
6. $S = 5752.6681$, or $5752.67
7. $S = 86{,}496.2599$, or $86,496.26
8. $S = 77{,}765.6860$, or $77,765.69
9. $t = 6.63$ years; round to 6 years, 9 months (the next full compounding period).
10. $t = 7.73$ years; round to 7 years, 9 months.
11. $t = 13.78$ years; round to 13 years, 10 months.
12. $t = 11.57$ years; round to 11 years, 9 months.
13. About 10.13%
14. About 22.17%
15. About 7.07%
16. About 4.19%
17. $t = 12.14$; round to 12 years, 3 months.
18. $t = 17.62$; round to 17 years, 8 months.
19. Doubles in about 7.7016 years; $48,217.82 after 15 years
20. Doubles in about 8.6643 years; $106,243.74 after 15 years
21. $r \approx 17.33\%$; $127,816.26 after 15 years
22. $r \approx 11.55\%$; $95,035.15 after 15 years
23. Doubles in about 17.4152 years, or 17 years, 6 months
24. Doubles in about 11.6389 years, or 11 years, 9 months
25. Doubles in about 8.7507 years, or 9 years (*almost* by 8 years, 9 months)
26. Doubles in about 7.0178 years, or 7 years, 3 months (*almost* by 7 years)
27. Doubles in about 10.2448 years, or 11 years
28. Doubles in about 9.9885 years, or 10 years
29. Doubles in about 9.9310 years, or 10 years
30. Doubles in about 9.9021 years
31. APY = 6.1364%
32. APY = 5.9180%
33. APY = 6.5027%
34. APY = 4.8026%
35. The APYs are 5.1162%, and 5.1984%; choose the latter.
36. The APYs are $5.125\% = 5\frac{1}{8}\%$ and 5.1271%; choose the latter.
37. $42,211.46
38. $80,367.73
39. $239.42 per month
40. $158.03 per month
41. $219.51 per month
42. $145.74 per month
43. $676.57 per month
44. $856.39 per month
45. **(a)** 14.3171 years, or 14 years, 4 months (with a smaller payment of $846.57 in the last month). **(b)** $360 \cdot \$884.61 = \$318{,}459.60$; $172 \cdot \$1050 = \$180{,}600$, a difference of $137,859.60 (the actual savings are $138,063.03, after accounting for the smaller payment).
46. **(a)** After 10 years, the remaining loan balance is $80,338.75; another 12.1330 years are required to pay off the loan, for a total of 22 years, 2 months (the final payment would only be $626.93). **(b)** $120 \cdot \$884.61 + 146 \cdot \$1050 = \$259{,}453.20$, a savings of $59,006.40 (the actual savings are $59,429.47).
47. **(b)** Model: $y = 124.2657(1.0874)^x$. Window: $[0, 5] \times [120, 200]$.

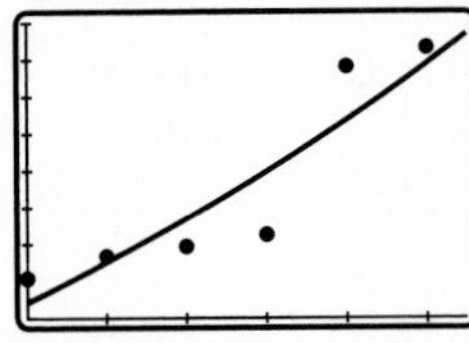

(c) When $x = 10$, $y = \$287.23$ million.

48. **(b)** Model: $y = 161.7282(1.0781)^x$. Window: $[-1, 5] \times [160, 240]$.

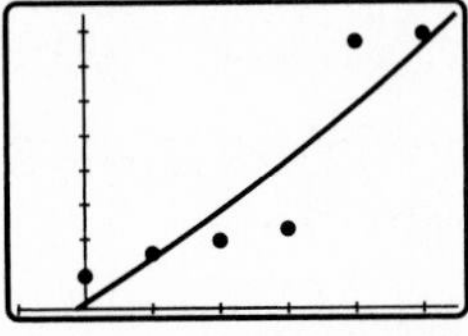

(c) When $x = 12$, $y = \$398.86$ million.

49. The APY is the percentage increase from the initial balance $S(0)$ to the end-of-year balance $S(1)$; specifically, it is $S(1)/S(0) - 1$. Multiplying the initial balance by P results in the end-of-year balance being multiplied by the same amount, so that the ratio remains unchanged. With a $1 initial investment—or a $1000 initial investment—APY = $\left(1 + \frac{r}{k}\right)^k$.
50. The APR will be lower than the APY (except under annual compounding), so the bank's offer looks more attractive when the APR is given. If we assume

monthly compounding, the APY is 4.594%.

51. Answers may vary.

52. **(a)** Steve's balance will always remain \$1000, since interest is not added to it; every year he receives 6% of that \$1000 in interest: 6% in the first year, then another 6% in the second year (for a total of $2 \cdot 6\% = 12\%$), then another 6% (totaling $3 \cdot 6\% = 18\%$), and so on; after t years, he has been paid $6t\%$ of the \$1000 investment, meaning that altogether he has $1000 + 1000 \cdot \frac{6t}{100} = 1000(1 + 0.06t)$. **(b)**

Year	Not Compounded	Compounded
0	\$1000.00	\$1000.00
1	1060.00	1060.00
2	1120.00	1123.60
3	1180.00	1191.02
4	1240.00	1262.48
5	1300.00	1338.23
6	1360.00	1418.52
7	1420.00	1503.63
8	1480.00	1593.85
9	1540.00	1689.48
10	1600.00	1790.85

53. **(a)** $r = 8\%$. **(b)** 12 payments per year. **(c)** Each payment is \$100.

54. **(a)** $r = 8\%$. **(b)** 12 payments per year. **(c)** Each payment is \$200.

Chapter 4 Review

1. $x = 3^5 = 243$

2. $x = 2^y$

3. $\frac{x}{y} = e^{-2}$

4. $\frac{a}{b} = 10^{-3} = 0.001$

5. $f(x) = 2^{-2x} + 3$; starting from 2^x, horizontally shrink by 1/2, reflect through the y-axis, and translate upward 3 units.

6. $f(x) = -2^{-2x}$; starting from 2^x, horizontally shrink by 1/2, reflect through the y-axis, and reflect through the x-axis.

7. $f(x) = -2^{-3x} - 3$; starting from 2^x, horizontally shrink by 1/3, reflect through the y-axis, reflect through the x-axis, and translate downward 3 units.

8. $f(x) = 2^{-3x} + 3$; starting from 2^x, horizontally shrink by 1/3, reflect through the y-axis, and translate upward 3 units.

9. Starting from e^x, reflect through the y-axis and translate right 4 units; or translate left 4 units and then reflect through the y-axis.

10. Starting from e^x, reflect through the y-axis and translate right 1 unit; or translate left 1 unit and reflect through the y-axis.

11. Starting from e^x, horizontally shrink by 1/2 and translate right 3/2 units; or translate right 3 units and horizontally shrink by 1/2.

12. Starting from e^x, horizontally shrink by 1/3 and translate right 4/3 units; or translate right 4 units and horizontally shrink by 1/3.

13. Window: $[-5, 5] \times [-5, 10]$.

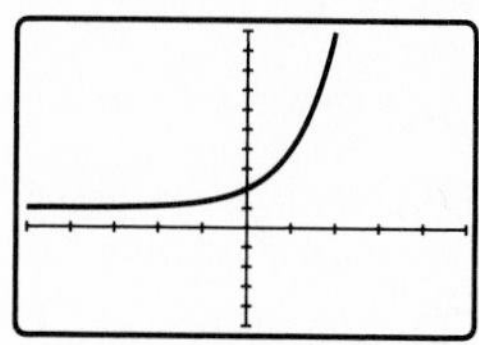

14. Window: $[-5, 5] \times [-5, 10]$.

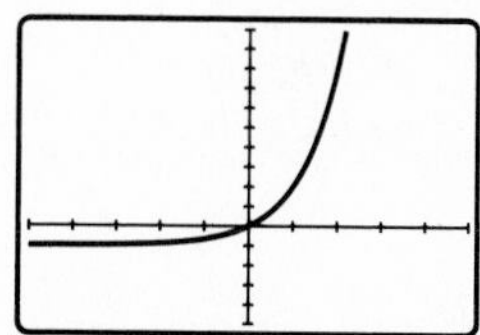

15. Window: $[-5, 10] \times [-5, 15]$.

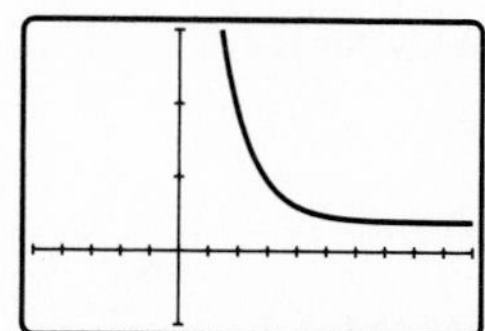

16. Window: $[-5, 10] \times [-5, 15]$.

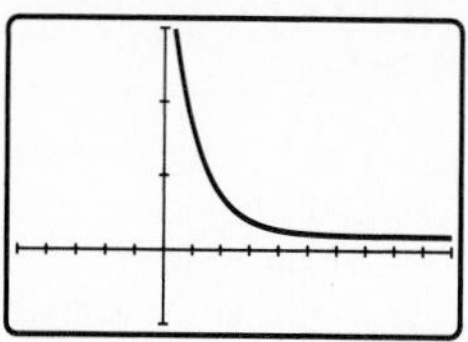

17. Window: $[-5, 5] \times [-5, 15]$.

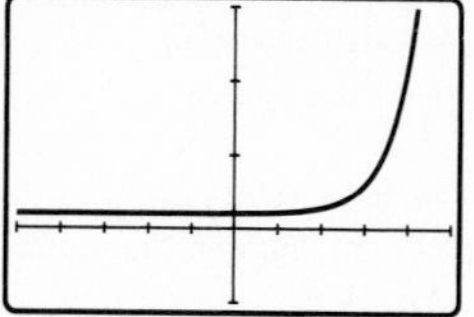

18. Window: $[-5, 5] \times [-10, 50]$.

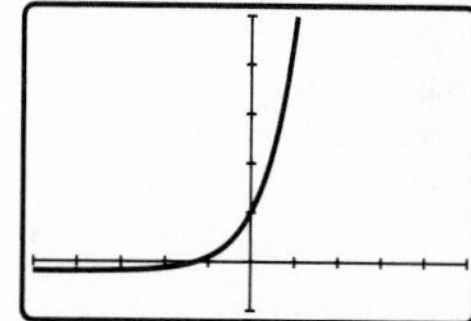

19. Translate left 4 units.

20. Translate right 5 units.

21. Reflect through the y-axis and translate right 4 units or translate left 4 units and then reflect through the y-axis.

22. Reflect through the y-axis and translate right 5 units; or translate left 5 units and reflect through the y-axis.

23. Translate right 1 unit, reflect through the x-axis, and translate upward 2 units.

24. Translate left 1 unit, reflect through the x-axis, and translate upward 4 units.

25. Decreasing on $(0, 0.3679]$, increasing on $[0.3679, \infty)$ (0.3679 is approximate). Minimum: approximately $(0.3679, -0.3679)$; no maximum. Domain: $(0, \infty)$; range: approximately $[-0.3679, \infty)$. $f(x) \to 0$ as $x \to 0^+$; $f(x) \to \infty$ as $x \to \infty$

26. Decreasing on $(0, 0.6065]$, increasing on $[0.6065, \infty)$ (0.6065 is approximate). Minimum: approximately $(0.6065, -0.1839)$; no maximum. Domain:

$(0, \infty)$; range: approximately $[-0.1839, \infty)$. $f(x) \to 0$ as $x \to 0^+$; $f(x) \to \infty$ as $x \to \infty$.

27. Decreasing on $(-\infty, -0.6065]$, increasing on $[-0.6065, 0)$, decreasing on $(0, 0.6065]$, increasing on $[0.6065, \infty)$ (endpoints are approximate). Minima: approximately $(-0.6065, -0.1839)$ and $(0.6065, -0.1839)$; no maximum. Domain: $(-\infty, 0) \cup (0, \infty)$; range: approximately $[-0.1839, \infty)$. $f(x) \to \infty$ as $x \to -\infty$; $f(x) \to \infty$ as $x \to \infty$

28. Increasing on $(0, e] \approx (0, 2.7183]$, decreasing on $[e, \infty)$. No minimum; maximum: $(e, \frac{1}{e})$. Domain: $(0, \infty)$; range: $(-\infty, \frac{1}{e}]$. $f(x) \to -\infty$ as $x \to 0^+$; $f(x) \to 0$ as $x \to \infty$.

29. 20

30. 25

31. 10

32. 12

33. e^{-6x}

34. e^{8x}

35. $x < 0: y_3 < y_2 < y_1$; $x > 0: y_1 < y_2 < y_3$

36. $x < 0: y_1 < y_2 < y_3$; $x > 0: y_3 < y_2 < y_1$

37. $y_2 = y_3$, since $2^{3x-3/2} = (2^3)^{x-1/2} = 8^{x-1/2}$

38. $y_1 = y_2$, since $2^{-(x-4)} = (2^{-1})^{x-4} = (0.5)^{x-4}$

39. $n = 6$ (in fact, $y_1 = y_2$ when $x \approx 5.147$); one possible window is $[0, 6] \times [0, 200]$

40. $n = 10$ (in fact, $y_1 = y_2$ when $x \approx 9.940$); one possible window is $[0, 11] \times [0, 1{,}200{,}000]$.

41. $x = \log 4 \approx 0.6021$

42. $x = \ln 0.25 \approx -1.3863$

43. $x \approx 22.5171$

44. $x \approx 15.7473$

45. $x = e^{5.4} \approx 221.4064$

46. $x = 10^{-7} = 0.0000001$

47. $x \approx 4.4650$

48. $x = -2$

49. $x = 4$

50. $x = 3^{7/2} = 27\sqrt{3} \approx 46.7654$

51. $x \approx 2.1049$

52. $x = \frac{1}{2} \ln \frac{6}{11} \approx -0.3031$

53. $x \approx 99.5112$

54. $x \approx -0.4915$

55. Increasing on $(-\infty, 0.9102]$, decreasing on $[0.9102, \infty)$ (endpoints are approximate). No minimum; maximum: about $(0.9102, 0.3349)$. Domain: $(-\infty, \infty)$; range: approximately $(-\infty, 0.3349]$. $f(x) \to -\infty$ as $x \to -\infty$; $f(x) \to 0$ as $x \to \infty$.

56. Increasing on $(-\infty, 0]$, decreasing on $[0, \infty)$. No minimum; maximum: $(0, e^4) \approx (0, 54.5982)$. Domain: $(-\infty, \infty)$; range: $(0, e^4]$. $g(x) \to 0$ as $x \to -\infty$; $g(x) \to 0$ as $x \to \infty$.

57. Increasing on $(-\infty, 0]$, decreasing on $[0, \infty)$. No minimum; maximum: $\left(0, \frac{1}{\sqrt{2\pi}}\right) \approx (0, 0.3989)$. Domain: $(-\infty, \infty)$; range: $\left(0, \frac{1}{\sqrt{2\pi}}\right]$. $f(x) \to 0$ as $x \to -\infty$; $f(x) \to 0$ as $x \to \infty$.

58. Vertically stretch by e^b the graph of $(e^a)^x$.

59. $k \approx -0.3054$

60. $k \approx -0.4055$

61. $k \approx 0.4055$

62. $k \approx 0.5493$

63. $k \approx -0.0841$

64. $k \approx -0.3365$

65. $y \approx 18.0652 + 13.4715 \ln x$. When $x = 23$, $y \approx 60.31$ metric tons.

66. $y \approx 48.4031 + 71.1024 \ln x$; when $x = 16$, $y \approx 245.54$ metric tons.

67. **(a)** $y \approx 3.9354(1.0167)^t$. **(b)** When $t = 40$, $y \approx 7.63$ million. **(c)** The model predicts $y = 7.88$ million when $t \approx 41.9626$, or about 2002.

68. **(a)** $y \approx 16.7765(1.0023)^t$. **(b)** When $t = 40$, $y \approx 18.40$ million. **(c)** The model predicts $y = 33.56$ million when $t = 299.6460$, in about 2260.

69. **(a)** $f(0) = 90$ units. **(b)** $f(2) = 32.8722$ units. **(c)** Window: $[0, 4] \times [0, 90]$.

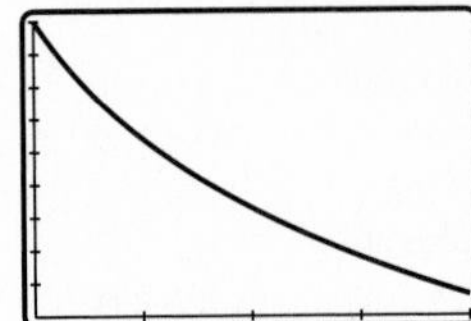

70. **(a)** $P(t) = 123{,}000(1 - 0.024)^t = 123{,}000(0.976)^t$. **(b)** When $t \approx 12.8588$, $P(t) = 90{,}000$, so in about 13 years.

71. **(a)** $P(t) = 89{,}000(1 - 0.018)^t = 89{,}000(0.982)^t$. **(b)** When $t \approx 31.7449$, $P(t) = 50{,}000$, so in about 32 years.

72. **(a)** $P(0) \approx 5.3959$ (5 or 6 students). **(b)** $P(3) \approx 80.6824$ (80 or 81 students). **(c)** $P(t) = 100$ when $t \approx 3.3069$, or sometime on the fourth day. **(d)** 300.

73. **(a)** $P(t) = 20 \cdot 2^t$, where t is time in months. (Or $20 \cdot 2^{12t}$ if t is in years, or $20 \cdot 2^{t/30}$ if t is in days). **(b)** $P(12) = 81{,}920$ rabbits present after 1 year. $P(60) \approx 2.3058 \times 10^{19}$ rabbits present after 5 years. **(c)** After about 8.9658 months, or 8 months and about 29 days.

74. **(a)** $P(t) = 4 \cdot 2^t = 2^{t+2}$, where t is time in days. **(b)** $P(4) = 64$ guppies present after 4 days. $P(7) = 512$ guppies present after 1 week. **(c)** After about 8.9658 days, or 8 days and about 23 hours.

75. **(a)** $S(t) = S_0 \cdot (\frac{1}{2})^{t/1.5}$, where t is time in seconds. **(b)** $S(1.5) = S_0/2$. $S(3) = S_0/4$. **(c)** If $S(60) = 1$ g, then $S_0 = 1 \cdot$

$2^{60/1.5} = 2^{40} \approx 1.0995 \times 10^{12}$ g = 1.0995×10^9 kg = 1,099,500 metric tons.

76. **(a)** $S(t) = S_0 \cdot \left(\frac{1}{2}\right)^{t/2.5}$, where t is time in seconds. **(b)** $S(2.5) = S_0/2$. $S(7.5) = S_0/8$. **(c)** If $S(60) = 1$ g, then $S_0 = 1 \cdot 2^{60/2.5} = 2^{24} = 16{,}777{,}216$ g = 16,777.216 kg.

77. 11.5678 years; round to 11 years, 9 months.

78. 12.2068 years

79. $t \approx 137.7940$, or about 11 years, 6 months

80. $t \approx 308.1550$, or about 25 years, 9 months

81. $r \approx 8.5692\%$

82. $r \approx 7.4655\%$

83. $I = 12 \cdot 10^{-0.0125 \cdot 25} \approx 5.84$ lumens

84. $\log_b x = \dfrac{\ln x}{\ln b}$; vertically stretch if $e^{-1} < b < e$ (so that $0 < |\ln b| < 1$), and vertically shrink if $b > e$ or $0 < b < e^{-1}$.

85. Vertically stretch if $0.1 < b < 10$ (so that $0 < |\log b| < 1$), and vertically shrink if $b > 10$ or $0 < b < 0.1$.

86. $g(x) = \ln [a \cdot b^x] = \ln a + \ln b^x = \ln a + x \ln b$; slope $\ln b$, y-intercept $\ln a$

87. **(a)** $P(0) = 16$. **(b)** $P(t) = 800$ when $t = 11.4878$, or about $11\frac{1}{2}$ days. **(c)** $P(t) = 400$ when $t = 8.7413$, or about 8–9 days.

88. **(a)** $P(0) = 12$. **(b)** $P(t) = 1000$ when $t = 15.5114$, or about $15\frac{1}{2}$ years. **(c)** As $t \to \infty$, $P(t) \to 1200$.

89. $k \approx 0.0655$, so $t \approx 41.54$ min

90. $k \approx 0.0188$, so $t \approx 105.17$ min

91. **(a)** $r = 9\%$. **(b)** 4 payments per year. **(c)** Each payment is \$100.

92. **(a)** $r = 11\%$. **(b)** 4 payments per year. **(c)** Each payment is \$200.

93. **(a)** Grace's balance will always remain at \$1000, since interest is not added to it. Every year she receives 5% of that \$1000 in interest; after t years, she has been paid $5t\%$ of the \$1000 investment, meaning that altogether she has $1000 + 1000 \cdot \frac{5t}{100} = 1000(1 + 0.05t)$.
(b)

Year	Not Compounded	Compounded
0	\$1000.00	\$1000.00
1	1050.00	1051.27
2	1100.00	1105.17
3	1150.00	1161.83
4	1200.00	1221.40
5	1250.00	1284.03
6	1300.00	1349.86
7	1350.00	1419.07
8	1400.00	1491.82
9	1450.00	1568.31
10	1500.00	1648.72

CHAPTER 5

QUICK REVIEW 5.1

1. $x = -4/3$

2. $x = 5/2$

3. $x = 3$ or $x = -4$

4. $x = 1$ or $x = -5$

5. $f(x) = (2x - 1)(x + 3)$, $x = -3$ or $x = 0.5$

6. $f(x) = (3x + 4)(x - 2)$, $x = -4/3$ or $x = 2$

7. $f(x) = (x + 2)(x - 2)$, $x = \pm 2$

8. $f(x) = (x + 1)(x - 1)$, $x = \pm 1$

9. $f(x) = (x - 1)(x^2 + x + 1)$, $x = 1$

10. No real zeros

11. Quotient 2, R 7

12. Quotient 2, R 5

13. Quotient 3, R -5

14. Quotient $\frac{5}{2}$, R -1

15. 37.5% acid

16. $66.\overline{6}\%$ (2/3) acid

17. $23.\overline{6}\%$ acid

18. 44% acid

SECTION EXERCISES 5.1

1. Yes: Both numerator and denominator are polynomials.

2. No: The denominator contains a square root expression.

3. No: The numerator contains a square root expression.

4. Yes: Both numerator and denominator are polynomials.

5. Can be written $f(x) = \dfrac{1 - 4x}{x}$; domain: $(-\infty, 0) \cup (0, \infty)$.

6. Can be written $f(x) = \dfrac{2x - 1}{x}$; domain: $(-\infty, 0) \cup (0, \infty)$.

7. Can be written $f(x) = \dfrac{4x + 5}{x + 1}$; domain: $(-\infty, -1) \cup (-1, \infty)$.

8. Can be written $f(x) = \dfrac{-2x - 4}{x + 3}$; domain: $(-\infty, -3) \cup (-3, \infty)$.

9. Can be written $f(x) = \dfrac{x^2 + 2x + 3}{x^2 + 3}$; domain: $(-\infty, 0) \cup (0, \infty)$.

10. Can be written $f(x) = \dfrac{3x^2 - x - 6}{x^2 - 2}$; domain: $(-\infty, -\sqrt{2}) \cup (-\sqrt{2}, 0) \cup (0, \sqrt{2}) \cup (\sqrt{2}, \infty)$.

11. ∞

12. $-\infty$

13. 0

14. 0

15. ∞

16. $-\infty$

17. 5

18. 5

19. $x = -1$; as $x \to -1^+, f(x) \to \infty$; as $x \to -1^-, f(x) \to -\infty$

20. $x = 2$; as $x \to 2^+, g(x) \to -\infty$; as $x \to 2^-, g(x) \to \infty$

21. $x = 1$; as $x \to 1^+, h(x) \to \infty$; as $x \to 1^-$, $h(x) \to -\infty$

22. $x = -4$; as $x \to -4^+, f(x) \to -\infty$; as $x \to -4^-, f(x) \to \infty$

23. $x = -1$ and $x = 1.5$; as $x \to -1^+$, $g(x) \to -\infty$; as $x \to -1^-$, $g(x) \to \infty$; as $x \to 1.5^+$, $g(x) \to \infty$; as $x \to 1.5^-$, $g(x) \to -\infty$

24. $x = -3$ and $x = -1$; as $x \to -3^+$, $h(x) \to -\infty$; as $x \to -3^-$, $h(x) \to \infty$; as $x \to -1^+$, $h(x) \to \infty$; as $x \to -1^-$, $h(x) \to -\infty$

25. $y = -3$; as $|x| \to \infty, f(x) \to -3$

26. $y = 4$; as $|x| \to \infty, f(x) \to 4$

27. $y = 2$; as $|x| \to \infty, f(x) \to 2$

28. $y = 3$; as $|x| \to \infty, f(x) \to 3$

29. $y = 0$; as $|x| \to \infty, f(x) \to 0$

30. $y = 0$; as $|x| \to \infty, f(x) \to 0$

31. $y = \frac{3}{5}$; as $|x| \to \infty, f(x) \to \frac{3}{5}$

32. $y = -\frac{7}{2}$; as $|x| \to \infty, f(x) \to -\frac{7}{2}$

33. $y = 7$; as $|x| \to \infty, f(x) \to 7$. Note that the numerator is not written in standard form.

34. $y = -5$; as $|x| \to \infty, f(x) \to -5$

35. $y = 3, x = -2$. Window: $[-15, 15] \times [-10, 15]$.

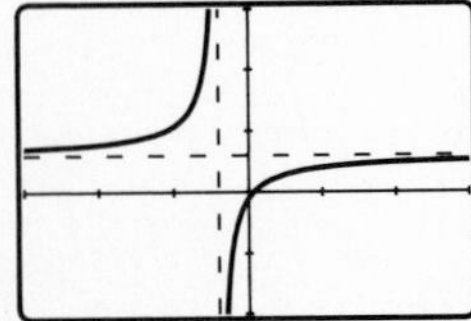

36. $y = 4, x = 2$. Window: $[-15, 15] \times [-10, 15]$.

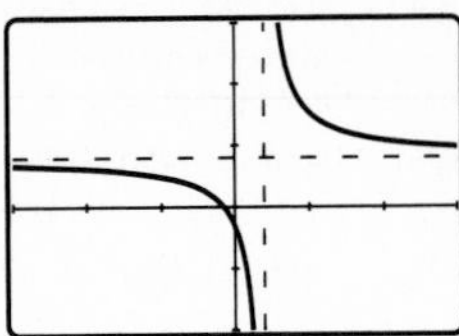

37. $y = 2, x = 3$. Window: $[-15, 15] \times [-10, 15]$.

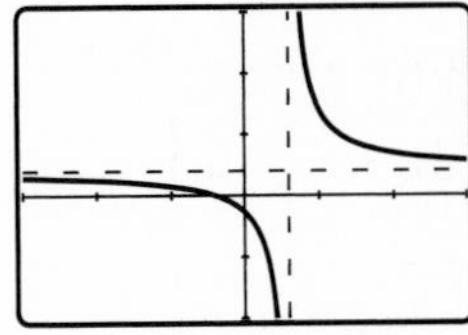

38. $y = \frac{1}{2}, x = -\frac{5}{2}$. Window: $[-10, 10] \times [-5, 5]$.

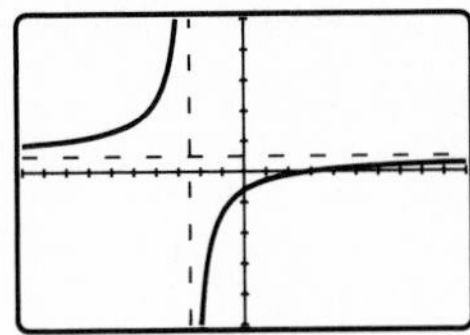

39. $y = 1, x = -4$. Window: $[-10, 10] \times [-10, 10]$.

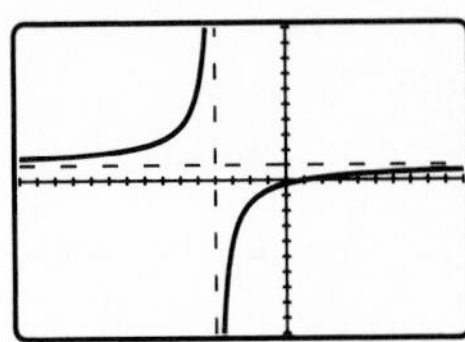

40. $y = 2, x = -2$. Window: $[-15, 10] \times [-10, 15]$.

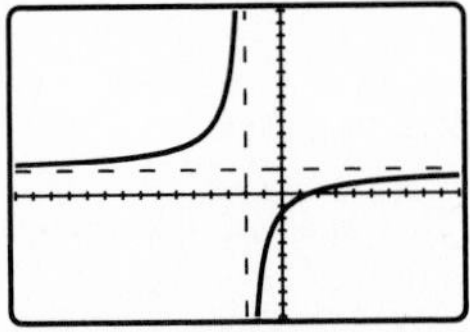

41. The graph in part (d); $X\text{min} = -2$, $X\text{max} = 8$, $X\text{scl} = 1$; $Y\text{min} = -3$, $Y\text{max} = 3$, $Y\text{scl} = 1$

42. The graph in part (b); $X\text{min} = -6$, $X\text{max} = 2$, $X\text{scl} = 1$; $Y\text{min} = -3$, $Y\text{max} = 3$, $Y\text{scl} = 1$

43. The graph in part (a); $X\text{min} = -3$, $X\text{max} = 5$, $X\text{scl} = 1$; $Y\text{min} = -5$, $Y\text{max} = 10$, $Y\text{scl} = 1$

44. The graph in part (f); $X\text{min} = -6$, $X\text{max} = 2$, $X\text{scl} = 1$; $Y\text{min} = -5$, $Y\text{max} = 5$, $Y\text{scl} = 1$

45. The graph in part (e); $X\text{min} = -2$, $X\text{max} = 8$, $X\text{scl} = 1$; $Y\text{min} = -3$, $Y\text{max} = 3$, $Y\text{scl} = 1$

46. The graph in part (c); $X\text{min} = -3$, $X\text{max} = 5$, $X\text{scl} = 1$; $Y\text{min} = -3$, $Y\text{max} = 8$, $Y\text{scl} = 1$

47. **(a)** The total amount of solution is $125 + x$ oz; of this, the amount of acid is x plus 60% of the original amount, or $x + 0.6(125)$. **(b)** $y = 0.83$. **(c)** $C(x) = \dfrac{x + 75}{x + 125} \geq 0.83$; $x \geq 169.12$ oz (approximately).

48. **(a)** $C(x) = \dfrac{x + 0.35(100)}{x + 100} = \dfrac{x + 35}{x + 100}$. **(b)** Graph $C(x)$ along with $y = 0.75$; observe where the first graph is below the second. Window: $[0, 250] \times [0, 1]$.

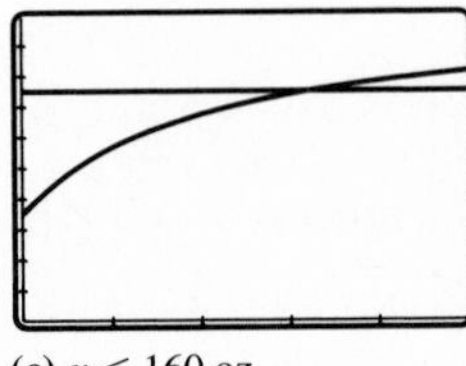

(c) $x < 160$ oz.

49. **(a)** $C(x) = \dfrac{3000 + 2.12x}{x}$. **(b)** A profit is realized if $C(x) < 2.75$; $x > 4761.9$, or 4762 hats per week. **(c)** 6350 (6349.2 . . .) hats.

50. **(a)** $P(10) = 200$, $P(40) = 350$, and $P(100) = 425$. **(b)** As $t \to \infty$, $P(t) \to 500$. **(c)** 500 bear

51. **(a)** Window: $[0, 10] \times [3, 14]$.

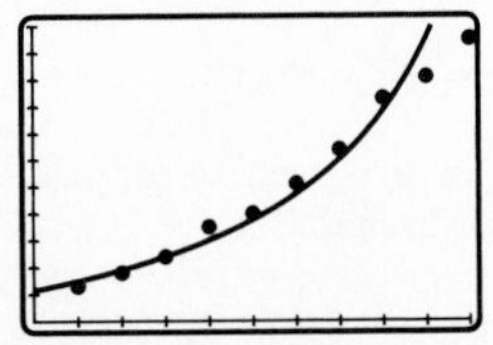

(b) $f(-26) = 1.38$, so the model predicts to \$1.38 billion.

52. **(a)** $x \approx 12.96$. **(b)** The model is realistic to about 1995 because it suggests that in about 1996, the expenditures grow without bound and then become large and negative.

53. $y = 9.25$; if the company sells a large number of T-shirts, the average cost is very close to the cost per shirt, and the \$2000 fixed cost becomes negligible.

54. **(a)** No: The domain of f is $(-\infty, 3) \cup (3, \infty)$; the domain of g is all real numbers. **(b)** No: Although not defined at 3, it does not tend toward $\pm\infty$ on either side. **(c)** Most viewing windows do not reveal that f is undefined at 3. **(d)** Almost, but not quite.

55. $f(x) = \dfrac{2x}{x - 3}$ is one possibility; any number (except -6) can be added to the numerator.

56. $f(x) = \dfrac{-3x}{x + 2}$ is one possibility; any number except -6 can be added to the numerator.

57. $f(x) = \dfrac{-1}{x - 2}$ is one possibility.

58. $f(x) = \dfrac{-1}{x + 4}$ is one possibility.

59. Horizontal asymptotes: $y = \pm 2$ and

$$h(x) = \begin{cases} \dfrac{2x - 3}{x + 2} & x \geq 0 \\ \dfrac{2x - 3}{-x + 2} & x < 0 \end{cases}$$

60. Horizontal asymptotes: $y = \pm 3$,

$$h(x) = \begin{cases} \dfrac{3x + 5}{x + 3} & x \geq 0 \\ \dfrac{3x + 5}{-x + 3} & x < 0 \end{cases}$$

61. Horizontal asymptotes: $y = \pm 3$;

$$f(x) = \begin{cases} \dfrac{5 - 3x}{x + 4} & x \geq 0 \\ \dfrac{5 - 3x}{-x + 4} & x < 0 \end{cases}$$

62. Horizontal asymptotes: $y = \pm 2$.

$$f(x) = \begin{cases} \dfrac{3 - 2x}{x + 1} & x \geq 0 \\ \dfrac{3 - 2x}{-x + 1} & x < 0 \end{cases}$$

63. The functions in Exercises 59 and 61 are not rational functions because of the absolute values.

64. By choosing x values close to 0 and making a table, this can be seen easily. Alternatively, consider that, if x is very close to zero, then $1/x$ is very large (in absolute value), so that $x - \frac{1}{x}$ is almost equal to $-\frac{1}{x}$; then $\dfrac{1}{x - 1/x} \approx \dfrac{1}{-1/x} = -x$, which is close to 0, so that $f(x) \approx 1$. Window: $[-4, 4] \times [-4, 4]$.

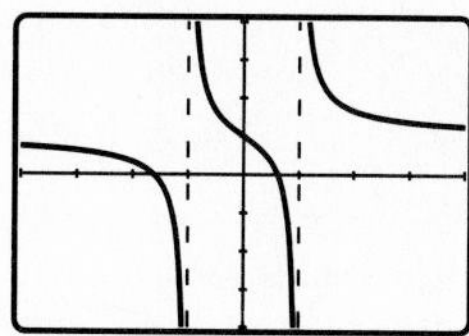

QUICK REVIEW 5.2

1. $(0, 3)$ and $(-1.5, 0)$

2. $(0, -6)$ and $\left(\frac{6}{7}, 0\right)$

3. $(0, -15)$, $(-2.5, 0)$, and $(3, 0)$

4. $(0, -28)$, $(-4, 0)$, and $\left(\frac{7}{3}, 0\right)$

5. $(0, 0)$ and $(2, 0)$

6. $(0, 0)$ and $(-1, 0)$

7. $(0, -1)$ and $\left(\frac{3}{4} \pm \frac{1}{4}\sqrt{17}, 0\right)$

8. $(0, -2)$ and $\left(\frac{1}{3} \pm \frac{1}{3}\sqrt{7}, 0\right)$

9. Translate down 4 units.

10. Translate left 4 units.

11. Reflect across the y-axis.

12. Reflect across the x-axis.

13. Translate left 2 units and reflect across the x-axis.

14. Vertically stretch by 2 and translate upward 3 units.

SECTION EXERCISES 5.2

1. Asymptotes: $x = -3$ and $x = 1$; intercepts: $\left(0, -\frac{2}{3}\right)$ and $(-2, 0)$

2. Asymptotes: $x = -1$ and $x = 2$; intercepts: $(0, 1.5)$ and $(3, 0)$

3. Asymptotes: $x = -3$ and $x = 3$; intercepts: $\left(0, \frac{2}{9}\right)$, $(-2, 0)$, and $(1, 0)$

4. Asymptotes: $x = -2$ and $x = 4$; intercepts: $(0, 0.25)$, $(-1, 0)$, and $(2, 0)$

5. **(a)**

x	0	-0.9	-0.99	-0.999	-0.9999
y	1.6667	15.13	150.13	1500.13	15,000.13

; as $x \to -1^+$, $y \to \infty$.

(b)

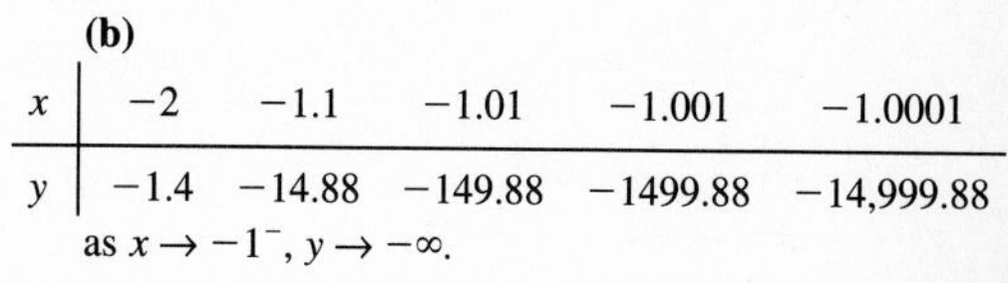

x	-2	-1.1	-1.01	-1.001	-1.0001
y	-1.4	-14.88	-149.88	-1499.88	$-14{,}999.88$

; as $x \to -1^-$, $y \to -\infty$.

6. **(a)**

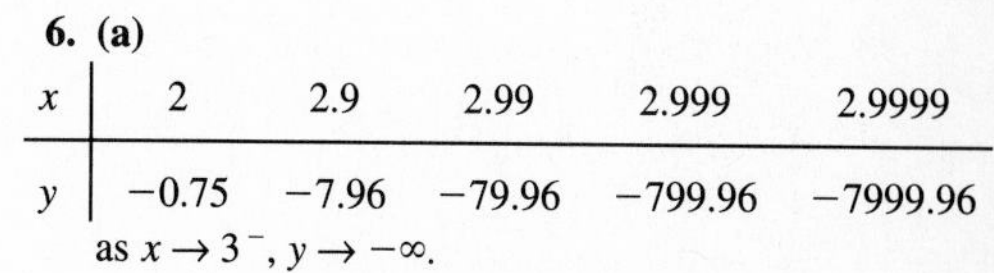

x	2	2.9	2.99	2.999	2.9999
y	-0.75	-7.96	-79.96	-799.96	-7999.96

; as $x \to 3^-$, $y \to -\infty$.

(b)

x	4	3.1	3.01	3.001	3.0001
y	0.83	8.04	80.04	800.04	8000.04

; as $x \to 3^+$, $y \to \infty$.

7. Vertical asymptotes: $x = 2$ and $x = 3$. As $x \to 2^-$, $f(x) \to \infty$; as $x \to 2^+$, $f(x) \to -\infty$. As $x \to 3^-$, $f(x) \to -\infty$; as $x \to 3^+$, $f(x) \to \infty$.

8. Vertical asymptotes: $x = 2$ and $x = 3$; as $x \to 2^-$, $f(x) \to \infty$; as $x \to 2^+$, $f(x) \to -\infty$. As $x \to 3^-$, $f(x) \to -\infty$; as $x \to 3^+$, $f(x) \to \infty$.

9. Vertical asymptotes: $x = -3$ and $x = 1$. As $x \to -3^-$, $f(x) \to \infty$; as $x \to -3^+$, $f(x) \to -\infty$. As $x \to 1^-$, $f(x) \to \infty$; as $x \to 1^+$, $f(x) \to -\infty$.

10. Vertical asymptotes: $x = -3$ and $x = 4$. As $x \to -3^-$, $f(x) \to \infty$; as $x \to -3^+$, $f(x) \to -\infty$. As $x \to 4^-$, $f(x) \to -\infty$; as $x \to 4^+$, $f(x) \to \infty$.

11. **(a)**

x	1	10	100	1000	10,000
y	1.5	2.1164	2.8677	2.9861	2.9986

; as $x \to \infty$, $y \to 3$.

(b)

x	-1	-10	-100	-1000	$-10{,}000$
y	-0.125	6.2826	3.1486	3.0141	3.0014

;

as $x \to -\infty$, $y \to 3$.

12. (a)

x	1	10	100	1000	$10{,}000$
y	-0.6	-0.9785	-1.4304	-1.4928	-1.4993

;

as $x \to \infty$, $y \to -1.5$.

(b)

x	-1	-10	-100	-1000	$-10{,}000$
y	0.4	-2.7669	-1.5756	-1.5073	-1.5007

;

as $x \to -\infty$, $y \to -1.5$.

13. Vertical asymptotes: $x = -1.5$ and $x = 1$. As $x \to -1.5^-$, $f(x) \to \infty$; as $x \to -1.5^+$, $f(x) \to -\infty$. As $x \to 1^-$, $f(x) \to -\infty$; as $x \to 1^+$, $f(x) \to \infty$. As $|x| \to \infty$, $f(x) \to 0$. Local maximum: approximately $(0.0726, -2.3252)$; intercepts $\left(\frac{7}{3}, 0\right)$ and $\left(0, -\frac{7}{3}\right)$.

14. Vertical asymptotes: $x = -1$ and $x = 2.5$. As $x \to -1^-$, $f(x) \to \infty$; as $x \to -1^+$, $f(x) \to -\infty$. As $x \to 2.5^-$, $f(x) \to -\infty$; as $x \to 2.5^+$, $f(x) \to \infty$. As $|x| \to \infty$, $f(x) \to 0$. Local maximum: approximately $(1.5858, -0.2991)$; intercepts $(3, 0)$ and $\left(0, -\frac{3}{5}\right)$.

15. Vertical asymptotes: $x = 1$ and $x = 3$. As $x \to 1^-$, $f(x) \to \infty$; as $x \to 1^+$, $f(x) \to -\infty$. As $x \to 3^-$, $f(x) \to -\infty$; as $x \to 3^+$, $f(x) \to \infty$. As $|x| \to \infty$, $f(x) \to 1$. Local maximum: approximately $(1.8030, -26.9164)$; intercepts $(-5, 0)$, $(-2, 0)$, and $\left(0, \frac{10}{3}\right)$.

16. Vertical asymptotes: $x = -0.5$ and $x = 1$. As $x \to -0.5^-$, $f(x) \to -\infty$; as $x \to -0.5^+$, $f(x) \to \infty$. As $x \to 1^-$, $f(x) \to \infty$; as $x \to 1^+$, $f(x) \to -\infty$. As $|x| \to \infty$, $f(x) \to 1.5$. Local maximum (hidden): approximately $(-2.9639, 1.8500)$; local minimum: approximately $(0.0750, 7.9277)$; intercepts $\left(1 \pm \sqrt{\frac{11}{3}}, 0\right)$, that is, about $(-0.91, 0)$ and $(2.91, 0)$, and $(0, 8)$.

17. Translate right 3 units; asymptotes: $x = 3$, $y = 0$.

18. Translate left 2 units and vertically stretch by 2; asymptotes: $x = -2$, $y = 0$.

19. Translate left 5 units, reflect across the x-axis, and vertically stretch by 2; asymptotes: $x = -5$, $y = 0$.

20. Translate left 3 units; asymptotes: $x = -3$, $y = 0$.

21. Translate left 1 unit and translate downward 3 units; asymptotes: $x = -1$, $y = -3$.

22. Translate left 3 units and translate downward 2 units; asymptotes: $x = -3$, $y = -2$.

23. Translate left 1 unit, reflect across the y-axis, vertically stretch by 5, and translate upward 2 units; asymptotes: $x = 1$, $y = 2$.

24. Translate left 1 unit, reflect through the x-axis, and translate downward 1 unit; asymptotes: $x = -1$, $y = -1$.

25. Translate left 3 units, reflect across the x-axis, vertically stretch by 7, and translate upward 2 units; asymptotes: $x = -3$, $y = 2$.

26. Translate right 1 unit and translate upward 3 units; asymptotes: $x = 1$, $y = 3$.

27. Translate left 4 units, vertically stretch by 13, and translate downward 2 units; asymptotes: $x = -4$, $y = -2$.

28. Translate right 5 units, reflect across the x-axis, vertically stretch by 11, and translate downward 3 units; asymptotes: $x = 5$, $y = -3$.

29. Intercept: $\left(0, -\frac{2}{3}\right)$; asymptotes: $x = 3$ and $y = 0$. Window: $[-3, 10] \times [-5, 5]$.

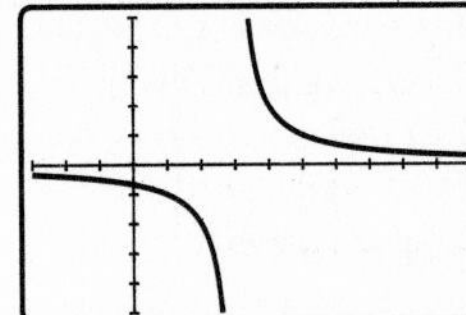

30. Intercept: $\left(0, -\frac{3}{2}\right)$; asymptotes: $x = -2$ and $y = 0$. Window: $[-7, 5] \times [-5, 5]$.

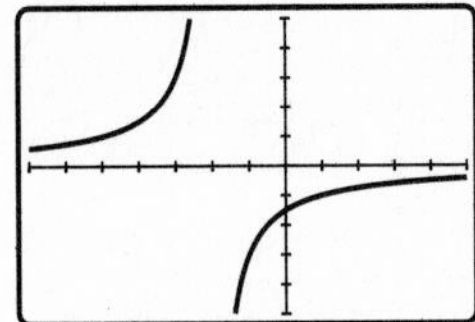

31. Intercepts: $\left(0, \frac{2}{3}\right)$ and $(2, 0)$; asymptotes: $x = -1$, $x = 3$, and $y = 0$. Window: $[-4, 6] \times [-5, 5]$.

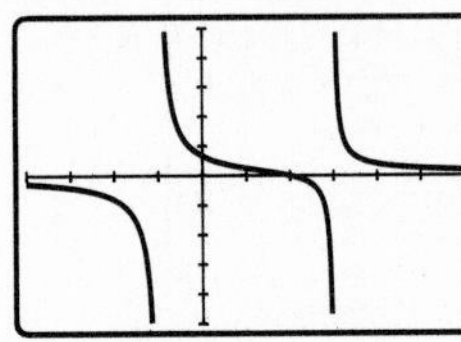

32. Intercepts: $\left(0, -\frac{2}{3}\right)$ and $(-2, 0)$; asymptotes: $x = -3$, $x = 1$ and $y = 0$. Window: $[-6, 4] \times [-5, 5]$.

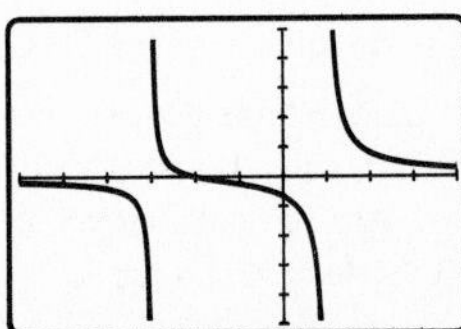

33. No intercepts; asymptotes: $x = -1$, $x = 0$, $x = 1$, and $y = 0$. Window: $[-3, 3] \times [-15, 15]$.

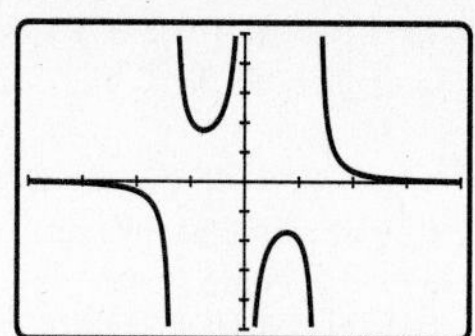

34. No intercepts; asymptotes: $x = -2$, $x = 0$, $x = 2$ and $y = 0$. Window: $[-4, 4] \times [-5, 5]$.

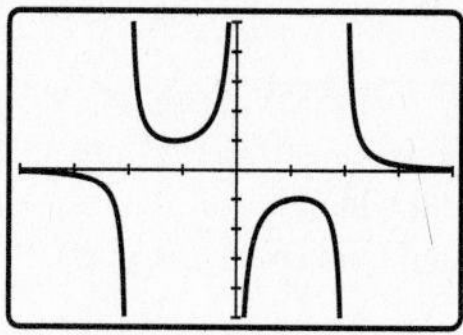

35. Intercepts: $(0, -\frac{1}{3})$ and $(1, 0)$; asymptote: $y = 0$. Window: $[-10, 10] \times [-1, 1]$.

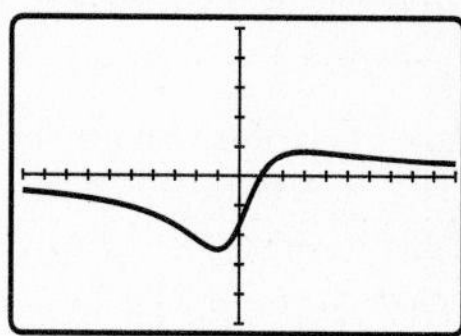

36. Intercepts: $(0, 3)$ and $(-3, 0)$; asymptote: $y = 0$. Window: $[-6, 6] \times [-1, 4]$.

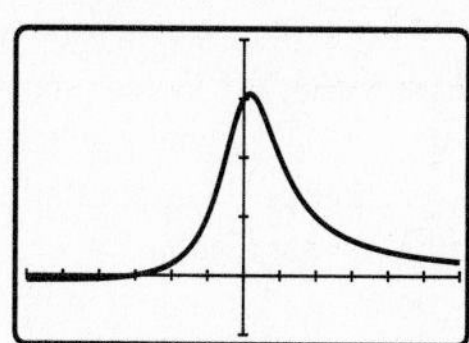

37. Domain: $(-\infty, h) \cup (h, \infty)$; range: $(-\infty, k) \cup (k, \infty)$

38. Domain: $(-\infty, -d/c) \cup (-d/c, \infty)$; range: $(-\infty, a/c) \cup (a/c, \infty)$

39. Translate right or left (as $h > 0$ or $h < 0$, respectively) $|h|$ units. If $r < 0$, reflect across the x-axis. Vertically stretch (or shrink) by $|r|$. Translate up (or down) $|k|$ units.

40. Translate right (or left) $|d/c|$ units. If $c < 0$, reflect across the y-axis; horizontally stretch (or shrink) by $1/|c|$. If $bc < ad/c$, reflect across the x-axis; vertically stretch (or shrink) by $|b - ad/c|$. Translate upward (or downward) $|a/c|$ units.

QUICK REVIEW 5.3

1. $(x + 5)(x - 3)$

2. $(2x - 3)(x + 4)$

3. $(2x + 3)(x - 1)$

4. $(2x + 1)(3x - 1)$

5. Quotient $x^2 - 3x - 5$, R -14

6. Quotient $2x^2 + 5x - 8$, R 4

7. Quotient $5x - 7$, R $-12x + 17$

8. Quotient $3x^2 - 5x + 10$, R $-19x + 31$

SECTION EXERCISES 5.3

1. (a)

x	1	10	100	1000	10,000
$f(x)$	−1	200.38	20,000.03	2,000,000.003	$\approx 2 \times 10^8$
$g(x)$	2	200	20,000	2,000,000	2×10^8

(b)

x	−1	−10	−100	−1000	−10,000
$f(x)$	1	199.75	19,999.97	1,999,999.997	$\approx 2 \times 10^8$
$g(x)$	2	200	20,000	2,000,000	2×10^8

When $|x|$ is large, $f(x)$ and $g(x)$ are almost exactly the same.

2. (a)

x	1	10	100	1000	10,000
$f(x)$	3	310.29	30,100.02	3,001,000.002	≈300,010,000
$g(x)$	4	310	30,100	3,001,001	300,010,000

(b)

x	−1	−10	−100	−1000	−10,000
$f(x)$	1.5	289.85	29,899.98	2,998,999.998	≈299,990,000
$g(x)$	2	290	29,900	2,999,000	299,990,000

When $|x|$ is large, $f(x)$ and $g(x)$ are almost exactly the same.

3. $y = 3$

4. $y = 4$

5. $y = x$

6. $y = x$

7. $y = x^3 - x^2 + x - 1$

8. $y = 2x^3 + 2x + 1$

9. $y = x^4 - 2x^3 + 4x^2 - 8x + 16$

10. $y = x^4 + x^3 + x^2 + x + 1$

11. (a); (c)

12. (a); (c)

13. (b); (d)

14. (b); (d)

15. **(a)** Window: $[-10, 10] \times [-50, 100]$.

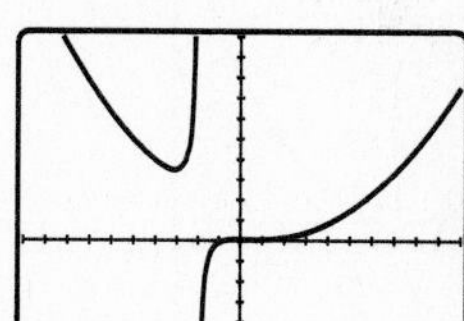

(b) Window: $[-40, 40] \times [-100, 1600]$.

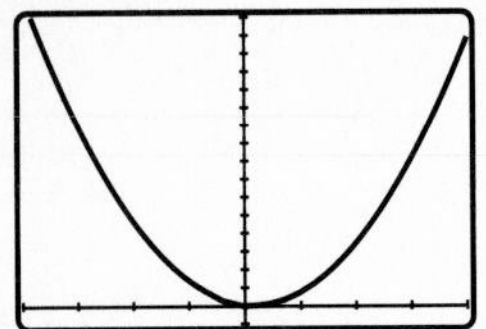

16. **(a)** Window: $[-10, 10] \times [-10, 30]$.

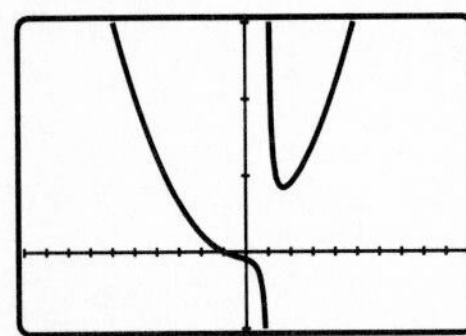

(b) Window: $[-40, 40] \times [-100, 1600]$.

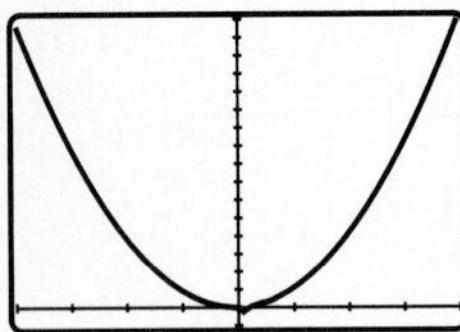

17. **(a)** Window: $[-10, 10] \times [-50, 150]$.

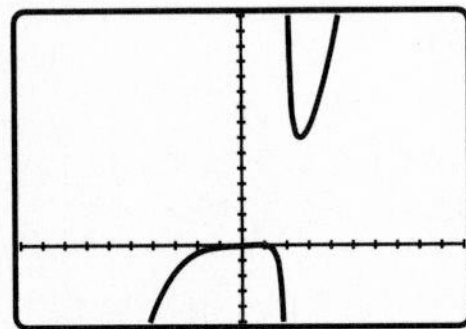

(b) Window: $[-15, 15] \times [-2000, 2000]$.

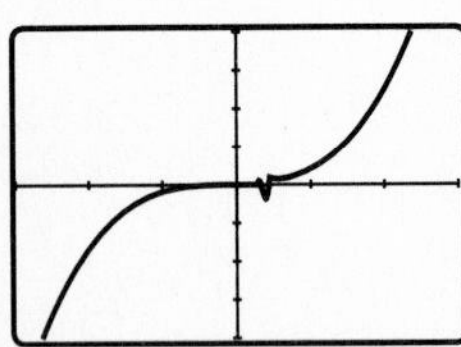

18. **(a)** No vertical asymptotes. **(b)** Window: $[-10, 10] \times [-1000, 1000]$.

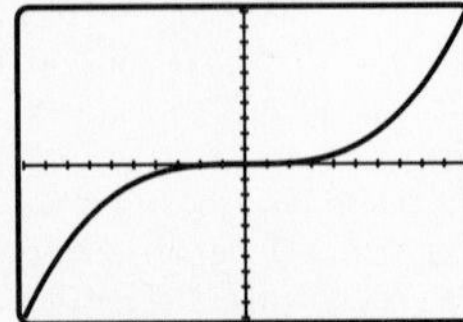

19. Horizontal asymptote: $y = 2$; vertical asymptotes: $x = \pm 1$; x-intercepts: approximately -1.28 and 0.78; y-intercept: $(0, 2)$. Intermediate behavior: $f(x) \to -\infty$ as $x \to -1^-$; $f(x) \to \infty$ as $x \to -1^+$; $f(x) \to -\infty$ as $x \to 1^-$; $f(x) \to \infty$ as $x \to 1^+$.

20. Horizontal asymptote: $y = -3$; vertical asymptotes: $x = \pm 2$; x-intercepts: approximately -1.84 and 2.17; y-intercept: $(0, -3)$. Intermediate behavior: $f(x) \to -\infty$ as $x \to -2^-$; $f(x) \to \infty$ as $x \to -2^+$; $f(x) \to -\infty$ as $x \to 2^-$; $f(x) \to \infty$ as $x \to 2^+$.

21. End behavior asymptote: $y = x - 4$; vertical asymptote: $x = -2$; x-intercepts: none; y-intercept: $\left(0, \frac{3}{2}\right)$. Intermediate behavior: $f(x) \to -\infty$ as $x \to -2^-$; $f(x) \to \infty$ as $x \to -2^+$.

22. End behavior asymptote: $y = x - 6$; vertical asymptote: $x = -3$; x-intercepts: approximately -1.54 and 4.54; y-intercept: $\left(0, -\frac{7}{3}\right)$. Intermediate behavior: $g(x) \to -\infty$ as $x \to -3^-$; $g(x) \to \infty$ as $x \to -3^+$.

23. End behavior asymptote: $y = \frac{1}{2}x^2 - \frac{3}{4}x + \frac{1}{8}$; vertical asymptote: $x = \frac{1}{2}$; x-intercept: approximately 1.75; y-intercept: $(0, 1)$. Intermediate behavior: $f(x) \to \infty$ as $x \to \frac{1}{2}^-$; $f(x) \to -\infty$ as $x \to \frac{1}{2}^+$.

24. End behavior asymptote: $y = 2x^2 + 2x + 3$; vertical asymptote: $x = 2$; x-intercept: approximately -1.19; y-intercept: $(0, -2.5)$. Intermediate behavior: $g(x) \to -\infty$ as $x \to 2^-$; $g(x) \to \infty$ as $x \to 2^+$.

25. End behavior asymptote: $y = x^3 + 2x^2 - 3x + 1$; vertical asymptote: $x = \frac{5}{2}$; x-intercepts: approximately -3.09 and 2.52; y-intercept: $(0, 1.2)$. Intermediate behavior: $f(x) \to \infty$ as $x \to \frac{5}{2}^-$; $f(x) \to -\infty$ as $x \to \frac{5}{2}^+$.

26. End behavior asymptote: $y = 2x^4 + 2x^3 - x^2 - x + 1$; vertical asymptotes: $x = 1$; x-intercept: approximately 1.33; y-intercept: $(0, 4)$. Intermediate behavior: $g(x) \to \infty$ as $x \to 1^-$; $g(x) \to -\infty$ as $x \to 1^+$.

27. **(a)** $P(x) = 2x + 364/x$. **(b)** $x = y \approx 13.49$ (a square); $P \approx 53.96$.

28. **(a)** $A(x) = (x + 1.75)\left(\frac{40}{x} + 2.5\right)$. **(b)** $x \approx 5.29$, so the dimensions are about 7.04 in. $\times$ 10.06 in.; $A \approx 70.8325$.

29. **(a)** $A(x) = (x + 4)\left(\frac{1000}{x} + 4\right)$. **(b)** $x = y = \sqrt{1000} \approx 31.62$, so the dimensions are about 35.62 ft $\times$ 35.62 ft; $A \approx 1268.98$.

30. **(a)** When $x = 10.3859$, in about 1996–1997. **(b)** It has a vertical asymptote at $x = 27.1644$, which makes it a bad model around (and after) that point.

31. **(a)** They are very similar for $0 < x < 7$, but they diverge after that. **(b)** g has no vertical asymptote; also, the rational model shows an increase for $x > 7$ that does not seem to be indicated by the data.

32. y_1 has a "hole" in it (that is, it is not defined) at $x = 1$; y_2 is the result of carrying out the division of y_1 (except that it does not include the restriction that $x \neq 1$). y_1 does not have a vertical asymptote at 1, since the function values do not tend to $\pm\infty$ around that point.

33. y_1 has a "hole" in it at $x = -1$; y_2 is the result of carrying out the division of y_1 (except that it does not include the restriction that $x \neq -1$); y_1 does not have a vertical asymptote at -1, since the function values do not tend to $\pm\infty$ around that point.

34. After canceling common factors in the numerator and denominator of y_1, we are left with y_2. y_1, though, is not defined at $x = -1$. The reason y_1 has a vertical asymptote at $x = 1$ in spite of the cancellation is that 1 is a zero of multiplicity 2 in the denominator but only multiplicity 1 in the numerator.

35. Canceling common factors in the numerator and denominator of y_1 leaves y_2, except that y_1 is not defined at $x = 1$; y_1 has a vertical asymptote at $x = -2$ because -2 is a zero of multiplicity 2 in the denominator but only multiplicity 1 in the numerator.

36. The multiplicity in the numerator must be smaller than the multiplicity in the denominator.

37. $x^2 + 2x + \dfrac{c}{x-2} = \dfrac{x^3 - 4x + c}{x - 2}$, where $c \neq 0$

38. $2x + 1 + \dfrac{c}{x-3} = \dfrac{2x^2 - 5x - 3 + c}{x - 3}$, where $c \neq 0$

39. $x^2 + 1 + \dfrac{cx + d}{x^2 - 1} = \dfrac{x^4 - 1 + cx + d}{x^2 - 1}$, where $cx + d \neq 0$ when $x = \pm 1$

40. $x^3 - x + \dfrac{cx + d}{x^2 + x - 6} = \dfrac{x^5 + x^4 - 7x^3 - x^2 + 6x + cx + d}{x^2 + x - 6}$, where $cx + d \neq 0$ when $x = -3$ or $x = 2$

41. Horizontal (end behavior) asymptote: $y = \frac{2}{3}$; vertical asymptote: $x \approx 0.7181$; zeros: $x = 1$ or $x = \pm\frac{1}{2}\sqrt{2}$. Local minimum: approximately $(0.1385, -1.0723)$. Decreases on $(-\infty, -1)$ and $(-1, 0.1385]$; Increases on $[0.1385, 0.7181)$ and $(0.7181, \infty)$; endpoints are approximate values.

42. End behavior asymptote: $y = x^3 - 2x^2 + x - 30$; vertical asymptote: $x = -4$; zeros: $x \approx -3.9923$ or $x \approx 3.8027$. Local minima: approximately $(-3.8725, -114.0949)$ and $(1.0193, -29.8004)$; local maxima: approximately $(-4.1209, -146.3362)$ and $(0.3074, -29.6204)$. Increasing on $(-\infty, -4.1209]$, $[-3.8725, 0.3074]$, and $[1.0193, \infty)$. Decreasing on $[-4.1209, 4)$, $(4, -3.8725]$, and $[0.3074, 1.0193]$. Endpoints are approximate.

QUICK REVIEW 5.4

1. $4.5x^2$

2. $4xz$

3. $3z$

4. $2yz^3$

5. $5(x - 1) = 5x - 5$

6. $2x(x + 4) = 2x^2 + 8x$

7. $(x + 1)^2 = x^2 + 2x + 1$

8. $(x + 4)(x - 1) = x^2 + 3x - 4$

9. $x = \frac{3}{4} \pm \frac{1}{4}\sqrt{17}$

10. $x = \frac{5}{4} \pm \frac{1}{4}\sqrt{33}$

11. $x = -\frac{1}{3} \pm \frac{1}{3}\sqrt{7}$

12. $x = \frac{3}{2} \pm \frac{3}{2}\sqrt{5}$

SECTION EXERCISES 5.4

1. $x = -3.5 = -7/2$

2. $x = 5.75$, or $\frac{23}{4}$

3. $x = 2$ or $x = 5$

4. $x = 3$ or $x = 4$

5. $x = -5$ or $x = 3$

6. $x = -7$ or $x = 2$

7. $x = \frac{11}{8} \pm \frac{1}{8}\sqrt{73}$

8. $x = \frac{13}{16} \pm \frac{1}{16}\sqrt{105}$

9. $-\frac{1}{3}$; extraneous: 2

10. $-\frac{3}{4}$; extraneous: 1

11. 5; extraneous: 0

12. 3; extraneous: 0

13. No solution; extraneous: -2 and 0

14. -1; extraneous: -3

15. $(-\infty, 2.75) \cup (3, \infty)$

16. $(-\infty, -6] \cup (-5, \infty)$

17. $(-\infty, 3.5) \cup \left(\frac{38}{9}, \infty\right)$

18. $(-6.5, -4)$

19. $(-\infty, -2) \cup (1, 2)$

20. $(-\infty, -3) \cup (-2, 3)$

21. $[-1, 1]$

22. $(-\infty, -2) \cup (2, \infty)$

23. $(-\infty, -4) \cup (3, \infty)$

24. $(-5, 2)$

25. $(-\infty, -2) \cup (-1, 5)$

26. $(-4, -3) \cup (3, \infty)$

27. $x = 3 \pm \sqrt{2}$

28. $x = -\frac{3}{2} \pm \frac{1}{2}\sqrt{31}$

29. $[-1, 0] \cup [1, \infty)$

30. $(-\infty, -2] \cup [0, 2]$

31. Approximately $x = -3.1004$, $x = 0.6611$, or $x = 2.4393$

32. Approximately $x = -2.3977$, $x = -0.4408$, or $x = 2.8385$

33. $(-\infty, 0) \cup (\sqrt[3]{2}, \infty)$

34. $(-\infty, -\sqrt[3]{4}] \cup (0, \infty)$

35. **(a)** $R = \dfrac{2.3x}{x + 2.3}$. **(b)** $x \approx 6.52$ ohms.

36. **(a)** $P = 2x + 400/x$. **(b)** 7.1922 m × 27.8078 m. **(c)** Both dimensions are between 7.1922 m and 27.8078 m.

37. **(a)** $S = 2\pi x^2 + 1000/x$. **(b)** Either $x \approx 1.12$ cm and $h \approx 126.66$ cm or $x \approx 11.37$ and $h \approx 1.23$.

38. **(a)** Approximately $1.12 < x < 11.37$ and $1.23 < h < 126.66$. **(b)** When $x \approx 4.30$ cm and $h \approx 8.60$, the minimum surface area, $S \approx 348.73$ cm^2, is achieved.

39. **(a)** $D = \frac{1}{4.75} + \frac{1}{t}$. **(b)** $t \approx 5.74$ h.

40. **(a)** $T = \frac{17}{x} + \dfrac{53}{x + 43}$. **(b)** $x \approx 20.45$ mph.

41. **(a)** $P = 2L + 1800/L$. **(b)** A 30 ft × 30 ft house has the minimum perimeter (120 ft).

42. When combined into a single rational expression, common factors in the numerator and denominator cancel to leave

$\frac{2x+1}{x-1}$, which gives the only *actual* solution, $-\frac{1}{2}$. The extraneous solution arises from the factor $(x-3)$, which is canceled during simplification.

43. **(a)** Rewriting leaves $\frac{x^3-20x+4}{x}=0$. The zeros of this rational function are zeros of the numerator. Using synthetic division, we can see that 5 (as well as many other numbers) is an upper bound on the zeros of the numerator. **(b)** As found in Example 2, the zeros are approximately -4.57, 0.20, and 4.37.

44. $x=-\frac{y-2}{y-1}$

45. $x=\frac{y}{y-1}$

46. $x=-\frac{y-1}{y-2}$

47. $x=\frac{2y-3}{y-2}$

48. Multiply both sides of $a<b$ by the positive number a to get $a^2<ab$; similarly, multiply both sides by b to get $ab<b^2$; then $a^2<ab<b^2$, which gives the desired inequality after dropping the middle.

49. Multiply both sides of $a<b$ by the positive number $\frac{1}{ab}$.

QUICK REVIEW 5.5

1. $q(x)=x-3$, $r(x)=2x+1$; $\frac{f(x)}{g(x)}=x-3+\frac{2x+1}{x^2+1}$

2. $q(x)=2x+5$, $r(x)=3$; $\frac{f(x)}{g(x)}=2x+5+\frac{3}{x^2-1}$

3. $q(x)=x^2-3$, $r(x)=2x^2+1$; $\frac{f(x)}{g(x)}=x^2-3+\frac{2x^2+1}{x^3-4x}$

4. $q(x)=2x+3$, $r(x)=-x^2+1$; $\frac{f(x)}{g(x)}=2x+3+\frac{-x^2+1}{x^3+x}$

5. $(x-2)(x^2+1)$

6. $(x+3)(x^2+4)$

7. $(2x+1)(x^2+1)$

8. $(3x-1)(x^2+4)$

9. $f(x)=x+1$

10. $f(x)=-5x-3$

11. $f(x)=3x^2+2$

12. $f(x)=-2x^2+5$

13. $f(x)=\frac{5}{9}x^2-\frac{20}{9}x+\frac{29}{9}$

14. $f(x)=-0.375x^2-2.25x+0.625$

SECTION EXERCISES 5.5

1. $\frac{1}{x-5}-\frac{1}{x-3}$

2. $\frac{1}{x+3}-\frac{1}{x+7}$

3. $\frac{2}{x-1}-\frac{2}{x+1}$

4. $\frac{1}{x-3}-\frac{1}{x+3}$

5. $\frac{1}{2x}-\frac{1/2}{x+2}$

6. $\frac{-2}{x-3}+\frac{2}{x}$

7. $\frac{1}{x-3}-\frac{2}{x+4}$

8. $\frac{4}{x-5}+\frac{3}{x+2}$

9. $\frac{-2}{x+3}+\frac{5}{2x-1}$

10. $\frac{3}{x+1}-\frac{2}{2x-3}$

11. $\frac{3}{(x^2+1)^2}+\frac{2}{x^2+1}$

12. $\frac{1}{(x^2+1)^2}+\frac{3}{x^2+1}$

13. $\frac{2}{(x-1)^2}-\frac{1}{x-1}+\frac{2}{x}$

14. $\frac{7/3}{(x-3)^2}-\frac{25/9}{x-3}+\frac{25/9}{x}$

15. $\frac{2}{x-3}-\frac{1}{x^2}+\frac{1}{x}$

16. $-\frac{1}{x^2}+\frac{2}{x}+\frac{3}{x+4}$

17. $-\frac{1}{(x^2+2)^2}+\frac{2x}{x^2+2}$

18. $-\frac{1}{(x^2+2)^2}+\frac{3x}{x^2+2}$

19. $\frac{2}{x-1}-\frac{x}{x^2+x+1}$

20. $\frac{3}{x+1}-\frac{x}{x^2-x+1}$

21. $\frac{2x^2+x+3}{x^2-1}=2+\frac{x+5}{x^2-1}$; $\frac{r(x)}{h(x)}=\frac{x+5}{x^2-1}=\frac{3}{x-1}-\frac{2}{x+1}$

22. $\frac{3x^2+2x}{x^2-4}=3+\frac{2x+12}{x^2-4}$; $\frac{r(x)}{h(x)}=\frac{2x+12}{x^2-4}=\frac{4}{x-2}-\frac{2}{x+2}$

23. $\frac{x^3-2}{x^2+x}=x-1+\frac{x-2}{x^2+x}$; $\frac{r(x)}{h(x)}=\frac{x-2}{x^2+x}=-\frac{2}{x}+\frac{3}{x+1}$

24. $\frac{x^3+2}{x^2-x}=x+1+\frac{x+2}{x^2-x}$; $\frac{r(x)}{h(x)}=\frac{x+2}{x^2-x}=\frac{3}{x-1}-\frac{2}{x}$

25. **(a)** $f(x)=2x-1+\frac{-x+10}{x^2-4}$. **(b)** $\frac{r(x)}{h(x)}=\frac{-x+10}{x^2-4}=\frac{2}{x-2}+-\frac{3}{x+2}$ **(c)** The end behavior asymptote $y=2x-1$ gives the overall appearance of the graph; the partial fractions $\frac{2}{x-2}$ and $-\frac{3}{x+2}$ indicate where the vertical asymptotes will be ($x = 2$ and $x = -2$); the graph can be sketched by adding the functions "manually" (as was done in Chapter 2).

26. **(a)** $f(x) = 2x^2 - 2 + \dfrac{2x^2 + 3x + 3}{x^3 + x}$. **(b)** $\dfrac{r(x)}{h(x)} = \dfrac{2x^2 + 3x + 3}{x^3 + x} = \dfrac{3}{x} + \dfrac{3 - x}{x^2 + 1}$. **(c)** The end behavior asymptote $y = 2x^2 - 2$ gives the overall appearance of the graph of f; the partial fractions $\dfrac{3}{x}$ and $\dfrac{3 - x}{x^2 + 1}$ indicate where the vertical asymptote will be ($x = 0$); the graph can be sketched by adding the functions "manually" (as was done in Chapter 2).

Chapter 5 Review

1. Domain: $(-\infty, -1) \cup (-1, \infty)$; asymptote: $x = -1$

2. Domain: $(-\infty, 3) \cup (3, \infty)$; asymptote: $x = 3$

3. Domain: $(-\infty, -1) \cup (-1, 1) \cup (1, \infty)$; asymptotes: $y = 1$, $x = -1$, and $x = 1$

4. Domain: $(-\infty, -3) \cup (-3, 2) \cup (2, \infty)$; asymptotes: $y = 2$, $x = -3$, and $x = 2$

5. Domain: $(-\infty, \infty)$; asymptotes: none

6. Domain: $(-\infty, \infty)$; asymptotes: none

7. End behavior (horizontal) asymptote: $y = -1$; vertical asymptote: $x = 1$; as $x \to 1^-, f(x) \to -\infty$; as $x \to 1^+, f(x) \to \infty$. Intercepts: approximately $(1.32, 0)$ and $(0, -1)$. Window: $[-5, 5] \times [-3, 3]$.

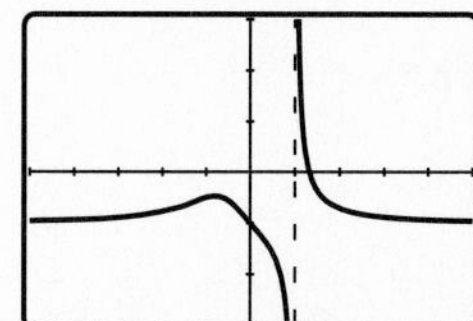

8. End behavior (horizontal) asymptote: $y = 1$; vertical asymptote: $x = -1$; as $x \to -1^-, f(x) \to -\infty$; as $x \to -1^+$, $f(x) \to \infty$. Intercepts: approximately $(-1.32, 0)$ and $(0, 1)$. Window: $[-5, 5] \times [-3, 3]$.

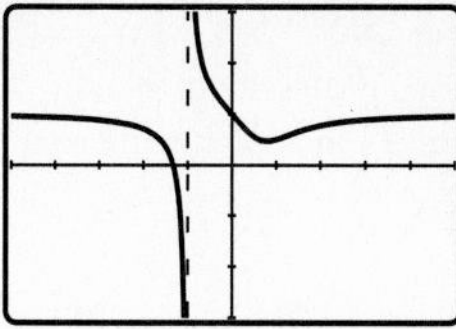

9. End behavior asymptote: $y = x - 7$; vertical asymptote: $x = -3$; as $x \to -3^-, f(x) \to -\infty$; as $x \to -3^+, f(x) \to \infty$. Intercept: $(0, \frac{5}{3})$. Window: $[-15, 10] \times [-30, 10]$.

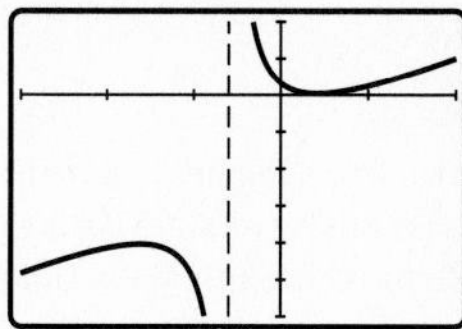

10. End behavior asymptote: $y = x - 6$; vertical asymptote: $x = -3$; as $x \to -3^-, f(x) \to -\infty$; as $x \to -3^+, f(x) \to \infty$. Intercepts: approximately $(-1.54, 0)$, $(4.54, 0)$, and $(0, -\frac{7}{3})$. Window: $[-20, 10] \times [-30, 10]$.

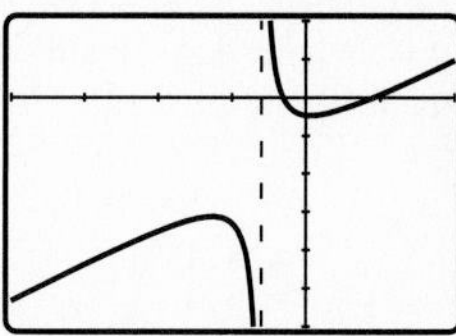

11. End behavior asymptote: $y = x$; vertical asymptote: none. Intercepts: $(-1, 0)$ and $(0, 1)$. Window: $[-3, 3] \times [-2, 2]$.

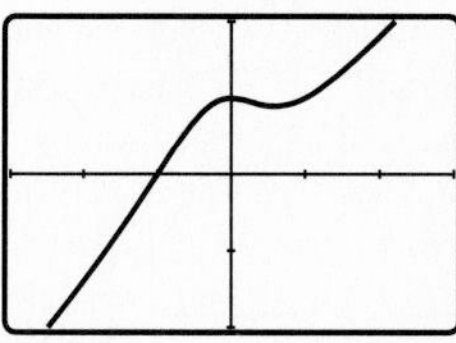

12. End behavior asymptote: $y = 2x$; vertical asymptote: none. Intercepts: approximately $(-1.37, 0)$, $(0.37, 0)$, $(1, 0)$, and $(0, \frac{1}{4})$. Window: $[-4, 4] \times [-4, 4]$.

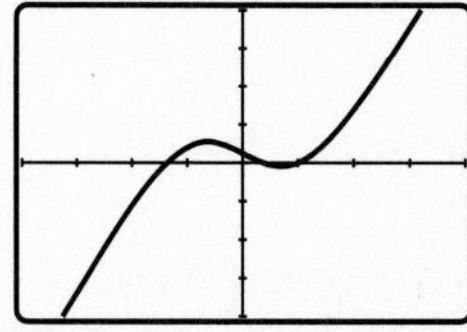

13. End behavior asymptote: $y = x^2 - x$; vertical asymptote: $x = 3$; as $x \to 3^-$, $f(x) \to -\infty$; as $x \to 3^+, f(x) \to \infty$. Intercepts: approximately $(-0.41, 0)$, $(2.41, 0)$, $(2, 0)$, and $(0, -\frac{2}{3})$. Window: $[-5, 5] \times [-5, 20]$.

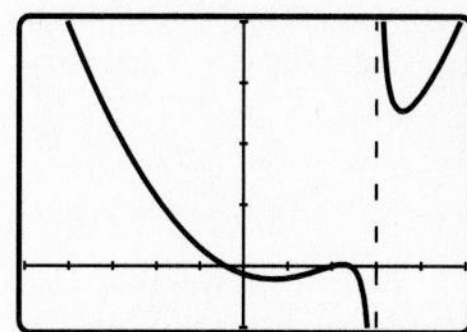

14. End behavior asymptote: $y = -x^2 + 3$; vertical asymptotes: $x = \pm 1$; as $x \to 1^+$, $f(x) \to -\infty$; as $x \to 1^-, f(x) \to \infty$, as $x \to -1^+, f(x) \to \infty$; as $x \to -1^-$, $f(x) \to -\infty$. y-intercept: $(0, 4)$, x-intercept: $(\pm\sqrt{2}, 0)$. Window: $[-3, 3] \times [-6, 6]$.

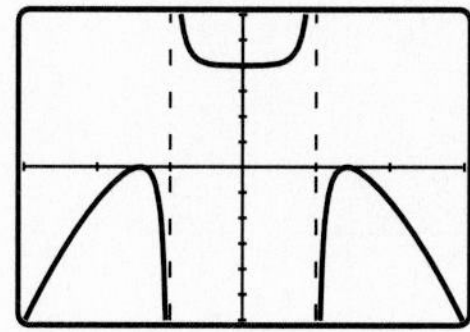

15. $f(x) = x^2 + 2x + \dfrac{c}{x - 2} = \dfrac{x^3 - 4x + c}{x - 2}$, where $c \neq 0$

16. $f(x) = 2x + 1 + \dfrac{c}{x - 3} = \dfrac{2x^2 - 5x - 3 + c}{x - 3}$, where $c \neq 0$

17. $f(x) = x^2 + 1 + \dfrac{c}{x^2 - 1} = \dfrac{x^4 - 1 + c}{x^2 - 1}$, where $c \neq 0$

18. $f(x) = x^3 - x + \dfrac{c}{(x + 3)(x - 2)} = \dfrac{x^5 + x^4 - 7x^3 - x^2 + 6x + c}{x^2 + x - 6}$, where $c \neq 0$

19. $f(x) = -1 + \dfrac{2}{x - 5}$; translate right 5 units and vertically stretch by 2 (either order) and then translate down 1 unit.

Horizontal asymptote: $y = -1$; vertical asymptote: $x = 5$.

20. $f(x) = 3 - \dfrac{1}{x+2}$; translate left 2 units and reflect across the x-axis (either order) and then translate upward 3 units. Horizontal asymptote: $y = 3$; vertical asymptote: $x = -2$.

21. $x = 1.5$ or $x = 4$

22. $x = -5$

23. $[-3, -2) \cup (2, \infty)$, or $-3 \le x < -2$ or $2 < x$

24. $(-\infty, 2) \cup (1, 3)$

25. $\dfrac{2}{x-4} + \dfrac{1}{x+1}$

26. $\dfrac{6}{x+2} - \dfrac{5}{x-1}$

27. $\dfrac{1}{x+1} + \dfrac{2}{(x+1)^2} - \dfrac{1}{x+2}$

28. $\dfrac{1}{x+2} - \dfrac{3}{(x+2)^2} + \dfrac{2}{x-1}$

29. $\dfrac{2}{x+1} + \dfrac{3x-4}{x^2+1}$

30. $\dfrac{1}{x+2} - \dfrac{2x+1}{x^2+4}$

31. **(a)** $C = 4.32 + 4000/x$. **(b)** $x \approx$ 12,903.23; round up to 12,904.

32. **(a)** $P(15) = 325$, $P(70) = 600$, and $P(100) = 648$. **(b)** $y = 800$. **(c)** 800 deer

33. **(a)** $P = 2x + 750/x$. **(b)** Approximately 19.36 ft × 19.36 ft; $P \approx 77.46$ ft.

34. **(a)** If x is the height of the print material, $A(x) = (x + 2.5)(\frac{40}{x} + 1.5)$; if x is the width of the print material, $A(x) = (\frac{40}{x} + 2.5)(x + 1.5)$. **(b)** The least area, approximately 68.24 in.^{2}, occurs when the page is about 10.66 in. (height) by 6.40 in. (width); note also that the printed material is about 8.16 in. × 4.9 in.

35. **(a)** $A = (x + 5)(\frac{1200}{x} + 5)$. **(b)** $x = y \approx 34.64$; the plot is about 39.64 ft × 39.64 ft and has an approximate area of 1571.41 ft^2.

36. **(a)** $P(x) = 2(x + \frac{150}{x}) = 2x + \frac{300}{x}$. **(b)** $P(x) = 60$ when $x = 15 \pm \sqrt{75}$, or approximately 6.34 or 23.66; these two answers yield the same rectangle: about 6.34 m × 23.66 m. **(c)** x must be between 5 and 30, or $5 < x < 30$; the other dimension of the rectangle has the same restriction.

37. $y = 1$; when a large amount of pure acid is added, the resulting mixture is almost pure acid, so the concentration is almost 100%, or 1.

38. **(a)** No; the domain of f is $(-\infty, -2) \cup (-2, \infty)$, or "$x \ne 2$", and the domain of g is $(-\infty, \infty)$, or "all real numbers." **(b)** No; although f is not defined at -2, it does not tend toward $\pm\infty$ on either side. **(c)** Most viewing windows do not reveal that f is undefined at -2. **(d)** Almost, but not quite.

39. **(a)** $R_2 = \dfrac{1.2x}{x - 1.2}$. **(b)** When $x = 3$, $R_2 = 2$ ohms.

40. **(a)** $C(x) = \dfrac{50}{50 + x}$. **(b)** Shown is the window $[0, 50] \times [0, 1]$, with the graphs of $y = C(x)$ and $y = 0.6$; the two graphs cross when $x \approx 33.33$ ounces of distilled water. Therefore, add more than 33.33 ounces of water.

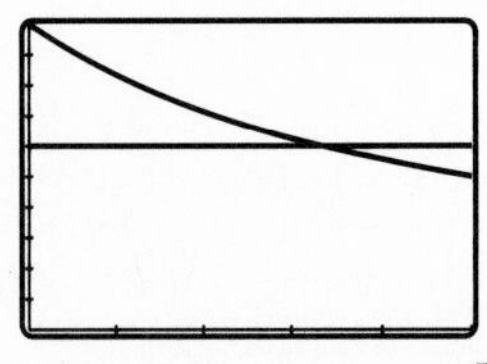

(c) Algebraic solution of $\dfrac{50}{50 + x} = 0.6$ leads to $0.6x = 20$, or $x = \frac{100}{3} = 33.\overline{3}$.

41. **(a)** $S = 2\pi x^2 + 2000/x$. **(b)** Either $x \approx$ 2.31 cm and $h \approx 59.75$ cm or $x \approx 10.65$ and $h \approx 2.81$. **(c)** Approximately $2.31 < x < 10.65$ and $2.81 < h < 59.75$.

42. **(a)** Add the area of the base to 4 times the area of one side; each side is a rectangle with dimensions $x \times \dfrac{1000}{x^2}$, so $S(x) = x^2 + 4x \cdot \dfrac{1000}{x^2} = x^2 + \dfrac{4000}{x}$. **(b)** Either $x = 20$, giving the dimensions 20 ft × 20 ft × 2.5 ft, or $x \approx 7.32$, giving the approximate dimensions 7.32 ft × 7.32 ft × 18.66 ft. **(c)** $7.32 < x < 20$ (lower bound approximate), so the height must be between 2.5 and about 18.66.

43. **(a)** $S = x^2 + 2400/x$. **(b)** $x \approx 5.06$ or $x \approx 19.40$; either about 5.06 ft × 5.06 ft × 23.44 ft or 19.40 ft × 19.40 ft × 1.59 ft. **(c)** Approximately $5.06 < x < 19.40$ and $1.59 < h < 23.44$. **(d)** When $x \approx 10.63$, $S \approx 338.77$ ft^2.

44. **(a)** $D = D(t) = \dfrac{1}{3} - \dfrac{1}{t} = \dfrac{t-3}{3t}$ of the pool each hour. **(b)** If $t = 5$, $D = \frac{1}{3} - \frac{1}{5} = \frac{2}{15}$, so drain B by itself can empty the pool in $\frac{15}{2} = 7.5$ hours.

45. **(a)** $t = \dfrac{10}{x - 40} + \dfrac{35}{x}$ **(b)** $x \approx 62.66$, so the car traveled at an average of about 62.66 mph.

CHAPTER 6

QUICK REVIEW 6.1

1. $C = 5\pi$ in.

2. $C = 9.2\pi$ m

3. $r = \frac{6}{\pi}$ m

4. $r = \frac{4}{\pi}$ ft

5. **(a)** $s = 47.52$ ft. **(b)** $s = 39.77$ km.

6. **(a)** $v = 26.1$ m/sec. **(b)** $v = 8.06$ ft/sec.

7. $r = \frac{26}{59} = 0.4406 \ldots$ cm

8. $r = \frac{939}{16} = 58.6875$ ft

9. $s = 3559.24$ km

10. $s = 11.2$ ft

11. 88 ft/sec

12. 66 ft/sec

13. 6 mph

14. 90 mph

15. Positive

16. Negative

SECTION EXERCISES 6.1

1. 400° and −320° are possible answers.

2. 330° and −390° are possible answers.

3. 203° and −517° are possible answers.

4. 97° and −263° are possible answers.

5. $\frac{\pi}{6}$ and $-\frac{23\pi}{6}$ are possible answers.

6. $\frac{5\pi}{4}$ and $-\frac{11\pi}{4}$ are possible answers.

7. $\frac{19\pi}{6}$ and $-\frac{5\pi}{6}$ are possible answers.

8. $\frac{7\pi}{2}$ and $-\frac{\pi}{2}$ are possible answers.

9.

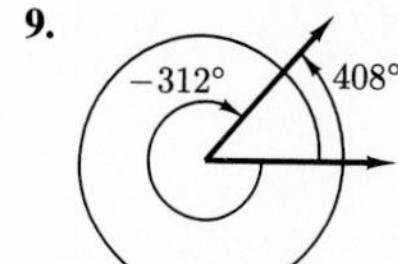

10.

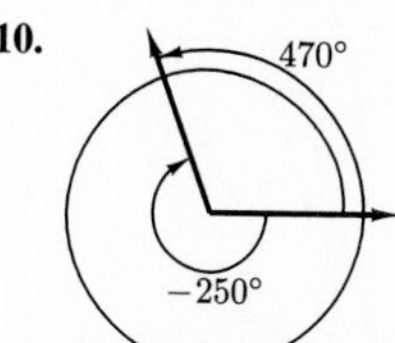

11.

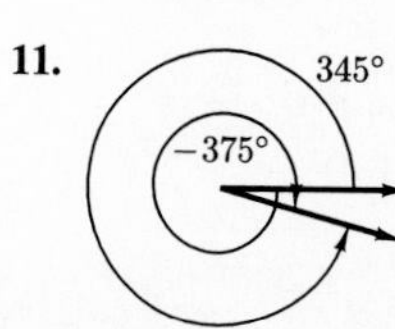

12.

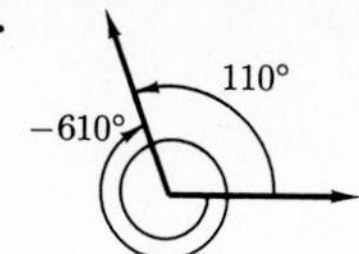

13.

14.

15. Complement: 55°; supplement: 145°

16. Complement: 67°; supplement: 157°

17. Complement: 22°; supplement: 112°

18. Complement: 78°; supplement: 168°

19. No complement or supplement

20. No complement or supplement

21. Complement: $\frac{\pi}{3}$; supplement: $\frac{5\pi}{6}$

22. Complement: $\frac{\pi}{6}$; supplement: $\frac{2\pi}{3}$

23. No complement or supplement

24. No complement or supplement

25. 23.2°

26. 35.4°

27. 118.7375°

28. 48.51°

29. 21°12′

30. 49°42′

31. 118°19′12″

32. 99°22′12″

33. $s = 50$ in.

34. $s = 70$ cm

35. $r = \frac{6}{\pi}$ ft

36. $r = \frac{7.5}{\pi}$ cm

37. $\theta = 3$ rad

38. $\theta = \frac{4}{7}$ rad

39. $\theta = \frac{9}{11}$ rad; $s_2 = 36$ cm

40. $\theta = 4.5$ rad; $r_2 = 16$ km

41. $\frac{\pi}{3}$

42. $\frac{\pi}{2}$

43. $\frac{3\pi}{2}$

44. $-\frac{5\pi}{6}$

45. Approximately 1.2518 rad

46. Approximately 0.2065 rad

47. Approximately −1.0716 rad

48. Approximately −1.3177 rad

49. 30°

50. 45°

51. 270°

52. −240°

53. Approximately −343.7747°

54. Approximately 286.4789°

55. Approximately 108.8620°

56. Approximately −309.3972°

57. **(a)** 45° **(b)** 22.5° **(c)** 247.5°

58. **(a)** 202.5° **(b)** 292.5° **(c)** 337.5°

59. ESE is 112.5°

60. SW is 225°

61. The angle between them is 9°42′ ≈ 0.1693 rad, so the distance is about 4.2324 mi.

62. $V \approx 52.36$ mph

63. Approximately 387.85 rpm

64. Approximately 12,566.3706 teeth per sec

65.

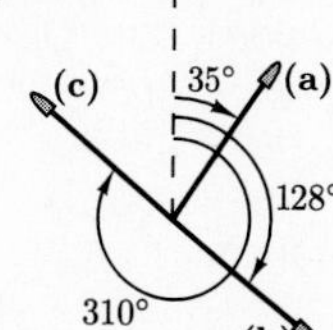

66.

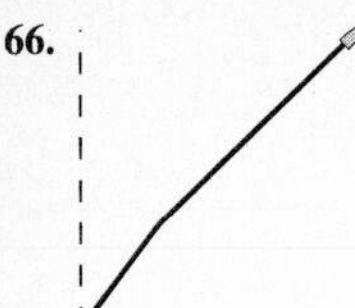

67. Approximately 295.7437 stat mi

68. Approximately 777.7511 naut mi

69. $\theta = 180° = \pi$

70. $\theta = 90° = \frac{\pi}{2}$

71. $\theta = 45° = \frac{\pi}{4}$

72. $\theta = 135° = \frac{3\pi}{4}$

73. $\theta = 315° = \frac{7\pi}{4}$

74. $\theta = 0° = 0$ rad

75. **(a)** $\pi = 3.141\ldots$ m
(b) $5\pi = 15.707\ldots$ m

76. **(a)** $16\pi = 50.265\ldots$ in., or $\frac{4}{3}\pi = 4.188\ldots$ ft **(b)** $2\pi = 6.283\ldots$ ft

77. $\frac{13}{45}\pi = 0.907\ldots$ ft

78. **(a)** 4π rad/sec **(b)** 28π cm/sec
(c) 7π rad/sec

79. **(a)** 4.5π rad/sec **(b)** 5.4π m/sec
(c) 2.7π m/sec

80. 38°03′

81. 43°12′

82. 5°38′

83. 42°09′

84. 149 naut mi

85. 591 naut mi

86. 898 naut mi

87. 517 naut mi

88. The area of the whole circle is πr^2; the sector with central angle θ accounts for $\frac{\theta}{2\pi}$ of that area, or $\frac{\theta}{2\pi} \cdot \pi r^2 = \frac{1}{2}\theta r^2$.

89. **(a)** $3.481\pi = 10.935\ldots \text{ ft}^2$
(b) 4.736 km^2

90.

37°
340°
B
60 mi
A

91. Bike wheels: about 56.5714 rad/sec; pedal sprocket: about 18.8571 rad/sec; wheel sprocket: about 56.5714 rad/sec.

QUICK REVIEW 6.2

1. $x = \sqrt{50} = 5\sqrt{2}$

2. $x = \sqrt{208} = 4\sqrt{13}$

3. $x = 6$

4. $x = \sqrt{12} = 2\sqrt{3}$

5. 43,824 ft

6. 100.8 in.

7. $\frac{47}{264} = 0.17\overline{803}$ mi

8. $63.\overline{3}$ ft

9. $a = 7.9152$ km

10. $b = 13.895\ldots$ ft

11. $\alpha = 1.0101\ldots$ (no units)

12. $\beta = 4.1899\ldots$ (no units)

13. $\frac{3\sqrt{7}}{7}$

14. $\frac{\sqrt{13}}{13}$

15. Opposite: 4; adjacent: 3

16. Opposite: 12; adjacent: 5

SECTION EXERCISES 6.2

1. $\sin\theta = \frac{4}{5}$, $\cos\theta = \frac{3}{5}$, $\tan\theta = \frac{4}{3}$; $\csc\theta = \frac{5}{4}$, $\sec\theta = \frac{5}{3}$, $\cot\theta = \frac{3}{4}$

2. $\sin\theta = \frac{8}{\sqrt{113}}$, $\cos\theta = \frac{7}{\sqrt{113}}$, $\tan\theta = \frac{8}{7}$; $\csc\theta = \frac{\sqrt{113}}{8}$, $\sec\theta = \frac{\sqrt{113}}{7}$, $\cot\theta = \frac{7}{8}$

3. $\sin\theta = \frac{12}{13}$, $\cos\theta = \frac{5}{13}$, $\tan\theta = \frac{12}{5}$; $\csc\theta = \frac{13}{12}$, $\sec\theta = \frac{13}{5}$, $\cot\theta = \frac{5}{12}$

4. $\sin\theta = \frac{8}{17}$, $\cos\theta = \frac{15}{17}$, $\tan\theta = \frac{8}{15}$; $\csc\theta = \frac{17}{8}$, $\sec\theta = \frac{17}{15}$, $\cot\theta = \frac{15}{8}$

5. $\sin\theta = \frac{7}{\sqrt{170}}$, $\cos\theta = \frac{11}{\sqrt{170}}$, $\tan\theta = \frac{7}{11}$; $\csc\theta = \frac{\sqrt{170}}{7}$, $\sec\theta = \frac{\sqrt{170}}{11}$, $\cot\theta = \frac{11}{7}$

6. $\sin\theta = \frac{\sqrt{7}}{4}$, $\cos\theta = \frac{3}{4}$, $\tan\theta = \frac{\sqrt{7}}{3}$; $\csc\theta = \frac{4}{\sqrt{7}}$, $\sec\theta = \frac{4}{3}$, $\cot\theta = \frac{3}{\sqrt{7}}$

7. $\sin\theta = \frac{\sqrt{57}}{11}$, $\cos\theta = \frac{8}{11}$, $\tan\theta = \frac{\sqrt{57}}{8}$; $\csc\theta = \frac{11}{\sqrt{57}}$, $\sec\theta = \frac{11}{8}$, $\cot\theta = \frac{8}{\sqrt{57}}$

8. $\sin\theta = \frac{9}{13}$, $\cos\theta = \frac{2\sqrt{22}}{13}$, $\tan\theta = \frac{9}{2\sqrt{22}}$; $\csc\theta = \frac{13}{9}$, $\sec\theta = \frac{13}{2\sqrt{22}}$, $\cot\theta = \frac{2\sqrt{22}}{9}$

9. $\sin\theta = \frac{3}{7}$, $\cos\theta = \frac{\sqrt{40}}{7}$, $\tan\theta = \frac{3}{\sqrt{40}}$; $\csc\theta = \frac{7}{3}$, $\sec\theta = \frac{7}{\sqrt{40}}$, $\cot\theta = \frac{\sqrt{40}}{3}$

10. $\sin\theta = \frac{2}{3}$, $\cos\theta = \frac{\sqrt{5}}{3}$, $\tan\theta = \frac{2}{\sqrt{5}}$; $\csc\theta = \frac{3}{2}$, $\sec\theta = \frac{3}{\sqrt{5}}$, $\cot\theta = \frac{\sqrt{5}}{2}$

11. $\sin\theta = \frac{4\sqrt{6}}{11}$, $\cos\theta = \frac{5}{11}$, $\tan\theta = \frac{4\sqrt{6}}{5}$; $\csc\theta = \frac{11}{4\sqrt{6}}$, $\sec\theta = \frac{11}{5}$, $\cot\theta = \frac{5}{4\sqrt{6}}$

12. $\sin\theta = \frac{\sqrt{39}}{8}$, $\cos\theta = \frac{5}{8}$, $\tan\theta = \frac{\sqrt{39}}{5}$; $\csc\theta = \frac{8}{\sqrt{39}}$, $\sec\theta = \frac{8}{5}$, $\cot\theta = \frac{5}{\sqrt{39}}$

13. $\sin\theta = \frac{5}{\sqrt{106}}$, $\cos\theta = \frac{9}{\sqrt{106}}$, $\tan\theta = \frac{5}{9}$; $\csc\theta = \frac{\sqrt{106}}{5}$, $\sec\theta = \frac{\sqrt{106}}{9}$, $\cot\theta = \frac{9}{5}$

14. $\sin\theta = \frac{12}{\sqrt{313}}$, $\cos\theta = \frac{13}{\sqrt{313}}$, $\tan\theta = \frac{12}{13}$; $\csc\theta = \frac{\sqrt{313}}{12}$, $\sec\theta = \frac{\sqrt{313}}{13}$, $\cot\theta = \frac{13}{12}$

15. $\sin\theta = \frac{3}{\sqrt{130}}$, $\cos\theta = \frac{11}{\sqrt{130}}$, $\tan\theta = \frac{3}{11}$; $\csc\theta = \frac{\sqrt{130}}{3}$, $\sec\theta = \frac{\sqrt{130}}{11}$, $\cot\theta = \frac{11}{3}$

16. $\sin\theta = \frac{5}{12}$, $\cos\theta = \frac{\sqrt{119}}{12}$, $\tan\theta = \frac{5}{\sqrt{119}}$; $\csc\theta = \frac{12}{5}$, $\sec\theta = \frac{12}{\sqrt{119}}$, $\cot\theta = \frac{\sqrt{119}}{5}$

17. $\sin\theta = \frac{9}{23}$, $\cos\theta = \frac{8\sqrt{7}}{23}$, $\tan\theta = \frac{9}{8\sqrt{7}}$; $\csc\theta = \frac{23}{9}$, $\sec\theta = \frac{23}{8\sqrt{7}}$, $\cot\theta = \frac{8\sqrt{7}}{9}$

18. $\sin\theta = \frac{2\sqrt{66}}{17}$, $\cos\theta = \frac{5}{17}$, $\tan\theta = \frac{2\sqrt{66}}{5}$; $\csc\theta = \frac{17}{2\sqrt{66}}$, $\sec\theta = \frac{17}{5}$, $\cot\theta = \frac{5}{2\sqrt{66}}$

19. $\frac{\sqrt{3}}{2}$

20. 1

21. $\sqrt{3}$

22. 2

23. $\frac{1}{\sqrt{2}} = \frac{\sqrt{2}}{2}$

24. $\frac{2}{\sqrt{3}} = \frac{2\sqrt{3}}{3}$

25. 0.96

26. 0.67

27. 0.14

28. 0.99

29. 0.94

30. 0.44

31. 0.42

32. 0.61

33. 0.27

34. 0.21

35. 1.52

36. 3.07

37. 0.51

38. 2.25

39. 3.08

40. 0.81

41. 2.63

42. 1.11

43. 2.41

44. 3.24

45. $\frac{a}{c} = \frac{1}{c/a}$

46. $\frac{a}{b} = \frac{1}{b/a}$

47. $\frac{c}{b} = \frac{1}{b/c}$

48. $\frac{a}{b} = \frac{a/c}{b/c}$

49. $1 + (\frac{a}{b})^2 = (\frac{c}{b})^2$, which simplifies to $b^2 + a^2 = c^2$

50. $1 + (\frac{b}{a})^2 = (\frac{c}{a})^2$, which simplifies to $a^2 + b^2 = c^2$

51. $\theta = 30° = \frac{\pi}{6}$

52. $\theta = 60° = \frac{\pi}{3}$

53. $\theta = 60° = \frac{\pi}{3}$

54. $\theta = 45° = \frac{\pi}{4}$

55. $\theta = 60° = \frac{\pi}{3}$

56. $\theta = 45° = \frac{\pi}{4}$

57. $\theta = 30° = \frac{\pi}{6}$

58. $\theta = 30° = \frac{\pi}{6}$

59. $\theta = 22.95°$

60. $\theta = 75.93°$

61. $\theta = 65.80°$

62. $\theta = 27.13°$

63. $\theta = 72.65°$

64. $\theta = 55.41°$

65. $\theta = 58.21°$

66. $\theta = 43.95°$

67. $x = \frac{15}{\sin 34°} \approx 26.82$

68. $z = \frac{23}{\cos 39°} \approx 29.60$

69. $y = \frac{32}{\tan 57°} \approx 20.78$

70. $x = 14 \sin 43° \approx 9.55$

71. $b \approx 33.79$, $c \approx 35.96$, $\beta = 70°$

72. $c = 5, \alpha \approx 36.87°, \beta \approx 53.13°$

73. $a \approx 6.56, b \approx 7.55, \beta = 49°$

74. $b \approx 22.25, c \approx 27.16, \alpha = 35°$

75. $a \approx 4.90, \alpha \approx 44.42°, \beta \approx 45.58°$

76. $b \approx 46.56, \alpha \approx 23.46°, \beta \approx 66.54°$

77. $c \approx 9.46, \alpha \approx 12.20°, \beta \approx 77.80°$

78. $b \approx 8.32, x \approx 9.71, \alpha = 31°$

79. $a \approx 12.58, b \approx 2.79, \alpha = 77.5°$

80. $a \approx 2.57, b \approx 14.27, \beta = 79.8°$

81. $h = 55 \tan 75° \approx 205.26$ ft

82. Angle with horizontal: 62.18°; angle with antenna: 27.82°; height: $8\sqrt{11}$

83. $A = 12 \cdot 5/\sin 54° \approx 74.16 \text{ ft}^2$

84. $h = 130 \tan 82.9° \approx 1043.70$ ft

85. $AC = 100 \tan 75°12'42'' \approx 378.80$ ft

86. $h = 5 + 120 \tan 8° \approx 21.86$ ft

87. The first computation was done in DEGREE mode, the second in RADIAN mode.

88. No; since it fails for $\alpha = \beta = 1$, it cannot be true in general.

89. $\sin\theta = \dfrac{\sqrt{2}}{2}$

90. $\cos\theta = \dfrac{1}{2}$

91. $\sin\theta = \dfrac{\sqrt{2}}{2}$

92. $\sin\theta = \dfrac{\sqrt{15}}{4}$

93. False; this is cot 56°.

94. True; $\cos\theta$ and $\sec\theta$ are reciprocals.

95. False; although $\cos 34° = \sin 56°$; since $|\cos\theta| \le 1$ and $|\sin\theta| \le 1$, we would need $\cos\theta = \sin\theta = \pm 1$ in order for this to work, but there is no such θ.

96. True; $\cos 19° = \sin 71°$, and $\sin\theta$ and $\csc\theta$ are reciprocals.

97. False; we know that $(\cos 43°)^2 + (\sin 43°)^2 = 1$ and that $\sin 47° \ne \sin 43°$.

98. False; $\cot 39° = \frac{\cos 39°}{\sin 39°} > \cos 39°$ and, similarly, $\tan 39° > \sin 39°$, so $(\cot 39°)^2 + (\tan 39°)^2 > (\cos 39°)^2 + (\sin 39°)^2 = 1$.

99. True; this is a rearrangement of one of the Pythagorean identities.

100. False; $1 + (\tan 39°)^2 = (\sec 39°)^2 \ne (\cot 39°)^2$

101. $PB = 5\left(\dfrac{5\cos 30° - 2}{5\cos 30°}\right) = 5 - \dfrac{2}{\cos 30°} \approx 2.69$ m

102. 18 in. to the right of C (or 12 in. to the left of D).

QUICK REVIEW 6.3

1. Quadrant II

2. Quadrant III

3. Quadrant IV

4. Quadrant III

5. Quadrant I

6. Quadrant III

7. $\dfrac{\sqrt{2}}{2}$

8. $\dfrac{\sqrt{3}}{2}$

9. $\dfrac{1}{\sqrt{3}} = \dfrac{\sqrt{3}}{3}$

10. 1

11. $\sqrt{2}$

12. 2

13. $\sin\theta = \frac{5}{13}, \cos\theta = \frac{12}{13}, \tan\theta = \frac{5}{12}$; $\csc\theta = \frac{13}{5}, \sec\theta = \frac{13}{12}, \cot\theta = \frac{12}{5}$

14. $\sin\theta = \frac{8}{17}, \cos\theta = \frac{15}{17}, \tan\theta = \frac{8}{15}$; $\csc\theta = \frac{17}{8}, \sec\theta = \frac{17}{15}, \cot\theta = \frac{15}{8}$

15. $\sin\theta = \frac{20}{29}, \cos\theta = \frac{21}{29}, \tan\theta = \frac{20}{21}$; $\csc\theta = \frac{29}{20}, \sec\theta = \frac{29}{21}, \cot\theta = \frac{21}{20}$

16. $\sin\theta = \frac{20}{29}, \cos\theta = \frac{21}{29}, \tan\theta = \frac{20}{21}$; $\csc\theta = \frac{29}{20}, \sec\theta = \frac{29}{21}, \cot\theta = \frac{21}{20}$

17. Graph $x = t, y = t^2 + 3$. Shown is the window $[-5, 5] \times [-5, 30]$, with Tmin $= -5$ and Tmax $= 5$.

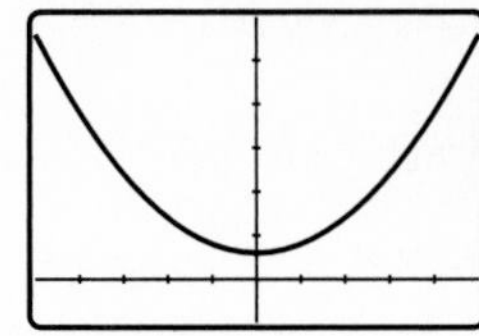

18. Graph $x = t, y = \sqrt{t}$; shown is the window $[-2, 30] \times [-1, 6]$, with Tmin $= 0$, Tmax $= 30$.

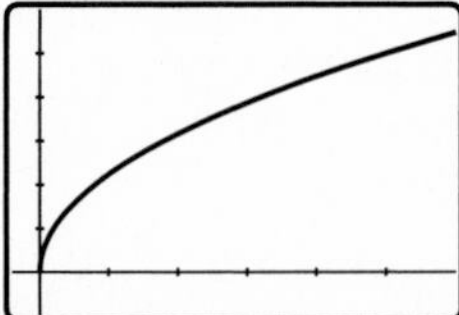

SECTION EXERCISES 6.3

1. $\sin\theta = \dfrac{2}{\sqrt{5}}, \cos\theta = -\dfrac{1}{\sqrt{5}}$, $\tan\theta = -2$; $\csc\theta = \dfrac{\sqrt{5}}{2}$, $\sec\theta = -\sqrt{5}, \cot\theta = -\dfrac{1}{2}$

2. $\sin\theta = -\dfrac{3}{5}, \cos\theta = \dfrac{4}{5}, \tan\theta = -\dfrac{3}{4}$; $\csc\theta = -\dfrac{5}{3}, \sec\theta = \dfrac{5}{4}, \cot\theta = -\dfrac{4}{3}$

3. $\sin\theta = -\dfrac{1}{\sqrt{2}}, \cos\theta = -\dfrac{1}{\sqrt{2}}$, $\tan\theta = 1$; $\csc\theta = -\sqrt{2}$, $\sec\theta = -\sqrt{2}, \cot\theta = 1$

4. $\sin\theta = -\dfrac{5}{\sqrt{34}}, \cos\theta = \dfrac{3}{\sqrt{34}}$,

$\tan\theta = -\frac{5}{3}$; $\csc\theta = -\frac{\sqrt{34}}{5}$,
$\sec\theta = \frac{\sqrt{34}}{3}$, $\cot\theta = -\frac{3}{5}$

5. $\sin\theta = \frac{2}{\sqrt{5}}$, $\cos\theta = \frac{1}{\sqrt{5}}$, $\tan\theta = 2$;
$\csc\theta = \frac{\sqrt{5}}{2}$, $\sec\theta = \sqrt{5}$, $\cot\theta = \frac{1}{2}$

6. $\sin\theta = \frac{4}{5}$, $\cos\theta = \frac{3}{5}$, $\tan\theta = \frac{4}{3}$;
$\csc\theta = \frac{5}{4}$, $\sec\theta = \frac{5}{3}$, $\cot\theta = \frac{3}{4}$

7. $\sin\theta = 0$, $\cos\theta = -1$, $\tan\theta = 0$;
$\csc\theta =$ undefined, $\sec\theta = -1$,
$\cot\theta =$ undefined

8. $\sin\theta = 1$, $\cos\theta = 0$, $\tan\theta =$ undefined;
$\csc\theta = 1$, $\sec\theta =$ undefined, $\cot\theta = 0$

9. $\sin\theta = -\frac{3}{\sqrt{13}}$, $\cos\theta = -\frac{2}{\sqrt{13}}$,
$\tan\theta = \frac{3}{2}$; $\csc\theta = -\frac{\sqrt{13}}{3}$,
$\sec\theta = -\frac{\sqrt{13}}{2}$, $\cot\theta = \frac{2}{3}$

10. $\sin\theta = -\frac{2}{\sqrt{29}}$, $\cos\theta = \frac{5}{\sqrt{29}}$,
$\tan\theta = -\frac{2}{5}$; $\csc\theta = -\frac{\sqrt{29}}{2}$,
$\sec\theta = \frac{\sqrt{29}}{5}$, $\cot\theta = -\frac{5}{2}$

11. $\sin\theta = -\frac{1}{\sqrt{2}}$, $\cos\theta = \frac{1}{\sqrt{2}}$,
$\tan\theta = -1$; $\csc\theta = -\sqrt{2}$,
$\sec\theta = \sqrt{2}$, $\cot\theta = -1$

12. $\sin\theta = -\frac{1}{\sqrt{65}}$, $\cos\theta = -\frac{8}{\sqrt{65}}$,
$\tan\theta = \frac{1}{8}$; $\csc\theta = -\sqrt{65}$,
$\sec\theta = -\frac{\sqrt{65}}{8}$, $\cot\theta = 8$

13. **(a)** + **(b)** + **(c)** +

14. **(a)** + **(b)** − **(c)** −

15. **(a)** − **(b)** − **(c)** +

16. **(a)** − **(b)** + **(c)** −

17. Negative

18. Positive

19. Negative

20. Negative

21. Negative

22. Negative

23. Positive

24. Positive

25. (a)

26. (b)

27. (a)

28. (b)

29. **(a)** -1 **(b)** 0 **(c)** Undefined

30. **(a)** 1 **(b)** 0 **(c)** Undefined

31. **(a)** 0 **(b)** -1 **(c)** 0

32. **(a)** -1 **(b)** 0 **(c)** Undefined

33. **(a)** 1 **(b)** 0 **(c)** Undefined

34. **(a)** 0 **(b)** 1 **(c)** 0

35. $-\frac{1}{2}$

36. $-\sqrt{3}$

37. 2

38. $\sqrt{2}$

39. $\frac{1}{2}$

40. $\frac{1}{2}$

41. 1

42. 1

43. $\sin\theta = \frac{\sqrt{5}}{3}$, $\tan\theta = \frac{\sqrt{5}}{2}$

44. $\cos\theta = -\frac{\sqrt{15}}{4}$, $\cot\theta = -\sqrt{15}$

45. $\tan\theta = -\frac{2}{\sqrt{21}}$, $\sec\theta = \frac{5}{\sqrt{21}}$

46. $\sin\theta = -\frac{7}{\sqrt{58}}$, $\cos\theta = -\frac{3}{\sqrt{58}}$

47. $\sec\theta = -\frac{5}{4}$, $\csc\theta = \frac{5}{3}$

48. $\csc\theta = \frac{5}{4}$, $\cot\theta = -\frac{3}{4}$

49. $\frac{\sqrt{3}}{2}$

50. $\frac{\sqrt{3}}{2}$

51. $\frac{\sqrt{2}}{2}$

52. $-\frac{\sqrt{3}}{2}$

53. $-\frac{\sqrt{3}}{2}$

54. $\frac{\sqrt{2}}{2}$

55. $\frac{1}{\sqrt{3}}$

56. $\sqrt{3}$

57. $(-\infty, \infty)$

58. $(-\infty, \infty)$

59. $(-\infty, 0) \cup (0, \infty)$

60. $(-\infty, 1) \cup (1, \infty)$

61. $(0, \infty)$

62. $\ldots(-3\pi, -2\pi) \cup (-2\pi, -\pi) \cup (-\pi, 0) \cup (0, \pi) \cup (\pi, 2\pi)\ldots$, $x \neq n\pi$ for all integers n.

63. $[-3, 3]$

64. $[-2, 2]$

65. $[-\frac{1}{2}, \frac{1}{2}]$

66. $[-\frac{1}{3}, \frac{1}{3}]$

67. $[-\sqrt{2}, \sqrt{2}]$

68. $[-\sqrt{5}, \sqrt{5}]$

69. Graph $x = 3\cos t$, $y = 3\sin t$ with $0 \le t \le 2\pi$ (in RADIAN mode).

70. Graph $x = 5\cos t$, $y = 5\sin t$ with $0 \le t \le 2\pi$ (in RADIAN mode).

71. $\sin t$ is the y-coordinate of the point on the unit circle t units counterclockwise from (1, 0); it repeats every 2π units, since the distance around the circle is 2π.

72. $\cos t$ is the x-coordinate of the point on the unit circle t units counterclockwise from (1, 0); It will repeat every 2π units, since the distance around the circle is 2π.

73. $\mu = \frac{\sin 83°}{\sin 36°} \approx 1.69$

74. $\sin\theta_2 = \frac{\sin 42°}{1.52} \approx 0.44$, so $\theta_2 \approx 26.12°$

75. **(a)** 0.4 in. **(b)** about 0.1852 in.

76. $t = 0$: $\theta = 0.25$ rad; $t = 2.5$: $\theta = -0.2002\ldots$

77. $d = 600/\sin\theta$. **(a)** $d \approx 848.53$ ft **(b)** $d \approx 600$ ft. **(c)** $d \approx 933.43$ ft

78. January: 103.25; April ($t = 4$): 125.83; June ($t = 6$): 72.4; October ($t = 10$): 18.97; December ($t = 12$): 72.4. June and December are the same; perhaps by June most people have suits for the summer, and by December they are beginning to purchase them for next summer (or as Christmas presents, or for midwinter vacations).

79. $\cos\theta = \pm\sqrt{1 - 0.36^2} = \pm 0.932\ldots$

80. $\tan\theta = \dfrac{\pm\sqrt{1-(-0.65)^2}}{-0.65} = \pm 1.169\ldots$

81. $\frac{5\pi}{6}$

82. $\frac{11\pi}{6}$

83. $\frac{7\pi}{4}$

84. $\frac{5\pi}{4}$

85. The two triangles are congruent: both have hypotenuse 1; the corresponding angles are congruent because the two acute angles in a right triangle sum to $\frac{\pi}{2}$.

86. These coordinates give the lengths of the legs of the triangles from Exercise 85, and these triangles are congruent. For example, the length of the horizontal leg of the triangle with vertex P is given by the (absolute value of the) x-coordinate of P, which must be the same as the (absolute value of the) y-coordinate of Q.

87. One possible answer: Starting from point (a, b) on the unit circle—at an angle of t so that $\cos t = a$—then measuring a quarter of the way around the circle (which corresponds to adding $\frac{\pi}{2}$ to the angle), we end at $(-b, a)$ so that $\sin(t + \frac{\pi}{2}) = a$. This result is shown in the figure for (a, b) in quadrant I; similar illustrations can be drawn for the other quadrants.

88. One possible answer: Starting from point (a, b) on the unit circle—at an angle of t, so that $\sin t = b$—then measuring a quarter of the way around the circle (which corresponds to adding $\frac{\pi}{2}$ to the angle), we end at $(-b, a)$ so that $\cos(t + \frac{\pi}{2}) = -b = -\sin t$. This result is shown in the figure for (a, b) in quadrant I; similar illustrations can be drawn for the other quadrants.

89. Starting from point (a, b) on the unit circle—at an angle of t so that $\cos t = a$—then measuring a quarter of the way around the circle (which corresponds to adding $\frac{\pi}{2}$ to the angle), we end at $(-b, a)$ so that $\sin(t + \frac{\pi}{2}) = a$. This outcome holds when (a, b) is in quadrant II, as it did for quadrant I.

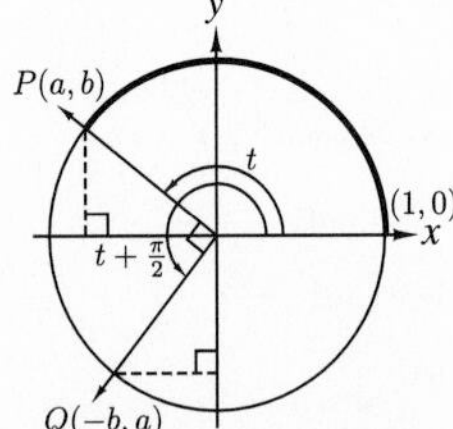

90. **(a)** Both triangles are right triangles with hypotenuse 1, and the angles at the origin are both t (for the triangle on the left, the angle is the supplement of $\pi - t$); therefore, the vertical legs are also congruent, their lengths corresponding to the sines of t and $\pi - t$. **(b)** The points P and Q are reflections of each other through the y-axis, so they are the same distance (but opposite directions) from the y-axis; alternatively, use the congruent triangles argument from part (a).

91. Seven decimal places are shown so that the slight differences can be seen. The magnitude of the relative error is less than 1% when $|\theta| < 0.2441$ (approximately).

θ	$\sin\theta$	$\sin\theta - \theta$	$\left\lvert\frac{\sin\theta - \theta}{\sin\theta}\right\rvert$
−0.03	−0.0299955	0.0000045	0.0001500
−0.02	−0.0199987	0.0000013	0.0000667
−0.01	−0.0099998	0.0000002	0.0000167
0	0	0	—
0.01	0.0099998	0.0000002	0.0000167
0.02	0.0199987	0.0000013	0.0000667
0.03	0.0299955	0.0000045	0.0001500

92. The Taylor polynomial is a better approximation for θ close to 0; it is slightly larger than $\sin\theta$ when $\theta < 0$ and slightly smaller when $\theta > 0$.

θ	$\sin\theta$	$\theta - \frac{1}{6}\theta^3$	$\sin\theta - (\theta - \frac{1}{6}\theta^3)$
−0.3	−0.2955202	−0.2955000	−0.0000202
−0.2	−0.1986693	−0.1986667	−0.0000027
−0.1	−0.0998334	−0.0998333	−0.0000001
0	0	0	0
0.1	0.0998334	0.0998333	0.0000001
0.2	0.1986693	0.1986667	0.0000027
0.3	0.2955202	0.2955000	0.0000202

93. The Taylor polynomial is a better approximation for θ close to 0; it is slightly larger than $\cos\theta$ when $\theta \ne 0$.

θ	$\cos\theta$	$1 - \frac{1}{2}\theta^2 + \frac{1}{24}\theta^4$	$\cos\theta - (1 - \frac{1}{2}\theta^2 + \frac{1}{24}\theta^4)$
−0.3	0.9553365	0.9553375	−0.0000010
−0.2	0.9800666	0.9800667	−0.0000001
−0.1	0.9950042	0.9950042	−0.0000000
0	1	1	0
0.1	0.9950042	0.9950042	−0.0000000
0.2	0.9800666	0.9800667	−0.0000001
0.3	0.9553365	0.9553375	−0.0000010

QUICK REVIEW 6.4

1. In order: +, +, −, −
2. In order: +, −, −, +
3. In order: +, −, +, −
4. In order: +, −, +, −
5. In order: +, −, −, +
6. In order: +, +, −, −
7. $\frac{3\pi}{4}$
8. $\frac{3\pi}{2}$
9. $-\frac{5\pi}{6}$
10. $-\frac{\pi}{6}$
11. $\frac{5\pi}{2}$
12. $\frac{17\pi}{6}$
13. Starting with the graph of y_1, shift left 2 units.
14. Starting with the graph of y_1, vertically stretch by 3.
15. Starting with the graph of y_1, reflect through the y-axis.
16. Starting with the graph of y_1, vertically shrink by 0.5.
17. Starting with the graph of y_1, shift downward 2 units.
18. Starting with the graph of y_1, reflect through the y-axis (or the x-axis).

SECTION EXERCISES 6.4

1. Period 2π; window: $[-\pi, \pi] \times [-3, 3]$.

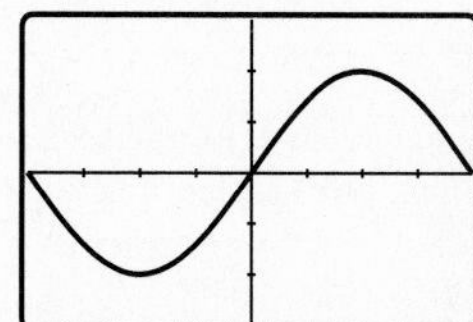

2. Period 2π; window: $[-\pi, \pi] \times [-3, 3]$.

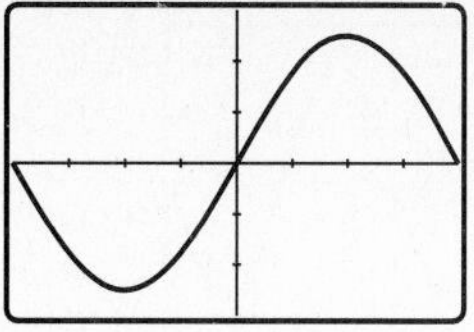

3. Period 2π; window: $[-\pi, \pi] \times [-3.1, 3.1]$.

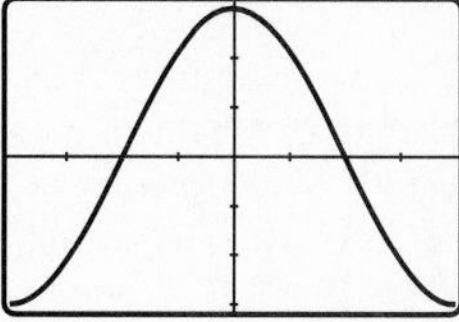

4. Period 2π; window: $[-\pi, \pi] \times [-3, 3]$.

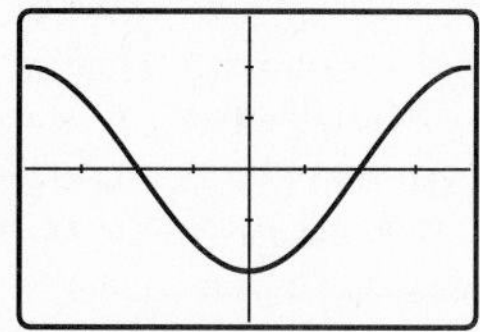

5. Period π; window: $[-1.5\pi, 1.5\pi] \times [-6, 6]$.

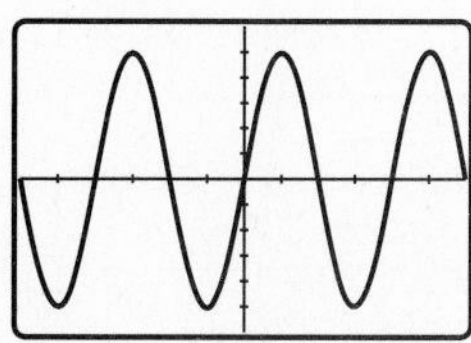

6. Period 4π; window: $[-6\pi, 6\pi] \times [-4, 4]$.

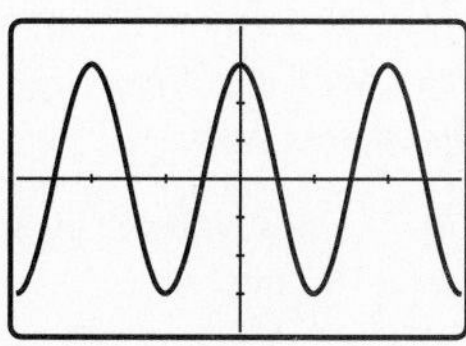

7. Period $\frac{2\pi}{3}$; window: $[-\pi, \pi] \times [-1, 1]$.

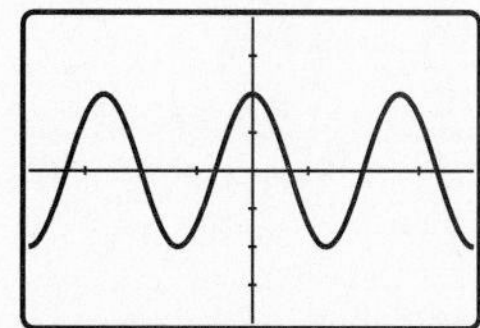

8. Period $\frac{\pi}{2}$; window: $[-0.75\pi, 0.75\pi] \times [-25, 25]$.

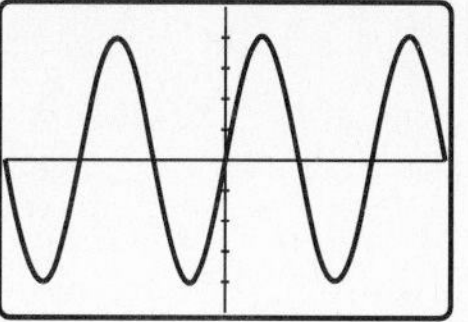

9. Period 8π; window: $[-12\pi, 12\pi] \times [-5, 5]$.

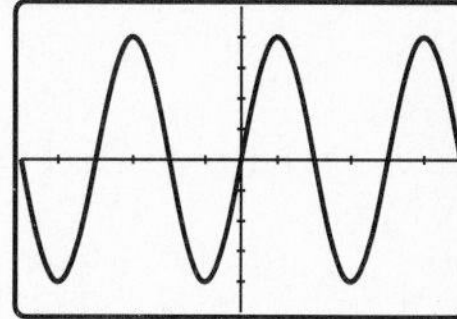

10. Period $\frac{2\pi}{5}$; window: $[-0.6\pi, 0.6\pi] \times [-10, 10]$.

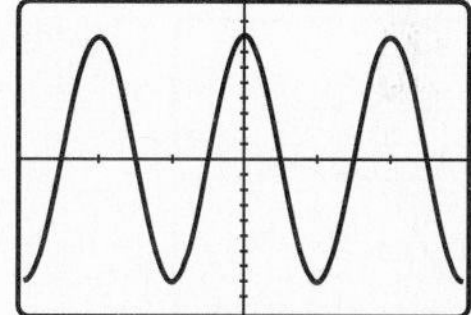

11. Period: π; amplitude: 1.5; window: $[-2\pi, 2\pi] \times [-2, 2]$.
12. Period: $\frac{2\pi}{3}$; amplitude: 2; window: $[-\frac{2\pi}{3}, \frac{2\pi}{3}] \times [-4, 4]$.
13. Period: π; amplitude: 3; window: $[-2\pi, 2\pi] \times [-4, 4]$.
14. Period: 4π; amplitude: 5; window: $[-4\pi, 4\pi] \times [-10, 10]$.
15. Period: 6; amplitude: 4; window: $[-3, 3] \times [-5, 5]$.
16. Period: 2; amplitude: 3; window: $[-4, 4] \times [-5, 5]$.
17. Maximum: 2 (at $-\frac{3\pi}{2}$ and $\frac{\pi}{2}$); minimum: -2 (at $-\frac{\pi}{2}$ and $\frac{3\pi}{2}$). Zeros: 0, $\pm\pi$, $\pm 2\pi$.
18. Maximum: 3 (at 0); minimum: -3 (at $\pm 2\pi$). Zeros: $\pm\pi$.
19. Maximum: 1 (at 0, $\pm\pi$, $\pm 2\pi$); minimum: -1 (at $\pm\frac{\pi}{2}$ and $\pm\frac{3\pi}{2}$). Zeros: $\pm\frac{\pi}{4}$, $\pm\frac{3\pi}{4}$, $\pm\frac{5\pi}{4}$, $\pm\frac{7\pi}{4}$.

20. Maximum: $\frac{1}{2}$ (at $-\frac{3\pi}{2}$ and $\frac{\pi}{2}$); minimum: $-\frac{1}{2}$ (at $-\frac{\pi}{2}$ and $\frac{3\pi}{2}$). Zeros: 0, $\pm\pi$, $\pm 2\pi$.

21. $x \approx 0.85$

22. $x \approx 0.84$

23. $x \approx 5.12$

24. $x \approx 2.50$

25. $x \approx 4.07$

26. $x \approx 1.77$

27. Starting from $y = \sin x$, horizontally shrink by $\frac{1}{3}$ and vertically shrink by 0.5; one possible window: $[-\frac{2\pi}{3}, \frac{2\pi}{3}] \times [-\frac{3}{4}, \frac{3}{4}]$.

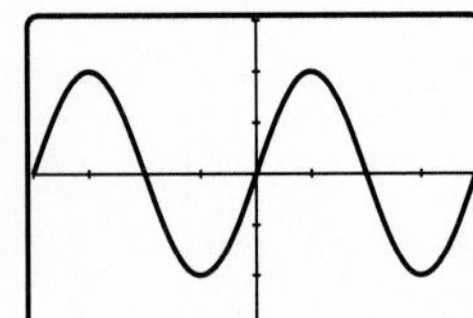

28. Starting from $y = \cos x$, horizontally shrink by $\frac{1}{4}$ and vertically stretch by 1.5; one possible window: $[-\frac{\pi}{2}, \frac{\pi}{2}] \times [-2, 2]$.

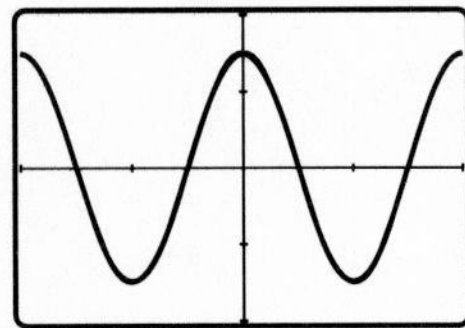

29. Starting from $y = \cos x$, horizontally stretch by 3, vertically shrink by $\frac{2}{3}$, and reflect through the x-axis; one possible window: $[-6\pi, 6\pi] \times [-1, 1]$.

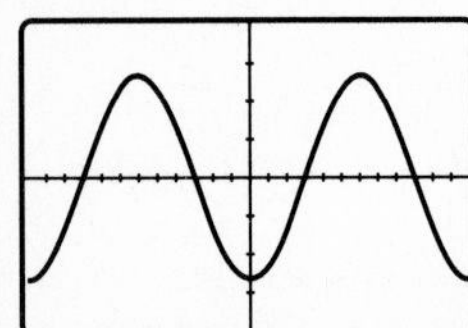

30. Starting from $y = \sin x$, horizontally stretch by 5 and vertically shrink by $\frac{3}{4}$; one possible window: $[-10\pi, 10\pi] \times [-1, 1]$.

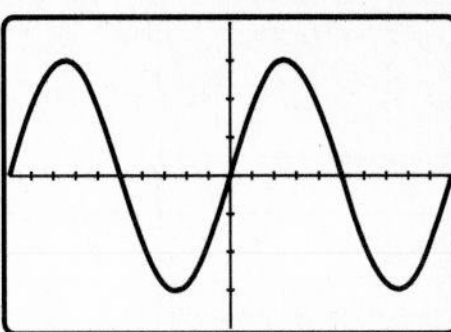

31. Starting from $y = \cos x$, horizontally shrink by $\frac{3}{2\pi}$ and vertically stretch by 3; one possible window: $[-3, 3] \times [-3.5, 3.5]$.

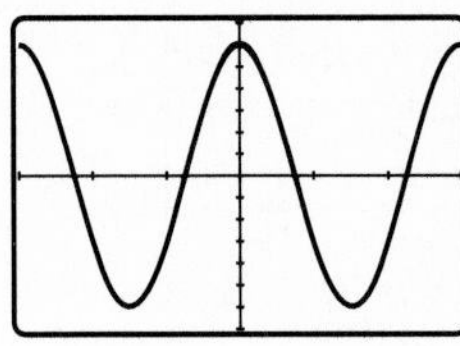

32. Starting from $y = \sin x$, horizontally stretch by $\frac{4}{\pi}$, vertically stretch by 2, and reflect through the x-axis; one possible window: $[-8, 8] \times [-3, 3]$.

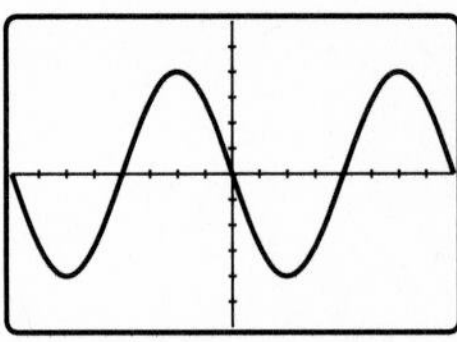

33. Starting with f, vertically stretch by 2.

34. Starting with f, vertically stretch by $\frac{5}{3}$.

35. Starting with f, shift left $\frac{\pi}{3}$ units.

36. Starting with f, shift right $\frac{\pi}{12}$ units and vertically shrink by $\frac{1}{2}$.

37. Starting with f, vertically shrink by $\frac{1}{3}$ and then shift down $\frac{4}{3}$ units.

38. Starting with f, vertically stretch by $\frac{5}{3}$ and shift upward $\frac{23}{3}$ units.

39. Starting with f, horizontally shrink by $\frac{1}{2}$.

40. Starting with f, horizontally stretch by 2 and vertically shrink by $\frac{2}{3}$.

41. (a) and (b)

42. (a) and (b)

43. (a) and (b)

44. (a) and (c)

45. Amplitude: 2; period: 2π; phase shift: $\frac{\pi}{4}$; vertical translation: 1

46. Amplitude: 3.5; period: π; phase shift: $\frac{\pi}{4}$; vertical translation: -1

47. Amplitude: 5; period: $\frac{2\pi}{3}$; phase shift: $\frac{\pi}{18}$; vertical translation: $\frac{1}{2}$

48. Amplitude: 3; period: 2π; phase shift: -3; vertical translation: -2

49. Amplitude: 2; period: 1; phase shift: 0; vertical translation: 1

50. Amplitude: 4; period: $\frac{2}{3}$; phase shift: 0; vertical translation: -2

51. Amplitude: $\frac{7}{3}$; period: 4π; phase shift: -5; vertical translation: -1

52. Amplitude: $\frac{2}{3}$; period: 8π; phase shift: 3; vertical translation: 1

53. $y = 2 \sin 2x \quad (a = 2, b = 2, h = 0, k = 0)$

54. $y = 3 \sin [2(x + 0.5)] \quad (a = 3, b = 2, h = -0.5, k = 0)$

55. **(a)** $\sin x$ is in $[-1, 1]$, so since $\sin [b(x - h)]$ only involves horizontal changes, it also is between -1 and $+1$. Then $a \sin [b(x - h)]$ varies from a minimum of $-|a|$ to a maximum of $+|a|$. The largest and smallest values of f are thus obtained by adding k (the vertical shift). **(b)** Using the expressions from (a), we find that $\max f - \min f = 2|a|$ and know that the amplitude is $|a|$.

56. The cosine graph is related to the sine graph by a shift $\frac{\pi}{2}$ units left; thus the cosine function has zeros $\frac{\pi}{2}$ units to the left of the zeros of the sine function's zeros, and so on.

57. **(a)** There are three points of intersection. **(b)** The coordinates are approximately (1.40, 0.34), (5.78, 1.75), and (6.62, 1.89).

58. **(a)** There are two points of intersection in that interval. **(b)** The coordinates are (0, 1) and $(2\pi, 1.3^{-2\pi})$, or approximately (6.28, 0.19).

59. $a = 4$ and $b = \frac{2\pi}{3.5}$, or $\frac{4\pi}{7}$

60. About 15.90 sec.

61. $L \approx 972{,}000$ ft

62. **(a)** 1 second. **(b)** Each peak corresponds to a heartbeat (60 per minute). **(c)** $[0, 10] \times [80, 160]$.

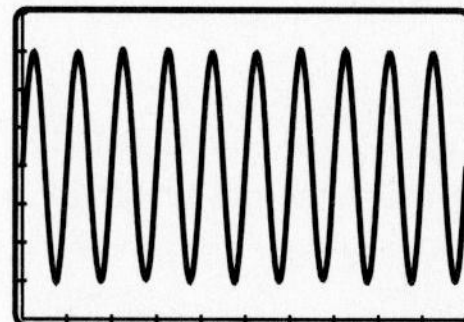

63. **(a)** 1:00 A.M. **(b)** At 4:00 A.M.: about 8.90 ft; at 9:00 P.M.: about 10.52 ft. **(c)** 4:06 A.M.

64. If we say that $A = 1$ corresponds to a full moon, $A = 0.5 + 0.5\cos(\frac{2\pi}{29.5}t)$. Window: $[0, 29.5] \times [-0.1, 1.1]$.

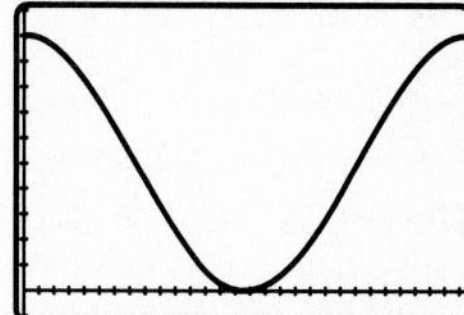

65.

(a) Window: $[-3.5, 3.5] \times [-1.2, 1.2]$. **(b)** $\cos x \approx 0.0246x^4 - (1.4396 \times 10^{-14})x^3 - 0.4410x^2 - (1.0058 \times 10^{-13})x + 0.9703 \approx 0.9703 - 0.4410x^2 + 0.0246x^4$. **(c)** Taylor polynomial $1 - \frac{1}{2}x^2 + \frac{1}{24}x^4$; their graphs are fairly similar for x near zero.

66.

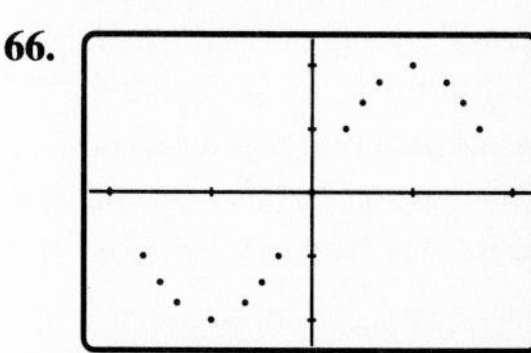

(a) Window: $[-3.5, 3.5] \times [-1.2, 1.2]$. **(b)** $\sin x \approx -0.0872x^3 - (3.5999 \times 10^{-15})x^2 + 0.8263x + (1.8845 \times 10^{-14}) \approx 0.8263x - 0.0872x^3$. **(c)** Taylor polynomial $x - \frac{1}{6}x^3$; their graphs are fairly similar for x near zero.

67. **(a)** $p = \frac{1}{262}$ sec. **(b)** $f = 262$ cycles per sec. **(c)** Window: $[0, 0.025] \times [-2, 2]$.

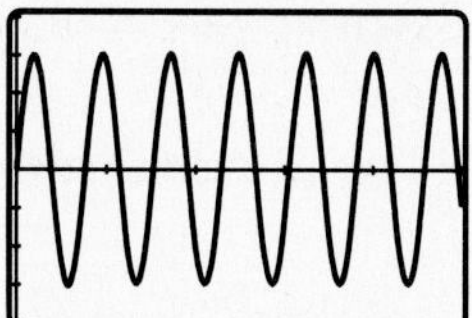

68. Since the cursor moves at a constant rate, its distance from the center must be made up of linear pieces as shown (the slope of the line is the rate of motion).

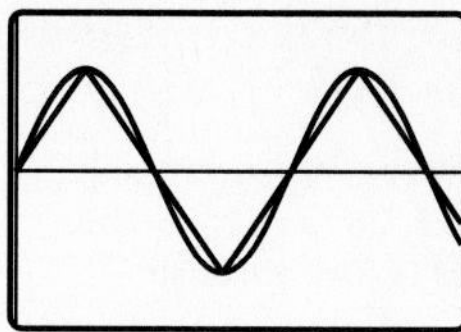

69. $B = (0, 3)$ and $C = (\frac{3\pi}{4}, 0)$

70. $B = (\frac{3\pi}{4}, 4.5)$ and $C = (\frac{9\pi}{4}, 0)$

71. $B = (\frac{\pi}{4}, 2)$ and $C = (\frac{3\pi}{4}, 0)$

72. $A = (\frac{\pi}{2}, 0)$, $B = (\frac{3\pi}{4}, 3)$, and $C = (\frac{3\pi}{2}, 0)$

73. **(a)** $a \sin[-B(x-h)] + k = -a\sin[B(x-h)] + k$, since $\sin(-\theta) = -\sin\theta$; any expression with a negative value of b can be rewritten as an expression of the same general form but with a positive coefficient. **(b)** Since $\cos\theta = \sin(\theta + \frac{\pi}{2})$, $\cos[b(x-h)] = \sin[b(x-h) + \frac{\pi}{2}] = \sin[b(x - h + \frac{\pi}{2b})]$; multiplying by a and adding k gives the desired result; thus $H = h - \frac{\pi}{2b}$. **(c)** Adding π to the angle corresponds to traveling halfway around the circle; this has the effect of negating both the x- and y-coordinates.

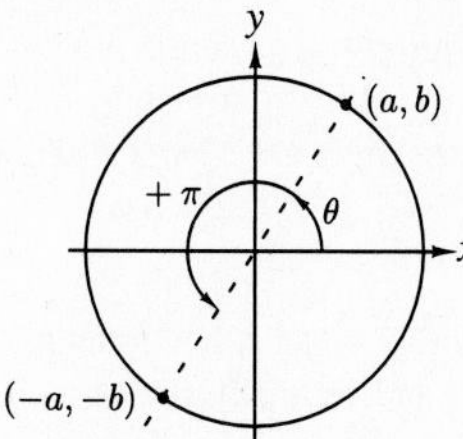

(d) Take $H = h - \frac{\pi}{b}$: $-a\sin[b(x-h)] + k = a\sin[b(x-h) + \pi] + k = a\sin[b(x - h + \frac{\pi}{b})] + k$. **(e)** From (a), we can assume $b > 0$; from (d), we can similarly choose h so that $a > 0$.

QUICK REVIEW 6.5

1. π

2. $\frac{2\pi}{3}$

3. 6π

4. 4π

5. Zero: 3; asymptote: $x = -4$

6. Zero: -5; asymptote: $x = 1$

7. Zero: -1; asymptotes: $x = 2$ and $x = -2$

8. Zero: -2; asymptotes: $x = 0$ and $x = 3$

9. $\sin\theta = \frac{3}{5}$, $\cos\theta = \frac{4}{5}$, $\tan\theta = \frac{3}{4}$; $\csc\theta = \frac{5}{3}$, $\sec\theta = \frac{5}{4}$, $\cot\theta = \frac{4}{3}$

10. $\sin\theta = \frac{5}{13}$, $\cos\theta = \frac{12}{13}$, $\tan\theta = \frac{5}{12}$; $\csc\theta = \frac{13}{5}$, $\sec\theta = \frac{13}{12}$, $\cot\theta = \frac{12}{5}$

11. Even

12. Neither

13. Odd

14. Even

15. Neither

16. Odd

SECTION EXERCISES 6.5

1. $y_1 = 2\csc x$ and $y_2 = \csc x$

2. $y_1 = 5\tan x$ and $y_2 = 0.5\tan x$

3. $y_1 = 3\csc 2x$ and $y_2 = \csc x$

4. $y_1 = \cot(x - 0.5) + 3$ and $y_2 = \cot x$

5. Possible window: $[-\frac{\pi}{2}, \frac{\pi}{2}] \times [-6, 6]$ (Most calculators will not properly show the vertical asymptotes.)

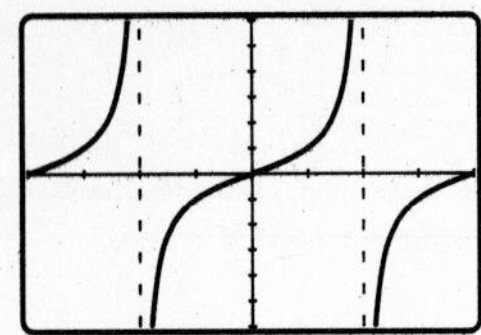

6. Possible window: $[-\frac{\pi}{3}, \frac{\pi}{3}] \times [-6, 6]$

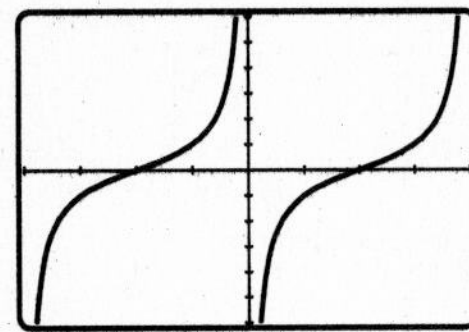

7. Possible window: $[-\frac{2\pi}{3}, \frac{2\pi}{3}] \times [-6, 6]$

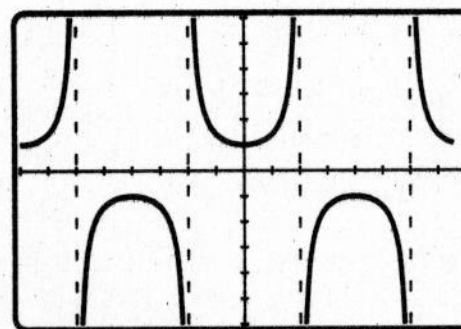

8. Possible window: $[-\pi, \pi] \times [-6, 6]$

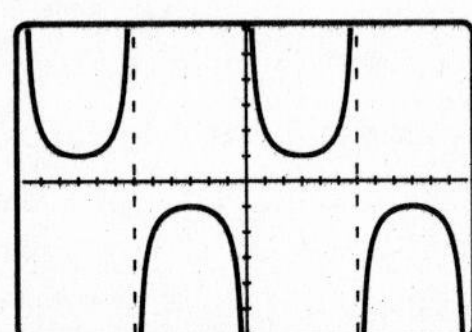

9. Possible window: $[-\frac{\pi}{2}, \frac{\pi}{2}] \times [-6, 6]$

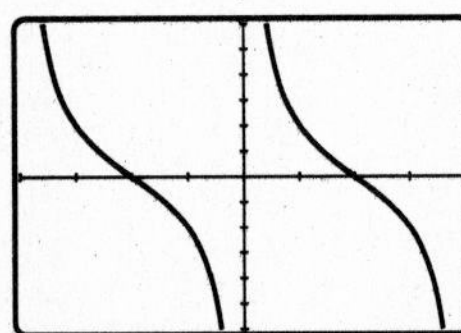

10. Possible window: $[-2\pi, 2\pi] \times [-15, 15]$

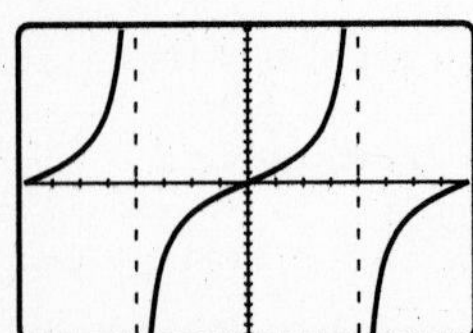

11. Possible window: $[-4\pi, 4\pi] \times [-6, 6]$

12. Possible window: $[-\frac{\pi}{2}, \frac{\pi}{2}] \times [-15, 15]$

13. The graph in part (a); Xmin $= -\pi$ and Xmax $= \pi$

14. The graph in part (d); Xmin $= -\pi$ and Xmax $= \pi$

15. The graph in part (c); Xmin $= -\pi$ and Xmax $= \pi$

16. The graph in part (b); Xmin $= -\pi$ and Xmax $= \pi$

17. Starting with $y = \tan x$, vertically stretch by 3; π; $x \neq (2n + 1)\frac{\pi}{2}$; $(-\infty, \infty)$; $n\pi$; $x = (2n + 1)\frac{\pi}{2}$.

18. Starting with $y = \tan x$, reflect across the x-axis; π; $x \neq (2n + 1)\frac{\pi}{2}$; $(-\infty, \infty)$; $n\pi$; $x = (2n + 1)\frac{\pi}{2}$.

19. Starting with $y = \sec x$, vertically shrink by 0.5; 2π; $x \neq (2n + 1)\frac{\pi}{2}$; $(-\infty, -0.5] \cup [0.5, \infty)$; none; $x = (2n + 1)\frac{\pi}{2}$.

20. Starting with $y = \sec x$, reflect across the y-axis; 2π; $x \neq (2n + 1)\frac{\pi}{2}$; $(-\infty, -1] \cup [1, \infty)$; none; $x = (2n + 1)\frac{\pi}{2}$.

21. Starting with $y = \csc x$, vertically stretch by 3; 2π; $x \neq n\pi$; $(-\infty, -3] \cup [3, \infty)$; none; $x = n\pi$.

22. Starting with $y = \tan x$, vertically stretch by 2; π; $(2n + 1)\frac{\pi}{2}$; $(-\infty, \infty)$; $n\pi$; $x = (2n + 1)\frac{\pi}{2}$.

23. Starting with $y = \cot x$, horizontally stretch by 2, vertically stretch by 3, and reflect across the x-axis; 2π; $x \neq 2n\pi$; $(-\infty, \infty)$; $x = (2n + 1)\pi$; $x = 2n\pi$.

24. Starting with $y = \cot x$, horizontally stretch by 2 and vertically stretch by 2; 2π; $x \neq 2n\pi$; $(-\infty, \infty)$; $x = (2n + 1)\pi$; $x = 2n\pi$.

25. Starting with $y = \csc x$, vertically stretch by 2; 2π; $x \neq n\pi$; $(-\infty, -2] \cup [2, \infty)$; none; $x = n\pi$.

26. Starting with $y = \sec x$, horizontally stretch by 2, vertically stretch by 2, and reflect across the x-axis; 2π; $x \neq (2n + 1)\pi$; $(-\infty, -2] \cup [2, \infty)$; none; $x = (2n + 1)\pi$.

27. Starting with $y = \tan x$, horizontally shrink by $\frac{1}{3}$ and vertically stretch by 2; $\frac{\pi}{3}$; $x \neq (2n + 1)\frac{\pi}{6}$; $(-\infty, \infty)$; $x = n\frac{\pi}{3}$; $x = (2n + 1)\frac{\pi}{6}$.

28. Starting with $y = \sec x$, horizontally stretch by 2 and reflect across the y-axis; 4π; $x \neq (2n + 1)\pi$; $(-\infty, -1] \cup [1, \infty)$; none; $x = (2n + 1)\pi$.

29. Starting with $y = \tan x$, horizontally shrink by $\frac{2}{\pi}$ and reflect across the x-axis; 2; $x \neq 2n + 1$; $(-\infty, \infty)$; $x = 2n$; $x = 2n + 1$.

30. Starting with $y = \tan x$, horizontally shrink by $\frac{1}{\pi}$ and vertically stretch by 2; 1; $x \neq \frac{2n+1}{2}$; $(-\infty, \infty)$; $x = n$; $x = \frac{2n+1}{2}$.

31. Starting with $y = \sec x$, horizontally shrink by $\frac{1}{2}$ and vertically stretch by 3; π; $x \neq (2n + 1)\frac{\pi}{4}$; $(-\infty, -3] \cup [3, \infty)$; none; $x = (2n + 1)\frac{\pi}{4}$.

32. Starting with $y = \csc x$, horizontally stretch by 3 and vertically stretch by 4; 6π; $x + 3n\pi$; $(-\infty, -4] \cup [4, \infty)$; none; $x = 3n\pi$.

33. **(a)** One explanation: If O is the origin, the right triangles with hypotenuses $\overline{OP_1}$ and $\overline{OP_2}$, and one leg (each) on the x-axis, are congruent, so the legs have the same lengths; these lengths give the magnitudes of the coordinates of P_1 and P_2; therefore these coordinates differ only in sign.

(b) $\tan t = \dfrac{\sin t}{\cos t} = \dfrac{b}{a}$.

(c) $\tan (t - \pi) = \dfrac{\sin (t - \pi)}{\cos (t - \pi)} = \dfrac{-b}{-a} = \dfrac{b}{a}$.

(d) From any point P_1 on the unit circle, the arguments of (a)–(c) indicate that coordinates of the point P_2 halfway around the circle are the opposites of those of P_1 and that the ratio of those coordinates is the tangent of the angle; this ratio is thus the same at P_1 and P_2. **(e)** The tangent function repeats every π units; therefore so does its reciprocal, the cotangent (see also Exercise 37).

34. The terminal side passes through (0, 0) and ($\cos x$, $\sin x$); the slope is therefore $m = \dfrac{\sin x - 0}{\cos x - 0} = \dfrac{\sin x}{\cos x} = \tan x$.

35. Use the definitions from Section 6.3: $\cos \theta = \frac{x}{r}$ and $\sin \theta = \frac{y}{r}$, and, since (x, y) is on the circle, $x^2 + y^2 = r^2$. Divide this equation by r^2 to get the equation

$\cos^2\theta + \sin^2\theta = 1$. Similarly, dividing the circle equation by x^2 gives $1 + (\frac{y}{x})^2 = (\frac{r}{x})^2$, which is equivalent to $1 + (\tan\theta)^2 = (\sec\theta)^2$. Finally, dividing by y^2 gives $(\frac{x}{y})^2 + 1 = (\frac{1}{y})^2$, so $(\cot\theta)^2 + 1 = (\csc\theta)^2$.

36. Use the definitions from Section 6.3: $\sin\theta = \frac{y}{r}$, $\csc\theta = \frac{r}{y} = \left(\frac{y}{r}\right)^{-1} = \frac{1}{\sin\theta}$, and $\sin\theta = \left(\frac{r}{y}\right)^{-1} = \frac{1}{\csc\theta}$. Likewise, $\cos\theta = \frac{x}{r}$, $\sec\theta = \frac{r}{x} = \left(\frac{x}{r}\right)^{-1} = \frac{1}{\cos\theta}$, and $\cos\theta = \left(\frac{r}{x}\right)^{-1} = \frac{1}{\sec\theta}$. Finally, $\tan\theta = \frac{y}{x} = \left(\frac{x}{y}\right)^{-1} = \frac{1}{\cot\theta}$ and $\cot\theta = \frac{x}{y} = \left(\frac{y}{x}\right)^{-1} = \frac{1}{\tan\theta}$. For the quotient identities, note that $\tan\theta = \frac{y}{x} = \frac{y/r}{y/x} = \frac{\sin\theta}{\cos\theta}$ and $\cot\theta = \frac{x}{y} = \frac{x/r}{y/r} = \frac{\cos\theta}{\sin\theta}$.

37. For any x, $\left(\frac{1}{f}\right)(x + p) = \frac{1}{f(x + p)} = \frac{1}{f(x)} = \left(\frac{1}{f}\right)(x)$. This result is not true for any smaller value of p, since this is the smallest value that works for f.

38. **(a)** and **(b)** Start at (1, 0), measure t units counterclockwise around the circle, and then measure π units farther; this second point is halfway around the circle from the first; the second x- and y-coordinates are the opposites of the original x- and y-coordinates, and these coordinates correspond to the cosine and sine of $t + \pi$. **(c)** $\tan(t + \pi) = \frac{\sin(t + \pi)}{\cos(t + \pi)} = \frac{-\sin t}{-\cos t} = \frac{\sin t}{\cos t} = \tan t$; in order to determine that the period of $\tan t$ is π, we would need to show that no $p < \pi$ satisfies $\tan(t + p) = \tan t$ for all t.

39. $x \approx 0.92$

40. $x \approx 1.14$

41. $x \approx 5.25$

42. $x \approx 3.87$

43. $x \approx 0.52$ or $x \approx 2.62$

44. $x \approx 0.29$ or $x \approx 3.43$

45. **(a)** $d = 350 \sec x$.
(b) $d = 16{,}831.1083....$ ft

46. **(a)** $x = 800 \cot y$.
(b) $x = 5{,}051.0012....$ ft **(c)** 9°.

47. $x \approx -0.90$ or $x \approx 0.90$

48. $x \approx 0.67$ or $x \approx 2.48$

49. $x \approx \pm 1.11$ or $x \approx \pm 2.03$

50. $x \approx 1.08$ or $x \approx 2.06$

51. On the interval $[-\pi, \pi]$, $f > g$ on about $(-0.44, 0) \cup (0.44, \pi)$

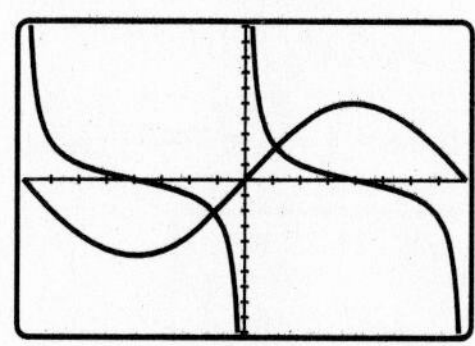

52. On the interval $[-\pi, \pi]$, $f > g$ on about $(-\pi, -2.24) \cup (-\frac{\pi}{2}, 0) \cup (\frac{\pi}{2}, 2.24)$

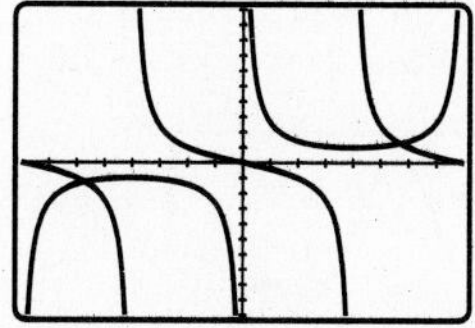

53. cot x is not defined at 0; the definition of "increasing on (a, b)" requires that the function be defined everywhere in (a, b); also, choosing $a = \frac{-\pi}{4}$ and $b = \frac{\pi}{4}$, we have $a < b$ but $f(a) = 1 > f(b) = -1$.

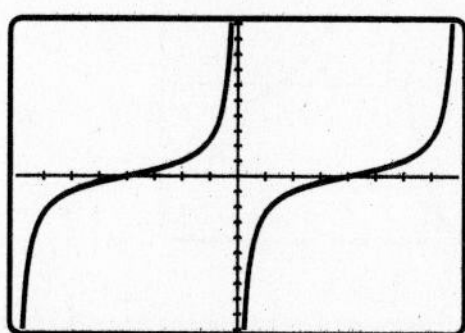

54. They look similar on this window, but they are noticeably different at the edges (near 0 and π); also, if f were equal to g, it would follow that $\frac{1}{f} = -\cos x = \frac{1}{g} = x - \frac{\pi}{2}$ on this interval, which we know to be false.

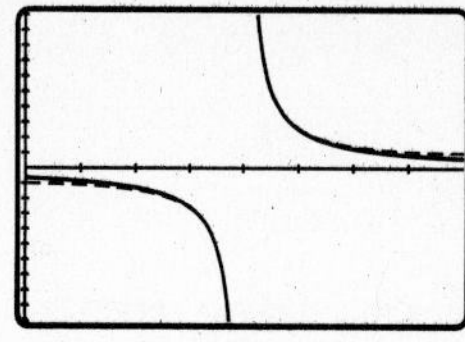

55. $d = 30 \sec x$; shown is the window $[-\frac{\pi}{2}, \frac{\pi}{2}] \times [0, 100]$

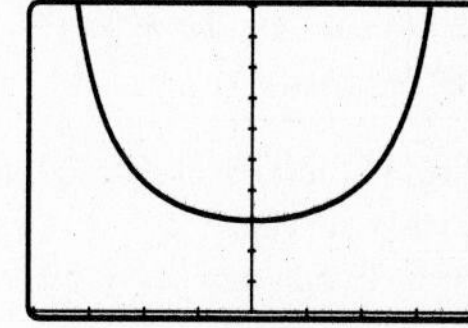

56. **(a)** For any acute angle θ, $\cos\theta = \sin(\frac{\pi}{2} - \theta)$, the sine of the complement of θ, which can be seen from the right-triangle definition of sine and cosine: If one of the acute angles is θ, then the other acute angle is $\frac{\pi}{2} - \theta$, since all three angles in a triangle must add to π; the side opposite the angle θ is the side adjacent to the other acute angle.
(b) For $\triangle ODE \sim \triangle OBD$, $DE = \frac{DE}{OD} = \frac{BD}{OB} = \frac{\sin t}{\cos t} = \tan t$ and $OE = \frac{OE}{OD} = \frac{OD}{OB} = \frac{1}{\cos t} = \sec t$. Also, since $\triangle ODC \sim \triangle OAD$, $DC = \frac{DC}{OD} = \frac{AD}{OA} = \cot t$ and $OC = \frac{OC}{OD} = \frac{OD}{OA} = \csc t$. **(c)** $\overline{DE}$ is a tangent segment (part of the tangent line); $\overline{OE}$ is a secant segment (part of a secant line, which crosses the circle at two points). The names "cotangent" and "cosecant" arise in the same way as "cosine": They are the tangent and secant (respectively) of the complement. That is, just as $\overline{DE}$ and $\overline{OE}$ go with $\angle BOD$ (which has measure t), $\overline{CD}$ and $\overline{OC}$ go with $\angle AOD$ (the complement of $\angle BOD$, with the measure $\frac{\pi}{2} - t$).

57. $0.058\text{ N/m} = \frac{1}{2}(1.5\text{ m})(1050\text{ kg/m}^3)\cdot(9.8\text{ m/sec}^2)(4.7 \times 10^{-6}\text{ m})\sec\phi$, so $\sec\phi = 1.5990\ldots$, and $\phi = 0.8951\ldots$ radians $\approx 51.29°$

QUICK REVIEW 6.6

1. Domain: $(-\infty, \infty)$; range: $[-3, 3]$

2. Domain: $(-\infty, \infty)$; range: $[-2, 2]$

3. Domain: $[1, \infty)$; range: $[0, \infty)$

4. Domain: $[0, \infty)$; range: $[0, \infty)$

5. Domain: $(-\infty, \infty)$; range: $[-2, \infty)$

6. Domain: $(-\infty, \infty)$; range: $[1, \infty)$

7. Since the sine function has period 2π, $\sin(x + 4\pi) = \sin(x + 2\pi + 2\pi) = \sin(x + 2\pi) = \sin x$.

8. Since the cosine function has period 2π, $\cos(x + 4\pi) = \cos(x + 2\pi + 2\pi) = \cos(x + 2\pi) = \cos x$.

9. Since the tangent function has period π, $\tan(2x + 3\pi) = \tan(2x + \pi + \pi + \pi) = \tan(2x + \pi + \pi) = \tan(2x + \pi) = \tan 2x$.

10. Since the sine function has period 2π, $\sin(4x + 6\pi) = \sin(4x + 2\pi + 2\pi + 2\pi) = \sin(4x + 2\pi + 2\pi) = \sin(4x + 2\pi) = \sin 4x$.

11. Take the portion of the graph of y_2 that lies below the x-axis (that is, $-2 < x < 2$) and reflect it across the x-axis; the result is the graph of y_1.

12. Take the portion of the graph of y_2 that lies below the x-axis (that is, $x < 1.5$) and reflect it across the x-axis; the result is the graph of y_1.

13. As $x \to -\infty$, $f(x) \to \infty$; as $x \to \infty$, $f(x) \to 0$

14. As $x \to -\infty$, $g(x) \to -\infty$; as $x \to \infty$, $g(x) \to 0$

15. As $x \to -\infty$, $f(x) \to -\infty$; as $x \to \infty$, $f(x) \to 0$

16. As $x \to -\infty$, $g(x) \to \infty$; as $x \to \infty$, $g(x) \to 0$

17. $f \circ g(x) = (\sqrt{x})^2 - 4 = x - 4$, domain: $[0, \infty)$; $g \circ f(x) = \sqrt{x^2 - 4}$, domain: $(-\infty, -2] \cup [2, \infty)$

18. $f \circ g(x) = (\cos x)^2 = \cos^2 x$, domain: $(-\infty, \infty)$; $g \circ f(x) = \cos(x^2)$, domain: $(-\infty, \infty)$

SECTION EXERCISES 6.6

1. Yes

2. Yes

3. No

4. Yes

5. No

6. No

7. $A \approx 2.24$, $b = 1$, and $h \approx -1.11$, so $f(x) \approx 2.24 \sin(x + 1.11)$

8. $A \approx 3.16$, $b = 2$, and $h \approx 0.16$, so $f(x) \approx 3.16 \sin[2(x - 0.16)]$

9. $A \approx 3.61$, $b = 2$, and $h \approx 0.49$, so $f(x) \approx 3.61 \sin[2(x - 0.49)]$

10. $A \approx 2.24$, $b = 3$, and $h \approx -0.15$, so $f(x) \approx 2.24 \sin[3(x + 0.15)]$

11. $A \approx 2.24$, $b = \pi$, and $h \approx 0.35$, so $f(x) \approx 2.24 \sin[\pi(x - 0.35)]$

12. $A \approx 3.16$, $b = 2\pi$, and $h \approx -0.05$, so $f(x) \approx 3.16 \sin[2\pi(x + 0.05)]$

13. Possible window: $[-\pi, \pi] \times [-3, 3]$

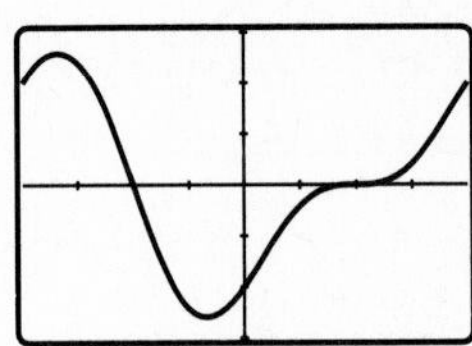

14. Possible window: $[-\pi, \pi] \times [-3.5, 3.5]$

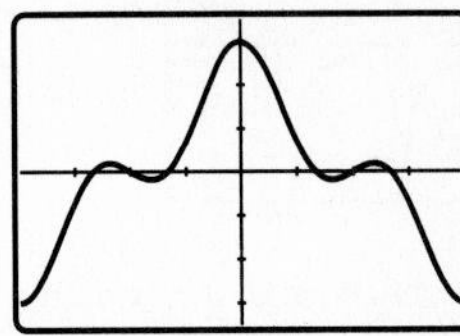

15. Possible window: $[-\pi, \pi] \times [-3, 3]$

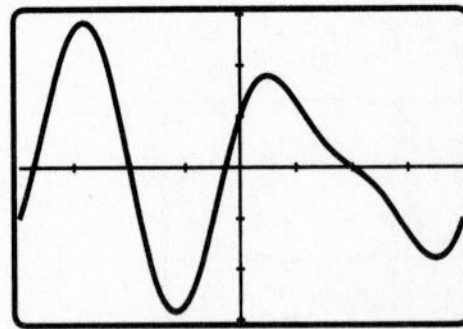

16. Possible window: $[-\pi, \pi] \times [-5, 5]$

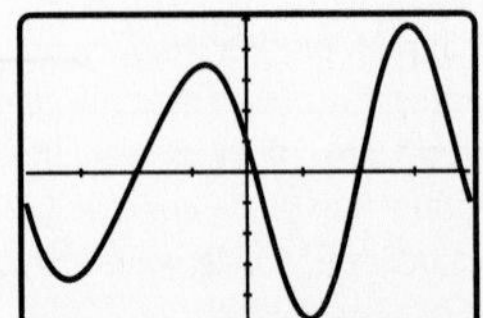

17. Possible window: $[-\pi, \pi] \times [-5, 5]$

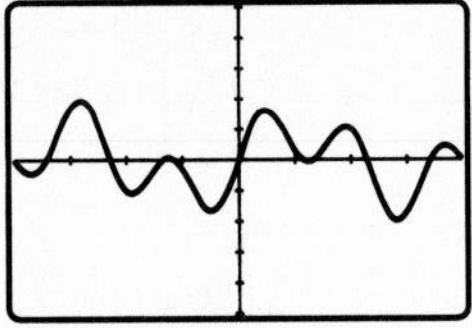

18. Possible window: $[-\pi, \pi] \times [-5, 5]$

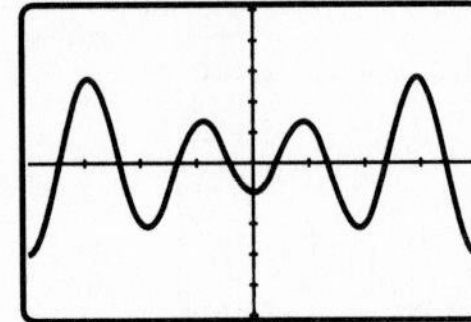

19. (a)

20. (d)

21. (c)

22. (b)

23. Possible window: $[-\pi, \pi] \times [-1.2, 1.2]$. Period is π because the cosine function has the property that $\cos(x + \pi) = -\cos x$ so that $[\cos(x + \pi)]^2 = (-\cos x)^2 = \cos^2 x$.

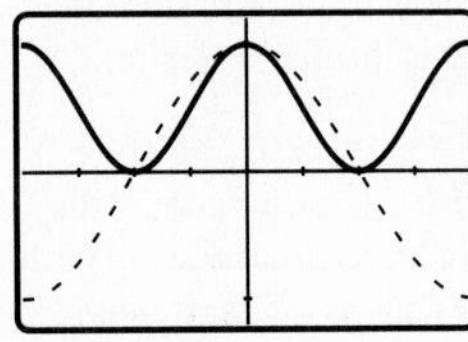

24. Possible window: $[-2\pi, 2\pi] \times [-1.2, 1.2]$; period 2π, the same as the cosine function: $[\cos(x + 2\pi)]^3 = [\cos x]^3 = \cos^3 x$.

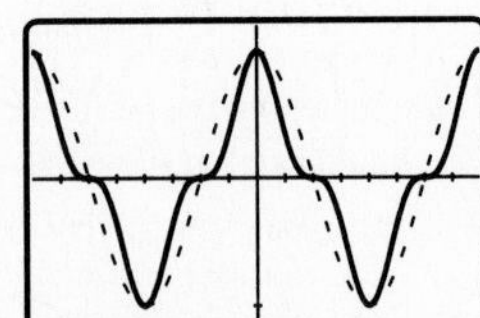

25. Possible window: $[-\pi, \pi] \times [-1.2, 1.2]$. Period is π because the cosine function has the property that $\cos(x + \pi) = -\cos x$ and this function is the same as $|\cos x|$.

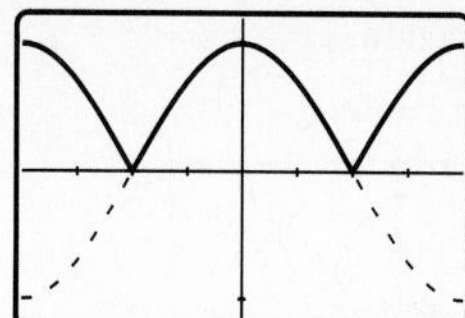

26. Possible window: $[-\pi, \pi] \times [-1.2, 1.2]$. Period is π because the cosine function has the property that $\cos(x + \pi) = -\cos x$ so that $|[\cos(x + \pi)]^3| = |(-\cos x)^3| = |-\cos^3 x| = |\cos^3 x|$.

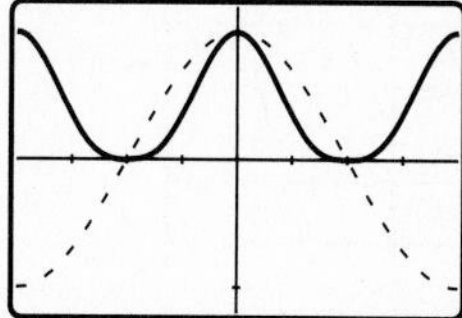

27. Domain: $(-\infty, \infty)$; range: $[0, 1]$; possible window: $[-2\pi, 2\pi] \times [-0.25, 1.25]$

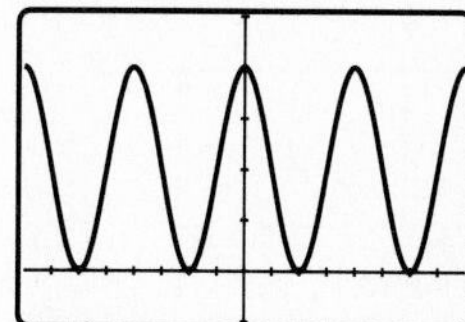

28. Domain: $(-\infty, \infty)$; range: $[0, 1]$; possible window: $[-2\pi, 2\pi] \times [-0.25, 1.25]$

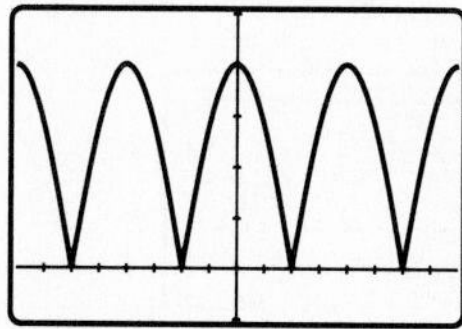

29. Domain: all $x \neq \pm n\pi$, n an integer; range: $[0, \infty)$; possible window: $[-2\pi, 2\pi] \times [-0.5, 4]$

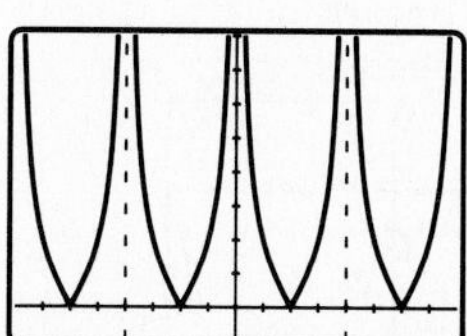

30. Domain: $(-\infty, \infty)$; range: $[-1, 1]$; possible window: $[-4\pi, 4\pi] \times [-1.2, 1.2]$

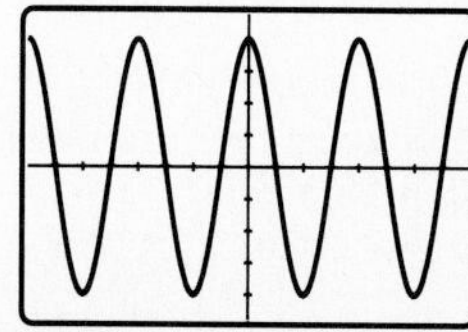

31. Domain: all $x \neq \frac{\pi}{2} \pm n\pi$, n an integer; range: $(-\infty, 0]$; possible window: $[-2\pi, 2\pi] \times [-10, 0.2]$

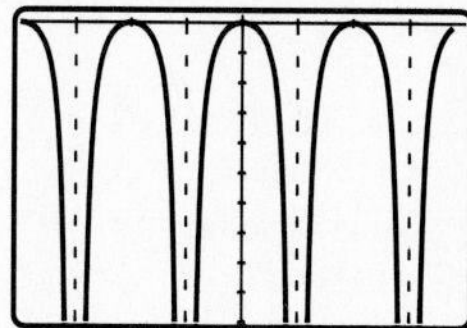

32. Domain: $(-\infty, \infty)$; range: $[-1, 0]$; possible window: $[-2\pi, 2\pi] \times [-1.25, 0.25]$

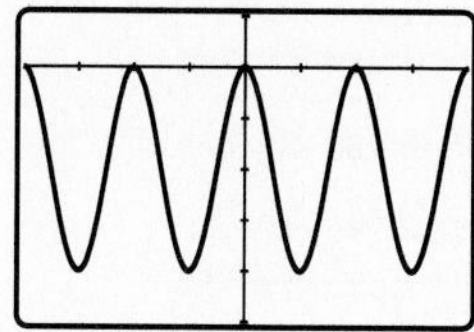

33. $2x - 1 \leq y \leq 2x + 1$; possible window: $[-10, 10] \times [-20, 20]$

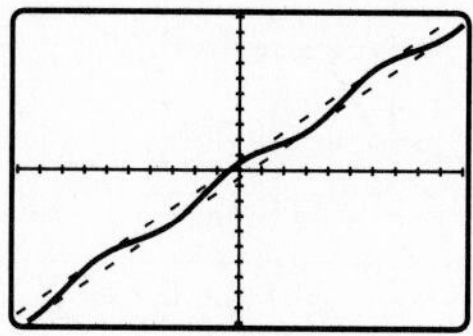

34. $-0.5x \leq y \leq 2 - 0.5x$; possible window: $[-10, 10] \times [-10, 10]$

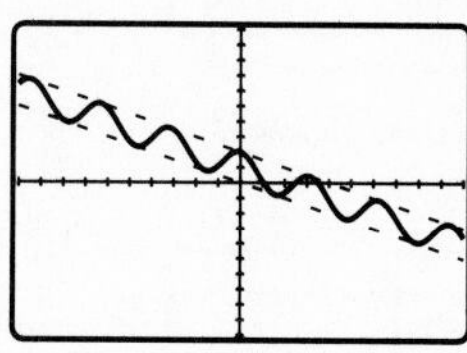

35. $1 - 0.3x \leq y \leq 3 - 0.3x$; possible window: $[-10, 10] \times [-4, 8]$

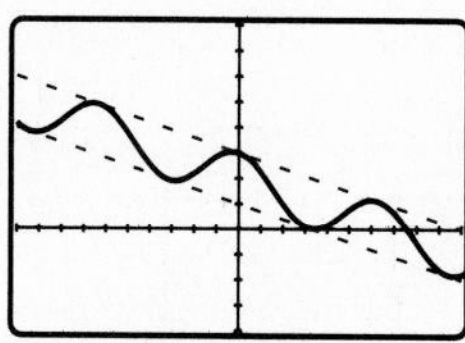

36. $x \leq y \leq x + 2$; possible window: $[-10, 10] \times [-10, 10]$

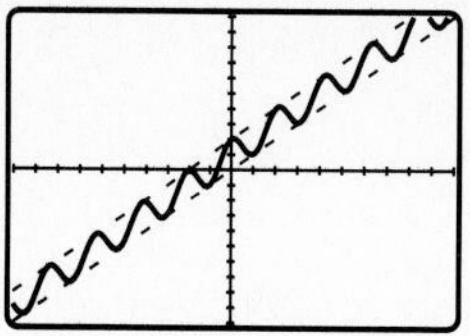

37. The graph of $f(x)$ lies between $y = x$ and $y = -x$ (in the vertical direction); possible window: $[-4\pi, 4\pi] \times [-4\pi, 4\pi]$.

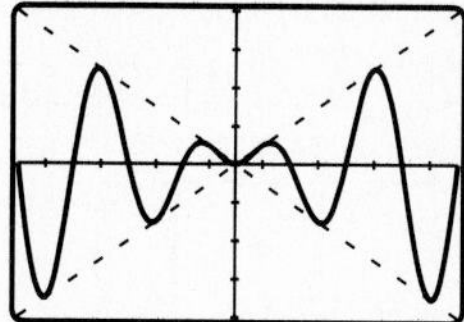

38. The graph of $f(x)$ lies (vertically) between $y = -x$ and the x-axis; possible window: $[-4\pi, 4\pi] \times [-4\pi, 4\pi]$.

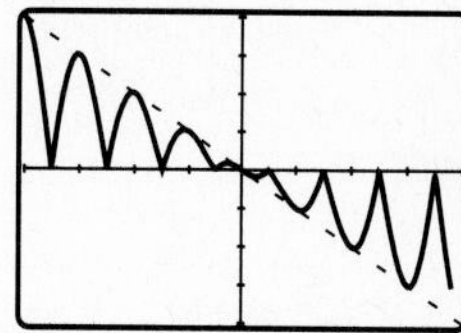

39. The graph of $f(x)$ lies above the x-axis and below the graphs of $y = -x$ (on the left) and $y = x$ (on the right); that is, it lies below the graph of $y = |x|$; possible window: $[-4\pi, 4\pi] \times [-4\pi, 4\pi]$.

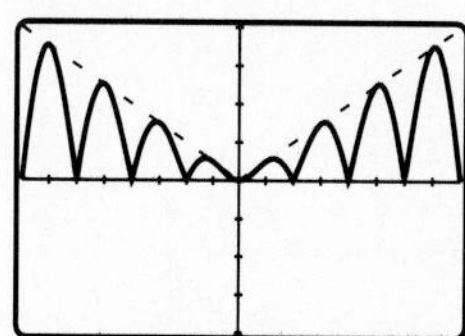

40. The graph of $f(x)$ lies between $y = x$ and $y = -x$ (in the vertical direction); possible window: $[-4\pi, 4\pi] \times [-4\pi, 4\pi]$.

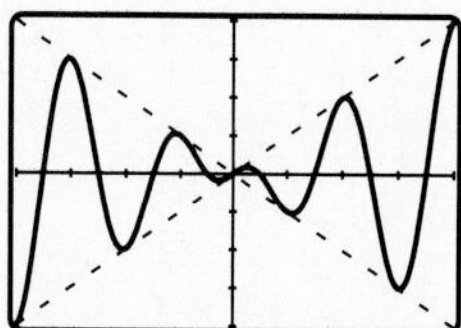

41. Shown is the window $[0, 4\pi] \times [-1, 1]$; f oscillates up and down between 1.2^{-x} and -1.2^{-x}; as $x \to \infty$, $f(x) \to 0$.

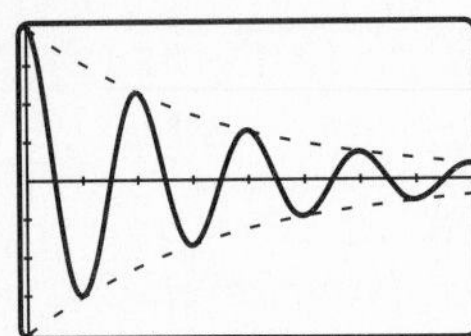

42. Shown is the window $[0, 2\pi] \times [-1, 1]$; f oscillates up and down between 2^{-x} and -2^{-x}; as $x \to \infty$, $f(x) \to 0$.

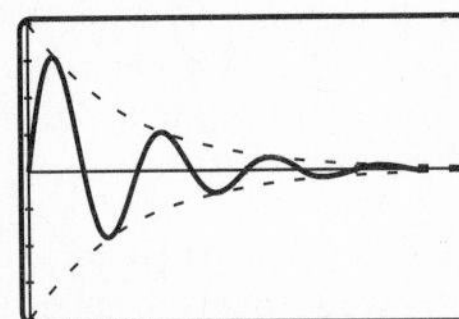

43. Shown is the window $[0, 4\pi] \times [-1.5, 1.5]$; f oscillates up and down between $\frac{1}{x}$ and $-\frac{1}{x}$; as $x \to \infty$, $f(x) \to 0$.

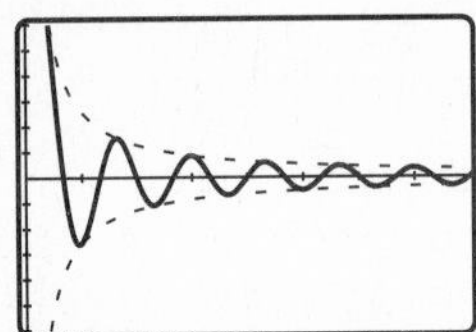

44. Shown is the window $[0, 1.5\pi] \times [-1, 1]$; f oscillates up and down between e^{-x} and $-e^{-x}$; as $x \to \infty$, $f(x) \to 0$.

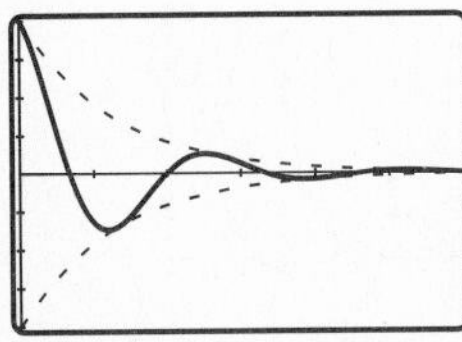

45. Period 2π; possible window: $[-2\pi, 2\pi] \times [-3.4, 2.8]$

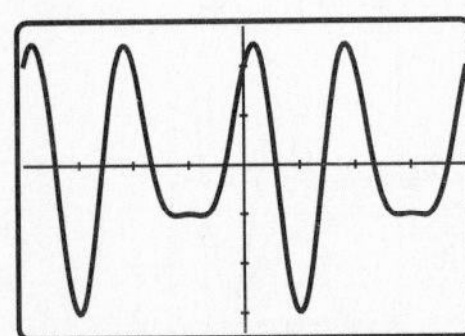

46. Period 2π; possible window: $[-2\pi, 2\pi] \times [-7, 6]$

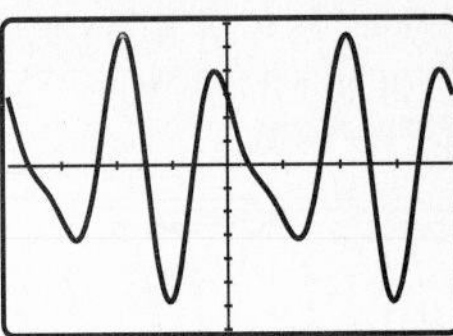

47. Period π; possible window: $[-\pi, \pi] \times [-1, 1]$

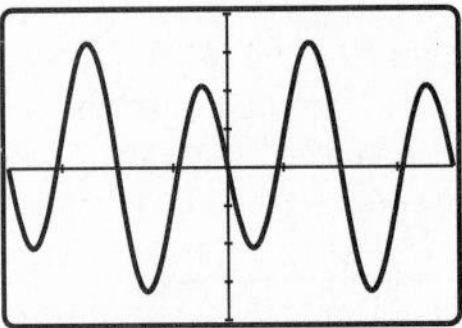

48. Period $\frac{2\pi}{3}$; possible window: $[-\frac{2\pi}{3}, \frac{2\pi}{3}] \times [-2, 2.5]$

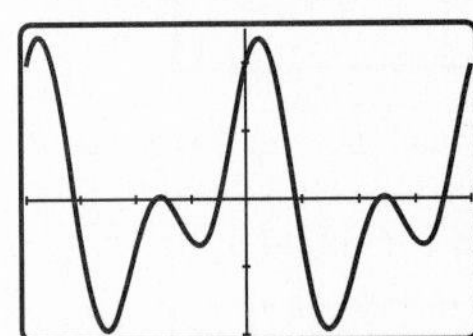

49. Period 2π; possible window: $[-2\pi, 2\pi] \times [-3, 3]$

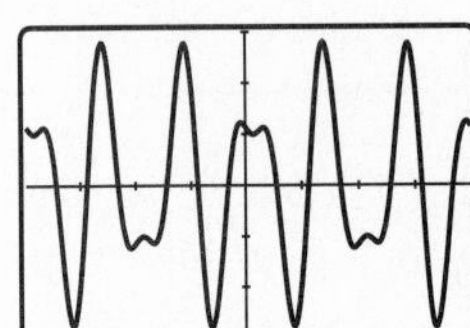

50. Period 2π; possible window: $[-2\pi, 2\pi] \times [-8, 7]$

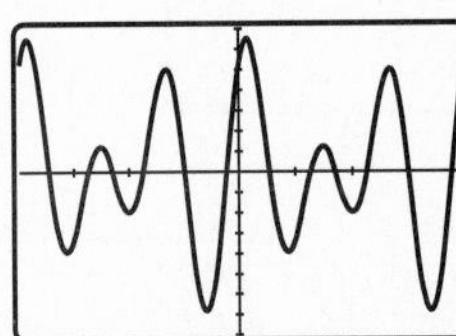

51. Period 2π; Ymin $= -1$ and Ymax $= 4$

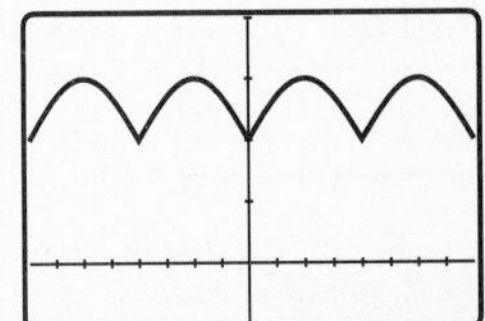

52. Not periodic; Ymin $= -50$ and Ymax $= 50$

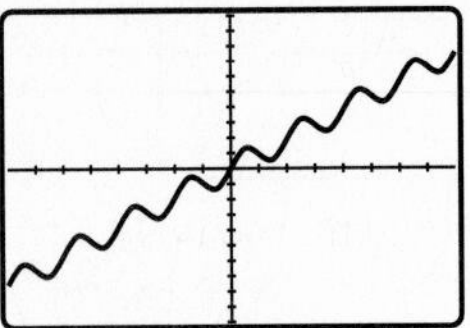

53. Not periodic; Ymin $= -13$ and Ymax $= 13$

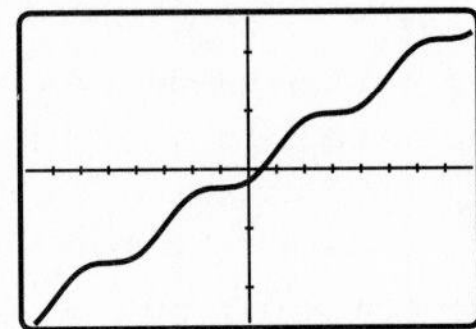

54. Not periodic; Ymin $= -13$ and Ymax $= 13$

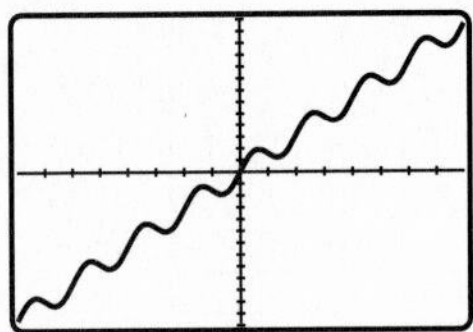

55. Not periodic; Ymin $= -7$ and Ymax $= 7$

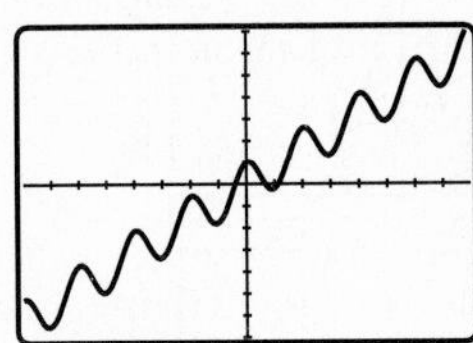

56. Not periodic; Ymin $= -10$ and Ymax $= 15$

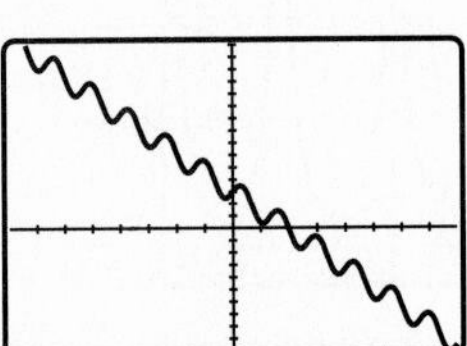

57. Domain: $(-\infty, \infty)$; range: $(-\infty, \infty)$

58. Domain: $(-\infty, \infty)$; range: $(-\infty, \infty)$

59. Domain: $(-\infty, \infty)$; range: $[1, \infty)$

60. Domain: $(-\infty, \infty)$; range: $(-\infty, \infty)$

61. Domain: $\ldots \cup [-2\pi, -\pi] \cup [0, \pi] \cup [2\pi, 3\pi] \cup \ldots$; that is, all x with $2n\pi \le x \le (2n+1)\pi$, n an integer; range: $[0, 1]$

62. Domain: $(-\infty, \infty)$; range: $[-1, 1]$

63. Domain: $(-\infty, \infty)$; range: $[0, 1]$

64. Domain: $\cdots \cup [-\frac{5\pi}{2}, -\frac{3\pi}{2}] \cup [-\frac{\pi}{2}, \frac{\pi}{2}] \cup [\frac{3\pi}{2}, \frac{5\pi}{2}] \cup \cdots$; that is, all x with $\frac{(4n-1)\pi}{2} \le x \le \frac{(4n+1)\pi}{2}$, n an integer; range: $[0, 1]$

65. **(a)** By definition of *period,* there is no number a with $0 < a < p$ such that $f(x+a) = f(x)$ for all x; hence $s > p$. **(b)** $f(x+s) = f(x+s-p+p) = f(x+s-p)$, since p is the period of f; similarly, $f(x+s-p) = f(x+s-2p) = f(x+s-3p) = \ldots$; in particular, $f(x+s) = f(x+s-pq) = f(x+r)$, and also $f(x+s) = f(x)$ by assumption, so $f(x+r) = f(x)$. **(c)** There is no number a with $0 < a < p$ such that $f(x+a) = f(x)$ for all x; from (b), we have $0 \le r < p$ and $f(x+r) = f(x)$ for all x, so r must be 0; therefore $s = pq + r = pq$, a multiple of p.

66. $f(x+2p) = f(x+p+p) = f(x+p)$, since p is the period of f, and $f(x+p) = f(x)$; more generally, $f(x+np) = f(x)$ for any integer n.

67. **(a)** Window: $[0, 15] \times [70, 140]$.

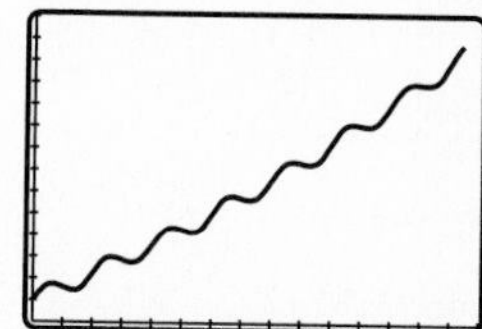

(b) The economic cycle lasts 2 years. **(c)** When $x = 6$, $y \approx 94.90$ million dollars.

68. **(a)** Window: $[0, 12] \times [-0.5, 0.5]$.

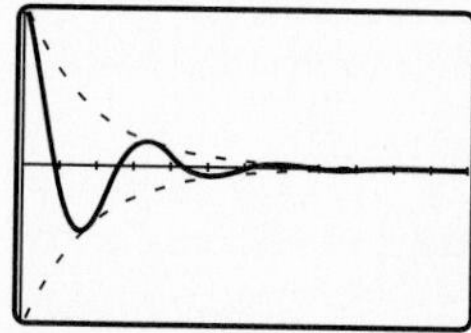

(b) For $t \ge 0.51$ (approximately).

69. No. Although the graph of $y = \sin x^3$ suggests this possibility, no other section of the graph looks like the section between -1 and 1. In particular, there is only one zero of the function in that interval (at $x = 0$); nowhere else can we find an interval this long with only one zero.

70. One explanation: The v-shaped section around $x = 0$ is unique; it does not appear anywhere else on the graph.

71. (a), which is obtained by adding x to all parts of the inequality $-1 \le \sin x \le 1$. In (b), after subtracting x from both sides, we are left with $-\sin x \le \sin x$, which is false when $\sin x$ is negative.

72. (b); (a) is impossible (even ignoring the middle part) if $x < 0$, since then $-x \not\le x$.

73. The graph of $x \sin x$ lies (vertically) between the graphs of $y = -|x|$ (the lower "v"-shaped graph) and $y = |x|$ (the upper "v"-shaped graph).

74. The graph of $3\frac{\sin x}{x}$ lies (vertically) between the graphs of $y = -|\frac{3}{x}|$ (the lower two curves) and $y = |\frac{3}{x}|$ (the upper two curves); note that this inequality is meaningless when $x = 0$.

75. The graph in part (d); window: $[-2\pi, 2\pi] \times [-4, 4]$

76. The graph in part (a); window $[-2\pi, 2\pi] \times [-4, 4]$

77. The graph in part (b); window: $[-2\pi, 2\pi] \times [-4, 4]$

78. The graph in part (c); window $[-2\pi, 2\pi] \times [-4, 4]$

79. Graph $y_1 = \sin x$ and $y_2 = \cos(\frac{\pi}{2} - x)$ and observe that the two graphs appear to be one.

80. Graph $y_1 = \cos x$ and $y_2 = \sin(\frac{\pi}{2} - x)$ and observe that the two graphs appear to be one.

81. Graph $y_1 = \tan x$ and $y_2 = \cot(\frac{\pi}{2} - x) = \frac{1}{\tan(\pi/2 - x)}$ and observe that the two graphs appear to be one.

82. Graph $y_1 = \cot x = \frac{1}{\tan x}$ and $y_2 = \tan(\frac{\pi}{2} - x)$ and observe that the two graphs appear to be one.

83. **(a)** Answers will vary. For a TI-81: $\frac{\pi}{47.5} = 0.0661\ldots \approx 0.07$; for a TI-82: $\frac{\pi}{47} = 0.0668\ldots \approx 0.07$; for a TI-85: $\frac{\pi}{63} = 0.0498\ldots \approx 0.05$; for a TI -92: $\frac{\pi}{119} = 0.263\ldots$. **(b)** Period: $p = \frac{\pi}{125} = 0.0251\ldots$. For any of the TI-8x calculators, there are about 1 or 3 cycles between each pair of pixels; the graphs produced are therefore inaccurate, since so much detail is lost.

84. Answers will vary but should be close to $186 \sin[\frac{2\pi}{365}(x - 80)] + 731$ minutes of daylight.

85. Domain: $(-\infty, \infty)$; range: $[-1, 1]$; horizontal asymptote: $y = 1$; zeros at $\ln(\frac{\pi}{2} + n\pi)$, n a nonnegative integer. Shown is the window $[-3, 3] \times [-1.2, 1.2]$.

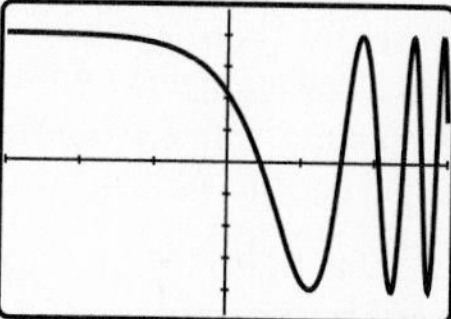

86. Period: π; domain: $x \ne \frac{\pi}{2} \pm n\pi$, n an integer; range: $(0, \infty)$; vertical asymptotes at missing points of domain: $x = \frac{\pi}{2} \pm n\pi$, n an integer. Shown is the window $[-\frac{5\pi}{2}, \frac{5\pi}{2}] \times [-0.2, 5]$.

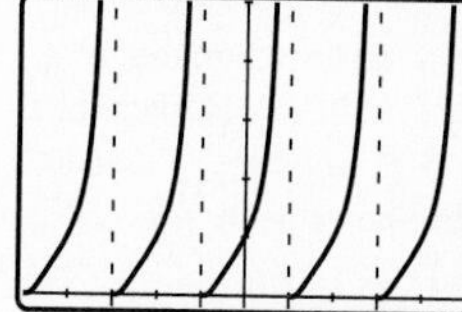

87. Domain: $[0, \infty)$; range: $(-\infty, \infty)$. Zeros at $n\pi$, n a nonnegative integer. Shown is the window $[-0.5, 4\pi] \times [-4, 4]$.

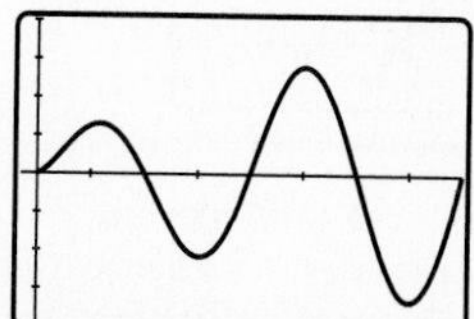

88. Domain: $[-2, 2]$; range: $[0, 2.94]$ (approximately); zeros at -2 and 2.

Shown is the window $[-2.5, 2.5] \times [-0.5, 3.5]$.

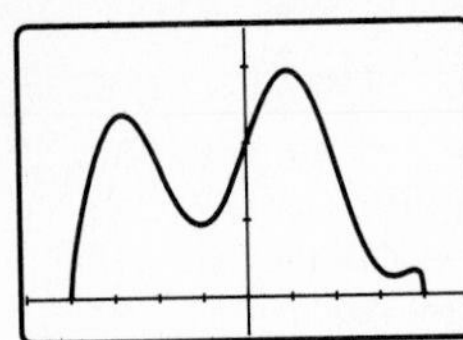

89. Domain: $(-\infty, 0) \cup (0, \infty)$; range: approximately $[-0.22, 1)$. Horizontal asymptote: $y = 0$; zeros at $n\pi$, n a nonzero integer. Shown is the window $[-5\pi, 5\pi] \times [-0.5, 1.2]$.

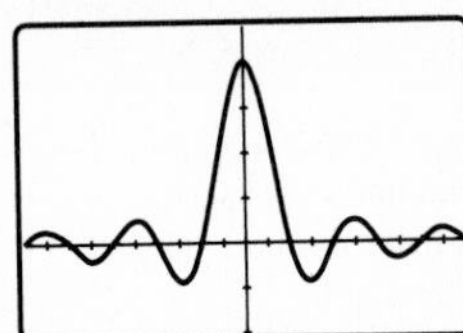

90. Domain: $(-\infty, 0) \cup (0, \infty)$; range: $(-\infty, \infty)$; horizontal asymptote: $y = 0$; vertical asymptote: $x = 0$; zeros at $n\pi$, n a nonzero integer. Shown is the window $[-4\pi, 4\pi] \times [-0.5, 0.5]$.

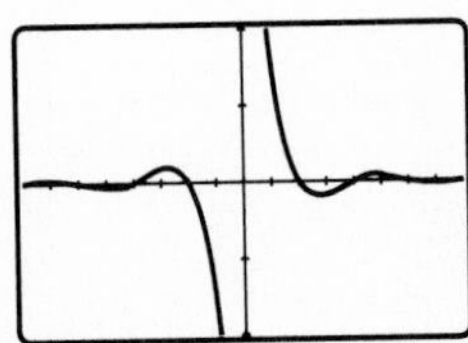

91. Domain: $(-\infty, 0) \cup (0, \infty)$; range: approximately $[-0.22, 1)$. Horizontal asymptote: $y = 1$; zeros at $\frac{1}{n\pi}$, n a nonzero integer. Shown is the window $[-\pi, \pi] \times [-0.3, 1.2]$.

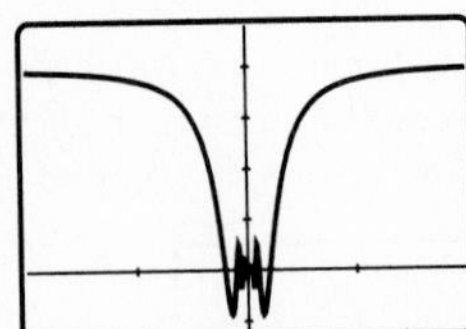

92. Domain: $(-\infty, 0) \cup (0, \infty)$; range: $(-\infty, \infty)$; zeros at $\frac{1}{n\pi}$, n a nonzero integer. Note: The graph also suggests the end-behavior asymptote $y = x$. Shown is the window $[-1, 1] \times [-1, 1]$.

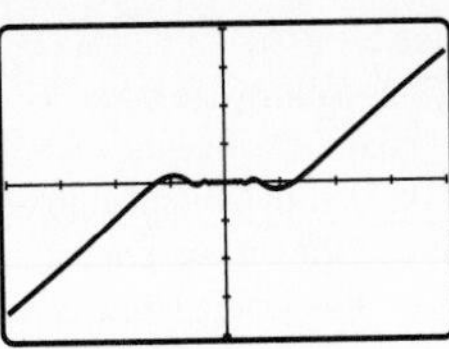

93. $\ln|\tan x| = \ln \dfrac{|\sin x|}{|\cos x|} =$ $\ln|\sin x| - \ln|\cos x|$, using properties of logarithms.

94. $\ln|\cot x| = \ln \dfrac{|\cos x|}{|\sin x|} =$ $\ln|\cos x| - \ln|\sin x|$, using properties of logarithms

95. $\log|\sec x| = \log \dfrac{1}{|\cos x|} =$ $\log 1 - \log|\cos x| = -\log|\cos x|$, using properties of logarithms.

96. $\log(\sin^2 x + \cos^2 x) = \log 1 = 0$, using the Pythagorean identity

QUICK REVIEW 6.7

1. sine: positive; cosine: positive; tangent: positive

2. sine: positive; cosine: negative; tangent: negative

3. sine: negative; cosine: negative; tangent: positive

4. sine: negative; cosine: positive; tangent: negative

5. $\frac{1}{2}$

6. 1

7. $-\frac{1}{2}$

8. $\dfrac{\sqrt{3}}{2}$

9. $-\frac{1}{2}$

10. $\frac{1}{2}$

11. $\dfrac{\sqrt{2}}{2}$

12. -1

13. 1

14. 0

15. Counterclockwise from the bottom of the circle: $-\frac{\pi}{2}, -\frac{\pi}{3}, -\frac{\pi}{4}, -\frac{\pi}{6}, 0, \frac{\pi}{6}, \frac{\pi}{4}, \frac{\pi}{3}, \frac{\pi}{2}, \frac{2\pi}{3}, \frac{3\pi}{4}, \frac{5\pi}{6}, \pi$

16. They must satisfy $x^2 + y^2 = 1$. For $(0, 1)$, we have $0^2 + 1^2 = 1$; for $(\frac{1}{2}, \frac{\sqrt{3}}{2})$, $(\frac{1}{2})^2 + (\frac{\sqrt{3}}{2})^2 = \frac{1}{4} + \frac{3}{4} = 1$; for $(\frac{\sqrt{2}}{2}, \frac{\sqrt{2}}{2})$, $(\frac{\sqrt{2}}{2})^2 + (\frac{\sqrt{2}}{2})^2 = \frac{2}{4} + \frac{2}{4} = 1$. All other points are essentially the same as one of these.

SECTION EXERCISES 6.7

1. $\frac{\pi}{3}$

2. $-\frac{\pi}{6}$

3. 0

4. 0

5. $\frac{\pi}{3}$

6. $\frac{\pi}{4}$

7. $-\frac{\pi}{4}$

8. $\frac{5\pi}{6}$

9. $-\frac{\pi}{4}$

10. $\frac{-\pi}{3}$

11. $\frac{\pi}{2}$

12. $\frac{\pi}{2}$

13. Approximately 21.22°

14. Approximately 42.07°

15. Approximately −85.43°

16. Approximately 103.30°

17. Approximately 87.59°

18. Approximately 80.21°

19. Approximately 1.172

20. Approximately 1.527

21. Approximately −0.478

22. Approximately 2.593

23. $\frac{\sqrt{3}}{2}$

24. $\frac{\sqrt{2}}{2}$

25. $\frac{\pi}{4}$

26. $\frac{\pi}{4}$

27. $\frac{1}{2}$

28. $-\frac{\sqrt{2}}{2}$

29. $\frac{\pi}{6}$

30. 0

31. Domain: $[-\frac{1}{2}, \frac{1}{2}]$; range: $[-\frac{\pi}{2}, \frac{\pi}{2}]$. Starting from $y = \sin^{-1} x$, horizontally shrink by $\frac{1}{2}$.

32. Domain: $[-\frac{1}{2}, \frac{1}{2}]$; range: $[0, 3\pi]$. Starting from $y = \cos^{-1} x$, horizontally shrink by $\frac{1}{2}$ and vertically stretch by 3 (either order).

33. Domain: $(-\infty, \infty)$; range: $(-\frac{5\pi}{2}, \frac{5\pi}{2})$. Starting from $y = \tan^{-1} x$, horizontally stretch by 2 and vertically stretch by 5 (either order).

34. Domain: $[-2, 0]$; range: $[-\frac{\pi}{2}, \frac{\pi}{2}]$. Starting from $y = \arcsin x = \sin^{-1} x$, translate left 1 unit.

35. Domain: $[-2, 2]$; range: $[0, 3\pi]$. Starting from $y = \arccos x = \cos^{-1} x$, horizontally stretch by 2 and vertically stretch by 3 (either order).

36. Domain: $(-\infty, \infty)$; range: $(-\frac{3\pi}{20}, \frac{3\pi}{20})$. Starting from $y = \arctan x = \tan^{-1} x$, horizontally shrink by $\frac{1}{2}$ and vertically shrink by 0.3 (either order).

37. $x = 1$

38. $x = 1 + 2n\pi$ or $x = -1 + 2n\pi$, for all integers n

39. $x = \sin \frac{1}{2} = 0.479\ldots$

40. $x = \tan(-1) = -1.557\ldots$

41. $\dfrac{x}{\sqrt{1 + x^2}}$

42. $\dfrac{1}{\sqrt{1 + x^2}}$

43. $\dfrac{x}{\sqrt{1 - x^2}}$

44. $\dfrac{x}{\sqrt{1 - x^2}}$

45. $\dfrac{1}{\sqrt{1 + 4x^2}}$

46. $\sqrt{1 - 9x^2}$

47. **(a)** If we call the smaller (unlabeled) angle in the lower left α, then $\tan \alpha = (\frac{2}{x})$, or $\alpha = \tan^{-1}(\frac{2}{x})$ (since α is acute); also, $\theta + \alpha$ is the measure of one acute angle in the right triangle formed by the floor and the wall; for this triangle, $\tan(\theta + \alpha) = \frac{14}{x}$; then $\theta + \alpha = \tan^{-1}(\frac{14}{x})$ (since $\theta + \alpha$ is acute), so $\theta = \tan^{-1}(\frac{14}{x}) - \alpha = \tan^{-1}(\frac{14}{x}) - \tan^{-1}(\frac{2}{x})$. **(b)** Graph is shown; the actual maximum occurs at $x \approx 5.29$ ft, where $\theta \approx 48.59°$.

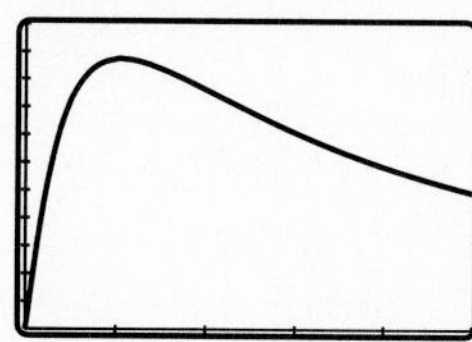

(c) Either $x \approx 1.83$ or $x \approx 15.31$—these round to 2 ft or 15 ft.

48. **(a)** θ is one acute angle in the right triangle with leg lengths x (opposite) and 3 (adjacent); thus $\tan \theta = \frac{x}{3}$, and $\theta = \tan^{-1}(\frac{x}{3})$ (since θ is acute). **(b)** Graph is shown (DEGREE mode); negative values of x correspond to the point P being "upshore" ("into" the picture) instead of downshore (as shown in the illustration); positive angles point downshore, and negative angles point upshore.

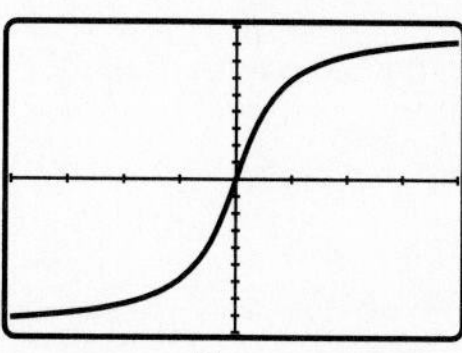

(c) $\theta = \tan^{-1} 5 \approx 78.69°$.

49. **(a)** $\theta = \tan^{-1}(\frac{s}{500})$. **(b)** As s changes from 10 to 20 ft, θ changes from about 1.1458° to 2.2906°, almost exactly doubling (a 99.92% increase); as s changes from 200 to 210 ft, θ changes from about 21.80° to 22.78°, an increase of less than 1°, and a very small relative change (only about 4.25%). **(c)** The x-axis represents the height and the y-axis represents the angle; the angle cannot grow past 90° (in fact, it *approaches* but never exactly equals 90°).

50. The restricted sine and inverse sine graphs are reflections of each other across the line $y = x$; window: *X*min and *X*max vary among calculators; [*Y*min, *Y*max] should be something similar to $[-\frac{\pi}{2}, \frac{\pi}{2}]$.

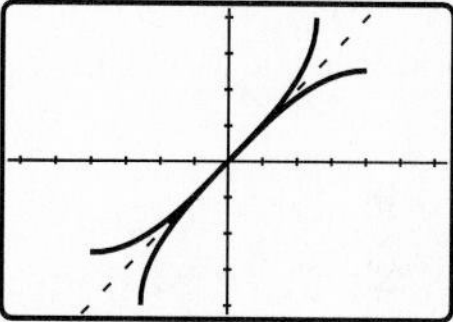

51. The restricted cosine and inverse cosine graphs are reflections of each other across the line $y = x$. Window: *X*min and *X*max vary among calculators; [*Y*min, *Y*max] should be something similar to $[-1, \pi]$.

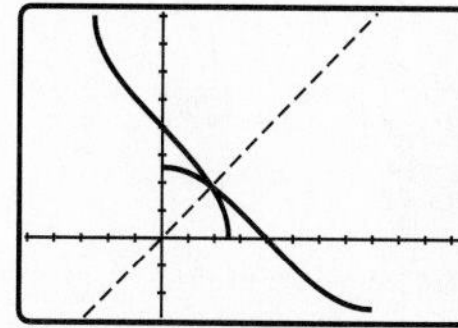

52. The restricted tangent and inverse tangent graphs are reflections of each other across the line $y = x$; the window shown has *Y*min = −4 and *Y*max = 4.

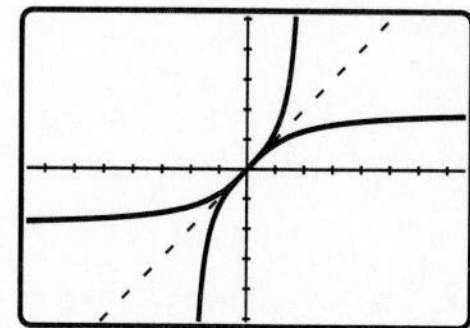

53. There are infinitely many solutions to (a); namely, $x = \frac{\pi}{6} \pm 2n\pi$ for integers n or $x = \frac{5\pi}{6} \pm 2n\pi$ for integers n. The expression in (b) evaluates to $\frac{\pi}{6}$, the one solution from (a) that falls in the restricted domain of the sine function used to define $\sin^{-1} x$. This answer potentially may be extended to obtain *all* solutions in (a).

54. $x = -\frac{\pi}{2}$, $0 \le x \le \frac{\pi}{2}$, or $\{-\frac{\pi}{2}\} \cup [0, \frac{\pi}{2}]$

55. Approximately $-1.42 \le x < 0$ or $0 < x \le 1.42$; that is, $[-1.42, 0) \cup (0, 1.42]$

56. Let $\theta = \sin^{-1}(-x)$; note that $-\frac{\pi}{2} \le \theta \le \frac{\pi}{2}$. Since sine is an odd function, $\sin(-\theta) = -\sin\theta = -(-x) = x$; thus $-\theta$ is an angle in $[-\frac{\pi}{2}, \frac{\pi}{2}]$ with $\sin(-\theta) = x$ and $\sin^{-1} x = -\theta$; that is, $\sin^{-1}(-x) = \theta = -\sin^{-1} x$.

57. Let $\theta = \tan^{-1} x$; note that $-\frac{\pi}{2} < \theta < \frac{\pi}{2}$. Since tangent is an odd function, $\tan(-\theta) = -\tan\theta = -x$. Thus $-\theta$ is an angle in $(-\frac{\pi}{2}, \frac{\pi}{2})$ with $\tan(-\theta) = -x$, and $\tan^{-1}(-x) = -\theta$; that is, $\tan^{-1} x = \theta = -\tan^{-1}(-x)$.

58. Let $\theta = \sin^{-1} x$. When $-1 < x < 1$, the inverse sine function returns θ in $(-\frac{\pi}{2}, \frac{\pi}{2})$; in that interval, $\cos\theta$ is positive; a Pythagorean identity gives $\cos\theta = \sqrt{1 - \sin^2\theta} = \sqrt{1 - x^2}$. Therefore $\tan\theta = \dfrac{\sin\theta}{\cos\theta} = \dfrac{x}{\sqrt{1 - x^2}}$; since θ is in $\left(-\dfrac{\pi}{2}, \dfrac{\pi}{2}\right)$, $\tan^{-1}\dfrac{x}{\sqrt{1 - x^2}} = \theta = \sin^{-1} x$.

59. Let $\theta = \cos^{-1} x$. When $0 < x < 1$, the inverse cosine function returns θ in $(0, \frac{\pi}{2})$; in that interval, $\sin\theta$ is positive. Using a Pythagorean identity, we get $\sin\theta = \sqrt{1 - \cos^2\theta} = \sqrt{1 - x^2}$, so $\tan\theta = \dfrac{\sin\theta}{\cos\theta} = \dfrac{\sqrt{1 - x^2}}{x}$; since θ is in $\left(0, \dfrac{\pi}{2}\right)$, $\tan^{-1}\dfrac{\sqrt{1 - x^2}}{x} = \theta = \cos^{-1} x$.

60. Let $\theta = \sin^{-1} x$. Now, since cosine is even, $\cos(\frac{\pi}{2} - \theta) = \cos(\theta - \frac{\pi}{2})$, which equals $\sin\theta$ by one of the horizontal translation identities. Therefore, $\cos(\frac{\pi}{2} - \theta) = \sin\theta = x$, and since $\frac{\pi}{2} - \theta$ is in $[0, \pi]$, it follows that $\cos^{-1} x = \frac{\pi}{2} - \theta$, so $\sin^{-1} x + \cos^{-1} x = \theta + \frac{\pi}{2} - \theta = \frac{\pi}{2}$.

61. From Exercises 59 and 60, we know that $\sin^{-1} x + \cos^{-1} x = \frac{\pi}{2}$ and that, when $0 < x \le 1$, $\tan^{-1}\dfrac{\sqrt{1 - x^2}}{x} = \cos^{-1} x$; combining these two facts gives the desired result.

62. Let $\theta = \sin^{-1} x$. Note that θ is in $[-\frac{\pi}{2}, \frac{\pi}{2}]$, and in that interval $\cos\theta$ is nonnegative; a Pythagorean identity yields $\cos(\sin^{-1} x) = \cos\theta = \sqrt{1 - \sin^2\theta} = \sqrt{1 - x^2}$.

63. Let $\theta = \cos^{-1} x$. Since $0 < x \le 1$, we have θ in $[0, \frac{\pi}{2})$, and $\sin\theta$ is nonnegative for all those values. Using a Pythagorean identity, we get $\sin\theta = \sqrt{1 - \cos^2\theta} = \sqrt{1 - x^2}$, and $\tan(\cos^{-1} x) = \tan\theta = \dfrac{\sin\theta}{\cos\theta} = \dfrac{\sqrt{1 - x^2}}{x}$.

QUICK REVIEW 6.8

1. $b = 15 \cot 31° = 24.964\ldots$, $c = 15 \csc 31° = 29.124\ldots$

2. $a = 25 \cos 68° = 9.365\ldots$, $b = 25 \sin 68° = 23.179\ldots$

3. $a = 28 \csc 44° = 40.307\ldots$, $b = 28 \cot 28° - 28 \cot 44° = 23.665\ldots$, $c = 28 \csc 28° = 59.641\ldots$

4. $a = 21 \csc 48° = 28.258\ldots$, $b = 21 \cot 31° - 21 \cot 48° = 16.041\ldots$, $c = 21 \csc 31° = 40.773\ldots$

5. Complement: 58°; supplement: 148°

6. Complement: 17°; supplement: 107°

7. Complement: none; supplement: 85°

8. Complement: none; supplement: 60°

9. 45°

10. 202.5°

11. Amplitude: 3; period: π

12. Amplitude: 4; period: $\frac{\pi}{2}$

SECTION EXERCISES 6.8

1. $h = 300 \tan 60° = 300\sqrt{3} = 519.615\ldots$ ft

2. $h = 100 \tan 34°13'12'' = 68.010\ldots$ ft

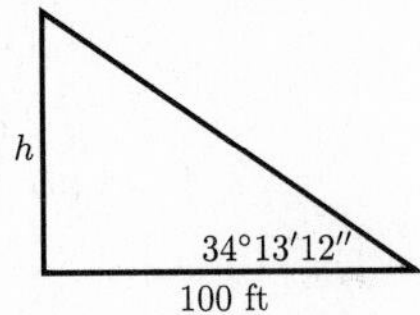

3. $d = 120 \cot 10° = 680.553\ldots$ ft

4. $d = 90 \cot 14° = 360.970\ldots$ ft

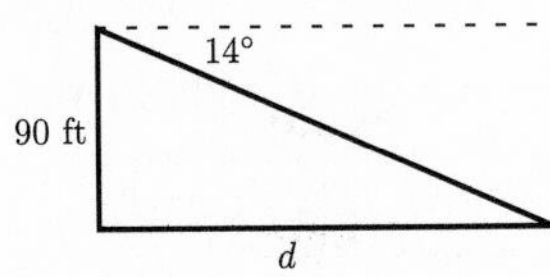

5. Wire length $5 \sec 80° = 28.793\ldots$ ft; tower height is $5 \tan 80° = 28.356\ldots$ ft.

6. tower height $16 \tan 62° = 30.091\ldots$ ft; wire length $16 \sec 62° = 34.080\ldots$ ft

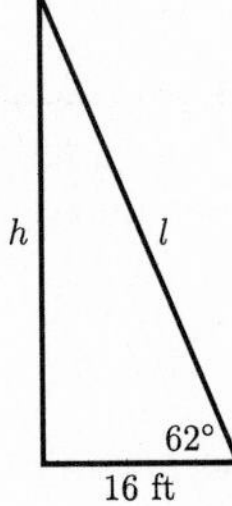

7. $h = 185 \tan 80°1'12'' = 1051.332\ldots$ ft

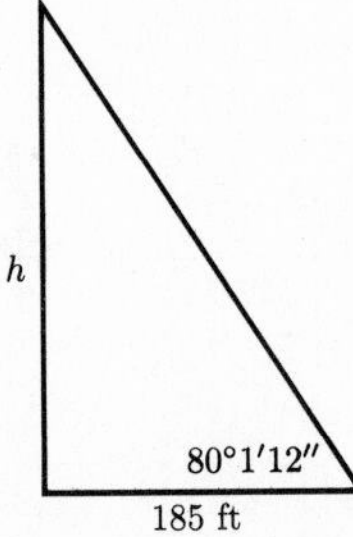

8. $h = 1580 \tan 38° = 1234.431\ldots$ ft

9. $h = 100 \tan 83°12' = 838.625\ldots$ ft

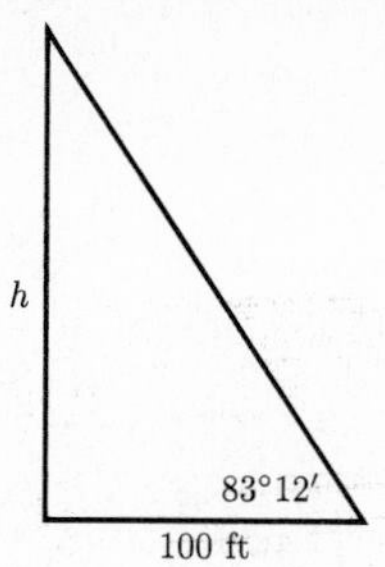

10. $\theta = \sin^{-1}\frac{32}{470} \approx 3.9°$

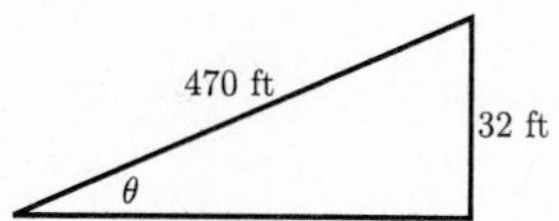

11. $h = 10 \tan 55° = 14.281\ldots$ m

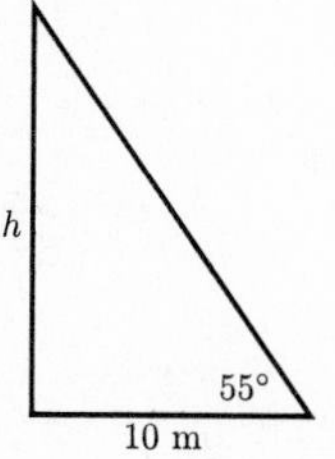

12. $h = 125 \tan 29°48' = 71.588\ldots$ ft

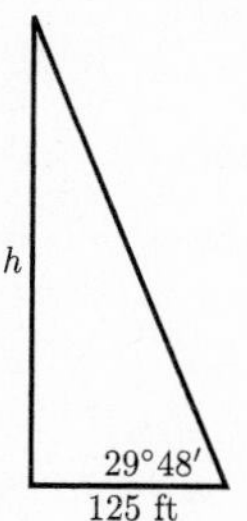

13. $LP = 4.25 \tan 35° = 2.975\ldots$ mi

14. $x = h \cot 35°$ and $x + 1000 = h \cot 30°$, so $h = \frac{1000}{\cot 30° - \cot 35°} = 3290.5\ldots$ ft

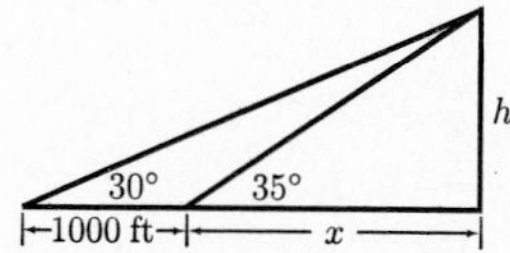

15. $h = 200(\tan 40° - \tan 30°) = 52.349\ldots$ ft

16. $d = 100(\cot 15° - \cot 33°) = 219.218\ldots$ ft

17. distance: $60\sqrt{2} = 84.852\ldots$ naut mi; bearing: 140°

18. Distance: $80\sqrt{5} = 178.885\ldots$ naut mi; bearing: $(65 + \tan^{-1} 2)° = 128.434\ldots°$

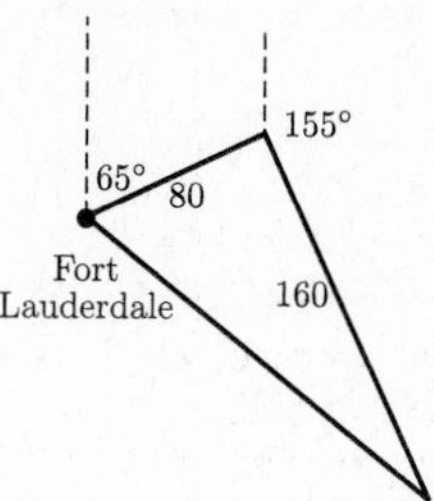

19. $w = 1097 \cot 19° = 3185.919\ldots$ ft

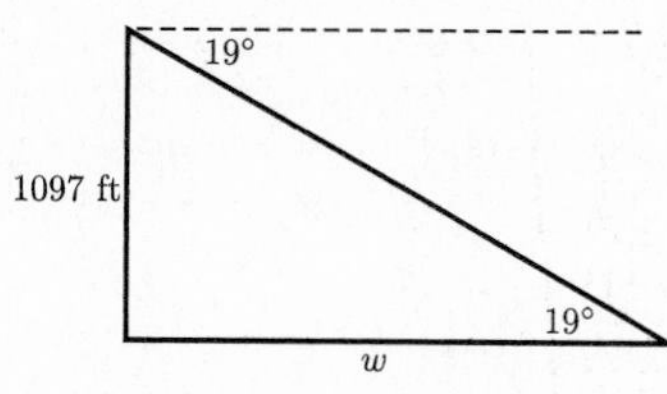

20. $d = 73 \cot 1°20' = 3136.377\ldots$ ft (from base of tower)

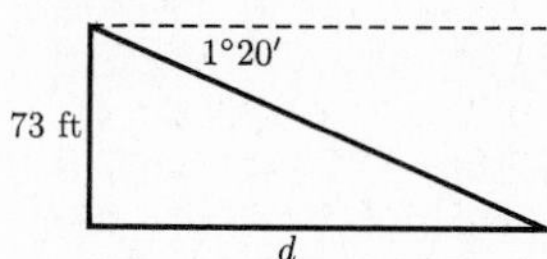

21. $l = 325 \tan 63° = 637.848\ldots$ ft

22. $h = 12 \tan 17° = 3.668\ldots$ mi

23. $h = 36.5 \tan 15° = 9.780\ldots$ ft

24. $d = 760 \cot 5.25° = 8271.020\ldots$ ft

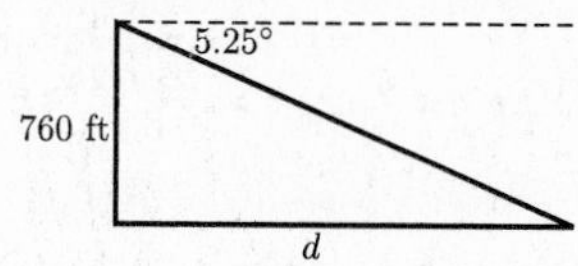

25. $d = \dfrac{550}{\cot 70° - \cot 80°} = 2931.093\ldots$ ft

26. If t is the time until the boats collide, the law enforcement boat travels $23t$ naut mi; during that same time, the smugglers' craft travels xt naut mi, where x is that craft's speed. These two distances are the legs of a right triangle (shown); then $x = 23 \tan 15° = 6.162\ldots$ knots.

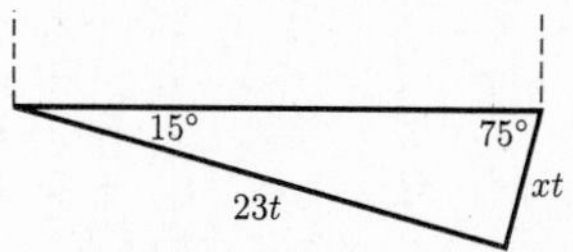

27. **(a)** 8 oscillations/sec. **(b)** $d = 6 \cos 16\pi t$ in. **(c)** When $t = 2.85$, $d = 1.854\ldots$, or about 4.1 in. left of the starting position.

28. **(a)** $\frac{1}{2}$ oscillation/sec. **(b)** $d = 18 \cos \pi t$ cm. **(c)** Since 1 oscillation takes 2 sec, there are 30 oscillations/min.

29. $d = 3 \cos 4\pi t$

30. $\omega = 1056\pi$ rad/sec

31. **(a)** $a = 25$ ft. **(b)** $k = 33$ ft. **(c)** $\omega = \pi/10$ rad/sec.

32. **(a)** One possibility is $h = -8 \cos \frac{\pi t}{10} + 9$ m. **(b)** Window: $[0, 30] \times [-1, 20]$. **(c)** $h(4) = 6.527\ldots$ m; $h(10) = 17$ m.

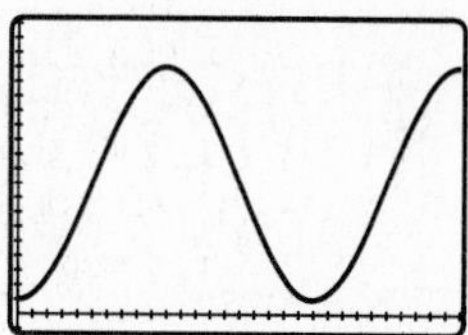

33. **(a)** $b = \frac{\pi}{6}$. **(b)** The amplitude is half this difference (the maximum minus the minimum): $a = 25$ (degrees). **(c)** $k = 55$. **(d)** $h = 5$ (halfway between the times of the minimum and maximum); $y = 25 \sin\left[\frac{\pi}{6}(t - 5)\right] + 55$. **(e)** Scatter plot and graph shown in $[0, 13] \times [0, 85]$:

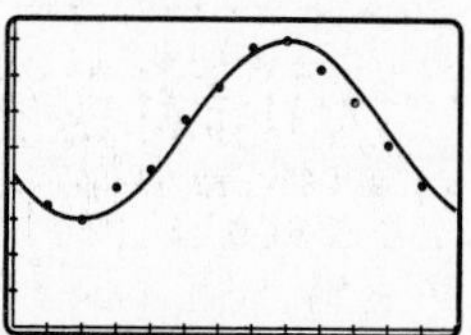

(f) $t \approx 6.23$ (about July 7) or $t \approx 9.77$ (about October 24).

34. (d); the others are incorrect in various ways; the closest of the others is (c), but it does not have the property that $h(0) = 8$; this is accomplished by the horizontal shift in (d).

35. **(a)** Scatter plot shown:

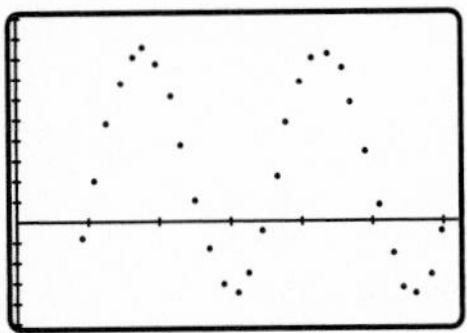

(b) The first is the best. **(c)** About $\frac{1232}{\pi} \approx 392$ oscillations/sec.

36. **(a)** Newborn: about 6 hours; 4-year-old: about 24 hours; adult: about 24 hours. **(b)** The adult sleep cycle is most like a sinusoid.

37. The 7-gon can be split into 14 congruent right triangles with a common vertex at the center; the legs of these triangles measure a and 2.5; the angle at the center is $\frac{2\pi}{14} = \frac{\pi}{7}$, so $a = 2.5 \cot \frac{\pi}{7} = 5.191\ldots.$

38. The 7-gon can be split into 14 congruent right triangles with a common vertex at the center; the legs of these triangles measure a and 2.5, and the hypotenuse has length r; the angle at the center is $\frac{2\pi}{14} = \frac{\pi}{7}$, so $r = 2.5 \csc \frac{\pi}{7} = 5.761\ldots.$

39. Choosing point E in the center of the rhombus, we have $\triangle AEB$ with right angle at E, and $m\angle EAB = 21°$; then $AE = 18 \cos 21°$ in. and $BE = 18 \sin 21°$ in., so $AC = 2AE = 33.608\ldots$ in. and $BD = 2BE = 12.901\ldots$ in.

40. **(a)** $BE = 20 \tan 50° = 23.835\ldots$ ft. **(b)** $CD = BE + 45 \tan 20° = 40.213\ldots$ ft. **(c)** $AE + ED = 20 \sec 50° + 45 \sec 20° = 79.002\ldots$ ft, so the total distance across the top of the roof is about 158 ft.

41. $\theta = \tan^{-1} 0.06 = 3.433\ldots°$

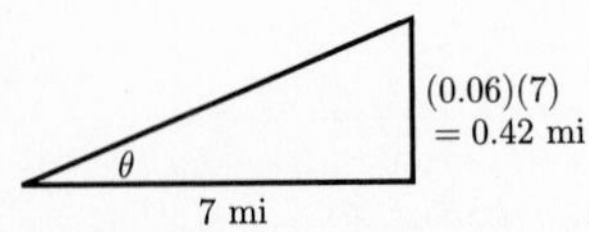

42. $\theta = 2 \cos^{-1} \frac{40}{41} = 25.360\ldots°$

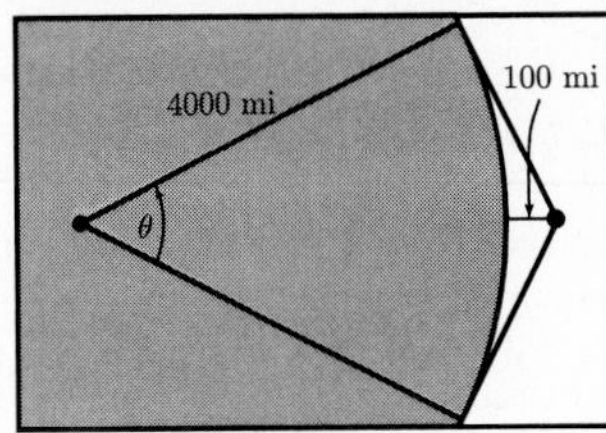

43. **(a)** Scatter plot shown; window: $[0, 0.0092] \times [-1.6, 1.6]$.

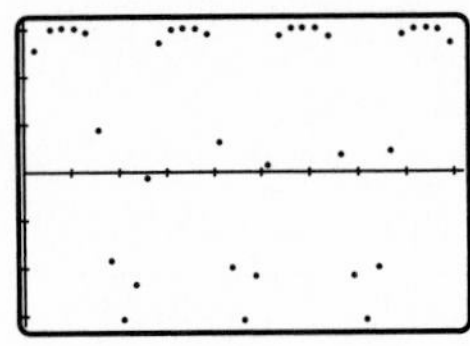

(b) One fairly good match is $y = 1.51971 \sin [2467(t - 0.0002)$; that is, $a = 1.51971$, $b = 2467$, and $h = 0.0002$. **(c)** The frequency is about 392 or 393 Hz, and the note appears to be a G. **(d)** This note also appears to be a G.

Chapter 6 Review

1. On the positive y-axis (between quadrants I and II); 450°

2. Quadrant II; 135°

3. Quadrant III; $-\frac{3\pi}{4}$

4. Quadrant IV; $-\frac{\pi}{4}$

5. Quadrant I; $\frac{13\pi}{30}$

6. Quadrant II; $\frac{28\pi}{45}$

7. Quadrant I; 15°

8. Quadrant II; 126°

9. 270°, or $\frac{3\pi}{2}$ radians

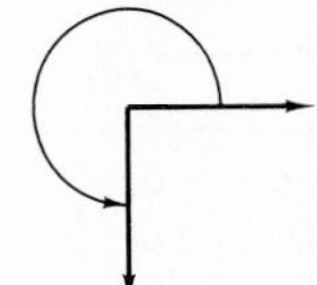

10. 900° or 5π radians

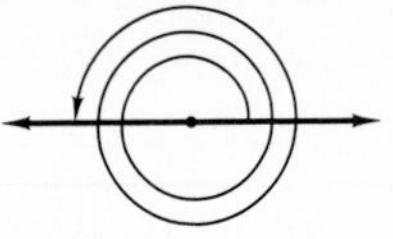

11. $\theta = \tan^{-1} 2 = 63.434\ldots° = 1.107\ldots$ radians

12. $\theta = 135° = \frac{3\pi}{4}$ radians

13. $\theta = 120° = \frac{2\pi}{3}$ radians

14. $\theta = 225° = \frac{5\pi}{4}$ radians

15. $\theta = 360° + \tan^{-1}(-2) = 296.565\ldots° = 5.176\ldots$ radians

16. $\theta = 30° = \frac{\pi}{6}$ radians

17. $\frac{1}{2}$

18. $\dfrac{\sqrt{3}}{2}$

19. 1

20. $-\sqrt{2}$

21. $\frac{1}{2}$

22. $\dfrac{2}{\sqrt{3}}$

23. 2

24. $\sqrt{3}$

25. -1

26. -1

27. undefined

28. 0

29. $\sin\theta = -\frac{1}{2}$, $\cos\theta = \dfrac{\sqrt{3}}{2}$, $\tan\theta = -\dfrac{1}{\sqrt{3}}$; $\csc\theta = -2$, $\sec\theta = \dfrac{2}{\sqrt{3}}$, $\cot\theta = -\sqrt{3}$

30. $\sin\theta = \dfrac{1}{\sqrt{2}}$, $\cos\theta = -\dfrac{1}{\sqrt{2}}$, $\tan\theta = -1$; $\csc\theta = \sqrt{2}$, $\sec\theta = -\sqrt{2}$, $\cot\theta = -1$

31. $\sin\theta = -\dfrac{1}{\sqrt{2}}$, $\cos\theta = -\dfrac{1}{\sqrt{2}}$, $\tan\theta = 1$; $\csc\theta = -\sqrt{2}$, $\sec\theta = -\sqrt{2}$, $\cot\theta = 1$

32. $\sin\theta = \dfrac{\sqrt{3}}{2}$, $\cos\theta = \frac{1}{2}$, $\tan\theta = \sqrt{3}$; $\csc\theta = \dfrac{2}{\sqrt{3}}$, $\sec\theta = 2$, $\cot\theta = \dfrac{1}{\sqrt{3}}$

33. $\sin\alpha = \frac{5}{13}$, $\cos\alpha = \frac{12}{13}$, $\tan\alpha = \frac{5}{12}$; $\csc\alpha = \frac{13}{5}$, $\sec\alpha = \frac{13}{12}$, $\cot\alpha = \frac{12}{5}$

34. $\sin\theta = \dfrac{2\sqrt{6}}{7}$, $\cos\theta = \frac{5}{7}$, $\tan\theta = \dfrac{2\sqrt{6}}{5}$; $\csc\theta = \dfrac{7}{2\sqrt{6}}$, $\sec\theta = \frac{7}{5}$, $\cot\theta = \dfrac{5}{2\sqrt{6}}$

35. $\sin\theta = \frac{15}{17}$, $\cos\theta = \frac{8}{17}$, $\tan\theta = \frac{15}{8}$; $\csc\theta = \frac{17}{15}$, $\sec\theta = \frac{17}{8}$, $\cot\theta = \frac{8}{15}$

36. $\theta = 64.623\ldots°$

37. $x = 4.074\ldots$ radians

38. $x = 0.219\ldots$ or $x = 2.921\ldots$ radians

39. $a = 8.603\ldots$, $b = 12.287\ldots$, $\beta = 55°$

40. $a = 6$, $\alpha = 36.869\ldots°$, $\beta = 53.130\ldots°$

41. $b = 7.774\ldots$, $c = 10.461\ldots$, $\alpha = 42°$

42. $a = 3.755\ldots$, $b = 7.063\ldots$, $\beta = 62°$

43. $a = 2\sqrt{6}$, $\alpha = 44.415\ldots°$, $\beta = 45.584\ldots°$

44. $c = \sqrt{59.54} = 7.716\ldots$, $\alpha = 18.904\ldots°$, $\beta = 71.095\ldots°$

45. Quadrant III

46. Quadrant II

47. Quadrant II

48. Quadrant II

49. $\tan\theta = \dfrac{\pm\sqrt{0.3276}}{0.82} = \pm 0.698\ldots$

50. $\tan\theta = \dfrac{-0.58}{\pm\sqrt{0.6636}} = \pm 0.711\ldots$

51. $\sin\theta = \dfrac{2}{\sqrt{5}}$, $\cos\theta = -\dfrac{1}{\sqrt{5}}$, $\tan\theta = -2$; $\csc\theta = \dfrac{\sqrt{5}}{2}$, $\sec\theta = -\sqrt{5}$, $\cot\theta = -\dfrac{1}{2}$

52. $\sin\theta = \dfrac{7}{\sqrt{193}}$, $\cos\theta = \dfrac{12}{\sqrt{193}}$, $\tan\theta = \dfrac{7}{12}$; $\csc\theta = \dfrac{\sqrt{193}}{7}$, $\sec\theta = \dfrac{\sqrt{193}}{12}$, $\cot\theta = \dfrac{12}{7}$

53. $\sin\theta = -\dfrac{3}{\sqrt{34}}$, $\cos\theta = -\dfrac{5}{\sqrt{34}}$, $\tan\theta = \dfrac{3}{5}$; $\csc\theta = -\dfrac{\sqrt{34}}{3}$, $\sec\theta = -\dfrac{\sqrt{34}}{5}$, $\cot\theta = \dfrac{5}{3}$

54. $\sin\theta = \dfrac{9}{\sqrt{97}}$, $\cos\theta = \dfrac{4}{\sqrt{97}}$, $\tan\theta = \dfrac{9}{4}$; $\csc\theta = \dfrac{\sqrt{97}}{9}$, $\sec\theta = \dfrac{\sqrt{97}}{4}$, $\cot\theta = \dfrac{4}{9}$

55. Starting from $y = \sin x$, translate left π units; possible window: $[-2\pi, 2\pi] \times [-1.2, 1.2]$.

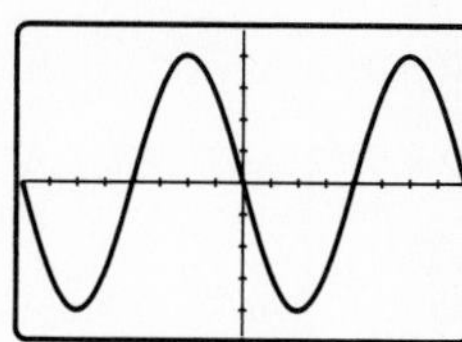

56. Starting from $y = \cos x$, vertically stretch by 2, translate upward 3 units; possible window: $[-2\pi, 2\pi] \times [-1, 6]$.

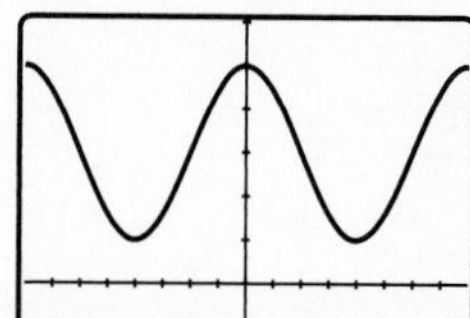

57. Starting from $y = \cos x$, translate left $\frac{\pi}{2}$ units, reflect through the x-axis, and translate upward 4 units; possible window: $[-2\pi, 2\pi] \times [-1, 6]$.

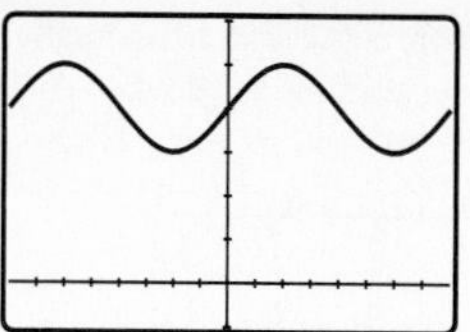

58. Starting from $y = \sin x$, translate right π units, vertically stretch by 3, reflect through the x-axis, and translate downward 2 units; possible window: $[-2\pi, 2\pi] \times [-6, 2]$.

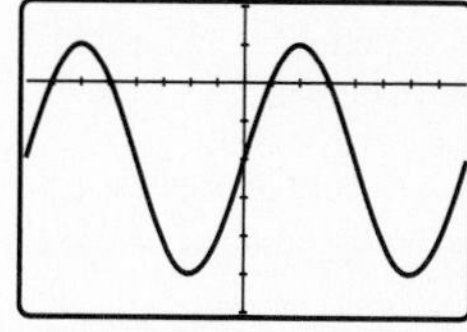

59. Starting from $y = \tan x$, horizontally shrink by $\frac{1}{2}$; possible window: $[-\frac{\pi}{2}, \frac{\pi}{2}] \times [-5, 5]$.

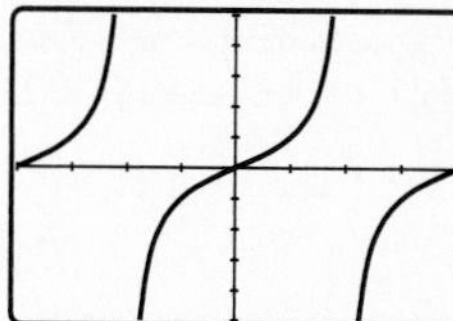

60. Starting from $y = \cot x$, horizontally shrink by $\frac{1}{3}$, vertically stretch by 2, and reflect through the x-axis (in any order); possible window: $[-\frac{\pi}{3}, \frac{\pi}{3}] \times [-10, 10]$.

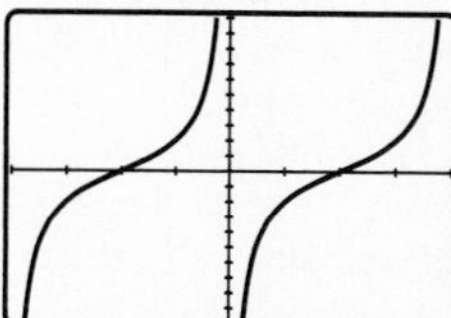

61. Starting from $y = \sec x$, horizontally stretch by 2, vertically stretch by 2, and reflect through the x-axis (in any order); possible window: $[-4\pi, 4\pi] \times [-8, 8]$.

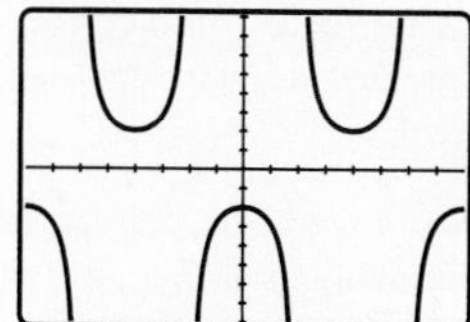

62. Starting from $y = \csc x$, horizontally shrink by $\frac{1}{\pi}$; possible window: $[-2, 2] \times [-8, 8]$.

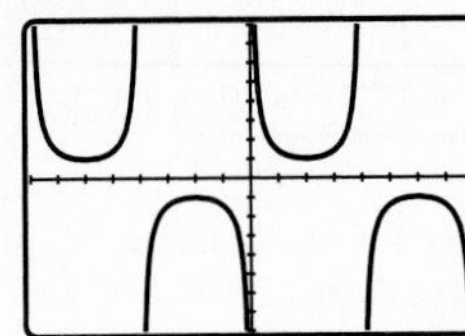

63. Amplitude: 2; period: $\frac{2\pi}{3}$; phase shift: 0; domain: $(-\infty, \infty)$; range: $[-2, 2]$

64. Amplitude: 3; period: $\frac{\pi}{2}$; phase shift: 0; domain: $(-\infty, \infty)$; range: $[-3, 3]$

65. Amplitude: 1.5; period: π; phase shift: $\frac{\pi}{8}$; domain: $(-\infty, \infty)$; range: $[-1.5, 1.5]$

66. Amplitude: 2; period: $\frac{2\pi}{3}$; phase shift: $\frac{\pi}{9}$; domain: $(-\infty, \infty)$; range: $[-2, 2]$

67. Amplitude: 4; period: π; phase shift: $\frac{1}{2}$; domain: $(-\infty, \infty)$; range: $[-4, 4]$

68. Amplitude: 2; period: $\frac{2\pi}{3}$; phase shift: $-\frac{1}{3}$; domain: $(-\infty, \infty)$; range: $[-2, 2]$

69. $A \approx 4.47$, $b = 1$, and $h \approx 1.11$, so $f(x) \approx 4.47 \sin(x - 1.11)$

70. $A \approx 3.61$, $b = 2$, and $h \approx -1.08$, so $f(x) \approx 3.61 \sin[2(x + 1.08)]$

71. $49.996\ldots°$, or $0.872\ldots$ rad

72. $61.379\ldots°$, or $1.071\ldots$ rad

73. $45°$, or $\frac{\pi}{4}$ rad

74. $60°$, or $\frac{\pi}{3}$ rad

75. Starting from $y = \sin^{-1} x$, horizontally shrink by $\frac{1}{3}$; domain: $[-\frac{1}{3}, \frac{1}{3}]$; range: $[-\frac{\pi}{2}, \frac{\pi}{2}]$.

76. Starting from $y = \tan^{-1} x$, horizontally shrink by $\frac{1}{2}$; domain: $(-\infty, \infty)$; range: $(-\frac{\pi}{2}, \frac{\pi}{2})$.

77. Starting from $y = \sin^{-1} x$, translate right 1 unit, horizontally shrink by $\frac{1}{3}$, and translate upward 2 units; domain: $[0, \frac{2}{3}]$; range: $[2 - \frac{\pi}{2}, 2 + \frac{\pi}{2}]$.

78. Starting from $y = \cos^{-1} x$, translate left 1 unit, horizontally shrink by $\frac{1}{2}$, translate downward 3 units; domain: $[-1, 0]$; range: $[-3, \pi - 3]$.

79. $x = \frac{\pi}{6}$ or $x = \frac{5\pi}{6}$

80. $x = \frac{\pi}{6}$

81. $x = \frac{3\pi}{4}$ or $x = \frac{7\pi}{4}$

82. $x = 2.366\ldots$

83. As $|x| \to \infty$, $\dfrac{\sin x}{x^2} \to 0$

84. As $x \to \infty$, $\frac{3}{5}e^{-x/12} \sin(2x - 3) \to 0$; as $x \to -\infty$, the function oscillates from positive to negative, and the oscillations become larger and larger.

85. 1

86. $\frac{\pi}{3}$

87. $\dfrac{0.75}{\sqrt{0.4375}} = 1.133\ldots$

88. $1.366\ldots$

89. Periodic; period π; domain: $x \neq \frac{\pi}{2} \pm n\pi$, n an integer; range: $[1, \infty)$

90. Not periodic; domain: $(-\infty, \infty)$; range: $[-1, 1]$

91. Not periodic; domain: $x \neq \frac{\pi}{2} \pm n\pi$, n an integer; range: $(-\infty, \infty)$

92. Periodic; period 2π; domain: $(-\infty, \infty)$; range: approximately $[-5, 4.65]$

93. $\frac{4\pi}{3}$

94. $\dfrac{\sqrt{1 - x^2}}{x}$

95. $h = 100 \tan 78° = 470.463\ldots$ m

96. $h = 51 \tan 25° = 23.781\ldots$ ft

97. $d = 150 \cot 18° - 150 \cot 42° = 295.060\ldots$ ft

98. $PQ = 4 \tan 22° = 1.616\ldots$ mi

99.

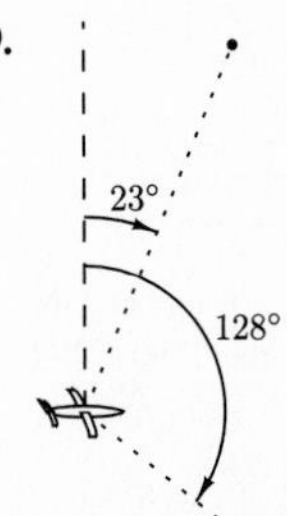

100. $d = \dfrac{855}{\tan 33° - \tan 25°} = 4669.581\ldots$ ft

101. $h = 62 \tan 72°24' = 195.448\ldots$ ft

102. **(a)** $h = 75 \sin 22° = 28.095\ldots$ ft.
(b) Now $h = 75 \sin 27° = 34.049\ldots$ ft.

103. $s = \frac{22\pi}{15} = 4.607\ldots$ in.

104. $A = \frac{11}{36}[\pi(20)^2 - \pi(4)^2] = \frac{352\pi}{3} = 368.613\ldots \text{ in}^2$

CHAPTER 7

QUICK REVIEW 7.1

1. $(a - b)^2$

2. $(2u + 1)^2$

3. $(2x + y)(x - 2y)$

4. $(2v + 1)(v - 3)$

5. $\dfrac{y - 2x}{xy}$

6. $\dfrac{ay + bx}{xy}$

7. xy

8. $\dfrac{x^2 + y^2}{x^2 - y^2}$

9. **(a)** This graph is *correct* but not very useful because very few of the interesting details are visible.

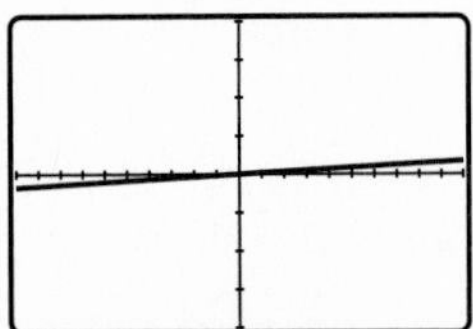

(b) This second view is much more useful because the two full periods of the function are visible.

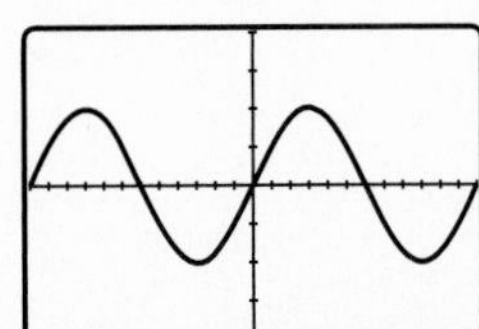

10. **(a)** The graph is not correct, because there should be about 115 periods in this window and there are only 20 visible.

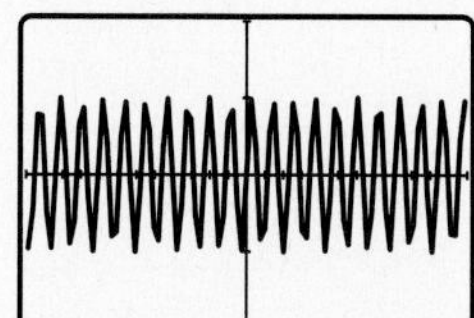

(b) This view is much more useful because only a few periods of the function are visible.

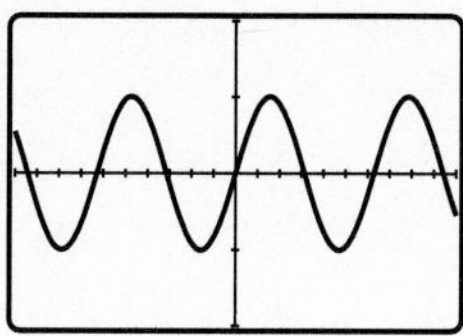

SECTION EXERCISES 7.1

1. $\sin x$
2. 1
3. 1
4. $\cos u$
5. $\tan^2 x$
6. $\sin \theta$
7. $(\cos x)(\sin^2 x)$
8. $\sec u$
9. -1
10. 1
11. 1
12. 1
13. 1
14. 1
15. $\sin x \tan x$
16. $\tan x$
17. 2
18. 1
19. $\dfrac{1}{\sin \alpha \cos \alpha}$
20. $2 \sin \theta$
21. $\dfrac{1 + 2 \sin y \cos y}{\sin^3 y \cos y}$
22. $\cos \theta - \sin \theta$
23. 1
24. $\cos x \sin x$
25. $2 \sec^2 x$
26. $2 \cot^2 x$
27. -1
28. $\cot x$
29. $2 \csc x$
30. $\tan x$
31. $\tan^2 \theta$
32. $\cos x + 1$
33. $(\cos x + 1)^2$
34. $(1 - \sin x)^2$
35. $(1 - \sin x)^2$
36. $(\sin x - 1)(\sin x + 2)$
37. $(2 \cos x - 1)(\cos x + 1)$
38. $(\sin x + 1)^2$
39. $(2 \tan x - 1)^2$
40. $(2 \sec x + 1)(\sec x - 1)$
41. $1 - \sin x$
42. $\tan \alpha - 1$
43. $1 - \cos x$
44. $\sec x - 1$
45. Yes
46. No. For a counterexample, take any x for which $\sin x < 0$, such as $-\frac{\pi}{2}$.
47. No. For a counterexample, take any x for which $\sec x < 0$, such as π.
48. Yes
49. Yes
50. No. Any $x \neq n\pi$ can be used.
51. Yes
52. No. $x = 0$ is the simplest counterexample.
53. $\sqrt{1 - \cos^2 \theta} = |\sin \theta| = \sin \theta$, since $0 \leq \theta \leq \frac{\pi}{2}$
54. $\sqrt{\tan^2 \theta + 1} = |\sec \theta| = \sec \theta$, since $0 \leq \theta \leq \frac{\pi}{2}$
55. $\sqrt{9 \sec^2 \theta - 9} = 3|\tan \theta| = 3 \tan \theta$, since $0 \leq \theta \leq \frac{\pi}{2}$
56. $\sqrt{36 - 36 \sin^2 \theta} = 6|\cos \theta| = 6 \cos \theta$, since $0 \leq \theta \leq \frac{\pi}{2}$
57. $\sqrt{81 \tan^2 \theta + 81} = 9|\sec \theta| = 9 \sec \theta$, since $0 \leq \theta \leq \frac{\pi}{2}$
58. $\sqrt{100 \sec^2 \theta - 100} = 10|\tan \theta| = 10 \tan \theta$, since $0 \leq \theta \leq \frac{\pi}{2}$
59. **(a)** True because $\cos^2 x = 1 - \sin^2 x$. **(b)** Since $\frac{\pi}{2} \leq x \leq \pi$, the cosine function is negative, so $\cos x = -\sqrt{1 - (\frac{3}{5})^2} = -\frac{4}{5}$. **(c)** Since $\frac{3\pi}{2} \leq x \leq 2\pi$, the cosine function is positive, so $\cos x = \sqrt{1 - (-\frac{3}{5})^2} = \frac{4}{5}$.
60. **(a)** True because $\sec^2 x = \tan^2 x + 1$. **(b)** Since $\pi \leq x \leq \frac{3\pi}{2}$, the secant function is negative, so $\sec x = -\sqrt{(\frac{5}{12})^2 + 1} = -\frac{13}{12}$. **(c)** Since $2\pi \leq x \leq \frac{5\pi}{2}$, the secant function is positive, so $\sec x = \sqrt{(\frac{5}{12})^2 + 1} = \frac{13}{12}$.
61. $\cos x = \pm\sqrt{1 - \sin^2 x}$, $\tan x = \pm\dfrac{\sin x}{\sqrt{1 - \sin^2 x}}$, $\csc x = \dfrac{1}{\sin x}$, $\sec x = \pm\dfrac{1}{\sqrt{1 - \sin^2 x}}$, $\cot x = \pm\dfrac{\sqrt{1 - \sin^2 x}}{\sin x}$

62. $\sin x = \pm\sqrt{1-\cos^2 x}$,
$\tan x = \pm\dfrac{\sqrt{1-\cos^2 x}}{\cos x}$,
$\csc x = \pm\dfrac{1}{\sqrt{1-\cos^2 x}}$,
$\sec x = \dfrac{1}{\cos x}$,
$\cot x = \pm\dfrac{\cos x}{\sqrt{1-\cos^2 x}}$

63. As defined in Section 6.3, $\csc\theta = \dfrac{r}{y}$ and $y = r\sin\theta$. Therefore $\csc\theta = \dfrac{r}{y} = \dfrac{r}{r\sin\theta} = \dfrac{1}{\sin\theta}$. Similarly, $\cot\theta = \dfrac{x}{y}$ and $x = r\cos\theta$. Therefore $\cot\theta = \dfrac{x}{y} = \dfrac{r\cos\theta}{r\sin\theta} = \dfrac{\cos\theta}{\sin\theta} = \dfrac{1}{\sin\theta/\cos\theta} = \dfrac{1}{\tan\theta}$.

QUICK REVIEW 7.2

1. $\dfrac{\sin x + \cos x}{\sin x\cos x}$

2. $\dfrac{1}{\sin x\cos x}$

3. $\dfrac{1}{\sin x\cos x}$

4. $\cos\theta - \sin\theta$, $x \neq n\frac{\pi}{2}$ and n an integer

5. 1, $x \neq n\frac{\pi}{2}$ and n an integer

6. 1, $x \neq n\frac{\pi}{2}$ and n an integer

7. No. For a counterexample, take any $x < 0$.

8. Yes. Cubing and cube-rooting do not affect the sign; that is, the cube of a positive number is positive, and the cube of a negative number is negative, and similarly for cube roots. These two functions undo one another.

9. No. For a counterexample, take any x for which $\sin x < 0$, such as $-\frac{\pi}{2}$.

10. No. For a counterexample, take any x for which $\tan x < 0$, such as $-\frac{\pi}{4}$.

11. Yes by properties of logarithms. Note that 0 is not in the domain of either function.

12. No because f can take any $x \neq 0$, but g requires that $x > 0$.

SECTION EXERCISES 7.2

1. No. There are many possible choices of x for counterexamples.

2. No. For a counterexample, take any $x \neq n\frac{\pi}{2}$, n an integer.

3. Yes

4. No. For a counterexample, take any $x \neq n\pi$, n an integer.

5. No. A simple counterexample is $x = 0$.

6. No. They are, in fact, opposites; take any $x \neq \frac{\pi}{4} + n\frac{\pi}{2}$, n an integer.

7. $(\cos x)(\tan x + \sin x\cot x) = \cos x\dfrac{\sin x}{\cos x} + \cos x\sin x\dfrac{\cos x}{\sin x} = \sin x + \cos^2 x$

8. $(\sin x)(\cot x + \cos x\tan x) = \sin x\dfrac{\cos x}{\sin x} + \sin x\cos x\dfrac{\sin x}{\cos x} = \cos x + \sin^2 x$

9. $(1-\tan x)^2 = 1 - 2\tan x + \tan^2 x = (1+\tan^2 x) - 2\tan x = \sec^2 x - 2\tan x$

10. $(\cos x - \sin x)^2 = \cos^2 x - 2\sin x\cdot\cos x + \sin^2 x = (\cos^2 x + \sin^2 x) - 2\sin x\cos x = 1 - 2\sin x\cos x$

11. $\tan x + \cot x = \dfrac{\sin x}{\cos x} + \dfrac{\cos x}{\sin x} = \dfrac{\sin^2 x + \cos^2 x}{\cos x\sin x} = \dfrac{1}{\cos x\sin x} = \sec x\csc x$

12. $\tan x + \sec x = \dfrac{\sin x}{\cos x} + \dfrac{1}{\cos x} = \dfrac{\sin x + 1}{\cos x} = \dfrac{\cos x(\sin x + 1)}{\cos^2 x} = \dfrac{\cos x(\sin x + 1)}{1 - \sin^2 x} = \dfrac{\cos x}{1 - \sin x}$

13. $\dfrac{\cos^2 x - 1}{\cos x} = \dfrac{-\sin^2 x}{\cos x} = -\dfrac{\sin x}{\cos x}\sin x = -\tan x\sin x$

14. $\dfrac{\sec^2\theta - 1}{\sin\theta} = \dfrac{\tan^2\theta}{\sin\theta} = \dfrac{1}{\sin\theta}\cdot\dfrac{\sin^2\theta}{\cos^2\theta} = \dfrac{\sin\theta}{\cos^2\theta} = \dfrac{\sin\theta}{1 - \sin^2\theta}$

15. Multiply out the expression.

16. $\dfrac{1}{1-\cos x} + \dfrac{1}{1+\cos x} = \dfrac{2}{(1-\cos x)(1+\cos x)} = \dfrac{2}{1-\cos^2 x} = \dfrac{2}{\sin^2 x} = 2\csc^2 x$

17. $(\cos t - \sin t)^2 + (\cos t + \sin t)^2 = \cos^2 t - 2\cos t\sin t + \sin^2 t + \cos^2 t + 2\cos t\sin t + \sin^2 t = \cos^2 t + \sin^2 t + \cos^2 t + \sin^2 t = 2$

18. $\sin^2\alpha - \cos^2\alpha = 1 - \cos^2\alpha - \cos^2\alpha = 1 - 2\cos^2\alpha$

19. $\dfrac{1+\tan^2 x}{\sin^2 x + \cos^2 x} = \dfrac{\sec^2 x}{1} = \sec^2 x$

20. $\dfrac{1}{\tan\beta} + \tan\beta = \dfrac{\cos\beta}{\sin\beta} + \dfrac{\sin\beta}{\cos\beta} = \dfrac{\sin^2\beta + \cos^2\beta}{\cos\beta\sin\beta} = \dfrac{1}{\cos\beta\sin\beta} = \sec\beta\csc\beta$

21. $\dfrac{\cos\beta}{1+\sin\beta} = \dfrac{\cos^2\beta}{\cos\beta(1+\sin\beta)} = \dfrac{1-\sin^2\beta}{\cos\beta(1+\sin\beta)} = \dfrac{1-\sin\beta}{\cos\beta}$

22. $\dfrac{\sec x + 1}{\tan x} = \dfrac{\tan x(1+\sec x)}{\tan^2 x} = \dfrac{\tan x(1+\sec x)}{\sec^2 x - 1} = \dfrac{\tan x}{\sec x - 1}$

23. $\dfrac{\tan^2 x}{\sec x + 1} = \dfrac{\sec^2 x - 1}{\sec x + 1} = \sec x - 1 = \dfrac{1}{\cos x} - 1 = \dfrac{1-\cos x}{\cos x}$

24. $\dfrac{\cot v - 1}{\cot v + 1} = \dfrac{\cot v - 1}{\cot v + 1}\cdot\dfrac{\tan v}{\tan v} = \dfrac{\cot v\tan v - \tan v}{\cot v\tan v + \tan v} = \dfrac{1-\tan v}{1+\tan v}$ since $\cot v\tan v = \dfrac{\cos v\sin v}{\sin v\cos v} = 1$.

25. $\cot^2 x - \cos^2 x = \left(\dfrac{\cos x}{\sin x}\right)^2 - \cos^2 x = \dfrac{\cos^2 x(1-\sin^2 x)}{\sin^2 x} = \dfrac{\cos^4 x}{\sin^2 x} = \cos^2 x\dfrac{\cos^2 x}{\sin^2 x} = \cos^2 x\cot^2 x$

26. $\tan^2\theta - \sin^2\theta = \left(\dfrac{\sin\theta}{\cos\theta}\right)^2 - \sin^2\theta = \dfrac{\sin^2\theta(1-\cos^2\theta)}{\cos^2\theta} = \dfrac{\sin^4\theta}{\cos^2\theta} = \sin^2\theta\dfrac{\sin^2\theta}{\cos^2\theta} = \sin^2\theta\tan^2\theta$

27. $\cos^4 x - \sin^4 x = (\cos^2 x + \sin^2 x)(\cos^2 x - \sin^2 x) = 1(\cos^2 x - \sin^2 x) = \cos^2 x - \sin^2 x$

28. $\tan^4 t + \tan^2 t = \tan^2 t(\tan^2 t + 1) = (\sec^2 t - 1)(\sec^2 t) = \sec^4 t - \sec^2 t$

29. $(x\sin\alpha + y\cos\alpha)^2 + (x\cos\alpha - y\cdot\sin\alpha)^2 = x^2\sin^2\alpha + 2xy\sin\alpha\cos\alpha + y^2\cos^2\alpha + x^2\cos^2\alpha - 2xy\cos\alpha\cdot\sin\alpha + y^2\sin^2\alpha = x^2\sin^2\alpha + y^2\cos^2\alpha + x^2\cos^2\alpha + y^2\sin^2\alpha = (x^2+y^2)(\sin^2\alpha + \cos^2\alpha) = x^2 + y^2$

30. $\dfrac{1-\cos\theta}{\sin\theta} = \dfrac{1-\cos^2\theta}{\sin\theta(1+\cos\theta)} = \dfrac{\sin^2\theta}{\sin\theta(1+\cos\theta)} = \dfrac{\sin\theta}{1+\cos\theta}$

31. $\dfrac{\tan x}{\sec x - 1} = \dfrac{\tan x(\sec x + 1)}{\sec^2 x - 1} = \dfrac{\tan x(\sec x + 1)}{\tan^2 x} = \dfrac{\sec x + 1}{\tan x}$

32. $\dfrac{\sin t}{1+\cos t} + \dfrac{1+\cos t}{\sin t} = \dfrac{\sin^2 t + (1+\cos t)^2}{(\sin t)(1+\cos t)} = \dfrac{\sin^2 t + 1 + 2\cos t + \cos^2 t}{(\sin t)(1+\cos t)} = \dfrac{2 + 2\cos t}{(\sin t)(1+\cos t)} = \dfrac{2}{\sin t} = 2\csc t$

33. $\dfrac{\sin x - \cos x}{\sin x + \cos x} = \dfrac{(\sin x - \cos x)(\sin x + \cos x)}{(\sin x + \cos x)^2} = \dfrac{\sin^2 x - \cos^2 x}{\sin^2 x + 2\sin x\cos x + \cos^2 x} = \dfrac{\sin^2 x - (1 - \sin^2 x)}{1 + 2\sin x\cos x} = \dfrac{2\sin^2 x - 1}{1 + 2\sin x\cos x}$

34. $\dfrac{1+\cos x}{1-\cos x} = \dfrac{1+\cos x}{1-\cos x}\cdot\dfrac{\sec x}{\sec x} = \dfrac{\sec x + \cos x\sec x}{\sec x - \cos x\sec x} = \dfrac{\sec x + 1}{\sec x - 1}$ since $\cos x\sec x = \cos x\dfrac{1}{\cos x} = 1.$

35. $\dfrac{\sin t}{1-\cos t} + \dfrac{1+\cos t}{\sin t} = \dfrac{\sin^2 t + (1+\cos t)(1-\cos t)}{(\sin t)(1-\cos t)} = \dfrac{\sin^2 t + 1 - \cos^2 t}{(\sin t)(1-\cos t)} = \dfrac{2(1-\cos^2 t)}{(\sin t)(1-\cos t)} = \dfrac{2(1+\cos t)}{\sin t}$

36. $\dfrac{\sin A\cos B + \cos A\sin B}{\cos A\cos B - \sin A\sin B} = \dfrac{1/\cos B\ 1/\cos A}{1/\cos B\ 1/\cos A}\cdot\dfrac{\sin A\cos B + \cos A\sin B}{\cos A\cos B - \sin A\sin B} = \dfrac{\sin A/\cos A + \sin B/\cos B}{1 - \sin A/\cos A\ \sin B/\cos B} = \dfrac{\tan A + \tan B}{1 - \tan A\tan B}$

37. $\sin^2 x\cos^3 x = \sin^2 x\cos^2 x\cos x = \sin^2 x(1-\sin^2 x)\cos x = (\sin^2 x - \sin^4 x)\cos x$

38. $\sin^5 x\cos^2 x = \sin^4 x\cos^2 x\sin x = (\sin^2 x)^2\cos^2 x\sin x = (1-\cos^2 x)^2\cdot\cos^2 x\sin x = (1 - 2\cos^2 x + \cos^4 x)\cdot\cos^2 x\sin x = (\cos^2 x - 2\cos^4 x + \cos^6 x)\sin x$

39. $\cos^5 x = \cos^4 x\cos x = (\cos^2 x)^2\cdot\cos x = (1 - \sin^2 x)^2\cos x = (1 - 2\sin^2 x + \sin^4 x)^2\cos x$

40. $\sin^3 x\cos^3 x = \sin^3 x\cos^2 x\cos x = \sin^3 x(1 - \sin^2 x)\cos x = (\sin^3 x - \sin^5 x)\cos x$

41. $\dfrac{\tan x}{1-\cot x} + \dfrac{\cot x}{1-\tan x} = \dfrac{\tan x}{1-\cot x}\cdot\dfrac{\sin x}{\sin x} + \dfrac{\cot x}{1-\tan x}\cdot\dfrac{\cos x}{\cos x} = \dfrac{\sin^2 x/\cos x}{\sin x - \cos x} + \dfrac{\cos^2 x/\sin x}{\cos x - \sin x} = \dfrac{\sin^2 x/\cos x - \cos^2 x/\sin x}{\sin x - \cos x} = \dfrac{\sin^3 x - \cos^3 x}{\sin x\cos x(\sin x - \cos x)} = \dfrac{(\sin x - \cos x)(\sin^2 x + \sin x\cos x + \cos^2 x)}{\sin x\cos x(\sin x - \cos x)} = \dfrac{1 + \sin x\cos x}{\sin x\cos x} = \dfrac{1}{\sin x\cos x} + 1 = \csc x\sec x + 1$

42. $\dfrac{\cos x}{1+\sin x} + \dfrac{\cos x}{1-\sin x} = \dfrac{(\cos x)(1-\sin x) + (\cos x)(1+\sin x)}{(1+\sin x)(1-\sin x)} = \dfrac{\cos x - \cos x\sin x + \cos x + \cos x\sin x}{1 - \sin^2 x} = \dfrac{2\cos x}{\cos^2 x} = 2\sec x$

43. $\dfrac{2\tan x}{1-\tan^2 x} + \dfrac{1}{2\cos^2 x - 1} = \dfrac{2\tan x}{1-\tan^2 x}\cdot\dfrac{\cos^2 x}{\cos^2 x} + \dfrac{1}{\cos^2 x - \sin^2 x} = \dfrac{2\sin x\cos x}{\cos^2 x - \sin^2 x} + \dfrac{\cos^2 x + \sin^2 x}{\cos^2 x - \sin^2 x} = \dfrac{2\sin x\cos x + \cos^2 x + \sin^2 x}{(\cos x - \sin x)(\cos x + \sin x)} = \dfrac{(\cos x + \sin x)^2}{(\cos x - \sin x)(\cos x + \sin x)} = \dfrac{\cos x + \sin x}{\cos x - \sin x}$

44. $\dfrac{1 - 3\cos x - 4\cos^2 x}{\sin^2 x} = \dfrac{(1+\cos x)(1 - 4\cos x)}{1 - \cos^2 x} = \dfrac{(1+\cos x)(1-4\cos x)}{(1+\cos x)(1-\cos x)} = \dfrac{1 - 4\cos x}{1 - \cos x}$

45. **(b)**; divide by $\cos x$: $\dfrac{1+\sin x}{\cos x} = \dfrac{1}{\cos x} + \dfrac{\sin x}{\cos x} = \sec x + \tan x.$

46. **(d)**; $(1 + \sec x)(1 - \cos x) = 1 - \cos x + \sec x - \sec x\cos x = 1 - \cos x + \dfrac{1}{\cos x} - 1 = \dfrac{1 - \cos^2 x}{\cos x} = \dfrac{\sin^2 x}{\cos x} = \dfrac{\sin x}{\cos x}\sin x = \tan x\sin x.$

47. **(a)**; put over a common denominator: $\sec^2 x + \csc^2 x = \left(\frac{1}{\cos x}\right)^2 + \left(\frac{1}{\sin x}\right)^2 = \frac{\sin^2 x + \cos^2 x}{\cos^2 x \sin^2 x} = \frac{1}{\cos^2 x \sin^2 x} = \left(\frac{1}{\cos x}\frac{1}{\sin x}\right)^2 = \sec^2 x \csc^2 x.$

48. **(c)**; put over a common denominator: $\frac{1}{1 + \sin x} + \frac{1}{1 - \sin x} = \frac{1 - \sin x + 1 + \sin x}{1 - \sin^2 x} = \frac{2}{\cos^2 x} = 2\sec^2 x.$

49. **(e)**; multiply and divide by $\sin x \cos x$: $\frac{1}{\tan x + \cot x} = \frac{\sin x \cos x}{(\sin x/\cos x + \cos x/\sin x)(\sin x \cos x)} = \frac{\sin x \cos x}{\sin^2 x + \cos^2 x} = \frac{\sin x \cos x}{1} = \sin x \cos x.$

50. **(b)**; multiply and divide by $\sec x + \tan x$: $\tan x$: $\frac{\sec x + \tan x}{(\sec x - \tan x)(\sec x + \tan x)} = \frac{\sec x + \tan x}{\sec^2 x - \tan^2 x} = \frac{\sec x + \tan x}{1}$ (using a Pythagorean identity).

51. $\cos x$; $\sin x \cot x = \sin x \frac{\cos x}{\sin x} = \cos x$

52. $\sin x$; $\cos x \tan x = \cos x \frac{\sin x}{\cos x} = \sin x$

53. 1; $\frac{\sin x}{\csc x} + \frac{\cos x}{\sec x} = \frac{\sin x}{1/\sin x} + \frac{\cos x}{1/\cos x} = \sin^2 x + \cos^2 x = 1$

54. 1; $\frac{\csc x}{\sin x} - \frac{\cot x \csc x}{\sec x} = \frac{1/\sin x}{\sin x} - \frac{\cos x/\sin^2 x}{1/\cos x} = \frac{1}{\sin^2 x} - \frac{\cos^2 x}{\sin^2 x} = \frac{1 - \cos^2 x}{\sin^2 x} = \frac{\sin^2 x}{\sin^2 x} = 1$

55. $\cos x$; $\frac{\sin x}{\tan x} = \frac{\sin x}{\sin x/\cos x} = \cos x$

56. 1; $\left(\frac{1}{\cos x}\right)^2(\cos^2 x) = 1$

57. Multiply and divide by $1 - \sin t$ under the radical: $\sqrt{\frac{1 - \sin t}{1 + \sin t}\cdot\frac{1 - \sin t}{1 - \sin t}} = \sqrt{\frac{(1 - \sin t)^2}{1 - \sin^2 t}} = \sqrt{\frac{(1 - \sin t)^2}{\cos^2 t}} = \frac{|1 - \sin t|}{|\cos t|}$, since $\sqrt{a^2} = |a|$. Now, since $1 - \sin t \ge 0$, we can dispense with the absolute value in the numerator, but it must stay in the denominator.

58. Multiply and divide by $1 + \cos t$ under the radical: $\sqrt{\frac{1 + \cos t}{1 - \cos t}\cdot\frac{1 + \cos t}{1 + \cos t}} = \sqrt{\frac{(1 + \cos t)^2}{1 - \cos^2 t}} = \sqrt{\frac{(1 + \cos t)^2}{\sin^2 t}} = \frac{|1 + \cos t|}{|\sin t|}$, since $\sqrt{a^2} = |a|$. Now, since $1 + \cos t \ge 0$, we can dispense with the absolute value in the numerator, but it must stay in the denominator.

59. $\sin^6 x + \cos^6 x = (\sin^2 x)^3 + \cos^6 x = (1 - \cos^2 x)^3 + \cos^6 x = 1 - 3\cos^2 x + 3\cos^4 x - \cos^6 x + \cos^6 x = 1 - 3\cos^2 x(1 - \cos^2 x) = 1 - 3\cos^2 x \sin^2 x$

60. $a^3 - b^3 = (a - b)(a^2 + ab + b^2)$; also note that $a^2 + ab + b^2 = a^2 + 2ab + b^2 - ab = (a + b)^2 - ab$. Taking $a = \cos^2 x$ and $b = \sin^2 x$, we have $\cos^6 x - \sin^6 x = (\cos^2 x - \sin^2 x)(\cos^4 x + \cos^2 x \sin^2 x + \sin^4 x) = (\cos^2 x - \sin^2 x)[(\cos^2 x + \sin^2 x)^2 - \cos^2 x \sin^2 x] = (\cos^2 x - \sin^2 x)\cdot(1 - \cos^2 x \sin^2 x).$

61. **(a)** They are not equal; shown is the window $[-2\pi, 2\pi] \times [-2, 2]$; graphing on nearly any viewing window does not show any apparent difference, but, using TRACE, we find that the y-coordinates are not identical; likewise, a table of values will show slight differences; for example, when $x = 1$, $y_1 = 0.53988\ldots$ and $y_2 = 0.54030\ldots$

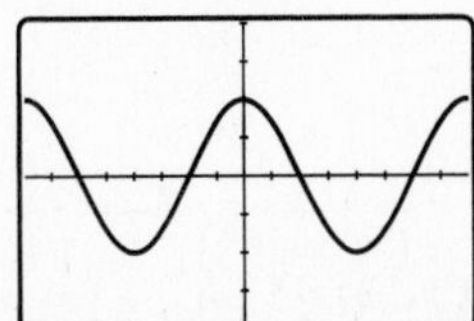

(b) One choice for h is 0.001, as shown; the function y_3 is a combination of three sinusoidal functions [$1000 \sin(x + 0.001)$, $-1000 \sin x$, and $-\cos x$], all with period 2π.

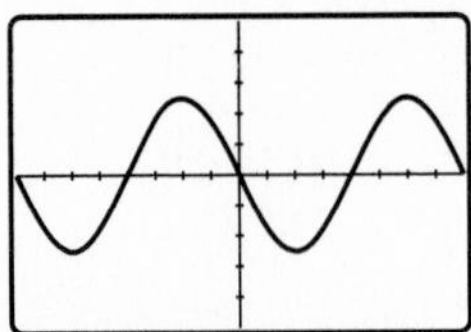

62. **(a)** $\cosh^2 x - \sinh^2 x = \frac{1}{4}(e^x + e^{-x})^2 - \frac{1}{4}(e^x - e^{-x})^2 = \frac{1}{4}[e^{2x} + 2 + e^{-2x} - (e^{2x} - 2 + e^{-2x})] = \frac{1}{4}(4) = 1.$

(b) $1 - \tanh^2 x = 1 - \frac{\sinh^2 x}{\cosh^2 x} = \frac{\cosh^2 x - \sinh^2 x}{\cosh^2 x} = \frac{1}{\cosh^2 x}$, using the result from **(a)**, which equals $\operatorname{sech}^2 x$.

(c) $\coth^2 x - 1 = \frac{\cosh^2 x}{\sinh^2 x} - 1 = \frac{\cosh^2 x - \sinh^2 x}{\sinh^2 x} = \frac{1}{\sinh^2 x}$, using the result from **(a)**, which equals $\operatorname{csch}^2 x$.

63. In the decimal window, the x-coordinates used to plot the graph on the calculator are, for example, 0, 0.1, 0.2, 0.3, and so on; that is, $x = \frac{n}{10}$, where n is an integer. Then $10\pi x = \pi n$, and the sine of integer multiples of π is 0; therefore $\cos x + \sin 10\pi x = \cos x + \sin \pi n = \cos x + 0 = \cos x$. However, for other choices of x, such as $x = \frac{1}{\pi}$, we have $\cos x + \sin 10\pi x = \cos x + \sin 10 \neq \cos x$.

QUICK REVIEW 7.3

1. 1.1760 . . . radians = 67.380 . . .°
2. 0.9272 . . . radians = 53.130 . . .°
3. 2.4980 . . . radians = 143.130 . . .°
4. −0.3947 . . . radians = −22.619 . . .°
5. $x = 0.9272\ldots$ or $x = 2.2142\ldots$
6. $x = 2.7468\ldots$ or $x = 3.5363\ldots$
7. No solutions

8. No solutions

9. $x = \frac{3\pi}{4}$ or $x = \frac{7\pi}{4}$

10. $x = \frac{2\pi}{3}$ or $x = \frac{4\pi}{3}$

11. $x = 0.2013\ldots$ or $x = 2.9402\ldots$

12. $x = 0.3217\ldots$ or $x = 3.4633\ldots$

13. No solutions

14. No solutions

15. $\frac{\pi}{6}, \frac{5\pi}{6}, \frac{3\pi}{2}$

16. $\frac{\pi}{2}, \frac{7\pi}{6}, \frac{11\pi}{6}$

17. $\frac{\pi}{8}, \frac{9\pi}{8}$

18. $\frac{\pi}{6}, \frac{2\pi}{3}, \frac{7\pi}{6}, \frac{5\pi}{3}$

SECTION EXERCISES 7.3

1. $x = \frac{\pi}{6} + 2n\pi$ or $x = \frac{5\pi}{6} + 2n\pi$, n an integer

2. $\theta = \frac{\pi}{6} + 2n\pi$ or $\theta = \frac{11\pi}{6} + 2n\pi$, n an integer; or, $\theta = \pm\frac{\pi}{6} + 2n\pi$, n an integer

3. $x = \frac{\pi}{3} + 2n\pi$ or $x = \frac{5\pi}{3} + 2n\pi$, n an integer; or, $x = \pm\frac{\pi}{3} + 2n\pi$, n an integer

4. $\alpha = \frac{\pi}{6} + n\pi$, n an integer

5. $x = \frac{\pi}{6} + 2n\pi$ or $x = \frac{5\pi}{6} + 2n\pi$, n an integer

6. No solutions

7. $x = -\frac{\pi}{3} + n\pi$, n an integer

8. $y = -\frac{\pi}{6} + 2n\pi$ or $y = \frac{7\pi}{6} + 2n\pi$, n an integer

9. $\cos x(2\sin x - 1) = 0$, so $x = \frac{\pi}{2} + n\pi$ or $x = \frac{\pi}{6} + 2n\pi$ or $x = \frac{5\pi}{6} + 2n\pi$, n an integer; window: $[-2\pi, 2\pi] \times [-2, 2]$

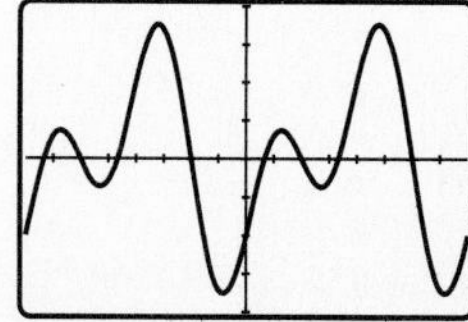

10. $\tan x(\sqrt{2}\cos x - 1) = 0$, so $x = n\pi$ or $x = \pm\frac{\pi}{4} + 2n\pi$, n an integer; window: $[-2\pi, 2\pi] \times [-4, 4]$

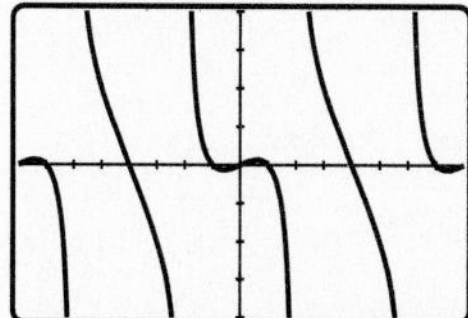

11. Either $\tan x = 0$ or $\sin^2 x = 1$, so $x = n\pi$ or $x = \frac{\pi}{2} + n\pi$, n an integer. But since $\tan x$ excludes $x = \frac{\pi}{2} + n\pi$, we have only $x = n\pi$, n an integer; window: $[-2\pi, 2\pi] \times [-1, 1]$

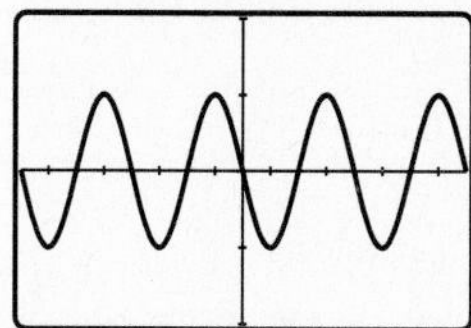

12. Either $\sin x = 0$ or $\tan^2 x = 1$, so $x = n\pi$ or $x = \frac{\pi}{4} + n\frac{\pi}{2}$, n an integer. Put another way, all multiples of $\frac{\pi}{4}$ except for $\pm\frac{\pi}{2}, \pm\frac{3\pi}{2}$, and so on. Window: $[-2\pi, 2\pi] \times [-4, 4]$

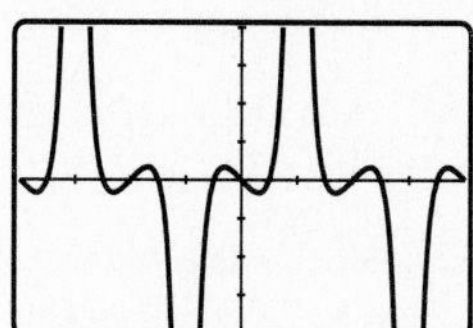

13. $\tan x = \pm\sqrt{3}$, so $x = \pm\frac{\pi}{3} + n\pi$, n an integer; window: $[-2\pi, 2\pi] \times [-1, 5]$

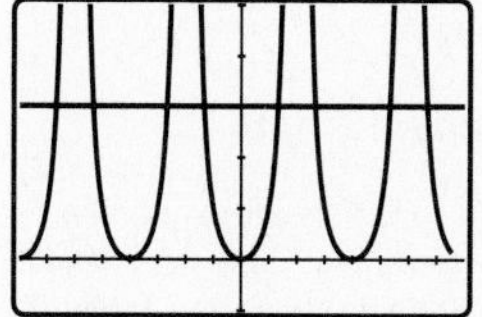

14. $\sin x = \pm\dfrac{1}{\sqrt{2}}$, so $x = \dfrac{\pi}{4} + n\dfrac{\pi}{2}$, n an integer; window: $[-2\pi, 2\pi] \times [-1, 3]$

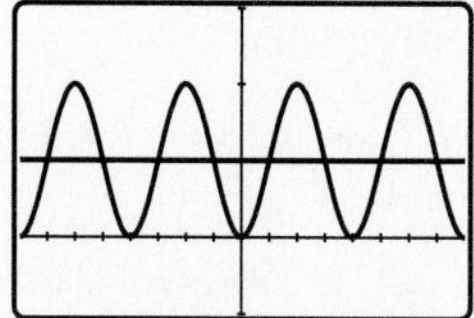

15. $(2\cos x - 1)^2 = 0$, so $\cos x = \frac{1}{2}$; therefore $x = \pm\frac{\pi}{3} + 2n\pi$, n an integer; window: $[-2\pi, 2\pi] \times [-1, 10]$

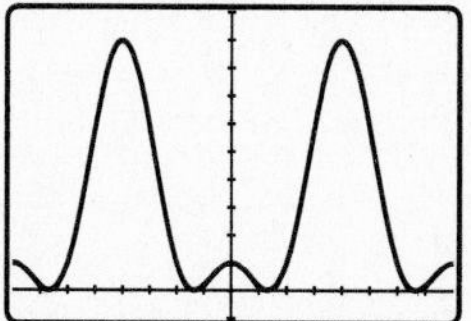

16. $(2\sin x + 1)(\sin x + 1) = 0$, so $\sin x = -\frac{1}{2}$ or $\sin x = -1$; therefore $x = -\frac{\pi}{6} + 2n\pi$, $x = -\frac{5\pi}{6} + 2n\pi$, or $x = -\frac{\pi}{2} + 2n\pi$, n an integer; window: $[-2\pi, 2\pi] \times [-1, 6]$

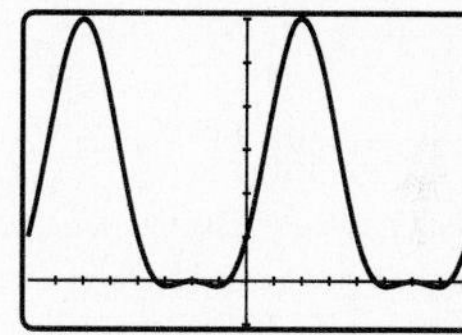

17. $(\sqrt{2}\cos x - 1)(2\sin x - 1) = 0$, so $x = \pm\frac{\pi}{4} + 2n\pi$ or $x = \frac{\pi}{6} + 2n\pi$ or $x = \frac{5\pi}{6} + 2n\pi$, n an integer; window: $[-2\pi, 2\pi] \times [-2, 5]$

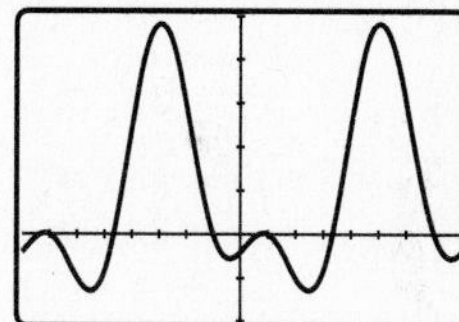

18. $(2\cos x + 1)(\tan x - 1) = 0$, so $x = \pm\frac{2\pi}{3} + 2n\pi$ or $x = \frac{\pi}{4} + n\pi$, n an integer; window: $[-2\pi, 2\pi] \times [-3, 3]$

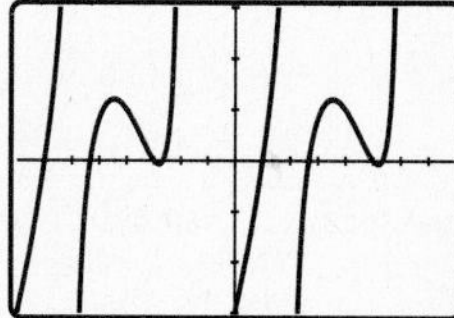

19. $x = \frac{\pi}{4}$ or $x = \frac{5\pi}{4}$

20. $t = \frac{\pi}{6}, t = \frac{5\pi}{6}, t = \frac{3\pi}{2}$

21. $t = \frac{\pi}{9}, t = \frac{5\pi}{9}, t = \frac{7\pi}{9}, t = \frac{11\pi}{9}, t = \frac{13\pi}{9}$, or $t = \frac{17\pi}{9}$

22. $x = \frac{\pi}{8}, x = \frac{5\pi}{8}, x = \frac{9\pi}{8}$, or $x = \frac{13\pi}{8}$

23. Let $\alpha = \frac{1}{2}\tan^{-1}\frac{1}{3} = 0.1608\ldots$. The solutions are $\theta = \alpha, \theta = \alpha + \frac{\pi}{2}$, $\theta = \alpha + \pi$, or $\theta = \alpha + \frac{3\pi}{2}$.

24. Let $\alpha = \frac{1}{2}\cos^{-1}(\frac{2}{3}) = 0.4205\ldots$ The solutions are $t = \alpha, t = \pi - \alpha, t = \alpha + \pi$, or $t = 2\pi - \alpha$.

25. $x = 0, x = \frac{\pi}{3}, x = \frac{2\pi}{3}, x = \pi, x = \frac{4\pi}{3}$, or $x = \frac{5\pi}{3}$

26. $x = \frac{\pi}{8}, x = \frac{3\pi}{8}, x = \frac{5\pi}{8}, x = \frac{7\pi}{8}, x = \frac{9\pi}{8}, x = \frac{11\pi}{8}, x = \frac{13\pi}{8}$, or $x = \frac{15\pi}{8}$

27. $x = 0, x = \frac{3\pi}{8}, x = \frac{\pi}{2}, x = \frac{7\pi}{8}, x = \pi, x = \frac{11\pi}{8}, x = \frac{3\pi}{2}$, or $x = \frac{15\pi}{8}$

28. $x = \frac{\pi}{4}, x = \frac{7\pi}{12}, x = \frac{3\pi}{4}, x = \frac{11\pi}{12}, x = \frac{5\pi}{4}, x = \frac{19\pi}{12}, x = \frac{7\pi}{4}$, or $x = \frac{23\pi}{12}$

29. $\cos x = \dfrac{3 \pm \sqrt{41}}{4}$; only the "minus" choice gives a number between -1 and 1; let $\alpha = \cos^{-1}\left(\dfrac{3 - \sqrt{41}}{4}\right) = 2.5882\ldots$. The solutions are $x = \alpha$ or $x = 2\pi - \alpha = 3.6949\ldots$.

30. $\sin x = \dfrac{\sqrt{2} \pm \sqrt{6}}{2}$; only the "minus" choice gives a number between -1 and 1. Let $\alpha = \sin^{-1}\left(\dfrac{\sqrt{2} - \sqrt{6}}{2}\right) = -0.5440\ldots$; the solutions are $x = \pi - \alpha = 3.6856\ldots$ or $x = 2\pi + \alpha = 5.7390\ldots$

31. $\tan 2x = 1 \pm \sqrt{6}$; let $\alpha = \frac{1}{2}\tan^{-1}(1 - \sqrt{6}) = -0.4834\ldots$ and $\beta = \frac{1}{2}\tan^{-1}(1 + \sqrt{6}) = 0.6443\ldots$. The solutions are $x = \beta, x = \alpha + \frac{\pi}{2} = 1.0873\ldots, x = \beta + \frac{\pi}{2} = 2.2151\ldots, x = \alpha + \pi = 2.6581\ldots, x = \beta + \pi = 3.7859\ldots, x = \alpha + \frac{3\pi}{2} = 4.2289\ldots, x = \beta + \frac{3\pi}{2} = 5.3567\ldots$, or $x = \alpha + 2\pi = 5.7997\ldots$

32. $\sin 2x = \dfrac{-1 \pm \sqrt{13}}{6}$; let $\alpha = \dfrac{1}{2}\sin^{-1}\left(\dfrac{-1 - \sqrt{13}}{6}\right) = -0.4375\ldots$ and $\beta = \dfrac{1}{2}\sin^{-1}\left(\dfrac{-1 + \sqrt{13}}{6}\right) = 0.2246\ldots$. The solutions are $x = \beta, x = \dfrac{\pi}{2} - \beta = 1.3461\ldots, x = \dfrac{\pi}{2} - \alpha = 2.0083\ldots, x = \alpha + \pi = 2.7040\ldots, x = \beta + \pi = 3.3662\ldots, x = \dfrac{3\pi}{2} - \beta = 4.4877\ldots, x = \dfrac{3\pi}{2} - \alpha = 5.1499\ldots$, or $x = \alpha + 2\pi = 5.8456\ldots$.

33. $\theta = n\pi$, n an integer

34. $t = \pm\dfrac{1}{2}\cos^{-1}\left(\dfrac{3}{8}\right) + n\pi = \pm 0.5931\ldots + n\pi$, n an integer

35. $\sin t = -2$ or $\sin t = \frac{1}{2}$; $t = \frac{\pi}{6} + 2n\pi$ or $t = \frac{5\pi}{6} + 2n\pi$, n an integer

36. $y = n\pi$ or $y = \pm\, 1.047\ldots + 2n\pi = \pm\frac{\pi}{3} + 2n\pi$, n an integer

37. $x = n\pi$, n an integer

38. $x = \pm\cos^{-1}[\tan^{-1}(\frac{3}{4})] + 2n\pi = \pm 0.8717\ldots + 2n\pi$, n an integer

39. $\sin x = -2$ or $\sin x = \frac{1}{2}$; $x = \frac{\pi}{6} + 2n\pi$ or $x = \frac{5\pi}{6} + 2n\pi$, n an integer

40. $\cos 2x = -1$ or $\cos 2x = \frac{1}{2}$; $x = \frac{\pi}{6} + n\frac{\pi}{3}$, n an integer

41. $x = \frac{\pi}{3} + n\frac{2\pi}{3}$, n an integer. (The three zeros in $[0, 2\pi)$ are $\pi/3$, π, and $5\pi/3$.)

42. There are no solutions: This factors to $(\cos x)(\tan^2 x + 5) = 0$; the second factor is always positive, and anywhere that $\cos x = 0$ is not allowed, since $\tan x$ is undefined there.

43. $x \approx -2.99$

44. $x \approx -0.64$ or $x \approx 1.41$

45. $x \approx -1.82, x \approx -0.93$, or $x \approx 0.44$

46. $x \approx 1.51$ or $x \approx 2.21$ or $x \approx 3.38$

47. About day 159 (158.70) and about day 226 (225.80)

48. Up to about day 96 (96.28) and after about day 288 (288.22)

49. Either $\theta \approx 0.15$ rad $(8.75°)$ or $\theta \approx 1.42$ rad $(81.25°)$

50. $\theta = \frac{\pi}{4}$ rad, or 45°; the maximum distance is 7812.5 ft

51. **(a)** $h(x) = 55\cos\left(\frac{\pi x}{210}\right) + 35$; that is, $a = 55$, $b = \frac{\pi}{210}$, and $k = 35$.
(b) $h(x) = 0$ when $x \approx 151.11$ ft

52. Falling 50 ft brings the front car to a height of 40 ft; $h(x) = 40$ when $x \approx 98.91$ ft.

53. $\theta \approx 0.583$ rad (about 33.41°), or $\theta \approx 0.988$ rad (about 56.59°)

54. $\theta \approx 0.595$ rad (about 34.06°), or $\theta \approx 0.976$ rad (about 55.94°)

55. $0 \le x < \frac{\pi}{6}$ or $\frac{5\pi}{6} < x \le 2\pi$; that is, $[0, \frac{\pi}{6}) \cup (\frac{5\pi}{6}, 2\pi]$

56. $0 \le x < \frac{\pi}{4}$ or $\frac{5\pi}{4} < x \le 2\pi$; that is, $[0, \frac{\pi}{4}) \cup (\frac{5\pi}{4}, 2\pi]$

57. $0 \le x < \frac{\pi}{4}$ or $\frac{3\pi}{4} < x < \frac{5\pi}{4}$ or $\frac{7\pi}{4} < x \le 2\pi$; that is, $[0, \frac{\pi}{4}) \cup (\frac{3\pi}{4}, \frac{5\pi}{4}) \cup (\frac{7\pi}{4}, 2\pi]$

58. $0 < x < \frac{\pi}{2}$ or $\pi < x < \frac{3\pi}{2}$; that is, $(0, \frac{\pi}{2}) \cup (\pi, \frac{3\pi}{2})$

59. $x = n\pi$, n a nonzero integer

60. $x = \frac{\pi}{2} + n\pi$, n an integer

61. $x = x_1 + np$, n an integer

62. $x = x_1 + np$ or $x = x_2 + np$, n an integer

63. Since $\sin x$ has period 2π, there are also two solutions to $\sin x = 0.5$ in the interval $[2\pi, 4\pi)$ and two more in $[4\pi, 6\pi)$—a total of six solutions in $[0, 6\pi)$. Starting from $y_1 = \sin x$, we obtain the graph of $y_2 = \sin 3x$ with a horizontal shrink by $\frac{1}{3}$; that is, the portion of the graph of y_1 with $0 \le x < 6\pi$ is shrunk to the interval $0 \le x < 2\pi$. Then the six solutions to $\sin x = 0$ from the interval $[0, 6\pi)$ become solutions to the equation $\sin 3x = 0$ in the interval $[0, 2\pi)$.

QUICK REVIEW 7.4

1. $45° - 30°$

2. $45° + 30°$

3. $3 \cdot 45° + 30° = 135° + 30°$, or $4 \cdot 30° + 45° = 120° + 45°$

4. $5 \cdot 45° + 30° = 225° + 30°$, or $7 \cdot 30° + 45° = 210° + 45°$

5. $2 - \frac{\pi}{6} - \frac{\pi}{4} = \frac{\pi}{3} - \frac{\pi}{4}$

6. $4 - \frac{\pi}{6} - \frac{\pi}{4} = \frac{2\pi}{3} - \frac{\pi}{4}$

7. $5 - \frac{\pi}{6} - \frac{\pi}{4} = \frac{5\pi}{6} - \frac{\pi}{4}$

8. $\frac{\pi}{4} - 2 - \frac{\pi}{6} = \frac{\pi}{4} - \frac{\pi}{3}$

9. Yes; amplitude: ≈ 3.61; period: π; phase shift: ≈ -0.29

10. Yes; amplitude ≈ 6.40; period $\frac{2\pi}{3}$; phase shift ≈ 0.22

11. No

12. No

SECTION EXERCISES 7.4

1. $\sin 15° = \sin(45° - 30°) = \sin 45° \cos 30° - \cos 45° \sin 30° = \frac{1}{\sqrt{2}}\frac{\sqrt{3}}{2} - \frac{1}{\sqrt{2}}\frac{1}{2} = \frac{\sqrt{3} - 1}{2\sqrt{2}} = \frac{\sqrt{6} - \sqrt{2}}{4}$

2. $\tan 15° = \tan(45° - 30°) = \frac{\tan 45° - \tan 30°}{1 + \tan 45° \tan 30°} = \frac{1 - \sqrt{3}/3}{1 + \sqrt{3}/3} = \frac{3 - \sqrt{3}}{3 + \sqrt{3}} = \frac{(3 - \sqrt{3})^2}{9 - 3} = 2 - \sqrt{3}$

3. $\sin 75° = \sin(45° + 30°) = \sin 45° \cos 30° + \cos 45° \sin 30° = \frac{1}{\sqrt{2}} \cdot \frac{\sqrt{3}}{2} + \frac{1}{\sqrt{2}} \cdot \frac{1}{2} = \frac{\sqrt{3} + 1}{2\sqrt{2}} = \frac{\sqrt{6} + \sqrt{2}}{4}$

4. $\cos 75° = \cos(45° + 30°) = \cos 45° \cos 30° - \sin 45° \sin 30° = \frac{1}{\sqrt{2}}\frac{\sqrt{3}}{2} - \frac{1}{\sqrt{2}}\frac{1}{2} = \frac{\sqrt{3} - 1}{2\sqrt{2}} = \frac{\sqrt{6} - \sqrt{2}}{4}$

5. $\cos\frac{\pi}{12} = \cos\left(\frac{\pi}{3} - \frac{\pi}{4}\right) = \cos\frac{\pi}{3}\cos\frac{\pi}{4} + \sin\frac{\pi}{3}\sin\frac{\pi}{4} = \frac{1}{2}\frac{1}{\sqrt{2}} + \frac{\sqrt{3}}{2}\frac{1}{\sqrt{2}} = \frac{1 + \sqrt{3}}{2\sqrt{2}} = \frac{\sqrt{2} + \sqrt{6}}{4}$

6. $\sin\frac{7\pi}{12} = \sin\left(\frac{\pi}{3} + \frac{\pi}{4}\right) = \sin\frac{\pi}{3}\cos\frac{\pi}{4} + \cos\frac{\pi}{3}\sin\frac{\pi}{4} = \frac{\sqrt{3}}{2}\frac{1}{\sqrt{2}} + \frac{1}{2}\frac{1}{\sqrt{2}} = \frac{\sqrt{3} + 1}{2\sqrt{2}} = \frac{\sqrt{6} + \sqrt{2}}{4}$

7. $\tan\frac{5\pi}{12} = \tan\left(\frac{2\pi}{3} - \frac{\pi}{4}\right) = \frac{\tan 2\pi/3 - \tan\pi/4}{1 + \tan 2\pi/3 \tan\pi/4} = \frac{-\sqrt{3} - 1}{1 - \sqrt{3}} = \frac{\sqrt{3} + 1}{\sqrt{3} - 1} = \frac{(\sqrt{3} + 1)^2}{3 - 1} = 2 + \sqrt{3}$

8. $\tan\frac{11\pi}{12} = \tan\left(\frac{2\pi}{3} + \frac{\pi}{4}\right) = \frac{\tan 2\pi/3 + \tan\pi/4}{1 - \tan 2\pi/3 \tan\pi/4} = \frac{-\sqrt{3} + 1}{1 + \sqrt{3}} = \frac{1 - \sqrt{3}}{1 + \sqrt{3}} = \frac{(1 - \sqrt{3})^2}{1 - 3} = \sqrt{3} - 2$

9. $\cos\frac{7\pi}{12} = \cos\left(\frac{5\pi}{6} - \frac{\pi}{4}\right) = \cos\frac{5\pi}{6}\cos\frac{\pi}{4} + \sin\frac{5\pi}{6}\sin\frac{\pi}{4} = -\frac{\sqrt{3}}{2}\frac{1}{\sqrt{2}} + \frac{1}{2}\frac{1}{\sqrt{2}} = \frac{1 - \sqrt{3}}{2\sqrt{2}} = \frac{\sqrt{2} - \sqrt{6}}{4}$

10. $\sin -\frac{\pi}{12} = \sin\left(\frac{\pi}{6} - \frac{\pi}{4}\right) = \sin\frac{\pi}{6}\cos\frac{\pi}{4} - \cos\frac{\pi}{6}\sin\frac{\pi}{4} = \frac{1}{2}\frac{1}{\sqrt{2}} - \frac{\sqrt{3}}{2}\frac{1}{\sqrt{2}} = \frac{1 - \sqrt{3}}{2\sqrt{2}} = \frac{\sqrt{2} - \sqrt{6}}{4}$

11. $\sin(42° - 17°) = \sin 25°$

12. $\cos(94° - 18°) = \cos 76°$

13. $\sin\left(\frac{\pi}{5} + \frac{\pi}{2}\right) = \sin\frac{7\pi}{10}$

14. $\sin\left(\frac{\pi}{3} - \frac{\pi}{7}\right) = \sin\frac{4\pi}{21}$

15. $\tan(19° + 47°) = \tan 66°$

16. $\tan\left(\frac{\pi}{5} - \frac{\pi}{3}\right) = \tan\frac{-2\pi}{15}$

17. $\cos\left(\frac{\pi}{7} - x\right) = \cos\left(x - \frac{\pi}{7}\right)$

18. $\cos\left(\frac{\pi}{7} + x\right) = \cos\left(x + \frac{\pi}{7}\right)$

19. $\sin(3x - x) = \sin 2x$

20. $\cos(7y + 3y) = \cos 10y$

21. $\tan(2y + 3x)$

22. $\tan(3\alpha - 2\beta)$

23. $\sin\left(x - \frac{\pi}{2}\right) = \sin x \cos\frac{\pi}{2} - \cos x \sin\frac{\pi}{2} = \sin x \cdot 0 - \cos x \cdot 1 = -\cos x$

24. Using the difference identity for the tangent function, we encounter $\tan\frac{\pi}{2}$, which is undefined; however, we can compute $\tan\left(x - \frac{\pi}{2}\right) = \frac{\sin(x - \pi/2)}{\cos(x - \pi/2)}$. From Exercise 23, $\sin\left(x - \frac{\pi}{2}\right) = -\cos x$; since the cosine function is even, $\cos\left(x - \frac{\pi}{2}\right) = \cos\left(\frac{\pi}{2} - x\right) = \sin x$ (Example 4); therefore this simplifies to $\frac{-\cos x}{\sin x} = -\cot x$.

25. $\cos\left(x - \frac{\pi}{2}\right) = \cos x \cos\frac{\pi}{2} + \sin x \sin\frac{\pi}{2} = \cos x \cdot 0 + \sin x \cdot 1 = \sin x$

26. Note that $\left(\frac{\pi}{2} - x\right) - y = \frac{\pi}{2} - x - y = \frac{\pi}{2} - (x + y)$ so that $\cos\left[\left(\frac{\pi}{2} - x\right) - y\right] = \cos\left[\frac{\pi}{2} - (x + y)\right]$. Now recall that $\cos(-\theta) = \cos\theta$ so that $\cos\left[\frac{\pi}{2} - (x + y)\right] = \cos\left[(x + y) - \frac{\pi}{2}\right]$, which equals $\sin(x + y)$ (from Exercise 25).

27. $\sin\left(x + \frac{\pi}{6}\right) = \sin x \cos\frac{\pi}{6} + \cos x \sin\frac{\pi}{6} = \sin x \cdot \frac{\sqrt{3}}{2} + \cos x \cdot \frac{1}{2}$

28. $\cos\left(x - \frac{\pi}{4}\right) = \cos x \cos\frac{\pi}{4} + \sin x \sin\frac{\pi}{4} = \cos x \frac{1}{\sqrt{2}} + \sin x \frac{1}{\sqrt{2}} = \frac{1}{\sqrt{2}}(\cos x + \sin x) = \frac{\sqrt{2}}{2}(\cos x + \sin x)$

29. $\tan\left(\theta + \frac{\pi}{4}\right) = \frac{\tan\theta + \tan\pi/4}{1 - \tan\theta \tan\pi/4} = \frac{\tan\theta + 1}{1 - \tan\theta \cdot 1} = \frac{1 + \tan\theta}{1 - \tan\theta}$

30. $\cos\left(\theta + \frac{\pi}{2}\right) = \cos\theta \cos\frac{\pi}{2} - \sin\theta \sin\frac{\pi}{2} = \cos\theta \cdot 0 - \sin\theta \cdot 1 = -\sin\theta$

31. (b) and (f)

32. (c) and (e)

33. (d) and (h)

34. (a) and (g)

35. Rewrite as $\sin 2x \cos x - \cos 2x \sin x = 0$; the left side equals $\sin(2x - x) = \sin x$, so $x = n\pi$, n an integer.

36. Rewrite as $\cos 3x \cos x - \sin 3x \sin x = 0$; the left side equals $\cos(3x + x) = \cos 4x$, so $x = \frac{\pi}{8} + n\frac{\pi}{4}$, n an integer.

37. $\sin(\frac{\pi}{2} - u) = \sin\frac{\pi}{2}\cos u - \cos\frac{\pi}{2}\sin u = 1 \cdot \cos u - 0 \cdot \sin u = \cos u$.

38. Using the difference identity for the tangent function, we encounter $\tan\frac{\pi}{2}$, which is undefined; however, we can compute $\tan(\frac{\pi}{2} - u) = \dfrac{\sin(\pi/2 - u)}{\cos(\pi/2 - u)}$; using the first two cofunction identities, we find that the numerator and denominator are $\cos u$ and $\sin u$, respectively so that this is $\cot u$; or use Exercise 24 and the fact that the tangent function is odd.

39. $\cot(\frac{\pi}{2} - u) = \dfrac{\cos(\pi/2 - u)}{\sin(\pi/2 - u)} = \dfrac{\sin u}{\cos u} = \tan u$, when we use the first two cofunction identities.

40. $\sec(\frac{\pi}{2} - u) = \dfrac{1}{\cos(\pi/2 - u)} = \dfrac{1}{\sin u} = \csc u$, when we use the first cofunction identity.

41. $\csc(\frac{\pi}{2} - u) = \dfrac{1}{\sin(\pi/2 - u)} = \dfrac{1}{\cos u} = \sec u$, when we use the second cofunction identity.

42. $\sin(u - v) = \cos[\frac{\pi}{2} - (u - v)] = \cos[(\frac{\pi}{2} - u) + v] = \cos(\frac{\pi}{2} - u)\cos v - \sin(\frac{\pi}{2} - u)\sin v = \sin u \cos v - \cos u \sin v$.

43. $\tan(u + v) = \dfrac{\sin(u + v)}{\cos(u + v)} = \dfrac{\sin u \cos v + \cos u \sin v}{\cos u \cos v - \sin u \sin v}$. Now divide both numerator and denominator by $\cos u \cos v$, leaving $\dfrac{\sin u/\cos u + \sin v/\cos v}{1 - \sin u/\cos u \sin v/\cos v} = \dfrac{\tan u + \tan v}{1 - \tan u \tan v}$.

44. $\tan(u - v) = \dfrac{\sin(u - v)}{\cos(u - v)} = \dfrac{\sin u \cos v - \cos u \sin v}{\cos u \cos v + \sin u \sin v}$; now divide both numerator and denominator by $\cos u \cos v$, leaving $\dfrac{\sin u/\cos u - \sin v/\cos v}{1 + \sin u/\cos u \sin v/\cos v} = \dfrac{\tan u - \tan v}{1 + \tan u \tan v}$.

45. $a = \sqrt{5}$, period $\frac{2\pi}{3}$, $h = \frac{1}{3}\tan^{-1}(-\frac{1}{2}) = -0.1545\ldots$

46. $|a| = \sqrt{13}$, period π, $h = \frac{1}{2}\tan^{-1}(\frac{3}{2}) = 0.4913\ldots$

47. $a = \sqrt{(3 + 5\cos 2)^2 + (5\sin 2)^2} = \sqrt{34 + 30\cos 2} = 4.6384\ldots$, period 2π, $h = \tan^{-1}\left(-\dfrac{5\sin 2}{3 + 5\cos 2}\right) = -1.3712\ldots$

48. $|a| = \sqrt{(3\sin 1 - 5\sin 3)^2 + (3\cos 1 + 5\cos 3)^2} = \sqrt{9 + 25 - 30\sin 1\sin 3 + 30\cos 1\cos 3} = \sqrt{34 + 30\cos(1 + 3)} = \sqrt{34 + 30\cos 4} = 3.7935\ldots$, period π, $h = \frac{1}{2}\tan^{-1}\left(\dfrac{3\sin 1 - 5\sin 3}{3\cos 1 + 5\cos 3}\right) = -0.2500\ldots$

49. $\dfrac{\cos(x + h) - \cos x}{h} = \dfrac{\cos x\cos h - \sin x\sin h - \cos x}{h} = \dfrac{\cos x(\cos h - 1) - \sin x\sin h}{h} = \cos x\left(\dfrac{\cos h - 1}{h}\right) - \sin x\left(\dfrac{\sin h}{h}\right)$

50. $\dfrac{\tan(x + h) - \tan x}{h} = \dfrac{1}{h}\left(\dfrac{\tan x + \tan h}{1 - \tan x\tan h} - \tan x\right) = \dfrac{\tan x + \tan h - \tan x(1 - \tan x\tan h)}{h(1 - \tan x\tan h)} = \dfrac{\tan h(1 + \tan^2 x)}{h(1 - \tan x\tan h)} = \dfrac{\tan h\sec^2 x}{h(1 - \tan x\tan h)} = \sec^2 x\dfrac{\sin h}{h\cos h(1 - \tan x\tan h)} = \sec^2 x\left(\dfrac{\sin h}{h}\right)\cdot\dfrac{1}{\cos h - \tan x\sin h}$

51. $\sin(x - y) + \sin(x + y) = (\sin x\cos y - \cos x\sin y) + (\sin x\cos y + \cos x \cdot \sin y) = 2\sin x\cos y$

52. $\cos(x - y) + \cos(x + y) = (\cos x \cdot \cos y + \sin x\sin y) + (\cos x\cos y - \sin x\sin y) = 2\cos x\cos y$

53. $\cos 3x = \cos[(x + x) + x] = \cos(x + x)\cos x - \sin(x + x)\sin x = (\cos x\cos x - \sin x\sin x)\cos x - (\sin x\cos x + \cos x\sin x)\sin x = \cos^3 x - \sin^2 x\cos x - \sin^2 x\cos x - \cos x\sin^2 x = \cos^3 x - 3\sin^2 x\cos x$

54. $\sin 3u = \sin[(u + u) + u] = \sin(u + u)\cos u + \cos(u + u)\sin u = (\sin u\cos u + \cos u\sin u)\cos u + (\cos u\cos u - \sin u\sin u)\sin u = 2\cos^2 u\sin u + \cos^2 u\sin u - \sin^3 u = 3\cos^2 u\sin u - \sin^3 u$

55. $\cos 3x + \cos x = \cos(2x + x) + \cos(2x - x)$; use Exercise 52 with x replaced with $2x$ and y replaced with x.

56. $\sin 4x + \sin 2x = \sin(3x + x) + \sin(3x - x)$; use Exercise 51, with x replaced with $3x$ and y replaced with x.

57. $\tan(x + y)\tan(x - y) = \dfrac{\tan x + \tan y}{1 - \tan x\tan y}\dfrac{\tan x - \tan y}{1 + \tan x\tan y} = \dfrac{\tan^2 x - \tan^2 y}{1 - \tan^2 x\tan^2 y}$, since both the top and bottom are factored forms for differences of squares.

58. $\tan 5u\tan 3u = \tan(4u + u) \cdot \tan(4u - u)$; use Exercise 57, with $x = 4u$ and $y = u$.

59. $\cos(x + y + z) = \cos(x + y)\cos z -$

$\sin(x + y)\sin z = (\cos x \cos y - \sin x \sin y)\cos z - (\sin x \cos y + \cos x \sin y)\sin z = \cos x \cos y \cos z - \sin x \sin y \cos z - \sin x \cos y \sin z - \cos x \sin y \sin z$

60. This equation is easier to deal with after rewriting it as $\cos 5x \cos 4x + \sin 5x \sin 4x = 0$. The left side of this equation is the expanded form of $\cos(5x - 4x)$, which of course equals $\cos x$; the graph shown is simply $y = \cos x$. The equation $\cos x = 0$ is easily solved on the interval $[-2\pi, 2\pi]$: $x = \pm\frac{\pi}{2}$ or $x = \pm\frac{3\pi}{2}$. The original graph is so crowded that we cannot see where crossings occur. Window: $[-2\pi, 2\pi] \times [-1.1, 1.1]$

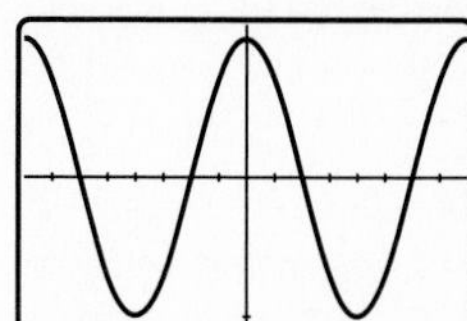

61. Rewrite as $\sin 3x \cos x - \cos 3x \sin x = 0$; the left side of this equation is the expanded form of $\sin(3x - x)$, which equals $\sin 2x$. Solving $\sin 2x = 0$ algebraically yields $x = n\frac{\pi}{2}$, n an integer. Solving this exercise graphically by the multigraph method (graphing $y_1 = \sin 3x \cos x$ and $y_2 = \cos 3x \sin x$) is difficult; shown is the window $[-2\pi, 2\pi] \times [-1.1, 1.1]$.

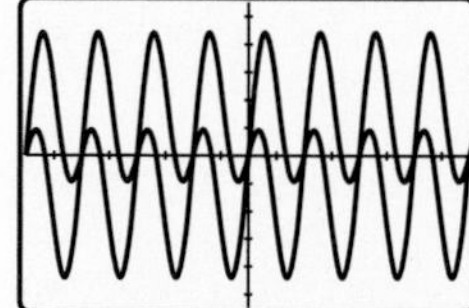

Graphing $y_1 = \sin 3x \cos x - \cos 3x \sin x$ and looking for x-axis crossings is easier and suggests the simplification found by solving algebraically; this graph is shown on the same window as the other graph.

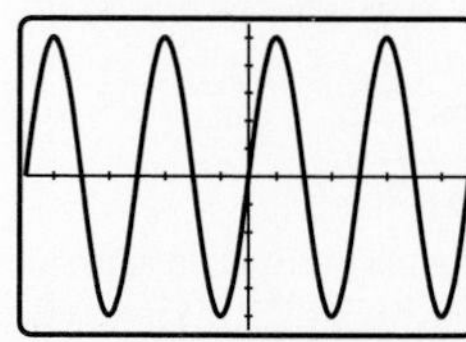

62. $x = a\cos\left(\frac{2\pi}{T}t + \delta\right) = a\left[\cos\left(\frac{2\pi}{T}t\right) \cdot \cos\delta - \sin\left(\frac{2\pi}{T}t\right)\sin\delta\right] = (a\cos\delta) \cdot \cos\left(\frac{2\pi}{T}t\right) + (-a\sin\delta)\sin\left(\frac{2\pi}{T}t\right)$

63. $B = \frac{E_0}{c}\cos\left(\omega t - \frac{\omega x}{c}\right) + \frac{E_0}{c}\cos\left(\omega t + \frac{\omega x}{c}\right) = \frac{E_0}{c}\left(\cos\omega t \cos\frac{\omega x}{c} + \sin\omega t \sin\frac{\omega x}{c} + \cos\omega t \cos\frac{\omega x}{c} - \sin\omega t \sin\frac{\omega x}{c}\right) = \frac{E_0}{c}\left(2\cos\omega t \cos\frac{\omega x}{c}\right) = 2\frac{E_0}{c}\cos\omega t \cos\frac{\omega x}{c}$

QUICK REVIEW 7.5

1. $x = \frac{\pi}{4} + n\pi$, n an integer
2. $x = -\frac{\pi}{4} + n\pi$, n an integer
3. $x = \frac{\pi}{2} + n\pi$, n an integer
4. $x = n\pi$, n an integer
5. $x = -\frac{\pi}{4} + n\pi$, n an integer
6. $x = \frac{\pi}{4} + n\pi$, n an integer
7. $x = \frac{\pi}{6} + 2n\pi$, $x = \frac{2\pi}{3} + 2n\pi$, $x = \frac{5\pi}{6} + 2n\pi$, or $x = \frac{4\pi}{3} + 2n\pi$, n an integer
8. $x = \pm\frac{\pi}{4} + 2n\pi$ or $x = \frac{3\pi}{2} + 2n\pi$, n an integer
9. $(2)(3) + \frac{1}{2}(1)(3) + \frac{1}{2}(2)(3) = 10.5$ square units
10. $(2)(5) + \frac{1}{2}(1)(5) + \frac{1}{2}(1)(5) = 15$ square units
11. $\sqrt{3^2 - 1^2} = \sqrt{8} = 2\sqrt{2}$

SECTION EXERCISES 7.5

1. $\cos 2u = \cos(u + u) = \cos u \cos u - \sin u \sin u = \cos^2 u - \sin^2 u$
2. $\cos 2u = \cos^2 u - \sin^2 u = \cos^2 u - (1 - \cos^2 u) = 2\cos^2 u - 1$
3. $\cos 2u = \cos^2 u - \sin^2 u = (1 - \sin^2 u) - \sin^2 u = 1 - 2\sin^2 u$
4. $\tan 2u = \tan(u + u) = \dfrac{\tan u + \tan u}{1 - \tan u \tan u} = \dfrac{2\tan u}{1 - \tan^2 u}$
5. $x = 0$ or $x = \pi$
6. $x = \frac{\pi}{6}$, $x = \frac{\pi}{2}$, $x = \frac{5\pi}{6}$ or $x = \frac{3\pi}{2}$
7. $x = \frac{\pi}{6}$, $x = \frac{5\pi}{6}$, $x = \frac{3\pi}{2}$
8. $x = 0$, $x = \frac{2\pi}{3}$ or $x = \frac{4\pi}{3}$
9. $x = 0$, $x = \frac{\pi}{4}$, $x = \frac{3\pi}{4}$, $x = \pi$, $x = \frac{5\pi}{4}$, or $x = \frac{7\pi}{4}$
10. $x = \cos^{-1}\left(\dfrac{1 - \sqrt{5}}{2}\right) = 2.2370\ldots$ or $x = 2\pi - \cos^{-1}\left(\dfrac{1 - \sqrt{5}}{2}\right) = 4.0461\ldots$
11. $2\sin\theta\cos\theta + \cos\theta = \cos\theta(2\sin\theta + 1)$
12. $2\sin\theta\cos\theta + \cos^2\theta - \sin^2\theta$
13. $2\sin\theta\cos\theta + \cos 2\theta\cos\theta - \sin 2\theta\sin\theta = 2\sin\theta\cos\theta + (\cos^2\theta - \sin^2\theta)\cos\theta - 2\sin^2\theta\cos\theta = 2\sin\theta \cdot \cos\theta + \cos^3\theta - 3\sin^2\theta\cos\theta = 2\sin\theta\cos\theta + 4\cos^3\theta - 3\cos\theta$
14. $\sin 2\theta\cos\theta + \cos 2\theta\sin\theta + \cos^2\theta - \sin^2\theta = 2\sin\theta\cos^2\theta + (\cos^2\theta - \sin^2\theta)\sin\theta + \cos^2\theta - \sin^2\theta = 3\sin\theta\cos^2\theta - \sin^3\theta + \cos^2\theta - \sin^2\theta$
15. $\sin 4x = \sin 2(2x) = 2\sin 2x\cos 2x$
16. $\cos 6x = \cos 2(3x) = 2\cos^2 3x - 1$
17. $2\csc 2x = \dfrac{2}{\sin 2x} = \dfrac{2}{2\sin x\cos x} = \dfrac{1}{\sin^2 x}\dfrac{\sin x}{\cos x} = \csc^2 x\tan x$
18. $2\cot 2x = \dfrac{2}{\tan 2x} = \dfrac{2(1 - \tan^2 x)}{2\tan x} = \dfrac{1}{\tan x} - \tan x = \cot x - \tan x$
19. $\sin 3x = \sin 2x\cos x + \cos 2x\sin x = 2\sin x\cos^2 x + (2\cos^2 x - 1)\sin x = \sin x(4\cos^2 x - 1)$
20. $\sin 3x = \sin 2x\cos x + \cos 2x\sin x = 2\sin x\cos^2 x + (1 - 2\sin^2 x)\sin x = \sin x(2\cos^2 x + 1 - 2\sin^2 x) = \sin x(3 - 4\sin^2 x)$
21. $\cos 4x = \cos 2(2x) = 1 - 2\sin^2 2x = 1 - 2(2\sin x\cos x)^2 = 1 - 8\sin^2 x\cos^2 x$

22. $\sin 4x = \sin 2(2x) = 2 \sin 2x \cos 2x = 2(2 \sin x \cos x)(2 \cos^2 x - 1) = 4 \sin x \cos x(2\cos^2 x - 1)$

23. $2 \cos^2 x + \cos x - 1 = 0$, so $\cos x = -1$ or $\cos x = \frac{1}{2}$; $x = \frac{\pi}{3}$, $x = \pi$, $x = \frac{5\pi}{3}$

24. $-2 \sin^2 x + 2 \sin x + 1 = 0$, so $\sin x = \dfrac{1 - \sqrt{3}}{2}$. Let $\alpha = \sin^{-1}\left(\dfrac{1 - \sqrt{3}}{2}\right) = -0.3747\ldots$; then $x = \pi - \alpha = 3.5163\ldots$ or $x = 2\pi + \alpha = 5.9084\ldots$.

25. The left side can be rewritten as $2 \cos x(1 - 2 \sin^2 x)$; which equals 0 in $[0, 2\pi)$ when $x = \frac{\pi}{4}$, $x = \frac{\pi}{2}$, $x = \frac{3\pi}{4}$, $x = \frac{5\pi}{4}$, $x = \frac{3\pi}{2}$, or $x = \frac{7\pi}{4}$.

26. If we use Exercise 19, this becomes $\sin x \cdot 4 \cos^2 x = 0$, so $x = 0$, $x = \frac{\pi}{2}$, $x = \pi$, or $x = \frac{3\pi}{2}$.

27. This can be rewritten as $\sin 2x(1 + 2 \cos 2x) = 0$; solutions in $[0, 2\pi)$ are $x = 0$, $x = \frac{\pi}{3}$, $x = \frac{\pi}{2}$, $x = \frac{2\pi}{3}$, $x = \pi$, $x = \frac{4\pi}{3}$, $x = \frac{3\pi}{2}$, or $x = \frac{5\pi}{3}$.

28. With $u = 2x$, this becomes $\cos u + \cos 2u = 0$, or the same as Exercise 23. This means that $u = \frac{\pi}{3}$, $u = \pi$, $u = \frac{5\pi}{3}$, and so on; then $x = \frac{\pi}{6}$, $x = \frac{\pi}{2}$, $x = \frac{5\pi}{6}$, $x = \frac{7\pi}{6}$, $x = \frac{3\pi}{2}$, or $x = \frac{11\pi}{6}$.

29. This can be rewritten as $\cos x(4 \sin^2 x + 2 \sin x - 1) = 0$; $\cos x = 0$ when $x = \frac{\pi}{2}$ or $x = \frac{3\pi}{2}$; the expression in parentheses equals zero when $\sin x = \frac{-1 \pm \sqrt{5}}{4}$; that means (as can be observed by taking, for example, $\sin^{-1} \frac{-1 \pm \sqrt{5}}{4}$) that $x = 0.1\pi$, $x = 0.9\pi$, $x = 1.3\pi$, or $x = 1.7\pi$.

30. Using Exercise 14, the left side can be rewritten as $3 \sin x \cos^2 x - \sin^3 x + \cos^2 x - \sin^2 x = 3 \sin x(1 - \sin^2 x) - \sin^3 x + 1 - 2 \sin^2 x = 3 \sin x - 4 \sin^3 x + 1 - 2 \sin^2 x = (\sin x + 1) \cdot (-4 \sin^2 x + 2 \sin x + 1)$. This equals zero when $x = \frac{3\pi}{2}$ (where $\sin x = -1$) and at four other places. Those other places can be found by using $\sin^{-1}(\,)$, the solutions to the quadratic equation $-4u^2 + 2u + 1 = 0$; these values turn out to be $x = 0.3\pi$, $x = 0.7\pi$, $x = 1.1\pi$, and $x = 1.9\pi$.

31. $\sin 15° = \pm\sqrt{\dfrac{1 - \cos 30°}{2}} = \pm\sqrt{\dfrac{1}{2}\left(1 - \dfrac{\sqrt{3}}{2}\right)} = \pm\dfrac{1}{2}\sqrt{2 - \sqrt{3}}$; $\sin 15°$ is positive, so take the positive square root. Note that this result is the same as $\dfrac{\sqrt{6} - \sqrt{2}}{4}$, the value found in Exercise 1, Section 7.4.

32. $\tan 195° = \dfrac{1 - \cos 390°}{\sin 390°} = \dfrac{1 - \sqrt{3}/2}{1/2} = 2 - \sqrt{3}$

33. $\cos 75° = \pm\sqrt{\dfrac{1 + \cos 150°}{2}} = \pm\sqrt{\dfrac{1}{2}\left(1 - \dfrac{\sqrt{3}}{2}\right)} = \pm\dfrac{1}{2}\sqrt{2 - \sqrt{3}}$; $\cos 75°$ is positive, so take the positive square root.

34. $\sin\dfrac{5\pi}{12} = \pm\sqrt{\dfrac{1 - \cos(5\pi/6)}{2}} = \pm\sqrt{\dfrac{1}{2}\left(1 + \dfrac{\sqrt{3}}{2}\right)} = \pm\dfrac{1}{2}\sqrt{2 + \sqrt{3}}$; $\sin\dfrac{5\pi}{12}$ is positive, so take the positive square root.

35. $\tan\dfrac{7\pi}{12} = \dfrac{1 - \cos(7\pi/6)}{\sin(7\pi/6)} = \dfrac{1 + \sqrt{3}/2}{-1/2} = -2 - \sqrt{3}$

36. $\cos\dfrac{\pi}{8} = \pm\sqrt{\dfrac{1 + \cos(\pi/4)}{2}} = \pm\sqrt{\dfrac{1}{2}\left(1 + \dfrac{\sqrt{2}}{2}\right)} = \pm\dfrac{1}{2}\sqrt{2 + \sqrt{2}}$; $\cos\dfrac{\pi}{8}$ is positive, so take the positive square root.

37. Starting from the right side: $\dfrac{1}{2}(1 - \cos 2u) = \dfrac{1}{2}[1 - (1 - 2 \sin^2 u)] = \dfrac{1}{2}(2 \sin^2 u) = \sin^2 u$.

38. Starting from the right side; $\frac{1}{2}(1 + \cos 2u) = \frac{1}{2}[1 + (2 \cos^2 u - 1)] = \frac{1}{2}(2 \cos^2 u) = \cos^2 u$

39. $\sin^4 x = (\sin^2 x)^2 = [\frac{1}{2}(1 - \cos 2x)]^2 = \frac{1}{4}(1 - 2 \cos 2x + \cos^2 2x) = \frac{1}{4}[1 - 2 \cos 2x + \frac{1}{2}(1 + \cos 4x)] = \frac{1}{8}(2 - 4 \cos 2x + 1 + \cos 4x) = \frac{1}{8}(3 - 4 \cos 2x + \cos 4x)$

40. $\cos^3 x = \cos x \cos^2 x = \cos x \cdot \frac{1}{2}(1 + \cos 2x) = \frac{1}{2}(\cos x)(1 + \cos 2x)$

41. $\sin^3 2x = \sin 2x \sin^2 2x = \sin 2x \cdot \frac{1}{2}(1 - \cos 4x) = \frac{1}{2} \sin 2x(1 - \cos 4x)$

42. $\sin^5 x = \sin x(\sin^2 x)^2 = \sin x[\frac{1}{2}(1 - \cos 2x)]^2 = \frac{1}{4} \sin x(1 - 2 \cos 2x + \cos^2 2x) = \frac{1}{4} \sin x[1 - 2 \cos 2x + \frac{1}{2}(1 + \cos 4x)] = \frac{1}{8} \sin x(2 - 4 \cos 2x + 1 + \cos 4x) = \frac{1}{8} \sin x(3 - 4 \cos 2x + \cos 4x)$

43. $\cos^2 x = \dfrac{1 - \cos x}{2}$, so $\cos x = -1$ or $\cos x = \frac{1}{2}$; in the interval $[0, 2\pi)$, $x = \frac{\pi}{3}$, $x = \pi$, or $x = \frac{5\pi}{3}$. The general solution is $x = \pm\frac{\pi}{3} + 2n\pi$ or $x = \pi + 2n\pi$, n an integer.

44. $1 - \cos^2 x = \dfrac{1 + \cos x}{2}$, so $\cos x = -1$ or $\cos x = \frac{1}{2}$. In the interval $[0, 2\pi)$, $x = \frac{\pi}{3}$, $x = \pi$, or $x = \frac{5\pi}{3}$. The general solution is $x = \pm\frac{\pi}{3} + 2n\pi$ or $x = \pi + 2n\pi$, n an integer.

45. The right side equals $\tan^2(\frac{x}{2})$; the only way that $\tan(\frac{x}{2}) = \tan^2(\frac{x}{2})$ is if either $\tan(\frac{x}{2}) = 0$ or $\tan(\frac{x}{2}) = 1$; in $[0, 2\pi)$, this happens when $x = 0$ or $x = \frac{\pi}{2}$. The general solution is $x = 2n\pi$ or $x = \frac{\pi}{2} + 2n\pi$, n an integer.

46. $\dfrac{1 - \cos x}{2} = 2 \cos^2 x - 1$, so $\cos x = -1$ or $\cos x = \frac{3}{4}$. Let $\alpha = \cos^{-1}(\frac{3}{4}) = 0.7227\ldots$; in the interval $[0, 2\pi)$, $x = \alpha$, $x = \pi$, or $x = 2\pi - \alpha$. The general solution is $x = \pm\alpha + 2n\pi$ or $x = \pi + 2n\pi$, n an integer.

47. (a) In the figure, the triangle with side lengths $\frac{x}{2}$ and R is a right triangle, since

R is given as the perpendicular distance; then the tangent of the angle $\frac{\theta}{2}$ is the ratio "opposite over adjacent," $\tan\frac{\theta}{2} = \frac{x/2}{R}$, and solving for x gives the equation desired; the central angle θ is $\frac{2\pi}{n}$, since one full revolution of 2π radians is divided evenly into n sections. **(b)** $5.87 \approx 2R\tan\frac{\theta}{2}$, where $\theta = \frac{2\pi}{11}$, so $R \approx \dfrac{5.87}{2\tan\left(\frac{\pi}{11}\right)} = 9.9956\ldots$; $R = 10$.

48. **(a)** Call the center of the rhombus E. Consider right $\triangle ABE$, with leg lengths $\frac{d_2}{2}$ and $\frac{d_1}{2}$ and hypotenuse length x. $\angle ABE$ has measure $\frac{\theta}{2}$ and, using "sine equals $\dfrac{\text{opp}}{\text{hyp}}$" and "cosine equals $\dfrac{\text{adj}}{\text{hyp}}$," we have $\cos\dfrac{\theta}{2} = \dfrac{d_2/2}{x} = \dfrac{d_2}{2x}$ and $\sin\dfrac{\theta}{2} = \dfrac{d_1/2}{x} = \dfrac{d_1}{2x}$. **(b)** Use the double angle formula for the sine function: $\sin\theta = \sin 2\left(\dfrac{\theta}{2}\right) = 2\sin\dfrac{\theta}{2}\cos\dfrac{\theta}{2} = 2\dfrac{d_1}{2x}\dfrac{d_2}{2x} = \dfrac{d_1 d_2}{2x^2}$.

49. The volume is 10 ft times the area of the end; the end is made up of two identical triangles [area $\frac{1}{2}(\sin\theta)(\cos\theta)$ each] and a rectangle [area $(1)(\cos\theta)$], so the total volume is $10\cdot(\sin\theta\cos\theta + \cos\theta) = 10\cos\theta(1 + \sin\theta)$. Considering only $\frac{\pi}{2} \le \theta \le \frac{\pi}{2}$, the maximum value occurs when $\theta \approx 0.52$ (in fact, it happens exactly at $\theta = \frac{\pi}{6}$) and is about 12.99 ft^3.

50. **(a)** The height of the tunnel is y, and its width is $2x$, so the area is $2xy$; the x- and y-coordinates of the vertex are $20\cos\theta$ and $20\sin\theta$, so the area is $2(20\cos\theta)(20\sin\theta) = 400(2\cos\theta\sin\theta) = 400\sin 2\theta$. **(b)** For $0 \le \theta \le \frac{\pi}{2}$, the maximum area occurs when $\theta = \frac{\pi}{4}$, or about 0.79; this gives $x = 20\cos\frac{\pi}{4} = 10\sqrt{2}$, or about 14.14, for a width of about 28.28 and a height of $y = 10\sqrt{2} \approx 14.14$.

51. $\csc 2u = \dfrac{1}{\sin 2u} = \dfrac{1}{2\sin u\cos u} = \dfrac{1}{2}\dfrac{1}{\sin u}\dfrac{1}{\cos u} = \dfrac{1}{2}\csc u\sec u$

52. $\cot 2u = \dfrac{1}{\tan 2u} = \dfrac{1}{2\tan u/(1 - \tan^2 u)} = \dfrac{1 - \tan^2 u}{2\tan u} = \dfrac{(1 - \tan^2 u)\cot^2 u}{2\tan u\cot^2 u} = \dfrac{\cot^2 u - 1}{2\cot u}$

53. $\sec 2u = \dfrac{1}{\cos 2u} = \dfrac{1}{1 - 2\sin^2 u} = \dfrac{1}{1 - 2\sin^2 u}\cdot\dfrac{\csc^2 u}{\csc^2 u} = \dfrac{\csc^2 u}{\csc^2 u - 2}$

54. $\sec 2u = \dfrac{1}{\cos 2u} = \dfrac{1}{2\cos^2 u - 1} = \dfrac{1}{2\cos^2 u - 1}\cdot\dfrac{\sec^2 u}{\sec^2 u} = \dfrac{\sec^2 u}{2 - \sec^2 u}$

55. $\sec 2u = \dfrac{1}{\cos 2u} = \dfrac{1}{\cos^2 u - \sin^2 u} = \dfrac{1}{\cos^2 u - \sin^2 u}\cdot\dfrac{\sec^2 u\csc^2 u}{\sec^2 u\csc^2 u} = \dfrac{\sec^2 u\csc^2 u}{\csc^2 u - \sec^2 u}$

56. The second equation cannot work for any values of x for which $\sin x < 0$, since the square root cannot be negative. The first is correct, since a double angle identity for the cosine gives $\cos 2x = 1 - 2\sin^2 x$; solving for $\sin x$ gives $\sin^2 x = \frac{1}{2}(1 - \cos 2x)$, so $\sin x = \pm\sqrt{\frac{1}{2}(1 - \cos 2x)}$. The absolute value on both sides removes the "$\pm$."

Chapter 7 Review

1. $\sin 200°$

2. $\tan 80°$

3. 1; the expression simplifies to $(\cos 2\theta)^2 + (2\sin\theta\cos\theta)^2 = (\cos 2\theta)^2 + (\sin 2\theta)^2 = 1$.

4. $\cos^2 2x$; $1 - (2\sin x\cos x)^2 = 1 - (\sin 2x)^2 = \cos^2 2x$

5. $\cos 3x = \cos(2x + x) = \cos 2x\cos x - \sin 2x\sin x = (\cos^2 x - \sin^2 x)\cos x - (2\sin x\cos x)\sin x = \cos^3 x - 3\sin^2 x\cdot\cos x = \cos^3 x - 3(1 - \cos^2 x)\cos x = \cos^3 x - 3\cos x + 3\cos^3 x = 4\cos^3 x - 3\cos x$

6. $\cos^2 2x - \cos^2 x = (1 - \sin^2 2x) - (1 - \sin^2 x) = \sin^2 x - \sin^2 2x$

7. $\tan^2 x - \sin^2 x = \sin^2 x\dfrac{1 - \cos^2 x}{\cos^2 x} = \sin^2 x\dfrac{\sin^2 x}{\cos^2 x} = \sin^2 x\tan^2 x$

8. $2\sin\theta\cos^3\theta + 2\sin^3\theta\cos\theta = 2\sin\theta\cos\theta(\cos^2\theta + \sin^2\theta) = 2\sin\theta\cos\theta = \sin 2\theta$

9. $\csc x - \cos x\cot x = \dfrac{1 - \cos^2 x}{\sin x} = \dfrac{\sin^2 x}{\sin x} = \sin x$

10. $\dfrac{\tan\theta + \sin\theta}{2\tan\theta} = \dfrac{1 + \cos\theta}{2} = \left(\pm\sqrt{\dfrac{1 + \cos\theta}{2}}\right)^2 = \left(\cos\dfrac{\theta}{2}\right)^2$

11. $\dfrac{1 + \tan\theta}{1 - \tan\theta} + \dfrac{1 + \cot\theta}{1 - \cot\theta} = \dfrac{(1 + \tan\theta)(1 - \cot\theta) + (1 + \cot\theta)(1 - \tan\theta)}{(1 - \tan\theta)(1 - \cot\theta)} = \dfrac{0}{(1 - \tan\theta)(1 - \cot\theta)} = 0$

12. $\sin 3\theta = \sin(2\theta + \theta) = \sin 2\theta\cos\theta + \cos 2\theta\sin\theta = 2\sin\theta\cos^2\theta + (\cos^2\theta - \sin^2\theta)\sin\theta = 3\sin\theta\cos^2\theta - \sin^3\theta$

13. $\cos^2\frac{t}{2} = \left(\pm\sqrt{\frac{1}{2}(1 + \cos t)}\right)^2 = \dfrac{1}{2}(1 + \cos t) = \dfrac{1 + \cos t}{2}\cdot\dfrac{\sec t}{\sec t} = \dfrac{\sec t + 1}{2\sec t}$

14. $\dfrac{\tan^3\gamma - \cot^3\gamma}{\tan^2\gamma + \csc^2\gamma} = \dfrac{(\tan\gamma - \cot\gamma)(\tan^2\gamma + \tan\gamma\cot\gamma + \cot^2\gamma)}{\tan^2\gamma + \csc^2\gamma} = \dfrac{(\tan\gamma - \cot\gamma)(\tan^2\gamma + 1 + \cot^2\gamma)}{\tan^2\gamma + \csc^2\gamma} = \dfrac{(\tan\gamma - \cot\gamma)(\tan^2\gamma + \csc^2\gamma)}{\tan^2\gamma + \csc^2\gamma} = \tan\gamma - \cot\gamma$

15. $\dfrac{\cos\phi}{1 - \tan\phi} + \dfrac{\sin\phi}{1 - \cot\phi} = \dfrac{\cos\phi}{1 - \tan\phi}\cdot\dfrac{\cos\phi}{\cos\phi} + \dfrac{\sin\phi}{1 - \cot\phi}\cdot$

$\dfrac{\sin \phi}{\sin \phi} = \dfrac{\cos^2 \phi}{\cos \phi - \sin \phi} + \dfrac{\sin^2 \phi}{\sin \phi - \cos \phi} = \dfrac{\cos^2 \phi - \sin^2 \phi}{\cos \phi - \sin \phi} = \cos \phi + \sin \phi$

16. $\dfrac{\cos(-z)}{\sec(-z) + \tan(-z)} = \dfrac{\cos(-z)}{[1 + \sin(-z)]/\cos(-z)} = \dfrac{\cos^2(-z)}{1 + \sin(-z)} = \dfrac{1 - \sin^2 z}{1 - \sin z} = 1 + \sin z$

17. $\sqrt{\dfrac{1 - \cos y}{1 + \cos y}} = \sqrt{\dfrac{(1 - \cos y)^2}{(1 + \cos y)(1 - \cos y)}} = \sqrt{\dfrac{(1 - \cos y)^2}{1 - \cos^2 y}} = \sqrt{\dfrac{(1 - \cos y)^2}{\sin^2 y}} = \dfrac{|1 - \cos y|}{|\sin y|} = \dfrac{1 - \cos y}{|\sin y|}$; since $1 - \cos y \geq 0$, we can drop that absolute value.

18. $\sqrt{\dfrac{1 - \sin \gamma}{1 + \sin \gamma}} = \sqrt{\dfrac{(1 - \sin \gamma)(1 + \sin \gamma)}{(1 + \sin \gamma)^2}} = \sqrt{\dfrac{1 - \sin^2 \gamma}{(1 + \sin \gamma)^2}} = \sqrt{\dfrac{\cos^2 \gamma}{(1 + \sin \gamma)^2}} = \dfrac{|\cos \gamma|}{|1 + \sin \gamma|} = \dfrac{|\cos \gamma|}{1 + \sin \gamma}$, since $1 + \sin \gamma \geq 0$, we can drop that absolute value.

19. $\tan\left(u + \frac{3\pi}{4}\right) = \dfrac{\tan u + \tan 3\pi/4}{1 - \tan u \tan 3\pi/4} = \dfrac{\tan u + (-1)}{1 - \tan u(-1)} = \dfrac{\tan u - 1}{1 + \tan u}$

20. $\frac{1}{4}\sin 4\gamma = \frac{1}{4}\sin 2(2\gamma) = \frac{1}{4}(2 \sin 2\gamma \cos 2\gamma) = \frac{1}{2}(2 \sin \gamma \cos \gamma);$ $(\cos^2 \gamma - \sin^2 \gamma) = \sin \gamma \cos^3 \gamma - \cos \gamma \sin^3 \gamma$

21. $\tan \frac{1}{2}\beta = \dfrac{1 - \cos \beta}{\sin \beta} = \dfrac{1}{\sin \beta} - \dfrac{\cos \beta}{\sin \beta} = \csc \beta - \cot \beta$

22. Let $\theta = \arctan t$ so that $\tan \theta = t$; then $\tan 2\theta = \dfrac{2 \tan \theta}{1 - \tan^2 \theta} = \dfrac{2t}{1 - t^2}$. Note also that, since $-1 < t < 1$, $-\frac{\pi}{4} < \theta < \frac{\pi}{4}$ and therefore $-\frac{\pi}{2} < 2\theta < \frac{\pi}{2}$. That means that 2θ is in the range of the arctan function, so $2\theta = \arctan \dfrac{2t}{1 - t^2}$, or equivalently $\theta = \dfrac{1}{2} \arctan \dfrac{2t}{1 - t^2}$; and of course, $\theta = \arctan t$.

23. Yes; $\sec x - \sin x \tan x = \dfrac{1}{\cos x} - \dfrac{\sin^2 x}{\cos x} = \dfrac{1 - \sin^2 x}{\cos x} = \dfrac{\cos^2 x}{\cos x} = \cos x$

24. Yes; $(\sin^2 \alpha - \cos^2 \alpha)(\tan^2 \alpha + 1) = (\sin^2 \alpha - \cos^2 \alpha) \sec^2 \alpha = \dfrac{\sin^2 \alpha - \cos^2 \alpha}{\cos^2 \alpha} = \dfrac{\sin^2 \alpha}{\cos^2 \alpha} - 1 = \tan^2 \alpha - 1$

25. Possible answer: $\sin 3x + \cos 3x = (3 \sin x - 4 \sin^3 x) + (4 \cos^3 x - 3 \cos x) = 3(\sin x - \cos x) - 4(\sin^3 x - \cos^3 x) = (\sin x - \cos x)[3 - 4(\sin^2 x + \sin x \cos x + \cos^2 x)] = (\sin x - \cos x)(3 - 4 - 4 \sin x \cos x) = (\cos x - \sin x)(1 + 4 \sin x \cos x)$.

26. Possible answer: $\sin 2x + \cos 3x = 2 \sin x \cos x + 4 \cos^3 x - 3 \cos x = \cos x(2 \sin x + 4 \cos^2 x - 3) = \cos x(2 \sin x + 1 - 4 \sin^2 x)$.

27. Possible answer: $\cos^2 2x - \sin 2x = 1 - \sin^2 2x - \sin 2x = 1 - 4 \sin^2 x \cos^2 x - 2 \sin x \cos x$.

28. Possible answer: $\sin 3x - 3 \sin 2x = 3 \cos^2 x \sin x - \sin^3 x - 6 \sin x \cos x = \sin x(3 \cos^2 x - \sin^2 x - 6 \cos x) = \sin x(4 \cos^2 x - 1 - 6 \cos x)$.

29. $x = \frac{\pi}{12} + n\pi$ or $x = \frac{5\pi}{12} + n\pi$, n an integer

30. $x = \pm\frac{\pi}{6} + 2n\pi$, n an integer

31. $x = -\frac{\pi}{4} + n\pi$, n an integer

32. $x = \sin^{-1} 0.7 + 2n\pi$ or $x = \pi - \sin^{-1} 0.7 + 2n\pi$, n an integer

33. $x = \pm\frac{1}{2}\cos^{-1} 0.13 + n\pi$, n an integer

34. $x = \cot^{-1} 1.5 + n\pi = \tan^{-1}\frac{2}{3} + n\pi$, n an integer

35. $x = \sin^{-1} 0.3 + 2n\pi$ or $x = \pi - \sin^{-1} 0.3 + 2n\pi$, n an integer

36. $x = \pm\frac{1}{3}\cos^{-1} 0.225 + n\frac{2\pi}{3}$, n an integer

37. $x = \sin \frac{\sqrt{2}}{2} = 0.6496\ldots$

38. $x = \tan 1 = 1.5574\ldots$

39. $x \approx \pm 1.12 + 2n\pi$, n an integer; window: $[-2\pi, 2\pi] \times [-4, 4]$.

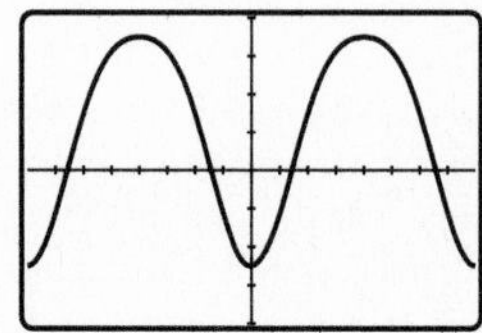

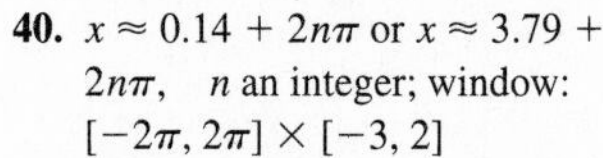

40. $x \approx 0.14 + 2n\pi$ or $x \approx 3.79 + 2n\pi$, n an integer; window: $[-2\pi, 2\pi] \times [-3, 2]$

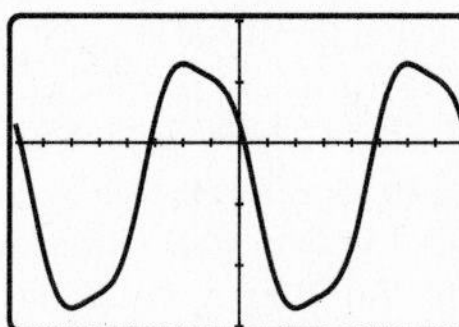

41. $x \approx \pm 1.15$; window: $[-2\pi, 2\pi] \times [-3, 10]$

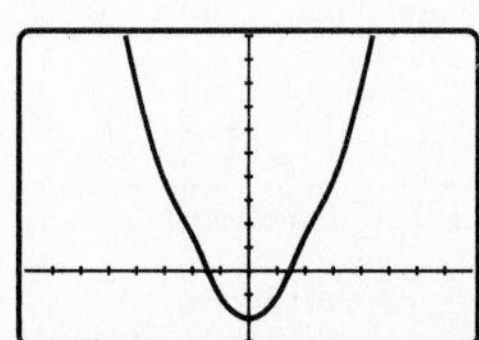

42. $x \approx -0.23$, $x \approx 1.85$ or $x \approx 3.59$; window: $[-\pi, 2\pi] \times [-6, 6]$

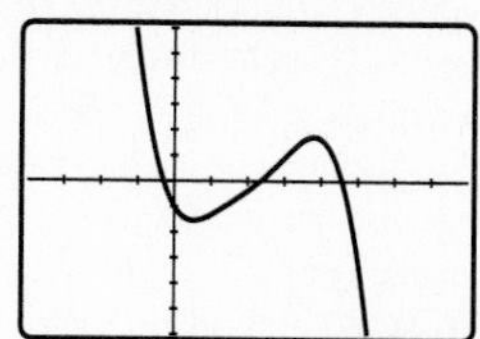

43. Let $\alpha = \frac{1}{2}\sin^{-1}\frac{1}{3} = 0.1699\ldots$; $x = \alpha$, $x = \frac{\pi}{2} - \alpha = 1.4008\ldots$, $x = \pi + \alpha = 3.3115\ldots$, or $x = \frac{3\pi}{2} - \alpha = 4.5424\ldots$

44. $x = \frac{\pi}{3}$ or $x = \frac{5\pi}{3}$

45. $x = 0$, $x = \frac{\pi}{4}$, $x = \frac{3\pi}{4}$, $x = \pi$, $x = \frac{5\pi}{4}$, or $x = \frac{7\pi}{4}$

46. $x = \frac{3\pi}{2}$

47. $t = 0$, $t = \frac{2\pi}{3}$, or $t = \frac{4\pi}{3}$

48. No solutions

49. $0 \le x < \frac{\pi}{6}$, $\frac{5\pi}{6} < x < \frac{7\pi}{6}$, or $\frac{11\pi}{6} < x < 2\pi$

50. $\frac{\pi}{2} < x < \frac{3\pi}{2}$

51. $\frac{\pi}{3} < x < \frac{5\pi}{3}$

52. $-\frac{\pi}{2} < x < 0$

53. $y = 5\sin[3(x + \frac{1}{3}\tan^{-1}\frac{4}{3})]$; that is, $a = 5$, $b = 3$, $h = -\frac{1}{3}\tan^{-1}\frac{4}{3} = -0.3090\ldots$

54. $y = -13\sin[2(x - \frac{1}{2}\tan^{-1}\frac{5}{12})]$; that is, $a = -13$, $b = 2$, $h = \frac{1}{2}\tan^{-1}\frac{5}{12} = 0.1973\ldots$

55. **(a)** $A(\theta) = \sin\theta + \sin\theta\cos\theta = \sin\theta(1 + \cos\theta)$. **(b)** $\theta = \frac{\pi}{3} = 60°$; maximum area $= \frac{3}{4}\sqrt{3} \approx 1.30$ square units.

56. **(a)** $S(\theta) = 6.825 + 0.63375(-\cot\theta + \sqrt{3}\cdot\csc\theta) = 6.825 + 0.63375\left(\frac{\sqrt{3} - \cos\theta}{\sin\theta}\right)$. Shown is the window $[-2\pi, 2\pi] \times [-5, 15]$.

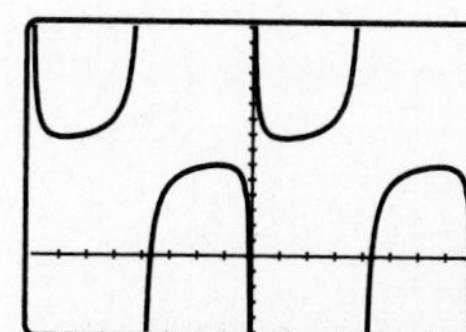

(b) Considering only angles between 0 and π, the minimum occurs when $\theta \approx 0.96 \approx 54.74°$. **(c)** The minimum value of S is approximately $S(0.96) \approx 7.72$ in.^{2}

57. **(a)** $h = 4000\sec\frac{\theta}{2} - 4000$ miles. **(b)** $\theta \approx 0.62 \approx 35.51°$.

58. Using the double angle sine formula and Review Exercise 12, we rewrite the left side as $\sin x - 2\sin x\cos x + 3\cos^2 x\sin x - \sin^3 x = \sin x(1 - 2\cos x + 3\cos^2 x - \sin^2 x) = \sin x(4\cos^2 x - 2\cos x) = 2\sin x\cos x(2\cos x - 1)$. This expression equals 0 when $x = n\frac{\pi}{2}$ or $x = \pm\frac{\pi}{3} + 2n\pi$, n an integer.

CHAPTER 8

QUICK REVIEW 8.1

1. $a = bc/d$

2. $b = ad/c$

3. $c = ad/b$

4. $d = bc/a$

5. 13.313 . . .

6. 31.888 . . .

7. $x = 17.457\ldots°$

8. $x = 162.542\ldots°$

9. $x = 224.427\ldots°$

10. $x = 315.572\ldots°$

SECTION EXERCISES 8.1

1. $a = 4.531\ldots$; $c = 5.054\ldots$; $\gamma = 75°$

2. $a = 13.880\ldots$; $b = 5.080\ldots$; $\alpha = 45°$

3. $b = 15.796\ldots$; $c = 12.813\ldots$; $\beta = 45°$

4. $a = 141.364\ldots$; $c = 122.683\ldots$; $\gamma = 59°$

5. $a = 12.855\ldots$; $c = 18.793\ldots$; $\gamma = 110°$

6. $b = 4.610\ldots$; $c = 4.841\ldots$; $\gamma = 68°$

7. $a = 4.057\ldots$; $c = 7.258\ldots$; $\gamma = 77°$

8. $a = 10.771\ldots$; $b = 3.394\ldots$; $\alpha = 61°$

9. $c = 25.297\ldots$; $\beta = 20.052\ldots°$; $\gamma = 127.947\ldots°$

10. $c = 42.399\ldots$; $\beta = 41.328\ldots°$; $\gamma = 89.671\ldots°$

11. $a = 14.235\ldots$; $\alpha = 72.836\ldots°$; $\gamma = 37.163\ldots°$

12. $a = 31.029\ldots$; $\alpha = 29.712\ldots°$; $\beta = 47.287\ldots°$

13. Zero

14. One

15. Two

16. Zero

17. One

18. Zero

19. $c = 12.199\ldots$; $\beta = 72.740\ldots°$; $\gamma = 43.259\ldots°$ or $c = 2.705\ldots$; $\beta = 107.259\ldots°$; $\gamma = 8.740\ldots°$

20. $a = 33.986\ldots$; $\alpha = 94.867\ldots°$; $\gamma = 47.132\ldots°$ or $a = 5.413\ldots$; $\alpha = 9.132\ldots$; $\gamma = 132.867\ldots°$

21. $b = 10.813\ldots$; $\alpha = 78.151\ldots°$; $\beta = 33.848\ldots°$ or $b = 3.421\ldots$; $\alpha = 101.848\ldots°$; $\beta = 10.151\ldots°$

22. $c = 9.850\ldots$; $\alpha = 67.299\ldots°$; $\gamma = 55.700\ldots°$ or $c = 2.131\ldots$; $\alpha = 112.700\ldots°$; $\gamma = 10.299\ldots°$

23. $h = 10\sin 42° = 6.691\ldots$. **(a)** $6.691\ldots < b < 10$. **(b)** $b = 6.691\ldots$ or $b \ge 10$. **(c)** $b < 6.691\ldots$.

24. $h = 12\sin 53° = 9.583\ldots$. **(a)** $9.583\ldots < c < 12$. **(b)** $c = 9.583\ldots$ or $c \ge 12$. **(c)** $c < 9.583\ldots$.

25. **(a)** No. This is an SAS case (which requires methods of Section 8.2). **(b)** No. Only two pieces of information were given.

26. **(a)** Yes. $b = 88.536\ldots$; $c = 146.127\ldots$; $\beta = 32°$. **(b)** No. This is an SAS case (which requires methods of Section 8.2).

27. No triangle is formed.

28. $c = 25.624\ldots$; $\alpha = 16.177\ldots°$; $\gamma = 116.822\ldots°$

29. No triangle is formed.

30. No triangle is formed.

31. $a = 28.250\ldots$; $b = 19.138\ldots$; $\alpha = 99°$

32. $a = 9.793\ldots$; $c = 27.480\ldots$; $\gamma = 114°$

33. $a = 20.671\ldots$; $\alpha = 24.581\ldots°$; $\beta = 80.418\ldots°$ or $a = 4.692\ldots$; $\alpha = 5.418\ldots°$; $\beta = 99.581\ldots°$

34. $c = 13.478\ldots$; $\beta = 68.984\ldots°$; $\gamma = 57.015\ldots°$ or $c = 4.154\ldots$; $\beta = 111.015\ldots°$; $\gamma = 14.984\ldots°$

35. Cannot be solved with law of sines.

36. Cannot be solved with law of sines.

37. **(a)** $AC = b = 54.597\ldots$ ft **(b)** $b \sin\alpha (= a\sin\beta) = 51.925\ldots$ ft

38. $a = 19.703\ldots$ mi; $b = 15.047\ldots$ mi; $h = b\sin\alpha = a\sin\beta = 11.857\ldots$ mi

39. $a = 24.925\ldots$ ft

40. Altitude $= b\sin\alpha = (1.540\ldots)\sin 28° = 0.723\ldots$ mi or $a\sin\beta = (1.201\ldots)\sin 37°$ mi

41. $4\sin 28° - (\tan 10°)(4\cos 28°) = 1.255\ldots$ ft

42. $r = \dfrac{7.75}{\sin 11.25°} \approx 39.7$ ft

43. $108.906\ldots$ ft

44. $61.732\ldots$ ft

45. $a = 36.581\ldots$ mi; $b = 28.936\ldots$ mi

46. **(a)** Since $\alpha = 90°$, $\sin\alpha = 1$; using the right-triangle definition of the sine function, we have $\sin\beta = \dfrac{b}{a}$ and $\sin\gamma = \dfrac{c}{a}$, so $\dfrac{\sin\alpha}{a} = \dfrac{1}{a} = \dfrac{\sin\beta}{b} = \dfrac{\sin\gamma}{c}$. **(b)** With A at the origin, B on positive x-axis, and C in the first quadrant, we have (just as for the obtuse case) $h = b\sin\alpha = a\sin\beta$; with A at the origin, C on positive x-axis, and B in the first quadrant, we have (just as for the obtuse case) $k = c\sin\alpha = a\sin\gamma$.

47. If α is obtuse (or even right), there will always be either zero or one possible triangles: The longest side of a triangle is always opposite the largest angle, and an obtuse (or right) angle *must* be the largest; therefore a should be greater than b (indicating one triangle), but if it is not, there can be no triangle.

48. **(a)** For any triangle with side lengths a, b, and c, the law of sines says that $\dfrac{\sin\alpha}{a} = \dfrac{\sin\beta}{b} = \dfrac{\sin\gamma}{c}$; we can also find another triangle (using ASA) with two angles the same as the first (in which case the third angle is also the same) and a different side length, say, a'; if $a' = ka$ for some constant k; for this triangle we have $\dfrac{\sin\alpha}{a'} = \dfrac{\sin\beta}{b'} = \dfrac{\sin\gamma}{c'}$; since $\dfrac{\sin\alpha}{a'} = \dfrac{\sin\alpha}{ka} = \dfrac{1}{k}\cdot\dfrac{\sin\alpha}{a}$, we see that $\dfrac{\sin\beta}{b'} = \dfrac{1}{k}\cdot\dfrac{\sin\beta}{b}$, so $b' = kb$, and similarly, $c' = kc$; thus for any choice of a positive constant k, we can create a triangle with angles α, β, and γ. **(b)** Possible answers are $a = 1$, $b = \sqrt{3}$, $c = 2$ (or any set of three numbers proportional to these). **(c)** Any set of three identical numbers.

49. $c = 3.850\ldots$; $\alpha = 29.107\ldots°$; $\beta = 128.892\ldots°$

50. $AC = b = 8.691\ldots$ mi; $BC = a = 12.232\ldots$ mi; height $= a\sin 25° = b\sin 36.5° = 5.169\ldots$ mi

51. **(a)** The notation AAS, etc., does not specify going around clockwise or counterclockwise: if we have been given α, β, and a, for example, this might be an AAS case viewed clockwise but an SAA case viewed counterclockwise. **(b)** If we have been given α, b, and a, for example, this might be an ASS case viewed clockwise but an SSA case viewed counterclockwise. **(c)** If we have three pieces of information, each of which is either "A" or "S," there are only $2^3 = 8$ possibilities (2 choices for the first piece of information, 2 for the second, 2 for the third). In (a) and (b) we showed that two pairs are equivalent, leaving only 6 distinct possibilities.

52. **(a)** Use the formula $A = \frac{1}{2}$(base width)(height); if we orient the triangle with $\overline{AB}$ as the base, the base width is c; the perpendicular segment from C to the line containing the other side gives the height; this segment is a leg on two right triangles; one with hypotenuse $\overline{AC}$ and one with hypotenuse $\overline{BC}$; the height can then be determined to be $b\sin\alpha$ or $a\sin\beta$, respectively, from these triangles; this gives both the first and last versions of the area formula, and using a triangle with $\overline{BC}$ or $\overline{AC}$ as the base gives the other. **(b)** Since these formulas all represent the area, they must be equal, and therefore, for example, $0.5ac\sin\beta = 0.5ab\sin\gamma$, so $c\sin\beta = b\sin\gamma$, or $\dfrac{\sin\beta}{b} = \dfrac{\sin\gamma}{c}$.

QUICK REVIEW 8.2

1. $\alpha = 53.130\ldots°$

2. $\beta = 51.317\ldots°$

3. $\gamma = 103.297\ldots°$

4. $\alpha = 132.843\ldots°$

5. $\beta = 46.369\ldots°$

6. $\gamma = 50.208\ldots°$

7. $\cos\alpha = \dfrac{81 - x^2 - y^2}{-2xy} = \dfrac{x^2 + y^2 - 81}{2xy}$; $\alpha = \cos^{-1}\left(\dfrac{x^2 + y^2 - 81}{2xy}\right)$

8. $\cos\alpha = \dfrac{y^2 - x^2 - 25}{-10} = \dfrac{x^2 - y^2 + 25}{10}$; $\alpha = \cos^{-1}\left(\dfrac{x^2 - y^2 + 25}{10}\right)$

9. One answer is $(x-1)(x-2) = x^2 - 3x + 2$.

10. One answer is $(x-1)(x+1) = x^2 - 1$.

11. One answer is $(x-i)(x+i) = x^2 + 1$.

12. One answer is $(x-1)^2 = x^2 - 2x + 1$.

SECTION EXERCISES 8.2

1. $b = 19.221\ldots$; $\alpha = 30.692\ldots°$; $\gamma = 18.307\ldots°$

2. $c = 9.502\ldots$; $\alpha = 80.331\ldots°$; $\beta = 57.668\ldots°$

3. $\alpha = 76.816\ldots°$; $\beta = 43.247\ldots°$; $\gamma = 59.935\ldots°$

4. $\alpha = 52.160\ldots°$; $\beta = 99.187\ldots°$; $\gamma = 28.651\ldots°$

5. $a = 9.830\ldots$; $\beta = 89.317\ldots°$; $\gamma = 35.682\ldots°$

6. $b = 29.521\ldots$; $\alpha = 123.336\ldots°$; $\gamma = 21.663\ldots°$

7. $c = 25.078\ldots$; $\alpha = 28.468\ldots°$; $\beta = 56.531\ldots°$

8. $a = 35.428\ldots$; $\beta = 37.946\ldots°$; $\gamma = 60.053\ldots°$

9. No triangles possible

10. No triangles possible

11. $\alpha = 24.558\ldots°$; $\beta = 99.215\ldots°$; $\gamma = 56.225\ldots°$

12. No triangles possible

13. $c = 9.487\ldots$; $\beta = 72.921\ldots°$; $\gamma = 65.078\ldots°$ or $c = 5.375\ldots$; $\beta = 107.078\ldots°$; $\gamma = 30.921\ldots°$

14. $c = 12.564\ldots$; $\beta = 49.678\ldots°$; $\gamma = 73.321\ldots°$

15. No triangles possible

16. $c = 7.446\ldots$; $\beta = 59.789\ldots°$; $\gamma = 49.210\ldots°$

17. $a = 23.573\ldots$, so area ≈ 222.33 ft^2

18. $a = 16.582\ldots$, so area ≈ 115.84 m^2

19. $b = 25.844\ldots$, so area ≈ 107.98 cm^2

20. $c = 6.010\ldots$, so area ≈ 4.26 in.^{2}

21. Area $= \sqrt{66.9375} = 8.181\ldots$

22. Area $= \sqrt{303.1875} = 17.412\ldots$

23. No triangle can be formed.

24. Area $= \sqrt{12960} = 36\sqrt{10} = 113.841\ldots$

25. Area $= \sqrt{46720.3464} = 216.148\ldots$

26. No triangle

27. Area $= \sqrt{98629.1856} = 314.052\ldots$

28. Area $= \sqrt{10269.224} = 101.337\ldots$

29. The largest angle is opposite the largest side; it has measure $82.819\ldots°$, or $1.445\ldots$ radians.

30. $16.510\ldots$ ft

31. $AB = 130.422\ldots$ ft

32. **(a)** Distance to second base: $90\sqrt{2} - 60.5 = 66.779\ldots$ ft; a bit more than $c = 63.717\ldots$ ft **(b)** $\beta = 92.824\ldots°$

33. **(a)** $c = 42.495\ldots$ ft **(b)** $60\sqrt{2} - 40 = 44.852\ldots$ ft **(c)** $\beta = 93.273\ldots°$

34. $AB = 841.215\ldots$ ft

35. **(a)** $\tan^{-1}\left(\frac{6}{18}\right) = 18.434\ldots°$ **(b)** $DF = 4.481\ldots$ ft **(c)** $EF = 7.626\ldots$ ft

36. $290.847\ldots$ mi

37. $AB = 12.504\ldots$ yd

38. $HB = 36.955\ldots$; $HC = 48.284\ldots$; $HD = 52.262\ldots$

39. $m\angle CAB = 37.874\ldots°$

40. $m\angle ABC = \sin^{-1}\left(\frac{1}{3}\right) = 19.471\ldots$

41. AAA: If $\alpha + \beta + \gamma = 180°$, there are infinitely many triangles.
ASA: Given angles α and β, if $\alpha + \beta < 180°$, there is a unique triangle.
SAA: Given angles α and β, if $\alpha + \beta < 180°$, there is a unique triangle.
SSA: Given α, a, b, there is at least one triangle (and possibly two) if $a \geq b \sin\alpha$ and α is acute. If α is obtuse or right, there is a unique triangle if $a > b$.
SAS: For any angle $\alpha < 180°$, there is exactly one triangle.
SSS: Given a, b, c, there is exactly one triangle, provided $a + b > c$, $a + c > b$, and $b + c > a$.

42. Law of sines: ASA, SAA, and SSA; law of cosines: SSA, SAS, and SSS

43. The missing side length is $162.922\ldots$ ft; the missing angles are $84.402\ldots°$ (lower left) and $66.597\ldots°$ (upper right).

44. Missing side: $114.332\ldots$ ft; missing angles (clockwise from the given angle): $121.850\ldots°$, $57.029\ldots°$, $91.119\ldots°$

45. $\cos 90° = 0$, so the formula reduces to $c^2 = a^2 + b^2$.

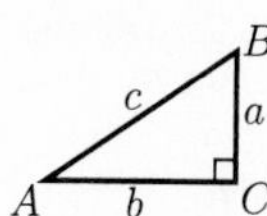

46. $\cos 0° = 1$, so the formula reduces to $c^2 = a^2 + b^2 - 2ab = (a-b)^2$. Then $|c| = |a - b|$, which gives $c = |a - b|$ since $c \geq 0$.

A B C (c, a, b) or B A C (c, b, a)

47. $\cos 180° = -1$, so the formula reduces to $c^2 = a^2 + b^2 + 2ab = (a+b)^2$: then $|c| = |a + b|$, which gives $c = a + b$, since a, b, and c are nonnegative.

B C A (a, b, c)

QUICK REVIEW 8.3

1. $x = \dfrac{9\sqrt{3}}{2}$, $y = 4.5$

2. $x = -7.5$, $y = \dfrac{15\sqrt{3}}{2}$

3. $x = 7\cos 220° = -5.362\ldots$, $y = 7\sin 220° = -4.499\ldots$

4. $x = 6\cos(-50°) = 3.856\ldots$, $y = 6\sin(-50°) = -4.596\ldots$

5. $\theta = 75.522\ldots°$

6. $\theta = 38.682\ldots°$

7. $\theta = 33.854\ldots°$

8. $\theta = 104.963\ldots°$

9. $\theta = 180° + \tan^{-1}(-\frac{4}{3}) = 126.869\ldots°$

10. $\theta = \tan^{-1}(\frac{9}{5}) = 60.945\ldots°$

11. $\theta = 360° + \tan^{-1}(-\frac{7}{5}) = 305.537\ldots°$

12. $\theta = 180° + \tan^{-1}(\frac{5}{2}) = 248.198\ldots°$

13. Distance: about 254.14 naut mi; bearing: approximately 95.40°.

SECTION EXERCISES 8.3

1. $\|\overrightarrow{RS}\| = \sqrt{3^2 + (-2)^2} = \sqrt{13} = \|\overrightarrow{OP}\|$; directions of both are right and down, with slope $\frac{-2}{3} = \frac{-2}{3}$.

2. $\|\overrightarrow{RS}\| = \sqrt{(-3)^2 + (-2)^2} = \sqrt{13} = \|\overrightarrow{OP}\|$; directions of both are left and down, with slope $\frac{-2}{-3} = \frac{2}{3}$.

3. $\|\overrightarrow{RS}\| = \sqrt{(-2)^2 + (-2)^2} = \sqrt{8} = 2\sqrt{2} = \|\overrightarrow{OP}\|$; directions of both are left and down, with slope $\frac{-2}{-2} = 1$.

4. $\|\overrightarrow{RS}\| = \sqrt{(4)^2 + (5)^2} = \sqrt{41} = \|\overrightarrow{OP}\|$; directions of both are right and up, with slope $\frac{-5}{-4} = 1.25$.

5. $\mathbf{v} = \langle 4, 2\rangle$; $\|\mathbf{v}\| = \sqrt{20} = 2\sqrt{5}$

6. $\mathbf{v} = \langle 4, -13\rangle$; $\|\mathbf{v}\| = \sqrt{185}$

7. $\mathbf{v} = \langle -5, 1\rangle$; $\|\mathbf{v}\| = \sqrt{26}$

8. $\mathbf{v} = \langle 3, -10\rangle$; $\|\mathbf{v}\| = \sqrt{109}$

9. $\mathbf{v} = \langle -2, -24\rangle$; $\|\mathbf{v}\| = \sqrt{580} = 2\sqrt{145}$

10. $\mathbf{v} = \langle -\sqrt{2}, 3\sqrt{2}\rangle$; $\|\mathbf{v}\| = \sqrt{20} = 2\sqrt{5}$

11. $\mathbf{v} = \langle -12, -7\rangle$; $\|\mathbf{v}\| = \sqrt{193}$

12. $\mathbf{v} = \langle -9, -16\rangle$; $\|\mathbf{v}\| = \sqrt{337}$

13. $\langle 1, 7\rangle$

14. $\langle -3, -1\rangle$

15. $\langle -3, 8\rangle$

16. $\langle 6, 12\rangle$

17. $\langle 4, -9\rangle$

18. $\langle -10, -10\rangle$

19. $\langle -4, -18\rangle$

20. $\langle -1, -7\rangle$

21. (a) $\left\langle \frac{2}{\sqrt{5}}, \frac{1}{\sqrt{5}}\right\rangle$. (b) $\frac{2}{\sqrt{5}}\mathbf{i} + \frac{1}{\sqrt{5}}\mathbf{j}$.

22. (a) $\left\langle -\frac{3}{\sqrt{13}}, \frac{2}{\sqrt{13}}\right\rangle$. (b) $-\frac{3}{\sqrt{13}}\mathbf{i} + \frac{2}{\sqrt{13}}\mathbf{j}$.

23. (a) $\left\langle -\frac{4}{\sqrt{41}}, \left(-\frac{5}{\sqrt{41}}\right)\right\rangle$. (b) $-\frac{4}{\sqrt{41}}\mathbf{i} + \left(-\frac{5}{\sqrt{41}}\right)\mathbf{j} = -\frac{4}{\sqrt{41}}\mathbf{i} - \frac{5}{\sqrt{41}}\mathbf{j}$.

24. (a) $\left\langle \frac{3}{5}, \left(-\frac{4}{5}\right)\right\rangle$. (b) $\frac{3}{5}\mathbf{i} + \left(-\frac{4}{5}\right)\mathbf{j} = \frac{3}{5}\mathbf{i} - \frac{4}{5}\mathbf{j}$.

25. $\mathbf{v} = \langle 18\cos 25°, 18\sin 25°\rangle = \langle 16.313\ldots, 7.607\ldots\rangle$

26. $\mathbf{v} = \langle 14\cos 55°, 14\sin 55°\rangle = \langle 8.030\ldots, 11.468\ldots\rangle$

27. $\mathbf{v} = \langle 47\cos 108°, 47\sin 108°\rangle = \langle -14.523\ldots, 44.699\ldots\rangle$

28. $\mathbf{v} = \langle 33\cos 136°, 33\sin 136°\rangle = \langle -23.738\ldots, 22.923\ldots\rangle$

29. (a) $\theta_u = \tan^{-1}(\frac{4}{3}) = 53.130\ldots°$; $\theta_v = \tan^{-1}0 = 0°$. (b) $\theta_u - \theta_v = 53.130\ldots°$.

30. (a) $\theta_u = 180° + \tan^{-1}(-2) = 116.565\ldots°$; $\theta_v = \tan^{-1}(\frac{2}{3}) = 33.690\ldots°$. (b) $\theta_u - \theta_v = 82.874\ldots°$.

31. (a) $\theta_u = 180° + \tan^{-1}(-2) = 116.565\ldots°$; $\theta_v = \tan^{-1}(-\frac{4}{3}) = -53.130\ldots°$. (b) $\theta_u - \theta_v = 169.695\ldots°$.

32. (a) $\theta_u = \tan^{-1}(-1.5) = -56.309\ldots°$; $\theta_v = 180° + \tan^{-1}(\frac{5}{3}) = 239.036\ldots°$ or $-120.963\ldots°$. (b) $\theta_u - \theta_v = 64.653\ldots°$.

33. $\mathbf{v} \approx \langle -223.99, 480.34\rangle$

34. $\mathbf{v} \approx \langle 79.88, -453.01\rangle$

35. (a) $\mathbf{v} \approx \langle -111.16, 305.40\rangle$. (b) Actual velocity vector: $\mathbf{v} + \langle -25.71, 30.64\rangle = \langle -136.87, 336.04\rangle$; speed: $\sqrt{136.87^2 + 336.04^2} \approx 362.84$ mph; bearing: approximately 337.84°.

36. (a) $\mathbf{v} \approx \langle 79.88, -453.01\rangle$. (b) Actual velocity vector: $\mathbf{v} + \langle -27.36, -75.18\rangle = \langle 52.52, -528.19\rangle$; speed: $\sqrt{52.52^2 + 528.19^2} \approx 530.79$ mph; bearing: approximately 174.32°.

37. (a) $\mathbf{v} \approx \langle 3.42, 9.40\rangle$. (b) The ball's horizontal speed is about 3.42 m/sec; the initial vertical speed is about 9.40 m/sec (although this will change due to gravity).

38. (a) $\mathbf{v} \approx \langle 2.41, 0.65\rangle$. (b) The box's horizontal force is about 2.41 lb; the vertical force is about 0.65 lb

39. We need to choose $\mathbf{w} = \langle a, b\rangle = k\langle\cos 33°, \sin 33°\rangle$ so that $k\cos(33° - 15°) = 2.5$; then $k \approx 2.63$ lb so that $\mathbf{w} \approx \langle 2.20, 1.43\rangle$.

40. Juana's force can be represented by $\langle 21.87, 7.11\rangle$, whereas Diego's force is $\langle 17.39, -4.66\rangle$; their total force is therefore $\langle 39.26, 2.45\rangle$, so Corporal must be pulling with an equal force in the opposite direction, $\langle -39.26, -2.45\rangle$, with a force of about 39.34 lb.

41. $\mathbf{F} = \langle 50 \cos 45°, 50 \sin 45° \rangle + \langle 75 \cos(-30°), 75 \sin(-30°) \rangle \approx \langle 100.31, -2.14 \rangle$, so $\|\mathbf{F}\| \approx 100.33$ lb and $\theta \approx -1.22°$

42. $\mathbf{F} = 100\langle \cos 50°, \sin 50° \rangle + 50\langle \cos 160°, \sin 160° \rangle + 80\langle \cos(-20°), \sin(-20°) \rangle \approx \langle 92.47, 66.34 \rangle$, so $\|\mathbf{F}\| \approx 113.81$ lb and $\theta \approx 35.66°$

43. **(a)** Suppose that $\mathbf{v} = \langle a, b \rangle$; then $\|\mathbf{v}\| = \sqrt{a^2 + b^2}$, so $\mathbf{u} = \left\langle \frac{a}{\sqrt{a^2+b^2}}, \frac{b}{\sqrt{a^2+b^2}} \right\rangle = \langle c, d \rangle$, say; therefore $\|\mathbf{u}\| = \sqrt{c^2 + d^2} = \sqrt{\frac{a^2}{a^2+b^2} + \frac{b^2}{a^2+b^2}} = \sqrt{\frac{a^2+b^2}{a^2+b^2}} = 1$. **(b)** $\mathbf{v}$ has direction angle θ_v, with $\tan \theta_v = \frac{b}{a}$; θ_v is also determined by the signs of a and b (which determine the quadrant of the terminal point when the vector is in standard position); similarly, $\mathbf{u}$ has direction angle θ_u, with $\tan \theta_u = \frac{d}{c}$; which equals $\frac{b}{a} = \tan \theta_v$; the signs of c and d are also the same as a and b, respectively, so $\theta_v = \theta_u$. **(c)** $k = \|\mathbf{w}\|$.

44. Let $\mathbf{u} = \langle u_1, u_2 \rangle$, $\mathbf{v} = \langle v_1, v_2 \rangle$, and $\mathbf{w} = \langle w_1, w_2 \rangle$.
(a) $\mathbf{u} + \mathbf{v} = \langle u_1, u_2 \rangle + \langle v_1, v_2 \rangle = \langle u_1 + v_1, u_2 + v_2 \rangle = \langle v_1 + u_1, v_2 + u_2 \rangle = \langle v_1, v_2 \rangle + \langle u_1, u_2 \rangle = \mathbf{v} + \mathbf{u}$.
(b) $(\mathbf{u} + \mathbf{v}) + \mathbf{w} = \langle u_1 + v_1, u_2 + v_2 \rangle + \langle w_1, w_2 \rangle = \langle u_1 + v_1 + w_1, u_2 + v_2 + w_2 \rangle = \langle u_1, u_2 \rangle + \langle v_1 + w_1, v_2 + w_2 \rangle = \mathbf{u} + (\mathbf{v} + \mathbf{w})$.
(c) $\mathbf{u} + \mathbf{0} = \langle u_1, u_2 \rangle + \langle 0, 0 \rangle = \langle u_1 + 0, u_2 + 0 \rangle = \langle u_1, u_2 \rangle = \mathbf{u}$.
(d) $\mathbf{u} + (-\mathbf{u}) = \langle u_1, u_2 \rangle + \langle -u_1, -u_2 \rangle = \langle u_1 + (-u_1), u_2 + (-u_2) \rangle = \langle 0, 0 \rangle = \mathbf{0}$.
(e) $a(\mathbf{u} + \mathbf{v}) = a(\langle u_1 + v_1, u_2 + v_2 \rangle) = \langle a(u_1 + v_1), a(u_2 + v_2) \rangle = \langle au_1 + av_1, au_2 + av_2 \rangle = \langle au_1, au_2 \rangle + \langle av_1, av_2 \rangle = a\langle u_1, u_2 \rangle + a\langle v_1, v_2 \rangle = a\mathbf{u} + a\mathbf{v}$.
(f) $(a + b)\mathbf{u} = \langle (a + b)u_1, (a + b)u_2 \rangle = \langle au_1 + bu_1, au_2 + bu_2 \rangle = \langle au_1, au_2 \rangle + \langle bu_1, bu_2 \rangle = a\langle u_1, u_2 \rangle + b\langle u_1, u_2 \rangle = a\mathbf{u} + b\mathbf{u}$.
(g) $(ab)\mathbf{u} = \langle (ab)u_1, (ab)u_2 \rangle = \langle a(bu_1), a(bu_2) \rangle = a\langle bu_1, bu_2 \rangle = a(b\mathbf{u})$.
(h) $a\mathbf{0} = a\langle 0, 0 \rangle = \langle a0, a0 \rangle = \langle 0, 0 \rangle = \mathbf{0}$. $0\mathbf{u} = 0\langle u_1, u_2 \rangle = \langle 0u_1, 0u_2 \rangle = \langle 0, 0 \rangle = \mathbf{0}$.
(i) $(1)\mathbf{u} = \langle (1)u_1, (1)u_2 \rangle = \langle u_1, u_2 \rangle = \mathbf{u}$. $(-1)\mathbf{u} = \langle (-1)u_1, (-1)u_2 \rangle = \langle -u_1, -u_2 \rangle = -\mathbf{u}$.
(j) $\|a\mathbf{u}\| = \|\langle au_1, au_2 \rangle\| = \sqrt{(au_1)^2 + (au_2)^2} = \sqrt{a^2u_1^2 + a^2u_2^2} = \sqrt{a^2(u_1^2 + u_2^2)} = |a|\sqrt{u_1^2 + u_2^2} = |a|\,\|\mathbf{u}\|$.

45. $\langle 5, 7 \rangle = 3.8r + 0.6s$; if you take a vector 3.8 times as long as $\mathbf{r}$ (in the same direction) and place it head-to-tail with a vector 0.6 times as long as s, the resulting vector (the third side of the resulting triangle) is $\langle 5, 7 \rangle$.

46. $\mathbf{v} = \langle x, y \rangle = a\mathbf{r} + b\mathbf{s}$, where $a = 0.2x + 0.4y = \frac{(x + 2y)}{5}$ and $b = 0.4x - 0.2y = \frac{(2x - y)}{5}$

47. This can be done a number of ways; one way is to consider the slopes. **(a)** Slope of $\mathbf{u}$ is $\frac{3}{2}$, and slope of $\mathbf{v}$ is $-\frac{2}{3}$; the product of these slopes is -1. **(b)** Slope of $\mathbf{u}$ is -1, and slope of $\mathbf{v}$ is 1; the product of these slopes is -1. **(c)** Slope of $\mathbf{u}$ is $-\frac{3}{5}$; and slope of $\mathbf{v}$ is $\frac{5}{3}$; the product of these slopes is -1.

48. **(a)** $\mathbf{u} \cdot \mathbf{v} = (2)(-3) + (3)(2) = 0$. **(b)** $\mathbf{u} \cdot \mathbf{v} = (1)(-1) + (-1)(-1) = 0$. **(c)** $\mathbf{u} \cdot \mathbf{v} = (5)(-3) + (-3)(-5) = 0$.

QUICK REVIEW 8.4

1. $6 - i$

2. $-2 - 2i$

3. $14 + 2i$

4. $4 + 8i$

5. $0.5 - 2.5i$

6. $1.6 + 0.2i$

7. $\theta = \frac{5\pi}{6}$

8. $\theta = \frac{7\pi}{4}$

9. $\theta = \frac{4\pi}{3}$

10. $\theta = \frac{11\pi}{6}$

11. $\theta = \frac{5\pi}{4}$

12. $\theta = \frac{2\pi}{3}$

13. **(a)** $\|\langle -1, 1 \rangle\| = \sqrt{2}$. **(b)** $d = \sqrt{2}$. **(c)** $r = \sqrt{2}, \theta = \frac{3\pi}{4}$.

14. **(a)** $\|\langle 2, -2 \rangle\| = 2\sqrt{2}$. **(b)** $d = 2\sqrt{2}$. **(c)** $r = 2\sqrt{2}, \theta = \frac{7\pi}{4}$.

15. **(a)** $\|\langle -\sqrt{3}, -1 \rangle\| = 2$. **(b)** $d = 2$. **(c)** $r = 2, \theta = \frac{7\pi}{6}$.

16. **(a)** $\|\langle 1, \sqrt{3} \rangle\| = 2$. **(b)** $d = 2$. **(c)** $r = 2, \theta = \frac{\pi}{3}$.

SECTION EXERCISES 8.4

1. $|2 + 7i| = \sqrt{53}$

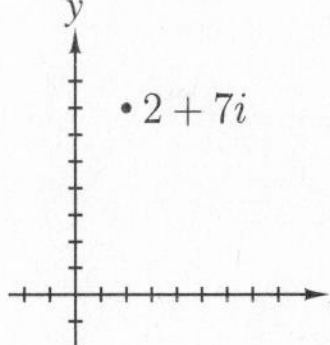

2. $|-4 + i| = \sqrt{17}$

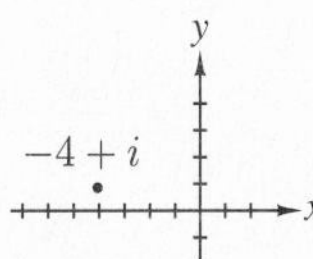

3. $|6 - 5i| = \sqrt{61}$

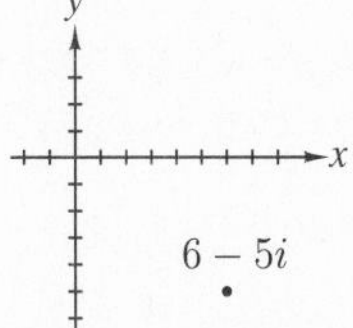

4. $|-1-7i| = \sqrt{50} = 5\sqrt{2}$

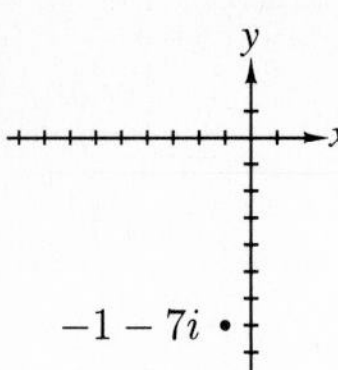

5. $|5-3i| = \sqrt{34}$

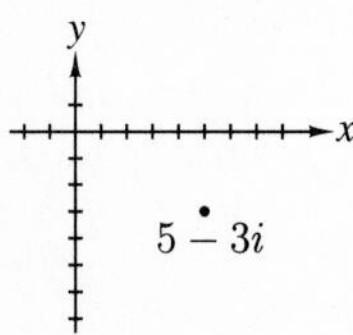

6. $|3+2i| = \sqrt{13}$

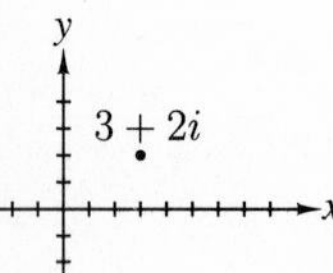

7. $3i = 3\left(\cos\frac{\pi}{2} + i\sin\frac{\pi}{2}\right)$

8. $-2i = 2\left(\cos\frac{3\pi}{2} + i\sin\frac{3\pi}{2}\right)$

9. $2+2i = 2\sqrt{2}\left(\cos\frac{\pi}{4} + i\sin\frac{\pi}{4}\right)$

10. $\sqrt{3}+i = 2\left(\cos\frac{\pi}{6} + i\sin\frac{\pi}{6}\right)$

11. $-2+2i\sqrt{3} = 4\left(\cos\frac{2\pi}{3} + i\sin\frac{2\pi}{3}\right)$

12. $3-3i = 3\sqrt{2}\left(\cos\frac{7\pi}{4} + i\sin\frac{7\pi}{4}\right)$

13. $\frac{1}{2} - \frac{\sqrt{3}}{2}i = 1\left(\cos\frac{5\pi}{3} + i\sin\frac{5\pi}{3}\right) = \cos\frac{5\pi}{3} + i\sin\frac{5\pi}{3}$

14. $-\frac{\sqrt{3}}{2} - \frac{1}{2}i = 1\left(\cos\frac{7\pi}{6} + i\sin\frac{7\pi}{6}\right) = \cos\frac{7\pi}{6} + i\sin\frac{7\pi}{6}$

15. $3+2i \approx 3.6(\cos 0.588 + i\sin 0.588)$

16. $4-7i \approx 8.06(\cos 5.232 + i\sin 5.232)$

17. $-1+3i \approx 3.16(\cos 1.893 + i\sin 1.893)$

18. $-9+i \approx 9.06(\cos 3.031 + i\sin 3.031)$

19. $13+8i \approx 15.26(\cos 0.552 + i\sin 0.552)$

20. $-11-17i \approx 20.25(\cos 4.138 + i\sin 4.138)$

21. $\sqrt{5} - i\sqrt{3} \approx 2.83(\cos 5.624 + i\sin 5.624)$

22. $-3\sqrt{2} - i\sqrt{5} \approx 4.80(\cos 3.627 + i\sin 3.627)$

23. $\dfrac{3\sqrt{3}}{2} - 1.5i$

24. $\sqrt{2} + i\sqrt{2}$

25. $-4\sqrt{3} - 4i$

26. $2.5 - 2.5\sqrt{3}\,i$

27. $1.5 + 1.5\sqrt{3}\,i$

28. $1 - \sqrt{3}\,i$

29. $2.5\sqrt{2} + 2.5\sqrt{2}\,i$

30. $2 - 2\sqrt{3}\,i$

31. $-0.5\sqrt{6} - 0.5\sqrt{2}\,i$

32. $3.5\sqrt{3} - 3.5i$

33. $-0.5\sqrt{6} - 0.5\sqrt{6}\,i$

34. $2.555\ldots + i\,0.684\ldots$

35. $14(\cos 155° + i\sin 155°)$

36. $0.5\sqrt{2}(\cos 99° + i\sin 99°)$

37. $15\left(\cos\frac{23\pi}{12} + i\sin\frac{23\pi}{12}\right)$

38. $\dfrac{1}{3}\sqrt{3}\left(\cos\dfrac{11\pi}{12} + i\sin\dfrac{11\pi}{12}\right) = \dfrac{1}{\sqrt{3}}\left(\cos\dfrac{11\pi}{12} + i\sin\dfrac{11\pi}{12}\right)$

39. $\frac{2}{3}(\cos(-30°) + i\sin(-30°)) = \frac{2}{3}(\cos 330° + i\sin 330°)$

40. $.5(\cos 105° + i\sin 105°)$

41. $2(\cos 3\pi + i\sin 3\pi) = 2(\cos\pi + i\sin\pi)$

42. $1\left(\cos\frac{\pi}{4} + i\sin\frac{\pi}{4}\right) = \cos\frac{\pi}{4} + i\sin\frac{\pi}{4}$

43. **(a)** $\dfrac{\sqrt{13}[\cos(5.695\ldots) + i\sin(5.695\ldots)]}{\sqrt{2}[\cos(\pi/4) + i\sin(\pi/4)]} = \sqrt{6.5}\left[\cos\left(5.695\ldots - \frac{\pi}{4}\right) + i\sin\left(5.695\ldots - \frac{\pi}{4}\right)\right] = 0.5 - 2.5i$

(b) $\dfrac{3-2i}{1+i} = \dfrac{3-2i}{1+i}\cdot\dfrac{1-i}{1-i} = \dfrac{1-5i}{2} = 0.5 - 2.5i$

44. **(a)** $\dfrac{\sqrt{2}[\cos(7\pi/4) + i\sin(7\pi/4)]}{2[\cos(\pi/6) + i\sin(\pi/6)]} = \dfrac{1}{\sqrt{2}}\left(\cos\dfrac{19\pi}{12} + i\sin\dfrac{19\pi}{12}\right) \approx 0.183 - 0.683i.$ **(b)** $\dfrac{1-i}{\sqrt{3}+i} = \dfrac{(1-i)(\sqrt{3}-i)}{4} = \dfrac{\sqrt{3}-1}{4} - \dfrac{\sqrt{3}+1}{4}i.$

45. **(a)**
$\dfrac{\sqrt{10}[\cos(0.321\ldots) + i\sin(0.321\ldots)]}{\sqrt{34}[\cos(-0.540\ldots) + i\sin(-0.540\ldots)]} = 0.542\ldots[\cos(0.321\ldots + 0.540\ldots) + i\sin(0.321\ldots + 0.540\ldots)] \approx 0.353 + 0.412i$

(b) $\dfrac{3+i}{5-3i} = \dfrac{(3+i)(5+3i)}{34} = \dfrac{1}{17}(6+7i)$

46. **(a)**
$\dfrac{\sqrt{13}[\cos(-0.982\ldots) + i\sin(-0.982\ldots)]}{2[\cos(-\pi/3) + i\sin(-\pi/3)]} = \dfrac{\sqrt{13}}{2}(\cos(0.064\ldots) + i\sin(0.064\ldots)) \approx 1.799 + 0.116i.$

(b)
$\dfrac{2-3i}{1-\sqrt{3}i} = \dfrac{(2-3i)(1+\sqrt{3}i)}{4} = \dfrac{1}{4}[2 + 3\sqrt{3} + (2\sqrt{3}-3)i].$

47.
$\dfrac{z_1}{z_2} = \dfrac{r_1(\cos\theta_1 + i\sin\theta_1)}{r_2(\cos\theta_2 + i\sin\theta_2)} = \dfrac{r_1}{r_2}\cdot \dfrac{\cos\theta_1 + i\sin\theta_1}{\cos\theta_2 + i\sin\theta_2}\cdot\dfrac{\cos\theta_2 - i\sin\theta_2}{\cos\theta_2 - i\sin\theta_2} = \dfrac{r_1}{r_2}\,\dfrac{\cos\theta_1\cos\theta_2 + \sin\theta_1\sin\theta_2 + i(\sin\theta_1\cos\theta_2 - \cos\theta_1\sin\theta_2)}{(\cos\theta_2)^2 + (\sin\theta_2)^2} = \dfrac{r_1}{r_2}[\cos\theta_1\cos\theta_2 + \sin\theta_1\sin\theta_2 + i(\sin\theta_1\cos\theta_2 - \cos\theta_1\sin\theta_2)]$; now use the angle difference formulas $\cos(\theta_1 - \theta_2) = \cos\theta_1\cos\theta_2 + \sin\theta_1\sin\theta_2$ and $\sin(\theta_1 - \theta_2) = \sin\theta_1\cos\theta_2 - \cos\theta_1\sin\theta_2$.

48. Since $\cos(-\theta) = \cos\theta$ and $\sin(-\theta) = -\sin\theta$, the suggested form equals $r(\cos\theta - i\sin\theta)$; the complex conjugate of $a + bi$ is $a - bi$, so this is the appropriate form for the complex conjugate of $r(\cos\theta + i\sin\theta)$.

Alternatively, multiplying a complex number $a + bi$ by its complex conjugate gives $a^2 + b^2$, so we just need to confirm that this multiplication results in r^2; since multiplying complex numbers in trigonometric form means adding the angles, we have $r(\cos\theta + i\sin\theta) \cdot r[\cos(-\theta) + i\sin(-\theta)] = r^2[\cos(\theta - \theta) + i\sin(\theta - \theta)] = r^2(\cos 0 + i\sin 0) = r^2$.

49. **(a)** Trigonometric is most convenient, since $r_1 = |z_1|$ and $r_2 = |z_2|$; the magnitude of the product $z_1 z_2$ is easily read from the trigonometric form of the multiplication. **(b)** $|z_1 z_2| = |r_1 r_2 \cos(\theta_1 + \theta_2) + i r_1 r_2 \sin(\theta_1 + \theta_2)| = \sqrt{(r_1 r_2)^2[\cos(\theta_1 + \theta_2)]^2 + (r_1 r_2)^2[\sin(\theta_1 + \theta_2)]^2} = r_1 r_2\sqrt{[\cos(\theta_1 + \theta_2)]^2 + [\sin(\theta_1 + \theta_2)]^2} = r_1 r_2$.

50. **(a)** A circle centered at the origin, with radius 1. **(b)** A circle centered at the origin, with radius 3.

51. Suppose that $z_1 = r_1(\cos\theta_1 + i\sin\theta_1)$ and $z_2 = r_2(\cos\theta_2 + i\sin\theta_2)$. Each triangle's angle at the origin has the same measure: For the smaller triangle, this angle has measure θ_1; for the larger triangle, the side from the origin to z_2 makes an angle of θ_2 with the x-axis, and the side from the origin to $z_1 z_2$ makes an angle of $\theta_1 + \theta_2$, so the angle between is θ_1 as well.

The corresponding side lengths for the sides adjacent to these angles have the same ratio: The two longest sides have lengths $|z_1| = r_1$ (for the smaller) and $|z_1 z_2| = r_1 r_2$, for a ratio of r_2; for the other two sides, the lengths are 1 and r_2, again giving a ratio of r_2; finally, the law of sines can be used to show that the remaining sides have that same ratio.

52. Construct an angle with vertex at z_2, and one ray from z_2 to 0, congruent to the angle formed by 0, 1, and z_1, with vertex 1; also, be sure that this new angle is oriented in the appropriate direction; for example, if z_1 is located "counterclockwise" from 1, the points on this new ray should also be located counterclockwise from z_2. Now similarly construct an angle with vertex at 0 and one ray from 0 to z_2, congruent to the angle formed by 1, 0, and z_1, with vertex 0; the intersection of the two newly constructed rays is $z_1 z_2$.

QUICK REVIEW 8.5

1. $5\{\cos[\tan^{-1}(-\frac{4}{3})] + i\sin[\tan^{-1}(-\frac{4}{3})]\} = 5[\cos(-0.927\ldots) + i\sin(-0.927\ldots)] = 5[\cos(5.355\ldots) + i\sin(5.355\ldots)]$

2. $2\sqrt{2}(\cos\frac{3\pi}{4} + i\sin\frac{3\pi}{4})$

3. $\sqrt{34}[\cos(\tan^{-1} 0.6) + i\sin(\tan^{-1} 0.6)] = \sqrt{34}[\cos(0.540\ldots) + i\sin(0.540\ldots)]$

4. $\sqrt{89}\{\cos[\tan^{-1}(-0.625)] + i\sin[\tan^{-1}(-0.625)]\} = \sqrt{89}[\cos(-0.558\ldots) + i\sin(-0.558\ldots)] = \sqrt{89}[\cos(5.724\ldots) + i\sin(5.724\ldots)]$

5. $2[\cos(-\frac{\pi}{2}) + i\sin(-\frac{\pi}{2})] = 2(\cos\frac{3\pi}{2} + i\sin\frac{3\pi}{2})$

6. $10(\cos\pi + i\sin\pi)$

7. $1 + 0i = 1$

8. $0 - i = -i$

9. $-4 - 4i$

10. $-4 + 0i = -4$

11. $x = 1$

12. $x = \pm 1$

13. $x = 1$

14. $x = \pm 1$

SECTION EXERCISES 8.5

1. $\cos\frac{3\pi}{4} + i\sin\frac{3\pi}{4}$

2. $\cos\frac{4\pi}{3} + i\sin\frac{4\pi}{3}$

3. $2(\cos 6\pi + i\sin 6\pi) = 2(\cos 0 + i\sin 0)$

4. $3(\cos\frac{15\pi}{2} + i\sin\frac{15\pi}{2}) = 3(\cos\frac{3\pi}{2} + i\sin\frac{3\pi}{2})$

5. $2(\cos\frac{9\pi}{4} + i\sin\frac{9\pi}{4}) = 2(\cos\frac{\pi}{4} + i\sin\frac{\pi}{4})$

6. $5(\cos\frac{10\pi}{3} + i\sin\frac{10\pi}{3}) = 5(\cos\frac{4\pi}{3} + i\sin\frac{4\pi}{3})$

7. $(\sqrt{2}^5)(\cos\frac{5\pi}{4} + i\sin\frac{5\pi}{4}) = 4\sqrt{2}(\cos\frac{5\pi}{4} + i\sin\frac{5\pi}{4})$

8. $5^{20}[\cos(20\tan^{-1}(\frac{4}{3})) + i\sin(20\tan^{-1}(\frac{4}{3}))] = 5^{20}[\cos(5.979\ldots) + i\sin(5.979\ldots)]$

9. $8(\cos 5\pi + i\sin 5\pi) = 8(\cos\pi + i\sin\pi)$

10. $\cos 7\pi + i\sin 7\pi = \cos(-\pi) + i\sin(-\pi) = \cos\pi + i\sin\pi$

11. $-1(\cos\pi + i\sin\pi) = 1(\cos 0 + i\sin 0)$

12. $\cos\pi + i\sin\pi$

13. $\sqrt[3]{2}\left(\cos\frac{2\pi(k+1)}{3} + i\sin\frac{2\pi(k+1)}{3}\right)$, $k = 0, 1, 2$; $\sqrt[3]{2}\left(\cos\frac{2\pi}{3} + i\sin\frac{2\pi}{3}\right)$, $\sqrt[3]{2}\left(\cos\frac{4\pi}{3} + i\sin\frac{4\pi}{3}\right)$, $\sqrt[3]{2}(\cos 2\pi + i\sin 2\pi) = \sqrt[3]{2}$

14. $\sqrt[3]{2}\left(\cos\frac{\pi(8k+1)}{12} + i\sin\frac{\pi(8k+1)}{12}\right)$, $k = 0, 1, 2$; $\sqrt[3]{2}\left(\cos\frac{\pi}{12} + i\sin\frac{\pi}{12}\right)$, $\sqrt[3]{2}\left(\cos\frac{3\pi}{4} + i\sin\frac{3\pi}{4}\right)$, $\sqrt[3]{2}\left(\cos\frac{17\pi}{12} + i\sin\frac{17\pi}{12}\right)$

15. $\sqrt[3]{3}\left(\cos\frac{2\pi(3k+2)}{9} + i\sin\frac{2\pi(3k+2)}{9}\right)$, $k = 0, 1, 2$; $\sqrt[3]{3}\left(\cos\frac{4\pi}{9} + i\sin\frac{4\pi}{9}\right)$,

$\sqrt[3]{3}\left(\cos\frac{10\pi}{9} + i\sin\frac{10\pi}{9}\right)$,

$\sqrt[3]{3}\left(\cos\frac{16\pi}{9} + i\sin\frac{16\pi}{9}\right)$

16. $3\left(\cos\frac{\pi(12k + 11)}{18} + i\sin\frac{\pi(12k + 11)}{18}\right)$, $k = 0, 1, 2$;

$3\left(\cos\frac{11\pi}{18} + i\sin\frac{11\pi}{18}\right)$,

$3\left(\cos\frac{23\pi}{18} + i\sin\frac{23\pi}{18}\right)$,

$3\left(\cos\frac{35\pi}{18} + i\sin\frac{35\pi}{18}\right)$

17. $\sqrt[3]{5}\left(\cos\frac{5.355\ldots + 2\pi k}{3} + i\sin\frac{5.355\ldots + 2\pi k}{3}\right)$, $k = 0, 1, 2$;

$\sqrt[3]{5}(\cos 1.785\ldots + i\sin 1.785\ldots)$,
$\sqrt[3]{5}(\cos 3.879\ldots + i\sin 3.879\ldots)$,
$\sqrt[3]{5}(\cos 5.974\ldots + i\sin 5.974\ldots)$

18. Note first that $\sqrt[3]{2\sqrt{2}} = \sqrt{2}$.

$\sqrt{2}\left(\cos\frac{\pi(8k + 3)}{12} + i\sin\frac{\pi(8k + 3)}{12}\right)$, $k = 0, 1, 2$;

$\sqrt{2}\left(\cos\frac{\pi}{4} + i\sin\frac{\pi}{4}\right)$,

$\sqrt{2}\left(\cos\frac{11\pi}{12} + i\sin\frac{11\pi}{12}\right)$,

$\sqrt{2}\left(\cos\frac{19\pi}{12} + i\sin\frac{19\pi}{12}\right)$

19. $\cos\frac{\pi(2k + 1)}{5} + i\sin\frac{\pi(2k + 1)}{5}$,

$k = 0, 1, 2, 3, 4$; $\cos\frac{\pi}{5} + i\sin\frac{\pi}{5}$,

$\cos\frac{3\pi}{5} + i\sin\frac{3\pi}{5}$, $\cos\pi + i\sin\pi$,

$\cos\frac{7\pi}{5} + i\sin\frac{7\pi}{5}$, $\cos\frac{9\pi}{5} + i\sin\frac{9\pi}{5}$

20. $2\left(\cos\frac{\pi(4k + 1)}{10} + i\sin\frac{\pi(4k + 1)}{10}\right)$,
$k = 0, 1, 2, 3, 4$;

$2\left(\cos\frac{\pi}{10} + i\sin\frac{\pi}{10}\right)$,

$2\left(\cos\frac{\pi}{2} + i\sin\frac{\pi}{2}\right)$,

$2\left(\cos\frac{9\pi}{10} + i\sin\frac{9\pi}{10}\right)$,

$2\left(\cos\frac{13\pi}{10} + i\sin\frac{13\pi}{10}\right)$,

$2\left(\cos\frac{17\pi}{10} + i\sin\frac{17\pi}{10}\right)$

21. $\sqrt[5]{2}\left(\cos\frac{\pi(12k + 1)}{30} + i\sin\frac{\pi(12k + 1)}{30}\right)$, $k = 0, 1, 2, 3, 4$;

$\sqrt[5]{2}\left(\cos\frac{\pi}{30} + i\sin\frac{\pi}{30}\right)$,

$\sqrt[5]{2}\left(\cos\frac{13\pi}{30} + i\sin\frac{13\pi}{30}\right)$,

$\sqrt[5]{2}\left(\cos\frac{5\pi}{6} + i\sin\frac{5\pi}{6}\right)$,

$\sqrt[5]{2}\left(\cos\frac{37\pi}{30} + i\sin\frac{37\pi}{30}\right)$,

$\sqrt[5]{2}\left(\cos\frac{49\pi}{30} + i\sin\frac{49\pi}{30}\right)$

22. $\sqrt[5]{2}\left(\cos\frac{\pi(8k + 1)}{20} + i\sin\frac{\pi(8k + 1)}{20}\right)$, $k = 0, 1, 2, 3, 4$;

$\sqrt[5]{2}\left(\cos\frac{\pi}{20} + i\sin\frac{\pi}{20}\right)$,

$\sqrt[5]{2}\left(\cos\frac{9\pi}{20} + i\sin\frac{9\pi}{20}\right)$,

$\sqrt[5]{2}\left(\cos\frac{17\pi}{20} + i\sin\frac{17\pi}{20}\right)$,

$\sqrt[5]{2}\left(\cos\frac{5\pi}{4} + i\sin\frac{5\pi}{4}\right)$,

$\sqrt[5]{2}\left(\cos\frac{33\pi}{20} + i\sin\frac{33\pi}{20}\right)$

23. $\sqrt[5]{2}\left(\cos\frac{\pi(4k + 1)}{10} + i\sin\frac{\pi(4k + 1)}{10}\right)$, $k = 0, 1, 2, 3, 4$;

$\sqrt[5]{2}\left(\cos\frac{\pi}{10} + i\sin\frac{\pi}{10}\right)$,

$\sqrt[5]{2}\left(\cos\frac{\pi}{2} + i\sin\frac{\pi}{2}\right)$,

$\sqrt[5]{2}\left(\cos\frac{9\pi}{10} + i\sin\frac{9\pi}{10}\right)$,

$\sqrt[5]{2}\left(\cos\frac{13\pi}{10} + i\sin\frac{13\pi}{10}\right)$,

$\sqrt[5]{2}\left(\cos\frac{17\pi}{10} + i\sin\frac{17\pi}{10}\right)$

24. $\sqrt[5]{2}\left(\cos\frac{\pi(6k + 1)}{15} + i\sin\frac{\pi(6k + 1)}{15}\right)$, $k = 0, 1, 2, 3, 4$;

$\sqrt[5]{2}\left(\cos\frac{\pi}{15} + i\sin\frac{\pi}{15}\right)$,

$\sqrt[5]{2}\left(\cos\frac{7\pi}{15} + i\sin\frac{7\pi}{15}\right)$,

$\sqrt[5]{2}\left(\cos\frac{13\pi}{15} + i\sin\frac{13\pi}{15}\right)$,

$\sqrt[5]{2}\left(\cos\frac{19\pi}{15} + i\sin\frac{19\pi}{15}\right)$,

$\sqrt[5]{2}\left(\cos\frac{5\pi}{3} + i\sin\frac{5\pi}{3}\right)$

25. $\sqrt[8]{2}\left(\cos\frac{\pi(8k + 1)}{16} + i\sin\frac{\pi(8k + 1)}{16}\right)$, $k = 0, 1, 2, 3$;

$\sqrt[8]{2}\left(\cos\frac{\pi}{16} + i\sin\frac{\pi}{16}\right)$,

$\sqrt[8]{2}\left(\cos\frac{9\pi}{16} + i\sin\frac{9\pi}{16}\right)$,

$\sqrt[8]{2}\left(\cos\frac{17\pi}{16} + i\sin\frac{17\pi}{16}\right)$,

$\sqrt[8]{2}\left(\cos\frac{25\pi}{16} + i\sin\frac{25\pi}{16}\right)$

26. $\sqrt[12]{2}\left(\cos\frac{\pi(8k + 7)}{24} + i\sin\frac{\pi(8k + 7)}{24}\right)$, $k = 0, 1, 2, 3, 4, 5$;

$\sqrt[12]{2}\left(\cos\frac{7\pi}{24} + i\sin\frac{7\pi}{24}\right)$,

$\sqrt[12]{2}\left(\cos\frac{5\pi}{8} + i\sin\frac{5\pi}{8}\right)$,

$\sqrt[12]{2}\left(\cos\frac{23\pi}{24} + i\sin\frac{23\pi}{24}\right)$,

$\sqrt[12]{2}\left(\cos\frac{31\pi}{24} + i\sin\frac{31\pi}{24}\right)$,

$\sqrt[12]{2}\left(\cos\frac{13\pi}{8} + i\sin\frac{13\pi}{8}\right)$,

$\sqrt[12]{2}\left(\cos\frac{47\pi}{24} + i\sin\frac{47\pi}{24}\right)$

27. $\sqrt{2}\left(\cos\frac{\pi(8k + 1)}{12} + i\sin\frac{\pi(8k + 1)}{12}\right)$, $k = 0, 1, 2$;

$\sqrt{2}\left(\cos\frac{\pi}{12} + i\sin\frac{\pi}{12}\right)$,

$\sqrt{2}\left(\cos\frac{3\pi}{4} + i\sin\frac{3\pi}{4}\right)$,

$\sqrt{2}\left(\cos\frac{17\pi}{12} + i\sin\frac{17\pi}{12}\right)$

28. $\sqrt[8]{8}\left(\cos\frac{\pi(8k+3)}{16} + i\sin\frac{\pi(8k+3)}{16}\right)$, $k = 0, 1, 2, 3$;

$\sqrt[8]{8}\left(\cos\frac{3\pi}{16} + i\sin\frac{3\pi}{16}\right)$,

$\sqrt[8]{8}\left(\cos\frac{11\pi}{16} + i\sin\frac{11\pi}{16}\right)$,

$\sqrt[8]{8}\left(\cos\frac{19\pi}{16} + i\sin\frac{19\pi}{16}\right)$,

$\sqrt[8]{8}\left(\cos\frac{27\pi}{16} + i\sin\frac{27\pi}{16}\right)$

29. $\sqrt[6]{2}\left(\cos\frac{\pi(4k+3)}{12} + i\sin\frac{\pi(4k+3)}{12}\right)$, $k = 0, 1, 2, 3, 4, 5$;

$\sqrt[6]{2}\left(\cos\frac{\pi}{4} + i\sin\frac{\pi}{4}\right)$,

$\sqrt[6]{2}\left(\cos\frac{7\pi}{12} + i\sin\frac{7\pi}{12}\right)$,

$\sqrt[6]{2}\left(\cos\frac{11\pi}{12} + i\sin\frac{11\pi}{12}\right)$,

$\sqrt[6]{2}\left(\cos\frac{5\pi}{4} + i\sin\frac{5\pi}{4}\right)$,

$\sqrt[6]{2}\left(\cos\frac{19\pi}{12} + i\sin\frac{19\pi}{12}\right)$,

$\sqrt[6]{2}\left(\cos\frac{23\pi}{12} + i\sin\frac{23\pi}{12}\right)$

30. $2\left(\cos\frac{2\pi k}{5} + i\sin\frac{2\pi k}{5}\right)$, $k = 0, 1, 2, 3, 4$; $2(\cos 0 + i\sin 0) = 2$,

$2\left(\cos\frac{2\pi}{5} + i\sin\frac{2\pi}{5}\right)$,

$2\left(\cos\frac{4\pi}{5} + i\sin\frac{4\pi}{5}\right)$,

$2\left(\cos\frac{6\pi}{5} + i\sin\frac{6\pi}{5}\right)$,

$2\left(\cos\frac{8\pi}{5} + i\sin\frac{8\pi}{5}\right)$

31. $1, -\frac{1}{2} \pm \frac{\sqrt{3}}{2}i$

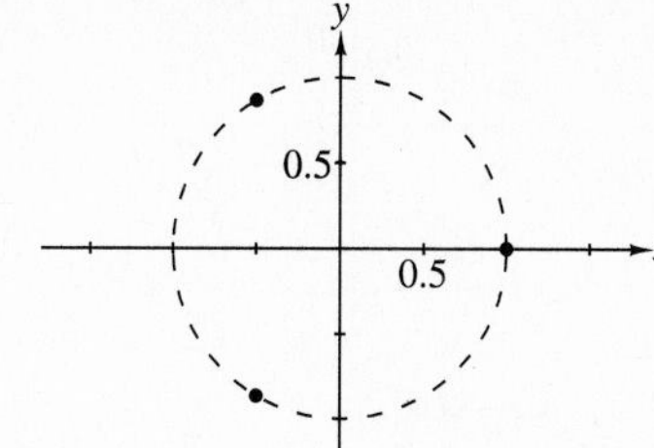

32. $\pm 1, \pm i$

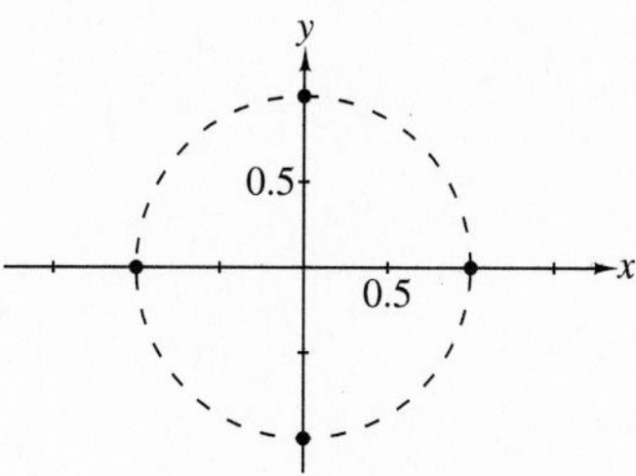

33. $\pm 1, \frac{1}{2} \pm \frac{\sqrt{3}}{2}i, -\frac{1}{2} \pm \frac{\sqrt{3}}{2}i$

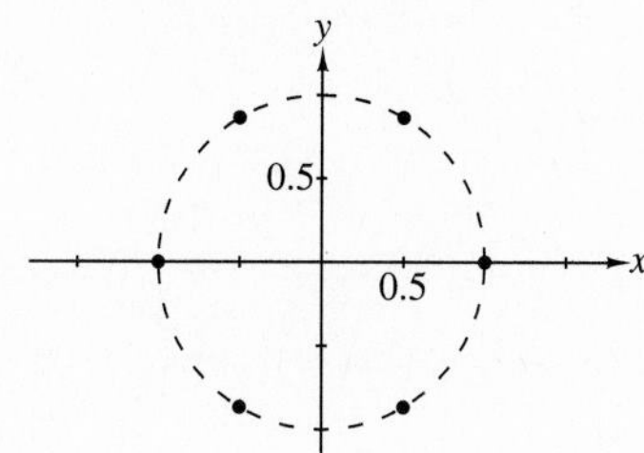

34. $1, -1$

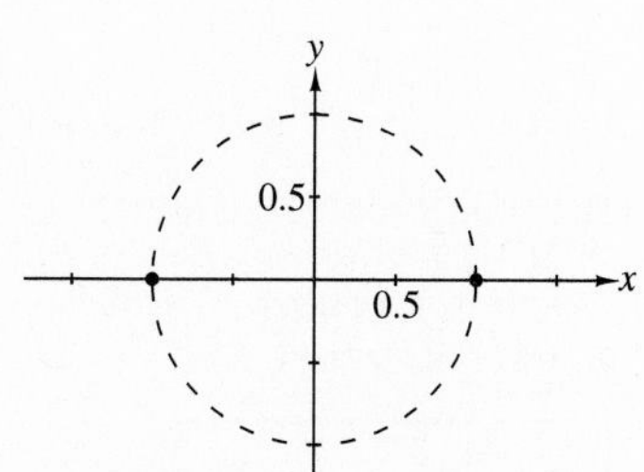

35. **(a)** $|s^n(\cos n\alpha + i\sin n\alpha)| = |s^n\cos n\alpha + i\, s^n \sin n\alpha| = \sqrt{s^{2n}\cos^2 n\alpha + s^{2n}\sin^2 n\alpha} = \sqrt{s^{2n}(\cos^2 n\alpha + \sin^2 n\alpha)} = \sqrt{s^{2n}(1)} = |s^n| = s^n$ (assuming that $s \geq 0$). **(b)** Follow essentially the steps of **(a)**. **(c)** If two complex numbers are equal, their absolute values must be equal, so $|s^n(\cos n\alpha + i\sin n\alpha)| = |r(\cos\theta + i\sin\theta)|$; then $s^n = r$, and since s and r are both real and nonnegative, $s = \sqrt[n]{r}$.

36. The n nth roots are given by

$\sqrt[n]{r}\left(\cos\frac{\theta}{n} + i\sin\frac{\theta}{n}\right)$,

$\sqrt[n]{r}\left(\cos\frac{\theta + 2\pi}{n} + i\sin\frac{\theta + 2\pi}{n}\right)$,

$\sqrt[n]{r}\left(\cos\frac{\theta + 4\pi}{n} + i\sin\frac{\theta + 4\pi}{n}\right)$,

...,

$\sqrt[n]{r}\left(\cos\frac{\theta + 2(n-1)\pi}{n} + i\sin\frac{\theta + 2(n-1)\pi}{n}\right)$; the angles between successive values in this list differ by $\frac{2\pi}{n}$ radians, while the first and last roots differ by $2\pi - \frac{2\pi}{n}$, which also makes the angle between them $\frac{2\pi}{n}$; also, the modulus of each root is $\sqrt[n]{r}$, placing it at that distance from the origin, on the circle with radius $\sqrt[n]{r}$.

37. In parametric mode, set $x_1 = \cos\left(\frac{2\pi t}{7}\right)$ and $y_1 = \sin\left(\frac{2\pi t}{7}\right)$, with $T\text{min} = 0$, $T\text{max} = 7$, and $T\text{step} = 1$; choose a viewing window similar to $[-2.3, 2.3] \times [-1.5, 1.5]$.

38. In parametric mode, set $x_1 = \cos\left(\frac{2\pi t}{5}\right)$ and $y_1 = \sin\left(\frac{2\pi t}{5}\right)$, with $T\text{min} = 0$, $T\text{max} = 5$, and $T\text{step} = 1$; choose a viewing window similar to $[-2.3, 2.3] \times [-1.5, 1.5]$.

39. Fourth roots of i:
$\cos\frac{\pi}{8} + i\sin\frac{\pi}{8}$, $\cos\frac{5\pi}{8} + i\sin\frac{5\pi}{8}$, $\cos\frac{9\pi}{8} + i\sin\frac{9\pi}{8}$, $\cos\frac{13\pi}{8} + i\sin\frac{13\pi}{8}$

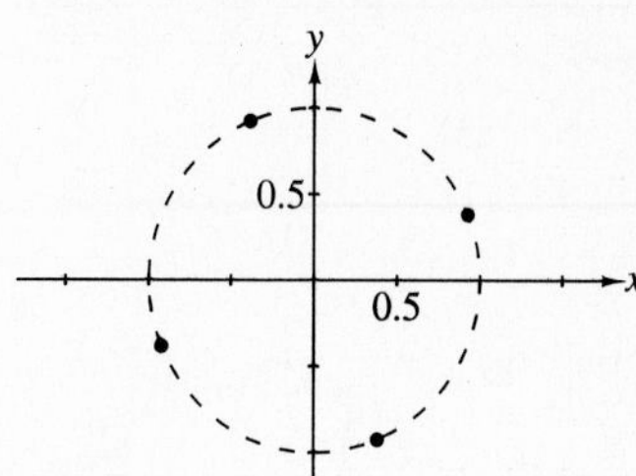

40. Fourth roots of $5 - 5i$: $\sqrt[8]{50}\left(\cos\frac{7\pi}{16} + i\sin\frac{7\pi}{16}\right)$, $\sqrt[8]{50}\left(\cos\frac{15\pi}{16} + i\sin\frac{15\pi}{16}\right)$, $\sqrt[8]{50}\left(\cos\frac{23\pi}{16} + i\sin\frac{23\pi}{16}\right)$, $\sqrt[8]{50}\left(\cos\frac{31\pi}{16} + i\sin\frac{31\pi}{16}\right)$

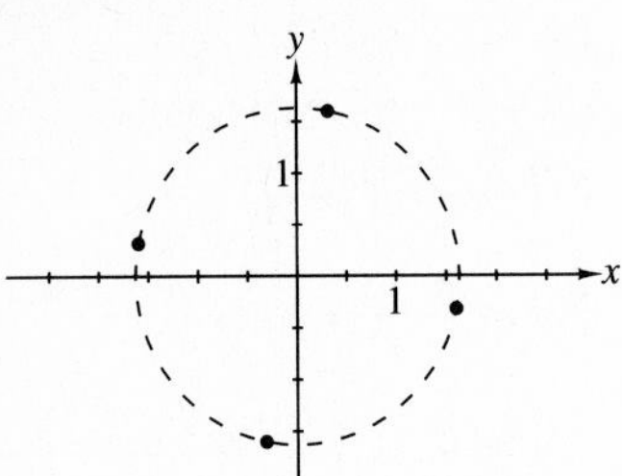

41. Sixth roots of $64i$: $2\left(\cos\frac{\pi}{12} + i\sin\frac{\pi}{12}\right)$, $2\left(\cos\frac{5\pi}{12} + i\sin\frac{5\pi}{12}\right)$, $2\left(\cos\frac{3\pi}{4} + i\sin\frac{3\pi}{4}\right)$, $2\left(\cos\frac{13\pi}{12} + i\sin\frac{13\pi}{12}\right)$, $2\left(\cos\frac{17\pi}{12} + i\sin\frac{17\pi}{12}\right)$, $2\left(\cos\frac{7\pi}{4} + i\sin\frac{7\pi}{4}\right)$

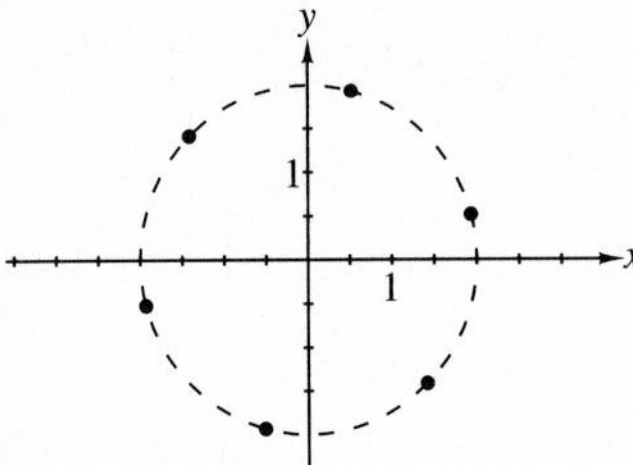

42. Sixth roots of $2i$: $\sqrt[6]{2}\left(\cos\frac{\pi}{12} + i\sin\frac{\pi}{12}\right)$, $\sqrt[6]{2}\left(\cos\frac{5\pi}{12} + i\sin\frac{5\pi}{12}\right)$, $\sqrt[6]{2}\left(\cos\frac{3\pi}{4} + i\sin\frac{3\pi}{4}\right)$, $\sqrt[6]{2}\left(\cos\frac{13\pi}{12} + i\sin\frac{13\pi}{12}\right)$, $\sqrt[6]{2}\left(\cos\frac{17\pi}{12} + i\sin\frac{17\pi}{12}\right)$, $\sqrt[6]{2}\left(\cos\frac{7\pi}{4} + i\sin\frac{7\pi}{4}\right)$

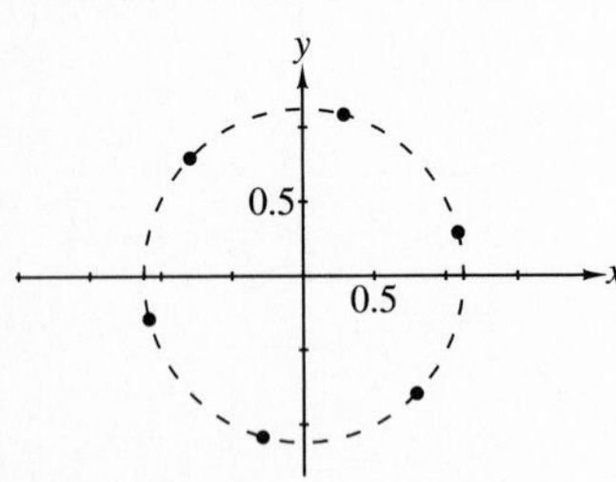

43. Cube roots of $-1 + i$: $\sqrt[6]{2}\left(\cos\frac{\pi}{4} + i\sin\frac{\pi}{4}\right)$, $\sqrt[6]{2}\left(\cos\frac{11\pi}{12} + i\sin\frac{11\pi}{12}\right)$, $\sqrt[6]{2}\left(\cos\frac{19\pi}{12} + i\sin\frac{19\pi}{12}\right)$

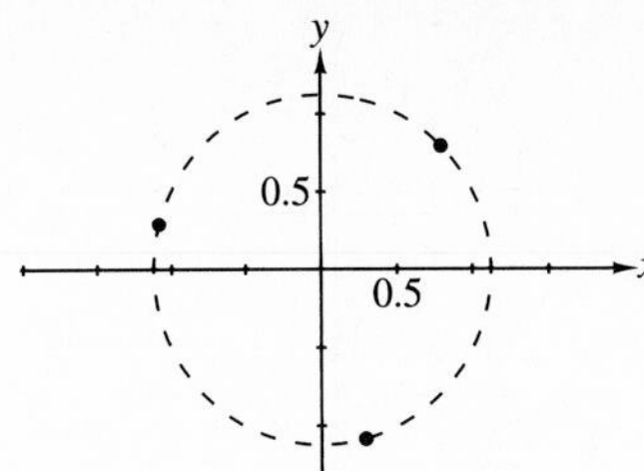

44. Cube roots of $\sqrt{3} - i$: $\sqrt[3]{2}\left(\cos\frac{11\pi}{18} + i\sin\frac{11\pi}{18}\right)$, $\sqrt[3]{2}\left(\cos\frac{23\pi}{18} + i\sin\frac{23\pi}{18}\right)$, $\sqrt[3]{2}\left(\cos\frac{35\pi}{18} + i\sin\frac{35\pi}{18}\right)$

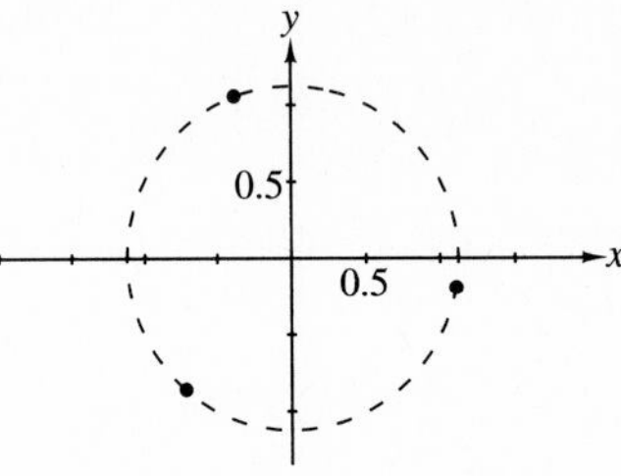

45. $z = -8$; the cube roots are -2 and $1 \pm \sqrt{3}\,i$

46. $z = -64$; the fourth roots are $2 \pm 2i$ and $-2 \pm 2i$

47. $-1 + i = \sqrt{2}\left(\cos\frac{3\pi}{4} + i\sin\frac{3\pi}{4}\right)$, so $(-1+i)^{12} = (\sqrt{2})^{12}(\cos 9\pi + i\sin 9\pi) = 2^6(\cos\pi + i\sin\pi) = -64$

48. $1 + \sqrt{3}\,i = 2\left(\cos\frac{\pi}{3} + i\sin\frac{\pi}{3}\right)$, so $(1 + \sqrt{3}\,i)^8 = 2^8\left(\cos\frac{8\pi}{3} + i\sin\frac{8\pi}{3}\right) = 256\left(\cos\frac{2\pi}{3} + i\sin\frac{2\pi}{3}\right) = 256\left(-\frac{1}{2} + \frac{\sqrt{3}}{2}i\right) = 128(-1 + \sqrt{3}\,i)$

49. Using $\tan^{-1}\left(\dfrac{1}{\sqrt{2}}\right) \approx 0.6155$, we have $x(t) = (\sqrt{3})^t\cos(0.6155t)$ and $y(t) = (\sqrt{3})^t\sin(0.6155t)$; use Tmin = 0, Tmax = 4, and Tstep = 1; shown is $[-8, 4] \times [-1, 7]$.

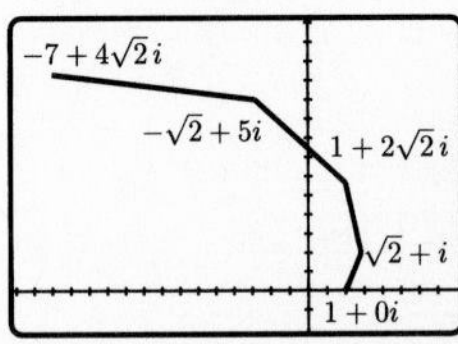

50. $x(t) = (\sqrt{2})^t\cos(0.75\pi t)$ and $y(t) = (\sqrt{2})^t\sin(0.75\pi t)$; use Tmin = 0, Tmax = 4, and Tstep = 1; shown is $[-5, 4] \times [-3, 3]$.

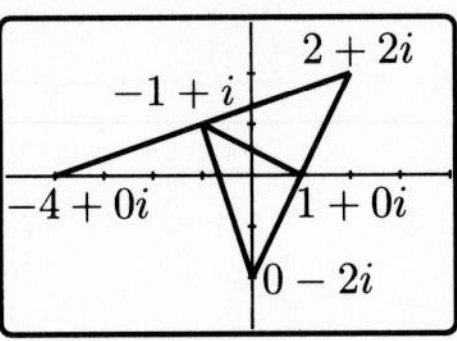

Chapter 8 Review

1. $b = 3.883\ldots$; $c = 6.611\ldots$; $\gamma = 68°$

2. $c = 4.764\ldots$; $\alpha = 35.966\ldots°$; $\gamma = 34.033\ldots°$

3. No triangle is formed.

4. $b = 26.595\ldots$; $c = 16.359\ldots$; $\beta = 117.7°$

5. $a = 2.939\ldots$; $b = 5.053\ldots$; $\gamma = 72°$

6. $\alpha = 19.452\ldots$; $b = 48.898\ldots$; $\beta = 102°$

7. $\alpha = 44.415\ldots°$; $\beta = 78.463\ldots°$; $\gamma = 57.121\ldots°$

8. $c = 4.834\ldots$; $\beta = 41.615\ldots°$; $\gamma = 53.384\ldots°$

9. Area $= \sqrt{56} = 7.483\ldots$

10. $c = 7.672\ldots$, so area $= 22.981\ldots$

11. $h = 12\sin 28° = 5.633\ldots$. **(a)** $5.633\ldots < b < 12$. **(b)** $b = 5.633\ldots$ or $b \geq 12$. **(c)** $b < 5.633\ldots$.

12. **(a)** $AC = b = 102.537\ldots$ ft. **(b)** $b\sin\alpha = a\sin\beta = 96.353\ldots$ ft.

13. $b\sin\alpha = a\sin\beta = 0.610\ldots$ mi

14. $AB = 849.768\ldots$ ft

15. The largest angle is opposite the longest side: It measures $71.790\ldots°$ or $1.252\ldots$ radians.

16. $15.793\ldots$ and $36.777\ldots$

17. $2 + 4i$

18. $-2 - 5i$

19. $5 - i$

20. $10 - 10i$

21. $-3 + i$

22. $-\frac{17}{58} - \frac{1}{58}i$

23.

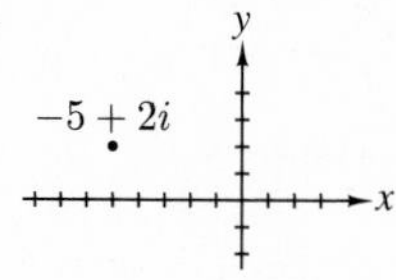

24.

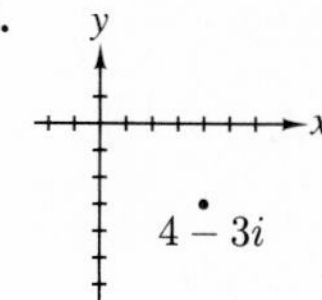

25.

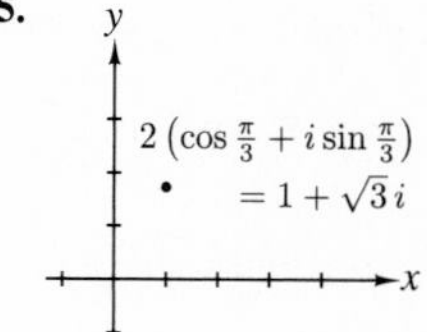

26.

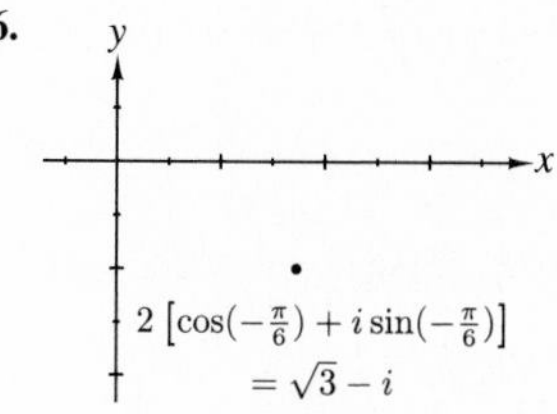

27. $a = -3$; $b = 4$; $|z_1| = 5$

28. $z_1 = 5[\cos(2.214\ldots) + i\sin(2.214\ldots)]$

29. $-4 + 0i = -4$

30. $23 + 14i$

31. $\sqrt{493}$

32. $-16 + 30i$

33. $3\sqrt{3} + 3i$

34. $-1.5\sqrt{3} + 1.5i$

35. $-1.25 - 1.25\sqrt{3}i$

36. $-3.204\ldots + 2.393\ldots i$

37. $3\sqrt{2}\left(\cos\frac{7\pi}{4} + i\sin\frac{7\pi}{4}\right)$. One possibility is to add 2π, 4π, and -2π to $7\pi/4$.

38. $\sqrt{3}[\cos(2.186\ldots) + i\sin(2.186\ldots)]$. One possibility is to add 2π, 4π, and -2π to $2.186\cdots$.

39. $\sqrt{34}[\cos(5.252\ldots) + i\sin(5.252\ldots)]$. One possibility is to add 2π, 4π, and -2π to $5.252\cdots$.

40. $2\sqrt{2}\left(\cos\frac{5\pi}{4} + i\sin\frac{5\pi}{4}\right)$. One possibility is to add 2π, 4π, and -2π to $5\pi/4$.

41. $z_1 z_2 = 12(\cos 90^\circ + i\sin 90^\circ) = 12i$; $\frac{z_1}{z_2} = 0.75[\cos(-30^\circ) + i\sin(-30^\circ)] = 0.75(\cos 330^\circ + i\sin 330^\circ)$

42. $z_1 z_2 = -10(\cos 65^\circ + i\sin 65^\circ) = 10(\cos 245^\circ + i\sin 245^\circ)$, $\frac{z_1}{z_2} = -2.5[\cos(-25^\circ) + i\sin(-25^\circ)] = 2.5(\cos 155^\circ + i\sin 155^\circ)$

43. $z_1 = 1.5\sqrt{3} + 1.5i$; $z_2 = 2 + 2\sqrt{3}i$; $\frac{z_1}{z_2} = \frac{(1.5\sqrt{3} + 1.5i)(2 - 2\sqrt{3}i)}{16} = \frac{6\sqrt{3} - 6i}{16} = \frac{3}{8}\sqrt{3} - \frac{3}{8}i$. This quotient agrees with the trigonometric form.

44. $z_1 = 4.698\ldots + 1.710\ldots i$; $z_2 = -\sqrt{2} - \sqrt{2}i$; $z_1 z_2 = (4.698\cdots + 1.710\cdots i)(-\sqrt{2} - \sqrt{2}i) = (-4.226\cdots) - (9.063\cdots)i$. This product agrees with the trigonometric form.

45. $3^5\left(\cos\frac{5\pi}{4} + i\sin\frac{5\pi}{4}\right) = 243\left(\cos\frac{5\pi}{4} + i\sin\frac{5\pi}{4}\right) = -\frac{243\sqrt{2}}{2} - \frac{243\sqrt{2}}{2}i$

46. $2^8\left(\cos\frac{2\pi}{3} + i\sin\frac{2\pi}{3}\right) = 256\left(\cos\frac{2\pi}{3} + i\sin\frac{2\pi}{3}\right) = -128 + 128i\sqrt{3}$

47. $5^3(\cos 5\pi + i\sin 5\pi) = 125(\cos\pi + i\sin\pi) = -125 + 0i = -125$

48. $7^6\left(\cos\frac{\pi}{4} + i\sin\frac{\pi}{4}\right) = 117{,}649\left(\cos\frac{\pi}{4} + i\sin\frac{\pi}{4}\right) = \frac{117{,}649\sqrt{2}}{2} + \frac{117{,}649\sqrt{2}}{2}i$

49. $\sqrt[8]{18}\left(\cos\frac{\pi(8k+1)}{16} + i\sin\frac{\pi(8k+1)}{16}\right)$; $k = 0, 1, 2, 3$; $\sqrt[8]{18}\left(\cos\frac{\pi}{16} + i\sin\frac{\pi}{16}\right)$, $\sqrt[8]{18}\left(\cos\frac{9\pi}{16} + i\sin\frac{9\pi}{16}\right)$, $\sqrt[8]{18}\left(\cos\frac{17\pi}{16} + i\sin\frac{17\pi}{16}\right)$, $\sqrt[8]{18}\left(\cos\frac{25\pi}{16} + i\sin\frac{25\pi}{16}\right)$

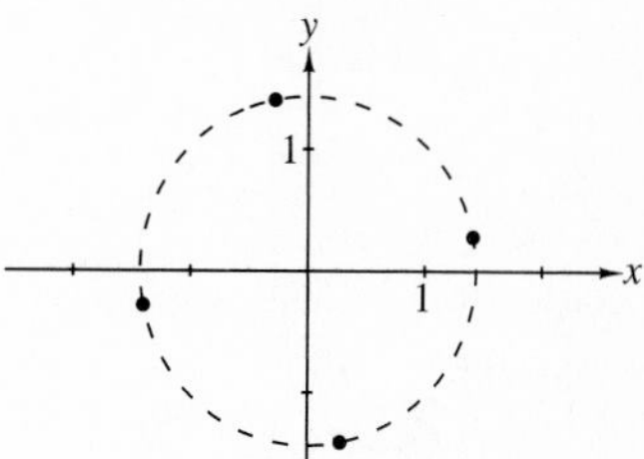

50. $2\left(\cos\frac{2\pi k}{3} + i\sin\frac{2\pi k}{3}\right)$, $k = 0, 1, 2$; $2(\cos 0 + i\sin 0) = 2$, $2\left(\cos\frac{2\pi}{3} + i\sin\frac{2\pi}{3}\right)$, $2\left(\cos\frac{4\pi}{3} + i\sin\frac{4\pi}{3}\right)$

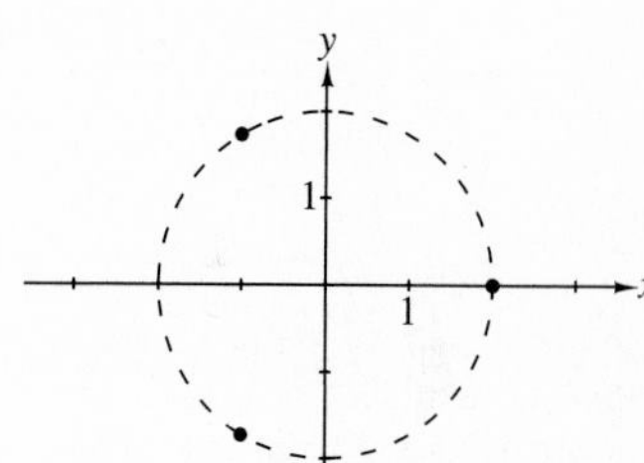

51. $\cos\frac{2\pi k}{5} + i\sin\frac{2\pi k}{5}$, $k = 0, 1, 2, 3, 4$; $\cos 0 + i\sin 0 = 1$, $\cos\frac{2\pi}{5} + i\sin\frac{2\pi}{5}$, $\cos\frac{4\pi}{5} + i\sin\frac{4\pi}{5}$, $\cos\frac{6\pi}{5} + i\sin\frac{6\pi}{5}$, $\cos\frac{8\pi}{5} + i\sin\frac{8\pi}{5}$

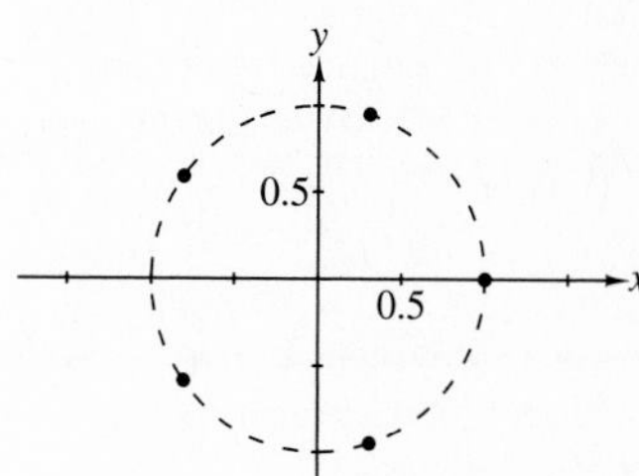

52. $\cos\frac{\pi(2k+1)}{6} + i\sin\frac{\pi(2k+1)}{6}$, $k = 0, 1, 2, 3, 4, 5$; $\cos\frac{\pi}{6} + i\sin\frac{\pi}{6}$, $\cos\frac{\pi}{2} +$

$i \sin \frac{\pi}{2}$, $\cos \frac{5\pi}{6} + i \sin \frac{5\pi}{6}$, $\cos \frac{7\pi}{6} + i \sin \frac{7\pi}{6}$, $\cos \frac{3\pi}{2} + i \sin \frac{3\pi}{2}$, $\cos \frac{11\pi}{6} + i \sin \frac{11\pi}{6}$

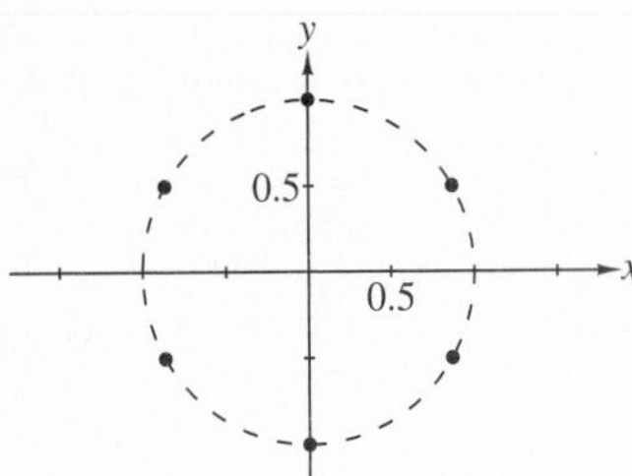

53. $\langle -2, -3 \rangle$

54. $\langle 1, 7 \rangle$

55. $\sqrt{37}$

56. $\sqrt{10}$

57. $\langle 3, 6 \rangle$; $\|3\overrightarrow{AB}\| = 3\sqrt{5}$

58. $\langle 6, -5 \rangle$; $\|\overrightarrow{AB} + \overrightarrow{CD}\| = \sqrt{61}$

59. $\langle -8, -3 \rangle$; $\|\overrightarrow{AC} + \overrightarrow{BD}\| = \sqrt{73}$

60. $\langle 4, -9 \rangle$; $\|\overrightarrow{CD} - \overrightarrow{AB}\| = \sqrt{97}$

61. $\overrightarrow{AB} = -2\mathbf{i} + 1\mathbf{j} = -2\mathbf{i} + \mathbf{j}$

62. $\overrightarrow{AB} = 2\mathbf{i} + 0\mathbf{j} = 2\mathbf{i}$

63. **(a)** $\theta_u = \tan^{-1} 0.75 = 0.643\ldots$; $\theta_v = \tan^{-1} 2.5 = 1.190\ldots$. **(b)** $\theta_v - \theta_u = 0.546\ldots$.

64. **(a)** $\theta_u = \pi + \tan^{-1}(-2) = 2.034\ldots$; $\theta_v = \tan^{-1}\left(\frac{2}{3}\right) = 0.588\ldots$. **(b)** $\theta_u - \theta_v = 1.446\ldots$

65. **(a)** $\mathbf{v} \approx \langle 531.80, 93.77 \rangle$. **(b)** Actual velocity vector: $\mathbf{v} + \langle 54.16, -9.55 \rangle = \langle 585.96, 84.22 \rangle$; speed: $\|\langle 585.96, 84.22 \rangle\| \approx 591.98$ mph; bearing: approximately 81.82°.

66. **(a)** $\mathbf{v} \approx \langle -463.64, 124.23 \rangle$. **(b)** Actual velocity vector: $\mathbf{v} + \langle -29.89, -2.61 \rangle = \langle -493.53, 121.62 \rangle$; speed: $\|\langle -493.53, 121.62 \rangle\| \approx 508.29$ mph; bearing: approximately 283.84°.

67. $\mathbf{F} = \langle 120 \cos 20°, 120 \sin 20° \rangle + \langle 300 \cos(-5°), 300 \sin(-5°) \rangle \approx \langle 411.62, 14.90 \rangle$, so $\|\mathbf{F}\| \approx 411.89$ lb and $\theta \approx 2.07°$

68. The hexagon's area is $384\sqrt{3}$ cm^2 and the radius of the circle is 16 cm, so the desired area is $256\pi - 384\sqrt{3} = 139.140\ldots$ cm^2.

69. The pentagon is made up of five triangles with base length 12 cm and height $\frac{6}{\tan 36°} = 8.258\ldots$ cm, so its area is $247.748\ldots$ cm^2; the radius of the circle is the height of those triangles, so the desired area is $247.748\ldots - \pi(8.258\ldots)^2 = 33.494\ldots$ cm^2.

70. The wheel of cheese has volume 405π cm^3; a 15° wedge would have fraction $\frac{15}{360} = \frac{1}{24}$ of that volume, or $\frac{405\pi}{24} \approx 53.01$ cm^3.

71. Solve $a\mathbf{r} + b\mathbf{s} = \mathbf{v}$, or $a + 2b = x$ and $2a - b = y$, for a and b; this gives $a = 0.2x + 0.4y$ and $b = 0.4x - 0.2y$, so $\mathbf{v} = (0.2x + 0.4y)\mathbf{r} + (0.4x - 0.2y)\mathbf{s}$

72. The difference $\mathbf{u} - \mathbf{v}$ can be found by placing the terminal points of the vectors together; $\mathbf{u} - \mathbf{v}$ is then the vector that goes from the initial point of $\mathbf{u}$ to the initial point of $\mathbf{v}$. Alternatively, "flip" the vector $\mathbf{v}$ around to point in the opposite direction (this gives $-\mathbf{v}$), and place its initial point at the terminal point of $\mathbf{u}$; then $\mathbf{u} - \mathbf{v}$ goes from the initial point of $\mathbf{u}$ to the terminal point of the flipped vector.

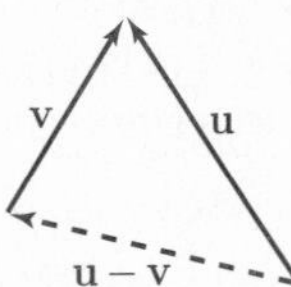

CHAPTER 9

QUICK REVIEW 9.1

1. $y + 2 = \frac{8}{7}(x + 3)$ or $y - 6 = \frac{8}{7}(x - 4)$

2. $y - 3 = -\frac{6}{5}(x + 1)$ or $y + 3 = -\frac{6}{5}(x - 4)$

3. Window: $[-3, 7] \times [-7, 7]$

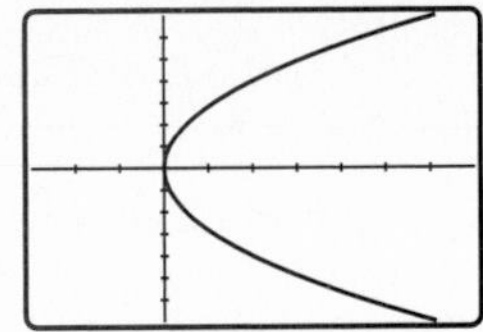

4. Window: $[-7, 2] \times [-7, 7]$

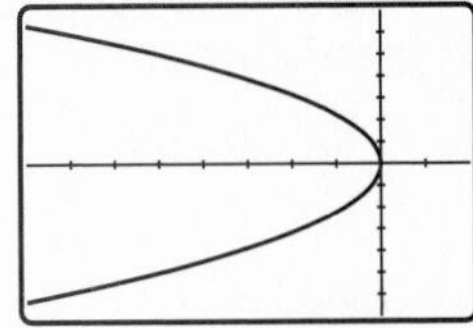

5. $x^2 + y^2 = 4$

6. $(x - 2)^2 + (y - 3)^2 = 16$

7. $x^2 + (y - 2)^2 = 25$

8. $(x + 2)^2 + (y - 5)^2 = 9$

9. 20π rad/sec

10. $\frac{70}{3}\pi$ rad/sec

11. $P(\cos 35°, \sin 35°) \approx (0.8192, 0.5736)$

12. $P(\cos 160°, \sin 160°) \approx (-0.9397, 0.3420)$

13. $P\left(\cos \frac{\pi}{6}, \sin \frac{\pi}{6}\right) = \left(\frac{\sqrt{3}}{2}, \frac{1}{2}\right)$

14. $P\left(\cos \frac{4\pi}{3}, \sin \frac{4\pi}{3}\right) = \left(-\frac{1}{2}, -\frac{\sqrt{3}}{2}\right)$

15. $a \approx 6.3442$, $b \approx 2.9583$

16. $a \approx 7.3724$, $b \approx 5.1622$

17. $a = -15$, $b = 15\sqrt{3} \approx 25.9808$

18. $a = -25\sqrt{3} \approx -43.3013$, $b = 25$

SECTION EXERCISES 9.1

1. The graph in part **(c)**; $[-5, 5] \times [-5, 5]$

2. The graph in part **(a)**; $[-2, 2] \times [-2, 2]$

3. The graph in part **(d)**; $[-10, 10] \times [-10, 10]$

4. The graph in part **(b)**; $[-15, 15] \times [-15, 15]$

5. Endpoints: $A(1, 0)$ and $B(6, 5)$; a line segment from A to B

6. Endpoints: $A(-2, -3)$ and $B(3, 2)$; a line segment from A to B

7. Endpoints: $A(5, 4)$ and $B(-7, 8)$; a line segment from A to B

8. Endpoints: $A(3, -3)$ and $B(7, -11)$; a line segment from A to B

9. Endpoints: $A(-8, -0.4)$ and $B(2, 0.4)$; a piece of the graph of $y = \frac{2}{x+3}$, with a vertical asymptote at $x = -3$

10. Endpoints: $A(-6, -0.5)$ and $B(10, 0.5)$; a piece of the graph of $y = \frac{2}{x-2}$, with a vertical asymptote at $x = 2$

11. Endpoints: $A(8, 3)$ and $B(-4, 7)$; $x + 3y = 17$

12. Endpoints: $A(8, 1)$ and $B(-4, 5)$; $x + 3y = 11$

13. Endpoints: $A(0, 2)$ and $B(-2, 0)$; $x^2 + y^2 = 4$

14. Endpoints: $A(3, 0)$ and $B(-3, 0)$; $x^2 + y^2 = 9$, or $y = \sqrt{9 - x^2}$

15. Endpoints: $A(-7, -2)$ and $B(9, 2)$; $y = \sqrt[3]{x - 1}$

16. Endpoints: $A(-1, -19)$ and $B(1, 13)$; $y = 16x^3 - 3$

17. One possibility is $x = t + 3$, $y = 4 - \frac{7}{3}t$, $0 \le t \le 3$.

18. One possibility is $x = 5 - t$, $y = 2 - \frac{6}{7}t$, $0 \le t \le 7$.

19. One possibility is $x = 5 + 3\cos t$, $y = 2 + 3\sin t$, $0 \le t \le 2\pi$.

20. One possibility is $x = -2 + 2\cos t$, $y = -4 + 2\sin t$, $0 \le t \le 2\pi$.

21. **(a)** This shows the motion along the horizontal line $y = 2$. **(b)** The motion is "spread out" vertically so that the changes in direction are visible. **(c)** This shows Kathy's distance to the left or right of the horizontal line's origin as an "ordinary" graph with time on the horizontal axis and distance on the vertical axis.

22. **(a)** This shows the motion along the horizontal line $y = 2$. **(b)** The motion is "spread out" vertically so that the changes in direction are visible. **(c)** This shows Pete's distance to the left or right of the horizontal line's origin as an "ordinary" graph with time on the horizontal axis and distance on the vertical axis.

23. $x = s(t)$, $y = 2$ and $x = s(t)$, $y = -2 - t$ are two possibilities; the particle begins at $+6$, moves left to -6.25 at $t = 5/2$, and then changes direction and returns to $+6$.

24. $x = s(t)$, $y = 2$ and $x = s(t)$, $y = -2 - t$ are two possibilities. The particle begins at $+7$, moves left to -9, and then changes direction and ends at -5.

25. $x = s(t)$, $y = 2$ and $x = s(t)$, $y = -3 - t$ are two possibilities; the particle begins at -10, moves right to $+2.25$, and then changes direction and ends at -4.

26. $x = s(t)$, $y = 2$ and $x = s(t)$, $y = -2 - t$ are two possibilities. The particle begins at -5, moves right to $+4$, and then changes direction and returns to -5.

27. $x = s(t)$, $y = 2$ and $x = s(t)$, $y = -2 - t$ are two possibilities; the particle begins at -5, moves right to about $+0.07$, changes direction and moves left to about -20.81, and then changes direction and ends at $+7$.

28. $x = s(t)$, $y = 2$ and $x = s(t)$, $y = -2 - t$ are two possibilities. The particle begins at -10, moves right to about $+0.88$, changes direction and moves left to about -6.06, and then changes direction and ends at $+20$.

29. Possible answers: **(a)** $0 < t < \frac{\pi}{2}$ (t in radians). **(b)** $0 < t < \pi$. **(c)** $\frac{\pi}{2} < t < \frac{3\pi}{2}$.

30. **(a)** Both pairs of equations can be changed to $x^2 + y^2 = 9$, a circle centered at the origin with radius 3; also, when we choose a point on this circle and swap the x- and y-coordinates, we obtain another point on the same circle. **(b)** The first begins at the right side (when $t = 0$) and traces the circle counterclockwise; the second begins at the top (when $t = 0$) and traces the circle clockwise.

31. **(a)** About 2.80 sec. **(b)** About 7.18 ft. **(c)** The outfielder's glove should be at or near the height of the ball as it approaches the wall; if hit at an angle of 20°, the ball would strike the wall about 19.74 ft up (after 2.84 sec), which the outfielder could not reach and could catch the ball only after it bounces off the wall.

32. **(a)** No. **(b)** The ball hits the wall about 24.59 ft up (at time $t \approx 3.37$); it isn't catchable (except perhaps after it hits the wall).

33. **(a)** Yes. **(b)** The ball clears the wall with about 1.59 ft to spare (at time $t \approx 3.21$).

34. The minimum distance is about 6.60 ft, at time $t \approx 1.206$ sec.

35. No. The dart lands (at time $t \approx 1.86$) about 19.11 ft from Tony, just over 10 in. short of the target.

36. Yes. The dart lands (at time $t \approx 1.45$) about 20.82 ft from Sue, inside the target.

37. If the wind is blowing at just over 10 ft/sec, the ball will hit near the top of the wall.

38. For a level course. **(a)** ≈ 506.25 ft. **(b)** ≈ 650.82 ft. **(c)** ≈ 775.62 ft. **(d)** ≈ 876.85 ft.

39. $x = 35\cos\left(\frac{\pi}{6}t\right)$ and $y = 50 + 35\sin\left(\frac{\pi}{6}t\right)$

40. The minimum distance, when $t \approx 2.19$ sec, is about 3.47 ft.

41. **(a)** When $t = \pi$ (or 3π, or 5π, etc.), $y = 2$, which corresponds to the highest points on the graph. **(b)** The x-intercepts occur where $y = 0$, which happens when $t = 0, 2\pi, 4\pi$, etc.; the x-coordinates at those times are $0, 2\pi, 4\pi$, etc., respectively, so they are 2π units apart.

42. **(a)** Shown on the window $[-4.7, 4.7] \times [-3.1, 3.1]$. **(b)** All 2's should be changed to 3's.

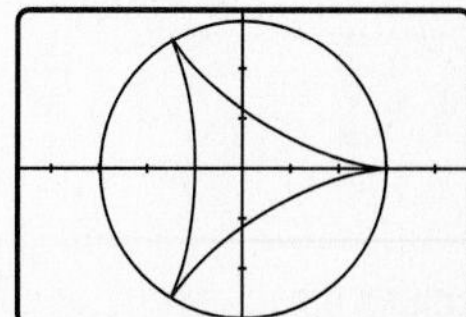

43. **(a)** $x(0) = 0c + (1 - 0)a = a$ and $y(0) = 0d + (1 - 0)b = b$. **(b)** $x(1) = 1c + (1 - 1)a = c$ and $y(1) = 1d + (1 - 1)b = d$.

44. $x(0.5) = 0.5c + (1 - 0.5)a = (a + c)/2$ and $y(0.5) = (b + d)/2$, the correct coordinates for the midpoint.

45. After $t \approx 21.50$ sec, the distance is about 4.11 ft.

46. After $t \approx 12.32$ sec, the distance is about 10.48 ft.

47. All graphs are part of the line $y = 3x - 1$. **(a)** The equations represent the entire line. **(b)** The equations represent the portion of the line to the right of the y-axis, since all the x-coordinates must be positive. **(c)** The equations represent the portion of the line with x-coordinates in the interval $[-1, 1]$.

48. In function mode, set $y_1 = \sqrt{[20\cos(\frac{\pi}{6}x) + 30x - 75]^2 + [20 + 20\sin(\frac{\pi}{6}x) + 16x^2 - (30\sqrt{3})x]^2}$ and graph on the window $[0, 4] \times [-5, 20]$ to obtain the view shown in Figure 9.12.

QUICK REVIEW 9.2

1. $\sqrt{13}$

2. $\sqrt{61}$

3. $\sqrt{(a - 2)^2 + (b + 3)^2}$

4. $\sqrt{(a + 3)^2 + (b + 4)^2}$

5. $y = (5 - 2x)/3 = -\frac{2}{3}x + \frac{5}{3}$

6. $y = (3x + 1)/2 = 1.5x + 0.5$

7. $y = \pm 2\sqrt{x}$

8. $y = \pm\sqrt{5x}$

9. $y = \pm 3\sqrt{1 - \frac{x^2}{4}} = \pm\frac{3}{2}\sqrt{4 - x^2}$

10. $y = \pm 4\sqrt{1 + \frac{x^2}{9}} = \pm\frac{4}{3}\sqrt{9 + x^2}$

SECTION EXERCISES 9.2

1. $x = -\frac{1}{12}y^2$

2. $y = \frac{1}{8}x^2$

3. $y = -\frac{1}{16}x^2$

4. $x = \frac{1}{8}y^2$

5. $y = \frac{1}{20}x^2$

6. $x = -\frac{1}{16}y^2$

7. Focus: $(0, \frac{1}{16})$; directrix: $y = -\frac{1}{16}$; line of symmetry: $x = 0$

8. Focus: $(0, -1.5)$; directrix: $y = 1.5$; line of symmetry: $x = 0$

9. Focus: $(-\frac{1}{32}, 0)$; directrix: $x = \frac{1}{32}$; line of symmetry: $y = 0$

10. Focus: $(\frac{1}{8}, 0)$; directrix: $x = -\frac{1}{8}$; line of symmetry: $y = 0$

11. Major axis from $(-8, 0)$ to $(8, 0)$; minor axis from $(0, -6)$ to $(0, 6)$. Window: $[-9.2, 9.2] \times [-6.1, 6.1]$.

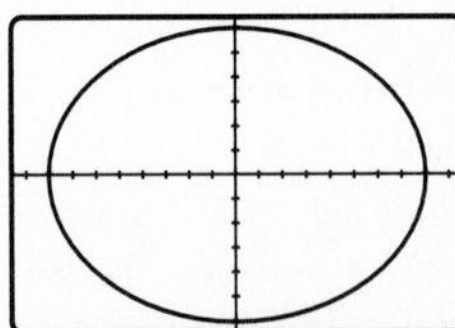

12. Major axis from $(-9, 0)$ to $(9, 0)$; minor axis from $(0, -5)$ to $(0, 5)$. Window: $[-9.1, 9.1] \times [-6, 6]$.

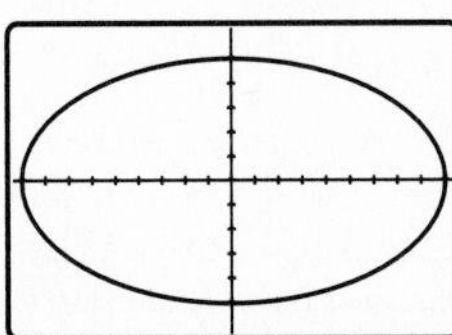

13. Major axis from $(-6, 0)$ to $(6, 0)$; minor axis from $(0, -4)$ to $(0, 4)$. Window: $[-9.1, 9.1] \times [-6, 6]$.

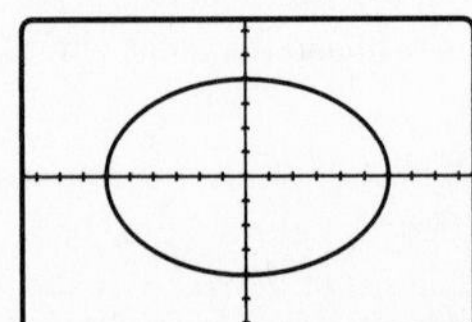

14. Major axis from $(0, -7)$ to $(0, 7)$; minor axis from $(-5, 0)$ to $(5, 0)$. Window: $[-10.8, 10.8] \times [-7.1, 7.1]$.

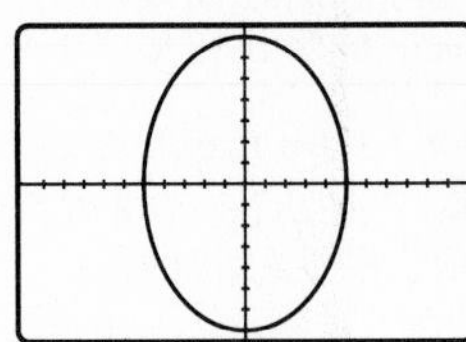

15. $\frac{x^2}{16} + \frac{y^2}{25} = 1$

16. $\frac{x^2}{49} + \frac{y^2}{16} = 1$

17. $\frac{x^2}{4} + \frac{y^2}{9} = 1$

18. $\frac{x^2}{49} + \frac{y^2}{25} = 1$

19. $\frac{x^2}{25} + \frac{y^2}{21} = 1$

20. $\frac{x^2}{16} + \frac{y^2}{25} = 1$

21. $\frac{x^2}{25} + \frac{y^2}{4} = 1$; foci $(\pm\sqrt{21}, 0)$

22. $\frac{x^2}{16} + \frac{y^2}{36} = 1$; foci $(0, \pm 2\sqrt{5})$

23. $\frac{x^2}{25} + \frac{y^2}{16} = 1$; foci $(\pm 3, 0)$

24. $\frac{x^2}{144} + \frac{y^2}{169} = 1$; foci $(0, \pm 5)$

25. Transverse axis from $(-7, 0)$ to $(7, 0)$; asymptotes: $y = \pm\frac{5}{7}x$; window: $[-20, 20] \times [-15, 15]$

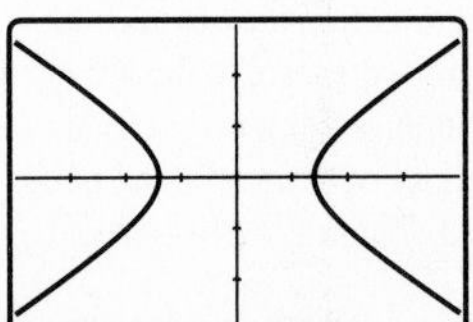

26. Transverse axis from $(0, -8)$ to $(0, 8)$; asymptotes: $y = \pm\frac{8}{5}x$; window: $[-20, 20] \times [-25, 25]$

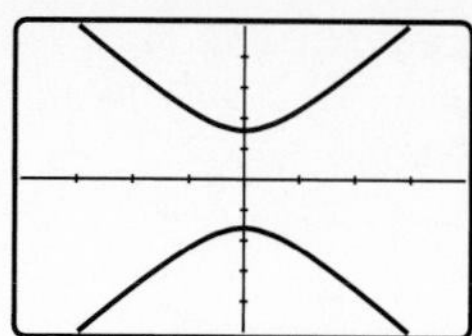

27. Transverse axis from $(0, -5)$ to $(0, 5)$; asymptotes: $y = \pm\frac{5}{4}x$; window: $[-20, 20] \times [-25, 25]$

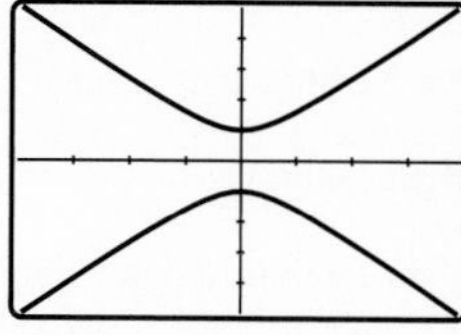

28. Transverse axis from $(-13, 0)$ to $(13, 0)$; asymptotes: $y = \pm\frac{12}{13}x$; window: $[-30, 30] \times [-25, 25]$

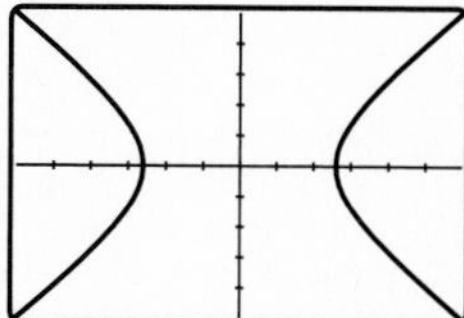

29. $\dfrac{x^2}{4} - \dfrac{y^2}{5} = 1$

30. $\dfrac{y^2}{4} - \dfrac{x^2}{5} = 1$

31. $\dfrac{y^2}{16} - \dfrac{x^2}{209} = 1$

32. $\dfrac{x^2}{2.25} - \dfrac{y^2}{22.75} = 1$, or $\dfrac{x^2}{9/4} - \dfrac{y^2}{91/4} = 1$

33. $y = -\dfrac{1}{4}x^2$

34. $x = \dfrac{1}{8}y^2$

35. $\dfrac{x^2}{9} - \dfrac{y^2}{16} = 1$

36. $\dfrac{x^2}{16} + \dfrac{y^2}{9} = 1$

37. Ellipse; major axis from $(-4, 0)$ to $(4, 0)$; minor axis from $(0, -\sqrt{5})$ to $(0, \sqrt{5})$; foci: $(\pm\sqrt{11}, 0)$

38. Hyperbola; transverse axis from $(-1, 0)$ to $(1, 0)$; foci: $(\pm\sqrt{5}, 0)$; asymptotes: $y = \pm 2x$.

39. Parabola; vertex: $(0, 0)$; directrix: $y = 1.5$; focus: $(0, -1.5)$

40. Ellipse; major axis from $(0, -3)$ to $(0, 3)$; minor axis from $(-\sqrt{7}, 0)$ to $(\sqrt{7}, 0)$; foci: $(0, \pm\sqrt{2})$

41. Hyperbola; transverse axis from $(0, -2)$ to $(0, 2)$; foci: $(0, \pm\sqrt{5})$; asymptotes: $y = \pm 2x$

42. Parabola; vertex: $(0, 0)$; directrix: $x = \frac{7}{4}$; focus: $(-\frac{7}{4}, 0)$

43. $y = \pm 15\sqrt{1 - \frac{1}{144}x^2} = \pm\frac{5}{4}\sqrt{144 - x^2}$; window: $[-25, 25] \times [-16.5, 16.5]$

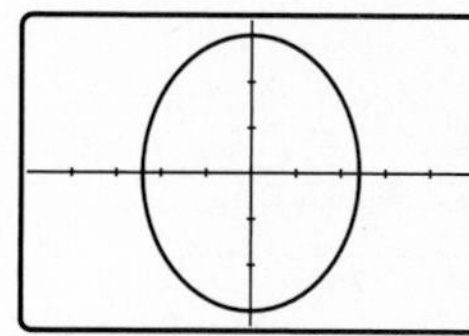

44. $y = \pm\sqrt{5x}$; window: $[-5, 45] \times [-15.1, 15.1]$

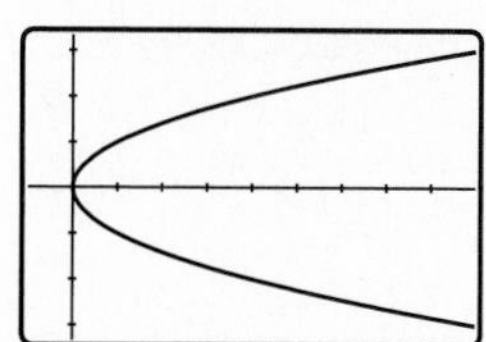

45. $y = \pm 3\sqrt{\frac{1}{4}x^2 - 1} = \pm\frac{3}{2}\sqrt{x^2 - 4}$; window: $[-15, 15] \times [-15, 15]$

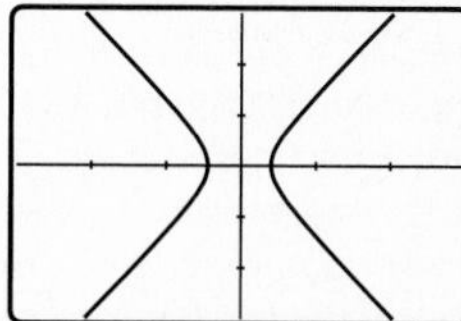

46. $y = \pm 10\sqrt{1 - \frac{1}{324}x^2} = \pm\frac{5}{9}\sqrt{324 - x^2}$; window: $[-25, 25] \times [-16.5, 16.5]$

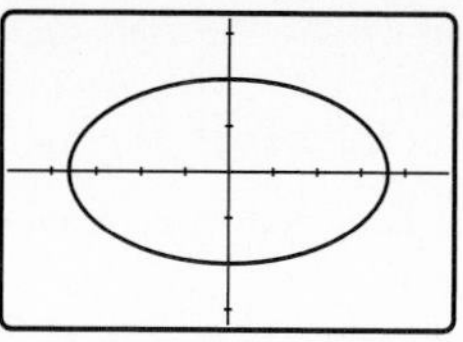

47. $y = \pm\sqrt{-3x}$; window: $[-25, 5] \times [-10, 10]$

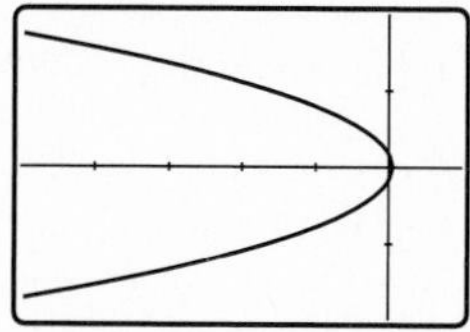

48. $y = \pm 4\sqrt{1 + \frac{1}{9}x^2} = \pm\frac{4}{3}\sqrt{9 + x^2}$; window: $[-10, 10] \times [-15, 15]$

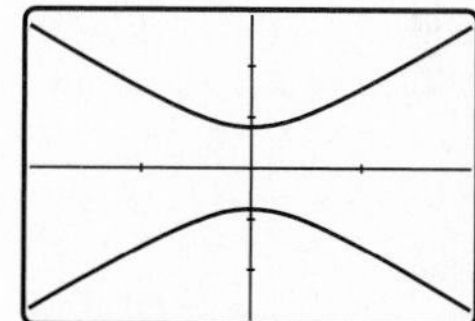

49. Multiply both sides of the equation by r^2 to get $x^2 + y^2 = r^2$ (since $r^2 = a^2 + b^2$).

50. For any point $P(x, y)$ on the parabola, the distance $PF = \sqrt{(x - p)^2 + y^2}$ must equal the distance from P to the directrix, which is $|x - (-p)| = |x + p|$; therefore $\sqrt{(x - p)^2 + y^2} = |x + p|$, so $(x - p)^2 + y^2 = (x + p)^2$. Expanding the squares gives $x^2 - 2px + p^2 + y^2 = x^2 + 2px + p^2$; the x^2 and p^2 terms cancel, leaving $y^2 - 2px = 2px$, or $y^2 = 4px$, so $x = \frac{1}{4p}y^2$.

51. Only the point $(x, y) = (0, 0)$ solves this equation.

52. Only the point $(x, y) = (0, 0)$ solves this equation.

53. Rewrite as $9x^2 = 4y^2$; the graph consists of the lines $y = \pm 1.5x$.

54. Rewrite as $b^2x^2 = a^2y^2$; the graph consists of the lines $y = \pm\frac{b}{a}x$.

55. The microphone should be placed at the focus, $(0, 2.5)$.

56. Place the light bulb at the focus, (0, 3).

57. $(\pm\sqrt{51.75}, 0) \approx (\pm 7.19, 0)$

58. At the foci, 12 in. from the center along the major axis

59. Let $d_1 = \sqrt{(x-c)^2 + y^2}$ (the distance to one focus) and $d_2 = \sqrt{(x+c)^2 + y^2}$ (the distance to the other focus); for an ellipse, we have $d_1 + d_2 = 2a$. Note for future reference that $(d_1 d_2)^2 = [(x-c)^2 + y^2][(x+c)^2 + y^2] = (x^2 + c^2 + y^2 - 2cx)(x^2 + c^2 + y^2 + 2cx) = (x^2 + y^2 + c^2)^2 - 4c^2x^2 = (x^2 + y^2)^2 + c^4 + 2c^2(x^2 + y^2) - 4c^2x^2 = (x^2 + y^2)^2 + c^4 - 2c^2(x^2 - y^2)$. Also, $d_1^2 + d_2^2 = (x-c)^2 + y^2 + (x+c)^2 + y^2 = x^2 - 2cx + c^2 + y^2 + x^2 + 2cx + c^2 + y^2 = 2(x^2 + y^2 + c^2)$.

Square both sides of $d_1 + d_2 = 2a$ to get $d_1^2 + 2d_1d_2 + d_2^2 = 4a^2$; this simplifies to $2(x^2 + y^2 + c^2) + 2d_1d_2 = 4a^2$ so that $d_1d_2 = 2a^2 - (x^2 + y^2 + c^2) = 2a^2 - c^2 - x^2 - y^2 = a^2 + b^2 - (x^2 + y^2)$; squaring both sides again gives $(x^2 + y^2)^2 + c^4 - 2c^2(x^2 - y^2)$ on the left and $(x^2 + y^2 - a^2 - b^2)^2 = (x^2 + y^2)^2 + (a^2 + b^2)^2 - 2(x^2 + y^2)(a^2 + b^2)$ on the right; then $(x^2 + y^2)^2 + c^4 - 2c^2(x^2 - y^2) = (x^2 + y^2)^2 + (a^2 + b^2)^2 - 2(x^2 + y^2)(a^2 + b^2)$, so $c^4 - 2c^2(x^2 - y^2) = (a^2 + b^2)^2 - 2(x^2 + y^2)(a^2 + b^2)$; collect all the x and y terms on the left and the others on the right: $2(x^2 + y^2)(a^2 + b^2) - 2c^2(x^2 - y^2) = (a^2 + b^2)^2 - c^4$ so that $2(a^2 + b^2 - c^2)x^2 + 2(a^2 + b^2 + c^2)y^2 = (a^2 + b^2)^2 - c^4$; now substitute b^2 for $a^2 - c^2$ and $a^2 - b^2$ for c^2 on the left, and replace c^4 with $(a^2 - b^2)^2$, which leaves $4b^2x^2 + 4a^2y^2 = (a^2 + b^2)^2 - (a^2 - b^2)^2 = 4a^2b^2$; dividing by $4a^2b^2$ finishes the task.

60. Let $d_1 = \sqrt{(x-c)^2 + y^2}$ (the distance to one focus) and $d_2 = \sqrt{(x+c)^2 + y^2}$ (the distance to the other focus); for a hyperbola, we have $|d_1 - d_2| = 2a$. Note for future reference that $(d_1 d_2)^2 = [(x-c)^2 + y^2][(x+c)^2 + y^2] = (x^2 + c^2 + y^2 - 2cx)(x^2 + c^2 + y^2 + 2cx) = (x^2 + y^2 + c^2)^2 - 4c^2x^2 = (x^2 + y^2)^2 + c^4 + 2c^2(x^2 + y^2) - 4c^2x^2 = (x^2 + y^2)^2 + c^4 - 2c^2(x^2 - y^2)$. Also, $d_1^2 + d_2^2 = (x-c)^2 + y^2 + (x+c)^2 + y^2 = 2(x^2 + y^2 + c^2)$.

Square boths sides of $|d_1 - d_2| = 2a$ to get $d_1^2 - 2d_1d_2 + d_2^2 = 4a^2$; this simplifies to $2(x^2 + y^2 + c^2) - 2d_1d_2 = 4a^2$ so that $d_1d_2 = x^2 + y^2 + c^2 - 2a^2 = x^2 + y^2 - (2a^2 - c^2) = x^2 + y^2 - (a^2 - b^2)$; squaring both sides again gives $(x^2 + y^2)^2 + c^4 - 2c^2(x^2 - y^2)$ on the left and $(x^2 + y^2 - a^2 + b^2)^2 = (x^2 + y^2)^2 + (a^2 - b^2)^2 - 2(x^2 + y^2)(a^2 - b^2)$ on the right; then $(x^2 + y^2)^2 + c^4 - 2c^2(x^2 - y^2) = (x^2 + y^2)^2 + (a^2 - b^2)^2 - 2(x^2 + y^2)\cdot(a^2 - b^2)$, so $c^4 - 2c^2(x^2 - y^2) = (a^2 - b^2)^2 - 2(x^2 + y^2)(a^2 - b^2)$; collect all the x and y terms on the left and the others on the right: $2(x^2 + y^2)\cdot(a^2 - b^2) - 2c^2(x^2 - y^2) = (a^2 - b^2)^2 - c^4$, so that $2(a^2 - b^2 - c^2)x^2 + 2(a^2 - b^2 + c^2)y^2 = (a^2 - b^2)^2 - c^4$; now substitute $a^2 + b^2$ for c^2 on the left, and replace c^4 with $(a^2 + b^2)^2$; this leaves $-4b^2x^2 + 4a^2y^2 = (a^2 - b^2)^2 - (a^2 + b^2)^2 = -4a^2b^2$; dividing by $-4a^2b^2$ finishes the task.

61. If $|x|$ is very large, $\frac{b^2}{a^2}x^2$ also will be large; as will $y^2 = \frac{b^2}{a^2}x^2 - b^2$ and (for large enough $|x|$) very close to $\frac{b^2}{a^2}x^2$; then, since $y^2 \approx \frac{b^2}{a^2}x^2$, we can conclude that $y \approx \pm\sqrt{\frac{b^2}{a^2}x^2} = \pm\frac{b}{a}|x|$ (assuming that a and b are positive).

62. The standard forms involve multiples of x, x^2, y, and y^2, as well as constants; therefore they can be rewritten in the general form $Ax^2 + Cy^2 + Dx + Ey + F = 0$ (none of the standard forms we have seen require a Bxy term). For example, rewrite $y = ax^2$ as $ax^2 - y = 0$; this is the general form with $A = a$ and $E = -1$, and all others 0. Similarly, the hyperbola $\frac{y^2}{b^2} - \frac{x^2}{a^2} = 1$ can be put in standard form with $A = -\frac{1}{a^2}$, $C = \frac{1}{b^2}$, $F = -1$, and $D = E = 0$.

QUICK REVIEW 9.3

1. $y = -5$ or $y = \frac{1}{2}$

2. $y = -4$ or $y = -\frac{1}{3}$

3. $y = \frac{-1 \pm \sqrt{5}}{3}$

4. $y = -\frac{1}{2} \pm \frac{\sqrt{3}}{4}$

5. Translate left 2 units and up 5 units.

6. Translate right 3 units and up 2 units.

7. Translate right 7 units and down 3 units.

8. Translate left 2 units and down 5 units.

9. $(x + 2)^2 - 5 = 0$, so $x = -2 \pm \sqrt{5}$

10. $(x - 1)^2 - 3 = 0$, so $x = 1 \pm \sqrt{3}$

11. $\left(x - \frac{3}{2}\right)^2 - \frac{15}{4} = 0$, so $x = \frac{3 \pm \sqrt{15}}{2}$

12. $(x + 1)^2 - \frac{7}{2} = 0$, so $x = -1 \pm \sqrt{\frac{7}{2}}$

SECTION EXERCISES 9.3

1. $x + 4 = \frac{1}{8}(y + 4)^2$

2. $y - 6 = -\frac{1}{12}(x + 5)^2$

3. $y - 2.5 = \frac{1}{6}(x - 3)^2$

4. $x - 3.5 = -\frac{1}{6}(y + 3)^2$

5. $x - 4 = -\frac{1}{8}(y - 3)^2$

6. $y - 5 = -\frac{1}{8}(x - 3)^2$

7. Vertex: $(3, -1)$; line of symmetry: $x = 3$; focus: $(3, 2)$; directrix: $y = -4$; window: $[-15, 25] \times [-10, 25]$

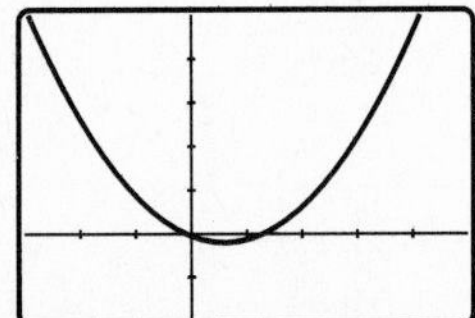

8. Vertex: $(-1, 3)$; line of symmetry: $x = -1$; focus: $(-1, 4.5)$; directrix: $y = 1.5$; window: $[-12, 10] \times [-2, 20]$

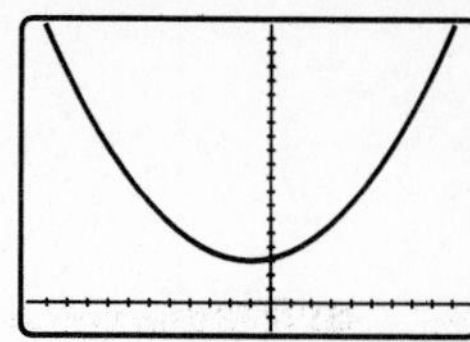

9. Vertex: (3, 2); line of symmetry: $x = 3$; focus: $\left(3, \frac{127}{64}\right)$; directrix: $y = \frac{129}{64}$; window: $[-1, 6] \times [-15, 5]$

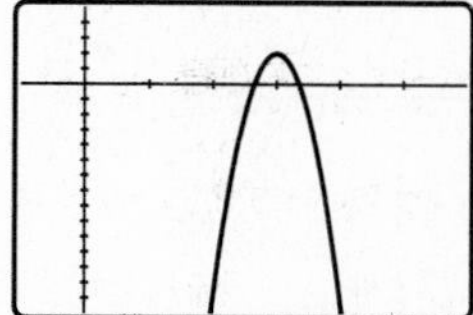

10. Vertex: $(-4, 1)$; line of symmetry: $x = -4$; focus: $(-4, -0.5)$; directrix: $y = 2.5$; window: $[-15, 5] \times [-15, 2]$

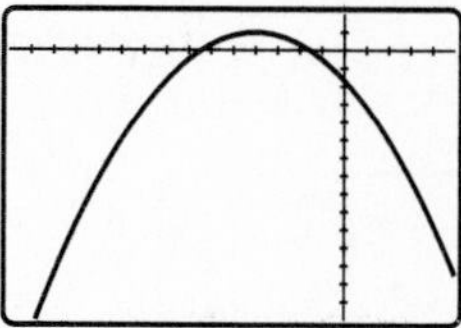

11. $y - 2 = (x + 1)^2$; vertex: $(-1, 2)$; focus: $(-1, 2.25)$; directrix: $y = 1.75$

12. $y - \frac{7}{6} = \frac{1}{2}(x - 1)^2$; vertex: $\left(1, \frac{7}{6}\right)$; focus: $\left(1, \frac{5}{3}\right)$; directrix: $y = \frac{2}{3}$

13. $x - 2 = \frac{1}{8}(y - 2)^2$; vertex: (2, 2); focus: (4, 2); directrix: $x = 0$

14. $x - 3.25 = -\frac{1}{4}(y - 1)^2$; vertex: (3.25, 1); focus: (2.25, 1); directrix: $x = 4.25$

15. $\frac{(x-1)^2}{16} + \frac{(y-2)^2}{36} = 1$

16. $\frac{(x+2)^2}{4} + \frac{(y-2)^2}{25} = 1$

17. $\frac{(x-3)^2}{9} + \frac{(y+4)^2}{5} = 1$

18. $\frac{(x+2)^2}{12} + \frac{(y-3)^2}{16} = 1$

19. $\frac{(x-3)^2}{9} + \frac{(y+2)^2}{25} = 1$

20. $\frac{(x+1)^2}{16} + \frac{(y-2)^2}{9} = 1$

21. $\frac{(x-2)^2}{16} + \frac{(y-3)^2}{9} = 1$

22. $\frac{(x+4)^2}{16} + \frac{(y-2)^2}{9} = 1$

23. Major axis endpoints: $(1, -1)$ and $(1, -5)$; minor axis endpoints: $(1 \pm \sqrt{2}, -3)$; center $(1, -3)$; foci: $(1, -3 \pm \sqrt{2})$; lines of symmetry: $x = 1$ and $y = -3$; window: $[-4, 5.4] \times [-6, 0.2]$

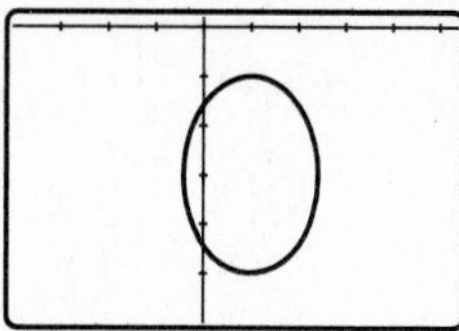

24. Major axis endpoints: $(-7, 1)$ and $(1, 1)$; minor axis endpoints: $(-3, -1)$ and $(-3, 3)$; center $(-3, 1)$; foci: $(-3 \pm \sqrt{12}, 1)$; lines of symmetry: $x = -3$ and $y = 1$; window: $[-7.2, 2.2] \times [-2.1, 4.1]$

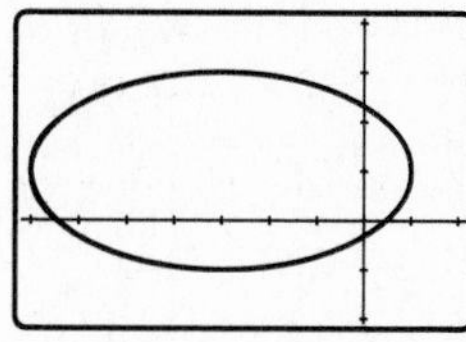

25. Major axis endpoints: $(-2 \pm \sqrt{5}, 1)$; minor axis endpoints: $(-2, 1 \pm \sqrt{0.5})$; center $(-2, 1)$; foci: $(-2 \pm \sqrt{4.5}, 1)$; lines of symmetry: $x = -2$ and $y = 1$; window: $[-7.2, 2.2] \times [-2.1, 4.1]$

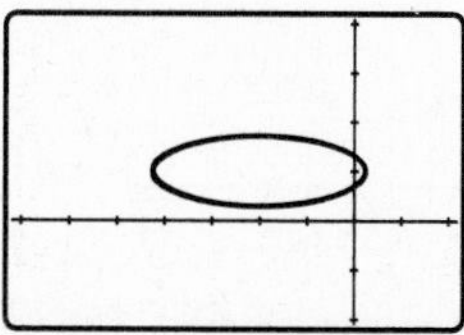

26. Major axis endpoints: $(4 \pm \sqrt{128}, -4)$; minor axis endpoints: $(4, -4 \pm \sqrt{0.5})$; center $(4, -4)$; foci: $(4 \pm \sqrt{127.5}, -4)$; lines of symmetry: $x = 4$ and $y = -4$; window: $[-8, 16] \times [-6, 1]$

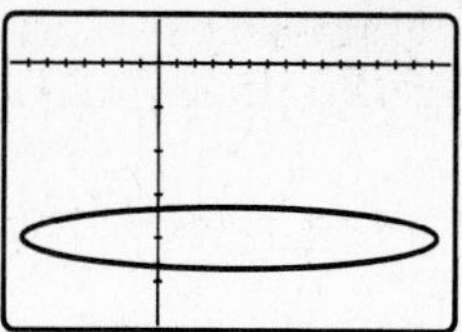

27. $\frac{(x-1)^2}{4} + \frac{(y+1)^2}{9} = 1$; major axis endpoints: $(1, -4)$ and $(1, 2)$; minor axis endpoints: $(-1, -1)$ and $(3, -1)$; foci: $(1, -1 \pm \sqrt{5})$

28. $\frac{(x-2)^2}{5} + \frac{(y+3)^2}{3} = 1$; major axis endpoints: $(2 \pm \sqrt{5}, -3)$; minor axis endpoints: $(2, -3 \pm \sqrt{3})$; foci: $(2 \pm \sqrt{2}, -3)$

29. $\frac{(x+3)^2}{16} + \frac{(y-1)^2}{9} = 1$; major axis endpoints: $(-7, 1)$ and $(1, 1)$; minor axis endpoints: $(-3, -2)$ and $(-3, 4)$; foci: $(-3 \pm \sqrt{7}, 1)$

30. $(x-4)^2 + \frac{(y+8)^2}{4} = 1$; major axis endpoints: $(4, -10)$ and $(4, -6)$; minor axis endpoints: $(3, -8)$ and $(5, -8)$; foci: $(4, -8 \pm \sqrt{3})$

31. $\frac{(x-2)^2}{9} - \frac{(y-3)^2}{16} = 1$

32. $\frac{(y-2.5)^2}{4.5^2} - \frac{(x+2)^2}{3.375^2} = 1$

33. $\frac{(x+1)^2}{4} - \frac{(y-2)^2}{5} = 1$

34. $\frac{(y+5.5)^2}{3.5^2} - \frac{(x+3)^2}{18} = 1$

35. Center: (0, 3); foci: $(\pm 3, 3)$; vertices (transverse axis endpoints): $(\pm 2, 3)$; lines of symmetry: $x = 0$ and $y = 3$; asymptotes: $y - 3 = \pm\frac{\sqrt{5}}{2}x$; window: $[-10, 10] \times [-10, 15]$

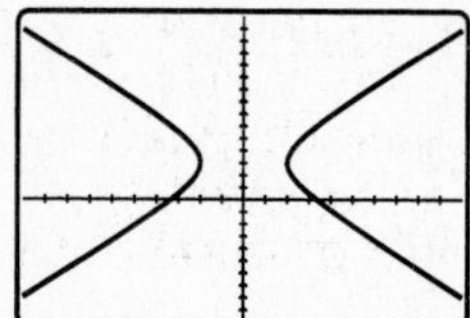

36. Center: $(-2, 3)$; foci: $(-2, 3 \pm \sqrt{13})$; vertices (transverse axis endpoints): $(-2, 0)$ and $(-2, 6)$; lines of symmetry: $x = -2$ and $y = 3$; asymptotes: $y - 3 = \pm\frac{3}{2}(x + 2)$, or $y = -\frac{3}{2}x$ and $y = \frac{3}{2}x + 6$; window: $[-12, 8] \times [-15, 20]$

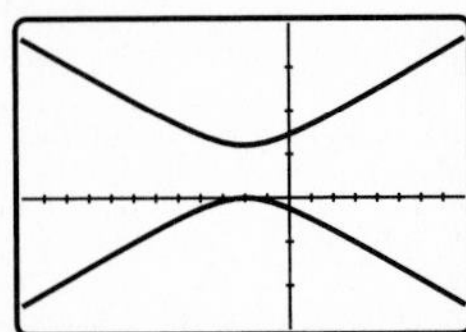

37. Center: $(3, 1)$; foci: $(3, 1 \pm \sqrt{13})$; vertices (transverse axis endpoints): $(3, -2)$ and $(3, 4)$; lines of symmetry: $x = 3$ and $y = 1$; asymptotes: $y - 1 = \pm\frac{3}{2}(x - 3)$, or $y = \frac{3}{2}x - \frac{7}{2}$ and $y = -\frac{3}{2}x + \frac{11}{2}$; window: $[-12, 18] \times [-15, 20]$

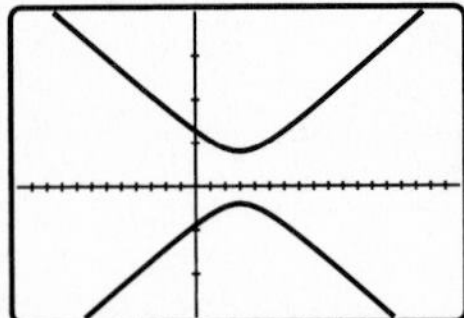

38. Center: $(2, -4)$; foci: $(2 \pm \frac{\sqrt{13}}{6}, -4)$; vertices (transverse axis endpoints): $(1.5, -4)$ and $(2.5, -4)$; lines of symmetry: $x = 2$ and $y = -4$; asymptotes: $y + 4 = \pm\frac{2}{3}(x - 2)$, or $y = \frac{2}{3}x - \frac{16}{3}$ and $y = -\frac{2}{3}x - \frac{8}{3}$; window: $[-10, 15] \times [-15, 5]$

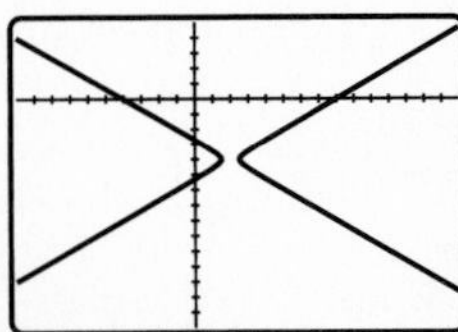

39. $\dfrac{(x-2)^2}{4} - \dfrac{(y-1)^2}{9} = 1$; transverse axis endpoints: $(0, 1)$ and $(4, 1)$; asymptotes: $y - 1 = \pm\frac{3}{2}(x - 2)$, or $y = \frac{3}{2}x - 2$ and $y = -\frac{3}{2}x + 4$

40. $\dfrac{(y-1)^2}{9} - \dfrac{(x+3)^2}{25} = 1$; transverse axis endpoints: $(-3, -2)$ and $(-3, 4)$; asymptotes: $y - 1 = \pm\frac{3}{5}(x + 3)$, or $y = 0.6x + 2.8$ and $y = -0.6x - 0.8$.

41. Hyperbola: $\dfrac{(y-1)^2}{9} - \dfrac{(x+1)^2}{4} = 1$; foci: $(-1, 1 \pm \sqrt{13})$; center $(-1, 1)$

42. Ellipse: $\dfrac{(x+3)^2}{3} + \dfrac{(y-4)^2}{2} = 1$; foci: $(-4, 4)$ and $(-2, 4)$; center $(-3, 4)$

43. Ellipse: $\dfrac{(x-1)^2}{4} + \dfrac{(y+2)^2}{9} = 1$; foci: $(1, -2 \pm \sqrt{5})$; center $(1, -2)$

44. Hyperbola: $\dfrac{(x-1)^2}{4} - \dfrac{(y+3)^2}{64} = 1$; foci: $(1 \pm \sqrt{68}, -3) = (1 \pm 2\sqrt{17}, -3)$; center $(1, -3)$

45. Ellipse: $\dfrac{(x-1)^2}{10} + \dfrac{(y-3)^2}{20} = 1$; foci: $(1, 3 \pm \sqrt{10})$; center $(1, 3)$

46. Hyperbola: $\dfrac{y^2}{4} - \dfrac{(x+1)^2}{2} = 1$; foci: $(-1, \pm\sqrt{6})$; center $(-1, 0)$

47. $y = -2 \pm \frac{1}{2}\sqrt{6(-x^2 + 2x + 1)}$, an ellipse; window: $[-4.7, 4.7] \times [-4.2, 2]$

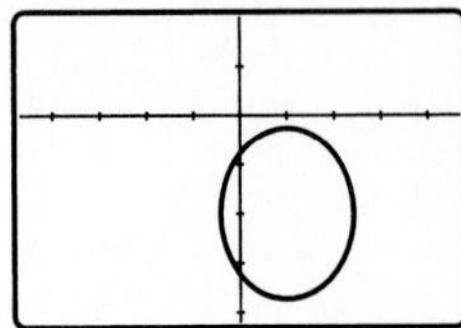

48. $y = 1 \pm \frac{1}{6}\sqrt{60x^2 - 24x + 144}$, a hyperbola; window: $[-6, 6] \times [-6, 8]$

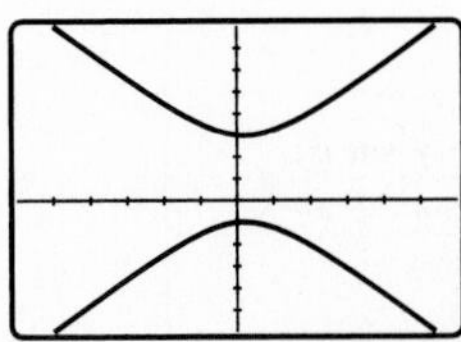

49. $y = \dfrac{x - 4 \pm \sqrt{-23x^2 + 28x + 88}}{6}$, an ellipse; $\alpha = \frac{\pi}{8}$; window: $[-4.7, 4.7] \times [-3.1, 3.1]$

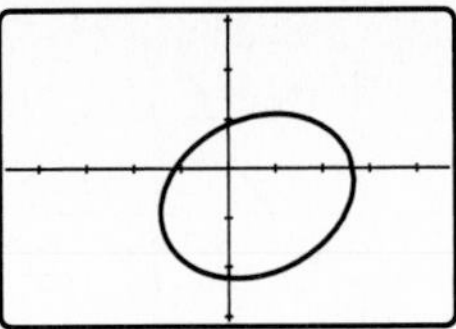

50. $y = \dfrac{10 - 3x \pm \sqrt{25x^2 + 20x + 420}}{8}$, a hyperbola; $\alpha = \frac{\pi}{2} + \frac{1}{2}\tan^{-1}(-\frac{3}{5}) = 1.300\ldots$; window: $[-10, 10] \times [-10, 10]$

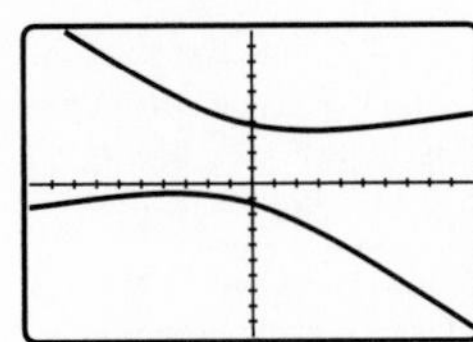

51. $y = \dfrac{x - 1 \pm \sqrt{3(-x^2 + 6x + 9)}}{4}$, an ellipse; $\alpha = \frac{1}{2}\tan^{-1}\frac{2}{3} = 0.294\ldots$; window: $[-2, 7.4] \times [-3.1, 3.1]$

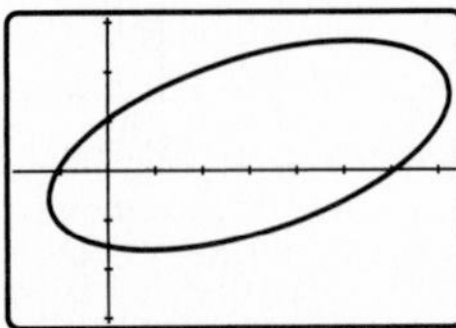

52. $y = x - \frac{3}{2} \pm \frac{1}{2}\sqrt{39 - 2x}$, a parabola; $\alpha = \frac{\pi}{4}$; window: $[-5, 20] \times [-5, 20]$

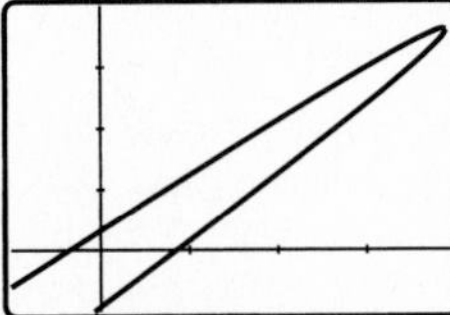

53. About 886.66 mi east and 1045.83 mi north, a bearing and distance of about 40.29° and 1371.11 mi, respectively.

54. About 11,714.3 ft (approximately 2.22 mi) east and 9,792.5 ft (approximately 1.85 mi) north, a bearing and distance of about 50.11° and 15,628.2 ft (approximately 2.89 mi), respectively.

55. Perihelion: $a - c = a - ea = a(1 - e) = 36 \cdot 0.79 = 28.44$ million miles; aphelion: $a + c = a(1 + e) = 36 \cdot 1.21 =$ 43.56 million miles.

56. Perihelion: $a - c = a - ea = a(1 - e) = 1680 \cdot 0.03 = 50.4$ million miles; aphelion: $a + c = a(1 + e) = 1680 \cdot 1.97 = 3309.6$ million miles.

57. Since $A = 0$, we can simply write $Cy^2 + Dx + Ey + F = 0$; divide the equation by C and complete the square for the terms involving y: $\left(y + \frac{E}{2C}\right)^2 - \frac{E^2}{4C^2} + \frac{D}{C}x + \frac{F}{C} = 0$, or $\left(y + \frac{E}{2C}\right)^2 = \frac{E^2}{4C^2} - \frac{D}{C}x - \frac{F}{C}$. If $D = 0$, there are 0, 1, or 2 (real) values of y that solve this equation (the degenerate case). If $D \neq 0$, rewrite the right side as $-\frac{D}{C}\left(x + \frac{F}{D} - \frac{E^2}{4CD}\right)$ and multiply both sides by $-\frac{C}{D}$; the equation is then in standard form for a parabola, with $k = -\frac{E}{2C}$, $h = \frac{E^2}{4CD} - \frac{F}{D}$, and $a = -\frac{C}{D}$.

58. Divide the equation by AC and complete the square for x and y: $\frac{1}{C}\left(x^2 + \frac{D}{A}x\right) + \frac{1}{A}\left(y^2 + \frac{E}{C}y\right) + \frac{F}{AC} = \frac{1}{C}\left(x + \frac{D}{2A}\right)^2 + \frac{1}{A}\left(y + \frac{E}{2C}\right)^2 - \frac{D^2}{4A^2C} - \frac{E^2}{4C^2A} + \frac{F}{AC} = 0$; move the constant terms to the right side and put them over a common denominator, which leaves $\frac{1}{C}\left(x + \frac{D}{2A}\right)^2 + \frac{1}{A}\left(y + \frac{E}{2C}\right)^2 = (CD^2 + AE^2 - 4ACF)/(4A^2C^2)$. Call the right side K; if $K = 0$, only $(x, y) = \left(-\frac{D}{2A}, -\frac{E}{2C}\right)$ solves the equation (a "degenerate" case); if $K \neq 0$, divide the equation by K, leaving $\frac{1}{CK}\left(x + \frac{D}{2A}\right)^2 + \frac{1}{AK}\left(y + \frac{E}{2C}\right)^2 = 1$.

Since $AC > 0$, both A and C have the same sign, and therefore both AK and CK have the same sign. If $AK < 0$ and $CK < 0$, there are no solutions, since the left side is negative but the right is positive; if AK and CK are positive, this is standard form for an ellipse, with $h = -\frac{D}{2A}$, $k = -\frac{E}{2C}$, $a^2 = CK = (CD^2 + AE^2 - 4ACF)/(4A^2C)$, and $b^2 = AK = (CD^2 + AE^2 - 4ACF)/(4AC^2)$.

59. Points (x, y) on the right branch have $x > 0$; with foci $F_1(-c, 0)$ and $F_2(c, 0)$, we have $d_1 = \sqrt{(x + c)^2 + y^2} = \sqrt{x^2 + 2cx + c^2 + y^2}$ and $d_2 = \sqrt{(x - c)^2 + y^2} = \sqrt{x^2 - 2cx + c^2 + y^2}$. Referring to Exercise 60, Section 9.2, we know that $d_1 - d_2 = \pm 2a$; since $c > 0$ and $x > 0$, $2cx > 0$ and $-2cx < 0$, so that the expression under the radical for d_1 is greater than that under the radical for d_2; then $d_1 > d_2$, so $d_1 - d_2 > 0$ and $d_1 - d_2 = 2a$.

Similarly, if (x, y) is on the left branch, $x < 0$, so $2cx < 0$ and $-2cx > 0$; therefore $d_1 < d_2$, and $d_1 - d_2 = -2a$.

60. Take the answer for Exercise 59 and reverse roles of x and y and of a and b.

61. **(a)** The segment from the origin to the point is the hypotenuse of two right triangles: one with one leg on the x-axis and the other with one leg on the x'-axis; the coordinates in the $x'y'$ system are given by the lengths of the legs of the latter triangle, which are determined by the length of the hypotenuse (r) and the sine and cosine of the angle at the origin; this angle is $\theta - \alpha$, so $x' = r\cos(\theta - \alpha)$ and $y' = r\sin(\theta - \alpha)$; the angle difference identities give the desired equations. **(b)** Referring to the first of the two right triangles, we have $x = r\cos\theta$ and $y = r\sin\theta$; making these substitutions gives the desired equations. **(c)** Using the first equation, we get $x = \dfrac{(x' - y\sin\alpha)}{\cos\alpha}$ (assuming that $\cos\alpha \neq 0$); substituting into the second equation leaves $y' = (y\sin\alpha - x')\tan\alpha + y\cos\alpha$, so $y = \dfrac{y' + x'\tan\alpha}{\sin\alpha\tan\alpha + \cos\alpha} = \dfrac{y'\cos\alpha + x'\sin\alpha}{\cos\alpha(\sin\alpha\tan\alpha + \cos\alpha)} = \dfrac{y'\cos\alpha + x'\sin\alpha}{\sin^2\alpha + \cos^2\alpha} = y'\cos\alpha + x'\sin\alpha$; making this substitution into $x = \dfrac{(x' - y\sin\alpha)}{\cos\alpha}$ gives $x = \dfrac{x' - (y'\cos\alpha + x'\sin\alpha)\sin\alpha}{\cos\alpha} = \dfrac{x'(1 - \sin^2\alpha) - y'\cos\alpha\sin\alpha}{\cos\alpha} = \dfrac{x'\cos^2\alpha - y'\cos\alpha\sin\alpha}{\cos\alpha} = x'\cos\alpha - y'\sin\alpha$.

QUICK REVIEW 9.4

1. Quadrant II
2. Quadrant III
3. Quadrant I
4. Quadrant III
5. One possible answer is $\frac{7\pi}{4}, -\frac{9\pi}{4}$.
6. One possible answer is $\frac{7\pi}{3}, -\frac{5\pi}{3}$.
7. One possible answer is 520°, −200°.
8. One possible answer is 240°, −480°.
9. 1.107 . . . rad, or 63.434 . . . °
10. $\frac{\pi}{6}$ rad or 30°
11. $\theta = 0.411\ldots$ or $\theta = 2.730\ldots$
12. $\theta = \frac{\pi}{4}$ or $\theta = \frac{5\pi}{4}$
13. $\sin\theta = \dfrac{2}{\sqrt{13}}$, $\cos\theta = \dfrac{3}{\sqrt{13}}$, $\tan\theta = \dfrac{2}{3}$; $\csc\theta = \dfrac{\sqrt{13}}{2}$, $\sec\theta = \dfrac{\sqrt{13}}{3}$, $\cot\theta = \dfrac{3}{2}$
14. $\sin\theta = \dfrac{\sqrt{55}}{8}$, $\cos\theta = \dfrac{3}{8}$, $\tan\theta = \dfrac{\sqrt{55}}{3}$; $\csc\theta = \dfrac{8}{\sqrt{55}}$, $\sec\theta = \dfrac{8}{3}$, $\cot\theta = \dfrac{3}{\sqrt{55}}$
15. $(x - 3)^2 + y^2 = 4$
16. $x^2 + (y + 4)^2 = 9$

SECTION EXERCISES 9.4

1.

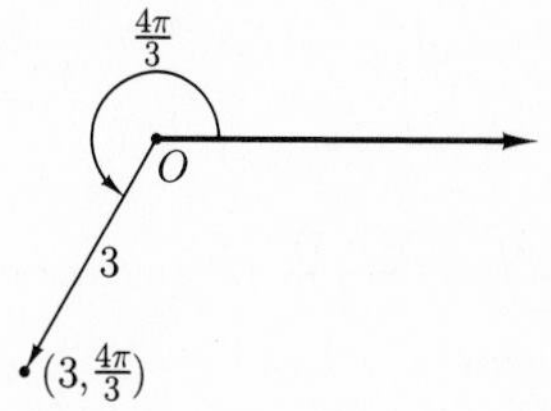

$\frac{4\pi}{3}$, O, 3, $\left(3, \frac{4\pi}{3}\right)$

2. $\left(2, \frac{5\pi}{6}\right)$

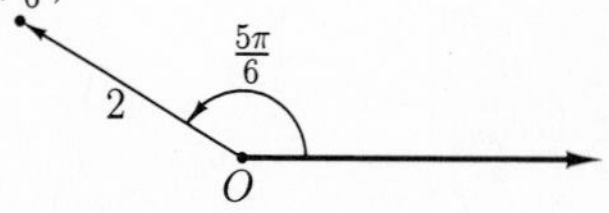

$\frac{5\pi}{6}$, 2, O

3. $\frac{2\pi}{5}$ O 1 $(-1, \frac{2\pi}{5})$

4. $(-3, \frac{17\pi}{10})$

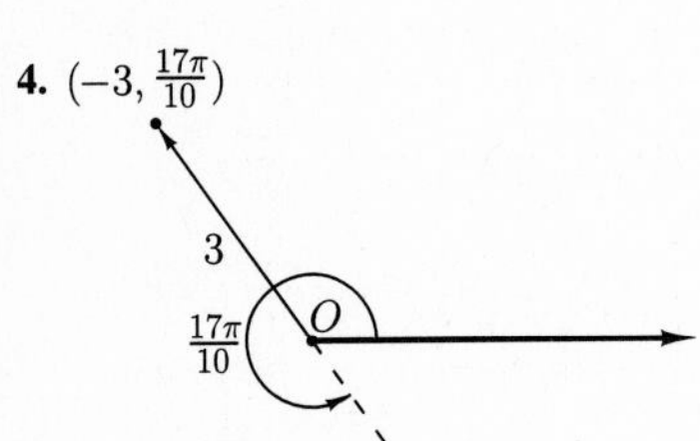

5.

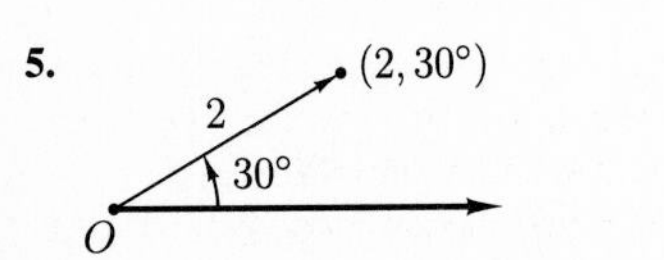

6.

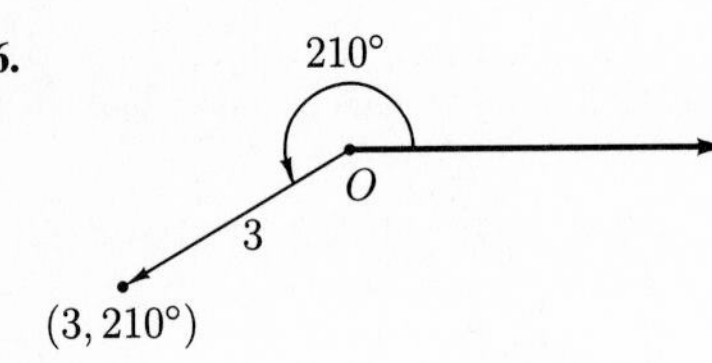

7.

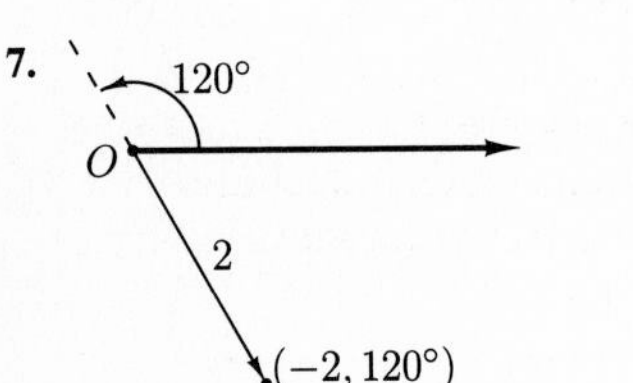

8.

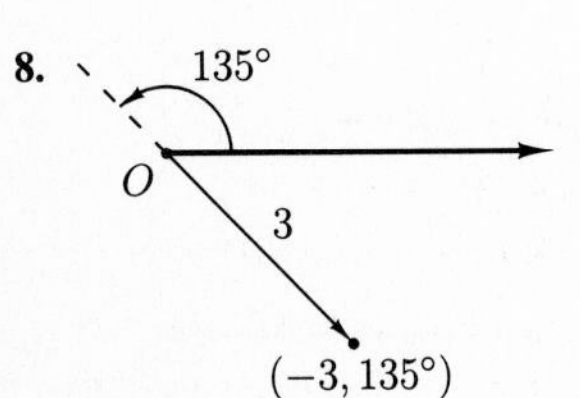

9. $(\frac{3}{4}, \frac{3\sqrt{3}}{4})$

10. $(\frac{5}{4}\sqrt{2}, \frac{5}{4}\sqrt{2})$

11. $(-2.702\ldots, 1.301\ldots)$

12. $(-1.618\ldots, 1.175\ldots)$

13. $(2, 0)$

14. $(0, 1)$

15. $(0, -2)$

16. $(-3, 0)$

17. $(2, \frac{\pi}{6} + 2n\pi)$ and $(-2, \frac{\pi}{6} + (2n+1)\pi)$, n an integer

18. $(1, -\frac{\pi}{4} + 2n\pi)$ and $(-1, -\frac{\pi}{4} + (2n+1)\pi)$, n an integer

19. $(1.5, -20° + 360n°)$ and $(-1.5, 160° + 360n°)$, n an integer

20. $(-2.5, 50° + 360n°)$ and $(2.5, 230° + 360n°)$, n an integer

21. **(a)** $(\sqrt{2}, \frac{\pi}{4})$ or $(-\sqrt{2}, \frac{5\pi}{4})$. **(b)** $(\sqrt{2}, \frac{\pi}{4})$ or $(-\sqrt{2}, -\frac{3\pi}{4})$. **(c)** The answers from (a) and $(\sqrt{2}, \frac{9\pi}{4})$ or $(-\sqrt{2}, \frac{13\pi}{4})$.

22. **(a)** $(\sqrt{10}, \tan^{-1} 3) = (\sqrt{10}, 1.249\ldots)$ or $(-\sqrt{10}, \tan^{-1} 3 + \pi) = (-\sqrt{10}, 4.390\ldots)$. **(b)** $(\sqrt{10}, \tan^{-1} 3) = (\sqrt{10}, 1.249\ldots)$ or $(-\sqrt{10}, \tan^{-1} 3 - \pi) = (-\sqrt{10}, -1.892\ldots)$. **(c)** The answers from (a) and $(\sqrt{10}, \tan^{-1} 3 + 2\pi) = (\sqrt{10}, 7.532\ldots)$ or $(-\sqrt{10}, \tan^{-1} 3 + 3\pi) = (-\sqrt{10}, 10.673\ldots)$.

23. **(a)** $(\sqrt{29}, \tan^{-1}(-2.5) + \pi) = (\sqrt{29}, 1.951\ldots)$ or $(-\sqrt{29}, \tan^{-1}(-2.5) + 2\pi) = (-\sqrt{29}, 5.092\ldots)$. **(b)** $(-\sqrt{29}, \tan^{-1}(-2.5)) = (-\sqrt{29}, -1.190\ldots)$ or $(\sqrt{29}, \tan^{-1}(-2.5) + \pi) = (\sqrt{29}, 1.951\ldots)$. **(c)** The answers from (a) and $(\sqrt{29}, \tan^{-1}(-2.5) + 3\pi) = (\sqrt{29}, 8.234\ldots)$ or $(-\sqrt{29}, \tan^{-1}(-2.5) + 4\pi) = (-\sqrt{29}, 11.376\ldots)$.

24. **(a)** $(-\sqrt{5}, \tan^{-1} 2) = (-\sqrt{5}, 1.107\ldots)$ or $(\sqrt{5}, \tan^{-1} 2 + \pi) = (\sqrt{5}, 4.248\ldots)$. **(b)** $(-\sqrt{5}, \tan^{-1} 2) = (-\sqrt{5}, 1.107\ldots)$ or $(\sqrt{5}, \tan^{-1} 2 - \pi) = (\sqrt{5}, -2.034\ldots)$. **(c)** The answers from (a) and $(-\sqrt{5}, \tan^{-1} 2 + 2\pi) = (-\sqrt{5}, 7.390\ldots)$ or $(\sqrt{5}, \tan^{-1} 2 + 3\pi) = (\sqrt{5}, 10.531\ldots)$.

25. (b)

26. (d)

27. (c)

28. (a)

29. $x = 3$, a vertical line

30. $y = -2$, a horizontal line

31. $r^2 + 3r\sin\theta = 0$, or $x^2 + y^2 + 3y = 0$; completing the square gives $x^2 + (y + \frac{3}{2})^2 = \frac{9}{4}$, a circle centered at $(0, -\frac{3}{2})$ with radius $\frac{3}{2}$.

32. $r^2 + 4r\cos\theta = 0$, or $x^2 + y^2 + 4x = 0$; completing the square gives $(x+2)^2 + y^2 = 4$; a circle centered at $(-2, 0)$ with radius 2.

33. $r^2 - r\sin\theta = 0$, or $x^2 + y^2 - y = 0$; completing the square gives $x^2 + (y - \frac{1}{2})^2 = \frac{1}{4}$, a circle centered at $(0, \frac{1}{2})$ with radius $\frac{1}{2}$.

34. $r^2 + 3r\cos\theta = 0$, or $x^2 + y^2 - 3x = 0$; completing the square gives $(x - \frac{3}{2})^2 + y^2 = \frac{9}{4}$, a circle centered at $(\frac{3}{2}, 0)$ with radius $\frac{3}{2}$.

35. $r\cos\theta - 2r\sin\theta = 5$, or $x - 2y = 5$, which is more familiar as $y = \frac{1}{2}x - \frac{5}{2}$, a line

36. $3r\cos\theta + r\sin\theta = 2$, or $3x + y = 2$, which is more familiar as $y = 2 - 3x$, a line

37. $r = \frac{2}{\cos\theta} = 2\sec\theta$

38. $r = \frac{5}{\cos\theta} = 5\sec\theta$

39. $r = \frac{5}{2\cos\theta - 3\sin\theta}$

40. $r = \frac{2}{3\cos\theta + 4\sin\theta}$

41. $r^2 - 6r\cos\theta = 0$, so $r = 6\cos\theta$

42. $r^2 - 2r\sin\theta = 0$, so $r = 2\sin\theta$

43. $r^2 = \dfrac{4}{\cos^2\theta - \sin^2\theta} = \dfrac{4}{\cos 2\theta}$, or $r = \dfrac{\pm 2}{\sqrt{\cos 2\theta}}$

44. $r^2 = \dfrac{1}{\sin^2\theta - \cos^2\theta} = \dfrac{-1}{\cos 2\theta}$, or
$r = \dfrac{\pm 1}{\sqrt{-\cos 2\theta}}$

45. $d = \sqrt{4^2 + 2^2 - 2(4)(2)\cos(12° - 72°)} = \sqrt{20 - 16\cos 60°} = \sqrt{20 - 8} = \sqrt{12} = 2\sqrt{3}$ mi

46. $d = \sqrt{3^2 + 5^2 - 2(3)(5)\cos(170° - 150°)} = \sqrt{34 - 30\cos 20°} = \sqrt{5.809\ldots} = 2.410\ldots$ mi

47. $\left(\dfrac{a}{\sqrt{2}}, \dfrac{\pi}{4}\right), \left(\dfrac{a}{\sqrt{2}}, \dfrac{3\pi}{4}\right), \left(\dfrac{a}{\sqrt{2}}, \dfrac{5\pi}{4}\right)$, and $\left(\dfrac{a}{\sqrt{2}}, \dfrac{7\pi}{4}\right)$; other polar coordinates for these points are possible, of course.

48. $(a, 0)$, $\left(a, \frac{2\pi}{5}\right)$, $\left(a, \frac{4\pi}{5}\right)$, $\left(a, \frac{6\pi}{5}\right)$, and $\left(a, \frac{8\pi}{5}\right)$; other polar coordinates for these points are possible, of course.

49. Draw a triangle with vertices $(0, 0)$, (r_1, θ_1), and (r_2, θ_2). To find the distance d between (r_1, θ_1) and (r_2, θ_2), use the law of cosines, noting that the angle at the origin is $\alpha = |\theta_1 - \theta_2|$; since the cosine function is even, $\cos\alpha = \cos(-\alpha)$, so you may assume that $\alpha = \theta_1 - \theta_2$.
Now $d^2 = r_1^2 + r_2^2 - 2r_1r_2 \cos(\theta_1 - \theta_2)$; taking square roots on both sides of the equation gives $d = \pm\sqrt{r_1^2 + r_2^2 - 2r_1r_2\cos(\theta_1 - \theta_2)}$; d is a distance, so we choose the positive root.

50. Multiply both sides by r to get $r^2 = 6r\cos\theta + 8r\sin\theta$; this becomes $x^2 + y^2 - 6x - 8y = 0$, or $(x - 3)^2 + (y - 4)^2 = 25$, which is a circle centered at $(3, 4)$ with radius 5.

51. Polar graphers can graph only equations that can be solved to give r alone on one side of the equation, with nothing involving r on the other side. (Note that, in the same way, function graphers can only graph those equations that can be written with y alone on one side of the equation). This equation does not fit that, since r does not appear in it anywhere; however, the set of all points with θ-coordinate $\frac{\pi}{6}$ is easily expressed in rectangular coordinates; starting with the polar point $\left(r, \frac{\pi}{6}\right)$, the corresponding rectangular coordinates are $x = r\cos\frac{\pi}{6} = \dfrac{\sqrt{3}}{2}r$ and $y = r\sin\dfrac{\pi}{6} = \dfrac{1}{2}r$; then $x = \sqrt{3}y$, or $y = \dfrac{1}{\sqrt{3}}x$, a line.

52. The final screen looks the same in either case: it is a circle centered at $(0, 0)$ with radius 3; both are drawn counterclockwise by the grapher, but the first begins on the positive x-axis (at "3 o'clock") and the other begins on the negative x-axis (at "9 o'clock").

QUICK REVIEW 9.5

1. $(x + 2)^2 + y^2 = 4$; center $(-2, 0)$; radius 2

2. $x^2 + \left(y - \frac{3}{2}\right)^2 = \frac{9}{4}$; center $\left(0, \frac{3}{2}\right)$ and radius $\frac{3}{2}$

3. $(x + 3)^2 + (y - 2)^2 = 13$; center $(-3, 2)$; radius $\sqrt{13}$

4. $(x - 4)^2 + (y + 1)^2 = 17$; center $(4, -1)$ and radius $\sqrt{17}$

5. $\sin(\pi - \theta) = \sin\theta$

6. $\cos(\pi - \theta) = -\cos\theta$

7. $\sin(-2\theta) = -2\sin\theta\cos\theta$

8. $\cos(-3\theta) = \cos 3\theta = \cos 2\theta\cos\theta - \sin 2\theta\sin\theta$; from here many answers are possible, including $\cos^3\theta - 3\sin^2\theta\cos\theta$ or $4\cos^3\theta - 3\cos\theta$.

9. $\cos 2(\pi + \theta) = \cos(2\pi + 2\theta) = \cos 2\theta$

10. $\sin 2(\pi + \theta) = \sin(2\pi + 2\theta) = \sin 2\theta$

SECTION EXERCISES 9.5

1. Yes. It is a three-petal rose; $0 \le \theta \le \pi$ is sufficient to show the entire shape. Window: $[-3, 3] \times [-2, 2]$.

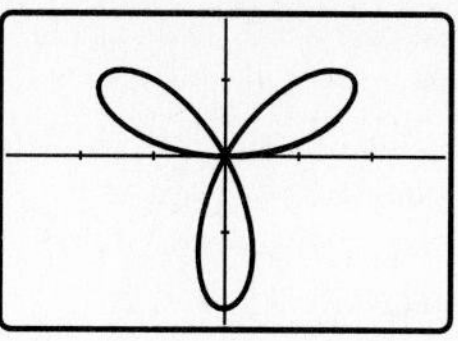

2. Yes. It is an eight-petal rose; $0 \le \theta \le 2\pi$ shows the entire shape. Window: $[-4.7, 4.7] \times [-3.1, 3.1]$.

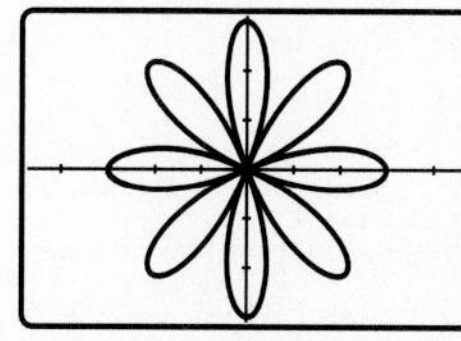

3. No. It is not the correct form (nor can it be transformed to the correct form); however, the graph does resemble a flower with different-sized petals. Window: $[-4, 4] \times [-2.5, 2.5]$.

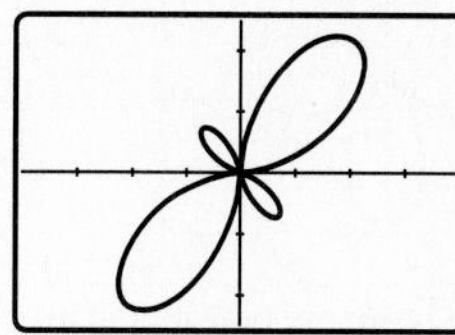

4. No. It is not the correct form (nor can it be transformed to the correct form); however, the graph does resemble a flower with different-sized petals. Window: $[-4.7, 4.7] \times [-2.5, 4.5]$.

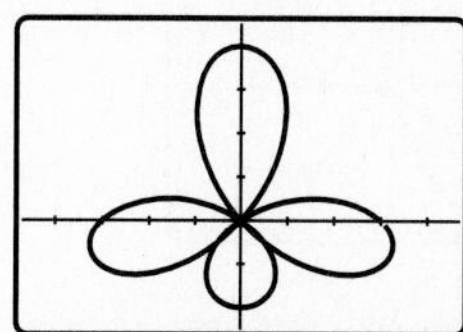

5. $k = \pi$; the petals are drawn as numbered; petals 1 and 3 are generated by positive r, whereas r is negative for petal 2. Window: $[-4.7, 4.7] \times [-3.7, 2.5]$.

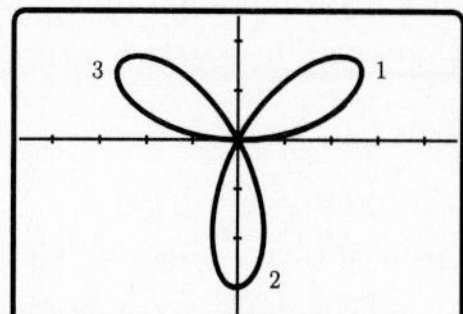

6. $k = 2\pi$; the petals are drawn as numbered (the bottom half of petal 1/5 was drawn first, the top half last); petals 2 and 4 are generated by positive r, whereas r is negative for petals 1/5 and 3. Window: $[-4.7, 4.7] \times [-3.1, 3.1]$.

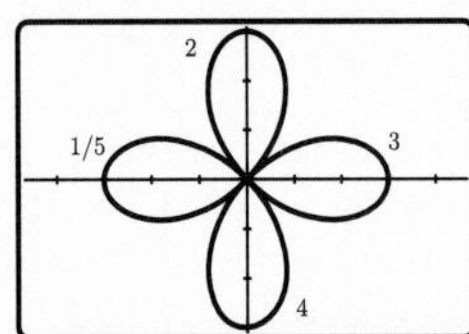

7. $k = 2\pi$; the petals are drawn as numbered (the top half of petal 1/5 was drawn first, the bottom half last); petals 1/5 and 3 are generated by positive r, whereas r is negative for petals 2 and 4. Window: $[-4.7, 4.7] \times [-3.1, 3.1]$.

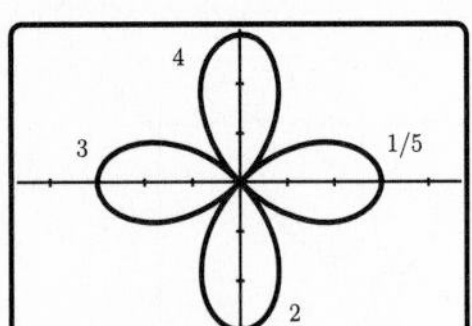

8. $k = \pi$; the petals are drawn as numbered; petals 1, 3, and 5 are generated by positive r, whereas r is negative for petals 2 and 4. Window: $[-4.7, 4.7] \times [-3.1, 3.1]$.

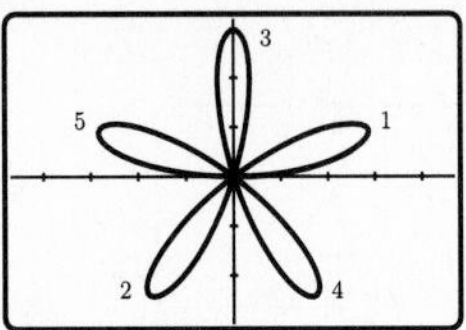

9. r_1 not shown (this is a 12-petal rose). r_2 not shown (this is a 16-petal rose). r_3 is the graph in part (b).

10. $6 \cos 2\theta \sin 2\theta = 3(2 \cos u \sin u)$, where $u = 2\theta$; this equals $3 \sin 2u = 3 \sin 4\theta$; $r = 3 \sin 4\theta$ is the equation for the eight-petal rose shown in the graph in part (a).

11. The graph in part (b) is $r = 2 - 2 \cos \theta$; taking $\theta = 0$ and $\theta = \frac{\pi}{2}$, we get $r = 2$ and $r = 4$ from the first equation, and $r = 0$ and $r = 2$ from the second equation. No graph matches the first of these (r, θ) pairs, but the graph in part (b) matches the latter (and any others we might choose).

12. The graph in part (c) is $r = 2 + 3 \cos \theta$; taking $\theta = 0$, we get $r = -1$ from the other equation, which matches nothing; any (r, θ) pair from the first equation matches the graph in part (c), however.

13. The graph in part (a) is $r = 2 - 2 \sin \theta$, when $\theta = \frac{\pi}{2}$, $2 + 2 \cos \theta = 2$, but $(2, \frac{\pi}{2})$ is clearly not on this graph; however, $2 - 2 \sin \frac{\pi}{2} = 0$, so $(0, \frac{\pi}{2})$ (the origin) is part of the graph in part (a).

14. The graph in part (d) is $r = 2 - 1.5 \sin \theta$, when $\theta = \frac{\pi}{2}$, $2 + 1.5 \cos \theta = 2$, but $(2, \frac{\pi}{2})$ is clearly not on the graph in part (d); however, $2 - 1.5 \sin \frac{\pi}{2} = 0.5$, so $(0.5, \frac{\pi}{2})$ is part of the graph in part (d).

15. r_1 and r_2 produce identical graphs: a limaçon with an inner loop; r_1 traces out most of the outer loop first, whereas r_2 traces a small part of the outer loop first then does the inner loop before completing the outer loop.

16. r_1 and r_3 produce identical graphs: a limaçon with an inner loop; r_1 traces out the top half of the outer loop first, then the inner loop, and finally the bottom half of the outer loop; meanwhile r_3 traces out the top half of the *inner* loop, then the outer loop, and finally the bottom half of the inner loop.

17. r_2 and r_3 produce identical graphs: a limaçon with an inner loop; r_2 traces out the bottom half of the inner loop first, then the outer loop, and finally the top half of the inner loop; meanwhile r_3 traces out the bottom half of the *outer* loop, then the inner loop, and finally the top half of the outer loop.

18. r_1 and r_2 produce identical graphs; a cardioid; r_1 traces out the "rounder" half of the cardioid first and then the "dimpled" part; r_2 traces them in reverse order.

19. Maximum r is 3 (along with -3), when $\theta = 2n\pi/3$, n an integer (or 0, $2\pi/3$, $4\pi/3$).

20. Maximum r is 4 (along with -4), when $\theta = n\pi/4$, n an odd integer (or $\pi/4$, $3\pi/4$, $5\pi/4$, $7\pi/4$).

21. Maximum r is 5, when $\theta = 0$.

22. Maximum r is 5 (actually, -5), when $\theta = \frac{3\pi}{2}$.

23. The two large petals are 6 units long, and the small ones are 2 units long.

24. The two large petals are 8 units long, and the small ones are 2 units long.

25. The large (outer) petals are 5 units long, and the small (inner) ones are 3 units long.

26. The large (outer) petals are 7 units long, and the small (inner) ones are 1 unit long.

27. Symmetric about the y-axis; replacing (r, θ) with $(r, \pi - \theta)$ gives the same equation, since $\sin(\pi - \theta) = \sin \theta$.

28. Symmetric about the polar (x-) axis; replacing (r, θ) with $(r, -\theta)$ gives the same equation, since $\cos(-\theta) = \cos \theta$.

29. Symmetric about the polar (x-) axis; replacing (r, θ) with $(r, -\theta)$ gives the same equation, since $\cos(-\theta) = \cos \theta$.

30. Symmetric about the y-axis; replacing (r, θ) with $(r, \pi - \theta)$ gives the same equation, since $\sin(\pi - \theta) = \sin \theta$.

31. All three symmetries. Polar axis: replacing (r, θ) with $(r, -\theta)$ gives the same equation, since $\cos(-2\theta) = \cos 2\theta$; y-axis: replacing (r, θ) with $(r, \pi - \theta)$ gives the same equation, since $\cos[2(\pi - \theta)] = \cos(2\pi - 2\theta) = \cos(-2\theta) = \cos 2\theta$; pole: replacing (r, θ) with $(r, \theta + \pi)$ gives the same equation, since $\cos[2(\theta + \pi)] = \cos(2\theta + 2\pi) = \cos 2\theta$.

32. Symmetric about the y-axis; replacing (r, θ) with $(-r, -\theta)$ gives the same equation, since $\sin(-3\theta) = -\sin 3\theta$.

33. Symmetric about the y-axis; replacing (r, θ) with $(r, \pi - \theta)$ gives the same equation, since $\sin(\pi - \theta) = \sin \theta$.

34. Symmetric about the polar (x-) axis; replacing (r, θ) with $(r, -\theta)$ gives the same equation, since $\cos(-\theta) = \cos\theta$.

35. y-axis symmetry. Window: $[-9, 9] \times [-2.5, 9.5]$.

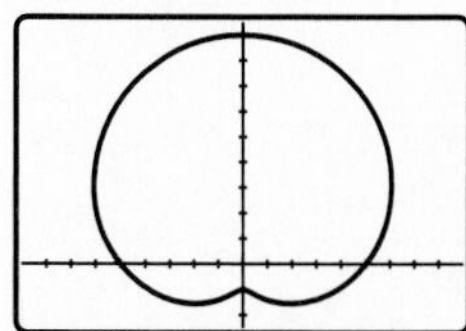

36. Polar (x-) axis symmetry. Window: $[-15, 9] \times [-8, 8]$.

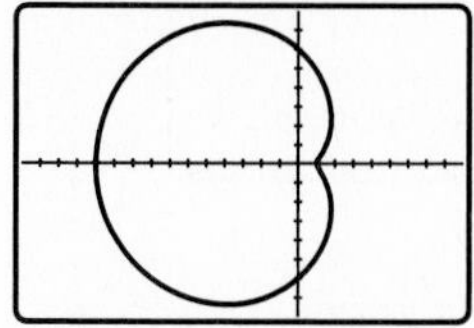

37. Polar (x-) axis symmetry. Window: $[-2.5, 6.5] \times [-3.1, 3.1]$.

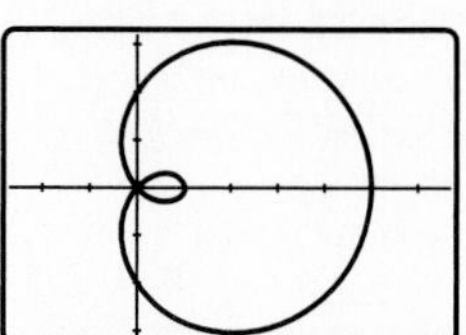

38. y-axis symmetry. Window: $[-5.5, 5.5] \times [-7.5, 1]$.

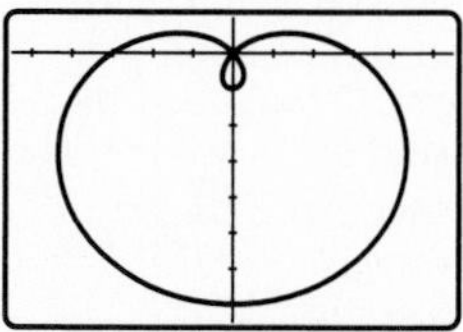

39. Polar (x-) axis symmetry. Window: $[-7, 5] \times [-4, 4]$.

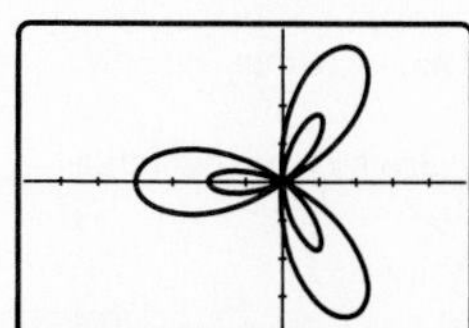

40. y-axis symmetry. Window: $[-6, 6] \times [-5, 3]$.

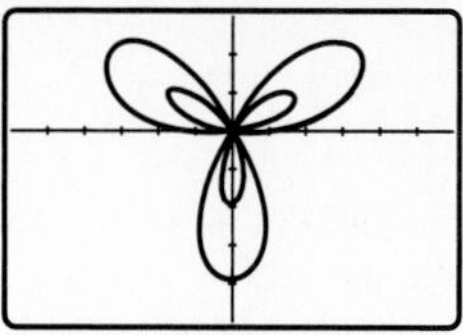

41. Starting with the graph of r_1, if we rotate clockwise (centered at the origin) by $\frac{\pi}{12}$ radians (15°), we get the graph of r_2; rotating r_1 clockwise by $\frac{\pi}{4}$ radians (45°) gives the graph of r_3. Window: $[-4.5, 4.5] \times [-3.1, 3.1]$.

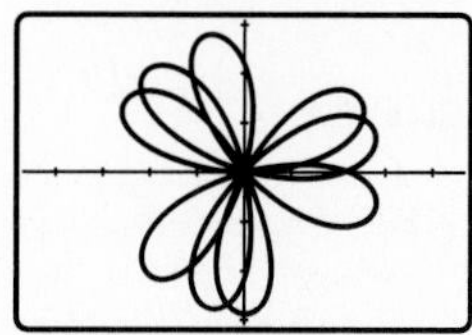

42. Starting with the graph of r_1, if we rotate counterclockwise (centered at the origin) by $\frac{\pi}{4}$ radians (45°), we get the graph of r_2; rotating r_1 counterclockwise by $\frac{\pi}{3}$ radians (60°) gives the graph of r_3. Window: $[-3, 7] \times [-2, 4]$.

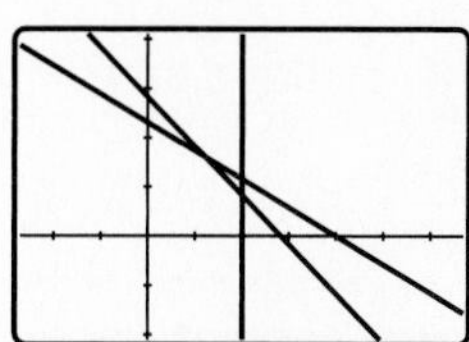

43. Starting with the graph of r_1, if we rotate clockwise (centered at the origin) by $\frac{\pi}{4}$ radians (45°), we get the graph of r_2; rotating r_1 clockwise by $\frac{\pi}{3}$ radians (60°) gives the graph of r_3. Window: $[-6.5, 4.5] \times [-3, 4.5]$.

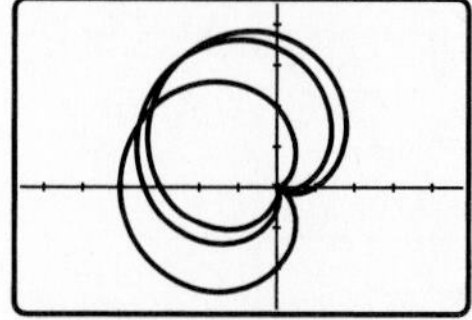

44. The second graph is the result of rotating the first graph clockwise (centered at the origin) through an angle of α; the third graph results from rotating the first graph counterclockwise through the same angle. This can be explained because the value of r, for example, when $\theta = 0$ in the first equation is obtained when $\theta = -\alpha$ for the second equation, and when $\theta = \alpha$ for the third equation.

45. **(a)** For the first graph; $0 \le \theta \le 4\pi$ (or any interval that is 4π units long); for the second graph, same answer. **(b)** First graph, 10 (overlapping) petals; second graph, 14 (overlapping) petals.

46. The period of $\sin n\theta$ is $2\pi/|n|$, so for $0 \le \theta \le 2\pi$ there are n periods of $\sin n\theta$. Considering $0 \le \theta \le \frac{2\pi}{n}$—one period—one petal (call it #1) is traced out for $0 \le \theta \le \frac{\pi}{n}$, and another petal (#2) is traced out for $\frac{\pi}{n} \le \theta < \frac{2\pi}{n}$; for every other period, two petals are drawn. None of these petals overlap. With two petals drawn for every period, we have $2n$ petals altogether.

47. The equation $r = a\sec\theta = \frac{a}{\cos\theta}$ is valid whenever $\cos\theta \ne 0$; in such cases we can multiply both sides by $\cos\theta$ and get $r\cos\theta = a$. The left side (in rectangular coordinates) is x; the graph of $x = a$ is a vertical line. (If $a = 0$, we have the equation $r = 0$, the graph of which is simply the origin.)

48. The amplitude of $y = a\sin x$ is $|a|$; it indicates that the largest and smallest y-values are a and $-a$. These are also the largest and smallest r-values for $r = a\sin\theta$; the points of the graph of $r = a\sin\theta$ are no more than $|a|$ units away from the origin.

49. The parametric equations $x = f(\theta)\cos\theta$ and $y = f(\theta)\sin\theta$ are simply restatements of the well-known relations $x = r\cos\theta$ and $y = r\sin\theta$, using the fact that $r = f(\theta)$. Note that these *are* parametric equations, even though the parameter is θ rather than the more familiar t.

50. This is a spiral (called a "logarithmic" or "equiangular" spiral); the exact appearance varies with the choice of θMin and θMax; shown is $-6\pi \le \theta \le 6\pi$ on the window $[-4.7, 4.7] \times [-3.1, 3.1]$; as $\theta \to -\infty$, $r \to 0$.

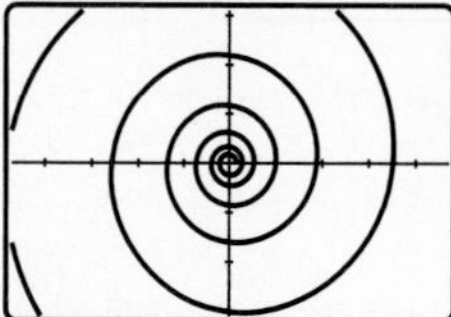

51. For $\theta \geq 1$, this is a spiral; when $0 < \theta < 1$, it is the curve under the negative x-axis extending off to the left (see the graph); the exact appearance varies with the choice of θMin and θMax; shown is $0.01 \leq \theta \leq 8\pi$ on the window $[-4.7, 4.7] \times [-3.1, 3.1]$. Note that $\theta > 0$.

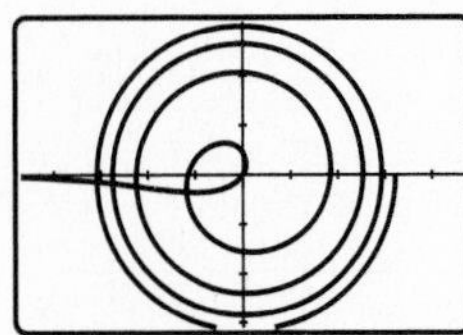

52. This graph is made up of two spirals, which are mirror images of each other through the y-axis: One is determined by $\theta > 0$, the other by $\theta < 0$. The exact appearance varies with the choice of θMin and θMax; shown is $-6\pi \leq \theta \leq 6\pi$ on the window $[-7.5, 7.5] \times [-2, 8]$; as $|\theta| \to \infty$, $r \to 0$.

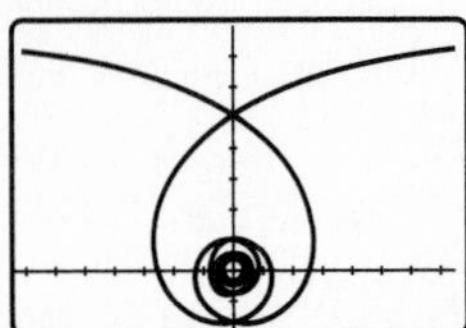

53. This graph (called a "lituus") is made up of two spirals: One is the graph of $r = \dfrac{10}{\sqrt{\theta}}$, and the other is $r = -\dfrac{10}{\sqrt{\theta}}$; the graph is symmetric about the pole; the exact appearance varies with the choice of θMin and θMax; shown is $0.01 \leq \theta \leq 12\pi$ on the window $[-13.5, 13.5] \times [-9, 9]$. Note that θ must be positive: As $\theta \to \infty$, $r \to 0$.

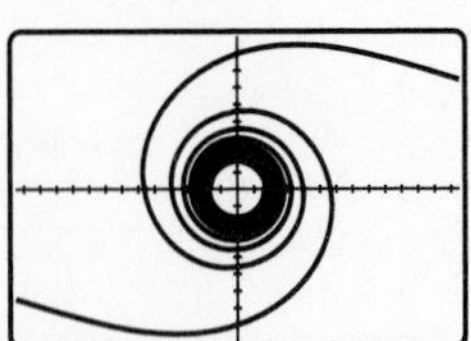

QUICK REVIEW 9.6

1. $(3, \theta) = (-3, \theta + \pi)$

2. $(-2, \theta) = (2, \theta + \pi)$

3. $\theta = -\frac{5\pi}{6}$ or $\theta = \frac{7\pi}{6}$

4. $\theta = -\frac{5\pi}{3}$ or $\theta = \frac{\pi}{3}$

5. Foci: $(\pm\sqrt{5}, 0)$; major axis endpoints: $(\pm 3, 0)$

6. Foci: $(0, \pm 4)$; major axis endpoints: $(0, \pm 5)$

7. Foci: $(\pm 5, 0)$; transverse axis endpoints: $(\pm 4, 0)$

8. Foci: $(0, \pm\sqrt{40}) = (0, \pm 2\sqrt{10})$; transverse axis endpoints: $(0, \pm 6)$

SECTION EXERCISES 9.6

1. The graph in part (b); $[-15, 5] \times [-10, 10]$

2. The graph in part (d); $[-5, 5] \times [-3, 3]$

3. The graph in part (f); $[-5, 5] \times [-3, 3]$

4. The graph in part (e); $[-5, 5] \times [-3, 5]$

5. The graph in part (c); $[-10, 10] \times [-5, 10]$

6. The graph in part (a); $[-3, 3] \times [-6, 6]$

7. $r = \dfrac{2}{1 - \cos\theta}$, a parabola; window: $[-10, 20] \times [-10, 10]$

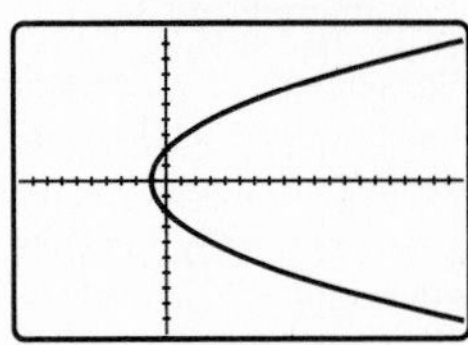

8. $r = \dfrac{5}{1 + (5/4)\cos\theta} = \dfrac{20}{4 + 5\cos\theta}$, a hyperbola; window: $[-20, 40] \times [-20, 20]$

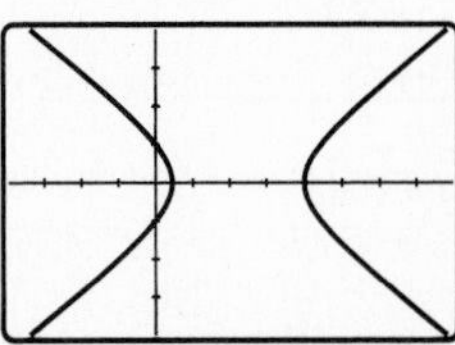

9. $r = \dfrac{12/5}{1 + (3/5)\sin\theta} = \dfrac{12}{5 + 3\sin\theta}$, an ellipse; window: $[-6, 6] \times [-6.3, 1.7]$

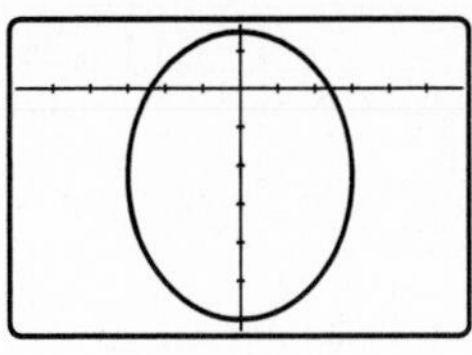

10. $r = \dfrac{2}{1 + \sin\theta}$, a parabola; window: $[-6, 6] \times [-6.3, 1.7]$

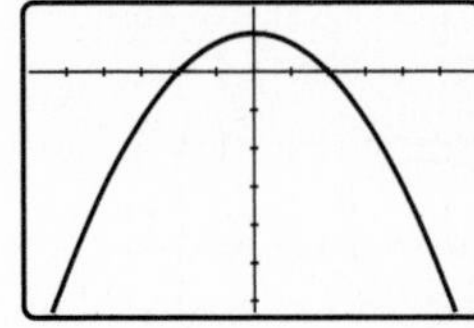

11. $r = \dfrac{7/3}{1 - (7/3)\sin\theta} = \dfrac{7}{3 - 7\sin\theta}$, a hyperbola; window: $[-6, 6] \times [-5.5, 2.5]$

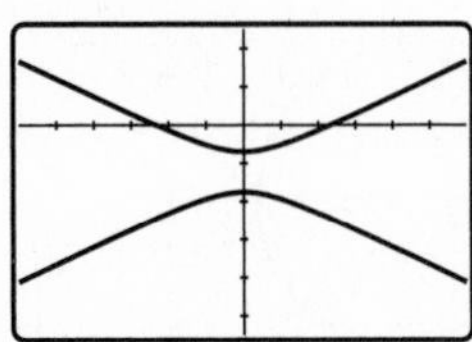

12. $r = \dfrac{10/3}{1 - (2/3)\cos\theta} = \dfrac{10}{3 - 2\cos\theta}$, an ellipse; window: $[-4, 11] \times [-5, 5]$

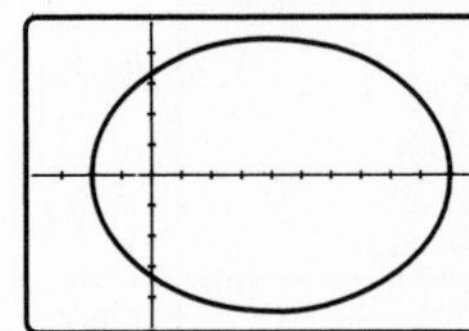

13. Parabola with $e = 1$ and directrix $x = 2$

14. Hyperbola with $e = 2$ and directrix $x = 3$

15. Parabola with $e = 1$ and directrix $y = -2.5$

16. Ellipse with $e = \frac{1}{4} = 0.25$ and directrix $x = -2$

17. Ellipse with $e = \frac{5}{6}$ and directrix $y = 4$

18. Hyperbola with $e = 3.5$ and directrix $y = -6$

19. Ellipse with $e = \frac{2}{5} = 0.4$ and directrix $x = 3$

20. Hyperbola with $e = 2.5$ and directrix $y = 4$

21. $r = \dfrac{2.4}{1 + (3/5)\cos\theta} = \dfrac{12}{5 + 3\cos\theta}$ ($e = \dfrac{3}{5} = 0.6$ and directrix $x = 4$)

22. $r = \dfrac{1.2}{1 - (1/5)\cos\theta} = \dfrac{6}{5 - \cos\theta}$ ($e = \frac{1}{5} = 0.2$ and directrix $x = -6$)

23. $r = \dfrac{1.5}{1 + (1/2)\sin\theta} = \dfrac{3}{2 + \sin\theta}$ ($e = \frac{1}{2} = 0.5$ and directrix $y = 3$)

24. $r = \dfrac{1.2}{1 - (3/5)\sin\theta} = \dfrac{6}{5 - 3\sin\theta}$ ($e = \frac{3}{5} = 0.6$ and directrix $y = -2$)

25. $r = \dfrac{7.5}{1 + (3/2)\cos\theta} = \dfrac{15}{2 + 3\cos\theta}$ ($e = \frac{3}{2} = 1.5$ and directrix $x = 5$)

26. $r = \dfrac{6}{1 - 3\cos\theta}$ ($e = 3$ and directrix $x = -2$)

27. $r = \dfrac{6}{1 + (3/2)\sin\theta} = \dfrac{12}{2 + 3\sin\theta}$ ($e = \frac{3}{2} = 1.5$ and directrix $y = 4$)

28. $r = \dfrac{6}{1 - 2\sin\theta}$ ($e = 2$ and directrix $y = -3$)

29. $r = \dfrac{1.2}{1 + (3/5)\cos\theta} = \dfrac{6}{5 + 3\cos\theta}$ ($e = \frac{3}{5} = 0.6$ and directrix $x = 2$)

30. $r = \dfrac{2}{1 + \sin\theta}$ ($e = 1$ and directrix $y = 2$)

31. Draw diagrams similar to those in Example 1. Let $P(r, \theta)$ be a point on the conic; if P and F are on the same side of the directrix (so that $x = r\cos\theta > -p$), assume that $r > 0$; then $PF = r$ and $PQ = p + x = p + r\cos\theta$. Since $e = \frac{PF}{PQ}$, $r = e(p + r\cos\theta)$ so that $r(1 - e\cos\theta) = ep$, and $r = \dfrac{ep}{1 - e\cos\theta}$.

If P and F are on opposite sides of the directrix, $x = r\cos\theta < -p < 0$; assume that $r < 0$ and $\cos\theta > 0$; then $PF = -r$ and $PQ = -x - p = -r\cos\theta - p$, and since $e = \frac{PF}{PQ}$, $-r = e(-r\cos\theta - p)$ so that $r = e(p + r\cos\theta)$ as before.

32. Draw diagrams similar to those in Example 1. Let $P(r, \theta)$ be a point on the conic; if P and F are on the same side of the directrix (so that $y = r\sin\theta < p$), assume that $r > 0$; then $PF = r$ and $PQ = p - y = p - r\sin\theta$. Since $e = \frac{PF}{PQ}$, $r = e(p - r\sin\theta)$ so that $r(1 + e\sin\theta) = ep$ and $r = \dfrac{ep}{1 + e\sin\theta}$.

If P and F are on opposite sides of the directrix, $y = r\sin\theta > p > 0$; assume that $r < 0$ and $\sin\theta < 0$; then $PF = -r$ and $PQ = y - p = r\sin\theta - p$, and since $e = \frac{PF}{PQ}$, $-r = e(r\sin\theta - p)$ so that $r = e(p - r\sin\theta)$ as before.

33. Let $P(r, \theta)$ be any point on the graph $r = \dfrac{ep}{1 - e\cos\theta}$; the distance PF equals $|r|$; PQ, the distance from P to the line $x = -p$, is the difference between the x-coordinates: $PQ = |r\cos\theta + p| = \left|\dfrac{ep\cos\theta}{1 - e\cos\theta} + p\right| = \left|\dfrac{ep\cos\theta + p(1 - e\cos\theta)}{1 - e\cos\theta}\right| = \left|\dfrac{p}{1 - e\cos\theta}\right| = \left|\dfrac{r}{e}\right|$; therefore the ratio $\dfrac{PF}{PQ} = \dfrac{|r|}{|r/e|} = e$, satisfying the polar definition of conics.

34. Let $P(r, \theta)$ be any point on the graph of $r = \dfrac{ep}{1 + e\sin\theta}$; the distance PF equals $|r|$; PQ, the distance from P to the line $y = p$, is the difference between the y-coordinates: $PQ = |r\sin\theta - p| = \left|\dfrac{ep\sin\theta}{1 + e\sin\theta} - p\right| = \left|\dfrac{ep\sin\theta - p(1 + e\sin\theta)}{1 + e\sin\theta}\right| = \left|\dfrac{-p}{1 + e\sin\theta}\right| = \left|\dfrac{r}{e}\right|$; therefore the ratio $\dfrac{PF}{PQ} = \dfrac{|r|}{|r/e|} = e$, satisfying the polar definition of conics.

35. $p \approx 102.36$, so the model for the comet's orbit is $r = \dfrac{99.288}{1 + 0.97\cos\theta}$; the perihelion, 50.4 million miles, occurs when $\theta = 0$; the aphelion, 3309.6 million miles, occurs when $\theta = \pi$.

36. $p \approx 163.87$, so the model for the planet's orbit is $r = \dfrac{34.4124}{1 + 0.21\cos\theta}$; the perihelion, 28.44 million miles, occurs when $\theta = 0$; the aphelion, 43.56 million miles, occurs when $\theta = \pi$.

37. **(a)** The total radius of the orbit is $r = 250 + 1740 = 1990$ km $= 1{,}990{,}000$ m; then $v \approx \sqrt{2{,}406{,}030} \approx 1551$ m/sec $= 1.551$ km/sec. **(b)** The circumference of one orbit is $2\pi r \approx 12503.5$ km; one orbit therefore takes about 8061 seconds, or about 2 hr 14 min.

38. The total radius of the orbit is $r = 1000 + 2100 = 3100$ miles; 1 mi is about 1.61 km, so $r \approx 4989$ km $= 4{,}989{,}000$ m; then $v \approx \sqrt{8{,}797{,}400} \approx 2966$ m/sec $= 2.966$ km/sec.

39. **(a)** Let $P(x, y)$ be a point on the ellipse; the distance to the focus $(c, 0)$ is $PF = \sqrt{(x - c)^2 + y^2} = \sqrt{x^2 - 2cx + c^2 + y^2}$; the horizontal distance from P to point $Q\left(\dfrac{a^2}{c}, y\right)$ on line L is $PQ = \dfrac{a^2}{c} - x$. To confirm that $\dfrac{PF}{PQ} = \dfrac{c}{a}$, cross-multiply to get $aPF = cPQ$; we need to confirm that $a\sqrt{x^2 - 2cx + c^2 + y^2} = a^2 - cx$. Square both sides: $a^2(x^2 - 2cx + c^2 + y^2) = a^4 - 2a^2cx + c^2x^2$; substitute $a^2 - b^2$ for c^2, multiply out both sides, and cancel out terms, leaving $a^2y^2 - a^2b^2 = -b^2x^2$; since P is on the ellipse, $\dfrac{x^2}{a^2} + \dfrac{y^2}{b^2} = 1$ or, equivalently,

$b^2x^2 + a^2y^2 = a^2b^2$; this confirms the equality. **(b)** According to the polar definition, the eccentricity is the ratio $\frac{PF}{PQ}$, which we found to be $\frac{c}{a}$ in (a).

(c) Since $e = \frac{c}{a}$, $\frac{a}{e} = \frac{a}{c/a} = \frac{a^2}{c}$ and $ae = c$; the distance from F to L is $\frac{a^2}{c} - c = \frac{a}{e} - ea$, as desired.

40. **(a)** Let $P(x, y)$ be a point on the hyperbola; the distance to the focus $(c, 0)$ is $PF = \sqrt{(x - c)^2 + y^2} = \sqrt{x^2 - 2cx + c^2 + y^2}$; the horizontal distance from P to point $Q\left(\frac{a^2}{c}, y\right)$ on line L is $PQ = \left|\frac{a^2}{c} - x\right|$. To confirm that $\frac{PF}{PQ} = \frac{c}{a}$, cross-multiply to get $aPF = cPQ$; we need to confirm that $a\sqrt{x^2 - 2cx + c^2 + y^2} = |a^2 - cx|$; square both sides: $a^2(x^2 - 2cx + c^2 + y^2) = a^4 - 2a^2cx + c^2x^2$; substitute $a^2 + b^2$ for c^2, multiply out both sides, and cancel out terms, leaving $a^2y^2 + a^2b^2 = b^2x^2$; since P is on the hyperbola, $\frac{x^2}{a^2} - \frac{y^2}{b^2} = 1$ or, equivalently, $b^2x^2 - a^2y^2 = a^2b^2$; this confirms the equality. **(b)** According to the polar definition, the eccentricity is the ratio $\frac{PF}{PQ}$, which we found to be $\frac{c}{a}$ in (a).

(c) Since $e = \frac{c}{a}$, $\frac{a}{e} = \frac{a}{c/a} = \frac{a^2}{c}$ and $ae = c$; the distance from F to L is $\frac{a^2}{c} - c = \frac{a}{e} - ea$, as desired.

41. If P and F are on opposite sides of the directrix and we use $(-r, \theta + \pi)$ for P, then $x = -r\cos(\theta + \pi) > p > 0$ with both $-r$ and $\cos(\theta + \pi)$ negative; since PF must be positive, $PF = |r| = r$; also $PQ = x - p = r\cos\theta - p$; since $e = \frac{PF}{PQ}$, $r = e(r\cos\theta - p)$ so that $r(1 - e\cos\theta) = -ep$, so $r = \frac{-ep}{1 - e\cos\theta}$. This is equivalent to $-r = \frac{ep}{1 + e\cos(\theta + \pi)}$, which shows that $P(-r, \theta + \pi)$ solves the correct equation.

Chapter 9 Review

1. The graph resembles $y = \sqrt{x + 1}$, but the endpoint $(-1, 0)$ is missing.

2. The parabola $x - 3 = 2(y + 1)^2$; no endpoints

3. A circle with center $(0, 0)$ and radius 3; the circle is traced twice, beginning and ending at $(3, 0)$.

4. The top half $(y \geq 0)$ of a circle with center $(0, 0)$ and radius 4, traced from left to right, beginning at $(-4, 0)$ and ending at $(4, 0)$

5.

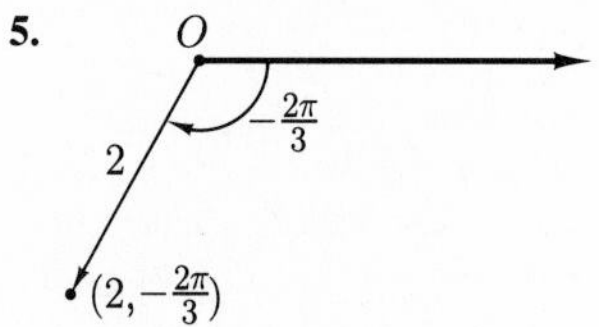

6.

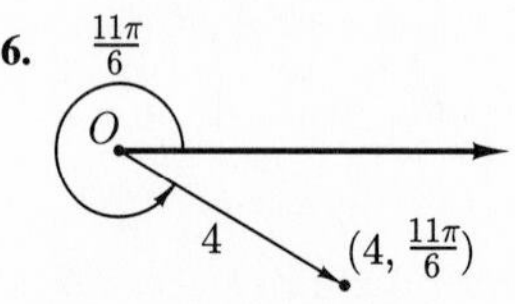

7.

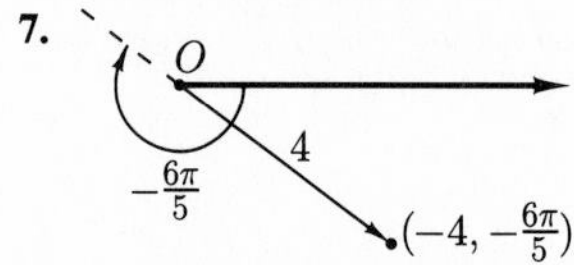

8.

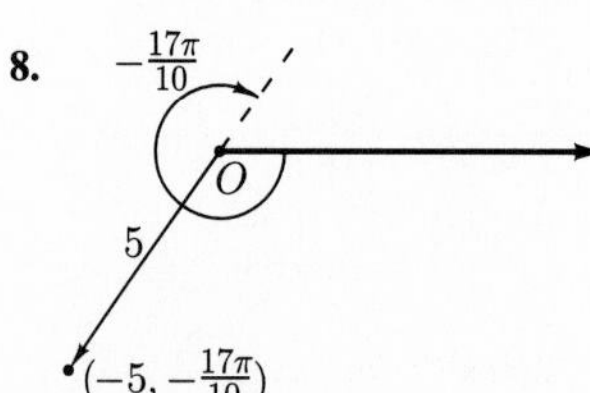

9. $(-2.5\cos 25°, -2.5\sin 25°) = (-2.265\ldots, -1.056\ldots)$

10. $(-3.1\cos 135°, -3.1\sin 135°) = (-1.55\sqrt{2}, -1.55\sqrt{2}) = (2.192\ldots, -2.192\ldots)$

11. $(2\cos(-\frac{\pi}{4}), 2\sin(-\frac{\pi}{4})) = (\sqrt{2}, -\sqrt{2})$

12. $(3.6\cos(\frac{3\pi}{4}), 3.6\sin(\frac{3\pi}{4})) = (-1.8\sqrt{2}, 1.8\sqrt{2}) = (-2.545\ldots, 2.545\ldots)$

13. $(1, -\frac{2\pi}{3} + (2n + 1)\pi)$ and $(-1, -\frac{2\pi}{3} + 2n\pi)$, n an integer

14. $(2, \frac{5\pi}{6} + (2n + 1)\pi)$ and $(-2, \frac{5\pi}{6} + 2n\pi)$, n an integer

15. **(a)** $(-\sqrt{13}, \pi + \tan^{-1}(-1.5)) = (-\sqrt{13}, 2.158\ldots)$ or $(\sqrt{13}, 2\pi + \tan^{-1}(-1.5)) = (\sqrt{13}, 5.300\ldots)$. **(b)** $(\sqrt{13}, \tan^{-1}(-1.5)) = (\sqrt{13}, -0.982\ldots)$ or $(-\sqrt{13}, \pi + \tan^{-1}(-1.5)) = (-\sqrt{13}, 2.158\ldots)$. **(c)** The answers from (a) and $(-\sqrt{13}, 3\pi + \tan^{-1}(-1.5)) = (-\sqrt{13}, 8.441\ldots)$ or $(\sqrt{13}, 4\pi + \tan^{-1}(-1.5)) = (\sqrt{13}, 11.583\ldots)$.

16. **(a)** $(-10, 0)$ or $(10, \pi)$ or $(-10, 2\pi)$. **(b)** $(10, -\pi)$ or $(-10, 0)$ or $(10, \pi)$. **(c)** The answers from (a) and $(10, 3\pi)$ or $(-10, 4\pi)$.

17. **(a)** $(5, 0)$ or $(-5, \pi)$ or $(5, 2\pi)$. **(b)** $(-5, -\pi)$ or $(5, 0)$ or $(-5, \pi)$. **(c)** The answers from **(a)** and $(-5, 3\pi)$ or $(5, 4\pi)$.

18. **(a)** $(-2, \frac{\pi}{2})$ or $(2, \frac{3\pi}{2})$. **(b)** $(2, -\frac{\pi}{2})$ or $(-2, \frac{\pi}{2})$. **(c)** The answers from (a) and $(-2, \frac{5\pi}{2})$ or $(2, \frac{7\pi}{2})$.

19. $3x + 5y = -11$; endpoints $A(13, -10)$ and $B(-12, 5)$

20. $5x + y = 12$, or $y = 12 - 5x$; endpoints $A(1, 7)$ and $B(9, -33)$

21. $x = 2\ln y^{1/4} = \frac{1}{2}\ln y$ or $y = e^{2x}$; endpoint $A(\ln 4, 16) = (1.386\ldots, 16)$; the parametric equation for x is undefined when $t = 0$.

22. $y = \ln x^{1/3} = \frac{1}{3}\ln x$; endpoint $A(8, \ln 2) = (8, 0.693\ldots)$; the parametric equation for y is undefined when $t = 0$.

23. For any point $P(x, y)$ on the parabola, the distance $PF = \sqrt{x^2 + (y - p)^2}$ must equal the distance from P to the directrix, which is $|y - (-p)| = |y + p|$; therefore $\sqrt{x^2 + (y - p)^2} = |y + p|$, so $x^2 + (y - p)^2 = (y + p)^2$. Expanding the squares gives $x^2 + y^2 - 2py + p^2 = y^2 + 2py + p^2$; the y^2 and p^2 terms cancel, leaving $x^2 - 2py = 2py$, or $x^2 = 4py$, so $y = \frac{1}{4p}x^2$.

24. The point $P(ay^2, y)$ is on the graph of $x = ay^2$; the distance from this point to $(\frac{1}{4a}, 0)$ is $\sqrt{\left(\frac{1}{4a} - ay^2\right)^2 + y^2} = \sqrt{\frac{1}{16a^2} - \frac{1}{2}y^2 + a^2y^4 + y^2} = \sqrt{\frac{1}{16a^2} + \frac{1}{2}y^2 + a^2y^4} = \sqrt{\left(\frac{1}{4a} + ay^2\right)^2} = |\frac{1}{4a} + ay^2|$. The distance from P to the line $x = -\frac{1}{4a}$ is the absolute value of the difference between the x-coordinates; $|-\frac{1}{4a} - ay^2| = |\frac{1}{4a} + ay^2|$, or the same.

25. $y = \dfrac{8x + 5 \pm \sqrt{-8x^2 + 200x - 455}}{12}$
Window: $[0, 25] \times [0, 20]$.

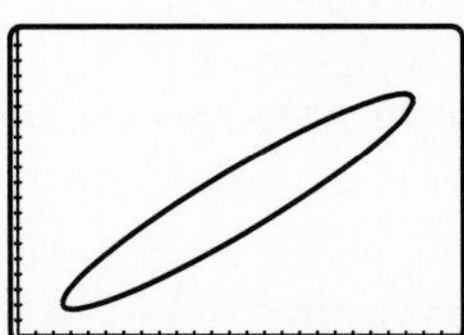

26. $y = \dfrac{8x + 5 \pm \sqrt{-176x^2 - 112x + 745}}{12}$
Window: $[-4.7, 4.7] \times [-3.1, 3.1]$.

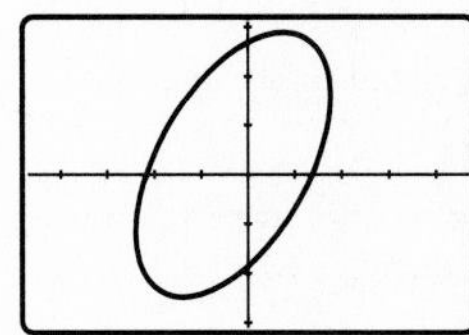

27. $y = \dfrac{3x^2 - 5x - 10}{2x - 6}$
Window: $[-8, 12] \times [-5, 15]$.

28. $y = \dfrac{5x - 17 \pm \sqrt{25x^2 + 70x + 769}}{12}$
Window: $[-15.2, 15.2] \times [-10, 10]$.

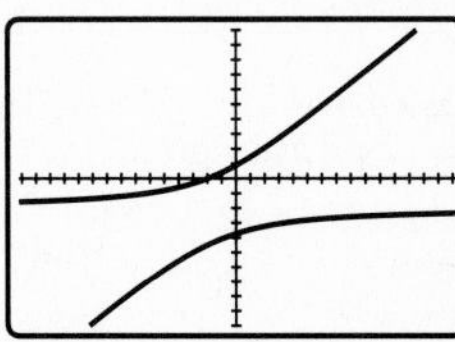

29. $y = \dfrac{7x + 20 \pm \sqrt{25x^2 + 272x + 280}}{4}$
Window: $[-22, 17] \times [-18, 8]$.

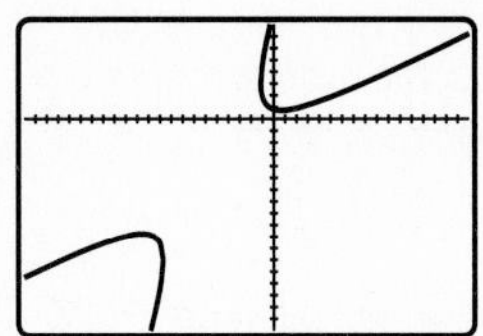

30. $y = \dfrac{7x + 3 \pm \sqrt{25x^2 + 26x - 71}}{4}$
Window: $[-20, 19] \times [-13, 13]$.

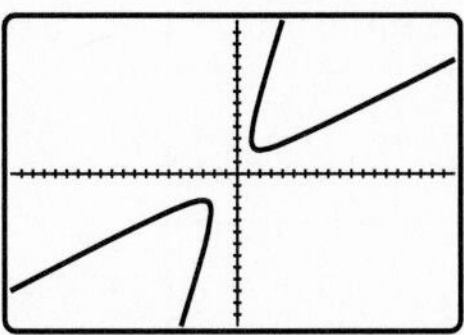

31. The graph in part (b)

32. Not shown

33. The graph in part (a)

34. Not shown

35. Not shown

36. The graph in part (d)

37. The graph in part (c)

38. Not shown

39. Ellipse; center (0, 0); foci: $(0, \pm\sqrt{3})$; major axis endpoints: $(0, \pm 2\sqrt{2})$; minor axis endpoints: $(\pm\sqrt{5}, 0)$. Window: $[-4.7, 4.7] \times [-3.1, 3.1]$.

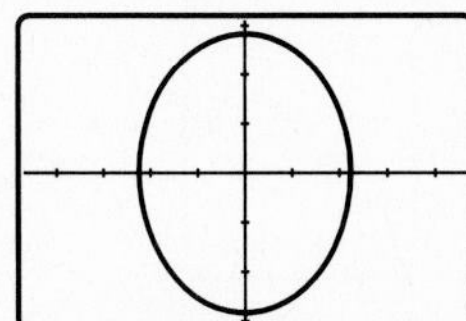

40. Hyperbola; center: (0, 0); foci: $(0, \pm\sqrt{65})$; transverse axis endpoints: $(0, \pm 4)$. Window: $[-18, 18] \times [-12, 12]$.

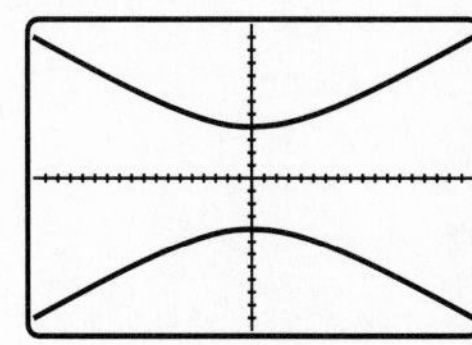

41. Hyperbola; center: $(-3, 5)$; foci: $(-3 \pm \sqrt{46}, 5)$; transverse axis endpoints: $(-3 \pm 3\sqrt{2}, 5)$. Window: $[-18, 12] \times [-10, 20]$.

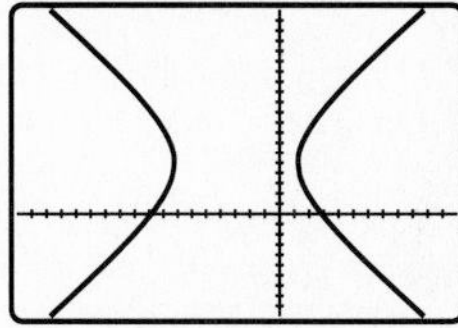

42. Hyperbola; center: (7, 3); foci: $(7, 3 \pm \sqrt{21})$; transverse axis endpoints: (7, 0) and (7, 6). Window: $[-8, 22] \times [-10, 16]$.

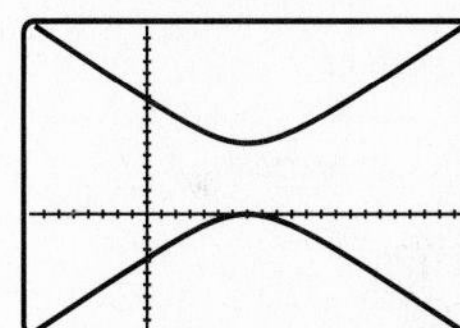

43. $\dfrac{(x - 2)^2}{4} + \dfrac{(y - 3)^2}{13} = 1$

44. $\dfrac{(x + 2)^2}{25} + \dfrac{(y + 3)^2}{16} = 1$

45. $\dfrac{(x - 1)^2}{4} + \dfrac{(y - 3)^2}{16} = 1$

46. $\dfrac{(x + 2)^2}{25} + \dfrac{(y + 1)^2}{9} = 1$

47. $\dfrac{x^2}{25} + \dfrac{y^2}{4} = 1$, an ellipse

48. $\dfrac{x^2}{16} + \dfrac{y^2}{36} = 1$, an ellipse

49. $(x + 2)^2 + (y - 4)^2 = 1$, an ellipse (a circle)

50. $\dfrac{(x-5)^2}{9}+\dfrac{(y+3)^2}{9}=1$, or $(x-5)^2+(y+3)^2=9$, which is an ellipse (a circle)

51. $\dfrac{x^2}{9}-\dfrac{y^2}{25}=1$, a hyperbola

52. $\dfrac{x^2}{16}-\dfrac{y^2}{9}=1$, a hyperbola

53. $x^2+y^2=4$, a circle with center (0, 0) and radius 2

54. $x^2+y^2=16$, a circle with center (0, 0) and radius 4

55. $y=x$, a line

56. $y=(-\sqrt{3})x$, a line

57. Multiply by r^2 to obtain $r^4 = 4(r\sin\theta)(r\cos\theta)$ so that $(x^2+y^2)^2 = 4xy$; this graph is called a lemniscate.

58. Multiply by r^2 to obtain $r^4 = 3(r^2\cos^2\theta - r^2\sin^2\theta)$ so that $(x^2+y^2)^2 = 3(x^2-y^2)$; this graph is called a lemniscate.

59. $xy=3$, a hyperbola

60. $x^2-y^2=9$, a hyperbola

61. $x-2y=5$, a line

62. $3x+y=2$, a line

63. $9x^2+4y^2=36$, or $\frac{x^2}{4}+\frac{y^2}{9}=1$, which is an ellipse

64. $9x^2+16y^2=144$, or $\frac{x^2}{16}+\frac{y^2}{9}=1$, which is an ellipse

65. Polar: $(2, 0), (4, \frac{\pi}{2}), (4, \frac{3\pi}{2}), (2, 2\pi)$; rectangular: $(2, 0), (0, 4), (0, -4)$

66. Polar: $(5, 0), (5, \pi), (2.5, \frac{3\pi}{2}), (5, 2\pi)$; rectangular: $(5, 0), (-5, 0), (0, -2.5)$

67. Polar: $(2, 0), (\frac{4}{3}, \frac{\pi}{2}), (1, \pi), (\frac{4}{3}, \frac{3\pi}{2})$, $(2, 2\pi)$; rectangular: $(2, 0), (0, \frac{4}{3})$, $(-1, 0), 0, -\frac{4}{3})$

68. Polar: $(0.75, 0), (0.6, \frac{\pi}{2}), (0.75, \pi)$, $(1, \frac{3\pi}{2}), (0.75, 2\pi)$; rectangular: $(0.75, 0), (0, 0.6), (-0.75, 0), (0, -1)$

69. Polar: $(17.5, 0), (-7, \frac{\pi}{2}), (17.5, \pi)$, $(\frac{35}{9}, \frac{3\pi}{2}), (17.5, 2\pi)$; rectangular: $(17.5, 0), (0, -7), (-17.5, 0)$, $(0, -35/9)$

70. Polar: $(\frac{15}{7}, 0), (7.5, \frac{\pi}{2}), (-5, \pi)$, $(7.5, \frac{3\pi}{2}), (\frac{15}{7}, 2\pi)$; rectangular: $(\frac{15}{7}, 0), (0, 7.5), (5, 0), (0, -7.5)$

71. $r = -\dfrac{4}{\sin\theta} = -4\csc\theta$
Window: $[-10, 10] \times [-10, 10]$.

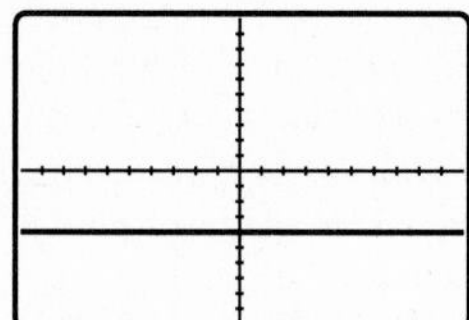

72. $r = \dfrac{3}{\sin\theta} = 3\csc\theta$
Window: $[-10, 10] \times [-10, 10]$.

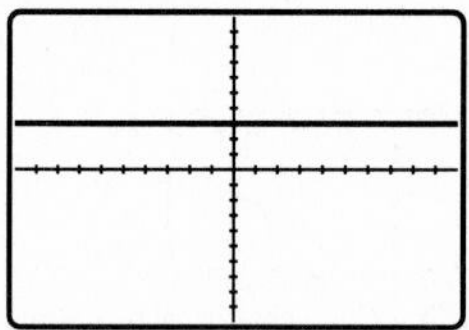

73. $r^2 = \dfrac{2}{\sin\theta\cos\theta} = 4\csc 2\theta$, so $r = \pm 2\sqrt{\csc 2\theta}$
Window: $[-9, 9] \times [-6, 6]$.

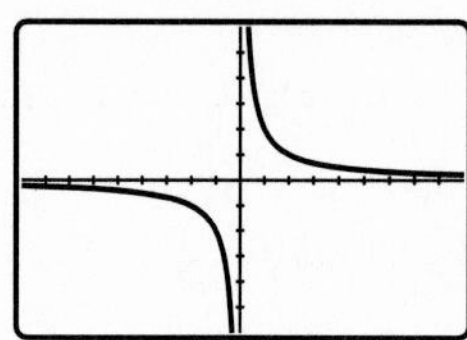

74. $r = -\dfrac{4}{\sin\theta\cos\theta} = -8\csc 2\theta$, so $r = \pm 2\sqrt{-2\csc 2\theta}$
Window: $[-9, 9] \times [-6, 6]$.

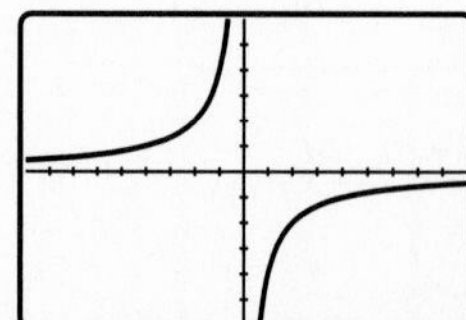

75. $r^2 = \dfrac{1}{1+\sin\theta\cos\theta}$, so $r = \dfrac{\pm 1}{\sqrt{1+\sin\theta\cos\theta}}$; the same graph is produced by both the (+) and (−) versions of this equation. Window: $[-3, 3] \times [-2, 2]$.

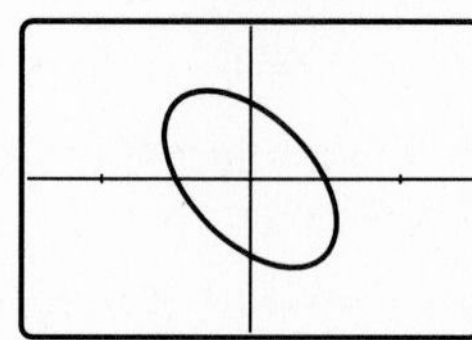

76. $r^2 = \dfrac{2}{1-\sin\theta\cos\theta} = \dfrac{4}{2-\sin 2\theta}$, so $r = \dfrac{\pm 2}{\sqrt{2-\sin 2\theta}}$; the same graph is produced by both the (+) and (−) versions of this equation. Window: $[-3, 3] \times [-2, 2]$.

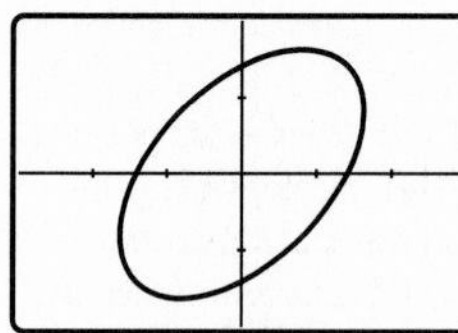

77. $r = 6\cos\theta - 2\sin\theta$. Window: $[-4, 11] \times [-6, 4]$.

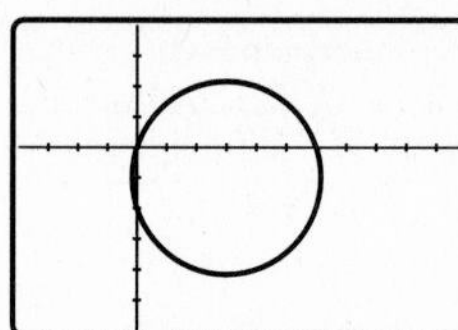

78. $r = -4\cos\theta + 2\sin\theta$. Window: $[-7, 3] \times [-2, 4]$.

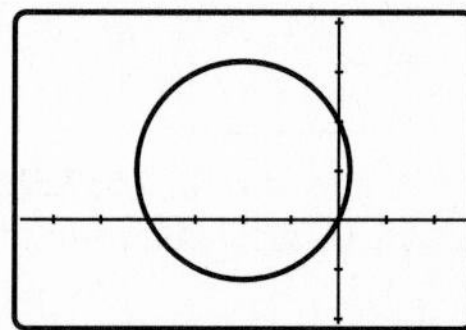

79. Center (1, 2.5); radius $\sqrt{7.25}$

80. Center (−0.5, 1.5); radius $\sqrt{2.5}$

81. Center (−1.5, −1); radius $\sqrt{3.25}$

82. Center (0.5, −1); radius $\sqrt{1.25}$

83. All are equivalent to $y = 3x - 1$. If t ranges over all real numbers, the graph for **(a)** has domain $(-\infty, \infty)$ and range $(-\infty, \infty)$; the graph for **(b)** has domain $(0, \infty)$ and range $(-1, \infty)$; the graph for **(c)** has domain $[-1, 1]$ and range $[-4, 2]$.

84. Since $\cos^2 t + \sin^2 t = 1, \dfrac{x^2}{a^2} + \dfrac{y^2}{b^2} = 1$; the center is $(0, 0)$, and the intercepts are $(\pm a, 0)$ and $(0, \pm b)$.

85. $x = 35 \cos\left(\frac{\pi}{10}t\right), y = 50 + 35 \sin\left(\frac{\pi}{10}t\right)$ if the wheel turns counterclockwise

86. $x = 40 \sin\left(\frac{2\pi}{15}t\right), y = 50 - 40 \cos\left(\frac{2\pi}{15}t\right)$ if the wheel turns counterclockwise

87. $x = -40 \sin\left(\frac{\pi}{9}t\right), y = 50 + 40 \cos\left(\frac{\pi}{9}t\right)$ if the wheel turns counterclockwise

88. $(0, 4.5)$

89. $(3.75, 0)$

90. **(a)** If the ball is closer to the pocket than the spot, the best strategy is to aim for the pocket. But if the ball is closer to the spot, the "shark" can aim for the spot rather than the hole; it is passes over the spot, the reflective properties tell us that it will bounce off the side and go to the pocket. **(b)** Let $a = 3$, $b = 2$, $c = \sqrt{5}$. If the center of the table is $(0, 0)$ then the foci are at $(-\sqrt{5}, 0)$ and $(\sqrt{5}, 0)$. If the pocket is at $(-\sqrt{5}, 0)$, the ball should be aimed at $(\sqrt{5}, 0)$.

91. $P_1P_t =$
$\sqrt{\{a - [tc + (1 - t)a]\}^2 + \{b - [td + (1 - t)b]\}^2} =$
$\sqrt{(-tc + ta)^2 + (-td + tb)^2} =$
$\sqrt{t^2(a - c)^2 + t^2(b - d)^2} =$
$t\sqrt{(a - c)^2 + (b - d)^2} = tP_1P_2$,
as desired

92. **(a)** Shown on the window $[-7.5, 7.5] \times [-5, 5]$. **(b)** All 4's should be changed to 5's.

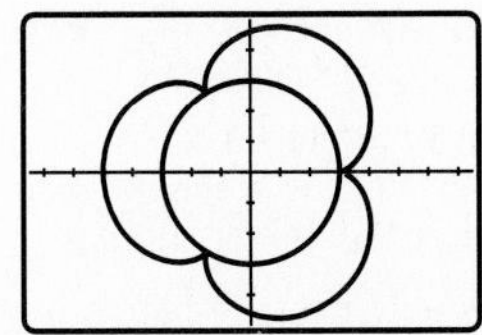

93. $d = \sqrt{5^2 + 6^2 - 2(5)(6)\cos 30°} = \sqrt{61 - 30\sqrt{3}} = 3.006\ldots$ mi

94. $d = \sqrt{7^2 + 9^2 - 2(7)(9)\cos 20°} = \sqrt{11.598\ldots} = 3.405\ldots$ mi

95. One petal is traced out in $-\frac{\pi}{2n} \le \theta \le \frac{\pi}{2n}$, and another in $\frac{\pi}{2n} \le \theta \le \frac{3\pi}{2n}$, and so forth. n petals are traced out in the interval $-\frac{\pi}{2n} \le \theta \le \frac{(2n-1)\pi}{2n}$ of length π. In the next interval of length π the n petals are retraced. We have n petals altogether.

96. The equation $r = a \csc \theta = \frac{a}{\sin \theta}$ is valid whenever $\sin \theta \neq 0$; in such cases we can multiply both sides by $\sin \theta$ and get $r \sin \theta = a$. The left side (in rectangular coordinates) is y; the graph of $y = a$ is a horizontal line. (If $a = 0$, we have the equation $r = 0$, the graph of which is simply the origin.)

97. A parabola, symmetric about the y-axis; replacing θ with $\pi - \theta$ produces the same equation. Window: $[-7.5, 7.5] \times [-5, 5]$.

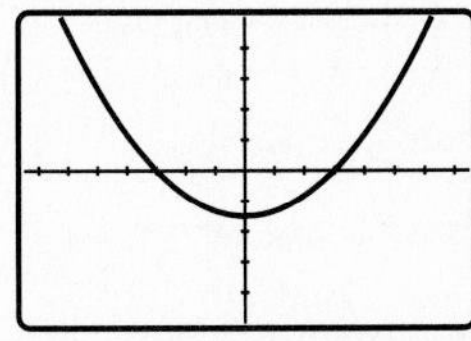

98. A parabola, symmetric about the polar (x-) axis; replacing θ with $-\theta$ produces the same equation. Window: $[-7.5, 7.5] \times [-5, 5]$.

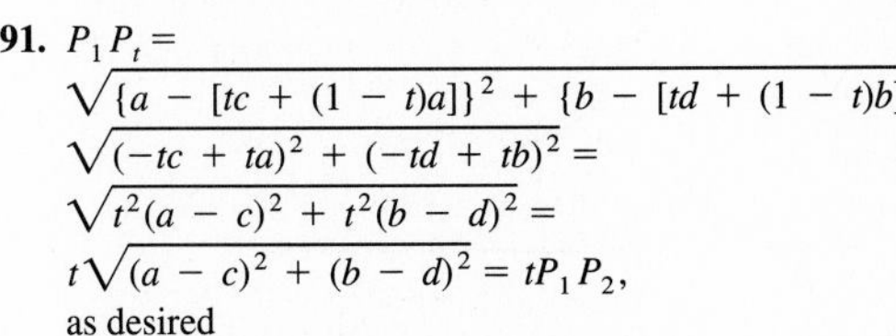

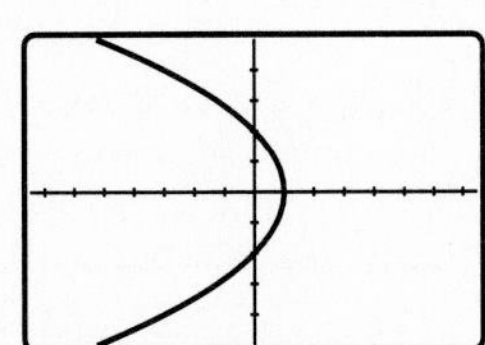

99. Symmetric about the pole; replacing θ with $\theta + \pi$ produces the same equation. Window: $[-4.7, 4.7] \times [-3.1, 3.1]$.

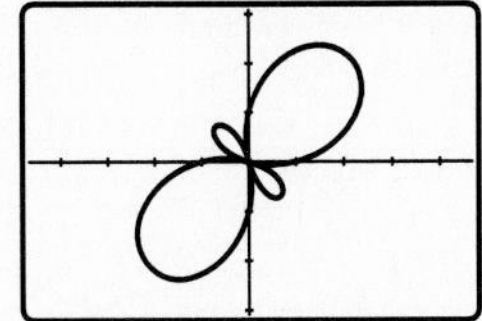

100. Symmetric about the pole, the polar (x-) axis, and the y-axis; replacing θ with $\theta + \pi$, $-\theta$, or with $\pi - \theta$, produces the same equation. Window: $[-6.5, 6.5] \times [-4.5, 4.5]$.

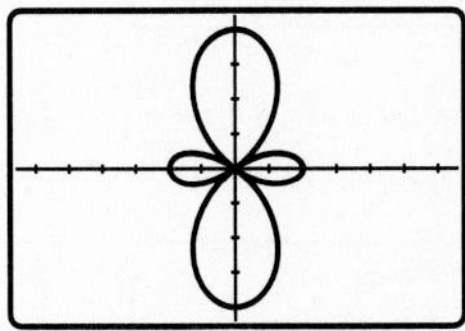

101. Symmetric about the polar (x-) axis, y-axis, and pole; replacing (r, θ) with $(r, -\theta)$, $(-r, -\theta)$, or $(-r, \theta)$ produces the same equation. Window: $[-3, 3] \times [-2, 2]$.

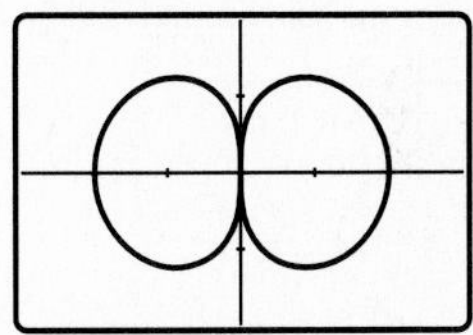

102. A four-petal rose, with all three symmetries; replacing (r, θ) with $(-r, \pi - \theta)$, $(-r, -\theta)$, or $(r, \theta + \pi)$ produces the same equation. Window: $[-4.7, 4.7] \times [-3.1, 3.1]$.

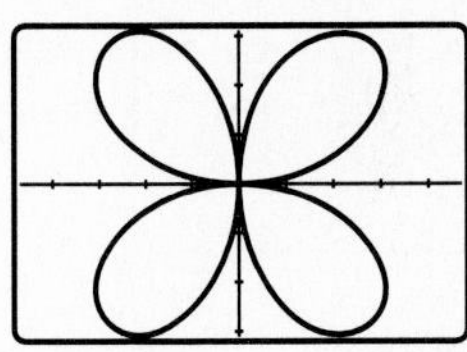

103. The graph of $y = x^3$, symmetric about the pole; replacing (r, θ) with $(-r, \theta)$ produces the same equation. Window: $[-4.7, 4.7] \times [-3.1, 3.1]$.

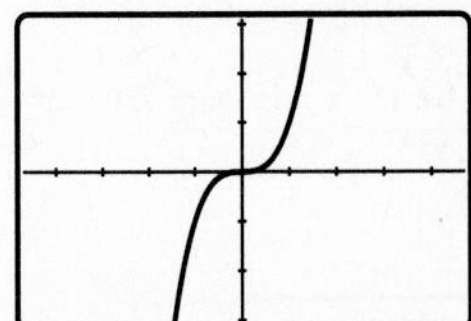

104. The graph of $y = \sqrt[3]{x}$, symmetric about the pole; replacing (r, θ) with $(-r, \theta)$ produces the same equation. Window: $[-4.7, 4.7] \times [-3.1, 3.1]$.

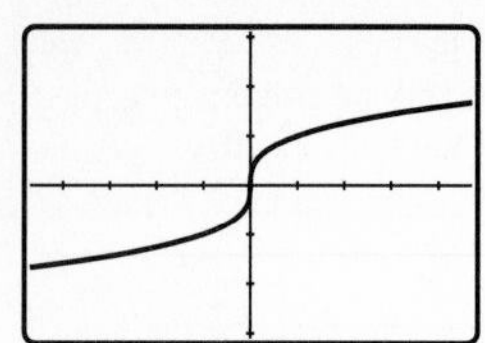

105. Shown on $[-9, 9] \times [-6, 6]$; $0 \le \theta \le 2\pi$ (or any interval of length 2π)

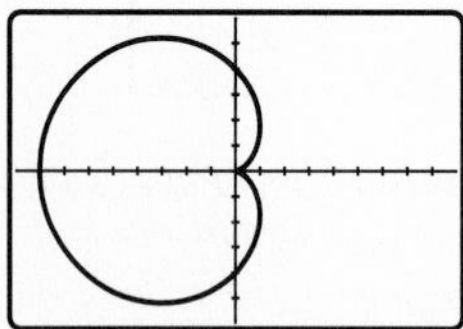

106. Shown on $[-9, 9] \times [-6, 6]$; $0 \le \theta \le 2\pi$ (or any interval of length 2π)

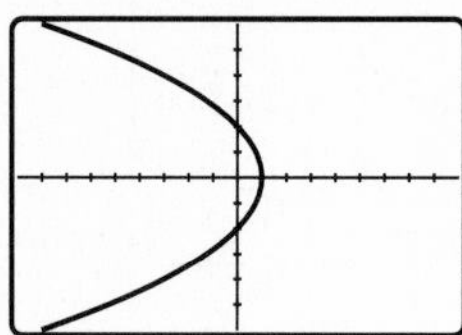

107. Shown on $[-6, 6] \times [-4, 4]$; $0 \le \theta \le \pi$ (or any interval of length π)

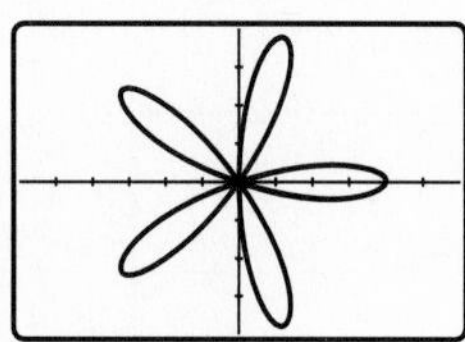

108. Shown on $[-3, 3] \times [-1.9, 2.1]$; $0 \le \theta \le \pi$ (or any interval of length π)

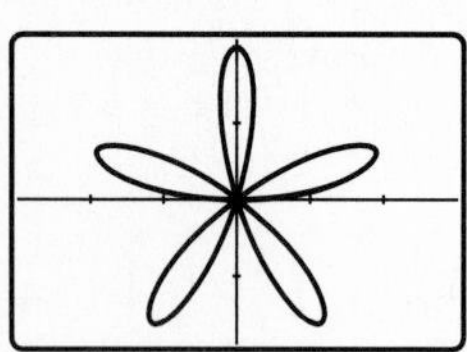

109. Shown on $[-9, 9] \times [-6, 6]$; $0 \le \theta \le 2\pi$ (or any interval of length 2π)

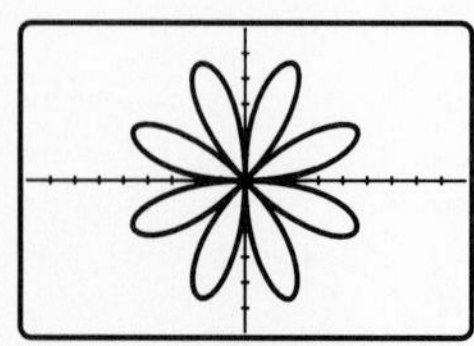

110. Shown on $[-4.7, 4.7] \times [-3.1, 3.1]$; $0 \le \theta \le 2\pi$ (or any interval of length 2π)

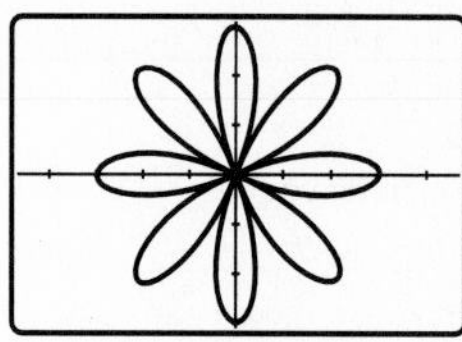

111. Shown on $[-4.7, 4.7] \times [-3.1, 3.1]$; $0 \le \theta \le 4\pi$ (or any interval of length 4π)

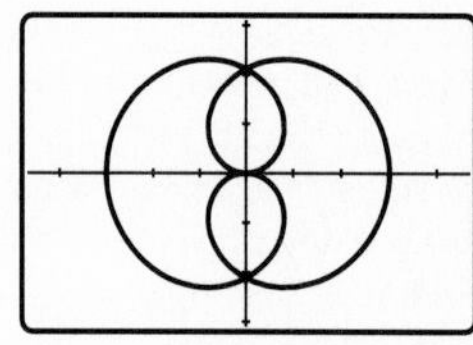

112. Shown on $[-4.7, 4.7] \times [-3.1, 3.1]$; $0 \le \theta \le 4\pi$ (or any interval of length 4π)

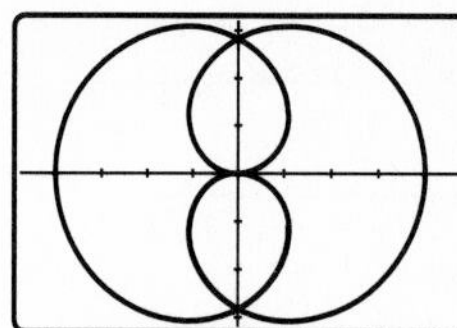

113. About 0.71 sec; about 46.75 ft

114. About 1.06 sec; about 68.65 ft

115. Yes. The ball has traveled 40 yd (120 ft) horizontally after about 2.42 sec, at which point it is about 25.96 ft above the ground.

116. No. The ball has traveled 40 yd (120 ft) horizontally after about 1.98 sec, at which point it is only about 6.59 ft above the ground.

117. Assume that the initial height is 0 ft. **(a)** about 77.59 ft. **(b)** about 4.40 sec.

118. Just over 125 ft/sec

119. Minimum distance: about 17.65 ft, after about 2.64 sec

120. No. The dart lands about 15.61 ft away.

121. No. The dart lands about 15.52 ft away.

122. **(a)** One mile is about 1.61 km, so 500 mi are about 805 km, or 805,000 m. The total radius of the orbit is $r = 805 + 6380 = 7185$ km, or 7,185,000 m; then $v \approx 7452$ m/sec $= 7.452$ km/sec. **(b)** About 6058 sec or 1 h and 41 min.

123. The major axis's length is 18,000 km, plus 170 km, plus the diameter of the earth, so $a \approx 15{,}465$ km $= 15{,}465{,}000$ m. At apogee, $r = 18{,}000 + 6380 = 24{,}380$ km, so $v \approx 2633$ m/sec; at perigee, $r = 6380 + 170 = 6550$ km, so $v \approx 9800$ m/sec.

124. Let $d_1 = \sqrt{x^2 + (y - c)^2}$ (the distance to one focus) and $d_2 = \sqrt{x^2 + (y + c)^2}$ (the distance to the other focus); for an ellipse, we have $|d_1 + d_2| = 2b$. Note for future reference that $(d_1 d_2)^2 = [x^2 + (y - c)^2] \cdot [x^2 + (y + c)^2] = (x^2 + c^2 + y^2 - 2cy)(x^2 + c^2 + y^2 + 2cy) = (x^2 + y^2 + c^2)^2 - 4c^2y^2 = (x^2 + y^2)^2 + c^4 + 2c^2(x^2 + y^2) - 4c^2y^2 = (x^2 + y^2)^2 + c^4 + 2c^2(x^2 - y^2)$. Also, $d_1^2 + d_2^2 = x^2 + (y - c)^2 + x^2 + (y + c)^2 = x^2 + y^2 - 2cy + c^2 + x^2 + y^2 + 2cy + c^2 = 2(x^2 + y^2 + c^2)$.

Square both sides of $d_1 + d_2 = 2b$ to get $d_1^2 + 2d_1d_2 + d_2^2 = 4b^2$. This simplifies to $2(x^2 + y^2 + c^2) + 2d_1d_2 = 4b^2$ so that $d_1d_2 = 2b^2 - (x^2 + y^2 + c^2) = 2b^2 - c^2 - x^2 - y^2 = a^2 + b^2 - (x^2 + y^2)$; squaring both sides again gives $(x^2 + y^2)^2 + c^4 + 2c^2(x^2 - y^2)$ on the left and $(x^2 + y^2 - a^2 - b^2)^2 = (x^2 + y^2)^2 + (a^2 + b^2)^2 - 2(x^2 + y^2)(a^2 + b^2)$ on the right; then $(x^2 + y^2)^2 + c^4 + 2c^2(x^2 - y^2) = (x^2 + y^2)^2 + (a^2 + b^2)^2 - 2(x^2 + y^2)(a^2 + b^2)$, so $c^4 + 2c^2(x^2 - y^2) = (a^2 + b^2)^2 - 2(x^2 + y^2)(a^2 + b^2)$; collect all the x and y terms on the left and the others on the right; $2(x^2 + y^2)(a^2 + b^2) + 2c^2(x^2 - y^2) = (a^2 + b^2)^2 - c^4$ so that $2(a^2 + b^2 + c^2)x^2 + 2(a^2 + b^2 - c^2)y^2 = (a^2 + b^2)^2 - c^4$; now substitute b^2 for $c^2 + a^2$ and a^2 for $b^2 - c^2$ on the left, and replace c^4 with $(b^2 - a^2)^2$; this leaves $4b^2x^2 + 4a^2y^2 = (a^2 + b^2)^2 - (b^2 - a^2)^2 = 4a^2b^2$; dividing by $4a^2b^2$ finishes the task.

125. Let $d_1 = \sqrt{x^2 + (y - c)^2}$ (the distance to one focus) and $d_2 =$

$\sqrt{x^2 + (y + c)^2}$ (the distance to the other focus); for a hyperbola, we have $|d_1 - d_2| = 2b$. Note for future reference that $(d_1 d_2)^2 = [x^2 + (y - c)^2] = [x^2 + (y + c)^2] = (x^2 + c^2 + y^2 - 2cy)(x^2 + c^2 + y^2 + 2cy) = (x^2 + y^2 + c^2)^2 - 4c^2y^2 = (x^2 + y^2)^2 + c^4 + 2c^2(x^2 + y^2) - 4c^2y^2 = (x^2 + y^2)^2 + c^4 + 2c^2(x^2 - y^2)$. Also, $d_1^2 + d_2^2 = x^2 + (y - c)^2 + x^2 + (y + c)^2 = x^2 + y^2 - 2cy + c^2 + x^2 + y^2 + 2cy + c^2 = 2(x^2 + y^2 + c^2)$.

Square both sides of $|d_1 - d_2| = 2b$ to get $d_1^2 - 2d_1d_2 + d_2^2 = 4b^2$; this simplifies to $2(x^2 + y^2 + c^2) - 2d_1d_2 = 4b^2$ so that $d_1d_2 = x^2 + y^2 + c^2 - 2b^2 = x^2 + y^2 - (2b^2 - c^2) = x^2 + y^2 - (b^2 - a^2)$; squaring both sides again gives $(x^2 + y^2)^2 + c^4 + 2c^2(x^2 - y^2)$ on the left and $(x^2 + y^2 - b^2 + a^2)^2 = (x^2 + y^2)^2 + (b^2 - a^2)^2 - 2(x^2 + y^2)(b^2 - a^2)$ on the right; then $(x^2 + y^2)^2 + c^4 + 2c^2(x^2 - y^2) = (x^2 + y^2)^2 + (b^2 - a^2)^2 - 2(x^2 + y^2)(b^2 - a^2)$, so $c^4 + 2c^2(x^2 - y^2) = (b^2 - a^2)^2 - 2(x^2 + y^2)(b^2 - a^2)$; collect all the x and y terms on the left and the others on the right, $2(x^2 + y^2)(b^2 - a^2) + 2c^2(x^2 - y^2) = (b^2 - a^2)^2 - c^4$ so that $2(b^2 - a^2 + c^2)x^2 + 2(b^2 - a^2 - c^2)y^2 = (b^2 - a^2)^2 - c^4$; now substitute b^2 for $c^2 - a^2$ and $-a^2$ for $b^2 - c^2$ on the left, and replace c^4 with $(a^2 + b^2)^2$, which leaves $4b^2x^2 - 4a^2y^2 = (a^2 - b^2)^2 - (a^2 + b^2)^2 = -4a^2b^2$; dividing by $-4a^2b^2$ finishes the task.

CHAPTER 10

QUICK REVIEW 10.1

1. **(a)** Yes **(b)** No
2. **(a)** No **(b)** Yes
3. $y = \dfrac{5 - 2x}{3}$
4. $y = \dfrac{4 - x}{x} = \dfrac{4}{x} - 1$
5. $\dfrac{1 \pm \sqrt{7}}{3}$
6. $x = \dfrac{-5 \pm \sqrt{105}}{4}$
7. $x = \frac{3}{2}$
8. $x = -\frac{4}{7}$
9. No solution
10. All real x

SECTION EXERCISES 10.1

1. **(a)** No **(b)** Yes **(c)** No
2. **(a)** Yes **(b)** No **(c)** Yes
3. $(9, -2)$
4. $(3, -17)$
5. $(\frac{50}{7}, -\frac{10}{7})$
6. $(-\frac{23}{5}, \frac{23}{5})$
7. $(-\frac{1}{2}, 2)$
8. $(-3, 2)$
9. No solution
10. Infinitely many solutions
11. $(3, 9), (-3, 9)$
12. $(0, -3)$ or $(4, 1)$
13. $(8, -2)$
14. $(3, 4)$
15. $(4, 2)$
16. $(-2, 3)$
17. Infinitely many solutions
18. No solution
19. $(0, 1), (3, -2)$
20. $(1.5, 1)$
21. No solution
22. $(0, -4)$ or $(\pm\sqrt{7}, 3)$
23. One
24. No solution
25. Infinitely many solutions
26. One solution
27. $(0.69, -0.37)$
28. $(1.13, 1.27)$
29. $(-2.32, -3.16); (0.47, -1.77); (1.85, -1.08)$
30. $(-0.70, -2.40)$ or $(5.70, 10.40)$
31. $(2.92, -0.46); (-1.48, 1.74)$
32. $(-1.48, -1.74)$ or $(2.92, 0.46)$
33. $(-3.10, 3.55); (2.10, 0.95)$
34. $(-3.10, -3.55)$ or $(2.10, -0.95)$
35. $(-1.04, -0.86); (1.04, -0.86); (-1.37, 0.73); (1.37, 0.73)$
36. $(-2.20, \pm 1.36)$ or $(2.20, \pm 1.36)$
37. $(\frac{1}{3}, \frac{2}{3}); (-\frac{3}{2}, \frac{27}{2})$
38. $(-4, 28)$ or $(2.5, 15)$
39. $(0, 0); (3, 18)$
40. $(0, 0)$ or $(-2, 4)$
41. $x = -0.1 \pm 3\sqrt{0.89}$, $y = 0.3 \pm \sqrt{0.89}$
42. $x = \dfrac{52 \pm 7\sqrt{871}}{65}$, $y = \dfrac{91 \mp 4\sqrt{871}}{65}$
43. $(3.75, 143.75)$
44. $x = 130, p = 5.9$
45. **(a)** Window: $[-1, 5] \times [0, 150]$.

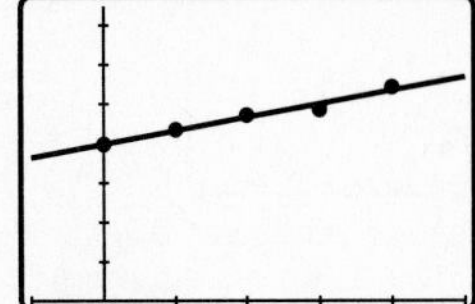

(b) Window: $[-1, 5] \times [0, 150]$.

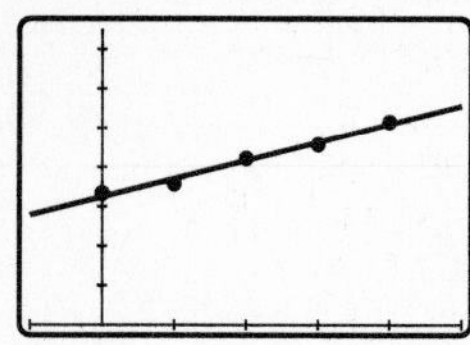

(c) Revenues: \$134.18 (billion); payments: \$138.78 (billion)

46. At the end of the year 2006 ($x = 16$), the model predicts that about \$4.9 billion is left in the surplus fund (ignoring any interest that might be earned on the fund); the next year (2007), the fund is depleted, and the year ends with a deficit of about \$21.57 billion.

47. 5.28 m × 94.72 m

48. 50 yd × 60 yd

49. Rowing speed: 3.56 mph; current speed: 1.06 mph

50. Airplane speed: $617.\overline{42}$ mph; wind speed: $49.\overline{24}$ mph

51. Medium: \$0.79; Large: \$0.95

52. $\frac{175}{57} \approx 3.07$ lb of peanuts and $\frac{110}{57} \approx 1.93$ lb of cashews

53. $a = \frac{2}{3}, b = \frac{14}{3}$

54. $a = 2.5, b = -3$

55. **(a)** $y = 3.59x^2 - 25.81x + 191.64$. Window: $[0, 35] \times [0, 5000]$.

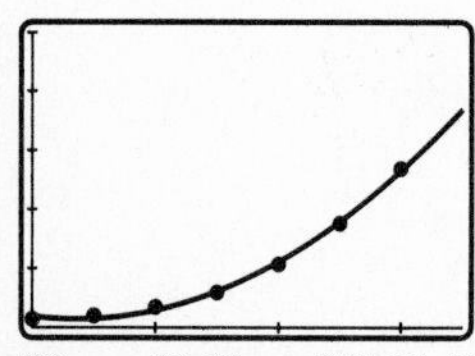

(b) $y = 81.89x - 257.11$. Window: $[0, 35] \times [0, 5000]$.

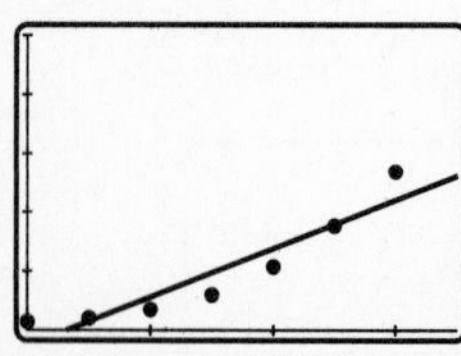

(c) Quadratic: \$3250; linear: \$2445.

(d) Answers will vary; one example: the growth in expenditures is not linear.

56. **(a)** $y \approx 534.92x - 4736.45$; Window $[19, 31] \times [5500, 12{,}000]$.

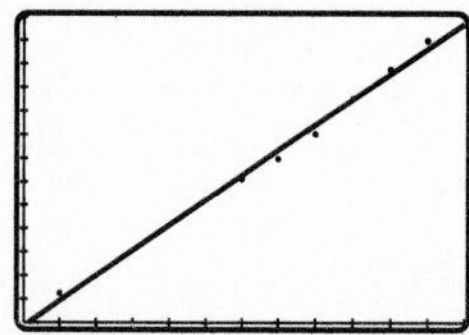

(b) Using the quadratic model from Exercise 55: When $x \approx 13.5$ (1973–74) and $x \approx 23.5$ (1983–84). Using the linear model from Exercise 55: When $x \approx 27.5$ (1987–88).

57. Zero, one, or two solutions

58. $x = \pm\sqrt{\frac{2}{3}}, y = \frac{10}{3}$

59. $(\pm 1, 0)$

60. 12.5 units

61. 5.16 units

QUICK REVIEW 10.2

1. 12.8 L

2. 8.4 milliliters

3. 38 L

4. 24 milliliters

5. (a) $(-1, 6)$

6. (a) $(0, -1)$

7. $x = 2 + z$

8. $y = 2w + 3$

9. $y = 1 + w - z$

10. $x = 2z - w + 3$

SECTION EXERCISES 10.2

1. (b)

2. None are solutions.

3. None are solutions.

4. All three are solutions.

5. $(1, 2, 1)$

6. $(\frac{5}{13}, \frac{10}{13}, \frac{74}{13})$

7. $(4.5, 3.5, 4, -7.5)$

8. $(2, 1, 0, -1)$

9. $(0, -10, 1)$

10. $(\frac{34}{71}, \frac{352}{71}, \frac{199}{142}) = (0.478\ldots, 4.957\ldots, 1.401\ldots)$

11. $(3, 3, -2, 0)$

12. $(-1, 2, 4, -1)$

13. $(2 - 1.5z, -4 - 0.5z, z)$

14. $(0.2z - 1, 0.6z - 2, z)$

15. $(-2w - 1, w + 1, -w, w)$

16. $(w, 1 - 2w, -w - 1, w)$

17. $(-2 - w, 0.5 - z, z, w)$

18. $(z - 3w + 1, 2w - 2z + 4, z, w)$

19. No solution

20. $(1 - 2w, w + 1, -2, w)$

21. 825 children, 410 adults, and 165 seniors

22. $\frac{160}{11} = 14.\overline{54}$ g of the 22% alloy; $\frac{320}{11} = 29.\overline{09}$ g of the 30% alloy; $\frac{400}{11} = 36.\overline{36}$ g of the 42% alloy

23. \$14,500 in CDs, \$5500 in bonds, and \$60,000 in growth fund

24. It cannot be determined: If z dollars are invested at 10% ($9000 \le z \le 14500$), then $z - 9000$ dollars invested at 6% and $29000 - 2z$ invested at 8% satisfy all conditions.

25. \$0 in CDs, \$38,983.05 in bonds, and \$11,016.95 in growth fund

26. 0 liters of 10%, 28.8 liters of 25%, and 11.2 liters of 50%

27. $f(x) = 2x^2 - 3x - 2$

28. $f(x) = 3x^3 - x^2 + 2x - 5$

29. $f(x) = (-c - 3)x^2 + x + c$, if $c = 3$, then $f(x)$ is a linear function.

30. $f(x) = ax^3 - x^2 + (4 - a)x - 1$, for any real number a

31. No. If there is exactly one solution, the triangular form resulting from Gaussian elimination would require one equation for each of the four variables, a total of four equations. This is impossible because there are only three equations.

32. A line can intersect a hyperbola (or a parabola or an ellipse) at 0, 1, or 2 points: If we solve the linear equation for one variable in terms of the other, and substitute into the quadratic (hyperbola) equation, the result is a quadratic equation in one variable, which has no more than two solutions.

33. $(2, -3)$

34. $(-3, -1)$

35. No solution

36. $(4, -1)$

37. Zero, one, or infinitely many solutions

38. A linear equation in three variables defines a plane in space; two linear equations may define the same plane, or two planes that intersect in a line, or two distinct, nonintersecting parallel planes; the intersection with the third plane will leave either a point, a line, a plane, or nothing.

39. Each vector's representative in standard position has its terminal point on a line through $(5, -2, 0)$ parallel to the vector $(-3, 2, 1)$

QUICK REVIEW 10.3

1. $-3x + 6y - 3w + 15$

2. $2x + y - w - 2$

3. $x + 8y - 2z - 1$

4. $10y - 3z - 10$

5. $\dfrac{3x - 5}{(x - 1)(x - 3)} = \dfrac{3x - 5}{x^2 - 4x + 3}$

6. $\dfrac{3x - 3}{(x + 4)(x + 1)} = \dfrac{3(x - 1)}{x^2 + 5x + 4}$

7. $\dfrac{4x^2 + 6x + 1}{x(x + 1)^2} = \dfrac{4x^2 + 6x + 1}{x^3 + 2x^2 + x}$

8. $\dfrac{3x^2 - x + 2}{(x^2 + 1)^2}$

9. $A = 3, B = -1, C = 1$

10. $A = -2, B = 2, C = -1, D = -5$

SECTION EXERCISES 10.3

1. 2×3; not square

2. 2×2; square

3. 3×2; not square

4. 1×3; not square

5. 3×1; not square

6. 1×1; square

7. $a_{13} = 3$

8. $a_{24} = -1$

9. $a_{32} = 4$

10. $a_{33} = -1$

11. R_{12}

12. $2R_2 + R_1$

13. $(-3)R_2 + R_3$

14. $\frac{1}{4}R_3$

15. $\begin{pmatrix} 2 & -6 & 4 \\ 1 & 2 & -3 \\ 0 & -8 & 4 \end{pmatrix}$

16. $\begin{pmatrix} 1 & -3 & 2 \\ 1 & 2 & -3 \\ -3 & 1 & -2 \end{pmatrix}$

17. $\begin{pmatrix} 0 & -10 & 10 \\ 1 & 2 & -3 \\ -3 & 1 & -2 \end{pmatrix}$

18. $\begin{pmatrix} 2 & -6 & 4 \\ 3 & -4 & 1 \\ -3 & 1 & -2 \end{pmatrix}$

19. (a)

20. (c)

21. (b)

22. (a)

23. Answers will vary; one example is $\begin{pmatrix} 1 & 3 & -1 \\ 0 & 1 & -1.2 \\ 0 & 0 & 1 \end{pmatrix}$

24. Answers will vary; one example is $\begin{pmatrix} 1 & 2 & -3 \\ 0 & 0 & 1 \\ 0 & 0 & 0 \end{pmatrix}$

25. Answers will vary; one example is $\begin{pmatrix} 1 & 2 & 3 & -4 \\ 0 & 1 & 0 & -0.6 \\ 0 & 0 & 1 & -9.2 \end{pmatrix}$

26. Anwers will vary; one example is $\begin{pmatrix} 1 & 2 & 3 & -2 \\ 0 & 1 & -1 & 1 \end{pmatrix}$

27. $\begin{pmatrix} 1 & 0 & 2 & 1 \\ 0 & 1 & -1 & 2 \\ 0 & 0 & 0 & 0 \end{pmatrix}$

28. $\begin{pmatrix} 1 & 0 & 0 & 0 & -19 \\ 0 & 1 & 0 & 0 & -4 \\ 0 & 0 & 1 & 0 & 7 \\ 0 & 0 & 0 & 1 & -2 \end{pmatrix}$

29. $\begin{pmatrix} 1 & 0 & -1 & 3 \\ 0 & 1 & 2 & -1 \end{pmatrix}$

30. $\begin{pmatrix} 1 & -2 & 1 & -1 \\ 0 & 0 & 0 & 0 \\ 0 & 0 & 0 & 0 \end{pmatrix}$

31. $\begin{pmatrix} 2 & -3 & 1 & 1 \\ -1 & 1 & -4 & -3 \\ 3 & 0 & -1 & 2 \end{pmatrix}$

32. $\begin{pmatrix} 3 & -4 & 1 & -1 & 1 \\ 1 & 0 & 1 & -2 & 4 \end{pmatrix}$

33. $\begin{pmatrix} 2 & -5 & 1 & -1 & -3 \\ 1 & 0 & -2 & 1 & 4 \\ 0 & 2 & -3 & -1 & 5 \end{pmatrix}$

34. $\begin{pmatrix} 3 & -2 & 5 \\ -1 & 5 & 7 \end{pmatrix}$

35. $\begin{aligned} 3x + 2y &= -1 \\ -4x + 5y &= 2 \end{aligned}$

36. $x - z + 2w = -3$
$2x + y - w = 4$
$-x + y + 2z = 0$

37. $2x + z = 3$
$-x + y = 2$
$2y - 3z = -1$

38. $2x + y - 2z = 4$
$-3x + 2z = -1$

39. $(2, -3)$

40. $(-1, -1)$

41. $(-2, 3, 1)$

42. $(7, 6, 3)$

43. No solution

44. $(z + 2, -z - 1, z)$

45. $(2 - z, 1 + z, z)$

46. $(z + 53, z - 26, z)$

47. No solution

48. $(1, 2)$

49. $\dfrac{-3}{x + 4} + \dfrac{4}{x - 2}$

50. $\dfrac{2}{x + 3} - \dfrac{1}{x}$

51. $\dfrac{3}{x^2 + 1} + \dfrac{2x - 1}{(x^2 + 1)^2}$

52. $\dfrac{1}{x} + \dfrac{2}{x^2} - \dfrac{1}{x + 2}$

53. $f(x) = \dfrac{2}{3x - 1}$

54. $f(x) = -3x^2 + 5x - 4$

55. 22 nickels, 35 dimes, and 17 quarters

56. 27 one-dollar bills, 18 fives, and 6 tens

57. Adding one row to another row is the same as multiplying that first row by 1 and then adding it to the other, so it falls into the category of the second type of elementary row operation.

58. Subtracting one row from another is the same as multiplying that first row by -1 and then adding it to the other, so it falls into the category of the second type of elementary row operation.

59. Answers will vary.

60. Starting with any matrix in row echelon form, you can perform the operation $kR_i + R_j$, for any constant k, with $i > j$, and obtain another matrix in row echelon form. For example, consider $\begin{pmatrix} 1 & 1 & 1 \\ 0 & 1 & 1 \end{pmatrix}$ and $\begin{pmatrix} 1 & 2 & 2 \\ 0 & 1 & 1 \end{pmatrix}$, two equivalent matrices (the second can be obtained from the first via $R_2 + R_1$), both of which are in row echelon form.

QUICK REVIEW 10.4

1. $x = \frac{1}{6}$

2. $x = \frac{1}{13}$

3. $x = -\frac{1}{11}$

4. $x = -\frac{1}{5}$

5. $x = 1$

6. $x = 1$

7. $x = -1$

8. $x = -1$

9. $a = \frac{1}{x}$

10. If $x \neq 1$, then $a = 0$. If $x = 1$, no conclusion can be drawn.

11. $f \circ I(x) = f(I(x)) = f(x)$; $I \circ f(x) = I(f(x)) = f(x)$ because $I(x) = x$

12. $f \circ g(x) = \dfrac{2\,g(x) - 1}{g(x) + 2} = x$;
$g \circ f(x) = -\dfrac{2\,f(x) + 1}{f(x) - 2} = x$

13. $f \circ g(x) = \dfrac{1}{g(x) - 3} = x$;
$g \circ f(x) = 3 + \dfrac{1}{f(x)} = x$

SECTION EXERCISES 10.4

1. **(a)** $\begin{pmatrix} -4 & -18 \\ -11 & -17 \end{pmatrix}$ **(b)** $\begin{pmatrix} 5 & -12 \\ 0 & -26 \end{pmatrix}$

2. **(a)** $AB = \begin{pmatrix} 13 & 13 \\ -2 & -16 \end{pmatrix}$
(b) $BA = \begin{pmatrix} 7 & -14 \\ -8 & -10 \end{pmatrix}$

3. **(a)** $\begin{pmatrix} 2 & 2 \\ -11 & 12 \end{pmatrix}$ **(b)** $\begin{pmatrix} 4 & 8 & -5 \\ -5 & 4 & -6 \\ -2 & -8 & 6 \end{pmatrix}$

4. **(a)** $AB = \begin{pmatrix} 19 & -1 \\ 2 & 10 \end{pmatrix}$
(b) $BA = \begin{pmatrix} 3 & -1 & -14 & 16 \\ 4 & 2 & 8 & -2 \\ 5 & 3 & 14 & -6 \\ 8 & 2 & 0 & 10 \end{pmatrix}$

5. **(a)** $\begin{pmatrix} 6 & -7 & -2 \\ 3 & 7 & 3 \\ 8 & -1 & -1 \end{pmatrix}$
(b) $\begin{pmatrix} 2 & 1 & 3 \\ 5 & 0 & 0 \\ -18 & -3 & 10 \end{pmatrix}$

6. **(a)** $AB = \begin{pmatrix} -8 & 8 & 5 \\ 0 & 7 & -8 \\ 11 & 4 & 11 \end{pmatrix}$
(b) $BA = \begin{pmatrix} -3 & 18 & -2 \\ 11 & 2 & 11 \\ 2 & -11 & 11 \end{pmatrix}$

7. **(a)** (-8) **(b)** $\begin{pmatrix} -10 & 5 & -15 \\ 8 & -4 & 12 \\ 4 & -2 & 6 \end{pmatrix}$

8. **(a)** $AB = \begin{pmatrix} 2 & -4 & -8 \\ -3 & 6 & 12 \\ 4 & -8 & -16 \end{pmatrix}$
(b) $BA = (-8)$

9. **(a)** AB is not possible. **(b)** $(18 \quad 14)$

10. **(a)** $AB = \begin{pmatrix} 1 & 15 \\ 2 & 3 \\ 5 & -6 \\ -17 & 15 \end{pmatrix}$
(b) BA is not possible.

11. **(a)** $\begin{pmatrix} -1 & 3 & 4 \\ 2 & 0 & 1 \\ 1 & 2 & 1 \end{pmatrix}$ **(b)** $\begin{pmatrix} 1 & 2 & 1 \\ 1 & 0 & 2 \\ 4 & 3 & -1 \end{pmatrix}$

12. **(a)** $AB = \begin{pmatrix} -3 & 2 & 1 & 3 \\ 2 & 1 & 0 & -1 \\ -1 & 2 & 3 & -4 \\ 4 & 0 & 2 & -1 \end{pmatrix}$

(b) $BA = \begin{pmatrix} 3 & 2 & -1 & -4 \\ 0 & 1 & 2 & -1 \\ 1 & 2 & -3 & 3 \\ 2 & 0 & 4 & -1 \end{pmatrix}$

13. $a = 5, b = 2$

14. $a = 3, b = -1$

15. $a = -2, b = 0$

16. $a = 1, b = 6$

17. Show that $AB = BA = I_2$.

18. Show that $AB = BA = I_3$.

19. $\begin{pmatrix} -1 & 1.5 \\ 1 & -1 \end{pmatrix}$

20. No inverse exists.

21. No inverse exists.

22. $\begin{pmatrix} 1 & 1 & -3 \\ -0.25 & -0.5 & 1.75 \\ 0.25 & 0.5 & -0.75 \end{pmatrix}$; to confirm, carry out the multiplication.

23. $\begin{pmatrix} 2 & 5 \\ 1 & -2 \end{pmatrix}\begin{pmatrix} x \\ y \end{pmatrix} = \begin{pmatrix} -3 \\ 1 \end{pmatrix}$

24. $\begin{pmatrix} 1 & -2 \\ 2 & -5 \end{pmatrix}\begin{pmatrix} x \\ y \end{pmatrix} = \begin{pmatrix} 1 \\ 3 \end{pmatrix}$

25. $\begin{pmatrix} 5 & -7 & 1 \\ 2 & -3 & -1 \\ 1 & 1 & 1 \end{pmatrix}\begin{pmatrix} x \\ y \\ z \end{pmatrix} = \begin{pmatrix} 2 \\ 3 \\ -3 \end{pmatrix}$

26. $\begin{pmatrix} 2 & 3 & -1 \\ 2 & -3 & 2 \\ -1 & -1 & 3 \end{pmatrix}\begin{pmatrix} x \\ y \\ z \end{pmatrix} = \begin{pmatrix} 2 \\ -1 \\ -4 \end{pmatrix}$

27. $3x - y = -1$
$2x + 4y = 3$

28. $2x + 4y = 5$
$-x - 2y = -2$

29. $x - 3z = 3$
$2x - y + 3z = -1$
$-2x + 3y - 4z = 2$

30. $x - y = 3$
$2x + y - 3z = -1$
$-x + y + 2z = 4$

31. $(-2, 3)$

32. $(1, -1.5)$

33. $(2, 0)$

34. $(-0.5, -4)$

35. $a = -\frac{14}{15}, b = \frac{41}{5}, c = -\frac{29}{3}$

36. $a = \frac{19}{84}, b = \frac{-257}{84}, c = \frac{185}{14}$

37. $a = 0.28, b = -4.8, c = 24.50, d = -29$

38. $a = -\frac{41}{720}, b = 0.75, c = -\frac{283}{144},$
$d = -3.75, e = \frac{316}{45}$

39. $(x, p) = (5.33, 73.33)$

40. $(x, p) = (\frac{10}{3}, 110)$

41. **(a)** $y = 462.2x + 6921.8$. Window: $[0, 20] \times [0, 20{,}000]$.

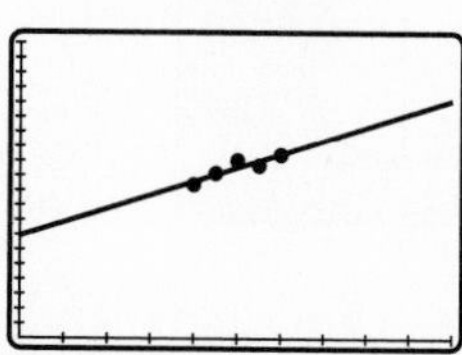

(b) $y = 155.6x + 11{,}333.20$. Window: $[0, 20] \times [0, 20{,}000]$.

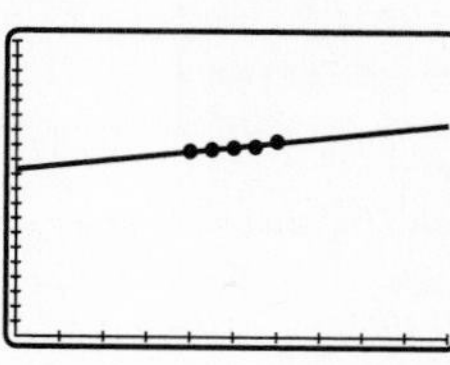

(c) At $x = 14.39$ (after 1994)

42. **(a)** $n = 2$:

$AI_2 = \begin{pmatrix} a_{11} & a_{12} \\ a_{21} & a_{22} \end{pmatrix}\begin{pmatrix} 1 & 0 \\ 0 & 1 \end{pmatrix} =$

$\begin{pmatrix} a_{11}(1) + a_{12}(0) & a_{11}(0) + a_{12}(1) \\ a_{21}(1) + a_{22}(0) & a_{21}(0) + a_{22}(1) \end{pmatrix} =$

$\begin{pmatrix} a_{11} & a_{12} \\ a_{21} & a_{22} \end{pmatrix} = A$; similarly, $I_2A = A$.

The technique is similar for $n = 3$ and 4. **(b)** The result of multiplying each entry in the ith row of A with the corresponding entry in the jth column of I_n will be all zeros except for the jth product, which will be $(a_{ij})(1) = a_{ij}$; therefore the entry in row i, column j of AI_n will be a_{ij}, the same as in row i, column j of A.

43. Answers will vary. (Note that $x = 1$ *and* $x = 0$ are not possible.)

44. Matrix B must have n rows; it can have any number of columns.

45. **(a)** $C(x) = 13 - 8x + x^2$ **(c)** 5.73, 2.27 **(d)** det $A = 13$; y-intercept is $(0, 13)$ **(e)** $a_{11} + a_{22} = 8$; sum of eigenvalues is 8.

46. **(a)** $C(x) = (2 - x)^2 - 5$ **(c)** $2 \pm \sqrt{5}$ **(d)** They are the same (-1). **(e)** They are the same (4).

QUICK REVIEW 10.5

1. $x < \frac{4}{3}$

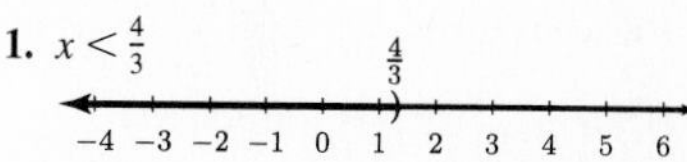

2. $x \leq 4$

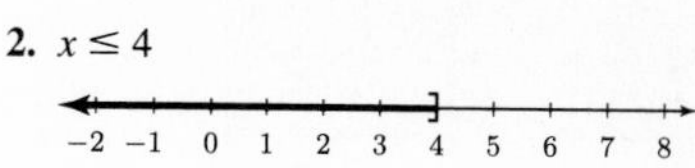

3. $x \leq 0$

4. $x < 0$

5. x-intercept: $(3, 0)$; y-intercept: $(0, -2)$.

6. x-intercept: $(6, 0)$; y-intercept: $(0, 3)$

7. $-2, 0, 2$

8. $x = 0$ or $x = \pm 1$

9. $-1, 1$

10. $x = \pm 2$

SECTION EXERCISES 10.5

1. The graph in part (c); boundary is included.

2. The graph in part (f); boundary is excluded.

3. The graph in part (b); boundary is included.

4. The graph in part (d); boundary is excluded.

5. The graph in part (e); boundary is included.

6. The graph in part (a); boundary is excluded.

7. Window: $[-10, 10] \times [-10, 10]$.

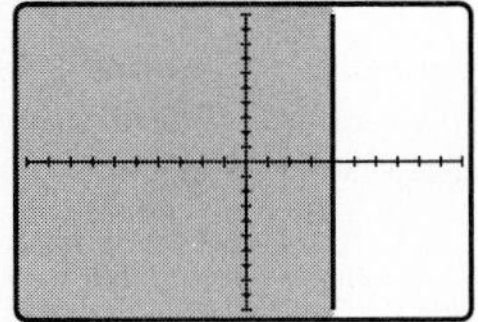

8. Window: $[-9, 9] \times [-6, 6]$

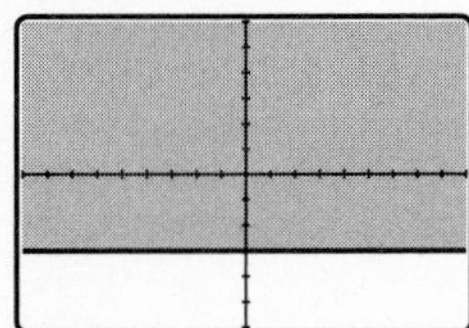

9. Window: $[-10, 10] \times [-10, 10]$.

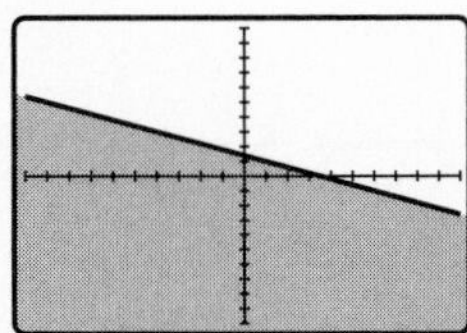

10. Window: $[-9, 9] \times [-6, 6]$

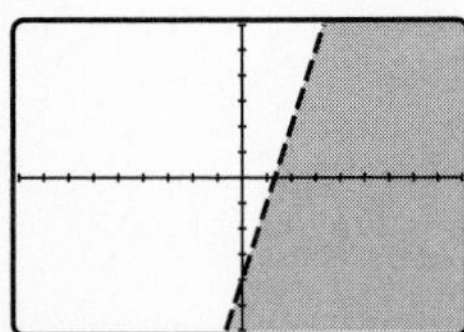

11. Window: $[-10, 10] \times [-10, 10]$.

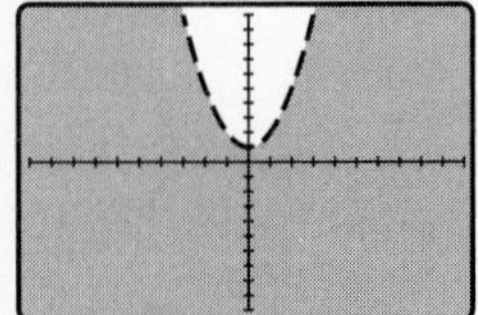

12. Window: $[-6, 6] \times [-4, 4]$

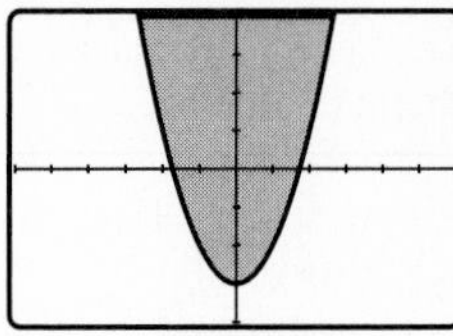

13. Window: $[-4.7, 4.7] \times [-3.1, 3.1]$.

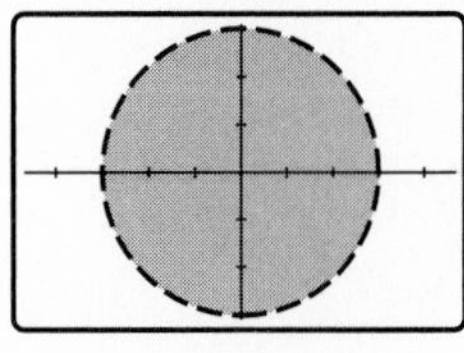

14. Window: $[-4.7, 4.7] \times [-3.1, 3.1]$.

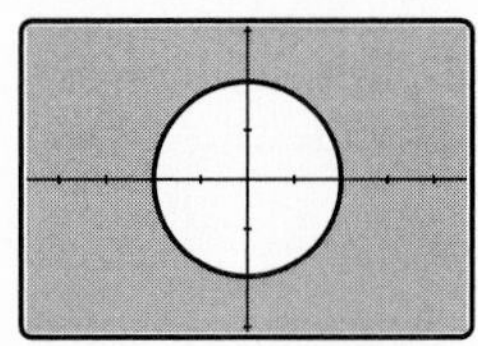

15. Window: $[-4.7, 4.7] \times [-3.1, 3.1]$.

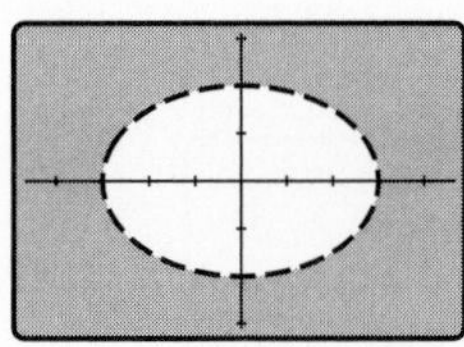

16. Window: $[-7.5, 7.5] \times [-5, 5]$

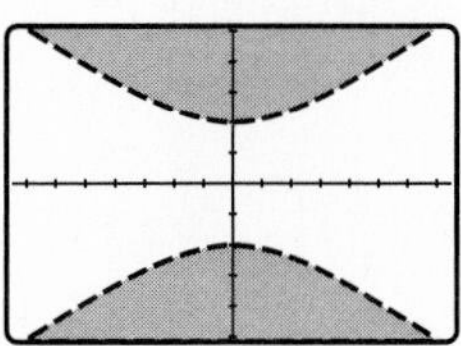

17. Window: $[-10, 10] \times [-10, 10]$; corner point: $(2, 3)$

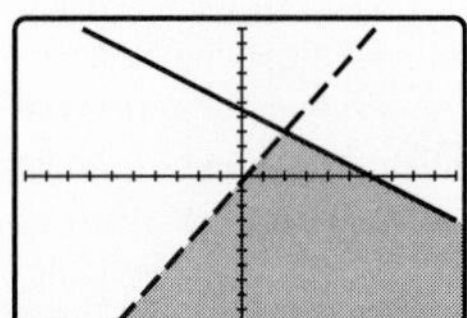

18. Window: $[-7, 2.4] \times [-2, 4.1]$; corner point: $(-3, 2)$

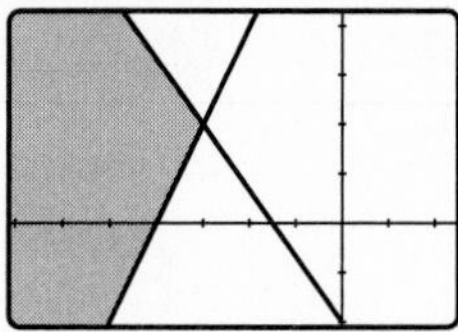

19. Window: $[-10, 10] \times [-5, 15]$; corner points: about $(-1.45, 0.10)$ and $(3.45, 9.90)$

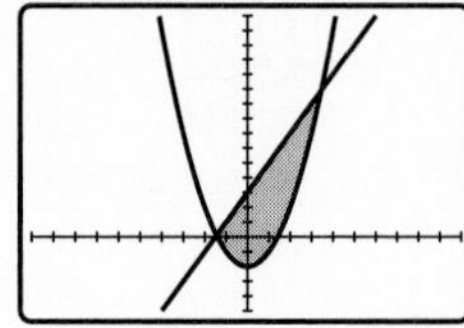

20. Window: $[-6, 6] \times [-4, 4]$; corner points: about $(-3.482, -3.161)$ and $(1.149, -1.617)$

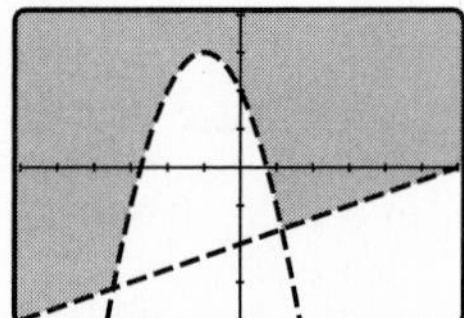

21. Window: $[-4.7, 4.7] \times [-3.1, 3.1]$; corner points: about $(-1.25, 1.56)$ and $(1.25, 1.56)$

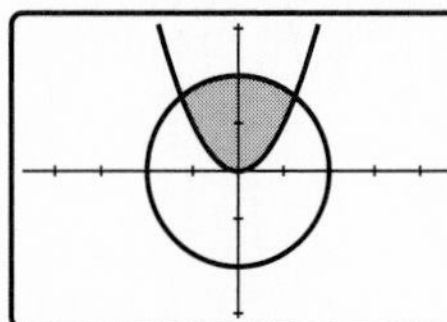

22. Window: $[-6, 6] \times [-4, 4]$; corner points: about $(-2.550, \pm 1.581)$ and $(2.550, \pm 1.581)$

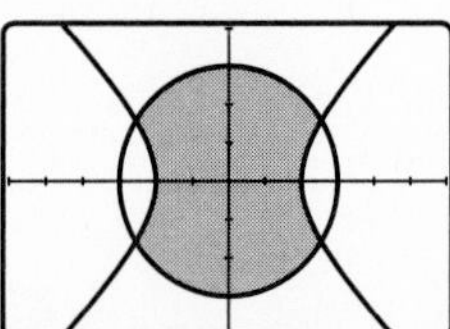

23. Window: $[0, 85] \times [0, 85]$; corner points: about $(0, 0)$, $(0, 40)$, $(26.67, 26.67)$, and $(40, 0)$

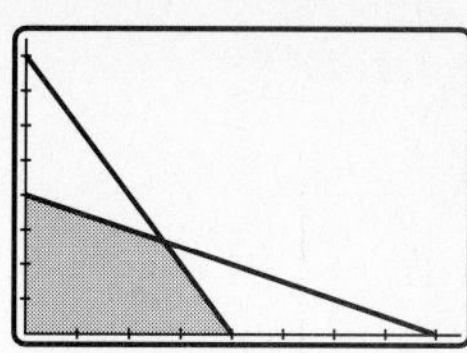

24. Window: $[-10, 110] \times [-10, 100]$; corner points: (0, 90), (32, 18) and (80, 0)

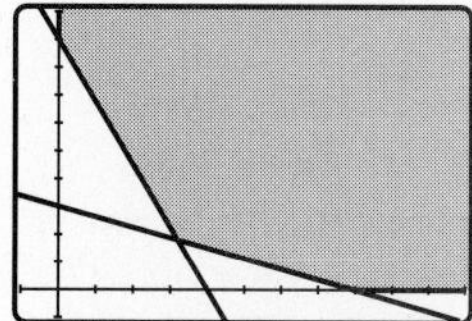

25. Window: $[0, 10] \times [0, 10]$; corner points: about (0, 2), (0, 6), (2.18, 4.55), (2, 0), and (4, 0)

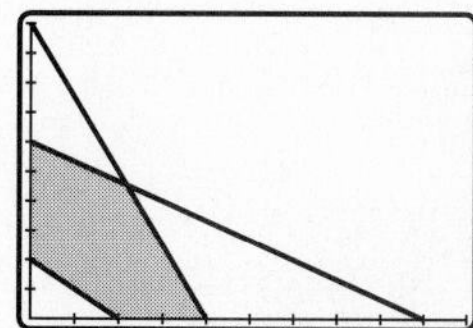

26. Window: $[-20, 60] \times [-10, 50]$; corner points: (0, 30), (21, 21) and (30, 0)

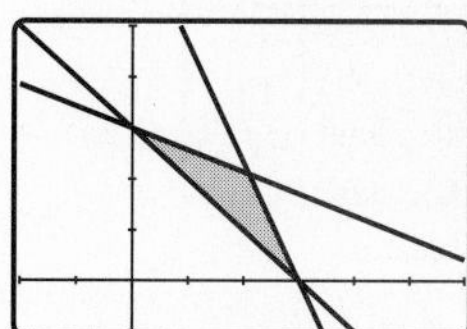

27. $x^2 + y^2 \leq 4$
$y \geq 1 - x^2$

28. $y \geq 0$
$x^2 + y^2 \leq 4$

29. $y \leq -0.5x + 5$
$y \leq -1.5x + 9$
$x \geq 0$
$y \geq 0$

30. $x \geq 0$
$y \geq 0$
$5x + 2y \geq 12$
$x + 3y \geq 5$

31. $C = 880/3$ at $(x, y) = (160/3, 80/3)$

32. $C = 967.5$ at $(x, y) = (22.5, 67.5)$

33. About 13.48 tons of R; 20.87 tons of S; minimum cost is $1926.09.

34. 1.6 units of A; 9.6 units of B; minimum cost is $10.88.

35. From 0 to 4 solutions: The two curves might not intersect or they might be (externally) tangent at a single point, externally tangent at two points, internally or externally tanget at one point and crossing at two other points, or crossing at two or four points.

36. Yes. A simple example: the ellipse (circle) $x^2 + y^2 = 1$ and any hyperbola with transverse axis endpoints $(-2, 0)$ and (1, 0); the latter is the only point of intersection.

37. Ellipses; corner points approximately: (0.21, 1.87), (1.87, 0.21), $(-2.75, -0.53)$, and $(-0.53, -2.75)$; window: $[-5, 5] \times [-5, 5]$; the solution is the shaded portion.

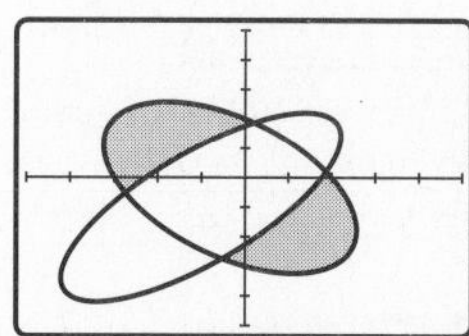

38. A hyperbola and an ellipse; corner points: approximately $(-3.27, -1.21)$, $(-1.91, -3.72)$, (2.06, 1.93), and (2.12, 0.85); window: $[-9, 9] \times [-6, 6]$; the solution is the shaded portion.

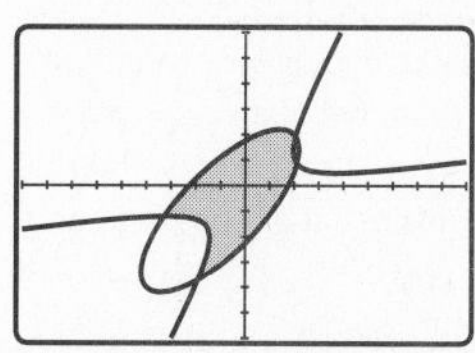

39. Hyperbolas; corner points: approximately $(-4.34, -0.69)$, $(-0.86, -3.51)$, (0.98, 3.06), and (3.71, 0.81); window: $[-8, 8] \times [-8, 8]$; the solution is the shaded portion.

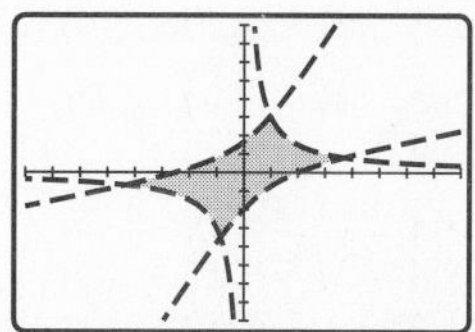

40. An ellipse and a circle (i.e., another ellipse); corner points: approximately $(-1.19, -0.43)$, $(-0.43, -1.72)$, (2.21, 1.88), and $(3.24, -0.02)$; window: $[-4.7, 4.7] \times [-3.1, 3.1]$; the solution is the shaded portion.

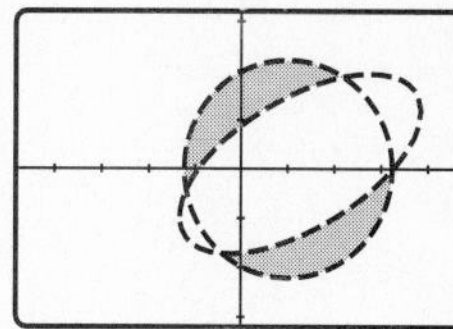

CHAPTER 10 REVIEW

1. (1, 2)

2. Yes; $(x, y) = (-3, -1)$

3. No solution

4. No solution

5. $\begin{pmatrix} 1 & 0 & 2 \\ 0 & 1 & -1 \\ 0 & 0 & 0 \end{pmatrix}$

6. $\begin{pmatrix} 1 & 0 & 0 & 1 \\ 0 & 1 & 0 & 2 \\ 0 & 0 & 1 & -3 \end{pmatrix}$

7. $\begin{pmatrix} 1 & 0 & 0 & 8 \\ 0 & 1 & 0 & -11 \\ 0 & 0 & 1 & 5 \end{pmatrix}$

8. $\begin{pmatrix} 1 & 0 & 0 & 2 \\ 0 & 1 & 0 & -1 \\ 0 & 0 & 1 & 1 \end{pmatrix}$

9. $(2 - z - w, w + 1, z, w)$

10. $(-w - 2, -z - w, z, w)$

11. No solution

12. $(0.25z + 0.75, 1.75z + 1.25, z)$

13. $(1 - 2z + w, 2 + z - w, z, w)$

14. $(-w + 2, -z - 1, z, w)$

15. $(x, p) = (-1.5 - 0.5\sqrt{329}, 15.5 - 1.5\sqrt{329}) = (-10.569\cdots, -11.707\cdots)$ or $(-1.5 + 0.5\sqrt{329}, 15.5 + 1.5\sqrt{329}) = (7.569\cdots, 42.707\cdots)$; only the latter makes sense in this case.

16. $(x, p) = (-20 - 5\sqrt{46}, -75 - 20\sqrt{46}) = (-53.911\ldots, -210.646\ldots)$ or $(-20 + 5\sqrt{46}, -75 + 20\sqrt{46})$ or $(13.911\ldots, 60.646\ldots)$; only the latter makes sense in this case.

17. $(-1, -2.5), (3, 1.5)$

18. $(\pm 2, 1)$

19. BA not possible;
$$AB = \begin{pmatrix} -3 & -7 & 11 \\ 0 & -12 & 24 \end{pmatrix}$$

20. AB not possible; $BA = \begin{pmatrix} 15 & -4 \\ 1 & 3 \\ -5 & -13 \end{pmatrix}$

21. $AB = (3 \quad 7)$; BA not possible

22. AB not possible; $BA = \begin{pmatrix} -3 & -1 \\ -1 & 3 \\ -3 & 4 \\ -1 & 2 \end{pmatrix}$

23. $AB = \begin{pmatrix} 1 & 2 & -3 \\ 2 & -3 & 4 \\ -2 & 1 & -1 \end{pmatrix}$;
$$BA = \begin{pmatrix} -3 & 2 & 4 \\ 2 & 1 & -3 \\ 1 & -2 & -1 \end{pmatrix}$$

24. $AB = \begin{pmatrix} 3 & 0 & 2 & 1 \\ -2 & 1 & 0 & 1 \\ 3 & -2 & 1 & 0 \\ -1 & 1 & 2 & -1 \end{pmatrix}$;
$$BA = \begin{pmatrix} 1 & -2 & 1 & 0 \\ 0 & 3 & 1 & 2 \\ 1 & -1 & -1 & 2 \\ -2 & 3 & 0 & 1 \end{pmatrix}$$

25. Show that $AB = I_4$ and $BA = I_4$.

26. Show that $AB = I_3$ and $BA = I_3$.

27. $\begin{pmatrix} -2 & -5 & 6 & -1 \\ 0 & -1 & 1 & 0 \\ 10 & 24 & -27 & 4 \\ -3 & -7 & 8 & -1 \end{pmatrix}$

28. $\begin{pmatrix} -0.4 & 0.2 & 0.2 \\ -0.2 & -0.4 & 0.6 \\ 0.6 & 0.2 & 0.2 \end{pmatrix}$

29. $(2.25, -0.75, -1.75)$

30. $(0.5, -2.5, -2.5)$

31. No solution

32. $(x, y, z, w) = (4.\overline{3}, -2.\overline{6}, -0.\overline{3}, 7.\overline{3})$

33. 380 student tickets; 72 nonstudent tickets

34. About \$21,333.33 at 7.5% and \$16,666.67 at 6%

35. **(a)** $y = 354.2x + 10967$. Window: $[0, 10] \times [0, 20{,}000]$.

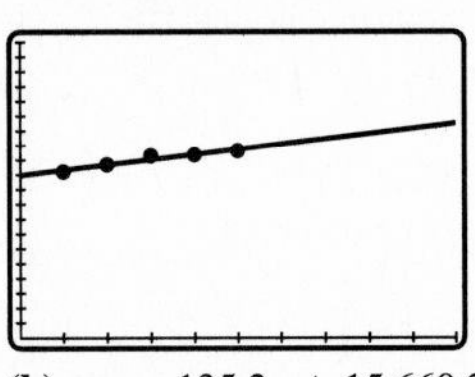

(b) $y = -135.2x + 15{,}660.2$. Window: $[0, 10] \times [0, 20{,}000]$.

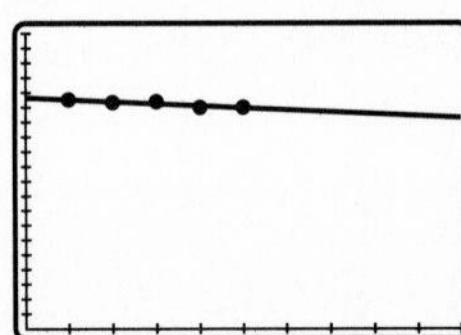

(c) At $x = 9.59$ (after 1996)

36. From 0 to 4 solutions: The two curves might not intersect, or they might be (externally) tangent at a single point, externally tangent at two points, internally or externally tangent at one point and crossing at two other points, or crossing at two or four points.

37. Answers will vary; one example is 2 vans, 4 small trucks, and 2 large trucks.

38. \$160,000 at 4%; \$170,000 at 6.5%; \$320,000 at 9%

39. Sue: $9\frac{1}{3}$ hours; Esther: 12 hours; Murphy: 16.8 hours

40. Pipe A: 15 hours; pipe B: $\frac{60}{11} = 5.\overline{45}$ hours; pipe C: 12 hours

41. $(2 - w, z + 3, z, w)$

42. $(2 - w, z + 3, z, w)$

43. $(-2, 1, 3, -1)$

44. $(1, -w - 3, w + 2, w)$

45. $a = \frac{17}{840}, b = -\frac{33}{280}, c = -\frac{571}{420}, d = \frac{386}{35}$

46. $a = \frac{19}{108}, b = -\frac{29}{18}, c = \frac{59}{36}, d = \frac{505}{54}, e = -\frac{68}{9}$

47. B is an $n \times n$ matrix. The number of columns of B is n because AB is defined, and the number of rows of B is n because BA is defined.

48. $n = p$; the number of columns in A is the same as the number of rows in B.

49. No solution

50. $(\pm 2, 0)$

51. Window: $[-10, 10] \times [-10, 10]$; no corner points

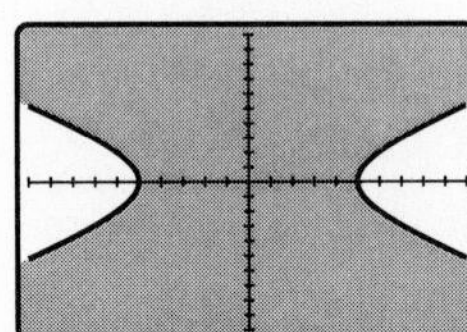

52. Window: $[-9, 9] \times [-6, 6]$; no corner points

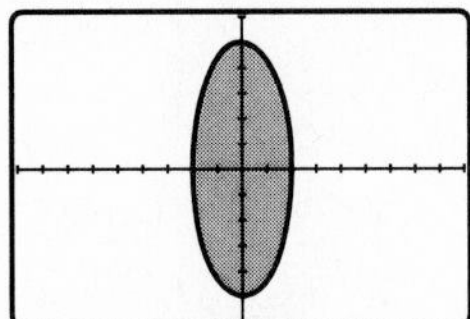

53. Window: $[-10, 10] \times [-5, 15]$; corner points: approximately $(0.92, 2.31)$ and $(5.41, 3.80)$

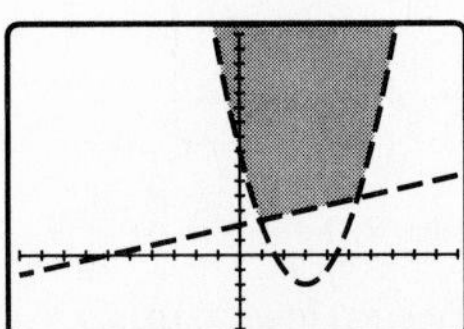

54. Window: $[-10, 10] \times [-10, 10]$; corner points: approximately $(-2.41, 3.20)$ and $(2.91, 0.55)$

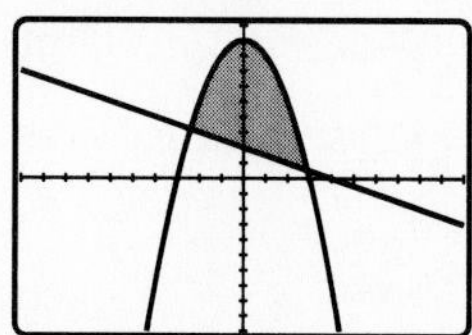

55. Window: $[-4.7, 4.7] \times [-3.1, 3.1]$; corner points: $(\pm 2, 0)$

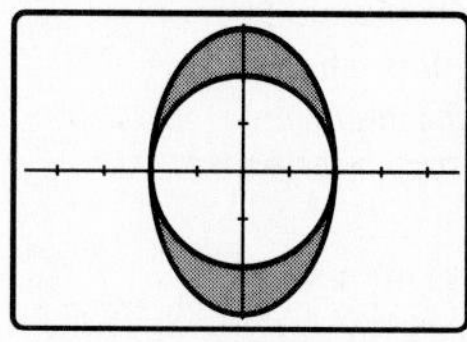

56. Window: $[-9, 9] \times [-3, 9]$; no corner points

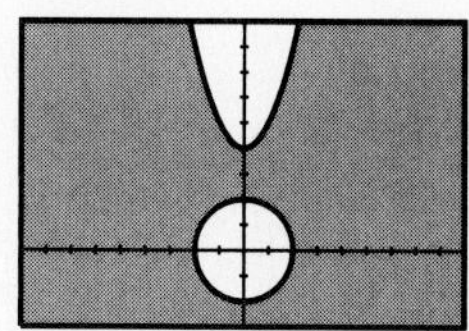

57. Window: $[-10, 100] \times [-10, 100]$; corner points: $(0, 90)$, $(27.69, 27.69)$, and $(90, 0)$

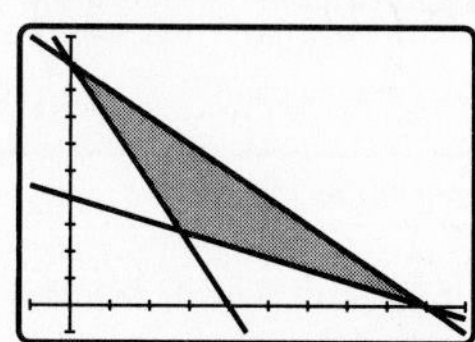

58. Window: $[-2, 10] \times [-1, 7]$; corner points: $(0, 3)$, $(0, 7)$, $(\frac{30}{13}, \frac{70}{13})$, $(3, 0)$, and $(5, 0)$

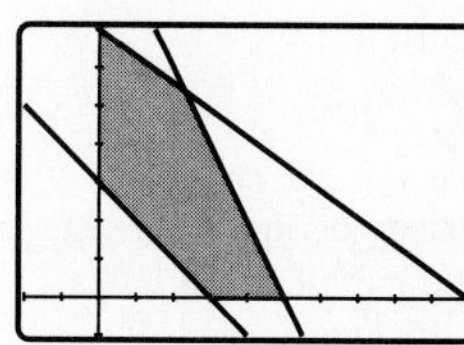

59. $C = 106$ when $(x, y) = (10, 6)$

60. $C = 138$ when $(x, y) = (8, 10)$

CHAPTER 11

QUICK REVIEW 11.1

1. -8

2. 16

3. 729

4. 256

5. -729

6. -256

7. **(a)** approximately 1422.10 **(b)** approximately 2411.71

8. **(a)** approximately 1508.30 **(b)** approximately 6623.41

9. **(a)** $f(a + b) = 3a + 3b + 2$ **(b)** $f(f(x)) = 9x + 8$

10. **(a)** $f(a + b) = 2(a + b) - 3$ **(b)** $f(f(x)) = 4x - 9$

11. -13

12. -1

13. -4

14. $-14.$

15. $[1, \infty)$

16. $(-\infty, -1) \cup (-1, 1) \cup (1, \infty)$.

17. $x = -1, y = 2$

18. $a = 2, b = -3$

SECTION EXERCISES 11.1

1. $2, \frac{3}{2}, \frac{4}{3}, \frac{5}{4}, \frac{6}{5}, \frac{7}{6}; \frac{101}{100}$

2. $\frac{4}{3}, 1, \frac{4}{5}, \frac{2}{3}, \frac{4}{7}, \frac{1}{2}; \frac{2}{51}$

3. $0, 6, 24, 60, 120, 210; 999{,}900$

4. $-4, -6, -6, -4, 0, 6; 9500$

5. $8, 4, 0, -4; -20$

6. $-3, 7, 17, 27; 67$

7. $2, 6, 18, 54; 4374$

8. $0.75, -1.5, 3, -6; -96$

9. $2, 4, 16, 256; \approx 3.40 \times 10^{38}$

10. $-2, 4, 16, 256; \approx 3.40 \times 10^{38}$

11. $2, -1, 1, 0; 3$

12. $-2, 3, 1, 4; 23$

13. The table in part (c)

14. The table in part (b)

15. The table in part (d)

16. The table in part (a)

17. $1, -3; (-3)^{99} \approx -1.72 \times 10^{47}$

18. $4, -8; (-2)^{101} \approx -2.54 \times 10^{30}$

19. $0.2, -\frac{1}{3}; \frac{(-2)^{99}}{104} \approx -6.09 \times 10^{27}$

20. $\frac{1}{4}, -\frac{3}{5}; \frac{(-3)^{99}}{103} \approx -1.67 \times 10^{45}$

21. **(a)** 4 **(b)** 42 **(c)** $a_n = a_{n-1} + 4$ **(d)** $a_n = 6 + (n - 1)4$

22. **(a)** 5 **(b)** 41 **(c)** $a_n = a_{n-1} + 5$ **(d)** $a_n = -4 + 5(n - 1)$

23. **(a)** 3 **(b)** 22 **(c)** $a_n = a_{n-1} + 3$ **(d)** $a_n = -5 + (n - 1)3$

24. **(a)** 11 **(b)** 92 **(c)** $a_n = a_{n-1} + 11$ **(d)** $a_n = -7 + 11(n - 1)$

25. **(a)** 3 **(b)** 4374 **(c)** $a_n = a_{n-1} \cdot 3$ **(d)** $a_n = 2(3)^{n-1}$

26. **(a)** 2 **(b)** 384 **(c)** $a_n = 2a_{n-1}$ **(d)** $a_n = 3 \cdot 2^{n-1}$

27. **(a)** -2 **(b)** -128 **(c)** $a_n = a_{n-1} \cdot (-2)$ **(d)** $a_n = (-2)^{n-1}$

28. **(a)** -1 **(b)** 2 **(c)** $a_n = -1a_{n-1}$ **(d)** $a_n = -2(-1)^{n-1}$

29. $a_1 = -20, a_n = a_{n-1} + 4$

30. $a_1 = 7, a_n = a_{n-1} - 3$

31. $a_1 = \frac{3}{2}$ or $-\frac{3}{2}$, $r = 2$ or -2, $a_n = \frac{3}{2}(2)^{n-1}$ or $a_n = (-\frac{3}{2})(-2)^{n-1}$

32. $a_1 = -3, r = 5, a_n = -3(5)^{n-1}$

33. Geometric, $r = 2$, $a_n = 5(2)^{n-1}$

34. Geometric, $r = -4$, $a_n = -0.25(-4)^{n-1} = (-4)^{n-2}$

35. Arithmetic, $d = 7$, $a_n = 7n - 23$

36. Geometric, $r = 1.01$, $a_n = 10.1(1.01)^{n-1}$

37. Geometric, $r = -\frac{1}{2}$, $a_n = -2(-\frac{1}{2})^{n-1}$

38. Neither arithmetic nor geometric

39. $a_n = 3 + (n - 1)2 = 2n + 1$

40. $b_n = -5 + 7(n - 1) = 7n - 12$

41. $u_n = 2(3)^{n-1}$

42. $v_n = 3(-2)^{n-1}$

43. $a_n = 8(\frac{1}{4})^{n-1}$, for $n = \{1, 2, 3, \ldots, 8\}$

44. $a_n = -7 + 3(n - 1) = 3n - 10$, for $n = \{1, 2, 3, \ldots, 115\}$

45. $a_n = \dfrac{(-1)^n}{3n}$, for $n = \{1, 2, 3, \ldots, 333\}$

46. $a_n = -1(-\frac{1}{5})^{n-1}$, for $n = \{1, 2, 3, \ldots, 12\}$

47. The graph in part (c)

48. The graph in part (d)

49. The graph in part (b)

50. The graph in part (a)

51. Window: $[0, 5] \times [-2, 5]$

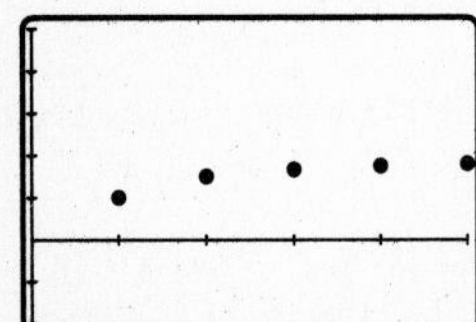

52. Window: $[0, 5] \times [-5, 5]$

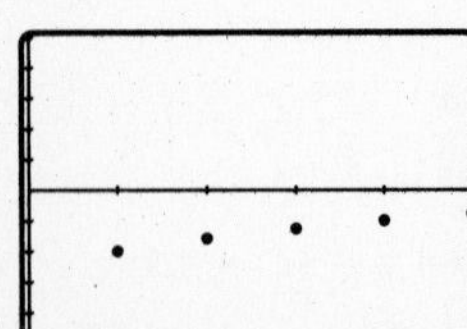

53. Window: $[0, 5] \times [-5, 20]$

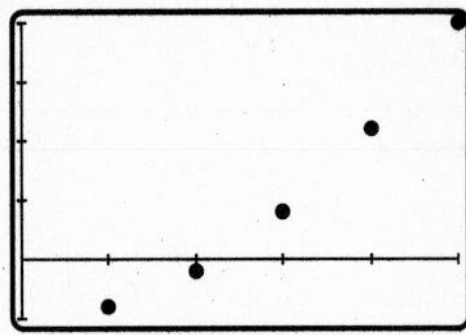

54. Window: $[0, 5] \times [-5, 15]$

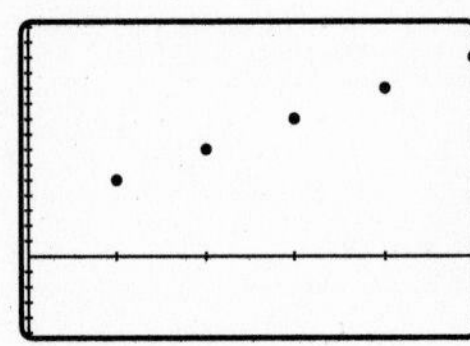

55. **(a)** Window: $[0, 5] \times [-1, 20]$

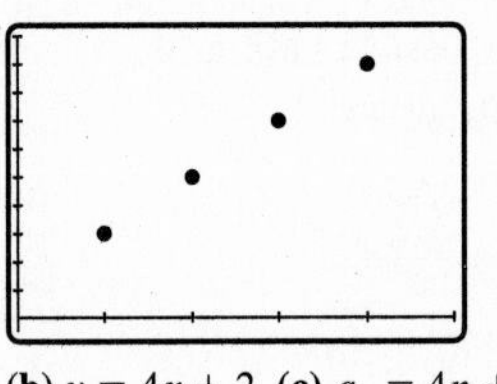

(b) $y = 4x + 2$ **(c)** $a_n = 4n + 2$

56. **(a)** Window: $[0, 5] \times [-5, 15]$

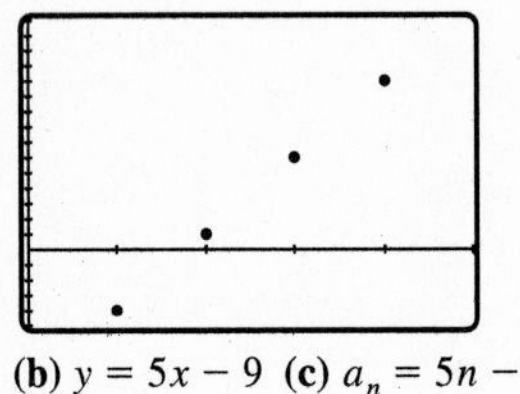

(b) $y = 5x - 9$ **(c)** $a_n = 5n - 9$

57. **(a)** Window: $[0, 5] \times [-7, 7]$

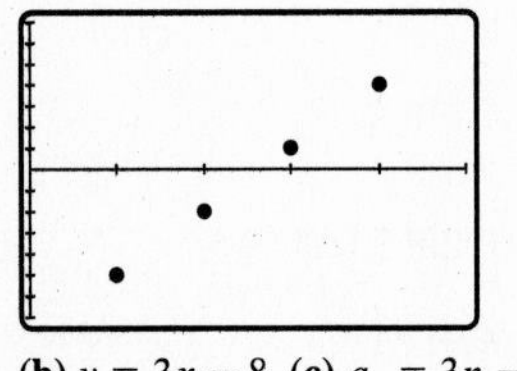

(b) $y = 3x - 8$ **(c)** $a_n = 3n - 8$

58. **(a)** Window: $[0, 5] \times [-10, 30]$

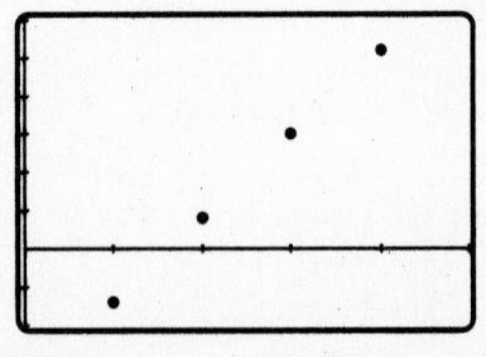

(b) $y = 11x - 18$ **(c)** $a_n = 11n - 18$

59. \$1712.61

60. About \$9126.56

61. 700, 702.3, 704.6, 706.9, ..., 815, 817.3

62. 2019: about \$2.30; 1969: about \$0.15

63.

Time (t) (in billion-year intervals)	Mass (g)
0	16
14	8
28	4
42	2
56	1

The first column is an arithmetic sequence with common difference 14, and the second column is a geometric sequence with common ratio $\frac{1}{2}$.

64. **(a)**

	Huntsville	Conroe
1985	25,854	22,314
1986	26,255	23,285
1987	26,662	24,298
1988	27,075	25,354
1989	27,495	26,457
1990	27,921	27,608
1991	28,354	28,809
1992	28,793	30,062
1993	29,239	31,370
1994	29,693	32,735
1995	30,153	34,159

(b) In about 1990 or 1991.

65. **(a)** $a_1 = 1$ because the colony begins with 1 male–female pair;
$a_2 = 1$ because it takes 1 month for the male–female pair to become fertile;
$a_3 = 2$ because the male–female pair produced one new male–female pair so there are now 2 male–female pairs.
(b) $a_4 = 3$, $a_5 = 5$, $a_6 = 8$, $a_7 = 13$, $a_8 = 21$, $a_9 = 34$, $a_{10} = 55$, $a_{11} = 89$, $a_{12} = 144$, $a_{13} = 233$
(c) a_2 represents the end of the first month, so a_{13} represents the end of 12 months. The rabbit colony is represented by the sequence $a_1, a_2, a_3, \ldots, a_{13}$.

66. $P_1 = 525{,}000$; $P_n = P_{n-1}(1.0175)$, $n \geq 2$

67. $y = -2x + 4$

68. $y = 3x - 7$

69. $y = x^2 - 4x + 1$

70. $y = -x^2 + 2x + 3$

71. **(a)**
$a_2 = a_1 + d;$
$a_3 = a_2 + d = (a_1 + d) + d = a_1 + 2d;$
$a_4 = a_3 + d = (a_1 + 2d) + d = a_1 + 3d;$
$a_5 = a_4 + d = (a_1 + 3d) + d = a_1 + 4d;$
$\vdots$
$a_n = a_{n-1} + d = (a_1 + (n-2)d) + d = a_1 + (n-1)d$
(b) $a_n = f(n)$
$= a_1 + (n-1)d$
$= a_1 + dn - d$
$= dn + (a_1 - d)$

This is of the form $f(x) = mx + b$ with variable n, slope $m = d$, and y-intercept $b = a_1 - d$.

72. The nth term is $a_n = a_{n-1} \cdot r = (a_{n-2} \cdot r) \cdot r = a_{n-2} \cdot r^2 = a_{n-3} \cdot r^3 = \cdots = a_1 \cdot r^{n-1}$.

73. $a_1 = 1$, $a_2 = 1$, $a_3 = 2$, $a_4 = 3$, $a_5 = 5$, $a_6 = 8$, and $a_7 = 13$; these values (1, 1, 2, 3, 5, 8, and 13) are the first seven terms of the Fibonacci sequence.

74. Consider the constant sequence $a_1, a_1, a_1, \ldots, a_1, \ldots$ **(a)** Then $a_n = a_1 + 0(n-1)$, which is an arithmetic sequence with common difference 0; also, $a_n = a_1 \cdot 1^{n-1}$, which is a geometric sequence with common ratio 1. **(b)** Recursive definitions would be $a_n = a_{n-1} + 0$ and $a_n = 1 \cdot a_{n-1}$; explicit definition would be $a_n = a_1$.

75. **(a)** Yes **(b)** Yes **(c)** No **(d)** No

76. $\{a_n\}$ is increasing if $a_1 \leq a_2 \leq a_3 \leq \ldots \leq a_n \leq \ldots$; $\{a_n\}$ is decreasing if $a_1 \geq a_2 \geq a_3 \geq \ldots \geq a_n \geq \ldots$.

77. **(a)** Form a right triangle in quadrant I of any figure; the measure of the angle formed is $\theta = \frac{2\pi}{2n} = \frac{\pi}{n}$, so $\sin\theta = \sin\frac{\pi}{n}$; let a_n = the perimeter of the polygon; then the length of a side of the polygon is $\frac{a_n}{n}$, and the length of the side in quadrant I is $\frac{1}{2}\left(\frac{a_n}{n}\right) = \frac{a_n}{2n}$; from the figure, $\sin\theta = \frac{a_n}{2n}$ or $\sin\left(\frac{\pi}{n}\right) = \frac{a_n}{2n}$ and $a_n = 2n\sin\left(\frac{\pi}{n}\right)$.

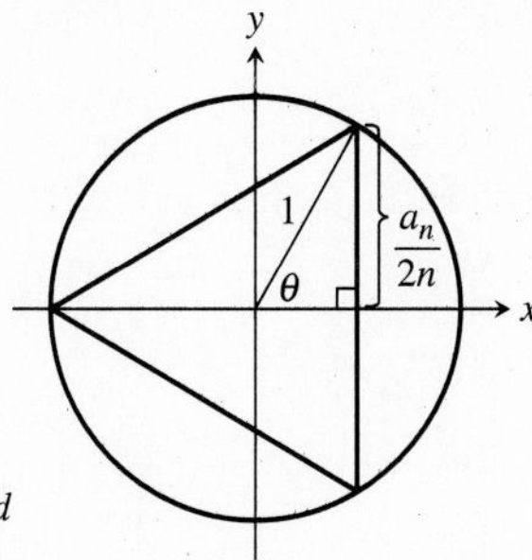

(b) $a_n \to 2\pi$ as $n \to \infty$

78. If $a_n = ra_{n-1}$, then $\log a_n = \log r + \log a_{n-1}$; an arithmetic sequence with common difference $\log r$ and first term $\log a_1$.

79. If $b_n = b_1 + (n-1)d$, then $10^{b_n} = 10^{b_1+(n-1)d} = [10^{b_1}][(10^d)^{n-1}]$; a geometric sequence with first term 10^{b_1} and common ratio 10^d.

80. $a_1 = 1$, $a_2 = 2 \cdot a_1 = 2 \cdot 1 = 2$, $a_3 = 3 \cdot a_2 = 3 \cdot 2 \cdot 1 = 6$, $a_4 = 4 \cdot a_3 = 4 \cdot 3 \cdot 2 \cdot 1 = 24$, $a_5 = 5 \cdot a_4 = 5 \cdot 4 \cdot 3 \cdot 2 \cdot 1 = 120$, $a_6 = 6 \cdot a_5 = 6 \cdot 5 \cdot 4 \cdot 3 \cdot 2 \cdot 1 = 720$

81. **(a)** $K = 2$; window is $[0, 10] \times [0, 4]$ with Xscl = 1 and Yscl = 1.

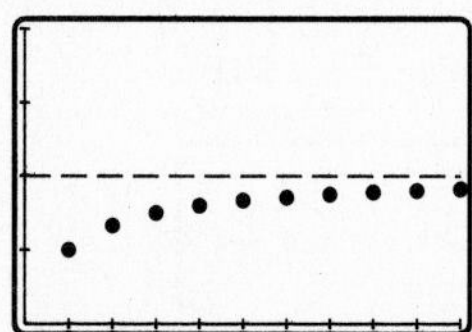

(b) $K \approx 1.0513$; window is $[0, 10] \times [1.04, 1.06]$ with Xscl = 1 and Yscl = 0.005.

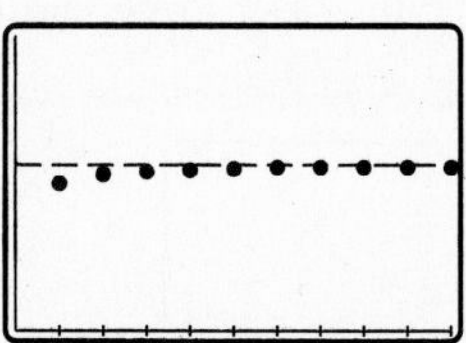

(c) $K = 3$; window is $[0, 10] \times [0, 5]$ with Xscl = 1 and Yscl = 1.

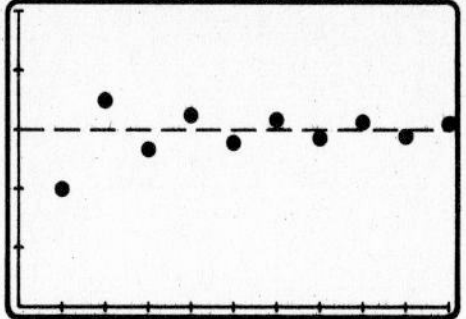

(d) $K \approx 1.571$; window is $[0, 10] \times [0, 2]$ with Xscl = 1 and Yscl = 0.5.

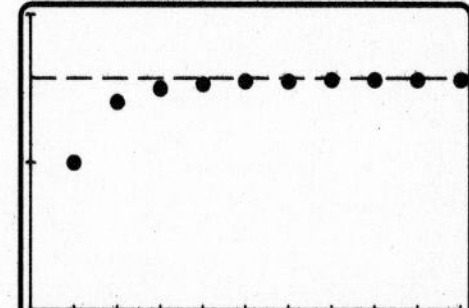

82. **(a)** $a_1 = 0.5$, $a_2 \approx 0.707$, $a_3 \approx 0.613$, $a_4 \approx 0.654$, $a_5 \approx 0.635$, $a_6 \approx 0.644$, $a_7 \approx 0.640$, $a_8 \approx 0.642$. **(b)** Approximately 0.641. **(c)** The solution is approximately 0.641; the solution and the limit of the sequence appear to be equal.

83. **(a)** $1, i, -1, -i, 1, i, -1, -i, 1$
(b) $2 + i, -1 + 2i, -2 - i, 1 - 2i, 2 + i, -1 + 2i, -2 - i, 1 - 2i, 2 + i$; the first four terms of the sequence are the endpoints of a square graphed in the complex plane:

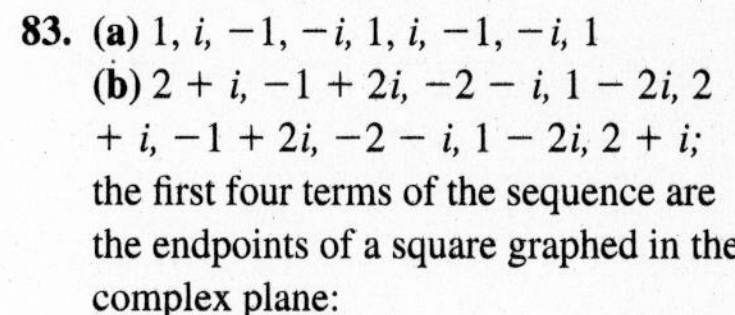

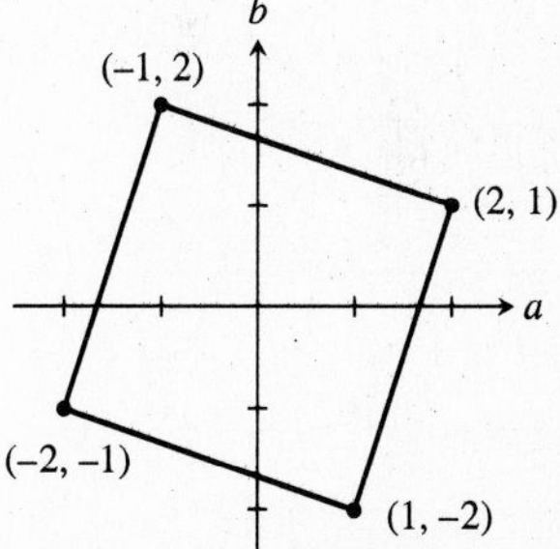

(c) The result is a complex number of the form $-b + ai$; for any complex number $a + bi$, $(a + bi)(i) = ai + bi^2 = ai + b(-1) = ai - b = -b + ai$.

84. $a_1 = (1\ 1)$, $a_2 = (1\ 2)$, $a_3 = (2\ 3)$, $a_4 = (3\ 5)$, $a_5 = (5\ 8)$, $a_6 = (8\ 13)$, $a_7 = (13\ 21)$; the terms of this sequence are successive pairs of terms from the Fibonacci sequence.

85. **(a)** Finite; $\{-2, 1, 4, 7, 10, 13, 16\}$
(b) Finite; $\{2, 10, 26, 58, 122, 250\}$

(c) Infinite; $\{4, 9, 14, 19, 24, 29, 34, \ldots\}$
(d) Infinite; $\{5, 17, 37, 65, 101, 145, \ldots\}$

86. **(a)** $-2, 43, 238, 763, 1870, 3883, 7198.$
(b) $58, 250, 1018, 4090, 16378, 65530.$

QUICK REVIEW 11.2

1. Terminates

2. Terminates

3. Repeats

4. Repeats

5. 15; 78

6. For $k = 6, \dfrac{k(k+3)}{2} = \dfrac{6(9)}{2} = 27$, and for $k = 11, \dfrac{k(k+3)}{2} = \dfrac{11(14)}{2} = 77$

7. 5.90625; 62.34375

8. For $r = \dfrac{1}{3}, \dfrac{4(1-r^5)}{1-r} \approx 5.975$, and for $r = \dfrac{3}{4}, \dfrac{4(1-r^5)}{1-r} \approx 12.203$

9. As $x \to \infty, \dfrac{x}{x+1} \to 1$

10. As $t \to \infty, \dfrac{5t}{t+8} \to 5$

11. As $x \to \infty, \dfrac{2x}{3x+1} \to \dfrac{2}{3}$

12. As $x \to \infty, \dfrac{3x}{6x+2} \to \dfrac{1}{2}$

13. As $x \to \infty, \dfrac{2x^3+1}{x^2-5} \to \infty$

14. As $x \to \infty, \dfrac{2x+1}{x^2-3x+1} \to 0$

15. As $x \to \infty, \dfrac{x^2+1}{2x^2-x-3} \to \dfrac{1}{2}$

16. As $x \to \infty, \dfrac{5x^2-2}{2x^2+3x-1} \to \dfrac{5}{2}$

17. $ab_1 + ab_2 + ab_3$

18. $1 - x^5$

SECTION EXERCISES 11.2

1. 18

2. 57

3. 3240

4. 1260

5. -120

6. -833

7. 975

8. 2001

9. 24,573

10. 147,620

11. $50.4 - 50.4(6^{-9}) \approx 50.4$

12. $36 - 6^{-8} \approx 36$

13. 155

14. -90

15. 1050

16. 576

17. $(\frac{8}{3})(1 - 2^{-12}) \approx 2.666$

18. $4(1 + 2^{-11}) \approx 4.002$

19. $-196{,}495{,}641$

20. 66,151,030

21. Window: $[-1, 12] \times [-2, 7]$. The sum is approaching $\frac{30}{7} \approx 4.285\ldots.$

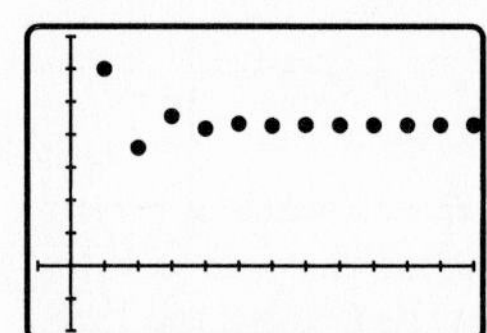

22. Window: $[0, 10] \times [-2, 6]$; the sum is approaching $\frac{20}{7} = 2.857\ldots.$

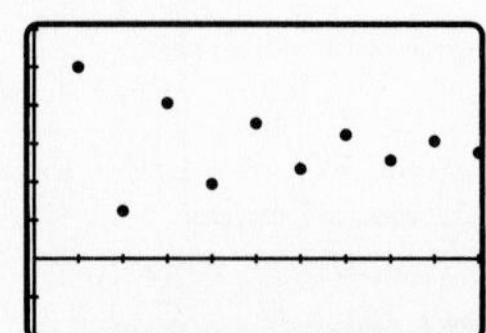

23. Converges; $S = 12$

24. Converges; $S = 6$

25. Diverges, since $r = 2$

26. Diverges, since $r = 3$

27. Converges; $S = 1$

28. Converges; $S = 10$

29. $\frac{707}{99}$

30. $\frac{588}{99} = \frac{196}{33}$

31. $-\frac{17{,}251}{999}$

32. $-\frac{12864}{999} = -\frac{4288}{333}$

33. $\sum_{j=1}^{n} (2j - j^2) = [2(1) - 1^2] + [2(2) - 2^2] + [2(3) - 3^2] + \cdots + (2n - n^2);$
$\sum_{k=1}^{n} (2k - k^2) = [2(1) - 1^2] + [2(2) - 2^2] + [2(3) - 3^2] + \cdots + (2n - n^2)$

34. $\sum_{k=1}^{n} (3\sqrt{k} - k) = (3\sqrt{1} - 1) + (3\sqrt{2} - 2) + (3\sqrt{3} - 3) + \cdots + (3\sqrt{n} - n);$
$\sum_{i=1}^{n} (3\sqrt{i} - i) = (3\sqrt{1} - 1) + (3\sqrt{2} - 2) + (3\sqrt{3} - 3) + \cdots + (3\sqrt{n} - n)$

35. $\sum_{k=1}^{7} (k^2 + 3k + 1)$
$= \sum_{k=1}^{7} k^2 + \sum_{k=1}^{7} 3k + \sum_{k=1}^{7} 1$
$= 140 + 84 + 7$
$= 231$

36. 250

37. $\sum_{j=1}^{\infty} \left(-\frac{1}{4}\right) = \left(-\frac{1}{4}\right)^1 + \left(-\frac{1}{4}\right)^2 + \left(-\frac{1}{4}\right)^3 + \left(-\frac{1}{4}\right)^4 + \left(-\frac{1}{4}\right)^5 + \cdots$
$= -\frac{1}{4} + \frac{1}{16} - \frac{1}{64} + \frac{1}{256} - \frac{1}{1024} + \cdots$
The sum is -0.2, or $-\frac{1}{5}$.

38. $-\frac{2}{7}$

39. $\sum_{r=1}^{n+2} r = 1 + 2 + 3 + 4 + 5 + \cdots + (n+2) = \dfrac{(n+2)(n+3)}{2}$

40. $\dfrac{(m+3)(m+4)}{2}$

41. $\sum_{k=1}^{11} (6k - 13)$

42. $\sum_{k=1}^{10} (3k - 1)$

43. $\sum_{k=0}^{n} (k+1)^2$

44. $\sum_{k=0}^{n} (k+1)^3$

45. $\sum_{k=1}^{\infty} 6(-2)^{k-1}$

46. $\sum_{k=1}^{\infty} 5(-3)^{k-1}$

47. No; the sum of the series $2 + 2.1 + 2.205 + 2.31525 + \cdots$ for a particular value of n can be found by evaluating $f(x) = 2\left(\dfrac{1 - 1.05^x}{1 - 1.05}\right)$ at n. Since the common ratio 1.05 is greater than 1, the series will diverge; this is shown by the graph of the function since it does not have a horizontal asymptote.

48. **(a)** In Seattle it rains 150 days per year; Los Angeles only has 37 rainy days.
(b) January is the rainiest month of the year; August is the least rainy.

49. $2 + 4\sum_{k=1}^{9} 0.9^k \approx 24.05$ m

50. \$86,001.96

51. The payment at the end of the first period earns $R(1+i)^{n-1}$, the payment at the end of the second period earns $R(1+i)^{n-2}$, and so on. The value V of an annuity after n periods is

$$V = R(1+i)^{n-1} + R(1+i)^{n-2} + \cdots + R$$
$$= R + R(1+i) + \cdots + R(1+i)^{n-1}$$
$$= \sum_{k=1}^{n} a_1 r^{k-1}$$

where $a_1 = R$ and $r = 1 + i$.

52. $\sum_{k=1}^{8} (k^2 - 2)$

53. **(a)** $\sum_{k=1}^{n} [a_1 + (k-1)d] = \dfrac{n(a_1 + a_n)}{2}$; $\sum_{k=1}^{n} [a_1 + (k-1)d] = \dfrac{n}{2}[2a_1 + (n-1)d]$

(b) $\sum_{k=1}^{\infty} [a_1 + (k-1)d]$

(c) $\sum_{k=1}^{n} a_1 r^{k-1} = \dfrac{a_1(1 - r^n)}{1 - r}$

(d) $\sum_{k=1}^{\infty} a_1 r^{k-1} = \dfrac{a_1}{1 - r}$

54. $T_{n-1} + T_n = \dfrac{(n-1)n}{2} + \dfrac{n(n+1)}{2} = \dfrac{n^2 - n + n^2 + n}{2} = n^2$

55.

n	F_n	S_n	$F_{n+2} - 1$
1	1	1	1
2	1	2	2
3	2	4	4
4	3	7	7
5	5	12	12
6	8	20	20
7	13	33	33
8	21	54	54
9	34	88	88

$$S_n = F_{n+2} - 1 = \sum_{k=1}^{n} F_k$$

56. Answers may vary.

57. **(a)** Heartland: 17,659,690 people
Southeast: 36,087,346 people
(b) Heartland: 517,825 mi^2
Southeast: 348,999 mi^2
(c) Heartland: $\dfrac{17{,}659{,}690}{517{,}825} \approx$ 34.10 people/mi^2
Southeast: $\dfrac{36{,}087{,}346}{348{,}999} \approx$ 103.40 people/mi^2

(d)

Heartland:			Southeast:		
Iowa	≈	49.34	Alabama	≈	78.15
Kansas	≈	30.11	Arkansas	≈	44.20
Minnesota	≈	51.84	Florida	≈	220.62
Missouri	≈	73.42	Georgia	≈	109.97
Nebraska	≈	20.40	Louisiana	≈	88.37
N. Dakota	≈	9.03	Mississippi	≈	53.96
S. Dakota	≈	9.03	S. Carolina	≈	112.07
Total	≈	243.17	Total	≈	707.34

$\dfrac{243.17}{7} \approx 34.74$ $\dfrac{707.34}{7} \approx 101.05$

The answers differ because $\dfrac{\Sigma \text{ populations}}{\Sigma \text{ area}} \neq$ average of $\dfrac{\text{populations}}{\text{areas}}$.

58. Yes: $\sum_{k=1}^{n} (2k-1) = \sum_{k=1}^{n} 2k - \sum_{k=1}^{n} 1 = 2\left(\sum_{k=1}^{n} k\right) - (1 + 1 + \cdots + 1) = n(n+1) - (n) = n^2$

59. **(a)** $\sum_{k=1}^{n} c = c + c + c + c + \cdots + c$ added n times, which is cn

(b) $\sum_{k=1}^{n} [a_1 + (k-1)d]$
$$= \sum_{k=1}^{n} a_1 + \sum_{k=1}^{n} (dk - d)$$
$$= \sum_{k=1}^{n} a_1 + \sum_{k=1}^{n} dk - \sum_{k=1}^{n} d$$
$$= \sum_{k=1}^{n} a_1 + d\sum_{k=1}^{n} k - \sum_{k=1}^{n} d$$
$$= \sum_{k=1}^{n} a_1 + d\left[\frac{n(n+1)}{2}\right] - \sum_{k=1}^{n} d$$
$$= na_1 + nd\left(\frac{n+1}{2}\right) - nd$$
$$= \frac{n}{2}(2a_1 + d(n+1) - 2d)$$
$$= \frac{n}{2}[2a_1 + nd + d - 2d]$$
$$= \frac{n}{2}[2a_1 + nd - d]$$
$$= \frac{n}{2}[2a_1 + (n-1)d]$$

60. Window: $[0, 10] \times [-2, 4]$; if $\sum_{k=1}^{n} \frac{1}{k} \geq \ln n$ for all n, the sum diverges, since as $n \to \infty$, $\ln n \to \infty$.

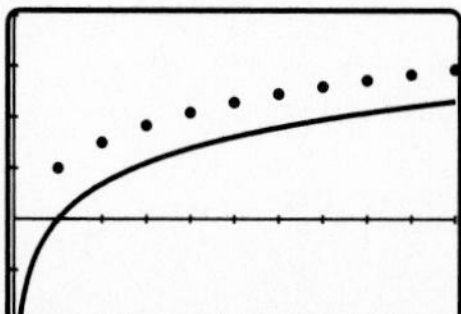

61. Let $\sum_{k=1}^{\infty} a_k = a_1 + a_2 + a_3 + \cdots + a_k + \cdots$ be an arithmetic series with $a_k = a_1 + (k-1)d$. The sum of the first n terms is $S_n = \frac{n}{2}(2a_1 + (n-1)d)$. Because $|S_n| \to \infty$ as $n \to \infty$, the series diverges.

QUICK REVIEW 11.3

1. $n^2 + 5n$

2. $n^2 - n - 6$

3. $k^3 + 3k^2 + 2k$

4. $k^3 + 6k^2 + 8k$

5. $(n+3)(n-1)$

6. $(k+4)(k+1)$

7. $(k+1)^3$

8. $(n-1)^3$

9. $f(1) = 5; f(t) = t + 4; f(t+1) = t + 5$

10. $f(1) = -2, f(t) = t - 3, f(t+1) = t - 2$

11. $f(1) = \frac{1}{2}; f(k) = \frac{k}{k+1};$ $f(k+1) = \frac{k+1}{k+2}$

12. $f(a) = \frac{a}{a-7}, f(k) = \frac{k}{k-7},$ $f(k+1) = \frac{k+1}{k-6}$

13. $P(1) = \frac{1}{2}; P(k) = \frac{2k}{3k+1};$ $P(k+1) = \frac{2k+2}{3k+4}$

14. $P(1) = \frac{3}{8}, P(k) = \frac{3k}{6k+2},$ $P(k+1) = \frac{3k+3}{6k+8}$

15. $P(1) = -2; P(k) = 2k^2 - k - 3;$ $P(k+1) = 2k^2 + 3k - 2$

16. $P(1) = 6, P(k) = 3k^2 + 4k - 1,$ $P(k+1) = 3k^2 + 10k + 6$

SECTION EXERCISES 11.3

1. $P_n: 2 + 4 + 6 + \cdots + 2n = n^2 + n.$ P_1 is true: it says that $2 = 1^2 + 1$. Now assume P_k is true, so that $2 + 4 + 6 + \cdots + 2k = k^2 + k$. Then:
$2 + 4 + 6 + \cdots + 2k + 2(k+1)$
$= k^2 + k + 2(k+1)$
$= k^2 + k + 2k + 2$
$= k^2 + 3k + 2$
$= (k^2 + 2k + 1) + (k+1)$
$= (k+1)(k+1) + (k+1)$
$= (k+1)^2 + (k+1),$
so P_{k+1} is true.
Therefore the statement is true for all $n \geq 1$.

2. $P_n: 8 + 10 + 12 + \cdots + (2n+6) = n^2 + 7n.$
P_1 is true: it says that $8 = 1^2 + 7 \cdot 1$. Now assume P_k is true, so that $8 + 10 + 12 + \cdots + (2k+6) = k^2 + 7k$. Add $2(k+1) + 6 = 2k + 8$ to both sides, and we have $8 + 10 + 12 + \cdots + (2k+6) + (2k+8) = k^2 + 7k + 2k + 8 = (k^2 + 2k + 1) + 7k + 7 = (k+1)^2 + 7(k+1)$, so P_{k+1} is true. Therefore the statement is true for all $n \geq 1$.

3. $P_n: 6 + 10 + 14 + \cdots + (4n+2) = n(2n+4).$
P_1 is true: it says that $6 = 1[2(1) + 4]$. Now assume P_k is true, so that $6 + 10 + 14 + \cdots + (4k+2) = k(2k+4)$.
Then:
$6 + 10 + 14 + \cdots + (4k+2)$
$\quad + [4(k+1) + 2]$
$= k(2k+4) + [4(k+1) + 2]$
$= 2k^2 + 4k + 4k + 6$
$= (2k^2 + 2k) + (6k + 6)$
$= 2k(k+1) + 6(k+1)$
$= (k+1)(2k+6)$
$= (k+1)(2k+2+4)$
$= (k+1)[2(k+1) + 4],$
so P_{k+1} is true.
Therefore the statement is true for all $n \geq 1$.

4. P_n: $14 + 18 + 22 + \cdots + (4n+10) = 2n(n+6)$.
P_1 is true: it says that $14 = 2 \cdot 1(1+6)$. Now assume P_k is true, so that $14 + 18 + 22 + \cdots + (4k+10) = 2k(k+6)$. Add $4(k+1) + 10 = 4k + 14$ to both sides, and we have $14 + 18 + 22 + \cdots + (4k+10) + (4k+14) = 2k(k+6) + (4k+14) = 2(k^2 + 8k + 7) = 2(k+1)(k+7) = 2(k+1)(k + 1 + 6)$, so P_{k+1} is true. Therefore the statement is true for all $n \geq 1$.

5. P_n: $a_n = 3 + (n-1)5$.
P_1 is true: it says that $3 = 3 + (1-1)5$. Now assume P_k is true, so that $a_k = 3 + (k-1)5$. Then:
$a_{k+1} = a_k + 5$
$= 3 + (k-1)5 + 5$
$= 3 + 5k - 5 + 5$
$= 3 + 5k$
$= 3 + ((k+1) - 1)5,$
so P_{k+1} is true.
Therefore the statement is true for all $n \geq 1$.

6. P_n: $a_n = 5 + 2n$. P_1 is true: it says that $a_1 = 5 + 2 \cdot 1 = 7$. Now assume P_k is true, so that $a_k = 5 + 2k$. To get a_{k+1}, add 2 to a_k; that is, $a_{k+1} = (5 + 2k) + 2 = 5 + 2(k+1)$. This shows that P_{k+1} is true. Therefore the statement is true for all $n \geq 1$.

7. P_n: $a_n = 2 \cdot 3^{n-1}$. P_1 is true: it says that $2 = 2(3^{1-1})$. Now assume P_k is true, so that $a_k = 2(3^{k-1})$. Then:
$a_{k+1} = 3a_k$
$= 3[2(3^{k-1})]$
$= 3\left[2\left(\frac{3^k}{3^1}\right)\right]$
$= \frac{3}{3}[2(3^k)]$
$= 2(3^k)$
$= 2\{3^{[(k+1)-1]}\},$
so P_{k+1} is true.
Therefore the statement is true for all $n \geq 1$.

8. P_n: $a_n = 3 \cdot 5^{n-1}$. P_1 is true: it says that $a_1 = 3 \cdot 5^{1-1} = 3 \cdot 5^0 = 3$. Now assume P_k is true, so that $a_k = 3 \cdot 5^{k-1}$. To get a_{k+1}, multiply a_k by 5; that is, $a_{k+1} = 5 \cdot 3 \cdot 5^{k-1} = 3 \cdot 5^k = 3 \cdot 5^{k+1-1}$. This shows that P_{k+1} is true. Therefore the statement is true for all $n \geq 1$.

9. $P_1: 1 = \frac{1(1+1)}{2}; P_k: 1 + 2 + \cdots + k = \frac{k(k+1)}{2}; P_{k+1}: 1 + 2 + \cdots + k + (k+1) = \frac{(k+1)(k+2)}{2}$

10. $P_1: 1^2 = \frac{1(1+1)(2+1)}{6}$; $P_k: 1^2 + 2^2 + \cdots + k^2 = \frac{k(k+1)(2k+1)}{6}$; $P_{k+1}: 1^2 + 2^2 + \cdots + k^2 + (k+1)^2 = \frac{(k+1)(k+2)(2k+3)}{6}$

11. $P_1: \frac{1}{1(2)} = \frac{1}{1+1}$; $P_k: \frac{1}{1(2)} + \frac{1}{2(3)} + \cdots + \frac{1}{k(k+1)} = \frac{k}{k+1}$; $P_{k+1}: \frac{1}{1(2)} + \frac{1}{2(3)} + \cdots + \frac{1}{k(k+1)} + \frac{1}{(k+1)(k+2)} = \frac{k+1}{k+2}$

12. $P_1: 1^4 = \frac{1(1+1)(2+1)(3+3-1)}{30}$; $P_k: 1^4 + 2^4 + \cdots + k^4 = \frac{k(k+1)(2k+1)(3k^2+3k-1)}{30}$; $P_{k+1}: 1^4 + 2^4 + \cdots + k^4 + (k+1)^4 = \frac{(k+1)(k+2)(2k+3)(3k^2+9k+5)}{30}$

13. $P_n: 1 + 5 + 9 + \cdots + (4n-3) = n(2n-1)$.
P_1 is true since $(4)1 - 3 = 1$ and $1[2(1) - 1] = 1(2-1) = 1(1) = 1$.
Assume that P_k is true: $1 + 5 + 9 + \cdots + (4k-3) = k(2k-1)$. Then:

$$\begin{aligned} 1 + 5 + 9 + \cdots + (4k-3) + [4(k+1) - 3] &= k(2k-1) + [4(k+1) - 3] \\ &= k(2k-1) + (4k+1) \\ &= 2k^2 - k + 4k + 1 \\ &= 2k^2 + 3k + 1 \\ &= (k+1)(2k+1) \\ &= (k+1)(2k+2-1) \\ &= (k+1)[2(k+1) - 1], \end{aligned}$$

so P_{k+1} is true.
Therefore P_n is true for all $n \geq 1$.

14. $P_n: 1 + 2 + 2^2 + \cdots + 2^{n-1} = 2^n - 1$.
P_1 is true since $1 = 2^1 - 1 = 2 - 1$.
Now assume P_k is true, so that $1 + 2 + 4 + \cdots + 2^{k-1} = 2^k - 1$. Add 2^k to both sides, and we have $1 + 2 + 4 + \cdots + 2^{k-1} + 2^k = 2^k - 1 + 2^k = 2 \cdot 2^k - 1 = 2^{k+1} - 1$, so P_{k+1} is true.
Therefore P_n is true for all $n \geq 1$.

15. $P_n: \frac{1}{1(2)} + \frac{1}{2(3)} + \cdots + \frac{1}{n(n+1)} = \frac{n}{n+1}$.
P_1 is true since $\frac{1}{1 \cdot 2} = \frac{1}{1+1}$.
Assume that P_k is true: $\frac{1}{1(2)} + \frac{1}{2(3)} + \cdots + \frac{1}{k(k+1)} = \frac{k}{k+1}$. Then:

$$\begin{aligned} &\frac{1}{1(2)} + \frac{1}{2(3)} + \cdots + \frac{1}{k(k+1)} + \frac{1}{(k+1)(k+2)} \\ &= \frac{k}{k+1} + \frac{1}{(k+1)(k+2)} \\ &= \frac{k(k+2)+1}{(k+1)(k+2)} \\ &= \frac{k^2+2k+1}{(k+1)(k+2)} \\ &= \frac{(k+1)(k+1)}{(k+1)(k+2)} \\ &= \frac{k+1}{k+2} \\ &= \frac{k+1}{(k+1)+1}, \end{aligned}$$

so P_{k+1} is true.
Therefore P_n is true for all $n \geq 1$.

16. $P_n: \frac{1}{1 \cdot 3} + \frac{1}{3 \cdot 5} + \cdots + \frac{1}{(2n-1)(2n+1)} = \frac{n}{2n+1}$.
P_1 is true since $\frac{1}{1 \cdot 3} = \frac{1}{2 \cdot 1 + 1}$.
Now assume P_k is true, so that $\frac{1}{1 \cdot 3} + \frac{1}{3 \cdot 5} + \cdots + \frac{1}{(2k-1)(2k+1)} = \frac{k}{2k+1}$. Add $\frac{1}{[2(k+1)-1][2(k+1)+1]} = \frac{1}{(2k+1)(2k+3)}$ to both sides, and we have

$$\begin{aligned} &\frac{1}{1 \cdot 3} + \frac{1}{3 \cdot 5} + \cdots + \frac{1}{(2k-1)(2k+1)} + \frac{1}{(2k+1)(2k+3)} \\ &= \frac{k}{2k+1} + \frac{1}{(2k+1)(2k+3)} \\ &= \frac{2k^2+3k+1}{(2k+1)(2k+3)} = \\ &\frac{(2k+1)(k+1)}{(2k+1)(2k+3)} = \frac{k+1}{2(k+1)+1}, \end{aligned}$$

so P_{k+1} is true.
Therefore P_n is true for all $n \geq 1$.

17. $P_n: 2^n \geq 2n$.
P_1 is true since $2^1 \geq 2(1)$.
Assume that P_k is true: $2^k \geq 2k$. Then, $2^{k+1} = 2^k(2) \geq 4k = 2k + 2k \geq 2k + 2 = 2(k+1)$, so P_{k+1} is true.
Therefore P_n is true for all $n \geq 1$.

18. $P_n: 3^n \geq 3n$.
P_1 is true since $3^1 \geq 3 \cdot 1$. Now assume P_k is true, so that $3^k \geq 3k$. Then $3^{k+1} = 3(3^k) \geq 3 \cdot 3k = 3(k + 2k) \geq 3(k+1)$, so P_{k+1} is true. Therefore P_n is true for all $n \geq 1$.

19. P_n: 3 is a factor of $n^3 + 2n$.
P_1 is true because 3 is a factor of $1^3 + 2(1) = 3$.
Assume that P_k is true: 3 is a factor of $k^3 + 2k$. Then $(k+1)^3 + 2(k+1)$
$= (k^3 + 3k^2 + 3k + 1) + (2k + 2)$
$= (k^3 + 2k) + 3(k^2 + k + 1)$
Therefore 3 is a factor of $(k+1)^3 + 2(k+1)$, P_{k+1} is true, and P_n is true for all $n \geq 1$.

20. P_n: 6 is a factor of $n(n+1)(n+2)$.
P_1 is true because 6 is a factor of $1(1+1)(1+2) = 6$.
Assume that P_k is true: 6 is a factor of $k(k+1)(k+2)$. Then $(k+1)(k+2) \cdot (k+3) = k(k+1)(k+2) + 3(k+1)(k+2)$. Either $(k+1)$ or $(k+2)$ is even, so 6 is a factor of $3(k+1)(k+2)$. Therefore 6 is a factor of $k(k+1)(k+2) + 3(k+1)(k+2) = (k+1)(k+2)(k+3)$, so P_{k+1} is true. Therefore, P_n is true for all $n \geq 1$.

21. $P_n: a_n = a_1 \cdot r^{n-1}$.
P_1 is true since $a_1 \cdot r^{1-1} = a_1(r^0) = a_1(1) = a_1$.
Assume that P_k is true: $a_k = a_1(r^{k-1})$. Then:

$$\begin{aligned} a_k r &= (a_1 r^{k-1})r \\ a_{k+1} &= a_1 \frac{r^k}{r} r \\ &= a_1 r^k \\ &= a_1 r^{[(k+1)-1]}, \end{aligned}$$

so P_{k+1} is true.
Therefore P_n is true for all $n \geq 1$.

22. $P_n: S_n = \frac{n}{2}[2a_1 + (n-1)d]$
First note by Example 2 that $a_n = a_1 + (n-1)d$. P_1 is true since $S_1 = \frac{1}{2}[2a_1 + (1-1)d] = \frac{1}{2}(2a_1) = a_1$. Now as-

sume P_k is true, so that $S_k = \frac{k}{2}[2a_1 + (k-1)d]$. Add $a_{k+1} = a_1 + kd$ to both sides, and observe that $S_k + a_{k+1} = S_{k+1}$. Then we have $S_{k+1} = \frac{k}{2}[2a_1 + (k-1)d] + a_1 + kd = ka_1 + \frac{1}{2}k(k-1)d + a_1 + kd = (k+1)a_1 + \frac{1}{2}k(k+1)d = \frac{k+1}{2}[2a_1 + (k+1-1)d]$, so P_{k+1} is true. Therefore P_n is true for all $n \geq 1$.

23. P_n: $\sum_{k=1}^{n} k = \frac{n(n+1)}{2}$.

P_1 is true since $\sum_{k=1}^{1} k = 1$ and $\frac{1(1+1)}{2} = \frac{2}{2} = 1$.

Assume that P_j is true:

$$\sum_{k=1}^{j} k = \frac{j(j+1)}{2}.$$

Then

$$\begin{aligned}
&1 + 2 + 3 + \cdots + j + (j+1) \\
&= \frac{j(j+1)}{2} + (j+1) \\
&= \frac{j^2+j}{2} + \frac{2j}{2} + \frac{2}{2} \\
&= \frac{j^2+3j+2}{2} \\
&= \frac{(j+1)(j+2)}{2} \\
&= \frac{(j+1)((j+1)+1)}{2},
\end{aligned}$$

so P_{j+1} is true.

Therefore P_n is true for all $n \geq 1$.

24. P_n: $\sum_{k=1}^{n} k^2 = \frac{n(n+1)(2n+1)}{6}$.

P_1 is true since $1^2 = \frac{1 \cdot 2 \cdot 3}{6}$. Now assume P_k is true, so that $\sum_{i=1}^{k} i^2 = \frac{k(k+1)(2k+1)}{6}$. Add $(k+1)^2$ to both sides, and we have $\sum_{i=1}^{k+1} i^2$

$$\begin{aligned}
&= \frac{k(k+1)(2k+1)}{6} + (k+1)^2 \\
&= \tfrac{1}{6}(k+1)[k(2k+1) + 6(k+1)] \\
&= \tfrac{1}{6}(k+1)(2k^2+7k+6) = \tfrac{1}{6}(k+1)(k+2)(2k+3) \\
&= \frac{(k+1)(k+1+1)[2(k+1)+1]}{6},
\end{aligned}$$

so P_{k+1} is true. Therefore P_n is true for all $n \geq 1$.

25. $\frac{(500)(501)}{2} = 125{,}250$

26. $\frac{(250)(251)(501)}{6} = 5{,}239{,}625$

27. $\frac{(n-3)(n+4)}{2}$

28. $\frac{(75^2)(76^2)}{4} = 8{,}122{,}500$

29. $2^{35} - 1 \approx 3.44 \times 10^{10}$

30. $\frac{(15^2)(16^2)}{4} = 14{,}400$

31. $\frac{n^3 - 3n^2 + 8n}{3}$

32. $\frac{n(4n^2 + 21n + 5)}{6}$

33. $\frac{n^4 + 2n^3 + n^2 - 4n}{4}$

34. $\frac{n(n^3 + 2n^2 + 9n - 12)}{4}$

35. P_n: $3n - 4 \geq n$ for $n \geq 2$.
P_2 is true since $3(2) - 4 \geq 2$.
Assume that P_k is true: $3k - 4 \geq k$.
Then $3(k+1) - 4 = 3k + 3 - 4 = (3k-4) + 3$, $(3k-4) + 3 \geq k + 3 \geq k + 1$, so P_{k+1} is true. Therefore P_n is true for all $n \geq 2$.

36. P_n: $2^n \geq n^2$, for all $n \geq 4$.
P_4 is true since $2^4 \geq 4^2$. Assume that P_k is true: $2^k \geq k^2$. Then $2^{k+1} = 2(2^k) \geq 2k^2 = k^2 + k^2 \geq k^2 + 3k \geq k^2 + 2k + 1 = (k+1)^2$, so P_{k+1} is true. Therefore P_n is true for all $n \geq 4$.

37. Let P_n be the statement that the sum of the measures of the exterior angles of any n-sided polygon is 360°, or $m\angle 1 + m\angle 2 + m\angle 3 + \cdots + m\angle n = 360°$ for $n \geq 3$.

1. If $n = 3$, the polygon is a triangle and the exterior angles are $\angle 1$, $\angle 2$, and $\angle 3$ as in the figure; then the angles of the triangle are $(180° - m\angle 1) + (180° - m\angle 2) + (180° - m\angle 3) = 180°$, $180°(3) - (m\angle 1 + m\angle 2 + m\angle 3) = 180°$, $180°(3) - 180° = m\angle 1 + m\angle 2 + m\angle 3$, and $360° = m\angle 1 + m\angle 2 + m\angle 3$.

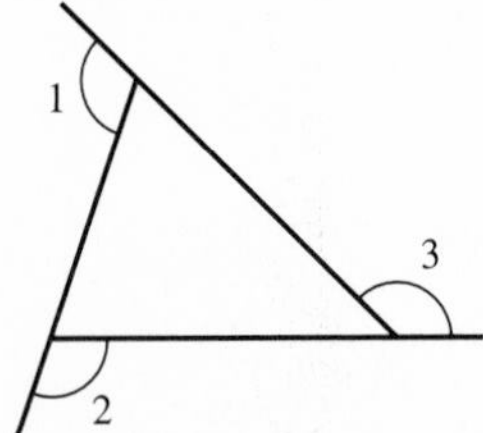

2. The proof for general n is similar to the case $n = 3$. By Example 6, the sum of the interior angles of an n-sided polygon is $180°(n-2)$. So, $(180° - m\angle 1) + (180° - m\angle 2) + \cdots + (180° - m\angle n) = 180°(n-2)$, $180°(n) - (m\angle 1 + m\angle 2 + \cdots + m\angle n) = 180°(n-2)$, and $m\angle 1 + m\angle 2 + \cdots + m\angle n = 360°$.

38. Use the formulas proved in Example 4 and Exercise 23.

39. P_n: $F_{n+2} - 1 = \sum_{k=1}^{n} F_k$.

P_1 is true since $F_{1+2} - 1 = F_3 - 1 = 2 - 1 = 1$ and $\sum_{k=1}^{1} F_k = F_1 = 1$.

Assume that P_k is true: $F_{k+2} - 1 = \sum_{j=1}^{k} F_j$. Then

$$\begin{aligned}
&F_{(k+1)+2} - 1 \\
&= F_{k+3} - 1 \\
&= F_{k+2} + F_{k+1} - 1 \\
&= (F_{k+2} - 1) + F_{k+1} \\
&= \left(\sum_{j=1}^{k} F_j\right) + F_{k+1} \\
&= \sum_{j=1}^{k+1} F_j, \text{ so } P_{k+1} \text{ is true.}
\end{aligned}$$

Therefore P_n is true for all integers $n \geq 1$.

40. P_1 is easy: $\sqrt{2} < 2$. Suppose that P_k: $a_k < 2$; note that $a_{k+1} = \sqrt{2 + a_k}$, so $a_{k+1}^2 = 2 + a_k < 2 + 2 = 4$; therefore $a_{k+1} < 2$.

QUICK REVIEW 11.4

1. 120
2. 110
3. 20
4. 55
5. $x^2 + 2xy + y^2$
6. $a^2 + 2ab + b^2$
7. $25x^2 - 10xy + y^2$
8. $a^2 - 6ab + 9b^2$
9. $9s^2 + 12st + 4t^2$
10. $9p^2 - 24pq + 16q^2$
11. $u^3 + 3u^2v + 3uv^2 + v^3$
12. $b^3 - 3b^2c + 3bc^2 - c^3$
13. $8x^3 - 36x^2y + 54xy^2 - 27y^3$
14. $64m^3 + 144m^2n + 108mn^2 + 27n^3$

SECTION EXERCISES 11.4

1. 1 8 28 56 70 56 28 8 1
2. 1 10 45 120 210 252 210 120 45 10 1
3. $a^5 + 5a^4b + 10a^3b^2 + 10a^2b^3 + 5ab^4 + b^5$
4. $x^6 + 6x^5y + 15x^4y^2 + 20x^3y^3 + 15x^2y^4 + 6xy^5 + y^6$
5. $x^7 - 7x^6y + 21x^5y^2 - 35x^4y^3 + 35x^3y^4 - 21x^2y^5 + 7xy^6 - y^7$
6. $a^6 - 6a^5b + 15a^4b^2 - 20a^3b^3 + 15a^2b^4 - 6ab^5 + b^6$
7. $81x^4 - 108x^3y + 54x^2y^2 - 12xy^3 + y^4$
8. $a^4 - 8a^3b^2 + 24a^2b^4 - 32ab^6 + 16b^8$
9. 35
10. 36
11. 55
12. 1365
13. 924
14. 1716
15. 1
16. 1
17. $\binom{n}{1} = \dfrac{n!}{1!(n-1)!} = \dfrac{n!}{(n-1)!} = \dfrac{n(n-1)!}{(n-1)!} = n;\ \binom{n}{n-1} = \dfrac{n!}{(n-1)!(n-(n-1))!} = \dfrac{n(n-1)!}{(n-1)!(1)!} = \dfrac{n(n-1)!}{(n-1)!} = n$
18. $\binom{n}{r} = \dfrac{n!}{r!(n-r)!} = \dfrac{n!}{(n-r)!r!} = \binom{n}{n-r}$
19. $(x-2)^5 = x^5 - 10x^4 + 40x^3 - 80x^2 + 80x - 32$
20. $(x+3)^6 = x^6 + 18x^5 + 135x^4 + 540x^3 + 1215x^2 + 1458x + 729$
21. $(2x-1)^7 = 128x^7 - 448x^6 + 672x^5 - 560x^4 + 280x^3 - 84x^2 + 14x - 1$
22. $(3x+4)^5 = 243x^5 + 1620x^4 + 4320x^3 + 5760x^2 + 3840x + 1024$
23. $16x^4 + 32x^3y + 24x^2y^2 + 8xy^3 + y^4$
24. $32y^5 - 240y^4x + 720y^3x^2 - 1080y^2x^3 + 810yx^4 - 243x^5$
25. $x^3 - 6x^{5/2}y^{1/2} + 15x^2y - 20x^{3/2}y^{3/2} + 15xy^2 - 6x^{1/2}y^{5/2} + y^3$
26. $x^2 + 4x\sqrt{3x} + 18x + 12\sqrt{3x} + 9$
27. $x^{-10} + 15x^{-8} + 90x^{-6} + 270x^{-4} + 405x^{-2} + 243$
28. $a^7 - 7a^6b^{-3} + 21a^5b^{-6} - 35a^4b^{-9} + 35a^3b^{-12} - 21a^2b^{-15} + 7ab^{-18} - b^{-21}$
29. $\binom{14}{3} = 364$
30. $\binom{13}{5} = \binom{13}{8} = 1287$
31. $\binom{12}{8}(-2)^8 = 126{,}720$
32. $(-3)^4\binom{11}{7} = 26{,}730$
33. 69 for the TI-82; answers will vary for other calculators.
34. $\binom{n}{2} + \binom{n+1}{2} = \dfrac{n!}{2!(n-2)!} + \dfrac{(n+1)!}{2!(n-1)!} = \dfrac{n(n-1)}{2} + \dfrac{(n+1)(n)}{2} = n^2$
35. **(a)** Answers may vary; for $n = 2$ and $m = 3$, $(2+3)! = 5! = 120$, but $2! + 3! = (2 \cdot 1) + (3 \cdot 2 \cdot 1) = 2 + 6 = 8$. **(b)** Answers may vary; for $n = 2$ and $m = 3$, $(2 \cdot 3)! = 6! = 720$, but $2! \cdot 3! = 2 \cdot 1 \cdot 3 \cdot 2 \cdot 1 = 12$.
36. P_n: $a_n = n!$
    P_0 is true because $0! = 1$. Assume P_k is true: $a_k = k!$ Then, $a_{k+1} = (k+1)a_k = (k+1)k! = (k+1)!$ so P_{k+1} is true. Therefore P_n is true for all $n \geq 0$.
37. **(a)** $\binom{0}{0} = 1$ and $2^0 = 1$; $\binom{1}{0} + \binom{1}{1} = 1 + 1 = 2$ and $2^1 = 2$; $\binom{2}{0} + \binom{2}{1} + \binom{2}{2} = 1 + 2 + 1 = 4$ and $2^2 = 4$; $\binom{3}{0} + \binom{3}{1} + \binom{3}{2} + \binom{3}{3} = 1 + 3 + 3 + 1 = 8$ and $2^3 = 8$; $\binom{4}{0} + \binom{4}{1} + \binom{4}{2} + \binom{4}{3} + \binom{4}{4} = 1 + 4 + 6 + 4 + 1 = 16$ and $2^4 = 16$; $\binom{5}{0} + \binom{5}{1} + \binom{5}{2} + \binom{5}{3} + \binom{5}{4} + \binom{5}{5} = 1 + 5 + 10 + 10 + 5 + 1 = 32$ and $2^5 = 32$.
    (b) $2^n = (1+1)^n = \binom{n}{0}(1)^n +$

$\binom{n}{1}(1)^{n-1}(1)+\binom{n}{2}(1)^{n-2}(1)^2+\cdots$
$+\binom{n}{n-1}(1)^1(1)^{n-1}+\binom{n}{n}(1)^n=$
$\binom{n}{0}+\binom{n}{1}+\binom{n}{2}+\cdots+\binom{n}{n-1}$
$+\binom{n}{n}$.

38. **(a)** $\dfrac{(n+1)!}{2!(n-1)!}=\dfrac{(n+1)(n)}{2}$ is a triangular number. **(b)** These numbers are located one entry in from the left on each row.

39. Window: $[-0.5, 9] \times [0, 3]$

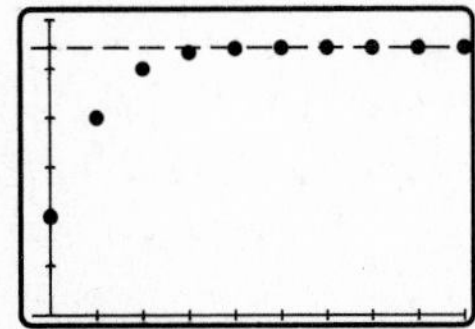

L1	L2	L3
0	1	2.7183
1	2	2.7183
2	2.5	2.7183
3	2.6667	2.7183
4	2.7083	2.7183
5	2.7167	2.7183
6	2.7181	2.7183

L1(1)=0

40. **(a)** The sequence of partial sums is 1, 1.18, 1.1935, 1.19404, ≈1.19405, ≈1.19405, ≈1.19405. The last number is, in fact, exactly equal to 1.03^6. **(b)** The sequence of partial sums is 1, 1.4, 1.464, 1.46912, ≈1.46932, ≈1.46933. The last number is, in fact, exactly equal to 1.08^5.

41. $\binom{n}{r}$ is a positive integer because the coefficients of $(1+x)^n$ are positive integers. In particular, $\binom{n}{3}$ is a positive integer.

$\binom{n}{3}=\dfrac{n!}{3!(n-3)!}=$
$\dfrac{n(n-1)(n-2)}{6}$

The product of any three consecutive integers can be put into the form $n(n-1)\cdot(n-2)$ for appropriate n, so is divisible by 6 since $n(n-1)(n-2)/6$ is a positive integer.

42. P_n: $(a+b)^n=\sum_{r=0}^{n}\binom{n}{r}a^{n-r}b^r$

P_1 is true because $(a+b)^1=\sum_{r=0}^{1}\binom{1}{r}a^{1-r}b^r=a+b$.

Assume that P_k is true:

$(a+b)^k=\sum_{r=0}^{k}\binom{k}{r}a^{k-r}b^r$. Then:

$(a+b)^{k+1}=a(a+b)^k+b(a+b)^k$
$=\sum_{r=0}^{k}\binom{k}{r}a^{k+1-r}b^r+$
$\sum_{r=0}^{k}\binom{k}{r}a^{k-r}b^{r+1}$
$=\sum_{r=0}^{k}\binom{k}{r}a^{k+1-r}b^r+$
$\sum_{s=1}^{k+1}\binom{k}{s-1}a^{k+1-s}b^s$

Let $s=r+1$

$=\binom{k}{0}a^{k+1}+\sum_{r=1}^{k}\binom{k}{r}a^{k+1-r}b^r+$
$\sum_{s=1}^{k}\binom{k}{s-1}a^{k+1-s}b^s+$
$\binom{k}{k}b^{k+1}$
$=\binom{k+1}{0}a^{k+1}+$
$\sum_{r=1}^{k}\left[\binom{k}{r}+\binom{k}{r-1}\right]\cdot$
$a^{k+1-r}b^r+\binom{k+1}{k+1}b^{k+1}$
$=\binom{k+1}{0}a^{k+1}+\sum_{r=1}^{k}\binom{k+1}{r}\cdot$
$a^{k+1-r}b^r+\binom{k+1}{k+1}b^{k+1}$
$=\sum_{r=0}^{k+1}\binom{k+1}{r}a^{k+1-r}b^r$,

so P_{k+1} is true.
Therefore, P_n is true for all $n \ge 1$.

43. Use the pattern $(a+b)^n=a^n+na^{n-1}b+\dfrac{n(n-1)}{2}a^{n-2}b^2+\dfrac{n(n-1)(n-2)}{2\cdot 3}a^{n-3}b^3+\cdots$ with $a=1$, $b=x$, and $n=\frac{1}{2}$ to obtain

$\sqrt{1+x}=1+\frac{1}{2}x+\dfrac{(\frac{1}{2})(\frac{1}{2}-1)}{2}x^2$
$+\dfrac{(\frac{1}{2})(\frac{1}{2}-1)(\frac{1}{2}-2)}{2\cdot 3}x^3+\cdots$
$=1+\frac{1}{2}x-\frac{1}{8}x^2+\frac{1}{16}x^3+\cdots$

44. **(a)** Let $k=n-r$, then

$(a+b)^n=\sum_{r=0}^{n}\binom{n}{r}a^{n-r}b^r$
$=\sum_{k=0}^{n}\binom{n}{n-k}a^kb^{n-k}$
$=\sum_{k=0}^{n}\binom{n}{k}a^kb^{n-k}$, because
$\binom{n}{n-k}=\binom{n}{k}$

(b) The series is the binomial theorem expansion of $[x+(1-x)]^n=1^n=1$.

QUICK REVIEW 11.5

1. 840

2. 3024

3. 330

4. 45

5. 3,628,800

6. 87,178,291,200

7. 840

8. 30

9. 10

10. 35

11. $\approx 4.959185329 \times 10^{14}$

12. 5,200,300

13. 20,708,500

14. $\approx 2.458 \times 10^{20}$

SECTION EXERCISES 11.5

1. 6; *ATX, AXT, XAT, XTA, TAX,* and *TXA*

2. 24

3. 12;

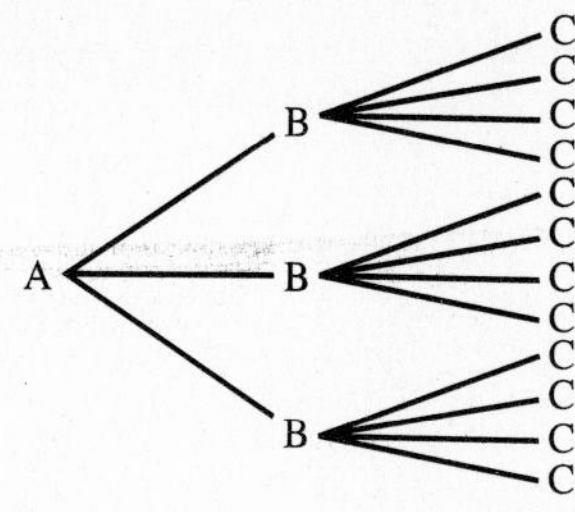

4. 144

5. 12; $K_1Q_1, K_1Q_2, K_1Q_3, K_1Q_4, K_2Q_1, K_2Q_2, K_2Q_3, K_2Q_4, K_3Q_1, K_3Q_2, K_3Q_3$, and K_3Q_4

6. 8

7. 8,000,000

8. 1,000,000,000

9. 19,656,000

10. 60,466,176

11. 36

12. 1024

13. 6720

14. 3,991,680

15. 13,366,080

16. 1,663,200

17. 2002

18. 28

19. 1,307,504

20. 43,758

21. 2300

22. 2,598,960

23. 17,296

24. 77,520

25. 37,353,738,800

26. 56

27. ${}_6C_3 + {}_6C_2 + {}_6C_1 = 41$

28. ${}_5C_3 + {}_5C_2 + {}_5C_1 = 25$

29. 7776

30. ${}_{20}C_8 = 125{,}970$

31. ${}_nC_n = \dfrac{n!}{n!(n-n)!} = \dfrac{n!}{n!0!} = \dfrac{1}{1 \cdot 1} = 1$ and ${}_nC_0 = \dfrac{n!}{0!(n-0)!} = \dfrac{n!}{0!n!} = \dfrac{1}{1 \cdot 1} = 1$

32. ${}_nC_1 = \dfrac{n!}{1!(n-1)!} = \dfrac{n!}{(n-1)!1!} = {}_nC_{n-1} = n$

33. ${}_nP_n = \dfrac{n!}{(n-n)!} = \dfrac{n!}{0!} = \dfrac{n!}{1} = n!$ and ${}_nP_{n-1} = \dfrac{n!}{[n-(n-1)]!} = \dfrac{n!}{1!} = n!$

34. ${}_nP_1 = \dfrac{n!}{(n-1)!} = \dfrac{n!}{1!(n-1)!} = {}_nC_1$

35. ${}_nC_r = \dfrac{n!}{r!(n-r)!}$ and ${}_nC_{n-r} = \dfrac{n!}{(n-r)!(n-(n-r))!} = \dfrac{n!}{(n-r)!(0+r)!} = \dfrac{n!}{r!(n-r)!}$

36. $\dbinom{n}{r} + \dbinom{n}{r+1} = \dfrac{n!}{r!(n-r)!} + \dfrac{n!}{(r+1)!(n-r-1)!} = \dfrac{n!(r+1)}{(r+1)!(n-r)!} + \dfrac{n!(n-r)}{(r+1)!(n-r)!} = \dfrac{n!(r+1+n-r)}{(r+1)!(n-r)!} = \dfrac{(n+1)!}{(r+1)!(n-r)!} = \dbinom{n+1}{r+1}$

37. There will be $5^{52} \approx 2.22 \times 10^{36}$ letters sent in one year, which exceeds the entire population. Somebody will have to receive another letter later during the year.

38. 6

39. 20,123 years (including leap years)

40. $792^2 = 627{,}264$

41. (a)

x	${}_{52}C_x$
0	1
1	52
2	1326
3	22,100
4	270,725
5	2.60×10^6
6	2.04×10^7
7	1.34×10^8
8	7.53×10^8
9	3.68×10^9
10	1.6×10^{10}
11	6.0×10^{10}
12	2.1×10^{11}
13	6.4×10^{11}
14	1.8×10^{12}
15	4.5×10^{12}

The table represents the number of hands possible when x cards are dealt.

(b)

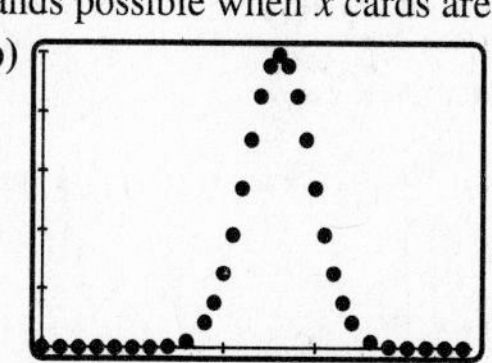

This is related to the binomial coefficients of $(a+b)^{52}$ since $(a+b)^{52} = \dbinom{52}{0}a^{52} + \dbinom{52}{1}a^{51}b + \dbinom{52}{2}a^{50}b^2 + \cdots + \dbinom{52}{50}a^2b^{50} + \dbinom{52}{51}ab^{51} + \dbinom{52}{52}b^{52}$. From Exercise 35, ${}_nC_r = {}_nC_{n-r}$, so ${}_{52}C_0 = {}_{52}C_{52}$, ${}_{52}C_1 = {}_{52}C_{51}$, etc., which explains the symmetry of the graph. You can also use the rows of Pascal's triangle to explain the symmetry of the graph.

QUICK REVIEW 11.6

1. $\frac{7}{18}$

2. $-\frac{17}{12}$

3. $\frac{1}{12}$

4. $\frac{1}{30}$

5. $\frac{5}{8}$

6. $\frac{1}{40}$

7. $\frac{1}{20}$

8. $\frac{1}{840}$

9. $\frac{3}{8}$

10. $\frac{2}{5}$

11. $\frac{7}{264}$

12. $\frac{1}{42}$

13. Approximately 0.78

14. Approximately 0.22

15. Approximately 0.67

16. Approximately 0.33

SECTION EXERCISES 11.6

1. $\{1, 2, 3, 4, 5, 6\}$; 6

2. $\{H, T\}$; 2

3. $\{HHH, HHT, HTH, HTT, THH, THT, TTH, TTT\}$; 8

4. $\{1, 2, 3, 4, 5, 6, 7, 8, 9, 10\}$; 10

5. $\{(1, 1), (1, 2), (1, 3), (1, 4), (1, 5), (2, 1), (2, 2), (2, 3), (2, 4), (2, 5), (3, 1), (3, 2), (3, 3), (3, 4), (3, 5), (4, 1), (4, 2), (4, 3), (4, 4), (4, 5), (5, 1), (5, 2), (5, 3), (5, 4), (5, 5)\}$; 25

6. There are 56 different outcomes: all ordered pairs (a, b) of integers between 1 and 8, with $a \neq b$.

7. $\{(3, 6), (4, 5), (5, 4), (6, 3)\}$; $\frac{1}{9}$

8. $\{(1, 1), (1, 3), (1, 5), (2, 2), (2, 4), (2, 6), (3, 1), (3, 3), (3, 5), (4, 2), (4, 4), (4, 6), (5, 1), (5, 3), (5, 5), (6, 2), (6, 4), (6, 6)\}$; $\frac{1}{2}$

9. $\{(2, 1), (3, 1), (3, 2), (4, 1), (4, 2), (4, 3), (5, 1), (5, 2), (5, 3), (5, 4), (6, 1), (6, 2), (6, 3), (6, 4), (6, 5)\}$; $\frac{5}{12}$

10. $\{(1, 1), (1, 2), (1, 3), (1, 4), (1, 5), (1, 6), (2, 1), (2, 2), (2, 3), (2, 4), (2, 5), (2, 6), (3, 1), (3, 2), (3, 3), (3, 4), (3, 5), (3, 6), (4, 1), (4, 2), (4, 3), (4, 4), (4, 5), (5, 1), (5, 2), (5, 3), (5, 4), (6, 1), (6, 2), (6, 3)\}$; $\frac{5}{6}$

11. $\{(1, 1), (1, 3), (1, 5), (3, 1), (3, 3), (3, 5), (5, 1), (5, 3), (5, 5)\}$; $\frac{1}{4}$

12. $\{(2, 2), (2, 4), (2, 6), (4, 2), (4, 4), (4, 6), (6, 2), (6, 4), (6, 6)\}$; $\frac{1}{4}$

13. 0.06048

14. ≈ 0.0036

15. 0.8

16. 0.6

17. 0.4

18. 0.4

19. 0.2

20. 0.8

21. 0.7

22. 0.6

23. 0.15

24. $\frac{1}{4} = 0.25$

25. 0.49

26. $\frac{1}{10} = 0.1$

27. 0.09

28. 0.01

29. 0.08

30. 0.06

31. 0.64

32. 0.72

33. $\frac{1}{134{,}596}$

34. $\frac{4}{134596} = \frac{1}{33649}$

35. $\frac{5}{3542}$

36. There are $\binom{2}{2}\binom{22}{4} = \binom{22}{4}$ ways to get both black jacks, and 4 "other" cards; similarly, there are $\binom{22}{4}$ ways to get both red jacks. These two numbers together count *twice* the $\binom{2}{2}\binom{2}{2}\binom{20}{2} = \binom{20}{2}$ to get all four jacks; therefore there are $2\binom{22}{4} - \binom{20}{2} = 14{,}440$ distinct ways to have both bowers, so $P(E) = \frac{14{,}440}{134{,}596} = \frac{190}{1771}$.

37. $\frac{1}{2^{10}}$

38. $\frac{1}{2^{10}}$

39. $\frac{1}{2^{10}}$

40. $\frac{10}{2^{10}} = \frac{5}{2^9}$

41. $\frac{45}{1024}$

42. $\frac{120}{2^{10}} = \frac{15}{2^7}$

43. $1 - \frac{1}{2^{10}}$

44. $1 - \frac{11}{2^{10}}$

45. $\frac{19}{30}$

46. ≈ 0.860

47. 0.5

48. **(a)** 0.67 **(b)** 0.33

49. $\frac{1}{1296}$

50. The sum of the probabilities is greater than 1—an impossibility, since the events are mutually exclusive.

51. **(a)** No; the events are not independent (assuming the first card is not replaced before drawing the second). **(b)** Yes; (1) there are 3 observations, so n is fixed; (2) each observation has exactly 2 outcomes; (3) the coins are independent; and (4) $P(H)$ is the same for each coin. **(c)** Yes; (1) there are 5 observations, so n is fixed; (2) each observation has exactly 2 outcomes; (3) the events are independent (assuming none are identical twins); and (4) P(brown eyes) is the same for each child. **(d)** No; since 4 bulbs are from one assembly process and 3 are from another, P(defect) may not be the same for all 7 bulbs.

52. The probability of at least one false positive is approximately 0.344.

53. **(a)** $\frac{86}{127}$ **(b)** $\frac{91}{127}$ **(c)** $\frac{62}{127}$

54. **(a)** $\frac{1}{16}$ **(b)** 0 **(c)** 1 **(d)** $\frac{11}{16}$

55. **(a)** $\frac{8568}{15504} = \frac{21}{38}$ **(b)** $\frac{6120}{15504} = \frac{15}{38}$ **(c)** 1

56. **(a)** $\frac{77}{3230}$ **(b)** $\frac{308}{969}$ **(c)** $\frac{176}{323}$

57. $\frac{1}{36}$

58. **(a)** $\frac{1365}{2^{15}}$ **(b)** $\frac{1941}{2^{15}}$

59. The statement could not be true. P(not cloudy all day) = 1 − P(cloudy all day); then P(not cloudy all day) must be 0.78, since 1 − 0.22 = 0.78; as P(at least 1 hour of sun) = 0.78, then P(at least 1 hour of sun) must equal P(not cloudy all day). However, the probabilities are not only measured in terms of hours since there is a probability for at least 30 minutes of rain, and P(at least 30 minutes of sun) must be greater than P(at least 1 hour of sun). P(at least 30 minutes of sun) + P (cloudy all day) > 1, so the statement could not be true.

60. **(a)** 0.15 **(b)** $(0.85)^4(0.15) = 0.0783\ldots$ **(c)** $1 - (0.85)^4 = 0.477\ldots$

61. **(a)** ≈0.34 **(b)** ≈0.0014

62. **(a)** 1.84, or about 2 **(b)** Yes, this would be an unusual sample. **(c)** Approximately 0.019

63.

n	Probability of n Children Having Blood Type A
0	$\frac{243}{1024} \approx 0.237$
1	$\frac{405}{1024} \approx 0.396$
2	$\frac{270}{1024} \approx 0.264$
3	$\frac{90}{1024} \approx 0.088$
4	$\frac{15}{1024} \approx 0.015$
5	$\frac{1}{1024} \approx 0.001$
Total	$\frac{1024}{1024} = 1$

64. **(a)** About 0.004. **(b)** Approximately 0.006. **(c)** The university's graduation rate seems to be exaggerated; at least, this particular class did not fare as well as the university claims.

65. **(a)** $\frac{7}{10}$ **(b)** $\frac{4}{13}$ **(c)** $\frac{5}{18}$

66. Let E_k be the event "H occurs exactly k times." Since $E_k, E_{k+1}, \ldots$ are all mutually exclusive, $P(E_k$ or E_{k+1} or . . . or $E_n)$ equals $P(E_k) + P(E_{k+1}) + \cdots + P(E_n)$, which, in summation notation, looks like the expression in the book.

67. Suppose that digits are selected at random; although you would expect no particular pattern to the order of the digits selected, you would expect about the same number of occurrences of each digit after a very large number of selections.

QUICK REVIEW 11.7

1. Approximately 15.48%

2. Approximately 20.94%

3. Approximately 14.44%

4. Approximately 27.22%

5. 1723

6. 9200

7. There are many possible graphs.

8. There are many possible graphs.

SECTION EXERCISES 11.7

1.

Stem	Leaf
5	7
6	8197237
7	2210

Most life expectancies for males in South American countries are in the 60s.

2.

Stem	Leaf
6	6788
7	58535276

Most life expectancies for females in South American countries are in the 70s.

3.

Female		Male
	5	7
6788	6	8197237
58535276	7	2210

The life expectancies for females in South American countries are greater than those for males.

4. Aaron was a more productive home-run hitter overall, even though Maris had a higher "record" year.

Aaron		Maris
	0	895
320	1	463
76490	2	836
0942984	3	93
40544470	4	
	5	
	6	1

5.

Interval	Frequency
55–59	1
60–64	3
65–69	4
70–74	4

6.

Interval	Frequency
65–69	4
70–74	2
75–79	6

7. Window: [50, 80] × [−2, 5]

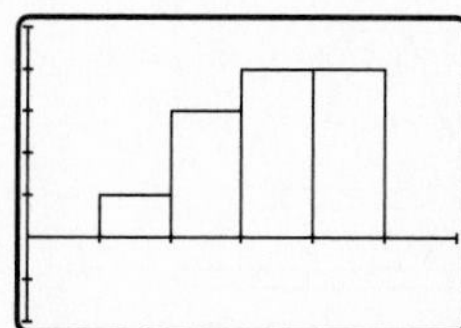

8. Window: [60, 85] × [−2, 8]

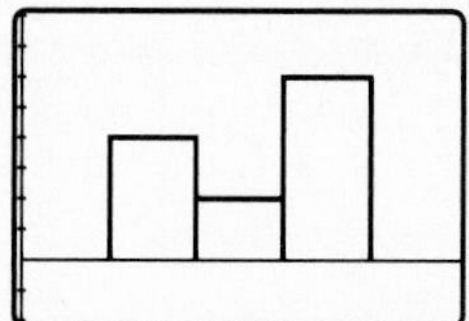

9. Window: $[35, 70] \times [-2, 8]$

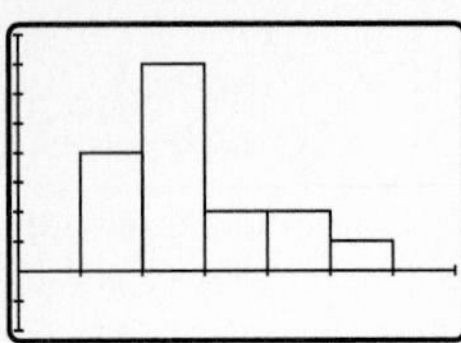

10. Window: $[40, 75] \times [-1, 6]$

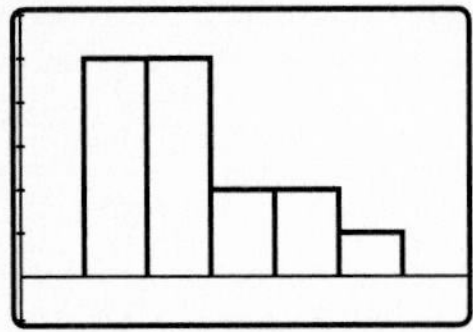

11. Window: $[0, 55] \times [-2, 5]$

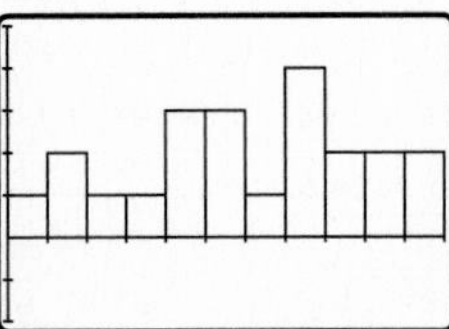

12. Window: $[5, 60] \times [-1, 6]$

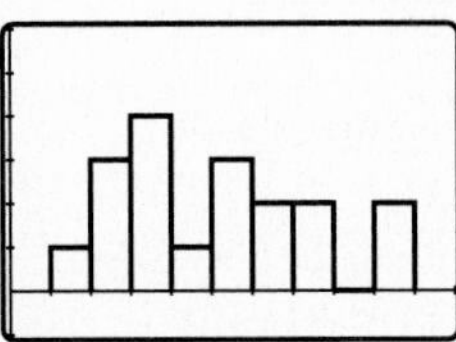

13. Window: $[1950, 1973] \times [-2, 60]$

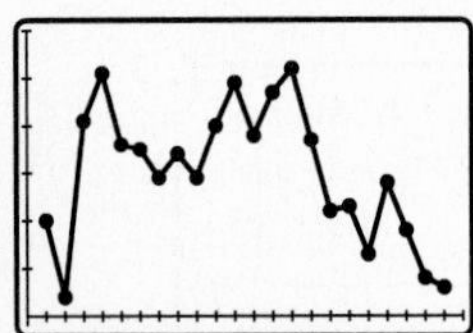

14. Window: $[1950, 1969] \times [-5, 60]$

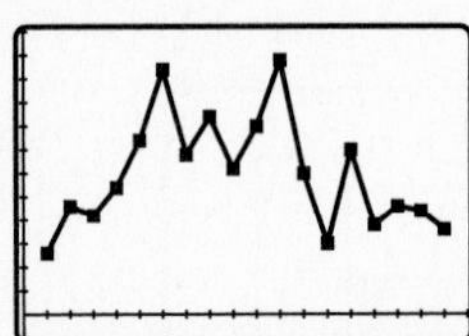

15. The immunization rates for Houston toddlers increased rapidly from 1991 to 1995.

16. Life expectancy for Russian men fell at an alarming rate from 1991 to 1994.

17. Window: $[1950, 1973] \times [-2, 60]$
● = Mays; + = Mantle

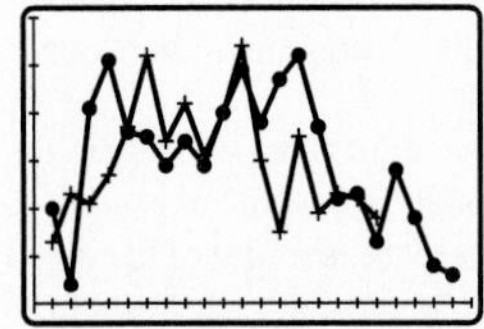

The general shape of both graphs indicates that both Mays and Mantle hit more home runs in the middle of their careers, and the length of the graphs indicates that Mays's career lasted longer than Mantle's.

18. Aaron was more consistent over the years; Mays faded after about 1965.

19. **(a)**

Stem	Leaf
28	2
29	37
30	
31	76
32	87
33	5855
34	882
35	334
36	73
37	
38	5

(b)

Interval	Frequency
25–29	3
30–34	11
35–39	6

(c) Window: $[20, 45] \times [-2, 15]$

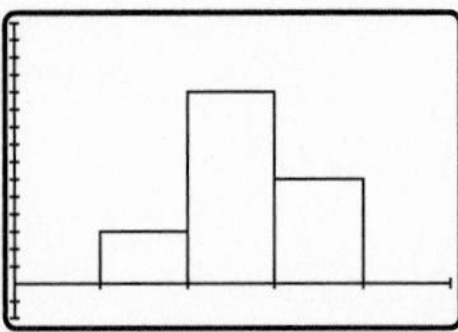

(d) Because the horizontal axis value is not time.

20. **(a)**

Stem	Leaf
6	9239
7	7999
8	70392879
9	0126330451703
10	23693766
11	2046
12	509

(b)

Interval	Frequency
6.0–6.9	4
7.0–7.9	4
8.0–8.9	8
9.0–9.9	13
10.0–10.9	8
11.0–11.9	4
12.0–12.9	3

(c) Window: $[5, 14] \times [-2, 15]$

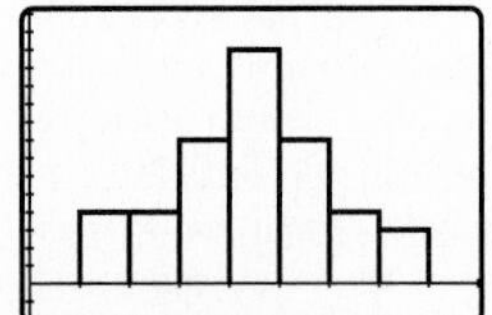

(d) Not categorical.

21. ● = California; + = New York; ■ = Texas
Window: $[1890, 2000] \times [-5, 35]$

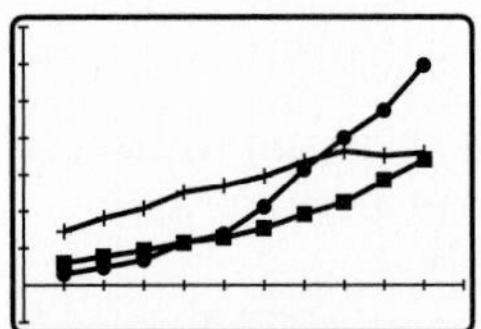

All three populations increased over time, but the population of California increased more than that of Texas and New York.

22. ■ = Florida; + = Illinois; ● = Pennsylvania
Window: $[1900, 2000] \times [-2, 14]$

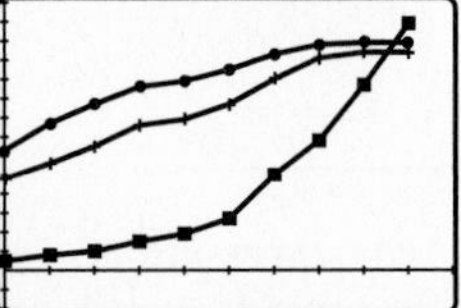

All three populations increased over time, but the population of Florida increased more than that of Illinois and Pennsylvania.

23. Table 11.7 is a good representation of "raw" data, which simply involves recording the data as they are obtained; the stem and leaf plot provides a good visualization of the most and least common salary ranges without requiring much reading; although similar to the stem plot, the frequency table shows the salary ranges more clearly; the histogram, because of its visual impact, probably will be remembered longest. The picture will likely be easier to recall than the lists of numbers used in the tabulations.

24. Responses will vary.

25.

Revenue Source	$ Millions	Percent
Individual income taxes	509,680	44.2
Corporation income taxes	117,520	10.2
Social insurance taxes	428,300	37.1
Other	97,675	8.5

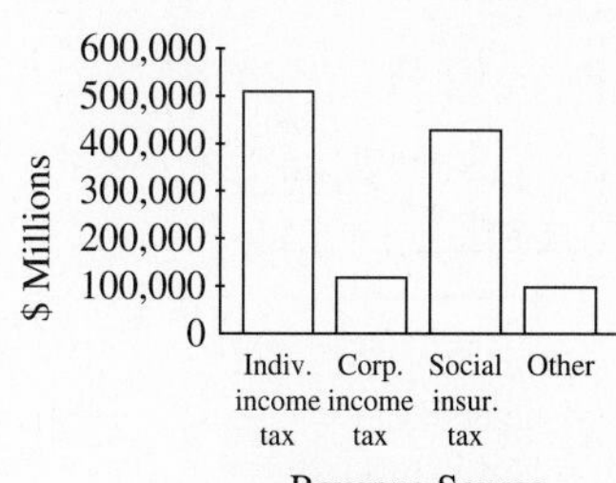

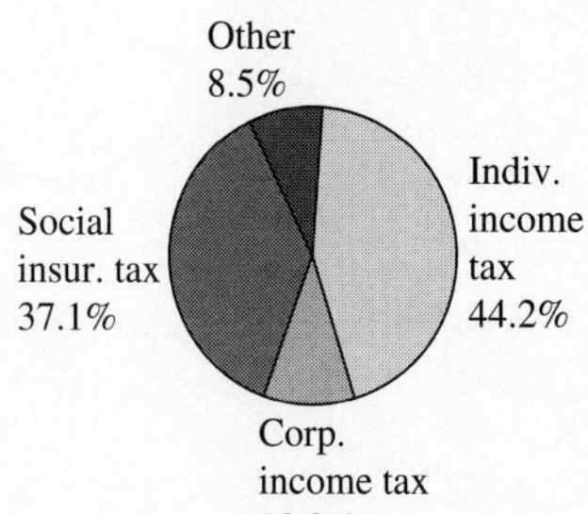

26.

Allocation	$ Million	Percentage
Defense	278,576	19.8
Health and Human Services	282,774	20.1
Interest	292,502	20.8
Other	554,270	39.4

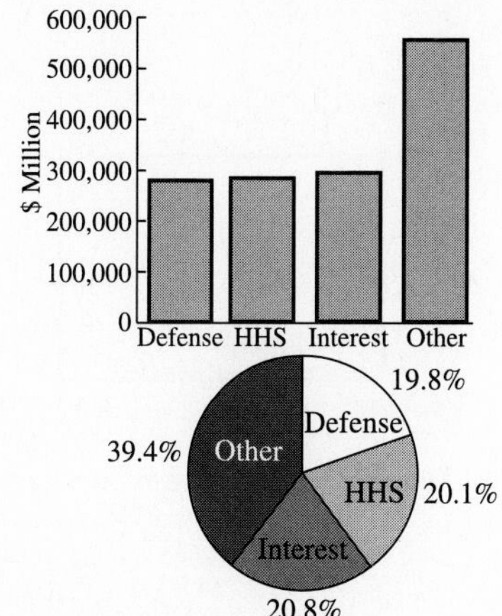

27. $254,947,000,000

28. Window: $[0, 13] \times [-10, 35]$

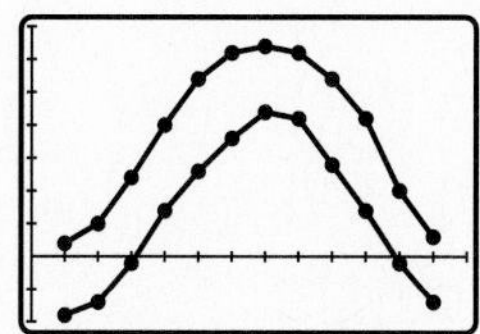

29. High: $f(t) \approx 17 + 15 \sin\left[\frac{\pi}{6}(t - 3.5)\right]$;
Low: $f(t) \approx 6.5 + 15.5 \sin\left[\frac{\pi}{6}(t - 3.5)\right]$

QUICK REVIEW 11.8

1. $x_1 + x_2 + \cdots + x_7$
2. $x_1 + x_2 + x_3 + x_4 + x_5 - 5\bar{x}$
3. $\frac{1}{7}(x_1 + x_2 + \cdots + x_7)$
4. $\frac{1}{5}(x_1 + x_2 + x_3 + x_4 + x_5) - \bar{x}$
5. $\frac{1}{5}[(x_1 - \bar{x})^2 + (x_2 - \bar{x})^2 + \cdots + (x_5 - \bar{x})^2]$
6. $\sqrt{\frac{1}{5}(x_1^2 + x_2^2 + x_3^2 + x_4^2 + x_5^2) - \bar{x}^2}$

SECTION EXERCISES 11.8

1. 26.8
2. 9.5
3. ≈ 60.12
4. $\frac{992}{35} \approx 28.34$
5. 3,900,000
6. ≈ 16.83
7. ≈ 6.78, or about 7 moons
8. 21,176,240 km^2
9. Mays: 30; Mantle: ≈ 29.78; Mays had the greater production rate.
10. State College: 2.4 houses/day; College Station: about 2.14 houses/day; the workers in State College worked faster.
11. What-Next Fashion
12. India: approximately \$352.38; Mexico: approximately \$2485.32; Mexico has the higher PCI.
13. Median: 10.5; modes: 9 and 11
14. Median: 2.5; mode: 6
15. Median: 10; mode: 10
16. Median: 13; modes: 13 and 16
17. ≈ 70.35, or about 70 years
18. ≈ 62.37, or about 62 years
19. **(a)** $\bar{x} \approx 18.42°$C **(b)** The weighted mean is $\bar{x} \approx 18.49°$C. **(c)** The weighted average is the better indicator.
20. **(a)** $\bar{x} \approx 6.42°$C **(b)** The weighted mean is $\bar{x} \approx 6.49°$C. **(c)** The weighted average is the better indicator.
21. In thousands of dollars: min = 21.4; $Q_1 = 24.5$; median = 28.05; $Q_3 = 29.6$; max = 31.7; range = 10.3; $IQR = 5.1$; outliers: none
22. Mantle: min = 13; $Q_1 = 21$; median = 28.5; $Q_3 = 37$; max = 54;

range = 41; IQR = 16; outliers: none
Mays: min = 4; Q_1 = 20; median = 31.5; Q_3 = 40; max = 52; range = 48; IQR = 20; outliers: none

23. Min = 6.2; Q_1 = 8.5; median = 9.25; Q_3 = 10.6; max = 12.9; range = 6.7; IQR = 2.1; outliers: none

24. Min = 28.2; Q_1 = 31.7; median = 33.5; Q_3 = 35.3; max = 38.5; range = 10.3; IQR = 3.6; outliers: none

25. Window: [20, 35] × [−1, 3]; modified box plot is the same

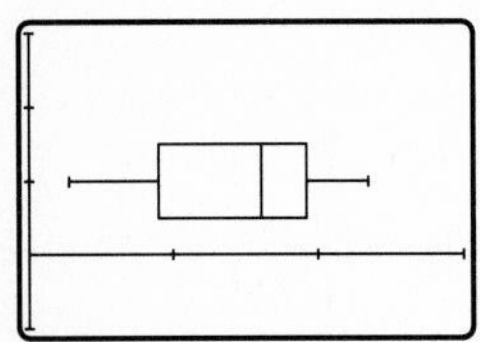

26. Window: [−1, 57] × [−1, 3]; modified boxplot is the same.

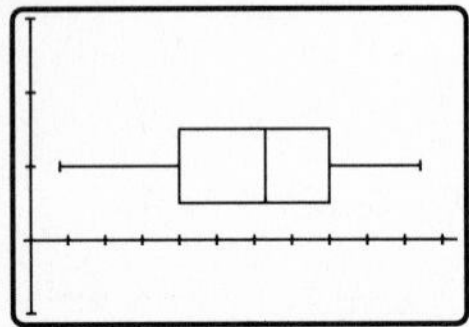

27. Window: [0, 165] × [−1, 3]
(a)

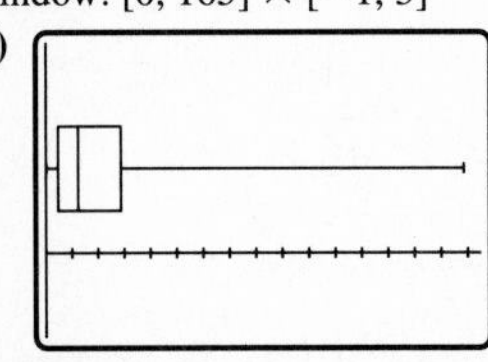

(b)

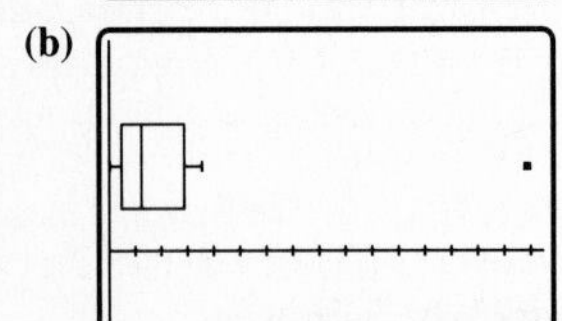

28. Window: [0, 100] × [−1, 3]
(a)

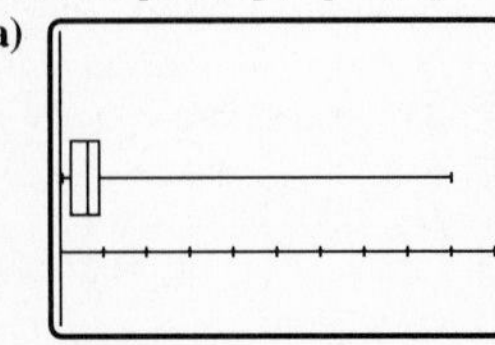

(b)

29. $\frac{12}{44} = \frac{3}{11}$

30. $\frac{40}{44} = \frac{10}{11}$

31. Window: [0, 60] × [0, 3]

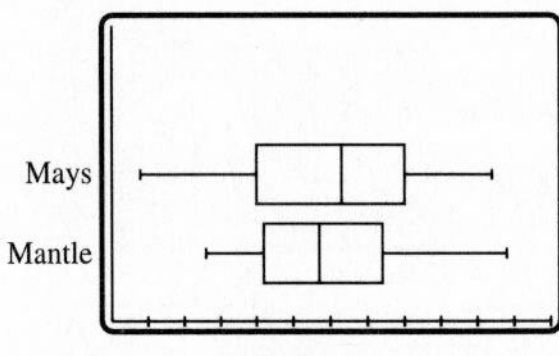

(a) Mays's data set **(b)** Mays's data set

32. Although Mays's best year was not as good as Mantle's best year, Mays was in some ways a better home-run hitter: His median season was 31.5 home runs, compared to 28.5 for Mantle. However, Mays did have three extremely bad years when he hit less than 10 home runs; these make him look worse by comparison.

33. $\sigma = 9.08$; $\sigma^2 = 82.5$

34. $\sigma \approx 23.99$, $\sigma^2 = 575.64$

35. $\sigma \approx 3.10$; $\sigma^2 \approx 9.585$

36. $\sigma \approx 11.61$, $\sigma^2 \approx 134.73$

37. $\sigma \approx 1.53$; $\sigma^2 \approx 2.34$

38. $\sigma \approx 2.60$, $\sigma^2 \approx 6.77$

39. Yes

40. There are many possible answers; the following are several. **(a)** {2, 2, 2, 3, 6, 8, 20} **(b)** {1, 2, 3, 4, 6, 48, 48} **(c)** {−20, 1, 1, 1, 2, 3, 4, 5, 6}

41. Answers may vary; examples are given. **(a)** {2, 4, 6, 8} **(b)** {1, 5, 5, 6, 6, 9} **(c)** {3, 3, 3, 9, 9, 9}

42. No: $(x_i - \bar{x})^2 \le (\max - \min)^2 = (\text{range})^2$, so $\sigma^2 = \frac{1}{n}\sum_{i=1}^{n}(x_i - \bar{x})^2 \le \frac{1}{n}\sum_{i=1}^{n}(\text{range})^2 = (\text{range})^2$; then $\sigma = \sqrt{\sigma^2} \le \sqrt{(\text{range})^2} = \text{range}$.

43. The mode does not have to be a numerical value because it is the data value that occurs most often; both the mean and the median are numerical only, since they both involve calculations of numerical data values.

44. Many answers are possible; examples are given.
(a) and **(c)** {1, 1, 2, 6, 7}
(b) {1, 6, 6, 6, 6, 10}

CHAPTER 11 REVIEW

1. 0, 1, 2, 3, 4, 5; 39

2. $-1, \frac{4}{3}, -2, 3.2, -\frac{16}{3}, \frac{64}{7}$; $\approx 2.68 \times 10^{10}$

3. −1, 2, 5, 8, 11, 14; 32

4. 5, 10, 20, 40, 80, 160; 10,240

5. −5, −3.5, −2, −0.5, 1, 2.5; 11.5

6. $3, 1, \frac{1}{3}, \frac{1}{9}, \frac{1}{27}, \frac{1}{81}; \frac{1}{3^{10}}$

7. −3, 1, −2, −1, −3, −4; −76

8. −3, 2, −1, 1, 0, 1; 13

9. $a_n = 12 + (n - 1)(-2.5)$; $d = -2.5$

10. $a_n = -5 + 4(n - 1) = 4n - 9$; $d = 4$

11. $a_n = 10(1.2)^{n-1}$; $r = 1.2$

12. $a_n = (-2)^{n-1}/8$; $r = -2$

13. $a_n = -11 + (n - 1)4.5$; $d = 4.5$

14. $b_n = 7(\frac{1}{4})^{n-1}$; $r = \frac{1}{4}$

15. $a_n = 3(-4)^{n-1}$; $r = -4$

16. $a_n = 21 - 3.5(n - 1) = 24.5 - 3.5n$; $d = -3.5$

17. −4

18. 7

19. −985.5

20. 775

21. $\frac{21}{8} = 2.625$

22. $-\frac{121}{27}$

23. 59,048

24. -5461

25. $\frac{29,524}{9} \approx 3280.44\ldots$

26. 805

27. Window: $[0, 50] \times [0, 2]$

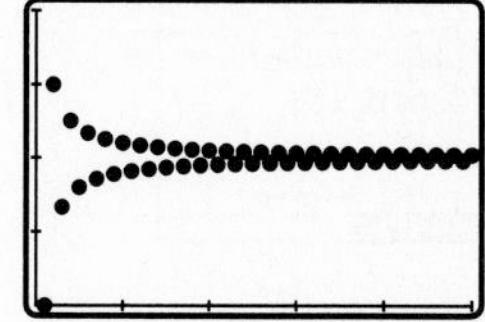

28. Window: $[0, 10] \times [-50, 200]$.

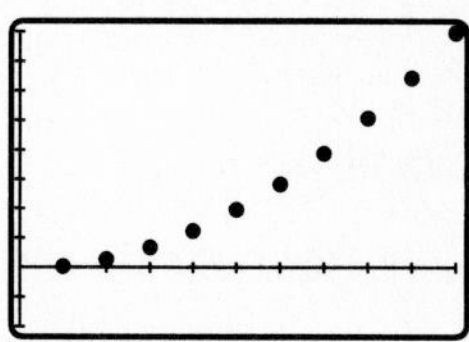

29. Converges; 6

30. Converges; $-\frac{1}{2}$

31. Diverges

32. Diverges

33. Converges; 3

34. Diverges

35. $\sum_{k=1}^{21} (-13 + 5k)$

36. $\sum_{k=1}^{10} (-2)^{k+1}$

37. $\sum_{k=1}^{\infty} (2k - 1)^2$

38. $\sum_{k=0}^{\infty} (\frac{1}{2})^k$

39. $\dfrac{3n^2 + 5n}{2}$

40. $\dfrac{n(n + 1)(2n + 1)}{2}$

41. 4650

42. 25,492,400

43. 5,328,575

44. 7,979,200

45. 792

46. 310,866

47. 18,564

48. 6,724,520

49. 3,991,680

50. 259,459,200

51. $32x^5 + 80x^4y + 80x^3y^2 + 40x^2y^3 + 10xy^4 + y^5$

52. $16384a^7 - 86016a^6b + 193536a^5b^2 - 241920a^4b^3 + 181440a^3b^4 - 81648a^2b^5 + 20412ab^6 - 2187b^7$

53. $243x^{10} + 405x^8y^3 + 270x^6y^6 + 90x^4y^9 + 15x^2y^{12} + y^{15}$

54. $1 + 6x^{-1} + 15x^{-2} + 20x^{-3} + 15x^{-4} + 6x^{-5} + x^{-6}$

55. $512a^{27} - 2304a^{24}b^2 + 4608a^{21}b^4 - 5376a^{18}b^6 + 4032a^{15}b^8 - 2016a^{12}b^{10} + 672a^9b^{12} - 144a^6b^{14} + 18a^3b^{16} - b^{18}$

56. $x^{-8} + 4x^{-6}y^{-1} + 6x^{-4}y^{-2} + 4x^{-2}y^{-3} + y^{-4}$

57. $P_n: 1 + 3 + 6 + \cdots + \dfrac{n(n + 1)}{2} = \dfrac{n(n + 1)(n + 2)}{6}$.

P_1 is true since $1 = 1$ and $\dfrac{1(1 + 1)(1 + 2)}{6} = \dfrac{1(2)(3)}{6} = \dfrac{6}{6} = 1$.

Assume that P_k is true: $1 + 3 + 6 + \cdots + \dfrac{k(k + 1)}{2} = \dfrac{k(k + 1)(k + 2)}{6}$.

Then P_{k+1} is $1 + 3 + 6 + \cdots + \dfrac{k(k + 1)}{2} + \dfrac{(k + 1)[(k + 1) + 1]}{2}$

$= \dfrac{k(k + 1)(k + 2)}{6} + \dfrac{(k + 1)[(k + 1) + 1]}{2}$

$= \dfrac{k(k + 1)(k + 2)}{6} + \dfrac{3(k + 1)(k + 2)}{6}$

$= \dfrac{k(k + 1)(k + 2) + 3(k + 1)(k + 2)}{6}$

$= \dfrac{(k + 1)(k + 2)(k + 3)}{6}$

$= \dfrac{(k + 1)[(k + 1) + 1][(k + 1) + 2]}{6}$,

so P_{k+1} is true. Therefore P_n is true for all $n \geq 1$.

58. $P_n: 1 \cdot 2 + 2 \cdot 3 + \cdots + n(n + 1) = \dfrac{n(n + 1)(n + 2)}{3}$.

P_1 is true since $1 \cdot 2 = \dfrac{1 \cdot 2 \cdot 3}{3}$.

Now assume P_k is true, so that $1 \cdot 2 + 2 \cdot 3 + 3 \cdot 4 + \cdots + k(k + 1) = \dfrac{k(k + 1)(k + 2)}{3}$. Add $(k + 1)(k + 2)$ to both sides, and we have $1 \cdot 2 + 2 \cdot 3 + 3 \cdot 4 + \cdots + k(k + 1) + (k + 1)(k + 2) = \dfrac{k(k + 1)(k + 2)}{3} + (k + 1)(k + 2) = (\frac{1}{3}k + 1)(k + 1)(k + 2) = \frac{1}{3}(k + 3)(k + 1)(k + 2) = \dfrac{(k + 1)(k + 1 + 1)(k + 1 + 2)}{3}$, so P_{k+1} is true. Therefore P_n is true for all $n \geq 1$.

59. $P_n: 2^{n-1} \leq n!$

P_1 is true since $2^{1-1} = 2^0 = 1$ and $1! = 1$ and $1 \leq 1$.

Assume that P_k is true: $2^{k-1} \leq k!$. Then $2 \cdot 2^{k-1} \leq 2 \cdot k! \leq (k + 1)k! = (k + 1)!$ so P_{k+1} is true. Therefore P_n is true for all $n \geq 1$.

60. $P_n: n^3 + 2n$ is divisible by 3.

P_1 is true because $1^3 + 2(1) = 3$ is divisible by 3. Assume P_k is true: $k^3 + 2k$ is divisible by 3. Then

$(k + 1)^3 + 2(k + 1) = k^3 + 3k^2 + 3k + 1 + 2k + 2$
$= (k^3 + 2k) + 3(k^2 + k + 1)$.

Since $(k^3 + 2k)$ and $3(k^2 + k + 1)$ are each divisible by 3, so is their sum $(k^3 + 2k) + 3(k^2 + k + 1) = (k + 1)^3 + 2(k + 1)$. Therefore P_{k+1} is true and P_n is true for all $n \geq 1$.

61. 43,670,016

62. 15

63. 14,508,000

64. 14,190

65. 8,217,822,536

66. 708,930,508

67. 26

68. 19,110

69. {1, 2, 3, 4, 5, 6}

70. {(1, 1), (1, 2), (1, 3), (1, 4), (1, 5), (1, 6), (2, 1), (2, 2), . . . , (6, 6)}

71. {13, 16, 31, 36, 61, 63}

72. {Defective, Nondefective}

73. {*HHH, HHT, HTH, HTT, THH, THT, TTH, TTT*}

74. {*HHT, HTH, THH, TTH, THT, HTT*}

75. {*HHH, TTT*}

76. $\frac{7}{8}$

77. $\frac{1}{2^6}$

78. $\frac{1}{4}$

79. $\frac{10}{2^5} = \frac{5}{16}$

80. $(0.997)^{10} \approx 0.97$

81. $\frac{1}{4} = 0.25$

82. We can redefine the words *success* and *failure* to mean the opposite of what they meant before; in this sense the experiment is symmetrical because $P(S) = P(F)$.

83. 0.24

84. 0.096

85. 0.64

86. Successes are less likely than failures, so the two are not interchangeable.

87. **(a)**

Stem	Leaf
48	5
49	9
50	72
51	06411077
52	136
53	3546122
54	9
55	1
56	
57	7
58	
59	5
60	70
61	
62	3

(b)

Intervals	Frequency
48	1
49	1
50	2
51	8
52	3
53	7
54	1
55	1
56	0
57	1
58	0
59	1
60	2
61	0
62	1

(c) Window: $[45, 65] \times [-2, 10]$

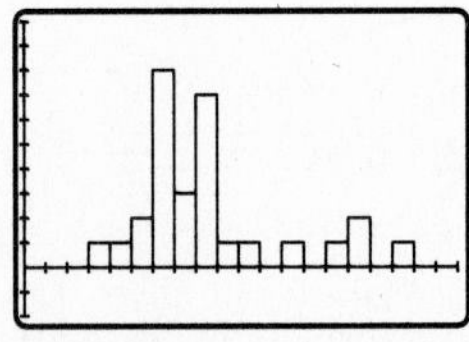

88. Round to nearest whole number first.

(a)

Stem	Leaf
4	799995779937
5	438000419412
6	006
7	21

(b)

Interval	Frequency
40–44.9	2
45–49.9	12
50–54.9	8
55–59.9	3
60–64.9	1
65–69.9	1
70–74.9	2

(c) Window: $[35, 80] \times [-1, 15]$

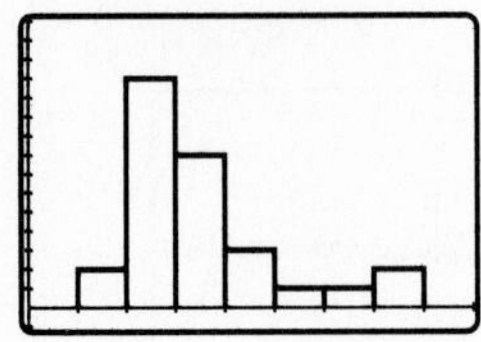

89. **(a)**

Stem	Leaf
0	56373856568424523572998821
1	1535730984399
2	73542
3	136
4	48
5	
6	71
7	
8	
9	
10	4

(b)

Interval	Frequency
0–4	10
5–9	16
10–14	6
15–19	7
20–24	3
25–29	2
30–34	2
35–39	1
40–44	1
45–49	1
50–54	0
55–59	0
60–64	1
65–69	1
70–74	0
75–79	0
80–84	0
85–89	0
90–94	0
95–99	0
100–104	1

(c) Window: $[0, 105] \times [-2, 20]$

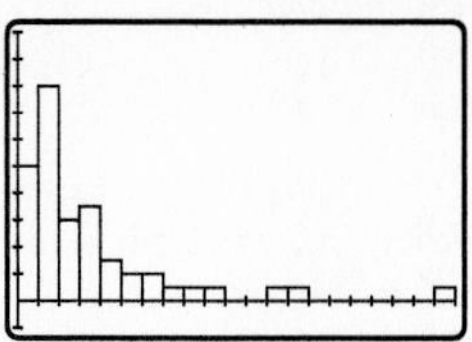

90. Round to nearest hundred yards first.

(a)

Stem	Leaf
1	1
2	47835
3	13565
4	0773
5	06

(b)

Interval	Frequency
10–19	1
20–29	5
30–39	5
40–49	4
50–59	2

(c) Window: $[0, 65] \times [-1, 6]$

91. Min = 48.5; $Q_1 = 51.1$; median = 52.6; $Q_3 = 54.25$; max = 62.3; range = 13.8; $IQR = 3.15$; $\sigma \approx 3.37$; $\sigma^2 \approx 11.36$; outliers: 59.5, 60.0, 60.7, and 62.3

92. Min = 42.8; $Q_1 = 48.7$; median = 50; $Q_3 = 56.2$; max = 71.7; range = 28.9; $IQR = 7.5$; $\sigma \approx 7.15$; $\sigma^2 \approx 51.17$; outliers: 71.3, 71.7

93. Min = 1; $Q_1 = 5$; median = 9.5; $Q_3 = 20.5$; max = 104; range = 103; $IQR = 15.5$; $\sigma \approx 18.99$; $\sigma^2 \approx 360.77$; outliers: 44, 48, 61, 67, and 104

94. Min = 1112; $Q_1 = 2615$; median = 3485; $Q_3 = 4476.5$; max = 5648; range = 4536; $IQR = 1861.5$; $\sigma \approx 1113$; $\sigma^2 \approx 1{,}238{,}765$; outliers: none

95. Window: $[45, 65] \times [-1, 3]$

(a) (b)

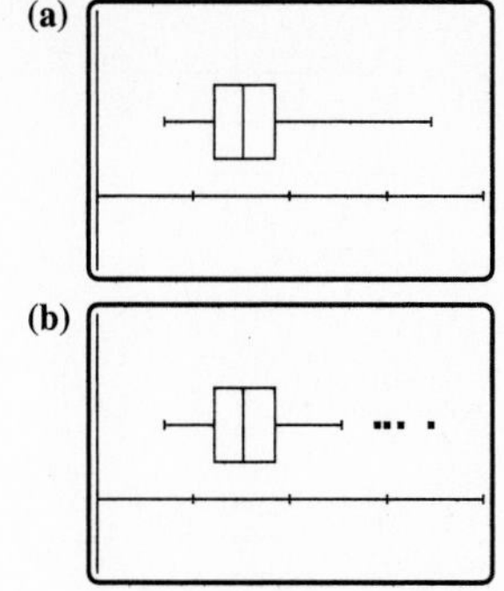

96. The modified boxplot is above the (unmodified) boxplot. Window: $[41, 73] \times [-1, 3]$

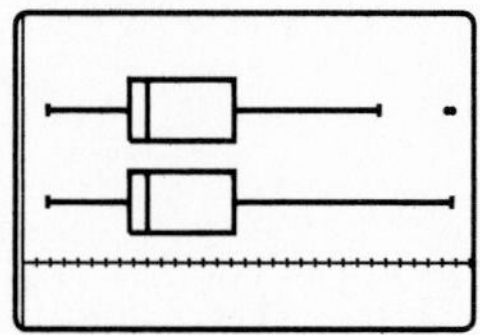

97. Window: $[0, 105] \times [-1, 3]$

(a) (b)

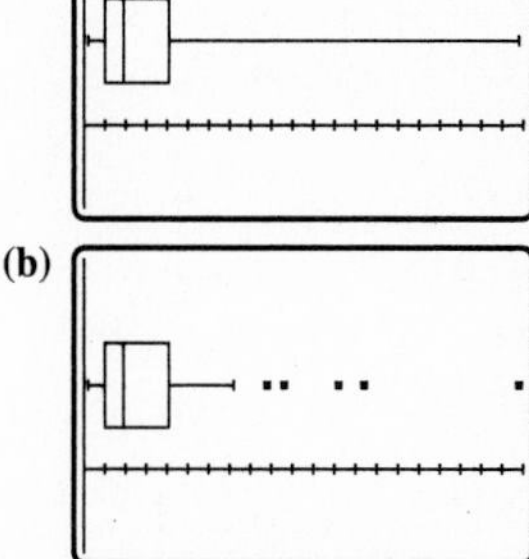

98. The modified boxplot looks the same. Window: $[0, 6000] \times [-1, 3]$

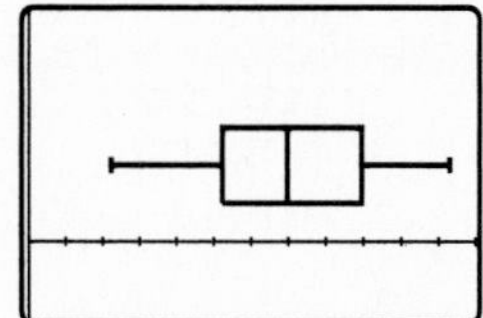

99.

Private		Public
8	42	
	43	
9	44	
	45	
76	46	
21	47	
8868	48	5
5742	49	9
08	50	72
38	51	06411077
6	52	136
58	53	3546122
0	54	9
	55	1
	56	
	57	7
46	58	
8	59	5
4	60	70
	61	
	62	3
	63	
	64	
	65	
3	66	
	67	
	68	
	69	
	70	
73	71	

The median salary appears to be higher for faculty in public institutions and the salary range is greater for the private institutions.

100. Both the range and the *IQR* are considerably larger for the private college data (on the bottom). Window: $[41, 73] \times [0, 3]$

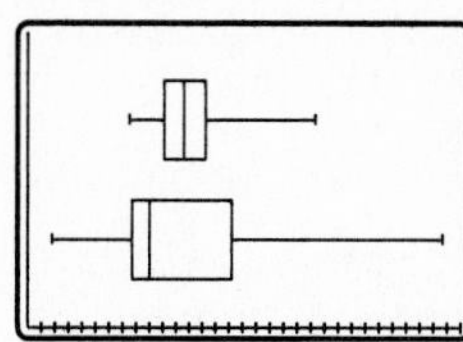

101. Window: $[0, 13] \times [-5, 40]$

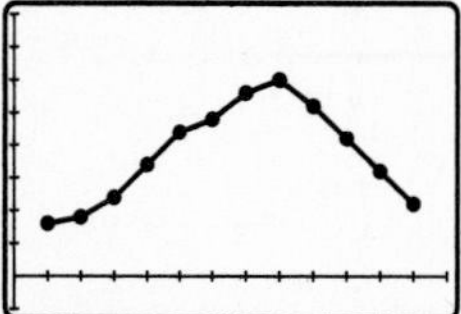

102. Window: [0, 13] × [−5, 25]

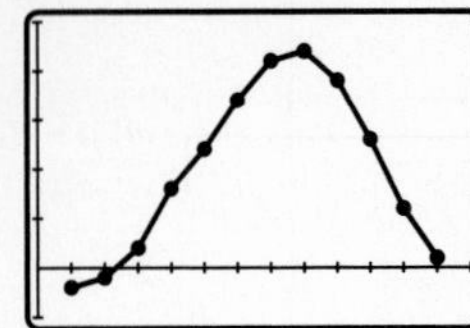

103. P_n: $\sum_{k=1}^{n} a_1 r^{k-1} = \dfrac{a_1(1 - r^n)}{1 - r}$.
P_1 is true since $a_1 = a_1$ and $\dfrac{a_1(1 - r^1)}{1 - r} = a_1$; assume that P_k is true; then:

$$\sum_{j=1}^{k} a_1 r^{j-1} = \frac{a_1(1 - r^k)}{1 - r}$$

and $a_{k+1} = a_k r$; so for P_{k+1},

$$\begin{aligned}\sum_{j=1}^{k+1} a_1 r^{j-1} &= \frac{a_1(1 - r^k)}{1 - r} + a_1 r^k \\ &= \frac{a_1 - a_1 r^k}{1 - r} + \frac{(1 - r)(a_1 r^k)}{1 - r} \\ &= \frac{a_1 - a_1 r^k + a_1 r^k - a_1 r^{k+1}}{1 - r} \\ &= \frac{a_1 - a_1 r^{k+1}}{1 - r} \\ &= \frac{a_1(1 - r^{k+1})}{1 - r}\end{aligned}$$

Therefore P_n is true for all $n \geq 1$.

104. 1 9 36 84 126 126 84 36 9 1

105.
$$\begin{aligned}{}_nP_k \times {}_{n-k}P_j &= \frac{n!}{(n - k)!} \cdot \frac{(n - k)!}{[(n - k) - j]!} \\ &= \frac{n!}{(n - k - j)!} \\ &= \frac{n!}{[n - (k + j)]!} \\ &= {}_nP_{k+j}\end{aligned}$$

106. **(a)** $(0.98)^4 \approx 0.922$
(b) $4(0.98)^3(0.02) \approx 0.075$

107. **(a)** ≈ 0.996 **(b)** $\approx 7.18 \times 10^{-6}$

108. Use a line graph (a time plot) because you are probably interested in how the weather changes over time.

INDEX